STANDARD 7: Concepts of Whole Number Operations

In grades K-4, the mathematics curriculum should include concepts of addition, subtraction, multiplication, and division of whole numbers so that students can—
- develop meaning for the operations by modeling and discussing a rich variety of problem situations;
- relate the mathematical language and symbolism of operations to problem situations and informal language;
- recognize that a wide variety of problem structures can be represented by a single operation;
- develop operation sense.

STANDARD 8: Whole Number Computation

In grades K-4, the mathematics curriculum should develop whole number computation so that students can—
- model, explain, and develop reasonable proficiency with basic facts and algorithms;
- use a variety of mental computation and estimation techniques;
- use calculators in appropriate computational situations;
- select and use computation techniques appropriate to specific problems and determine whether the results are reasonable.

STANDARD 9: Geometry and Spatial Sense

In grades K-4, the mathematics curriculum should include two- and three-dimensional geometry so that students can—
- describe, model, draw, and classify shapes;
- investigate and predict the results of combining, subdividing, and changing shapes;
- develop spatial sense;
- relate geometric ideas to number and measurement ideas;
- recognize and appreciate geometry in their world.

STANDARD 10: Measurement

In grades K-4, the mathematical curriculum should include measurement so that students can—
- understand the attributes of length, capacity, weight, area, volume, time, temperature, and angle;
- develop the process of measuring and concepts related to units of measurement;
- make and use estimates of measurement;
- make and use measurements in problems and everyday situations.

STANDARD 11: Statistics and Probability

In grades K-4, the mathematics curriculum should include experiences with data analysis and probability so that students can—
- collect, organize, and describe data;
- construct, read, and interpret displays of data;
- formulate and solve problems that involve collecting and analyzing data;
- explore concepts of chance.

STANDARD 12: Fractions and Decimals

In grades K-4, the mathematics curriculum should include fractions and decimals so that students can—
- develop concepts of fractions, mixed numbers, and decimals;
- develop number sense for fractions and decimals;
- use models to relate fractions to decimals and to find equivalent fractions;
- use models to explore operations on fractions and decimals;
- apply fractions and decimals to problem situations.

STANDARD 13: Patterns and Relationships

In grades K-4, the mathematics curriculum should include the study of patterns and relationships so that students can—
- recognize, describe, extend, and create a wide variety of patterns;
- represent and describe mathematical relationships;
- explore the use of variables and open sentences to express relationships.

MATHEMATICAL REASONING
for Elementary Teachers

MATHEMATICAL REASONING
for Elementary Teachers

Calvin T. Long
Washington State University

Duane W. DeTemple
Washington State University

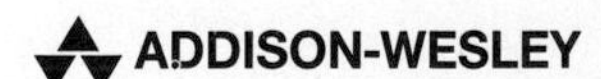

An imprint of Addison Wesley Longman, Inc.

Reading, Massachusetts • Menlo Park, California • New York • Harlow, England
Don Mills, Ontario • Sydney • Mexico City • Madrid • Amsterdam

To my wife and constant helpmate, Jean. *C.T.L.*
To my wife, Janet, and my daughters, Jill and Rachel. *D.W.D.*

Sponsoring Editor: George Duda
Developmental Editor: Margaret Malde-Arnosti
Project Editor: Ann Buesing/Ginny Guerrant
Design Administrator: Jess Schaal
Text Design: Kay D. Fulton
Cover Design: Kay D. Fulton
Cover Photo: Paul Lundquist
Photo Researcher: Rosemary Hunter
Production Administrator: Randee Wire
Compositor: Interactive Composition Corporation
Printer and Binder: R. R. Donnelley & Sons Company
Cover Printer: The Lehigh Press, Inc.

We want to extend our appreciation to the following individuals who contributed elements for the cover: Jay Graening, University of Arkansas, Fayetteville; Stanley Salzman, American River College; Nicholas Richmond, Willard School; Koehler Wider, St. Genevieve DuBois School.

Library of Congress Cataloging-in-Publication Data

De Temple, Duane W.
Mathematical reasoning for elementary teachers / Calvin T. Long, Duane W. DeTemple.
p. cm.
Includes index.
ISBN 0-673-46483-0
1. Mathematics—Study and teaching (Elementary) I. Long, Calvin T. II. Title.
QA135.5.D477 1995
510—dc20 94-20863
CIP

Reprinted with corrections November, 1996. ISBN 0-321-01330-1

97 98 9 8 7 6 5 4

Contents

14 Coordinate Geometry 897

Preface

This text is for use in mathematics content courses for prospective mathematics teachers. We assume that students enrolled in these courses will have completed two years of high school algebra and one year of high school geometry. We do not assume that the students will be highly proficient in algebra and geometry, but that they have a basic knowledge of these subjects and reasonable arithmetic skills. Typically, students bring widely varying backgrounds to these courses, and this text is written to accommodate this diversity.

GUIDING PHILOSOPHY AND APPROACH

The content and processes of mathematics are presented in an appealing yet logically sound approach with these important goals:

- that students come to view mathematics as a fascinating and stimulating intellectual endeavor that provides skills, insights, and modes of thinking that are essential in modern life;
- that students become confident problem solvers, able to think critically and creatively about quantitative, spatial, and logical situations;
- that students develop effective communication skills, with an ability to convey detailed information with clarity and accuracy, and with the capacity to construct well-reasoned explanations;
- that students see the connections between the various parts of mathematics and between mathematics and other subjects, and appreciate the importance of applications of mathematics to the solution of real-world problems.

In short, our goals seek to prepare future teachers to implement the recommendations set forth in the *Curriculum and Evaluation Standards for School Mathematics* (the *Standards*), a landmark document calling for reforms in mathematics education published by the National Council of Teachers of Mathematics (NCTM) in 1989.

Our approach to achieving these goals reflects the recommendations set forth in the NCTM's *Professional Standards for Teaching Mathematics,* issued in 1991 as a continuation of the reforms initiated with the *Standards*. Since teachers often pattern their own teaching after the ways they have been taught, this text models effective teaching by emphasizing activities, manipulatives, investigations, written projects and discussion questions, use of visualization and physical modeling, and—above all else—solving problems. These emphases will prepare future teachers to orient their classroom environment in the five ways called for in the *Professional Teaching Standards*:

- toward classrooms as mathematical communities
  —away from classrooms as simply a collection of individuals;
- toward logic and mathematical evidence as verification
  —away from the teacher as the sole authority for right answers;
- toward mathematical reasoning
  —away from merely memorizing procedures;
- toward conjecturing, inventing, and problem solving
  —away from an emphasis on mechanistic answer finding;
- toward connecting mathematics, its ideas, and its applications
  —away from treating mathematics as a body of isolated concepts and procedures

EMPHASES OF THIS TEXT

Problem Solving and Critical Thinking

Problem solving is the principal vehicle for learning mathematics, and should be the heart of every mathematics content course for prospective teachers. This is echoed in the *Standards,* which lists *mathematics as problem solving* as the first standard for both grades K–4 and grades 5–8. The primary emphasis of this text is the development of skill in problem solving and competence in critical thinking. This is accomplished in a variety of ways.

- An unusually comprehensive initial chapter presents an extensive list of problem-solving strategies, complete with attractive and carefully worked out examples to fully explain Pólya's four-step process. Exercises in the problem sets invite the student to become thoughtfully engaged in doing mathematics and in thinking mathematically. In our own course, this chapter has been especially effective, not only in enhancing reasoning skills but also in promoting student interest in mathematics—often to their admitted surprise! The chapter closes with a section on problem solving in the real world that shows how these strategies apply to problems in the real world.
- A class activity, titled **Hands On,** opens each chapter with a cooperative problem-solving activity, leading naturally into the subject matter of the chapter.
- Concepts are illustrated and often introduced in a problem-solving mode, with many examples worked out in the Pólya four-step process.
- Exercises in the extensive problem sets that follow each section are carefully designed to interest and engage the student. The problems are categorized by type, including the category **Thinking Critically,** whose purpose is to develop higher-level problem-solving skills.
- The enjoyment of mathematical problem solving is encouraged by the **Just for Fun** and **Classic Conundrum** problems found in every chapter.

- Cooperative learning is emphasized in the **Thinking Cooperatively** category of the problem sets. Moreover, extended, often open-ended problems for group exploration and discussion are found in numerous **Cooperative Investigations** throughout the text. In addition, most of the **Thinking Critically** problems are suited to study in a cooperative setting.

Communicating, Reasoning, and Connections

Beyond the first NCTM standard of *mathematics as problem solving,* the next three standards call for emphasis on *mathematics as communication, mathematics as reasoning,* and *mathematical connections*. To prepare teachers to implement these standards in their own classrooms, this text asks students to do the following.

- Write about mathematical situations, critique solutions, and see mathematics as a valuable language in its own right. For example, see the **Communicating** category of the problem sets.
- Translate everyday language into its equivalent mathematical form, using appropriate symbolism.
- Recognize and describe patterns and relationships, and make conjectures and generalizations based on these observations.
- Draw logical conclusions based on accepted facts, and explain the reasoning that supports these conclusions.
- Understand the role of deduction and the utility of counterexamples to delineate the boundaries of what is known to be true and what is known to be false.
- Be aware of the value of mathematics in everyday life.
- Appreciate the historical evolution of mathematical concepts, demonstrating that culture is both a cause and an effect of this evolution.

Use of Technology

The NCTM *Standards* recommends that the elementary school curriculum "should make appropriate and ongoing use of calculators and computers." To enable future teachers to improve children's learning with the incorporation of appropriate technology, the use of calculators and computers is integrated thoroughly into the text.

- Most problem sets have a category titled **Using a Calculator,** marked with the symbol shown. Indeed, many problems in other categories are also approached best with a calculator in hand, and in many cases a calculator symbol is placed beside such problems to emphasize that calculators are a legitimate tool for problem solving.
- The special features of a calculator are highlighted in Section 3.6, "Getting the Most Out of Your Calculator." We particularly recommend the Texas Instrument's Math Explorer, but the algorithms presented can usually be adapted to any algebraic-entry calculator.
- A computer disk Numeric Surmiser/Surpriser (by Max Gerling and Nancy Taitt) comes free for adopters only. It contains user-friendly programs that allow a student to explore and conjecture, experiencing the computer as a powerful tool for creating and discovering new mathematics. The disk asks "what if" questions that extend the notion being investigated. Problems that use a program on the disk are indicated with a disk symbol as shown.

- **Logo** computer graphics are described in Appendix C. In addition, several sections in Chapters 10, 13, and 14 contain Logo examples and exercises that illustrate how the concepts of the section can be investigated in a Logo environment. Problems that use Logo are indicated by the symbol shown, and are found within the **Using a Computer** category of the problem sets.
- **The Geometer's Sketchpad** is introduced in Appendix D. The appendix shows how Sketchpad can create a dynamic range of geometric figures and how the properties of the figures can be explored wiith measurements, manipulations, and transformations. The appendix is specific to Sketchpad, but users of other computer geometry software can modify the procedures to fit the examples shown. Many of the examples, concepts, and problems of the geometry chapters are ideally suited for computer geometry software, but problems of special interest are found in the **Using a Computer** category of most problem sets in Chapters 10, 11, 12, and 13.
- The **Stat Explorer** is capable of producing pie charts, box and whisker diagrams, histograms, and other graphical displays of data, and also can compute the mean, standard deviation, and other statistical values. Problems in Chapter 8 that can take advantage of this or comparable software are marked with a computer symbol.

Cultural and Gender Diversity

In *A Call for Change: Recommendations for the Mathematical Preparation of Teachers of Mathematics,* the Mathematical Association of America (MAA) Committee on the Mathematical Education of Teachers recommends that all prospective teachers

> *"develop an appreciation of the contributions made by various cultures to the growth and development of mathematical ideas; investigate the contributions made by individuals, both female and male, and from a variety of cultures, in the development of ancient, modern, and current mathematical topics; [and] gain an understanding of the historical development of major school mathematical concepts."*

The examples and exercises in this text reflect ethnic, gender, and cultural diversity in both historic and contemporary mathematics. In particular, we have

- featured the contributions of people from around the world, both men and women, in the feature **Highlights from History;**
- used examples from the Chinese, Arabian, Egyptian, African, Maori, and other nonwestern cultures;
- included a variety of international names in exercises and examples, reminding future teachers of the growing ethnic and language diversity they will encounter in their classrooms.

CONTENT FEATURES

Problem Solving

The first section of Chapter 1, titled Some Surprising Tidbits, contains a variety of arithmetical and geometric patterns and tricks with unexpected results that our students have found both pleasing and nonthreatening. These are continued in the following problem set with exercises that students can do easily and with pleasure. The idea is to allow the students immediately to experience both success and enjoyment in doing mathematics. This trend then continues throughout the chapter as the students learn increasingly sophisticated problem-solving strategies and skills.

Number Systems

Chapters 2, 3, 5, 6, and 7 on the various number systems are replete with discussions of manipulatives and pictorial and graphical representations that promote understanding of the systems, their properties, and the various modes of computation. Exercises in the accompanying problem sets not only include routine exercises to provide conventional drill but also contain problems that provide continued practice in both individual and cooperative problem solving, reasoning, and communication.

Number Theory

Chapter 4 contains much material that is new and interesting to students. Notions of divisibility, divisors, multiples, greatest common divisors, and least common multiples are first developed via informative diagrams and then through the use of manipulatives, sets, prime factor representations, and the Euclidean algorithm. A final section deals with clock arithmetic (modular arithmetic) and contains interesting applications to secret codes and other codes such as the zip code and the book number codes in common use.

Statistics

Chapter 8 on statistics is designed not only to give the students an appreciation of the basic measures but also of the uses and misuses of statistics. This last issue is particularly important since we are confronted daily with a stream of "facts and figures" seductively intended to influence our thinking. The informed citizen needs not only to be aware of the legitimate predictive and descriptive power of statistics, but also to be wary of the way statistics can mislead.

Probability

In Chapter 9, we first study empirical probability—probability based on experience and repeated trials. This prepares the way for the subsequent study of the elements of theoretical probability of an event based on counting and other *a priori* considerations. The great surprise for the students is how closely the results agree, particularly when the number of trials is large. Of course, it is necessary to consider various methods of counting in order to compute theoretical probabilities. At the same time counting is an important topic in its own right, and we have made it accessible through the use of tree diagrams, Venn diagrams, and careful explanation of the use of the words *or* and *and.*

Geometry

The creative and inductive nature of geometric discovery is emphasized in Chapters 10, 11, 13, and 14, and students who acquired a distaste for the subject from a heavily axiomatic and deductive course will often see geometry in a positive new way. The text's approach to geometry is constructive and visual. Students are often asked to draw, cut, fold, paste, count, and so on, making geometry an experimental science. While the traditional construction and measurement tools continue to have a place, the visual and dynamic scope of geometry is enhanced with Logo and computer geometry software. Problem solving and applications permeate the geometry chapters, and sections on tilings and symmetry provide an opportunity to highlight the aesthetic and artistic aspects of geometry.

A VISUAL GUIDE

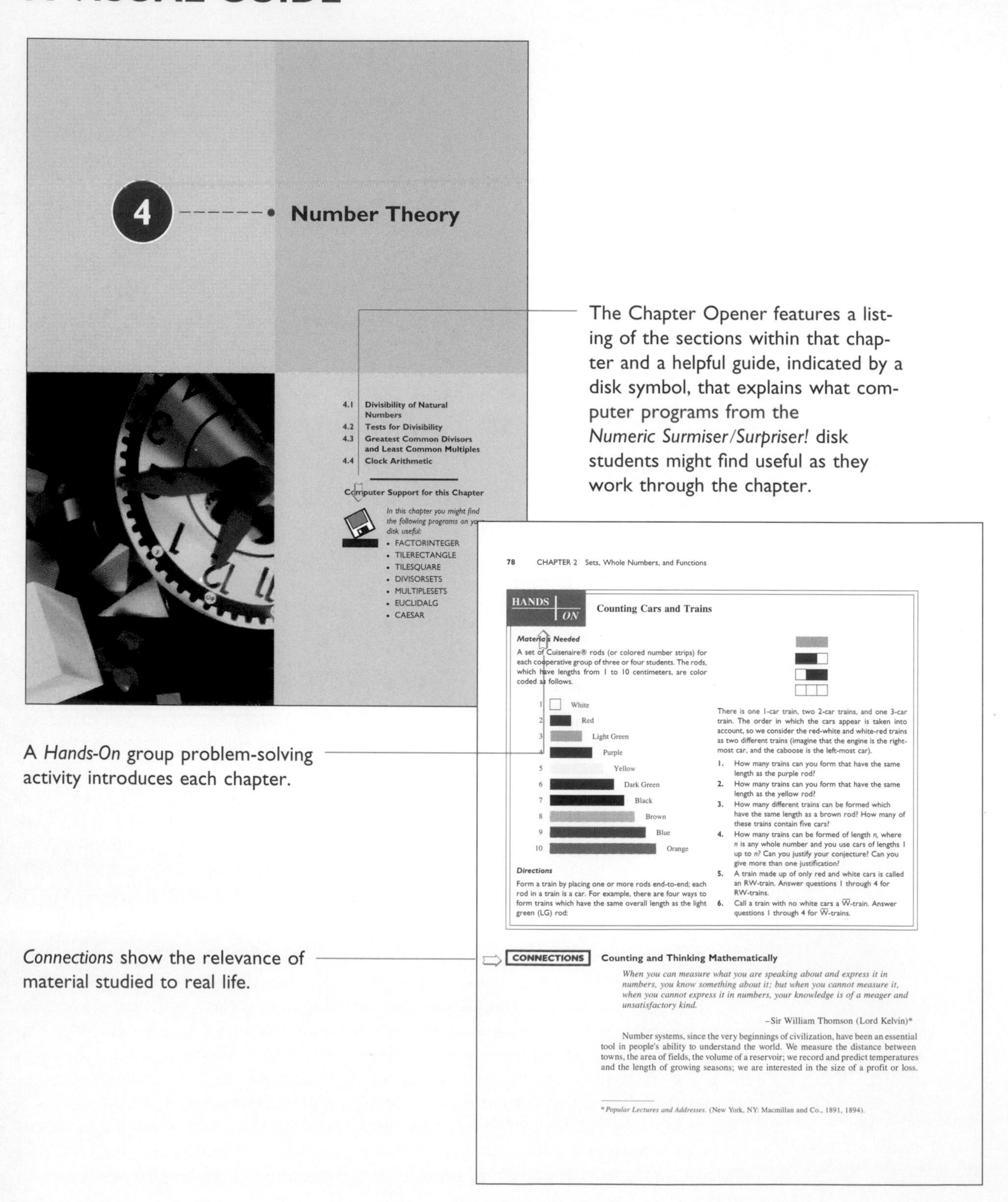

4
Number Theory
4.1 Divisibility of Natural Numbers
4.2 Tests for Divisibility
4.3 Greatest Common Divisors and Least Common Multiples
4.4 Clock Arithmetic
Computer Support for this Chapter
In this chapter you might find the following programs on your disk useful:
• FACTORINTEGER
• TILERECTANGLE
• TILESQUARE
• DIVISORSETS
• MULTIPLESETS
• EUCLIDALG
• CAESAR
78 CHAPTER 2 Sets, Whole Numbers, and Functions
HANDS ON
Counting Cars and Trains
Materials Needed
A set of Cuisenaire® rods (or colored number strips) for each cooperative group of three or four students. The rods, which have lengths from 1 to 10 centimeters, are color coded as follows.
1 White
2 Red
3 Light Green
4 Purple
5 Yellow
6 Dark Green
7 Black
8 Brown
9 Blue
10 Orange
Directions
Form a train by placing one or more rods end-to-end; each rod in a train is a car. For example, there are four ways to form trains which have the same overall length as the light green (LG) rod:
There is one 1-car train, two 2-car trains, and one 3-car train. The order in which the cars appear is taken into account, so we consider the red-white and white-red trains as two different trains (imagine that the engine is the right-most car, and the caboose is the left-most car).
1. How many trains can you form that have the same length as the purple rod?
2. How many trains can you form that have the same length as the yellow rod?
3. How many different trains can be formed which have the same length as a brown rod? How many of these trains contain five cars?
4. How many trains can be formed of length n, where n is any whole number and you use cars of lengths 1 up to n? Can you justify your conjecture? Can you give more than one justification?
5. A train made up of only red and white cars is called an RW-train. Answer questions 1 through 4 for RW-trains.
6. Call a train with no white cars a W̄-train. Answer questions 1 through 4 for W̄-trains.
CONNECTIONS
Counting and Thinking Mathematically
When you can measure what you are speaking about and express it in numbers, you know something about it; but when you cannot measure it, when you cannot express it in numbers, your knowledge is of a meager and unsatisfactory kind.
–Sir William Thomson (Lord Kelvin)*
Number systems, since the very beginnings of civilization, have been an essential tool in people's ability to understand the world. We measure the distance between towns, the area of fields, the volume of a reservoir; we record and predict temperatures and the length of growing seasons; we are interested in the size of a profit or loss.
* Popular Lectures and Addresses. (New York, NY: Macmillan and Co., 1891, 1894).

The Chapter Opener features a listing of the sections within that chapter and a helpful guide, indicated by a disk symbol, that explains what computer programs from the *Numeric Surmiser/Surpriser!* disk students might find useful as they work through the chapter.

A *Hands-On* group problem-solving activity introduces each chapter.

Connections show the relevance of material studied to real life.

Highlight From History vignettes illustrate the contributions that men and women have made to mathematics, providing cultural, historical, and personal perspectives on the development of mathematical concepts and thought.

HIGHLIGHT FROM HISTORY
Three Impossible Construction Problems

The straightedge allows us to draw a line of indefinite length through any two given points, and the compass* allows us to draw a circle with a given point as its center and passing through any given distinct second point. It then becomes a challenge to find procedures to construct a figure using only these simple tools. Many important contributions to geometry were inspired by attempts to solve the following famous problems, each of which arose in antiquity.

1. *The trisection of an angle:* Divide an arbitrary given angle into three congruent angles.

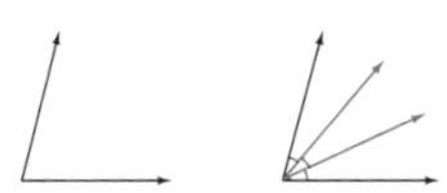

2. *The duplication of the cube:* Given a cube, construct a cube with twice the volume.

3. *The squaring of the circle:* Given a circle, construct a square of the same area as the circle.

Extensive efforts for over 2000 years failed to solve any of these problems. It was not until the early 1800s that it was shown that these problems were impossible to solve when only the straightedge and compass were allowed. It is interesting to note that methods of algebra were used to prove the impossibility of these geometric construction problems.

* The usage "the compass" is common, but some texts and authors still prefer "compasses" or even "a pair of compasses."

From The NCTM Standards features extensive excerpts from the NCTM standards to help future teachers comprehend the timeliness and relevance of mathematical ideas.

FROM THE NCTM STANDARDS **Constructing and Using Geometric Models**

Geometry gives children a different view of mathematics. As they explore patterns and relationships with models, blocks, geoboards, and graph paper, they learn about the properties of shapes and sharpen their intuitions and awareness of spatial concepts. Children's geometric ideas can be developed by having them sort and classify models of plane and solid figures, construct models from straws, make drawings, and create and manipulate shapes on a computer screen. Folding paper cutouts or using mirrors to investigate lines of symmetry are other ways for children to observe figures in a variety of positions, become aware of their important properties, and compare and contrast them. Related experiences help children avoid simplistic and misleading ideas about shapes, such as that implied by one child's observation, "This is an upside-down triangle."

SOURCE: From *Curriculum and Evaluation Standards for School Mathematics Grades K–4*, pp. 48–49. Copyright © 1989 by the National Council of Teachers of Mathematics, Inc. Reprinted by permission.

SOLUTION

Understand the problem • • • •

The numbers V, E, and F are not independent of one another. Th[e] is to uncover a formula which relates the three numbers corresponding polyhedron.

Devise a plan • • • •

Formulas are often revealed by seeing a pattern in special cases. making a table of values of V, F, and E we have a better chance to see this pattern may be. To be confident that the pattern holds for all polyhedra, we need to examine polyhedra of varied kinds.

Carry out the plan • • • •

A pentagonal pyramid, a hexagonal prism, a "house", and a truncated icosahedron are shown below. The truncated icosahedron, formed by slicing off the corners of an icosahedron to form pentagons, may look familiar; it is a common pattern on soccer balls.

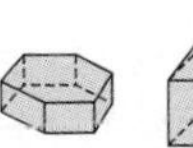
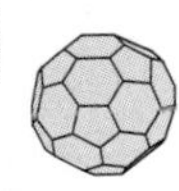

The following table lists the number of vertices, edges, and faces for these polyhedra as well as for some of the regular polyhedra depicted in Table 10.4.

Polyhedron	V	F	E
Pentagonal Pyramid	6	6	10
Hexagonal Prism	12	8	18
"House"	10	9	17
Cube	8	6	12
Tetrahedron	4	4	6
Octahedron	6	8	12
Truncated Icosahedron	60	32	90

The table reveals that the sum of the number of vertices and faces is 2 larger than the number of edges. That is, $V + F = E + 2$.

Look back • • • •

Knowing any two values of V, F, and E determine the remaining value, since the three numbers must satisfy the Euler formula $V + F = E + 2$. For example, the dodecahedron has 12 pentagonal faces. The product $5 \cdot 12$ counts

Many examples are presented in a problem-solving fashion, asking the student to independently obtain a solution that can be compared with the solution presented in the text. Solutions are frequently structured in the Pólya four-step format for easy comprehension.

868 CHAPTER 13 Geometric Transformations

WORTH READING

Symmetries of Culture

In this book we demonstrate how to use the geometric principles of crystallography to develop a descriptive classification of patterned design. Just as specific chemical assays permit objective analysis and comparison of objects, so too the description of designs by their geometric symmetries makes possible systematic study of their function and meaning within cultural contexts.

This particular type of analysis classifies the underlying structure of decorated forms; that is, the way the parts (elements, motifs, design units) are arranged in the whole design by the geometrical symmetries which repeat them. The classification emphasizes the way the design elements are repeated, not the nature of the elements themselves. The symmetry classes which this method yields, also called motion classes, can be used to describe any design whose parts are repeated in a regular fashion. On most decorated forms such repeated design, properly called pattern, is either planar or can be flattened (e.g., unrolled), so that these repeated designs can be described either as bands or strips (one-dimensional infinite) or as overall patterns (two-dimensional infinite) in a plane.

These excerpts are from the introduction to *Symmetries of Culture: Theory and Practice of Plane Pattern Analysis*, by Dorothy K. Washburn and Donald W. Crow. Nearly every page of *Symmetries of Culture* is

graced by beautiful photographs and drawings that illustrate the principles of symmetry discovered and utilized by contemporary and historic cultures from around the world.

SOURCE: From *Symmetries of Culture: Theory and Practice of Plane Pattern Analysis* by Dorothy K. Washburn and Donald W. Crow, Page ix. Copyright © 1988 by The University of Washington Press. Reprinted by permission.

border). Any glide reflection necessarily has a horizontal slide arrow and a horizontal line of reflection. However, not all combinations of these symmetries can exist simultaneously. For example, if a border pattern has both a horizontal and a vertical line of symmetry, then it necessarily has a 180° rotation symmetry (why?).

It has been shown that border patterns have just seven types of symmetry. The International Crystallographic Union designates each symmetry type with a two-symbol notation, as shown in Figure 13.17. The symbols can be assigned by the following procedure, where we view the border pattern in a horizontal position.

First symbol: m, if there is a vertical line of symmetry; 1, otherwise.
Second symbol: m, if there is a horizontal line of symmetry; g, if there is a glide reflection (but no horizontal symmetry line);
2, if there is a half-turn symmetry (but no horizontal reflection or glide reflection); 1, otherwise.

The seven symmetry types are mm, mg, $m1$, $1m$, $1g$, 12, and 11. The m refers to *mirror*, g to *glide*, and 2 to a *half-turn*.

The seven types of border patterns, and their corresponding Crystallographic Union symbol, are shown in Figure 13.17.

Worth Reading excerpts are taken from the kinds of mathematical literature that have special meaning to the classroom teacher.

Into The Classroom features give students insight on the use of ideas now being learned in classes they will subsequently teach.

640 CHAPTER 10 Geometric Figures

INTO THE CLASSROOM

Activity-based Learning and the van Hiele Levels

From kindergarten onward, geometry is learned best through hands-on activities. A successful teacher will make good advantage of the enjoyment children experience when working with colored paper, straws, string, crayons, toothpicks, and other tangible materials. Children learn geometry by doing geometry as they construct two- and three-dimensional shapes, combine their shapes to create attractive patterns, and build interesting space figures using plane shapes. By its nature, informal geometry provides unlimited opportunities to construct shapes, designs, and structures that capture a child's interest.

According to pioneering research of the van Hieles in the late 1950s, the knowledge children construct for themselves through hands-on activities is essential to learning geometry. Dr. Pierre van Hiele and his late wife Dr. Dieke van Hiele-Geldof, both former mathematics teachers in the Netherlands, theorized that learning geometry progresses through five "levels," which can be described briefly as follows.

Level 0–Recognition of shape

Children recognize shapes "holistically." Only the overall appearance of a figure is observed, with no attention given to the component parts of the figure. For example, a figure with three curved sides would likely be identified as a traingle by a child at Level 0. Similarly, a square tilted point downward may not be recognized as a square.

Level 1–Analysis of single shapes

Children at Level 1 are cognizant of the component parts of certain figures. For example, a rectangle has four straight sides which meet at "square" corners. However, at Level 1 the interrelationships of figures and properties is not understood.

Level 2–Relationships among shapes

At Level 2 children understand how common properties create abstract relationships among figures. For example, a square is both a rhombus and a rectangle. Also, simple deductions can be made about figures, using the analytic abilities acquired at Level 1.

Level 3–Deductive reasoning

The student at Level 3 views geometry as a formal mathematical system, and can write deductive proofs.

Level 4–Geometry as an axiomatic system

This is the abstract level, reached only in high level university courses. The focus is on the axiomatic foundations of a geometry, and no dependence is placed on concrete or pictorial models.

Ongoing research supports the thesis that students learn geometry by progressing through the van Hiele levels. This text–by means of hands-on activities, and examples and problems that require constructions and drawings–promotes the spirit of the van Hiele approach. However, it is the elementary classroom teacher who must bring geometry to life for his or her students, creating activities that support each child's progression through the first three van Hiele levels.

Figure 10.3
A segment $\overline{AB}$, its length AB, congruent segments, and the midpoint M of $\overline{AB}$

With their engaging puzzles and teasers that involve an element of surprise and humor, *Just For Fun* sections show the lighter side of mathematical problem solving.

JUST FOR FUN

Arranging Points at Integer Distances

It is easy to find threee points A, B, C, not all on one line, that are at integer distance from one another. The equilateral triangle with three sides of length one is the smallest example. Four points F, O, U, R, no three of which are collinear, determine the six segments as shown.

Can you arrange the four points so that all six segments are of integer length? Try experimenting with six straws, cut to lengths of 2″, 2″, 3″, 4″, 4″, 4″. The following arrangement of six points is remarkable because all 15 segments between pairs of points have integer length.

COOPERATIVE *Investigation* ***Diffy***

The process in this activity is sometimes called *Diffy*. The name comes from the process of taking successive differences of whole numbers, and the activity provides an interesting setting for practicing skills in subtraction. The activity can be done by hand, with a calculator, or by calling up DIFFY on your computer disc.

Directions

Step 1. Make an array of circles as shown and choose four whole numbers to place in the top four circles.

Step 2. In the first three circles of the second row write the differences of the numbers above and to the right and left of the circle in question, always being careful to subtract the smaller of these two numbers from the larger. In the fourth circle of the second row place the difference of the numbers in the first and fourth circles in the preceding row, again always subtracting the smaller number from the larger.

Step 3. Repeat Step 1 to fill in successive rows of circles in the diagram. You may stop if you obtain a row of zeros.

Step 4. Repeat steps 1, 2, and 3 several times, and each time, start with different numbers.

Questions

1. Do you think the process will always stop?
2. Can you find four numbers such that the process terminates at the first step? The second step? The third step? Try several sets of starting numbers.
3. On the basis of your work so far, what do you guess is the largest number of steps needed for the process to stop?
4. Can you find four starting numbers such that the process requires eight steps to reach termination?
5. Try DIFFY with the starting numbers 17, 32, 58, and 107.

Stressing the importance and practicality of cooperative learning, these *Cooperative Investigations* provide activities and open-ended problems for small group work.

PROBLEM SET 2.3

Understanding Concepts

1. Let $A = \{\text{apple, berry, peach}\}$, $B = \{\text{lemon, lime}\}$, $C = \{\text{lemon, berry, prune}\}$.
   (a) Find (i) $n(A \cup B)$, (ii) $n(A \cup C)$, (iii) $n(B \cup C)$.
   (b) In which case is the number of elements in the union *not* the sum of the elements in the individual sets?
2. Let $n(A) = 5$, $n(B) = 8$, and $n(A \cup B) = 10$. What can you say about $n(A \cap B)$?
3. Let $n(A) = 4$ and $n(A \cup B) = 8$.
   (a) What are the possible values of $n(B)$?
   (b) If $A \cap B = \emptyset$, what is the only possible value of $n(B)$?
4. Draw number strips (or rods) to illustrate
   (a) $4 + 6 = 10$;
   (b) $2 + 8 = 8 + 2$; and
   (c) $3 + (2 + 5) = (3 + 2) + 5$.
5. Make up a word problem which uses the set model of addition to illustrate $30 + 28$.
6. Make up a word problem which uses the measurement model of addition to illustrate $18 + 25$.

970 Appendix B

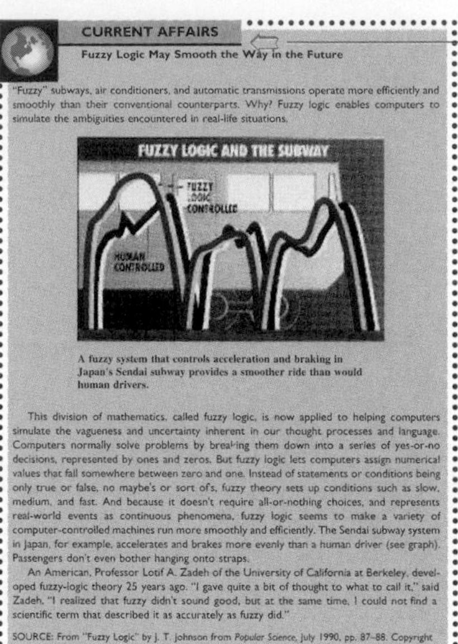
CURRENT AFFAIRS

Fuzzy Logic May Smooth the Way in the Future

"Fuzzy" subways, air conditioners, and automatic transmissions operate more efficiently and smoothly than their conventional counterparts. Why? Fuzzy logic enables computers to simulate the ambiguities encountered in real-life situations.

A fuzzy system that controls acceleration and braking in Japan's Sendai subway provides a smoother ride than would human drivers.

This division of mathematics, called fuzzy logic, is now applied to helping computers simulate the vagueness and uncertainty inherent in our thought processes and language. Computers normally solve problems by breaking them down into a series of yes-or-no decisions, represented by ones and zeros. But fuzzy logic lets computers assign numerical values that fall somewhere between zero and one. Instead of statements or conditions being only true or false, no maybe's or sort of's, fuzzy theory sets up conditions such as slow, medium, and fast. And because it doesn't require all-or-nothing choices, and represents real-world events as continuous phenomena, fuzzy logic seems to make a variety of computer-controlled machines run more smoothly and efficiently. The Sendai subway system in Japan, for example, accelerates and brakes more evenly than a human driver (see graph). Passengers don't even bother hanging onto straps.

An American, Professor Lotif A. Zadeh of the University of California at Berkeley, developed fuzzy-logic theory 25 years ago. "I gave quite a bit of thought to what to call it," said Zadeh. "I realized that fuzzy didn't sound good, but at the same time, I could not find a scientific term that described it as accurately as fuzzy did."

SOURCE: From "Fuzzy Logic" by J. T. Johnson from *Popular Science*, July 1990, pp. 87–88. Copyright © 1990 Times Mirror Magazines Inc. Distributed by L. A. Times Syndicate. Reprinted with permission from Popular Science Magazine.

If-then Statements

Another way to combine two statements p and q is to form the **conditional statement** "if p, then q," symbolized by $p \rightarrow q$. The statement is true if q *must* be true when p is true. Statement p is called the **hypothesis** of the conditional statement $p \rightarrow q$, and q is called the **conclusion.**

Current Affairs are examples of recent advancements in mathematics, often demonstrating the significant role mathematics plays in today's world and even in our personal lives.

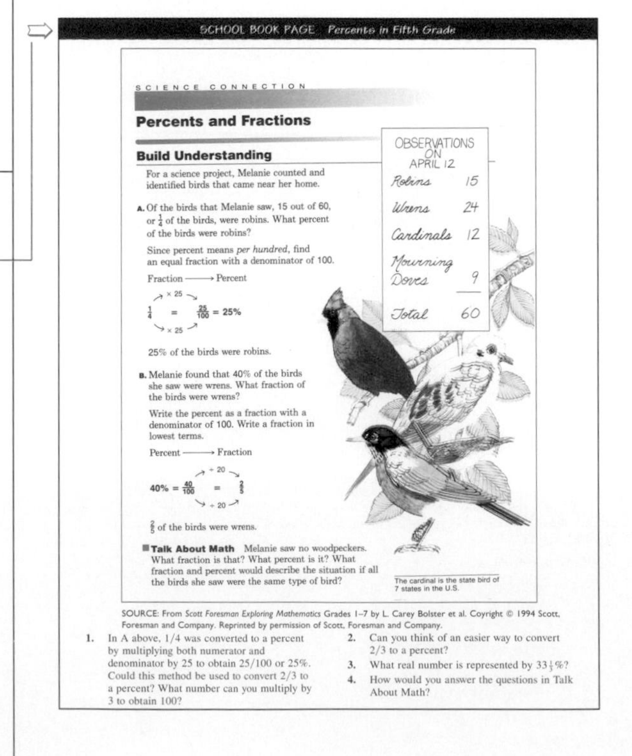
476 CHAPTER 7 Decimals and Real Numbers

SCHOOL BOOK PAGE *Percents in Fifth Grade*

SCIENCE CONNECTION

Percents and Fractions

Build Understanding

For a science project, Melanie counted and identified birds that came near her home.

A. Of the birds that Melanie saw, 15 out of 60, or $\frac{1}{4}$ of the birds, were robins. What percent of the birds were robins?

Since percent means *per hundred*, find an equal fraction with a denominator of 100.

Fraction ⟶ Percent

$$\frac{1}{4} \underset{\times 25}{\overset{\times 25}{=}} \frac{25}{100} = 25\%$$

25% of the birds were robins.

B. Melanie found that 40% of the birds she saw were wrens. What fraction of the birds were wrens?

Write the percent as a fraction with a denominator of 100. Write a fraction in lowest terms.

Percent ⟶ Fraction

$$40\% = \frac{40}{100} \underset{\div 20}{\overset{\div 20}{=}} \frac{2}{5}$$

$\frac{2}{5}$ of the birds were wrens.

Talk About Math Melanie saw no woodpeckers. What fraction is that? What percent is it? What fraction and percent would describe the situation if all the birds she saw were the same type of bird?

The cardinal is the state bird of 7 states in the U.S.

SOURCE: From *Scott Foresman Exploring Mathematics* Grades 1–7 by L. Carey Bolster et al. Copyright © 1994 Scott, Foresman and Company. Reprinted by permission of Scott, Foresman and Company.

1. In A above, 1/4 was converted to a percent by multiplying both numerator and denominator by 25 to obtain 25/100 or 25%. Could this method be used to convert 2/3 to a percent? What number can you multiply by 3 to obtain 100?
2. Can you think of an easier way to convert 2/3 to a percent?
3. What real number is represented by $33\frac{1}{3}\%$?
4. How would you answer the questions in Talk About Math?

Taken from actual elementary textbooks, *School Book Page* features show how the topics presented in this text are made meaningful to school children. They also show that the topics in the text are central to the elementary school curriculum. These pages are also followed by stimulating, thought-provoking questions.

The *Epilogue* consists of a brief concluding essay that discusses the importance of the material just covered and features a summarizing overview.

The *Classic Conundrum* is a final, stimulating puzzle that closes each chapter, helping to sustain student interest.

Chapter Summaries enumerate and review the chapter's key concepts.

Chapter Review Exercises help students to review chapter material; all are answered at the back of the text.

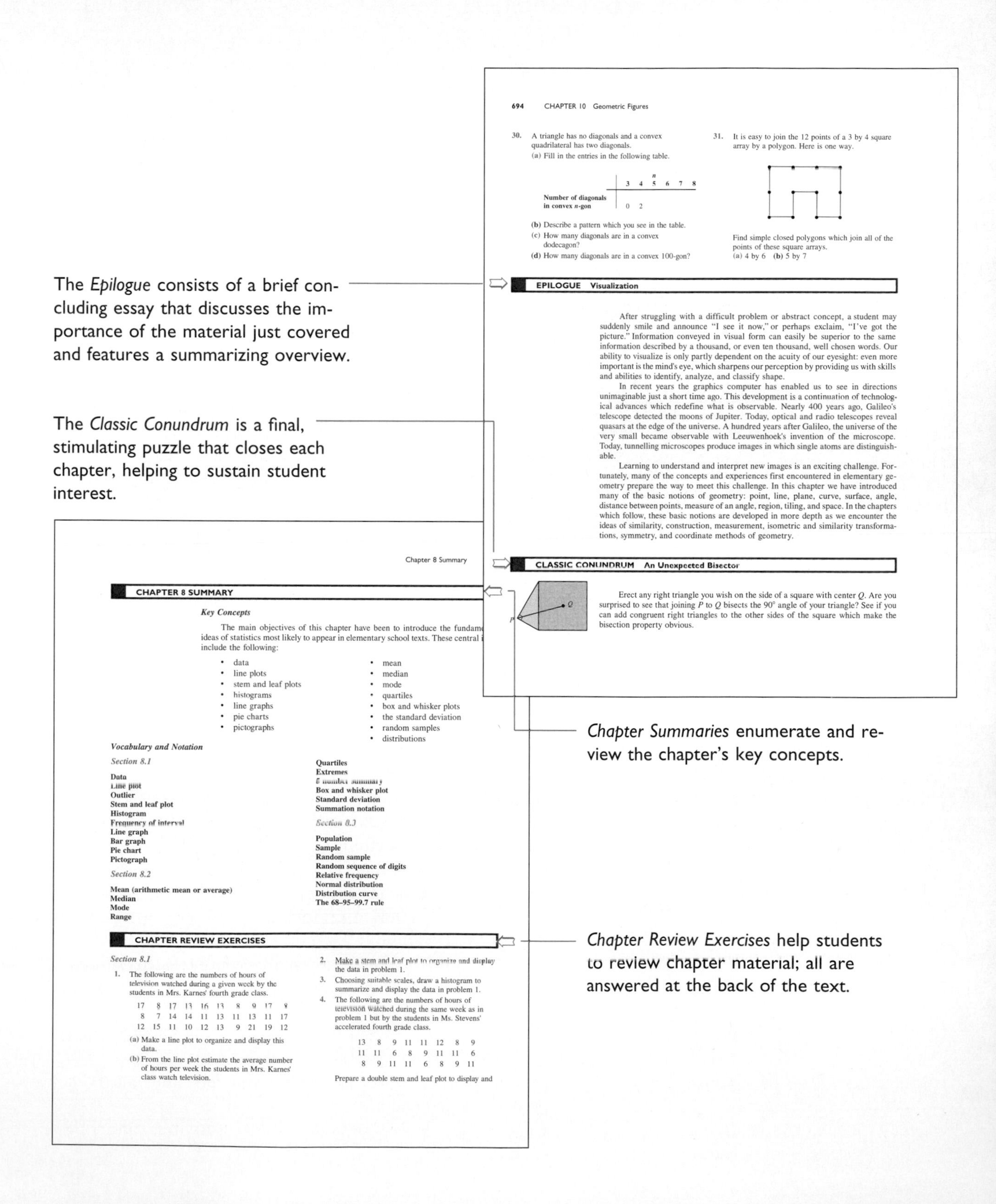

694 CHAPTER 10 Geometric Figures

30. A triangle has no diagonals and a convex quadrilateral has two diagonals.
(a) Fill in the entries in the following table.

n	3	4	5	6	7	8
Number of diagonals in convex n-gon	0	2				

(b) Describe a pattern which you see in the table.
(c) How many diagonals are in a convex dodecagon?
(d) How many diagonals are in a convex 100-gon?

31. It is easy to join the 12 points of a 3 by 4 square array by a polygon. Here is one way.

Find simple closed polygons which join all of the points of these square arrays.
(a) 4 by 6 (b) 5 by 7

EPILOGUE Visualization

After struggling with a difficult problem or abstract concept, a student may suddenly smile and announce "I see it now," or perhaps exclaim, "I've got the picture." Information conveyed in visual form can easily be superior to the same information described by a thousand, or even ten thousand, well chosen words. Our ability to visualize is only partly dependent on the acuity of our eyesight: even more important is the mind's eye, which sharpens our perception by providing us with skills and abilities to identify, analyze, and classify shape.

In recent years the graphics computer has enabled us to see in directions unimaginable just a short time ago. This development is a continuation of technological advances which redefine what is observable. Nearly 400 years ago, Galileo's telescope detected the moons of Jupiter. Today, optical and radio telescopes reveal quasars at the edge of the universe. A hundred years after Galileo, the universe of the very small became observable with Leeuwenhoek's invention of the microscope. Today, tunnelling microscopes produce images in which single atoms are distinguishable.

Learning to understand and interpret new images is an exciting challenge. Fortunately, many of the concepts and experiences first encountered in elementary geometry prepare the way to meet this challenge. In this chapter we have introduced many of the basic notions of geometry: point, line, plane, curve, surface, angle, distance between points, measure of an angle, region, tiling, and space. In the chapters which follow, these basic notions are developed in more depth as we encounter the ideas of similarity, construction, measurement, isometric and similarity transformations, symmetry, and coordinate methods of geometry.

CLASSIC CONUNDRUM An Unexpected Bisector

Erect any right triangle you wish on the side of a square with center Q. Are you surprised to see that joining P to Q bisects the 90° angle of your triangle? See if you can add congruent right triangles to the other sides of the square which make the bisection property obvious.

Chapter 8 Summary

CHAPTER 8 SUMMARY

Key Concepts

The main objectives of this chapter have been to introduce the fundam[ental] ideas of statistics most likely to appear in elementary school texts. These central i[deas] include the following:

- data
- line plots
- stem and leaf plots
- histograms
- line graphs
- pie charts
- pictographs
- mean
- median
- mode
- quartiles
- box and whisker plots
- the standard deviation
- random samples
- distributions

Vocabulary and Notation

Section 8.1

Data
Line plot
Outlier
Stem and leaf plot
Histogram
Frequency of interval
Line graph
Bar graph
Pie chart
Pictograph

Section 8.2

Mean (arithmetic mean or average)
Median
Mode
Range
Quartiles
Extremes
5 number summary
Box and whisker plot
Standard deviation
Summation notation

Section 8.3

Population
Sample
Random sample
Random sequence of digits
Relative frequency
Normal distribution
Distribution curve
The 68–95–99.7 rule

CHAPTER REVIEW EXERCISES

Section 8.1

1. The following are the numbers of hours of television watched during a given week by the students in Mrs. Karnes' fourth grade class.

17	8	17	13	16	13	8	9	17	8
8	7	14	14	11	13	11	13	11	17
12	15	11	10	12	13	9	21	19	12

(a) Make a line plot to organize and display this data.
(b) From the line plot estimate the average number of hours per week the students in Mrs. Karnes' class watch television.

2. Make a stem and leaf plot to organize and display the data in problem 1.
3. Choosing suitable scales, draw a histogram to summarize and display the data in problem 1.
4. The following are the numbers of hours of television watched during the same week as in problem 1 but by the students in Ms. Stevens' accelerated fourth grade class.

13	8	9	11	11	12	8	9	
11	11	6	8	9	11	11	6	
8	9	11	11	6	8	9	11	

Prepare a double stem and leaf plot to display and

SPECIAL PEDAGOGICAL FEATURES

The teacher of mathematics should be aware of the historical development of mathematics, have some knowledge of the principal contributors to mathematics, and realize that mathematics continues to be a lively area of research. Moreover, the teacher must always be alert to ways to foster mathematical power in the elementary classroom. This text contains a number of features that prospective teachers will find as valuable resources.

- **Highlights from History** are vignettes that illustrate the contributions individual men and women have made to mathematics, providing a cultural, historical, and personal perspective on the development of mathematical concepts and thought.
- **Into the Classroom** provides insights on teaching the topics in this text to elementary school children; often these tips come from current elementary school teachers.
- **Schoolbook Pages,** taken from actual elementary textbooks, show how topics from this text are made meaningful to schoolchildren. They also show that the topics in the text are central to the elementary school curriculum.
- **Current Affairs** are examples of recent advancements in mathematics, often demonstrating the important role mathematics has in today's world and even in our personal lives.
- **Worth Reading** presents excerpts from mathematical literature that have relevance to the classroom teacher.
- **Just for Fun** shows the lighter side of mathematical problem solving, with engaging puzzles and teasers that have an element of surprise and humor.
- **Classic Conundrum** contains a final, stimulating puzzle at the end of each chapter to help sustain student interest.

Chapter Structure

Each chapter is structured consistently and meaningfully according to the following pattern:

Chapter opener	A **Hands On** class activity introduces the topic of the chapter. This is followed by **Connections,** a preview of the topics and main ideas in the sections that follow.
Examples	Examples are presented in a problem-solving mode, asking the reader to independently obtain a solution that can be compared with the solution presented in the text. Solutions are frequently structured in the Pólya four-step format.
Figures and tables	A large number of figures visually reinforce the concepts, problems, and solutions. Many figures, and all tables, include informative captions.
From the NCTM Standards	Extensive excerpts from the *Standards* help students understand the timeliness and relevance of topics.
Think clouds	Marginal notes serve as quick reminders and clarify key points in a deduction.
Cooperative learning	**Cooperative Investigations** provide activities and open-ended problems for small group work.

Categorized problem sets	Section problems are grouped in the following categories: *Understanding Concepts* Provides drill and reinforces basic concepts. *Thinking Critically* Provides problem-solving practice related to the topic of the section. *Thinking Cooperatively* Provides problem-solving experiences for a cooperative small group; most of the problems in the *Thinking Critically* category are also suited to group attack. *Making Connections* Provides problems that apply the concepts of the section to the solution of real-life problems. *Communicating* Provides students with opportunities to write mathematics and investigate mathematics as a language. *Using a Calculator* Provides problems that are best solved with a calculator. *Using a Computer* Provides experiences with the programs on the disk that accompanies this text and with other available software. *For Review* Provides problems that review earlier material.
Epilogue	A brief concluding essay discusses the importance of the material just covered, and provides a summarizing overview.
End-of-Chapter Material	Each chapter closes with a Chapter Summary, Key Concepts, Vocabulary and Notation, Chapter Review Problems, and a Chapter Test.
Answers to Selected Problems	Problems with a colored problem number are answered at the back of the text. The answered problems help students check their understanding and provide an extensive collection of worked examples.

Lexicon, Pronunciation Guide, and Index

- Mathematical Lexicon — A brief mathematical lexicon is placed just before the index.
- Index and Pronunciation Guide — As an aid to the correct pronunciation of some mathematical terms and names of historical figures, the index includes a phonetic pronunciation guide.

COURSE FLEXIBILITY

Course Options

This text contains ample material for at least two semester courses. At Washington State University, elementary education majors are required to take two three-semester-hour courses, with an elective third course available particularly suited to the needs of upper elementary and middle school teachers. This text is used in all three courses, with some augmentation in the third course. The suggestions below are for semester courses, but instructors should have little difficulty selecting material that fits the coverage needed for courses in a quarter system.

- A First Course: Problem Solving and Number Systems
  Cover Chapters 1 through 7, with optional coverage of Appendix B:

Logic and Mathematical Reasoning. Our own first course devotes at least five weeks to Chapter 1. The problem-solving skills and enthusiasm developed in this chapter make it possible to move through most of the topics in Chapters 2 through 6 more quickly than usual. There is considerable latitude in which topics an instructor chooses to give lighter or heavier emphasis.

- A Second Course: Statistics, Probability, and Basic Geometry
  Cover Chapters 8 through 12, with optional inclusion of Logo (Appendix C) and/or computer geometry software. (Appendix D gives a brief introduction to *The Geometer's Sketchpad*.)
- An Alternative Second Course: Informal Geometry
  Cover Chapters 10 through 14, with optional inclusion of Logo and/or computer geometry software.

Once the basic notions and symbolism of geometry have been covered in Sections 10.1 and 10.2, and at least some of the fundamental ideas of congruence and similarity have been introduced from Chapter 11, the remaining chapters in geometry can be taken up in any order. Section 10.5 on Figures in Space should receive some coverage before taking up surface area and volume in Section 12.4.

SUPPLEMENTS

For the Instructor

In addition to the items mentioned below, a number of videotapes are provided for instructors. Core manipulative kits, calculators, and other valuable learning tools including a number of activity books from the well-organized *Good Year Books* will be available with an adoption. Special arrangements can also be made with your local HarperCollins College representative to obtain pertinent Scott Foresman educational materials. For information contact your local HarperCollins College representative.

- **Instructor's Complete Solutions Manual** includes solutions to all of the exercises and features that include exercises.
- **Instructor's Guide** includes teaching tips, objectives, chapter-by-chapter comments, and blackline masters.
- **Printed Test Bank and Forms** are available, with three forms per chapter, including true/false, multiple-choice, and open-ended questions.
- **HarperCollins Test Generator/Editor for Mathematics with QuizMaster** Available in IBM and Macintosh versions, both the Test Generator and the Editor are fully networkable. The Test Generator can be used to select questions by section or chapter for ready-made tests. The Editor enables instructors to edit any preexisting data or to create their own questions. The software is algorithm driven, allowing different constants to be generated while maintaining a specific problem type. This feature provides a large number of available test or quiz items in either multiple-choice or free-response formats. The system furnishes printed graphs and correct mathematical symbols. **QuizMaster** enables instructors to create tests and quizzes from the Test Generator/Editor and to save them on disks so they can be used by students on a stand-alone desk-top computer or a network. QuizMaster then grades the test or quiz to allow instructors

to create reports on either an individual or class basis. **IBM** 0–673–55818–5; **Macintosh** 0–673–55819–3.

- **GraphExplorer** With this sophisticated software students can graph rectangular, conic, polar, and parametric equations; zoom; transform functions; and experiment with families of equations quickly and easily.
- **GeoExplorer** With this computer disk, free to adopters, students can draw, measure, and transform geometric figures to illustrate and explore geometric postulates, definitions, and theorems. Teachers will appreciate GeoExplorer when it is time to demonstrate complex figures, theorems, and properties.
- **StatExplorer** A reproducible disk free to adopters, it enables students to enter data in a spreadsheet format, then build bar and circle charts, line and scatter graphs, histograms, and box plots. They can perform general statistical analysis on any data set or group of data sets. Other features include mathematical modeling, simulation, and frequency tables.

For the Student

- **Hands On Activity Manual** contains cooperative activities, calculator activities, data collection projects, writing assignments, drawing assignments, additional exercises, computer activities, and about fifty material cards.
- A **Student's Solution Manual** (by Charlotte Lewis and Allen Davis) contains solutions to the problems that are answered at the back of the book.
- **"Numeric Surmiser Surpriser!" Utility Disk** (by Max Gerling and Nancy Taitt) contains fifteen programs that step students through a process in a way that promotes understanding and encourages conjecturing.
- **GraphExplorer, GeoExplorer, StatExplorer** Students will be given an opportunity to use these programs with a department's adoption of *Mathematical Reasoning for Elementary Teachers*. Contact your local HarperCollins College representative for details. See description given above.

ACKNOWLEDGMENTS

We want to thank several individuals for their help in bringing this text into being. We are indebted to Carol Benson at Illinois State University for producing the Logo appendix and the Logo discussions and exercises in Chapters 10, 11, 13, and 14. Similarly, we are indebted to Nancy Taitt and Max Gerling at Eastern Illinois University for programming the innovative interactive computer disk, ***Numeric Surmiser/Surpriser.*** A special thanks goes to LuJane Herbert, who patiently converted our handwritten copy to typescript. We also thank two of our graduate students, Lani Shipley and Dorothy Kerzel, who worked out answers to the nearly 2000 problems. We especially want to thank the staff at HarperCollins for their many helpful suggestions, for their attention to detail, and for their constant encouragement. We especially thank George Duda, our acquisitions editor; Laurie Golson and Meg Malde-Arnosti, our developmental editors; Ginny Guerrant, our project editor; and Linda Youngman, director of development. All these individuals have materially and continuously contributed unstintingly of their imaginations, time, and talent to help make this a better text. Finally, we want to thank all of the reviewers listed below who carefully read the various drafts of the manuscript, catching errors and making numerous suggestions that have greatly improved our effort.

Richard Anderson-Sprecher
University of Wyoming

James E. Arnold
University of Wisconsin-Milwaukee

James K. Bidwell
Central Michigan University

James R. Boone
Texas A & M University

Peter Braunfeld
University of Illinois-Urbana

Louis J. Chatterley
Brigham Young University

Phyllis Chinn
Humboldt State University

Lynn Cleary
University of Maryland

Lynn D. Darragh
San Juan College

Allen Davis
Eastern Illinois University

Gary A. Deatsman
West Chester University

Stephen Drake
Northwestern Michigan College

Joseph C. Ferrar
Ohio State University

Marjorie A. Fitting
San Jose State University

Grace Peterson Foster
Beaufort County Community College

Fay Jester
Penn State University

Wilburn C. Jones
Western Kentucky University

Charlotte K. Lewis
University of New Orleans

Jennifer Luebeck
Sheridan College

Eldon L. Miller
University of Mississippi

F. A. Norman
University of North Carolina-Charlotte

Anthony Piccolino
Montclair State College

Jane M. Rood
Eastern Illinois University

Lisa M. Scheuerman
Eastern Illinois University

Carol J. Steiner
Kent State University

Many of the best aspects of the book are due to others; we assume all responsibility for the weaknesses and errors that may unfortunately remain. We hope these are few in number, and earnestly request that users of this text write or call with suggestions for improvements and corrections of errors.

We can be reached at the Department of Pure and Applied Mathematics, Washington State University, Pullman, WA 99164-3113; 509-335-3134 or 509-335-3161. Our e-mail addresses are CTL@ODIN.MATH.NAU.edu and DETEMPLE @DELTA.MATH.WSU.edu.

C.T.L.
D.W.D.

ABOUT THE AUTHORS

Both Calvin Long and Duane DeTemple have been extensively involved in mathematics education throughout their careers. They have taught the content courses for elementary teachers at Washington State University for many years. In addition, they have served as invited speakers at international, national, and regional mathematics education meetings; conducted extensive in-service programs for elementary, middle school, and high school teachers; taught in and directed numerous summer institutes designed for teachers; served as educational consultants to publishers, to the National Science Foundation, to the National Assessment of Educational Progress, to the State Superintendent of Public Instruction, and other organizations; and served on numerous committees of the National Council of Teachers of Mathematics and the Mathematical Association of America. Calvin Long also served on NCTM and MAA committees that drew up guidelines for the preparation of teachers of mathematics.

To The Student

You may be wondering what to expect from a college course in mathematics for prospective elementary school teachers. Will this course simply repeat arithmetic that you already know, or will the subject matter be new and interesting? Will any attention be given to methods of teaching mathematics to elementary school children? We'll try to answer some of your questions here, and give an orientation to the text that we hope you find useful.

Since you are preparing to teach elementary school, you must be prepared to teach all of the subject areas of the curriculum. This includes teaching mathematics, and the purpose of this text is to prepare you to meet this challenge. Some of you may think that you already know enough mathematics to do the job, but you should be aware that ideas about both the content and methods of teaching school mathematics have changed dramatically in recent years. Here we are concerned primarily with mathematics content, but the material is presented in a way that suggests effective approaches to teaching in your own classroom. Often you will discover methods that differ markedly from what you may have encountered in your school experience.

The NCTM Standards

In 1989 the National Council of Teachers of Mathematics (NCTM) published a document entitled *Curriculum and Evaluation Standards for School Mathematics* (the *Standards*), which was followed in 1991 by the publication of *Professional Standards for Teaching Mathematics*. These two documents have charted a new direction for school mathematics that directly affects what and how you will be expected to teach in the years ahead. In particular, this text is written as a direct response to the *Standards* to enable you to be as up to date as possible when you enter your professional career.

The first four of the *Standards* emphasize that the mathematics you should teach is much more than just imparting skills in arithmetic. These standards are as follows:

- Mathematics as Problem Solving
- Mathematics as Communication

- Mathematics as Reasoning
- Mathematics as Connections

Children need to develop skills in thinking critically and in writing and in speaking knowledgeably about quantitative and logical situations. They also need to understand the connections between the various parts of mathematics and between mathematics and the real world. Of course, if children are to be taught these skills they must be taught by teachers possessing the same understandings, and this is the primary emphasis of this text.

Problem Solving and Mathematical Reasoning

The entire first chapter of this text is devoted to developing skills in problem solving and critical thinking, and this theme is then continued throughout the entire book. At first, problem solving may seem daunting, but as you gain experience and begin to acquire an arsenal of strategies, you will become increasingly comfortable and will begin to find the challenge of solving a unique problem stimulating and even fun. Quite often, and much to their surprise, this has been the experience of students in our classes as they successfully match wits with challenging problems and gain insight that leads to even more success.

You should not expect to see instantly to the heart of a problem, or to know immediately how it can be solved. This text contains many exercises that check your understanding of basic concepts and build basic skills, but you will continually encounter problems characterized by the following couplet.

Problems worthy of attack
prove their worth by hitting back.

—*Piet Hein*

These problems are not unreasonably hard (indeed, many would be suitable with only a minor modification for use in classes you will subsequently teach). However, these problems do require thought. Expect to try a variety of approaches, be willing to discuss possibilities with your classmates, and form a study group to engage in cooperative problem solving. Don't be afraid to try and perhaps fail, but then try again. This is the way mathematics is done, even by professionals, and as you gain experience you will increasingly experience the real pleasure of success. Also, you will greatly improve your thinking and problem-solving skills if you take time to write carefully worded solutions that explain your methods and reasoning. Similarly, it will help to engage in verbal mathematical discourse with your instructor and with other students. Finally, remember as little as possible, but be able to figure out as much as possible. Mechanical skills learned by rote without understanding are soon forgotten and guarantee failure, both for you now, and for your students later. Conversely, the ability to think creatively makes it more likely that the task can be successfully completed.

How to Read This Book

Mathematics is not a spectator sport.

Learning is an inside job.

No mathematics textbook can be read passively. To understand the concepts and to benefit from the examples, you must be an active participant in a dialog with the text.

Often this means you need to check a calculation, make a drawing, take a measurement, construct a model, or use a calculator or computer. If you first attempt to answer questions raised in the examples on your own, the solutions written in the text will be more meaningful and useful than they would be without your personal involvement.

Many of the problems are fully or partially answered in the back of the book, including all of the Chapter Review problems and Chapter Test problems. This gives you an additional source of worked examples, but again you will benefit most fully by attempting to solve the problems on your own (or in a study group) before you check your reasoning by looking up the answer provided in the text. Other special features are described earlier in the Preface and the Visual Guide for this book.

Why Do I Need to Know This?

Teachers at any grade level, including the primary grades, must be well prepared in the mathematics content taught throughout the K–8 curriculum. The mathematics concepts of even the early grades are in reality more sophisticated than many prospective teachers realize, and the concepts must be taught in a sound manner that prepares children for what follows in the later grades. Moreover, the knowledgeable teacher can better respond to the individual needs of each student, and within a single classroom students are sure to exhibit a wide range of mathematics achievement. Finally, elementary school teachers are generally certified to teach all of the K–8 grades, and it is often the case that some initial assignments and subsequent reassignments will be at various grades.

The Enjoyment of Mathematics

In his *Scrap Book of Elementary Mathematics,* W.F. White aptly penned the following:

> *The beautiful has its place in mathematics. For here are beautiful theorems, proofs, and processes whose perfection of form has made them classic. He must be a practical man who can see no poetry in mathematics!*

There is a poetry and beauty in mathematics and it seems to us that every student deserves to be taught by a person who shares that point of view. Students taught by knowledgeable and enthusiastic teachers are far more apt to experience success than when taught by teachers who dislike the subject. This text contains a multitude of unexpected and interesting results designed to pique your curiosity and to tickle your fancy. As you use this text, it is our hope that you will often be struck by an unexpected and remarkable pattern or result that gives you a moment of pleasure and perhaps even an occasion to smile.

1 Thinking Critically

Computer Support for this Chapter

In this chapter you might find the following programs on your disk useful:

- SQDIGSUM
- PALINDROME
- KAPREKAR
- COLLATZ

HANDS ON

One Sided Paper

Materials Needed

1. Two strips of adding machine tape about 2 feet long for each student.
2. Scotch tape and scissors for each student.

Directions

Step 1. Give each strip a half twist and tape its two ends together to form a loop as shown. These two loops are called **Möbius strips.**

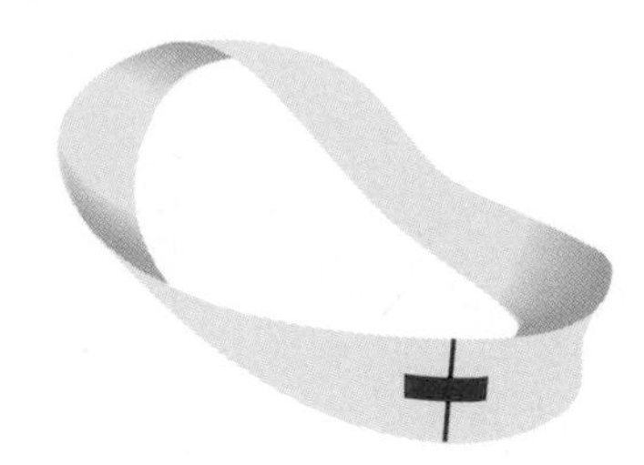

Step 2. Draw a line down the middle of one of the strips stopping only when there is a unique and compelling reason to do so. How many sides does the strip have?

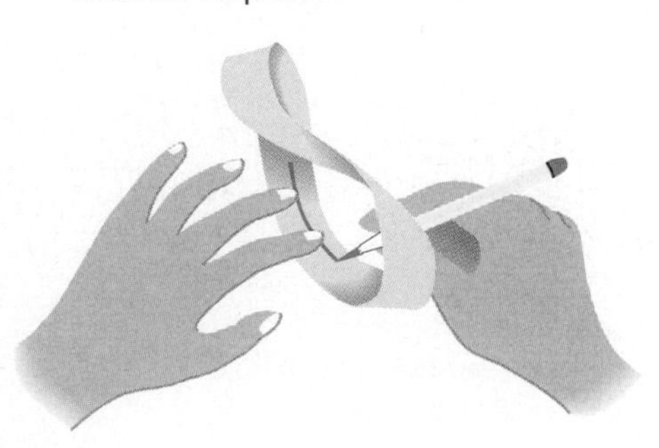

Step 3. Cut a notch out of one edge of the strip used in Step 2 and two notches out of the other edge as shown. Decide if these notches are really cut out of opposite edges of the strip. (*Hint:* Think of what was done in Step 2.)

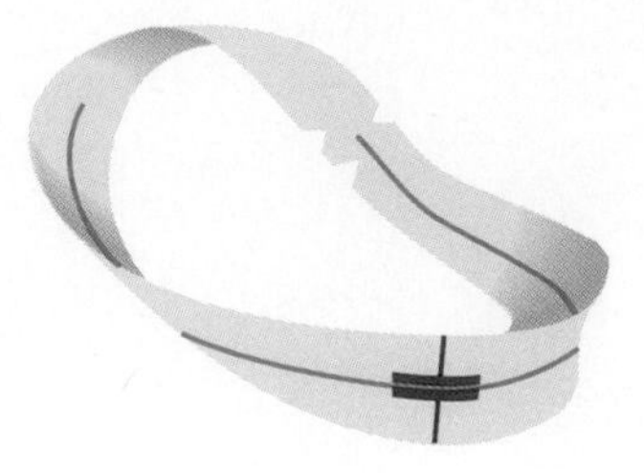

Step 4. Use the scissors to cut the strip of Step 3 in half lengthwise as shown. Are you surprised at the result?

Step 5. Predict what will happen if you cut the second Möbius strip lengthwise as in Step 4 but one-third of the way in from one edge. Now make the cut. Are you surprised again?

Step 6. In Step 5 you obtained two separate but intertwined loops. Predict what will happen if you cut the smaller of these loops in half lengthwise as in Step 4. Now make the cut. Was your guess correct?

Step 7. After cutting the small loop in Step 6 in half, cut it away from the large loop and discard it. Now cut the large loop in half lengthwise after first predicting what will happen if you do so. Was your prediction correct?

Step 8. Möbius strips are not just curiosities. If you had a belt twisted to form a Möbius strip running over pulleys as shown, would it last longer than an untwisted belt? Why or why not?

CONNECTIONS

Mathematics Is Problem Solving

In 1980, the National Council of Teachers of Mathematics (NCTM) published an important document titled *An Agenda for Action: Recommendations for School Mathematics of the 1980's.* This document has had important consequences, but none is more far-reaching than its first recommendation that:

- problem solving be the focus of school mathematics in the 1980's.

That this thrust continues to dominate thinking in mathematics education is borne out by Standard 1 from NCTM's 1989 document, *Curriculum and Evaluation Standards for School Mathematics,* which continues to receive broad national support. The conviction is that children need to learn to *think* about quantitative situations in insightful and imaginative ways, and that mere rote memorization of seemingly arbitrary rules for computation is largely unproductive.

FROM THE NCTM STANDARDS

Standard 1: Mathematics as Problem Solving

In grades K–4, the study of mathematics should emphasize problem solving so that students can—

- *use problem-solving approaches to investigate and understand mathematical content;*
- *formulate problems from everyday and mathematical situations;*
- *develop and apply strategies to solve a wide variety of problems;*
- *verify and interpret results with respect to the original problem;*
- *acquire confidence in using mathematics meaningfully.*

Focus

Problem solving should be the central focus of the mathematics curriculum. As such, it is a primary goal of all mathematics instruction and an integral part of all mathematical activity. Problem solving is not a distinct topic but a process that should permeate the entire program and provide the context in which concepts and skills can be learned.

This standard emphasizes a comprehensive and rich approach to problem solving in a classroom climate that encourages and supports problem-solving efforts. Ideally, students should share their thinking and approaches with other students and with teachers, and they should learn several ways of representing problems and strategies for solving them. In addition, they should learn to value the process of solving problems as much as they value the solutions. Students should have many experiences in creating problems from real-world activities, from organized data, and from equations.

In the early years of the K–4 program, most problem situations will arise from school and other everyday experiences. When mathematics evolves naturally from problem situations that have meaning to children and are regularly related to their environment, it becomes relevant and helps children link their knowledge to many kinds of situations. As children progress through the grades, they should encounter more diverse and complex types of problems that arise from both real-world and mathematical contexts.

When problem solving becomes an integral part of classroom instruction and children experience success in solving problems, they gain confidence in doing mathematics and develop persevering and inquiring minds. They also grow in their ability to communicate mathematically and use higher-level thinking processes.

SOURCE: From *Curriculum and Evaluation Standards for School Mathematics Grades K–4*, p. 23.

Of course, if children are to learn problem solving, their teachers must themselves be competent problem solvers and teachers of problem solving. Professor Robert Davis, a prominent mathematics educator, once said, "All too often we involve our students in a rhetoric of conclusions when, in fact, we ought to be involving them in a rhetoric of inquiry." This is true not only of elementary and secondary school students, but of college students as well. Thus, the purpose of this chapter, and indeed of this entire book, is to help you to think more critically and thoughtfully about mathematics, and to be more comfortable with mathematical reasoning and discourse.

We begin with some "mathematical tidbits" that we hope you will find interesting and surprising. Depending on the grade level at which you ultimately will teach, you may also find these tidbits useful in your own classroom.

After the tidbits, we spend the remainder of the chapter studying problem-solving strategies, and how to use them to solve problems. Strategies for problem solving are legion, and we cannot begin to discuss them all in a single chapter. However, we can introduce the most basic strategies, which you will be able to use and refine as you continue your study of this book.

1.1 Some Surprising Tidbits

Tidbit Number 1

Follow these directions.

For example, May would be 5.

Step 1. Write down the number of the month in which you were born.
Step 2. Double the number in Step 1.
Step 3. Add 5 to the result of Step 2.
Step 4. Multiply the result of Step 3 by 50.
Step 5. Subtract 250 from the present year.
Step 6. Add the result from Step 5 to the result from Step 4.
Step 7. Subtract the year of your birth from the result from Step 6.
Step 8. Circle the last two digits of the result from Step 7. The circled number should give your age on your birthday this year!
Step 9. The uncircled part of the number from Step 8 should be the number of your birth month!

Did the process work? Are you somewhat puzzled? Surprised? Pleased? Do you think your future students might find this trick interesting? Would this work if you were born after the year 1999?

Tidbit Number 2

Let's try another one.

A digit is one of 0, 1, 2, 3, 4, 5, 6, 7, 8, or 9.

Step 1. Write down your favorite digit.
Step 2. Multiply your favorite digit by 101.
Step 3. Multiply 110011 by the result from Step 2. (Use a calculator if you like.)

Are you surprised at the result? Do you think this will always work? Perhaps you should try a different favorite digit. How could you show that the trick always works? Clearly 101 and 110011 are "magic" numbers. Are 73 and 152207 "magic"?

Tidbit Number 3

See how quickly you can perform each of these multiplications. (Use a calculator if you like.)

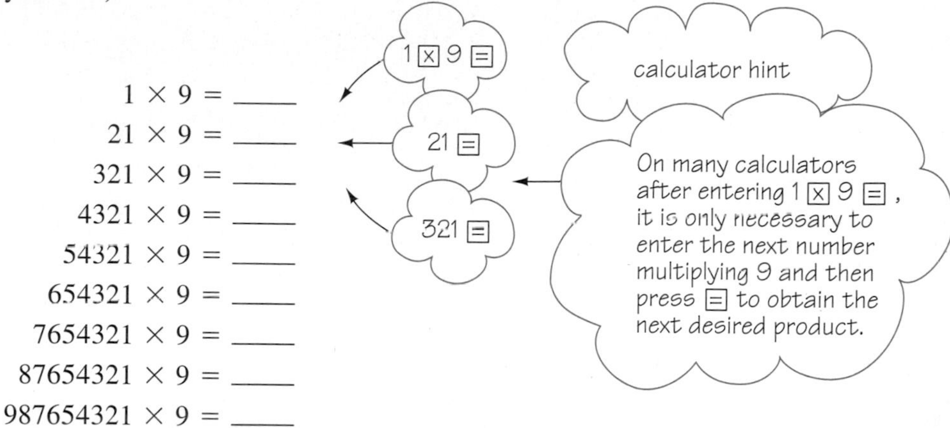

Neat, isn't it? All you have to do is complete the first three or four multiplications and notice the emerging pattern. But the pattern only suggests, it may not continue as you expect. You may want to check your guess by calculating the last product. Can you do this on your calculator? (Some calculators have too little display space.) Is the result what you expected? How about 10987654321×9?

Tidbit Number 4

1. **(a)** Draw two nonparallel nonintersecting line segments and three points on each segment labeled as shown.

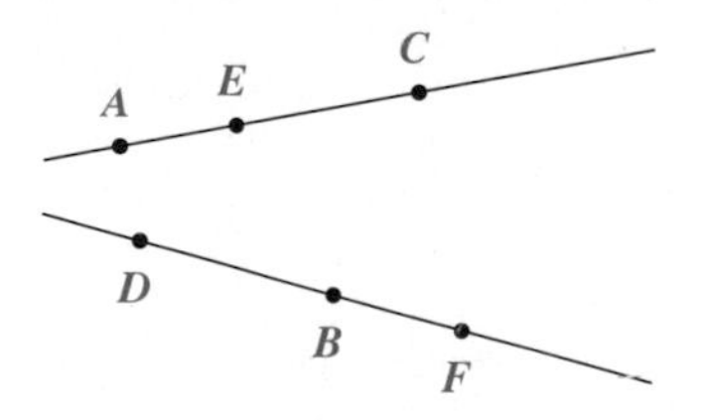

   (b) Draw the segments $\overline{AB}$, $\overline{BC}$, $\overline{CD}$, $\overline{DE}$, $\overline{EF}$, and $\overline{FA}$. Let P, Q, and R be the points of intersection of $\overline{AB}$ and $\overline{DE}$, $\overline{BC}$ and $\overline{EF}$, and $\overline{CD}$ and $\overline{FA}$ respectively. What appears to be the case about points P, Q, and R?

2. Try the construction again with the points labeled as shown here. To keep the drawing small put B to the right of A and F to the right of E. Also, extend the line segments as needed to get intersection points. What can be said about P, Q, and R this time? Do you suppose this will always be true no matter where A, B, C, and D, E, F are placed on the two segments respectively?

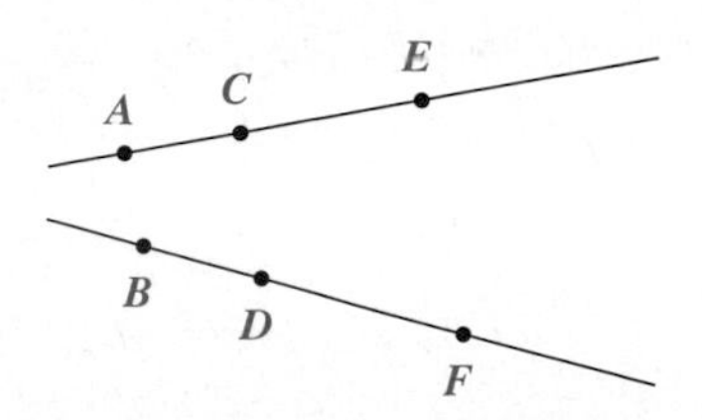

CURRENT AFFAIRS

The Status Quo Is not Good Enough

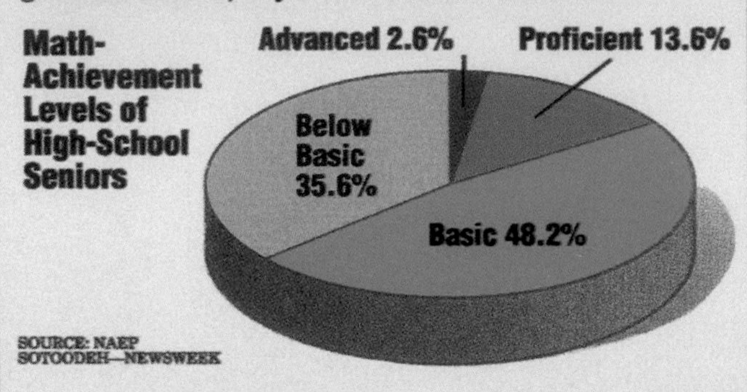

SOURCE: Chart, "Formula for Failure" by Sotoodah from *Newsweek,* October 14, 1991, p. 54.

At first glance, it might seem like good news about the nation's public schools: American youngsters today are just as good at math, science and reading as students were in 1970. That was the conclusion last week of a federal report summarizing 20 years of national testing. But, as he released the National Assessment of Educational Progress (NAEP), Education Secretary Lamar Alexander immediately warned against complacency. "What we did in 1970," he said, "is not nearly good enough in 1990."

Over the past two decades, other countries—particularly Japan and Germany—have developed curricula emphasizing advanced math and science in order to give their students the skills necessary to compete in an increasingly technological global economy. During the same period, American schools have swung from one extreme to another, from the open classrooms of the '70s to the back-to-basics movement of the 1980s. Says the University of Wisconsin's Thomas Romberg, an expert in worldwide math education, "We're acting as if America is still an agricultural society, and that all the mothers are home with their children." Now, many educators say, the highly politicized debate over the best way to teach has meant that American kids have been running in place while other countries trained for a marathon. "The world has changed, and what we expect and what the workplace expects is a much higher level of skills," says Diane Ravitch, assistant secretary of education and coauthor of "What Do Our 17-Year-Olds Know?"

The situation is especially alarming in math, a subject in which American students rank way below their counterparts in most industrialized countries. According to the Education Department, only one in five eighth graders has achieved competence for his or her age level. That news will probably come as a shock to many parents, who figure that as long as their kids can add up a bill or balance a checkbook, they're doing fine. But Iris Carle, president of the National Council of Teachers of Mathematics, says that without more sophisticated skills, "we'll be a nation of people completely out of contention for jobs. We'll be a nation of unemployed people because most entry-level positions today require some proficiency in algebra and high-level mathematics." Only 2.6 percent of high-school seniors were capable of doing advanced 12th-grade work such as calculus or sophisticated problem solving.

While minority students made some gains over the 20 years covered in the NAEP report, they still have a long way to go. Although the gap in achievement has narrowed at all ages tested, the average performance of blacks on the math and science tests was still significantly below that of white students tested in 1990.

Educators say parents should push their children—and the schools—to do more advanced work in technical subjects. Illiteracy is socially unacceptable, but lots of people still shrug their shoulders at math incompetence. "Too many parents feel that what counts in math is not effort, but natural ability," says Ravitch. There is a feeling that "my Jane is not good at math, so why should she work at it?" The answer, Ravitch and others say, is that if Jane doesn't "work at it" now, she may not be working at anything in 20 years.

SOURCE: "A Nation Running in Place" by Barbara Kantrowitz and Pat Wingert from *Newsweek,* October 14, 1991, p. 54.

The exercise set that follows contains more surprising examples and others that are easily developed. Elementary students enjoy problems like these and teachers report that using such activities in their classes help to create interest and enthusiasm for doing mathematics that were missing before.

This is not to say that mathematics is all "fun and games." It requires diligence and perseverance, but the effort need not be distasteful. Mathematics is full of unexpected and pleasing results that will continually excite and interest both you and your students. As a student in this course, you are preparing to teach. You want to teach well and with success. Experience and research have shown that you are much more likely to succeed if you are confident in the subject matter you teach and actually find it pleasing and personally satisfying.

PROBLEM SET 1.1

In doing these problems the use of calculators is encouraged. Problem numbers in color indicate that the answer to this problem can be found at the back of the book.

1. (a) Compute the products:

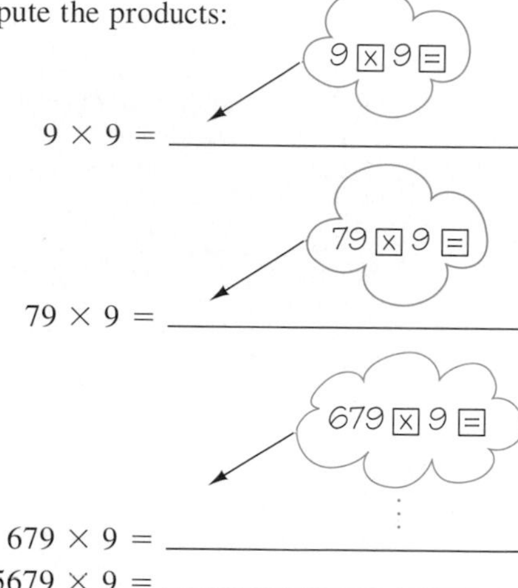

$9 \times 9 =$ ________

$79 \times 9 =$ ________

$679 \times 9 =$ ________

$5679 \times 9 =$ ________

$45679 \times 9 =$ ________

$345679 \times 9 =$ ________

$2345679 \times 9 =$ ________

$12345679 \times 9 =$ ________

(b) Did you have to complete all the multiplications in part (a) to be pretty sure you knew what all the answers would be? Explain in one *carefully* written sentence.

2. (a) Perform these computations being careful to multiply first and then add.

$1 \times 9 + 2 =$ ________

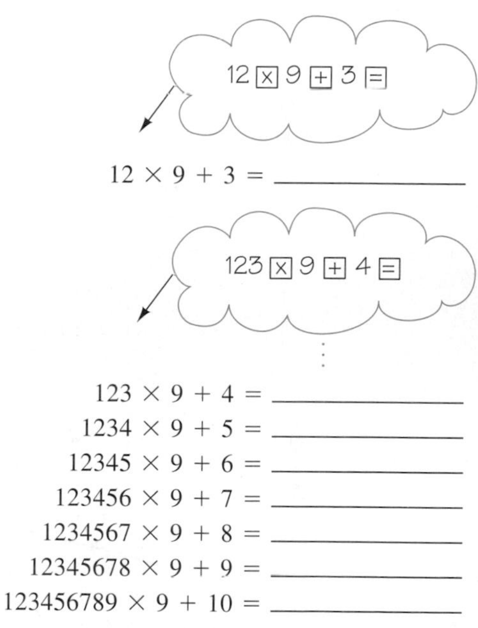

$12 \times 9 + 3 =$ ________

$123 \times 9 + 4 =$ ________

$1234 \times 9 + 5 =$ ________

$12345 \times 9 + 6 =$ ________

$123456 \times 9 + 7 =$ ________

$1234567 \times 9 + 8 =$ ________

$12345678 \times 9 + 9 =$ ________

$123456789 \times 9 + 10 =$ ________

(b) Did you have to complete all the work in part (a) to be pretty sure you knew all the answers? Explain in one *carefully* written sentence.

3. (a) Perform these computations:

$1 \times 8 + 1 =$ ________

$12 \times 8 + 2 =$ ________

$123 \times 8 + 3 =$ ________

(b) Guess the results of these computations:

$1234 \times 8 + 4 =$ ________

$123456789 \times 8 + 9 =$ ________

(c) Check the last answer to part (b) by some means.

4. (a) The numbers 3 and 37,037 are "magic." Pick your favorite digit from among 1, 2, . . . , 9 and multiply it by 3. Now multiply the result by 37,037. What is the result?

(b) Are 13 and 8547 "magic"?

(c) Find two other pairs of "magic" numbers. (*Hint:* Compute 3 × 37037 and 13 × 8547.)

5. (a) Compute these products:

$67 \times 67 =$ ______________

$667 \times 667 =$ ______________

$6667 \times 6667 =$ ______________

(b) Guess the result of multiplying 6666667 by itself. Are you sure your guess is correct? Explain in one *carefully* written sentence.

6. **(a)** This is a magic square. Compute the sums of the numbers in each row, column, and diagonal of the square and write your answers in the appropriate circles.

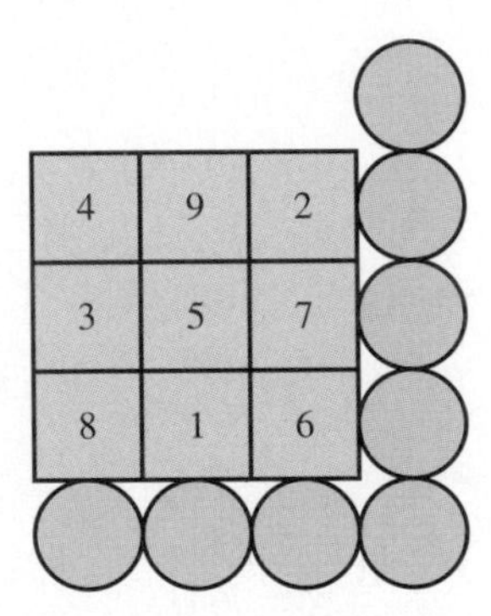

(b) Interchange the 2 and 8 and the 4 and 6 in the array in part (a) to create this magic *subtraction* square. Add the two end entries and subtract the middle entry from this sum for each row, column, and diagonal.

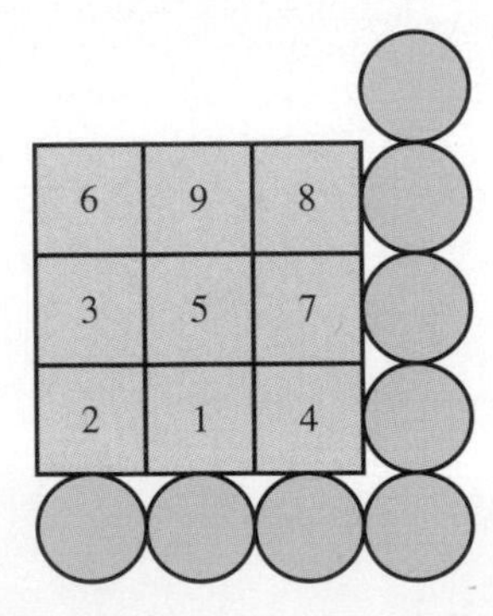

7. **(a)** Write the digits 0, 1, 2, 3, 4, 5, 6, 7, and 8 in the small squares to create another magic square. (*Hint:* Relate this to problem 6. Also, you may want to write these digits on nine small squares that you can move around easily to check various possibilities.)

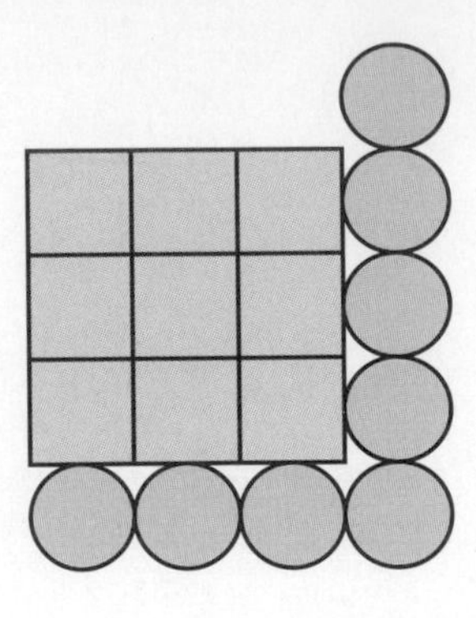

(b) Make a magic subtraction square using the numbers 0, 1, 2, 3, 4, 5, 6, 7, 8.

8. Flow charts are frequently used in computer science since they make it possible to chart a sequence of operations or events in a visual format that is easily followed and understood.

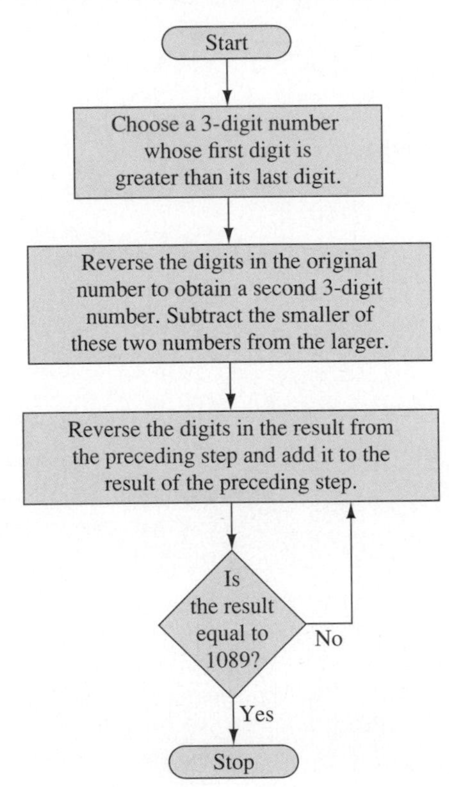

(a) Follow the steps indicated by this flow chart. All the work can be done on your calculator without writing anything down, though you may find it interesting to record the result of each step. Repeat the process with several different starting numbers.

(b) Do you believe that the process always stops? Explain briefly.

(c) Compute these products:

$1 \times 1089 =$ ____________
$2 \times 1089 =$ ____________
$3 \times 1089 =$ ____________
$4 \times 1089 =$ ____________
$5 \times 1089 =$ ____________
$6 \times 1089 =$ ____________
$7 \times 1089 =$ ____________
$8 \times 1089 =$ ____________
$9 \times 1089 =$ ____________

(d) Did you have to compute all the products in part (c) to be pretty sure you knew what all the answers would be? Explain briefly.

(e) Do you see any other interesting patterns in part (c)? Explain briefly.

9. (a) Compute these products:

$1 \times 142857 =$ ____________
$2 \times 142857 =$ ____________
$3 \times 142857 =$ ____________
$4 \times 142857 =$ ____________
$5 \times 142857 =$ ____________

(b) Predict the product of 6 and 142857. Now calculate the product and see if your prediction was correct.

(c) Predict the result of multiplying 7 times 142857, then compute this product.

(d) What does part (c) suggest about apparent patterns? Explain.

10. **(a)** Compute these products:

1×76923	5×76923	9×76923
2×76923	6×76923	10×76923
3×76923	7×76923	11×76923
4×76923	8×76923	12×76923

(b) Compute 13×76923.

(c) What patterns do you see in part (a)? Explain briefly but carefully.

11. (a) Draw a circle and place six points on it labeled as shown here. Draw the chords $\overline{AB}$, $\overline{BC}$, $\overline{CD}$, $\overline{DE}$, $\overline{EF}$, and $\overline{FA}$. Let P, Q, and R be the points of intersection of $\overline{AB}$ and $\overline{DE}$, $\overline{BC}$ and $\overline{EF}$, and $\overline{CD}$ and $\overline{FA}$ respectively. What seems to be true of the points P, Q, and R?

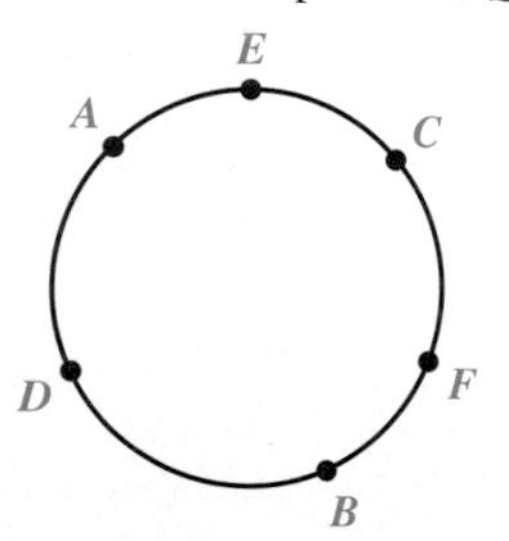

(b) Would the result of part (a) still hold if the six points were labeled in a different order? Try some cases being careful to locate the points so that the line segments, or their extensions if needed, intersect on the paper on which you are drawing.

12. **(a)** Draw three line segments l, m, and n from a common point O and draw two triangles $\triangle ABC$ and $\triangle A'B'C'$ with corresponding vertices on l, m, and n respectively as shown. Let P, Q, and R be the points where the lines $\overleftrightarrow{AB}$ and $\overleftrightarrow{A'B'}$, $\overleftrightarrow{AC}$ and $\overleftrightarrow{A'C'}$, and $\overleftrightarrow{BC}$ and $\overleftrightarrow{B'C'}$ respectively intersect. What seems to be true about P, Q, and R?

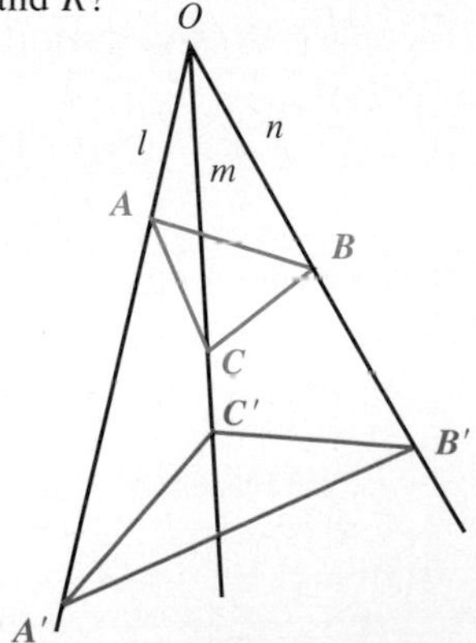

(b) Try this again for a different placement of A, B, C, and A', B', C'. What do these examples suggest?

13. **(a)** Draw a circle and use a ruler or straightedge to carefully draw six tangent lines l, m, n, o, p, and q that form a hexagon around the circle, as shown below. Let A, B, C, D, E, and F respectively be the points where the consecutive tangent lines l and m, m and n, n and o, o and p, p and q, and q and l respectively intersect. What appears to be true about the line segments $\overline{AD}$, $\overline{BE}$, and $\overline{CF}$?

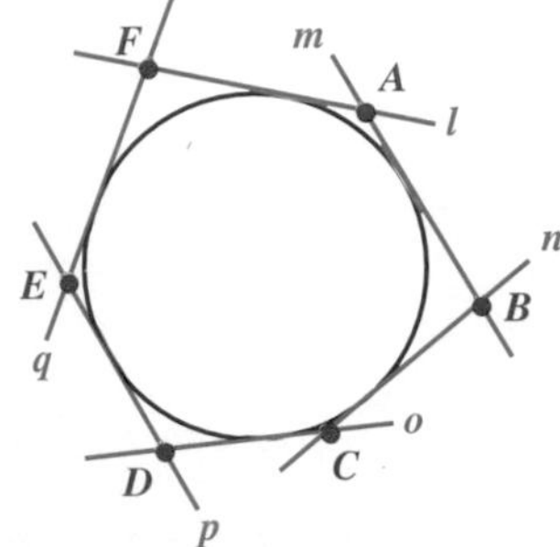

(b) Try this again with a different circle and a different set of tangent lines. Does the result still appear to be true?

14. **(a)** Draw three circles of different sizes and carefully draw the pairs of external tangent lines to each pair of circles as shown. Let P, Q, and R be the points where the pairs of external tangents intersect. What appears to be true about P, Q, and R?

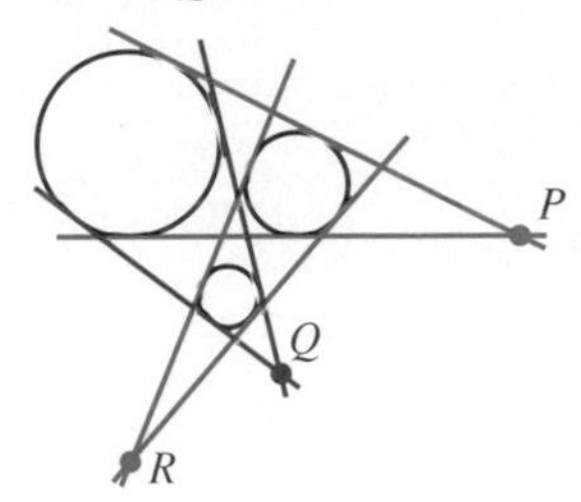

(b) Draw a second figure with three different circles. Does your conjecture of part (a) still appear to hold?

15. **(a)** **Sums of Squares of Digits.** Carry out the activities explained in the flow chart below by using your computer disk and calling up the program SQDIGSUM. You can also do this by hand or on a calculator but, if you do, be sure to record your result at each step as shown here. Try this several times. Do you think the process always stops no matter what whole number you start with?

(b) It takes six steps for the process to stop if you start with 98. How many steps are required if you start with 248? with 999? with 9999?

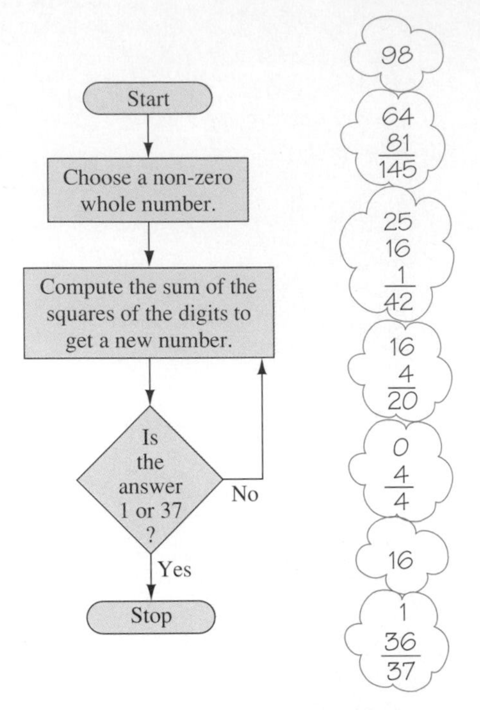

16. **(a)** **Palindromes.** A number palindrome is a number like 242 or 3113 that reads the same both forward and backward. A famous palindrome in words, attributed to Napoleon, is, "Able was I ere I saw Elba." Use your computer disc and call up PALINDROME to complete the activity of this flow chart. This can also be done with a calculator, but if you use your calculator, be sure to record your results at each step as shown here. Try this several times.

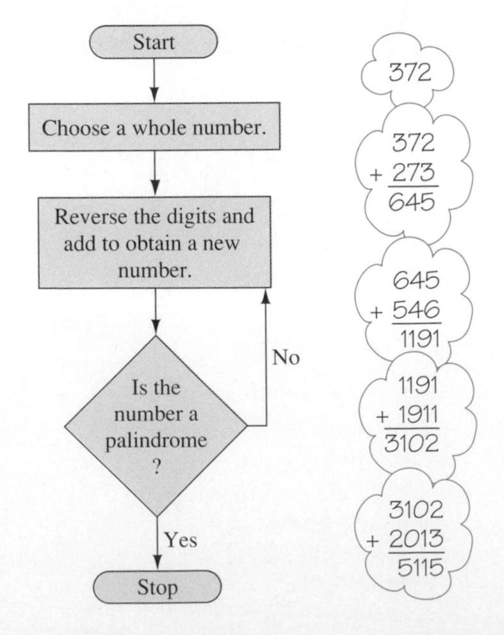

(b) It takes four steps for the process to stop if we start with 372. Will it stop if we start with 98? If so, in how many steps?

(c) Do you think the process will always stop?

17. **(a) Kaprekar's Number.** Using the computer, call up KAPREKAR to complete this activity. Alternatively, use your calculator being sure to record the intermediate steps as shown here.

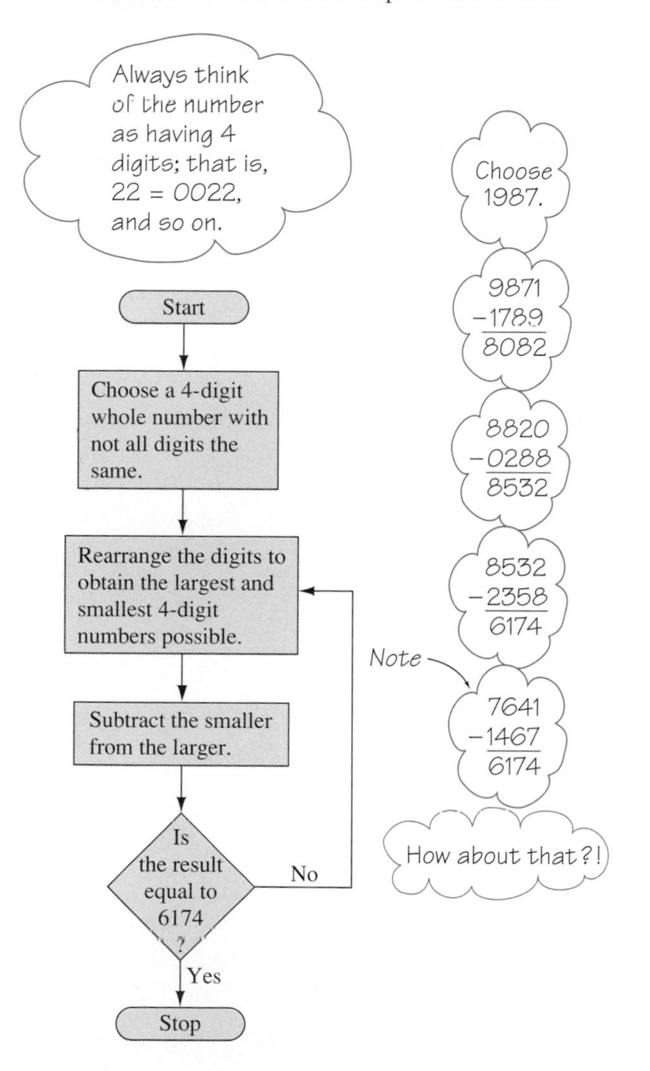

(b) Will the process stop if you start with 1996? If so, in how many steps?

(c) Starting with any 4-digit number, do you think the process will always stop?

(d) What happens if you start with a 3-digit number? A 5-digit number? Explain briefly.

18. (a) **Collatz's Problem**. Using the computer, call up COLLATZ on your computer disc to complete this activity. This can also be done easily on a calculator, but, if you do, be sure to write down all the steps shown as here. Try this for a number of different starting values. Do you think the process will always stop?

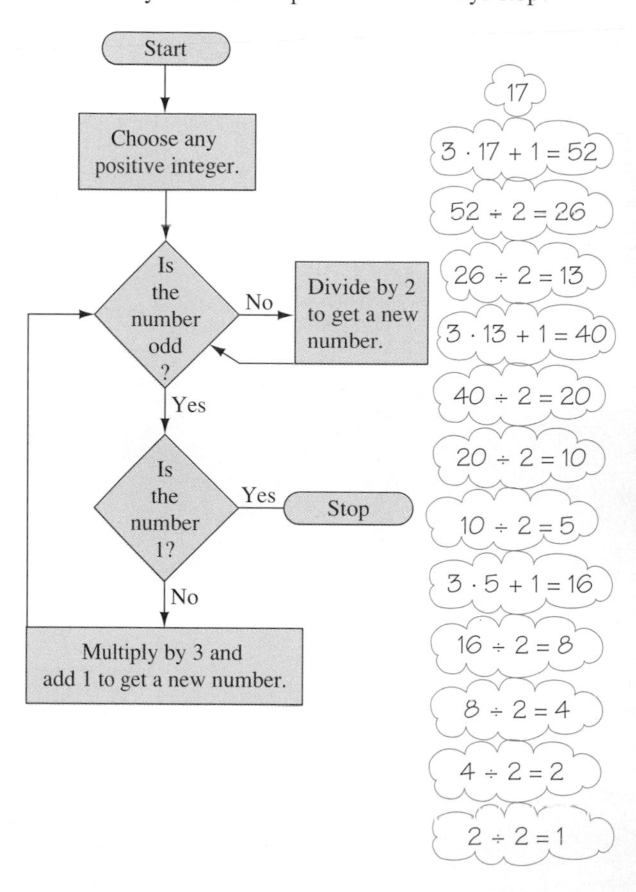

(b) How many steps are required if you start with 9?

1.2 An Introduction to Problem Solving

When the children arrived in Frank Capek's fifth grade class one day, this "special" problem was on the blackboard.

> Old MacDonald had a total of 37 chickens and pigs on his farm. All together they had 98 feet. How many chickens were there and how many pigs?

After organizing the children into problem-solving teams, Mr. Capek asked them to solve the problem. "Special" problems were always fun and the children got right to work. Let's listen in on the group with Mary, Joe, Carlos, and Sue.

"I'll bet there were 20 chickens and 17 pigs," said Mary.

"Let's see," said Joe. "If you're right there are 2 × 20 or 40 chicken feet and 4 × 17 or 68 pigs feet. This gives 108 feet. That's too many feet."

"Let's try 30 chickens and 7 pigs," said Sue. "This should give us less feet."

"Hey," said Carlos. "With Mary's guess we got 108 feet and Sue's guess gives us 88 feet. Since 108 is 10 too much and 88 is 10 too few, I'll bet we should guess 25 chickens—just half way between Mary's and Sue's guesses!"

These children are using a **guess and check** strategy. If their guess gives an answer that is too large or too small they adjust the guess to get a smaller or larger answer as needed. This can be a very effective strategy. By the way, is Carlos's guess right?

Let's look in on another group.

"Let's make a table," says Nandita. "We've had good luck that way before."

"Right, Nani," responded Ann. "Let's see. If we start with 20 chickens and 17 pigs, we have 2 × 20 or 40 chicken feet and 4 × 17 or 68 pig feet. If we have 21 chickens . . . "

This is a powerful refinement of guess and check.

Chickens	Pigs	Chicken Feet	Pig Feet	Total
20	17	40	68	108
21	16	42	64	106
22	15	44	60	104
•	•	•	•	•
•	•	•	•	•
•	•	•	•	•

Making a table to look for a pattern is often an excellent strategy. Do you think the group with Nandita and Ann will soon find a solution? How many more rows of the table will they have to fill in? Can you think of a shortcut?

Mike says, "Let's draw a picture. We can draw 37 circles for heads and put two lines under each circle to represent feet. Then we can add two extra feet under enough circles to make 98. That should do it."

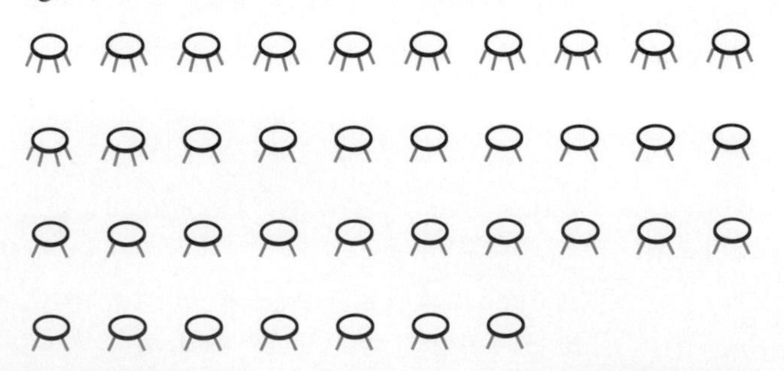

Drawing a picture is often a good strategy. Does it work in this case?

JUST FOR FUN

For Careful Readers

1. Two engineers were standing on a street corner. The first engineer was the second engineer's father but the second engineer was not the first engineer's son. How could this be?
2. If an electric train is traveling 40 miles an hour due west and a wind of 30 miles per hour is blowing due east, which way is the smoke from the train blowing?
3. At noon a rope ladder with rungs 1 foot apart is hanging over the side of the ship and the 12th rung down is even with the water surface. Later, after the tide has risen 3 feet, which rung of the ladder is just even with the surface of the water?

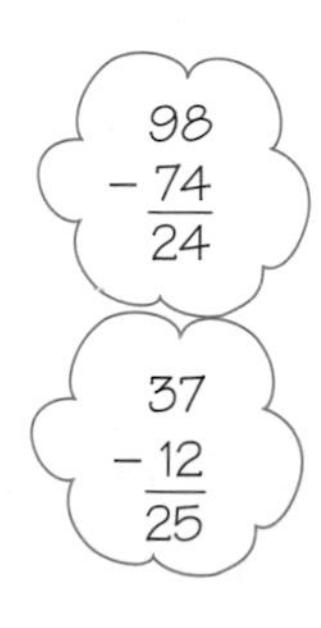

"Oh! The problem is easy," says Jennifer. "If we have all the pigs stand on their hind legs then there are 2 × 37 or 74 feet touching the ground. That means the pigs must be holding 24 front feet up in the air. This means there must be 12 pigs and 25 chickens!"

It helps if you can be ingenious like Jennifer, but it is not essential, and children *can* be taught strategies like

Guess and check
Make a table
Look for a pattern
Draw a picture

Other useful strategies will be discussed later but for now let's try some problems on our own.*

EXAMPLE 1.1

Guessing Toni's Number

Toni is thinking of a number. If you double the number and add 11, the result is 39. What number is Toni thinking of?

SOLUTION 1 GUESSING AND CHECKING

Guess 10. $2 \cdot 10 + 11 = 20 + 11 = 31.$ This is too small.
Guess 20. $2 \cdot 20 + 11 = 40 + 11 = 51.$ This is too large.
Guess 15. $2 \cdot 15 + 11 = 30 + 11 = 41.$ This is a bit large.
Guess 14. $2 \cdot 14 + 11 = 28 + 11 = 39.$ This checks!
Toni's number must be 14.

SOLUTION 2 MAKING A TABLE AND LOOKING FOR A PATTERN

Trial Number	Result Using Toni's Rule	
5	$2 \cdot 5 + 11 = 21$	
		2 larger
6	$2 \cdot 6 + 11 = 23$	
		2 larger
7	$2 \cdot 7 + 11 = 25$	
		2 larger
8	$2 \cdot 8 + 11 = 27$	
⋮	⋮	⋮

*Note that algebra students might solve this problem by solving the equation $2x + 4(37 - x) = 98$ where x denotes the number of chickens. But this approach is not available to fifth graders and it is certainly not as quick as Jennifer's solution!

We need to get to 39 and we jump by 2 each time we take a step of 1. Therefore, we need to take

$$\frac{39 - 27}{2} = \frac{12}{2} = 6$$

more steps; we should guess $8 + 6 = 14$ as Toni's number as before. ■

EXAMPLE 1.2 **Guessing and Checking**

(a) Place the digits 1, 2, 3, 4, and 5 in these circles so that the sums across and vertically are the same. Is there more than one solution?

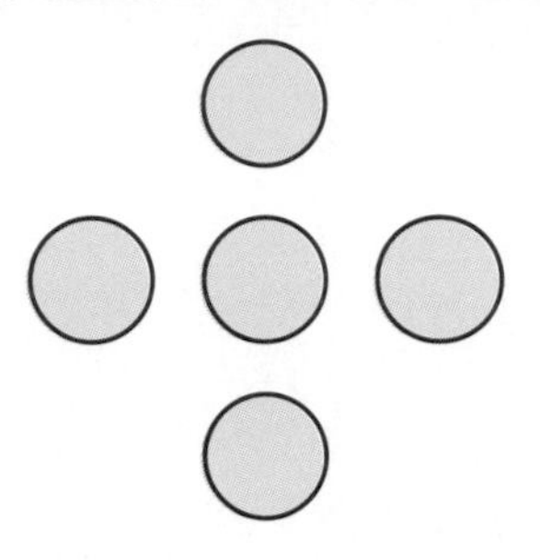

(b) Can part (a) be accomplished if 2 is placed in the center? Why or why not?

SOLUTION

(a) Using the guess and check strategy, suppose we put the 3 in the center circle. Since the sums across and down must be the same, we must pair the remaining numbers so that they have equal sums. But this is easy since $1 + 5 = 2 + 4$. Thus, one solution to the problem is as shown here.

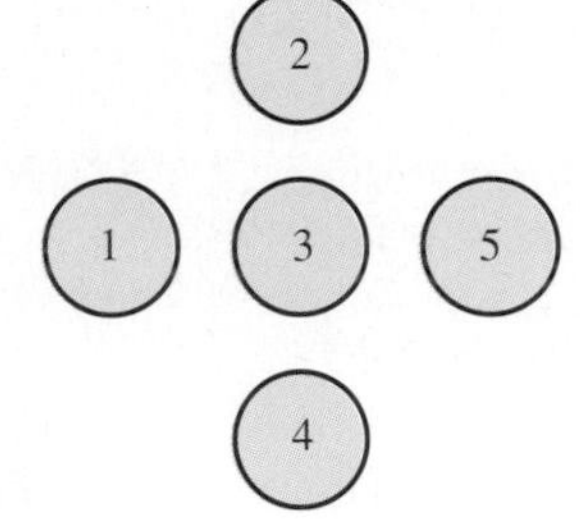

Checking further, we find other solutions like these.

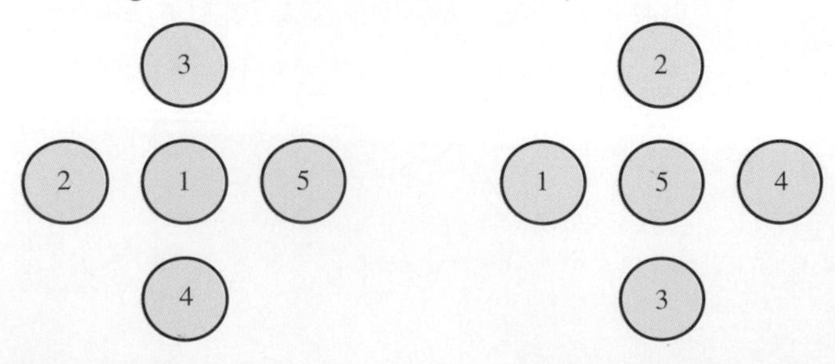

(b) What about putting 2 in the center? The remaining digits are 1, 3, 4, and 5 and these cannot be grouped into two pairs with equal sums since one sum is necessarily odd and the other even. Therefore, there is no solution with 2 in the center circle. ■

PROBLEM SET 1.2

1. Levinson's Hardware has a number of bikes and trikes for sale. There are 27 seats and 60 wheels all told. Determine how many bikes there are and how many trikes.
   (a) Use the guess and check strategy to find a solution.
   (b) Complete this table to find a solution.

Bikes	Trikes	Bike Wheels	Trike Wheels	Total
17	10	34	30	64
18	9	36	27	63
•	•	•	•	•
•	•	•	•	•
•	•	•	•	•

   (c) Find a solution by completing this diagram.

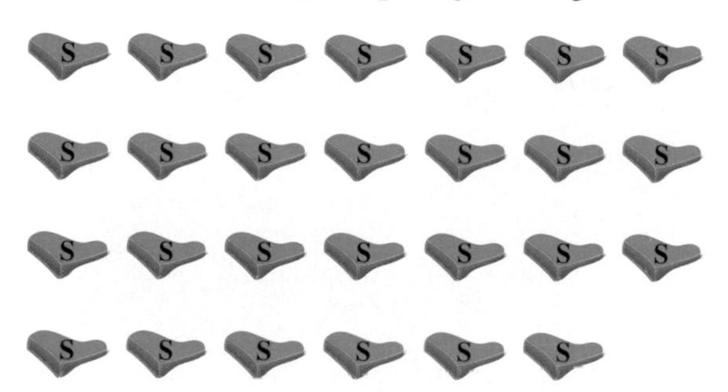

   (d) Would Jennifer's method work for this problem? Explain briefly.

2. **(a)** Mr. Aiken has 32 18-cent and 27-cent stamps all told. The stamps are worth $7.65. How many of each kind of stamp does he have?
   (b) Summarize your solution method in one or two *carefully* written sentences.
3. Make up a problem similar to problems 1 and 2.
4. Who am I? If you multiply me by 5 and subtract 8, the result is 52.
5. Who am I? If you multiply me by 15 and add 28, the result is 103.
6. Make up a problem like problems 4 and 5.
7. (a) Place the digits 4, 6, 7, 8, and 9 in the circles to make the sum across and vertically equal 19.
   (b) Is there more than one answer to part (a)? Explain briefly.

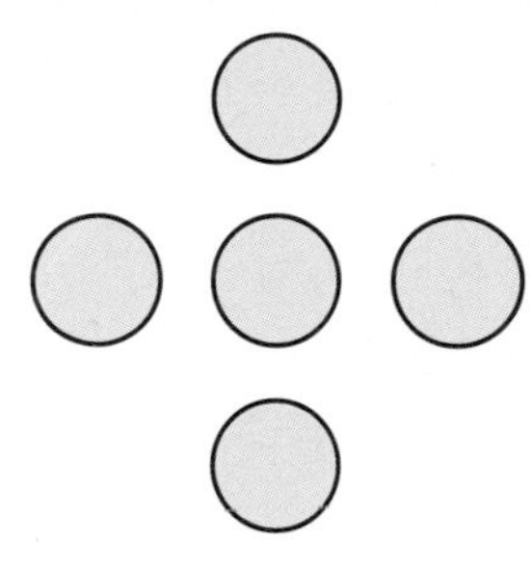

8. Melissa Dietz has nine coins with a total value of 48 cents. What coins does Melissa have?
9. (a) Using each of 1, 2, 3, 4, 5, and 6 once and only once, fill in the circles so that the sums of the numbers on each of the three sides of the triangle are equal.

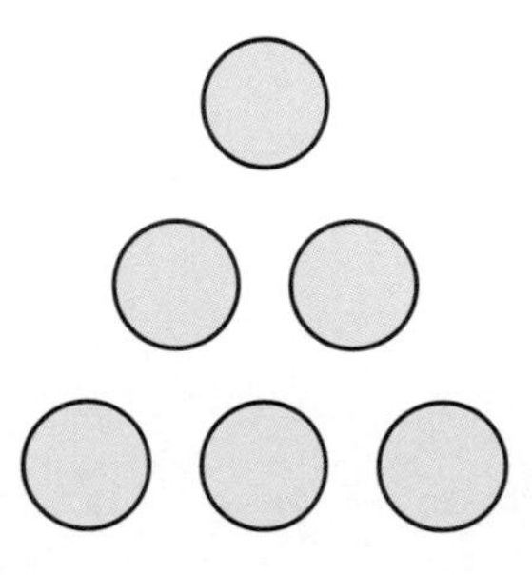

   (b) Does part (a) have more than one solution?
   (c) Write up a brief but careful description of the thought process you used in solving this problem.
10. In this diagram, the sum of any two horizontally adjacent numbers is the number immediately below and between them. Using the same rule of formation, complete these arrays.

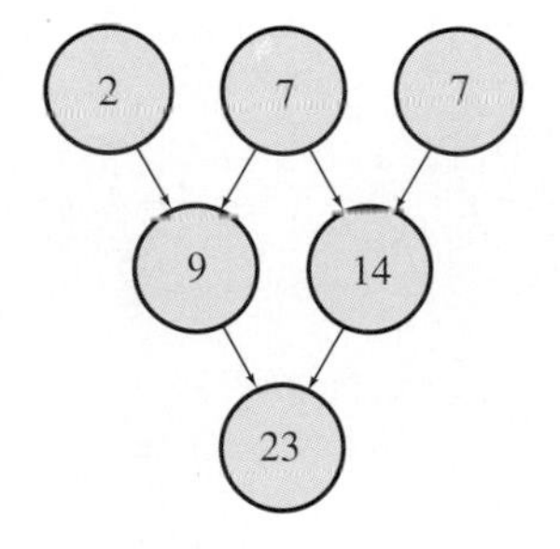

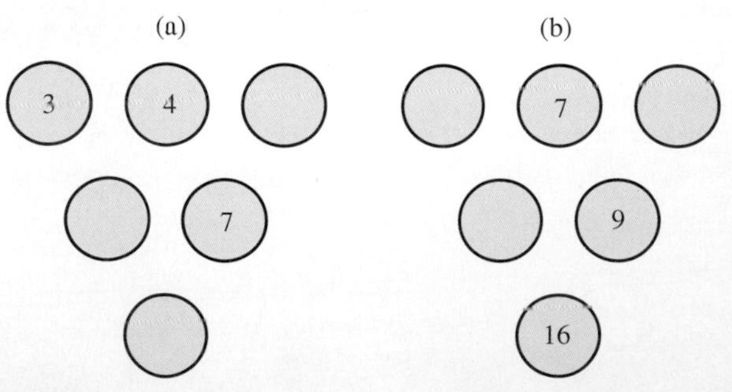

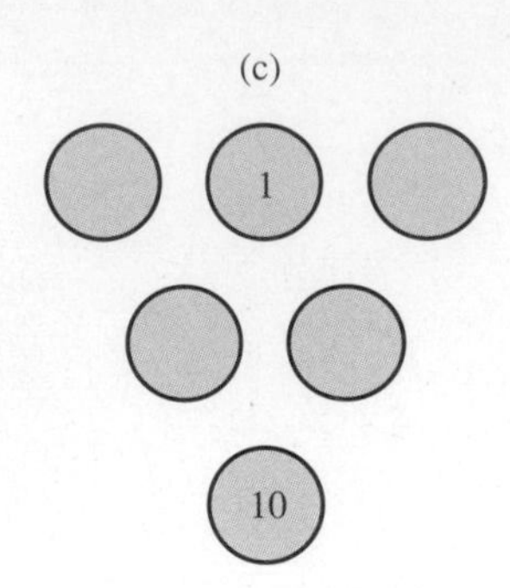

(d) Is there more than one solution to parts (a), (b), or (c)?

11. Study the sample diagram. Note that

$$2 + 8 = 10,$$

$$5 + 3 = 8,$$

$$2 + 5 = 7,$$

$$3 + 8 = 11.$$

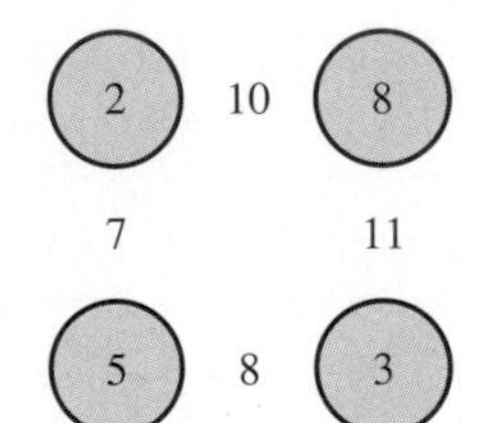

If possible, complete each of these diagrams so that the same pattern holds.

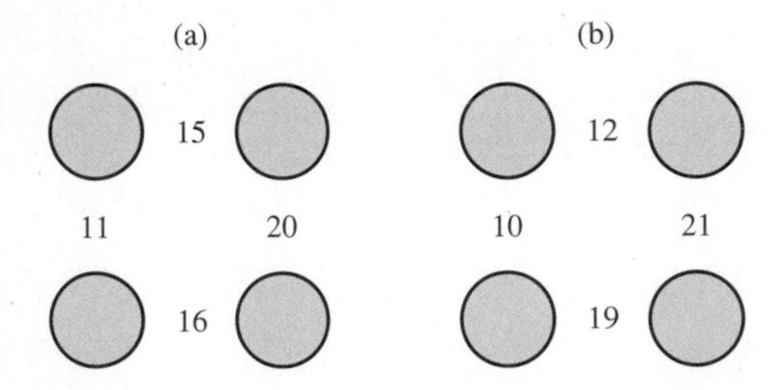

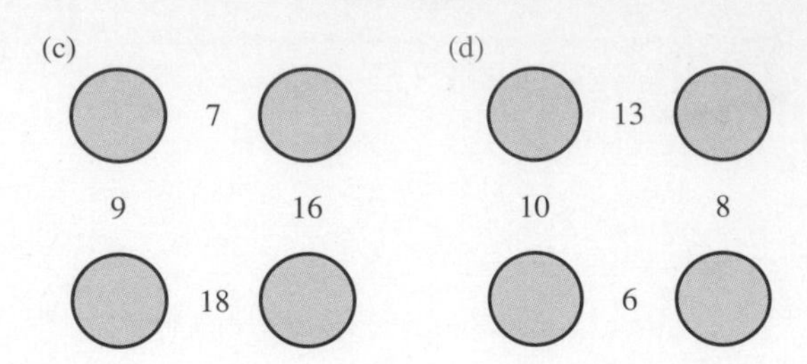

12. Study this sequence of numbers: 3, 4, 7, 11, 18, 29, 47, 76. Note that $3 + 4 = 7$, $4 + 7 = 11$, $11 + 18 = 29$, and so on. Use the same rule to complete these sequences.

(a) 1, 2, 3, ______, ______, ______, ______

(b) 2, ______, 8, ______, ______, ______, ______

(c) 3, ______, ______, 13, ______, ______, ______

(d) 2, ______, ______, ______, ______, 26

(e) 2, ______, ______, ______, ______, 11

13. (a) Use each of the numbers 2, 3, 4, 5, and 6 once and only once to fill in the circles so that the sum of the numbers in the three horizontal circles equals the sum of the numbers in the three vertical circles.

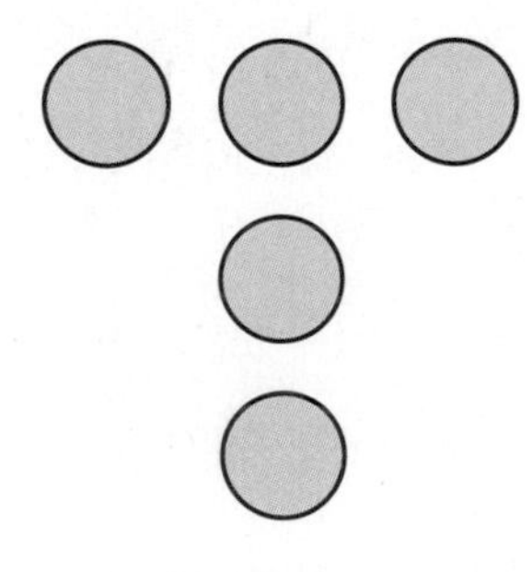

(b) Can you find more than one solution?

(c) Can you have a solution with 3 in the middle of the top row? Explain in two *carefully* written sentences.

14. Make up a guess and check problem of your own and solve it.

1.3 Pólya's Problem-Solving Principles

STRATEGIES

- ***Guess and check.***
- ***Make an orderly list.***
- ***Draw a diagram.***

In *How to Solve It**, George Pólya identifies four principles that form the basis for any serious attempt at problem solving. He then proceeds to develop an extensive list of questions that teachers should ask students who need help in solving a problem, questions students can and should ask themselves as they seek solutions to problems.

* George Pólya, *How to Solve it* (Garden City, NY: Doubleday and Co., Inc., 1957.)

HIGHLIGHT FROM HISTORY
George Pólya (1887–1985)

How does one most efficiently proceed to solve a problem? Can the art of problem solving be taught or is it a talent possessed by only a select few? Over the years, many have thought about these questions but none so effectively and definitively as the late George Pólya, and he maintained that the skill of problem solving can be taught.

Pólya was born in Hungary in 1887 and received his Ph.D. in mathematics from the University of Budapest. He taught for many years at the Swiss Federal Institute of Technology in Zurich and would no doubt have continued to do so but for the advent of Nazism in Germany. Deeply concerned by this threat to civilization, Pólya moved to the United States in 1940 and taught briefly at Brown University and then, for the remainder of his life, at Stanford University. He was extraordinarily capable both as a mathematician and as a teacher. He also maintained a life-long interest in studying the thought processes that are productive in both learning and doing mathematics. Indeed, among the numerous books that he wrote he seemed most proud of *How to Solve It* (1945), which has sold nearly one million copies and has been translated into 17 languages. This book, along with his two two-volume treatises, *Mathematics and Plausible Reasoning* (1954) and *Mathematical Discovery* (1962), form the definitive basis for the current thinking in mathematics education and are as timely and important today as when they were written.

Pólya's First Principle: Understand the problem

This principle seems so obvious that it need not be mentioned. However, students are often stymied in their efforts to solve a problem because they don't understand it fully, or even in part. Teachers should ask students such questions as:

- Do you understand all the words used in stating the problem? If not, look them up in the index, in a dictionary, or wherever they can be found.
- What are you asked to find or show?
- Can you restate the problem in your own words?
- Is there yet another way to state the problem?
- What does (key word) really mean?
- Could you work out some numerical examples that would help make the problem clear?
- Could you think of a picture or diagram that might help you understand the problem?
- Is there enough information to enable you to find a solution?
- Is there extraneous information?
- What do you really need to know to find a solution?

Pólya's Second Principle: Devise a plan • • • •

Devising a plan for solving a problem once it is fully understood may still require substantial effort. But don't be afraid to make a start—you may be on the right track. There are often many reasonable ways to try to solve a problem, and the successful idea may emerge only gradually after several unsuccessful trials. A partial list of strategies include:

- guess and check
- make an orderly list
- think of the problem as partially solved
- eliminate possibilities
- solve an equivalent problem
- use symmetry
- consider special cases
- use direct reasoning
- solve an equation
- look for a pattern
- draw a picture
- think of a similar problem already solved
- solve a simpler problem
- solve an analogous problem
- use a model
- work backward
- use a formula
- be ingenious!

Skill at choosing an appropriate strategy is best learned by solving many problems. As you gain experience, you will find choosing a strategy increasingly easy—and the satisfaction of making the right choice and having it work is considerable! Again, teachers can turn the above list of strategies into appropriate questions to ask students in helping them learn the art of problem solving.

Pólya's Third Principle: Carry out the plan • • • •

Carrying out the plan is usually easier than devising the plan. In general, all you need is care and patience, given that you have the necessary skills. If a plan does not work immediately, be persistent. If it still doesn't work, discard it and try a new strategy. Don't be misled, this is the way mathematics is done, even by professionals.

Pólya's Fourth Principle: Look back • • • •

Much can be gained by looking back at a completed solution to analyze your thinking and ascertain just what was the key to solving the problem. This is how we gain "mathematical power," the ability to come up with good ideas for solving problems never encountered before. The French mathematician and philosopher, Henri Poincaré (1854–1912), put this rather strongly when he wrote

> *Suppose I apply myself to a complicated calculation and with much difficulty arrive at a result. I shall have gained nothing by my trouble if it has not enabled me to forsee the results of other analogous calculations, and to direct them with certainty, avoiding the blind groping with which I had to be content the first time.*

Clearly, Poincaré felt that merely solving a problem was essentially meaningless if he did not also gain experience and insight that increased his "mathematical power." Often the connection between very dissimilar problems is tenuous at best. Yet, in working on a problem, there is something lurking in the back of your mind from a previous effort that says, "I'll bet if . . . ," and the plan does indeed work!

Also, looking back, you can often see an easier or more powerful strategy that further enhances your ability to solve problems. Looking back is an often overlooked but extremely important step in developing problem-solving skills.

Let's now look at some examples. □

Guess and Check

PROBLEM-SOLVING STRATEGY 1 • Guess and Check

Make a guess and check to see if it satisfies the demands of the problem. If it doesn't, alter the guess appropriately and check again. When the guess finally checks, a solution has been found.

Students often feel that it is not "proper" to solve a problem by guessing. And they are right if the guess is not accompanied by a check. However, a process of guessing, checking, altering the guess if it does not check, guessing again in light of the preceding check, and so on, is a legitimate and effective strategy. When a guess finally checks, there can be no doubt that a solution has been found. If we can be sure that there is only one solution, *the* solution has been found. Moreover, the process is often quite efficient and it may be the only approach available.

EXAMPLE 1.3

Guess and Check

In the first diagram the numbers in the big circles are found by adding the numbers in the two adjacent smaller circles as shown. Complete the second diagram so that the same pattern holds.

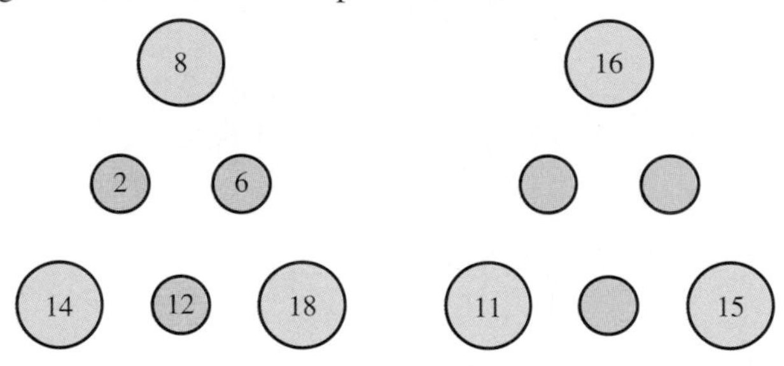

SOLUTION

Understand the problem • • • •

Considering the example, it is pretty clear that we must find three numbers—a, b, and c—such that

$$a + b = 16,$$
$$a + c = 11,$$
$$b + c = 15.$$

How should we proceed?*

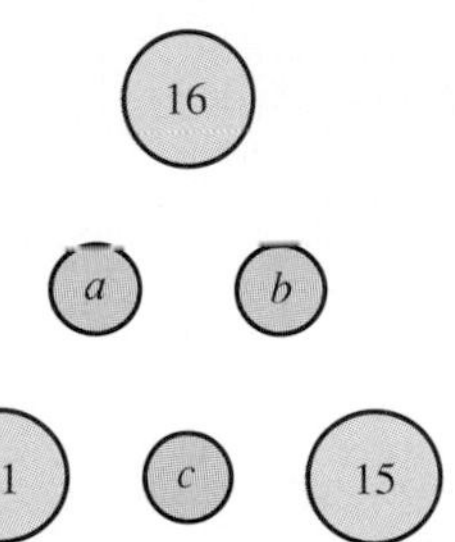

*Students who know algebra could solve this system of simultaneous equations, but elementary school students don't know algebra.

Devise a plan • • • •

Let's try the *guess and check* strategy. It worked on several problems somewhat like this in the last problem set. Also, even if the strategy fails, it may at least suggest an approach that will work.

Carry out the plan • • • •

We start by guessing a value for a. Suppose we guess that a is 10. Then, since $a + b$ must be 16, b must be 6. Similarly, since $b + c$ must be 15, c must be 9. But then $a + c$ is 19 instead of 11 as it is supposed to be. This does not check.

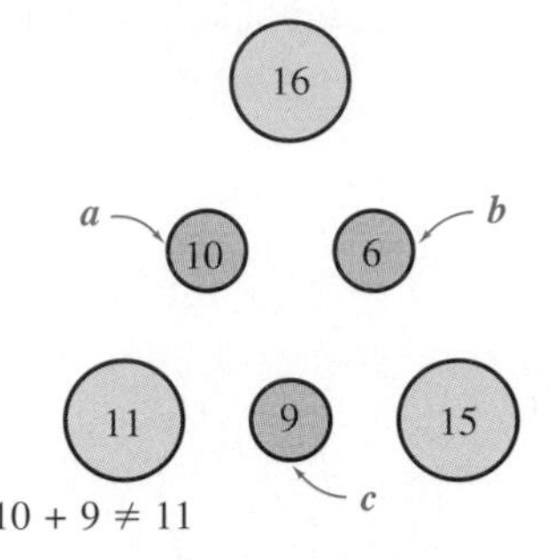

Since 19 is too large, we try again with a smaller guess for a. Guess that a is 5. Then, as above, this implies that b is 11 and that c is 4. But then $a + c$ is 9 and this is too small, but by just a little bit. We should guess that a is just a bit larger than 5.

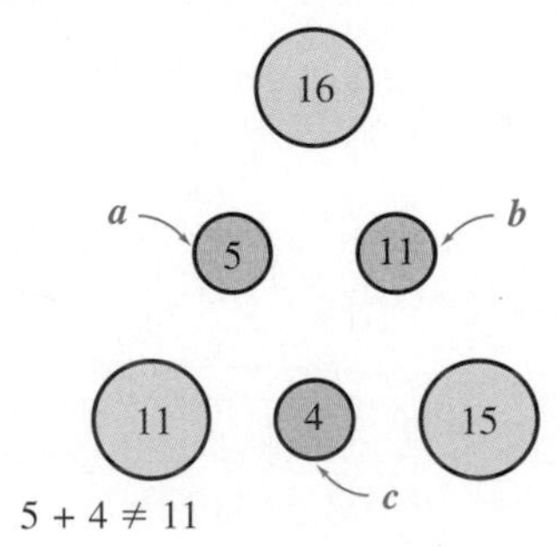

Guess that $a = 6$. Again, as above, this implies that b is 10 and c is 5. Now $a + b$ is 11 as desired and we have the solution.

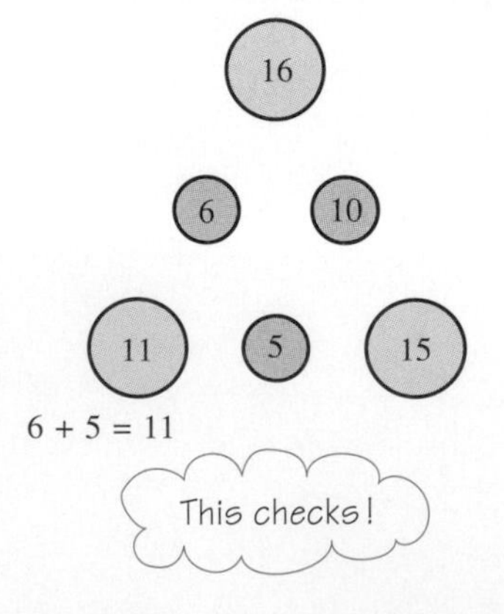

Look back • • • •

Guess and check worked fine. Our first choice of 10 for a was too large so we chose a smaller value. Our second choice of 5 was too small, but quite close. Choosing $a = 6$, which was between 10 and 5 but quite near 5, we obtained a solution that checked. Surely this approach would work equally well on other similar problems.

But wait. Have we fully understood this problem? Might there be an easier solution?

Look back at the initial example and also at the completed solution to the problem. Do you see any special relationship between the numbers in the large circles and those in the small circles?

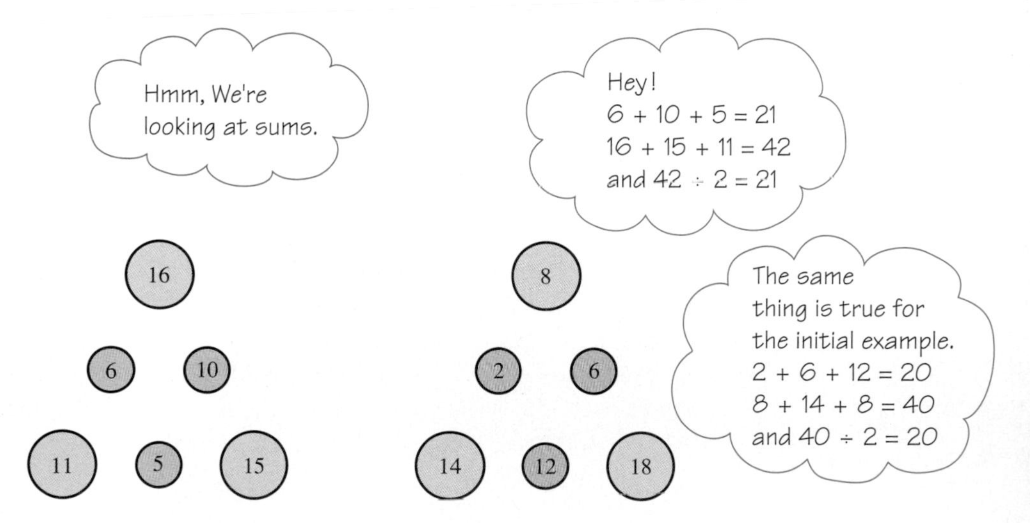

That's interesting; the sum of the numbers in the small circles in each case is just half the sum of the numbers in the large circles. Could we use this to find another solution method?

Sure! Since $16 + 15 + 11 = 42$ and $a + b + c$ is half as much, $a + b + c = 21$. But $a + b = 16$, so c must equal 5; that is,

$$c = 21 - 16 = 5$$
$$b = 21 - 11 = 10, \quad \text{and}$$
$$a = 21 - 15 = 6.$$

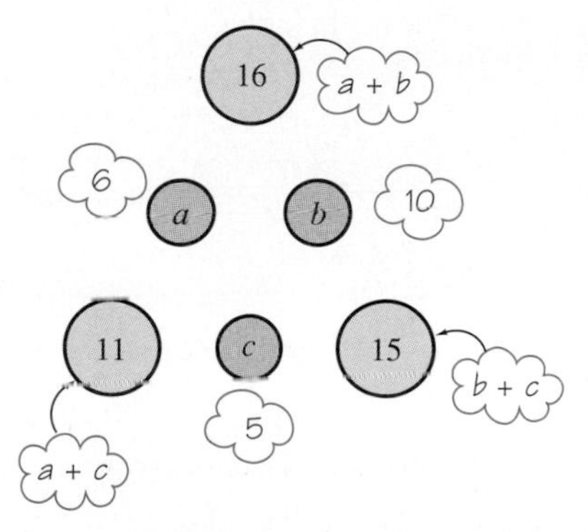

This is much easier than our first solution and, for that matter, the algebraic solution. Quickly now, does it work on this diagram? Try it.

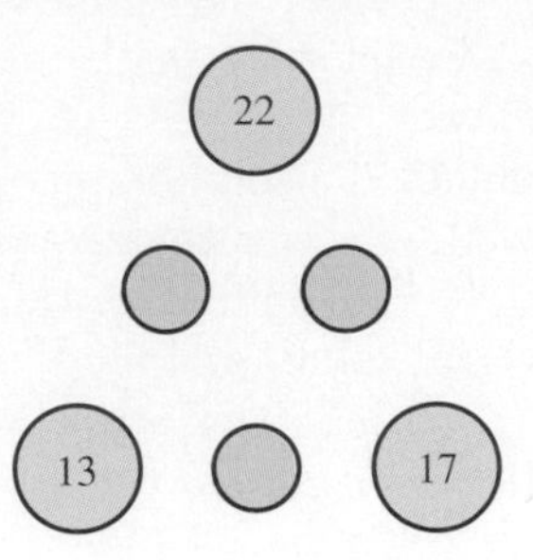

But there's one more thing. Do you understand *why* the sum of the numbers in the little circles equals half the sum of the numbers in the big circles? This diagram might help to make it clear.

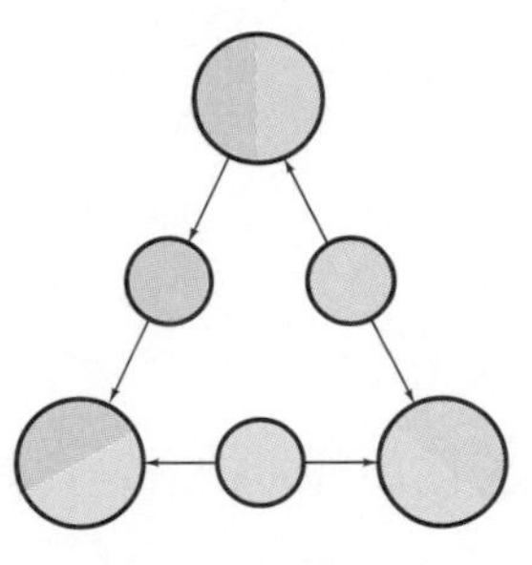

Make an Orderly List

PROBLEM-SOLVING STRATEGY 2 • Make an Orderly List

For problems that require consideration of many possibilities, make an orderly list or a table to make sure that no possibilities are missed.

Sometimes a problem is sufficiently involved that the task of sorting out all the possibilities seems quite forbidding. Often these problems can be solved by making a carefully structured list so that you can be sure that all of the data and all of the cases have been considered. Consider the next example.

EXAMPLE 1.4

Make an Orderly List

How many different total scores could you make if you hit the dart board shown with three darts?

1
5
10

SOLUTION

Understand the problem • • • •

Three darts hit the dart board and each scores a 1, 5, or 10. The total score is the sum of the scores for the three darts. There could be three 1s, two 1s and a 5, one 5 and two 10s, and so on. The fact that we are told to find the total score when throwing three darts at a dart board is just a way of asking what sums can be made using three numbers, each of which is either 1, 5, or 10.

Devise a plan • • • •

If we just write down sums hit or miss, we will almost surely overlook some of the possibilities. Using an orderly scheme instead, we can make sure that we obtain all possible scores. Let's make such a list. We first list the score if we have three 1s, then two 1s and one 5, then two 1s and no 5s, and so on. In this way, we can be sure that no score is missed.

Carry out the plan • • • •

Number of 1s	Number of 5s	Number of 10s	Total score
3	0	0	3
2	1	0	7
2	0	1	12
1	2	0	11
1	1	1	16
1	0	2	21
0	3	0	15
0	2	1	20
0	1	2	25
0	0	3	30

The possible total scores are listed.

Look back • • • •

Here the key to the solution was in being very systematic. We were careful first to obtain all possible scores with three 1s, then two 1s, then no 1s. With two 1s there could be either a 5 or a 10 as shown. For one 1 the only possibilities are two 5s and no 10s, one 5 and one 10, or no 5s and two 10s. Constructing the table in this orderly way makes it clear that we have not missed any possibilities. ■

Draw a Diagram

PROBLEM-SOLVING STRATEGY 3 • Draw a Diagram

Draw a diagram or picture that represents the data of a problem as accurately as possible.

JUST FOR FUN

How Many Heaps?

Into the bright and refreshing outskirts of a forest, which was full of numerous trees with their branches bent down with the weight of flowers and fruits—trees such as jambu, lime, plantains, areca palms, jack trees, date palms, hintala, palmyra, punnags, and mangos—a number of weary travelers entered with joy. Sixty-three heaps of plantain fruits were put together and combined with seven more plantain fruits, and these were equally distributed among 23 travelers so as to have no remainder. If there were the same number of plantain fruits in each heap, what is the least number of plantain fruits each traveler could have received?

—Brahmagupta (d. A.D. 660)

The aphorism "a picture is worth a thousand words" is certainly applicable to solving many problems. Language used to describe situations and state problems often can be clarified by drawing a suitable diagram, and unforseen relationships and properties often become clear. As with the problem of the pigs and chickens on Old MacDonald's farm, even problems that do not appear to have pictorial relationships can sometimes be solved using this technique. Would you immediately draw a picture in attempting to solve the problem in the next example? Some would and some wouldn't, but it's surely the most efficient approach.

EXAMPLE 1.5

Draw a Diagram

In a stock car race the first five finishers in some order were a Ford, a Pontiac, a Chevrolet, a Buick, and a Dodge.

(a) The Ford finished seven seconds before the Chevrolet.
(b) The Pontiac finished six seconds after the Buick.
(c) The Dodge finished eight seconds after the Buick.
(d) The Chevrolet finished two seconds before the Pontiac.

In what order did the cars finish the race?

SOLUTION

Devise a plan

Imagine the cars in a line as they race toward the finish. If they do not pass one another, this is the order in which they will finish the race. We can draw a line to represent the track at the finish of the race and place the cars on it according to the conditions of the problem. Mark the line off in time intervals of one second. Then, using the first letter of each car's name to represent the car, see if we can line up B, C, D, F, and P according to the given information.

Carry out the plan

Here is a line with equally spaced points to represent one second time intervals. Pick some point and label it C to represent the Chevrolet's finishing position.

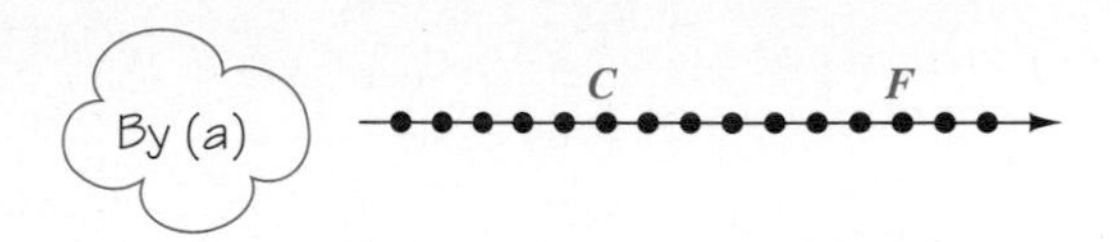

Then F is seven seconds ahead of C by condition (a) as shown above. Conditions (b) and (c) cannot yet be used since they do not relate to the positions of either C or F. However, (d) allows us to place P two seconds behind (to the left) of C as shown.

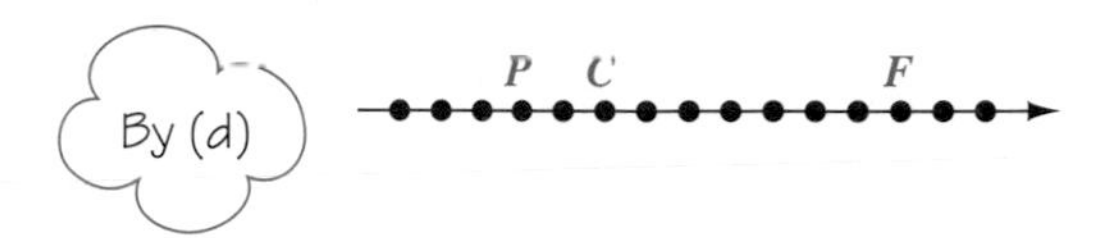

Since (b) relates the finishing position of B to P, we place B six seconds ahead of P.

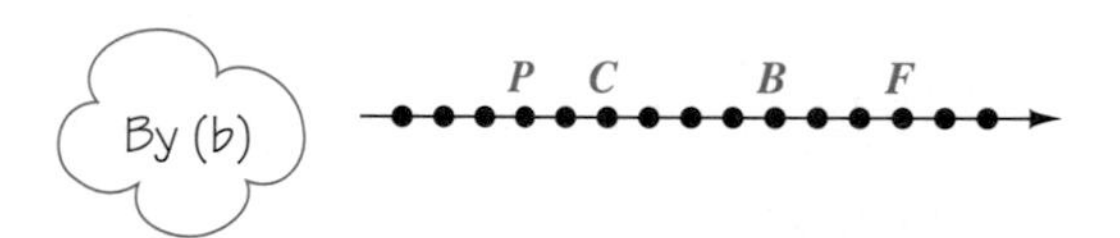

Similarly, (c) relates the finishing positions of D and B and allows us to place D eight seconds behind (to the left) of B. Since this accounts for all the cars, a glance at the last diagram reveals the order in which the cars finished the race. We repeatedly drew the line for pedagogical purposes in order to show the placement of the cars as each new condition was used. Ordinarily, all the work would be done on a single line since it is not necessary to show what happens at each stage as we did here.

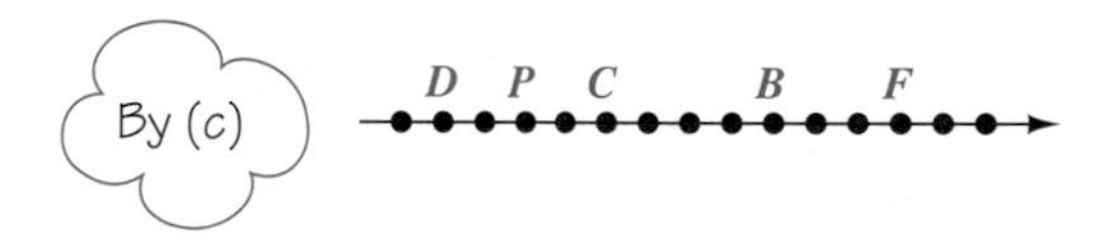

Look back • • • •

Like the problem of the pigs and chickens on Old MacDonald's farm, this problem did not immediately suggest drawing a picture. However, having seen pictures used to solve these problems will help you to see how pictures can be used to solve other even vaguely related problems. ■

PROBLEM SET 1.3

1. Libby Zeitler is thinking of a number. If you multiply it by 5 and add 13 you get 48. Could Libby's number be 10? Why or why not?
2. John Young is thinking of a number. If you divide it by 2 and add 16, you get 28. What number is John thinking of?
3. Lisa Wunderle is thinking of a number. If you multiply it by 7 and subtract 4, you get 17. What is the number?
4. Vicky Valerin is thinking of a number. Twice the number increased by 1 is 5 less than three times the number. What is the number? (*Hint:* For each guess, compute two numbers and compare.)
5. In Mrs. Garcia's class they sometimes play a game called **Guess My Rule.** The student who is IT makes up a rule for changing one number into another. The other students then call out numbers and the person who is IT tells what number the

rule gives back. The first person in the class to guess the rule then becomes IT and gets to make up a new rule.

(a) For Juan's rule the results were:

Numbers chosen	2	5	4	0	8
Numbers Juan gave back	7	22	17	−3	37

Could Juan's rule have been, "multiply the chosen number by 5 and subtract 3?" Could it have been, "reduce the chosen number by 1, multiply the result by 5 and then add 2?" Are these rules really different? Discuss briefly.

(b) For Mary's rule, the results were:

Numbers chosen	3	7	1	0	9
Numbers Mary gave back	10	50	2	1	82

What is Mary's rule?

(c) For Peter's rule, the results were:

Numbers chosen	0	1	2	3	4
Numbers Peter gave back	7	10	13	16	19

Observe that the students began to choose the numbers in order starting with 0. Why is that a good idea? What is Peter's rule?

6. Put the numbers 2, 3, 4, 5, and 6 in the circles to make the sum across and the sum down equal 12.

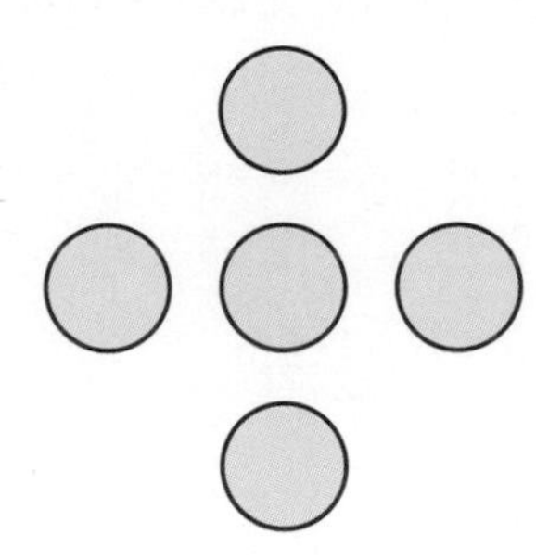

7. As in Example 1.3, the numbers in the big circles are the sums of the numbers in the two small adjacent circles. Place numbers in the empty circles in each of these arrays so that the same scheme holds.

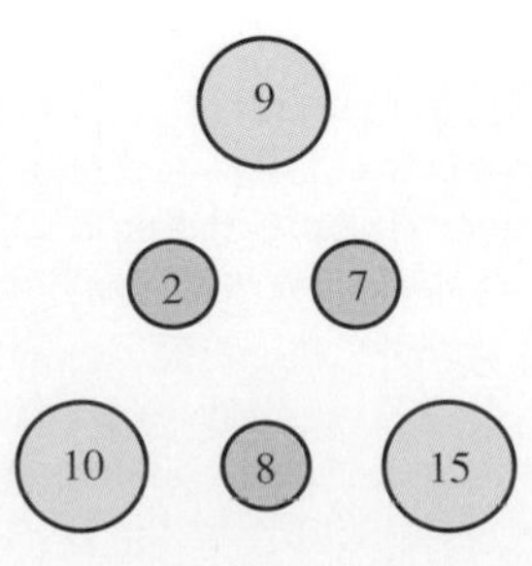

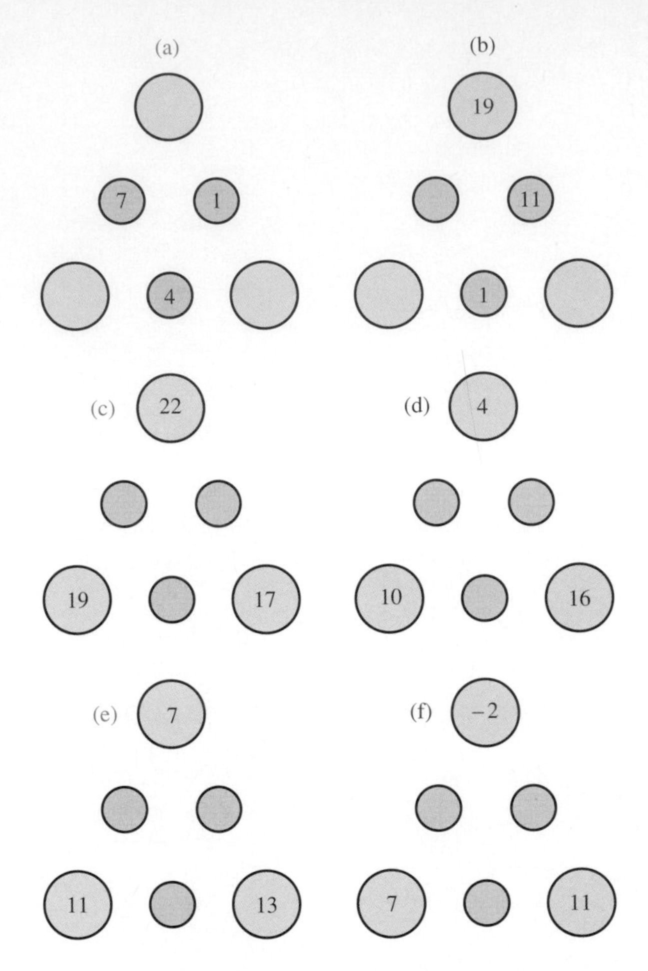

8. How many different amounts of money can you pay if you use four coins including only nickels, dimes, and quarters?

9. How many different ways can you make change for a 50 cent coin using quarters, nickels, dimes, and pennies?

10. List the 3-digit numbers that can be written using each of the digits 2, 5, and 8 once and only once.

11. List the 4-digit numbers that can be written using each of 1, 3, 5, and 7 once and only once.

12. When Anita Virgilio made a purchase she gave the clerk a dollar and received 21 cents in change. Complete this table to show what Anita's change could have been.

Number of Dimes	Number of Nickels	Number of Pennies
2	0	1

13. Julie Rislov has 25 pearls. She put them in three velvet bags with an odd number of pearls in each bag. What are the possibilities?

14. A rectangle has an area of 120 cm^2. Its length and width are whole numbers.

    (a) What are the possibilities for the two numbers?

    (b) Which possibility gives the smallest perimeter?

15. The product of two whole numbers is 96 and their sum is less than 30. What are the possibilities for the two numbers?

16. Peter and Jill each worked a different number of days but earned the same amount of money. Use these clues to determine how many days each worked:

    Peter earned $20 a day.
    Jill earned $30 a day.
    Peter worked five more days than Jill.

17. Bob Straub can cut through a log in one minute. How long will it take Bob to cut a 20 foot log into 2-foot sections? (*Hint:* Draw a diagram.)

18. How many posts does it take to support a straight fence 200 yards long if a post is placed every 20 yards?

19. How many posts does it take to support a fence around a square field measuring 200 yards on a side if posts are placed every 20 yards?

20. Albright, Badgett, Chalmers, Dawkins, and Ertl all entered the primary to seek election to the city council. Albright received 2000 more votes than Badgett and 4000 less than Chalmers. Ertl received 2000 votes less than Dawkins and 5000 votes more than Badgett. In what order did each person finish in the balloting?

21. Nine square tiles are laid out on a table so that they make a solid pattern. Each tile must touch at least one other tile along an entire edge. The squares all have sides of length one.

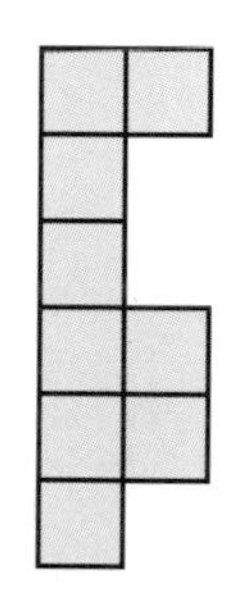

    (a) What are the possible perimeters of the figures that can be formed?

    (b) Which figure has the least perimeter?

22. A 9 meter by 12 meter rectangular lawn has a concrete walk one meter wide all around it. What is the area of the walk?

1.4 More Problem-Solving Strategies

STRATEGIES

- ***Look for a pattern.***
- ***Make a table.***
- ***Use a variable.***
- ***Consider special cases.***
- ***Solve an equivalent problem.***
- ***Solve an easier, simpler problem.***
- ***Argue from special cases.***

Look for a Pattern

PROBLEM-SOLVING STRATEGY 4 • Look for a Pattern

Consider an ordered sequence of particular examples of the general situation described in a problem. Then carefully scrutinize these results, looking for a pattern that may be the key to the problem.

It is no overstatement to assert that this strategy is the most important of all problem-solving strategies. In fact, mathematics is often characterized as the study of patterns, and patterns occur in some form in almost all problem-solving situations. Think about the problems we have already considered and you will see patterns everywhere—numerical patterns, geometrical patterns, counting patterns, listing patterns, rhetorical patterns—patterns of all kinds.

Some problems, like those in the next examples, are plainly pattern problems, but looking for a pattern is almost never a bad way to start solving a problem.

EXAMPLE 1.6 Look for Patterns in Numerical Sequences

Continue these numerical sequences. Fill in the next three blanks in each part.

(a) 1, 4, 7, 10, 13, ______, ______, ______

(b) 19, 20, 22, 25, 29, ______, ______, ______

(c) 1, 1, 2, 3, 5, ______, ______, ______

(d) 1, 4, 9, 16, 25, ______, ______, ______

SOLUTION

Understand the problem • • • •

In each case we are asked to discover a reasonable pattern suggested by the first five numbers and then to continue the pattern for three more terms.

Devise a plan • • • •

Questions we might ask ourselves and answer in search of a pattern include: Are the numbers growing steadily larger? Steadily smaller? How is each number related to its predecessor? Is it perhaps the case that a particular term depends on its two predecessors? On its three predecessors? Perhaps each term depends in a special way on the number of the term in the sequence; can we notice any such a dependence? Are the numbers in the sequence somehow special numbers that we recognize? This is rather like playing *Guess My Rule.* Let's see how successful we can be.*

Carry out the plan • • • •

In each of (a), (b), (c), and (d) the numbers grow steadily larger. How is each term related to the preceding term or terms in each case? Are the terms related to their numbered place in the sequence? Do they have a special form we can recognize?

(a) For the sequence in part (a), each number listed is three larger than its predecessor. If this pattern continues, the next three entries will be 16, 19, and 22.

(b) Here the numbers increase by 1, by 2, by 3, and by 4. If we continue this scheme, the next three numbers will be 5 more, 6 more, and 7 more than their predecessors. This would give 34, 40, and 47.

(c) If we use the idea of (a) and (b) for this sequence, we should check how much larger each entry is than its predecessor. The numbers that must be added are *0*, *1*, *1*, and *2*; that is, $0 + 1 = 1$, $1 + 1 = 2$, $1 + 2 = 3$, $2 + 3 = 5$. This just amounts to adding any two consecutive terms of the sequence to obtain the next term. The next three terms must be 8, 13, and 21.

---

*Actually, there is a touchy point here. To be strictly accurate, **any** three numbers you choose in each case can be considered correct. There are actually infinitely many different rules that will give you any first five numbers followed by any three other numbers. What we seek here are relatively simple rules that apply to the given numbers and tell how to obtain the next three in each case.

SCHOOL BOOK PAGE *Looking for Patterns in Fifth Grade*

Practice

For More Practice, see Set B, pages 302–303.

Draw the next three figures in each pattern.
Copy and complete each table.

5.

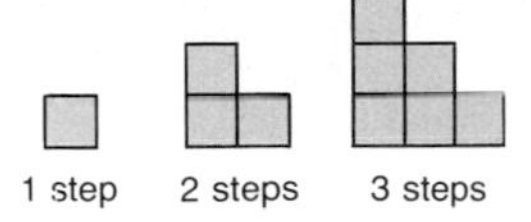

6. Number of

Steps	1	2	3	4	5	6
Squares	1	3	6			

7. How many squares are in 10 steps?

8.

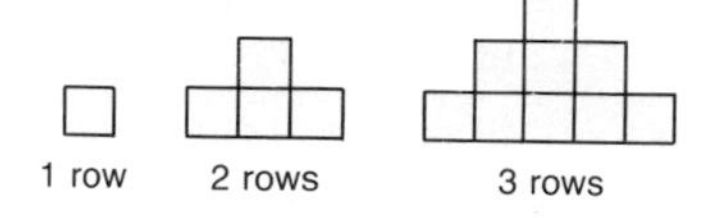

9. Number of

Rows	1	2	3	4	5	6
Squares	1	4	9			

10. How many squares are in 10 rows?

11.

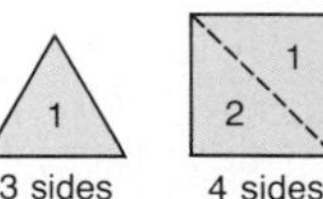

5 sides

12. Number of

Sides	3	4	5	6	7	8
Triangles	1	2	3			

13. How many triangles are in a figure with 10 sides?

Problem Solving

In the school cafeteria, 4 people can sit together at 1 table. If 2 tables are placed together, 6 people can sit together.

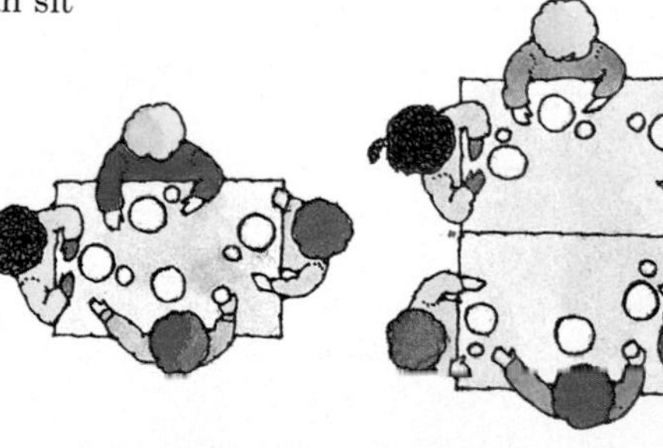

How many tables must be placed together in a row to seat

14. 10 people? **15.** 20 people? **16.** 30 people?

If the tables are placed together in a row, how many people can be seated using

17. 10 tables? **18.** 15 tables? **19.** 20 tables?

1. Do problems 5 through 12 above.
2. Do you think problems 11 through 13 are clearly stated? Explain briefly.
3. How many tables must be placed in a row as in the Problem Solving section above to seat 100 people?
4. If 50 tables were placed in a row as in the Problem Solving section above, how many people could be seated?

(d) Here 4 is 3 larger than 1, 9 is 5 larger than 4, 16 is 7 larger than 9, and 25 is 9 larger than 16. The terms seem to be increasing by the next largest **odd** number each time. Thus, the next three numbers should probably be $25 + 11 = 36$, $36 + 13 = 49$, and $49 + 15 = 64$. Alternatively, in this case, we may recognize that the numbers 1, 4, 9, 16, and 25 are special numbers. Thus, $1 = 1^2$, $4 = 2^2$, $9 = 3^2$, $16 = 4^2$, and $25 = 5^2$. The sequence appears to be just the sequence of **square numbers.** The 6th, 7th, and 8th terms are just $6^2 = 36$, $7^2 = 49$, and $8^2 = 64$ as before.

Look back • • • •

In all four sequences, we checked to see how much larger each number was than its predecessor. In each case we were able to discover a pattern that allowed us to write the next three terms of the sequence. In part (d), we also noted that the first term was 1^2, the second term was 2^2, the third term was 3^2, and so on. Thus, it was reasonable to guess that each term was the square of the number of its position in the sequence. This allowed us to write the next few terms with the same result as before. ■

We have already seen in earlier examples how making a table is often an excellent strategy, particularly when combined with the strategy of looking for a pattern. Like drawing a picture or making a diagram, making a table often reveals unexpected patterns and relationships that help to solve a problem.

Make a Table

PROBLEM-SOLVING STRATEGY 5 • Make a Table

Make a table reflecting the data in a problem. If done in an orderly way, such a table will often reveal patterns and relationships that suggest how the problem can be solved.

EXAMPLE 1.7 **Make a Table**

(a) Draw the next two diagrams to continue this dot sequence.

(b) How many dots are in each figure?

______, ______, ______, ______, ______, ______

(c) How many dots would be in the one hundredth figure?

(d) How many dots would be in the one millionth figure?

SOLUTION

Understand the problem • • • •

What is given?

In part (a), we are given an ordered sequence of arrays of dots. We are asked to recognize how the arrays are being formed and to continue the pattern for two more steps. In part (b), we are asked to record the number of dots in each array in part (a). In parts (c) and (d) we are asked to determine specific numerical terms in the sequence of part (b).

Devise a plan • • • •

In part (a), we are asked to continue the pattern of a sequence of arrays of dots. As with numerical sequences, our strategy will be to see how each array relates to its predecessor or predecessors hoping to discern a pattern that we can extend 2 more steps.

In part (b), we will simply count and record the numbers of dots in the successive arrays in part (a).

In parts (c) and (d), we will study the numerical sequence of part (b) just as we did in Example 1.6 hoping to discern a pattern and understand it sufficiently well that we can determine its one hundredth and one millionth terms.

Carry out the plan • • • •

In part (a), we observe that the arrays of dots are similar but that each one has a column of 2 more dots than its predecessor. Thus, the next two arrays are

```
  • • • •           • • • • •
• • • • •   and   • • • • • •
```

For part (b), we count the dots in each array of part (a) to obtain

$$1, 3, 5, 7, 9, 11, \cdots$$

These are just the odd numbers, and we could write out the first 1 million odd numbers and so answer parts (c) and (d). But surely there's an easier way.

Let's review how the successive terms were obtained. A table may help.

Number of Entry	Entry
1	$1 = 1$
2	$3 = 1 + 2$
3	$5 = 1 + 2 + 2 = 1 + 2 \times 2$
4	$7 = 1 + 2 + 2 + 2 = 1 + 3 \times 2$
5	$9 = 1 + 2 + 2 + 2 + 2 = 1 + 4 \times 2$

Reviewing this carefully, we finally experience an Aha!

The *second* term is $1 + 1 \times 2$ ($2-1$)

The *third* term is $1 + 2 \times 2$ ($3-1$)

The *fourth* term is $1 + 3 \times 2$ ($4-1$)

The number of twos added is one less than the number of the term. Therefore, the one hundredth term is

$$1 + 99 \times 2 = 199 \quad (100-1)$$

and the one millionth term is

$$1 + (1000000 - 1) \times 2 = 1 + 999999 \times 2 = 1999999.$$

Look back • • • •

The basic observation was that each diagram could be obtained by adding an additional column of two dots to its predecessor. Hence, the successive terms in the numerical sequence were obtained by adding 2 to each entry to get the next entry. Using this notion, we examined the successive terms and discovered that any entry could be found by subtracting 1 from the number of the entry, doubling the result, and adding 1. But this last sentence is rather cumbersome and we have already seen that using *symbols* can make it easier to make mathematical statements. If we use n for the number of the term, the above sentence can be translated into this mathematical sentence:

$$e_n = (n - 1) \times 2 + 1 = 2n - 2 + 1 = 2n - 1.*$$

Here e_n, read "e sub n," is the formula that gives the nth entry in the sequence of part (b). All we have to do is replace n by 1, 2, 100, and so on, to find the 1st entry, the 2nd entry, the one hundredth entry, and so on. Thus,

$$e_1 = 2 \times 1 - 1 = 1,$$
$$e_2 = 2 \times 2 - 1 = 3,$$
$$e_{100} = 2 \times 100 - 1 = 199,$$

and so on. ■

$2n - 1$ is the nth odd number.

Calculator Note

Many electronic calculators have a constant function. Pressing the keys 1 [+] 2 [=] [=] [=] causes the calculator to add 2 to 1 three times. With such a calculator, the successive terms of the sequence of part (b) in the preceding example can be obtained by pressing keys 1 [+] 2 and then pressing [=] for each new term. Try this to show that 49 is the twenty-fifth odd number.

Use a Variable

In the preceding example, and even earlier, we saw how using symbols or variables often makes it easier to express mathematical ideas and so to solve problems.

* The formula $e_n = 2n - 1$ is an example of a function. Functions are of considerable importance in mathematics and will be discussed in detail in Chapter 2.

HIGHLIGHT FROM HISTORY
Carl Friedrich Gauss (1777–1855)

Sometimes called the "prince of mathematicians" and always rated as one of the three greatest mathematicians who ever lived, Carl Friedrich Gauss was born of poor parents in Braunschweig, Germany in 1777. His father was a self-righteous lout; his mother an intelligent but barely literate housewife. Gauss, on the other hand was a genius who began making significant mathematical discoveries while still in his teens. Gauss contributed many original ideas across the spectrum of science and engineering, but he characterized mathematics as the "queen of the sciences."

EXAMPLE 1.8 **Use a Variable—Gauss's Trick**

Carl Gauss (1777–1855) is generally acknowledged as one of the three greatest mathematicians of all time. When just a young school boy, the teacher instructed the students in his class to add all the numbers from 1 to 100 expecting this to take a long time. To the teacher's surprise, young Gauss completed the task in about half a minute.

SOLUTION

Gauss's strategy was to use a variable. If

$$s = 1 + 2 + 3 + \cdots + 100,$$

then

$$s = 100 + 99 + 98 + \cdots + 1.$$

Therefore,

$$2s = 101 + 101 + 101 + \cdots + 101,$$
$$2s = 100 \times 101,$$

a sum with 100 terms

and

$$s = \frac{100 \times 101}{2} = 5050.$$

■

DEFINITION A Variable

A **variable** is a letter that can represent any of the numbers of some set of numbers.

Sometimes we want to use a variable in representing the general term in a sequence. Thus, as we saw in Example 1.7, $2n - 1$ is the nth odd number. In the expression $2n - 1$, n is the variable and it can be replaced by any natural number.

Other times we want to find the numerical replacement for a variable that makes a statement true. For example, if we want to know which odd number 85 is, we need to determine n such that

$$2n - 1 = 85.$$

This implies that $2n = 86$ and so $n = 43$. Hence, 85 is the 43rd odd number.

PROBLEM-SOLVING STRATEGY 6 • Use a Variable

Often a problem requires that a number be determined. Represent the number by a variable and use the conditions of the problem to set up an equation that can be solved to ascertain the desired number.

EXAMPLE 1.9

Use a Variable to Determine a General Formula

Look at these corresponding geometrical and numerical sequences.

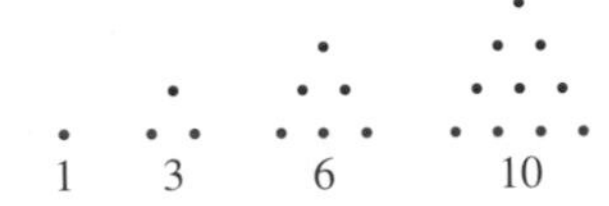

For fairly obvious reasons, the numbers 1, 3, 6, 10 are called **triangular numbers** (the diagram with a single dot is considered a *degenerate triangle*). The numbers 1, 3, 6, and 10 are the first four triangular numbers. Find a formula for the nth triangular number.

SOLUTION

Understand the problem

Having just gone through a similar problem in Example 1.7, we understand that we are to find a formula for t_n, the nth triangular number.

Devise a plan

The geometrical and numerical sequences in the statement of the problem suggest that we look for a pattern. How is each diagram related to its predecessor? How is each triangular number related to its predecessor?

Carry out the plan

We add a diagonal of 2 dots to the first diagram to obtain the second, a diagonal of 3 dots to the second diagram to obtain the third, and so on. Thus, the next three diagrams should be as shown here. Numerically, we add 2 to the first triangular number to obtain the second, 3 to the second triangular number to obtain the third, and so on. To make this even more clear, we can construct the following table.

Number of Entry	Entry
1	$t_1 = 1$
2	$t_2 = 1 + 2 = 3$
3	$t_3 = 1 + 2 + 3 = 6$
4	$t_4 = 1 + 2 + 3 + 4 = 10$
5	$t_5 = 1 + 2 + 3 + 4 + 5 = 15$

Indeed, it appears that

$$t_n = 1 + 2 + 3 + \cdots + n;$$

but we can still do better. Let

$$t_n = 1 + 2 + 3 + \cdots + n.$$

Then, using Gauss's trick,

$$t_n = n + (n - 1) + (n - 2) + \cdots + 1.$$

So,

$$\begin{aligned} 2t_n &= (n + 1) + (n + 1) + (n + 1) + \cdots + (n + 1) \\ &= n(n + 1) \end{aligned}$$

$1 + n = n + 1$
$2 + (n - 1) = n + 1$
$3 + (n - 2) = n + 1$
$\vdots$

n terms

and

$$t_n = \frac{n(n + 1)}{2}$$

as required.

Look back • • • •

Looking back, the key to our solution lay in considering the sequence of special cases t_1, t_2, t_3, t_4, and t_5 and in looking for a pattern. This approach deserves special recognition as a problem-solving strategy. We call it **considering special cases.** ■

Consider Special Cases

PROBLEM-SOLVING STRATEGY 7 • Consider Special Cases

In trying to solve a complex problem, consider a sequence of special cases. This will often show how to proceed naturally from case to case until one arrives at the case in question. Alternatively, the special cases may reveal a pattern that makes it possible to solve the problem.

Pascal's Triangle

One of the most interesting and useful patterns in all of mathematics is the numerical array called Pascal's triangle.

Consider the problem of finding how many different paths there are from A to P on the grid shown in Figure 1.1 if you can only move **down** along edges in the grid. If we start to trace out paths willy-nilly, our chances of finding all possibilities are not good. A better approach is to notice that any path from A to P must contain four moves downward and to the left along the edges of small squares in the grid and six moves downward and to the right. Thus, the question becomes one of finding the number of different ways we can arrange 4 Ls (for left) and 6 Rs (for right) in order. This, by the way, illustrates another important problem-solving strategy, the strategy of **solving an equivalent problem.** The idea is to find a problem that is equivalent to the original problem that may be easier to solve. Here, finding how many ways you can put four Ls and six Rs in order using the strategy *make an orderly list* is easier than the original problem of finding the number of paths from A to P on the grid. Even this approach with four Ls and six Rs is sufficiently complicated that it is probably better to see if we can find yet another strategy.

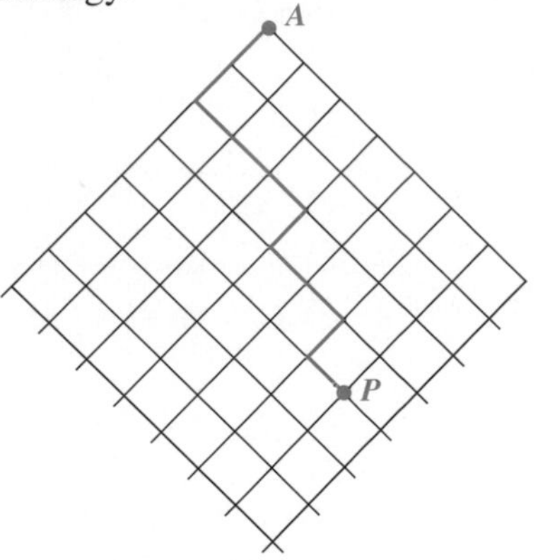

Figure 1.1
A path from A to P

A strategy that is often helpful is to **solve an easier similar problem.** It would certainly be easier if P were not so far down in the grid. Consider the easier similar problem of finding the number of paths from A to E in Figure 1.2.

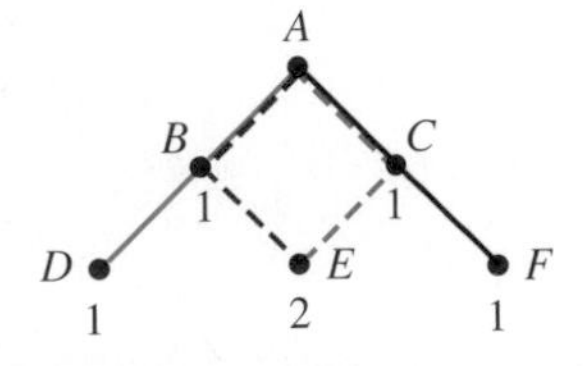

Figure 1.2
Solving an easier similar problem

Or, consider solving a similar *set* of easier problems all at once. How many different paths are there from A to each of B, C, D, E, and F? (Note that this is an example of considering a series of special cases.) Clearly, there is only one way to go from A to each of B and C, and we indicate this by the 1s under B and C in Figure 1.2. Also, the only route to D is through B, so there is only one path from A to D as indicated. For the same reason, there is one path from A to F. On the other hand, there are two ways to go from A to E—one route through B and one through C. We indicate this by placing a 2 under E on the diagram. This certainly doesn't solve the original problem, but it gives us a start and even suggests how we might proceed. Consider the diagram of Figure 1.3.

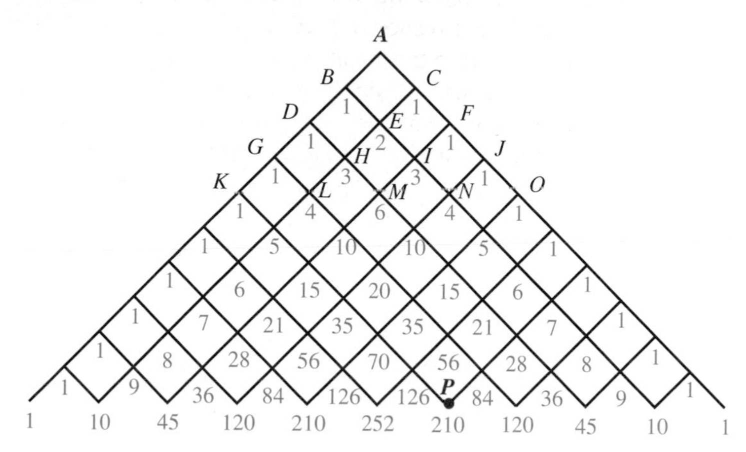

Figure 1.3
The number of paths from A to P

Having determined the number of paths from A to each of B, C, D, E, and F, could we perhaps determine the number of paths to G, H, I, and J and then continue on down the grid to eventually solve the original problem?

1. Always moving downward, the only way to get to G from A is via D. But there is only one path to D and only one path from D to G. Thus, there is only one path from A to G and we enter a 1 under G on the diagram as shown.
2. The only way to get to H from A is via D or E. Since there is only one path from A to D and one from D to H, there is only one path from A to H via D. However, since there are two paths from A to E and one path from E to H, there are two paths from A to H via E. The number of paths from A to H is the number via D plus the number via E for $1 + 2 = 3$ paths, and we enter 3 under H on the diagram as shown.
3. The arguments for I and J are the same as for H and G, so we enter 3 and 1 under I and J on the diagram.
4. But this reveals a very nice pattern that enables us to solve the original problem with ease. There can be only one path to any edge vertex on the grid since we have to go straight down the edge to get to such a vertex. For interior points on the grid, however, we can always reach the grid by paths through the points immediately above and to the left and right of such a point. Since there is only one path from each of these points to the point in question, the total number of paths to this point is the *sum* of the number of paths to these preceding two points. Thus, we easily generate the number of paths from A to any given point in the grid by simple addition. In particular, there are 210 different paths from A to P as we initially set out to determine.

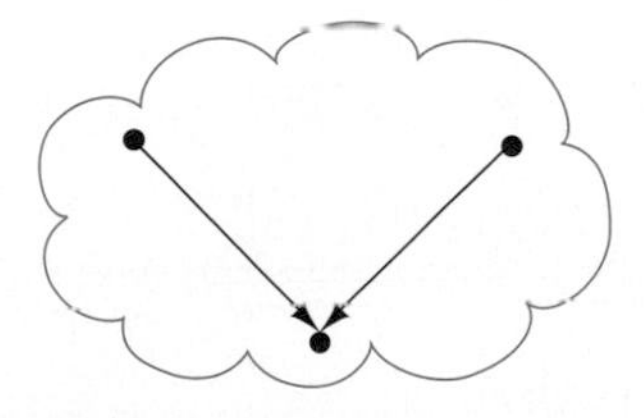

Without the grid and with an additional 1 at the top to complete a triangle, the number array of Figure 1.4 is called **Pascal's triangle.**

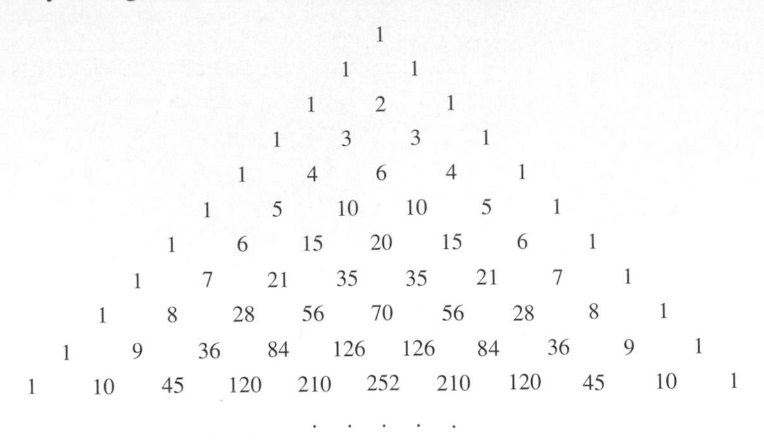

Figure 1.4
Pascal's triangle

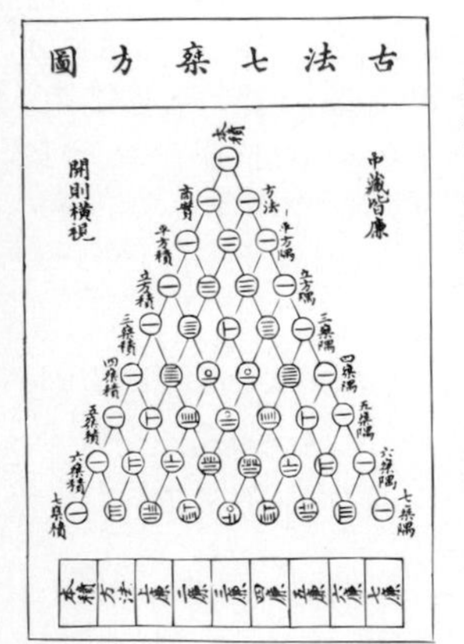

Figure 1.5
Pascal's triangle from Chu Shih—Chieh's *Ssu Yuan Yü Chien*, A.D. 1303

The array is named after the French mathematician Blaise Pascal (1623–1662) who showed that these numbers play an important role in the theory of probability. However, the triangle was certainly known in China as early as the twelfth century. An interesting and clear depiction of the famous triangle from a fourteenth century manuscript is shown in Figure 1.5.

Pascal's triangle is rich with remarkable patterns and is also extremely useful.

Before discussing the patterns, we observe that it is customary to call the single 1 at the top of the triangle the 0th row (since, for example, in the path counting problem discussed above, this 1 would represent a path of length 0). For consistency, we will also call the initial 1 in any row the 0th element in the row, and the initial diagonal of 1s the 0th diagonal. (See Figure 1.6.) Thus, 1 is the zeroth element in the fourth row, 4 is the first element, 6 the second element, and so on.

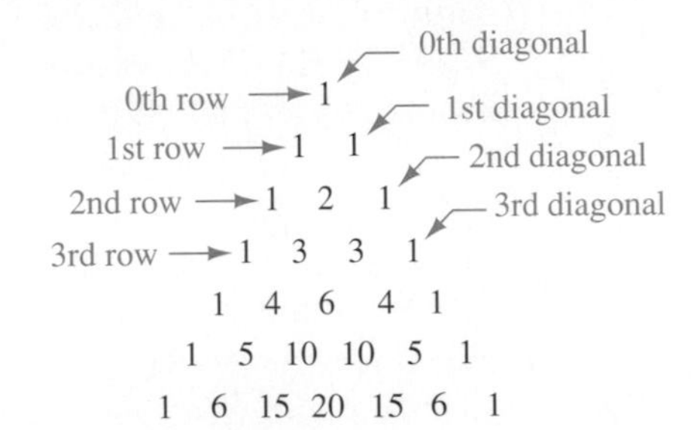

Figure 1.6
Numbered rows and diagonals in Pascal's triangle

EXAMPLE 1.10 Find a Pattern in the Row Sums of Pascal's Triangle

(a) Compute the sum of the elements in each of rows zero through four of Pascal's triangle.

(b) Look for a pattern in the results of part (a) and guess a general rule.

(c) Use Figure 1.4 to check your guess for rows five through eight.

(d) Give a convincing argument that your guess in part (b) is correct.

SOLUTION

Understand the problem • • • •

For part (a), we must add the elements in the indicated rows. For part (b), we are asked to discover a pattern in the numbers generated in part (a). For part (c), we must compute the sums for four more rows and see if the results obtained continue the pattern guessed in part (b). In part (d), we are asked to argue convincingly that our guess in part (b) is correct.

Devise a plan • • • •

Part (a) is certainly straightforward; we must compute the desired sums. To find the pattern requested in part (b) we should ask the question, "Have we ever seen a similar problem before?" The answer, of course, is a resounding yes—all the problems in this section, but particularly Example 1.6, have involved looking for patterns. Surely, the techniques that succeeded earlier should be tried here. Appropriate questions to ask and answer include: "How are the successive numbers related to their predecessors?" Are the numbers special numbers that we can easily recognize?" "Can we relate the successive numbers to their numbered location in the sequence of numbers being generated?" Answering these questions should help us make the desired guess. For part (c) we will compute the sums of the elements in rows five through eight to see if these numbers agree with our guess in part (b). If they *don't* agree, we will go back and modify our guess. If they *do* agree we will proceed to part (d) and try to make a convincing argument that our guess is correct. About all we have to go on is the fact that the initial and terminal elements in each row are 1s and that the sum of any two consecutive elements in a row is the element between these two elements but in the next row down.

Carry out the plan • • • •

(a) $1 = 1$
$1 + 1 = 2$
$1 + 2 + 1 = 4$
$1 + 3 + 3 + 1 = 8$
$1 + 4 + 6 + 4 + 1 = 16$

(b) It appears that each number in part (a) is just twice its predecessor. The numbers are just

$$1, 2 \cdot 1 = 2, 2 \cdot 2 = 2^2, 2 \cdot 2^2 = 2^3, 2 \cdot 2^3 = 2^4.$$

It appears that the sum of the elements of

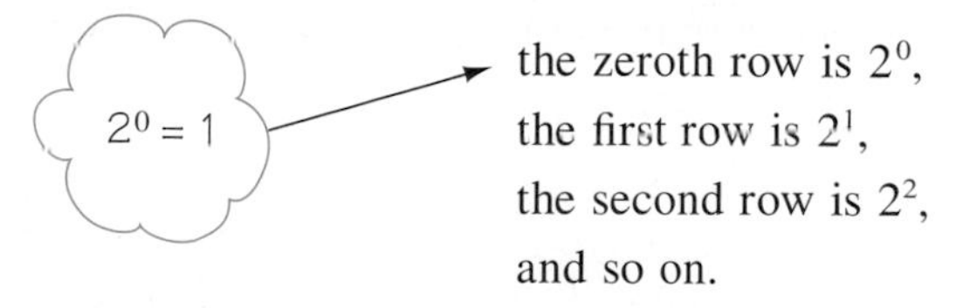

the zeroth row is 2^0,
the first row is 2^1,
the second row is 2^2,
and so on.

Our guess is that the sum of the elements in the nth row is 2^n.

(c) Computing these sums for the next four rows, we have

5th row	$1 + 5 + 10 + 10 + 5 + 1 = 32 = 2^5$
6th row	$1 + 6 + 15 + 20 + 15 + 6 + 1 = 64 = 2^6$
7th row	$1 + 7 + 21 + 35 + 21 + 7 + 1 = 128 = 2^7$
8th row	$1 + 8 + 28 + 56 + 70 + 56 + 28 + 8 + 1 = 256 = 2^8$

Since these results do not contradict our guess, we proceed to try to make a convincing argument that our guess is correct.

(d) What happens as we go from one row to the next? How is the next row obtained? Consider the third and fourth rows shown here. The arrows show how the fourth row is obtained from the third, and we see that each of the 1, 3, 3, and 1 in the third row appears *twice* in the sum of the elements in the fourth row. Since this argument would hold for any two consecutive rows, the sum of the numbers in any row is just twice the sum of the numbers in the preceding row. Hence, from above, the sum of the numbers in the 11th row must be $2 \cdot 2^{10} = 2^{11}$, in the 12th row it must be $2 \cdot 2^{11} = 2^{12}$, and so on. Thus, the result is true in general as claimed.

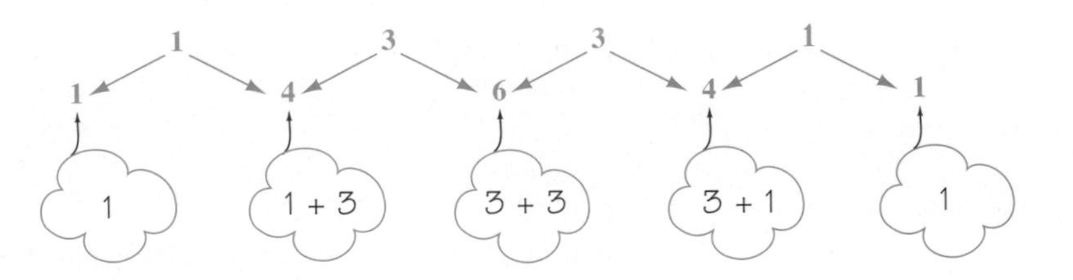

Look back ••••

Several aspects of our solution merit special comment. Beginning with the strategies of considering special cases and looking for a pattern, we were led to guess that the sum of the elements in the nth row of the triangle is 2^n. In an attempt to argue that this guess was correct, we considered the special case of obtaining the fourth row from the third. This showed that the sum of the elements in the fourth row was twice the sum of the elements in the third row. *Since the argument did not depend on the actual numbers appearing in the third and fourth rows but only on the general rule of formation of the triangle,* it would hold for any two consecutive rows and so actually proves that our conjecture was correct. This type of argument is called **arguing from a special case,** and is an important problem-solving strategy. ■

WORTH READING

Aha!

Experimental psychologists like to tell a story about a professor who investigated the ability of chimpanzees to solve problems. A banana was suspended from the center of the ceiling, at a height that the chimp could not reach by jumping. The room was bare of all objects except several packing crates placed around the room at random. The test was to see whether a lady chimp would think of first stacking the crates in the center of the room, and then of climbing on top of the crates to get the banana.

The chimp sat quietly in a corner, watching the psychologist arrange the crates. She waited patiently until the professor crossed the middle of the room. When he was directly below the fruit, the chimp suddenly jumped on his shoulder, then leaped into the air and grabbed the banana.

The moral of this anecdote is: A problem that seems difficult may have a simple, unexpected solution. In this case the chimp may have been doing no more than following her instincts or past experience, but the point is that the chimp solved the problem in a direct way that the professor had failed to anticipate.

This quotation is from the introduction to Martin Gardner's delightful book, *Aha! Insight* (Scientific American, Inc./W. H. Freeman and Company, 1978). It is full of interesting, thought-provoking problems as well as insightful solutions and analyses of the problem-solving process.

In considering Pascal's triangle we found it helpful to use three additional problem-solving strategies that are well worth highlighting.

PROBLEM-SOLVING STRATEGY 8 • Solve an Equivalent Problem

Try to recast the problem in totally different terms that change it into a new, different, but completely equivalent problem. Solving the equivalent problem also solves the original problem, but it may be easier.

PROBLEM-SOLVING STRATEGY 9 • Solve an Easier Similar Problem

Instead of attempting immediately to solve a problem in general or for a reasonably large value like 10 or 20, try first to solve it for small values like 2 or 3, or even 1, 2, and 3. This may show how to solve the larger problem.

PROBLEM-SOLVING STRATEGY 10 • Argue from Special Cases

A convincing argument for a general case can often be made by discussing a special case but only using those features of the special case that are typical of the general case.

PROBLEM SET 1.4

1. Look for a pattern and fill in the next three blanks with the most likely choices for each sequence.
   (a) 2, 5, 8, 11, ____, ____, ____
   (b) $-5, -3, -1, 1$, ____, ____, ____
   (c) 1, 1, 3, 3, 6, 6, 10, ____, ____, ____
   (d) 1, 3, 4, 7, 11, ____, ____, ____
   (e) 2, 6, 18, 54, ____, ____, ____.

2. (a) Draw three diagrams to continue this dot sequence.
   (b) What number sequence corresponds to the pattern of part (a)?
   (c) What is the tenth term in the sequence of part (b)? The one hundredth term?

(d) Which even number is $2n$?

(e) What term in the sequence is 2402?

(f) Compute the sums $2 + 4 + 6 + \cdots + 2402$. (*Hint:* Use Gauss's trick.)

3. **(a)** Fill in the blanks to continue this dot sequence in the most likely way.

```
        •      • •      • • •
•     • •    • • •    • • • •
•,    • •,   • • •,   • • • •,    ____, ____
```

(b) What number sequence corresponds to the sequence of dot patterns of part (a)?

(c) What is the tenth term in the sequence of part (b)? The one hundredth term?

(d) Which term in the sequence is 101? (*Hint:* How many 3s must be added to 2 to get 101?)

(e) Compute the sum $2 + 5 + 8 + \cdots + 101$. (*Hint:* Use Gauss's trick and the result of part (d).)

4. Sequences like 2, 5, 8, $\cdots$, where each term is greater (or less) than its predecessor by a constant amount, are called **arithmetic** (a-rith-me´tic) **progressions.** Find the number of terms in each of these arithmetic progressions.

(a) 5, 7, 9, $\cdots$, 35

(b) -4, 1, 6, $\cdots$, 46

(c) 3, 7, 11, $\cdots$, 67

5. Compute the sum of each of these arithmetic progressions.

(a) $5 + 7 + 9 + \cdots + 35$

(b) $-4 + 1 + 6 + \cdots + 46$

(c) $3 + 7 + 11 + \cdots + 67$

(d) $1 + 7 + 13 + \cdots + 73$

6. Consider the arithmetic progression: 2, 9, 16, 23, . . . , 86.

(a) How many 7s must be added to 2 to obtain 86?

(b) Compute the sum $2 + 9 + 16 + \cdots + 86$.

(c) What is the nth number in the progression? (*Hint:* Show that $2 + 7(n - 1) = 7n - 5$.)

(d) Compute the sum $2 + 9 + 16 + \cdots + (7n - 5)$.

7. (a) Fill in the blanks to continue this sequence of equations.

$$1 = 1$$
$$1 + 2 + 1 = 4$$
$$1 + 2 + 3 + 2 + 1 = 9$$
$$1 + 2 + 3 + 4 + 3 + 2 + 1 = 16$$

________________________ = ______

________________________ = ______

(b) Compute this sum.

$1 + 2 + 3 + \cdots + 99 + 100 + 99 + \cdots + 3 + 2 + 1 =$ ________

(c) Fill in the blank to complete this equation.

$1 + 2 + 3 + \cdots + (n - 1) + n + (n - 1) + \cdots + 3 + 2 + 1 =$ ________

8. **(a)** Fill in the blanks to continue this sequence of equations.

$$1 = 0 + 1$$
$$1 + 3 + 1 = 1 + 4$$
$$1 + 3 + 5 + 3 + 1 = 4 + 9$$

________________ = ______

________________ = ______

(b) What expression, suggested by part (a), should be placed in the blank to complete this equation?

$1 + 3 + 5 + \cdots + (2n - 3) + (2n - 1) + (2n - 3) + \cdots + 5 + 3 + 1 =$ ________

(*Hint:* The number preceding n is $n - 1$.)

9. Here is the start of a 100 chart.

1	2	3	4	5	6	7	8	9	10
11	12	13	14	15	16	17	18	19	20
21	22	23	24	25	26	27	28	29	30

Shown below are parts of the chart. Without extending the chart, determine which numbers should go in the lavender squares.

(a) (b)

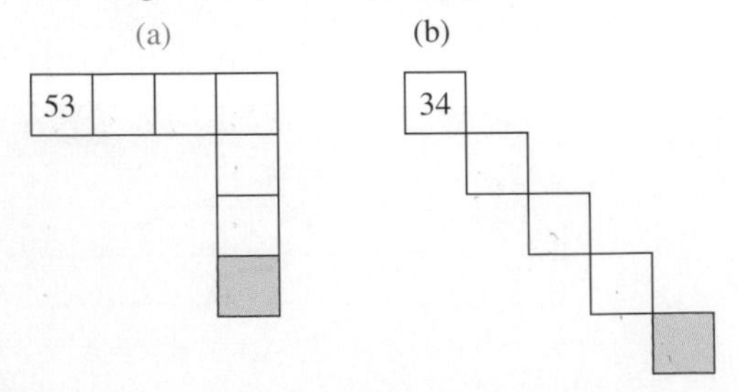

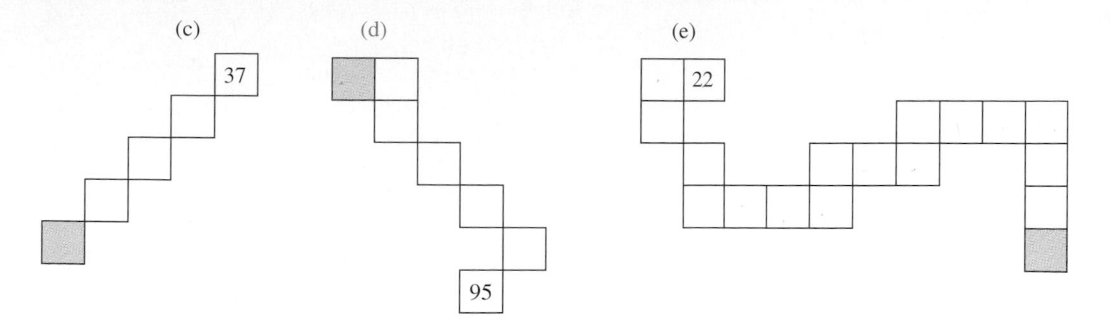

10. Five blue and five red discs are lined up as shown.

Switching just two adjacent discs at a time, what is the least number of moves you can make to achieve the blue, red, blue, red, . . . , arrangement shown here?

(*Hint:* Start with one chip of each color, then two chips of each color, then three chips of each color, and so on.)

11. (a) Complete the next two of this sequence of equations.

$$1 = 1$$
$$1 - 4 = -3$$
$$1 - 4 + 9 = 6$$
$$1 - 4 + 9 - 16 = -10$$

________________ = ________________

________________ = ________________

(b) Write the seventh and eighth equations in the sequence of equations of part (a). (*Hint:* Have you encountered the number sequence 1, 3, 6, 10, . . . before?)

(c) Write general equations suggested by parts (a) and (b) for even n and for odd n where n is the number of the equation.

12. (a) How many rectangles are there in each of these figures? (*Note:* Rectangles may measure 1 by 1, 1 by 2, 1 by 3, and so on.)

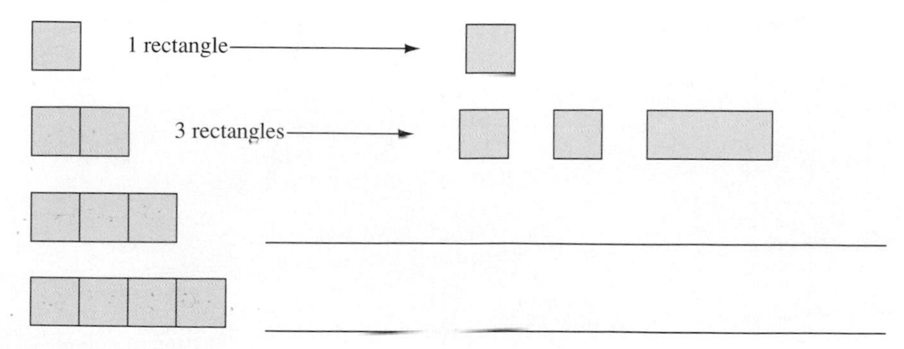

(b) How many rectangles are in this figure? ________________

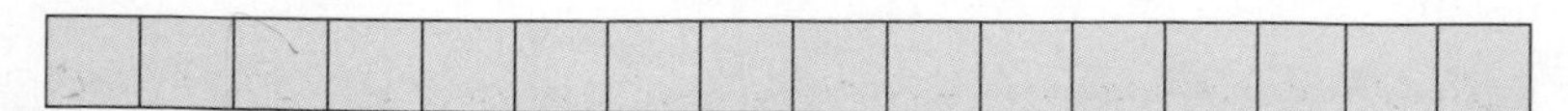

(c) How many rectangles are in a $1 \times n$ strip?

(d) Argue that your guess in part (c) is correct, giving a lucid and careful write-up. (*Hint:* How many of each type of rectangle begin with each small square?)

13. (a) In how many ways can you exactly cover this diagram with "dominoes" that are just the size of two small squares?

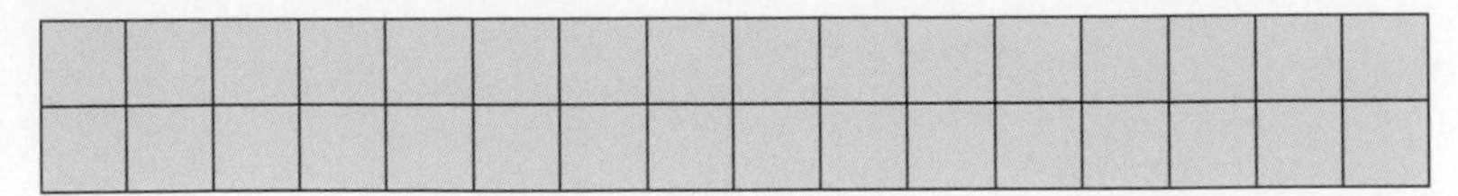

(*Hint:* This is too complicated as is. Consider this sequence of simpler arrays and look for a pattern.)

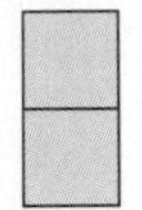
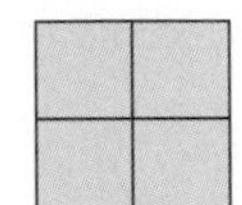
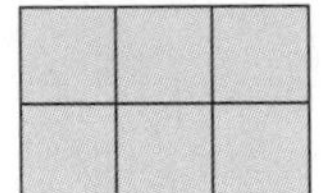
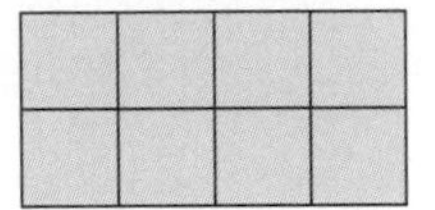

(b) Argue carefully that your solution in part (a) is correct. (*Hint:* A covering must start with one vertical domino or two horizontal dominoes.)

14. How many line segments are determined by joining dots on a circle if there are

(a) four dots? (b) ten dots?

(c) 100 dots? (d) n dots?

(e) Argue that your solution to part (d) is correct.

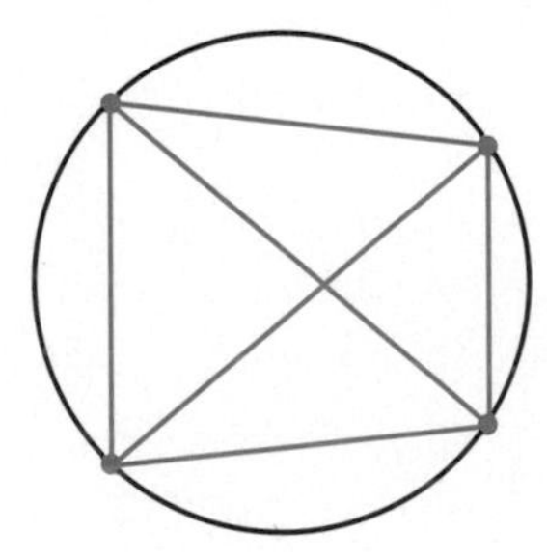

15. (a) How many games are played in a round robin tournament with ten teams if every team plays every other team once?

(b) How many games are played if there are 11 teams?

(c) Is this problem related to problem 14? If so how?

16. Here is an addition table.

+	0	1	2	3	4	5	6	7	8	9
0	0	1	2	3	4	5	6	7	8	9
1	1	2	3	4	5	6	7	8	9	10
2	2	3	4	5	6	7	8	9	10	11
3	3	4	5	6	7	8	9	10	11	12
4	4	5	6	7	8	9	10	11	12	13
5	5	6	7	8	9	10	11	12	13	14
6	6	7	8	9	10	11	12	13	14	15
7	7	8	9	10	11	12	13	14	15	16
8	8	9	10	11	12	13	14	15	16	17
9	9	10	11	12	13	14	15	16	17	18

(a) Find the sum of the entries in these squares of entries from the addition table.

2	3
3	4

5	6
6	7

11	12
12	13

15	16
16	17

Look for a pattern and write a clear and simple rule for finding such sums almost at a glance.

(b) Find the sum of the entries in these squares of entries from the table.

4	5	6
5	6	7
6	7	8

10	11	12
11	12	13
12	13	14

14	15	16
15	16	17
16	17	18

(c) Write a clear and simple rule for computing these sums.

(d) Write a clear and simple rule for computing the sum of the entries in any square of entries from the addition table.

17. We have already considered the triangular numbers:

and the square numbers:

1 4 9 16 . . .

$s_n = n^2$

(a) Draw the next two figures to continue this sequence of dot patterns.

(b) List the sequence of numbers that corresponds to the sequence of part (a). These are called **pentagonal numbers.**

(c) Complete this list of equations suggested by parts (a) and (b).

$$1 = 1$$
$$1 + 4 = 5$$
$$1 + 4 + 7 = 12$$
$$1 + 4 + 7 + 10 = 22$$
$$\_\_\_\_\_\_\_\_\_\_ = \_\_\_\_\_\_$$
$$\_\_\_\_\_\_\_\_\_\_ = \_\_\_\_\_\_$$

Observe that each pentagonal number is the sum of an arithmetic progression.

(d) Compute the 10th term in the arithmetic progression 1, 4, 7, 10, $\cdots$.

(e) Compute the 10th pentagonal number.

(f) Determine the nth term in the arithmetic progression 1, 4, 7, 10, $\cdots$.

(g) Compute the nth pentagonal number, p_n.

18. (a) The **hexagonal numbers** are associated with this sequence of dot patterns. Complete the next two diagrams in the sequence.

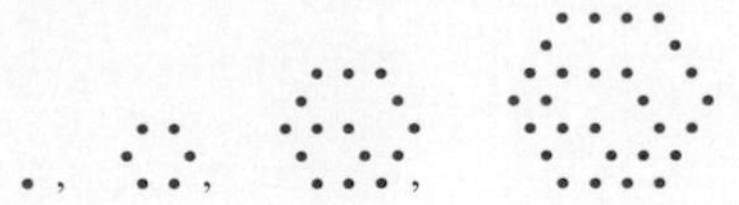

(b) Write the first five hexagonal numbers.

(c) What is the tenth hexagonal number?

(d) Compute a formula for h_n, the nth hexagonal number.

19. **Figurate numbers** are numbers associated with polygons—heptagons, octagons, nonagons, decagons, and so on—in a manner *analogous* to the definitions of triangular, square, pentagonal, and hexagonal numbers of problems 17 and 18. *What if we generalize* our study to include all these numbers? We could proceed step by step as before, but let's make a table of results already obtained and see if we can spot any useful patterns.

Reason by analogy.
Ask "what if?"
Seek to generalize.
Make a table.
Look for a pattern.

(a) Use data from problems 17 and 18, and use patterns to complete this table of figurate numbers.

Figurate Numbers n	1	2	3	4	5	6	General Formula
Triangular	1	3	6	10	15		$t_n = n(n+1)/2$
Square	1	4	9	16	25		$s_n = n^2$
Pentagonal	1	5	12	22			$p_n =$
Hexagonal	1	6	15	28			$h_n =$
Heptagonal	1	7					$H_n =$
Octagonal							$O_n =$
Nonagonal							$N_n =$
Decagonal							$D_n =$

(b) How might you define *linear numbers*. Would they fit the pattern of this table? Explain briefly.

20. Compute the sum of the numbers in the "handle" of each "hockey stick" in Pascal's triangle. Explain what you observe. Does the pattern always appear to hold?

21. Remembering that the elements in each row of Pascal's triangle are numbered counting from 0,

(a) compute the sum of the odd numbered elements in each of rows 1 through 8.

(b) compute the sum of the even numbered elements in rows 1 through 8.

(c) Guess what these sums would be in the nth row.

22. (a) Compute the sum of the squares of the entries in each of the zeroth, first, second, third, and fourth rows of Pascal's triangle.
(b) Carefully compare your answers to part (a) with Pascal's triangle. Can you spot an interesting pattern? Explain briefly.
(c) Describe, *but do not compute*, the sum of the squares of the entries in the eighth row of Pascal's triangle.
(d) Identify the sum of the squares of the elements of the entries in the nth row of Pascal's triangle. What number in the triangle is this?

23. (a) Compute the square root of the product of the six elements surrounding an element in Pascal's triangle. In particular, do this for the six entries surrounding each of 4, 15, and 35.
(b) Does the limited amount of data from part (a) suggest a general conjecture? What appears to be true in general?

1.5 Still More Problem-Solving Strategies

STRATEGIES

- ***Work backward.***
- ***Eliminate possibilities.***
- ***Use the pigeonhole principle.***

Working Backward

PROBLEM-SOLVING STRATEGY 11 • Work Backward

Start from the desired result and work backward step-by-step until the initial conditions of the problem are achieved.

Many problems require that a sequence of events occur that result in a desired final outcome. These problems at first seem obscure and intractable, and one is tempted to try a guess and check approach. However, it is often easier to work backward from the end result to see how the process would have to start to achieve the desired end. To make this more clear, consider the following example.

EXAMPLE 1.11

Working Backward—The Gold Coin Game

This is a two person game. Place 15 golden coins (markers) on a desk top. The players play in turn and, on each play, can remove one, two, or three coins from the desk top. The player who takes the last coin wins the game. Can one player or the other devise a strategy that guarantees a win?

SOLUTION

Understand the problem

The assertion that the game is played with gold coins is just so much window dressing. What is important is that the players start with 15 objects, that they can remove one, two, or three objects on each play, and that the player who takes the last object wins the game. The question is, how can one play in such a way that he or she is sure of winning?

Devise a plan

Since it is not clear how to begin to play or how to continue as the play proceeds, we turn the problem around to see how the game must end. We then work backwards step-by-step to see how we can guarantee that the game ends as we desire.

Carry out the plan • • • •

We carry out the plan by presenting an imaginary dialogue that you could have with yourself to arrive finally at the solution to the problem.

Q. What must be the case just before the last person wins the game?
A. There must be 1, 2, or 3 markers on the desk.
Q. So how can I avoid leaving this arrangement for my opponent?
A. Clearly, I must leave at least 4 markers on the desk in my next to last move. Indeed, if I leave precisely 4 markers, my opponent must take 1, 2, or 3 leaving me with 3, 2, or 1. I can remove all of these on my last play to win the game.
Q. So how can I be sure to leave precisely 4 markers on my next to last play?
A. If I leave 5, 6, or 7 markers on my previous play, my opponent can leave *me* with 4 markers and he or she can then win. Thus, I must be sure to leave my opponent 8 markers on the previous play.
Q. All right. So how can I be sure to leave 8 markers on the previous play?
A. Well, I can't leave 9, 10, or 11 markers on the next previous play or my opponent can take 1, 2, or 3 markers as necessary and so leave me with 8 markers. But then, as just seen, my opponent can be sure to win the game. Therefore, at this point, I must leave 12 markers on the desk.
Q. Can I be sure of doing this?
A. Only if I play first and remove 3 markers the first time. Otherwise, I have to be lucky, and hope that my opponent will make a mistake and still allow me to leave 12, 8, or 4 markers at the end of one of my plays. The following outlines the play if I get to play first.

- I take 3 markers, leaving 12.
- My opponent takes 1, 2, or 3 markers, leaving 11, 10, or 9.
- I take 3, 2, or 1 markers as needed to make sure that I leave 8.
- My opponent takes 1, 2, or 3 markers, leaving 7, 6, or 5.
- I take 3, 2, or 1 markers as needed to assure leaving 4 markers on the desk.
- My opponent takes 1, 2, or 3 markers, leaving 3, 2, or 1.
- I take the remaining markers and win the game!

Look back • • • •

In mathematics it is critical that you keep in mind the goal you want to achieve. Asking yourself (and answering) such questions as:

- What am I asked to find?
- What is desired?
- What am I asked to show?
- What would I have to know to determine what I am asked to find?

Working backward is a must in many problem-solving situation, and particularly so with a problem like this. The desired strategy to win the game described is not at all clear. Here the strategy of working backward is somewhat similar to considering special cases and looking for a pattern. It is as if we started with just a few counters where the strategy was more apparent and gradually increased the number of counters until we reached the given number of 15. Working backward is a powerful strategy that ought to be in every problem solver's repertoire. ■

JUST FOR FUN

How many pages in the book?

If it takes 867 digits to number the pages of a book starting with page 1, how many pages are in the book?

Eliminate Possibilities

One way of determining what must happen in a given situation is to determine what the possibilities are and then to eliminate them one by one. If you can eliminate all but one possibility in this way, then that possibility must, in fact, prevail. Suppose that either John, Jim, or Yuri is singing in the shower. Suppose also that you are able to recognize both John's voice and Yuri's voice but that you do not recognize the voice of the person singing in the shower. Then the person in the shower must be Jim. This is another important problem-solving strategy that should not be overlooked.

PROBLEM-SOLVING STRATEGY 12 • Eliminate Possibilities

Suppose you are guaranteed that a problem has a solution. Use the data of the problem to decide which outcomes are impossible. Then at least one of the possibilities not ruled out must prevail. If all but one possibility can be ruled out, then it must prevail.

Of course, if you use this strategy on a problem and **all** possibilities can be correctly ruled out, the problem has no solution. Don't be misled. It is certainly possible to have problems with no solution! Consider the problem of finding a number such that three more than twice the number is 15 and six more than four times the number is 34. This problem has no solution since the first condition is only satisfied by 6 and the second is only satisfied by 7. Yet $6 \neq 7$.

However, if we know that a problem has a solution, it is sometimes easier to determine what can't be true than what must be true. In this approach to problem solving one eliminates possibilities until the only one left must yield the desired solution.

Consider the following problem.

EXAMPLE 1.12

Eliminate Possibilities

Yam, Bam, Uam, Iam, and Gam are aliens on a space ship.

(a) Yam is younger than Uam.

(b) Yam is not the youngest in the group.

(c) Only one alien is older than Gam.

(d) Gam is younger than Bam.

Arrange Yam, Bam, Uam, Iam, and Gam in order of increasing age.

SOLUTION

Understand the problem • • • •

The important part of this problem concerns the relative ages of the individuals on the space ship. Given this information we are to decide who is youngest, who is next oldest, and so on.

Devise a plan • • • •

One's first reaction is that the problem is quite confusing. Perhaps a table can be constructed that will bring some order out of the chaos and make it possible to draw conclusions from the given information.

Carry out the plan • • • •

While one table would suffice, for pedagogical purposes we exhibit a sequence of tables showing step-by-step how we are able to X out possibilities on the basis of information given. Steps a and b will be combined and we will use X_a, X_b, and so on, to indicate impossibilities due to statement (a), statement (b), and so on. Make a table of your own to see how all the work can be done on a single table.

	youngest → oldest				
Yam	X_b				X_a
Uam	X_a				
Bam					
Iam					
Gam					

Statements a, b

	youngest → oldest				
Yam	X_b			X_c	X_a
Uam	X_a			X_c	
Bam				X_c	
Iam				X_c	
Gam	X_c	X_c	X_c	P_c	X_c

Statement c

(a) Statement (a) says that Yam is younger than Uam. It follows that Yam is not the oldest alien and that Uam is not the youngest. This, justifies the X_a in the last box of the Yam row and the first box of the Uam row.

(b) Similarly, statement (b) says that Yam is not the youngest. This justifies the X_b in the first box of the Yam row.

(c) Statement (c) says that only one alien is older than Gam. Thus, Gam is the next to oldest alien and we put P_c, for "possibility justified by c," in the next to the last box of the Gam row. But, if Gam is the second oldest alien, then no one else is. This justifies placing an X_c in every other box in the row and column containing P_c as shown.

	youngest → oldest				
Yam	X_b			X_c	X_a
Uam	X_a			X_c	$\mathbf{X_d}$
Bam	$\mathbf{X_d}$	$\mathbf{X_d}$	$\mathbf{X_d}$	X_c	$\mathbf{P_d}$
Iam	$\mathbf{P_d}$	$\mathbf{X_d}$	$\mathbf{X_d}$	X_c	$\mathbf{X_d}$
Gam	X_c	X_c	X_c	P_c	X_c

Statement d

	youngest → oldest				
Yam	X_b	P_a	X_a	X_c	X_a
Uam	X_a	X_a	P_a	X_c	$\mathbf{X_d}$
Bam	$\mathbf{X_d}$	$\mathbf{X_d}$	$\mathbf{X_d}$	X_c	$\mathbf{P_d}$
Iam	$\mathbf{P_d}$	$\mathbf{X_d}$	$\mathbf{X_d}$	X_c	$\mathbf{X_d}$
Gam	X_c	X_c	X_c	P_c	X_c

Statement a

(d) Statement (d) affirms that Gam is younger than Bam. Since we just discovered that Gam is the second oldest, this forces us to place P_d in the last box of the Bam row and X_d in the remaining empty boxes in

SCHOOL BOOK PAGE *Problem Solving in Fifth Grade*

Use Logical Reasoning

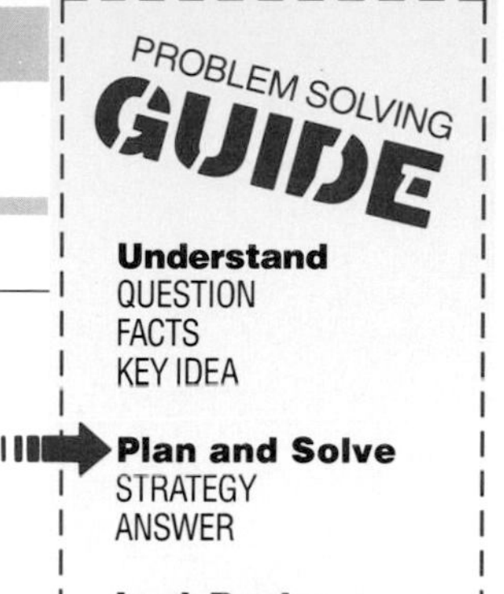

Build Understanding

West School has teams only in volleyball, swimming, soccer, and basketball. Erica, Justin, Molly, and Dave each play a different sport. Justin's sport does not use a ball. Molly is older than the volleyball player. Neither Molly nor Dave plays soccer. Who plays volleyball?

Understand Only one student plays volleyball. You know a ball is not used in swimming and people cannot be older than themselves.

Plan and Solve STRATEGY Logical reasoning is thinking in a sensible and orderly way to draw conclusions using known facts. Putting the facts in a chart can help.

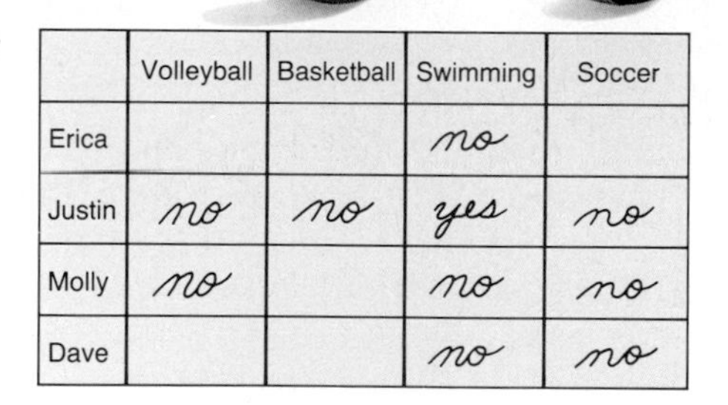

	Volleyball	Basketball	Swimming	Soccer
Erica			no	
Justin	no	no	yes	no
Molly	no		no	no
Dave			no	no

Justin is the swimmer since his sport does not use a ball. Write yes for swimming and no for the other sports. Also, write no for swimming for the other students. Molly is not older than herself. Write no for her for volleyball. Since neither Molly nor Dave plays soccer, write no for them for soccer. The chart shows only Erica can play soccer. So she does not play volleyball.

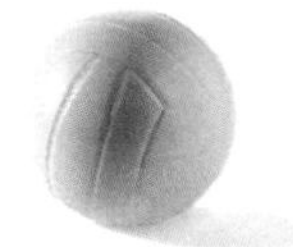

ANSWER Dave plays volleyball.

Look Back Check that the answer fits all the facts.

■ **Talk About Math** Who plays basketball? How does using facts together help you learn new facts?

Problem-solving situations should always be kept open-ended.

1. The discussion in the *Plan and Solve* shows that Justin swims and Erica plays soccer. Who plays volleyball and who plays basketball?

2. What could be concluded if the condition that Molly is older than the volleyball player is left out? Does the problem have a unique solution in this case?

this row and column as shown. At this point the only vacant box in the first column is in the Iam row so this must contain P_d as well and every other empty box in this row must contain X_d.

(e) We now return to statement (a) which affirms that Yam is younger than Uam. Since there are only two possibilities left for Yam and Uam, this forces us to place P_a in the second box in the Yam row and in the third box of the Uam row. It also forces us to put X_a in the two remaining boxes. Finally, the aliens listed from youngest to oldest, as indicated in the last table, are Iam, Yam, Uam, Gam, and Bam.

Look back • • • •

In this problem, we were confronted with a mass of data that was simply too extensive to be analyzed without some organizational scheme. To bring order into this chaos, it seemed reasonable to make a table allowing for all possibilities. We then used the statements of the problem one at a time to see what they permitted us to say about the table. In this way we were able to *X* out certain possibilities and, step-by-step, draw necessary conclusions until the only remaining possibility completed the solution. ■

Eliminating possibilities is often a successful approach to solving a problem.

The Pigeonhole Principle

If 101 guests are staying at a hotel with 100 rooms, can we make any conclusion about how many people there are in a room? It is probable that a number of the rooms are empty since some of the guests probably include married couples, families with children, and friends staying together to save money. But suppose most of the hotel's guests desire single rooms? How many such persons could the hotel possibly accommodate? Putting just one person per room, all 100 rooms would be occupied with one person left over. Thus, if *all* 101 guests are to be accommodated, there must be at least two persons in one of the rooms. To summarize:

> If 101 guests are staying in a hotel with 100 guest rooms, then at least one of the rooms must be occupied by at least two guests.

This reasoning is essentially trivial, but it is also suprisingly powerful. Indeed, it is so often useful that we name it **the pigeonhole principle,** which is stated here.

PROBLEM-SOLVING STRATEGY 13 • The Pigeonhole Principle

If m pigeons are placed in n pigeonholes and $m > n$, then there must be at least two pigeons in one pigeonhole.

For example, if we place three pigeons in two pigeonholes, then there must be at least two pigeons in one pigeonhole. To make this quite clear, consider all possibilities as shown here:

Pigeonhole Number 1	Pigeonhole Number 2
3 pigeons	0 pigeons
2 pigeons	1 pigeon
1 pigeon	2 pigeons
0 pigeons	3 pigeons

In every case there are at least two pigeons in one of the pigeonholes.

A second useful way to understand this reasoning is to try to avoid the conclusion by spreading out the pigeons as much as possible. Suppose we start by placing one pigeon in each pigeonhole as indicated below. Then we have one more pigeon to put in a pigeonhole, and it must go in either hole number one or hole number two. In either case, one of the holes must contain a second pigeon and the conclusion follows.

Pigeonhole Number 1	Pigeonhole Number 2
1	1

EXAMPLE 1.13

Using the Pigeonhole Principle

An electrician working in a tight space in an attic can barely reach a box containing twelve 15-amp fuses and twelve 20-amp fuses. If her position is such that she cannot see into the box, how many fuses must she select to be sure that she has at least two fuses of the same strength?

SOLUTION

Understand the problem • • • •

The box contains twelve 15-amp fuses and twelve 20-amp fuses. The electrician is in a tight spot and can barely reach the box into which she cannot see. In one attempt, she wants to select enough fuses to be sure that she has at least two fuses of the same strength. We must determine how many fuses she must select to insure the desired result.

Devise a plan • • • •

Let's consider possibilities. To make sure we don't miss one, we make an orderly list.

Carry out the plan • • • •

If the electrician chooses two fuses she must have:

two 15-amp fuses	and	zero 20-amp fuses, or
one 15-amp fuse	and	one 20-amp fuse, or
zero 15-amp fuses	and	two 20-amp fuses.

Two fuses are **not** enough; she might get one of each kind. But if she selects a third fuse, it must either be a third 15-amp fuse, a second 15-amp fuse, a second 20-amp fuse, or a third 20-amp fuse. In any case, she has two fuses of the same strength and the condition of the problem is satisfied. Therefore, she only needs to select three fuses from the box.

Look back • • • •

We certainly solved the problem considering possibilities. But might there be an easier solution? Choosing fuses of two kinds is much like putting pigeons into two pigeonholes. Thus, if we select three fuses, at least two must be the same kind by the pigeonhole principle, and we have the same result as before. ■

Observe that the number 12 in the statement of the preceding problem is misleading (only two fuses of each kind in the box are really needed), and it causes many students to respond that the answer is 13. This is incorrect as we have just seen, but the good teacher will make use of this error to stimulate further study. For example, one might ask, "What question can I ask about the electrician for which 13 **is** the correct answer? What question might I ask for which 14 is the correct answer? How many fuses must the electrician select to be sure that she has twelve 20 amp fuses?" Also, one might repeat the problem with twelve 15 amp fuses, twelve 20 amp fuses, and twelve 30 amp fuses. The possibilities are almost limitless.

PROBLEM SET 1.5

1. Play this game with a partner. The first player marks down 1, 2, 3, or 4 tallies on a sheet of paper. The second player then adds to this by marking down 1, 2, 3, or 4 more tallies. The first player to exceed a total of 30 loses the game. Can one player or the other devise a surefire winning strategy? Explain carefully.
2. Consider this mathematical machine.

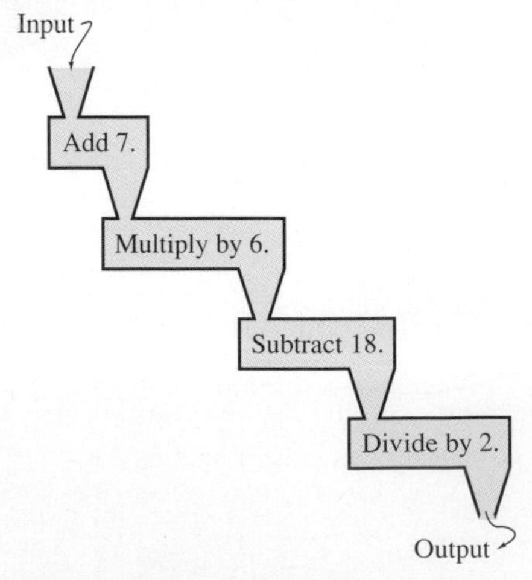

   (a) What number would you have to use as input if you wanted 39 as the output?

   (b) What would you have to input to obtain an output of 57?

   (c) Describe a strategy for attacking this problem different from the one you used for parts (a) and (b).
3. Josh wanted to buy a bicycle but didn't have enough money. After telling his troubles to Sam Slick, Sam said, "I can fix that. See that fence? Each time you jump that fence, I'll double your money. There's one small thing though. You must give me \$32 each time for the privilege of jumping." Josh agreed, jumped the fence, received his payment from Sam Slick and paid him \$32. Repeating the routine twice more, Josh was distressed to find that, on the last jump after Sam had made his payment to Josh, Josh had only \$32 with which to pay Sam and so had nothing left. Sam, of course, went merrily on his way leaving Josh wishing that he had known a little more about mathematics.

   (a) How much did Josh have before he made his deal with Sam?

   (b) Suppose the problem is the same but this time Josh jumps the fence five times before running out of money. How much did Josh start with this time?

4. A stack of ten cards numbered 0, 1, 2, . . . , 9 in some order lies face up on a desk. Form a new stack as follows: Place the top card face up in your hand, place the second card face up **under** the first card, place the third card face up **on top** of the new stack, place the fourth card face up **on the bottom** of the new stack, and so on. Arrange the cards in the original stack so that they appear in the new stack numbered in increasing order from the top down.

5. You have five markers, with three red colored and two green colored, arranged on a strip as shown.

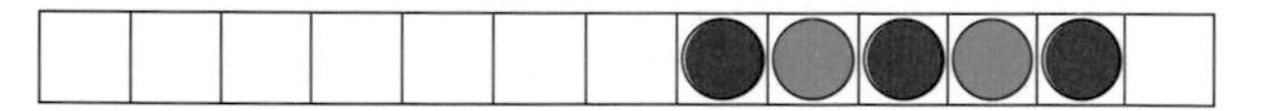

You are allowed to move any pair of adjacent red and green colored markers to any position determined by the markings on the strip. The object is to arrange the markers in the order red, red, red, green, green as shown here in any five contiguous squares along the strip.

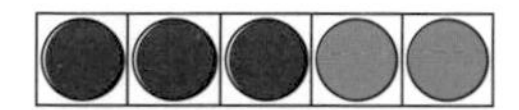

(*Hint:* Mark a strip as indicated across the middle of a piece of paper and make five colored markers to move back and forth in the prescribed manner. Finally, work backward; it's much easier!) (This problem is from Wayne A. Wickelgren's book *How to Solve Problems,* W. H. Freeman and Co., San Francisco, 1974.)

6. Moe, Joe, and Hiram are brothers. One day, in some haste, they left home with each one wearing the hat and coat of one of the others. Joe was wearing Moe's coat and Hiram's hat. Whose hat and coat was each one wearing?

7. Lisa Kosh-Granger likes to play number games with the students in her class since it improves their skill at both mental arithmetic and critical thinking. Solve each of these number riddles she gave to her class.

(a) I'm thinking of a number.
The number is odd.
It is more than 1 but less than 100.
It is greater than 20.
It is less than $5 \cdot 7$.
The sum of its digits is 7.
It is evenly divisible by 5.
What is the number? Was all the information needed?

(b) I'm thinking of a number.
The number is not even.
The sum of its digits is divisible by 2.
The number is a multiple of 11.
It is greater than $4 \cdot 5$.
It is a multiple of 3.
It is less than $7 \cdot 8 + 23$.
What is the number? Is more than one answer possible?

(c) I am thinking of a number.
The number is even.
It is not divisible by 3.
It is not divisible by 4.
It is not greater than 9^2.
It is not less than 8^2.
What is the number? Is more than one answer possible?

8. Beth, Jane, and Mitzi play on the basketball team. Their positions are forward, center, and guard.
Beth and the guard bought a milk shake for Mitzi.
Beth is not the forward.
Who plays each position?

9. Four married couples belong to a bridge club. The wives' names are Kitty, Sarah, Josie, and Anne. Their husbands' names (in some order) are David, Will, Gus, and Floyd.
Will is Josie's brother.
Josie and Floyd dated some, but then Floyd met his present wife.
Kitty is married to Gus.
Anne has two brothers.
Anne's husband is an only child.

Use this table to sort out who is married to whom.

	Kitty	Sarah	Josie	Anne
David				
Will				
Floyd				
Gus				

10. **(a)** Kathy Konradt chose one of the numbers 1, 2, 3, · · · , 1024 and challenged Sherrie Firavich to determine the number by asking no more than 10 questions to which Kathy would respond truthfully either 'yes' or 'no.' Determine the number she chose if the questions and answers are as follows:

Questions	*Answers*
Is the number greater than 512?	no
Is the number greater than 256?	no
Is the number greater than 128?	yes
Is the number greater than 192?	yes
Is the number greater than 224?	no
Is the number greater than 208?	no
Is the number greater than 200?	yes
Is the number greater than 204?	no
Is the number greater than 202?	no
Is the number 202?	no

(b) In part (a), we were able to dispose of 1023 possibilities by asking just ten questions. How many questions would Sherrie have to ask to determine Kathy's number if it is one of 1, 2, 3, · · · , 8192? If it is one of 1, 2, 3, · · · , 8000? Explain briefly but clearly. (*Hint:* Determine the differences between 512, 256, 128, 192, and so on.)

(c) How many possibilities might be disposed of with 20 questions?

11. If 9 is not placed in the center, is it possible to place the digits 5, 6, 7, 8, 9 in the circles so that the vertical sum and the horizontal sum in the cross are the same? Explain briefly.

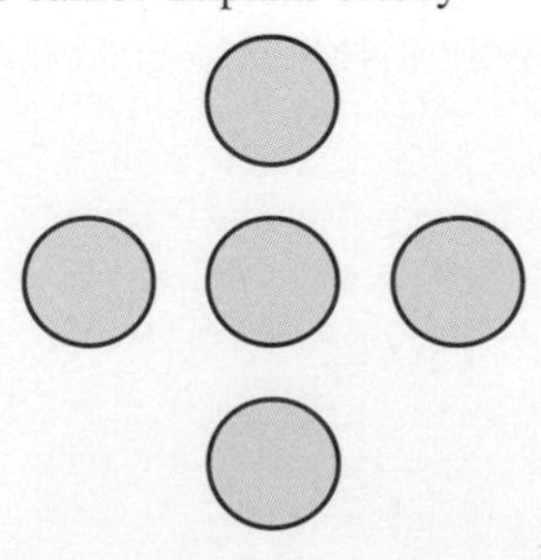

12. **(a)** How many students must be in a room to be sure that at least two are of the same sex?

(b) How many students must be in a room to be sure that at least six are boys or at least six are girls?

13. **(a)** How many people must be in a room to be sure that at least two people in the room have the same birthday (not birth date)? Assume that there are 365 days in a year.

(b) How many people must be in a room to be sure that at least three have the same birthday?

14. In any collection of 11 natural numbers, show that there must be at least two whose difference is evenly divisible by ten. Helpful question: When is the difference of two natural numbers divisible by ten?

15. **(a)** In any collection of seven natural numbers, show that there must be two whose sum or difference is divisible by ten. (*Hint:* Try a number of particular cases. Try to choose numbers that show that the conclusion is false. What must be the case if the sum of two natural numbers is divisible by ten?)

(b) Find six numbers for which the conclusion of part (a) is false.

16. Show that if five points are chosen in or on the boundary of a square one unit on a side, at least two of them must be no more than $\sqrt{2}/2$ units apart. (*Hint:* Consider this figure.)

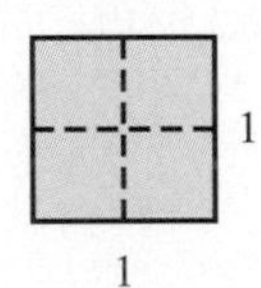

17. Show that if five points are chosen in or on the boundary of an equilateral triangle with sides one unit long, at least two of them must be no more than $1/2$ unit apart.

18. Think of ten cups with one marble in the first cup, two marbles in the second cup, three marbles in the third cup, and so on. If the cups are arranged in a circle in any order whatsoever, show that some three adjacent cups in the circle must contain a total of at least 17 marbles.

19. A fruit grower packs apples in boxes. Each box contains at least 240 apples and at most 250 apples. How many boxes must be selected to be certain that at least three boxes contain the same number of apples?

20. Show that at a party of 20 people, there are at least two people with the same number of friends at the party. Presume that the friendship is mutual. (*Hint:* Consider the following three cases! (i) Everyone has at least one friend at the party; (ii) Precisely one person has no friends at the party; (iii) At least two people have no friends at the party.)

21. Argue convincingly that at least two people in New York have precisely the same number of hairs on their heads. (*Hint:* You may need to determine a reasonable figure for the number of hairs on a human head.)

1.6 Problem Solving in the Real World

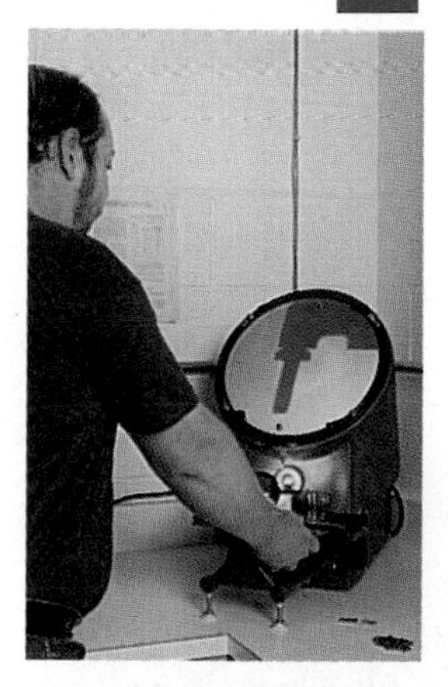

For most people it is the practical aspect of mathematics that makes it an important part of their educational background. Many of the advances in science and technology have depended on contributions from mathematics and, in contemporary society, applications of mathematics have become increasingly numerous, complex, and essential. In the workplace, many new jobs require advanced education and even so-called "blue-collar" jobs require substantial mathematical expertise. In personal life, decisions we face as consumers, financial planners, environmental caretakers, and informed voters can be approached more confidently using mathematical methods and skills.

Many problems of the "real world" (that is, the world beyond the school walls) involve large and complex systems. To solve such problems, a company or governmental agency will often turn to the mathematical scientist, a person who can bring the most advanced methods and latest technology to bear. Even so, the professional uses strategies of problem solving which are no different from those which school children should learn in their mathematics classes.

The examples that follow are simplified but otherwise representative of real problems encountered in today's world.

Two Problems on Routes of Shortest Time

In a world of limited resources and finite available time, companies are understandably interested in operating at maximal efficiency. Accomplishing a task in the shortest time and at a minimum cost brings success, whereas inefficient designs and procedures send business to competitors. The mathematics of **optimization** is concerned with the study of doing the most for the least expenditure of time, space, or money.

EXAMPLE 1.14 **Solving a Minimum-Time Routing Problem**

Four city blocks are shown in the figure. The minutes required to travel each block are shown; the times vary because of different amounts of traffic and congestion on each block. If you must head either north, N, or east, E, at each corner, what route from point A will enable you to reach point B in the shortest time?

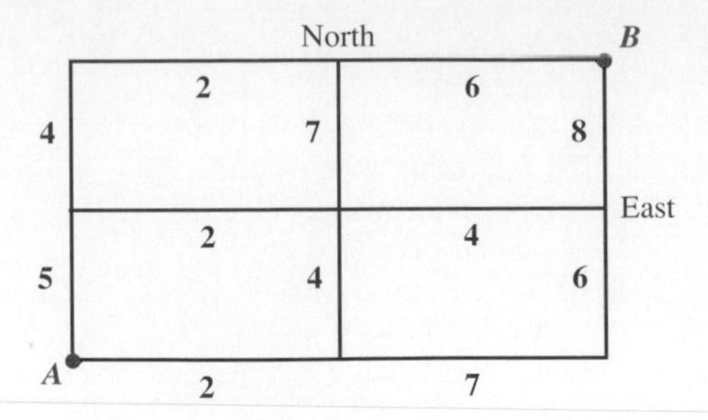

SOLUTION

Understand the problem • • • •

Each route from A to B must cover blocks headed north or east; no block can be traveled heading west or south. The travel time for each route is computed by adding the times to cover each block. We need to find a route that reaches B in the shortest possible time.

Devise a plan • • • •

Plan 1. Direct search One approach is to make an orderly list of all routes and times. An objection to this plan is that it requires lots of computation. Also, most of the routes examined are far from being the best so much of the work is wasted. If we had 30 blocks instead of just four, a direct search might be overwhelming to carry out.

Plan 2. Work backward Starting at B, the point to the west of B is 6 minutes from B and the point to the south of B is 8 minutes from B. The circled numbers give the time to reach B from the corresponding corners.

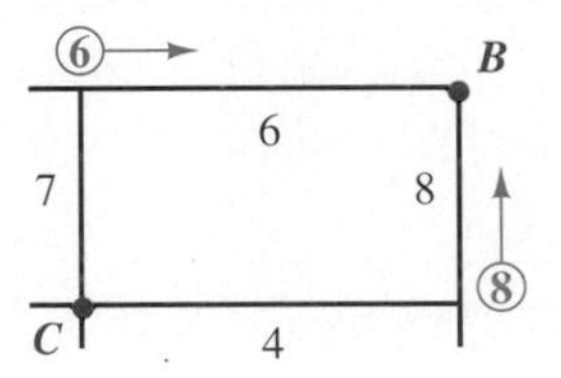

At corner C, heading north 7 minutes takes you to a point 6 minutes from B, and so C is $7 + 6 = 13$ minutes from B along the north-east route. However, the east-north route takes $4 + 8 = 12$ minutes. If you find yourself at C then it is 12 minutes from B, and the shortest route heads east. The shortest time and best direction are shown by a circled arrow at corner C. Plan 2 is to continue to work backward to discover a shortest path from A to B.

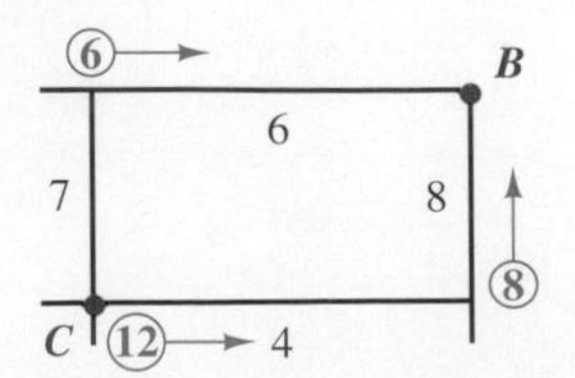

HIGHLIGHT FROM HISTORY
Henri Poincaré (1854–1912)

The most remarkable mathematician at the beginning of the twentieth century was Henri Poincaré, born in Nancy, France, on August 29, 1854. Poincaré was the last person to take all of mathematics as his province. It is certainly true today that this would be impossible for any single individual and, even in 1880, it was generally believed that Gauss was the last mathematician of whom this could be said. But it was true of Poincaré whose work was monumental—including over 500 landmark papers and some 30 important books.

Among Poincaré's many interests was the psychology of mathematical invention and discovery, or, as we have phrased it in this text, the psychology of problem solving. Poincaré thought and wrote extensively about this subject. One idea he expressed is akin to Pólya's look back principle. One learns to solve problems by solving problems and each time remembering the key to the solution so that it can be utilized again and again to solve other problems. As in any other endeavor, the key to success in problem solving is practice.

Carry out the plan

Plan 1. The six routes and their times are easy to list (since there are just four blocks!). There is one route, NNEE, which connects A to B in 17 minutes. The other routes all require more time.

Route	Time in Minutes
EENN	23
ENEN	18
ENNE	19
NEEN	19
NENE	20
NNEE	17

Plan 2. The work backward plan, when fully carried out as indicated below, also shows that point A is 17 minutes from B. By following the route corresponding to the arrows, the optimal route NNEE is again discovered as shown.

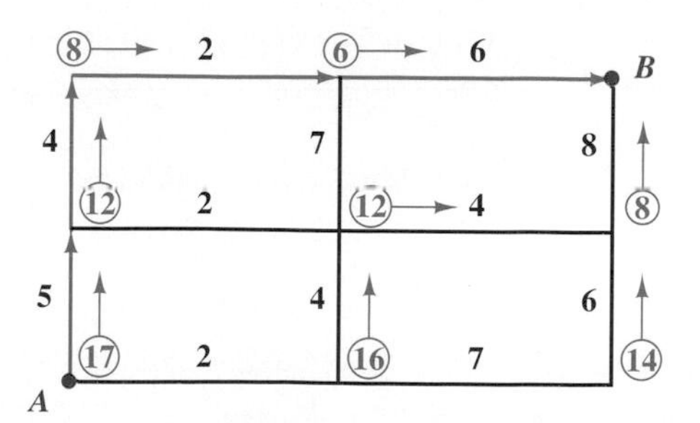

Look back • • • •

Plan 1 required 18 additions (three for each of the six routes from *A* to *B*). Then it was necessary to compare the six travel times to pick out the least. Plan 2 required only 10 additions and four comparisons. For a larger problem, the working backward method offers the better hope for success. ■

EXAMPLE 1.15 Solving a Larger Minimum-Time Routing Problem

Find the route of least time from *A* to *B* for the 3 by 4 block system shown in the figure.

SOLUTION

There are 35 routes from *A* to *B*. To compute the time of one route takes six additions, so there would be 210 additions all together. Let's try the work backward solution shown below instead. The circled values and arrows in the figure take only 29 additions and 12 comparisons between pairs of numbers.

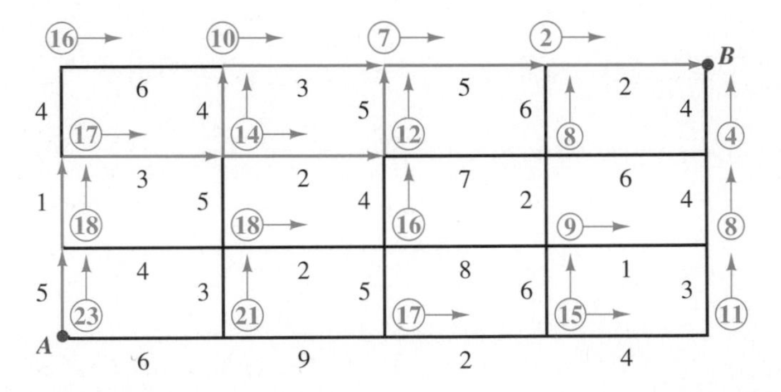

We quickly find that the shortest time from *A* to *B* is 20 minutes. There is one route which has the smallest possible travel time, NNENEEE. ■

The work backward solution is a special case of **dynamic programming,** a branch of mathematics created about 1950 by the mathematician Richard Bellman. Problems of shortest routes, quickest routes, or cheapest routes are real problems faced in today's world. A telecommunications network should direct its signal over routes with the smallest total rental fees. A power company will try to deliver service to its customers with the shortest number of miles of transmission lines. It is easy to add other examples from the airline industry, shipping companies and so forth.

Two Fair Division Problems

An interesting application of elementary mathematics is the problem of **fair division.**

EXAMPLE 1.16

Fairly Dividing a Gift

When Joyce joined the army, she decided to give her boom box to Ron and Jim, her two high-school age brothers. After she heard Ron and Jim quarrel over using the boom box, Joyce told them they had better figure out a fair method of division so that one of them would get the stereo and the other would get a fair payment. Determine a method that is fair to Ron and Jim.

SOLUTION

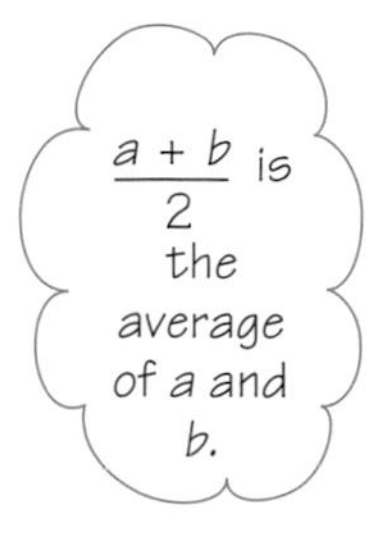
$\frac{a+b}{2}$ is
the
average
of a and
b.

The following method allows for the possibility that, while both Ron and Jim want the boom box, they have different assessments of its worth. Independently Ron and Jim are asked to write down what they think the stereo is worth. Let's say that Jim thinks it's worth \$90, but Ron thinks it's worth \$120. Since Ron places a higher value on the stereo, it should be his, and the question now is—how much should Ron pay Jim?

The answer is \$52.50. To see why this is fair, recall that each brother feels he should get half of the value of Joyce's gift. Thus, Ron will be happy with a \$60 value, and Jim will feel he is fairly treated if he gets \$45. The average of the two values is \$52.50. For \$52.50 Ron gets \$60 in stereo value. At the same time Jim, who expects his half of the \$90 gift should be \$45, is also getting \$52.50. Each receives \$7.50 more value than expected. ■

The solution just described works equally well when several people must divide any number of items among themselves. The method is a legally accepted process in many states for settling estate claims.

EXAMPLE 1.17

Settling an Estate Claim

Alice, Bill, and Carl have jointly inherited a piano, a car, a boat, and \$20,000 in cash. The lawyer has asked each of them to bid on the value of the four items. Their bids are shown in this table. Decide which person receives each item, and what cash payments are made so that Alice, Bill and Carl all feel they have received at least what each considers to be his or her fair share of the inheritance.

Bids Alice, Bill, and Carl have Made on the Four Items of Their Inheritance

	Alice	Bill	Carl
Piano	\$1900	1500	2000
Car	5000	5200	6000
Boat	2500	1800	2000
Cash	20,000	20,000	20,000
Total	\$29,400	\$28,500	\$30,000
Fair share	9800	9500	10,000

SOLUTION

Each person expects to receive one-third of what they assess as the total value of the estate. For example, Carl feels he should receive some combination of cash and items which he believes are worth $10,000. Let's first award each item to its highest bidder, and then award enough of the cash to bring each person up to his or her "fair share" amount as shown in this table.

A Preliminary Settlement: Items and Cash Are Distributed so that Each Heir Receives what They Perceive as One-Third of the Estate Value

	Alice	Bill	Carl
"Fair share"	9800	9500	10,000
Assigned items	boat	—	piano, car
Value of items	2500	0	8000
Cash to bring to "fair share"	7300	9500	2000

This uses $7300 + $9500 + $2000 = $18,800 of the $20,000 in cash, so there is $1200 cash yet to be distributed. Dividing this equally among the three heirs, we can give *each* of Alice, Bill, and Carl $400 more than they expect for a fair settlement. The final settlement is as shown in this table.

	Alice	Bill	Carl
Final settlement	boat + $7700	$9900	piano + car + $2400

■

Mathematics has made many contributions to the general notion of fairness. The equitable allocation of resources and funding is one example. Other examples relate to the democratic process—how many representatives is each state entitled to have congress? How can voters fairly rank candidates in a race contested by more than two contenders for office? As one can see, mathematics reaches into arenas of application far beyond the natural sciences.

A Problem in Correcting Errors*

Making errors is sometimes viewed as a uniquely human attribute, but computers, memory banks, and communications equipment are also susceptible to error. For example, the telemetry data sent by a spacecraft from the far reaches of the solar system will surely be affected by background radiation.

Most signals are coded into strings of pulses that can be viewed as strings of zeros and ones. At first glance this may appear to be overly confining, but even as complex a signal as music can be coded in this way—this is why the compact disc is "digital." Although longer strings are needed for most applications, even strings of four 0s and 1s—0000, 0001, 0010, . . . , 1111—provide a vocabulary of 16 words. The use of 0s and 1s to store, transmit, and receive data is called **binary coding.** Each 0 or 1 is a **binary digit,** or **bit** in the shorthand parlance of computing.

*Additional information on binary coding and error-corrections can be found in Chapter 10 of *For All Practical Purposes,* 3rd ed., (New York: W. H. Freeman and Company, 1994).

It is hoped that errors will be infrequent. But even when a single 0 or 1 is misread as a 1 or 0 the intended word is replaced by a different word. In 1948, Richard Hamming, a mathematician at Bell Laboratories, proposed an efficient method to detect errors. As an added bonus, the **Hamming code** can correct the error if just one 0 or 1 has been replaced by a 1 or 0.

A simple way to describe Hamming's idea is with the diagrams shown in Figure 1.7. A 4-bit word, say 1001, is stored in the 4 regions of intersection of three circles as shown.

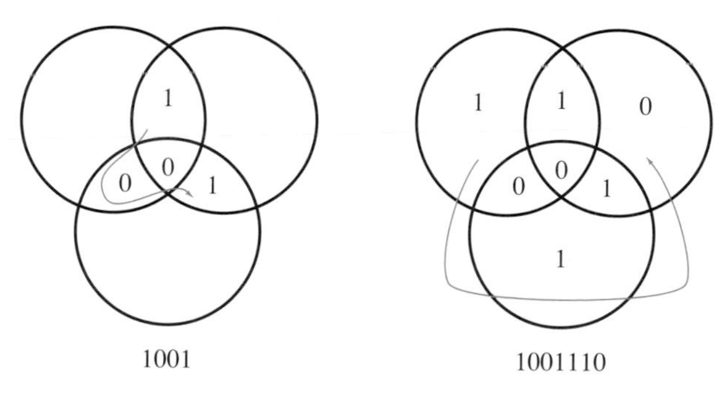

Figure 1.7
Extending the 4-bit "word" 1001 to the Hamming 7-bit 1001110

Next, starting with the upper left hand circle and proceeding counterclockwise the empty region of each circle is assigned a 0 or 1 so that each circle contains an even number of 1s. The 4-bit word 1001 becomes, in the Hamming code, the 7-bit word 1001110.

More generally, any 4-bit word abcd can be completed to its 7-bit Hamming code abcdefg as shown in Figure 1.8.

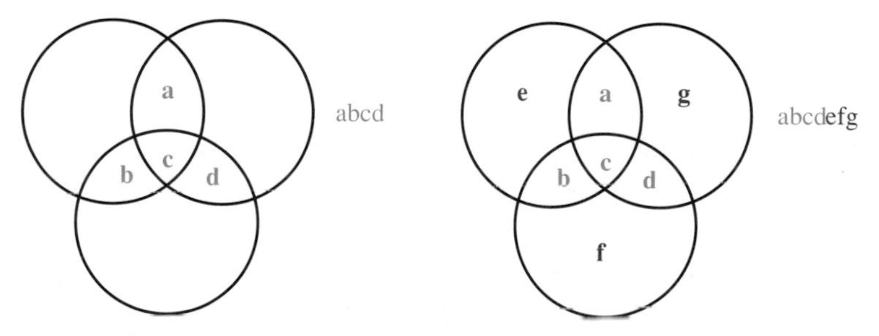

Figure 1.8
Hamming's scheme for extending a 4-bit code word to a 7-bit error-correcting word.

The redundancy of the 7-bit representation provides an **error-correcting code,** since any single error of one of the seven bits can be corrected. The following example shows how this works.

EXAMPLE 1.18 Correcting Errors with the Hamming Code

The following 7-bit representations have *at most* a single bit in error. Determine if an error is present and, if so, correct it.

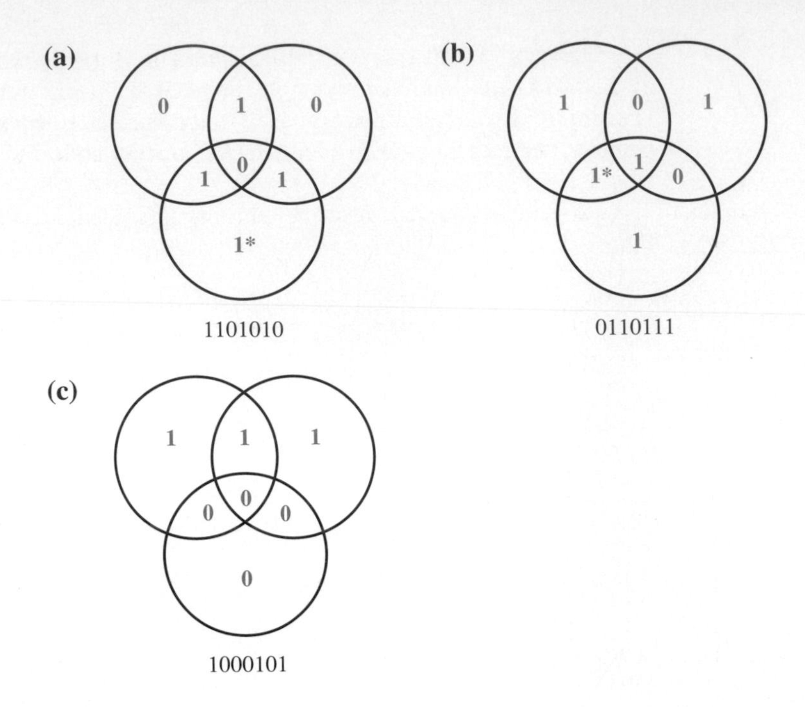

SOLUTION

(a) The upper two circles contain two 1s and are "good," but the lower circle is "bad" since it contains three 1s. Thus, one of the four bits in the lowest circle is in error. If the 0 is changed to a 1, the "good" circles become "bad." For the same reason, neither of the uppermost 1s in the lowest circle can be replaced by a 0. This leaves just the lowest 1 (marked with an asterisk) that can be changed. Thus, the corrected word is 1101000.

(b) The upper right circle is good but the remaining circles are bad. Because both bad circles must be changed to good, but we don't want the good circle to become bad, it follows that the 1 marked by the asterisk is the single bit in error. The corrected word is 0010111.

(c) Each of the three circles has evenly many 1s so we assume the word 1000101 is correct. ■

If two errors occur in a 7-bit word, the error-correction procedure can actually make things worse by introducing yet a third error. Fortunately, if single errors are relatively infrequent, the chances of two errors in the same word are quite rare.

In situations where an error may be critical—imagine the flight control system of an airplane, for example—redundant computing units may be required to reduce the chance of error to an acceptably low value. In more critical applications, more complex error-correcting codes are used. Many compact disc players can correct up to 14,000 consecutive errors. Even so, the ratio of error-correcting bits to sound bits is just 1 to 3, so there is a modest 33 percent overhead to provide error-correction capability.

Coded information—zip code codes, uniform price codes, book number codes, and others—plays an increasingly important role in what has become the "age of information." Much of the mathematics involved is accessible and exciting and provides opportunities for problem solving in the real world.

PROBLEM SET 1.6

1. Find the least amount of time to go from A to B. The times are in minutes. One must travel east or north from each corner.

2. How many paths are there from A to B in problem 1? Compare the number of calculations required to solve problem 1 by the direct search method and work backward (dynamic programming) method.

3. Find the quickest route from A to B in this diagram. You must head north or east at each corner.

4. The rental costs of natural gas pipelines are shown (in thousands of dollars per month). What is the cheapest route from A to B? The direction of flow is indicated by the arrows.

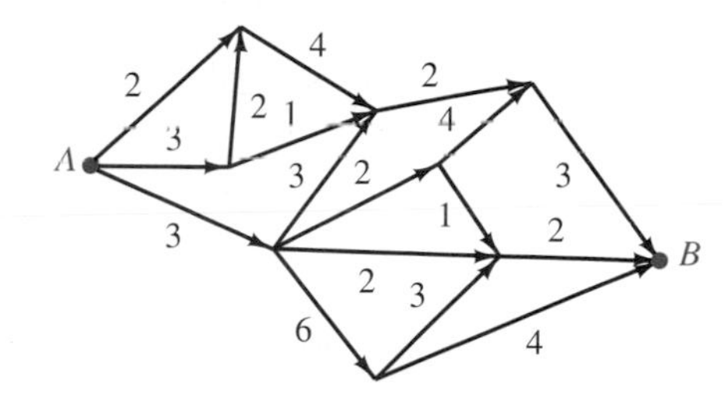

5. Sara and Ken are given a tennis racket by their uncle. Sara values the racket at $25 and Ken thinks the racket is worth $20. Who should get the racket? What payment should be made?

6. Suppose Ken, in problem 5, already has a tennis racket and is not interested in the racket from his uncle.
   - **(a)** Assuming Ken would like his sister to pay him as much as possible, what strategy should he follow when making his bid? Should he bid at all?
   - **(b)** If he guesses that Sara thinks the racket is worth $20 to $30, what bid would be best?
   - **(c)** What can Sara do, knowing that Ken isn't much interested in the racket?

7. La'Tiece, Mike, and Nancy wish to divide fairly a bike, a stereo, and a computer game. They have submitted bids as shown. Who should receive each item and what payments do they need to make so that each feels they received a fair share?

	La'Tiece	Mike	Nancy
Bike	$40	$50	$45
Stereo	$75	$60	$40
Game	$44	$40	$50

8. Suppose La'Tiece learns that Mike only bid $60 for the stereo in problem 7. If La'Tiece also knows that Nancy isn't very interested in the stereo, can she adjust her bid to take advantage of her knowledge? How does this affect Mike and Nancy? Discuss briefly.

9. At most one bit is in error in the following 7-bit Hamming code representations. Determine if the representation is correct; if not, correct the error.

(a)

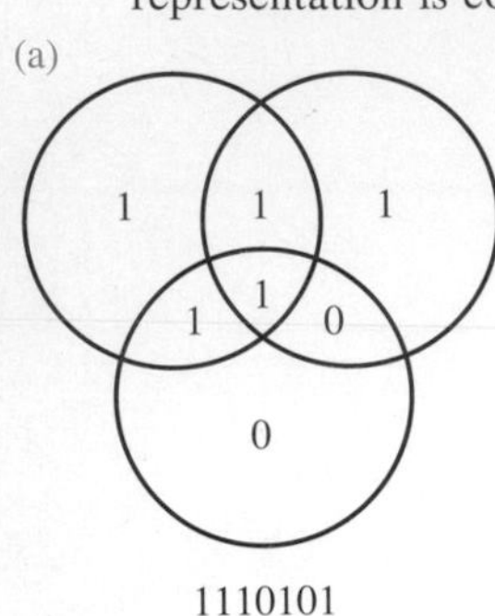

1110101

(b)

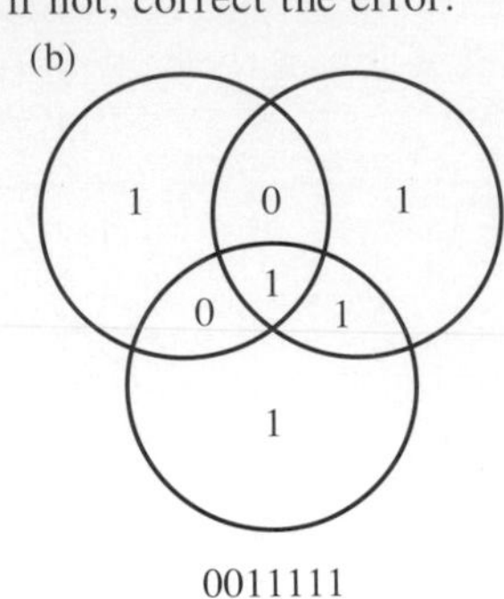

0011111

(c)

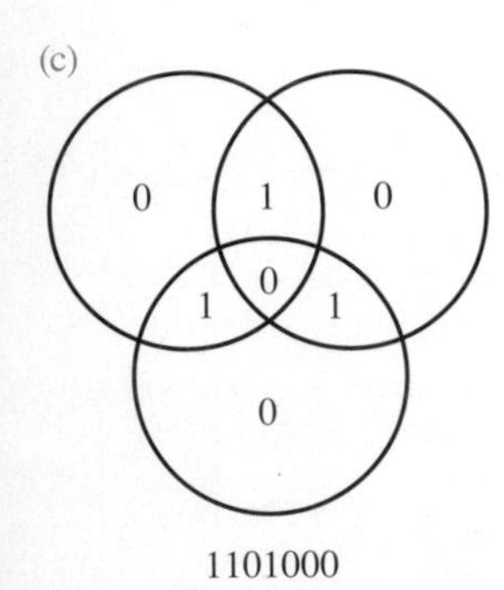

1101000

(d)

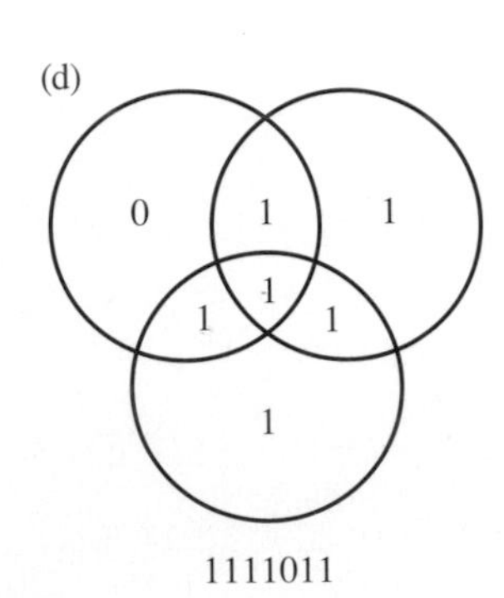

1111011

10. (a) Correct the single error in the following 7-bit Hamming code representation.

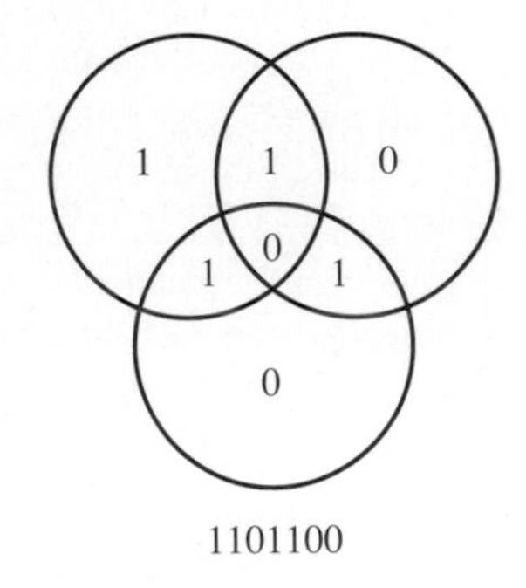

1101100

(b) If all three circles are "bad" and there is an error in a single bit, why is it easy to correct the error?

11. Suppose the bits marked by an asterisk are incorrect. Show that the error-correction procedure makes things worse by introducing a third error changing just one bit.

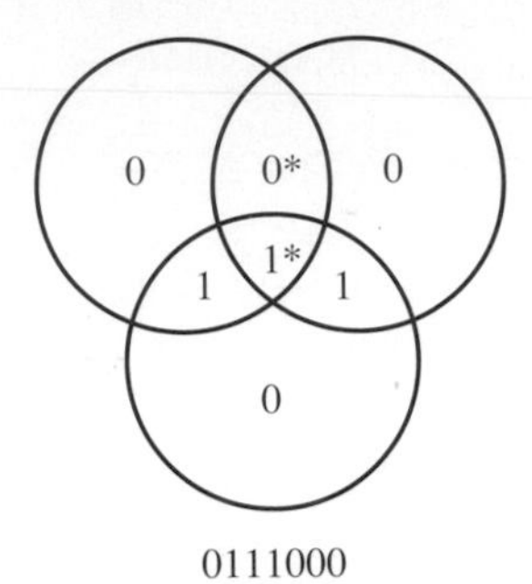

0111000

12. There are four 2-bit words: 00, 01, 10, 11. A *Gray code* is a circular ordering of the four words with the property that adjacent words differ by just one bit.

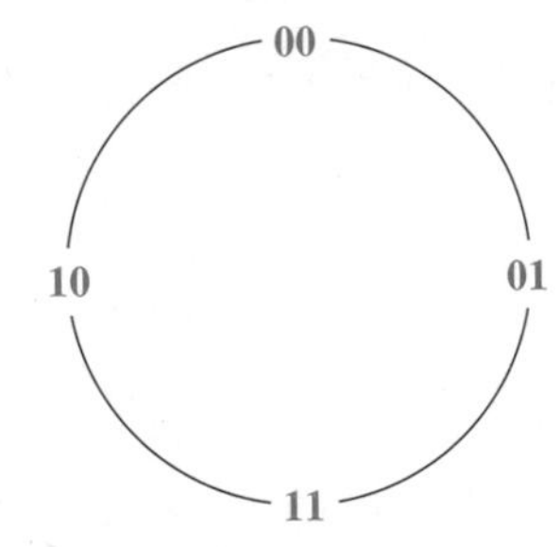

(a) Find a Gray code for the eight 3-bit words. (*Hint:* Try adding a 0 on the left of each of the 2-bit words. Then add a 1 on the left of each 2-bit word and see if the two sequences of 3-bit words can be properly joined together.)

(b) Find a Gray code for the sixteen 4-bit words.

---

13. Burger Boy has three stores along Division Street, all doing the same volume of business. Store A is 4 miles from store B, and store C is another 2 miles beyond B.

(a) Burger Boy would like to build a warehouse somewhere along Division Street to supply all three stores. Where is the best location? (*Hint:* Assume that the cost is proportional to the distance traveled in delivering goods from the warehouse to the stores, use the guess and check strategy, and use a variable.)

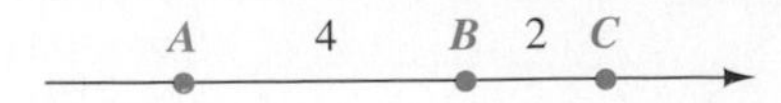

(b) Burger Boy has now decided to open a fourth store D which is 3 miles beyond store C. Now where is the best location for the warehouse?

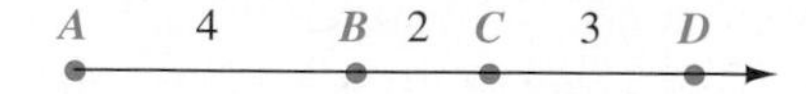

(c) Suppose store A does twice the volume of each of the other three stores in part (b). Now where is the best location for the warehouse?

(d) Describe the problem-solving strategies you have used.

14. The Acme Yardstick Company frequently sends boxes of yardsticks measuring $\frac{1}{4}$ by $1\frac{1}{2}$ by 36 inches to distribution centers. Naturally the boxes are a yard long, but postal regulations limit the girth of the box to 36 inches. The girth is the perimeter of the box when viewed from an end. Acme has been using boxes 6 inches high and 12 inches wide, but wonders if a different shaped box might hold more of their rulers. What shape do you believe may be better? (*Hint:* Complete this table.)

Height h (Inches)	Width w (Inches)	Area = hw (Square Inches)
6	12	72
7		
8		
9		
10		
11		

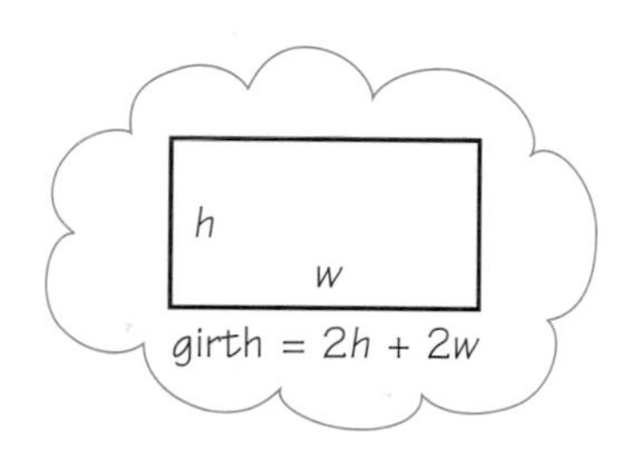

INTO THE CLASSROOM
Teaching Problem Solving

Here are some bits of advice for teaching problem solving.

- Set out now to develop a store of interesting and challenging problems appropriate to the skill levels of the students in your class. This process should continue over your professional lifetime. A good start is to keep your present textbook and to use many of the problems you find here as is or modified to be more accessible to your students.
- Really listen to your students. They often have good ideas. They also frequently find it difficult to express their ideas clearly. Listening carefully and helping your students to communicate clearly are skills you should continually seek to improve.
- Don't be afraid of problems you don't already know how to solve. Say, "Well, I don't know. Let's work together and see what we can come up with." You needn't be the oracle who "knows all." Indeed, many students will be excited and motivated by the prospect of "working with my teacher" to solve a problem. They also learn not to be afraid to tackle the unknown, and to make mistakes and yet persevere until a solution is finally achieved.
- Don't be too quick to help your students or to simply tell them how to solve a problem.
- Develop a long list of leading questions you can ask of your students, and that they can ask themselves, to help clarify their thinking and eventually arrive at a solution.
- Don't be afraid of having students in your class who are brighter and better at solving problems than you are. Just be happy to have such students; give them all the encouragement you can, and use them to help teach the others. Your ego should not be on the line—we have all taught students who are brighter than we are!

EPILOGUE Fascination with Mathematics—Past and Present

The fascination of individuals with mathematical problems, both real world and fanciful, goes back at least as far as recorded history. The reason for the interest in real-world problems, then and now, is obvious. The successful conduct of many human activities requires mathematical understanding. Thus, much of ancient mathematics was developed to meet particular needs and to enable individuals to accomplish desired tasks. The not so simple matter of making a reliable calendar, for example, requires reasonably sophisticated understanding of mathematics, and the Babylonians had this ability by approximately 4700 B.C. Similarly, many practical notions from geometry, including formulas for areas and volumes of geometrical figures and the properties of right triangles, were known at least 1000 years before Pythagoras and Euclid.

But it is also true that much of mathematics was studied for its own sake, simply because it was interesting. It is even true that many of the problems still popular today are of extraordinarily ancient origin. One of the most interesting of ancient mathematical documents is the Rhind papyrus purchased by the Scottish Egyptologist, Henry Rhind, in a small shop in Egypt in 1858. The scroll dates from approximately 1650 B.C. and is something of a mathematical handbook. However, it also contains a number of fanciful problems including this cryptic set of data.

Estate		
Houses	7	7^1
Cats	49	7^2
Mice	343	7^3
Heads of wheat	2401	7^4
Hekat measures	16807	7^5

The problem is not explained, but one can guess that the challenge was to compute the sum

$$7 + 7^2 + 7^3 + 7^4 + 7^5.$$

This problem reappears in A.D. 1202 in *Liber abaci* by the celebrated thirteenth century mathematician, Leonardo of Pisa or Fibonacci. In translation, Fibonacci's rendition of the problem runs as follows

> There are seven old women on the road to Rome. Each woman has seven mules; each mule carries seven sacks; each sack contains seven loaves; with each loaf are seven knives and sheaths. How many are there in all on the road to Rome?

The modern version of the problem is the well-known nursery rhyme

As I was going to St. Ives,
I met a man with seven wives,
Each wife had seven sacks,
Each sack had seven cats,
Each cat had seven kits.
Kits, cats, sacks, and wives,
How many were going to St. Ives?

It has been conjectured that the original problem in the Rhind papyrus might have been: "An estate consisted of seven houses; each house had seven cats; each cat ate seven mice; each mouse ate seven heads of wheat; each head of wheat when planted would produce seven hekats of grain. How many of all these items were in the estate?" However that may be, here is a problem, still popular in children's literature, that was already ancient when Fibonacci copied it nearly 800 years ago. One even wonders if the O'Henry twist in the nursery rhyme version of the problem might not have existed in some form in the ancient Egyptian version.

In any case, mathematical problems and puzzles have piqued the curiosity and challenged the ingenuity of individuals for millennia and this is no less true today. Introducing problem solving into the curriculum improves students' skills and their ability to think creatively and carefully; it also greatly enhances both student and teacher enjoyment of the entire educational process.

Finally, listing all possible strategies for solving problems is not possible. Those presented in this chapter are some of the most basic. But all have variants that we have not mentioned, all can be mixed and matched in many ways, and other methods, approaches, ideas, and schemes will occur both to you and to your students over the years.

CLASSIC CONUNDRUM 10 = 9?

Ten weary, footsore travelers,
All in a woeful plight,
Sought shelter at a wayside inn
One dark and stormy night.

"Nine rooms, no more," the landlord said,
"Have I to offer you.
To each of eight a single bed.
But the ninth must serve for two."

A din arose. The troubled host
Could only scratch his head,
For of those tired men no two
Would occupy one bed.

The puzzled host was soon at ease—
He was a clever man—
And so to please his guests devised
This most ingenious plan.

In room marked A two men were placed,
The third was lodged in B,
The fourth to C was then assigned,
The fifth retired to D.

In E the sixth he tucked away,
In F the seventh man,
The eighth and ninth in G and H,
And then to A he ran,

Wherein the host, as I have said,
Had laid two travelers by;
Then taking one—the tenth and last—
He lodged him safe in I.

Nine single rooms—a room for each—
Were made to serve for ten;
And this it is that puzzles me
And many wiser men.

CHAPTER 1 SUMMARY

Key Concepts

The thrust of this chapter has been to introduce the basic ideas of problem solving or critical mathematical thinking. Fundamental to this entire process are Pólya's Principles:

- Understand the problem
- Devise a plan
- Carry out the plan
- Look back

Understand the problem • • • •

Students are often stymied because they don't understand a problem fully or even in part. To help, teachers should ask such questions as:

- Do you understand all the words in the problem? If not, look them up in your book, in a dictionary, and so on.
- What are you given?
- What are you asked to find or show?
- Can you restate the problem in your own words?
- What does (key word) really mean?
- Could you work out some numerical problems that might clarify the problem at hand?
- Would a picture or diagram help?
- If you knew (key word) could you solve the problem?
- Could you determine (key word)?

As a teacher, your entire professional life should be spent adding to and refining this list. Students can't solve a problem unless they understand it.

Devise a plan • • • •

Devising a plan involves seeking an appropriate strategy that will ultimately lead to a solution. A list of such strategies includes:

- guess and check
- make an orderly list
- think of the problem as partially solved
- eliminate possibilities
- solve an equivalent problem
- use symmetry
- consider special cases
- work backward
- use direct reasoning
- be ingenious
- look for a pattern
- draw a picture
- think of a similar problem already solved
- solve a simpler problem
- solve an analogous problem
- solve an equation
- use a diagram or model
- use a formula (function)
- use indirect reasoning

Not all of these strategies have been discussed in this chapter. But problem solving pervades this text, and these strategies will occur and be discussed further in examples from time to time.

Carry out the plan • • • •

Having decided on a strategy, try to carry it out. Don't give up too soon. But if your strategy doesn't work, give it up and try another strategy. With practice, you are increasingly apt to choose a successful strategy the first time.

Look back • • • •

An important part of problem solving is looking back at a completed solution to see what the key ideas were, what ultimately led you to a solution, and how your approach might be modified to solve other even remotely related problems. Solving problems begets problem-solving skill. If we do not learn by our successes and *failures,* we are not apt to make much progress as problem solvers. Conversely, success at solving problems improves our chances of solving additional problems more easily.

Problem-solving ability is learned by solving problems! □

Vocabulary and Notation

Hands On

Möbius strip

Section 1.1

Palindrome
Kaprekar's number
Collatz's problem

Sections 1.2 and 1.3

Guess and check
Make a table
Look for a pattern
Make an orderly list
Draw a picture or diagram
Guess my rule

Section 1.4

Look for a pattern
Make a table
Variable
Use a variable
Triangular numbers
Consider special cases
Pascal's triangle
Solve an equivalent problem
Solve an easier similar problem
Argue from special cases

Section 1.5

Work backward
Eliminate possibilities
The pigeonhole principle

Section 1.6

Optimization
Dynamic programming
Fair division
Binary code
Binary digit
Hamming code
Error-correcting code

CHAPTER REVIEW EXERCISES

Section 1.1

1. (a) See how quickly you can compute these products. You may use a calculator if you like.
   $1 \cdot 8 =$ ________
   $21 \cdot 8 =$ ________
   $321 \cdot 8 =$ ________
   $4321 \cdot 8 =$ ________
   $54321 \cdot 8 =$ ________
   $654321 \cdot 8 =$ ________
   $7654321 \cdot 8 =$ ________
   $87654321 \cdot 8 =$ ________
   $987654321 \cdot 8 =$ ________

   (b) Did you have to perform all the multiplications to complete the problem?

   (c) Does the pattern you noticed hold for the next to the last product?

   (d) Does the pattern you noticed hold for the last product? What general principle about guessing does this suggest?

2. (a) Multiply your favorite digit by 239. Now multiply the product by 4649. The numbers 239 and 4649 are "magic" numbers as discussed in section 1.1, Tidbit Number 2.

   (b) To see why 239 and 4649 are "magic," compute $239 \cdot 4649$.

   (c) Use the program FACTORINTEGER to show that there are no other magic numbers whose product is a 7-digit integer.

3. (a) Form a magic addition square using the numbers 1, 7, 13, 31, 37, 43, 61, 67, and 73.

   (b) Form a magic subtraction square using the numbers in part (a).

4. Draw three circles with different diameters roughly as shown. Also, draw internal tangents T_1 and T_2 to one pair of the circles, T_3 and T_4 to another pair of the circles, and external tangents T_5 and T_6 to the other pair of circles as shown below. Let P, Q, and R be the points of intersection of T_1 and T_2, T_3 and T_4, and T_5 and T_6 respectively. Make a conjecture about P, Q, and R.

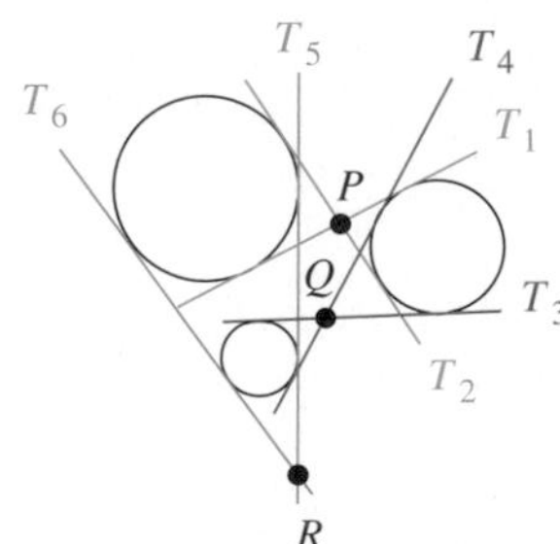

5. Generate a palindrome starting with the number 87. (See Problem 16 in Problem Set 1.1.)

Sections 1.2 and 1.3

6. Standard Lumber has 8′ and 10′ two-by-fours. If Mr. Zimmermann bought 90 two-by-fours with a total length of 844 feet, how many were eight feet long? Give two solutions, (a) and (b), using different strategies. (*Note:* Zimmermann is the German word for carpenter. Have you ever known anyone with this last name?)

7. (a) Using each of 1, 2, 3, 4, 5, 6, 7, 8, and 9 once and only once fill in the circles in this diagram so that the sum of the three digit numbers formed is 999.

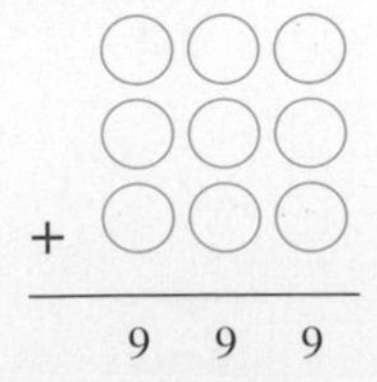

(b) Is there more than one solution to this problem? Explain briefly.

(c) Is there a solution to this problem with the digit 1 not in the hundreds column? Explain briefly.

8. Bill Cornett's purchases at the store cost $4.79. In how many ways can Bill receive change if he pays with a five dollar bill?

9. How many 3-letter code words can be made using the letters a, e, i, o, and u at most once each time?

10. A flower bed measuring 8′ by 10′ is bordered by a concrete walk two feet wide. What is the area of the concrete walk?

11. Karen LaBonte is thinking of a number, if you double it and subtract 7 you obtain 11. What is Karen's number?

12. (a) Chanty is IT in a game of *Guess My Rule.* If you give her a number, she uses her rule to determine another number. The numbers the other students gave Chanty and her responses are as shown. Can you guess her rule?

Student Input	Chanty's Responses
2	8
7	33
4	18
0	−2
3	13
⋮	⋮

(b) Can you suggest a better strategy for the students to use in attempting to determine Chanty's rule? Explain briefly.

Section 1.4

13. Study this sequence: 2, 6, 18, 54, 162, ... Each number is obtained by multiplying the preceding number by 3. The sequences in (a) through (e) are formed in the same way but with a different multiplier. Complete each sequence.

(a) 3, 6, 12, ______, ______, ______

(b) 4, ______, 16, ______, ______, ______

(c) 1, ______, ______, 216, ______, ______

(d) 2, ______, ______, ______, 1250, ______

(e) 7, ______, ______, ______, ______, 7

14. Because of the high cost of living, Kimberly, Terry, and Otis each holds down two jobs, but no two have the same occupation. The occupations are doctor, engineer, teacher, lawyer, writer, and painter. Given the following information, determine the occupations of each individual.

(a) The doctor had lunch with the teacher.

(b) The teacher went fishing with Kimberly, who is not the writer.

(c) The painter is related to the engineer.

(d) The doctor hired the painter to do a job.

(e) Terry lives next door to the writer.

(f) Otis beat Terry and the painter at tennis.

(g) Otis is not the doctor.

15. (a) Write down the next three rows to continue this sequence of equations.

$$1 = 1 = 1^3$$
$$3 + 5 = 8 = 2^3$$
$$7 + 9 + 11 = 27 = 3^3$$
$$13 + 15 + 17 + 19 = 64 = 4^3$$

______

______

______

(b) Consider the sequence 1, 3, 7, 13, . . . of the first terms in the sums of part (a). Write the first ten terms of this sequence.

(c) Write out the tenth row in the pattern established in part (a).

16. (a) Write down the next three rows to continue this sequence of equations.

$$2 = 1^3 + 1$$
$$4 + 6 = 2^3 + 2$$
$$8 + 10 + 12 = 3^3 + 3$$

______

______

______

(b) Write down the tenth row in the sequence in part (a).

17. (a) How many terms are in the arithmetic progression 7, 10, 13, 16, . . . , 79?

(b) Compute the sum of the terms in part (a).

18. (a) A **geometric progression** is a sequence of numbers where each term is a constant multiple of the preceding term. Thus, 3, 6, 12, 24, . . . , 3072 is a geometric progression since each term is twice its predecessor.

Analyzing these terms, we have

$$\begin{aligned} 3 &= 3 \\ 6 &= 2^1 \cdot 3, \\ 12 &= 2 \cdot 6 = 2(2^1 \cdot 3) = 2^2 \cdot 3, \\ 24 &= 2 \cdot 12 = 2(2^2 \cdot 3) = 2^3 \cdot 3, \\ &\vdots \end{aligned}$$

Which term in the sequence is 3072?

(b) Let $S = 3 + 6 + 12 + 24 + \cdots + 1536 + 3072$ denote the sum of the progression. Compute S.

(c) Note that $2S = 6 + 12 + 24 + \cdots + 3072 + 6144$.

(d) Note that $2S - S = S$ and use (b) and (c) to compute S a second time.

19. Compute the sum of this geometric progression:

$$5, 15, 45, \ldots, 295{,}245$$

20. Consider a circle divided by n chords in such a way that every chord intersects every other chord interior to the circle and no three chords intersect in a common point. Complete this table and answer these questions.

(a) Into how many regions is the circle divided by the chords?

(b) How many points of intersection are there?

(c) Into how many segments do the chords divide one another?

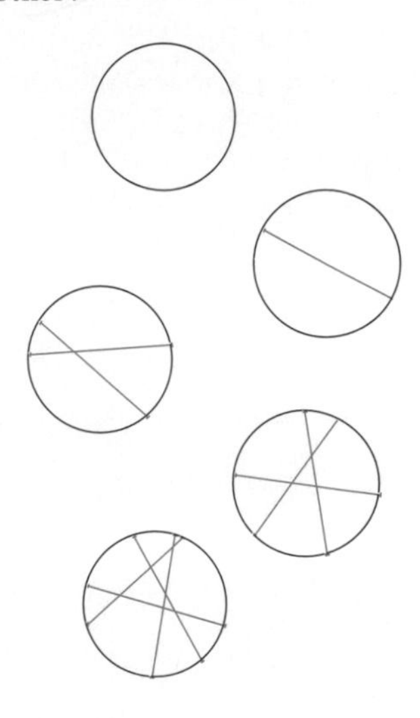

Number of Chords	Number of Regions	Number of Intersections	Number of Segments
0	1	0	0
1	2	0	1
2	4	1	4
3			
4			
5			
6			
⋮			
n			

21. Consider Pascal's triangle.

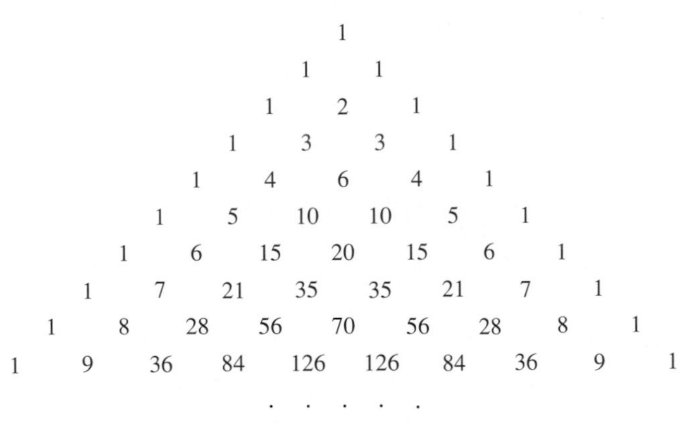

The numbers form a triangular grid which could be represented by an array of dots as shown. Consider the pattern of circled and squared dots in the diagram. Compute the product of the circled dots (entries in Pascal's triangle) and the product of the squared dots for several placements of this pattern in the triangle. Does your work suggest a plausible conjecture? Explain briefly.

22. Compute the following sums associated with Pascal's triangle.

(a) $1 + 1 \cdot 2$

(b) $1 + 2 \cdot 2 + 1 \cdot 2^2$

(c) $1 + 3 \cdot 2 + 3 \cdot 2^2 + 1 \cdot 2^3$

(d) What do these sums suggest? Explain briefly.

(e) Compute the sums

$$1 + 1 \cdot 3$$
$$1 + 2 \cdot 3 + 1 \cdot 3^2$$
$$1 + 3 \cdot 3 + 3 \cdot 3^2 + 1 \cdot 3^3$$

(f) What do (a), (b), (c), (d), and (e) together suggest? What might you do to check your guess further? Explain briefly.

Section 1.5

23. How many cards must be drawn from a standard deck of 52 playing cards to be sure that:

(a) at least two are of the same suit?

(b) at least three are of the same suit?

(c) at least two are aces?

24. How many books must you choose from among a collection of 7 mathematics books, 18 books of short stories, 12 chemistry books, and 11 physics books to be certain that you have at least 5 books of the same type?

Section 1.6

25. The numbers on the railroad network shown give the cost (in hundreds of dollars) of moving a trainload of coal along various links between junctions in the system.

(a) Determine the least possible cost of moving a load of coal from A to M.

(b) Starting with A, list the junctions through which the least cost route of part (a) passes.

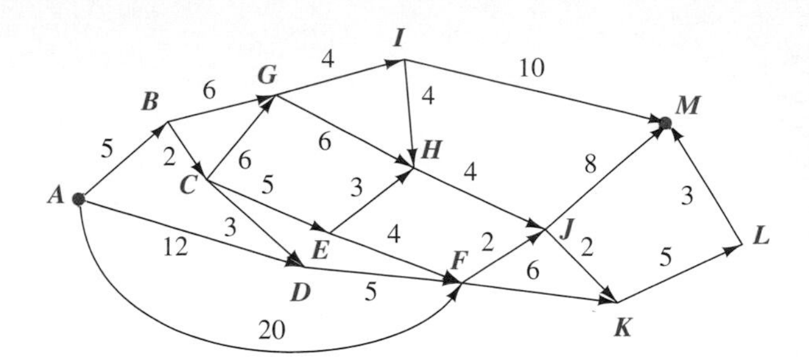

26. Judy, John, JoAnn, and Joshua are bequeathed a motor home, a car, a house, and a painting by a wealthy uncle. Since the items are clearly of unequal value and the executor of the estate wants to fairly divide the bequest, she asks the four siblings to submit statements of the value they place on the items. Given the list of values shown, who should inherit which items and what financial adjustments should be made so that each inheritee receives what he or she considers at least a fair share of the bequest?

	Judy	John	Joshua	JoAnn
Motor home	27,000	32,000	35,000	30,000
Automobile	19,000	18,000	23,000	24,000
House	250,000	200,000	220,000	230,000
Painting	900,000	800,000	1,000,000	700,000

27. Write the 7-bit Hamming code representation for each of these 4-bit binary words.

(a) 1101 (b) 1000 (c) 1010

28. The following 7-bit Hamming code words contain at most one error. Determine if an error is present and, if so, correct it.

(a) 1110001 (b) 0101101 (c) 1000010

CHAPTER TEST

1. Perform these multiplications as quickly as possible.

$2345679 \times 9 =$ ____________

$1345679 \times 9 =$ ____________

$1245679 \times 9 =$ ____________

$1235679 \times 9 =$ ____________

$1234679 \times 9 =$ ____________

$1234579 \times 9 =$ ____________

$1234569 \times 9 =$ ____________

$1234568 \times 9 =$ ____________

2. Consider the following equations.

$$1 = 0 + 1 = 1 - 0$$
$$2 + 3 + 4 = 1 + 8 = 9 - 0$$
$$5 + 6 + 7 + 8 + 9 = 8 + 27 = 36 - 1$$
$$10 + 11 + 12 + 13 + 14 + 15 + 16 = 27 + 64 = 100 - 9$$

(a) Continue this sequence for two more equations.

(b) What is the tenth row in the sequence?

(c) What is the nth row in the sequence?

3. While three watchmen were guarding an orchard, a thief slipped in and stole some apples. On his way out, he met the three watchmen one after another, and to each in turn he gave half the apples he had and two besides. In this way he managed to escape with one apple. How many had he stolen originally?

4. If five pigeons are placed in two pigeonholes, what is the least possible number of pigeons in the pigeonhole with the greatest number of pigeons?

5. (a) Make a magic square using each of the numbers 2, 7, 12, 27, 32, 37, 52, 57, 62 once and only once.

   (b) Make a magic subtraction square using each of the numbers in part (a) once and only once.

6. A frog is in a well 12 feet deep. Each day he climbs up 3 feet and each night he slips back 2 feet. How many days will it take the frog to get out of the well?

7. (a) Write the next two lines in this sequence of equations.

$$2 = 2 = 0^2 + 2 \cdot 1^2$$
$$2 + 5 + 2 = 9 = 1^2 + 2 \cdot 2^2$$
$$2 + 5 + 8 + 5 + 2 = 22 = 2^2 + 2 \cdot 3^2$$

____

____

   (b) Write the tenth line in the sequence of part (a).

8. Consider these equations.

$$s_1 = 1 = 1$$
$$s_2 = 1 - 3 = -2$$
$$s_3 = 1 - 3 + 6 = 4$$
$$s_4 = 1 - 3 + 6 - 10 = -6$$
$$s_5 = 1 - 3 + 6 - 10 + 15 = 9$$
$$s_6 = 1 - 3 + 6 - 10 + 15 - 21 = -12$$

   (a) Guess the values of s_{20} and s_{21}.

   (b) Guess the value of s_n. (*Hint:* You may want to consider n even and n odd separately.)

9. (a) Write the Hamming 7-bit code word for the 4-bit word 1101.

   (b) Correct the error in the Hamming code word 0101111.

10. Find the cost of the shortest path from A to B on the following network if you can only move east or north.

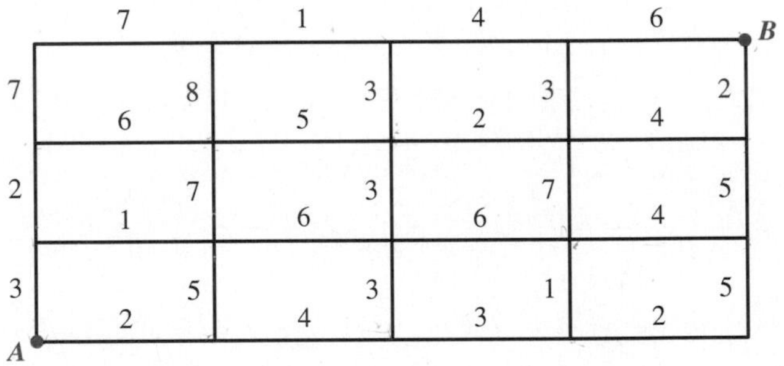

7 1 4 6 B
7 8 3 3 2
6 5 2 4
2 7 3 7 5
1 6 6 4
3 5 3 1 5
2 4 3 2
A

2 Sets, Whole Numbers, and Functions

Computer Support for this Chapter

In this chapter you might find the following program on your disk useful:

- DIFFY

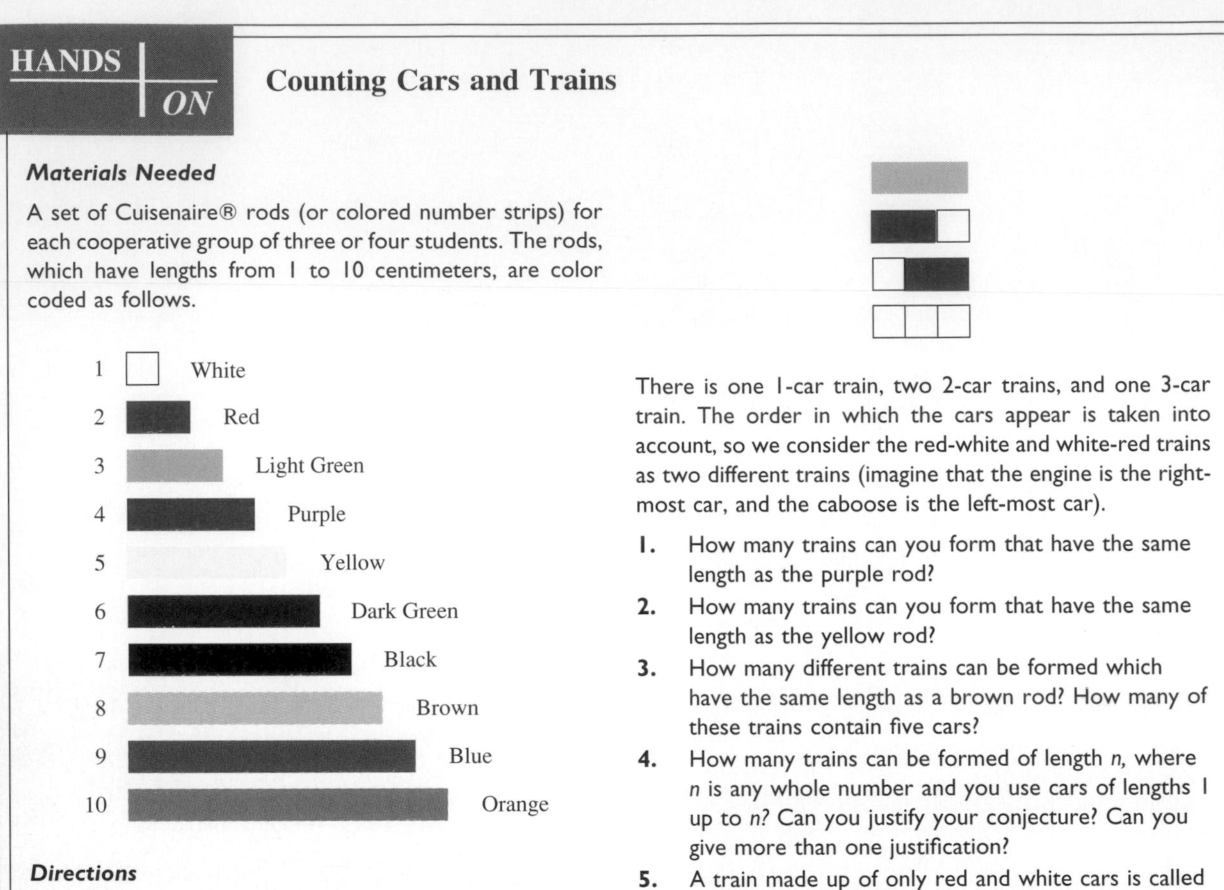

HANDS ON — Counting Cars and Trains

Materials Needed

A set of Cuisenaire® rods (or colored number strips) for each cooperative group of three or four students. The rods, which have lengths from 1 to 10 centimeters, are color coded as follows.

Directions

Form a train by placing one or more rods end-to-end; each rod in a train is a car. For example, there are four ways to form trains which have the same overall length as the light green (LG) rod:

There is one 1-car train, two 2-car trains, and one 3-car train. The order in which the cars appear is taken into account, so we consider the red-white and white-red trains as two different trains (imagine that the engine is the right-most car, and the caboose is the left-most car).

1. How many trains can you form that have the same length as the purple rod?

2. How many trains can you form that have the same length as the yellow rod?

3. How many different trains can be formed which have the same length as a brown rod? How many of these trains contain five cars?

4. How many trains can be formed of length *n*, where *n* is any whole number and you use cars of lengths 1 up to *n*? Can you justify your conjecture? Can you give more than one justification?

5. A train made up of only red and white cars is called an RW-train. Answer questions 1 through 4 for RW-trains.

6. Call a train with no white cars a $\overline{W}$-train. Answer questions 1 through 4 for $\overline{W}$-trains.

CONNECTIONS

Counting and Thinking Mathematically

When you can measure what you are speaking about and express it in numbers, you know something about it; but when you cannot measure it, when you cannot express it in numbers, your knowledge is of a meager and unsatisfactory kind.

—Sir William Thomson (Lord Kelvin)*

Number systems, since the very beginnings of civilization, have been an essential tool in people's ability to understand the world. We measure the distance between towns, the area of fields, the volume of a reservoir; we record and predict temperatures and the length of growing seasons; we are interested in the size of a profit or loss.

---

* *Popular Lectures and Addresses.* (New York, NY: Macmillan and Co., 1891, 1894).

HIGHLIGHT FROM HISTORY
The Early Origins of Mathematics

"The history of mathematics should really be the kernel of the history of culture."

—George Sarton

The transition from simply gathering food to actually producing it occurred some 10,000 years ago, marking the change from the Paleolithic to the Neolithic age. Agriculture required farmers to stay in one place for long periods, so people designed and constructed permanent dwellings. Laying out fields, tending flocks, measuring the amount of grain that should be stored over the winter, knowing whether excess grain could be traded to neighboring villages—all gave rise to problems dealing with quantity and form. Thus, the two important branches of mathematics—number and geometry—have origins concurrent with the dawn of civilization.

Until about 600 B.C., mathematics was pursued primarily for its practical, decorative, and religious values. Problems in land apportionment, interest payments, and tax rates required solution for the development of commerce. Many problems of a mathematical nature were successfully solved during the construction of irrigation canals, temples, and pyramids. Interest in astronomy grew out of the need for calendars sufficiently accurate to forecast flood and growing seasons. Number systems and notations were developed, and even some empirically derived formulas from algebra and geometry were known. However, little use was made of symbolism, scant attention was given to abstraction and general methods, and nowhere was the notion of proof or even informal justification to be found.

A portion of the Rhind papyrus that discusses the measurement of the area of a triangle and the slopes of pyramids. The papyrus, which contains 85 mathematical problems, was copied c. 1575 B.C. by the scribe Ahmes from a work written almost three centuries earlier.

The Incas recorded and communicated quantitative data with quipus. A quipu is an assemblage of cords, with numerical values determined by the colors of the cords, the way the cords are spaced and interconnected, and the type and placement of knots tied in the cords. The quipu shown is in the Museo Nacional de Antropologia y Arqueologia, Lima, Peru.

Perhaps the most basic mathematical questions of all times are these: How many? How much?

In this chapter we introduce the concept of **sets**, discuss how sets may be related to each other, and define **operations** on sets by which two or more sets can be combined to form new sets. The **whole numbers**—0, 1, 2, 3, and so on—are then defined in terms of sets. The relations and operations on sets give meaning to the order of the whole numbers and the definitions and properties of the basic arithmetic operations of addition, subtraction, multiplication, and division. In the concluding section, we explore functions and relations.

2.1 Sets and Operations on Sets

The notion of a set originated in the last half of the nineteenth century, which makes it a latecomer in the history of mathematics. Even so, sets have become indispensable to nearly every branch of mathematics. Sets make it possible to organize, classify, describe, and communicate. For example, each number system that we will investigate—the whole numbers, the integers, the rationals, and the real numbers—is best viewed as a set together with a list of operations and properties that the numbers in the system possess.

And just what is a set? The great German mathematician Georg Cantor (1845–1918), whose work on "aggregates" pioneered set theory, described a set as

> *a bringing together into a whole of definite well-distinguished objects of our perception or thought—which are to be called the elements of the set.*

Intuitively, a set is a collection of objects. An object which belongs to the collection is called an **element** or **member** of the set. Words like *collection, family,* or *class* are frequently used interchangeably with set. In fact, a set of dishes, a collection of stamps, and the class of 1995 at Harvard University are all examples of sets.

Cantor requires that a set be **well-defined.** This means two things. First, there is a **universe** of objects which are allowed into consideration. Second, any object in the universe is either an element of the set or it is not an element of the set.

To be well-defined, the description of a set can leave no ambiguity about what is intended as the universe and which objects in the universe are members of the set. For example, "the first few presidents of the United States" does not well-define the set, since "few" is a matter of varying opinion. On the other hand, "the first three presidents of the United States" does provide an adequate verbal description of a set, with the understanding that the universe is all people who have ever lived.

Besides a verbal description, there are two symbolic descriptions of sets:

Listing in Braces: {George Washington, Thomas Jefferson, John Adams}
Set Builder Notation: $\{x \mid x$ is one of the first three presidents of the United States$\}$

The order in which elements in a set are listed is arbitrary, so listing Jefferson, the third president, before Adams is permissible. Set builder notation, when verbalized, is read "the set of all x such that x is . . . ." The letter x used in set builder notation can be replaced with any convenient letter.

HIGHLIGHT FROM HISTORY
Georg Cantor (1845–1918)

Georg Cantor was born in St. Petersburg, Russia, but at age 12 moved with his family to Germany. Cantor excelled in mathematics, and in 1867 completed his doctorate from the prestigious University of Berlin. His research soon led him to sets and the comparison of the size of various infinite sets. Today, this work is regarded as fundamental to mathematical thought, but at the time of its development it generated considerable controversy. In particular, Leopold Kronecker exhibited an unpleasant animosity toward Cantor and his work. This stifled Cantor's dream of a professorship at the University of Berlin, and may also have contributed to the onset of the mental breakdowns that plagued Cantor from age 40 until his death at age 73 in a mental hospital. Eventually, Cantor's work was widely recognized and appreciated. The mathematician David Hilbert proclaimed that "no one shall expel us from the paradise which Cantor has created for us."

Capital letters, A, B, C . . . are generally used to denote sets. Membership is symbolized by $\in$, so that if P designates the above set of three U.S. presidents then John Adams $\in P$ and James Monroe $\notin P$, where $\in$ is read "is a member of" and $\notin$ is read "is not a member of." It is sometimes useful to choose letters that suggest the set being designated. For example, the set of **natural,** or **counting,** numbers will be written

$$N = \{1, \quad 2, \quad 3, \quad \ldots\}$$

where ". . ." indicates "and so on."

EXAMPLE 2.1

Describing Sets

Each set below is taken from the universe N of the natural numbers, and has been described either verbally, by listing in braces, or with set builder notation. Provide the two remaining types of description of each set.

(a) The set of natural numbers greater than 12 and less than 17.

(b) $\{x \mid x = 2n \text{ and } n = 1, 2, 3, 4, 5\}$

(c) $\{3, 6, 9, 12, \ldots\}$

(d) The set of the first ten odd natural numbers.

(e) $\{1, 3, 5, 7, \ldots\}$

(f) $\{x \mid x = n^2 \text{ and } n \in N\}$

SOLUTION

(a) $\{13, 14, 15, 16\}$
$\{n \mid n \in N \text{ and } 12 < n < 17\}$

(b) $\{2, 4, 6, 8, 10\}$
The set of the first five even natural numbers.

(c) The set of natural numbers that are multiples of 3.
$\{x \mid x = 3n \text{ and } n \in N\}$

(d) $\{1, 3, 5, 7, 9, 11, 13, 15, 17, 19\}$
$\{x \mid x = 2n - 1 \text{ and } n = 1, 2, \ldots, 10\}$

(e) The set of the odd natural numbers
$\{x \mid x = 2n - 1 \text{ and } n \in N\}$

(f) $\{1, 4, 9, 16, 25, \ldots\}$
The set of the squares of the natural numbers. ■

Venn Diagrams

Sets can be represented pictorially by **Venn diagrams**, named for the English logician John Venn (1834–1923). The universal set, which we denote by U, is represented by a rectangle. Any set within the universe is represented by a closed loop lying within the rectangle. The region inside the loop is associated with the elements in the set. An example is given in Figure 2.1, which shows the Venn diagram for the set of vowels $V = \{\text{a, e, i, o, u}\}$ in the universe $U = \{\text{a, b, c}, \ldots, \text{z}\}$.

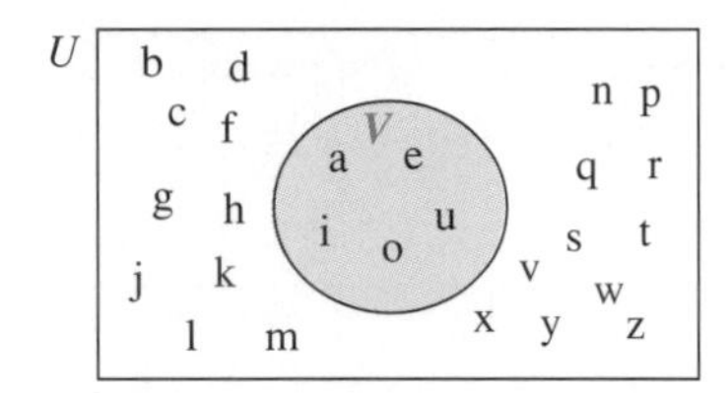

Figure 2.1
The Venn diagram showing the set of vowels in the universe of the 26 letter alphabet

Coloring, shading, or cross-hatching are also useful devices to distinguish the sets in a Venn diagram. For example, set A is red in Figure 2.2 and the set of elements of the universe that do not belong to A is colored blue.

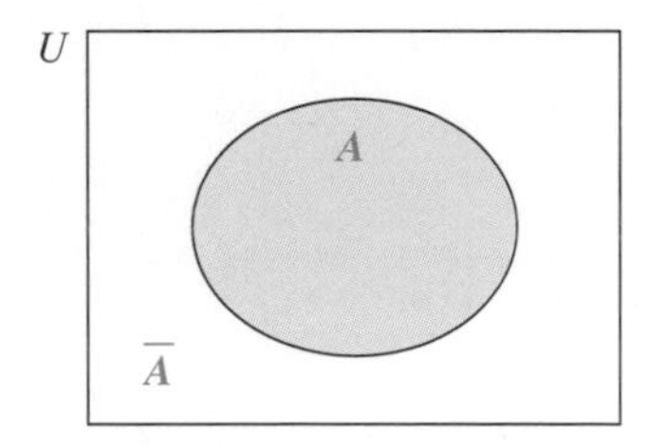

Figure 2.2
The Venn diagram of set A (shown in blue), its complement $\overline{A}$ (shown in white), and the universal set U (the region within the rectangle)

The elements in the universe that are not in set A form a set called the **complement** of A, which is written $\overline{A}$.

DEFINITION The Complement of a Set A

The **complement of set A,** written $\overline{A}$, is the set of elements in the universal set U which are not elements of A. That is,

$$\overline{A} = \{x \mid x \in U \quad \text{and} \quad x \notin A\}.$$

EXAMPLE 2.2

Finding Set Complements

Let $U = N = \{1, 2, 3, \ldots\}$ be the set of natural numbers. For the sets E and F given below, find the complementary sets $\overline{E}$ and $\overline{F}$.

(a) $E = \{2, 4, 6, \ldots\}$
(b) $F = \{n \mid n > 10\}$

SOLUTION

(a) $\overline{E} = \{1, 3, 5, \ldots\}$. That is, the complement of the set E of even natural numbers is the set $\overline{E}$ of odd natural numbers.
(b) $\overline{F} = \{1, 2, 3, 4, 5, 6, 7, 8, 9, 10\}$. ■

Relationships and Operations on Sets

Consider several sets, labeled $A, B, C, D, \ldots$, whose members all belong to the same universal set U. It is useful to understand how sets may be related to one another, and how two or more sets can be used to define new sets.

> **DEFINITION Subset**
>
> The set A is a **subset** of B, written $A \subseteq B$, if, and only if, every element of A is also an element of B.

If A is a subset of B, every element of A also belongs to B. In this case it is useful in the Venn diagram to place the loop for representing A within the loop representing set B, as shown in Figure 2.3.

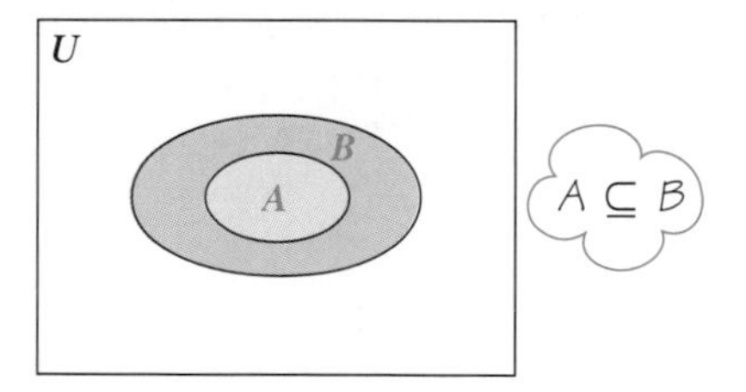

Figure 2.3
A Venn diagram when it is known that A is a subset of B

If two sets A and B have precisely the same elements then they are **equal** and we write $A = B$. If $A \subseteq B$ but $A \neq B$, we say that A is a **proper subset** of B and write $A \subset B$. If $A \subset B$, there must be some element of B which is not also an element of A; that is, there is some x for which $x \in B$ and $x \notin A$.

> **DEFINITION Intersection of Sets**
>
> The **intersection** of two sets A and B, written $A \cap B$, is the set of elements common to both A and B. That is,
>
> $$A \cap B = \{x \mid x \in A \quad \text{and} \quad x \in B\}.$$

For example, $\{a, b, c, d\} \cap \{a, d, e, f\} = \{a, d\}$. The symbol $\cap$ is a special mathematics symbol called a **cap.**

Two sets with no element in common are said to be **disjoint.** The intersection of disjoint sets is then the set with no members, which is called the **empty set.** The empty set is given the special mathematical symbol $\emptyset$, which should not be mistaken for the Greek letter ϕ (phi). In symbols, two sets C and D are disjoint if and only if $C \cap D = \emptyset$. See Figure 2.4.

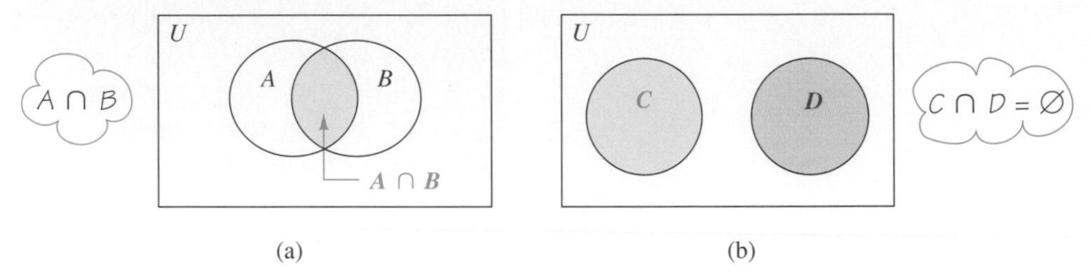

Figure 2.4
(a) A Venn diagram whose shaded region shows the intersection $A \cap B$.
(b) A Venn diagram for sets C and D that are disjoint. That is, $C \cap D = \emptyset$.

DEFINITION Union of Sets

The **union** of sets A and B, written $A \cup B$, is the set of all elements that are in A or B. That is,

$$A \cup B = \{x \mid x \in A \quad \text{or} \quad x \in B\}.$$

For example, if $A = \{\text{a, e, i, o, u}\}$ and $B = \{\text{a, b, c, d, e}\}$ then $A \cup B = \{\text{a, b, c, d, e, i, o, u}\}$. Elements such as a and e that belong to both A and B are listed just once in $A \cup B$. The word *or* in the definition of union is taken in the inclusive sense of and/or. The symbol for union is the **cup,** $\cup$. The cup symbol must be carefully distinguished from the letter U used to denote the universal set. See Figure 2.5.

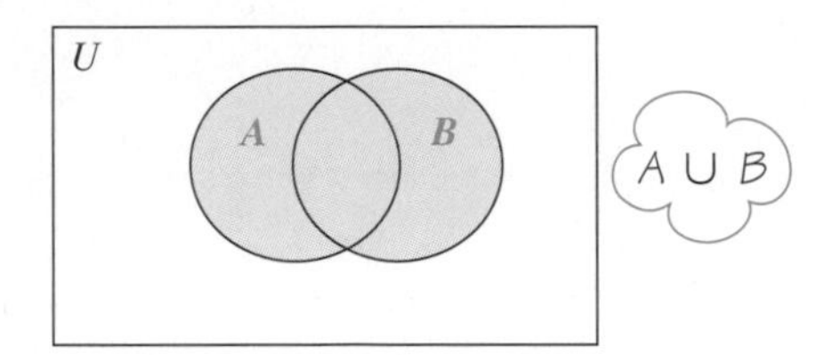

Figure 2.5
The shaded region corresponds to the union, $A \cup B$, of sets A and B

DEFINITION Difference of Sets

The **difference** of sets A and B, written $A - B$, is the set of all elements of A that are not elements of B. That is,

$$A - B = \{x \mid x \in A \quad \text{and} \quad x \notin B\}.$$

For example,

$$\{\text{a, b, c, d, e}\} - \{\text{a, c, e, h}\} = \{\text{b, d}\}.$$

The operation of set difference $A - B$ can be viewed as removing those elements from A that belong to B. Figure 2.6 shows how $A - B$ is visualized with a Venn diagram. We see that $A - B \subseteq A$.

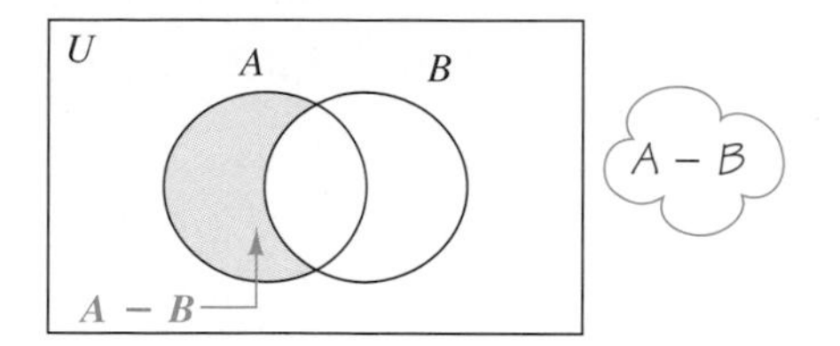

Figure 2.6
The shaded region corresponds to the set difference $A - B$

EXAMPLE 2.3

Performing Operations on Sets

Let $U = \{p, q, r, s, t, u, v, w, x, y\}$ be the universe, and let $A = \{p, q, r\}$, $B = \{q, r, s, t, u\}$, $C = \{r, u, w, y\}$. Locate all ten elements of U in a three loop Venn diagram, and then find the following sets:

(a) $A \cup C$ **(b)** $A \cap C$ **(c)** $A - C$ **(d)** $\overline{C}$
(e) $A \cap \overline{C}$ **(f)** $C - A$ **(g)** $A \cup B$ **(h)** $A \cap B$

SOLUTION

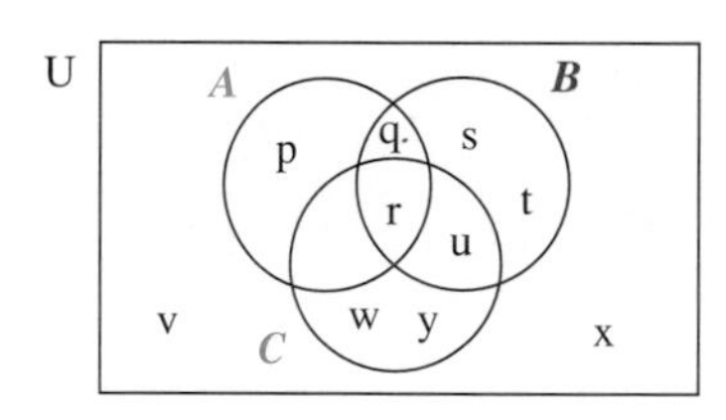

(a) $A \cup C = \{p, q, r, u, w, y\}$ **(b)** $A \cap C = \{r\}$
(c) $A - C = \{p, q\}$ **(d)** $\overline{C} = \{p, q, s, t, v, x\}$
(e) $A \cap \overline{C} = \{p, q\}$ **(f)** $C - A = \{u, w, y\}$
(g) $A \cup B = \{p, q, s, t, u\}$ **(h)** $A \cap B = \{q, r\}$

■

The Cartesian Product

Another way to form a new set from two other sets A and B is to form their **Cartesian product.** The name honors the French philosopher and mathematician René Descartes (1596–1650). To define the Cartesian product, recall that an **ordered pair** is an ordered list of two objects. An ordered pair is written with parentheses in the form (a, b), where the comma separates the first **component**, a, from the second component, b. Two ordered pairs are equal if, and only if, their respective components are equal. For example, $(2, 3) = (4 - 2,\ 1 + 2)$ and $(2, 3) \neq (3, 2)$.

We can now state the definition of the Cartesian product.

DEFINITION Cartesian Product of Sets

The **Cartesian product** of sets A and B, written $A \times B$, is the set of all ordered pairs whose first component is an element of set A and whose second component is an element of set B. That is,

$$A \times B = \{(a, b) \mid a \in A \quad \text{and} \quad b \in B\}.$$

The cross symbol $\times$ (not to be mistaken for the letter x) denotes the Cartesian product, so $A \times B$ is usually read "A cross B." If $A = \{a, b\}$ and $B = \{a, c, d\}$, then

$$A \times B = \{(a, a), (a, c), (a, d), (b, a), (b, c), (b, d)\}.$$

It is useful to see that the elements in $A \times B$ can be arranged in a rectangular array, as shown in Table 2.1.

TABLE 2.1 The Array of Entries in the Cartesian Product $A \times B$, Where $A = \{a, b\}$ and $B = \{a, c, d\}$

		B		
		a	c	d
A	a	(a, a)	(a, c)	(a, d)
	b	(b, a)	(b, c)	(b, d)

EXAMPLE 2.4 Applying the Cartesian Product

Firavich's Frozen Yogurt offers four flavors of yogurt—vanilla, strawberry, mint, raspberry—and three toppings—walnuts, caramel, fudge. Discuss how the Cartesian product can be used to express a customer's possible choices of yogurt with one topping.

SOLUTION

Let $F = \{v, s, m, r\}$ denote the set of flavors of yogurt and $T = \{w, c, f\}$ denote the set of toppings. Then

$$F \times T = \{(v, w), (v, c), (v, f), (s, w), (s, c), (s, f), (m, w), (m, c), (m, f), (r, w), (r, c), (r, f)\}.$$

This gives all possible combinations of yogurt flavors and toppings. ■

The fact that four flavors and three toppings results in 12 combinations corresponds to the multiplication $4 \cdot 3 = 12$. In Section 2.4 we will see that the Cartesian product of sets provides one way to define multiplication.

Using Sets for Problem Solving

The notions of sets and their operations are often used to understand a problem and communicate its solution. Moreover, Venn diagrams provide a means of visual understanding and representation.

EXAMPLE 2.5 Using Sets to Solve a Problem in Color Graphics

The cathode ray tube (CRT) on a color monitor uses three types of phosphors, each of which when excited by an electron beam produces one of the three primary colors—red, blue, or green. By exciting different combinations of phosphors a wider range of colors is possible. Use a Venn diagram to show what combinations are possible.

SOLUTION

Introduce three loops in a Venn diagram—one for each of the primary colors. As shown below, eight colors can be achieved. Most computers also allow the intensity of each color to be specified. In this way even more colors can be obtained.

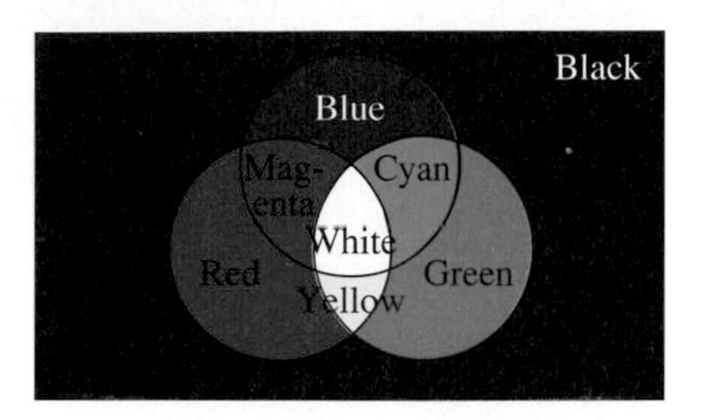

■

Notice that if red and green are excited, then including blue will produce white. On the other hand, if red is excited then including both green and blue will also produce white. In symbols, this becomes

$$(R \cap G) \cap B = R \cap (G \cap B).$$

Since the result is independent of where the parentheses are placed, this illustrates what is called the **associative property** of the intersection operation. Since the order in which the intersections are taken has no effect on the outcome, we do not need parentheses and it is meaningful to write $R \cap G \cap B$.

The associative property is just one example of a number of useful properties that hold for set operations and relations. The properties listed in the following theorem can be proved by reasoning directly from the definitions given earlier. They can also be justified by considering appropriately shaded Venn diagrams.

THEOREM Properties of Set Operations and Relations

1. Transitivity of inclusion

$$\text{If } A \subseteq B \text{and } B \subseteq C \text{ then } A \subseteq C.$$

2. Commutativity of union and intersection

$$A \cup B = B \cup A$$
$$A \cap B = B \cap A$$

3. Associativity of union and intersection

$$A \cup (B \cup C) = (A \cup B) \cup C$$
$$A \cap (B \cap C) = (A \cap B) \cap C$$

4. Properties of the empty set

$$A \cup \emptyset = \emptyset \cup A = A$$
$$A \cap \emptyset = \emptyset \cap A = \emptyset$$
$$A \times \emptyset = \emptyset \times A = \emptyset$$

5. Distributive properties of union and intersection

$$A \cap (B \cup C) = (A \cap B) \cup (A \cap C)$$
$$A \cup (B \cap C) = (A \cup B) \cap (A \cup C)$$

The reason $\emptyset \times A = \emptyset$ is that no element exists for a first component of any ordered pair. The distributive properties show how the operations of union and intersection interact.

EXAMPLE 2.6 **Verifying Properties with Venn diagrams**

(a) Verify the property $A \cap (B \cup C) = (A \cap B) \cup (A \cap C)$.

(b) Show that set difference is not commutative.

(c) Show that $A \cup B \cap C$ is not meaningful without parentheses.

SOLUTION

(a) To shade the region corresponding to $A \cap (B \cup C)$, we intersect the A loop with the "figure 8" loop of $B \cup C$.

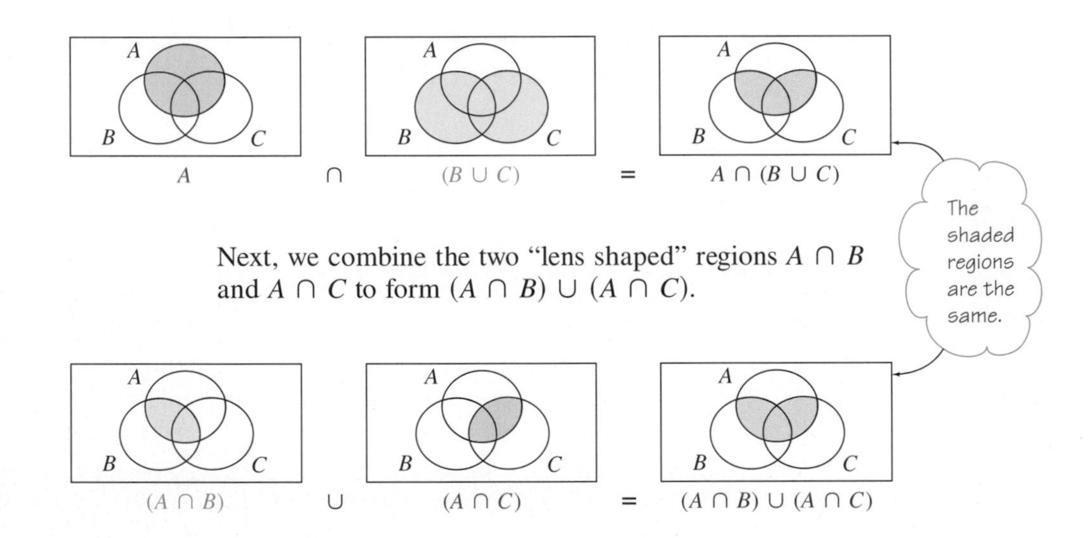

Next, we combine the two "lens shaped" regions $A \cap B$ and $A \cap C$ to form $(A \cap B) \cup (A \cap C)$.

The shaded regions at the right agree, which verifies that intersection distributes over union; that is, that

$A \cap (B \cup C) = (A \cap B) \cup (A \cap C)$.

(b)

If $A = \{a\}$ and $B = \{b\}$ then $A - B = \{a\}$ and $= B - A = \{b\}$. This shows that it is not generally true that $A - B = B - A$. An example like this that shows that something is *not true* is called a **counterexample.**

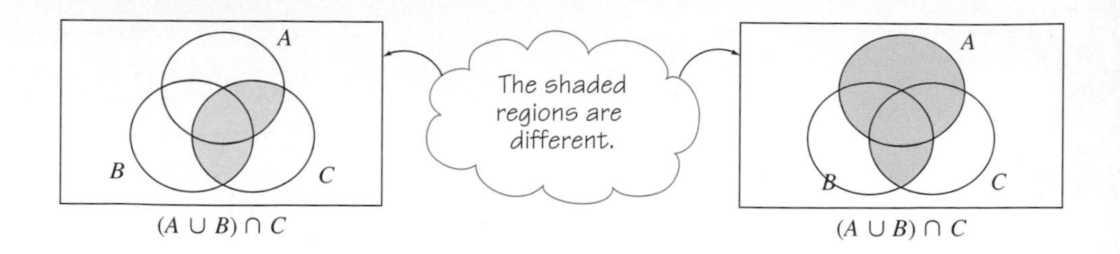

(c) A counterexample is provided by $A = \{a\}$, $B = \{a, b\}$ and $C = \{b, c\}$, since $(A \cup B) \cap C = \{b\}$ and $A \cup \{B \cap C\} = \{a, b\}$. ■

The next example considers a type of problem which Lewis Carroll (the pen name of Oxford University mathematician Charles Dodgson) included in his brilliant but eccentric book *Symbolic Logic,* published in 1896. The book is subtitled "A Fascinating Mental Recreation for the Young." It contains a large number of curious puzzles, including this one from Part One.

HIGHLIGHT FROM HISTORY
Lewis Carroll (Charles Dodgson, 1832–1898)

The Reverend Charles Dodgson, more widely known by his pen name Lewis Carroll, was a well-known mathematician and logician at Oxford University in England. Dodgson wrote several mathematics books that were used in many schools during the last century. He also delighted in thinking up challenging mathematical problems to amuse his friends and educate his pupils. In 1882, he was coaxed by the young daughter of a friend into telling a story.

Later, Charles Dodgson, now as Lewis Carroll, wrote and published this story under the title *Alice's Adventures in Wonderland.* Six years later Lewis Carroll published another story about Alice called *Through the Looking Glass.* Although these books were written for children, many adults enjoyed analyzing the clever wit and twisted logic found in them as Alice tries to make sense out of the nonsense world she finds herself in. Both books contain many subtle allusions to mathematical concepts.

EXAMPLE 2.7 **Solving a Lewis Carroll Puzzle**

(1) All the old articles in this cupboard are cracked.
(2) No jug in this cupboard is new.
(3) Nothing in this cupboard, that is cracked, will hold water.

Is there a jug in this cupboard that will hold water?

SOLUTION

Understand the problem • • • •

Information has been given about objects in a cupboard. We must determine if there is a jug in the cupboard that can hold water, keeping in mind that we must be consistent with the conditions given.

Devise a plan • • • •

Since we are dealing with a collection of objects, it seems natural to introduce sets. The objects in the cupboard which share a common attribute will define various sets. Using set operations and Venn diagrams, it may be possible to decide if the cupboard contains a jug that can hold water.

Carry out the plan • • • •

Looking over the statement of the problem, it seems useful to define the following sets:

$$U = \{x \mid x \text{ is an article in this cupboard}\}$$
$$W = \{x \mid x \text{ is able to hold water}\}$$
$$C = \{x \mid x \text{ is cracked}\}$$
$$J = \{x \mid x \text{ is a jug}\}$$
$$D = \{x \mid x \text{ is old}\}$$

The three conditions of the problem can now be expressed in the language of sets as follows:

$D \subseteq C$ from condition (1)
$J \subseteq D$ from condition (2)
$C \cap W = \emptyset$ from condition (3)

Now draw the Venn diagram shown in the figure. It is evident that J and W are disjoint sets, so this cupboard cannot contain a jug that is able to hold water.

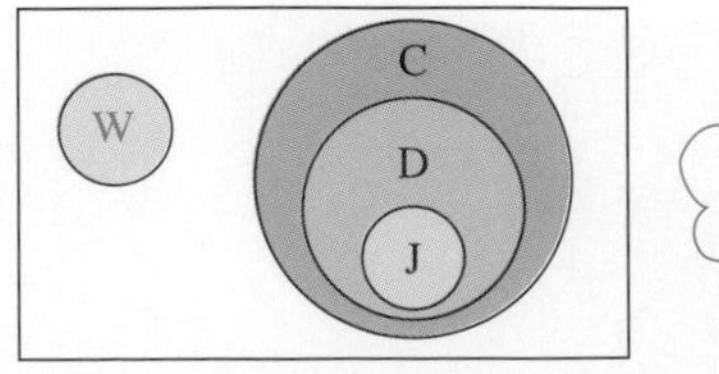

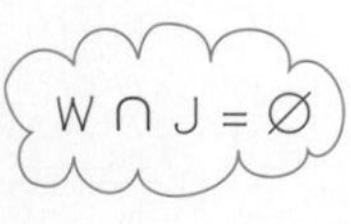

Look back

This Lewis Carroll puzzle is typical of problems that ask us to organize and understand information about objects according to the presence or absence of various attributes. The method of sets, and the visualization offered by Venn diagrams, should be helpful for other problems of this type. ■

PROBLEM SET 2.1

Understanding Concepts

Problem numbers in color indicate that the answer to this problem can be found at the back of the book.

1. Write the following sets by listing their elements.
   (a) The set of states in the U.S. which border Nevada.
   (b) The set of states in the U.S. whose names begin with the letter M.
   (c) The set of states in the U.S. whose names contain the letter Z.
2. Write the following sets.
   (a) The set of letters used in the sentence "list the elements in a set only once."
   (b) The set of letters that are needed to spell these words: team, meat, mate, tame.
3. Let $U = \{1, 2, 3, \ldots, 20\}$. Write these sets by listing the elements in braces.
   (a) $\{x \in U \mid 7 < x \leq 12\}$
   (b) $\{x \in U \mid x \text{ is odd and } 4 \leq x \leq 13\}$
   (c) $\{x \in U \mid x \text{ is divisible by } 3\}$
   (d) $\{x \in U \mid x = 3n \text{ for some } n \in N\}$
   (e) $\{x \in U \mid x = n^2 \text{ for some } n \in N\}$
4. Write these sets in set builder notation, where $U = \{1, 2, \ldots, 20\}$.
   (a) $\{11, 12, 13, 14\}$
   (b) $\{6, 8, 10, 12, 16\}$
   (c) $\{4, 8, 12, 16, 20\}$
   (d) $\{2, 5, 10, 17\}$
5. Use set builder notation to write the following subsets of the natural numbers.
   (a) The even natural numbers larger than 12.
   (b) The squares of the odd numbers larger than or equal to 25.
   (c) The natural numbers divisible by 3.
6. Discuss whether or not the following verbal descriptions describe well-defined sets:
   (a) Moscow, Lima, Paris, Duluth
   (b) Moscow ID, Lima MT, Paris TX, Duluth GA
   (c) The smart students in my math class.
   (d) The students in my class with a 3.5 GPA.

Trucker rolls to right city, but wrong state

DULUTH, Minn. (AP)
A truck driver made a longer trip than he bargained for when he ended up in the right town – but the wrong state.

SOURCE: From "Trucker rolls to right city, but wrong state" from *The Daily Evergreen*, Vol. 97, No. 92, January 21, 1991. Reprinted by permission of The Daily Evergreen.

7. Decide if the following set relationships are *true* or *false*.
   (a) $\{t, w, e, n, t, y, o, n, e\} = \{t, w, e, n, t, y, t, w, o\}$
   (b) $\{2\} \subset \{1, 2, 3\}$
   (c) $\{1, 2, 3\} = \{3, 1, 2\}$
   (d) $\{2\} \subseteq \{1, 2, 3\}$
   (e) $\{1, 2, 3\} \subseteq \{3, 2, 1\}$
   (f) $\{3\} \subset \{2\}$
8. Let $U = \{a, b, c, d, e, f, g, h\}$, $A = \{a, b, c, d, e\}$, $B = \{a, b, c\}$, and $C = \{a, b, h\}$. Locate all eight elements of U in a three loop Venn diagram, and then list the elements in the following sets.
   (a) $B \cup C$ **(b)** $A \cap B$ (c) $A - B$
   (d) $B - A$ (e) $\overline{A}$ **(f)** $A \cap C$
   (g) $A \cup (B \cap C)$
9. Let $L = \{6, 12, 18, 24, \ldots\}$ be the set of multiples of 6 and let $M = \{45, 90, 135, \ldots\}$ be the set of multiples of 45.
   (a) Use your calculator to extend the list of elements in M to include 4 more elements.
   (b) Describe $L \cap M$.
   (c) What is the smallest element in $L \cap M$?

10. Let $G = \{n \mid n \text{ divides } 90\}$ and $D = \{n \mid n \text{ divides } 144\}$. In listed form $G = \{1, 2, 3, 5, 6, 9, 10, 15, 18, 30, 45, 90\}$.

(a) Find the listed form of the set D.

(b) Find $G \cap D$.

(c) Which element of $G \cap D$ is largest?

11. Draw and shade Venn diagrams that correspond to the following sets.

(a) $A \cap B \cap C$ **(b)** $A \cup (B \cap \overline{C})$
(c) $(A \cap B) \cup C$ **(d)** $\overline{A} \cup (B \cap C)$
(e) $A \cup B \cup C$ (f) $\overline{A} \cap B \cap C$

12. For each part, draw a Venn diagram that clearly shows that the conditions listed must hold.

(a) $A \subseteq C, \quad B \subseteq C, \quad A \cap B = \emptyset$

(b) $C \subseteq (A \cap B)$

(c) $(A \cap B) \subseteq C$

13. If $A \cup B = A \cup C$, is it necessarily true that $B = C$? Explain briefly.

14. The English mathematician Augustus DeMorgan (1806–1871) showed that

$$\overline{A \cap B} = \overline{A} \cup \overline{B} \qquad \text{and} \qquad \overline{A \cup B} = \overline{A} \cap \overline{B}.$$

These identities are now called *DeMorgan's Laws.* On four separate Venn diagrams, shade in regions that represent each of the sets $\overline{A \cap B}$, $\overline{A} \cup \overline{B}$, $\overline{A \cup B}$, and $\overline{A} \cap \overline{B}$. Two pairs of your shaded diagrams should be identical. This justifies DeMorgan's Laws.

15. Let U be the set of natural numbers 1 through 20, let A be the set of even numbers in U, and let B be the set of numbers in U that are divisible by 3.

(a) Find $\overline{A \cap B}, \overline{A} \cup \overline{B}, \overline{A \cup B}, \overline{A} \cap \overline{B}$.

(b) Verify that DeMorgan's Laws (see problem 14) hold.

16. Find these Cartesian products of sets.

(a) $\{1, 2\} \times \{a, b\}$

(b) $\{x, y, z\} \times \{\square, \star\}$

(c) $\{6\} \times \{t\}$

(d) $\{1, 2, 3\} \times \{2, 3\}$

17. Write each of these sets as a Cartesian product.

(a) $\{(d, 4), (e, 6), (e, 4), (d, 6)\}$

(b) $\{(0, 1), (0, 0), (1, 0), (2, 0), (2, 1), (1, 1)\}$

(c) $\{(m, n) \mid m \in N \text{ and } n \in N\}$

(d) $\{(x, y)\}$

(e) $\emptyset$ [Give at least three representations for this one.]

18. The Cartesian product $A \times B \times C$ of three sets A, B, and C is defined as the set of all ordered triples of the form (a, b, c), where $a \in A$, $b \in B$, and $c \in C$.

(a) Let $A = \{a, b\}$, $B = \{1, 2\}$, $C = \{r, s, t\}$. Find $A \times B \times C$. (*Hint:* One element of $A \times B \times C$ is $(a, 1, r)$.)

(b) At Perulli's Pizza Parlor, pizzas come in various sizes (small, medium, large), crust types (thin, pan) and toppings (black olives, green pepper, onions, pepperoni, anchovies). Let $S = \{s, m, l\}$, $C = \{t, p\}$, $T = \{b, g, o, pe, a\}$. Then $S \times C \times T$ is the set of ways to order a pizza with one topping. Describe these pizzas: (s, t, b), (m, p, pe), (l, t, a).

(c) How many ways can you order a one topping pizza at Perulli's?

19. (a) Explain why the Cartesian product is not a commutative operation.

(b) Under what conditions on sets C and D is it possible that $C \times D = D \times C$? (*Hint:* Don't overlook the empty set.)

20. The 12 shapes in the Venn diagram below are described by the following attributes.

Shape:	circle, hexagon, or triangle
Size:	small or large
Color:	red or blue

Let C, H, and T denote the respective sets of circular, hexagonal, and triangular shapes. Similarly, let S, L, R, and B denote the sets of small, large, red, and blue shapes, respectively. The set A contains the shapes that are small and not a triangle, so $A = S \cap \overline{T}$.

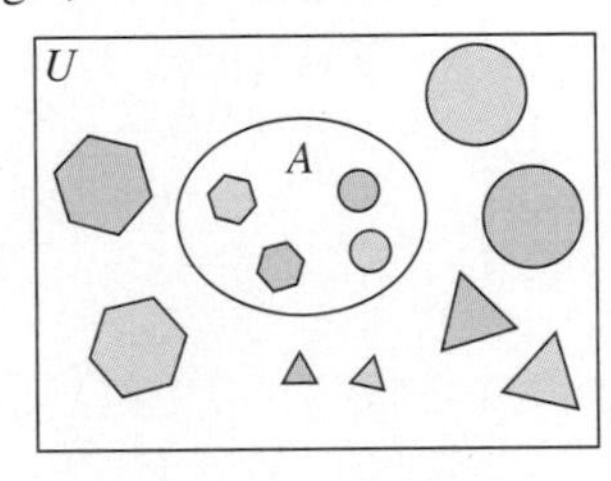

Draw the shapes found in these sets.

(a) $R \cap C$ **(b)** $L \cap H$ (c) $T \cup H$
(d) $L \cap T$ (e) $B - C$ **(f)** $H \cap S \cap R$

21. Let C, H, T, S, L, R, and B denote the sets of attribute shapes described in problem 20. The set of small hexagons can be written $S \cap H$ in symbolic form. Express these sets in symbolic form.

(a) The large triangles.

(b) The blue polygonal (that is, noncircular) shapes.

(c) The small or triangular shapes.

(d) The shapes that are either blue circles or are red.

Thinking Critically

22. The set {a} has two subsets, ∅ and {a}. The set {a, b} has four subsets: ∅, {a}, {b}, {a, b}.
    (a) How many subsets are there of {a, b, c}? (*Hint:* Make an orderly list.)
    (b) How many subsets are there of {a, b, c, d}?
    (c) How many of the subsets of {a, b, c, d} do not contain the element d?
    (d) If a set has n elements, guess a formula for the number of subsets. Explain carefully, on the basis of consideration of parts (a), (b), and (c).
    (e) How many subsets does the set of 26 letters {a, b, . . . , z} of the alphabet have?
    (f) A set has exactly 524,288 subsets. How many elements does it have?

23. Circular loops in a Venn diagram divide the universe U into distinct regions.

No sets

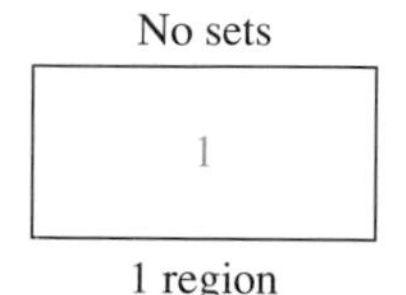

1 region

1 set

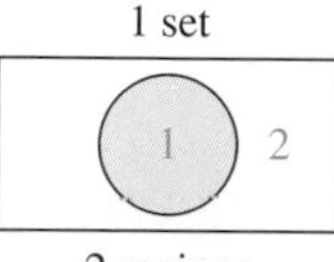

2 regions

2 sets

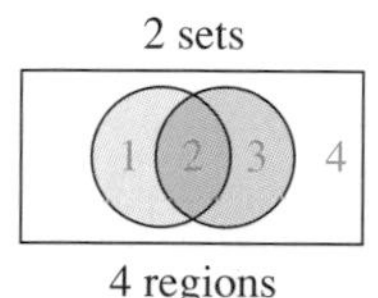

4 regions

    (a) Draw a diagram for three circles that gives the largest number of regions.
    (b) How many regions do the four circles define in this figure?

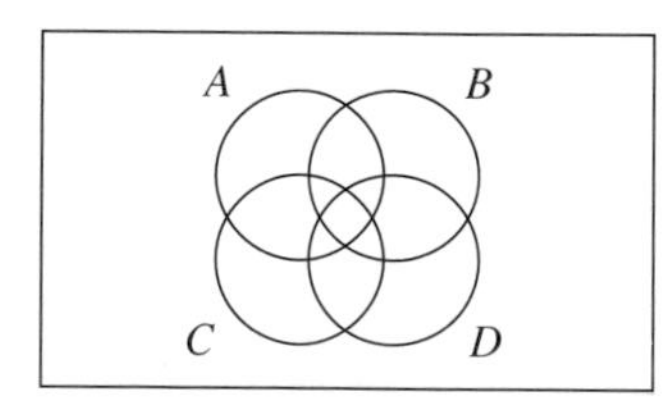

    (c) Verify that the four circle diagram is missing a region corresponding to $A \cap \overline{B} \cap \overline{C} \cap D$. What other set has no corresponding region?
    (d) Will this Venn diagram allow for all possible combinations of four sets? Explain briefly but carefully.

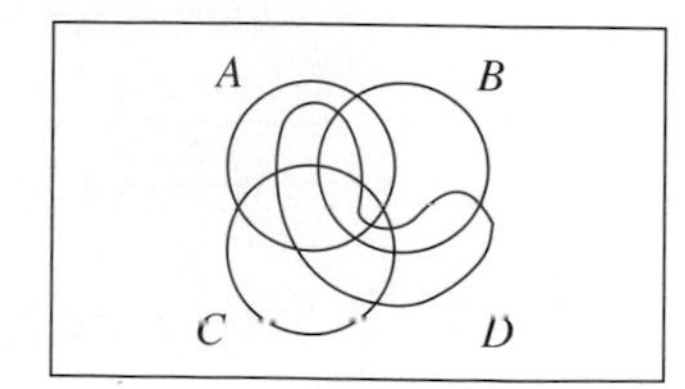

24. Solve this Lewis Carroll puzzle.
    (1) No birds, except ostriches, are 9 feet high;
    (2) There are no birds in this aviary that belong to anyone but me;
    (3) No ostrich lives on mince pies;
    (4) I have no birds less than 9 feet high.

    Do any birds in this aviary live on mince pies? (*Hint:* Let U be the set of birds, A the set of birds in this aviary, L the birds living on mince pies, M be my birds, N be birds 9 feet high, and S be ostriches.)

25. Solve this Lewis Carroll puzzle.
    (1) All writers, who understand human nature are clever;
    (2) No one is a true poet unless he can stir the hearts of men;
    (3) Shakespeare wrote "Hamlet";
    (4) No writer, who does not understand human nature, can stir the hearts of men;
    (5) None but a true poet could have written "Hamlet."

    Show that Shakespeare was clever. (*Hint:* Let U be the set of writers, A the set of writers able to stir hearts of men, C the set of clever writers, P the set of true poets, N the set of writers who understand human nature, H the set of writers of "Hamlet," and S = {Shakespeare}.)

26. A set is *partitioned* when it is written as a union of subsets, no two of which have elements in common. For example, set C = {a, b, c} can be partitioned in these five ways, counting C itself.

$$\{a, b, c\} = \{a\} \cup \{b, c\} = \{b\} \cup \{a, c\}$$
$$= \{c\} \cup \{a, b\} = \{a\} \cup \{b\} \cup \{c\}$$

    (a) How many ways can the set D = {a, b, c, d} be partitioned?
    (b) How many ways can the set E = {a, b, c, d, e} be partitioned? (*Hint:* There are ten subsets of E which contain two elements.)

Thinking Cooperatively

27. Let C, H, T, S, L, R, and B denote the sets of attribute shapes described in problem 20. Each loop in the Venn diagrams below corresponds to one of the above sets. Just some of the twelve attribute pieces are depicted in each figure. Determine how to label the loops. Is there more than one correct way to label each diagram?
    (a)

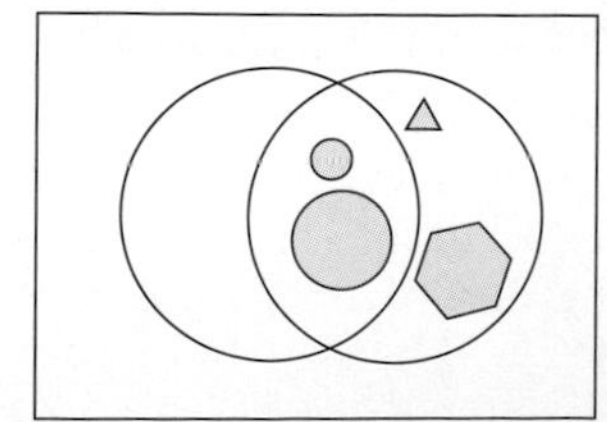

(b)

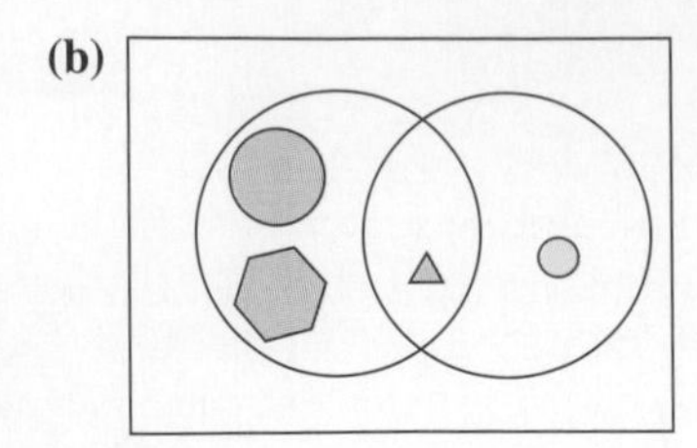

(c)

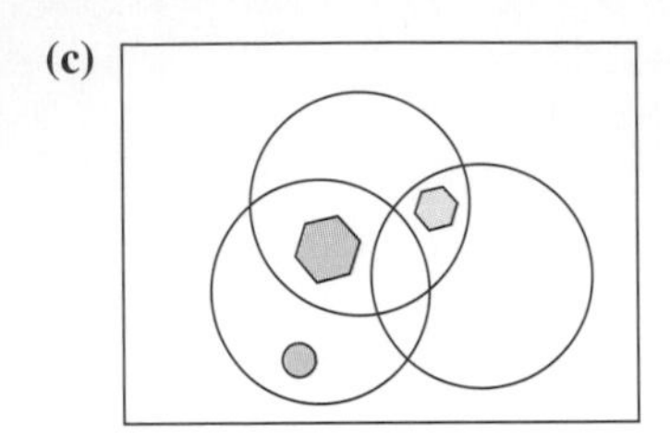

(d) Invent your own "missing label" puzzles similar to those shown. Trade the puzzles with classmates and search for their solutions.

28. Each of the objects depicted in problem 20 is described by its attributes: shape (three choices), size (two choices), and color (two choices). The number of pieces in a full set of shapes can be varied by changing the number of attributes and the number of possible choices of an attribute. How many pieces are contained in these attribute sets?
    (a) Shape: circle, hexagon, equilateral triangle, isosceles right triangle, rectangle, square
    Size: large, small
    Color: red, yellow, blue
    (b) As in part (a), but also include this attribute:
    Thickness: thick, thin

29. Attribute cards can be made by drawing various shapes on cards, varying the figure shown, the color used, and so on. Here are some examples of a few attribute cards.

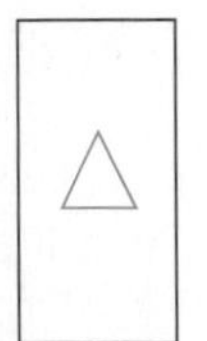 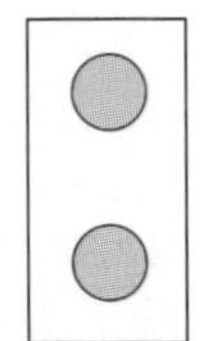 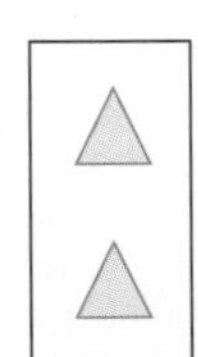 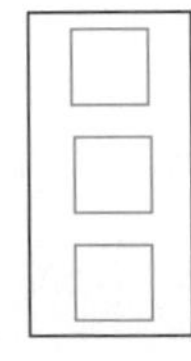 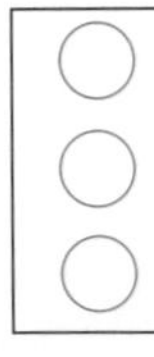

(a) Conjecture how many cards make a full deck. Explain how you get your answer.
(b) Design and make your own set of attribute cards, drawing simple figures on small rectangles of cardstock with colored pens.
(c) Exchange decks of attribute cards among cooperative groups. Shuffle each deck and turn over just a few of the cards. Can you predict the number of cards in the complete deck?

Making Connections

30. **The ABO System of Blood Typing.** Until the beginning of the twentieth century, it was assumed that all human blood was identical. About 1900, however, the Austrian–American pathologist Karl Landsteiner discovered that blood could be classified into four groups according to the presence of proteins called antigens. This discovery made it possible to transfuse blood safely. A person with the antigens A, or B, or both A and B, has the respective blood type A, B, or AB. If neither antigen is present the type is O. Draw and label a Venn diagram which illustrates the ABO system.

31. **The Rh System of Blood Typing.** In 1940, Karl Landsteiner (see problem 30) and the American pathologist Alexander Wiener discovered another protein which coats the red blood cells of some persons. Since the initial research was on rhesus monkeys, a person with the protein is classified as Rh positive (Rh+), and a person whose blood cells lack the protein is Rh negative (Rh−). Draw and label a Venn diagram which illustrates the classification of blood in the eight major types A+, A−, B+, B−, AB+, AB−, O+, O−.

Communicating

32. The word "set" has been given precise mathematical meaning. The same word is also used as part of our ordinary language. For example, you may own a set of golf clubs. Make a list of other examples where "set," or "family," "aggregate," "class," or "collection" are used, and discuss to what extent their usage corresponds to the mathematical concept of set.

33. Many of the terms introduced in this section also have meaning in nonmathematical contents. For example, "complement," "union," and "intersection" are used in ordinary speech. Other terms, such as "transitive," "commutative," "associative," and "distributive" are rare in everyday usage but there are closely related words which are quite common, such as "transit," "commute," "associate," and "distribute." Discuss the differences and similarities in meanings of the words, and word roots, of the terms introduced in this section.

For Review

34. Show that the natural numbers 1, 2, . . . , 15 can be arranged in a list so that the sum of each adjacent pair is a square number. Write your solution to illustrate Pólya's four steps of problem solving.

35. A power company intends to number its power poles from 1 to 10,000. Each numeral is formed by gluing stamped-metal digits to the poles. How many metal 2s should the company order? Write your solution to illustrate Pólya's four steps of problem solving.

2.2 Sets, Counting, and the Whole Numbers

If you attended a student raffle, you might hear the following announcement when the entry forms are drawn:

> *"The student with identification number 50768–973 has just won second prize—four tickets to the big game this Saturday."*

There are three types of numbers.

This sentence contains three numbers, each of a different type, and each serving a different purpose.

First, a number can be an **identification,** or **nominal number.** A nominal number is a sequence of digits used as a name or label. Telephone numbers, social security numbers, account numbers with stores and banks, serial numbers, and driver's license numbers are just a few examples of the use of number for identification and naming. The role such numbers play in contemporary society has expanded rapidly with the advent of computers.

The next type of number used by the raffle announcer is an **ordinal number.** The words first, second, third, fourth, and so on, are used to describe the relative position of the objects in an ordered sequence. Thus, ordinal numbers communicate location in an ordered collection. First class, second rate, third base, Fourth of July, page 5, volume 6, and twenty-first century are all familiar examples of ordinal numbers.

The final use of number by the raffle announcer was to tell how many tickets had been won. That is, the prize is a set of tickets and *four* tells us *how many* tickets are in the set. More generally, a **cardinal number** of a set is the number of objects in the set. Thus cardinal numbers help communicate the basic notion of "how many."

It should be noticed that numbers, of whatever type, can be expressed verbally (in a language) or symbolically (in a numeration system). For example, the number of moons of Mars is "two" in English, "zwei" in German, and "dos" in Spanish. Symbolically, we could write 2 in the Hindu-Arabic system or II in Roman numerals. Numeration, as a system of symbolic representation of number, is closely related to algorithms for computation. Numeration systems, both historic and contemporary, are described in Chapter 3.

In the remainder of this section, we explore the notion of cardinal number more fully. Of special importance is the set of whole numbers, which can be viewed as the set of cardinal numbers of finite sets.

One-to-one Correspondence and Equivalent Sets

Suppose, when you come to your classroom, each student is seated at his or her desk. You would immediately know, *without counting,* that the number of children and the number of occupied desks are the same. Similarly, by placing the tips of the fingers on the left hand against the tips of the fingers on the right hand, most people can affirm that each of their hands has the same number of fingers. These two examples illustrate the concept of a **one-to-one correspondence** between sets in which each element of one set is paired with exactly one element of the second set, and each element of the two sets belongs to exactly one of the pairs.

> **DEFINITION One-to-One Correspondence**
>
> A **one-to-one correspondence** between sets A and B is a pairing of the elements of A with the elements of B in such a way that each element of A and B belongs to one, and only one, pair.

Figure 2.7 illustrates a one-to-one correspondence between the sets of board members and officers of the Math Club.

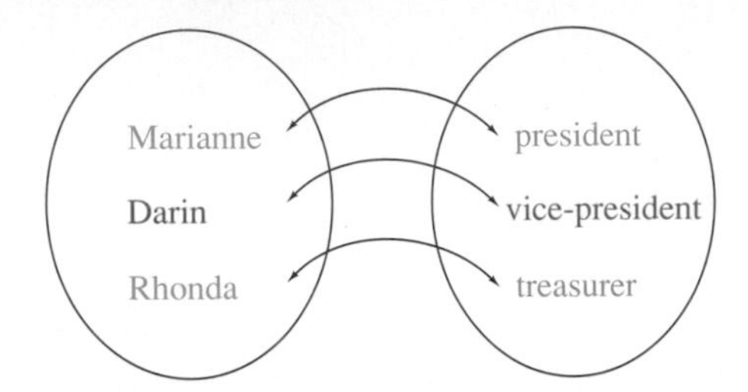

Figure 2.7
A one-to-one correspondence between the set of board members {Marianne, Darin, Rhonda} and offices {president, vice-president, treasurer} of the Math Club

When a one-to-one correspondence can be established between sets {m, d, r} and {p, v, t}, we say that they are **equivalent sets** and write {m, d, r} $\sim$ {p, v, t}. More generally we have the following definition.

> **DEFINITION Equivalent Sets**
>
> Sets A and B are **equivalent** if there is a one-to-one correspondence between A and B. When A and B are equivalent, we write $A \sim B$. We also say that equivalent sets **match.** If A and B are not equivalent we write $A \nsim B$.

It is easy to see that equal sets match. To see why, suppose that $A = B$. Since each x in A is also in B, the natural matching $x \leftrightarrow x$ is a one-to-one correspondence between A and B. Thus $A \sim B$. On the other hand, there is no reason to believe equivalent sets must be equal. For example, $\{\square, \star\} \sim \{1, 2\}$ but $\{\square, \star\} \neq \{1, 2\}$.

EXAMPLE 2.8

Investigating Sets for Equivalence

Let $A = \{x \mid x \text{ is a moon of Mars}\}$
$B = \{x \mid x \text{ is a former U.S. president whose last name is Adams}\}$
$C = \{x \mid x$ is one of the Bronte sisters of nineteenth century literary fame$\}$
$D = \{x \mid x \text{ is a satellite of the fourth closest planet to the sun}\}$

Which of these relationships, $=, \neq, \sim, \nsim$, holds between distinct pairs of the four sets?

SOLUTION

$A \neq B, A \neq C, A = D, B \neq C, B \neq D, A \sim B, A \nsim C, A \sim C, C \neq D, B \nsim C, B \sim C, C \nsim D.$

It may be useful to write the sets in listed form:

$A = D =$ {Diemos, Phobos}, $B =$ {John Adams, John Quincy Adams},
$C =$ {Ann, Charlotte, Emily} ■

Finite and Infinite Sets

Consider the set of natural numbers $N = \{1, 2, 3, \ldots\}$ and the set of squares of natural numbers $S = \{1, 4, 9, \ldots\}$. If you are asked to compare the "number of elements" (that is, the cardinality) of N and S, you may well be quite perplexed as to how to answer. Since S is a *proper* subset of N, it seems reasonable to feel that there are more elements in N than in S. On the other hand, the elements in N can be paired with the elements of S to form a one-to-one correspondence:

$$\begin{array}{cccccccc} N & 1 & 2 & 3 & \ldots & n & \ldots \\ & \updownarrow & \updownarrow & \updownarrow & & \updownarrow & \\ S & 1 & 4 & 9 & \ldots & n^2 & \ldots \end{array}$$

This matching tells us that the cardinalities of the two sets are equal.

The difficulty arises because N and S have an infinite number of elements. Infinity has fascinated individuals since the time of the ancient Greek philosophers. For thousands of years, mathematicians have attempted to understand the idea of infinity and resolve the many paradoxical situations which arise in the presence of infinity. Fortunately, the notion of equivalence between sets can be used to give a precise and useful definition which distinguishes an infinite set from a finite set. The definition was formulated in 1887 by the German mathematician Richard Dedekind (1831–1916).

DEFINITION Infinite and Finite Sets

A set is **infinite** if it can be put in one-to-one correspondence to one of its proper subsets. A set that is not infinite is **finite.**

It is not necessary to use the set of squares as the proper subset to show that N is an infinite set. For example, we could also use the set $\{2, 3, 4, \ldots\}$. This is a proper subset of N and it matches N by the following one-to-one correspondence:

$$\begin{array}{cccccc} 1 & 2 & 3 & \ldots & n & \ldots \\ \updownarrow & \updownarrow & \updownarrow & & \updownarrow & \\ 2 & 3 & 4 & \ldots & n+1 & \ldots \end{array}$$

The Whole Numbers

The sets {a, b, c}, {□, ○, △}, {Mercury, Venus, Earth}, {Larry, Moe, Curly} are distinct, but they do share the property of "threeness." The English word "three," and the Hindu-Arabic numeral 3, are used to identify this common property of all sets that are equivalent to {1, 2, 3}. In a similar way, we use the word "two," and the symbol 2, to convey the idea that all of the sets that are equivalent to the set {1, 2} have the same cardinality.

The whole numbers, as defined below, allow us to classify any finite set according to how many elements the set contains. It is useful to adopt the symbol $n(A)$ to

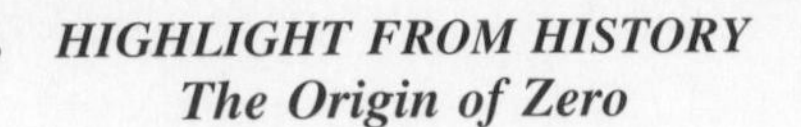

The Origin of Zero

It required thousands of years of mathematical thought for the concept of zero to emerge and its importance be recognized. Zero, as is true of any whole number, expresses a quantity. Thus zero *is* something, and it should not be mistaken as synonymous with nothing. (Ask any student who gets a zero on a test if it means nothing to his or her grade!) The contemplation of the void was an important aspect of ancient Eastern philosophy, and it is possible to trace the idea, and even the word zero, to this tradition. The Hindu word *sunya,* meaning "void" or "empty," was translated into Arabic as *sifr.* In its Latinized form sifr became *zephirum,* from which the word *zero* ultimately evolved.

represent that cardinality of the finite set A. If A is not the empty set, then $n(A)$ is a counting number. The cardinality of the empty set, however, requires a new name and symbol: we let **zero** designate the cardinality of the empty set, and write $0 = n(\emptyset)$.

A common error is made by omitting the $n(\quad)$ symbol: be sure not to write "$A = 4$" when your intention is to state that $n(A) = 4$.

DEFINITION The Whole Numbers

The **whole numbers** are the cardinal numbers of finite sets; that is, the numbers of elements in finite sets. If $A \sim \{1, 2, 3, \ldots, m\}$ then $n(A) = m$, and $n(\emptyset) = 0$, where $n(A)$ denotes the cardinality of set A. The set of whole numbers is written $W = \{0, 1, 2, 3, \ldots\}$.

EXAMPLE 2.9 **Determining Whole Numbers**

For each set find the whole number that gives the number of elements in the set.

(a) $M = \{x \mid x \text{ is a month of the year}\}$
(b) $A = \{\text{a, b, c, . . . , z}\}$
(c) $B = \{n \in N \mid n \text{ is a square number smaller than } 200\}$
(d) $Z = \{n \in N \mid n \text{ is a square number between 70 and } 80\}$
(e) $S = \{0\}$

SOLUTION

(a) $n(M) = 12$ since $M \sim \{1, 2, 3, \ldots, 12\}$, and this amounts to counting the elements in M.
(b) $n(A) = 26$
(c) $n(B) = 14$ since $B = \{1, 4, 9, 16, 25, 36, 49, 64, 81, 100, 121, 144, 169, 196\}$ and contains 14 elements.
(d) $n(Z) = 0$ since there are no square numbers between 70 and 80 and therefore $Z = \emptyset$.
(e) $n(S) = 1$ since the set $\{0\}$ contains one element. This shows that zero is *not* the same as "nothing"! ■

EXAMPLE 2.10 Problem Solving with Whole Numbers and Venn Diagrams

In a recent survey, the 60 students living in Harris Hall were asked about their enrollments in science, engineering, and humanities classes. The results were as follows:

24 are taking a science class
22 are taking an engineering class
17 are taking a humanities class
5 are taking both science and engineering classes
4 are taking both science and humanities classes
3 are taking both engineering and humanities classes
2 are taking classes in all three areas.

How many students are not taking classes in any of the three areas? How many students are taking a class in just one area? Use a Venn diagram, indicating the number of students in each region of the diagram.

SOLUTION

Let S, E, and H denote the set of students in science, engineering, and humanities classes, respectively. Since $n(S \cap E \cap H) = 2$, begin by placing the 2 in the region of the Venn diagram corresponding to the subset $S \cap E \cap H$. Then, by comparing $n(S \cap E \cap H) = 2$ and $n(E \cap H) = 3$, we conclude there is one student who is taking both engineering and humanities, but not a science class. This allows us to fill in the 1 in the Venn diagram. Analogous reasoning leads to the entries within the loops of the following Venn diagram.

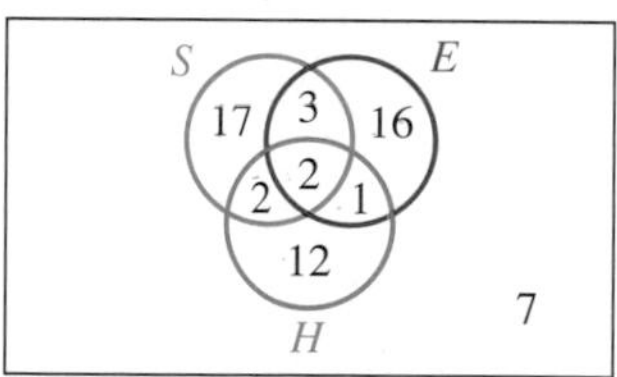

The values within the loops account for 53 of the 60 students, so it follows that 7 students are not taking classes in any of the three areas. Also, since $17 + 16 + 12 = 45$, we conclude that 45 students are taking a class in just one area. ■

Disc Discoveries

Consider the set of three discs shown here.

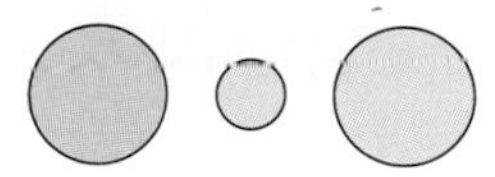

Which disc is most like each of the remaining two discs? If you were told that the three discs shown were originally part of a set of four discs, what do you suppose the fourth disc looks like?

Physical and Pictorial Representations for Whole Numbers

Sets provide a basis for defining the whole numbers. As we will see shortly, sets also provide a useful conceptual model through which the properties and operations of whole numbers can be defined and understood. Also many problem-solving situations and activities with whole numbers are best interpreted from the viewpoint of sets. At the same time, it is not appropriate to use the *abstract* notion of sets to teach young children. Instead, the teacher should use actual collections of objects to illustrate numerical concepts. Colored chips, attribute pieces, tiles, beans and so on, all become useful concrete embodiments of elements that model sets—sets that can be seen and touched. The older child can later move successfully to pictorial representations. Later still, the student will become comfortable dealing with abstract models, represented entirely in words and symbols.

Here are just a few physical and pictorial representations useful for explaining whole and natural number concepts. It should be noticed that we have not yet introduced representations that require grouping, number base, or place value. In the next chapter, additional representations are introduced that help illustrate connections to numeration and algorithms for the operations of arithmetic.

Tiles Tiles are congruent squares, each about 2 centimeters (3/4 inch) on a side. They should be sufficiently thick to be easily picked up and moved about. Colored plastic tiles are available from suppliers, but tiles are easily homemade from vinyl tile or cardboard. Of course, beans, circular discs, and other objects can be used as well. However, square tiles can be arranged into rectangular patterns, and such patterns reveal many of the fundamental properties of the whole numbers. See Figure 2.8.

Figure 2.8
Some representations of six with square tiles

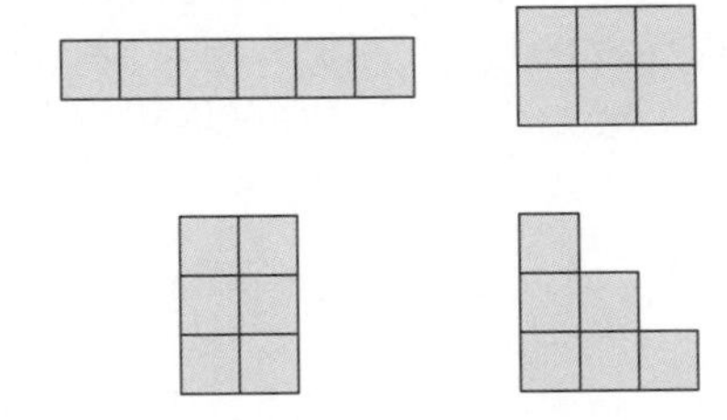

Cubes Cubes are much like tiles, but they can form both three-dimensional and two-dimensional patterns. Several attractive versions are commercially available. Unifix™ Cubes can be snapped together to form linear groupings. Multilink™ Cubes permit planar and spatial patterns as we see in Figure 2.9.

Figure 2.9
Some representations of twelve with cubes

Number Strips and Rods Colored strips of cardboard or heavy paper, divided into squares, can be used to demonstrate and reinforce whole number properties and operations. The squares should be ruled off, and colors can be used to visually identify the number of squares in a strip, as indicated in Figure 2.10. Number strips can be used nearly interchangeably with Cuisenaire® rods, which have been used effectively for many years. The colors shown in the figure correspond to those of Cuisenaire® rods.

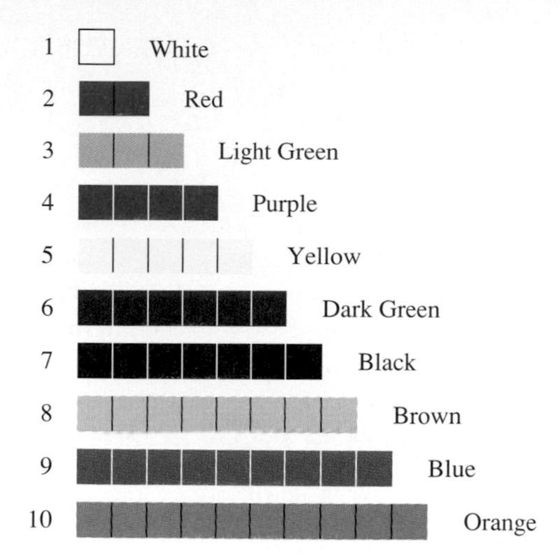

Figure 2.10
Number strips for the natural numbers 1 through 10

Whole numbers larger than ten are illustrated by placing strips, or rods, end-to-end to form a "train."

Number Line The number line is a pictorial model in which two distinct points on a line are labeled 0 and 1, and then the remaining whole numbers are laid off in succession with the same spacing (see Figure 2.11). Any whole number is interpreted as a distance from 0, which can be shown by an arrow. The number line model is particularly important because it can also be used to visualize the number systems that are developed later. In this way, it is easier to see how the whole numbers are extended to the integers, and how the integers are related to the rational and real number systems.

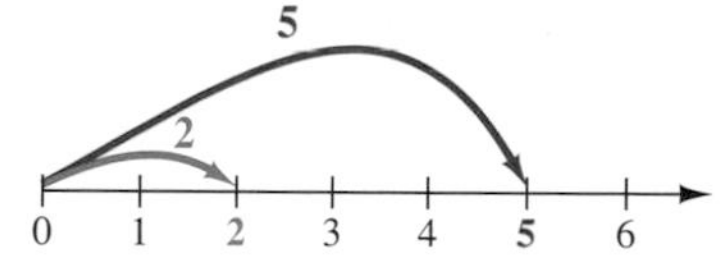

Figure 2.11
Illustrating two and five on the number line model of the whole numbers

Order the Whole Numbers

We often wish to relate the number of elements in two given sets. For example, if each child in the class is given one cupcake, and there are still some cupcakes left, we would know that there are more cupcakes for the party than children. Notice that children have been matched to a *proper* subset of the set of cupcakes.

The order of the whole numbers can be defined in the following way.

> **DEFINITION Ordering the Whole Numbers**
>
> Let $a = n(A)$ and $b = n(B)$ be whole numbers, where A and B are finite sets. If A matches a *proper* subset of B, we say that ***a* is less than *b*** and write $a < b$.

The expression $b > a$ is read "b is greater than a," and is equivalent to $a < b$. Also, $a \leq b$ means "a is less than or equal to b." The sets A and B used in checking the definition are arbitrary, and it is sometimes useful to choose $A \subset B$.

HIGHLIGHT FROM HISTORY
The Equality and Inequality Symbols

Many of the mathematical symbols we commonly use today originated in Renaissance times or shortly thereafter. The sign = was introduced in 1557 by Robert Recorde (1510–1558) of Cambridge. It appeared in *The Whetstone of Witte,* the first English book on algebra. According to Recorde, no two things are so alike as two parallel lines, and hence the symbol of equality. The symbols > and <, for "greater" and "less," first appeared in 1631 in a posthumous publication of Thomas Harriot (1560–1621). The equality sign in Harriot's book was very long, looking like this: ═══════

EXAMPLE 2.11 Showing the Order of Whole Numbers

Show that $4 < 7$ using (a) sets; (b) tiles; (c) rods; and (d) the number line.

SOLUTION

(a) The following diagram shows that a set with 4 elements matches a proper subset of a set with 7 elements. Therefore, $4 < 7$.

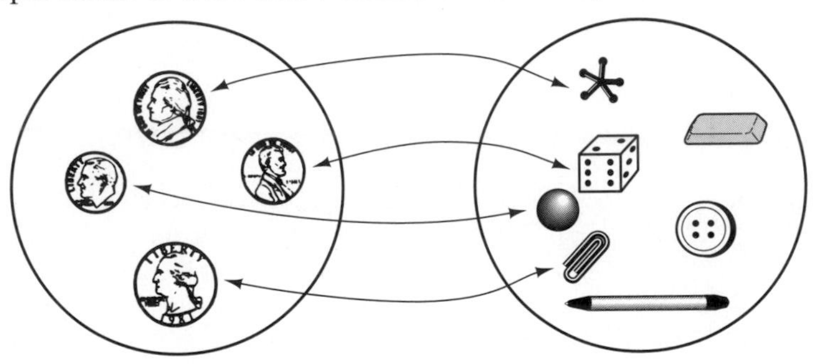

(b) By setting tiles side by side, it is seen that the 4 colored tiles match a proper subset of the 7 uncolored tiles.

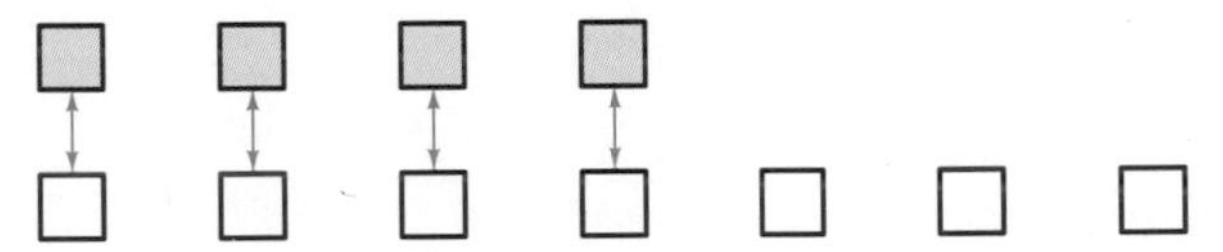

(c) With rods, the order of whole numbers is interpreted by comparing the lengths of the rods.

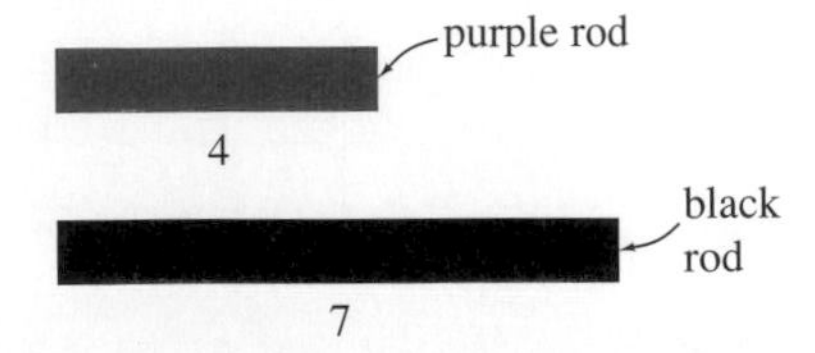

Numbers larger than ten are compared by forming side by side trains of rods.

(d) On the number line, $4 < 7$ because 4 is to the left of 7.

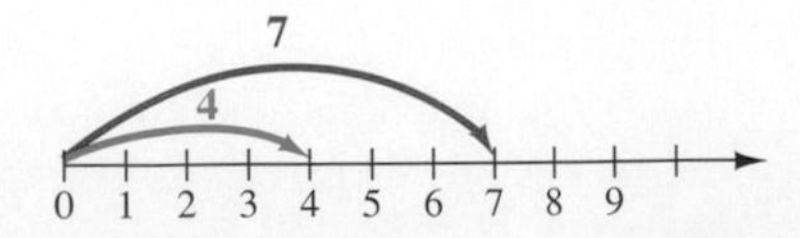

■

In Section 2.1 we learned that if $A \subset B$ and $B \subset C$ then $A \subset C$. When A, B, and C are each finite sets, we see that the whole numbers have a corresponding transitive property.

> **PROPERTY Transitivity of "Less Than"**
>
> If a, b, and c are whole numbers with $a < b$ and $b < c$, then $a < c$.

PROBLEM SET 2.2

Understanding Concepts

1. Classify by type—cardinal, ordinal, or nominal—the numbers that appear in these sentences.
   (a) On June 13, Mary was promoted to first vice-president.
   (b) On hole eleven, Joe's second shot with a 6-iron went 160 yards.
   (c) Erin's bowling partner left the 7 pin in frame six, which added 9 to her score.
2. Let $A = \{\text{h, d, l}\}$ and $B = \{x \mid x \text{ is a nephew of Donald Duck}\}$. Show $A \sim B$ by giving two different one-to-one correspondences between A and B.
3. Kelly plans to attend the concert at Smith Auditorium. The concert is free so Kelly is surprised to learn that a ticket must be obtained for admission. Why do you think this is so?
4. (a) Establish a one-to-one correspondence between the sets {a, b, c} and {□, ★, △}.
   (b) Establish a one-to-one correspondence between the sets {△, □, ★} and {6, ⊕, y}.
   (c) Explain how the correspondences you chose in parts (a) and (b) can be combined to give a one-to-one correspondence between {a, b, c} and {6, ⊕, y}.
5. Use counting to determine the whole number that corresponds to the cardinality of these sets.
   (a) $A = \{x \mid x \in N \text{ and } 20 \leq x < 35\}$
   (b) $B = \{x \mid x \in N \text{ and } x + 1 = x\}$
   (c) $C = \{x \mid x \in N \text{ and } (x - 3)(x - 8) = 0\}$
   (d) $D = \{x \mid x \in N, 1 \leq x \leq 100, x \text{ is divisible by both 4 and 6}\}$
6. Let $A = \{n \mid n \text{ is a cube of a natural number and } 1 \leq n \leq 100\}$
   $B = \{s \mid s \text{ is a state in the U.S. which borders Mexico}\}$
   Is $A \sim B$?
7. Let $T = \{n \mid n \text{ is a natural number divisible by 3}\}$ and
   $F = \{n \mid n \text{ is a natural number divisible by 5}\}$
   Is $T \sim F$? If so, describe a one-to-one correspondence between T and F.
8. Decide which of the following sets are finite.
   (a) {grains of sand on all the world's beaches}
   (b) {whole numbers divisible by 46182970138}
   (c) {points on a line segment that is one inch long}
9. Show that the following sets are infinite.
   (a) $\{n \mid n = m^3 \text{ for some } m \in N\}$
   (b) $\{1, 10, 10^2, \ldots\}$
10. The figure below shows two line segments, L_1 and L_2. The rays through point P give, geometrically, a one-to-one correspondence between the points on L_1 and the points on L_2; for example, $Q_1 \longleftrightarrow Q_2$.

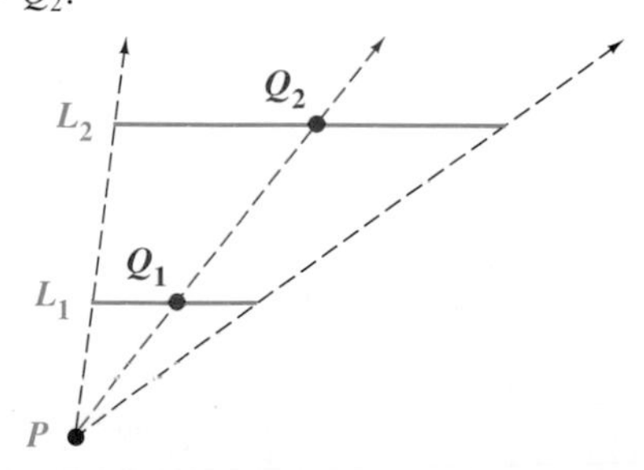

Use similar geometric diagrams to show that the following figures are equivalent sets of points.

(a) Two concentric circles

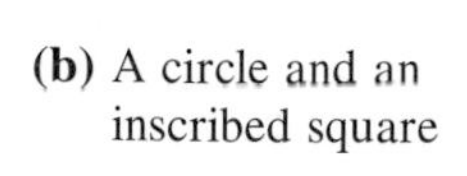

(b) A circle and an inscribed square

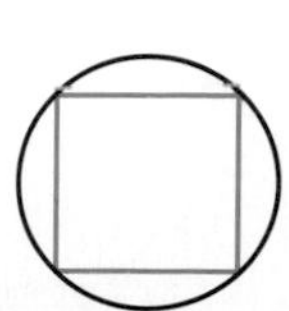

(c) A triangle and its circumscribing circle

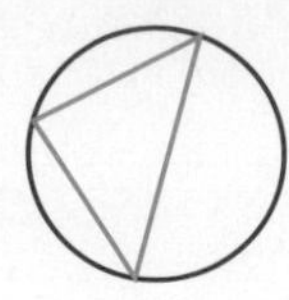

(d) A semicircle and its diameter

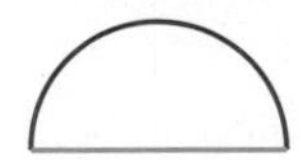

11. Decide whether or not the following statements are true or false. If false, give a counterexample; that is, give two sets which satisfy the hypothesis but not the conclusion of the statement. A and B designate finite sets.
   (a) If $A \subseteq B$ then $n(A) \leq n(B)$
   (b) If $n(A) < n(B)$ then $A \subset B$
   (c) If $n(A - B) = n(A)$ then $A \cap B = \varnothing$
   (d) If $n(A - B) = 0$ then $A = B$.
12. Let $A = \{a, b, c, d\}$ and $B = \{\star, \square\}$. Find $A \times B$ and $n(A \times B)$.
13. Suppose $n(A \times B) = 12$. What are the possible values of $n(A)$ and $n(B)$?
14. Let A and B be finite sets.
   (a) Explain why $n(A \cap B) \leq n(A)$.
   (b) Explain why $n(A) \leq n(A \cup B)$.
   (c) Suppose $n(A \cap B) = n(A \cup B)$. What more can be said about A and B?
15. The Cartesian product of three sets A, B, and C, written $A \times B \times C$, is the set of ordered triples $\{(a, b, c) \mid a \in A, b \in B, c \in C\}$.
   (a) If $n(A) = 4$, $n(B) = 2$, and $n(C) = 5$, explain why $n(A \times B \times C) = 40$.
   (b) If $n(E \times F \times G) = 6$, what are the possible values of $n(E)$?
16. Finish labeling the number of elements in the regions in the Venn diagram shown, where the subsets A, B, and C of the universe U satisfy the conditions listed.

   $n(U) = 100$
   $n(A) = 40$
   $n(B) = 50$
   $n(C) = 30$
   $n(A \cap B) = 17$
   $n(B \cap C) = 12$
   $n(A \cap C) = 15$
   $n(A \cap B \cap C) = 7$

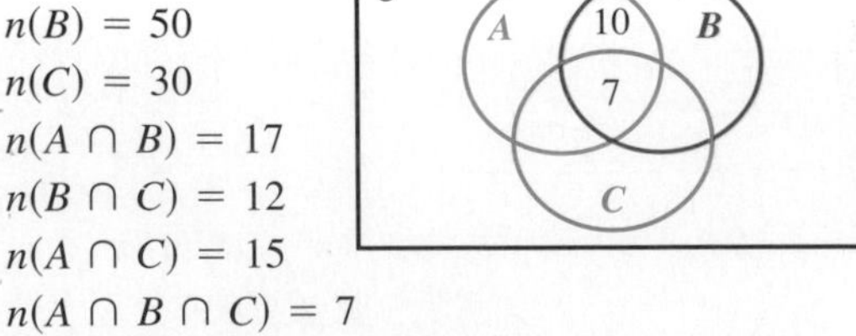

17. Evelyn's Electronics Emporium hired Sloppy Survey Services (SSS) to poll 100 households at random. Evelyn's report from SSS contained the following data on ownership of a TV, VCR, or stereo:

TV only	8
TV and VCR	70
TV, VCR, and stereo	65
Stereo only	3
Stereo and TV	74
No TV, VCR, or stereo	4

   Evelyn, who assumes anyone with a VCR also has a TV, is wondering if she should believe if the figures are accurate. Should she?
18. Number tiles can be arranged to form patterns which relate to properties of the numbers.
   (a) Arrange number tiles to show why 1, 4, 9, 16, 25, . . . are called the "square" numbers.
   (b) Arrange number tiles to show why 1, 3, 6, 10, 15, . . . are called "triangular" numbers.
19. Suppose the square number tiles are replaced by regular hexagons. Instead of building rectangular patterns, it is natural to build "hex" patterns with 6-fold symmetry. The first three **hex** numbers are 1, 7, and 19.

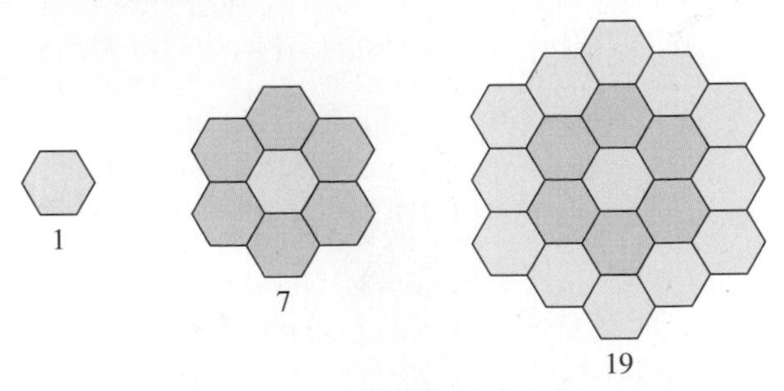

   (a) Find the next two hex numbers.
   (b) Use number cubes to form corner shapes, as shown. The two walls and the floor are the thickness of one cube. What connection do you see to hex numbers?

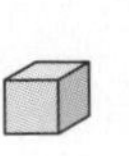
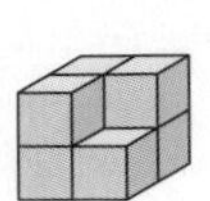
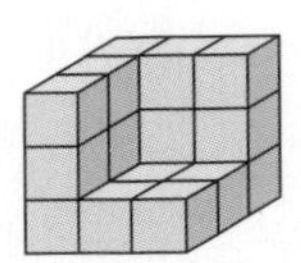

   (c) Draw the next corner shape.

Thinking Critically

20. Joe Freespender makes \$15 per week, but his extravagant lifestyle requires \$50 per week. After ten weeks, Joe has earned \$150 and so he can pay off the bills he accumulated during the first three weeks. Fortunately Joe lives forever.

(a) Will he have any bills which he can never pay?

(b) When can Joe pay off the bills for the weeks 100, 101, and 102?

21. The diagram shows the set of dots in a square array that extends upward and rightward without end. This set is equivalent to the set N of natural numbers, since a one-to-one correspondence is obtained by counting the dots along successive downward diagonals.

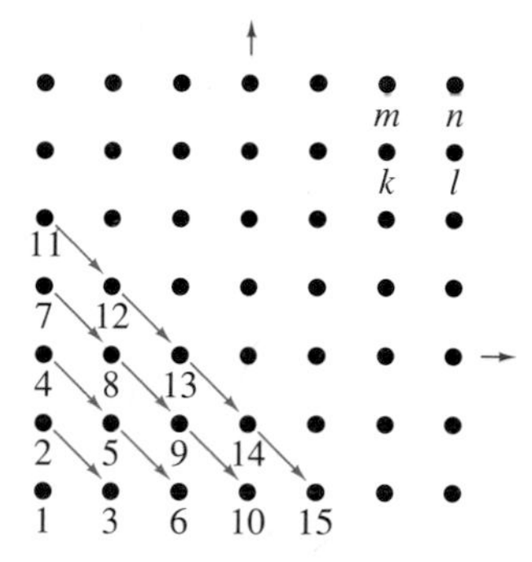

(a) Without extending the array, look for patterns that allow you to determine k, l, m, and n in the diagram. Carefully describe the patterns and how they are used.

(b) The dot in the first row and one hundredth column corresponds to the one hundredth triangular number, which is $(100)(101)/2 = 5050$. Use this fact to determine which row and column contains the dot numbered 5000.

(c) What number is assigned to the dot in the twentieth row and the thirtieth column of the array?

22. Show that set equivalence satisfies the following properties. Carefully describe the one-to-one correspondences you choose.

(a) The reflexive property: $A \sim A$

(b) The symmetric property: If $A \sim B$, then $B \sim A$.

(c) The transitive property: If $A \sim B$ and $B \sim C$, then $A \sim C$.

23. A political polling organization sent out a questionnaire that asked the following question: "Which taxes—income, sales, excise—would you be willing to have raised?" Sixty voters' opinions were tallied by the office clerk:

Tax	Number Willing to Raise the Tax
Income	20
Sales	28
Excise	29
Income and sales	7
Income and excise	8
Sales and excise	10
Unwilling to raise any tax	5

The clerk neglected to count how many, if any, of the 60 voters are willing to raise all three of the taxes. Can you help the pollsters? Explain how.

24. There are 40 students in the Travel Club. They discovered that 17 members have visited Mexico, 28 have visited Canada, 10 have been to England, 12 have visited both Mexico and Canada, 3 have been only to England, and 4 have been only to Mexico. Some club members have not been to any of the three foreign countries and, curiously, an equal number have been to all three countries.

(a) How many students have been to all three countries?

(b) How many students have been only to Canada?

25. The 12 white tiles are surrounded by a border of 18 blue tiles.

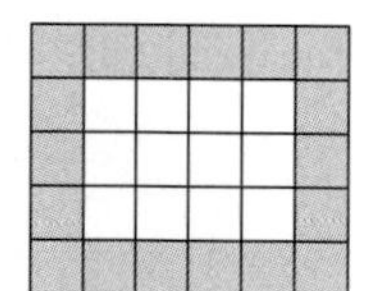

Find all possible rectangles of white tiles which have a border of blue tiles and use equally many white and blue tiles. (*Hint:* Match as many blue tiles to an adjacent white tile as you can, and determine how many blue tiles must be matched to nonadjacent white tiles.)

Thinking Cooperatively

26. There are two ways to match the sets {a, b} and {1, 2}, namely by the one-to-one correspondences a ↔ 1, b ↔ 2 and a ↔ 2, b ↔ 1.

(a) Find the six different one-to-one correspondences between the equivalent sets {a, b, c} and {1, 2, 3}.

(b) How many one-to-one correspondences are there between {p, q, r, s} and {1, 2, 3, 4}?

(c) Complete the following table.

$n(A)$	1	2	3	4	5
Number of one-to-one correspondences of set A to $\{1, \ldots, m\}$, where $m = n(A)$					

Look for a pattern which allows you to fill in as many entries as you wish.

(d) If $m \in N$, then $m!$ (read as "m factorial") is defined by $m! = m \times (m - 1) \times (m - 2) \times \cdots \times 3 \times 2 \times 1$. Thus $1! = 1$, $2! = 2 \times 1 = 2$, $3! = 3 \times 2 \times 1 = 6$, and so on. If $n(A) = m$, how many ways can A be matched to set $\{1, 2, \ldots, m\}$? Explain how you arrived at your answer.

(e) How many different ways can you put a dozen eggs back in the carton?

(f) A set of history books can be placed on the shelf in 362,880 ways. How many volumes are there?

27. To show that $r \leq n$, where r and n are whole numbers, the set method requires us to choose a subset with r elements from a set with n elements. For example, if $C = \{a, b, c\}$ so that $n = 3$, the subsets of C with r elements are as follows:

r	0	1	2	3	4
Subsets with r elements	$\emptyset$	{a}, {b}, {c}	{a, b}, {a, c}, {b, c}	{a, b, c}	none
Number of subsets with r elements	1	3	3	1	0

Thus, for example, there are three ways to choose two elements from a set with three elements.

(a) Fill in the following table of subsets of $D = \{a, b, c, d\}$, arranged by the number of elements in the subsets.

r	0	1	2	3	4	5
Subsets with r elements						
Number of subsets with r elements						

(b) Fill in the table below. The value in the (n, r) position should be the number of ways to choose r elements from a set with n elements. The row for $n = 3$ has already been filled in. Since there is no way to form a subset of $r = 4$ elements from a set with $n = 3$ elements, there is a zero in the (3, 4) position of the table.

n \ r	0	1	2	3	4	5
0						
1						
2						
3	1	3	3	1	0	0
4						

(c) What general pattern do you see in the table of part (b)? Describe it, and then complete additional rows in the table.

(d) A steering committee for the class play consists of Amy, Byron, Clea, Don, Edie, and Franco. How many ways can they choose a 3-member subcommittee to arrange publicity?

Making Connections

28. In the late twentieth century, modern society has entered what some observers have called "the information age." People are now accustomed to being identified by number as often as by name. Make a list of your own identification numbers—social security, credit cards, telephone, and so on. It may be helpful to look through your billfold!

29. Blood tests of 100 people showed that 45 had the A antigen and 14 had the B antigen (see problem 30 of Section 2.1). Another 45 had neither antigen, and so are of type O. How many people are of type AB, having both the A and B antigens? Draw and label a Venn diagram that shows the number of people with blood type A, B, AB, and O.

Communicating

30. The English words for the whole numbers are zero, one, two, three, and so on. Make similar lists in other languages. For example, your list in German would begin *Null, eins, zwei, drei, ...* Do you notice any common roots of the words?

31. Describe how you might teach a very young child the idea of "color" and the words—"blue" or "red" for example—that describe particular colors.

(a) Do you see any similarities in teaching the idea of "number" and such words as "two" or "five"?

(b) Why do you think "zero" is a difficult idea?

32. Imagine yourself teaching a third-grader the transitive property of "less than." Write an imagined dialogue with the student, using number strips (or colored rods) as a manipulative.

For Review

33. Let $U = \{0, 1, 2, \ldots, 25\}$, $A = \{n \mid n \text{ is even}\}$, $B = \{m \mid m = n^2 \text{ for some } n \in U\}$, $C = \{p \mid p \in U \text{ and } p \text{ divides into 48 with no remainder}\}$. Make a Venn diagram that shows the location of all the elements of U.

34. The elements of a universal set $U = \{a, b, c, d, e, f, g\}$ with two subsets A and B are shown in the following Venn diagram.

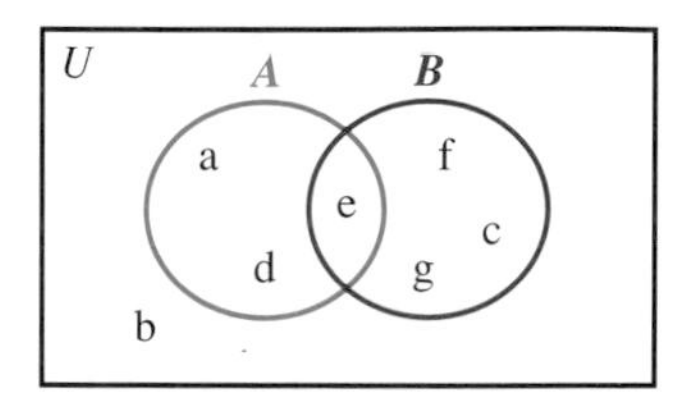

List the elements in the following sets.

(a) $\overline{A}$ **(b)** $A \cap B$
(c) $A \cup \overline{B}$ **(d)** $A - B$
(e) $\overline{A} \cap \overline{B}$ **(f)** $\overline{A \cup B}$
(g) $A \times B$ **(h)** $(A \cup B) - (A \cap B)$

35. Draw a three-loop Venn diagram for each set given, shading the region corresponding to the set.

(a) $(A - B) \cup C$
(b) $(A \cup B \cup C)$
(c) $(A \cup B \cup C) - (A \cap B \cap C)$

2.3 Addition and Subtraction of Whole Numbers

In this section we introduce the operations of addition and subtraction on the set of whole numbers $W = \{0, 1, 2, 3, \ldots\}$. In each operation, two whole numbers are combined to form another whole number. Because *two* whole numbers are added to form the sum, addition is called a **binary operation.** Similarly, subtraction is defined on a pair of numbers, so subtraction is also a binary operation.

The definitions of addition and subtraction are accompanied by a variety of conceptual models that give the operations both intuitive and practical meaning. A corresponding variety of physical and pictorial models are described which reinforce these conceptual notions at the concrete and visual levels. It is vitally important that children be able to interpret and express the operations and their properties through manipulatives and visualization. These activities prepare them to understand the algorithms of computation and build confidence in their ability to select the appropriate operations for problem solving.

The Set Model of Whole Number Addition

The whole numbers answer the basic question "how many?" For example, if Alok collects baseball cards and A is the set of cards in his collection, then $a = n(A)$ is the number of cards he owns. Suppose his friend Barbara also has a collection of baseball cards, forming a set B with $b = n(B)$ cards. If Alok and Barbara decide to combine their collections, the new collection would be the set $A \cup B$ and would contain $n(A \cup B)$ cards. The **addition,** or **sum,** of a and b can be defined as the number of cards in the combined collection. That is, the sum of two whole numbers a and b is given by $a + b = n(A \cup B)$, where A and B are disjoint sets and $a = n(A)$, $b = n(B)$.

Addition answers the question "how many elements are in the union of two disjoint sets?" Figure 2.12 shows how the set model is used to show that $3 + 5 = 8$. First, disjoint sets A and B are found, with $n(A) = 3$ and $n(B) = 5$. Since $n(A \cup B) = 8$, we have shown that $3 + 5 = 8$.

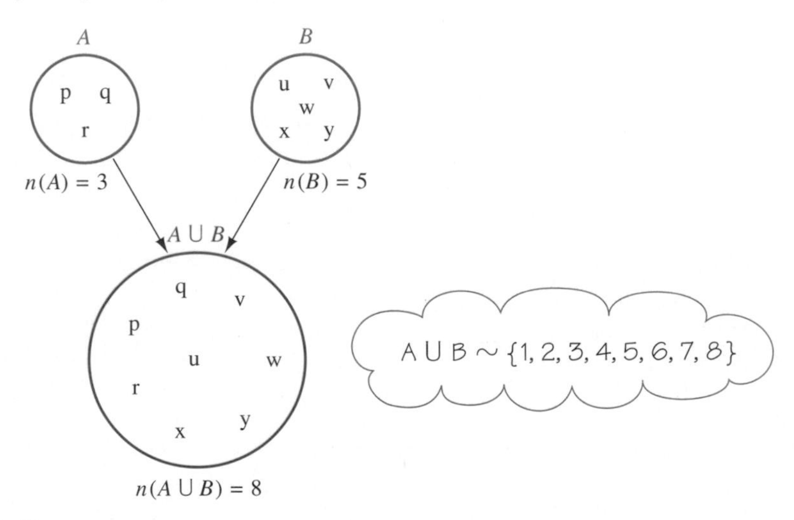

Figure 2.12
Showing 3 + 5 = 8 with the set model of addition

Here is the general definition.

> **DEFINITION The Addition of Whole Numbers**
>
> Let a and b be any two whole numbers. If A and B are any two disjoint sets for which $a = n(A)$ and $b = n(B)$, then the **sum of *a* and *b*,** written $a + b$, is given by $a + b = n(A \cup B)$.

The expression $a + b$ is read "a plus b" and a and b are called the **addends** or **summands.**

$$\mathbf{a + b}$$
↑ ↑ addends

EXAMPLE 2.12 **Using the Set Model of Addition**

Let $F = \{a, b, c, d\}$, $G = \{e, x\}$, $H = \{e, f, g\}$.

(a) Can these sets be used to find the sum $4 + 3$?

(b) Can these sets be used to find the sum $2 + 3$?

SOLUTION

(a) Since $n(F) = 4$, $n(H) = 3$, and $F \cap H = \emptyset$, then $4 + 3 = n(F \cup G) = n(\{a, b, c, d, e, f, g\}) = 7$.

(b) $n(G) = 2$ and $n(H) = 3$. However, G and H are not disjoint, since $G \cap H = \{e\}$. Thus, we cannot say that $2 + 3$ is given by $n(G \cup H)$. For a proper choice of sets we can show $2 + 3 = 5$. However, $n(G \cup H) = n(\{e, x, f, g\}) = 4$. ■

The set model of addition can be illustrated with manipulatives such as those described in the last section. Many addition facts and patterns become evident when they are discovered and visualized by means of concrete embodiments. An example is the triangular numbers $t_1 = 1$, $t_2 = 3$, $t_3 = 6, \ldots$ introduced in Chapter 1. Recall that t_n denotes that nth triangular number, where the name refers to the triangular pattern which can be formed with t_n objects. By using number tiles, as shown in Figure 2.13, we see that the sum of any two successive triangular numbers forms a square number. Indeed, we have the general formula:

$$t_{n-1} + t_n = n^2, \quad n = 1, 2, \ldots.$$

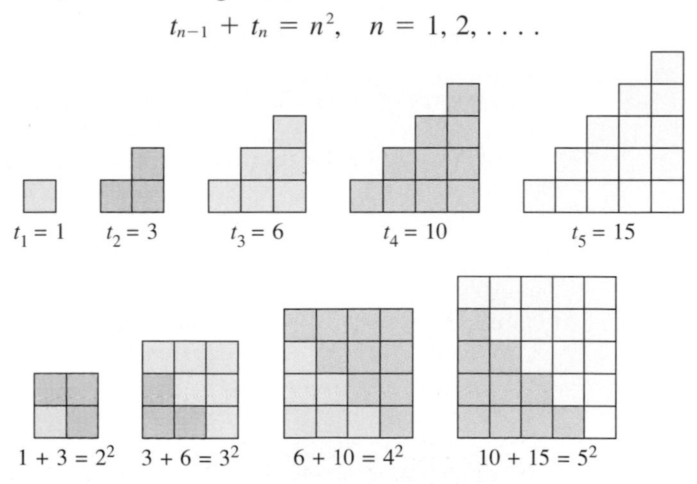

Figure 2.13
The sum of two successive triangular numbers is a square number

The Measurement Model of Addition

On the number line, whole numbers are geometrically interpreted as distances. Addition can be visualized as combining two distances to get a total distance. Right pointing arrows are used to indicate distances. Figure 2.14 illustrates $3 + 5$ on the number line. It is important to notice that the two distances are not overlapping, and the tail of the arrow representing 5 is placed at the head of the arrow representing 3.

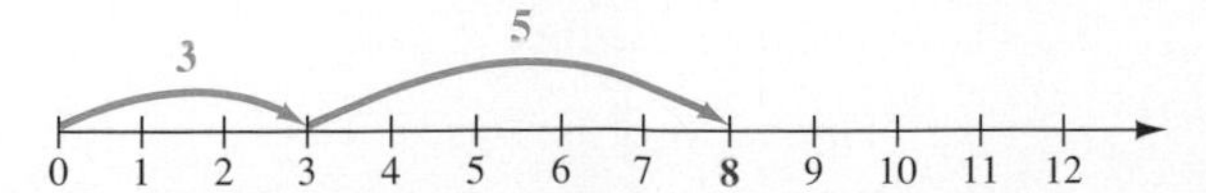

Figure 2.14
Illustrating 3 + 5 on the number line

Properties of Whole Number Addition

The sum of any two whole numbers is also a whole number, so we say that the set of whole numbers has the **closure property** under addition. This property is so obvious for addition it may seem unnecessary to mention it. However, the set of whole numbers is not closed under subtraction or division. Even under addition, many subsets of whole numbers do not have the closure property. For example, if $D = \{1, 3, 5, \ldots\}$ denotes the subset of the odd whole numbers, then D does *not* have the closure property under addition. For example, the sum of 1 and 3 is not in D. On the other hand, the set of even whole numbers, $E = \{0, 2, 4, 6, \ldots\}$, is closed under addition; this is so because the sum of any two even whole numbers is also an even whole number.

Some other important properties of whole number addition correspond to properties of operations on finite sets. For example, the commutative property of union, $A \cup B = B \cup A$, proves that $a + b = b + a$ for all whole numbers a and b. Similarly, the associative property $A \cup (B \cup C) = (A \cup B) \cup C$ tells us that $a + (b + c) = (a + b) + c$, and $A \cup \emptyset = \emptyset \cup A = A$ gives us the additive identity property, $a + 0 = 0 + a = a$, of zero.

THEOREM Properties of Whole Number Addition

Closure Property	If a and b are any two whole numbers, then $a + b$ is a unique whole number.
Commutative Property	If a and b are any two whole numbers, then $a + b = b + a$.
Associative Property	If a, b, and c are any three whole numbers, then $a + (b + c) = (a + b) + c$.
Additive Identity Property of Zero	If a is any whole number, then $a + 0 = 0 + a = a$.

The properties of whole number addition can also be illustrated with any of the physical and pictorial models of the whole numbers. Figure 2.15 shows how the associative property can be demonstrated with number tiles.

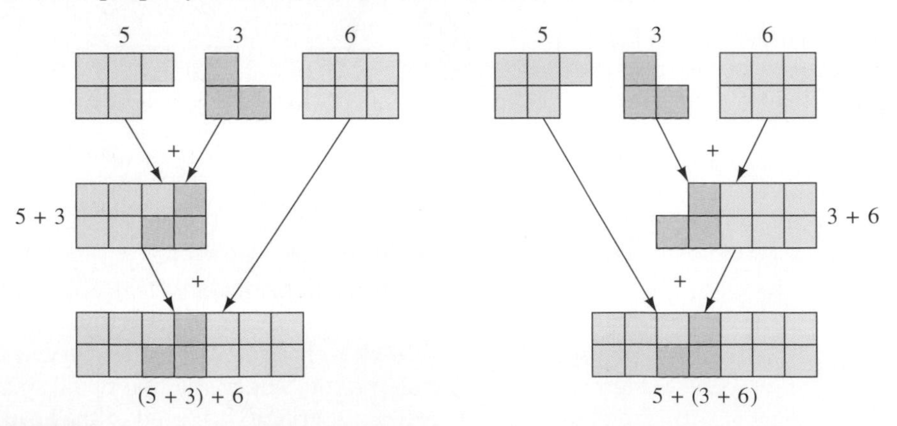

Figure 2.15
Illustrating the associative property (5 + 3) + 6 = 5 + (3 + 6) with number tiles

The addition properties are very useful when adding several whole numbers, since we are permitted to rearrange the order of the addends and the order in which pairs of addends are summed.

INTO THE CLASSROOM
Using Addition Properties to Learn Addition Facts

As children learn addition, the properties of whole number addition should become a habit of thought from the very beginning. Indeed, the properties are useful even for learning and recalling the sums of one digit numbers, as found in the addition table shown. The commutative property $a + b = b + a$ means the entries above the diagonal repeat the entries below the diagonal. Also, the first column of addition by zero is easy by the additive identity property. The next four columns, giving addition by 1, 2, 3, and 4, can be learned by "counting on." For example, 8 + 3 is viewed as "8 plus 1 makes 9, plus 1 more makes 10, plus 1 more makes 11." The diagonal entries $1 + 1 = 2$, $2 + 2 = 4, \ldots$, are the "doubles," which are readily learned by knowing how to count by twos; 2, 4, . . . , 18. The remaining entries in the table can be obtained by combining "doubles" and "counting on." For example, 6 + 8 is viewed as $6 + (6 + 2)$, which is $(6 + 6) + 2$; knowing that 6 doubled is 12 and counting on 2 gives the answer 14.

Other effective strategies include:

Making tens: For example, $8 + 6 = (8 + 2) + 4 = 10 + 4 = 14$.

Counting back: For example, 9 is 1 less than 10, so 6 + 9 is 1 less than $6 + 10 = 16$, giving 15. In symbols, this amounts to
$6 + 9 = (6 + 10) - 1 = 16 - 1 = 15$.

+	0	1	2	3	4	5	6	7	8	9
0	0	1	2	3	4	5	6	7	8	9
1	1	2	3	4	5	6	7	8	9	10
2	2	3	4	5	6	7	8	9	10	11
3	3	4	5	6	7	8	9	10	11	12
4	4	5	6	7	8	9	10	11	12	13
5	5	6	7	8	9	10	11	12	13	14
6	6	7	8	9	10	11	12	13	14	15
7	7	8	9	10	11	12	13	14	15	16
8	8	9	10	11	12	13	14	15	16	17
9	9	10	11	12	13	14	15	16	17	18

Look for patterns. They also help learn facts!

EXAMPLE 2.13

Using the Properties of Whole Number Addition

(a) Which property justifies each of the following statements?
(i) $8 + 3 = 3 + 8$
(ii) $(7 + 5) + 8 = 7 + (5 + 8)$
(iii) A million plus a quintillion is not infinite.

(b) Justify each equality below:

$$\begin{aligned}(20 + 2) + (30 + 8) &= 20 + [2 + (30 + 8)] && \text{(i)}\\ &= 20 + [(30 + 8) + 2] && \text{(ii)}\\ &= 20 + [30 + (8 + 2)] && \text{(iii)}\\ &= (20 + 30) + (8 + 2) && \text{(iv)}\end{aligned}$$

SOLUTION

(a) (i) Commutative property, (ii) associative property, (iii) the sum is a whole number by the closure property, and is therefore a finite value.

(b) (i) associative property, (ii) commutative property, (iii) associative property, (iv) associative property. ■

With practice, the associative and commutative properties become nearly automatic. To sum 22 + 38 mentally, the formal justification in Example 2.13(b) is replaced by

$$22 + 38 = (20 + 30) + (8 + 2) = 50 + 10 = 60.$$

Mental arithmetic is explored in more depth in Chapter 3.

EXAMPLE 2.14

Illustrating Properties on the Number Line

What properties of whole number addition are shown below?

(a)

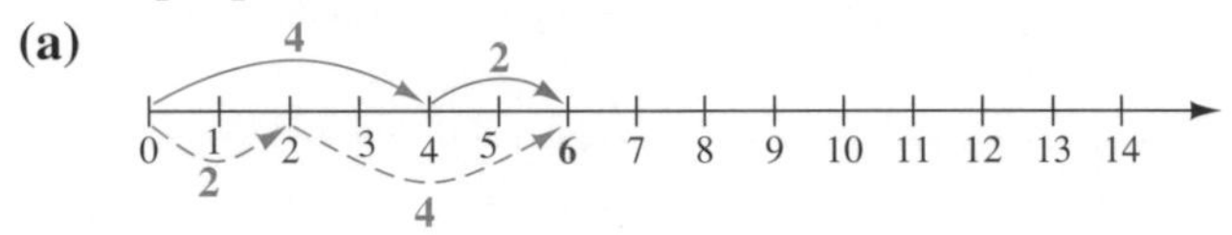

(b)

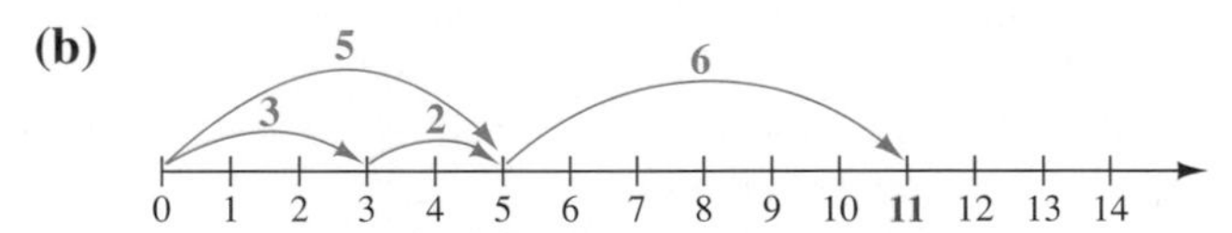

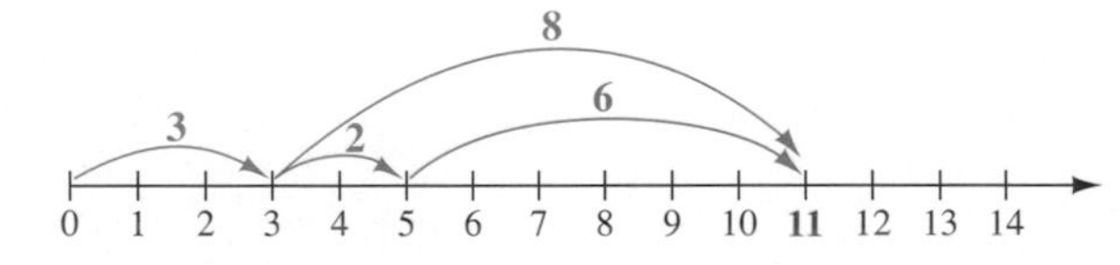

(c)

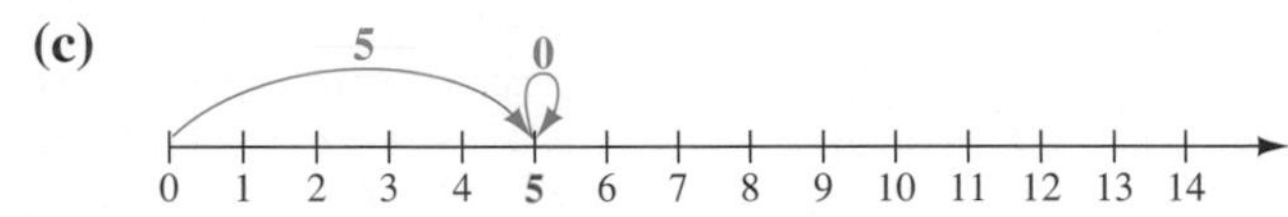

SOLUTION

(a) The commutative property: $4 + 2 = 2 + 4$.

(b) The associative property: $(3 + 2) + 6 = 3 + (2 + 6)$.

(c) The additive identity property: $5 + 0 = 5$. ■

Subtraction of Whole Numbers

Subtraction can be defined in a brief statement.

DEFINITION Subtraction of Whole Numbers

Let a and b be whole numbers. The **difference,** written $a - b$, is the unique whole number c such that $a = b + c$. That is, $a - b = c$ if, and only if, there is a whole number c such that $a = b + c$.

The expression $a - b$ is read " a minus b. " The a is called the **minuend** and b is the **subtrahend.**

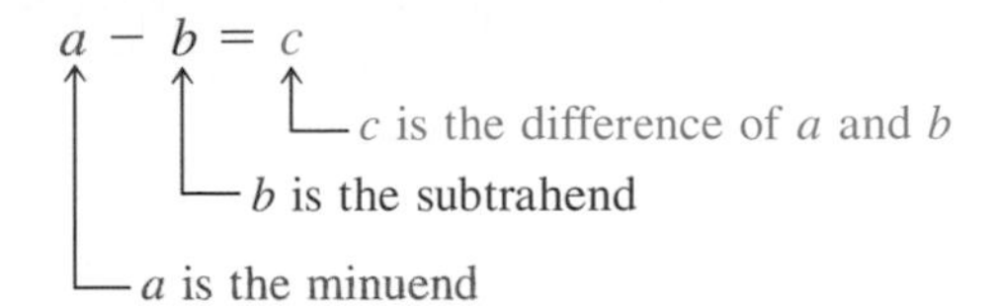

This definition relates subtraction to addition, and is the definition most easily extended to the integers, rationals, and real numbers. An alternative definition of subtraction of whole numbers can be given in terms of sets, as explored in problem 14.

Since $8 = 5 + 3$ the definition tells us that $8 - 5 = 3$. However,the practical use of subtraction is not revealed in the definition above. To understand the nature and value of subtraction, we will introduce four conceptual models: **take-away, missing addend, comparison,** and **number-line** (or **measurement**). The following four problems illustrate each of the four conceptual models.

Take-away:
Eroll has \$8 and spent \$5 for a ticket to the movies. How much money does Eroll have left?
Missing addend:
Alice has read 5 chapters of her book. If there are 8 chapters in all, how many more chapters must she read to finish the book?
Comparison:
Georgia has 8 mice and Tonya has 5 mice. How many more mice does Georgia have than Tonya?
Number-line:
Mike hiked up the mountain trail 8 miles. Five of these miles were hiked after lunch. How many miles did Mike hike before lunch?

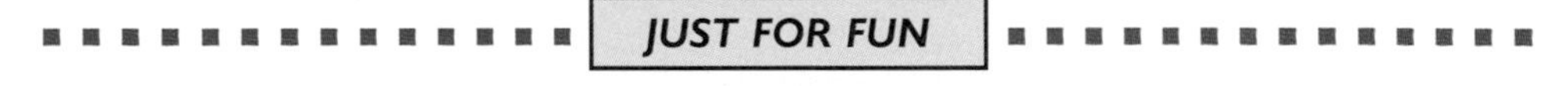

Paper Clip Comparison

In full view of a group of people, show them that you have ten paper clips on the table. Still in plain sight, divide the paper clips into two groups of five and put a piece of cardboard on the table which hides one group. Next move one clip from the hidden pile in the back, to the pile in sight in the front. Now ask, "How many more paper clips are in the front than in the back?" Don't be surprised if more than just a few people say there is just one more in front!

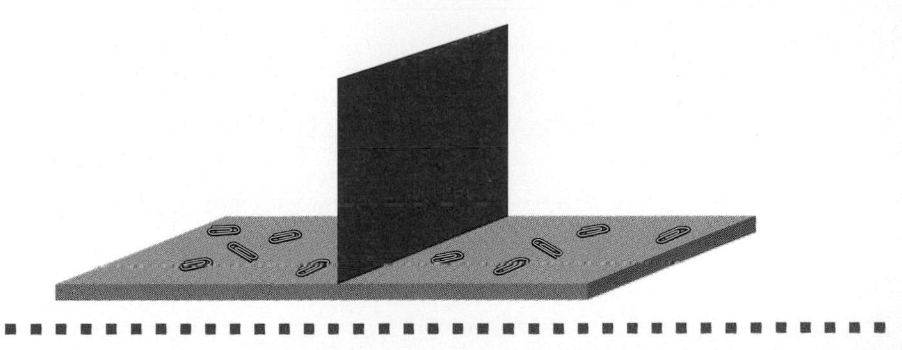

In all four problems, the answer is 3 because we know that $8 = 5 + 3$. For example, Eroll's \$8, when written as \$5 + \$3, shows that a \$5 movie ticket and the \$3 still in his pocket account for all of the \$8 he had originally. The cashier at the box office "took away" 5 of Eroll's 8 dollars, leaving him with \$3. Thus the problem is an example of the **take-away model** of subtraction. Alice, having read 5 chapters, wants to know how many more chapters she must read; the emphasis now is on what number is to be added to 5 to get 8, so the second problem illustrates the **missing addend model** of subtraction. Similarly, the remaining problems illustrate the **comparison model** and the **number-line** (or **measurement**) **model** of subtraction.

The four basic conceptual models of the subtraction $8 - 5$ are visualized as follows.

Take-away model

1. Start with 8 objects.

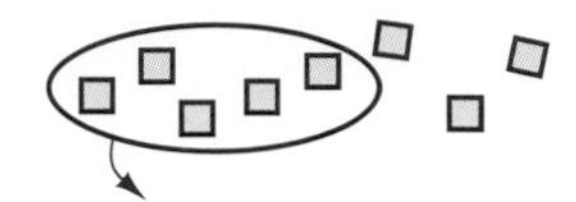

2. Take away 5 objects.
3. How many objects are left?

Missing addend model

1. Start with 5 objects.
2. How many more objects are needed to give a total of 8 objects?

Comparison model

1. Start with two collections, with 8 objects in one collection and 5 in the other.

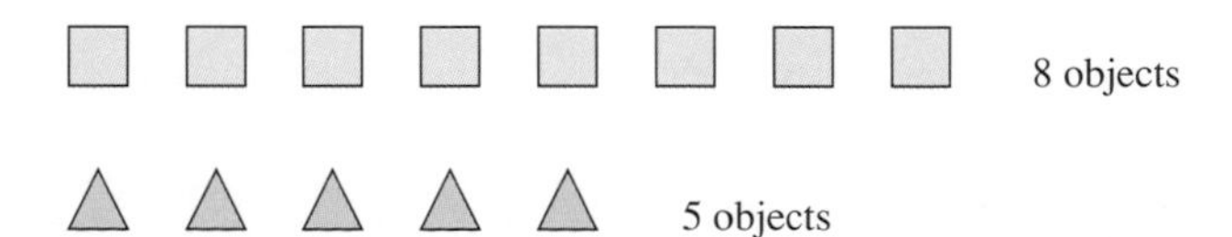

2. How many more objects are in the larger collection?

Number-line Model

1. Move forward (to the right) 8 units.

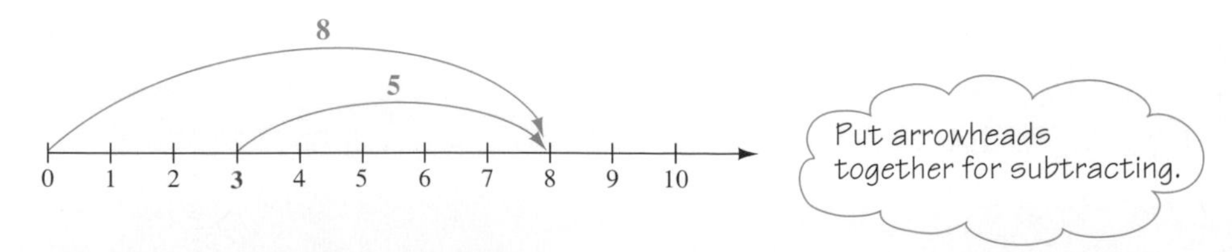

2. Remove a jump of 5 units.
3. What is the distance from 0?

Notice that the head of the arrow representing 5 is positioned at the head of the arrow representing 8.

EXAMPLE 2.15 Identifying Conceptual Models of Subtraction

Identify the conceptual model of subtraction that best fits these problems.

(a) Mary got 43 pieces of candy trick-or-treating on Halloween. Karen got 36 pieces. How many more pieces of candy does Mary have than Karen?

(b) Mary gave 20 pieces of her candy to her sick brother Jon. How many pieces of candy does Mary have left?

(c) Karen's older brother Ken collected 53 pieces of candy. How many more pieces of candy would Karen need to have as many as Ken?

(d) Ken left home and walked 10 blocks east trick-or-treating on one side of the street. He then turned around and rang doorbells on the other side of the street covering 4 blocks. How far is Ken from home?

SOLUTION

(a) Comparison model

(b) Take-away model

(c) Missing addend model

(d) Number-line model ■

The set of whole numbers is not closed under subtraction. For example, $2 - 5$ is undefined since there is no whole number n which satisfies $2 = 5 + n$. Similar reasoning shows subtraction is not commutative, so the order in which a and b are taken is important. Neither is subtraction associative, which means that parentheses must be placed with care in expressions involving subtractions. For example,

$$5 - (3 - 1) = 5 - 2 = 3 \quad \text{but} \quad (5 - 3) - 1 = 2 - 1 = 1.$$

HIGHLIGHT FROM HISTORY

The Origin of the + and − Symbols

The + and − symbols first appeared in an arithmetic book by Johannes Widman, published in Leipzig in 1489. The signs were not used as symbols of operation, but simply to indicate excess or deficiency. Widman states, "What is −, that is minus, what is +, that is more." The + symbol is thought to be a contraction of the Latin word *et,* which often indicated addition. The minus symbol may be a contraction of the abbreviation − for minus. The + and − symbols were used to designate algebraic operations by the Dutch mathematician van der Hoelse in 1514.

COOPERATIVE Investigation

Diffy

The process in this activity is sometimes called *Diffy*. The name comes from the process of taking successive differences of whole numbers, and the activity provides an interesting setting for practicing skills in subtraction. The activity can be done by hand, with a calculator, or by calling up DIFFY on your computer disc.

Directions

3 18 7 100
15 11 93 97
4 82 4 82

Step 1. Make an array of circles as shown and choose four whole numbers to place in the top four circles.

Step 2. In the first three circles of the second row write the differences of the numbers above and to the right and left of the circle in question, always being careful to subtract the smaller of these two numbers from the larger. In the fourth circle of the second row place the difference of the numbers in the first and fourth circles in the preceding row, again always subtracting the smaller number from the larger.

Step 3. Repeat Step 1 to fill in successive rows of circles in the diagram. You may stop if you obtain a row of zeros.

Step 4. Repeat steps 1, 2, and 3 several times, and each time, start with different numbers.

Questions

1. Do you think the process will always stop?
2. Can you find four numbers such that the process terminates at the first step? The second step? The third step? Try several sets of starting numbers.
3. On the basis of your work so far, what do you guess is the largest number of steps needed for the process to stop?
4. Can you find four starting numbers such that the process requires eight steps to reach termination?
5. Try DIFFY with the starting numbers 17, 32, 58, and 107.

PROBLEM SET 2.3

Understanding Concepts

1. Let $A = \{\text{apple, berry, peach}\}$, $B = \{\text{lemon, lime}\}$, $C = \{\text{lemon, berry, prune}\}$.
   (a) Find (i) $n(A \cup B)$, (ii) $n(A \cup C)$, (iii) $n(B \cup C)$,
   (b) In which case is the number of elements in the union *not* the sum of the number of elements in the individual sets?
2. Let $n(A) = 5$, $n(B) = 8$, and $n(A \cup B) = 10$. What can you say about $n(A \cap B)$?
3. Let $n(A) = 4$ and $n(A \cup B) = 8$.
   (a) What are the possible values of $n(B)$?
   (b) If $A \cap B = \emptyset$, what is the only possible value of $n(B)$?
4. Draw number strips (or rods) to illustrate
   (a) $4 + 6 = 10$;
   (b) $2 + 8 = 8 + 2$; and
   (c) $3 + (2 + 5) = (3 + 2) + 5$.
5. Make up a word problem which uses the set model of addition to illustrate $30 + 28$.
6. Make up a word problem which uses the measurement model of addition to illustrate $18 + 25$.

7. Which of the following sets of whole numbers are closed under addition? If the set is not closed, give an example of two elements from the set whose sum is not in the set.
   (a) $\{10, 15, 20, 25, 30, 35, 40, \ldots\}$
   (b) $\{1, 2, 3, \ldots, 1000\}$
   (c) $\{0\}$ (d) $\{1, 5, 6, 11, 17, 28, \ldots\}$
   (e) $\{n \in N: n \geq 19\}$
   (f) $\{0, 3, 6, 9, 12, 15, 18, \ldots\}$
8. What properties of addition are used in these equalities?
   (a) $14 + 18 = 18 + 14$
   (b) $12345678 + 97865342$ is a whole number
   (c) $18 + 0 = 18$
   (d) $(17 + 14) + 13 = 30 + 14$
   (e) $(12 + 15) + (5 + 38) = 50 + 20$
9. Carefully explain why the parentheses on the left side of this expression can be overlooked and the addends rearranged as on the right side:
   $$(8 + (3 + (4 + 2))) + (7 + 6) = (8 + 2) + (3 + 7) + (4 + 6)$$
10. An easy way to add $1 + 2 + 3 + 4 + 5 + 6 + 7 + 8 + 9 + 10$ is to write the sum as $(1 + 10) + (2 + 9) + (3 + 8) + (4 + 7) + (5 + 6) = 11 + 11 + 11 + 11 + 11 = 55$.
(a) Compute $1 + 2 + 3 + \cdots + 20$. Describe your procedure.
(b) What properties of addition are you using to justify why your procedure works?
11. If it is known that $5 + 9 = 14$, then it follows that $9 + 5 = 14$, $14 - 9 = 5$ and $14 - 5 = 9$. Many elementary texts call such a group of four basic facts a "fact family."
   (a) What is the fact family that contains $5 + 7 = 12$?
   (b) What is the fact family that contains $12 - 4 = 8$?
12. For each subtraction statement below, write the equivalent missing addend addition statement.
   (a) $627 - x = 419$
   (b) $1289 - 613 = y$
13. For each statement find all whole numbers, if any, which make the statement true:
   (a) $\square + 6 = 15$ (b) $18 - \square - 9$
   (c) $\square + 2 = 2$ (d) $7 - \square = 0$
   (e) $\square + 2 \leq 7$ (f) $\square + 3 \leq 1$
   (g) $\square - 5 \geq 3$ (h) $2 - \square \leq 1$
14. The take-away model of subtraction suggests that subtraction can be defined in terms of set operations as follows: Let $B \subseteq A$. If $a = n(A)$ and $b = n(B)$, then $a - b = n(A - B)$.
   (a) Choose sets A and B which illustrate $8 - 3$ with this definition.
   (b) Why is it important that B be a subset of A?
15. Here are some subtraction problems found in *Exploring Mathematics, Grade 2* (From *Scott Foresman Exploring Mathematics* Grades 1-7 by L. Carey Bolster et al. Copyright © 1994 Scott, Foresman and Company. Reprinted by permission of Scott, Foresman and Company.). Which subtraction model—take-away, missing addend, comparison, or measurement—corresponds best to each problem?
   (a) The second graders have 38 more books than the first graders. The second graders have 87 books. How many books do the first graders have?
   (b) Jodi's book has 75 pages. On Monday she read 32 pages. How many pages did she have left to read?
   (c) Lee took 65¢ to the swimming pool. He spent 33¢ for a waterslide ticket. How much did Lee have left?
   (d) Amy's book has 124 pages. Steve's book has 133 pages. How many more pages are in Steve's book than Amy's?
16. Make up a word problem which corresponds well to each of the four conceptual models for subtraction.
   (a) Take-away (b) Missing addend
   (c) Comparison (d) Measurement (number-line)
17. Jeff must read the last chapter of his book. It begins on the top of page 241 and ends at the bottom of page 257. How many pages must he read?
18. Notice that $(6 + (8 - 5)) - 2 = 7$, but a different placement of parentheses on the left side would give the statement $(6 + 8) - (5 - 2) = 11$. Place parentheses to turn these into *true* statements.
   (a) $8 - 5 - 2 - 1 = 2$
   (b) $8 - 5 - 2 - 1 = 4$
   (c) $8 - 5 - 2 - 1 = 0$
   (d) $8 + 5 \quad 2 + 1 = 12$
   (e) $8 + 5 - 2 + 1 = 10$
19. Each circled number is the sum of the adjacent row, column, or diagonal of the numbers in the square array.

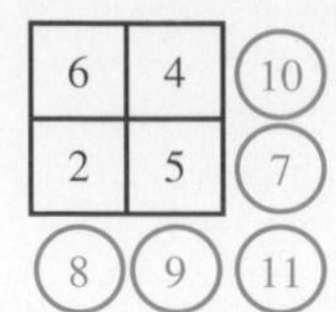

Fill in the missing entries in these patterns.

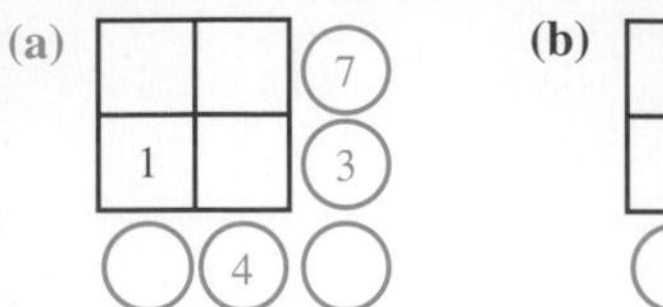

20. The first figure below shows that the numbers 1, 2, 3, 4, 5, 6 can be placed around a triangle in such a way that the three numbers along any side sum to 9, which is shown circled. Arrange the numbers 1, 2, 3, 4, 5, 6 to give the sums circled in the next three figures.

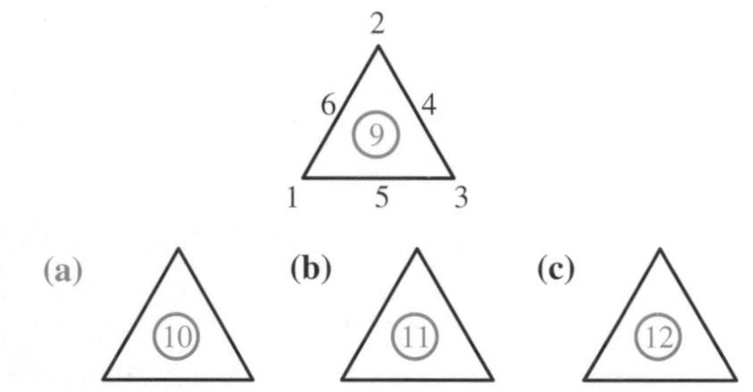

21. Nearly a month earlier, Glyn had saved $68 and Wade had saved $17. Both have paper routes paying $6 a day, but Wade also has a dishwashing job paying $3 a day. While watching TV together after work on a Friday evening, Wade discovered his savings had caught up with Glyn's. How far behind was Wade on the previous Wednesday morning of that same week?

Thinking Critically

22. If $A = \{a, b, c, d\}$ and $B = \{c, d, e, f, g\}$, then $n(A \cup B) = n(\{a, b, c, d, e, f, g\}) = 7$, $n(A) = 4$, $n(B) = 5$, and $n(A \cap B) = n(\{c, d,\}) = 2$. Since $7 = 4 + 5 - 2$, this suggests that

$$n(A \cup B) = n(A) + n(B) - n(A \cap B).$$

Use Venn diagrams to justify this formula for arbitrary finite sets A and B.

23. Since $12 \times 16 = 192$ and $5 \times 40 = 200$, it follows that among the first 200 natural numbers $\{1, 2, \ldots, 200\}$ there are 16 that are multiples of 12 and 40 that are multiples of 5. Just 3 are multiples of *both* 5 and 12, namely 60, 120, and 180. Use the formula of problem 22 to find how many natural numbers in the set $\{1, 2, \ldots, 200\}$ are divisible by *either* 12 or 5 or both.

24. In a "double-six" set of dominoes, each square half of the domino has from 0 to 6 spots, and no two dominoes are alike. Here is the beginning of an arrangement of the dominoes into a triangular pattern.

(a) By completing the pattern, explain why there are 28 dominoes in a "double-six" set. Observe that 28 is the seventh triangular number.

(b) Dominoes also come in "double-nine" sets, where the number of spots in each half is between 0 and 9. How many dominoes are in a double-nine set?

25. The odd whole numbers 1, 3, 5, 7, . . . can be represented by number tiles arranged in an angle shape. For example, the first four odd numbers correspond to these shapes.

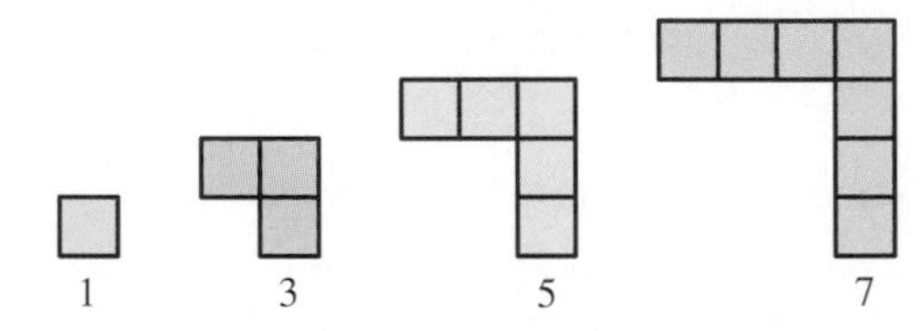

(a) Draw number tile arrangements to represent the odd whole numbers 9, 11, and 13.

(b) Use your angular shapes to show that the sum of the first five odd whole numbers, $1 + 3 + 5 + 7 + 9$, is the square number 25.

(c) The one-hundredth odd whole number is 199. What is the sum $1 + 3 + 5 + \cdots + 197 + 199$?

(d) Find a formula for $1 + 3 + 5 + \cdots + (2n - 1)$, which is the sum of the first n odd whole numbers.

26. Let $d_1 = 1$, $d_2 = 8$, $d_3 = 16$, . . . , be the "donut" numbers, as shown below.

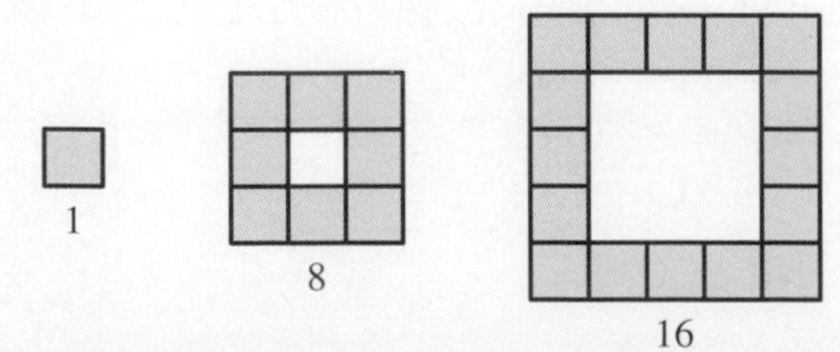

(a) Find d_4 and d_5.

(b) Find a formula for d_n expressed with the variable n.

(c) Notice that $1 = 1^2$, $1 + 8 = 3^2$, $1 + 8 + 16 = 5^2$. Why should it be visually apparent that the sum of the first n donut numbers is the square of the nth odd whole number; that is, that $d_1 + d_2 + \cdots + d_n = (2n - 1)^2$?

27. Begin with a single small cube. The original small cube can be covered with 26 more small cubes like the first one to form a larger solid cube with three small cubes on each edge as shown.

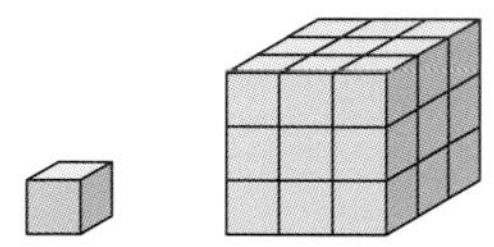

(a) Now cover the larger cube with a layer of small cubes to form a solid cube with five small cubes on each edge. How many small cubes must be added to form this next larger cube?

(b) If the large cube of part (a) is now covered by a layer of cubes to form a solid cube with seven cubes on each edge, how many small cubes must be added?

(c) How many small cubes must be added next to obtain a solid cube with nine small cubes on an edge?

28. There is a nonempty subset of the whole numbers that is closed under subtraction. Find this subset.

29. The set C contains 2 and 3 and is closed under addition.

(a) What whole numbers must be in C?

(b) What whole numbers may not be in C?

(c) Are there any whole numbers definitely not in C?

(d) How would your answers to (a), (b), and (c) change if 2 and 4 were contained in C instead of 2 and 3?

Thinking Cooperatively

30. The triangular numbers $t_1 = 1$, $t_2 = 3$, $t_3 = 6, \ldots$, are shown in Figure 2.13.

(a) Complete the following table of the first fifteen triangular numbers:

n	1	2	3	4	5	6	7	8	9	10	11	12	13	14	15
t_n	1	3	6	10											

(b) The first ten natural numbers can be expressed as sums of triangular numbers. For example, $1 = 1$, $2 = 1 + 1$, $3 = 3$, $4 = 1 + 3$, $5 = 1 + 1 + 3$, $6 = 6$, $7 = 1 + 6$, $8 = 1 + 1 + 6$, $9 = 3 + 6$, $10 = 10$. Show that the natural numbers 11 through 25 can be written as a sum of triangular numbers. Use as few triangular numbers as possible each time.

(c) Choose five more numbers at random (don't look at your table!) between 26 and 120 and write each of them as a sum of as few triangular numbers as possible. What is the largest number of triangular numbers needed?

(d) On July 10, 1796, the 19 year old Carl Friedrich Gauss wrote in his notebook "EUREKA! NUM $= \triangle + \triangle + \triangle$." What theorem do you think Gauss had proved?

31. The following table represents a partially complete scrambled addition table. The rows and columns have both been mixed up. See if you can complete the table.

+	5					2				3
3										
					18					
				12						
			5			6				
							0			
			8						14	
5										
						3				
8								16		

32. Make up your own incomplete scrambled addition table puzzle, similar to the one in problem 31. It may be helpful to use two colors, to separate the entries which are given from the entries which are determined by what is given. Trade puzzles among groups and see which presents the most challenge.

Making Connections

33. Addition and subtraction problems arise frequently in everyday life. Consider such activities as scheduling time, making budgets, sewing clothes, making home repairs, planning finances, modifying recipes, and making purchases. Make a list of five addition and five subtraction problems you have encountered at home or in your work. Which conceptual model of addition or subtraction corresponds best to each problem?

For Review

34. The Venn diagram shows the number of elements in each region. Find the numbers of elements in the sets indicated.

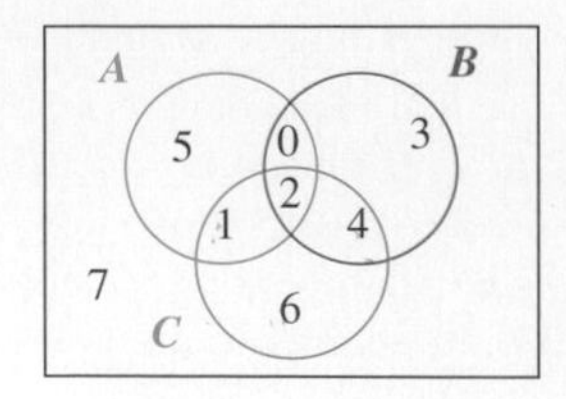

(a) $n(A \cap B)$ **(b)** $n(B \cup C)$
(c) $n(\overline{A} \cap C)$ **(d)** $n(B \cap C)$
(e) $n(A - C)$ **(f)** $n(\overline{A \cup B} \cup C)$

35. Let $A = \{\text{red, blue, green}\}$, $B = \{\text{2-door, 4-door}\}$.
(a) Find $A \times B$. **(b)** Find $n(A \times B)$.

36. Maria was told there are fifteen ways to get from Abbottsville to Central City which go through Brownsport. Maria knows there are three routes from Abbottsville to Brownsport. How many choices does she have at Brownsport to complete her trip?

2.4 Multiplication and Division of Whole Numbers

Multiplication as Repeated Addition

Misha has an after school job at a local bike factory. Each day she has a 3 mile round trip walk to the factory. At her job she assembles 4 hubs and wheels. How many hubs and wheels does she assemble in 5 afternoons? How many miles does she walk to and from her job each week?

These two problems can be answered by repeated additions. Misha assembles

$$4 + 4 + 4 + 4 + 4 = 20$$

5 fours, written $5 \cdot 4$

hubs and wheels. This can be illustrated with the set model as shown in Figure 2.16

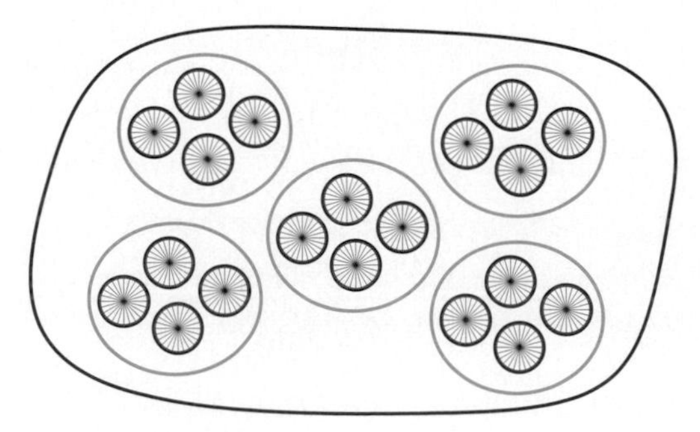

Figure 2.16
Set model to show 5 times 4 is 20

Misha walks

$$3 + 3 + 3 + 3 + 3 = 15$$

5 threes, written $5 \cdot 3$

miles each week. This can be illustrated by the measurement model as in Figure 2.17.

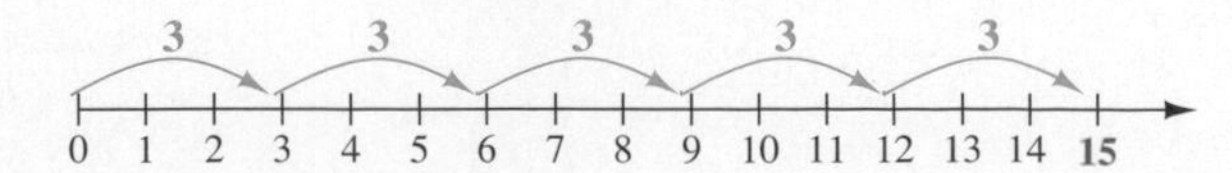

Figure 2.17
Measurement model to show 5 times 3 is 15

If Misha only worked one day in the week, she would assemble $1 \cdot 4 = 4$ wheels. If she were sick all week and missed work entirely, she would not assemble any wheels; therefore $0 \cdot 4 = 0$.

DEFINITION Multiplication of Whole Numbers as Repeated Additions

Let a and b be any two whole numbers. Then the **product** of a and b, written $a \cdot b$, is defined by:

$$a \cdot b = \underbrace{b + b + \cdots + b}_{a \text{ addends}} \quad \text{when} \quad a \neq 0;$$

and by:

$$0 \cdot b = 0.$$

Since multiplication is defined for all pairs of whole numbers, and the outcome is also a whole number, we see that multiplication is a binary operation on the set of whole numbers. The dot symbol for multiplication is often replaced by a cross $\times$ (not to be mistaken for the letter x or the cross product of sets) or by a star (asterisk)*, the symbol computers use most often. Sometimes no symbol at all is used, or parentheses are placed around the factors. Thus, the expressions

$$a \cdot b, \quad a \times b, \quad a{*}b, \quad ab, \quad (a)(b)$$

all denote the multiplication of a and b. Each whole number, a and b, is a **factor** of the product $a \cdot b$, and often $a \cdot b$ is read "a times b."

In addition to the set model and the measurement model, there are two other useful models of multiplication, the *array model* and the *Cartesian product model.*

The Array Model for Multiplication

Suppose Lida, as part of her biology research, planted 5 rows of bean seeds and each row contains 8 seeds. How many seeds did she plant in her rectangular plot?

The 40 seeds Lida planted form a 5 by 8 rectangular array, as shown in Figure 2.18. The arrays can also be drawn with squares, giving an important connection with the measurement model of the whole numbers. This measurement model for multiplication is especially important because it extends from the whole numbers to the rational numbers, the integers, and even the real number system.

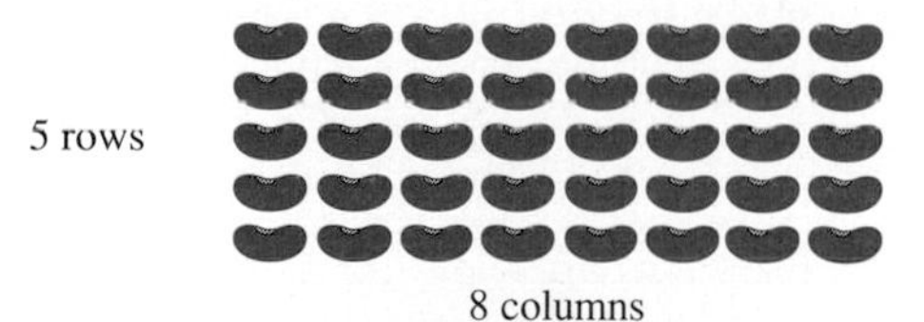

Figure 2.18
Array model to show 5 · 8 = 40

The Cartesian Product Model of Multiplication

At Sonya's Ice Cream Shop, one can order either a sugar or waffle cone and one of four flavors—vanilla, chocolate, mint, or raspberry. If C = {sugar, waffle} is the set of types of cones and F = {vanilla, chocolate, mint, raspberry} is the set of flavors, then the Cartesian product $C \times F$ gives the ways an ice cream cone can be ordered at Sonya's. A rectangular table of entries provides a visualization, and shows why the Cartesian product model is closely associated with the rectangular array model.

		Flavor			
		v	c	m	r
Type of cone	s	(s, v)	(s, c)	(s, m)	(s, r)
	w	(w, v)	(w, c)	(w, m)	(w, r)

More generally we can express an alternative, but equivalent, definition of multiplication as follows.

ALTERNATIVE DEFINITION Multiplication of Whole Numbers via the Cartesian Product

Let a and b be whole numbers, and A and B be any sets for which $a = n(A)$ and $b = n(B)$. Then $a \cdot b = n(A \times B)$.

Since $\emptyset \times B = \emptyset$ and $0 = n(\emptyset)$, this definition of multiplication is consistent with the earlier definition of multiplication by 0; that is, $0 \cdot b = b \cdot 0 = 0$.

EXAMPLE 2.16

Using the Cartesian Product Model of Multiplication

To get to work, Juan either walks, bikes, or takes a cab from his house to downtown. From downtown, he either continues on the bus the rest of the way to work or catches the train to his place of business. How many ways can Juan get to work?

SOLUTION

If $A = \{\text{w, b, c}\}$ is the set of possibilities for the first leg of his trip, and $B = \{\text{b, t}\}$ is the next set of choices, then Juan has altogether

$$3 \cdot 2 = n(A \times B) = n(\{(\text{w, b}), (\text{w, t}), (\text{b, b}), (\text{b, t}), (\text{c, b}), (\text{c, t})\}) = 6$$

ways to commute to work. We notice that it doesn't matter whether or not A and B have an element in common. ■

Properties of Whole Number Multiplication

It follows from the definition that the set of whole numbers is **closed under multiplication:** the product of any two whole numbers is a unique whole number. We have also observed earlier that $0 \cdot b = 0$ and $1 \cdot b = b$ for all whole numbers b.

By rotating a rectangular array through 90°, we interchange the number of rows and the number of columns in the array, but we do not change the total number of objects in the array. Thus $a \cdot b = b \cdot a$, which demonstrates the **commutative property of multiplication.** An example is shown in Figure 2.19.

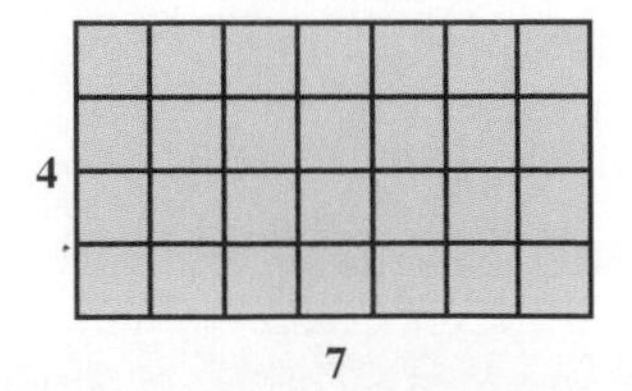

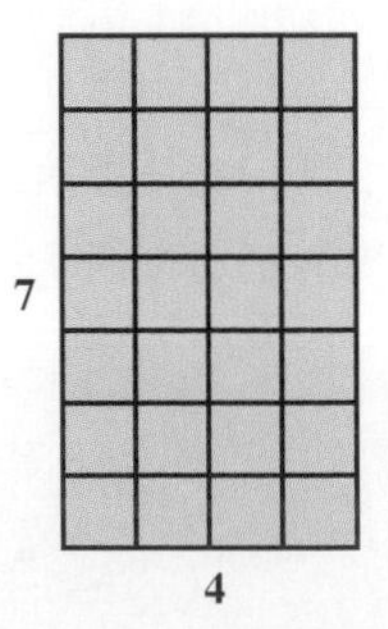

Figure 2.19 **Multiplication is commutative: $4 \cdot 7 = 7 \cdot 4$**

Since we already know that $1 \cdot b = b$ for any whole number b, the commutative property tells us $1 \cdot b = b \cdot 1 = b$. This makes 1 a **multiplicative identity.**

A three-dimensional model, as shown in Figure 2.20, shows that multiplication is **associative,** so that $a \cdot (b \cdot c) = (a \cdot b) \cdot (c)$ This means that an expression such as $4 \cdot 3 \cdot 5$ is meaningful without parentheses: the product is the same for both ways in which parentheses can be placed.

Figure 2.20
Multiplication is associative:
$4 \cdot (3 \cdot 5) = (4 \cdot 3) \cdot 5$

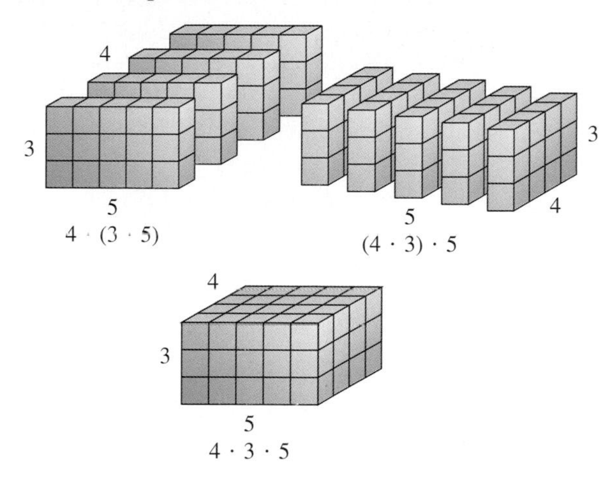

There is one more important property, the **distributive property,** which relates multiplication and addition. This property is the basis for the multiplication algorithm discussed in the next chapter. The distributive property can be nicely visualized by the array model. In Figure 2.21, it is seen that $4 \cdot (6 + 3) = (4 \cdot 6) + (4 \cdot 3)$; that is, the factor 4 *distributes* itself over each term in the sum $6 + 3$.

Figure 2.21
Multiplication is distributive over addition:
$4 \cdot (6 + 3) = (4 \cdot 6) + (4 \cdot 3)$

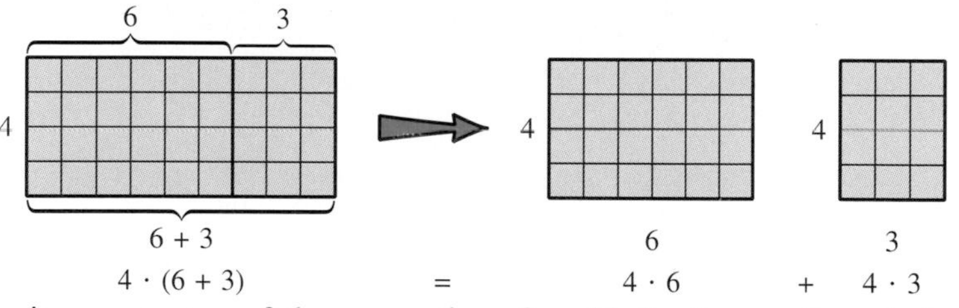

Here is a summary of the properties of multiplication on the whole numbers.

PROPERTIES Whole Number Multiplication

Closure Property If a and b are any two whole numbers, then $a \cdot b$ is a unique whole number.

Commutative Property If a and b are any two whole numbers, then $a \cdot b = b \cdot a$

Associative Property If a, b, and c are any three whole numbers, then $a \cdot (b \cdot c) = (a \cdot b) \cdot (c)$

Multiplicative Identity Property of One The number 1 is the unique whole number for which $b \cdot 1 = 1 \cdot b = b$ holds for all whole numbers b.

Multiplication by Zero Property For all whole numbers b, $0 \cdot b = b \cdot 0 = 0$.

Distributive Property of Multiplication over Addition If a, b, and c are any three whole numbers, then $a \cdot (b + c) = (a \cdot b) + (a \cdot c)$ and $(a + b) \cdot c = (a \cdot c) + (b \cdot c)$.

FROM THE NCTM STANDARDS

Developing the Operations on the Whole Numbers

Connecting problem structures to operations should be emphasized throughout grades K–4 for both one-step and appropriate two-step problems. For example, multiplication is most commonly linked to the process of combining equal groups. Children also need to see that it relates to array, "times as many," and "combination" (e.g., three blouses, four skirts, how many outfits?) situations. An example of combination situations is finding the number of outfits that can be made with two blouses and three skirts.

The language of basic operations, such as the terms *addend, sum, difference, factor, multiple, product,* and *quotient,* can be introduced and used informally in work with operations. The notions of factors and multiples can prompt interesting explorations. Children can find the factors of a number using tiles or graph paper. This can lead to an investigation of numbers that have only two factors (prime numbers) and numbers with two equal factors (square numbers).

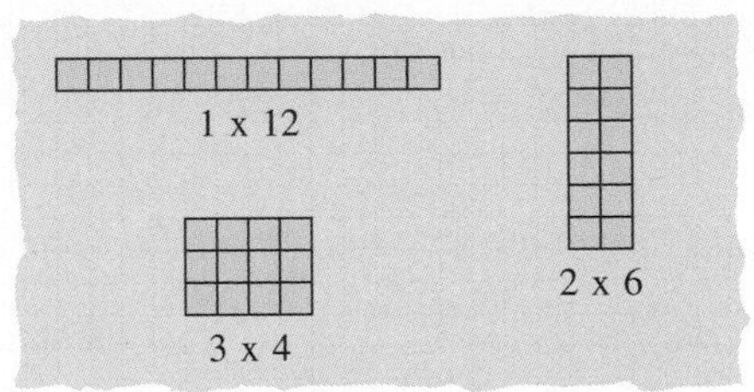

Factors of 12

SOURCE: Excerpted from the discussion of Standard 7: Concepts of Whole Number Operations. From *Curriculum and Evaluation Standards for School Mathematics Grades K–4,* p. 41.

EXAMPLE 2.17

Multiplying Two Binomial Expressions

(a) Use the properties of multiplication to justify the formula $(a + b)(c + d) = ac + ad + bc + bd$. The factor $(a + b)$ is called a **binomial** since it has two terms.

(b) Visualize the expansion $(a + b)(c + d) = ac + ad + bc + bd$ with the array model of multiplication.

SOLUTION

(a)

$(a + b)(c + d) = (a + b)c + (a + b)d$	Distributive property
$= ac + bc + ad + bd$	Distributive property
$= ac + ad + bc + bd$	Commutative property of addition

(b)

a	$a \cdot c$	$a \cdot d$
b	$b \cdot c$	$b \cdot d$
	c	d

■

Division of Whole Numbers

There are three conceptual models for the division $a \div b$ of a whole number a by a nonzero whole number b: the **repeated-subtraction** model, the **partition** model and the **missing-factor** model.

The Repeated-Subtraction Model of Division

Ms. Rislov has 28 students in her class, that she wishes to divide into cooperative learning groups of 4 students per group. If each group requires a set of Cuisenaire® rods, how many sets of rods must Ms. Rislov have available? The answer, 7, is pictured in Figure 2.22, and is obtained by counting how many times groups of 4 can be formed, starting with 28. Thus, $28 \div 4 = 7$. The repeated subtraction model can be realized easily with physical objects; the process is called **division by grouping.**

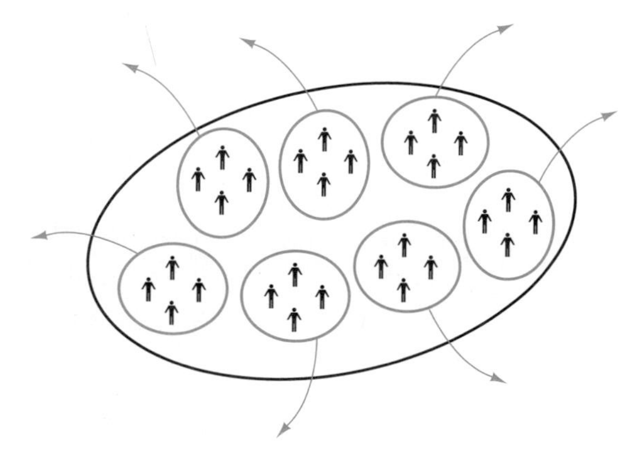

Figure 2.22
Division as repeated subtraction: $28 \div 4 = 7$ because seven 4s can be subtracted from 28.

The Partition Model of Division

When Ms. Rislov checked her supply cupboard, she discovered she had only 4 sets of Cuisenaire® rods to use with the 28 students in her class. How many students must she assign to each set of rods? The answer, 7 students in each group, is depicted in Figure 2.23. The partition model is also realized easily with physical objects, in which case the process is called **division by sharing.**

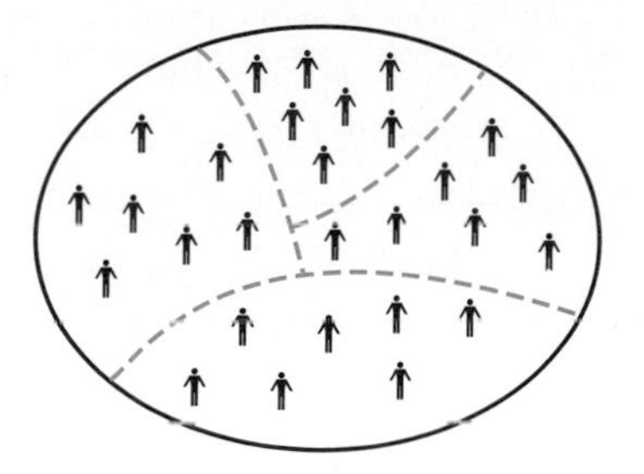

Figure 2.23
Division as a partition: $28 \div 4 = 7$ because when 28 objects are partitioned into 4 equal sized groups, there are 7 objects in each group.

The Missing-Factor Model of Division

In the repeated-subtraction model, $28 \div 4 = 7$ because $28 = 4 + 4 + 4 + 4 + 4 + 4 + 4 = 7 \cdot 4$. However, in the partition model, $28 \div 4 = 7$ because $28 = 7 + 7 + 7 + 7 = 4 \cdot 7$. In both cases, the division $28 \div 4$ can be viewed as finding the factor c for which $28 = 4 \cdot c$ or $28 = c \cdot 4$. The missing-factor model is usually the concept adopted to define division formally.

DEFINITION Division in Whole Numbers

Let a and b be whole numbers with $b \neq 0$. Then $a \div b = c$ if, and only if, $a = b \cdot c$ for a unique whole number c.

The symbol $a \div b$ is read "a divided by b," where a is the **dividend** and b is the **divisor.** If $a \div b = c$, then we say that ***b* divides *a*** or is a **divisor** of a, and c is called the **quotient.**

$$\mathbf{a} \div \mathbf{b} = \mathbf{c}$$

dividend (a), divisor (b), quotient (c)

Division is also symbolized by a/b or $\frac{a}{b}$; the backslash notation is used by most computers.

Any multiplication fact with nonzero factors, say $3 \cdot 5 = 15$, is related to four equivalent facts which form a **fact family** such as

$$3 \cdot 5 = 15 \qquad 15 \div 3 = 5$$
$$5 \cdot 3 = 15 \qquad 15 \div 5 = 3$$

With the array model in mind, a fact family is associated with rectangular arrays of 15 objects:

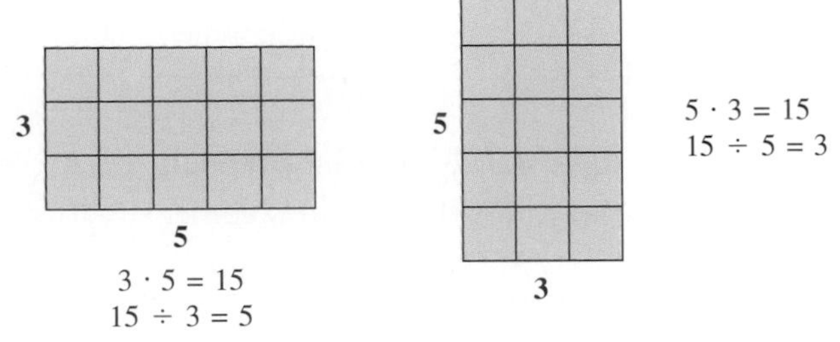

EXAMPLE 2.18 Computing Quotients with Manipulatives

Suppose you have 78 number tiles. Describe how to illustrate $78 \div 13$ with the tiles, using each of the three basic conceptual models for division.

SOLUTION

(a) **Repeated-subtraction.** Remove groups of 13 tiles each. Since 6 groups are formed, $78 \div 13 = 6$.

(b) **Partition.** Partition the tiles into 13 groups. Since each group contains exactly 6 tiles, $78 \div 13 = 6$.

(c) **Missing factor.** Use the 78 tiles to form a rectangle with 13 rows. Since it turns out there are 6 columns in the rectangle, then $78 \div 13 = 6$. ■

SCHOOL BOOK PAGE *Arrays and Families of Facts*

Families of Facts

Build Understanding

Draw 3 rows of Xs with the same number of Xs in each row. Now write two multiplication sentences that describe your picture.

Nancy drew the picture at the right.

Nancy wrote two multiplication sentences about her picture.

3 × 4 = 12 Think: 3 rows of 4 is 12.

4 × 3 = 12 Think: 4 columns of 3 is 12.

She can also write two division sentences about her picture.

12 ÷ 4 = 3 Think: 12 divided into rows of 4 is 3 rows.

12 ÷ 3 = 4 Think: 12 divided into columns of 3 is 4 columns.

These four number sentences form a ***family of facts.***

Write About Math Write the complete family of facts for the picture you drew.

Check Understanding

For another example, see Set B, pages 206–207.

Draw a picture for Exercises 1–2.

1. Draw eight rows of Xs with the same number of Xs in each row. Write the family of facts.
2. Write a family of facts that has only two number sentences. Draw a picture to show the facts.

1. What fact family is associated with a carton of eggs?
2. What fact family is associated with a case of 24 cans of cola?
3. Go on a rectangular array search. Report on arrays you find across campus, around town, in packaging of products, in landscape designs, and so on.

Division by Zero Is Undefined

The definition $a \div b$ does not allow b to be zero. The reason for this can best be explained by taking the cases $a \neq 0$ and $a = 0$ separately.

Case 1: $a \neq 0$. In this case, "$a \div 0$" would be equivalent to finding the missing factor c that makes $a = c \cdot 0$. But $c \cdot 0 = 0$ for all whole numbers c, so there is no solution when $a \neq 0$. Thus $a \div 0$ is undefined for $a \neq 0$.

Case 2: $a = 0$. In this case "$0 \div 0$" is equivalent to finding a unique whole number c for which $0 = 0 \cdot c$. This equation is satisfied by every choice of the factor c. Since no *unique* factor c exists, the division of 0 by 0 is also undefined.

In the division $a \div b$, there is no restriction on the dividend a. For all $b \neq 0$, we have that $0 \div b = 0$ since $0 = b \cdot 0$.

The multiplicative indentity 1 has two simple but useful relationships to division:

$$\frac{b}{b} = 1 \qquad \text{for all } b, \quad \text{where } b \neq 0;$$

$$\frac{a}{1} = a \qquad \text{for all } a.$$

Division with Remainders

Consider the division problem $27 \div 6$. There is no whole number c that satisfies $c \cdot 6 = 27$, so $27 \div 6$ is not defined in the whole numbers. That is, the set of whole numbers is *not closed* under division.

By allowing the possibility of a **remainder**, we can extend the division operation. Consider $27 \div 6$, where division is viewed as repeated subtraction. Four groups of 6 can be removed from 27. This leaves 3, which is too few to form another group of 6. This can be written

$$27 = 4 \cdot 6 + 3.$$

Here 4 is called the **quotient** and 3 is the **remainder**. This information is also written

$$27 \div 6 = 4 \text{ R } 3,$$

where R separates the quotient from the remainder. A visualization of division with a remainder is shown in Figure 2.24.

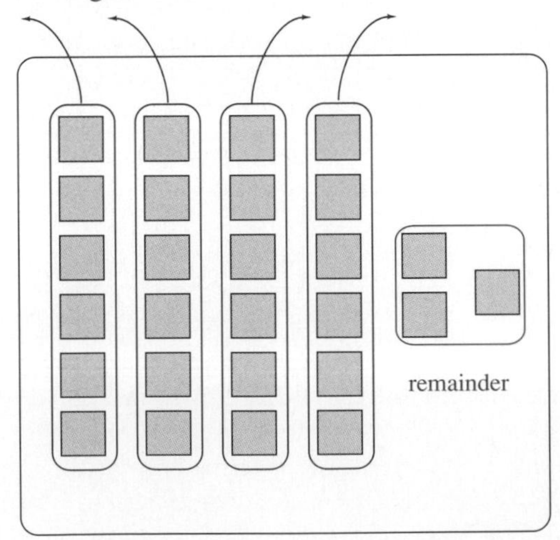

Figure 2.24
Visualizing division with a remainder: 27 ÷ 6 = 4 R 3

In general, we have the following result.

THEOREM The Division Algorithm

Let a and b be whole numbers with $b \neq 0$. Then there is a unique whole number q called the **quotient** and a unique whole number r, called the **remainder** such that,

$$a = q \cdot b + r, \quad 0 \leq r < b.$$

It is common to write

$$a \div b = q \text{ R } r$$

if $a = q \cdot b + r, 0 \leq r < b$. The quotient q is the largest whole number of groups of b objects that can be formed from a objects, and the remainder r is the number of objects that are left over. The remainder is 0 if and only if b divides a according to the definition of division in whole numbers. This important case is explored fully in Chapter 4.

EXAMPLE 2.19

Using the Division Algorithm to Solve the Marching Band Problem

Mr. Garza was happy to see that so many students in the school band had turned out for the parade. He had them form into rows of 6, but it turned out that just one tuba player was in the back row. To his dismay, when he reformed the band into rows of 5, there was still a lone tuba player in the back row. In desperation, Mr. Garza had the band reassemble into rows of 7. To his relief, every row was filled! How many students are marching in the parade band?

SOLUTION

If we let n denote the number of students marching in the band, then we know that when n is divided by 6 the remainder is 1. Thus $n = 6q + 1$ for some whole number $q = 0, 1, 2, \ldots$. This means n is somewhere in the list:

1, 7, 13, 19, 25, 31, 37, 43, 49, 55, 61, 67, 73, 79, 85, 91, 97, 103, ...

Similarly, $n \div 5$ has a remainder of 1, so n is also a number in the list:

1, 6, 11, 16, 21, 26, 31, 36, 41, 46, 51, 56, 61, 66, 71, 76, 81, 86, 91, 96, 101, ...

Finally, 7 is a divisor of n so n is one of the numbers:

7, 14, 21, 28, 35, 42, 49, 56, 63, 70, 77, 84, 91, 98, 105, ...

Comparing the three lists, we find that 91 is the smallest number common to all three lists, so it is possible there are 91 band members. By extending the lists (it helps to notice that the numbers 1, 31, 61, 91, . . . common to the first two lists jump ahead by 30 each step), we find that the next smallest number in all three lists is 301, since $301 = 50 \cdot 6 + 1 = 60 \cdot 5 + 1 = 43 \cdot 7$. It seems pretty unlikely the band is this large, so we conclude that there are 91 members of the band. Mr. Garza has arranged them in 13 rows, since $91 \div 7 = 13$. ■

Exponents and the Power Operation

Multiplication is a shortcut for repeated addition with the same addend. A similar shortcut has been found useful for repeated multiplication with the same factor. For example, instead of writing $3 \cdot 3 \cdot 3 \cdot 3 \cdot 3$ we can write 3^5. This operation is called "taking 3 to the fifth power." The general definition is described as follows.

DEFINITION The Power Operation for Whole Numbers

Let a and m be whole numbers, where $m \neq 0$. Then ***a* to the *m*th power,** written a^m, is defined by:

$$a^1 = a, \quad \text{if } m = 1,$$

and

$$a^m = \overbrace{a \cdot a \cdot \cdots \cdot a}^{m \text{ factors}}, \quad \text{if } m > 1.$$

The number a is called the **base**, m is called the **exponent** or **power**, and a^m is called an **exponential.** Special cases include squares and cubes. For example, 7^2 is read "7 squared," and 10^3 is read "10 cubed." On most computers, 7^2 and 10^3 would be typed in as 7^2 and 10^3, where the circumflex ^ separates the base from the exponent.

EXAMPLE 2.20 **Working with Exponents**

Compute the following product and powers, expressing your answers in the form of a single exponential a^m.

(a) $7^4 \cdot 7^2$ **(b)** $6^5 \cdot 6^3$ **(c)** $2^3 \cdot 5^3$
(d) $3^2 \cdot 5^2 \cdot 4^2$ **(e)** $(3^2)^5$ **(f)** $(4^2)^3$

SOLUTION

(a) $7^4 \cdot 7^2 = (7 \cdot 7 \cdot 7 \cdot 7) \cdot (7 \cdot 7) = 7 \cdot 7 \cdot 7 \cdot 7 \cdot 7 \cdot 7 = 7^6$

(b) $6^3 \cdot 6^5 = (6 \cdot 6 \cdot 6) \cdot (6 \cdot 6 \cdot 6 \cdot 6 \cdot 6)$
$= 6 \cdot 6 \cdot 6 \cdot 6 \cdot 6 \cdot 6 \cdot 6 \cdot 6 = 6^8$

(c) $2^3 \cdot 5^3 = (2 \cdot 2 \cdot 2) \cdot (5 \cdot 5 \cdot 5) = (2 \cdot 5) \cdot (2 \cdot 5) \cdot (2 \cdot 5)$
$= (2 \cdot 5)^3 = 10^3$

(d) $3^2 \cdot 5^2 \cdot 4^2 = (3 \cdot 3) \cdot (5 \cdot 5) \cdot (4 \cdot 4) = (3 \cdot 5 \cdot 4) \cdot (3 \cdot 5 \cdot 4)$
$= (3 \cdot 5 \cdot 4)^2 = 60^2$

(e) $(3^2)^5 = (3)^2 \cdot (3)^2 \cdot (3)^2 \cdot (3)^2 \cdot (3)^2$
$= (3 \cdot 3) \cdot (3 \cdot 3) \cdot (3 \cdot 3) \cdot (3 \cdot 3) \cdot (3 \cdot 3)$
$= 3 \cdot 3 \cdot 3 \cdot 3 \cdot 3 \cdot 3 \cdot 3 \cdot 3 \cdot 3 \cdot 3 = 3^{10}$

(f) $(4^2)^3 = (4^2) \cdot (4^2) \cdot (4^2) = (4 \cdot 4) \cdot (4 \cdot 4) \cdot (4 \cdot 4)$
$= 4 \cdot 4 \cdot 4 \cdot 4 \cdot 4 \cdot 4 = 4^6$ ■

Example 2.20 reveals that multiplying exponentials follows useful patterns that can be used to shorten calculations. For example, $7^4 \cdot 7^2 = 7^{4+2}$ and $6^5 \cdot 6^3 = 6^{5+3}$ are two special cases of the general rule $a^m \cdot a^n = a^{m+n}$ for multiplying exponentials with the same base. Similarly, $2^3 \cdot 5^3 = (2 \cdot 5)^3$ is a special case of $a^n \cdot b^n = (a \cdot b)^n$, and $(3^2)^5 = 3^{2 \cdot 5}$ is a special case of $(a^m)^n = a^{m \cdot n}$.

HIGHLIGHT FROM HISTORY
Symbols for Multiplication, Division, and Power

William Oughtred introduced the symbol × for multiplication in 1631. Gottfried Leibniz objected to the ×, writing to Bernoulli in 1698 that it ". . . is easily confounded with *x*." Leibniz preferred the dot · symbol which Thomas Harriot had introduced in the same year 1631. The symbol ÷ was introduced in 1659 by the Swiss mathematician J. H. Rohn. It was adopted into the English speaking countries by John Wallis and others, but on the European continent the colon symbol a : b of Leibniz was the symbol of regular choice. The slanted line /, and the horizontal line, gradually became more common. Exponential notation became common with its systematic use by Descartes (1596–1650) in *La Geometrie.* For use with computers, where it is best to keep expressions on one line, the circumflex ^ is used: thus 4 ^ 7 denotes 4^7.

THEOREM Multiplication Rules of Exponentials

Let a, b, m, n be whole numbers, where $m \neq 0$ and $n \neq 0$. Then:

(i) $a^m \cdot a^n = a^{m+n}$

(ii) $a^m \cdot b^m = (a \cdot b)^m$

(iii) $(a^m)^n = a^{m \cdot n}$.

PROOF OF (I): $a^m \cdot a^n = a^{m+n}$

$$a^m \cdot a^n = \underbrace{a \cdot a \cdot \cdots \cdot a}_{m \text{ factors}} \cdot \underbrace{a \cdot a \cdot \cdots \cdot a}_{n \text{ factors}} = \underbrace{a \cdot a \cdot \cdots \cdot a}_{m + n \text{ factors}} = a^{m+n}.$$

The proofs of (ii) and (iii) follow similarly. ■

If the formula $a^m \cdot a^n = a^{m+n}$ were extended to allow $m = 0$, it would state that $a^0 \cdot a^n = a^{0+n} = a^n$. This suggests that it is reasonable to define $a^0 = 1$ when $a \neq 0$.

DEFINITION Zero as an Exponent

Let a be any whole number, $a \neq 0$. Define $a^0 = 1$.

To see why 0^0 is not defined, notice that there are two conflicting patterns:

$$3^0 = 1, \quad 2^0 = 1, \quad 1^0 = 1, \quad 0^0 = ?$$
$$0^3 = 0, \quad 0^2 = 0, \quad 0^1 = 0, \quad 0^0 = ?$$

The multiplication formula $a^m \cdot a^n = a^{m+n}$ can also be converted to a corresponding division fact. For example,

$$a^{5-3} \cdot a^3 = a^{(5-3)+3} = a^5$$

so

$$a^5/a^3 = a^{5-3}.$$

In general, we have the following theorem.

THEOREM Rules for Division of Exponentials

Let a, b, m, n be whole numbers, where $m \geq n > 0$, $b \neq 0$, and $a \div b$ is defined. Then

(i) $$b^m/b^n = b^{m-n}$$

and

(iii) $$(a^m/b^m) = (a/b)^m.$$

EXAMPLE 2.21

Working with Exponents

Rewrite these expressions in the form a^m of exponentials.

(a) $5^{12} \cdot 5^8$ **(b)** $7^{14}/7^5$ **(c)** $3^2 \cdot 3^5 \cdot 3^8$
(d) $8^7/4^7$ **(e)** 2^{5-5} **(f)** $(3^5)^2/3^4$

SOLUTION

(a) $5^{12+8} = 5^{20}$ **(b)** $7^{14-5} = 7^9$ **(c)** $3^{2+5+8} = 3^{15}$
(d) $(8/4)^7 = 2^7$ **(e)** $2^0 = 1$ **(f)** $3^{5 \cdot 2 - 4} = 3^6$ ■

COOPERATIVE *Investigation*

The Krypto Game

The Krypto game, which is available commercially or which can be homemade, consists of 25 cards numbered 1 through 25. Each player, or cooperative group, is dealt a hand of 5 cards plus a sixth card which designates the target number. Here is an example:

Hand: 21 | 5 | 17 | 8 | 3   Target: 12

The object of the game is to combine the cards in the hand, using any of the operations $+, -, \times, \div$ and using each card once and only once, to obtain the target number. Here is one solution:

$$21 - 5 = 16$$
$$16 \div 8 = 2$$
$$2 + 3 = 5$$
$$17 - 5 = 12 \leftarrow \text{Target}$$

That is, $17 - (((21 - 5) \div 8) + 3) = 12$.

In this game guess and check and working backwards will be useful problem-solving strategies. Parentheses will also be needed to show the order in which operations are performed.

PROBLEM SET 2.4

Understanding Concepts

1. What multiplication fact is illustrated in each of these diagrams?

   (a)

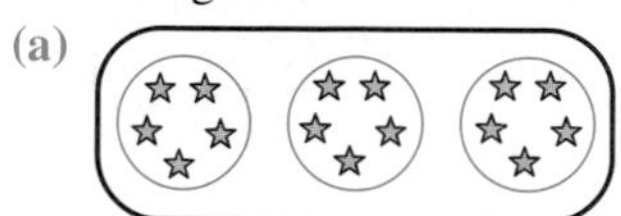

   (b)

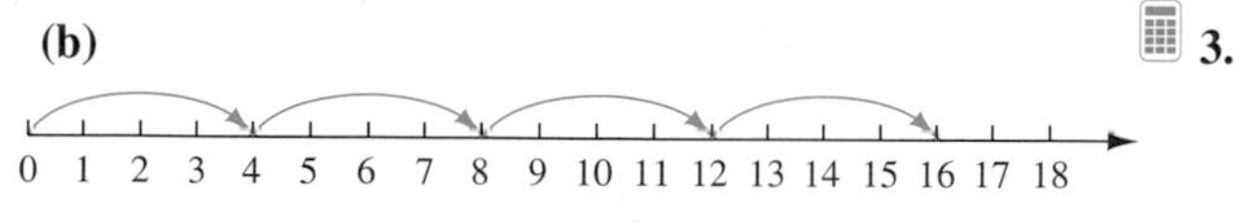

   (c)

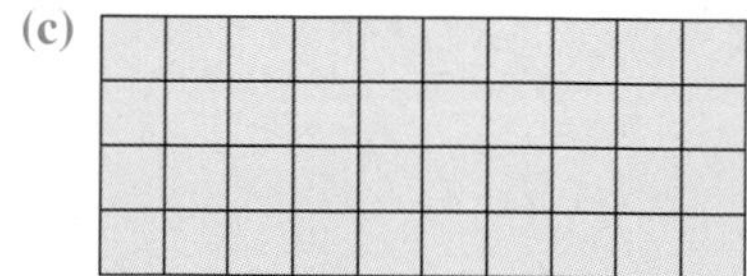

   (d)

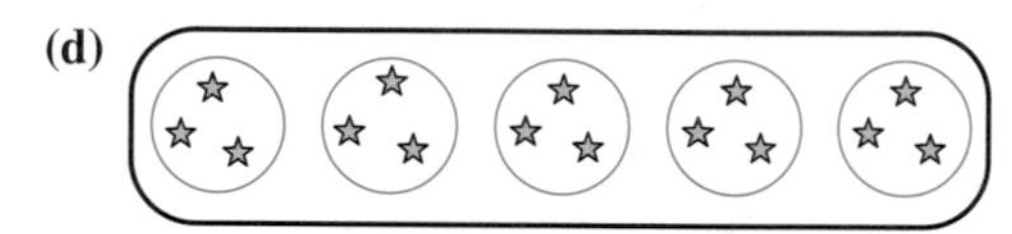

   (e)

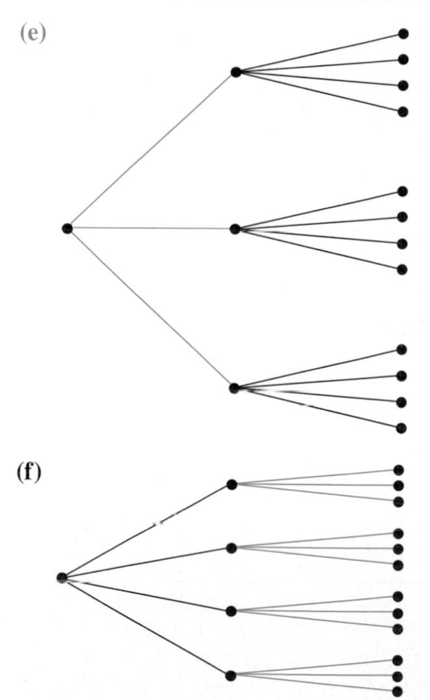

   (f)

2. Discuss which model of multiplication best fits the following problems:

   (a) A set of dominoes came in a box containing eleven stacks of five dominoes each. How many dominoes are in the set?

   (b) Marja has 3 skirts which she can "mix or match" with 6 blouses. How many outfits does she have to wear?

   (c) Harold hiked 10 miles each day, and crossed the mountains after 5 days. How long was his hike?

   (d) Ace Widgit Company makes 35 widgits a day. How many widgits are made in a 5-day workweek?

3. Multiplication as repeated-addition can be illustrated on most calculators. For example, $4 \cdot 7$ is computed by 7 [+] 7 [+] 7 [+] 7 [=]. Each press of the [+] key completes any pending addition and sets up the next one, so the intermediate products $2 \cdot 7 = 14$ and $3 \cdot 7 = 21$ are displayed along the way. Many calculators have a constant feature, which avoids having to reenter the same addend over and over. For example, [+] 734 [=] [=] [=], or 734 [+] [+] [+], may compute $3 \cdot 734$; it all depends on how your particular calculator operates.

   (a) Explain carefully how repeated addition is best accomplished on your calculator.

   (b) Use repeated additions on your calculator to compute these products. Check your result by using the [×] button.

   (i) $4 \cdot 9$ (ii) 7×536
   (iii) 6×47819 (iv) 56108×6 (What property may help?)

4. The Cartesian product of finite and nonempty sets can be illustrated by the intersections of a crossing-line pattern, as shown.

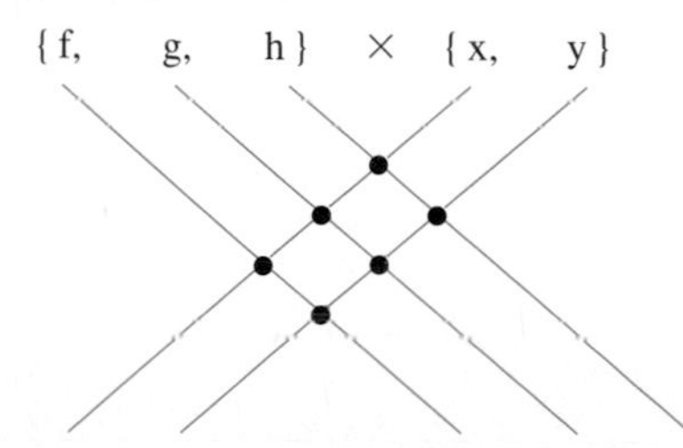

   (a) Explain why the number of intersection points in the crossing-line pattern for $A \times B$ is $a \cdot b$, where $a = n(A)$ and $b = n(B)$.

   (b) Draw the pattern for {□, △} × {♥, ♦, ♣, ♠}

5. Which of the following sets of whole numbers are closed under multiplication? Explain your reasoning.

   (a) {1, 2} (b) {0, 1} (c) {0, 2, 4}
   (d) {0, 2, 4, . . .} (the even whole numbers)
   (e) {1, 3, 5, . . .} (the odd whole numbers)
   (f) $\{1, 2, 2^2, 2^3\}$ (g) $\{1, 2, 2^2, 2^3, \ldots\}$
   (h) $\{1, 7, 7^2, 7^3, \ldots\}$

6. Which of the following subsets of the whole numbers $W = \{0, 1, 2, \ldots\}$ are closed under multiplication? Explain carefully.

(a) $W - \{5\}$ (that is, the whole numbers except for 5)

(b) $W - \{6\}$

(c) $W - \{2, 3\}$

7. What properties of whole number multiplication justify these equalities?

(a) $4 \cdot 9 = 9 \cdot 4$

(b) $4 \cdot (6 + 2) = 4 \cdot 6 + 4 \cdot 2$

(c) $0 \cdot 439 = 0$ **(d)** $7 \cdot 3 + 7 \cdot 8 = 7 \cdot (3 + 8)$

(e) $5 \cdot (9 \cdot 11) = (5 \cdot 9) \cdot 11$ **(f)** $1 \cdot 12 = 12$

8. What property of multiplication is illustrated in each of these diagrams?

(a)

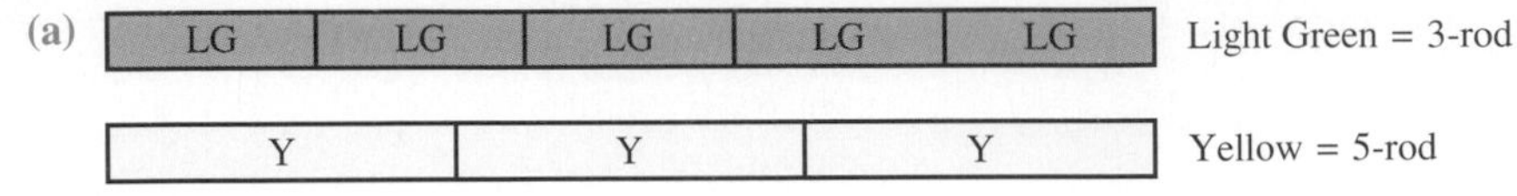

(b)

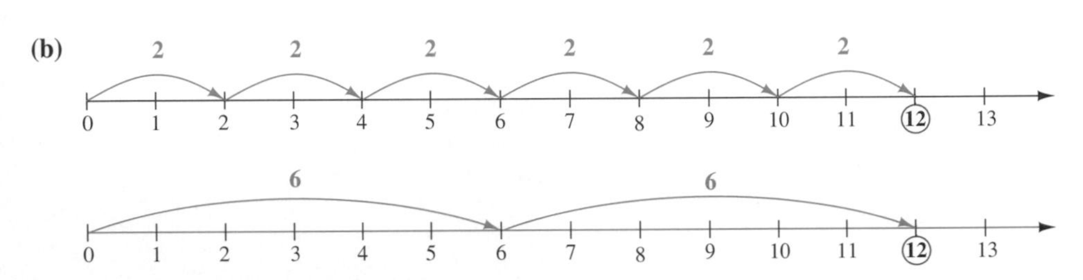

9. Use the rectangular array model to illustrate each of the following statements. Make drawings similar to Figure 2.21.

(a) $(2 + 5) \cdot 3 = 2 \cdot 3 + 5 \cdot 3$

(b) $3 \cdot (2 + 5 + 1) = 3 \cdot 2 + 3 \cdot 5 + 3 \cdot 1$

(c) $(3 + 2) \cdot (4 + 3) = 3 \cdot 4 + 3 \cdot 3 + 2 \cdot 4 + 2 \cdot 3$.

10. Use the figure below to show that the product of trinomials, $(a + b + c) \cdot (d + e + f)$, can be written as a sum of nine products.

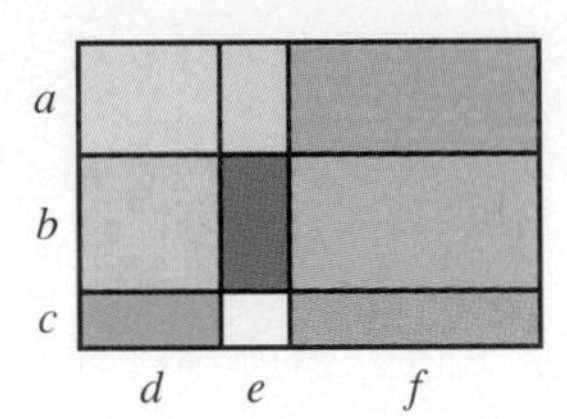

11. Modify Figure 2.20 to show how the associative property $3 \cdot (2 \cdot 4)$ can be illustrated with arrays of cubes in 3-dimensional space.

12. Shannon bought 18 nuts and 18 bolts. The bolts were 86¢ each, and the nuts were 14¢ each. The store clerk computed the bill as shown below. How did Shannon already know the answer by simple mental math?

$$\begin{array}{r} {}^{4} \\ 86 \\ \times\ 18 \\ \hline 688 \\ 86 \\ \hline 1548 \end{array} \qquad \begin{array}{r} {}^{3} \\ 14 \\ \times\ 18 \\ \hline 112 \\ 14 \\ \hline 252 \end{array} \qquad \begin{array}{r} {}^{1\ 1} \\ 15.48 \\ +\ 2.52 \\ \hline 18.00 \end{array}$$

13. What properties of multiplication make it easy to compute these values mentally?

(a) $7 \cdot 19 + 3 \cdot 19$

(b) $24 \cdot 17 + 24 \cdot 3$

(c) $36 \cdot 15 - 12 \cdot 45$

14. What division facts are illustrated below?

(a) **(b)**

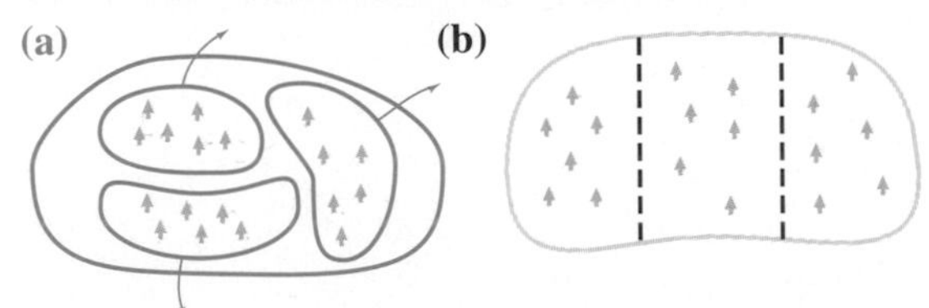

15. A 2 by 3 rectangular array is associated with the fact family $2 \cdot 3 = 6$, $3 \cdot 2 = 6$, $6 \div 2 = 3$, $6 \div 3 = 2$. What fact family is associated with each of these rectangular arrays?

(a) 4 by 8 **(b)** 6 by 5

16. Discuss which of three conceptual models of division—repeated-subtraction, partition, missing-factor—best corresponds to the following problems. More than one model may fit.

(a) Preston owes \$3200 on his car. If his payments are \$200 a month, how many more months will Preston make car payments?

(b) An estate of \$76,000 is to be split among 4 heirs. How much can each heir expect to inherit?

(c) Anita was given a grant of \$375 to cover expenses on her trip. She expects that it will cost her \$75 a day. How many days can she plan to be gone?

17. Use repeated subtraction on your calculator to compute the following division problems, where remainders are possible. Be sure to take advantage of the constant feature of your calculator.
(a) $78 \div 13$ **(b)** $832 \div 52$
(c) $96 \div 14$ **(d)** $548245 \div 45687$

18. Show that the following statements are not always true, where a, b, and c represent nonzero whole numbers.
(a) $a \div b$ is a whole number
(b) $a \div b = b \div a$
(c) $(a \div b) \div c = a \div (b \div c)$

19. (a) Show that if a/b and d/b are defined, then $(a + d)/b$ is defined and $(a + d)/b = (a/b) + (d/b)$.
(b) Show by an example that division is *not* distributive over addition from the left; that is, it is not always true that $a/(b + c) = (a/b) + (a/c)$.

20. Solve for the unknown whole number in the following expressions.
(a) $y \div 5 = 5$ R 4
(b) $20 \div x = 3$ R 2

21. Rewrite the following in the form of a single exponential.
(a) $3^{20} \cdot 3^{15}$ **(b)** $4^8 \cdot 7^8$ (c) $(3^2)^5$
(d) $x^7 \cdot x^9$ (e) $y^3 \cdot z^3$ **(f)** $(t^3)^4$

22. Write the following as 2^m for some whole number m.
(a) 8 **(b)** $4 \cdot 8$ (c) 1024 **(d)** 8^4

23. (a) Show that $2^4 = 4^2$. Does this mean that the power operation is commutative?
(b) Is the power operation associative on the nonzero whole numbers? That is, is $(a^b)^c = a^{(b^c)}$ for *all* a, b, and c?
(c) Are there *certain* values for which $(a^b)^c = a^{(b^c)}$?

24. Find the exponents that make the following equations true.
(a) $3^m = 81$ **(b)** $3^n = 531{,}441$
(c) $4^p = 1{,}048{,}576$ **(d)** $2^q = 1{,}048{,}576$

Thinking Critically

25. Let a and b be whole numbers. Explain why the square array shown below illustrates the formula $(a + b)^2 = a^2 + 2ab + b^2$.

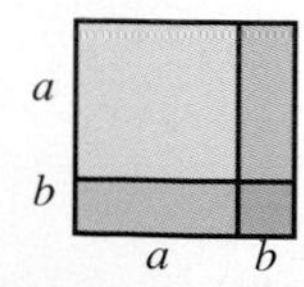

26. Let a and b be whole numbers, with $a > b$. Explain how the square array shown below illustrates the formula $(a + b)^2 - (a - b)^2 = 4ab$.

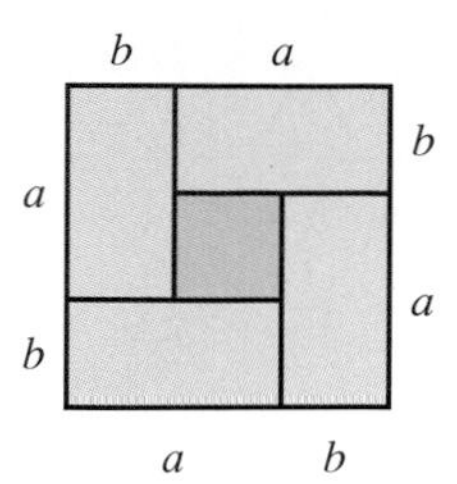

27. Let a and b be whole numbers with $a > b$. Explain how the square array shown below illustrates the formula $(a + b)^2 + (a - b)^2 = 2a^2 + 2b^2$.

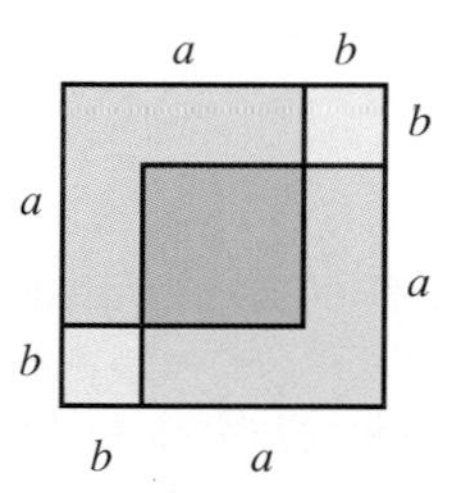

28. Let a and b be positive whole numbers with $a > b$. Use the figure below to illustrate and justify the formula:

$$a^2 - b^2 = (a - b)(a + b).$$

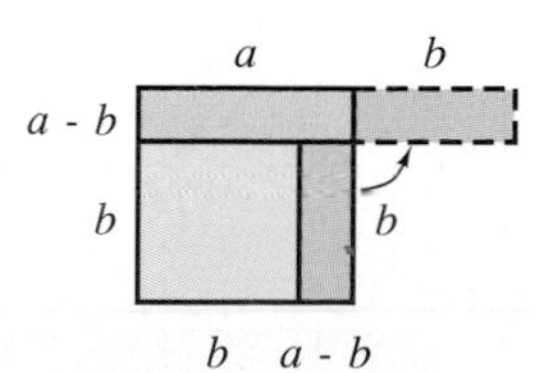

29. (a) Verify that $4 \cdot (5 - 2) = 4 \cdot 5 - 4 \cdot 2$.
(b) Prove that multiplication distributes over subtraction. That is, for all whole number a, b, and c, with $b \geq c$, show that $a \cdot (b - c) = a \cdot b - a \cdot c$.

30. **Adams' Magic Hexagon.** In 1957 Clifford W. Adams discovered a magic hexagon, in which the sum of the numbers in any "row" is 38. Fill in the empty cells of the partially completed hexagon shown, using the whole numbers 6, 7, . . . , 15, to recreate Adams' discovery. When completed, each cell is assigned one of the numbers 1, 2, . . . , 19.

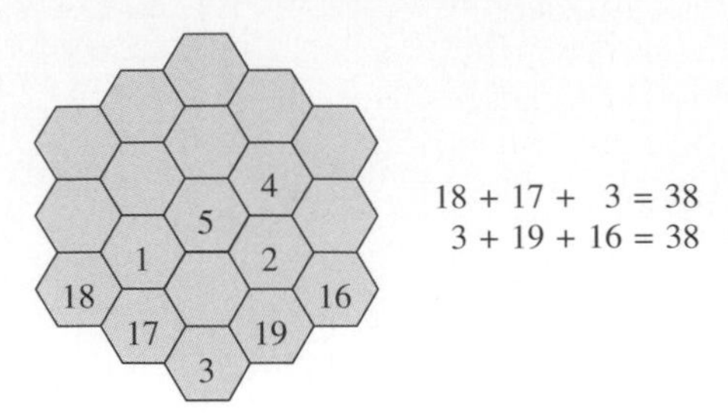

31. A clock chimes on the hour, once at 1 o'clock, twice at 2 o'clock and so on. How many times does it chime every 24 hour period?

32. How many total spots are on the 28 dominoes of a "double-six" domino set? (See problem 24 of Problem Set 2.3.) (*Hint:* Suppose you have *two* sets of dominoes. Match each domino of one set with its "complement" in the second set, where the complement of the domino with m and n spots has $6 - m$ and $6 - n$ spots.) For example, the following dominoes are complementary:

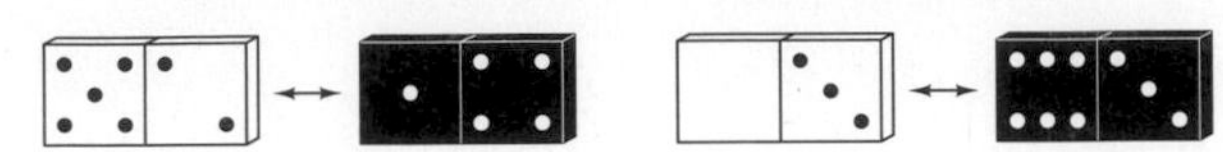

Thinking Cooperatively

33. Numbers of the form $1 \cdot 2, 2 \cdot 3, \ldots, n \cdot (n + 1), \ldots$ are called **oblong** numbers.

(a) Complete the following table of the first fifteen oblong numbers.

n	1	2	3	4	5	6	7	8	9	10	11	12	13	14	15
nth oblong number	2	6	12	20											

(b) Recall that $t_n = 1 + 2 + 3 + \cdots + n$ is called the nth triangular number. Add a new row to the table in part (a) that lists the first fifteen triangular numbers.

(c) Describe a pattern in your table that suggests the formula $t_n = n(n + 1)/2$.

(d) Draw rectangular arrays of squares of size 4 by 5 and 5 by 6, and then color them to show why each array is composed of two identical triangular arrays. Here is how to draw the 3 by 4 case.

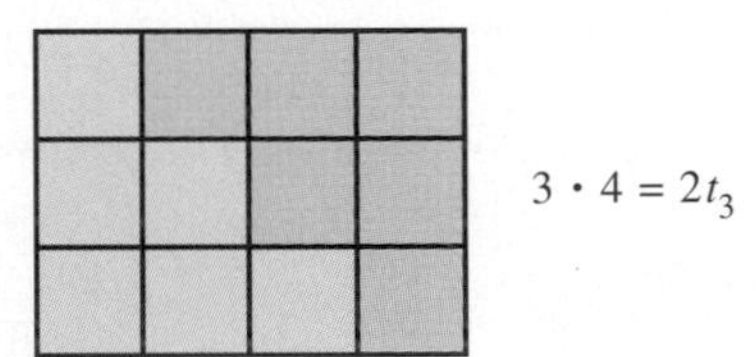

34. This diagram shows that 8 triangular numbers $t_2 = 3$, added to 1, sum to the third *odd* square 5^2. That is, $8 \cdot t_2 + 1 = 5^2$.

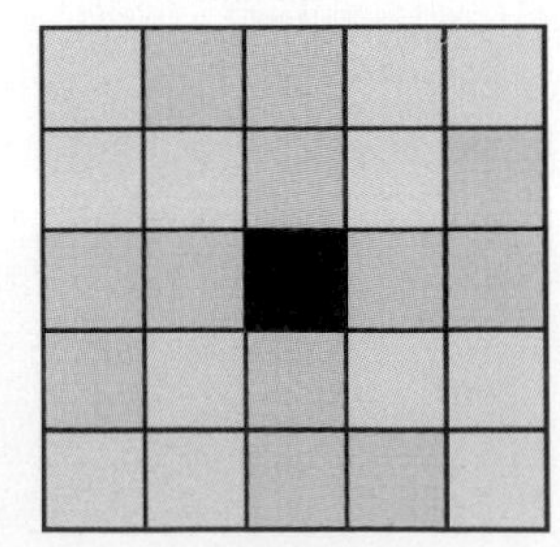

(a) Shade a 7×7 square to show that $8 \cdot t_3 + 1 = 7^2$.

(b) Verify $8 \cdot t_4 + 1$ is a square number. Draw a diagram to show your result.

(c) Guess a formula for $8 \cdot t_n + 1$.

35. Let T_n denote the sum of first n triangular numbers, so that

$$T_n = 1 + 3 + 6 + 10 + \cdots + t_n,$$

where

$$t_n = 1 + 2 + 3 + \cdots + n.$$

Now observe that

$$T_1 = 1 = 1 \cdot 2 \cdot 3/6$$
$$T_2 = 1 + 3 = 4 = 2 \cdot 3 \cdot 4/6$$
$$T_3 = 1 + 3 + 6 = 10 = 3 \cdot 4 \cdot 5/6$$

(a) See if the pattern continues to hold for T_4, $T_5, \ldots, T_{12}$.

(b) What is T_{100}?

36. Consider the squares of whole numbers which end in the digit 5:

$$15^2 = 225 = (1 \cdot 2)100 + 25$$
$$25^2 = 625 = (2 \cdot 3)100 + 25$$
$$35^2 = 1225 = (3 \cdot 4)100 + 25$$

(a) Verify that the pattern continues to hold for $45^2, 55^2, \ldots, 95^2$.

(b) Show that $(10a + 5)^2 = a \cdot (a + 1) \cdot 100 + 25$, and then use the formula to explain why the trick works.

(c) Use mental math to compute these squares, and check your answers with a calculator: 105^2, 195^2, 1005^2.

(d) Give the square roots of these numbers: 5625, 42,025, 99,900,025.

Using A Calculator

37. The eighth triangular number is $t_8 = (8 \cdot 9)/2 = 36$, which is also a square number since $6^2 = 36$. Verify that the triangular numbers t_{49}, t_{288}, and t_{1681} are also square numbers. Recall that $t_n = n(n + 1)/2$ (see problem 33).

For Review

38. What addition-subtraction fact family contains $2 + 7 = 9$?

39. Let $n(A) = 2$, $n(B) = 7$, and $n(A \cup B) = 9$. What is $n(A \cap B)$?

40. What basic facts are illustrated by these number strip trains?

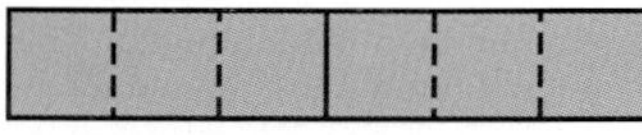

2.5 Relations and Functions

The word *relation* usually suggests a family connection between two people. For example "is a sister of" and "is a grandparent of" easily come to mind. We might also relate people to countries by "is a citizen of," and relate a person to a number by "is of age." In this section we study some important relations that help compare and classify mathematical objects, such as numbers, sets, and geometric figures. A particular type of relation called a function is of special importance.

Relations

In mathematics, relations are used to describe how one object compares, or is connected, to another object. Many relations are already familiar from our study of sets and whole numbers. Several other relations may be recalled from basic geometry. The following examples will help motivate the formal definition of relation that is given below.

- *Some relations on the set of whole numbers*

  less than $<$ — The whole number a is related to the whole number b if, and only if, $a < b$. Thus $5 < 8$, but $8 \not< 5$.

  divides $|$ — The natural number m is related to the whole number n if, and only if, m divides n. Using a vertical bar $|$ to symbolize "divides," we see that $7 \mid 35$ and $13 \nmid 25$.

- *Some relations useful in set theory*

  membership $\in$ — An element x is related to set A, if and only if, $x \in A$. Thus $4 \in N$ and $0 \notin N$, where $N = \{1, 2, \ldots\}$ denotes the set of natural numbers.

  set equivalence $\sim$ — Two sets A and B are related, if and only if, there is a one-to-one correspondence between A and B, which is symbolized by $A \sim B$. Thus

$$\{a, b, c\} \sim \{x, y, z\}.$$

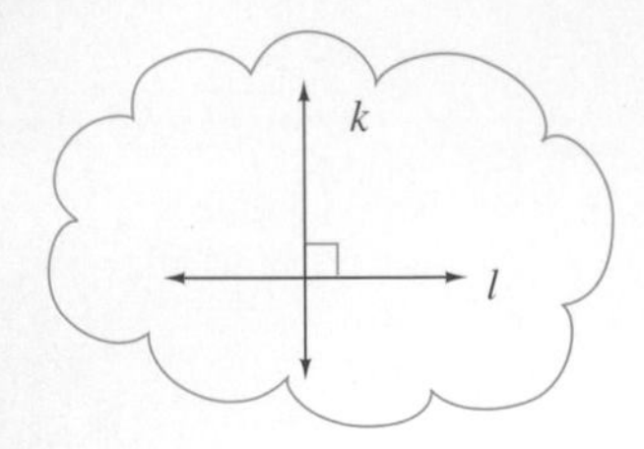

- *Some relations from geometry*

  perpendicularity $\perp$ Two lines k and l are related if, and only if, they intersect in a right angle, in which case we write $k \perp l$.

  congruence $\cong$ Two geometric figures F and G are related by congruence if, and only if, they have the same size and shape, in which case we write that $F \cong G$.

It would not be hard to extend the list of familiar mathematical relations: "greater than or equal," "subset of," "similar to," and "parallel to" may come to mind, with the respective symbols $\geq$, $\subseteq$, $\sim$, and $\|$.

With the examples listed above in mind, we give a general definition of a relation, R, from a set S to a set T. When an object a, $a \in S$ is related to object b, $b \in T$, the ordered pair (a, b) is listed in the relation R.

DEFINITION Relation from S to T

A **relation R from set S to set T** is a subset of the Cartesian product $S \times T$. That is, $R \subseteq S \times T$. If $S = T$, so that $R \subseteq S \times S$, then we say R is a **relation on S.**

As our examples illustrate, it is more common to write a symbol between two related objects than it is to use the ordered pair notation. Let's agree to use a box, $\square$, to symbolize a relation. We will write $a \square b$ to indicate that a is related to b by relation $\square$. The $\square$ can stand for any of the usual symbols $<, \leq, >, \geq, \subseteq, \in, \sim$, and so on. More generally, $\square$ can symbolize any relation of interest to us.

Arrow Diagrams

Relations are sometimes visualized by drawing **arrow diagrams.** A relation $\square$ on S to T is pictured by drawing an arrow from element a to element b whenever $a \square b$. If $S = T$, either two copies of S are drawn or else just one copy of S is drawn with arrows drawn between the related members of S. Examples of arrow diagrams are shown in Figure 2.25. In (a) we see that $a \square 1$, $b \square 1$, $b \square 2$, and $c \square 3$. In (b) and (c) we see the two alternative ways to show that $a \square b$, $b \square b$, $c \square c$, $c \square d$, and $d \square b$.

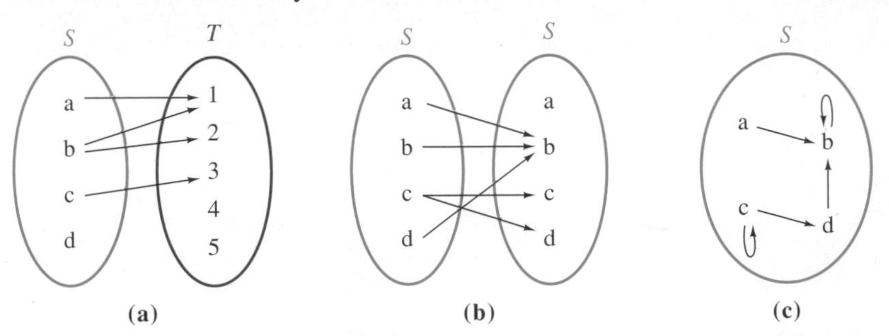

Figure 2.25
Relations can be visualized with arrow diagrams. Both (b) and (c) depict the same relation on *S*.

EXAMPLE 2.22 **Finding and Describing Relations**

Let $S = \{2, 3, 4\}$ and $T = \{8, 9, 10, 11, 12\}$.

(a) Give a rule that describes the relation $\square$ on $S \times T$, where $2 \square 8$, $2 \square 10$, $2 \square 12$, $3 \square 9$, $3 \square 12$, $4 \square 8$, $4 \square 12$.

(b) List the pairs of elements in the relation □ on S to T defined by "added to 9 equals." That is, $a \ \square \ b$ when $a + 9 = b$.

(c) Draw the arrow diagrams of the relations in parts (a) and (b).

SOLUTION

(a) The □ denotes the "divides" relation, usually symbolized by a vertical bar $|$. That is, $2 \mid 8$, $2 \mid 10$, and so on.

(b) $2 \ \square \ 11$ and $3 \ \square \ 12$.

(c)

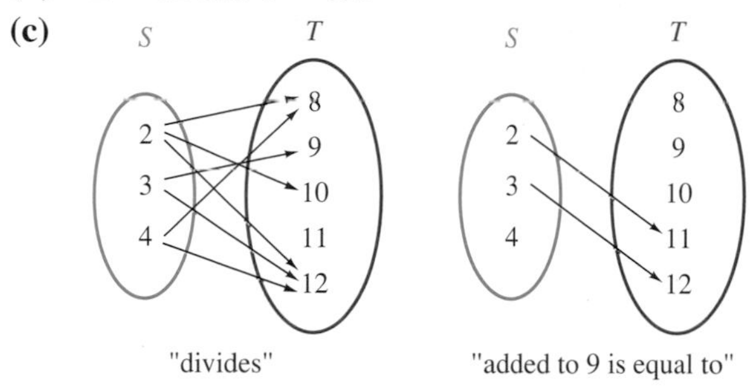

"divides" "added to 9 is equal to" ■

The Reflexive, Symmetric, and Transitive Properties

Let's consider a relation □ on some set S. There are three useful properties that a relation may have, which we now define and illustrate in turn.

> **DEFINITION The Reflexive Property**
>
> A relation □ on set S is **reflexive** if, and only if, $a \ \square \ a$ for all $a \in S$.

Relations mentioned earlier that are reflexive include "is equal to," "is set equivalent to," "divides," and "is congruent to." On the other hand, "less than," "is a proper subset of," and "is perpendicular to" are not reflexive relations.

> **DEFINITION The Symmetric Property**
>
> A relation □ on set S is **symmetric** if, and only if, for all elements $a, b \in S$, whenever $a \sqcap b$ then $b \ \square \ a$.

For example, equality is symmetric since if $a = b$ then $b = a$. Likewise, if k and l are lines and $k \perp l$, then $l \perp k$ as well; this means perpendicularity is a symmetric relation on the set of lines in the plane. On the other hand, $2 < 5$ but $5 \nless 2$. Thus, "less than" is not symmetric.

> **DEFINITION The Transitive Property**
>
> A relation □ on S is **transitive** if, and only if, for all elements $a, b, c \in S$, whenever $a \ \square \ b$ and $b \ \square \ c$, then $a \ \square \ c$.

WORTH READING

"A Provocative Perspective on a Disturbing Modern Problem"

In the following excerpt, writer and mathematician John Allen Paulos delivers a compelling and passionate treatise on the phenomenon of "innumeracy" in the modern world.

"Although few students get past elementary school without knowing their arithmetic tables, many do pass through without understanding that if one drives at 35 mph for four hours, one will have driven 140 miles; that if peanuts cost 40 cents an ounce and a bag of them costs $2.20, then there are 5.5 ounces of peanuts in the bag; that if 1/4 of the world's population is Chinese and 1/5 of the remainder is Indian, then 3/20 or 15 percent of the world is Indian. This sort of understanding is, of course, not the same as simply knowing that $35 \times 4 = 140$; that $(2.2)/(.4) = 5.5$; that $1/5 \times (1 - 1/4) = 3/20 = .15 = 15$ percent. And since it doesn't come naturally to many elementary students, it must be furthered by doing numerous problems, some practical, some more fanciful. . . .

"Some of the blame for the generally poor instruction in elementary schools must ultimately lie with teachers who . . . too often have little interest in or appreciation of mathematics. In turn, some of the blame for that lies, I think, with schools of education in colleges and universities which place little or no emphasis on mathematics in their teacher training courses . . .

"If mathematics education communicated [the] playful aspect of the subject, formally at the elementary, secondary, or college level or informally via popular books, I don't think innumeracy would be as widespread as it is."

SOURCE: John Allen Paulos, *Innumeracy: Mathematical Illiteracy and Its Consequences*. New York: Hill and Wang, pp. 73–74, 75 and 77.

For example, the "greater than or equal" relation of the whole numbers is transitive; that is, if $a \geq b$ and $b \geq c$, then $a \geq c$, where $a, b, c \in W = \{0, 1, 2, \ldots\}$. The relations $=, \leq, <, \subseteq, \subset$, and $\sim$ are also transitive. On the other hand, perpendicularity is not a transitive relation. If k, l, and m are lines in the place with $k \perp l$ and $l \perp m$, it is *not* true that k is perpendicular to m.

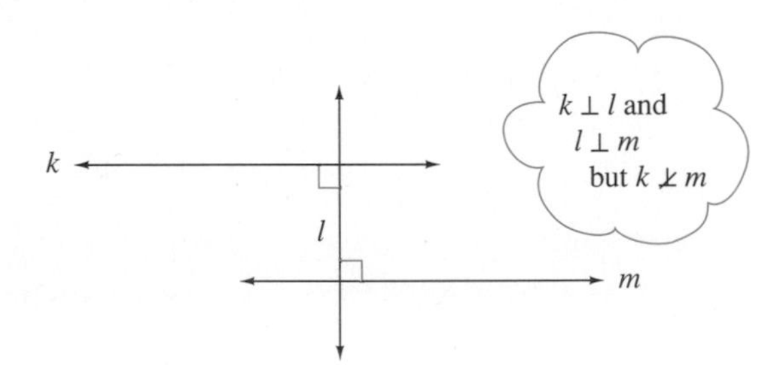

EXAMPLE 2.23

Checking for Properties of Relations

Decide whether or not the following relations on the set of all people are reflexive, symmetric, or transitive.

(a) "Is an ancestor of"
(b) "Is a different age than"
(c) "Has the same annual income as"
(d) "Knows the telephone number for"

SOLUTION

The answers are given in this table.

Relation	Reflexive	Symmetric	Transitive
(a) "Is an ancestor of"	no	no	yes
(b) "Is a different age than"	no	yes	no
(c) "Has the same income as"	yes	yes	yes
(d) "Knows a telephone number for"	no	no	no

It may be surprising that (d) is not reflexive, but there are many people without a telephone. ■

Equivalence Relations

Many of the most basic relations in mathematics—equality, set equivalence, congruence, similarity, to name a few—are reflexive, symmetric, *and* transitive. When all three properties are satisfied, a relation is called an **equivalence relation.**

DEFINITION Equivalence Relation

An **equivalence relation** on a set S is a relation that is reflexive, symmetric, and transitive.

An equivalence relation on a set S partitions S into disjoint subsets each consisting of the related elements of S. For example, consider the relation "has the same parity (evenness of oddness) as" on the set of whole numbers $W = \{0, 1, 2, \ldots\}$. This equivalence relation (verify that it is!) partitions W into two disjoint sets, the evens

JUST FOR FUN

Red and Green Jellybeans

Caralee has two jars of jellybeans, labeled R and G. The jellybeans in one jar R are all red, and those in the other jar G are all green. Caralee removed 20 red jellybeans, mixed them up with the green jellybeans in jar G and then without looking grabbed 20 jellybeans from the mixed jar and put them back in jar R. How does the number of green jellybeans in jar R compare to the number of red jellybeans in jar G?

and the odds:

$$W = \{0, 2, 4, \ldots\} \cup \{1, 3, 5, \ldots\}.$$

The disjoint sets into which an equivalence relation partitions S are called **equivalence classes.** For example, the relation of set equivalence partitions the finite sets into equivalence classes which each contain the same whole number of elements. In Example 2.23, only "has the same income as" is an equivalence relation. A typical equivalence class would consist of everyone earning, say, $25,000 per year.

Functions

The area of a circle of radius 5 is $\pi \cdot (5)^2$, or 25π. Likewise, the area of a circle of radius 10 is $\pi \cdot (10)^2$, or 100π. Thus, radius 5 is related to area 25π, and radius 10 is related to 100π. Indeed, the formula $A = \pi r^2$ for the area A of a circle of radius r defines a relation between every positive real number r and the unique number A given by $A = \pi r^2$. This means that the area of a circle is a **function** of its radius, since we have the following definition.

DEFINITION Function

A relation from S to T is a **function** if each element $a \in S$ is related to precisely one element $b \in T$. S is called the **domain** of the function.

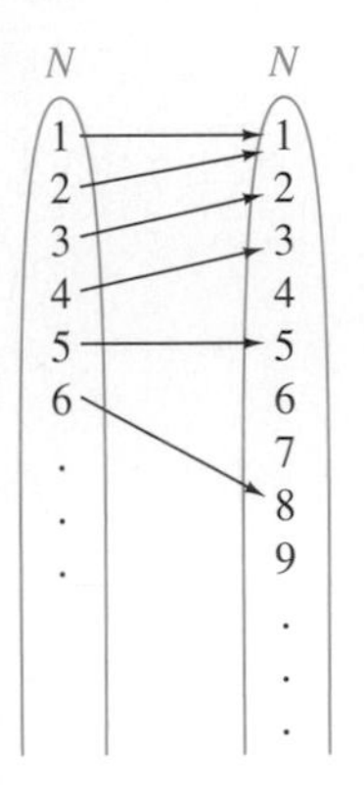

Figure 2.26
The arrow diagram of the Fibonacci sequence

A function is often denoted by a letter such as f, and we write $f(x) = y$ when $x \in S$ is related to $y \in T$. Often the element y is called the **image,** or **value,** of f at x. For example, if f is the function relating the radius of a circle to the area of the circle, then $f(5) = 25\pi$, $f(10) = 100\pi$ and in general $f(r) = \pi r^2$ for all $r > 0$.

The requirement that each x in the domain of a function has a unique image is clearly a reasonable condition for the area of a circle function. It would be strange if a circle of radius 5 were to have two different areas.

When an arrow diagram is drawn for a function precisely one arrow emanates from each element of the domain and terminates at the corresponding image point. Figure 2.26 shows an arrow diagram for the function F defined on the natural numbers which relates $n \in N$ to the nth Fibonacci number. A function defined on the natural numbers is also called a **sequence,** and it is common to write F_n in place of $F(n)$.

If f is a function from domain S to set T, the set of image points in T is called the **range** of f.

DEFINITION Range of a Function

The **range** of a function f on S to T is the set of images of f. That is,

$$\text{range } f = \{y \in T \mid y = f(x) \text{ for some } x \in S\}.$$

For example, the Fibonacci function shown in Figure 2.26 has the range $\{1, 2, 3, 5, 8, \ldots\}$

HIGHLIGHT FROM HISTORY
***Hypatia* (A.D. 370–415)**

The early history of mathematics mentions few women's names. Indirect evidence, however, suggests that at least some women of ancient times had access to mathematical knowledge, and likely made contributions to it. For example, the misnamed "brotherhood" of Pythagoreans, at the insistence of Pythagoras himself, included women in the Order both as teachers and scholars. Indeed, Theano, the wife and former student of Pythagoras, assumed leadership of the school at the death of her husband. Theano wrote several treatises on mathematics, physics, medicine, and child psychology.

Hypatia was the first woman to attain lasting prominence in mathematics history. Her father, Theon, was a professor of mathematics at the Alexandrian Museum. Theon took extraordinary interest in his daughter, and saw to it that she was thoroughly educated in arts, literature, science, philosophy, and of course mathematics. Hypatia's fame as a mathematician was secured in Athens, where she studied with Plutarch the Younger and his daughter Asclepigenia. Later she returned to the university at Alexandria, where she lectured on Diophantus' *Arithmetica* and Apollonius' *Conic Sections* and wrote several treatises of her own. The account of Hypatia's life of accomplishments in mathematics, astronomy, and teaching ends on a tragic note, for in March of 415 she was seized by a mob of religious fanatics and brutally murdered.

Describing Functions

There are several useful ways to describe and visualize function.

- **Functions as formulas.** For all numbers in a specified domain, some algebraic expression in a variable tells us how to compute the image value. For example, $f(r) = 2\pi r$ gives us the circumference of a circle for each positive value r of the radius. As we saw in Chapter 1, the sequence of odd numbers 1, 3, 5, . . . is given by the formula $h(n) = 2n - 1$, where $n \in \{0, 1, 2, \ldots\}$. Thus, the fiftieth odd number is

$$h(50) = 2(50) - 1 = 99.$$

- **Functions as tables.** The following table gives the grades of three students on an essay question.

Student	Grade
Raygene	8
Sergei	7
Leticia	10

No algebraic formula connects the student to the grade. Even so, as long as the table assigns one value on the right to each domain element on the left, then a function is completely described.

- **Functions as machines.** Viewing a function as a machine gives students an attractive dynamic visual model. The machine has an input hopper, which accepts any domain element x, and an output chute which gives the image $y = f(x)$. A few machines are shown in Figure 2.27.

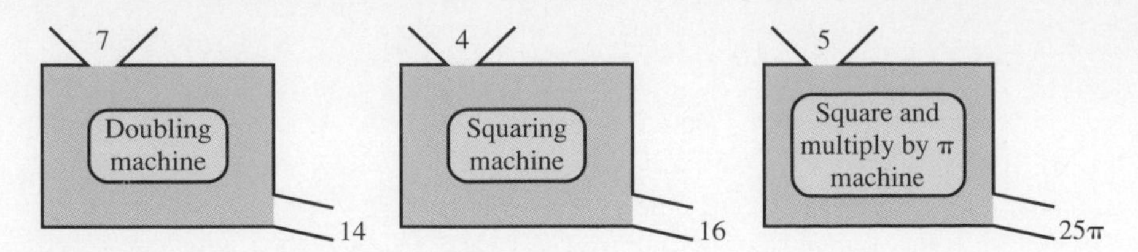

Figure 2.27
Three function machines

- **Functions as graphs.** A function whose domain and range are sets of numbers can be graphed on a two-dimensional coordinate system: if $f(x) = y$, plot the point (x, y). The Fibonacci sequence and the doubling function are plotted in Figure 2.28.

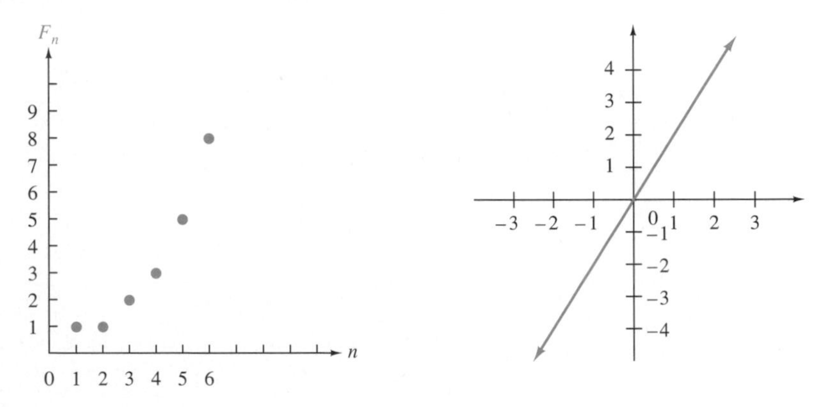

Figure 2.28
Graphs of (a) the Fibonacci sequence and (b) the doubling function

Additional information on graphs of functions may be found in Chapter 14.

The following example illustrates a function game that elementary students enjoy.

EXAMPLE 2.24 **Guessing Erica's rule**

Erica is "it" in a game of Guess My Rule. As the children pick an input number, Erica tells what number her rule gives back as shown in this table. Can you guess Erica's rule?

Children's Choice	Result of Erica's Rule (function)
2	−1
5	8
6	11
0	−7
1	−4

SOLUTION

Understand the problem • • • •

The rule Erica has chosen is a function—given an input number, she uses her function to determine the output number that she reveals to the class. We must guess her function. We will express the formula in terms of a variable x or n as in the above examples.

Devise a plan • • • •

The children's choices are somewhat random. Perhaps a pattern will become more apparent if we arrange their input numbers in order of increasing size. We anticipate that Erica's function is given by a formula.

Carry out the plan • • • •

Rearranging the children's choices in order of increasing size, we have the following table.

Children's Choice	Result of Erica's Rule (function)
0	−7
1	−4
2	−1
5	8
6	11

When 0 is input the formula yields −7. This suggests that −7 is a separate term in the formula; when $n = 0$ all the other terms are zero. Also, we observe that when n increases by 1 from 0 to 1, from 1 to 2, and from 5 to 6, the ouput number increases by 3. This suggests that the formula also contains the term $3n$, since this quantity increases by 3 each time n increases by 1. Combining these observations, we guess that Erica's function (rule), E, is given by the formula:

$$E(n) = 3n - 7.$$

Checking, we see that $E(0) = 3 \cdot 0 - 7 = -7$, $E(1) = 3 \cdot 1 - 7 = -4$, $E(2) = -1$, $E(5) = 8$, and $E(6) = 11$ as in Erica's table. When challenged, Erica reveals that we have guessed correctly.

Look back • • • •

Erica's rule is really a function—given a single input, her rule returns a single output. We guessed the rule by arranging the data in a more orderly way, noting that her rule associated −7 with 0, and that the output number increased by 3 each time the input number increased by 1. Thus, we correctly guessed Erica's function to be $E(n) = 3n - 7$. ■

CURRENT AFFAIRS

Mathematics in the Cradle

To most of us, newborn babies are little more than limp conglomerations of needs, passively soaking in stimuli from the outside world. Until recently, even the standard psychological-development model reflected this assumption. But in the past decade psychologists have learned that even very young babies are active participants in the world: they can discriminate objects by shape, size, and color; they can discern whether a person's lip movements are in sync with his or her speech; and this past year a researcher at the University of Arizona has found that baby brains can perform basic arithmetic.

In a landmark study hailed as a "notable event" in the field, developmental psychologist Karen Wynn tested the mathematical skills of infants by using Mickey Mouse dolls. A captive audience of 64 subjects, each five months old, watched as Wynn demonstrated simple additions and subtractions with the dolls. To add one plus one, for example, Wynn first put a doll on a stage, then hid it behind a small screen. In view of the infant, she then moved a second doll behind the screen. The screen was then lowered to reveal either the correct "answer" (two dolls) or an impossible outcome (one doll). The infants were shown both answers three times.

To find out whether the babies knew the correct answer, Wynn used a technique that, after 20 years of development, has finally become accepted as a tried-and-true method of accurately assessing what goes on in the minds of babes: they stare longer at something unexpected than they do at something that is predicted or familiar. By measuring how long each baby looked at the revealed answers, Wynn was able to determine if the infant found the results surprising or not.

Wynn found that the infants stared significantly longer at the incorrect answer than they did at the correct, presumably expected, answer. A set of clever controls assured that the infants were calculating rather than simply responding to a change in the number of dolls. She found similar results for the subtraction problem two minus one.

"This shows that infants have a much richer understanding of the relationship between numbers than we gave them credit for," Wynn says. "Considering the results from studies with chimps and other animals, it's now looking more and more like there is an innate mental mechanism for determining numbers." Wynn speculates that infants even younger than five months might have similar math ability.

SOURCE: "Counting In The Cradle" by David J. Fishman from *Discover*, January 1993, Volume 14, Number 1, p. 92.

PROBLEM SET 2.5

Understanding Concepts

1. Indicate which of these relations on the set of people in the United States is reflexive. If a relation is not reflexive, explain why.
   (a) "is a mutual friend of"
   (b) "is the sister of"
   (c) "is the father of"
   (d) "is in the same grade at school as"
   (e) "lives in the same state as"
   (*Hint for (e):* Consider the President of the United States.)
2. Which of the relations in problem 1 are symmetric? If a relation is not symmetric, explain why.
3. Which of the relations in problem 1 are transitive? If a relation is not transitive, explain why.
4. Which of these relations are reflexive? If the relation is not reflexive, give an example of an element of the set not related to itself.
   (a) The relation "has the same parity as" on the set of natural numbers. (Two natural numbers have the same parity when both are even or both are odd.)
   (b) The relation "divides" on the set of natural numbers.
   (c) The relation □ on the whole numbers defined by $a \,\square\, b$ if and only if $ab > 0$.
5. Which of the relations in problem 4 are symmetric? If a relation is not symmetric, explain why.
6. Which of the relations in problem 4 is transitive? If a relation is not transitive, explain why.

7. The traditional Japanese children's game *Jan-Kem-Po* is also known as Rock-Paper-Scissors. After one player chants *Jan-Kem-Po,* each of the two players display hand signs at the same time. The winner is shown by the arrows: Rock breaks scissors, scissors cut paper, and paper covers rock.

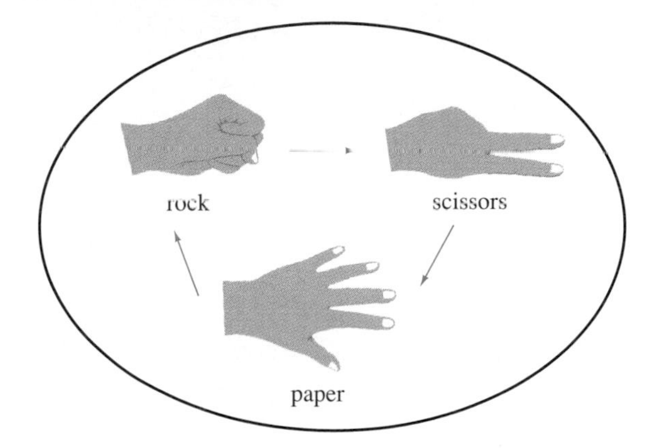

(a) What properties does the relation have, if any?

(b) Suppose the rules are changed to rock breaks paper, thereby reversing the arrow between rock and paper. Why is the game no longer interesting?

(c) What are the properties of the relation as modified in part (b)?

8. Determine if the relations given by the following arrow diagrams are reflexive, symmetric, or transitive.

(a)

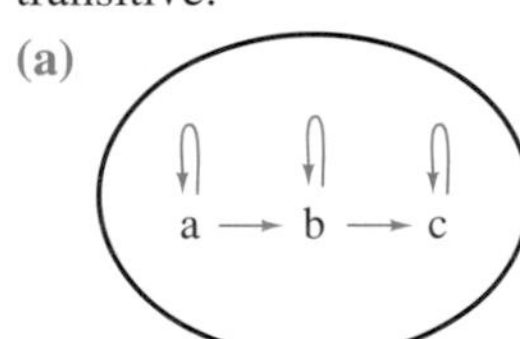

(b)

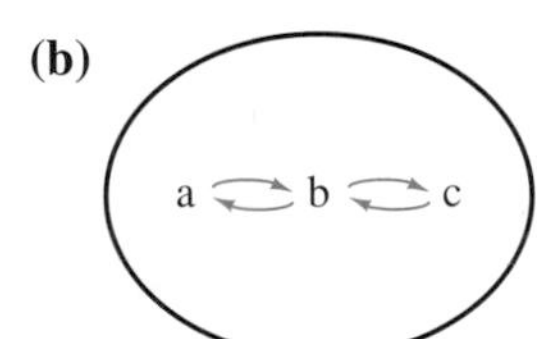

(c)

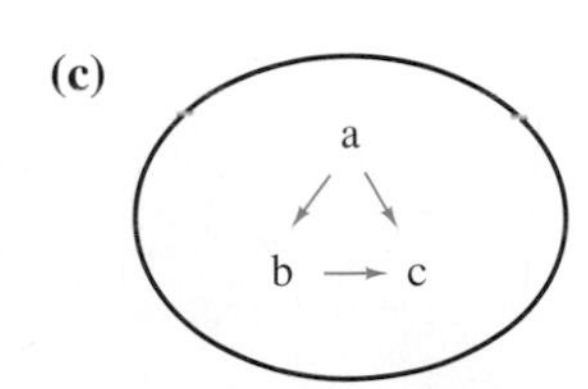

9. Draw an arrow diagram (similar to those in problem 8) that represents a relation that is

(a) reflexive and symmetric, but not transitive.

(b) reflexive and transitive, but not symmetric.

(c) symmetric and transitive, but not reflexive.

10. Which of these relations are equivalence relations? If a relation is not an equivalence relation, explain why.

(a) The less than or equal relation, $\leq$, on the set of whole numbers.

(b) The congruence relation, $\cong$, on the set of geometric figures.

(c) The subset relation, $\subseteq$, for sets.

(d) The relation $\square$ on the set of natural numbers defined by $a \square b$ if, and only if, $a^2 + b^2$ is even. (For example, $10 \square 2$ since $10^2 + 2^2 = 104$ is even.)

11. Let g be the function from $S = \{0, 1, 2, 3, 4\}$ to the whole numbers W given by the formula $g(x) = 5 - 2x + x^2$.

(a) Find $g(0)$, $g(1)$, $g(2)$, $g(3)$, $g(4)$.

(b) What is the range of g?

12. Let h be the function defined by $h(x) = x^2 - 1$, where the domain is the set of real numbers (that is, the set of positive and negative decimal numbers).

(a) Find $h(2)$

(b) Find $h(-2)$

(c) If $h(t) = 15$, what are the possible values of t?

(d) Find $h(7.32)$

13. Consider the "double-square" rectangles of width x and length $2x$, as shown.

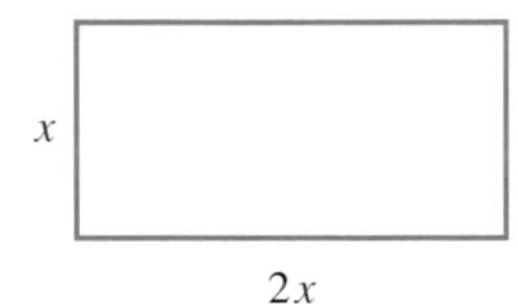

(a) Find a formula that gives the area A of a double-square of width x.

(b) Find a formula that gives the perimeter P of a double-square of width x.

(c) Use the Pythagorean theorem to obtain a formula for the diagonal D of a double-square of width x.

---

14. Che is "it" in a game of Guess My Rule. The students' inputs, and Che's outputs, are shown in the table below.

Input given to Che	5	2	4	0	−2	−5
Output reported by Che	24	3	15	−1	3	24

(a) Guess Che's rule (function), $y = C(x)$.

(b) Antonio says that Che's rule is $C(x) = x^2 - 1$, but Claudette claims that it is $C(x) = (x + 1)(x - 1)$. Who is correct, Antonio or Claudette? Explain.

15. (a) Draw the next two "picket fence" arrangements of dots to continue the pattern shown.

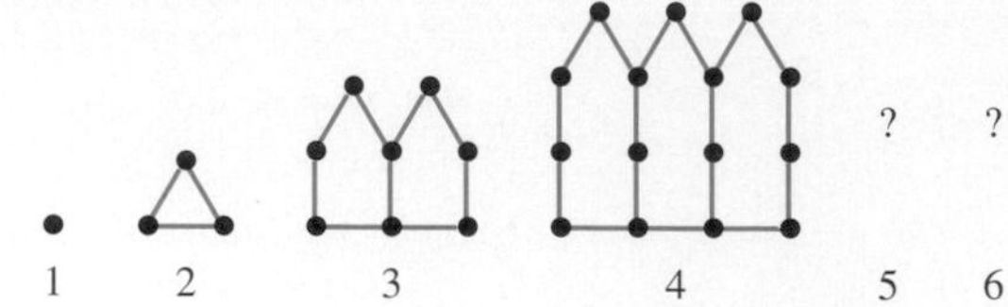

(b) Count the number of dots in each "picket-fence" arrangement. Then guess how many dots are in the seventh and eighth arrangements.

(c) Guess a formula $d = f(n)$ that gives the number of dots, d, in the nth "picket-fence."

Thinking Critically

16. Let □ denote a symmetric and transitive relation on a set S.

(a) If $a \square b$, explain why you also know that $a \square a$ and $b \square b$.

(b) Why doesn't it follow from part (a) that □ is necessarily reflexive? You may want to review your answer to problems 4(c) and 9(c).

17. Define the relation ≡ on the set of whole numbers, W, by the condition $m \equiv n$ if, and only if, both m and n have the same remainder when divided by 3.

(a) Check that $19 \equiv 31$ and $28 \not\equiv 15$.

(b) Show that $E_0 = \{n \mid n = 3k \text{ and } k \in W\} = \{0, 3, 6, \ldots\}$ is an equivalence class of the relation ≡.

(c) Show that $E_1 = \{n \mid n = 3k + 1 \text{ and } k \in W\} = \{1, 4, 7, \ldots\}$ is an equivalence class of the relation ≡.

(d) There is one more equivalence class of ≡. What is it?

18. Define the relation □ on the set $S = W \times W$ by

$$(m, n) \square (p, q) \text{ if, and only if, } m + q = n + p,$$

where m, n, p, and q are whole numbers.

(a) Check that $(5, 8) \square (11, 14)$ and $(4, 9) \not\square (6, 17)$.

(b) Verify that □ is an equivalence relation on $W \times W$.

(c) Show that $E_0 = \{(0, 0), (1, 1), (2, 2), \ldots, (m, m), \ldots\}$ is an equivalence class of □.

(d) Describe the equivalence class E_1 that contains the element $(1, 0)$.

19. Define the relation △ on the set $S = N \times N$ by

$$(m, n) \triangle (p, q) \text{ if, and only if, } mq = np,$$

where m, n, p, and q are natural numbers.

(a) Check that $(12, 3) \triangle (16, 4)$ and $(5, 11) \not\triangle (6, 12)$.

(b) Verify that △ is an equivalence relation on $N \times N$.

(c) Describe the equivalence class that contains $(1, 2)$.

(d) Describe the equivalence class that contains $(3, 1)$.

Thinking Cooperatively

20. Beginning with $F_1 = 1$ and $F_2 = 1$, the successive terms in the Fibonacci sequence are defined by the recursive formula $F_{n+2} = F_{n+1} + F_n$, $n = 1, 2, \ldots$. Investigate the following "tribonacci" sequence: $T_1 = 1$, $T_2 = 2$, $T_3 = 4$, $T_{n+3} = T_{n+2} + T_{n+1} + T_n$. For example, $T_4 = T_3 + T_2 + T_1 = 4 + 2 + 1 = 7$.

(a) Complete this table of values of T_n.

n	1	2	3	4	5	6	7	8	9	10
T_n	1	2	4	7						

(b) Find the missing values in this table, *without* finding $T_{11}, \ldots, T_{19}$. Explain how you obtain the function values.

n	20	21	22	23	24
T_n	121415	223317		755476	

21. In the HANDS ON activity introducing this chapter, trains of length n were formed with Cuisenaire® rods. You may have discovered that the Fibonacci sequence helps answer this question: *How many trains of length* n *can be formed from white (length 1) and red (length 2) cars?* The answer is the $(n + 1)$st Fibonacci number F_{n+1}. For example, trains of length $n = 4$ can be formed in $F_{4+1} = F_5 = 5$ ways, namely WWWW, WWR, WRW, RWW, and RR. Now suppose light green cars of length 3 are also allowed. Let T_n denote the number of ways to form trains of length n using W (white), R (red), and G (light green) cars. For example, WWW, RW, WR, and G are all trains of length $n = 3$.

(a) Use Cuisenaire® rods (or form letter strings such as GWWR) to fill in this table.

Length of train, n	1	2	3	4	5
Number of trains, T_n	1	2			

(b) Explain why $T_{n+3} = T_{n+2} + T_{n+1} + T_n$. (*Hint:* Add white, red, and green cabooses to the respective trains of length $n + 2$, $n + 1$ and n.)

(c) Find T_6, T_7, and T_8.

22. Shaquita noticed the following pattern when summing the squares of the successive natural numbers.

$S_1 = 1^2 = 1 = 2 \cdot 3 \cdot 4/24$
$S_2 = 1^2 + 2^2 = 5 = 4 \cdot 5 \cdot 6/24$
$S_3 = 1^2 + 2^2 + 3^2 = 14 = 6 \cdot 7 \cdot 8/24$

(a) Does Shaquita's pattern hold for S_5 and S_6?

(b) Find a formula that gives Shaquita's function S_n.

23. The 3×3 square array shown contains nine unit 1×1 squares, four 2×2 squares, and one 3×3 square, making $1 + 4 + 9 = 14$ squares in all.

(a) Fill in this partially completed table counting the squares of various sizes in a given square array.

Size of the Array	Number of Squares of Size 1×1	2×2	3×3	4×4	5×5	Total Number of Squares
1×1	1					1
2×2	4	1				5
3×3	9	4	1			14
4×4						
5×5						

(b) Make a conjecture about the number of squares, S_6, that can be found in a 6×6 array.

(c) Give a formula for S_n, the number of squares in an $n \times n$ array.

24. A **cevian** of a triangle (named for the Italian geometer Giovanni Ceva, ca. 1647–1736) is any line segment extending from a vertex to the opposite side.

(a) If p cevians are drawn from vertex A of triangle ABC, how many regions are formed inside the triangle? Express your answer as a function of the variable p.

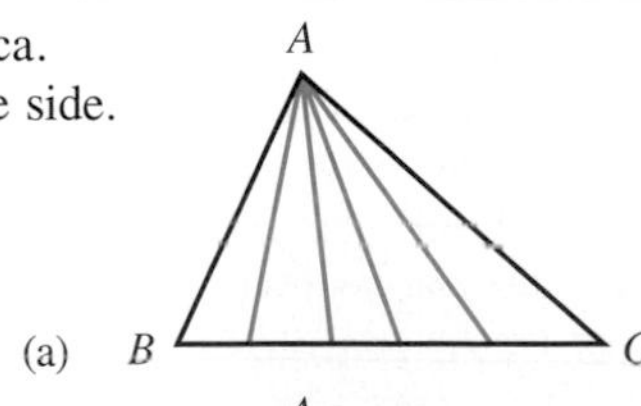

(b) Suppose triangle ABC is divided into pieces by p cevians from A and q cevians from B. How many regions are formed inside the triangle? Express your answer as a formula in the variables p and q.

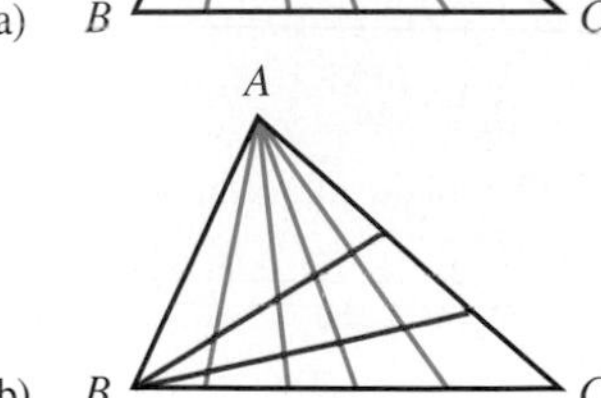

(c) Suppose p, q, and r cevians are drawn from the respective vertices A, B, and C of triangle ABC. Guess a formula in the three variables p, q, and r that gives the number of regions the cevians form inside the triangle. Assume that no three cevians intersect at the same point inside the triangle. (*Hint:* How many *additional* regions are created by each cevian drawn from vertex C?) (This is a problem of I. M. Yaglom.)

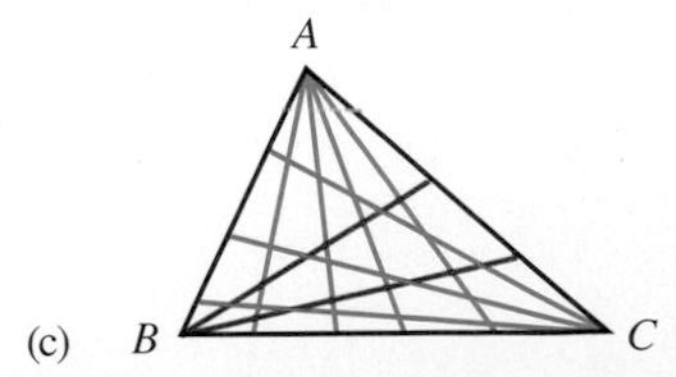

Making Connections

Linear functions. *A function f is called* ***linear*** *if there are two numbers m and b so that* $f(x) = mx + b$ *for all x in the domain of the function. Some useful linear functions are investigated in problems 25 through 29.*

25. **Hooke's Law.** If a (small) weight w is suspended from a spring, the length L of the stretched spring is a linear function of w. Find m and b in the formula $L = mw + b$ if the unstretched spring has length 10″ and a weight of 2 pounds stretches it to 14″.

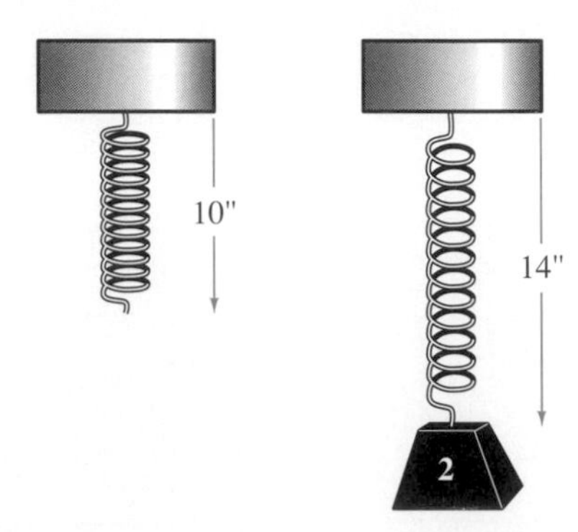

26. **Temperature Conversion.** The temperature at which water freezes is 32 degrees Fahrenheit and 0 degrees Celsius, and the temperature at which water boils is 212 degrees Fahrenheit and 100 degrees Celsius. Find the constants m and b in the formula $F = mC + b$ to express the Fahrenheit temperature F as a function of the Celsius temperature C.

27. **Speed of a Dropped Object.** Neglecting the resistance of air, the (downward) speed of an object t seconds after being dropped is given by $v = 32t$ feet per second. How many seconds does it take an object to reach a downward speed of 100 miles per hour? Recall that there are 5280 feet in one mile, since you'll need to express 100 mph in feet per second.

28. **Straight Line Depreciation.** Suppose a car originally valued at \$18,500 is worth \$10,400 after 5 years. Express the value V of the car as a linear function of the age t of the car, where t is measured in years. Use the formula to determine the value of the car after three years of service. (*Hint:* m is negative.)

29. **The Lightning Distance Function.** The speed of sound is about 760 miles per hour. Assuming the lightning flash takes no appreciable time to be seen, and the peal of thunder is heard after t seconds, show that $d = t/5$ gives the approximate distance d (in miles) to the strike.

30. **Taxi Fare.** The Evergreen Taxi Company in Pullman, Washington, charges \$2.15 for the first mile (or fraction thereof), and \$1.25 for each additional mile. What is the charge for these trips?
   (a) From Pullman to the airport, which is 4.3 miles.
   (b) From Pullman to Moscow, Idaho, a 7.8 mile trip.

31. The U.S. Post Office calls the perimeter of a cross-section of a package the "girth." The combined girth plus length of a package cannot exceed 108 inches.
   (a) Find a formula for the girth, g, of a rectangular box whose cross-section is of width w and height h.
   (b) What is the maximum length of a rectangular box whose cross-section is 10″ by 18″ that can be mailed at the U.S. Post Office?

32. **First Class Postage.** First class postage in the United States is 32 cents for the first ounce, and 23 cents for each additional ounce or fraction of an ounce. What does it cost to mail a package weighing (a) 1 1/2 ounces? (b) half a pound? (c) 2 ounces?

33. **Tax Tables.** A portion of the 1992 Tax Table is shown below.

(a) Demitrius had a taxable income of $35,241 in 1992. If Demitrius is single, what is his tax?

(b) How many of the last $100 of Demitrius' 1992 earnings were paid in income tax?

(c) Kay and Andres filed a joint return on $35,830 of taxable income. What is their tax?

(d) How much of Andres and Kay's last $100 of income went to taxes?

(e) Suppose Kay discovers an arithmetic error, and her joint return with Andres must really report a taxable income of $35,930. How much of the additional $100 of income will go to income tax?

If line 37 (taxable income) is—		And you are—			
At least	But less than	Single	Married filing jointly *	Married filing sepa-rately	Head of a house-hold
		Your tax is—			
35,000					
35,000	35,050	7,019	5,254	7,480	6,070
35,050	35,100	7,033	5,261	7,494	6,084
35,100	35,150	7,047	5,269	7,508	6,098
35,150	35,200	7,061	5,276	7,522	6,112
35,200	35,250	7,075	5,284	7,536	6,126
35,250	35,300	7,089	5,291	7,550	6,140
35,300	35,350	7,103	5,299	7,564	6,154
35,350	35,400	7,117	5,306	7,578	6,168
35,400	35,450	7,131	5,314	7,592	6,182
35,450	35,500	7,145	5,321	7,606	6,196
35,500	35,550	7,159	5,329	7,620	6,210
35,550	35,600	7,173	5,336	7,634	6,224
35,600	35,650	7,187	5,344	7,648	6,238
35,650	35,700	7,201	5,351	7,662	6,252
35,700	35,750	7,215	5,359	7,676	6,266
35,750	35,800	7,229	5,366	7,690	6,280
35,800	35,850	7,243	5,377	7,704	6,294
35,850	35,900	7,257	5,391	7,718	6,308
35,900	35,950	7,271	5,405	7,732	6,322
35,950	36,000	7,285	5,419	7,746	6,336
36,000					

Using a Calculator

34. **Tax rate schedules.** Tax Tables (use problem 33) apply to incomes less than $100,000. Those with higher taxable incomes determine their tax by a Tax Rate Schedule. For example, a single person would use Schedule X below for 1992. Notice that the schedule is shown for *all* levels of income, so that all taxpayers can see the rate that applies to them.

Schedule X—Use if your filing status is **Single**

If the amount on Form 1040, line 37, is: Over—	But not over—	Enter on Form 1040, line 38	of the amount over—
$0	$21,450	.......15%	$0
21,450	51,900	$3,217.50 + 28%	21,450
51,900	....	11,743.50 + 31%	51,900

(a) If Demitrius, a single person with $35,241 of taxable income, could use Schedule X, what would his tax be?

(b) Cyndi is a single person whose 1992 taxable income is $134,520. What is her income tax?

35. **The Windchill Index.** In windy cold weather, the increased rate of heat loss makes the temperature feel colder than the actual temperature. To describe an equivalent temperature that more closely matches how it "feels," weather reports often give a windchill index, WCI. The WCI is a function of both the temperature F (in degrees Fahrenheit) and the wind speed v (in miles per hour). For wind speeds v between 4 and 45 miles per hour, the WCI is given by the formula

$$WCI = 91.4 - \frac{(10.45 + 6.69\sqrt{v} - 0.447v)(91.4 - F)}{22}$$

(a) Verify that a temperature of 10°F in a wind of 20 miles per hour produces a windchill index of −25°F.

(b) A weather forecaster claims that a wind of 36 miles per hour has resulted in a WCI of −50°. What is the actual temperature to the nearest degree?

For Review

36. In each of 10,000 boxes of Cracker Smacks one of the following prizes is placed: a ring, a top, a ball, or a marble. How many boxes must be purchased to be absolutely certain you receive a prize of each type? Assume there are equally many prizes of each type, one per box.

37. Let $A = \{a, b, c, d\}$, $B = \{b, d, e, f\}$, and $C = \{a, b, e, g, h\}$. Draw a Venn diagram of A, B, and C in the universe of the first half of the alphabet, $U = \{a, b, \ldots, l, m\}$.

38. For the sets listed in problem 37, give
(a) $n(A \cup B)$ (b) $n(B \cap C)$
(c) $n(\overline{B})$ (d) $n(C - A)$.

39. What fact family contains $42 \div 7 = 6$?

EPILOGUE The Pythagorean View of Number

No figure of Ionian science was more influential than Pythagoras, born about 560 B.C. on the island of Samos. As a young man he studied mathematics with Thales and then continued his studies during his travels to Egypt and Babylonia. Later he founded the famous Pythagoras school at Croton (now Crotona, in southern Italy), dedicated to mathematical, ethical, political, and philosophical deliberations.

To the Pythagoreans, whose mathematics was a curious blend of fact and fantasy, it soon became the case that "all things are numbers." In his *Metaphysics,* Aristotle writes that the Pythagoreans not only advanced mathematics, ". . . but saturated with it, they fancied that the principle of mathematics was the principle of all things."

In the mystical numerology of the Pythagoreans, the odd numbers were male and the even numbers female. Unity and omnipotence were associated with 1, diversity was associated with 2, and 3 combined both unity and diversity. Justice was associated with 4. Marriage corresponded to 5, combining the male and female numbers, 2 and 3. The number 6 was considered perfect since it was the sum of all its divisors except for 6 itself. That is,

$$6 = 1 + 2 + 3.$$

The next perfect number is 28 since

$$28 = 1 + 2 + 4 + 7 + 14.$$

Numbers a and b were "amicable" or friendly if the sum of the divisors of a, excluding a, was equal to b and the sum of the divisors of b, excluding b, was equal to a. Thus, 220 and 284 were amicable since the divisors of 220 excluding 220 are 1, 2, 4, 5, 10, 11, 20, 22, 44, 55, and 110 which sum to 284, and the divisors of 284 excluding 284 are 1, 2, 4, 71, and 142 which sum to 220.

In spite of the mystical nature of much of the mathematics of the Pythagoreans, they did much to enhance our understanding of numbers and of mathematics generally.

CLASSIC CONUNDRUM A Problem from the Dark Ages

Alcuin of York compiled *Problems for the Quickening of the Mind* in the eighth century. Can you solve this problem from his collection?

Thirty flasks, ten full, ten half-empty, and ten entirely empty are to be divided equally among three sons so that flasks and contents should be shared equally. How can this be done?

CHAPTER 2 SUMMARY

Key Concepts

1. Sets and operations on sets
   - **(a)** A set is a collection of objects from a stated universe. A set can be described either verbally, by a list, or with set builder notation.
   - **(b)** Sets can be visualized with Venn diagrams, in which the universe is a rectangle and closed loops correspond to sets. Elements of the set are associated with points within the loop.
   - **(c)** Sets are combined and related by set complement $\overline{A}$, subset $A \subseteq B$, proper subset $A \subset B$, intersection $A \cap B$, union $A \cup B$, set difference $A - B$, and the Cartesian product $A \times B$.
2. Sets, counting, and the whole numbers
   - **(a)** Numbers are used in three ways: to name (nominal numbers), to indicate position in an order (ordinal numbers), and to indicate how many elements are in a set (cardinal numbers).
   - **(b)** Two sets that have a one-to-one correspondence of their elements are equivalent sets. A set that is equivalent to any of its proper subsets is an infinite set; sets which are not infinite are finite sets.
   - **(c)** The whole numbers are the cardinal numbers of the finite sets, with zero the cardinal number of the empty set. The whole numbers can be represented and visualized by a variety of manipulatives and diagrams, including tiles, cubes, number strips, rods, and the number line.
   - **(d)** The whole numbers are ordered, so that m is less than n if a set with n elements has a proper set which matches a set with m elements.
3. Addition and subtraction of whole numbers
   - **(a)** Addition of whole numbers is defined in the set model by $a + b = n(A \cup B)$, where $a = n(A)$, $b = n(B)$ and A and B are disjoint finite sets. Addition can also be visualized on the number line with the measurement model.
   - **(b)** Addition of whole numbers is a closed binary operation. Addition satisfies the commutative property $a + b = b + a$ and the associative property $a + (b + c) = (a + b) + c$. Zero is an additive identity: $a + 0 = 0 + a = a$.
   - **(c)** Subtraction of whole numbers is defined by $a - b = c$, where c is the unique whole number for which $a = b + c$. There are four conceptual models: take-away, missing addend, comparison and number-line.

4. Multiplication and division of whole numbers.

(a) Multiplication is defined as a repeated addition, so that $a \cdot b = b + b + \cdots + b$, where there are a addends. Alternatively, multiplication is defined by the Cartesian product, $a \cdot b = n(A \times B)$, and can be visualized as the number of objects in an a row by b column rectangular array.

(b) Multiplication is closed, commutative $a \cdot b = b \cdot a$, associative $a \cdot (b \cdot c) = (a \cdot b) \cdot c$, distributes over addition, $a \cdot (b + c) = a \cdot b + a \cdot c$; has 1 as a multiplicative identity, $1 \cdot a = a \cdot 1 = a$; and satisfies the multiplication by zero property $0 \cdot a = a \cdot 0 = 0$.

(c) Division $a \div b$, where $b \neq 0$, is defined in whole numbers if, and only if, $a = b \cdot c$ for a unique whole number c. Division is modeled as repeated-subtraction (grouping), a partition (sharing), and as the missing factor c in the equation $a = b \cdot c$.

(d) The division algorithm extends the division operation to all nonzero whole number divisors by allowing remainders. By the division algorithm, given whole numbers a and b, $b \neq 0$, there is a quotient q and remainder r so that $a = b \cdot q + r$, $0 \leq r < b$.

5. Relations and functions

(a) A relation from set S to set T is a subset of ordered pairs from $S \times T$. Relations are commonly denoted by a symbol, such as $\square$, $<$, $\leq$, $\subseteq$, $\subset$, $=$, $\sim$. Arrow diagrams provide a visualization of a relation.

(b) A relation $\square$ on a set S is reflexive if $a \square a$ for all $a \in S$, is symmetric if $a \square b$ implies that $b \square a$, and is transitive if $a \square b$ and $b \square c$ implies that $a \square c$.

(c) A relation that is reflexive, symmetric, and transitive is an equivalence relation. An equivalence relation partitions S into disjoint subsets called equivalence classes.

(d) A function f from S to T is a relation which assigns to each element $x \in S$ precisely one element $f(x)$ of T. Functions are often described by formulas and tables, and can be visualized with graphs.

Vocabulary and Notation

Section 2.1

Universe U
Element of $\in$
List in braces $\{a, b, \ldots\}$
Set builder notation $\{x \mid x$ is (condition)$\}$
Venn diagram
Complement $\overline{A}$
Subset $A \subseteq B$
Equal sets $A = B$
Proper subset $A \subset B$
Intersection $A \cap B$
Disjoint sets $A \cap B = \emptyset$
Empty set $\emptyset$
Union $A \cup B$
Difference $A - B$
Cartesian product $A \times B$

Section 2.2

Nominal (naming, identification) number
Ordinal number
Cardinal number
One-to-one correspondence
Equivalent sets $A \sim B$
Finite set
Infinite set
Whole number $n(A)$
Zero $0 = n(\emptyset)$
Number tiles, cubes, strips, rods
Number line
Less than: $a < b$, greater than: $a > b$

Section 2.3

Binary operation
Addition (sum) $a + b$

Addend
Summand
Set model: $a + b = n(A \cup B)$ for $A \cap B = \emptyset$
Measurement (number-line) model
Properties of addition
Closure, commutative, associative, additive identity property of zero
Subtraction (difference) $a - b$
Minuend
Subtraend
Take-away model
Missing addend model
Comparison model
Measurement (number-line) model
Addition-subtraction fact family

Section 2.4

Multiplication (product) $a \cdot b$
Factor
Repeated addition model
Array model
Cartesian product model
Properties of multiplication
Closure, commutative, associative property
Multiplicative identity property of 1
Multiplication by zero property
Distributivity over addition

Division $a \div b$ (or a/b)
Repeated-subtraction (grouping) model
Partition (sharing) model
Missing Factor model
Dividend
Divisor
Multiplication-division fact family
Division with remainders
Division algorithm $a \div b = q$ R r or $a = bq + r, 0 \le r < b$
Power operation
Base and exponent a^m

Section 2.5

Relation, $\square$
Arrow diagram
Reflexive property
Symmetric property
Transitive property
Equivalence relation
Equivalence class
Function, f
Image (or value) of f at x, $f(x)$
Domain of a function
Range of a function

CHAPTER REVIEW EXERCISES

Section 2.1

1. Let $U = \{n \mid n$ is a whole number and $2 \le n \le 25\}$
$S = \{n \mid n \in U$ and n is a square number$\}$
$P = \{n \mid n \in U$ and n is a prime number$\}$
$T = \{n \mid n \in U$ and n is a power of $2\}$
(a) Write S, P, T in listed form.
(b) Find the following sets: $\overline{P}$, $S \cap T$, $S \cup T$, $S - T$
2. Draw a Venn diagram of the sets S, P, T in problem 1.
3. File boxes come in two sizes—standard and legal—and four colors—grey, tan, blue, yellow. Use the Cartesian product of sets to explain what choices are available.

Section 2.2

4. Let $S = \{$s, e, t$\}$ and $T = \{$t, h, e, o, r, y$\}$. Find $n(S)$, $n(T)$, $n(S \cup T)$, $n(S \cap T)$, $n(S - T)$, $n(T - S)$.
5. Show that the set of square natural numbers less than 101 is in one-to-one correspondence with the set $\{$a, b, c, d, e, f, g, h, i, j$\}$.
6. Show that the set of cubes $\{1, 8, 27, \ldots\}$ is an infinite set.

Section 2.3

7. Explain how to illustrate $5 + 2$ with
(a) the set model of addition;
(b) the number line (measurement) model of addition.
8. What properties of whole number addition are illustrated on the number lines below?

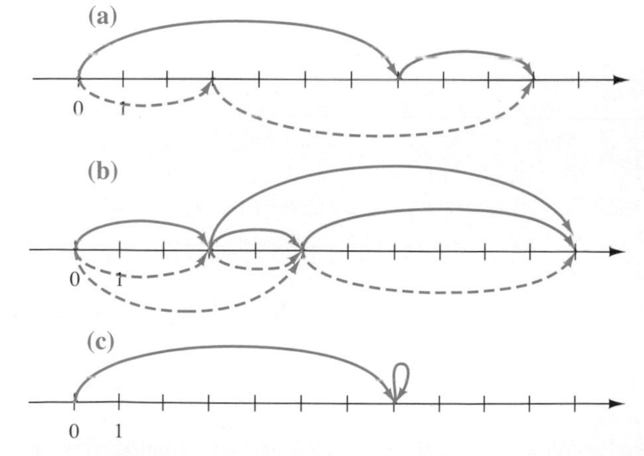

9. Draw figures that illustrate $6 - 4$
(a) using sets
(b) using the number line.

Section 2.4

10. Whiffle balls 2″ in diameter are packed individually in cubical boxes and then in cartons of 3 dozen balls. What are the dimensions of a suitable rectangular carton?
11. A drill sergeant lined up 92 soldiers in rows of 12, except for a partial row in the back. How many rows are formed? How many soldiers are in the back row?
12. Draw figures that illustrate the division problem $15 \div 3$, using these models:
    (a) repeated-subtraction (grouping objects in a set)
    (b) partition (sharing objects)
    (c) missing factor (rectangular array)

Section 2.5

13. Let S be the set of all people in the world today. Describe the properties of the following relations on S:
    (a) "Is an ancestor of"
    (b) "Lives in the same country as"
    (c) "Speaks a common language with"
14. Let $S = A \cup B \cup C$, where A, B, C are nonempty subsets of S and each element $x \in S$ belongs to precisely one of A, B, or C. Define the relation $\square$ on S by $x \square y$ if x and y belong to the same subset—A, B, or C—of S. Show $\square$ is an equivalence relation on S, and list the equivalence classes.
15. Let f be a function defined by the formula $f(x) = 2x(x - 3)$.
    (a) Find $f(3), f(0.5), f(-2)$
    (b) If $f(x) = 0$, what are the possible values of x?
16. When light passes into two face-to-face layers of glass, it is either reflected or transmitted when it strikes the glass surfaces. There is 1 path with no reflections, 2 paths with 1 reflection, and 3 paths with 2 reflections.

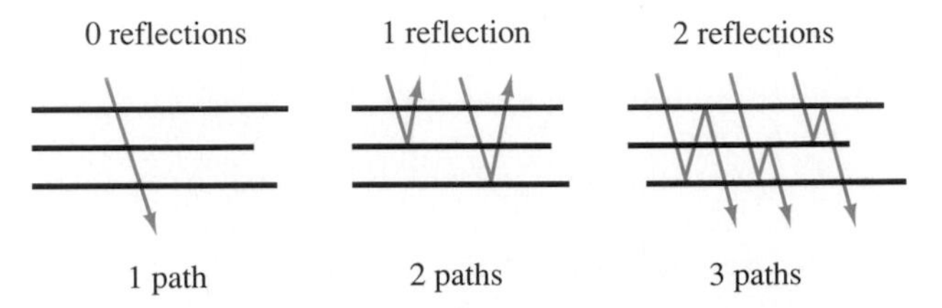

    (a) Show that there are 5 paths with 3 internal reflections.
    (b) Conjecture a formula which gives the number of paths which have n internal reflections.

CHAPTER TEST

1. The following sentence contains three types of numbers:
   "On the *15th* of April, Joe Taxpayer sent in form *1040* and a money order for *\$253*."
   Name and describe the three kinds of numbers.
2. Shade regions in Venn diagrams which correspond to the following sets.
   (a) $A \cap (B \cup C)$
   (b) $A \cup B$
   (c) $(A - B) \cup (B - A)$
3. Suppose $A \cap B = A$. What can you say about $A - B$?
4. Let $A = \{w, h, o, l, e\}$, $B = \{n, u, m, b, e, r\}$, $C = \{z, e, r, o\}$. Find:
   (a) $n(A \cup B)$ (b) $n(B - C)$
   (c) $n(A \cap C)$ (d) $n(A \times C)$
5. What property of whole numbers justifies the following equalities?
   (a) $4 + (6 + 2) = (4 + 6) + 2$
   (b) $8 \cdot (4 + x) = 8 \cdot 4 + 8x$
   (c) $3 + 0 = 0 + 3$
   (d) $2 \cdot (8 \cdot 5) = (2 \cdot 8) \cdot 5$
6. Althea made 5 gallons of root beer and wishes to bottle it in 10 ounce bottles.
   (a) How many bottles does she need? Remember that a quart contains 32 ounces.
   (b) Discuss which model of division—grouping or sharing—corresponds best to your answer for part (a).
7. Let A and B be two sets in the universe $U = \{a, b, c, \ldots, z\}$. If $n(A) = 12$, $n(B) = 14$, and $n(A \cup B) = 21$, find $n(A \cap B)$ and $n(\overline{A \cap B})$.
8. Let $n(A \times B) = 21$. What are the possible number of elements in the set A?

9. What operation on whole numbers is being illustrated in the following diagrams?
   (a)

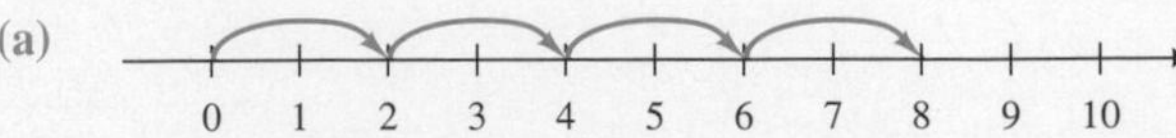

(b)

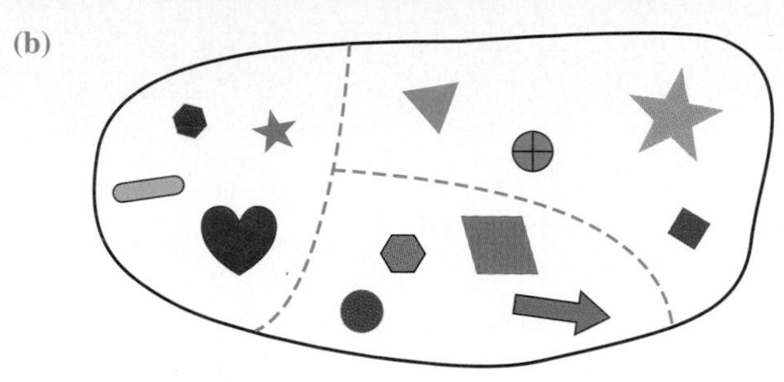

(c)

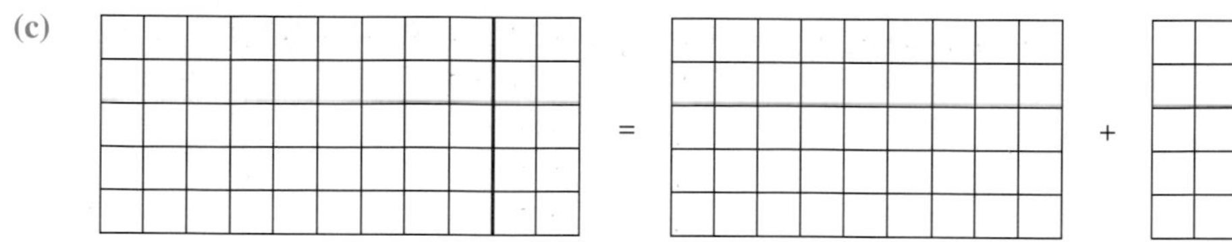

(d)

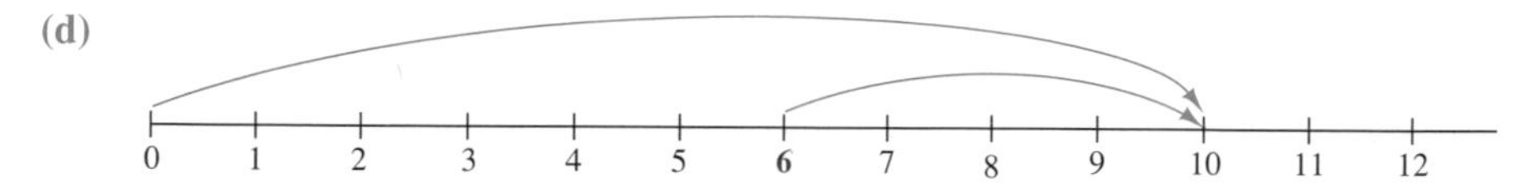

10. Let $S = \{1, 2, 4, 8, 16, \ldots\}$ denote the set of powers of 2.
    (a) Is S closed under multiplication?
    (b) Is S closed under addition?
    Explain why you answered as you did.
11. Show how to illustrate the following properties of whole number multiplication with the rectangular array model.
    (a) $2 \cdot (4 + 3) = 2 \cdot 4 + 2 \cdot 3$
    (b) $2 \cdot 5 = 5 \cdot 2$
12. Discuss which conceptual model of subtraction you feel best corresponds to each of the following problems.
    (a) On Monday, Roberto hiked 11 miles to the lake. On Friday at noon he had hiked 6 miles back down the trail. How much farther does Roberto have to hike to get back to the trailhead?
    (b) Kerri has 56 customers on her paper route, but her manager only left her 48 papers. How many extra papers should she have brought over before starting her delivery?
    (c) All but 7 of the 3 dozen picnic plates were used. How many people came to the picnic?
13. Let S be a set of points in the plane, with an origin chosen as point O. Let two points P, Q in the plane S be related if they are the same distance from the origin O. That is, $P \square Q$ if $OP = OQ$, where OP denotes the distance from O to P.
    (a) Show $\square$ is an equivalence relation on S.
    (b) Describe all of the equivalence classes of $\square$.
14. Let S be the set of cities in North America. Define the relation $\square$ by $a \square b$ if a "is no farther from the equator than" b. Describe the properties of $\square$ on S.
15. An arrow diagram on the set {a, b, c, d} is shown below.

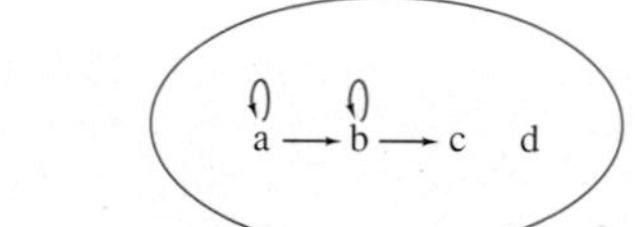

Using as few new arrows as possible, append new arrows so the modified diagram represents
    (a) a reflexive relation; (b) a symmetric relation;
    (c) a transitive relation; (d) an equivalence relation.
    Show separate diagrams for each part.
16. Let f be the function given by $f(x) = x^2 - 4x + 5$ on the domain {1, 2, 3, 4, 5}. Give the range of the function.
17. A bee starts in cell 1, and moves by crawling over cell walls. Assume that the bee always moves to the right, from one cell to one of the two adjacent cells.

There is one path to cell 2, two paths to cell 3, and three paths to cell 4.
    (a) How many paths are there to cell 5?
    (b) Let F be the function for which $F(n)$ is the number of paths from cell 1 to cell n. Explain why $F(n + 2) = F(n + 1) + F(n)$ for $n = 1, 2, 3, \ldots$.
    (c) How many paths can the bee follow to cell 12?

3 Numeration and Computation

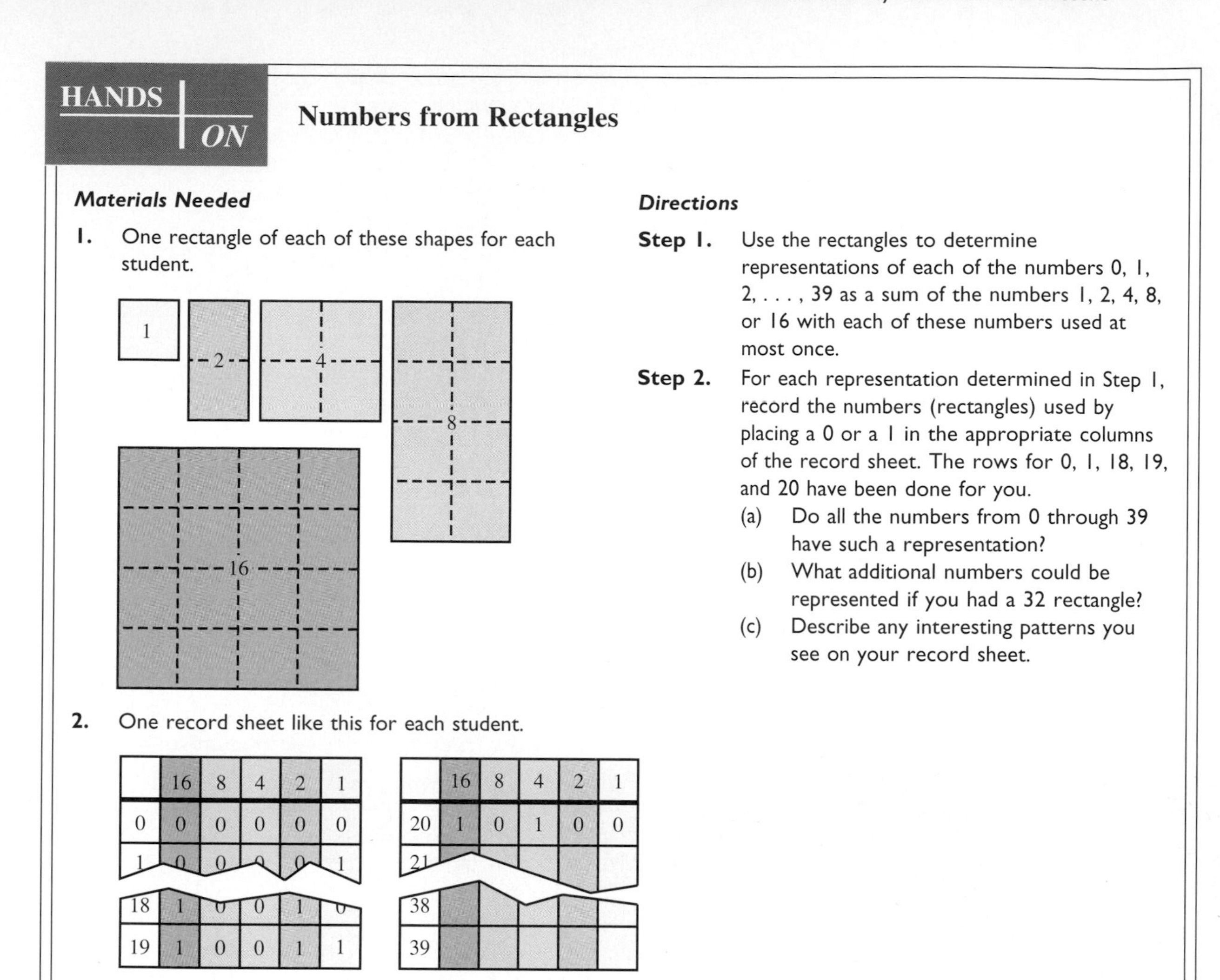

HANDS ON

Numbers from Rectangles

Materials Needed

1. One rectangle of each of these shapes for each student.

2. One record sheet like this for each student.

Directions

Step 1. Use the rectangles to determine representations of each of the numbers 0, 1, 2, . . . , 39 as a sum of the numbers 1, 2, 4, 8, or 16 with each of these numbers used at most once.

Step 2. For each representation determined in Step 1, record the numbers (rectangles) used by placing a 0 or a 1 in the appropriate columns of the record sheet. The rows for 0, 1, 18, 19, and 20 have been done for you.

(a) Do all the numbers from 0 through 39 have such a representation?

(b) What additional numbers could be represented if you had a 32 rectangle?

(c) Describe any interesting patterns you see on your record sheet.

CONNECTIONS

Counting and Calculating

In Chapter 2, we introduced whole numbers as the cardinal numbers of finite sets. For example, if A is the set shown, then $n(A) = 5$; that is, the number five is an abstract idea which represents the count of the elements in A and in each set equivalent to A.

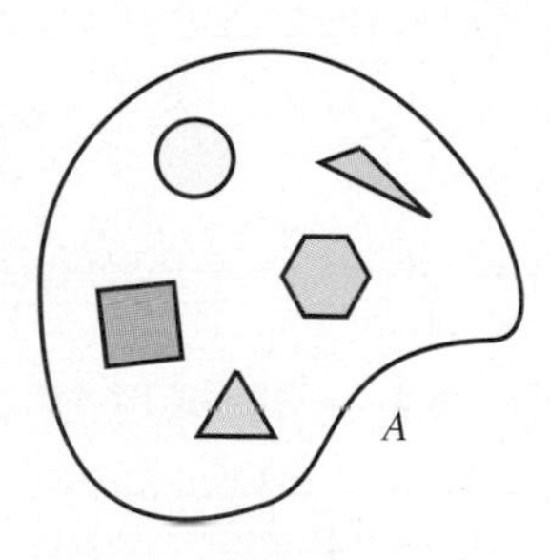

Historically, the notion of number developed over many years from people's need to count objects—sheep, goats, arrows, warriors, beads, and so on. Indeed, the need to count collections of concrete objects to develop an understanding of numbers was not just true historically; it remains equally true for individuals today. Research has shown that it is essentially impossible for children to understand the abstract notion of "five" without counting many sets of five objects—five beans, five fingers, five pennies, five concrete objects of any kind. Thus, the need to use a variety of manipulative devices in the elementary school classroom to help children develop an understanding of numerousness—or number—cannot be ignored.

In this chapter we consider how we write numbers and how this is reflected in the methods used to perform calculations both by hand and with calculators and computers. We begin with a brief look at some of the earlier numeration systems and then consider the Hindu-Arabic system we use today.

It is essential for children's understanding that these notions be introduced with extensive use of hands-on manipulative devices and copious pictorial representations. To do otherwise is to condemn students to rote memorization and meaningless manipulation of symbols. Put positively, students who are allowed to work with appropriate devices—to handle, to manipulate, and to experience—are much more likely to develop real understanding of concepts and skills, and also to develop enthusiasm for the study of mathematics.

3.1 Numeration Systems Past and Present

To appreciate the power of the Hindu-Arabic numeration system, or decimal system as we call it, it is important to know something about numeration systems of the past. Just as the idea of number historically arose from the need to determine "how many," the demands of commerce in an increasingly sophisticated society stimulated the development of convenient symbolism for writing numbers and methods for calculating. The symbols for writing numbers are called **numerals** and the methods for calculating are called **algorithms.** Taken together, any particular system of numerals and algorithms is called a **numeration system.**

The earliest means of recording numbers consisted of creating a set of tallies—marks on stone, stones in a bag, notches in a stick—one-for-one for each item being counted. Indeed, the original meaning of the word tally was a stick with notches cut into it to record debts owed or paid. Often such a stick was split in half with one half going to the debtor and the other half to the creditor. It is still common practice today to keep count by making tallies or marks with the minor but useful refinement of marking off the tallies in groups of five. Thus,

is much easier to read as twenty-three than

But such systems for recording numbers were much too simplistic for large numbers and for calculating.

The Egyptian System

As early as 3400 B.C., the Egyptians developed a system for recording numbers on stone tablets using hieroglyphics. This system was based on 10, as is our modern system, and probably for the same reason. That is, we humans come with a built-in "digital" calculator, as it were, with 10 convenient keys. Of course, for the same reason, other systems were often based on 5 or 20. The French word for 80, for example, is *quatre-vingts* which literally means "four twenties" and suggests the early use of a system based on twenty. Other systems were based on two, three, and even sixty.

The Egyptians had symbols for one and the first few powers of ten, and then combined symbols to represent other numbers.

TABLE 3.1 Egyptian Symbols for One Through Nine

Egyptian Symbol	I	II	III	IIII	III II	III III	IIII III	IIII IIII	IIIII IIII
Modern Equivalent	1	2	3	4	5	6	7	8	9

TABLE 3.2 Egyptian Symbols for Powers of Ten

Power of 10	Egyptian Symbol	Description
$10^0 = 1$	𓏺	a vertical staff
$10^1 = 10$	𓎆	a heel bone or yoke
$10^2 = 100$	𓍢	a scroll or coil of rope
$10^3 = 1{,}000$	𓆼	a lotus flower
$10^4 = 10{,}000$	𓂭	a pointing finger
$10^5 = 100{,}000$	𓆛	a fish
$10^6 = 1{,}000{,}000$	𓁨	an amazed person

The Egyptian system was an *additive* system since the values of the various symbols were simply added together to obtain the desired number as the following example shows.

EXAMPLE 3.1

Writing a Number in Egyptian Notation

Write 4293 using Egyptian notation.

SOLUTION

Since this is an additive system, it suffices to write the symbols for four thousands, two hundreds, nine tens, and three ones. Referring to Tables 3.1 and 3.2, we see that 4293 is written in Egyptian symbols as

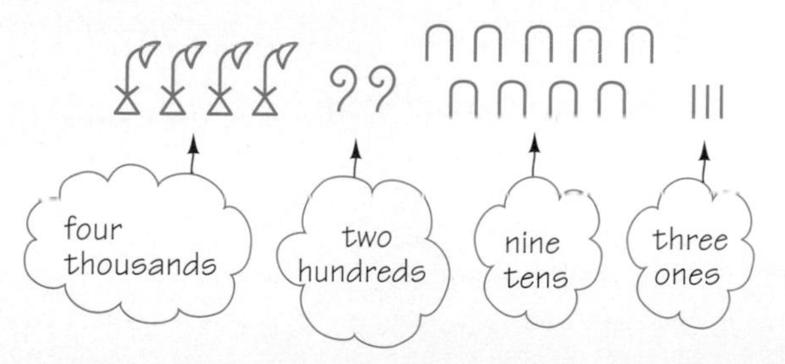

■

EXAMPLE 3.2

Converting from Egyptian to Standard Notation

What number does this Egyptian numeral represent?

SOLUTION

From Tables 3.1 and 3.2 we see that

represents 10,000

represents 2000

represents 300

represents 20

represents 7

and this all adds up to 12,327. ■

Arithmetic in the Egyptian system was cumbersome but, at least for addition and subtraction, not conceptually difficult. Since it was an additive system, to add two numbers all one had to do was to draw all the numerals used to represent both numbers and then simplify using the fact that ten 10s give 100, ten 100s give 1000, and so on. Subtraction could be done in much the same way except that it was necessary to "take away" the symbols needed to represent the number being subtracted. Of course, as with our present system, this might first require some exchanging or regrouping; that is, writing 100 as ten 10s, and so on.

EXAMPLE 3.3

Adding in the Egyptian System

Add to using Egyptian notation only.

SOLUTION

336
+125
461

■

EXAMPLE 3.4 **Subtracting in the Egyptian System**

Subtract 𓍢𓍢𓎆𓎆𓎆𓎆𓎆||| from 𓍢𓍢𓍢𓎆||||||| using Egyptian notation only.

SOLUTION

𓍢𓍢𓍢 𓎆 ||||||| − 𓍢𓍢 𓎆𓎆𓎆𓎆𓎆 |||

317
−253
64

𓍢 = 𓎆𓎆𓎆𓎆𓎆𓎆𓎆𓎆𓎆𓎆

= 𓍢𓍢 𓎆𓎆𓎆𓎆𓎆𓎆𓎆𓎆𓎆𓎆𓎆 ||||||| − 𓍢𓍢 𓎆𓎆𓎆𓎆𓎆 |||

= 𓎆𓎆𓎆𓎆𓎆𓎆 |||| ■

Multiplication and division in the Egyptian system were quite difficult since these operations had to be treated as repeated addition and repeated subtraction. This would normally involve considerable simplifying and combining of symbols and substantial mental arithmetic.

The Roman System

The Roman system of numeration is already somewhat familiar from its current usage on the faces of analog watches and clocks, on cornerstones, and on the façades of buildings to record when they were built. Originally, the system was completely additive, like the Egyptian system, with the familiar symbols as shown in Table 3.3.

TABLE 3.3 Roman Numerals and Their Modern Equivalents

Roman Symbols	Modern Equivalent
I	1
V	5
X	10
L	50
C	100
D	500
M	1000

Using these symbols, 1959 originally was written as

M D CCCC L V IIII

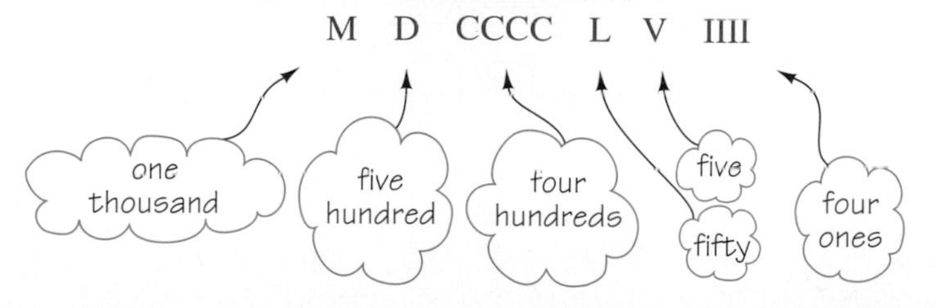

Later a subtractive principle was introduced to shorten the notation. If a single symbol for a lesser number was written to the left of a symbol for a greater number than the lesser number was to be *subtracted* from the greater number. In particular, the common representations were as shown here.

IIII was written as IV — $4 = 5 - 1$

VIIII was written as IX — $9 = 10 - 1$

XXXX was written as XL — $40 = 50 - 10$

CCCC was written as CD — $400 = 500 - 100$

DCCCC was written as CM — $900 = 1000 - 100$

Using this principle, 1959 could then be written more succinctly as

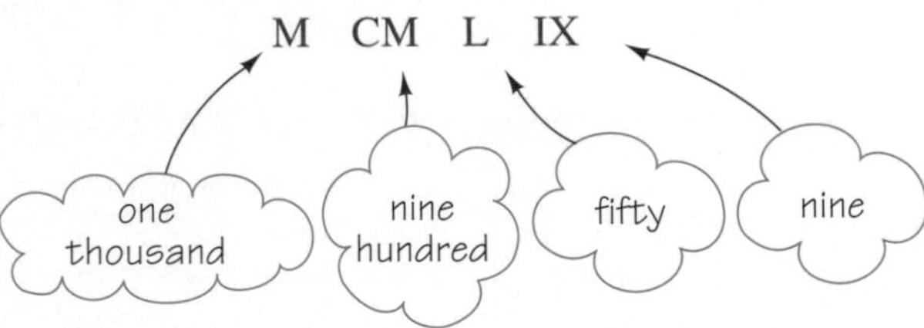

As with the Egyptian system, addition and subtraction in the Roman system was cumbersome but not conceptually difficult as these examples show.

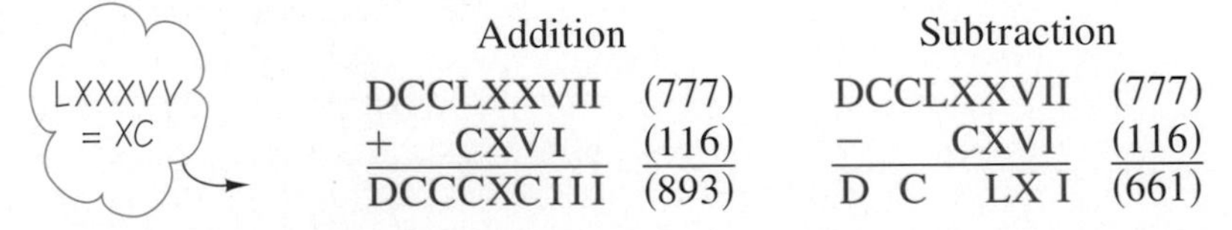

Addition		Subtraction	
DCCLXXVII	(777)	DCCLXXVII	(777)
+ CXVI	(116)	− CXVI	(116)
DCCCXCIII	(893)	D C LX I	(661)

As with the Egyptian system, however, multiplication and division were quite cumbersome. For this reason, much of the commercial calculation of the time was performed on devices like abacuses, counting boards, sand trays, and the like.

EXAMPLE 3.5

Writing a Number in the Roman System

Write 1998 in Roman numerals.

SOLUTION

One could write

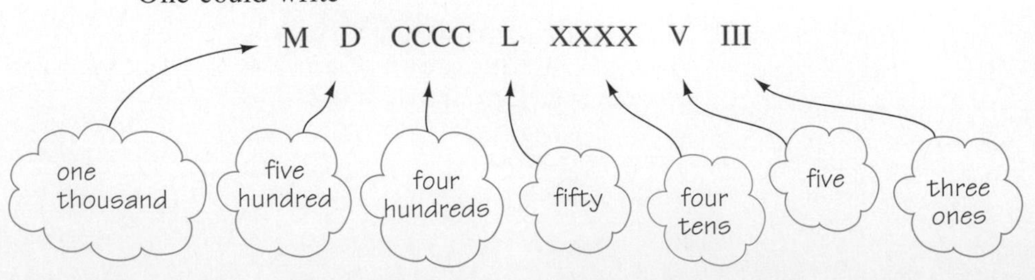

or, using the subtraction principle,

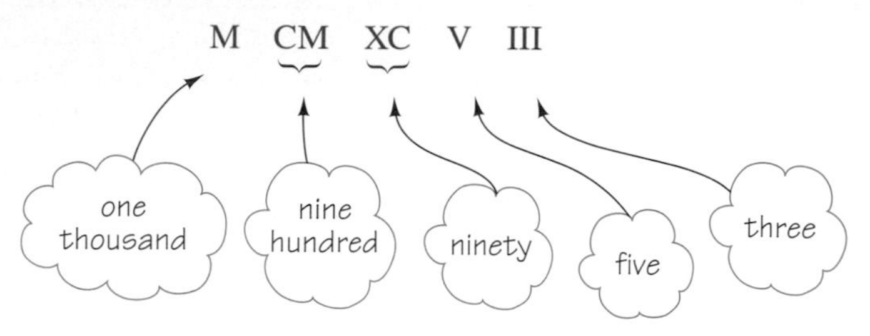

■

EXAMPLE 3.6 **Converting from Roman to Standard Notation**

Write MDCXLIII using modern notation.

SOLUTION

M	represents	1000
D	represents	500
C	represents	100
XL	represents	40
III	represents	3

Adding, we obtain 1,643. ■

The Mayan System

One of the most interesting of the ancient systems of numeration was developed by the Mayans in the region now known as the Yucatan Peninsula in southeastern Mexico. As early as A.D. 200, these resourceful people had developed a remarkably advanced society. They were the first Native Americans to develop a system of writing and to manufacture paper and books. Their learned scholars knew more about astronomy than was known at that time anywhere else in the world. Their calendar was very accurate with a 365 day year and a leap year every fourth year. In short, many scholars believe that the Mayans very early developed the most sophisticated society ever attained by early residents of the Western Hemisphere.

Like their other achievements, the Mayan system of numeration was remarkably advanced. It contained a symbol for zero which did not appear elsewhere until about A.D. 800. Also, it was a positional system of notation, like our current system, which made it possible to write very large numbers with relatively few symbols. Unlike our present system, the Mayan system was essentially a *vigesimal* or base twenty system, although the positions from the third on were for $18 \cdot 20$, $18 \cdot 20^2$, $18 \cdot 20^3$, and so on, rather than 20^2, 20^3, 20^4, and so on, as might be expected.* As with any fully developed positional system, the Mayans needed symbols for the numbers 0 through 19, the base minus one. It was then possible to write any whole number in positional notation in one and only one way using only these 20 symbols. Their symbols for 0 through 19 were as shown in Table 3.4. You will no doubt agree that these symbols are simple and natural.

* It is conjectured that this was because $18 \cdot 20 = 360$, approximately the number of days in a year.

TABLE 3.4 Mayan Numerals for 0 Through 19

Mayan Symbol	Modern Equivalent	Mayan Symbol	Modern Equivalent
	0		10
	1		11
	2		12
	3		13
	4		14
	5		15
	6		16
	7		17
	8		18
	9		19

The Mayans wrote their numerals in a vertical style with the unit's position on the bottom. For example, the number 43,487 would appear as

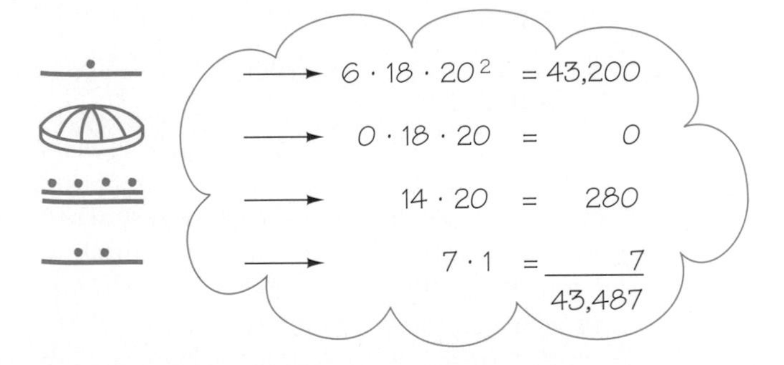

6 · 18 · 20² = 43,200
0 · 18 · 20 = 0
14 · 20 = 280
7 · 1 = 7
43,487

All that is needed to convert a Mayan numeral to modern notation is a knowledge of the digits and the value of the position each digit occupies as shown in Table 3.5.

TABLE 3.5 Position Values in the Mayan System

Position Level	Position Value
.	.
.	.
.	.
fifth	$18 \cdot 20^3 = 144{,}000$
fourth	$18 \cdot 20^2 = 7200$
third	$18 \cdot 20 = 360$
second	20
first or lowest	1

EXAMPLE 3.7 **Writing a Number Using Mayan Notation**

Write 27,408 in Mayan notation.

SOLUTION

We will not need the fifth or any higher position since $18 \cdot 20^3 = 144{,}000$ is already greater than 27,408. How many 7200s, 360s, 20s, and 1s are needed? This can be answered by repeated subtraction or, more simply, by division. From the arithmetic shown,

$$\begin{array}{r} 3 \\ 7200\overline{)27{,}408} \\ \underline{21{,}600} \\ 5\,808 \end{array} \qquad \begin{array}{r} 16 \\ 360\overline{)5808} \\ \underline{360} \\ 2208 \\ \underline{2160} \\ 48 \end{array} \qquad \begin{array}{r} 2 \\ 20\overline{)48} \\ \underline{40} \\ 8 \end{array} \qquad \begin{array}{r} 8 \\ 1\overline{)8} \\ \underline{8} \\ 0 \end{array}$$

it follows that we need three 7200s, sixteen 360s, two 20s and eight 1s. Thus, in Mayan notation, 27,408 appears as

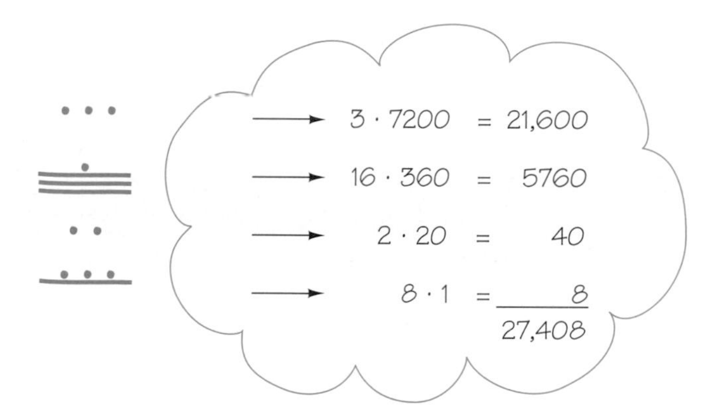

■

The ingenuity of this system, particularly since it involves the 18s, is easy to overlook. It turns out that such a positional system *will not work* unless the value of each position is a number that evenly divides the value of the next higher position. This fact is not at all obvious!

The Hindu-Arabic System

Today the most universally used system of numeration is the **Hindu-Arabic,** or **decimal system.** The system was named jointly for the Hindus, who invented it in India at least as early as 800 B.C., and for the Arabs who transmitted it to the Western world. Like the Mayan, it is a positional system. Since its **base** is ten, it requires special symbols for the numbers zero through nine. Over the years various notational choices have been made as shown in Table 3.6.

The common symbols 0, 1, 2, 3, 4, 5, 6, 7, 8, and 9 are called **digits** as are our fingers and toes, and it is easy to imagine the historical significance of this terminology. With these ten symbols and the idea of positional notation, all that is needed to write the numeral for any whole number is the value of each digit and the value of the position the digit occupies in the numeral. In the decimal system, the positional values as shown in Table 3.7 are well known.

JUST FOR FUN

For Careful Readers

1. How much dirt is there in a hole 10 meters long, 5 meters wide, and 3 meters deep?
2. The outer track of a phonograph record is a circle of radius 6 inches. The unused center of the disk is a circle of radius 3 inches. If the record has 20 grooves per inch, how far, to the nearest inch, does the needle travel while the record plays from beginning to end?
3. Joni has two U.S. coins in her pocket. One of the coins is not a quarter. The total value of the coins is 26¢. What are the two coins?

TABLE 3.6 Symbols for Zero Through Nine, Ancient and Modern

Tenth century Hindu	0	[illegible]	[illegible]	[illegible]	[illegible]	[illegible]	[illegible]	[illegible]	[illegible]	[illegible]
Current Arabic	٠	١	٢	٣	٤	٥	٦	٧	٨	٩
Fifteenth century European	0	/	2	3	[illegible]	[illegible]	[illegible]	Λ	8	9
Modern cursive	0	1	2	3	4	5	6	7	8	9
Modern print	0	1	2	3	4	5	6	7	8	9
Lighted scoreboard	0	1	2	3	4	5	6	7	8	9
Calculator display	0	1	2	3	4	5	6	7	8	9
Machine readable	0	1	2	3	4	5	6	7	8	9

TABLE 3.7 Positional Values in Base Ten

Position Names	...	Hundred Thousands	Ten Thousands	Thousands	Hundreds	Tens	Units
Decimal Form	...	100,000	10,000	1000	100	10	1
Powers of Ten	...	10^5	10^4	10^3	10^2	10^1	10^0

As our number words suggest, the symbol 2572 means two thousands plus five hundreds plus seven tens plus two ones. In so-called **expanded notation** we write

2000 + 50 + 7 + 2

$$\begin{aligned} 2572 &= 2 \cdot 1000 + 5 \cdot 10 + 7 \cdot 10 + 2 \cdot 1 \\ &= 2 \cdot 10^3 + 5 \cdot 10^2 + 7 \cdot 10^1 + 2 \cdot 10^0 \end{aligned}$$

The amount each digit contributes to the number is the value of the digit times the value of the position the digit occupies in the representation.

Physical Models for Positional Systems

The classroom abacus The idea for the Hindu-Arabic scheme for representing numbers derived historically from the use of counting boards and abacuses of various types to facilitate computations in commercial transactions. Such devices are still useful today in helping children to understand the basic concepts. One such device, commercially available for classroom use, is shown in Figure 3.1.

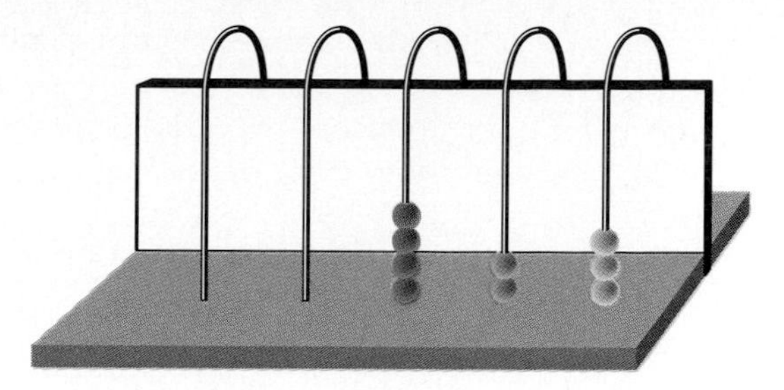

Figure 3.1
A classroom demonstration abacus

The device consists of a series of wire loops fixed into a wooden base and with a vertical shield affixed to the center of the base under the wire loops so that the back is hidden from the students. On each wire loop are 19 or 20 beads that can be moved from the back of the shield to the front and *vice versa.* Also, each wire is equipped with a clip to separate off any number of beads depending on what is to be demonstrated. For our present purpose we set off ten beads on each wire behind the shield.

To demonstrate counting and positional notation one begins by moving beads from the back of the shield to the front on the wire to the students' right (the instructor's left) counting 1, 2, 3, and so on, as the beads are moved to the front of the shield. Once all 10 beads set off on the first wire are counted, all are moved back behind the shield and the fact that the beads on this wire have all been counted once is recorded by moving a single bead to the front of the shield on the second wire. The count 11, 12, 13, and so on, then continues with a single bead on the first wire again moved forward with each count. When the count reaches 20, all the beads on the first wire will have been counted a second time and this is recorded by moving a second bead forward on the second wire and moving all the beads to the back again on the first wire. Then the count continues. When the count reaches 34, for example, the

HIGHLIGHT FROM HISTORY
Fibonacci

The most talented mathematician of the Middle Ages was Leonardo of Pisa (ca. 1170–1250), the son of a Pisan merchant named Bonaccio. The Latin, *Leonardo Filius Bonaccio* (Leonardo son of Bonaccio) was soon contracted to Leonardo Fibonacci. This was further shortened simply to Fibonacci which is still popularly used today. The young Fibonacci was brought up in Bougie—still an active port in modern Algeria—where his father served for many years as customs manager. It was here and on numerous trips throughout the Mediterranean region with his father that Fibonacci became acquainted with the Hindu-Arabic numerals and the algorithms for computing with them which we still use today. Fibonacci did outstanding original work in geometry and number theory and is best known today for the remarkable sequence

$$1, 1, 2, 3, 5, 8, 13, 21, 34, 55, 89, 144, \ldots$$

which bears his name. However, his most important contribution to Western civilization remains his popularization of the Hindu-Arabic numeration system in his book *Liber abaci* written in 1202. This book so effectively illustrated the vast superiority of this system over the other systems then in use that it soon was widely adopted not only for use in commerce but also in serious mathematical studies. Its use so simplified computational procedures that its effect on the rapid growth of mathematics during the Renaissance and beyond can only be characterized as profound.

children will see the arrangement of beads shown schematically in Figure 3.2 with three beads showing on the second wire and four beads showing on the first wire. A natural way to record the result is to write 34; that is, three tens and four ones, or 34. Note that we never leave ten beads up on any wire.

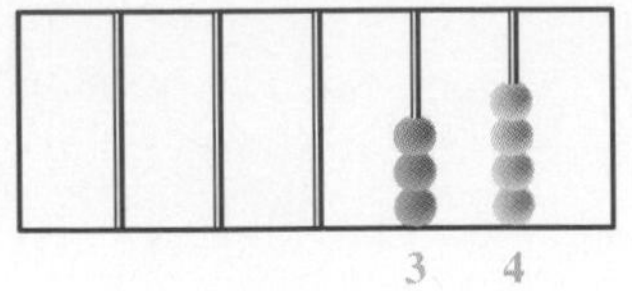

Figure 3.2
Thirty-four on the classroom abacus

What happens if we count to 100; that is, if we count all ten beads on the first wire ten times? Since we move one bead on the second wire forward each time we count all the beads on the first wire once, we will have ten beads showing on the second wire; that is, we will have counted *all* the beads on the second wire once. But we record this on the abacus by moving one bead forward on the third wire and moving all beads on the second wire back behind the shield. Thus, each bead showing

INTO THE CLASSROOM

Mary Cavanaugh Discusses the Use of Manipulatives

Teaching for meaning in mathematics is a major focus of math educators. The work of Piaget and others suggests that students begin their exploration of mathematical concepts through hands-on experiences with manipulatives. Manipulatives appeal to several senses and are used to physically involve students in a learning situation. Students manipulate the objects with actions such as forming, ordering, comparing, tracing, joining, or separating groups. By these actions they gain an understanding of the meanings of number and various operations in mathematics.

Research points to multi-digit numeration as the pivotal concept that students must learn with the help of manipulatives before they can succeed in learning computational algorithms. A comfortable progression through representing numbers, trading, and computing with manipulatives is developmentally appropriate for providing a solid foundation for understanding arithmetic concepts. Premature introduction of paper-and-pencil procedures pushes students into memorizing a complex sequence of mathematical acts before the acts have meaning.

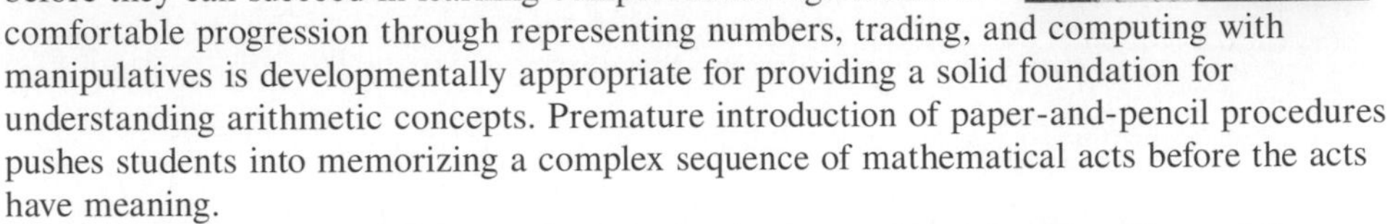

Manipulating objects allows students to develop internal images of number and the relative magnitude of number that can be recalled when needed. Pictures of concrete materials that follow manipulative experiences form a connection between the internal images, the pictures, and abstract symbols. The guidance of teachers is needed to help bridge the gap between what is learned from manipulatives and what is written using mathematical symbols.

A research recommendation, easily realized with manipulatives, is that teachers can and should expect students to enjoy the learning of mathematics. Concrete materials appeal to a variety of senses and motivate students. The frequent use of manipulatives in the classroom helps make the expectation of enjoyment evident.

SOURCE: From *ScottForesman Exploring Mathematics* Grades 1–7 by L. Carey Bolster et al. Copyright © 1994 Scott, Foresman and Company. Reprinted by permission of Scott, Foresman and Company. Mary Cavanaugh is the coordinator of "Parenting Through Math, Science, and Beyond" and "Matemáticas y Ciencia Limites" for Solana Beach School District, Solana Beach, California.

on the first wire counts for 1, each bead showing on the second wire counts for 10, each bead showing on the third wire counts for 100, and so on. If we count to 423, for example, the arrangement of beads is as illustrated earlier in Figure 3.1 and the count is naturally recorded as 423.

Approached in this way, the notion of place value becomes much more concrete. One can actually experience the fact that each bead on the second wire counts for 10, each bead on the third wire counts for 100, and so on—particularly if allowed to handle the device and move the beads while counting. One can also see that the counting process on the abacus can proceed as far as desired though more and more wire loops would have to be added to the device. It follows that any whole number can be represented in one and only one way in our modern system using only the digits 0, 1, 2, 3, 4, 5, 6, 7, 8, and 9.

Other physical devices can also be used to illustrate positional notation and many are even more concrete than the historical abacus mimicked above.

Sticks in bundles One simple idea for introducing positional notation is to use bundles of small sticks which can be purchased inexpensively at almost any craft store. Single sticks are ones. Ten sticks can be bound together in a bundle to represent ten. Ten bundles can be banded together to represent 100, and so on. Thus, 34 would be represented as shown in Figure 3.3.

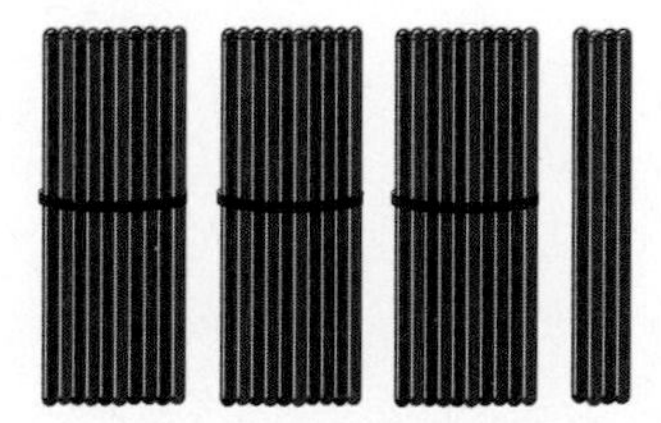

Figure 3.3
Representing 34 with sticks and bundles of sticks

Unifix™ cubes Here single cubes represent ones. Ten cubes snapped together to form a stick represent ten. Ten sticks bound together represent 100, and so on. Again, 34 would be represented as shown in Figure 3.4.

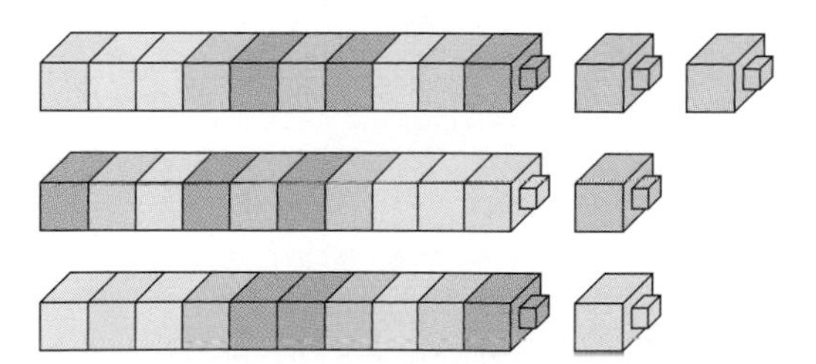

Figure 3.4
Representing 34 with Unifix™ cubes

Units, strips, and mats Pieces for this concrete realization of the decimal system are easily cut from graph paper with reasonably large squares. A unit is a single square, a strip is a strip of ten squares, and a mat is a square ten units on a side as illustrated in Figure 3.5. To make the units, strips, and mats more substantial and hence easier to manipulate, the graph paper can be pasted to a substantial tag board or it can be copied on reasonably heavy stock. Using units, strips, and mats, 254 would be represented as shown in Figure 3.6.

Base ten blocks These commercially prepared materials include single cubes that represent ones, sticks called "longs" made up of ten cubes that represent ten, "flats" made up of ten longs that represent 100, and "blocks" made up of ten flats that represent 1000 as shown in Figure 3.7.

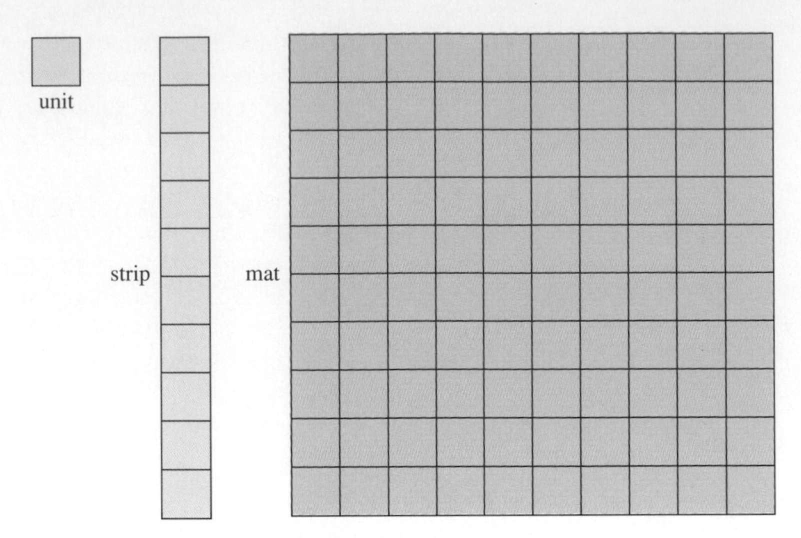

Figure 3.5 **A unit, a strip, and a mat**

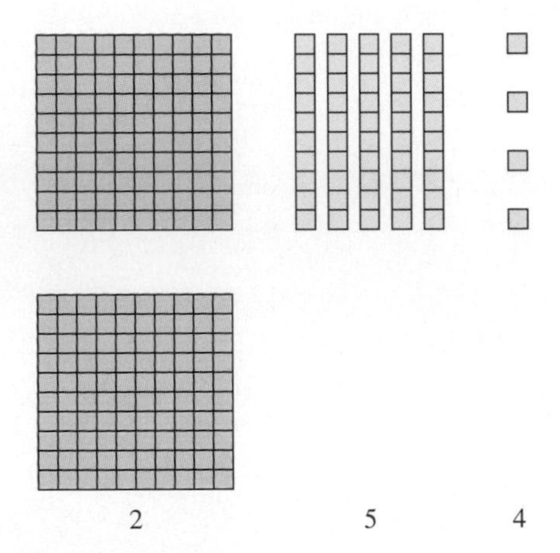

Figure 3.6 **Representing 254 with units, strips, and mats**

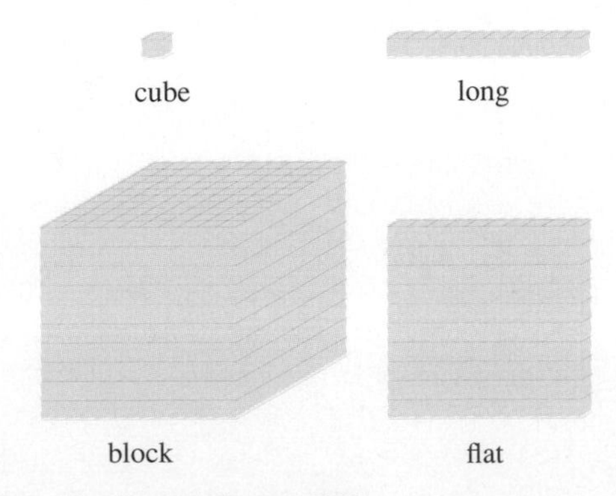

Figure 3.7 **Base ten blocks for 1, 10, 100, and 1000**

These materials are quite neat and have the advantage of giving a good representation of 1000. However, they are expensive. Most schools will not want to purchase more than one set per classroom. Also, since they come ready made, the youngsters do not have the advantage of the very real learning experienced in cutting a strip into ten units, a mat into ten strips, and so on.

PROBLEM SET 3.1

Understanding Concepts

1. Write the Hindu-Arabic equivalent of each of the following.

   (a) 

   (b)

   (c)

   (d) MDCCXXIX

   (e) DCXCVII

   (f) CMLXXXIV

   (g) (h)

   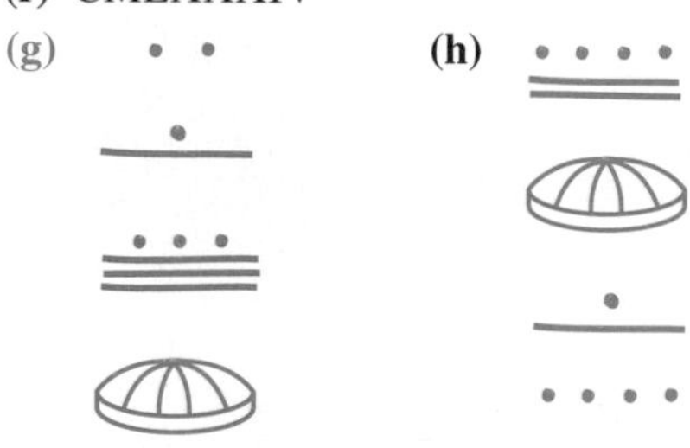

2. Write the following in Egyptian notation.

   (a) 11 (b) 597 (c) 1949

3. Write the following in Roman notation using the subtraction principle as appropriate.

   (a) 9 (b) 486 (c) 1945

4. Write these numbers in Mayan notation.

   (a) 12 (b) 584 (c) 12473

5. Add

   and

   using only Egyptian notation. Write your answer as simply as possible.

6. Add MDCCCXVI and MCCCLXIV using only Roman notation. Write your answer as simply as possible.

7. Add and using only Mayan notation. Write your answer as simply as possible.

8. Subtract

   from

   using only Egyptian notation.

9. Subtract MCCXCV from MDCCCXVI using only Roman notation.

10. Subtract from using only Mayan notation.

11. Draw a sketch to illustrate 452 using units, strips, and mats. Use dots to represent units, vertical line segments to represent strips, and squares to represent mats.

12. Draw a sketch to illustrate 234 using base ten blocks.

13. Draw a sketch of the exposed side of a classroom abacus to illustrate 2475.

14. Write 24,872 and 3071 in expanded notation.

15. Suppose you have 3 mats, 24 strips, and 13 units for a total count of 553. Briefly describe the exchanges you would make to keep the same total count but have the smallest possible number of manipulative pieces.

16. Suppose you have 2 mats, 7 strips, and 6 units in one hand and 4 mats, 5 strips, and 9 units in the other.

    (a) Put all these pieces together and make the necessary exchanges to keep the same total count but with the smallest possible number of manipulative pieces.

    (b) Explain briefly but clearly what mathematics the manipulation in part (a) represents.

17. Suppose you have 3 mats and 6 units on your desk and want to remove a count represented by 3 strips and 8 units.

(a) Describe briefly but clearly the exchange that must take place to accomplish the desired task.

(b) After removing the 3 strips and 8 units, what manipulative pieces are left on your desk?

(c) What mathematics does the manipulation in part (a) respresent?

Thinking Critically

18. Think of the classroom abacus of Figure 3.1 with the clips inserted behind the shield and below the fifth bead on each wire. Count as before moving one bead forward on the units wire each time the count increases by one. This time, however, when *five* beads have been moved forward on any wire, move them back behind the shield and move one bead forward on the next wire. Thus, when a count is completed, at most four beads on any wire will show in front of the shield. Finally, as the count proceeds, record the number of beads showing on all wires in positions corresponding to the positions of the wires and omitting the zeros corresponding to all wires to the left of the last wire with any beads. Thus, the count of one hundred thirteen would be shown as in Figure 3.1. It would be recorded as 423 (read "four-two-three" and *not* as "four hundred twenty-three").

What numbers (counts) are represented if the beads on the abacus described are as shown in these diagrams?

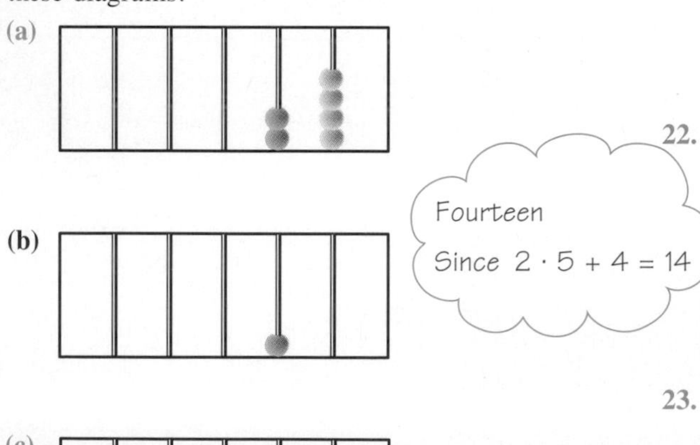

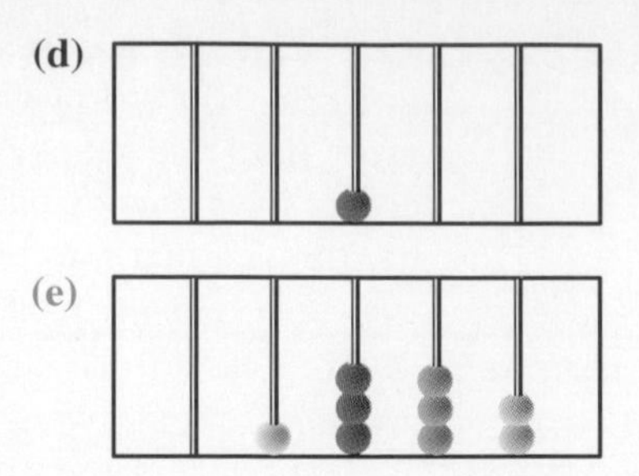

19. The count shown in the diagram of problem 18, part (b) would be recorded as 10 ("one-zero"). How would the counts for parts (a), (c), (d), and (e) naturally be recorded?

Thinking Cooperatively

20. Suppose the abacus of problem 18 is configured as shown here.

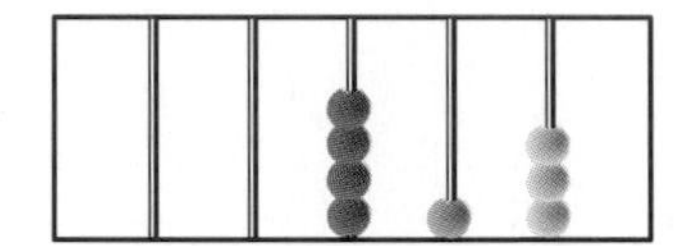

(a) How would this count naturally be recorded?

(b) Suppose the count shown in part (a) continues for seven more units. Draw a diagram to show how the abacus will then be configured.

21. Suppose the abacus of problem 18 is as shown here.

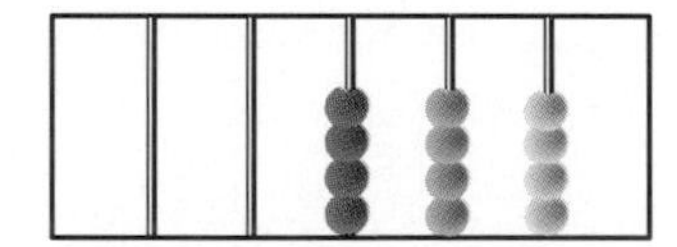

How will it appear if the count is increased by 1? Draw a suitable diagram.

22. Draw diagrams of the abacus of problem 18 for each of these counts.

(a) five **(b)** seven

(c) ten **(d)** twenty-five

(e) thirty-two **(f)** one hundred-four

(g) one hundred twenty-five

(h) one hundred forty-seven

23. The right-most wire on the abacus of problem 18 is called the units wire. What would be appropriate names for

(a) the second wire from the right?

(b) the third wire from the right?

24. If the count on the abacus of problem 18 is recorded as 132, what is the count?

Communicating

25. Suppose the abacus of problem 18 had been invented by a race of one-handed, five-fingered aliens.
    (a) Could they represent any number, however large, on such an abacus with suitably many wires? Explain.
    (b) Carefully describe the numeration system that the aliens of part (a) might use.
26. Write a two-page paper describing one of these ancient numeration systems: Greek, Babylonian, or Chinese.

For Review

27. Given sets $U = \{1, 2, 3, 4, 5, 6, 7, 8, 9\}$, $A = \{1, 3, 5, 7, 9\}$, $B = \{1, 2, 3, 4\}$, and $C = \{2, 4, 6, 8\}$, complete the following.
    (a) $A \cup B =$ ______ (b) $A \cap B =$ ______
    (c) $\overline{A} =$ ______ (d) $A \cap C =$ ______
    (e) $A \cap (B \cup C) =$ ______
    (f) $(A \cap B) \cup (A \cap C) =$ ______
    (g) $n(A) =$ ______, $n(C) =$ ______, $n(A \cup C) =$ ______
    (h) $n(A) =$ ______, $n(B) =$ ______, $n(A \cup B) =$ ______, $n(A) + n(B) - n(A \cap B) =$ ______
    (i) Explain in two carefully written sentences the reason for the difference between the results of parts (g) and (h).
28. Name the property that justifies each of the following.
    (a) $3x + 5x = (3 + 5)x$
    (b) $4\pi + 7\pi = 7\pi + 4\pi$
    (c) $3x + (4y + z) = 3x + (z + 4y)$
    (d) $21 + 7 = 20 + (1 + 7)$
29. Write two subtraction equations corresponding to each of these equations.
    (a) $3 + 7 = 10$ (b) $11 + 5 = 16$
    (c) $21 + 19 = 40$
30. Write one addition equation and a second subtraction equation that corresponds to each of these equations.
    (a) $17 - 8 = 9$ (b) $23 - 15 = 8$
    (c) $23 - 5 = 18$

3.2 Nondecimal Positional Systems

In the preceding section we discussed numeration systems including the decimal system in common use today. As already mentioned, the decimal system is a positional system based on ten. This is probably so because we have ten fingers and, as now, people often counted on their fingers. Suppose that there are little green aliens on Mars who have only one hand with five fingers. If Martians were to go through essentially the same process of developing a number system as the Hindus did, how would the system work?

Base Five Notation

Problems 18–24 in the preceding section begin to suggest a positional system with base five. In fact, the idea can be extended similarly to any integer base b, with b greater than one. Let's look briefly at base five.

For our purposes, perhaps the quickest route to understanding **base five notation** is to reconsider the abacus of Figure 3.1. This time, however, we suppose that we put the clips on the wires so that only five beads can be moved forward on each wire. As before, we start with the wire to the students' right (our left) and move one bead forward each time as we count one, two, three, and so on. When we reach five, we have counted the first wire once and we record this on the abacus by moving one bead forward on the second wire while moving all the beads on the first wire to the back. Continuing to count, when the count reaches ten we will have counted all the beads on the first wire a second time. We move a second bead forward on the second wire and again move all the beads on the first wire to the back as indicated in Figure 3.8.

HIGHLIGHT FROM HISTORY
Benjamin Banneker (1731–1806)

Born in Maryland before the American Revolutionary War, the son of free black parents, Benjamin Banneker owed much of his education to his Quaker neighbors. As a teenager, he developed an interest in mathematics, science, astronomy, and mechanics. His diverse accomplishments included a mathematical study of the 17-year locust, the accurate prediction of solar eclipses, and an extraordinary wooden clock that is accepted as the first clock made entirely in America. Like another American inventor and scientist, Benjamin Franklin, Banneker published a popular almanac. It included antislavery essays, information on crops and tides, and astronomical data. He was one of three members of a commission that planned and surveyed the site for the federal capital at Washington. When he died in 1806, Thomas Jefferson occupied the White House in the city he had helped to design.

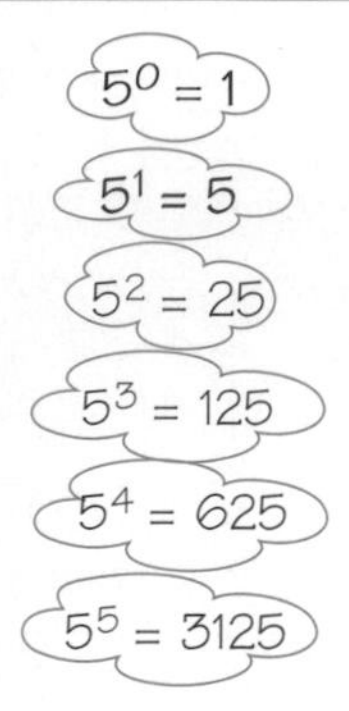

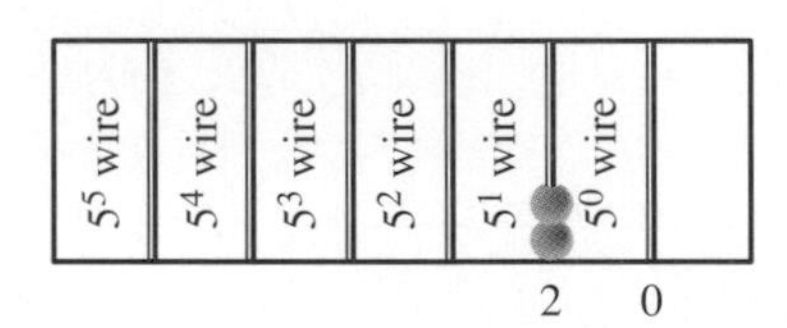

Figure 3.8 A count of ten on a 5-bead abacus

Thus, each bead on the second wire counts for five; that is, the second wire is the "fives" wire. Continuing the count, when it reaches 19, the abacus will appear as in Figure 3.2 and

$$19 = 3 \cdot 5 + 4.$$

If we continue the count to 25, all the beads on the first wire will have been counted five times. This means that we have moved all five beads to the front on the *second* wire. This is recorded on the abacus by returning these beads to the back of the abacus and moving one bead forward on the *third* wire. Thus, the third wire becomes the $25 = 5^2$ wire. The count can be continued in this way, and it is apparent that any whole number will eventually be counted and can be recorded on the abacus with only 0, 1, 2, 3, or 4 beads per wire showing at the front; that is, we need only the digits 0, 1, 2, 3, and 4 in base five. For example, if we continue the counting process up to 113, the abacus will be configured as shown in Figure 3.1 and

$$113 = 4 \cdot 5^2 + 2 \cdot 5 + 3.$$

Just as we shorten $4 \cdot 10^2 + 2 \cdot 10 + 3$ to 423, the little green Martian might also shorten $4 \cdot 5^2 + 2 \cdot 5 + 3$ to 423. Thus, the three-digit sequence 423 can represent many different numbers depending on which base is chosen. To the average American, who doesn't even think about it, it means "four hundred twenty-three." To our Martian friend it means *flug globs zeit-tab;* that is, Martian for $4 \cdot 5^2 + 2 \cdot 5 + 3$. Note that our Martian friend certainly would **not** say "one hundred thirteen" since that is decimal, or base ten, language. To avoid confusion in talking about base five numeration, we agree that we will *not* say "four hundred twenty-three" when we read 423 as a base five numeral. Instead, we will say "four two three base five" and will write 423_{five} where the subscript five indicates the base. In base six, we understand that

$423_{\text{six}} = 159_{\text{ten}}$ since
$4 \cdot 6^2 + 2 \cdot 6 + 3 = 159_{\text{ten}}$

$$423_{\text{six}} = 4 \cdot 6^2 + 2 \cdot 6 + 3$$

and read the numeral as "four two three base six." As an additional example, 423_{twelve} should be read "four two three base twelve," and in expanded form, we have

$$423_{\text{twelve}} = 4 \cdot 12^2 + 2 \cdot 12 + 3 = 603_{\text{ten}}.$$

Unless expressly stated to the contrary a numeral written *without* a subscript should be read as a base ten numeral.

Observe that, in base five, we need the digits 0, 1, 2, 3, and 4 and we need to know the values of the positions in base five. In base six, the digits are 0, 1, 2, 3, 4, and 5 and we need to know the positional values in base six. In base twelve, we use the digits 0, 1, 2, 3, 4, 5, 6, 7, 8, 9, T, and E where T and E are, respectively, the digits for ten and eleven. For these bases, the positional values are given in Tables 3.8, 3.9, and 3.10.

TABLE 3.8 Positional Values in Base Five

Position Names	. . .	Three Thousand One Hundred-twenty-fives	Six Hundred Twenty-fives	One Hundred Twenty-fives	Twenty-fives	Fives	Units
Decimal Form	. . .	3125	625	125	25	5	1
Powers of Five	. . .	5^5	5^4	5^3	5^2	5^1	5^0

TABLE 3.9 Positional Values in Base Six

Position Names	. . .	Seven Thousand Seven Hundred Seventy-sixes	One Thousand Two Hundred Ninety-sixes	Two Hundred Sixteens	Thirty-sixes	Sixes	Units
Decimal Form	. . .	7776	1296	216	36	6	1
Powers of Six	. . .	6^5	6^4	6^3	6^2	6^1	6^0

TABLE 3.10 Positional Values in Base Twelve

Position Names	...	Twenty Thousand Seven Hundred Thirty-sixes	One Thousand Seven Hundred Twenty-eights	One Hundred Forty-fours	Twelves	Units
Decimal Form	...	20736	1728	144	12	1
Powers of Twelve	...	12^4	12^3	12^2	12^1	12^0

EXAMPLE 3.8

Converting from Base Five to Base Ten Notation

Write the base ten representation of 3214_{five}.

SOLUTION

We use the place values of Table 3.8 along with the digit values. Thus,

$$\begin{aligned} 3214_{\text{five}} &= 3 \cdot 5^3 + 2 \cdot 5^2 + 1 \cdot 5^1 + 4 \cdot 5^0 \\ &= 3 \cdot 125 + 2 \cdot 25 + 1 \cdot 5 + 4 \cdot 1 \\ &= 375 + 50 + 5 + 4 \\ &= 434. \end{aligned}$$

Therefore, three two one four base five is four hundred thirty-four. We just need to understand the expanded form of 3214_{five} as shown here. ■

EXAMPLE 3.9

Converting from Base Ten to Base Five Notation

Write the base five representation of 344.

SOLUTION

Remember that 344 without a subscript has its usual meaning as a base ten numeral. Thus, we need to determine how many of each position value in base five are required to obtain 344. This can be determined by successive divisions as in Example 3.7, in Section 3.1. Referring to Table 3.8 for position values, we see that 625 is too big. Thus, we begin with 125 and divide successive remainders by successively lower position values.

$$\begin{array}{r} 2 \\ 125\overline{)344} \\ \underline{250} \\ 94 \end{array} \qquad \begin{array}{r} 3 \\ 25\overline{)94} \\ \underline{75} \\ 19 \end{array} \qquad \begin{array}{r} 3 \\ 5\overline{)19} \\ \underline{15} \\ 4 \end{array} \qquad \begin{array}{r} 4 \\ 1\overline{)4} \\ \underline{4} \\ 0 \end{array}$$

These divisions reveal that we need 2 125s, 3 25s, 3 5s, and 4 1s or, in tabular form,

125s	25s	5s	1s
2	3	3	4

Thus,

$$344_{\text{ten}} = 2334_{\text{five}}.$$

As a check, we note that

$$\begin{aligned} 2334_{\text{five}} &= 2 \cdot 125 + 3 \cdot 25 + 3 \cdot 5 + 4 \\ &= 250 + 75 + 15 + 4 \\ &= 344_{\text{ten}}. \end{aligned}$$

■

CURRENT AFFAIRS

The Utility of Other Bases

At first thought it might appear that positional numeration systems in bases other than base ten are merely interesting diversions. Quite the contrary, they are extremely important in today's society. This is particularly true of the base two or binary system, the base eight or octal system, and the base sixteen or hexadecimal system. The degree to which calculators and computers have affected modern life is simply enormous, and the basic notion that allows these devices to operate is base two arithmetic. A switch is either on or off, a spot on a magnetic grid is either magnetized or not magnetized, a spot on a compact disc is activated or it is not. All of these devices are capable of recording two states—either 0 or 1—and so can be programmed to do base two arithmetic and to record other data in code "words" consisting of strings of 0s and 1s. A drawback of base two notation is that numerals for relatively small numbers become quite long. Thus,

$$60_{ten} = 111100_{two}$$

and this makes programming a computer somewhat cumbersome. The notation is greatly simplified by using octal or hexadecimal notation; since these are intimately related to binary notation. Thus, triples of binary digits become single octal digits and vice versa and, quadruples of digits in binary notation correspond to single digits in hexadecimal. For example, since

$$111_{two} = 7 \quad \text{and} \quad 100_{two} = 4,$$

it follows that

$$60_{ten} = 111100_{two} = 74_{eight}.$$

For this reason, computer programmers work in octal or hexadecimal notation.

PROBLEM SET 3.2

Understanding Concepts

1. In a long column write the base five numerals for the numbers from zero through twenty-five.
2. Briefly describe the pattern or patterns you observe in the list of numerals in problem 1.
3. Here are the base six representations of the numbers from zero through thirty-five arranged in a square array. Briefly describe any patterns you observe in this array.

0	1	2	3	4	5
10	11	12	13	14	15
20	21	22	23	24	25
30	31	32	33	34	35
40	41	42	43	44	45
50	51	52	53	54	55

4. What would be the next entry in the table of problem 3?
5. Write the base ten respresentations of each of the following.
   (a) 413_{five} **(b)** 2004_{five} (c) 10_{five}
   (d) 100_{five} (e) 1000_{five} **(f)** 2134_{five}
6. Write the base ten respresentations of each of the following.
   (a) 413_{six} **(b)** 2004_{six} (c) 10_{six}
   (d) 100_{six} (e) 1000_{six} **(f)** 2134_{six}
7. Write the base ten respresentations of each of these. Remember that in base twelve the symbols for the digits ten and eleven are T and E.
   (a) 413_{twelve} **(b)** 2004_{twelve} (c) 10_{twelve}
   (d) 100_{twelve} (e) 1000_{twelve} **(f)** $2TE4_{twelve}$
8. Determine the base five representation for each of the following. Remember that a numeral with no subscript is understood to be in base ten.
   (a) 362 **(b)** 27 (c) 5 **(d)** 25

9. Determine the base six representation for the following.
(a) 342 **(b)** 21 (c) 6 **(d)** 216

10. Determine the base five representation for each of these numbers. (*Hint:* First find the base ten representation.)
(a) 342_{six} **(b)** 21_{six} (c) 41_{six} **(d)** TE_{twelve}

11. Determine the base twelve representation for each of the following.
(a) 2743 **(b)** 563 (c) 144 **(d)** 1584

12. Base two is a very useful base. Since it only requires two digits, 0 and 1, it is the system on which all calculators and computers are based.
(a) Make a table of position values for base two up as far as $2^{10} = 1024$.
(b) Write each of these in base ten notation.
(i) 1101_{two} (ii) 111_{two} (iii) 1000_{two}
(iv) 10101_{two}
(c) Write each of these in base two notation.
(i) 24 (ii) 18 (iii) 2 (iv) 8
(d) Write the numbers from zero to thirty-one in base two notation in a vertical column and discuss any pattern you observe in a short paragraph.
(e) In three or four sentences compare part (d) of this problem with the *Hands On* activity at the beginning of the chapter.

Thinking Critically

13. Here's an interesting trick. Consider these pictures of cards you might make for use in your class.

1	3	5	7	9	11
13	15	17	19	21	
23	25	27	29	31	

2	3	6	7	10	11
14	15	18	19	22	
23	26	27	30	31	

4	5	6	7	12	13
14	15	20	21	22	
23	28	29	30	31	

8	9	10	11	12	13
14	15	24	25	26	
27	28	29	30	31	

16	17	18	19	20	21
22	23	24	25	26	
27	28	29	30	31	

(a) Record the first number on each card that shows the day of the month on which you were born.
(b) Add the numbers in part (a).
(c) Surprised? Our experience is that elementary school students are too and that they immediately want to know how the trick works. See if you can discover the secret by carefully comparing the cards with your answer to problem 12, part (d).

14. (a) Add $11{,}111{,}111_{two}$ and 1_{two} in base 2.
(b) What are the base two numerals for 2^n and $2^n - 1$? Explain briefly, describing a pattern.

15. It is a remarkable fact that the number of odd entries in the row of Pascal's triangle whose second entry is n is 2^f where f is the number of 1s in the base 2 representation of n. For example, the row of the triangle whose second entry is 5 is 1, 5, 10, 10, 5, 1 and four of these are odd. Also, $5 = 101_{two}$ with two 1s in the base 2 representation. Finally, $2^2 = 4$ and this gives the number of odd entries in this row of Pascal's triangle as claimed. Check that this result is true for the first eight rows of Pascal's triangle.

Thinking Cooperatively

16. The two 1-digit sequences of 0s and 1s are 0 and 1. The four 2-digit sequences of 0s and 1s are 00, 10, 01, and 11.

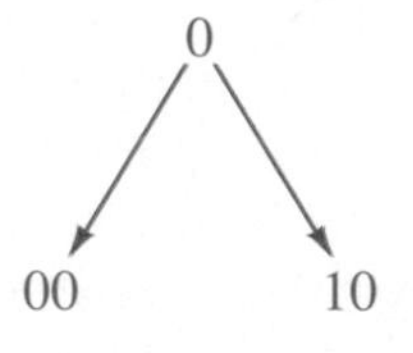

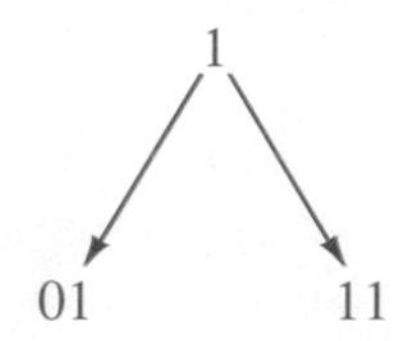

(a) Write down all of the 3-digit sequences of 0s and 1s. Can you think of an easy way to do this? Explain.
(b) Write down all of the 4-digit sequences of 0s and 1s. Can you think of an easy way to do this? Explain.
(c) How many 5-digit sequences of 0s and 1s are there?
(d) How many n-digit sequences of 0s and 1s are there? Explain why this is so.

17. (a) Think of each 3-digit sequence of 0s and 1s in problem 16(a) as a base two numeral. What are the base ten equivalents of these numerals?

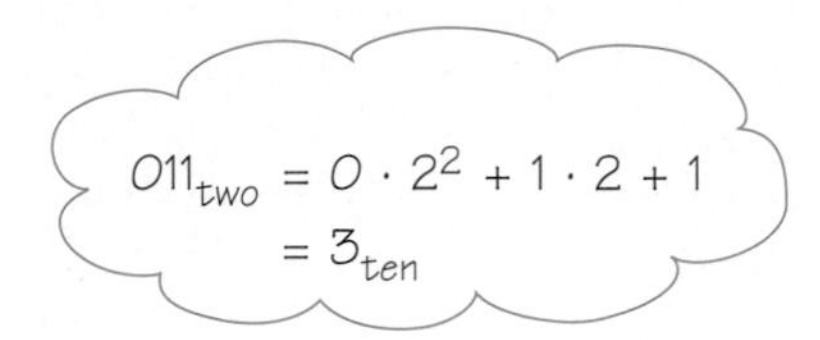
$011_{two} = 0 \cdot 2^2 + 1 \cdot 2 + 1$
$= 3_{ten}$

(b) Think of each 4-digit sequence of 0s and 1s in problem 16 (b) as a base two numeral. What are the base ten equivalents of these numerals?

(c) Think of each n-digit sequence of 0s and 1s as a base two numeral. What do you think are the base ten equivalents of these numerals? Explain.

18. (a) Set up a one-to-one correspondence between the 3-digit base two numerals (including numerals like 010 and 011) and the subsets of the set {a, b, c,} as follows. Include a in a subset if the left most digit in the corresponding numeral is 1 and exclude it otherwise. Make similar decisions about b and c depending on the middle and right digits respectively. Complete the following table of subsets of {a, b, c}, base two numerals, and their base ten equivalents.

Subset	Base Two Numeral	Base Ten Equivalent
∅	000	0
{a}	100	4
{b}	010	2
____	001	____
____	110	____
{a, c}	____	____
____		____
____	____	____

(b) Could a one-to-one correspondence as in part (a) be set up between the 4-digit base two numerals (including those with 0s on the left) and the subsets of {a, b, c, d}? Explain briefly.

(c) Since sets in one-to-one correspondence have the same number of elements, how many subsets of {a, b, c, d} are there?

(d) How many subsets are there of an n-element set? Note problem 16 (d) above. Observe that this very nearly constitutes a proof of your answer to problem 16, part (d), in Problem Set 2.1.

Communicating

19. Read the article "Counting" by L. L. Conant in *The World of Mathematics*, vol. 1 (New York: Simon and Schuster, 1956), and write a brief two page summary.

For Review

20. Write two division equations corresponding to each of these multiplication equations.
(a) $3 \cdot 17 = 51$ (b) $11 \cdot 91 = 1001$
(c) $9 \cdot 121 = 1089$

21. Write a multiplication equation and a second division equation corresponding to each of these division equations.
(a) $341 \div 11 = 31$ (b) $455 \div 65 = 7$
(c) $124857 \div 13873 = 9$

22. Lida Lee has three sweaters, four blouses, and two pairs of slacks that mix and match beautifully. How many different outfits can she wear using these nine items of clothing?

23. (a) If n represents Shane Stagg's age six years ago, how old is Shane now?
(b) If m represents Shane's age now, how old was Shane six years ago?

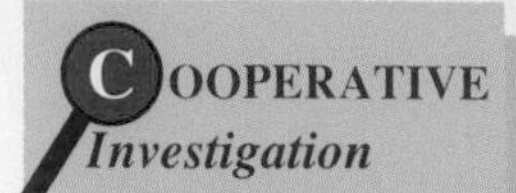

A Remarkable Base Three Trick

Mathematical magic certainly has a place in the classroom. Students find it interesting, fun, and highly motivational. Consider the following problem: You are given 12 coins that appear identical but one of the coins is false and is either heavy or light, you don't know which. You have a balance scale and are to find the false coin and to determine whether it is heavy or light in just three weighings. This is a difficult problem that is usually solved by considering a whole series of cases. However, if you know base three arithmetic you can determine the false coin so quickly right in front of your class that you appear to be a real wizard! Here's how it can be done.

1. Number the coins from 1 through 12.
2. With your back turned, ask your class to agree on the number of the coin they want to be false and to decide whether it is to be heavy or light.
3. Indicate that you are going to make these three weighings and ask the class what the movement of the *left hand* pan on each weighing will be. If the left hand pan goes up, record a 2. If it balances, record a 1. If it goes down, record a 0. Now use the results of these weighings to form a 3-digit base three numeral. The first weighing determines the *nines* digit, the second weighing determines the *threes* digit, and the third weighing determines the *units* digit. If the base ten number determined in this way is less than 13, it names the false coin. If the number is more than 13, then 26 minus the number generated names the false coin. You can tell, by knowing the false coin, whether it is heavy or light by noting how the left hand pan moved on a weighing involving the false coin.

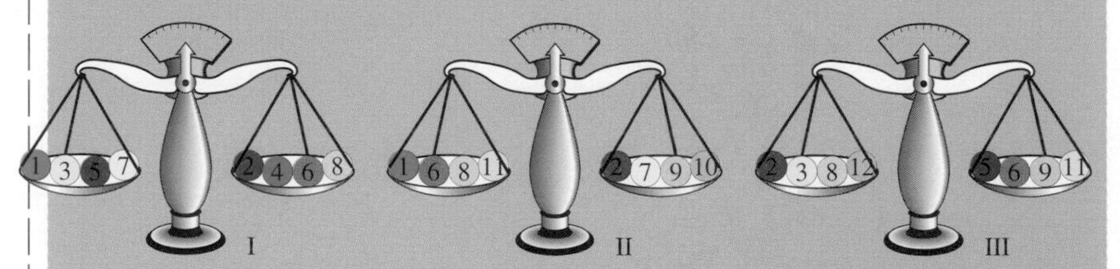

For example, suppose the class chooses coin 7 as the false coin and decides that it should be heavy. Then on the three weighings the left hand pan goes down, goes up, and balances and we generate $021_{\text{three}} = 7$, the number of the false coin. Moreover, since coin 7 was in the left hand pan, which went down on the first weighing, coin 7 must be heavy. On the other hand, suppose that coin 7 is light. Then on the three weighings the left hand pan goes up, goes down, and balances, and we generate $201_{\text{three}} = 19$. But then $26 - 19 = 7$, so coin 7 is the false coin and must be light since the left hand pan went up on the first weighing.

Our experience is that children like the trick very much and are strongly motivated to learn base three notation in order to pull the trick on their friends and parents. Note also that to do the trick well students must be able to perform mental calculations quickly—itself a worthy goal.

3.3 Algorithms for Adding and Subtracting Whole Numbers

Today's predominant opinion, as summarized in NCTM's *Curriculum and Evaluation Standards for School Mathematics,* is that we spend much more time than necessary trying to teach youngsters the intricacies of pencil and paper calculations. The time would be much better spent helping students learn to think critically about quantitative situations, to develop thoughtful mathematical behavior as opposed to rote memorization of rules, and to learn effective techniques for solving thought provoking problems while handling the arithmetic details by appropriate use of a calculator.

At the same time, for students to be able to interpret calculator output and to judge its correctness, it is important that the basics of the decimal system and the algorithms used in calculation be understood. In particular, this involves developing skill at mental arithmetic and in estimation.

When using the calculator, was an incorrect key pressed by mistake? Does the displayed answer make sense? The critical aphorism for calculating or computing is *garbage in, garbage out.* The very speed and neatness with which calculators and computers process data and return answers causes some unquestioningly to accept the results as correct. But this is frequently not so, and students must be able to ascertain with skill the reasonableness of results they obtain.

The calculation skills and essential knowledge for school mathematics include:

- knowledge of the one-digit arithmetic facts,
- the meaning of the arithmetic operations,
- the properties of numbers,
- the meaning of positional notation,
- the ability to perform mental arithmetic,
- the ability to calculate correctly and efficiently with a calculator,
- the ability to estimate the results of calculations.

We address these essentials here.

The Addition Algorithm

Consider the addition shown.

$$\begin{array}{r} {}^{1} \\ 28 \\ +\ 45 \\ \hline 73 \end{array}$$

This process depends entirely on our positional system of notation. But why does it work this way? Why "carry" the "one"? Why add by columns? To many young people, these procedures remain a great mystery—you add this way because you were told to—it's simply done by rote with no understanding.

As noted earlier, children learn abstract notions by first experiencing them concretely with devices they can actually see, touch, and manipulate. Thus, one should introduce the addition algorithm with manipulatives like base ten blocks, sticks and bundles of sticks; units, strips, and mats; abacuses; or, better yet, by using a number of these devices. For illustrative purposes here we use units, strips, and mats.

After cutting out their strips and mats, students will be aware that there are ten units on a strip and ten strips or 100 units on a mat, and they will be aware that, in manipulating these materials, they can make these exchanges back and forth as needed.

EXAMPLE 3.10 **Making Exchanges with Units, Strips, and Mats**

Suppose a number is represented by 15 units, 11 strips and 2 mats. What exchanges must be made in order to represent the same number but with the smallest number of manipulative pieces?

SOLUTION

Understand the problem • • • •

We are given the mats, strips, and units indicated in the statement of the problem and are asked to make exchanges that reduce the number of loose pieces of apparatus while keeping the same number of units in all.

Devise a plan • • • •

Since we know that 10 units form a strip and 10 strips form a flat, we can reduce the number of loose pieces by making these exchanges.

Carry out the plan • • • •

If we actually had in hand the pieces described we would physically make the desired exchanges. Here in the text, we illustrate the exchanges pictorially.

We reduce the number of pieces by replacing 10 units by 1 strip and 10 strips by 1 mat. This gives 3 mats, 2 strips, and 5 units for 325 units as shown.

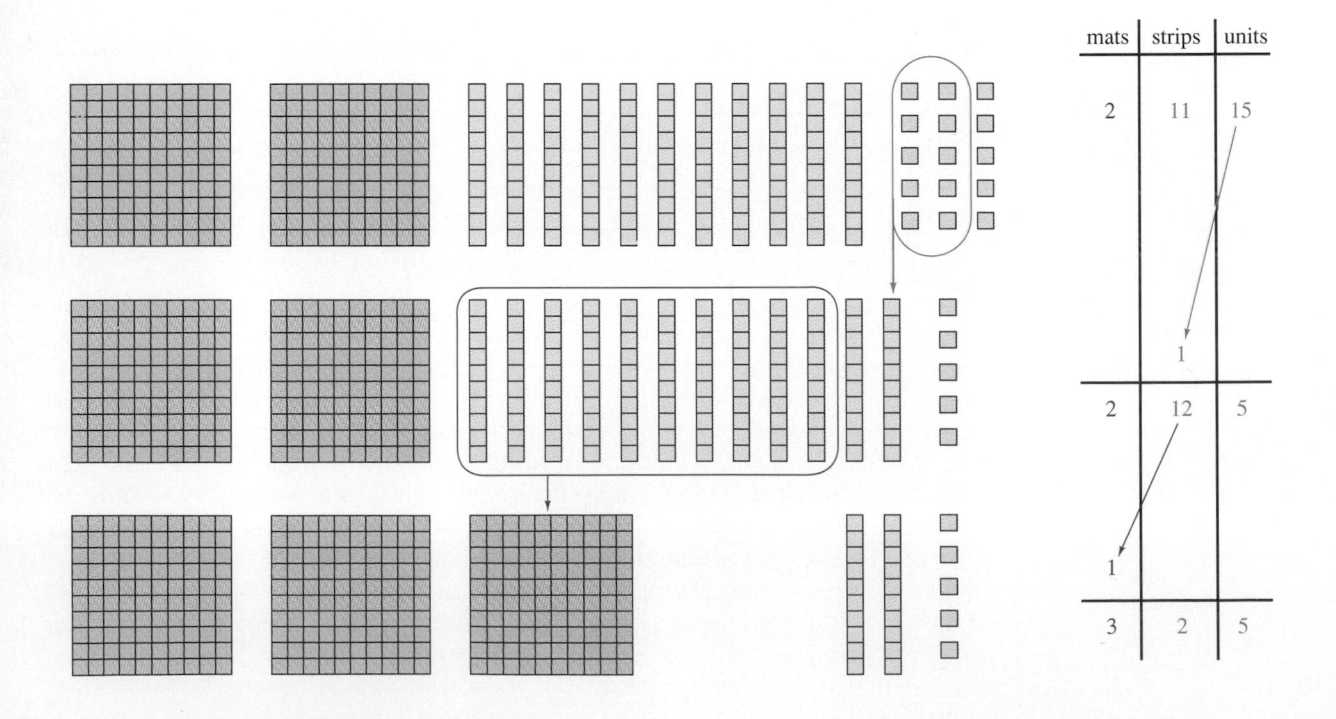

Look back • • • •

No more combining can take place because it takes 10 units to make a strip and 10 strips to make a mat. Thus, we finish with 3 mats, 2 strips, and 5 units for a total of 325 units. Moreover, we have not changed the total number of units since

$$2 \cdot 100 + 11 \cdot 10 + 15 = 200 + 110 + 15 = 325.$$

■

EXAMPLE 3.11

Developing the Addition Algorithm

Find the sum of 135 and 243.

SOLUTION

With units, strips, and mats

One hundred thirty-five is represented by 1 mat, 3 strips and 5 units, and 243 is represented by 2 mats, 4 strips and 3 units, as shown. All told this gives a total of 3 mats, 7 strips, and 8 units. Therefore, since no exchanges are possible, the sum is 378. Note how this illustrates the column by column addition algorithm typically used in pencil and paper calculation.

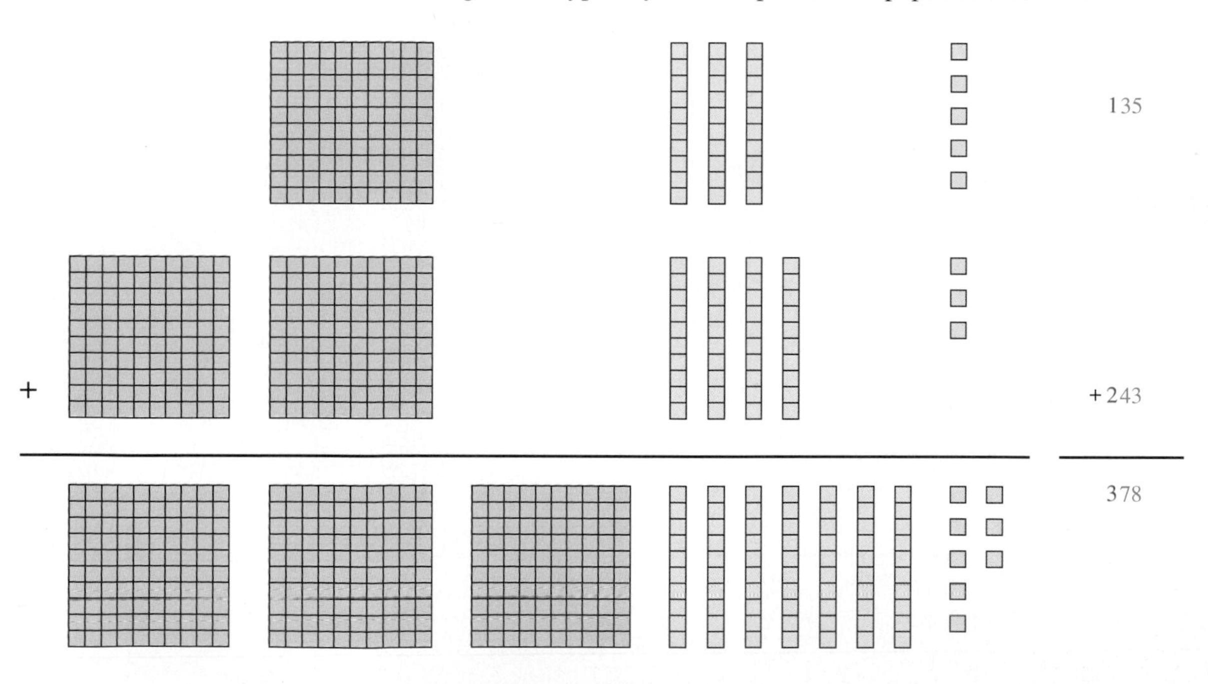

With place value cards

A somewhat more abstract approach to this problem is to use **place value cards;** that is, cards marked off in squares labeled 1s, 10s, and 100s right to left, and with the appropriate number of markers placed in each square to represent the desired number. A marker on the second square is worth 10 markers on the first square; a marker on the third square is worth

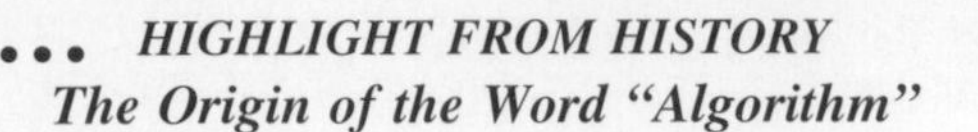

HIGHLIGHT FROM HISTORY
The Origin of the Word "Algorithm"

The word algorithm derives from al-Khowarizmi (al-ko-wár-izmi), the name of the eighth century Arab scribe who wrote two books on arithmetic and algebra giving a very careful account of the Hindu system of numeration and methods of calculation. Though al-Khowarizmi made no claim to inventing the system, careless readers of Latin translations of his book began to attribute the system to him and to call the new methods of calculation al-Khowarizmi or, carelessly, algorismi. Over time the word became algorithm, and came to mean any orderly, repetitive scheme.

10 markers on the second square, and so on. the addition of this example is illustrated as follows.

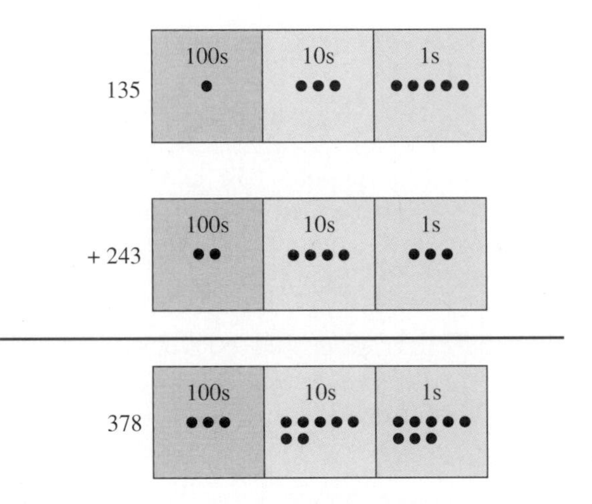

With place value diagrams and instructional algorithms

An even more abstract approach leading finally to the usual algorithm is provided by the following place value diagrams and instructional algorithms.

Place Value Diagram

	100s	10s	1s
	1	3	5
+	2	4	3
	3	7	8

Instructional Algorithm I

$$\begin{array}{r} 135 \\ +\ 243 \\ \hline 8 \\ 70 \\ 300 \\ \hline 378 \end{array}$$

8 ← 5 + 3
70 ← 30 + 40
300 ← 100 + 200

Instructional Algorithm II

$$\begin{array}{r} 135 \\ +\ 243 \\ \hline 8 \\ 7 \\ 3 \\ \hline 378 \end{array}$$

Suppress the zeros on the right.

Final Algorithm

$$\begin{array}{r} 135 \\ +\ 243 \\ \hline 378 \end{array}$$

Compress the result to a single line.

Notice how the degree of abstraction steadily increases as we move through the various solutions. If the elementary school teacher goes directly to the final algorithm, many students will be lost along the way. The approach from the concrete gradually moving toward the abstract is more likely to impart the desired understanding. ■

SCHOOL BOOK PAGE *Adding with Manipulatives*

CONSUMER CONNECTION

Exploring Three-Digit Addition

Build Understanding

Birthday Money
Materials: Play money
Groups: 3–5 students

A. Jim and Joanne put their money together to buy a birthday present for their mother. How much did they have altogether?

	$	d	p
Jim's money →	2	8	7
Joanne's money →	1	7	5

a. Use your play money to show each amount. Put the money together.

b. Trade 10 pennies for 1 dime.

c. Trade 10 dimes for 1 dollar.

d. Show the final amount: 4 dollars, 6 dimes, and 2 pennies.

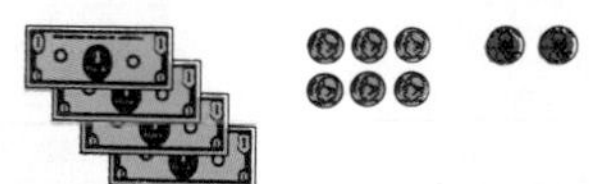

Jim and Joanne had $4.62 in all.

1. Here the manipulatives being used to illustrate the addition algorithm are pennies, dimes, and dollars. Do these seem particularly appropriate for use with elementary school students? Explain.
2. On this sample page the word "trade" is used instead of exchange or carry. Why is it better to use words like trade or exchange, rather than carry and phrases like "carry 1 to the next column?" Explain.
3. Illustrate this addition using a place value diagram.

EXAMPLE 3.12 Adding with Exchanging

Find the sum of 357 and 274.

SOLUTION

With units, strips, and mats

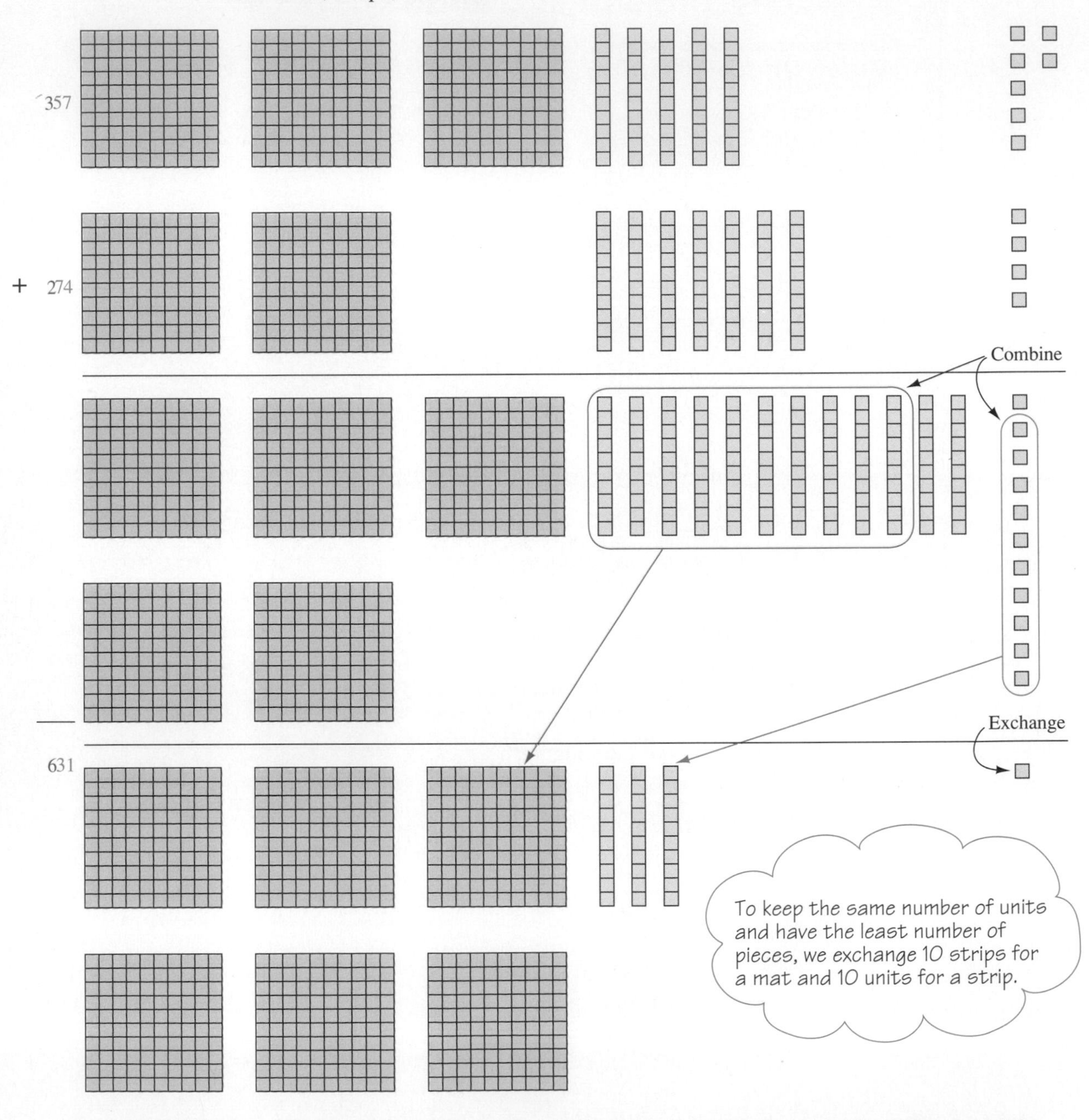
357
+ 274
Combine
631
Exchange
To keep the same number of units and have the least number of pieces, we exchange 10 strips for a mat and 10 units for a strip.

```
 1 1
 357
+274
 631
```

Note how the above not only illustrates the usual column by column addition of pencil and paper arithmetic, but also "carrying" or, more appropriately, "exchanging." The 1 "carried" from the first column to the second is really 1 ten (represented by a strip) and the one "carried" from the second column to the third is really 1 hundred (represented by a mat).

With place value cards

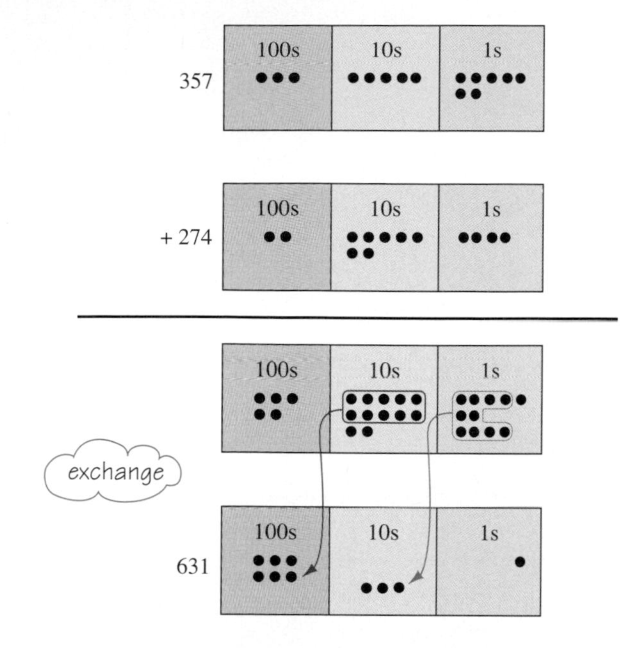

With place value diagrams and instructional algorithms

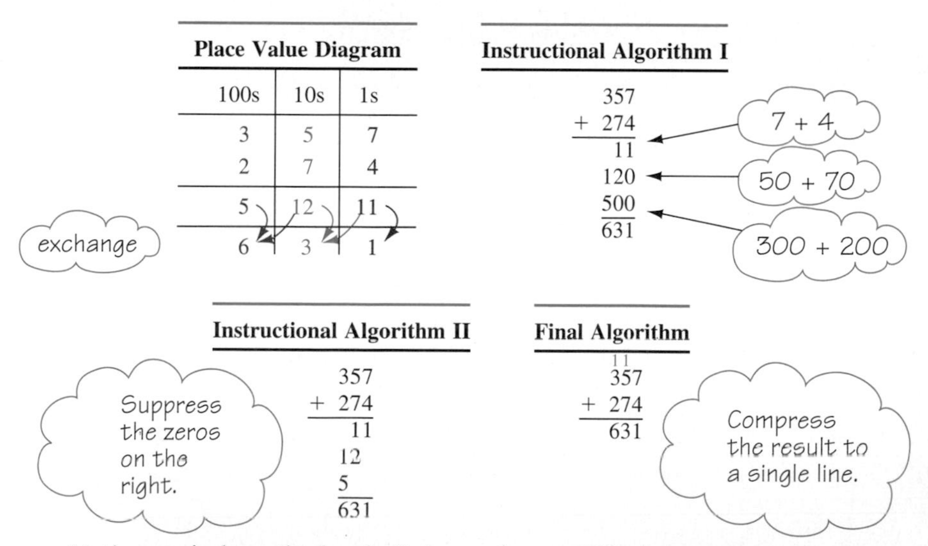

Notice again how the level of abstraction steadily increases as we move through these solutions. This example, and the various solutions should make it clear to students that at no point are they "carrying a one." It is always the case of exchanging 10 ones from the ones column to the tens column, 10 tens from the tens column to the hundreds column, and so on. ■

The Subtraction Algorithm

We can illustrate the subtraction algorithm in much the same way as the addition algorithm. For primary children, the idea of subtraction is often understood in terms of "take away." Thus, if you have 9 apples and I take away 5, you have 4 left as in Figure 3.9.

Figure 3.9
Subtraction as "take away"

EXAMPLE 3.13

Subtracting without Exchanging

Subtract 243 from 375.

SOLUTION

With units, strips, and mats

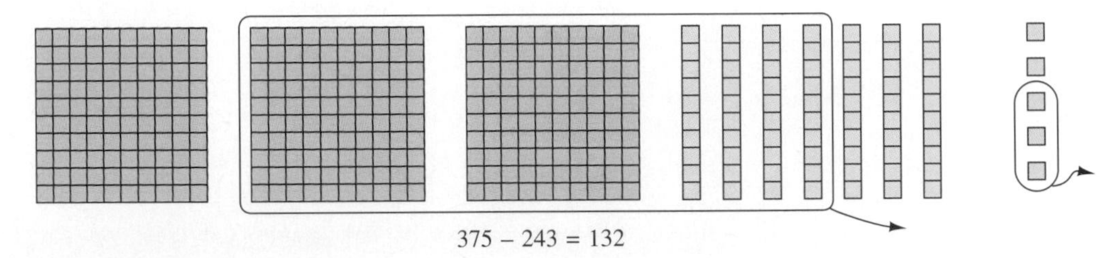

With place value cards

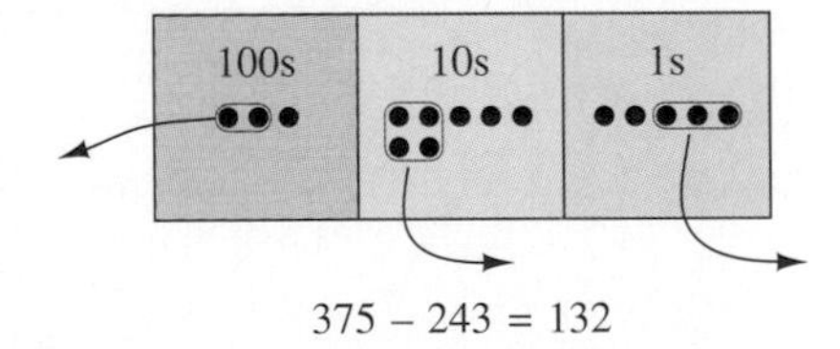

With place value diagrams and instructional algorithms

Place Value Diagram

	100s	10s	1s
	3	7	5
−	2	4	3
	1	3	2

Instructional Algorithm I

$$\begin{array}{r} 375 \\ -\ 243 \\ \hline 2 \\ 30 \\ 100 \\ \hline 132 \end{array}$$

5 − 3 (for 2)
70 − 40 (for 30)
300 − 200 (for 100)

Instructional Algorithm II

$$\begin{array}{r} 375 \\ -\ 243 \\ \hline 2 \\ 3 \\ 1 \\ \hline 132 \end{array}$$

Suppress the zeros on the right.

Final Algorithm

$$\begin{array}{r} 375 \\ -\ 243 \\ \hline 132 \end{array}$$

Compress the result to a single line. ■

EXAMPLE 3.14 **Subtracting with "Exchanging"**

Subtract 185 from 362.

SOLUTION

With units, strips, and mats

We start with 3 mats, 6 strips, and 2 units.

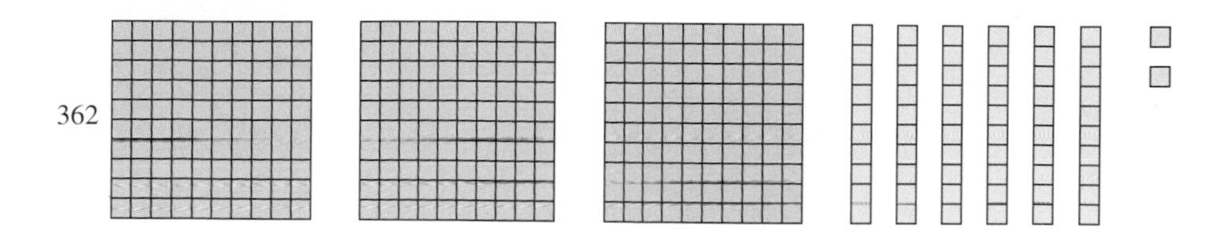

We want to take away 1 mat, 8 strips, and 5 units. Since we cannot pick up 5 units from our present arrangement, we exchange a strip for 10 units to obtain 3 mats, 5 strips and 12 units.

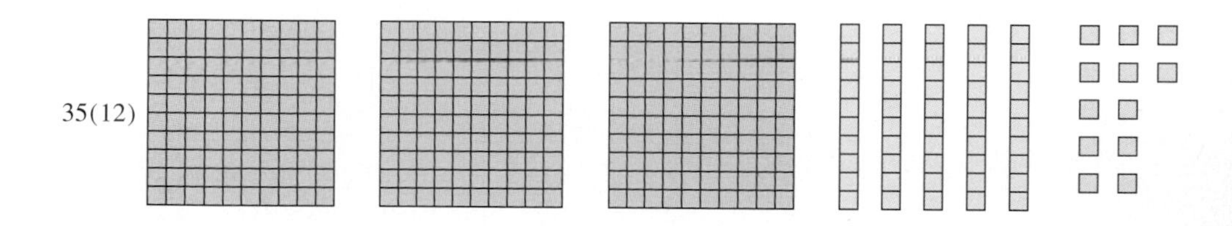

We can now take away 5 units, but we still cannot pick up 8 strips. Therefore, we exchange a mat for 10 strips to obtain 2 mats, 15 strips, and 12 units. Finally we are able to take away 1 mat, 8 strips, and 5 units (that is, 185 units) as shown. This leaves 1 mat, 7 strips, and 7 units. So

$$362 - 185 = 177.$$

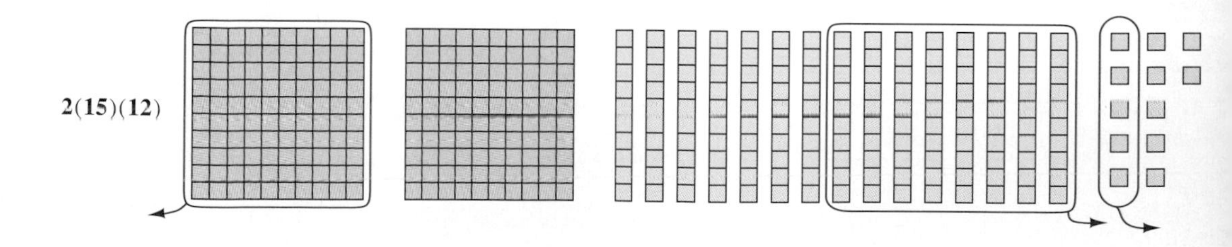

With place value cards

We must take away 1 marker from the 100s square, 8 markers from the 10s square, and 5 markers from the 1s square. To make this possible we trade 1 marker from the 10s square for 10 markers on the 1s square and 1 marker on the 100s square for 10 markers on the 10s square. Now, taking away the desired markers, we have 1 marker left on the 100s square, 7 markers left on the 10s square, and 7 markers left on the 1s square for 177.

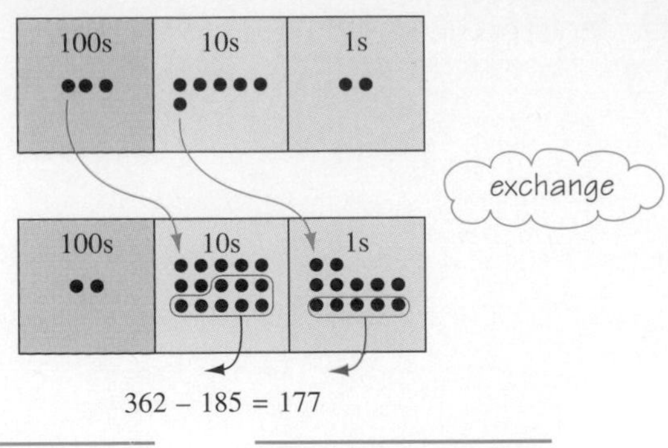

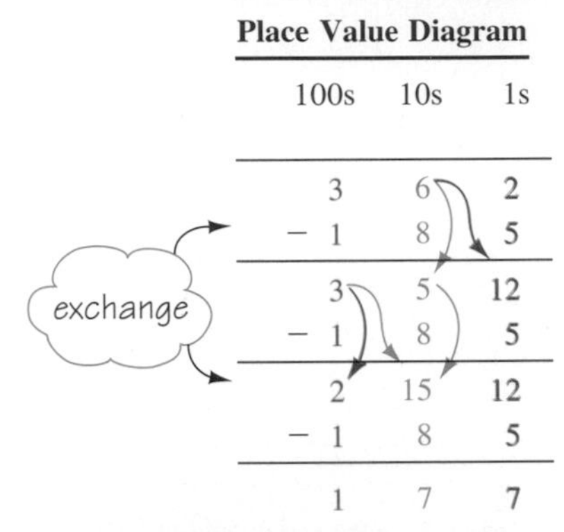

Instructional Algorithm

3	6	2
− 1	8	5
3	5	(12)
− 1	8	5
2	(15)	(12)
− 1	8	5
1	7	7

exchange

Final Algorithm

~~3~~ 2	~~6~~ [1]5	[1]2
− 1	8	5
1	7	7

Arithmetic in Other Bases

It is interesting to note that the algorithms for addition and subtraction are just as valid in base five or any other base as they are in base ten, and that the notions can be developed with manipulatives just as for base ten. Indeed, it is reasonable to consider other bases right along with base ten. Studied together this way, students gain a greater understanding of the whole idea of positional notation and the related algorithms. In base five, for example, the place value cards would have a units or 1s square, a 5s square, a 5^2 or 25s square, and so on. With sticks, we could use loose sticks, bundles of 5 sticks, bundles of 5 bundles of sticks, and so on. With units, strips, and mats, we would have units, strips with 5 units each, and mats with 5 strips per mat.

JUST FOR FUN

What's the Difference?

Perform these subtractions.

91	95	42	62	74
− 19	− 59	− 24	− 26	− 47
61	82	81	32	54
− 16	− 28	− 18	− 23	− 45

(a) What do you notice about the answer in each case?

(b) Can you predict the answers to all problems like this just by looking at the two digits involved?

(c) What seems to be true about the sums of the digits in the answers to all such 2-digit problems?

Adding in Base Five

The next example shows us how the addition algorithm works in base five. Skipping the concrete realization for brevity, let's see how the process works using place value cards.

EXAMPLE 3.15

Adding in Base Five

Compute the sum of 143_{five} and 234_{five} in base five notation.

SOLUTION

With place value cards

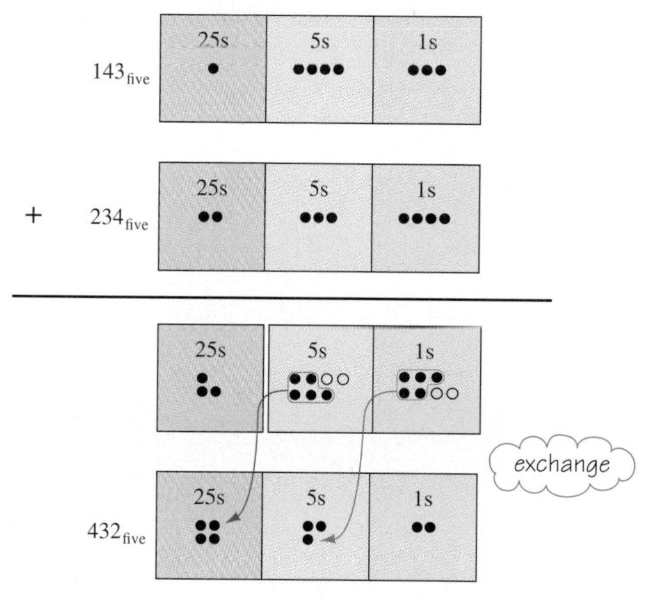

With place value diagrams and instructional algorithms

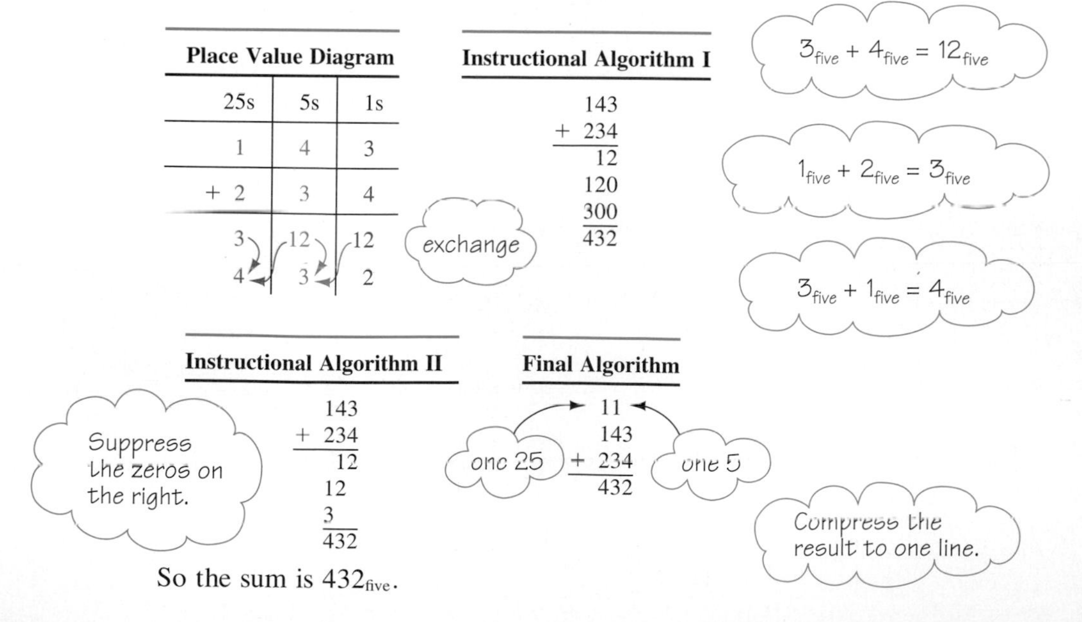

Place Value Diagram

25s	5s	1s
1	4	3
+ 2	3	4
3 / 4	12 / 3	12 / 2

So the sum is 432_{five}.

To check the addition, we convert everything to base ten where we feel comfortable.

$$\begin{array}{rcr} 143_{\text{five}} &= 1 \cdot 25 + 4 \cdot 5 + 3 \cdot 1 =& 48_{\text{ten}} \\ \underline{234_{\text{five}}} &\underline{= 2 \cdot 25 + 3 \cdot 5 + 4 \cdot 1 =}& \underline{69_{\text{ten}}} \\ 432_{\text{five}} &= 4 \cdot 25 + 3 \cdot 5 + 2 \cdot 1 =& 117_{\text{ten}} \end{array}$$

The result is confirmed since $48_{\text{ten}} + 69_{\text{ten}} = 117_{\text{ten}}$. ■

The difficulty in doing base five arithmetic is unfamiliarity with the meaning of the symbols—we do not recognize at a glance, for example, that $8_{\text{ten}} = 13_{\text{five}}$. On the other hand, thinking of units, strips with 5 units per strip, and so on, it is not hard to think of 1 strip and 3 units as 8. In fact, doing arithmetic in another base gives good practice in mental arithmetic which is a desirable end in itself. Of course, elementary school children memorize the addition and multiplication tables in base ten so that the needed symbols are readily recalled and we could do the same thing here. The addition table in base five, for example, is as shown.

+	0	1	2	3	4
0	0	1	2	3	4
1	1	2	3	4	10
2	2	3	4	10	11
3	3	4	10	11	12
4	4	10	11	12	13

These numerals are in base five.

This could be used to make the above addition easier and more immediate. From the table we see that $3 + 4 = 12_{\text{five}}$ so we write down the 2 and exchange the five 1s for a five in the fives column as shown above. Next, $4 + 3 = 12_{\text{five}}$ and the 1 from the exchange gives 13_{five}. Thus, we write down the 3 and exchange the five 5s for one 25 in the next column. Finally, $1 + 2 = 3$ and 1 from the exchange makes 4, so we obtain 423_{five} as before. However, there is no merit in memorizing the addition table in base five; it is far better just to think carefully about what is going on.

Before we finish this section, let's look at subtraction.

Subtracting in Base Five

EXAMPLE 3.16 **Subtracting in Base Five**

Subtract 143_{five} from 234_{five} in base five notation.

SOLUTION

With place value cards

There is no problem in taking away 3 markers from the 1s square, but we cannot remove 4 markers from the 5s square without exchanging 1 marker on the 25s square for 5 markers on the 5s square. Taking away the desired markers, we are left with zero 25s, four 5s, and one 1 for 41_{five}.

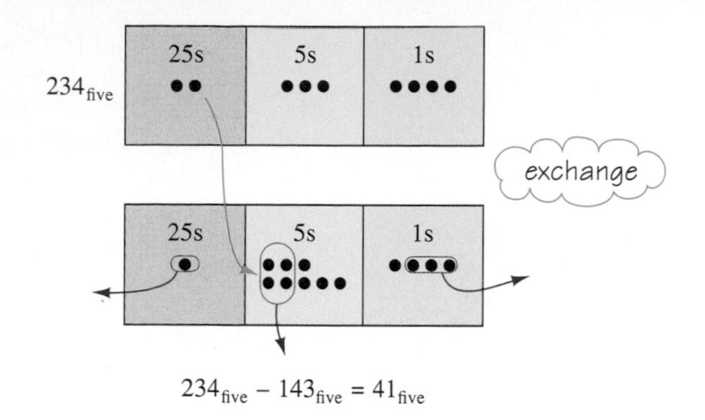

With place value diagrams and instructional algorithms

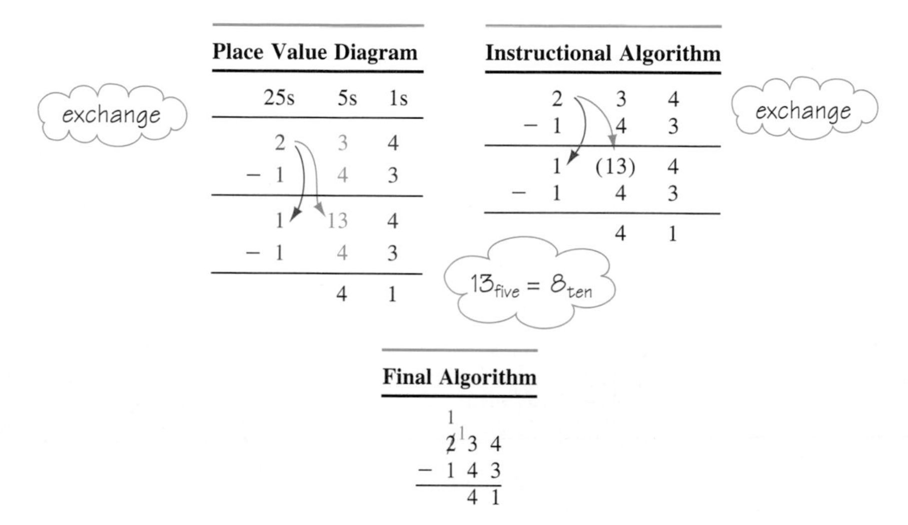

So the answer is 41_{five}.

As a check, we already know that $243_{five} = 69_{ten}$ and $143_{five} = 48_{ten}$. Since $69 - 48 = 21$ and $41_{five} = 4 \cdot 5 + 1 \cdot 1 = 21$, the calculation checks. ■

PROBLEM SET 3.3

Understanding Concepts

1. Sketch the solution to 36 + 75 using:
   (a) mats, strips, and units. Draw a square for a mat, a vertical line segment for a strip, and a dot for a unit.
   (b) place value cards marked 1s, 10s, and 100s from right to left.
2. Use the addition Instructional Algorithm I to perform the following additions:
   (a) 23 + 44 **(b)** 57 + 84 **(c)** 324 + 78
3. The hand calculation of the sum 279 and 84 involves two exchanges and might appear as follows.

$$\begin{array}{r} {\scriptstyle 11} \\ 279 \\ 84 \\ \hline 363 \end{array}$$

   Carefully describe each of the exchanges.
4. While Sylvia was trying to balance her father's checkbook, her calculator battery went dead. When she added up the outstanding checks by hand, her work looked like this. Is the addition correct? Discuss each of the exchanges shown. Is this how you would have proceeded? How else might you have done it?

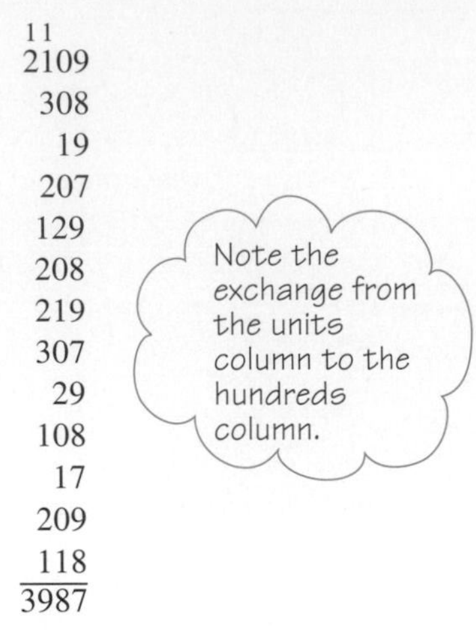

11
2109
308
19
207
129
208
219
307
29
108
17
209
118
3987
Note the exchange from the units column to the hundreds column.

5. Sketch the solution to 275 − 136 using:
   (a) mats, strips, and units.
   (b) place value cards marked 1s, 10s, and 100s from right to left.
6. Use the subtraction Instructional Algorithm I to perform these subtractions:
   (a) 78 − 35 **(b)** 75 − 38 **(c)** 414 − 175
7. The calculation of the difference 523 − 247 might look like this.

```
   4 11
   5̸ 2̸¹3
 − 2 4 7
   2 7 6
```

Carefully discuss the exchanges indicated.
8. The calculation of the difference 30007 − 1098 might look like this.

```
  2  9  9  9
  3̸¹0̸¹0̸¹0̸¹7
 −   1 0 9 8
   2 8 9 0 9
```

Carefully discuss the exchanges indicated.
9. The column by column addition of numbers can be justified as follows. State the property of the whole numbers that justifies each of these steps. We begin with expanded notation.

$36 + 52 = (3 \cdot 10 + 6) + (5 \cdot 10 + 2)$	expanded notation
$= 3 \cdot 10 + [6 + (5 \cdot 10 + 2)]$	(a) ______
$= 3 \cdot 10 + [(6 + 5 \cdot 10) + 2]$	(b) ______
$= 3 \cdot 10 + [(5 \cdot 10 + 6) + 2]$	(c) ______
$= 3 \cdot 10 + [5 \cdot 10 + (6 + 2)]$	(d) ______
$= (3 \cdot 10 + 5 \cdot 10) + (6 + 2)$	(e) ______
$= (3 + 5) \cdot 10 + (6 + 2)$	(f) ______
$= 8 \cdot 10 + 8$	addition facts
$= 88$	expanded notation

10. (a) In a single column, write the base four representations of the numbers from 0 to 15 inclusive.
    (b) Briefly discuss any pattern you noticed in part (a).
11. Complete the addition table for base four arithmetic shown by placing the sum of $a + b$ at the intersection of the ath row and the bth column. The subscript four may be omitted here.

+	**0**	**1**	**2**	**3**
0				
1				
2				
3				

12. Complete the following computations in base four notation. The numerals are written in base four and the subscript indicating base four may be omitted from your work. Check your work by converting to base ten notation.

```
(a)   231   (b)   303   (c)   1223
    + 121       +  33       +  231

(d)   333   (e)    32   (f)    302
    + 101       +  13       −  103

(g)   212   (h)  3102
    −  33      − 1033
```

Thinking Critically

13. On his way to school, Peter dropped his arithmetic paper in a puddle of water blotting out some of his work. What digits should go under the blots on these problems? (The base is ten.)

```
(a)   6■■3   (b)   77■   (c)    8■
    + ■51■        + ■■2       + 362
     ■2282          871        ■■43

(d)   248■   (e)   4■2   (f)   34■5
    − 1■22       − 1843      − ■748
      ■1■9        ■15■         ■2■
```

14. Find the missing digits in each of these base ten addition problems.

```
(a)   _437   (b)   _721
       2_1          901_
    + 347_       + 71_3
      6_94         _0_26

(c)   38_1   (d)    5_4
       24_3         612_
    + 512_       +  8_1
      __5_9        76_6
```

15. Fill in the missing digits in each of these base ten subtraction problems.

(a)
$$\begin{array}{r} -3- \\ -\ 2-1 \\ \hline 594 \end{array}$$

(b)
$$\begin{array}{r} 3--4 \\ -\ 346 \\ \hline 175- \end{array}$$

(c)
$$\begin{array}{r} 7-4- \\ -\ -5-4 \\ \hline 808 \end{array}$$

(d)
$$\begin{array}{r} 63--4 \\ -\ 2-12- \\ \hline -6209 \end{array}$$

16. These additions and subtractions are written in different bases. Determine the base used in each case. (*Hint:* There may be more than one correct answer.)

(a)
$$\begin{array}{r} 231 \\ +\ 414 \\ \hline 1200 \end{array}$$

(b)
$$\begin{array}{r} 231 \\ +\ 414 \\ \hline 1045 \end{array}$$

(c)
$$\begin{array}{r} 231 \\ +\ 414 \\ \hline 645 \end{array}$$

(d)
$$\begin{array}{r} 344 \\ +\ 143 \\ \hline 1042 \end{array}$$

(e)
$$\begin{array}{r} 523 \\ -\ 254 \\ \hline 236 \end{array}$$

(f)
$$\begin{array}{r} 523 \\ -\ 254 \\ \hline 247 \end{array}$$

(g)
$$\begin{array}{r} 523 \\ -\ 254 \\ \hline 28E \end{array}$$

(h)
$$\begin{array}{r} 1020 \\ -\ 203 \\ \hline 312 \end{array}$$

17. There is a rather interesting addition algorithm called the **scratch** method that proceeds as follows. Consider this sum.

$$\begin{array}{r} 2\ 834 \\ \not{5}\not{7}\not{6} \\ 4\ \not{8}35 \\ \not{2}\ 743 \\ \hline 10{,}988 \end{array}$$

Begin by adding from the top down in the units column. When you add a digit that makes your sum 10 or more, scratch out the digit as shown and make a mental note of the units digit of your present sum. Start with the digit noted and continue adding and scratching until you have completed the units column, writing down the units digit of the last sum as the units digit of the answer as shown. Now, count the number of scratches in the units column and, starting with this number, add on down the tens column repeating the scratch process as you go. Continue the entire process until all the columns have been added. This gives the desired answer. Explain why the algorithm works.

Thinking Cooperatively

18. Form two 4-digit numbers using each of 1, 2, 3, 4, 5, 6, 7, and 8 once, and only once, so that

(a) the sum of the two numbers is as large as possible.

(b) the sum of the two numbers is as small as possible.

(c) You can do better than guess and check on parts (a) and (b). Explain your solution strategy briefly.

(d) Is there only one answer to each of parts (a) and (b)? Explain in two sentences.

19. (a) Let s denote the sum of the numbers in problem 18, parts (a) and (b). Use your calculator to compute these products.

$1 \cdot s =$ ________

$2 \cdot s =$ ________

$3 \cdot s =$ ________

$4 \cdot s =$ ________

$5 \cdot s =$ ________

$6 \cdot s =$ ________

$7 \cdot s =$ ________

$8 \cdot s =$ ________

$9 \cdot s =$ ________

$10 \cdot s =$ ________

(b) Briefly discuss any patterns you notice in part (a).

(c) See if you can predict the product $41 \cdot s$ and then check the result with a calculator.

(d) Compute $14 \cdot s$, $23 \cdot s$, and $32 \cdot s$.

(e) Compute $6 \cdot s$, $17 \cdot s$, $28 \cdot s$, and $39 \cdot s$.

(f) What kind of sequences are 14, 23, 32, 41 and 6, 17, 28, 39?

20. Form two 4-digit numbers using each of the digits 1, 2, 3, 4, 5, 6, 7, and 8 precisely once so that:

(a) the difference of the two numbers is a natural number that is as small as possible.

(b) the difference of the two numbers is as large as possible.

(c) Briefly explain the strategy you used in solving this problem.

(d) Is there more than one solution to parts (a) and (b)? Why or why not?

Using a Calculator

21. When doing calculations with very large numbers, calculators must often make approximations. However, exact calculations with large numbers can be accomplished using a combination of pencil and paper and calculator techniques.

Suppose that you want to add 3,472,859,467,283 to 23,147,289,846,782. Use your calculator and organize the work as follows.

1	
3,472,859	467,283
+ 23,147,289	846,782
26,620,149	314,065

note the exchange to the millions column

Thus the answer is 26,620,149,314,065 as desired. Perform each of the following additions accurately by the above method.

(a) 34,270,185,934,875 + 25,071,400,283,468

(b) 692,878,467,958,871 + 728,765,483,924,877

22. Exact subtraction of very large numbers can be carried out in much the same way as the additions in problem 23. However, it is important to break the numbers into segments in such a way that negative numbers are avoided. For example, in subtracting 223,471,425, 235 from 873,751,234,612, if we break the numbers into 6-digit groups, we obtain this array. Then the subtraction gives

873,751	234,612
− 223,471	425,235
650,280	−190,623

The negative number here causes the problem.

the result shown and the two parts can only be combined with further tedious arithmetic. However, if we break the numbers one column further right, we have

8737512	34612
− 2234714	25235
6502798	09377

so the desired answer is 650,279,809,377. Perform each of these subtractions by this method.

(a) 65,421,534,784 − 24,131,243,677

(b) 5,378,246,102 − 2,157,327,081

For Review

23. Write two division equations and a second multiplication equation corresponding to each of these multiplication equations.

(a) $11 \cdot 91 = 1001$ **(b)** $7 \cdot 84 = 588$

(c) $13 \cdot 77 = 1001$

24. Write a multiplication equation and a second division equation corresponding to each of these division equations.

(a) $1001 \div 7 = 143$ **(b)** $323 \div 19 = 17$

(c) $899 \div 31 = 29$

25. If Jennifer Hurwitz had four sweaters, three blouses, and two pairs of slacks that all mixed and matched, how many different outfits could she wear?

26. Let $A = \{1, 2, 3, 4, 5\}$, $B = \{2, 4, 6, 8\}$, and $C = \{3, 4, 5, 6, 7\}$.

(a) Determine $A \cup B \cup C$, $A \cap B$, $A \cap C$, $B \cap C$, and $A \cap B \cap C$.

(b) Determine $n(A \cup B \cup C)$.

(c) Compute $n(A) + n(B) + n(C) - n(A \cap B) - n(A \cap C) - n(B \cap C) + n(A \cap B \cap C)$.

27. **(a)** Repeat problem 26 with three different sets A, B, and C of your own choosing.

(b) What general result do problems 26 and 27 suggest?

(c) Write a paragraph of argument that the result guessed in part (b) is true. (*Hint:* Consider an element x not in any of A, B, or C; y in A but not B or C, and so on.)

3.4 Algorithms for Multiplication and Division of Whole Numbers

Multiplication and division in most ancient numeration systems were quite complicated. The decimal system makes these processes much easier, but many people still find them confusing. At least part of the difficulty is that the ideas are often presented as a collection of rules to be learned by rote, with little or no effort made to impart understanding. In this section we endeavor to strip away some of the mystery.

Multiplication Algorithms

Multiplication is repeated addition. Thus, $2 \cdot 9$ means $9 + 9$, $3 \cdot 9$ means $9 + 9 + 9$, and so on. But repeated addition is slow and tedious and easier algorithms exist. As with addition and subtraction, these should be introduced starting with concrete approaches and gradually becoming more and more abstract. The development should proceed through units, strips, and mats; place value cards; classroom abacuses; and so on. Let's consider the product $9 \cdot 3$.

Multiplication Using Units, Strips, and Mats

Since $9 \cdot 3 = 3 + 3 + 3 + 3 + 3 + 3 + 3 + 3 + 3$, we can illustrate this as in Figure 3.10 with 9 rows of 3 units each. Simplifying the original array by appropriately exchanging units for strips, we eventually have 2 strips and 7 units which is recorded as 27. Of course, elementary school children should actually handle the materials, making the necessary exchanges of units for strips.

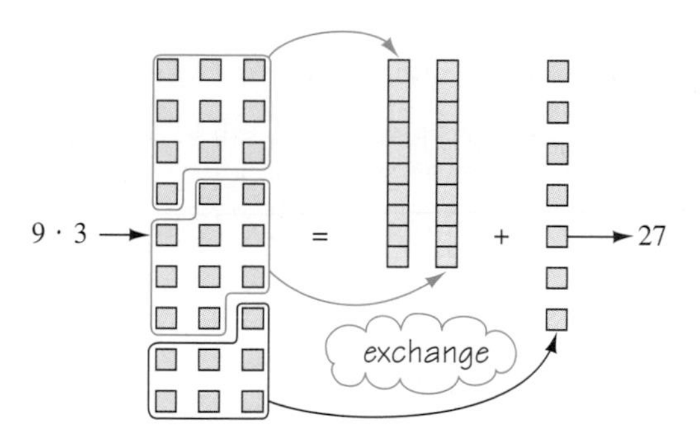

Figure 3.10
Illustrating 9 · 3 with units, strips, and mats

Multiplication Using Place Value Cards

With place value cards students should start with 9 rows of 3 markers each of the 1s square of their place value cards as shown in Figure 3.11. They then exchange 10 markers on the 1s square for 1 marker on the 10s square as many times as possible, and record the fact that this gives 2 tens plus 7 units or 27.

Figure 3.11
Illustrating 9 · 3 with place value cards

Using manipulatives as illustrated, children can experience, learn and actually *understand* all the one-digit multiplication facts. It should not be the case that $9 \cdot 3 = 27$ is a string of meaningless symbols memorized by rote. This basic fact must be understood.

Once the one-digit facts are thoroughly understood *and memorized,* as they must be, even for intelligent use of a calculator, one can move on to more complicated problems. Consider, for example, 3 · 213. This could be illustrated with units, strips, and mats but, for brevity, we will go directly to place value cards, expanded notation, and then to an algorithm as illustrated in Figure 3.12. Again we make use of the understanding that 3 · 213 = 213 + 213 + 213.

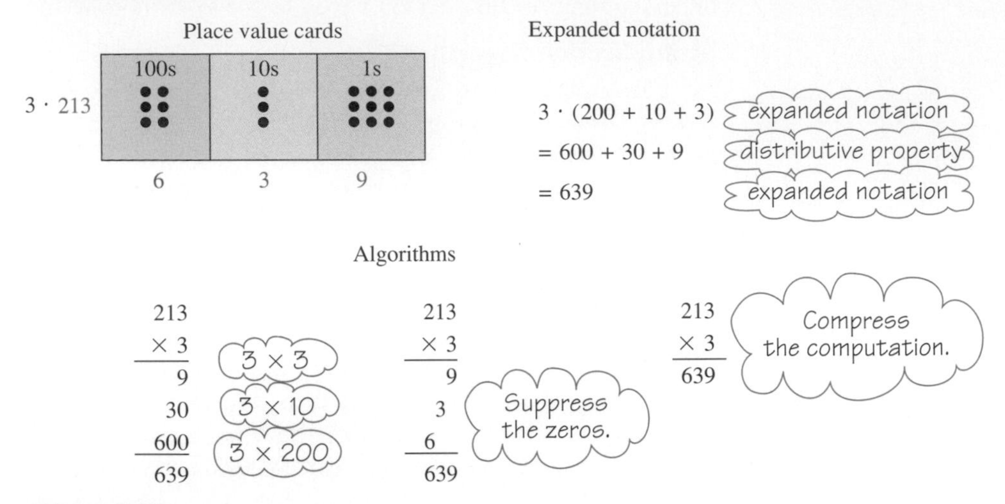

Figure 3.12
Developing 3 · 213 from concrete representation to final algorithm

The product 3 · 213 did not require exchanging. Consider the product 4 · 243. Presentation of this product proceeds from the concrete representation to the final algorithm as shown in Figure 3.13.

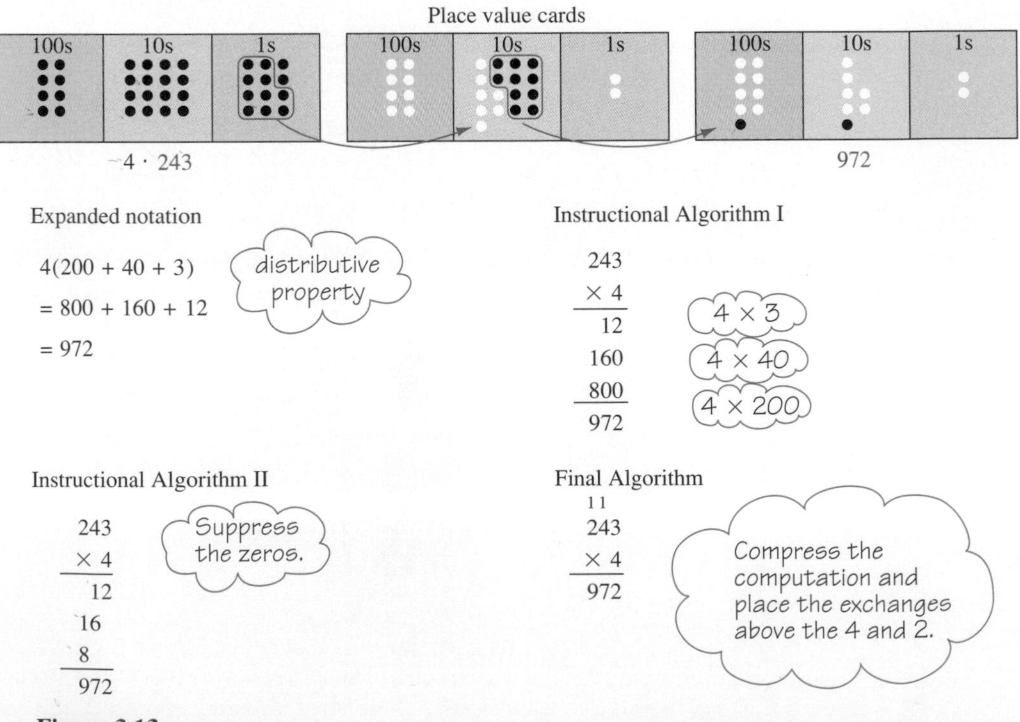

Figure 3.13
Developing the product 4 · 243 from concrete realization to final algorithm with increasing levels of abstraction

Of course, this is not intended to represent one day's lesson. Though the various demonstrations should clearly be tied together, enough time should be spent to assure that each level of the chain of reasoning is understood before proceeding to the next.

Finally, consider the product 24 · 324. Unquestionably, the most efficient algorithm is that provided by the calculator. The pencil and paper algorithm, however, is not unusually difficult as the following series of calculations show.

Expanded notation

$$
\begin{aligned}
24 \cdot 324 &= (20 + 4) \cdot 324 && \text{distributive property}\\
&= 20 \cdot 324 + 4 \cdot 324\\
&= 20(300 + 20 + 4) + 4(300 + 20 + 4) && \text{expanded notation}\\
&= 6000 + 400 + 80 + 1200 + 80 + 16 && \text{distributive property}\\
&= 7776
\end{aligned}
$$

Instructional algorithm I

$$
\begin{array}{rrl}
 & 324 & \\
 & \times\ \ 24 & \\
\hline
 & 16 & 4 \cdot 4\\
4 \cdot 20 & 80 & \\
 & 1200 & 4 \cdot 300\\
20 \cdot 4 & 80 & \\
 & 400 & 20 \cdot 20\\
20 \cdot 300 & 6000 & \\
\hline
 & 7776 &
\end{array}
$$

Instructional algorithm II

$$
\begin{array}{rl}
324 & \\
\times\ \ 24 & \\
\hline
16 & \\
8\ & \\
12\ \ & \text{Suppress the zeros on the right.}\\
8\ & \\
4\ \ & \\
6\ \ \ & \\
\hline
7776 &
\end{array}
$$

Final algorithm

$$
\begin{array}{r}
3\overset{1}{2}4\\
\times\ \ 24\\
\hline
1296\\
648\ \\
\hline
7776
\end{array}
$$

Division Algorithms

An approach to division discussed in Chapter 2 was repeated subtraction. This ultimately led to the so-called **division algorithm** which we restate here for easy reference.

THEOREM The Division Algorithm

If a and b are whole numbers with b not zero, there exists precisely one pair of whole numbers q and r with $0 \leq r < b$ such that $a = bq + r$.

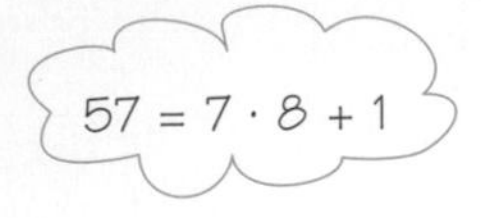

The Long Division Algorithm

Suppose we want to divide 942 by 7. Doing this by repeated subtraction will take a long time even with a calculator. But, suppose we subtract several sevens at a time and keep track of the number we subtract each time. Indeed, since it is so easy to multiply a number by 10, 100, 1000, and so on, let's subtract hundreds of sevens, tens of sevens, and so on. The work might be organized like this.

```
7)941
  700   Subtract 100 sevens
  241
   70   Subtract  10 sevens
  171
   70   Subtract  10 sevens
  101
   70   Subtract  10 sevens
   31
   28   Subtract   4 sevens
    3            134 number of sevens subtracted
```

Since $3 < 7$, the process stops and we see that 941 divided by 7 gives a quotient of 134 and a remainder of 3. As a check we note that $941 = 7 \cdot 134 + 3$. The above work could have been shortened if we had subtracted the three tens of sevens all at once and then the four sevens all at once like this.

```
7)941
  700    100
  241
  210     30
   31
   28      4
    3    134
```

A slightly different form of this algorithm, sometimes called the **scaffold** algorithm, is obtained by writing the 100, 30, and 4 above the divide symbol like this.

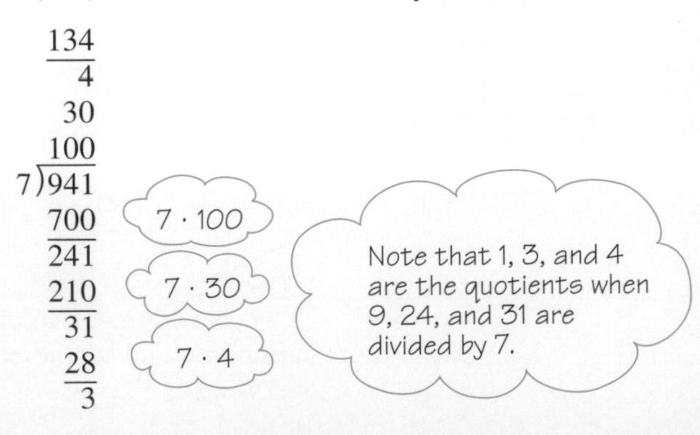

134
4
30
100
7)941
700
241
210
31
28
3
7 · 100
7 · 30
7 · 4
Note that 1, 3, and 4 are the quotients when 9, 24, and 31 are divided by 7.

Finally, suppressing the zeros on the right and in the 100 and 30, we obtain the usual long division algorithm

$$
\begin{array}{r}
134 \\
7\overline{)941} \\
\underline{7} \\
24 \\
\underline{21} \\
31 \\
\underline{28} \\
3
\end{array}
$$

which focuses our attention on the 9, 24, and 32 as noted above.

A further example with a larger divisor may be helpful.

EXAMPLE 3.17

Using the Long Division Algorithm

Divide 28,762 by 307.

SOLUTION

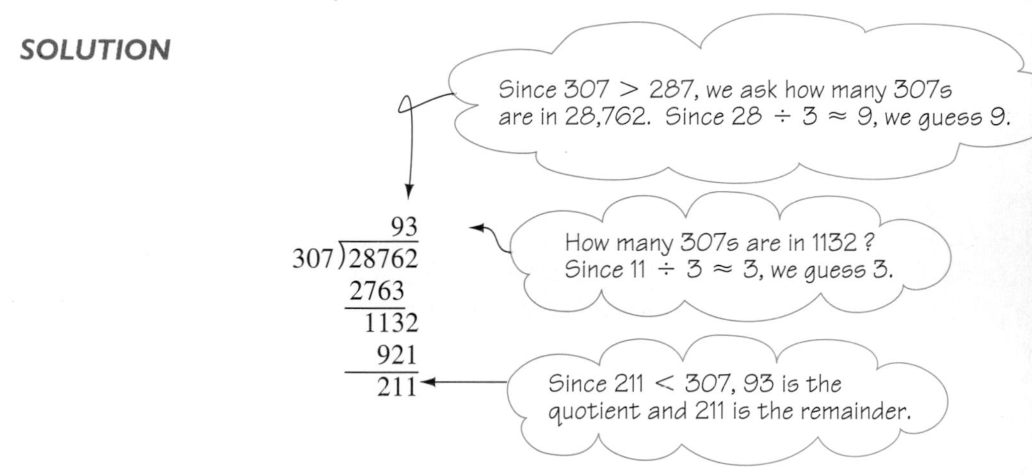

Check: $28{,}762 = 307 \cdot 93 + 211$. ■

JUST FOR FUN

What's the Sum?

Perform these additions.

$$
\begin{array}{ccccc}
\begin{array}{r} 91 \\ +\ 19 \\ \hline \end{array} &
\begin{array}{r} 95 \\ +\ 59 \\ \hline \end{array} &
\begin{array}{r} 42 \\ +\ 24 \\ \hline \end{array} &
\begin{array}{r} 62 \\ +\ 26 \\ \hline \end{array} &
\begin{array}{r} 74 \\ +\ 47 \\ \hline \end{array} \\
\begin{array}{r} 61 \\ +\ 16 \\ \hline \end{array} &
\begin{array}{r} 82 \\ +\ 28 \\ \hline \end{array} &
\begin{array}{r} 81 \\ +\ 18 \\ \hline \end{array} &
\begin{array}{r} 32 \\ +\ 23 \\ \hline \end{array} &
\begin{array}{r} 54 \\ +\ 45 \\ \hline \end{array}
\end{array}
$$

(a) Investigate the results of dividing the answers to the above additions by 11.

(b) Can you predict the answers to all problems just by looking at the two digits involved?

The Short Division Algorithm

A division algorithm that is quite useful, even in this calculator age, is the **short division** algorithm. This is a much simplified version of the long division algorithm and is quite useful and quick when the divisor is a single digit. This can be developed using the scaffold method as above. Also, it follows directly from the long division algorithm if that is already known.

EXAMPLE 3.18

Using the Short Division Algorithm

Divide 2834 by 3 and check your answer.

SOLUTION

Consider the following divisions which show how the short division algorithm derives from the long division algorithm.

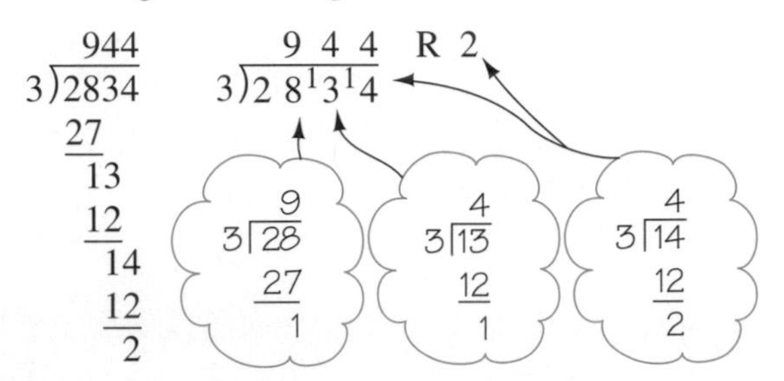

■

Multiplication and Division in Other Bases

As with addition and subtraction, the multiplication and division algorithms depend on the idea of positional notation but are independent of the base. The arithmetic is more awkward since we do not think in these bases as we do in base ten, but the ideas are the same. Moreover, working in other bases not only enhances understanding of base ten but also provides an interesting activity that helps to improve mental arithmetic skills. Let's consider some examples.

EXAMPLE 3.19

Multiplying in Base Seven

Compute the product $3_{\text{seven}} \cdot 3614_{\text{seven}}$.

SOLUTION

We suppress the subscript since all the work will be done in base seven. Moreover, while we naturally think in base ten, the results are quickly converted to base seven if we think of units, strips with 7 units per strip, and so on.

$$\begin{array}{r} {\scriptstyle 1\,2\,1} \\ 3614 \\ \times \quad 3 \\ \hline 14445 \end{array}$$

Think

3 · 4 = twelve; 1 strip and 5 units. Write 5 below the line and exchange seven 1s for one 7.

$7^0 = 1$
$7^1 = 7$
$7^2 = 49$
$7^3 = 343$
$7^4 = 2401$

3 · 1 + 1 = four; 4 strips. Write 4 below the line in the 7s column.

3 · 6 = eighteen; 2 strips and 4 units. Write 4 below the line in the 49s column and exchange 2 sevens of 49s for 2 343s.

3 · 3 + 2 = eleven; 1 strip and 4 units. Write 4 below the line in the 343s column and exchange 1 seven of 343s for 1 2401.

To check we convert all numerals to base ten. Thus,

$$\begin{aligned}
3614_{\text{seven}} &= 3 \cdot 7^3 + 6 \cdot 7^2 + 1 \cdot 7^1 + 4 \cdot 7^0 \\
&= 3 \cdot 343 + 6 \cdot 49 + 1 \cdot 7 + 4 \cdot 1 \\
&= 1334_{\text{ten}} \\
3_{\text{seven}} &= 3_{\text{ten}} \\
14445_{\text{seven}} &= 1 \cdot 4^4 + 1 \cdot 7^3 + 4 \cdot 7^2 + 4 \cdot 7 + 5 \\
&= 1 \cdot 2401 + 4 \cdot 343 + 4 \cdot 49 + 4 \cdot 7 + 5 \cdot 1 \\
&= 4002_{\text{ten}}
\end{aligned}$$

and

$$3_{\text{ten}} \cdot 1334_{\text{ten}} = 4002_{\text{ten}}$$

so the multiplication checks. ■

EXAMPLE 3.20 **More Multiplication in Base Seven**

Compute the product $24_{\text{seven}} \cdot 364_{\text{seven}}$ in base seven.

SOLUTION

Suppressing the subscripts and using mental arithmetic we have the following.

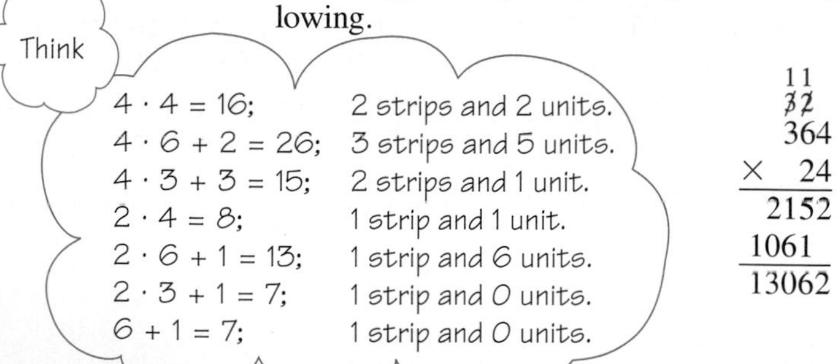

Think
4 · 4 = 16; 2 strips and 2 units.
4 · 6 + 2 = 26; 3 strips and 5 units.
4 · 3 + 3 = 15; 2 strips and 1 unit.
2 · 4 = 8; 1 strip and 1 unit.
2 · 6 + 1 = 13; 1 strip and 6 units.
2 · 3 + 1 = 7; 1 strip and 0 units.
6 + 1 = 7; 1 strip and 0 units.
11
32
364
× 24
2152
1061
13062

To check we convert all numerals to base ten. Thus,

$$
\begin{aligned}
364_{\text{seven}} &= 3 \cdot 7^2 + 6 \cdot 7^1 + 4 \cdot 7^0 \\
&= 3 \cdot 49 + 6 \cdot 7 + 4 \cdot 1 \\
&= 193_{\text{ten}} \\
24_{\text{seven}} &= 2 \cdot 7^1 + 4 \cdot 7^0 \\
&= 2 \cdot 7 + 4 \\
&= 18_{\text{ten}} \\
13062_{\text{seven}} &= 1 \cdot 7^4 + 3 \cdot 7^3 + 0 \cdot 7^2 + 6 \cdot 7^1 + 2 \cdot 7^0 \\
&= 1 \cdot 2401 + 1 \cdot 343 + 6 \cdot 7 + 2 \cdot 1 \\
&= 3474_{\text{ten}}
\end{aligned}
$$

and $18_{\text{ten}} \cdot 193_{\text{ten}} = 3474_{\text{ten}}$. ■

Division in another base is more difficult than the other operations. Here is an example.

EXAMPLE 3.21 **Dividing in Base Seven**

Find the quotient and remainder when 6324_{seven} is divided by 421_{seven}.

SOLUTION

The algorithm works just as it does in base ten. We suppress the subscripts and do all the work in base seven.

```
       13  R 221
421)6324
    421
    2114
    1563
     221
```

Think

The 1st trial divisor is certainly 1. What's the 2nd trial divisor? Since $4 \cdot 4 = 16$; that is, 2 strips and 2 or 22_{seven}, this is too much. So we try 3.

To check we compute $13 \cdot 421 + 221$ in base seven. Again, the subscript indicating the base is suppressed.

```
   421       6103
 ×  13     +  221
  1563       6324
  421
  6103
```

The base seven arithmetic checks. ■

PROBLEM SET 3.4

Understanding Concepts

(Note: Problems 1 through 11 are all base ten problems.)

1. (a) Make a suitable drawing of units and strips to illustrate the product $4 \cdot 8 = 32$.

   (b) Make a suitable sketch of place value cards to illustrate the product $4 \cdot 8 = 32$.

2. Make a suitable sketch of place value cards to illustrate the product $3 \cdot 254 = 762$.

3. (a) In the product shown, what does the colored 2 actually represent?

```
     2
   274
 ×  34
  1 096
  8 22
 10,316
```

   (b) In the product shown in part (a) when multiplying $4 \cdot 7$ one "exchanges" a 2. What is actually being exchanged?

4. The diagram shown illustrates the product $27 \cdot 32$. Discuss how this is related to finding the product by Instructional Algorithm I.

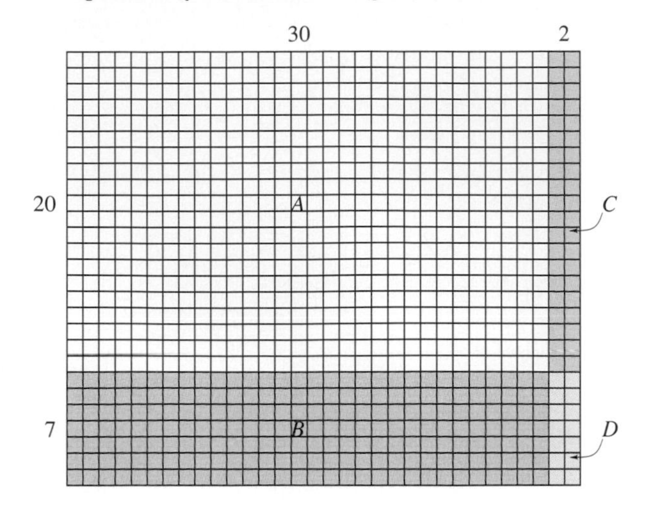

5. What property of the whole numbers justifies each step in this calculation?

$17 \cdot 4$	
$= (10 + 7) \cdot 4$	expanded notation
$= 10 \cdot 4 + 7 \cdot 4$	(a) ________
$= 10 \cdot 4 + 28$	1-digit multiplication fact
$= 10 \cdot 4 + (2 \cdot 10 + 8)$	expanded notation
$= 4 \cdot 10 + (2 \cdot 10 + 8)$	(b) ________
$= (4 \cdot 10 + 2 \cdot 10) + 8$	(c) ________
$= (4 + 2) \cdot 10 + 8$	(d) ________
$= 6 \cdot 10 + 8$	1-digit addition fact
$= 68$	expanded notation

6. Draw a sequence of sketches of units, strips, and mats to illustrate dividing 429 by 3.

7. What calculation does this sequence of sketches illustrate? Explain briefly.

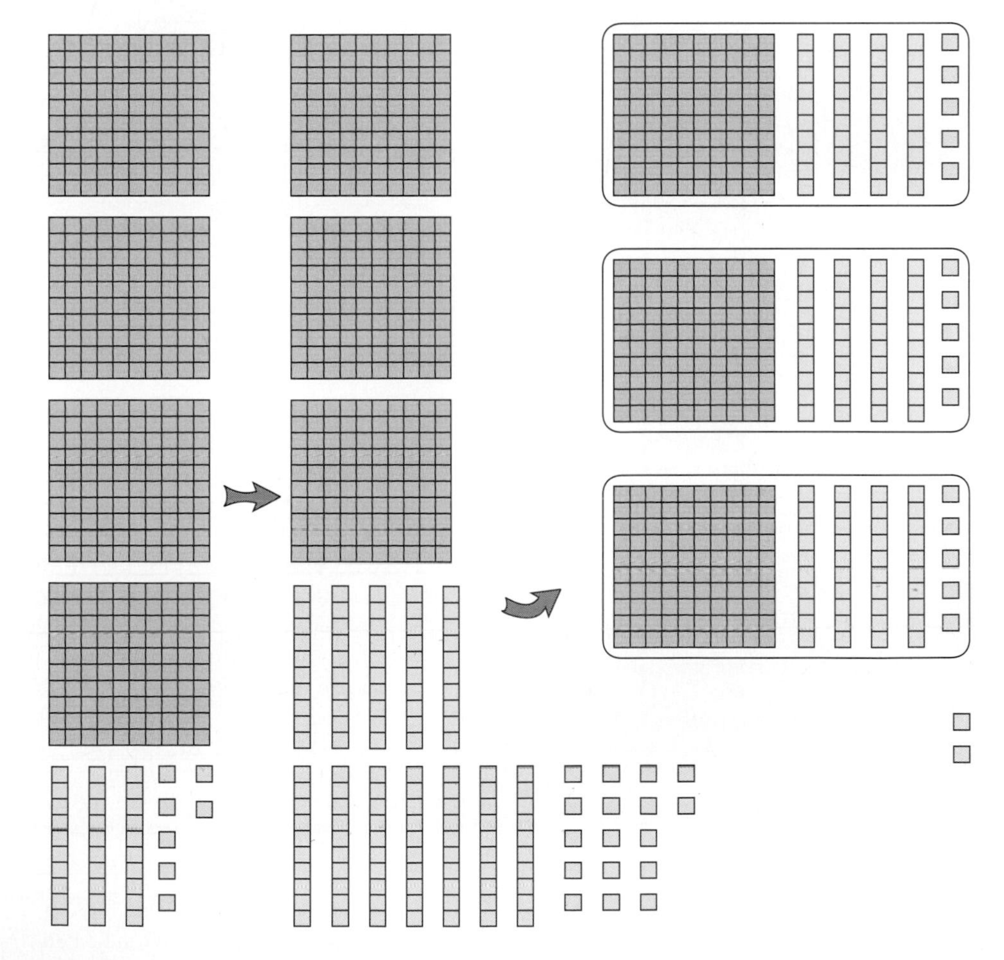

8. Multiply 352 by 27 using Instructional Algorithm I.

9. Find the quotient q and a remainder r when a is divided by b and write the result in the form $a = bq + r$ of the division algorithm for each of these choices of a and b.

(a) $a = 27, \quad b = 4$ **(b)** $a = 354, \quad b = 29$
(c) $a = 871, \quad b = 17$

10. Perform each of the following divisions by the scaffold method. In each case check your results by using the equation of the division algorithm.

(a) $351\overline{)7425}$ **(b)** $23\overline{)6814}$ **(c)** $213\overline{)3175}$

11. Use short division to find the quotient and remainder for each of these. Check each result.

(a) $5\overline{)873}$ **(b)** $7\overline{)2432}$ **(c)** $8\overline{)10,095}$

(Note: Problems 12 through 15 are in bases other than ten.)

12. Construct base five addition and multiplication tables.

13. What is being illustrated by the following sequences of sketches of place value cards? Explain briefly.

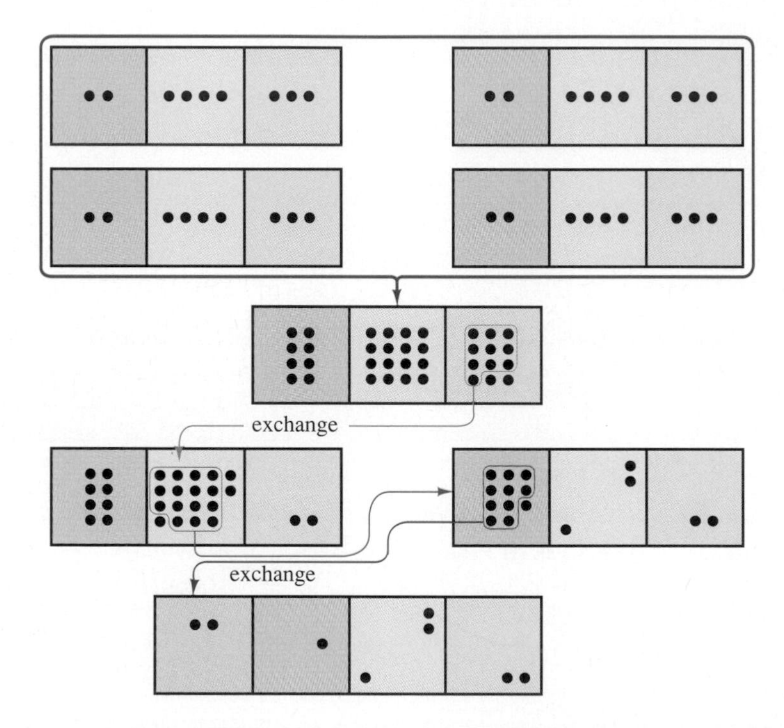

14. Carry out these multiplications using base five notation. All the numerals are already written in base five so that no subscript is needed.

(a) $\begin{array}{r} 23 \\ \times\ 3 \\ \hline \end{array}$ **(b)** $\begin{array}{r} 342 \\ \times\ 41 \\ \hline \end{array}$ **(c)** $\begin{array}{r} 2013 \\ \times\ 23 \\ \hline \end{array}$

(d) Convert the numerals in parts (a), (b), and (c) to base ten and check the results of your base five computations.

15. Carry out these divisions using base five notation. All numerals are already in base five so that the subscripts are omitted.

(a) $4\overline{)231}$ **(b)** $32\overline{)2342}$ **(c)** $213\overline{)34122}$

(d) Convert the numerals in parts (a), (b), and (c) to base ten and check the results of your base five computations.

Thinking Critically

16. The **Egyptian algorithm** for multiplication was one of the interesting subjects explained in the Rhind papyrus discussed in Chapter 2. We will explain the algorithm by giving the following example. Suppose we want to compute 19 times 35. Successively doubling 35 we obtain this list.

$$\begin{aligned} \longrightarrow \quad 1 \cdot 35 &= 35 \\ \longrightarrow \quad 2 \cdot 35 &= 70 \\ 4 \cdot 35 &= 140 \\ 8 \cdot 35 &= 280 \\ \longrightarrow \quad 16 \cdot 35 &= 560 \end{aligned}$$

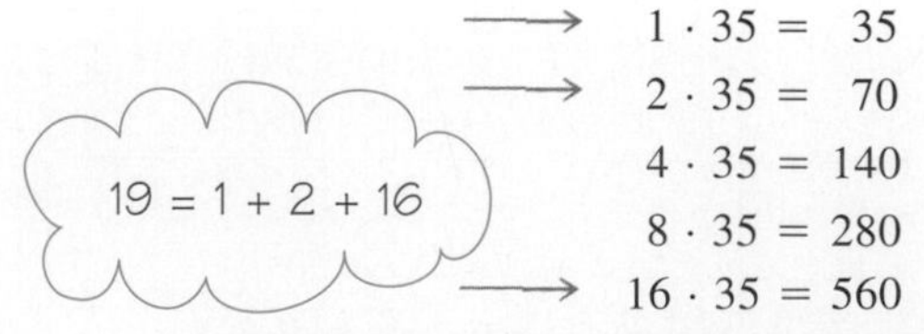

Adding the results in the indicated rows gives us 665 as the desired product.

(a) After carefully considering the above computation, write a short paragraph explaining how and why the process always works.

(b) This scheme is also known as the **duplation algorithm.** Use duplation to find the product of 24 and 71.

17. The **Russian peasant algorithm** for multiplication is similar to the duplation algorithm described in problem 16. To find the product of 34 and 54, for example, successively divide the 34 by 2 (ignoring remainders if they occur) and successively multiply 54 by 2. This gives the following lists.

~~34~~	~~54~~
17	108
~~8~~	~~216~~
~~4~~	~~432~~
~~2~~	~~864~~
1	1728
	1836

Now cross out the even numbers in the left-hand column and the companion numbers in the right-hand column. Add the remaining numbers in the right-hand column to obtain the desired product. To see why the process works, consider the products $34 \cdot 54 = 1836$ and $17 \cdot 108 = 1836$. Also, consider $8 \cdot 216$, $4 \cdot 432$, and $2 \cdot 864$.

(a) Why are $34 \cdot 54$ and $17 \cdot 108$ the same?

(b) Why are $17 \cdot 108$ and $8 \cdot 216$ different? How much do they differ?

(c) Why are $8 \cdot 216$, $4 \cdot 432$, $2 \cdot 864$, and $1 \cdot 1728$ all the same?

(d) Write a short paragraph explaining why the Russian peasant algorithm works.

(e) Use the Russian peasant algorithm to compute $29 \cdot 81$ and $11 \cdot 243$.

18. Another multiplication algorithm is the **lattice algorithm.** Suppose, for example, you want to multiply 324 by 73. Form a two by three rectangular array of boxes with the 3, 2, and 4 across the top and the 7 and 3 down the right side as shown. Now compute the products $3 \cdot 7 = 21$, $2 \cdot 7 = 14$, $4 \cdot 7 = 28$, $3 \cdot 3 = 9$, $2 \cdot 3 = 6$, and $4 \cdot 3 = 12$. Placing the products in the appropriate boxes ($3 \cdot 7$ is in the 3 column and the 7 row) with the units digit of the product below the diagonal in each box and the tens digit (if there is one) above the diagonal. Now add down the diagonals and add any "exchanges" to the sum in the next diagonal. The result of 23,652 is the desired product.

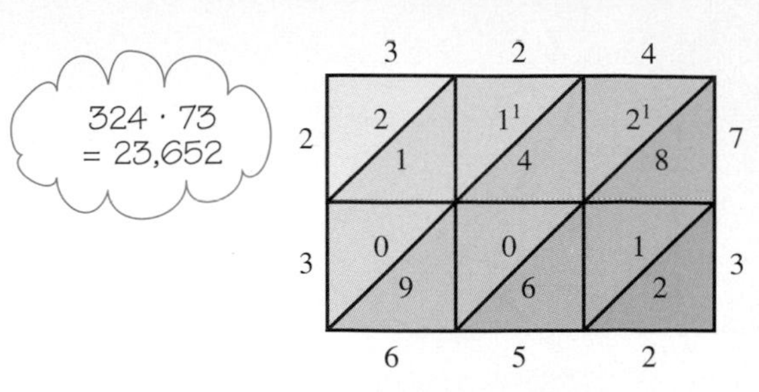

(a) Multiply 374 by 215 using the lattice algorithm.

(b) Write a short paragraph comparing the lattice algorithm with the standard pencil and paper algorithm.

19. Consider the following computation.

```
  374
×  23
  748
 1122
 8602
```

(a) Is the algorithm correct? Explain briefly.

(b) Multiply 285 by 362 by this method.

20. Earlier in this chapter we showed how to convert a base ten numeral for a number into a numeral in another base. The process was described as it was to impart proper understanding. However, the conversion can be completed much more easily. The trouble with showing this new scheme (at least in the beginning) is that it can be carried out with absolutely no understanding. Using the earlier method, we find that $583_{ten} = 4313_{five}$. Now let's see how the new scheme works. Divide 583 by 5 to obtain a quotient and a remainder, then divide the quotient by 5 and record the next remainder, and so on like this:

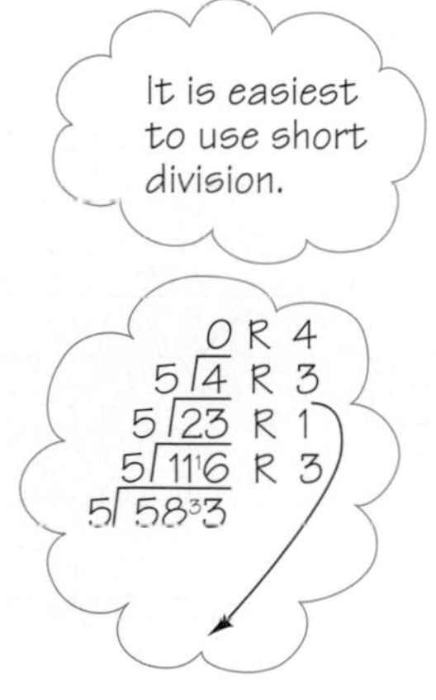

Remarkably, $4313_{five} = 583_{ten}$ as we have just seen. But why does the method work? If we write each of the above divisions in the form of the division algorithm, we have

$$585 = 5 \cdot 116 + 3$$
$$116 = 5 \cdot 23 + 1$$

and

$$23 = 5 \cdot 4 + 3.$$

Combining these, we obtain

$$\begin{aligned} 585 &= 5 \cdot 116 + 3 \\ &= 5(5 \cdot 23 + 1) + 3 \\ &= 5^2 \cdot 23 + 5 \cdot 1 + 3 \\ &= 5^2(5 \cdot 4 + 3) + 5 \cdot 1 + 3 \\ &= 5^3 \cdot 4 + 5^2 \cdot 3 + 5 \cdot 1 + 3 \end{aligned}$$

and this is the expanded form of 4313_{five}. Thus, $585_{\text{ten}} = 4313_{\text{five}}$ as noted. Write 482_{ten} in each of these bases and check by reconverting the numeral obtained to base ten.

(a) base seven (b) base four
(c) base two (d) base twelve

Thinking Cooperatively

21. Use each of 1, 3, 5, 7, and 9 once, and only once, in the boxes to obtain the largest possible product in each case.

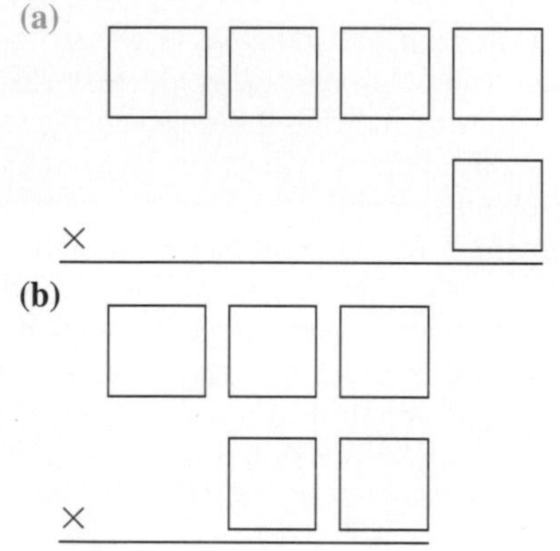

22. Use each of 1, 3, 5, 7, and 9 once and only once in the boxes to obtain the smallest possible product in each case.

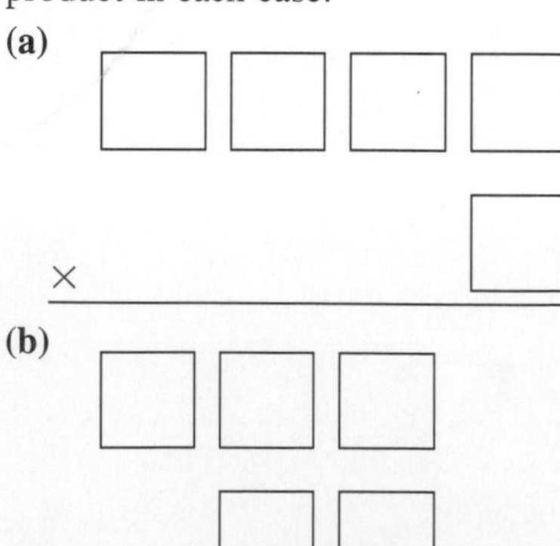

23. A druggist has a balance scale to weigh objects. She also has two 1 gram weights, two 3 gram weights, two 9 gram weights, and two 27 gram weights. She places an object to be weighed in one pan and her weights in the other.

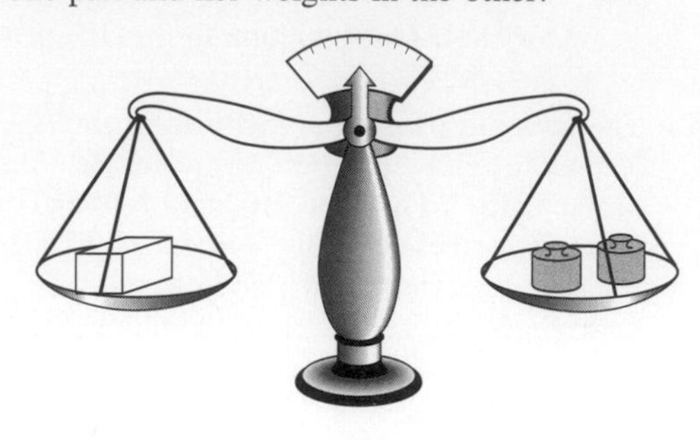

(a) Listed below are the number and sizes of weights she used to weigh objects A, B, C, D, and E. How much did each object weigh?

	27g	9g	3g	1g	Total Weight
A	0	1	0	2	
B	0	2	1	0	
C	1	0	0	2	
D	0	1	2	2	
E	2	1	0	1	

(b) Complete this table to show how to weigh objects weighing the given amounts.

Weight of Object	27g	9g	3g	1g
35g				
11g	0	1	0	2
74g				
26g				
42g				

(c) What weights of objects could you possibly weigh in this way?

(d) What mathematics would this be illustrating to students if you did this in one of your classes?

(e) Suppose you had only one of each type of weight and that you could put weights in both scales, some along with the object being weighed. This table shows how the druggist weighed objects H, I, J, K, and L in this way where the numerals in red indicate weights that were placed in the pan with the object being weighed and those in black indicate weights that were placed in the pan not containing the object. How much did each of H, I, J, K, and L weigh?

	27g	9g	3g	1g	Weight of Object
H	0	1	**1**	1	
I	1	0	0	**1**	
J	0	1	1	0	
K	1	**1**	**1**	**1**	
L	1	**0**	1	1	

(f) Complete this table showing how you could weigh objects of the indicated weights.

Weight of Object	27g	9g	3g	1g
2 grams				
13 grams				
8 grams				
40 grams				
33 grams				

(g) What weights of objects could you possibly weigh in this way?

(h) Could you "weigh" a balloon filled with helium that was tied to one of the pans of the scales and exerted an upward pull of 11 grams on the pan? What balloons could you "weigh" in this way?

Using a Calculator

24. (a) Compute 375×2432.

(b) Examine the following calculation where certain zeros are suppressed.

		24	32	
×		3	75	
	1	1 24	00	$75 \cdot 32$
	18	00		$75 \cdot 2400$
		96		$300 \cdot 32$
	72			$300 \cdot 2400$
	91	20	00	

Note that this agrees with your answer to part (a). Use the same idea to compute the following product which would ordinarily exceed the capacity of your calculator.

	3748	2325
×	263	1473

25. The *Math Explorer* calculator manufactured by Texas Instruments for elementary classroom use has an *integer divide key*. For example, 374821 [INT÷] 357 [=] gives a quotient of 1049 and a remainder of 328. However, most calculators give 1049.9188. Here the quotient is 1049 and the remainder r satisfies $r \doteq 357 \cdot (0.9188)$. Thus, $r \doteq 357 \cdot (0.9188) = 328.116$. Since r is a whole number, we conclude that $r = 328$. (The error is the rounded-off error in the machine; indeed 0.9188 is the rounded-off value of 0.9187675 . . . .) To check, note that

$$374{,}821 = 1049 \cdot 357 + 328$$

Without using the divide with remainder capability of a *Math Explorer,* use your calculator to compute the quotient and remainder when

(a) 276,523 is divided by 511.

(b) 347,285 is divided by 87.

For Review

26. An alternative to using expanded form in explaining addition and subtraction algorithms is to write out in words what numerals mean. For example, we could write

$$232 = 2 \text{ hundreds} + 3 \text{ tens} + 2 \text{ ones}$$

which we might call **word expanded form.** Then, we could write

$$\begin{aligned} 232 &= 2 \text{ hundreds} + 3 \text{ tens} + 2 \text{ ones} \\ +\ 465 &= 4 \text{ hundreds} + 6 \text{ tens} + 5 \text{ ones} \\ \hline &\ 6 \text{ hundreds} + 9 \text{ tens} + 7 \text{ ones} \\ &= 697 \end{aligned}$$

Use word expanded form to perform these additions and subtractions.

(a) $\begin{array}{r} 634 \\ +\ 163 \\ \hline \end{array}$ **(b)** $\begin{array}{r} 247 \\ +\ 332 \\ \hline \end{array}$ (c) $\begin{array}{r} 783 \\ +\ 532 \\ \hline \end{array}$

(d) $\begin{array}{r} 674 \\ -\ 122 \\ \hline \end{array}$ (e) $\begin{array}{r} 725 \\ -\ 413 \\ \hline \end{array}$ **(f)** $\begin{array}{r} 544 \\ -\ 432 \\ \hline \end{array}$

27. Note that

2 hundreds + 14 tens + 5 ones
= 2 hundreds + 10 tens + 4 tens + 5 ones
= 2 hundreds + 1 hundred + 4 tens + 5 ones
= 3 hundreds + 4 tens + 5 ones

Now perform these additions and subtractions using word expanded form.

(a) $374 + 483$ **(b)** $264 + 327$ (c) $724 + 532$

(d) $418 - 237$ (e) $367 - 249$ **(f)** $642 - 246$

28. In base five, the word expanded form of 231 is 2 twenty-fives + 3 fives + 1 one. Use the base five word expanded form to perform each of these additions and subtractions.

(a) $213 + 131$ **(b)** $332 + 12$ (c) $142 + 123$

(d) $231 - 130$ (e) $344 - 232$ **(f)** $342 - 104$

29. Write 495_{ten} and 7821_{ten} in base six

(a) using the positional value method and Table 3.9, in Section 3.2.

(b) using the short division with remainder method of problem 22 above.

30. Perform these computations entirely in base five. The numerals are already written in base five, so the subscripts are omitted.

(a) $34 + 23$ **(b)** $243 + 22$ (c) $312 - 21$

(d) $423 - 234$ (e) 32×4 **(f)** 241×22

(g) $23\overline{)344}$ **(h)** $32\overline{)2341}$

Calvin and Hobbes by Bill Watterson

COOPERATIVE *Investigation*

Magic in Base Three

Materials Needed

A deck of cards.

Discussion

This is an extremely effective card trick that surprises and excites students. They immediately want to know how the trick is done and this provides great motivation for learning. Doing the trick well in front of a group not only requires knowing base three, but also the ability to do mental arithmetic accurately and swiftly while keeping up a steady stream of chatter about mind reading and other psychic nonsense. The activity is a great skill and understanding builder that has many times the payoff of a dull sheet of drill problems.

To pull this trick on your class, proceed as follows.

Step 1. Ask your class to select a number n between 1 and 27.

Step 2. Represent $n - 1$ in base 3, say $n - 1 = abc_{three}$.

Step 3. Lay out 27 **face up** cards in the order shown. Note that the numbers only indicate the way the cards are to be distributed. Thus, 1 might be the queen of spades, 2 the ace of diamonds, and so on,

Continued

Suppose they choose 16?

$16 - 1 = 15$
$= 120_{\text{three}}$

To be effective, Step 2 should be done in your head at the same time you are doing Step 3 and keeping up a stream of chatter. It just takes practice.

Step 4. Now ask your class to agree secretly on one of the cards and to tell you in which column the card lies.

Step 5. Pick up the cards by columns and place them **face up** in your palm. But this must be done in a very special way. Think of the three possible positions for placing the columns in your palm. A column can be placed in

the 0 position—next to your palm,
the 1 position—in the middle,
the 2 position—away from your palm, as indicated in the diagram.

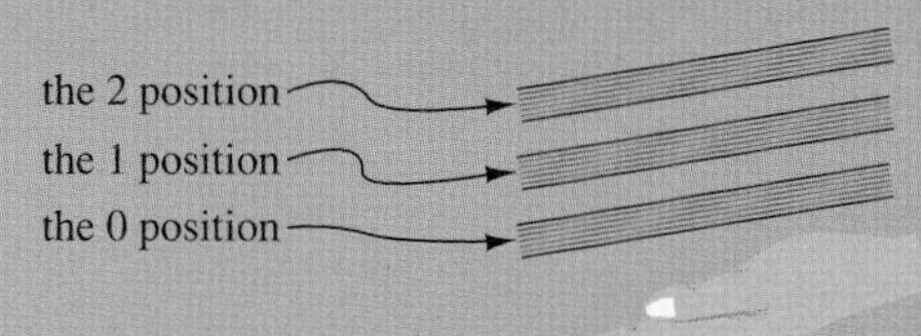

Recalling that $n - 1 = abc_{\text{three}}$ with each of a, b and c one of 0, 1, or 2. This first time the cards are picked up make sure the column with the selected card is placed in the ***c*** position in your palm.

Step 6. Turning the assembled cards **face down** in your hand, redistribute them in the order shown above and again ask the students to indicate the column in which their card now appears.

Step 7. Again pick up the cards by columns, this time placing the column with the selected card in the ***b*** position in your palm.

Step 8. Finally, repeat Step 6 and again pick up the cards, this time placing the column with the selected card in the ***a*** position in your palm.

Step 9. All that remains is to place the assembled deck of 27 cards **face down** in your palm. Counting out the cards from the top down, the class will be surprised and impressed when the nth card turned over is the card they selected.

Step 10. But there is more. What if you had 9 cards? 81 cards? 243 cards? Could the trick be modified so that it would still work? Could a similar trick be worked out for other bases? If so, how would it work? A little experimentation would suggest answers to these questions.

3.5 Mental Arithmetic and Estimation

The ability to make accurate estimates and do mental arithmetic is increasingly important in today's society. When buying several items in a store it is helpful to know before you go to the cashier that you have enough money to pay for your purchase. It is also worthwhile to keep a mental check on the cashier to make sure that you are charged correctly. More importantly, as we increase our use of calculators and computers, it has become essential that we be able to tell if an answer is "about right." Because of their quickness and the neatness with which they display answers, it is tempting to accept as true whatever answer your calculator or computer gives. It is quite easy to enter an incorrect number or press an incorrect operation key and so obtain an incorrect answer. Care must be exercised, and this requires estimation skills.

The One-Digit Facts

It is essential that the basic addition and multiplication facts be memorized since all other numerical calculations and estimations depend on this foundation. At the same time, this should not be rote memorization of symbols. Using a variety of concrete objects students should actually *experience* the fact that $8 + 7 = 15$, that $9 \cdot 7 = 63$, and so on. Moreover, rather than having children simply memorize the addition and multiplication tables, these should be learned by the frequent and long-term use of manipulatives, games, puzzles, oral activities, and appropriate problem-solving activities. In the same way, children learn the basic properties of the whole numbers which, in turn, can be used to recall some momentarily forgotten arithmetic fact. For example, $7 + 8$ can be recalled as $7 + 7 + 1$, $6 \cdot 9$ can be recalled as $5 \cdot 9 + 9$, and so on. In the same way, the properties of whole numbers along with the one-digit facts form the basis for mental calculation. Here are several strategies for mental calculation.

Easy Combinations

Always look for **easy combinations** in doing mental calculations. The next example shows how this works.

WORTH READING

Calculating Prodigies of the Past

When Zerah Colburn in London in 1812 was asked to compute

$$8 \times 8 \times 8 \times 8 \times 8 \times 8 \times 8 \times 8 \times 8 \times 8 \times 8 \times 8 \times 8 \times 8 \times 8 \times 8 = 8^{16}$$

he answered "promptly and with facility,"

281,474,976,710,656

which you can show to be correct with a combination of hand and calculator techniques. Another such prodigy was Truman Henry Safford (1836–1901). According to a Reverend H. W. Adams, when Safford was asked to compute the product

365,365,365,365,365,365 · 365,365,365,365,365,365

"he flew around the room like a top, pulled his pantaloons over the tops of his boots, bit his hands, rolled his eyes in their sockets, sometimes smiling and talking, and then seeming to be in agony, until, in not more than one minute, said he,

133,491,850,208,566,925,016,658,299,941,583,225."

An electronic computer would do the job somewhat more quickly, but it would also be much less impressive!

An interesting account of child prodigies along with a delightful collection of puzzles, paradoxes, problems, magic squares, and the like, can be found in W. W. Rouse Ball and H. S. M. Coxeter, *Mathematical Recreations and Essays,* 12th ed. (Toronto: University of Toronto Press, 1974).

EXAMPLE 3.22 **Using Easy Combinations**

Use mental processes to perform these calculations.

(a) $35 + 7 + 15$ **(b)** $8 + 3 + 4 + 6 + 7 + 12 + 4 + 3 + 6 + 3$
(c) $25 \cdot 8$ **(d)** $4 \cdot 99$ **(e)** $57 - 25$ **(f)** $47 \cdot 5$

SOLUTION

(a) Using the commutative and associative properties, we have

$$35 + 7 + 15 = 35 + 5 + 10 + 7 = 40 + 10 + 7 = 50 + 7 = 57$$

Think 35, 40, 50, 57 The answer is 57.

(b) Note numbers that add to 10 or multiples of 10.

$$8 + 3 + 4 + 6 + 7 + 12 + 4 + 3 + 6 + 3 = 56$$

Think 20, 30, 40, 50, 53, 56 The answer is 56.

(c) $25 \cdot 8 = 25 \cdot 4 \cdot 2 = 100 \cdot 2 = 200.$

Think
25, 100, 200
The answer is 200.

(d) $4 \cdot 99 = 4(100 - 1) = 400 - 4 = 396.$

Think $400 - 4 = 396$

(e) $57 - 25 = 50 - 25 + 7 = 25 + 7 = 32.$

Think $50 = 2 \cdot 25$, so $50 - 25$ plus 7 gives 32.

Think Two quarters are worth 50¢.

(f) $47 \cdot 5 = 47 \cdot 10 \div 2 = 470 \div 2 = 235.$

Think
47, 470, 235

■

Adjustment

In parts (d) and (e) of Example 3.22, we made use of the fact that 99 and 57 were close to 100 and 50 respectively. This is an example of adjustment. **Adjustment** simply means that we modify numbers in a calculation to minimize the mental effort required.

EXAMPLE 3.23 **Using Adjustment in Mental Calculation**

Use mental processes to perform these calculations.

(a) $57 + 84$ **(b)** $83 - 48$ **(c)** $286 + 347$

(d) $4931 \cdot 7$ **(e)** $2646 \div 9$ **(f)** $639 \div 7$

SOLUTION

(a)
$$\begin{aligned} 57 + 84 &= (57 + 3) + (84 - 3) \\ &= 60 + 81 = 60 + 80 + 1 \\ &= 140 + 1 = 141 \end{aligned}$$

Think: 57 + 84, 60 + 81, 140, 141

(b) $83 - 48 = (83 + 2) - (48 + 2) = 85 - 50 = 35.$

Think: 83 – 48, 85 – 50, 35

(c)
$$\begin{aligned} 286 + 347 &= (286 + 14) + (347 - 14) \\ &= 300 + 300 + 47 - 14 \\ &= 600 + 33 = 633 \end{aligned}$$

Think: 300, 647 – 14, 633

(d) $493 \cdot 7 = (500 - 7) \cdot 7 = 3500 - 49 = 3451.$

Think: (500 – 7) • 7, 3500 – 49, 3451

(e) $2646 \div 9 = (2700 - 54) \div 9 = 300 - 6 = 294.$

Think: ÷ 9, 2700 – 54, 300 – 6, 294

(f) $639 \div 7 = (630 + 7 + 2) \div 7 = 90 + 1\ R\ 2 = 91\ R\ 2.$

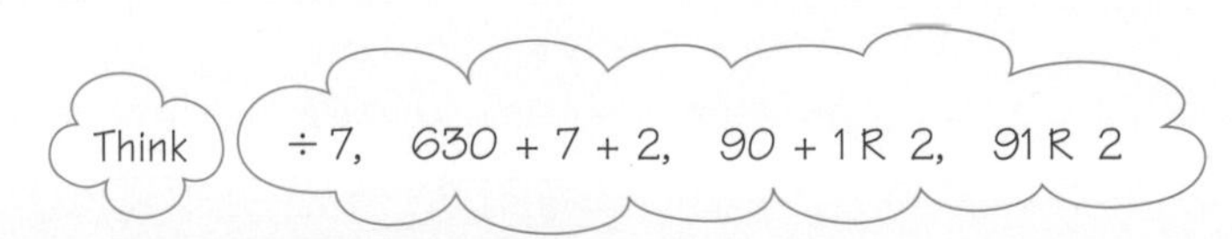

■

Working from Left to Right

Because it tends to reduce the amount one has to remember, many expert mental calculators **work from left to right** rather than the other way around as in most of our standard algorithms.

EXAMPLE 3.24 **Working from Left to Right**

Use mental processes to perform these calculations.

(a) $352 + 647$ **(b)** $739 - 224$ **(c)** $4 \cdot 235$

SOLUTION

(a)
$$\begin{aligned}352 + 647 &= (300 + 50 + 2) + (600 + 40 + 7)\\ &= (300 + 600) + (50 + 40) + (2 + 7)\\ &= 900 + 90 + 9 = 999\end{aligned}$$

(b)
$$\begin{aligned}739 - 224 &= (700 + 30 + 9) - (200 + 20 + 4)\\ &= (700 - 200) + (30 - 20) + (9 - 4)\\ &= 500 + 10 + 5 = 515\end{aligned}$$

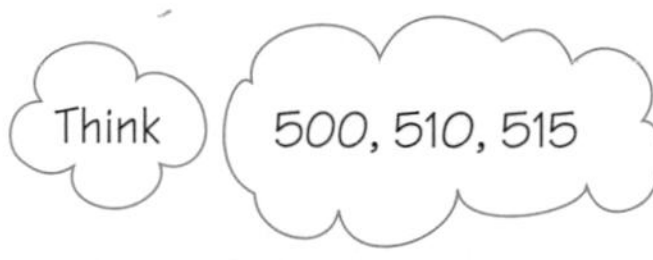

(c)
$$\begin{aligned}4 \cdot 235 &= 4(200 + 30 + 5)\\ &= 800 + 120 + 20\\ &= 920 + 20\\ &= 940\end{aligned}$$

■

Left to right methods often combine nicely with an understanding of positional notation to simplify mental calculation. Since $4200 = 42 \cdot 100$, for example, we might compute the sum

$$\begin{array}{r} 3700 \\ 900 \\ 2800 \\ +\ 5600 \\ \hline \end{array}$$

by thinking of

$$\begin{array}{r} 37 \\ 9 \\ 28 \\ +\ 56 \\ \hline \end{array}$$

Then, working from left to right, we think

30, 50, 100, 107, 116, 124, 130
times 100. The answer is 13,000.

Rounding

Often we are **not** interested in exact values. This is certainly true when *estimating* the results of numerical calculations, and it is often the case that exact values are actually unobtainable. What does it mean, for example, to say that the population of California in 1990 was 25,874,293? Even if this is supposed to be the actual count on a given day, it is almost surely in error because of the sheer difficulty in conducting a census. How many illegal immigrants were not counted? How many homeless people? How many transients? In gross terms, it is probably accurate to say that the population of California was approximately 26,000,000 or 26 million people. To obtain this figure we **round** to the nearest million. This is accomplished by considering the digit in the hundred thousands position. If this digit is 5 or more, we increase the digit in the millions position by one and replace all the digits to the right of this position by zeros. If the hundred thousands digit is 4 or less, we leave the millions digit unchanged and replace all the digits to its right by zeros.

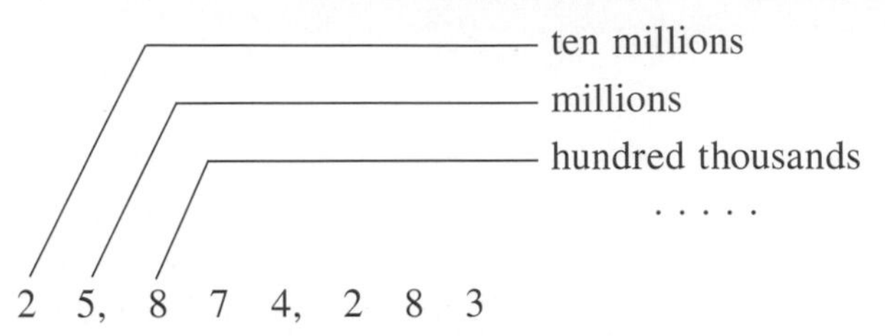

Rounding Using the 5-up Rule

1. Determine to which position you are rounding.
2. If the digit to the right of this position is 5 or more, add one to the digit in the position to which you are rounding. Otherwise leave the digit unchanged.
3. Replace by zeros all digits to the right of the position to which you are rounding.

EXAMPLE 3.25 **Using the 5-up Rule to Round Whole Numbers**

Round 27,250 to the position indicated.

(a) the nearest ten thousand **(b)** the nearest thousand
(c) the nearest hundred **(d)** the nearest ten

SOLUTION

(a) The digit in the ten thousand position is 2. Since the digit to its right is seven and $7 > 5$, we add 1 to 2 and replace all digits to the right of the rounded digit by zeros. This gives 30,000.

(b) This time 7 is the critical digit and 2 is the digit to its right. Thus, to the nearest thousand, 27,250 is rounded to 27,000.

(c) Here 2 is the digit in the hundreds position and 5 is to its right. Thus 27,250 is rounded to 27,300 to the nearest hundred.

(d) This time 5 is the critical digit and zero is the digit on its right. Thus, no change need be made and 27,250 is already rounded to the nearest ten. ■

Estimation

The ever increasing use of calculators and computers makes it essential that students develop skill at estimation. How large an answer should I expect? Is this about

HIGHLIGHT FROM HISTORY
Emmy Noether (1882–1935)

Emmy Noether was born in Erlangen, Germany, to a family noted for mathematical talent. Much of her life was spent at the University of Göttingen, exploring, teaching, and writing about algebra. This university—where Carl Gauss had taught a century earlier—was the first in Germany to grant a doctoral degree to a woman. Yet Noether met with frustrating discrimination there. For many years she was denied appointment to the faculty; finally she was given an impressive title as "extraordinary professor"—with no salary. But her abilities overcame the obstacles that daunted many other women in mathematics. Her work in the 1920s brought invitations to lecture throughout Europe and in Moscow. In 1933, as the Nazi party came to power in Germany, Emmy Noether met with persecution not only as a woman but as an intellectual, a Jew, a pacifist, and a political liberal. She fled to the United States, where she taught and lectured at Bryn Mawr and Princeton until her death in 1935.

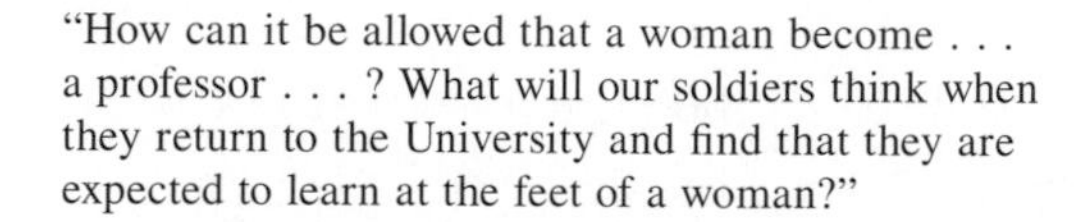
"How can it be allowed that a woman become . . . a professor . . . ? What will our soldiers think when they return to the University and find that they are expected to learn at the feet of a woman?"

—Faculty member at Göttingen, in 1918

". . . for two of the most significant sides of the theory of relativity, she gave at that time [1919] the genuine and universal mathematical formulation."

—Hermann Weyl, colleague at Göttingen

"In the judgement of the most competent living mathematicians, Fraulein Noether was the most significant creative mathematical genius thus far produced since the higher education of women began. In the realm of algebra . . . she discovered methods which have proved of enormous importance . . . ."

—Albert Einstein, 1935

"She was the most creative abstract algebraist in the world."

—Eric Temple Bell in *Men of Mathematics*

SOURCE: Biographical information is from Lynn Osen, *Women in Mathematics* (MIT Press, 1974). Quotations are cited in that source, original references including: Einstein, *New York Times,* May 4, 1935; Weyl, *Scripta Mathematica,* Vol. 3, 1935; *Men of Mathematics,* p. 261. Simon and Schuster, 1965; anonymous faculty member, Constance Reid, *Hilbert* (Springer-Verlag, 1970, p. 143). From Mathematics in Modules, *Intermediate Algebra,* A5, Teachers Edition. Reprinted by permission.

the right answer? These are questions students should ask and be able to answer. And they can be answered reasonably effectively on the basis of a good understanding of one-digit arithmetic facts and positional notation. To be effective, the estimator must also be adept at mental arithmetic.

Ranges for Answers

Ranges in which answers lie can usually be obtained by replacing the numbers in a computation by one-digit approximations using the left-most digit. For example, since $400 < 456 < 500$ and $200 < 278 < 300$, it follows that

$$600 < 456 + 278 < 800.$$

It also follows that

$$100 < 456 - 278 < 300.$$

Finally, it follows that

$$80{,}000 < 456 \cdot 278 < 150{,}000.$$

Also, since $40 < 47 < 50$ and $7000 < 7826 < 8000$ it follows that

$$140 < 7826 \div 47 < 200.$$

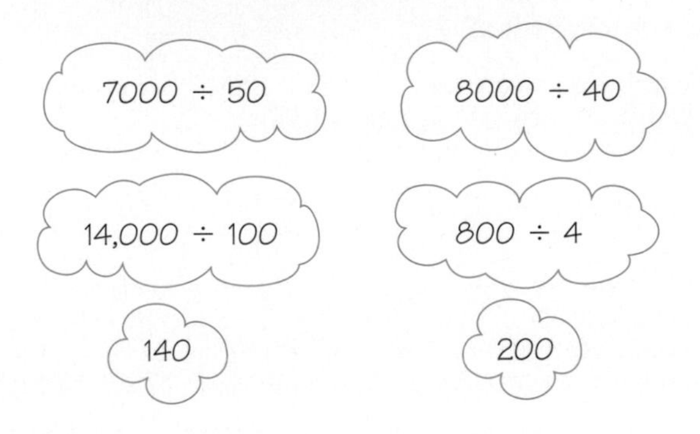

In each case, the range was found using one-digit facts and positional notation.

$r < a < s$
$u < b < v$

THEOREM Ranges of Sums, Products, Differences, and Quotients

Let a and b be natural numbers, let r and s be lower and upper 1-digit approximations of a, and let u and v be lower and upper 1-digit approximations of b. Then the following hold.

(i) $r + u < a + b < s + v$
(ii) $r \cdot u < a \cdot b < s \cdot v$
(iii) $r - v < a - b < s - u$
(iv) $r \div v < a \div b < s \div u$

EXAMPLE 3.26 **Finding Ranges for Answers**

Find a range in which the answers to each of the following must lie using left digit approximations.

(a) 681 + 241 **(b)** 681 − 241
(c) 681 · 241 **(d)** 57,801 ÷ 336

SOLUTION

For parts (a), (b) and (c) we observe that 600 < 681 < 700 and 200 < 241 < 300. Thus,

(a) 800 < 681 + 241 < 1000

(b) 300 < 681 − 241 < 500

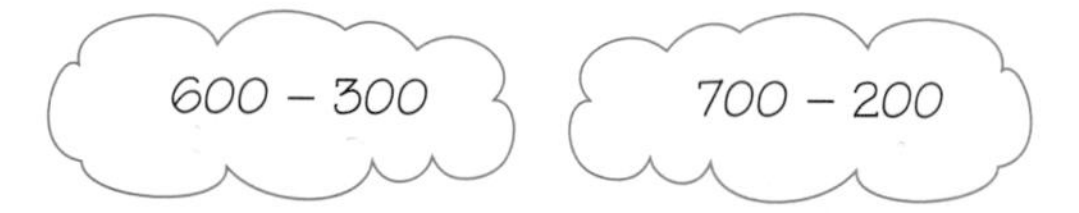

(c) 120,000 < 681 · 241 < 210,000

(d) Here the ranges for 57,801 and 336 are 50,000 < 57,801 < 60,000 and 300 < 336 < 400, and it follows that

125 < 57,801 ÷ 336 < 200.

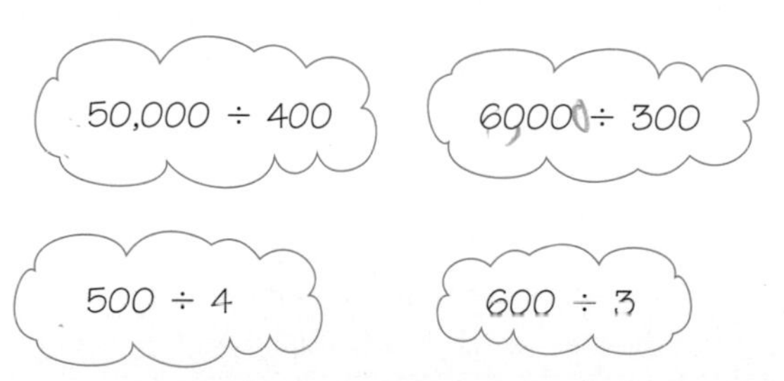

A somewhat weaker but more easily computed lower limit on the range of this division can be found by replacing 50,000 by the even smaller 40,000. This gives a lower limit of 40,000 ÷ 400 = 100, which is much more easily computed mentally and is still useful. ■

Approximating by Rounding

Rounding is often used in finding estimates. The advantage of **approximating by rounding** is that it gives a single estimate that is reasonably close to the desired answer. The idea is to round the numbers involved in a calculation to the position of the left-most digit or two digits and to use these rounded numbers in making the estimate. For example, for 467 + 221 we would obtain the range

$$600 < 467 + 221 < 800.$$

400 + 200
500 + 300

Rounding to the nearest hundreds, we have

$$467 \approx 500 \quad \text{and} \quad 221 \approx 200,$$

where we use the symbol ≈ to mean is estimated by. Thus, we obtain the estimate

$$500 + 200 \approx 700.$$

The actual answer is 688, so the approximation is reasonably good. Rounding to the nearest ten usually gives an even closer approximation if it is needed. Thus,

$$467 \approx 470, \quad 221 \approx 220,$$

and 470 + 220 gives the very close approximation 690.

EXAMPLE 3.27

Approximating by Rounding

Round to the left-most digit to find approximate answers to each of these. Note that the numbers are the same as in Example 3.26 and compare the results. Also, compute the exact answer in each case.

(a) 681 + 241 **(b)** 681 − 241
(c) 681 · 241 **(d)** 57,801 ÷ 336

SOLUTION

To the nearest hundred 681 ≈ 700 and 241 ≈ 200. Also, 57,801 ≈ 60,000 and 336 ≈ 300. Using these, we obtain the approximations shown which are consistent with the results of Example 3.26.

	Approximation	*Exact Answer*
(a)	681 + 241 ≈ 700 + 200 = 900	681 + 241 = 922
(b)	681 − 241 ≈ 700 − 200 = 500	681 − 241 = 440
(c)	681 · 241 ≈ 700 · 200 = 140,000	681 · 241 = 164,121
(d)	57,801 ÷ 336 ≈ 60,000 ÷ 300 = 200	57,801 ÷ 336 = 172 R 9 ■

Here the quotient
is approximately
200.

EXAMPLE 3.28

Approximating by Rounding

Approximate the results of these computations by rounding to the position of the left digit. Also, compute the exact answer in each instance.

(a) 545 + 376 **(b)** 545 − 376
(c) 545 · 376 **(d)** 54,376 ÷ 38

SCHOOLBOOK PAGE *Using Estimates*

Adding Larger Numbers

Build Understanding

A. The observation level of Toronto's CN Tower is 1,136 feet above the ground. The tower with its antenna continues upward for another 679 feet. What is the height of the CN Tower?

Find 1,136 + 679.
Estimate: 1,100 + 700 = 1,800

Paper and Pencil

Calculator

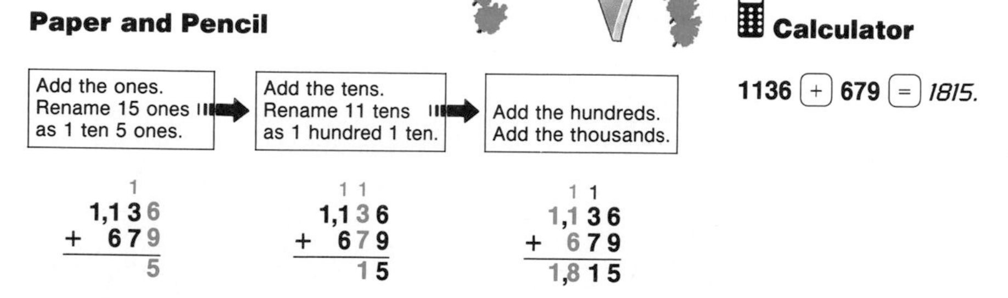

Add the ones.
Rename 15 ones as 1 ten 5 ones.
Add the tens.
Rename 11 tens as 1 hundred 1 ten.
Add the hundreds.
Add the thousands.
1
1,136
+ 679
5
1 1
1,136
+ 679
15
1 1
1,136
+ 679
1,815
1136 + 679 = 1815.

The height of the CN Tower is 1,815 feet. The estimate shows that the answer is reasonable.

B. Estimation Use ***front-end digits with adjusting.***

Use front-end digits.

273 + 412 + 148
↓ ↓
200 + 400 + 100 = 700

Then adjust the estimate.

73 + 12 + 48 > 100

The sum is more than 800.

C. Mental Math Use ***compensation.***

Change one number to make it easy to add. Then change the other number.

29 + 1 = 30 45 − 1 = 44 30 + 44
29 + 45 = 74

Change both numbers to make them easy to add. Then change the answer.

98 + 2 = 100 97 + 3 = 100 200 − 5
98 + 97 = 195

■ **Talk About Math** Explain how you would use mental math to find $3.25 + $2.60 + $4.75 + $1.45.

1. The above estimate for 1136 + 679 was obtained by rounding each number to the nearest 100 as opposed to rounding to the left-most digit. Which method seems preferable here? Explain.
2. Discuss the merits of using estimates, pencil and paper methods, and a calculator to do arithmetic all in one lesson as here.
3. What is the accurate value of the sum in part A?
4. If you estimate the sum in Talk about Math above by rounding to the nearest dollar, how close is your estimate to the actual total?

SOLUTION

Rounding gives the approximate values $545 \approx 500$, $376 \approx 400$, $54{,}376 \approx 50{,}000$, and $37 \approx 40$. Using these values, we obtain the approximations shown.

	Approximation	*Exact Answer*
(a)	$545 + 376 \approx 500 + 400 = 900$	$545 + 376 = 921$
(b)	$545 - 376 \approx 500 - 400 = 100$	$545 - 376 = 169$
(c)	$545 \cdot 376 \approx 500 \cdot 400 = 200{,}000$	$545 \cdot 376 = 204{,}920$
(d)	$54{,}376 \div 38 \approx 50{,}000 \div 40 = 1250$	$54{,}376 \div 38 = 1440$ R 16 ■

PROBLEM SET 3.5

Understanding Concepts

1. Calculate mentally, using easy combinations. Write a sequence of numbers indicating intermediate steps in your thought process. The first one is done for you.

   (a) $7 + 11 + 5 + 3 + 9 + 16 + 4 + 3$.

   Think 10, 30, 50, 55, 58

   (b) $6 + 9 + 17 + 5 + 8 + 12 + 3 + 6$
   (c) $27 + 42 + 23$ **(d)** $47 - 23$
   (e) $48 \cdot 5$ **(f)** $21{,}600 \div 50$

2. Calculate mentally using adjustment. Write down a sequence of numbers indicating intermediate steps in your thought process. The first one is done for you.

   (a) $78 + 64$

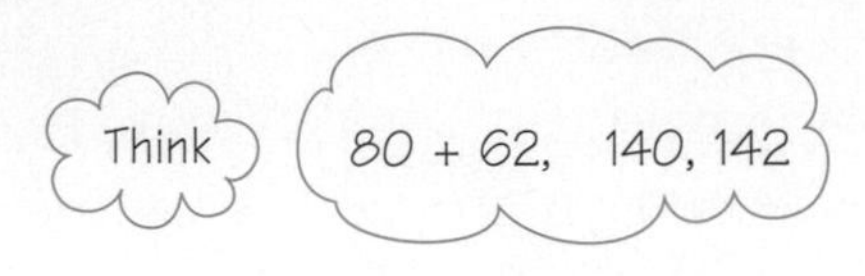
Think
80 + 62, 140, 142

   (b) $294 + 177$ (c) $306 - 168$ **(d)** $294 - 102$
   (e) $479 + 97$ **(f)** $3493 \div 7$ **(g)** $412 \cdot 7$

3. Perform these calculations mentally from left to right. Write down a sequence of numbers indicating intermediate steps in your thought process.

   (a) $425 + 362$ **(b)** $363 + 274$ **(c)** $572 - 251$
   (d) $764 - 282$ (e) $3 \cdot 342$
   (f) $47 + 32 + 71 + 9 + 26 + 32$

4. Round 235,476 to the
   (a) nearest ten thousand.
   (b) nearest thousand.
   (c) nearest hundred.

5. Round each of these to the position indicated.
   (a) 947 to the nearest 100.
   (b) 850 to the nearest 100.
   (c) 27,462,312 to the nearest million.
   (d) 2461 to the nearest thousand.

6. Find a range (upper and lower estimates) for the answers to these computations using one-digit estimates. The first one is done for you.

   (a) $478 + 631$

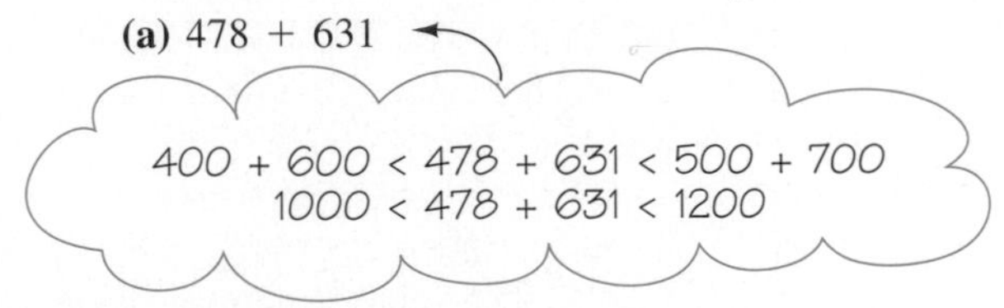
400 + 600 < 478 + 631 < 500 + 700
1000 < 478 + 631 < 1200

   (b) $782 + 346$ (c) $678 - 431$ **(d)** $257 \cdot 364$
   (e) $7403 \cdot 28$ **(f)** $28{,}329 \div 43$
   (g) $71{,}908 \div 824$

7. **(a)** Rounding to the left-most digit, calculate approximate values to the computations in problem 6.
   (b) Compute exact answers to parts (a) through (g) of problem 6.

8. (a) Rounding to the nearest 10, use mental processes to determine the approximate value of this sum.

   $$284 + 3046$$

   (b) Use your calculator to determine the exact value of the sum in part (a).

9. Rounding to the nearest thousands and using mental arithmetic, estimate each of these sums and differences.

(a)		(b)	
	17,281		2734
	6 564		3541
	12,147		2284
	2 481		3478
	+ 13,671		+ 7124

(c)
$$\begin{array}{r} 28{,}341 \\ 942 \\ 2\,431 \\ 4\,716 \\ +\ 12{,}472 \\ \hline \end{array}$$
(d)
$$\begin{array}{r} 4720 \\ -\ 1324 \\ \hline \end{array}$$

(e)
$$\begin{array}{r} 21{,}243 \\ -\ \ 7\,824 \\ \hline \end{array}$$
(f)
$$\begin{array}{r} 37{,}481 \\ -\ 16{,}249 \\ \hline \end{array}$$

(g) Use your calculator to compute the exact value of the answer to parts (a) through (f).

10. Using rounding to the left-most digit, estimate these products.

(a) $2748 \cdot 31$ **(b)** $4781 \cdot 342$ **(c)** $23{,}247 \cdot 357$

(d) Use your calculator to determine the exact value of the products in parts (a) through (c).

11. Use rounding to the left-most digit to estimate the quotient in each of the following.

(a) $29{,}342 \div 42$ **(b)** $7431 \div 37$

(c) $79{,}287 \div 429$

(d) Use your calculator to determine the exact value of the quotients in each of parts (a) through (c).

Thinking Critically

12. Theresa DePalo and Fontaine Elvado each used their calculators to compute $357 + 492$. Fontaine's answer was 749 and Theresa's was 849. Who was most likely correct? In two brief sentences tell how estimation can help you decide whose answer was probably correct.

13. Use rounding to estimate the results of each of the following.

(a) $\dfrac{452 + 371}{281}$ **(b)** $\dfrac{3 \cdot 271 + 465}{74 + 9}$

(c) $\dfrac{845 \cdot 215}{416}$

(d) Use your calculator to determine the exact answer to each problem in parts (a) through (c).

14. Sometimes the last digits of numbers can help you decide if calculator computations are correct.

(a) Given that one of 27,453; 27,587; or 27,451 is the correct result of multiplying 283 by 97, which answer is correct?

(b) In two brief sentences, tell how consideration of last digits helped you answer part (a).

15. Since 25,781; 24,323; 26,012; and 25,243 are all about the same size, about how large is their sum? Explain briefly.

Thinking Cooperatively

16. Note that $(2)(4678) = 9356$. Place parentheses in each of the following strings of digits to make the equality true.

(a) 2 4 6 7 8 = 16,272

(b) 2 4 6 7 8 = 19,188

(c) 2 4 6 7 8 = 19,736

(d) 2 4 6 7 8 = 11,232

17. Place parentheses and plus signs in each string of digits to make these equalities true. (*Hint:* $(88) + (88) + (88) + (88) = 352$.)

(a) 8 8 8 8 8 8 8 8 = 136

(b) 8 8 8 8 8 8 8 8 = 17,776

(c) 8 8 8 8 8 8 8 8 = 928

(d) 8 8 8 8 8 8 8 8 = 9064

(e) 8 8 8 8 8 8 8 8 = 8920

18. Place parentheses and divide signs in each of these strings of digits so that the equalities are true. Remember that it is **not** generally the case that $(a \div b) \div c = a \div (b \div c)$. (*Hint:* $(844 \div (4 \div 2)) \div (2 \div 1) = 211$.)

(a) 8 4 4 4 2 2 1 = 844,422

(b) 8 4 4 4 2 2 1 = 42,221

(c) 8 4 4 4 2 2 1 = 42

(d) 8 4 4 4 2 2 1 = 38 R 46

Connections

19. While grocery shopping with $40, you buy the following items at the price listed:

2 gallons of milk	$2.29 a gallon
1 dozen eggs	$1.63 per dozen
2 rolls of paper towels	$1.21 per roll
1 five pound pork roast	$1.47 per pound
2 boxes of breakfast cereal	$3.19 each
1 azalea	$9.95 each

(a) About how much will all of this cost?

(b) If you don't buy the azalea, about how much change should you receive?

For Review

20. In a college mathematics class all the students are also taking anthropology, history or psychology and some of the students are taking two or even all three of these courses. If (i) forty students are taking anthropology, (ii) eleven students are taking history, (iii) twelve students are taking psychology, (iv) three students are taking all three courses, (v) six students are taking anthropology and history, and (vi) six students are taking psychology and anthropology,

(a) how many students are taking only anthropology?

(b) how many students are taking anthropology or history?

(c) how many students are taking history and anthropology but not psychology?

21. Fill in the missing digits in each of these addition problems.

(a)	(b)	(c)
–742	2341	–21–
41–	4–30	–0–
69–3	1–––	41–
+ 2–18	+ 3–18	+ 771–
–2,818	–3,100	9666

22. Fill in the missing digits in each of these subtraction problems.

(a)	(b)	(c)
27–4	7–01	–22–
– –64–	– 192–	– 2333
91	–8–9	1––9

23. Fill in the missing digits in each of these multiplication and division problems.

(a)	(b)
34–	3–––7
× ––	× 1–
6–4	––– 296
–––8	–7–––
––,57–	–6–,6––

(c)

```
        1–
–61)3 – 24
   – 6 –
    1 2 1 –
    – – – –
        – – 0
```

COOPERATIVE Investigation — *Multiplication Tic-Tac-Toe*

Materials Needed

1. A calculator for each student.
2. A multiplication Tic-Tac-Toe handout for each pair of students.

Procedure

This is like regular tic-tac-toe except that on each play the player chooses two numbers from the list and places his or her mark (X or O) on the square containing the product of the numbers chosen. Thus, in part (a), if the first player's symbol is X, and 11 and 23 are chosen on the first play, an X is placed on 253 on the diagram and also the notation 11 · 23. As usual, the player to get three Xs or Os in a row wins the game. If a player chooses a product not in a square, (s)he loses that turn.

(a) 11, 12, 15, 19, 23

345	132	285
228	209	437
X 253 (11 · 23)	276	180

(b) 13, 14, 17, 19, 21

221	266	273
247	238	182
323	399	357

(c) 9, 13, 17, 23, 25

425	153	299
117	391	325
225	575	207

Continued

(d) 7, 23, 341, 2706, 4123

94,829	2,387	11,156,838
1,405,943	7,843	18,942
28,861	62,238	922,746

Parts (e) and (f) are the same as parts (a) through (d) except that the numbers in the squares are the approximations of the true products obtained by rounding each number to the left position.

(e) 23, 27, 36, 47, 55

600 $\approx 23 \cdot 27$ (crossed out)	800	1800
2000	1500	1200
1000	2400	3000

(f) 143, 254, 361, 2391, 2511

40,000	800,000	300,000
600,000	200,000	1,200,000
120,000	6,000,000	900,000

(g) For parts (a) through (d), discuss how looking at the last digits in the numerals can help you play the game.

(h) For part (d), discuss how estimation can help you play the game.

3.6 Getting the Most Out of Your Calculator

STRATEGY

- ***Use algorithmic thinking.***

It's a safe bet that almost everyone who reads these words is the owner of at least one electronic calculator. It's also a safe bet that almost every reader knows how to add, subtract, multiply, and divide with his or her calculator. However, most calculators have useful features that are not well understood and are not employed by some users. In this section, we discuss some of these special features.

Figure 3.14
The Math Explorer Calculator

First, we note that there are three common types of logic that determine how your calculator operates—arithmetic (á-rith-mé-tic), algebraic, and reverse Polish. **Reverse Polish notation** is a powerful system used on sophisticated scientific calculators and is not suitable for elementary school students. Machines utilizing **arithmetic logic** are too simplistic and again are not recommended for classroom use. Machines using **algebraic logic** are the most appropriate for student use and have built-in features that make calculations easier and more natural. Of course, your instruction manual is the basic source of information about the operation of your particular calculator. We will discuss the features of the *Math Explorer* calculator. This machine, shown in Figure 3.14, is made by Texas Instruments especially for use by elementary school students. The *Math Explorer* uses algebraic logic, does arithmetic both with fractions and decimals, and also does integer division with remainders. Most of the discussion here applies equally well to other algebraic machines, but your instruction manual is the final arbiter.

FROM THE NCTM STANDARDS

Calculators and Computers

The K–4 curriculum should make appropriate and ongoing use of calculators and computers. Calculators must be accepted at the K–4 level as valuable tools for learning mathematics. Calculators enable children to explore number ideas and patterns, to have valuable concept-development experiences, to focus on problem-solving processes, and to investigate realistic applications. The thoughtful use of calculators can increase the quality of the curriculum as well as the quality of children's learning.

Calculators do not replace the need to learn basic facts, to compute mentally, or to do reasonable paper-and-pencil computation. Classroom experience indicates that young children take a common sense view about calculators and recognize the importance of not relying on them when it is more appropriate to compute in other ways. The availability of calculators means, however, that educators must develop a broader view of the various ways computation can be carried out and must place less emphasis on complex paper-and-pencil computation. Calculators also highlight the importance of teaching children to recognize whether computed results are reasonable.

The power of computers also needs to be used in contemporary mathematics programs. Computer languages that are geometric in nature help young children become familiar with important geometric ideas. Computer simulations of mathematical ideas, such as modeling the renaming of numbers, are an important aid in helping children identify the key features of mathematics. Many software programs provide interesting problem-solving situations and applications.

The thoughtful and creative use of technology can greatly improve both the quality of the curriculum and the quality of children's learning. Integrating calculators and computers into school mathematics programs is critical in meeting the goals of a redefined curriculum.

SOURCE: From *Curriculum and Evaluation Standards For School Mathematics Grades K–4,* p. 19.

Priority of Operations

Suppose you enter

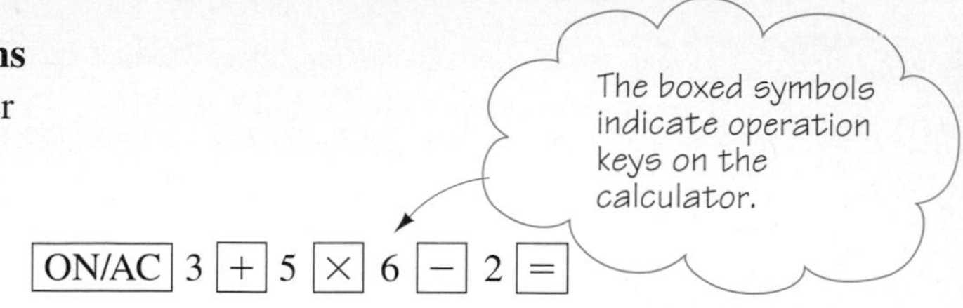

into your calculator. A calculator with arithmetic logic would perform each operation in exactly the order entered and would give the answer 46; that is,

$$3 + 5 = 8$$
$$8 \times 6 = 48$$
$$48 - 2 = 46.$$

The boxed symbols indicate operation keys on the calculator.
ON/AC 3 + 5 × 6 − 2 =

In contrast, a calculator with algebraic logic multiplies and divides *before* adding and subtracting. Also, if there are pending operations of equal priority, these calculators execute them from left to right. Since these priorities are those generally accepted in mathematics, algebraic machines correspond well to the standard rules and properties of arithmetic. Thus, for the above sequence of entries an algebraic calculator gives the answer 31 obtained as follows:

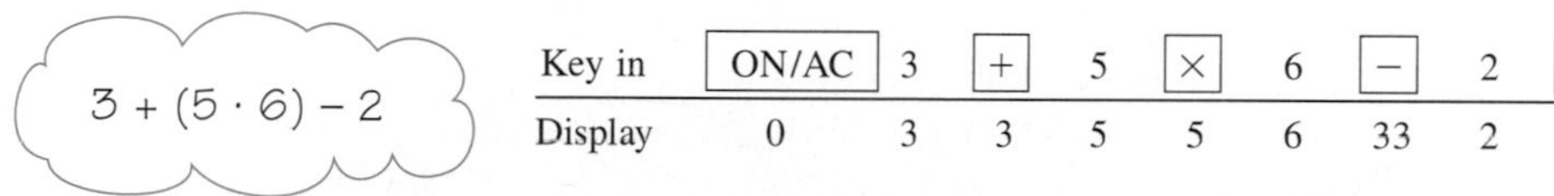

Key in	ON/AC	3	+	5	×	6	−	2	=
Display	0	3	3	5	5	6	33	2	31

When the − was pressed, the calculator *first* multiplied 5 × 6 to get 30, *then* completed the pending addition to 3 giving 33, and then subtracted the 2 to give 31.

While we have not yet discussed the use of all of these keys, the priority of operations for machines with algebraic logic is as shown in Table 3.11.

TABLE 3.11 Priority of Operations on Calculators with Algebraic Logic

Priority	Keys	Explanation
1	( )	Operations in parentheses are performed before other operations.
2	x^2 √ 10^n 1/x %	Operations performed on a single number.
3	y^x	Exponentiation.
4	× ÷	Multiplications and divisions are completed before additions and subtractions.
5	+ −	Additions and subtractions are completed last.
6	=	Terminates a calculation.

EXAMPLE 3.29 Understanding the Order of Calculator Operations

(a) Indicate what should be entered into a calculator with algebraic notation to compute

$$27 \div 3 + 24 \cdot 4.$$

(b) Actually key the sequence in part (a) into your calculator and complete the computation.

(c) If you entered the same sequence into a calculator with arithmetic logic, what would the result be?

SOLUTION

(a) Because of the priority of operations, it is only necessary to key in the following:

$$\boxed{\text{ON/AC}}\ 27\ \boxed{\div}\ 3\ \boxed{+}\ 24\ \boxed{\times}\ 4\ \boxed{=}.$$

(b) On a machine with algebraic logic, the calculator makes the following sequence of calculations:

$$27 \div 3 = 9, \quad 24 \cdot 4 = 96, \quad 9 + 96 = 105.$$

(c) A machine with arithmetic logic would compute

$$27 \div 3 = 9, \quad 9 + 24 = 33, \quad 33 \cdot 4 = 132.$$ ■

It is possible to override the priority of operations built into the calculator by use of the parentheses keys. Suppose you want to compute

$$(789 + 364) \cdot (863 + 939).$$

This is accomplished by keying the sequence

$$\boxed{\text{ON/AC}}\ \boxed{(}\ 789\ \boxed{+}\ 364\ \boxed{)}\ \boxed{\times}\ \boxed{(}\ 863\ \boxed{+}\ 939\ \boxed{)}\ \boxed{=}$$

into the calculator. The desired answer is 2,077,706. If we keyed in

$$\boxed{\text{ON/AC}}\ 789\ \boxed{+}\ 364\ \boxed{\times}\ 863\ \boxed{+}\ 939\ \boxed{=}$$

(the same entries but omitting the parentheses) the calculator would show 315,860 determined by computing the product of 364 and 863 and then adding 789 and 939 in that order.

EXAMPLE 3.30 **Using Parentheses**

Perform this computation on your calculator.

$$216 \div (3 + 24) \cdot 4$$

SOLUTION

Key in ON/AC 216 ÷ (3 + 24) × 4 = to obtain the answer 32. Remember that divide and multiply are operations of the same priority. Thus, the calculator first computes the sum in parentheses, divides 216 by that sum, and multiplies the result by 4. Using clearer notation, what the calculator is computing is

$$(216 \div (3 + 24)) \cdot 4$$

but the built-in priority of operations makes it unnecessary to key the expression into the calculator this way. ■

Before proceeding, several observations should be made.

1. It is always a good idea to begin each calculation by pressing the ON/AC key. This key turns the calculator on. Equally importantly, it clears all preceding data from all parts of the calculator so that present work will not be rendered incorrect by the presence of unexpected and unwanted information from a preceding calculation. As a reminder, in all our examples we indicate starting by depressing this key.
2. Pressing the = key causes the calculator to complete all entered calculations up to that point. This can be used on occasion to simplify calculation. For example, to compute $(29 + 37) \div 11$, one could use parentheses or alternatively enter

   ON/AC 29 + 37 = ÷ 11 =

   to obtain the correct answer of 6. As a check, key in

   ON/AC (29 + 37) ÷ 11 =

   to see that you obtain the same answer.
3. Parentheses must always be entered in pairs; that is, for each left parenthesis entered a right parenthesis must be entered later. Otherwise, when = is entered, the expression *Error P* will appear in the display to inform you of an error in entering parentheses. Other common error messages are:

   Error A Arithmetic error. For example, attempting to divide by zero.

   Error O Overflow error. The number is too large for the calculator.

   Error U Underflow error. The number is too small for the calculator.

   Error 4 You have entered four or more pending operations. The limit for the *Math Explorer* is three.

If, in the midst of a calculation, you obtain an error message of any kind, you must press the ON/AC key to clear the machine. Then repeat the calculation being careful to restructure your procedure to avoid the previous error.

It may be useful to consider one more example concerning priority of operations.

EXAMPLE 3.31

Prioritizing Operations

Compute $\dfrac{323 - 4 \cdot 38}{19}$.

SOLUTION

If you enter

[ON/AC] 323 [−] 4 [×] 38 [÷] 19 [=]

into the calculator, you obtain the incorrect answer 315. Because of the priority of operations, this sequence of commands computes $4 \cdot 38 \div 19$ and subtracts this from 323. However, the problem requires that the entire quantity $323 - 4 \cdot 38$ be divided by 19. This can be accomplished in several ways but perhaps the following are easiest. Using parentheses

[ON/AC] [(] 323 [−] 4 [×] 38 [)] [÷] 19 [=]

Using [=] twice

[ON/AC] 323 [−] 4 [×] 38 [=] [÷] 19 [=]

Key in each of these sequences and note that each gives the correct answer of 9. ■

Using the [x^2], [$\sqrt{\ }$], [10^n], and [$1/x$] Keys

All of these keys cause the calculator to perform an operation on a single number, and all but 10^n operate in the same way. To compute

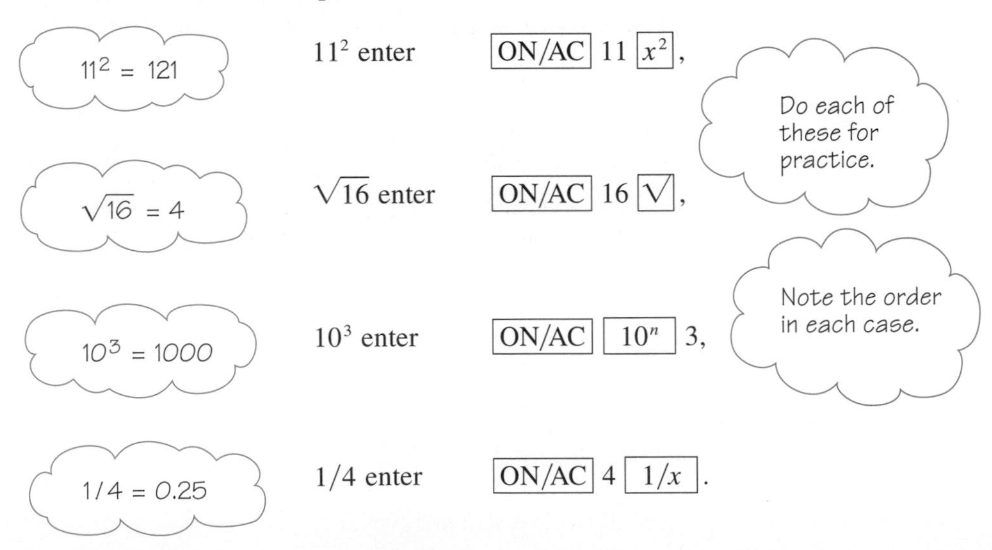

11² = 121
11² enter [ON/AC] 11 [x²],
√16 = 4
√16 enter [ON/AC] 16 [√],
10³ = 1000
10³ enter [ON/AC] [10ⁿ] 3,
1/4 = 0.25
1/4 enter [ON/AC] 4 [1/x].
Do each of these for practice.
Note the order in each case.

In each case, the operation is carried out immediately; it is not necessary to use the [=] key to complete the computation.

Using the [y^x] Key

To compute 10^3 on the calculator you enter the string [ON/AC] [10^n] 3. To compute 5^3 we proceed similarly but with the [y^x] key. Here we have to tell the calculator what number we want to raise to a power and what power to raise it to. For

5^3, $y = 5$, $x = 3$, and we enter 5 [y^x] 3 [=] to obtain 125. Check to see that the entry strings

[ON/AC] 2 [y^x] 5 [=]

and

[ON/AC] 2 [×] 2 [×] 2 [×] 2 [×] 2 [=]

both give 32.

Using the Memory Keys—[M+], [M−], [MR], and [x⇄M]

The *Math Explorer* has a memory that is accessed with the [M+], [M−], [MR], and [x⇄M] keys. Other good algebraic calculators will have similar capabilities though the keys may bear different symbols like [M], [RCL], and so on.

Using the [M+] or [M−] keys adds or subtracts respectively the number in the calculator display to the number in the memory. If nothing has been placed in the memory, it will be assumed to contain a zero. Also, when you place a number in the memory, the display will show a small M to remind you that the memory is not empty. The [MR] key recalls what is in the memory and places it in the display. The [x⇄M] key exchanges the number in the display, x, with the number in the memory, M. These capabilities are useful in elementary ways but, with a little imagination, they are also useful in more surprising and powerful ways.

EXAMPLE 3.32 **Using the [M+] and [MR] Keys**

Compute the quotient

$$\frac{\sqrt{158{,}404} - 200}{6019 - 5986}$$

without using parentheses.

SOLUTION

The idea is to compute 6019 − 5986 and store it in the memory. Then compute $\sqrt{158{,}404} - 200$ and divide the result by the number stored in the memory using the divide ÷ and recall memory, [MR], keys. We enter the following to obtain the correct answer of 6.

[ON/AC] 6019 [−] 5986 [=] [M+] 158404 [√] [−] 200 [=] [÷] [MR] [=] ■

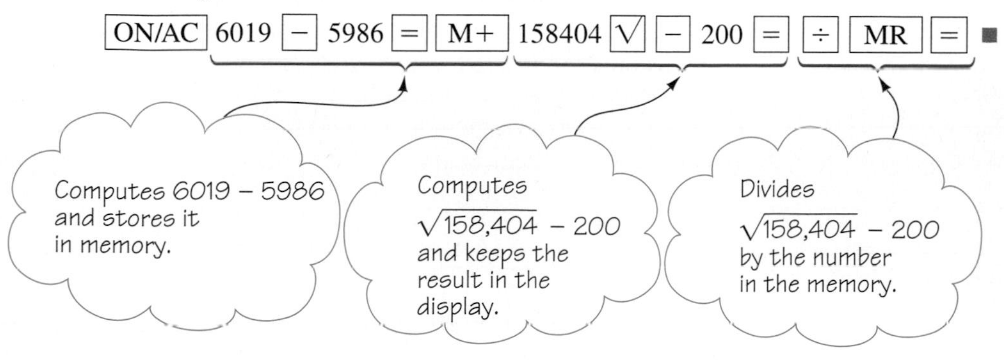

EXAMPLE 3.33 **Using the [M+] and [M−] Keys**

Compute

$$329 \cdot 742 - 3 \cdot 8914 + 2 \cdot 2175.$$

SOLUTION

The idea is to compute the products and add them to or subtract them from the contents of the memory. We then obtain the answer by using the [MR] key. The desired string entered in the calculator is

[ON/AC] 329 [×] 742 [=] [M+] 3 [×] 8914 [=] [M−] 2 [×] 2175 [=] [M+] [MR].

This gives the correct answer of 221,726. However, in this simple case, it is easier to enter the string

[ON/AC] 329 [×] 742 [−] 3 [×] 8914 [+] 2 [×] 2175 [=]

to obtain 221,726 as before. Be sure to check both entry sequences to see that they give this result. ■

EXAMPLE 3.34

Using the [x⟳M] Key

Compute this quotient.

$$\frac{2927 + 8976 - 3417}{3154 + 1089}$$

SOLUTION

The idea is to compute the numerator and place it in memory using the [M+] key. Then compute the denominator which will be in the display. Use [x⟳M] to exchange the two numbers. Then use [÷] [RM] to perform the desired division. The entry string

[ON/AC] 2927 [+] 8976 [−] 3417 [=] [M+] 3154 [+] 1089 [=] [x⟳M] [÷] [RM] [=]

yields the answer 2. Check to see that the entry string

[ON/AC] [(] 2927 [+] 8976 [−] 3417 [)] [÷] [(] 3154 [+] 1089 [)] [=]

with parentheses gives the same answer. ■

Using the [INT÷] Key to Compute Integer Division with Remainders

Since the *Math Explorer* calculator is designed for use in elementary classrooms, it was thought appropriate to give it the capability of dividing one natural number by another and recording the quotient as well as the remainder. Because of limited space in the display, there is a limitation on the size of the numbers that can be used. If either the quotient or the remainder is more than a four-digit number, the Error O (overflow) message will appear in the calculator display and the problem will have to be completed by some other method.

EXAMPLE 3.35

Using the [INT÷] Key

(a) Using the [INT÷] key, compute the quotient and the remainder when 89,765 is divided by 78.

(b) Using the [INT÷] key, compute the quotient and the remainder when 897,654 is divided by 81.

SOLUTION

(a) Entering [ON/AC] 89765 [INT÷] 78 [=] into your calculator, you see the quotient, 1150, and the remainder, 65, in the display. Use your calculator to check this result by showing that

$$1150 \cdot 78 + 65 = 89765.$$

(b) Entering [ON/AC] 897654 [INT÷] 81 [=] into your calculator, the display reads *Error O* since the quotient is a five-digit number. ■

Using the Built-in Constant Function

Consider the arithmetic progression

$$3, 7, 11, 15, 19, \ldots, 67.$$

It is easy to use your calculator to compute all the terms in this progression. Simply enter [ON/AC] 3 [+] 4 and then repeatedly press the [=] key to repeatedly add 4. This utilizes the built-in constant function of most calculators with algebraic logic.

EXAMPLE 3.36 Finding the Sum of an Arithmetic Progression

Compute the sum $3 + 7 + 11 + \cdots + 67$ of the terms in the arithmetic progression above.

SOLUTION

There are 17 terms since you must add 16 4s to 3 to obtain 67.

1. If we remember young Gauss's trick, then

$$\begin{aligned} S &= 3 + 7 + 11 + \cdots + 67 \\ S &= 67 + 63 + 59 + \cdots + 3 \\ \hline 2S &= 70 + 70 + 70 + \cdots + 70 = 17 \cdot 70 \end{aligned}$$

so

$$S = \frac{17 \cdot 70}{2} = 595.$$

2. An alternative approach to this problem is to use the [M+] and the constant function of your calculator. Thus, we enter

[ON/AC] 3 [M+] [+] 4 [=] [M+] [=] [M+] $\cdots$,

repeating the [=] [M+] sequence until we reach 67. We then press [M+] once more and [MR] to again obtain 595 as the answer. ■

EXAMPLE 3.37 Finding the Sum of a Geometric Progression

Find the sum of the first 15 terms of the geometric progression whose first 4 terms are 4, 12, 36, 108.

SOLUTION

The consecutive terms of the progression can be found on your calculator by entering 4 [×] 3 [=] and then pressing [=] repeatedly.

As with the arithmetic progression, this problem can also be solved on algebraic notation machines with a built-in constant. Thus, if we enter

[ON/AC] 4 [M+] [×] 3 [=] [M+] [=] [M+] [=] · · · [=] [M+] [MR]

where we use the [=] key 14 times (to add 15 terms), we obtain the desired answer of 28,697,812. Check this on your calculator. ■

Using the [+⟲−] Key

The [+⟲−] key is used to change the sign of a number in the calculator's display. Thus, if 23 is in the display, pressing [+⟲−] causes −23 to appear. Similarly, if −23 is in the display, pressing [+⟲−] causes 23 to appear.

EXAMPLE 3.38

Using the [+⟲−] Key

Compute $281 - \sqrt{133 \div 19 + 7914}$.

SOLUTION

Here we compute the square root first and then subtract it from 281. This could be done using the memory, but suppose there is a number in the memory that we want to use later. Using the fact that

$$281 - \sqrt{133 \div 19 + 7914} = -(\sqrt{133 \div 19 + 7914} - 281)$$

we enter the following command string.

[ON/AC] 133 ÷ 19 [+] 7914 [=] [√] [−] 281 [=] [+⟲−]

to obtain the desired answer of 192.

Check that the entry string

[ON/AC] 133 [÷] 19 [+] 7914 [=] [√] [M+] 281 [−] [MR] [=],

which performs the subtraction in the reverse order, gives the same answer. ■

Algorithmic Thinking

An approach to doing mathematics that is particularly important when working with calculators and computers is **algorithmic thinking**—the doing of mathematical tasks by means of a sequential and often repetitive set of steps. This was illustrated modestly in the preceding examples explaining the repeated use of the [=] and [M+] keys. But much more can be done with your calculator to illustrate this approach to problem solving. To make the idea more clear consider the following example.

EXAMPLE 3.39

Generating the Fibonacci Sequence Algorithmically

Develop an efficient algorithm for generating successive terms of the Fibonacci sequence using the special capabilities of your calculator.

SOLUTION

Recall that the Fibonacci numbers are $F_1 = 1$, $F_2 = 1$, $F_3 = 2$, $F_4 = 3$, $F_5 = 5, \ldots$, where we start with 1 and 1 and add any two successive terms to obtain the next term in the sequence. Obviously, this can be accomplished with the straightforward use of the [+] and [=] keys on your calculator. But this requires repeatedly entering the proper numbers. More efficient algorithmic approaches can be devised which only require entering one or two numbers initially and then repetitively using the special keys on your calculator to complete the task. **Note:** Since not all calculators with algebraic logic operate exactly the same, great care must be exercised in devising an algorithm suitable for your machine. Here we offer two alternatives depending on the special characteristics of different calculators.

Algorithm 1

For an algebraic calculator with [M+] and [x⟳M] or equivalent keys, the following algorithm, given in a vertical format that shows what is entered, what is in the display, x, and what is in the memory, M, at each step, generates the Fibonacci numbers. These numbers are shown in red.

Entry	x	M	
[ON/AC]	0	0	
1	**1**	0	Enters 1 in the display; 0 is in M.
[M+]	1	1	Adds the 1 to the 0 in M; keeps 1 in the display.
[x⟳M]	**1**	1	Interchanges the 1s in x and M.
[M+]	1	2	Adds the 1 in x to the 1 in M; keeps 1 in x.
[x⟳M]	**2**	1	Interchanges the 2 in x and the 1 in M.
[M+]	2	3	Adds the 2 in x to the 1 in M; keeps 2 in x.
[x⟳M]	**3**	2	Interchanges the 2 in x and the 3 in M.
[M+]	3	5	Adds the 3 in x to the 2 in M; keeps the 3 in x.
[x⟳M]	**5**	3	Interchanges the 3 in x and the 5 in M.
			•
			•
			•
[M+]	5	8	
[x⟳M]	**8**	5	
.	.	.	
.	.	.	
.	.	.	

The algorithm continues by repeating the entry sequence [M+] [x⟳M].

Algorithm 2

For algebraic calculators without an [x⟳M] key the following algorithm will efficiently generate the Fibonacci sequence:

[ON/AC] 1 [M+] [+] [M+] [MR] [+] [M+] [MR] [+] · · · [M+] [MR] [+] · · ·.

This algorithm causes the calculator to display these successive numbers. Here we highlight the Fibonacci numbers in red.

0, 1, **1**, **1**, 1, **2**, **3**, 3, **5**, **8**, 8, **13**, **21**, 21, ... ■

EXAMPLE 3.40 **Computing $1 + 2 + \cdots + n$ and $1^2 + 2^2 + \cdots + n^2$ Simultaneously**

Write an algorithm that simultaneously computes

$$t_{10} = 1 + 2 + 3 + 4 + 5 + 6 + 7 + 8 + 9 + 10$$

and

$$s_{10} = 1^2 + 2^2 + 3^2 + 4^2 + 5^2 + 6^2 + 7^2 + 8^2 + 9^2 + 10^2.$$

SOLUTION

The entry string

[ON/AC] 1 [M+] [x^2] [+] 2 [M+] [x^2] [+] 3 [M+] [x^2] [+] · · · [+] 10 [M+] [x^2] [=] [MR]

does the job quite nicely. Check to see that when [=] is pressed you obtain $s_{10} = 385$, and, when [MR] is pressed, you obtain $t_{10} = 55$. ■

PROBLEM-SOLVING STRATEGY 14 • Use Algorithmic Thinking

Sometimes a problem or set of problems can be solved by devising a set of operations that can be carried out repetitively on a calculator or computer. Doing the problem by hand may be prohibitively time-consuming.

PROBLEM SET 3.6

Understanding Concepts

1. Use your calculator to compute each of the following.
   (a) $284 + 357$ (b) $357 - 284$ (c) $284 \cdot 357$
   (d) $284 \div 71$ (e) $781 - 35 + 24$
   (f) $781 - (35 + 24)$ (g) $781 - (35 - 24)$
   (h) $861 - 423 - 201$ (i) $861 - (423 + 201)$
2. Compute the following using your calculator.
   (a) $271 \cdot 365$ (b) $1183 \div 91$
   (c) $1024 \div 16 \div 2$ (d) $1024 \div (16 \div 2)$
3. Use your calculator to calculate each of these division problems.
   (a) $\dfrac{420 + 315}{15}$ (b) $423 + 315 \div 15$
   (c) $\dfrac{4441 + 2332}{220 + 301}$ (d) $\dfrac{16{,}157 + 17 \cdot 13}{722 - 291}$
4. Evaluate the following using your calculator.
   (a) 29^2 (b) $\sqrt{1849}$
   (c) $\sqrt{2569 - 1480}$ (d) $\sqrt{1444} - \sqrt{784}$
5. Write out an entry string to compute
   $$\frac{\sqrt{784} - 91 \div 13}{8 \cdot 49 - 11 \cdot 35}$$
   (a) using parentheses.
   (b) not using parentheses.
6. Use the [M+], [M−], and [MR] keys to compute
   $$\frac{4041 + 1237}{91} + \frac{3381 + 2331}{84} - \frac{2113 + 2993}{46}$$
   (*Note:* You should be able to do this entirely with your calculator. Nothing need be (should be) written down but the answer.)
7. (a) Write out the expression you are evaluating if you enter [ON/AC] 1831 [−] 17 [×] 28 [+] 34 [=] into a calculator with algebraic logic.
   (b) Write the numerical answer to part (a).
   (c) What would the answer be if you entered the string in part (a) into a calculator with arithmetic logic?

8. **(a)** Write out the expression you are evaluating if you enter

[ON/AC] 42 [×] 34 [−] 14 [×] 6 [=] [÷] 28 [=]

into your calculator.

(b) What expression does this string evaluate?

[ON/AC] 42 [×] 34 [−] 14 [×] 6 [x^2] [=] [÷] 28 [=]

9. **(a)** Make your calculator count by 2s by entering

[ON/AC] 0 [+] 2 [=] [=] [=] . . . .

(b) Make your calculator generate the odd numbers by entering

[ON/AC] 1 [+] 2 [=] [=] [=] . . . .

10. **(a)** Make your calculator count by 17s by entering

[ON/AC] 0 [+] 17 [=] [=] [=] . . . .

(b) If you count by 17s do you ever get to 323? If so when?

11. Use the constant function to generate the first 10 terms of the arithmetic progression whose first 4 terms are 2, 6, 10, 14; that is, use the [=] key repeatedly.

12. Use the constant function to generate the first 10 terms of the geometric progression whose first 4 terms are 3, 15, 75, 375.

13. Use the [=] and [M+] keys to compute the sum of the arithmetic progression

$$5 + 8 + 11 + \cdots + 47.$$

14. Use the [=] and [M+] keys to compute the sum of the geometric progression

$$3 + 15 + 75 + \cdots + 234{,}375.$$

15. Compute 2^{20} using the built-in constant and mentally keeping track of the times you multiply by 2. (*Hint*: Enter [ON/AC] 2 [×] 2 [=] [=] . . . , repeating the [=] as long as necessary.)

Thinking Critically

16. **(a)** Write out the expression your calculator will evaluate if you enter the string

[ON/AC] 5 [√] [+] 1 [=] [÷] 2 [=] [M+] [y^x] 3 [=] [÷] 5 [√] [=]

(b) Compute the quantities indicated for the successive values of n and fill in the following table. Note that $(1 + \sqrt{5})/2$ can be computed and stored in [M+] to be recalled repeatedly for use in the successive calculations.

n	The integer nearest $\dfrac{\left(\frac{1+\sqrt{5}}{2}\right)^n}{\sqrt{5}}$
1	
2	
3	
4	
5	
6	
7	
8	
9	
10	

(c) What pattern do you notice in the table in part (b)? Have you seen these numbers before? Guess a theorem on the basis of this table.

17. Use your calculator to compute the following. (*Hint:* Let $G = (1 + \sqrt{5})/2$ and notice you are computing $(G^n - (1 - G)^n)/\sqrt{5}$ for $n = 1, 2, 3,$ 4 and 5. Calculate G and store it in memory for repeated use. In this form be sure to use parentheses.)

(a) $$\frac{\left(\frac{1+\sqrt{5}}{2}\right) - \left(\frac{1-\sqrt{5}}{2}\right)}{\sqrt{5}}$$

(b) $$\frac{\left(\frac{1+\sqrt{5}}{2}\right)^2 - \left(\frac{1-\sqrt{5}}{2}\right)^2}{\sqrt{5}}$$

(c) $$\frac{\left(\frac{1+\sqrt{5}}{2}\right)^3 - \left(\frac{1-\sqrt{5}}{2}\right)^3}{\sqrt{5}}$$

(d) $$\frac{\left(\frac{1+\sqrt{5}}{2}\right)^4 - \left(\frac{1-\sqrt{5}}{2}\right)^4}{\sqrt{5}}$$

(e) Predict the result of calculating

$$\frac{\left(\frac{1+\sqrt{5}}{2}\right)^5 - \left(\frac{1-\sqrt{5}}{2}\right)^5}{\sqrt{5}}.$$

(f) Predict the result of calculating

$$\frac{\left(\frac{1+\sqrt{5}}{2}\right)^n - \left(\frac{1-\sqrt{5}}{2}\right)^n}{\sqrt{5}}$$

for any natural number n.

18. Use your calculator to compute the following. (*Hint:* Compute and store $(1 + \sqrt{5})/2$ in memory as you did for problem 17.)

(a) $\left(\frac{1+\sqrt{5}}{2}\right) + \left(\frac{1-\sqrt{5}}{2}\right)$

(b) $\left(\frac{1+\sqrt{5}}{2}\right)^2 + \left(\frac{1-\sqrt{5}}{2}\right)^2$

(c) $\left(\frac{1+\sqrt{5}}{2}\right)^3 + \left(\frac{1-\sqrt{5}}{2}\right)^3$

(d) $\left(\frac{1+\sqrt{5}}{2}\right)^4 + \left(\frac{1-\sqrt{5}}{2}\right)^4$

(e) Predict the result of calculating

$$\left(\frac{1+\sqrt{5}}{2}\right)^5 + \left(\frac{1-\sqrt{5}}{2}\right)^5.$$

(f) Describe the sequence of numbers $L_1, L_2, L_3 \ldots$ where

$$L_n = \left(\frac{1+\sqrt{5}}{2}\right)^n + \left(\frac{1-\sqrt{5}}{2}\right)^n$$

for any natural number n.

Thinking Cooperatively

19. In Example 3.39, algorithms were developed to compute successive Fibonacci numbers. The Lucas numbers are $L_1 = 1$, $L_2 = 3$, $L_3 = 4$, $L_4 = 7, \ldots$ where we start with 1 and 3 and then add any two consecutive numbers in the sequence to obtain the next entry. Develop an algorithm similar to one of those in Example 3.39 for computing the successive Lucas numbers on your calculator. (*Hint:* Apart from the initial numerical entries you only need use the [M+] and [x⟲M] keys, otherwise use your own memory and the [+] and [=] keys.)

20. (a) Use the definition of the Fibonacci and Lucas numbers (see problem 19) to complete this table.

n	1	2	3	4	5	6	7	8	9
F_n	1	1							
L_n	1	3							

(b) Compute the sums $F_1 + F_3$, $F_2 + F_4$, $F_3 + F_5, \ldots$. What result do these sums suggest?

(c) Compute the sums $L_1 + L_3$, $L_2 + L_4$, $L_3 + L_5, \ldots$. What result do these sums suggest?

(d) Compute the sums $F_1 + L_1$, $F_2 + L_2$, $F_3 + L_3, \ldots$. What result do these sums suggest?

21. For $n \geq 2$, the nth Lucas number is the integer nearest to $((1 + \sqrt{5})/2)^n$. The entry string

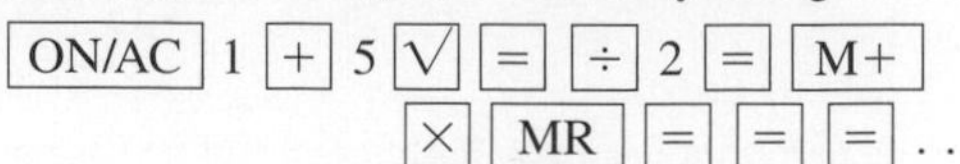

will calculate the consecutive Lucas numbers from the second on.

(a) Calculate L_{38}.

(b) Can you calculate L_{39}? Why or why not?

22. (a) Simultaneously compute $t_6 = 1 + 2 + 3 + 4 + 5 + 6$ and $c_6 = 1^3 + 2^3 + 3^3 + 4^3 + 5^3 + 6^3$.

(b) Repeat part (a) for t_{10} and c_{10}.

(c) Make a guess relating t_n and c_n for any n.

(d) Do your examples prove that your guess is correct? Explain briefly.

For Review

23. Write a second addition equation and two subtraction equations equivalent to $18 + 17 = 35$.

24. Write two addition equations and a second subtraction equation equivalent to $27 - 9 = 18$.

25. Write a second multiplication equation and two division equations equivalent to $27 \cdot 11 = 297$.

26. Write two multiplication equations and a second division equation equivalent to $96 \div 12 = 8$.

27. Draw a rectangular array to illustrate the product $5 \cdot 7$.

28. Draw an appropriate diagram to illustrate the equation $5(3 + 4) = 5 \cdot 3 + 5 \cdot 4$; that is, to illustrate the distributive property for whole numbers.

EPILOGUE Calculating Today

In this chapter we have considered the art of writing numbers and performing calculations, with methods extending from ancient to modern times. Our algorithms range from pencil and paper procedures to the use of modern electronic calculators and computers. History shows that the development of the art of calculation has been a tortuous task extending over several thousands of years, but that it has finally reached a stage of extraordinary speed and accuracy. Electronic calculators perform ordinary arithmetic at the touch of a few buttons and the speed of the latest high-speed electronic computing machines is measured in nano* seconds.

But speed is not the purpose of machines alone. History is replete with the names of calculating prodigies who could perform the most astounding feats of mental arithmetic quickly and accurately. Be that as it may, with command of the ideas discussed in this chapter, every child can become a calculating prodigy in his or her own right.

These important notions include the following:

- the basic number facts,
- positional notation in base 10 and others bases,
- the basic algorithms,
- estimation and approximation,
- the use of a calculator with algebraic notation.

This puts great arithmetic power within easy reach of everyone and makes it possible for students to spend a great deal more time thinking about and doing more significant and meaningful mathematics. We consider some of these more important ideas, starting in the next chapter with notions from number theory.

CLASSIC CONUNDRUM A Cryptarithm

Cryptarithms are classic arithmetic problems with letters in place of digits. Each different letter stands for a different digit. This puzzle is meant to be solved using knowledge of the basic arithmetic facts and the addition algorithm. Solve this cryptarithm.

$$\begin{array}{r} \text{FOOD} \\ +\quad \text{FAD} \\ \hline \text{DIETS} \end{array}$$

Is more than one answer possible?

CHAPTER 3 SUMMARY

Key Concepts

1. Numeration systems
   (a) Ancient systems like the Egyptian and Roman were additive systems that made it difficult to record large numbers and to calculate.
   (b) Positional systems like the Mayan and our modern Hindu-Arabic, or decimal, system make it possible to record large numbers with only a few symbols and to devise relatively easy algorithms for addition, subtraction, multiplication, and division.
   (c) Studying positional systems in bases other than base ten make positional notation more clear and also help to clarify computational algorithms.
2. Mental arithmetic and estimation
   (a) It is important that students develop skill at mental arithmetic. Chil-

*A nano second is one billionth of a second.

dren find mental arithmetic exciting (in games, contests, and so on), and it promotes greater confidence in doing mathematics generally.

(b) With calculators and computers giving quick and easy solutions, it is crucial that students be able to make good estimates so that they can reliably assess the correctness of results obtained. This not only requires a complete understanding of one-digit facts but also a good understanding of positional notation.

3. Getting the most out of your calculator

(a) Calculators take the drudgery out of everyday arithmetic.

(b) Calculators free students to spend considerably more time on problem solving and critical mathematical thinking.

(c) Beyond simple arithmetic, it is possible to use special capabilities of calculators to devise ingenious methods or algorithms for acomplishing numerous mathematical tasks.

Vocabulary and Notation

Section 3.1

- **Additive systems**
  - **The Egyptian system**
  - **The Roman system**
- **Positional systems**
  - **The Mayan system**
  - **The Hindu-Arabic (decimal) system**
    - **Base**
    - **Expanded notation**
    - **Models for positional notation**
      - **The classroom abacus**
      - **Sticks in bundles**
      - **Unifix™ cubes**
      - **Units, strips, and mats**
      - **Base 10 blocks**

Section 3.2

- **Positional systems in bases other than ten**
  - **Base five notation**
  - **Converting from base five to base ten notation**
  - **Converting from base ten to base five notation**

Section 3.3

- **Algorithms for addition**
  - **Using units, strips, and mats**
  - **Using place value cards**
  - **Using pencil and paper**
- **Algorithms for subtraction**
  - **Using units, strips, and mats**
  - **Using place value cards**
  - **Using pencil and paper**
- **Algorithms in other bases**
  - **Addition in base five**
  - **Subtraction in base five**

Section 3.4

- **Algorithms for multiplication**
  - **Using units, strips, and mats**
  - **Using place value cards**
  - **Using pencil and paper**
  - **Using expanded notation**
- **Algorithms for division**
  - **The division algorithm**
  - **The long division algorithm**
  - **The short division algorithm**
- **Multiplication and division in other bases**
  - **Multiplication in base seven**
  - **Division in base seven**

Section 3.5

- **Mental arithmetic**
  - **One-digit facts**
  - **Easy combinations**
  - **Adjustment**
  - **Working from left to right**
  - **Rounding**
  - **The 5-up rule**
- **Estimation**
  - **Ranges for answers**
  - **Approximating by rounding**

Section 3.6

- **Types of calculators**
  - **Reverse Polish notation**
  - **Arithmetic logic**
  - **Algebraic logic**
- **Priority of operations**
  - **Order for machines using arithmetic logic**
  - **Order for machines using algebraic logic**
  - **Use of parentheses**

Special capabilities
- **Use of the ON/AC key**
- **Use of the = key**
- **Pairing of parentheses**
- **Error messages**
- **Using the x^2, $\sqrt{\ }$, 10^n, and $1/x$ keys**
- **Using the y^x key**

Using the M+, M−, MR, and $x \rightleftarrows M$ keys
Using the INT÷ key
Using the built-in constant function
- **The sum of an arithmetic progression**
- **The sum of a geometric progression**

Using the $+ \rightleftarrows -$ key
Algorithmic thinking

CHAPTER REVIEW EXERCISES

Section 3.1

1. Write the Hindu-Arabic equivalent of each of these

(a)

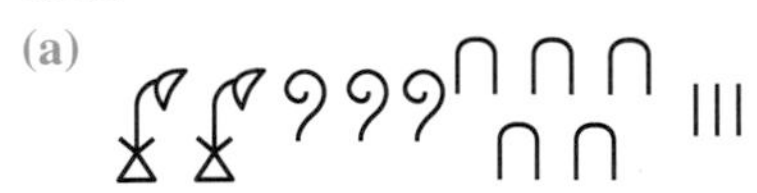

(b)

(c) MCMXCVIII

2. Write 234,572 in Mayan notation.
3. Suppose you have 5 mats, 27 strips, and 32 units for a total count of 802. Briefly describe the exchanges that must be made to represent this number with the smallest possible number of manipulative pieces. How many mats, strips, and units result?

Section 3.2

4. Find the base ten equivalent of each of the following.
   (a) 101101_{two} (b) 346_{seven} (c) $2T9_{\text{twelve}}$
5. Write 287_{ten} as a numeral in each base indicated.
   (a) base five (b) base two (c) base seven

Section 3.3

6. Sketch the solution to 47 + 25 using mats, strips, and units. Draw a square for each mat, a vertical line segment for each strip, and a dot for each unit.
7. Use Instructional Addition Algorithm I to perform the following additions.
   (a) 42 + 54 (b) 47 + 35 (c) 59 + 63
8. Use sketches of place value cards to illustrate each of these subtractions.
   (a) 487 − 275 (b) 547 − 152
9. Perform the following calculations in base five notation. Assume that the numerals are already written in base five.
   (a) $\begin{array}{r} 2433 \\ +\ 141 \\ \hline \end{array}$ (b) $\begin{array}{r} 2433 \\ -\ 141 \\ \hline \end{array}$ (c) $\begin{array}{r} 243 \\ \times\ 42 \\ \hline \end{array}$

Section 3.4

10. Perform these multiplications using Instructional Multiplication Algorithm I.
    (a) 4×357 (b) 27×642
11. Use the scaffold method to perform each of these divisions.
    (a) $7\overline{)895}$ (b) $347\overline{)27483}$
12. Use the short division algorithm to perform each of these divisions.
    (a) $5\overline{)27436}$ (b) $8\overline{)39584}$
13. Carry out each of these multiplications in base five. Assume that the numerals are already written in base five.
    (a) $23 \cdot 42$ (b) $2413 \cdot 332$
14. Use the Russian peasant method to compute the product $42 \cdot 35$.

Section 3.5

15. Round 274,535
    (a) to the nearest one hundred thousand.
    (b) to the nearest ten thousand.
    (c) to the nearest thousand.
16. Compute upper and lower limits for the answers to each of these.
    (a) 657 + 439 (b) 657 − 439
    (c) $657 \cdot 439$ (d) $1657 \div 23$
17. Rounding to the left-most digit, compute approximations to the computations in problem 16.

Section 3.6

18. Compute the following using your calculator.

$$\frac{\sqrt{1444} - 152 \div 19}{2874 - 2859}$$

19. (a) Compute the sum of this arithmetic progression.

$$4 + 11 + 18 + \cdots + 333$$

(b) Compute the sum of this geometric progression.

$$3 + 6 + 12 + 24 + \cdots + 24576$$

20. (a) Compute each of the following. (*Suggestion*: Compute $(1 + \sqrt{5})/2$ and store it in the memory for repeated use.)

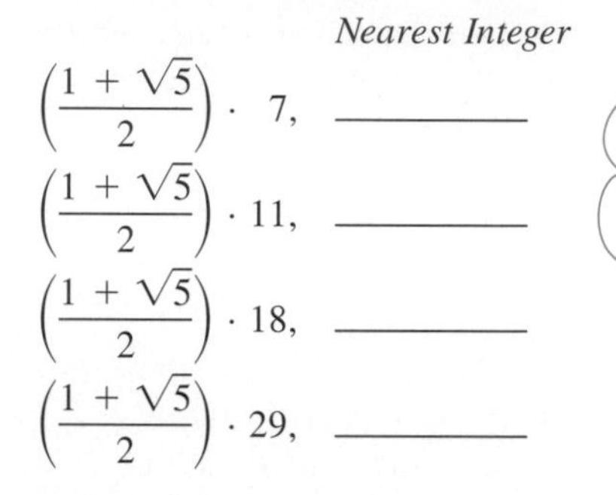

	Nearest Integer
$\left(\frac{1+\sqrt{5}}{2}\right) \cdot 7,$	______
$\left(\frac{1+\sqrt{5}}{2}\right) \cdot 11,$	______
$\left(\frac{1+\sqrt{5}}{2}\right) \cdot 18,$	______
$\left(\frac{1+\sqrt{5}}{2}\right) \cdot 29,$	______

Remember: The Lucas numbers are 1, 3, 4, 7, 11, 18, 29, ...

(b) Make a conjecture on the basis of part (a).

(c) Does your conjecture hold for all Lucas numbers? How about L_1, L_2, and L_3?

(d) How might you modify your conjecture in view of part (c)? Note that $L_{34} = 12{,}752{,}043$ and $L_{35} = 20{,}633{,}239$.

CHAPTER TEST

1. Write the base ten equivalents of each of the following.

   (a) 21022_{three} (b) 317_{eight} (c) 4213_{five}

2. Write 281_{ten} as a numeral in each of these bases.

   (a) base five (b) base two (c) base twelve

3. Perform each of the following calculations entirely in base five. The numerals are already written in base five.

   (a) $\begin{array}{r} 242 \\ +\ 43 \\ \hline \end{array}$ (b) $\begin{array}{r} 242 \\ -\ 43 \\ \hline \end{array}$ (c) $\begin{array}{r} 242 \\ \times\ 43 \\ \hline \end{array}$

4. Make a schematic drawing using mats, strips, and units to illustrate the addition of 74 and 48. Draw squares for mats, straight line segments for strips, and dots for units.

5. Write 39,485 in Mayan notation.

6. Fill in the missing digits in this addition problem.

$$\begin{array}{r} 2\text{–}37 \\ +\ \text{–}22\text{–} \\ \hline \text{–}0{,}0\text{–}1 \end{array}$$

7. Fill in the missing digits in this subtraction problem.

$$\begin{array}{r} \text{–}23\text{–} \\ -\ 35\text{–}2 \\ \hline 4\text{–}94 \end{array}$$

8. Round 3,376,500 to the

   (a) nearest million.
   (b) nearest one hundred thousand.
   (c) nearest ten thousand.
   (d) nearest one thousand.

9. Round each number to the left-most digit to find an estimate for the sum

$$378 + 64 + 291 + 39 + 3871.$$

10. Place the digits 1, 3, 5, 7, 9, in the proper boxes to achieve the maximum product.

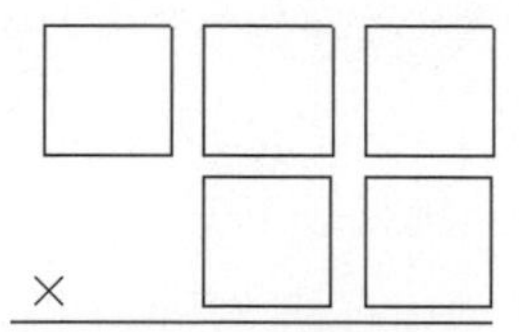

11. Place the digits 0, 2, 4, 6, 8 in the proper boxes to obtain the least product given that 0 cannot be placed in either of the left-hand boxes.

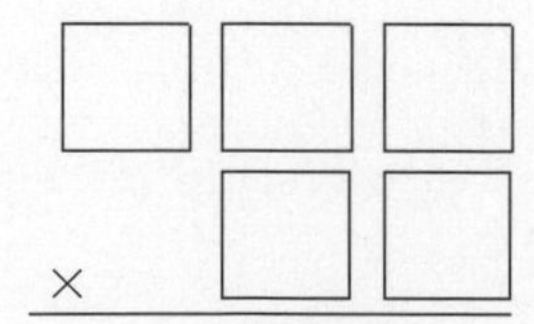

12. (a) Compute the integers nearest to each of

$$f_1 = \frac{1 + \sqrt{2}}{\sqrt{8}}, \quad f_2 = \frac{(1 + \sqrt{2})^2}{\sqrt{8}},$$
$$f_3 = \frac{(1 + \sqrt{2})^3}{\sqrt{8}}, \quad f_4 = \frac{(1 + \sqrt{2})^4}{\sqrt{8}}.$$

(b) Guess the most likely choices for f_5 and f_6.

(c) Guess what rule (other than that of part (a)) might be used to obtain the successive values of f_n.

13. (a) Compute the quantities:

1^3	and	1^2
$1^3 + 2^3$	and	$(1 + 2)^2$
$1^3 + 2^3 + 3^3$	and	$(1 + 2 + 3)^2$
$1^3 + 2^3 + 3^3 + 4^3$	and	$(1 + 2 + 3 + 4)^2$

(b) Compute the square roots of the answers to the computations of part (a).

(c) Guess a formula for $1^3 + 2^3 + \cdots + n^3$ and for $(1 + 2 + 3 + \cdots + n)^2$.

14. Compute the sum of this arithmetic progression.

$$3 + 8 + 13 + \cdots + 123$$

15. Compute the sum of this geometric progression.

$$3 + 15 + 75 + \cdots + 1{,}171{,}875$$

16. Write an algorithm to generate the sequence

$$2, 5, 7, 12, 19, \ldots$$

where we start with 2 and 5 and add any two consecutive terms to obtain the next term.

4 Number Theory

Computer Support for this Chapter

In this chapter you might find the following programs on your disk useful:

- FACTORINTEGER
- TILERECTANGLE
- TILESQUARE
- DIVISORSETS
- MULTIPLESETS
- EUCLIDALG
- CAESAR

Primes and Composites via Rectangular Arrays

Materials Needed

1. Twenty-five small cubes or number tiles for each student or small group of students.
2. One record sheet like this for each student.

Values of n	Dimensions of Rectangles	Number of Rectangles	
1			
2			
3			
*4	1 × 4, 2 × 2, 4 × 1	3	1, 2, 4
5			
6			
7			
8			
9			
10			
11			
12			
19			
20			
21			
22			
23			
24			
25			

Directions

1. For each value of n, make up all possible rectangular arrays of n tiles. Then record on your record sheet the dimensions of each rectangle and the number of rectangles. For $n = 4$, we have the rectangles shown here and we fill in the fourth row of the record sheet as shown above.

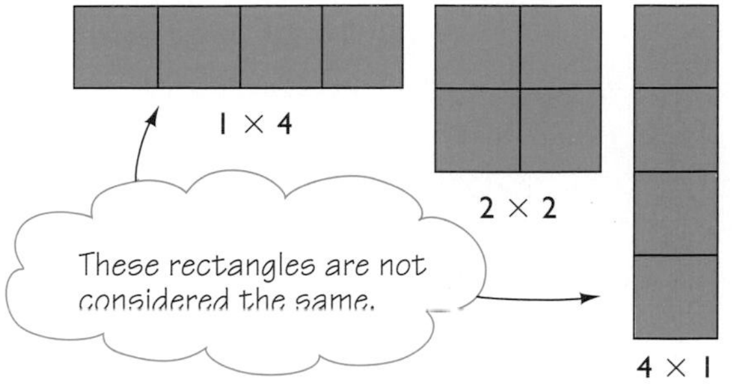

2. In Chapter 2, we used diagrams like this to illustrate the product $3 \cdot 5 = 15$.

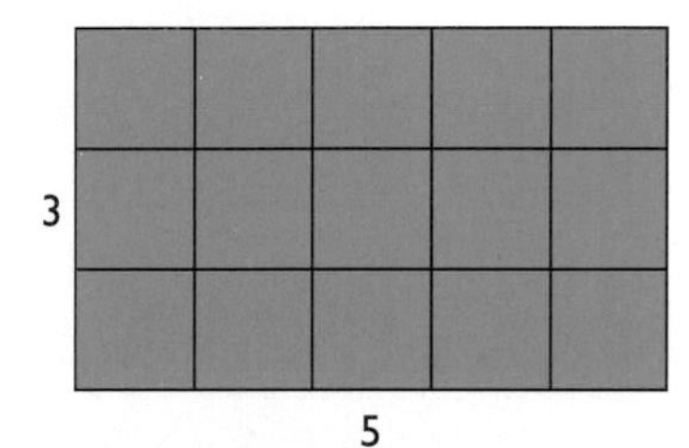

We call 3 and 5 **factors** of 15. Label the last column of your record sheet *Factors of n* and list the factors in increasing order for each value of n. The row for $n = 4$ is done for you.

3. Natural numbers that have exactly two factors are called **prime numbers.** Place a P along the left side of your record sheet next to each prime number.
4. Numbers with more than two factors are called **composite numbers.** Place a C along the left hand side of your record sheet next to each composite number.
5. Is 1 a prime or composite number or neither? Why?
6. Put an asterisk just to the left of each n that has an odd number of factors. Do these numbers seem to share some other property? State a guess (conjecture) about natural numbers with an odd number of factors.
7. Carefully considering all the data on your record sheet, see if you can guess a number with just 7 factors. Check to see that your guess is correct.

FROM THE NCTM STANDARDS

Number Theory

Challenging but accessible problems from number theory can be easily formulated and explored by students. For example, building rectangular arrays with a set of tiles can stimulate questions about divisibility and prime, composite, square, even, and odd numbers.

This activity and others can be extended to investigate other interesting topics, such as abundant, deficient, or perfect numbers; triangular and square numbers; cubes; palindromes; factorials; and Fibonacci numbers. The development of various procedures for finding the greatest common factor of two numbers can foreshadow important topics in the 9–12 curriculum, as students compare the advantages, disadvantages, and efficiency of various algorithms. String art and explorations with star polygons can relate number theory to geometry.

Another example from number theory involves making connections between the prime structure of a number and the number of its factors:

Find five examples of numbers that have exactly three factors. Repeat for four factors, then five factors. What can you say about the numbers in each of your lists?

Students might give 4, 9, 25, 49, and 121 as examples of numbers with exactly three factors. Each of these numbers is the square of a prime.

Without an understanding of number systems and number theory, mathematics is a mysterious collection of facts. With such an understanding, mathematics is seen as a beautiful, cohesive whole.

Only one rectangle can be made with seven tiles, so 7 is prime.

More than one rectangle can be made with eight tiles, so 8 is a composite.

SOURCE: From *Curriculum and Evaluation Standards for School Mathematics Grades 5–8,* p. 93. Fig. 6.1, "Tile explorations" from *Curriculum and Evaluation Standards for School Mathematics Grades 5–8,* p. 93.

CONNECTIONS

The Fascination with Numbers

In Chapter 2, we considered the whole numbers, operations with whole numbers, and some of their properties. In Chapter 3, we studied various systems for writing whole numbers and a variety of algorithms for calculating. Initially, these ideas arose in response to people's needs—the need to count, the need to record counts, and the need of merchants to calculate in the course of doing business. However, very early on, people began to be fascinated with numbers themselves and their many interesting properties. For example, the Greeks called 6 a **perfect number** since

$$6 = 1 + 2 + 3$$

and 1, 2, and 3 are all the natural numbers that divide 6 evenly except for 6 itself.

$1 \cdot 6 = 6$
$2 \cdot 3 = 6$

It turns out that 28 also has this property since

$$28 = 1 + 2 + 4 + 7 + 14$$

1 · 28 = 28
2 · 14 = 28
4 · 7 = 28

and 1, 2, 4, 7, and 14 are all the natural numbers that divide 28 evenly except for 28 itself. The next two perfect numbers are 496 and 8128. As of this writing, 33 perfect numbers are known. It is conjectured that there are infinitely many perfect numbers, but there is no proof that this is so.

In this chapter, we consider divisibility properties of the natural numbers. These ideas are necessary and useful (in working with fractions, for example), and are a rich source of interesting problems and puzzles that can be used to heighten student interest in learning mathematics.

4.1 Divisibility of Natural Numbers

Divides, Divisors, Factors, Multiples

In Chapter 3 we considered the division algorithm. If a and b are whole numbers with b not zero, when we divide a by b we obtain a unique quotient q and remainder r such that $a = bq + r$ *and* $0 \leq r < b$. Thus, the division.

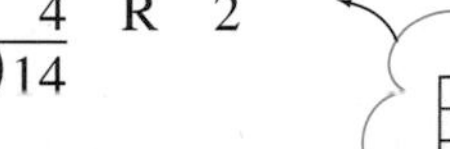

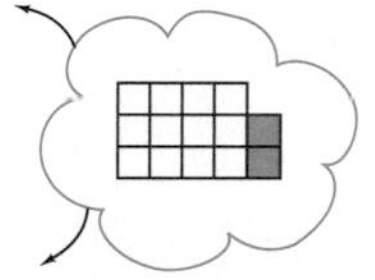

is equivalent to the equation

$$14 = 3 \cdot 4 + 2.$$

Of special interest in this chapter is the case when the remainder r is zero. Then $a = bq$ and we say that **b divides a evenly** or, more simply, **b divides a.** This is expressed in other terminology as indicated here.

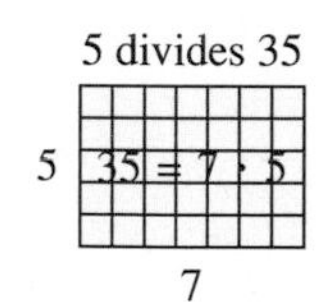

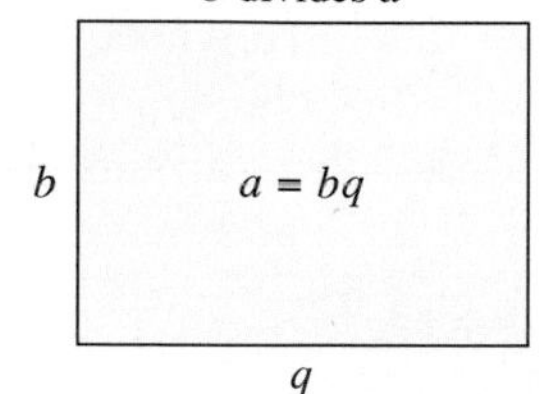

DEFINITION Divides, Divisor, Factor, Multiple

If a and b are whole numbers with $b \neq 0$ and there is a whole number q such that $a = bq$, we say that b **divides** a. We also say that b is a **factor** of a or a **divisor** of a and that a is a **multiple** of b. If b divides a and b is less than a, it is called a **proper divisor** of a.

It is also convenient to introduce some special notation so that we can write about divisibility more easily.

NOTATION The Divides Sign and Its Negation

Let a and b be whole numbers with $b \neq 0$. If b divides a, write $b \mid a$. If b does not divide a, write $b \nmid a$.

The notation $b \mid a$ is read "b divides a" or "b is a factor of a." Also, the divides sign is strictly vertical; it is not slanted as in a fraction. Indeed, $b \mid a$ if, and only if, the fraction a/b is a whole number.

EXAMPLE 4.1 **The Multiples of 2**

Write the multiples of 2.

SOLUTION

According to the above definition, the multiples of 2 are the numbers of the form $2q$ where q is a whole number. Taking $q = 0, 1, 2, 3, \ldots$, we obtain the multiples

$$0, 2, 4, 6, \ldots$$

which are just the **even** whole numbers. These multiples could also be obtained by starting with 0 and counting by 2s. Also, except for 0, the multiples of 2 can be illustrated by this series of rectangles

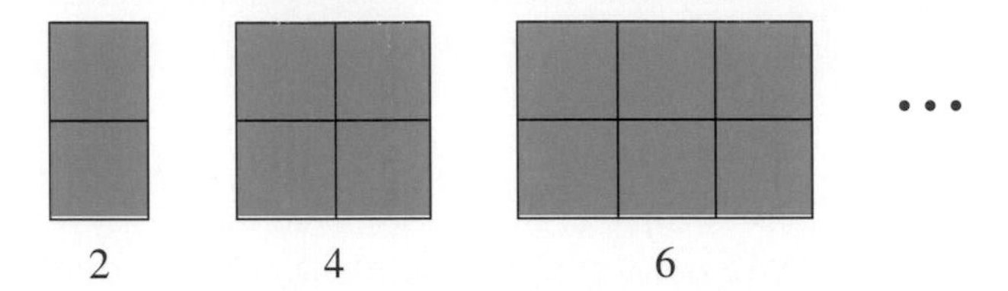

As in the preceding example, the multiples of 3 are

$$0, 3, 6, 9, \ldots$$

which can be obtained by starting with 0 and counting by 3s. In general, the multiples of m are

$$0, m, 2m, 3m, \ldots$$

obtained by starting with 0 and counting by ms.

EXAMPLE 4.2 **The Divisors or Factors of 6**

List all the factors of 6.

SOLUTION

We must determine all natural numbers b for which there is a whole number q such that $6 = bq$. This means that we must find all rectangular arrays with six small squares. By trial and error, we find that there are only four such rectangles and that the desired factors of 6 are 1, 2, 3, and 6. Note, by the way, that as b runs through the divisors of 6—1, 2, 3, 6—so does the quotient q but in the reverse order—6, 3, 2, 1. Finally, it is correct to say that 1, 2, 3, and 6 are factors of 6 and that 6 is a multiple of each of 1, 2, 3, and 6. It is also correct to write $1 \mid 6$, $2 \mid 6$, $3 \mid 6$, and $6 \mid 6$.

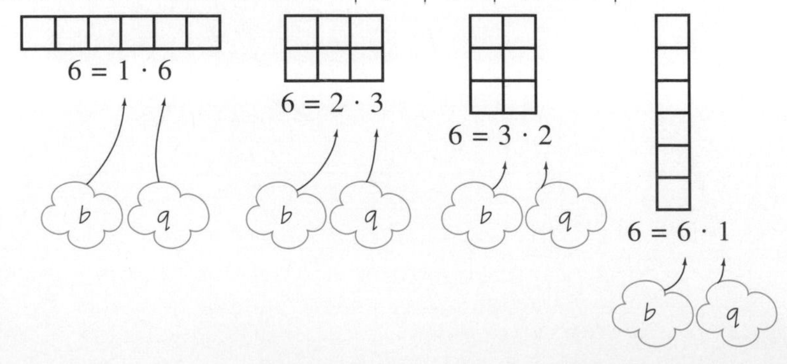

SCHOOL BOOK PAGE *Prime and Composites*

Prime and Composite Numbers

Build Understanding

A. Can a Square Be a Rectangle?
Materials: Small squares, one size
Groups: 3 or 4 students

a. Use 12 squares and form all possible rectangles. Copy the following table. Record the dimensions of each rectangle and the number of unique rectangles.

Dimensions 3 × 4 and 4 × 3 have the same shape, so they are counted as one rectangle.

Number of Squares	Dimensions of Rectangles	Number of Unique Rectangles
12	1 × 12, 12 × 1 2 × 6, 6 × 2 3 × 4, 4 × 3	3
13	1 × 13, 13 × 1	1
14	1 × 14, 14 × 1 2 × 7, 7 × 2	2

b. Repeat this activity with 14, 15, 16, 17, 18, and 19 squares. Record your results.

c. With which numbers of squares can you form only one unique rectangle?

d. With which numbers of squares can you form 2 or more unique rectangles?

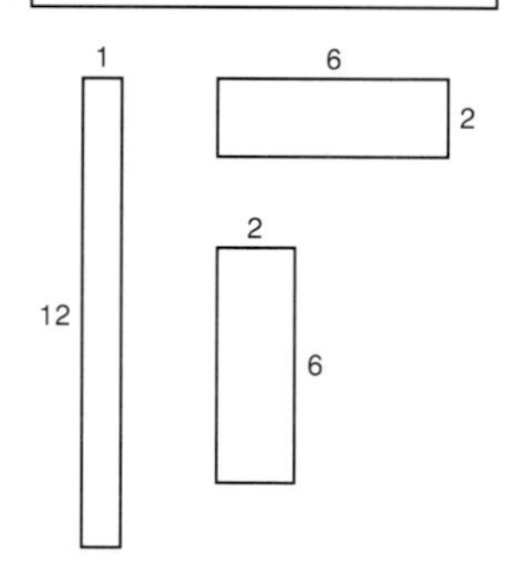

B. Any whole number greater than 1 with exactly 2 factors is a ***prime number***. Any whole number greater than 1 with more than 2 factors is a ***composite number***. The numbers 0 and 1 are neither prime nor composite.

Is 351 a prime number?

Two factors of 351 are 1 and 351. By the divisibility rules, you know that 2, 5, and 10 are not factors of 351. Since 3 + 5 + 1 = 9, the number 351 is divisible by 3 and by 9. So 351 is not a prime number.

1. Discuss the author's use of the word "unique" in the discussion above. Are the 1 by 12 and 12 by 1 rectangles the same or not?
2. What product does the 1 by 12 rectangle represent? The 12 by 1 rectangle?
3. Why do you think 1 is not considered a prime number? (*Hint:* Consider the fundamental theorem of arithmetic.)
4. What would you say to a student who places a rectangle diagonally on his or her desk? Is the student wrong?

Prime and Composite Numbers

Since $1 \cdot a = a$, 1 and a are *always* factors of a for every natural number a. For this reason, 1 and a are often called trivial factors of a and $1 \cdot a$ and $a \cdot 1$ are trivial factorings. Some numbers, like 2, 3, 5, and 7, have only trivial factorings. Other numbers, like 6, have nontrivial factorings. The number 1 stands alone since it has only one factor, 1 itself. All this is summarized in this definition.

DEFINITION Units, Primes, and Composite Numbers

A natural number that possesses only two factors, itself and 1, is called a **prime number.** A natural number that possesses more than two factors is called a **composite number.** The number 1 is called a **unit;** it is neither prime nor composite.

The primes are sometimes called the building blocks of the natural numbers since every natural number other than 1 is either a prime or a product of primes. For example, consider a number like 180. This is composite since, for example, we can write $180 = 10 \cdot 18$. Moreover, 10 and 18 are both composite since $10 = 2 \cdot 5$ and $18 = 2 \cdot 9$. Now 2 and 5 are both primes and cannot be factored further. But $9 = 3 \cdot 3$ and 3 is a prime. We simply continue to factor a composite number into smaller and smaller factors and stop when this can proceed no further; that is, when the factors are all primes. In the case of 180, we see that

$$\begin{aligned} 180 &= 10 \cdot 18 \\ &= 2 \cdot 5 \cdot 2 \cdot 9 \\ &= 2 \cdot 5 \cdot 2 \cdot 3 \cdot 3, \end{aligned}$$

and this is a product of primes.

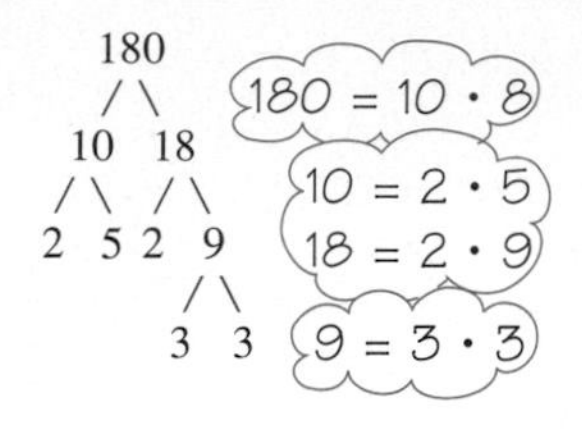

180 = 2 · 5 · 2 · 3 · 3

Figure 4.1
A factor tree for 180

A convenient way of organizing this work is to develop a **factor tree,** as shown in Figure 4.1, to keep track of each step in the process. But there are other ways to factor 180 as these factor trees show.

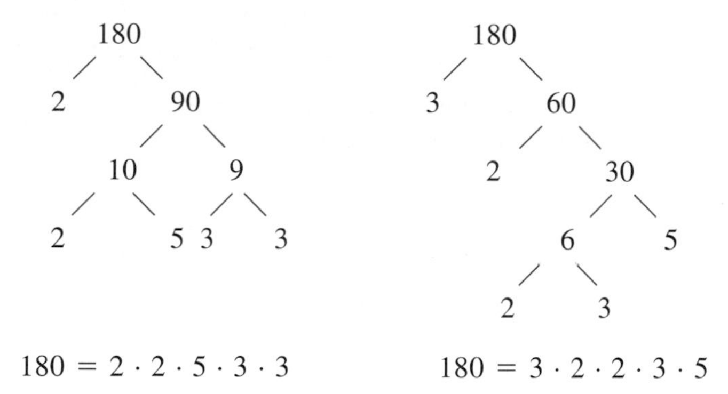

Another, perhaps even more convenient, scheme for finding the prime factors of a number should be mentioned as well. Repeatedly use prime divisors and short division until arriving at a quotient that is a prime. For 180 we might have these divisions.

$$\begin{array}{r} 5 \\ 3\overline{)15} \\ 3\overline{)45} \\ 2\overline{)90} \\ 2\overline{)180} \end{array} \qquad \begin{array}{r} 3 \\ 3\overline{)9} \\ 2\overline{)18} \\ 2\overline{)36} \\ 5\overline{)180} \end{array} \qquad \begin{array}{r} 5 \\ 2\overline{)10} \\ 2\overline{)20} \\ 3\overline{)60} \\ 3\overline{)180} \end{array}$$

$180 = 2 \cdot 2 \cdot 3 \cdot 3 \cdot 5$ $\quad$ $180 = 5 \cdot 2 \cdot 2 \cdot 3 \cdot 3$ $\quad$ $180 = 3 \cdot 3 \cdot 2 \cdot 2 \cdot 5$

In making a factor tree most of the arithmetic is done mentally. The short division scheme makes for a bit more writing while still leaving a written record of the results. Thus, some find the short division approach more helpful.

The most important thing about all the factorings shown for 180 is that, no matter what scheme is used and no matter how it is carried out, the *same prime factors always* result. That this is always the case is stated here without proof.

THEOREM Simple Product Form of the Fundamental Theorem of Arithmetic

Every natural number greater than 1 is a prime or can be expressed as a product of primes in one and only one way apart from order.

EXAMPLE 4.3 **The Prime Factors of 600**

Represent 600 as a product of prime factors.

SOLUTION

Using a factor tree: Using short division:

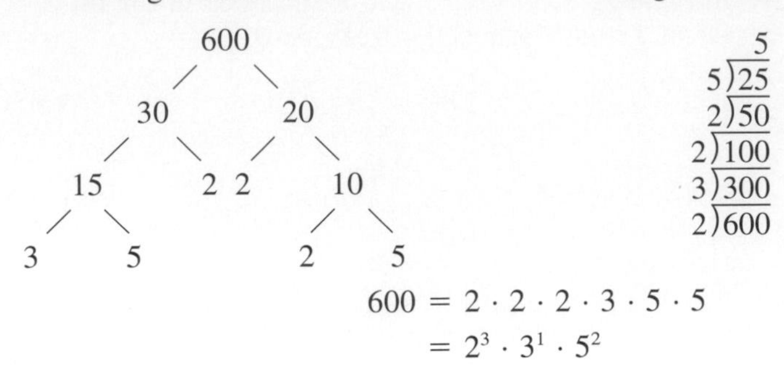

$$600 = 2 \cdot 2 \cdot 2 \cdot 3 \cdot 5 \cdot 5$$
$$= 2^3 \cdot 3^1 \cdot 5^2$$ ■

In Example 4.3, we simplified the representation of 600 as a product of primes by collecting like primes together and writing their products as powers. Since this could be done for any natural number, it is useful to give an alternative version of the **fundamental theorem of arithmetic.**

THEOREM Prime Power Form of the Fundamental Theorem of Arithmetic

Every natural number n greater than 1 is a power of a prime or it can be expressed as a product of powers of primes in one and only one way apart from order. This representation is called the **prime power representation of n.**

EXAMPLE 4.4 **The Prime Power Representation of 675**

Determine the prime power representation of 675.

SOLUTION

$$\begin{array}{r} 3 \\ 3\overline{)9} \\ 3\overline{)27} \\ 5\overline{)135} \\ 5\overline{)675} \end{array}$$

Using the short division method we see that $675 = 3^3 \cdot 5^2$. ■

The Divisors of a Natural Number

An important feature of the prime power representations of numbers is that they make it possible to tell at a glance if one number divides another. In Example 4.4, we saw that $675 = 3^3 \cdot 5^2$. Suppose that $r \mid 675$, then there is an integer s such that

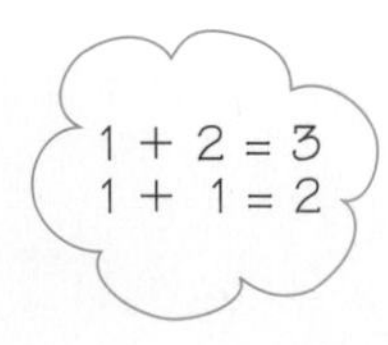

$$rs = 675 = 3^3 \cdot 5^2.$$

Then those primes that divide r must be among the primes that divide 675 and they must appear to no higher power. Thus, we see that $r = 15 = 3^1 \cdot 5^1$ divides 675. Moreover, the quotient s must be $3^2 \cdot 5^1 = 45$ so that the product of r and s contains three 3s and two 5s. In fact, all the divisors of 675 can be written down by making a systematic list as shown here.

$$\begin{array}{rcl} 1 = 3^0 \cdot 5^0 & 5 = 3^0 \cdot 5^1 & 25 = 3^0 \cdot 5^2 \\ 3 = 3^1 \cdot 5^0 & 15 = 3^1 \cdot 5^1 & 75 = 3^1 \cdot 5^2 \\ 9 = 3^2 \cdot 5^0 & 45 = 3^2 \cdot 5^1 & 225 = 3^2 \cdot 5^2 \\ 27 = 3^3 \cdot 5^0 & 135 = 3^3 \cdot 5^1 & 675 = 3^3 \cdot 5^2 \end{array}$$

In determining r above, there are four choices for the exponent on 3—0, 1, 2, or 3—and three choices for the exponent on 5—0, 1, or 2. Thus, there are $3 \cdot 4 = 12$ possible factors of 675 as the above list shows. This argument can be repeated in general and the result stated as a theorem that tells which numbers divide a given number and also gives the **number of divisors of a number.**

THEOREM The Divisors of a Natural Number

Let $n = p_1^{a_1} p_2^{a_2} \ldots p_r^{a_r}$ be the prime power representation of n. Then $m \mid n$ if, and only if, $m = p_1^{b_1} p_2^{b_2} \ldots p_r^{b_r}$ where $0 \leq b_1 \leq a_1$, $0 \leq b_2 \leq a_2, \ldots, 0 \leq b_r \leq a_r$. Moreover, the number of factors or divisors of a is given by $N = (a_1 + 1)(a_2 + 1) \ldots (a_r + 1)$.

EXAMPLE 4.5

The Divisors of 600

(a) Compute the number of divisors of 600.

(b) List all the divisors of 600.

SOLUTION

(a) If we complete part (b) first and list all the divisors of 600, we can count them and answer part (a). However, by the theorem, this number can be determined separately. From Example 4.3,

$$600 = 2^3 \cdot 3^1 \cdot 5^2.$$

Therefore, there must be $N = (3 + 1)(1 + 1)(2 + 1) = 24$ divisors of 600.

(b) The actual divisors must be all the numbers of the form $2^r \cdot 3^s \cdot 5^t$ with $r \in \{0, 1, 2, 3\}$, $s \in \{0, 1\}$, and $t \in \{0, 1, 2\}$. We make a systematic list as follows.

$s = 0, \quad t = 0$	$s = 0, \quad t = 1$	$s = 0, \quad t = 2$
$r = 0, 1, 2, 3$	$r = 0, 1, 2, 3$	$r = 0, 1, 2, 3$
$1 = 2^0 \cdot 3^0 \cdot 5^0$	$5 = 2^0 \cdot 3^0 \cdot 5^1$	$25 = 2^0 \cdot 3^0 \cdot 5^2$
$2 = 2^1 \cdot 3^0 \cdot 5^0$	$10 = 2^1 \cdot 3^0 \cdot 5^1$	$50 = 2^1 \cdot 3^0 \cdot 5^2$
$4 = 2^2 \cdot 3^0 \cdot 5^0$	$20 = 2^2 \cdot 3^0 \cdot 5^1$	$100 = 2^2 \cdot 3^0 \cdot 5^2$
$8 = 2^3 \cdot 3^0 \cdot 5^0$	$40 = 2^3 \cdot 3^0 \cdot 5^1$	$200 = 2^3 \cdot 3^0 \cdot 5^2$

$s = 1,\ t = 0$	$s = 1,\ t = 1$	$s = 1,\ t = 2$
$r = 0, 1, 2, 3$	$r = 0, 1, 2, 3$	$r = 0, 1, 2, 3$
$3 = 2^0 \cdot 3^1 \cdot 5^0$	$15 = 2^0 \cdot 3^1 \cdot 5^1$	$75 = 2^0 \cdot 3^1 \cdot 5^2$
$6 = 2^1 \cdot 3^1 \cdot 5^0$	$30 = 2^1 \cdot 3^1 \cdot 5^1$	$150 = 2^1 \cdot 3^1 \cdot 5^2$
$12 = 2^2 \cdot 3^1 \cdot 5^0$	$60 = 2^2 \cdot 3^1 \cdot 5^1$	$300 = 2^2 \cdot 3^1 \cdot 5^2$
$24 = 2^3 \cdot 3^1 \cdot 5^0$	$120 = 2^3 \cdot 3^1 \cdot 5^1$	$600 = 2^3 \cdot 3^1 \cdot 5^2$ ■

Two Questions About Primes

Since we have just seen how important the primes are as "building blocks" for the natural numbers, it is reasonable to ask at least two questions:

- How many primes are there?
- How does one determine if a given number is a prime?

We answer these questions in the order asked.

There Are Infinitely Many Primes

The answer to the first of the questions just posed is that there are infinitely many primes. To see this we describe a step-by-step process that determines at least one new prime at each step. Since the process can be continued without end, it follows that the set of primes is infinite. The argument proceeds as follows.

Since 1 is the only natural number less than 2, the only factors of 2 are 1 and 2 itself. Thus, 2 is a prime. Consider $3 = 2 + 1$. Since $2 \nmid 3$, the only divisors of 3 are 1 and 3. So 3 is also a prime. Consider $7 = 2 \cdot 3 + 1$. Recalling the division algorithm, this says that, if you divide 7 by either 2 or 3, there is a remainder of 1.

HIGHLIGHT FROM HISTORY
Euclid of Alexandria

One of the great mathematicians of the ancient world was Euclid of Alexandria. Little is known of Euclid's life. The dates of his birth and death are not even known (though he lived about 300 B.C.) nor is his birth place or any of the other little details one might find of interest. What is known is that he was a first rate teacher/scholar at the great university at Alexandria under Ptolemy II and that he authored some dozen books on mathematics. The most important of these is his *Elements (Stoichia)* which set a new standard of rigor for mathematical thought that persists to this day. It is almost certain that no book, save the Bible, has been more used, studied, or edited than the *Elements* and none has had a more profound effect on scientific thought.

This monumental work is divided into 13 books and is more popularly known for its remarkable treatment of geometry—an approach that set the pattern for what is taught in schools even today. However, books VII, VIII, and IX deal with number theory and contain many interesting results. Euclid was the first to prove that there are infinitely many primes, and his proof is essentially the one given in this text. He also invented the Euclidean algorithm for determining the greatest common divisor of two natural numbers which we consider shortly. Another fascinating result is Euclid's formula for even perfect numbers. Recall that a number is considered **perfect** if it is the sum of its proper divisors. Thus, as observed earlier,

$$6 = 1 + 2 + 3 \quad \text{and} \quad 28 = 1 + 2 + 4 + 7 + 14$$

are the first two perfect numbers. The reader may find it interesting, perhaps with the help of FACTORINTEGER, to show that 496, 8128, and 33,550,336 are the next three perfect numbers. That Euclid was able to devise a formula for all even perfect numbers was truly remarkable.

So impressed was she by Euclid's work that the American poet, Edna St. Vincent Millay, was moved to write, "Euclid alone has looked at beauty bare!"

Thus, $2 \nmid 7$ and $3 \nmid 7$. But, by the fundamental theorem of arithmetic, 7 must have a prime divisor. Therefore, 7 has a prime factor different from both 2 and 3. In fact, 7 is itself a prime. Similarly, $43 = 2 \cdot 3 \cdot 7 + 1$ leaves a remainder of 1 when divided by 2, 3, or 7. Thus, 43 is divisible by a prime different from 2, 3, and 7. In fact, 43 is also a prime. We next consider $1807 = 2 \cdot 3 \cdot 7 \cdot 43 + 1$. As before, 1807 is not divisible by any of 2, 3, 7, or 43 (since all leave a remainder of 1) so there must exist a prime other than these primes. In this case, $1807 = 13 \cdot 139$ and both 13 and 139 are primes. In any event, if we multiply all the known primes together and add 1, we obtain a number that must be divisible by at least one new prime. Since this process can be continued *ad infinitum,* it follows that there is no end to the list of primes; that is, there are infinitely many primes.

THEOREM The Number of Primes

There are infinitely many primes.

EXAMPLE 4.6

A Prime Different from 2, 3, 7, 13, and 43

Using the memory function of your calculator and trial and error or the computer program FACTORINTEGER, find a new prime different from 2, 3, 7, 13, and 43 that divides $23{,}479 = 2 \cdot 3 \cdot 7 \cdot 13 \cdot 43 + 1$.

SOLUTION

With persistence and the calculator, or using the program FACTORINTEGER, we find that $23{,}479 = 53 \cdot 443$ where 53 and 443 are both primes. Thus, we have found two new primes different from 2, 3, 7, 13, and 43. ■

Determining if a Given Integer Is a Prime

We now answer the second question posed above. Observe that if n is composite it must have at least one nontrivial factor (and hence at least one prime factor) not greater than its own square root. For, suppose $n = bc$ with $b > \sqrt{n}$ and $c > \sqrt{n}$. Then,

$$n = bc > \sqrt{n} \cdot \sqrt{n} = n.$$

But this says that $n > n$ which is nonsense. Thus, either $b \leq \sqrt{n}$ or $c \leq \sqrt{n}$ and we have this theorem.

THEOREM Prime Divisors of *n*

If n is composite, then there is a prime p such that $p \mid n$ and $p \leq \sqrt{n}$ (that is, $p^2 \leq n$).

To use this theorem, we still need to know all the primes less than or equal to $\sqrt{n}$. Remarkably, a systematic method for determining all the primes up to a given limit was devised by the Greek mathematician, Eratosthenes (276–195 B.C.). Aptly called the **sieve of Eratosthenes,** the method depends only on counting. Suppose we want to determine all the primes up to 100. Since 1 is neither prime nor composite, we write down all the whole numbers from 2 to 100. Note that 2 is a prime since its only possible factors are itself and 1. However, every second number after 2 is a multiple of 2 and so is composite. Thus, we delete or "sieve out," these numbers from our list. The next number not deleted is 3. It must be a prime since it is not a multiple of the only smaller prime. We now "sieve out" all multiples of 3 after 3 itself; that is, we strike from the list every third number after 3 whether it has been struck out before or not. The next number not already deleted must also be a prime since it is not a multiple of 2 or 3, the only smaller primes. Thus, 5 is a prime and every fifth number after 5 must be deleted as a multiple of 5. In the same way, we determine that 7 is a prime and delete every seventh number after 7. Similarly, 11 is the next prime, but, when we delete every eleventh number after 11, no new numbers are deleted. In fact, all the remaining numbers at this point are primes and the sieve, as shown in Figure 4.2, is complete.

	(2)	(3)	~~4~~	(5)	~~6~~	(7)	~~8~~	~~9~~	~~10~~
(11)	~~12~~	(13)	~~14~~	~~15~~	~~16~~	(17)	~~18~~	(19)	~~20~~
~~21~~	~~22~~	(23)	~~24~~	~~25~~	~~26~~	~~27~~	~~28~~	(29)	~~30~~
(31)	~~32~~	~~33~~	~~34~~	~~35~~	~~36~~	(37)	~~38~~	~~39~~	~~40~~
(41)	~~42~~	(43)	~~44~~	~~45~~	~~46~~	(47)	~~48~~	~~49~~	~~50~~
~~51~~	~~52~~	(53)	~~54~~	~~55~~	~~56~~	~~57~~	~~58~~	(59)	~~60~~
(61)	~~62~~	~~63~~	~~64~~	~~65~~	~~66~~	(67)	~~68~~	~~69~~	~~70~~
(71)	~~72~~	(73)	~~74~~	~~75~~	~~76~~	~~77~~	~~78~~	(79)	~~80~~
~~81~~	~~82~~	(83)	~~84~~	~~85~~	~~86~~	~~87~~	~~88~~	(89)	~~90~~
~~91~~	~~92~~	~~93~~	~~94~~	~~95~~	~~96~~	(97)	~~98~~	~~99~~	~~100~~

Figure 4.2
The sieve of Eratosthenes for $n = 100$

The primes have been circled to make them stand out. To see why the sieve is complete at this point, note that if $n < 100$, then $\sqrt{n} < \sqrt{100} = 10$. Hence, if n is composite and is in the list, it must have a prime factor less than or equal to 10 by the preceding theorem. Thus, the sieve is complete by the time we have deleted all multiples of 2, 3, 5, and 7.

EXAMPLE 4.7 Determining the Primality of 443

Show that 443 is a prime.

SOLUTION

If we used FACTORINTEGER for the solution of Example 4.6, we already know that 443 is a prime. However, this is also easily determined with your calculator in just a couple of minutes. Since $\sqrt{443} \doteq 21$, we see from the sieve of Eratosthenes that, if 443 is composite, it is a multiple of one of 2, 3, 5, 7, 11, 13, 17, or 19. Putting 443 in the memory of our calculator, we easily complete the necessary divisions and determine that 443 is a prime since it is not divisible by 2, 3, 5, 7, 11, 13, 17, or 19. ■

Divisibility of Sums and Differences

Because it is so useful in answering divisibility questions, we close this section by proving a result about divisibility of sums and differences. Suppose that a, b, and c are natural numbers and that $a \mid b$ and $a \mid c$. Does a also divide the sum $b + c$ and the difference $b - c$? Since $a \mid b$ and $a \mid c$, there exist natural numbers r and s such that $ar = b$ and $as = c$. But then

$$b + c = ar + as = a(r + s) \qquad \text{and} \qquad b - c = ar - as = a(r - s).$$

Hence, $a \mid (b + c)$ and $a \mid (b - c)$.

> **THEOREM Divisibility of Sums and Differences**
>
> If a, b, and c are natural numbers and $a \mid b$ and $a \mid c$, then $a \mid (b + c)$ and $a \mid (b - c)$.

EXAMPLE 4.8

An Application to Fractions

If a, b, and $\frac{1}{a} + \frac{1}{b}$ are natural numbers, show that $a = b$. (*Hint:* Show that $a \mid b$ and $b \mid a$.)

SOLUTION

Understand the problem • • • •

We are told that a, b, and $\frac{1}{a} + \frac{1}{b}$ are natural numbers. We must show that $a = b$.

Restate the problem.

Devise a plan • • • •

We don't have much to go on. Could we restate the problem? What does it mean to say that $\frac{1}{a} + \frac{1}{b}$ is a natural number? It means that there is some natural number, r, such that

Use a variable.

$$\frac{1}{a} + \frac{1}{b} = r.$$

Perhaps we can use this equation and some divisibility notions as suggested in the hint.

Carry out the plan • • • •

The equation

$$\frac{1}{a} + \frac{1}{b} = r$$

can be cleared of fractions by multiplying both sides by ab. This gives

$$b + a = abr.$$

Differently put, this says that

$$a = abr - b \qquad \text{and} \qquad b = abr - a.$$

But then, $b \mid b$ and $b \mid abr$. So by the preceding theorem, $b \mid a$; that is $bu = a$ for some natural number u. Similarly, $a \mid a$ and $a \mid abr$, so $a \mid b$, say $av = b$ for some natural number v. Hence,

$$a = bu = (av)u = a(vu)$$

and this implies that $vu = 1$. But then $v = u = 1$, and so $a = b$ as was to be shown.

Look back

The basic strategy was to restate the problem by introducing a new variable. This gave the equation

$$\frac{1}{a} + \frac{1}{b} = r$$

and there was little that could be done with this except clear of fractions. But then we could use divisibility properties of integers to show that $a = b$ as required. ■

JUST FOR FUN

A Problem of Punctuation

Insert two commas, a semicolon, and a period in the following so that it makes a meaningful sentence.

That that is is that that is not is not

PROBLEM SET 4.1

Understanding Concepts

1. Draw diagrams to show that
   (a) 4 is a factor of 36.
   (b) 6 is a factor of 36.
2. Draw diagrams to illustrate all the factorings of 35 taking order into account; that is, think of $1 \cdot 35$ as different from $35 \cdot 1$.
3. (a) List the first ten multiples of 8 starting with $0 \cdot 8 = 0$.
   (b) List the first ten multiples of 6 starting with $0 \cdot 6 = 0$.
   (c) Use parts (a) and (b) to determine the least natural number that is a multiple of both 8 and 6.
4. Complete this table of all factors of 18 and their corresponding quotients.

Factors of 18	1	2						
Corresponding quotients	18	9						

5. Construct factor trees for each of these numbers.
   (a) 72 (b) 126 (c) 264 (d) 550
6. Use short division to find all the prime factors of each of these numbers.
   (a) 700 (b) 198 (c) 450 (d) 528
7. (a) List all the factors of 48.
   (b) List all the factors of 54.
   (c) Use parts (a) and (b) to find the largest common factor of 48 and 54.
8. Determine the prime power representation of each of these numbers.
   (a) 48 (b) 108 (c) 2250 (d) 24,750
9. Let $a = 2^3 \cdot 3^1 \cdot 7^2$.
   (a) Is $2^2 \cdot 7^1 = 28$ a factor of a? Why or why not?
   (b) Is $2^1 \cdot 3^2 \cdot 7^1 = 126$ a factor of a? Why or why not?
   (c) One factor of a is $b = 2^2 \cdot 3^1$. What is the quotient when a is divided by b?
   (d) How many different factors does a possess?
   (e) Make an orderly list of all of the factors of a.

10. To determine if 599 is a prime, which primes must you check as possible divisors?

11. Use your calculator and information from the sieve of Eratosthenes in Figure 4.2 to determine if 1139 is prime. If it is not prime, give its prime factors.

12. If n is composite, is it true that all prime factors of n must not exceed $\sqrt{n}$? Explain briefly.

Thinking Critically

13. Which of the following are *true* or *false*? Justify your answer in each case.
   (a) $n \mid 0$ for every natural number n.
   (b) $0 \mid n$ for every natural number n.
   (c) $1 \mid n$ for every natural number n.
   (d) $n \mid n$ for every natural number n.
   (e) $0 \mid 0$

14. Assume that $a \mid b$ and $b \mid c$ where a, b, and c are natural numbers. Give a carefully written three sentence argument showing that $a \mid c$.

15. If p is a prime, b and c are natural numbers, and $p \mid bc$, is it necessarily the case that $p \mid b$ or $p \mid c$? Justify your answer.

16. If n, b, and c are natural numbers and $n \mid bc$, is it necessarily the case that $n \mid b$ or $n \mid c$? Justify your answer.

17. Assume that p and q are different primes and that n is a natural number. If $p \mid n$ and $q \mid n$, argue briefly that $pq \mid n$.

18. If m and n are natural numbers and $m \mid n$ and $n \mid m$, show that $m = n$. (*Hint:* What does it mean to say that $m \mid n$? That $n \mid m$?)

19. If a, b, and c are natural numbers and

$$c = \frac{1}{a} + \frac{1}{b},$$

prove that $a = b = 1$ or $a = b = 2$.

20. Consider the statement, "If $a \mid b$ and $a \nmid c$, then $a \nmid (b + c)$." Is the statement *true* or *false?* Justify your answer.

21. Consider the statement, "If $a \mid (b + c)$, then $a \mid b$ and $a \mid c$." Is the statement necessarily true? Justify your answer.

22. Consider the statement, "If $a \nmid b$ and $a \nmid c$, then $a \nmid (b + c)$." Is the statement *true* or *false*? Justify your answer.

Thinking Cooperatively

23. Let N_n be the natural number whose decimal representation consists of n consecutive 1s. For example, $N_2 = 11$, $N_7 = 1{,}111{,}111$, and $N_9 = 111{,}111{,}111$.
   (a) Show that $N_2 \mid N_4$, $N_2 \mid N_6$, and $N_2 \mid N_8$.
   (b) Guess the quotient when N_{18} is divided by N_2 and check your guess by multiplication by hand.

24. Let N_n be as in problem 23.
   (a) Show that $N_3 \mid N_6$, $N_3 \mid N_9$, and $N_3 \mid N_{12}$.
   (b) Guess the quotient when N_{18} is divided by N_3 and check your guess by multiplication by hand.

25. Let N_n be as in problems 23 and 24.
   (a) Does $N_3 \mid N_5$? (b) Does $N_3 \mid N_7$?
   (c) Does $N_3 \mid N_{15}$?
   (d) Guess conditions on m and n that guarantee that $N_m \mid N_n$.

26. Recall that the Fibonacci numbers are the numbers in the sequence 1, 1, 2, 3, 5, 8, 13, 21, 34, 55, . . . which starts with two 1s and where $F_{n+2} = F_{n+1} + F_n$ for all $n \geq 1$ where F_n denotes the nth Fibonacci number.
   (a) The above list contains the first ten Fibonacci numbers. Compute F_n for $n = 11, 12, 13, \ldots, 20$.
   (b) Which Fibonacci numbers are divisible by F_3? Justify your answer in two or three sentences.
   (c) Guess which Fibonacci numbers are divisible by F_4? By F_5?
   (d) Guess which Fibonacci numbers are divisible by F_m.

27. Draw a square measuring 10 centimeters on a side. Draw vertical and horizontal line segments dividing the square into rectangles of areas 12, 18, 28, and 42 square centimeters respectively. Where should A, B, C, and D be located?

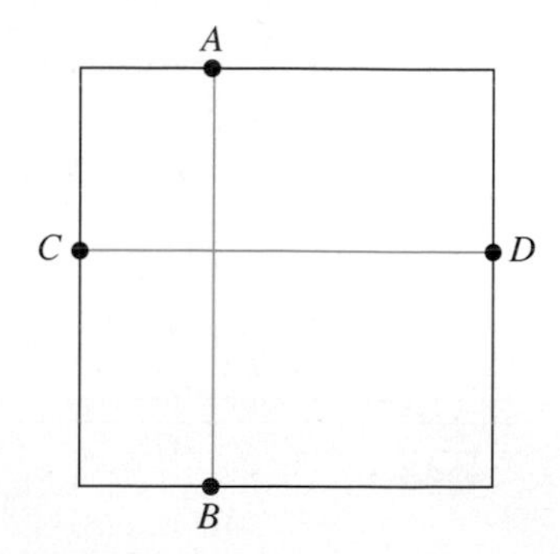

28. When the students arrived in Mr. Gowing's classroom one day there were 28 three by five cards neatly propped up in the chalk tray. They were numbered from 1 through 28 and arranged in order with the numbers facing the chalkboard. After the students were seated Mr. Gowing asked the first student in the first row of seats to go to the board and turn each card so that they all faced the classroom. He then asked the second student to go

to the board and turn every second card back facing the board. The third student was to turn every third card whether it had been turned before or not. Thus, for example, the sixth card would have been turned toward the class by the first student, toward the board by the second, and toward the class again by the third student. The process continues until the twenty-eighth student turns the twenty-eighth card for the last time.

(a) When the process is complete, which numbers are left facing the students?

(b) Show why the answer to part (a) is as it is. (*Hint:* You may also want to make a table to exhibit what is happening.)

Using a Computer

29. Use FACTORINTEGER to determine the prime power representation of each of these numbers.

(a) 548 **(b)** 936 (c) 274 **(d)** 45,864

(e) Use the results of parts (a), (b), (c), and (d) to determine which, if any, of these numbers divides another of these numbers.

(f) Which of the numbers in part (a), (b), (c), and (d) has precisely six divisors?

(g) How many divisors does 45,864 have? Explain briefly.

30. Use FACTORINTEGER to determine the prime power representation of each of these integers.

(a) 894,348 **(b)** 245,025 (c) 1,265,625

(d) Which of the numbers in parts (a), (b), and (c) are squares?

(e) What can you say about the prime power representation of a square? Explain briefly.

(f) Guess how you might see from a glance at its prime power representation that 93,576,664 is the cube of a natural number. Explain.

For Review

31. Let $S = \{1, 3, 5, \ldots\}$ be the set of all odd natural numbers. Which of the following are *true* and which are *false*? Justify your answer in each case.

(a) S is closed with respect to addition.

(b) S is closed with respect to multiplication.

(c) The commutative property for addition holds for elements of S.

(d) The commutative property for multiplication holds for elements of S.

(e) The associative property for addition holds for elements of S.

(f) The associative property for multiplication holds for elements of S.

(g) The distributive property for multiplication over addition holds for elements of S.

(h) S possesses an additive identity.

(i) S possesses a multiplicative identity.

32. (a) Compute the one hundredth term in the arithmetic sequence 2, 5, 8, 11, ...

(b) Compute the sum of the first one hundred terms in the sequence of part (a).

33. Write the most likely choices for the next three terms in each of these sequences.

(a) 3, 7, 11, 15, ____, ____, ____

(b) 1, 3, 7, 17, 41, ____, ____, ____

(c) 1, 3, 6, 10, ____, ____, ____

(d) 1, 0, 1, 0, 2, 0, 3, 0, 5, ____, ____, ____

(e) 3, 6, 12, 24, ____, ____, ____

(f) 2, 9, 13, 31, 57, 119, ____, ____, ____

34. Compute the sum of the terms in each of these arithmetic progressions.

(a) 2, 5, 8, . . . , 155

(b) 7, 12, 17, . . . , 152

(c) 3, 10, 17, . . . , 689

4.2 Tests for Divisibility

It is convenient for both teachers and students to have simple tests that show when one number is divisible by another. Relatively easy tests exist for the primes 2, 3, 5, 7, 11, and 13, for 9, 10, and powers of 2, and for products of numbers in the preceding list. Indeed, tests for divisibility by any prime exist but they become quite cumbersome and are not useful. In general, for larger primes, we are reduced to using technology and/or trial and error. Also, all the tests depend on using base ten notation.

Consider a number like 76. Since

$$76 = 7 \cdot 10 + 6$$

COOPERATIVE *Investigation*

A Neat Fibonacci Trick

Materials Needed

1. A calculator for each student.
2. A Fibonacci Sum record sheet for each student as below.

Directions

1. Start the activity by placing any two natural numbers in the first two rows of column one of the record sheet and then complete the column by adding any two consecutive entries to obtain the next entry in the Fibonacci manner. Finally, add the ten entries obtained and divide the sum by 11.
2. Repeat the process to complete all but the last column of the record sheet. Then look for a pattern and make a conjecture. Lastly, prove that your conjecture is correct by placing a and b in the first and second positions in the last column and then repeating the process as before.

Fibonacci Sum Record Sheet

	1	**2**	**3**	**4**	**5**	**General Case**
1						a
2						b
3						
4						
5						
6						
7						
8						
9						
10						
Sum						
Sum ÷ 11						

and $2 \cdot 5 = 10$ and $2 \cdot 3 = 6$, we can write

$$\begin{aligned} 76 &= 7 \cdot 2 \cdot 5 + 2 \cdot 3 \\ &= 2(7 \cdot 5 + 3) \end{aligned}$$

and it follows that $2 \mid 76$. Conversely, since $76 = 2 \cdot 38$, we can write

$$\begin{aligned} 6 &= 76 - 7 \cdot 10 \\ &= 2 \cdot 38 - 7 \cdot 2 \cdot 5 \\ &= 2 \cdot (38 - 7 \cdot 5) \end{aligned}$$

and it follows that $2 \mid 6$. Thus, 76 is divisible by 2 since $2 \mid 6$, and 6 is divisible by 2 since $2 \mid 76$. Moreover, the same argument holds if 6 is replaced by any of 0, 2, 4, or

8. In general, a number is divisible by 2 if, and only if, its units digit is divisible by 2. Since a similar argument also holds for divisibility by 5, we have the following theorem.

THEOREM Divisibility by 2 and 5

Let n be a natural number. Then n is divisible by 2 if, and only if, its units digit is 0, 2, 4, 6, or 8. Similarly, n is divisible by 5 if, and only if, its units digit is 0 or 5.

THEOREM Divisibility by 10

Let n be a natural number. Then n is divisible by 10 if, and only if, 2 and 5 divide n; that is, if, and only if, the units digit of n is 0.

PROOF By definition, $10 \mid n$ if, and only if, $10q = n$ for some q. But, since $2 \cdot 5 = 10$, this is so if, and only if, $2(5q) = n$ and $5(2q) = n$; that is, if, and only if, $2 \mid n$ and $5 \mid n$. Thus, the divisibility tests for both 2 and 5 must be satisfied, and this is so if, and only if, the units digit of n is 0. ■

It is important to note that the preceding theorem is a special case of a much more general result.

THEOREM Divisibility by Products

Let a and b be natural numbers with no common factor other than 1. Then, if $a \mid c$ and $b \mid c$, it follows that $ab \mid c$.

PROOF From Section 4.1, all the prime factors of a must appear in c and to at least as high a power as they appear in a. Similarly, all prime factors of b must appear in c and to at least as high a power as they appear in b. But, since a and b have no factors in common, they can have no prime factors in common and this implies that all primes appearing in either a or b must appear in c and to at least as high a power as they appear in ab. Therefore, $ab \mid c$ as claimed. ■

EXAMPLE 4.9 **Divisibility by Products**

Using the fundamental theorem of arithmetic, show that $18 \cdot 25 \mid 22{,}050$.

SOLUTION

Since $18 = 2^1 \cdot 3^2$, $25 = 5^2$, and $22{,}050 = 2^1 \cdot 3^2 \cdot 5^2 \cdot 7^2$, it is clear that $18 \mid 22{,}050$ and that $25 \mid 22{,}050$. Moreover, since 18 and 25 have no common factor other than 1, it follows from the preceding theorem that $(18 \cdot 25) \mid 22{,}050$. In fact, $22{,}050 \div (18 \cdot 25) = 49$. ■

HIGHLIGHT FROM HISTORY
Eratosthenes of Cyrene (276–195 B.C.)

Much has been written about the remarkable scholarly accomplishments of the ancient Greeks. Such accounts are sprinkled with names like Pythagoras, Thales, Euclid, Archimedes, Aristarchus, Eudoxus, Hypatia, Theano, and Apollonius. Another extraordinarily able contributor to Greek thought was Eratosthenes, who was remarkably gifted not only in mathematics but also in poetry, astronomy, geography, history, philosophy, and even athletics. Eratosthenes was born in Cyrene on the south shore of the Mediterranean, lived most of his early life in Athens, and, later, was invited to Alexandria by Ptolemy III of Egypt to tutor his son and to serve as librarian of the great Alexandrian University. We have already mentioned his remarkable sieve for determining all the prime numbers up to any given limit, but he is perhaps best known for his unusually accurate determination that the circumference of the earth was very nearly 25,000 miles—a figure much nearer the actual value than that of any of the ancients either before or after. In old age, Eratosthenes became almost totally blind and it is said that he died of voluntary starvation in about 195 B.C.

EXAMPLE 4.10 Nondivisibility by Products

Using the fundamental theorem of arithmetic, show that 22,050 is not divisible by $18 \cdot 15$.

SOLUTION

Here $18 = 2^1 \cdot 3^2$, $15 = 3^1 \cdot 5^1$ and $22{,}050 = 2^1 \cdot 3^2 \cdot 5^2 \cdot 7^2$. Thus, $18 \mid 22{,}050$ and $15 \mid 22{,}050$ as above. But this time 18 and 15 have a common factor other than 1. Moreover, $18 \cdot 15 = 2^1 \cdot 3^3 \cdot 5^1$. Since this product involves 3^3 and the prime power representation of 22,050 only contains 3^2, it follows that $(18 \cdot 15) \nmid 22{,}050$. In fact, using a calculator, $22{,}050 \div (18 \cdot 25) = 81$ R 180. ■

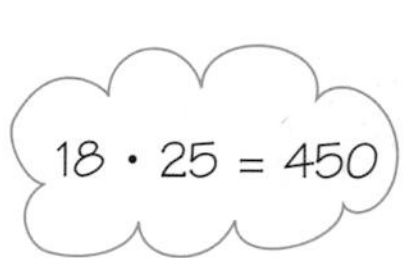

THEOREM Tests for Divisibility by 3 and 9

A natural number is divisible by 3, if and only if, the sum of its digits is divisible by 3. Similarly, a natural number is divisible by 9 if, and only if, the sum of its digits is divisible by 9.

Instead of a formal proof, we use an example to reveal why these tests hold. Consider $n = 27{,}435$ and let $s = 2 + 7 + 4 + 3 + 5$ denote the sum of the digits of n. Then

$$\begin{aligned} n - s &= 27{,}435 - (2 + 7 + 4 + 3 + 5) \\ &= (20{,}000 - 2) + (7{,}000 - 7) + (400 - 4) + (30 - 3) \\ &= 2(10{,}000 - 1) + 7(1{,}000 - 1) + 4(100 - 1) + 3(10 - 1) \\ &= 2 \cdot 9999 + 7 \cdot 999 + 4 \cdot 99 + 3 \cdot 9 \\ &= 9(2 \cdot 1111 + 7 \cdot 111 + 4 \cdot 11 + 3 \cdot 1) \\ &= 9q. \end{aligned}$$

$q = 2 \cdot 1111 + 7 \cdot 111 + 4 \cdot 11 + 3 \cdot 1$

Therefore, by the theorem on divisibility of sums and differences, $9 \mid n$ if and only if $9 \mid s$. Since $9 = 3 \cdot 3$, the same argument holds for divisibility by 3.

SCHOOL BOOK PAGE *Divisibility of Whole Numbers*

Divisibility

Build Understanding

If the remainder is zero when a whole number is divided by another whole number, the first number is ***divisible*** by the second number.

The numbers that are multiplied to give a product are called ***factors*** of that product.

Divisibility Rules

A number is divisible by 2 if its ones digit is 0, 2, 4, 6, or 8.
A number is divisible by 3 if the sum of its digits is divisible by 3.
A number is divisible by 5 if its ones digit is 0 or 5.
A number is divisible by 9 if the sum of its digits is divisible by 9.
A number is divisible by 10 if its ones digit is 0.

All whole numbers that are divisible by 2 are ***even*** numbers. Whole numbers that are not divisible by 2 are ***odd*** numbers.

Winston has 225 muffins to package for the bake sale. He wants to put the same number of muffins in each bag and have none left over. What can he do?

Winston can use the divisibility rules to see if 225 is divisible by 2, 3, 5, 9, or 10.

225 is not divisible by 2 because its ones digit is not 0, 2, 4, 6, or 8.

225 is divisible by 3 because the sum 2 + 2 + 5 is divisible by 3.

225 is divisible by 5 because its ones digit is 5.

225 is divisible by 9 because the sum 2 + 2 + 5 is divisible by 9.

225 is not divisible by 10 because its ones digit is not 0.

Winston can package 225 muffins in bags of 3, 5, or 9.

$225 \div 3 = 75$

$225 \div 5 = 45$

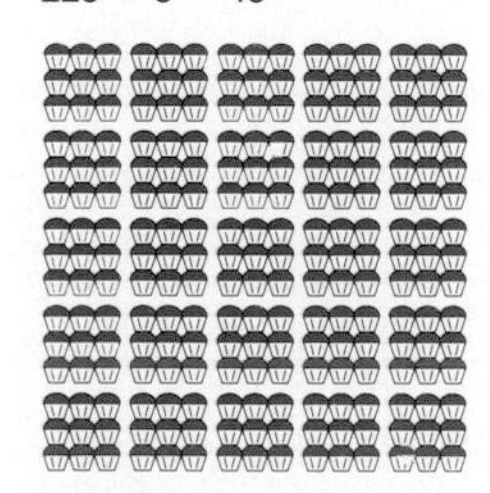

$225 \div 9 = 25$

■ **Talk About Math** Give an example of a 4-digit number that is divisible by 2, 3, and 5.

Check Understanding

For another example, see Set A, pages 180–181.

1. Is 51 divisible by 3? by 5?
2. Is 657 divisible by 5? by 9?
3. Is 198 divisible by 2 *and* 3?
4. If a number is divisible by 10, is it also divisible by 2? Explain your answer.

1. From the above divisibility rules, what must be the case if a number is divisible by 6? By 18?
2. If the sum of the digits of a number is divisible by 6 is the number divisible by 6?
3. If the sum of the digits of a number is divisible by 7 is the number divisible by 7?
4. Give an example of a 10-digit number that is divisible by 9.

EXAMPLE 4.11 **Divisibility by 2, 3, 5, 9, and 10**

State whether each of the following is *true* or *false* without actually dividing. Give a reason for your answer.

(a) $2 \mid 43{,}826$ **(b)** $3 \mid 111{,}111$ **(c)** $10 \mid 26{,}785$
(d) $9 \mid 10{,}020{,}006$ **(e)** $6 \mid 111{,}111$ **(f)** $5 \mid (287 + 78)$

SOLUTION

(a) True, since the units digit of 43,826 is 6 (that is, even).
(b) True, since $1 + 1 + 1 + 1 + 1 + 1 = 6$ and $3 \mid 6$.
(c) False, since 10 divides a number if, and only if, its units digit is 0. Here it is 5.
(d) True, since $1 + 0 + 0 + 2 + 0 + 0 + 0 + 6 = 9$ and $9 \mid 9$.
(e) False. If $6 \mid 111{,}111$, then $2 \mid 111{,}111$ since $2 \mid 6$. But $2 \nmid 111{,}111$ since the units digit is not even.
(f) True. $5 \nmid 287$ and $5 \nmid 78$. Yet $287 + 78 = 365$, and $5 \mid 365$ since the units digit is 5. ■

We now state without proof a test for divisibility by 11.

> **THEOREM Divisibility Test for 11**
>
> A natural number is divisible by 11 if, and only if, the difference of the sums of the digits in the even and odd positions in the number is divisible by 11.

EXAMPLE 4.12 **Testing for Divisibility by 11**

Show that $n = 8{,}193{,}246{,}781{,}053{,}476{,}109$ is divisible by 11 and that $m = 76{,}124{,}738{,}465{,}372{,}103$ is not divisible by 11.

SOLUTION

First consider n. Using mental arithmetic, we find that

$$8 + 9 + 2 + 6 + 8 + 0 + 3 + 7 + 1 + 9 = 53$$

and

$$1 + 3 + 4 + 7 + 1 + 5 + 4 + 6 + 0 = 31.$$

Since $53 - 31 = 22$ and $11 \mid 22$, it follows that $11 \mid n$. In fact, you can use mental arithmetic and short division to show that $8{,}193{,}246{,}781{,}053{,}476{,}109 \div 11 = 744{,}840{,}616{,}459{,}406{,}919$.

Now consider m. Using mental arithmetic, we find that

$$7 + 1 + 4 + 3 + 4 + 5 + 7 + 1 + 3 = 35$$

and

$$6 + 2 + 7 + 8 + 6 + 3 + 2 + 0 = 34.$$

Since $35 - 34 = 1$ and $11 \nmid 1$, it follows that $11 \nmid m$. Again, using mental arithmetic and short division you can show that

$$76{,}124{,}738{,}465{,}372{,}103 \div 11 = 6{,}920{,}430{,}769{,}579{,}282 \text{ R } 1.$$ ■

A Combined Test for Divisibility by 7, 11, and 13

We now describe a test for divisibility by 7, 11, and 13 all at the same time. Consider a number like

$$n = 35{,}253{,}663{,}472{,}614{,}159{,}218{,}828.$$

Break n up into the series of 3-digit numbers

$$035,\ 253,\ 663,\ 472,\ 614,\ 159,\ 218,\ 828$$

determined by the 3-digit groups starting from the right in n. Now, as in the test for divisibility by 11, compute the sum of the numbers in the odd-numbered positions in the above list and the sum of the numbers in the even-numbered positions. This gives

$$\begin{array}{r} 035 \\ 663 \\ 614 \\ +\ 218 \\ \hline 1530 \end{array} \quad \text{and} \quad \begin{array}{r} 253 \\ 472 \\ 159 \\ 828 \\ \hline 1712 \end{array}$$

as shown. The difference in these sums is

$$1712 - 1530 = 182$$

and n will be divisible

by 7 if, and only if, 182 is divisible by 7,

by 11 if, and only if, 182 is divisible by 11, and

by 13 if, and only if, 182 is divisible by 13.

Since $182 = 7 \cdot 26$ and $182 = 13 \cdot 14$, it follows that $7 \mid n$ and $13 \mid n$. On the other hand, $11 \nmid 182$ and so $11 \nmid n$. Use short division and mental processes to find the quotients when n is divided by 7 and by 13 and also to find the quotient and remainder when n is divided by 11.

EXAMPLE 4.13 **Divisibility by 7, 11, and 13**

Test to see if $n = 8{,}346{,}261{,}059{,}482{,}647$ is divisible by 7, 11, or 13.

SOLUTION

Breaking n up into 3-digit numbers and computing the sums of those in even positions and odd positions we have

$$\begin{array}{r} 008 \\ 261 \\ 482 \\ \hline 751 \end{array} \quad \text{and} \quad \begin{array}{r} 346 \\ 59 \\ 647 \\ \hline 1052 \end{array}$$

The test number is $1052 - 751 = 301$. Since $7 \cdot 43 = 301$, $7 \mid 301$ and hence $7 \mid n$. However,

$$301 = 11 \cdot 27 + 4 \quad \text{and} \quad 301 = 13 \cdot 23 + 2$$

so $11 \nmid 301$ and $13 \nmid 301$. ■

A Test for Divisibility by Powers of Two

> **THEOREM A Test for Divisibility by 4, 8, and Other Powers of 2**
>
> Let n be a natural number. Then $4 \mid n$ if, and only if, 4 divides the number named by the last two digits of n. Similarly, $8 \mid n$ if, and only if, 8 divides the number named by the last three digits of n. In general, $2^r \mid n$ if, and only if, 2^r divides the number named by the last r digits of n.

This theorem can be understood by looking at an example. Consider a number like 28,476,324. This can be written in the form

$$\begin{aligned} 28{,}476{,}324 &= 284{,}763 \cdot 100 + 24 \\ &= 4 \cdot 25 \cdot 284{,}763 + 24. \end{aligned}$$

Thus, by the theorem on divisibility of sums and differences, $4 \mid 28{,}476{,}324$ if, and only if, $4 \mid 24$. Therefore, 4 does indeed divide 28,476,324.

In general, when we divide any natural number n by 100 we obtain a quotient q and a remainder r such that

$$n = 100q + r = 4(25q) + r$$

and by the argument above $4 \mid n$ if, and only if, $4 \mid r$.

A similar argument holds for divisibility by 8 since $n = 1000q + s$ where s is the number represented by the last three digits of n and $8 \mid 1000$. For higher powers of 2, the test depends on the fact that $16 \mid 10{,}000$. $32 \mid 100{,}000$, and so on.

EXAMPLE 4.14

Testing for Divisibility by 4 and 8

Test the following for divisibility by 4 and 8.

(a) 2452 **(b)** 3849 **(c)** 7672 **(d)** 39,000

SOLUTION

Since $2 \mid 4$, $4 \mid 8$, $8 \mid 16$, and so on, divisibility of a number by any power of 2 guarantees divisibility by any lower power. Thus, it follows that if $2 \nmid n$, then $2^k \nmid n$ for any k.

(a) Since $452 = 8 \cdot 56 + 4$, $8 \nmid 452$. Thus $8 \nmid 2452$. However, $4 \mid 52$ and so $4 \mid 2452$ and $2 \mid 2452$.

(b) Since $2 \nmid 9$, it follows that $2 \nmid 3849$, $4 \nmid 3849$, and $8 \nmid 3849$.

(c) Since $672 = 8 \cdot 84$, it follows that $8 \mid 7672$, $4 \mid 7672$, and $2 \mid 7672$.

(d) Since $8 \mid 1000$, it follows that $8 \mid 39{,}000$, $4 \mid 39{,}000$, and $2 \mid 39{,}000$. ■

PROBLEM SET 4.2

Understanding Concepts

1. Test each number for divisibility by each of 2, 3, and 5. Do the work mentally.
   (a) 1554 **(b)** 1999
   (c) 805 **(d)** 2450

2. Use the results of problem 1 to decide which, if any, of the numbers in problem 1 are divisible by
   (a) 6 **(b)** 10
   (c) 15 **(d)** 30

3. Test each of these for divisibility by 7, 11, and 13.
   (a) 539 **(b)** 253,799
   (c) 834,197 **(d)** 1,960,511
4. Use the results of problem 3 to decide which of the numbers in problem 3 are divisible by
   (a) 77 **(b)** 91
   (c) 143 **(d)** 1001
5. Is 1,927,643,001,548 divisible by 11? Explain briefly.
6. Let $a = 2^3 \cdot 3^1 \cdot 5^2$, $b = 2^2 \cdot 3^2 \cdot 5^1 \cdot 7^1$, $c = 7^1 \cdot 11^2$, and $d = 2^4 \cdot 3^5 \cdot 5^3 \cdot 7^2 \cdot 11^2$. Determine which of these statements are *true* and which are *false*. Justify your answer in each case.
   (a) $a \mid d$ **(b)** $b \mid d$ (c) $c \mid d$
   (d) $ac \mid d$ (e) $ab \mid d$
   (f) Does it necessarily follow that if $a \mid n$ and $b \mid n$, then $ab \mid n$? Explain.
7. **(a)** Fill in the missing digit so that 897,650,243,28__ is divisible by 6. Can this be done in more than one way?
   (b) Fill in the missing digit so that the number in part (a) is divisible by 11.
8. A palindrome is a number like 2,743,472 that reads the same forward and backward.
   (a) Give a clear but brief argument to show that every palindrome with an even number of digits is divisible by 11.
   (b) Is it possible for a palindrome with an odd number of digits to be divisible by 11? Explain.

Thinking Critically

9. Show that every number whose decimal representation has the form *abc*, *abc* is divisible by 7, 11, and 13.
10. **(a)** Reverse the digits in a two-digit number to obtain a second number and then subtract the smaller of these two numbers from the larger. What possible numbers can result? Can you tell at a glance what number will result? Explain carefully but do not attempt to prove.
    (b) Repeat part (a) for 3-digit numbers.
11. A common error in banking is to make an interchange or transposition of some of the digits in a number involved in a transaction. For example, a teller may pay out $43.34 on a check actually written for $34.43 and so be short by $8.91 at the end of the day. Show that such a mistake always causes the teller's balance sheet to show an error in pennies, here 891, that is divisible by 9. (*Hint:* Recall that, in the proof of the divisibility test for 9, we showed that every number differs from the sum of its digits by a multiple of 9.)
12. If a teller's record of the day's work is out of balance by an amount, in pennies, that is a multiple of 9, is it necessarily the case that he or she has made a transposition error in the course of the day's work? Explain.
13. **(a)** Consider a 6-digit number like 142,857 that is divisible by 27. Show by actually dividing that each of 428,571; 285,714; 857,142; 571,428; and 714,285 is also divisible by 27.
    (b) 769,230 and 153,846 are two other numbers divisible by 27. See if the pattern of part (a) holds for these numbers.
    (c) What general property is suggested by parts (a) and (b) of this problem? State it carefully.
    (d) Find another 6-digit number that is divisible by 27. Does it have the property you guessed in part (c)?
    (e) Check to see that all the numbers in parts (a) and (b) are also divisible by 11. Is this true of all 6-digit numbers divisible by 27?

Thinking Cooperatively

14. Compute the quotients for each of these sequences of division problems.
    (a) $126 \div 9 =$
    $10{,}206 \div 9 =$
    $1{,}002{,}006 \div 9 =$
    $100{,}020{,}006 \div 9 =$
    $10{,}000{,}200{,}006 \div 9 =$
    $1{,}000{,}002{,}000{,}006 \div 9 =$
    (b) $252 \div 9 =$
    $20{,}502 \div 9 =$
    $2{,}005{,}002 \div 9 =$
    $200{,}050{,}002 \div 9 =$
    $20{,}000{,}500{,}002 \div 9 =$
    $2{,}000{,}005{,}000{,}002 \div 9 =$
    (c) $432 \div 9 =$
    $40{,}302 \div 9 =$
    $4{,}003{,}002 \div 9 =$
    $400{,}030{,}002 \div 9 =$
    $40{,}000{,}300{,}002 \div 9 =$
    $4{,}000{,}003{,}000{,}002 \div 9 =$
    (d) Given that $342 \div 9 = 38$, predict the quotient in the division $3{,}000{,}004{,}000{,}002 \div 9$.
    (e) Check your guess for part (d) by pencil and paper multiplication.

15. Compute the quotients in each of these sequences of division problems.

(a) $5401 \div 11 =$
$540{,}001 \div 11 =$
$54{,}000{,}001 \div 11 =$
$5{,}400{,}000{,}001 \div 11 =$
$540{,}000{,}000{,}001 \div 11 =$
$54{,}000{,}000{,}000{,}001 \div 11 =$

(b) $2013 \div 11 =$
$201{,}003 \div 11 =$
$20{,}100{,}003 \div 11 =$
$2{,}010{,}000{,}003 \div 11 =$
$201{,}000{,}000{,}003 \div 11 =$
$20{,}100{,}000{,}000{,}003 \div 11 =$

(c) Given that $7117 \div 11 = 647$ and $711{,}007 \div 11 = 64{,}637$, predict the quotient in the division $711{,}000{,}000{,}007 \div 11$.

(d) Check your guess in part (c) by pencil and paper multiplication.

16. Let a, b, c, and d be natural numbers with $a = b + c$. If d divides any two of a, b, and c, show that it also divides the third.

17. Let b, c, and d be natural numbers. If $d \mid (b + c)$ show that d divides both b and c or neither b nor c.

For Review

18. Use short division to determine the prime power representation of these natural numbers.
(a) 8064 **(b)** 2700 **(c)** 19,602

19. Construct a factor tree for each of these natural numbers and write their prime power representations.
(a) 7000 **(b)** 6174 **(c)** 6237

20. How many factors does each number in problem 19 possess? Be sure to show your work on this problem.

21. What is the quotient when a is divided by b if $a = 2^3 \cdot 3^4 \cdot 5^1 \cdot 7^2$ and $b = 2^1 \cdot 3^2 \cdot 7^1$?

22. **(a)** Is $2^4 \cdot 3^2 \cdot 5^1 \cdot 7^2$ divisible by $2^3 \cdot 3^1 \cdot 7^2$? Why or why not?
(b) Is $2^4 \cdot 3^2 \cdot 5^1 \cdot 7^2$ divisible by $2^2 \cdot 5^2 \cdot 7^2$? Why or why not?

4.3 Greatest Common Divisors and Least Common Multiples

Let m and n be any two natural numbers. In this section we consider finding the greatest common divisor and least common multiple of m and n, at least partly because these are useful in the arithmetic of fractions. We begin with a pictorial scheme.

The Greatest Common Divisor

Visualizing the greatest common divisor by cutting squares off rectangles Consider the 18 by 24 rectangle divided into 2 by 2 squares as shown in Figure 4.3. Since the rectangle is nine 2 by 2 squares high and twelve 2 by 2 squares wide, we know that $9 \cdot 2 = 18$ and $12 \cdot 2 = 24$. Thus, $2 \mid 18$ and $2 \mid 24$, so 2 is a common divisor of 18 and 24. Are there other common divisors of 18 and 24? Reflecting on the division of the rectangle into 2 by 2 squares makes it clear that answering this question is equivalent to finding other disections of the rectangle into squares. Such a disection into squares is called a **tiling** by squares, so we are looking for other such tilings. Trial and error shows that the 18 by 24 rectangle can be tiled by 1 by 1, 3 by 3, and 6 by

Making a Chain

Here are six sections of chain, each containing four links. What is the *least* number of links you can open and close to join these sections into a single chain of 24 links?

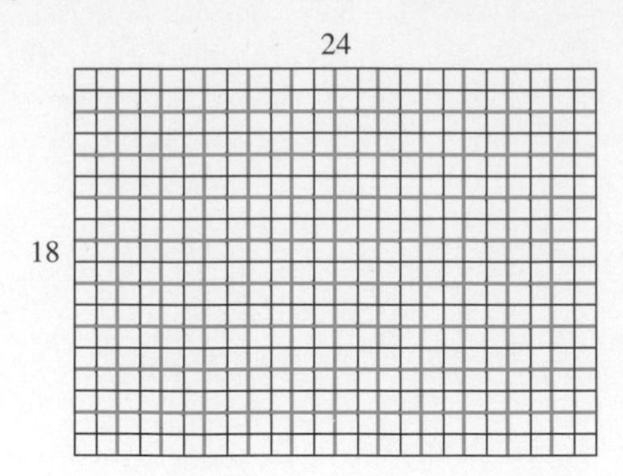

Figure 4.3
Tiling an 18 by 24 rectangle by 2 by 2 squares. 2 is a common divisor of 18 and 24.

Solve an equivalent problem.

6 squares, as shown in Figures 4.4, 4.5, and 4.6, in addition to the 2 by 2 tiling already discussed. Moreover, one becomes aware that to find such a tiling it is necessary to find a number d that divides both 18 and 24; that is, that finding a tiling by squares and finding a common divisor d are equivalent problems.

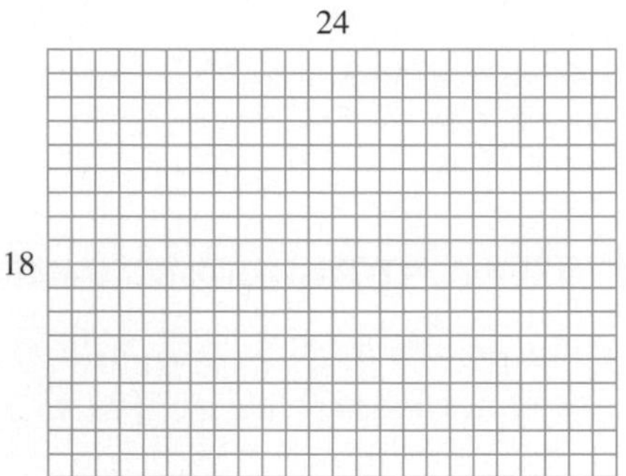

Figure 4.4
Tiling an 18 by 24 rectangle by 1 by 1 squares. 1 is a common divisor of 18 and 24.

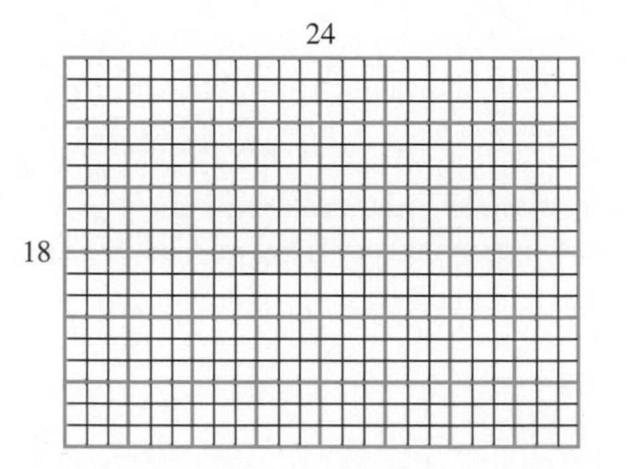

Figure 4.5
Tiling an 18 by 24 rectangle by 3 by 3 squares. 3 is a common divisor of 18 and 24.

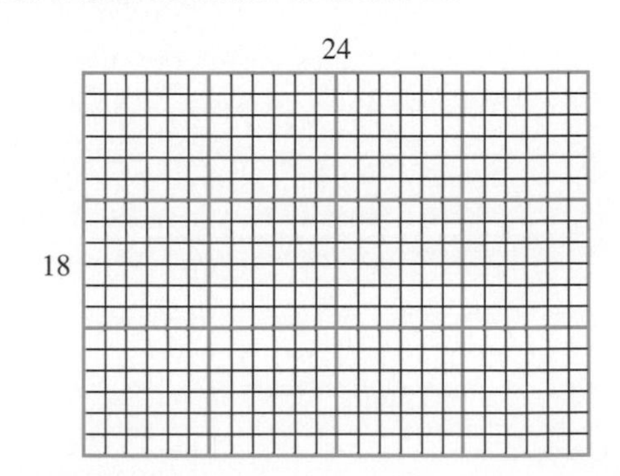

Figure 4.6
Tiling an 18 by 24 rectangle by 6 by 6 squares. 6 is a common divisor of 18 and 24.

Since the 6 by 6 square is the largest square that tiles the 18 by 24 rectangle, it is apparent that 6 is the greatest common divisor of 18 and 24.

DEFINITION Greatest Common Divisor

Let m and n be natural numbers. The greatest natural number d that divides both m and n is called their **greatest common divisor** and we write $d = \text{GCD}(m, n)$.

In the preceding discussion, we determined the greatest common divisor of 18 and 24 by a trial and error process of finding all squares that tile the 18 by 24 rectangle. Is there a more systematic way in which to proceed? What happens if you cut an 18 by 18 square off the end of the 18 by 24 rectangle leaving an 18 by 6 rectangle? The above figures show that the 18 by 6 rectangle can be tiled by squares of a given size if, and only if, the 18 by 24 rectangle can be tiled by squares of the same size. Since this is equally true for any two rectangles where one is obtained from the other by cutting a square off the end of the larger rectangle, we have a theorem.

THEOREM Tiling Rectangles that Differ by Squares

Let R and T be two rectangles where R is obtained from T by cutting a square off the end of T. Then R can be tiled by t by t squares if, and only if, T can be tiled by t by t squares.

In view of the preceding theorem, we can reduce the problem of tiling the 18 by 24 rectangle to the problem of tiling the 6 by 18 rectangle. But the argument can be repeated and we cut off a 6 by 6 square from the 6 by 18 rectangle to obtain a 6 by 12 rectangle, and then again cut off a 6 by 6 rectangle to obtain a 6 by 6 rectangle. Since the divisors remain the same at each step and since the last rectangle is a square that tiles itself, the greatest common divisor of 18 and 24 is 6; that is, GCD(18, 24) = 6. In general this gives us a method of **finding greatest common divisors by tiling.** What emerges is a method of **finding greatest common divisors by cutting squares off rectangles.** The sequence of cuts for the 18 by 24 rectangle is illustrated in Figure 4.7.

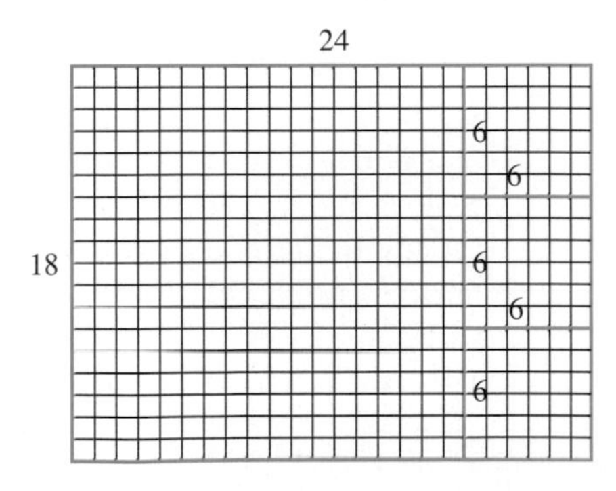

Figure 4.7
Finding the greatest common divisor of 18 and 24 by cutting squares off rectangles. GCD (18, 24) = 6.

EXAMPLE 4.15

Finding the Greatest Common Divisor of 28 and 44

Use the method of tiling rectangles and cutting off squares to compute GCD(28, 44).

SOLUTION

The needed sequence of cuts of the 28 by 44 rectangle are as shown here.

JUST FOR FUN

A Weighty Matter

If a basketball weighs 21 ounces plus half its own weight, how much does it weigh?

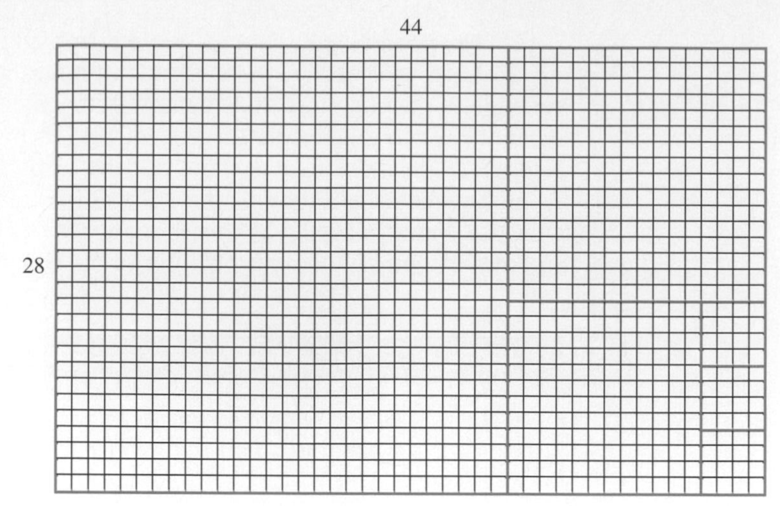

Since the process terminates with a 4 by 4 square, the greatest common divisor of 28 and 44 is 4. ■

Greatest common divisors by intersection of sets This method works rather well when the numbers involved are small and all the divisors of both numbers are easily written down. By way of explanation, it is probably best to do an example.

EXAMPLE 4.16

Finding the Greatest Common Divisor by Intersection of Sets

Find the greatest common divisor of 18 and 45.

SOLUTION

Let D_{18} and D_{45} denote the sets of divisors of 18 and 45. Since

$$D_{18} = \{1, 2, 3, 6, 9, 18\} \qquad \text{and} \qquad D_{45} = \{1, 3, 5, 9, 15, 45\},$$

$D_{18} \cap D_{45} = \{1, 3, 9\}$ is the set of common divisors of 18 and 45. Thus, $\text{GCD}(18, 45) = 9$. ■

Greatest common divisors from prime power representations The greatest common divisor of two numbers can also be found by using their prime power representations. Consider

$$54 = 2^1 \cdot 3^3 \qquad \text{and} \qquad 45 = 3^2 \cdot 5^1.$$

Since $2^0 = 5^0 = 1$, we can write these representations so that they *appear* to be products of powers of the same primes. Thus,

$$54 = 2^1 \cdot 3^3 \cdot 5^0 \qquad \text{and} \qquad 45 = 2^0 \cdot 3^2 \cdot 5^1.$$

As seen earlier, the divisors of 54 are numbers of the form

$$d = 2^a \cdot 3^b \cdot 5^c \qquad \begin{cases} 0 \le a \le 1 \\ 0 \le b \le 3 \\ 0 \le c \le 0 \end{cases}$$

and the divisors of 45 are numbers of the form

$$d = 2^a \cdot 3^b \cdot 5^c. \qquad \begin{cases} 0 \le a \le 0 \\ 0 \le b \le 2 \\ 0 \le c \le 1 \end{cases}$$

Thus, we obtain the largest common divisor by choosing a, b, and c as large as possible while still satisfying both the above sets of inequalities.

Hence,

$$\text{GCD}(54, 45) = 2^0 \cdot 3^2 \cdot 5^0 = 9.$$

Zero is the smaller of 1 and 0.
Two is the smaller of 3 and 2.
Zero is the smaller of 0 and 1.

THEOREM GCDs from Prime Power Representations

Let

$$a = p^{a_1}p^{a_2} \cdots p_r^{a_r} \qquad \text{and} \qquad b = p_1^{b_1}p_2^{b_2} \cdots p_r^{b_r}$$

be the prime power representations of a and b. Then

$$\text{GCD}(a, b) = p_1^{c_1}p_2^{c_2} \cdots p_r^{c_r}$$

where c_1 = the smaller of (a_1, b_1), c_2 = the smaller of (a_2, b_2), . . . , and c_r = the smaller of (a_r, b_r).

EXAMPLE 4.17

Finding the Greatest Common Divisor by Using Prime Power Representations

Compute the greatest common divisor of 504 and 3675.

SOLUTION

We first find the prime power representation of each number.

3	3
3)9	7)21
7)63	7)147
2)126	5)735
2)252	5)3675
2)504	

$$504 = 2^3 \cdot 3^2 \cdot 5^0 \cdot 7^1 \qquad 3675 = 2^0 \cdot 3^1 \cdot 5^2 \cdot 7^2$$

Choosing the smaller of the exponents on each prime, we see that

$$\text{GCD}(504, 3675) = 2^0 \cdot 3^1 \cdot 5^0 \cdot 7^1 = 21.$$

As a check, we note that $504 \div 21 = 24$, that $3675 \div 21 = 175$, and that 24 and 175 have no common factor other than 1. ■

The Euclidean algorithm This method for finding greatest common divisors is found in Book IV of Euclid's *Elements* written in about 300 B.C. It has the distinct advantage of working unfailingly no matter how large the two numbers are or how complicated the arithmetic. It all depends on the division algorithm.

Dividing a by b, we obtain a quotient q and a remainder r such that

$$a = bq + r, \quad 0 \leq r < b.$$

But then

$$r = a - bq.$$

Therefore, by the theorem on divisors of sums and differences, $d \mid a$ and $d \mid b$ if, and only if, $d \mid b$ and $d \mid r$. Thus, a and b must have the same set of divisors as b and r and, hence, the same greatest common divisor.

HIGHLIGHT FROM HISTORY
Julia Robinson

The long struggle for women's rights is certainly reflected in mathematics. From Hypatia (370?–A.D. 415) to the twentieth century, few women worked in mathematics—long considered a male domain. Though strides still need to be made, many women now ply the trade and there is even an Association of Women in Mathematics (AWM). One of the most successful women mathematicians of our time was Professor Julia Bowman Robinson, solver of the tenth in David Hilbert's famous list of problems. Professor Robinson was born in St. Louis in 1919. Her graduate work in logic was done at the University of California where she received her Ph.D. in 1948. It was also at Berkeley that she met her husband to be, number theorist Raphael Robinson. After solving Hilbert's tenth problem in 1970, she was made a member of the prestigious National Academy of Sciences as well as the American Academy of Arts and Sciences and was promptly awarded the rank of Professor at UC Berkeley. Professor Robinson served as President of the American Mathematical Society and was also active in the Association of Women in Mathematics. She died on July 30, 1985, hoping that she would not be remembered "as the first woman this or that, I would prefer to be remembered, as a mathematician should, simply for the theorems I have proved and the problems I have solved." There is little doubt but that her hope will be realized.

THEOREM The GCD and the Division Algorithm

Let a and b be any two natural numbers and let q and r be determined by the division algorithm. Thus,

$$a = bq + r, \quad 0 \leq r < b.$$

Then, $\text{GCD}(a, b) = \text{GCD}(b, r)$.

This theorem is the basis of the **Euclidean algorithm** which we illustrate by determining GCD(1539, 3144). We begin by dividing 3144 by 1539 and then continue by dividing successive divisors by successive remainders. It turns out that the last nonzero remainder is the desired greatest common divisor. Of course, these divisions can be performed by hand or with a calculator. (With the *Math Explorer* calculator they are particularly easy using the [INT÷] key as discussed in Section 3.6.) We obtain these divisions.

$$1539\overline{)3144}^{\,2\text{ R }66} \quad 66\overline{)1539}^{\,23\text{ R }21} \quad 21\overline{)66}^{\,3\text{ R }3} \quad 3\overline{)21}^{\,7\text{ R }0}$$

In the form of the division algorithm these give

$$\begin{aligned} 3144 &= 2 \cdot 1539 + 66 \\ 1539 &= 23 \cdot 66 + 21 \\ 66 &= 3 \cdot 21 + 3 \\ 21 &= 3 \cdot 7. \end{aligned}$$

Hence, from the preceding theorem,

$$\text{GCD}(1539, 3144) = \text{GCD}(66, 1539) = \text{GCD}(21, 66) = \text{GCD}(3, 21) = 3$$

since $3 \mid 21$. Thus, the last nonzero remainder is the greatest common divisor of 1539 and 3144 as asserted.

THEOREM The Euclidean Algorithm

Let a and b be any two natural numbers. Using the division algorithm, determine natural numbers $q_1, q_2, \ldots, q_s$ and $r_1, r_2, \ldots, r_{s-1}$ such that

$$\begin{aligned}
a &= bq_1 + r_1, & 0 &< r_1 < b\\
b &= r_1q_2 + r_2, & 0 &\le r_2 < r_1\\
r_1 &= r_2q_3 + r_3, & 0 &\le r_3 < r_2\\
&\vdots\\
r_{s-3} &= r_{s-2}q_{s-1} + r_{s-1}, & 0 &\le r_{s-1} < r_{s-2}\\
r_{s-2} &= r_{s-1}\, q_s.
\end{aligned}$$

Then, $\text{GCD}(a, b) = r_{s-1}$.

EXAMPLE 4.18

Using the Euclidean Algorithm

Compute GCD(18,411, 1649) using the Euclidean algorithm.

SOLUTION

Using the Euclidean algorithm, we have

$$1649\overline{)18{,}411}^{\;11\text{ R }272} \qquad 272\overline{)1649}^{\;6\text{ R }17} \qquad 17\overline{)272}^{\;16\text{ R }0}$$

and it follows that GCD(18,411, 1649) = 17. ■

The Least Common Multiple

The least common multiple by tiling squares by rectangles Again we begin with a pictorial approach. Consider a 4 by 6 rectangle and ask which is the smallest square that can be tiled by such rectangles.

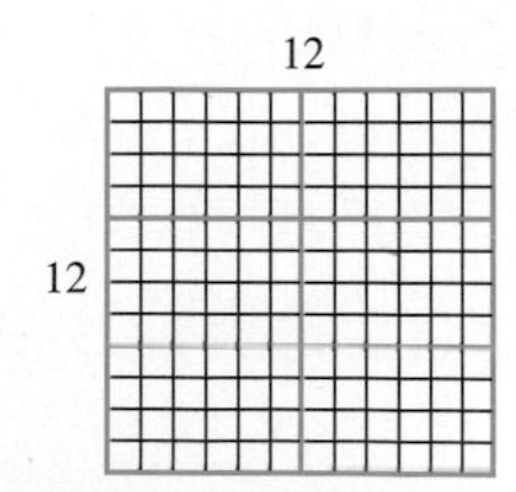

Figure 4.8
Tiling a 12 by 12 square by 4 by 6 rectangles

Given a supply of 4 by 6 rectangles, they can be moved around and it will shortly be discovered that the smallest square that can be tiled by such rectangles is the 12 by 12 square. It is clear from Figure 4.8 that $4 \mid 12$ and $6 \mid 12$. Hence, 12 is a common multiple of both 4 and 6. Since this is the smallest square tilable by 4 by 6 rectangles, 12 is the least common multiple of 4 and 6.

DEFINITION Least Common Multiple

Let a and b be natural numbers. The least natural number m that is a multiple of both a and b is called their **least common multiple** and we write $m = \text{LCM}(a, b)$.

In general, if a rectangle tiles a square, then the length of each side of the rectangle divides the length of the side of the square. Thus, finding the smallest square tilable by a rectangle with dimensions m by n, is equivalent to finding the least common multiple of m and n, and we have the following theorem.

THEOREM Finding Least Common Multiples by Tiling Squares by Rectangles

Let m and n be natural numbers. The least common multiple of m and n is the length of the side of the smallest square that can be tiled by a rectangle with dimensions m and n.

EXAMPLE 4.19 Finding a Least Common Multiple by Tiling

Determine the least common multiple of 9 and 12 by finding the smallest square that can be tiled by a 9 by 12 rectangle.

SOLUTION

Cut out a number of 9 by 12 rectangles and manipulate them by trial and error to determine that the 36 by 36 square is the smallest square obtainable. Thus, $36 = \text{LCM}(9, 12)$.

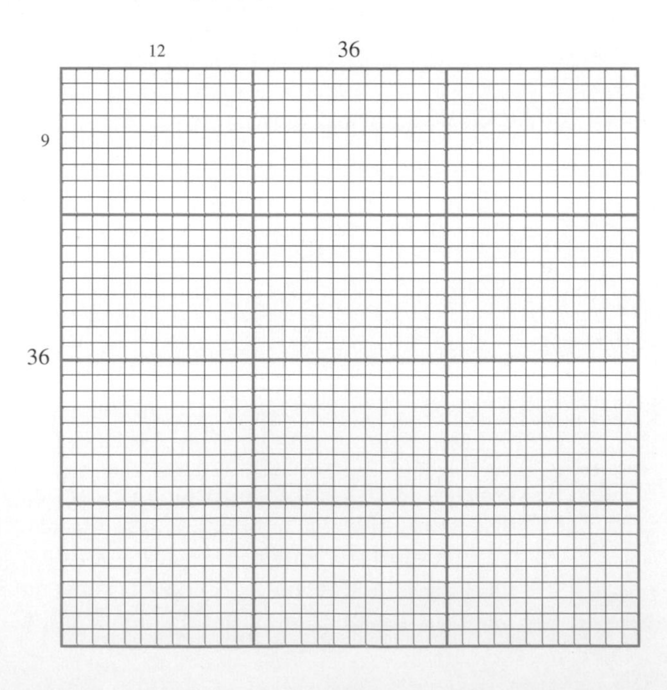

■

But we ought not have to rely on trial and error. How else might we proceed to find the smallest square tilable by an m by n rectangle?

Consider the 2 by 3 rectangle of Figure 4.9. Since $3 \cdot 2 = 2 \cdot 3 = 6$, it follows that six of these rectangles can be formed into a 6 by 6 square. The height of the rectangle in squares gives the width of the large square in rectangles and the width of the rectangle in squares gives the height of the large square in rectangles and *vice versa*. Trial and error suffices to show that no smaller square can be tiled by the 2 by 3 rectangle.

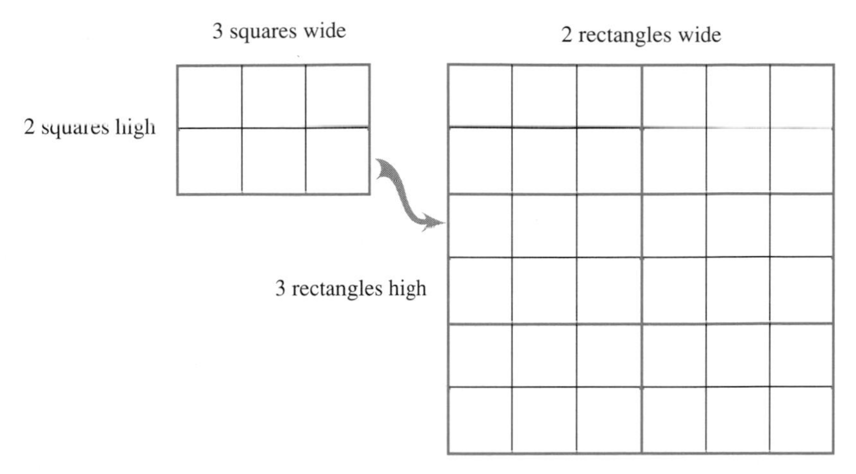

Figure 4.9
Six 2 by 3 rectangles form a 6 by 6 square

But consider a 4 by 6 rectangle. As shown in Figure 4.10, the same process works. An array 4 rectangles wide and 6 rectangles high is indeed a square. However, it is not the *smallest* such square obtainable. This is because the 4 by 6 rectangle can be tiled by larger than 1 by 1 squares. In fact, it can be tiled by 2 by 2 squares (the largest possible) and, in 2 by 2 units, the 4 by 6 rectangle becomes a 2 by 3 rectangle. Thus, the construction that yielded Figure 4.9 here leads to a smaller large square that is only 2 rectangles wide and 3 rectangles high. This could be argued in general and gives us the following theorem.

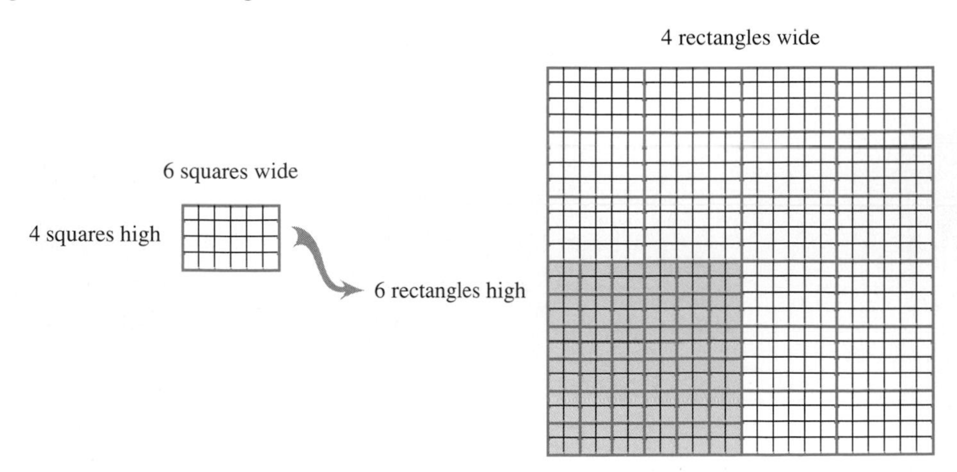

Figure 4.10
Tiling a square by 4 by 6 rectangles

THEOREM Finding the Smallest Square Tileable by a Rectangle

1. Tile the rectangle by the largest possible squares.
2. In terms of this tiling let m denote the height of the rectangle and let n denote its width.
3. The array of rectangles that is m rectangles wide and n rectangles high is the smallest square tilable by the given rectangle.

EXAMPLE 4.20

Finding the Smallest Square Tileable by a Rectangle

Find the smallest square that can be tiled by a 6 by 8 rectangle.

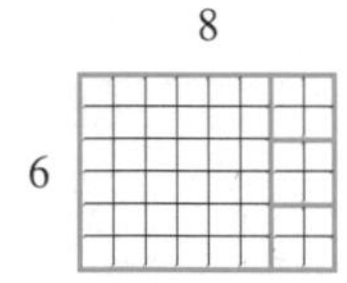

SOLUTION

To find the largest square that will tile the 6 by 8 rectangle, we use the process of repeatedly cutting off squares as in Example 4.15. This shows that the 2 by 2 square is the largest square that will tile the 6 by 8 rectangle (and hence that GCD(6, 8) = 2). Carrying out this tiling, we see that, in terms of 2 by 2 squares, the 6 by 8 rectangle is 3 squares high and 4 squares wide. Thus, the desired large square is 3 rectangles wide and 4 rectangles high as shown. The 24 by 24 square is the smallest square tileable by the 6 by 8 rectangle. Also, from above, it follows that LCM(6, 8) = 24.

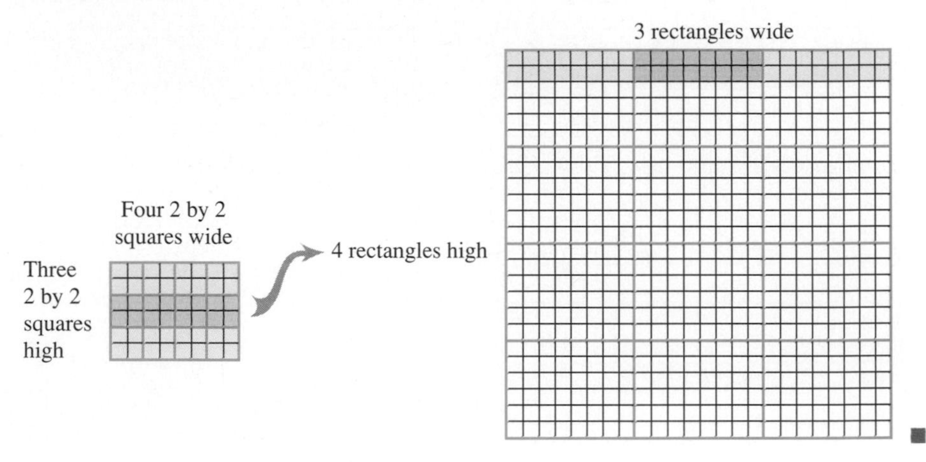

■

Before moving on, it is profitable to look back at the solution of the preceding example. The diagrams show that

$$\text{GCD}(6, 8) = 2 \qquad \text{and} \qquad \text{LCM}(6, 8) = 24.$$

Moreover, the height of the rectangle in 2 by 2 squares determined the width of the large square in rectangles. Since the rectangle is three 2 by 2 squares high and four 2 by 2 squares wide, it follows that the square is twelve 2 by 2 squares wide. Hence the twelve 2 by 2 squares in the rectangle exactly tile the 2 by 24 strip along the top of the square. *Thus, the area of the rectangle and the area of the 2 by 24 strip at the top of the square are the same.* In numbers, this gives the equality

$$6 \cdot 8 = 2 \cdot 24 = \text{GCD}(6, 8) \cdot \text{LCM}(6, 8).$$

This argument could be repeated in general to obtain the next theorem.

CURRENT AFFAIRS

Eureka!

"At last, shout of 'Eureka!' in age-old math mystery," so read the headline in the New York Times on June 24, 1993. The following article then proceeded to detail the announcement of the proof of Fermat's last theorem, a problem that has plagued mathematicians for over 350 years. It has been known since the time of the Babylonians (ca. 1600 B.C.) that there are infinitely many triples of natural numbers x, y, and z for which $x^2 + y^2 = z^2$. However, Pierre de Fermat, the greatest French mathematician of the seventeenth century, in 1637 noted in the margin of his copy of Diophantus's *Arithmetica,* "that the equation $x^n + y^n = z^n$ has **no** solution in natural numbers if n is greater than two." And he went on to write, ". . . and I have assuredly found an admirable proof of this, but the margin is too narrow to contain it." This produced a frenzy on the part of others to find the proof and much first rate mathematics has been produced in the process. However, the proof resisted all efforts for over three centuries until Dr. Andrew Wiles, a British mathematician at Princeton University, using the most sophisticated of the accumulated methods, finally produced a proof. In view of the great effort required to construct it, it is seriously doubted that Fermat actually had a valid proof. Nevertheless, he had the insight to guess the result, and the efforts to provide a proof have greatly enriched mathematics.

THEOREM ab = GCD(a, b,) · LCM(a, b)

If a and b are any two natural numbers, then $ab = \text{GCD}(a, b) \cdot \text{LCM}(a, b)$.

Least common multiples by intersection of sets Let a and b be any two natural numbers. The multiples of a and b are the sets

$$M_a = \{0, a, 2a, 3a, \ldots\} \quad \text{and} \quad M_b = \{0, b, 2b, 3b, \ldots\}.$$

Therefore, $M_a \cap M_b$ is the set of all common multiples of a and b, and the least natural number in this set is the least common multiple of a and b.

EXAMPLE 4.21 **Finding a Least Common Multiple by Set Intersection**

Find the least common multiple of 9 and 15.

SOLUTION

$$M_9 = \{0, 9, 27, 36, 45, 54, 63, 72, 81, 90, \ldots\}$$
$$M_{15} = \{0, 15, 30, 45, 60, 75, 90, 105, \ldots\}$$
$$M_9 \cap M_{15} = \{0, 45, 90, \ldots\}$$

Since, $M_9 \cap M_{15}$ is the set of all common multiples of 9 and 15, the least nonzero element of this set is LCM(9, 15). Therefore, LCM(9, 15) = 45. Notice that GCD(9, 15) = 3 and $9 \cdot 15 = 135 = 3 \cdot 45 =$ GCD(9, 15) $\cdot$ LCM(9, 15). ■

Least common multiples from prime power representations Suppose we want to find the least common multiple of 54 and 45. Computing the prime power representations of these numbers, we obtain

$$54 = 2^1 \cdot 3^3 \cdot 5^0 \qquad \text{and} \qquad 45 = 2^0 \cdot 3^2 \cdot 5^1$$

where we use the zero exponents to make it *appear* that 54 and 45 are products of powers of the same primes. By the theorem just preceding Example 4.5, the multiples of 54 are numbers of the form

$$m = 2^a \cdot 3^b \cdot 5^c \qquad \begin{cases} a \geq 1 \\ b \geq 3 \\ c \geq 0 \end{cases}$$

and the multiples of 45 are numbers of the form

$$m = 2^a \cdot 3^b \cdot 5^c \qquad \begin{cases} a \geq 0 \\ b \geq 2 \\ c \geq 1. \end{cases}$$

We obtain the least common multiple of 54 and 45 by choosing a, b, and c as small as possible while still satisfying both the above sets of inequalities. It follows that we must choose a to be the larger of 1 and 0, b the larger of 3 and 2, and c the larger of 0 and 1. Hence,

$$m = 2^1 \cdot 3^3 \cdot 5^1 = 270 = \text{LCM}(54, 45).$$

Since this argument could be repeated in general, we have the following theorem.

THEOREM LCMs from Prime Power Representations

Let

$$a = p_1^{a_1} p_2^{a_2} \cdots p_s^{a_s} \qquad \text{and} \qquad b = p_1^{b_1} p_2^{b_2} \cdots p_s^{b_s}$$

be the prime power representations of a and b. Then

$$\text{LCM}(a, b) = p_1^{d_1} p_2^{d_2} \cdots p_s^{d_s}$$

where d_1 = larger of (a_1, b_1), d_2 = larger of (a_2, b_2), . . . , and d_s = larger of (a_s, b_s).

EXAMPLE 4.22 Finding LCMs and GCDs Using Prime Power Representations

Compute the least common multiple and greatest common divisor of $r = 2^2 \cdot 3^4 \cdot 7^1$ and $s = 3^2 \cdot 5^2 \cdot 7^3$.

$5^\circ = 1$
$2^\circ = 1$

SOLUTION

Both the least common multiple and the greatest common divisor will be of the form $2^a \cdot 3^b \cdot 5^c \cdot 7^d$. For the greatest common divisor we choose the smaller of the two exponents with which each prime appears in r and s, and for the least common multiple we choose the larger of each pair of exponents. Thus, since

$$r = 2^2 \cdot 3^4 \cdot 5^0 \cdot 7^1 = 2268 \quad \text{and} \quad s = 2^0 \cdot 3^2 \cdot 5^2 \cdot 7^3 = 77{,}175,$$

we have that

$$\text{GCD}(r, s) = 2^0 \cdot 3^2 \cdot 5^0 \cdot 7^1 = 63$$

and

$$\text{LCM}(r, s) = 2^2 \cdot 3^4 \cdot 5^2 \cdot 7^3 = 2{,}778{,}300.$$

Since both exponents on each prime were used in finding GCD(r, s) and LCM(r, s), it follows that

$$rs = \text{GCD}(r, s) \cdot \text{LCM}(r, s).$$

Thus,

$$rs = 2268 \cdot 77{,}175 = 175{,}032{,}900$$

and

$$\text{GCD}(r, s) \cdot \text{LCM}(r, s) = 63 \cdot 2{,}778{,}300 = 175{,}032{,}900. \quad \blacksquare$$

Least common multiples using the Euclidean algorithm As noted above, if a and b are natural numbers, then GCD(a, b) · LCM(a, b) = ab. Thus,

$$\text{LCM}(a, b) = \frac{ab}{\text{GCD}(a, b)}$$

and we have already learned that GCD(a, b) can always be found using the Euclidean algorithm.

EXAMPLE 4.23 Finding a Least Common Multiple by Use of the Euclidean Algorithm

Find the least common multiple of 2268 and 77,175 by using the Euclidean algorithm.

SOLUTION

These are the same two numbers treated in Example 4.22. However, here our procedure is totally different. Using the Euclidean algorithm we have these divisions.

$$\begin{array}{r} 34 \text{ R } 63 \\ 2268 \overline{)77{,}175} \end{array} \qquad \begin{array}{r} 36 \text{ R } 0 \\ 63 \overline{)2268} \end{array}$$

Since the last nonzero remainder is 63, GCD(2268, 77,175) = 63 and

$$\text{LCM}(2268, 77{,}175) = \frac{2268 \cdot 77{,}175}{63} = 2{,}778{,}300$$

as before. ■

PROBLEM SET 4.3

Understanding Concepts

1. (a) Use the method of cutting off squares to find the largest square that will tile a 9 by 15 rectangle. Make a diagram to illustrate the process.
   (b) Exhibit a tiling of the 9 by 15 rectangle by the largest possible square.
2. Use the method of cutting squares off rectangles to determine the greatest common divisors of each of these pairs of numbers. Make a diagram in each case.
   (a) 8 and 22 **(b)** 33 and 42 **(c)** 21 and 34
3. Find the least common multiple of 4 and 7 by finding the smallest square that can be tiled by a 4 by 7 rectangle. Draw a diagram to illustrate the process.
4. Find the least common multiple of 8 and 12 by finding the smallest square that can be tiled by an 8 by 12 rectangle. Draw a diagram to illustrate the process.
5. Find the greatest common divisor of each of these pairs of numbers by the method of intersection of sets of divisors.
   (a) 24 and 27 **(b)** 14 and 22 **(c)** 48 and 72
6. Find the least common multiple of each of these pairs of numbers by the method of intersection of sets of multiples.
   (a) 24 and 27 **(b)** 14 and 22 **(b)** 48 and 72.
7. Use the results of problems 5 and 6 to show that:
   (a) $24 \cdot 27 = \text{GCD}(24, 27) \cdot \text{LCM}(24, 27)$
   (b) $14 \cdot 22 = \text{GCD}(14, 22) \cdot \text{LCM}(14, 22)$
   (c) $48 \cdot 72 = \text{GCD}(48, 72) \cdot \text{LCM}(48, 72)$
8. Use the method based on prime power representations to find the greatest common divisor and least common multiple of each of these pairs of numbers.
   (a) $r = 2^2 \cdot 3^1 \cdot 5^3$ and $s = 2^1 \cdot 3^3 \cdot 5^2$
   (b) $u = 5^1 \cdot 7^2 \cdot 11^1$ and $v = 2^2 \cdot 5^3 \cdot 7^1$
   (c) $w = 2^2 \cdot 3^3 \cdot 5^2$ and $x = 2^1 \cdot 5^3 \cdot 7^2$
9. Use the Euclidean algorithm to find each of the following.
   (a) GCD(3500, 550) and LCM(3500, 550)
   (b) GCD(3915, 825) and LCM(3915, 825)
   (c) GCD(624, 1044) and LCM(624, 1044)

Thinking Critically

10. The notions of the greatest common divisor and the least common multiple extend naturally to more than two numbers. Moreover, the prime power method extends naturally to finding GCD(a, b, c,) and LCM(a, b, c).
    (a) If $a = 2^2 \cdot 3^1 \cdot 5^2$, $b = 2^1 \cdot 3^3 \cdot 5^1$ and $c = 3^2 \cdot 5^3 \cdot 7^1$, compute GCD($a$, b, c) and LCM(a, b, c).
    (b) Is it necessarily true that $\text{GCD}(a, b, c) \cdot \text{LCM}(a, b, c) = abc$?
    (c) Find numbers r, s, and t such that $\text{GCD}(r, s, t) \cdot \text{LCM}(r, s, t) = rst$.
11. Use the method of intersection of sets to compute the following.
    (a) GCD(18, 24, 12) and LCM(18, 24, 12)
    (b) GCD(8, 20, 14) and LCM(8, 20, 14)
    (c) Is it true that $\text{GCD}(8, 20, 14) \cdot \text{LCM}(8, 20, 14) = 8 \cdot 20 \cdot 14$?
12. **(a)** Compute GCD(6, 35, 143) and LCM(6, 35, 143).
    (b) Is it true that $\text{GCD}(6, 35, 143) \cdot \text{LCM}(6, 35, 143) = 6 \cdot 35 \cdot 143$?
    (c) Guess under what conditions $\text{GCD}(a, b, c) \cdot \text{LCM}(a, b, c) = abc$.
13. (a) Compute GCD(GCD(24, 18), 12) and LCM(LCM(24, 18), 12) and compare with the results of problem 11(a).
    (b) Compute GCD(GCD(8, 20), 14) and LCM(LCM(8, 20), 14) and compare the results with those of problem 11(b).
    (c) In one or two written sentences, state a conjecture based on parts (a) and (b).

14. Cuisenaire® rods are colored rods one centimeter square and of lengths 1 cm, 2 cm, 3 cm, . . . , 10 cm as shown here.

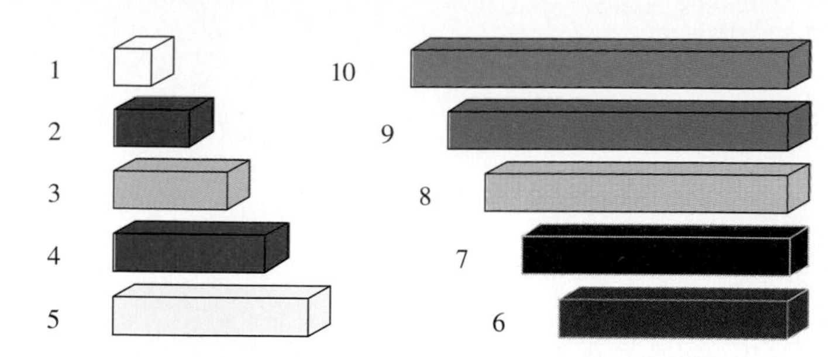

We say that a smaller rod *measures* a longer rod if a number of the smaller rods placed end to end are the same length as the given rod.

(a) Which rods will measure the 9 rod?

(b) Which rods will measure the 6 rod?

(c) What is the greatest common divisor of 9 and 6?

(d) Place the 10 rod and the 8 rod end to end to form an 18 train as in the HANDS ON in Chapter 2. Which rods or trains will measure the "18 train"?

(e) Place the 7 rod and the 5 rod end to end to form a "12 train." Which rods or trains will measure the "12 train"?

(f) What is the greatest common divisor of 12 and 18?

(g) Briefly describe how you might use Cuisenaire® rods to demonstrate the notion of greatest common divisor to children.

15. Cuisenaire® rods are described in the preceding problem.

(a) What is the shortest train or length that can be measured by both 4-rods and 6-rods?

(b) What is the least common multiple of 4 and 6?

(c) What is the shortest train or length that can be measured by both 6-rods and 9-rods?

(d) What is the least common multiple of 6 and 9?

(e) Briefly describe how you could use Cuisenaire® rods to demonstrate the notion of least common multiple to children.

16. **(a)** Indicate how you could use a number line to illustrate the notion of greatest common divisor to children.

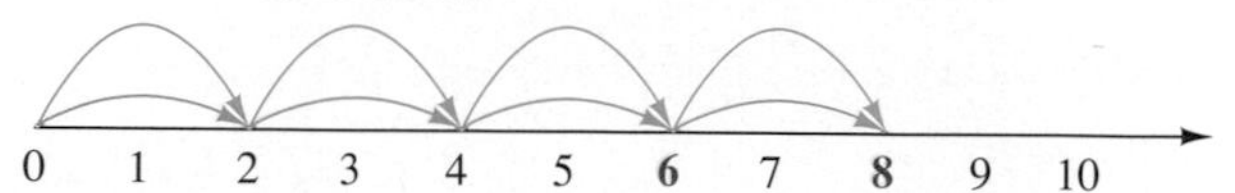

(b) Indicate how you could use a number line to illustrate the notion of least common multiple to children.

Thinking Cooperatively

17. The greatest common divisor machine of Andres Zavrotsky (U.S. patent number 2 978 816, April 11, 1961) is described in Martin Gardner's, *The Sixth Book of Mathematical Games from Scientific American* (San Francisco: W. H. Freeman, 1971). For example, build a 6 by 8 "billiard table" out of mirrors and shine a light at 45° from a corner, P. The beam will bounce around the "table" and will eventually be absorbed at a corner.
Let Q be the point on the long side of the "table" closest to P where the light beam "bounces." Then

$$\text{GCD}(6, 8) = (1/2) \cdot PQ = (1/2) \cdot 4 = 2.$$

Use graph paper and draw a similar diagram to determine GCD(9, 15).

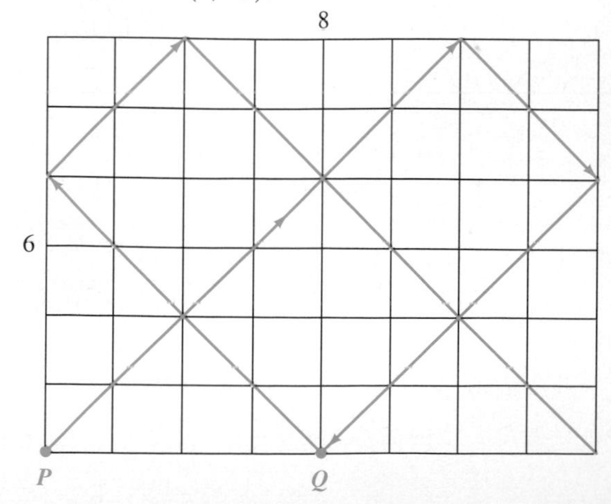

18. A simple graphical scheme for finding the greatest common divisor of two numbers is illustrated below. Suppose again that we want to compute GCD(6, 8). Draw a 6 by 8 rectangle and draw a diagonal from corner to corner.

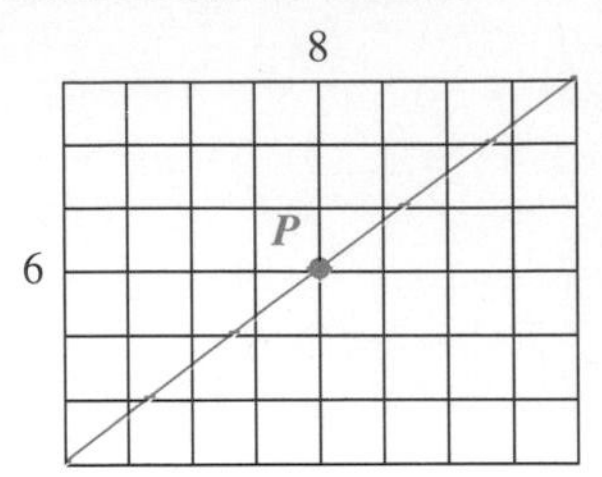

In this case, the diagonal passes through just one point P that is a corner of squares on the graph paper. P divides the diagonal into 2 parts and $2 = \text{GCD}(6, 8)$. Use graph paper and draw a similar diagram to determine GCD(15, 25).

---

19. Recall the Fibonacci sequence defined in problem 26 of Problem Set 4.1. For easy reference, the first 30 numbers in the sequence are listed here. Note that we use F_n to denote the nth Fibonacci number

n	1	2	3	4	5	6	7	8	9	10
F_n	1	1	2	3	5	8	13	21	34	55

n	11	12	13	14	15	16	17	18	19	20
F_n	89	144	233	377	610	987	1597	2584	4181	6765

n	21	22	23	24	25	26	27	28	29	30
F_n	10,946	17,711	28,657	46,368	75,025	121,393	196,418	317,811	514,229	832,040

(a) In problem 26 of Problem Set 4.1, you should have guessed that, for $n > 2$, $F_n \mid F_m$ if, and only if, $n \mid m$. Compute each of these quotients: $F_{12} \div F_6$, $F_{18} \div F_9$, $F_{30} \div F_{15}$.

(b) If m is composite must F_m be composite? Explain briefly.

(c) If n is a prime must F_n be a prime? Explain briefly.

(d) Complete the following list.

$\text{GCD}(F_6, F_9) = \text{GCD}(8, 34) = 2 = F_3$	$\text{GCD}(6, 9) = 3$
$\text{GCD}(F_{14}, F_{21}) = \text{GCD}(377, 10{,}946) = 13 = F_7$	$\text{GCD}(14, 21) = 7$
$\text{GCD}(F_{10}, F_{15}) =$	$\text{GCD}(10, 15) =$
$\text{GCD}(F_{20}, F_{30}) =$	$\text{GCD}(20, 30) =$
$\text{GCD}(F_{16}, F_{24}) =$	$\text{GCD}(16, 24) =$
$\text{GCD}(F_{12}, F_{18}) =$	$\text{GCD}(12, 18) =$

(e) On the basis of the calculations in part (c), make a conjecture about $\text{GCD}(F_m, F_n)$.

(f) Test your conjecture by computing $\text{GCD}(F_{24}, F_{28})$.

(g) Does the result of part (f) prove that your conjecture in part (e) is correct?

(h) If $\text{GCD}(F_{16}, F_{20})$ were to equal 4, what could you conclude about your conjecture in part (e)?

(i) Actually compute $\text{GCD}(F_{16}, F_{20})$.

Making Connections

20. The front wheel of a tricycle has a circumference of 54 inches and the back wheels have a circumference of 36 inches.

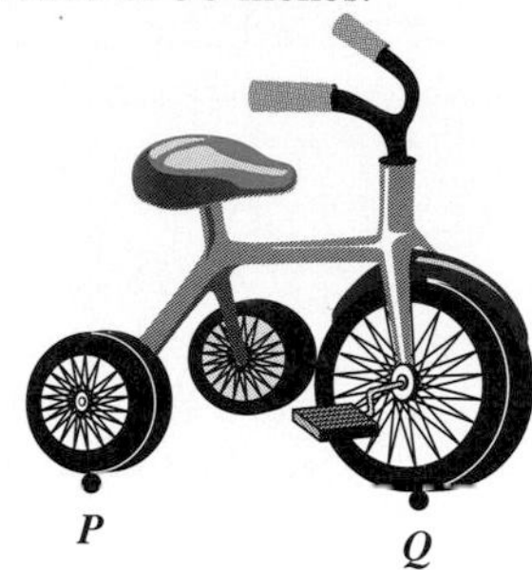

If points P and Q are both touching the sidewalk when Marja starts to ride, how far will she have ridden when P and Q first touch the sidewalk at the same time again?

21. Sarah Speed and Hi Velocity are racing cars around a track. If Sarah can make a complete circuit in 72 seconds and Hi completes a circuit in 68 seconds,

(a) how many seconds will it take for Hi to first pass Sarah at the starting line?

(b) how many laps will Sarah have made when Hi laps her the first time?

22. In a machine, a gear with 45 teeth is engaged with one with 96 teeth, with teeth A and B on the small and large gears respectively in contact as shown. How many more revolutions will the small gear have to make before A and B are again in the same position?

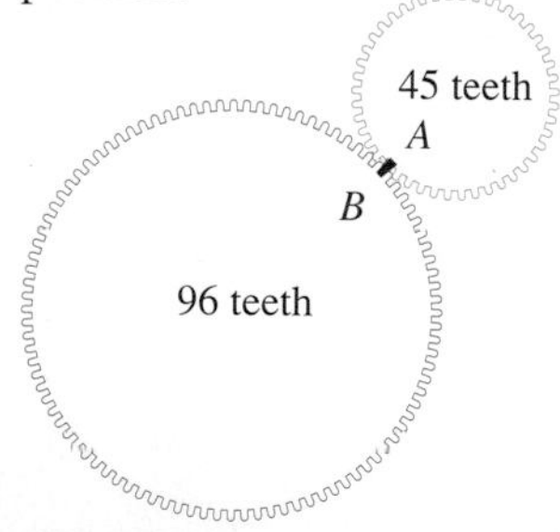

23. The musical notes A and E just below and above middle C have fundamental frequencies of 220 and 330 cycles per second (cps) respectively. Whenever these notes are sounded, overtones whose frequencies are multiples of the fundamental frequencies are also generated. A and E harmonize because many of their overtones have the same frequencies (660, 1320, 1980, . . .). The closer the least common multiple of the fundamental frequencies is to the fundamental frequencies, the better the harmony.

(a) A, C, and C sharp have frequencies of 220, 264, and 275 cps respectively. Compute LCM(220, 264), LCM(220, 275), and LCM(264, 275).

(b) On the basis of part (a), which pair of these three notes will produce the most pleasing harmony?

(c) Compute GCD(220, 264), GCD(220, 275), and GCD(264, 275).

(d) How can the results of part (c) tell you if two notes harmonize nicely?

Using a Computer

24. Use TILERECTANGLE to compute each of the following.

(a) GCD(189, 294) **(b)** GCD(117, 195)

25. Use TILESQUARE to compute each of these numbers.

(a) LCM(189, 294) **(b)** LCM(117, 195)

26. Compute the following using DIVISORSETS.

(a) GCD(45, 48) **(b)** GCD(48, 54)

27. Use MULTIPLESETS to compute each of these numbers.

(a) LCM(45, 48) **(b)** LCM(48, 54)

28. Use FACTORINTEGER to compute each of the following.

(a) The prime power representation of 205,800, 31,460, and 25,840.

(b) GCD(31,460, 205,800)

(c) LCM(31,460, 205,800)

(d) GCD(205,800, 31,460, 25,840)

(e) LCM(205,800, 31,460, 25,840)

29. Compute the following using EUCLIDALG.

(a) GCD(36,461, 33,269)

(b) GCD(16,583, 16,377)

(c) LCM(36,461, 33,269)

(d) LCM(16,583, 16,377)

For Review

30. Let $A = \{a, b, c, d, e, f, g\}$, $B = \{a, c, d, e, g\}$, $C = \{a, b, c, d\}$, and $U = \{a, b, c, \ldots, z\}$.

(a) Show that $A \cap (B \cup C) = (A \cap B) \cup (A \cap C)$.

(b) Show that $A \cup (B \cap C) = (A \cup B) \cap (A \cup C)$.

(c) Show that $\overline{A \cup B} = \overline{A} \cap \overline{B}$.

(d) Show that $\overline{A \cap B} = \overline{A} \cup \overline{B}$.

31. (a) Check to see if $n^2 - 81n + 1681$ is prime for $n = 1, 2, 3, 4$, and 5.
(b) On the basis of part (a), make a conjecture about the values of $n^2 - 81n + 1681$.
(c) Does the fact that your conjecture is true for $n = 1, 2, 3, 4$, and 5 mean that the conjecture is always true? Explain briefly.
(d) Check your conjecture for $n = 6, 7$, and 8. What do you conclude? Explain briefly.
(e) Check your conjecture for $n = 80$. How does this affect your belief in your conjecture? Explain.
(f) Check your conjecture for $n = 81$. What do you conclude? Explain briefly.

32. Let $n = 2 \cdot 5 \cdot 7 + 1$.
(a) Could 2, 5, or 7 divide n evenly? Why or why not?
(b) Is n prime or composite?

33. Let $n = 2 \cdot 3 \cdot 5 + 7 \cdot 11 \cdot 13$.
(a) Do any of 2, 3, 5, 7, 11, or 13 divide n evenly? Explain briefly.
(b) Must n have a prime divisor different from 2, 3, 5, 7, 11, or 13? Explain briefly.
(c) Is n prime or composite? (*Hint:* Use FACTORINTEGER.)

34. Let $n = 2 \cdot 3 \cdot 5 + 7 \cdot 11 \cdot 13 + 17 \cdot 19 \cdot 23$.
(a) Do any of 2, 3, 5, 7, 11, 13, or 23 divide n evenly? Explain briefly.
(b) Can you argue as in problem 33(b) that n must have a prime divisor different from 2, 3, 5, 7, 11, 13, 17, 19, and 23?
(c) Find the prime power representation of n using FACTORINTEGER.

COOPERATIVE *Investigation*

The All Star Game

To give her students more experience with greatest common divisors and least common multiples, Leslie Laposka frequently uses the All Star Game as an activity. She gives each pair of students four All Star Game sheets. The first one is illustrated on the next page and the others are the same except that the second sheet has 7 circles with 8 points per circle, the third has 8 circles with 9 points per circle, and the fourth has 11 circles with 12 points per circle. The rules are as follows:

1. The first student to play writes his or her name in the space marked with a cross on the record sheet and then chooses a "star number," n, and marks it with an X. He or she then draws a "star" shaped polygon by joining the **highest** numbered point on the circle by a straight line segment to the point n spaces counterclockwise around the circle, then joining that point to the next point n spaces beyond that, and so on, continuing until the highest numbered point again is reached. The student then writes down the number of points on the star obtained in the space provided. The "star" for "star number" 6 has been drawn for you on the record sheet and the number of star points has been entered in the table.
2. The second student now enters his or her name in the space marked with a circle on the record sheet, chooses a "star number" and circles it in the table, draws the corresponding star, and finally records the number of star points in the proper place on the record sheet.
3. The play continues until all the star numbers have been used.
4. The player with the greatest total number of star points wins the game.
5. The play then continues on the next three All Star Game sheets with each player going first every other time.

For the first record sheet can you guess how to choose a "star number" that will always result in a 15 pointed star? Write an explanation agreed upon by both players on the back of this record sheet. Write a similar discussion on the back of each of the other three sheets.

Since the first to play probably has an advantage, how can the two players fairly decide who should go first?

Continued

All Star Game Sheet

Name X: ________
Name O: ________

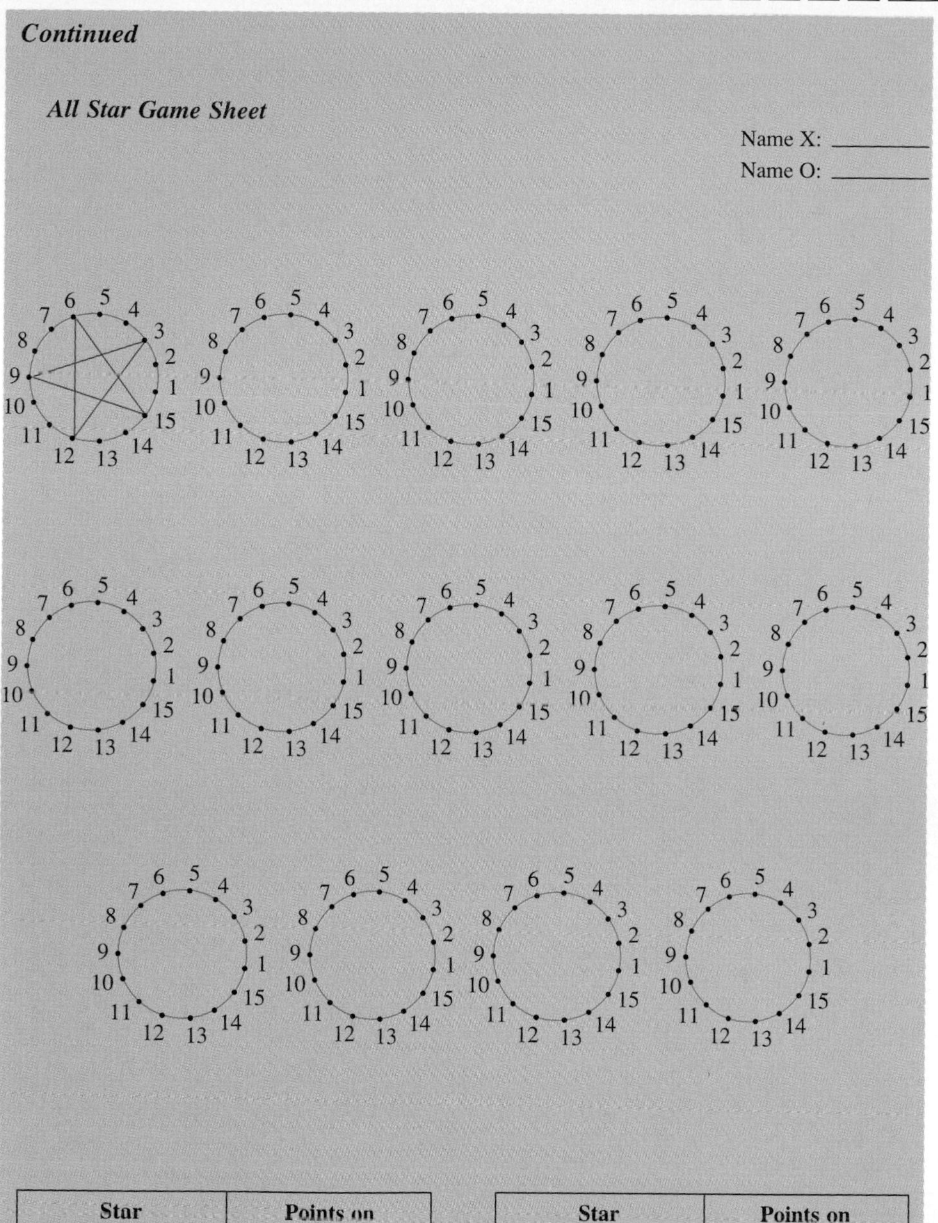

Star Number	Points on Star
1	
2	
3	
4	
5	
~~6~~	5
7	

Star Number	Points on Star
8	
9	
10	
11	
12	
13	
14	

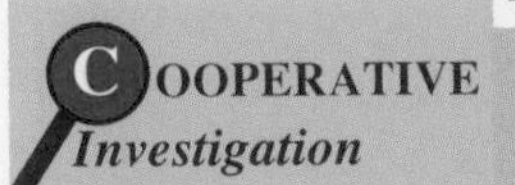

Second Day of the All Star Game

After checking to see that she has enough correctly completed sets of data sheets from the first day of this activity, Ms. Laposka begins the second day by dividing her class into groups of three or four students each, gives each group a set of record sheets from the first day, and gives each student four new record sheets, as shown below, to be filled out with data from the preceding day's work—one sheet for each of the first day's sheets. For ease in communication, we let

- m denote the number of points on the circle,
- n denote the "star number,"
- r denote the number of points on a star, and
- s denote the number of spaces between star points.

Suppose also that a bug crawls around each star. We let

- b denote the number of times the bug goes around the center of the circle.

Note that to count the number of bug circuits for a given star, it is helpful to draw a line segment from the center of the circle to the highest numbered point on the circle. Then, as the bug crawls around the star, each time it crosses this line and the last time it touches this line, it completes a circuit.

1. Complete the second day All Star Game sheet for $n = 15, 8, 9$, and 12.
2. Guess how you can determine r given n and m.
3. Guess how you can determine s given n and m.
4. Guess how you can determine b given n and m.

Second Day All Star Game Sheet

Name ________

m	n	r	s	b	GCD(m, n)	LCM(m, n)	n/GCD(m, n)	LCM$(m, n)/m$	LCM$(m, n)/n$

4.4 Clock Arithmetic

Mathematics is all around us. Making use of such naturally occurring mathematical ideas is an important and useful teaching strategy. What, for example, is the mathematics of an ordinary clock as depicted in Figure 4.11?

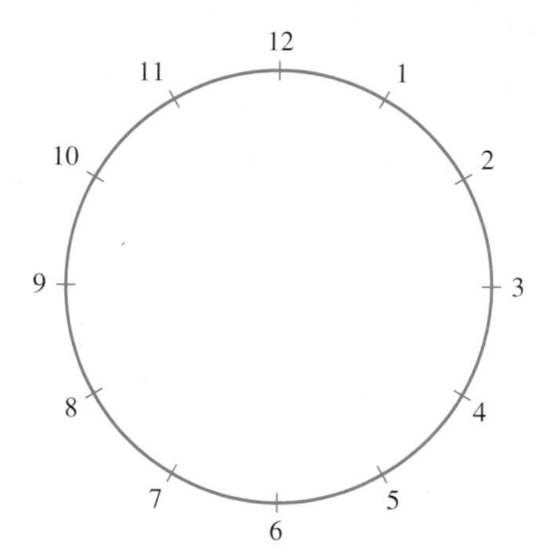

Figure 4.11
A 12-hour clock

Clock Addition

What is the point on the number line 9 units to the right of 7?

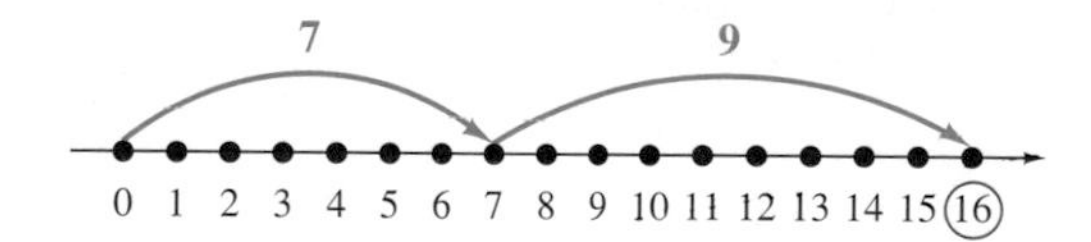

This is an addition problem that can be solved on the number line by "counting on." Starting at 0, we first count out to 7 and then count on 9 more to 16. Thus, $7 + 9 = 16$. Now, thinking of the clock, we might ask "what time is it 9 hours after 7 o'clock?" Again this can be answered by counting, but this time we count clockwise around the face of a clock. Starting at 12 on the clock diagrammed in Figure 4.11, we count to 7 and then count on 9 more to arrive at 4. Thus, in **12-hour clock arithmetic,** it is reasonable to say that 7 plus 9 is 4 and to write

$$7 +_{12} 9 = 4$$

where we indicate 12-hour clock addition by the symbol $+_{12}$.

EXAMPLE 4.24 **Computing Sums in 12-hour Clock Arithmetic**

Compute these in 12-hour clock arithmetic.

(a) $6 +_{12} 8$ **(b)** $3 +_{12} 7$ **(c)** $8 +_{12} 10$ **(d)** $9 +_{12} 12$

SOLUTION

We use the "counting on" method just discussed.

(a) $6 +_{12} 8 = 2$
(b) $3 +_{12} 7 = 10$
(c) $8 +_{12} 10 = 6$
(d) $9 +_{12} 12 = 9$ ■

Alternatively, to add in 12-hour clock arithmetic, we can add in the normal way. Then, if the answer obtained is 12 or less, it is also the answer in 12-hour clock arithmetic. If the answer is more than 12, subtract 12 to obtain the answer in 12-hour clock arithmetic. Differently put, if we add in the usual way and divide the answer by 12, the nonzero remainder obtained is the answer in 12-hour clock arithmetic. If the remainder is 0, the 12-hour clock answer is 12. Thus, since $11 + 8 + 9 = 28$ and $28 = 2 \cdot 12 + 4$, then $11 +_{12} 8 +_{12} 9 = 4$. Also, since $7 + 11 + 6 = 24$ and $24 = 2 \cdot 12 + 0$, $7 +_{12} 11 +_{12} 6 = 12$.

DEFINITION **Computing 12-hour Clock "Sums"**

Let $T = \{1, 2, 3, 4, 5, 6, 7, 8, 9, 10, 11, 12\}$. To compute the "sum" $a +_{12} b$ for a and b in T, proceed as follows.

1. Compute $a + b$.
2. Divide $a + b$ by 12 so that $a + b = 12q + r$ with $0 \le r < 12$.
3. If $r > 0$, set $a +_{12} b = r$.
4. If $r = 0$, set $a +_{12} b = 12$.

CURRENT AFFAIRS

Public Key Encryption

Number theory is usually considered a part of pure mathematics; that is, mathematics studied for its own sake with no thought that it might be, or even could be, applied to real world problems. Who would have guessed that the notions of primality and factoring would turn out to provide the basis for a new but simple and remarkably secure method of sending secret messages in code? The idea is to determine two, 100-digit primes, p and q, and to publish the product pq for all to see. Using this "key" anyone can send you a message, but no one can read the message unless they know p and q; that is, unless they are able to factor the 200-digit product, pq. While it is a three or four minute job to determine two 100-digit primes on one of today's fastest computers, it would take one of these same machines on the order of 10^9 years (that is, 1 billion years), to factor the product pq. Thus, unless someone has discovered a way to break the code without determining p and q—a very unlikely circumstance—it is effectively unbreakable. The importance of the existence of such a code in our modern society, with its need to keep masses of data (financial records, industrial secrets, computer data bases of all sorts) secret, cannot be overstated.

Interestingly, it follows from the preceding theorem that all addition properties for the whole numbers also hold for 12-hour clock arithmetic. For example,

$$(a + b) + c = a + (b + c)$$

in ordinary whole number arithmetic. Since we obtain clock sums by dividing by 12 and recording the remainder or 12 as appropriate, it follows that

$$(a +_{12} b) +_{12} c = a +_{12} (b +_{12} c).$$

One minor difference, is that the additive identity in 12-hour clock arithmetic is 12 rather than 0; that is,

$0 + n = n + 0 = n$ for all whole numbers n, and

$12 +_{12} n = n +_{12} 12 = n$ for all n in 12-hour clock arithmetic.

THEOREM 12-hour Clock Addition Properties

Let $T = \{1, 2, 3, \ldots, 12\}$ as above. Then, for all a and b in T:

Closure Property	$a +_{12} b \in T$
Commutative Property	$a +_{12} b = b +_{12} a$
Associative Property	$(a +_{12} b) +_{12} c = a +_{12} (b +_{12} c)$
Additive Identity	$12 +_{12} a = a +_{12} 12 = a$

Clock Multiplication

Like ordinary multiplication, clock multiplication, denoted by $\times_{12}$, is understood as repeated addition. Thus, we would have

$$5 \times_{12} 8 = 8 +_{12} 8 +_{12} 8 +_{12} 8 +_{12} 8,$$

and it is easily determined by counting that $5 \times_{12} 8 = 4$. Alternatively, $5 \times_{12} 8$ is the remainder obtained when 5×8 is divided by 12. Since

$$5 \cdot 8 = 40 = 3 \cdot 12 + 4,$$

it again follows that

$$5 \times_{12} 8 = 4.$$

DEFINITION Computing 12-hour Clock Products

Let $T = \{1, 2, 3, \ldots, 12\}$. To compute $a \times_{12} b$ for $a \in T$ and $b \in T$, proceed as follows.

1. Compute $a \cdot b$.
2. Divide $a \cdot b$ by 12 so that $a \cdot b = 12q + r$ with $0 \le r < 12$.
3. If $r > 0$, set $a \times_{12} b = r$.
4. If $r = 0$, set $a \times_{12} b = 12$.

EXAMPLE 4.25 **Computing Products in 12-hour Clock Arithmetic**

Compute each of these products.

(a) $7 \times_{12} 9$ **(b)** $8 \times_{12} 12$ **(c)** $4 \times_{12} 8$ **(d)** $4 \times_{12} 2$

SOLUTION

(a) $7 \cdot 9 = 63 = 5 \cdot 12 + 3$. Therefore, $7 \times_{12} 9 = 3$.
(b) $8 \cdot 12 = 96 = 8 \cdot 12 + 0$. Therefore, $8 \times_{12} 12 = 12$.
(c) $4 \cdot 8 = 32 = 2 \cdot 12 + 8$. Therefore, $4 \times_{12} 8 = 8$.
(d) $4 \cdot 2 = 8 = 0 \cdot 12 + 8$. Therefore, $4 \times_{12} 2 = 8$. ■

As with addition, the usual multiplication properties follow from the same properties from ordinary whole number arithmetic. Even the multiplicative property for 0 holds although it looks different in 12-hour clock arithmetic. Recall that 0 is the additive identity for whole number arithmetic; that is,

$$0 + a = a + 0 = a$$

for every whole number a and that

$$a \cdot 0 = 0 \cdot a = 0.$$

As noted above, 12 is the additive identity in 12-hour clock arithmetic, and, since $12 \cdot a = a \cdot 12$ always leaves a remainder of 0 when divided by 12, it follows that

$$12 \times_{12} a = a \times_{12} 12 = 12$$

for all a in 12-hour clock arithmetic. In both whole number arithmetic and clock arithmetic it might be better to call this the multiplicative property of the additive identity. Then the apparent difference disappears.

THEOREM 12-hour Clock Multiplication Properties

Let $T = \{1, 2, 3, \ldots, 12\}$. For all a, b, and c in T:

Closure Property	$a \times_{12} b \in T$
Commutative Property	$a \times_{12} b = b \times_{12} a$
Associative Property	$(a \times_{12} b) \times_{12} c = a \times_{12} (b \times_{12} c)$
Multiplicative Property of 12	$a \times_{12} 12 = 12 \times_{12} a = 12$
Distributive Property	$a \times_{12} (b +_{12} c) = (a \times_{12} b) +_{12} (a \times_{12} c)$

Clock Subtraction

Like whole number subtraction, subtraction in clock arithmetic is defined in terms of addition. In whole number subtraction, $a - b = c$ if, and only if, $a = b + c$. In like manner, we have this definition.

DEFINITION Clock Subtraction

Let $T = \{1, 2, 3, \ldots, 12\}$. For all $a \in T$ and $b \in T$, $a -_{12} b = c$ if, and only if, $a = b +_{12} c$.

EXAMPLE 4.26 **12-hour Clock Subtraction**

Compute the following clock "differences."

(a) $11 -_{12} 4$ **(b)** $4 -_{12} 7$ **(c)** $12 -_{12} 5$ **(d)** $5 -_{12} 12$

SOLUTION

(a) Since $4 +_{12} 7 = 11$, it follows that $11 -_{12} 4 = 7$.

(b) Since $7 +_{12} 9 = 4$, it follows that $4 -_{12} 7 = 9$.

(c) Here $12 = 5 +_{12} 7$ and so $12 -_{12} 5 = 7$.

(d) Since $12 +_{12} 5 = 5$, it follows that $5 -_{12} 12 = 5$. This is analogous to $5 - 0 = 5$ in ordinary arithmetic.

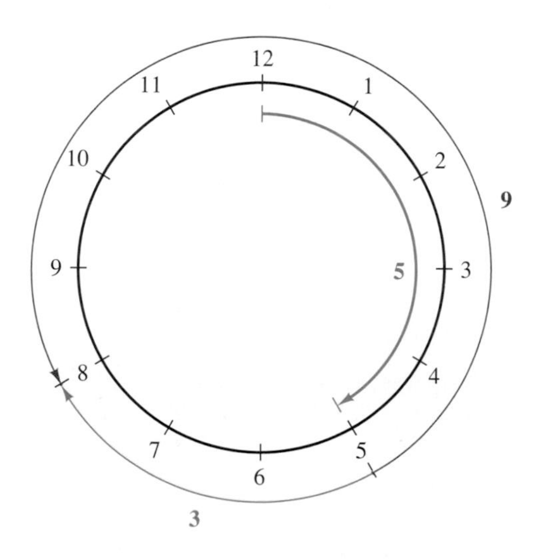

$5 -_{12} 9 = 5 +_3 3 = 8$ ■

Since any point on the clock can be reached from any other point by counting in either the clockwise direction or the counterclockwise direction, clock subtraction can always be performed as clock addition. For example, $5 -_{12} 9 = 8$ since, starting at 12, if we count 5 in the clockwise direction and 9 in the counterclockwise direction, we arrive at 8. However, 8 can also be reached by starting at 12 and counting 5 in the clockwise direction and then 3 more in the clockwise direction. Thus,

$$5 -_{12} 9 = 5 +_{12} 3 = 8.$$

Indeed, since the counts of 9 and 3 have to reach all the way around the circle, it is clear that $9 + 3 = 12$. This is always the case, and we have the following theorem.

THEOREM Clock Subtraction as Clock Addition

Let $T = \{1, 2, 3, \ldots, 12\}$. For all a and b in T, $a -_{12} b = a +_{12} (12 - b)$.

WORTH READING

Fascinating Fibonaccis

"For many years I have been captivated and intrigued by Fibonacci numbers. It has been enormously satisfying to share what I know about them with my young . . . students. Their responses have run the gamut from profound disbelief to patronizing good humor, sprinkled with scientific inquisitiveness. Invariably they ask for more information that they might take home to share with family and friends, demonstrating a confidence in others' interest in the subject and thus revealing their own. Unfortunately the information they seek has been deeply buried in scientific journals or aging mathematical literature, occasionally surfacing as one perfunctory page in a textbook or as a scholarly article in a popular periodical.

"It seemed to me that the time had come to collect and sort out what is currently known about this fascinating subject, to make it understandable and to excite the curiosity of believers and skeptics alike. What is so special about these numbers? Where did they come from? Why do they keep popping up in unlikely, unrelated places? Where might they be that no one has yet thought to look? What answers might they hold for the world, if not the universe?"

This is the preface to the book by middle school teacher Trudi Hammel Garland, *Fascinating Fibonaccis: Mystery and Magic in Numbers*. The book brims with great ideas that will not only stimulate your imagination but will do the same for your students.

SOURCE: From *Fascinating Fibonaccis: Mystery and Magic in Numbers* by Trudi Hammel Garland. Copyright © 1987 by Dale Seymour Publications, Palo Alto, CA 94303. Reprinted by permission.

EXAMPLE 4.27

Subtracting by Adding in Clock Arithmetic

Perform the following subtractions as additions.

(a) $3 -_{12} 8$ **(b)** $11 -_{12} 7$ **(c)** $9 -_{12} 11$ **(d)** $2 -_{12} 6$

SOLUTION

(a) Since $12 - 8 = 4$, $3 -_{12} 8 = 3 +_{12} 4 = 7$. Check: $8 +_{12} 7 = 3$

(b) Since $12 - 7 = 5$, $11 -_{12} 7 = 11 +_{12} 5 = 4$. Check: $7 +_{12} 4 = 11$

(c) Since $12 - 11 = 1$, $9 -_{12} 11 = 9 +_{12} 1 = 10$. Check: $11 +_{12} 10 = 9$

(d) Since $12 - 6 = 6$, $2 -_{12} 6 = 2 +_{12} 6 = 8$. Check: $6 +_{12} 8 = 2$ ■

In the preceding theorem and example, we saw that we can subtract by b in 12-hour clock arithmetic by *adding* $12 - b$. Indeed, we note that

$$b +_{12} (12 - b) = 12$$

and that 12 is the additive identity in T. We call $12 - b$ the **additive inverse** of b and note that we can subtract b from a by *adding the additive inverse of b to a.*

DEFINITION The Additive Inverse

Let $T = \{1, 2, \ldots, 12\}$. If $a \in T$ and $a +_{12} b = b +_{12} a = 12$, then b is called the **additive inverse** of a. Also, a is the **additive inverse** of b.

Using this terminology, the preceding theorem can now be written as follows.

THEOREM Clock Subtraction as Clock Addition

To subtract b from a in clock arithmetic, add the additive inverse of b to a.

Since clock subtraction can always be performed as clock addition and clock arithmetic is closed under clock addition, it follows that clock arithmetic is also closed under clock subtraction. In this respect, clock arithmetic is like integer arithmetic rather than whole number arithmetic. At the same time, like whole number and integer arithmetic, clock arithmetic is not commutative or associative under subtraction.

THEOREM 12-hour Clock Subtraction Properties

1. Clock arithmetic is closed under subtraction.
2. Subtraction is *not* commutative in clock arithmetic.
3. Subtraction is *not* associative in clock arithmetic.

Division in Clock Arithmetic

Division without remainder is defined in clock arithmetic just as it is for whole numbers.

DEFINITION Clock Division

Let $T = \{1, 2, 3, \ldots, 12\}$. For $a \in T$ and $b \in T$, we say that ***b* divides *a*** in clock arithmetic, and write $a \div_{12} b = c$, if, and only if, there exists a *unique* $c \in T$ such that $a = b \times_{12} c$.

Division in clock arithmetic is greatly facilitated if one has a complete multiplication table as shown in Table 4.1.

TABLE 4.1 12-hour Clock Multiplication Table

$\times_{12}$	**1**	**2**	**3**	**4**	**5**	**6**	**7**	**8**	**9**	**10**	**11**	**12**
1	1	2	3	4	5	6	7	8	9	10	11	12
2	2	4	6	8	10	12	2	4	6	8	10	12
3	3	6	9	12	3	6	9	12	3	6	9	12
4	4	8	12	4	8	12	4	8	12	4	8	12
5	5	10	3	8	1	6	11	4	9	2	7	12
6	6	12	6	12	6	12	6	12	6	12	6	12
7	7	2	9	4	11	6	1	8	3	10	5	12
8	8	4	12	8	4	12	8	4	12	8	4	12
9	9	6	3	12	9	6	3	12	9	6	3	12
10	10	8	6	4	2	12	10	8	6	4	2	12
11	11	10	9	8	7	6	5	4	3	2	1	12
12	12	12	12	12	12	12	12	12	12	12	12	12

EXAMPLE 4.28 **12-hour Clock Division**

Perform these divisions if possible.

(a) $8 \div_{12} 5$ **(b)** $7 \div_{12} 8$ **(c)** $4 \div_{12} 10$ **(d)** $7 \div_{12} 11$

SOLUTION

(a) From the table, 4 is the only element a in T such that $8 = 5 \times_{12} a$. Therefore, $8 \div_{12} 5 = 4$.

(b) From the table, there is no $c \in T$ such that $7 = 8 \times_{12} c$. Therefore, this division is not defined.

(c) From the table, $10 \times_{12} 4 = 10 \times_{12} 10 = 4$. Since there is *more than one number* $c \in T$ such that $10 \times_{12} c = 4$, this division is *not* defined.

(d) From the table, the only number $c \in T$ for which $11 \times_{12} c = 7$ is 5. Therefore, $7 \div_{12} 11 = 5$. ■

EXAMPLE 4.29 **Division on the 12-hour Clock**

(a) For which numbers in 12-hour clock arithmetic is division always defined?

(b) For which numbers in 12-hour clock arithmetic is division never defined?

SOLUTION

(a) Scanning Table 4.1, we see that every entry in T appears in each row (and column) numbered 1, 5, 7, and 11. Thus, $a = b \times_{12} c$ is satisfied by precisely one $c \in T$ for every a for each of $b = 1, 5, 7$, or 11. Therefore, $a \div_{12} b$ is defined for each of these values of b.

(b) Again from Table 4.1, we see that no entry appears only once and some entries in T never appear in rows numbered 2, 3, 4, 6, 8, 9, 10, and 12. Therefore, division by these numbers is never defined in 12-hour clock arithmetic. ■

THEOREM 12-hour Clock Division Properties

1. Clock arithmetic is *not* closed under division.
2. The commutative law for division does *not* hold in 12-hour clock arithmetic.
3. The associative law for division does *not* hold in 12-hour clock arithmetic.

A Post Office Application of Clock Arithmetic

Suppose we had a "10-hour" clock as shown in Figure 4.12 instead of a 12-hour clock. The resulting arithmetic is like 12-hour clock arithmetic except that to find the 10-hour sum or product of two elements of

$$T = \{1, 2, 3, 4, 5, 6, 7, 8, 9, 10\},$$

we find the sum as in ordinary arithmetic, divide the answer by 10, and use the resulting remainder, r, if $r \neq 0$ and 10 if $r = 0$.

Subtraction and division are then defined in terms of addition and of multiplication as above.

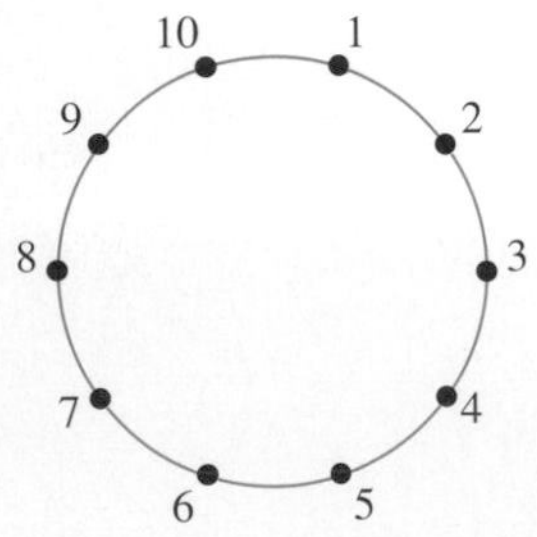

Figure 4.12
A ten-hour clock

INTO THE CLASSROOM
Clock Arithmetic

Clock arithmetic should be considered a worthwhile enrichment topic for the elementary classroom. Our experience is that children like mathematical ideas drawn from their immediate surroundings, and that they are intrigued and interested by this arithmetic generated by an ordinary clock; an arithmetic that is both similar to yet different from, ordinary arithmetic. As usual, the arithmetic should be introduced via manipulatives (in this case an ordinary clock and, later, diagrams of n-hour clocks for other values of n), and then students should be led to develop the resulting arithmetic pretty much on their own. The teacher should ask only occasional pertinent questions and make occasional suggestions as the development proceeds. Of special importance is the difference between the arithmetics generated by n-hour clocks when n is prime and when n is composite. In particular, n-hour clock arithmetic where n is a prime number is closed under division except for division by zero and, as we have seen, this is not so when n is composite. Other interesting questions and properties of clock arithmetic that can be turned into classroom activities will be seen in Problem Set 4.4.

Also, this material should not be presented to students in a vacuum. Classroom activities involving zip codes and Caesar ciphers are attractive to students as are other codes, like ISBN numbers in books, that rely at least partly on clock arithmetic.

EXAMPLE 4.30

Calculating in 10-hour Clock Arithmetic

Perform each of these 10-hour clock computations.

(a) $6 +_{10} 9$ **(b)** $6 -_{10} 9$ **(c)** $6 \times_{10} 9$ **(d)** $6 \div_{10} 9$

SOLUTION

(a) Since $6 + 9 = 15 = 1 \cdot 10 + 5$, $6 +_{10} 9 = 5$.

(b) Here we must find $c \in T$ such that $6 = 9 +_{10} c$. Since the remainder on dividing a number by 10 is just its units digit and $9 + 7 = 16$, it follows that $c = 7$.

(c) Since $6 \cdot 9 = 54 = 5 \cdot 10 + 4$, $6 \times_{10} 9 = 4$.

(d) Here we must find $c \in T$ such that $9 \times_{10} c = 6$. Since $4 \cdot 9 = 36 = 3 \cdot 10 + 6$, it follows that $c = 4$. Thus, $6 \div_{10} 9 = 4$. ■

Have you ever wondered what all the strange marks mean that appear so often on mail (particularly junk mail) that you regularly receive? Consider the following problem.

EXAMPLE 4.31

The Bar Code on Envelopes and Postal Cards

Discover the meaning of bar codes like those appearing on the postal cards shown on the next page.

SOLUTION

Understand the problem • • • •

Surely the markings are placed on the cards and envelopes for a purpose. They must mean something; they must convey some sort of information.

That means the string of marks must be a code of some sort. But what is secret about business reply cards and other pieces of mail? Why should there be coded information on each such card? Since it is unlikely that the information is intended to be kept secret, it is no doubt coded for some other purpose. Do you suppose that it is encoded in this way so that it is machine readable? Actually, given the ever increasing use of automation coupled with the steady increase in the amount of mail being processed, it is quite likely that this has something to do with the automatic sorting of mail.

Devise a plan • • • •

The first bit of information needed to deliver a piece of mail is the post office to which the mail must be sent, and this is determined by the 5-digit zip code. Actually, we all have a 9-digit zip code as on the post cards shown, with the first two of the last four digits indicating a portion of a delivery area (for example, a specific delivery route) and the last two digits indicating a specific business or organization, a specific building, or a specific portion (say a part of a street) of a delivery route. Do you suppose the bar code in question gives the zip code in machine readable form? Let's analyze the bar codes and see if we can associate them with the zip codes.

Carry out the plan • • • •

To simplify the work, we repeat here the two zip codes and what we guess are the associated bar codes.

07713-0001

20077-9964

Looking for patterns and trying to associate the numbers with the bar code, we note that each string begins and ends with a long bar. In fact, this is always true as you can easily check by finding other examples in your mail box. These two bars probably tell the machine when the code starts and stops. Doing a count, we find that there are 52 bars in each code. Thus, deleting the start and stop bars, there are 50 bars to determine the zip code. Since there are nine numbers and a dash in the zip code, this may mean that all code groups are of length five. Thus, we break the bar code up into 5-bar groups as shown below and try to associate the successive numbers in the zip code with the successive groups. As a check, we can use the fact that several digits are repeated both within and between the two codes. For the first code we obtain the pairings shown. But this can't make sense because then 0 and the dash correspond to the same code group.

0 7 7 1 3 – 0 0 0 1

Indeed, given the four 0s in this code, it appears that no code group corresponds to the dash and that ||ııı corresponds to zero. This means that the last two code groups both correspond to 1s so there must be additional information in the code. Let's look again at both codes, this time leaving out the dash. We obtain these pairings.

0 7 7 1 3 0 0 0 1 1

2 0 0 7 7 9 9 6 4 6

Note that each of 0, 1, and 7 appear more than once in these codes and that each time they appear, they have the same representation in bar code! This is very strong evidence that we are correct, that the bar code represents the zip code, and that the code representations are as shown here. Of course, the last 1 and 6 in the above diagrams were determined by the earlier appearance of the same 5-bar patterns in each zip code. Finally, it is not clear from the illustration what code group should correspond to each of 5 and 8, though it might be guessed with careful study of the following patterns. Otherwise, these can be determined by looking at other pieces of mail whose zip codes contain these digits.

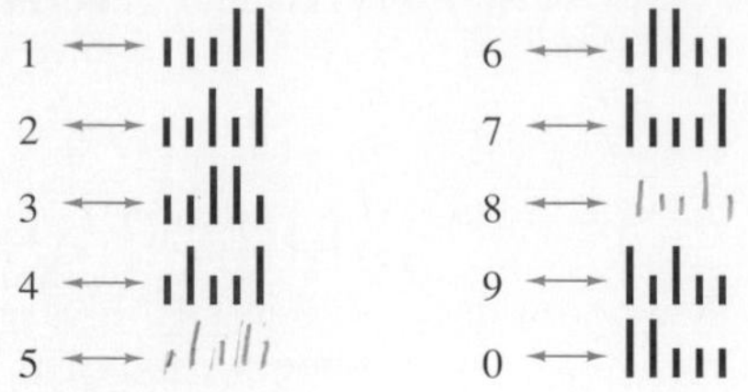

Look back • • • •

We first guessed that the bar code must indeed be a code and that it likely gave the zip code in machine readable form. We then noticed that both codes had 52 bars beginning and ending with a long bar. Ignoring the first and last bars, we had a bar code with 50 bars. Our guess was that each five consecutive bars represented a digit which, in turn, could be determined by looking at the printed zip code.

This gave us 10 digits and a question remains regarding the tenth digit. Trying a variety of possibilities, we find that

$$0 + 7 + 7 + 1 + 3 + 0 + 0 + 0 + 1 + 1 = 20$$

and

$$2 + 0 + 0 + 7 + 7 + 9 + 9 + 6 + 4 + 6 = 50.$$

It appears that the tenth digit in each case, here 1 and 6 respectively, has been chosen so that the sum of all the digits is a multiple of 10; that is, 10 in 10-hour clock arithmetic. This suggests that the tenth digit is a *check digit* and that it is included so that incorrect codes can be detected by machine. In fact, this is the case, and, if an error is detected in this way, the piece of mail is automatically shunted aside for human inspection and correction of the zip code. ■

An Application to Secret Codes

One of the first methods of encoding was devised by Julius Caesar for communicating with his generals in the Gallic wars during the first century B.C. Caesar's idea was simply to replace each letter in the alphabet by the letter three places to its right with the stipulation that x be replaced by a, y by b, and z by c. Thus the plaintext

REINFORCEMENTS COMING

would be sent in code (or ciphertext) as

ULHQIRUFHPHQWV FRPLQJ.

The original text is then easily recovered by replacing each letter in the ciphertext by the letter three places to its left in the alphabet with a being replaced by x, b by y, and c by z.

This method is easily computerized as follows:

1. Number the letters from 1 to 26 as shown.

A	B	C	D	E	F	G	H	I	J	K	L	M	N	O	P	Q	R	S	T	U	V	W	X	Y	Z
1	2	3	4	5	6	7	8	9	10	11	12	13	14	15	16	17	18	19	20	21	22	23	24	25	26

2. Let C denote the numerical equivalent of a ciphertext or codetext letter and let P denote the numerical equivalent of a plaintext letter. Then, using 26-hour clock arithmetic,

$$C = P +_{26} 3$$

is used for enciphering and

$$P = C -_{26} 3 = C +_{26} 23$$

is used for deciphering a message. The above message would first be converted to the numerical plaintext,

18 5 9 14 6 15 18 3 5 13 5 14 20 19
3 15 13 9 14 7.

Then, using $C = P +_{26} 3$, it would be transformed into the numerical ciphertext

21 8 12 17 9 18 21 6 8 16 8 17 23 22
6 18 16 12 17 10.

To disguise word length, the message would actually be transmitted in blocks of length five as

21 8 12 17 9 18 21 6 8 16 8 17 23 22 6
18 16 12 17 10.

Using $P = C +_{26} 23$, the recipient of the message would transform it back to the numerical plaintext,

18 5 9 14 6 15 18 3 5 13 5 14 20 19 3
15 13 9 14 7.

and then to the alphabetic text,

REINF ORCEM ENTSC OMING.

Lastly, this is rewritten from recognition of the words as

REINFORCEMENTS COMING.

A generalized Caesar cipher would employ the transformations

$$C = P +_{26} k \quad \text{and} \quad P = C +_{26} (26 - k) = C -_{26} k$$

where k is called the **shift constant.**

The difficulty with Caesar ciphers is that they are easily broken by cryptanalysts. Since the substitution of letters is one for one, the letters that appear most often in the ciphertext are the letters that appear most frequently in the plaintext. For written English, studies have shown that E appears with a frequency of 13%, T with a frequency of 9%, and so on, to I, N, and R at 8%, and so on. Thus, the most frequently appearing letter in a coded message is most likely the substitute for E and this permits the determination of k and hence the reading of the message.

EXAMPLE 4.32 **Breaking a Caesar Cipher**

Assuming that it was enciphered with a Caesar transformation with constant k, decipher the message

BPMKW LMJZM ISMZA

SOLUTION

A frequency count of the letters in the message reveals that M appears most often. Thus, M is likely the ciphertext equivalent of E. In numbers, 13 is likely the numerical ciphertext equivalent of 5. Since

$$C = P +_{26} k,$$

this means that it is quite likely that

$$13 = 5 +_{26} k,$$

so $k = 8$. But then

$$P = C -_{26} 8 = C +_{26} (26 - 8) = C +_{26} 18.$$

Converting the ciphertext into numerical form, the message becomes

2 16 13 11 23 12 13 10 26 13 9 19 13 26 1

and, using $P = C +_{26} 18$, this gives the numerical plaintext

20 8 5 3 15 4 5 2 18 5 1 11 5 18 19.

Finally, converting to alphabetic form, we have

THECO DEBRE AKERS

which we quickly regroup into THE CODEBREAKERS. This, by the way, is the name of a huge but interesting tome on cryptography by David Kahn (New York: Macmillan Pub. Co., 1967). ■

Of course, in any given message, E need not be the most frequently occurring letter. Thus, in breaking a Caesar cipher or any other substitution cipher, we would try the next most frequently occurring letters in order until the message made sense.

PROBLEM SET 4.4

Understanding Concepts

1. Compute these 12-hour clock sums.
   (a) $5 +_{12} 9$ **(b)** $12 +_{12} 7$ (c) $8 +_{12} 4$
   (d) $4 +_{12} 7$ (e) $10 +_{12} 10$ **(f)** $7 +_{12} 8$
2. Compute these n-hour clock sums, where $+_n$ denotes n-hour clock addition; that is, $5 +_7 3$ indicates the 7-hour clock addition of 5 and 3.
   (a) $3 +_5 4$ **(b)** $17 +_{26} 13$ (c) $2 +_{10} 7$
   (d) $7 +_9 4$ (e) $12 +_{16} 10$ **(f)** $2 +_7 7$
3. In 12-hour clock arithmetic, $a -_{12} b$ can be computed by starting at a and counting b steps *counterclockwise* around the clock. Use this method to compute each difference.
   (a) $9 -_{12} 7$ **(b)** $8 -_{12} 11$ (c) $5 -_{12} 9$
   (d) $8 -_{12} 12$ (e) $2 -_{12} 11$ **(f)** $8 -_{12} 8$
4. Recall that 12 is the additive identity in 12-hour clock arithmetic and that b is the additive inverse of a if, and only if, $a +_{12} b = b +_{12} a = 12$. Compute the additive inverses of each of these numbers in 12-hour clock arithmetic.
   (a) 7 **(b)** 11 (c) 9 **(d)** 12

5. Compute each of these differences by adding. Be sure to show what you are adding each time.
   (a) $9 -_{12} 7$ (b) $8 -_{12} 11$ (c) $5 -_{12} 9$
   (d) $8 -_{12} 12$ (e) $2 -_{12} 11$ (f) $8 -_{12} 8$
6. Compute these products in 12-hour clock arithmetic.
   (a) $5 \times_{12} 7$ (b) $9 \times_{12} 11$ (c) $12 \times_{12} 5$
   (d) $8 \times_{12} 6$ (e) $4 \times_{12} 6$ (f) $4 \times_{12} 9$
7. Perform these divisions if they are defined. (*Suggestion:* See Table 4.1.)
   (a) $5 \div_{12} 7$ (b) $7 \div_{12} 10$ (c) $8 \div_{12} 4$
   (d) $8 \div_{12} 5$ (e) $9 \div_{12} 5$ (f) $6 \div_{12} 11$
8. Two numbers are said to be **relatively prime** if their greatest common divisor is 1.
   (a) List the numbers in $T = \{1, 2, \ldots, 12\}$ that are relatively prime to 12.
   (b) List the numbers in $T = \{1, 2, \ldots, 12\}$ that are not relatively prime to 12.
   (c) Compare the results of parts (a) and (b) with the results of Example 4.29. What conjecture does this comparison suggest?
9. Construct complete addition and multiplication tables for 5-hour clock arithmetic.
10. Perform these computations in 5-hour clock arithmetic.
    (a) $3 +_5 4$ (b) $2 +_5 5$ (c) $4 +_5 4$
    (d) $3 \times_5 4$ (e) $2 \times_5 5$ (f) $4 \times_5 4$
    (g) $3 -_5 4$ (h) $2 -_5 5$ (i) $4 -_5 4$
    (j) $3 \div_5 4$ (k) $2 \div_5 5$ (l) $4 \div_5 4$
11. (a) What is the additive identity in 5-hour clock arithmetic? Why?
    (b) What is the multiplicative identity in 5-hour clock arithmetic? Why?
12. If $a +_5 b = b +_5 a = 5$, then **b is the additive inverse of a** and **a is the additive inverse of b** in 5-hour clock arithmetic; that is, $a = 5 -_5 b$ and $b = 5 -_5 a$.
    (a) Compute the additive inverse of each of 1, 2, 3, 4, and 5 in 5-hour clock arithmetic.

    As in 12-hour clock arithmetic, we can subtract b from a in 5-hour clock arithmetic by *adding* the additive inverse of b to a. Perform each of these subtractions in two ways—(i) by counting backwards on a 5-hour clock and (ii) by adding the additive inverse of the number being subtracted.
    (b) $2 -_5 4$ (c) $3 -_5 2$ (d) $1 -_5 5$
13. If $a \times_5 b = b \times_5 a = 1$, then **$a$ is called the multiplicative inverse of b** and **b is called the multiplicative inverse of a** in 5-hour clock arithmetic; that is, we write $a = b^{-1}$ and $b = a^{-1}$.
    (a) Which numbers in 5-hour clock arithmetic possess multiplicative inverses? (*Hint:* Check the results of problem 9 above.)
    (b) Which numbers in 12-hour clock arithmetic possess multiplicative inverses? (*Hint:* Check Table 4.1.)
14. Just as we can subtract in clock arithmetic by adding the additive inverse, we can divide by multiplying by the multiplicative inverse. Use the definition of clock division to compute each of the following.
    (a) $4 \div_{12} 7$ (b) $3 \div_{12} 11$ (c) $3 \div_5 2$
    (d) $2 \div_5 2$ (e) $2 \div_5 4$ (f) $4 \div_5 3$

    Compute each of the quantities in parts (g) through (l) by multiplying by the multiplicative inverse; that is, compute each of these products.
    (g) $4 \times_{12} 7^{-1}$ (h) $3 \times_{12} 11^{-1}$ (i) $3 \times_5 2^{-1}$
    (j) $2 \times_5 2^{-1}$ (k) $2 \times_5 4^{-1}$ (l) $4 \times_5 3^{-1}$
15. Using Table 4.1 solve the following equations in 12-hour clock arithmetic. The first one is done for you.
    (a) $(3 \times_{12} y) +_{12} 7 = 4$
    (b) $(7 \times_{12} y) -_{12} 4 = 8$
    (c) $(y +_{12} 2) \div_{12} 11 = 3$
    (d) $(2 \div_{12} y) -_{12} 4 = 3$

    Solution:
    $3 \times_{12} y = 4 +_{12} 5$
    $3 \times_{12} y = 9$
    $y = 3, 7, \text{ or } 11$
16. Write the bar code for each of these zip codes.
    (a) 99164–3113 (b) 18374–2147
    (c) 38423–1747

---

17. Write the zip code given by each of these bar codes unless the code is necessarily incorrect. If it is necessarily incorrect, tell why.
    (a)
    (b)
    (c)

18. Encipher the message

FOUR SCORE AND SEVEN YEARS AGO

using a Caesar cipher with shift constant $k = 20$.

19. Assuming that it was enciphered with a Caesar cipher with shift constant k, determine k and decipher the message

YKIAH APQON AWOKJ PKCAP DANMP.

Thinking Critically

20. Powers in *n*-hour Clock Arithmetic. Since $a^s = a \cdot a \cdots a$ with s factors of a, we can compute a^s in n-hour clock arithmetic in the usual way. That is, $a^s = r$ in n-hour clock arithmetic where $r = n$ if $n \mid a^s$ and r is the remainder when a^s is divided by n if $n \nmid a^s$. Moreover, it follows that the usual rules

$$a^s a^t = a^{s+t} \qquad \text{and} \qquad (a^s)^t = a^{st}$$

hold in clock arithmetic just as they do in ordinary arithmetic. Compute these in 12-hour clock arithmetic. (*Note:* To compute 3^5, for example, using your calculator, enter the string 3 [y^x] 5 [=]. Then reenter the answer and use [INT÷] to determine the result in 12-hour clock arithmetic.)

(a) 3^3 **(b)** 3^4 (c) $3^3 \cdot 3^4$
(d) 3^{3+4} (e) 4^2 **(f)** $(4^2)^5$
(g) $4^{2 \cdot 5}$ **(h)** 4^{10}

21. Computing Large Powers in Clock Arithmetic. The preceding example shows in principle how to compute powers in clock arithmetic. However, if the exponent is quite large, the answer will exceed the capacity of your calculator and an error message will result. An efficient method for computing such powers is sometimes called the **Russian peasant method for finding powers in clock arithmetic.** Suppose you want to compute 3^{54} in 12-hour clock arithmetic. Since

$$54 = 110110_{\text{two}} = 2^5 + 2^4 + 2^2 + 2^1 = 32 + 16 + 4 + 2,$$
$$3^{54} = 3^{2+4+16+32} = 3^2 \cdot 3^4 \cdot 3^{16} \cdot 3^{32}.$$

Moreover, the factors in this last product can be selected from the following list of successive squares

$$3, 3^2, 3^4 = (3^2)^2, 3^8 = (3^4)^2, 3^{16} = (3^8)^2, 3^{32} = (3^{16})^2$$

which, when computed in 12-hour clock arithmetic will *not* exceed the capacity of your calculator. Since the work of converting 54_{ten} to base two can be accomplished by successively dividing by 2 and noting the remainders as we have already seen, the work for computing 3^{54} in 12-hour clock arithmetic might reasonably be arranged as follows:

12-hour clock arithmetic

0 R 1 $3^{32} = 9$
2)1 R 1 $3^{16} = 9$
2)3 ~~R 0 $3^8 = 9$~~
2)6 R 1 $3^4 = 9$
2)13 R 1 $3^2 = 9$
2)27 ~~R 0 3 = 3~~
2)54

$3^2 \times_{12} 3^4 = 9$

$3^2 \times_{12} 3^4 \times_{12} 3^{16} = 9$

$3^2 \times_{12} 3^4 \times_{12} 3^{16} \times_{12} 3^{32} = 9$

$9^2 = 81 = 6 \cdot 12 + 9$ so $9^2 = 9$ in 12-hour clock arithmetic.

Therefore, $3^{54} = 9$ in 12-hour clock arithmetic. This means that $3^{54} = 12q + 9$ for some whole number q. Also $3^{54} - 9 = 12q$ and so $12 \mid (3^{54} - 9)$. Use this method to compute each of the following.

(a) 5^{21} in 12-hour clock arithmetic **(b)** 2^{151} in 31-hour clock arithmetic

22. Mersenne Primes. The largest prime known as of this writing is an enormous giant whose decimal expansion contains 258,716 digits! This number is of the form $2^n - 1$ and is called a Mersenne number after the French monk, Father Marin Mersenne (1588–1648), who first studied such numbers. Show that $2^{37} - 1$ is *not* a prime using 223-hour clock arithmetic; that is, show that $223 \mid 2^{37} - 1$. (*Hint:* Use the Russian peasant method for computing 2^{37} in 223-hour clock arithmetic.)

23. **Fermat Numbers.** The great French amateur mathematician, Pierre de Fermat (1601–1665), noted that

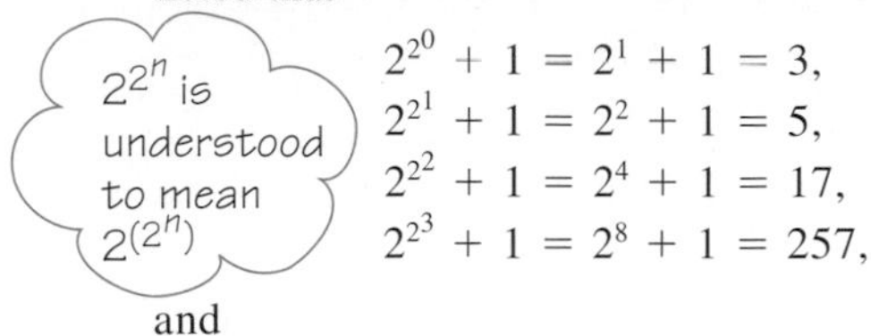

$$2^{2^0} + 1 = 2^1 + 1 = 3,$$
$$2^{2^1} + 1 = 2^2 + 1 = 5,$$
$$2^{2^2} + 1 = 2^4 + 1 = 17,$$
$$2^{2^3} + 1 = 2^8 + 1 = 257,$$

and

$$2^{2^4} + 1 = 2^{16} + 1 = 65{,}537$$

were all primes and so conjectured that all numbers of the form $2^{2^n} + 1$ were primes. Show that this is *not* generally true by computing $2^{2^5} + 1 = 2^{32} + 1$ in 641-hour arithmetic. (Indeed, no Fermat primes have been found other than those noted by Fermat and it is now generally believed that no others exist. The smallest Fermat number of unknown primality status is $2^{2^{22}} + 1$.)

Thinking Cooperatively

24. Fill in the following table for a number of choices of a and b in $T = \{1, 2, 3, \ldots, 12\}$ and endeavor to discover when $a \times_{12} y = b$ is solvable and how many solutions this equation has. (*Hint:* Use Table 4.1.)

a	b	GCD $(a, 12)$	b/GCD $(a, 12)$	12/GCD $(a, 12)$	No. of solutions of $a \times_{12} y = b$

25. In n-hour clock arithmetic, a number s is said to be a **divisor of zero** if $s \neq n$ and there is a number t in the arithmetic such that $t \neq n$ yet $s \times_n t = n$.

(a) List all the divisors of zero in 12-hour clock arithmetic.

(b) List the numbers in 12-hour clock arithmetic for which division is not defined.

(c) Compare (a) and (b) and make a conjecture.

(d) List all the divisors of zero in 5-hour clock arithmetic. (*Hint:* See problem 9 above.)

26. In ordinary arithmetic there are no divisors of zero; that is, $ab = 0$ if, and only if, at least one of a or b is zero. This is useful in solving equations since, for example $(y - 2)(y - 3) = 0$ if, and only if, $y = 2$ or $y = 3$. Find all solutions to $(y -_{12} 2) \times_{12} (y -_{12} 3) = 12$. As noted earlier, 12 plays the role in 12-hour clock arithmetic that 0 does in ordinary arithmetic.

Making Connections

27. Since about 1972 all books published anywhere in the world have been given an identifying number called an International Standard Book Number (ISBN). These numbers greatly facilitate buying and selling books, inventory control, and so on. A typical ISBN number is

0–13–257502–7

and, somewhat like the zip code, the last digit is a check digit. It works like this. There are 10 digits in the code and the check digit is chosen so that 10 times the first digit plus 9 times the second digit, plus 8 times the third digit, . . . , plus 1 times the check digit is 11 in 11-hour clock arithmetic. Thus, for the above, $(10 \times_{11} 0) +_{11} (9 \times_{11} 1) +_{11} (8 \times_{11} 3) +_{11} (7 \times_{11} 2) +_{11} (6 \times_{11} 5) +_{11} (5 \times_{11} 7) +_{11} (4 \times_{11} 5) +_{11} (3 \times_{11} 0) +_{11} (2 \times_{11} 2) +_{11} (1 \times_{11} 7) = 11$ since the result in ordinary arithmetic is 143 and $11 \mid 143$. The check digit may be an X, denoting a ten.

(a) Which of these ISBN numbers is correct?

(i) 0–70–808228–7

(ii) 0–201–30722–7

(b) Supply the check digit to complete each of these correct ISBN numbers.

(i) 5–648–00738–

(ii) 3–540–11200–

28. Supply the correct check digit for each of these zip codes.

(a) 24763–8117–

(b) 35992–1712–

Using a Computer

29. Use CAESAR with shift constant $k = 7$ to encrypt the first phrase of the Gettysburg address.

(a) Give the numerical form of the encrypted message.

(b) Give the alphabetic form of the encrypted message.

30. Use CAESAR to determine k and decrypt this message, given that it was encrypted with a CAESAR cipher with shift constant k.

YDHUS UDJJY CUJXU KIUEV SQBSK
BQJEH IQDTS ECFKJ UHIXQ IRUSE
CUIEF UHLQI YLUJX QJULU DJXUU
BUCUD JQHOI SXEEB CQJXU CQJYS
ISKHH YSKBK CCKIJ SXQDW UJEHU
VBUSJ JXYIH UQBYJ OQOCO

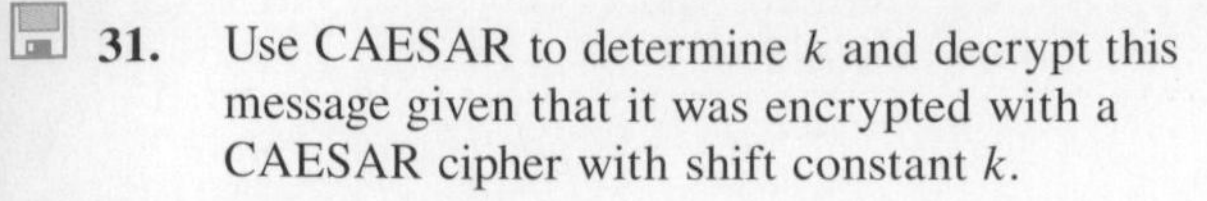

31. Use CAESAR to determine k and decrypt this message given that it was encrypted with a CAESAR cipher with shift constant k.

FGESL ZWESL AUASF USFTW SUGEH
DWLWE SLZWE SLAUA SFOZG AKFGL
SDKGK GEWLZ AFYGX SHGWL

For Review

32. Draw rectangular diagrams to illustrate all the factorings of 12 taking order into account; that is, think of $1 \cdot 12$ as different from $12 \cdot 1$.

33. Use the method of intersection of sets to determine the following.
(a) GCD(60, 150) **(b)** LCM(60, 150)

34. (a) Use a factor tree to determine all the prime divisors of 540.
(b) Determine the number of different divisors of 540.
(c) How can you tell at a glance if 540 is or is not divisible by 9?

35. (a) Write the prime power representations of 540 and 600.
(b) Use part (a) to determine GCD(540, 600).
(c) Use part (a) to determine LCM(540, 600).

36. Does $2^3 \cdot 3^5 \cdot 7^2 \cdot 11^6$ evenly divide $2^4 \cdot 3^7 \cdot 5^2 \cdot 7^1 \cdot 11^8$? Explain briefly.

37. To determine if 427 is a prime, which primes must you check as possible divisors?

EPILOGUE Number Theory, the Queen of Mathematics

Almost 200 years ago Carl Gauss wrote, "Mathematics is the queen of the sciences, but number theory is the queen of mathematics." What Gauss was really saying is that our modern technology, and work in science generally, depends so heavily on mathematics that progress in these areas would be essentially impossible without mathematical skill. This assertion is increasingly true in the social sciences as well. Beyond that, Gauss was saying that mathematics is also intrinsically interesting and intellectually satisfying and that this is particularly true of number theory.

In this chapter we have introduced the basic number theoretic notions of divisibility, factoring, factors and multiples, primes and composite numbers, least common multiples and greatest common divisors, and related ideas from clock arithmetic. These ideas are not only useful in other parts of mathematics and in disciplines like computer science but they also provide interesting and stimulating motivational material for the elementary mathematics classroom. In particular, number theory is replete with interesting and challenging problems which provide additional opportunities to further develop problem-solving skills. It is increasingly the case that number theoretic notions appear in elementary school texts and that these ideas must be understood by teachers.

CLASSIC CONUNDRUM The Chinese Remainder Problem

Find the least positive integer having remainders of 2, 3, and 4 when divided by 5, 7, and 9 respectively. What is the next larger integer having this property? Problems like these were studied by Chinese mathematicians like Sun-Tse as early as the first century A.D.

CHAPTER 4 SUMMARY

Key Concepts

The main objectives of this chapter have been to introduce the fundamental ideas of number theory and clock arithmetic most likely to appear in modern elementary school texts. The central ideas are these.

- divisibility
- composite numbers
- multiples
- least common multiples
- the fundamental theorem of arithmetic
- primes
- factors or divisors
- greatest common divisors
- clock arithmetic
- the Euclidean algorithm

Vocabulary and Notation

Hands On

Factor
Prime
Composite

Section 4.1

Divides
Divisor, proper divisor
Factor
Multiple
$a \mid b, a \nmid b$
Prime number
Composite number
Unit
Factor tree
Fundamental theorem of arithmetic
Prime power representation of n
Number of divisors of a number

Section 4.2

Divisibility by 2 and 5
Divisibility by 10
Divisibility by products
Divisibility by 3 and 9
Divisibility by 11
Divisibility by 7, 11, and 13
Divisibility by 4, 8, and other powers of 2

Section 4.3

Tiling rectangles by squares and common divisors
Greatest common divisor
$\text{GCD}(a, b)$
Finding GCDs by tiling
Finding GCDs by cutting squares off rectangles
Finding GCDs by intersection of sets
Finding GCDs from prime power representations
Finding the GCD by the Euclidean algorithm
Least common multiple
Finding the LCM by tiling squares by rectangles
$\text{GCD}(a, b) \cdot \text{LCM}(a, b) = ab$
Finding least common multiples by intersection of sets
Finding LCMs from prime power representations
Finding LCMs using the Euclidean algorithm

Section 4.4

12-hour clock arithmetic
12-hour clock addition
12-hour clock multiplication
12-hour clock subtraction
Additive inverse
12-hour clock subtraction by 12-hour clock addition
12-hour clock division
n**-hour clock arithmetic**
10-hour clock arithmetic
Zip codes, bar codes, and 10-hour clock arithmetic
Caesar ciphers and 26-hour clock arithmetic

CHAPTER REVIEW EXERCISES

Section 4.1

1. Draw rectangular diagrams to illustrate the factorings of 15 taking order into account; that is, think of $1 \cdot 15$ as different from $15 \cdot 1$.
2. Construct a factor tree for 96.
3. (a) Determine the set D_{60} of all divisors of 60.
   (b) Determine the set D_{72} of all divisors of 72.
   (c) Use $D_{60} \cap D_{72}$ to determine GCD(60, 72).
4. (a) Determine the prime power representation of 1200.
   (b) Determine the prime power representation of 2940.
   (c) Use parts (a) and (b) to determine GCD(1200, 2940) and LCM(1200, 2940).
5. Use information from the sieve of Eratosthenes in Figure 4.2 to determine if 847 is prime or composite.
6. (a) Determine a composite natural number n with a prime factor greater than $\sqrt{n}$.
   (b) Does the n in part (a) have a prime divisor less than or equal to $\sqrt{n}$? If so what is it?
7. Determine natural numbers r, s, and m such that $r \mid m$ and $s \mid m$, but $rs \nmid m$.
8. Use the number $n = 3 \cdot 5 \cdot 7 + 11 \cdot 13 \cdot 17$ to determine a prime different from 3, 5, 7, 11, 13, or 17.

Section 4.2

9. Using mental methods, test each number for divisibility by 2, 3, 5, and 11.
   (a) 9310 (b) 2079
   (c) 5635 (d) 5665
10. Test each number for divisibility by 7, 11, and 13.
   (a) 10,197 (b) 9373 (c) 36,751
11. Use the results of problem 9 to decide which of these are *true.*
   (a) $15 \mid 9310$ (b) $33 \mid 2079$
   (c) $55 \mid 5635$ (d) $55 \mid 5665$
12. Let $m = 3^4 \cdot 7^2$.
   (a) How many divisors does m have?
   (b) List all the divisors of m.
13. Determine d so that $2{,}765{,}301{,}2d3$ is divisible by 11.
14. (a) Determine the least natural number divisible by both q and m if $q = 2^3 \cdot 3^5 \cdot 7^2 \cdot 11^1$ and $m = 2^1 \cdot 7^3 \cdot 11^3 \cdot 13^1$.
   (b) Determine the largest number less than the q of part (a) that divides q.

Section 4.3

15. (a) Use the method of cutting off squares to determine the largest square that will tile a 56 by 84 rectangle.
   (b) What is the greatest common divisor of 56 and 84?
16. (a) Determine the smallest square that can be tiled by a 56 by 84 rectangle.
   (b) What is the least common multiple of 56 and 84?
17. (a) Find the greatest common divisor of 63 and 91 by the method of intersection of sets of divisors.
   (b) Find the least common multiple of 63 and 91 by the method of intersection of sets of multiples.
   (c) Demonstrate that $\text{GCD}(63, 91) \cdot \text{LCM}(63, 91) = 63 \cdot 91$.
18. If $r = 2^1 \cdot 3^2 \cdot 5^1 \cdot 11^3$, $s = 2^2 \cdot 5^2 \cdot 11^2$, and $t = 2^3 \cdot 3^1 \cdot 7^1 \cdot 11^3$, determine each of the following.
   (a) GCD(r, s, t) (b) LCM(r, s, t)
19. Determine each of the following using the Euclidean algorithm.
   (a) GCD(119,790, 12,100)
   (b) LCM (119,790, 12,100)
20. Seventeen year locusts and thirteen year locusts both emerged in 1971. When will these insect's descendants next emerge in the same year?

Section 4.4

21. Perform the indicated clock calculations if they are defined.
   (a) $4 +_{12} 9$ (b) $9 -_{12} 4$ (c) $4 \times_{12} 9$
   (d) $4 \div_{12} 9$ (e) $9 +_{12} 8$ (f) $9 \times_{12} 12$
   (g) $4 \div_{12} 12$ (h) $9 \div_{12} 7$ (i) $9 \times_{12} 7$
22. Perform these clock calculations.
   (a) $5 +_7 6$ (b) $6 -_7 5$
   (c) $6 \times_7 5$ (d) $6 \div_7 5$
23. List the numbers in 10-hour clock arithmetic for which 10-hour clock division is *not* defined.
24. Determine the check digit for each of these zip codes.
   (a) 87243–1772 (b) 22001–8941

25. Write out the zip code named by each of these bar codes. Which one, if either, is incorrect?

(a)

(b)

26. Use the Russian peasant method to compute each of these in 23-hour clock arithmetic.

(a) 3^{45} (b) 21^{200} (c) 5^{181}

27. Use a Caesar cipher with shift constant 7 to encode this message.

I SURRENDER

28. Assuming the message below was encoded using a Caesar cipher with shift constant k, determine k and decipher the message.

DRVIO OJWZA MZZST

CHAPTER TEST

1. Indicate whether each of these is *always true* (T) or *not always true* (F).
   (a) If $a \mid c$ and $b \mid c$, then $ab \mid c$.
   (b) If $r \mid s$ and $s \mid t$, then $r \mid t$.
   (c) If $a \mid b$ and $a \mid c$, then $a \mid (b + c)$.
   (d) If $a \nmid b$ and $a \nmid c$, then $a \nmid (b + c)$.
2. (a) Make a factor tree for 8532.
   (b) Write the prime power representation of 8532.
   (c) Name the largest natural number smaller than 8532 that divides 8532.
   (d) Name the smallest number larger than 8532 that is divisible by 8532.
3. Using mental methods, test each of these for divisibility by 2, 3, 9, 11, and 33.
   (a) 62,418 (b) 222,789
4. Use the Euclidean algorithm to determine each of the following.
   (a) GCD(13,534, 997,476)
   (b) LCM(13,534, 997,476)
5. Let $m = 2^3 \cdot 5^2 \cdot 7^1 \cdot 11^4$ and $n = 2^2 \cdot 7^2 \cdot 11^3$.
   (a) Does $r \mid m$ if $r = 2^2 \cdot 5^1 \cdot 7^2 \cdot 11^3$? Why or why not?
   (b) How many divisors does m have?
   (c) Determine GCD(m, n).
   (d) Determine LCM(m, n).
6. Draw diagrams of Cuisenaire® rods to illustrate all the divisors of 21.
7. Perform the indicated clock calculations.
   (a) $7 +_8 5$ (b) $7 +_{12} 5$ (c) $5 -_7 7$
   (d) $7 \times_8 5$ (e) $7 \div_8 5$
   (f) 7^5 (in 8-hour clock arithmetic)
8. Use a Caesar cipher with shift constant 5 to encipher VICTORY.
9. Assuming that it was enciphered using a Caesar cipher with shift constant k, determine k and decipher

   DAZZG ADIZS

5 Integers

Black–Red Game

Materials Needed

1. Twenty-five cards for each pair of students—two of each type shown plus three with 2 red discs and two with 3 black discs.

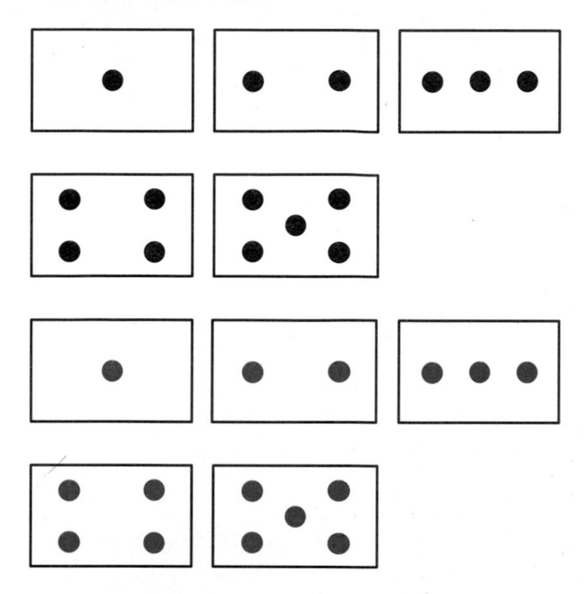

2. Five record sheets for each pair of students as shown here.

Record Sheet

Player 1					Player 2			
# R	# B	Score	Cumulative Score		# R	# B	Score	Cumulative Score
				Winners Name				

Directions

This is a 2-person game played as follows:

1. The cards are shuffled and then laid out facedown in a 5 by 5 array.
2. A player plays by picking up two cards and recording the score using the rule that each red disc cancels a black disc and vice versa. Thus, if a player turns up a red 4 and a black 5, the score is 1B. If, on the other hand, a red 4 and a black 2 are turned up, the score is 2R. The player records his or her score and also the cumulative score obtained by combining the score on a play with the cumulative score on the preceding play. (On the first play for each player the cumulative score is the same as the score.) After a player's score is recorded, the play shifts to the other player.
3. To start the play, each player turns over a card and the player with the greatest black score, or, if neither player has a black score, the player with the least red score, goes first.
4. Play continues until the record sheet is full. The player with the greatest cumulative black score, or, if neither has a cumulative black score, the player with the least cumulative red score, is the winner. The privilege of going first alternates between the two players from game to game. Also, the cards are shuffled and redistributed after each game.

CONNECTIONS

A Further Extension of the Number System

In Chapter 2 we discussed the whole numbers and operations with whole numbers. In particular, we observed that the whole numbers and their operations were developed as a direct result of people's need to count. But a modern society has many quantitative needs aside from counting and these often require numbers other than whole numbers.

In this chapter, we consider the set I of integers which consists of:

- the **natural numbers** or **positive integers** denoted by 1, 2, 3, . . . ,
- the number **zero,** 0, and
- the **negative integers** denoted by -1, -2, -3, ...

We note immediately that

$$N \subset W \subset I.$$

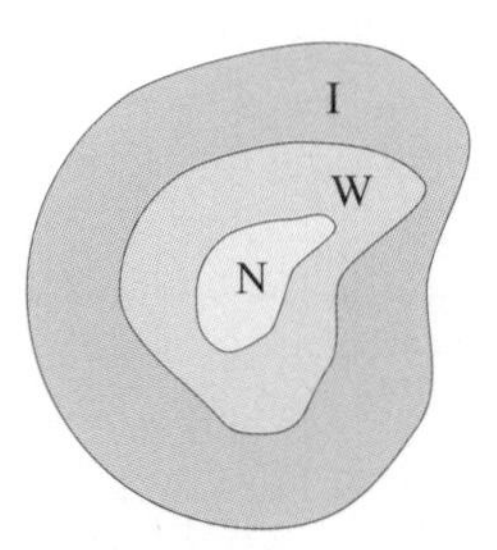

In today's society, these numbers are used to record debits and credits, profits and losses, changes in prices in the stock market, degrees above zero and degrees below zero in measuring temperature, yards lost and yards gained in football, points won or points "in the hole" in many card and board games, and so on.

As usual, we introduce the set of integers by means of manipulatives, pictures, and diagrams with steadily increasing levels of abstraction to suggest, in turn, how these ideas might be presented to elementary school children.

5.1 Representations of Integers

Business people of all kinds regularly use the phrases "in the black" and "in the red" to indicate whether a given business has experienced a profit or a loss or to indicate whether a bank account still has money available or if it has been overdrawn. Since accountants often note these states of affairs with black and red ink, we adopt that same convention here.

Representing Integers by "Drops" of Colored Counters

Mark Garza gives each student in his class a collection of about 25 counters colored black on one side and red on the other. These are easily made by duplicating the desired shapes on red construction paper and then having the students cut them out and color one side of each counter black.

Mr. Garza has each student drop several counters on his or her desk top, match the black and red counters, and record as the score for the drop the number of unmatched black or red counters. This could be viewed as recording the results of playing a game against two opponents and winning points (black counters) from one opponent and losing points (red counters) to the other opponent. The score is then the net gain or loss on a given play, and this can be used to represent positive or negative integers. For example, Figure 5.1 shows a drop of 6 black and 4 red counters for a score of 2B. We interpret this as representation of *positive two* and write 2. Similarly, we interpret a score of 2R as *negative two* and write -2.

Figure 5.1
A drop of 6 black and 4 red counters. The score is 2B and represents positive two.

It is important to observe that different drops can result in the same score. For example, the two diagrams in Figure 5.2 both represent -2. Moreover, -2 could also be represented by a diagram with only 2 red counters. A drop with counters of only one color is the simplest of all such representations and we will have occasion to use such drops in what follows. Mr. Garza discusses this at length with his students and asks particularly how to increase or decrease the number of counters in a drop without changing the score. A moment's reflection makes it clear that adding or taking away the same number of each of black and red counters does not change the score of a drop. All this is summarized in this theorem.

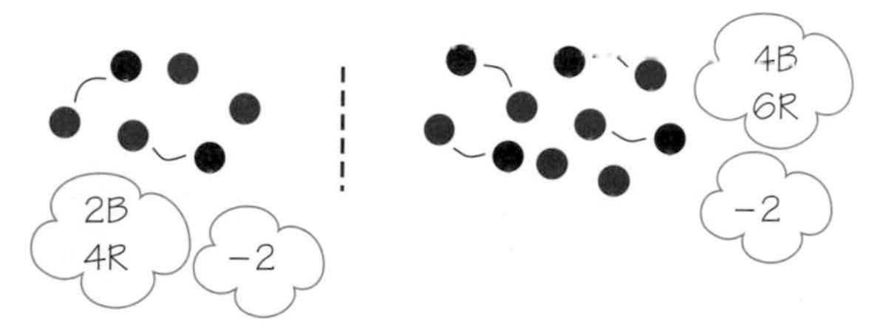

Figure 5.2
Two drops representing -2

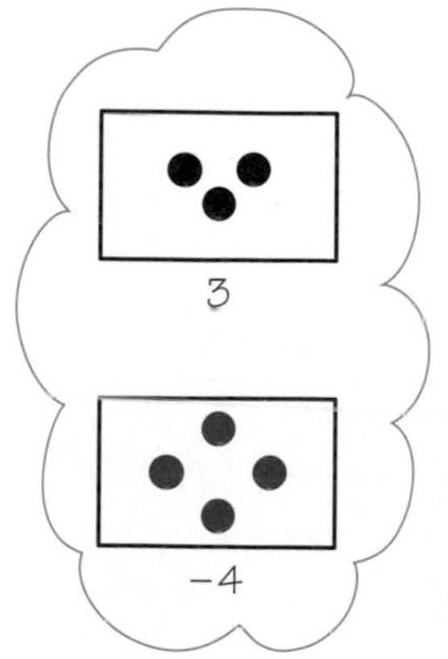

THEOREM Equivalence of Representation of Integers by Drops of Colored Counters

Two drops of colored counters are **equivalent** and represent the same integer if one can be obtained from the other by adding (or deleting) the same number of each of black and red counters. The simplest representation of the positive integer n is a drop of n black counters. The simplest representation of the negative integer $-n$ is a drop of n red counters.

A drop consisting of an equal number of black and red counters is of special interest. Since no counters remain unmatched, such a drop represents 0 which is therefore *neither positive nor negative.* Moreover, as we have just seen, adding such a drop to an existing drop does not change its score and this corresponds numerically to the fact that

$$0 + n = n + 0 = n$$

for *any* integer n—positive, negative, or zero. Finally, any such drop can be viewed as a combination of two drops—one of r black counters and one of r red counters for some natural number r. Separately these drops would score rB and rR and represent r and $-r$ respectively. Since together they score 0, we have the following definition.

DEFINITION The Integers

The **positive integers** are the natural numbers.

The **negative integers** are the numbers $-1, -2, -3, \ldots$, where $-r$ is defined by the equality

$$r + (-r) = (-r) + r = 0.$$

The integer 0 is neither positive nor negative and has the property

$$0 + n = n + 0 = n$$

for every integer n. The **integers** consist of the positive integers, the negative integers, and zero.

A Comment on Notation

Observe that it is standard to use the same sign in writing $8 - 6$ and -5. The expression $8 - 6$ is often read "8 minus 6" and the sign, $-$, is often called the *minus* sign. However, there is a duplicity in the usage that must be clearly understood. In $8 - 6$, the sign is used to indicate the *operation of subtraction,* and, in writing -5, it is used to indicate the **negative** or **additive inverse** of 5. This distinction is made on your calculator with two separate keys, one for subtraction and one for negation. To compute $8 - 6$ on the calculator one keys in the sequence

8 [−] 6 [=]

and to enter -5 in the display one keys in the sequence

[5] [+⇄−].

Indeed, keying in the sequence

5 [+⇄−] [+⇄−] [+⇄−] [+⇄−]

shows 5, -5, 5, -5, and 5 in the calculator display since the [+⇄−] key changes the sign of the number in the display each time it is pressed. This suggests that the negative

FROM THE NCTM STANDARDS **Number Systems**

In grades 5–8, the mathematics curriculum should include the study of number systems and number theory so that students can—

- *understand and appreciate the need for numbers beyond the whole numbers;*
- *develop and use order relations for whole numbers, fractions, decimals, integers, and rational numbers;*
- *extend their understanding of whole number operations to fractions, decimals, integers, and rational numbers;*
- *understand how the basic arithmetic operations are related to one another;*
- *develop and apply number theory concepts (e.g, primes, factors, and multiples) in real-world and mathematical problem situations.*

SOURCE: From *Curriculum and Evaluation Standards for School Mathematics Grades 5–8*, p. 91.

of negative 5 is 5 (that $-(-5) = 5$), and we will see later that this is so. Indeed, as we will show, $-(-n) = n$ for every integer n.

As a final note on this double use of the minus sign, we observe that many texts for elementary school try to avoid the difficulty by using a raised minus sign to indicate the negative of a number. Thus, they write $^{-}5$ in place of -5. And some even write $^{+}2$ in place of 2. However, these texts invariably ultimately change to the standard notation of -5 and 2 used here. We feel that it causes less confusion to do this at the outset, stressing that the context makes it clear when "subtract" is meant as opposed to "the negative of."

EXAMPLE 5.1

Scoring Drops of Colored Counters

Determine the score and the integer represented by each of these drops.

(a) 

(b) 

(c)

(d) 

SOLUTION

(a) Matching the two red counters with two of the black counters, we see that the score is 3B representing 3.

(b) Matching the red counter with a black counter, we see that the score is 2B representing 2.

(c) Matching the 3 red with the 3 black counters, we see that the score is 0; we have no red or black counters unmatched.

(d) Since there are no black counters, the score is 3R and the integer represented is -3. ■

EXAMPLE 5.2

Drops for Given Integers

Illustrate two different drops for each of these integers.

(a) 2 **(b)** -3 **(c)** 0 **(d)** 5

SOLUTION

(a) Here the drops must have 2 more black counters than red counters. Two possibilities are

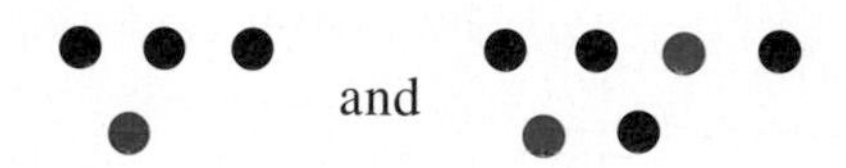

(b) Here the drop must have 3 less black counters than red counters. Possibilities include

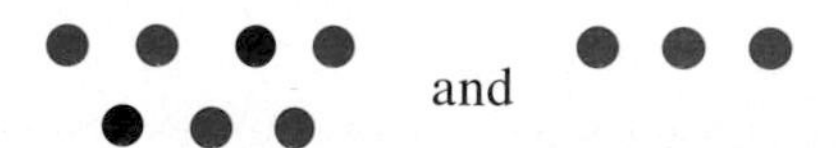

(c) Here the drop must contain the same number of red and black counters or no counters at all. Possibilities include

and .

(d) Here the drop must have 5 more black than red counters. Possibilities include

and . ■

Mail-time Representations of Integers

Integers can also be represented in other real-life situations. The following examples illustrate how the mail delivery of a check or a bill affects your overall net worth, or the value of your assets at any given time.

At mail time suppose that you are delivered a check for \$20. What happens to your net worth? Answer: It goes up by \$20.

At mail time you are delivered a bill for \$35. What happens to your net worth? Answer: It goes down by \$35.

At mail time, you received a check for \$10 and a bill for \$10. What happens to your net worth? Answer: It stays the same.

EXAMPLE 5.3

Interpreting Mail-time Situations

(a) At mail time you are delivered a check for \$27. What happens to your net worth?

(b) At mail time you are delivered a bill for \$36. Are you richer or poorer? By how much?

SOLUTION

(a) Your net worth goes up \$27.

(b) Poorer. Your net worth goes down by \$36. ■

EXAMPLE 5.4

Describing Mail-time Situations for Given Integers

Describe a mail-time situation corresponding to each of these integers.

(a) -42 **(b)** 75 **(c)** 0

SOLUTION

(a) At mail time the letter carrier brought you a bill for \$42. Are you richer or poorer and by how much?

(b) At mail time you were delivered a check for \$75. What happens to your net worth?

(c) Quite to your surprise, at mail time the mail carrier skipped your house so you received no checks and no bills. Are you richer or poorer and by how much? ■

Number Line Representations of Integers

We have already used a number line to illustrate whole numbers, and it can be used equally effectively to represent integers. Choose an arbitrary point on the number

HIGHLIGHT FROM HISTORY
Edward A. Bouchet (1852–1918)

Edward Bouchet's list of firsts is impressive; the first black man to attend and graduate from Yale University; the first elected to Phi Beta Kappa, the oldest collegiate honor society; the first, in 1876, to earn a doctoral degree. His Ph.D. was earned for an application of mathematics to physics, "Measuring Refractive Indices." Instead of resting on his scholarly laurels, Dr. Bouchet moved to Philadelphia to spend 26 years teaching physics, chemistry, and mathematics at the Institute for Colored Youth. This quaker school, now known as Cheyney State College, was founded by another eminent black mathematician, Charles Reason. Bouchet's example has been followed by many black scholars, such as Dr. Elbert Cox, the first to earn a doctorate in pure mathematics (1925), who devoted his efforts to encouraging graduate work by students at Howard University. As a young student in his hometown, New Haven, Connecticut, as high school valedictorian, and as teacher, high school principal, and scholar, Bouchet's accomplishments and warm personality were a source of inspiration to countless young men and women, black and white.

SOURCE: From *Mathematics in Modules, Algebra, A4,* Teacher's Edition. Reprinted by permission.

line for 0. Then successively measure out unit distances on each side of 0 and label successive points on the right of 0 with successive positive integers and points on the left with successive negative integers as shown in Figure 5.3.

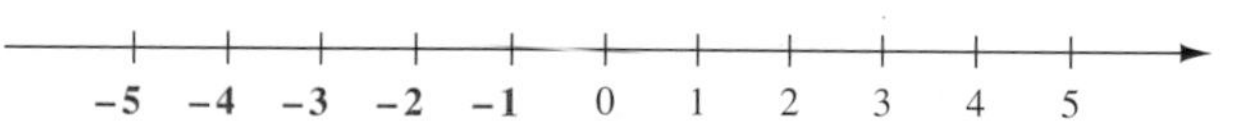

Figure 5.3
Representing integers on a number line

This corresponds nicely to marking thermometers with degrees above zero and degrees below zero, and with the practice in most parts of the world (with the notable exceptions of North America and Russia) of numbering floors above ground and below ground in a skyscraper. It also corresponds to the countdown of the seconds to lift-off and beyond in a space shuttle launch. The count 9, 8, 7, 6, 5, 4, 3, 2, 1, lift-off!, 1, 2, 3, . . . is not really counting backward and then forward but forward all the time. The count is actually

9 seconds before lift-off	-9
8 seconds before lift-off	-8
.	.
.	.
.	.
1 second before lift-off	-1
Lift-off!	0
1 second after lift-off	1
2 seconds after lift-off	2

and so on. This is very real to space-age children and helps to make positive and negative numbers real and understandable.

In this connection it is also often helpful to represent positive and negative integers by curved arrows. For example, an arrow from any point to a point 5 units to the right represents 5, and an arrow from any point to a point 5 units to the left represents -5 as illustrated in Figure 5.4.

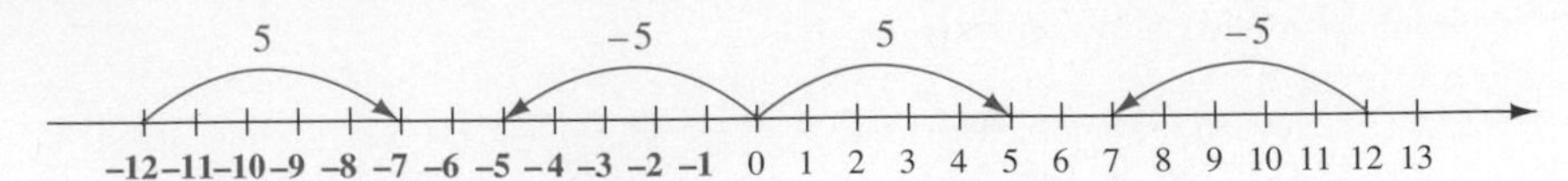

Figure 5.4
Using arrows to represent integers

Thus, the arrow for -5 has the same length as the arrow for 5 but is directed in the opposite direction and vice versa. This is indicated in some elementary school texts by writing **opp 5** to indicate **-5.** Since opp means that you simply change the direction of the arrow, it follows that opp (opp n) $= n$ for any integer n. In our notation, this would be written $-(-n) = n$ for any integer n.

THEOREM The Negative of the Negative of an Integer

For any integer n, $-(-n) = n$.

Absolute Value of an Integer

While discussing number line representations it is convenient to introduce the notion of absolute value of an integer. We have just observed that the integer 5 can be illustrated by the point numbered 5 on the number line. Similarly, -5 is represented by the point numbered -5. On the other hand both 5 and -5 are five units from 0 on the number line as shown in Figure 5.5. The **absolute value** of an integer n is defined to be the distance of the corresponding point on the number line from 0. We indicate the absolute value of n by writing $|n|$.

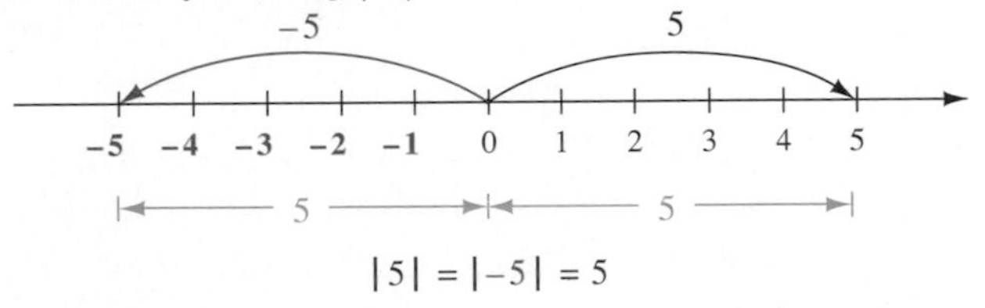

Figure 5.5
The absolute value of both 5 and -5 is 5.

EXAMPLE 5.5 **Finding Absolute Values**

Find the absolute values of these integers.

(a) -11 **(b)** 13 **(c)** 0 **(d)** -9

SOLUTION

We plot the numbers on the number line and determine the distance of the points from 0.

(a) $|-11| = 11 = -(-11)$ since -11 is 11 units from 0.
(b) $|13| = 13$ since 13 is 13 units from 0.
(c) $|0| = 0$ since 0 is 0 units from 0.
(d) $|-9| = 9 = -(-9)$ since -9 is 9 units from 0.

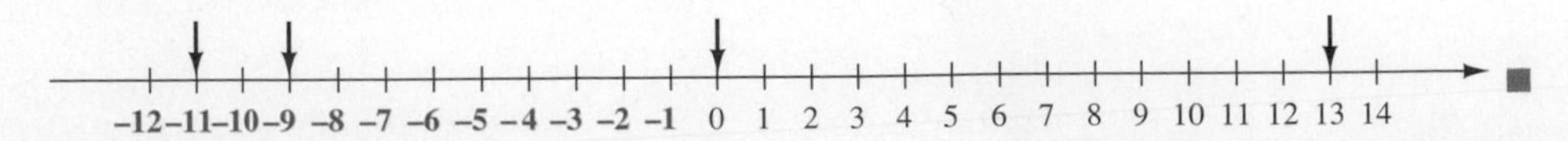

■

SCHOOL BOOK PAGE *Integers on a Number Line*

Enrichment

Numbers Greater Than and Less Than Zero

Most thermometers show temperatures both above and below 0°. Numbers above zero are greater than 0°. Numbers below zero are less than 0° and are written with a negative sign. A temperature of 5 degrees below zero is written as −5°.

By counting degrees on the thermometer, you can find the new temperature after it rises or falls.

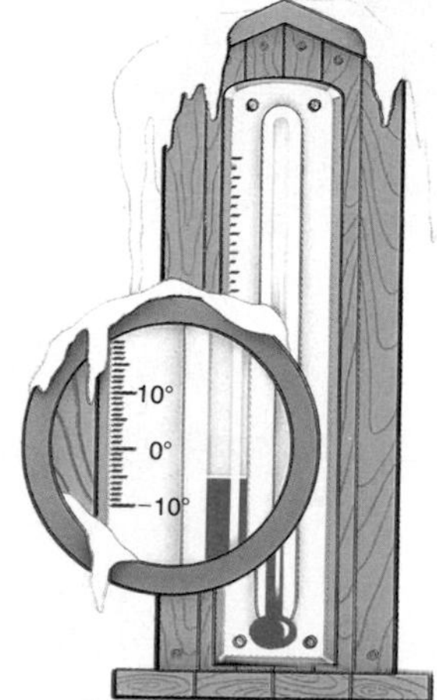

Temperature at 7:00 A.M. was −5°. It rose 11° during the day. Count up 11 degrees from −5°. The new temperature is 6°.

Temperature at 7:00 P.M. was 3°. It fell 7° during the night. Count down 7 degrees from 3°. What is the new temperature?

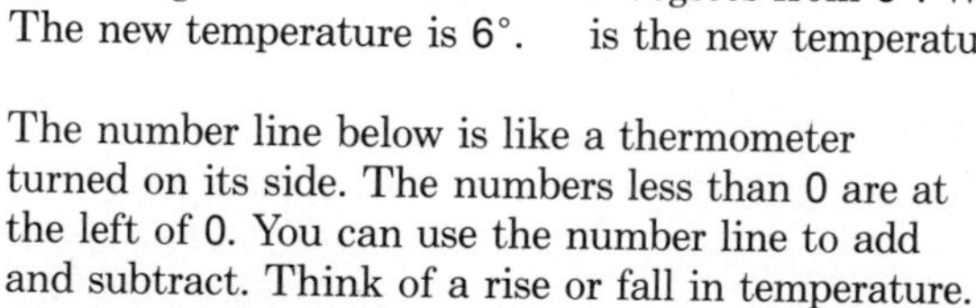

The number line below is like a thermometer turned on its side. The numbers less than 0 are at the left of 0. You can use the number line to add and subtract. Think of a rise or fall in temperature.

Find −7 + 3.

Add 3 to −7.
Move 3 units to the right from −7.
$-7 + 3 = -4$

Find 6 − 9.

Subtract 9 from 6.
Move 9 units to the left from 6.
$6 - 9 = -3$

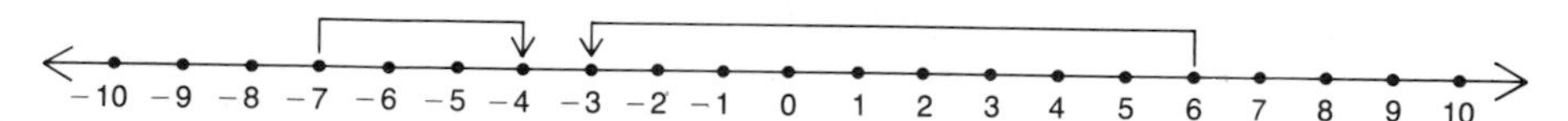

The temperature at 6:00 A.M. was −2°. Find the new temperature if it

1. rises 1°. **2.** falls 3°. **3.** rises 6°. **4.** falls 8°.

Use the number line to add or subtract.

5. $-6 + 2$ **6.** $-3 + 4$ **7.** $-4 + 4$ **8.** $-8 + 0$ **9.** $-1 + 7$

10. $4 - 7$ **11.** $1 - 5$ **12.** $0 - 2$ **13.** $-1 - 5$ **14.** $-6 - 4$

1. The lesson shown uses the thermometer to introduce negative numbers and the number line. Do you think this is a good idea? Explain.
2. In some children's games you can either gain or lose points during play. Would discussing such games be a good way to introduce negative numbers and adding or subtracting these numbers? Explain.
3. The subtraction $6 - 9$ is illustrated using the number line above. How would you illustrate $6 - (-9)$ on the number line? Explain.

Careful examination of the results of Example 5.5 suggests an alternative definition that makes it possible to avoid drawing diagrams to find absolute values.

$|-2| = -(-2)$
$= 2$

DEFINITION Absolute Value of an Integer

If a is an integer, then

$$|a| = \begin{cases} a \text{ if } a \text{ is positive or zero} \\ -a \text{ if } a \text{ is negative.} \end{cases}$$

EXAMPLE 5.6 **Finding Absolute Values**

Determine the absolute values of these integers.

(a) -71 **(b)** 29 **(c)** 0 **(d)** -852

SOLUTION

(a) Since -71 is negative, $|-71| = -(-71) = 71$.
(b) Since 29 is positive, $|29| = 29$.
(c) Since 0 is 0, $|0| = 0$.
(d) Since -852 is negative, $|-852| = -(-852) = 852$. ■

PROBLEM SET 5.1

Understanding Concepts

1. Draw two colored counter diagrams to represent each of these scores or integers.
   (a) 5B **(b)** 2R (c) 0 **(d)** -3 **(e)** 3
2. Draw a colored counter diagram with the least number of counters to represent each of the following.
   (a) 3 **(b)** -4 (c) 0 **(d)** 2
3. (a) Draw a colored counter diagram to represent 5.
   (b) Draw what you would see if you turned over all the colored counters in your diagram for part (a). What integer would this new diagram represent?
   (c) We could quite reasonably call the diagram of part (b) the opposite of the diagram of part (a). Thus, we might reasonably call -5 the opposite of 5 and write opp 5. As noted earlier, some elementary texts actually use this terminology. What would these texts write in place of -17?
   (d) Using the opp idea with colored counters, write an argument that $-(-5) = 5$.
   (e) Use the colored counter model to argue convincingly that $-(-n) = n$ for any integer n.
4. (a) Describe a mail-time situation that illustrates 14.
   (b) Describe a mail-time situation that illustrates -27.
5. At mail time you are delivered a check for \$48 and a bill for \$31.
   (a) Are you richer or poorer and by how much?
   (b) What integer does this situation illustrate?
6. (a) At mail time you are delivered a check for \$27 and a bill for \$42. What integer does this situation illustrate?
   (b) Describe a different mail-time situation that illustrates the same integer as in part (a).
7. Draw a number line and plot the points representing these integers.
   (a) 0 **(b)** 4 (c) -4 **(d)** 8 (e) $(4 + 8)/2$
   (f) Where is $(4 + 8)/2$ relative to 4 and 8?

8. What integers are represented by the curved arrow on each of these number line diagrams?

(a)

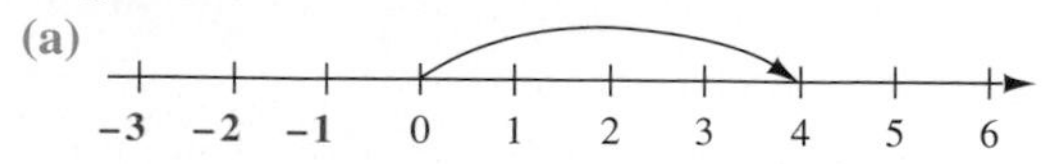

(b)

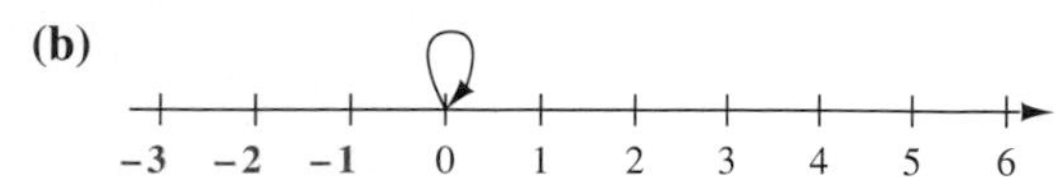

(c)

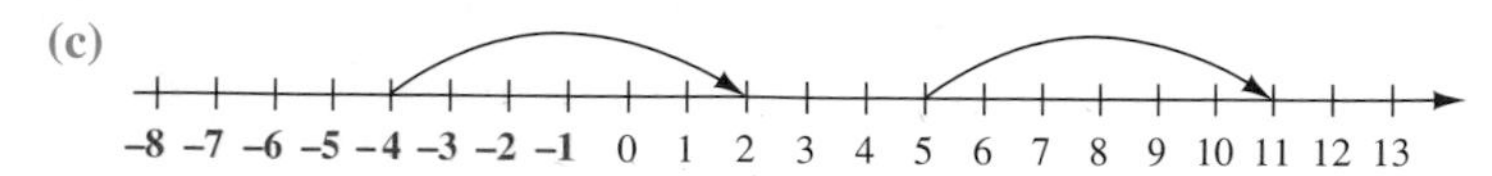

(d)

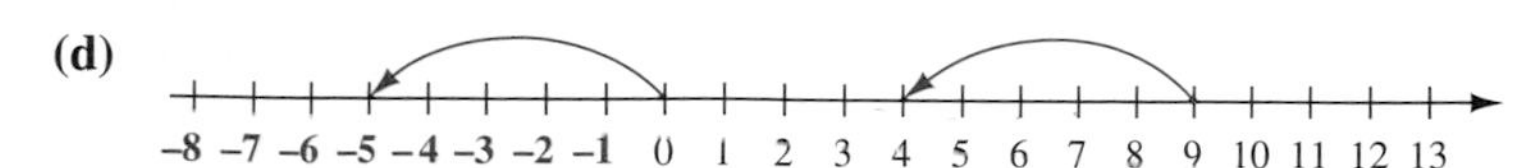

9. Draw number line diagrams to represent each of these integers.

(a) 7 **(b)** 0 (c) -9 **(d)** 9

10. In Europe, the floor of a building at ground level is called the ground floor. What in America is called the second floor is called the first floor in Europe and so on.

(a) If an elevator in a tall building in Paris, France, starts on the fifth basement level below ground, B5, and goes up 27 floors, on which numbered floor does it stop?

(b) What would the answer to part (a) be if the building were located in New York?

11. If an elevator starts on basement level B3 and goes down to B6, how far down has it gone?

12. Find the absolute values of these quantities.

(a) 34 **(b)** $4 - 4$ (c) -76 **(d)** $5 - 17$

(e) How far is it between 5 and 17 on the number line?

13. For what values of x are these equations true?

(a) $|x| = 13$ **(b)** $|x| + 1 = 2$ **(c)** $|x| + 5 = 0$

Thinking Critically

14. (a) What colored counters would have to be added to this array in order to represent -3?

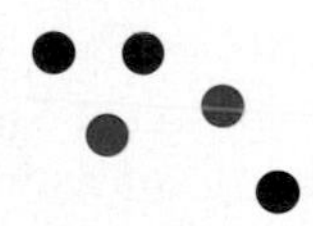

(b) Could the question in part (a) be answered in more than one way?

(c) How many different representations of -3 can be made with 20 or fewer counters?

15. If all the counters are used each time, list all the integers that can be represented using

(a) 12 counters. **(b)** 11 counters.

16. **(a)** If all the counters are used each time, describe the set of integers that can be represented using n counters.

(b) How many different integers are representable using n counters as in part (a)?

17. If some or all of the counters are used each time, describe the set of integers that can be represented

(a) using 12 counters. **(b)** using 11 counters.

(c) using n counters.

Thinking Cooperatively

18. (a) How many different appearing rows of counters can you make using all of 20 counters each time? Remember that each counter has a black side and a red side. Two possibilities with only 4 counters are shown here.

(b) How many different appearing rows of counters can you make with at least 1 and at most 20 counters?

(c) How many different appearing rows of counters can you make with n counters (all used each time) if precisely 2 of the counters show red and the others show black? (*Suggestion:* Consider the special cases $n = 2$, 3, 4, and 5.)

19. (a) How many black and how many red counters are there in a triangular array like this but with 20 rows?

You should not actually need to make a diagram with 20 rows in order to answer this question.

(b) Write a brief explanation of your solution to part (a).

(c) Repeat part (a) but with a triangular array with 21 rows.

(d) What integers are represented by the triangular arrays in parts (a) and (c)?

(e) Make a table of integers represented by triangular arrays like those in parts (a) and (c) but with n rows for $n = 1, 2, 3, 4, 5, 6, 7,$ and 8.

(f) Carefully considering the table of part (e), conjecture what integer is represented in a triangular array like those in parts (a) and (c) but with n rows where n is any natural number. (*Suggestion:* Consider n odd and n even separately.)

For Review

20. How many different factors does each of these numbers have?

(a) $2310 = 2 \cdot 3 \cdot 5 \cdot 7 \cdot 11$

(b) $5{,}336{,}100 = 2^2 \cdot 3^2 \cdot 5^2 \cdot 7^2 \cdot 11^2$

21. If $a = 2^1 \cdot 3^3 \cdot 5^2$ and $b = 2^3 \cdot 3^2 \cdot 5^1 \cdot 7^1$, compute each of the following.

(a) GCD(a, b) **(b)** LCM(a, b)
(c) GCD$(a, b) \cdot$ LCM(a, b) **(d)** $a \cdot b$

22. (a) Does $c \mid a$ if a is as in problem 21 and $c = 2^1 \cdot 3^4 \cdot 5^1$? Why or why not?

(b) Is d a multiple of b if b is as in problem 21 and $d = 2^4 \cdot 3^2 \cdot 5^2 \cdot 7^1$? Why or why not?

23. Determine the prime power representation of each of these numbers.

(a) 1400 **(b)** 5445 **(c)** 4554

24. Determine each of the following using the Euclidean algorithm.

(a) GCD(4554, 5445) **(b)** LCM(4554, 5445)

5.2 Addition and Subtraction of Integers

Addition of Integers

In the preceding section we introduced the integers using devices like colored counters, mail-time stories, and number lines. In this section we consider adding and subtracting integers and it is helpful to use these same devices to illustrate how these operations should be performed in this enlarged number system.

Addition of Integers Using Sets of Colored Counters

We discussed integers in the preceding section in terms of sets of colored counters. Also, in Chapter 2, addition of whole numbers was defined in terms of sets. If $a = n(A)$, $b = n(B)$, and $A \cap B = \emptyset$, then $a + b$ was defined as $n(A \cup B)$.

This idea works equally well for integers using **sets of colored counters.** Notice that when working with actual sets of counters, the counters are necessarily different so that any two distinct sets A and B clearly satisfy the condition $A \cap B = \emptyset$. In what follows, we presume that the diagrams show actual physical sets of counters so that the condition $A \cap B = \emptyset$ is automatically satisfied. Moreover, we avoid using set notation by drawing a loop around two sets we wish to combine into a single set. Suppose, for example, that we wish to illustrate the addition of 8 and −3. We draw the diagram shown in Figure 5.6 and interpret this as a drop consisting of all the counters in the combined set.

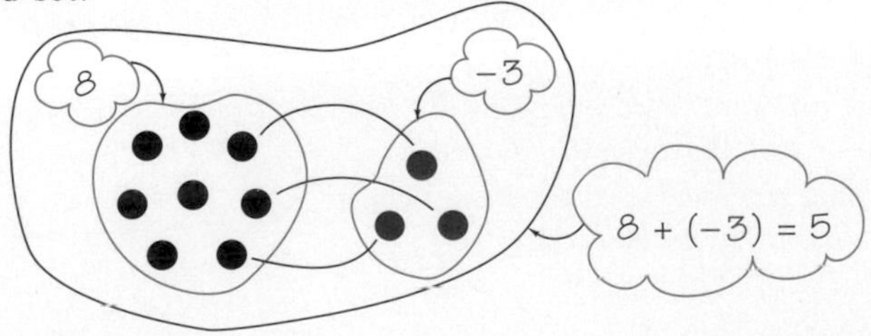

Figure 5.6
Diagram of colored counters illustrating 8 + (−3) = 5

The score of this drop, 5B, illustrates that the desired sum is 5 as shown. Of course, in Figure 5.6 we used the simplest possible representation of 8 and -3. However, as shown in Figure 5.7, the result is the same if we use other, equivalent, representations for 8 and -3. In both Figures 5.6 and 5.7 the total number of black counters exceeds the total number of red counters by 5. Thus, in each case, the score of the combined set is 5B representing 5.

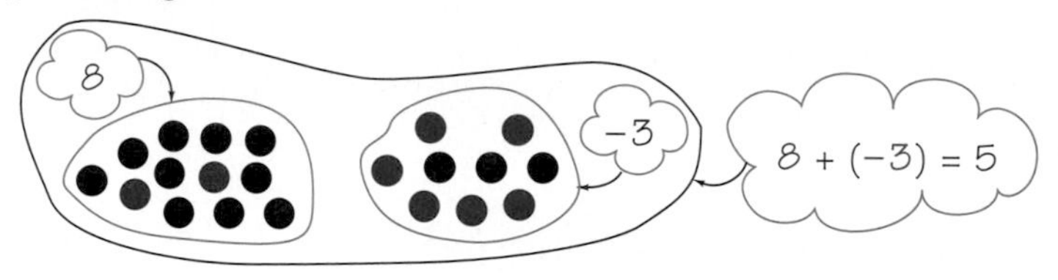

Figure 5.7
Another representation of $8 + (-3) = 5$ using sets of colored counters

EXAMPLE 5.7

Representing Sums of Integers Using Colored Counter Diagrams

Draw appropriate diagrams of colored counters to illustrate each of these sums.

(a) $(-3) + 5$ **(b)** $(-2) + (-4)$ **(c)** $5 + (-7)$ **(d)** $4 + (-4)$

SOLUTION

(a) Using the simplest representations of -3 and 5, we draw this diagram.

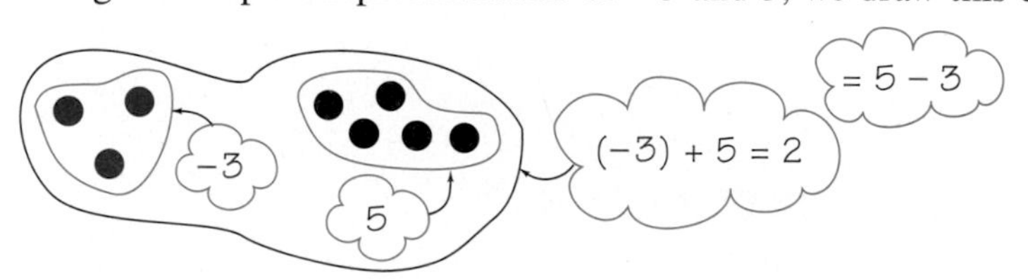

Since the combined set has a score of 2B, this represents 2. Thus, $(-3) + 5 = 2$ and we note that $2 = 5 - 3$ as well. Hence, $(-3) + 5 = 5 - 3 = 2$.

(b) This sum can be represented as shown.

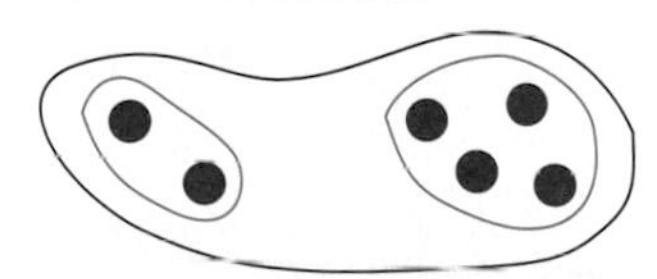

Since the combined set has a score of 6R, this illustrates the sum

$$(-2) + (-4) = -6,$$

and we note that $-6 = -(2 + 4)$. Thus,

$$(-2) + (-4) = -(2 + 4) = -6.$$

(c) $5 + (-7) = -2 = -(7 - 5)$

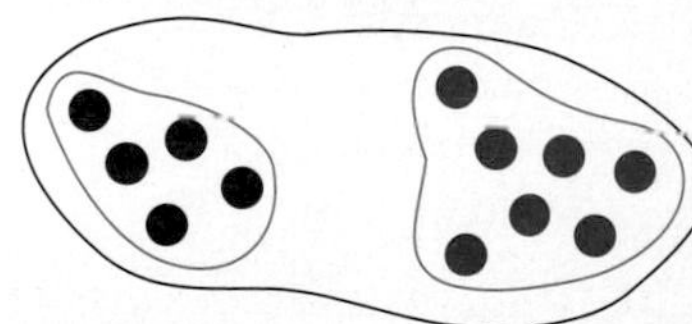

(d) $4 + (-4) = 0 = 4 - 4$

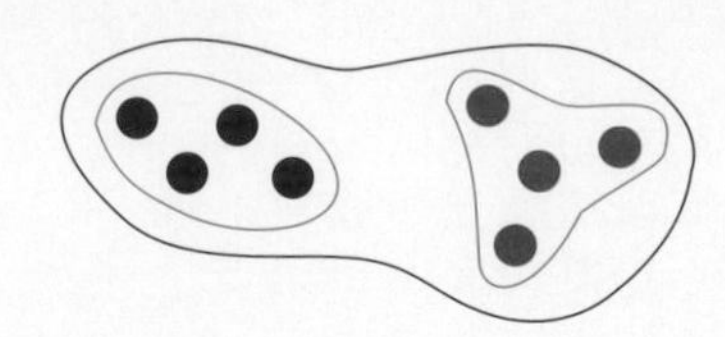

■

The results of Example 5.7 are entirely typical and we state them here as a theorem.

THEOREM Adding Integers

Let m and n be positive integers so that $-m$ and $-n$ are negative. Then the following are true:

- $(-m) + (-n) = -(m + n)$
- If $m > n$, then $m + (-n) = m - n$.
- If $m < n$, then $m + (-n) = -(n - m)$.
- $n + (-n) = (-n) + n = 0$

EXAMPLE 5.8

Adding Integers

Compute these sums.

(a) $7 + 11$ **(b)** $(-6) + (-5)$ **(c)** $7 + (-3)$
(d) $4 + (-9)$ **(e)** $6 + (-6)$ **(f)** $(-8) + 3$

SOLUTION

(a) $7 + 11 = 18$
(b) $(-6) + (-5) = -(6 + 5) = -11$
(c) Since $7 > 3$, $7 + (-3) = 7 - 3 = 4$.
(d) Since $4 < 9$, $4 + (-9) = -(9 - 4) = -5$
(e) $6 + (-6) = 0$
(f) $(-8) + 3 = 3 + (-8)$ and since $8 > 3$,
$3 + (-8) = -(8 - 3) = -5$. Therefore, $(-8) + 3 = -5$. ■

As with the properties of whole number addition, except for the existence of additive inverses, the properties of integer addition derive from the corresponding properties of sets.

THEOREM Properties of the Addition of Integers

Let m, n, and r be integers. Then the following hold.

Closure Property	$m + n$ is an integer
Commutative Property	$m + n = n + m$
Associative Property	$m + (n + r) = (m + n) + r$
Additive identity	$0 + m = m + 0 = m$
Additive inverse	$(-m) + m = m + (-m) = 0$

HIGHLIGHT FROM HISTORY
Samuel Morse (1791–1872)

As a young scholar at Yale, Samuel Morse learned what little was known of electricity in the early nineteenth century. There had been experiments (like those of Benjamin Franklin) and some theories had been formed. The idea that electricity could be used to send messages sparked his imagination. Although Morse made his living as a portrait painter (his subjects included President James Monroe and the author James Fenimore Cooper), he is known today as the inventor of the electromagnetic telegraph and the Morse code—a system of dots and dashes that is one of the earliest examples of a binary code.

SOURCE: From *Mathematics in Modules, Basic Math Skills, MS I* Teacher's Edition. Reprinted by permission.

PROOF Since we have defined integers in terms of unions of sets of colored counters, the first four properties in the theorem follow from the fact that, for any sets M, N, and R,

$$M \cup N \text{ is a set,}$$
$$M \cup N = N \cup M,$$
$$M \cup (N \cup R) = (M \cup N) \cup R, \text{ and}$$
$$\emptyset \cup M = M \cup \emptyset = M.$$

The existence of the additive inverse of m for every integer m follows as a generalization of part (d) of Example 5.7. ■

Bringing something *to* you is adding.

Addition of Integers Using Mail-time Stories

A second useful approach to addition of integers is by means of **mail-time stories.**

At mail time suppose you receive a check for \$13 and another check for \$6. Are you richer or poorer and by how much? Answer: Richer by \$19. This illustrates $13 + 6 = 19$.

EXAMPLE 5.9 **Adding Integers Using Mail-time Stories**

Write the addition equation illustrated by each of these stories.

(a) At mail time you receive a check for \$3 and a check for \$5. Are you richer or poorer and by how much?

(b) At mail time you receive a bill for \$2 and another bill for \$4. Are you richer or poorer and by how much?

(c) At mail time you receive a check for \$5 and a bill for \$7. Are you richer or poorer and by how much?

(d) At mail time, you receive a check for \$4 and a bill for \$4. Are you richer or poorer and by how much?

SOLUTION

(a) Receiving a check for \$3 and a check for \$5 makes you \$8 richer. This illustrates $3 + 5 = 8$.

(b) Receiving a bill for \$2 and another bill for \$4 makes you \$6 poorer. This illustrates $(-2) + (-4) = -6$.

(c) Receiving a check for \$5 makes you richer by \$5, but receiving a bill for \$7 makes you \$7 poorer. The net effect is that you are \$2 poorer. This illustrates $5 + (-7) = -2$.

(d) Receiving a \$4 check and a \$4 bill exactly balances out and you are neither richer nor poorer. This illustrates $4 + (-4) = 0$. ■

Note that these results are exactly the results of Example 5.7 above. Moreover, the arguments hold in general and we are again led to the theorem immediately preceding Example 5.8.

Addition of Integers Using a Number Line

Suppose we want to illustrate $5 + 4$ on a number line. The addition can be thought of as starting at 0 and counting five units to the right (in the positive direction on the number line) and then counting on 4 more units to the right. Figure 5.8 shows that this is the same as counting 9 units to the right from 0 straight away. Thus, $5 + 4 = 9$.

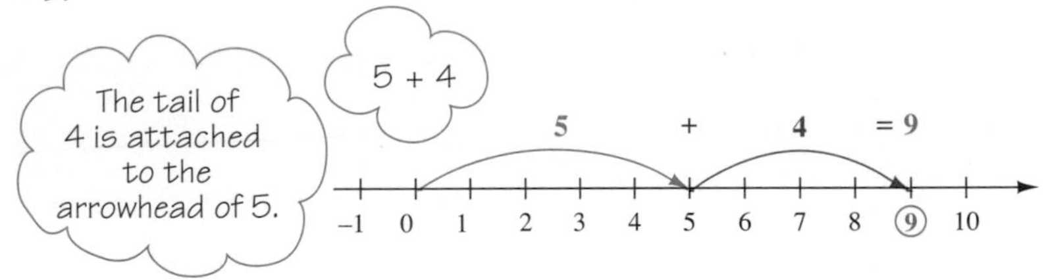

Figure 5.8
Illustrating $5 + 4 = 9$ on a number line

If 4 is depicted by counting four units to the right, then -4 should be depicted by counting four units to the left. Thus, the addition $5 + (-4)$ is depicted on the number line as in Figure 5.9.

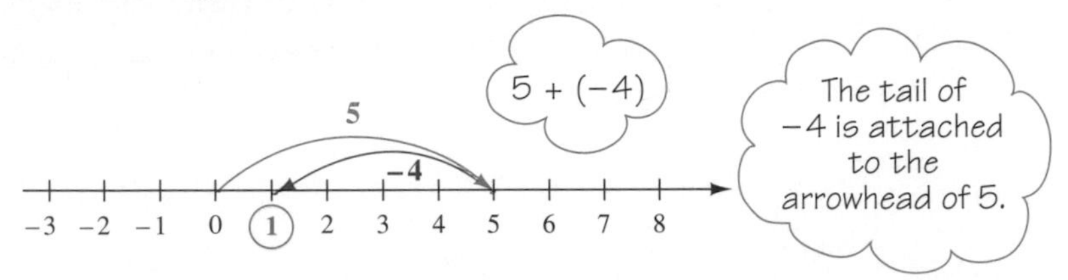

Figure 5.9
Illustrating $5 + (-4) = 1$ on a number line

We count five units to the right from 0 and then count "on" 4 units to the left in the direction of the -4 arrow. As seen on the diagram this justifies $5 + (-4) = 1$.

EXAMPLE 5.10 **Adding Integers on a Number Line**

What addition fact is illustrated by each of these diagrams?

(a)

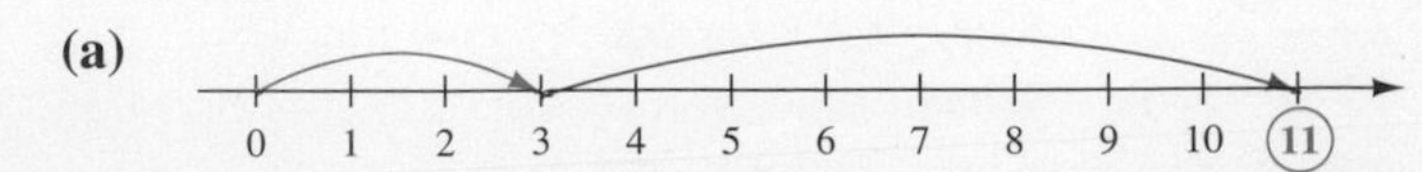

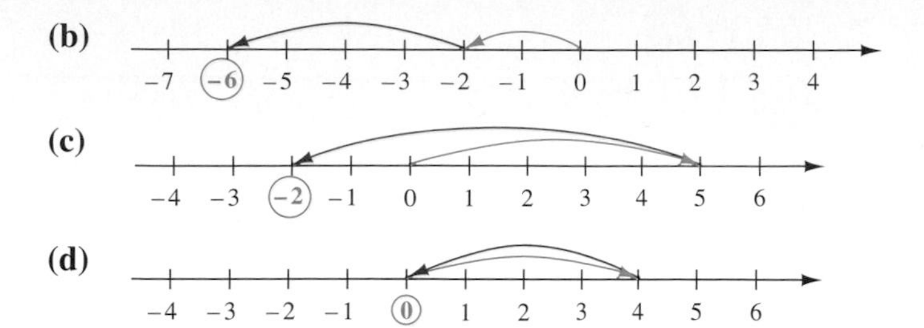

SOLUTION

Since counting in the direction indicated by an arrow means adding, these diagrams represent the following sums.

(a) $3 + 8 = 11$ **(b)** $(-2) + (-4) = -6$

(c) $5 + (-7) = -2$ **(d)** $4 + (-4) = 0$ ■

Note that these are again precisely the same results as in Example 5.7.

EXAMPLE 5.11

Drawing a Number Line Diagram for a Given Sum

Draw a number line diagram to illustrate $(-7) + 4$.

SOLUTION

Since -7 is indicated by counting 7 units to the left from 0 (in the direction indicated by the minus sign on the -7), and adding 4 is indicated by counting on to the right (in the positive direction since 4 is a positive integer), we have this diagram.

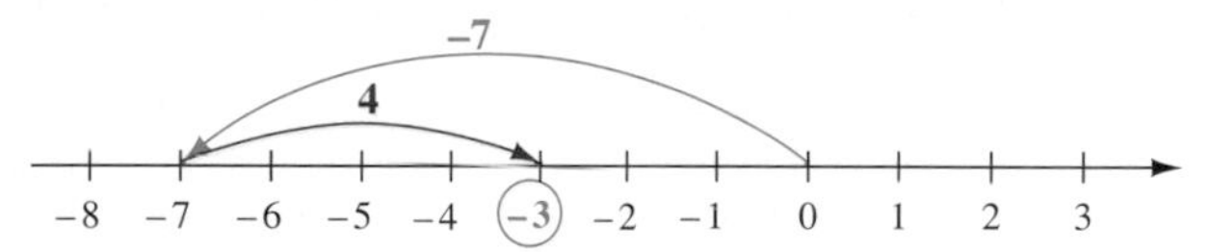

Thus, $(-7) + 4 = -3 = -(7 - 4)$. ■

Ordering the Set of Integers

Since the set with 3 black counters as shown in Figure 5.10 contains fewer counters than the set with 7 black counters, we say that **3 is less than 7** and write $3 < 7$. We also observe that $3 + 4 = 7$ and say that 7 is 4 more than 3.

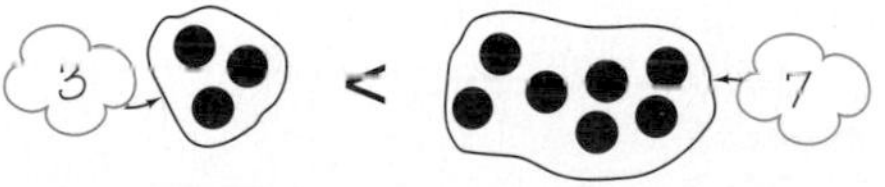

Figure 5.10
Comparing 3 and 7; $3 < 7$

This idea is also easily illustrated on a number line as shown in Figure 5.11. In particular, we note again that $3 + 4 = 7$ and this implies that 3 *is to the left of* 7 on a number line. With this in mind, we extend the notion of the less than to the set of all integers. In particular, if a is to the left of b on a number line, then there is a positive integer c such that $a + c = b$ and so $a < b$.

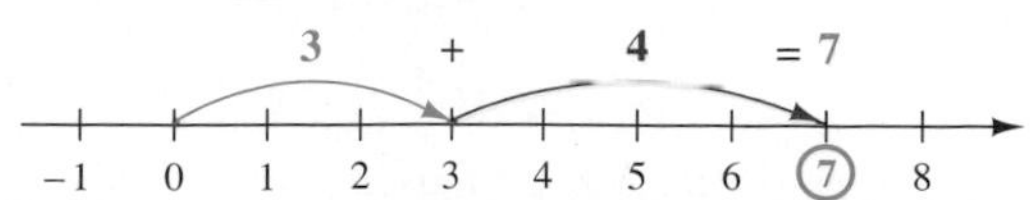

Figure 5.11
Comparing 3 and 7 on a number line; $3 < 7$

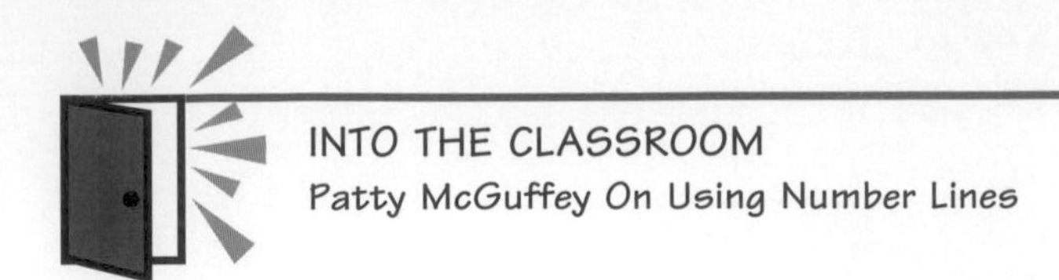

INTO THE CLASSROOM
Patty McGuffey On Using Number Lines

In many cases, learning mathematics has become a passive experience for students. I have found that involving students in developing rules for addition, subtraction, multiplication, and division of integers significantly improves learning.

In introducing integers, I establish concepts before having students become involved in making up rules. I explain that a number is made up of two components: the sign, which tells direction, and the digit or digits (absolute value) which indicates the number of units or distance to move. To illustrate direction, I demonstrate on a number line that a negative indicates movement to the left and a positive indicates movement to the right. Using these ideas of direction and distance, students use the number line to solve problems involving addition of two positive addends.

Once students understand ideas of direction and distance, it is easy for them to move on to problems in which both addends are negative, or one addend is positive and the other is negative. I put at least five problems on the board and have students use the number line to determine the sum. It is helpful to have students work in pairs, and I have them answer questions such as:

- Were all the answers positive? Were all the answers negative?
- In what situations were the answers positive? Negative?
- Can you see any pattern in your problems and answers?
- Can we make a generalization?

The next step is to write and test the rules that students have stated.

Using the number line to work on several examples, students gradually conclude that subtraction is the same as adding the opposite.

When developing a rule for multiplying a positive factor and a negative factor, students use the commutative property and the idea that multiplication is repeated addition. I have my students copy a pattern similar to the one in *Lesson 4.* Then I draw a number line to the right of it if they are working with a problem that involves two negative factors. The visualization of the pattern of products on the number line is helpful.

To begin the investigation of division, I ask, *How do you check the answer in a division problem?* Students quickly see the relationship between multiplication and division and determine a set of rules for division of integers.

By using "discovery learning," students develop problem-solving strategies and use higher-level strategies as they actively participate in their own learning.

SOURCE: *ScottForesman Exploring Mathematics,* Grades 1–7 by L. Carey Bolster et al. Copyright © 1994 Scott, Foresman and Company. Reprinted by permission of Scott, Foresman and Company. Patty McGuffey teaches at Central Intermediate School in Brownsville, TX.

DEFINITION Less Than and Greater Than for the Set of Integers

Let a and b be integers. We say that ***a* is less than *b*,** and write $a < b$, if, and only if, there is a positive integer c such that $a + c = b$. We say that ***b* is greater than *a*,** and write $b > a$, if, and only if, $a < b$.

Notions similar to less than and greater than are **less than or equal** and **greater than or equal,** and these mean just what they say. That is, we say that a is less than or equal to b and write $a \leq b$ if, and only if, $a < b$ or $a = b$. Similarly, we say that a is greater than or equal to b and write $a \geq b$ if, and only if $a > b$ or $a = b$. Moreover, it is important to note that if a and b are any two points on a number line, either a is to the left of b, or $a = b$, or a is to the right of b. Thus, the integers satisfy the so-called law of trichotomy.

THEOREM The Law of Trichotomy

If a and b are any two integers, then precisely one of these three possibilities must hold:

$$a < b \quad \text{or} \quad a = b \quad \text{or} \quad a > b.$$

Since the law of trichotomy holds for the integers, they are said to be **ordered;** that is, they can be lined up on a number line in order of increasing size.

EXAMPLE 5.12

Ordering Pairs of Integers

Place a less than or greater than sign in the circle as appropriate.

(a) 17 ◯ 121 **(b)** 2 ◯ −7 **(c)** −7 ◯ −27 **(d)** 0 ◯ −6

SOLUTION

(a) $17 < 121$ since $17 + 104 = 121$.
(b) $2 > -7$ since $-7 + 9 = 2$.
(c) $-7 > -27$ since $-27 + 20 = -7$.
(d) $0 > -6$ since $-6 + 6 = 0$.

104, 9, 20, and 6 are all positive.

■

EXAMPLE 5.13

Ordering a Set of Integers

Plot each of these integers on a number line and then list them in increasing order: −5, −9, 7, 0, 12, −8.

SOLUTION

−9 −8 −5 0 7 12

−10 −9 −8 −7 −6 −5 −4 −3 −2 −1 0 1 2 3 4 5 6 7 8 9 10 11 12 13 14 15 16

Reading from left to right, we see that

$$-9 < -8 < -5 < 0 < 7 < 12.$$

■

EXAMPLE 5.14

Operating with Inequalities

If a, b, and c are integers and $a < b$, prove that $a + c < b + c$.

JUST FOR FUN

Choosing the Right Box

Three boxes are on a shelf so high that you can reach into but not see into them. One box contains only apples, one contains only oranges, and one contains both apples and oranges. Also, the boxes are labeled "apples," "oranges," and "apples and oranges" but each box is mislabeled. How can you select a single piece of fruit from just one of the boxes and then proceed to label all the boxes correctly?

SOLUTION

Understand the problem • • • •

We are told that a, b, and c are integers with $a < b$. We're asked to show that $a + c < b + c$.

Devise a plan • • • •

Imagine the following internal dialogue, that reveals how a plan emerges:

Perhaps I should ask and answer some questions of myself to see if I can come up with a plan.

Say it in another way.

Q. What am I given?
A. That $a < b$ where a and b are integers.
Q. What must I show?
A. That $a + c < b + c$ where c is also an integer.
Q. What does it mean to say that $a + c < b + c$? Can I say this in another way?
A. I guess so. According to the definition, $a + c < b + c$ if, and only if, some natural number added to $a + c$ gives $b + c$.
Q. Okay. So how could I find such a natural number? Do I have anying to go on?
A. Well. I'm given that $a < b$. This means that there is some natural number r such that $a + r = b$. Perhaps I can use this.

Carry out the plan • • • •

Since $a < b$, there is some natural number r such that $a + r = b$. Therefore, add c to both sides,

$$a + r + c = b + c.$$

Then, by the commutative property for addition,

$$a + c + r = b + c.$$

This shows that adding the natural number r to $a + c$ gives $b + c$ and that means that $a + c < b + c$ as was to be shown.

Look back • • • •

Basically, we obtained the solution just by making sure we understood the problem. Key questions answered were: What is given? What must be shown? Can we say all this in a different way (that is, what does $a < b$ mean, and so on)? With answers to these questions clearly in mind, a little arithmetic finished the job. ■

Subtraction of Integers

Subtraction of Integers Using Sets of Colored Counters

As with the subtraction of whole numbers, one approach to the subtraction of integers is the notion of "take away." Consider the subtraction

$$7 - (-3).$$

Modeling 7 with counters, we must "take away" a representation of -3. Since any representation of -3 must have at least 3 red counters, we must use a representation of 7 with at least 3 red counters. The simplest representation of this subtraction is shown in Figure 5.12.

Figure 5.12
Colored counter representation of $7 - (-3) = 10$

Thus, $7 - (-3) = 10$. Also, the result does not change if we use another representation of 7 with at least three red counters since equivalent representations are obtained by adding (or deleting) the same number of counters of each color from a given representation. A second representation of $7 - (-3) = 10$ is shown in Figure 5.13.

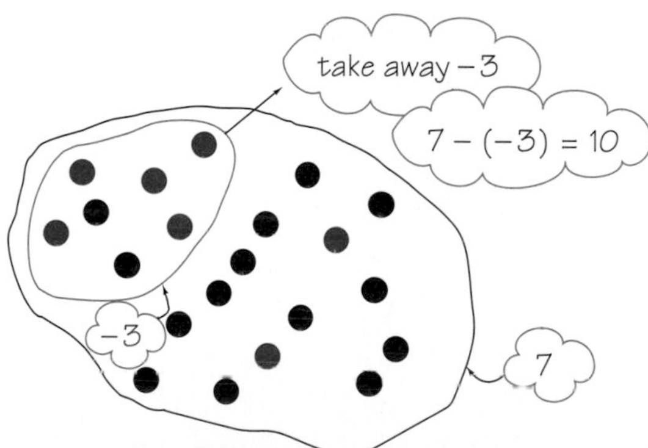

Figure 5.13
A second colored counter representation of $7 - (-3) = 10$

EXAMPLE 5.15 **Subtracting Using Colored Counters**

Write out the subtraction equation illustrated by each of these diagrams.

(a)

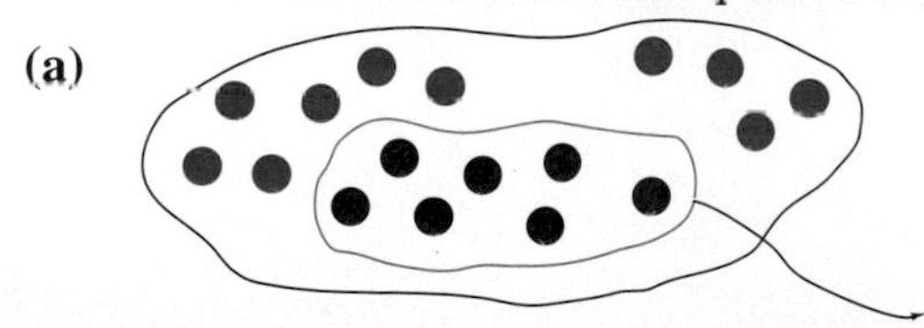

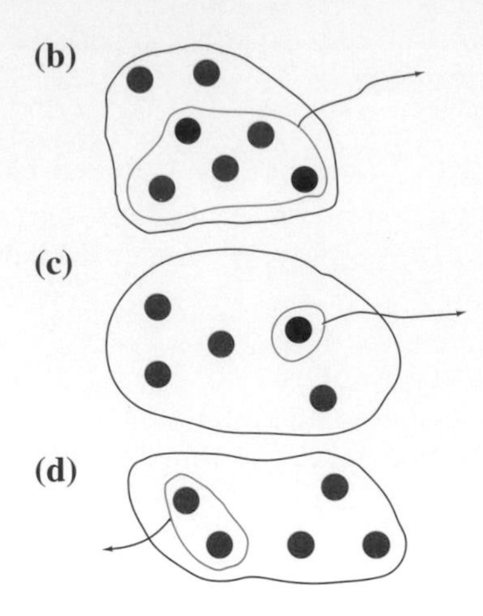

SOLUTION

(a) Since the large loop contains 10 red counters and 7 black counters, it represents -3. The small loop contains 7 black counters and so represents 7. If we remove the small loop of counters as indicated, we are left with 10 red markers representing -10. Thus, this diagram illustrates the subtraction $(-3) - 7 = -10$.

(b) This diagram represents the subtraction $(-3) - (-1) = -2$.

(c) This diagram represents the subtraction $(-3) - 1 = -4$.

(d) This diagram represents the subtraction $(-5) - (-2) = -3$. ■

EXAMPLE 5.16

Drawing Diagrams for Given Subtraction Problems

Draw a diagram of colored counters to illustrate each of these subtractions and determine the result in each case.

(a) $7 - 3$ **(b)** $(-7) - (-3)$ **(c)** $7 - (-3)$ **(d)** $(-7) - 3$

SOLUTION

(a) Many different diagrams could be drawn, but the simplest is shown here.

(b) As in part (a) we can use a diagram with counters of only one color.

(c) This is illustrated in Figures 5.12 and 5.13.

(d) Here, in order to remove 3 black counters (that is, subtract 3), the representation for -7 must have at least 3 black counters. The simplest diagram is as shown.

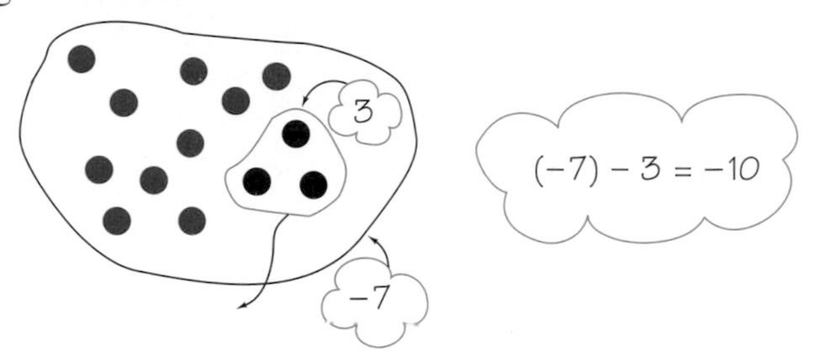

■

Notice that we can interpret any subtraction diagram as an addition diagram and *vice versa*. For example, if we interpret the diagram of Figure 5.14 as taking away the four red counters, the diagram represents the subtraction

$$5 - (-4) = 9.$$

On the other hand, if we view it as a diagram showing the combining of the set of black counters with the set of red counters, it illustrates the addition

$$5 = (-4) + 9.$$

Figure 5.14
Diagram illustrating both $5 - (-4) = 9$ and $5 = (-4) + 9$

Since all the subtraction diagrams can be interpreted in this way, we have the following definition.

DEFINITION Subtraction of Integers

If a, b, and c are integers, then

$$a - b = c$$

if, and only if, $a = b + c$.

This was also the definition of subtraction of whole numbers. Thus, as before, we have a family of equivalent facts; that is,

$$a - b = c, \qquad a = b + c, \qquad a = c + b, \qquad \text{and} \qquad a - c = b$$

all express essentially the same relationship among the integers a, b, and c. If we know that any one of these equations is true, then all are true.

Another important fact about subtraction of integers is illustrated in Figure 5.15.

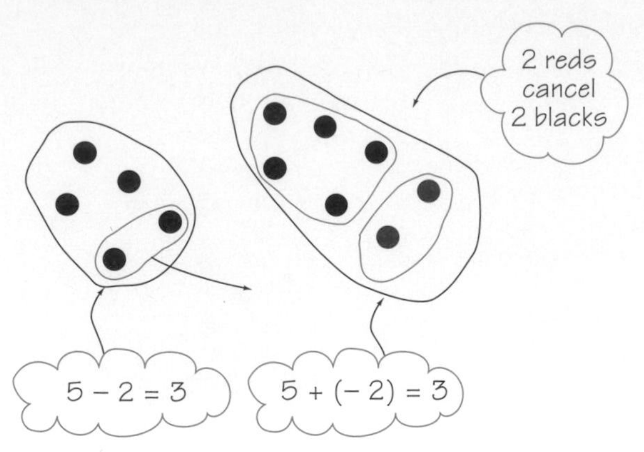

Figure 5.15
Subtracting by adding the negative

Note that the diagram on the left illustrates the subtraction

$$5 - 2 = 3$$

while the diagram on the right illustrates the addition

$$5 + (-2) = 3.$$

Since the effect of adding the negative of 2 to 5 is the same as subtracting 2 from 5 we see that

$$5 - 2 = 5 + (-2).$$

Moreover, since the opposite of an integer represented by a set of counters is always represented by the set with all the counters turned over, this proposition is true in general and can be expressed as a theorem.

To subtract, add the negative.

THEOREM Subtracting by Adding the Negative

Let a and b be any integers. Then

$$a - b = a + (-b).$$

Finally, since we have already seen that the set of integers is closed under addition, an immediate consequence of this theorem is that, unlike the set of whole numbers, *the set of integers is closed under subtraction.*

THEOREM Closure Property for the Subtraction of Integers

The set of integers is closed under subtraction.

EXAMPLE 5.17 **Subtracting by Adding**

Perform each of these subtractions as additions.

(a) $7 - 3$ **(b)** $(-7) - (-3)$ **(c)** $7 - (-3)$ **(d)** $(-7) - 3$

SOLUTION

We make use of the theorem stating that $a - b = a + (-b)$ for any integers a and b.

(a) $7 - 3 = 7 + (-3) = 4$

(b) $(-7) - (-3) = (-7) + 3 = -4$

(c) $7 - (-3) = 7 + 3 = 10$

(d) $(-7) - 3 = -7 + (-3) = -10$ ■

Subtraction of Integers Using Mail-time Stories

For this model of subtraction to work we must imagine a situation where checks and bills are immediately credited or debited to your account as soon as they are delivered whether or not they are really intended for you. If an error has been made by the mail carrier, he or she must return and reclaim delivered mail and take it to the intended recipient. Thus,

bringing a check adds a positive number,
bringing a bill adds a negative number,
taking away a check subtracts a positive number, and
taking away a bill subtracts a negative number.

EXAMPLE 5.18 Subtraction Facts from Mail-time Stories

Indicate the subtraction facts illustrated by each of these mail-time stories.

(a) The mail carrier brings you a check for \$7 and takes away a check for \$3. Are you richer or poorer and by how much?

(b) The mail carrier brings you a bill for \$7 and takes away a bill for \$3. Are you richer or poorer and by how much?

(c) The mail carrier brings you a check for \$7 and takes away a bill for \$3. Are you richer or poorer and by how much?

(d) The mail carrier brings you a bill for \$7 and takes away a check for \$3. Are you richer or poorer and by how much?

SOLUTION

(a) You are \$4 richer. This illustrates $7 - 3 = 4$.

(b) You are \$4 poorer. This illustrates $(-7) - (-3) = -4$.

(c) You are \$10 richer. This illustrates $7 - (-3) = 10$.

(d) You are \$10 poorer. This illustrates $(-7) - 3 = -10$. ■

Note that these are precisely the same subtractions as illustrated in Example 5.16 with diagrams of colored counters and in Example 5.17 by adding additive inverses.

JUST FOR FUN

Fun with a Flow Chart

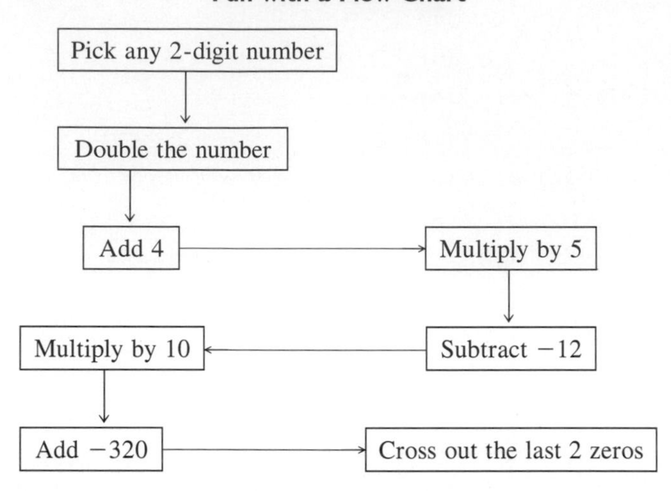

What number results?

Subtraction of Integers Using the Number Line

The addition $5 + 3 = 8$ is illustrated on a number line in Figure 5.16.

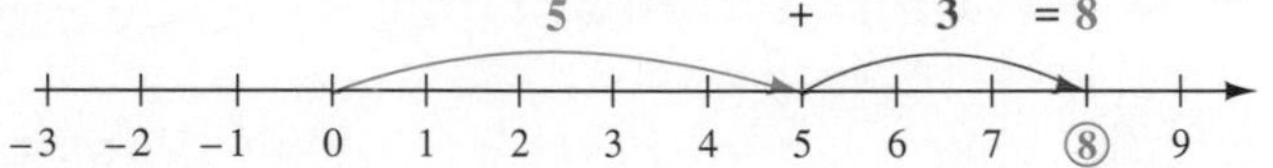

Figure 5.16
The sum $5 + 3 = 8$ on the number line

Now consider the subtraction

$$5 - 3.$$

This time we start at 0 and count 5 units to the right as before and then count *backwards* 3 units in the direction *opposite from that indicated by the 3 arrow.* Figure 5.17 shows us that the result is then 2 as we already know from the subtraction of whole numbers.

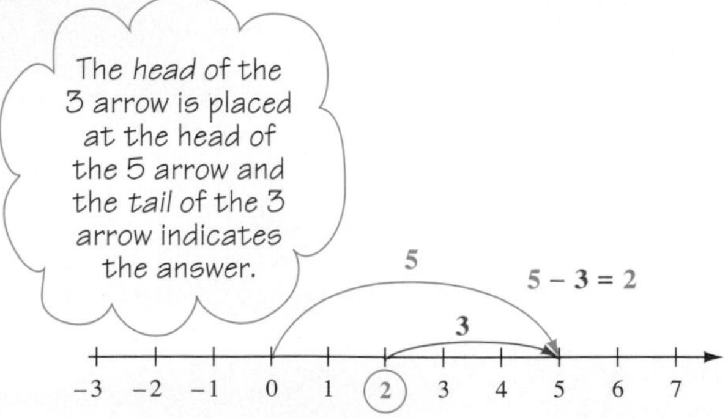

Figure 5.17
The subtraction $5 - 3 = 2$ on the number line

EXAMPLE 5.19 Drawing Diagrams to Illustrate Subtraction on the Number Line

Illustrate each of these subtractions on a number line and give the result in each case.

(a) $7 - 3$ **(b)** $(-7) - (-3)$ **(c)** $7 - (-3)$ **(d)** $(-7) - 3$

SCHOOL BOOK PAGE *Subtracting Integers in Elementary School*

ALGEBRA

Subtracting Integers

Build Understanding

A. Look for a Pattern
Groups: With a partner

a. Study these equations. Then give the next three equations that continue the pattern.

10 − 2 = 8
10 − 1 = 9
10 − 0 = 10
10 − (−1) = 11
10 − (−2) = 12
10 − (−3) = 13

b. Give the numbers that complete the equations that match the subtraction equations in part a.

10 − 2 = 8 **10 + (−2) = 8**	**10 − 1 = 9** **10 + ▒ = 9**
10 − 0 = 10 **10 + ▒ = 10**	**10 − (−1) = 11** **10 + ▒ = 11**
10 − (−2) = 12 **10 + ▒ = 12**	**10 − (−3) = 13** **10 + ▒ = 13**

c. Study the pattern in part b and describe a way for using addition to subtract two integers.

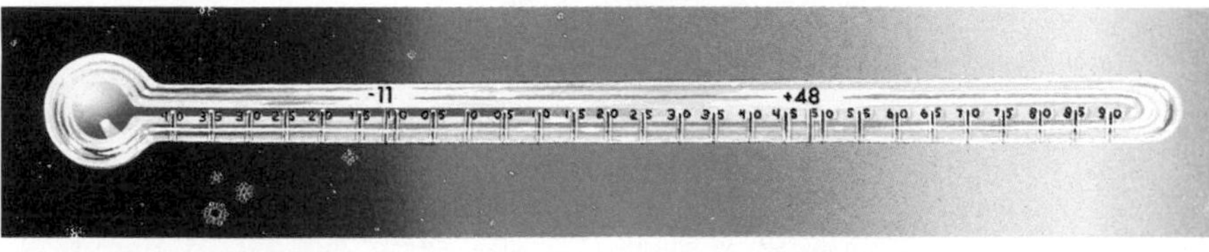

B. One day the high temperature was 48°; the low was −11°. How many degrees did it fall? Compare your rule to the computation shown. Does your rule give the same answer?

48 − (−11)
48 + 11 = +59

The temperature had dropped 59°.

Talk About Math If you subtract a positive number from a negative number, will the answer be positive or negative? Explain.

Check Understanding

For another example, see Set C, pages 482–483

Find the missing numbers.

1. 5 − (−2) = 5 + ▒ = ▒

2. −8 − 3 = −8 + ▒ = ▒

3. −6 − (−1) = −6 + ▒ = ▒

4. 9 − 17 = 9 + ▒ = ▒

1. In part (a) of A above, subtraction of negative integers is introduced by considering a pattern. Do you think this is a good approach for elementary school? Discuss briefly.
2. What point is being made in part (b) of A above? How would you respond to part (c)?
3. How would you respond to the above question in Talk About Math?
4. Write down at least two other questions you might discuss with your class when considering this lesson.

SOLUTION

(a)

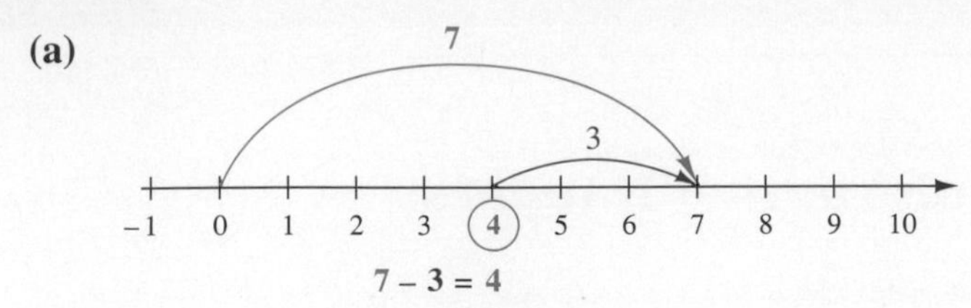

(b)

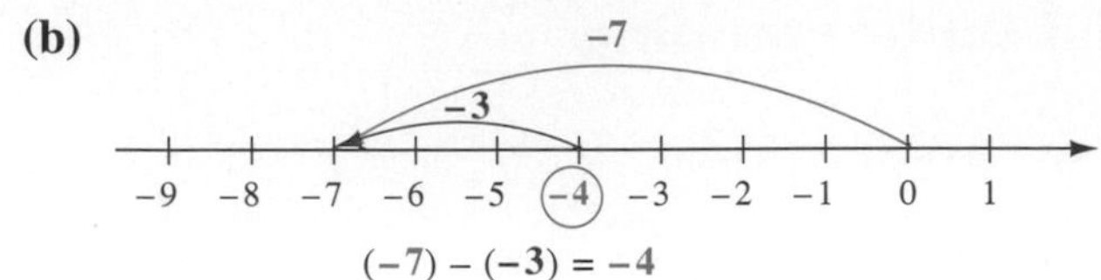

(c)

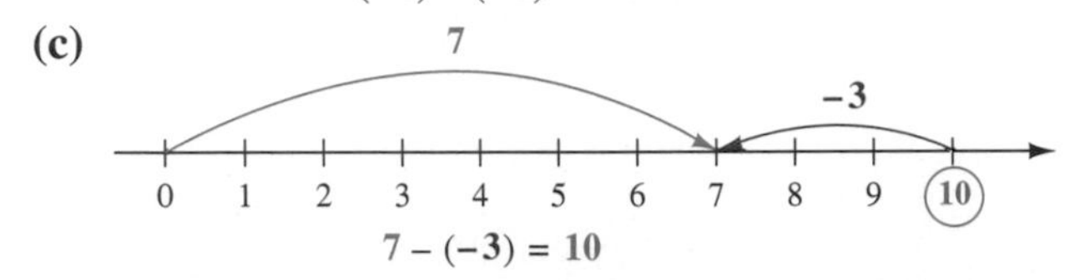

(d)

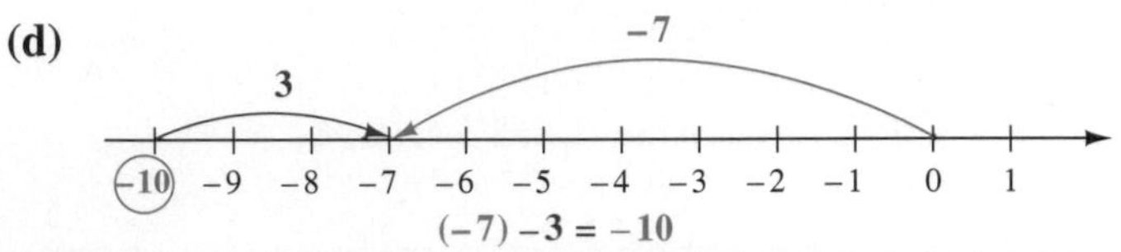

■

The number line is particularly useful in demonstrating the result

$$a - b = a + (-b)$$

noted earlier. The diagram illustrating that $(-7) - 3 = -10$ of part (d) in the preceding example is essentially the same as that in Figure 5.18 which illustrates $(-7) + (-3) = -10$. Thus, as in the theorem,

$$(-7) - 3 = (-7) + (-3).$$

(−7) + (−3) = −10 −7

−3

−11 −10 −9 −8 −7 −6 −5 −4 −3 −2 −1 0 1

Figure 5.18
The sum $(-7) + (-3) = -10$ on the number line

Adding and Subtracting Integers with a Calculator

Calculator addition and subtraction of natural numbers has already been discussed. Integers can also be added and subtracted on a calculator but care must be taken since integers are both positive and negative.

Using the Change of Sign Key, [+ ⟲ −]

We observed earlier that to enter -5 into the display of your calculator you enter

5 [+ ⟲ −].

Thus, $27 + (-83)$ can be found by keying into your calculator the string

[ON/AC] 27 [+] 83 [+ ⟲ −] [=]

and the result is -56. Similarly, entering the string

[ON/AC] 71 [+ ⟳ −] [−] 93 [+ ⟳ −] [=]

performs the subtraction

$$(-71) - (-93) = 22.$$

EXAMPLE 5.20 **Adding and Subtracting Integers Using the [+ ⟳ −] Key**

Perform these computations using the [+ ⟳ −] key.

(a) $(-27) + (-95)$ **(b)** $(-27) - (-95)$ **(c)** $3250 + (-4729)$

SOLUTION

In each case we indicate the key string entered into the calculator and the resulting answer.

(a) [ON/AC] 27 [+ ⟳ −] [+] 95 [+ ⟳ −] [=], -122
(b) [ON/AC] 27 [+ ⟳ −] [−] 95 [+ ⟳ −] [=], 68
(c) [ON/AC] 3250 [+] 4729 [+ ⟳ −] [=], -1479

Alternatively, one can use the theorems in this section to write

$$(-27) + (-95) = -(27 + 95),$$
$$(-27) - (-95) = (-27) + 95 = 95 - 27, \text{ and}$$
$$3250 + (-4729) = 3250 - 4729.$$

Then these can be calculated as follows.

(a) [ON/AC] 27 [+] 95 [=] [+ ⟳ −], -122
(b) [ON/AC] 95 [−] 27 [=], 68
(c) [ON/AC] 3250 [−] 4729 [=], -1479 ■

PROBLEM SET 5.2

Understanding Concepts

1. Draw diagrams of colored counters to illustrate these computations and state the answer in each case.
   (a) $8 + (-3)$ **(b)** $(-8) + 3$ (c) $(-8) - (-3)$
   (d) $8 - (-3)$ (e) $9 + 4$ **(f)** $9 + (-4)$
   (g) $(-9) + 4$ **(h)** $(-9) - (-4)$
2. Describe mail-time situations that illustrate each computation and state the answer in each case.
   (a) $(-27) + (-13)$ **(b)** $(-27) - 13$
   (c) $27 + 13$ **(d)** $27 - 13$
   (e) $(-41) + 13$ **(f)** $(-41) - 13$
   (g) $(-13) + 41$ **(h)** $13 - 41$
3. Draw number line diagrams that illustrate each computation and state the answer in each case.
   (a) $8 + (-3)$ **(b)** $8 - (-3)$ (c) $(-8) + 3$
   (d) $(-8) - (-3)$ (e) $4 + (-7)$ **(f)** $4 - (-7)$
   (g) $(-4) + 7$ **(h)** $(-4) - (-7)$
4. Write each of these subtractions as an addition.
   (a) $13 - 7$ **(b)** $13 - (-7)$
   (c) $(-13) - 7$ **(d)** $(-13) - (-7)$
   (e) $3 - 8$ **(f)** $8 - (-3)$
   (g) $(-8) - 13$ **(h)** $(-8) - (-13)$
5. Perform each of these computations.
   (a) $27 - (-13)$ **(b)** $12 + (-24)$
   (c) $(-13) - 14$ **(d)** $-81 + 54$
   (e) $(-81) - 54$ **(f)** $(-81) - (-54)$
   (g) $(-81) + (-54)$ **(h)** $27 + (-13)$
   (i) $(-27) - 13$

6. By 2 P.M. the temperature in Cutbank, Montana had risen 31° from a nighttime low of 41° below zero.
   (a) What was the temperature at 2 P.M.?
   (b) What computation does part (a) illustrate?

7. **(a)** If the high temperature on a given day was 2° above zero and the morning's low was 27° below zero, how much did the temperature rise during the day?
   (b) What computation does part (a) illustrate?

8. **(a)** If the high temperature for a certain day was 8° above zero and that night's low temperature was 27° below zero how much did the temperature fall?
   (b) What computation does part (a) illustrate?

9. During the day Sam's Soda Shop took in \$314. That same day Sam paid a total of \$208 in bills.
   (a) Was Sam's net worth more or less at the end of the day? By how much?
   (b) What computation does part (a) illustrate?

10. During the day Sam's Soda Shop took in \$284. Also, Sam received a check in the mail for \$191 as a refund for several bills that he had inadvertently paid twice.
    (a) Was Sam's net worth more or less at the end of the day? By how much?
    (b) What computation does part (a) illustrate?
    (c) If you think of the \$191 check as removing or taking away the bills previously paid, what computation does this represent? Explain.

11. Place a less than or greater than sign in each circle to make a *true* statement.
    (a) $-117 \bigcirc -24$ **(b)** $0 \bigcirc -4$ (c) $18 \bigcirc 12$
    (d) $18 \bigcirc -12$ (e) $-5 \bigcirc 1$ **(f)** $-5 \bigcirc -9$

12. List these numbers in increasing order from least to greatest: $-5, 27, 5, -2, 0, 3, -17$.

Thinking Critically

13. Which of the following are *true*?
    (a) $3 < 12$ **(b)** $-3 < -12$
    (c) $-3 < 12$ **(d)** $3 < -12$

14. If $a < b$, is $a \le b$ true? Explain.

15. If $a \ge b$, is $a > b$ true? Explain.

16. Is $2 \ge 2$ true? Explain.

17. For which integers x is it true that $|x| < 7$?

18. For which integers x is it true that $|x| > 99$?

19. (a) Compute each of these absolute values.
    (i) $|5 - 11|$ (ii) $|(-4) - (-10)|$
    (iii) $|8-(-7)|$ (iv) $|(-9) - 2|$
    (b) Draw a number line and determine the distance between the points on the number line for each of these pairs of integers.
    (i) 5 and 11 (ii) -4 and -10
    (iii) 8 and -7 (iv) -9 and 2
    (c) Since parts (a) and (b) are completely representative of the corresponding general cases, state a general theorem summarizing these results.

20. (a) Compute each of these pairs of expressions.
    (i) $|7 + 2|$ and $|7| + |2|$
    (ii) $|(-8) + 5|$ and $|-8| + |5|$
    (iii) $|7 + (-6)|$ and $|7| + |-6|$
    (iv) $|(-9) + (-5)|$ and $|-9| + |-5|$
    (v) $|6 + 0|$ and $|6| + |0|$
    (vi) $|0 + (-7)|$ and $|0| + |-7|$
    (b) Since the results of part (a) are completely typical, place one of the signs $>$, $<$, $\ge$, or $\le$ in the circle to make the following a *true* statement.
    For any integers a and b,
    $$|a + b| \bigcirc |a| + |b|.$$

21. Let a and b be positive integers, with $a < b$. If c is a negative integer, prove that $ac > bc$. (*Suggestion:* Try using specific numbers first.)

Thinking Cooperatively

22. (a) Make a magic square using the numbers $-4, -3, -2, -1, 0, 1, 2, 3, 4$.

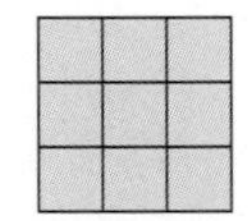

    (b) Make a magic subtraction square using the numbers $-4, -3, -2, -1, 0, 1, 2, 3, 4$.

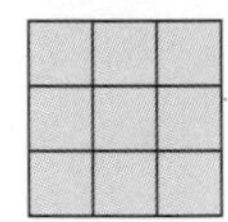

23. Place the numbers $-2, -1, 0, 1, 2$, in the circles in the diagram so that the sum of the numbers in each direction is the same.

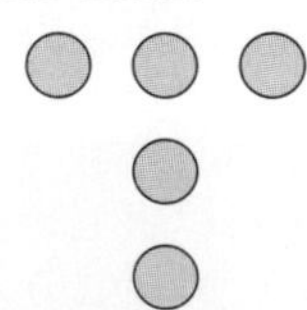

    (a) Can this be done with 0 in the middle of the top row? If so, show how. If not, why not?
    (b) Can this be done with 2 in the middle of the top row? If so, show how. If not, why not?

(c) Can this be done with -2 in the middle of the top row? If so, show how. If not, why not?

(d) Can this be done with 1 or -1 in the middle of the top row? If so, show how. If not, why not?

24. Perform these pairs of computations.

(a) $7 - (-3)$ and $(-3) - 7$

(b) $(-2) - (-5)$ and $(-5) - (-2)$

(c) Does the commutative law for subtraction hold for the set of integers? Explain briefly.

25. Does the associative law for subtraction hold for the set of integers (that is, is $a - (b - c) = (a - b) - c$ true for all integers a, b, and c)? If so, explain why. If not, give a counter example.

26. The equation $a(b - c) = ab - ac$ expresses the distributive law for multiplication over subtraction. Is it true for all integers a, b, and c? If so, explain why. If not, give a counter example.

Using A Calculator

27. If we write $a - b + c - d$, it is understood to mean $(((a - b) + c) - d)$. The operations are performed from left to right as on a calculator with algebraic logic.

(a) Compute $1 - 2 + 3 - \cdots + 99$.

(b) Could you think of an easy way to complete part (a) without a calculator?

(c) Compute $1 - 2 + 3 - \cdots + 99 - 100$.

28. Use your calculator to perform these calculations.

(a) $3742 + (-2167)$ **(b)** $(-2751) + (-3157)$

(c) $(-2167) - 3742$ **(d)** $(-3157) - (-2751)$

(e) $-(3571 - 5624)$ **(f)** $-[49{,}002 + (-37{,}621)]$

---

29. Recall that the triangular numbers are the numbers 1, 3, 6, 10, 15, . . . generated by the formula

$$t_n = \frac{n(n + 1)}{2}.$$

(a) Fill in the blanks to complete these equations and to extend the pattern.

$1 =$ ________

$1 - 3 =$ ________

$1 - 3 + 6 =$ ________

$1 - 3 + 6 - 10 =$ ________

________ $=$ ________

________ $=$ ________

(b) What would the nth equation in this pattern be? (*Hint:* You may have to consider two cases—n odd and n even.)

30. Recall that the Fibonacci numbers are the numbers $F_1 = 1$, $F_2 = 1$, $F_3 = 2$, $F_4 = 3$, $F_5 = 5$, . . . where we start with 1 and 1 and then add any two successive terms to obtain the next term in the sequence.

(a) Fill in the blanks to continue this pattern.

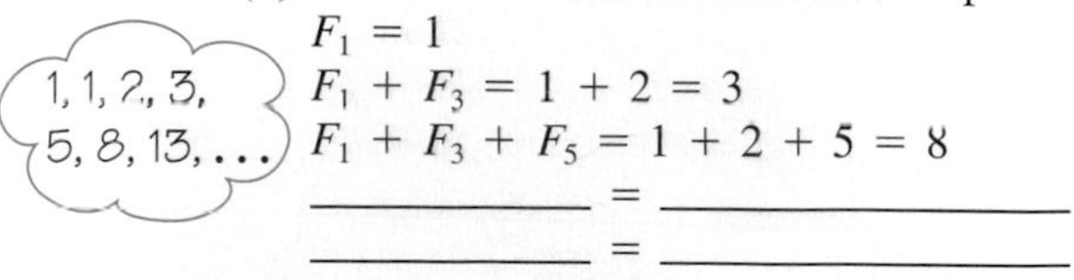

$F_1 = 1$

$F_1 + F_3 = 1 + 2 = 3$

$F_1 + F_3 + F_5 = 1 + 2 + 5 = 8$

________ $=$ ________

________ $=$ ________

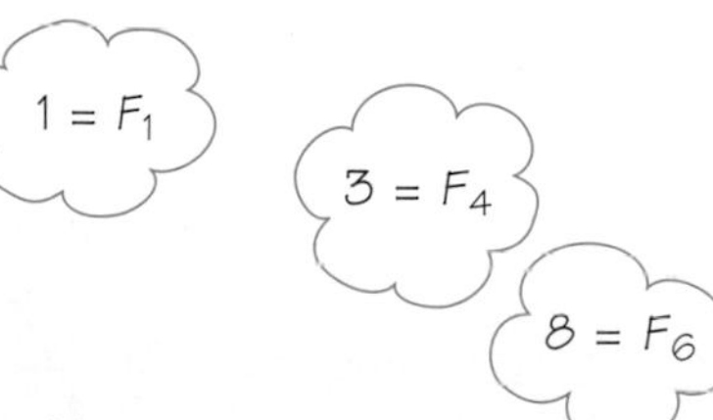

(b) Guess a formula for the general result suggested by part (a).

31. Continue each of these patterns two more steps and guess a general result in each case.

(a) $F_2 = 1$

$F_2 + F_4 = 1 + 3 = 4$

$F_2 + F_4 + F_6 = 1 + 3 + 8 = 12$

________ $=$ ________

________ $=$ ________

General result: ________________

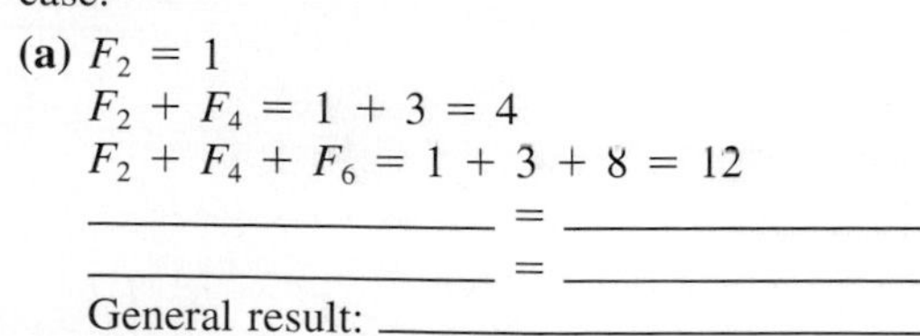

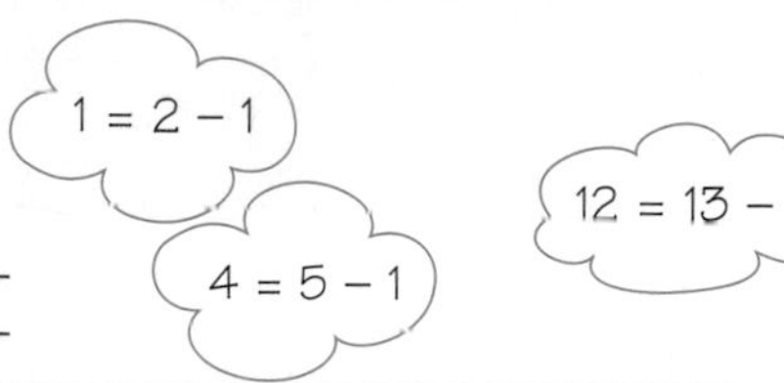

(b) $F_1 = 1$
$F_1 + F_2 = 1 + 1 = 2$
$F_1 + F_2 + F_3 = 1 + 1 + 2 = 4$
________________ = ________________
________________ = ________________
General result: ________________________________

32. Note that the successive powers $(-1)^1 = -1$, $(-1)^2 = 1$, $(-1)^3 = -1$, $(-1)^4 = 1$ alternate between 1 and -1. Thus, $(-1)^n = 1$ if n is even and -1 if n is odd. Using this idea, continue each of these patterns two more steps and guess a general result in each case.

(a) $F_1 = 1$
$F_1 - F_3 = 1 - 2 = -1$
$F_1 - F_3 + F_5 = 1 - 2 + 5 = 4$
________________ = ________________
________________ = ________________
General result: ________________________________

(b) $F_2 = 1$
$F_2 - F_4 = 1 - 3 = -2$
$F_2 - F_4 + F_6 = 1 - 3 + 8 = 6$
________________ = ________________
________________ = ________________
General result: ________________________________

(c) $1 = 1$
$1 - 1 = 0$
$1 - 1 + 2 = 2$
$1 - 1 + 2 - 3 = -1$
$1 - 1 + 2 - 3 + 5 = 4$
$1 - 1 + 2 - 3 + 5 - 8 = -4$
$1 - 1 + 2 - 3 + 5 - 8 + 13 = 9$
________________________________ = ________________
________________________________ = ________________
General result: ________________________________
(*Suggestion:* To guess the general result, think very carefully about the last three equations in part (c).)

33. **(a)** The general result guessed in problem 32, part (c) does not hold for $n = 1$ unless we define F_0 to be 0. Would this be consistent with the definition of the Fibonacci numbers as given in problem 30? Explain.

(b) Determine F_{-1}, F_{-2}, F_{-3}, and F_{-4} so that the pattern established by the definition of problem 30 still holds.

(c) Determine F_{-n} so that the definition of problem 30 holds for all integer values of n. (*Hint:* Use $(-1)^n$ as in problem 32.)

(d) Recall that the Lucas numbers are the numbers $L_1 = 1$, $L_2 = 3$, $L_3 = 4, \ldots$ where we start with 1 and 3 and then add any two successive terms to obtain the next term in the sequence. As in part (c), determine L_0 and L_{-n} so that this definition holds for all integer values of n.

Making Connections

34. One day Anne Felsted had the flu. At 8 A.M. her temperature was 101°. By noon her temperature had increased by 3° and then it fell 5° by six in the evening.

(a) Write a single addition equation to determine Anne's temperature at noon.

(b) Write a single equation using both addition and subtraction to determine Anne's temperature at 6 P.M.

35. Vicky Valerin was 12 years old on her birthday today.

(a) How old was Vicky on her birthday 7 years ago?

(b) How old will Vicky be on her birthday 7 years from now?

(c) Write addition equations that answer both parts (a) and (b) of this problem.

36. Greg Odjakjian's bank balance was $4500. During the month he wrote checks for $510, $87, $212, and $725. He also made deposits of $600 and $350. What was his balance at the end of the month?

37. A ball is thrown upward from the top of a building 144 feet high. Let h denote the height of the ball above the top of the building t seconds after it was thrown. It can be shown that $h = -16t^2 + 96t$ feet.

(a) Complete this table of values of h.

t	h
0	0
1	80
2	
3	
4	
5	
6	
7	

(b) Give a carefully worded plausibility argument (not a proof) that the greatest value of h in the table is the greatest height the ball reaches.

(c) Carefully interpret (explain) the meaning of the value of h when $t = 7$.

38. The speed of the ball in problem 37 in feet per second is given by the equation $s = -32t + 96$.

(a) Complete the table of values of s shown below.

(b) Carefully interpret (explain) the value of s when $t = 0$.

(c) Interpret the value of s when $t = 3$.

t	s
0	96
1	64
2	
3	
4	
5	
6	
7	

(d) Carefully interpret the meaning of the values of s for $t = 4, 5, 6,$ and 7.

(e) Compare the values of s for $t = 0$ and 6, 1, and 5, and 2 and 4. What do these values tell you about the motion of the ball?

39. A company manufacturing dolls calculates that the profit earned in making x dolls is $10{,}000 - (x - 100)^2$ dollars.

(a) How many dolls should the company make for maximum profit?

(b) Discuss what happens if the company produces 200 dolls.

(c) What is the result if the company produces 250 dolls? Explain.

Communicating

40. In playing Simon Says, children line up along a line on the floor, sidewalk, or ground and move steps forward or backward as directed by the person who is "it."

(a) How could you use this game to introduce the notion of a negative integer to children?

(b) Describe how you could use your version of Simon Says to help students understand addition and subtraction of positive and negative integers.

For Review

41. Test the following numbers for divisibility by each of 2, 3, 5, and 11.

(a) 214,221 **(b)** 106,090 **(c)** 1,092,315

42. Test each of the following for divisibility by 7, 11, and 13.

(a) 965,419 **(b)** 1,140,997 **(c)** 816,893

43. Test each of the following for divisibility by 4, 6, and 8.

(a) 62,418 **(b)** 83,224 **(c)** 244,824

44. Show that each of 271,134; 427,113; 342,711; 134,271; 113,427; and 711,342 is divisible by 27.

45. Let r, s, t, and u be natural numbers. If $r = s + t$, $u \mid s$ and $u \nmid t$, use the method of proof by contradiction to prove that $u \nmid r$.

5.3 Multiplication and Division of Integers

Multiplication of Integers

In this section we consider multiplication and division of integers. As in the preceding section, it is helpful to introduce these operations with integers by considering colored counters, Cartesian products of sets, mail-time stories, and number lines.

Multiplication of Integers Using Sets of Colored Counters

Multiplication of natural numbers was defined in Chapter 2 as repeated addition. Thus, $4 \cdot 5$ was shorthand for $5 + 5 + 5 + 5$, and this was represented diagrammatically by the array of Figure 5.19.

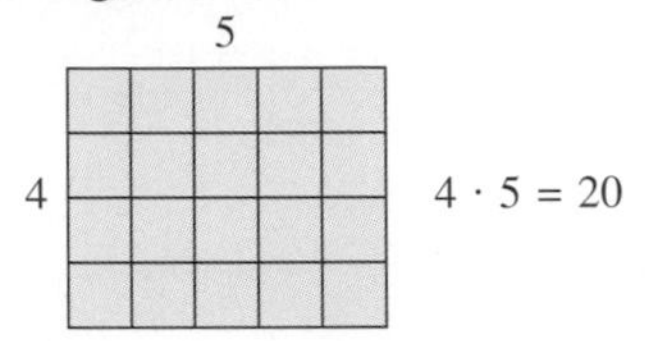

Figure 5.19
Array representation of $4 \cdot 5 = 20$

Using colored counters, this same concept would be represented by the diagram in Figure 5.20 with 4 rows of five black counters each. Similarly $4 \cdot (-5)$ is shorthand for $(-5) + (-5) + (-5) + (-5)$ and this can be represented as in Figure 5.20 by a diagram of 4 rows with 5 red counters per row.

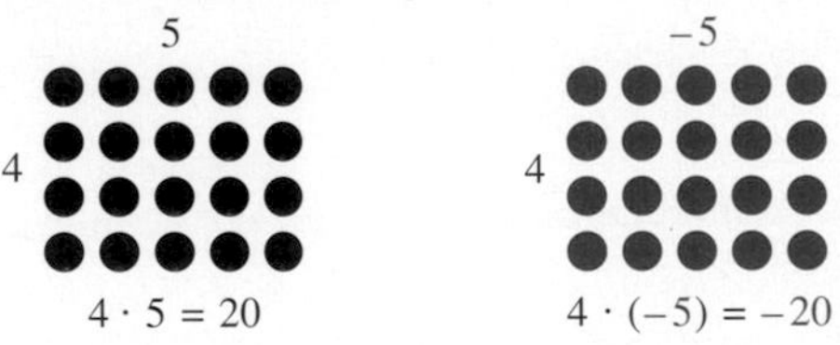

Figure 5.20
Representing $4 \cdot 5 = 20$ and $4 \cdot (-5) = -20$ with arrays of colored counters

If we rotate each diagram in Figure 5.20, we obtain the diagrams in Figure 5.21 which illustrate the commutative property for multiplication of integers. In particular, it follows from Figures 5.20 and 5.21, that

$$4 \cdot (-5) = (-5) \cdot 4 = -20$$

positive times negative gives negative

and, more generally,

$$a \cdot (-b) = (-b) \cdot a = -ab$$

for any positive integers a and b.

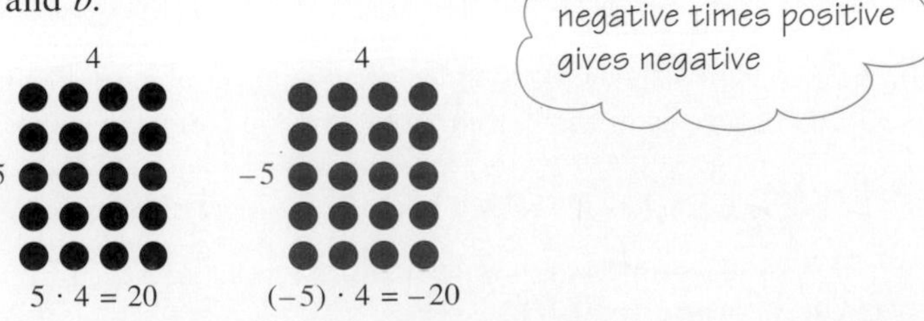

Figure 5.21
Representing $5 \cdot 4 = 20$ and $(-5) \cdot 4 = -20$ with arrays of colored counters

It is useful to think of the diagrams in Figures 5.20 and 5.21 as **Cartesian products of sets.** If we identify the counter pair (●, ●) with the single counter ● and the pairs (●, ●) and (●, ●) with the single counter ● then Figure 5.20 appears as the Cartesian product in Figure 5.22 and Figure 5.21 appears as the Cartesian product in Figure 5.23.

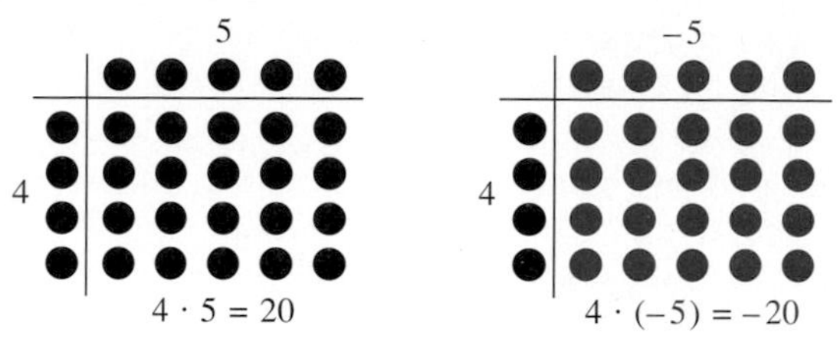

Figure 5.22
Representing 4 · 5 = 20 and 4 · (−5) = −20 by Cartesian products

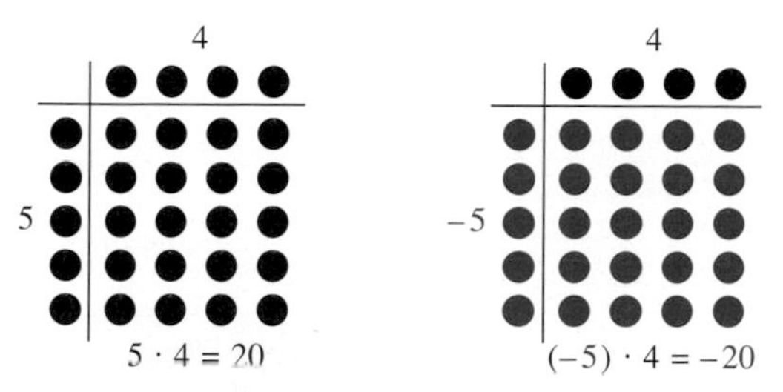

Figure 5.23
Representing 5 · 4 = 20 and (−5) · 4 = −20 by Cartesian products

The Cartesian product of any set with the empty set is the empty set, so

$$a \cdot 0 = 0 \cdot a = 0$$

for any integer a. But 0 has many representations as a set of colored counters (e.g., 0 is represented by the set of 4 black and 4 red counters), and the Cartesian product of such a set with any other set of counters must also represent 0. Consider the Cartesian product representing $(-5) \cdot 0$. We form the representation in Figure 5.24 but necessarily leave it incomplete since we have not determined which colored counter should be used as a replacement for the counter pair (●, ●). In Figure 5.24, the completed portion of the diagram corresponds to the product $(-5) \cdot 4 = -20$.

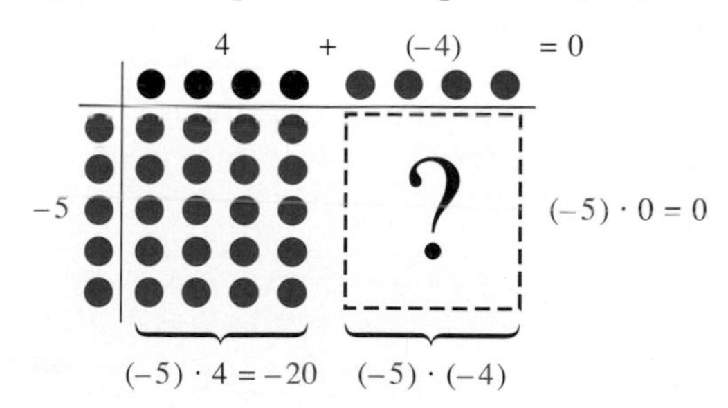

Figure 5.24
Incomplete Cartesian product representation of (−5) · 0 = 0

Since the entire representation must be a representation of 0, it follows that the remaining portion of the diagram should be a representation of 20. Thus, the Cartesian product should be completed by replacing each red-red counter pair, (●, ●), by a black counter, ●, as shown in Figure 5.25.

HIGHLIGHT FROM HISTORY
Charlotte Angas Scott (1858–1931)

Charlotte Scott is an example of a woman who has overcome great obstacles in the field of mathematics. Born in England, she attended Girton College of Cambridge University at a time when women were barred from receiving degrees and from even attempting the examinations for honors. She took the exams "informally," however, and besides two first places won eighth place in math, a field "too difficult for women." As the honors lists were read, shouts for "*Scott of Girton!*" cheered her achievement. Scott received her doctoral degree from the University of London and was called to the United States to become the only woman on the founding faculty of Bryn Mawr College in Pennsylvania. A long record of scholarship, writing, and contributions to the field of analytic geometry followed. Scott's students found her classes exciting, her mathematical style "elegant," and her own gift for clear explanation combined nicely with her quick understanding of their "stupidity." In later years, her record at Cambridge opened new doors for women; and many of her women students joined her in the lists of distinguished mathematicians and teachers.

SOURCE: From *Mathematics in Modules, Algebra, A3,* Teacher's Edition. Reprinted by permission.

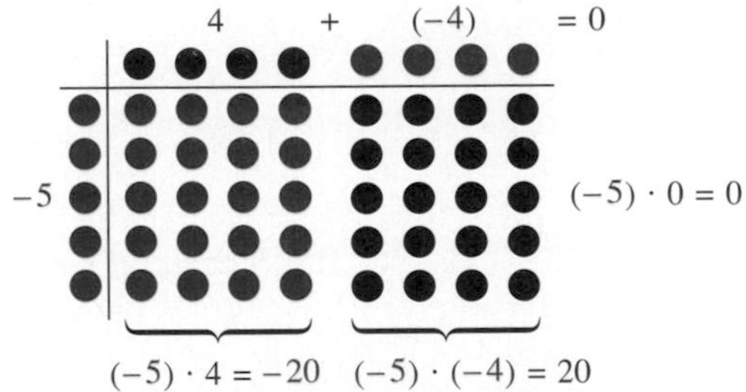

Figure 5.25
Cartesian product representation of $(-5) \cdot 0$. This implies that $(-5) \cdot (-4) = 20$

We conclude that $(-5) \cdot (-4) = 20$. Similarly, and in general, we see that

$$(-a) \cdot (-b) = ab$$

Negative times negative gives positive.

for any positive integers a and b.

The preceding results are frequently called the rule of signs, and we collect them here as a theorem.

THEOREM The Rule of Signs

Let m and n be positive integers so that $-m$ and $-n$ are negative integers. Then the following are true:

- $m \cdot (-n) = -(mn)$
- $(-m) \cdot n = -(mn)$
- $(-m) \cdot (-n) = mn$

and,

- $a \cdot 0 = 0 \cdot a = 0$

for any integer a.

EXAMPLE 5.21 **Multiplying Integers**

Compute these products.

(a) $7 \cdot 8$ **(b)** $(-6) \cdot (-5)$ **(c)** $3 \cdot (-7)$
(d) $7 \cdot (-3)$ **(e)** $0 \cdot (-5)$ **(f)** $(-4) \cdot 9$
(g) $5 \cdot 0$ **(h)** $9 \cdot (5 - 7)$

SOLUTION

(a) $7 \cdot 8 = 56$ **(b)** $(-6) \cdot (-5) = 6 \cdot 5 = 30$
(c) $3 \cdot (-7) = -(3 \cdot 7) = -21$ **(d)** $7 \cdot (-3) = -(7 \cdot 3) = -21$
(e) $0 \cdot (-5) = 0$ **(f)** $(-4) \cdot 9 = -(4 \cdot 9) = -36$
(g) $5 \cdot 0 = 0$ **(h)** $9 \cdot (5 - 7) = 9 \cdot (-2) = -(9 \cdot 2) = -18$ ■

EXAMPLE 5.22 **Illustrating Multiplication of Integers Using Cartesian Products**

Depict each of these products using a Cartesian product diagram. State the result of the indicated calculation in each case.

(a) $3 \cdot (-5)$ **(b)** $(-5) \cdot 3$ **(c)** $(-3) \cdot (-5)$
(d) $[(-3) \cdot (-2)] \cdot 4$ **(e)** $(-3) \cdot [(-2) \cdot 4]$

SOLUTION

To make a Cartesian product of two sets of colored counters, recall that we replace the counter pairs (●, ●) and (●, ●) by a single black counter, ●, and the pairs (●, ●) and (●, ●) by a single red counter, ●.

(a)

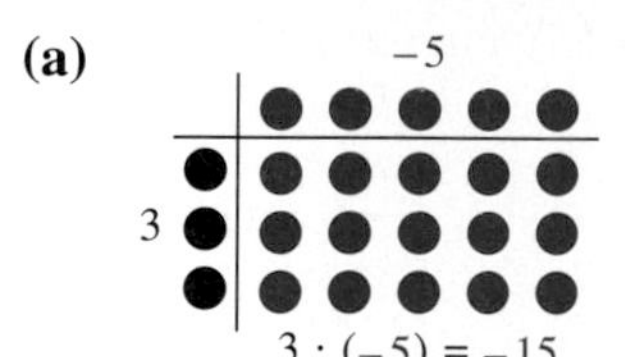

(b)

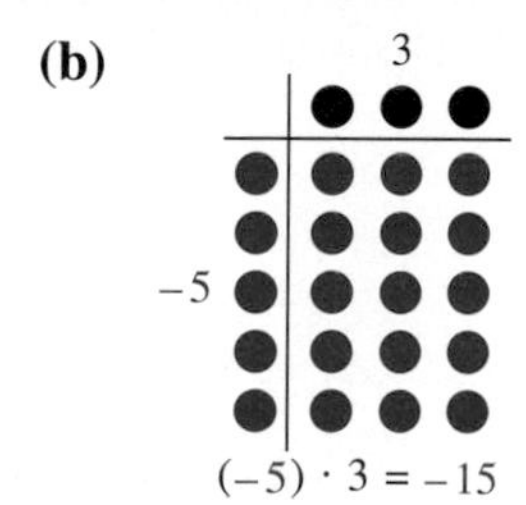

(c)

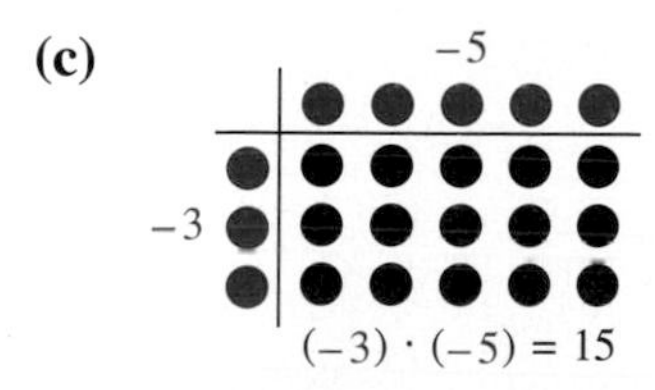

(d)

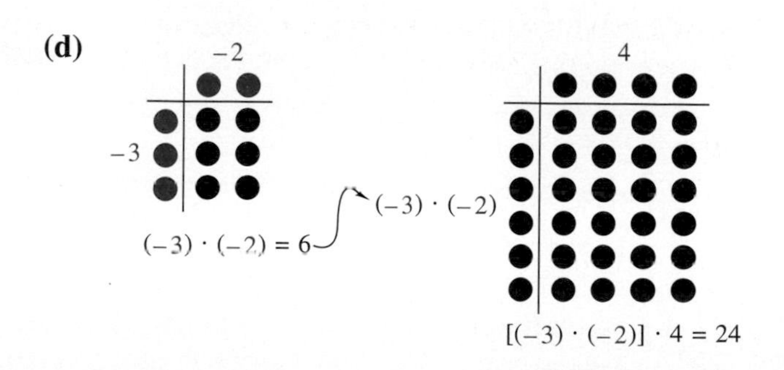

(e)

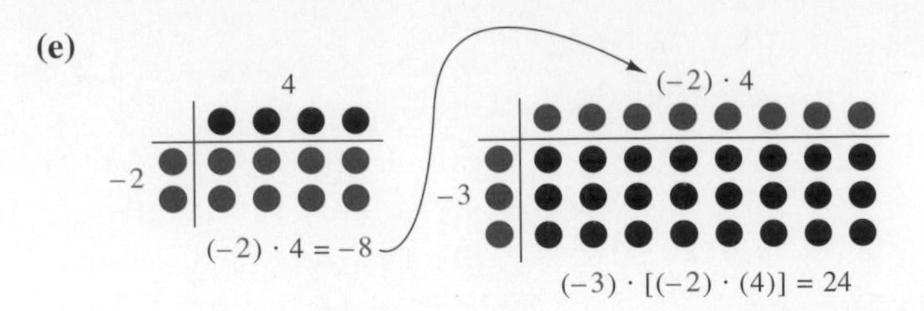

Parts (d) and (e) illustrate that the associative law for multiplication holds for the integers. ■

Since these examples are completely typical, we list the following properties of multiplication of integers.

THEOREM Multiplication Properties of Integers

Let r, s, and t be any integers.

Closure Property	rs is an integer.
Commutative Property	$rs = sr$.
Associative Property	$r(st) = (rs)t$.
Distributive Property	$r(s + t) = rs + rt$.

Multiplication of Integers Using Mail-time Stories

Recall that in mail-time stories the mail carrier bringing checks and bills corresponds to adding positive and negative numbers respectively. Similarly, taking away checks and bills corresponds to subtracting positive and negative numbers.

Suppose the letter carrier brings 5 bills for \$11 each. Are you richer or poorer and by how much? Answer: Poorer by \$55. Since this is repeated addition, this illustrates the product

$$5 \cdot (-11) = -55.$$

Suppose the mail carrier takes away 4 bills for \$13 each. Are you richer or poorer and by how much? Answer: Richer by \$52. This illustrates the product

$$(-4) \cdot (-13) = 52$$

since 4 bills for the same amount are *taken away*.

EXAMPLE 5.23 **Writing Mail-time Stories for Multiplication of Integers**

Write a mail-time story to illustrate each of these products.

(a) $(-4) \cdot 16$ **(b)** $(-4) \cdot (-16)$ **(c)** $4 \cdot (-16)$ **(d)** $4 \cdot 16$

SOLUTION

(a) The letter carrier takes away 4 checks for \$16 each. Are you richer or poorer and by how much? Answer: \$64 poorer, $(-4) \cdot 16 = -64$.

(b) The letter carrier takes away 4 bills for \$16 each. Are you richer or poorer and by how much? Answer: \$64 richer, $(-4) \cdot (-16) = 64$.

JUST FOR FUN

Three on a Bike

Three teenagers arrived at a rental agency each desiring to rent a bicycle for the day. There was only one bicycle left, but the teenagers finally agreed to share equally the \$30 charge for the one day and also to share equally in the bicycle's use. Later the manager of the agency decided that his charge was excessive and sent his helper to return \$5 to the teenagers. Being dishonest, the helper pocketed \$2 and only returned \$3 to the teenagers. Thus, each paid \$9 for the use of the bicycle. But \$27 plus the \$2 the helper took makes only \$29. What happened to the other dollar?

(c) The letter carrier brings 4 bills for \$16 each. Are you richer or poorer and by how much? Answer: Poorer by \$64, $4 \cdot (-16) = -64$.

(d) The letter carrier brings you 4 checks for \$16 each. Are you richer or poorer and by how much? Answer: Richer by \$64, $4 \cdot 16 = 64$. ■

Multiplication of Integers Using a Number Line

Think of $3 \cdot 4$ as $4 + 4 + 4$,
$3 \cdot (-4)$ as $(-4) + (-4) + (-4)$,
$(-3) \cdot (4)$ as $-4 - 4 - 4$, and
$(-3) \cdot (-4)$ as $-(-4) - (-4) - (-4)$.

These can be illustrated on the number line as follows.

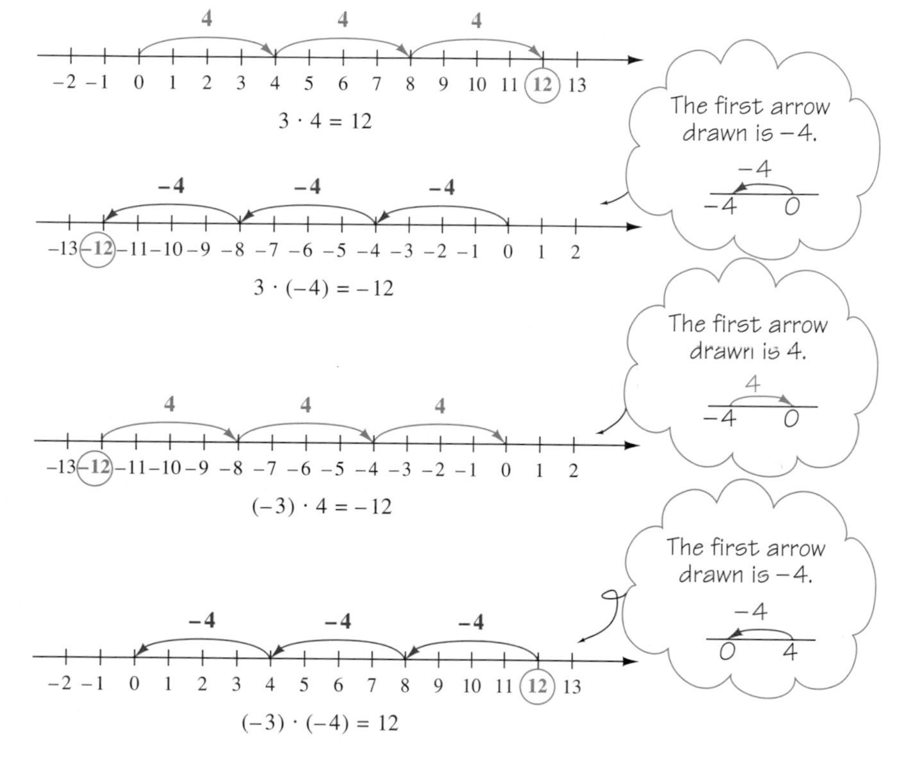

EXAMPLE 5.24

Multiplying Integers Using a Number Line

What products do each of these diagrams illustrate?

(a)

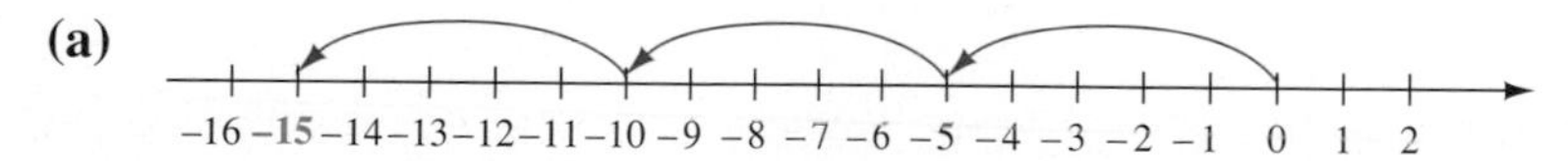

(b)

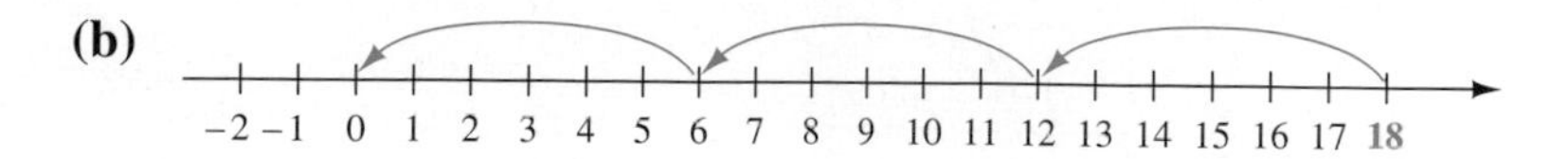

SOLUTION

(a) Starting with the tail of the right most arrow at 0, we move in the direction of the arrows 5 units to the left, then 5 more units to the left, and finally five more units to the left. Thus, we are *adding* -5 to -5 to -5 and the result is $3 \cdot (-5) = -15$.

(b) Here we start with the head of the left more arrow at 0 and move in the direction *opposite* to the direction of the arrows 6 units to the right, 6 more units to the right, and finally 6 more units to the right. Since the arrows are directed from right to left, we are *subtracting* -6 and -6 and -6. The result is $(-3) \cdot (-6) = 18$. ■

Division of Integers

Division of Integers Using Sets of Colored Counters

When discussing division of natural numbers, we considered families of facts. Thus, the equations

$$12 = 3 \cdot 4 \qquad 12 \div 3 = 4$$
$$12 = 4 \cdot 3 \qquad 12 \div 4 = 3$$

all express the same relationship between the numbers 12, 3, and 4. Indeed, one definition of division of whole numbers was the following:

> If a, b, and c are whole numbers with $b \neq 0$, then $a \div b = c$ if, and only if, $a = bc$.

Thus, to find the quotient $143 \div 11$, we determine c such that $143 = 11 \cdot c$. Since $c = 13$ satisfies this equation, $143 \div 11 = 13$.

Now we extend this definition to integers.

DEFINITION Division of Integers

If a, b, and c are integers with $b \neq 0$, then $a \div b = c$ if, and only if, $a = b \cdot c$.

Earlier in this section we discussed multiplication of integers in terms of Cartesian products of sets of colored counters. Turning the process around, we can use much the same diagrams to illustrate division of integers. Suppose, for example, that we wish to illustrate $-15 \div 3$. Since this is equivalent to finding an integer c such that $-15 = 3 \cdot c$, -15 is the Cartesian product of a set with 3 black counters and a set with c counters as shown in Figure 5.26.

WORTH READING

On the Shoulders of Giants

Mathematics can be exciting for students—and great fun for younger children—if educators and parents adopt a fresh perspective on mathematical concepts and how they are presented in our schools. "On the Shoulders of Giants," published in 1990 by the Mathematical Sciences Education Board and edited by Lynn Arthur Steen, contains five stimulating essays written by experts in mathematics and mathematics education. These essays, each exploring a different concept, present mathematics as the language and science of patterns, offering numerous imaginative ways mathematical ideas can be developed in the classroom.

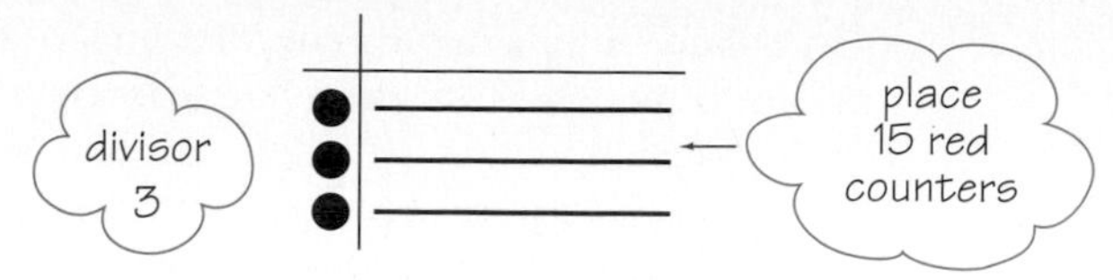

Figure 5.26
Incomplete Cartesian product diagram for (−15) ÷ 3

Since −15 corresponds to 15 red counters in the body of the diagram and each such counter corresponds to a (black, red) or (red, black) pair, the quotient set must consist of red counters and their number is determined by distributing the 15 red counters equally along the 3 black lines to complete a **Cartesian product diagram.** The only way this can be done is to place 5 red counters along each of the black lines. Thus, the quotient is a set of 5 red counters; that is, $c = -5$. The completed diagram is as shown in Figure 5.27.

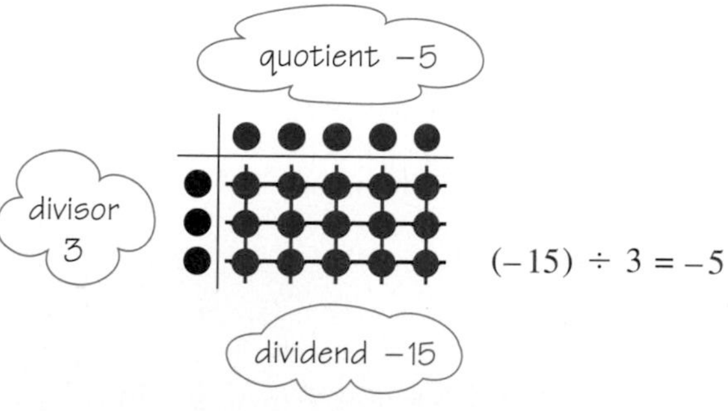

Figure 5.27
Cartesian product diagram for (−15) ÷ 3 = −5

EXAMPLE 5.25

Illustrating Division of Integers by Cartesian Product Diagrams

Draw Cartesian product diagrams to illustrate each of these quotients and state the result in each case.

(a) $(-15) \div (-3)$ **(b)** $20 \div (-4)$

SOLUTION

(a) Since we are dividing −15 by −3, our diagram must start with 3 red counters on the left side of the diagram with three red lines extending to the right from each counter, and we must apportion 15 red counters equally along these lines. Since a red counter in the body of the diagram corresponds to a (red, black) pair, the quotient set must consist of black counters, and we complete the diagram as shown. Thus,

$$(-15) \div (-3) = 5.$$

(b) This time, since we are dividing 20 by -4, we start with 4 red counters on the left with a red line extending to the right from each counter. We must apportion 20 black counters equally along these four lines. Since these counters are black and a single black counter in the body of the diagram corresponds to a (red, red) pair, it follows that the quotient set must consist of red counters. Thus, we complete the diagram as shown and conclude that

$$20 \div (-4) = -5.$$

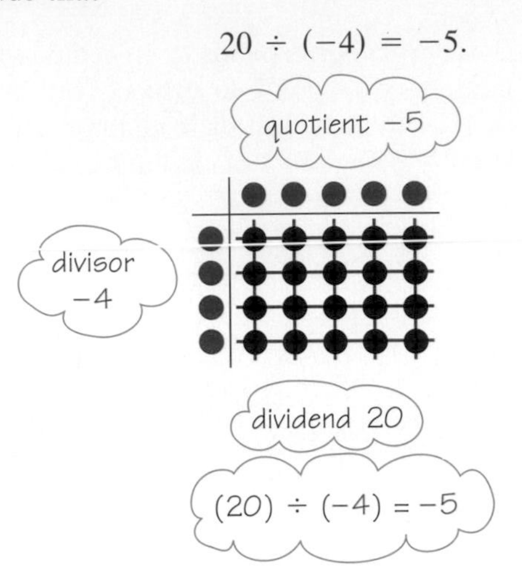

■

EXAMPLE 5.26 **Completing a Division Diagram**

Complete this division diagram given that there are 18 red counters in the body of the diagram. Also, indicate what division fact it illustrates.

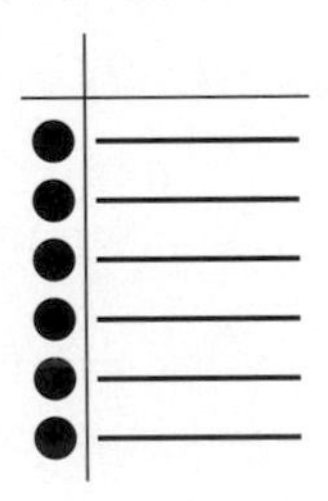

SOLUTION

Since the 18 red counters are divided into 6 groups of 3 counters each along the 6 black lines and a red counter in the body of the diagram corresponds to a (black, red) pair, the quotient set must consist of 3 red counters. Thus, we complete the diagram as shown and conclude that

$$(-18) \div 6 = -3.$$

■

Since these results are entirely typical, we state here the rule of signs for division.

THEOREM Rule of Signs for Division of Integers

Let m and n be positive integers so that $-m$ and $-n$ are negative integers and suppose that $n \mid m$. Then the following are true:

- $m \div (-n) = -(m \div n)$
- $(-m) \div n = -(m \div n)$
- $(-m) \div (-n) = m \div n$

Thus, *given that n divides m*, we see that:

- a positive integer divided by a negative integer is a negative integer,
- a negative integer divided by a positive integer is a negative integer,
- a negative integer divided by a negative integer is a positive integer, and
- a positive integer divided by a positive integer is a positive integer.

EXAMPLE 5.27

Performing Division of Integers

If possible, compute each of these quotients.

(a) $(-24) \div (-8)$ **(b)** $24 \div (-8)$ **(c)** $48 \div 12$
(d) $(-48) \div 12$ **(e)** $(-57) \div 19$ **(f)** $(-12) \div 0$
(g) $(-51) \div (-17)$ **(h)** $28 \div (9 - 5)$ **(i)** $(27 + 9) \div (-4)$

SOLUTION

We use the preceding theorem and the definition of division of integers.

(a) $(-24) \div (-8) = 3$. Check: $-24 = (-8) \cdot 3$
(b) $24 \div (-8) = -3$. Check: $24 = (-8) \cdot (-3)$
(c) $48 \div 12 = 4$. Check: $48 = 12 \cdot 4$
(d) $(-48) \div 12 = -4$. Check: $-48 = 12 \cdot (-4)$
(e) $(-57) \div 19 = -3$. Check: $-57 = 19 \cdot (-3)$
(f) $(-12) \div 0$ is not defined since there is no number c such that $-12 = 0 \cdot c = 0$.
(g) $(-51) \div (-17) = 3$. Check: $-51 = (-17) \cdot 3$
(h) $28 \div (9 - 5) = 28 \div 4 = 7$. Check: $28 = 4 \cdot 7$
(i) $(27 + 9) \div (-4) = 36 \div (-4) = -9$. Check: $36 = (-4) \cdot (-9)$. ■

PROBLEM SET 5.3

Understanding Concepts

1. Perform these multiplications.
   (a) $7 \cdot 11$ **(b)** $7 \cdot (-11)$
   (c) $(-7) \cdot 11$ **(d)** $(-7) \cdot (-11)$
   (e) $12 \cdot 9$ **(f)** $12 \cdot (-9)$
   (g) $(-12) \cdot 9$ **(h)** $(-12) \cdot (-9)$
   (i) $(-12) \cdot 0$
2. Perform these divisions.
   (a) $36 \div 9$ **(b)** $(-36) \div 9$
   (c) $36 \div (-9)$ **(d)** $(-36) \div (-9)$
   (e) $(-143) \div 11$ **(f)** $165 \div (-11)$
   (g) $(-144) \div (-9)$ **(h)** $275 \div 11$
   (i) $72 \div (21 - 19)$
3. Write another multiplication equation and two division equations that are equivalent to $(-11) \cdot (-25{,}753) = 283{,}283$.
4. Write two multiplication equations and another division equation that are equivalent to $(-1001) \div 13 = -91$.
5. (a) Draw a Cartesian product diagram of colored counters to illustrate the product $4 \cdot 6 = 24$.
   (b) If you turn over each counter in the bottom four rows of the diagram of part (a) including the counters to the left of the vertical line, what product is represented? Be sure to draw the appropriate diagram.
   (c) Draw the diagram obtained if you turn over each counter in the six columns on the right of the array in part (a) including those above the horizontal line. What product does this new diagram represent?
   (d) Draw the diagram obtained from the diagram of part (c) by turning over each counter in the bottom four rows of the diagram including those to the left of the vertical line. What product does this new diagram represent?
6. What computation does this Cartesian product diagram represent?

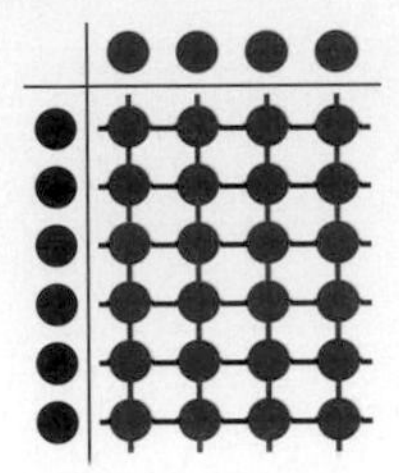

7. Draw a Cartesian product diagram with colored counters to illustrate the product $(-4) \cdot (-4)$. State the result of the multiplication.
8. Complete this Cartesian product diagram and indicate what division the diagram illustrates.

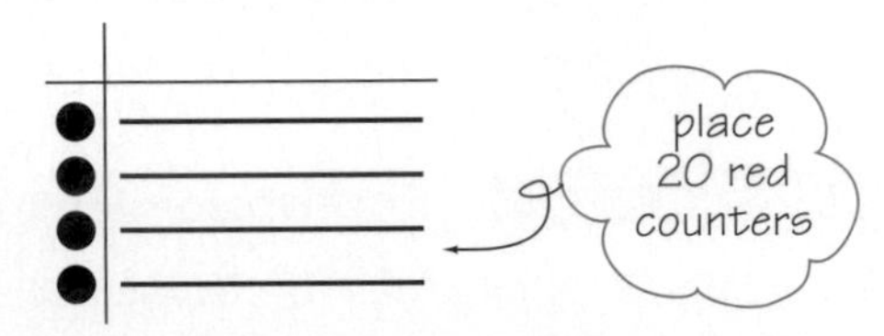

9. What computation does each of these mail-time stories illustrate?
   (a) The mail carrier brings you 6 checks for \$13 each. Are you richer or poorer? By how much?
   (b) The mail carrier brings you 4 bills for \$23 each. Are you richer or poorer? By how much?
   (c) The mail carrier takes away 3 bills for \$17 each. Are you richer or poorer? By how much?
   (d) The mail carrier takes away 5 checks for \$20 each. Are you richer or poorer? By how much?

10. What computation does each of these number line diagrams represent?
    (a)

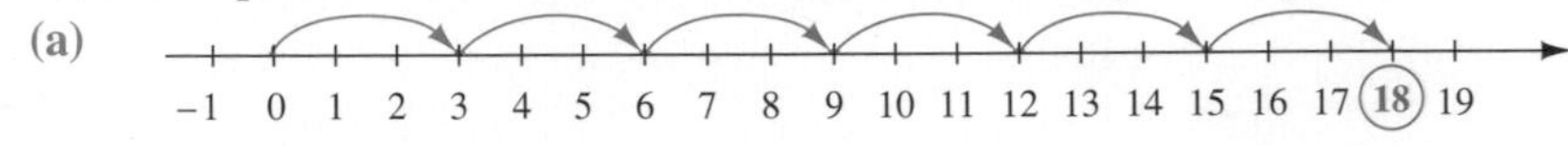

(b)

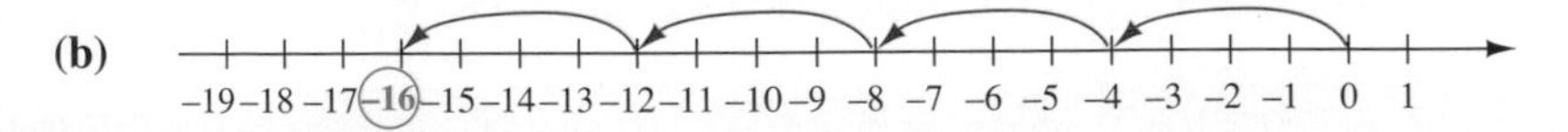

(c)

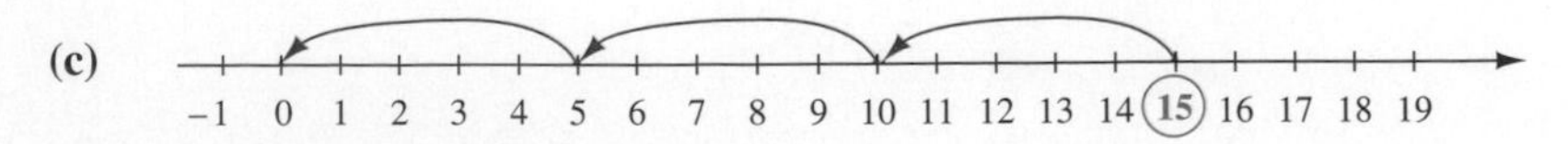

11. Draw a number line diagram to illustrate $(-4) \cdot 3 = -12$.

Thinking Critically

12. (a) Fill in the blanks to continue this pattern.

$$3 \cdot 3 = 9$$
$$3 \cdot 2 = 6$$
$$3 \cdot 1 = 3$$
$$3 \cdot 0 = 0$$
$$3 \cdot (-1) = \_\_\_\_\_\_$$
$$3 \cdot (-2) = \_\_\_\_\_\_$$
$$3 \cdot (-3) = \_\_\_\_\_\_$$

(b) What general result is suggested by this pattern?

13. Fill in the blanks to continue each of these patterns. What general result is suggested by the pattern in each case?

(a) Using problem 12, part (b), we have

$$3 \cdot (-3) = -9$$
$$2 \cdot (-3) = -6$$
$$1 \cdot (-3) = -3$$
$$0 \cdot (-3) = \_\_\_\_\_\_$$
$$(-1) \cdot (-3) = \_\_\_\_\_\_$$
$$(-2) \cdot (-3) = \_\_\_\_\_\_$$
$$(-3) \cdot (-3) = \_\_\_\_\_\_$$

General result suggested: __________

(b) $(-3) \cdot (-3) = 9$
$(-3) \cdot (-2) = 6$
$(-3) \cdot (-1) = 3$
$(-3) \cdot 0 = \_\_\_\_\_\_$
$(-3) \cdot 1 = \_\_\_\_\_\_$
$(-3) \cdot 2 = \_\_\_\_\_\_$
$(-3) \cdot 3 = \_\_\_\_\_\_$
General result suggested: __________

(c) Patterns do not prove general results, but they certainly suggest them. Do you think that use of such patterns might be helpful in teaching elementary school students that $a \cdot (-b) = -(ab)$, $(-a) \cdot (-b) = ab$, and $(-a) \cdot (b) = -(ab)$? Explain.

(d) Could you use this scheme to help students understand that the multiplication of integers is commutative? Explain briefly.

---

14. If we assume that the integers obey the distributive property as illustrated earlier in this section, then a proof of the rule of signs is available. Fill in the blanks to justify steps of these proofs.

(a)

$3 \cdot 0 = 0$	__________
$3 \cdot [5 + (-5)] = 0$	Definition of additive inverse
$3 \cdot 5 + 3 \cdot (-5) = 0$	__________
$15 + 3 \cdot (-5) = 0$	Arithmetic fact
$3 \cdot (-5) = -15$	__________

(b) If a and b are any positive integers, then

$a \cdot 0 = 0$	__________
$a[b + (-b)] = 0$	__________
$ab + a(-b) = 0$	__________
$a \cdot (-b) = -(ab)$	__________

(c)

$0 \cdot 5 = 0$	__________
$[3 + (-3)] \cdot 5 = 0$	Definition of additive inverse
$3 \cdot 5 + (-3) \cdot 5 = 0$	__________
$15 + (-3) \cdot 5 = 0$	Arithmetic fact
$(-3) \cdot 5 = -15$	__________

(d) If a and b are any positive integers, then

$0 \cdot b = 0$	__________
$[a + (-a)] \cdot b = 0$	__________
$ab + (-a)b = 0$	__________
$(-a)b = -(ab)$	__________

(e)

$(-3) \cdot 0 = 0$	__________
$(-3) \cdot [5 + (-5)] = 0$	__________
$(-3) \cdot 5 + (-3) \cdot (-5) = 0$	__________
$-15 + (-3)(-5) = 0$	By part (c)
$(-3) \cdot (-5) = 15$	__________
$(-3) \cdot (-5) = 3 \cdot 5$	__________

(f) $(-a) \cdot 0 = 0$ ______________________
$(-a) \cdot [b + (-b)] = 0$ ______________________
$(-a)b + (-a)(-b) = 0$ ______________________
$-(ab) + (-a)(-b) = 0$ ______________________
$(-a)(-b) = ab$ ______________________

(g) Would parts (b), (d), and (f) be appropriate for an elementary school classroom? Why or why not?

(h) Might parts (a), (c), and (e) be appropriate for an upper elementary or middle school classroom? Discuss.

Thinking Cooperatively

15. In the diagram below the sum of the numbers in any two small circles is the number in the large circle between them. Complete the following so that the same pattern holds in each case.

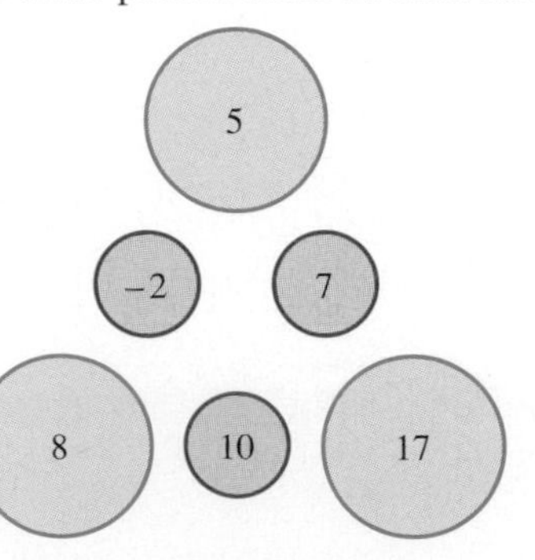

(a)

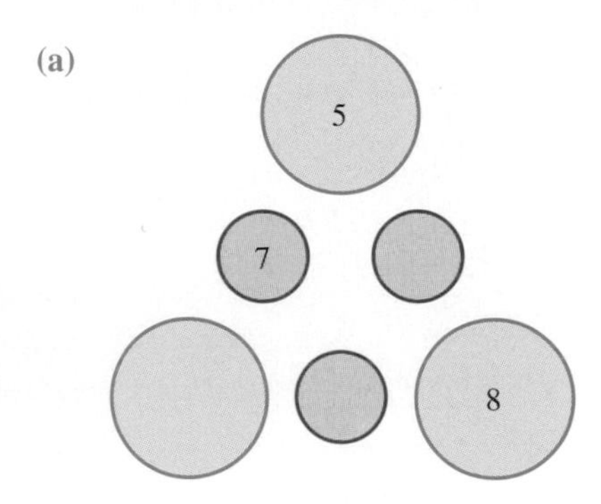

(b)

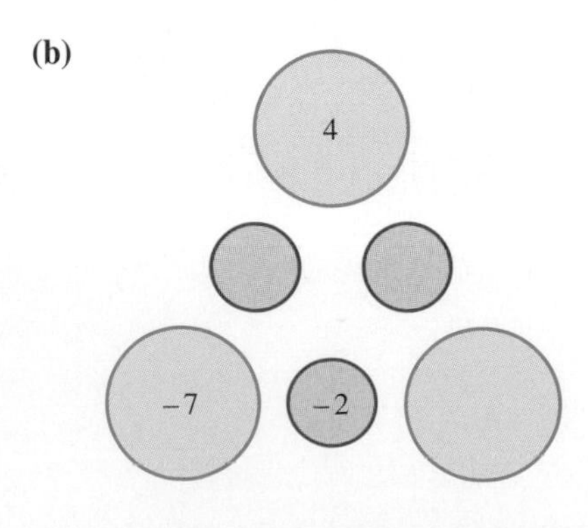

(c)

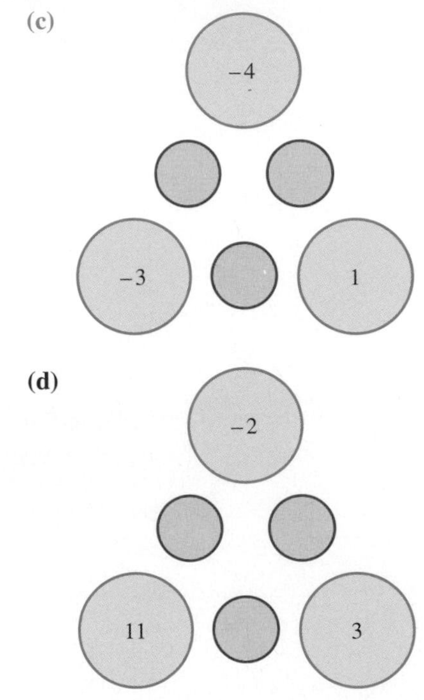

(d)

16. This problem is like the preceding one except that there are 5 large and 5 small circles. Again the sum of the numbers in two small circles is the number in the large circle between them.

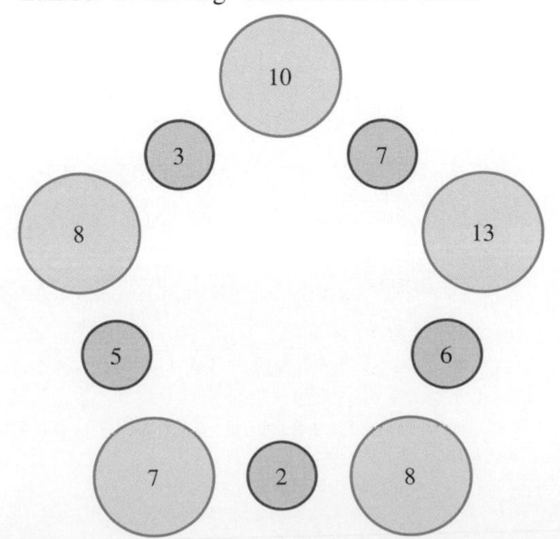

Complete each of the following so that the same pattern holds in each case. Indicate if the solution to each of these is unique.

(a)

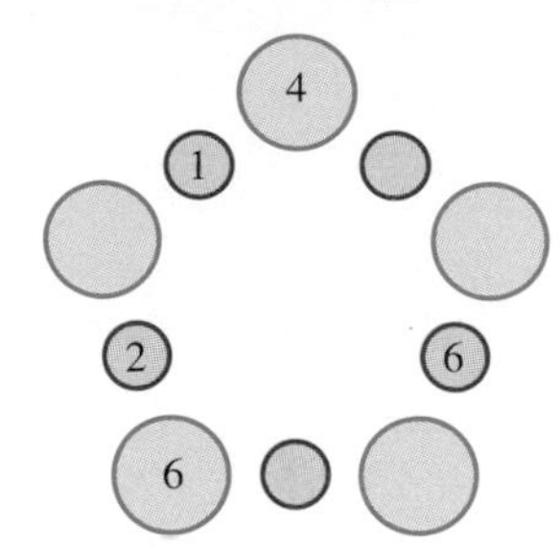

(b)

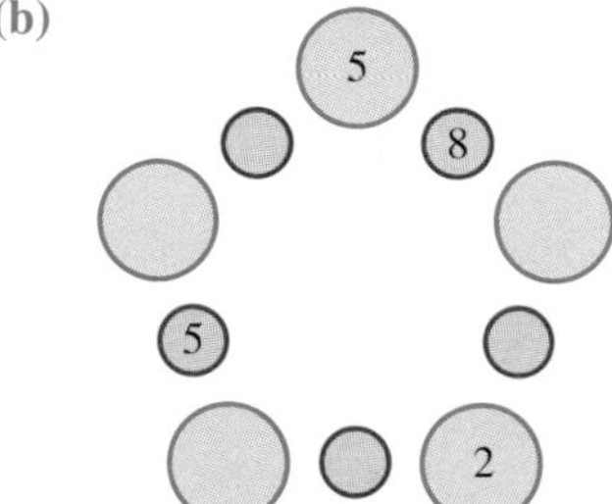

(c)

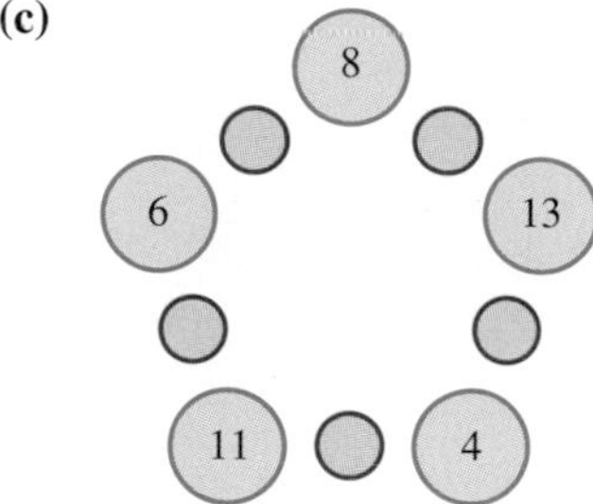

Use variables

(d) Is a puzzle like these always solvable? Explain. (*Hint:* Let u, v, w, x, and y be the numbers in the small circles.)

17. Consider a four large circle version of problem 16 as shown below.

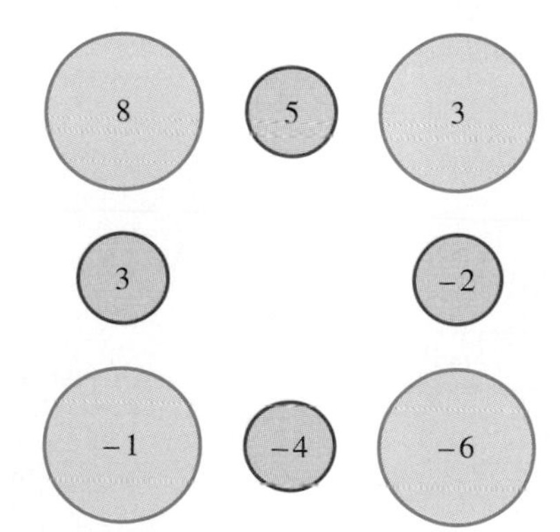

(a) Complete the diagram below so that the same relationships hold.

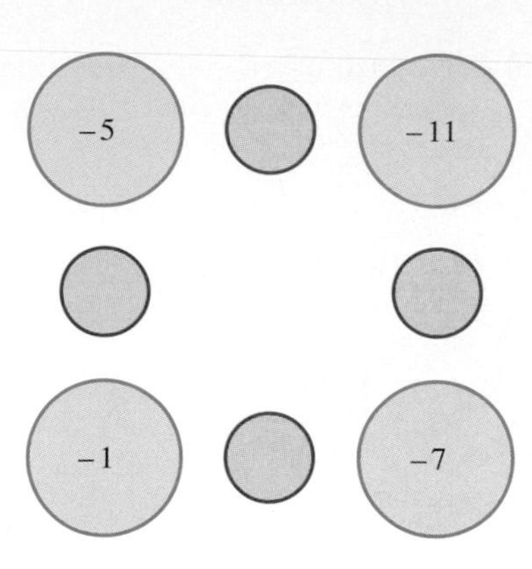

(b) Is a puzzle like this always solvable? Explain. (*Hint:* Let x, y, z, and w be the numbers in the small circles.)

(c) What must be the case to ensure that a puzzle with 6 large circles is solvable? Explain carefully.

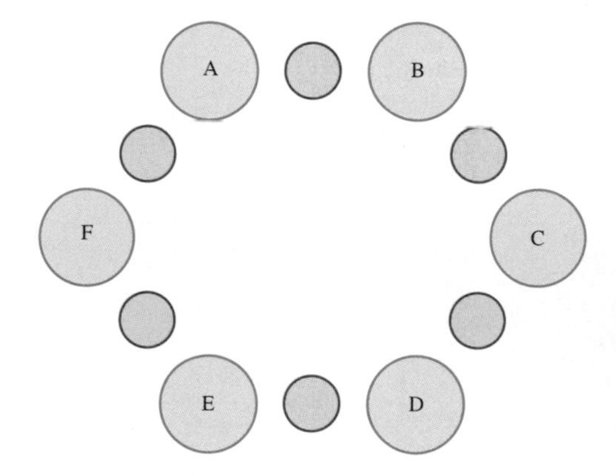

18. Use the numbers $-8, -6, -4, -2, 0, 2, 4, 6, 8$ to make a magic square. What should the sum in each row, column, and diagonal be? What should the middle number be?

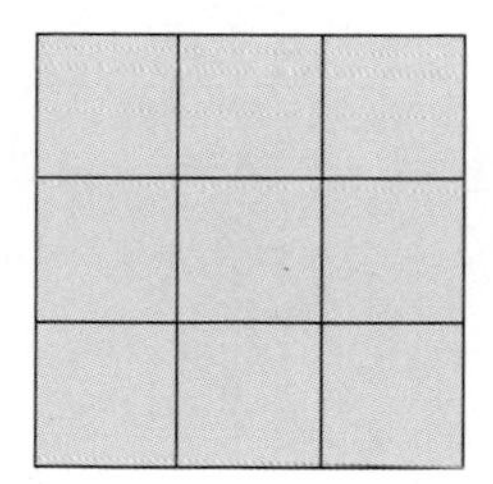

19. Use the numbers $-7, -6, -5, -1, 0, 1, 5, 6, 7$ to make a magic square.

20. **(a)** Using each of -3, -1, 1, and 3 at most once, what sums can be found? For example, three such sums are

$$3 + (-1) = 2, (-3) + (-1) = -4, \text{ and } 1 = 1.$$

(b) What strategy did you use to solve part (a)? Explain.

Making Connections

21. The Pep Club tried to raise money by raffling off a pig, agreeing that if they actually lost money on the enterprise the members would share the loss equally.

(a) If they lost $105 and there are 15 members in the club, how much did each club member have to pay?

(b) What arithmetic might be used to illustrate this story? Explain.

22. It cost $39 each to buy sweatshirts for members of the Pep Club.

(a) If there are 15 members in the Pep Club, what was the total cost?

(b) What arithmetic might be used to illustrate this story? Explain.

Using a Calculator

23. If you enter 13 in the display of your calculator and then press [+ ⇄ −], you read -13 in the display. If you press [+ ⇄ −] again what do you see in the display? What does the calculator have to say about $-(-13)$?

24. Calculate each of these without using the [M+], [M−], or [MR] keys on your calculator.

(a) $31 - 47 + 88 + 16 - 5$

(b) $57 + 165 \div (-11) + 17$

(c) $(47 + 81 - 56 + 9) \div (67 - 31 - 9)$

25. (a) Use the [M+], [M−], and [MR] keys to simultaneously calculate

$$a_n = 1^2 - 2^2 + 3^2 - \cdots + (-1)^{n-1}n^2$$

and

$$s_n = 1^2 + 2^2 + 3^2 + \cdots + n^2$$

for $n = 1, 2, 3, 4, 5,$ and 6.

(b) Compute $3s_n \div a_n$ from part (a) for $n = 1, 2, 3, 4, 5,$ and 6.

(c) Make a conjecture based on the results of part (b).

---

26. **(a)** Complete this table for all values of r and s such that $r + s = 3, 4, 5, 6, 7,$ and 8 and with $r > s > 0$.

r	s	$r + s$	$x = 2rs$	$y = r^2 - s^2$	$z = r^2 + s^2$	$x^2 + y^2$	z^2
2	1	3					
3	1	4					
4	1	5					
3	2	5					
5	1	6					
4	2	6					
6	1	7					
5	2	7					
4	3	7					
7	1	8					
6	2	8					
5	3	8					

(b) Make a conjecture on the basis of the data in the table of part (a).

For Review

27. Use the constant function on a calculator to compute the twentieth term in each of these sequences.

(a) 5, 8, 11, 14, 17, . . .

(b) 5, 10, 20, 40, 80, . . .

28. Let $a_1 = 1$, $a_2 = 2$, and $a_{n+2} = 2a_{n+1} + a_n$ for $n \geq 1$.

(a) Compute a_3, a_4, a_5, and a_6.

(b) Recall that x refers to the number in the display of your calculator and M refers to the number in memory. Complete this table using symbols and the facts from part (a)—that $a_3 = 2a_2 + a_1$, $a_4 = 2a_3 + a_2$, $a_5 = 2a_4 + a_3$, and so on. Notice that, from step 8 on, the indicated operations in this algorithm occur in a repeating pattern.

Entry	x	M	*Action*
ON/AC	0	0	Turns on and/or clears the calculator.
a_1	a_1	0	Places a_1 in the display; 0 is in memory.
x ⇄ M	0	a_1	Interchanges the a_1 in x and the 0 in M.
a_2	a_2	a_1	Places a_2 in x; keeps a_1 in M.
× 2 =	$2a_2$	a_1	. . .
M+	$2a_2$	a_3	$2a_2 + a_1 = a_3$
÷ 2 =	a_2	a_3	
x ⇄ M			
× 2 =			
M+			
÷ 2 =			
x ⇄ M			
× 2 =			
M+			
÷ 2 =			
x ⇄ M			
× 2 =			
M+			
÷ 2 =			
x ⇄ M			
⋮			

(c) List the entries that appear in the display, x, each time you press x ⇄ M after the first time.

(d) Carry out the above algorithm starting with $a_1 = 1$ and $a_2 = 2$. List the first six terms of the sequence being generated.

(e) Start with $b_1 = 2$ and $b_2 = 6$ and compute the sequence b_1, b_2, b_3, b_4, b_5, and b_6 using the algorithm of part (a).

(f) Compute $a_1 + a_3$, $a_2 + a_4$, $a_3 + a_5$, and $a_4 + a_6$ and compare with the results of part (e). Make a conjecture based on your observations.

(g) Compute $(b_1 + b_3) \div 8$, $(b_2 + b_4) \div 8$, $(b_3 + b_5) \div 8$, and $(b_4 + b_6) \div 8$. Make a conjecture based on these computations.

(h) Compute $b_1 + 2a_1$, $b_2 + 2a_2$, $b_3 + 2a_3$, $b_4 + 2a_4$, $b_5 + 2a_5$, and $b_6 + 2a_6$. Make a conjecture based on these computations.

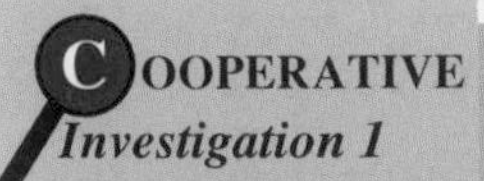

Multiplication Diagram for Integers

Consider the two crossed number lines m and n as shown.

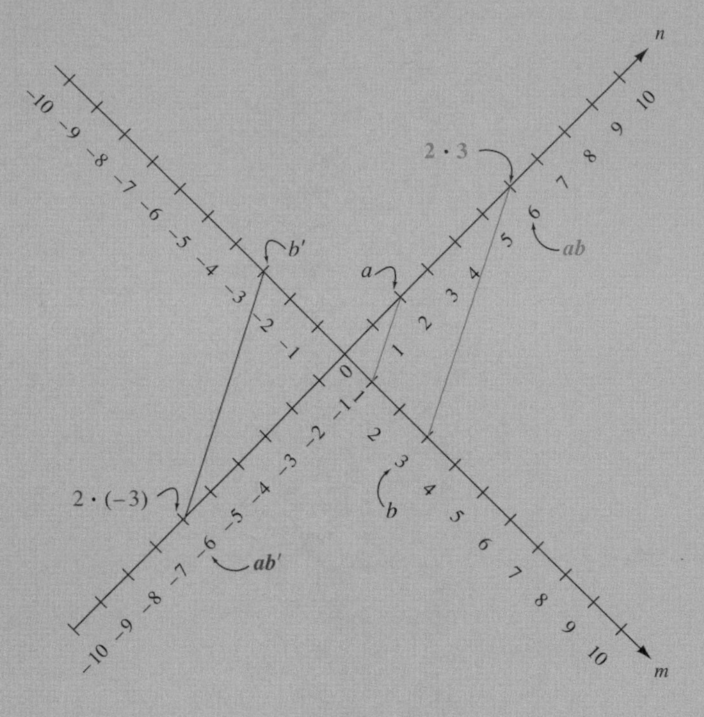

1. To multiply a by b on the crossed number lines, join 1 on m to a on n. Then draw a line through b on m parallel to the line joining 1 to a. Where this line crosses n is the product of a and b. Note that parallel lines are easily drawn by holding one draftsman's triangle fixed and sliding another such triangle along it to the desired position.

2. As shown, the drawing illustrates the products $2 \cdot 3$ and $2 \cdot (-3)$.
3. Copy this diagram and draw the lines necessary to illustrate the products $(-2) \cdot 3$, and $(-2) \cdot (-3)$.

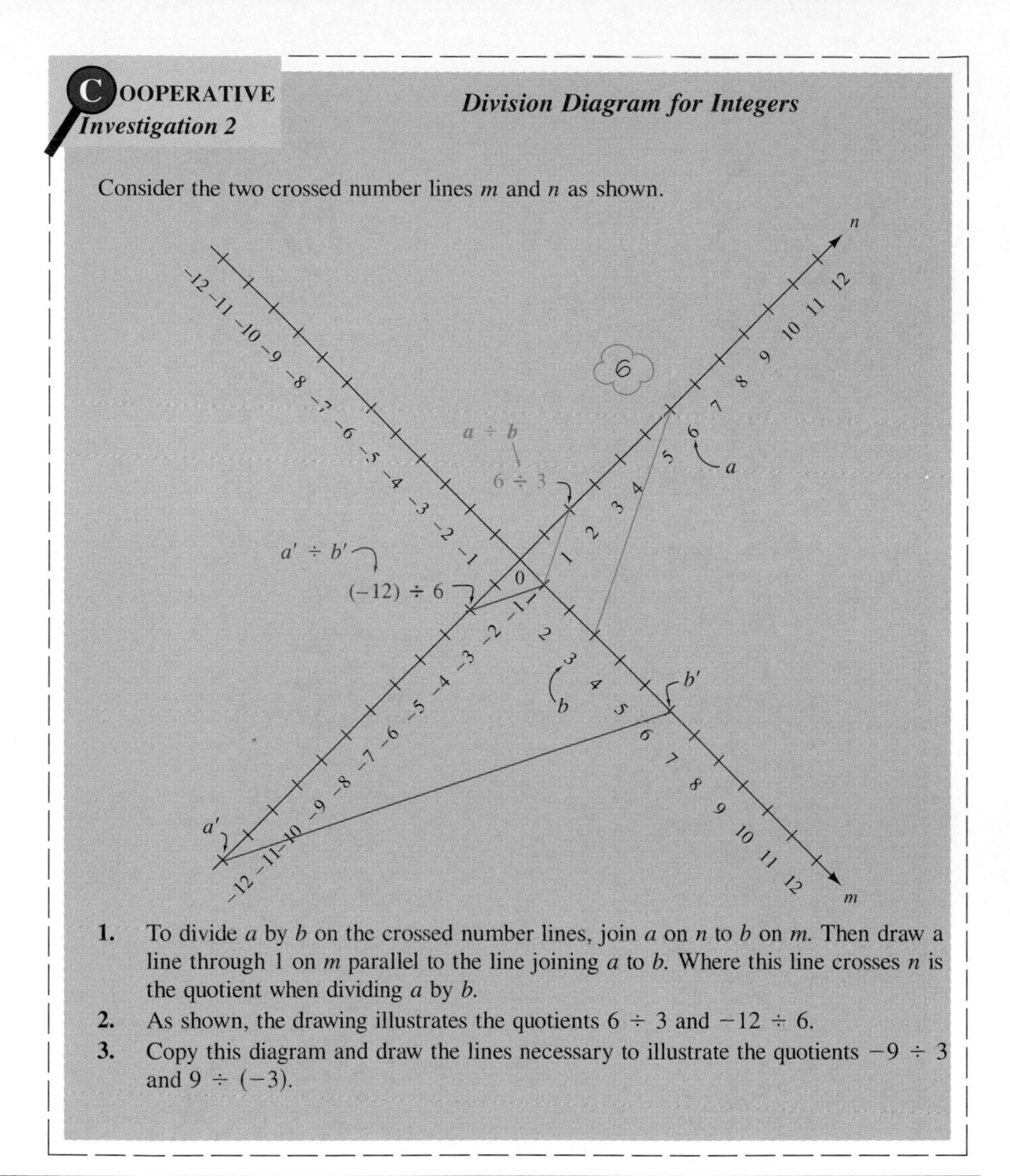

COOPERATIVE *Investigation 2*

Division Diagram for Integers

Consider the two crossed number lines m and n as shown.

1. To divide a by b on the crossed number lines, join a on n to b on m. Then draw a line through 1 on m parallel to the line joining a to b. Where this line crosses n is the quotient when dividing a by b.
2. As shown, the drawing illustrates the quotients $6 \div 3$ and $-12 \div 6$.
3. Copy this diagram and draw the lines necessary to illustrate the quotients $-9 \div 3$ and $9 \div (-3)$.

EPILOGUE Developing the Number System

In the struggle to develop numbers, history shows that it took a very long time for people to develop the natural numbers and even longer to develop the notion of zero. Even then, zero was first introduced only as a placeholder in positional systems like the Mayan, and it was not until much later that it was considered to be a number.

Somewhat surprisingly, negative numbers made their appearance on the mathematical scene before zero did. No evidence of the recognition of negative numbers, as distinct from subtrahends, appears in ancient Egyptian, Babylonian, Hindu, Chinese, or Greek mathematics. Still, the rules of signs, considered at length in this chapter, were established early on by considering such products as $(8 - 4) \cdot (7 - 5)$. The Chinese made use of such subtractions at least as early as 200 B.C., but the rules of signs in Chinese mathematics were not stated explicitly until A.D. 1299. The first mention of negative numbers in western mathematics occurred in *Arithmetica* by the Greek mathematician, Diophantus, in about A.D. 275, though spoken of there in disparaging terms. Diophantus called the equation $4x + 20 = 4$ *absurd* since it would require that $x = -4$. The first substantial use of negative numbers occurred in the work of the Hindu mathematician, Brahmagupta, in about A.D. 628, and after that time they appear in all Indian works on the subject.

In this chapter, we have used various devices to introduce the notions of negative numbers and the rules for operating with them. The fact that these numbers were considered absurd or nonnumbers by early mathematicians notwithstanding, negative numbers, like all the other numbers, are simply ideas in our minds. They are in constant use to solve problems that occur daily in modern technological society.

CLASSIC CONUNDRUM On Stealing Apples

While three watchmen were guarding an orchard, a thief slipped in and stole some apples. On his way out he met the three watchmen one after the other. To each one in turn he gave half the apples he then had and two more besides. Thus, he managed to escape with just one apple. How many did he originally steal?

CHAPTER 5 SUMMARY

Key Concepts

The thrust of this chapter has been to introduce the notion of negative numbers and the rules for operating with these numbers. Various physical and diagrammatic devices were used. The essential feature of negative numbers is that, if a is a natural number, then $-a$, is the number having the property

$$a + (-a) = (-a) + a = 0;$$

that is, $-a$ was introduced primarily as the additive inverse of a.

Properties of the Integers

The integers satisfy these properties:

- The closure properties for addition, subtraction, and multiplication

  If r and s are any integers, then
  $r + s$, $r - s$, and rs are integers.

- The commutative properties for addition and multiplication

  If r and s are any integers, then
  $r + s = s + r$, and $rs = sr$.

- The associative properties for addition and multiplication

  If r, s, and t are any integers, then
  $(r + s) + t = r + (s + t)$ and $(rs)t = r(st)$.

- The distributive property for multiplication over addition and subtraction

  If r, s, and t are any integers, then
  $r(s + t) = rs + rt$ and
  $(s + t)r = sr + tr$.
  Also, $r(s - t) = rs - st$
  and $(s - t)r = sr - tr$.

- 0 is the additive identity for the integers.

  If r is any integer,
  then $0 + r = r + 0 = r$.

- 1 is the multiplicative identity for the integers.

  If r is any integer,
  then $1 \cdot r = r \cdot 1 = r$.

- $r \cdot 0 = 0 \cdot r = 0$ for any integer r.

Rules for Integer Computation

If r, s, and t are positive integers so that $-r$, $-s$, and $-t$ are negative, then:

- $r + (-s) = r - s$ if $r > s$,
- $r + (-s) = -(s - r)$ if $r < s$,
- $(-r) + (-s) = -(r + s)$,
- $r \cdot (-s) = -(rs)$,
- $(-r) \cdot s = -(rs)$,
- $(-r) \cdot (-s) = rs$.

Also, if $s \mid r$, then:

- $r \div (-s) = -(r \div s)$,
- $(-r) \div s = -(r \div s)$,
- $(-r) \div (-s) = r \div s$.

Vocabulary and Notation

Section 5.1

Representing integers using colored counters
Representing integers using mail-time stories
Representing integers on the number line
Absolute value
The set of integers consists of
—the natural numbers or positive integers 1, 2, 3, . . .
—the number 0
—the negative integers −1, −2, −3, . . .
Representing integers by drops of colored counters

Section 5.2

Addition of integers
—using sets of colored counters
—using mail-time stories
—using the number line
Ordering the set of integers
—less than
—greater than
—less than or equal to
—greater than or equal to
—the law of trichotomy

Subtraction of integers
—using sets of colored counters
—using mail-time stories
—using the number line

Section 5.3

Multiplication of integers
—using sets of colored counters
—using Cartesian products of sets
—the rule of signs
—using mail-time stories
—using a number line

Division of integers
—using sets of colored counters
—using Cartesian product diagrams

CHAPTER REVIEW EXERCISES

Section 5.1

1. You have 15 counters colored black on one side and red on the other.
   (a) If you drop them on your desk top and 7 come up black and 8 come up red, what integer is represented?
   (b) If you drop them on your desk top and twice as many come up black as red, what number is being represented?
   (c) What numbers are represented by all possible drops of the 15 counters?
2. (a) If the mail carrier brings you a check for \$12 are you richer or poorer and by how much? What integer does this situation illustrate?
   (b) If the mail carrier brings you a bill for \$37 are you richer or poorer and by how much? What integer does this situation illustrate?
3. (a) 12° above zero illustrates what integer?
   (b) 24° below zero illustrates what integer?
4. (a) List five different "drops" of colored counters that represent the integer -5.
   (b) List five different drops of colored counters that represent the integer 6.
5. (a) Give a mail-time story that illustrates -85.
   (b) Give a mail-time story that illustrates 47.
6. (a) What number must you add to 44 to obtain 0?
   (b) What number must you add to -61 to obtain 0?

Section 5.2

7. What addition is represented by this diagram?

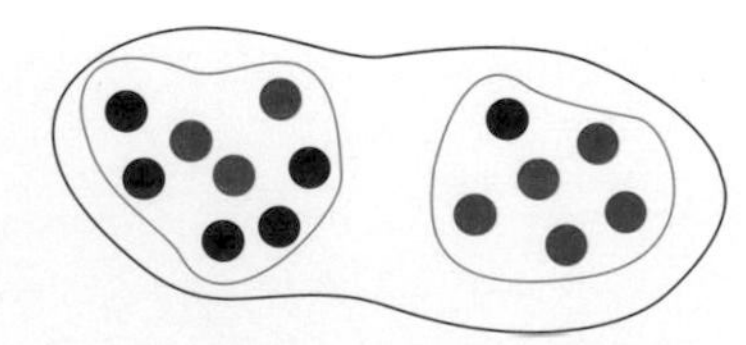

8. What subtraction is represented by this diagram?

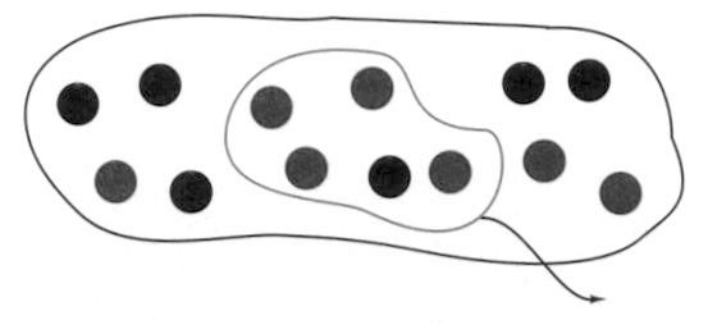

9. What additions and subtractions are represented by these mail-time stories?
   (a) At mail time the letter carrier brings you a check for \$45 and a bill for \$68. Are you richer or poorer and by how much?
   (b) At mail time the letter carrier brings you a check for \$45 and takes away a bill for \$68 left previously. Are you richer or poorer and by how much?
10. Perform these additions and subtractions.
    (a) $5 + (-7)$ (b) $(-27) - (-5)$
    (c) $(-27) + (-5)$ (d) $5 - (-7)$
    (e) $8 - (-12)$ (f) $8 - 12$
11. (a) If it is 15° below zero and the temperature falls 12°, what temperature is it?
    (b) What arithmetic does this situation illustrate?
12. (a) Dina's bank account was overdrawn by \$12. What was her balance after she deposited \$37 she earned working at a local pizza parlor?
    (b) What arithmetic does this situation illustrate?

Section 5.3

13. What multiplications are represented by these two diagrams of colored counters?
    (a)

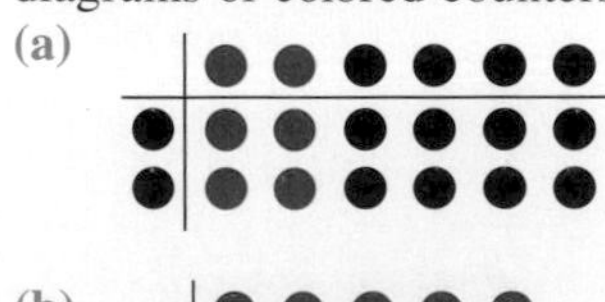

    (b)

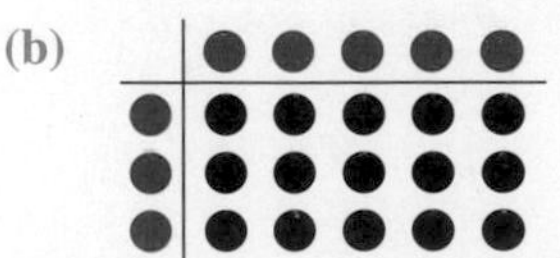

14. (a) Complete this diagram so that it represents the product $7 \cdot (-5) = -35$. How many rows must the diagram have?

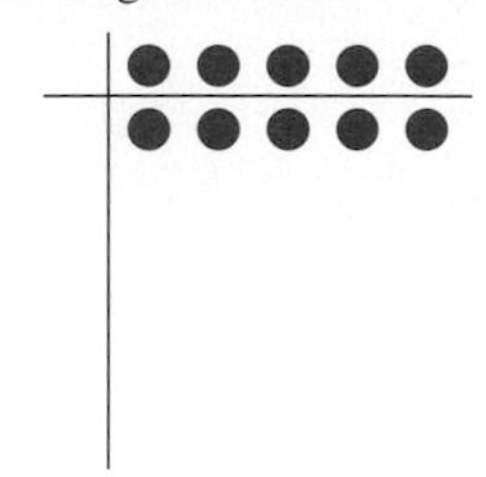

(b) Asking how many rows there should be in the diagram of part (a) is the same as asking for what number r it is true that $r \cdot (-5) = -35$; that is, asking for the determination of $r = (-35) \div (-5)$. Thus, the diagram of part (a) can also be viewed as a division diagram. Draw a similar diagram to illustrate the quotient $(-28) \div (-4)$.

15. Draw the Cartesian product diagram to represent each of these operations and give the result in each case.

(a) $(-8) \cdot (-4)$ (b) $(-15) \div 5$

16. Draw the Cartesian product diagram to illustrate each of these quotients and give the result in each case.

(a) $54 \div (-9)$ (b) $(-54) \div (-9)$

17. Perform each of these computations

(a) $(-8) \cdot (-7)$ (b) $8 \cdot (-7)$
(c) $(-8) \cdot 7$ (d) $84 \div (-12)$
(e) $(-84) \div 7$ (f) $(-84) \div (-7)$

18. Write a mail-time story to illustrate each of these products.

(a) $7 \cdot (12)$ (b) $(-7) \cdot (13)$ (c) $(-7) \cdot (-13)$

19. If a and b are integers, the greatest common divisor of a and b is the largest positive integer dividing both a and b. Compute each of the following.

(a) GCD(255, -39) (b) GCD(-1001, 2651)

20. If n is an integer not divisible by 2 or 3, show that $n^2 - 1$ is divisible by 24. (*Hint:* By the division algorithm, n must be of one of these forms: $6q$, $6q + 1$, $6q + 2$, $6q + 3$, $6q + 4$, or $6q + 5$.)

CHAPTER TEST

1. Complete this Cartesian product division diagram and tell what quotient it illustrates.

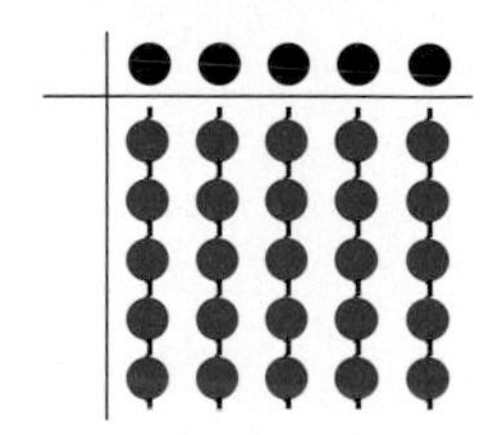

2. Perform each of these computations.

(a) $(-7) + (-19)$ (b) $(-7) - (-19)$
(c) $7 - (-19)$ (d) $7 + (-19)$
(e) $(-6859) \div 19$ (f) $(-24) \cdot 17$
(g) $36 \cdot (-24)$ (h) $(-1155) \div (-11)$
(i) $0 \div (-27)$

3. The least common multiple of integers a and b is the least positive integer divisible by both a and b. Compute LCM(-240, 54)

4. At mail time, if the mail carrier took away 5 bills for \$27 each are you richer or poorer and by how much? What calculation does this illustrate?

5. Draw a number line diagram to illustrate each of these calculations

(a) $(-7) + 10$ (b) $10 - (-7)$ (c) $7 \cdot (-5)$

6. Draw a diagram of colored counters to illustrate the subtraction $7 - (-4) = 11$.

7. Mary Lou's checkbook balance was \$129. What was it after she deposited \$341 and then wrote checks for \$13, \$47, and \$29? What arithmetic does this illustrate?

8. Tammie, Jody, and Nora formed a small club. After a party celebrating the first anniversary of the club's existence they owed the local pizzeria \$27. The bill was paid and shared equally by the three girls. Was each one richer or poorer and by how much? What arithmetic does this illustrate?

9. The Fibonacci sequence is formed by adding any two consecutive numbers in the sequence to obtain the next. If the same rule is followed in each of these, correctly fill in the blanks.

(a) -5, -3, ______, ______, ______, ______

(b) 7, ______, 2, ______, ______, ______

(c) 6, ______, ______, ______, -12, ______

10. (a) What sums can be obtained using only the numbers -10, -5, -2, -1, 1, 2, 5, and 10, each at most once and without using any number with its double?

(b) Do the representations in part (a) appear to be unique?

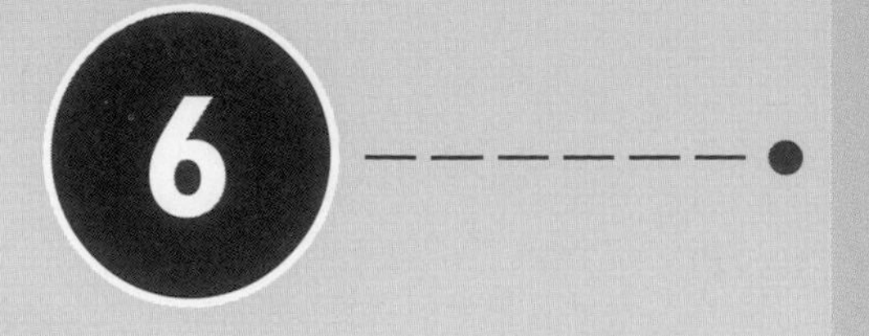

Fractions and Rational Numbers

Computer Support for this Chapter

In this chapter you might find the following programs on your computer disk useful:

- DIFFY
- DIVVY

HANDS ON Subdividing Segments

Suppose you are given a line segment $\overline{AB}$, where A and B are endpoints.

A B

If you could divide it into four equal pieces, you could find the points $\frac{1}{4}$, $\frac{2}{4}$, and $\frac{3}{4}$ of the way along the segment. A ruler is not too helpful, since the length of the segment can only be measured approximately.

Here is a convenient method for subdividing a segment into any number of parts of precisely equal length.

Materials Needed

A sheet of lined notebook paper and a 3″ × 5″ or 4″ × 6″ file card. Number the lines of the sheet of notebook paper this way, with the bottom line numbered 0.

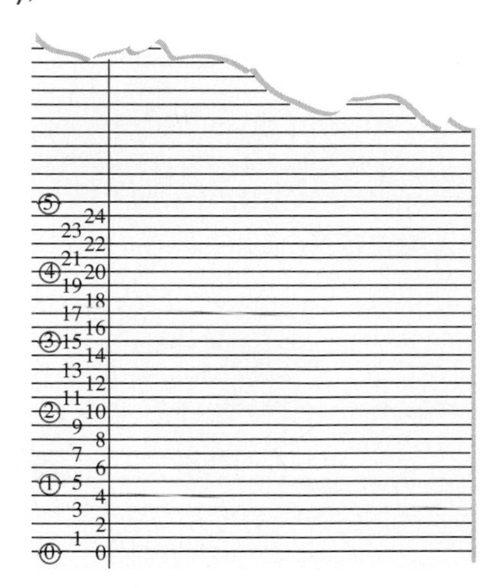

Procedure

Lay the edge of the file card against the given segment $\overline{AB}$ and transfer its length to the card. Let A' and B' denote the points on the file card corresponding to A and B.

Now put A' on the 0 line of the lined paper and rotate the card until point B' falls on the 4 line. The 1 line touches the edge of the file card $\frac{1}{4}$ of the way from A' toward B'. Similarly, the 2 and 3 lines correspond to $\frac{2}{4}$ and $\frac{3}{4}$ of the way. The 5 line is $\frac{5}{4}$ of the way; that is, it is a full length plus an additional $\frac{1}{4}$ of the length of $\overline{A'B'}$. If $\overline{AB}$ is a short segment, or a larger denominator is used, the lines with uncircled numbers can replace the lines with circled numbers.

Once the file card is marked, the distances can be transferred back to the original segment $\overline{AB}$.

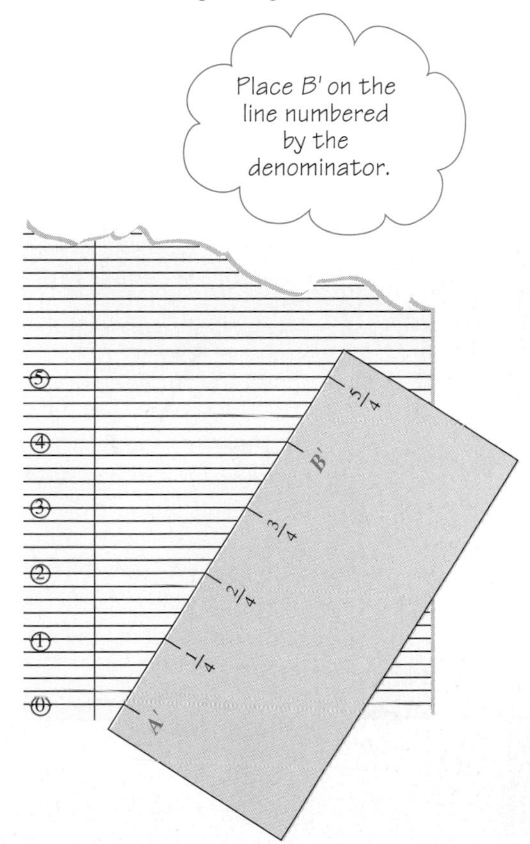

Activities

In 1 through 8 below, a segment and its length is given. Use the procedure just described to find the point with the desired distance from the left endpoint. In 6, 7, and 8, you may need to use a two-step procedure.

1. A 1 unit B Find C so that $AC = 2/3$.

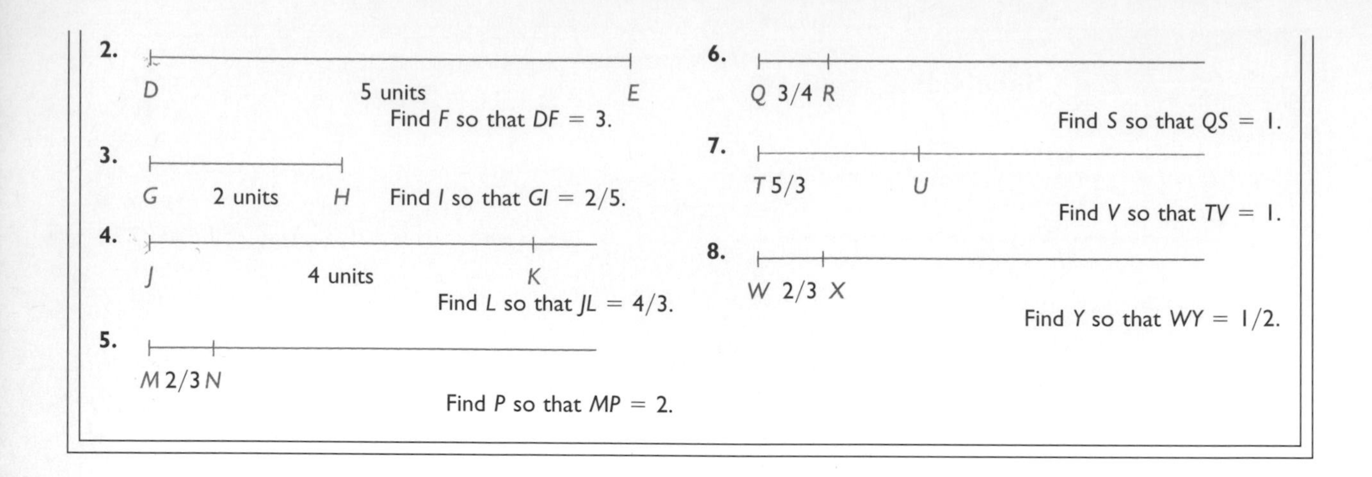

2.
D
5 units
E
Find F so that DF = 3.
3.
G
2 units
H
Find I so that GI = 2/5.
4.
J
4 units
K
Find L so that JL = 4/3.
5.
M 2/3 N
Find P so that MP = 2.
6.
Q 3/4 R
Find S so that QS = 1.
7.
T 5/3
U
Find V so that TV = 1.
8.
W 2/3 X
Find Y so that WY = 1/2.

CONNECTIONS

Numbers for Parts and Pieces

In Chapter 5 the set of whole numbers, $W = \{0, 1, 2, \ldots\}$ was extended to the set of integers $I = \{\ldots, -2, -1, 0, 1, 2, 3, \ldots\}$. In the set of integers it becomes possible to solve problems that are more difficult or even impossible to solve in the whole number system. In particular, the integers are closed under subtraction. Therefore equation $x + a = b$ has the unique solution $x = b - a$ for any integers a and b.

In this chapter the set of integers I is extended to a number system called the rational numbers, in which each number is represented by a pair of integers called a fraction. Section 6.1 introduces the basic concepts of fractions and rational numbers. A fraction is an ordered pair of integers a and b, $b \neq 0$, written in the form $\frac{a}{b}$. Pictorial and physical models are used to illustrate the basic concepts of equivalent fractions, fractions in simplest form, common denominators, and inequality. In Section 6.2, the arithmetic operations—addition, subtraction, multiplication, and division—are defined for rational numbers by introducing corresponding operations for fractions. The properties of the rational number system are explored in the concluding Section 6.3. Of special importance is the existence of reciprocals of nonzero fractions, which means that the rational number system is closed under division. The concluding section also discusses estimations and mental arithmetic, the use of calculators, and practical applications of the rational numbers.

6.1 The Basic Concepts of Fractions and Rational Numbers

The whole numbers, as suggested by the word "whole," arise most often in *counting* problems, where the units or objects being counted cannot be subdivided into smaller parts. For example, Earth has one moon and Mars has two moons, but no planet can have some number of moons between one and two. If Earth's moon were to split into two parts, we would have two moons! In many situations, however, the objects or quantities of interest can be meaningfully subdivided. For example, when just one cookie remains, two young children may well display a good understanding of what "one half" of a cookie is all about. Similarly, two pizzas can be shared equally among three people.

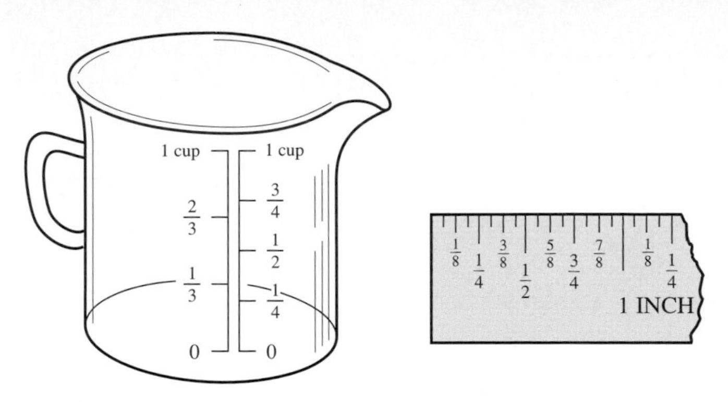

Figure 6.1
Fractions are used on measuring cups and rulers to indicate subdivisions of the basic unit.

Fractions were first introduced in *measurement* problems, to express a quantity that is less than a whole unit. Indeed, the word "fraction" comes from the Latin word *fractio,* meaning "the act of breaking into pieces." Figure 6.1 shows a measuring cup and a ruler, two common items on which fractions appear.

Fractions indicate amounts or distances in which the basic unit is subdivided into a whole number of equal parts. For example, if a pizza is cut into 8 equal pieces and 3 of these are eaten, then the fraction $\frac{3}{8}$ expresses the amount of a pizza consumed as shown in Figure 6.2. Similarly we say that $\frac{5}{8}$ of the pizza remains. In this example, the unit of measurement is one pizza.

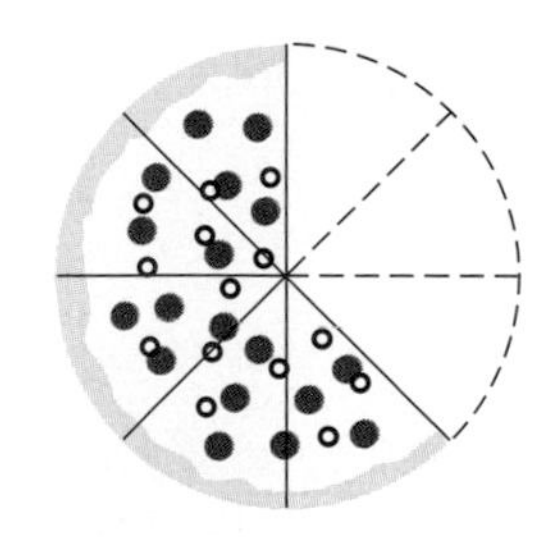

Figure 6.2
$\frac{3}{8}$ of the pizza has been eaten and $\frac{5}{8}$ remains

More generally expressions of the form $\frac{3}{8}, \frac{-4}{12}, \frac{31}{-4}, \frac{0}{1}$ and so on, are examples of **fractions.**

"WHAT I DO IS BELIEVE 1/5 OF WHAT THE CONVICTED PERJURERS SAY, 2/5 OF WHAT THE SUSPECTED PERJURERS SAY, AND 3/5 OF WHAT ANYONE ELSE SAYS."

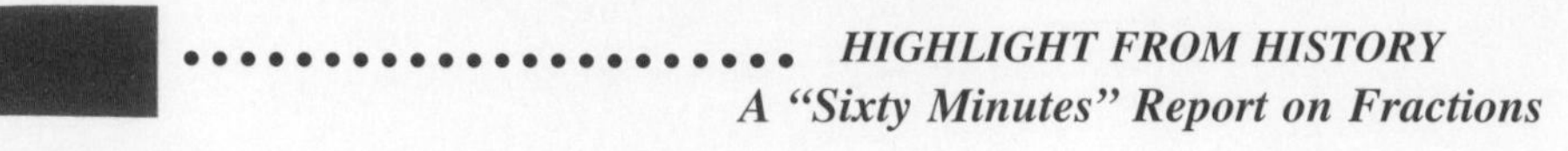

HIGHLIGHT FROM HISTORY
A "Sixty Minutes" Report on Fractions

The division of an hour into 60 minutes can be traced back several thousand years to the Babylonians, who developed a positional base sixty numeration system. It is not clear why 60 came into use, but one theory suggests that fractions were responsible. Suppose a system of weight measurement is based on a unit which is subdivided into thirds and halves and grouped into tens. Next, suppose a second system of measurement whose unit is 60 times larger is required, for political or social reasons perhaps, to be merged with the first system, In base sixty, $1/3$, $1/2$, and $2/3$ of the large unit would be easily expressed as multiples of the smaller unit, or even multiples of a group of ten small units.

DEFINITION Fractions

A **fraction** is an ordered pair of integers a and b, $b \neq 0$, written $\frac{a}{b}$ or a/b. The integer a is called the **numerator** of the fraction, and the integer b is called the **denominator** of the fraction.

Any integer m is viewed as the fraction $\frac{m}{1}$. Usually the denominator 1 is not written explicitly. For example, we write 3 instead of $\frac{3}{1}$.

Models for Fractions

The "pie" diagram shown in Figure 6.2 is a representation of a fraction by a pictorial model. In a pictorial or physical model of, say $\frac{3}{5}$, a "whole" unit is subdivided into 5 congruent parts, and 3 of the small parts are distinguished by shading or coloring. Colored regions, sets, fraction strips, and the fraction number line are useful models for the classroom. In each model it is essential that the unit be clearly identified, and it must be apparent how the unit has been subdivided.

Colored Regions

A shape is chosen to represent the unit and is then subdivided into subregions of equal size. A fraction is visualized by coloring some of the subregions. Three examples are shown in Figure 6.3. Colored region models are sometimes called **area** models, although area is treated quite informally.

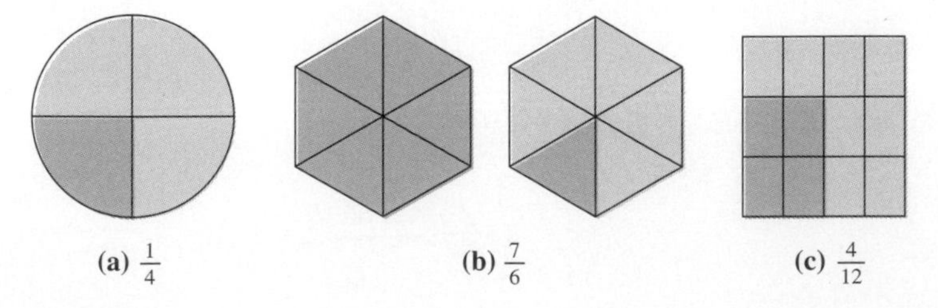

Figure 6.3
Some colored region models for fractions

The Set Model

The unit is a finite set of objects, U. Each subset A of U corresponds to the fraction $\frac{n(A)}{n(U)}$. For example, the set of 10 apples shown in Figure 6.4 contains a subset

Figure 6.4
The set model depicts that $\frac{3}{10}$ of the apples are wormy.

of 3 which are wormy. Therefore we would say that $\frac{3}{10}$ of the apples are wormy. In Chapter 9 we will see that the set model of fractions is particularly useful in probability.

Fraction Strips

Here the unit is defined by a rectangular strip of cardstock. A fraction, such as $\frac{3}{6}$, is modeled by shading 3 of 6 equally sized subrectangles of the card. Sample fraction strips are shown in Figure 6.5. A set of fraction strips typically contains strips for the denominators 1, 2, 3, 4, 6, 8, and 12.

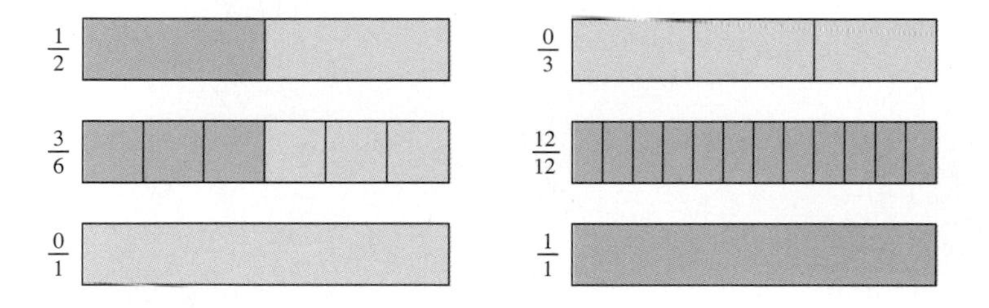

Figure 6.5
Examples of fractions modeled by fraction strips

The Number Line Model

Once the points corresponding to 0 and 1 are assigned on a number line, all of the points corresponding to the integers are determined. A fraction such as $\frac{5}{4}$ is assigned to a point along the number line by subdividing the unit interval into 4 equal parts, and then counting off 5 of these lengths to the right of 0. Typical fractions are shown in Figure 6.6. It should be noticed that the same distance from 0 can be named by different fraction expressions. For example $\frac{1}{2}$ and $\frac{2}{4}$ both correspond to the same distance. The number line model has the advantage of allowing negative fractions such as $-\frac{3}{4}$ to be visualized.

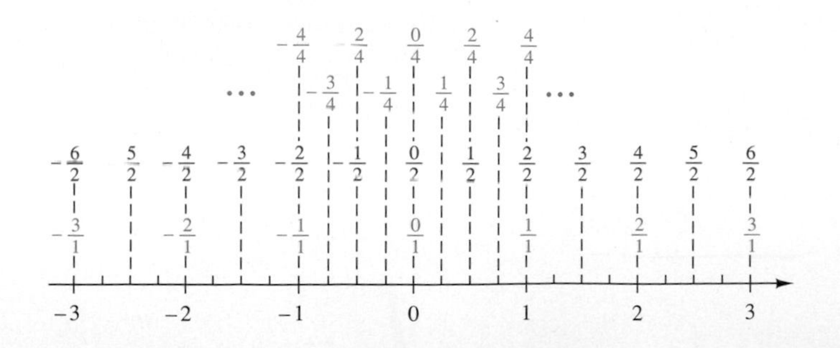

Figure 6.6
Fractions on the number line model

FROM THE NCTM STANDARDS **Fractions**

All work at the K–4 level should involve fractions that are useful in everyday life, that is, fractions that can be easily modeled. Initial work with fractions should draw on children's experiences in sharing, such as asking four children to share a candy bar. The concept of a unit and its subdivision into equal parts is fundamental to understanding fractions and decimals, whether the quantity to be divided is a rectangular candy bar, a handful of jelly beans, or a piece of licorice. Initial instruction needs to emphasize oral language (one-fourth, two-thirds) and connect it to the models. Many productive activities can be used for initial instruction, such as folding paper strips into equal parts and describing the kind of parts (e.g., fifths) and the amount being considered (e.g., two-fifths).

SOURCE: From *Curriculum and Evaluation Standards for School Mathematics Grades K–4,* p. 57.

Equivalent Fractions

In Figure 6.7, the fraction strip representing $\frac{2}{3}$ is further subdivided by the vertical dashed lines to show that $\frac{4}{6}$, $\frac{6}{9}$, and $\frac{8}{12}$ are other fractions which express the *same* shaded portion of a whole strip.

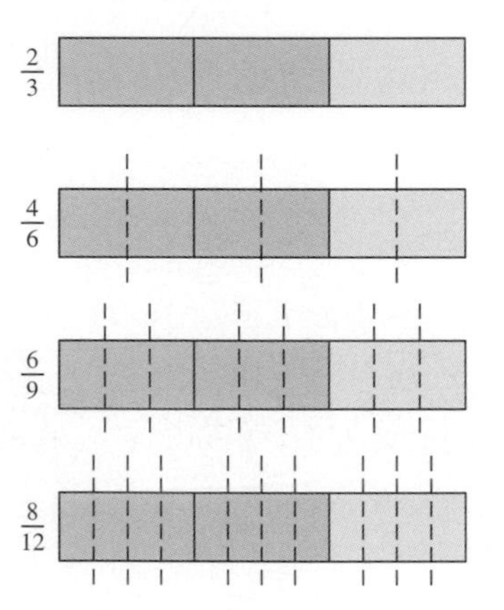

Figure 6.7
The fraction strip model showing that $\frac{2}{3}$, $\frac{4}{6}$, $\frac{6}{9}$, and $\frac{8}{12}$ are equivalent fractions

Fractions that express the same quantity are called **equivalent fractions.** The equality symbol, =, is used to signify that fractions are equivalent, so we write

$$\frac{2}{3} = \frac{4}{6} = \frac{6}{9} = \frac{8}{12}.$$

The number of additional dashed lines between each vertical pair of solid lines shown in Figure 6.6 can be increased arbitrarily; therefore,

$$\frac{2}{3} = \frac{2 \cdot n}{3 \cdot n}$$

for $n = 1, 2, \ldots$. In this way, beginning with one fraction $\frac{a}{b}$, an infinite list, $\frac{2a}{2b}$, $\frac{3a}{3b}$, $\frac{4a}{4b}$, . . . , of equivalent fractions is obtained.

Equivalent fractions can also be obtained by dividing both the numerator and denominator by a common divisor. For example, the numerator and denominator of $\frac{35}{21}$ are each divisible by 7, so

$$\frac{35}{21} = \frac{35 \div 7}{21 \div 7} = \frac{5}{3}$$

In summary, we have the following important property.

PROPERTY The Fundamental Law of Fractions

Let $\frac{a}{b}$ be a fraction. Then

$$\frac{a}{b} = \frac{an}{bn}, \text{ for any integer } n \neq 0;$$

and

$$\frac{a}{b} = \frac{a \div d}{b \div d}, \text{ for any common divisor } d \text{ of } a \text{ and } b.$$

Now suppose we are given two fractions, say $\frac{3}{12}$ and $\frac{2}{8}$, and wish to know if they are equivalent. From the fundamental law of fractions we know that

$$\frac{3}{12} = \frac{3 \cdot 8}{12 \cdot 8} = \frac{24}{96} \text{ and } \frac{2}{8} = \frac{2 \cdot 12}{8 \cdot 12} = \frac{24}{96}.$$

Thus, $\frac{3}{12}$ and $\frac{2}{8}$ are equivalent fractions. This illustrates the following theorem.

THEOREM A Characterization of Equivalent Fractions

The fractions $\frac{a}{b}$ and $\frac{c}{d}$ are **equivalent** if, and only if, $ad = bc$.

EXAMPLE 6.1

Showing the Equivalence of Fractions

(a) Show that $\frac{6}{14} = \frac{9}{21}$.

(b) Show that $\frac{-23}{47} = \frac{2231}{-4559}$.

(c) Find m if $\frac{m}{6} = \frac{10}{15}$.

(d) Find *all* fractions that are equivalent to $\frac{2}{3}$.

SOLUTION

(a) Since $6 \cdot 21 = 126 = 9 \cdot 14$, $\frac{6}{14} = \frac{9}{21}$ by the preceding theorem.

(b) Using a calculator (or paper and pencil), $(-23) \cdot (-4559) = 104{,}857$ and $2231 \cdot 47 = 104{,}857$.

(c) $\frac{m}{6} = \frac{10}{15}$ if, and only if, $m \cdot 15 = 10 \cdot 6$. Thus, $m = 60 \div 15 = 4$.

(d) Suppose $\frac{a}{b} = \frac{2}{3}$, so that $3a = 2b$. Since 3 divides the left side, then 3 divides the right side; that is, $3 \mid (2b)$. But 3 does not divide 2, so $3 \mid b$. Thus, for some integer n we have $3n = b$. Since $b \neq 0$, we know that $n \neq 0$. Thus, $3a = 2b = 2(3n)$, so $a = 2n$. It follows that every fraction $\frac{a}{b}$ equivalent to $\frac{2}{3}$ can be written as $\frac{2 \cdot n}{3 \cdot n}$ for some integer n. Therefore, the set of all fractions equivalent to $\frac{2}{3}$ is

$$\left\{ \ldots, \frac{-6}{-9}, \frac{-4}{-6}, \frac{-2}{-3}, \frac{2}{3}, \frac{4}{6}, \frac{6}{9}, \ldots \right\}.$$ ■

Fractions in Simplest Form

Often it is preferable to use the simplest equivalent form of a fraction instead of a more complicated fraction. For example, consider the fraction $\frac{34}{-51}$. By the fundamental law of fractions

$$\frac{34}{-51} = \frac{34 \cdot (-1)}{(-51) \cdot (-1)} = \frac{-34}{51}.$$

This gives us an equivalent fraction with a positive denominator. Next, 17 is a common divisor factor of -34 and 51, since $-34 = -2 \cdot 17$ and $51 = 3 \cdot 17$. The fundamental law of fractions now shows that

$$\frac{-34}{51} = \frac{-2 \cdot 17}{3 \cdot 17} = \frac{-2}{3}.$$

Since -2 and 3 have no common divisor larger than 1, no further simplification is possible. We say that $\frac{-2}{3}$ has **simplest** or **reduced form** or is in **lowest terms.** Here is the general definition.

DEFINITION Fractions in Simplest Form

A fraction $\frac{a}{b}$ is in **simplest form** if a and b have no common divisor larger than 1 and b is positive.

There are several ways to determine the simplest form of a fraction $\frac{a}{b}$.

Method 1. Divide successively by common factors. Suppose we want to write $\frac{560}{960}$ in simplest form. Using the fundamental law of fractions, we successively divide both numerator and denominator by common factors. Since 560 and 960 are both divisible by 10, it follows that

$$\frac{560}{960} = \frac{56}{96}.$$

But both 56 and 96 are even (that is, divisible by 2), so

$56 \div 2 = 28$
$96 \div 2 = 48$

$$\frac{56}{96} = \frac{28}{48}.$$

Again, 28 and 48 are easily seen to be divisible by 4 so that

$28 \div 4 = 7$
$48 \div 4 = 12$

$$\frac{28}{48} = \frac{7}{12}.$$

Finally, since 7 and 12 have no common factor other than 1, $\frac{7}{12}$ is the simplest form of $\frac{560}{960}$. Indeed, one might quickly and efficiently carry out this simplification as here

$$\frac{\overset{\overset{7}{\cancel{28}}}{\cancel{560}}}{\underset{\underset{12}{\cancel{48}}}{\cancel{960}}} \quad \text{or} \quad \frac{560}{960} = \frac{56}{96} = \frac{28}{48} = \frac{7}{12}$$

to obtain the desired result.

Method 2. Divide *a* and *b* by GCD(*a*, *b*). Using the ideas of Chapter 4, we determine that GCD(560, 960) = 80. Therefore, $\frac{560}{960} = \frac{560 \div 80}{960 \div 80} = \frac{7}{12}$.

Method 3. Divide by the common factors in the prime factorizations of *a* and *b*. Using this method,

$$\frac{560}{960} = \frac{2^4 \cdot 5 \cdot 7}{2^6 \cdot 3 \cdot 5} = \frac{7}{2^2 \cdot 3} = \frac{7}{12}.$$

Method 4. Use a fraction calculator. On the *Math Explorer,* the fraction $\frac{560}{960}$ can be entered by pressing 560 [/] 960. To simplify, pressing [Simp] [=] gives 280/480 in the display. Pressing [x ⇄ y] displays the common factor 2 that was divided into both 560 and 960. Pressing [x ⇄ y] again returns 280/480 to the display, and [Simp] [=] gives 140/240, where a second factor of 2 has been divided out. Continuing the sequence [Simp] [=] three more steps gives the final result of 7/12. By making a list of the common factors, 2, 2, 2, 2, 5, the *Math Explorer* makes it easy to compute GCD(560, 960) = 2 · 2 · 2 · 2 · 5 = 80.

EXAMPLE 6.2

Simplifying Fractions

Find the simplest form of each fraction.

(a) $\frac{240}{72}$ **(b)** $\frac{-450}{1500}$ **(c)** $\frac{294}{-84}$ **(d)** $\frac{399}{483}$.

SOLUTION

(a) By Method 1, $\frac{240}{72} = \frac{120}{36} = \frac{60}{18} = \frac{10}{3}$, where the successive common factors 2, 2, and 6 were divided into the numerator and denominator.

(b) Since $1500 = 3 \cdot 450 + 150$ and $450 = 3 \cdot 150 + 0$, the Euclidean algorithm shows that GCD(450, 1500) = 150. Thus, by Method 2,

$$\frac{-450}{1500} = \frac{-450 \div 150}{1500 \div 150} = \frac{-3}{10}.$$

(c) Using Method 3 this time, we find

$$\frac{294}{-84} = \frac{2 \cdot 3 \cdot 7^2}{(-2) \cdot 2 \cdot 3 \cdot 7} = \frac{7}{-2} = \frac{-7}{2}.$$

(d) If 399 [/] 483 is entered on the *Math Explorer,* then pressing [Simp] [=] repeatedly gives first 133/161 and next 19/23. Therefore $\frac{399}{483} = \frac{19}{23}$. The common factors are 3 and 7, so GCD(399, 483) = $3 \cdot 7 = 21$. ■

Common Denominators

When working with two fractions, it is important to know how they can be replaced with equivalent fractions with the *same* denominator. For example, $\frac{5}{8}$ and $\frac{7}{10}$ can each be replaced by equivalent fractions with the **common denominator** $8 \cdot 10 = 80$ so that

$$\frac{5}{8} = \frac{5 \cdot 10}{8 \cdot 10} = \frac{50}{80} \quad \text{and} \quad \frac{7}{10} = \frac{7 \cdot 8}{10 \cdot 8} = \frac{56}{80}.$$

In the same way any two fractions $\frac{a}{b}$ and $\frac{c}{d}$ can be rewritten with the common denominator $b \cdot d$, since $\frac{a}{b} = \frac{a \cdot d}{b \cdot d}$ and $\frac{c}{d} = \frac{c \cdot b}{d \cdot b}$.

It is sometimes worthwhile to find the common positive denominator which is as small as possible. Assuming $\frac{a}{b}$ and $\frac{c}{d}$ are in simplest form we require a common denominator that is a multiple of both b and d. The least such common multiple is

JUST FOR FUN

Suspicious Simplifications

Dividing the numerator and denominator by a common factor will simplify a fraction. What do you think of the following "shortcut" method?

$$\frac{1\not{6}}{\not{6}4} = \frac{1}{4}, \quad \frac{1\not{9}}{\not{9}5} = \frac{1}{5}, \quad \frac{4\not{9}}{\not{9}8} = \frac{4}{8}$$

Does the shortcut always work? Can you find any other fraction where this shortcut works?

called the **least common denominator** and is the least common multiple of b and d. For the example $\frac{5}{8}$ and $\frac{7}{10}$, we would calculate LCM(8, 10) = 40, so 40 is the least common denominator. Therefore,

$$\frac{5}{8} = \frac{5 \cdot 5}{8 \cdot 5} = \frac{25}{40} \quad \text{and} \quad \frac{7}{10} = \frac{7 \cdot 4}{10 \cdot 4} = \frac{28}{40}.$$

EXAMPLE 6.3

Finding Common Denominators

Find equivalent fractions with a common denominator.

(a) $\frac{7}{12}$ and $\frac{1}{4}$

(b) $\frac{9}{8}$ and $\frac{-12}{7}$

(c) $\frac{14}{-16}$ and $\frac{11}{12}$

(d) $\frac{3}{4}$, $\frac{5}{8}$, and $\frac{2}{3}$

SOLUTION

(a) Since $4 \mid 12$, only the denominator in the second fraction must be changed: $\frac{1}{4} = \frac{1 \cdot 3}{4 \cdot 3} = \frac{3}{12}$.

(b) Since 8 and 7 have no common factors other than 1, $8 \cdot 7 = 56$ is the least common denominator:

$$\frac{9}{8} = \frac{9 \cdot 7}{8 \cdot 7} = \frac{63}{56}, \quad \frac{-12}{7} = \frac{-12 \cdot 8}{7 \cdot 8} = \frac{-96}{56}.$$

(c) First we simplify $\frac{14}{-16}$ to $\frac{-7}{8}$. Since 8 and 12 have 4 as their greatest common divisor, we see that LCM(8, 12) = $8 \cdot 12$/GCD(8, 12) = 96/4 = 24. Thus,

$$\frac{14}{-16} = \frac{-7}{8} = \frac{-7 \cdot 3}{8 \cdot 3} = \frac{-21}{24}, \quad \frac{11}{12} = \frac{11 \cdot 2}{12 \cdot 2} = \frac{22}{24}.$$

The LCM could have been computed directly, by finding the smallest common element in the sequences of positive multiples 8, 16, 24, 32, . . . and 12, 24, 36, ...

(d) With three fractions, it is still possible to use the product of all of the denominators as a common denominator. Since $4 \cdot 8 \cdot 3 = 96$, we have

$$\frac{3}{4} = \frac{3 \cdot 8 \cdot 3}{4 \cdot 8 \cdot 3} = \frac{72}{96}, \quad \frac{5}{8} = \frac{5 \cdot 4 \cdot 3}{8 \cdot 4 \cdot 3} = \frac{60}{96}, \quad \frac{2}{3} = \frac{2 \cdot 4 \cdot 8}{3 \cdot 4 \cdot 8} = \frac{64}{96}.$$

Alternatively, LCM(4, 8, 3) = 24, so we have

$$\frac{3}{4} = \frac{3 \cdot 6}{4 \cdot 6} = \frac{18}{24}, \quad \frac{5}{8} = \frac{5 \cdot 3}{8 \cdot 3} = \frac{15}{24}, \quad \frac{2}{3} = \frac{2 \cdot 8}{3 \cdot 8} = \frac{16}{24}.$$

This expresses the three fractions as equivalent fractions with the least common denominator. ■

Rational Numbers

We have seen that different fractions can nevertheless express the same amount or correspond to the same point on a number line. For example, the fraction strips in Figure 6.7 each show that two thirds of a whole strip has been shaded, but the fraction which represents this amount can be any one of the choices $2/3$, $4/6$, $6/9$, ... Similarly, in Figure 6.6 the single point on the number line at a distance of one half of a unit to the left of the origin can be expressed by any of the equivalent fractions $-1/2$, $1/-2$, $-2/4$, $2/-4$, ...

Numbers such as "two-thirds" and "negative one half" which can be represented by fractions are examples of **rational numbers.** The following definition is suitable for use in the elementary school classroom.*

DEFINITION Rational Numbers

A **rational number** is a number that can be represented by a fraction a/b, where a and b are integers, $b \neq 0$. Two rational numbers are **equal** if, and only if, they can be represented by equivalent fractions.

The set of rational numbers is denoted by Q. A rational number such as $3/4$ can also be represented by $6/8$, $30/40$, or any other fraction that is equivalent to $3/4$.

EXAMPLE 6.4

Representing Rational Numbers

How many different rational numbers are given in this list?

$$2/5, \quad 3, \quad -4/-10, \quad 39/13, \quad 7/4$$

SOLUTION

Since $2/5 = -4/-10$ and $3 = 39/13$, there are three different rational numbers: $2/5$, 3, and $7/4$. ■

Ordering Fractions and Rational Numbers

By placing the fraction strips for $\frac{3}{4}$ and $\frac{5}{6}$ side by side, as shown in Figure 6.8(**a**), it becomes geometrically apparent that $\frac{3}{4}$ represents a smaller shaded portion of a whole strip than $\frac{5}{6}$. The comparison can be made even clearer by replacing the strips $\frac{3}{4}$ and $\frac{5}{6}$ by the equivalent strips $\frac{9}{12}$ and $\frac{10}{12}$, as shown in Figure 6.8(**b**). Since $9 < 10$, we see that $\frac{9}{12}$ is less than $\frac{10}{12}$, and therefore why $\frac{3}{4}$ is less than $\frac{5}{6}$.

More generally, let two rational numbers be represented by the fractions $\frac{a}{b}$ and

* In advanced mathematics a **rational number** is defined to be a set containing all of the fractions that are equivalent to some given fraction. Thus, a rational number is an equivalence class of fractions, where the concept of an equivalence class was discussed in Section 5 of Chapter 2. The informal definition given above conveys the proper intuitive meaning of rational number however.

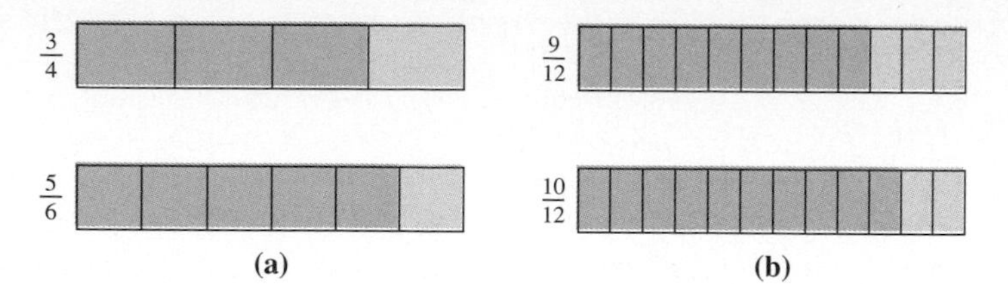

Figure 6.8
Showing $\frac{3}{4} < \frac{5}{6}$ and $\frac{9}{12} < \frac{10}{12}$ with fraction strips

$\frac{c}{d}$ with positive denominators b and d. These numbers can be compared by first rewriting them in equivalent form $\frac{ad}{bd}$ and $\frac{bc}{bd}$. Then $\frac{ad}{bd} < \frac{bc}{bd}$ if, and only if, $ad < bc$. This leads us to the following definition.

DEFINITION Inequality of Rational Numbers

Let two rational numbers be represented by fractions $\frac{a}{b}$ and $\frac{c}{d}$, with b and d positive. Then $\frac{a}{b}$ is **less than** $\frac{c}{d}$, written $\frac{a}{b} < \frac{c}{d}$, if, and only if, $ad < bc$.

The corresponding relations, less than or equal, $\leq$, greater than, $>$, and greater than or equal, $\geq$, are defined in the usual way.

EXAMPLE 6.5

Comparing Rational Numbers

Replace the box by the proper relation $<$, $=$, $>$ for each pair of rational numbers.

(a) $\frac{3}{4} \square \frac{2}{5}$ **(b)** $\frac{15}{29} \square \frac{6}{11}$

(c) $\frac{2106}{7047} \square \frac{234}{783}$ **(d)** $\frac{-10}{13} \square \frac{22}{-29}$

SOLUTION

(a) Since $3 \cdot 5 = 15 > 8 = 2 \cdot 4$, we have $\frac{3}{4} > \frac{2}{5}$.

(b) Since $15 \cdot 11 = 165 < 174 = 6 \cdot 29$, we have $\frac{15}{29} < \frac{6}{11}$.

(c) Using a calculator, $2106 \cdot 783 = 1648998 = 7047 \cdot 234$. Thus, the two fractions are equivalent: $\frac{2106}{7047} = \frac{234}{783}$.

(d) First write $\frac{22}{-29}$ as $\frac{-22}{29}$ so that its denominator is positive. Then, since $-10 \cdot 29 = -290 < -286 = -22 \cdot 13$, we conclude that $\frac{-10}{13} < \frac{22}{-29}$. ■

PROBLEM SET 6.1

Understanding Concepts

1. What fraction is represented by the darker shaded portion of the following figures?

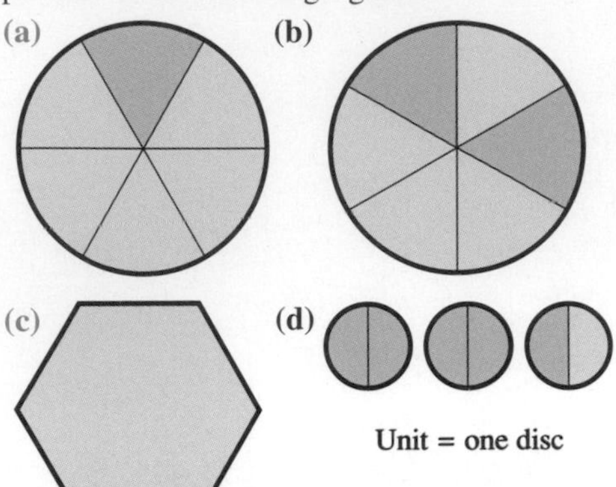

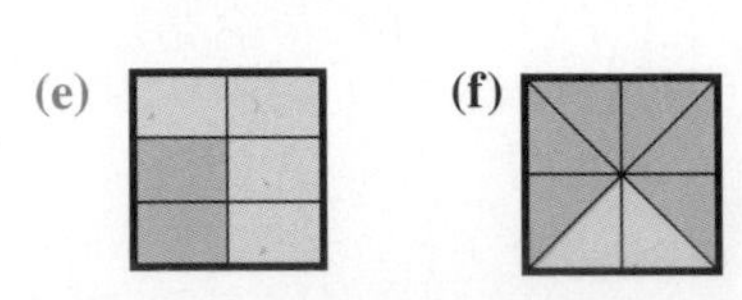

2. Subdivide and shade the unit octagons shown to represent the given fraction.

(a) $\frac{1}{8}$ (b) $\frac{6}{8}$ (c) $\frac{3}{4}$ (d) $\frac{11}{8}$

3. For each lettered point on the number lines below, express its position by a corresponding fraction.

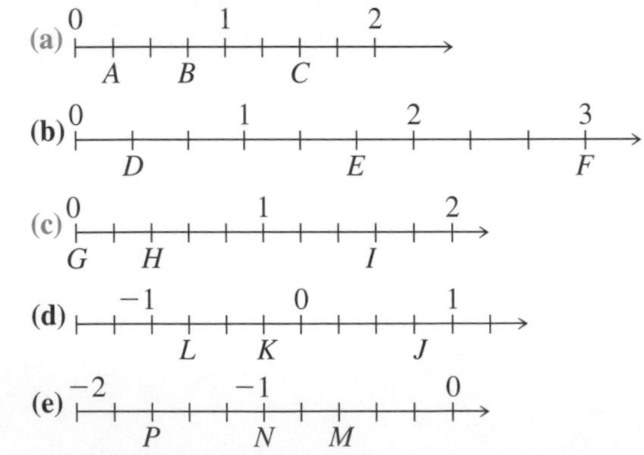

4. Depict the fraction $\frac{4}{6}$ with the following models.

(a) Colored region model (b) Set model
(c) Fraction strip model (d) Number line model

5. Express the following quantities by a fraction placed in the blank space.

(a) 20 minutes is _____ of an hour.
(b) 30 seconds is _____ of a minute.
(c) 5 days is _____ of a week.
(d) 25 years is _____ of a century.
(e) A quarter is _____ of a dollar.
(f) 3 eggs is _____ of a dozen.
(g) 2 feet is _____ of a yard.
(h) 3 cups is _____ of a quart.

6. In Figure 6.7, fraction strips show that $\frac{2}{3}, \frac{4}{6}, \frac{6}{9}$, and $\frac{8}{12}$ are equivalent fractions. Use a similar drawing of fraction strips to show that $\frac{3}{4}, \frac{6}{8}$, and $\frac{9}{12}$ are equivalent fractions.

7. What equivalence of fractions is shown in these pairs of colored region models?

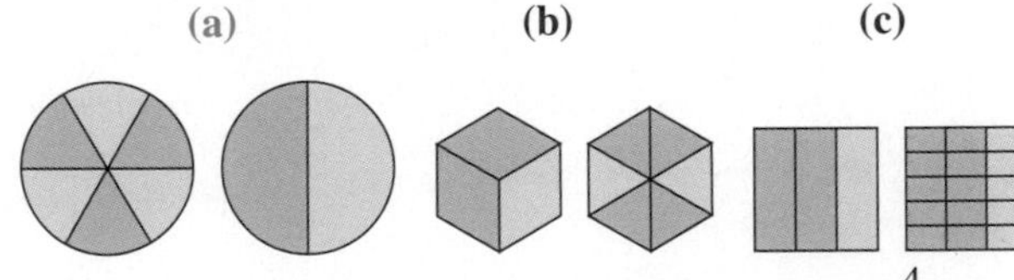

8. Find four different fractions equivalent to $\frac{4}{9}$.

9. Subdivide and mark the unit square on the right to illustrate that the given fractions are equivalent.

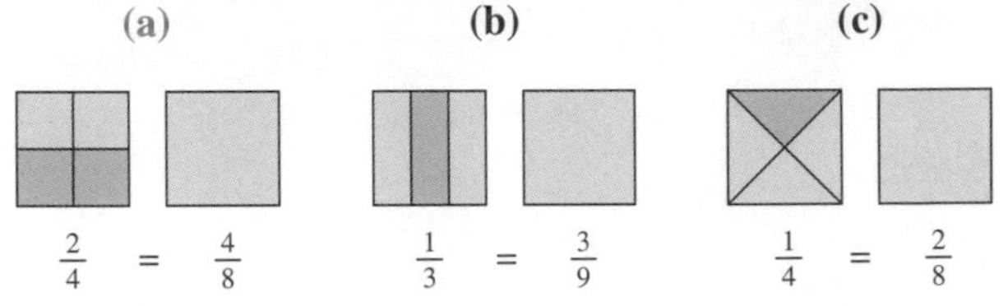

10. Fill in the missing integer to make the fractions equivalent.

(a) $\frac{4}{5} = \frac{}{30}$ (b) $\frac{6}{9} = \frac{2}{}$

(c) $\frac{-7}{25} = \frac{}{500}$ (d) $\frac{18}{3} = \frac{-6}{}$

11. Determine if each set of two fractions is equivalent by calculating equivalent fractions with a common denominator.

(a) $\frac{18}{42}$ and $\frac{3}{7}$ (b) $\frac{18}{49}$ and $\frac{5}{14}$

(c) $\frac{9}{25}$ and $\frac{140}{500}$ (d) $\frac{24}{144}$ and $\frac{32}{96}$

12. Determine which of these pairs of fractions are equivalent.

(a) $\frac{78}{24}$ and $\frac{546}{168}$ **(b)** $\frac{243}{317}$ and $\frac{2673}{3487}$

(c) $\frac{412}{-864}$ and $\frac{-308}{616}$

13. (a) Is it true that $\frac{4 \cdot 3}{9 \cdot 3} = \frac{12}{27}$?

(b) Is it true that $\frac{4 \cdot 3}{9 \cdot 3} = \frac{4}{9}$?

(c) Is it true that $\frac{4 + 3}{9 + 3} = \frac{7}{12}$?

(d) Is it true that $\frac{4 + 3}{9 + 3} = \frac{4}{9}$?

14. Rewrite the following fractions in simplest form.

(a) $\frac{84}{144}$ **(b)** $\frac{208}{272}$ (c) $\frac{-930}{1290}$ **(d)** $\frac{325}{231}$

15. Find the prime factorizations of the numerators and denominators of these fractions and use them to express the fractions in simplest form.

(a) $\frac{96}{288}$ **(b)** $\frac{247}{-75}$ **(c)** $\frac{2520}{378}$

16. For each of these sets of fractions, determine equivalent fractions with a common denominator.

(a) $\frac{3}{11}$ and $\frac{2}{5}$ **(b)** $\frac{5}{12}$ and $\frac{2}{3}$

(c) $\frac{4}{3}$, $\frac{5}{8}$, and $\frac{1}{6}$ **(d)** $\frac{1}{125}$ and $\frac{-3}{500}$

17. For each of these determine equivalent fractions with the least common denominator.

(a) $\frac{3}{8}$ and $\frac{5}{6}$ **(b)** $\frac{1}{7}$, $\frac{4}{5}$, and $\frac{2}{3}$

(c) $\frac{17}{12}$ and $\frac{7}{32}$ **(d)** $\frac{17}{51}$ and $\frac{56}{42}$

18. Order the rational numbers from least to greatest in each part.

(a) $\frac{2}{3}$, $\frac{7}{12}$ **(b)** $\frac{2}{3}$, $\frac{5}{6}$ (c) $\frac{5}{6}$, $\frac{29}{36}$

(d) $\frac{-5}{6}$, $\frac{-8}{9}$ **(e)** $\frac{2}{3}$, $\frac{5}{6}$, $\frac{29}{36}$, $\frac{8}{9}$

19. Decide if each statement is *true* or *false*. Explain your reasoning in a brief paragraph.

(a) There are infinitely many ways to replace two fractions with two equivalent fractions which have a common denominator.

(b) There is a unique least common denominator for a given pair of fractions.

(c) There is a least positive fraction.

(d) There are infinitely many fractions between 0 and 1.

20. For each, determine the set of all fractions that are equivalent to the given fraction.

(a) $\frac{3}{5}$ **(b)** $\frac{-7}{4}$ (c) 0 **(d)** $\frac{39}{65}$

21. How many different rational numbers are given by this list:

$\frac{27}{36}$, 4, $\frac{21}{28}$, $\frac{24}{6}$, $\frac{3}{4}$, $\frac{-8}{-2}$.

Thinking Critically

22. What fraction represents the part of the whole region which has been shaded? Draw additional lines to make your answer visually clear. For example, $\frac{2}{6}$ of the regular hexagon on the left is shaded, since the entire hexagon can be subdivided into six congruent regions as shown on the right.

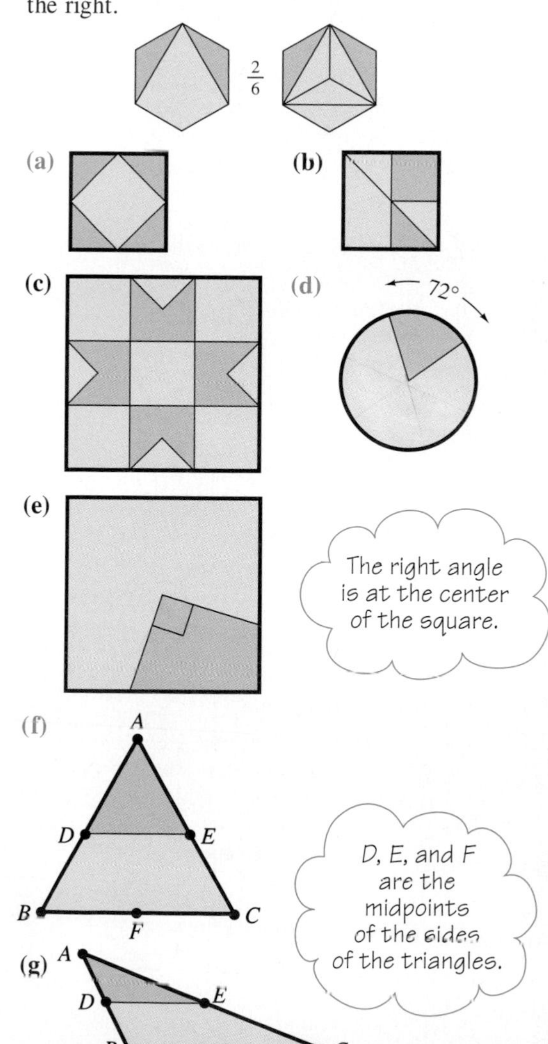

In (f) and (g) D, E, and F are midpoints of two sides of triangle ACB.

23. Recall the Fibonacci sequence $F_1 = 1, F_2 = 1, F_3 = 2, F_4 = 3, F_5 = 5, F_6 = 8, \ldots$.

(a) Arrange the fractions $\frac{1}{1}, \frac{2}{1}, \frac{3}{2}, \frac{5}{3}, \frac{8}{5}$ in increasing order. Notice that each fraction has the form F_{n+1}/F_n.

(b) Guess where the next fraction $\frac{13}{8}$ will be placed in the ordered list you made in part (a). Then check that your guess is correct.

(c) Repeat parts (a) and (b), but replace the Fibonacci sequence with the Lucas sequence $L_1 = 1, L_2 = 3, L_3 = 4, L_4 = 7, L_5 = 11, L_6 = 18, \ldots$. Again, order the first six fractions from the list $\frac{3}{1}, \frac{4}{3}, \frac{7}{4}, \ldots$.

(d) Make up your own Fibonacci-like sequence by starting with any two integers of your choosing. For example, you could start with 2 and 5 to form the sequence 2, 5, 7, 12, 19, 31, ... Repeat (a) and (b) for your sequence.

24. The **mediant** of a pair of fractions $\frac{a}{b}$ and $\frac{c}{d}$ with b and d positive is defined by $\frac{a+c}{b+d}$. For example, the mediant of $\frac{2}{5}$ and $\frac{3}{4}$ is $\frac{2+3}{5+4} = \frac{5}{9}$, which is *between* $\frac{2}{5}$ and $\frac{3}{4}$; that is,

$$\frac{2}{5} < \frac{2+3}{5+4} < \frac{3}{4}$$

(a) Verify that the mediant of $\frac{3}{4}$ and $\frac{4}{5}$ is between $\frac{3}{4}$ and $\frac{4}{5}$.

(b) Prove the following general result on the mediant:

$$\textit{If } \frac{a}{b} < \frac{c}{d} \textit{ then } \frac{a}{b} < \frac{a+c}{b+d} < \frac{c}{d}.$$

(c) What is the mediant of $\frac{4}{10}$ and $\frac{9}{12}$?

25. (a) Let $\frac{a}{b} < \frac{c}{d}$, where b and d are positive, and let m be any natural number. Show that $\frac{a+mc}{b+md}$ is a rational number between $\frac{a}{b}$ and $\frac{c}{d}$.

(b) Use part (a) to explain why there are *infinitely* many rational numbers between any two distinct rational numbers.

26. Consider fractions for which both numerator and denominator are positive integers.

(a) Show that "increasing the numerator" increases a positive fraction; that is, if $a < c$, prove that

$$\frac{a}{b} < \frac{c}{b}.$$

(b) Show that "decreasing the denominator" increases a positive fraction. That is, if $b > d$ prove that $\frac{a}{b} < \frac{a}{d}$.

(c) Use parts (a) and (b) to explain how the following inequalities can be verified by mental math.

$$\frac{4}{9} < \frac{17}{36}, \quad \frac{7}{4} < \frac{70}{37}, \quad \frac{2489}{503} < \frac{5}{1}$$

Thinking Cooperatively

27. The regular hexagon of side length 3 can be tiled in many different ways with 27 rhombuses of unit side length. Here are two partially completed tilings.

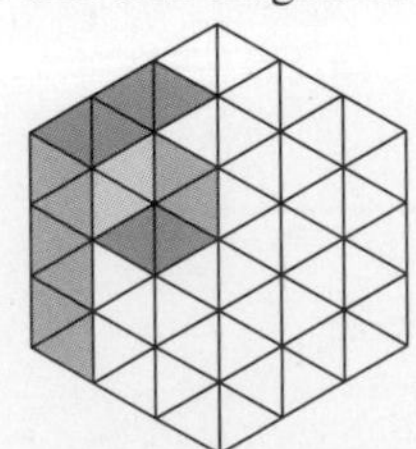

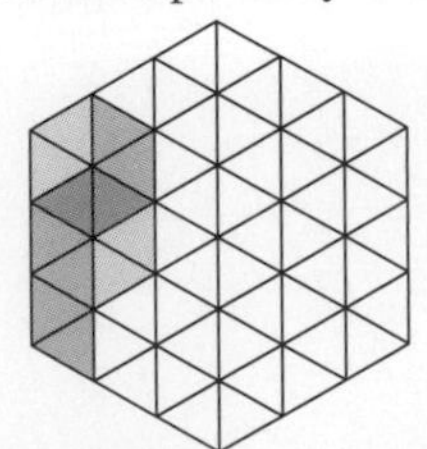

We see that the longer diagonal of the rhombuses have three distinct orientations: horizontal, upward to the right, upward to the left.

(a) Trace each hexagon above and complete the tiling by coloring in rhombuses according to their orientation.

(b) What fraction of the 27 tiles are horizontal in each of the two tilings you completed in part (a)? What fractions are upward to the right? To the left?

(c) Use three marking pens of different colors to tile tracings of the following hexagons with rhombuses. Color each rhombus according to its orientation, as in part (a).

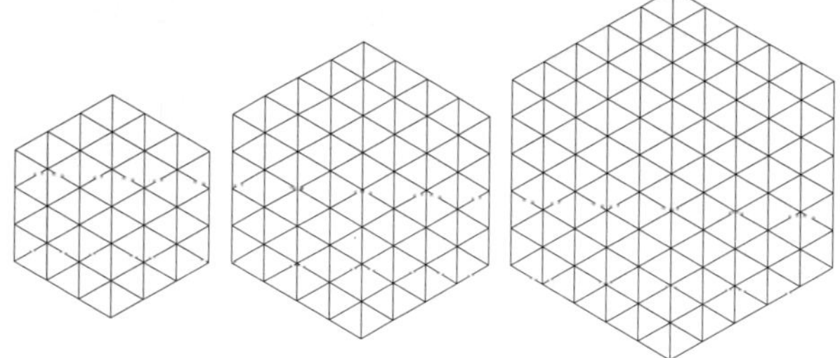

What fraction of rhombuses are in each orientation in your tilings?

(d) How many rhombuses are required to tile a regular hexagon of side length n? How many are in each orientation? Can you justify your answers? (*Hint:* Visualize cubes stacked in a corner; what would you see from different directions of sight?)

28. Let $\mathcal{F}_n$ be the sequence of fractions listed in increasing order from $\frac{0}{1}$ to $\frac{1}{1}$ which, in simplest form, have denominator no larger than n. $\mathcal{F}_n$ is called the ***n*th Farey sequence.** For example, the third and fourth Farey sequences are

$$\mathcal{F}_3: \frac{0}{1}, \frac{1}{3}, \frac{1}{2}, \frac{2}{3}, \frac{1}{1}$$

$$\mathcal{F}_4: \frac{0}{1}, \frac{1}{4}, \frac{1}{3}, \frac{1}{2}, \frac{2}{3}, \frac{3}{4}, \frac{1}{1}.$$

(a) Find the fifth and sixth Farey sequences.

(b) The three fractions $\frac{1}{4}, \frac{1}{3}, \frac{1}{2}$ occur in succession in $\mathcal{F}_4$. If the numerators and denominators of the two outside fractions are added, the resulting fraction is equivalent to the middle fraction:

$$\frac{1 + 1}{4 + 2} = \frac{2}{6} = \frac{1}{3}$$

Show that this property holds for every three consecutive terms of $\mathcal{F}_4$. This is called the *mediant property* (see problem 24) of $\mathcal{F}_4$.

(c) Examine the three term sequences $\frac{a}{b}, \frac{e}{f}, \frac{c}{d}$ in $\mathcal{F}_5$ and $\mathcal{F}_6$, and check that the mediant property $\frac{a + c}{b + d} = \frac{e}{f}$ continues to hold in these Farey sequences.

Making Connections

29. Francisco's pickup truck has a 24 gallon gas tank and an accurate fuel gauge. Estimate the number of gallons in the tank at these readings.

(a)

(b)

(c)

30. If 153 of the 307 graduating seniors go on to college, it is likely that a principal would claim that $\frac{1}{2}$ of the class is collegebound. Give simpler convenient fractions that approximately express the data in these situations.

(a) Esteban is on page 310 of a 498 page novel. He has read _____ of the book.

(b) Myra has saved \$73 toward the purchase of a \$215 plane ticket. She has saved _____ of the amount she needs.

(c) Nine students in Ms. Evaldo's class of 35 students did perfect work on the quiz. _____ of the class scored 100% on the quiz.

(d) The Math Club has sold 1623 of the 2400 raffle tickets. They have sold _____ of the available tickets.

31. Lakeside High won 19 of their season's 25 baseball games. Rival Shorecrest High School lost just 5 of the 21 games they played. Can Shorecrest claim to have had the better season? Explain.

32. **Basketball Math.** If a basketball player makes m free throws in n attempts, the rational number $\frac{m}{n}$ gives a measure of the player's skill at the foul line. Suppose, going into the playoffs, Carol and Joleen have the following records over the first and second halves of the regular season. For example, Carol made 38 free throws in 50 attempts during the first half of the season.

$\frac{m}{n}$	Carol	Joleen
Nov.–Jan.	$\frac{38}{50}$	$\frac{30}{40}$
Jan.–Mar.	$\frac{42}{70}$	$\frac{14}{24}$
Entire Season		

(a) Who was the more successful free throw shooter during the first half of the regular season? During the second half?

(b) Combine the data for each half-season. For example, Carol hit 80 free throws in 120 attempts. Which player had the higher success rate over the entire season?

(c) Suppose the data in the table represents the success rate of two drugs: drug C versus drug J. Which drug is best in each half of the 5 month trial period? Which drug looks most promising overall?

33. **Fractions in Probability.** If a card is picked at random from an ordinary deck of 52 playing cards, there are 4 ways it can be an ace, since it could be the ace of hearts, diamonds, clubs, or spades. To measure the chances of drawing an ace, it is common to give the probability as the rational number $\frac{4}{52}$. In general, if n equally likely outcomes are possible and m of these outcomes are successful for an event to occur, then the probability of the event is $\frac{m}{n}$. As another example, $\frac{5}{6}$ is the probability of rolling a single die and having more than one spot appear. Give fractions which express the probability of the following events:

(a) Getting a head in the flip of a fair coin.

(b) Drawing a face card from a deck of cards.

(c) Rolling an even number on a single die.

(d) Drawing a green marble from a bag which contains 20 red, 30 blue, and 25 green marbles.

(e) Drawing either a red or a blue marble from the bag of marbles described in part (d).

Communicating

34. **Fraction squares** are another popular manipulative for teaching the basic concepts of fractions. The unit is represented by a square. The unit can be divided into vertical or horizontal strips, or a combination of vertical and horizontal divisions are used to partition the unit square into congruent rectangles. Coloring or shading indicates the corresponding fraction. Here are some examples.

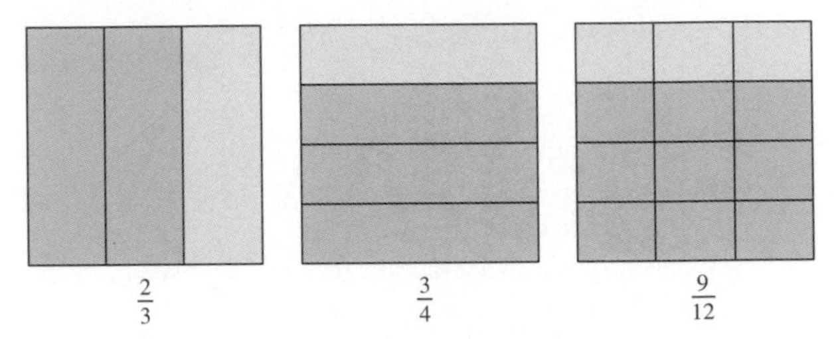

$\frac{2}{3}$ $\frac{3}{4}$ $\frac{9}{12}$

Write a report outlining how fraction squares can be used to illustrate these basic concepts:

(a) equivalence of fractions,

(b) common denominator,

(c) the fundamental law of fractions.

For Review

35. The Fibonacci sequence 1, 1, 2, 3, 5, . . . , is formed by adding the two successive terms to obtain the next one, beginning with 1 and 1. Suppose subtraction is used in place of addition, to get

$$1, 1, 0, 1, -1, 2, -3, \ldots$$

(a) Obtain the next ten terms of this sequence.

(b) If you know $F_{24} = 46{,}368$ and $F_{25} = 75{,}025$ are the twenty-fourth and twenty-fifth Fibonacci numbers, what do you think the twenty-seventh and twenty-eighth numbers in the subtractive sequence are? Explain your reasoning.

36. (a) Find integers m and n which solve the equation $8m - 11n = 1$. It may be helpful to compare a list of multiples of 8 to a list of multiples of 11.

(b) Explain why it is not possible to find integers x and y which satisfy the equation $8x - 12y = 1$. (*Hint:* Is there an integer which divides the left side of the equation but cannot divide the right side?)

37. Find the digit D so that $49{,}D84$ is divisible by 24. (*Hint:* Use divisibility tests for 3 and 8.)

6.2 The Arithmetic of Rational Numbers

In this section we consider addition, subtraction, multiplication, and division in the set of rational numbers. The geometric and physical models of fractions serve to motivate the definitions we adopt for each operation. Moreover, the models convey intuitive, visual, and practical meaning for each operation.

Addition of Rational Numbers

The sum of $\frac{3}{8}$ and $\frac{2}{8}$ is illustrated in Figure 6.9. The colored region and number-line models both show that $\frac{3}{8} + \frac{2}{8} = \frac{5}{8}$. The colored region visualization corresponds to the set model of addition, and the number line visualization corresponds to the measurement model of addition.

Figure 6.9
Showing $\frac{3}{8} + \frac{2}{8} = \frac{5}{8}$ with the colored region and number-line models

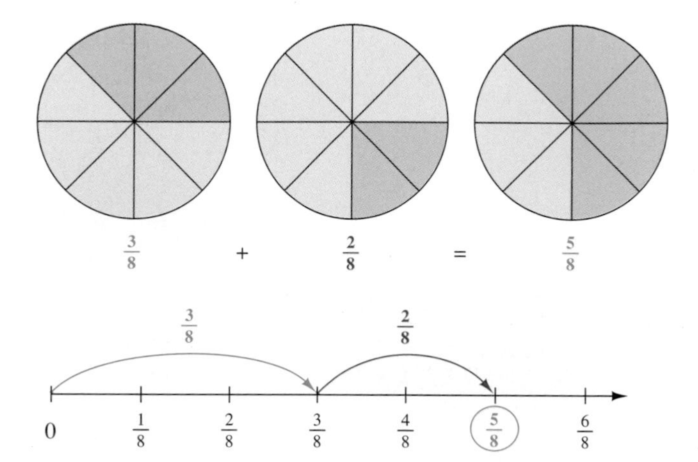

The models suggest that the sum of two rational numbers represented by fractions with a common denominator should be found by adding the two numerators. This motivates the following definition.

DEFINITION Addition of Rational Numbers

Let two rational numbers be represented by fractions $\frac{a}{b}$ and $\frac{c}{b}$ with a common denominator. Then their **sum** is the rational number given by

$$\frac{a}{b} + \frac{c}{b} = \frac{a + c}{b}.$$

To add rational numbers represented by fractions with unlike denominators, we first rewrite the fractions with a common denominator. For example, to add $\frac{1}{4}$ and $\frac{2}{3}$, we rewrite them with 12 as a common denominator:

$$\frac{1}{4} = \frac{1 \cdot 3}{4 \cdot 3} = \frac{3}{12}, \quad \frac{2}{3} = \frac{2 \cdot 4}{3 \cdot 4} = \frac{8}{12}.$$

According to the definition above, we then have

$$\frac{1}{4}+\frac{2}{3}=\frac{1\cdot 3}{4\cdot 3}+\frac{2\cdot 4}{3\cdot 4}=\frac{1\cdot 3+2\cdot 4}{4\cdot 3}=\frac{3+8}{12}=\frac{11}{12}.$$

The procedure just followed can be modeled nicely with fraction strips as shown in Figure 6.10.

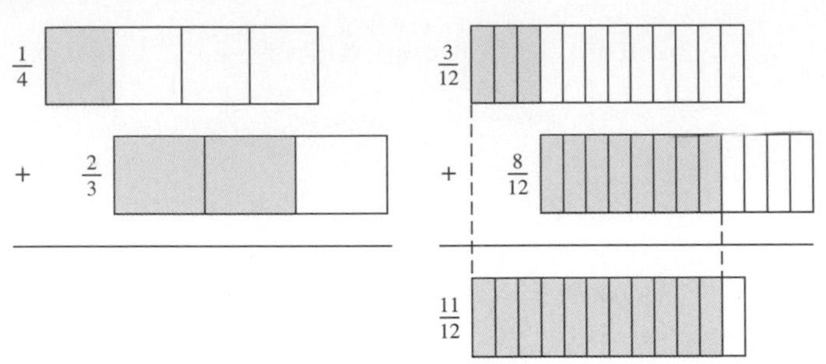

Figure 6.10
The fraction strip model showing $\frac{1}{4}+\frac{2}{3}=\frac{3}{12}+\frac{8}{12}=\frac{11}{12}$

The same procedure can be followed to add any two rational numbers $\frac{a}{b}$ and $\frac{c}{d}$.

We have

$$\frac{a}{b}+\frac{c}{d}=\frac{a\cdot d}{b\cdot d}+\frac{c\cdot b}{d\cdot b}=\frac{ad+bc}{bd}.$$

Rewrite the fractions with a common denominator.

Add the numerators, retain the common denominator.

This proves the following theorem.

THEOREM The Rational Numbers Addition Formula

Let two rational numbers be represented by the fractions $\frac{a}{b}$ and $\frac{c}{d}$. Then the sum of the rational numbers is given by

$$\frac{a}{b}+\frac{c}{d}=\frac{ad+bc}{bd}.$$

In many cases the denominator bd is quite large, and it is better to use a smaller common denominator. For example, it is more convenient to write $\frac{11}{26}+\frac{7}{52}=\frac{22}{52}+\frac{7}{52}=\frac{29}{52}$ than $\frac{11}{26}+\frac{7}{52}=\frac{11\cdot 52+26\cdot 7}{26\cdot 52}=\frac{754}{1352}$. Here we used the least common denominator LCM(26, 52) = 52.

EXAMPLE 6.6

Adding Rational Numbers

Compute these sums.

(a) $\frac{3}{10}+\frac{4}{7}$ **(b)** $\frac{3}{8}+\frac{-7}{24}$

(c) $\left(\frac{3}{4}+\frac{5}{6}\right)+\frac{2}{3}$ **(d)** $\frac{3}{4}+\left(\frac{5}{6}+\frac{2}{3}\right)$

SOLUTION

(a) $\frac{3}{10} + \frac{4}{7} = \frac{3 \cdot 7}{10 \cdot 7} + \frac{4 \cdot 10}{7 \cdot 10} = \frac{21 + 40}{70} = \frac{61}{70}$

(b) $\frac{3}{8} + \frac{-7}{24} = \frac{3 \cdot 24 + 8 \cdot (-7)}{8 \cdot 24} = \frac{72 - 56}{192} = \frac{16}{192}$

A better method is to use the least common denominator. Thus, $\frac{3}{8} + \frac{-7}{24} = \frac{3 \cdot 3}{8 \cdot 3} + \frac{-7}{24} = \frac{9}{24} + \frac{-7}{24} = \frac{2}{24}$. To see that the two answers are equivalent we rewrite each fraction in simplest form:

$$\frac{16}{192} = \frac{8}{96} = \frac{4}{48} = \frac{1}{12}, \quad \frac{2}{24} = \frac{1}{12}.$$

(c) The parentheses tell us to compute first

$$\frac{3}{4} + \frac{5}{6} = \frac{3 \cdot 6 + 4 \cdot 5}{4 \cdot 6} = \frac{18 + 20}{24} = \frac{38}{24} = \frac{19}{12}$$

where the last step was taken to simplify the result. Then,

$$\left(\frac{3}{4} + \frac{5}{6}\right) + \frac{2}{3} = \frac{19}{12} + \frac{2}{3} = \frac{19}{12} + \frac{8}{12} = \frac{19 + 8}{12} = \frac{27}{12},$$

which simplifies to $\frac{9}{4}$ when written in simplest form.

(d) The sum in parentheses is

$$\frac{5}{6} + \frac{2}{3} = \frac{5}{6} + \frac{4}{6} = \frac{9}{6} = \frac{3}{2}.$$

Then $\frac{3}{4} + \left(\frac{5}{6} + \frac{2}{3}\right) = \frac{3}{4} + \frac{3}{2} = \frac{3}{4} + \frac{6}{4} = \frac{9}{4}$. ■

Parts (c) and (d) of Example 6.6 show that

$$\left(\frac{3}{4} + \frac{5}{6}\right) + \frac{2}{3} = \frac{3}{4} + \left(\frac{5}{6} + \frac{2}{3}\right),$$

since each side represents the rational number $\frac{9}{4}$. This result is a particular example of the associative property for the addition of rational numbers. The properties of the arithmetic operations on the rational numbers will be explored in the next section.

> **JUST FOR FUN**
>
> **The Sultan's Estate**
>
> When the sultan died, his will divided his possessions among his three sons. The sultan's eldest son was to receive one-half of his horses, his second son was to receive one-third of the horses, and his youngest son was to receive one-ninth of them. The sultan left 17 horses, and after struggling for some time trying to decide how to carry out the instructions of the will, the perplexed sons decided to turn to their wise uncle who had just ridden up on his own horse. The uncle, after a moment's reflection said, "Let me contribute my own horse, making 18 all together." The sons saw now that the eldest received $\frac{1}{2}$ of 18, or 9 horses. The middle son inherited 6 horses, since that is $\frac{1}{3}$ of 18, and the youngest son was given 2 horses, since $\frac{1}{9}$ of 18 is 2. Once the 9 + 6 + 2 = 17 horses were presented to the sons, the uncle returned home mounted on his own horse!

Proper Fractions and Mixed Numbers

The sum of a natural number and a fraction is most often written as a **mixed number.** For example, $3 + \frac{1}{7}$ would be written as $3\frac{1}{7}$ and would be read "three and one seventh." It is important to realize that it is the addition symbol + which is suppressed, since the common notation xy for multiplication might suggest, incorrectly, that $3\frac{1}{7}$ is $3 \cdot \frac{1}{7}$. Thus, $3\frac{1}{7} = 3 + \frac{1}{7}$, not $\frac{3}{7}$.

A mixed number can always be rewritten in the standard form $\frac{a}{b}$ of a fraction.

For example,

$$3\frac{1}{7} = 3 + \frac{1}{7} = \frac{3}{1} + \frac{1}{7} = \frac{3 \cdot 7 + 1 \cdot 1}{1 \cdot 7} = \frac{21 + 1}{7} = \frac{22}{7}$$

and

$$-4\frac{3}{5} = (-4) + \left(\frac{-3}{5}\right) = \frac{-4}{1} + \frac{-3}{5} = \frac{(-4)(5) + (1)(-3)}{1 \cdot 5}$$
$$= \frac{(-20) + (-3)}{5} = \frac{-23}{5}.$$

A fraction $\frac{a}{b}$ for which $0 \leq |a| < b$ is called a **proper fraction.** For example, $2/3$ is a proper fraction, but $\frac{3}{2}$, $\frac{-8}{5}$, and $\frac{6}{6}$ are not proper fractions. It is common, though not necessary, to rewrite fractions which are not proper as mixed numbers. For example, to express $\frac{439}{19}$ as a mixed number we first use the division algorithm (or division with remainder) to find that $439 = 23 \cdot 19 + 2$. Then

$$\frac{439}{19} = \frac{23 \cdot 19 + 2}{19} = \frac{23}{1} + \frac{2}{19} = 23 + \frac{2}{19} = 23\frac{2}{19}.$$

In mixed number form, it is obvious that $23\frac{2}{19}$ is just slightly larger than 23; this fact was not as evident in the original fraction form $\frac{439}{19}$. Nevertheless, it is perfectly acceptable to express rational numbers as "improper" fractions. In general, the fractional form $\frac{a}{b}$ is the more convenient form for arithmetic and algebra, and the mixed number form is easiest to understand for practical applications. For example, it would be more common to buy $2\frac{1}{4}$ yards of material than to request $\frac{9}{4}$ yards.

EXAMPLE 6.7

Working with Mixed Numbers

(a) Give an improper fraction for $3\frac{17}{120}$.

(b) Give a mixed number for $\frac{355}{133}$.

(c) Give a mixed number for $\frac{-15}{4}$.

(d) Compute $2\frac{3}{4} + 4\frac{2}{5}$.

SOLUTION

(a) $3\frac{17}{120} = \frac{3}{1} + \frac{17}{120} = \frac{3 \cdot 120 + 1 \cdot 17}{120} = \frac{360 + 17}{120} = \frac{377}{120}$.
This rational number was given by Claudius Ptolemy around A.D. 150 to approximate π, the ratio of the circumference of a circle to its diameter.

It has better accuracy than $3\frac{1}{7}$, the value proposed by Archimedes in about 240 B.C.

(b) Using the division algorithm, $355 = 3 \cdot 113 + 16$. Therefore,

$$\frac{355}{113} = \frac{3 \cdot 113 + 16}{113} = \frac{3}{1} + \frac{16}{113} = 3\frac{16}{113}$$

which corresponds to a point somewhat to the right of 3 on the number line.

The value $\frac{355}{113}$ was used around A.D. 480 in China to approximate π; as a decimal it is correct to six places!

(c) $\frac{-15}{4} = \frac{-(3 \cdot 4 + 3)}{4} = -\left(3 + \frac{3}{4}\right) = -3\frac{3}{4}.$

(d) $2\frac{3}{4} + 4\frac{2}{5} = 2 + 4 + \frac{3}{4} + \frac{2}{5} = 6 + \frac{15}{20} + \frac{8}{20} = 6 + \frac{23}{20} = 7\frac{3}{20}.$ ■

Subtraction of Rational Numbers

Figure 6.11 shows how the take-away, measurement, and missing-addend conceptual models of the subtraction operation can be illustrated with colored regions, the number line, and fraction strips. In each case we see that $\frac{7}{6} - \frac{3}{6} = \frac{4}{6}$.

Figure 6.11
Models that show $\frac{7}{6} - \frac{3}{6} = \frac{4}{6}$

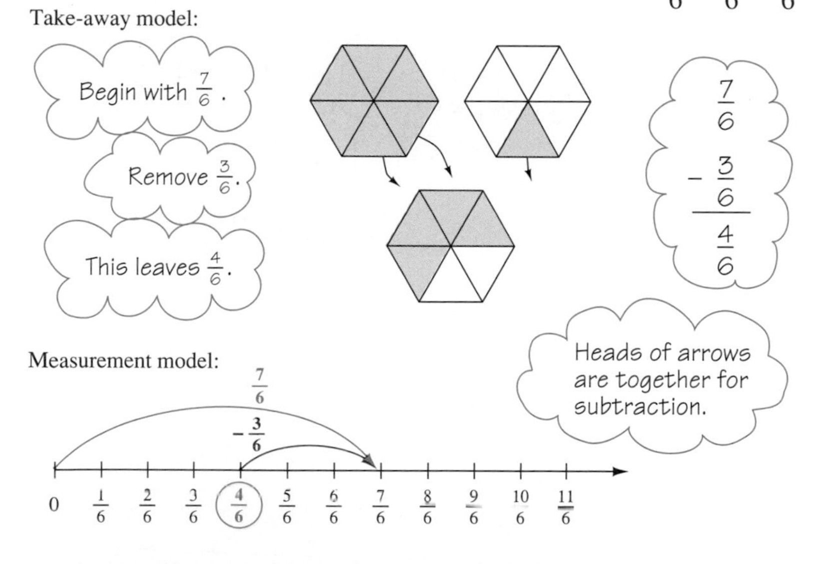

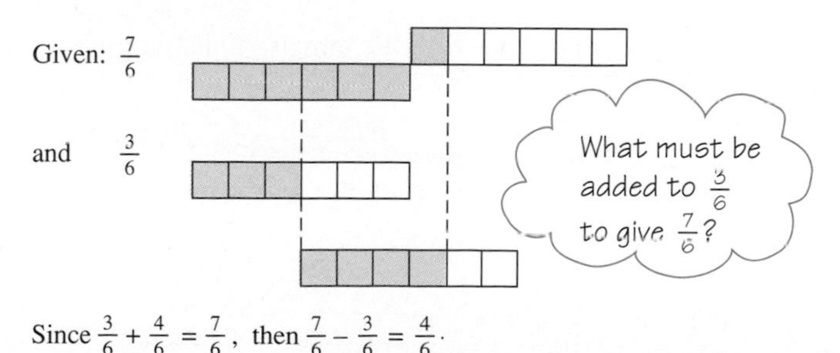

HIGHLIGHT FROM HISTORY
Egyptian Fractions

The hieroglyphic numerals of ancient Egypt were described in Chapter 3. Fractions were likewise expressed with hieroglyphs. For example, the symbol ⬭ originally indicated $\frac{1}{320}$ of a bushel, but later it came to denote a fraction in general. Here are some hieroglyphic fractions and their modern equivalents:

$\frac{1}{5}$ $\frac{1}{10}$ $\frac{1}{15}$

Some common fractions were denoted by special symbols:

$\frac{1}{2}$ $\frac{2}{3}$ $\frac{1}{4}$

With $\frac{2}{3}$ as an exception, Egyptian fractions were written as sums of fractions with numerator 1 and distinct denominators, though no summation sign appeared. For example, $\frac{2}{5}$ would be written

$\frac{1}{3}+\frac{1}{15}$

With more effort the Egyptians expressed $\frac{7}{29}$ as $\frac{1}{6}+\frac{1}{24}+\frac{1}{58}+\frac{1}{87}+\frac{1}{232}$. The Egyptians assumed every fraction could be written as a sum of unit numerator fractions, but no record has been found that justifies this assumption. In A.D. 1202, Leonardo of Pisa—otherwise known as Fibonacci—gave the first proof of this theorem: *Every fraction between 0 and 1 can be written as a sum of distinct fractions with numerator 1.*

Subtraction of whole numbers and integers was defined on the basis of the missing-addend approach, which emphasizes that subtraction is the inverse operation to addition. Subtraction of rational numbers can be defined in the same way.

DEFINITION Subtraction of Rational Numbers

Let $\frac{a}{b}$ and $\frac{c}{d}$ be rational numbers. Then $\frac{a}{b}-\frac{c}{d}=\frac{e}{f}$ if, and only if, $\frac{a}{b}=\frac{c}{d}+\frac{e}{f}$.

The following easily verified theorem gives algorithms for calculating differences. There are two cases, depending on whether the fractions representing the rational numbers have like or unlike denominators.

THEOREM Calculating Differences of Rational Numbers

Rational numbers with a common denominator:

$$\frac{a}{b}-\frac{c}{b}=\frac{a-c}{b}$$

Rational numbers with unlike denominators:

$$\frac{a}{b}-\frac{c}{d}=\frac{ad-bc}{bd}$$

JUST FOR FUN

Bookworm Math

A bookworm is on page 1 of Volume 1 of a set of encyclopedias. He (or she—how do you tell?) decides to eat straight to the last page of Volume 2. How far does the bookworm travel? The covers of each volume are $\frac{1}{8}''$ thick, the pages in each volume are, in total, $\frac{3}{4}''$ thick, and the volumes are neatly in place side by side on a shelf.

EXAMPLE 6.8 **Subtracting Rational Numbers**

Find the following differences

(a) $\frac{4}{5} - \frac{2}{3}$ (b) $\frac{103}{24} - \frac{-35}{16}$ (c) $4\frac{1}{4} - 2\frac{2}{3}$

SOLUTION

(a) $\frac{4}{5} - \frac{2}{3} = \frac{4 \cdot 3 - 5 \cdot 2}{5 \cdot 3} = \frac{12 - 10}{15} = \frac{2}{15}$

(b) $\frac{103}{24} - \frac{-35}{16} = \frac{103 \cdot 16 - 24 \cdot (-35)}{24 \cdot 16} = \frac{1648 + 840}{384} = \frac{2488}{384}$

Alternatively, since LCM(24, 16) = 48, the least common denominator, 48, can be used to give

$$\frac{103}{24} - \frac{-35}{16} = \frac{206}{48} - \frac{-105}{48} = \frac{206 - (-105)}{48} = \frac{311}{48}.$$

Since $\frac{2488}{384} = \frac{8 \cdot 311}{8 \cdot 48} = \frac{311}{48}$, the first answer is equivalent to the second.

(c) $4\frac{1}{4} - 2\frac{2}{3} = \frac{17}{4} - \frac{8}{3} = \frac{17 \cdot 3 - 4 \cdot 8}{4 \cdot 3} = \frac{51 - 32}{12} = \frac{19}{12} = 1\frac{7}{12}$

Alternatively, subtraction of mixed numbers can follow the familiar regrouping algorithm:

$$\begin{array}{r} 4\frac{1}{4} \\ -\ 2\frac{2}{3} \\ \hline \end{array} \Rightarrow \begin{array}{r} 4\frac{3}{12} \\ -\ 2\frac{8}{12} \\ \hline \end{array} \Rightarrow \begin{array}{r} 3\frac{15}{12} \\ -\ 2\frac{8}{12} \\ \hline 1\frac{7}{12} \end{array}$$

$4\frac{3}{12} = 3 + \frac{12}{12} + \frac{3}{12}$

■

The regrouping method should always be followed when the integer parts of the mixed numbers are large; for example,

$$248\frac{1}{4} - 187\frac{2}{3} = 247\frac{5}{4} - 187\frac{2}{3} = (247 - 187) + \left(\frac{15}{12} - \frac{8}{12}\right) = 60\frac{7}{12}.$$

Multiplication of Rational Numbers

In earlier chapters, the array model provided a useful visualization of multiplication for both the whole numbers and the integers. The array model also serves well to motivate the definition of multiplication of rational numbers.

In Figure 6.12, we see that $2 \cdot 3 = 6$, since shading a 2 by 3 rectangle covers exactly 6 units (that is, 1 by 1) squares. In (b), the vertical dashed lines divide the unit squares into congruent halves, and $2 \cdot \frac{3}{2} = \frac{6}{2}$ since 6 half-units form the shaded rectangle. We also see that $2 \cdot \frac{3}{2} = \frac{3}{2} + \frac{3}{2} = \frac{6}{2}$, and so multiplication as repeated addition is retained when a whole number multiplies a rational number. Parts (c) and (d) of Figure 6.12 show that we can just as well multiply a rational number times either a whole number or another fraction.

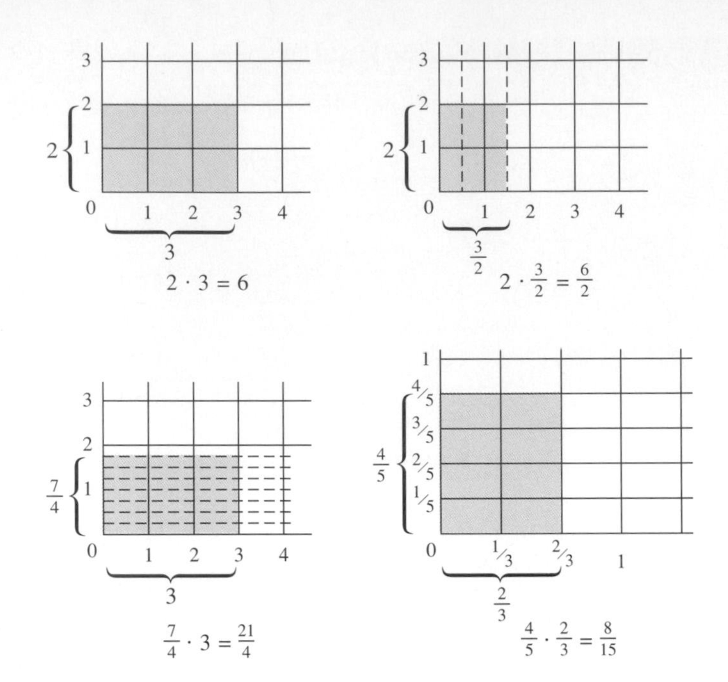

Figure 6.12
Extending the array model of multiplication to fractions

In general, if $\frac{a}{b}$ and $\frac{c}{d}$ are any two fractions, each unit square is divided into bd congruent rectangles and ac of the small rectangles are shaded. This shows that $\frac{ac}{bd}$ units are shaded. Thus, extending the array model of multiplication to fractions leads to the following definition.

DEFINITION Multiplication of Rational Numbers

Let $\frac{a}{b}$ and $\frac{c}{d}$ be rational numbers. Then their **product** is given by

$$\frac{a}{b} \cdot \frac{c}{d} = \frac{ac}{bd}.$$

Often a product, say $\frac{4}{5} \cdot \frac{2}{3}$, is read as four-fifths "of" two-thirds. The association between "of" and "times" is natural for multiplication by whole numbers and extends naturally to multiplication by rational numbers. For example, "I'll buy three *of* the half-gallon size bottles" is equivalent to buying $3 \cdot \frac{1}{2} = \frac{3}{2} = 1\frac{1}{2}$ gallons.

EXAMPLE 6.9 Calculating Products of Rational Numbers

Compute each product, expressing the answer in simplest form.

(a) $\frac{5}{8} \cdot \frac{2}{3}$ **(b)** $\frac{56}{88} \cdot \frac{-4}{7}$ **(c)** $3\frac{1}{7} \cdot 5\frac{1}{4}$

SCHOOL BOOK PAGE *Using Fraction Models to Build Understanding*

Subtracting Mixed Numbers

Build Understanding

A. Shades of Subtraction
Materials: Fraction pieces
Groups: Four students

Find $1\frac{1}{2} - \frac{5}{8}$ using fraction pieces.

a. Place fraction pieces representing $1\frac{1}{2}$ on your desk. Trace around each piece. What pieces did you use?

b. Next use fraction pieces to help you shade a part of your drawing that represents $\frac{5}{8}$. What pieces did you use?

c. What fraction is represented by the unshaded part of your drawing? Use fraction pieces, if needed, to find out. What is $1\frac{1}{2} - \frac{5}{8}$?

d. Now find $1\frac{2}{5} - \frac{1}{2}$ by tracing around fraction pieces. Describe the steps you used to solve the problem.

B. Rename $4\frac{2}{5}$ as $3\frac{\square}{5}$.

$4\frac{2}{5} = 4 + \frac{2}{5}$

$= 3 + 1 + \frac{2}{5}$

$= 3 + \frac{5}{5} + \frac{2}{5}$

$= 3\frac{7}{5}$

1. A student claims that if $\frac{a}{b} = \frac{c}{d}$, then $\frac{a+b}{b} = \frac{c+d}{d}$. For example, since $\frac{2}{3} = \frac{4}{6}$ then $\frac{2+3}{3} = \frac{4+6}{6}$. Is the student correct? Verify the example using fraction models.

2. Yet another student noticed that $\frac{3}{4} = \frac{9}{12}$ and $\frac{3+9}{4+12} = \frac{9}{12}$. Help this student investigate if $\frac{a}{b} = \frac{c}{d}$ implies that $\frac{a+c}{b+d} = \frac{c}{d}$.

SOLUTION

It is useful to look for common factors in the numerator and denominator before doing any multiplications.

(a) $\frac{5}{8} \cdot \frac{2}{3} = \frac{5 \cdot 2}{8 \cdot 3} = \frac{5 \cdot 2}{2 \cdot 4 \cdot 3} = \frac{5}{12}$

(b) $\frac{56}{88} \cdot \frac{-4}{7} = \frac{56 \cdot (-4)}{88 \cdot 7} = \frac{7 \cdot 8 \cdot (-4)}{8 \cdot 11 \cdot 7} = \frac{-4}{11}$

(c) $3\frac{1}{7} \cdot 5\frac{1}{4} = \frac{22}{7} \cdot \frac{21}{4} = \frac{22 \cdot 21}{7 \cdot 4} = \frac{2 \cdot 11 \cdot 3 \cdot 7}{7 \cdot 2 \cdot 2} = \frac{33}{2}.$ ■

EXAMPLE 6.10

Computing the Area of a Carpet

The hallway in the Bateks' house is a rectangle 4 feet wide and 20 feet long; that is, in yards it measures $\frac{4}{3}$ yards by $\frac{20}{3}$ yards. What is the area of the carpet in square yards? Mrs. Batek wants to know, since carpet is priced by the square yard.

SOLUTION

Since $\frac{4}{3} \cdot \frac{20}{3} = \frac{80}{9} = 8\frac{8}{9}$, the area is $8\frac{8}{9}$ square yards, or nearly 9 square yards. This can be seen in the diagram below, which shows the hallway divided into six full square yards, six $\frac{1}{3}$-square yard rectangular regions, and eight square regions which are each $\frac{1}{9}$ of a square yard. This gives the total area of $6 + \frac{6}{3} + \frac{8}{9} = 8\frac{8}{9}$ square yards.

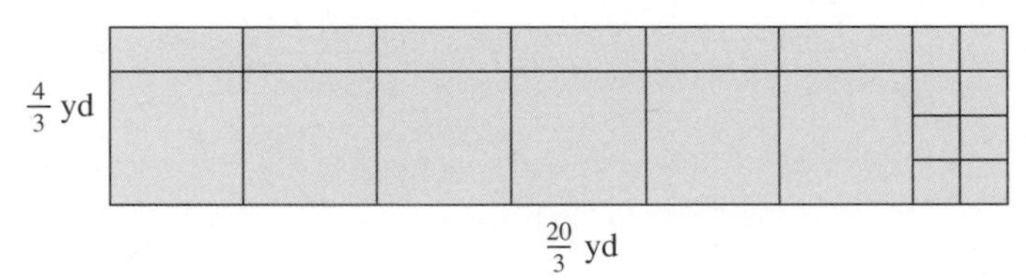

■

Division of Rational Numbers

In the whole number system, the division operation was made meaningful by associating it with conceptual models, such as repeated subtraction and determination of a missing factor. These models are equally useful for understanding division of rational numbers.

In the following problem, both the repeated-subtraction and missing-factor models are used to solve a real-life problem.

EXAMPLE 6.11

Bottling Root Beer: A Division Problem

Ari is making homemade root beer. The recipe he followed nearly fills a 5-gallon glass jug, and he estimates it contains $4\frac{3}{4}$ gallons of root beer. He is now ready to bottle his root beer. How many $\frac{1}{2}$ gallon bottles can he fill?

INTO THE CLASSROOM

Margaret Foster Discusses Multiplication and Division of Fractions

In learning about multiplication and division of fractions, students can more easily visualize the concepts if they participate in a cooperative-learning activity. The activity should use readily available and inexpensive manipulatives and should include ideas to which students can relate.

To begin the activity, I gather some of the materials I have acquired, including tops from margarine tubs, whipped-cream containers, and freezer tubs of various shapes. I have cut each lid into fraction parts: halves, fourths, and so on. For this activity, I use sets of lid fractions and simple granola-bar recipes which list a number of fractions.

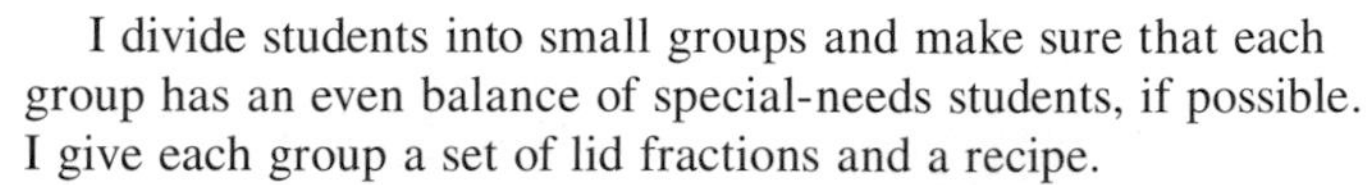

I divide students into small groups and make sure that each group has an even balance of special-needs students, if possible. I give each group a set of lid fractions and a recipe.

Then I present each group with three problems to solve using the lid fractions. First, the group members must properly assemble all necessary fractions for the recipe. Second, they must multiply the recipe portions by 2 to show how twice as many granola bars could be made. Third, they divide the original quantities by 2 to show how the recipe could be modified to produce half the number of granola bars.

When they are finished, I ask them to write the recipe that doubles the ingredients and the recipe that halves the ingredients.

This activity encourages student communication, exploration of concepts, discovery learning, application of concepts to a real-life situation, and "writing across the curriculum."

SOURCE: From *ScottForesman Exploring Mathematics* Grades 1–7 by L. Carey Bolster et al. Margaret Foster is Chairperson at Colleyville Middle School, Colleyville, TX.

SOLUTION

The repeated-subtraction approach. In the language of the repeated-subtraction model, the division problem $4\frac{3}{4} \div \frac{1}{2}$ is phrased in the form of a question: "How many $\frac{1}{2}$ gallons are in $4\frac{3}{4}$ gallons?" Since

$$4\frac{3}{4} = \left(\frac{1}{2} + \frac{1}{2} + \frac{1}{2} + \frac{1}{2} + \frac{1}{2} + \frac{1}{2} + \frac{1}{2} + \frac{1}{2} + \frac{1}{2}\right) + \left(\frac{1}{2}\right) \cdot \frac{1}{2} = \left(9 + \frac{1}{2}\right) \cdot \frac{1}{2},$$

we see that there are $9\frac{1}{2}$ "halves" in $4\frac{3}{4}$. Thus, Ari can fill 9 half-gallon bottles, and half of another one. This is shown in Figure 6.13. Ari will need 9 half-gallon bottles, and he will probably see if he can also find a quart bottle to use.

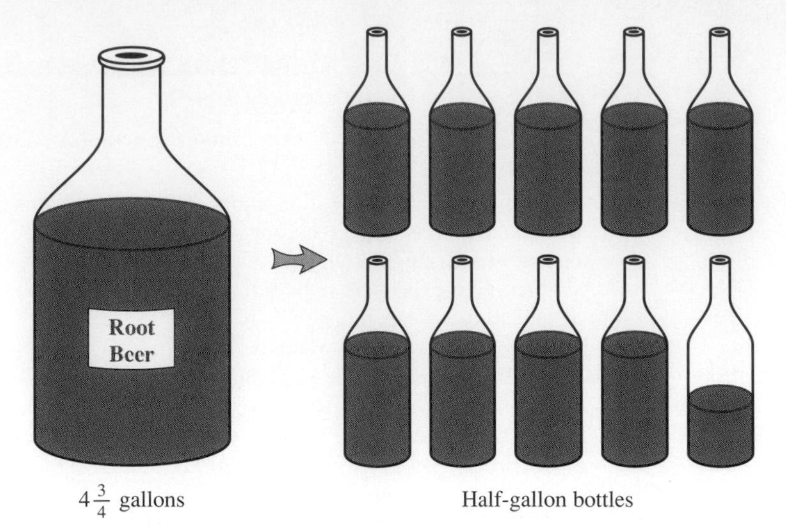

Figure 6.13

Since $4\frac{3}{4}$ gallons will fill $9\frac{1}{2}$ half-gallon bottles, $4\frac{3}{4} \div \frac{1}{2} = 9\frac{1}{2}$. That is, $\frac{19}{4} \div \frac{1}{2} = \frac{19}{2}$.

The missing-factor approach. Let x denote the number of half-gallon bottles required, where x will be allowed to be a fraction since we expect some bottle may be only partially filled. We must then solve the equation

$$x \cdot \frac{1}{2} = 4\frac{3}{4},$$

since this is the missing-factor problem that is equivalent to the division problem $4\frac{3}{4} \div \frac{1}{2}$.

Since

$$4\frac{3}{4} = \frac{19}{4} = \frac{19}{2} \cdot \frac{1}{2},$$

we see that the missing factor is $x = \frac{19}{2} = 9\frac{1}{2}$. ■

As with division in whole numbers and integers, the missing-factor model is the basis for the definition of division of rational numbers.

DEFINITION Division of Rational Numbers

Let $\frac{a}{b}$ and $\frac{c}{d}$ be rational numbers, where $\frac{c}{d}$ is not zero. Then $\frac{a}{b} \div \frac{c}{d} = \frac{e}{f}$ if, and only if, $\frac{a}{b} = \frac{c}{d} \cdot \frac{e}{f}$.

The definition gives us an equation the quotient must satisfy, but it does not tell us how to compute the quotient. Let's see if we can develop a formula for the quotient $\frac{2}{5} \div \frac{3}{7}$. According to the definition above, we need to solve the equation

$$\frac{2}{5} = \frac{3}{7} \cdot \frac{e}{f}$$

for $\frac{e}{f}$. By the definition of multiplication of rational numbers, we have

$$\frac{2}{5} = \frac{3 \cdot e}{7 \cdot f}.$$

To solve for $\frac{e}{f}$, we will multiply both sides of the equation by $\frac{7}{3}$, finding

$$\frac{2}{5} \cdot \frac{7}{3} = \frac{3 \cdot e}{7 \cdot f} \cdot \frac{7}{3}.$$

The right hand side of the equation simplifies nicely, since

$$\frac{3 \cdot e}{7 \cdot f} \cdot \frac{7}{3} = \frac{e}{f}.$$

We have found that if $\frac{e}{f} = \frac{2}{5} \div \frac{3}{7}$ then $\frac{e}{f} = \frac{2}{5} \cdot \frac{7}{3}$.

The same procedure proves a general formula for the division of rational numbers.

THEOREM An Algorithm for Division of Rational Numbers

Let $\frac{a}{b}$ and $\frac{c}{d}$ be rational numbers where $\frac{c}{d}$ is not zero. Then $\frac{a}{b} \div \frac{c}{d} = \frac{a}{b} \cdot \frac{d}{c}$.

For obvious reasons the algorithm is called the **invert the divisor and multiply rule.** To "invert" $\frac{c}{d}$ means to form the rational number $\frac{d}{c}$, which is called the **reciprocal** of $\frac{c}{d}$.

DEFINITION Reciprocal of a Rational Number

The **reciprocal** of a nonzero rational number $\frac{c}{d}$ is $\frac{d}{c}$.

Since

$$\frac{c}{d} \cdot \frac{d}{c} = \frac{c \cdot d}{d \cdot c} = 1,$$

$\frac{d}{c}$ is also known as the **multiplicative inverse** of $\frac{c}{d}$. The importance of the multiplicative inverse is investigated in the next section.

Sometimes the invert and multiply rule is mistaken for a definition of division of fractions. This is unfortunate; the algorithm does not relate directly to any conceptual model of division, and division becomes a meaningless and unpleasant process of following abstract and seemingly arbitrary rules.

The algorithm for division tells us that the rational numbers are closed under division, since every quotient $\frac{a}{b} \div \frac{c}{d}$ with $c \neq 0$ defines the unique fraction $\frac{ad}{bc}$. The whole numbers are not closed under division. For example, 3 is not divisible by 8. On the other hand, if we view 3 and 8 as rational numbers, then we see that

$$3 \div 8 = \frac{3}{1} \div \frac{8}{1} = \frac{3 \cdot 1}{1 \cdot 8} = \frac{3}{8}.$$

In general, if a and b are any integers, with $b \neq 0$, then interpreting a and b as *rational numbers* we find that

$$a \div b = \frac{a}{b}.$$

This shows that extending the set of integers to the rational numbers gives us a set of numbers for which division is closed.

EXAMPLE 6.12

Dividing Rational Numbers

Find the following quotients.

(a) $\frac{3}{4} \div \frac{1}{8}$ **(b)** $\frac{-7}{4} \div \frac{2}{3}$ **(c)** $3 \div \frac{4}{3}$

(d) $39 \div 13$ **(e)** $13 \div 39$ **(f)** $4\frac{1}{6} \div 2\frac{1}{3}$

SOLUTION

(a) $\frac{3}{4} \div \frac{1}{8} = \frac{3}{4} \cdot \frac{8}{1} = \frac{24}{4} = 6$ **(b)** $\frac{-7}{4} \div \frac{2}{3} = \frac{-7}{4} \cdot \frac{3}{2} = \frac{-21}{8}$

(c) $3 \div \frac{4}{3} = \frac{3}{1} \cdot \frac{3}{4} = \frac{9}{4}$ **(d)** $39 \div 13 = \frac{39}{1} \cdot \frac{1}{13} = \frac{39}{13} = 3$

(e) $13 \div 39 = \frac{13}{1} \cdot \frac{1}{39} = \frac{13}{39} = \frac{1}{3}$

(f) $4\frac{1}{6} \div 2\frac{1}{3} = \frac{25}{6} \div \frac{7}{3} = \frac{25}{6} \cdot \frac{3}{7} = \frac{25}{14}$ ■

EXAMPLE 6.13

Scale of a Map: Illustrating Division of Rational Numbers with a Measurement Model

Colton and Uniontown are $2\frac{1}{8}''$ apart on the map. The map is scaled so that $\frac{3}{4}''$ on the map corresponds to 1 mile of actual distance. How far is it between the two towns?

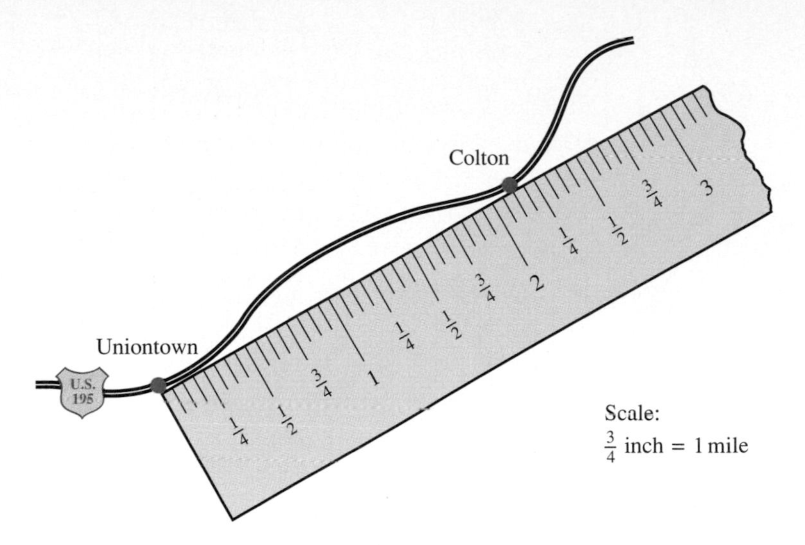

SOLUTION

Each $\frac{3}{4}''$ on the map represents 1 mile of distance. To find the number of miles between Uniontown and Colton, we must know how many segments of length $\frac{3}{4}''$ lie between the two points on the map.

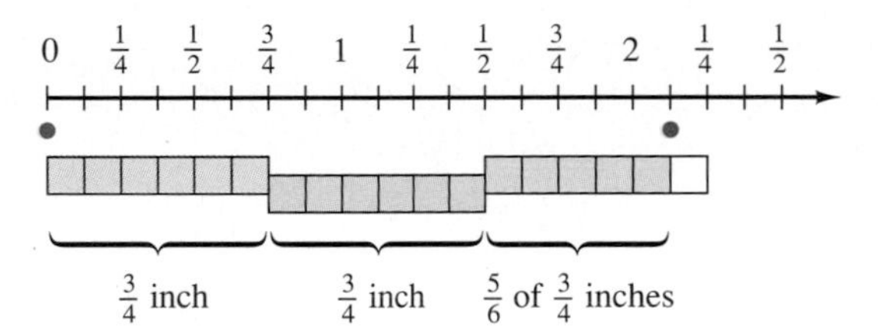

1 mile + 1 mile + $\frac{5}{6}$ of 1 mile = $2\frac{5}{6}$ miles

The figure shows that there are $1 + 1 + \frac{5}{6} = 2\frac{5}{6}$ segments of length $\frac{3}{4}''$ in a segment $2\frac{1}{8}''$ long. Thus it is $2\frac{5}{6}$ miles between Uniontown and Colton. Instead of relying on the figure, we can use the algorithm for division of rational numbers:

$$2\frac{1}{8} \div \frac{3}{4} = \frac{17}{8} \cdot \frac{4}{3} = \frac{17}{6} = 2\frac{5}{6}.$$

■

Using a Fraction Calculator

Some calculators, such as the Texas Instrument *Math Explorer,* are capable of performing "rational arithmetic." That is, fractions can be added, subtracted, multiplied, and divided, with the result written as a fraction. For example, consider the expression

$$\left(2\frac{3}{4} + \frac{2}{3}\right) \div \left(\frac{4}{5} - \frac{1}{2}\right).$$

The entry string

ON/AC (2 Unit 3 / 4 + 2 / 3) ÷ (4 / 5 − 1 / 2) =

yields the fractional answer 410/36. Pressing Simp = expresses the answer in simplest form, 205/18. Pressing Ab/c converts the fraction to mixed number form, 11⌴7/18, where ⌴ separates the units to the left from the fractional part at the right. Pressing F ⇆ D gives the decimal number 11.388889 that approximates the fraction $11\frac{7}{18}$. The F ⇆ D key switches the display from fraction to decimal form, and vice versa.

PROBLEM SET 6.2

Understanding Concepts

1. (a) What addition fact is illustrated by the following fraction strip model?

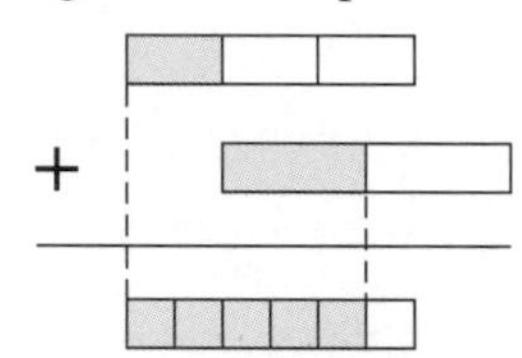

(b) Illustrate $\frac{1}{6} + \frac{1}{4}$ with the fraction strip model.

(c) Illustrate $\frac{2}{3} + \frac{3}{4}$ with the fraction strip model. (The sum will require two strips.)

2. Use the colored region model to illustrate these sums.

(a) $\frac{2}{5} + \frac{6}{5}$ (b) $\frac{2}{3} + \frac{1}{4}$.

3. Use the number-line model to illustrate these sums.

(a) $\frac{3}{4} + \frac{-2}{4}$ (b) $\frac{-3}{4} + \frac{2}{4}$ (c) $\frac{-3}{4} + \frac{-1}{4}$.

4. Perform these additions. Express each answer in simplest form.

(a) $\frac{2}{7} + \frac{3}{7}$ (b) $\frac{6}{5} + \frac{4}{5}$ (c) $\frac{3}{8} + \frac{11}{24}$

(d) $\frac{6}{13} + \frac{2}{5}$ (e) $\frac{5}{12} + \frac{17}{20}$ (f) $\frac{6}{8} + \frac{-25}{100}$

(g) $\frac{-57}{100} + \frac{13}{10}$ (h) $\frac{213}{450} + \frac{12}{50}$

5. Express these fractions as mixed numbers.

(a) $\frac{9}{4}$ (b) $\frac{17}{3}$

(c) $\frac{111}{23}$ (d) $\frac{3571}{-100}$

6. Express these mixed numbers as fractions.

(a) $2\frac{3}{8}$ (b) $15\frac{2}{3}$

(c) $111\frac{2}{5}$ (d) $-10\frac{7}{9}$.

7. (a) What subtraction fact is illustrated by this fraction strip model?

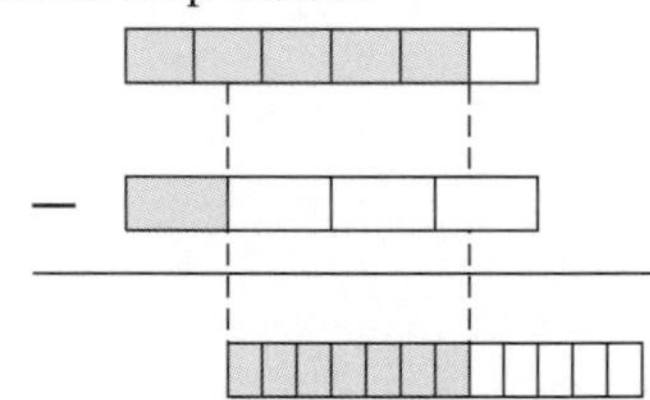

(b) Use the fraction strip, colored region, and number-line models to illustrate $\frac{2}{3} - \frac{1}{4}$.

8. Compute these differences, expressing each answer in simplest form.

(a) $\frac{5}{8} - \frac{2}{8}$ (b) $\frac{3}{5} - \frac{2}{4}$ (c) $2\frac{2}{3} - 1\frac{1}{3}$

(d) $4\frac{1}{4} - 3\frac{1}{3}$ (e) $\frac{6}{8} - \frac{5}{12}$ (f) $\frac{1}{4} - \frac{14}{56}$

(g) $\frac{137}{214} - \frac{-1}{3}$ (h) $\frac{-23}{100} - \frac{198}{1000}$

9. What multiplication facts are illustrated by these colored rectangular region models?

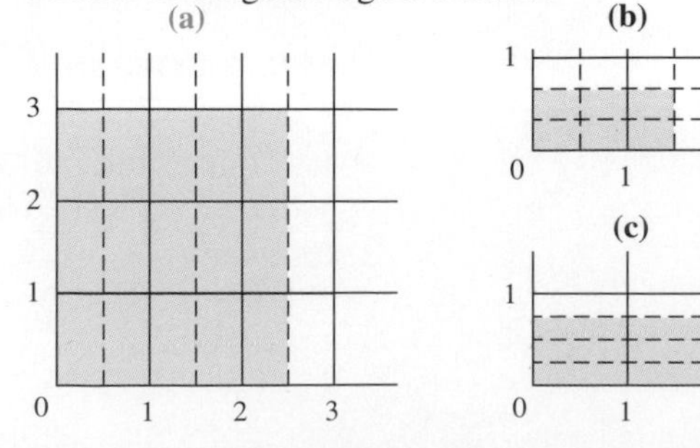

10. Illustrate these multiplications with the model used in problem 9.

(a) $2 \times \frac{3}{5}$ (b) $\frac{3}{2} \times \frac{3}{4}$ (c) $1\frac{2}{3} \times 2\frac{1}{4}$

11. A rectangular plot of land is $2\frac{1}{4}$ miles wide and $3\frac{1}{2}$ miles long. What is the area of the plot, in square miles? Draw a sketch that verifies your answer.

12. Find the reciprocals of the following rational numbers.

(a) $\frac{3}{8}$ (b) $\frac{4}{3}$ (c) $2\frac{1}{4}$

(d) $\frac{1}{8}$ (e) 5 (f) 1

13. If $\frac{a}{b}$ is between 0 and 1, what can you say about the reciprocal $\frac{b}{a}$? If $\frac{c}{d}$ is nearly 1, what can you say about $\frac{d}{c}$?

14. If the positive rational numbers $\frac{a}{b}$ and $\frac{c}{d}$ satisfy $\frac{a}{b} < \frac{c}{d}$, how are their reciprocals ordered?

15. Compute these divisions, expressing each answer in simplest form.

(a) $\frac{2}{5} \div \frac{3}{4}$ (b) $\frac{6}{11} \div \frac{4}{3}$ (c) $\frac{100}{33} \div \frac{10}{3}$

(d) $2\frac{3}{8} \div 5$ (e) $3 \div 5\frac{1}{4}$ (f) $\frac{21}{25} \div \frac{7}{25}$

16. A sign on a roll-end of canvas says it contains 42 square yards. The width of the canvas, which is easily measured without unrolling, is 14 feet $\left(\text{or } 4\frac{2}{3} \text{ yards}\right)$. What is the length of the piece of canvas, in yards?

17. Complicated expressions such as

$$\frac{\frac{1}{2} - \frac{1}{5}}{\frac{3}{4} + \frac{2}{3}}$$

are meaningful when interpreted as $\left(\frac{1}{2} - \frac{1}{5}\right) \div \left(\frac{3}{4} + \frac{2}{3}\right)$. These expressions are called *complex fractions*. In the example above, the complex fraction simplifies to

$$\begin{aligned}\left(\frac{1}{2} - \frac{1}{5}\right) \div \left(\frac{3}{4} + \frac{2}{3}\right) &= \left(\frac{5}{10} - \frac{2}{10}\right) \\ &\quad \div \left(\frac{9}{12} + \frac{8}{12}\right) \\ &= \frac{3}{10} \div \frac{17}{12} \\ &= \frac{3}{10} \cdot \frac{12}{17} \\ &= \frac{18}{85}.\end{aligned}$$

Simplify these complex fractions.

(a) $\dfrac{\frac{1}{2} - \left(\frac{2}{3} \cdot \frac{3}{8}\right)}{\frac{2}{7} - \frac{1}{14}}$ (b) $\dfrac{3\frac{1}{3} - 2\frac{1}{2}}{4\frac{1}{4} + 5\frac{1}{5}}$

18. If $\frac{a}{b} < \frac{c}{d}$, there is a positive rational number $\frac{e}{f}$ $\left(\text{namely } \frac{c}{d} - \frac{a}{b}\right)$ which satisfies $\frac{a}{b} + \frac{e}{f} = \frac{c}{d}$. An alternative, but equivalent, definition of rational number inequality is the following:

$\frac{a}{b} < \frac{c}{d}$ if, and only if, there is a positive rational number $\frac{e}{f}$ such that $\frac{a}{b} + \frac{e}{f} = \frac{c}{d}$.

Use the alternative definition to verify these inequalities.

(a) $\frac{2}{3} < \frac{3}{4}$ (b) $\frac{4}{5} < \frac{14}{17}$ (c) $\frac{19}{10} < \frac{99}{50}$

Thinking Critically

19. Reread the Just For Fun, "The Sultan's Estate" in Section 6.2. Can you determine the sum $\frac{1}{2} + \frac{1}{3} + \frac{1}{9}$ just by the outcome of the story, with no need to do the calculations explicitly? Now check your answer by actually computing the sum.

20. (a) Find the missing fractions in the Magic Fraction Square below, so that the entries in every row, column, and diagonal add to 1.

$\frac{1}{2}$	$\frac{1}{12}$	
	$\frac{1}{3}$	

(b) Here is a Magic Square in the distinct whole numbers 1, 2, . . . , 9 on the left, and a partially completed Magic Fraction Square on the right. How was the left square converted to the corresponding fraction version on the right?

8	3	4
1	5	9
6	7	2

	$\frac{1}{3}$	$\frac{3}{5}$
		$\frac{2}{15}$

(c) Here is another partially completed Magic Fraction Square, where again each row, column, and diagonal adds to 1. Is it related to the whole number square of part (b)? If so, describe how.

$\frac{1}{6}$	$\frac{1}{3}$	
$\frac{3}{8}$		

21. Consider the sums of the reciprocals of the natural numbers: $S_1 = \frac{1}{1}$, $S_2 = \frac{1}{1} + \frac{1}{2} = \frac{3}{2}$, $S_3 = \frac{1}{1} + \frac{1}{2} + \frac{1}{3} = \frac{11}{6}$, $S_4 = \frac{25}{12}, \ldots$.

(a) Find S_5, S_6, S_7, S_8, where $S_n = \frac{1}{1} + \frac{1}{2} + \cdots + \frac{1}{n}$, writing each in lowest terms.

(b) Verify that each fraction representing $S_2, \ldots, S_8$, when in lowest terms, has an odd numerator and an even denominator.

(c) It can be shown that each fraction S_n, $n = 2, 3, \ldots$, in lowest terms, has an odd numerator and an even denominator. Use this fact to explain carefully why S_n is not a whole number for any $n \geq 2$.

22. Let A, b, c be natural numbers. Carefully *prove* that the mixed number $A\frac{b}{c}$ has the fraction form $\frac{Ac + b}{c}$. Explain your steps and reasoning.

23. **Other Algorithms for Division.** The "invert and multiply" algorithm, $\frac{a}{b} \div \frac{c}{d} = \frac{a}{b} \cdot \frac{d}{c}$, transforms a division of rational numbers into a multiplication. Here are two other useful algorithms:

Common-denominators: $\frac{a}{b} \div \frac{c}{b} = \frac{a}{c}$

Divide-numerators and denominators:

$$\frac{a}{b} \div \frac{c}{d} = \frac{a \div c}{b \div d}$$

Thus,

$$\frac{3}{19} \div \frac{15}{19} = \frac{3}{15} = \frac{1}{5} \quad \text{and} \quad \frac{24}{35} \div \frac{6}{5} = \frac{24 \div 6}{35 \div 5} = \frac{4}{7}.$$

(a) Choose one of the two new algorithms to perform these calculations:

$$\frac{7}{12} \div \frac{11}{12}, \quad \frac{4}{15} \div \frac{2}{3}, \quad \frac{19}{210} \div \frac{19}{70}$$

(b) Verify the correctness of the two new algorithms.

Thinking Cooperatively

24. With the exception of $\frac{2}{3}$, which was given the hieroglyph [hieroglyph], the ancient Egyptians attempted to express all rational numbers as a sum of *unit* fractions, fractions with 1 as the numerator.

(a) Verify that $\frac{23}{25} = \frac{1}{2} + \frac{1}{3} + \frac{1}{15} + \frac{1}{50}$.

(b) Verify that $\frac{7}{29} = \frac{1}{6} + \frac{1}{24} + \frac{1}{58} + \frac{1}{87} + \frac{1}{232}$.

(c) Verify that $\frac{7}{29} = \frac{1}{5} + \frac{1}{29} + \frac{1}{145}$.

(d) If [hieroglyph] represents $1\frac{1}{2}$, and [hieroglyph] is interpreted as "one over," does the symbol [hieroglyph] seem reasonable for $\frac{2}{3}$?

25. The Rhind papyrus was copied by the scribe Ahmes about 1650 B.C. from earlier hieratic writing. The papyrus was purchased in Egypt in 1858 by Scottish Egyptologist Henry Rhind and is one of the few surviving mathematical documents of ancient times in Egypt. The papyrus opens with the words "Directions for Obtaining the Knowledge of All Dark Things" and goes on to state and solve 85 problems. The papyrus also contains a table for expressing fractions with numerator 2 and odd denominators 5 through 101 as a sum of fractions with numerator 1. For example:

$$\frac{2}{5} = \frac{1}{3} + \frac{1}{15} \quad \text{and} \quad \frac{2}{7} = \frac{1}{4} + \frac{1}{28}.$$

(a) Express $\frac{2}{9}$ and $\frac{2}{11}$ as sums of fractions with numerator 1. (*Hint:* Notice that 3 is half of $5 + 1$ and 4 is half of $7 + 1$ in the two examples given.)

(b) Look for a pattern which allows you to write any fraction $\frac{2}{3}, \frac{2}{5}, \frac{2}{7}, \frac{2}{9}, \frac{2}{11}, \frac{2}{13}, \ldots,$ $\frac{2}{2n-1}, \ldots$ as a sum of fractions with numerator 1. Describe your conjecture carefully and check it on several examples. Try to prove your conjecture by deriving a formula which is true for any n, $n = 2, 3, \ldots$.

26. Let $a_1 = \frac{1}{2}$, $a_2 = \frac{1}{2} + \frac{1}{4}$, $a_3 = \frac{1}{2} + \frac{1}{4} + \frac{1}{8}, \ldots,$

$$a_n = \frac{1}{2} + \frac{1}{4} + \cdots + \frac{1}{2^n}.$$

(a) Show that $a_1 = 1 - \frac{1}{2}$, $a_2 = 1 - \frac{1}{4}$, and $a_3 = 1 - \frac{1}{8}$.

(b) Guess a formula for a_n. Use the subdivided unit square below to give a visual justification of your conjecture.

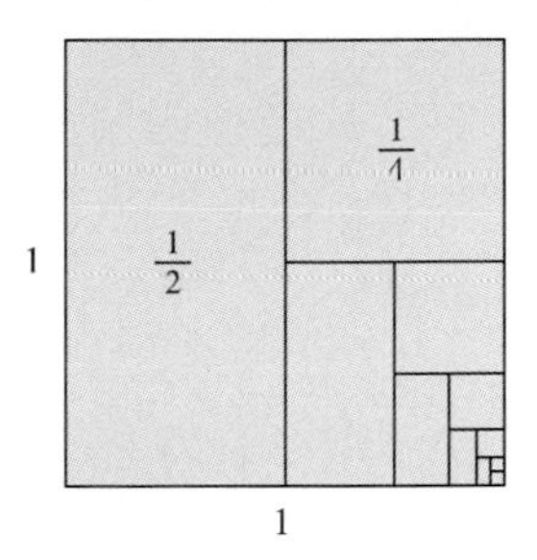

Using a Calculator

27. **DIFFY with Fractions.** The DIFFY game, described in Section 2.3, was played with whole numbers. Here are the beginning lines of DIFFY where the entries are fractions. A new line is formed by subtracting the smaller fraction from the larger. The DIFFY computer program can also be used for fraction entries.

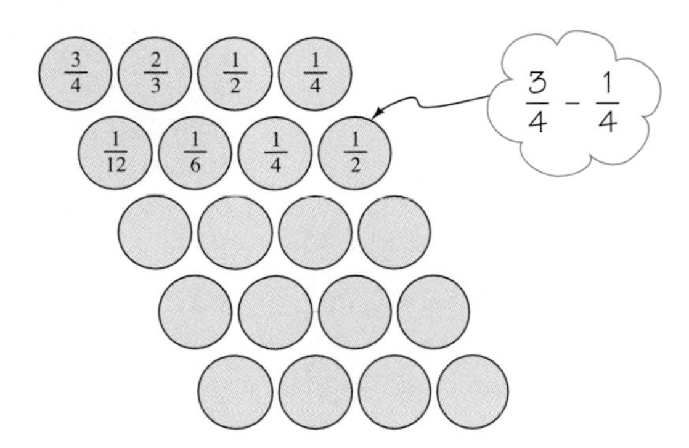

(a) Fill in additional lines of the DIFFY array started above. Does it terminate?

(b) Try fraction DIFFY with these fractions in your first row: $\frac{2}{7}, \frac{4}{5}, \frac{3}{2}, \frac{5}{6}$.

(c) Suppose that you know DIFFY with whole number entries always terminates with 0, 0, 0, 0. Does it necessarily follow that DIFFY with fractions must terminate? Explain your reasoning carefully.

28. **DIVVY.** The process called DIVVY is like DIFFY (see problem 27) except that the larger fraction is *divided* by the smaller. The first few rows of DIVVY are shown.

(a) Continue to fill in additional rows, using a calculator if you like, or the DIFFY computer program. Does the process terminate? How?

(b) Try DIVVY with $\frac{2}{7}, \frac{4}{5}, \frac{3}{2}, \frac{5}{6}$ in the first row. Don't let complicated fractions put you off! Things should get better if you persist.

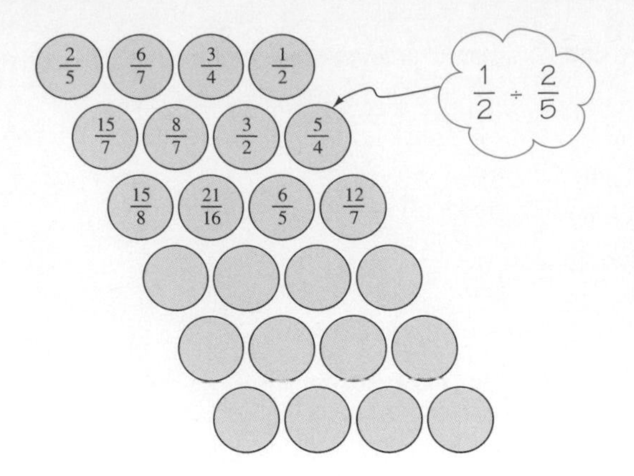

Making Connections

29. At a certain university, a student's senior thesis is acceptable if at least $\frac{3}{4}$ of the student's committee votes in its favor. What is the smallest number of favorable votes needed to accept a thesis if the committee has 3 members? 4 members? 5 members? 6 members? 7 members? 8 members?

30. Tongue-and-groove decking boards are each $2\frac{1}{4}''$ wide. How many boards must be placed side by side to do a deck 14 feet in width?

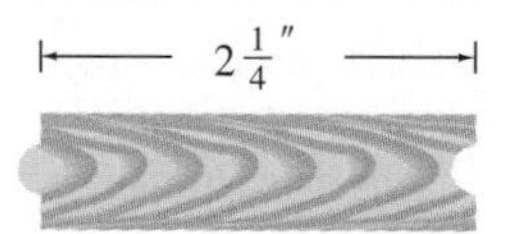

31. Six bows can be made from $1\frac{1}{2}$ yards of ribbon. How many bows can be made from $5\frac{3}{4}$ yards of ribbon?

32. Andre has 35 yards of material available to make aprons. Each apron requires $\frac{3}{4}$ yard. How many aprons can Andre make?

33. Gisela paid \$28 for a shirt that was "$\frac{1}{3}$ off." What was the original price of the skirt?

34. A soup recipe calls for $2\frac{3}{4}$ cups of chicken broth, and will make enough to serve 8 people. How much broth is required if the recipe is modified to serve 6 people?

Communicating

35. For each fraction operation below, make up a realistic word problem whose solution requires the computation shown. Try to create an interesting and original situation. Include a carefully drawn diagram.

(a) $\frac{19}{32} + \frac{1}{4}$ **(b)** $3\frac{1}{4} - 1\frac{1}{16}$

(c) $\frac{4}{5} \times \frac{7}{8}$ **(d)** $\frac{9}{10} \div \frac{3}{5}$

36. Respond to a student who asks "should fractions always be written in simplest form, or are there situations where it would be better not to simplify?"

For Review

37. Represent these rational numbers by fractions in simplest form.

(a) $\frac{168}{48}$ **(b)** $\frac{945}{3780}$

38. Arrange these rational numbers in increasing order:

$$\frac{31}{90}, \frac{1}{3}, \frac{19}{60}, \frac{4}{13}$$

39. Find the least common denominator of $\frac{4}{9}, \frac{5}{12}, \frac{4}{15}$ and then find the sum of the three numbers.

6.3 The Rational Number System

This section explores the properties of the rational numbers. Many properties will be familiar, since the integers have corresponding properties. However, we will also discover some important new properties of rational numbers which have no counterpart in the integers. This section also gives techniques for estimation and computation, and presents additional examples of the application of rational numbers to the solution of practical problems.

Properties of Addition and Subtraction

To add two rational numbers, the definition introduced in the preceding section tells us to "represent the rational numbers by fractions with a common denominator, and then add the two numerators." For example, to add $\frac{5}{6}$ and $\frac{3}{10}$ we could use the common denominator 30; thus

$$\frac{5}{6} + \frac{3}{10} = \frac{25}{30} + \frac{9}{30} = \frac{34}{30}.$$

We could also have used the common denominator 60 to find,

$$\frac{5}{6} + \frac{3}{10} = \frac{50}{60} + \frac{18}{60} = \frac{68}{60}.$$

The two answers, namely $\frac{34}{30}$ and $\frac{68}{60}$, are different fractions. However, they are equivalent fractions, and both represent the *same* rational number, $\frac{17}{15}$, when expressed by a fraction in simplest form. More generally *any* two rational numbers have a unique rational number which is their sum. That is, the rational numbers are closed under addition.

It is also straightforward to check that addition in the rational numbers is commutative, associative, and has 0 as an additive identity. For example, $\frac{5}{6} + \frac{3}{10} = \frac{3}{10} + \frac{5}{6}$, $\frac{3}{4} + \left(\frac{-1}{3} + \frac{2}{5}\right) = \left(\frac{3}{4} + \frac{-1}{3}\right) + \frac{2}{5}$, and $\frac{7}{9} + 0 = \frac{7}{9}$ illustrate these respective properties.

The rational numbers share one more property with the integers, namely the existence of **negatives,** or **additive inverses.** For example, the negative of $\frac{4}{7}$ is $\frac{-4}{7}$ since

$$\frac{4}{7} + \frac{-4}{7} = \frac{4 + (-4)}{7} = \frac{0}{7} = 0.$$

We have the following definition.

DEFINITION Negative or Additive Inverse

Let $\frac{a}{b}$ be a rational number. Its **negative,** or **additive inverse,** written $-\frac{a}{b}$, is the rational number $\frac{-a}{b}$.

For example, $-\frac{2}{3} = \frac{-2}{3}$. Since $\frac{-2}{3} = \frac{2}{-3}$, we see that $-\frac{2}{3} = \frac{-2}{3} = \frac{2}{-3}$. There are *three* places for the negative sign, and all are useful: $-\frac{2}{3}$ denotes the negative of the rational number $\frac{2}{3}$; $\frac{-2}{3}$ denotes that the negative of the numerator is taken; $\frac{2}{-3}$ denotes that the negative of the denominator is taken. Similarly, $-\frac{-5}{7} = \frac{-(-5)}{7} = \frac{5}{7}$, showing that $\frac{5}{7}$ is the negative of $\frac{-5}{7}$.

A negative is also called an **opposite.** This term describes how a rational number and its negative are positioned on the number line: $\frac{-a}{b}$ is on the opposite side of 0 from $\frac{a}{b}$.

The properties of addition on the rational numbers are listed in the following theorem.

THEOREM Properties of Addition of Rational Numbers

Let $\frac{a}{b}$, $\frac{c}{d}$, and $\frac{e}{f}$ be rational numbers. The following properties hold.

Closure Property	$\frac{a}{b} + \frac{c}{d}$ is a rational number.
Commutative Property	$\frac{a}{b} + \frac{c}{d} = \frac{c}{d} + \frac{a}{b}$
Associative Property	$\left(\frac{a}{b} + \frac{c}{d}\right) + \frac{e}{f} = \frac{a}{b} + \left(\frac{c}{d} + \frac{e}{f}\right)$
Zero is an Additive Identity	$\frac{a}{b} + 0 = \frac{a}{b}$
Existence of Additive Inverses	$\frac{a}{b} + \left(-\frac{a}{b}\right) = 0$, where $-\frac{a}{b} = \frac{-a}{b}$.

In the integers we discovered that subtraction was equivalent to the addition of the negative. The same result holds for the rational numbers. Since the rational numbers are closed under addition and every rational number has a negative, this means that the rational numbers are closed under subtraction.

THEOREM Formulas for Subtraction of Rational Numbers

Let $\frac{a}{b}$ and $\frac{c}{d}$ be rational numbers. Then $\frac{a}{b} - \frac{c}{d} = \frac{a}{b} + \left(-\frac{c}{d}\right) = \frac{ad - bc}{bd}$.

Subtraction is neither commutative nor associative, as examples such as $\frac{1}{2} - \frac{1}{4} \neq \frac{1}{4} - \frac{1}{2}$ and $1 - \left(\frac{1}{2} - \frac{1}{4}\right) \neq \left(1 - \frac{1}{2}\right) - \frac{1}{4}$ show. This means that subtraction requires that careful attention be given to the order of the terms and the placement of parentheses.

EXAMPLE 6.14

Subtracting Rational Numbers

Compute the following differences.

(a) $\frac{3}{4} - \frac{7}{6}$ **(b)** $\frac{2}{3} - \frac{-9}{8}$ **(c)** $\left(-2\frac{1}{4}\right) - \left(4\frac{2}{3}\right)$

SOLUTION

(a) $\frac{3}{4} - \frac{7}{6} = \frac{3 \cdot 6 - 4 \cdot 7}{4 \cdot 6} = \frac{18 - 28}{24} = \frac{-10}{24} = \frac{-5}{12} = -\frac{5}{12}.$

(b) $\frac{2}{3} - \frac{-9}{8} = \frac{2 \cdot 8 - 3 \cdot (-9)}{3 \cdot 8} = \frac{16 - (-27)}{24} = \frac{43}{24}.$

(c) $\left(-2\frac{1}{4}\right) - \left(4\frac{2}{3}\right) = (-2 - 4) - \left(\frac{1}{4} + \frac{2}{3}\right) = -6 - \left(\frac{3}{12} + \frac{8}{12}\right)$

$= -6\frac{11}{12}.$ ■

Properties of Multiplication and Division

Multiplication of rational numbers includes all of the properties of multiplication for the integers. For example, let's investigate the distributive property of multiplication over addition by considering a specific case:

$$\frac{2}{5} \cdot \left(\frac{3}{4} + \frac{7}{8}\right) = \frac{2}{5} \cdot \left(\frac{6}{8} + \frac{7}{8}\right) = \frac{2}{5} \cdot \frac{13}{8} = \frac{26}{40}$$

Add first, then multiply.

$$\frac{2}{5} \cdot \frac{3}{4} + \frac{2}{5} \cdot \frac{7}{8} = \frac{6}{20} + \frac{14}{40} = \frac{12}{40} + \frac{14}{40} = \frac{26}{40}$$

Multiply first, then add.

These computations show that multiplication by $\frac{2}{5}$ distributes over the sum $\frac{3}{4} + \frac{7}{8}$. A similar procedure proves the general distributive property $\frac{a}{b} \cdot \left(\frac{c}{d} + \frac{e}{f}\right) = \frac{a}{b} \cdot \frac{c}{d} + \frac{a}{b} \cdot \frac{e}{f}$.

There is one important new property of multiplication of rational numbers which is *not* true for the integers: the existence of multiplicative inverses. For example, the nonzero rational number $\frac{5}{8}$ has the multiplicative inverse $\frac{8}{5}$, since

$$\frac{5}{8} \cdot \frac{8}{5} = 1.$$

This property does not hold in the integers. For example, the integer 2 does not have a multiplicative inverse in the set of integers since there is no integer m for which $2 \cdot m = 1$.

THEOREM Properties of Multiplication of Rational Numbers

Let $\frac{a}{b}$, $\frac{c}{d}$, and $\frac{e}{f}$ be rational numbers. The following properties hold.

Closure Property $\frac{a}{b} \cdot \frac{c}{d}$ is a rational number.

Commutative Property $\frac{a}{b} \cdot \frac{c}{d} = \frac{c}{d} \cdot \frac{a}{b}$

Associative Property $\left(\frac{a}{b} \cdot \frac{c}{d}\right) \cdot \frac{e}{f} = \frac{a}{b} \cdot \left(\frac{c}{d} \cdot \frac{e}{f}\right)$

Distributive Property of Multiplication over Addition and Subtraction

$\frac{a}{b} \cdot \left(\frac{c}{d} + \frac{e}{f}\right) = \frac{a}{b} \cdot \frac{c}{d} + \frac{a}{b} \cdot \frac{e}{f}$ and $\frac{a}{b} \cdot \left(\frac{c}{d} - \frac{e}{f}\right) = \frac{a}{b} \cdot \frac{c}{d} - \frac{a}{b} \cdot \frac{e}{f}$

Multiplication by Zero $0 \cdot \frac{a}{b} = 0$

One is a Multiplicative Identity $1 \cdot \frac{a}{b} = \frac{a}{b}$

Existence of Multiplicative Inverse If $\frac{a}{b} \neq 0$, then there is a unique rational number, namely $\frac{b}{a}$, for which $\frac{a}{b} \cdot \frac{b}{a} = 1$.

EXAMPLE 6.15

Solving an Equation with the Multiplicative Inverse

Consuela paid \$36 for a pair of shoes at the "one fourth off" sale. What was the price of the shoes before the sale?

SOLUTION

Let x be the original price of the shoes, in dollars. Consuela paid $\frac{3}{4}$ of this price, so $\frac{3}{4} \cdot x = 36$. To solve this equation for the unknown x, multiply both sides by $\frac{4}{3}$ to get

$$\frac{4}{3} \cdot \frac{3}{4} \cdot x = \frac{4}{3} \cdot 36.$$

But $\frac{4}{3} \cdot \frac{3}{4} = 1$ by the multiplicative inverse property, so

$$1 \cdot x = \frac{4 \cdot 36}{3} = 48.$$

Since 1 is a multiplicative identity we find that $1 \cdot x = x = 48$. Therefore, the original price of the pair of shoes was \$48. ■

CURRENT AFFAIRS

Simpson's Paradox

In baseball, a player's "batting average" is given by the fraction of hits obtained in the total times at bat. For example, Dave Justice had 12 hits in 51 at bats in 1989, giving him a batting average of $\frac{12}{51}$. That same year, Andy Van Slyke had a batting average of $\frac{113}{476}$. Since $\frac{113}{476} > \frac{12}{51}$ Andy out hit Dave in 1989.

	Dave Justice			Andy Van Slyke		
	Hits	At bats	Batting Average	Hits	At bats	Batting Average
1989	12	51	$\frac{12}{51} \doteq .235$	113	476	$\frac{113}{476} \doteq .237$
1990	124	439	$\frac{124}{439} \doteq .282$	140	493	$\frac{140}{493} \doteq .284$
1989–1990 combined	136	490	$\frac{136}{490} \doteq .278$	253	969	$\frac{253}{969} \doteq .261$

As the table* shows, Andy also outhit Dave in 1990, since $\frac{140}{493} > \frac{124}{439}$. By combining the data for 1989 and 1990, we see that Dave's two year average is $\frac{136}{490}$, and Andy's two year average is $\frac{253}{969}$. But $\frac{136}{490} > \frac{253}{969}$! This means that Dave outhit Andy over the two year period, even though Andy had the better record in each individual year.

This situation is known as *Simpson's paradox,* since it was first described in a paper by E. H. Simpson that appeared in 1951. The paradox itself is much older, and continues to arise in real-life cases. Several examples are known in baseball statistics, but Simpson's paradox also occurs in more serious situations. In one well-known case, the combined data suggested that graduate admissions at the University of California at Berkeley were gender biased in favor of men. However, when the admissions were examined department by department, it was discovered that there was a slight bias in *favor* of women applicants.

*SOURCE: From figure from "Ol' Abner Has Done It Again" by Richard J. Friedlander from *The American Mathematical Monthly,* Volume 99, Number 9, November 1992, p. 845.

Division is also not commutative. For example, $\frac{1}{3} \div \frac{1}{2} = \frac{2}{3}$ but $\frac{1}{2} \div \frac{1}{3} = \frac{3}{2}$. Nor is division associative. For example, the expression $2/3/4$ is ambiguous without parentheses; it could either be $\left(\frac{2}{3}\right)\Big/4 = \frac{2}{12} = \frac{1}{6}$ or $2\Big/\left(\frac{3}{4}\right) = \frac{8}{3}$.

Properties of the Order Relation

The order relation has many useful properties which are not difficult to prove.

THEOREM Properties of the Order Relation on the Rational Numbers

Let $\frac{a}{b}, \frac{c}{d}, \frac{e}{f}$ be rational numbers.

Transitive Property

$$\text{If } \frac{a}{b} < \frac{c}{d} \text{ and } \frac{c}{d} < \frac{e}{f}, \text{ then } \frac{a}{b} < \frac{e}{f}.$$

Addition Property

$$\text{If } \frac{a}{b} < \frac{c}{d}, \text{ then } \frac{a}{b} + \frac{e}{f} < \frac{c}{d} + \frac{e}{f}.$$

Multiplication Property

$$\text{If } \frac{a}{b} < \frac{c}{d} \text{ and } \frac{e}{f} > 0, \text{ then } \frac{a}{b} \cdot \frac{e}{f} < \frac{c}{d} \cdot \frac{e}{f}.$$

$$\text{If } \frac{a}{b} < \frac{c}{d} \text{ and } \frac{e}{f} < 0, \text{ then } \frac{a}{b} \cdot \frac{e}{f} > \frac{c}{d} \cdot \frac{e}{f}.$$

Trichotomy Property Exactly one of the following holds:

$$\frac{a}{b} < \frac{c}{d}, \quad \frac{a}{b} = \frac{c}{d}, \quad \text{or} \quad \frac{a}{b} > \frac{c}{d}.$$

The Density Property of Rational Numbers

By the definition of inequality, we know that

$$\frac{1}{2} < \frac{2}{3}.$$

Alternatively, using 6 as a common denominator, we have

$$\frac{3}{6} < \frac{4}{6}$$

and with 12 as a common denominator we have

$$\frac{6}{12} < \frac{8}{12}.$$

In the last form, we see that $\frac{7}{12}$ is a rational number which is between $\frac{6}{12}$ and $\frac{8}{12}$. That is,

$$\frac{1}{2} < \frac{7}{12} < \frac{2}{3}$$

as shown in Figure 6.14.

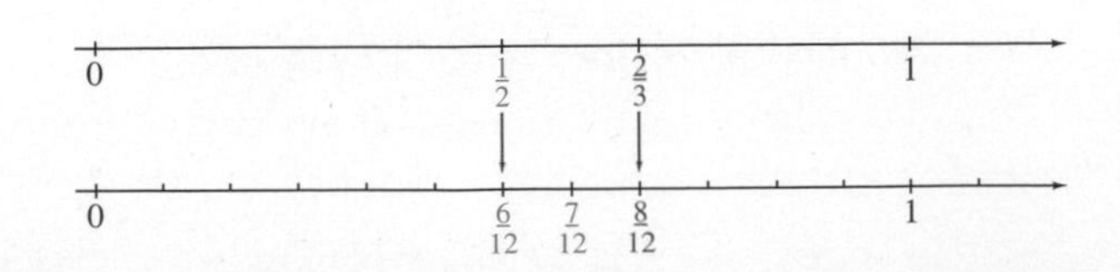

Figure 6.14

The rational number $\frac{7}{12}$ is between $\frac{1}{2}$ and $\frac{2}{3}$

The idea used to find a rational number that is between $\frac{1}{2}$ and $\frac{2}{3}$ can be extended to show that between *any* two rational numbers there is some other rational number. This interesting fact is called the **density property** of the rational numbers. The analogous property does not hold in the integers. For example, there is no integer between 1 and 2.

THEOREM The Density Property of Rational Numbers

Let $\frac{a}{b}$ and $\frac{b}{c}$ be any two rational numbers, with $\frac{a}{b} < \frac{c}{d}$. Then there is a rational number $\frac{e}{f}$ between $\frac{a}{b}$ and $\frac{c}{d}$; that is, $\frac{a}{b} < \frac{e}{f} < \frac{c}{d}$.

PROOF We can assume b and d are positive. Using $2bd$ as a common denominator, we then rewrite $\frac{a}{b} < \frac{c}{d}$ as $\frac{2ad}{2bd} < \frac{2bc}{2bd}$. Therefore $2ad < 2bc$. Let e be any integer, such as the *odd* integer $2ad + 1$, that lies between the *even* integers $2ad$ and $2bc$. Since $2ad < e < 2bc$, we have $\frac{2ad}{2bd} < \frac{e}{2bd} < \frac{2bc}{2bd}$. Therefore $\frac{e}{f}$, where $f = 2bd$, is a rational number between $\frac{a}{b}$ and $\frac{c}{d}$. ■

EXAMPLE 6.16

Finding Rational Numbers Between Two Rational Numbers

Find a rational number between the two given fractions.

(a) $\frac{2}{3}$ and $\frac{3}{4}$ **(b)** $\frac{5}{12}$ and $\frac{3}{8}$

SOLUTION

(a) Using $2 \cdot 3 \cdot 4 = 24$ as a common denominator, we have $\frac{2}{3} = \frac{16}{24}$ and $\frac{3}{4} = \frac{18}{24}$. Since $\frac{16}{24} < \frac{17}{24} < \frac{18}{24}$, this shows that $\frac{17}{24}$ is one answer.

(b) If we follow the idea shown in the proof of the theorem, we would use $2 \cdot 12 \cdot 8 = 192$ as a common denominator to write $\frac{5}{12} = \frac{80}{192}$ and $\frac{3}{8} = \frac{72}{192}$. Therefore $\frac{73}{192}, \frac{74}{192}, \ldots, \frac{79}{192}$ are all fractions between $\frac{3}{8}$ and $\frac{5}{12}$. If 48 is taken as a common denominator, so that $\frac{5}{12} = \frac{20}{48}$ and $\frac{3}{8} = \frac{18}{48}$, we see that $\frac{19}{48}$ is also a solution. Using the least common denominator 24 is not helpful because $\frac{3}{8} = \frac{9}{24}$ and $\frac{5}{12} = \frac{10}{24}$ and there is no fraction between $\frac{3}{8}$ and $\frac{5}{12}$ whose denominator is 24. ■

WORTH READING

The Importance of Generalization

Extending a domain by introducing new symbols in such a way that the laws which hold in the original domain continue to hold in the larger domain is one aspect of the characteristic mathematical process of *generalization*. The generalization from the natural to the rational numbers satisfies both the theoretical need for removing the restrictions on subtraction and division, and the practical need for numbers to express the results of measurement. It is the fact that the rational numbers fill this two-fold need that gives them their true significance. As we have seen, this extension of the number concept was made possible by the creation of new numbers in the form of abstract symbols like 0, −2, and 3/4. Today, when we deal with such numbers as a matter of course, it is hard to believe that as late as the seventeenth century they were not generally credited with the same legitimacy as the positive integers, and that they were used, when necessary, with a certain amount of doubt and trepidation. The inherent human tendency to cling to the "concrete," as exemplified by the natural numbers, was responsible for this slowness in taking an inevitable step. Only in the realm of the abstract can a satisfactory system of arithmetic be created.

SOURCE: Richard Courant and Herbert Robbins, *What Is Mathematics? An Elementary Approach to Ideas and Methods* (New York: Oxford University Press, 1941), pg. 56.

Computations with Rational Numbers

To work confidently with rational numbers, it is important to develop skills in estimation, mental arithmetic, efficient paper-and-pencil computation, and the use of the calculator.

Estimations

In many applications, the exact fractional value can be rounded off to the nearest integer value; if more precision is required, values can be rounded to the nearest half or perhaps even the nearest quarter.

EXAMPLE 6.17

Estimating Playing Time of a Recording

Laura wishes to tape a compact disc. The time of the eight selections are given as follows, in minutes and seconds.

1. 5′02″	**5.** 4′21″
2. 4′35″	**6.** 6′46″
3. 7′10″	**7.** 5′29″
4. 4′51″	**8.** 6′56″

To select the right length of tape, Laura wants to know, at least approximately, the total playing time of the CD.

SOLUTION

Here are some techniques Laura might try.

Range estimation: Since $5'02'' = 5\frac{2}{60}$ is between 5 and 6 minutes, and the second selection satisfies $4' < 4\frac{35}{60} < 5'$, and so on, the total playing time is between $5 + 4 + 7 + \cdots + 5 + 6 = 41$ minutes and $6 + 5 + 8 + \cdots + 6 + 7 = 49$ minutes.

Rounding to the nearest integer: Rounding down if the number of seconds is under 30, and up otherwise, the total time is estimated by $5 + 5 + 7 + 5 + 4 + 7 + 5 + 7 = 45$ minutes.

Rounding to the nearest half: The sum of the approximate times is $5 + 4\frac{1}{2} + 7 + 5 + 4\frac{1}{2} + 7 + 5\frac{1}{2} + 7 = 45\frac{1}{2}$ minutes, which can be computed mentally by adding the whole numbers and then adding the halves. This is in fact quite a good estimate, since the exact sum is 45′10″. ■

HIGHLIGHT FROM HISTORY

Rational Numbers and the Pythagorean Musical Scale

One of the earliest applications of mathematics was made to music. The Pythagoreans (c. 500 B.C.) noticed that two tones plucked on stretched strings sounded pleasing together if the length of the shorter string was a simple fraction of the length of the longer string. For example, if the longer string, say of length 1, is the note C (the middle white key on a piano), then a string of length $\frac{1}{2}$ sounds C′, the note an octave higher than C. Similarly, a string of length $\frac{2}{3}$ produces the note G, which is the fifth note in the scale C-D-E-F-G. Thus, to raise the pitch by a "fifth," the length of the string is multiplied by $\frac{2}{3}$. In particular, a "fifth" above G is the note D′ sounded by a string of length $\frac{2}{3} \cdot \frac{2}{3} = \frac{4}{9}$. Doubling this length to $\frac{8}{9}$ gives D, the pitch an octave below D′. By ascending by fifths $\left(\text{multiplying by } \frac{2}{3}\right)$ and lowering by octaves (multiplying by 2), the lengths of the strings which sound the white key pitches C, G, D′, D, A, E′, E, and B are obtained.

A string of length $\frac{3}{4}$ produces the pitch of F, the fourth note of the scale. A "fourth" above F is B♭ (B flat), a black key on the piano sounding the pitch of a string of length $\frac{3}{4} \cdot \frac{3}{4} = \frac{9}{16}$. By ascending by fourths $\left(\text{multiplying by } \frac{3}{4}\right)$ and lowering by octaves, the pitches C, F, B♭, E♭′, E♭, A♭, D♭′, D♭, G♭, C♭ are obtained. The chart below shows how the Pythagoreans filled in the notes of their scale.

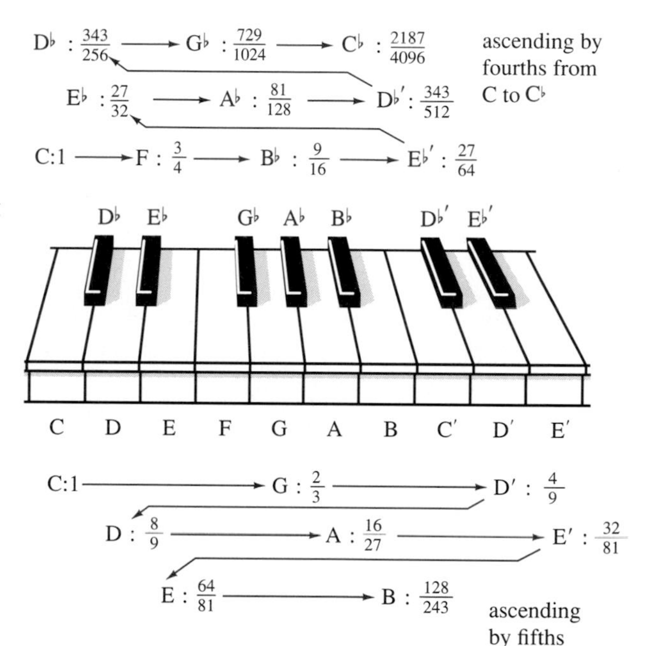

Ascending by fifths eventually reaches the note B: $\frac{128}{243}$, and ascending by fourths reaches C♭: $\frac{2187}{4096}$. These notes are assumed to be the same on a piano, but in fact C♭ is a slightly lower pitch than B, as the following calculation shows.

$$\frac{2187}{4096} \div \frac{128}{243} = \frac{531441}{524288} = 1.0136\ldots.$$

Amazingly, the Pythagoreans were aware of this difference in pitch; it is called the *Pythagorean Comma.* To avoid the comma, a number of different tunings have been used throughout musical history. Modern pianos are tuned to the "equal tempered" scale, in which the intervals from one pitch to the next are made equal.

Mental Arithmetic

By taking advantage of the properties, formulas, and algorithms associated with the operations, it is often possible to simplify the computational process. Some useful strategies are demonstrated in the following example.

EXAMPLE 6.18 Computational Strategies for Rational Number Arithmetic

Perform these computations mentally.

(a) $53 - 29\frac{3}{5}$ **(b)** $\left(2\frac{1}{8} - 4\frac{2}{3}\right) + 7\frac{7}{8}$

(c) $\frac{7}{15} \times 90$ **(d)** $\frac{3}{8} \times 14 + \frac{3}{4} \times 25$

(e) $4\frac{1}{6} \times 18$ **(f)** $\frac{5}{8} \times \left(\frac{7}{10} \times \frac{24}{49}\right)$

SOLUTION

(a) Adding $\frac{2}{5}$ to each term gives $53 - 29\frac{3}{5} = 53\frac{2}{5} - 30 = 23\frac{2}{5}$.

(b) $$\left(2\frac{1}{8} - 4\frac{2}{3}\right) + 7\frac{7}{8} = \left(2 + 7 + \frac{1}{8} + \frac{7}{8}\right) - 4\frac{2}{3}$$
$$= 10 - 4\frac{2}{3} = 10\frac{1}{3} - 5 = 5\frac{1}{3}$$

(c) $$\frac{7}{15} \times 90 = 7 \times \frac{90}{15} = 7 \times 6 = 42$$

(d) $$\frac{3}{8} \times 14 + \frac{3}{4} \times 25 = \frac{3}{8} \times 14 + \frac{3}{8} \times 50$$
$$= \frac{3}{8} \times (14 + 50) = \frac{3}{8} \times 64$$
$$= 3 \times \frac{64}{8} = 3 \times 8 = 24$$

(e) $$4\frac{1}{6} \times 18 = \left(4 + \frac{1}{6}\right) \times 18 = 4 \times 18 + \frac{1}{6} \times 18 = 72 + 3 = 75$$

(f) $$\frac{5}{8} \times \left(\frac{7}{10} \times \frac{24}{49}\right) = \frac{5}{10} \times \frac{7}{49} \times \frac{24}{8} = \frac{1}{2} \times \frac{1}{7} \times 3 = \frac{3}{14}$$ ■

Rational Numbers on a Calculator

Some special purpose calculators, such as the *Math Explorer,* can perform arithmetic on fractional values and display the result as a fraction. Most calculators, however, perform calculations only in decimal form, and it takes special care to represent the final answer as a fraction.

EXAMPLE 6.19 Computing with Rational Numbers on a Calculator

Use a calculator to find $\frac{3}{8} \times \left(\frac{-7}{12} + \frac{9}{5}\right)$.

SOLUTION

On a fraction calculator. On the *Math Explorer* the steps are

[ON/AC] 3 [/] 8 [×] [(] 7 [+ ⇄ −] [/] 12 [+] 9 [/] 5 [)] [=]

which gives 219/480 in the display. Pressing [Simp] [=] gives the simplified form 73/160.

On a decimal calculator. We can first obtain a decimal answer by pressing

[ON/AC] 3 [÷] 8 [×] [(] 7 [+ ⇄ −] [÷] 12 [+] 9 [÷] 5 [)] [=],

obtaining 0.45625. To express this as a fraction, we first observe that 8, 12, and 5 are the denominators of the terms which appear in the problem. Thus, the answer can be written as a fraction with $8 \cdot 12 \cdot 5 = 480$ as a denominator. Multiplying by 480 we get $0.45625 \times 480 = 219$, so the decimal answer 0.45625 is equivalent to the fraction 219/480; in simplest form this is 73/160, since GCD(219, 480) = 3 and $(219 \div 3)/(480 \div 3) = 73/160$. ■

We conclude with a real-life application of the use of rational numbers and their arithmetic.

EXAMPLE 6.20

Designing Wooden Stairs

A deck is 4′2″ above the surface of the patio. How many steps and risers should there be in a stairway which connects the deck to the patio? The decking is $1\frac{1}{2}$″ thick, the stair treads are $1\frac{1}{8}$″ thick, and the steps should each rise the same distance, from one to the next. Calculate the vertical dimension of each riser.

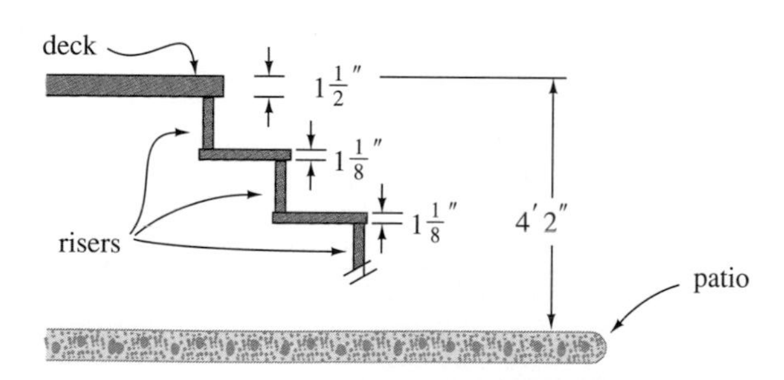

SOLUTION

Understand the problem • • • •

We have been given some important dimensions, including the thickness of the treads and the deck. We have *not* been told what dimension to cut the risers. This, we see from the figure, depends on the number of steps we choose, with a lower rise corresponding to a greater number of steps. We must choose a number of steps which feels natural to walk on. We must also be sure each step rises the same amount.

Devise a plan

If we know what vertical rise from step to step is customary, we can first get a reasonable estimate of the number of steps to use. Once it is agreed what number of steps to incorporate in the design, we can calculate the height of each riser. The riser meeting the deck must be adjusted to account for the decking being thicker than the stair tread.

Carry out the plan

A brief survey of existing stairways shows that most steps rise 5″ to 7″ from one to the next. Since $4'2'' = 50''$ and $\frac{50}{5} = 10$ and $\frac{50}{7} = 7\frac{1}{7}$, this suggests that 8, or perhaps 9, steps will work well, including the step onto the deck itself. Let's choose 8. Then $\frac{50}{8}'' = 6\frac{2}{8}'' = 6\frac{1}{4}''$; that is, each combination of riser plus tread is to be $6\frac{1}{4}''$. Since the treads are $1\frac{1}{8}''$ thick, the first seven risers require the boards to be cut $6\frac{1}{4}'' - 1\frac{1}{8}'' = 5\frac{1}{8}''$ wide. Since the decking is $1\frac{1}{2}''$ thick, the uppermost riser is $6\frac{1}{4}'' - 1\frac{1}{2}'' = 4\frac{3}{4}''$ wide.

Look back

If we were concerned that the steps will take up too much room on the patio, we might use a steeper 7 step design. Each riser plus tread would rise $7\frac{1}{7}''$. Experienced carpenters would think of this as a "hair" more than $7\frac{1}{8}''$, and cut the lower six risers from a board 8″ wide for a total rise of $7\frac{1}{8}''$. The last riser is cut to make any final small adjustment. ■

PROBLEM SET 6.3

Understanding Concepts

1. Explain what properties of addition of rational numbers can be used to make this sum very easy to compute.

$$\left(3\frac{1}{5} + 2\frac{2}{5}\right) + 8\frac{1}{5}$$

2. What properties can you use to make these computations easy?

   (a) $\frac{2}{5} + \left(\frac{3}{5} + \frac{2}{3}\right)$ **(b)** $\frac{1}{4} + \left(\frac{2}{5} + \frac{3}{4}\right)$

   (c) $\frac{2}{3} \cdot \frac{1}{8} + \frac{2}{3} \cdot \frac{7}{8}$ **(d)** $\frac{3}{4} \cdot \left(\frac{5}{9} \cdot \frac{4}{3}\right)$

3. Find the negatives (that is, the additive inverses) of the following rational numbers. Show each number and its negative on the number line.

   (a) $\frac{4}{5}$ **(b)** $\frac{-3}{2}$ (c) $\frac{8}{-3}$ **(d)** $\frac{4}{2}$

4. Compute these sums of rational numbers. Explain what properties of rational numbers you find useful. Express your answers in simplest form.

   (a) $\frac{1}{6} + \frac{2}{-3}$ **(b)** $\frac{-4}{5} + \frac{3}{2}$

   (c) $\frac{9}{4} + \frac{-7}{8}$ **(d)** $\frac{3}{4} + \frac{-5}{8} + \frac{7}{-12}$

5. Compute these differences of rational numbers. Explain what properties you find useful. Express your answers in simplest form.
(a) $\frac{2}{5} - \frac{3}{4}$ **(b)** $\frac{-6}{7} - \frac{4}{7}$ (c) $\frac{3}{8} - \frac{1}{12}$
(d) $3\frac{2}{5} - \frac{7}{10}$ (e) $2\frac{1}{3} - 5\frac{3}{4}$ **(f)** $-4\frac{2}{3} - \frac{-19}{6}$

6. Calculate these products of rational numbers. Explain what properties of rational numbers you find useful. Express your answers in simplest form.
(a) $\frac{3}{5} \cdot \frac{7}{8} \cdot \left(\frac{5}{3}\right)$ **(b)** $\frac{-2}{7} \cdot \frac{3}{4}$
(c) $\frac{-4}{3} \cdot \frac{6}{-16}$ **(d)** $3\frac{1}{8} \cdot 2\frac{1}{5} \cdot (40)$
(e) $\frac{14}{15} \cdot \frac{60}{7}$ **(f)** $\left(\frac{4}{11} \cdot \frac{22}{7}\right) \cdot \left(\frac{-3}{8}\right)$

7. Find the reciprocals (that is, the multiplicative inverses) of these rational numbers. Show each number and its reciprocal on the number line.
(a) $\frac{3}{2}$ **(b)** $\frac{4}{-9}$ (c) $\frac{-4}{-11}$
(d) 5 (e) -2 **(f)** $2\frac{1}{2}$

8. Use the properties of the operations of rational number arithmetic to perform these calculations. Express your answers in simplest form.
(a) $\frac{2}{3} \cdot \frac{4}{7} + \frac{2}{3} \cdot \frac{3}{7}$ **(b)** $\frac{4}{5} \cdot \frac{2}{3} - \frac{3}{10} \cdot \frac{2}{3}$
(c) $\frac{4}{7} \cdot \frac{3}{2} - \frac{4}{7} \cdot \frac{6}{4}$ **(d)** $\left(\frac{4}{7} \cdot \frac{2}{5}\right) \div \frac{2}{7}$

9. Justify each step in the following proof of the distributive property of multiplication over addition. The two addends have a common denominator, but this is always possible by using a common denominator for two fractions.

$$\frac{a}{b} \cdot \left(\frac{c}{d} + \frac{e}{d}\right) = \frac{a}{b} \cdot \frac{c+e}{d} \quad \text{(a) Why?}$$

$$= \frac{a \cdot (c+e)}{b \cdot d} \quad \text{(b) Why?}$$

$$= \frac{a \cdot c + a \cdot e}{b \cdot d} \quad \text{(c) Why?}$$

$$= \frac{a \cdot c}{b \cdot d} + \frac{a \cdot e}{b \cdot d} \quad \text{(d) Why?}$$

$$= \frac{a}{b} \cdot \frac{c}{d} + \frac{a}{b} \cdot \frac{e}{d}. \quad \text{(e) Why?}$$

10. If $\frac{a}{b} \cdot \frac{4}{7} = \frac{2}{3}$, what is $\frac{a}{b}$? Carefully explain how you obtain your answer. What properties do you use?

11. Solve each equation for the rational number x. Show your steps, and explain what property justifies each step.
(a) $4x + 3 = 0$ **(b)** $x + \frac{3}{4} = \frac{7}{8}$
(c) $\frac{2}{3}x + \frac{4}{5} = 0$ **(d)** $3\left(x + \frac{1}{8}\right) = -\frac{2}{3}$

12. The Fahrenheit and Celsius temperature scales are related by the formula $F = \frac{9}{5}C + 32$. For example, a temperature of 20° Celsius corresponds to 68° Fahrenheit since

$$\frac{9}{5} \cdot 20 + 32 = 36 + 32 = 68.$$

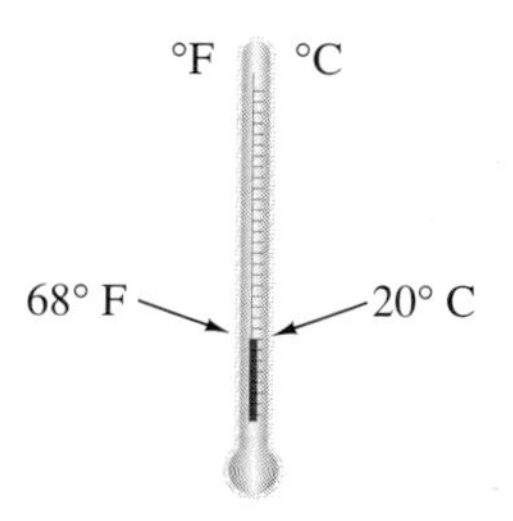

(a) Explain what properties of rational number arithmetic permit you to deduce the equivalent formula $C = \frac{5}{9}(F - 32)$ relating the two temperature scales.
(b) Fill in the missing entries in this table.

°C	−40°		0°	10°	20°		
°F		−13°			68°	104°	212°

(c) Electronic signboards frequently give the temperature in both Fahrenheit and Celsius. What is the temperature if both temperatures are the same (that is, F = C)? The negative of one another (that is, F = −C)? Answer the latter question with an exact rational number and the approximate integer that would be seen on the signboard.

13. Arrange each group of rational numbers in increasing order. Show, at least approximately, the numbers on the number line.
(a) $\frac{4}{5}, -\frac{1}{5}, \frac{2}{5}$ **(b)** $\frac{-3}{7}, \frac{4}{7}, \frac{-5}{7}$
(c) $\frac{3}{8}, \frac{1}{2}, \frac{3}{4}$ **(d)** $\frac{-7}{12}, \frac{-2}{3}, \frac{3}{-4}$

14. Verify these inequalities.
(a) $\frac{-4}{5} < \frac{-3}{4}$ **(b)** $\frac{1}{10} > -\frac{1}{4}$ **(c)** $-\frac{19}{60} > \frac{-1}{3}$

15. The properties of the order relation can be used to solve inequalities. For example, if $-\frac{2}{3}x + \frac{1}{4} < -\frac{1}{2}$, then

$$-\frac{2}{3}x + \frac{1}{4} + \left(-\frac{1}{4}\right) < -\frac{1}{2} + \left(-\frac{1}{4}\right)$$

Add $-\frac{1}{4}$ to both sides of the inequality.

$$-\frac{2}{3}x < -\frac{3}{4}$$

$$\left(-\frac{3}{2}\right) \cdot \left(-\frac{2}{3}\right)x > \left(-\frac{3}{2}\right) \cdot \left(-\frac{3}{4}\right)$$

Note that multiplying by the negative number $-\frac{3}{2}$ reversed the direction of inequality.

$$1 \cdot x > \frac{9}{8}$$

That is, all rational numbers x greater than $\frac{9}{8}$, and only these, satisfy the given inequality. Solve these inequalities. Show all of your steps.

(a) $x + \frac{2}{3} > -\frac{1}{3}$ **(b)** $x - \left(-\frac{3}{4}\right) < \frac{1}{4}$

(c) $\frac{3}{4}x < -\frac{1}{2}$ **(d)** $-\frac{2}{5}x + \frac{1}{5} > -1.$

16. Find a rational number which is between the two given rational numbers.

(a) $\frac{4}{9}$ and $\frac{6}{11}$ **(b)** $\frac{1}{9}$ and $\frac{1}{10}$

(c) $\frac{14}{23}$ and $\frac{7}{12}$ **(d)** $\frac{141}{568}$ and $\frac{183}{737}$

17. Find three rational numbers between $\frac{1}{4}$ and $\frac{2}{5}$.

18. For each given rational number, choose the best estimate from the list provided.

(a) $\frac{104}{391}$ is approximately $\frac{1}{3}$, $\frac{1}{4}$, $\frac{1}{2}$.

(b) $\frac{217}{340}$ is approximately $\frac{1}{3}$, $\frac{1}{2}$, $\frac{2}{3}$.

(c) $\frac{-193}{211}$ is approximately $-\frac{1}{2}$, -1, 1, $\frac{1}{2}$.

(d) $\frac{453}{307}$ is approximately $\frac{3}{4}$, 1, $1\frac{1}{3}$, $1\frac{1}{2}$.

19. Use estimations to choose the best approximation of the following expressions.

(a) $3\frac{19}{40} + 5\frac{11}{19}$ is approximately 8, $8\frac{1}{2}$, 9, $9\frac{1}{2}$.

(b) $2\frac{6}{19} + 5\frac{1}{3} - 4\frac{7}{20}$ is approximately 3, $3\frac{1}{3}$, $3\frac{1}{2}$, 4.

(c) $17\frac{8}{9} \div 5\frac{10}{11}$ is approximately 2, 3, $3\frac{1}{2}$, 4.

20. Do the following calculations mentally.

(a) $\frac{1}{2} + \frac{1}{4} + \frac{3}{4}$ **(b)** $\frac{5}{2} \cdot \left(\frac{2}{5} - \frac{2}{10}\right)$

(c) $\frac{3}{4} \cdot \frac{12}{15}$ **(d)** $\frac{2}{9} \div \frac{1}{3}$

(e) $2\frac{2}{3} \times 15$ **(f)** $3\frac{1}{5} - 1\frac{1}{4} + 7\frac{4}{5}$

(g) $6\frac{1}{8} - 8\frac{1}{4}$ **(h)** $\frac{2}{3} \cdot \frac{7}{4} - \frac{2}{3} \cdot \frac{1}{4}$

Using a Calculator

21. In the following problems first mentally estimate an answer accurate to a nearby integer. Next, use a calculator to perform these computations. Express your answers in simplest form. Was your guess accurate?

(a) $\frac{119}{346} + \frac{86}{121}$ **(b)** $\frac{-155}{47} + \frac{58}{13}$

(c) $\frac{104}{160} \cdot \frac{-65}{39}$ **(d)** $\frac{47}{11} \div \frac{21}{19}$

22. Use a calculator to evaluate the following expressions. Express your answers in simplest form.

(a) $\frac{4}{5} \times \left(\frac{18}{25} - \frac{3}{4}\right)$ **(b)** $\left(\frac{-7}{3} + \frac{4}{9}\right) \div \frac{1}{5}$

Thinking Critically

23. Problem 31 of the Rhind papyrus, if translated literally, reads: "A quantity, its $\frac{2}{3}$, its $\frac{1}{2}$, its $\frac{1}{7}$, its whole, amount to 33." That is, in modern notation, $\frac{2}{3}x + \frac{1}{2}x + \frac{1}{7}x + x = 33$. What is the quantity?

24. The ancient Egyptians measured steepness of a slope by the fraction $\frac{x}{y}$, where x is the number of hands of horizontal "run" and y is the number of cubits of vertical "rise." Seven hands form a

cubit. Problem 56 of the Ahmes Papyrus asks for the steepness of the face of a pyramid 250 cubits high and having a square base 360 cubits on a side. The papyrus gives the answer $5\frac{1}{25}$. Show why this is correct.

25. **The Law of the Lever.** One of Archimedes (c. 225 B.C.) great achievements was the law of the lever. The law can be described in terms of a condition under which weights hung from a beam pivoting at 0 will be in balance. For example, the two beams shown below are in balance.

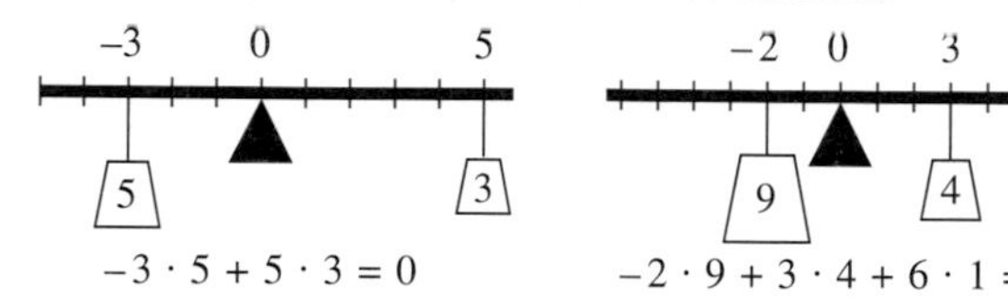

$-3 \cdot 5 + 5 \cdot 3 = 0$ $\qquad$ $-2 \cdot 9 + 3 \cdot 4 + 6 \cdot 1 = 0$

In general, if weights $W_1, W_2, \ldots, W_n$ are hung at positions $x_1, x_2, \ldots, x_n$, then the beam is balanced if, and only if, $x_1 \cdot W_1 + x_2 \cdot W_2 + \cdots + x_n \cdot W_n = 0$.

For each diagram below, find the missing weight W or unknown position x which will balance the beam. You can expect rational number answers.

(a) **(b)**

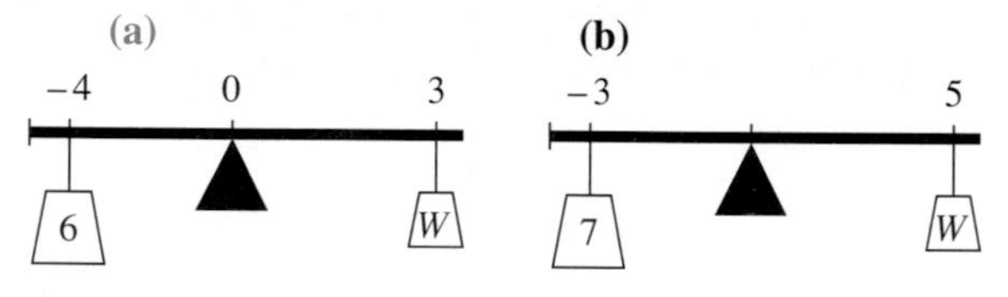

(c) **(d)**

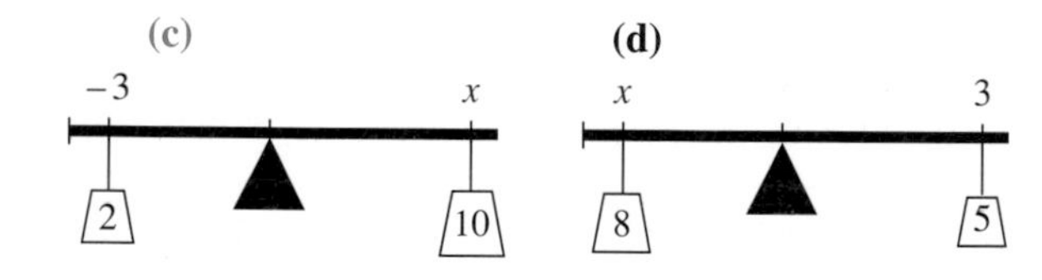

(e) **(f)**

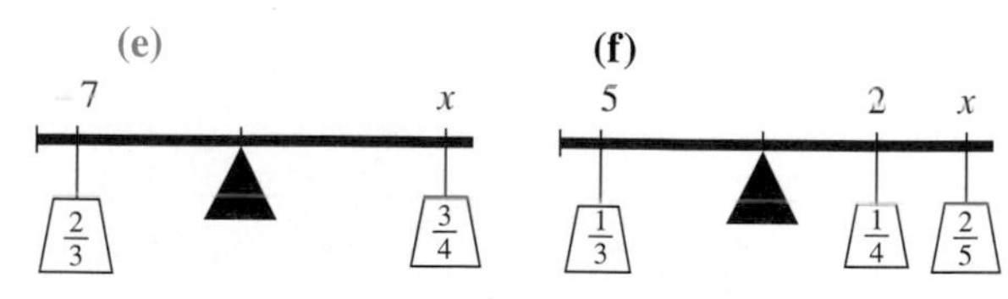

26. Let $s_1 = \dfrac{1}{1 \cdot 2}$, $s_2 = \dfrac{1}{1 \cdot 2} + \dfrac{1}{2 \cdot 3}$,

$$s_3 = \frac{1}{1 \cdot 2} + \frac{1}{2 \cdot 3} + \frac{1}{3 \cdot 4}, \ldots,$$

$$s_n = \frac{1}{1 \cdot 2} + \frac{1}{2 \cdot 3} + \cdots + \frac{1}{n \cdot (n+1)}.$$

(a) Show that $s_1 = 1 - \dfrac{1}{2}$, $s_2 = \left(1 - \dfrac{1}{2}\right) + \left(\dfrac{1}{2} - \dfrac{1}{3}\right)$, and $s_3 = \left(1 - \dfrac{1}{2}\right) + \left(\dfrac{1}{2} - \dfrac{1}{3}\right) + \left(\dfrac{1}{3} - \dfrac{1}{4}\right)$.

(b) Use part (a) to explain why $s_2 = \dfrac{2}{3}$ and $s_3 = \dfrac{3}{4}$.

(c) What pattern do you see? Can you guess values for s_9? s_{99}? s_n?

27. Prove the transitive property of order. That is, if $\dfrac{a}{b}, \dfrac{c}{d}$, and $\dfrac{e}{f}$ are rational numbers which satisfy $\dfrac{a}{b} < \dfrac{c}{d}$ and $\dfrac{c}{d} < \dfrac{e}{f}$, show that $\dfrac{a}{b} < \dfrac{e}{f}$.

28. **(a)** Let x, y, and z be arbitrary rational numbers. Show that the sums of the three entries in every row, column, and diagonal in the Magic Square are the same.

$x - z$	$x - y + z$	$x + y$
$x + y + z$	x	$x - y - z$
$x - y$	$x + y - z$	$x + z$

(b) Let $x = \dfrac{1}{2}$, $y = \dfrac{1}{3}$, $z = \dfrac{1}{4}$. Find the corresponding Magic Square.

(c) Here is a partial Magic Square that corresponds to the form shown in part (a). Find x, y, z, and then fill in the remaining entries of the square.

$-\frac{5}{12}$		1
$-\frac{1}{3}$		

(*Hint:* $2x = (x + y) + (x - y)$.)

Thinking Cooperatively

29. The definition of fraction multiplication, namely $\frac{a}{b} \cdot \frac{c}{d} = \frac{ac}{bd}$, suggests that addition might be defined by the rule $\frac{a}{b} \oplus \frac{c}{d} = \frac{a+c}{b+d}$, instead of the more complicated rule $\frac{a}{b} + \frac{c}{d} = \frac{ad+bc}{bd}$. The new symbol $\oplus$ distinguishes the *mediant* $\frac{a}{b} \oplus \frac{c}{d}$, from the ordinary sum.

(a) The mediant of one-half and two-thirds is $\frac{3}{5}$; at least, this is so if one-half is represented by the fraction $\frac{1}{2}$ and two-thirds is represented by $\frac{2}{3}$. What happens to the mediant if $\frac{2}{4}$ and $\frac{2}{3}$ are chosen as the fractions representing one-half and two-thirds?

(b) Is the inequality $\frac{2}{3} < \frac{2}{3} + \frac{1}{2}$ still true if the ordinary sum + on the right is replaced by mediant, $\oplus$?

(c) The mediant operation $\oplus$ has some uses (that is, see problems 24, 25, 28, and 32 of Section 6.1), but its algebraic properties are different from those of ordinary addition. Explore and discuss properties of the $\oplus$ operation. Which properties hold? Which do not?

Making Connections

30. The ruler shown below divides each unit into quarters. It is also fitted with a sliding scale which divides each unit into fifths. Without the sliding scale we would only be able to say that the length L to be measured is between $1\frac{3}{4}$ and 2. With the slide L can be found to the nearest $\frac{1}{20}$ of a unit. To see how, observe that the 3 marked on the slide is aligned with the $2\frac{1}{2}$ mark on the fixed scale. Therefore $y = \frac{3}{5}$ and $z = \frac{3}{4}$, so $x = z - y = \frac{3}{4} - \frac{3}{5} = \frac{3}{20}$, where x is the distance to the left edge of the slide from the mark at $1\frac{3}{4}$. Therefore $L = 1\frac{3}{4} + x = 1\frac{3}{4} + \frac{3}{20} = 1\frac{18}{20}$.

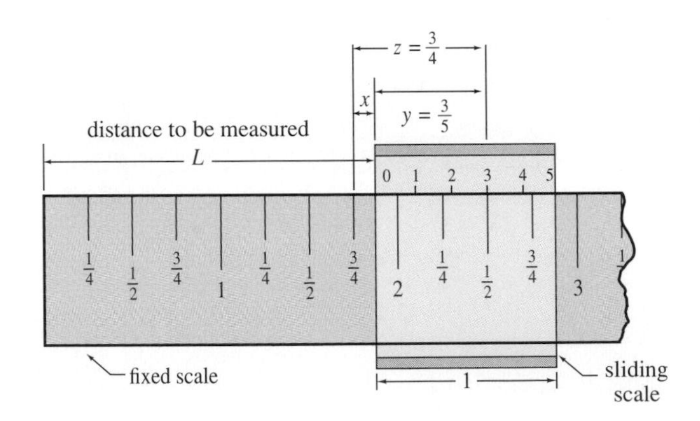

This simple yet effective device to improve the accuracy of a measurement is from the French mathematician and inventor Pierre Vernier (1580–1637). The sliding scale is usually called the *vernier,* in his honor. For parts (a) and (b) depicted below, find the distances L, x, y, and z.

(a)

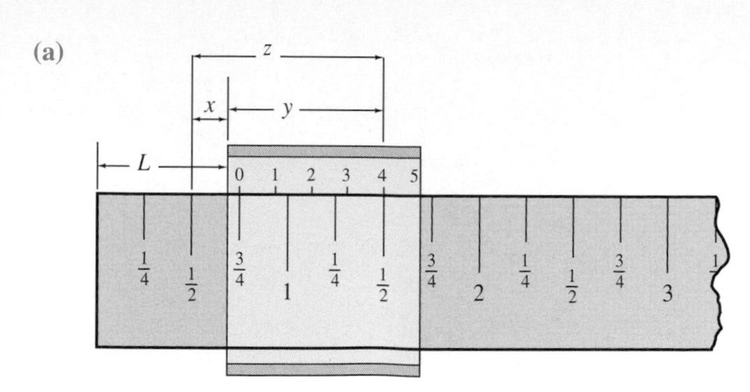

(b)

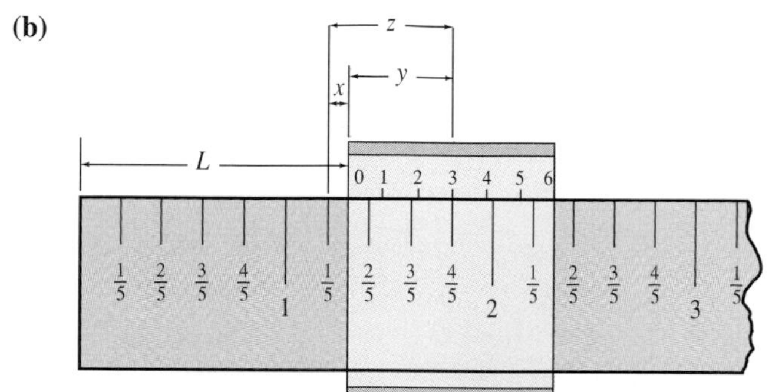

(c) Suppose you have a ruler marked at intervals of $\frac{1}{8}$ inch. Describe how you could make a vernier scale for your ruler. How accurately can you measure with the vernier?

31. Stock prices are quoted in dollars and fractions of a dollar. For example, AAR Corporation was $13\frac{7}{8}$ at the close of the New York Stock Exchange on Wednesday, meaning that shares last sold for $\$13\frac{7}{8}$ each. On Thursday, the closing price increased to 14, which is reported as a change of $+\frac{1}{8}$. Similarly the closing prices of ACMin dropped from $11\frac{7}{8}$ to $11\frac{3}{4}$, for a change of $11\frac{3}{4} - 11\frac{7}{8} = -\frac{1}{8}$.

Wednesday Close			Thursday Close		
AAR	$13\frac{7}{8}$		AAR	14	$+\frac{1}{8}$
ACELIn	$30\frac{3}{8}$	$+\frac{1}{8}$	ACELIn	$31\frac{1}{2}$	$+1\frac{1}{8}$
ACMin	$11\frac{7}{8}$	$+\frac{1}{8}$	ACMin	$11\frac{3}{4}$	$-\frac{1}{8}$
ACMOp	$9\frac{5}{8}$		ACMOp	$9\frac{5}{8}$	
ACM Sc1	$11\frac{5}{8}$		ACM Sc1	$11\frac{5}{8}$	

Fill in the missing entries in this table.

Company	Wed. Close	Change	Thurs. Close	Change
(a) Chrysler	$55\frac{1}{4}$	$+\frac{3}{8}$	$54\frac{1}{2}$	?
(b) Compaq	68	$+2\frac{1}{4}$	?	$+\frac{5}{8}$
(c) Dupont	?	$-\frac{1}{8}$	$46\frac{1}{2}$	$-\frac{1}{2}$
(d) Ford	$61\frac{3}{4}$	$+\frac{7}{8}$	?	$1\frac{1}{8}$

32. Design a stairway which connects the patio to the deck (see Example 6.20).

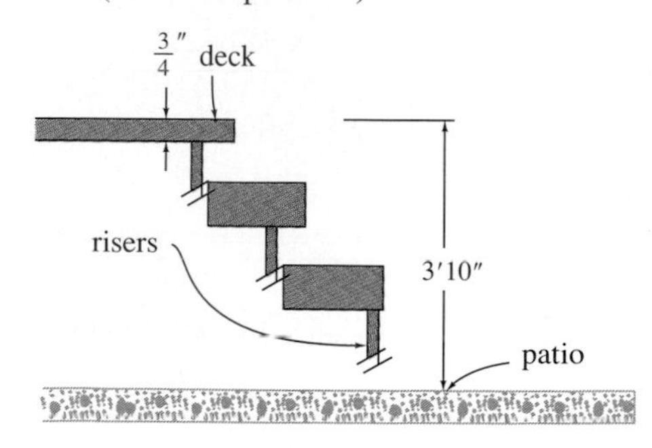

Assume the tread of each stair is $1\frac{1}{2}''$ thick.

33. Anja has a box of photographic print paper. Each sheet measures 8 by 10 inches. She could easily cut a sheet into four 4 by 5 inch rectangles. However, to save money, she wants to get six prints, all the same size, from each 8 by 10 inch sheet. Describe how she can cut the paper.

34. Krystoff has an 8′ piece of picture frame molding, shown in cross section below.

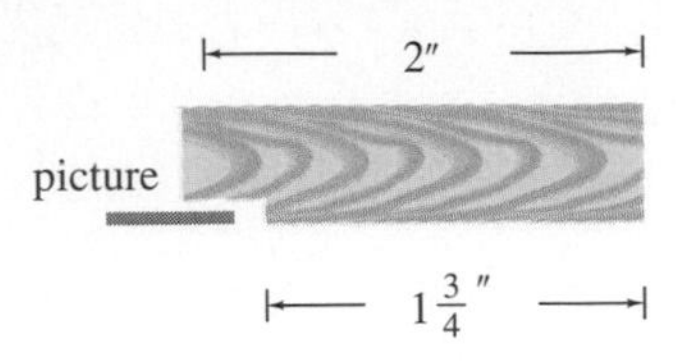

Is this 8′ piece sufficient to frame a 16″ by 20″ picture? Allow for saw cuts and some extra space to ensure that the picture fits easily into the frame.

For Review

35. (a) The Ahmes papyrus is about 6 yards long and $\frac{1}{3}$ of a yard wide. What is its area in square yards?

(b) The Moscow papyrus, another source of mathematics of ancient Egypt, is about the same length as the Ahmes papyrus, but has $\frac{1}{4}$ the area. What is the width of the Moscow papyrus?

36. A TV, regularly priced at \$345, is offered on sale at "up to one-third off." What range of prices would you expect to pay?

37. The triangular shape shown to the right is $\frac{1}{4}$ of a whole figure. What does the whole figure look like? Give several answers.

EPILOGUE Fractions and Technology

The equation $x + m = n$ cannot always be solved in the whole numbers. However, by extending the whole numbers to the integers, this equation always has a unique solution. Computationally, going from whole number to integer arithmetic comes at a modest cost, since the algorithms used to perform calculations on integers are essentially the same as those for whole numbers; the only added step is determining if the answer is positive or negative.

In the rational number system, the greatest gain is that the equation $rx + p = q$ can always be solved uniquely for the unknown x, whatever values are given for the rational numbers r, p, and q (although r must be nonzero).

However, going to rational number arithmetic comes with a high cost in the complexity of the computations. For example, adding two rationals $\frac{a}{b} + \frac{c}{d}$ requires three integer multiplications and an integer addition to produce the answer $\frac{ad + bc}{bd}$. Even then, the answer may not be in simplest form. It is scarcely surprising that school children (and adults) often find fractions and the arithmetic of fractions difficult.

When calculators and computers first became widely available, there was some talk of diminishing, or even removing, fractions from the curriculum. After all, these machines were based on decimal fractions, not on common fractions. There are many good reasons this did not and should not occur, including:

- fractions, as expressions of ratios and rates, convey a dynamic sense of number which is absent in decimal fractions;
- the algebraic concepts used in rational number arithmetic provide important readiness skills for learning algebra.

At the present time, technology is moving in a direction which supports teaching, learning, and using fractions. For example, inexpensive calculators with fraction arithmetic capabilities are now becoming common in the classroom. Also, computer algebra systems for both calculators and computers can now perform rational number arithmetic in the blink of an eye to unlimited precision. With these devices, problem solving and critical thinking in the realm of rational numbers can actually be a source of pleasure.

In this chapter, we have discussed the properties and arithmetic of fractions and rational numbers, both from a concrete and pictorial point of view and from a more theoretical standpoint. We have also seen how the operations with fractions often depend on notions from number theory. Many of the examples illustrated the practical side of fractions, in which the solution of real-life problems led naturally to operations with rational numbers.

CLASSIC CONUNDRUM The Slice is Right?

Divide 100 loaves among five men in such a way that the shares received shall be in arithmetic progression and that one seventh of the sum of the largest three shares shall be equal to the sum of the smallest two. (This problem from the Rhind papyrus shows that the ancient Egyptians had an advanced knowledge of fractional quantities.)

CHAPTER 6 SUMMARY

Key Concepts

1. The Basic Concepts of Fractions and Rational Numbers

(a) A fraction is an ordered pair of integers a and b, $b \neq 0$, written $\frac{a}{b}$ or a/b. Fractions are physically and pictorially modeled with colored regions, sets, fraction strips, and the number line.

(b) Two fractions that express the same quantity, or correspond to the same point on a number line, are called equivalent fractions. In particular $\frac{a}{b} = \frac{a \cdot n}{b \cdot n}$ for all integers n, $n \neq 0$ and $\frac{a}{b} = \frac{a \div d}{b \div d}$ if d divides a and b (the fundamental law of fractions), and $\frac{a}{b} = \frac{c}{d}$ if, and only if, $ad = bc$.

(c) Any fraction is equivalent to a fraction in simplest form. Two or more fractions can always be replaced by equivalent fractions with a common denominator.

(d) A rational number is a number represented by a common fraction $\frac{a}{b}$. The same rational number can also be represented by any fraction equivalent to $\frac{a}{b}$.

(e) If two rational numbers are represented by $\frac{a}{b}$ and $\frac{c}{d}$, with $b > 0$ and $d > 0$, then $\frac{a}{b} < \frac{c}{d}$ if, and only if, $ad < bc$.

2. The Arithmetic of Rational Numbers

(a) The sum of two rational numbers represented by fractions $\frac{a}{b}$ and $\frac{c}{b}$ with a common denominator is defined by $\frac{a}{b} + \frac{c}{b} = \frac{a+c}{b}$. From this it follows that $\frac{a}{b} + \frac{c}{d} = \frac{ad+bc}{bd}$. The definition is motivated by colored region and number-line models.

(b) Subtraction is defined by the missing-addend approach: $\frac{a}{b} - \frac{c}{d} = \frac{e}{f}$ if, and only if, $\frac{a}{b} = \frac{c}{d} + \frac{e}{f}$. The subtraction formula $\frac{a}{b} - \frac{c}{d} = \frac{ad-bc}{bd}$ follows from the definition of subtraction.

(c) Multiplication is defined by $\frac{a}{b} \cdot \frac{c}{d} = \frac{ac}{bd}$. This definition is motivated by extending the rectangular array model of multiplication.

(d) Division is defined by the missing-factor approach: $\frac{a}{b} \div \frac{c}{d} = \frac{e}{f}$ if, and only if, $\frac{a}{b} = \frac{c}{d} \cdot \frac{e}{f}$. The "invert the divisor and multiply" algorithm follows from this definition, so that $\frac{a}{b} \div \frac{c}{d} = \frac{a}{b} \cdot \frac{d}{c}$.

(e) A nonzero rational number, $\frac{a}{b}$, has a unique multiplicative inverse, the reciprocal $\frac{b}{a}$, which when multiplied by $\frac{a}{b}$ gives the product 1. That is, $\frac{a}{b} \cdot \frac{b}{a} = 1$.

3. The Rational Number System

(a) The rational numbers are closed under addition. Addition is commutative, associative, and zero is the additive identity.

(b) Each rational number, $\frac{a}{b}$, has a unique negative, $-\frac{a}{b}$, given by $\frac{-a}{b}$, which is the additive inverse of $\frac{a}{b}$. Subtraction is equivalent to adding the negative, so $\frac{a}{b} - \frac{b}{d} = \frac{a}{b} + \left(-\frac{b}{d}\right)$.

(c) Multiplication is closed, commutative, associative, one is the multiplicative identity, and multiplication distributes over addition and subtraction.

(d) Each nonzero rational number $\frac{a}{b}$ has a unique multiplicative inverse given by the reciprocal $\frac{b}{a}$. Division is equivalent to multiplication by the multiplicative inverse of the divisor, so $\frac{a}{b} \div \frac{c}{d} = \frac{a}{b} \cdot \frac{d}{c}$.

(e) The rational numbers have the density property; that is, there is a rational number between any two rational numbers.

(f) Computational skills—estimations, mental arithmetic, paper and pencil and electronic calculations—are as useful and necessary for work with the rational numbers as for any other number system.

Vocabulary and Notation

Section 6.1

Fraction
Numerator
Denominator
Models for fractions:
- **Colored regions**
- **Sets**
- **Fraction strips**
- **Number line**

Equivalent fraction
Fundamental law of fractions
Simplest (or reduced) form
Common denominator
Least common denominator
Rational number
Set of rational numbers, Q

Section 6.2

Operations on fractions:
- **Addition** $\frac{a}{b} + \frac{c}{d}$
- **Subtraction** $\frac{a}{b} - \frac{c}{d}$
- **Multiplication** $\frac{a}{b} \cdot \frac{c}{d}$ **or** $\frac{a}{b} \times \frac{c}{d}$
- **Division** $\frac{a}{b} \div \frac{c}{d}$ **or** $\frac{a}{b} \Big/ \frac{c}{d}$.

Mixed number
Proper fraction
Reciprocal, or multiplicative inverse
Invert the divisor and multiply algorithm

Section 6.3

Negative (or opposite, or additive inverse)
Transitive property of order
Density property

CHAPTER REVIEW EXERCISES

Section 6.1

1. What fraction is represented in each of the colored region models shown. In (d) the unit is the region inside one circle.

(a)
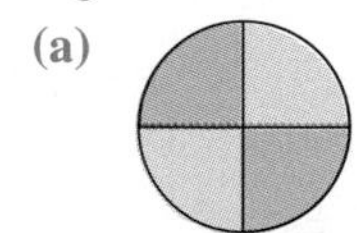

(b)
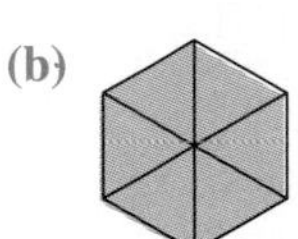

(c)
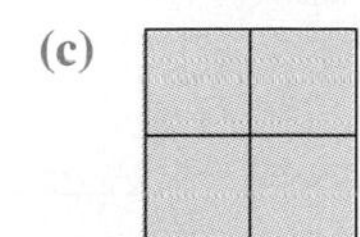

(d)
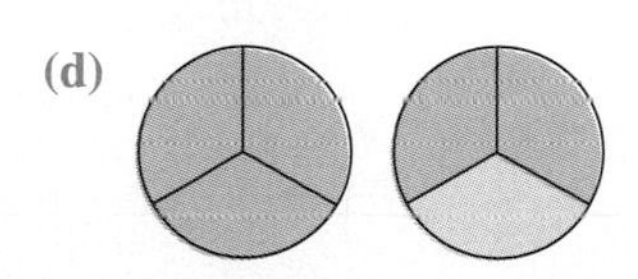

2. Label the points on the number line which correspond to these rational numbers:

(a) $\frac{3}{4}$ (b) $\frac{12}{8}$

(c) 1 (d) $2\frac{3}{8}$

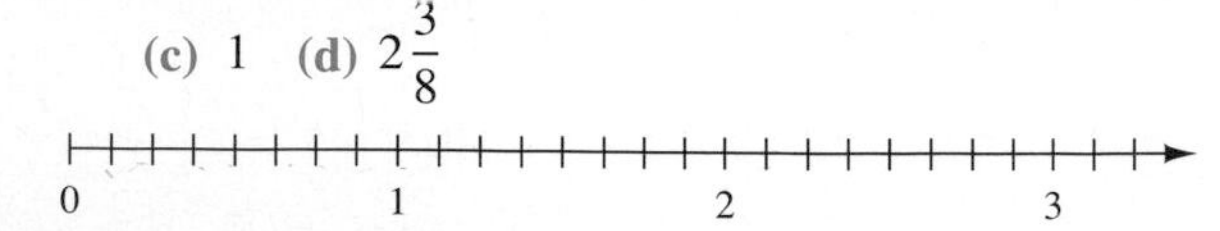

3. Express each rational by a fraction in simplest form.

(a) $\frac{27}{81}$ (b) $\frac{100}{825}$

(c) $\frac{378}{72}$ (d) $\frac{3^5 \cdot 7^2 \cdot 11^3}{3^2 \cdot 7^3 \cdot 11^2}$

4. Order these fractions from smallest to largest.

$$\frac{1}{2}, \frac{13}{27}, \frac{25}{49}, \frac{13}{30}, \frac{26}{49}$$

5. Describe the set of all fractions equivalent to the given fraction.

(a) $\frac{5}{8}$ (b) -4 (c) $\frac{18}{24}$.

Section 6.2

6. Illustrate $\frac{3}{4} + \frac{7}{8}$ on the number line.

7. Illustrate $\frac{3}{4} - \frac{1}{3}$ with fraction strips.

8. Compute these sums and differences.
(a) $\frac{3}{8} + \frac{1}{4}$ (b) $\frac{2}{9} + \frac{-5}{12}$
(c) $\frac{4}{5} - \frac{2}{3}$ (d) $5\frac{1}{4} - 1\frac{5}{6}$

9. Illustrate these products by labeling and coloring appropriate rectangular regions.
(a) $3 \times \frac{1}{3}$ (b) $\frac{2}{3} \times 4$ (c) $\frac{5}{6} \times \frac{3}{2}$

10. On a map, it is $7\frac{1''}{8}$ from Arlington to Banks. If the scale of the map is $1\frac{1''}{4}$ per mile, how far is it between the two towns?

Section 6.3

11. Perform these calculations, expressing your answers in simplest form.
(a) $\frac{-3}{4} + \frac{5}{8}$ (b) $\frac{4}{5} - \frac{-7}{10}$
(c) $\left(\frac{3}{8} \cdot \frac{-4}{27}\right) \div \frac{1}{9}$ (d) $\frac{2}{5} \cdot \left(\frac{3}{4} - \frac{5}{2}\right)$

12. Solve each equation. Give the rational number x as a fraction in simplest form.
(a) $3x + 5 = 11$ (b) $x + \frac{2}{3} = \frac{1}{2}$
(c) $\frac{3}{5}x + \frac{1}{2} = \frac{2}{3}$ (d) $-\frac{4}{3}x + 1 = \frac{1}{4}$

13. Find two rational numbers between $\frac{5}{6}$ and $\frac{10}{11}$.

14. Do the following calculations mentally. Explain your method.
(a) $1\frac{1}{3} + 2\frac{5}{12} + \frac{1}{4}$ (b) $\frac{6}{7} \cdot \frac{28}{3} \cdot \frac{5}{8}$ (c) $\frac{36}{5} \div \frac{9}{25}$

CHAPTER TEST

1. Illustrate $\frac{2}{3}$ (a) on the number line, (b) with a fraction strip, (c) with a colored region model, and (d) with the set model.

2. Give three different fractions each equivalent to $-\frac{3}{4}$.

3. Order these rational numbers from least to greatest. $\frac{16}{5}$, $\frac{2}{3}$, $-1\frac{1}{2}$, 0, $\frac{5}{8}$, 3, -3.

4. Perform these calculations:
(a) $\frac{1}{3} + \frac{5}{8} - \frac{5}{6}$ (b) $\left(\frac{2}{3} - \frac{5}{4}\right) \div \frac{3}{4}$
(c) $\frac{4}{7} \cdot \left(\frac{35}{4} + \frac{-42}{12}\right)$ (d) $\frac{123}{369} \div \frac{1}{3}$

5. (a) Define division in the rational number system.
(b) Justify the invert and multiply algorithm; that is, prove that $\frac{a}{b} \div \frac{c}{d} = \frac{a}{b} \cdot \frac{d}{c}$, where $d \neq 0$.

6. (a) Invent a realistic problem whose solution requires the calculation $\frac{4}{5} \cdot \frac{2}{3}$.
(b) Make up a realistic problem which leads to $\frac{3}{8} \div \frac{3}{10}$.

7. Solve the following equations and inequalities for all possible rational numbers x. Show all of your steps.
(a) $2x + 3 > 0$ (b) $\frac{3}{4}x + \frac{1}{2} = \frac{1}{3}$
(c) $\frac{5}{4}x > -\frac{1}{3}$ (d) $\frac{1}{2} < 4x + \frac{5}{6}$

8. Carefully explain why $\frac{a+b}{b} = \frac{c+d}{d}$ if, and only if, $\frac{a}{b} = \frac{c}{d}$.

9. An acre is $\frac{1}{640}$ of a square mile.
(a) How many acres are in a rectangular plot of ground $\frac{1}{8}$ mile wide by $\frac{1}{2}$ mile long?
(b) A rectangular piece of property contains 80 acres and is $\frac{1}{2}$ mile long. What is the width of the property?

10. (a) What is the density property of the rational numbers?
(b) Find a rational number between $\frac{3}{5}$ and $\frac{2}{3}$.

11. What is the best approximate answer listed for each of these problems?

(a) $2\frac{1}{48} + 3\frac{1}{99} + 6\frac{13}{25}$ is approximately

$11, \quad 11\frac{1}{2}, \quad 12, \quad 12\frac{1}{4}.$

(b) $8 \cdot \left(2\frac{1}{2} + 3\frac{7}{15}\right)$ is approximately

$40, \quad 44, \quad 48, \quad 56.$

(c) $11\frac{9}{10} \div \frac{21}{40}$ is approximately

$20, \quad 23, \quad 26, \quad 30.$

12. (a) Define *additive inverse.*

(b) Find the additive inverses of the following rational numbers: $\frac{3}{4}, \frac{-7}{4}, \frac{8}{-2}$.

13. (a) Define *multiplicative inverse.*

(b) Find the multiplicative inverses of these rational numbers: $\frac{3}{2}, \frac{-4}{5}, -5.$

14. Describe how the following calculations can be performed efficiently with "mental math."

(a) $\frac{19}{111} \cdot \left(\frac{2}{3} + \frac{-4}{6}\right)$ (b) $\frac{5}{6} \cdot \frac{36}{15}$

(c) $\frac{5}{8} \cdot \left(\frac{9}{5} - \frac{1}{5}\right)$ (d) $\frac{2}{3} \cdot \frac{3}{4} \cdot \frac{4}{5} \cdot \frac{5}{6}$

7 Decimals and Real Numbers

Computer Support for this Chapter

In this chapter you might find the following programs on your computer disk useful:

- DIFFY
- DIVVY

HANDS ON — On Triangles and Squares

Materials Needed

One 3″ × 5″ card, scissors, a sharp pencil, and one sheet of plain white paper.

Directions

Step 1. Cut a triangle with unequal sides off the corner of the card and denote the lengths of the three sides by a, b, and c as shown.

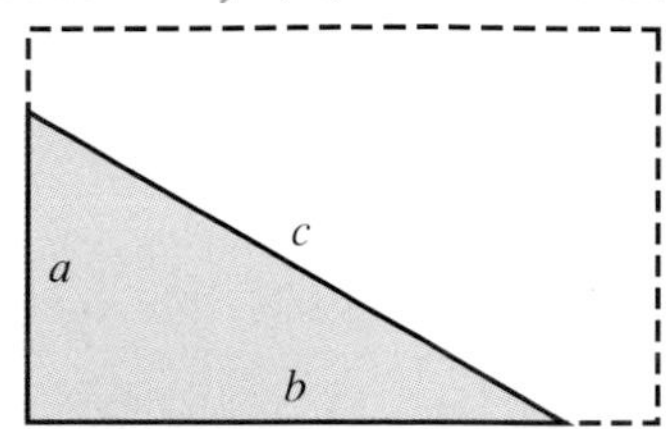

Step 2. With the lettering on the triangle always facing up, carefully trace around the triangle four times to form a square as shown here.

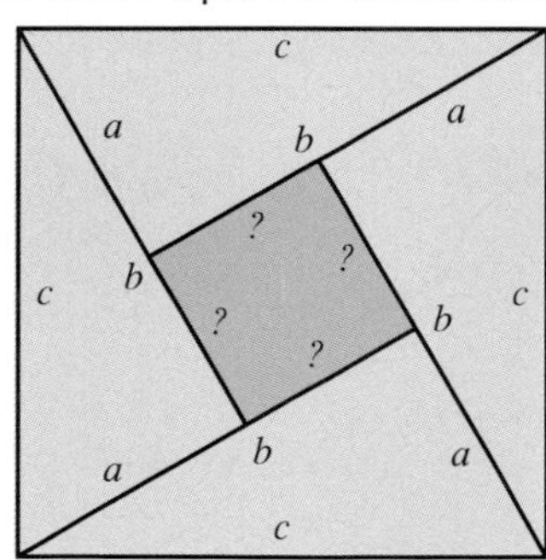

Step 3. Determine the dimensions of the small square in the center and then find its area. Remember that, using the distributive property of multiplication over subtraction and the commutative property,

$$\begin{aligned}(x-y)(x-y) &= (x-y)x-(x-y)y\\ &= x^2-yx-xy+y^2\\ &= x^2-2xy+y^2.\end{aligned}$$

Step 4. The area of the large square is clearly four times the area of the triangle plus the area of the small square; it is also equal to c^2. Express these two ways of finding the area of the large square by an equation and simplify it as much as possible.

Step 5. Relate the equation in Step 4 to the diagram shown here. What does the equation reveal about triangles with one square corner (that is, one right angle)? Explain briefly.

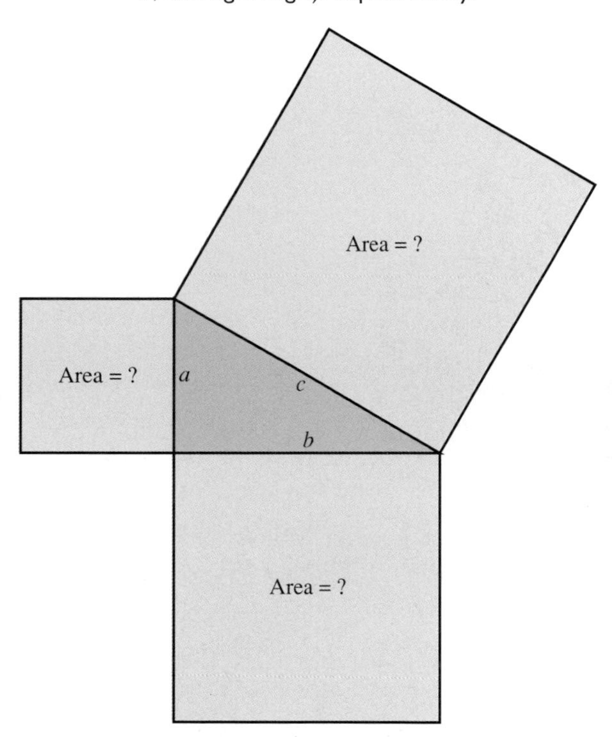

Step 6. If c is the length of the diagonal of a square one unit on a side, then the formula of Step 4 gives $c^2 = 2$. Remarkably, as we will see later, this number c is not a rational number.

CONNECTIONS — Further Enlarging the Number System

In the preceding chapters we have considered:

- the set of natural numbers, N,
- the set of whole numbers, W,
- the set of integers, I, and
- the set of rational numbers, Q.

The fact that each succeeding set is an extension of the immediately preceding set can be expressed in a Venn diagram as shown in Figure 7.1.

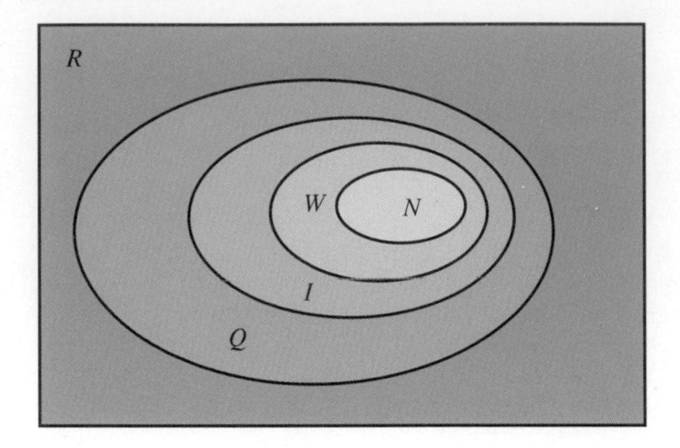

Figure 7.1
Venn diagram for the natural numbers, whole numbers, integers, and rational numbers

The following statements illustrate the relationships between these sets:

- The number 2 is a natural number, a whole number, an integer, and a rational number.
- The number 0 is a whole number, an integer, and a rational number; it is *not* a natural number.
- The number -5 is an integer and a rational number; it is not a whole number or a natural number.
- The number $3/4$ is a rational number; it is not an integer, a whole number, or a natural number.

From a practical point of view, each number system was created to meet a specific need:

- The natural numbers came into being as a result of people's need to count.
- The number 0, which with the natural numbers comprises the set of whole numbers, was first introduced simply as a place holder in positional number systems, but it also allows us to count the number of elements in the empty set and has other useful arithmetic properties.
- The integers allow us to keep track of debits and credits, gains and losses, degrees above zero and degrees below zero, and so on.
- The rational numbers are needed to make accurate measurements of lengths, areas, volumes, and other quantities.

From a more mathematical point of view the natural numbers allow us to solve equations like

$$x - 3 = 0 \quad \text{and} \quad 7x - 21 = 0.$$

The whole numbers allow us to solve equations like

$$3x = 0 \quad \text{and} \quad (x - 5)(x - 7) = 0.$$

Extending the number system to include all integers allows us to solve equations like

$$x + 4 = 0 \quad \text{and} \quad 9x + 36 = 0.$$

Extending the number system to include the rational numbers allows us to solve equations like

$$3x - 4 = 0 \quad \text{and} \quad (2x + 1)(5x - 4) = 0.$$

In summary, the rational number system permits us to deal with a wide range of practical and theoretical problems. However, remarkably, there remain simple questions that ought to have simple answers but do not if we do not extend the number system beyond the rationals. For example, if we stopped with the rationals, we would be unable to answer such a simple question as, "How long is the diagonal of a square measuring 1 unit on a side?" Thus, it is necessary to extend the number system at least once more. See Figure 7.2. In this chapter, we consider the system of numbers called the real numbers.* We do so by means of decimals.

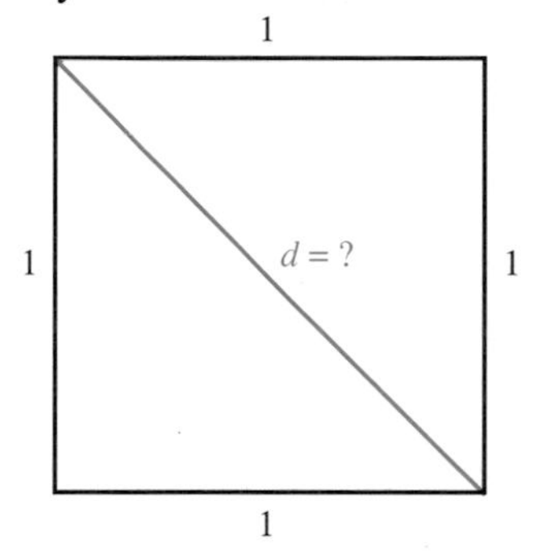

Figure 7.2
The length of the diagonal of a square 1 unit on a side is *not* a rational number.

7.1 Decimals

Since it is a positional system based on ten, we have used the term decimal system (from the Latin *decimus* meaning tenth) to refer generally to the Hindu-Arabic system of numeration in common use today. However, more colloquially, people often refer to expressions like 0.235 or 2.7142 as **decimals** as opposed to 24, 98, 0 or 2478, which they more often speak of as whole numbers or integers. In fact, both are part and parcel of the same system. Just as the **expanded form** of 2478 is

$$\begin{aligned} 2478 &= 2 \cdot 10^3 + 4 \cdot 10^2 + 7 \cdot 10^1 + 8 \cdot 10^0 \\ &= 2000 + 400 + 70 + 8, \end{aligned}$$

the expanded form of 0.235 is

$$\begin{aligned} 0.235 &= 2 \cdot \frac{1}{10^1} + 3 \cdot \frac{1}{10^2} + 5 \cdot \frac{1}{10^3} \\ &= \frac{2}{10} + \frac{3}{100} + \frac{5}{1000}, \end{aligned}$$

and the expanded form of 23.47 is

$$\begin{aligned} 23.47 &= 2 \cdot 10^1 + 3 \cdot 10^0 + 4 \cdot \frac{1}{10^1} + 7 \cdot \frac{1}{10^2} \\ &= 20 + 3 + \frac{4}{10} + \frac{7}{100}. \end{aligned}$$

* The word *real,* when applied to numbers, is used to distinguish these numbers from those called *imaginary* in a final extension of the number system to the system of *complex numbers.* We do not consider the complex number system in this book.

FROM THE NCTM STANDARDS **Fractions and Decimals**

Fractions and decimals represent a significant extension of children's knowledge about numbers. When children possess a sound understanding of fraction and decimal concepts, they can use this knowledge to describe real-world phenomena and apply it to problems involving measurement, probability, and statistics. An understanding of fractions and decimals broadens students' awareness of the usefulness and power of numbers and extends their knowledge of the number system. It is critical in grades K–4 to develop concepts and relationships that will serve as a foundation for more advanced concepts and skills.

SOURCE: From *Curriculum and Evaluation Standards for School Mathematics Grades K–4,* page 57.

Each digit of a decimal numeral contributes an amount to the number being represented that depends both on the digit and on its position in the number as shown in Table 7.1.

TABLE 7.1 Positional Values in the Decimal System

Position Names	...	Hundreds	Tens	Units	Tenths	Hundredths	Thousandths	Ten Thousandths	...
Decimal Form	...	100	10	1	0.1	0.01	0.001	0.0001	...
Rational Form	...	100	10	1	$\frac{1}{10}$	$\frac{1}{100}$	$\frac{1}{1000}$	$\frac{1}{10,000}$	...
Powers of 10	...	10^2	10^1	10^0	10^{-1}	10^{-2}	10^{-3}	10^{-4}	...

The pattern in expanded notation becomes even more clear if we use negative exponents.

Negative Exponents and Expanded Exponential Form

We have already observed that, for a natural number n,

$$a^n \text{ is shorthand for the product } \overbrace{a \cdot a \cdots a}^{n \text{ factors}}$$

and therefore

$$a^n \cdot a^m = \overbrace{a \cdot a \cdots a}^{n \text{ factors}} \cdot \overbrace{a \cdot a \cdots a}^{m \text{ factors}} = a^{m+n}.$$

Moreover, if this rule is to hold for $m = 0$, then

$$a^n \cdot a^0 = a^{n+0} = a^n$$

and we must define $a^0 = 1$ as we did in Chapter 2. Now suppose that n is a positive integer so that $-n$ is negative. What shall we mean by a^{-n}? If we require that the above rule of exponents holds for **negative exponents** as well as for 0, then

$$a^n \cdot a^{-n} = a^{n+(-n)} = a^0 = 1.$$

$r \cdot \frac{1}{r} = 1$

This implies that we should define a^{-n} to be $1/a^n$.

DEFINITION Negative Numbers and Zero as Exponents

If n is a positive integer and $a \neq 0$, then $a^0 = 1$ and $a^{-n} = \dfrac{1}{a^n}$.

Using this definition, we complete the last row of Table 7.1 as shown. We also write the decimals at the beginning of this section in **expanded exponential form** as follows:

$$
\begin{aligned}
0.235 &= 0 + 2 \cdot \frac{1}{10^1} + 3 \cdot \frac{1}{10^2} + 5 \cdot \frac{1}{10^3} \\
&= 2 \cdot 10^{-1} + 3 \cdot 10^{-2} + 5 \cdot 10^{-3} \\
23.47 &= 2 \cdot 10^1 + 3 \cdot 10^0 + 4 \cdot \frac{1}{10^1} + 7 \cdot \frac{1}{10^2} \\
&= 2 \cdot 10^1 + 3 \cdot 10^0 + 4 \cdot 10^{-1} + 7 \cdot 10^{-2}
\end{aligned}
$$

In this form, the pattern of decreasing exponents is both clear and neat.

EXAMPLE 7.1

Writing Decimals in Expanded Exponential Form

Write each of these in expanded exponential form.

(a) 234.72 **(b)** 30.0012

SOLUTION

(a) $234.72 = 2 \cdot 10^2 + 3 \cdot 10^1 + 4 \cdot 10^0 + 7 \cdot 10^{-1} + 2 \cdot 10^{-2}$

(b) $30.0012 = 3 \cdot 10^1 + 0 \cdot 10^0 + 0 \cdot 10^{-1} + 0 \cdot 10^{-2} + 1 \cdot 10^{-3} + 2 \cdot 10^{-4}$ ■

Multiplying and Dividing Decimals by Powers of 10

Consider the decimal

$$25.723 = 2 \cdot 10^1 + 5 \cdot 10^0 + 7 \cdot 10^{-1} + 2 \cdot 10^{-2} + 3 \cdot 10^{-3}.$$

If we multiply by 10^2 using the distributive property and the above rule for multiplying powers, we obtain

$$
\begin{aligned}
(10^2)(25.723) &= (10^2)(2 \cdot 10^1 + 5 \cdot 10^0 + 7 \cdot 10^{-1} + 2 \cdot 10^{-2} + 3 \cdot 10^{-3}) \\
&= 2 \cdot 10^{2+1} + 5 \cdot 10^{2+0} + 7 \cdot 10^{2+(-1)} + 2 \cdot 10^{2+(-2)} + 3 \cdot 10^{2+(-3)} \\
&= 2 \cdot 10^3 + 5 \cdot 10^2 + 7 \cdot 10^1 + 2 \cdot 10^0 + 3 \cdot 10^{-1} \\
&= 2572.3
\end{aligned}
$$

and the notational effect is to move the decimal point 2 places to the right. Note that 2 is the exponent of the power of 10 we are multiplying by and also the number of zeros in 100.

$10^2 = 100$

The result is analogous if we divide 25.723 by 10^2 except that the notational effect is to move the decimal point two places *to the left*. To see this, recall that we can divide by multiplying by the multiplicative inverse. Thus,

$10^{-2} = \frac{1}{10^2}$

$$
\begin{aligned}
(25.723) \div 10^2 &= (25.723) \cdot (1/10^2) \\
&= (25.723) \cdot 10^{-2} \\
&= (2 \cdot 10^1 + 5 \cdot 10^0 + 7 \cdot 10^{-1} + 2 \cdot 10^{-2} + 3 \cdot 10^{-3}) \cdot (10^{-2}) \\
&= 2 \cdot 10^{1+(-2)} + 5 \cdot 10^{0+(-2)} + 7 \cdot 10^{(-1)+(-2)} + 2 \cdot 10^{(-2)+(-2)} \\
&\quad + 3 \cdot 10^{(-3)+(-2)} \\
&= 2 \cdot 10^{-1} + 5 \cdot 10^{-2} + 7 \cdot 10^{-3} + 2 \cdot 10^{-4} + 3 \cdot 10^{-5} \\
&= 0.25723.
\end{aligned}
$$

These results are typical of the general case which we state here as a theorem.

THEOREM Multiplying and Dividing Decimals by Powers of 10

The effect of multiplying a decimal by 10^r is to move the decimal point r places to the right. The effect of dividing a decimal by 10^r (that is, multiplying by 10^{-r}) is to move the decimal point r places to the left.

EXAMPLE 7.2 Multiplying and Dividing Decimals by Powers of 10

Compute each of the following.

(a) $(10^3)(253.26)$ **(b)** $(253.26) \div 10^3$
(c) $(100)(34.764)$ **(d)** $(34.764) \div 10{,}000$

SOLUTION

The preceding theorem gives the desired results.

(a) $(10^3)(253.26) = 253{,}260$
(b) $(253.26) \div 10^3 = 0.25326$
(c) $(100)(34.764) = (10^2)(34.764) = 3476.4$
(d) $(34.764) \div 10{,}000 = 34.764 \div (10^4) = 0.0034764$ ■

Terminating Decimals as Fractions

From grade school we know that

$$\frac{1}{3} = 0.333 \cdots$$

where the ellipsis dots indicate that the string of 3s continues without end. Such a decimal is called a **nonterminating decimal.** A decimal like 21.00357 which has only finitely many digits is called a **terminating decimal.** Using the preceding ideas we show that every terminating decimal represents a rational number; that is, any terminating decimal can be represented by a fraction with integers in the numerator and denominator. For example, using expanded notation

$$
\begin{aligned}
24.357 &= 20 + 4 + \frac{3}{10} + \frac{5}{100} + \frac{7}{1000} \\
&= \frac{20{,}000}{1000} + \frac{4000}{1000} + \frac{300}{1000} + \frac{50}{1000} + \frac{7}{1000} \\
&= \frac{24{,}357}{1000}.
\end{aligned}
$$

Alternatively,

$$
\begin{aligned}
24.357 &= 24.357 \cdot \frac{1000}{1000} \\
&= \frac{24{,}357}{1000}
\end{aligned}
$$

$a \cdot \frac{b}{c} = \frac{ab}{c}$

as before. In each case, the denominator is determined by the position of the right-most digit (in this case, the 7 is in the thousandths position).

Also, numbers like the above are often written as mixed numbers. Returning to the above

$$
\begin{aligned}
24.357 &= 20 + 4 + \frac{3}{10} + \frac{5}{100} + \frac{7}{1000} \\
&= 24 + \frac{300}{1000} + \frac{50}{1000} + \frac{7}{1000} \\
&= 24 + \frac{357}{1000} \\
&= 24\frac{357}{1000},
\end{aligned}
$$

which is read "twenty-four and three hundred fifty-seven thousandths." Either form of the fraction is acceptable. The form chosen depends on how the result is to be used.

The preceding discussion shows how to convert a terminating decimal into a ratio of two integers, that is, into a fraction. Not all rational numbers have finite decimal expansions but those that do can be converted to decimal form as follows. Consider

$$\frac{17}{40} = \frac{17}{2^3 \cdot 5^1}.$$

Since the prime factor representation of the denominator contains only 2s and 5s, the fraction can be written so that the denominator is a power of 10. Thus,

Multiply numerator and denominator by 5^2.

$$\frac{17}{40} = \frac{17}{2^3 \cdot 5^1} = \frac{17 \cdot 5^2}{2^3 \cdot 5^3} = \frac{17 \cdot 25}{10^3} = \frac{425}{1000} = 0.425.$$

HIGHLIGHT FROM HISTORY
Simon Stevin (1548–1620)

The most influential mathematician from the Low Countries in the sixteenth century was the Flemish mathematician, Simon Stevin. Stevin served in the Dutch army as quartermaster and was something of an expert on military engineering and the building of fortifications. His most significant contribution to mathematics was the extension of the use of the positional Hindu-Arabic system of numeration to the rational and real numbers, thus greatly facilitating numerical calculation. To the local Dutch he was most famous for his invention of a carriage powered by sails and capable of carrying more than a score of people at speeds up to twenty miles per hour.

Similarly,

$$\begin{aligned}\frac{2473}{1250} &= \frac{2473}{2^1 \cdot 5^4}\\ &= \frac{2^3 \cdot 2473}{2^4 \cdot 5^4}\\ &= \frac{19{,}784}{10^4}\\ &= 1.9784.\end{aligned}$$

Since the numbers in the preceding discussion are typical of the general case, the results can be summarized as a theorem.

THEOREM Terminating Decimals and Rational Numbers

If a and b are integers with $b \neq 0$, if a/b is in simplest form, and if the prime factor representation of b contains only 2s and 5s, then a/b can be represented as a terminating decimal and conversely.

EXAMPLE 7.3 **Writing a Terminating Decimal as a Ratio of Two Integers**

Express each of these in the form a/b where the fraction is in simplest form.

(a) 31.75 **(b)** 4.112 **(c)** −0.035

SOLUTION

(a) $31.75 = 31.75 \cdot \frac{100}{100} = \frac{3175}{100} = \frac{127}{4}$

(b) $4.112 = 4.112 \cdot \frac{1000}{1000} = \frac{4112}{1000} = \frac{514}{125}$

(c) $-0.035 = -0.035 \cdot \frac{1000}{1000} = -\frac{35}{1000} = \frac{7}{200}$ ■

We just saw how to write 17/40 and 2473/1250 as finite decimals by multiplying both numerator and denominator by powers of 2 or 5. But the task can also be accomplished by division. Thus, using a calculator, we have that

$$\frac{17}{40} = 0.425 \quad \text{and} \quad \frac{2473}{1250} = 1.9784.$$

$\frac{17}{40} = 17 \div 40$

$\frac{2473}{1250} = 2473 \div 1250$

EXAMPLE 7.4

Converting Certain Fractions to Decimals

Convert each of these fractions to decimals by writing each as an equivalent fraction whose denominator is a power of 10. Check by dividing using a calculator.

(a) $\frac{37}{40}$ **(b)** $-\frac{29}{200}$

SOLUTION

(a)

$$\begin{aligned}\frac{37}{40} &= \frac{37}{2^3 \cdot 5} = \frac{37 \cdot 5^2}{2^3 \cdot 5^3}\\ &= \frac{37 \cdot 25}{10^3}\\ &= \frac{925}{1000}\\ &= \frac{900 + 20 + 5}{1000}\\ &= \frac{900}{1000} + \frac{20}{1000} + \frac{5}{1000}\\ &= \frac{9}{10} + \frac{2}{100} + \frac{5}{1000}\\ &= 9 \cdot 10^{-1} + 2 \cdot 10^{-2} + 5 \cdot 10^{-3} = 0.925\end{aligned}$$

$5^1 \cdot 5^2 = 5^3$

$5^2 = 25,\ 37 \cdot 25 = 925$

Also, by calculator, $37 \div 40 = 0.925$.

(b)

$$\begin{aligned}-\frac{29}{200} &= -\frac{29}{2^3 \cdot 5^2} = -\frac{29 \cdot 5}{2^3 \cdot 5^3}\\ &= -\frac{145}{10^3}\\ &= -\frac{145}{1000}\\ &= -\left(\frac{100 + 40 + 5}{1000}\right)\\ &= -\left(\frac{100}{1000} + \frac{40}{1000} + \frac{5}{1000}\right)\\ &= -\left(\frac{1}{10} + \frac{4}{100} + \frac{5}{1000}\right)\\ &= -(1 \cdot 10^{-1} + 4 \cdot 10^{-2} + 5 \cdot 10^{-3}) = -0.145.\end{aligned}$$

$\frac{a + b + c}{d} = \frac{a}{d} + \frac{b}{d} + \frac{c}{d}$

Also, by calculator, $-(29 \div 200) = -0.145$. ■

Nonterminating Decimals and Rational Numbers

Somewhat surprisingly not all rational numbers have decimal expansions that terminate. For example, it is well-known that

$$\frac{1}{3} = 0.333\ldots = 0.\overline{3}$$

where the three dots indicate that the decimal continues *ad infinitum* and the bar over the 3 indicates the digit or group of digits that repeats. To see why this is so, set

$$x = 0.333\ldots.^*$$

Then, using the preceding theorem,

$$10x = 3.333\ldots$$

so that

$$10x - x = 9x = 3.$$

3.333...
− 0.333...
3.000...

But this implies that

$$x = \frac{3}{9} = \frac{1}{3}.$$

Notice that the decimal expansion of 1/3 is a nonterminating but *repeating* decimal; that is, the stream of 3s repeats without end. In general, we have the following definition.

DEFINITION A Repeating Decimal

A nonterminating decimal that has the property that a digit or group of digits repeats *ad infinitum* from some point on is called a **periodic** or **repeating decimal.** The number of digits in the repeating group is called the **length of the period.**

Just as $0.333\cdots = .\overline{3}$ represents the rational number 1/3, so every repeating decimal represents a rational number.

EXAMPLE 7.5

Repeating Decimals as Rational Numbers

Write each of these repeating decimals in the form a/b where a and b are integers and the fraction is in simplest form. Check by dividing a by b with your calculator.

(a) $0.242424\ldots = 0.\overline{24}$ **(b)** $3.14555\ldots = 3.14\overline{5}$

SOLUTION

(a) Let $x = 0.242424\ldots.$
Then,

a decimal of period 2

Multiply by $10^2 = 100$ to move the decimal point two places to the right.

$$100x = 24.242424\ldots$$

---

*Actually there is a touchy point here that we gloss over. The question is whether $0.333\cdots = 3/10 + 3/100 + 3/1000 + \cdots$ means anything at all since it is the sum of an *infinite* number of numbers. That the answer is yes really depends on ideas from calculus!

and

$$100x - x = 99x = 24.$$

24.242424...
− 0.242424...
24.000000...

But then,

$$x = \frac{24}{99} = \frac{8}{33}.$$

Also, by calculator, $8 \div 33 \doteq 0.2424242$.

Here the answer is only approximate since only finitely many digits can appear in the calculator display.

(b) Let $x = 3.14555\ldots.$

Multiply by 10^2 to move the decimal point two places to the right where the period starts.

Then,

$$100x = 314.555\ldots$$

a decimal of period 1

and

$$1000x = 3145.555\ldots.$$

$10 \cdot 100 = 1000$

Multiply by 10 again to move the decimal point one more place (the length of the period) to the right.

Then,

$$1000x - 100x = 900x = 2831$$

3145.555...
− 314.555...
2831.000...

and

$$x = \frac{2831}{900}.$$

Also, by calculator, $2831 \div 900 \doteq 3.1455556$. Explain why the decimal is finite and ends with a 6. ■

Example 7.5 is typical of the general case and so we have the following theorem.

THEOREM Rational Numbers and Periodic Decimals

Every repeating decimal represents a rational number a/b. If a/b is in simplest form, b must contain a prime factor other than 2 or 5. Conversely, if a/b is such a rational number its decimal representation must be repeating.

SCHOOL BOOK PAGE *Terminating and Repeating Decimals*

Independent Study ENRICHMENT

Enrichment

Terminating and Repeating Decimals

Byron was using his calculator to change the fractions on page 329 to decimals. After he finished with those fractions, he thought he would try some others. Here are the fractions he tried, along with the results:

$\frac{5}{12}$ **Press:** 5 ÷ 12 = **Display:** *0.4166666*

$\frac{6}{11}$ **Press:** 6 ÷ 11 = **Display:** *0.5454545*

$\frac{7}{32}$ **Press:** 7 ÷ 32 = **Display:** *0.21875*

The decimals for $\frac{5}{12}$ and $\frac{6}{11}$ are ***repeating decimals***. A repeating decimal has a digit or pattern of digits that repeats endlessly. To write a repeating decimal, put a bar over the digit or digits that repeat.

$0.4166666\ldots = 0.41\overline{6}$ $\quad$ $0.5454545\ldots = 0.\overline{54}$

The decimal for $\frac{7}{32}$ is a ***terminating decimal***. Terminating decimals have digits in an exact number of decimal places.

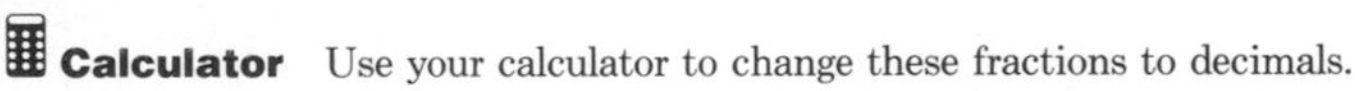

Calculator Use your calculator to change these fractions to decimals.

1. $\frac{3}{4}$ **2.** $\frac{5}{8}$ **3.** $\frac{4}{9}$ **4.** $\frac{5}{6}$ **5.** $\frac{17}{20}$ **6.** $\frac{5}{24}$

7. $\frac{11}{32}$ **8.** $\frac{8}{15}$ **9.** $\frac{13}{16}$ **10.** $\frac{29}{40}$ **11.** $\frac{9}{11}$ **12.** $\frac{7}{18}$

13. For each fraction, write the prime factorization of the denominator.

14. What are the prime factors of denominators in fractions that become terminating decimals?

15. What are the prime factors of denominators in fractions that become repeating decimals?

16. How can you tell if a fraction will be a terminating or a repeating decimal by looking at the denominator?

1. When you use your calculator to divide 4 by 9, the display shows 0.4444444. Does it follow that $4/9 = 0.\overline{4}$? Briefly describe how you would explain this to a class of fifth graders.

2. When you use your calculator to divide 7 by 18, the display shows 0.3888889. Is it true that $7/18 = 0.3888889$? Explain briefly.

3. How would you show a class of fifth graders that $7/18 = 0.3\overline{8}$?

"I THOUGHT IT WAS A BREAKTHROUGH, BUT IT WAS ONLY A MISPLACED DECIMAL."

Ordering Decimals

Ordering decimals is much like ordering integers. For example, to determine the larger of 247,761 and 2,326,447 write both numerals as if they had the same number of digits; that is, write

$$0{,}247{,}761 \qquad \text{and} \qquad 2{,}326{,}447.$$

Then determine the first place from the *left* where the digits differ. It follows from the idea of positional notation that the larger integer is the integer with the larger of these two different digits. In the present case, the first digits differ and so

$$0{,}247{,}761 < 2{,}326{,}447. \qquad (0 < 2)$$

In a similar example,

$$34{,}716 < 34{,}723 \qquad (1 < 2)$$

since the first pair of corresponding digits that differ are the 1 and the 2 and $1 < 2$.

Now consider the decimals,

$$0.2346612359 \quad \text{and} \quad 0.2348999.$$

Suppose we multiply both numbers by 10^{10} so that both become integers. We obtain

$$2{,}346{,}612{,}359 \quad \text{and} \quad 2{,}348{,}999{,}000.$$

We order these as integers in the manner just discussed to obtain

$$2{,}346{,}612{,}359 < 2{,}348{,}999{,}000;$$

6 < 8

that is,

$$(10^{10})(0.2346612359) < (10^{10})(0.2348999000).$$

But this implies that

$$0.2346612359 < 0.2348999$$

Divide both sides by 10^{10}.

and we are done.

These arguments could be repeated in general with the following result.

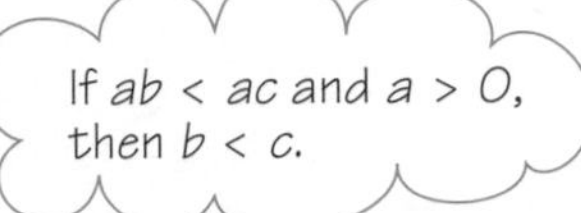

> **THEOREM Ordering Decimals**
>
> To order two positive decimals, add zeros on the left if necessary so that there are the same number of digits to the left of the decimal point, and then determine the first digits from the left that differ. The decimal with the lesser of these two digits is the lesser decimal.

EXAMPLE 7.6

Ordering Decimals

In each case, decide which of the decimals represents the lesser number.

(a) 2.35714 and 2.35709 **(b)** 23.45 and $23.4\overline{5}$

SOLUTION

(a) Here the first digits from the left that differ are 1 and 0. Since $0 < 1$, it follows that $2.35709 < 2.35714$.

(b) Since $23.4\overline{5} = 23.4545\cdots$, the first digits from the left that differ are 0 and 5. Since $0 < 5$, it follows that $23.45 < 23.4\overline{5}$ ■

EXAMPLE 7.7

Ordering Decimals and Fractions

Arrange these numbers in order from least to greatest.

$$\frac{11}{24},\quad \frac{3}{8},\quad 0.37,\quad 0.4584,\quad 0.37666\cdots,\quad 0.4583$$

SOLUTION

The easiest approach is to write all these numbers as decimals. Since

$$\frac{11}{24} = 0.458333\cdots \qquad \text{and} \qquad \frac{3}{8} = 0.375$$

it follows that

0.375   0.458333 . . .

$$0.37 < \frac{3}{8} < 0.37666\cdots < 0.4583 < \frac{11}{24} < 0.4584.$$ ■

The Set of Real Numbers

Earlier we showed that

the decimal expansion of a rational number either terminates or is nonterminating and repeating.

This raises an interesting question. What kind of numbers are represented by nonterminating nonrepeating decimals like 0.101001000100001 . . . ? Such numbers are called **irrational numbers** to distinguish them from the rational numbers considered in Chapter 6. The set consisting of all rational and irrational numbers is called **the set of real numbers.**

> **DEFINITION Irrational and Real Numbers**
>
> Numbers represented by nonterminating nonperiodic decimals are called **irrational numbers.** The set R consisting of all rational numbers and all irrational numbers is called the set of **real numbers.**

Thus, the set of real numbers is another extension of the number system. This number system contains all the earlier systems as illustrated by the Venn diagram in Figure 7.1.

Irrationality of $\sqrt{2}$

We complete this section by proving that the length of the diagonal of a square measuring 1 unit on a side is not a rational number. Recall that the equation $c^2 = a^2 + b^2$, relates the lengths of the long side of a triangle with one square corner to the lengths of the other two sides as discovered in the HANDS ON* at the beginning of this chapter. If c is the length of the diagonal of a square 1 unit on a side, then it follows that

$$c^2 = 1^2 + 1^2 = 2.$$

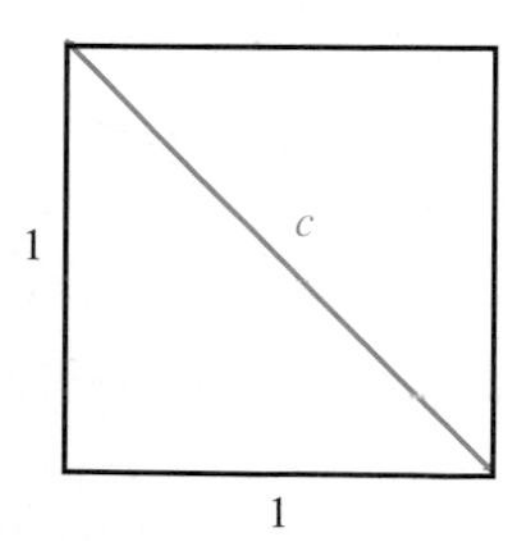

Recall that if $c^2 = 2$, then c is called the square root of 2 and we write $c = \sqrt{2}$. We now show that $\sqrt{2}$ is not a rational number.

* A triangle with one square corner (that is, a corner forming a 90° angle) is called a **right** triangle. The fact that $c^2 = a^2 + b^2$ for such a triangle is called the **Pythagorean theorem.**

HIGHLIGHT FROM HISTORY
Pythagoras and Irrational Numbers

The great Greek mathematician Pythagoras (c. 585–497 B.C.) is best known because of the Pythagorean theorem considered in the Hands On at the beginning of this chapter. Actually, it is doubtful that Pythagoras discovered this theorem or even proved the result true. At best, the theorem may be due to one of the members of the religious, philosophical, and scientific society founded by Pythagoras and called the Pythagorean School. The Pythagoreans said that "all is number" (and by numbers they meant whole numbers) and made numbers the basis for their entire system of thought. It is true that the Pythagoreans considered ratios and were able to deal with situations like measurement that required rational numbers. However, the discovery by Hippasus of Metapotum that the diagonal and side of a square are incommensurable (that is, have no common unit of measure), which amounts to showing that $\sqrt{2}$ is not rational and so cannot be expressed as a ratio of integers, struck a serious blow to the Pythagorean system. The story goes that Hippasus made his discovery while on a sea voyage with his fellow Pythagoreans who were so distressed that Hippasus was promptly thrown overboard and drowned.

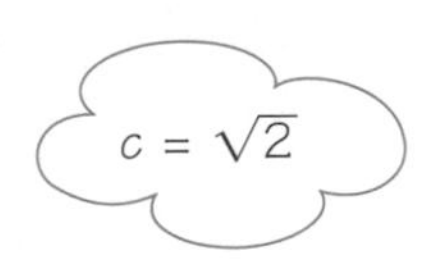

THEOREM Irrationality of $\sqrt{2}$

If $c^2 = 2$, then c is not a rational number.

PROOF (BY CONTRADICTION) Suppose that c is a rational number. Then

$$\sqrt{c} = \frac{r}{s}$$

where r and s are integers. But every rational number can be represented by a fraction in simplest form. Thus,

$$\sqrt{c} = \frac{u}{v}$$

where u and v have no common factor other than 1. But $c^2 = 2$, so

$$2 = \frac{u^2}{v^2}.$$

Thus,

$$2v^2 = u^2.$$

This says that 2 is a prime factor of u^2. Hence, by the fundamental theorem of arithmetic, 2 is a prime factor of u. But then $u = 2k$ for some integer k and

$$2v^2 = u^2 = (2k)^2 = 4k^2.$$

This implies that

$$v^2 = 2k^2$$

and so 2 is also a factor of v^2 and hence of v. But then u and v have a factor of 2 in common and u/v is not in simplest form. This contradicts the fact that *u/v is in lowest terms*. In view of this contradiction, the assumption that c was rational that started this chain of reasoning must be false. Therefore, c is irrational as was to be proved. ■

EXAMPLE 7.8

Proving Numbers Irrational

Show that $r + \sqrt{2}$ is irrational for any rational number r.

SOLUTION

Understand the problem • • • •

We have just seen that $\sqrt{2}$ is irrational; that is, is not rational, and we are told that r is rational. We must show that $r + \sqrt{2}$ is irrational.

Devise a plan • • • •

Does the assertion even make sense? Is it possible that $r + \sqrt{2}$ is rational? If so, then $r + \sqrt{2} = s$ where s is some rational number. Perhaps we can use this equation, and the fact that we already know that $\sqrt{2}$ is irrational, to arrive at the desired conclusion.

Carry out the plan • • • •

Since $r + \sqrt{2} = s$, it follows that $\sqrt{2} = s - r$. But r and s are rational and we know that the rational numbers are closed under subtraction. This implies that $\sqrt{2}$ is rational, and we know that that is not so. Therefore, the assumption that $r + \sqrt{2} = s$, where s is rational must be false. Hence, $r + \sqrt{2}$ is irrational as was to be shown.

Look back • • • •

Initially, it was not clear that the assertion of the problem had to be true, so we assumed briefly that it was not true. But this led directly to the contradiction that $\sqrt{2}$ was rational and so the assumption that r + $\sqrt{2}$ was rational had to be false; that is, $r + \sqrt{2}$ had to be irrational as we were to prove. ■

Since there are infinitely many rational numbers the preceding example shows that there are infinitely many irrational numbers. Moreover, in a very precise sense, which we will not discuss here, there are many more irrational numbers than rational numbers. One such number which is well-known but not usually known to be irrational is the number

$$\pi = 3.14159265\ldots$$

occurring in the formula $A = \pi r^2$ for the area of a circle of radius r or the formula $C = 2\pi r$ for the circumference of a circle. Moreover, if m the natural number is not a perfect square, it can be shown that $\sqrt{m}$ is also irrational. Indeed, if m is not a perfect nth power, $\sqrt[n]{m}$ is irrational, where by $\sqrt[n]{m}$ we mean a number c such that $c^n = m$.

Real Numbers and the Number Line

When we discussed the length of the diagonal of the square measuring 1 unit on a side and proved that $\sqrt{2}$ is irrational, we made a tacit assumption that is easily overlooked. Simply put, the assumption was that to every possible length there corresponds exactly one real number that names the length in question. Figure 7.3 illustrates how we can accurately plot the length $\sqrt{2}$ on the number line.

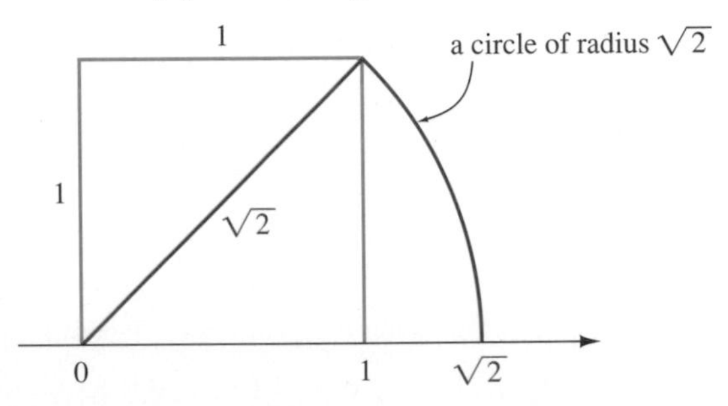

Figure 7.3
$\sqrt{2}$ on the number line

Also, the decimal expansion of $\sqrt{2}$ can be found step-by-step as follows. Using a calculator and trial and error, we find that

$$1^2 = 1 < 2 < 4 = 2^2,$$
$$1.4^2 = 1.96 < 2 < 2.25 = 1.5^2,$$
$$1.41^2 = 1.9881 < 2 < 2.0164 = 1.42^2,$$
$$1.414^2 = 1.999396 < 2 < 2.002225 = 1.415^2,$$

and so on. Thus, the decimal expansion of $\sqrt{2}$ is

$$\sqrt{2} = 1.414\ldots$$

Note that this is determined by the lower estimates above.

where successive digits are found by catching $\sqrt{2}$ between successive units, between successive tenths, between successive hundredths, and so on. To show that this can be done for any real number is somewhat tricky but the following theorem can be proved.

> **THEOREM Real Numbers and the Number Line**
>
> There is a one-to-one correspondence between the set of real numbers and the set of points on a number line. The absolute value of the number associated with any given point gives the distance of the point to the right or left of the 0.

Just as it can be shown that between any two rational numbers there is another rational number, it can also be shown that between any two irrational numbers there is an irrational number. Indeed, between any two rational numbers there is an irrational number and between any two irrational numbers there is a rational number.

This seems to suggest that the two infinite sets—the set of rational numbers and the set of irrational numbers—are about the same size. However, this is not so. As noted earlier, there are, imprecisely put, infinitely many more irrational numbers than rational numbers. Fortunately, when we compute with numbers, we use only truncated decimals to as many places as we need, so, computationally, there is no difficulty.

PROBLEM SET 7.1

Understanding Concepts

1. Write the following decimals in expanded form and in expanded exponential form.
 (a) 273.412 **(b)** 0.000723 **(c)** 0.20305
2. Write these decimals as fractions in lowest terms and determine the prime factorization of the denominator in each case.
 (a) 0.324 **(b)** 0.028 **(c)** 4.25
3. Write these fractions as terminating decimals.
 (a) $\frac{7}{20}$ **(b)** $\frac{7}{16}$ (c) $\frac{3}{75}$ **(d)** $\frac{18}{2^2 \cdot 5^4}$
4. Determine the rational fraction in lowest terms represented by each of these periodic decimals.
 (a) $0.321321\ldots = 0.\overline{321}$
 (b) $0.12414141\ldots = 0.12\overline{41}$
 (c) $3.262626\ldots = 3.\overline{26}$
 (d) $0.999\ldots = 0.\overline{9}$
 (e) $0.24999\ldots = 0.24\overline{9}$
 (f) $0.666\ldots = 0.\overline{6}$
 (g) $0.142857142857142857\ldots = 0.\overline{142857}$
 (h) $0.153846153846153846\ldots = 0.\overline{153846}$
5. (a) Give an example of a fraction whose decimal expansion is repeating and the length of the period is 3.
 (b) Use your calculator to check your answer to part (a). Is this check really complete? Explain.
 (c) Check your answer to part (a) by some other means.
6. In each of these, write the numbers in order of increasing size (from least to greatest).
 (a) $0.017, 0.007, 0.01\overline{7}, 0.027$
 (b) $25.412, 25.312, 24.999, 25.41\overline{2}$
 (c) $\frac{9}{25}, 0.35, 0.36, \frac{10}{25}, 0.\overline{35}$
 (d) $\frac{1}{4}, \frac{5}{24}, \frac{1}{3}, \frac{1}{6}$
7. Show that $\sqrt{3}$ is irrational. (*Hint:* Have you ever seen a similar result proved?)
8. Show that $3 - \sqrt{2}$ is irrational.
9. Show that $2\sqrt{2}$ is irrational.
10. Show that $1/\sqrt{2}$ is irrational.

Thinking Critically

11. If the fraction a/b is in simplest form with $0 < a < b$ and $b = 2 \cdot 5^4$, how many zeros appear to the right of the decimal point in the decimal expansion of a/b? Explain briefly.
12. In problem 4, parts (d) and (e) you may have been surprised to discover that some rational numbers have two *different* decimal representations.
 (a) What rational number is represented by the periodic decimal $0.0999\ldots = 0.0\overline{9}$?
 (b) Write two other decimals that are different but represent the same rational number.
 (c) What rational numbers possess two different decimal expansions? Explain briefly.
13. Compute the decimal expansions of each of the following.
 (a) $\frac{1}{11}$ **(b)** $\frac{1}{111}$ (c) $\frac{1}{1111}$
 (d) Guess the decimal expansion of $1/11111$.
 (e) Check your guess in part (d) by converting the decimal of your guess back into a ratio of two integers.
 (f) Carefully describe the decimal expansion of $1/N_n$ where $N_n = 111\ldots1$ with n 1s in its representation.
14. What rational number corresponds to each of these periodic decimals? (Refer to Example 7.5 if you need help.)
 (a) $0.747474\ldots = 0.\overline{74}$ **(b)** $0.777\ldots = 0.\overline{7}$
 (c) $0.235235235\ldots = 0.\overline{235}$
 (d) If a, b, and c are digits, what rational numbers are represented by each of these periodic decimals? You should be able to guess these results on the basis of patterns observed in parts (a), (b), and (c).
 (i) $0.aaa\ldots = 0.\overline{a}$
 (ii) $0.ababab\ldots = 0.\overline{ab}$
 (iii) $0.abcabcabc\ldots = 0.\overline{abc}$

15. Write the decimal representing each of these rational numbers without doing any calculation or at most doing only mental calculation.

(a) $\frac{5}{9}$ **(b)** $\frac{22}{99}$ **(c)** $\frac{317}{999}$

(d) $\frac{17}{33}$ **(e)** $\frac{14}{11}$

(f) Check the answers in parts (a) through (e) using your calculator. Is this check foolproof? Explain.

Thinking Cooperatively

16. (a) Give an example that shows that the sum of two irrational numbers is sometimes rational.

(b) Give an example that shows that the sum of two irrational numbers is sometimes irrational.

17. Give an example that shows that the product of two irrational numbers is sometimes rational.

18. Is $2/\sqrt{2}$ rational or irrational? Explain.

19. Give an example that shows that the quotient of two irrational numbers is sometimes rational.

20. What real numbers are represented by points A, B, and C in this diagram? Explain briefly.

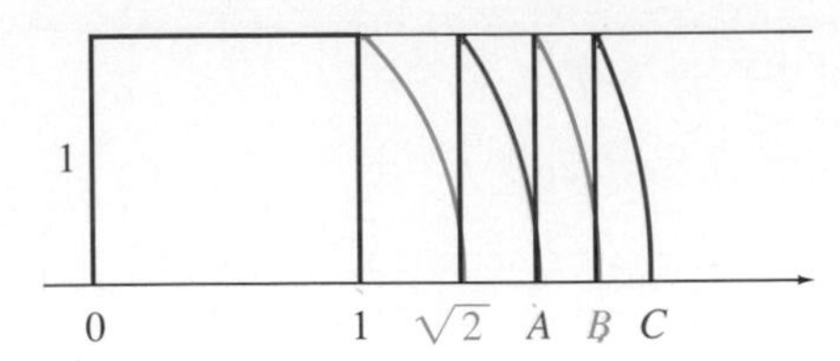

21. Find a rational number between the irrational numbers π and $2 + \sqrt{2}$.

---

22. **(a)** If a is an integer, what are the possibilities for the last digit of the decimal representation of a^2?

Last digit of a	0	1	2	3	4	5	6	7	8	9
Last digit of a^2										

(b) If b is an integer, what are the possibilities for the last digit of the decimal representation of $2b^2$?

Last digit of b	0	1	2	3	4	5	6	7	8	9
Last digit of $2b^2$										

(c) Using the results of parts (a) and (b), make a careful argument that $\sqrt{2} = a/b$, with a and b integers, is impossible.

Communicating

23. Write a proof that $\sqrt{6}$ is irrational.

24. If r is rational and $r \neq 0$, write a proof that $r \cdot \sqrt{2}$ is irrational.

Using A Calculator

25. (a) Choose any two positive numbers a and b and compute the first 20 terms of the sequence

$$f_1 = a, f_2 = b, f_3 = a + b, f_4 = a + 2b, \ldots,$$

where every term in the sequence after the first two is the sum of its two predecessors. (*Hint:* Use a calculator algorithm as in Chapter 3.)

(b) Compute the decimal expansion of each of

$$\frac{f_2}{f_1}, \frac{f_3}{f_2}, \frac{f_4}{f_3}, \ldots, \frac{f_{20}}{f_{19}}.$$

(c) Examine the decimal expansions of part (b). What appears to be the case? Explain.

Compare your results with a friend in the class who started with different choices for a and b.

(d) Compute the decimal expansion of $(1 + \sqrt{5})/2$. Comparing this to the results of part (b), what seems to be the case? Explain.

For Review

26. Arrange these fractions in order from least to greatest: $3/5$, $3/4$, $-2/3$, $27/29$, $7/8$, $1/2$

27. The *mediant* of a/b and c/d is the fraction $(a + c)/(b + d)$. Compute the mediant of the following and then plot each of the three numbers on a number line.

(a) $\frac{3}{8}$ and $\frac{17}{24}$ **(b)** $\frac{1}{3}$ and $\frac{11}{12}$

(c) Guess a general result on the basis of parts (a) and (b).

(d) Give a convincing argument that your guess in part (c) is correct for natural numbers a, b, c, and d.

28. Perform the following computations leaving all answers as fractions in reduced form.

(a) $\frac{1}{2}+\frac{2}{3}$ (b) $\frac{1}{2}-\frac{2}{3}$

(c) $\frac{1}{2}\cdot\frac{2}{3}$ (d) $\frac{1}{2}\div\frac{2}{3}$

(e) $\frac{3}{4}\cdot\left(1+\frac{3}{5}\right)$ (f) $\frac{3}{4}\div\left(1-\frac{3}{5}\right)$

29. Find two rational numbers between 1/2 and 1/3.

30. In a certain population, 2/3 of the men are married and 1/2 of the women are married. What fraction of the adult population is unmarried?

7.2 Computations with Decimals

Computations with decimals are similar to computations with integers except for the complication of dealing with the decimal point.

Rounding Decimals

In practical situations, we are almost never interested in all the digits in decimals like 2.3254071 or 0.0015243; the accuracy they suggest is usually not warranted and not useful. As with integers with many digits we almost invariably round such decimals to fewer places. Moreover, the process of rounding here is precisely the same as it was for integers. Thus, 2.3254071 rounded

- to the nearest integer is 2,
- to the nearest tenth is 2.3,
- to the nearest hundredth is 2.33,

and so on. As before, we are using the 5-up rule.

> **RULE The 5-up Rule for Rounding Decimals**
>
> To round a decimal to a given place, consider the digit in the next place to the right. If this digit is less than 5, leave the digit in the place under consideration unchanged and replace all digits to its right by 0s. If the digit to the right of the place in question is 5 or more, increase the digit in the place under consideration by 1 and replace all digits to the right by 0s.

EXAMPLE 7.9 Rounding Decimals

Round each of these to the indicated position.

(a) 23.2047 to the nearest integer

(b) 3.6147 to the nearest tenth.

(c) 0.015 to the nearest hundredth.

SOLUTION

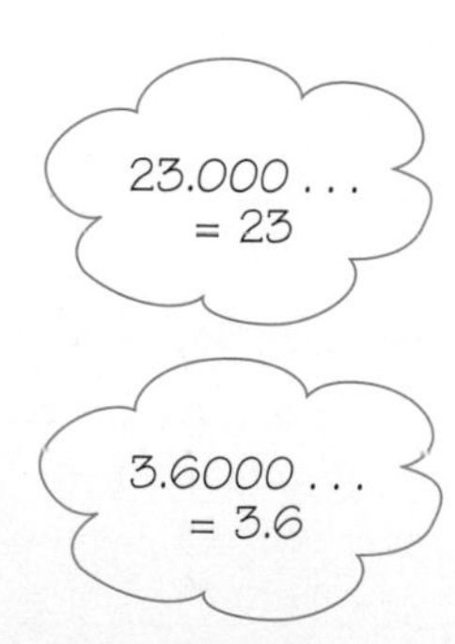

(a) Since we are asked to round 23.2047 to the nearest integer, we consider the digit to the right of 3. Since this digit is 2 and $2 < 5$, we leave the 3 unchanged and replace the digits to its right by 0s. Thus, 23.2047 rounded to the nearest integer is 23.

(b) In rounding 3.6147 to the nearest tenth, we note that the digit to the right of the 6 is 1. Since $1 < 5$, we leave the 6 unchanged and replace the digits to its right by 0s. Thus, 3.6147 rounded to the nearest tenth is 3.6.

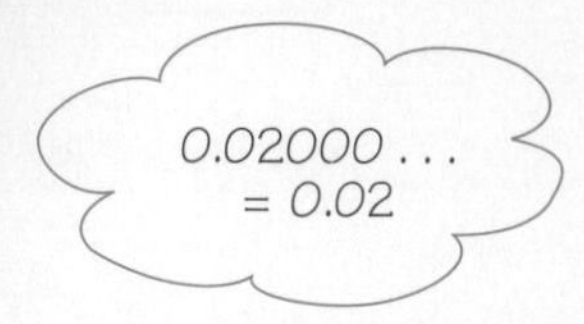

(c) Here the digit in question is 1 and the digit to its right is 5. Thus, we increase the 1 by 1 and replace the digits to its right by 0s. Hence, 0.015 rounded by the nearest hundredth is 0.02. ■

Adding and Subtracting Decimals

Suppose we wish to add 2.71 and 32.762. Converting these decimals to fractions we have

$$2.71 = \frac{271}{100} = \frac{2710}{1000} \quad \text{and} \quad 37.762 = \frac{37{,}762}{1000}.$$

Thus,

$$\begin{aligned} 2.71 + 37.762 &= \frac{2710}{1000} + \frac{37{,}762}{1000} \\ &= \frac{40{,}472}{1000} \\ &= 40.472. \end{aligned}$$

$$\begin{array}{r} 2710 \\ +\ 37{,}762 \\ \hline 40{,}472 \end{array} \qquad \begin{array}{r} 02.710 \\ +\ 37.762 \\ \hline 40.472 \end{array} \qquad \begin{array}{r} 2.71 \\ +\ 37.762 \\ \hline 40.472 \end{array}$$

To add decimals by hand, write the numbers in vertical style lining up the decimal points and then add essentially just as we add integers. With a calculator, we enter the string

[ON/AC] 2.71 [+] 37.762 [=]

and automatically obtain the desired sum 40.472. The important thing with a calculator is to recognize that we are adding approximately 3 to approximately 38 so the sum should be approximately 41. The estimation and mental arithmetic should proceed right along with the calculator manipulation to be sure that we recognize if we have inadvertently made a calculator error.

HIGHLIGHT FROM HISTORY
The Decimal Point

The development of mathematics over time has been paralleled by the development of good mathematical notation that facilitates doing mathematics. One of these developments was that of the Hindu-Arabic or decimal system of notation as already observed. Related to this is the use of a period (the decimal point) to separate the integer part and the fractional part of a numeral. Thus,

$$25.423 = 25\frac{423}{1000}.$$

As we have seen, use of the decimal point makes possible the easy extension of the algorithms for integer computation to decimal computation. Though it seems natural to us now, the decimal point was developed rather late in the history of mathematics and even today its use is not completely standardized. To illustrate the development, we list here various notations of the past and present.

3.4813	modern American
3 · 4813	modern English
3,4813	modern continental Europe
3|4813	Rudolph, 1530
34813	F. Vieta, 1579
3 ⓪ 4 ① 8 ② 1 ③ 3 ④	S. Stevin, 1585
$\overset{o}{3} . \overset{i}{4} \overset{ii}{8} \overset{iii}{1} \overset{iv}{3}$	J. Beyer, 1603
3 4813	J. Beyer, 1603
$3^{(0)}4^{(1)}8^{(2)}1^{(3)}3^{(4)}$	R. Norton, 1608
34813 ④	W. Kalcheim, 1629

Suppose now that we want to subtract 2.71 from 37.762. The calculation is much the same as above and we have

$$\begin{aligned} 37.762 - 2.71 &= \frac{37{,}762}{1000} - \frac{2710}{1000} \\ &= \frac{35{,}052}{1000} \\ &= 35.052 \end{aligned}$$

37,762
− 2710
35,052

37.762
− 2.710
35.052

As before, in hand calculation, we write the problem in vertical style lining up the decimal points and then subtract essentially as we subtract integers. Again, however, this is more easily handled on a calculator by entering the string

$$\boxed{\text{ON/AC}}\ 37.762\ \boxed{-}\ 2.71\ \boxed{=}$$

to obtain 35.052 automatically. Again, we should estimate and perform mental calculation as we manipulate the calculator. Thus, we think

approximately 38 minus approximately 3 gives approximately 35

and thus avoid gross calculator errors.

EXAMPLE 7.10

Adding and Subtracting Decimals

Make a mental calculation to determine an approximate answer and then use your calculator to determine the accurate result of each of these calculations.

(a) 23.47 + 7.81 **(b)** 351.42 − 417.815

SOLUTION

(a) Approximately 23 plus approximately 8 should give approximately 31.

$$\boxed{\text{ON/AC}}\ 23.47\ \boxed{+}\ 7.81\ \boxed{=}\ 31.28$$

(b) Approximately 350 minus approximately 400 should give approximately −50.

$$\boxed{\text{ON/AC}}\ 351.42\ \boxed{-}\ 417.815\ \boxed{=}\ -66.395$$

■

Multiplying Decimals

Suppose we wish to compute the product (31.76) · (4.6). Converting these decimals to fractions, we have

$$31.76 = \frac{3176}{100} \quad \text{and} \quad 4.6 = \frac{46}{10}.$$

Thus,

$$\begin{aligned} (31.76) \cdot (4.6) &= \frac{3176}{100} \cdot \frac{46}{10} \\ &= \frac{3176 \cdot 46}{100 \cdot 10} \\ &= \frac{146096}{1000} \\ &= 146.096 \end{aligned}$$

3176
× 46
19053
12704
146,096

three digits to the right of the decimal point since we are dividing by 1000 = 10^3

Or, by hand,

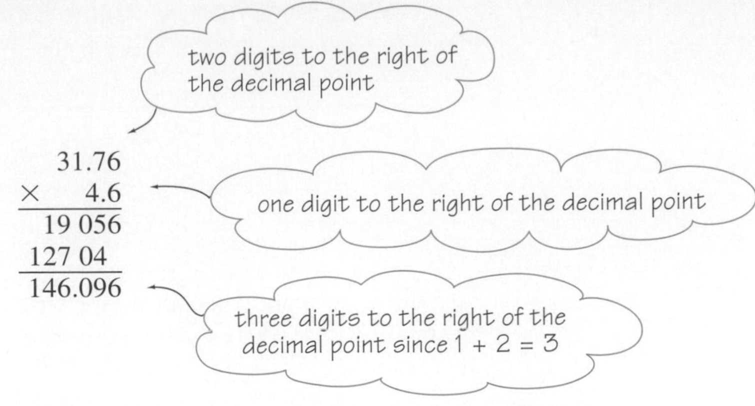
two digits to the right of the decimal point
31.76
× 4.6
19 056
127 04
146.096
one digit to the right of the decimal point
three digits to the right of the decimal point since 1 + 2 = 3

Since this is typical of the general case, we summarize it as a theorem.

THEOREM Multiplying Decimals

To multiply two decimals:

1. Multiply as with integers.
2. Count the number of digits to the right of the decimal point in each number in the product, add these numbers, and call their sum t.
3. Finally, place the decimal point in the product obtained so that there are t digits to the right of the decimal point.

As before the easier and more reasonable thing to do in computing the above product is to use your calculator. Entering the string

[ON/AC] 31.76 [×] 4.6 [=]

yields the answer 146.096 automatically. One should also work mentally to avoid a gross error.

Think
approximately 30 times approximately 5 gives approximately 150

As a further quick check, one can observe that each number in the product ends in a 6, and, since $6 \cdot 6 = 36$, the product must end in a 6. Thus, 146.096 is very likely correct.

EXAMPLE 7.11 **Multiplying Decimals**

Compute these products by estimating, by calculator, and by hand.

(a) $(471.2) \cdot (2.3)$ **(b)** $(36.34) \cdot (1.02)$

SOLUTION

(a) By estimating: Approximately 500 times approximately 2 gives approximately 1000.

By calculator: [ON/AC] 471.2 [×] 2.3 [=] 1083.76

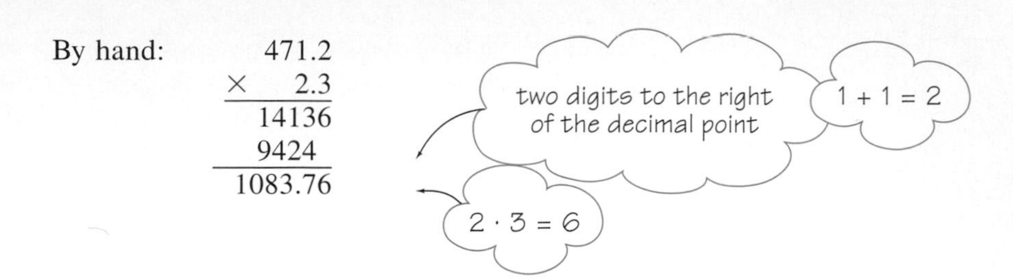

(b) By estimating: Approximately 40 times approximately 1 gives approximately 40.

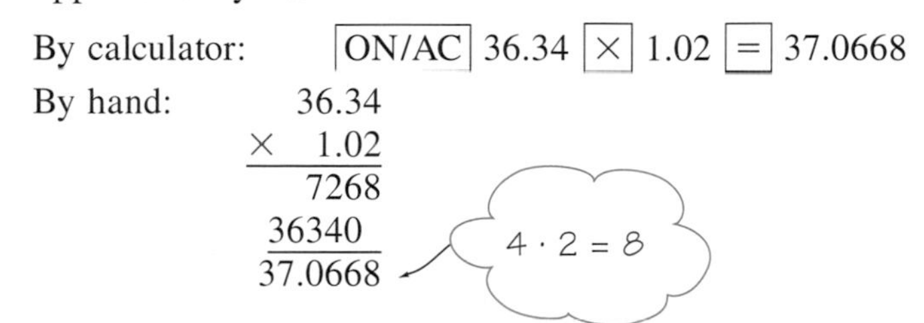

By calculator: ON/AC 36.34 × 1.02 = 37.0668
By hand:
36.34
× 1.02
7268
36340
37.0668
4 · 2 = 8

■

Dividing Decimals

Suppose we want to divide 537.6 by 2.56. Converting these decimals to fractions, we have

$$
\begin{aligned}
537.6 \div 2.56 &= \frac{5376}{10} \div \frac{256}{100} \\
&= \frac{5376}{10} \cdot \frac{100}{256} \\
&= \frac{537{,}600}{2560} \\
&= \frac{53{,}760}{256}
\end{aligned}
$$

and the problem is reduced to that of dividing 53,760 by 256; that is, to dividing integers. Recall that, when confronted by a division like

$$2.56\overline{)537.6},$$

students are often told to "move the decimal point in both the divisor and the dividend 2 places to the right so that the divisor becomes an integer." The preceding calculation with fractions justifies this rule and, by hand, we have

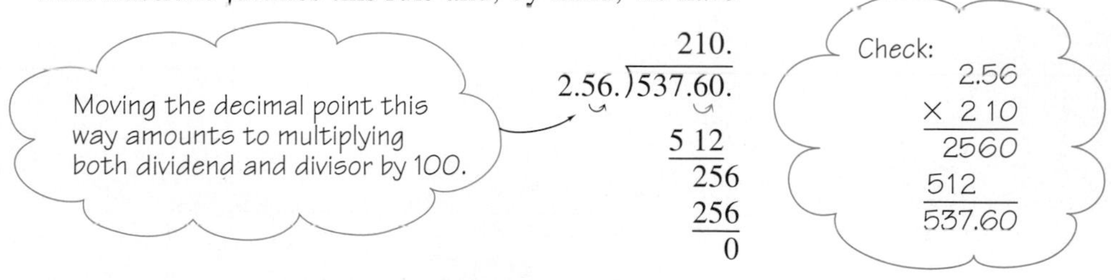

Moving the decimal point this way amounts to multiplying both dividend and divisor by 100.
210.
2.56.)537.60.
5 12
256
256
0
Check:
2.56
× 2 10
2560
512
537.60

An easier approach is to use a calculator while doing some mental calculation to see what answer is reasonable. Entering

ON/AC 537.6 ÷ 2.56 =

into the calculator, we immediately obtain 210 as the answer, and this is probably correct since we are dividing approximately 500 by approximately 2.

Earlier we claimed that the decimal expansion of a reduced fraction a/b where b has prime factors other than 2 and 5, is necessarily nonterminating but repeating. We are now in a position to give a justification and we do so by example.

EXAMPLE 7.12

The Decimal Expansion of 3/7

Use long division to find the decimal expansion of 3/7.

SOLUTION

Since $3/7 = 3 \div 7$, we obtain the desired decimal expansion by dividing 3 by 7. We have

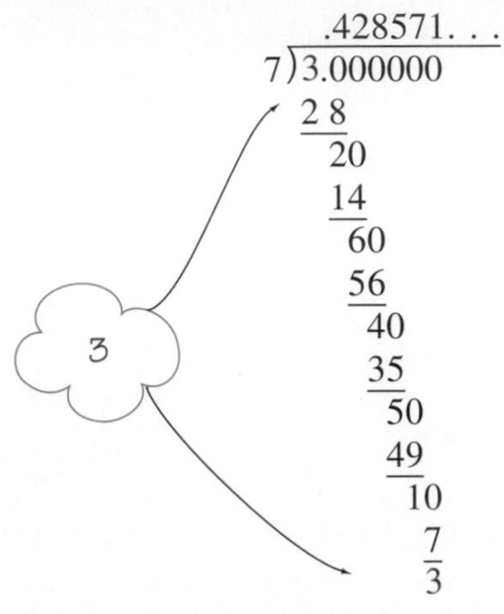

and, since the remainder at this stage is 3, the number with which we began, the division will continue *ad infinitum* with this repeating pattern. Thus,

$$\frac{3}{7} = 0.428571428571428571\ldots = 0.\overline{428571}.$$

Moreover, since we are dividing by 7, the only possible remainders are 0, 1, 2, 3, 4, 5, and 6. Thus, if the division does not terminate, it must repeat after at most 6 steps. This is true in general and so essentially demonstrates the theorem mentioned.

As a check of the preceding result, we compute the rational number determined by the decimal

$$x = 0.428571428571428571\ldots = 0.\overline{428571}.$$

It follows that

$$1{,}000{,}000x - x = 999{,}999x = 428{,}571.$$

So

$$x = \frac{428{,}571}{999{,}999} = \frac{3}{7}$$

as claimed.

Divide both numerator and denominator by 142,857.

■

JUST FOR FUN

The Decimal Expansion of 1/89

Recall that the Fibonacci numbers are the numbers 0, 1, 1, 2, 3, 5, 8, 13, 21, 34, 55, 89, 144, 233, ... Computing the decimal expansion of 1/89, we obtain

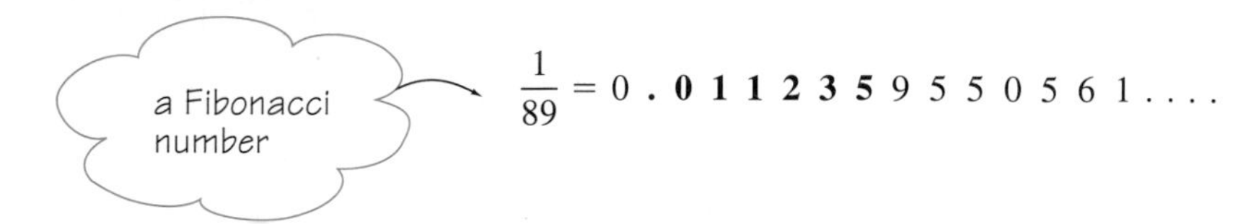

a Fibonacci number
1/89 = 0 . 0 1 1 2 3 5 9 5 5 0 5 6 1 . . . .

After starting with the digits 0, 1, 1, 2, 3, 5, we expect the next digit to be 8, the next Fibonacci number. But it is 9. Have we made an error? No. But all is not lost. By hand, compute this sum.

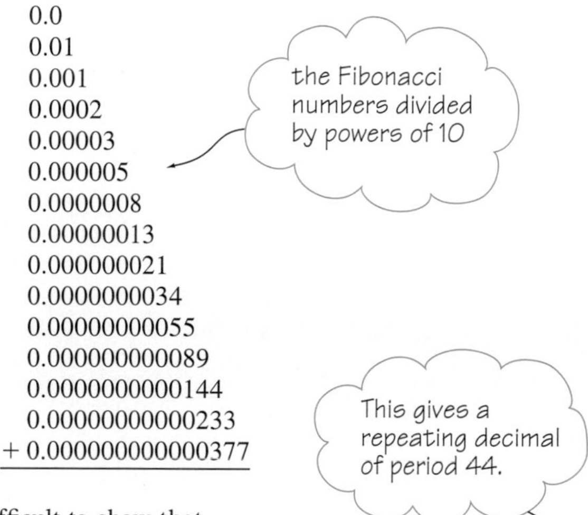

0.0
0.01
0.001
0.0002
0.00003
0.000005
0.0000008
0.00000013
0.000000021
0.0000000034
0.00000000055
0.000000000089
0.0000000000144
0.00000000000233
+ 0.000000000000377
the Fibonacci numbers divided by powers of 10
This gives a repeating decimal of period 44.

Surprised?! In fact, it is not too difficult to show that

$$\frac{1}{89} = \frac{0}{10^1} + \frac{1}{10^2} + \frac{1}{10^3} + \frac{2}{10^4} + \frac{3}{10^5} + \frac{5}{10^6} + \frac{8}{10^7} + \frac{13}{10^8} + \cdots + \frac{F_n}{10^{n+1}} + \cdots$$

just as the nonterminating sum

$$\frac{3}{10^1} + \frac{3}{10^2} + \frac{3}{10^3} + \frac{3}{10^4} + \cdots = .3333\ldots = \frac{1}{3}.$$

Scientific Notation

Some products and quotients cannot be calculated directly on a nonscientific calculator because the results are either too large or too small. For example, attempting to calculate either

$$876592 \cdot 7654 \qquad \text{or} \qquad 0.0011 \div 65536$$

results in an error message. On scientific calculators, the display for these two calculations shows something like

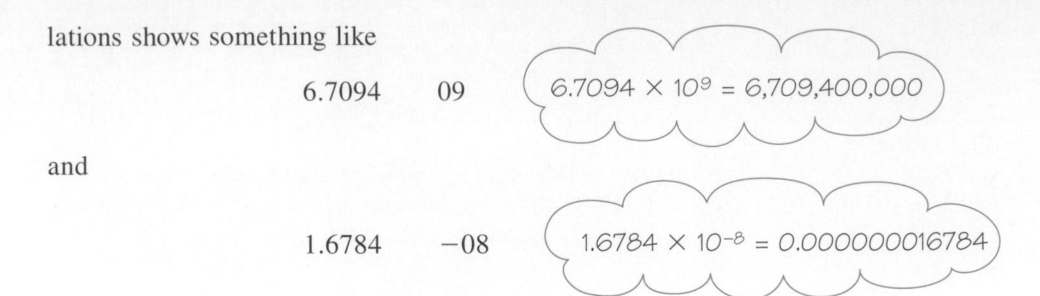

respectively and these are understood to mean

$$6.7094 \times 10^9 \qquad \text{and} \qquad 1.6784 \times 10^{-8}.$$

What these calculators are doing is giving an **approximate** answer in each case, using a form of what is called **scientific notation.*** That is, they give the correct first few digits of the answer in each case as a number between 1 and 10 and then tell what power of 10 to multiply this by, to correctly place the decimal point.

In computing answers like

$$6709435168 \qquad \text{and} \qquad 0.000000016784667968 75\ldots,$$

it is highly unlikely that we are interested in, or, because of inaccuracy of measurements, for example, that we are very sure of the last several digits of each result. Thus, the first answer is much more informative if we understand it as essentially 6.7 billion or as 6.7×10^9 and the second answer is more readily appreciated as 1.7×10^{-8}. We usually *round off* the number either to the number of digits we are interested in or the number of digits we are sure of and then use scientific notation. The digits we write before writing the power of ten are called **significant digits.** Thus, we might write the first answer above as

$$6.7 \times 10^9 \qquad \text{to 2 significant digits}$$

or

$$6.7094 \times 10^9 \qquad \text{to 5 significant digits.}$$

Similarly, the second answer would be written as

$$1.7 \times 10^{-8} \qquad \text{to 2 significant digits}$$

and

$$1.6784 \times 10^{-8} \qquad \text{to 5 significant digits.}$$

DEFINITION Scientific Notation

To write a number in **scientific notation,** write it as the product of a number whose absolute value is greater than or equal to 1 and less than 10 times the appropriate power of 10 to correctly place the decimal point. The digits in the number multiplying the power of 10 are called **significant digits.**

* The actual answers respectively are 6709435168 and 0.0000000167846679687 5 . . . .

EXAMPLE 7.13

Writing Numbers in Scientific Notation

Write each of these in scientific notation using the number of significant digits indicated.

(a) 93000000 using 2 significant digits
(b) 93000000 using 3 significant digits
(c) 0.000027841 using 2 significant digits
(d) 0.000027841 using 3 significant digits

SOLUTION

(a) First write $93000000 = 9.3000000 \times 10^7$. Then round this to 9.3×10^7 to obtain the answer to 2 significant digits.
(b) This is the same as part (a) and we proceed as before except that we round the answer to 9.30×10^7 to obtain 3 significant digits.
(c) Here $0.000027841 = 2.7841 \times 10^{-5}$ and this is rounded to 2.8×10^{-5} to obtain the answer correct to 2 significant digits.
(d) This is the same as part (c) but 2.7841×10^{-5} is rounded to 2.78×10^{-5} to obtain the answer correct to 3 significant digits. ■

Some calculators have a key marked [EXP] or [E] or [EE], and one can calculate in exponential notation using this key. For example, to calculate the product

$$(8.77 \times 10^5) \cdot (7.65 \times 10^3)$$

and write the answer to 3 significant figures, enter the string

[ON/AC] 8.77 [EXP] 5 [×] 7.65 [EXP] 3 [=].

In the display we read something like 6.709 09 and write this to three significant figures as $6.71 \cdot 10^9$.

EXAMPLE 7.14

Calculating Using Scientific Notation

Compute each of these using scientific notation to 3 significant figures.

(a) $(2.47 \times 10^{-5}) \cdot (8.15 \times 10^{-9})$
(b) $(2.47 \times 10^{-5}) \div (8.15 \times 10^{-9})$

SOLUTION

(a) Enter [ON/AC] 2.47 [EXP] 5 [+⇄−] [×] 8.15 [EXP] 9 [+⇄−] [=]. Read 2.013 − 13 in the display and record $2.01 \cdot 10^{-13}$.
(b) Enter [ON/AC] 2.47 [EXP] 5 [+⇄−] [÷] 8.15 [EXP] 9 [+⇄−] [=]. Read 3030.6748 in the display and record 3.03×10^3. ■

Note too, how the use of scientific notation can assist in mental approximation of calculator answers. For example, consider Example 7.14, part (a). Using only left digit approximations and rules for multiplication of exponents, think

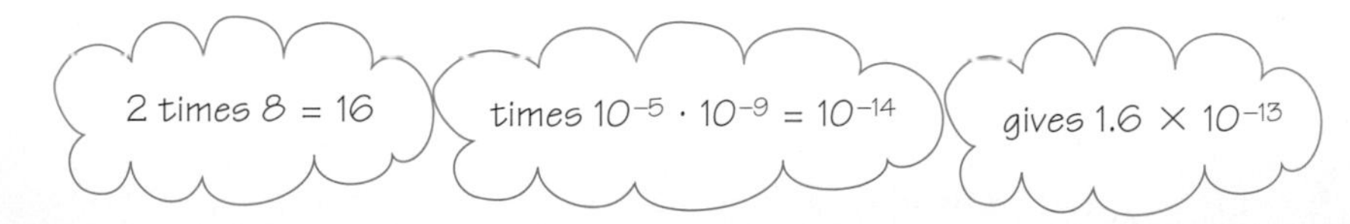
2 times 8 = 16
times 10⁻⁵ · 10⁻⁹ = 10⁻¹⁴
gives 1.6 × 10⁻¹³

so the above answer is probably correct. An even closer approximation can be obtained by using a 2-digit approximation on the first number and a one digit approximation on the second. Think

and the approximation is quite close. Similarly, if we use the same approximation on part (b), we think

2.4 divided by 8 = 0.3 = 3 · 10^{-1}

times $10^{-5} \cdot 10^9 = 10^4$

gives 3 × 10^3

and the approximation is again quite good.

JUST FOR FUN

The Decimal Expansion of 199/9899

Recall that if we define $L_0 = 2$, then the Lucas numbers are 2, 1, 3, 4, 7, 11, 18, 29, 47, 76, 123, 199, 322, 521, 843, 1364, . . . where we add any two consecutive numbers to obtain the next number in the sequence. Note in particular that $L_{11} = 199$ and $19 \cdot L_{13} = 9899$. Computing the decimal of 199/9899, we obtain

$$\frac{199}{9899} = \mathbf{0.020103040711182947}7774674 \cdots .$$

After the decimal starts with the numbers 02, 01, 03, 04, 07, 11, 18, 29, and 47, we expect the next two digits to name the next number, 76. But it is 77. Have we made an error? No. But all is not lost. By hand, compute this sum.

This gives a repeating decimal of period 63.

$$\begin{array}{r@{}l} & 0.02 \\ & 0.0001 \\ & 0.000003 \\ & 0.00000004 \\ & 0.0000000007 \\ & 0.000000000011 \\ & 0.00000000000018 \\ & 0.0000000000000029 \\ & 0.000000000000000047 \\ & 0.00000000000000000076 \\ & 0.0000000000000000000123 \\ & 0.000000000000000000000199 \\ & 0.00000000000000000000000322 \\ + & 0.0000000000000000000000000521 \\ \hline \end{array}$$

Surprised? In fact, it is not too difficult to show that

$$\frac{199}{9899} = \frac{2}{100} + \frac{1}{100^2} + \frac{3}{100^3} + \frac{4}{100^4} + \frac{7}{100^5} + \frac{11}{100^6} + \frac{18}{100^8} + \cdots + \frac{L_{n-1}}{100^n} + \cdots$$

just as the nonterminating sum

$$\frac{33}{100} + \frac{33}{100^2} + \frac{33}{100^4} + \frac{33}{100^6} + \cdots = 0.\overline{3} = \frac{1}{3}.$$

PROBLEM SET 7.2

Understanding Concepts

1. Perform these additions and subtractions by hand.
   (a) 32.174 + 371.5 **(b)** 371.5 − 32.174
   (c) 0.057 + 1.08 **(d)** 0.057 − 1.08
2. Perform these multiplications and divisions by hand.
   (a) (37.1) · (4.7) **(b)** (3.71) · (0.47)
   (c) 14.664 ÷ 4.7 **(d)** 1466.4 ÷ 47
3. Estimate the result of each of these computations mentally and then perform the calculations using a calculator.
   (a) 4.112 + 31.3 **(b)** 31.3 − 4.112
   (c) (4.112) · (31.3) **(d)** 31.3 ÷ 4.112
4. By long division determine the decimal expansion of each of these fractions.
   (a) 5/6 **(b)** 3/11 **(c)** 2/27
5. Write each of these in scientific notation with the indicated number of significant digits.
   (a) 276543421 to 3 significant digits
   (b) 0.000005341 to 2 significant digits
   (c) 376712.543248 to 2 significant digits
6. Calculate each of these with a suitable calculator and write the answer to 3 significant digits. Note that the *Math Explorer* will simply give an error message.
   (a) 0.0000127 × 0.000008235
   (b) 98613428 × 5746312
   (c) 0.0000127 ÷ 98613428
   (d) 98613428 ÷ 0.000008234
7. For each of these, estimate the answer and then calculate the result to 3 significant digits on a suitable calculator.
   (a) $(7.123 \times 10^5) \cdot (2.142 \times 10^4)$
   (b) $(7.123 \times 10^5) \div (2.142 \times 10^4)$
   (c) $(7.123 \times 10^5) \cdot (2.142 \times 10^{-9})$
   (d) $(7.123 \times 10^{-2}) \div (2.142 \times 10^8)$

Thinking Critically

8. Use these numbers to make a magic square. See problem 6 in Problem Set 1.1.
   0.123, 0.246, 0.369, 0.492, 0.615, 0.738, 0.861, 0.984, 1.107
9. Use the numbers in problem 8 to form a magic subtraction square. See problem 6 in Problem Set 1.1.
10. Use these numbers to make a magic square.
    7.02, 16.38, 11.70, 18.72, 2.34, 9.36, 4.68, 21.06, 14.04
11. The sum of the numbers in any two adjacent blanks is the number immediately below and between these two numbers. Complete each of these so that the same pattern holds. The first one has been completed for you.
    (a) 2.107 1.3 4.26
    3.407 5.568
    8.967
    (b) 21.06 3.21 ______
    ______ 5.00
    ______
    (c) ______ 0.041 ______
    2.415 ______
    7.723
    (d) ______ 1.414 ______
    ______ ______
    3.142
    (e) Can any of these be completed in more than one way? Explain.
12. Fill in the blanks so that each of these is an arithmetic progression.
    (a) 3.4, 4.3, 5.2, ______, ______, ______
    (b) −31.56, ______, −21.10, ______, ______, ______
    (c) 0.0114, ______, ______, 0.3204, ______, ______
    (d) 1.07, ______, ______, ______, 8.78, ______
13. Fill in the blanks so that each of these is a geometric progression.
    (a) 2.11, 2.327, ______, ______, ______
    (b) 35.1, ______, 2.835, ______, ______
    (c) 6.01, ______, ______, 0.75125, ______

Thinking Cooperatively

14. The numbers 3.447, 2.821, 5.764, 3.2, 2.351, 3.001, and 4.2444 are placed in a circle in some order. Show that the sum of some three consecutive numbers must exceed 10.6. (*Hint:* Have you ever seen a similar problem before?)

15. Place numbers in the circles in these diagrams so that the numbers in the large circles are the sums of the numbers in the two adjacent smaller circles.

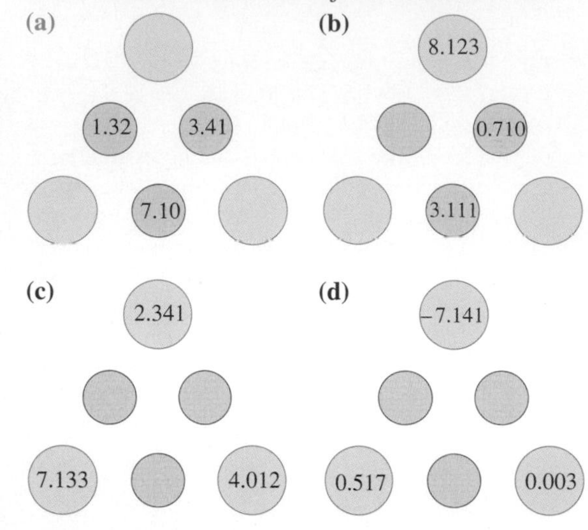

16. If possible, place numbers in the circles in these diagrams so that the numbers in the large circles exactly equal the sum of the numbers in the adjacent small circles.

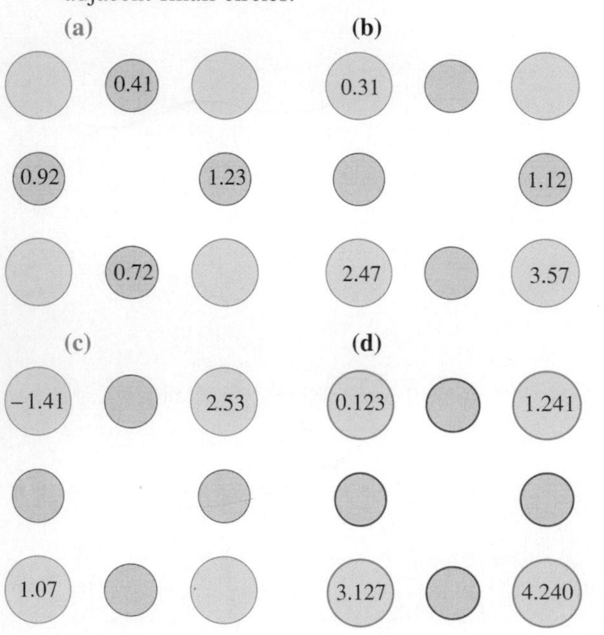

(e) What must be the case for these problems to be solvable? Explain.

(f) Is there more than one solution to all or any of these?

Making Connections

17. Kristina bought pairs of gloves as Christmas presents for 3 of her best friends. If the gloves cost \$9.72 a pair, how much did she spend for these presents?

18. Yolanda also bought identical pairs of gloves for each one of her four best friends. If her total bill was \$44.92, how much did each pair of gloves cost?

19. Dante cashed a check for \$74.29 and then bought a walkman for \$42.91 and a special pair of ear phones for \$17.02. After paying for his purchases, how much did he have left?

20. A picture frame 2.25 inches wide surrounds a picture 17.5 inches wide by 24.75 inches high.

(a) What is the area of the picture frame?

(b) What is the area of the picture?

21. What is the area of a square if it measures $\sqrt{3}$ inches on each side?

Communicating

22. In Chapter 3 we saw how to use positional notation to bases other than ten to represent integers. The same ideas can be used to represent rational and real numbers. Write a two or three page paper explaining how this would work. Be sure to include examples.

Using A Calculator

23. Among the cryptic notes of the Indian mathematician Srinivasa Ramanujan is the equation

$$\pi^4 = 97.409091 \cdots.$$

This suggests that $\pi^4 \doteq 97.4\overline{09}$.

(a) Show that $97.4\overline{09} = 97\frac{1}{2} - \frac{1}{11}$.

(b) Use your calculator to calculate $\pi^4 - \left(97\frac{1}{2} - \frac{1}{11}\right)$ and determine just how good Ramanujan's approximation is. (If you use a *Math Explorer,* how do you interpret its display?)

Using A Computer

24. (a) In Chapter 2 you were introduced to the process called DIFFY. Use DIFFY to complete this array. Remember that the first, second, and third circles in any row contain the differences (greatest minus least) of the numbers in the preceding row and the fourth circle contains the difference of the elements in the first and fourth circles of the preceding row.

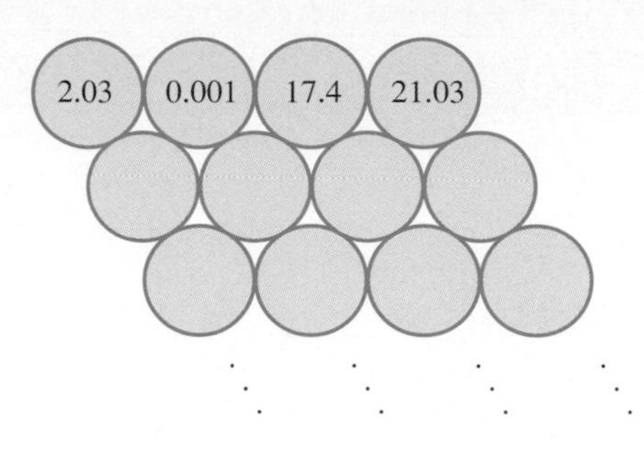

(b) Choose four more numbers and complete a second Diffy array.

(c) Do you think the process will always terminate?

(d) Can you find four numbers that cause the process to continue for at least eight steps?

(e) Use 17.34, 31.62, 58.14, 107.1 and complete a Diffy process. How many steps does the process continue?

(f) Use 1, 1.839286, 3.382973, and 6.222549 and complete the process using DIFFY.

(g) Does the result of part (f) affect your belief in your answer to part (c)? Explain.

25. (a) For this problem use DIVVY to complete this array.

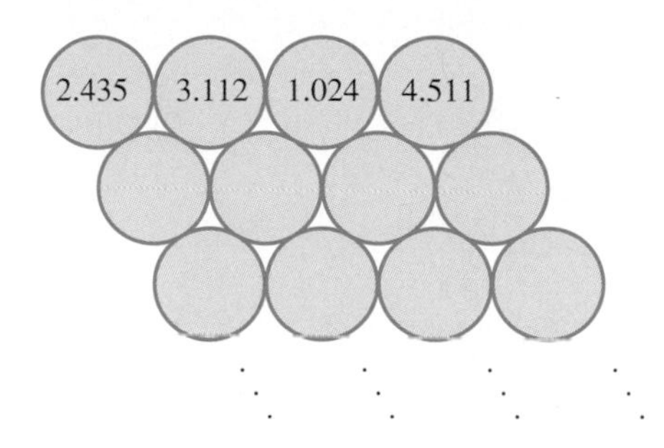

(b) Choose four positive real numbers and use DIVVY to complete the Divvy process.

(c) Do you think the process will always terminate? Explain.

(d) Use 10, 69.07, 2415.3, and 1669513 and complete the process using DIVVY.

(e) How does the result of part (d) affect your confidence in your guess in part (c)?

For Review

26. Write each of these in Roman numerals.
(a) 24 (b) 219 (c) 1935

27. Write each of these as base 10 numerals.
(a) XXIX (b) MCMXXXII (c) DCLXXXVI

28. Write each of these in Mayan notation.
(a) 231 (b) 15,278 (c) 7142

29. Write each of these (now written in Mayan notation) in base ten notation.

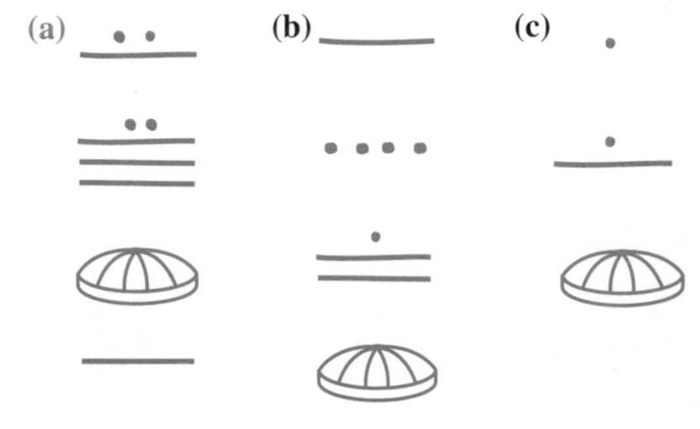

30. Write each of these using Egyptian numerals.
(a) 547 (b) 2486 (c) 854

31. Convert each of these Egyptian numerals to their base ten equivalents.

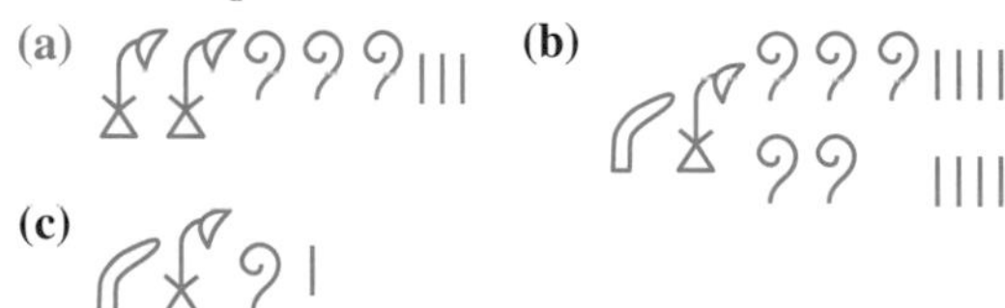

7.3 Ratio and Proportion

At basketball practice, Caralee missed 18 free throws out of 45 attempts. Since she made 27 free throws, we say that the **ratio** of the number missed to the number made was 18 to 27. This can be expressed by the fraction 18/27 or, somewhat archaically, by the notation 18 : 27. We will always use the fraction notation in what follows.

Other ratios from Caralee's basketball practice are:

- the ratio of the number of shots made to the number attempted—27/45,
- the ratio of the number of shots missed to the number attempted—18/45,
- the ratio of the number of shots made to the number missed—27/18.

DEFINITION Ratio

If a and b are real numbers with $b \neq 0$, the **ratio of a to b** is the quotient a/b.

Ratios occur with great frequency in everyday life. If you use 10.4 gallons of gasoline in driving 400.4 miles, the efficiency of your car is measured in miles per gallon given by the ratio 400.4/10.4 or 38.5 miles per gallon. If Lincoln Grade School has 405 students and 15 teachers, the student teacher ratio is the quotient 405/15. If Jose Varga got 56 hits in 181 times at bat, his batting average is the ratio 56/181. The number of examples that could be cited is almost endless.

EXAMPLE 7.15 Determining Ratios

Determine these ratios.

(a) The ratio of the number of boys to the number of girls in Martin Luther King High School if there are 285 boys and 228 girls.

(b) The ratio of the number of boys to the number of students in part (a).

SOLUTION

(a) The desired ratio is 285/228.

(b) Since the total number of students is 285 + 228 (that is, 513), the desired ratio is 285/513. ■

The ratio of the number of boys to the number of students in Martin Luther King High School was shown to be 285/513 or 285 to 513. This is certainly correct, but it is not nearly as informative as it would be if the ratio were written in simplest form.

WORTH READING

A Classic Pearl of Mathematical Exposition

The eminent mathematician Edward Kasner teamed with his former pupil, James Newman, more than fifty years ago to write this easily readable account of the nature of mathematics. Illuminating, witty, and insightful, the book remains one of the best in its class. An excerpt follows.

> Here, then, in mathematics we have a universal language, valid, useful, intelligible everywhere in place and in time—in banks and insurance companies, on the parchments of the architects who raised the Temple of Solomon, and on the blueprints of the engineers who, with their calculus of chaos, master the winds. Here is a discipline of a hundred branches, fabulously rich, literally without limit in its sphere of application, laden with honors for an unbroken record of magnificent accomplishment. Here is a creation of the mind, both mystic and pragmatic in appeal. Austere and imperious as logic, it is still sufficiently sensitive and flexible to meet each new need. Yet this vast edifice rests on the simplest and most primitive foundations, is wrought by imagination and logic out of a handful of childish rules. Even though no definition thus far has encompassed either its scope or its nature, can it be that the question "What is mathematics?" must go unanswered?

SOURCE: Edward Kasner and James Newman, *Mathematics and the Imagination* (New York: Simon and Schuster, 1940), p. 358.

Thus,

$$\frac{285}{513} = \frac{5}{9}$$

and this says that 5/9 (or a little more than 1/2) of the students in Martin Luther King High School are boys. Expressing a ratio given by a fraction to simplest form is often useful and informative.

EXAMPLE 7.16

Expressing Ratios in Simplest Form

Express these ratios in simplest form.

(a) The ratio of 385 to 440
(b) The ratio of 858 to 792
(c) 432/504

SOLUTION

(a) The ratio of 385 to 440 is the quotient 385/440. Expressing this in simplest form, we have

$$\frac{385}{440} = \frac{7}{8}.$$

Thus, in simplest form, the ratio is 7 to 8.

(b) The ratio of 858 to 792 is the quotient 858/792. Expressed in simplest form, we obtain

$$\frac{858}{792} = \frac{13}{12}.$$

Thus, in simplest form, the ratio is 13 to 12.

(c) Expressing the quotient in simplest form,

$$\frac{432}{504} = \frac{6}{7}.$$

Thus, in simplest form, the ratio is 6/7 or 6 to 7. ■

EXAMPLE 7.17

Determining a Less Obvious Ratio

If one seventh of the students at Garfield High are nonswimmers, what is the ratio of nonswimmers to swimmers?

SOLUTION

The desired ratio is the number of nonswimmers divided by the number of swimmers. Can we determine these numbers from the information given? Actually, no; but the problem can still be solved. Suppose there are n students in the school. Then $\frac{n}{7}$ are nonswimmers and $\frac{6n}{7}$ are swimmers. Thus, the desired ratio is

$$\frac{\frac{n}{7}}{\frac{6n}{7}} = \frac{n}{7} \cdot \frac{7}{6n} = \frac{1}{6}.$$ ■

Proportion

Ratios allow us to make clear comparisons when actual numbers sometimes make them more obscure. For example, at basketball practice, Caralee made 27 of 45 free throws attempted and Sonja made 24 of 40 attempts. Which player appears to be the better foul shot shooter? For Caralee, saying that the ratio of shots made to shots tried is 27/45 amounts to saying that she made 3/5 of her shots. That is,

$$\frac{27}{45} = \frac{3}{5}.$$

$\frac{27}{45}$ and $\frac{3}{5}$ are equivalent fractions.

Similarly, for Sonja the ratio of shots made to shots attempted is

$$\frac{24}{40} = \frac{3}{5},$$

$\frac{27}{45}$, $\frac{24}{40}$, and $\frac{3}{5}$ are all equivalent fractions.

and this suggests that the two girls are equally capable at shooting foul shots. Because of its importance in such comparisons, the equality of two ratios is called a **proportion.**

DEFINITION Proportion

If a/b and c/d are two ratios and

$$\frac{a}{b} = \frac{c}{d},$$

this equality is called a **proportion.**

From Chapter 6, we know that

$$\frac{a}{b} = \frac{c}{d}$$

for integers a, b, c, and d, if, and only if, $ad = bc$. But essentially the same argument holds if a, b, c, and d are real numbers. This leads to the next theorem.

THEOREM Conditions for a Proportion

The equality

$$\frac{a}{b} = \frac{c}{d},$$

is a proportion if, and only if, $ad = bc$.

EXAMPLE 7.18

Determining Proportions

In each of these, determine x so that the equality is a proportion.

(a) $\dfrac{28}{49} = \dfrac{x}{21}$ **(b)** $\dfrac{35}{63} = \dfrac{20}{x}$ **(c)** $\dfrac{2.11}{3.49} = \dfrac{1.7}{x}$

SOLUTION

We use the preceding theorem which amounts to multiplying both sides of the equality by the product of the denominators or "cross multiplying" as we often say.

(a)
$$\begin{aligned} \frac{28}{49} &= \frac{x}{21} \\ 28 \cdot 21 &= 49x \\ \frac{28 \cdot 21}{49} &= x \\ 12 &= x \end{aligned}$$

(b)
$$\begin{aligned} \frac{35}{63} &= \frac{20}{x} \\ 35x &= 63 \cdot 20 \\ x &= \frac{63 \cdot 20}{35} \\ x &= 36 \end{aligned}$$

(c)
$$\begin{aligned} \frac{2.11}{3.49} &= \frac{1.7}{x} \\ 2.11x &= (1.7)(3.49) \\ x &= \frac{(1.7)(3.49)}{2.11} \\ x &\doteq 2.8 \end{aligned}$$

■

EXAMPLE 7.19

Proving a Property of Proportions

If

$$\frac{a}{b} = \frac{c}{d},$$

prove that

$$\frac{a+b}{b} = \frac{c+d}{d}.$$

SOLUTION I

Understand the problem • • • •

We are given that $\dfrac{a}{b} = \dfrac{c}{d}$ is a proportion and are asked to show that $\dfrac{a+b}{b} = \dfrac{c+d}{d}$ is also a proportion.

Devise a strategy • • • •

Since it is not immediately clear what to do, we ask what it means to say that $\frac{a}{b} = \frac{a}{d}$ and $\frac{a + b}{b} = \frac{c + d}{d}$ are proportions. Perhaps this will put the problem in a form that is easier to understand and to work on. By the preceding theorem,

$$\frac{a}{b} = \frac{c}{d} \text{ if, and only if, } ad = bc$$

and

$$\frac{a + b}{b} = \frac{c + d}{d}$$

if, and only if, $(a + b)d = b(c + d)$. Perhaps we can use the first of these equations to prove the second.

Carry out the plan • • • •

We want to show that $(a + b)d = b(c + d)$; that is, using the distributive property,

$$ad + bd = bc + bd.$$

Say it in a different way.

But we know that

$$ad = bc$$

and adding bd to both sides of this equation gives

$$ad + bd = bc + bd$$

and hence

$$\frac{a + b}{b} = \frac{c + d}{d}$$

as was to be shown.

Look back • • • •

Here our principal strategy was simply to ask, "What does it mean to say that $\frac{a}{b} = \frac{c}{d}$ and $\frac{a + b}{b} = \frac{c + d}{d}$ are proportions?" Answering this question allowed us to "say it in a different way"; that is, to state an equivalent problem that proved to be quite easy to solve. The strategy, **say it in a different way,** is often very useful.

SOLUTION 2

There is a second easy solution to this problem, but it is perhaps less obvious. We are given

$$\frac{a}{b} = \frac{c}{d}.$$

Therefore,

$$\frac{a}{b} + 1 = \frac{c}{d} + 1,$$ (add 1 to both sides)

$$\frac{a}{b} + \frac{b}{b} = \frac{c}{d} + \frac{d}{d},$$ (since $\frac{b}{b} = 1 = \frac{d}{d}$)

and

$$\frac{a+b}{b} = \frac{c+d}{d}.$$ (add fractions) ■

Applications of Proportions

Suppose that a car is traveling at a constant rate of 55 miles per hour. Table 7.2 gives the distances the car will travel in different time periods.

TABLE 7.2 Distance Traveled in t Hours at 55 Miles per Hour

t = Time	d = Distance	t = Time	d = Distance
1	55	5	275
2	110	6	330
3	165	7	385
4	220	8	440

The ratios d/t are all equal for the various time periods shown. That is,

$$\frac{55}{1} = \frac{110}{2} = \frac{165}{3} = \frac{220}{4} = \frac{275}{5} = \frac{330}{6}$$

and so on. Thus, each pair of ratios from the list form a proportion. Indeed, $d/t = 55$ for every pair d and t. This is also expressed by saying that the distance traveled at a constant rate is proportional to the elapsed time. In the above instance

$$d = 55t$$

for every pair d and t. The number 55 is called the **constant of proportionality.**

> **DEFINITION y Proportional to x**
>
> If the variables x and y are related by the equation
>
> $$y = kx,$$ ($\frac{y}{x} = k$)
>
> then **y is said to be proportional to x** and k is called the **constant of proportionality.**

JUST FOR FUN

Who Shaves Francisco?

Francisco lives in Seville and shaves those and only those in Seville who do not shave themselves. Who shaves Francisco?

This situation is extremely common in everyday life. Gasoline consumed by your car is proportional to the miles traveled. The cost of pencils purchased is proportional to the number of pencils purchased. Income from the school raffle is proportional to the number of tickets sold, and so on.

EXAMPLE 7.20

Income from the School Raffle

The sixth grade class at Jefferson Middle School is raffling off a turkey as a money making project. If the turkey cost \$22 and raffle tickets are sold for 75¢ each, how many tickets will have to be sold for the class to break even? How many tickets will have to be sold if the class is to make a profit of \$20?

SOLUTION

If I represents income in dollars and N represents the number of tickets sold, then $I = 0.75N$. To break even, the class must sell enough tickets that

$$22 = 0.75N;$$

that is,

$$N = 22 \div 0.75 = 29.\overline{3}.$$

To break even the class will have to sell at least 30 tickets. In order to make a profit of \$20, enough tickets must be sold so that

$$42 = 0.75N;$$

that is,

$$N = 42 \div 0.75 = 56.$$

To make a profit of \$20, 56 tickets must be sold. ■

EXAMPLE 7.21

Computing the Cost of 25 Golf Balls

If Kourash paid \$10 for 4 golf balls, how much would he have to pay for 25 golf balls at the same price per ball?

SOLUTION 1

If Kourash paid \$10 for 4 balls, then each ball cost

$$\frac{\$10}{4} = \$2.50.$$

Therefore, 25 balls would cost

$$(\$2.50) \cdot 25 = \$62.50.$$

SOLUTION 2

The cost of a number of balls is proportional to the number of balls purchased. Thus, if c is the cost of 25 balls,

$$\frac{c}{25} = \frac{10}{4}.$$

Therefore,

$$c = 25 \cdot \frac{10}{4} = \frac{250}{4} = 62.50$$

The 25 balls cost \$62.50. ■

Recall that in geometry, two figures are said to be similar if they are the same shape but not necessarily the same size; i.e., one is a magnification of the other.

EXAMPLE 7.22

Computing the Height of a Tree

Ms. Gulley-Pavey's fifth grade class had been studying ratio and proportion. One afternoon she took her students outside and challenged them to find the height of a tree in the school yard. After a lively discussion the students decided to measure the lengths of the shadow cast by a yardstick and that cast by the tree, arguing that these should be proportional. To help convince the class that this was so, Omari drew the picture shown. If the lengths of the shadows are 4′7″ and 18′9″, complete the calculation to determine the height of the tree.

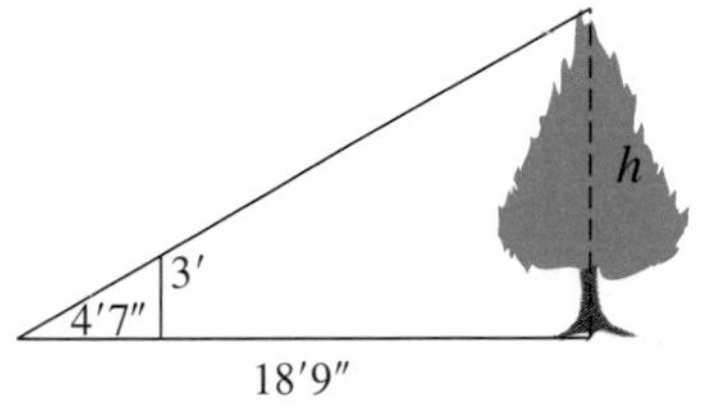

SOLUTION

Since the two triangles shown in the diagram are similar, the lengths of the sides are proportional. Also, 7″ = 7/12 feet and 9″ = 9/12 feet. Therefore, we have the ratios

$$\frac{h}{3} = \frac{18\frac{9}{12}}{4\frac{7}{12}} = \frac{\frac{225}{12}}{\frac{55}{12}} = \frac{225}{12} \cdot \frac{12}{55}$$

The units of both ratios must be the same.

and it follows that

$$h = \frac{3 \cdot 225 \cdot 12}{12 \cdot 55} \doteq 12.27' \doteq 12'3''$$

$0.27 \times 12 \doteq 3$

Thus, the tree was approximately 12′3″ tall. ■

EXAMPLE 7.23

Finding a Distance from a Map

On a map in an atlas, the scale shows that one centimeter represents 250 miles. If the straight line distance from Chicago to St. Louis measures 1.1 cm on the map, what is the approximate airline distance from Chicago to St. Louis?

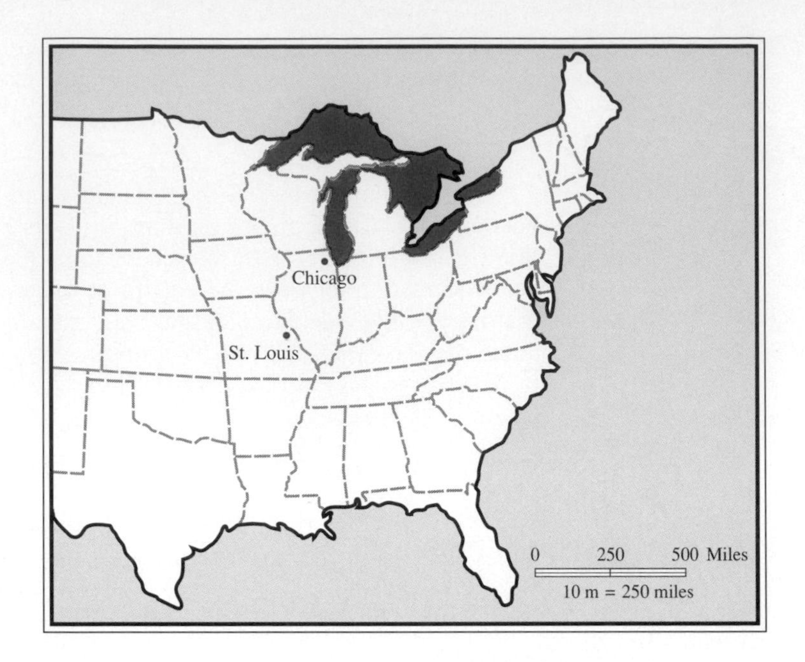

SOLUTION 1

Since the map is drawn to scale, the actual distance is proportional to the measured distance. Thus,

$$\frac{250}{1} \doteq \frac{d}{1.1}$$

	Scale	Actual
distance in miles	250	d
distance in centimeters	1	1.1

is a proportion. Therefore,

$$250(1.1) \doteq d$$

and $d \doteq 275$ miles.

SOLUTION 2

Since 1 centimeter represents 250 miles, approximately 1.1 centimeters equals (1.1)250 or approximately 275 miles. ■

PROBLEM SET 7.3

Understanding Concepts

1. There are 10 girls and 14 boys in Mr. Tilden's fifth grade class. What is the ratio of
   (a) boys to girls? **(b)** girls to students?
   (c) boys to students? **(d)** girls to boys?
   (e) students to girls? **(f)** students to boys?

2. Determine which of these are proportions.
   (a) $\frac{2}{3} = \frac{8}{12}$ **(b)** $\frac{21}{28} = \frac{27}{36}$
   (c) $\frac{7}{28} = \frac{8}{31}$ **(d)** $\frac{51}{85} = \frac{57}{95}$
   (e) $\frac{14}{49} = \frac{18}{60}$ **(f)** $\frac{20}{35} = \frac{28}{48}$

3. Determine values of r, s, and t so that each of these is a proportion.
   (a) $\frac{6}{14} = \frac{r}{21}$ (b) $\frac{8}{12} = \frac{10}{r}$
   (c) $\frac{51}{t} = \frac{85}{95}$ (d) $\frac{47}{3.2} = \frac{s}{7.8}$
4. Express each of these ratios in simplest form.
   (a) A ratio of 24 to 16
   (b) A ratio of 296 to 111
   (c) A ratio of 248 to 372
   (d) A ratio of 209 to 341
5. Collene had an after-school job at Taco Time at \$5.50 per hour.
   (a) How much did she earn on Monday if she worked $3\frac{1}{2}$ hours?
   (b) On Tuesday she earned \$27.50. How long did she work?
   (c) Show that the ratio of the time worked to the amount earned on Monday is equal to the ratio of the time worked to the amount earned on Tuesday; that is, show that these two ratios form a proportion.
   (d) Show that the ratio of the time worked on Monday to the time worked on Tuesday equals the ratio of the amount earned on Monday to the amount earned on Tuesday; that is, show that these two ratios also form a proportion.
6. David Horwitz bought four sweatshirts for \$107.92. How much would it cost him to buy nine sweatshirts at the same price per sweatshirt?
7. If s is proportional to t and $s = 62.5$ when $t = 7$, what is s when $t = 10$?
8. The flag pole at Sunnyside Elementary School casts a shadow 9′8″ long at the same time Mr. Schaal's shadow is 3′2″ long. If Mr. Schaal is 6′3″ tall, how tall is the flag pole to the nearest foot?
9. A kilometer is a bit more than six tenths of a mile. If the speed limit along a stretch of highway in Canada is 90 kilometers per hour, about how fast can you travel in miles per hour and still not break the speed limit?

Thinking Critically

10. If a is to b as c is to d; that is, if
$$\frac{a}{b} = \frac{c}{d},$$
   (a) show that b is to a as d is to c.
   (b) show that $a - b$ is to b as $c - d$ is to d.
   (c) show that a is to $a + b$ as c is to $c + d$.
   (d) show that $a + b$ is to $c + d$ as b is to d.
11. (a) If y is proportional to x^2 and $y = 27$ when $x = 6$, determine y when $x = 12$.
   (b) Determine the ratio of the y-values in part (a).
   (c) If y and x are related as in part (a), what happens to the value of y if the value of x is doubled? Explain.
12. (a) If y is proportional to x^3 and $y = 32$ when $x = 12$, determine the value of y when $x = 6$.
   (b) Determine the ratio of the y-values in part (a).
   (c) If y and x are related as in part (a), what happens to the value of y if the value of x is doubled? Tripled? Quadrupled? Explain.
13. If y is proportional to $1/x$ and $y = 3.5$ when $x = 84$, determine y when $x = 14$. (*Hint*: $y = k(1/x)$.)

Thinking Cooperatively

14. (a) You stand in front of a mirror on a wall and can just barely see your entire reflection. If your height is h and H is the vertical dimension of the mirror, determine the ratio h to H. (*Hint:* Make a suitable drawing.)
   (b) Where should the mirror of minimum height H in part (a) be located on the wall?
   (c) Does the distance you stand from the wall make a difference in your answers to parts (a) and (b)? Explain.

Making Connections

15. Celeste Neal won a one-hundred meter race with a time of 11.6 seconds and Michelle Beese came in second with a time of 11.8 seconds. Given that each girl ran at a constant rate throughout the race, determine the ratio of Celeste's speed to Michelle's in simplest terms.
16. On a trip of 320 miles, Sunao's truck averaged 9.2 miles per gallon. At the same rate, how much gasoline would his truck use on a trip of 440 miles?
17. Which is the best buy in each case?
   (a) 32 ounces of cheese for 90¢ or 40 ounces of cheese for \$1.20.
   (b) A gallon of milk for \$2.21 or two half gallons at \$1.11 per half gallon.
   (c) A 16 ounce box of bran flakes at \$3.85 per box or a 12 ounce box at \$2.94 per box.
18. The ratio of Dexter's salary to Claudine's is 4 to 5. If Claudine earns \$3200 per month, how much of a raise will Dexter have to receive to make the ratio of his salary to Claudine's 5 to 6?
19. The ratio of boys to girls in Ms. Zombo's class is 3 to 2. In Mr. Stolarski's class it is 4 to 3. If there are 30 students in Ms. Zombo's class and 28 students in Mr. Stolarski's class, what is the ratio of boys to girls in the combined classes?

Communicating

20. Write a two page summary of "On Being the Right Size" by J. B. S. Haldane, on pages 952–57 of *The World of Mathematics,* James R. Newman, ed. (New York: Simon and Schuster, 1956). In particular, make clear how ratio and proportion play a role in this study.

Using A Calculator

21. Use your calculator to determine if these equalities are proportions.

(a) $\frac{2.52}{7.56} = \frac{6.31}{18.93}$ **(b)** $\frac{3.71}{6.81} = \frac{8.24}{24.72}$

22. Use your calculator to determine x so that this is a proportion.

$$\frac{2.7}{3.5} = \frac{6.3}{x}$$

23. If a 12″ pepperoni pizza from Ricco's costs \$9.56, what should a 14″ pizza from Ricco's cost?

24. Suppose your car uses 8.7 gallons of gas traveling 192 miles. Determine approximately how many gallons it would use traveling a distance of 305 miles.

25. If it takes $1\frac{1}{3}$ cups of sugar to make a batch of cookies, how much sugar would be required to make four batches?

26. If three equally priced shirts cost a total of \$59.97, how much would seven shirts cost at the same price per shirt?

27. If 12 erasers cost \$8.04 and Mrs. Orton bought \$14.07 worth of erasers, how many erasers did she buy?

28. One day Kenji Okubo took his class of sixth graders outside and challenged them to find the distance between two rocks that could easily be seen one above the other on the vertical face of a bluff near the school. After some discussion the children decided to hold a rod in a vertical position at a point 100′ from the base of the cliff.

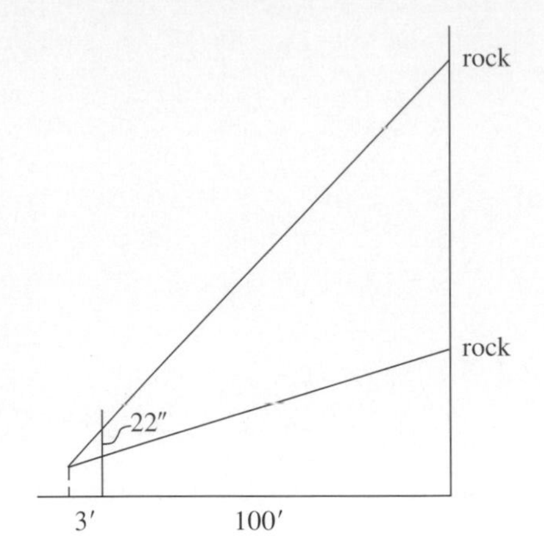

They also decided to mark the points on the rod where the lines of sight of a student standing 3′ further from the cliff and looking at the rocks cut the rod. If the marks on the rod were 22″ apart, what was the distance between the two rocks? (*Caution:* Convert all measurements to feet.)

For Review

29. Perform these computations.

(a) $4 + (-7)$ **(b)** $8 - (-5)$
(c) $(-5) + (-7)$ **(d)** $(-8) + (-5)$
(e) $8 - (-4)$ **(f)** $(-3)(-5)$
(g) $12 \div (-3)$ **(h)** $(-28) \div 4$
(i) $(-28) \div (-4)$ **(j)** $(-495) \div 11$
(k) $57 \div (-19)$ **(l)** $[4 + (-15)] \div 11$

30. Find the prime power representation of 521,752.

31. If $a = 3^5 \cdot 5^2 \cdot 11^3$, how many divisors does a have?

32. Does $b = 3^2 \cdot 5 \cdot 11^3$ divide a of problem 31? Explain.

33. Find the smallest integer greater than b of problem 32 that is divisible by b.

34. Find the largest integer less than b of problem 32 that divides b.

7.4 Percent

One of the most important uses of ratios in school mathematics is the notion of **percent** (from the Latin *per centum*—per hundred). Thus, 50% is the ratio 50/100, and this is quickly reduced to the fraction 1/2 or written as the decimal 0.50. Thus, if I have \$98 and give you 50% of what I have, I give you

$$\frac{1}{2} \cdot \$98 = \$49 \quad \text{or} \quad 0.5 \times \$98 = \$49.$$

The "of" in the preceding sentence translates into "times." Thus,

$$50\% \text{ of} \quad \text{means} \quad 50\%\times,$$

$$\frac{1}{2} \text{ of} \quad \text{means} \quad \frac{1}{2}\times,$$

and

$$0.5 \text{ of} \quad \text{means} \quad 0.5\times.$$

DEFINITION Percent

If r is any nonnegative real number, then r percent, written $r\%$, is the ratio

$$\frac{r}{100}.$$

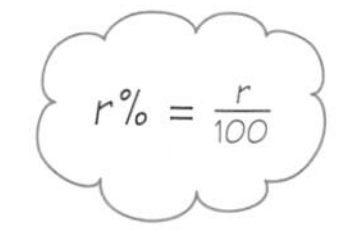

Since $r\%$ is defined as the ratio $r/100$ and dividing by 100 moves the decimal point 2 places to the left, it is easy to write a given percent as a decimal. For example, $12\% = 0.12$, $25\% = 0.25$, $130\% = 1.3$, and so on. Conversely, to write a decimal as a percent, we need only move the decimal point 2 places to the right. Thus, $0.125 = 12.5\%$ or $12\frac{1}{2}\%$, $0.10 = 10\%$, $1.50 = 150\%$, and so on.

EXAMPLE 7.24

Expressing Decimals as Percents

Express these decimals as percents.

(a) 0.25 **(b)** 0.333 . . . **(c)** 2.15

SOLUTION

(a) $0.25 = 25\%$
(b) $0.33\ldots = 33.333\ldots\%$ $= 33\frac{1}{3}\%$
(c) $1.255 = 125.5\%$

■

EXAMPLE 7.25

Expressing Percents as Decimals

Express these percents as decimals.

(a) 40% **(b)** 12% **(c)** 127%

SOLUTION

(a) $40\% = 0.40$ $40\% = \frac{40}{100} = 0.40$
(b) $12\% = 0.12$
(c) $127\% = 1.27$

■

SCHOOL BOOK PAGE *Percents in Fifth Grade*

SCIENCE CONNECTION

Percents and Fractions

Build Understanding

For a science project, Melanie counted and identified birds that came near her home.

A. Of the birds that Melanie saw, 15 out of 60, or $\frac{1}{4}$ of the birds, were robins. What percent of the birds were robins?

Since percent means *per hundred*, find an equal fraction with a denominator of 100.

Fraction ⟶ Percent

$$\frac{1}{4} \xrightarrow{\times 25} = \frac{25}{100} = 25\%$$

25% of the birds were robins.

B. Melanie found that 40% of the birds she saw were wrens. What fraction of the birds were wrens?

Write the percent as a fraction with a denominator of 100. Write a fraction in lowest terms.

Percent ⟶ Fraction

$$40\% = \frac{40}{100} \xrightarrow{\div 20} = \frac{2}{5}$$

$\frac{2}{5}$ of the birds were wrens.

Talk About Math Melanie saw no woodpeckers. What fraction is that? What percent is it? What fraction and percent would describe the situation if all the birds she saw were the same type of bird?

OBSERVATIONS ON APRIL 12

Robins	15
Wrens	24
Cardinals	12
Mourning Doves	9
Total	60

The cardinal is the state bird of 7 states in the U.S.

1. In A above, 1/4 was converted to a percent by multiplying both numerator and denominator by 25 to obtain 25/100 or 25%. Could this method be used to convert 2/3 to a percent? What number can you multiply by 3 to obtain 100?
2. Can you think of an easier way to convert 2/3 to a percent?
3. What real number is represented by $33\frac{1}{3}\%$?
4. How would you answer the questions in Talk About Math?

EXAMPLE 7.26 **Expressing Percents as Fractions**

Express each of these percents as fractions in lowest terms.

(a) 60% **(b)** $37\frac{1}{2}\%$ **(c)** $66\frac{2}{3}\%$ **(d)** 125%

SOLUTION

(a) By definition 60% means 60/100. Therefore,

$$60\% = \frac{60}{100} = \frac{3}{5}.$$

(b) Here

$$37\frac{1}{2}\% = \frac{37\frac{1}{2}}{100} = \frac{\frac{65}{2}}{100} = \frac{65}{200} = \frac{13}{40}.$$

(c) $66\frac{2}{3}\% = \dfrac{66\frac{2}{3}}{100} = \dfrac{\frac{200}{3}}{100} = \dfrac{2}{3}.$

(d) $125\% = \dfrac{125}{100} = \dfrac{5}{4}$ or $1\dfrac{1}{4}.$ ■

EXAMPLE 7.27 **Expressing Fractions as Percents**

Express these fractions as percents.

(a) $\frac{1}{8}$ **(b)** $\frac{1}{3}$ **(c)** $\frac{4}{9}$ **(d)** $\frac{16}{5}$

SOLUTION I (Using proportions)

Since percents are ratios, we can use variables to determine the desired percents.

(a) Suppose $1/8 = r\% = r/100$. Then

$$r = 100 \cdot \frac{1}{8} = 12.5$$

and

$$r\% = 12.5\%$$

(b) Let $1/3 = s\% = s/100$. Then

$$s = \frac{100}{3} = 33\frac{1}{3} = 33.\overline{3}$$

and

$$s\% = 33.\overline{3}\%.$$

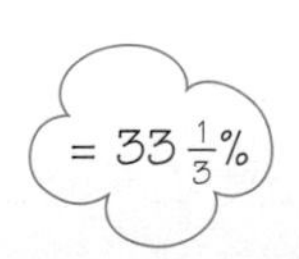

(c) Let $4/9 = t\% = t/100$. Then

$$t = \frac{4}{9} \cdot 100 = 44.\overline{4}$$

and

$$t\% = 44.\overline{4}\%.$$

$= 44\frac{4}{9}\%$

(d) Let $16/5 = u\% = u/100$. Then

$$u = \frac{16}{5} \cdot 100 = 320$$

and

$$u\% = 320\%.$$

SOLUTION 2 (Using decimals)

Here we write the fractions as decimals and then as percents.

(a) By division

$$\frac{1}{8} = 0.125 = 12.5\%.$$

(b) Here

$$\frac{1}{3} = 0.333\ldots = 33.\overline{3}\% = 33\frac{1}{3}\%.$$

(c) $\frac{4}{9} = 0.444\cdots = 44.\overline{4}\% = 44\frac{4}{9}\%.$

(d) $\frac{16}{5} = 3.2 = 320\%.$ ■

Applications of Percent

Use of percents is commonplace. Three of the most common types of usages are illustrated in the next three examples.

EXAMPLE 7.28

Calculating a Percentage of a Number

The Smetanas bought a house for \$175,000. If a 15% down payment was required, how much was the down payment?

SOLUTION 1 (Using an equation)

The down payment is 15% of the cost of the house. Thus, if d is the down payment,

$$\begin{aligned} d &= 15\% \times \$175{,}000 \\ &= 0.15 \times \$175{,}000 \\ &= \$26{,}250 \end{aligned}$$

SOLUTION 2 *(Using ratio and proportion)*

The size of the down payment is proportional to the cost of the house; that is, $d/175{,}000$ is a ratio that is equivalent to the ratio 15%. Thus,

$$\frac{d}{\$175{,}000} = \frac{15}{100} = 0.15$$

and

$$\begin{aligned} d &= 0.15 \times \$175{,}000 \\ &= \$26{,}250. \end{aligned}$$

■

EXAMPLE 7.29 Calculating a Number of Which a Given Number Is a Given Percentage

Soo Ling scored 92% on her last test. If she got 23 questions right, how many problems were on the test?

SOLUTION 1 *(Using an equation)*

Let n denote the number of questions on the test. Since Soo Ling got 23 questions correct for a score of 92%, we know that 23 is 92% of n; that is,

$$23 = 92\% \times n = 0.92n.$$

Hence,

$$n = \frac{23}{0.92} = 25.$$

SOLUTION 2 *(Using ratio and proportion)*

The ratio of 23 to the number of questions on the test must be the same ratio as 92%. So,

$$\frac{23}{n} = \frac{92}{100} = 0.92.$$

Thus,

$$23 = 0.92 \times n$$

and

$$n = 23 \div 0.92 = 25$$

as before.

■

EXAMPLE 7.30 Calculating What Percentage One Number Is of Another

Tara got 28 out of 35 possible points on her last math test. What percentage score did the teacher record in her grade book for Tara?

SOLUTION 1 (Using the definition)

Tara got twenty-eight thirty-fifths of the test right. Since

$$\frac{28}{35} = 0.80 = 80\%,$$

the teacher recorded 80% in her grade book.

SOLUTION 2 (Using ratio and proportion)

Let x be the desired percentage, then

$$\frac{x}{100} = \frac{28}{35}$$

and

$$x = \frac{28 \cdot 100}{35} = 80\%. \quad \blacksquare$$

INTO THE CLASSROOM

Theresa Norris Discusses Percents and Decimals

The newspaper is an excellent source of problems and activities that can be used to teach percent.

In one of my favorite activities, students work in small groups and use a calculator and department store advertisements to find the "best buy" and the percent of savings for popular items such as jeans and athletic shoes. Students search the newspapers to find ads that give either the original price and the sale price or the original price and the percent of discount. If the original price and the sale price are given, I have students calculate the percent of discount. If the original price and the percent of discount are given, I have them find the sale price.

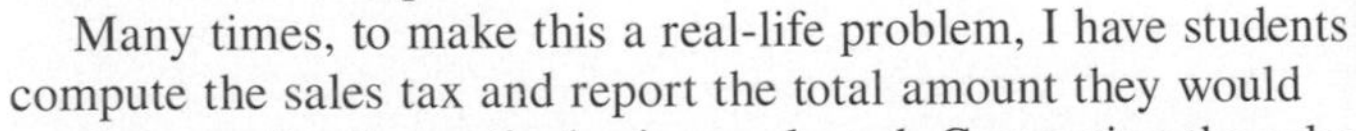

Many times, to make this a real-life problem, I have students compute the sales tax and report the total amount they would need to pay for the particular item selected. Computing the sales tax not only gives students additional practice in computing percents, but also helps develop students' consumer awareness.

To integrate the study of percents with statistics, I ask students to vote on their favorite brands of jeans, athletic shoes, or other popular items. We then show the results, as percentages, by using bar graphs and circle graphs.

I use many other newspaper activities to practice percents, decimals, fractions, and graphs. I have students change headlines by substituting percents for fractions or decimals and vice-versa and have them convert sports statistics to fractions or percents. I also ask students to compare the number of pages in one section of the newspaper to the number of pages in another section and write the comparison as a fraction, decimal, and percent and to change bar graphs to circle graphs and circle graphs to bar graphs. All of these activities can be completed in cooperative groups or as homework assignments.

SOURCE: From *Scott Foresman Exploring Mathematics* Grades 1–7 by L. Carey Bolster et al. Theresa Norris formerly taught at Quebec Heights CAMAS (Cincinnati Academy of Math and Science) and is currently teaching at Cresthill Middle School, Cresthill, Ohio.

Compound Interest

If you keep money in a savings account at a bank, the bank pays you interest at a fixed rate (percentage) for the privilege of using your money. For example, suppose you invest \$5000 for a year at 7% interest. How much is your investment worth at the end of the year? Since the interest earned is 7% of \$5000, the interest earned is

$$7\% \times \$5000 = 0.07 \times \$5000 = \$350$$

and the value of your investment at the end of the year is

$$\begin{aligned} \$5000 + \$350 &= \$5000 + 0.07 \times \$5000 \\ &= \$5000 \cdot (1.07) \\ &= \$5350. \end{aligned}$$

If you leave the total investment in the bank, its value at the end of the second year is

$$\begin{aligned} \$5350 + 0.07 \times \$5350 &= \$5350 \cdot (1.07) \\ &= \$5000 \cdot (1.07)(1.07) \\ &= \$5000 \cdot (1.07)^2 \\ &= \$5724.50. \end{aligned}$$

from above

Similarly, at the end of the third year, your investment would be worth

$$\$5000 \cdot (1.07)^3 = \$6125.22$$

to the nearest penny. In general, it would be worth

$$\$5000 \cdot (1.07)^n$$

at the end of n years. This is an example of **compound interest** where the term "compound" implies that each year you earn interest on all the interest earned in preceding years as well as on the original amount invested (the **principal**).

Usually, these days, interest is compounded more than once a year. Suppose the \$5000 investment just discussed was made in a bank that paid interest at the rate of 7% compounded semi-annually; that is, twice a year. Since the rate for a year is 7%, the rate for half a year is 3.5%. Thus, the value of the investment at the end of the year (that is, at the end of *two* interest periods) is

$$\$5000(1.035)^2 = \$5356.13$$

and the values at the end of 2 years and 3 years respectively are

$$\$5000(1.035)^4 = \$5737.62$$

and

$$\$5000(1.035)^6 = \$6146.28.$$

These can be easily calculated using the constant function of your calculator.

Compounding more and more frequently is to your advantage and, to attract customers, some banks are now compounding monthly or even daily.

The above calculations are typical, and are summarized in this theorem.

$\frac{r}{t}$ is the rate per interest period and nt is the number of interest periods.

THEOREM Calculating Compound Interest

The value of an investment of P dollars at the end of n years if interest is paid at the annual rate of $r\%$ compounded t times a year is

$$P\left(1 + \frac{r/t}{100}\right)^{nt}.$$

EXAMPLE 7.31

Computing the Cost of Debt

Many credit card companies charge 12% interest compounded monthly on unpaid balances. Suppose your card was "maxed out" at your credit limit of \$2000 and that you were unable to make any payments for two years. Aside from penalties, how much debt would you owe based on compound interest alone?

SOLUTION

Since the interest is computed at 12% compounded monthly, the effective rate per month is $12\%/12 = 1\%$ and the number of interest periods in 2 years is $12 \cdot 2 = 24$. Thus, your debt to the nearest penny would be

$$\$2000(1.01)^{24} = \$2{,}539.47.$$

If the debt went unpaid for 6 years, you would owe

$$\$2000(1.01)^{72} = \$4094.20.$$

This is more than double what you originally owed! The above calculations can be made by repeatedly multiplying \$2000 by 1.01 using the constant feature of your calculator. Even more easily, you can compute $\$2000(1.01)^{24}$ directly by entering the following string into your calculator.

[ON/AC] 2000 [×] 1.01 [y^x] 24 [=] ■

The Mathematics of Growth

Population growth occurs in exactly the same way that an investment grows if it is earning compound interest. Suppose, for example, that the population in the Puget Sound region in northwest Washington is approximately 2.2 million and that it is growing at the rate of 5% per year. In one year the population will be approximately

$$2.2(1.05) = 2.3$$

million. If the growth continues unabated, in 14 years it will be approximately

$$2.2(1.05)^{14} = 4.4$$

million or almost exactly twice what it is today. Given the fact that the area has already experienced several years of water shortages, these figures are cause for concern among officials in the area.

Like population growth, prices of commodities also rise with inflation as an investment grows drawing compound interest.

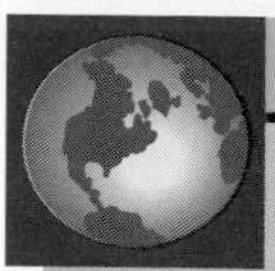

CURRENT AFFAIRS

It's Only $1000

The accompanying article appeared in the August 9, 1993 issue of *Newsweek*. This is a dramatic example of the power of compound interest—particularly at a rate as high as 24%. By the way, is the Wilson's calculation correct? If the bond was purchased on January 1, 1865, compute its value on January 1, 1993.

If Nevada officials are welshing, it's understandable. In April, Allen and Kathy Wilson of Montello, Nev., sued the state to cash in a $1,000 state bond issued in 1865 with a yearly interest rate of 24 percent. Why wouldn't the state pay up? With interest compounded, according to the couple's calculations, the bond is now worth $657 *trillion*. The Wilsons, who collect old documents like stock certificates and investigate if they're redeemable, were willing to settle for a lesser amount: say, a paltry $54 million. Last week a state judge ruled the bond had to have been cashed by 1872. "I'm glad the people who brought this absurd law-suit won't get a plugged nickel," said State Treasurer Bob Seale. The Wilsons plan an appeal.

EXAMPLE 7.32 **Pricing a Car**

If the economy were to experience a steady inflation rate of 2.5% per year, what would be the price of a new car in five years if the same quality car sells today for $18,400?

SOLUTION

Using the same formula as in computing compound interest, the price of the car five years from now, would be approximately

$$\$18{,}400(1.025)^5 = \$20{,}818.$$

■

PROBLEM SET 7.4

Understanding Concepts

1. Write each of these ratios as percents.
   (a) $\frac{3}{16}$ (b) $\frac{7}{25}$ (c) $\frac{37}{40}$
   (d) $\frac{5}{6}$ (e) $\frac{3.24}{8.91}$ (f) $\frac{7.801}{23.015}$
   (g) $\frac{1.6}{7}$ (h) $\frac{\sqrt{2}}{\sqrt{6}}$
2. Write each of these as percents.
   (a) 0.19 (b) 0.015 (c) 2.15 (d) 3
3. Write each of these as fractions in simplest form.
   (a) 10% (b) 25% (c) 62.5% (d) 137.5%
4. Compute each of the following.
   (a) 70% of 280 (b) 120% of 84
   (c) 38% of 751 (d) $7\frac{1}{2}$% of $20,000
   (e) .02% of 27,481 (f) 1.05% of 845
5. Compute each of these mentally.
   (a) 50% of 840 (b) 10% of 2480
   (c) 12.5% of 48 (d) 125% of 24
   (e) 200% of 56 (f) 110% of 180

6. Mentally convert each of these to a percent.

   (a) $\frac{7}{28}$ (b) $\frac{11}{33}$

   (c) $\frac{72}{144}$ (d) $\frac{44}{66}$

7. Mentally estimate the number that should go in the blank to make each of these *true*.

   (a) 27% of ________ equals 16

   (b) 4 is ________ % of 7.5

   (c) 41% of 120 = ________

Thinking Critically

8. In a given population of men and women, 40% of the men are married and 30% of the women are married. What percentage of the adult population is married?

9. A garage advertised car repairs at 10% off the usual price—5% on parts and 5% on labor. The Consumer Protection Agency chided the garage for false advertising. Was the CPA correct? Explain.

10. After declining to do a "shady" job for Mayor Pigg for 25% of the take, Chester Slocum finally agreed to the deal when the Mayor offered him 25% of 25% of the take. What does this say about Chester's understanding of arithmetic? Explain briefly.

11. During the first half of a basketball game, the basketball team at Red Cloud High School made 60% of their 40 field goal attempts. During the second half, they scored on only 25% of 44 attempts from the field. To the nearest 1%, what was their field goal shooting percentage for the entire game?

12. During the first half of a basketball game Skeeter Thoreson missed all five of her field goal attempts. During the second half she hit 75% of her 16 attempts from the field. What was her field goal shooting percentage for the game?

Thinking Cooperatively

13. Tabata's Furniture calculates the retail price of furniture they sell by marking up their cost at wholesale a full 100%.

    (a) If they had a sale with all items marked down 20% what percentage profit did they actually make on each item sold during the sale?

    (b) If they had a sale with all items marked down 50%, what percentage profit did they make on each item sold during the sale?

14. Arrange the following in order of increasing size.

    $$\frac{19}{25},\quad 0.\overline{7},\quad 77\%,\quad 0.777,\quad \text{and}\quad \frac{15}{19}$$

Making Connections

15. The mortgage company requires an 11% down payment on houses they finance. If the Sumis bought a house for $158,000, how much down payment did they have to make?

16. Mr. Swierkos invested $25,000 in a mutual fund. If his broker deducted a 6% commission before turning the rest of the money over to the mutual fund and the value of each share increased by 18% during the year, what percentage return on his investment did Mr. Swierkos realize at the end of the first year?

17. Kneblemann's Fine Clothes was having a sale with all merchandise marked down 20%.

    (a) How much would Bethany have to pay for a dress originally priced at $78?

    (b) If Natasha paid $84 for a suit at Kneblemann's sale, what was the original price of the suit?

    (c) Explain a quick and easy way to calculate the sale price of items for Kneblemann's sale using a calculator.

18. Show that the sale price of items marked down 15% is the same as 85% of the original retail price.

19. Find the value of each of these investments at the end of the time period specified.

    (a) $2500 invested at $5\frac{1}{4}$% interest compounded annually for 7 years.

    (b) $8000 invested at 7% interest compounded semi-annually for 10 years.

Communicating

20. When asked about his performance in an upset victory in a football game, the quarterback said that he gave 110% of effort. Briefly discuss the reasonableness of this assertion.

21. In a stockholder's meeting, the chief executive officer said that the company had earned 110% of the previous year's profits. Briefly discuss the reasonableness of this assertion.

Using A Calculator

22. How much would you have to invest at 6% interest compounded annually in order to have $16,000 at the end of 5 years? (*Note:* If P is the amount invested, then $16{,}000 = P \cdot (1.06)^5$.)

23. Suppose you place \$1000 in a bank account on January 1 of each year for nine years. If the bank pays interest on such accounts at the rate of 5.3% interest compounded annually, how much will your investment be worth on January 1 of the tenth year?

24. Successive powers of $1 + \frac{r}{100}$ are easily computed by using the constant feature of your calculator. Since the value of an investment at $r\%$ interest compounded annually for n years is $P(1 + \frac{r}{100})^n$, determine the least number of years for the value of an investment to double at each of these rates.

(a) 5% **(b)** 7% (c) 14% **(d)** 20%

(e) The "rule of 72" states that you divide 72 by the interest rate to obtain the doubling time. This rule is used by many bankers to give a crude approximation to the time for the value an investment to double at a given rate. Does it seem like a reasonable rule? Explain briefly.

25. (a) If the population of Oregon in 1990 was 2,900,000 and it is increasing at the rate of 4% per year, what will the population be in the year 2010? Give your answer correct to the nearest one hundred thousand.

(b) Using the same assumptions as in part (a), what was the population of Oregon in 1988? (*Hint:* Use the constant feature of your calculator to divide repeatedly by 1.04.)

26. A merchant obtains the retail price of an item by adding 20% to the wholesale price. Later he has a sale and marks every item down 20% from the retail price. Is the sale price the same as the wholesale price? Explain briefly. (*Hint:* Consider a specific item whose wholesale price is \$100.)

27. A store purchased bicycles from a factory at \$140 each. If the store marked up the price 40% (that is, added 40% of the original cost to the original cost) to determine the selling price, what was the selling price?

For Review

28. Write each of these fractions in simplest form.

(a) $\frac{51}{69}$ **(b)** $\frac{143}{1001}$ **(c)** $-\frac{38}{57}$

29. Perform each of these additions and subtractions. Be sure to write your answer in simplest form.

(a) $\frac{1}{12} + \frac{1}{4}$ **(b)** $\frac{1}{5} + \frac{1}{12}$

(c) $\frac{8}{15} - \frac{17}{45}$ **(d)** $\frac{3}{143} - \frac{1}{91}$

(e) $\frac{11}{84} + \frac{5}{96}$ **(f)** $\frac{21}{60} - \frac{7}{75}$

30. Perform each of these multiplications and divisions leaving your answers as fractions in simplest form.

(a) $\frac{27}{44} \cdot \frac{22}{81}$ **(b)** $\frac{176}{247} \cdot \frac{95}{99}$

(c) $\frac{33}{35} \div \frac{22}{63}$ **(d)** $\frac{159}{286} \div \frac{102}{253}$

(e) $\frac{25}{44} \cdot \frac{68}{75} \cdot \frac{11}{16}$ **(f)** $\frac{169}{289} \div \frac{104}{221}$

31. Determine r, s, and t to make each of these equalities *true*.

(a) $\frac{25}{65} = \frac{r}{26}$ **(b)** $\frac{86}{99} = \frac{129}{s}$ **(c)** $\frac{4}{3} \div t = \frac{7}{6}$

EPILOGUE The Number Systems of Arithmetic

With this chapter we have completed our study of the number systems of arithmetic. There is one more extension—to the complex numbers—that is needed for the study of algebra and other more advanced courses, but it is inappropriate for teaching on the elementary school level.

Table 7.3 provides a summary of the number systems we have considered and the properties that each system satisfies.

It is important to note that:

- The commutative property for neither subtraction nor division holds in any of these systems.
- Only the rational numbers, the real numbers, and the prime hour clock arithmetic have multiplicative inverses for all nonzero (nonadditive identity) elements. Hence, only these systems are closed under division except for division by zero.

TABLE 7.3 Number Systems and Their Properties

Properties	Natural numbers	Whole numbers	Integers	Rational numbers	Real numbers	n-hour clock arithmetic, n composite	n-hour clock arithmetic, n prime
Closure property for addition	•	•	•	•	•	•	•
Closure property for multiplication	•	•	•	•	•	•	•
Closure property for subtraction			•	•	•	•	•
Closure property for division except for division by zero				•	•		•
Commutative property for addition	•	•	•	•	•	•	•
Commutative property for multiplication	•	•	•	•	•	•	•
Commutative property for subtraction							
Commutative property for division							
Associative property for addition	•	•	•	•	•	•	•
Associative property for multiplication	•	•	•	•	•	•	•
Associative property for subtraction							
Associative property for division							
Distributive property for multiplication over addition	•	•	•	•	•	•	•
Distributive property for multiplication over subtraction	•	•	•	•	•	•	•
Contains the additive identity; 0		•	•	•	•	• (n)	• (n)
Contains the multiplicative identity; 1	•	•	•	•	•	•	•
The multiplication property for 0 holds		•	•	•	•	•	•
Each element possesses an additive inverse			•	•	•	•	•
Each element except 0 possesses a multiplicative inverse				•	•		•

This table is convenient, but just knowing the properties is *not* enough. It is at least as important that teachers know conceptual and physical models appropriate to illustrate the properties and operations for each number system. They must also know how to use the systems to model and solve problems, including real-world problems. Understanding and critical thinking must be the goal of mathematical instruction—not memorizing properties and procedures by rote!

CLASSIC CONUNDRUM Some Curious Fractions

Compute the improper fraction equivalent to each of these complex fractions.

$$1, \quad 1+\frac{1}{1}, \quad 1+\cfrac{1}{1+\cfrac{1}{1}}, \quad 1+\cfrac{1}{1+\cfrac{1}{1+\cfrac{1}{1}}},$$

$$1+\cfrac{1}{1+\cfrac{1}{1+\cfrac{1}{1+\cfrac{1}{1}}}}, \quad 1+\cfrac{1}{1+\cfrac{1}{1+\cfrac{1}{1+\cfrac{1}{1}}}}$$

What pattern do you observe in the above results? What do you think would be the improper fraction corresponding to the twentieth complex fraction in this list which involves thirty-nine 1s in all? (*Hint:* Simplify by starting at the bottom each time. For example,

$$1+\cfrac{1}{1+\cfrac{1}{1+\cfrac{1}{1}}} = 1+\cfrac{1}{1+\cfrac{1}{2}}$$

$$= 1+\cfrac{1}{\cfrac{3}{2}}$$

$$= 1+\frac{2}{3}$$

$$= \frac{5}{3}.$$

CHAPTER 7 SUMMARY

Key Concepts

The primary thrust of this chapter has been to extend the notion of decimal expansion to numbers other than integers and, thereby, to extend the number system to the system of real numbers. The rational numbers have decimal expansions that are either terminating or are nonterminating but periodic. The nonterminating and nonperiodic decimals represent numbers that are not rational; that is, irrational numbers. The set of real numbers is the set of all rational and irrational numbers. Equivalently, it is the set of all decimals—terminating, nonterminating but periodic, and nonterminating and nonperiodic.

The discussion of decimals included a review of the methods of computing with decimals. In addition to the basic arithmetic operations, this included the notions of rounding and estimation, and scientific notation and calculator computations using this notation.

From computation with decimals, the study turned to the important notions of ratio and proportion and the meaning of a phrase like "y is proportional to x." These notions are particularly important in everyday applications:

- distance traveled at a constant rate is proportional to the time of travel;
- total cost of items of the same price is proportional to the number of items purchased;
- revenue from ticket sales (at the same price per ticket) is proportional to the number of tickets sold;

and so on, with applications too numerous to mention.

Finally, the chapter closed with a study of percent—perhaps the most important ratio in everyday discourse. This discussion was extended to such applications as markups and discounts (sales) in business, compound interest, population growth, and the effects of inflation or prices.

Vocabulary and Notation

Section 7.1

Real numbers
Decimals
Negative exponents
Multiplying and dividing decimals by powers of 10
Nonterminating decimals
Terminating decimals
Repeating decimals
Ordering decimals
Irrational numbers
The set of real numbers

Section 7.2

Rounding decimals
Adding and subtracting decimals
Multiplying decimals
Dividing decimals
Scientific notation

Section 7.3

Ratio
Proportion
Conditions for a proportion
Proportional to
Constant of proportionality

Section 7.4

Percent
Expressing decimals as percents
Expressing percents as decimals
Expressing percents as fractions
Expressing fractions as percents
Calculating a percentage of a number
Calculating a number that is a given percent of another number
Calculating the percentage one number is of another
Compound interest
Mathematics of growth

CHAPTER REVIEW EXERCISES

Section 7.1

1. Write these decimals in expanded exponential form.
(a) 273.425 (b) 0.000354
2. Write these fractions in decimal form.
(a) $\frac{7}{125}$ (b) $\frac{6}{75}$ (c) $\frac{11}{80}$
3. Write these decimals as fractions in simplest form.
(a) 0.315 (b) 1.206 (c) 0.2001
4. Arrange these numbers in order from least to greatest. $\frac{4}{12}$, 0.33, 0.3334, $\frac{5}{13}$, $\frac{2}{66}$
5. Write these numbers as fractions in simplest form.
(a) $10.\overline{363}$ (b) $2.1\overline{42}$
6. Suppose $a = 0.20200200020000200000 2\ldots$ continuing in this way with one more zero between each successive pair of 2s. Is this number rational or irrational? Explain briefly.
7. Using only mental arithmetic determine the numbers represented by these base 10 numerals as fractions in simplest form.
(a) $0.222\ldots = 0.\overline{2}$ (b) $0.363636\ldots = 0.\overline{36}$

Section 7.2

8. Perform these computations by hand.
(a) 21.734 + 3.2145 + 71.24 (b) 23.471 − 2.89
(c) 35.4 × 2.37 (d) 24.15 ÷ 3.45
9. Compute the following using a calculator.
(a) 31.47 + 3.471 + 0.0027
(b) 31.47 − 3.471
(c) 31.47 × 3.471
(d) 138.87 ÷ 23.145
10. Write estimates of the results of these calculations, then do the computing accurately with a calculator.
(a) 47.25 + 13.134 (b) 52.914 − 13.101
(c) 47.25 × 13.134 (d) 47.25 ÷ 13.134
11. Write each of these in scientific notation using 4 significant digits.
(a) 24,732,654 (b) 0.000012473
12. Using an appropriate calculator and scientific notation perform each of these calculations and write the results using scientific notation with 3 significant digits.
(a) $(2.74 \times 10^5) \cdot (3.11 \times 10^4)$
(b) $(2.74 \times 10^5) \div (3.11 \times 10^{-4})$
13. Show that $3 - \sqrt{2}$ is irrational.
14. Show that the sum of two irrational numbers can be rational. (*Hint:* Consider problem 13.)
15. What can you say about the decimal expansion of an irrational number?
16. (a) A wall measures 8.25 feet by 112.5 feet. What is the area of the wall?
(b) If it takes 1 quart of paint to cover 110 square feet, how many quarts of paint must be purchased to paint the wall in part (a)?
17. Give an example of a fraction whose decimal is repeating and has a period of length 4.
18. Use your calculator to determine the periodic decimal expansions of the following numbers. Remember that the calculator will necessarily round off decimals, so don't be mislead by the last digit in the display if it seems to break a pattern.
$$\frac{5}{18}, \frac{41}{333}, \frac{11}{36}, \frac{7}{45}, \frac{13}{80}$$
(a) Determine where the period starts in each case.
(b) See if you can guess a rule for determining when the period of the decimal form of a fraction a/b begins.

Section 7.3

19. Maria made 11 out of 20 free throw attempts during a basketball game. What was the ratio of her successes to failures on free throws during the game?
20. Determine which of these are proportions.
(a) $\frac{775}{125} = \frac{155}{25}$ (b) $\frac{31}{64} = \frac{15}{32}$ (c) $\frac{9}{24} = \frac{12}{32}$
21. If Che bought 2 pounds of candy for \$3.15, how much would it cost him to buy 5 pounds of candy at the same price per pound?
22. It took Donnell 7.5 gallons of gas to drive 173 miles. Assuming that he gets the same mileage per gallon, how much gasoline will he need to travel 300 miles?
23. If y is proportional to x and $y = 7$ when $x = 3$, determine y when $x = 5$.
24. If a flag pole cast a shadow 12′ long when a yardstick cast a shadow 10″ long, how tall is the flag pole?

Section 7.4

25. Convert each of these to percents.

    (a) $\frac{5}{8}$ (b) 2.115 (c) 0.015

26. Convert each of these percents to decimals.

    (a) 28% (b) 1.05% (c) $33\frac{1}{3}\%$

27. If the sales tax is calculated at 7.2%, how much tax is due on a $49 purchase?

28. If a tax of $6.75 is charged on an $84.37 purchase, what is the sales tax rate?

29. Referring to problem 19 above, what percent of free throws attempted did Maria make during the game?

30. If you invest $3000 at 8% interest compounded every 3 months (quarterly), how much is your investment worth at the end of two years?

CHAPTER TEST

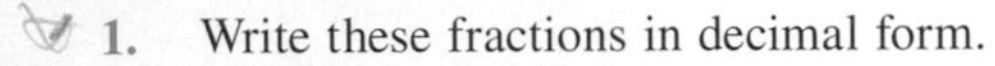

1. Write these fractions in decimal form.

   (a) $\frac{84}{175}$ (b) $\frac{24}{99}$ (c) $\frac{7}{11}$

2. Write each of these decimals as a fraction in reduced form.

   (a) $0.454545\cdots = 0.\overline{45}$
   (b) $31.5555\cdots = 31.\overline{5}$
   (c) $0.34999\cdots = 0.34\overline{9}$

3. Without doing the hand or calculator calculation, determine how many digits should appear to the right of the decimal point in the product 21.432×3.41.

4. Compute the product $(2.34 \times 10^6) \cdot (3.12 \times 10^{-19})$ using your calculator. Write your answer correct to three significant digits.

5. Give an example of a fraction whose decimal expansion is repeating of period 3.

6. The Pirates won 17 of their 32 hockey games.

   (a) What was the ratio of their wins to losses?

   (b) What percentage of their games did they win?

7. Mr. Spence paid $1425 down on a car selling for $9500. What percent of the purchase price did the dealer require as down payment?

8. If you invest $2000 now in a bank paying 4.2% interest compounded semi-annually, what is the least whole number of years you must leave your investment in the bank in order to withdraw at least $4000?

9. When did you invest $5000 in a bank paying $5\frac{3}{4}\%$ interest compounded annually if it is worth $6612.60 now?

10. Suppose that you borrow $1000 now at 9% compounded monthly. If you make no payments in the meantime, how much will you owe at the end of two years?

8 Statistics: The Interpretation of Data

Computer Support for this Chapter

In this chapter you may find a statistical package like Stat Explorer from your computer laboratory useful.

How Many Are Marked?

The choice of r must be the same for each container and should be unknown to the students.

Materials Needed

1. For each group of 10 students, a container with 100 white beans with r of the beans marked with a red dot where r is determined by the instructor.
2. A record sheet for each student as shown.

Name ______________

1. Number of marked beans in the sample. ________
2. Percentage of the sample consisting of marked beans. ________
3. Estimate of the percentage of marked beans in the container. ________
4. Total number of marked beans in all samples in the class. ________
5. Percentage of marked beans in all samples (the combined sample) in the class. ________
6. Estimate of the percentage of marked beans in each container based on the result of Step 5. ________
7. Statement of your opinion of the accuracy of the estimates in Steps 3 and 6 and why you believe this to be so. ________

____
____
____
____

Directions

Step 1. In turn, each student:

- thoroughly mixes the beans in the container;
- with eyes closed, selects a sample of 10 beans from the container;
- notes the number of marked beans in his or her sample;
- returns the beans to the container and passes it on to the next student;
- completes the first three lines on his or her record sheet and records on the chalkboard the number of marked beans in his or her sample.

Step 2. When all students have completed Step 1, each should use the data on the chalkboard and complete the remainder of his or her record sheet.

Step 3. Ask the instructor what the actual percentage of marked beans in the containers was and discuss with the class the results of the activity.

CONNECTIONS

The Onslaught of Information

Every day Americans are confronted by a deluge of "facts" and figures as shown in Figure 8.1. The public media is replete with assertions like—

> *"Fully 18% of Americans currently live below the poverty line."*
> *"Studies show that 62% of teenagers in America are sexually active."*
> *"According to the most recent Gallup poll, only 39% of Americans approve of the way the President is handling his job."*
> *"Eight out of ten dentists surveyed prefer Whito Toothpaste."*

How are such figures obtained? Are they accurate? Are they reliable? Are they misleading? Do polling organizations actually check with every American before issuing such statements? These are serious questions since the figures quoted often form the basis, not only for individual decisions, but for decisions made by government—decisions that affect us all. The fact is that many such statements are reliable while others are questionable at best.

In this chapter, we consider how assertions like these are generated and with what degree of confidence they can be believed. Without such understanding it is necessarily the case that much of what goes on in daily life must remain a mystery to be accepted or rejected on the basis of whim, or impression, or faith—a situation that is clearly less than satisfactory.

As a foundation for understanding, we must first know how data are collected, interpreted, and presented to the public.

8.1 The Graphical Representation of Data

Line Plots

In a class for prospective elementary school teachers the final examination scores for the students were as shown in Table 8.1. These are the **data** simply recorded in a list.

Is the Bush administration doing as much as it can to reduce crime at the local level?

	Whites	Blacks
Yes	24%	18%
No	65%	76%

Is the Bush administration doing as much as it can to guarantee equal justice for black Americans?

	Whites	Blacks
Yes	37%	17%
No	49%	78%

Figure 8.1
Polls in the news
Source: From the *Newsweek* Poll of April 30 May 1, 1992, "Bush administration/crime?" from *Newsweek,* May 11, 1992, p. 42.

TABLE 8.1 Final Examination Scores in Mathematics for Elementary School Teachers, Section 1

79	78	79	65	95	77	49
91	63	58	78	96	74	68
71	86	91	94	79	69	86
62	78	77	88	67	78	84
69	53	79	75	64	89	77

Just scanning the data gives some idea how the class did, but it is more revealing to organize the data by representing each score by an × placed above a number line as in Figure 8.2. Data depicted in this way is called a **line plot.** The line plot makes it possible to see at a glance that the scores ranged from 49 through 96, that most scores were between 60 and 80 with a large group between 75 and 80, and that the "typical" score was probably about 77 or 78. It also reveals that a score like 49 is quite atypical. Such a score is called an **outlier** since it is very unlike the other scores as a whole. Data organized and displayed on a line plot are much easier to interpret than raw data.

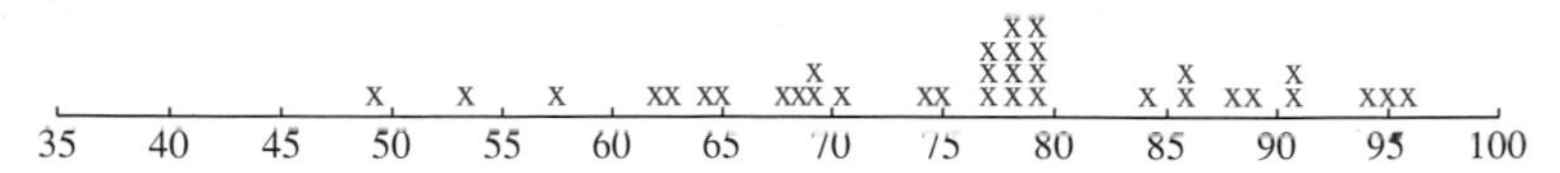

Figure 8.2
Line plot for final examination scores in Mathematics for Elementary School Teachers, Section 1

FROM THE NCTM STANDARDS **Statistics and Probability in the Classroom**

Collecting, organizing, describing, displaying, and interpreting data, as well as making decisions and predictions on the basis of that information, are skills that are increasingly important in a society based on technology and communication. These processes are particularly appropriate for young children because they can be used to solve problems that often are inherently interesting, represent significant applications of mathematics to practical questions, and offer rich opportunities for mathematical inquiry. The study of statistics and probability highlights the importance of questioning, conjecturing, and searching for relationships when formulating and solving real-world problems.

A spirit of investigation and exploration should permeate statistics instruction. Children's questions about the physical world can often be answered by collecting and analyzing data. After generating questions, they decide what information is appropriate and how it can be collected, displayed, and interpreted to answer their questions. The analysis and evaluation that occur as children attempt to draw conclusions about the original problem often lead to new conjectures and productive investigations. This entire process broadens children's views of mathematics and its usefulness.

Statistics and probability are important links to other content areas, such as social studies and science. They also can reinforce communication skills as children discuss and write about their activities and their conclusions. Within mathematics, these topics regularly involve the uses of number, measurement, estimation, and problem solving.

SOURCE: From *Curriculum and Evaluation Standards for School Mathematics Grades K–4,* p. 54.

Stem and Leaf Plots

Stem and leaf plots for displaying data are very similar to line plots and are particularly useful for comparison between two sets of data.

To draw a stem and leaf plot for the data in Table 8.1 we let the tens digits of the scores be the stems and let the units digits be the leaves. Thus, the scores 79, 78, and 79 are represented by

stem leaves

7 | 8 9 9

The completed plot appears in Figure 8.3.

```
4 | 9
5 | 3 8
6 | 2 3 4 5 7 8 9 9
7 | 1 4 5 7 7 7 8 8 8 8 9 9 9 9
8 | 4 6 6 8 9
9 | 1 1 4 5 6
```

Figure 8.3
Stem and leaf plot of the final examination scores in Mathematics for Elementary School Teachers, Section 1

The stem and leaf plot gives much the same visual impression as the line plot and allows a similar interpretation.

To compare two sets of similar data, it is useful to construct stem and leaf plots on the same stem. Figure 8.4 shows such a plot for final examination scores in Sections 1 and 2 of Mathematics for Elementary School Teachers.

Figure 8.4
Stem and leaf plots for final examination scores in Mathematics for Elementary School Teachers, Sections 1 and 2

Section 2		Section 1
3	4	9
9 8 7 5	5	3 8
8 8 5 5 5 5 3 1	6	2 3 4 5 7 8 9 9
5 5 4 4 3 0	7	1 4 5 7 7 7 8 8 8 8 9 9 9 9
9 7 6 4 4 2 0 0 0	8	4 6 6 8 9
6 5 5 0	9	1 1 4 5 6
0 0 0	10	

In this figure it is easy to see that, while the two classes are quite comparable, Section 2 had a wider range of scores with one lower and several higher than those in Section 1.

Histograms

Another common tool for organizing and summarizing data is a **histogram.** A histogram for the data in Table 8.1 is shown in Figure 8.5. In a histogram, scores are grouped in intervals and the number of scores in each interval is indicated by the height of the rectangle constructed above the interval. The number of times any particular data value occurs is called its **frequency.** Similarly, the number of data values in any interval is the **frequency of the interval.** Thus, the vertical axis of a histogram indicates frequency and the horizontal axis indicates data values or ranges of data values.

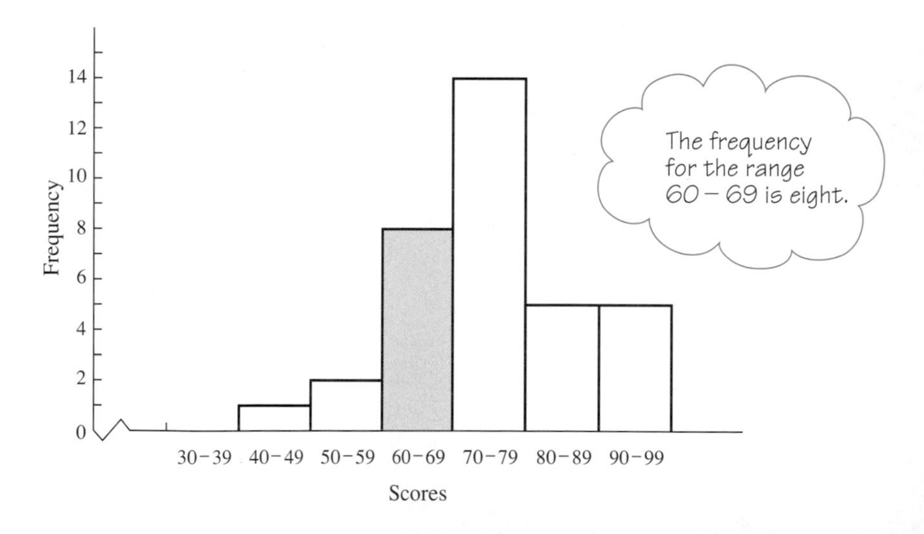

Figure 8.5
Histogram of final examination scores in Mathematics for Elementary School Teachers, Section 1

We lose some detail with the histogram. For example, the above histogram does not show how many students scored exactly 71. However, it has the advantage of giving a compact and accurate summary that is particularly useful with large collections of data that could not be conveniently represented using line or stem and leaf plots.

Histograms are useful for giving visual summaries of data that are either discrete or that vary continuously. For example, peoples' heights vary continuously. You may say that your height is 5′4″ (or 64″), but that is not exactly true. Instead, it is true only *to the nearest inch.* Consider the data in Table 8.2 that gives the heights of the 80 boys at Eisenhower High School. The heights were measured to the nearest inch so the numbers given already represent *grouped data;* that is, a measurement of 66 was recorded if a boy's height was judged to be between 65.5 and 66.5 inches. Even a

JUST FOR FUN

A Matter of Speed

In driving the 300 miles from Spokane to Seattle, Liz drove 50 miles per hour for the first 250 miles. How fast should she drive for the last 50 miles in order to average 60 miles per hour for the trip?

TABLE 8.2 Heights to the Nearest Inch of Boys at Eisenhower High School Arranged in Increasing Order

64	67	68	69	69	70	71	72
65	67	68	69	69	70	71	72
66	68	68	69	69	70	71	72
66	68	68	69	69	70	71	72
66	68	68	69	69	70	71	72
67	68	68	69	69	70	71	72
67	68	68	69	70	70	71	73
67	68	69	69	70	70	71	73
67	68	69	69	70	70	71	74
67	68	69	69	70	70	71	74

height of almost exactly 66.5 inches was grouped into the 66 or 67 inch class as deemed most appropriate by the person doing the measuring.

A histogram representing this data is shown in Figure 8.6. Note that, in both Figure 8.5 and Figure 8.6, the sum of the heights of the rectangles gives the number of data values.

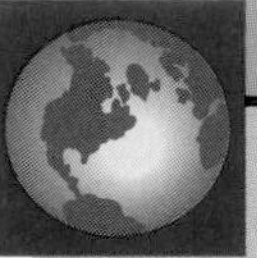

CURRENT AFFAIRS

All Polls Are Not Created Equal

The news media are obsessed with polls; almost every major newspaper and television station conducts its own polls. Unfortunately, the increase in quantity has not produced better quality. In the past, media polling was criticized because simplistic and sensational tones often led to unenlightened coverage. But a new and more fundamental problem has arisen—a polling credibility gap.

The 1990 cycle was among the first with an explosion in methodologically unsound media surveys conducted by so-called pollsters without sophisticated research training or campaign expertise. Too many election-eve media polls were well beyond the margin of error, and some projected the wrong winner. The Newark *Star-Ledger,* for instance, bestowed a 17-percentage-point lead on New Jersey Sen. Bill Bradley just seven days before he barely won with 3 points. And media pollsters have had several embarrassing failures this year. In New Hampshire, most surveys exaggerated Paul Tsongas's lead over Bill Clinton and underestimated support for Pat Buchanan.

Meanwhile, campaign-conducted polling has become a highly precise science. Most voters are unfamiliar with the differences between candidate and media polling. Media polling is designed primarily to predict an outcome. Campaign polling is designed primarily to prepare a strategy to affect an outcome. As a result, sampling techniques are different, wording is different and the analysis is different. If campaign pollsters had the same record as the average media polling outfit, they would quickly be out of business. Campaign pollsters live or die by their precision, particularly in the final days of a race. Media pollsters live by the audience they garner for their clients, not by accuracy. The more outlandish the predictions or polling result, the greater the public interest.

SOURCE: From "All polls are not created equal" from *U.S. News and World Report,* September 28, 1992, p. 24.

Grouping data into classes and displaying the data in a histogram is a useful visualization of the characteristics of the data set. In drawing a histogram, the scales should be chosen so that all the data can be represented. Also, the number of classes into which the data is grouped should not be so few that it hides too much information and not so numerous that one loses the visual advantage of constructing the diagram in the first place.

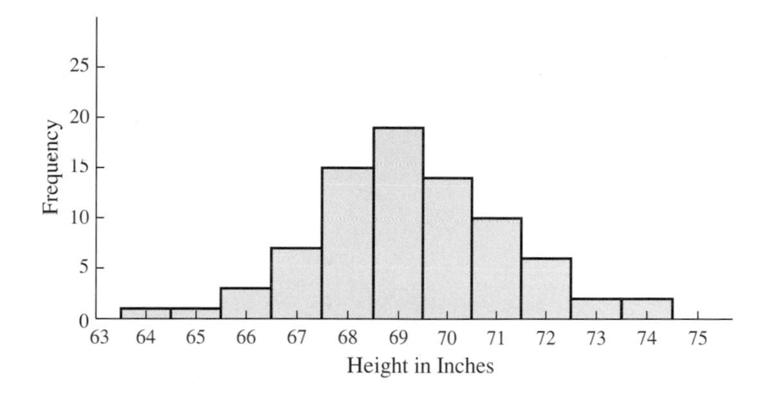

Figure 8.6
Histogram of heights of boys in Eisenhower High School

Line Graphs

A **line graph** for the data in Figure 8.6 is constructed by joining the midpoints of the tops of the rectangles in the figure by line segments (see Figure 8.7). Without the rectangles, which would not ordinarily be drawn, the line graph appears as in Figure 8.8. Since the vertical axis represents the frequency with which measurements occur, the line graph of a set of data like this is often called a **frequency polygon.**

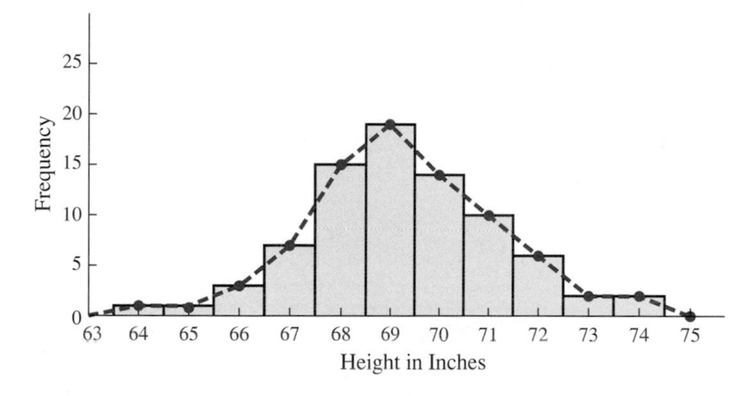

Figure 8.7
Histogram and line graph for the heights of boys at Eisenhower High School

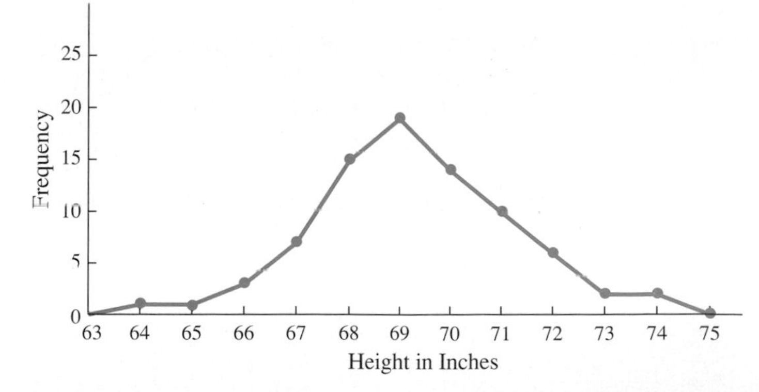

Figure 8.8
Line graph (frequency polygon) for the heights of boys at Eisenhower High School

A line graph or frequency polygon is particularly appropriate for representing data that vary continuously since this is strongly suggested by the sloping line segments. The rectangles in the histogram of Figure 8.6 tend to obscure the fact that the heights of boys represented by the rectangle centered at 67 range from 66.5″ to 67.5″. On the other hand, if the data really is discrete, it is probably better represented by a histogram than a frequency polygon. Note that the area under the frequency polygon or line graph now gives the approximate number of data values and the area under the graph and above a given interval now gives approximately the number of boys whose heights fall in that interval.

Line graphs are particularly effective when they are used to indicate trends over periods of years—trends in the stock market, trends in the consumption of electrical energy, and so on. For example, consider the data in Table 8.3 that gives the yearly consumer expenditure for food in the United States at five-year intervals from 1950 through 1990. This is represented visually by the line graph in Figure 8.9.

TABLE 8.3 Consumer Expenditure for Food in the United States in Billions of Dollars

Year	1950	1955	1960	1965	1970	1975	1980	1985	1990
Expenditure	44.0	53.1	66.9	81.1	110.6	167.0	264.4	345.4	440.8

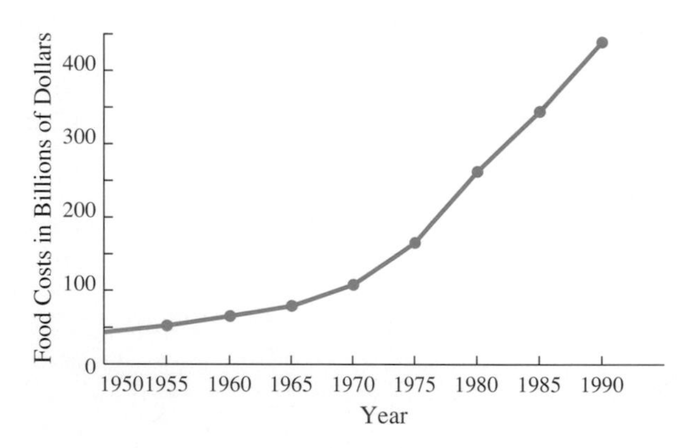

Figure 8.9
U.S. consumer expenditure for food in billions of dollars

One advantage of a line graph is that it makes it possible to estimate data values not explicitly given otherwise.

EXAMPLE 8.1 Estimating Data Values from a Line Graph

Using the graph in Figure 8.9, estimate the U.S. consumer expenditure for food in 1972.

SOLUTION

Understand the problem • • • •

The graph gives the expenditure at five-year intervals. We are asked to estimate the expenditure for 1972.

Devise a plan • • • •

Having Figure 8.9 already simplifies our task. The graph suggests that the total expenditure grows steadily each year and, while the growth is almost surely not "straight line growth" between data points as indicated by the diagram, the straight line joining the data points for 1970 and 1975 surely approximates the actual growth. If we draw a vertical line from the point representing 1972 on the horizontal axis, the height of the line segment should give us the approximate expenditure for 1972.

Carry out the plan • • • •

The point on the horizontal axis representing 1972 is two-fifths of the way from 1970 to 1975. Draw a vertical line from this point to the line graph. Then draw a horizontal line from the point where this line cuts the line graph to the vertical axis. This determines the point on the vertical axis that gives approximately $120 billion as the value of U.S. consumer expenditure for food in 1972. Note that these measurements are most easily carried out with a metric ruler since such a ruler is marked off in tenths of centimeters (millimeters) and hence lends itself to determination of numbers in base ten notation.

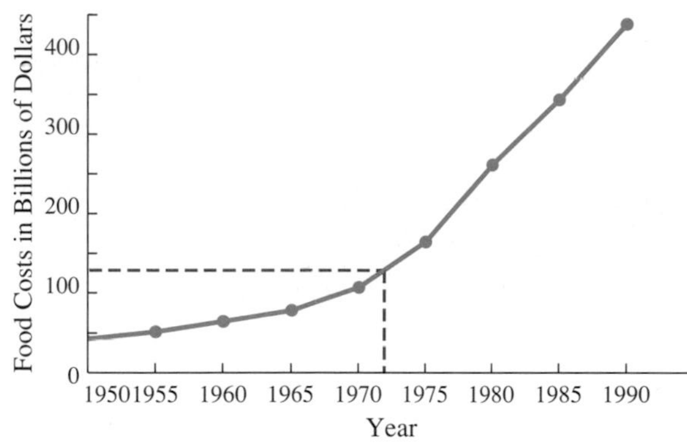

Look back • • • •

The solution was achieved by noting that the line graph, which gives the appropriate values of expenditures every five years, suggests that the expenditures steadily increase and that the actual values for intervening years no doubt lie relatively close to the straight line segments joining the given data points. Indeed, it appears that a curved line through the data points might give an even better approximation somewhat less than $120 billion. However, as an estimate, $120 billion is reasonably accurate. ■

Bar Graphs

Bar graphs, similar to histograms, are often useful in conveying information about so-called categorical data where the horizontal scale represents some nonnumerical attribute. For example, consider the final examination scores for Mathematics for Elementary School Teachers, Section 1, as listed in Table 8.1. Suppose that the instructor determines grades as indicated in the accompanying table. Then members of the class were awarded 5 As, 5 Bs, 14 Cs, 8 Ds, and 3 Fs. If we indicate grades on the horizontal scale and frequency on the vertical scale, we can construct the bar graph shown in Figure 8.10. In general, the rectangles in a bar graph do not abut and

Score	Grade
90–100	A
80–89	B
70–79	C
60–69	D
0–59	F

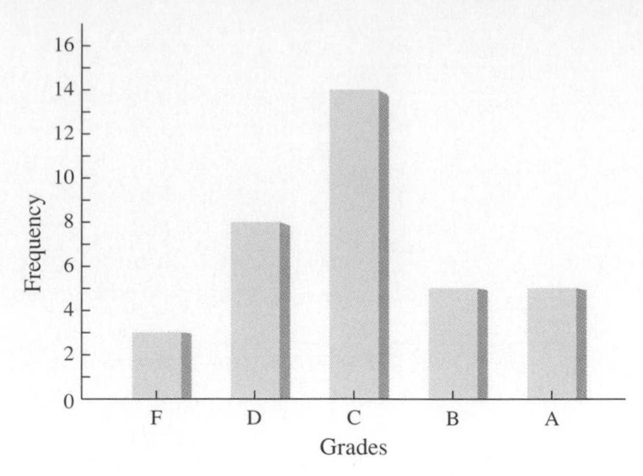

Figure 8.10
Bar graph of the final examination scores in Mathematics for Elementary School Teachers, Section 1

the horizontal scale may be designated by any attribute—grade in class, year, country, city, and so on. As usual, however, the vertical scale will denote frequency—the number of items in the given class.

Bar graphs are useful in displaying data concerning nonnumerical items. As the following example shows, they are also useful in comparing data concerning two or more similar groups of items.

EXAMPLE 8.2

Comparing Grades in Two Mathematics Classes by Means of a Bar Graph

Draw a suitable bar graph to make a comparison of the grades in Mathematics for Elementary School Teachers, Sections 1 and 2.

SOLUTION

The desired bar graph can be obtained by drawing two adjacent bars (rectangles) for each letter grade with a suitable indication of which bars to associate with each section. If we use blue bars for Section 1 and red bars for Section 2, a suitable graph might look like this. (See Figure 8.4 for the scores in Section 2. These scores merit 7 As, 9 Bs, 6 Cs, 8 Ds, and 5 Fs, using the same scale as for Section 1 above.)

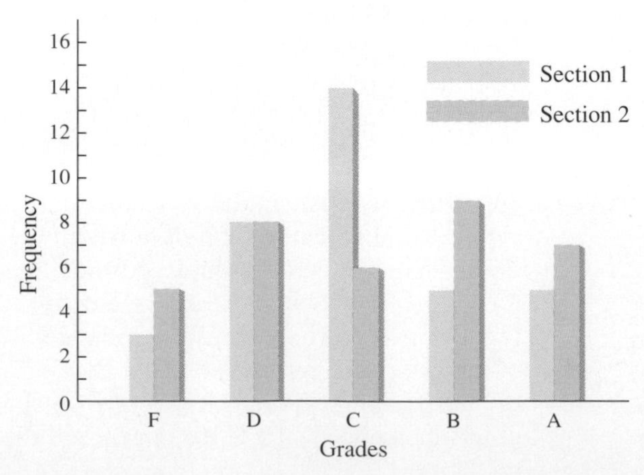

■

INTO THE CLASSROOM

Clem Boyer Comments on Statistics in Elementary School

In learning statistics, graphing, and probability, students are learning about the real world. When they work with tables, charts, and graphs and use the language and notation of graphing mathematics, they are developing important real-life skills in reading, interpreting, and communicating information. Working with statistics prepares students to deal with the endless number of statistics in today's world. In using probability to predict outcomes, they are learning how to cope mathematically with the uncertainties in the real world. Students should plan and carry out the collection and organization of data to satisfy their curiosity about everyday living. They need to construct, read, and interpret simple maps, tables, charts, and graphs. In doing these things, they find out how to present information about the numerical data.

In order for students to manage statistics in this age of technology, it is important for them to learn to find measures of central tendency (mean, median, and mode) and measures of dispersion (range and deviation). Further, students need to recognize the basic uses and misuses of statistical representation and inference in order for them to be wise consumers.

All of these skills in working with data improve students' ability to interpret the data they read and hear about every day. Being able to use the terminology when displaying data will help students communicate their findings.

Learning probability has applications in the real world too. Students find out how to identify situations in which immediate past experience does not affect the likelihood of future events. Their lives are enriched when they can see how mathematics is used to make predictions regarding election results, business forecasts, and sporting events.

SOURCE: From *ScottForesman Exploring Mathematics* Grades 1–7 by L. Carey Bolster et al. Copyright © 1994 Scott, Foresman and Company. Reprinted by permission of Scott, Foresman and Company. Clem Boyer is the former Coordinator of Mathematics, K–12, for the District School Board of Seminole County in Sanford, Florida.

Maintenance 11%
Misc. 3%
Administration 11%
Teachers' salaries 78%

Figure 8.11
Percent of each tax dollar expended by Mile High School District by category

Pie Charts

Another pictorial method for conveying information is a **pie chart.** For example, the pie chart in Figure 8.11 shows the parts of the budget of Mile High School District used for various purposes. The number of degrees in the angular measure of each part of the chart is the appropriate fraction or percentage of 360°. Thus, the angular sector for the portion representing administration measures

$$0.11 \times 360^\circ = 39.6^\circ$$

and so on. As shown here, pie charts are most often used to show how a whole (total budget, total revenues, total sources of oil, and so on) is divided up.

EXAMPLE 8.3 **Making a Pie Chart**

Ajax Steel Fabricators had a gross income of \$10,895,000 for the 1993–94 fiscal year. The expenses for the year were: labor—\$5,120,650;

materials—$4,064,450; new equipment—$329,550; and plant maintenance—$549,250; leaving $871,600 profit. Draw a pie chart illustrating how the income for Ajax was spent that year.

SOLUTION

Draw a circle and divide it into sectors whose central angles are appropriate fractions of 360°. To the nearest degree the angles for the various classes of expenditures are:

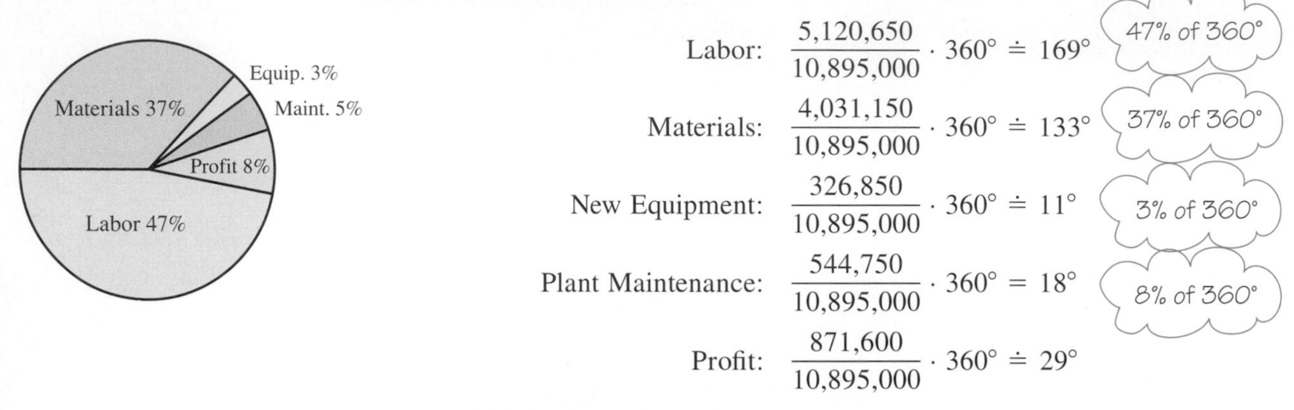

$$\text{Labor: } \frac{5{,}120{,}650}{10{,}895{,}000} \cdot 360° \doteq 169°$$

$$\text{Materials: } \frac{4{,}031{,}150}{10{,}895{,}000} \cdot 360° \doteq 133°$$

$$\text{New Equipment: } \frac{326{,}850}{10{,}895{,}000} \cdot 360° \doteq 11°$$

$$\text{Plant Maintenance: } \frac{544{,}750}{10{,}895{,}000} \cdot 360° = 18°$$

$$\text{Profit: } \frac{871{,}600}{10{,}895{,}000} \cdot 360° \doteq 29°$$

Using these as the central angles for the sectors, we obtain the pie chart shown. Such a diagram gives a quick mental image of the relative amounts of the budget spent in each category. ■

Increasingly, pie charts and other pictorial representations of data are drawn by computer and colored to give a more pleasing effect to the eye. If the pie chart is drawn in perspective, as if seen from an angle as in Figure 8.12, the central angles are no longer completely accurate. However, the pie chart still gives a good visual understanding of the apportionment of the whole being discussed. Also, in Figure 8.12, the pieces of the pie are separated slightly to produce a more pleasing visual effect.

Where the Income Came From:

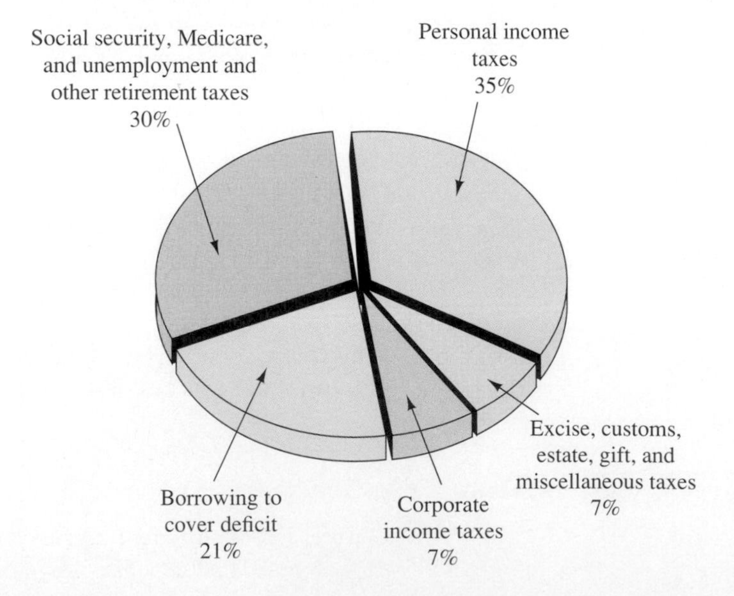

Figure 8.12
Pie chart showing U.S. government sources of revenue for fiscal year 1991. Source: 1992 IRS Form 1040 instruction booklet

Pictographs

A **pictograph** is a picture or set of pictures used to represent data and often to represent trends. Usually the picture is suggestively related to the data being represented. Consider the pictograph presentation of past and projected growth in world population in Figure 8.13. The pictograph accurately indicates that the world population more than doubled over the 40 year period from 1950 to 1990. It also suggests that, while the rate of increase is expected to diminish, the population will double again in the next 60 years.

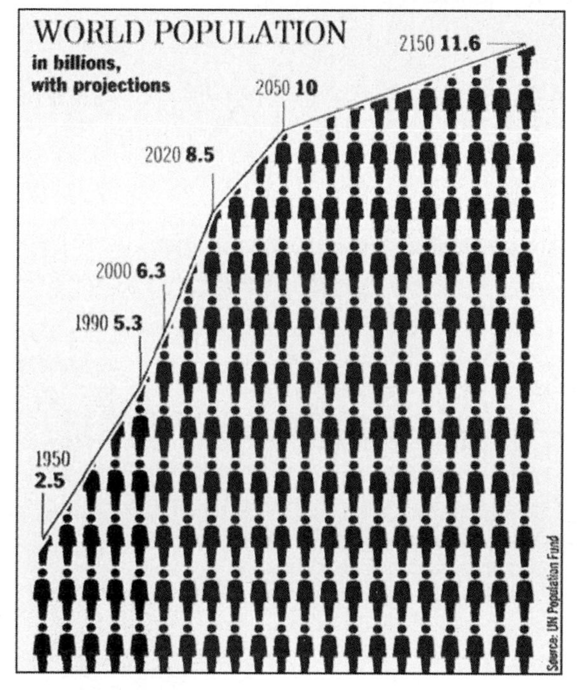

Figure 8.13
Pictograph of world population growth
SOURCE: Graph, "World Population" from *Time*, June 1, 1992, p. 54. Copyright © 1992 by Time Inc. Reprinted by permission.

Choosing Good Visualizations

Each of the graphical representations discussed is appropriate to summarize and present data so that the reader can visualize frequencies and determine trends. The various representations are more appropriate in some instances than others and most are subject to serious distortion if the intent is to mislead the reader. For example, the pictograph in Figure 8.14 represents the oil consumption in the United States for the years 1982 and 1992.

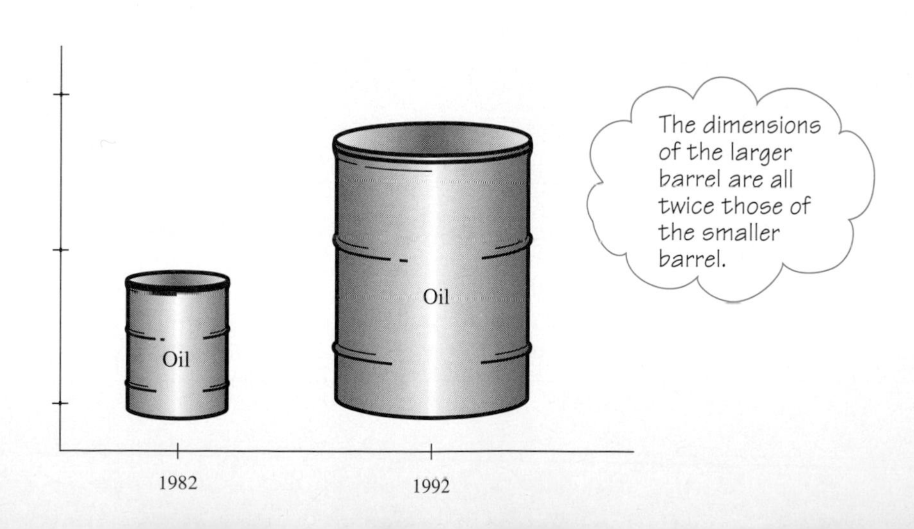

Figure 8.14
Oil consumption in the United States in 1982 and 1992

SCHOOL BOOK PAGE *Statistics in Fifth Grade*

SOCIAL STUDIES CONNECTION

Organizing Data

Build Understanding

What kind of job would you like after you finish school?

A. Aki collected data for a social studies report on new jobs expected to be available in the year 2000. With her data, she prepared a table to display the information.

After looking at the table, Aki saw that she could display her information better if she put it into a ***bar graph.*** A bar graph is one way to display information using numbers. The display helps you to compare quantities easily. How would you find the number of new jobs in food services from the graph?

Find the bar for food services. Move down from the end of the bar to the number on the horizontal scale.

What Kind of Job...

Type of Job	Number of New Jobs
Health Offices	1,400,000
Personnel	832,000
Food Services	2,500,000
Construction	890,000
Nursing	852,000

Where the New Jobs Will Be

Health Offices
Personnel
Food Services
Construction
Nursing

0 0.5 1 1.5 2 2.5

Number of Jobs (in millions)

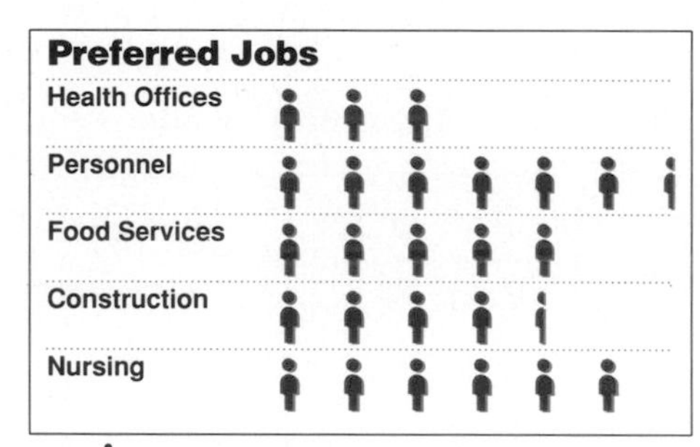

Each [figure] stands for two students.

B. Stella asked 50 students which of these jobs they preferred. The ***pictograph*** shows the results. A pictograph is another way to display quantities. It uses pictures or symbols.

How many students prefer personnel work?

Since [figure] stands for two students, [half figure] stands for one student.

There were 13 students who preferred personnel work.

1. The bar graph above has horizontal rather than vertical bars. Is there any reason to prefer one orientation over the other?
2. Write three questions you might ask students on the basis of the bar graph shown above.
3. What problems might students have in interpreting the pictograph above?
4. What problems might students have in drawing a pictograph? What help might you need to give?

While the vertical scale honestly indicates that approximately twice as much oil was used in 1992 as in 1982, the pictograph is misleading since the volume of the larger barrel is *eight times* the volume of the smaller barrel. The casual reader is quite likely to get a badly distorted idea of the relative amount of oil used in the two years. Of course, that may be exactly what the person who constructed the pictograph intended, particularly if the vertical scale is omitted as it is here. Have you ever seen such distortions on television? In the newspaper? In advertisements? Be observant the next time you see such a diagram.

Particularly suitable uses for the various visual representations of data described in this section are summarized here.

- Line plots: summarizing relatively small sets of data—grades in a class, heights of students in a class, birth months of students in a class, and so on.
- Stem and leaf plots: for essentially the same purposes as line plots; especially useful in comparing small data sets.
- Histograms: summarizing information in large sets of data that can be naturally grouped into intervals.
- Line graphs: summarizing trends over time.
- Pie charts: representing relative amounts of a whole.
- Pictographs: summarizing relative amounts, trends, and small data sets; useful in comparing quantities.

PROBLEM SET 8.1

Understanding Concepts

1. The scores below were obtained on the final examination in an introductory mathematics class of forty students.

98	80	98	76	79	94	71	45	89	71
62	61	95	77	83	49	65	58	56	89
66	87	74	64	75	58	72	75	48	88
75	51	84	76	95	69	61	69	33	86

(a) What is the highest score?

(b) Scanning the data, what do you think the "typical" or "average" score is?

(c) Make a line plot to organize this data.

(d) Looking at the line plot, what seems to be the "typical" score?

(e) Do you identify any scores that seem to be outliers for this data set? Explain.

(f) Write a two or three sentence description of the results of the final examination.

2. Make a stem and leaf plot of the data in problem 1.

3. Make a histogram for the data in problem 1 using the ranges 20–29, 30–39, . . . , 90–99 on the horizontal scale.

4. At the same time heights of the boys at Eisenhower High School were studied, heights of the girls were also studied.

(a) Draw a histogram to summarize this data that gives the heights to the nearest inch of the 75 girls at Eisenhower High School. Use intervals one unit wide centered at the whole number values 56, 57, . . . , 74.

57	62	63	64	66
58	62	63	64	66
60	62	63	64	66
60	62	63	65	66
61	62	63	65	66
61	62	63	65	66
61	62	63	65	66
61	63	64	65	66
61	63	64	65	66
61	63	64	65	66
62	63	64	65	67
62	63	64	65	67
62	63	64	65	70
62	63	64	65	70
62	63	64	66	73

(b) Write two or three sentences describing the distribution of the heights of the girls.

5. **(a)** Draw a frequency polygon for the data in problem 4 by joining the midpoints of the tops of the rectangles of the histogram by straight line segments.
   (b) What does the area under the frequency polygon and between the scores 60.5 and 64.5 represent? Explain briefly.

6. The scores on the first, second, and third hour tests given in a class in educational statistics as the term progressed are shown.

First hour test:	92,	80,	73,	74,	93,	75,	76,	68,	61,	76,
	83,	94,	63,	74,	76,	86,	82,	70,	65,	74,
	83,	87,	98,	77,	67,	64,	87,	96,	62,	64
Second hour test:	52,	65,	84,	91,	86,	76,	73,	52,	68,	79,
	88,	94,	98,	84,	53,	59,	63,	66,	77,	81,
	94,	81,	64,	56,	96,	58,	64,	57,	83,	87
Third hour test:	97,	91,	61,	67,	72,	81,	63,	56,	53,	59,
	43,	56,	64,	78,	93,	99,	84,	84,	61,	56,
	73,	77,	57,	46,	93,	87,	93,	78,	46,	87

   (a) Draw three separate but parallel line plots for these three sets of scores.
   (b) Write a three or four sentence analysis of these line plots suggesting what happened during the term to account for the changing distribution of scores.

7. The United States Department of Agriculture keeps careful records on agricultural production each year and uses them as a basis for many decisions on farm policy. This table gives data on milk production over the ten year period 1980–1989.

	1980	1981	1982	1983	1984	1985	1986	1987	1988	1989
Number of Milk Cows in the U.S. in millions	10.8	10.9	11.0	11.1	10.8	11.0	10.8	10.3	10.3	10.1
Milk Production in the U.S. in Millions of Tons	64.2	66.4	67.8	67.8	67.8	71.5	71.6	71.4	72.6	72.6

SOURCE: Agricultural Statistics, 1990

   (a) Using a vertical scale in *millions* and a horizontal scale indicating *years*, draw two line graphs on the same set of axes labeling one "number of milk cows in the United States" and the other "milk production in the United States in tons."
   (b) Determine the average number of tons of milk produced per cow during 1980 and during 1989.
   (c) How do you account for the increase in milk production over the ten year period?

8. A certain type of amoeba grows to maturity and divides into two once every minute. Using a vertical scale in *100s of amoeba* and a horizontal scale in *minutes*, draw a line graph to represent the number of amoeba in a culture at times $t = 0, 1, 2, \ldots, 10$ minutes. Assume that there is just 1 amoeba at time $t = 0$. (*Suggestion:* Mark your scales in units one centimeter long.)

9. Together the Smiths earn $64,000 per year which they spend as shown.

Taxes	$21,000
Rent	$10,800
Food	$ 5,000
Clothes	$ 2,000
Car payments	$ 4,800
Insurance	$ 5,200
Charity	$ 7,000
Savings	$ 6,000
Misc.	$ 2,200

Draw a pie chart to show how the Smiths spend their yearly income.

10. This pie chart indicates how the City of Metropolis allocates its revenues each year.

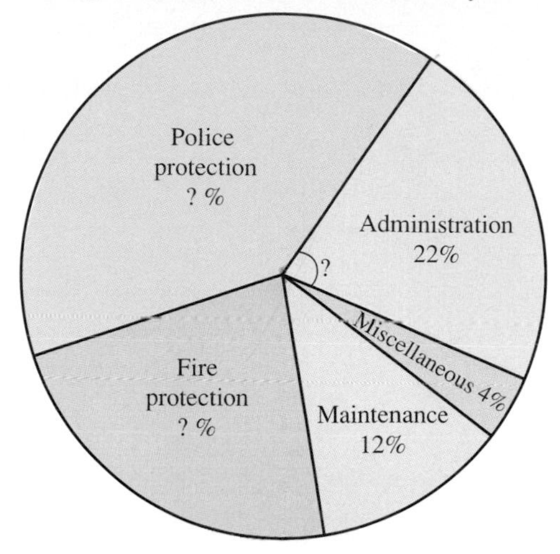

(a) What is the measurement of the central angle of the sector representing administrative expense?

(b) Using a protractor to measure the angle, determine what percent of the city budget goes for police protection.

(c) How does the city's expenditure for maintenance compare with its expenditure for police protection?

(d) How do the expenditures for administration and fire protection compare?

11. This bar graph shows the distribution of grades on the final examination in a class in English literature.

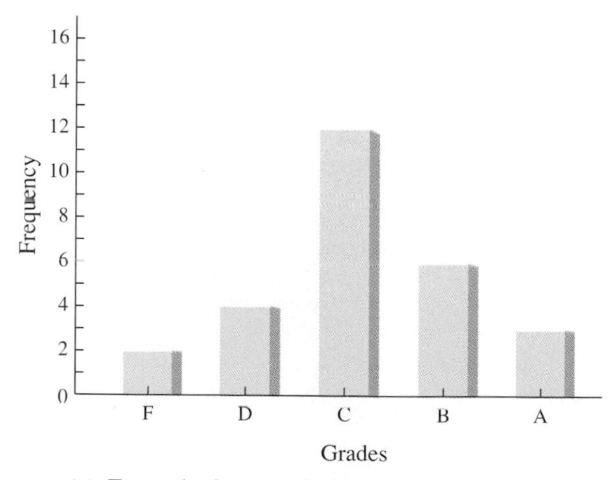

(a) From the bar graph determine how many students in the class got Cs.

(b) How many more students got Bs than Ds?

(c) What percent of the students earned As?

12. Bar graphs can summarize a variety of data all at once. Here is a bar graph showing Indonesia's export earnings from 1974 through 1986. The height of each bar shows total export earnings for each year. The green and tan portions of each bar differentiate earnings from the export of oil and gas products from all other exports.

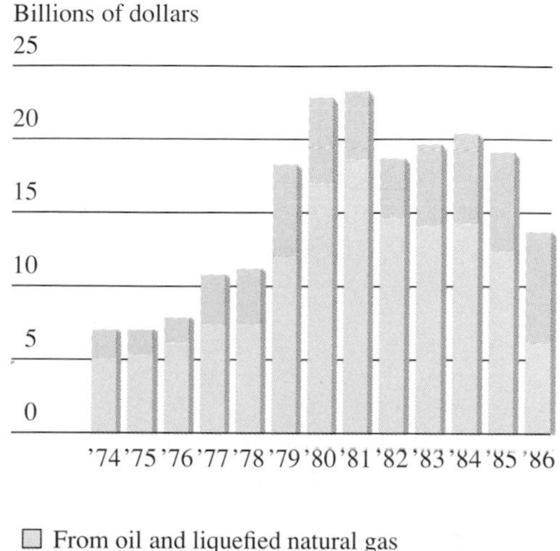

SOURCE: Graph, "Box figure 2.3B Indonesia's export earnings, 1974–86" from *World Development Report 1987*, page 25.

(a) Roughly what percent of the export earnings for 1974 came from the sale of oil and natural gas?

(b) Roughly what percent of the earnings in 1986 came from the sale of oil and gas?

(c) What was the approximate value of all the exports in 1980?

13. **(a)** Go to a busy campus parking lot and record the number of cars that are predominately white, black, red, gray, green, and "other."

(b) Make a bar graph to display and summarize your data.

(c) Based on part (a), if you were to stand on a busy street corner and watch 200 cars go by, how many would you expect to be predominately white?

14. **(a)** Roll two dice 50 times and record the total score for each roll. Make a bar graph with the vertical scale indicating frequency and the horizontal scale indicating the totals obtained on the various rolls.

(b) Estimate how often you would obtain each possible score if you were to repeat the experiment rolling the dice 1000 times.

15. Buy a small package of M&Ms with mixed colors. Open the package and pour out the M&Ms.
    (a) How many M&Ms of each color are in the package?
    (b) Make a bar graph of the data in part (a).

16. Using the figure, do the following:

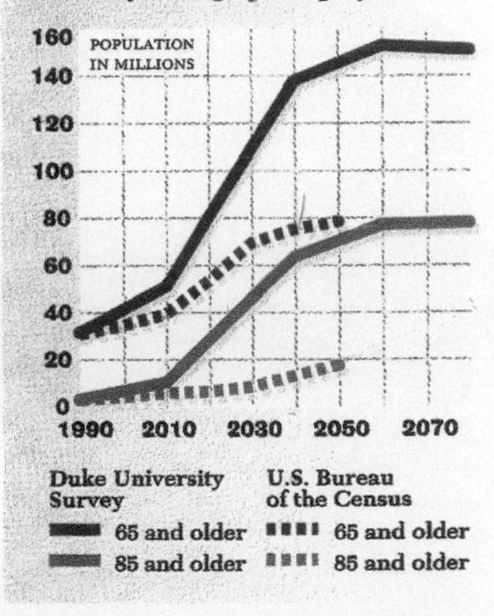

The Census Bureau's estimate of the growth of the older population by mid-century falls far short of what Duke University demographers project.

SOURCE: Chart, "How Long Will You Live?" from *Newsweek,* October 4, 1993, p. 66.

(a) estimate the U.S. population aged 65 and older in 2040 according to the Duke University survey.
(b) estimate the U.S. population aged 65 and older in 2040 as projected by the U.S. Bureau of the Census.
(c) estimate the U.S. population aged 85 and older in 2090 using the Duke University projections.
(d) estimate the U.S. population aged 85 and older in 2090 using the U.S. Bureau of the Census projections.

17. Consider the graph shown.
    (a) About how many arrests for violent crimes per 100,000 juveniles were made in 1983? Explain.
    (b) Estimate the number of arrests for violent crimes per 100,000 juveniles in 1995. Explain.

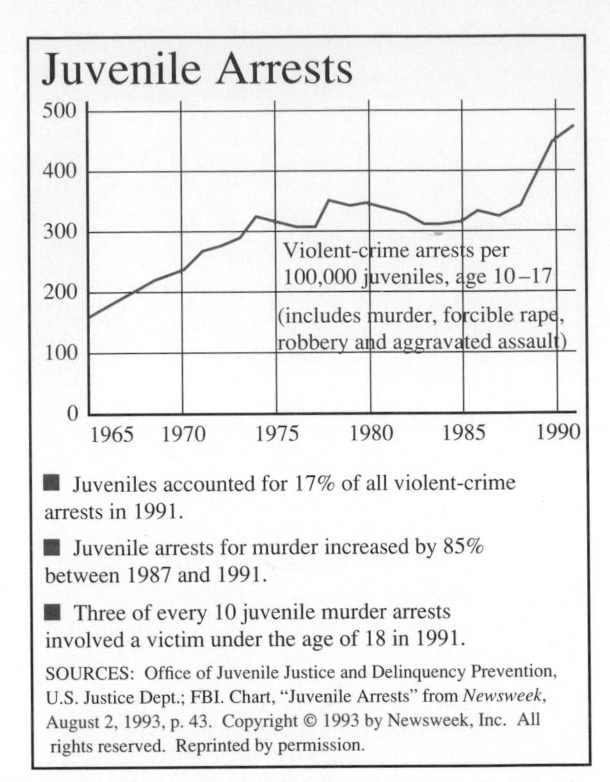

Juvenile Arrests
500
400
300
200
100
0
1965 1970 1975 1980 1985 1990
Violent-crime arrests per 100,000 juveniles, age 10–17
(includes murder, forcible rape, robbery and aggravated assault)
■ Juveniles accounted for 17% of all violent-crime arrests in 1991.
■ Juvenile arrests for murder increased by 85% between 1987 and 1991.
■ Three of every 10 juvenile murder arrests involved a victim under the age of 18 in 1991.

SOURCES: Office of Juvenile Justice and Delinquency Prevention, U.S. Justice Dept.; FBI. Chart, "Juvenile Arrests" from *Newsweek,* August 2, 1993, p. 43.

18. This bar graph/pictograph appeared in *Newsweek* in June, 1993 shortly after the war mounted to thwart Iraq's conquest of Kuwait.

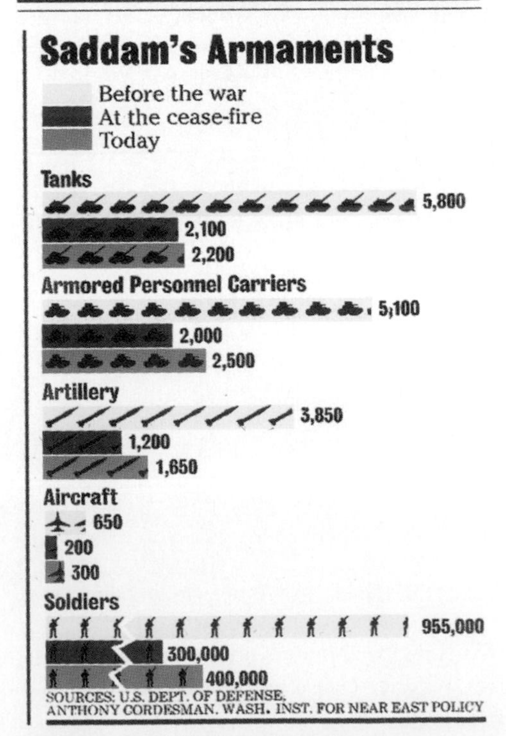

Armored Personnel Carriers

Artillery

Aircraft

SOURCES: U.S. DEPT. OF DEFENSE, ANTHONY CORDESMAN, WASH. INST. FOR NEAR EAST POLICY

SOURCE: Chart, "Saddam's Armaments" from *Newsweek,* July 12, 1993, p. 31.

(a) How many tanks does each tank icon represent in the pictograph?

(b) Do the icons in the other categories of armaments represent the same numbers of pieces of equipment as the tank icons in part (a)?

(c) Why are the bars in the soldiers category broken? Do the bar lengths as shown accurately represent the relative numbers of Iraqi soldiers before the war, after the cease fire, and at the time the graph was printed? Explain.

Thinking Critically

Data are often presented in a way that confuses or even purposely misleads the viewer.

19. Consider these two histograms for the same data. Briefly compare the impressions they convey. Which one do you feel most accurately or clearly describes the data?

(a)

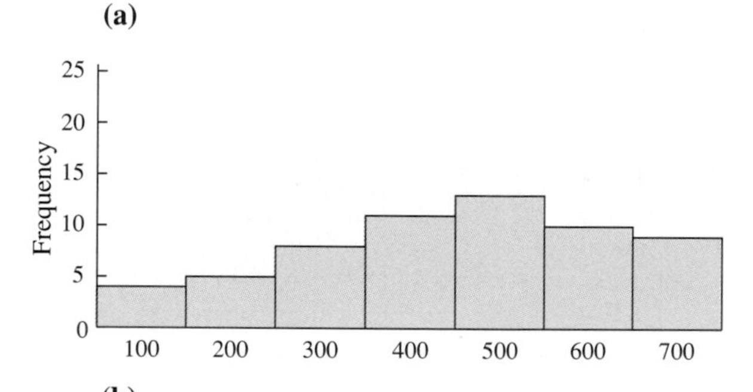

(b)

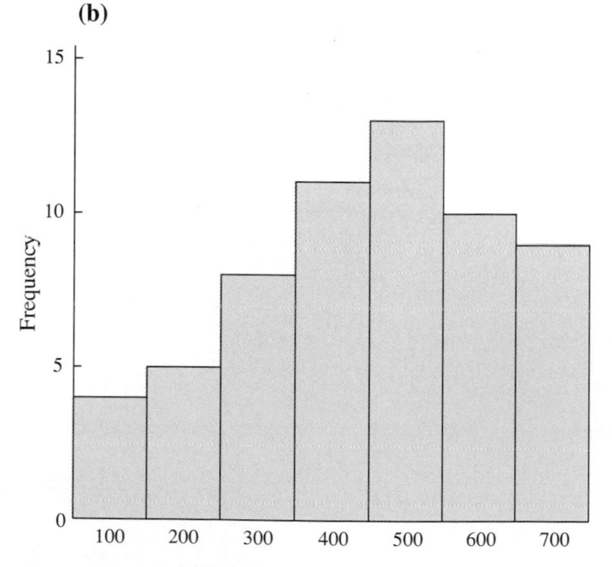

20. The bar graph shown gives the number of housing starts in the United States from November, 1992 through October, 1993.

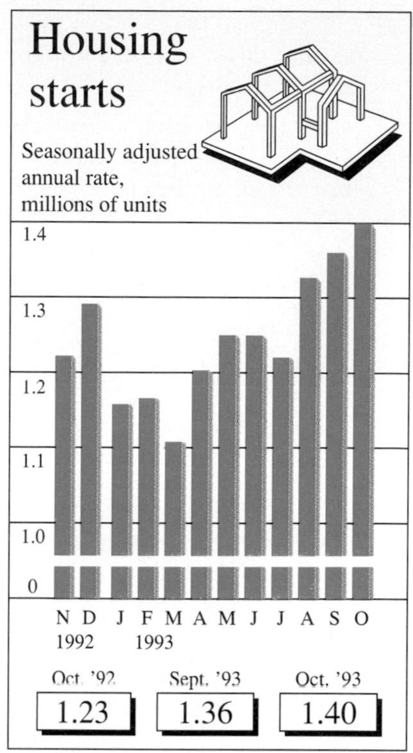

SOURCE: Graph, "Housing starts" from *The Moscow/Pullman Daily News*, November 17, 1993. Copyright © 1993 News Review Publishing Co. Reprinted by permission of The Moscow/Pullman Daily News.

(a) What does the gap in each bar between 0 and 1 on the vertical axis represent?

(b) Using data from this graph, draw a similar but complete graph with the vertical scale marked off in centimeters numbered 0, 0.1, 0.2, 0.3, . . . , 1.4.

(c) Is the graph in part (a) somewhat misleading? Explain, comparing the graph of part (a) with the graph you drew in part (b)

21. (a) Discuss briefly why the television evening news might show histogram (A) below rather than (B) in reporting stock market activity for the last seven days. Is one of these histograms misleading? Why or why not?

(b) What was the percentage drop in the Dow Jones average from the fourth to the fifth day as shown in the following histograms? As an investor should I worry very much about this 36 point drop in the market?

(c) Was the Dow Jones average on day five approximately half what it was on day four as suggested by histogram (A)?

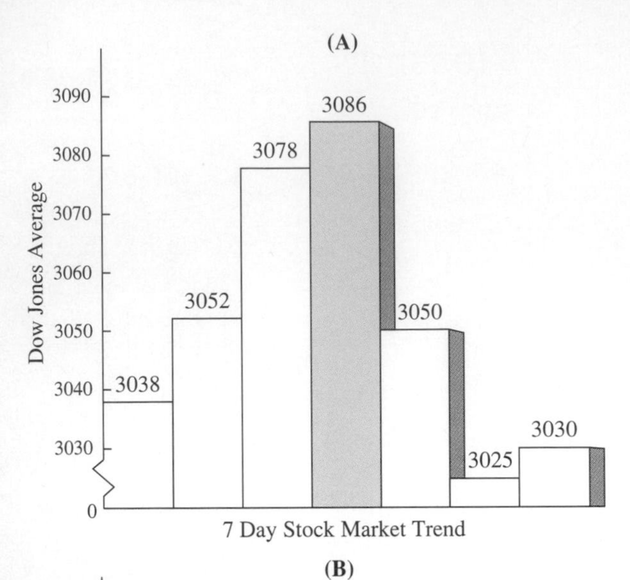

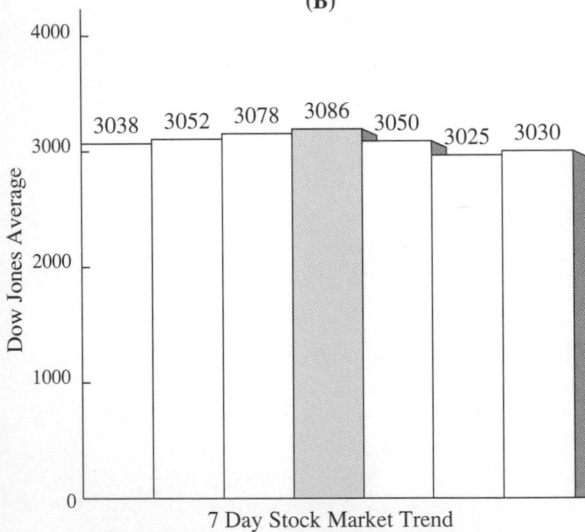

22. Longlife Insurance Company printed a brochure with the following pictographs showing the growth in company assets over the ten year period 1981–1990.

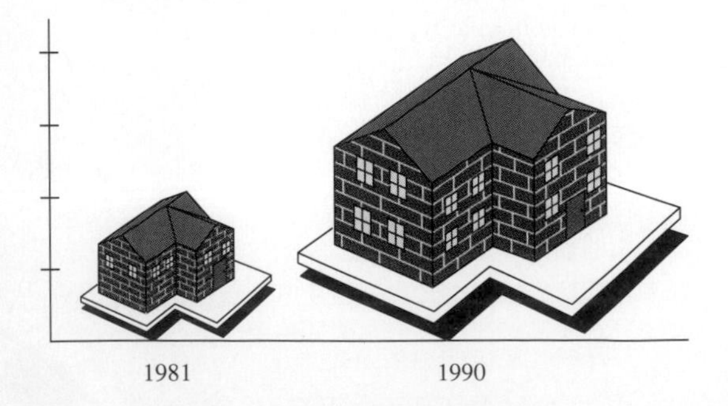

(a) Do the pictographs accurately indicate that the assets were 1 billion dollars in 1981 and 2 billion dollars in 1990, or might one assume from the pictographs that the assets were actually much greater? Explain briefly.

(b) The larger building shown is just twice the height of the smaller and the two buildings are similar as geometrical drawings. Does this accurately convey the impression that the assets of Longlife Insurance Company just doubled during the ten year period? Explain your reasoning. What is the ratio of the volume of the large building to the volume of the small building? (*Suggestion:* Suppose both buildings were rectangular boxes with the dimensions of the second just twice those of the first.)

(c) Would it have been more helpful (or honest) to print the actual asset value for each year on the front of each building?

23. The ability of some graphs and charts to distort data depends on the psychological reaction of individuals to stimuli. Consider these diagrams and answer each question both before and after checking.

(a) Which is longer, the vertical or horizontal line?

(b) Are lines l and m straight and parallel?

(c) Which line segment is longer: $\overline{AB}$ or $\overline{BC}$?

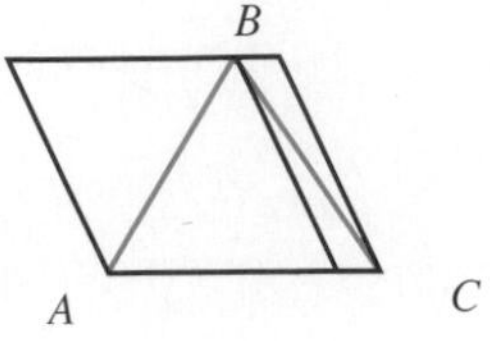

(d) How many sides does the "polygon" have? Is it a square?

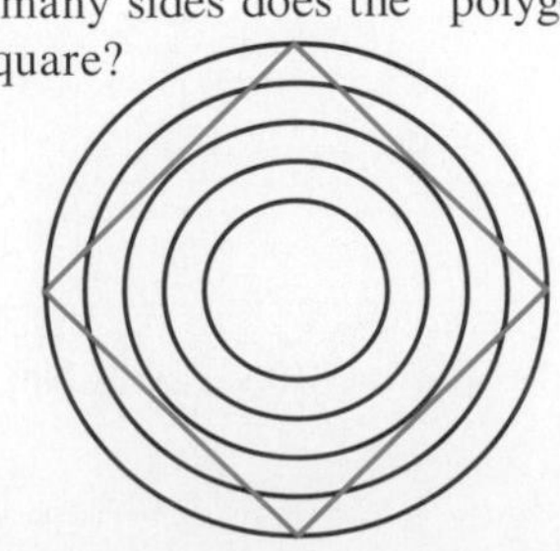

(e) Stare at the diagram below. Can you see four large posts rising up out of the paper? Stare some more and see if you can see four small posts.

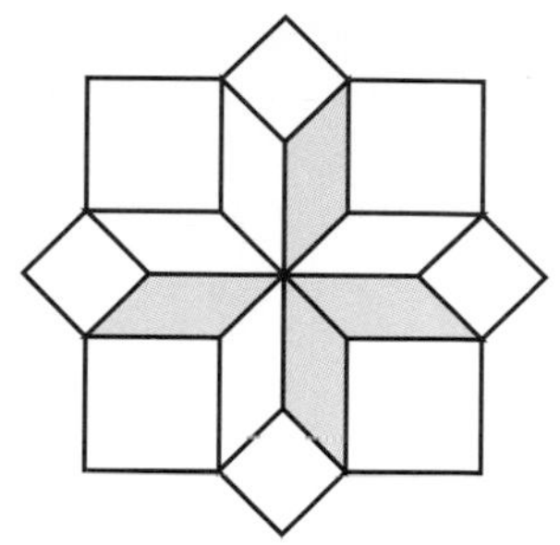

(c) Which shape of can do you see more often in the grocery store? Why do you suppose this is so?

Recall that the volume of a cylinder is given by the formula $V = \pi r^2 h$ where r is the radius and h is the height.

24. Some merchandisers take advantage of optical illusions just as some pollsters, advertisers, and others do.

(a) Which of the cans depicted here seems to have the greater volume? Note that the diagrams are drawn to scale.

(b) Actually compute the volumes of the cans.

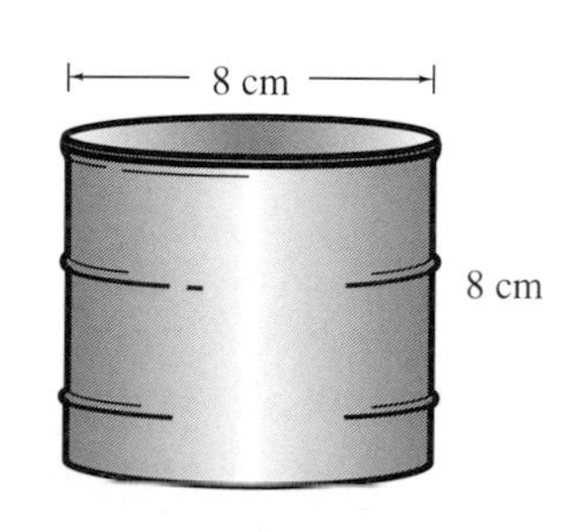

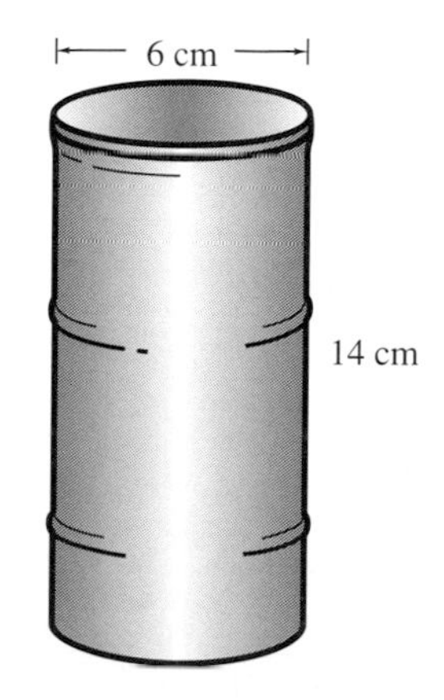

Making Connections

25. Read the brief article and study the line graph shown below.

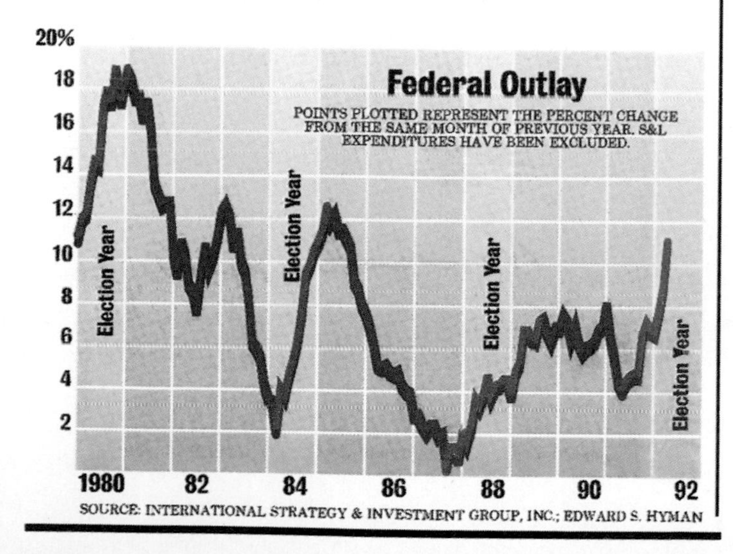

VITAL STATISTICS

Election-Year Spending Sprees

George Bush may talk about budget cutting, but in the end he's following the same pattern as his predecessor Ronald Reagan. When an election year approaches, the government shamelessly turns on the federal money spigots to boost the economy and attract voters. Here's a look at the spending pattern since 1980:

SOURCE: INTERNATIONAL STRATEGY & INVESTMENT GROUP, INC.; EDWARD S. HYMAN

(a) Does it effectively and fairly show that incumbent presidents endeavor to improve their chances of reelection by boosting government spending during election years? Discuss.

(b) By what percent did President Reagan increase federal spending in 1984 when he was running for his second term?

(c) Why do you suppose there was less increase in federal spending during 1988 when Vice President George Bush was running for president than during 1984?

(d) How do you account for the substantial increase in spending in 1992 when President Bush was running for reelection? Compare this increase to that of 1988.

26. The graphs shown here contrast enrollment in elementary and secondary schools and in institutions of higher education from the 1960–61 academic year through the 1990–91 academic year with expenditures for education during the same period.

Enrollment and total expenditures in current and constant dollars, by level of education: 1960–61 to 1990–91

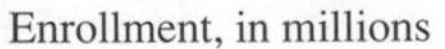

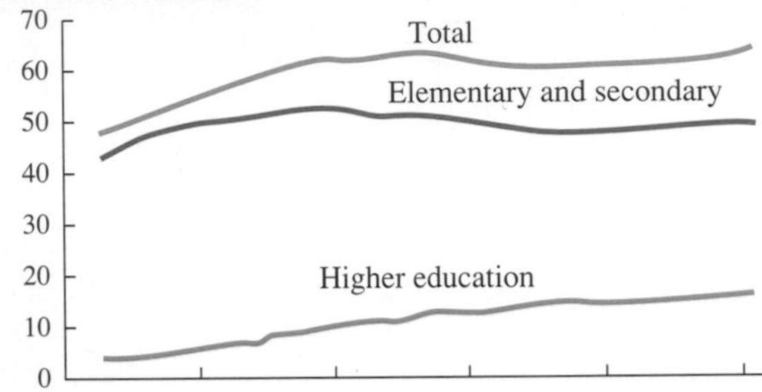

Expenditures, in billions of current dollars

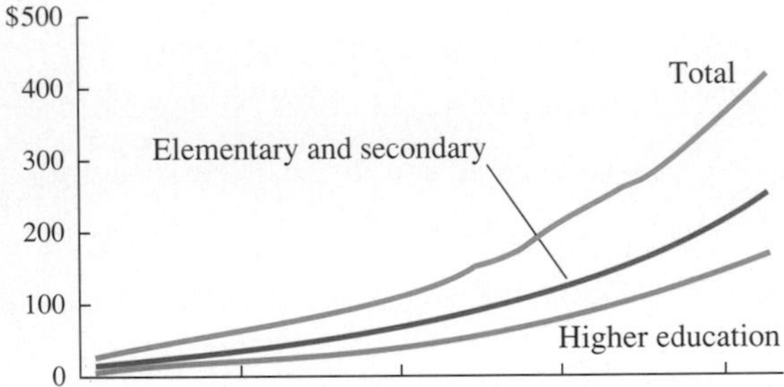

Expenditures, in billions of constant 1990-91 dollars

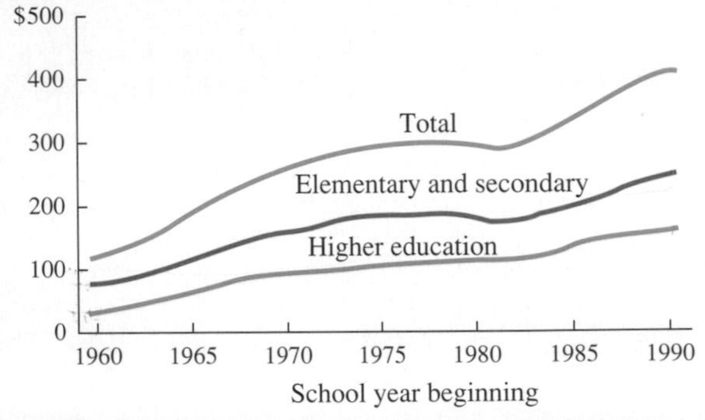

SOURCE: U.S. Department of Education, National Center for Education Statistics, *Statistics of State School Systems; Statistics of Public Elementary and Secondary School Systems; Statistics of Nonpublic Secondary School Systems; Statistics of Nonpublic Elementary and Secondary Schools; Revenues and Expenditures for Public Elementary and Secondary Education; Fall Enrollment in Institutions of Higher Education; Financial Statistics of Institutions of Higher Education;* Common Core of Data surveys; and Integrated Postsecondary Education Data System surveys. Figure 2 from *Digest of Education Statistics*, 1991, p.8.

(a) Estimate the average expenditure in 1990–91 dollars per elementary or secondary student in 1960 and in 1990.

(b) Estimate the average expenditure in 1990–91 dollars per higher education student in 1960 and in 1990–91.

(c) Discuss what factors may have contributed to the increased dollar cost per student indicated in parts (a) and (b).

(d) Explain the difference in the two line graphs of expenditures (in current dollars and in constant 1990–91 dollars). Which one of these graphs more accurately portrays the relationship between cost and student population size?

27. This pie chart from the *Digest of Education Statistics, 1991,* shows the percentage of persons 18 years and older in 1987 that attained various levels of education.

Highest degree earned by persons 18 years old and over: 1987

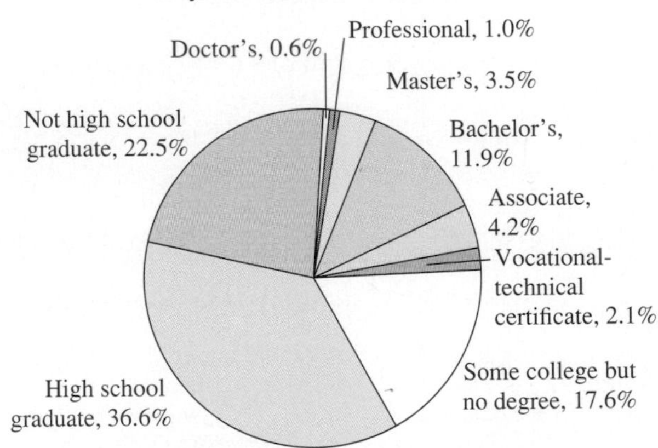

Total persons age 18 and over = 176 million

SOURCE: U.S. Department of Commerce, Bureau of the Census, *Current Population Reports,* Series P–70, No. 11, Educational Background and Economic Status: Spring 1987. Figure 5 from *Digest of Education Statistics*, 1991, p. 10.

(a) Assuming that these figures remain relatively constant from year to year, what percentage of high school graduates go to college but fail to earn a bachelor's degree?

(b) What percentage of high school graduates earn a bachelor's degree? (*Hint:* Assume that people who earn master's, professional, and doctor's degrees must first earn a bachelor's degree.)

(c) What percentage of the people who earn bachelor's degrees eventually earn a doctor's degree?

28. This table from the *Digest of Education Statistics, 1991,* summarizes perceptions of school problems in 1988.

Rating of School Problems by Teachers and Students: 1988

Selected Problems	Percent of Teachers who Say Problem Is "Very Serious" at Their School						Percent of Students Who Say They Know 10 or more Students Involved in Each Problem
	Total	Inner City	Urban	Suburban	Small Town	Rural	
1	2	3	4	5	6	7	8
The number of students requiring constant discipline	14	27	19	11	12	10	30
The number of students who lack basic skills (Students item: can't read)	16	38	16	12	14	13	5
The number of teenage pregnancies[1]	12	28	9	5	13	12	9
The number of students drinking alcohol[2]	33	32	20	24	38	38	47
The number of students using drugs[2]	14	26	11	14	12	15	25
The number of incidents involving violence in school[2]	4	10	4	3	5	—	—
Have threatened or become violent with other students	—	—	—	—	—	—	23
Have threatened or become violent with teachers	—	—	—	—	—	—	5
The number of dropouts	9	30	11	6	6	9	9

[1] Asked of junior high and high school students and teachers only.
[2] Asked of all students and junior high and high school teachers.
—Data not available.
SOURCE: Metropolitan Life/Louis Harris and Associates. Inc., *The American Teacher*. 1988. (This table was prepared May 1989.) From *Digest of Education Statistics,* 1991, p. 30.

(a) According to this study, what is the most prevalent problem overall faced by students and teachers?

(b) In which group (inner city, urban, suburban, and so on) is teenage drug abuse the most prevalent problem?

(c) In which category of schools does the total number of problems appear to be least?

29. Using data from problem 28, draw a double bar graph to compare the numbers of teachers who view drug abuse and alcohol consumption as serious problems in the various categories of schools. The vertical scale should show *percent* and the horizontal scale should indicate the *type of school.*

30. The line graph below compares salaries for elementary and secondary school teachers. Draw a double bar graph to make this same comparison for the years 1969–1970, 1974–1975, 1979–1980, 1984–1985, and 1989–1990.

Average annual salary for public elementary and secondary school teachers: 1969-1970 to 1990-1991 [In constant 1990-1991 dollars]

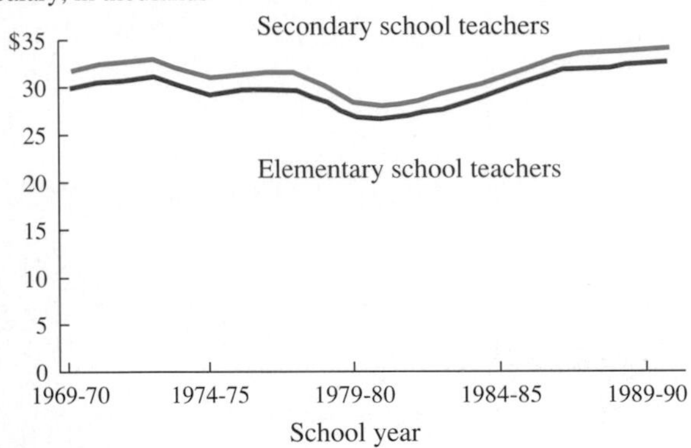

SOURCE: National Education Association, *Estimates of School Statistics:* and unpublished data. (Latest edition 1990–91. Copyright © 1991 by the National Education Association. All rights reserved.) Figure 10 from *Digest of Education Statistics*, 1991, p. 45.

Using A Computer

For the following problems, use a statistical package from your computer laboratory. Not all institutions will have such a facility and these problems will then be skipped. Note that programs for these problems are *not* found on your computer disk.

31. Consider the SAT scores recorded here.

Student	Verbal Score	Math Score
Dina	502	444
Carlos	590	520
Rosette	585	621
Broz	487	493
Coleen	585	602
Deiter	481	572
Karin	605	599
Luana	547	499

(a) Make a histogram to summarize the above data for verbal scores.

(b) Make a double bar graph to summarize the above data with one of each pair of bars for verbal scores and one for math scores.

32. For fiscal year 1991, federal expenditures were divided as follows:

Social programs—14%
Physical, human, and community development—14%
Net interest on debt—14%
Defense, veterans, and foreign affairs—24%
Social security, medicare, and other retirement—32%
Law enforcement and general government—2%

(a) Make a pie chart to reflect this data.

(b) Make a bar graph that reflects this data.

33. The scores in an elementary mathematics class were as follows:

83	86	86	69	78	88
74	89	63	82	60	79
64	87	79	75	75	89
70	69	66	71	84	92
64	84	97	74	91	57

(a) Make a histogram to summarize this data.

(b) Make a stem and leaf plot to summarize this data.

34. Make a line graph to graphically display this data.

Population in Washington in Millions	2.10	2.32	2.63	2.83	3.12	3.45	3.80
Year	1960	1965	1970	1975	1980	1985	1990

35. Use the computer to draw a pie chart for the data from problem 9 above and measure the angles to see how it compares with your drawing for problem 9.

For Review

36. Write these fractions in decimal form.

(a) $\frac{1}{8}$ **(b)** $\frac{7}{40}$ (c) $\frac{17}{250}$ **(d)** $\frac{7}{20}$

37. Write these fractions in decimal form.

(a) $\frac{1}{9}$ **(b)** $\frac{3}{7}$ (c) $\frac{3}{14}$ **(d)** $\frac{7}{15}$

38. Write these decimals as fractions in reduced form.

(a) 0.375 **(b)** 0.3125 (c) 0.444 **(d)** 0.33

39. Write these decimals as fractions in reduced form.

(a) $3.7444\ldots = 3.7\overline{4}$ **(b)** $.\overline{02}$ (c) $.0\overline{2}$
(d) $31.\overline{72}$ (e) $4.7\overline{314}$ **(f)** $.431\overline{23}$

8.2 Measures of Central Tendency and Variability

Measures of Central Tendency

Consider the data of Table 8.1 and the line plot of Figure 8.2 (both on page 493). The line plot is a considerable improvement over the unorganized raw data for the purpose of assessing the performance of the class. But even more precise information might be desired. For example, "What is the 'average' or 'typical' grade in the class?" As you might expect, there are several possible answers.

The Mean

One measure of the "average" or "typical" value for a collection of data is the **mean,** frequently called the **arithmetic mean** or **average,** of all the values.

DEFINITION The Mean of a Set of Data

The **mean** or **average** of a collection of values is $\bar{x} = S/n$ where S is the sum of the values and n is the number of values.

For the data in Table 8.1, we compute the mean by adding all the scores and dividing by 35, the number of scores. Thus,

$$\begin{aligned}\bar{x} = {} & (79 + 78 + 79 + 65 + 95 + 77 + 49 + 91 + 63 + 58 + 78 \\ & + 96 + 74 + 68 + 71 + 86 + 91 + 94 + 79 + 69 + 86 + 62 \\ & + 78 + 77 + 88 + 67 + 78 + 84 + 69 + 53 + 79 + 75 + 64 \\ & + 89 + 77)/35 \\ \doteq {} & 76.2\end{aligned}$$

and this is easily computed with a hand-held calculator. Observe that this agrees with our earlier informal assessment: 76.2 falls in the middle of the clump of grades between 75 and 80 and, more generally, roughly in the middle of the entire collection of grades arranged in order of increasing size. Actually, 76.2 is somewhat less than one might expect and this shows that the mean is sensitive to outliers. That is, the low grades of 49 and 57 that are clearly not typical of the grades in the class tend to cause the mean to be somewhat less than might be expected. However, the mean is often the most representative value of a set of data.

The Median

The **median** of a collection of values is the middle value in the collection arranged in order of increasing size or the average of the two middle values in case the number of values is even.

> **DEFINITION The Median of a Set of Data**
>
> Let a collection of n data values be written in order of increasing size. If n is odd, the **median,** denoted by $\hat{x}$, is the middle value in the list. If n is even, $\hat{x}$ is the average of the two middle values.

EXAMPLE 8.4

Determining a Median

Determine the median of the data in Table 8.1.

SOLUTION

The scores in Table 8.1 are arranged in order in the line plot of Figure 8.2. Since there are 35 scores, the median is the eighteenth score. Thus, counting, we see that $\hat{x} = 78$. ■

Note that the median in the preceding example not only closely approximates the mean, but also agrees with our intuitive idea of the typical value of the collection of scores.

Also, it follows from the definition, that the median is a data value if the number of data values is odd and is *not* necessarily a data value if the number of values is even. Thus, the median of the 9 scores

$$24, 25, 25, 27, 29, 31, 32, 34, 37$$

is 29, the fifth score; while the median of the 10 scores

$$42, 42, 43, 44, 44, 46, 47, 47, 47, 49$$

is 45, the average of the two middle scores.

The Mode

Another value often taken as "typical" of a set of data is the value occurring most frequently. This value is called the **mode**. From the definition, it is clear that there may be more than one mode. For example, the data in Table 8.1 has two modes, 78 and 79. Even so, it may be the case that a mode gives a better indication of the typical value of the data than either the median or the mean. Moreover, unlike the mean, neither the median nor the mode is affected by the existence of outliers.

> **DEFINITION A Mode of a Set of Data**
>
> A **mode** of a collection of values is a value that occurs at least as often as any other value. If two or more values occur equally often and more frequently than all other values, there are two or more modes.

JUST FOR FUN

How Many Are Single?

In a certain population $3/5$ of the men are married and $4/7$ of the women are married. What fraction of the adults in the population are single?

EXAMPLE 8.5

Determining Means, Medians, and Modes

Determine the mean, median, and mode for the grades in Mathematics for Elementary School Teachers, Section 2 from the data displayed in the stem and leaf plot below.

Section 2		Section 1
3	4	9
9 8 7 5	5	3 8
8 8 5 5 5 5 3 1	6	2 3 4 5 7 8 9 9
5 5 4 4 3 0	7	1 4 5 7 7 7 8 8 8 8 9 9 9 9
9 7 6 4 4 2 0 0 0	8	4 6 6 8 9
6 5 5 0	9	1 1 4 5 6
0 0 0	10	

SOLUTION

The mean is the sum of the data values divided by the number of data values. Thus,

$$\begin{aligned}\overline{x} = {} & (43 + 55 + 57 + 58 + 59 + 61 + 63 + 65 + 65 + 65 + 65 \\ & + 68 + 68 + 70 + 73 + 74 + 74 + 75 + 75 + 80 + 80 + 80 \\ & + 82 + 84 + 84 + 86 + 87 + 89 + 90 + 95 + 95 + 96 + 100 \\ & + 100 + 100)/35 \\ \doteq {} & 76.0\end{aligned}$$

Counting to find the eighteenth of the 35 scores, we see that the median is

$$\hat{x} = 75.$$

The mode is 65 since this score occurs four times and is thus the most frequently occurring score. Does the mode seem to be a good "typical" value of the scores in this case? ■

EXAMPLE 8.6

Determining an Average

All 12 players on the Uni Hi basketball team played in their 78 to 65 win over Lincoln. Jon Highpockets, Uni High's best player, scored 23 points in the game. How many points did each of the other players average?

SOLUTION

Understand the problem • • • •

The problem is to determine averages when we are not explicity given the data values. What we do know is that Uni Hi scored 78 points, that Jon Highpockets scored 23 of the points, and that 12 players are on the team.

Devise a plan • • • •

Since the average score for each of the 11 players other than Jon is the sum of their scores divided by 11, we must determine the sum of their scores.

Carry out the plan • • • •

Since Uni Hi scored a total of 78 points and Jon scored 23, the total for the rest of the team must have been $78 - 23 = 55$ points. Therefore, the average number of points for these players is $55 \div 11 = 5$.

Look back • • • •

The solution depended on knowing the definition of average. The real question was how many points were scored by all the players on Uni Hi's team other than Jon and how many such players there were. But those figures, and hence the solution to the problem, were easily obtained by subtraction. ■

EXAMPLE 8.7

Determining a Typical Value for a Set of Data

The owner/manager of a factory earned \$850,000 last year. The assistant manager earned \$48,000. Three secretaries earned \$18,000 each, and the other 16 employees each earned \$27,000.

(a) Prepare a line plot of the salaries of those deriving their income from the factory.

(b) Compute the mean, median, and mode of the salaries of those deriving their income from the factory.

(c) Which is most typical of the salaries of those associated with the factory—the mean, median, or mode?

SOLUTION

(a) The line plot is shown here.

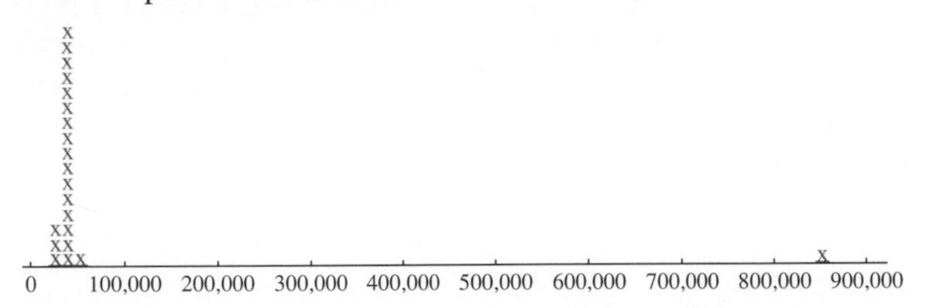

(b) The mean is

$$\bar{x} = \frac{18{,}000 + 18{,}000 + 18{,}000 + 27{,}000 + \cdots + 27{,}000 + 48{,}000 + 850{,}000}{21}$$

$$\doteq \$65{,}905.$$

The median $\hat{x}$, is the eleventh in the ordered list of salaries. Thus,

$$\hat{x} = 27{,}000 \text{ dollars.}$$

The mode is the most frequently occurring salary. Thus,

$$\text{mode} = 27{,}000 \text{ dollars.}$$

(c) The mean is clearly not typical of the salary most workers at the factory earn. The value of $\bar{x}$ is unduly affected by the huge salary earned by the owner/manager. Here the median and mode are the same and are more typical of salaries of those deriving their income from the factory since \$27,000 is the salary of 16 of the 21 people. Note that this last sentence is really an argument that, in this case, the most typical value is the mode. That the mode and median here are equal is incidental. ■

While the mean is the most commonly used indicator of the typical value of a data set, the preceding example makes it clear that this choice can be quite misleading. As will be seen in the problem set, it is easy to construct examples where the median is the most typical value and other examples, like the preceding, where the mode is most typical.

Measures of Variability

The most useful analysis of data would reveal both the center (typical value) and the *spread,* or *variability,* of the data. We now consider how the spread of data is determined. The simplest measure is the **range,** the difference between the smallest and largest data values. This certainly tells something about how the data occurs, but it is often misleading, particularly if there are outliers. A better understanding is obtained by determining **quartiles**.

DEFINITION Upper and Lower Quartiles

Consider a set of data arranged in order of increasing size. Let the number of data values, n, be written $n = 2r$ for n even or $n = 2r + 1$ for n odd for some integer r. In either case the **lower quartile,** denoted by Q_L, is the median of the first r data values. Also, the **upper quartile,** denoted by Q_U, is the median of the last r data values.

It follows from the definition that approximately 25% of the data values are less than or equal to Q_L, approximately 25% lie between Q_L and $\hat{x}$, approximately 25% lie between $\hat{x}$ and Q_U, and approximately 25% are greater than or equal to Q_U.

EXAMPLE 8.8 **Calculating Quartiles**

Determine Q_L and Q_U for the data in Table 8.1.

SOLUTION

The 35 data values are ordered in Figure 8.2 and $2r + 1 = 35$, so $r = 17$, Q_L is the median of the first 17 values, and Q_U is the median of the last 17 values. Thus, $Q_L = 68$ and $Q_U = 86$.

Symbolically, if the 35 points represent ordered data values, Q_L, $\hat{x}$, and Q_U are as illustrated here.

• •

49 ↑ $Q_L = 68$ ↑ $\hat{x} = 78$ ↑ $Q_U = 86$ 96

■

Box and Whisker Plots

The least and greatest scores, the **extremes**, along with the lower and upper quartiles and the median give a concise numerical summary, called the **5-number summary** of a set of data. Since the median of the data in Example 8.8 is 78 and the extremes are 49 and 96, the 5-number summary is 49 − 68 − 78 − 86 − 96. Moreover, a **box and whisker plot** gives a vivid graphical representation of the 5-number summary.

This is a pictorial representation of the 5-number summary.

DEFINITION Box and Whisker Plot

A **box and whisker plot** consists of a central box extending from the lower to the upper quartile with a line marking the median and line segments, or whiskers, extending outward from the box to the extremes.

For example, the box and whisker plot for Example 8.8 is shown in Figure 8.15.

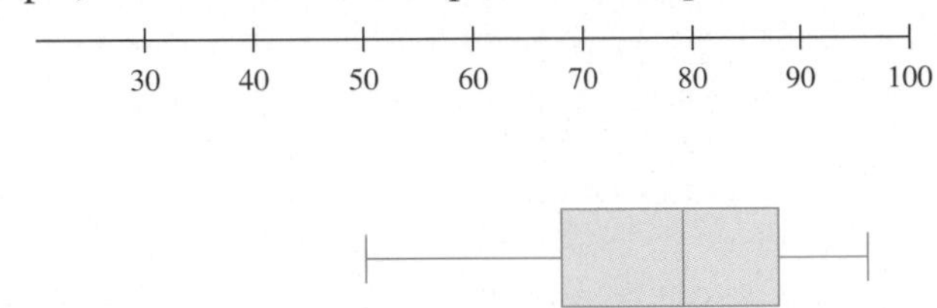

Figure 8.15
Box and whisker plot for the final examination scores in Mathematics for Elementary School Teachers, Section 1

In comparing sets of similar data it is often helpful to draw side-by-side box and whisker plots. For example, it is informative to compare side-by-side plots for the grades in Sections 1 and 2 of Mathematics for Elementary School Teachers. The 5-number summary for Section 1 was 49 − 68 − 78 − 86 − 96. From the data in Figure 8.4, the summary for Section 2 can be determined to be

$$43 - 65 - 75 - 87 - 100.$$

The side-by-side box and whisker plots are shown in Figure 8.16. These plots make it clear that the spread of scores in Section 2 was greater than in Section 1, that the extreme scores were both higher and lower, and that the median score in Section 2 was actually lower than in Section 1.

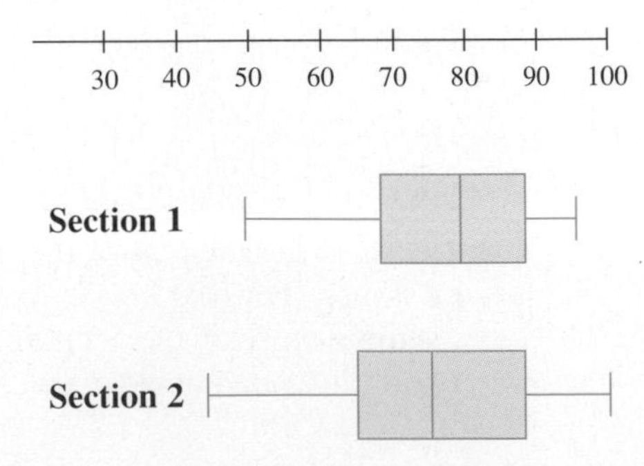

Figure 8.16
Box and whisker plots for final examination scores in Mathematics for Elementary School Teachers, Sections 1 and 2

SCHOOL BOOK PAGE *Displaying Data in the Third Grade*

CONSUMER CONNECTION

Pictographs

Build Understanding

Type of Movie	Tally	Number
Westerns	𝍸	5
Space	𝍸 𝍸 𝍸 𝍸	20
Comedy	𝍸 𝍸 𝍸	15
Adventure	𝍸 𝍸	10

Pinetree School students asked 50 children what their favorite kinds of movies are. Students worked in groups to make pictographs from their data.

Each group wrote a title.

Then each group listed the kinds of movies.

Each group used a face to stand for a certain number of children. They wrote that at the bottom.

They drew faces in the graph for the numbers on the tally chart.

Below are pictographs made by the groups of students. Think about how each group showed the same data.

Favorite Movies of Children	
Westerns	
Space	
Comedy	
Adventure	
Each ☺ stands for 5 children.	

Talk About Math Do you think a pictograph is more helpful than a tally chart in showing this kind of information? Why or why not?

A.

Favorite Movies of Children	
Westerns	☻
Space	☻ ☻ ☻ ☻
Comedy	☻ ☻ ☻
Adventure	☻ ☻
Each ☻ stands for 5 children.	

B.

Favorite Movies	
Westerns	◖
Adventure	☺
Comody	☺ ◖
Space	☺ ☺
Each ☺ stands for 10 children.	

Here children are involved in collecting and displaying data.

1. What points would you try to make if you were discussing the question in Talk About Math with your third grade class?
2. How would you compare the effectiveness and ease of construction of the pictographs in parts A and B in Talk About Math? Suppose one category of movie had 17 students listing it as their favorite?

An additional advantage of box and whisker plots is that they make it possible to make useful comparisons between data sets containing widely differing numbers of values. This is made clear in the next example.

EXAMPLE 8.9 Making Box and Whisker Plots for Comparisons

The data below are the final scores of men and women students in Calculus I. Draw box and whisker plots to compare the distribution of women's scores with the distribution of men's scores.

Women's scores: 95, 79, 53, 78, 71, 88, 77, 80, 79, 79

Men's scores: 84, 85, 53, 77, 66, 81, 79, 59, 65, 61, 81,
68, 68, 80, 76, 87, 85, 74, 92, 76, 70, 85,
55, 79, 74, 80, 73, 48, 66, 83, 48, 60, 87,
58, 64, 78, 82, 69, 76, 83, 94, 86, 73, 85,
75, 69, 49, 52, 59, 68, 65, 75, 31, 69, 73,
56, 95

SOLUTION

To make the plots, we need the 5-number summaries. First arrange the scores in order of increasing size.

Women's scores: 53, 71, 77, 78, 79, 79, 79, 80, 88, 95

Men's scores: 31, 48, 48, 49, 52, 53, 55, 56, 58, 59, 59,
60, 61, 64, 65, 65, 66, 66, 68, 68, 68, 69,
69, 70, 73, 73, 73, 74, 74, 75, 75, 76, 76,
76, 77, 78, 79, 79, 80, 80, 81, 81, 82, 83,
83, 84, 85, 85, 85, 85, 86, 87, 87, 92, 95

For the women, the extreme scores are 53 and 95 and the median is 79, the average of the fifth and sixth scores. The lower quartile is 77, the median of the first five scores. The upper quartile is 80, the median of the last five women's scores. Thus, the 5-number summary for the women's scores is

$$53 - 77 - 79 - 80 - 95.$$

Similarly, the 5-number summary of the men's scores is

$$31 - 64 - 74 - 81 - 95.$$

These give the box and whisker plots shown.

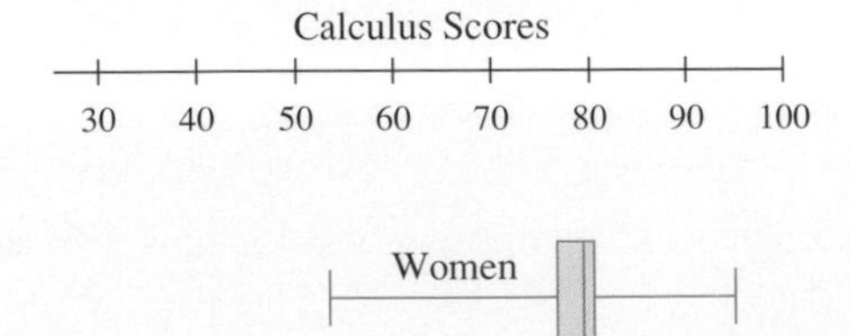

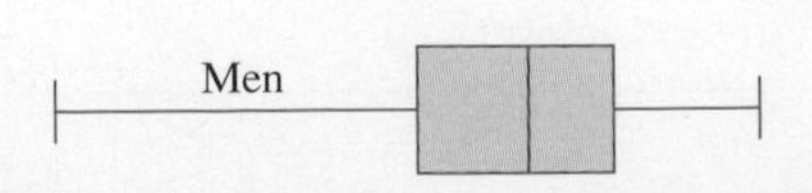

■

The box and whisker plots of Example 8.9 give a precise visual comparison of the performances of women and men students in calculus even though the numbers of students are quite different. One might reasonably speculate as to why the distributions of grades differ as they do. (For example, it is almost invariably the case that larger data sets have both smaller and larger extreme values as suggested here.) Nevertheless, one rather clear indication is that the women students in the class were no less able than the men.

The Standard Deviation

We have already observed that the range is one measure of the spread of a data set. It is not a very precise measure, however, since it depends only on the extreme data values which may differ markedly from the bulk of the data. This deficiency is largely remedied by the 5-number summary and its visualization by a box and whisker plot. However, an even better measure of variability is the **standard deviation.**

DEFINITION The Standard Deviation of a Set of Data

Let $x_1, x_2, x_3 \cdots, x_n$ be the values in a set of data and let $\overline{x}$ denote their mean. Then

$$s = \sqrt{\frac{(\overline{x} - x_1)^2 + (\overline{x} - x_2)^2 + \cdots + (\overline{x} - x_n)^2}{n}}$$

is the **standard deviation.**

Just as the mean is an indication of the typical value of a set of data, the standard deviation* is a measure of the typical deviation of the values from the mean. If the standard deviation is large, the data are more spread out; if it is small, the data are more concentrated near the mean. For example, the data represented by the two histograms in Figure 8.17 are equally numerous and have essentially the same mean. However, the standard deviation for the data in histogram (a) is 2.43 and the standard deviation for the data in histogram (b) is 1.43. Correspondingly, the data in histogram (b) is much more closely concentrated near the mean than it is for histogram (a).

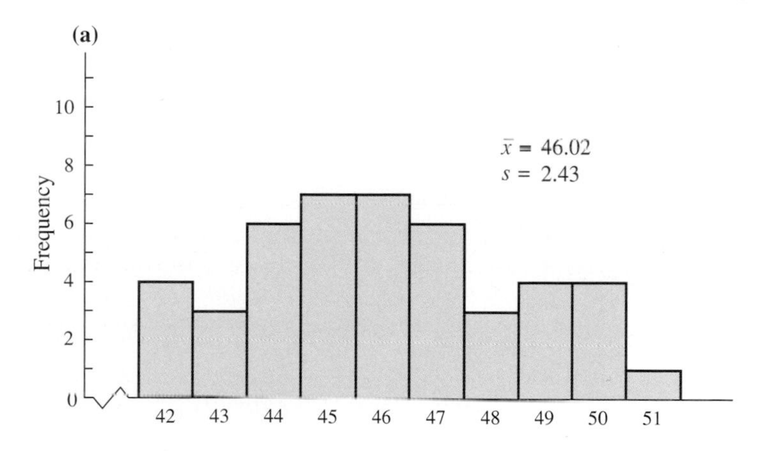

* The **variance,** $v = s^2$, of a set of data is also a good measure of the variability of the data. However, the more commonly used measure is the standard deviation.

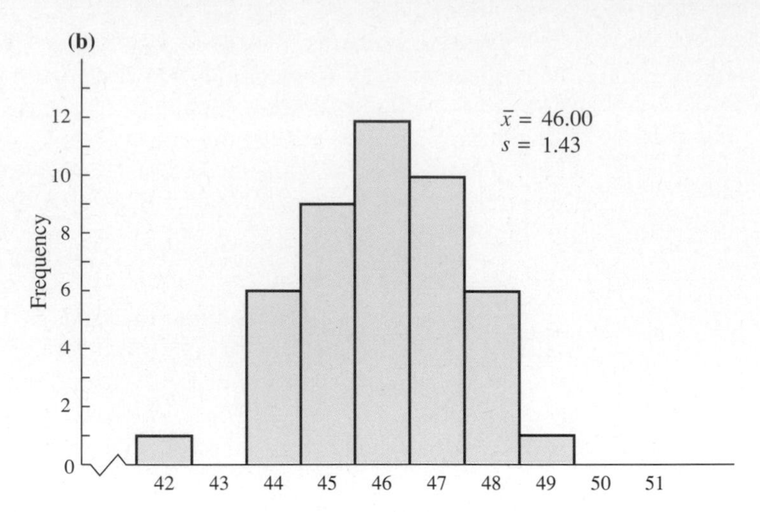

Figure 8.17
Smaller standard deviations imply less spread to data

EXAMPLE 8.10

Computing a Standard Deviation

Compute the mean and standard deviation for this set of data.

35	42	61	29	39
47	55	54	50	41
40	34	37	51	38

SOLUTION

$$\begin{aligned}\bar{x} &= (35 + 42 + 61 + 29 + 39 + 47 + 55 + 54 + 50 + 41 + 40 \\ &\quad + 34 + 37 + 51 + 38)/15 \\ &\doteq 43.5 \\ s^2 &= [(43.5 - 35)^2 + (43.5 - 42)^2 + (43.5 - 61)^2 + (43.5 - 29)^2 \\ &\quad + (43.5 - 39)^2 + (43.5 - 47)^2 + (43.5 - 55)^2 + (43.5 - 54)^2 \\ &\quad + (43.5 - 50)^2 + (43.5 - 41)^2 + (43.5 - 40)^2 + (43.5 - 34)^2 \\ &\quad + (43.5 - 37)^2 + (43.5 - 51)^2 + (43.5 - 38)^2]/15 \\ &\doteq 76.4\end{aligned}$$

Thus,

$$s \doteq \sqrt{76.4} \doteq 8.7.$$

■

The mean, $\bar{x}$, in the preceding example is easily computed on a calculator. The standard deviation, s, is also easily calculated on a machine with parentheses, M+ and MR keys. In the present instance, entering the following string does the job nicely.

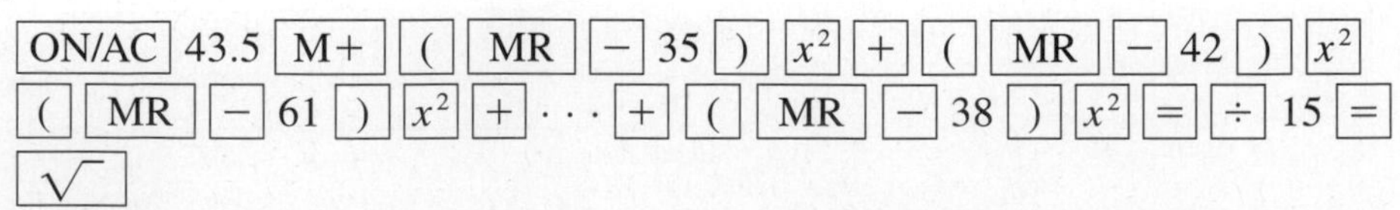

Scientific and business calculators frequently have built in statistics routines that make the calculation of means and standard deviations even simpler. Alternatively, these computations can be performed automatically on a computer using the *Stat Explorer* or other comparable software.

and so on. The seven terms in the sum are generated by the quantity $2 + i \cdot 3$, or $2 + 3i$, where i starts at 0 and increases to 6 by steps of 1. Thus,

$$2 + 5 + 8 + 11 + 14 + 17 + 20 = \sum_{i=0}^{6} (2 + 3i)$$

as desired.

Look back • • • •

The solution depended on finding the pattern in the sequence 2, 5, 8, 11, 14, 17, 20 and devising a formula for the general term. The initial term was $2 + 0 \cdot 3$, the next term was $2 + 1 \cdot 3$, and so on. Thus, the general term was $2 + i \cdot 3$, or $2 + 3i$, with i starting at 0 and continuing to 6. In earlier chapters, we argued that the nth term in such a sequence was $2 + (n-1) \cdot 3$ or $3n - 1$ where n started from 1. Had we made this choice instead of $2 + 3i$, the answer would have been $\Sigma_{n=1}^{7} (3n-1)$. Writing out this last sum gives the same result as above. ■

Using summation notation, the definitions for the mean and standard deviation of a collection of values is greatly simplified.

DEFINITION The Mean of a Set of Values

The mean of the values $x_1, x_2, \cdots, x_n$ is $\bar{x}$ where

$$\bar{x} = \frac{\sum_{i=1}^{n} x_i}{n}.$$

DEFINITION The Standard Deviation of a Set of Values

The standard deviation of $x_1, x_2, \cdots, x_n$ is s where

$$s = \sqrt{\frac{\sum_{i=1}^{n} (\bar{x} - x_i)^2}{n}}$$

and $\bar{x}$ is the mean.

PROBLEM SET 8.2

Understanding Concepts

1. Determine the mean, median, and mode for this set of data.

18 27 17 19 21 24 18 15
23 18 17 14 22 19 27 30

2. **(a)** Compute the mean, median, and mode for this set of data.

69 81 77 69 64 85 81 73 79
74 70 78 86 80 71 79 77 70
67 70 79 80 71 67 69 79 81

(b) Draw a line plot for the data in part (a).

(c) Does either the mean, median, or mode seem typical of the data in part (a)?

(d) Might it be reasonable to suspect that the data in part (a) actually comes from two essentially different populations (say daily incomes from two entirely different companies, for example)? Explain your reasoning.

3. (a) Compute the quartiles for the data in problem 2.

(b) Give the 5-number summary for the data in problem 2.

(c) Draw a box and whisker plot for the data in problem 2.

4. **(a)** Draw side-by-side box and whisker plots to compare students' performances in class A and class B if the final grades are as shown here.

Class A: 91, 63, 65, 73, 65, 86, 96, 75 75, 79, 84, 72, 80

Class B: 87, 72, 95, 89, 69, 79, 56, 64, 66, 67, 89, 47

(b) Briefly compare the performances in the two classes on the basis of the box and whisker plots in part (a).

5. Use the data in problem 1 to answer the following.

(a) Compute the mean.

(b) Compute the standard deviation.

(c) What percent of the data are within one standard deviation of the mean?

(d) What percent of the data are within two standard deviations of the mean?

(e) What percent of the data are within three standard deviations of the mean?

6. **(a)** Choose an appropriate scale and draw a line plot for this set of measurements of the heights in centimeters of 2-year-old ponderosa pine trees.

22.2 23.5 22.5 22.6 23.0 22.8
22.4 22.2 23.0 23.3 23.9 22.7

(b) Compute the mean and standard deviation for this data.

(c) What percent of the data are within one standard deviation of the mean?

(d) What percent of the data are within two standard deviations of the mean?

(e) What percent of the data are within three standard deviations of the mean?

7. Write out each of these sums.

(a) $\sum_{i=3}^{8} i^2$ **(b)** $\sum_{k=1}^{5} \frac{k}{k+1}$ **(c)** $\sum_{i=1}^{7} 3 \cdot 2^i$

8. Compute each of these sums, expressing the answer as a fraction in simplest form.

(a) $\sum_{i=1}^{1} \frac{1}{i(i+1)}$ (b) $\sum_{i=1}^{2} \frac{1}{i(i+1)}$ **(c)** $\sum_{i=1}^{3} \frac{1}{i(i+1)}$

(d) Conjecture the result of computing the sum

$$\sum_{i=1}^{n} \frac{1}{i(i+1)}$$

where n is a positive integer.

9. **(a)** Compute $(\Sigma_{i=1}^{5} i)^2$.

(b) Compute $\Sigma_{i=1}^{5} i^3$.

(c) Make a conjecture on the basis of parts (a) and (b) and test your conjecture when the upper limit is 10.

10. Write each of these sums using summation notation. F_n denotes the nth Fibonacci number.

(a) $3 + 5 + 7 + 9 + 11$ **(b)** $1 + 5 + 9 + 13$

(c) $2 + 6 + 18 + 54 + 162 + 486 + 1458$

(d) $1 + 1 + 2 + 3 + 5 + 8 + 13 + 21$

Thinking Critically

11. On June 1, 1993, the average age of the 33 employees at Acme Cement was 47 years. On June 1, 1994, three of the staff aged 65, 58, and 62, retired and were replaced by four employees aged 24, 31, 26, and 28. What was the average age of the employees at Acme Cement on June 1, 1994?

12. (a) Compute the mean and standard deviation for this data.

28 34 41 19 17 23

(b) Add 5 to each of the values in part (a) to obtain 33, 39, 46, 24, 22, and 28. Compute the mean and the standard deviation for this new set of values.

(c) What properties of the mean and standard deviation are suggested by parts (a) and (b)?

13. Compute the mean and standard deviation for the data represented by each of these two histograms.

(a)

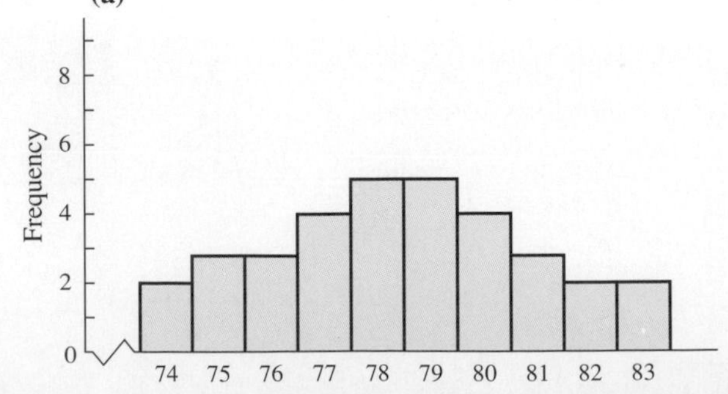

(b)

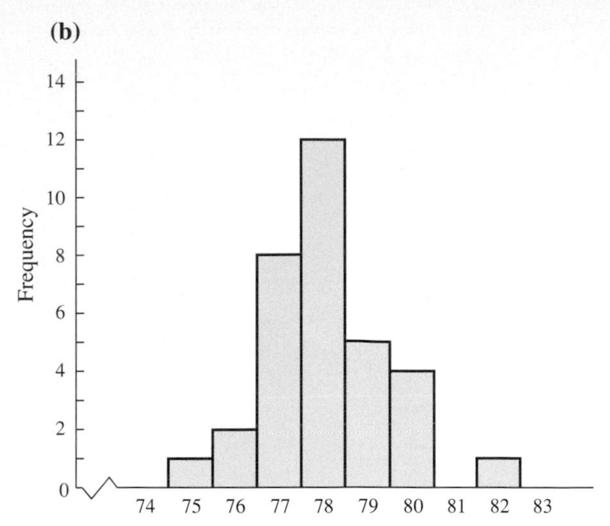

(c) Briefly explain why the standard deviation for the data of part (b) is less than that for part (a).

14. Does the mean, median, or mode seem to be the most typical value for this set of data? Explain briefly. (*Suggestion:* Draw a line plot.)

42 47 38 16 45 41 16 48 44

15. (a) Determine the mean, median, and mode of the data in the line plot shown.

(b) Does the mean, median, or mode seem to be the most typical of this data? Explain briefly.

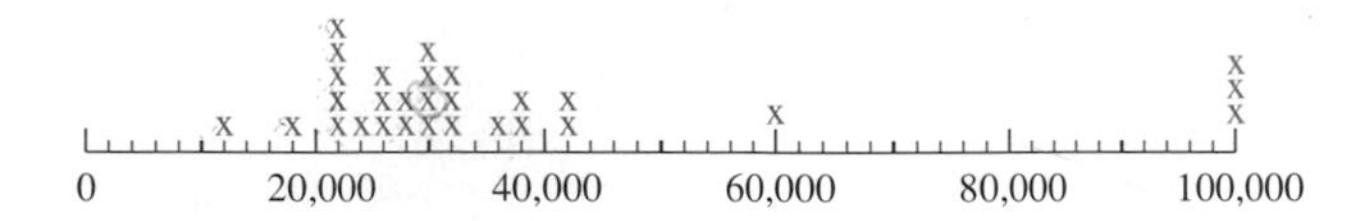

16. Produce sets of data that satisfy these conditions.

(a) mean = median < mode

(b) mean = mode < median

(c) median = mode < mean

17. (a) What can you conclude if the standard deviation of a set of data is zero? Explain.

(b) What can be said about the standard deviation of a set of data if the values all lie very near the mean? Explain.

18. Let Q_L, $\hat{x}$, and Q_U denote the lower quartile, median, and upper quartile of a set of data.

(a) Create a set of data with the property that 25% of the data lie in each of these ranges.

$$x < Q_L,\ Q_L < x < \hat{x},\ \hat{x} < x < Q_U,\ Q_U < x$$

(b) Create a set of data for which it is not true that 25% of the data lie in the ranges specified in part (a).

19. A collection of data contains 10 values consisting of a mix of ones, twos, and threes.

(a) If $\overline{x} = 3$ what is the data set?

(b) If $\overline{x} = 2$, what are the possibilities for the data set?

(c) If $\overline{x} = 1$ what is the data set?

(d) Could $\overline{x} = 1$ and $s \neq 0$ for this data set? Explain.

20. Compute the mean of each of these collections of data.

(a) A = {27, 38, 25, 29, 41}

(b) B = {27, 38, 25, 29, 41, 32}

(c) C = {27, 38, 25, 29, 41, 32, 32}

(d) D = {27, 38, 25, 29, 41, 32, 32, 32, 32, 32, 32}

(e) What conclusion is suggested by the calculations in parts (a) through (d)?

(f) Guess the mean of this set of data and then do the calculation to see if your guess is correct.

$$E = \{27, 38, 25, 29, 41, 60, 4, 60, 4\}$$

(g) What general result does the calculation in part (f) suggest?

21. (a) The mean of each of these collections of data is 45.

R = {45, 35, 55, 25, 65, 20, 70}
S = {45, 35, 55, 25, 65, 20, 70, 45, 45}
T = {45, 35, 55, 25, 65, 20, 70, 80, 10}

Which of R and S has the smaller standard deviation? No computation is needed; justify your response with a single sentence.

(b) Like the means of R and S in part (a), the mean of T is 45. Is the standard deviation for this set the same as that for S in part (a)? Note that both these sets have the same number of entries. Explain your conclusion.

22. According to Garrison Keillor, all the children in Lake Wobegon are above average. Is this assertion just a joke or is there a sense in which it could be true?

23. If the mean of A = $\{a_1, a_2, \cdots, a_{30}\}$ is 45 and the mean of B = $\{b_1, b_2, \cdots, b_{40}\}$ is 65, compute the mean of the combined data set. (*Hint:* The answer is not 55.)

24. (a) List the shoe size of ten of the women students in your class.

(b) Determine the mean, median, and mode of the data of part (a). Does the mean, median, or mode seem to be the most representative of the data of part (a)?

(c) Compute the standard deviation of the data of part (a).

(d) What percentage of the data of part (a) lies within one standard deviation of the mean? Two standard deviations? Three standard deviations?

25. Repeat problem 24 for a sample of 20 male students selected from classmates in various classes.

26. **(a)** Determine the amount of change ten students in your class have with them.

(b) Determine the mean, median, and mode for the data in part (a).

(c) Does the mean, median, or mode, seem most representative of the data in part (a)?

(d) Is it possible that the set of data could have no mode?

(e) Compute the standard deviation of the data collected in part (a).

(f) What percentage of the data collected in part (a) lies within one standard deviation of the mean? Two standard deviations? Three standard deviations?

Making Connections

27. **(a)** Use the data represented by the bar graph shown to determine as accurately as you can the average (mean) monthly trade deficit for the period from October, 1992 through September, 1993.

(b) What percentage of the data lies within one standard deviation of the mean? Two standard deviations?

U.S. merchandise trade deficit

Billions of dollars, seasonally adjusted; import figures exclude shipping and insurance.

13
12
11
10
9
8
7
6
5
4
3
2
1
0

O N D J F M A M J J A S
1992 1993

Sept. '92 8.31 Aug. '93 10.05 **Sept. '93 10.90**

SOURCE: Graph, "U.S. merchandise trade deficit" from *The Moscow/Pullman Daily News*, November 19, 1993. Copyright © 1993 News Review Publishing Co. Reprinted by permission of The Moscow/Pullman Daily News.

28. **(a)** Use data from the line graph shown to determine as accurately as possible the average (mean) number of homicides committed per year from 1976 through 1985.

(b) Repeat part (a) for the years 1986 through 1992.

(c) Do these means suggest a significant trend in homicide by hand guns? Explain briefly.

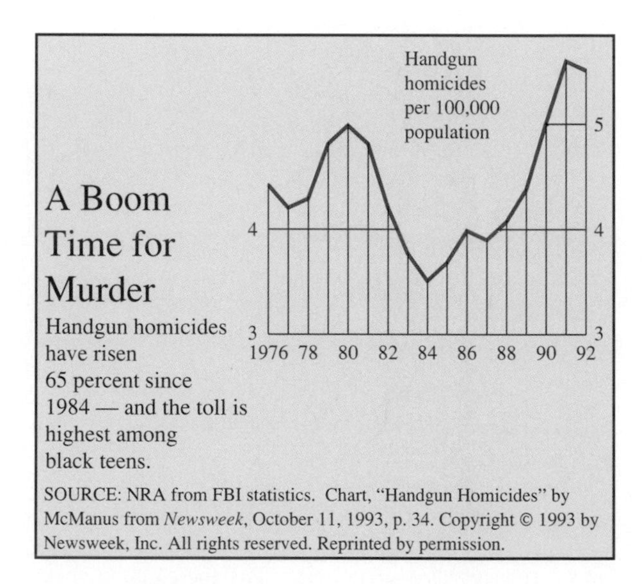

SOURCE: NRA from FBI statistics. Chart, "Handgun Homicides" by McManus from *Newsweek*, October 11, 1993, p. 34. Copyright © 1993 by Newsweek, Inc. All rights reserved. Reprinted by permission.

Using A Computer

29. Use the *Stat Explorer* or other available software to determine the following for the data in Table 8.1.

(a) $\overline{x}$

(b) $\hat{x}$

(c) the mode

(d) s

(e) Q_L

(f) Q_U

30. This table from *Statistical Abstract of the United States, 1992,* gives air pollutant data for the United States. Use the *Stat Explorer* or other available software to determine the mean and standard deviation of the data for the period 1980–1990 for each of these emissions.

(a) particulate matter **(b)** sulfur oxides

(c) nitrogen oxides **(d)** volatile organic compounds

(e) carbon monoxide **(f)** lead

(g) Do these data show that emission of air pollutants changed much during this eleven year period?

(h) How do the results of (a) through (f) above compare with the figures for the four years 1940, 1950, 1960, and 1970?

National Air Pollutant Emissions: 1940 to 1990

(In millions of metric tons, except lead in thousands of metric tons. Metric ton = 1.1023 short tons. PM = Particulate matter, SO_x = Sulfur oxides, NO_x = Nitrogen oxides, VOC = volatile organic compound, CO = Carbon monoxide, Pb = Lead. NA Not available)

	Emissions						Percent Change from Prior Year					
Year	**PM**	**SO_x**	**NO_x**	**VOC**	**CO**	**Pb**	**PM**	**SO_x**	**NO_x**	**VOC**	**CO**	**Pb**
1940	23.1	17.6	6.9	15.2	82.6	(NA)	(NA)	(NA)	(NA)	(NA)	(NA)	(NA)
1950	24.9	19.8	9.4	18.1	87.6	(NA)	7.8	12.5	36.2	19.1	6.1	(NA)
1960	21.6	19.7	13.0	21.0	89.7	(NA)	−13.3	−0.5	38.3	16.0	2.4	(NA)
1970	18.5	28.3	18.5	25.0	101.4	203.8	−14.4	43.7	42.3	19.0	13.0	(NA)
1980	8.5	23.4	20.9	21.1	79.6	70.6	−54.1	−17.3	13.0	−15.6	−21.5	−65.4
1981	8.0	22.6	20.9	19.8	77.4	56.4	−5.9	−3.4	0.0	−6.2	−2.8	−20.1
1982	7.1	21.4	20.0	18.4	72.4	54.4	−11.3	−5.3	−4.3	−7.1	−6.5	−3.5
1983	7.1	20.7	19.3	19.3	74.5	46.4	0.0	−3.3	−3.5	4.9	2.9	−14.7
1984	7.4	21.5	19.8	20.3	71.8	40.1	4.2	3.9	2.6	5.2	−3.6	−13.6
1985	7.2	21.1	19.9	20.1	68.7	20.1	−2.7	−1.9	0.5	−1.0	−4.3	−49.9
1986	6.7	20.9	19.1	19.0	63.2	8.4	−6.9	−0.9	−4.0	−5.5	−8.0	−58.2
1987	6.9	20.5	19.4	19.3	63.4	8.0	3.0	−1.9	1.6	1.6	0.3	−4.8
1988	7.5	20.6	20.0	19.4	64.7	7.6	8.7	0.5	3.1	0.5	−2.1	−5.0
1989	7.2	20.8	19.8	18.5	60.4	7.2	−4.0	1.0	−1.0	−4.6	−6.6	−5.3
1990	7.5	21.2	19.6	18.7	60.1	7.1	4.2	1.9	−1.0	1.1	−0.5	−1.4

SOURCE: *Statistical Abstract of the United States,* 1992, p. 213.

For Review

31. Find two rational numbers between 5/8 and 6/8.

32. Write the following real numbers in order from the least to the greatest. Be sure to identify the two numbers that are equal.

$$\frac{3}{4}, 0.74\overline{9}, 0.7\overline{49}, 0.74\overline{09}, 0.749, 0.74949$$

33. Given that $\sqrt{3}$ is irrational and r is rational, prove that $\sqrt{3} + r$ is irrational.

34. Given that $\sqrt{3}$ is irrational and r is rational, prove that $\sqrt{3} \div r$ is irrational.

8.3 Statistical Inference

Populations and Samples

In statistics a **population** is a particular set of things or operations about which one desires information. If the desire is to determine the average yearly income of all adults in the United States, the population is the set of *all* adults in the United States. If one is only concerned about the average yearly income of adults in Nevada, the

HIGHLIGHT FROM HISTORY
Dewey Wins!

As voters left the voting booths on election day in 1948 they were asked by representatives of the *Chicago Tribune* if they had voted for Harry Truman or Thomas Dewey for president of the United States. On the basis of this poll, which indicated that Dewey would win by a comfortable margin, they prepared and printed a story for their morning edition with the headline, *DEWEY DEFEATS TRUMAN.* To their chagrin Truman in fact won the election and other papers gleefully printed front page stories prominently displaying a picture of Truman triumphantly holding aloft a copy of the *Tribune* with its erroneous headline.

population is the set of *all* adults in Nevada. Other examples of populations are:

- all boys in Eisenhower High School in Yakima, Washington,
- all light bulbs manufactured on a given day by Acme Electric Company,
- all employees of AT&T,

and so on. One might want to determine:

- the average height of boys in Eisenhower High School,
- the average life of light bulbs produced by Acme Electric company, or
- the average cost of medical care for employees of AT&T.

Since it is often impractical or impossible to check each member of a population, the idea of statistics is to study a **sample** or subset of the population, and to make inferences about the entire population on the basis of the study of the sample.

If the goal of accurate estimation of population characteristics is to be achieved, the population must be carefully identified and the sample appropriately chosen. Indeed, lack of care in identifying populations and choosing samples was precisely the source of some of the most spectacular errors of the past in the use of statistics (see Highlight from History, Dewey Wins!). When a sample does not accurately reflect the composition of the entire population, it is not surprising that attributes of the sample do not accurately indicate attributes of the population.

For example, suppose a study of the heights of boys at Eisenhower High School is desired. Instead of measuring each boy in the school, it is decided to study a sample of just 20 of the boys. If the sample were to be selected by choosing every fourth name out of an alphabetical listing of all 80 boys, would it likely be representative of the entire population? Probably, since there is likely little or no connection between last names and heights. However, a sample consisting of the members of the basketball team is clearly not representative since basketball players tend to be unusually tall. It turns out that the best approach to sampling is to use a **random sample.**

BIZARRE SEQUENCE OF COMPUTER-GENERATED RANDOM NUMBERS

DEFINITION A Random Sample

A **random sample** of size r is a subset of r individuals from the population chosen in such a way that every such subset has an equal chance of being chosen.

For example, suppose an urn contains a mixture of red and white beans and you want to estimate what fraction of the beans are red by selecting a sample of 20 beans. You proceed by mixing the beans thoroughly and then, with your eyes closed, selecting 20 beans. Since each subset of 20 beans has an equal chance of being selected, the sample is indeed random.

Other schemes also work well.

Suppose AT&T wishes to study the employees at one of its plants by selecting a random sample of 20 employees and asking them to respond to a questionnaire. One way to obtain a random sample would be to put the names of all the employees on tags, place the tags in a large container, mix the tags thoroughly, and then have someone close his or her eyes and select a sample of 20 tags. The employees whose tags are chosen constitute the random sample.

Another way to obtain the sample is to use a sequence of digits chosen in such a way that each digit is equally likely to be any one of the ten possibilities and the choice of each digit is independent of the choice of every other digit. Such a sequence would be a **random sequence of digits.** One way to select such a random sequence is to construct a simple spinner* with ten 36° sectors numbered 0, 1, 2, 3, 4, 5, 6, 7, 8, and 9 as shown in Figure 8.18. Spinning the spinner repeatedly produces a string of

* For a description of a superior and more accurate spinner, see Figure 9.2.

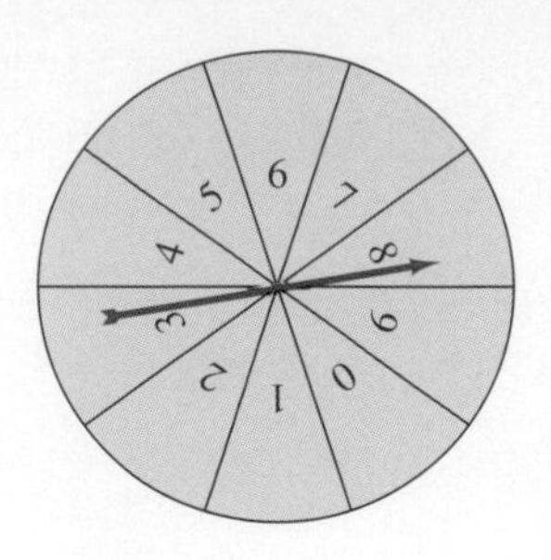

Figure 8.18
A simple random digit generator

digits. Since the result of each spin is independent of the result of every other spin, each digit is equally likely to be selected and the digit sequence is random.

The desired sample of AT&T employees can now be obtained as follows (suppose the plant in question has 9762 employees):

1. Assign each employee a 4-digit number from among 0001, 0002 · · · , 9762.
2. Generate a random sequence of 4-digit numbers by repeatedly spinning the spinner four times. Since each digit is equally likely to appear on any spin, each 4-digit number is equally likely to appear on any four spins. If the spinning process generates 0000, a 4-digit number greater than 9762, or any number already obtained, simply ignore it and continue to generate more 4-digit numbers. When 20 appropriate numbers have been generated, they can be used to identify the 20 employees to be included in the sample.

JUST FOR FUN

A Matter of Sums

Insert plus signs among the digits

1 2 3 4 5 6 7 8 9

to obtain a sum of 99. Here, for example, is a failed attempt:

1 + 23 + 4 + 5 + 67 + 8 + 9 = 117

Finally, as you might expect, many calculators have built-in statistics routines that will generate random numbers and samples, and computer software exists for the same purpose. On your computer disc, the program for this purpose is RANDOM-NUMBER.

Population Means and Standard Deviations

Professional statisticians find it helpful to use different symbols for means and standard deviations for populations and for samples. For populations the mean and standard deviation are denoted by the Greek letters μ and σ (mu and sigma). Thus, for a population of size N, the population mean and population standard deviation are given by

$$\mu = \frac{\sum_{i=1}^{N} x_1}{N} \quad \text{and} \quad \sigma = \sqrt{\frac{\sum_{i=1}^{N} (x_i - \mu)^2}{N}}.$$

EXAMPLE 8.14

Computing a Population Mean and Standard Deviation

Consider a population that consists of the numbers shown.

64 65 68 67 59 66 63 66 64 62 65 66 63 66 66 63
63 64 62 67 63 61 60 64 64 63 63 65 64 65 66 63
65 62 63 65 61 64 63 64 62 69 65 65 64 64 63 64
66 65 64 64 63 64 66 67 69 63 65 63 64 64 64 65
67 68 64 62 66 62 64 61 65 62 65 62 65 62 66 63

Use the *Stat Explorer* or other suitable software on your computer, or the built-in statistics routine on a suitable calculator to compute μ and σ for this population.

SOLUTION

Since there are 80 numbers in the population,

$$\mu = \frac{64 + 64 + \cdots + 63}{80} \doteq 64.2$$

and

$$\sigma = \sqrt{\frac{(64 - 64.3)^2 + (64 - 64.3)^2 + \cdots + (63 - 64.3)^2}{80}} \doteq 1.9. \quad \blacksquare$$

Estimating Population Means and Standard Deviations

Suppose we wish to know the mean and standard deviation of some large or inaccessible population. Since these statistics, μ and σ, may be difficult or impossible to compute, we estimate them with the sample mean

$$\bar{x} = \frac{\sum_{i=1}^{n} x_i}{n}$$

and the sample standard deviation

$$s = \sqrt{\frac{\sum_{i=1}^{n} (x_i - \bar{x})^2}{n}}$$

of a suitably chosen sample, $x_1, x_2, \ldots, x_n$, of size n.

EXAMPLE 8.15

Estimating a Population Mean and Standard Deviation

(a) Estimate the mean and standard deviation of the population in Example 8.14 by using a spinner as illustrated in Figure 8.18 to select a random sample of size ten.

(b) Compare the results of part (a) with the results obtained in Example 8.14.

SOLUTION

(a) Suppose your spinner generates the digit sequence 6, 2, 2, 9, 0, 6, 4, 4, 6, 4, 2, 7, 1, 2, 4, 6, 7, 0, 1, 8. Using these two at a time we obtain the 2-digit numbers

62 29 06 44 64
27 12 46 70 18.

The sixty-second number in the ordered data set is 66, and so on.

Since all are different, these determine the random sample shown here.

66 64 62 64 66
64 62 64 66 63

Thus, the mean of the numbers in the sample is

$$\bar{x} = (66 + 64 + 62 + 64 + 66 + 64 + 62 + 64 + 66 + 63)/10 \doteq 64.1.$$

The variance is

$$\begin{aligned} v &= s^2 \\ &= [(64.1 - 66)^2 + (64.1 - 64)^2 + (64.1 - 62)^2 \\ &\quad + (64.1 - 64)^2 + (64.1 - 66)^2 + (64.1 - 64)^2 \\ &\quad + (64.1 - 64)^2 + (64.1 - 62)^2 + (64.1 - 66)^2 \\ &\quad + (64.1 - 63)^2]/10 \doteq 2.09, \end{aligned}$$

and the sample standard deviation is

$$s = \sqrt{v} \doteq 1.45.$$

(b) We observe that $\bar{x}$ and s for the sample are reasonable approximations for μ and σ for the population as determined in Example 8.14. ■

Suppose we were to repeat the preceding example but with a random sample of size 15. For the resulting sample, $\bar{x}$ and s should be slightly better approximations to μ and σ from Example 8.14 than obtained above. In fact, it is generally true that larger samples tend to yield better approximations to population characteristics.

HIGHLIGHT FROM HISTORY
R. A. Fisher (1890–1962)

Statistics, a relative newcomer to the mathematical scene, was largely developed during the last one hundred years. One of the most influential personalities in this development was Sir Ronald A. Fisher, a British geneticist and statistician. To remove the effect of differences in soil fertility of different plots of ground used in testing the effects of various fertilizers on plant growth, Fisher introduced the idea of selecting plots by a random process and showed how to correctly compare results from such randomized experiments. Fisher also contributed other new ideas to statistics and his books and professional papers did much to shape statistics into an organized science.

Distributions

We return now to the data of Table 8.2 where the population consisted of all boys in Eisenhower High School. A histogram of the boys heights to the nearest inch appears in Figure 8.6 and a line graph for the same data is shown in Figure 8.8.

Since the heights of the columns in the histogram represent the number or frequency of the measurements in each range (63.5—64.5, 64.5—65.5, and so on) and the width of each column is one, the total area of all the columns in the histogram is 80, the total number of boys in the population.

The **relative frequency** of the measurements in each range in Figure 8.6 is the fraction (expressed as a decimal) of the total number of boys represented in that range. If the heights of the column in the histogram are determined by relative frequency, the diagram remains the same except for the designation on the vertical scale as shown in Figure 8.19. Also, since the width of each column is one, the *area* of each column gives the fraction of the population whose heights fall in that range. Moreover, the area of the first three columns gives the fraction of the population with heights ranging from 63.5 to 66.5 inches and the total area of the histogram is one.

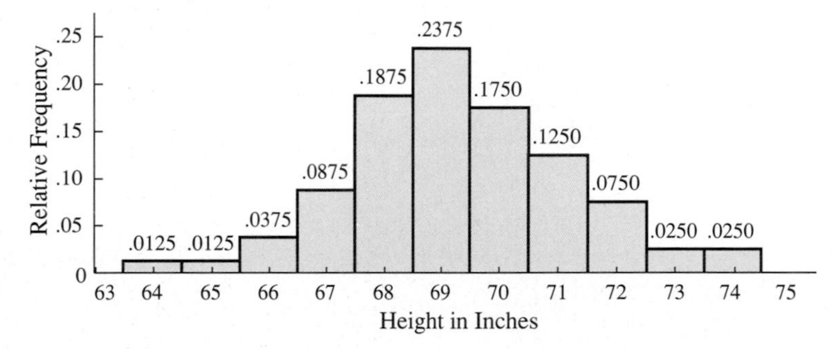

Figure 8.19
Histogram of Figure 8.16 but with the vertical scale denoting relative frequency

As noted earlier, histograms often are representations of grouped data that make it appear that all the data values in a given range are the same. Of course, as in the present case, this is frequently not so. The boys heights are listed to the nearest inch whereas people's heights actually vary continuously. Thus, a truer representation of such data is provided by a line graph or frequency polygon as in Figure 8.8. Again, if the vertical scale represented relative frequency rather than frequency the graph would appear unchanged as shown in Figure 8.20. Also, since such a diagram can be obtained from a histogram by deleting and adding small triangles of equal area, the area under the **relative frequency polygon** is still one and the area of that portion of

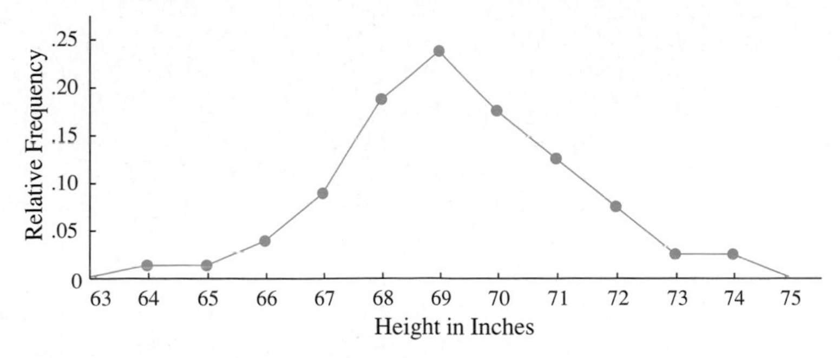

Figure 8.20
Relative frequency polygon of heights of boys in Eisenhower High School

the polygon from say 63.5 to 66.5 closely approximates the fraction of the population of boys whose heights lie in this range.

Additionally, had the measurements been taken more and more closely and the ranges in the histogram made narrower and narrower, the tops of the columns in the histogram and also the corresponding frequency polygon would have more and more closely approximated a smooth bell-shaped curve called a **normal distribution** as shown in Figure 8.21. Here also the area under the curve and above the interval between a and b indicates the relative frequency or fraction of the boys measured having heights between a and b. This fraction indicates the **likelihood** or **probability** that a boy chosen at random from the sample would have a height in the given range.

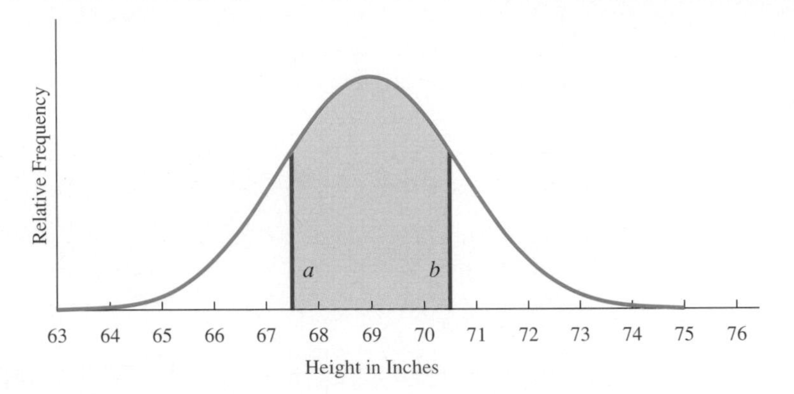

Figure 8.21
Normal distribution of the heights of boys in Eisenhower High School

For many populations the distribution of the measurements of the property will be a continuous (and often normal) curve as in Figure 8.21. However, in other cases the observations are not continuous, are not normal, or are discrete. If the observations are discrete the distribution "curve" is just a histogram.

DEFINITION A Distribution Curve

A curve or histogram that shows the relative frequency of the measurements of a characteristic of a population that lies in any given range is a **distribution curve.** The area under such a curve or histogram is always 1.

Knowing the distribution of a population frequently allows one to say with some precision what the average value is and what percentage of the population lie within different ranges. In particular, the normal distribution has been studied in great detail and it can be shown that very nearly 68% of the data lie within one standard deviation of the mean, very nearly 95% of the data lie within two standard deviations of the mean, and very nearly 99.7% (or virtually **all**) of the data lie within three standard deviations of the mean as illustrated in Figure 8.22. Using the language of probability, we would say that the probability that a given data value lies within one standard deviation of the mean is 0.68, the probability that a given data value lies within two standard deviations of the mean is 0.95, and the probability that any given data value lies within three standard deviations of the mean is 0.997 (virtually 100%). Considerations like these are what make it possible for very carefully designed polls and other studies to claim that their results are accurate to within a given tolerance, say 3%.

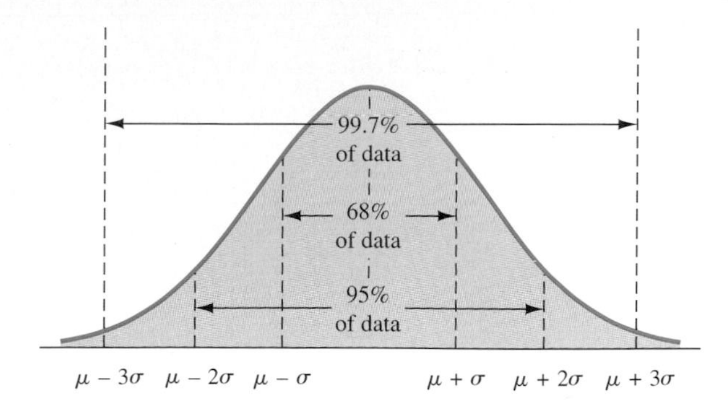

Figure 8.22
Percent of data within one, two, and three standard deviations of the mean of a normal distribution

THEOREM The 68–95–99.7 Rule for Normal Distributions

For a population that has a normal disribution, about 68% falls within one standard deviation of the mean, about 95% falls within two standard deviations of the mean, and about 99.7% falls within three standard deviations of the mean.

It turns out that many populations are normally distributed or approximately so. Thus, the 68–95–99.7 rule is approximately true for these populations and also for samples from these populations. For samples, the approximation is increasingly accurate for increasingly large sample sizes.

HIGHLIGHT FROM HISTORY
John von Neuman and Game Theory

Archeology suggests that game playing predates human history. In a larger sense, any competitive situation (war, business, labor relations) can be thought of as a game, though it may be played with deadly seriousness. Modern game theory, the mathematical study of games, is largely due to the remarkable John von Neuman at the Institute for Advanced Study who, with Princeton University economist Oskar Morgenstern, published the definitive *Theory of Games and Economic Behavior* in 1944. This book discusses such topics as games with perfect information with no randomness allowed; mixed games, which allow for randomness; cooperative games (that is, military alliances, business cartels, voting blocs), zero-sum games where one player's loss equals the opponent's gain, multi-sum games, and so on.

In recognition of the importance of game theory a medal (one of 18 that commemorate 18 stages in the development of Western consciousness) was struck for the Paris Musee de la Monnaie in 1971.

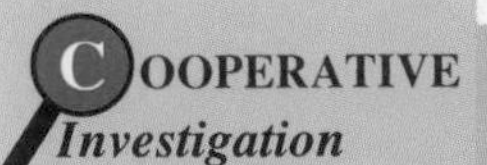

How Much Change do Students Carry?

Materials Needed

A calculator with a built-in statistics package.

Directions

1. Divide the students in the class into groups of approximately eight students each.
2. In each group, each student counts the change in his or her pocket or purse and lists the amount on the chalkboard.
3. In each group, the students compute the average amount of change per student for their group.
4. Have several students in the class individually use the data on the chalkboard and compute the average amount of change per student in the entire class. If their answers differ, someone has made an error and they should check their work.
5. Briefly conduct a class discussion on how well the averages for the various groups approximate the average amount of change per student in the class. Do the students feel that their groups are essentially random samples?
6. Have someone (perhaps the instructor) use a calculator with the statistics routine to determine the standard deviation of the data on the chalkboard.
7. Have the students compute what percentage of the data on the chalkboard lie within one standard deviation of the mean. Two standard deviations of the mean. Three standard deviations of the mean. Discuss how well these results compare with the 68–95–99.7 rule.

PROBLEM SET 8.3

Understanding Concepts

1. Describe the population that should be sampled to determine each of these.
   (a) The percentage of freshmen in U.S. colleges and universities in 1995 who earn baccalaureate degrees within 10 years.
   (b) The percentage of U.S. college and university football players in 1995 who earn baccalaureate degrees within 10 years.
   (c) The fraction of the people in the U.S. who feel that they have adequate police protection.
   (d) The fraction of the people in Los Angeles who feel that they have adequate police protection.
   (e) Would you include children in the population you describe in parts (c) and (d)? People in mental institutions? Known criminals?
2. Polls are often conducted by telephone. Might such a technique bias the results of the poll? Explain briefly.
3. Suppose a poll is conducted by face-to-face interviews but the names of the interviewees are selected at random from names listed in the telephone book. Would such a poll yield valid results? Discuss briefly.
4. The registrar at State University wants to determine the percentages of students (a) who live at home, (b) who live in apartments, and (c) who live in dormitories. There are 25,000 students in the university and the registrar proposes to select a sample of 100 students by choosing every 250th name from the list of all students arranged in alphabetical order.
   (a) What is the population?
   (b) Is the sample random? Explain.
5. In performing a study of college and university faculty attitudes in the United States, investigators first divided the population of all colleges and universities into groups according to size—25,000 or more students, 10,000 to 24,999 students, 3000

to 9999 students, and less than 3000 students. Using their judgment they then chose two schools from each group and asked each school to identify a random sample of 100 of their faculty.

(a) Was this a good way to obtain a statistically reliable (that is, random) sample of faculty? Why or why not?

(b) What is the population? Are there four distinct populations? Discuss briefly.

6. To determine the average life of light bulbs they manufacture, a company chooses a sample of the bulbs produced on a given day by selecting and testing to failure every 100th bulb.

(a) What is the population?

(b) Is this a good way to select a sample? Why or why not?

(c) Is the sample random? Why or why not?

7. Choose a representative sample of 20 students in your college or university and ask how many hours each person in your sample watches television each week.

(a) Describe how you chose your sample to assure that it was representative of your entire student body.

(b) On the basis of your sample, estimate how many hours of television most students on your campus watch each week.

(c) Combine your data with that of all the other students in your class and determine a revised estimate of the number of hours per week each student in your college or university watches television.

8. Describe and perform a study to determine how many of the students at your college or university have seen the movie *Animal House*.

9. A container holds a large number of beans. To estimate the number of beans in the container carry out the following steps:

Remove 25 beans and mark them with a red marker.

Return the 25 marked beans to the container and mix the beans thoroughly.

Now remove a handful of beans from the container and determine what fraction of the handful of beans is marked.

(a) Using the fraction of marked beans in your sample, determine how many beans were in the container originally.

(b) Repeat the procedure above but remove, mark, and return 50 beans to the container. Again estimate the original number of beans in the container. Compare this estimate to that of part (a). Which would you expect to be the better estimate? Explain.

(c) If the fraction of marked beans in the handful removed in part (a) is a/b, about how many beans were in the container originally? Explain briefly.

10. For a population with a normal distribution with mean 24.5 and standard deviation 2.7

(a) about 68% of the population lie between what limits?

(b) about 95% of the population lie between what limits?

(c) about 99.7% of the population lie between what limits?

Thinking Critically

11. Suppose you generate a sequence of 0s and 1s by repeatedly rolling a die and recording a 0 each time an even number comes up and a 1 each time an odd number comes up. Is this a random sequence of 0s and 1s? Explain.

12. **(a)** If you repeatedly tossed a single die would this produce a random sequence of the digits 1, 2, 3, 4, 5, and 6? Why or why not?

(b) If you repeatedly tossed a pair of dice would this produce a random sequence of the numbers 2, 3, 4, . . . , 12? Why or why not?

13. A TV ad proclaims that eight out of ten dentists surveyed prefer Whito Toothpaste. How could they possibly make such a claim if, in fact, only 1 dentist out of 10 actually prefers Whito?

14. Two sociologists mailed out questionnaires to 20,000 high school biology teachers. On the basis of the 200 responses they received, they claimed that fully 72% of high school biology teachers in the United States believe the biblical account of creation. Is their claim justified by this survey? Explain.

15. Consider the numbers 1, 2, 4, 8, and 16.

(a) Compute their mean, sometimes called the arithmetic mean.

(b) Compute their **geometric mean**; that is, compute $\sqrt[5]{1 \cdot 2 \cdot 4 \cdot 8 \cdot 16}$.

(c) Does the arithmetic mean or the geometric mean seem more typical of this sequence of numbers?

16. Katja can paddle a canoe 3 miles per hour in still water. The water in a stream flows at the rate of 1 mile per hour. Thus, going upstream in the canoe Katja's effective speed is 2 miles per hour, and going downstream it is 4 miles per hour.

(a) What is Katja's average speed (total distance ÷ total time) if she paddles 4 miles upstream and back?

(b) Compute the harmonic mean of 2 and 4—Katja's effective speeds upstream and downstream. The **harmonic mean** of several numbers is the reciprocal of the mean of their reciprocals.

(c) Compute the mean or arithmetic mean of 2 and 4.

(d) Discuss whether the arithmetic or harmonic mean seems the more appropriate "average" in this problem.

17. In the shoe business, which average of foot sizes is most important—the mean, median, or mode?

Making Connections

18. A fish biologist is studying the effect of recent management practices on the size of the cutthroat trout population in Idaho's Lochsa River. The biologist and her helpers first catch, tag, and release 200 such trout on a given day. Two weeks later, the team catches 150 cutthroat trout and determines that only three of these fish were tagged two weeks earlier. Estimate the size of the cutthroat population in the Lochsa River. (*Hint:* Determine the fraction of tagged cutthroat in the river in two different ways if there are x cutthroat in the river.)

Communicating

19. United States Representative, the Honorable J. J. Wacaser, recently sent a questionnaire to his constituents to determine their opinion on several bills being considered by the House of Representatives. Discuss how representative of the voters in his district the responses to his poll are likely to be.

20. Discuss how Representative Wacaser (see problem 19) might actually choose a random sample of voters in his district.

21. A large university was charged with sexual bias in admitting students to graduate school. Admissions were by departments and the figures are as shown.

	Men		Women	
Department	**Number of Applicants**	**Number Admitted**	**Number of Applicants**	**Number Admitted**
1	373	22	341	24
2	560	353	25	17
3	325	120	593	202
4	191	53	393	94
5	417	138	375	131
6	825	512	108	89
Totals	2691	1198	1835	557

(a) Compute the percentages of men and women applicants admitted by the school as a whole.

(b) Do the figures in part (a) suggest that sexual bias affected admission of students?

(c) Compute the percentages of men and women applicants admitted by each department.

(d) Do the figures in part (c) suggest that sexual bias affected admission to the various departments?

(e) Carefully but briefly explain this apparent contradiction. (*Hint:* See the Just for Fun in Section 9.3.)

22. (a) Carefully describe how you might reasonably estimate, without spending a whole evening, what fraction of those attending a movie on a given night purchase popcorn.

(b) Discuss what factors might complicate your sampling procedure in part (a) and how you might deal with them.

For Review

23. The following data are yields in pounds of hops.

3.4 4.4 4.8 4.5 5.1
4.8 5.5 4.7 3.5 3.6

(a) Compute the mean for this data.

(b) Compute the standard deviation for this data.

24. The 12 players on Uni Hi's basketball team averaged 5 points each during the first half of a game against Roosevelt. They averaged 7 points each for the entire game.

(a) How many points did Uni Hi score during the game?

(b) What was the average score for each player during the second half?

25. (a) What is the median of the data in problem 23?

(b) Draw a box and whisker plot for the data in problem 23.

26. Determine the mode for the data in problem 23.

27. (a) Compute the mean of 21, 25, 27, 20, 22.

(b) What is the mean of 21, 25, 27, 20, 22, 23?

(c) Were you sure of the answer to part (b) before doing the calculation? Explain.

(d) What is the mean of 21, 25, 27, 20, 22, 20, 26?

(e) Were you sure of the answer to part (d) before doing the calculation? Explain.

COOPERATIVE *Investigation*

Using Samples to Approximate Characteristics of Populations

The chart shown contains 100 integers (the population) displayed in such a way that they can be represented by a number pair (a, b). For example, entry (2, 7) is the integer 24 and entry (7, 3) is 26.

	0	1	2	3	4	5	6	7	8	9
0	21	22	20	24	22	29	25	21	27	17
1	25	12	28	21	22	17	28	18	18	26
2	19	17	23	29	19	16	24	24	25	19
3	22	17	26	11	31	19	14	20	23	17
4	26	13	30	26	18	23	37	24	27	28
5	14	15	25	20	24	18	20	30	35	21
6	18	30	22	20	20	23	27	26	33	13
7	24	21	23	26	28	19	28	29	31	23
8	21	27	22	25	21	16	23	27	16	25
9	23	22	24	22	16	15	19	24	25	20

(a) Use a spinner as in Figure 8.18 or the program RANDOMNUMBER to generate five number pairs (a, b) to determine a sample of five numbers from the above table. Compute the mean and standard deviation of your 5-number sample.

(b) Repeat part (a) but with a sample of size 10.

(c) On the basis of parts (a) and (b) give two estimates for each of the population mean and standard deviation of the population.

(d) Record your means for parts (a) and (b) on the chalkboard.

(e) Determine $\bar{x}_5$, the mean of the means of the samples of size 5 and $\bar{x}_{10}$, the mean of the means of samples of size 10 from the chalkboard. Also compute s_5 and s_{10}, the standard deviations of the means of the samples of size 5 and size 10.

(f) Using the result of part (e) again estimate the population mean and standard deviation.

(g) Using *Stat Explorer* or other suitable software, calculate the actual population mean and standard deviation for the entire population. Alternatively, these figures can be given to the class by the instructor.

(h) Briefly discuss the results of parts (a) through (g).

EPILOGUE The Information Age

In today's world we are often confronted with masses of data that must be organized and summarized to be understood. We are also constantly bombarded with numerical information prepared by other people—pollsters, politicians, special interest groups, federal and state governments, foundations, testing agencies, and others—which we must be able to evaluate if we are to be informed citizens. As H.G. Wells once wrote, "Statistical thinking will one day be as necessary for efficient citizenship as the ability to read and write." Statistical thinking requires that we be able to read graphs and charts of all kinds, understand diagrams, and properly understand that, while sample results vary from sample to sample, a random sample of the appropriate size can yield very accurate information about a population. At the same time, it is equally important that we be able to discern how graphs, charts, and statistics can misinform, either intentionally or unintentionally.

In this chapter we have discussed many aspects of statistics—the organization, interpretation, and display of data; the meaning of the mean, median, and mode as indications of a "typical" value of a set of data; notions like the range, box and whisker plots, and the standard deviation as indications of the spread of a set of data; and the notion of a population and how samples can be used to provide information about populations. Statistics is a powerful and useful tool, but it must be used with understanding and with care.

CLASSIC CONUNDRUM A Magic Magic Magic Square

Consider the magic magic magic square shown.

8	11	14	1
13	2	7	12
3	16	9	6
10	5	4	15

Compute the sums of the numbers indicated by the colored squares in these diagrams.

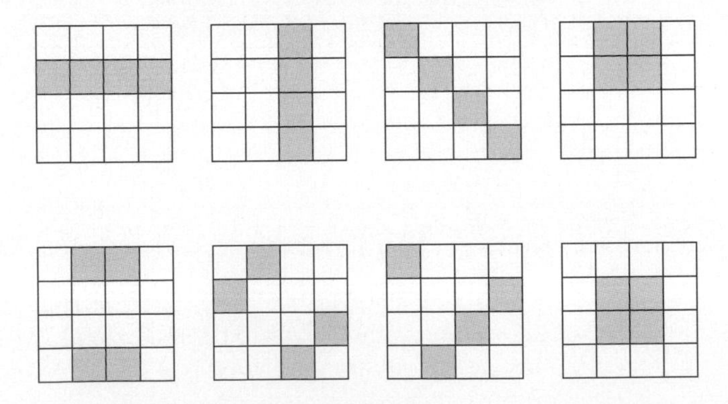

Can you find other interesting patterns of sums? Draw diagrams to show interesting patterns you find.

CHAPTER 8 SUMMARY

Key Concepts

The main objectives of this chapter have been to introduce the fundamental ideas of statistics most likely to appear in elementary school texts. These central ideas include the following:

- data
- line plots
- stem and leaf plots
- histograms
- line graphs
- pie charts
- pictographs
- mean
- median
- mode
- quartiles
- box and whisker plots
- the standard deviation
- random samples
- distributions

Vocabulary and Notation

Section 8.1

Data
Line plot
Outlier
Stem and leaf plot
Histogram
Frequency of interval
Line graph
Bar graph
Pie chart
Pictograph

Section 8.2

Mean (arithmetic mean or average)
Median
Mode
Range
Quartiles
Extremes
5-number summary
Box and whisker plot
Standard deviation
Summation notation

Section 8.3

Population
Sample
Random sample
Random sequence of digits
Relative frequency
Normal distribution
Distribution curve
The 68–95–99.7 rule

CHAPTER REVIEW EXERCISES

Section 8.1

1. The following are the numbers of hours of television watched during a given week by the students in Mrs. Karnes' fourth grade class.

17	8	17	13	16	13	8	9	17	7
8	7	14	14	11	13	11	13	11	17
12	15	11	10	12	13	9	21	19	12

(a) Make a line plot to organize and display this data.

(b) From the line plot estimate the average number of hours per week the students in Mrs. Karnes' class watch television.

2. Make a stem and leaf plot to organize and display the data in problem 1.

3. Choosing suitable scales, draw a histogram to summarize and display the data in problem 1.

4. The following are the numbers of hours of television watched during the same week as in problem 1 but by the students in Ms. Stevens' accelerated fourth grade class.

13	8	9	11	11	12	8	9
11	11	6	8	9	11	11	6
8	9	11	11	6	8	9	11

Prepare a double stem and leaf plot to display and

compare the number of hours of television watched by Mrs. Karnes' and Ms. Stevens' classes during the given week.

5. (a) Draw a line graph to show the trend in the retail price index of farm products as shown in this table.

1950	1955	1960	1965	1970	1975	1980	1985	1990
30	31	34	35	42	64	88	104	134

(b) Using part (a), estimate the retail price index for farm products in 1972.

(c) Using part (a), estimate the retail price index for farm products in 2000.

6. Find five numbers with four of the numbers less than the mean of all five.

7. Draw a pie chart to accurately illustrate how the State Department of Highways spends its budget if the figures are as shown. Administration—12%; New Construction—36%; Repairs—48%; Miscellaneous—4%.

8. (a) Criticize this pictograph designed to suggest that the administrative expenses for Cold Steel Metal appear to be less than double in 1994 than in 1993 though, in fact, the administrative expense did double.

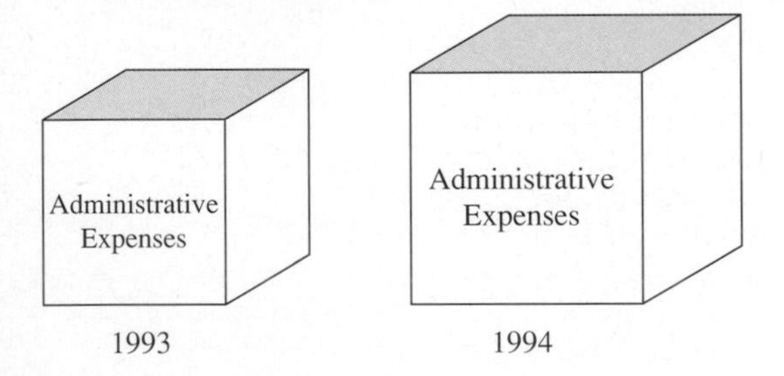

(b) If the administrators are challenged by the stock holders can they honestly defend the pictographs? (*Hint:* Measure the cubes very carefully with a metric ruler and compute their volumes.)

9. What would you need to know in order to take this statement seriously? "A survey shows that the average medical doctor in the United States earns $185,000 annually."

Section 8.2

10. Compute the mean, median, mode, and standard deviation for the data in problem 1.

11. (a) Compute the quartiles for the data in problem 1.

(b) Give the 5-number summary for the data in problem 1.

(c) Compute the quartiles for the data in problem 4.

(d) Give the 5-number summary for the data in problem 4.

(e) Using the same scales, draw side-by-side box and whisker plots to compare the data in problems 1 and 4.

12. Compute the mean and standard deviation for the data represented in these two histograms.

(a)

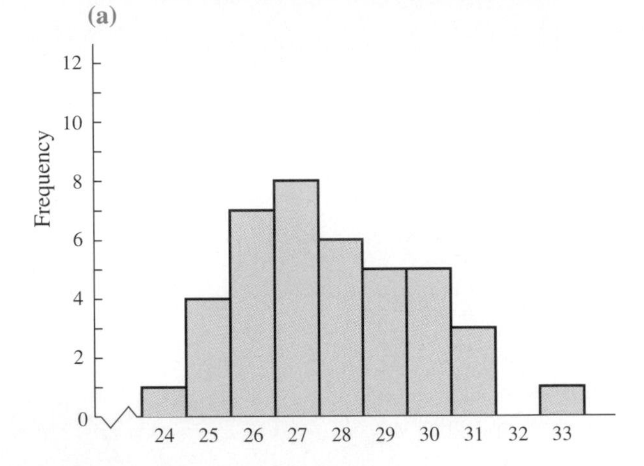

(b)

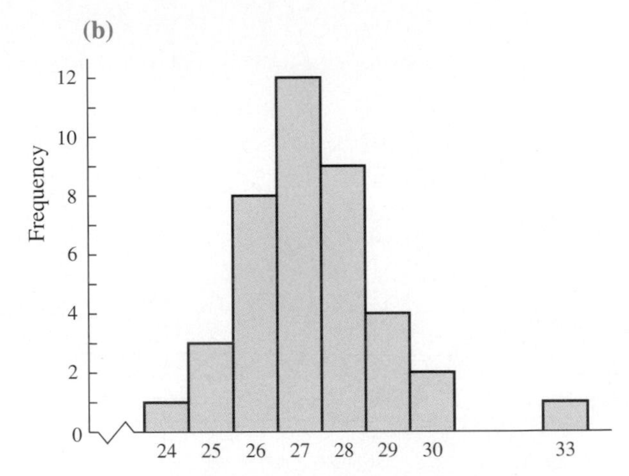

(c) Briefly explain the results of your computations in parts (a) and (b).

13. Three students were absent when the remaining 21 students took a test on which their average score was 77. When the three students took the test their scores were 69, 62, and 91. Taking these grades into account, what was the new average of all the test scores?

14. Mr. Renfro's second period Algebra I class of 27 students averaged 75 on a test and his fourth period class of 30 students averaged 78 on the same test. What was the average of all the second and fourth period test scores?

Section 8.3

15. In a study of drug use by college students in the United States, the investigators chose a sample of 200 students from State University. Was this an appropriate choice for the study? Explain.
16. Suppose you want to ascertain by a sampling procedure what percentage of the people in the United States are unemployed. How might you describe the population that should be sampled? Should every person residing in the U.S. be included in the population? Discuss briefly.
17. Discuss briefly the biases that are inherent in telephone polls.
18. Discuss briefly the biases that are inherent in samples obtained by voluntary responses to questionnaires like those sent out by members of Congress to their constituents.
19. Describe two different ways in which a random sample of 100 of the 10,000 students at State University can be obtained.
20. Suppose that only 1 dentist out of 10 actually prefers Whito Toothpaste over all other brands. By taking many random samples of size 10, might it be possible eventually to obtain a sample in which 8 out of 10 dentists in the sample preferred Whito? Explain.

CHAPTER TEST

1. (a) Prepare a line plot to summarize and visually display this data.

   42 86 80 90 74 84 86 80 63 92
   93 81 95 78 70 41 66 76 87 88
   75 88 87 78 89 85 77 87 81 57

   (b) Would you say that this data contains any outliers? Explain.
2. Compute the following for the data in problem 1:
   (a) the mean (b) the median (c) the mode
   (d) the standard deviation.
3. Prepare a stem and leaf plot to summarize the data in problem 1.
4. Sketch a box and whisker plot for the data in problem 1.
5. Suppose the average American spends 40% of his or her income paying taxes. If this were to be shown with a pie chart, what should the central angle be for this portion of the chart?
6. What does it mean to say that a sample is a random sample? Be brief but lucid.
7. Suppose that Nanda obtains scores of 77%, 79% and 72% on her first three tests in French. What total score must she earn on her last two tests in order to average at least 80% in the course?
8. The 68–95–99.7 rule does not apply to all populations but it does to a great many. What does the rule state?
9. Describe how you could choose a random sample of 200 students from your college or university.
10. In making inferences based on samples, why is it important to choose random samples?

9 Probability

Computer Support for this Chapter

In this chapter you might find the following program on your disk useful:

- COINTOSS

Are Pennies Fair?

Materials Needed

1. One penny, preferably new, for each student.
2. Tables or level topped desks for the students to work upon.

Directions

Each student stands his or her penny on edge on the table and raps the table top sharply to cause the penny to fall. This is repeated five times, noting each time whether the penny falls heads up or tails up. When all the students have completed this task, record the data on the chalkboard and compute the fraction of heads obtained on all trials.

Is the result surprising? What does it suggest about the likelihood that a penny set on edge on a table will fall heads up if the table is rapped sufficiently hard that the penny falls?

CONNECTIONS

From Polls to Probability

In the preceding chapter we considered the idea of estimating a characteristic of a population by determining the same characteristic of a sample randomly chosen from the population. Thus, if the result of a poll of 1500 randomly chosen residents of the United States shows that 61% of the sample choose blue as their favorite color, the pollster will claim that this true of the entire population with an error of say ±3%. This amounts to asserting that if you choose a person at random from the entire population, there is about a 61% chance that the person will hold the stated view. Alternatively, and perhaps more descriptively, it asserts that if you ask 100 randomly chosen residents of the United States, approximately 61 of them will choose blue as their favorite color. We could also say that the probability that a person will choose blue as his or her favorite color is about 0.61.

In this chapter, we study the notion of probability. As usual, we use concrete devices and experiences to gain an intuitive feel for the notion and to set the stage for a more mathematical approach considered later. Indeed, our approach is more than just pedagogical since there are two different but valid views of probability that correspond to the approaches taken here.

Empirical Probability

The first view of probability is that it is a measure of what happens in the long run. In the HANDS ON activity described above, unless something very unusual occurred, you discover that pennies stood on edge on a table top do not fall heads up and tails up equally often. Rather the penny falls heads up approximately nine-tenths of the time. In terms of probability we say that pennies stood on edge on a table top fall heads up with probability 9/10. This is a figure based on experience. There is no way to determine the probability on *a priori* grounds. Indeed, our intuition suggests that heads and tails should come up equally often. But experience shows that this is not so and this justifies the following definition.

DEFINITION Empirical Probability

Suppose an experiment with a number of possible outcomes is performed repeatedly—say n times, and that a specific outcome E occurs r times. The **empirical** (or experimental or experiential) **probability,** denoted by $P_e(E)$, that E will occur on any given trial of the experiment is given by

$$P_e(E) = \frac{r}{n}.$$

Theoretical Probability

Another view of probability is that, in many cases, the probability of an event can be determined by carefully analyzing the experiment to be performed and determining an *a priori* number that says what should happen if the experiment were to be performed. For example, quality dice are made in such a way, that, when a die is rolled, any one of the six sides is equally likely to come up. Since this is the case, it is reasonable to expect that if a die is rolled many times, any particular face will come up approximately one-sixth of the time. This type of reasoning justifies the second definition of probability. We refer to this second notion as **theoretical probability** or **mathematical probability,** or simply as **probability.**

DEFINITION Theoretical Probability of an Event, Sample Space

Let S, called the **sample space,** denote the set of *equally likely* outcomes of an experiment and let E be an **event** consisting of a subset of outcomes of the experiment. Let $n(S)$ and $n(E)$ denote the number of outcomes in S and E respectively. The **probability** of E, denoted by $P(E)$, is given by

$$P(E) = \frac{n(E)}{n(S)}.$$

Section 9.2 is devoted to the study of methods of counting needed to calculate theoretical probabilities.

9.1 Empirical Probability

Intuitively, most people feel that when a coin is tossed, there is a 50–50 chance that it will land heads up; that is, that

$$P(H) = \frac{1}{2},$$

where $P(H)$ is the probability of a head. However, for the same reasons, most people feel that $P(H) = 1/2$ if a penny is stood on edge on a table top and the table is jostled to cause the penny to fall. But experience shows that, in the latter instance, $P(H)$ is approximately 0.9. Even if our intuition is correct, the result of experimentation,

FROM THE NCTM STANDARDS **Probability in Elementary School**

An understanding of probability and the related area of statistics is essential to being an informed citizen. Often we read statements such as, "There is a 20 percent chance of rain or snow today." "The odds are three to two that the Cats will win the championship." "The probability of winning the grand prize in the state lottery is 1 in 7,240,000." Students in the middle grades have an intense interest in the notions of fairness and the chances of winning games. The study of probability develops concepts and methods for investigating such situations. These methods allow students to make predictions when uncertainty exists and to make sense of claims that they see and hear.

The study of probability in the grades should not focus on developing formulas or computing the likelihood of events pictured in texts. Students should actively explore situations by experimenting and simulating probability models. Such investigations should embody a variety of realistic problems, from questions about sports events to whether it will rain on the day of the school picnic. Students should talk about their ideas and use the results of their experiments to model situations or predict events. Probability is rich in interesting problems that can fascinate students and provide settings for developing or applying such concepts as ratios, fractions, percents, and decimals.

SOURCE: From *Curriculum and Evaluation Standards for School Mathematics, Grades 5–8,* p. 109.

particularly if the number of trials is small, may differ widely from what is expected. An important fact of probability, however, is the law of large numbers which we state here without proof.

THEOREM The Law of Large Numbers

If an experiment is performed repeatedly, the empirical probability of a particular outcome more and more closely approximates a fixed number as the number of trials increases.

A striking example of the law of large numbers was provided by John Kerrich, an English mathematician interned by the Germans during the Second World War. To while away the time, Kerrich decided to toss a coin 10,000 times and to recompute $P_e(H)$, the empirical probability of getting a head, after each toss. His first 10 tosses yielded 4 heads so $P_e(H)$ equalled 0.4 at that point. After 30 tosses $P_e(H)$ was 0.57, after 100 tosses it was 0.44, after 1000 tosses it was 0.49, and after that it varied up and down slightly but stayed very close to 0.5. After 10,000 tosses the number of heads obtained was 5067 for an empirical probability of $P_e(H) \doteq 0.51$. During the experiment, relatively long sequences of consecutive heads and also of consecutive tails occurred. Nevertheless, in the long run, the law of large numbers prevailed and the intuitive guess of $1/2$ was borne out. The variation in $P_e(H)$ is effectively illustrated in Figure 9.1.

HIGHLIGHT FROM HISTORY
The Mathematical Bernoullis

Like talent of any kind, mathematical talent of the highest order appears in individuals with relative infrequency. This makes it most remarkable that such talent appeared in three generations of the Bernoulli family of Basel, Switzerland. Nicolaus Bernoulli (1623–1708) was a merchant and not a mathematician. However, two of his sons, four of his grandsons, and two of his great grandsons were mathematicians of the highest order. The Bernoullis were fiercely competitive and often quarreled among themselves about priority in mathematical discoveries. Nevertheless, they contributed enormously to the development of mathematics including contributions to the theory of probability by Jakob and Johann, sons of the merchant Nicolaus, and by Daniel, one of Johann's three mathematically talented sons. In particular, Jakob was the first individual to clearly conceptualize and prove the law of large numbers, the basis for the notion of empirical probability.

Jakob Bernoulli

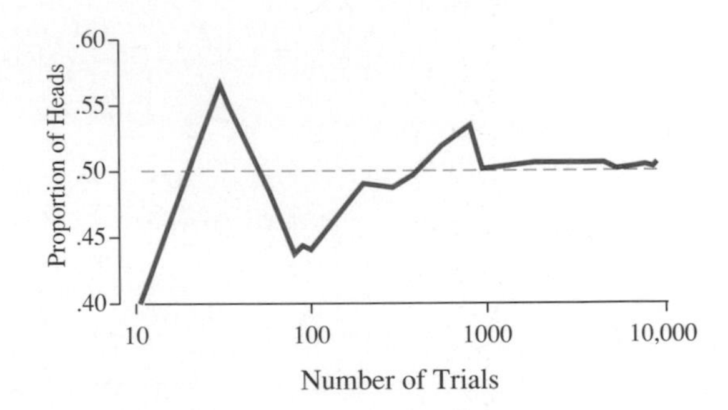

Figure 9.1
Ratio of the number of heads to the number of tosses in Kerrich's coin tossing experiment. (Adaptation of Figure 2 from *Statistics*, Second Edition by David Freedman, Robert Pisani, Roger Purves, and Ani Adhikari, p. 250. Copyright © 1991 by W.W. Norton & Company, Inc., copyright © 1978 by David Freedman, Robert Pisani, Roger Purves, and Ani Adhikari. Reprinted by permission of W.W. Norton & Company Inc.)

EXAMPLE 9.1 Determining an Empirical Probability

Open a book at random and note the number of the right hand page. Do this 20 times and determine the empirical probability that the number of the right hand page is divisible by 3.

SOLUTION

Since every third right hand page is numbered with a number that is divisible by 3, we would expect that the empirical probability should be about 1/3. Here are results obtained on checking 20 pages chosen at random.

349	69	267	407	133
395	269	123	331	373
155	235	187	273	401
297	83	852	263	303

Since a number is divisible by 3 if, and only if, the sum of its digits is divisible by 3, we determine that seven of these page numbers are divisible by 3. Hence,

$$P_e(\text{a right hand page number is divisible by 3}) = \frac{7}{20} = 0.35,$$

a very close approximation to 1/3. ■

Connections with Statistics

Most of what was said about statistics in Chapter 8 can be rephrased in terms of empirical probability. For example, saying that 28% of the items in a sample possess a certain property is the same as saying that the empirical probability that an item in the sample possesses the property is $P_e = 0.28$. Moreover, since population properties are estimated by properties of samples, one would go on to say that the empirical probability that an individual in the population possesses the property is also 0.28. Results of surveys, polls, summaries of data, and averages of all kinds also can be interpreted as empirical probabilities. Batting averages are empirical probabilities. Percentages of shots made in basketball are empirical probabilities. The entire insurance industry is based on empirical probabilities derived from **mortality tables** like that in Table 9.1. This table records what actually happened to 10,000,000 people; that is, the number alive at the beginning of each year and the number dying during that year, over a 100-year period. Empirical probability impinges on our lives almost daily by affecting, either consciously or unconsciously, many of our decisions. We should be aware that it does and also how it does.

TABLE 9.1 Commissioners 1958 Standard Ordinary Mortality Table

Age	Living at Beginning of Year	Dying During Year	Deaths Per 1,000	Age	Living at Beginning of Year	Dying During Year	Deaths Per 1,000
0	10,000,000	70,800	7.08	50	8,762,306	72,902	8.32
1	9,929,200	17,475	1.76	51	8,689,404	79,160	9.11
2	9,911,725	15,066	1.52	52	8,610,244	85,758	9.96
3	9,896,659	14,449	1.46	53	8,524,486	92,832	10.89
4	9,882,210	13,835	1.40	54	8,431,654	100,337	11.90
5	9,868,375	13,322	1.35	55	8,331,317	108,307	13.00
6	9,855,053	12,812	1.30	56	8,223,010	116,849	14.21
7	9,842,241	12,401	1.26	57	8,106,161	125,970	15.54
8	9,829,840	12,091	1.23	58	7,980,191	135,663	17.00
9	9,817,749	11,879	1.21	59	7,844,528	145,830	18.59
10	9,805,870	11,865	1.21	60	7,698,698	156,592	20.34
11	9,794,005	12,047	1.23	61	7,542,106	167,736	22.24
12	9,781,958	12,325	1.26	62	7,374,370	179,271	24.31
13	9,769,633	12,896	1.32	63	7,195,099	191,174	26.57
14	9,756,737	13,562	1.39	64	7,003,925	203,394	29.04
15	9,743,175	14,225	1.46	65	6,800,531	215,917	31.75
16	9,728,950	14,983	1.54	66	6,584,614	228,749	34.74
17	9,713,967	15,737	1.62	67	6,355,865	241,777	38.04
18	9,698,230	16,390	1.69	68	6,114,088	254,835	41.68

TABLE 9.1 (Contd.)

Age	Living at Beginning of Year	Dying During Year	Deaths Per 1,000	Age	Living at Beginning of Year	Dying During Year	Deaths Per 1,000
19	9,681,840	16,846	1.74	69	5,859,253	267,241	45.61
20	9,664,994	17,300	1.79	70	5,592,012	278,426	49.79
21	9,647,694	17,655	1.83	71	5,313,586	287,731	54.15
22	9,630,039	17,912	1.86	72	5,025,855	294,766	58.65
23	9,612,127	18,167	1.89	73	4,731,089	299,289	63.26
24	9,593,960	18,324	1.91	74	4,431,800	301,894	68.12
25	9,575,636	18,481	1.93	75	4,129,906	303,011	73.37
26	9,557,155	18,732	1.96	76	3,826,895	303,014	79.18
27	9,538,423	18,981	1.99	77	3,523,881	301,997	85.70
28	9,519,422	19,324	2.03	78	3,221,884	299,829	93.06
29	9,500,118	19,760	2.08	79	2,922,055	295,683	101.19
30	9,480,358	29,193	2.13	80	2,626,372	288,848	109.98
31	9,460,165	20,718	2.19	81	2,337,524	278,983	119.35
32	9,439,447	21,239	2.25	82	2,058,541	265,902	129.17
33	9,418,208	21,850	2.32	83	1,792,639	249,858	139.38
34	9,396,358	22,551	2.40	84	1,542,781	231,433	150.01
35	9,373,807	23,528	2.51	85	1,311,348	211,311	161.14
36	9,350,279	24,685	2.64	86	1,100,037	190,108	172.82
37	9,325,594	26,112	2.80	87	909,929	168,455	185.13
38	9,299,482	27,991	3.01	88	741,474	146,997	198.25
39	9,271,491	30,132	3.25	89	594,477	126,303	212.46
40	9,241,359	32,622	3.53	90	468,174	106,809	228.14
41	9,208,737	35,362	3.84	91	361,365	88,813	245.77
42	9,173,375	38,253	4.17	92	272,552	72,480	265.93
43	9,135,122	41,382	4.53	93	200,072	57,881	289.30
44	9,093,740	44,741	4.92	94	142,191	45,026	316.66
45	9,048,999	48,412	5.35	95	97,165	34,128	351.24
46	9,000,587	52,473	5.83	96	63,037	25,250	400.56
47	8,948,114	56,910	6.36	97	37,787	18,456	488.42
48	8,891,204	61,794	6.95	98	19,331	12,916	668.15
49	8,829,410	67,104	7.60	99	6,415	6,415	1,000.00

SOURCE: Denenberg, et al, *Risk and Insurance,* Prentice Hall, Englewood Cliffs, NJ, 1964. Reprinted with permission.

Computing Empirical Probabilities

Suppose you are on the fifth floor of a very tall building waiting for an elevator. When the next elevator comes is it more likely to be going up or going down? Given that the building is quite tall, the next elevator is probably above you and hence will be heading down when it reaches your floor.

EXAMPLE 9.2

Computing an Empirical Probability

Go to the second floor of a building with at least four floors and at least one elevator. Spend 15 minutes collecting data and determine the probability that an elevator arriving at the second floor will be headed down.

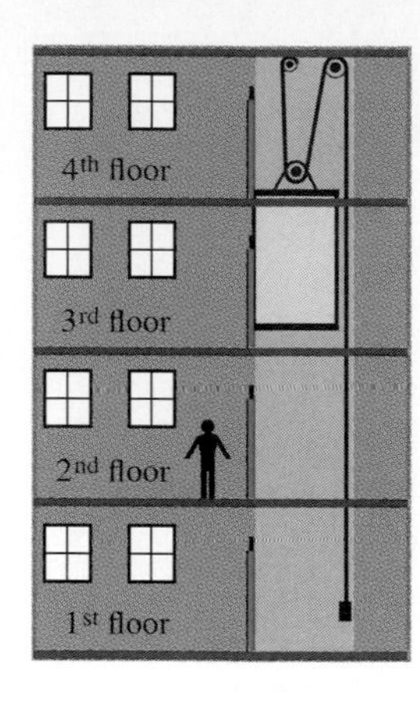

SOLUTION

Suppose that, during the 15 minute period an elevator arrives on the second floor 12 times and seven of those times it is heading down. Then the empirical probability that an elevator arriving on the second floor is headed down is 7/12.

(a) Do this exercise for the tallest building on your campus and determine the desired probability.

(b) Use the combined data for the entire class to again determine the empirical probability. Compare this figure with your result and with that of other members of the class.

(c) If the building were four stories tall and the floors were equally busy you might expect the elevator to be above you about twice as often as it is below as suggested by the diagram. For such a building, would the above figure of 7/12 seem reasonable? Discuss this with other members of the class. ■

EXAMPLE 9.3

Computing Empirical Probability from a Histogram

Suppose you wanted to study the heights of high school boys in the United States and decided to use Eisenhower High School as typical. Use Figure 8.19 to determine the empirical probability that the height in inches of a high school boy is in the range $67.5 < h < 70.5$.

SOLUTION

The percentages of boys with heights in the ranges 67.5–68.5, 68.5–69.5, and 69.5–70.5 are, respectively, 18.75%, 23.75%, and 17.5%. Thus, the total percentage of boys with heights in the range 67.5–70.5 is 18.75 + 23.75 + 17.5, or 60%. Therefore, $60\% = 0.60 = r/n$ where r is the number of boys with heights in the desired range and n is the number of boys in the high school. Since P_e also equals r/n, it follows that $P_e = 0.60$. ■

EXAMPLE 9.4

Computing Empirical Probability from Data

The final examination scores of students in a precalculus class are as shown. Compute the empirical probability that a student chosen at random from the class had a score in the 70s.

67	56	76	84	36	50	84
47	59	54	79	100	48	80
60	100	100	68	79	95	81
98	76	33	73	83	77	67

SOLUTION

Since five of the 28 students scored in the 70s, the empirical probability that a student chosen at random had a score in the 70s is $5/28 \doteq 0.18$. ■

EXAMPLE 9.5 **Computing Empirical Probability from an Experiment**

Roll a pair of dice 50 times and compute these empirical probabilities.

(a) $P_e(7)$ **(b)** $P_e(11)$ **(c)** $P_e(7 \text{ or } 11)$
(d) Show that $P_e(7 \text{ or } 11) = P_e(7) + P_e(11)$

SOLUTION

Actually performing the experiment, we obtained these results. From the data, the desired empirical probabilities are as shown.

2	3	4	5	6	7	8	9	10	11	12
I		III	IIII	~~IIII~~ ~~IIII~~	~~IIII~~ ~~IIII~~ II	~~IIII~~ III	IIII	IIII	II	II

(a) $P_e(7) = 12/50 = 0.24$
(b) $P_e(11) = 2/50 = 0.04$
(c) $P_e(7 \text{ or } 11) = 14/50 = 0.28$
(d) $P_e(7 \text{ or } 11) = 0.28 = 0.24 + 0.04 = P_e(7) + P_e(11)$. ■

Repeat this experiment on your own or with a friend and see what probabilities you obtain. Also, combine all the data for the entire class and recompute the empirical probabilities. Again compare the new empirical probabilities with your results and with those of other individuals in your class. Discuss the comparisons and, in particular, answer the question, "Do the results obtained for parts (c) and (d) suggest a general rule relating probability and the word "or"?

Before you put too much faith in your conjecture from the preceding example, perform the following experiment either by yourself or as a class activity. Toss two coins (say a penny and a dime) simultaneously and compute the empirical probability that the dime comes up heads or the two coins both come up the same. Also compute separately the probability that the dime comes up heads and the probability that the

WORTH READING

The Magic Numbers of Dr. Matrix

. . . When a friend of mine suggested in late December 1959 that I get in touch with a New York numerologist who called himself Dr. Matrix, I could hardly have been less interested.

"But you'll find him very amusing," my friend insisted. "He claims to be a reincarnation of Pythagoras, and he really does seem to know something about mathematics. For example, he pointed out to me that 1960 had to be an unusual year because 1,960 can be expressed as the sum of two squares—14^2 and 42^2—and 14 and 42 are multiples of the mystic number 7."

I made a quick check with pencil and paper. "By Plato, he's right!" I exclaimed. "He might be worth talking to at that."

So opens the delightful book *The Magic Numbers of Dr. Matrix* by Martin Gardner (New York: Prometheus Books, 1985). For years Martin Gardner delighted readers of his column, Mathematical Games, in the *Scientific American,* which from time to time contained essays about the enigmatic Dr. Matrix. Of course, Dr. Matrix is a fictional character. But, like Sherlock Holmes, he was brought to life in a remarkable way by Gardner's inspired writing.

This book collects all of Gardner's Dr. Matrix stories into one volume that is sure to delight and surprise its readers with its intriguing mixture of fact, fiction, and coincidence.

SOURCE: Martin Gardner, *The Magic Numbers of Dr. Matrix.* Buffalo, N.Y., Prometheus Books, 1985, p. 7.

two coins come up the same. Does this tend to modify your conjecture above? Discuss this as a class or with others in the class. Questions to consider are: Can the two outcomes occur simultaneously in Example 9.5? In the experiment just described?

EXAMPLE 9.6

Computing Empirical Probability and the Word "And"

Determine the empirical probability of obtaining two heads on a single toss of two coins; that is, the probability of obtaining a head on the first coin **and** a head on the second coin. Does the probability turn out to be about what you would expect? So that you can tell coins apart, use a penny and a dime.

SOLUTION

It is physically apparent that the fall of one coin has no effect on the fall of the other. Thus, the outcome of obtaining a head on the first coin is independent of the outcome of obtaining a head on the second coin. Let (H, H) indicate the outcome consisting of a head on the first coin and a head on the second coin; (H, T) is the outcome consisting of a head on the first coin and a tail on the second, and so on. Also, let $n(H, H)$ be the number of occurrences of the outcome (H, H) and so on. Toss the pair of coins 20 times and record the results in a table like this. Since $P_e(\text{H, H}) = n(\text{H, H})/20$, we easily complete the solution. The answer should approximate 0.25 since there are four outcomes and each should appear about equally often. Also, we have already seen from Kerrich's experiment that $P_e(\text{H}) \doteq 0.5$ where $P_e(\text{H})$ denotes the empirical probability of obtaining a head on a single throw of a coin. Since

$n(\text{H, H})$	
$n(\text{H, T})$	
$n(\text{T, H})$	
$n(\text{T, T})$	

$$(0.5)(0.5) = 0.25,$$

this suggests that

$$P_e(\text{H, H}) \doteq P_e(\text{H})P_e(\text{H}).$$

■

Combine all the data in the class from the preceding example and recompute the empirical probability. Does this strengthen your conjecture that, at least for large samples, $P_e(\text{H, H})$ is approximately equal to $P_e(\text{H})P_e(\text{H})$? At the same time, do some of the smaller samples suggest that the approximation may be very bad for small samples? Discuss this with your classmates.

EXAMPLE 9.7

Computing Empirical Probability from a Mortality Table

Use the mortality table, Table 9.1, to determine

(a) the empirical probability that a newborn baby will live to age 70.

(b) the empirical probability that a person who has just turned 60 will die before turning 70.

SOLUTION

(a) According to the table, of 10,000,000 babies born in year zero, 5,592,012 were still alive at age 70. Thus, the empirical probability that a newborn baby will live to age 70 is

$$P_e = \frac{5{,}592{,}012}{10{,}000{,}000} \doteq 0.56.$$

(b) According to the table, 7,698,698 people lived to be 60 years old. Of these, 5,592,012 were still alive at age 70. By subtraction, we learn that 2,106,686 of those alive at age 60 died before age 70. Hence, the empirical probability that a person age 60 will die before reaching age 70 is

$$P_e = \frac{2{,}106{,}686}{7{,}698{,}698} \doteq 0.27.$$

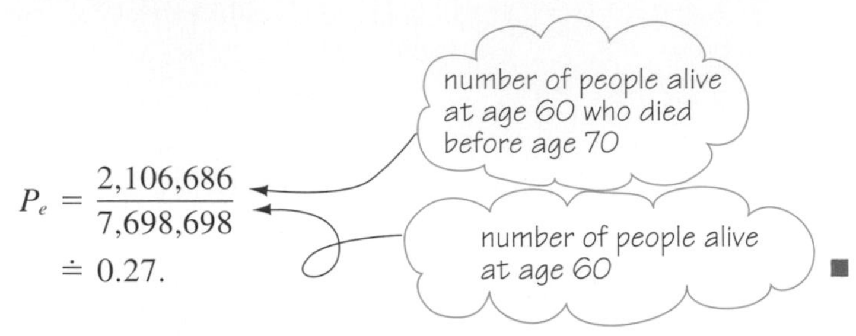

■

Empirical Probability and Geometry

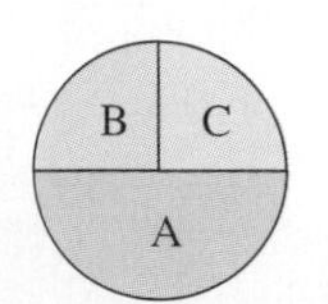

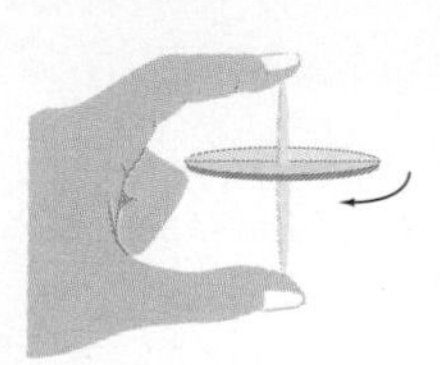

Figure 9.2
Making a spinner

An excellent spinner like the one shown in Figure 9.2 can be made by gluing an appropriately marked paper disc on top of a metal rimmed price tag readily obtainable at any stationery store. A round toothpick is forced through the center to act as an axle. Good results can be obtained as follows. Hold the spinner upside down and not too tightly in the left hand and up at eye level. Spin the spinner by twisting briskly between the thumb and forefinger of the right hand. Stop the spinner while it is still spinning rapidly by pinching it between the thumb and forefinger of the right hand with the forefinger on top. Before spinning the spinner, mark a vertical line on the tip of your right thumb with a pen. Then, when the spinner is pinched, let go with the left hand, turn the spinner over with the right hand, and note and record in which region the mark on your thumb lies. One can also use a standard spinner as illustrated in Figure 8.18 but such spinners are much less accurate and effective.

Three-Card Monte

You are shown three cards: one black on both sides, one white on both sides, and one black on one side and white on the other. One card is selected and shown to be black on one side.

(a) Considering the possibilities, what do you think is the likelihood or probability that the selected card is black on the other side as well?

(b) Conduct the experiment just described by selecting at random one of the three cards and looking at **one side only** of the card selected. If it is white, ignore it. If it is black record it and also record the color of the other side of the card. Repeat the experiment 10 times, carefully shuffling the cards between experiments. Compute the empirical probability that the second side of the card is black given that the first side is black. Does your experiment tend to confirm or refute your guess in part (a)?

EXAMPLE 9.8

Determining Empirical Probability Geometrically

The spinner shown is spun and stopped as just described.

(a) What do you intuitively feel the probability is that your thumb mark will fall in region A?

(b) Spin the spinner 20 times and record the number of times your thumb mark falls in each region. Then compute $P_e(\mathrm{A})$, $P_e(\mathrm{B})$ and $P_e(\mathrm{C})$, the empirical probabilities that the mark falls in regions A, B, and C respectively.

SOLUTION

(a) Since the arc length associated with regions B and C are one-quarter of the circumference of the spinner and that of region A is one-half the circumference, it is reasonable to guess that $P_e(\mathrm{A}) \doteq 1/2$ and $P_e(\mathrm{B}) \doteq P_e(\mathrm{c}) \doteq 1/4$.

(b) Actually spinning the spinner, denote the number of times the mark falls in A, B, and C respectively by $n(\mathrm{A})$, $n(\mathrm{B})$, and $n(\mathrm{C})$. Then

$$P_e(\mathrm{A}) = \frac{n(\mathrm{A})}{20}, \quad P_e(\mathrm{B}) = \frac{n(\mathrm{B})}{20}, \quad \text{and} \quad P_e(\mathrm{C}) = \frac{n(\mathrm{C})}{20}.$$

As guessed in part (a), these ratios should approximate 1/2, 1/4, and 1/4 or 0.5, 0.25, and 0.25 respectively. ■

EXAMPLE 9.9

Determining Geometry from Empirical Probability

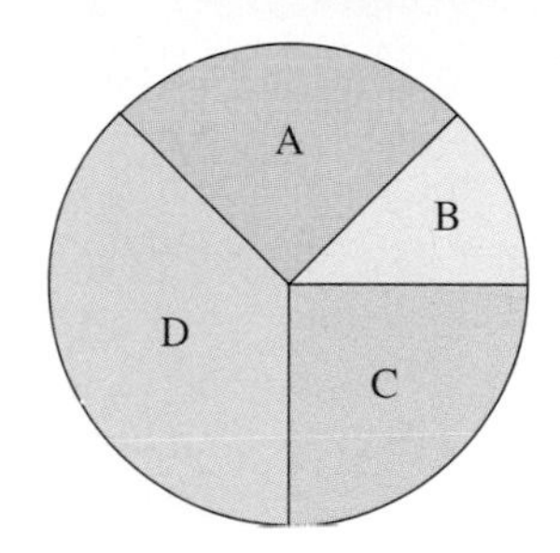

Make a spinner like the one shown here.

(a) Spin and stop the spinner 20 times and compute $P_e(\mathrm{A})$, $P_e(\mathrm{B})$, $P_e(\mathrm{C})$, and $P_e(\mathrm{D})$, the empirical probabilities that the mark on your thumb lands in region A, B, C, or D respectively.

(b) Compute $P_e(\mathrm{A}) \cdot 360$, $P_e(\mathrm{B}) \cdot 360$, $P_e(\mathrm{C}) \cdot 360$, and $P_e(\mathrm{D}) \cdot 360$.

(c) Measure the angles defining regions A, B, C, and D and compare the measurements with the results of part (b).

SOLUTION

A	~~IIII~~ I
B	III
C	IIII
D	~~IIII~~ II

(a) On 20 spins, we obtained the results shown. Thus,

$$P_e(\mathrm{A}) = \frac{6}{20} = 0.30,$$

$$P_e(\mathrm{B}) = \frac{3}{20} = 0.15,$$

$$P_e(\mathrm{C}) = \frac{4}{20} = 0.20, \text{ and}$$

$$P_e(\mathrm{D}) = \frac{7}{20} = 0.35.$$

(b) Multiplying these probabilities times 360° yields the approximate measures $m(\mathrm{A})$, $m(\mathrm{B})$, $m(\mathrm{C})$, and $m(\mathrm{D})$ of the central angles of the corresponding sectors. Thus,

$$m(A) \doteq 108^\circ \qquad m(B) \doteq 54^\circ$$
$$m(C) \doteq 72^\circ \qquad m(D) \doteq 126^\circ$$

(d) Finally, the actual measures are given by

$$m(A) \doteq 91^\circ \qquad m(B) \doteq 38^\circ$$
$$m(C) = 89^\circ \qquad m(D) \doteq 142^\circ$$

Thus, the measures correspond to the approximate values in part (b) only fairly well. According to the law of large numbers, the results would be much closer for 100 or even 1000 trials—not an unreasonable task for a team of two to carry out with one spinning and the other recording. ■

Simulation

Simulation is a method for determining answers to real problems by running experiments whose outcomes are analogous to the outcomes of the real problem. Suppose, for example, you want to know how many children a couple might expect to have if they want to be sure to have at least one boy and at least one girl.

To solve this problem by simulation, we must make some assumptions about the real problem and then design an experiment that takes these assumptions into account. In this case, we assume:

- that the birth of either a boy or a girl is equally likely, and
- that the sex of one child is completely independent of the sex of any other child.

These assumptions suggest tossing a coin since the occurrence of a head or a tail is equally likely and what happens on one toss of the coin is completely independent of what happens on any other toss. Of course, if we toss a coin repeatedly, it is quite possible that we could obtain 19 consecutive heads and then one tail, suggesting that the couple might expect to have 20 children! But is this likely? Not really. What we should do is perform the experiment repeatedly, recording the number of tosses needed to get at least one head and one tail, and then compute the average number of tosses required to obtain both a head and a tail. Here are the results of ten actual trials.

TH	TH
HHHHHHT	HT
TTH	HHHT
TH	HHT
HHT	HT

The average number of tosses per string is $30/10 = 3$. Thus, we estimate that the couple should expect to have three children. Finally, two observations should be made:

1. The fact that the above average was an integer was sheer luck. The average might well have been 2.9 or 3.2. But we would still guess that the couple should expect to have three children since the number of children must be an integer.
2. We would have greater confidence in the result of our simulation if we had repeated the experiment more than just ten times.

EXAMPLE 9.10 Using Simulation to Determine Empirical Probability

Use simulation to determine the empirical probability that a family with three children contains at least one boy and at least one girl.

SOLUTION

Using the same assumptions as above, we can simulate the real problem by repeatedly tossing three coins. Here are the results of such an experiment.

TTH	TTH	HHT	HHT	TTH
TTT	HTT	HTT	HHT	HHT
HHH	HHT	HHT	HHT	HHH
TTT	HHH	HTT	HHT	HHT
HHT	TTH	TTH	HHT	TTH

The empirical probability based on these results is

$$P_e = 20/25 = 0.80.$$

■

The simulation of the preceding problem could also have been made by sequentially selecting and replacing three balls one at a time from an urn containing an equal number of pink and blue balls. A pink ball represents a baby girl and a blue ball represents a baby boy. One could also use a sequence of random digits where an even digit represents a girl and an odd digit represents a boy. All of these methods are slow and tedious when compared to using a computer which can generate random digits and hence simulate rolling dice, tossing coins, selecting balls from an urn, and so on, with thousands or even millions of repetitions. The computer can even conduct the entire experiment and print out only the results! Use COINTOSS on your disk to repeat the experiments of Example 9.10 one thousand times each.

PROBLEM SET 9.1

Understanding Concepts

1. Refer to Table 8.1 and determine the empirical probability that a student in the class obtains a grade of 79.
2. **(a)** Refer to Table 8.2 and determine the empirical probability that a boy in Eisenhower High School is between 69.5 and 70.5 inches tall.

   (b) What is the empirical probability that one of the boys in Eisenhower High School is between 70.5 and 73.5 inches tall?
3. Prepare a 3″ × 5″ card by writing the numbers 1, 2, 3, and 4 on it as shown. Show the card to 20 college or university students, and ask them to choose a number and tell you their choice. Record the results on the back of the card and then compute $P_e(3)$, the empirical probability that a person chooses 3. Are you surprised at the result? Explain briefly.

1 2 3 4

4. **(a)** On a 3″ × 5″ card write the digits 1, 2, 3, 4, 5 as shown. Show the card to 20 different students, ask them to select a digit on the card, and then tell you which digit they selected. Record the results on the back of the card and compute the probabilities $P_e(1)$, $P_e(2)$, $P_e(3)$, $P_e(4)$, and $P_e(5)$.

1 2 3 4 5

   (b) Compute $P_e(1) + P_e(2), + P_e(3), + P_e(4), + P_e(5)$.

   (c) Did you need to perform the actual calculations in part (b) to be sure what the answer would be? Explain briefly.

(d) If you were to repeat part (a) with 100 different students, how many do you think would select the digit 4?

5. **(a)** Toss three coins 20 times and determine the empirical probability of obtaining three heads.

(b) Using the data from part (a), determine the empirical probability of *not* obtaining three heads.

(c) Using the data from part (a), determine the empirical probability of obtaining two heads and a tail.

(d) Could part (b) of this question be easily determined from your answer to part (a)? Explain.

6. **(a)** Make an orderly list of all possible outcomes resulting from tossing three coins. (*Hint:* Think of tossing a penny, a nickel, and a dime.)

(b) Does your listing in part (a) give you reason to believe that the result of problem 5(a) is about as expected? Explain briefly.

7. A computer is programmed to simulate experiments and to compute empirical probabilities. Match at least one of the computed probabilities with each of the descriptive sentences listed.

(a) $P_e(A) = 0$ **(b)** $P_e(B) = 0.5$
(c) $P_e(C) = -0.5$ **(d)** $P_e(D) = 1$
(e) $P_e(E) = 1.7$ **(f)** $P_e(F) = 0.9$

(i) This event occurred every time.
(ii) There was a bug in the program.
(iii) This event occurred often, but not every time.
(iv) This event never occurred.
(v) This event occurred half the time.

8. **(a)** Drop five thumbtacks on your desk top and determine the empirical probability that a tack dropped on a desk top will land point up.

(b) Repeat part (a) but with 20 thumbtacks.

(c) Give your best estimate of the number of thumbtacks that would land point up if 100 tacks were dropped on the desk top.

9. **(a)** Roll two dice 20 times and compute the empirical probability that you obtain a score of 8 on a single roll of two dice.

(b) List the ways you can obtain a score of 8 on a roll of two dice. (*Hint:* Think of rolling a red die and a green die.)

(c) Does the result of part (b) suggest that the result of part (a) is about right? (*Hint:* How many ways can a red die and a green die come up on a single roll?)

10. Construct a spinner with three regions A, B, and C as in Figure 9.2 but such that you would expect $P_e(A) \doteq 1/2$, $P_e(B) \doteq 1/3$, and $P_e(C) \doteq 1/6$, on say 18 spins. What size angles determine the regions A, B, and C?

11. Construct a spinner as in Figure 9.2 but with the circle marked as shown here.

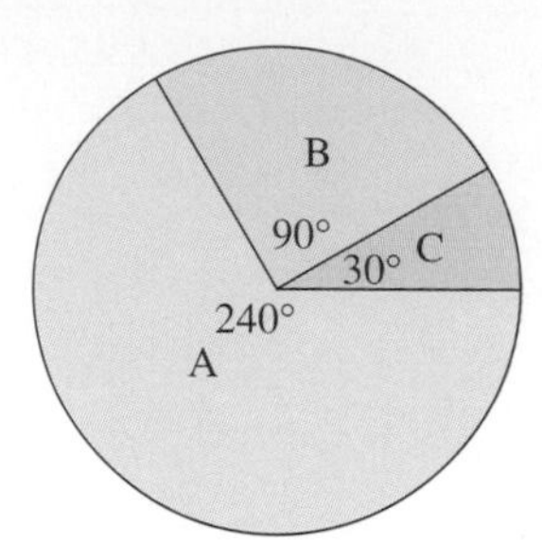

(a) Spin the spinner 20 times and compute $P_e(A)$, $P_e(B)$, and $P_e(C)$.

(b) Are the results of part (a) about as you expected? Explain briefly.

12. **(a)** A die is rolled repeatedly until a 6 is obtained. Repeat this experiment ten times and record the results. Estimate how many rolls it should take to obtain a 6.

(b) Might it take ten rolls to obtain a 6? Explain briefly.

(c) Might it take 100 rolls to obtain a 6? Why or why not?

13. From the data in problem 12, part (a), compute the empirical probability that a 6 first appears on the fourth roll of the die.

Thinking Critically

14. Suppose that an experiment is conducted 100 times.

(a) If event A never occurs, what is $P_e(A)$? Explain briefly.

(b) If event A occurs every time, what is $P_e(A)$? Explain briefly.

(c) What range of values are possible for $P_e(A)$? Explain briefly.

15. Roll a pair of dice 20 times and count the number of times you get a 7 and the number of times you get an 8.

(a) Compute $P_e(7)$.

(b) Compute $P_e(8)$.

(c) Compute $P_e(7 \text{ or } 8)$.

(d) Compute $P_e(7) + P_e(8)$.

(e) Explain why the results of parts (c) and (d) are as they are.

16. Consider the experiment of shuffling a deck of playing cards and selecting a card at random. Repeat this experiment 20 times and note the result each time. Let $P_e(R)$ denote the empirical probability that a card is red, let $P_e(F)$ denote the empirical probability that a card is a face card, let $P_e(R \text{ or } F)$ denote the empirical probability that a card is red or is a face card, and let $P_e(R \text{ and } F)$ denote the empirical probability that a card is red and is a face card. Compute these probabilities.
    (a) $P_e(R)$ (b) $P_e(F)$
    (c) $P_e(R \text{ or } F)$ (d) $P_e(R \text{ and } F)$
    (e) $P_e(R) + P_e(F) - P_e(R \text{ and } F)$
    (f) Compare the results of parts (c) and (e). Do these results suggest a general property? Explain.
    (g) Contrast the conclusions you drew in part (e) of problem 15 with those you drew in part (f) of this problem. Can you suggest why the results seem to differ?

17. **(a)** Perform an experiment similar to that of problem 16 and compute the empirical probability, $P_e(\text{A or } 9)$, that a single card chosen at random from the deck is an ace or a 9.
    (b) What fraction of cards in the deck are aces or 9s?
    (c) Briefly discuss the results of parts (a) and (b).

18. Consider an experiment of simultaneously tossing a single coin and rolling a single die. Repeat the experiment 20 times and compute these empirical probabilities.
    (a) $P_e(\text{H})$, the empirical probability that a head occurs.
    (b) $P_e(5)$, the empirical probability that a 5 occurs.
    (c) $P_e(\text{H and } 5)$, the empirical probability that a head and a 5 occur simultaneously.
    (d) $P_e(\text{H}) \cdot P_e(5)$
    (e) Should the results of parts (c) and (d) be about the same? Explain briefly.

19. Consider the experiment of shuffling a deck of playing cards and selecting a card at random. Perform the experiment 20 times and record the results.
    (a) Compute $P_e(\text{A})$, the empirical probability of selecting an ace.
    (b) Compute $P_e(\text{S})$, the empirical probability of selecting a spade.
    (c) Compute $P_e(\text{A and S})$, the empirical probability of selecting a card that is both an ace and a spade; that is, the ace of spades.
    (d) Compute $P_e(\text{A}) \cdot P_e(\text{S})$.
    (e) Should the results of parts (c) and (d) be about the same? Why or why not?

20. Two dice are rolled 20 times. Compute $P_e(13)$, the empirical probability of obtaining a score of 13. Explain the result very briefly.

21. Two dice are rolled 20 times. Compute

$$P_e(2) + P_e(3) + P_e(4) + \cdots + P_e(12).$$

Explain the result very briefly.

22. **(a)** Construct a spinner as in Figure 9.2 but with the circle marked as shown here. Spin the spinner 20 times and compute $P_e(\text{A})$, $P_e(\text{B})$, and $P_e(\text{C})$.

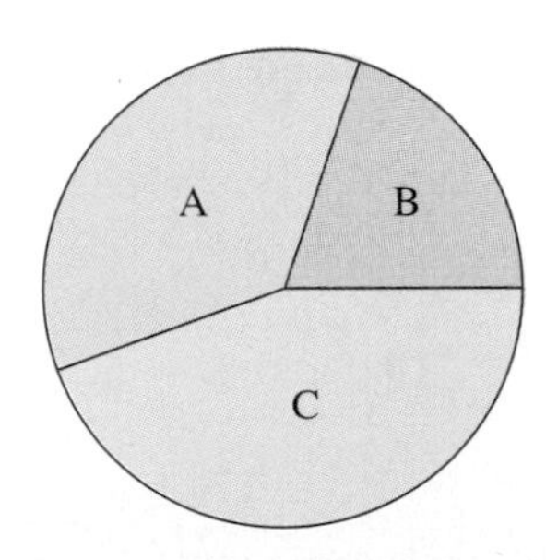

(b) Measure the central angles of this spinner and divide each measure by 360.
(c) Briefly discuss the relationship between the answers you obtained to parts (a) and (b).

23. Any point in a square one unit on a side can be determined by a number pair (a, b) that gives the point's horizontal and vertical distances from the lower left hand corner of the square. Thus, the point (0.4, 0.7) is located as shown.

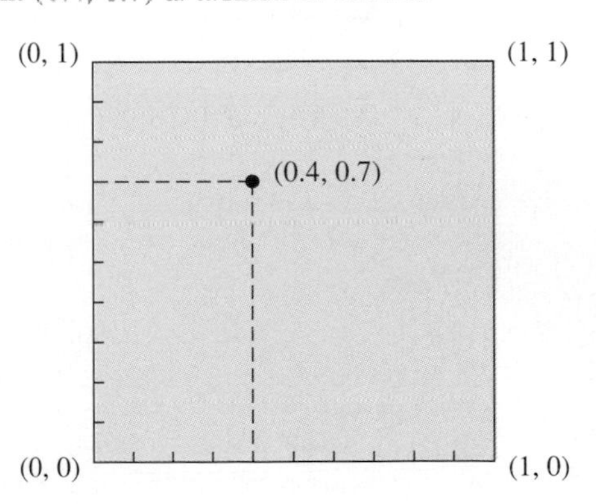

Consider next the square one unit on a side with regions A, B, C, and D as shown.

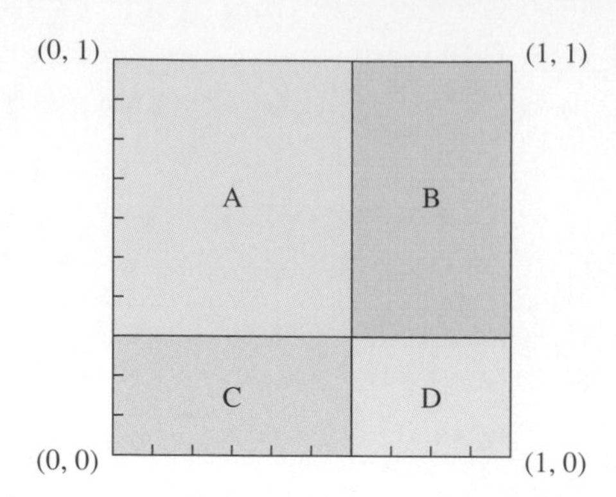

Random points in the square can be determined using a sequence of random numbers generated by a suitable spinner. For example, if you generate the sequence 0, 6, 4, 4, 6, 4, 2, 7, . . . the first two digits can be interpreted as 0.06 and each set of four successive entries from the sequence determine a point in the square. The first and second points would be (0.06, 0.44) and (0.64, 0.27). Continue in this way to determine 20 random points and determine these empirical probabilities. For example, $P_e(\text{A})$ is the empirical probability that a point lies in region A. If a point lies on a line, simply skip it and go on to the next point.

(a) $P_e(\text{A})$ **(b)** $P_e(\text{B})$ **(c)** $P_e(\text{C})$ **(d)** $P_e(\text{D})$

(e) Compute the areas of regions A, B, C, and D.

(f) Briefly discuss the relationship between these results.

Making Connections

24. Problem 23 suggests that the areas of subregions of the square one unit on a side can be approximated by the empirical probability that a point lies in the subregion. This is called a **Monte Carlo method,** an important technique in applied mathematics. Use the 20 points determined in problem 23 to make an estimate of the area in each of the figures shown here.

(a)

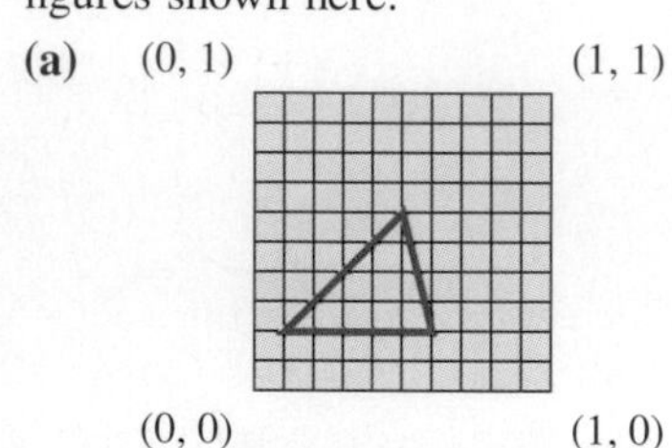

(b)

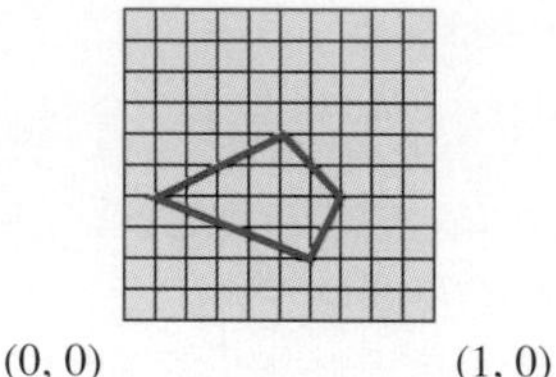

(c)

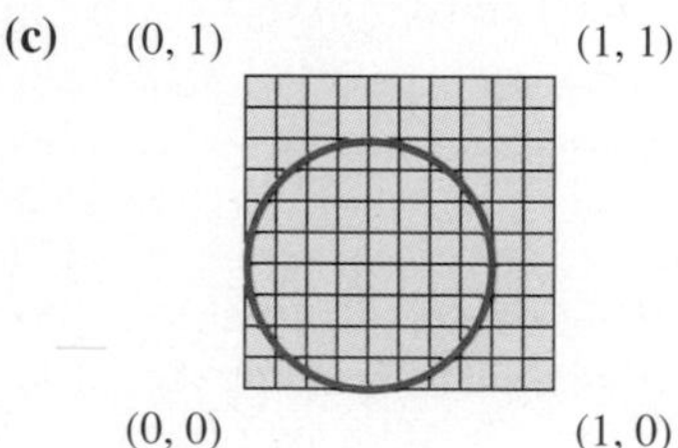

(d)

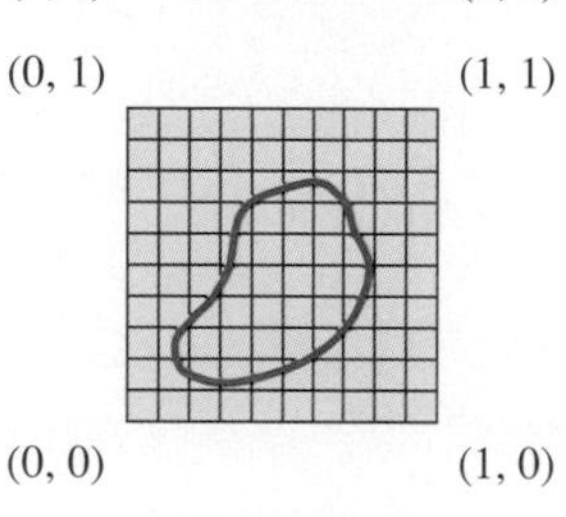

(e) Compute the areas of the regions in parts (a), (b), and (c) by standard means and compare the results to your answers to parts (a), (b), and (c) above.

(f) Does your answer to part (d) above seem reasonable? (*Hint:* How else might you roughly approximate the area of the curved region in part (d)?)

---

25. Use Table 9.1 to compute these empirical probabilities.
 (a) The empirical probability that a person 20 years old will live to be 80 years old.
 (b) The empirical probability that a person 65 years old will die before becoming 80 years old.

26. Suppose Honest Abe Insurance Company currently has insured the lives of a total of 95 sixty-five year old men for an average of $120,000 each. About how much money should the company expect to pay in death benefits to the beneficiaries of those men who die between the ages of 80 through 84? Explain briefly.

Using A Computer

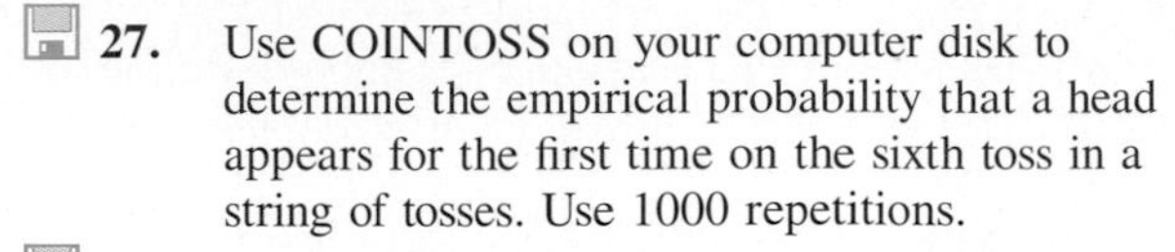

27. Use COINTOSS on your computer disk to determine the empirical probability that a head appears for the first time on the sixth toss in a string of tosses. Use 1000 repetitions.

28. Consider the experiment of tossing ten coins n times. Use COINTOSS on your computer disk to simulate this experiment. Also use *Stat Explorer* or other software you have available to draw a histogram of the results for these values of n.

(a) $n = 10$ **(b)** $n = 100$ **(c)** $n = 1000$

(d) Briefly discuss the results of (a), (b), and (c).

For Review

29. Compute the mean, median, and mode for these values.

30 28 34 33 29 28 27 31

30. Compute the standard deviation for the data in problem 29.

31. What percent of the data in problem 29 lie within one standard deviation of the mean? Two standard deviations? Three standard deviations?

32. If a population with a normal distribution has mean 28 and standard deviation 8.4, what is the probability that an individual from the population falls in the range $11.2 \leq x \leq 44.8$?

9.2 Principles of Counting

Kerrich's experiment, discussed in the last section, showed that the empirical probability of obtaining a head on a single toss of a coin was very nearly $1/2$. This is to be expected since there are only two ways a tossed coin can land and either heads or tails seems equally likely to come up. Similarly, if you repeatedly rolled a single die and computed the empirical probability of obtaining a 5 it would closely approximate $1/6 \doteq 0.17$. Again, this is expected since there are six faces on a die and these are very nearly equally likely to come up. In Example 9.6, the empirical probability of obtaining two heads on a single throw of two coins roughly approximated $1/4$ and HH is one of four equally likely possibilities: HH, HT, TH, TT.

These, and many of the examples and problems in the preceding problem set were designed to show that the empirical probability of an event could be predicted on the basis of a straightforward analysis of the possible outcomes of an experiment. Thus, if we desire the probability that two heads and a tail appear when three coins are tossed, we can list the possible outcomes—

HHH HHT HTH THH
HTT THT TTH TTT

—and observe that in three of the eight equally likely outcomes two heads and a tail appear in some order. Hence, we would expect the empirical probability to be about $3/8 = 0.375$ and, even for only moderately large numbers of trials, the experiment showed that our expectation was fulfilled. (See problem 5c, Problem Set 9.1.)

Considerations like these show that there is a close correlation between empirical and theoretical probability. Indeed, it is simply an extension of the law of large numbers to assert that the empirical probability of an event more and more closely approximates the theoretical probability provided the theoretical probability is correctly determined. In particular, it is important, as demonstrated in Example 9.6, that the equally likely outcomes of an experiment be properly identified. If this is not done successfully, the theoretical probabilities will almost surely be incorrect.

In Section 9.3 we consider theoretical probability in some detail. First, however, it is important to develop some principles of counting. We will not want to have to list all of the equally likely outcomes each time we solve a problem. Also, since sample spaces are *sets* and events are *subsets*, much of the discussion will be couched in terms

JUST FOR FUN

Social Security Numbers

Check and record the fifth digit of the Social Security number of 20 of your fellow students. Compute the empirical probability that this digit is even. Surprised?

of sets and subsets. In particular, if A is any set, then $n(A)$ will denote the number of elements of A as in Chapter 2.

The words "or" and "and" play key roles in counting and a proper understanding of these roles makes solving counting problems much easier. We begin by considering the role of "or."

Counting and the Word "Or"

Consider the question, "How many diamonds or face cards are in an ordinary deck of playing cards?" Of course, we can simply count the 13 diamonds and then proceed to count the nine face cards (kings, queens, and jacks) that are not diamonds. Thus, the answer of 22 is easily found. But there is another way to determine this sum. It seems more involved but reveals an important pattern that greatly facilitates solving more complex problems. Let D denote the set of diamonds in the deck of cards and let F denote the set of face cards. Then, since there are 13 cards in each suit and a total of 12 face cards with three in each suit,

$$n(D) = 13, n(F) = 12, \quad \text{and} \quad n(D \cap F) = 3$$

Moreover,

$$n(D) + n(F) - n(D \cap F) = 13 + 12 - 3 = 22,$$

the number of diamonds or face cards as before. That is,

$$n(D \text{ or } F) = n(D) + n(F) - n(D \cap F).$$

The elements in D or in F are in $D \cup F$ and conversely. $n(D \text{ or } F) = n(D \cup F)$

Moreover, it is easy to see why the preceding equation holds. Both $n(D \cup F)$ and $n(D) + n(F) - n(D \cap F)$ count:

- elements in D and not in F just once,
- elements in F and not in D just once, and
- elements in $D \cap F$ just once since $n(D \cap F)$ counts elements in $D \cap F$ once and $n(D) + n(F)$ counts them twice

as illustrated in Figure 9.3.

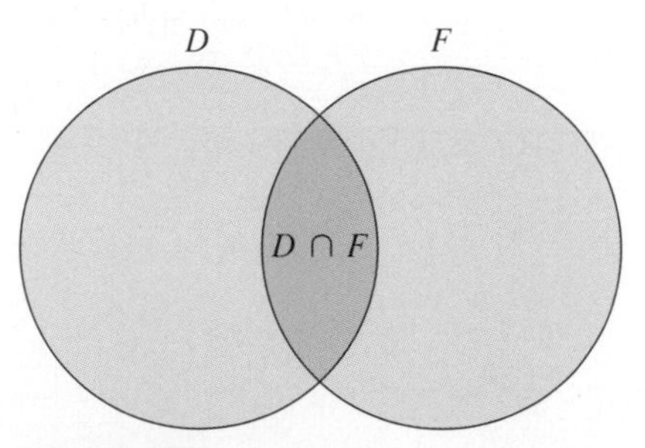

Figure 9.3
$n(D \cup F) = n(D) + n(F) - n(D \cap F)$

This argument is essentially general and yields the following theorem.

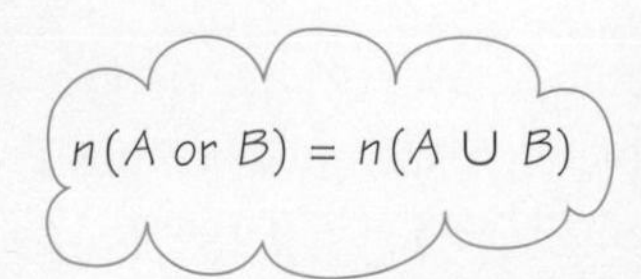

THEOREM The First Principle of Counting

If A and B are events, then $n(A \textbf{ or } B) = n(A) + n(B) - n(A \cap B)$.

EXAMPLE 9.11 **Counting and "Or"**

In how many ways can you select a red card or an ace from an ordinary deck of playing cards?

SOLUTION 1

As above, there are 26 red cards in a deck (including two aces) as well as two black aces. Thus, the desired answer is 28.

SOLUTION 2

Let R denote the set of red cards and let A denote the set of aces. Then $n(R) = 26$, $n(A) = 4$, $n(R \cap A) = 2$, and

$$\begin{aligned} n(R \text{ or } A) &= n(R) + n(A) - n(R \cap A) \\ &= 26 + 4 - 2 \\ &= 28 \end{aligned}$$

as before. ■

A very important special case of the first principle of counting is illustrated by the following example.

EXAMPLE 9.12 **Counting and "Or" When Events Are Mutually Exclusive**

Determine the number of ways of obtaining a score of 7 **or** 11 on a single roll of two dice.

SOLUTION 1

One way to solve this problem is simply to list all possible outcomes when rolling two dice and then to count those that are favorable. Thinking of rolling a red die and a green die makes it clear that there are 36 possible outcomes as shown below and that eight are favorable—six outcomes yield a score of 7 and two yield a score of 11. Thus,

$$n(7 \text{ or } 11) = 8 = 6 + 2 = n(7) + n(8).$$

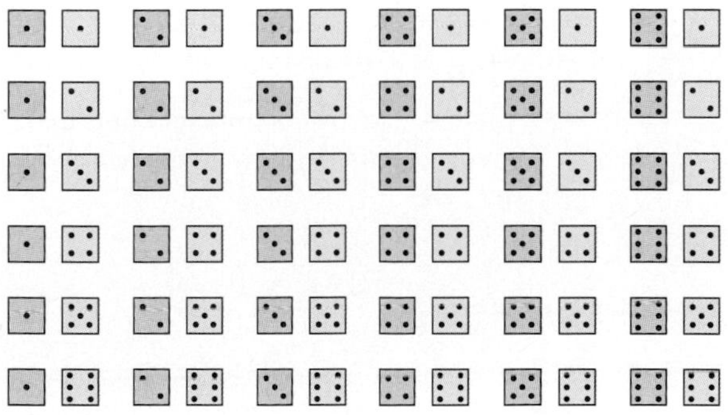

SOLUTION 2

Let F be the set of favorable outcomes when rolling the dice, let D be the set of outcomes yielding 7, and let E be the set of outcomes yielding 11.

Since $F = D \cup E$, we may use the first principle of counting to obtain:

$$\begin{aligned} n(F) &= n(D \cup E) \\ &= n(D \text{ or } E) \\ &= n(D) + n\,(E) - n(D \cap E). \end{aligned}$$

But $n(D \cap E) = 0$ since $D \cap E = \emptyset$. Thus,

$$n(F) = n(D) + n(E) = 6 + 2 = 8$$

as before. ■

The preceding solution showed that

$$n(D \text{ or } F) = n(D \cup F) = n(D) + n(F)$$

when $n(D \cap F) = 0$. This result is general and can be formalized as follows.

DEFINITION Mutually Exclusive Events

Events A and B are said to be **mutually exclusive** provided the occurrence of A precludes the occurrence of B and vice versa; that is, provided $A \cap B = \emptyset$.

THEOREM The Second Principle of Counting

If A and B are mutually exclusive events, then $n(A \textbf{ or } B) = n(A) + n(B)$.

Note that in both the first and second principles of counting, the word "or" suggests adding, though with a minor adjustment in the first principle. If problems like this are sufficiently complex, these principles are helpful. Otherwise one can simply list all possible outcomes and count the favorable ones.

EXAMPLE 9.13 Choosing a Chocolate

A box of chocolates contains 14 cremes, 16 caramels, and 10 chocolate covered nuts. In how many ways can you select a creme or a caramel from the box?

SOLUTION

Let C denote the set of cremes and let C^* denote the set of caramels. Then $C \cap C^* = \emptyset$ and

$$n(C \text{ or } C^*) = 14 + 16 = 30.$$

Thus, the number of ways of choosing a creme or a caramel is 30. ■

EXAMPLE 9.14

Determining How Many Are on an Airplane

On an airplane from Frankfurt to Paris, all the people speak only French or German. If 71 speak French, 85 speak German, and 29 speak both French and German, how many people are on the plane?

SOLUTION 1

Let F denote the set of French speakers on the plane and let G denote the set of German speakers. Of course, a person who speaks both French and German belongs to *both* sets F and G. Since all people on the plane speak only French or German, it follows that the number of persons on the plane is $n(F \cup G)$. But, by the second principle of counting,

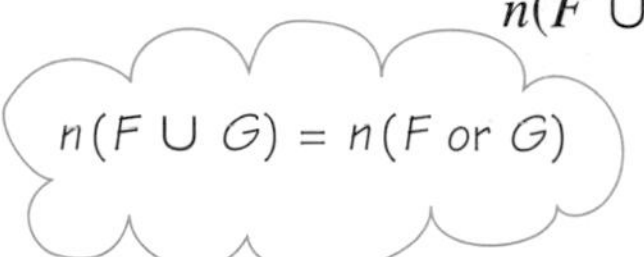

$$\begin{aligned} n(F \cup G) &= n(F) + n(G) - n(F \cap G) \\ &= 71 + 85 - 29 = 127. \end{aligned}$$

Thus, 127 people are on the plane.

SOLUTION 2

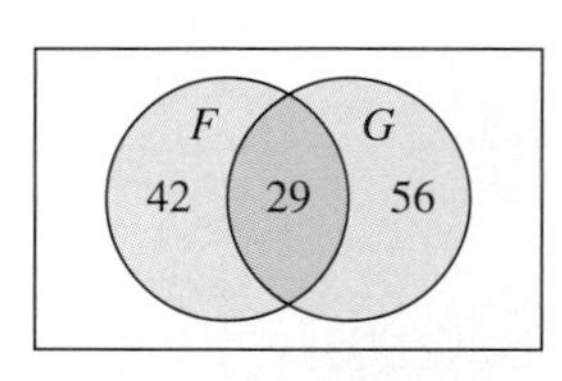

Let F and G be as above and consider the Venn diagram shown. Starting with the innermost region of the diagram, one fills in the appropriate numbers. Since 29 people fall in the set $F \cap G$, the common region of the two circles, and 71 are in the F circle, then $71 - 29 = 42$ people must fall inside the F circle but outside the common region as shown. Similarly, $85 - 29 = 56$ people must fall inside the G circle but outside the common region. Thus, finally,

$$n(F \cup G) = 42 + 29 + 56 = 127$$

as before. ■

Counting and the Word "And"

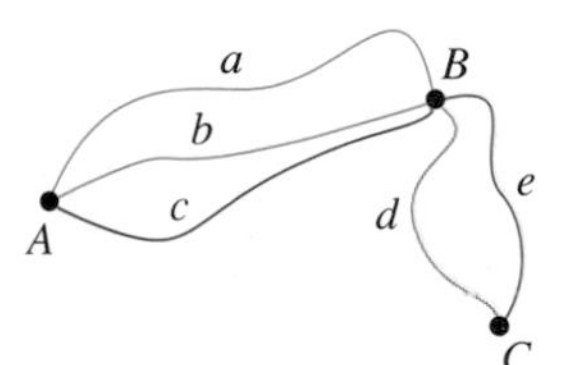

Figure 9.4
Routes from *A* to *B* to *C*

Suppose A, B, and C are towns with roads a, b, and c connecting A and B, and roads d and e connecting B and C as illustrated in Figure 9.4. How many different routes are there from A to C if two routes are considered different if any parts of the two routes are different? One way to solve this problem is to **make an orderly list.** Thus, the possibilities are

ae	*be*	*ce*
ad	*bd*	*cd*

and there are six possible routes from A to C. For each of the three available routes from A to B there are two ways to complete the trip to C, so there are

$$2 + 2 + 2 = 3 \cdot 2 = 6$$

routes in all.

In more complicated problems, it is more difficult to be sure that one has listed all possibilities and it is often useful to make a **possibility tree** as shown in Figure 9.5. (Curiously, trees are often drawn so that they appear upside down or sideways, but that is irrelevant to the listing and counting.) To determine the possible routes from A to C, the root of the tree is labeled A and the three "limbs" from A that correspond to the three routes—a, b, and c—from A to B are labeled a, b, and c.

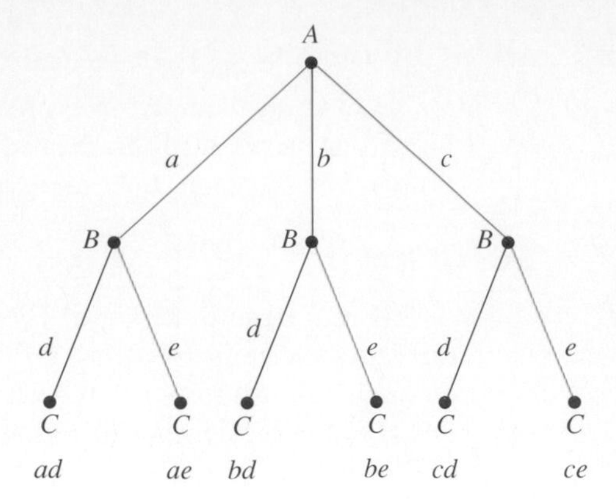

Figure 9.5
Tree of possible routes from A to B to C

INTO THE CLASSROOM

Jean Reherman Comments on Teaching Probability in Her Classroom

The probability activities I do in my classroom may be used with students of all ability levels. The activities require manipulatives that are both inexpensive and readily available. They can be used to improve several mathematical skills: estimating, predicting, gathering of data, and recording of probability outcomes.

For the first activity, I have students divide a sheet of paper into two equal halves by drawing a line down the center. Each student then uses a penny, a plastic chip, or a piece of poster board one-inch square to discover and explore the laws of probability and probable outcomes.

Each student drops his or her object several times from approximately the same height and same position onto the prepared sheet. Students observe where the object lands each time. After a given number of pretrial drops, we discuss students' predictions of probable results and the three possible outcomes.

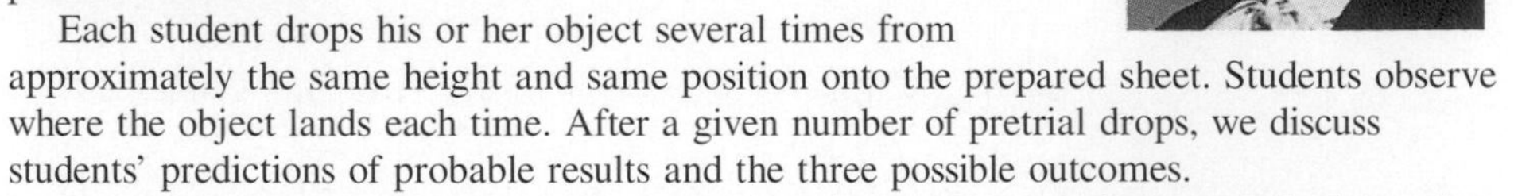

We also talk about how to use the chart shown below. First, students are asked to estimate the number of times, out of twelve drops, that the object will fall on the right half, on the left half, and on the center line of the sheet. Then students conduct the experiment twelve times and tally the results on their charts. After totaling the tally marks and writing the number in the Frequency column, students compare the actual results with their own predictions. Finally, students write the ratio of the frequency of each outcome to the total number of trials.

Outcomes	Estimate	Tally	Frequency	Ratio
Right half				
Left half				
Center line				

The second probability activity that I do has students drop a paper cup from a given height and predict the probability of three possible outcomes: the cup landing on its top, bottom, or side. The accuracy of student's estimates are then tested, as in the first activity, by keeping a record on a prepared chart. As students do their testing, I emphasize that the greater the number of trials they have made, the more reliable the probabilities will be as predictors of future events.

SOURCE: *Exploring Mathematics, Grade 7* (Glenview, IL: Scott, Foresman and Co., 1994), p. 418b. Jean Reherman teaches at Taft Middle School in Oklahoma City, Oklahoma.

But from B there are two routes to C and these are shown as "leaves"—d and e—two from each B on to each C. Starting down from A and proceeding to C along all possible limbs and leaves again yields the six possible routes

$$\begin{matrix} ad & bd & cd \\ ae & be & ce \end{matrix}$$

$3 \cdot 2 = 6$

as before.

This is illustrative of a general principle.

THEOREM Third Principle of Counting

Let A and B be events and let $n\ (A \mid B)$ be the number of ways A can occur given that B has already occurred. Then

$$n(A \text{ and } B) = n(B)\, n(A \mid B).$$

This theorem asserts that the number of ways A *and* B can occur simultaneously is the number of ways B can occur *times* the number of ways A can occur given that B has already occurred. Thus, in many counting problems, "and" is associated with multiplication.

EXAMPLE 9.15

Determining the Number of Code "Words"

How many 5-letter code "words" can be formed

(a) if repetition of letters is allowed?

(b) if repetition of letters is not allowed?

SOLUTION

Understand the problem • • • •

Since we are talking about words in code, a word need not look like a word. Thus, *arefg* is a perfectly acceptable code word. Moreover, to determine a 5-letter code word, we must choose a first letter *and* a second letter *and* a third letter *and* a fourth letter *and* a fifth letter. The question then is, "in how many ways can we do all these things?"

Devise a plan • • • •

Since we have to do a first thing *and* a second thing *and* a third thing, . . . , *and* a fifth thing, the word "and" suggests that we use the third principle of counting.

Carry out the plan • • • •

(a) We must choose letters to place in these blanks.

____ ____ ____ ____ ____

If repetition of letters is allowed, then the choice of any letter does not effect the choice of any other letter. Thus, there are 26 ways to fill each blank and so

$$\begin{aligned} &n(\text{5-letter code words with repetition allowed}) \\ &\quad = 26 \cdot 26 \cdot 26 \cdot 26 \cdot 26 \\ &\quad = 26^5 \\ &\quad = 11{,}881{,}376. \end{aligned}$$

(b) If repetition is not allowed, there are still 26 choices for the first letter, but only 25 for the second, 24 for the third, 23 for the fourth, and 22 for the fifth. Thus, the number of code words in this case is

$$\begin{aligned} &n(\text{5-letter code words without repetition}) \\ &\quad = 26 \cdot 25 \cdot 24 \cdot 23 \cdot 22 \\ &\quad = 7{,}893{,}600. \end{aligned}$$

Look back • • • •

The key to working this problem was the use of the word "and" in describing what actually must be done to accomplish the desired task; that is, choose a first letter *and* choose a second letter, *and* . . . , *and* choose a fifth letter. This suggests repeated use of the third principle of counting. Thus,

$$\begin{aligned} &n(\text{ways to choose a 5-letter code word}) = \\ &n(\text{ways to choose 1st letter}) \\ &\cdot n(\text{ways to choose 2nd letter} \mid \text{first letter}) \\ &\cdot n(\text{ways to choose 3rd letter} \mid \text{first 2 letters}) \\ &\cdot n(\text{ways to choose 4th letter} \mid \text{first 3 letters}) \\ &\cdot n(\text{ways to choose 5th letter} \mid \text{first 4 letters}). \end{aligned}$$

■

If $n(A \mid B) = n(A)$ and $n(B \mid A) = n(B)$, the occurrence of A does not effect the occurrence of B and conversely. In this case, A and B are said to be **independent events.** Moreover, it follows from the third principle of counting, that then

$$n(A \text{ and } B) = n(B)\, n(A \mid B) = n(B)\, n(A).$$

EXAMPLE 9.16 Determining the Number of Meals

At Fred's Fast Foods the menu board reads as follows.

Dinners		Drinks		Desserts	
Double Burger	1.95	Cola	.45	Berry pie	1.20
Cheese Burger	2.20	Uncola	.45	Apple pie	1.20
Burger	.99	Milk	.65	Ice cream	1.00
Fish and chips	2.65	Coffee	.40		
Chicken strips	2.50	Tea	.40		

(a) If you order dinner and a drink, how many different combinations can you choose?

(b) If you order dinner, a drink, and a dessert, how many different combinations can you choose?

HIGHLIGHT FROM HISTORY
Leonhard Euler (1707–1783)

Another of the titans of mathematics was the Swiss mathematician Leonhard Euler (pronounced Oiler), born in Basel in 1707 and a contemporary of the second generation of the remarkable Bernoulli family. Euler is the most prolific mathematician who ever lived, publishing 886 papers and books during his lifetime and posthumously for an average of approximately 800 pages of new mathematics per year. He was dubbed "Anaysis Incarnate" by the French academician Francois Arago who declared that Euler could produce mathematics without apparent effort "just as men breathe and eagles sustain themselves in the air." Euler went blind in 1766 to the considerable distress of his many friends and colleagues. However, aware that the condition was coming on, Euler taught himself to write his complicated formulas on a large slate with his eyes closed. Then with a scribe to write down the explanations of his formulas and with all the facts and formulas of the then known mathematics safely tucked away in his memory, his work continued unabated until his death.

On September 13, 1783, having earlier outlined on his slate the calculation of the orbit of the newly discovered planet Uranus, he called for his grandson to be brought in. While playing with the child he suffered a stroke and, with the words "I die," he quietly passed away.

SOLUTION

(a) Since there are five dinners and five drinks, and the choices are independent, the number of combinations is $5 \cdot 5 = 25$.

(b) Again, since the choices are independent the number of combinations is $5 \cdot 5 \cdot 3 = 75$. ■

Combinations and Permutations

In Example 9.15, we were required to find the number of 5-letter code words under certain conditions. Thus, *abcde* and *acdbe* are both acceptable code words and are different since the order of the letters is different even though the sets {a, b, c, d, e} and {a, c, d, b, e} are the same. These distinctions are formalized in the following definition.

DEFINITION Combinations and Permutations

Let U be a set of objects. A subset of U with r objects is called a **combination of r objects of U.** An ordered sequence of r objects from U is called a **permutation of r objects of U.**

EXAMPLE 9.17 **Determining Combinations and Permutations**

(a) Write all combinations of two elements of the set {a, b, c, d}; that is, write all subsets of {a, b, c, d} that contain two letters.

(b) Write all permutations of the set {a, b, c, d} that contain two letters.

SOLUTION

(a) We make an orderly list and write

$$\{a, b\} \qquad \{a, c,\} \qquad \{a, d\}$$
$$\{b, c,\} \qquad \{b, d\} \qquad \{c, d\}$$

This gives six different combinations of the four letters a, b, c, and d taken two at a time.

(b) Each combination in part (a) generates two permutations since we can write the two letters in either order. Thus, the permutations of a, b, c, and d taken two at a time are

ab ba ac ca ad da
bc cb bd db cd dc. ■

In counting problems, the actual permutations or combinations of a set of elements taken r at a time are not important but their number is. In this context the important questions are:

- How many permutations of n things taken r at a time are there?
- How many combinations of n things taken r at a time are there?

Before answering these questions it will be helpful to introduce just a bit of new notation. As in Example 9.15, part (b), products like $26 \cdot 25 \cdot 24 \cdot 23 \cdot 22$ or $7 \cdot 6 \cdot 5 \cdot 4 \cdot 3 \cdot 2 \cdot 1$ frequently appear. Since these are somewhat tedious to write out, we use the shorthand

$$7! = 7 \cdot 6 \cdot 5 \cdot 4 \cdot 3 \cdot 2 \cdot 1$$

where 7! is read "**7 factorial.**" Thus, for example,

$$5! = 5 \cdot 4 \cdot 3 \cdot 2 \cdot 1$$

and

$$7 \cdot 6 \cdot 5 = \frac{7 \cdot 6 \cdot 5 \cdot 4 \cdot 3 \cdot 2 \cdot 1}{4 \cdot 3 \cdot 2 \cdot 1} = \frac{7!}{4!}.$$

In general, we define $n!$ for every integer $n \geq 0$.

DEFINITION The Factorial, *n*!

Let n be a whole number. Then

$$n! = n \cdot (n - 1) \cdot (n - 2) \cdots 1 \quad \text{for} \quad n \geq 1$$

and

$$0! = 1.$$

We read $n!$ as ***n* factorial.**

That 0! is defined to be 1 may seem strange but we justify this part of the definition shortly.

EXAMPLE 9.18

Manipulating Factorials

Compute each of these expressions.

(a) $1!, 2!, 3!, 4!$ **(b)** $4 \cdot 3!$ **(c)** $(4 \cdot 3)!$
(d) $4! + 3!$ **(e)** $4! - 3!$ **(f)** $26! \div 24!$

SOLUTION

(a) $1! = 1, 2! = 2 \cdot 1 = 2, 3! = 3 \cdot 2 \cdot 1 = 6, 4! = 4 \cdot 3 \cdot 2 \cdot 1 = 24$
(b) $4 \cdot 3! = 4 \cdot (3 \cdot 2 \cdot 1) = 4! = 24$
(c) $(4 \cdot 3)! = 12! \doteq 4.79 \times 10^8$ (See below.)
(d) $4! + 3! = 4 \cdot 3! + 3! = 5 \cdot 3! = 30$
(e) $4! - 3! = 4 \cdot 3! - 3! = 3 \cdot 3! = 18$
(f) $26! \div 24! = \dfrac{26 \cdot 25 \cdot 24 \cdot 23 \cdot 22 \cdots 1}{24 \cdot 23 \cdot 22 \cdots 1} = 26 \cdot 25 = 650.$ ■

Many calculators have an [n!] key that provides for the immediate computation of $n!$ For example, to compute $4! + 3!$, enter this string into a suitable calculator.

[ON/AC] 4 [$n!$] [+] 3 [$n!$] [=]

Factorials are often very large even when n is relatively small, so a calculator may give only an approximate result expressed in scientific notation. For example,

26 [$n!$] gives 4.03×10^{26}

and

24 [$n!$] gives 6.20×10^{23}.

Nevertheless, on some calculators, entering the string

[ON/AC] 26 [$n!$] [÷] 24 [$n!$] [=]

gives the accurate answer, 650, as above, since the machine carries more accuracy internally than it displays.

With the factorial notation in hand, we now return to the problem of determining the number of permutations of n things taken r at a time and the number of combinations of n things taken r at a time.

NOTATION ***P*(*n*, *r*) and *C*(*n*, *r*)**

$P(n, r)$ denotes the **number of permutations of *n* things taken *r* at a time.**
$C(n, r)$ denote the **number of combinations of *n* things taken *r* at a time.**

THEOREM Formulas for $P(n, r)$ and $C(n, r)$

Let n and r be natural numbers with $0 < r \leq n$. Then

$$P(n, r) = n(n-1)(n-2) \cdots (n - r + 1),$$
$$P(n, n) = n!,$$

and

$$C(n, r) = \frac{n(n - 1)(n - 2) \cdots (n - r + 1)}{r!}.$$

Instead of giving a proof using abstract symbolism, we argue that the preceding theorem is true by considering representative special cases. The same argument could be repeated in general terms to give a proof.

Consider again the problem of determining the number of different 5-letter code words with repetition not allowed as in Example 9.15, part (b). Since the order in which letters appear in a code word certainly matters, this is precisely the problem of determining the number, $P(26, 5)$. Since there are 26 choices for the first letter, 25 choices for the second letter, and so on, the answer is

$$P(26,5) = 26 \cdot 25 \cdot 24 \cdot 23 \cdot 22.$$

five factors since we must choose five letters

Repeating the argument in general we would have

$$P(n, r) = n(n - 1)(n - 2) \cdots (n - r + 1)$$

Here there are r factors since we are choosing r letters.

as claimed. Note also that

$$P(n, r) = \frac{n(n - 1) \cdots (n - r + 1) \cdot (n - r) \cdots 1}{(n - r)(n - r - 1) \cdots 1}$$
$$= \frac{n!}{(n - r)!}.$$

Setting $r = n$ in the above two formulas, we obtain

$$P(n, n) = n! \qquad \text{and} \qquad P(n, n) = \frac{n!}{(n - n)!} = \frac{n!}{0!}.$$

Since these must be the same, it follows that we should define $0!$ to be 1 as above.

Suppose we wanted 26-letter code words without repetition; that is, suppose we wanted to compute $P(26, 26)$. Repeating the above argument we have that:

- the first letter can be chosen in 26 ways,
- the second letter can be chosen in 25 ways,
- the third letter can be chosen in 24 ways,

$$\vdots$$

- the last letter can be chosen in 1 way.

Since the product of these numbers gives the desired result, the number of permutations of 26 things taken all at a time is

$$P(26, 26) = 26 \cdot 25 \cdot 24 \cdots 1 = 26!$$

Repeating the argument for a general n, we have

$$P(n, n) = n!$$

Lastly, consider again the problem of determining the number of permutations of 26 things taken five at a time. One way to determine such a permutation is to:

- choose the five letters to appear in the permutation *and*
- determine the order in which the letters are to appear.

We can choose the five letters in $C(26, 5)$ ways and the five objects can be put in order in 5! ways. Thus, by the third principle of counting,

$$P(26, 5) = C(26, 5) \cdot 5!$$

Dividing both sides of the equation by 5!, we obtain

$$C(26, 5) = \frac{P(26, 5)}{5!} = \frac{26 \cdot 25 \cdot 24 \cdot 23 \cdot 22}{5!}.$$

five factors in both numerator and denominator

This argument could be repeated in general to give

$$C(n, r) = \frac{n(n-1) \cdots (n-r+1)}{r!}.$$

r factors in both numerator and denominator

EXAMPLE 9.19

Computing $P(n, r)$ and $C(n, r)$

Compute each of the following.

(a)	$P(7, 2)$	**(b)**	$P(8, 8)$	**(c)**	$P(25, 2)$
(d)	$C(7, 2)$	**(e)**	$C(8, 8)$	**(f)**	$C(25, 2)$

SOLUTION

(a) $P(7, 2) = 7 \cdot 6 = 42$

(b) $P(8, 8) = 8! = 40{,}320$

(c) $P(25, 2) = 25 \cdot 24 = 600$

(d) $C(7, 2) = \frac{7 \cdot 6}{2!} = 21$

(e) $C(8, 8) = \frac{8!}{8!} = 1$

(f) $C(25, 2) - \frac{25 \cdot 24}{1 \cdot 2} = 300$ ■

SCHOOL BOOK PAGE *Probability in Grade Three*

Practice

For More Practice, see Set G, pages 500-501.

Answer the following questions. You may want to do one of these activities with your partner.

Experiment A

a. Pick one card from the bag without looking.

b. Record the number. Return the card to the bag.

5. Is it possible to get an 8?

6. Is it possible to get a 12?

7. Do you have an equal chance to get a 3 or a 5?

8. Are you more likely to get a 4 than an 8?

9. How many times do you think you would get a 6 in 8 picks?

10. How many times do you think you would get a 6 in 80 picks?

Experiment B

a. Pick a marble from the jar without looking.

b. Record the color. Return the marble to the jar.

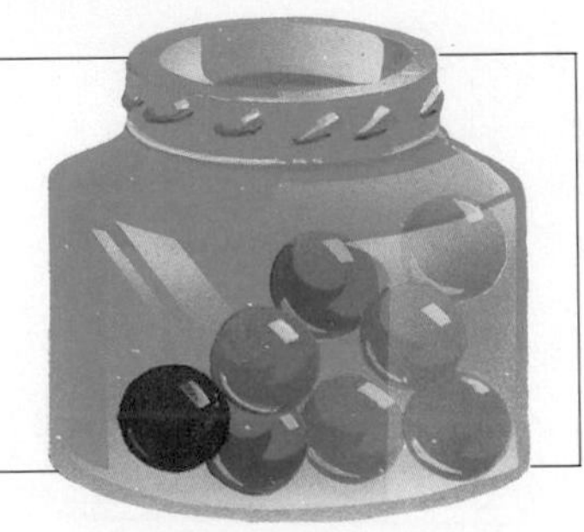

11. Do you have an equal chance to get a blue or a red marble?

12. Are you more likely to get a green marble than a red one?

13. How many blue marbles do you think you would get in 8 picks?

14. How many blue marbles do you think you would get in 80 picks?

1. Are problems 9, 10, 13, and 14 properly stated? If your students were confused by these problems, what would you say to help them out?

EXAMPLE 9.20

Permutations of Four Red Balls, Three White Balls, and Two Green Balls

How many ways can you put four red balls, three white balls, and two green balls in order in a line if the balls are indistinguishable except for color?

SOLUTION

Understand the problem • • • •

We must put four red balls, three white balls, and two green balls in order. Since the balls are indistinguishable except for color, interchanging balls of the same color will not make any difference; that is, the arrangement

R R W G W G R W R

will not change if we interchange red balls among themselves, or white balls among themselves, or green balls among themselves. Apparently, the only way to obtain a different arrangement is to choose different locations in which to place the red, white, and green balls.

Devise a plan • • • •

How can we determine in how many ways we can choose the four places for red balls, the three places for white balls, and the two places for green balls? This is just the number of ways we can choose four of the nine places to receive red balls *and* three of the remaining five places to receive white balls *and* two of the remaining two places to receive green balls. The words, "and," in the preceding sentence are the key. They suggest that we use the third principle of counting.

Carry out the plan • • • •

1. The number of ways we can choose four of the nine places to receive red balls is $C(9, 4)$, the number of combinations of nine things (spaces) taken four at a time.
2. Having chosen the four places to receive red balls, we must now choose three of the remaining five places to receive white balls and this can be done in $C(5, 3)$ ways.
3. This leaves two places from which we must choose the places to receive the two green balls and this can be done in $C(2, 2)$ ways.

But we have to do step 1 *and* step 2 *and* step 3 and so, by the third principle of counting, the desired answer is just the product of the number of ways we can complete each step. This is

$$\begin{aligned} C(9, 4) \cdot C(5, 3) \cdot C(2, 2) &= \frac{9 \cdot 8 \cdot 7 \cdot 6}{4!} \cdot \frac{5 \cdot 4 \cdot 3}{3!} \cdot \frac{2 \cdot 1}{2!} \\ &= \frac{9!}{4!\,3!\,2!}. \end{aligned}$$

Look back • • • •

We observed that a permutation of identical red, white, and green balls is not changed by interchanging balls of the same color among themselves. Hence, the number of permutations of four red, three white, and two green balls was the number of ways we could choose four out of the nine spaces to receive red balls *and* then three of the remaining five spaces to receive white balls, *and* then place the two remaining green balls in the two remaining spaces. Differently put, the desired number was the number of ways to choose four of the nine places to receive red balls *times* the number of ways to choose three of the remaining five places to receive white balls *times* the number of ways to choose the two of the remaining two places to receive the green balls. ■

This argument could be repeated in general, so the result can be stated as a theorem.

THEOREM Permutations of Like Things

The number of permutations of n things with a_1 things alike, a_2 other things alike, . . . , and a_r other things alike is given by

$$\frac{n!}{a_1!\,a_2!\,a_3!\cdots a_r!}.$$

EXAMPLE 9.21 Determining the Number of Zip Code Groups

As seen in Example 4.31, the bar code commonly seen under the address on pieces of mail is the zip code in machine readable form. When broken down, it turns out that each code group consists of two long bars and three short bars; like this, ||||| . How many different code groups can be formed in this way?

SOLUTION

This is just the number of ways one can put two long bars and three short bars in order. By the preceding theorem, the result is

$$\frac{5!}{2!\,3!} = \frac{5\cdot 4\cdot 3\cdot 2\cdot 1}{2\cdot 1\cdot 3\cdot 2\cdot 1} = 10.$$

Observe that the result is opportune since the ten code groups are just adequate to represent the ten digits—0, 1, 2, · · · , 9—needed to express a zip code. ■

PROBLEM SET 9.2

Understanding Concepts

1. Two dice are thrown. Determine the number of ways a score of 4 or 6 can be obtained. For example $4 = 3 + 1 = 1 + 3 = 2 + 2$, and so on.
2. Two dice are thrown. Determine the number of ways to obtain a score of at least 4. (*Hint:* In how many ways can you fail to obtain a score of at least 4? How many outcomes are possible all told?)
3. A coin is tossed and a die is rolled.
   (a) In how many ways can the outcome consist of a head and an even number?
   (b) In how many ways can the outcome consist of a head or an even number?
4. There are 26 students in Mrs. Pietz's fifth grade class at the International School of Tokyo. All of the students speak either English or Japanese and some speak both languages. If 18 of the students can speak English and 14 can speak Japanese,
   (a) how many speak both English and Japanese?
   (b) how many speak English but not Japanese? (*Hint:* Draw a Venn diagram.)
5. All of the 24 students in Mr. Walcott's fourth grade class at the International School of Tokyo speak either English or Japanese. If 11 of the students speak only English and nine of the students speak only Japanese, how many speak both languages?
6. In how many ways can you draw a club or a face card from an ordinary deck of playing cards?
7. In how many ways can you select a red face card or a black ace from an ordinary deck of playing cards?
8. (a) How many four digit natural numbers can be named using the digits 1, 2, 3, 4, 5, or 6 at most once?
   (b) How many of the numbers in part (a) begin with an odd digit? (*Hint:* Choose the first digit first.)
   (c) How many of the numbers in part (b) end with an odd digit? (*Hint:* Choose the first digit first and the last digit second.)
9. (a) If repetition of digits is not allowed, how many three digits numbers can be formed using the digits 1, 2, 3, 4, 5?
   (b) How many of the numbers in part (a) begin with either of the digits 2 or 3?
   (c) How many of the numbers in part (a) are even?
10. Construct a possibility tree to determine all 3-letter code words using only the letters a, b, and c without repetition.
11. If four coins are tossed, construct a possibility tree to determine in how many ways one can obtain two heads and two tails.
12. Evaluate each of these expressions.
   (a) $7!$ (b) $9! - 7!$ (c) $9! \div 7!$
   (d) $9! + 7!$ (e) $7 \cdot 7!$ (f) $0!$
13. Evaluate each of the following.
   (a) $P(13, 8)$ (b) $P(15, 15)$ (c) $P(15, 2)$
   (d) $C(13, 8)$ (e) $C(15, 15)$ (f) $C(15, 2)$
14. How many 4-letter code words can be formed using a standard 26 letter alphabet
   (a) if repetition is allowed?
   (b) if repetition is not allowed?
15. How many 5-digit numbers can be formed with the first three digits odd and the last two digits even
   (a) if repetition of digits is allowed?
   (b) if repetition of digits is not allowed?
16. How many different signals can be sent up on a flag pole if each signal requires three blue and three yellow flags and the flags are identical except for color?

17. (a) How many different arrangements are there of the letters in TOOT?
   (b) How many different arrangements are there of the letters in TESTERS?

Thinking Critically

18. Two dice are thrown.
   (a) In how many ways can the score obtained be even?
   (b) In how many ways can the score be a multiple of 5?
   (c) In how many ways can the score be a multiple of 3?
   (*Hint:* All possibilities when two dice are rolled are shown in Example 9.12.)

19. In how many ways can you arrange the nine letters a, a, a, a, b, b, b, c, c in a row?

20. How many of the arrangements in problem 19 start with a and end with b?

21. (a) In how many of the 120 arrangements of a, b, c, d, and e does b immediately follow a? (*Hint:* Think of ab as a single symbol.)

(b) In how many of the 120 possible arrangements of a, b, c, d, and e are a and b adjacent?

(c) In how many arrangements of a, b, c, d, and e does a precede e?

22. Sets of six cards are selected without replacement from an ordinary deck of playing cards.

(a) How many ways can you choose a set of 6 hearts?

(b) How many ways can you select a set of 3 hearts and 3 spades?

(c) How many ways can you select a set of 6 hearts or 6 spades?

23. How many different sequences of ten flips of a coin result in 5 heads and 5 tails?

24. Compute the values of $C(n, r)$ in this table and extend the table two more rows. Do you need the formula to do this last? Explain.

$$C(0, 0)$$
$$C(1, 0) \quad C(1, 1)$$
$$C(2, 0) \quad C(2, 1) \quad C(2, 2)$$
$$C(3, 0) \quad C(3, 1) \quad C(3, 2) \quad C(3, 3)$$
$$C(4, 0) \quad C(4, 1) \quad C(4, 2) \quad C(4, 3) \quad C(4, 4)$$

Making Connections

25. An electrician must connect a red, a white, and a black wire to a yellow, a blue, and a green wire in some order. How many different connections are possible?

26. (a) Mrs. Ruiz has 13 boys and 11 girls in her class. In how many ways can she select a committee to organize a class party if the committee must contain three boys and three girls?

(b) Lourdes, a girl, and Andy always fight. How many ways can Mrs. Ruiz select the committee of part (a) if she does not want both Lourdes and Andy on the committee? Note that Lourdes can be on the committee and Andy not on the committee **or** vice versa.

27. In the state of Washington each automobile license plate shows three letters followed by three digits or three digits followed by three letters. How many different license plates can be made

(a) if repetition of digits and letters is allowed?

(b) if repetition of digits and letters is not allowed?

Communicating

28. Write out a careful argument showing that

$$C(n, r) = C(n, n - r)$$

for whole numbers n and r with $0 \le r \le n$.

For Review

29. Toss four coins 20 times and determine the empirical probability of obtaining two heads and two tails.

30. Toss two dice 20 times and determine the empirical probability of obtaining a score of 4.

31. Argue carefully that the empirical probability of A or B satisfies the equation

$$P_e(A \text{ or } B) = P_e(A) + P_e(B) - P_e(A \text{ and } B).$$

32. Francine has lost 20 times in a row while playing roulette. She feels that her luck is bound to change soon and so begins to bet heavily. Briefly discuss Francine's reasoning.

33. Rashonda is 12 years old and gets a \$10 allowance each week. She typically spends \$5.50 for entertainment, \$3.50 for snacks, and \$1 for miscellaneous purchases. Draw a pie chart to show how Rashonda spends her allowance.

34. The final scores in Professor Kane's Ed Psych class were 53, 77, 82, 82, 86, 67, 77, 64, 72, 68, 60, 74, 56, 57, 82, 81, 88, and 90.

(a) Display this data using a line plot.

(b) Determine the mean, median, and mode for this data.

(c) Compute the 5-number summary for this data.

(d) Display the result of part (c) using a box and whisker plot.

9.3 Theoretical Probability

We now turn our attention to the study of **theoretical probability** which we refer to simply as probability. For convenience, we repeat here our earlier definition. It is convenient to couch this definition in terms of an experiment that may culminate in a variety of results each of which is as likely to occur as any other. Such results are said to be **equally likely.**

DEFINITION Theoretical Probability of an Event, Sample Space

Let S, called the **sample space,** denote the set of **equally likely outcomes** of an experiment and let E be an **event** consisting of a subset of outcomes of the experiment. Let $n(S)$ and $n(E)$ denote the number of outcomes and S and E respectively. The **probability** of E, denoted by $P(E)$, is given by

$$P(E) = \frac{n(E)}{n(S)}.$$

Be aware that the phrase **equally likely** is of critical importance in the above definitions. If the outcomes in the sample space are not equally likely, the definition of probability simply doesn't make sense. For example, suppose the experiment consists of rolling two dice. It would not do to think of the sample space as the eleven outcomes $\{2, 3, 4, \ldots, 12\}$ since these outcomes are not equally likely. The difficulty is that there is one way to obtain a 2, two ways to obtain a 3, three ways to obtain a 4, . . . , six ways to obtain a 7, . . . , and only one way to obtain a 12. Thus, the event of obtaining a 3 is twice as likely as the event of obtaining a 2, and so on. However, if we mistakenly thought of the sample space as the above set with each of 2, 3, 4, · · · , 11, and 12 equally likely, we would obtain

$$P(2) = P(3) = \cdots = P(12) = \frac{1}{11},$$

a manifest absurdity. In fact, as we have seen before, there are 36 equally likely ways two dice can come up so that

$$P(2) = \frac{1}{36},\ P(3) = \frac{2}{36},\ \ldots,\ P(12) = \frac{1}{36}.$$

Moreover, these probabilities are closely approximated by the corresponding empirical probabilities.

The definition of probability presumes that we can count outcomes. This implies that the sample space is finite. Thus, our discussion of theoretical probability will be largely restricted to the finite case. It is possible to treat the infinite case as well as with geometric probability, but the general theory is beyond the scope of this course. The ideas are perhaps best understood by considering a series of examples.

EXAMPLE 9.22 **Determining the Probability of Rolling Eleven with Two Dice**

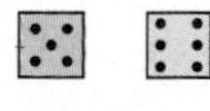

Compute the probability of obtaining a score of 11 on a single roll of two dice.

SOLUTION

Here the sample space, S, is the set of all 36 outcomes illustrated in Example 9.12. Let E denote the event of rolling a score of 11. Since only two of the outcomes in the sample space result in 11,

$$P(11) = \frac{n(E)}{n(S)} = \frac{2}{36} = \frac{1}{18}.$$

■

EXAMPLE 9.23 Computing the Probability of 5 or 8

Determine the probability of rolling a 5 *or* an 8 on a single roll of two dice.

SOLUTION 1

As we have just seen, the sample space, S, is the set of all 36 ways two dice can come up. The favorable outcomes are as shown here where F is the event of rolling a 5 and E is the event of rolling an 8.

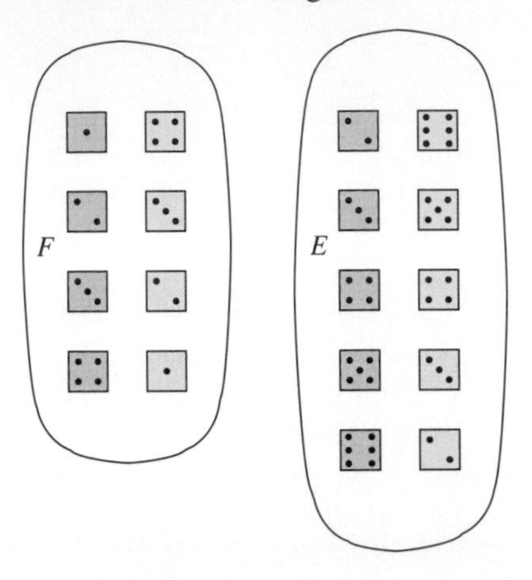

It follows that

$$P(5 \text{ or } 8) = \frac{4+5}{36} = \frac{1}{4}.$$

SOLUTION 2

Note that if 5 is rolled then 8 is not and conversely. Thus, rolling 5 and rolling 8 are mutually exclusive events and the **or** in the statement of the problem reminds us that we can use the second principle of counting. Thus,

$$\begin{aligned} P(5 \text{ or } 8) &= \frac{n(5 \text{ or } 8)}{n(S)} \\ &= \frac{n(5)+n(8)}{n(S)} \\ &= \frac{4+5}{36} \\ &= \frac{1}{4} \end{aligned}$$

as before. ■

EXAMPLE 9.24 Computing the Probability of Obtaining a Face Card or a Diamond

Determine the probability of obtaining a face card *or* a diamond if a card is drawn at random from an ordinary deck of playing cards.

> **JUST FOR FUN**
>
> **It's a Girl!**
>
> A family has two children including at least one boy. What is the probability that the other child is a girl? The answer is not 1/2!

SOLUTION 1

The sample space, S, is the set of all 52 cards in the deck. Let D denote the event of selecting a diamond and let F denote the event of selecting a face card. Since there are 13 diamonds (including face cards) and 9 nondiamond face cards,

$$P(D \text{ or } F) = \frac{13 + 9}{52} = \frac{22}{52} = \frac{11}{26}.$$

SOLUTION 2

Using the first principle of counting, we know that

$$\begin{aligned} P(D \text{ or } F) &= \frac{n(D \text{ or } F)}{n(S)} \\ &= \frac{n(D) + n(F) - n(D \cap F)}{n(S)} \\ &= \frac{13 + 12 - 3}{52} \\ &= \frac{22}{52} = \frac{11}{26} \end{aligned}$$

Three face cards are diamonds.

as before. ■

EXAMPLE 9.25 Determining Probabilities with Restrictive Conditions

All 24 students in Mr. Henry's fifth grade class are either brunette or blonde as shown in this table. A student is selected at random.

	Brunettes	Blondes
Boys	8	3
Girls	6	7

(a) What is the probability that the student is a brunette?

(b) What is the probability that the student is a brunette given that a boy was selected?

SOLUTION

Let B denote the set of boys, B_r thc sct of brunettes and S the set of all students in the class.

(a) $P(B_r) = \dfrac{n(B_r)}{n(S)} = \dfrac{14}{24}.$

(b) If we know that a boy was selected, then the sample space is not the set of all students in the class but *all boys in the class.* Similarly, the set of favorable outcomes is the set of all boys in the class who are brunettes. Thus, the desired probability is

$$P(B \cap B_r) = \frac{8}{11}.$$

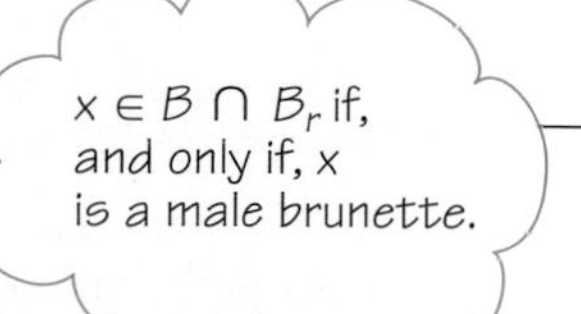

$x \in B \cap B_r$ if, and only if, x is a male brunette.

Actually, this is an example of so-called **conditional probability** and the standard notation is $P(B_r \mid B)$, read "the probability of B_r given that B has occurred." In this case $P(B_r \mid B)$ is the probability of selecting a brunette given that a boy has been selected. ■

EXAMPLE 9.26 **Using the Third Principle of Counting**

A red die and a green die are rolled. What is the probability of obtaining an even number on the red die *and* a multiple of 3 on the green die?

SOLUTION

The sets of favorable outcomes on the red and green dice respectively are $R = \{2, 4, 6\}$ and $G = \{3, 6\}$ and the sample space is the set of all 36 ways the two dice can come up. The word "and" is a broad hint to use the third principle of counting. Thus, the number of favorable outcomes is

$$n(R \text{ and } G) = n(R) \cdot n(G) = 2 \cdot 3 = 6$$

R and G are clearly independent.

Hence, the desired probability is $P = 6/36 = 1/6$. ■

JUST FOR FUN

A Probability Paradox

Consider the hats containing colored balls on the tables shown. The balls in the hats on tables *A* and *B* are combined and placed in the hats on table *C*. Let $P(R \mid G)$ denote the probability of randomly drawing a red ball from a grey hat and let $P(R \mid B)$ denote the probability of randomly drawing a red ball from a brown hat. Compute and compare $P(R \mid G)$ and $P(R \mid B)$ for each of tables *A*, *B*, and *C*. Is the result surprising?

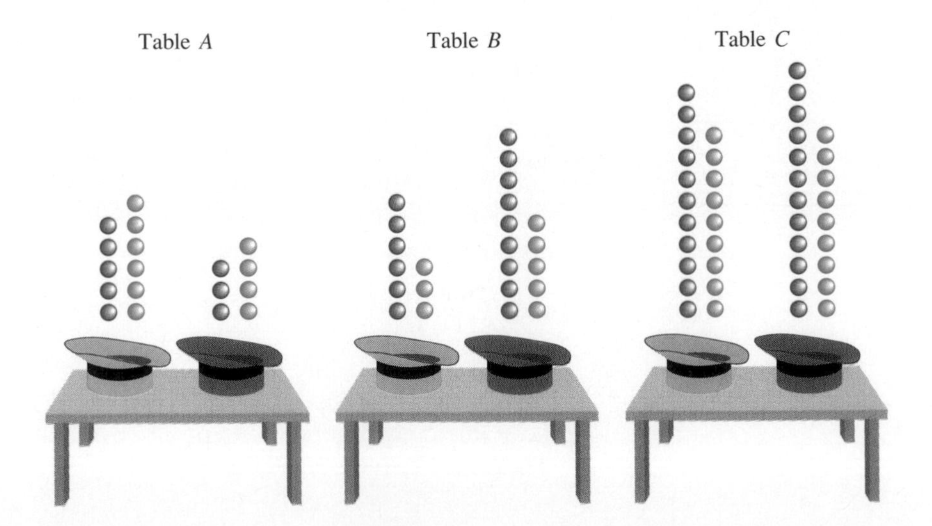

SOURCE: Adapted illustration by Ilil Arbel from "Mathematical Games" by Martin Gardner from *Scientific American*, March 1976, page 120.

EXAMPLE 9.27 Selecting Balls without Replacement

An urn contains three identical red and two identical white balls. Two balls are drawn one after the other without replacement.

(a) What is the probability that the first ball is red?

(b) What is the probability that the second ball is red given that the first ball is red?

(c) What is the probability that both balls are red?

SOLUTION 1

(a) Since there are initially five balls and three are red, $P(\text{1st ball is red}) = 3/5$.

(b) Having already selected a red ball on the first draw, the counts change for the selection of the second ball. For the second selection there remain four balls of which two are red. Thus,

given that

$$P(\text{2nd ball is red} \mid \text{1st ball is red}) = \frac{2}{4}.$$

(c) In order to select two red balls, it is necessary to select a red ball on the first draw *and* a red ball on the second draw. By the third principle of counting, the number of possible selections of two balls is $5 \cdot 4 = 20$ and the number of possible selections of two red balls is $3 \cdot 2 = 6$. Hence,

$$\begin{aligned} P(\text{2 red balls}) &= \frac{6}{20} \\ &= \frac{3 \cdot 2}{5 \cdot 4} \\ &= \frac{3}{5} \cdot \frac{2}{4} \\ &= P(\text{1st ball is red}) \cdot P(\text{2nd ball is red} \mid \text{1st ball is red}). \end{aligned}$$

SOLUTION 2

A second way to solve this problem is to label the balls r_1, r_2, r_3, w_1, and w_2 so that we can identify the different red and white balls. With the balls identified in this way we make an orderly list of all possible selections of two balls. In this listing, for example, the entry (r_1, r_2) indicates that red ball number 1 was chosen first and red ball number 2 was chosen second.

(r_1, r_2)	(r_1, r_3)	(r_1, w_1)	(r_1, w_2)
(r_2, r_1)	(r_2, r_3)	(r_2, w_1)	(r_2, w_2)
(r_3, r_1)	(r_3, r_2)	(r_3, w_1)	(r_3, w_2)
(w_1, r_1)	(w_1, r_2)	(w_1, r_3)	(w_1, w_2)
(w_2, r_1)	(w_2, r_2)	(w_2, r_3)	(w_2, w_1)

all selections with r_1 first

all selections with r_2 first

and so on

HIGHLIGHT FROM HISTORY
Srinivasa Ramanujan (1887–1920)

Perhaps the most exotic and mysterious of all mathematicians was Srinivasa Ramanujan, born to a high caste family of modest means in Kumbakonam in Southern India in 1887. Largely self-taught, Ramanujan, in 1913, wrote a letter to the eminent English mathematician, G. H. Hardy, at Cambridge University in which he included a list of formulas he had discovered. Of the formulas Hardy wrote "(such formulas) defeated me completely . . . a single look at them is enough to show that they were written by a mathematician of the highest class. They must be true because, if they were not true, no one would have had the imagination to invent them. Finally, (you must remember that I knew nothing about Ramanujan, and had to think of every possibility), the writer must be completely honest, because great mathematicians are commoner than thieves and humbugs of such incredible skill." In any event, Hardy arranged for Ramanujan to come to England in 1914 and the two collaborated intensively for the next three years with Ramanujan using his unorthodox methods and fantastic intuition to come up with deep and totally unexpected results which Hardy, with his considerable intellectual power and formal training, then proved. Ramanujan fell ill in 1917 and returned to India where he died in 1920.

Littlewood once remarked that every positive integer was one of Ramanujan's friends. Once, during Ramanujan's illness, Hardy visited him in the hospital in Putney. Trying to find a way to begin the conversation, Hardy remarked that he had come to the hospital in cab number 1729 and that he could not imagine a more uninteresting number. To which Ramanujan replied, "No, it is a very interesting number; it is the smallest number that can be written as the sum of two cubes in two different ways!"*

*$1729 = 9^3 + 10^3 = 1^3 + 12^3$

Counting possibilities, we have

(a) $P(\text{1st ball is red}) = \frac{12}{20} = \frac{3}{5}$,

(b) $P(\text{2nd ball is red} \mid \text{1st ball is red}) = \frac{6}{12} = \frac{2}{4}$, and

(c) $$P(\text{2 red balls}) = \frac{6}{20} = P(\text{1st ball red}) \cdot P(\text{2nd ball red} \mid \text{1st ball red})$$

as before. ■

EXAMPLE 9.28 Combinations and Probability

There are 10 boys and 13 girls in Mr. Fleck's fourth grade class and 12 boys and 11 girls in Mrs. Patero's fourth grade class. A picnic committee of six people is selected at random from the total group of students in both classes.

(a) What is the probability that all the committee members are girls?

(b) What is the probability that all the committee members are girls given that all come from Mr. Fleck's class?

(c) What is the probability that the committee has three girls and three boys?

(d) What is the probability that the committee has three girls and three boys given that Mary Akers and Ann-Marie Harborth are on the committee?

SOLUTION

(a) Since there are six committee members and 46 students in all, the total number of possible committees is $C(46, 6) = 46 \cdot 45 \cdot 44 \cdot 43 \cdot 42 \cdot 41/6!$. Since there are 24 girls, the number of committees with six girls is $C(24, 6) = 24 \cdot 23 \cdot 22 \cdot 21 \cdot 20 \cdot 19/6!$. Therefore,

$$
\begin{aligned}
P(\text{committee has six girls}) &= \frac{24 \cdot 23 \cdot 22 \cdot 21 \cdot 20 \cdot 19/6!}{46 \cdot 45 \cdot 44 \cdot 43 \cdot 42 \cdot 41/6!} \\
&= \frac{24 \cdot 23 \cdot 22 \cdot 21 \cdot 20 \cdot 19}{46 \cdot 45 \cdot 44 \cdot 43 \cdot 42 \cdot 41} \\
&= \frac{76}{5289} \\
&\doteq 0.014.
\end{aligned}
$$

(b) Here the population is the set of students in Mr. Fleck's class. Thus. P(six girls on committee | all committee members are chosen from

$$
\begin{aligned}
\text{Mr. Fleck's class}) &= \frac{C(13, 6)}{C(23, 6)} \\
&= \frac{13 \cdot 12 \cdot 11 \cdot 10 \cdot 9 \cdot 8/6!}{23 \cdot 22 \cdot 21 \cdot 20 \cdot 19 \cdot 18/6!} \\
&= \frac{13 \cdot 12 \cdot 11 \cdot 10 \cdot 9 \cdot 8}{23 \cdot 22 \cdot 21 \cdot 20 \cdot 19 \cdot 18} \\
&= \frac{52}{3059} \\
&\doteq 0.017.
\end{aligned}
$$

(c) The committee is chosen from all the students so the sample space contains $C(46, 6)$ possible committees. Since the three boys must be chosen from among the 22 boys in both classes *and* the three girls must be chosen from among the 24 girls in both classes, we have by the third principle of counting, that the number of committees with three girls *and* three boys is $C(22, 3) \cdot C(24, 3)$

Thus,

$$\frac{1}{C(46, 6)} = \frac{1 \cdot 2 \cdot 3 \cdot 4 \cdot 5 \cdot 6}{46 \cdot 45 \cdot 44 \cdot 43 \cdot 42 \cdot 41}$$

P(three girls and three boys on the committee)

$$
\begin{aligned}
&= \frac{C(22, 3)C(24, 3)}{C(46, 6)} \\
&= \frac{22 \cdot 21 \cdot 20}{3 \cdot 2 \cdot 1} \cdot \frac{24 \cdot 23 \cdot 22}{3 \cdot 2 \cdot 1} \cdot \frac{6 \cdot 5 \cdot 4 \cdot 3 \cdot 2 \cdot 1}{46 \cdot 45 \cdot 44 \cdot 43 \cdot 42 \cdot 41} \\
&= \frac{1760}{5289} \doteq 0.333.
\end{aligned}
$$

(d) Since Ann-Marie and Mary are on the committee, choosing the committee only requires choosing four more students. Thus, the sample space consists of the $C(44, 4)$ students chosen from among the 44 children other than Ann-Marie and Mary. Also, the number of favorable cases is found by selecting one more girl from among the 22 other girls in $C(22, 1)$ ways *and* selecting the three boys in $C(22, 3)$ ways. Thus,

P(committee has three girls and three boys | Mary and Ann-Marie are on the committee)

$$= \frac{C(22, 1)C(22, 3)}{C(44, 4)}$$

$$= \frac{22}{1} \cdot \frac{22 \cdot 21 \cdot 20}{3 \cdot 2 \cdot 1} \cdot \frac{4 \cdot 3 \cdot 2 \cdot 1}{44 \cdot 43 \cdot 42 \cdot 41}$$

$$= \frac{440}{1763} \doteq 0.250.$$

$$\frac{1}{C(44, 4)} = \frac{4 \cdot 3 \cdot 2 \cdot 1}{44 \cdot 43 \cdot 42 \cdot 41}$$

■

Complementary Events

If A and $\bar{A}$ are events such that $A \cup \bar{A} = S$ and $A \cap \bar{A} = \emptyset$, A and $\bar{A}$ are called **complementary events.** Moreover, $n(A) + n(\bar{A}) = n(S)$ and this implies that

$$P(A) + P(\bar{A}) = 1.$$

$$\frac{n(A)}{n(S)} + \frac{n(\bar{A})}{n(S)} = \frac{n(S)}{n(S)}$$

$$P(A) + P(\bar{A}) = 1$$

Alternatively, this implies that

$$P(A) = 1 - P(\bar{A}).$$

In ordinary language, suppose obtaining A is considered success. Then obtaining $\bar{A}$ is failure and we have that

$$P(\text{success}) = 1 - P(\text{failure}).$$

THEOREM P(success) = 1 − P(failure)

Let $A \cup \bar{A} = S$ and $A \cap \bar{A} = \emptyset$. Then

$$P(A) = 1 - P(\bar{A});$$

that is, the probability of success in an experiment is 1 minus the probability of failure.

The preceding theorem is often useful as this example shows.

EXAMPLE 9.29

Using Complementary Probability

Compute the probability of obtaining a score of at least 4 on a single roll of 2 dice.

SOLUTION

Here success is obtaining a 4 or a 5 or a 6 or . . . or a 12. The probability of doing this is the sum of all the individual probabilities and determining these requires considerable computation. However, failure occurs if we obtain either a sum of 2 or 3 and the probability of doing this is much easier to compute. Since S contains 36 equally likely outcomes and we fail by rolling two 1s, a 1 and a 2, or a 2 and a 1 in order, the probability of failure is $3/36 = 1/12$. Thus, the desired probability is $1 - 1/12 = 11/12$. ■

Odds

When someone speaks of the **odds** in favor of an event, E, they are comparing the likelihood that the event will happen to the likelihood that it will not happen. Consider an urn containing four blue balls and one yellow ball. If a ball is chosen at random, what are the odds that the ball is blue? Since a blue ball is four times as likely to be selected as a yellow ball, it is typical to say that the odds are 4 to 1. The odds are actually the ratio 4/1 but, when quoting odds, one usually writes 4 : 1 which is read "four to one." This is the basis for the following definition.

DEFINITION Odds

Let A be an event and let $\overline{A}$ be the complementary event. Then the **odds in favor** of A are $n(A)$ to $n(\overline{A})$ and the **odds against** A are $n(\overline{A})$ to $n(A)$.

EXAMPLE 9.30

Determining the Odds in Favor of 7 or 11

In the game of craps one wins on the first roll of the pair of dice if a 7 or 11 is thrown. What are the odds of winning on the first roll?

SOLUTION

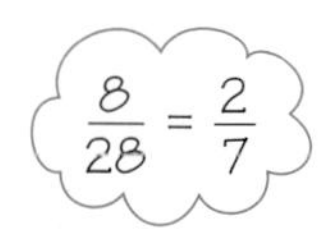

Let W be the set of outcomes that result in 7 or 11. Since $W = \{(1, 6), (2, 5), (3, 4), (4, 3), (5, 2), (6, 1), (1, 10), (10, 1)\}$ and there are 36 ways two dice can come up, $n(W) = 8$, and $n(\overline{W}) = 36 - 8 = 28$. Thus, the odds in favor of W are 8 to 28 or, more simply, 2 to 7. ■

EXAMPLE 9.31

Determining Probabilities from Odds

If the odds in favor of event E are 5 to 4, compute $P(E)$ and $P(\overline{E})$.

SOLUTION

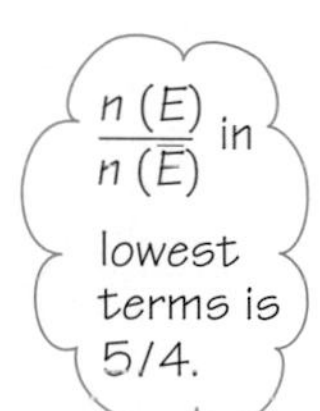

Since E and $\overline{E}$ are complementary, $n(S) = n(E) + n(\overline{E})$. Since the odds in favor of E are 5 to 4, $n(E) = 5k$ and $n(\overline{E}) = 4k$ for some integer k. Therefore,

$$P(E) = \frac{n(E)}{n(S)} = \frac{n(E)}{n(E) + n(\overline{E})} = \frac{5k}{5k + 4k} = \frac{5}{9}$$

and

$$P(\overline{E}) = \frac{n(\overline{E})}{n(S)} = \frac{n(\overline{E})}{n(E) + n(\overline{E})} = \frac{4k}{5k + 4k} = \frac{4}{9}.$$ ■

SCHOOL BOOK PAGE *Probability in Fifth Grade*

Practice

For More Practice, see Set F, pages 446–447.

Thom placed these cards face down and mixed them up. Then he drew a card and recorded the number on it.

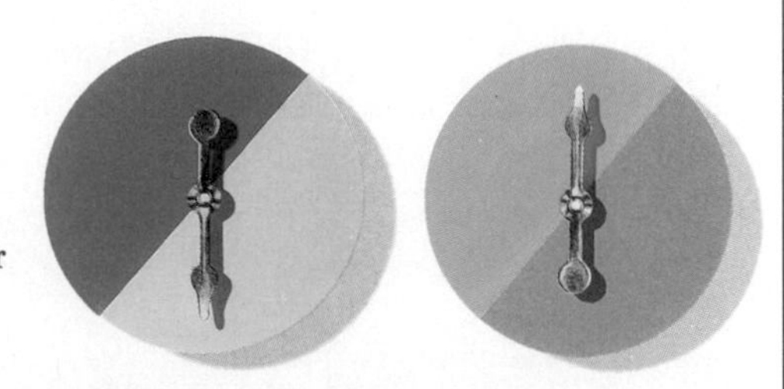

5. List all the possible outcomes.
6. Is Thom more likely to draw a 6 or to draw a 3?
7. Order all the possible outcomes from least likely to most likely.
8. Is Thom more likely to draw an odd number or an even number?
9. Gabe spins each spinner once. What are the possible outcomes for the colors the two spinners will land on?

Problem Solving

10. **Critical Thinking** In a game, Kimi spins the spinner shown. She needs to get a vowel to win. She thinks, "All the outcomes on this spinner are equally likely, so I have as good a chance to get a vowel as a consonant." Is Kimi correct about her chances? Why or why not?

Choose a _____ Strategy

In a certain game, you spin the spinner shown with Problem 10. Then you flip a coin and get heads or tails.

11. What are all the possible outcomes? How many possible outcomes are there?
12. Jan wins if she gets M and tails on her next turn. She loses if she gets any other outcome. Is Jan more likely to win or to lose?

1. In problem 8, how *much* more likely is Thom to draw an odd number than an even number? Give the answer in terms of odds and also in terms of probability.

EXAMPLE 9.32 **Odds from Probabilities**

Given $P(E)$, determine the odds in favor of E and the odds against E.

SOLUTION

The odds in favor of E are

$$\frac{n(E)}{n(\overline{E})} = \frac{n(E)/n(S)}{n(\overline{E})/n(S)} = \frac{P(E)}{P(\overline{E})} = \frac{P(E)}{1 - P(E)}.$$

$P(\overline{E}) = 1 - P(E)$

This last ratio would be expressed as a ratio of integers a/b in lowest terms and the odds quoted as a to b.

The odds against E are given by the reciprocal of the ratio giving the odds in favor of E. Thus, the odds against E are

$$\frac{1 - P(E)}{P(E)} = \frac{b}{a}$$

$= \frac{P(\overline{E})}{P(E)}$

and are quoted as b to a. ■

Expected Value

At a carnival you are offered the chance to play a game that consists of rolling a single die just once. If you play, you win the amount in dollars shown on the die. If you play the game several times how much would you expect to win? Of course, you may be lucky and win \$6 on each of a series of rolls. However, because of the law of large numbers, you would *expect* to roll a 6 only about 1/6 of the time. Since this is true for each of the numbers on the die, you should expect to win on average approximately

$$\begin{aligned} E &= \frac{1}{6} \cdot 1 + \frac{1}{6} \cdot 2 + \frac{1}{6} \cdot 3 + \frac{1}{6} \cdot 4 + \frac{1}{6} \cdot 5 + \frac{1}{6} \cdot 6 \\ &= \frac{1}{6} \cdot (1 + 2 + 3 + 4 + 5 + 6) \\ &= \frac{1}{6} \cdot 21 = \$3.50 \end{aligned}$$

per roll. If it costs you \$4 to play the game, the carnival confidently expects players to *lose* 50¢ per game on the average. Thus, the carnival stands to make a handsome profit if a large number of patrons play the game each night.

The preceding discussion introduces the notion of **expected value.**

DEFINITION Expected Value of an Experiment

Let the outcomes of an experiment be a sequence of real numbers (values), $v_1, v_2, \ldots, v_n$, and suppose the outcomes have respective probabilities $p_1, p_2, \ldots, p_n$. Then the **expected value** of the experiment is

$$e = v_1p_1 + v_2p_2 + \cdots + v_np_n.$$

EXAMPLE 9.33

Winning at Roulette

An American roulette wheel has 38 compartments around its rim. Two of these are colored green and are numbered 0 and 00. The remaining compartments are numbered from 1 to 36 and are alternately colored black and red. When the wheel is spun in one direction, a small ivory ball is rolled in the opposite direction around the rim. When the wheel and the ball slow down, the ball eventually falls in any one of the compartments with equal likelihood if the wheel is fair. One way to play is to bet on whether the ball will fall in a red slot or a black slot. If you bet on red for example, you win the amount of the bet if the ball lands in a red slot; otherwise you lose. What is the expected win if you consistently bet $5 on red?

SOLUTION

Since the probability of winning on any given try is 18/38 and the probability of losing is 20/38, your expected win is

$$\frac{18}{38} \cdot 5 + \frac{20}{38} \cdot (-5) = \frac{90 - 100}{38}$$
$$\doteq -0.26.$$

On average you should expect to lose 26¢ per play. Is it any wonder that casinos consistently make a handsome profit? ■

In the preceding example, it was pretty clear that you should expect to lose slightly more often than win. This next example is less clear.

EXAMPLE 9.34

Determining the Expected Value of an Unusual Game

Suppose you are offered the opportunity to play a game that consists of a single toss of three coins. It costs you $21 to play the game and you win $100 if you toss three heads, $20 if you toss two heads and a tail, and nothing if you toss more than one tail. Would you play the game?

SOLUTION

Many people would play the game hoping to "get lucky" and roll HHH frequently. But is this reasonable? What is your expected return? The expected value of the game is

$$\frac{1}{8} \cdot 100 + \frac{3}{8} \cdot 20 + \frac{3}{8} \cdot 0 + \frac{1}{8} \cdot 0 = \frac{160}{8} = \$20.$$

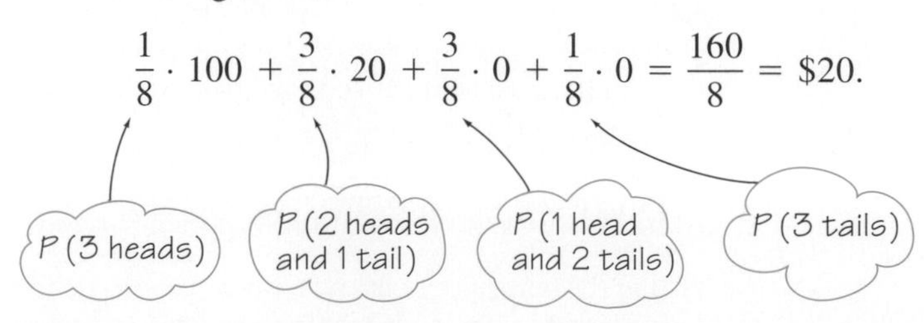

Thus, on average you expect to win $1 less than it costs you to play the game each time. Unless the excitement is worth at least $1, you should not play the game. ■

CURRENT AFFAIRS

State Lotteries

State lotteries are becoming increasingly numerous and popular. Over $20 billion worth of tickets were sold in 1990 and the figure continues to grow. The big draw is the chance to become an instant multimillionaire, but the chance of doing so is extremely small. Your expected return for each one dollar bet is only about 50 cents. In fact, your chances of winning are much better in Las Vegas or Atlantic City where the expected return on a one dollar bet is 85 to 95 cents. Indeed, the probabilities are constant in the big casinos and knowledgeable gamblers rely on the law of large numbers to predict with considerable accuracy their average winnings over the long term. But the probabilities are constantly changing in a lottery so that this calculation is quite complex. Indeed, it can be shown that, as the lottery jackpot increases and thousands of people rush to buy tickets, the probability of any ticket winning drops so much that the expected value of a dollar bet actually falls below 50 cents. State lotteries constantly produce substantial revenues and the only return for almost all players is the pleasure of imagining themselves becoming instantly wealthy.

Geometric Probability

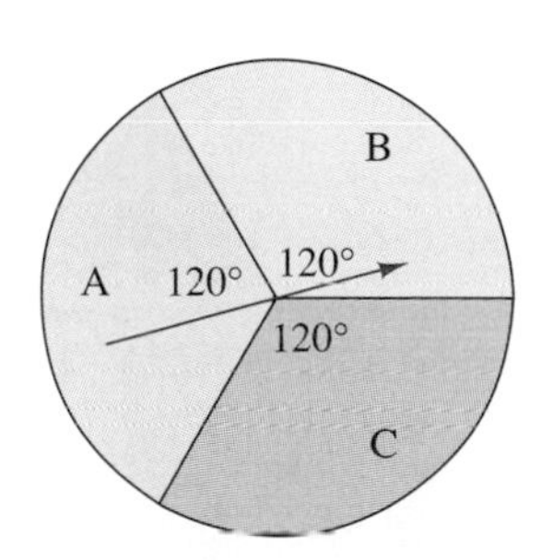

Figure 9.6
A spinner with equally likely outcomes

In Section 9.1, we considered the empirical probability of a spinner marked like that in Figure 9.6 stopping at any particular place. Here $P_e(A)$, $P_e(B)$, and $P_e(C)$ all turn out to be approximately equal to $1/3 = 0.\overline{3}$. This is not surprising since the three arcs bordering regions A, B, and C are equally long. More generally, we would define the theoretical **geometric probability** of a region on the spinner to be the ratio of its corresponding arc length to the circumference of the circle. Equivalently, the probability of stopping a spinner on a sector is the ratio of the measure of the central angle of the sector to 360°.

Similarly, consider the diagram of Figure 9.7. The probability that a point chosen at random in the square (say by using a sequence of random numbers) will belong to region A should be $1/4$ since one-fourth of the *area* of the region lies in A.

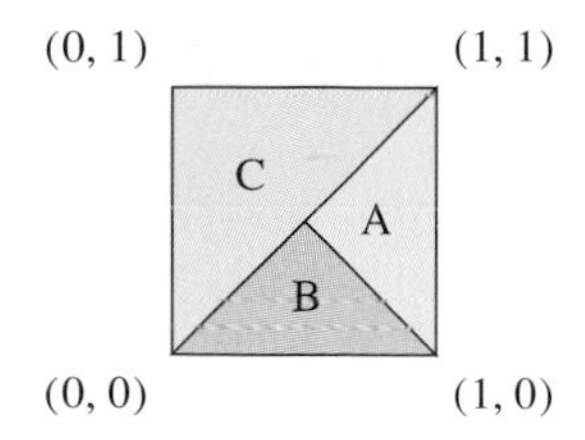

Figure 9.7
Probability as a ratio of areas

EXAMPLE 9.35 **Determining Geometric Probability**

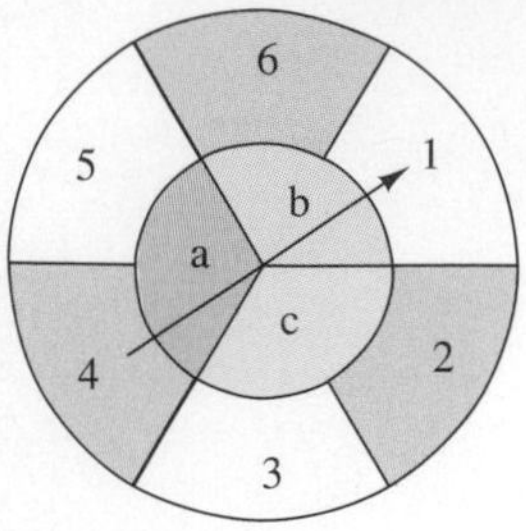

Consider the spinner shown here.

(a) If it is spun and stopped at random, what is the probability that your mark will land on b? That is, compute $P(\text{b})$.

(b) What is the probability, $P(1 \mid \text{b})$, that your mark will land on 1 given that it lands on b?

(c) Actually construct the spinner. Spin it 20 times and compute the empirical probabilities for parts (a) and (b).

(d) Recolor part of the top of the spinner so that, if the spinner is spun, the probability your mark will land in the colored region is 5/6.

SOLUTION

The central angles for the numbered regions all equal $360° \div 6 = 60°$ and the central angles for the lettered regions all equal 120°. Thus,

(a) $P(\text{b}) = \dfrac{120}{360} = \dfrac{1}{3}$, and

(b) $P(1 \mid \text{b}) = \dfrac{60}{120} = \dfrac{1}{2}$.

(c) $P_e(\text{b})$ and $P_e(1 \mid \text{b})$ should turn out to be approximately 1/3 and 1/2. The more trials one makes, the closer the approximation will be by the law of large numbers.

(d) There are several ways to color the spinner properly. One way would be to color regions a, b, and 2 since the sum of the central angles for these regions is $120 + 120 + 60 = 300$ and $300/360 = 5/6$. ■

Properties of Probability

Finally, we collect into a single theorem the properties of probability for finite sample spaces. The preceding examples are typical of general results and so suggest this theorem.

THEOREM Properties of Probability

1. $P(A) = 0$ if, and only if, A cannot occur.
2. $P(A) = 1$ if, and only if, A always occurs.
3. For any event A, $0 \le P(A) \le 1$.
4. For any events A and B,
$$P(A \text{ or } B) = P(A) + P(B) - P(A \text{ and } B).$$
5. If A and B are mutually exclusive events, then
$$P(A \text{ or } B) = P(A) + P(B).$$
6. For any events A and B,
$$P(A \text{ and } B) = P(A)P(B \mid A).$$
7. If A and B are independent events, then
$$P(A \text{ and } B) = P(A)P(B).$$
8. If E and $\overline{E}$ are complementary events, then $P(E) + P(\overline{E}) = 1$.

PROOF Let S be the sample space.

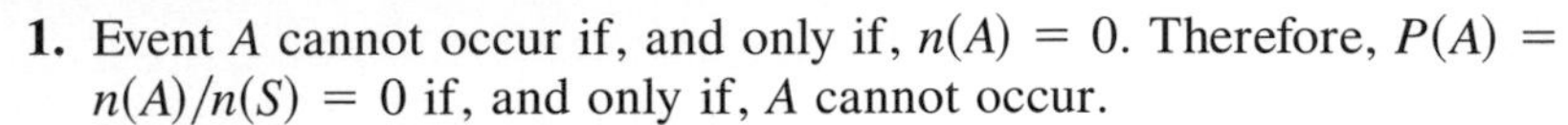

1. Event A cannot occur if, and only if, $n(A) = 0$. Therefore, $P(A) = n(A)/n(S) = 0$ if, and only if, A cannot occur.

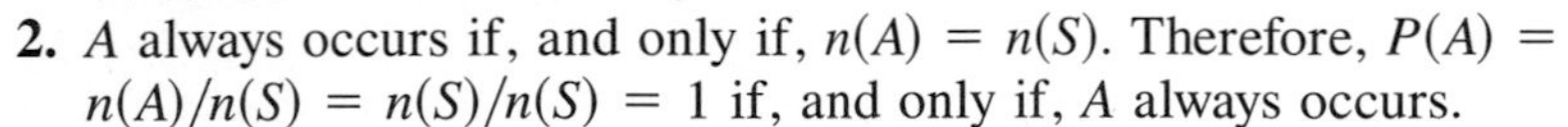

2. A always occurs if, and only if, $n(A) = n(S)$. Therefore, $P(A) = n(A)/n(S) = n(S)/n(S) = 1$ if, and only if, A always occurs.
3. An event A never occurs, sometimes occurs, or always occurs. Therefore, $0 \le n(A) \le n(S)$ and, dividing by $n(S)$, we obtain

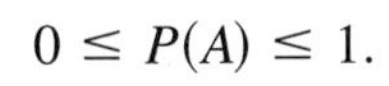

$$0 \le P(A) \le 1.$$

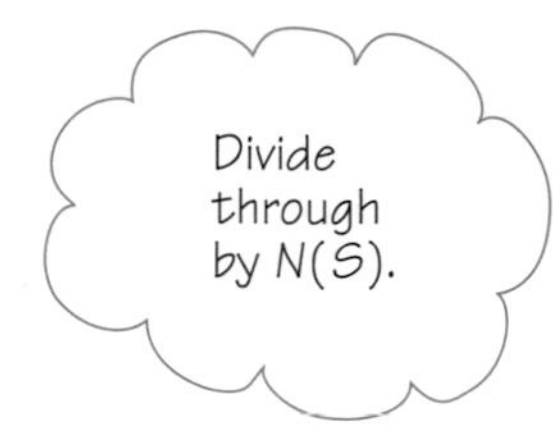

4–7. These *are* general results from earlier representative examples.

8. Since $E \cup \overline{E} = S$ and $E \cap \overline{E} = \emptyset$, it follows that $n(S) = n(E) + n(\overline{E})$ and hence that $1 = P(E) + P(\overline{E})$. ■

PROBLEM SET 9.3

Understanding Concepts

1. (a) Explicitly list all the outcomes for the experiment of tossing a penny, a nickel, a dime, and a quarter.
   (b) Determine the probability, $P(\text{HHTT})$, of obtaining a head on each of the penny and nickel and a tail on each of the dime and quarter in the experiment of part (a).
   (c) Determine the probability of obtaining two heads and two tails in the experiment of part (a).
   (d) Determine the probability of obtaining at least one head in the experiment of part (a). (*Hint:* Note that the complementary event consists of obtaining four tails.)
2. Determine the probability of obtaining a total score of 3 or 4 on a single throw of two dice.
3. Acme Auto Rental has three red Fords, four white Fords, and two black Fords. Acme also has six red Hondas, two white Hondas, and five black Hondas. If a car is selected at random for rental to a customer,
   (a) what is the probability that it is a white Ford?
   (b) what is the probability that it is a Ford?
   (c) what is the probability that it is white?
   (d) what is the probability that it is white given that the customer demands a Ford?
4. Five black balls numbered 1, 2, 3, 4, and 5 and seven white balls numbered 1, 2, 3, 4, 5, 6, and 7 are placed in an urn. If one is chosen at random
   (a) what is the probability it is numbered 1 or 2?
   (b) what is the probability that it is numbered 5 or that it is white?
   (c) what is the probability that it is numbered 5 given that it is white?
5. Mrs. Ricco has seven brown-eyed and two blue-eyed brunettes in her fifth grade class. She also has eight blue-eyed and three brown-eyed blondes. A child is selected at random.
   (a) What is the probability that the child is a brown-eyed brunette?
   (b) What is the probability that the child has brown eyes or is a brunette?
   (c) What is the probability that the child has brown eyes given that it is a brunette?
6. Suppose that you randomly select a 2-digit number (that is, one of 00, 01, 02, . . . , 99) from a sequence of random numbers obtained by repeatedly spinning a spinner. What is the probability that the number selected
   (a) is greater than 80?
   (b) is less than 10?
   (c) is a multiple of 3?
   (d) is even or is less than 50?
   (e) is even and is less than 50?
   (f) is even given that it is less than 50?
7. In a certain card game, you are dealt two cards face up. You then bet on whether a third card dealt is between the other two cards. (A 10 is between a 9 and a queen, and so on.) What is the probability of winning your bet if you are dealt
   (a) a 5 and a 7?
   (b) a jack and a queen?
   (c) a pair of 9s?
   (d) a 5 and a queen?
8. In playing draw poker, a flush is a hand with five cards all in one suit. You are dealt five cards and can throw away any of these and be dealt more cards to replace them. If you are dealt four hearts and a spade, what is the probability that you can

discard the spade and be dealt a heart to fill out your flush?

9. You are dealt three cards at random from an ordinary deck of playing cards. What is the probability that all three are hearts? (*Hint:* How many ways can you select three cards at random from the deck? How many ways can you select three cards at random from among the hearts?)

10. An urn contains 35 white beans. Red dots are placed on 24 of the beans.

(a) Perform the experiment of simultaneously selecting three beans at random 20 times and compute the empirical probability of obtaining one marked and two plain white beans.

(b) Compute the theoretical probability of selecting one marked *and* two plain white beans from the urn when three beans are selected simultaneously.

11. Consider a spinner made as in Figure 9.2 but marked and shaded as indicated here. The spinner is spun and grasped between your thumb and forefinger while it is still spinning rapidly.

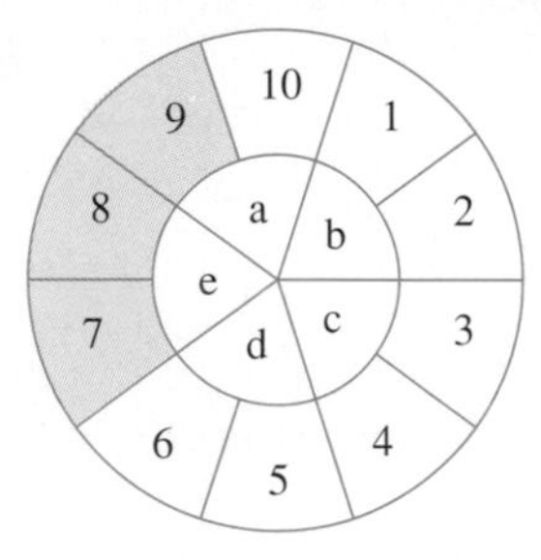

(a) What is the probability that the mark on your thumb is on the shaded area?

(b) What is the probability that your mark falls in regions 1 or 2?

(c) What is the probability that your mark falls in regions 10 or 6?

(d) What is the probability that your mark lands between the two radii that determine region e?

(e) What is the probability that the mark falls in region 8 given that it falls in the shaded area?

(f) What is the probability that it falls in a region marked by a vowel given that it falls in an odd numbered region?

(g) What is the probability that it falls in a region marked by a vowel or an odd number?

(h) Record the results of spinning the spinner in this problem 100 times and compute the empirical probabilities for each of (a) through (g) above.

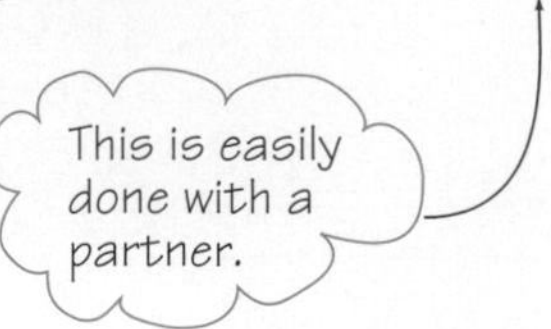

12. A dart board is marked as shown.

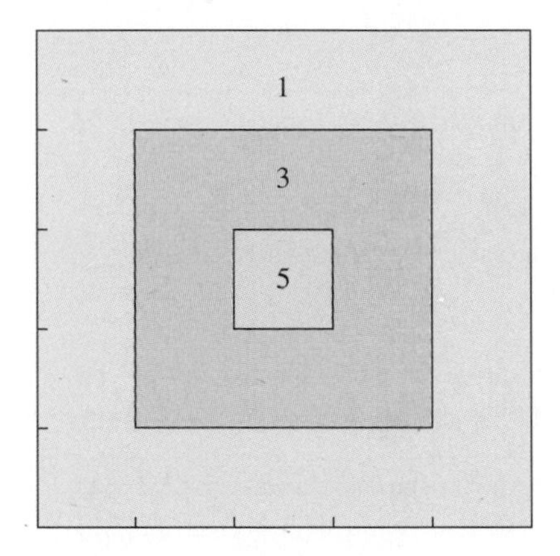

Josie is good enough that she always hits the dart board with her darts but, beyond that, the darts hit in random locations. If a single dart is thrown, compute these probabilities.

(a) $P(1)$ **(b)** $P(3)$ (c) $P(5)$

(d) If Josie wins the number of dollars indicated by the number of the region in which her dart falls, how much is her expected win (the expected value)?

(e) Suppose it costs Josie $2 each time she throws a dart. Should she play darts as in part (d)? Explain.

13. A dart board is marked like the spinner in problem 11. The radius of the inner circle is 1 and that of the outer circle is 2. What is the probability that a dart hitting the board at random

(a) hits in region b?

(b) hits in the inner circle but not in region b?

(c) hits in the inner circle but not in region b?

(d) hits in region b given that it hits in a, b, or c?

14. Two dice are thrown.

(a) What are the odds in favor of getting a score of 6?

(b) What are the odds against getting a 6?

(c) What is the probability of getting a 6?

15. If $P(A) = 2/5$, compute the odds in favor of A resulting from a single trial of an experiment.

16. If $P(A) = 1/2$, $P(B) = 1/3$, $P(C) = 1/6$, and A, B, and C are mutually exclusive, compute the odds in favor of A or C resulting from a single trial of an experiment.

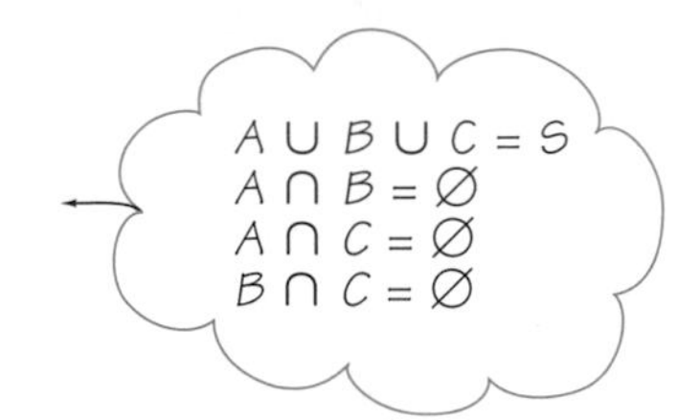

17. Compute the expected value of the score when rolling two dice.

---

18. A game consists of rolling a pair of dice. You win the amounts shown for rolling the score shown.

Roll	2	3	4	5	6	7	8	9	10	11	12
$ Won	4	6	8	10	20	40	20	10	8	6	4

Compute the expected value of the game.

Thinking Critically

19. Two balls are drawn at random from an urn containing six white and eight red balls.
(a) Compute the probability that both balls are white using combinations. Recall that

$$C(n, r) = \frac{n(n-1)(n-2)\cdots(n-r+1)}{r!}.$$

(b) Compute the probability that both balls are red.

20. An urn contains eight red, five white, and six green balls. Four balls are drawn at random.
(a) Compute P(all four are red).
(b) Compute P(exactly two are red and exactly two are green).
(c) Compute P(exactly two are red or exactly two are green).

21. Consider the set of all 5-letter code words without repetition of letters. (Recall that $P(n, r) = n(n-1)(n-2)\cdots(n-r+1)$.)
(a) What is the probability that a code word begins with the letter a?
(b) What is the probability that, in a code word, c is immediately followed by d?
(c) What is the probability that a code word starts with a vowel and ends with a consonant?
(d) In how many of the original set of 5-letter code words are c and d adjacent?

22. Six dice are rolled. What is the probability that all six numbers come up. (*Hint:* This can happen in more than one way. For example, 1, 2, 3, 4, 5, 6, and 6, 5, 1, 2, 3, 4 are just two of the possibilities.)

23. What is the probability that the six volumes of Churchill's *Second World War* appear in the correct order if they are randomly placed on a shelf?

24. What is the probability that a randomly dealt 5-card hand from a deck of playing cards will contain
(a) exactly two aces?
(b) at least two aces?

25. If seven dice are tossed, what is the probability that every number will appear? (*Hint:* In how many ways can the number that appears twice be chosen?)

26. (a) Show that $P(B \mid A) = \dfrac{P(B)P(A \mid B)}{P(A)}$
(*Hint:* Think of $P(A \text{ and } B)$.)
(b) Give a careful argument that

$$P(A) = P(B)P(A \mid B) + P(\overline{B})P(A \mid \overline{B}).$$

(*Hint:* Consider the Venn diagram shown.)

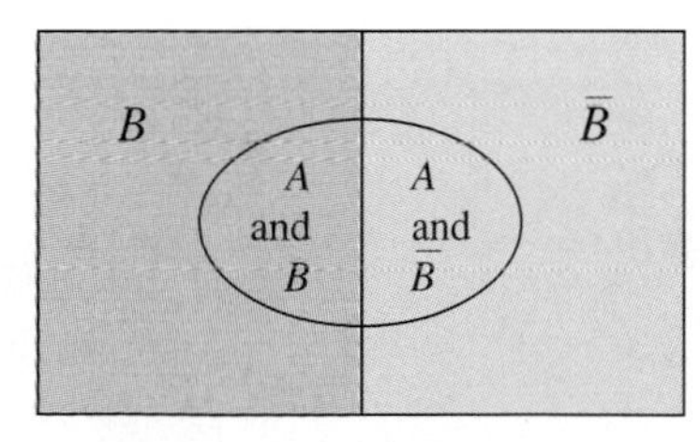

(c) Give a careful argument that

$$P(B \mid A) = \frac{P(B)P(A \mid B)}{P(B)P(A \mid B) + P(\overline{B})P(A \mid \overline{B})}.$$

You may know $P(A \mid B)$ but want to know $P(B \mid A)$. The results in (a) and (c) are two different forms of **Bayes' theorem** which makes it possible to compute reverse conditional probabilities.

Making Connections

27. On the basis of records kept in recent years it has been determined that the probability that a given person has AIDS is 0.015. Moreover, a test for detecting the disease has been developed that is 95% effective. That is, if T indicates a positive test, and $\overline{T}$ indicates a negative test, D indicates that a person has the disease, and $\overline{D}$ indicates that a person does not have the disease, then

$$P(T \mid D) = 0.95 \quad \text{and} \quad P(\overline{T} \mid \overline{D}) = 0.95.$$

Bayes' theorem, in problem 26, makes it possible to compute the important reverse conditional probabilities

$$P(D \mid T) \qquad \text{and} \qquad P(\overline{D} \mid \overline{T}),$$

the probability that a person who tests positive actually has the disease and the probability that a person who tests negative does not have the disease.

(a) Compute $P(D \mid T)$. (*Hint:* Identify T and D here with B and A respectively in problem 26, part (c).)

(b) Compute $P(\overline{D} \mid \overline{T})$. Recall that $\overline{\overline{T}} = T$ and $\overline{\overline{D}} = D$.

Communicating

28. After checking several sources in the library, write a concise two page essay on Sir R. A. Fisher's experiment on Ladies Tasting Tea. In particular, be sure to discuss Fisher's (then) new ideas of type 1 and type 2 errors.

For Review

29. In how many ways can you choose four marbles of the same color from an urn containing seven yellow and eight blue marbles?

30. In how many ways can you choose two yellow and three blue marbles from the urn of problem 29?

31. How many 5-letter code words can be made without repetition of letters if vowels and consonants must alternate?

32. How many different permutations are there of the letters a, a, a, a, b, b, c, c, c, c, c?

EPILOGUE Two Views of Probability

A seventeenth century Frenchman, the Chevalier de Méré, thought that the event of obtaining at least one 1 on four rolls of a die and the event of getting at least two 1s on 24 rolls of a pair of dice were equally likely. He reasoned as follows:

1st event:

- On one roll of a die, there is a $1/6$ chance of getting a 1.
- Therefore, on four rolls of a die, the chance of getting at least one 1 is $4 \cdot (1/6) = 2/3$.

2nd event:

- On one roll of a pair of dice the chance of a 2 is $1/36$.
- Therefore, on 24 rolls of a pair of dice the chance of at least one 2 is $24 \cdot (1/36) = 2/3$.

In each case, the chance was two-thirds. However, long experience with many trials showed that the first event was slightly more likely than the second and this difficulty became known as the Paradox of the Chevalier de Méré. De Méré asked his friend, the mathematician and philosopher, Blaise Pascal, about the problem and Pascal in turn sought the help of the jurist and amateur mathematician, Pierre de Fermat. Together, Pascal and Fermat were able to solve the problem showing that the first event occurs with probability 0.518 and that the second occurs with probability 0.491.

In fact, the study of probability substantially began with the French and in direct response to questions about games of chance, even though it had been considered briefly in the sixteenth century by such mathematicians as the Italian, Cardano. Today, probability and statistics play a critical role in society—in public opinion polls, in evaluating experimental data of all sorts, in quality control in manufacturing, in studying the behavior of atoms and of subatomic particles, and in countless other ways.

In this chapter we have considered the basic facts about probability. As revealed in the little vignette about the Chevalier de Méré, we have seen that there are two different but related notions of probability—empirical probability and theoretical probability. The first view of probability is that chance can be measured on the basis of long run experience. It was the experience of the Chevalier that event number one was slightly more likely than event number two even though his analysis suggested otherwise. The assertion based on experience was an expression of empirical probability—namely

$$P_e(\text{event number one}) > P_e(\text{event number two}).$$

The paradox was cleared up by Pascal and Fermat using strict mathematical principles that formed the genesis of theoretical probability. Much of the thrust of this chapter has been to show that these two notions are closely intertwined. Recall that the law of large numbers asserts that the empirical probability of an event more and more closely approaches the theoretical probability of the event as the number of trials is made larger and larger.

We have given attention to developing the ideas of probability in both senses and have seen in Chapter 8 how the interpretation of data, based on probability principles, can lead to surprisingly accurate conclusions. Without this kind of careful analysis, however, action based on collected data can lead to disastrous results.

CLASSIC CONUNDRUM Should the Contestant Switch?

On a popular TV game show the contestant is asked to select one of three doors. Behind one door is a very rich prize. Behind one door is a so-so, but not very valuable, prize. Behind the third door is a joke prize. After the contestant makes a selection but before the prize is revealed, the show host always opens a different door which reveals one of the two lesser prizes. He then asks the contestant if he or she would like to switch from the door originally chosen and pick the remaining unopened door instead. Should the contestant switch?

CHAPTER 9 SUMMARY

Key Concepts

The thrust of this chapter has been to develop the basic ideas of empirical probability, methods of counting, and theoretical probability.

It is convenient to think of an event as the result of an experiment:

- The empirical probability of an event is the fraction of times it occurs among a large number of trials.
- The theoretical probability of an event is the ratio of the number of times the event occurs in a sample space of equally likely outcomes to the number of outcomes in the sample space.
- The law of large numbers asserts that the empirical probability of an event more and more closely approximates the theoretical probability as the number of repetitions of the experiment becomes larger and larger.

Computing the theoretical probability of an event necessitated developing certain methods of counting:

- $n(A \text{ or } B) = n(A) + n(B) - n(A \cap B)$
  If $A \cap B = \emptyset$, $n(A \text{ or } B) = n(A) + n(B)$.
- If A is a set and B is a subset of r elements of A, then B is called a combination of r elements of A.
- If A is a set and B is an ordered subset of r elements of A, then B is called a permutation of r elements of A.
- For $n \geq 1$, the notation $n!$, read n factorial, is defined by $n! = n(n-1)(n-2) \cdots 1$. Also, $0! = 1$.
- The number of combinations of n things taken r at a time is
$$C(n, r) = \frac{n(n-1)(n-2) \cdots (n-r+1)}{r!}.$$
- The number of permutations of n things taken r at a time is
$$P(n, r) = n(n-1)(n-2) \cdots (n-r+1).$$
- If $n = a_1 + a_2 + \cdots + a_r$ and a_1 things are alike, a_2 other things are alike, . . . , and a_r other things are alike, then the number of permutations of all these things is
$$\frac{n!}{a_1!\, a_2! \cdots a_r!}.$$

The properties of theoretical probability correspond closely to the properties of counting. We consider only a finite sample space S of equally likely outcomes:

- $P(A) = 0$ if, and only if, A cannot occur.
- $P(A) = 1$ if, and only if, A must occur.
- For any A, $0 \leq P(A) \leq 1$.
- If A and $\overline{A}$ are complementary events so that $A \cup \overline{A} = S$ and $A \cap \overline{A} = \emptyset$, then $P(A) = 1 - P(\overline{A})$.
- $P(A \text{ or } B) = P(A) + P(B) - P(A \cap B)$. If A and B are mutually exclusive, then $P(A \text{ or } B) = P(A) + P(B)$.
- $P(A \text{ and } B) = P(B)P(A \mid B)$. If A and B are independent, then $P(A \text{ and } B) = P(A) \cdot P(B)$.

Vocabulary and Notation

Section 9.1

Empirical probability, P_e
Theoretical probability, mathematical probability, probability
Sample space, S
The law of large numbers
Mortality tables
Empirical probability and geometry
Simulation

Section 9.2

First principle of counting
Mututally exclusive events
Second principle of counting
Possibility tree
Third principle of counting
Combinations
Permutations
Permutations of like things

Section 9.3

Theoretical probability
Sample space, S
Outcome
Event, E
n factorial, $n!$
Probability of an event, $P(E)$
Equally likely
Conditional probability, $P(A \mid B)$
Complementary events, E and $\overline{E}$
Odds
Expected value
Geometric probability

CHAPTER REVIEW EXERCISES

Section 9.1

1. Toss four coins 20 times and determine the empirical probability of obtaining three heads and one tail.
2. (a) Roll three dice 20 times and determine the empirical probability of obtaining a total score of 3 or 4.
   (b) From the data in part (a) determine the empirical probability of obtaining a score of at least 5.
3. From the data of problem 2, determine
   $P_e(5 \text{ or } 6 \text{ or } 7 \mid 5 \text{ or } 6 \text{ or } 7 \text{ or } 8 \text{ or } 9)$.
4. Conduct a survey of 20 randomly chosen college students at your college or university and determine the empirical probability that chocolate is the favorite flavor of ice cream.
5. (a) Drop five thumbtacks on a table top 20 times and determine the empirical probability that precisely three of the tacks land point up.
   (b) From the data of part (a) determine the empirical probability that two or three of the five tacks in part (a) land point up.
6. (a) A die is rolled repeatedly until a 5 or a 6 appears. Perform this experiment ten times and estimate the number of rolls required.
   (b) From the data in part (a), compute the empirical probability that it takes precisely five rolls to obtain a 5 or 6 for the first time.
7. Shuffle a deck of cards and select a card at random. Return the card to the deck, shuffle, and draw again for a total of 20 trials.
   (a) Compute $P_e(\text{ace or heart})$.
   (b) Compute $P_e(\text{ace and heart})$.
   (c) Compute $P_e(\text{ace} \mid \text{heart})$.
8. Using the mortality table, Table 9.1, compute P_e(a 20 year old will live to be 90 years old).

Section 9.2

9. (a) Three coins are tossed. Make an orderly list of all possible outcomes.
   (b) In how many ways can you obtain two heads and one tail?
10. (a) How many ways can you select nine players for a baseball team from among 15 players if any player can play any position?
    (b) How many ways can you select the team in part (a) if only two players can pitch and only three others can catch? Note that these five players can also play all other positions.
11. (a) All of the 90 students in Ferry Hall speak at least one of French, English, or German. If 38 speak English, 24 speak French and English, 27 speak German and English, and 17 speak German, French, and English, how many speak German or French?
    (b) How many of the students in part (a) speak French and English but not German?
12. (a) How many ways can you select two clubs from an ordinary deck of playing cards?
    (b) How many ways can you select two face cards from an ordinary deck of playing cards?
    (c) How many ways can you select two clubs or two face cards from an ordinary deck of playing cards?
13. (a) How many 5-letter code words can be made if repetition of letters is not allowed?
    (b) How many of the 5-letter code words in part (a) start and end with vowels?
    (c) How many of the code words of part (a) contain the 3-letter sequence aef?
14. (a) In how many ways can the letters in STREETS be placed in recognizably different orders?
    (b) In how many of the orderings of part (a) are the two Es adjacent?
    (c) How many of the orderings in part (a) begin with a T?

Section 9.3

15. (a) Explicitly list all elements in the sample space if two coins and a die are tossed.
    (b) Compute $P(\text{T, T, 5})$, the probability of getting two tails on the coins and 5 on the die in part (a).
16. Compute $P(5 \mid \text{T, T})$, the probability of obtaining 5 on the die given that both coins came up tails in problem 15.
17. Compute the probability of obtaining a sum of at most 11 on a single roll of two dice.
18. An urn contains five white, six red, and four black balls. Two balls are chosen at random.
    (a) What is the probability that they are the same color?
    (b) What is the probability that both are white?
    (c) What is the probability that both are white given that they are the same color?
19. Consider a spinner made as shown in Figure 9.2 but marked as indicated here. Compute these probabilities.

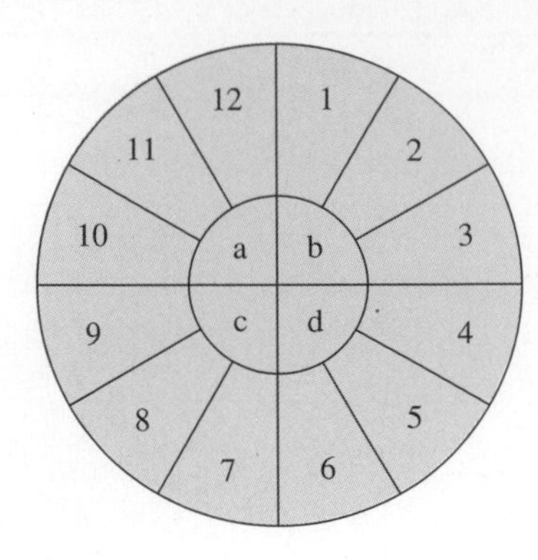

(a) $P(\text{b and } 8)$ (b) $P(\text{b or } 8)$ (c) $P(\text{b} \mid 8)$
(d) $P(\text{b and } 2)$ (e) $P(\text{b or } 2)$ (f) $P(2 \mid \text{b})$

20. Three coins are tossed.
    (a) What are the odds in favor of getting two heads and one tail?
    (b) What are the odds in favor of getting three heads?

21. (a) If $P(A) = 0.85$, what are the odds in favor of A occurring on any given trial?
    (b) If the odds in favor of A are 17 to 8, determine $P(A)$.

22. You play a game where you win the amount shown with the probability shown.

    $P(\$5) = 0.50 \quad P(\$10) = 0.25 \quad P(\$20) = 0.10$

    (a) What is the expected value of the game?
    (b) If it costs you $10 to play the game of part (a), is it wise to play? Explain.

23. A census taker was told by a neighbor that a family of five lived in the next house—two parents and three children. When the census taker visited the house, he was greeted by a girl. What is the probability that the other two children were both boys? Explain briefly.

CHAPTER TEST

1. We claim that the probability is about 0.6 that a person shown a card with the numbers 1, 2, 3, 4 printed on it will, when asked, choose the number 3. How can such a probability be calculated? Explain briefly.

1 2 3 4

2. What kind of probability is used when an assertion like, "the probability that penicillin will cure a case of strep throat is 0.9" is made? Explain briefly.

3. Calculate each of the following.
   (a) $7!$ (b) $\dfrac{9!}{6!}$ (c) $\dfrac{8!}{(8-8)!}$ (d) $7 \cdot 6!$
   (e) $P(8, 5)$ (f) $P(8, 8)$ (g) $C(9, 3)$
   (h) $C(9, 9)$

4. In Mrs. Spangler's calculus class all of the students are also studying one or more foreign languages. If
   27 students study French,
   29 students study German,
   17 students study Chinese,
   12 students study German and French,
   3 students study German and Chinese,
   2 students study French and Chinese,
   1 student studies French, German, and Chinese
   (a) How many students are in Mrs. Spangler's class?
   (b) How many students in the class study Chinese only?
   (c) How many students study French and German but not Chinese?

5. An urn contains ten distinct yellow, four distinct blue, and eight distinct green marbles. Five marbles are selected.
   (a) In how many ways can one select five green marbles?
   (b) In how many ways can one select five yellow and five green marbles?
   (c) In how many ways can one select five yellow or five green marbles?

6. If you select five marbles from the urn in problem 5, what is the probability that two are yellow given that three are green?

7. What are the odds in favor of selecting a yellow marble if a single marble is drawn from the urn described in problem 5?

8. If $P(E) = 0.35$, what are the odds of obtaining E on a single trial of an experiment?

9. How many 4-letter code words can be made using the letters a, b, c, d, e, f, and g
   (a) with repetition allowed?
   (b) with repetition not allowed?

10. How many of the code words in problem 9, part (b),
    (a) begin with a vowel?
    (b) have c and d adjacent?

10 Geometric Figures

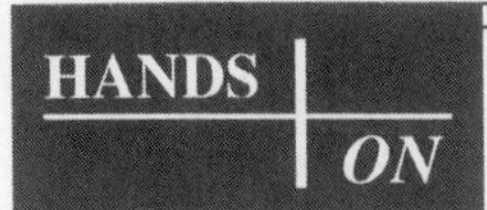

Exploring Polygons

Materials Needed

1. Colored pencils, ruler (in millimeters), protractor, unlined paper.
2. A set of 16 equilateral triangles and squares cut from patterns photocopied onto card stock. Each set consists of two of each of the following shapes. The segments drawn on the figures meet the sides at their midpoints. A small hole is punched at the center of each shape where the lines cross.

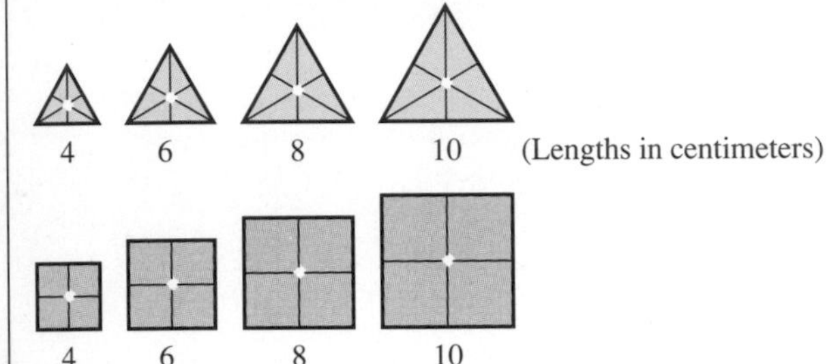

Directions

The cardstock triangles and squares will be arranged on a sheet of blank paper to form a new figure. Your goal is to discover, describe, and explore any orderly patterns and unexpected relationships that you see in the figures.

Exploring Quadrilaterals

Choose any four card stock squares and arrange them corner to corner, as on the left below. Place pencil points at the corners *A, B, C, D*, at the midpoints *M, N, P, Q*, and at the centers *W, X, Y, Z*, of the squares. Then connect the points with line segments as shown in the figure at the right.

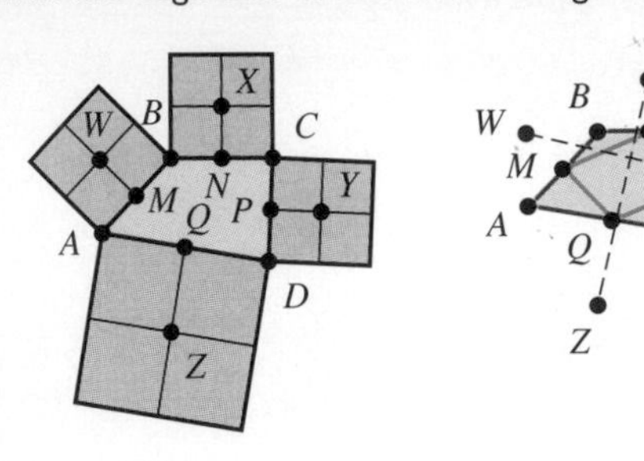

1. Does anything seem special about the quadrilateral (four-sided polygon) *MNPQ*?
2. Compare the line segment $\overline{XZ}$ (the one with endpoints *X* and *Z*) with the line segment $\overline{YW}$.
3. Suppose that the squares are chosen so that sides $\overline{AB}$ and $\overline{CD}$ are the same length, and $\overline{BC}$ and $\overline{AD}$ are the same length. What seems to be special about *WXYZ*?

Exploring Triangles

Choose any three card stock triangles, place them corner to corner, and locate the points forming the right hand figure.

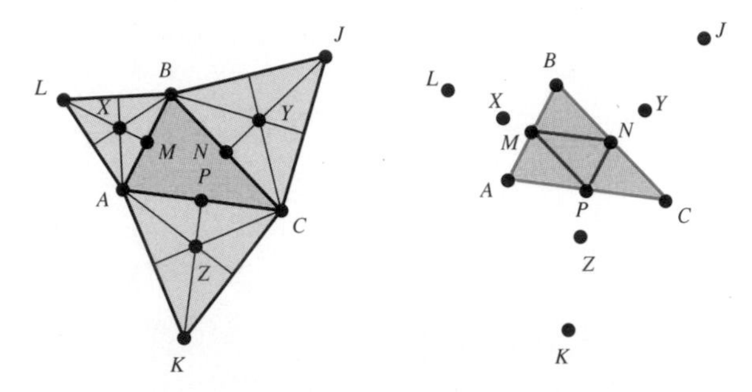

1. Compare the shapes of the triangles *ABC, AMP, BMN, CPN*, and *MNP*.
2. What is special about triangle *XYZ*?
3. Draw the line segments $\overline{AN}$, $\overline{BP}$, and $\overline{CM}$. What happens?
4. Draw the line segments $\overline{AJ}$, $\overline{BK}$, and $\overline{CL}$. What happens?
5. Draw the line segments $\overline{AY}$, $\overline{BZ}$, and $\overline{CX}$. What happens?

CONNECTIONS

The Nature of Informal Geometry

This chapter is the first of five dealing with topics in geometry which have importance to the teacher of elementary school mathematics. To understand the approach in these chapters it is necessary to know what we mean by geometry—and what we do not mean. Geometry for us means the informal study of shape. The three words in this description—*informal, shape,* and *study*—warrant some discussion.

Informal. Until about 600 B.C. geometry was pursued in response to practical, artistic, and religious needs. The pyramids of ancient Egypt (c. 2500 B.C.) and the Stonehenge observatory (c. 2800 B.C.) provide evidence that complex problems of form and measurement were solved even in Neolithic times. Over the centuries people built and interacted with a variety of patterns, objects, and structures. The shapes that recurred most often were named and some of their properties were discovered. Considerable knowledge of geometry was accumulated, but mathematics was not yet an organized and independent discipline, and the ideas of proof and deduction were still absent.

In the period 600–300 B.C., Thales, Pythagoras, Zeno, Eudoxus, Euclid and others organized this accumulated knowledge and experience, and transformed geometry into a theoretical science. Utilitarian considerations gave way to abstraction and general methods. With Euclid's *Elements,* geometry became a formal axiomatic system in which geometric theorems were deduced logically from a list of axioms. To many people, geometry is restricted to a Euclidean formalism in which exacting standards of proof and logical development must be met.

In this text, however, we return to learning by trusting our intuition and experience. Geometric facts are discovered by explorations with pictorial representations and physical models. Little attention will be given to the overall logical structure. However there will be many opportunities to verify patterns and conjectures by examining the consequences of properties and facts that have already been accepted.

Shape. Marjorie Senechal* points out that "shape" is an undefinable term, partly because we must leave room for new shapes as they are discovered. For example, fractals and CAT scan images are shapes of current interest and importance, made possible by a combination of mathematics and computers. For us, the shapes of interest will most often be figures such as polygons and curves, cubes and spheres, which are familiar in classical geometry.

Study. In common with geometers of ancient times, our goals are: to recognize differences and similarities among shapes; to analyze the properties of a shape or class of shapes; and, to model, construct, and draw shapes in a variety of ways. These goals are inseparably intertwined, but it will be seen that the discussion follows three threads of development: *classification, analysis,* and *representation.*

Materials for Explorations

Many examples will be presented in the form of an *exploration.* First, you will represent a shape, perhaps with a drawing or physical model, that satisfies the stated conditions. You are next asked to discover, analyze, and describe the properties of the shape. Often you will not want to read further until you have followed the directions and made some discoveries for yourself; only then should you read on to see if the patterns and relationships you have uncovered agree with those discussed in the text.

The following tools and materials will be useful to draw, construct, or create the shapes you will explore:

- colored pencils
- ruler (best if marked in millimeters)
- compass (be sure it is of good quality)
- tape
- glue
- unlined paper

---

* See "Shape," a chapter of *On the Shoulders of Giants: New Approaches to Numeracy,* L. A. Steen, ed., National Academy Press, Washington D. C., 1990.

- protractor
- drafting triangles (30°–60°–90° and 45°–45°–90°)
- dot paper in both square and triangular patterns
- scissors
- graph paper

A variety of manipulatives are available from commercial suppliers and are of great value in the study of geometry. Hopefully, you will have access to such items as:

- geoboards
- tangrams
- pattern blocks
- geometric solids (wood or plastic)
- pentominoes
- Mira®

Geometry with Computers

The computer has become an increasingly useful adjunct to the study of geometry. An introduction to Logo turtle graphics is found in Appendix C, and provides sufficient background to pursue the examples, activities, and problems that are found in many of the geometry sections. A more recently developed type of

FROM THE NCTM STANDARDS

Geometry and Spatial Sense in K–4

Geometry is an important component of the K–4 mathematics curriculum because geometric knowledge, relationships, and insights are useful in everyday situations and are connected to other mathematical topics and school subjects. Geometry helps us represent and describe in an orderly manner the world in which we live. Children are naturally interested in geometry and find it intriguing and motivating; their spatial capabilities frequently exceed their numerical skills, and tapping these strengths can foster an interest in mathematics and improve number understandings and skills.

Spatial understandings are necessary for interpreting, understanding, and appreciating our inherently geometric world. Insights and intuitions about two- and three-dimensional shapes and their characteristics, the interrelationships of shapes, and the effects of changes to shapes are important aspects of spatial sense. Children who develop a strong sense of spatial relationships and who master the concepts and language of geometry are better prepared to learn number and measurement ideas, as well as other advanced mathematical topics.

In learning geometry, children need to investigate, experiment, and explore with everyday objects and other physical materials. Exercises that ask children to visualize, draw, and compare shapes in various positions will help develop their spatial sense. Although a facility with the language of geometry is important, it should not be the focus of the geometry program but rather should grow naturally from exploration and experience. Explorations can range from simple activities to challenging problem-solving situations that develop useful mathematical thinking skills.

Evidence suggests that the development of geometric ideas progresses through a hierarchy of levels. Students first learn to recognize whole shapes and then to analyze the relevant properties of a shape. Later they can see relationships between shapes and make simple deductions. Curriculum development and instruction must consider this hierarchy because although learning can occur at several levels simultaneously, the learning of more complex concepts and strategies requires a firm foundation of basis skills.

SOURCE: From *Curriculum and Evaluation Standards for School Mathematics,* Grades K–4, p. 48.

software called an *automatic drawer* is described in Appendix D. If available, many of the construction and measurement problems can be pursued with an automatic drawer. Doing so adds a lively dynamic aspect to the study of shapes and their properties. Problems especially suitable for computer graphics are included in the *Using a Computer* section of the problem sets.

10.1 Figures in the Plane

The shapes in the Picture Gallery of Plane (p. 610) and Space Figures (p. 678) are each highly complex when viewed as a whole, but underlying this complexity is an orderly arrangement of simpler parts. In this section we consider the most basic shapes of geometry: points, lines, segments, rays, and angles.

Points and Lines

A point on paper is represented by a dot. A point on a television monitor is represented by a small rectangle of phosphors called a pixel that glows when excited by a beam of electrons. Neither a dot nor a pixel is an exact representation of a geometric point. In the mind's eye, dots and pixels are decreased in size until they become ideal **points,** that is, just locations in space. On paper we still draw dots to represent points, and we label the points with upper case letters as shown.

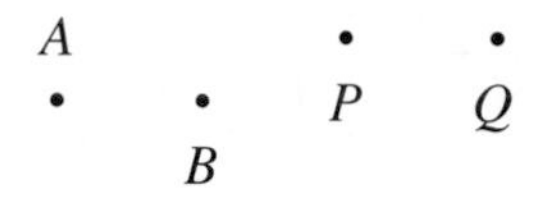

Line, like point, is undefined but its meaning is suggested by a tightly stretched thread, or a laser beam, or the edge of ruler. We assume that any two points determine one and only one line which contains the two points. Lines will often be denoted with lower case letters such as l and m. If A and B are two points then the line through A and B is denoted by $\overleftrightarrow{AB}$. On paper, lines can be drawn with either a ruler or a **straightedge.** A straightedge is like a ruler but without any marks on it.

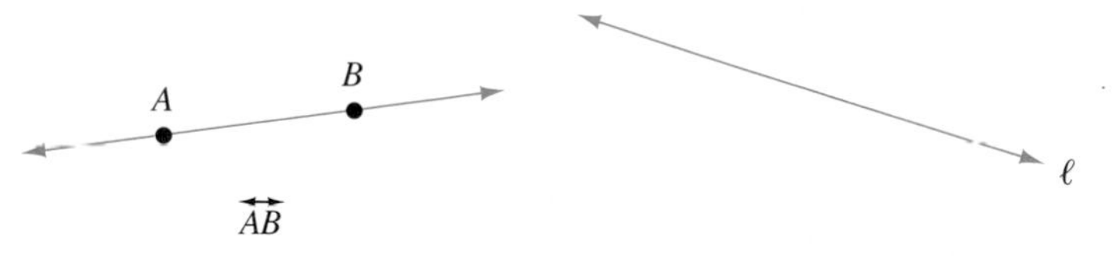

The arrows in the drawings and in the notation $\overleftrightarrow{AB}$ indicate that lines extend infinitely far in two directions.

Three or more points usually determine several lines, but if they lie on just one line then we say the points are **collinear** as shown in Figure 10.1.

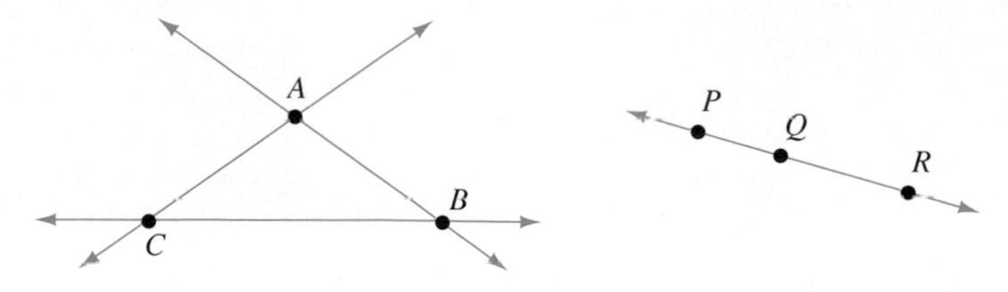

Figure 10.1
Three points determine either three lines or one line.

A Picture Gallery of Plane Figures

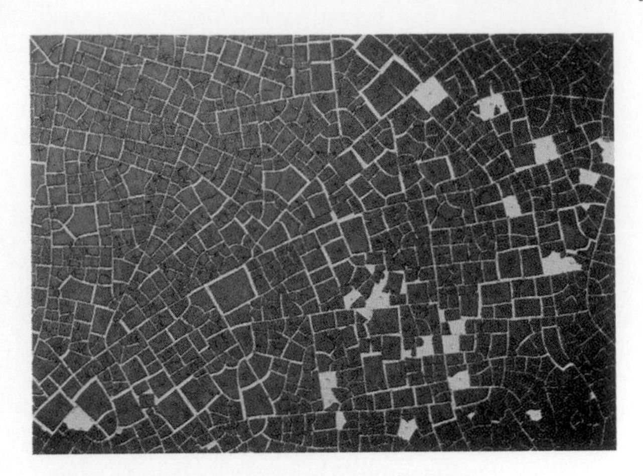

Fissures in a gelatinous preparation of tin oil

Butterfly wings

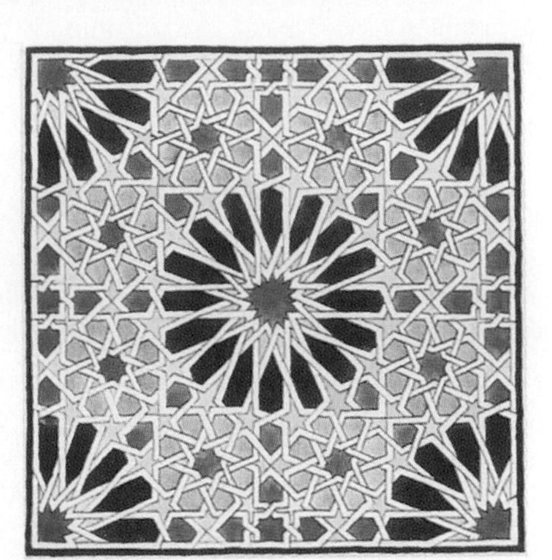

M. C. Escher's sketch of a wall mosaic in the Alhambra

A snow crystal

A fractal, an example of a complex beautiful image created with the computer.

The pattern of seeds in the head of a sunflower.

Three noncollinear points determine a **plane,** which is yet another undefined term used to describe a set of points which idealize a flat space such as a table top. In this section we consider only sets of points which belong to a single plane. Subsets of a plane are called **plane figures** or **plane shapes.** In a later section we explore solid shapes in which not all of the points belong to a single plane.

Two lines in the plane which do not have a point in common are called **parallel.** We write $l \parallel m$ if l and m are parallel lines. Two distinct lines p and q which are not parallel must have a single point in common, called their **point of intersection,** and we write $p \nparallel q$. Three lines can intersect at 0, 1, 2, or 3 points. In the case of one point of intersection, the three lines are said to be **concurrent.** A line which intersects two other lines is called a **transversal.** The different ways in which two or three lines can be arranged in the plane are shown in Figure 10.2.

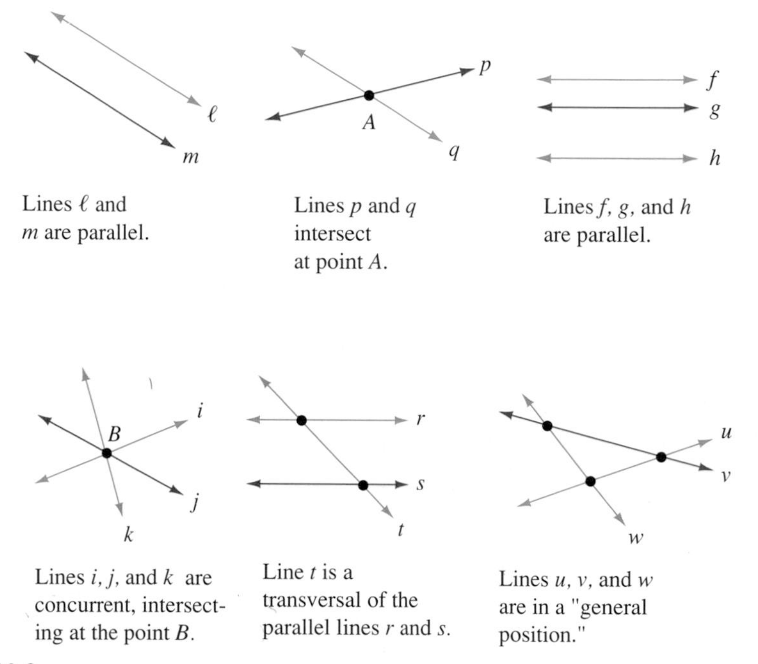

Figure 10.2
The possible arrangements of two and three lines in the plane.

EXAMPLE 10.1 Exploring Collinearity and Concurrency

(a) Draw three circles, C_1, C_2, and C_3 of different size, with no circle containing or intersecting another circle. Then draw two lines externally tangent to the circles C_1 and C_2 as shown, and let P denote the point of intersection.

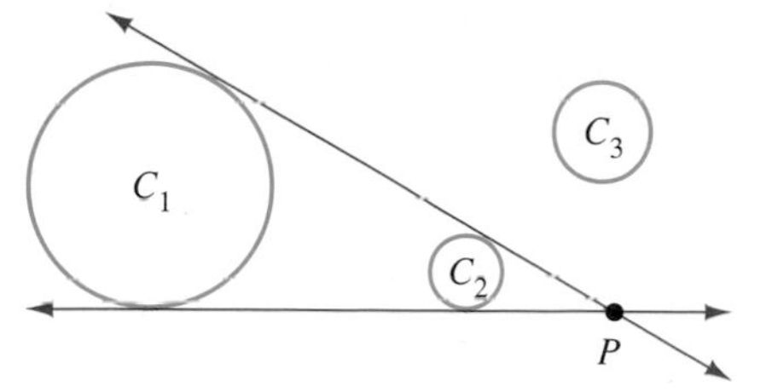

In the same way use your ruler to draw the two lines externally tangent to C_2 and C_3, and let Q be their point of intersection. Finally, draw the external tangents of C_1 and C_3, and let R be their point of intersection. What conjecture do you have concerning P, Q, and R?

(b) Draw a circle. Then draw any three lines that are tangent to the circle, at points labeled X, Y, and Z, and which intersect in pairs at the points labeled A, B, and C. Finally, draw the lines $\overleftrightarrow{AX}$, $\overleftrightarrow{BY}$, and $\overleftrightarrow{CZ}$. What conjecture can you make about $\overleftrightarrow{AX}$, $\overleftrightarrow{BY}$, and $\overleftrightarrow{CZ}$?

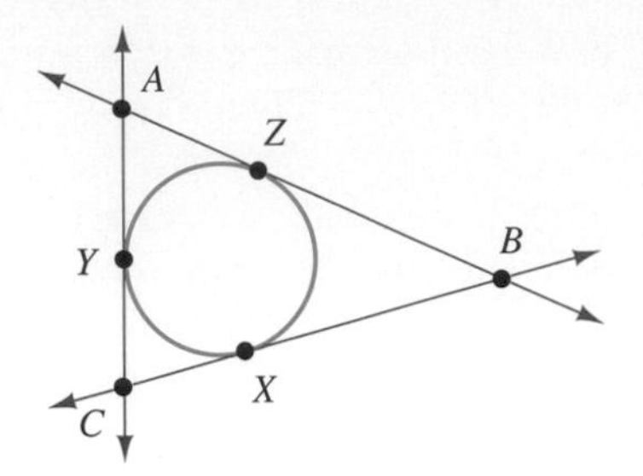

SOLUTION

(a) P, Q, and R are collinear

(b) $\overleftrightarrow{AX}$, $\overleftrightarrow{BY}$, $\overleftrightarrow{CZ}$ are concurrent. ■

Line Segments and the Distance Between Points

Let A and B be any two points. The line $\overleftrightarrow{AB}$ will be viewed as a copy of the number line. That is, every point on $\overleftrightarrow{AB}$ corresponds to a unique real number and every real number corresponds to a unique point on $\overleftrightarrow{AB}$. If A and B correspond to the real numbers x and y respectively, then the absolute value, $|x - y|$, gives the **distance** between A and B. We denote this distance by AB.

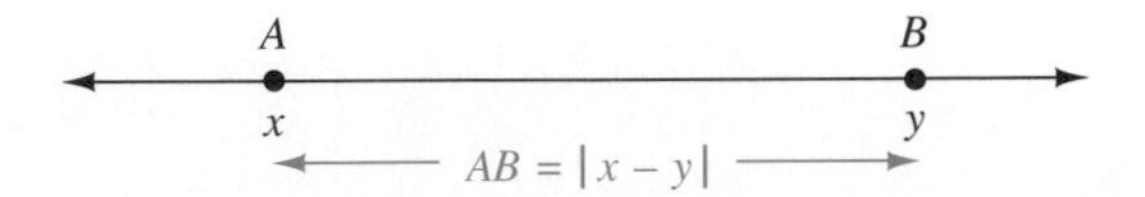

The points on the line $\overleftrightarrow{AB}$ which are between A and B, together with A and B themselves, form the **line segment** $\overline{AB}$. Points A and B are called the **endpoints** of $\overline{AB}$ and the distance AB is the **length** of $\overline{AB}$. It is important to see that the overbar used in the notation distinguishes the real number AB from the line segment $\overline{AB}$.

Two segments $\overline{AB}$ and $\overline{CD}$ are said to be **congruent** if they have the same length. This is symbolized by writing $\overline{AB} \cong \overline{CD}$. Thus, $\overline{AB} \cong \overline{CD}$ if, and only if, $AB = CD$. The point M in $\overline{AB}$ which is the same distance from A and B is called the **midpoint** of $\overline{AB}$. This is summarized in Figure 10.3.

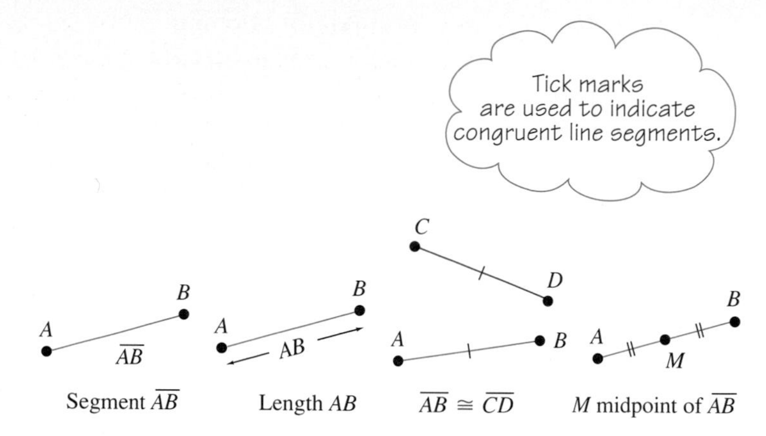

Figure 10.3
A segment $\overline{AB}$, its length AB, congruent segments, and the midpoint M of $\overline{AB}$

JUST FOR FUN

Arranging Points at Integer Distances

It is easy to find three points A, B, C, not all on one line, that are at integer distance from one another. The equilateral triangle with three sides of length one is the smallest example. Four points F, O, U, R, no three of which are collinear, determine the six segments as shown.

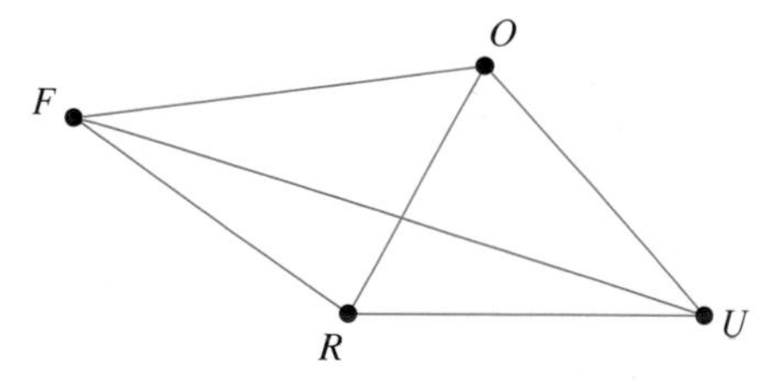

Can you arrange the four points so that all six segments are of integer length? Try experimenting with six straws, cut to lengths of 2″, 2″, 3″, 4″, 4″, 4″. The following arrangement of six points is remarkable because all 15 segments between pairs of points have integer length.

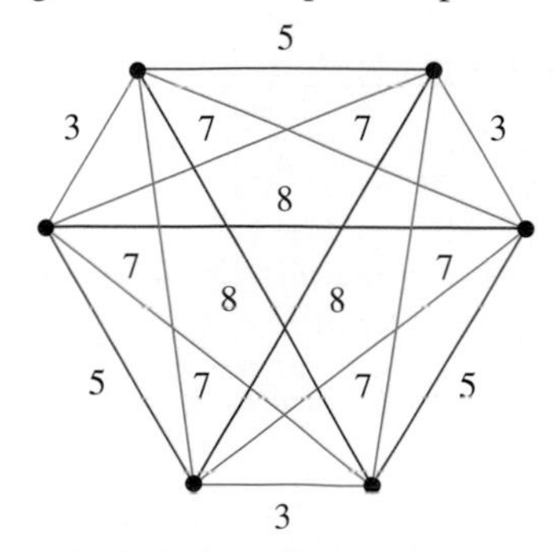

Rays, Angles, and Angle Measure

A **ray** is a subset of a line which contains a point P, the **endpoint** of the ray, and all points on the line lying to one side of P. If Q is any point of the ray other than P, then $\overrightarrow{PQ}$ denotes the ray. The union of two rays with a common endpoint is an **angle.** If the rays are $\overrightarrow{AB}$ and $\overrightarrow{AC}$ then the angle is denoted by $\angle BAC$. The common endpoint of the two rays is called the **vertex** of the angle and is the middle letter in the symbol for angle. The points B and C not at the vertex can be written in either order, so that $\angle CAB$ denotes the same angle so $\angle BAC$. The rays $\overrightarrow{AB}$ and $\overrightarrow{AC}$ are called the **sides** of the angle. See Figure 10.4.

An angle whose sides are not on the same line separates the remaining points of the plane into two parts, the **interior** and the **exterior** of the angle. The points along a line segment which joins an endpoint on side $\overrightarrow{AB}$ to an endpoint on $\overrightarrow{AC}$ are all interior points of $\angle BAC$.

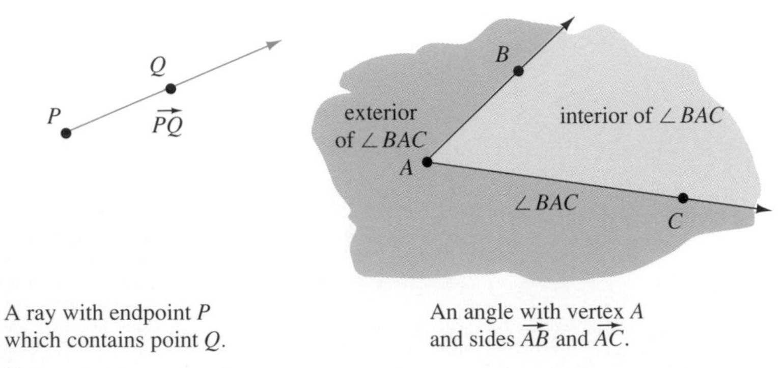

A ray with endpoint P which contains point Q.

An angle with vertex A and sides $\overrightarrow{AB}$ and $\overrightarrow{AC}$.

Figure 10.4
A ray $\overrightarrow{PQ}$ and an angle $\angle BAC$

If $\angle BAC$ is the only angle with its vertex at A it is common to write $\angle A$ in place of $\angle BAC$. When more than one angle has a vertex at A it is essential to use the full three letter symbol. Sometimes it is useful to number the angles which appear in a drawing and refer to $\angle 1$, $\angle 2$, $\angle 3$, and so on.

The size of an angle is measured by the amount of rotation required to turn one side of the angle to the other by pivoting about the vertex. The **measure of an angle** is generally given in **degrees,** where there are 360° in a full revolution. The measure

WORTH READING

A First-person Account of Life on a Plane

"I call our world Flatland, not because we call it so, but to make its nature clearer to you, my happy readers, who are privileged to live in Space.

Imagine a vast sheet of paper on which straight Lines, Triangles, Squares, Pentagons, Hexagons, and other figures, instead of remaining fixed in their places, move freely about, on or in the surface, but without the power of rising above or sinking below it, very much like shadows — only hard and with luminous edges —and you will then have a pretty correct notion of my country and countrymen. Alas, a few years ago, I should have said "my universe": but now my mind has been opened to a higher view of things."

More than a century ago, scholar and theologian Edwin Abbott anonymously published this delightful, satiric tale of the inhabitants of a two-dimensional world. Isaac Asimov called it "to this day, the best introduction one can find into the manner of perceiving dimensions."

SOURCE: Edwin Abbott Abbott, *Flatland: A Romance of Many Dimensions.* New York: Barnes & Noble, Inc., 1983, p. 3.

of $\angle A$ is denoted by $m(\angle A)$. If the rotation is imagined to pass through the interior of the angle, the measure is a number between 0° and 180°. Unless stated otherwise, $m(\angle A)$ is the measure of $\angle A$ not larger than 180°. In some applications the measure of interest corresponds to the rotation through the exterior of the angle and is therefore a number between 180° and 360°.

An angle with measure greater than 180° but less than 360° is called a **reflex angle.** An angle of measure 180° is a **straight angle,** an angle of 90° measure is a **right angle,** and an angle of measure 0° is a **zero angle.** Angles measuring between 0° and 90° are **acute,** and angles measuring between 90° and 180° are **obtuse.** The classification of angles according to their measure is summarized in Figure 10.5.

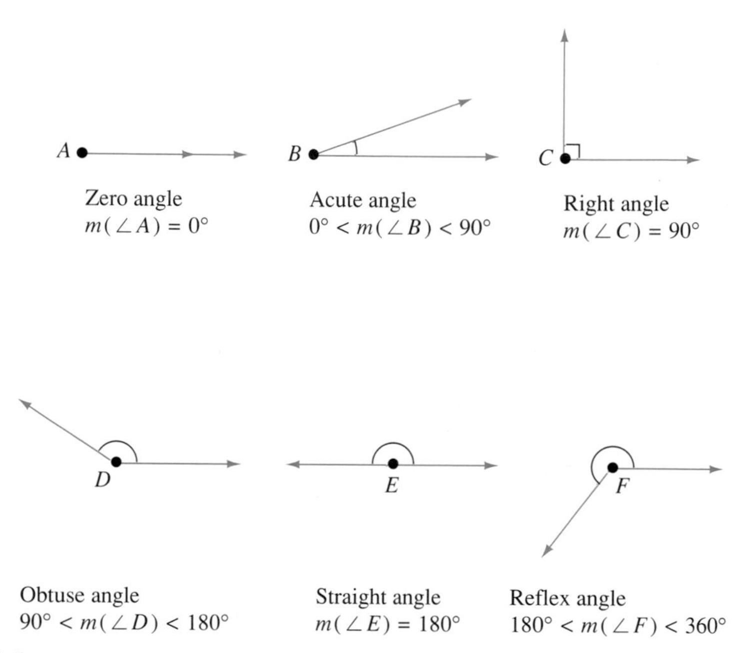

Figure 10.5
The classification of angles by their measure

Right angles in drawings are indicated by a small square placed at the vertex. A circular arc is required to indicate reflex angles.

Two lines l and m that intersect at right angles are called **perpendicular** lines. This is indicated in writing by $l \perp m$. Similarly two rays, or two segments, or a segment and a ray, are perpendicular if they are contained in perpendicular lines.

Two angles are **congruent** if, and only if, they have the same measure. The symbol $\cong$ is used to denote the congruence of angles. Thus,

$$\angle P \cong \angle Q \text{ if, and only if, } m(\angle P) = m(\angle Q).$$

The **protractor** is used both to measure angles and to draw angles having a given measure. The protractor and other traditional tools useful for drawing and measuring are shown in Figure 10.6. Increasingly these tools are being supplemented and replaced by the availability of automatic drawing software for the computer.

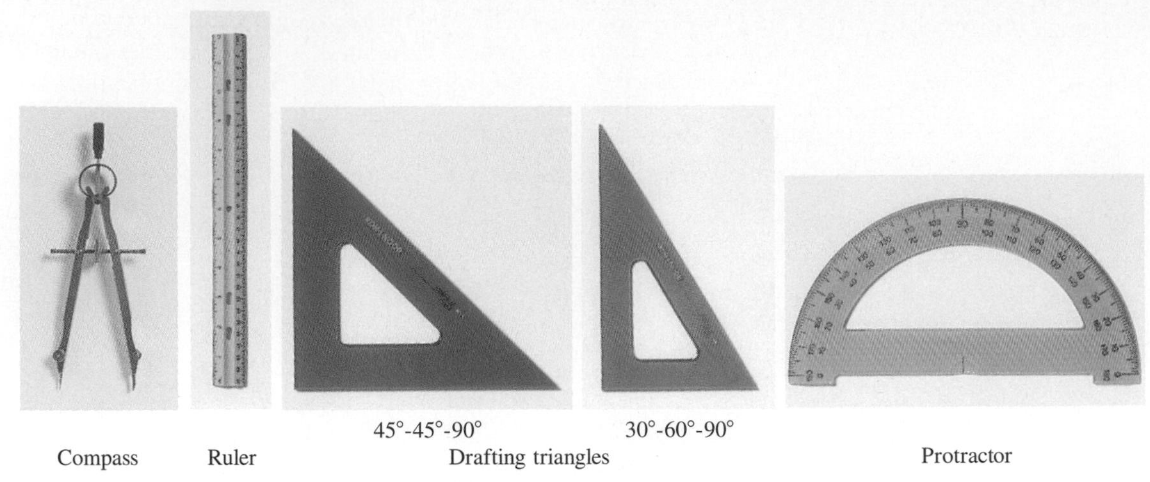

Figure 10.6
Some useful tools for measuring and drawing geometric figures

EXAMPLE 10.2

Exploring Angles and Distances in a Circle

(a) Draw a circle and choose any point P, other than the center, inside the circle. Any line l through P intersects the circle in two points, say A and B. Measure the distances PA and PB (to the nearest millimeter) and compute the product $PA \cdot PB$. Draw several lines through P and measure the distances of the two segments. Which line through P makes the product of distances, $AP \cdot PB$, as large as possible?

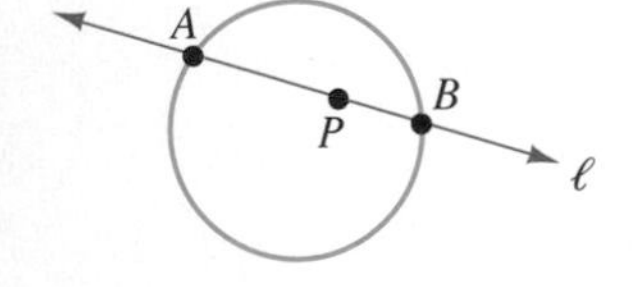

(b) Draw a circle with center C. Draw a line through C, and let A and B denote its intersections with the circle. Choose any three points P, Q, and R on the circle other than A or B. Use a protractor to measure $\angle APB$, $\angle AQB$, and $\angle ARB$. What general result does this suggest?

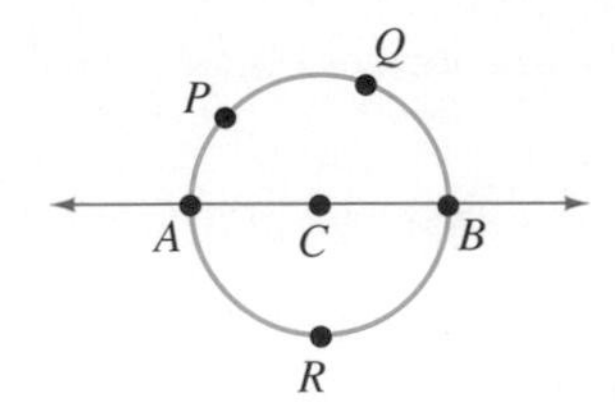

SOLUTION

(a) For every choice of l, the product $AP \cdot PB$ is the same. Therefore no line through P gives a larger product than any other line.

(b) Each angle is a right angle. This is one of geometry's earliest theorems, attributed to Thales of Miletus (c. 600 B.C.). ■

Pairs of Angles and the Corresponding Angles Theorem

Let two angles have measures m_1 and m_2 as shown in Figure 10.7 on page 618. The angles are said to be **complementary** if, and only if, $m_1 + m_2 = 90°$. The angles are **supplementary** if, and only if, $m_1 + m_2 = 180°$.

SCHOOL BOOK PAGE *Angles and Angle Measurement*

Angles and Angle Measurement

Build Understanding

ACTIVITY

A. Folding Angles
Materials: paper

a. Start with a square sheet of paper. Fold it in half so that the left edge meets the right edge. Next fold the paper in half so that the bottom edge meets the top edge. Finally, fold the paper in half diagonally so that the bottom edge meets the left edge.

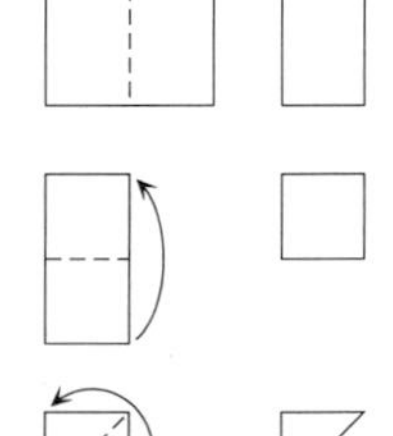

b. Open the paper and draw eight rays with a common endpoint (the center). Label the rays as shown in the diagram.

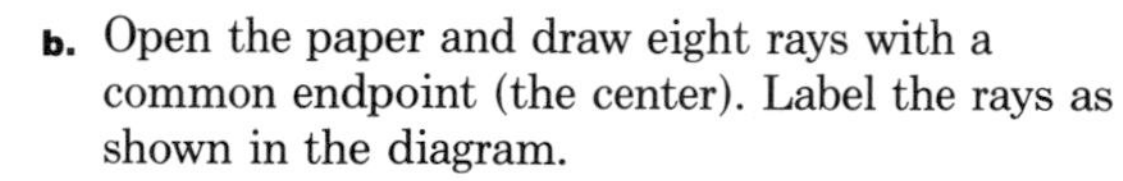

c. An ***angle*** is formed by two rays that have the same endpoint. I is the endpoint of ray IH and ray IA. The endpoint is the ***vertex*** of the angle. The rays are the ***sides*** of the angle.

Name as many other angles as you can find in the diagram.

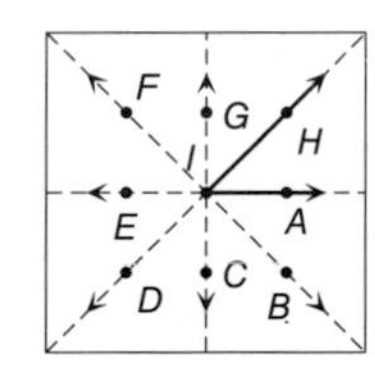

The angle formed by rays *IH* and *IA* is ∠*HIA*, read "angle *HIA*." The vertex is always the middle letter.

B. A ***protractor*** is a tool for measuring angles. It is divided into units called ***degrees***.

To measure an angle, place the center mark of the protractor on the vertex of the angle and the zero mark on a side of the angle.

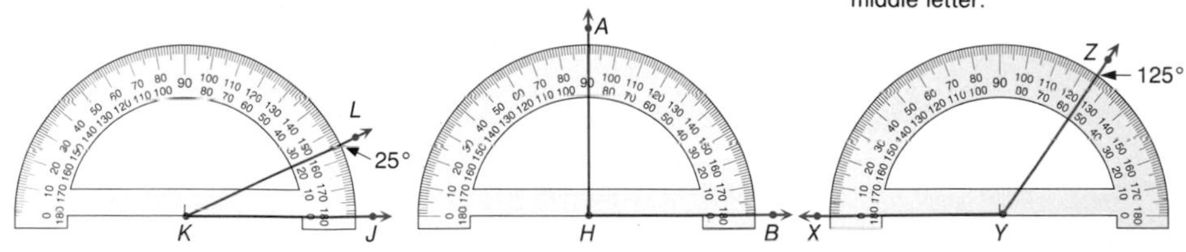

An ***acute angle*** is an angle with a measure less than 90°.

A ***right angle*** is an angle with a measure of 90°. The symbol ⅂ is used to show a right angle. Rays *HA* and *HB* are ***perpendicular.*** The symbol ⊥ is read "is perpendicular to."

An ***obtuse angle*** is an angle with a measure greater than 90° and less than 180°.

1. The ray with vertex I containing H is called "ray IH." How does this vary from the notation followed in this text?
2. A student wants to know if paper can be folded to create angles measuring 30° and 60°. Can you be of help?
3. A student wonders why there are two numerical scales on the protractor. Can you explain why? How are the two scales related?

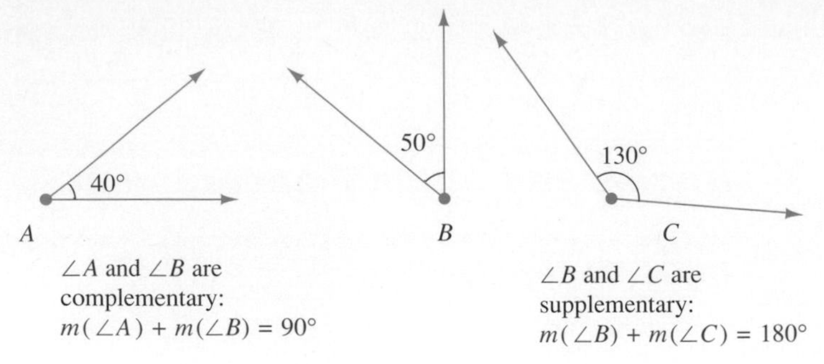

Figure 10.7
Examples of complementary and supplementary angles

Two angles that have a common side and nonoverlapping interiors are called **adjacent** angles. Supplementary and complementary angles frequently occur as adjacent angles, as shown in Figure 10.8.

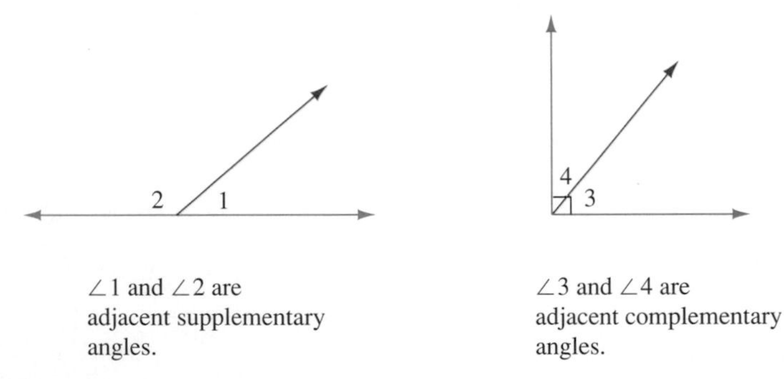

2
1
4
3
∠1 and ∠2 are adjacent supplementary angles.
∠3 and ∠4 are adjacent complementary angles.

Figure 10.8
Adjacent supplementary and complementary angles

HIGHLIGHT FROM HISTORY
Thales of Miletus (c. 640–546 B.C.)

Classical Greek mathematics developed in a succession of schools where a collection of scholars worked with a great leader. The first of these was the Ionian school, founded by Thales of Miletus. Thales traveled extensively and resided for some time in Egypt, where he learned Egyptian mathematics. He compared the length of the shadow cast by a pyramid to that cast by a vertically held rod and was able to calculate the height of the pyramid. Thales is credited with several propositions in geometry, and was instrumental in introducing the notions of abstraction and proof into mathematics.

Thales was known as a statesman, engineer, businessman, philosopher, astronomer, and mathematician. There are a number of anecdotes told about Thales. For example, there is the story of the recalcitrant mule which, when carrying a pack of salt, would lay down in the river to dissolve, and lighten, his load—Thales cleverly broke the mule of this habit by loading him with sponges. In Aristophanes' play, *The Clouds,* the wisdom of one of the characters is made clear when he is referred to as "a veritable Thales."

Two angles are called **vertical angles** when their four sides form two intersecting lines. In particular, a pair of intersecting lines forms two pairs of vertical angles, as shown in Figure 10.9. Since $\angle 1$ and $\angle 2$ are supplementary we know that $m(\angle 1) + m(\angle 2) = 180°$. Likewise, $\angle 2$ and $\angle 3$ are supplementary and we also have $m(\angle 2) + m(\angle 3) = 180°$. Comparing these two equations shows that $m(\angle 1) = m(\angle 3)$. This proves another theorem of Thales.

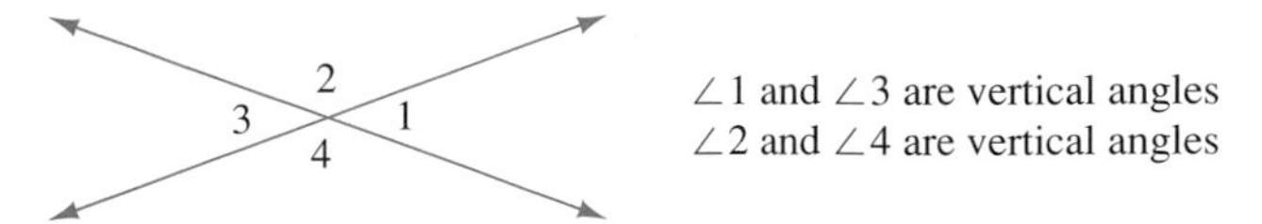

2
3
1
4
∠1 and ∠3 are vertical angles
∠2 and ∠4 are vertical angles

Figure 10.9
Intersecting lines form two pairs of vertical angles.

THEOREM Vertical Angles Theorem

Vertical angles have the same measure.

Now consider the angles formed when two lines l and m are intersected at two points by a transversal t. There are eight angles formed, in four pairs of **corresponding angles.**

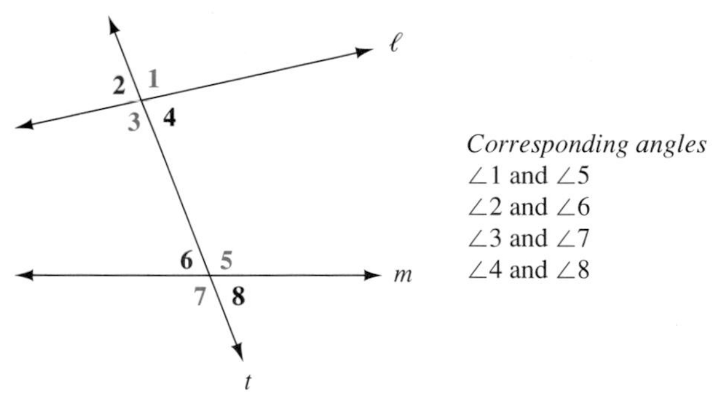

A case of special importance occurs when l and m are parallel lines, as shown in Figure 10.10. It would appear that each pair of corresponding angles is a pair of congruent angles. Conversely, if any one pair of corresponding angles is a congruent

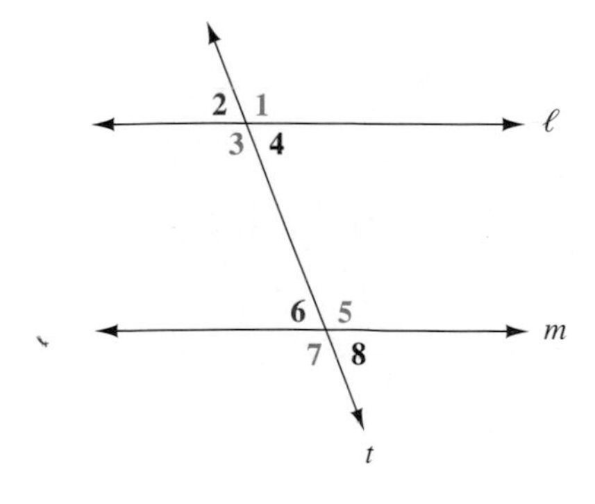

Figure 10.10
Line l and m are parallel if, and only if, the angles in some corresponding pair have the same measure.

pair of angles then the lines l and m appear to be parallel. We will accept the truth of these observations, giving us the corresponding angles property. Many formal treatments of Euclidean geometry introduce the corresponding angles property as an axiom.

PROPERTY Corresponding Angles Property

- If two parallel lines are cut by a transversal, then corresponding angles have the same measure.
- If two lines in the plane are cut by a transversal and some pair of corresponding angles has the same measure, then the lines are parallel.

EXAMPLE 10.3

Using the Corresponding Angles Property

(a) Lines l and m are parallel and $(\angle 6) = 35°$. Find the measures of the remaining seven angles.

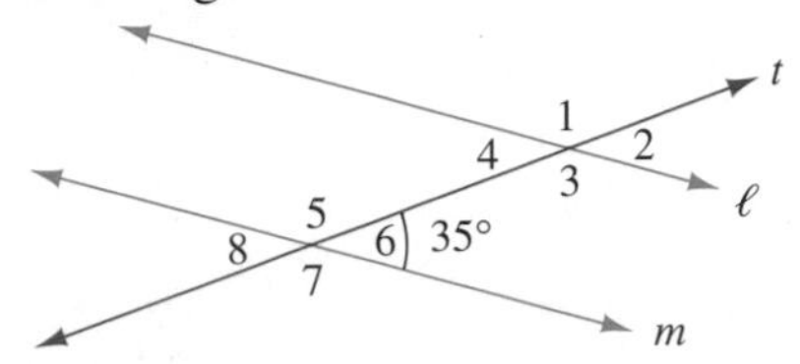

(b) Lines t and j intersect at P and form an angle measuring 122°. Describe how to use a protractor and straightedge to draw a line k through Q that is parallel to line j.

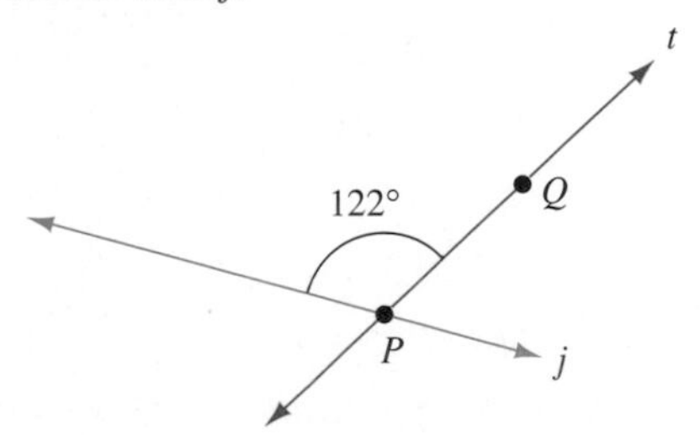

SOLUTION

(a) Since $\angle 8$ and $\angle 6$ are vertical angles, $m(\angle 8) = 35°$. Also $\angle 5$ and $\angle 7$ are supplements of $\angle 6$ and so $m(\angle 5) = m(\angle 7) = 180° - 35° = 145°$. By the corresponding angles property, $m(\angle 1) = m(\angle 5) = 145°$, $m(\angle 2) = m(\angle 6) = 35°$, $m(\angle 3) = m(\angle 7) = 145°$, and $m(\angle 4) = m(\angle 8) = 35°$.

(b) Use the protractor to form the corresponding angle measuring 122° at point Q.

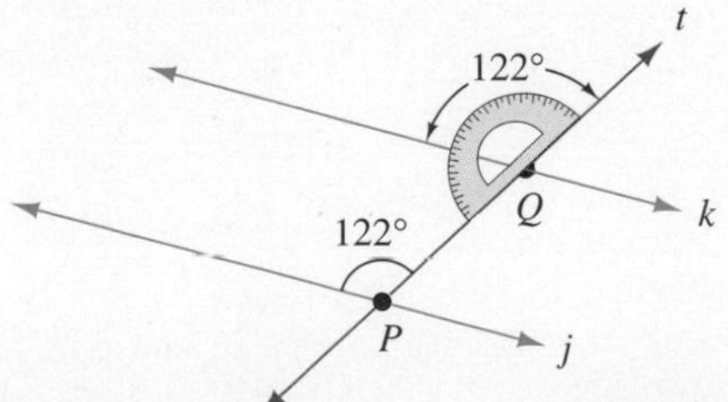

■

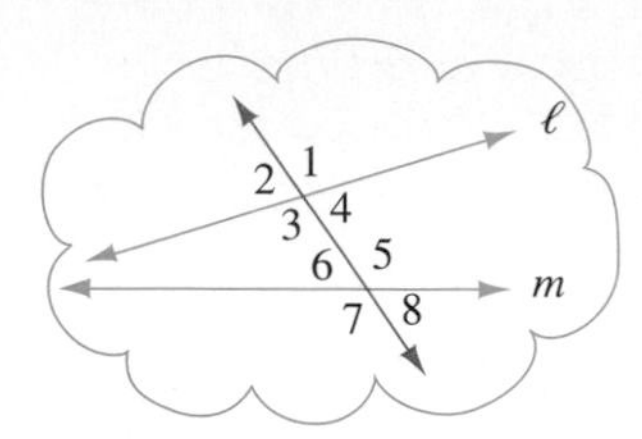

The pair of angles, $\angle 4$ and $\angle 6$, between l and m but on opposite sides of the transversal t, are called **alternate interior angles.** Since $\angle 2$ and $\angle 4$ are vertical angles, they are congruent by the vertical angles theorem. Thus, the corresponding angles $\angle 2$ and $\angle 6$ are congruent if, and only if, the alternate interior angles $\angle 4$ and $\angle 6$ are congruent. This gives the following consequence of the corresponding angles property.

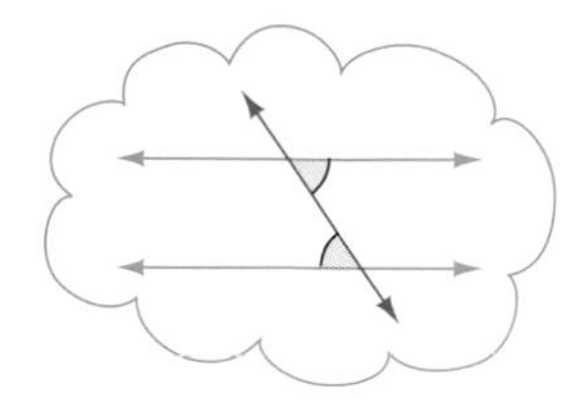

THEOREM Alternate Interior Angles Theorem

Two lines cut by a transversal are parallel if, and only if, a pair of alternate interior angles are congruent.

The Measure of Angles in Triangles

If a triangle ABC is cut from paper and its three corners are torn off, it is soon discovered that the three pieces will form a straight angle along a line l (see Figure 10.11). Thus $m(\angle 1) + m(\angle 2) + m(\angle 3) = 180°$ and we have a physical demonstration that the sum of the measures of the angles of a triangle is 180°.

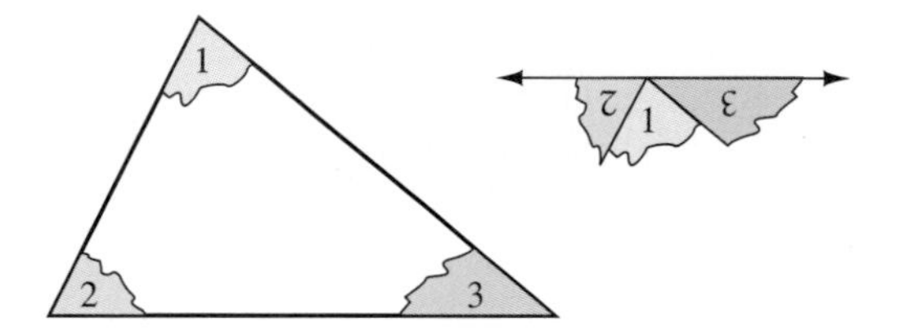

Figure 10.11
The torn corners of a triangle can be placed along a line to show that $m(\angle 1) + m(\angle 2) + m(\angle 3) = 180°$.

The alternate interior angles theorem can be used to give another demonstration of the result.

HIGHLIGHT FROM HISTORY
The Babylonian Origins of Degree Measure

There is no doubt that degree measure, with 360 degrees comprising a full turn, had its origin in ancient Babylonia. At one time it was suggested the Babylonians thought a year was 360 days, which would mean the sun would advance one degree per day as it revolved in its circular orbit about the sun. This explanation must be dismissed however, since the early Babylonians were fully aware that a year exceeded 360 days.

A more plausible explanation has been suggested by Otto Neugebauer, an authority on Babylonian mathematics and science. In the early Sumerian period, time was often measured by how long it took to travel a Babylonian mile (a Babylonian mile was about 7 modern miles). In particular, a day turned out to be the time required to travel 12 Babylonian miles. Since the Babylonian mile, being quite long, had been subdivided into 30 equal parts for convenience, there were $(12)(30) = 360$ parts in a day's journey. The complete turn of the earth each day was therefore divided into 360 parts, giving rise to degree measure of an angle.

THEOREM Sum of Angle Measures in a Triangle

The sum of the measures of the angles in a triangle is 180°.

PROOF Consider the line l through point A that is parallel to the line $m = \overleftrightarrow{BC}$.

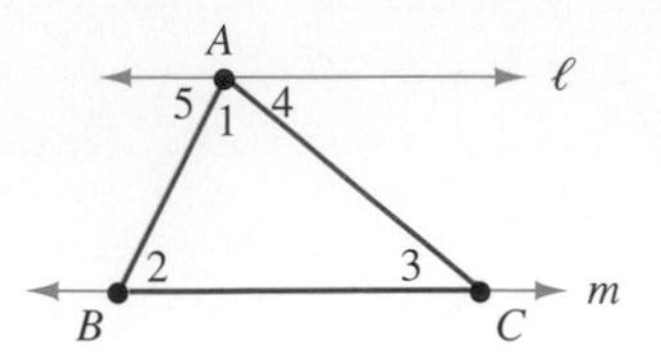

Line $\overleftrightarrow{AB}$ is a transversal to l and m for which $\angle 5$ and $\angle 2$ are alternate interior angles. Thus, $m(\angle 5) = m(\angle 2)$ by the alternate interior angles theorem. Similarly, $\angle 4$ and $\angle 3$ are alternate interior angles for the transversal $\overleftrightarrow{AC}$, so $m(\angle 4) = m(\angle 3)$. Since $\angle 5$, $\angle 1$, and $\angle 4$ form a straight angle at vertex A, we know that $m(\angle 5) + m(\angle 1) + m(\angle 4) = 180°$. Thus, by substitution, $m(\angle 2) + m(\angle 1) + m(\angle 3) = 180°$. ■

EXAMPLE 10.4

Measuring an Opposite Exterior Angle of a Triangle

In the figure below $\angle 4$ is called an **exterior angle** of triangle PQR, and $\angle 1$ and $\angle 2$ are its **opposite interior angles.** Show that the measure of the exterior angle is equal to the sum of the measures of the opposite interior angles; that is, show that $m(\angle 4) = m(\angle 1) + m(\angle 2)$.

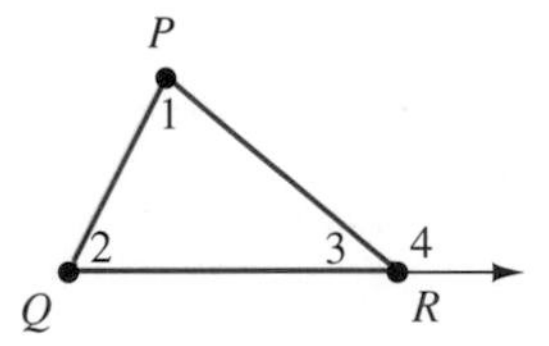

SOLUTION

By the preceding theorem, we have $m(\angle 1) + m(\angle 2) + m(\angle 3) = 180°$. Also, $\angle 3$ and $\angle 4$ are supplementary, so that $m(\angle 3) + m(\angle 4) = 180°$. Therefore $m(\angle 1) + m(\angle 2) + m(\angle 3) = m(\angle 3) + m(\angle 4)$. Subtracting $m(\angle 3)$ from both sides of this equation gives $m(\angle 1) + m(\angle 2) = m(\angle 4)$. ■

Directed Angles

Until now angles have been measured without regard to the *direction*—clockwise or counterclockwise—one side rotates until it coincides with the second side. Often it is useful to specify one side as the **initial side** and the other side as the **terminal side.** Angles are then measured by specifying the number of degrees to rotate the initial side to the terminal side. Mathematicians usually assign a positive number to counterclockwise turns, and a negative number to clockwise turns. Angles which specify an initial and final side and a direction of turn are called **directed angles.** Some examples are shown in Figure 10.12, where arrows on the circular arcs indicate the direction of turn. Notice that an angle measure of $-90°$ could also be assigned the measure $+270°$.

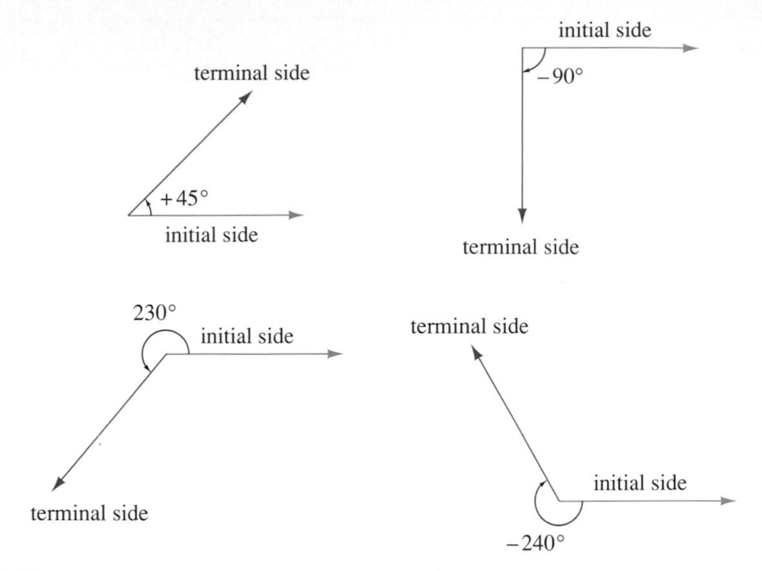

Figure 10.12
Directed angles, measured positively for counterclockwise turns

Directed angles are used in Logo turtle graphics by instructing the turtle to turn either to the right (clockwise) or to the left (counterclockwise). Automatic drawing software gives the user a choice of using directed angle measurements or not; for example, in the directed angle mode, if $m(\angle ABC) = 53°$ then $m(\angle CBA) = -53°$.

EXAMPLE 10.5

Measuring Directed Angles

Patty Pathfinder's trip through the woods to grandmother's house started and ended in an easterly direction, but zig-zagged through Wolf Woods to avoid trouble. The first two angles Patty turned through are 45° and −60°, as shown. Use a protractor to measure the three remaining turns. What is the sum of all five directed angles? Explain your surprise or lack of surprise.

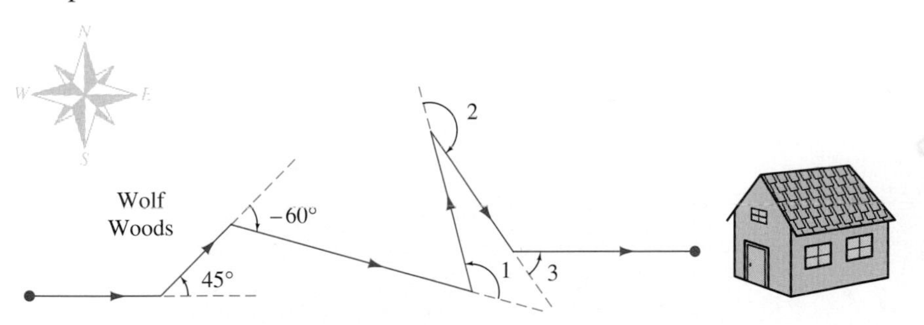

SOLUTION

$m(\angle 1) = 120°$, $m(\angle 2) = -155°$, $m(\angle 3) = 50°$. The sum of all five directed angles is $45° - 60° + 120° - 155° + 50° = 0°$. This is not surprising since Patty's path started and stopped in the same direction and her path didn't make any loops. ■

PROBLEM SET 10.1

Understanding Concepts

1. A *9-pin dot-matrix printer* for a computer might use a grid with 9 rows and 8 columns. Letters are printed by placing dots in particular cells. For example, a capital "A" and lower case "j" can be printed this way.

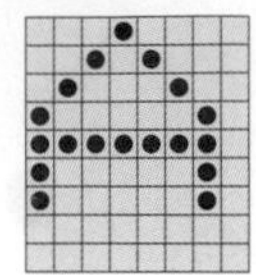

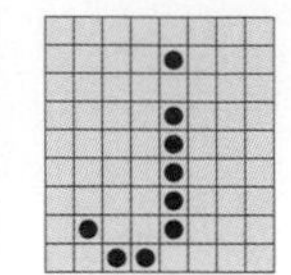

   (a) Create a dot-matrix capital letter O and lower case p in a 9 row 8 column grid.
   (b) Create a numeral for zero that can be distinguished from your capital O.
   (c) Discuss what advantages a 24-pin dot-matrix printer has over a 9-pin printer. Are there any disadvantages?

2. The points E, U, C, L, I, D are shown below.

E • • U
D • • C
I • • L

   Draw the following figures:
   (a) $\overleftrightarrow{EU}$ **(b)** $\overrightarrow{CL}$ **(c)** $\overline{ID}$

3. Trace the 5 by 5 square lattice shown, and draw the line segment $\overline{AB}$.

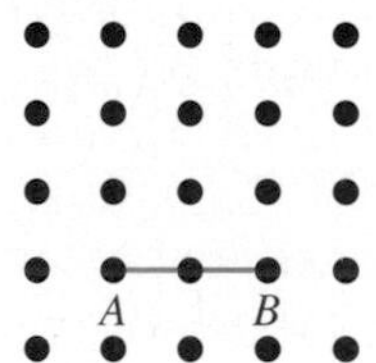

   Use colored pencils to circle all of the points C of the lattice that make $\angle BAC$ (a) a right angle, **(b)** an acute angle, **(c)** an obtuse angle, **(d)** a straight angle, and **(e)** a zero angle.

4. (a) How many nonzero angles are shown in the following figure? Give the three letter symbol, such as $\angle APB$, for each angle.

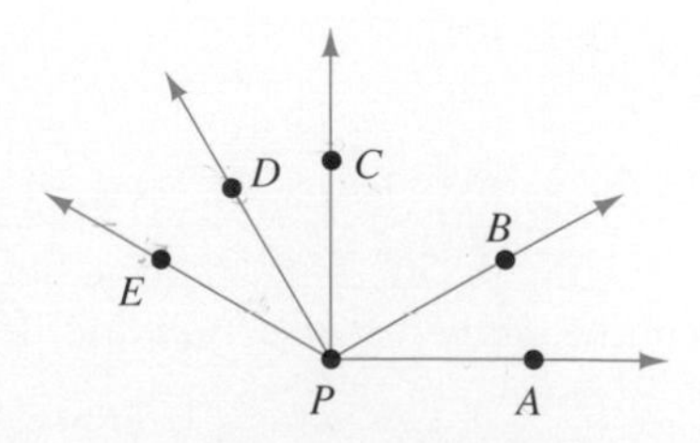

   (b) Measure each angle you identified in part (a), and classify it as acute, right, or obtuse.

5. In the figure shown, $\angle BXD$ is a right angle and $\angle AXE$ is a straight angle.

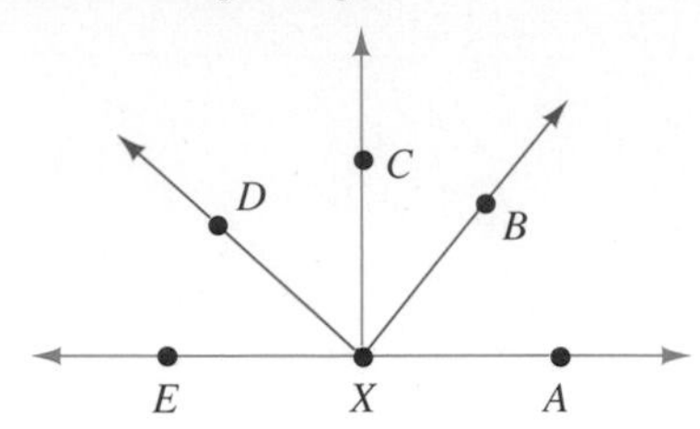

   If $m(\angle BXC) = 35°$ and $m(\angle BXE) = 132°$, explain how you can determine the measures of $\angle AXB$, $\angle CXD$, and $\angle DXE$ without using a protractor.

6. The point P shown below is the intersection of the two ***external* tangent** lines to a pair of circles. The point Q is the intersection of the two ***internal* tangent** lines to two circles.

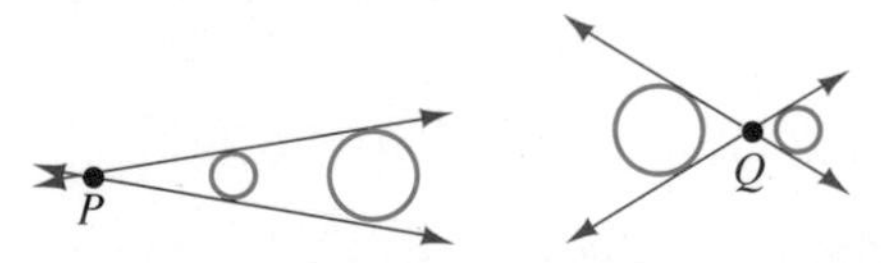

   Draw three circles, C_1, C_2, and C_3, no circle containing or intersecting either of the other circles. Let P be the intersection of the external tangent lines of C_1 and C_2, and let Q and R be the respective intersection of the internal tangent lines of the pairs of circles C_2, C_3, and C_1, C_3. What conclusion is suggested by your drawing?

7. Two intersecting circles determine a line, as shown below on the left. Draw three circles, where each circle intersects the other two circles as in the example shown on the right. Next draw the three lines determined by each pair of intersecting circles. What conclusion is suggested by your drawing?

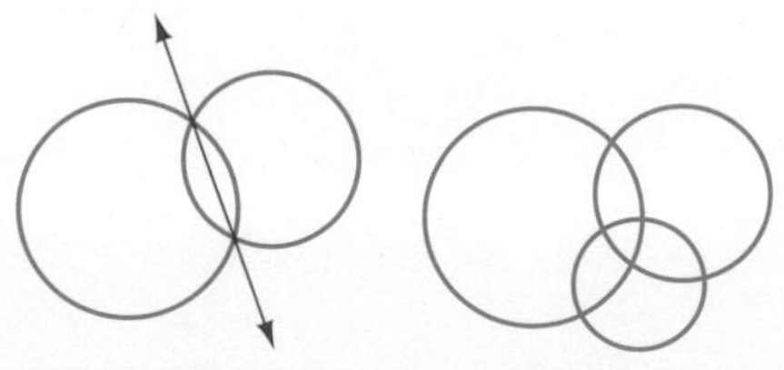

8. Draw a circle at a point A and a second circle of the same radius at a point B, where the radius is large enough to cause the circles to intersect at two points C and D.
   Let M be the point of intersection of $\overline{CD}$ and $\overline{AB}$.

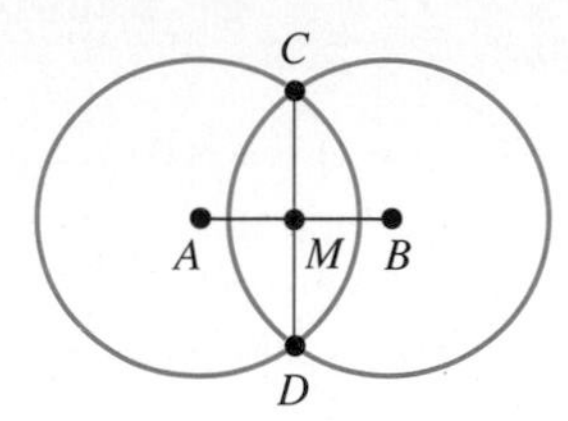

(a) Use a protractor to measure the angles at M. What can you say about how $\overline{CD}$ and $\overline{AB}$ intersect?

(b) Use a ruler to measure $\overline{MA}$ and $\overline{MB}$. What can you say about point M?

9. Draw two circles, and locate four points on each circle as shown.

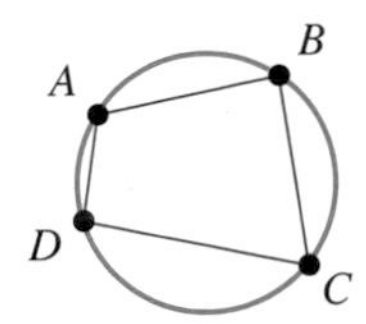

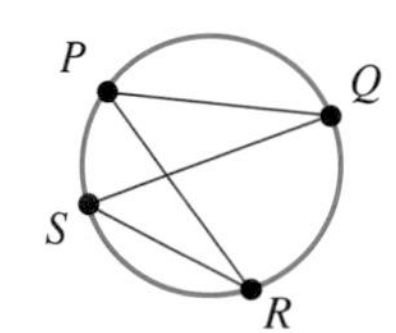

(a) Carefully measure the angles at A, B, C, and D with a protractor. What relationships do you see on the basis of your measurements?

(b) Measure the angles at P, Q, R, and S, and discuss what relationships hold between the angles in that figure.

10. Draw a circle, labeling its center as O. Draw an angle whose vertex is at O and whose sides intersect the circle at points A and B.

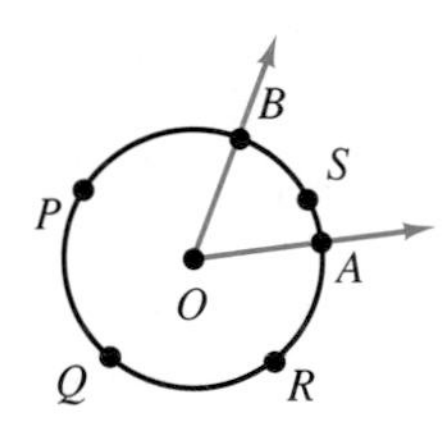

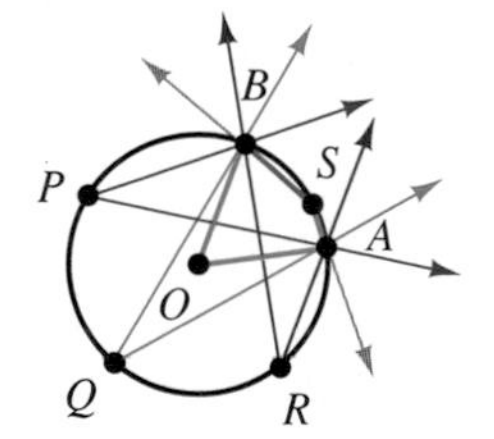

Let P, Q, and R be points on the circle which are in the exterior of $\angle AOB$. Let S be a point on the circle which is in the interior of $\angle AOB$.

(a) Use a protractor to find $m(\angle AOB)$, $m(\angle APB)$, $m(\angle AQB)$, and $m(\angle ARB)$. Describe how the angles compare to one another.

(b) Use a protractor to measure $\angle ASB$ where S is on the smaller arc between A and B. How is $\angle ASB$ related to the angles you considered in part (a)?

11. Machinists, engineers, astronomers and others often require angle measurements that are accurate to a fraction of a degree. Sometimes decimal fractions are used, such as 38.24°. It is also common to follow the historic idea of first subdividing a degree into 60 equal parts called **minutes** (from the medieval Latin *pars minuta prima,* meaning first minute [mī-noot′] part) and next subdividing a minute into 60 equal parts called **seconds** (*partes minutae secundae,* meaning second minute part). For example 24°13′46″ is read 24 degrees 13 minutes and 46 seconds. The following computation shows how to convert to decimal degrees, using the facts that $1' = \dfrac{1^\circ}{60}$ and $1'' = \dfrac{1^\circ}{3600}$.

$$
\begin{aligned}
24^\circ 13' 46'' &= 24^\circ + \left(\frac{13}{60}\right)^\circ + \left(\frac{46}{3600}\right)^\circ \\
&\doteq (24 + 0.217 + 0.013)^\circ = 24.230^\circ
\end{aligned}
$$

Here is how to convert to degrees-minutes-seconds from a decimal measure:

$$
\begin{aligned}
38.24^\circ &= 38^\circ + (0.24)(60)' = 38^\circ + 14.4' \\
&= 38^\circ + 14' + (0.4)(60)'' \\
&= 38^\circ\, 14' 24''.
\end{aligned}
$$

Use your calculator to convert the following angle measures from decimal to degrees-minutes-seconds, or the reverse.

(a) 58° 36′ 45″ **(b)** 141° 50′ 03″
(c) 71.32° **(d)** 0.913°

12. The hour and minute hands of a clock form a zero angle at noon and midnight. Between noon and midnight, how many times do the hands again form a zero angle?

13. How many degrees does the minute hand of a clock turn through (a) in 60 minutes? **(b)** in 5 minutes? **(c)** in one minute? How many degrees does the hour hand of a clock turn through (d) in 60 minutes? **(e)** in 5 minutes?

14. Find the angle formed by the minute and hour hands of a clock at these times.

(a) 3 o'clock **(b)** 6 o'clock (c) 4:30
(d) 10:20

15. The lines l and m are parallel. Find the measures of the numbered angles shown.

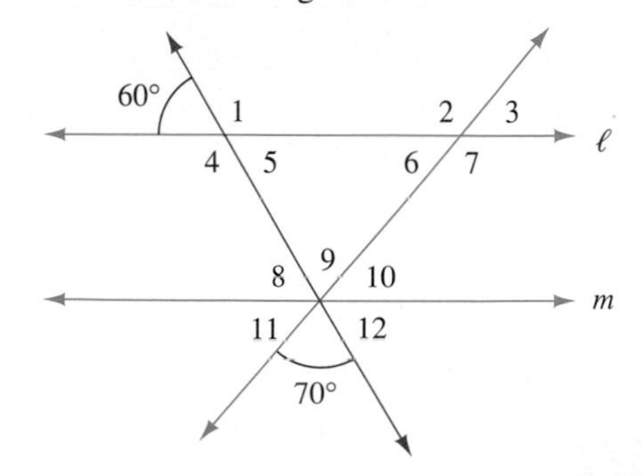

16. Determine the measure of $\angle P$ if $\overrightarrow{AB}$ and $\overrightarrow{CD}$ are parallel.

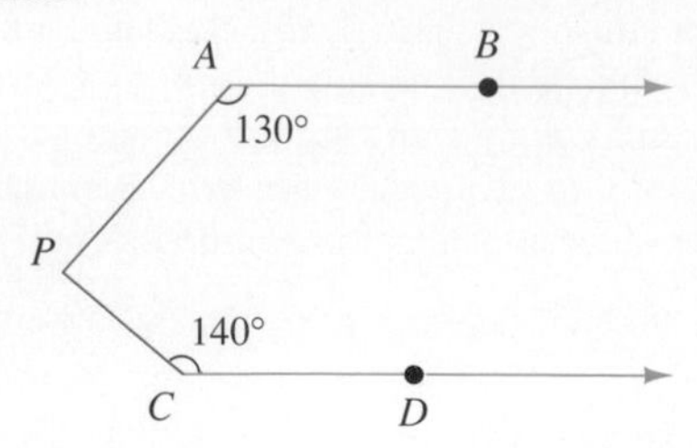

17. Find the measures of the numbered angles in the triangles shown.

(a)

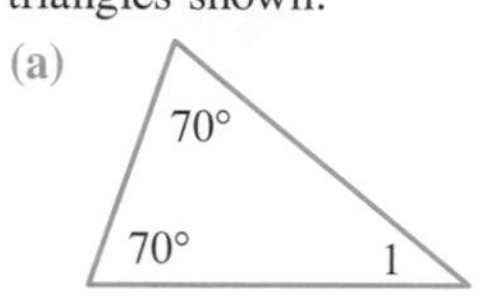

(b)

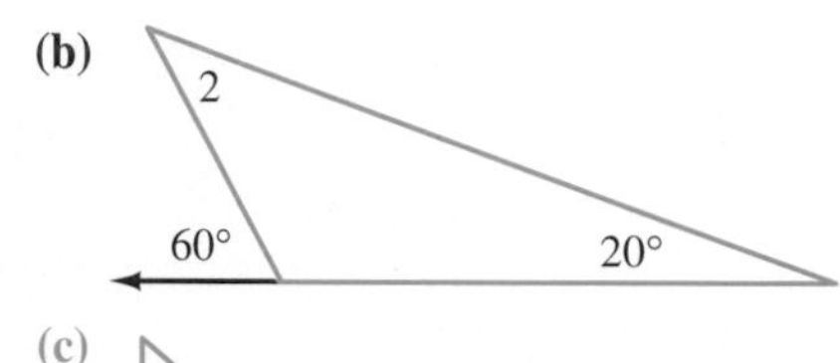

(c)

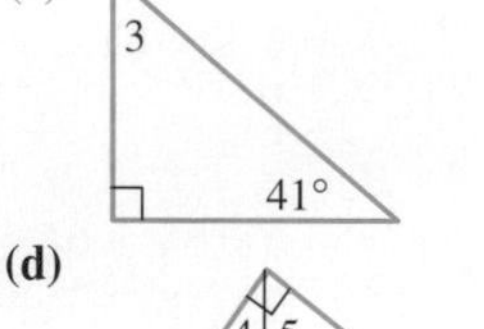

(d)

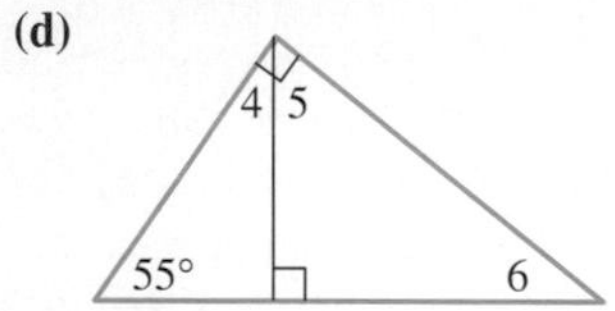

18. Find the measures of the interior angles of the following triangles.

(a)

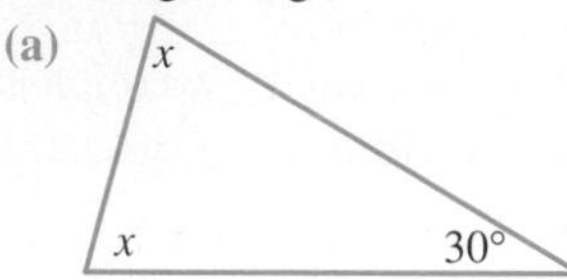

(b)

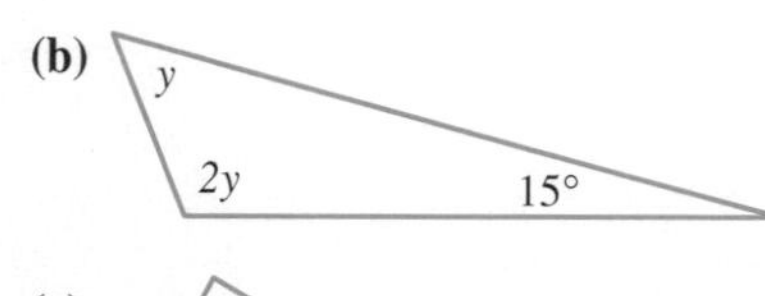

(c)

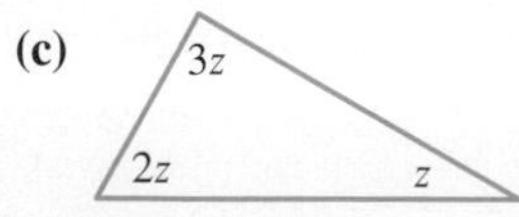

19. (a) Can a triangle have two obtuse angles? Why?
(b) Can a triangle have two right angles? Why?
(c) Suppose no angle of a triangle measures more than 60°. What do you know about the triangle?

Thinking Critically

20. Five lines are drawn in the plane.
(a) What is the smallest number of points of intersection of the five lines?
(b) What is the largest number of points of intersection?
(c) If m is an integer between the largest and smallest number of intersection points, can you arrange the lines to have m points of intersection?

21. Three noncollinear points determine three lines, as was shown in Figure 10.1.
(a) How many lines are determined by four points, no three of which are collinear?
(b) How many lines are determined by five points, no three of which are collinear?
(c) How many lines are determined by n points? Assume that no three points are collinear.

22. How many pairs of adjacent nonzero angles occur in this configuration of ten rays?

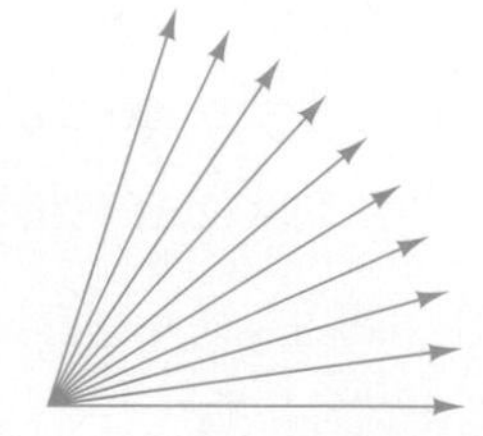

23. A boat B is anchored at points A and C. The lengths of the anchor lines are a and b. Find all of the possible positions of point B. (*Hint:* What are the positions of B if the anchor line to point C is cut? If the anchor line to point A is cut?)

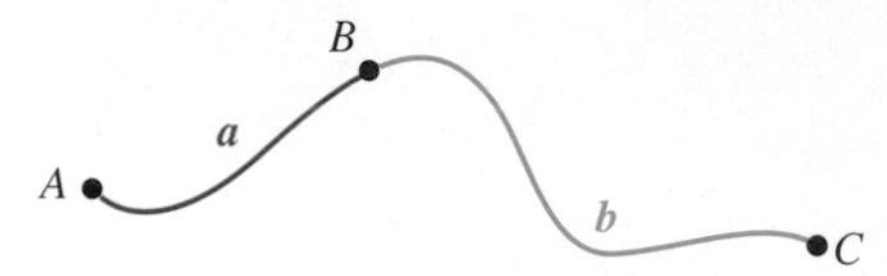

24. What is being demonstrated in this sequence of drawings? Explain in a short paragraph.

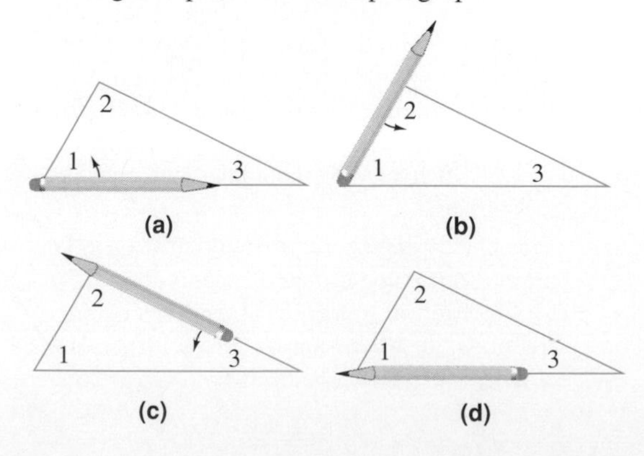

25. In "Taxicab" geometry the "points" are the corners of a square grid of "city blocks" in the plane. In the figure below the shortest trip from A to B must cover 5 blocks and so the **taxi-distance** from A to B is 5. A "taxi segment" is the set of points on a path of shortest taxi-distance from A to B, and so $\{A, W, X, Y, Z, B\}$ is a taxi segment from A to B.

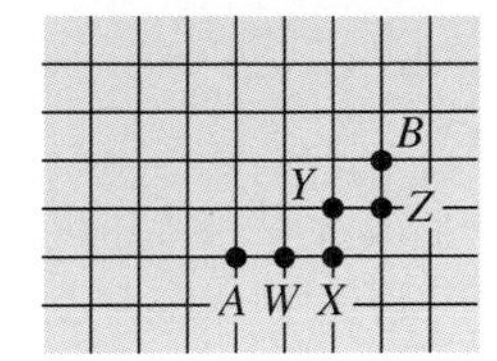

(a) How many taxi segments join A and B?

(b) Find all points which are at a taxi-distance of 5 from A. Does your "taxi circle" look like a circle drawn with a compass?

(c) Use pencils of different colors to draw the concentric taxi circles of taxi radius 1, 2, 3, 4, 5, 6. Describe the pattern you see.

Thinking Cooperatively

26. Searchlights are placed at points A, B, C, D in the plane. Each light can illuminate the sides and interior of a right angle and can be turned in any direction. If A, B, C, D were all the same point it would be easy to illuminate the entire plane.

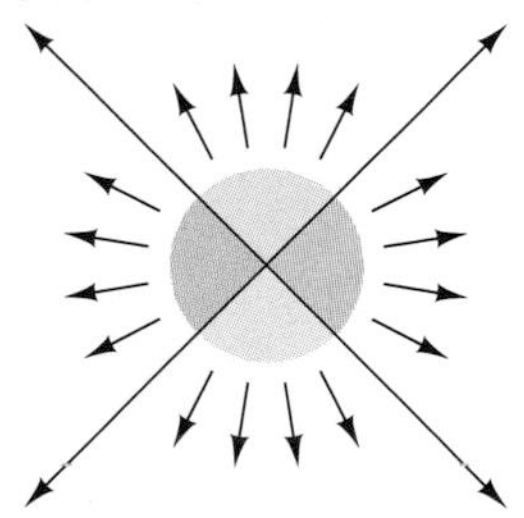

Are there positions for A, B, C, D from which it is impossible to illuminate the entire plane, or can the lights always be turned to illuminate the whole plane no matter how A, B, C, and D are located? Give a careful explanation of your conclusion. (This is a problem of Arthur Engel.)

27. The incident (incoming) and reflected (outgoing) rays of a light beam make congruent angles with the flat mirror.

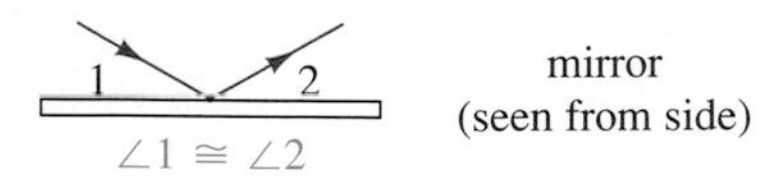

Suppose two mirrors are perpendicular to one another. Show that after the second reflection the outgoing ray is parallel to the incoming ray. (*Hint:* Show that $\angle 5$ and $\angle 6$ are supplementary. Why does this imply that the rays are parallel?)

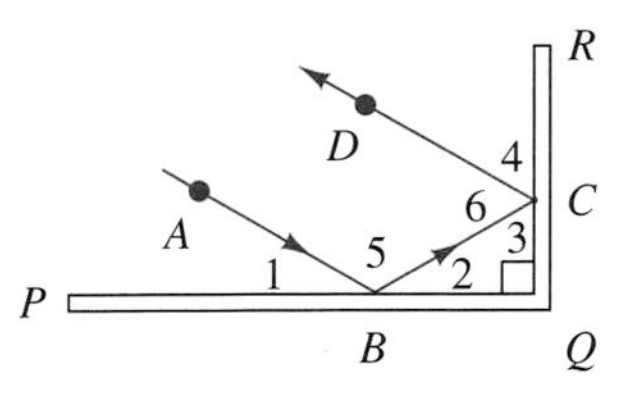

Making Connections

28. An explorer made the following trip from base camp.

First leg:	north 2 miles
Second leg:	southeast 5 miles
Third leg:	west 6 miles
Fourth leg:	south 1 mile

(a) Make a scale drawing of the trip.

(b) Show the angle the explorer turned through to go from one leg of the journey to the next.

(c) Estimate the compass heading and approximate distance the explorer needs to follow to return most directly to base camp.

29. A plumb bob suspended from the center of a protractor can be used to measure the angle of elevation of a tree top. If the string crosses the protractor's scale at the angle marked P, what is the measure of the angle of elevation?

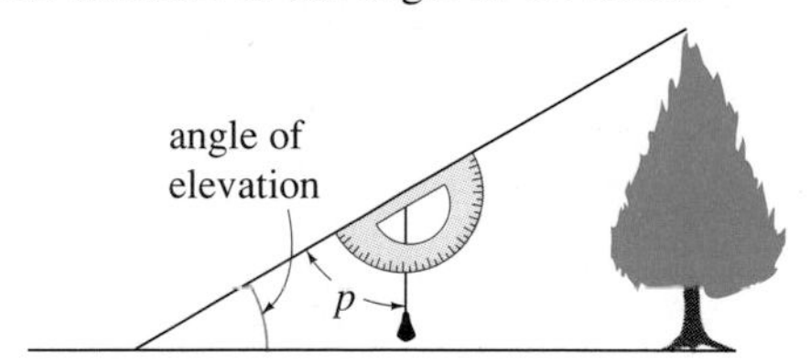

30. (a) How many degrees does the earth turn in one hour?

(b) How many degrees does the earth turn in one minute?

(c) On a clear night with a full moon it can be observed that the earth's rotation makes it appear that the moon moves a diameter in 2 minutes of time. What angle does a diameter of the moon make as seen from the earth?

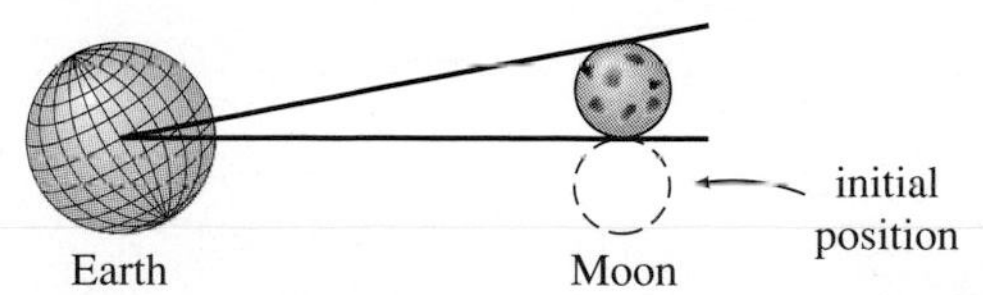

31. Suppose Polaris (the "pole star") is at an angle of elevation of 37° above the horizon. Explain how this information can be used to estimate your latitude, which is m($\angle EOP$) on the diagram below. N is the north pole, O is the earth's center, E is the point on the equator directly south of your position P, H is a point on the horizon to the north; $\overrightarrow{NS_1}$ and $\overrightarrow{PS_2}$ are parallel rays to the distant star Polaris.

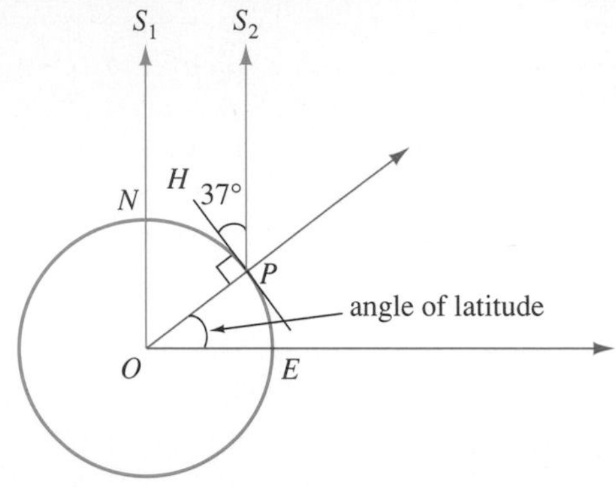

Using a Computer
Using an Automatic Drawer

32. Draw a quadrilateral $ABCD$ (that is, draw a four-sided polygon). Choose any point P on the diagonal line $\overleftrightarrow{AC}$, and draw two lines from P that intersect the sides of the quadrilateral at W, X, Y, and Z as shown. Next draw the diagonal line $\overleftrightarrow{BD}$ and the lines $\overleftrightarrow{XY}$ and $\overleftrightarrow{WZ}$. What conclusion is suggested by your drawing? If your software permits, drag selected vertices and edges and observe what effect it makes.

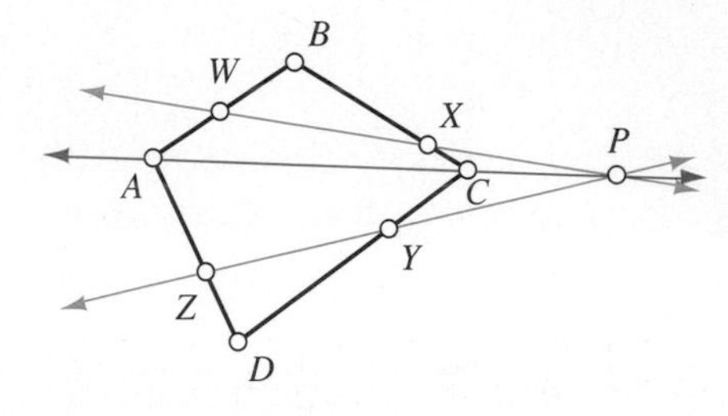

33. Draw a circle. From a point P outside of your circle draw three lines which each intersect the circle at two points. Let A, B, C, A', B', C' be the intersection points as shown. The segments $\overline{AB'}$ and $\overline{A'B}$ intersect to determine a point Q. Similarly, let $\overline{BC'}$ and $\overline{B'C}$, and $\overline{AC'}$ and $\overline{A'C}$, determine the respective points R and S.

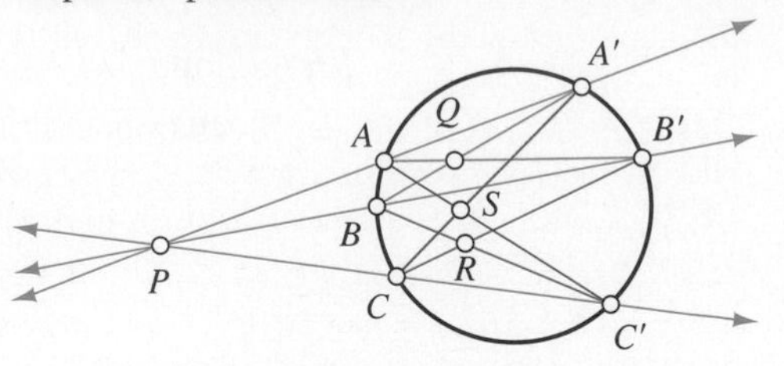

(a) What conjecture can you make concerning Q, R, and S? Drag P and the lines to investigate your conjecture.

(b) Discuss how the line $\overleftrightarrow{QR}$ can be used to construct the rays from P that are tangent to the circle.

34. Draw two squares that share a common vertex at A. Label the vertices of the squares $ABCD$ and $AB'C'D'$ in counterclockwise order.

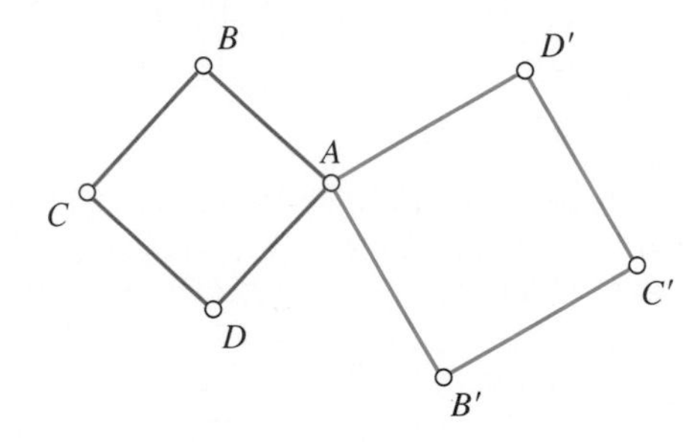

(a) Draw the lines $\overleftrightarrow{BB'}$ and $\overleftrightarrow{DD'}$, and let P denote their intersection point. Conjecture how the lines cross.

(b) Draw the line $\overleftrightarrow{CC'}$. Discuss how this line crosses the two lines drawn in part (a).

(c) Draw the line $\overleftrightarrow{AP}$. At what angle does it intersect the lines drawn before?

Communicating

35. The word "acute" has a nonmathematical meaning, as in *acute* appendicitis. Similarly, "obtuse" might be used to say someone's argument is "obtuse." Look these words up in a dictionary and comment on why their mathematical and nonmathematical meanings are related.

10.2 Curves and Polygons in the Plane

Curves and Regions

A **curve** in the plane can be described informally as a set of points that a pencil can trace without lifting until all points in the set are covered. A more precise definition is required for advanced mathematics but this intuitive idea of curve meets our present needs. If the pencil never touches a point more than once then the curve is **simple.** If the pencil is lifted at the same point at which it started tracing the curve then the curve is **closed.** If the common initial and final point of a closed curve is the only point touched more than once in tracing the curve then the curve is a **simple closed curve.** We require that the curve have both an initial and a final point, and so lines, rays, and angles are *not* curves for us.

Several examples of curves are shown in Figure 10.13.

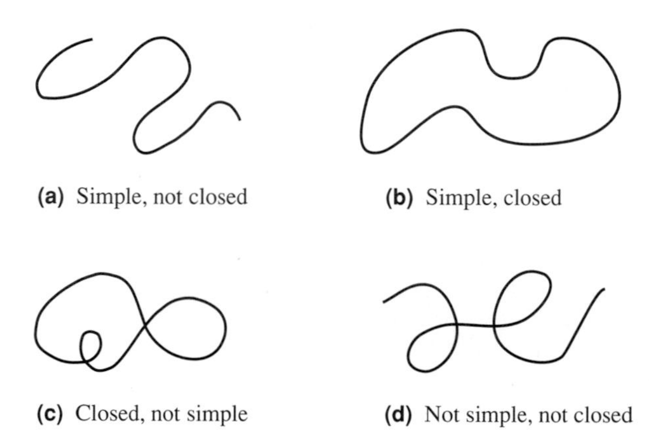

(a) Simple, not closed
(b) Simple, closed
(c) Closed, not simple
(d) Not simple, not closed

Figure 10.13
The classification of curves

Any simple closed curve separates the points of the plane into three disjoint pieces: the curve itself, the interior, and the exterior, as shown in Figure 10.14. The separation property of a simple closed curve may seem very obvious, but in fact it is an important theorem of mathematics.

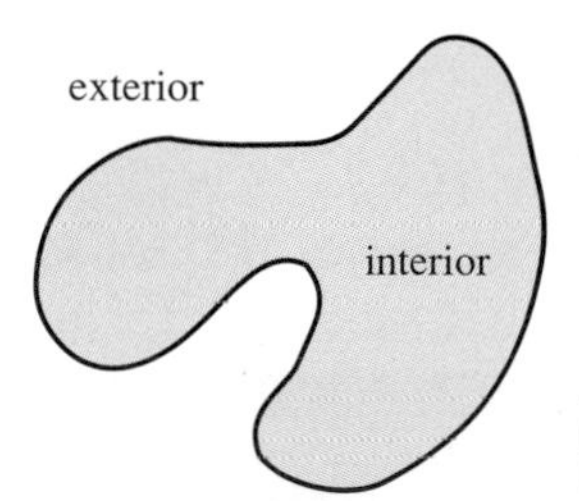

Figure 10.14
A simple closed curve and its interior and exterior

THEOREM Jordan Curve Theorem

A simple closed plane curve separates the plane into three disjoint subsets: the curve itself, the interior of the curve, and the exterior of the curve.

The French mathematician Camille Jordan (1838–1922) was the first to recognize that such an ''obvious'' result needed proof. To see why the theorem is difficult to prove (even Jordan's own proof was incorrect!) try to determine if the points G and H are inside or outside the very crinkly, but still simple, closed curve shown in Figure 10.15.

Figure 10.15
Is G in the interior or exterior of this simple closed curve? What about H?

EXAMPLE 10.6

Determining the Interior Points of a Simple Closed Curve in the Plane

Devise a method to determine if a given point is in the interior or the exterior of a given simple closed curve.

SOLUTION

Think of the curve as a fence. If we jump over the fence we either go from the interior to the exterior of the curve, or vice versa. Now draw any ray from the given point.

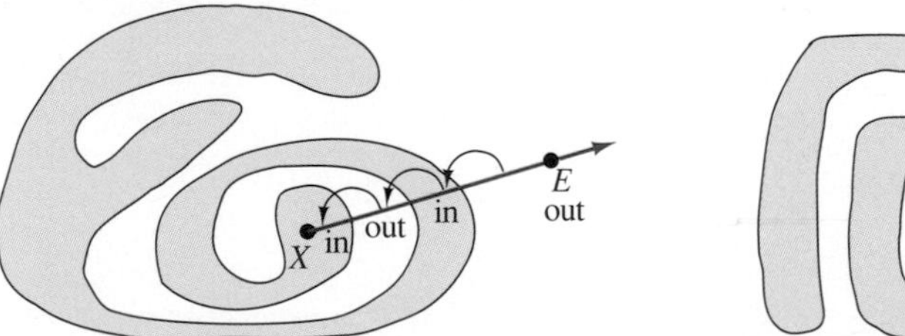

X is an interior point

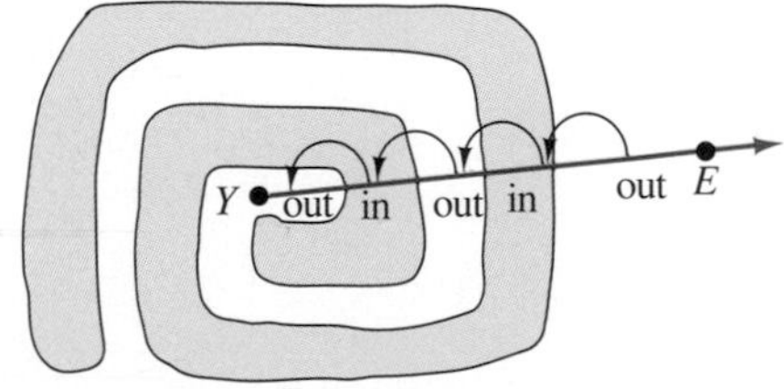

Y is an exterior point

Start at an exterior point E on the ray and see how many times the fence is crossed as you move to the endpoint of the ray. If the curve is crossed an odd number of times to reach the point X from the exterior point E, then the given point X is an interior point. An even number of crossings from E to reach a point Y means that Y is an exterior point. It should now be simple to verify that G is exterior and H is interior to the curve in Figure 10.15. ■

The interior and exterior of a simple closed curve are also called **regions.** More generally the complement of some system of lines, rays, and curves will be composed of one or more regions. For example, a line separates the plane into two regions called half planes. An angle, if not zero or straight, separates the plane into two regions called the interior and the exterior of the angle.

EXAMPLE 10.7

Counting Regions in the Plane

Count the number of regions into which the plane is separated by the following shapes.

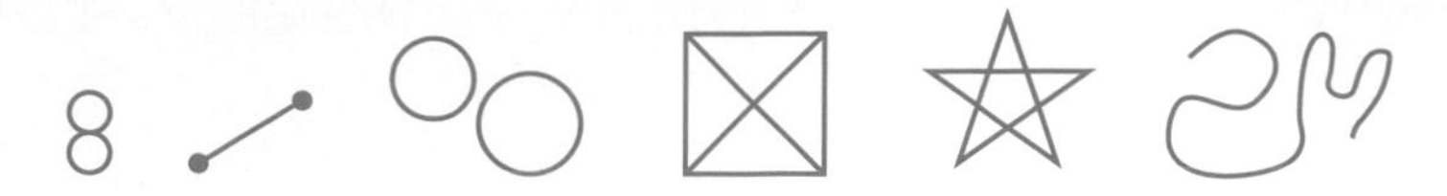

(a) a figure 8 **(b)** a segment **(c)** two nonintersecting circles
(d) a square and its two diagonals **(e)** a pentagram
(f) any simple nonclosed curve

SOLUTION

(a) 3 **(b)** 1 **(c)** 3 **(d)** 5 **(e)** 7 **(f)** 1

Notice that regions are subsets of the points *not* part of the shape itself. ■

Convex Curves and Figures

The interior of an angle has the property that the segment between any two interior points does not leave the interior. This means that the interior of an angle is a convex figure according to the following definition.

> **DEFINITION Convex Figures**
>
> A figure is **convex** if, and only if, it contains the segment $\overline{PQ}$ for each pair of points P and Q contained in the figure.

Several convex and nonconvex shapes are shown in Figure 10.16. To show that a figure is nonconvex, it is enough to find two points P and Q within the figure whose corresponding line segment $\overline{PQ}$ contains at least one point not in the figure. A nonconvex figure is sometimes called a **concave** figure.

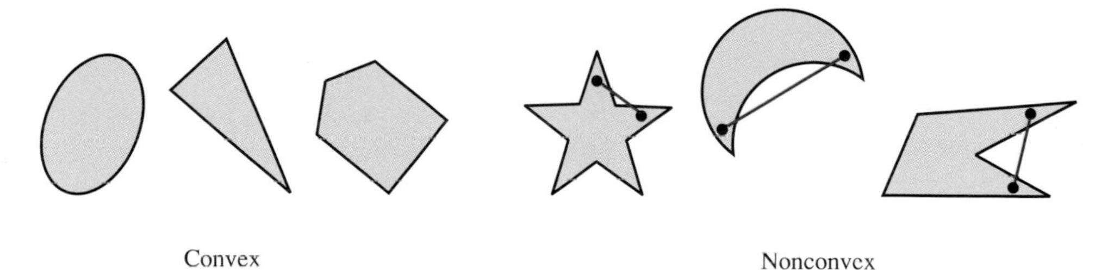

Figure 10.16
Convex and nonconvex plane figures

Polygonal Curves and Polygons

A curve that consists of a union of finitely many line segments is called **polygonal curve.** The endpoints of the segments are called **vertices,** and the segments are the **sides** of the polygonal curve. A **polygon** is a simple closed polygonal curve. The interior of a polygon is called a **polygonal region.** A **convex polygon** is a polygon whose interior is convex. Figure 10.17 illustrates examples of polygonal curves.

Polygons are named according to the number of sides or vertices they have. For example a polygon of seventeen sides is sometimes called a *heptadecagon* (hepta = seven, deca = ten). With more directness it can also be called a 17-gon. The common names of polygons are shown in Table 10.1.

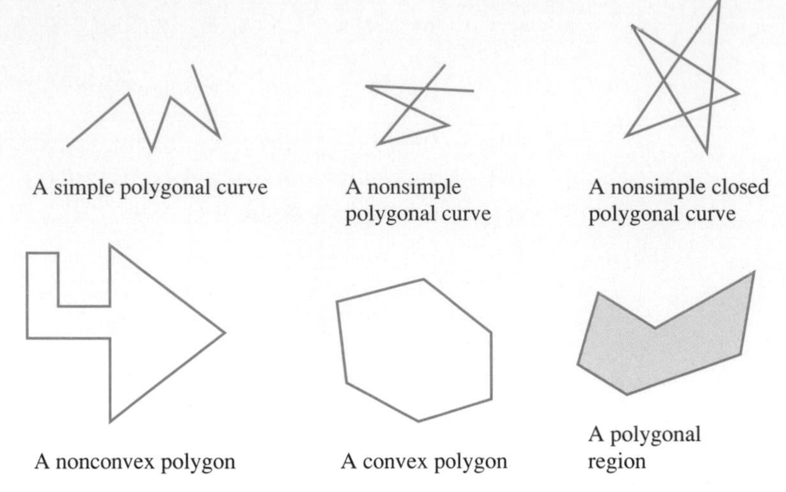

Figure 10.17
The classification of polygonal curves

TABLE 10.1 Names of Polygons

Polygon	Number of Sides	Example
Triangle	3	
Quadrilateral	4	
Pentagon	5	
Hexagon	6	
Heptagon	7	
Octagon	8	
Nonagon (or enneagon)	9	
Decagon	10	
n-gon	n	

JUST FOR FUN

Triangle Pick-Up-Sticks

The 16 matchsticks shown below form eight triangular regions. Remove a certain number of matches and attempt to leave behind two triangular regions in each case. There can be no matches remaining that do not border one of the two triangles. Which case gives you the most trouble?

a. Remove 6 matchsticks
b. Remove 7 matchsticks
c. Remove 8 matchsticks
d. Remove 9 matchsticks
e. Remove 10 matchsticks
f. Remove 11 matchsticks

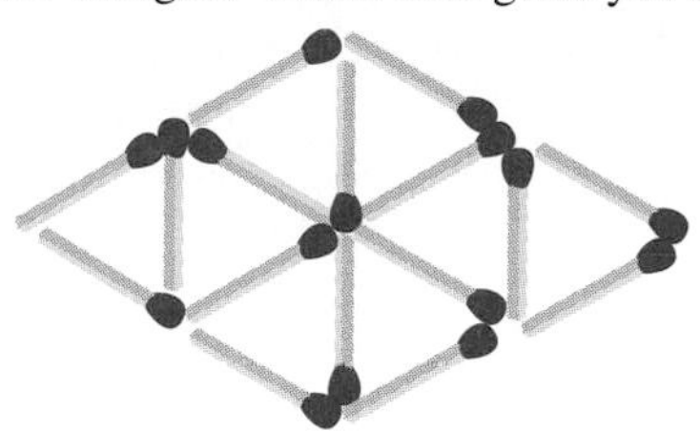

The rays containing two sides at a common vertex determine an **angle of the polygon.** For a convex polygon the interiors of these angles include the interior of the polygon. The angles are also called **interior angles,** as shown in Figure 10.18. An angle formed by replacing one of these rays with its opposite ray is an **exterior** angle of the polygon. The two exterior angles at a vertex are congruent by the vertical angles theorem. The interior angle and either of its adjacent exterior angles are supplements.

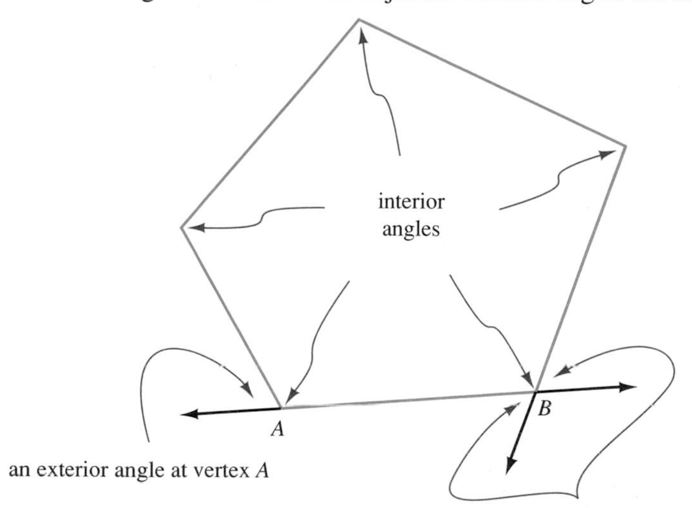

Figure 10.18
Angles in a convex polygon

In the following theorem we consider the interior angle and *one* of its supplementary exterior angles at each vertex of a convex polygon.

THEOREM Sums of the Angle Measures in a Convex Polygon

(a) The sum of the measures of the exterior angles of a convex polygon is 360°.
(b) The sum of the measures of the interior angles of a convex n-gon is $(n - 2)\,180°$.

PROOF

a. Imagine a walk completely about a polygon. At each vertex, we must turn through an exterior angle. At the conclusion of the walk we are heading in the same direction as we began, so our total turn is through 360°. If $\angle 1'$, $\angle 2', \ldots, \angle n'$ denote the exterior angles, this shows that $m(\angle 1') + m(\angle 2') + \cdots + m(\angle n') = 360°$.

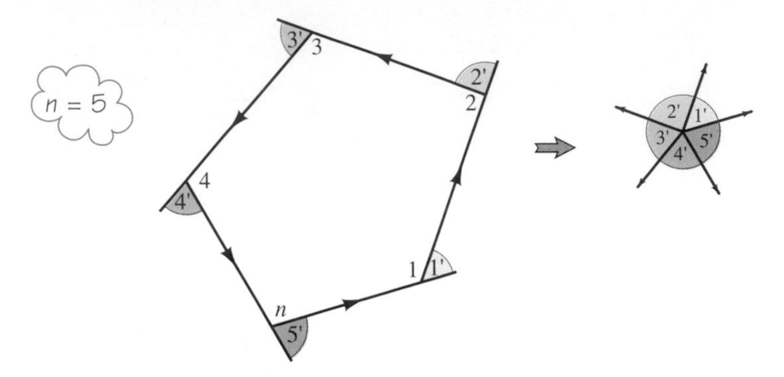

b. Since an interior and exterior angle at a vertex are supplementary, we have the equations $m(\angle 1) = 180° - m(\angle 1')$, $m(\angle 2) = 180° - m(\angle 2')$, . . . , $m(\angle n) = 180° - m(\angle n')$. Adding these n equations gives us

$$m(\angle 1) + m(\angle 2) + \cdots + m(\angle n) = n \cdot 180° - (m(\angle 1') + m(\angle 2') + \cdots + m(\angle n')), = n \cdot 180° - 360° = (n - 2) \cdot 180°,$$

where we need the result of part (a) in the second equality. ■

EXAMPLE 10.8 **Finding the Angles in a Pentagonal Arch**

Find the measures $3x$, $8x$, y, and z, of the interior and exterior angles of pentagon *PENTA* shown.

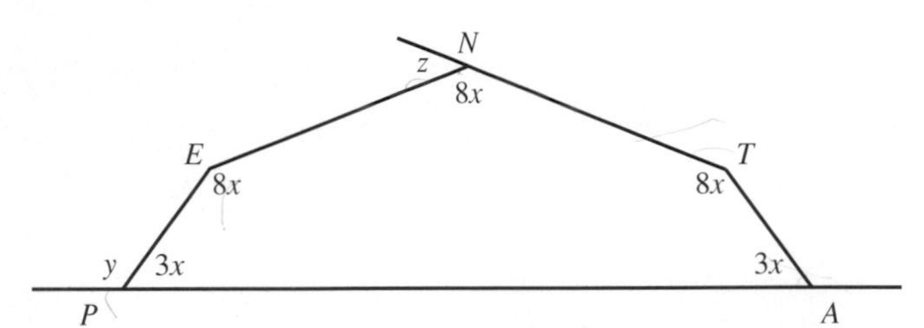

SOLUTION

By the theorem just proved we know that the sum of the measures of the five interior angles is

$$3x + 8x + 8x + 8x + 3x = (5 - 2)\ 180°.$$

That is, $30x = 3 \cdot 180°$ and so $x = (3 \cdot 180°/30) = 180°$. Thus, the interior angles at P and A measure $3 \cdot 18° = 54°$ and at E, N, and T measure $8 \cdot 18° = 144°$. The measures of the exterior angles are $y = 180° - 54° = 126°$ and $z = 180° - 144° = 36°$. ■

In a nonconvex polygon some of the interior angles are reflex angles, with measures larger than 180°. Nevertheless, it can be proved that the sum of the measures of the n interior angles is still given by $(n - 2)180°$.

THEOREM Sum of Interior Angle Measures of a General Polygon

The sum of the measures of the interior angles of any n-gon is $(n - 2)\ 180°$.

An example of a nonconvex heptagon is shown in Figure 10.19.

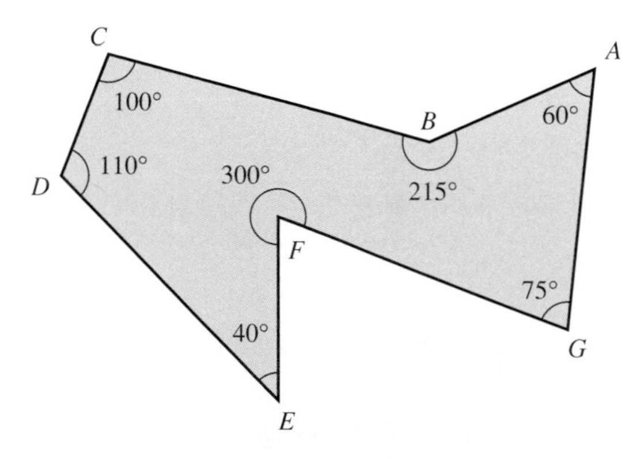

Figure 10.19
The sum of the interior angle measures of the 7-gon is $60° + 215° + 100° + 110° + 40° + 300° + 75° = 900° = (7 - 2) \cdot 180°$.

EXAMPLE 10.9

Measuring the Angles in a Five-pointed Star

The angles in each of the five "inward" points of the star shown have twice the measure of the angles of the "outward" points. What is the measure of the angle at each point of the star?

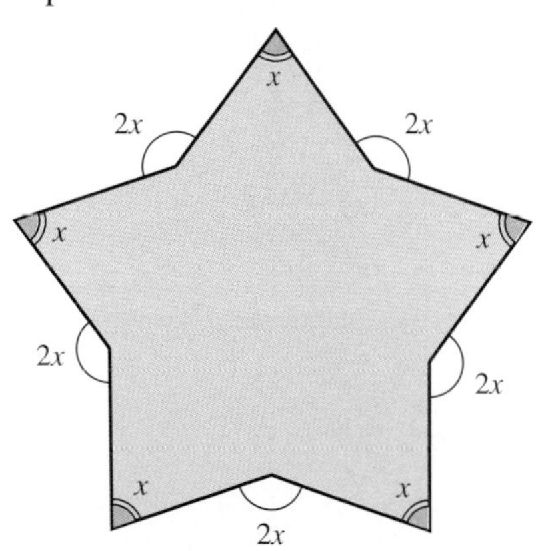

SOLUTION

The interior reflex angle at each "inward" point measures $360° - 2x$. Thus the sum of the measures of all ten interior angles of the star is $5x + 5(360° - 2x)$, or $1800° - 5x$. Since the star is a decagon (or 10-gon) the sum must equal $(10 - 2) \cdot 180° = 1440°$. This gives us the equation $1440° = 1800° - 5x$. Solving for x shows that $x = (1800° - 1440°) \div 5 = 72°$. Thus, each acute interior angle measures 72°, and the angles at the inward points each measure $2x = 144°$. ■

A walk about any closed polygonal curve, simple or nonsimple, which returns to the starting point at the same heading as the walk began must have turned through some integer multiple of 360°. In Logo computer graphics, a figure is drawn by instructing a "turtle" to walk a polygonal path. For the turtle to return to its initial position and heading, the amount of total turn in the instructions must be an integer multiple of 360°. This result is now known widely as the "Total Turtle Trip Theorem," a nice name which reminds us how to prove the theorem.

THEOREM The Total Turtle Trip Theorem

The total turning around any closed polygonal curve is an integral multiple of 360°.

More generally the total turn about any closed curve must be a multiple of 360°.

THEOREM The Total Turn Theorem

The total turn around any closed curve is an integral multiple of 360°.

EXAMPLE 10.10

Finding Total Turns

Find the total turn for each of the following closed curves. Trace the curve in the direction of the arrows, and assign positive values to turns which are counterclockwise (to the left), and assign negative values to turns which are clockwise (to the right).

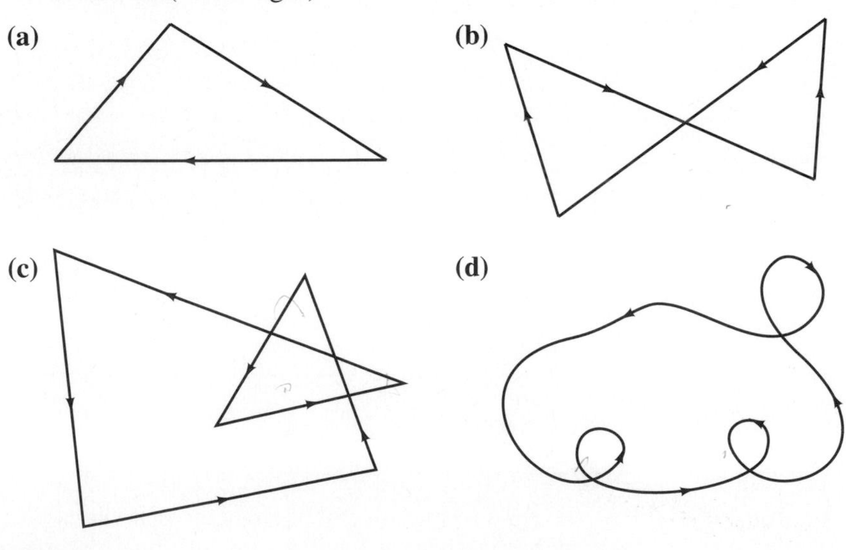

SOLUTION

(a) −360° **(b)** 0° **(c)** 720° **(d)** 720°. ■

Triangles

Triangles are classified by the measures of their angles or sides, as shown in Table 10.2.

TABLE 10.2 The Classification of Triangles

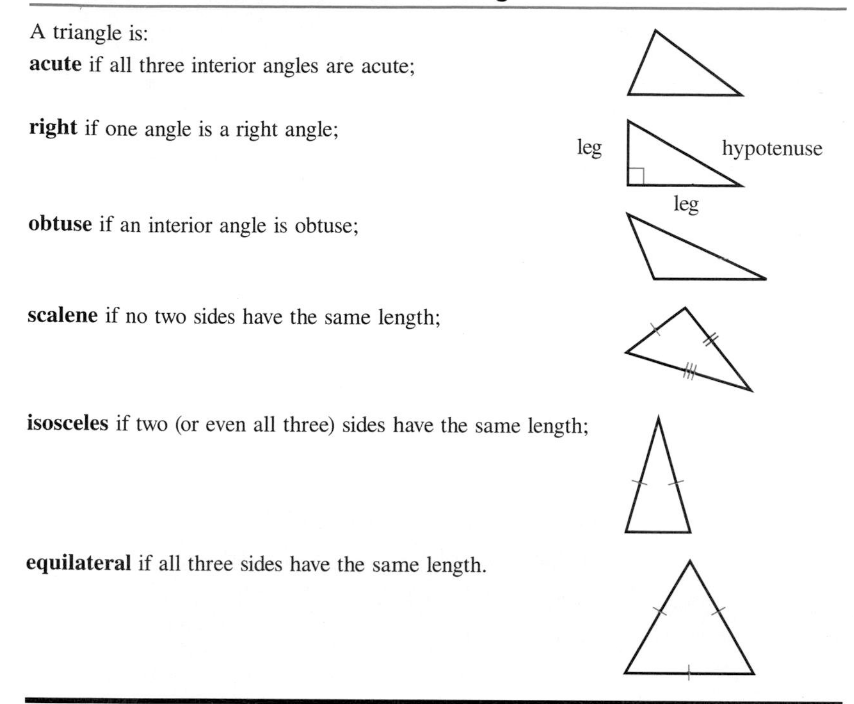

A triangle is:

acute if all three interior angles are acute;

right if one angle is a right angle;

obtuse if an interior angle is obtuse;

scalene if no two sides have the same length;

isosceles if two (or even all three) sides have the same length;

equilateral if all three sides have the same length.

EXAMPLE 10.11

Exploring Angle Trisectors in a Triangle

Draw any triangle, $\triangle ABC$. Use a protractor to measure each interior angle and draw in the rays which trisect each interior angle. Let the trisecting rays nearest side $\overline{AB}$ intersect at point P as shown. Similarly let Q be the point at which the rays nearest $\overline{BC}$ intersect, and let R be the intersection of the rays nearest $\overline{AC}$. What kind of triangle is PQR?

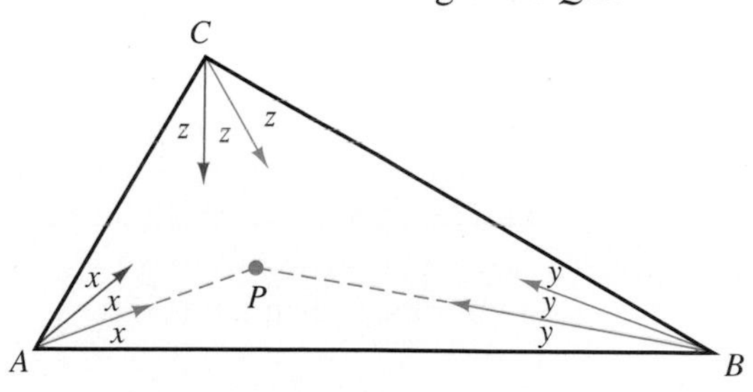

SOLUTION

A careful drawing indicates that $\triangle PQR$ is an equilateral triangle. This remarkable theorem was discovered about 1899 by Frank Morley, the father of the British novelist Christopher Morley. ■

Quadrilaterals

The convex four-sided polygons are classified as shown in Table 10.3. This classification allows a parallelogram to be described as a trapezoid, and similarly a square is a rectangle, and a rectangle is a parallelogram. Arranging figures

TABLE 10.3 The Classification of Convex Quadrilaterals

a. A **kite** is a quadrilateral with two distinct pairs of congruent adjacent sides.

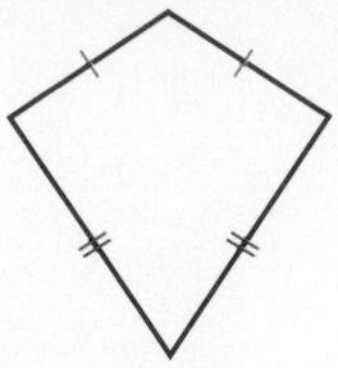

b. A **trapezoid** is a quadrilateral with at least one pair of parallel sides. *(Note: Some dictionaries and texts require a trapezoid to have exactly one pair of parallel sides.)*

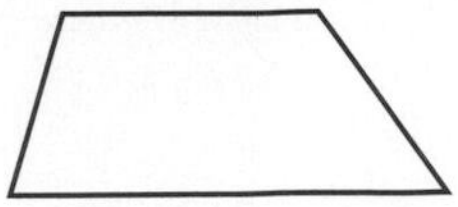

c. An **isosceles trapezoid** is a trapezoid with a pair of congruent angles along one of the parallel sides.

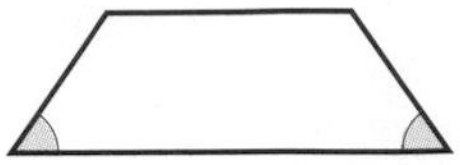

d. A **parallelogram** is a quadrilateral in which each pair of opposite sides is parallel.

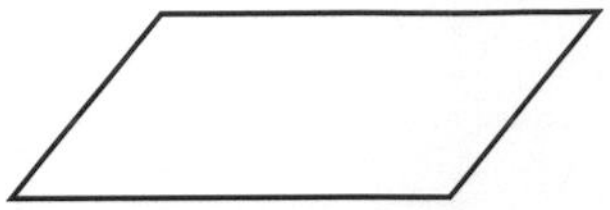

e. A **rhombus** is a parallelogram with all of its sides the same length.

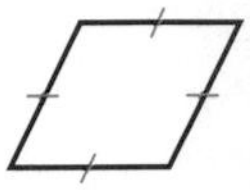

f. A **rectangle** is a parallelogram with a right angle.

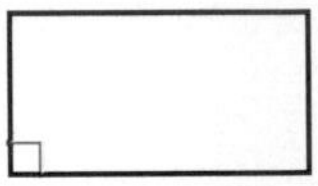

g. A **square** is a rectangle with all sides of equal length.

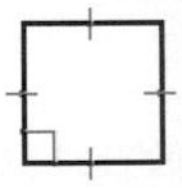

Hierarchy:

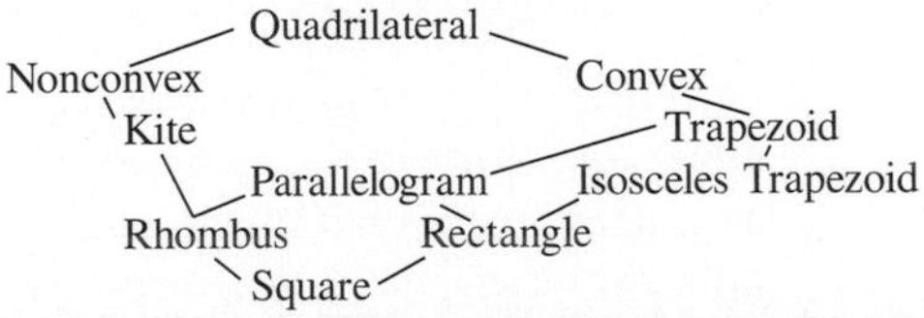

Quadrilateral
Nonconvex
Convex
Kite
Trapezoid
Parallelogram
Isosceles Trapezoid
Rhombus
Rectangle
Square

in classes which are subsets of one another can be very useful. For example, suppose we wish to show that the points A, B, C, D, are the vertices of a square. Step 1 may be to show that one pair of sides is parallel, telling us that $ABCD$ is a trapezoid. Step 2 may show that the remaining pair of opposite sides is parallel, and now we know $ABCD$ is a parallelogram. If Step 3 shows that $ABCD$ is a kite, and Step 4 shows that $\angle A$ is a right angle, then we correctly deduce that $ABCD$ is a square.

EXAMPLE 10.12

Exploring Quadrilaterals

Draw a convex quadrilateral $ABCD$. Use a compass to erect equilateral triangles on each side, alternatively pointing to the interior and exterior of the quadrilateral. Two such triangles are shown here, determining points P and Q. The equilateral triangles on the remaining sides determine points R and S.

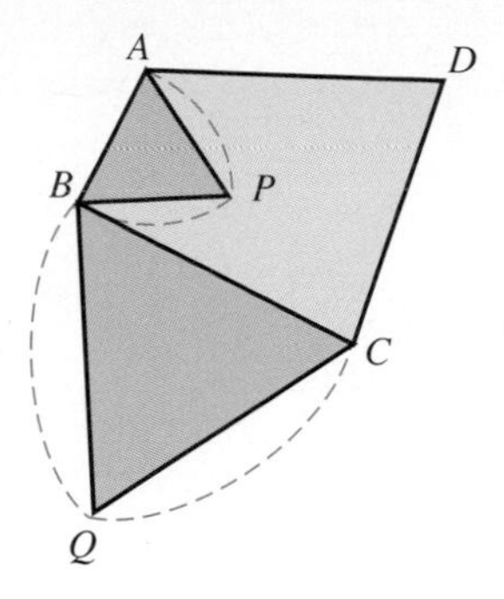

What can you say about quadrilateral $PQRS$? Support your conjecture by drawing another quadrilateral and its system of equilateral triangles. Use a ruler and protractor to measure the lengths of sides and the measure of the angles of $PQRS$.

SOLUTION

You should discover that $PQRS$ is a parallelogram. ■

Regular Polygons

A polygon with all of its sides congruent to one another is **equilateral** (that is, "equal sided"). Similarly, a convex polygon whose interior angles are all congruent is **equiangular** (that is, "equal angled"). A convex polygon which is both equilateral and equiangular is **regular.** Some hexagonal examples are shown in Figure 10.20.

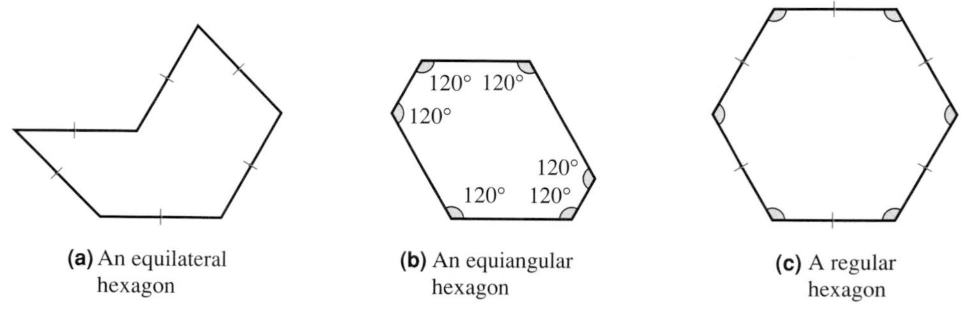

(a) An equilateral hexagon

(b) An equiangular hexagon

(c) A regular hexagon

Figure 10.20
Hexagons which are equilateral, equiangular, and regular

Since the six congruent interior angles in an equiangular hexagon have measures which add up to $(6 - 2) \cdot 180° = 720°$, each interior angle measures $720°/6 = 120°$. Similarly the measures of the interior angles of an equiangular n-gon add up to $(n - 2) \cdot 180°$, so each of the n congruent angles measures $(n - 2) \cdot 180°/n$.

In a regular n-gon, any angle with vertex at the center of a regular polygon and sides containing adjacent vertices of the polygon is called a **central angle** of the polygon. The following formulas give the measures of the exterior, interior, and central angles of a regular polygon.

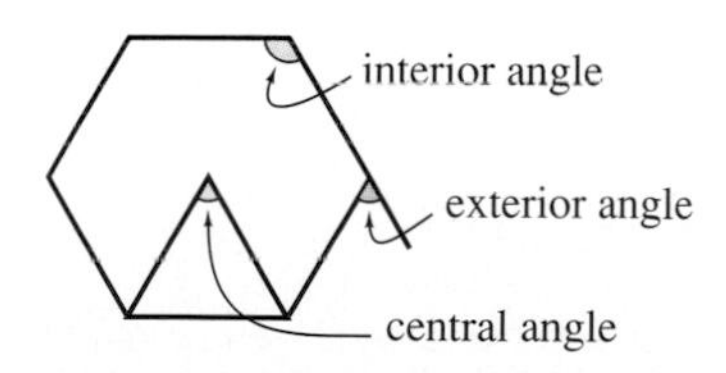

THEOREM Angle Measure in a Regular n-gon

In a regular n-gon:

- each interior angle has measure $(n - 2) \cdot 180°/n$
- each exterior angle has measure $360°/n$
- each central angle has measure $360°/n$

INTO THE CLASSROOM

Activity-based Learning and the van Hiele Levels

From kindergarten onward, geometry is learned best through hands-on activities. A successful teacher will make good advantage of the enjoyment children experience when working with colored paper, straws, string, crayons, toothpicks, and other tangible materials. Children learn geometry by doing geometry as they construct two- and three-dimensional shapes, combine their shapes to create attractive patterns, and build interesting space figures using plane shapes. By its nature, informal geometry provides unlimited opportunities to construct shapes, designs, and structures that capture a child's interest.

According to pioneering research of the van Hieles in the late 1950s, the knowledge children construct for themselves through hands-on activities is essential to learning geometry. Dr. Pierre van Hiele and his late wife Dr. Dieke van Hiele-Geldof, both former mathematics teachers in the Netherlands, theorized that learning geometry progresses through five "levels," which can be described briefly as follows.

Level 0—Recognition of shape

Children recognize shapes "holistically." Only the overall appearance of a figure is observed, with no attention given to the component parts of the figure. For example, a figure with three curved sides would likely be identified as a traingle by a child at Level 0. Similarly, a square tilted point downward may not be recognized as a square.

Level 1—Analysis of single shapes

Children at Level 1 are cognizant of the component parts of certain figures. For example, a rectangle has four straight sides which meet at "square" corners. However, at Level 1 the interrelationships of figures and properties is not understood.

Level 2—Relationships among shapes

At Level 2 children understand how common properties create abstract relationships among figures. For example, a square is both a rhombus and a rectangle. Also, simple deductions can be made about figures, using the analytic abilities acquired at Level 1.

Level 3—Deductive reasoning

The student at Level 3 views geometry as a formal mathematical system, and can write deductive proofs.

Level 4—Geometry as an axiomatic system

This is the abstract level, reached only in high level university courses. The focus is on the axiomatic foundations of a geometry, and no dependence is placed on concrete or pictorial models.

Ongoing research supports the thesis that students learn geometry by progressing through the van Hiele levels. This text—by means of hands-on activities, and examples and problems that require constructions and drawings—promotes the spirit of the van Hiele approach. However, it is the elementary classroom teacher who must bring geometry to life for his or her students, creating activities that support each child's progression through the first three van Hiele levels.

EXAMPLE 10.13 **Working with Angles in Regular *n*-gons**

(a) The Baha'i House of Worship in Wilmette, Illinois, has the unusual floor plan shown below. What are the measures of $\angle ABC$ and $\angle AOB$?

(b) Suppose an archeologist found a broken piece of pottery as shown on the right below. If the angle measures 160° and it is assumed the plate had the form of a regular polygon, how many sides would the complete plate have had?

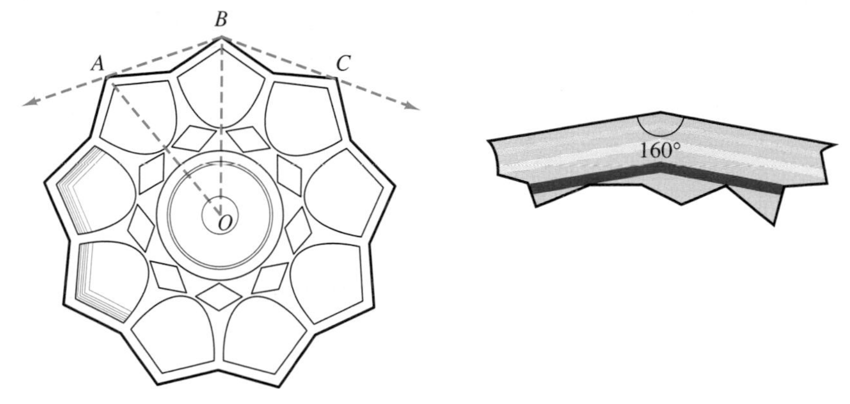

SOLUTION

(a) $\angle ABC$ is the interior angle of a regular 9-gon and it has measure $(9 - 2) \cdot 180°/9 = 140°$. $\angle AOB$ is a central angle of a regular 9-gon, so its measure is $360°/9 = 40°$.

(b) The corresponding exterior angle measures 20°. Since $20° = 360°/n$ it is seen that $n = 360°/20° = 18$. Under the assumptions stated the plate would have had 18 sides. ■

Circles

As n becomes larger a regular n-gon becomes increasingly circular. In fact, a circle in Logo is represented by drawing an approximating regular n-gon; a choice of, say, $n = 360$ will give a good "circle" when the length of each side is not too long. Since by definition a circle is the set of all points in the plane that are at a fixed distance—the **radius**—from a given point—the **center,** it is clear why a circle can be drawn easily with a compass. The **chord, diameter, tangent line, arc, sector,** and **segment** of a circle are shown in Figure 10.21. The word radius is used in two ways:

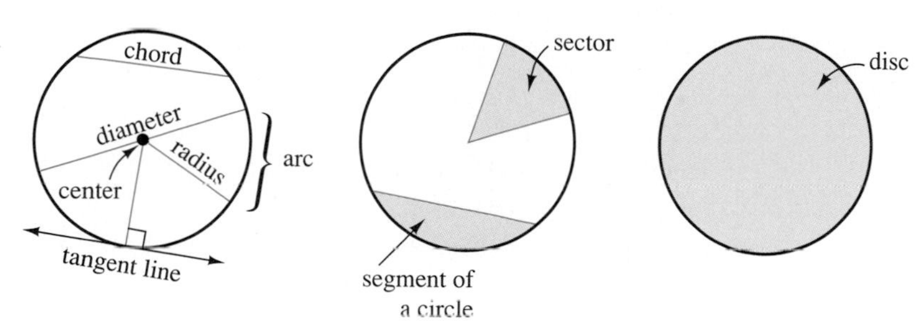

Figure 10.21
The parts of a circle

it is both a segment from the center to a point on the circle and the length of such a segment. Likewise diameter is both a segment and a length. The interior of a circle is a **disc.**

Logo Polygons and Stars

Logo can be used to draw regular polygons. In writing a Logo procedure, the turtle turns by rotating through the exterior angle at each vertex of the polygon. The exterior angles of a square are 90°. Therefore a procedure to draw a square of side length 40 is:

```
to square1
   repeat 4 [fd 40 rt 90]
end
```

Now we will write a procedure to draw an equilateral triangle with side length 40. First, we need to determine the measure of the exterior angles of an equilateral triangle. Since the turtle makes one full turn of 360° in a complete trip about any polygon, we find the measure of one exterior angle in a regular polygon by dividing 360° by the number of turns. There are three turns in a triangle, so each exterior angle measures $360°/3 = 120°$. Thus, the procedure to draw an equilateral triangle with side length 40 is:

```
to triangle1
   repeat 3 [fd 40 rt 120]
end
```

Logo permits us to use meaningful words as variables. For example, the word **:side** can represent the variable length of the side of a regular polygon. (See Appendix C for more instruction on the use of variables in Logo.) A procedure to draw a square with variable side length is:

```
to square2 :side
   repeat 4 [fd :side rt 90]
end
```

To draw a square with side length 40, the user would type in

```
square2 40 .
```

Similarly, we can write procedures to draw regular pentagons, hexagons, and other regular polygons with variable length sides. We can, using our knowledge of turtle turns, extend the procedures above to create a general procedure that will draw any regular polygon by the input of the variable number of sides and the variable length of each side. Choosing **:nsides** as the variable that represents the variable number of sides, our procedure to draw a regular polygon is:

```
to polygon :nsides :side
   repeat :sides [fd :side rt 360 / :nsides]
end
```

The figures below are regular polygons drawn using this procedure.

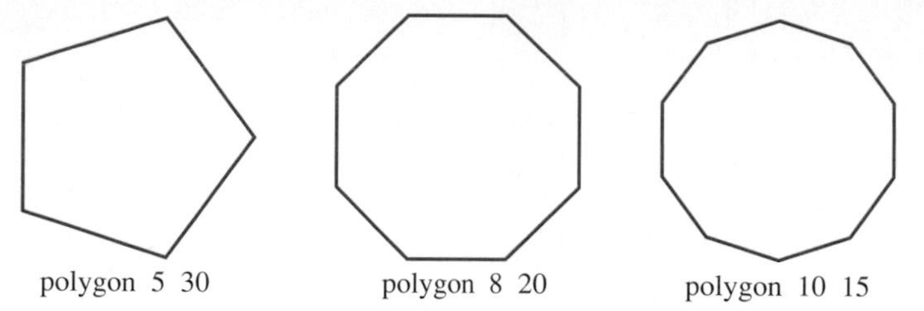

Stars are nearly as easy to draw with Logo as regular polygons. The open star below is drawn by the following Logo procedure.

```
to Ostar6
   repeat 6 [fd 30 rt 120 fd 30 lt 60]
end
```

The 120° right turn that the turtle makes at each point of the star is found by subtracting the interior angle, 60°, from 180°. The 60° left turn is found similarly, by subtracting the desired angle from 180°. This process of forward, right turn, forward, left turn is repeated for the six points of the star, forming a star-shaped dodecagon.

Drawing closed stars is as simple with Logo as drawing open stars. However, there is a restriction on the number of points in the star (this is investigated in problem 39). To draw a closed star, the sum of the measures of the turtle turn must add up to a multiple of 360°, but the measure itself must not divide 360 evenly (otherwise a regular polygon will be drawn instead of a star). Three procedures and the closed stars they draw are shown below. Two of the stars have seven points, but the angle of turn is different and so the shape of the stars is also different.

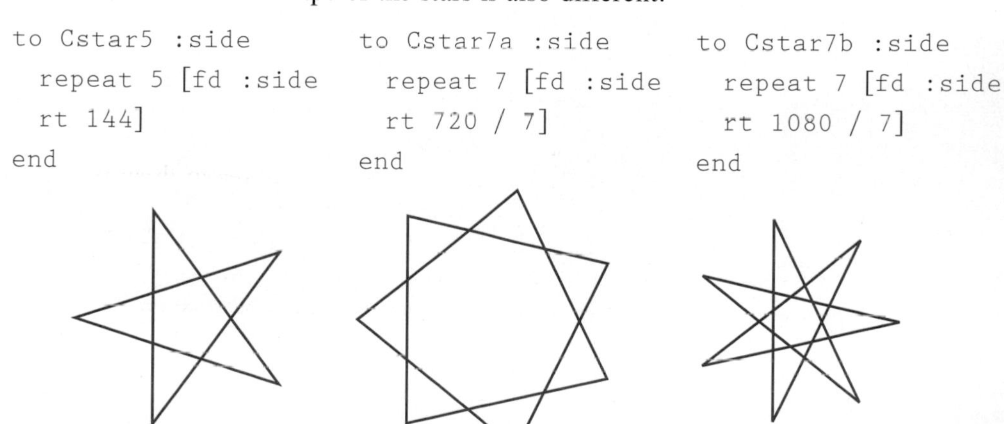

PROBLEM SET 10.2

Understanding Concepts

1. If the figure shown has the property listed, place a check in the table below.

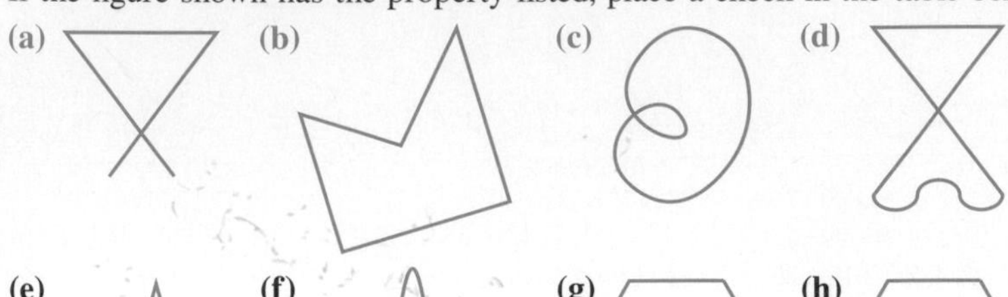

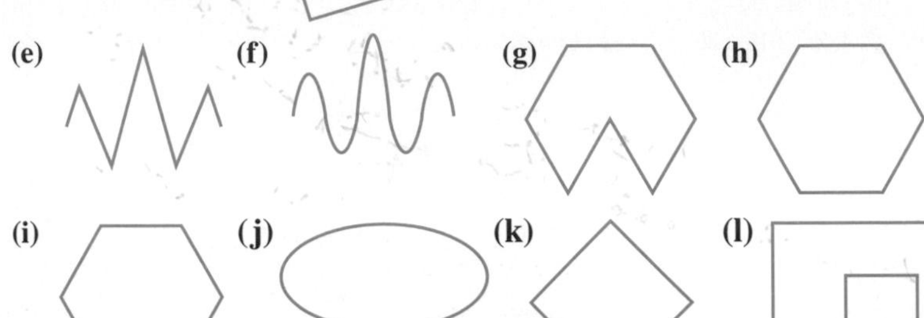

	(a)	(b)	(c)	(d)	(e)	(f)	(g)	(h)	(i)	(j)	(k)	(l)
Simple curve												
Closed curve												
Polygonal curve												
Polygon												

2. Draw figures which satisfy the given description.
   (a) a nonsimple closed four-sided polygonal curve
   (b) a nonconvex pentagon
   (c) an equiangular quadrilateral
   (d) a convex octagon
3. Can a line cross a simple closed plane curve 99 times? Explain why or why not.
4. Classify each region as convex or nonconvex.

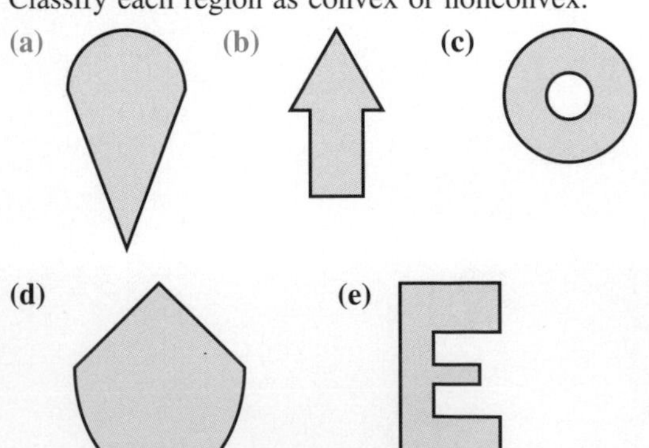

5. Imagine stretching a rubber band tightly about each figure shown. Shade the region within the band with a colored pencil. Is the shaded region always convex?

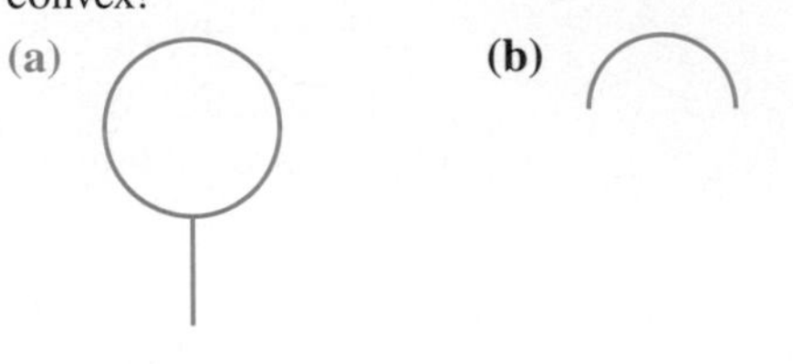

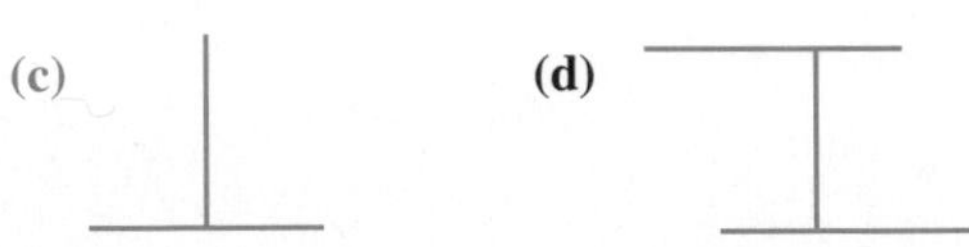

6. How many different regions in the plane are determined by these figures?

(a) 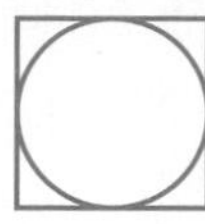 **(b)**

(c)  **(d)**

7. Determine the measures of the interior angles of this polygon.

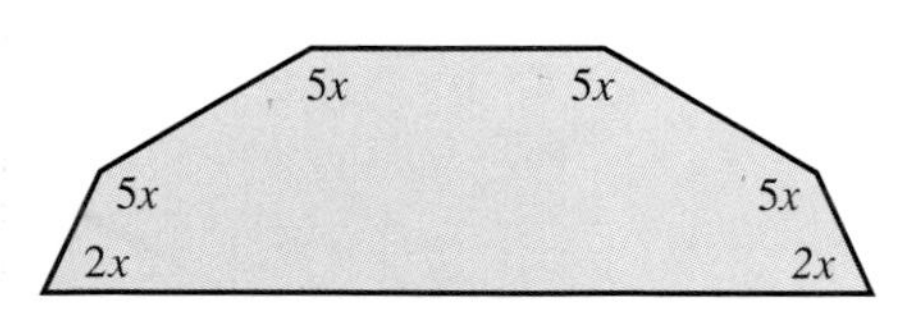

8. Calculate the measures of the angles in this nonconvex polygon.

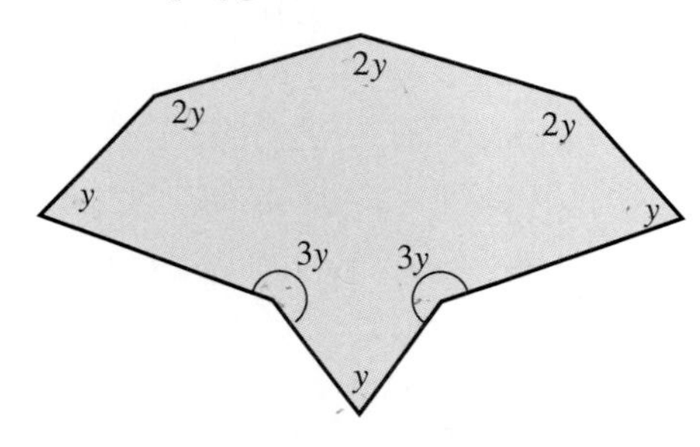

9. A **lattice polygon** is formed by a rubber band stretched over the nails of a geoboard. Find the sum of the measures of the interior angles of the following lattice polygons.

(a) **(b)**

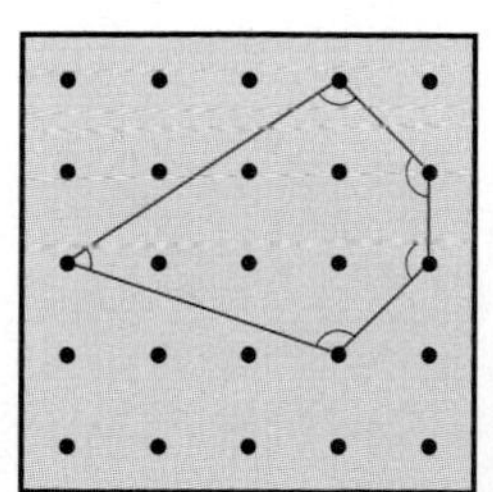

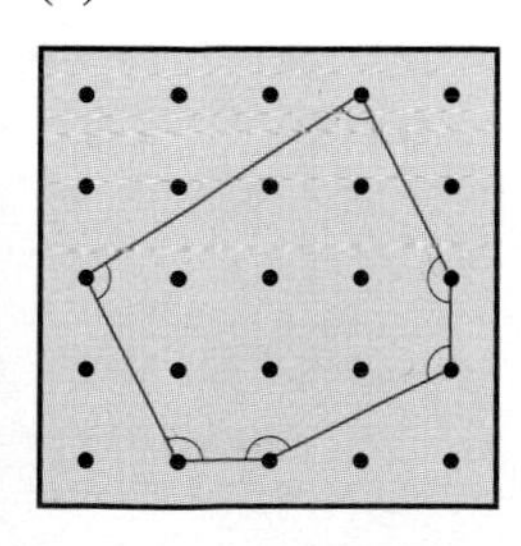

(c) **(d)**

10. Draw lattice polygons (see problem 9) on squared dot paper whose interior angles have the given sum of their measures. Place arcs which indicate the interior angles, as in problem 9.

(a) 180° **(b)** 1080° (c) 1440° **(d)** 1800°

11. The interior angles of an n-gon have an average measure of 175°.

(a) What is n?

(b) Suppose the polygon has flexible joints at the vertices. As the polygon is flexed to take on new shapes, what happens to the average measure of the interior angles? Explain your reasoning.

12. What is the amount of total turn for the following closed curves? Assign a positive measure to counterclockwise turning.

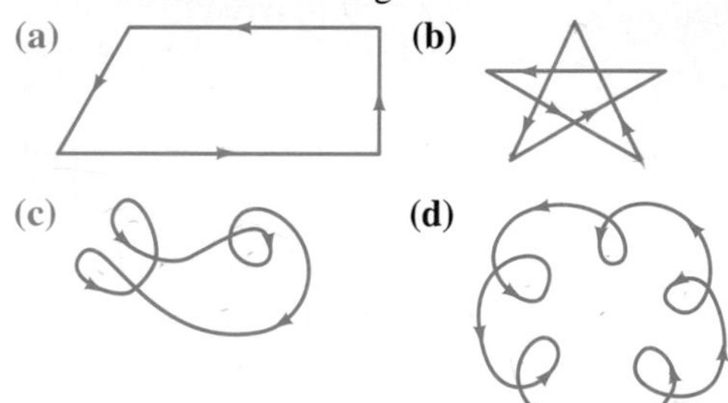

13. Suppose you walk north 10 paces, turn left through 24°, walk 10 paces, turn left through 24°, walk 10 paces, and so on.

(a) Will you return to your starting point?

(b) What is the shape of the path?

14. For each regular n-gon shown below, give the measures of the interior, exterior, and central angle.

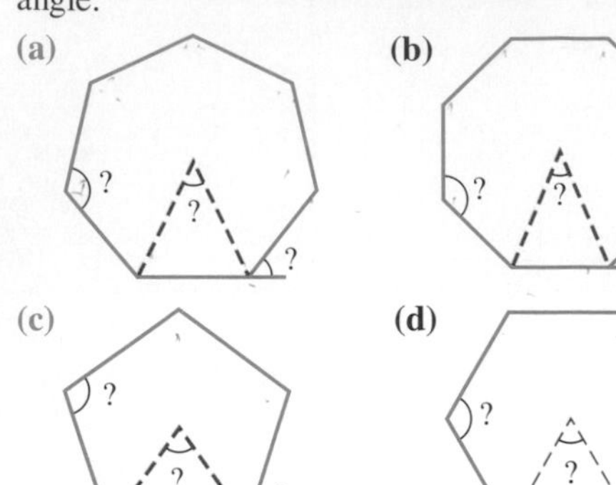

15. Use a protractor and a ruler to draw a regular heptagon whose sides are each 2 centimeters long. Describe your procedure.

16. (a) A regular n-gon has exterior angles of measure 15°. What is n?

    (b) A regular n-gon has interior angles each measuring $172\frac{1}{2}^\circ$. What is n?

Thinking Critically

17. The two rectangles shown below separate the plane into four regions. Draw similar pairs of rectangles that show it is possible to separate the plane into 3, 5, 6, 7, 8, 9, and 10 regions with two rectangles.

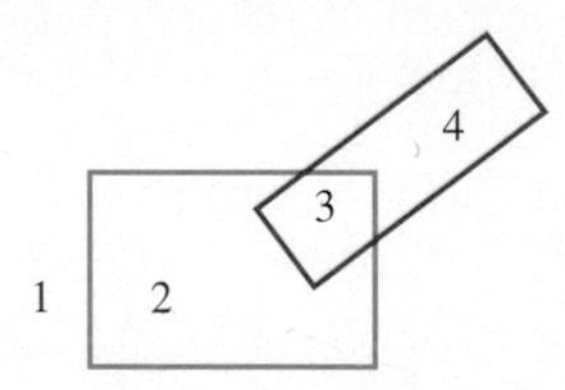

18. (a) How many regions in the plane can be formed with a system of three circles?

    (b) How many regions can be formed with four circles?

19. The segment $\overline{AB}$ is to be completed to become a side of a triangle ABC.

Describe, in words and sketches, the set of points C so that:

(a) $\triangle ABC$ is a right triangle and $\overline{AB}$ is a leg;

(b) $\triangle ABC$ is a right triangle and $\overline{AB}$ is a hypotenuse;

(c) $\triangle ABC$ is an acute triangle;

(d) $\triangle ABC$ is an obtuse triangle.

(*Hint:* For (b), proceed experimentally using the corner of a sheet of paper: C)

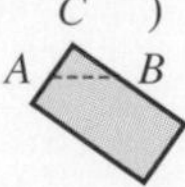

20. If the interior angle of a polygon has measure m, then $360° - m$ is called the measure of the **conjugate** angle at that vertex. Find a formula which gives the sum of the measures of the conjugate angles of an n-gon, and give a justification for your formula. As an example, the measures of the conjugate angles in this pentagon add up to 1260°.

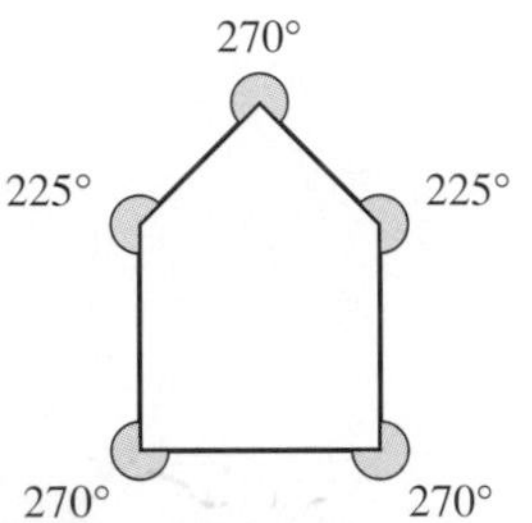

21. The heptagonal region shown on the left has been broken into five triangular regions by drawing four nonintersecting diagonals across the interior of the polygon, as shown on the right. In this way we say that the polygon is triangulated by diagonals.

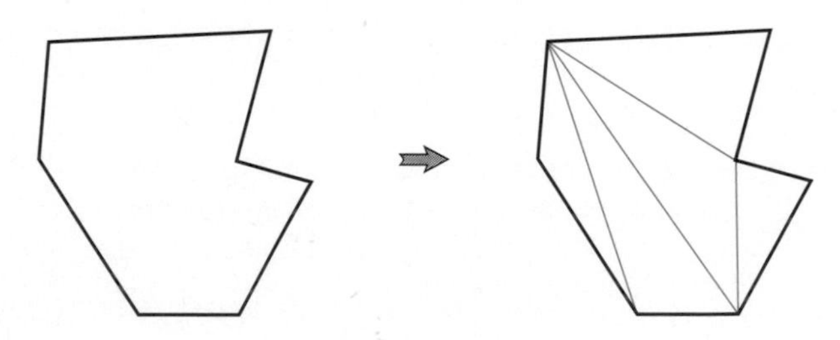

(a) Investigate how many diagonals are required to triangulate any n-gon.

(b) How many triangles are in any triangulation of an n-gon by diagonals?

(c) Explain how a triangulation by diagonals can give a new derivation of the formula $(n - 2) \cdot 180°$ for the sum of the measures of the interior angles of any n-gon.

22. The polygon on the left on the next page contains a point S in its interior which can be joined to any vertex by a segment which remains inside the polygon.

Drawing all such segments produces a triangulation of the interior of the polygon (compare to problem 21).

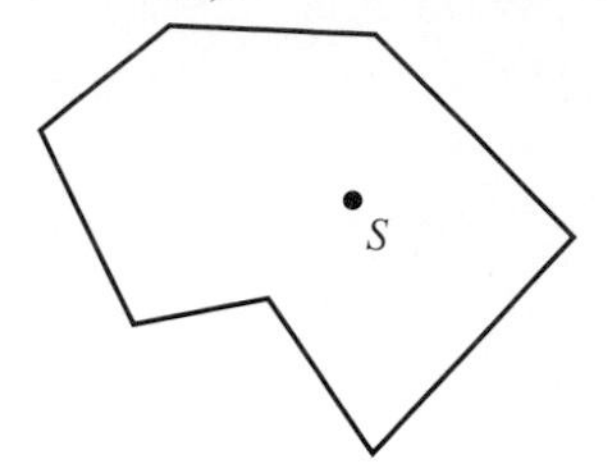

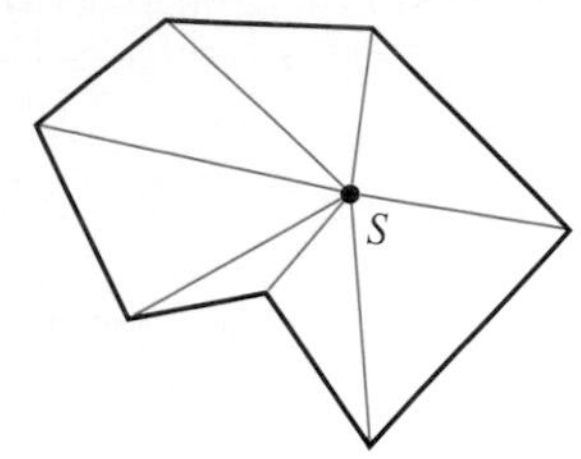

If an n-gon contains such a point S, explain how the corresponding triangulation can be used to derive the $(n - 2) \cdot 180°$ formula for the sum of interior angle measures.

23. (a) Find the sum of the angle measures in the 5-pointed star shown on the left below. Explain how the Total Turtle Trip Theorem can be used to obtain your answer.

(b) What is the measure of the angle in each point of the pentagram shown on the right?

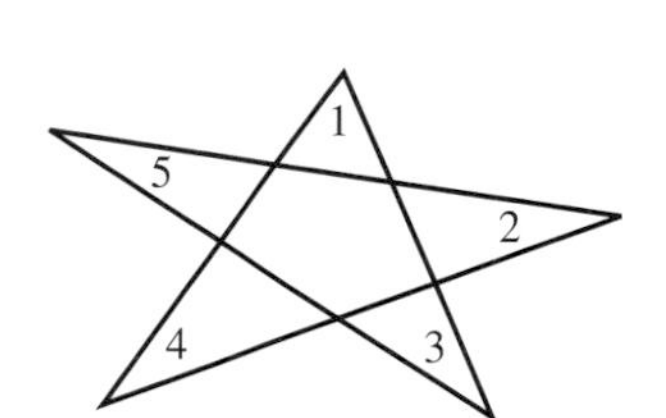

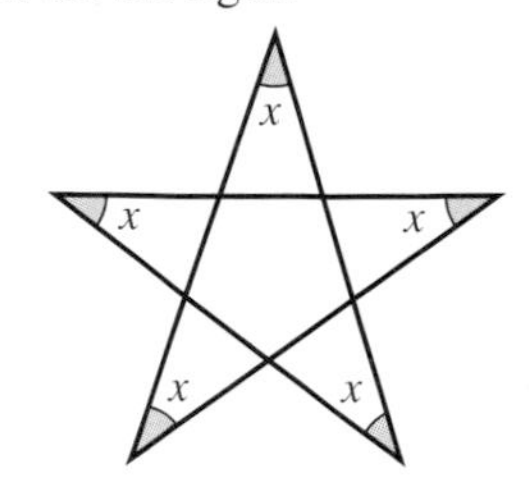

24. One-dimensional space consists of the points on a single line. Removing one point separates the line into two disjoint parts, and removing two points separates the line into three disjoint parts. In two-dimensional space (that is, a plane), removing one line separates the plane into two disjoint regions called half-planes. Removing two nonparallel lines separates the plane into four disjoint regions. The entries shown in the table below are easy to verify.

	Number of points removed from the Line or Number of Lines Removed from the Plane									
	0	**1**	**2**	**3**	**4**	**5**	**6**	**7**	...	n
Number of Pieces of Line Remaining	1	2	3							
Number of Pieces of Plane Remaining	1	2	4							

(a) Complete the upper missing row of the table.

(b) Draw three, four, or five lines in the plane which have general position, so that no two lines are parallel and no three lines are concurrent. Count the regions they determine and fill in the next three entries in the lower row of the table.

(c) Guess how many regions are determined by ten lines in general position (*Hint:* Look for a pattern in the table.)

(d) Find a formula for the number of regions determined by n lines in general position.

25. What is the largest number of regions you can form with a system of ten circles?

26. Regions can be formed in a circle by drawing chords, no three of which are concurrent. If C chords are drawn and they intersect in I points, determine a formula for the number of pieces, P (that is, the regions) that are formed inside the circle.

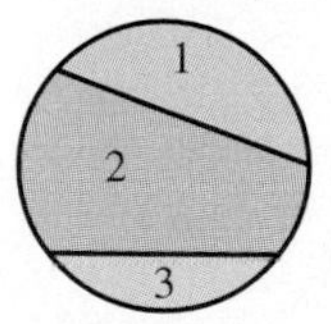

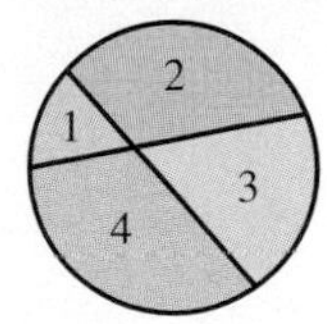

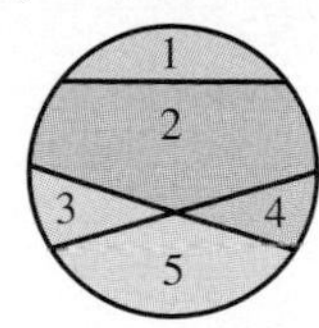

P	C	I
3	2	0
4	2	1
5	3	1

27. Show that for $n = 4$, 5, and 6 it is possible to draw n points in the plane so that every choice of three of the points are the vertices of an isoceles triangle.

28. Let $ABCD$ be a parallelogram.

(a) Prove that $\angle A$ and $\angle B$ are supplementary.

(b) Prove that $\angle A \cong \angle C$ and $\angle B \cong \angle D$.

29. Let $PQRS$ be a convex quadrilateral for which $\angle P \cong \angle R$ and $\angle Q \cong \angle S$.
(a) Prove that $\angle P$ and $\angle Q$ are supplementary.
(b) Prove that $PQRS$ is a parallelogram.

30. Let F and K be convex figures.
(a) Prove that $F \cap K$ is also a convex figure.
(b) Is $F \cup K$ necessarily a convex figure? Explain.

Thinking Cooperatively

31. With a black pencil draw a closed curve of any shape. With a red pencil draw a second closed curve which never passes through any point of intersection that the black curve has with itself or a previously drawn intersection of the red and black curve. Circle the points at which the black curve is crossed by the red curve.

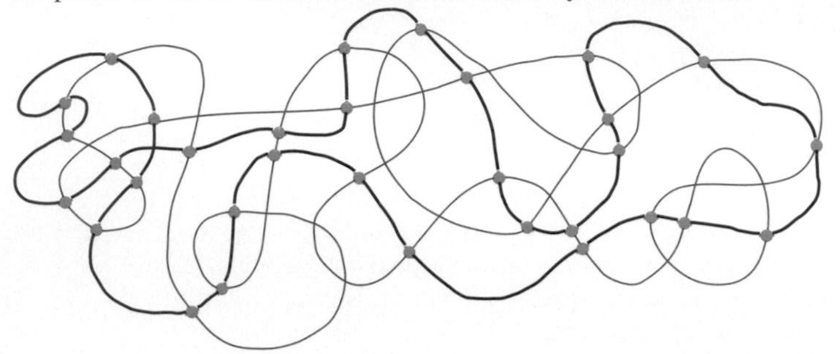

The example shows 32 points at which the red curve crosses the black one.
(a) Draw several examples of your own, and count the number of circled points of crossing.
(b) Explain why there are always an even number of crossing points. (*Hint:* Consider the regions determined by the black curve.) (This problem is due to Martin Gardner.)

32. Draw a simple closed curve C_1 in the plane with a black pencil. Next use a red pencil to draw a second simple closed curve C_2, where C_2 crosses C_1 several times as shown in the example below. The two curves separate the plane into regions of four types:
✓ : interior to both C_1 and C_2;
× : exterior to both C_1 and C_2;
□ : interior to C_1 and exterior to C_2;
■ : interior to C_2 and exterior to C_1.

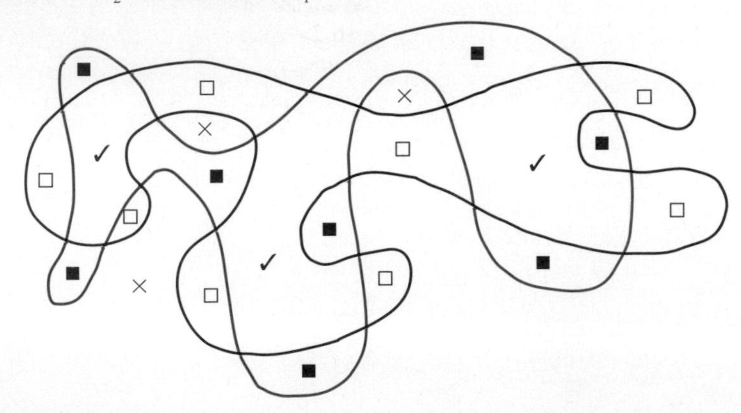

(a) Draw three (or more!) examples of curve systems as described. Label *all* of the regions with a ✓, ×, □, or ■ according to its type.
(b) Make a conjecture about the number of ✓ and × type regions on the basis of your examples.
(c) Make a conjecture about the number of □ and ■ type regions.

Making Connections

33. Access to underground utility cables, water pipes, and storm drains is usually provided by circular holes covered by heavy metal circular covers. What unsafe condition would be present if a square shape were used instead of the circular one?

34. The valve stems on fire hydrants are usually triangular or pentagonal. Fire trucks carry a special wrench with a triangular or pentagonal hole that fits over the valve stem.

(a) Why are squares and regular hexagons not used? (*Hint:* What is the shape of the jaws of ordinary adjustable wrenches?)

(b) Why are squares and regular hexagons the standard shape found in bolt heads and nuts?

Using a Computer
Using an Automatic Drawer

35. Draw two squares $ABCD$ and $AB'C'D'$ which share a common vertex at A. The labeling of the vertices is in the same direction (say counterclockwise) about the centers Q and E. Draw the midpoints R and S of $\overline{BD'}$ and $\overline{B'D}$.

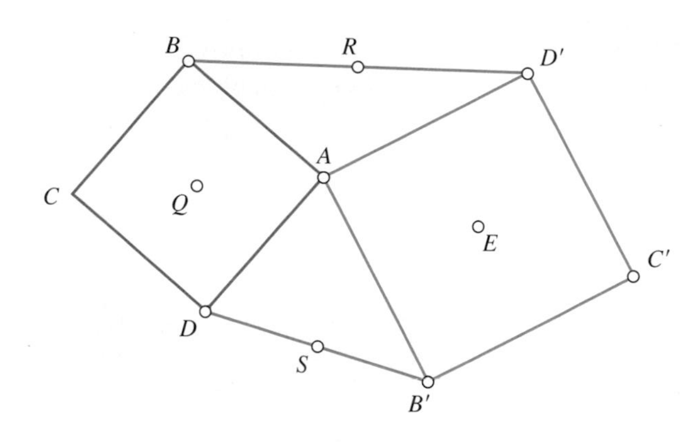

(a) What conclusion can you make about the quadrilateral $SQRE$?

(b) Does it make any difference if the squares overlap?

36. Draw a circle, and inscribe in it a convex quadrilateral $ABCD$ as shown. Next draw each pair of lines containing the opposite sides of $ABCD$. Let P and Q be the points of intersection of these pairs of lines. Finally, draw the rays which bisect $\angle APB$ and $\angle BQC$, and let R, S, T, and U be the points at which the rays intersect the quadrilateral $ABCD$. What kind of quadrilateral does $RSTU$ appear to be? Back up your guess with measurements of $RSTU$.

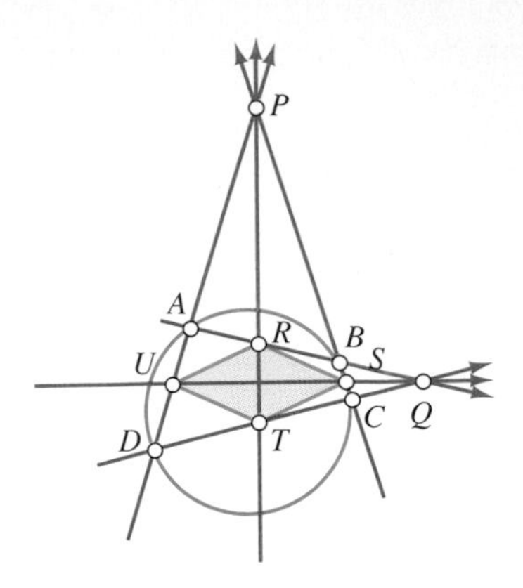

Using Logo

37. Write a Logo procedure to use to draw the following polygons.

(a) a regular pentagon (b) a regular hexagon
(c) a regular octagon (d) a regular nonagon
(e) a regular 20-gon (f) a regular 30-gon
(g) a regular 60-gon (h) a regular 360-gon

(i) How would you write a procedure to draw a circle?

(j) Images on a computer screen are obtained by lighting a finite number of pixels. Explain the difference between a circle on a computer screen and a circle in geometry.

38. (a) Write a procedure to draw the open star shown in Example 10.9, forming a decagon with five star points.

(b) Write a procedure to draw an open star with 16 sides and eight star points.

(c) Find a formula to determine the measures of the angles in the procedure for an open star with $2n$ sides and n star points, where $n \geq 3$ is an integer.

39. Fill in the table below with the missing information to draw closed stars. Write "no star possible" if it is not possible to draw a star with the specified number of points. Try to list all possible angle turns that give stars of different shapes.

Number of vertices	5	6	7	8	11	22	25
Angle turn	144°		720/7 or 1080/7				
Star							

For Review

40. Find two triangles, one acute and one obtuse, whose interior angle measures are each an integer multiple of 36°. Give drawings of each type of triangle.

41. In the figure below $\angle APC$ and $\angle BPD$ are right angles. Show that $\angle 1 \cong \angle 3$.

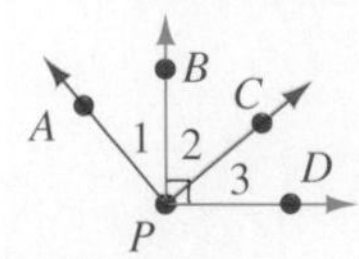

42. Prove that the two acute angles in a right triangle are complementary.

43. Let $\overline{PQ} \parallel \overline{AB}$ and $\overline{RQ} \parallel \overline{AC}$. Find the measures of $\angle 1, \angle 2, \ldots, \angle 8$. Explain how you find your answers.

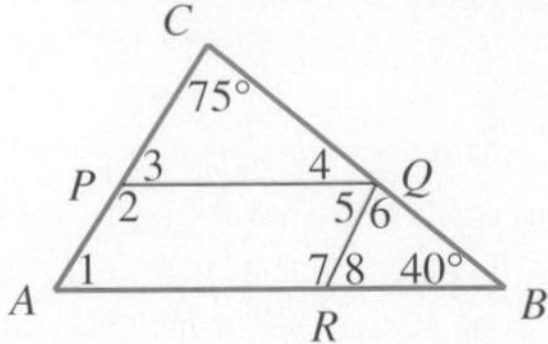

10.3 Tiling with Polygons

This section explores patterns in the plane which are formed by combining polygonal shapes. The art of tiling has a history as old as civilization itself. In virtually every ancient culture the artisan's choices of color and shape was guided as strongly by aesthetic urges as by structural or functional requirements. Imaginative and intricate patterns decorated baskets, pottery, fabrics, wall coverings, and weapons. Some examples of ornamental patterns from different cultures are shown in Figure 10.22.

In recent times the interest in tiling patterns has gone beyond its decorative value. For example, metallurgists and crystallographers wish to know how atoms can arrange themselves in a periodic array. Similarly, architects hope to know how simple structural components can be combined to create large building complexes, and computer engineers hope to integrate simple circuit patterns into powerful processors called neural networks. The mathematical analysis of tiling patterns is a response to these contemporary needs. At the same time the creation and exploration of tilings provides an inherently interesting setting for geometric discovery and problem solving in the elementary and middle school classroom.

Tiles and Tilings

The concepts introduced in the preceding section make it easy to give a precise definition.

> **DEFINITION Tiles and Tiling**
>
> A simple closed curve, together with its interior, is a **tile.** A set of tiles forms a **tiling** of a figure if the figure is completely covered by the tiles without overlapping any interior points of the tiles.

Since all points in the figure are covered there can be no gaps between tiles. Tilings are also known as **tessellations,** since the small square tiles in ancient Roman mosaics were called *tessella* in Latin.

Our main interest is when each tile is a polygonal shape. In most cases the figure to be covered by tiles is the entire plane.

Regular Tilings of the Plane

Each tiling shown in Figure 10.23 is a **regular tiling:** the tiles are regular polygons of one shape and they are joined edge-to-edge.

Figure 10.22
Tiling patterns from varied cultures

Pre-Inca fabric from Peru

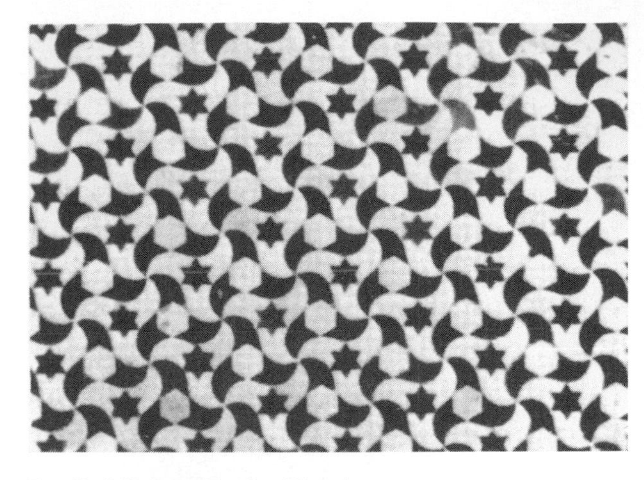

Detail of tiled wall in the Alhambra

Window of a fourteenth century mosque in Cairo

Mosaic floor of the fourteenth and fifteenth century in the Basilica of Saint Marks Cathedral, Venice

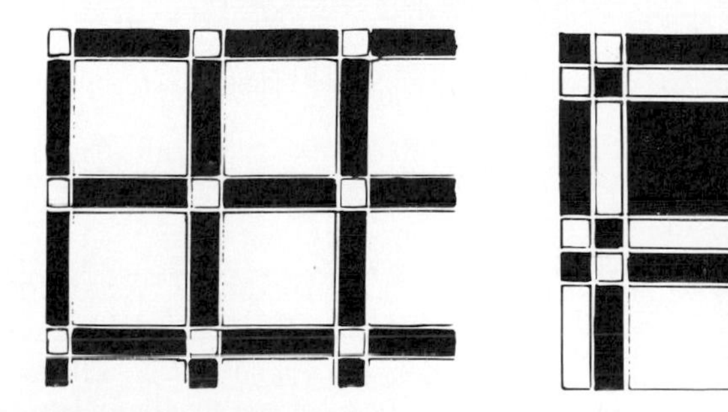

Tilings from Portugal, fifteenth to sixteenth centuries

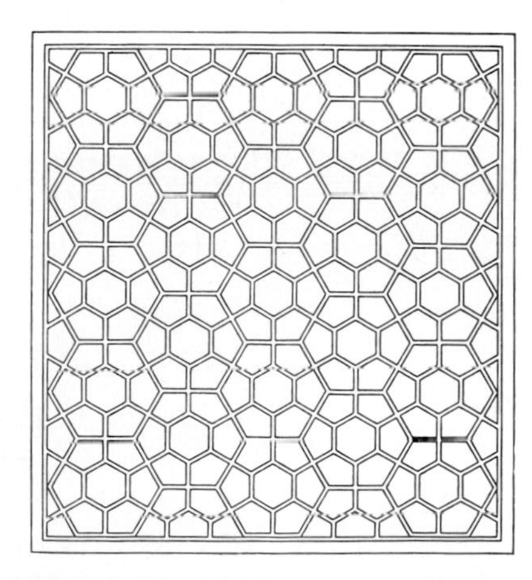

Chinese lattice work, used to support paper windows

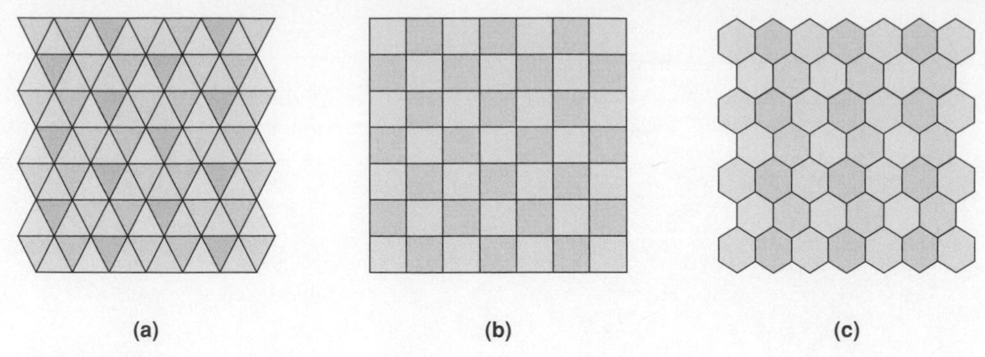

Figure 10.23
The three regular tilings of the plane

Any arrangement of nonoverlapping polygonal tiles surrounding a common vertex is called the **vertex figure.** Thus, four squares form each vertex figure of the regular square tiling, and three regular hexagons form each vertex figure of the hexagonal tiling. The measures of the interior angles meeting at a vertex figure must add to 360°. For example, in the square tiling $90° + 90° + 90° + 90° = 360°$.

Suppose we attempt to form a vertex figure with regular pentagons as shown in Figure 10.24. The interior angles of a regular pentagon each measure $(5 - 2) \cdot 180°/5 = 108°$. Thus, three regular pentagons fill in $3 \cdot 108° = 324°$ and leave a 36° gap. On the other hand four regular pentagons create an overlap since $4 \cdot 108° = 432° > 360°$. Since a vertex figure cannot be formed, no tiling of the plane by regular pentagons is possible.

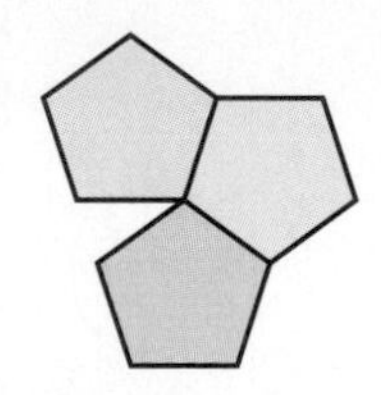
Three pentagons leave a gap.

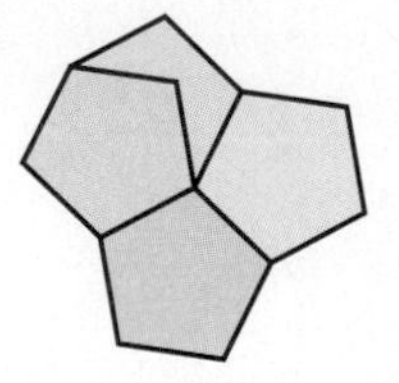
Four pentagons overlap.

Figure 10.24
Regular pentagons do not tile the plane.

Similarly a regular polygon of seven or more sides has an interior angle larger than 120°. Thus two such polygons leave a gap but three overlap. We have the following result.

> **THEOREM The Regular Tilings of the Plane**
>
> There are exactly three regular tilings of the plane: (a) by equilateral triangles, (b) by squares, and (c) by regular hexagons.

Semiregular Tilings of the Plane

A regular tiling uses congruent polygons of one type to tile the plane. What if regular polygons of several types are allowed? An edge-to-edge tiling of the plane with more than one type of regular polygon *and* with identical vertex figures is called a **semiregular tiling.** It is important to understand the restriction made about the vertex figures—*the same types of polygons must surround each vertex, and they must occur in the same order*. The two vertex figures in Figure 10.25 are not identical since the two triangles and the two hexagons on the left figure are adjacent but on the right side the triangles and the hexagons alternate with one another.

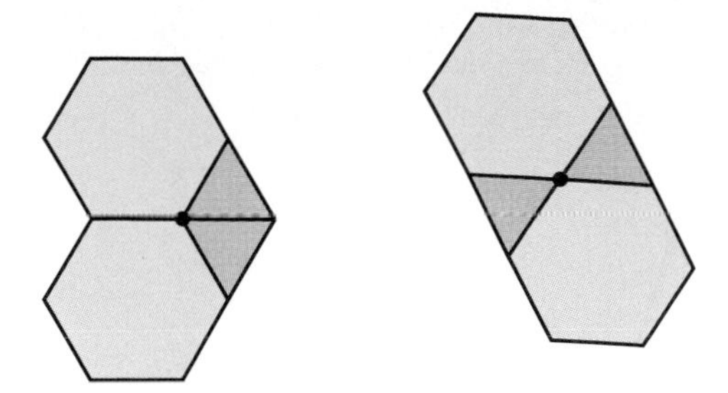

Figure 10.25
There are two distinct types of vertex figures formed by two equilateral triangles and two regular hexagons.

To see if the vertex figures in Figure 10.25 can be extended to form a semiregular tiling we must check to see if the pattern can be completed to make *all* of the vertex figures match the one shown. It is soon discovered that the pattern with adjacent triangles cannot be extended (try it!). On the other hand, the vertex figure of alternating triangles and hexagons extends to a semiregular tiling. You should be able to find it in Figure 10.26.

It can be shown that there are 18 ways to form a vertex figure with regular polygons of two or more types. Of these, eight extend to a semiregular tiling, shown in Figure 10.26.

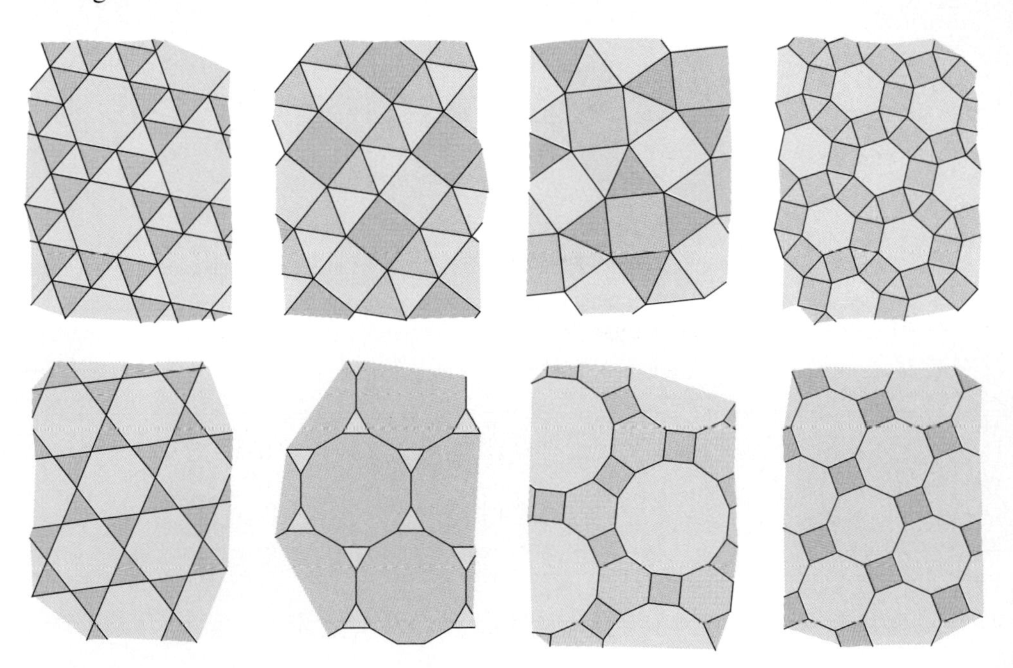

Figure 10.26
The eight semiregular tilings

HIGHLIGHT FROM HISTORY
Johannes Kepler and Tiling Patterns

The astronomer Johannes Kepler (1571–1630) is celebrated in scientific history for his identification of the elliptical shape of the orbits of the planets about the sun. Less known is Kepler's contribution to the theory of tiling. Here are some drawings from Kepler's book *Harmonice Mundi,* which he published in 1619.

SOURCE: Illustration from *Tilings and Patterns* by Branko Grünbaum and G.C. Shephard. Copyright © 1987 by W. H. Freeman and Company. Reprinted by permission.

Tilings with Irregular Polygons

In the following example it is helpful to cut tile patterns from cardboard or heavy card stock. You can then trace around them to form, if possible, a tiling of the plane.

EXAMPLE 10.14

Exploring Tilings with Irregular Polygons

Which of the polygons below tile the plane?

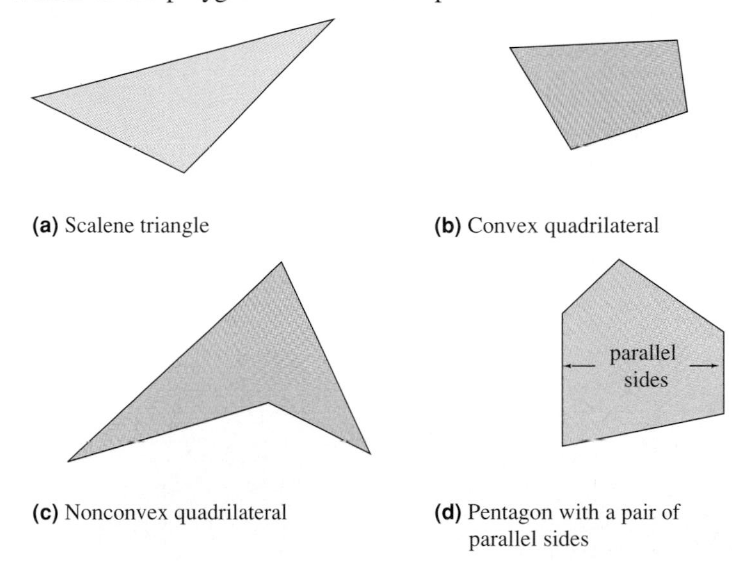

(a) Scalene triangle

(b) Convex quadrilateral

(c) Nonconvex quadrilateral

(d) Pentagon with a pair of parallel sides

SOLUTION

(a) Two triangles of identical size and shape can be joined along a corresponding edge to form a parallelogram. Since it is evident that parallelograms tile the plane, it follows that *any triangle will tile the plane.*

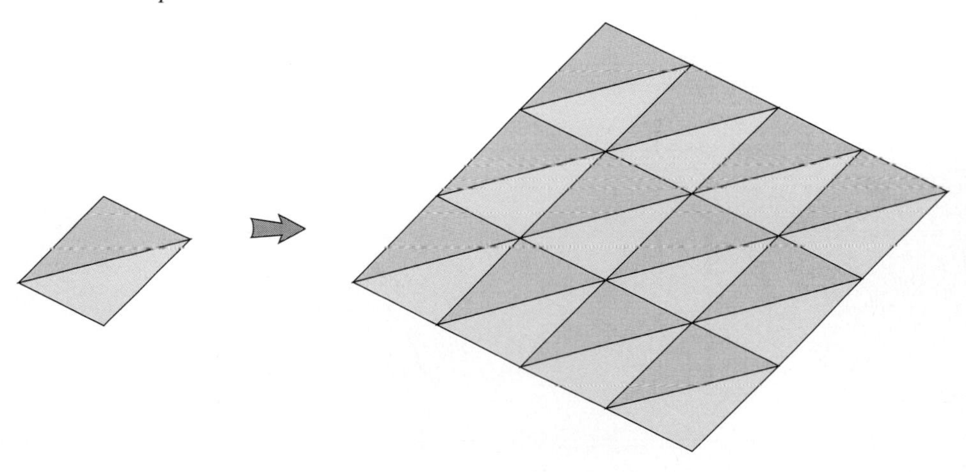

(b) **(c)** As illustrated below, any quadrilateral will tile the plane. A 180° turn about the midpoint of a side rotates the quadrilateral from one position to an adjacent position. Notice that each vertex of the tiling is surrounded by angles congruent to the four angles of the quadrilateral, whose measures add up to 360°.

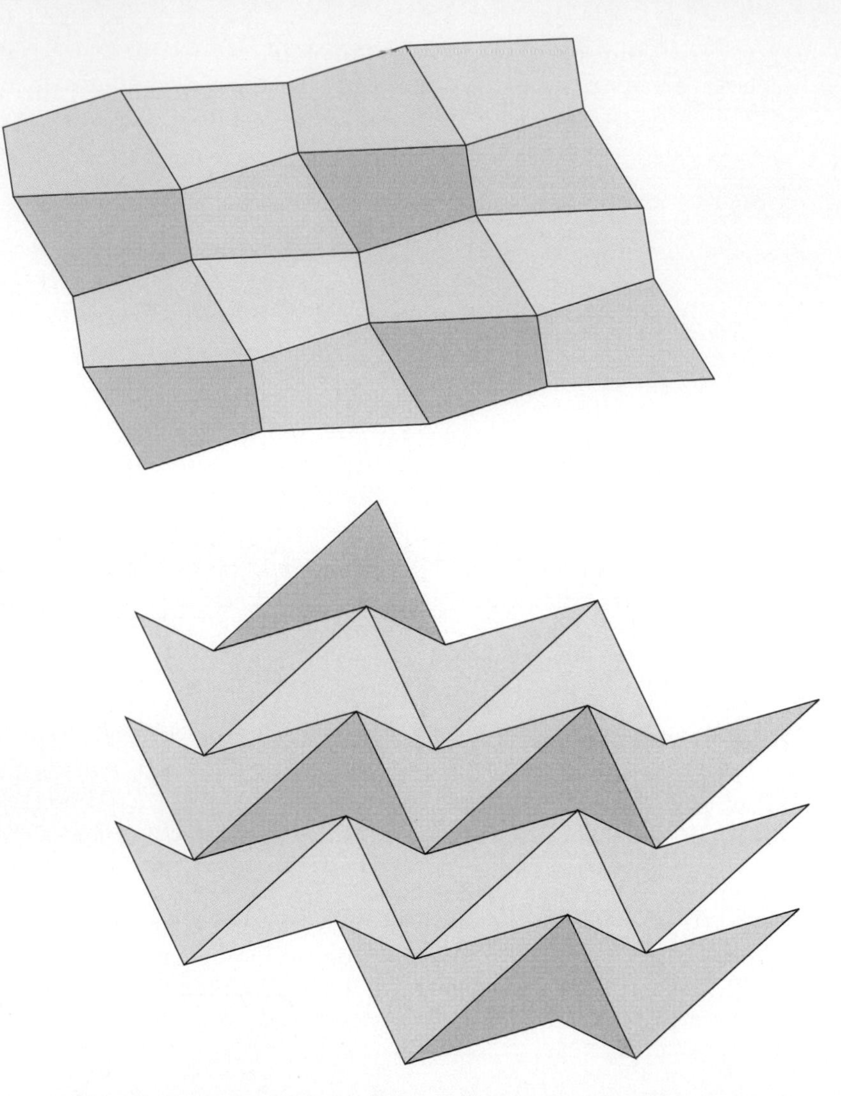

(d) A pentagonal tile with two parallel sides can always tile the plane. If the parallel edges are congruent the tiling will be edge-to-edge.

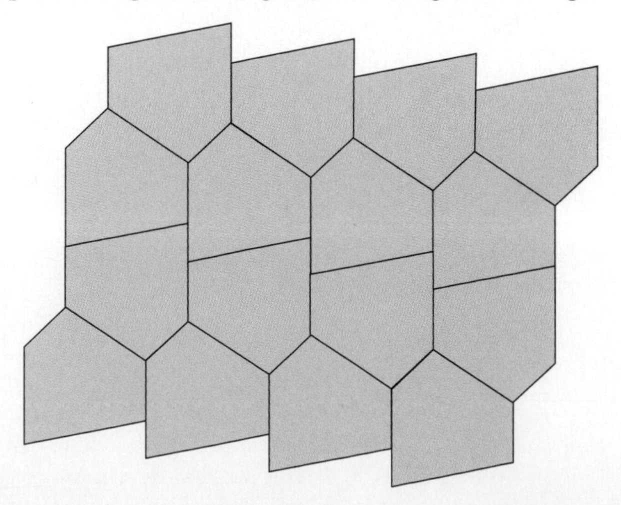

■

Much like the quadrilateral tiling shown on the previous page, a convex hexagon will tile the plane if it has a pair of opposite sides which are parallel and of the same length (see problem 12). If all three pairs of opposite sides are congruent and parallel, it is not even necessary to rotate the tile from one position to any other (see problem 11). It has been shown that no convex polygon with seven or more sides can tile the plane. The following theorem summarizes our discoveries.

THEOREM Tiling the Plane with Polygonal Tiles of the Same Shape

The plane can be tiled by:

- any triangular tile;
- any quadrangular tile, convex or not;
- certain pentagonal tiles (for example, those with two parallel sides);
- certain hexagonal tiles (for example, those with two opposite parallel sides of the same length).

The plane cannot be tiled by any convex tile with seven or more sides.

JUST FOR FUN

Dissection Delights

Trace the Star of David below and cut along the lines to obtain five pieces. E.B. Escott discovered that these same five pieces will form a square. Assemble a square from your pieces, and then sketch how the square is formed.

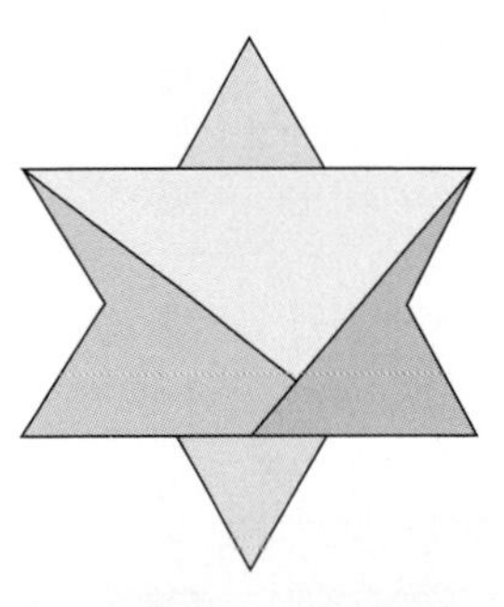

Harry Lindgren, of the Australian Patent Office, is an expert on dissection puzzles of this type. Show that the six pieces in his dissection of a regular dodecagon (12-gon) can also be arranged to form a square. A remarkable collection of dissections can be found in Lindgren's book *Recreational Problems in Geometric Dissections and How to Solve Them,* Dover Publications, 1972.

Although no convex polygon of seven or more sides can tile the plane, there are many interesting examples of nonconvex polygons that tile. Figure 10.27 shows a striking example of a spiral tiling by 9-gons (nonagons) created by Heinz Voderberg in 1936.

Figure 10.27
Heinz Voderberg's spiral tiling with nonagons

Tilings of Escher Type

The Dutch artist Maurits Cornelius Escher (1898–1972) has created a large number of artistic tilings. His designs have great appeal to the artist and general public and have also captured the interest of professional geometers. Escher's graphic work is most often based on modifications of known tiling patterns. However, he also discovered new principles of pattern formation which mathematicians had overlooked.

To see how Escher created his print of the birds on the left side of Figure 10.28, we begin by identifying the underlying grid of parallelograms shown on the right. The

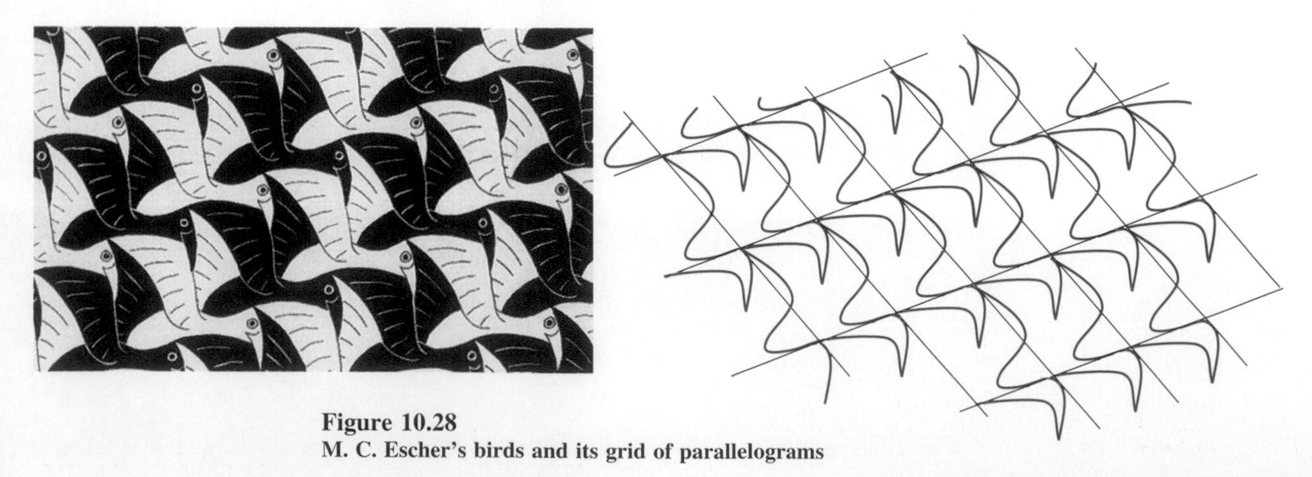

Figure 10.28
M. C. Escher's birds and its grid of parallelograms

modification of the upper left side of a parallelogram to form the V-shape between the wings is matched by an identical modification of the opposite lower right side to form the head and beak. The remaining pair of opposite sides of the parallelogram is modified in a like manner to complete the bird-shaped tile.

WORTH READING

Escher's Method of Artistic Tilings

"How did he do it?"

The work of M.C. Escher provokes that irrepressible question. A recurring theme as well as a device in his work from 1937 onward was, in his words, the "regular division of the plane." We see the jigsaw puzzle-like interlocking of birds, fish, lizards, or other creatures in his work; their rigid paving is usually just a fragment, a pause in a transition from two to three dimensions, a springboard from lockstep order to freedom. Escher confesses that the subject is for him a passion.

In his 1958 book *Regelmatige Vlakverdeling (Regular Division of the Plane),* he tells us much about why he uses regular division, explains some of the geometric elements of regular division, addresses the central question of figure and ground, and leads us through the development of a metamorphosis of form. But when we persist in asking "How did he do it?"—that is, how did he make those interlocking creatures—we do not find answers. We do find a few tantalizing hints in the book that Escher did study some technical papers and that he worked out his own theory:

> At first I had no idea at all of the possibility of systematically building up my figures. I did not know any "ground rules" and tried, almost without knowing what I was doing, to fit together congruent shapes that I attempted to give the form of animals. Gradually, designing new motifs became easier as a result of my study of literature on the subject, as far as this was possible for someone untrained in mathematics, and especially through the formulation of my own layman's theory, which forced me to think through the possibilites. It remains an extremely absorbing activity, a real mania to which I have become addicted, and from which I sometimes find it hard to tear myself away.

Professor Schattschneider's book is an account of Escher's discovery of the world of geometry and how he used his knowledge to create his intriguing interlocking figures.

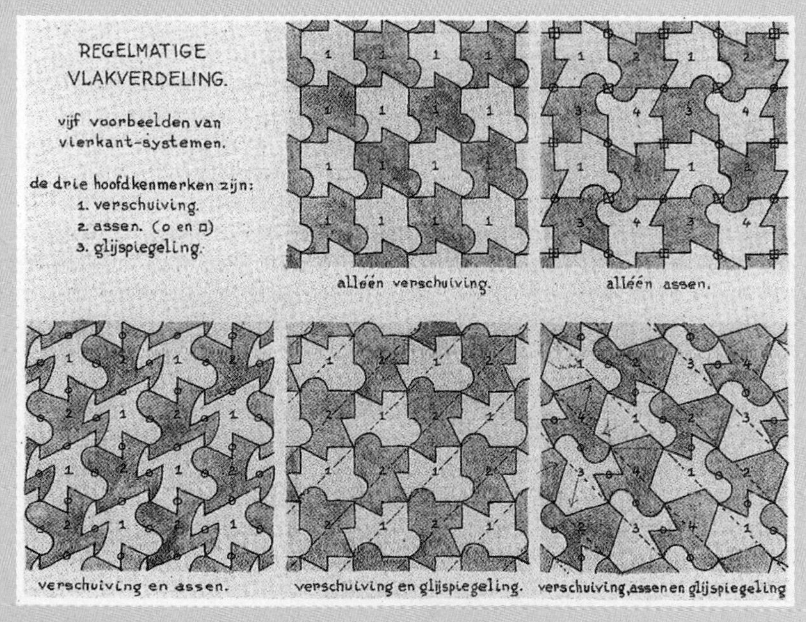

SOURCE: From *M. C. Escher Visions of Symmetry* by Doris Schattschneider.

Similar procedures will transform any polygonal tiling to an Escher-like tiling. For example, sixth grade teacher Nancy Putnam modified each of three pairs of opposite parallel congruent sides of hexagon *ABCDEF* to create the whale tiling shown in Figure 10.29.

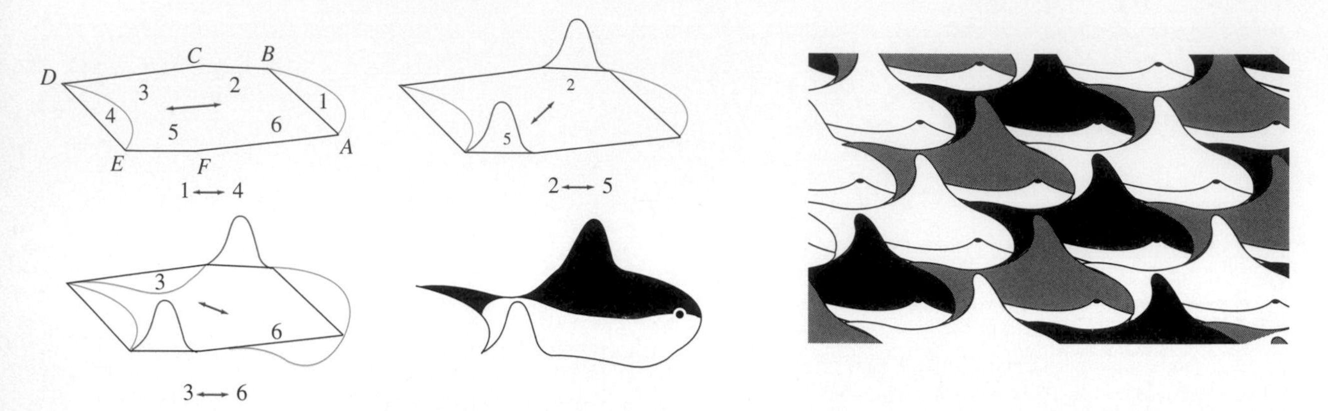

Figure 10.29
Sixth grade teacher Nancy Putnam modified a hexagon with opposite parallel congruent sides to create an Escher-type tiling.

Logo Tilings

Earlier in this section we discussed the three regular tilings of the plane. It is an interesting task to create these tilings on the computer screen with Logo procedures. We will consider the triangular tiling in the next example, leaving the square and hexagonal tilings as exercises.

Tiling the Plane with Equilateral Triangles

Create a Logo procedure to tile the screen with equilateral triangles.

SOLUTION

We approach the problem by dividing the task into simpler parts, using a bottom-up design of the procedure. We will begin with the simplest case—draw one triangle with a horizontal lower side. The procedure used in Section 10.2 to draw an equilateral triangle with a vertical left-hand side is easy to modify to draw the triangle with a horizontal lower side.

```
to flat.tri :side
   rt 30
   repeat 3 [fd :side rt 120]
   lt 30
end
```

Next we write a procedure to draw a row of flat triangles, introducing an additional variable to represent the number of triangles in the row. This number will vary, based on the length of the side of the triangles. The procedure consists of drawing a flat triangle, turning the turtle, moving it forward along the row, turning the turtle again, and then repeating this process.

```
to row.tri :num :side
   repeat :num [flat.tri :side rt 90 fd :side lt 90]
end
```

Next, we write a procedure to draw a double row of triangles. The turtle moves to create an additional triangle in each row that will help to fill out the screen with triangles.

```
to doublerow.tri :num :side
   row.tri :num :side
   rt 90 bk :num * :side
   lt 120 fd :side rt 30
   row.tri :num :side
   rt 30 fd :side
   rt 60 bk :num * :side lt 90
end
```

The tiling with triangles is created by the following procedure which repeats drawing double rows of triangles.

```
to tile.tri :num :side
   repeat 3 [doublerow.tri :num :side]
end
```

The tiling can be centered on the screen by first picking up the pen, moving the turtle to the lower left of the screen, and then putting the pen down and running the **tile.tri** procedure. ■

Next we will investigate tiling with an irregular polygon.

LOGO EXAMPLE 10.B

Tiling the Plane with a Trapezoid

Create a procedure to tile the plane with the trapezoidal tile shown.

```
to trap
   rt 30 fd 20 rt 60 fd 15 rt 60
   fd 20 rt 120 fd 35 rt 90
end
```

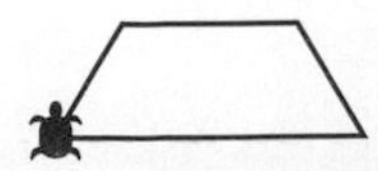

SOLUTION

We will proceed much like we did for the triangular tiling. The following procedure will draw one row of the isosceles trapezoids, leaving spaces to be filled in with downward facing trapezoids.

```
to row.trap :num
   repeat :num [trap pu
   rt 90 fd 50 lt 90 pd]
end
```

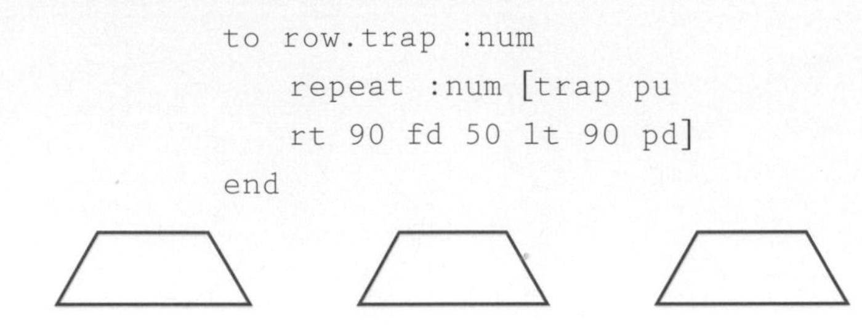

Next we will fill in the spaces in the row by turning the turtle downward and drawing the trapezoids that fit in the spaces.

```
to fullrow.trap :num
   row.trap :num
   rt 30 fd 20 rt 150
   row.trap :num
   pu lt 90 bk 10 lt 90 pd
end
```

We may now write the procedure to tile a portion of the plane with the specified trapezoids. As with the triangle tiling, the turtle can first be moved to the lower left of the screen to give a more centered tiling.

```
to tile.trap :num
   repeat 5 [ fullrow.trap :num]
end
```

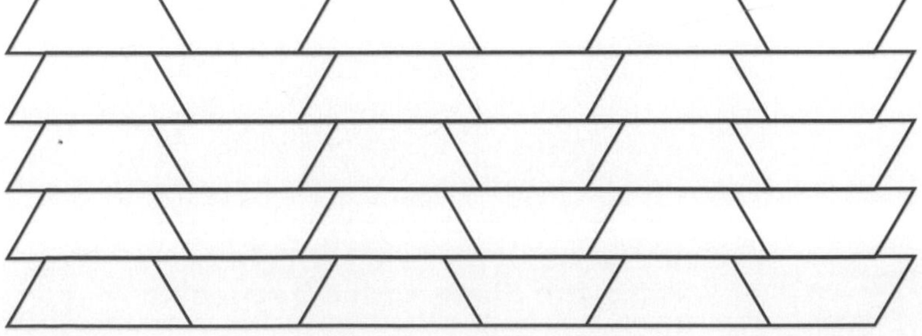

■

The methods illustrated in the examples can be adapted to create many other interesting and often beautiful tilings with Logo.

PROBLEM SET 10.3

Understanding Concepts

1. On dot paper arranged in a square grid show that the given shape will tile the plane.

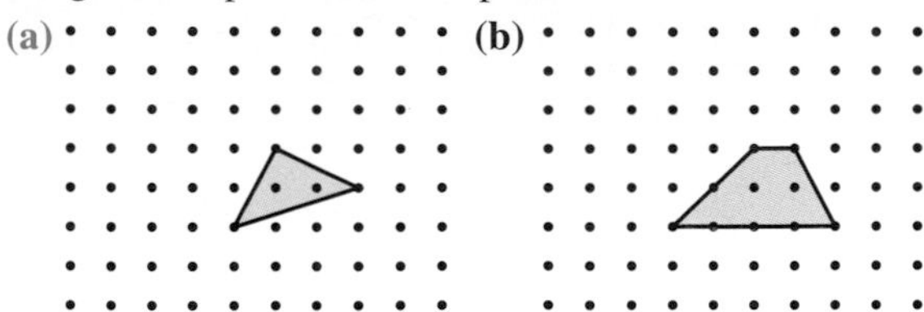

2. On "isometric" dot paper (arranged in a grid of equilateral triangles) show that the given shape will tile the plane.

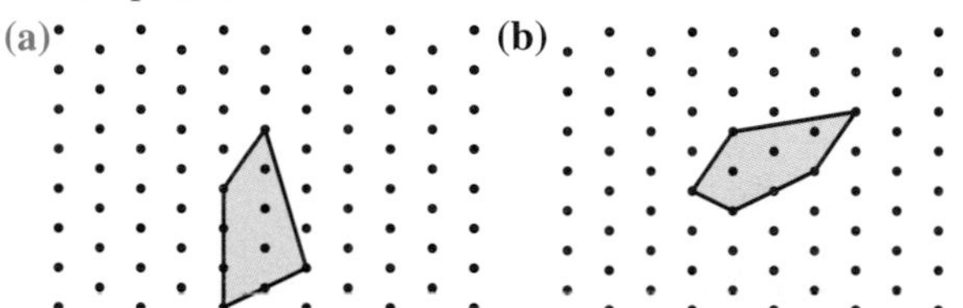

3. Branko Grünbaum and G. C. Shephard (*Tilings and Patterns,* W. H. Freeman and Co. 1987) discovered the tiling shown below in the children's coloring book *Altair Design* (E. Holiday, London: Pantheon, 1970).

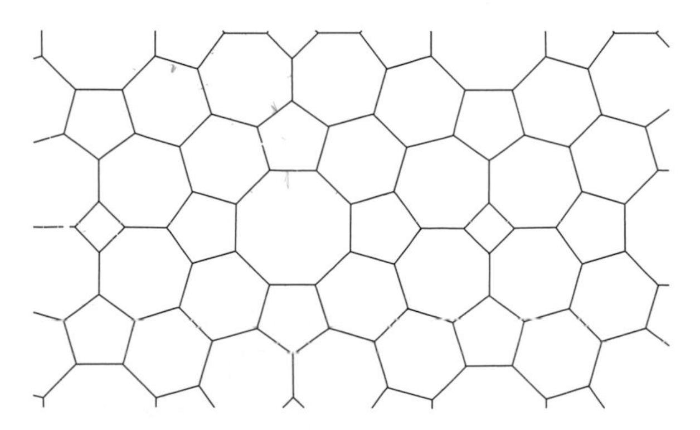

(a) What kinds of polygons appear?

(b) Grünbaum and Shephard claim this is a "fake" tiling by regular polygons. Explain why.

4. A vertex figure of regular polygons is shown.

(a) Find the angle measures of each polygon and directly verify that they add up to 360°.

(b) Explain why the vertex figure does not extend to form a semiregular tiling.

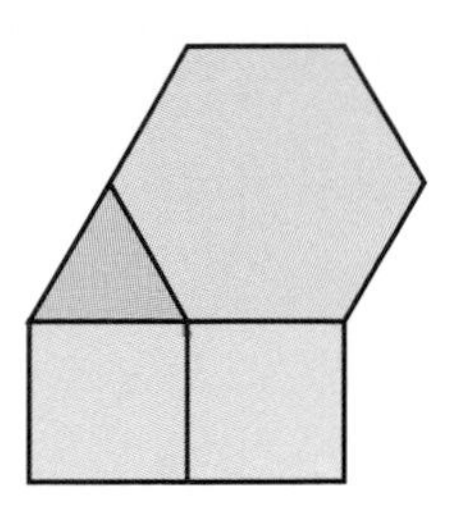

5. Consider the vertex figure formed by a square, a regular pentagon, and a regular 20-gon. Find the measures of the interior angle of each polygon and show that these three measures add up to 360°.

6. One of the most interesting sets of tiles are the seven **tangram** pieces, which originated in ancient China. As shown below, there are five triangles, a square, and a parallelogram. A serviceable set can be cut from a square of cardboard, although plastic and wooden sets are easy to buy or make.

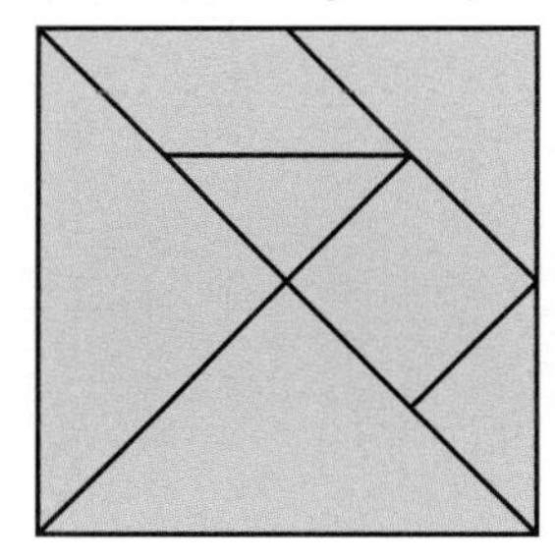

The most common activity is the Chinese ***tangram puzzle.*** A figure is shown in outline, and the challenge is to tile the figure using *all* 7 tangram pieces. Form the following animals (taken from the Multicultural Poster Set, National Council of Teachers of Mathematics, 1984).

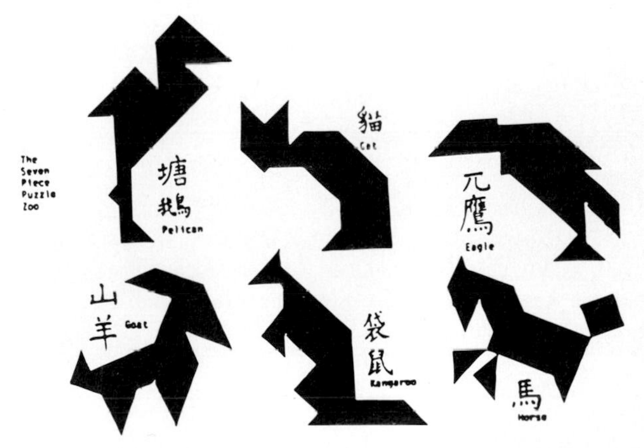

7. Some "letters" of the alphabet will tile the plane. For each letter shown, create an interesting tiling on square dot paper. Look for different patterns which use the same tile.

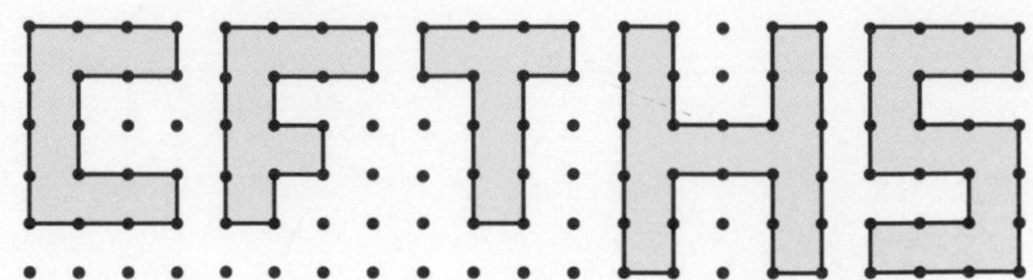

8. Tiles formed by joining five congruent squares edge-to-edge are called **pentominoes.** Pentominoes, and the more general class of polyominoes, were invented by mathematician and electrical engineer Solomon Golomb in 1953 in a talk to the Harvard mathematics club. There are 12 distinctly shaped pentominoes, shown labeled by letters which they somewhat resemble.

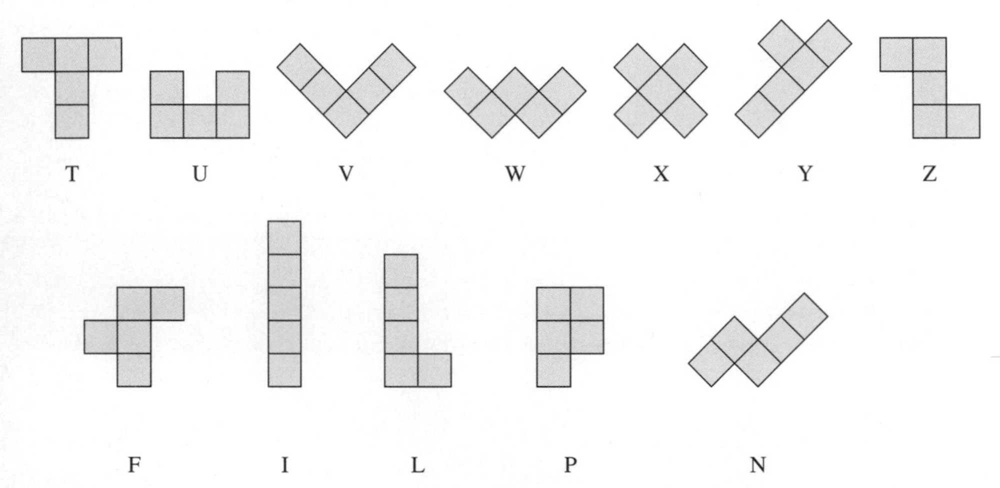

Source: Figure 10, "The pentominoes" from *Polyominoes* by Solomon W. Golomb, page 23. (Charles Scribner's Sons, 1965, Revised Edition, Princeton University Press, 1994).

Use squared paper (or dot paper) to decide which pentominoes tile the plane.

9. A tetromino is a tile formed by joining four congruent squares, where adjacent squares must share a common edge.

(a) Find the five differently shaped tetrominoes.

(b) Will the five tetrominoes combine to tile a 4 by 5 rectangle?

10. A hexagonal tile can be formed by rotating any convex quadrilateral 180° about the midpoint of any side. Since any of the four sides can be chosen, there are four hexagonal tiles which can be considered. Trace around patterns cut from card stock to show that each of the four hexagons tiles the plane.

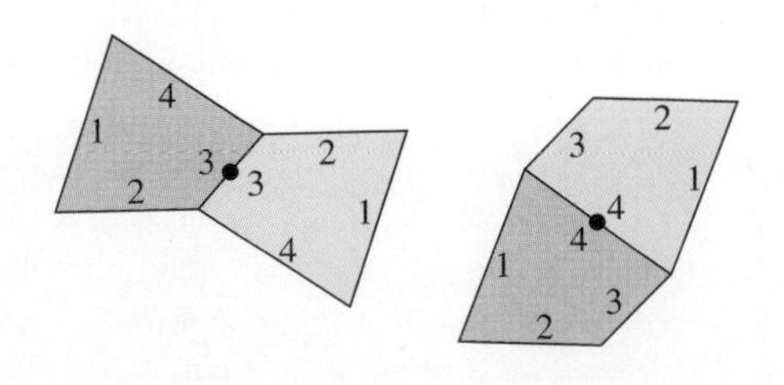

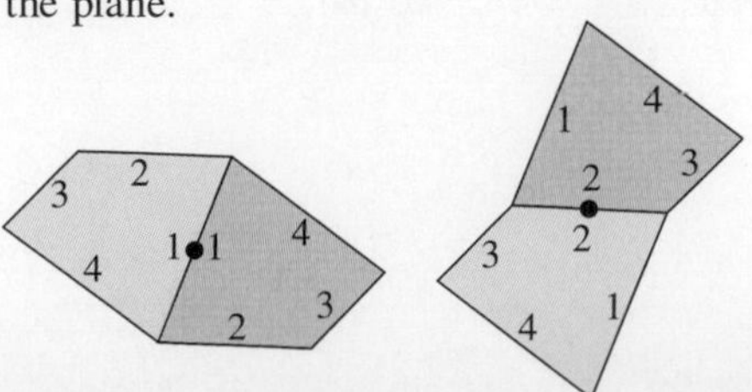

11. On heavy paper or cardboard, draw a convex nonregular hexagon with each side congruent and parallel to its opposite side. Cut out the shape with scissors, and use the pattern to show that the hexagon tiles the plane. Is it necessary to rotate the hexagon to move it to a new position in the tiling?

12. Repeat problem 11, but for a convex hexagon with just one pair of parallel and congruent opposite sides. (*Hint:* It will help to rotate the tile pattern 180° about the midpoint of the nonparallel sides.)

Thinking Critically

13. If overhead slides of the semiregular tilings of Figure 10.26 are made, in most cases it would not make any difference if the slide were placed on the projector upside down. There is one exception however. Which semiregular tiling is not identical to its mirror image? Carefully describe how the two images differ.

14. Suppose a vertex figure of regular polygons includes a regular octagon. Show that the figure must include another octagon and a square.

15. Create an Escher-like tiling by using the method of translational modification on a tiling by parallelograms.

16. Create an Escher-like tiling based on the translational modification of a hexagonal tiling such as shown in Figure 10.29.

17. For any integer n, $n \geq 3$, show that there is some n-gon that tiles the plane.

Thinking Cooperatively

18. An equilateral triangle and a parallelogram are each examples of "reptiles," short for "repeating tile." In each case copies of the tile can be arranged to form its own similar shape.

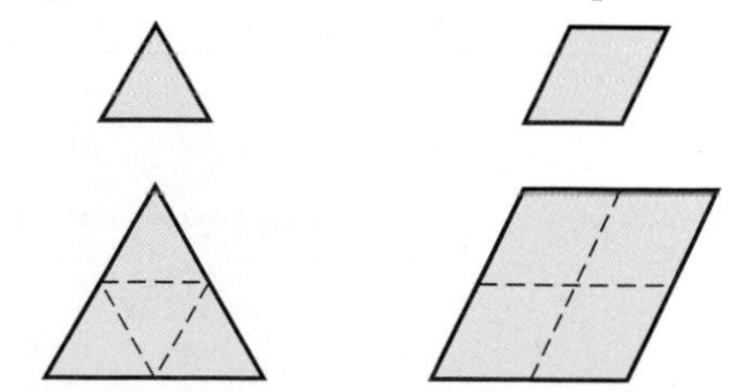

Use square dot paper to show that each of these shapes is a "reptile."

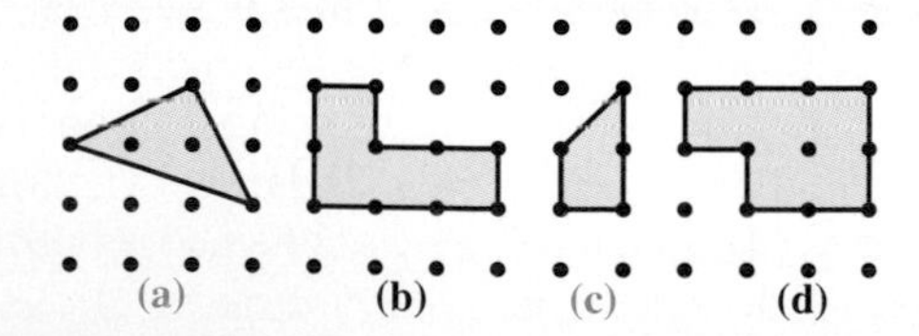

19. A hexiamond is formed from six congruent equilateral triangles. There are 12 different hexiamonds, including the Sphinx, Chevron, and Lobster shown below. Find the remaining nine hexiamonds and see if you can match their shape to their names: Hexagon, Crook, Crown, Hook, Snake, Yacht, Bar, Signpost, Butterfly.

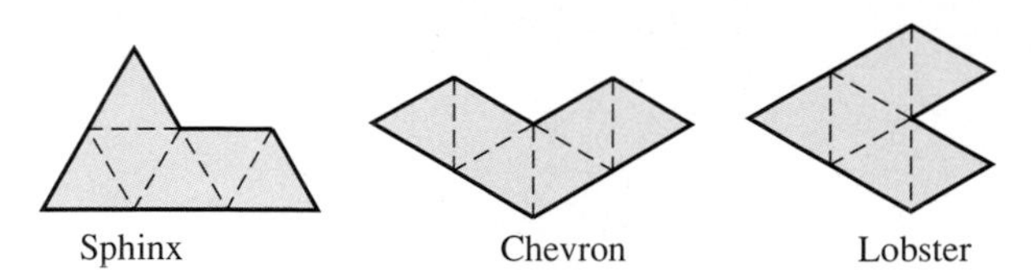

20. (a) Show that the Sphinx is a reptile. (See problems 18 and 19 for the definition of a reptile and a diagram of the Sphinx.)
   (b) It requires four copies of the Sphinx to form a second generation Sphinx. How many copies of the original Sphinx are required to form a third generation Sphinx? Explain your reasoning and provide a sketch.
   (c) Explain why any reptile provides a tiling of the plane.

21. (a) Show that the Chevron is a reptile. (See problems 18 and 19.)
   (b) Show that the Lobster is a reptile in *two* ways, one way using 36 Lobsters and a second way using only 25 Lobsters.

22. Find all of the different convex figures which can be tiled by tangrams (see problem 6). Be sure to use all seven tangram pieces in each of the convex figures. There are 13 noncongruent figures in all, and most of the figures can be tiled in several ways.

Using a Computer
Using an Automatic Drawer

23. Draw two nonoverlapping squares (of possibly different size) with a common vertex. Then complete it to an octagonal tile by drawing a parallelogram in each of the spaces between the squares, as shown below. Finally, draw parallel translates of the octagon to tile the plane, as shown on the next page.
   (a) What do you notice about the grid of points formed by the translates of the point V inside each octagon?
   (b) What do you notice about the grid of points formed by the corners A, C, E, G of the octagon?

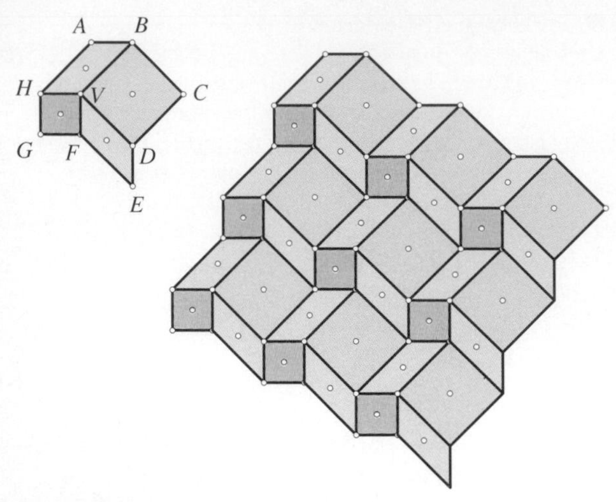

Using Logo

24. Use Logo Examples 10.A and B as a guide to write procedures to tile the plane with **(a)** squares and **(b)** regular hexagons.

25. **(a)** Write a procedure to create a tiling with the parallelogram shown.

```
to parallelogram
    repeat 2 [ fd 30 rt 110
    fd 40 rt 70 ]
end
```

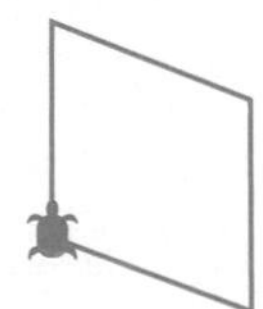

(b) Create your own parallelogram and write a procedure to tile the plane with it.

26. Write a Logo procedure to tile the plane with the triangular tile drawn by the following procedure. (*Hint*: First create a parallelogram with two copies of the triangle.)

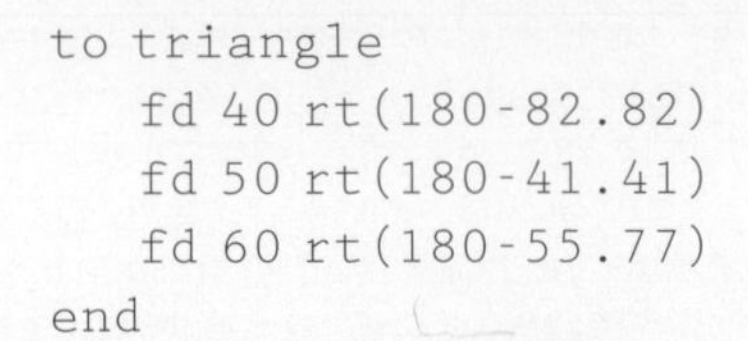

```
to triangle
    fd 40 rt(180-82.82)
    fd 50 rt(180-41.41)
    fd 60 rt(180-55.77)
end
```

27. Write a procedure for a rectangle and use it to create a brickwork tiling resembling that shown below.

For Review

28. What rule is used to separate the letters of the alphabet in the following arrangement?

A EF HI KLMN T VWXYZ
―――――――――――――――――
BCD G J OPQRS U

29. Find four points, A, B, C, D, in the plane that satisfy these conditions: $\overline{AD} \perp \overline{BC}$, $\overline{CD} \perp \overline{AB}$, $\overline{AC} \perp \overline{BD}$.

30. Show how to draw this closed polygonal curve in such a way that no side is retraced and the pencil is not lifted until the drawing is completed.

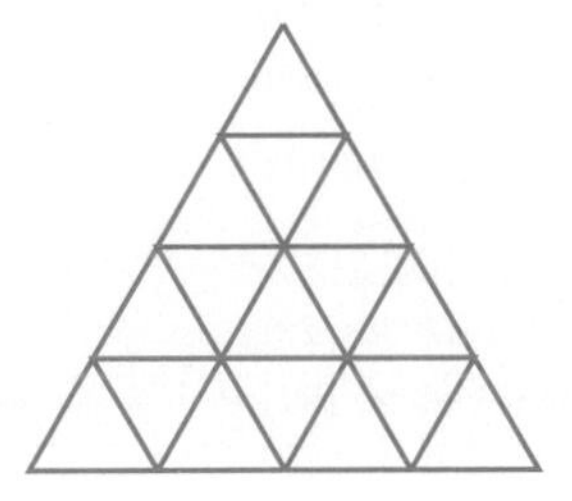

10.4 Networks

The Königsberg Bridge Problem

The great Swiss mathematician Leonhard Euler (1707–1783) (recall the pronunciation "oiler") lived for a short time in the East Prussian city of Königsberg, now called Kaliningrad in the Russian Federation. The Pregel River flows through the city, forming two islands. At the time of Euler there were seven bridges connecting the islands to one another and to the two sides of the river. Euler's own drawing is reproduced in Figure 10.30. The islands are labeled A and D, the shores of the river are labeled B and C, and the bridges are labeled a, b, c, d, e, f, and g.

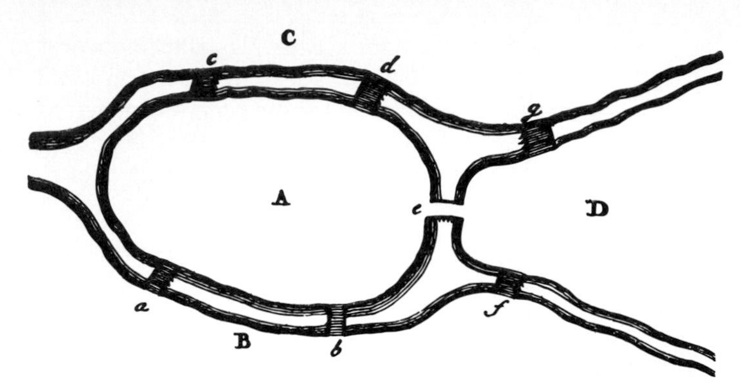

Figure 10.30
The seven bridges of Königsberg in the early 1700s

It was common for Königsbergers to take Sunday walks, and people wondered if it was possible to walk over all seven bridges without crossing any bridge more than one time. Euler solved the now famous Königsberg Bridge Problem to illustrate the ideas of what he called "the geometry of position," and what today is called topology. His most important step was to associate each of the four land masses with a point—A, B, C, or D. For each bridge joining one land mass to another he drew a curve from one point to the other corresponding point. The result, shown in Figure 10.31, is a system of points and curves known as a **network.**

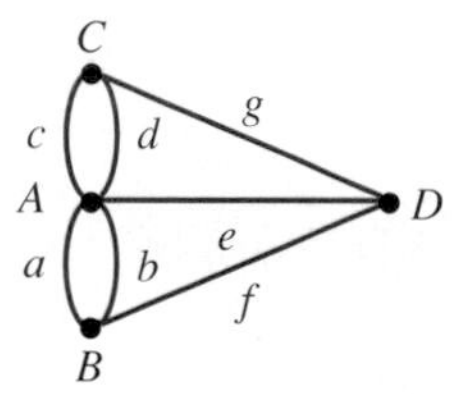

Figure 10.31
The network corresponding to the Königsberg Bridge Problem

The distances between points and the precise shape of the curves joining points are of no importance; what is important is that there are two bridges between A and B, that there is no bridge connecting B and C, and so on. Because the network contains all of the problem's relevant information, Euler was able to phrase the Königsberg Bridge Problem in a new way.

*Without lifting your pencil can you trace over **all** edges of the network exactly once?*

Euler realized that a deeper understanding of the problem would be gained if the question were asked for general networks, not just the one corresponding to the Königsberg Bridges.

DEFINITION Network

A **network** consists of two finite sets:

- a set of **vertices,** represented by a set of points in the plane,

and

- a set of **edges** that join some of the pairs of distinct vertices, represented by joining the corresponding points in the plane by a curve.

Some additional examples of the networks are shown in Figure 10.32. Any edge of a network must always connect two different vertices and there is no other vertex between the endpoints of the edge. In particular, a point at which two edges cross one another is not a vertex of the network. For example, network (2) of Figure 10.32 has just the six vertices shown by the large dots.

Figure 10.32
Four examples of networks

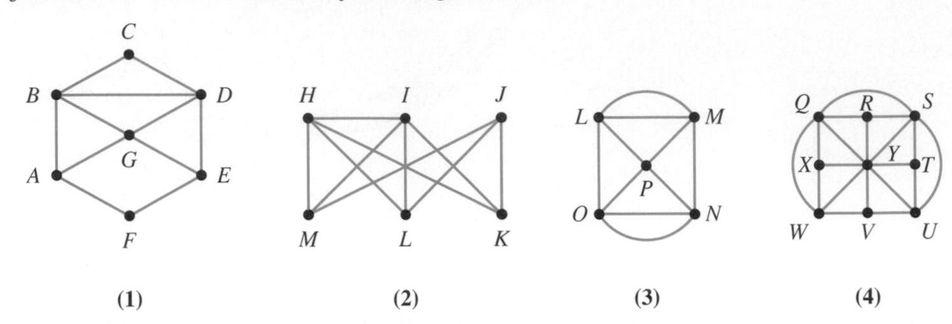

A **path** in a network is a curve traced by following a sequence of edges in the network, where no edge is retraced but vertices can be revisited. If every pair of vertices of a network can be joined by some path, then we say that the network is **connected.** For example, each of the four networks in Figure 10.32 is connected. The Konigsberg Bridge Problem is then equivalent to asking if there is a path that covers each edge of a network once and only once.

DEFINITION Euler Paths and Traversable Networks

An **Euler path** in a network is a path that traverses each edge once and only once. A network is **traversable** if, and only if, it has an Euler path.

In the next example, you will discover that classifying the vertices of a network as even or odd is most helpful in a search for an Euler path.

DEFINITION Degree, Even and Odd Vertices

- The **degree** of a vertex is the number of edges emanating from the vertex.
- A vertex is **odd** if it has odd degree.
- A vertex is **even** if it has even degree.

For example, vertices A and E of the network (1) of Figure 10.32 are odd since they have degree 3. The vertices B and C are even with respective degrees 4 and 2.

EXAMPLE 10.15

When Is a Network Traversable?

(a) Which of the networks in Figure 10.32 are traversable?
Experiment by tracing a path on a sheet of paper laid over the network.

(b) What is the number of odd vertices in each of the networks?

(c) Do you see any connection between the traversability of a network and the number of odd vertices?

SOLUTION

(a) Network (1) is traversable; one path is *ABCDEFAGBDGE*.
Network (2) is not traversable.
Network (3) is traversable; one path is *LMNOLMPNOPL*.
Network (4) is not traversable.

(b) Network (1) has 2 odd vertices: *A* and *E*.
Network (2) has 4 odd vertices: *J*, *K*, *L*, *M*.
Network (3) has 0 odd vertices.
Network (4) has 6 odd vertices: *Q*, *R*, *S*, *T*, *V*, *X*.

(c) The networks with 0 or 2 odd vertices are traversable.
The networks with 4 or 6 odd vertices are not traversable. ■

Euler observed that each time a path passes through a vertex it uses two edges: one to enter the vertex and another to exit. Except for the beginning and ending vertices all of the other vertices of a traversable network must therefore be even vertices, and it is impossible to traverse a network with more than two odd vertices. If there are two odd vertices, these are necessarily the endpoints of the Euler path. If there are no odd vertices the Euler path must terminate at the same vertex as it started, and the path forms a closed curve passing over each edge exactly one time.

This reasoning shows that if a network is traversable then it has 0 or 2 odd vertices. With more effort, the converse can also be shown: any connected network with 0 or 2 odd vertices is traversable. This is a celebrated result of Euler.

THEOREM Euler's Traversability Theorem

A connected network is traversable if, and only if, it has either no odd vertices or two odd vertices. If it has no odd vertices, any Euler path is a closed curve which ends at the same vertex it started from. If the network has two odd vertices, these vertices are the endpoints of any Euler path.

Since the Königsberg Bridge network has four odd vertices, Königsbergers must recross some bridge on their walk.

The Königsberg Bridge Problem and the traversability of networks may appear to have no practical importance, but in fact there are many useful applications to real problems. Here are two examples:

- What route can a telephone company inspection crew follow to check all of its lines without having to go over any section twice?
- What route can a street sweeper follow to clean all of the city streets and not have to travel over any blocks that have already been cleaned?

Counting Vertices, Edges, and Regions in Planar Networks

DEFINITION Planar Network

A network is **planar** if it is drawn in the plane without any intersection points of its edges other than endpoints.

For example, the network with four vertices and edges between each pair of distinct vertices is planar, as shown in Figure 10.33. There are no intersections of the edges. The network on five vertices with edges between each pair of distinct vertices is not planar, since it is impossible to arrange all of the edges so that no two of them intersect. By removing just one edge you can arrange the remaining nine edges between five vertices to form a planar network. Try it!

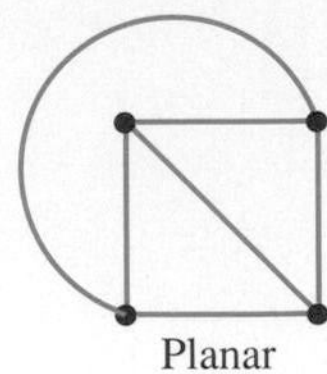

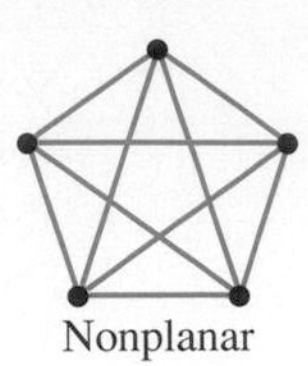

Figure 10.33
A planar and nonplanar network

Any connected planar network separates the plane into disjoint regions. It is interesting to count the number of vertices V, the number of edges E, and the number of regions R which correspond to a connected network.

EXAMPLE 10.16 Counting Vertices, Edges, and Regions

The connected network below has $V = 7$ vertices, $E = 9$ edges, and separates the plane into $R = 4$ regions.

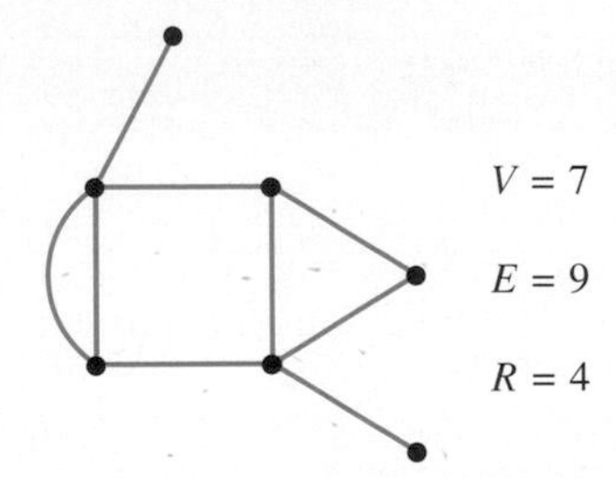

Count V, E, and R for each of the following networks.

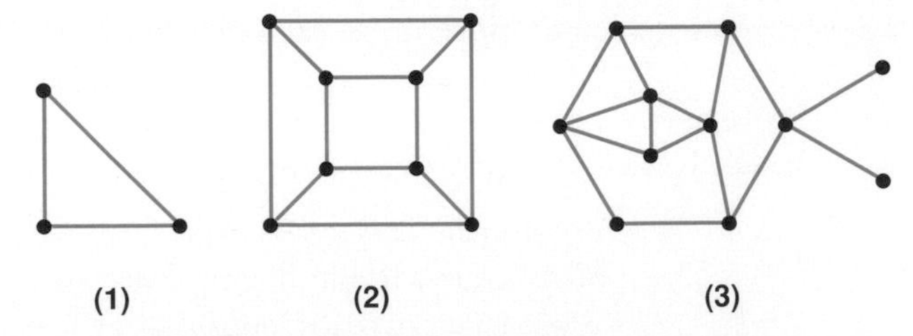

	V	R	E
	7	4	9
(1)			
(2)			
(3)			

Fill in the table of values of V, R, E. Do you see a pattern? Draw more examples to check your conjecture.

SOLUTION

You should discover that the sum of the number of vertices and the number of regions is 2 more than the number of edges. ■

The exploration in Example 10.16 leads to the conjecture that the formula $V + R = E + 2$ holds for every connected planar network. An example is shown in Figure 10.34.

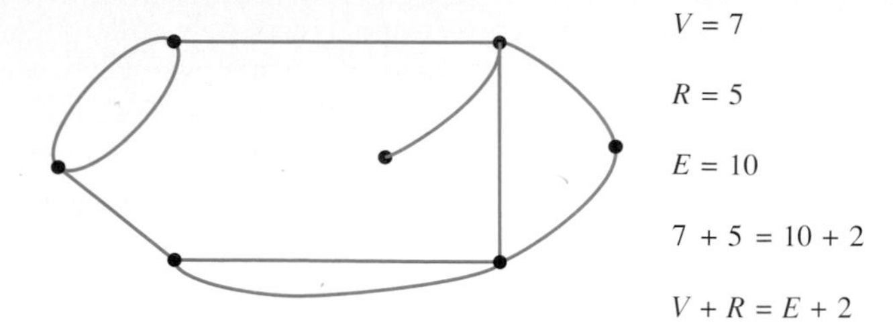

Figure 10.34
Euler's formula $V + R = E + 2$ holds for any connected planar network with V vertices, R regions, and E edges.

THEOREM Euler's Formula for Connected Planar Networks

Let V be the number of vertices, R the number of regions, and E the number of edges of a connected planar network. Then V, R, and E satisfy **Euler's formula:**

$$V + R = E + 2$$

PROOF* An informal justification for Euler's formula rests on the ideas of beginning with a single edge and then appending one edge at a time until any given network is completely drawn. When just one edge is in place we have two vertices and one region as shown. Thus $V_1 + R_1 = E_1 + 2$, since $V_1 = 2$, $R_1 = 1$, and $E_1 = 1$, which makes both sides of the equation 3.

$E_1 = 1, \quad V_1 = 2, \quad R_1 = 1$

There are two ways to draw the next edge, depending on whether we draw a new vertex or else join two already existing vertices.

Append a new edge and a new vertex

The new edge and new vertex combination does not create a new region, and so the number of regions R_2 equals the previous number of regions R_1. Since $R_2 = R_1$, $V_2 = V_1 + 1$, $E_2 = E_1 + 1$, we see that adding 1 to both sides of the formula $V_1 + R_1 = E_1 + 2$ gives $(V_1 + 1) + R_1 = (E_1 + 1) + 2$, which is the same as $V_2 + R_2 = E_2 + 2$ by substitution.

Append a new edge between existing vertices

* Optional

The new edge creates one new region, so $R_3 = R_2 + 1$. The number of vertices is not changed, so that $V_3 = V_2$. Since we have added one edge, we also have that $E_3 = E_2 + 1$. Again we add 1 to each side of the previous formula $V_2 + R_2 = E_2 + 2$ to get $V_2 + (R_2 + 1) = (E_2 + 1) + 2$. This is equivalent to $V_3 + R_3 = E_3 + 2$ by substitution, showing that Euler's formula holds for the partially drawn network.

Continuing to append new edges in either of the two ways just described will eventually produce any given connected planar network. Since Euler's formula holds at the first step and continues to hold from one step to the next, it holds for the completed network as well. ■

EXAMPLE 10.17 **Solving the Pizza Problem**

Suppose C cuts are made across a circular pizza, and there are I points of intersection of pairs of cuts. Assume no two cuts intersect on the bounding circle, and no three cuts intersect at the same point inside the pizza. How many pieces, P, of pizza are there?

SOLUTION

Understand the problem • • • •

It helps to examine a particular case, such as the one shown here. The cuts are drawn to satisfy the conditions of the problem. In the drawing we can see that 4 cuts and 3 intersections of cuts result in 8 pieces of pizza. Our goal is to see if we can find a formula that gives of the number P in terms of the variables C and I.

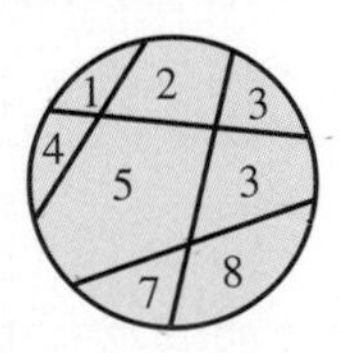

C = 4 cuts
I = 3 intersections of cuts
P = 8 pieces of pizza

Devise a plan • • • •

The cut up pizza can be viewed as a connected planar network. If we can determine the numbers V and E of vertices and edges, Euler's formula will allow us to solve for the number R of regions of the network inside the circle. Thus our plan is to relate V and E to the numbers C and I. Since all but the one region outside the circle corresponds to a piece of pizza, $P = R - 1$ will give us the formula we seek.

Carry out the plan • • • •

Each cut forms 2 vertices on the circle bounding the pizza, and each intersection of cuts gives one vertex of the network inside the circle. Altogether this gives $V = 2C + I$ vertices in the network. To count the number of edges in the network, let's first suppose that there are no intersecting cuts. Each cut then forms two edges on the circle and is itself an edge, giving $3C$ edges of the network. If we next suppose that some of the cuts intersect, it is seen that each intersection creates 2 additional edges of the network not yet counted. Altogether then, there are $E = 3C + 2I$ edges in the network. Solving for R in Euler's formula $V + R = E + 2$ we get $R = E - V + 2 = (3C + 2I) - (2C + I) + 2 = C + I + 2$. Since $P = R - 1$, we arrive at the final formula:

$$P = C + I + 1$$

Look back • • • •

In the example drawn on the previous page, there are 4 cuts intersecting in 3 points. Since $4 + 3 + 1 = 8$, we now see why we counted 8 pieces of pizza, and obtain a check that our formula is correct. ■

COOPERATIVE *Investigation*

The Game of Sprouts

The English mathematicians John Conway and Michael Paterson invented a game called Sprouts in 1967 in which two players take turns drawing new edges to an evolving network on n given initial vertices. It is permissible to draw an edge that returns to the starting vertex; such an edge is called a loop. There are just four simple rules:

1. Each new edge forms either a loop or joins two different vertices.
2. A new vertex must be created along the new edge.
3. The new edge cannot cross itself or any previously drawn edge, or pass through a vertex.
4. No vertex can have a degree larger than 3.

The first five moves of a Sprouts game on $n = 2$ initial vertices are shown below. The player who made the fifth move wins, since the other player cannot make a legal move.

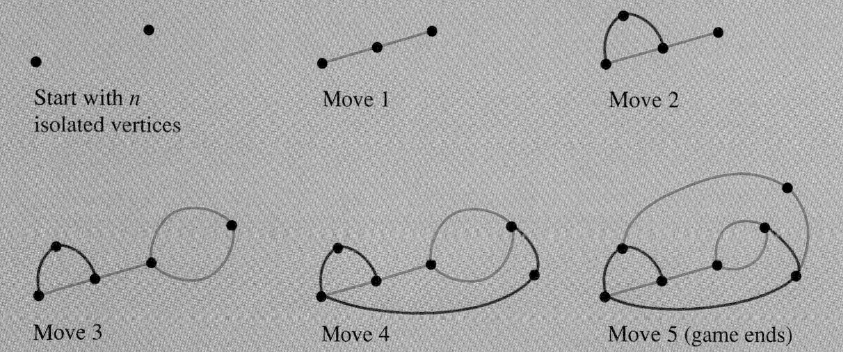

(a) Play several games on $n = 2$ initial vertices. At most how many moves do the games take?

(b) Play several games on $n = 3$ initial vertices. What is the largest number of moves possible before the games end?

(c) Look for a pattern on the largest number of moves possible in a Sprouts game on n initial vertices.

(d) What is the smallest number of moves to end a game on n initial vertices?

SOURCE: Sprouts is described in Martin Gardner's book *Mathematical Carnival,* New York: Knopf, 1975.

PROBLEM SET 10.4

Understanding Concepts

1. Decide which of the following networks are traversable. If it is traversable give an Euler path.

(a)

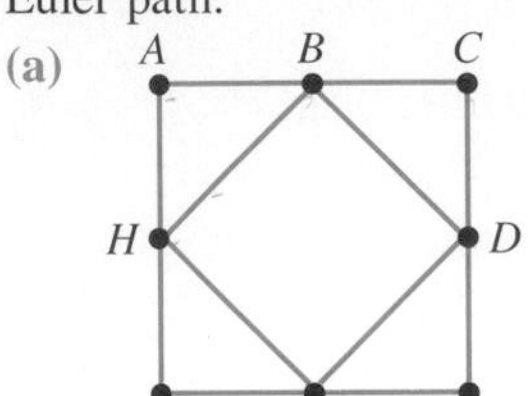

(b)

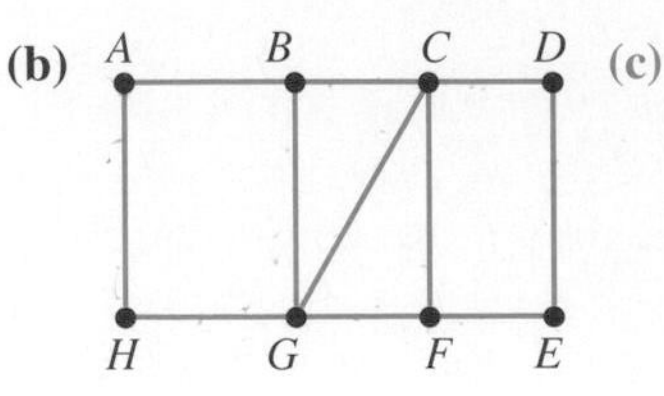

(c)

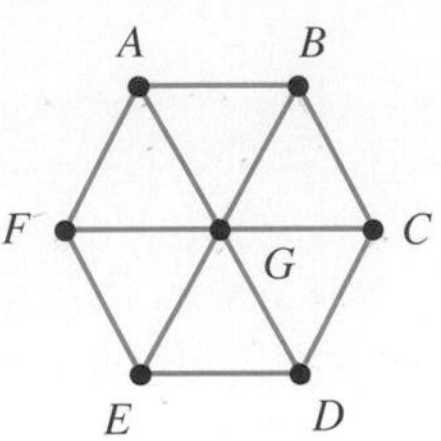

(d)

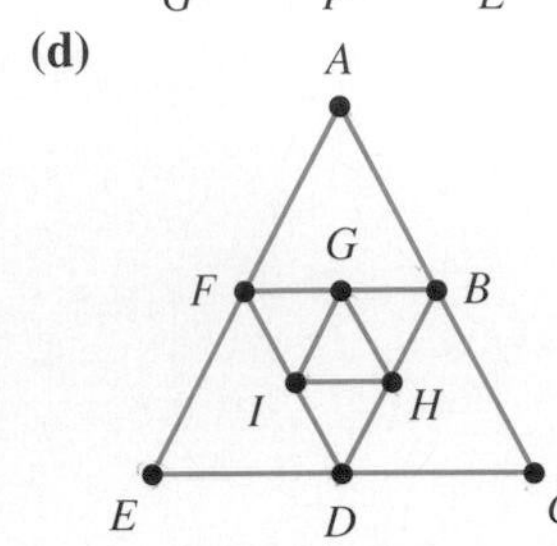

(e)

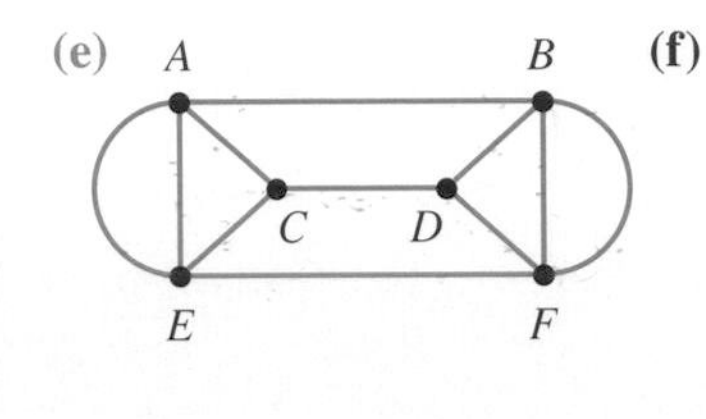

(f)

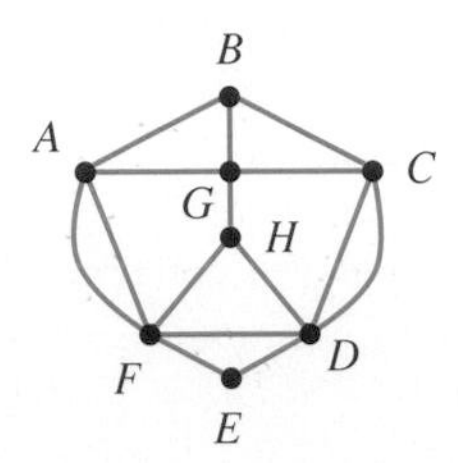

2. (a) The **total degree** D of a network is the sum of the degrees of all of the vertices. For example, the network in problem 1(a) has total degree $D = 2 + 4 + 2 + 4 + 2 + 4 + 2 + 4 = 24$. This network also has $E = 12$ edges. Find D and E for the remaining networks shown in problem 1.

   (b) Guess how the total degree D is related to the number of edges E in any network. Test your conjecture on several networks of your own choosing.

3. A connected network has the following degrees at its vertices: 2, 2, 4, 8, 3, 6, 6, 1.

   (a) Is the network traversable?

   (b) How many edges does the network have? (*Hint:* See problem 2(b).)

4. Here are two more examples from Euler's paper on the Königsberg Bridge Problem.

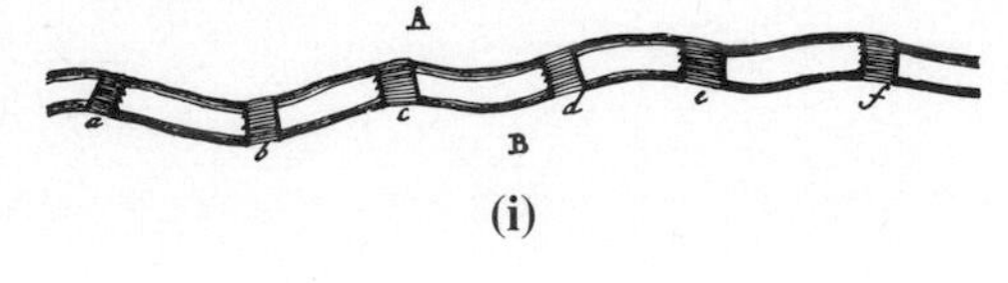

(i)

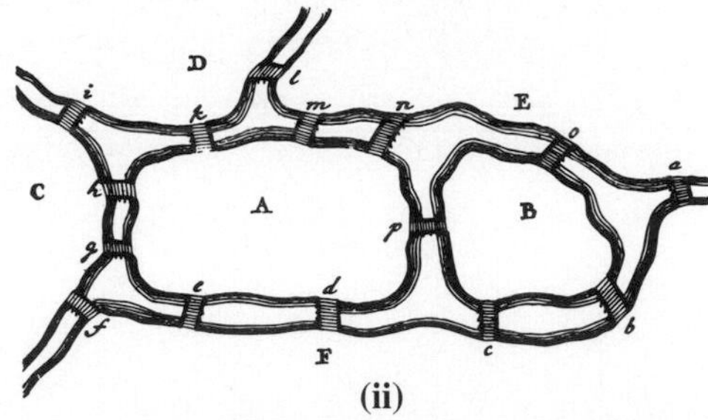

(ii)

   (a) Draw the networks corresponding to **(i)** and **(ii).**

   (b) Is network **(i)** traversable? Why?

   (c) Is network **(ii)** traversable? Explain your reasoning.

5. (a) Draw the network that corresponds to the following system of bridges and land masses.

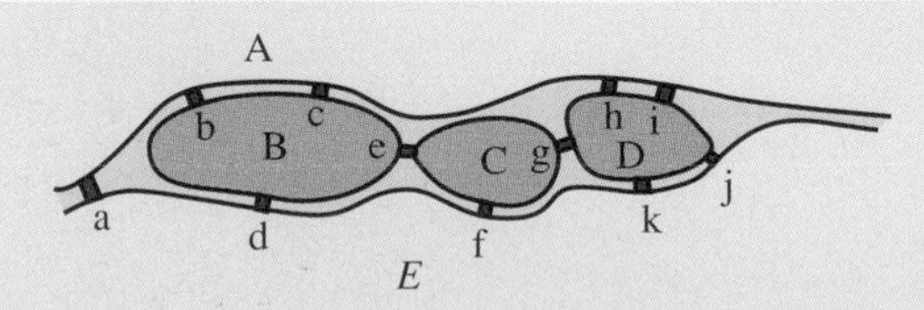

   (b) Explain why the network is not traversable.

   (c) What is the smallest number of new bridges required to form a traversable network? Where should the bridge(s) be placed?

6. A floor plan of a house is shown below. There are five rooms A, B, C, D, E and the outside O, connected by the doorways a, b, c, d, e, f, g.

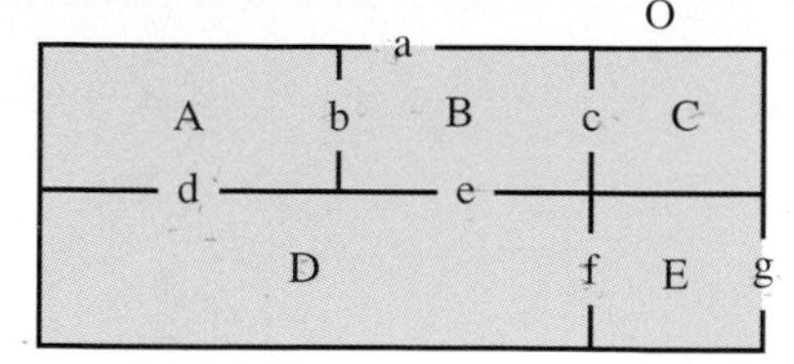

(a) Draw a network whose vertices are labeled A, B, C, D, E, O and whose edges correspond to the doorways.

(b) Is it possible to walk through each doorway exactly once? If so, where must you begin and end your walk?

7. A connected planar network is shown below.

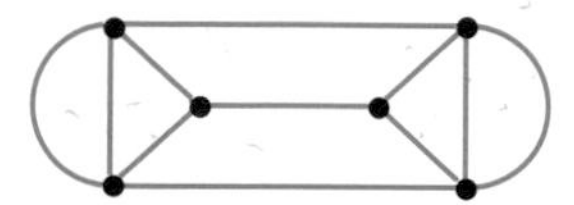

(a) What is the number of vertices V, regions R, and edges E?

(b) Does Euler's formula hold for this network?

8. (a) Draw a connected planar network whose 10 edges separate the plane into 6 regions.

(b) Can you draw a connected planar network with 10 edges that separates the plane into 12 regions? Explain.

Thinking Critically

9. The two networks shown below look different but are really equivalent to one another. There is a one-to-one correspondence between the vertices, namely $A \leftrightarrow 1$, $B \leftrightarrow 2$, $C \leftrightarrow 3$, $D \leftrightarrow 4$, $E \leftrightarrow 5$, and to every pair of vertices joined by an edge in one network the corresponding pair of vertices in the other network is also joined by an edge. For example, the edge joining B and D corresponds to the edge joining vertex 2 to vertex 4.

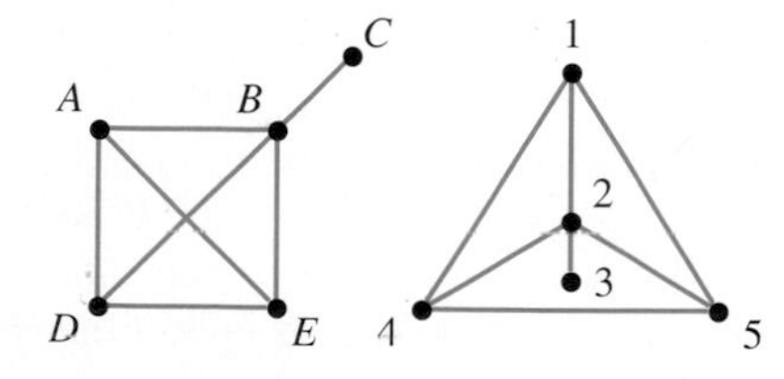

Two networks which are equivalent in this sense are called **isomorphic,** meaning *same form.* For each pair of networks below, explain whether they are isomorphic or not.

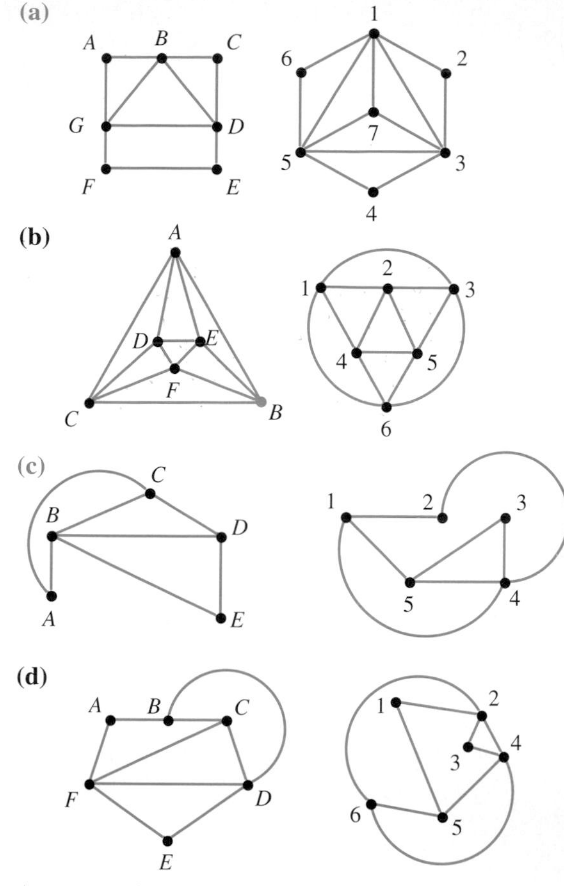

10. A connected network which does not contain a closed path of distinct edges is called a **tree.**

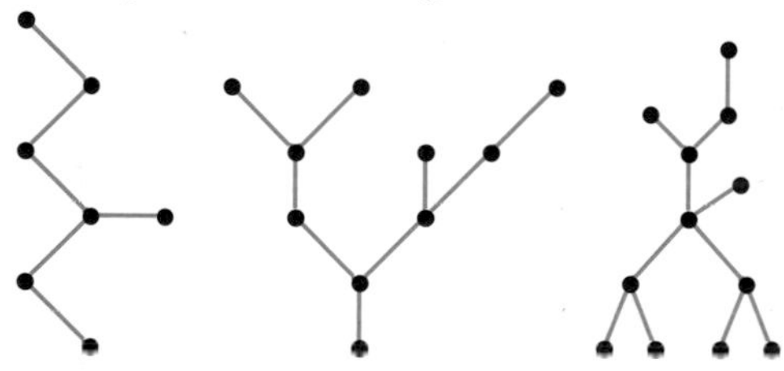

Trees

Not trees

(a) Suppose a network contains two vertices that can be joined by two different paths with no edges common to both paths. Why is it impossible for this network to be a tree?

(b) The **diameter** of a tree is the largest number of edges required in a path which joins any two vertices. For example, the leftmost tree above has diameter 5. Find the diameters of the two other trees drawn above.

(c) Let V and E denote the respective number of vertices and edges of a tree. What formula relates V and E? Prove your result.

11. What trees are traversable?

12. In problem 2 you discovered that the total degree D (the sum of degrees of all the vertices of the network) is given by $D = 2E$, where E is the number of edges. Suppose the degrees at the even vertices are $e_1, e_2, \ldots, e_m$, and the degrees at the odd vertices are $d_1, d_2, \ldots, d_n$. Thus $2E = D = e_1 + e_2 + \cdots + e_m + d_1 + d_2 + \cdots + d_n$.

(a) Explain why $d_1 + d_2 + \cdots + d_n$ is an even integer.

(b) Since $d_1 + d_2 + \cdots + d_n$ is even, explain why n is even. Since n is the number of odd vertices in the network, you've proved the following result:

The number of odd vertices in a network is always even.

13. Prove that the number of people who have shaken hands an odd number of times is an even number. (*Hint:* See problem 12.)

14. The connected network below has 6 odd vertices, so it cannot be traced without lifting the pencil.

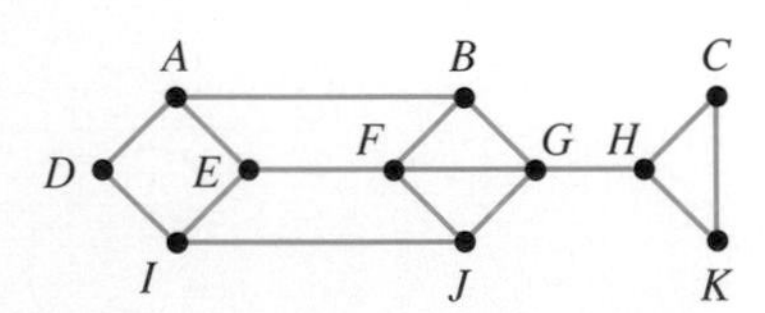

(a) Show that the network can be traced in three strokes; that is, the pencil is lifted twice and then placed at a different vertex.

(b) If a connected network has $2m$ odd vertices, with $m > 0$, explain why its edges can be traced in m strokes. (*Hint:* What is a useful way to add, temporarily, $m - 1$ new edges to the network?)

15. Suppose that you are asked to trace a connected network, without lifting your pencil, so that each edge is traced exactly twice. Is this always possible regardless of the number of odd vertices? Explain why or why not.

16. The planar network shown below on the vertices A, B, C, D, F, G, H, I, J, K, is not connected; indeed, it is made up of $P = 3$ connected pieces.

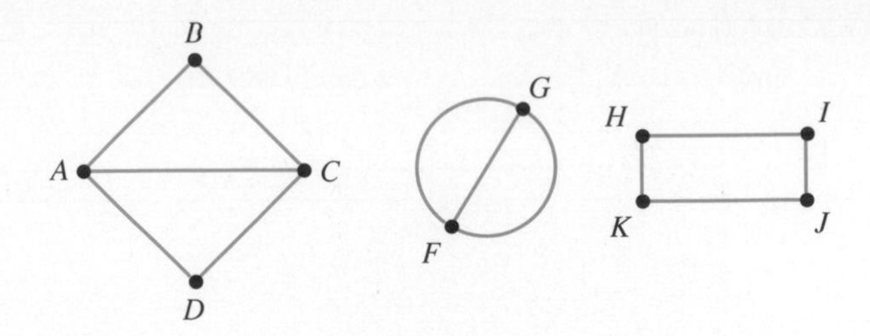

For this network $V = 10$, $E = 12$, $R = 6$ and we see that $V + R = 16$ but $E + 2 = 14$.

(a) Draw several more examples of disconnected planar networks. For each network count V, R, E and P, where P is the number of connected pieces in your network. Can you guess an Euler formula which relates V, R, E, and P?

(b) Prove your conjecture stated in part (a). (*Hint:* How many new edges does it take to connect a network with P pieces?)

17. One circle separates the plane into 2 regions, 2 circles which cross one another in 2 points separate the plane into 4 regions, and 3 circles form 8 regions when each pair of circles cross at 2 points and no more than 2 circles intersect at any point.

(a) What is the number of regions determined by 4 circles? Assume that each pair of circles intersect at 2 points, and no 3 circles intersect at any one point.

(b) How many regions are formed by n intersecting circles? (*Hint:* There are $\frac{1}{2}n(n - 1)$ pairs of circles. Now use Euler's formula and mimic the solution of Example 10.17.)

Thinking Cooperatively

18. There are two nonisomorphic trees on 4 vertices (See problems 9 and 10 for definitions of isomorphic and trees.)

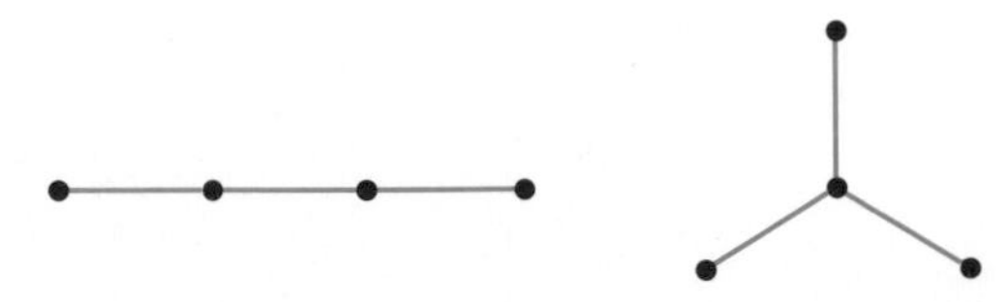

One way to see that these trees are nonisomorphic is to notice that the degrees of the vertices are 1, 2, 2, 1 on the left and 1, 1, 1, 3 on the right. Another way is to observe that the diameter of the tree on the left is 3 and that on the right is 2.

(a) Find the 3 nonisomorphic trees on 5 vertices.

(b) Find the 6 nonisomorphic trees on 6 vertices.

(c) Find the 11 nonisomorphic trees on 7 vertices.

Making Connections

19. An 8 pin connecting terminal is wired as shown at the left. An electrical engineer claims that 5 of the 12 connecting wires can be eliminated. On the right-hand diagram draw 7 of the 12 original wires which provide for the same current flows as the original circuit.

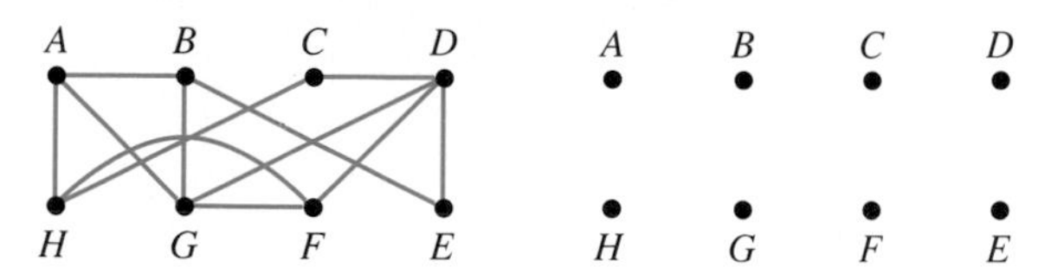

(This problem is found in *A Sourcebook of Applications of School Mathematics* by Donald Bushaw, Max Bell, Henry O. Pollak, Maynard Thompson, and Zalman Usiskin, NCTM, 1980).

20. The Big Eleven Conference has five teams—*A*, *B*, *C*, *D*, *E*— in the northern division and six teams—*U*, *V*, *W*, *X*, *Y*, *Z*—in the southern division. At a Conference Director's meeting, it was proposed that each team in the northern division must schedule exactly four opponents from the southern division, and each team from the southern division must play exactly four teams from the northern division.

(a) Attempt to make a suitable schedule by drawing a network in which each edge represents a game between two teams from opposite divisions.

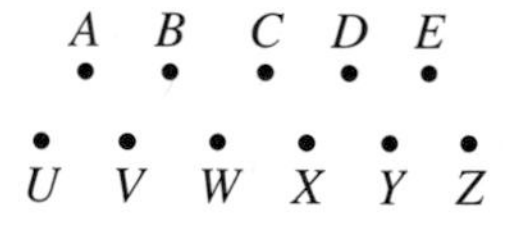

(b) If you cannot make a schedule with $k = 4$ games, try other values of k, such as $k = 1, 2,$ or 3 games.

(c) Is there any number of k games, $k \geq 1$, that can be required?

For Review

21. Draw a parallelogram, and then draw outward facing squares along each edge of your parallelogram. What type of quadrilateral is formed by joining the successive center points of the squares?

22. The following equilateral nonagon (9-gon) tiles the plane in several intersecting ways. Find the measures of all of the interior angles.

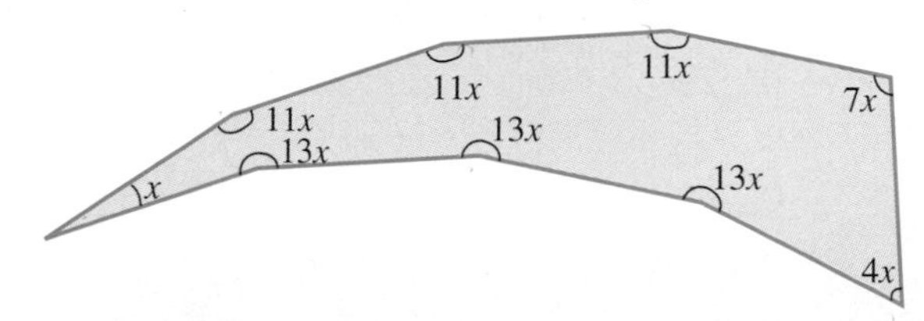

23. In the figure shown below lines l and m are parallel. Find the measures x, y, z of the indicated angles.

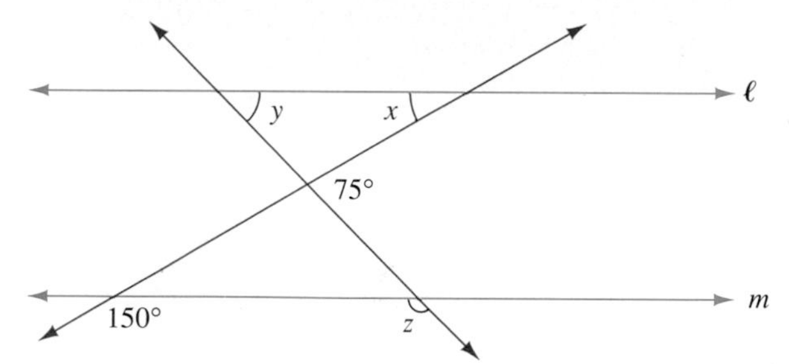

10.5 Figures in Space

The Picture Gallery of Space Figures on the next page shows several interesting examples of shapes whose points do not belong to a single plane. Intuitively, we think of space as three-dimensional. For example, the shape of a shoe box requires us to know not just width and length but height as well. In this section we classify, analyze, and represent some of the basic figures in space.

Planes and Lines in Space

There are infinitely many planes in space. Each plane separates the points of space into three disjoint sets: the plane itself and two regions called **half-spaces.** Two planes are either **parallel** or intersect in a line, as shown in Figure 10.35.

The angle between two intersecting planes is called the **dihedral angle** (*di* = two, *hedral* = face of a geometrical form). A dihedral angle is measured by measuring an angle whose sides lie in the planes and are perpendicular to the line of intersection of the planes. Some examples of dihedral angles and their measures are shown in Figure 10.36.

A Picture Gallery of Space Figures

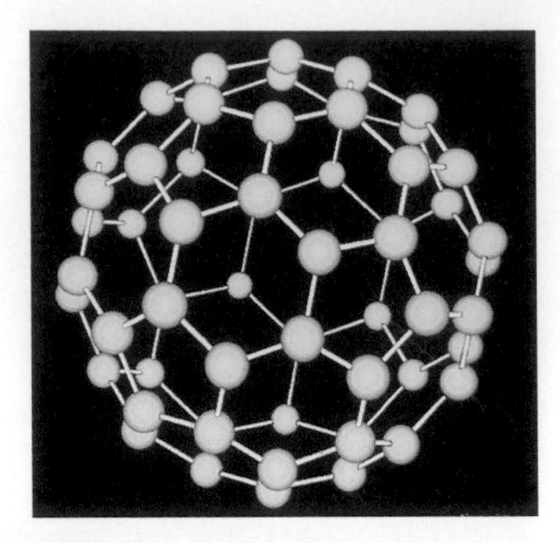

A Buckyball (named for Buckminster Fuller), the third form of pure carbon

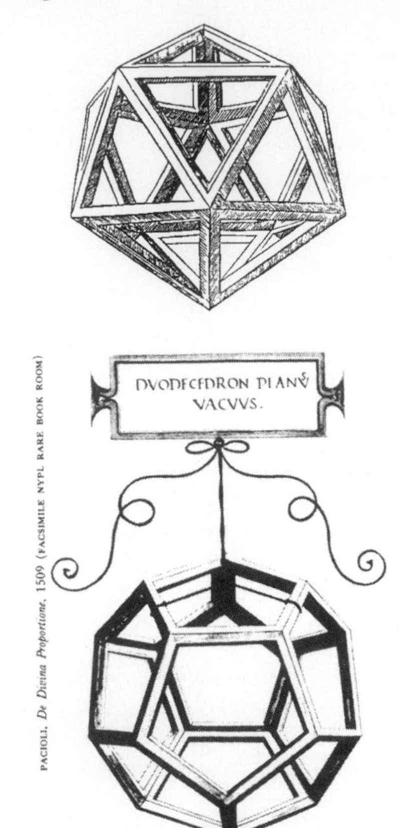

Leonardo da Vinci's drawings of an icosahedron and a dodecahedron for Fra Luca Pacioli's *Di Divina Proportione*

A seashell (left) and a computer drawn ideal representation

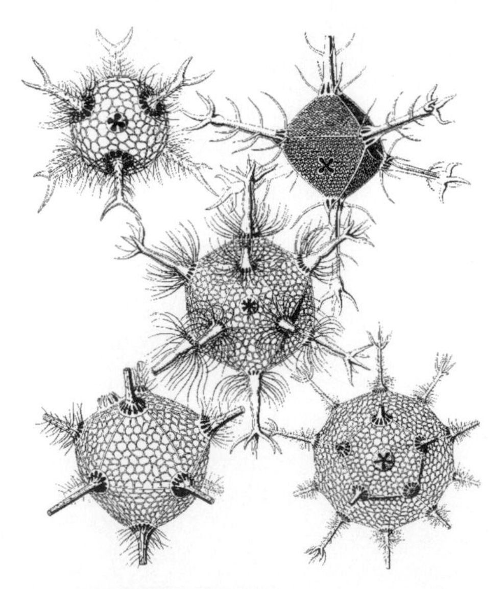

Skeletons of microscopic radiolaria

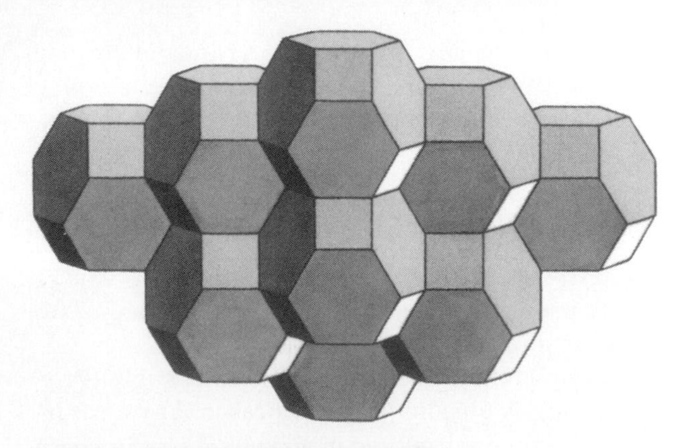

A filling of space by truncated octahedra

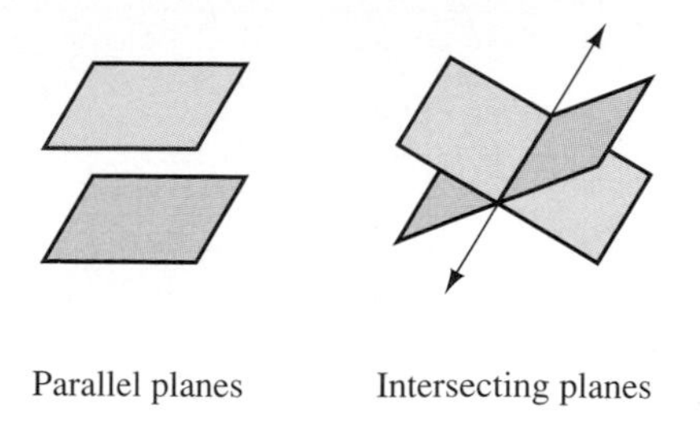

Figure 10.35
Parallel and intersecting planes

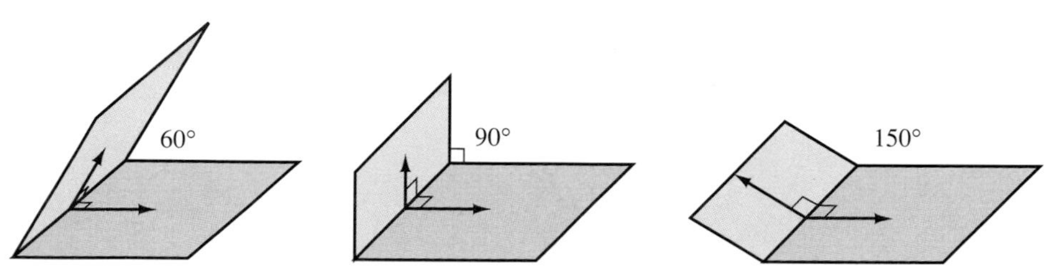

Figure 10.36
Dihedral angles and their measures

Two nonintersecting lines in space are **parallel lines** if they belong to a common plane. Two nonintersecting lines which do not belong to a common plane are called **skew lines.** A line l that does not intersect a plane P is said to be **parallel to the plane.** A line m is **perpendicular to a plane** Q at point A if every line in the plane through A intersects m at a right angle. Figures illustrating these terms are shown in Figure 10.37.

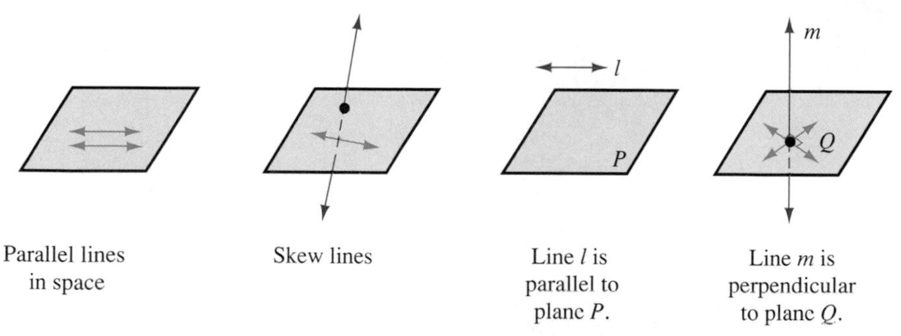

Figure 10.37
Lines and planes in space

Curves, Surfaces, and Solids

The intuitive concept of a curve can be extended from the plane to space by imagining figures drawn with a "magic" pencil whose point leaves a visible trace in the air. Two examples are shown in Figure 10.38, a helix (corkscrew) and a space octagon whose sides are 8 of the 12 edges of a cube.

No space curve will separate space. On the other hand, a **sphere,** which is the set of points at a constant distance from a single point called the **center,** does separate the remaining points of space into two disjoint regions, namely the points inside the

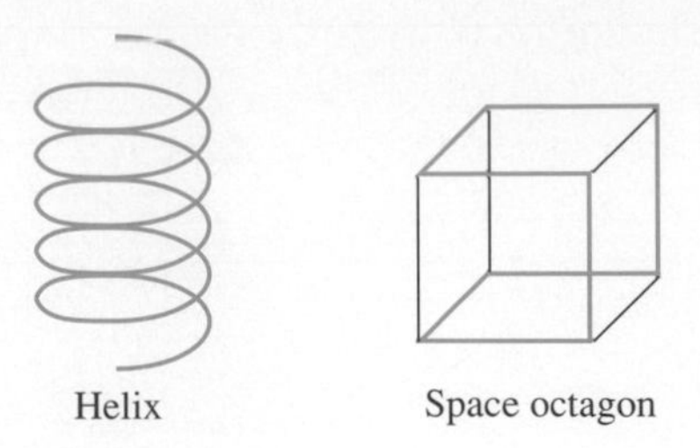

Figure 10.38
Two curves in space

sphere and the points outside the sphere. Any surface without holes and which encloses a hollow region—its interior—is called a **simple closed surface.** An easy check to see if a figure is a simple closed surface is to imagine what shape it would take if it were made of stretchy rubber: if it can be "blown up" into a sphere, then it is a simple closed surface.

The union of all points on a simple closed surface and all points in its interior form a space figure called a **solid.** For example, the shell of a hardboiled egg can be viewed, overlooking its thickness, as a simple closed surface; the shell together with the white and yolk of the egg form a solid.

A simple closed surface is **convex** if the line segment which joins any two of its points contains no point which is in the region exterior to the surface; that is, the solid bounded by the surface is a convex set in space. The sphere, soup can, and box shown in Figure 10.39 are all convex. The potato skin surface shown in the figure is not convex however, since it is possible to find two points on this surface for which the line segment contains points in the exterior region.

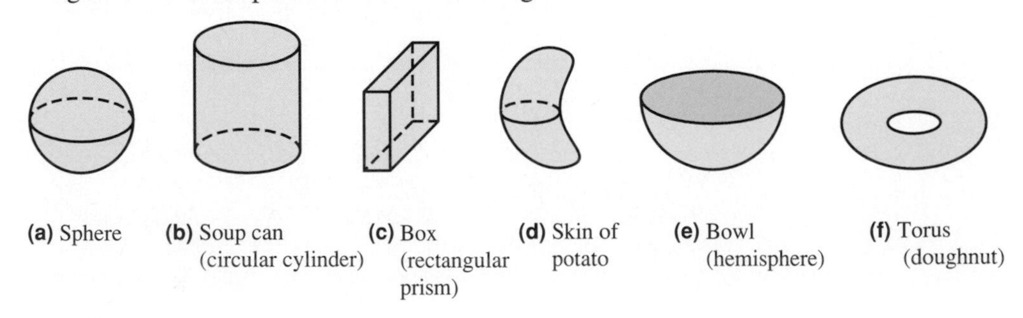

Figure 10.39
(a), (b), (c), and (d) are simple closed surfaces; (e) is a nonclosed surface; and (f) is closed but not simple

Polyhedra

Joining plane polygonal regions edge-to-edge forms a simple closed surface called a **polyhedron.**

DEFINITION Polyhedron

A **polyhedron** is a simple closed surface formed from planar polygonal regions. Each polygonal region is called a **face** of the polyhedron. The vertices and edges of the polygonal regions are called the **vertices** and **edges** of the polyhedron.

Polyhedra (*polyhedra* is the plural of polyhedron) are named according to the number of faces. For example, a **tetrahedron** has 4 faces, a **pentahedron** has 5 faces, a **hexahedron** has 6 faces, and so on. Several polyhedra are pictured in Figure 10.40.

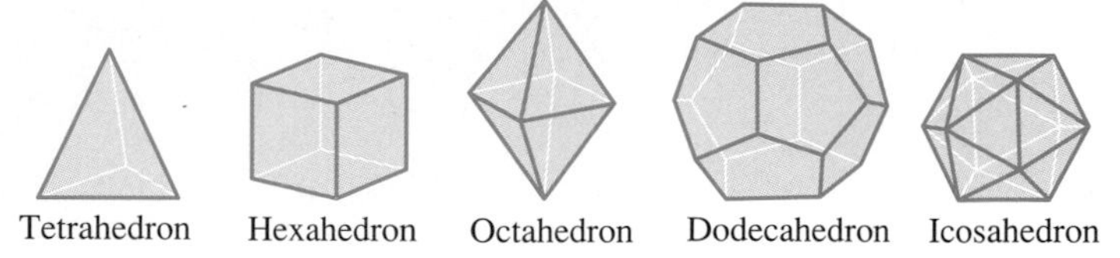

Figure 10.40
Examples of polyhedra

The most spectacular polyhedral shapes on Earth are the Egyptian and Mayan pyramids. Egyptian pyramids have a square base and four congruent triangular faces which meet at a common vertex. The Mayan pyramids have a stepped form. In geometry, a **pyramid** can have any polygonal region as a base. Triangular faces rise from the base edges to meet at a common vertex, called the **apex** of the pyramid, a point that is not in the plane of the base. Examples of pyramids and their names are shown in Figure 10.41.

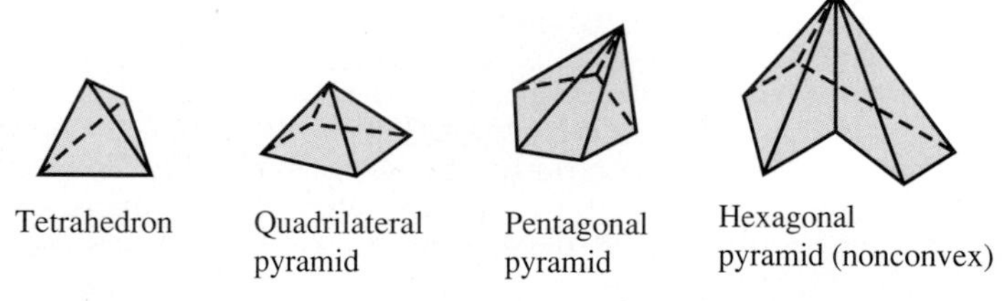

Figure 10.41
Pyramids and their names

Another commonly occurring shape is a **prism.** A prism has two **bases** which are congruent polygonal regions lying in parallel planes; the lateral faces joining the bases are all parallelograms. If the lateral faces of a prism are all rectangles it is a **right prism;** otherwise, it is an **oblique prism** and the lateral edges are not perpendicular to the plane of the base. Examples of prisms and their names are shown in Figure 10.42.

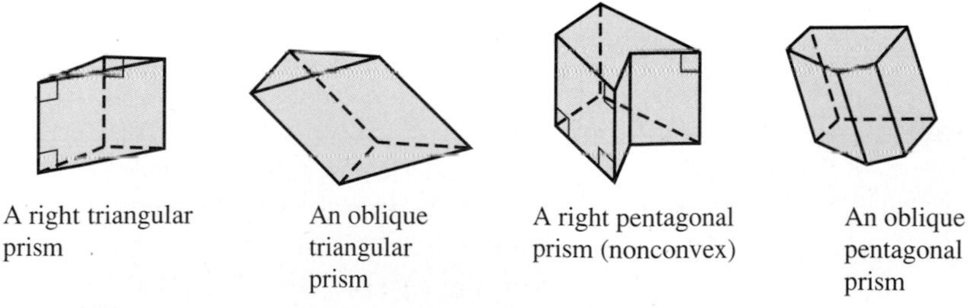

Figure 10.42
Right and oblique prisms

EXAMPLE 10.18

Determining Angles and Planes in a Hexagonal Prism

The bases of the right prism shown below are regular hexagonal regions:

(a) What are the measures of the dihedral angles at which the faces intersect?

(b) How many pairs of parallel planes contain the faces of this prism?

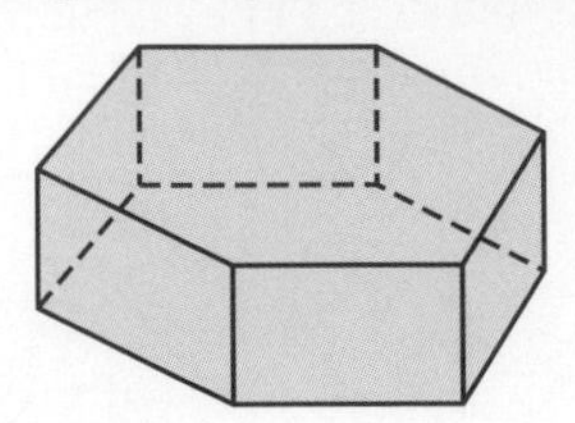

SOLUTION

(a) Since the prism is a right prism, the dihedral angles made by either base to any lateral face measures 90°. The lateral faces meet at 120° angles, since this is the measure of the interior angles of a regular hexagon.

(b) The opposite sides of a regular hexagon are parallel, so there are three pairs of parallel planes containing the opposite lateral faces of the prism. A fourth pair of parallel planes contains the hexagonal bases of the prism. ■

Regular Polyhedra

DEFINITION Regular Polyhedron

A **regular polyhedron** is a polyhedron with these properties:

- the surface is convex;
- the faces are congruent regular polygonal regions;
- the same number of faces meet at each vertex of the polyhedron.

The most common regular polyhedron is the cube: the six faces are congruent squares, and three squares meet at each of the eight vertices. The cube is the only regular polyhedron with square faces, since if we attempted to put four squares about a single vertex, their interior angle measures add up to 360°. That is, four edge-to-edge squares with a common vertex lie in a common plane and therefore cannot form a "corner" figure of a regular polyhedron.

Similar reasoning with equilateral triangles shows that corner figures in space can be formed with either 3, 4, or 5 congruent copies of the triangle. However, a convex corner cannot be formed with six or more equilateral triangles. Likewise, there is just one way to form a corner with congruent regular pentagons, and it is impossible to form a corner figure from regular n-gons for any $n \geq 6$. The possible corner figures of regular polyhedra are shown in Figure 10.43.

Surprisingly, each of the five corner figures depicted in Figure 10.43 can be completed to form a regular polyhedron. These are shown in Table 10.4, which also shows patterns called **nets.** Models of the polyhedra can be made by cutting the net from heavy paper, folding, and gluing. It helps to include flaps on every other outside edge of the net; these are then coated with glue and tucked under the adjoining face, forming a sturdy model.

JUST FOR FUN

Space Out for Success!

These five matches form two triangles. Show that six matches, touching only at their ends. can form four triangles.

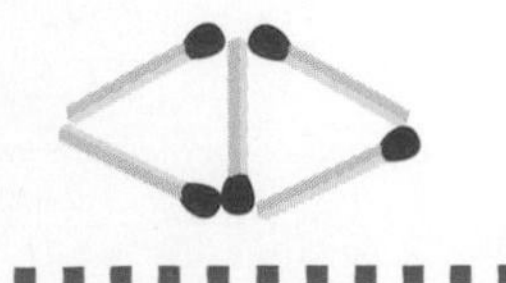

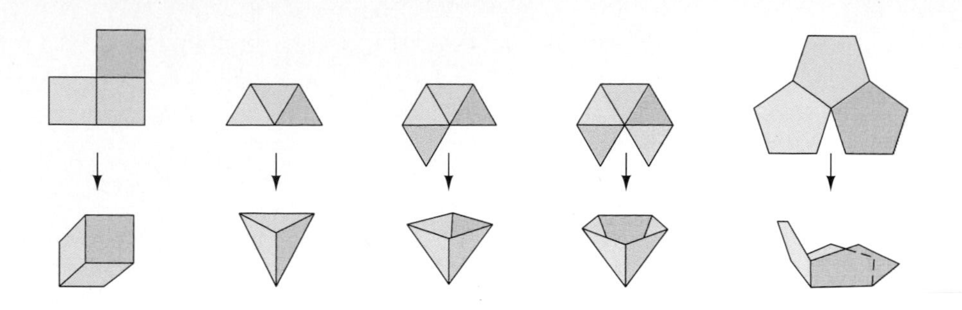

Figure 10.43
The five ways to form corner figures with congruent regular polygons

TABLE 10.4 The Five Regular Polyhedra

Polyhedron Name	Face Polygons	Net	Model
Cube	6 squares		
Tetrahedron	4 triangles		
Octahedron	8 triangles		
Icosahedron	20 triangles		
Dodecahedron	12 pentagons		

The five regular polyhedra were known to the ancient Greeks. They are described in Plato's book *The Republic* and so the shapes are often referred to as the **Platonic solids.** Theaetetas (c. 415–369 B.C.), a member of the Platonic school, is credited with the first proof that there are no regular polyhedra other than the five known to Plato.

Euler's Formula for Polyhedra

The name given to a polyhedron usually indicates its number of faces. For example, an octahedron has eight faces. A more complete description of a polyhedron may include its number of vertices and edges. The following notation will be useful:

F = the number of faces of a polyhedron

V = the number of vertices of a polyhedron

E = the number of edges of a polyhedron

For a regular octahedron we have $F = 8$, $V = 6$, and $E = 12$.

CURRENT AFFAIRS

A Flexible Polyhedron

When a paper model of a polyhedron is constructed by cutting, folding, and gluing a pattern net such as shown in Figure 10.43, the partially completed model is quite flexible. However, when the last face is glued in place no flexibility remains. The French mathematician Augustin-Louis Cauchy (1789–1857) conjectured that all polyhedra were rigid, and in 1813 proved that all *convex* polyhedra were indeed rigid. For over 150 years no one could show that nonconvex polyhedra must also be rigid. In 1978 the issue was resolved in an unexpected direction: Robert Connelly of Cornell University constructed a polyhedron from 18 triangular faces which is noticeably flexible! Several other flexible polyhedra are now known. The simplest one, with just 14 faces, was found by Klaus Steffen. It is pictured below, together with its pattern. All of the flexible polyhedra discovered so far share a remarkable property: as the surface flexes, the volume of the enclosed region remains the same!

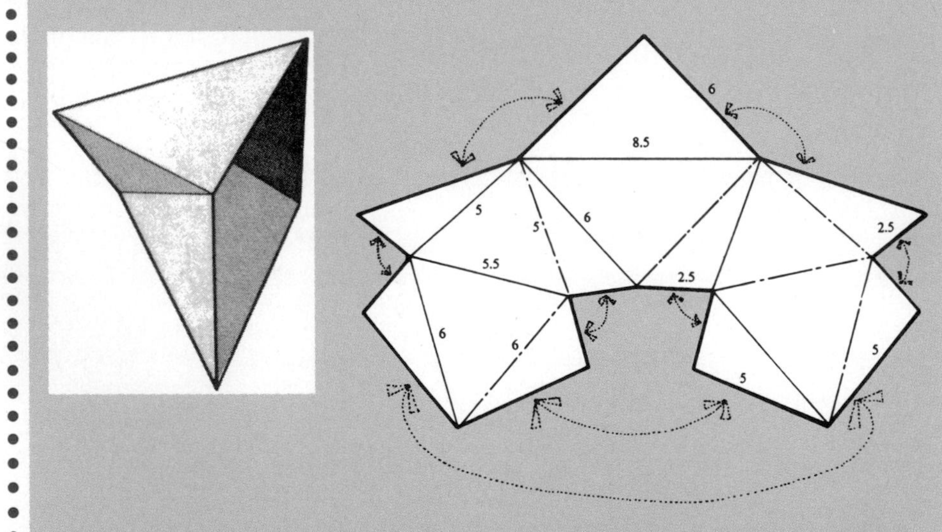

SOURCE: Figure 21 and Figure 22 from *Mathematics Magazine,* Volume 52, Number 5, November 1979, p. 281. Copyright © 1979 by The Mathematical Association of America. Reprinted by permission.

In 1752 the great Swiss mathematician Leonhard Euler discovered that the number of faces F, the number of vertices V, and the number edges E are related to one another. Euler was unaware that he had rediscovered a formula found about 1635 by the French mathematician-philosopher Rene Descartes.

EXAMPLE 10.19 **Discovering Euler's Formula**

Let V, F, and E denote the respective number of vertices, faces, and edges of a polyhedron. What relationship holds between V, F, and E?

SOLUTION

Understand the problem

The numbers V, E, and F are not independent of one another. The goal is to uncover a formula which relates the three numbers corresponding to *any* polyhedron.

Devise a plan

Formulas are often revealed by seeing a pattern in special cases. By making a table of values of V, F, and E we have a better chance to see what this pattern may be. To be confident that the pattern holds for all polyhedra, we need to examine polyhedra of varied kinds.

Carry out the plan

A pentagonal pyramid, a hexagonal prism, a "house", and a truncated icosahedron are shown below. The truncated icosahedron, formed by slicing off the corners of an icosahedron to form pentagons, may look familiar; it is a common pattern on soccer balls.

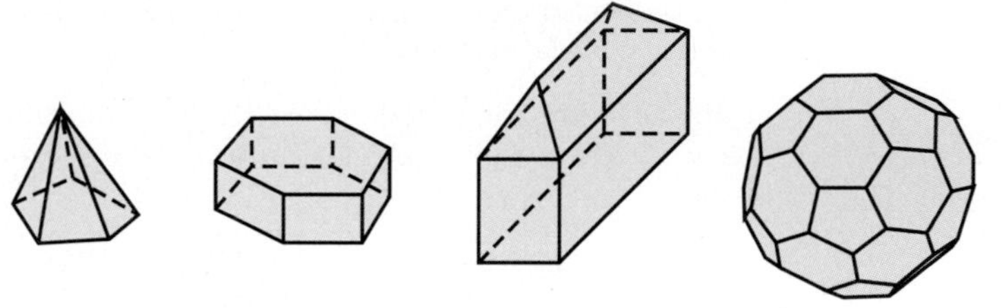

The following table lists the number of vertices, edges, and faces for these polyhedra as well as for some of the regular polyhedra depicted in Table 10.4.

Polyhedron	V	F	E
Pentagonal Pyramid	6	6	10
Hexagonal Prism	12	8	18
"House"	10	9	17
Cube	8	6	12
Tetrahedron	4	4	6
Octahedron	6	8	12
Truncated Icosahedron	60	32	90

The table reveals that the sum of the number of vertices and faces is 2 larger than the number of edges. That is, $V + F = E + 2$.

Look back

Knowing any two values of V, F, and E determine the remaining value, since the three numbers must satisfy the Euler formula $V + F = E + 2$. For example, the dodecahedron has 12 pentagonal faces. The product $5 \cdot 12$ counts *twice* the number of edges, since each edge borders two of the pentagonal faces. Thus, $F = 12$ and $E = 5 \cdot 12/2 = 30$ for the dodecahedron. Euler's formula can now be used to compute the number of vertices. Solving for V we get $V = E + 2 - F = 30 + 2 - 12 = 20$, so a dodecahedron has 20 vertices. ■

Euler's formula $V + F = E + 2$ for polyhedra follows directly from Euler's formula $V + R = E + 2$ for connected planar networks, which was proved in Section 10.4. To see this, replace each edge of a polyhedron with a segment of a rubber band. Next, stretch and flatten the rubber-band skeleton of the polyhedron to form a planar network of V vertices and E edges. Figure 10.44 shows this for a cube.

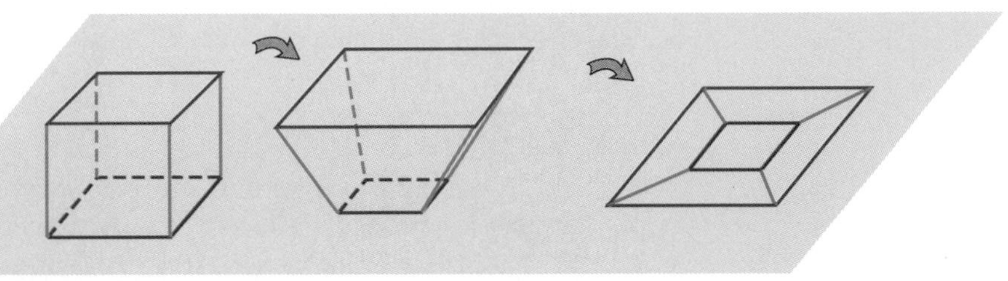

Figure 10.44
The skeleton of edges of any polyhedron, such as the cube shown, can be stretched and flattened to form a planar network.

Since each region of the network corresponds to a face of the polyhedron, we see that $R = F$. Thus, the two Euler formulas are really equivalent.

> **THEOREM Euler's Formula for Polyhedra**
>
> Let V, E, F denote the respective number of vertices, edges, and faces of a polyhedron. Then
>
> $$V + F = E + 2.$$

EXAMPLE 10.20

Searching for Polyhedra with Hexagonal Faces

The Epcot Center in Florida (below, left) is the site of one of the world's largest geodesic domes. The surface of the 165 foot diameter structure is covered with both hexagons and pentagons. Similarly the microscopic frame of the radiolarian (below, right) is covered by pentagons and hexagons. These shapes suggest the following question: *Can all the faces of a polyhedron be hexagonal?* Show that this is not possible, even if the hexagons need not all be congruent and are permitted to be irregular. Assume that three hexagons meet at each vertex.

SOLUTION

Suppose, to the contrary, that hexagonal faces can form a certain polyhedron. As usual, let F, E, and V denote the number of faces, edges, and vertices. Since each face is bordered by 6 edges and each edge touches 2 faces we obtain the formula $6F = 2E$. Thus we have $E = 3F$. Similarly,

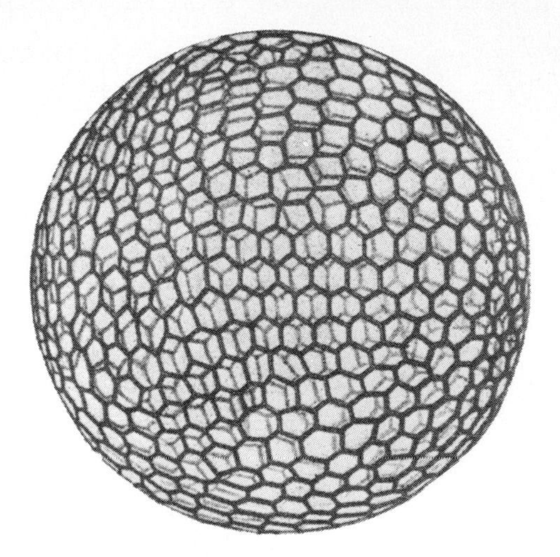

three hexagons surround each of the V vertices, and each of the F hexagonal faces touches 6 vertices. This gives us $3V = 6F$, which simplifies to $V = 2F$. Adding F to both sides of this equality gives $V + F = 3F$. But we have also shown that $3F = E$, so $V + F = E$. Since this contradicts Euler's formula, $V + F = E + 2$, we conclude that there is no polyhedron with each face hexagonal. ■

Cones and Cylinders

Cones and cylinders are simple closed surfaces which generalize pyramids and prisms respectively. A **cone** has a **base** that is any region bounded by a simple closed curve in a plane. The curved **lateral surface** is generated by the line segments which join one point—the **apex** (or **vertex**)—not in the plane of the base to the points of the curve bounding the base. A right circular cone, an oblique circular cone, and a general cone are shown in Figure 10.45. The line segment $\overline{AB}$ through the apex A of a cone that intersects the plane of the base perpendicularly at B is called the **altitude** of the cone.

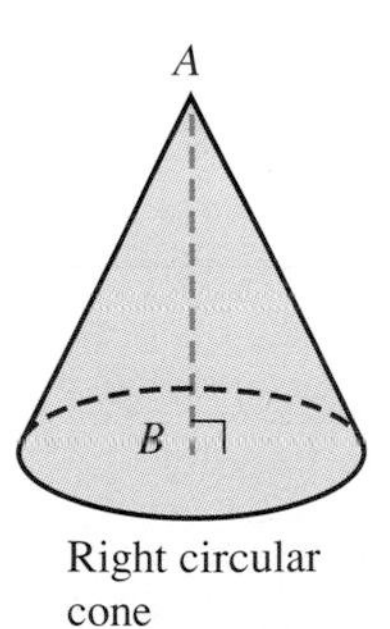

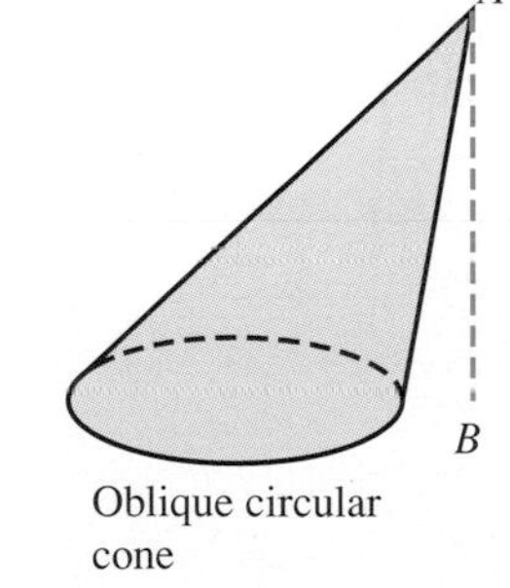

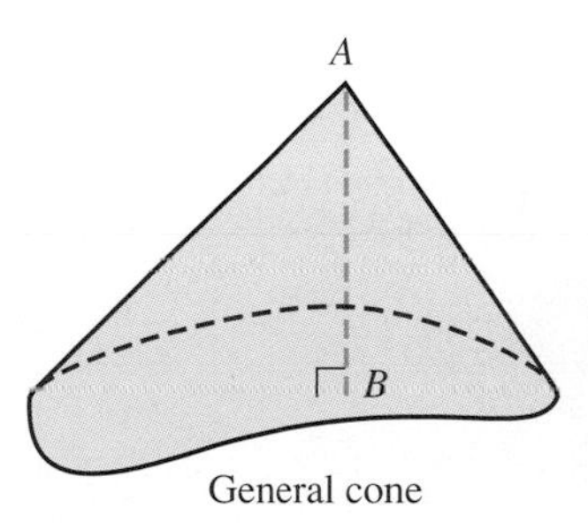

Figure 10.45
A right circular cone, an oblique circular cone, and a general cone

A **cylinder** is the surface generated by translating the points of a simple closed region in one plane to a parallel plane. Examples are shown in Figure 10.46. The points joining corresponding points on the curves bounding the bases form the **lateral surface.** If the line segments joining corresponding points in the two bases are perpendicular to the planes of the bases, it is a **right cylinder.** Cylinders that are not right cylinders are **oblique cylinders.**

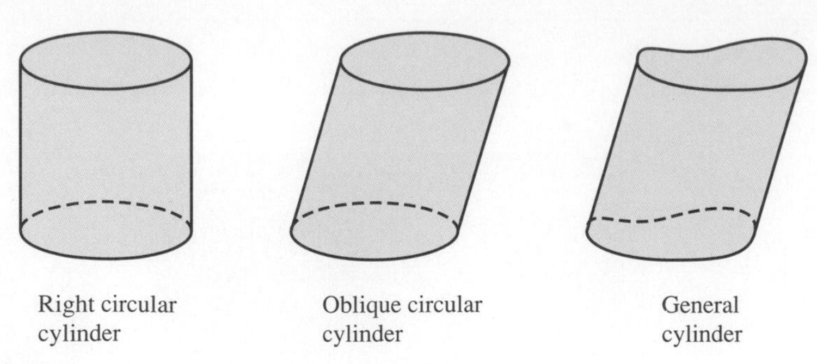

Figure 10.46
A right circular, oblique circular, and general cylinder.

COOPERATIVE *Investigation*

The Envelope Tetrahedron Model

Diagrams and photos of polyhedra are certainly useful, but physical models that can be seen and touched are much better. The construction of models of polyhedra is a worthwhile classroom activity; useful geometric principles are learned as students create beautiful and interesting shapes. Skeletal models are formed easily from drinking straws joined by thin string run through the straws and tied at the vertices. Paper models, in which a carefully drawn net of the polyhedron is cut, folded, and glued, can be colored in interesting ways.

Here is a quick way to construct a regular tetrahedron from an ordinary envelope.

1. Glue the flap of the envelope down.
2. Fold the envelope in half lengthwise, forming a crease $\overline{AB}$ along the centerline.
3. Fold a corner point C upward from corner D, so that C determines point E on the centerline. Once E is found, flatten out the fold.
4. Fold the envelope straight across at E, and then cut the envelope off at the height of E.
5. Make sharp folds along $\overline{DE}$ and $\overline{CE}$.
6. Open up the envelope by pulling the two sides of the envelope at E apart; a regular tetrahedron should appear!

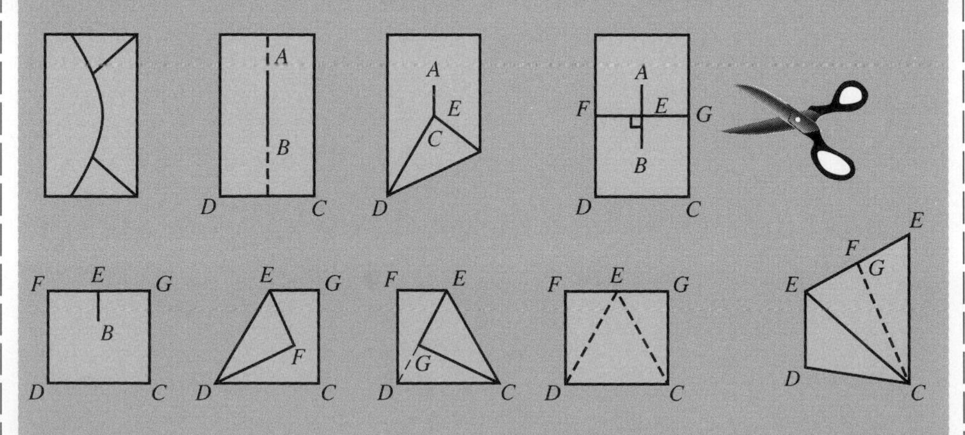

After completing your model, justify the construction procedure.

PROBLEM SET 10.5

Understanding Concepts

1. Which of the following figures are polyhedra?

(a)

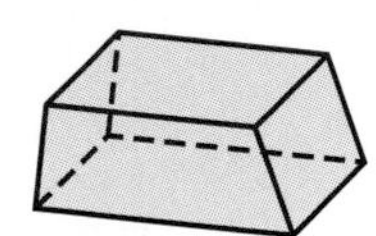

(b)

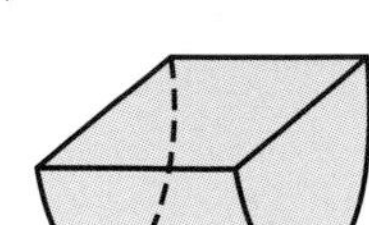

(c)

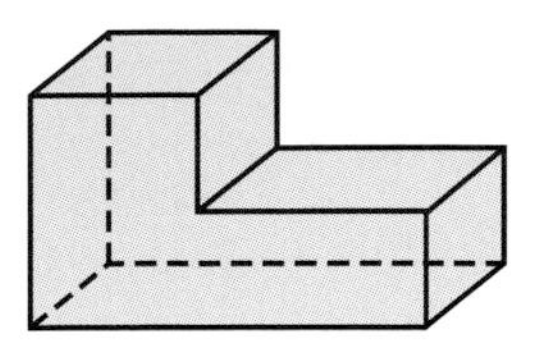

(d)

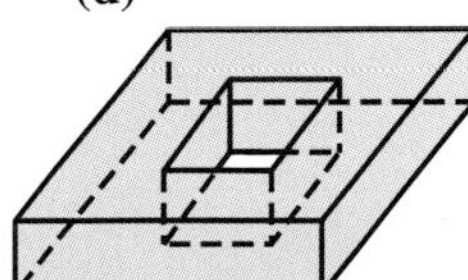

(e)

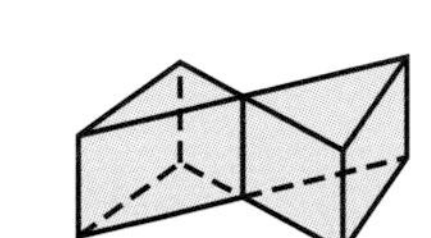

(f)

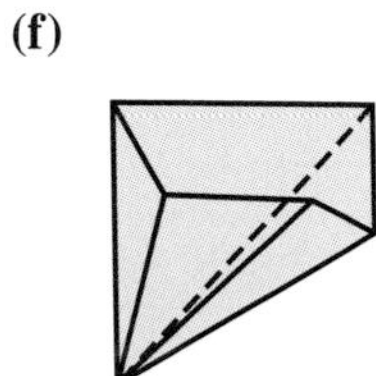

2. Name each of these surfaces.

(a)

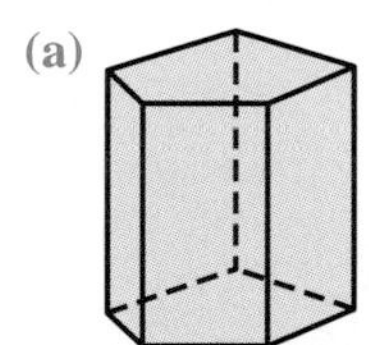

(b)

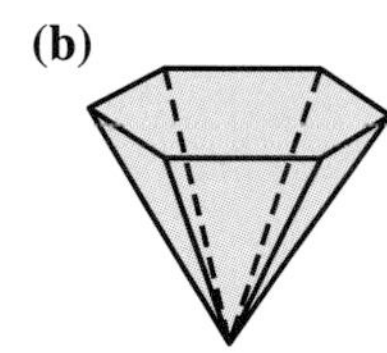

(c)

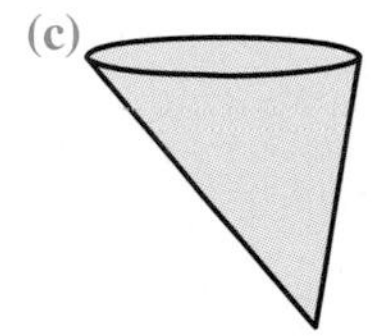

(d)

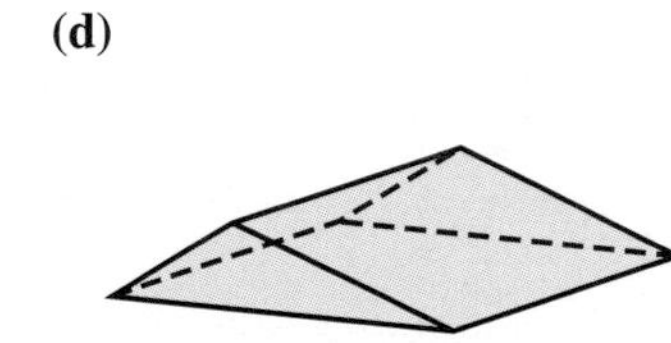

(e)

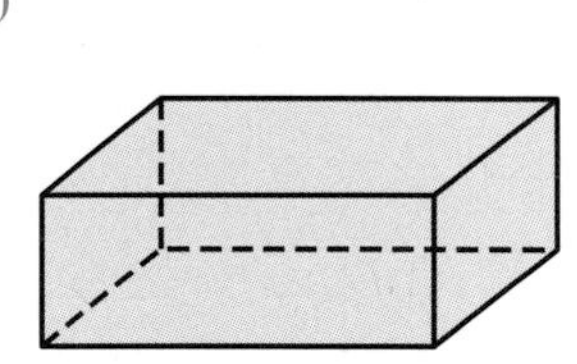

(f)

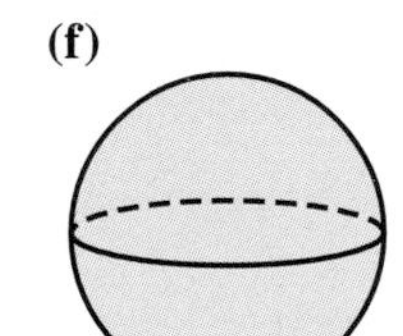

(g)

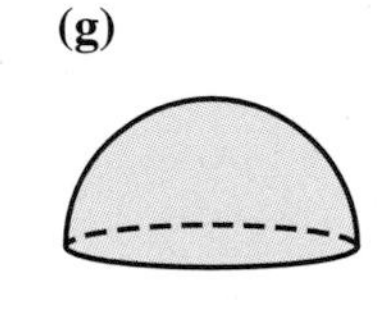

3. A tetrahedron is shown below.
   (a) How many planes contain its faces?
   (b) Name all of the edges.
   (c) Name all of the vertices.
   (d) Name all of the faces.

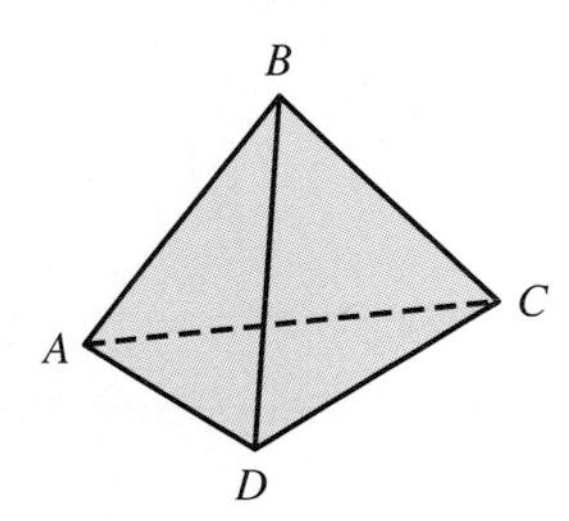

4. Draw free-hand pictures of the following figures, using dashed lines to indicate hidden edges. Don't just copy—transfer the image in your mind to paper.
   (a) cube **(b)** tetrahedron **(c)** square pyramid
   (d) pentagonal right cylinder
   (e) oblique hexagonal prism
   (f) octahedron **(g)** right circular cone

5. A cube with vertices *A*, *B*, *C*, *D*, *E*, *F*, *G*, *H* is shown below. Vertices *D*, *E*, *G*, *H* are the vertices of a tetrahedron inscribed in the cube.

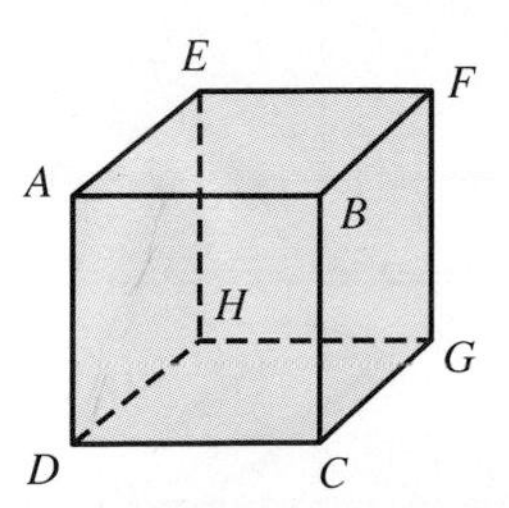

   (a) Trace the cube in one color and then draw the tetrahedron *DEGH* in a different color.

(b) Three of the faces of the tetrahedron *DEGH* are subsets of the faces of the cube. Find a tetrahedron inscribed in the cube which has none of its faces in the planes of the faces of the cube. Sketch your tetrahedron and the surrounding cube.

6. The pattern shown at the left folds up to form the cube on the right.

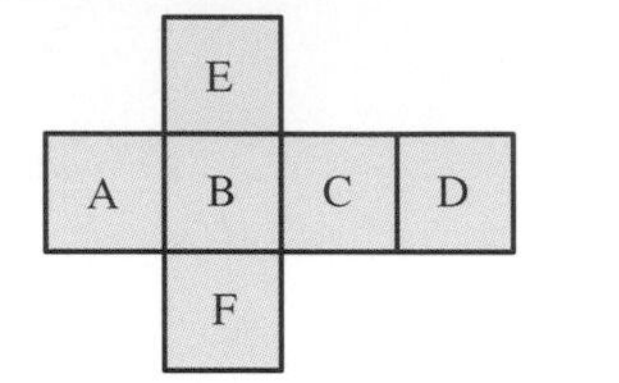

Sketch the letter, in its correct orientation, that should appear on each blank face shown below, where the same pattern is used.

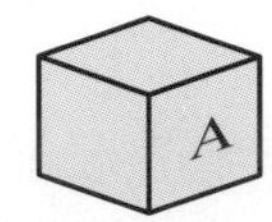

7. The apex of the pyramid shown below is at the center of the cube shown by the dashed lines. What is the dihedral angle which each lateral face of the pyramid makes (a) to the square base? (b) to an adjacent lateral face? (*Hint for (b):* It will help to imagine that the cube is filled with six congruent pyramids whose apexes coincide at the cube's center.)

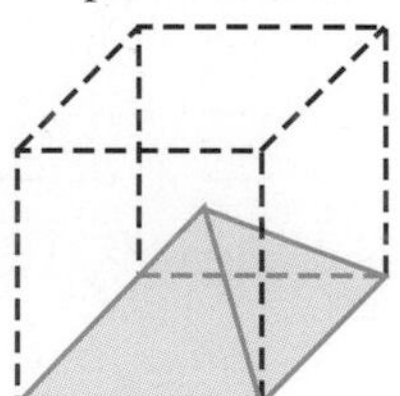

8. The figure shown below is a right prism whose bases are regular pentagons.

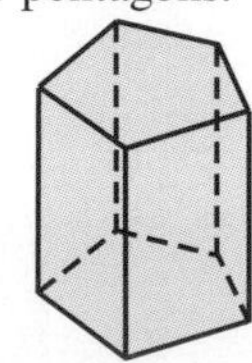

(a) What is the measure of the dihedral angle between each lateral face and a base?

(b) What is the measure of the dihedral angle between adjacent lateral faces?

9. The intersection of a plane and a three-dimensional figure is called a **cross section.** For example he cross section of the sphere shown below is a circle.

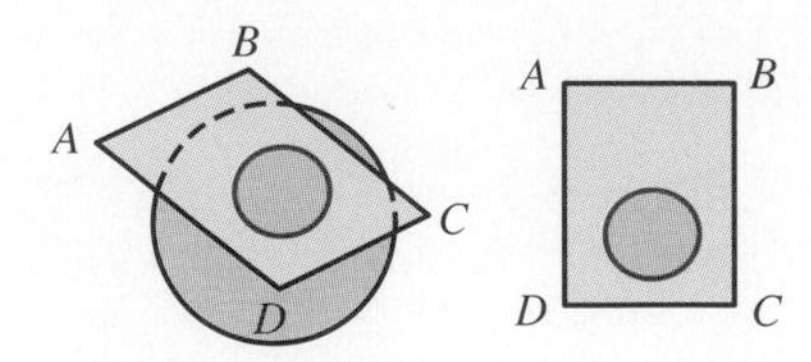

Describe each of the following cross sections, and sketch its shape.

(a)

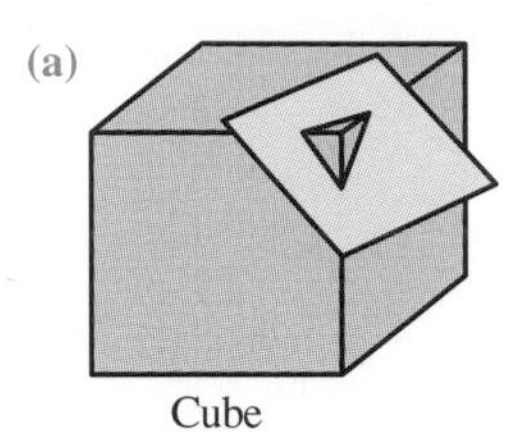

Cube

(b)

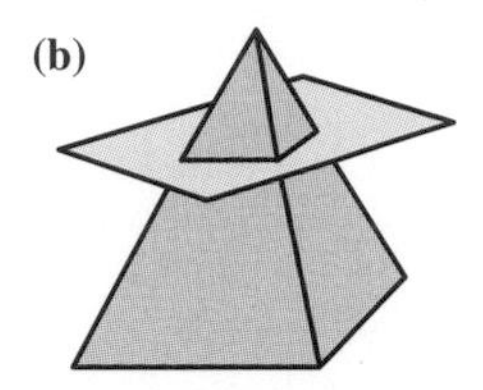

Regular tetrahedron, plane parallel to base

(c)

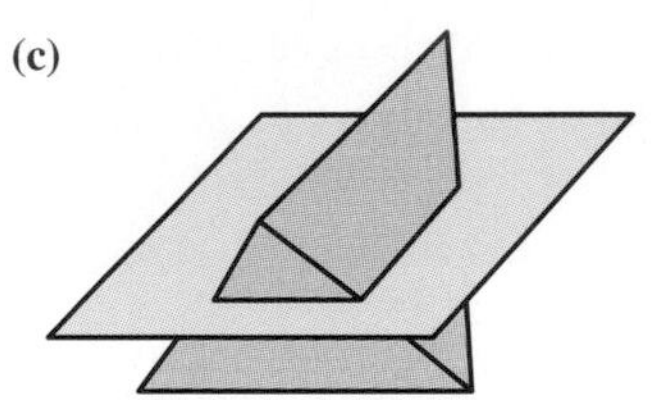

Regular tetrahedron, plane through midpoints of four sides

10. In problem 9(a), a triangle can be formed as some cross section of a cube. Which of the following polygons can be formed as a cross section of a cube? If the cross sectional shape is possible, explain how in words and show a suitable sketch. If the cross sectional shape is not possible, explain why not.

(a) an equilateral triangle

(b) a square

(c) a rectangle which is not a square

(d) a pentagon

(e) a hexagon

(f) a heptagon

11. Are any of the regular polyhedra (a) prisms? (b) pyramids?

12. Complete the table below using Euler's formula. The latter four polyhedra are representative of a class of 13 polyhedra discovered by Archimedes.

Polyhedron	Number of Vertices, Faces, and Edges V	F	E
Hexagonal pyramid	_____	_____	_____
Octagonal prism	_____	_____	_____
Icosahedron	_____	20	30
Truncated tetrahedron	12	8	_____
Truncated dodecahedron	_____	32	90
Snub cube	24	_____	60
Great rhombicosi-dodecahedron	120	62	_____

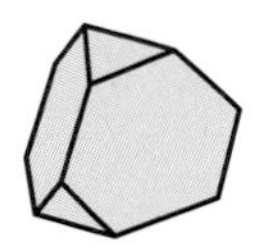

Truncated tetrahedron

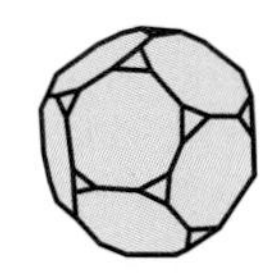

Truncated dodecahedron

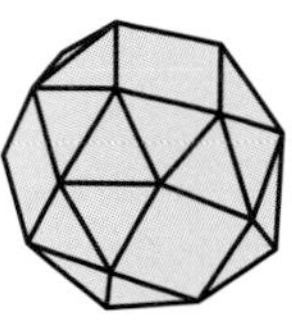

Snub cube

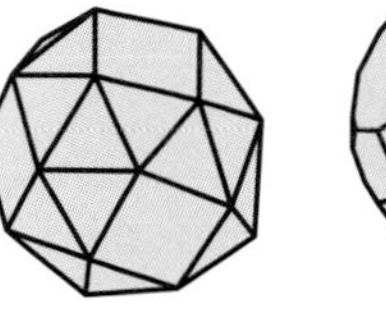

Great rhombicosidodecahedron

13. A **double pyramid** is a polyhedron with triangular faces arranged about a simple plane polygon and extending to a lower apex and an upper apex. An **antiprism,** like a prism, has two polygonal bases but the lateral faces are triangular. A pentagonal double pyramid and a hexagonal antiprism are shown.

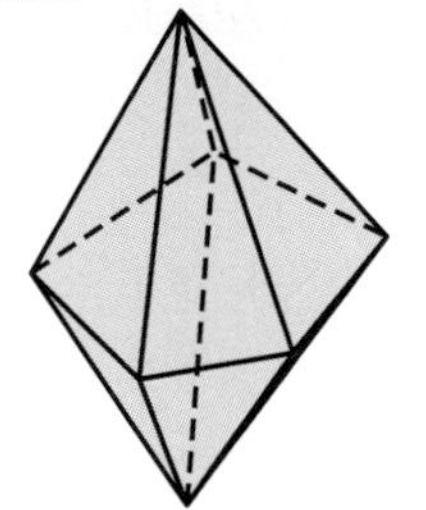

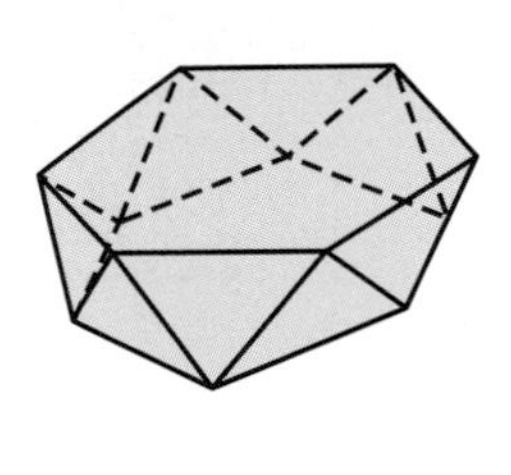

(a) Find the number of vertices, faces, and edges for the pentagonal double pyramid and the hexagonal antiprism. Then verify that Euler's formula holds for each.

(b) Repeat part (a) for the double pyramid whose lateral faces are built off a 20-gon.

(c) Repeat part (a) for an antiprism whose bases each have 23 sides.

14. Draw nets corresponding to the following polyhedra.

(a) A square pyramid with equilateral triangular faces.

(b) A truncated tetrahedron. (See the figure in problem 12.)

15. Check whether Euler's formula holds for the figures below. If not explain what assumption required for Euler's formula is not met.

(a)

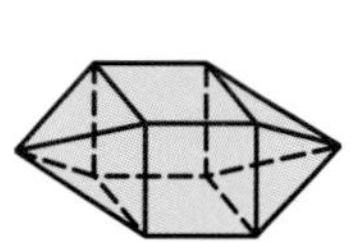

(b)

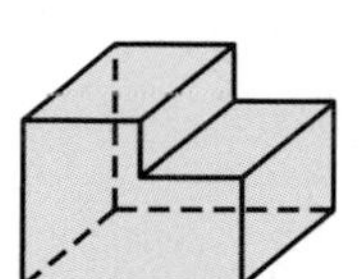

(c)

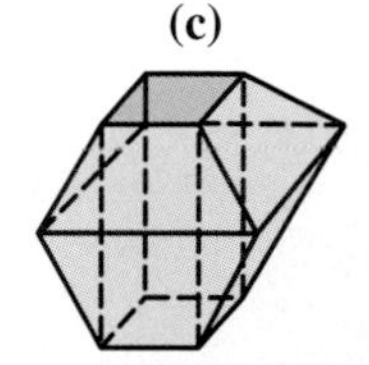

(An octahedron with a square hole)

Thinking Critically

16. Is it possible for a simple closed pentagon in space to look like this? The diagram shows which edge is closer to the viewer.

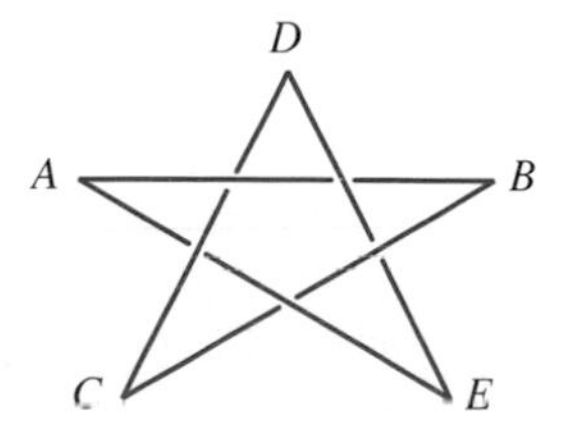

17. A simple closed polygonal curve is a regular polygon if all of its sides have the same length and every pair of adjacent sides intersect at the same angle.

(a) Describe and sketch a regular hexagon *in space* which does not lie in any plane.

(b) Describe regular polygons in space with 8, 10, 12, . . . sides which do not lie in a plane. (It is possible, but much more difficult, to find regular nonplanar 7–, 9–, 11–, . . . gons!)

18. Removing one point from a line leaves two pieces, and more generally removing n points from a line leaves $n + 1$ pieces. Similarly, (see problem 24 of Section 10.2) removing one line from the plane leaves two pieces, and removing two nonparallel lines leaves the plane in four pieces. Investigate how many pieces of space remain when n planes "in general position" (that is, no two planes parallel, and no three planes intersecting in a single line) are removed. Verify the entries in the partially completed table below, and see if you can discover a pattern which allows you to guess how the table can be filled out.

Number of Remaining Pieces in the	Number of Points Removed from the Line, or Lines Removed from the Plane, or Planes Removed from Space 0	1	2	3	4	5
Line	1	2	3	4	5	6
Plane	1	2				
Space	1	2				

19. Let V, E, F denote the number of vertices, edges, and faces of a polyhedron.

(a) Explain why $2E \geq 3F$. (*Hint:* Every face has at least three sides, and every edge borders two faces.)

(b) Explain why $2E \geq 3V$. (*Hint:* Every vertex is shared by at least three faces.)

(c) Show that every polyhedron has at least six edges.
(*Hint:* Add the inequalities of parts (a) and (b), and use Euler's formula.)

(d) Use (a) and (b) to prove that no polyhedron can have seven edges. (*Hint:* Use Euler's formula.)

(e) Show that there are polyhedra with 6, 8, 9, 10, . . . edges.

20. A convex polyhedron with five faces is called a **pentahedron.**

(a) Find and sketch two different types of pentahedra. (*Hint:* One is "easy as pie.")

(b) Prove that the two pentahedra described in part (a) are the only two possible types.

21. Suppose a skeletal model of a convex polyhedron is made, outlining just the edges of the polyhedron. If the model is viewed in perspective from a position just outside the center of a face, the edges of that face form a bounding polygon inside which the remaining edges are seen. The resulting pattern of edges is called a **Schlegel diagram,** named for the German mathematician Viktor Schlegel who invented the diagram in 1883. Schlegel diagrams for the cube and dodecahedron are shown on the right. Draw Schlegel diagrams for:

(a) the tetrahedron, (b) the octahedron, (c) the icosahedron. (*Hint:* Start by drawing a large equilateral triangle, and keep in mind each vertex must have degree five.)

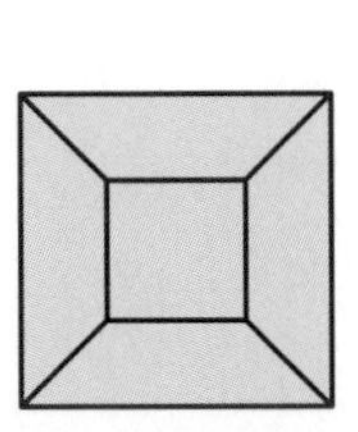

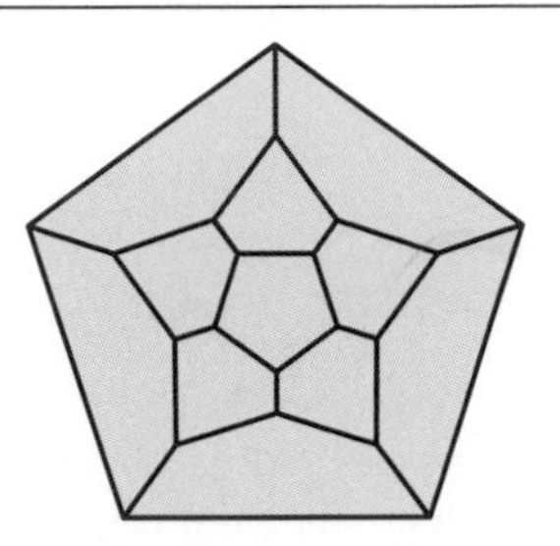

22. The dihedral angles of the regular polyhedra are given in the following table.

Regular Polyhedron	Measure of Dihedral Angle (in degrees and minutes)
Cube	90°
Tetrahedron	70° 32′
Octahedron	109° 28′
Dodecahedron	116° 34′
Icosahedron	138° 11′

(a) How many tetrahedra can be placed about a common edge without overlap? What is the size of the gap that remains?

(b) Why is the cube the only regular polyhedron that will fill space?

Thinking Cooperatively

23. A polyhedron with six faces is a **hexahedron.** For example, a cube is a hexahedron.

(a) Draw a pyramid that is a hexahedron.

(b) Draw a double pyramid (see problem 13) that is a hexahedron.

(c) Can you find other types of hexahedra? It has been shown there are seven distinct types.

24. A **hexomino** is formed by joining six congruent squares along edges. There are 35 different hexominoes in all, including the following.

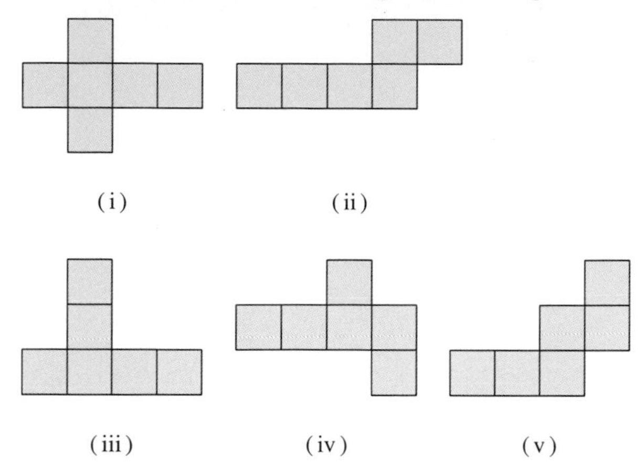

(a) Hexomino **(i)** is a net for a cube. Which of the other four hexominoes shown can be folded to form a cube?

(b) There are 11 hexominoes that can be folded to form a cube. Try to find all 11 shapes. Be careful not to repeat a shape; two congruent shapes may at first appear to be different when one is rotated or flipped over.

25. A net for a square pyramid is shown below on the left. A net for a more general quadrilateral pyramid is shown to the right. The dot at P in each net locates the point in the plane of the base directly beneath the apex of the pyramid.

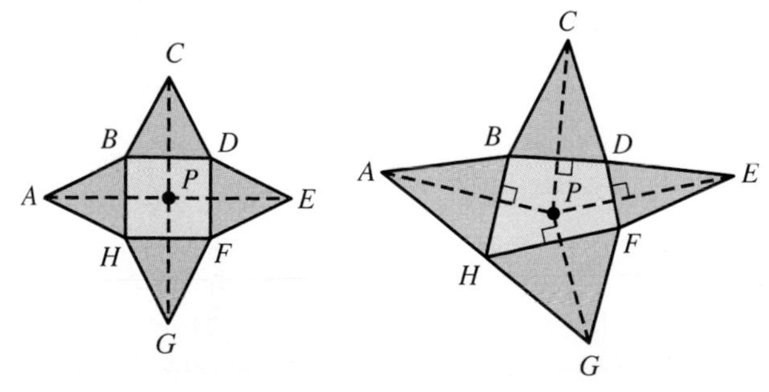

(a) Explain why $AB = BC$, $CD = DE$, . . . , $GH = HA$ in the nets and why the dashed lines from P are perpendicular to the sides of the base polygon.

(b) A net for a pentagonal pyramid has been started below.

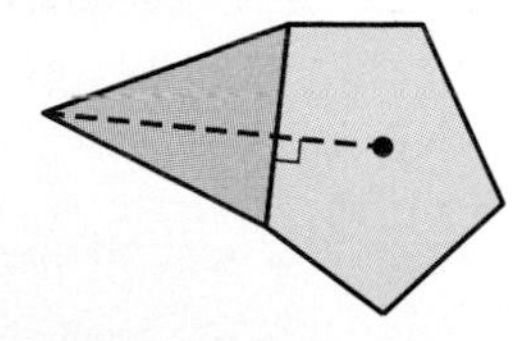

Trace the net on a piece of paper and complete the net. Cut and fold the pattern to see the shape which results.

Making Connections

26. The round door shown below has a problem. What facts of space geometry create a difficulty?

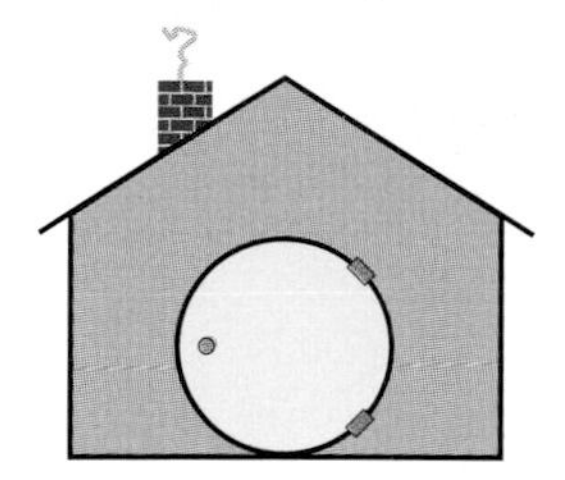

What is important about the placement of door hinges?

27. The ancient Greeks divided physical space into five parts: the universe, earth, air, fire, and water. Each of these was associated with one of the five regular polyhedra. Investigate what correspondence was made.

28. Biologists and physical scientists frequently become involved with the analysis of shape and form. For example, many viruses have an icosahedral structure and the crystalline structure of minerals are often polyhedral forms of considerable beauty. An example of a pyrite crystal is shown here.

Browse through your school library and see what three-dimensional shapes are receiving interest and attention. Report on your findings.

For Review

29. Find the measures of x and y in the following figure, where $l \parallel m$.

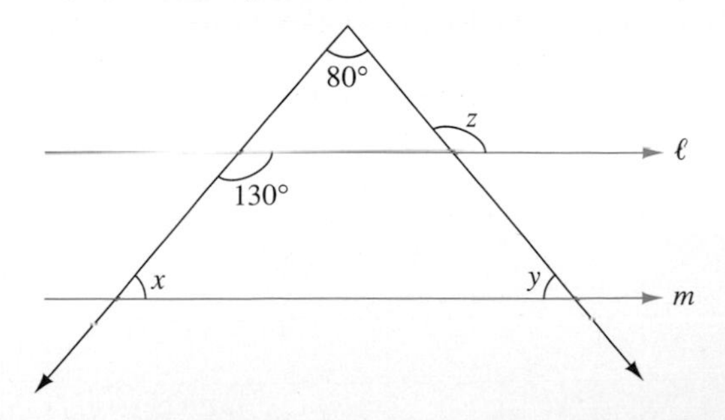

30. A triangle has no diagonals and a convex quadrilateral has two diagonals.
(a) Fill in the entries in the following table.

	n					
	3	4	5	6	7	8
Number of diagonals in convex n-gon	0	2				

(b) Describe a pattern which you see in the table.
(c) How many diagonals are in a convex dodecagon?
(d) How many diagonals are in a convex 100-gon?

31. It is easy to join the 12 points of a 3 by 4 square array by a polygon. Here is one way.

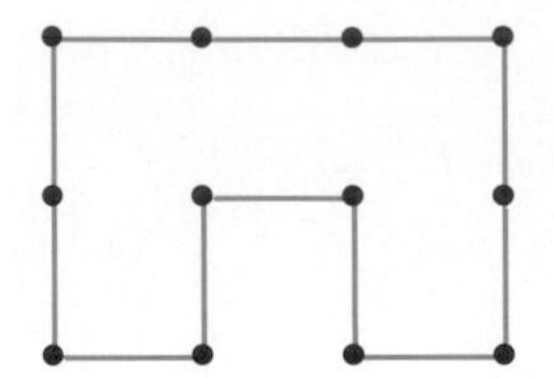

Find polygons which join all of the points of these square arrays.
(a) 4 by 6 **(b)** 5 by 7

EPILOGUE Visualization

After struggling with a difficult problem or abstract concept, a student may suddenly smile and announce "I see it now," or perhaps exclaim, "I've got the picture." Information conveyed in visual form can easily be superior to the same information described by a thousand, or even ten thousand, well chosen words. Our ability to visualize is only partly dependent on the acuity of our eyesight: even more important is the mind's eye, which sharpens our perception by providing us with skills and abilities to identify, analyze, and classify shape.

In recent years the graphics computer has enabled us to see in directions unimaginable just a short time ago. This development is a continuation of technological advances which redefine what is observable. Nearly 400 years ago, Galileo's telescope detected the moons of Jupiter. Today, optical and radio telescopes reveal quasars at the edge of the universe. A hundred years after Galileo, the universe of the very small became observable with Leeuwenhoek's invention of the microscope. Today, tunnelling microscopes produce images in which single atoms are distinguishable.

Learning to understand and interpret new images is an exciting challenge. Fortunately, many of the concepts and experiences first encountered in elementary geometry prepare the way to meet this challenge. In this chapter we have introduced many of the basic notions of geometry: point, line, plane, curve, surface, angle, distance between points, measure of an angle, region, tiling, and space. In the chapters which follow, these basic notions are developed in more depth as we encounter the ideas of similarity, construction, measurement, isometric and similarity transformations, symmetry, and coordinate methods of geometry.

CLASSIC CONUNDRUM An Unexpected Bisector

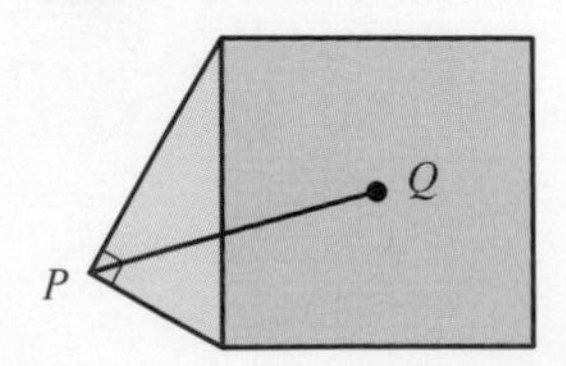

Erect any right triangle you wish on the side of a square with center Q. Are you surprised to see that joining P to Q bisects the 90° angle of your triangle? See if you can add congruent right triangles to the other sides of the square which make the bisection property obvious.

CHAPTER 10 SUMMARY

Key Concepts

This chapter introduced the basic figures of the plane and of space. Figures were represented by drawings or by physical models. Properties of figures were described and analyzed, and figures were classified according to their properties.

1. *Plane Figures*

   Two points A and B lie on a unique line $\overleftrightarrow{AB}$. The distance between A and B is AB, which is also the length of the line segment $\overline{AB}$. Two rays, $\overrightarrow{AB}$ and $\overrightarrow{AC}$, with a common endpoint A form an angle, $\angle BAC$, which has measure $m(\angle BAC)$. Angles are classified by their measures as zero, acute, right, obtuse, straight, or reflex. Two angles are complementary or supplementary if their measures add up respectively to 90° or 180°. The corresponding angles property, and the alternate interior angles and vertical angles theorems provided relationships among the angles formed when lines intersect. The measures of the interior angles of a triangle add up to 180°.

2. *Curves and Polygons in the Plane*

   Curves are classified as simple, closed, convex, and/or polygonal. A simple closed polygonal curve is a polygon. Triangles are classified as scalene, isosceles, equilateral, acute, right, and obtuse. Convex quadrilaterals are classified as trapezoids, parallelograms, rhombuses, rectangles, and squares. By the Jordan curve theorem, a simple closed curve in the plane separates the plane into two regions, the exterior and the interior of the curve. The measures of the interior angles of an n-sided polygon add up to $(n - 2)180°$. In a regular n-gon, each interior angle has measure $(n - 2)180°/n$ and each exterior angle measures $360°/n$. The total turn about any closed plane curve is an integral multiple of 360°.

3. *Tiling with Polygons*

   A tile is a simple closed curve in the plane together with its interior. There are three regular tilings of the plane—by equilateral triangles, squares, and regular hexagons—and eight semiregular tilings. Arbitrary triangles and quadrilaterals tile the plane. Some pentagons and hexagons tile the plane, but no convex n-gons tile the plane if $n \geq 7$.

4. *Networks*

   A network is a set of vertices and a set of edges joining some of the pairs of distinct vertices. A network is traversable if there is an Euler path covering each edge of the network once and only once. Euler showed that a connected network is traversable if, and only if, the network has 0 or 2 vertices of odd degree. A network is planar if no two edges cross one another. Euler's formula $V + R = E + 2$ relates the number of vertices, regions, and edges of a connected planar network.

5. *Figures in Space*

   A simple closed surface is a surface without holes which encloses a region called its interior. The simple closed surfaces formed by polygonal regions are called polyhedra, and include prisms, pyramids, and the five regular polyhedra. The number of vertices, faces, and edges of any polyhedron satisfy Euler's formula $V + F = E + 2$. Spheres, cones, and cylinders are examples of curved surfaces.

Vocabulary and Notation

Section 10.1

Point, A, B, . . .
Line l, $\overleftrightarrow{AB}$
Collinear points
Plane figure
Parallel, $l \parallel m$
Concurrent lines
Distance between points, AB
Line segment, $\overline{AB}$
Congruent line segments, $\overline{AB} \cong \overline{CD}$
Measure of an angle, $m(\angle ABC)$
Ray, $\overrightarrow{AB}$
Angle, $\angle ABC$
Vertex of an angle
Zero, acute, right, obtuse, straight, reflex angles
Perpendicular, $l \perp m$
Congruent angles, $\angle A \cong \angle B$
Complementary angles
Supplementary angles
Vertical angles
Corresponding angles
Alternate interior angles
Interior and exterior angles of a triangle
Directed angle
Initial and terminal side

Section 10.2

Curve
Simple, closed, simple closed
Jordan curve theorem
Interior and exterior region of a simple closed curve
Convex figure
Concave figure
Polygonal curve
Polygon
Pentagon, hexagon, heptagon, octagon, nonagon (or enneagon), decagon, . . . , n-gon
Interior, exterior angles of a polygon
Triangle: acute, right, obtuse, scalene, isosceles, equilateral
Quadrilateral: kite, trapezoid, isosceles trapezoid, parallelogram, rhombus, rectangle, square
Circle
Diameter, radius, center, chord, tangent line, arc, sector, segment

Section 10.3

Tile
Tiling (or tessellation)
Regular tiling of the plane
Vertex figure
Semiregular tiling of the plane

Section 10.4

Network
Vertices, edges, degree of a vertex, odd/even vertex, connected network
Traversable network
Euler path
Planar network

Section 10.5

Plane
Parallel and intersecting planes, dihedral angle, half-space, skew lines, perpendicular planes
Sphere
Simple closed surface
Polyhedron
Vertex, face, edge
Pyramid
Prism, right prism, oblique prism
Regular polyhedron
Cone
Cylinder

CHAPTER REVIEW EXERCISES

Section 10.1

1. Let $ABCD$ be the quadrilateral shown below.

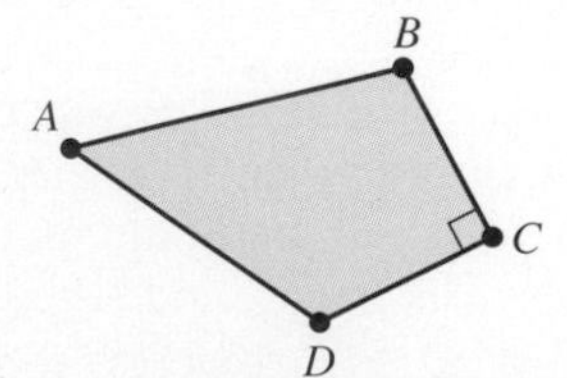

Give symbols for the following:

(a) The line containing the diagonal through C.
(b) The diagonal containing B.
(c) The length of the side containing A and D.
(d) The angle *not* containing D.
(e) The measure of the interior angle at C.
(f) The ray which has vertex at D and is perpendicular to a side of the quadrilateral.

2. For the quadrilateral shown in problem 1, which angle(s) appear to be:
(a) acute? (b) right? (c) obtuse?

3. An angle measures 37°. What is the measure of
(a) its supplementary angle?
(b) its complementary angle?

4. Lines l and m below are parallel. Find the measures p, q, r, and s of the angles shown.

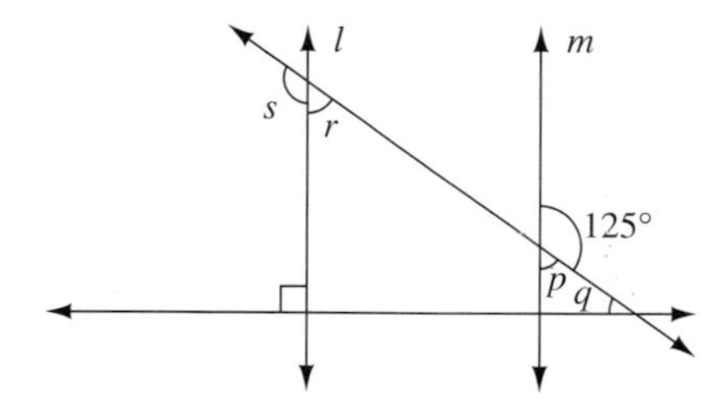

5. Find the measures x, y, and z of the angles in the following figure.

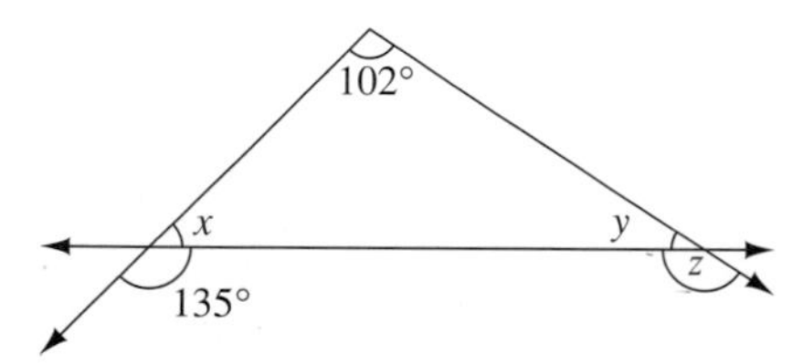

Section 10.2

6. Match each curve to one of the descriptions below.

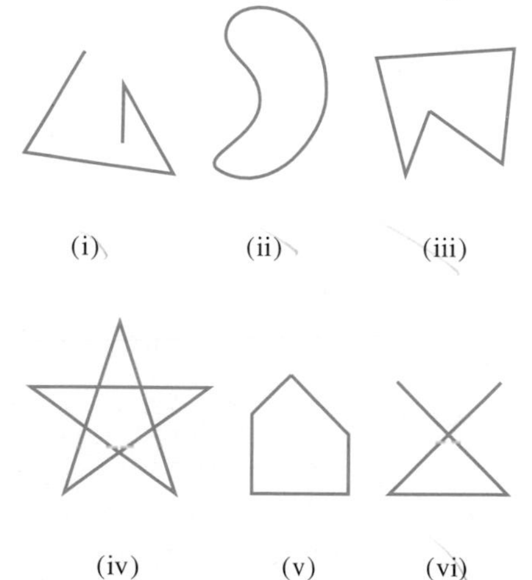

(a) nonconvex nonsimple polygonal curve
(b) nonclosed simple curve
(c) nonsimple nonclosed polygonal curve
(d) convex polygon
(e) simple closed nonconvex nonpolygonal curve
(f) nonconvex polygon

7. (a) Can a triangle have two obtuse angles?
(b) Can a convex quadrilateral have three obtuse interior angles?
(c) Is there an "acute" quadrilateral (that is, a quadrilateral whose angles are all acute)?
For each part, carefully explain your reasoning.

8. Find the measures of the interior angles of this polygon.

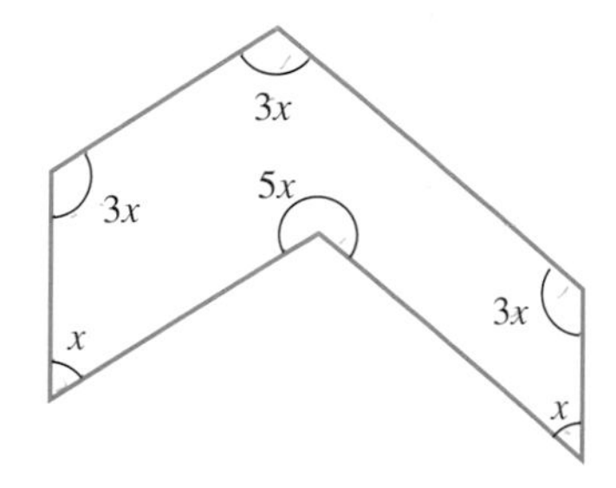

9. A turtle walks along the path $ABCDEFGA$ in the direction of the arrows, returning to the starting point and initial heading. What total angle does the turtle turn through?

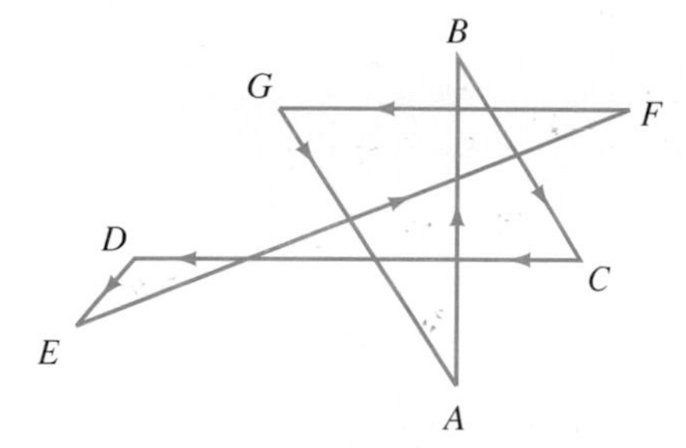

Section 10.3

10. Four regular polygons form a vertex figure in a tiling of the plane. Three of the polygons are a triangle, a square, and a hexagon. What is the fourth polygon?

11. Draw two different vertex figures which each incorporate three equilateral triangles and two squares.

12. Show that the shape below will tile the plane.

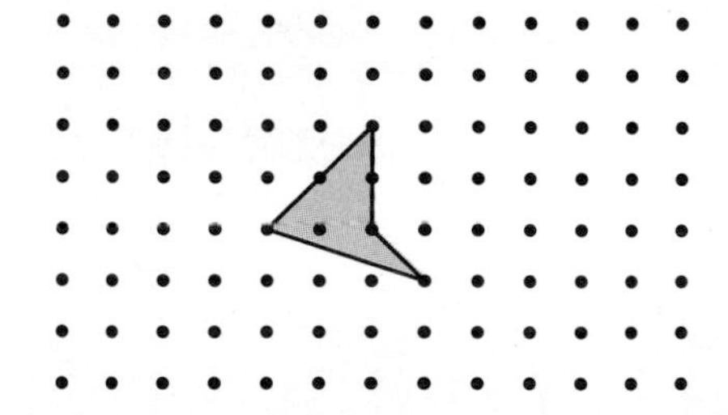

Section 10.4

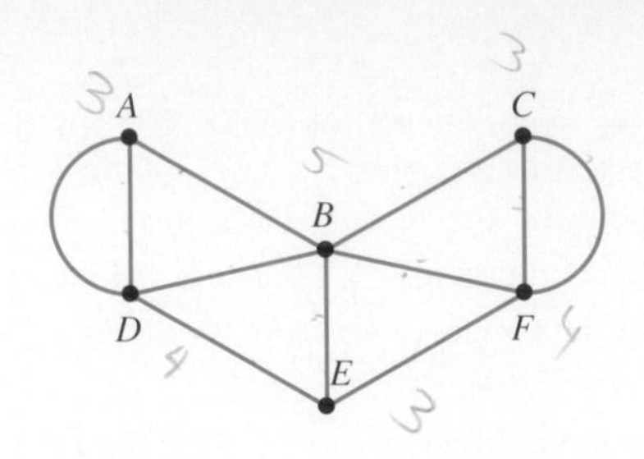

13. (a) Explain why the network shown is not traversable.

(b) Name two vertices which, if connected by a new edge, would make the resulting network traversable. Then list the vertices of an Euler path.

14. Is there a walk which crosses each of the bridges shown below exactly once? Explain your reasoning, using an appropriate network.

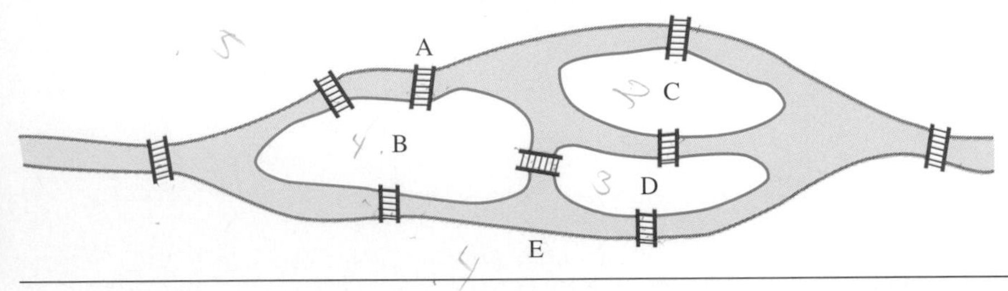

15. Count the number of vertices, regions, and edges in the following network, and then verify that Euler's formula holds.

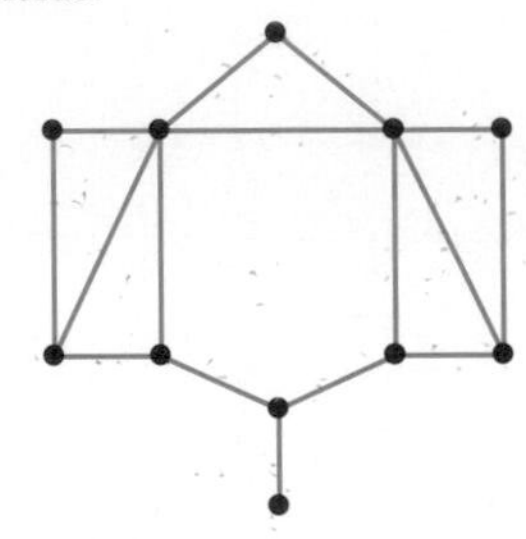

Section 10.5

16. Let *ABCDEFGH* be the vertices of a cube, as shown below.

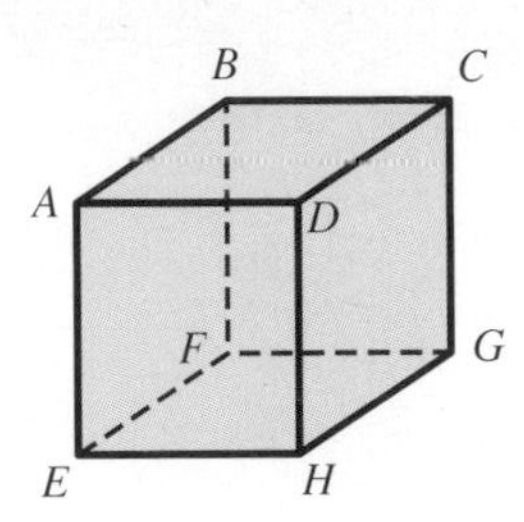

(a) How many planes are determined by the faces of the cube?

(b) Which edges of the cube are parallel to side $\overline{AB}$?

(c) Which edges of the cube are contained in lines which are skew to the line $\overleftrightarrow{AB}$?

(d) What is the measure of the dihedral angle between the plane containing *ABCD* and the plane containing *ABGH*?

17. Name the following surfaces in space.

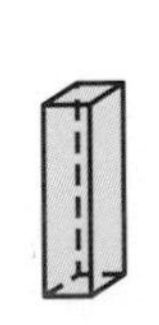 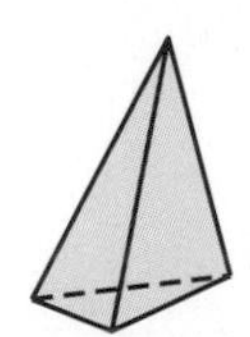   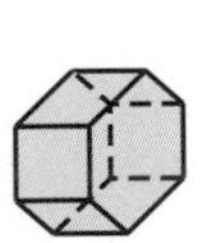

18. Draw the following shapes.

(a) a right circular cone

(b) a pentagonal prism

(c) a nonconvex quadrilateral pyramid

19. (a) Draw a regular octahedron.

(b) Using your drawing in part (a) count the number of vertices, faces, and edges of the octahedron, and then verify that Euler's formula for polyhedra holds.

20. A polyhedron has 14 faces and 24 edges. How many vertices does it have?

CHAPTER TEST

1. Let P, Q, R, S, T be the points shown. Draw and label the following figures.

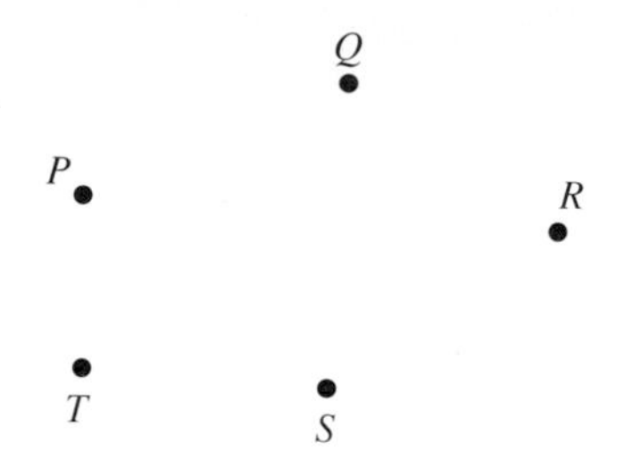

(a) $\overrightarrow{PQ}$ (b) $\overleftrightarrow{PR}$ (c) $\overline{SQ}$ (d) $\angle PST$

2. At which vertices of the polygon does the interior angle appear to be (a) acute? (b) right? (c) obtuse? (d) straight? (e) reflex?

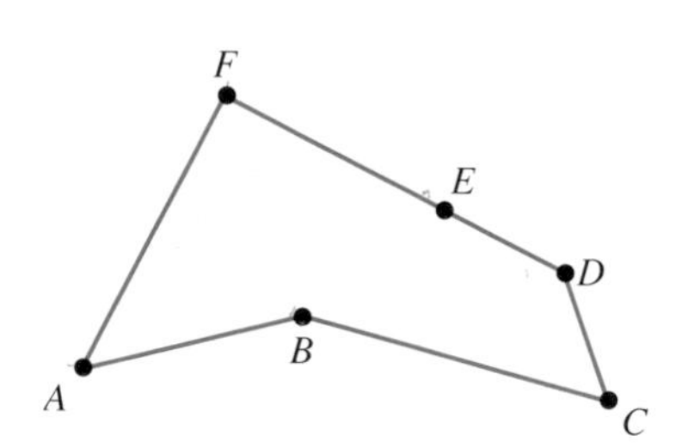

3. Sketch an example of each of the following types of curves.
   (a) a simple closed curve
   (b) a convex heptagon
   (c) a nonclosed simple polygonal curve
   (d) a closed nonsimple polygonal curve
4. Decide whether each statement below is *true* or *false*.
   (a) Every square is a rhombus.
   (b) Some right triangles are obtuse.
   (c) All equilateral triangles are isosceles.
   (d) All squares are kites.
5. For each part, name the regular polygon with the stated property.
   (a) The central angle has measure 36°.
   (b) The exterior angles each measure 45°.
   (c) The interior angles each measure 140°.
6. Find the angle measures r, s, t in the figure below, where lines l and m are parallel.

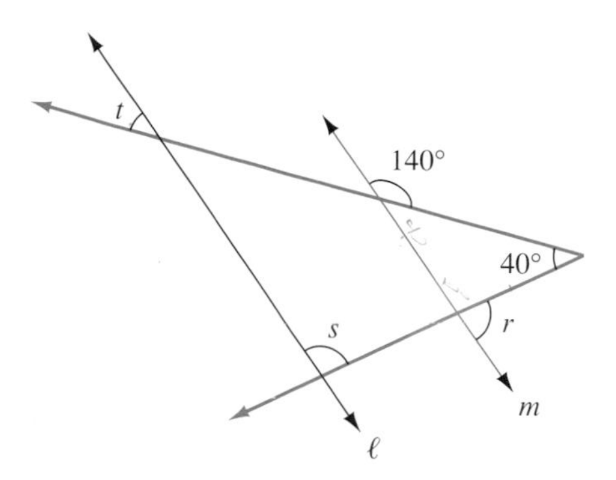

7. Which of the following shapes will tile the plane?

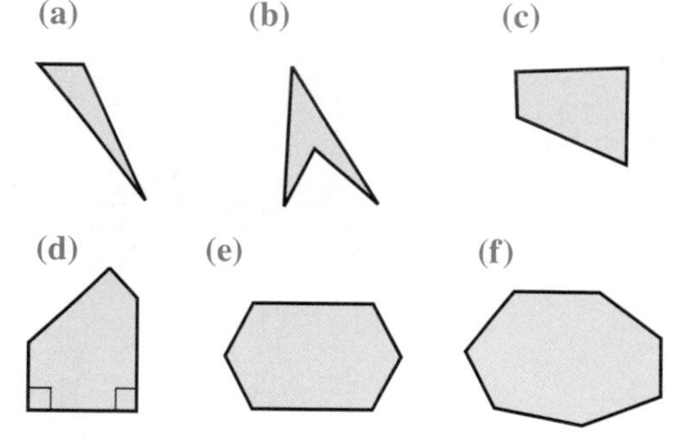

8. Sketch a portion of the semiregular tiling of the plane which uses square and octagonal tiles with a common side length.
9. What is the measure of the angles in the points of this symmetric 8-pointed star? Explain how you found your answer.

10. The average interior angle measure of a convex polygon is 174°. What is the number of sides of the polygon? Explain how you found your answer.

11. Consider the following networks.

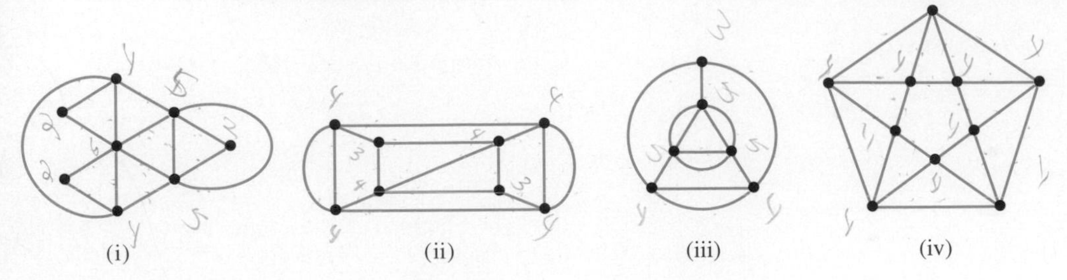

(a) Which networks are not traversable?
(b) Which networks are traversable starting at an arbitrary vertex?
(c) Which network is traversable starting at some, but not every, vertex?
(d) Which network becomes traversable when one additional edge is added to the network? Describe the new edge.

12. (a) A connected planar network with 11 edges separates the plane into 7 regions. How many vertices does this network have?
(b) Draw a connected planar network with 11 edges and 7 regions.

13. A right prism has bases bounded by regular pentagons. What is the dihedral angle at which two adjacent lateral faces meet?

14. A cube octahedron has eight triangular faces and six square faces.

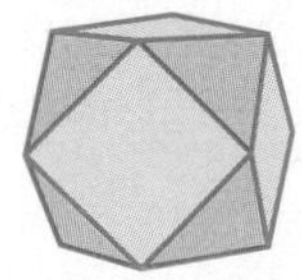

(a) What is the number of edges of a cube octahedron? Explain how you do your counting.
(b) What is the number of vertices of a cube octahedron? Explain how you found your answer.

15. Pyramids are erected outward to both bases of a heptagonal prism.
(a) Sketch the surface that is described.
(b) Directly count the number of vertices, faces, and edges of the resulting polyhedron.
(c) Verify that Euler's formula is satisfied.

Congruence, Constructions, and Similarity

Exploring Toothpick Triangles

Materials Needed

Toothpicks of equal length (20 per person)

Directions

Three toothpicks, placed end-to-end, form a triangle. Four toothpicks do not form a triangle. Five and six toothpicks each form just one triangle. Two different (that is, noncongruent) triangles can be formed with seven toothpicks.

Number of:										
Toothpicks, n	3	4	5	6	7	8	9	10	11	12
Triangles, $T(n)$	1	0	1	1	2					

Explore how many different triangles you can form with 8, 9, 10, 11, or 12 toothpicks. Organize your results in a table as shown above.

Questions for Consideration

1. How many isosceles toothpick triangles are there for which the two sides of equal length each use four toothpicks?
2. One side of a toothpick triangle uses three toothpicks and a second side uses five toothpicks. What are the possible numbers of toothpicks in the third side?
3. If two sides of a toothpick triangle together use 11 toothpicks, what is the largest number of toothpicks that can be used in the third side?
4. Suppose toothpicks form a triangle with p, q, and r toothpicks on its three sides. What can you say about the integer r in terms of the numbers p and q?
5. In your table of the number of triangles, suppose $T(n)$ is the number of different toothpick triangles formed from n toothpicks. For n odd, compare $T(n)$ to $T(n + 3)$. For example, compare $T(3)$ to $T(6)$, and compare $T(5)$ to $T(8)$. What pattern do you observe?

CONNECTIONS

Creating and Relating Geometric Figures

This chapter explores three key ideas of geometry—congruence, the construction of geometric figures, and similarity.

Two figures that are the same shape and size are **congruent**. A child is introduced to congruence when solving a jigsaw puzzle or copying a paper pattern onto cloth to make doll clothing. As adults we rely on congruence to provide affordable mass-produced consumer goods.

In the traditional sense of Euclid, a construction is a figure produced by the use of just the straightedge and compass—and pencil, although no marks are permitted on the straightedge. In this chapter we review many of the standard Euclidean constructions, but we also explore procedures which use non-Euclidean tools such as a ruler, a protractor, a Mira, or a computer.

Two figures of the same shape but of different size are **similar** to one another. Model ships, maps, a globe of the earth, and a floor plan of a house are familiar examples of similarity.

11.1 Congruent Triangles

Cookie cutters, candy molds, rubber stamps, and copy machines allow us to make multiple shapes identical to one another in both size and shape. In general, two figures F and G are called **congruent** if, and only if, they have the same size and shape. We write $F \cong G$ to indicate that F and G are congruent figures.

Two line segments are congruent if, and only if, they have equal length. Given a line segment $\overline{AB}$, the compass and straightedge allow us to construct a congruent segment $\overline{PQ}$, as shown in Figure 11.1.

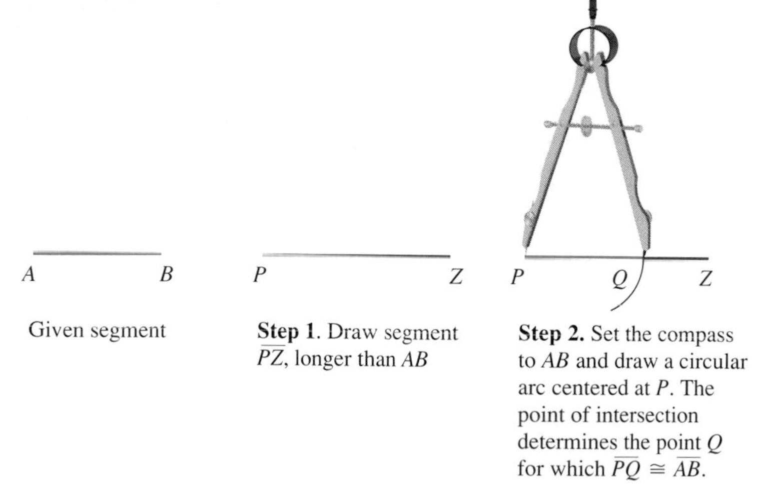

A
B
Given segment
P
Z
Step 1. Draw segment $\overline{PZ}$, longer than AB
P
Q
Z
Step 2. Set the compass to AB and draw a circular arc centered at P. The point of intersection determines the point Q for which $\overline{PQ} \cong \overline{AB}$.

Figure 11.1
Constructing a line segment $\overline{PQ}$ that is congruent to a given line segment $\overline{AB}$.

The size and shape of a triangle is described completely if we specify the **six parts of a triangle,** namely the three sides $\overline{AB}$, $\overline{BC}$, $\overline{CA}$ and the three angles $\angle A$, $\angle B$, and $\angle C$. A second triangle, PQR, is congruent to triangle ABC if there is a matching of vertices $A \leftrightarrow P$, $B \leftrightarrow Q$, $C \leftrightarrow R$ under which *all six* parts of triangle ABC are congruent to the corresponding six parts of triangle PQR. This is illustrated in Figure 11.2.

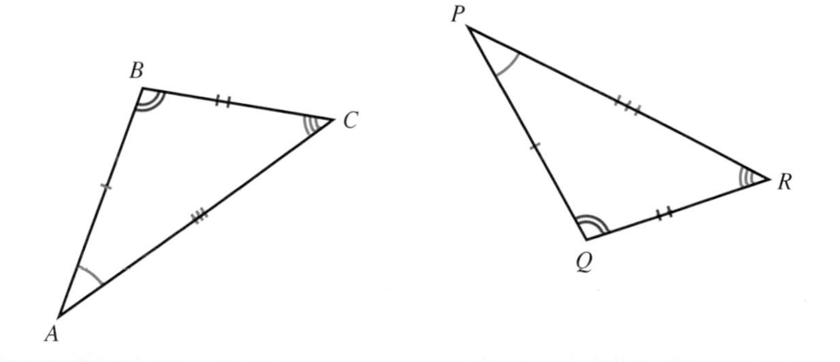

Figure 11.2
Triangles ABC and PQR are congruent under the vertex correspondence $A \leftrightarrow P$, $B \leftrightarrow Q$, $C \leftrightarrow R$ if, and only if, $\overline{AB} \cong \overline{PQ}$, $\overline{BC} \cong \overline{QR}$, $\overline{CA} \cong \overline{RP}$, and $\angle A \cong \angle P$, $\angle B \cong \angle Q$, and $\angle C \cong \angle R$.

DEFINITION Congruent Triangles

Two triangles are **congruent** if, and only if, there is a correspondence of vertices of the triangles such that the corresponding sides and corresponding angles are congruent.

The notation $\triangle ABC \cong \triangle PQR$ is read "triangle ABC is congruent to triangle PQR," and conveys the following information:

- The vertex correspondence is $A \leftrightarrow P$, $B \leftrightarrow Q$, $C \leftrightarrow R$.
- The corresponding sides are congruent: $\overline{AB} \cong \overline{PQ}$, $\overline{BC} \cong \overline{QR}$, $\overline{CA} \cong \overline{RP}$.
- The corresponding angles are congruent: $\angle A \cong \angle P$, $\angle B \cong \angle Q$, $\angle C \cong \angle R$.

SCHOOL BOOK PAGE *Introducing Congruent Figures*

Name

Congruence

These are the same shape and the same size.

These are the same shape but not the same size.

Color the one that is the same shape and size.

1.

2.

3.

Notes for Home Children identify shapes that have the same shape and size.

SOURCE: From *Scott Foresman Exploring Mathematics* Grades 1–7 by L. Carey Bolster et al.

1. Picture puzzles are familiar to most children and clearly involve the concept of congruence. Describe several other examples of congruence in a child's world.

2. How would you make it clear to children that two shapes may be congruent even if one is rotated, or even flipped over?

It is important to notice that the order in which the vertices are listed specifies the vertex correspondence. For the triangles depicted in Figure 11.2, we see that $\triangle ABC \not\cong \triangle QRP$ ($\not\cong$ is read "is not congruent to"). However, it is correct to say that $\triangle BCA \cong \triangle QRP$.

EXAMPLE 11.1

Exploring the Congruence Relation

Use a ruler and protractor to find the two pairs of congruent triangles among the six triangles shown. State the two congruences in the symbolic form $\triangle$ ____ $\cong \triangle$ ____.

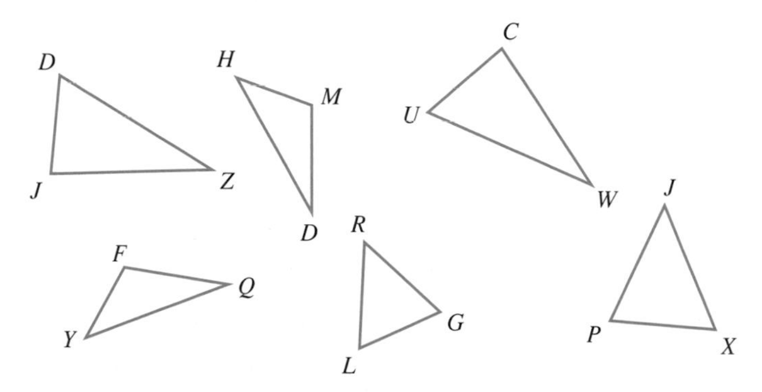

SOLUTION

$\triangle DJZ \cong \triangle UCW$ and $\triangle HMD \cong \triangle YFQ$. The order of the vertices can be permuted in each statement. For example, it would also be correct to express the first congruence as $\triangle ZDJ \cong \triangle WUC$. ■

Suppose that we have only partial information about a triangle ABC. For example, suppose that we know the lengths AB, BC, and CA of the three sides, but we are not given any information about the angles. Or, suppose we are given a side $\overline{AB}$ and two angles, $\angle A$ and $\angle B$. Is the information we have sufficient to construct a triangle PQR that is necessarily congruent to $\triangle ABC$? It is of interest to explore these questions constructively; that is, attempt to use a compass and straightedge to construct a triangle PQR that must be congruent to $\triangle ABC$.

The Side-Side-Side (SSS) Property

EXAMPLE 11.2

Exploring the Side-Side-Side Property

The three sides of triangle ABC are given as shown. Construct a triangle PQR that is congruent to $\triangle ABC$.

A ——— x ——— B   B —— y —— C   A —— z —— C

SOLUTION

Step 1 Construct a segment $\overline{PQ}$ of length $x = AB$.

P ——— x ——— Q

Step 2 Set the compass to radius $y = BC$ and draw a circle of radius y centered at Q.

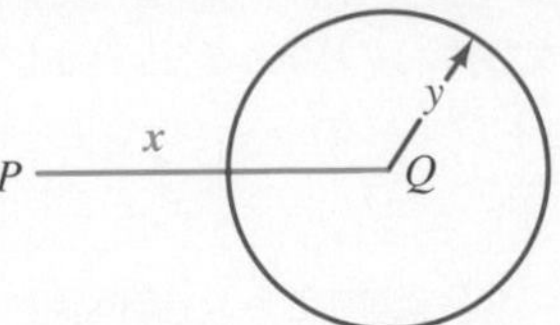

Step 3 Set the compass to radius $z = AC$ and draw a circular arc of radius z centered at P. Let R be either point of intersection with the circle drawn in Step 2.

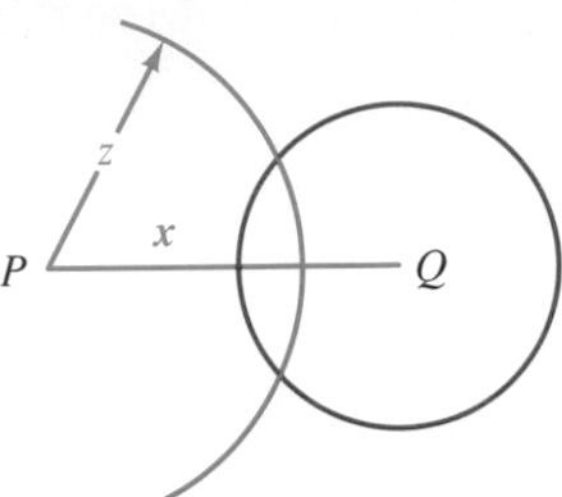

Step 4 Draw the segments $\overline{PR}$ and $\overline{PQ}$. Then $\triangle PQR \cong \triangle ABC$.

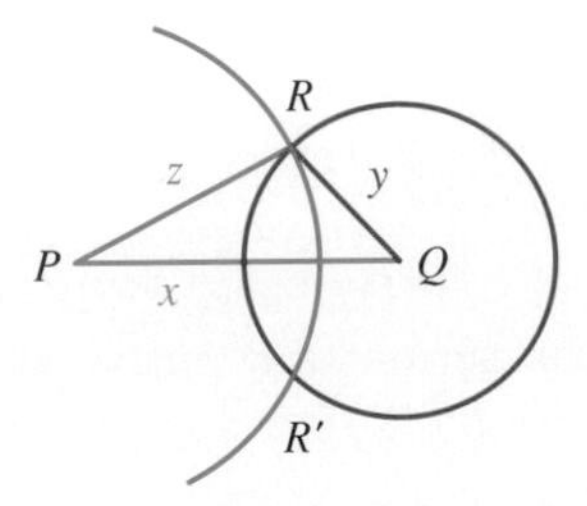

■

The size and shape of $\triangle PQR$ is uniquely determined. Even if we had chosen the second point of intersection R', we see that $\triangle PQR'$ has the same size and shape as $\triangle PQR$. Therefore we are lead to the following basic property.

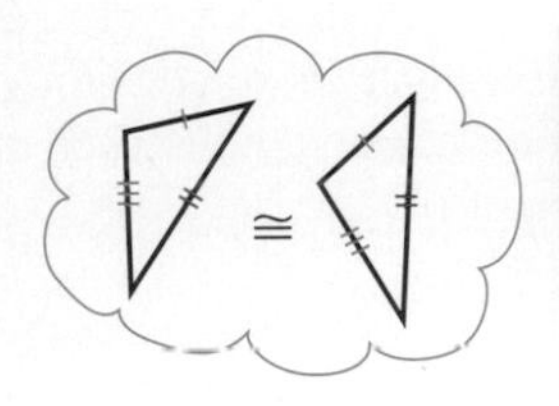

PROPERTY Side-Side-Side (SSS)

If the three sides of one triangle are respectively congruent to three sides of another triangle, then the two triangles are congruent.

In most formal treatments of Euclidean geometry, the SSS property is adopted as a postulate. That is, SSS is true by assumption, not by proof.

EXAMPLE 11.3

Using the SSS Property

Let $ABCD$ be a quadrilateral with opposite sides of equal length: $AB = CD$ and $AD = BC$. Show that $ABCD$ is a parallelogram.

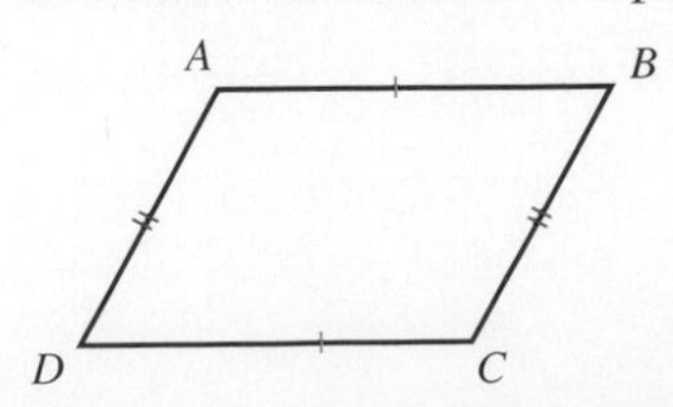

SOLUTION

Many problems in geometry are solved by constructing additional lines or arcs to reveal features of the original figure that would otherwise remain hidden. In this case we construct the diagonal $\overline{AC}$.

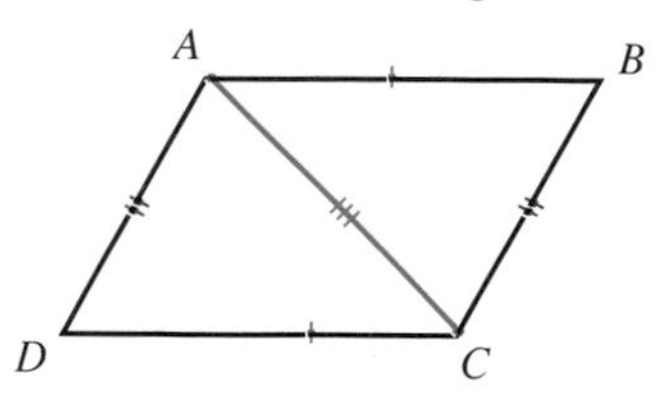

Since $\overline{AC}$ is congruent to itself we see that $\triangle ABC \cong \triangle CDA$ by the SSS property. Thus the corresponding angles $\angle BAC$ and $\angle DCA$ are congruent. By the alternate interior angles theorem of Chapter 10, we conclude that $\overline{AB} \parallel \overline{DC}$. Similarly, the congruence $\angle BCA \cong \angle DAC$ shows that $\overline{AD} \parallel \overline{BC}$. ■

An important consequence of the SSS property is the following construction of a congruent angle.

1 CONSTRUCTION Construct a Congruent Angle

Construct an angle that is congruent to a given angle, $\angle D$ as shown.

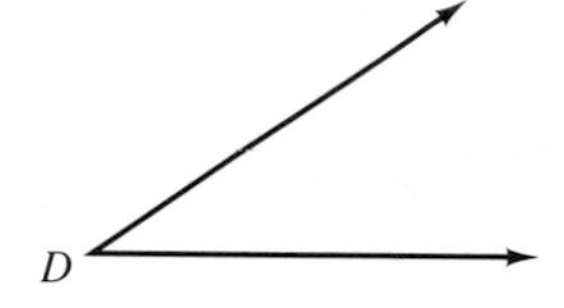

Procedure

Step 1	Step 2	Step 3	Step 4
Use the straightedge to draw a line segment $\overline{QY}$.	Draw arcs of the same radius centered at D and Q; let E and F be the points at which the arc intersects the sides of the given angle, and let R be the point at which the arc intersects $\overline{QY}$.	Place the point of the compass at E and adjust it to draw an arc through F. Draw an arc of the same radius centered at R to locate S.	Use the straightedge to draw a line segment from Q through S; then $\angle Q \cong \angle D$.

The construction procedure shows that $DE = QR$, $DF = QS$, and $EF = RS$. Therefore $\triangle DEF \cong \triangle QRS$ by the SSS property, and we see that the corresponding angles, $\angle D$ and $\angle Q$, are congruent.

CURRENT AFFAIRS

You Can't Listen for Congruence

In 1966 Mark Kac of Rockefeller University asked an apparently simple question, "Can you hear the shape of a drum?" A drum, for Kac, can have any shape. All that's required is that it be a two-dimensional figure having an interior (the drumhead) and a boundary (the rim). The shape of the boundary determines an infinite set of characteristic frequencies at which the interior "drum head" will vibrate. Two drums of the same shape generate the same set of frequencies, but Kac wanted to know if the converse is true: If you hear the same set of frequencies from two drums, are the drums necessarily the same shape?

In 1991 Carolyn Gordon and David Webb at Washington University and Scott Wolpert at the University of Maryland showed that you cannot hear the shape of a drum. They did so by finding two noncongruent shapes which, if made into drums with drumheads of the same material stretched at the same tension, would vibrate at exactly the same frequencies.

SOURCE: Adapted from Barry Cipra, "You Can't Hear the Shape of a Drum," *Science*, Vol. 255, (March 27, 1992): 1642.

The Triangle Inequality

The SSS property guarantees that two triangles are congruent if they have corresponding sides of the same length. However, not every triple of given lengths corresponds to a triangle, since the length of any side must be less than the sum of the lengths of the other two sides. An example is shown in Figure 11.3.

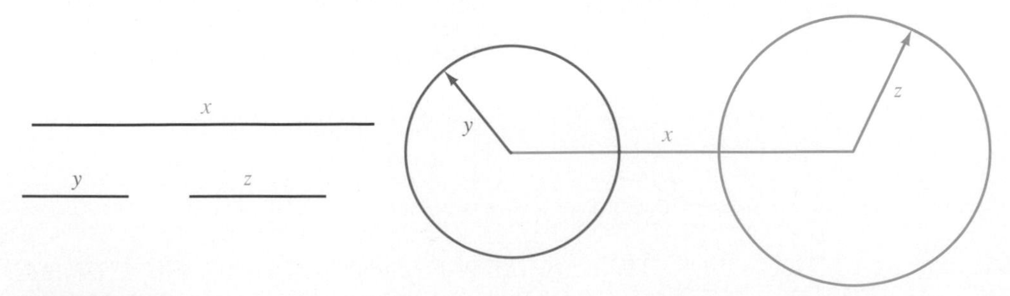

Figure 11.3
If $x \geq y + z$ there is no triangle with sides of length x, y, and z.

The lengths of the sides of a triangle must satisfy the following property.

PROPERTY Triangle Inequality

The sum of the lengths of any two sides of a triangle must be greater than the length of the third side.

A triangle gives rise to three inequalities, as shown in Figure 11.4.

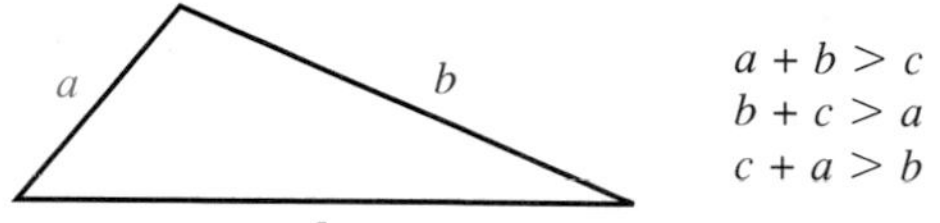

Figure 11.4
The lengths of the sides of any triangle satisfy the triangle inequalities.

EXAMPLE 11.4

Applying the Triangle Inequality

The four towns of Abbott, Brownsville, Connell, and Davis are building a new power generating plant that will serve all four communities. To keep the costs of the power lines at a minimum, the plant is to be located so that the sum of the distances from the plant to the four towns is as small as possible. An engineer recommended locating the plant at point E. A mathematician, seeing that the four towns formed a convex quadrilateral as shown, recommended that the plant be built at the point M at which the diagonals of the quadrilateral intersect. Why is location M better than E?

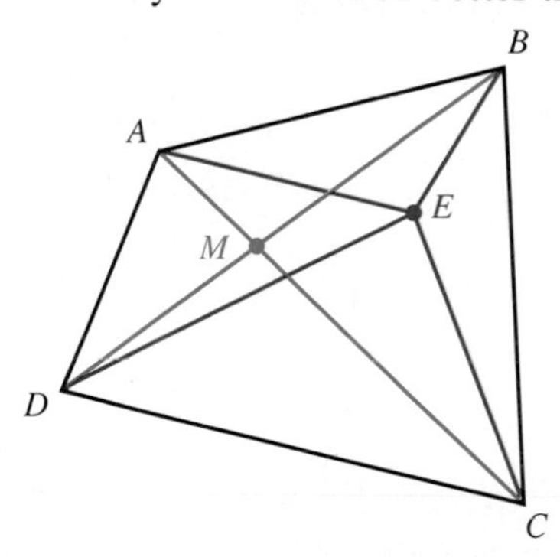

SOLUTION

The triangle inequality, applied to $\triangle ACE$, gives $EA + EC > AC$. Since M is on the diagonal $\overline{AC}$ we also know $AC = MA + MC$, and therefore

$$EA + EC > MA + MC.$$

If we apply the triangle inequality to $\triangle BDE$, the same reasoning gives us the inequality

$$EB + ED > MB + MC.$$

Adding these two inequalities gives us

$$EA + EB + EC + ED > MA + MB + MC + MD.$$

This shows that the sum of the distances to the towns from point E is greater than the sum of the distances to the towns from point M. ■

HIGHLIGHT FROM HISTORY
Christine Ladd-Franklin (1847–1930)

The "Metaphysical Club" of the Johns Hopkins University met for the first time in October 1879. The Club's founder, the great logician C. S. Pierce, read the paper "Non-Euclidean Space." What was unusual was that the paper's author, seated in the audience, was a newly admitted graduate student named Christine Ladd—and John Hopkins University did not admit women! Some earlier background explains why an exception was made.

Following her 1869 B.A. from Vassar, Christine had hoped to continue studies in physics but was denied access to laboratories largely because of her gender. She therefore turned to mathematics, which she studied on her own as she took a succession of positions teaching science. Shortly after the founding of Johns Hopkins University in 1876, Christine's application to admission to graduate studies was given a favorable report by Fabian Franklin, a young member of the mathematics department who was impressed by several articles and problem solutions she had published in some English periodicals.

Admitted "on a special status" in 1879, Christine Ladd studied mathematics, philosophy, and psychology, and wrote papers in logic, algebra, geometry and other subjects. By 1882 she had married Fabian Franklin and completed the requirements for a Ph.D. but, as a woman, could not be given the degree (this was finally rectified in 1924, when she received the degree at age 76).

Christine Ladd-Franklin continued to write papers in logic throughout her life, but after 1882 turned her attention again to experimental science, especially the psychology of vision. Her lifetime work includes about twenty papers in mathematics, another twenty in logic, and about fifty papers and a book on the theory of vision. Near the end of her life she was described as "the most distinguished woman scientist America had produced."

The Side-Angle-Side (SAS) Property

The next example explores how to construct a triangle with two given sides and the angle included between the given sides.

EXAMPLE 11.5 **Exploring the Side-Angle-Side Condition**

Two sides, $\overline{AB}$ and $\overline{AC}$, and the angle $\angle A$ included by these sides, are given for $\triangle ABC$ as shown. Show that a triangle PQR can be constructed for which $\triangle PQR \cong \triangle ABC$.

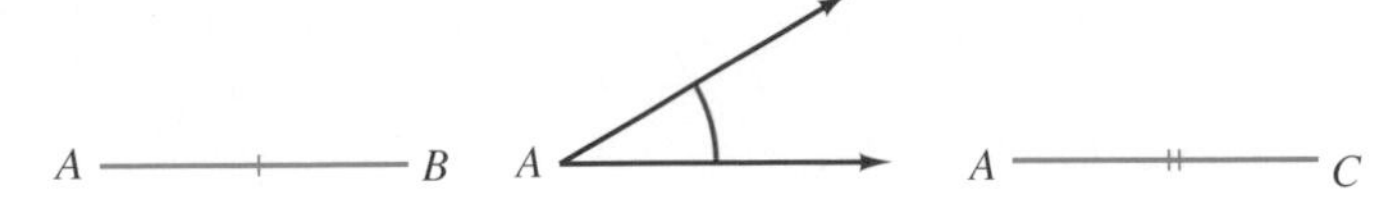

SOLUTION

Step 1 Construct an angle congruent to $\angle A$; let P denote its vertex.

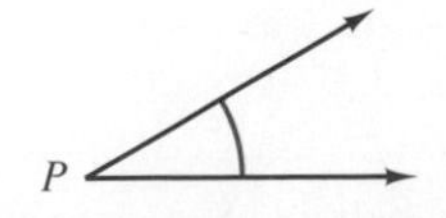

Step 2 Construct segments $\overline{PQ}$ and $\overline{PR}$ along the sides of $\angle P$ which are, respectively, of length AB and AC.

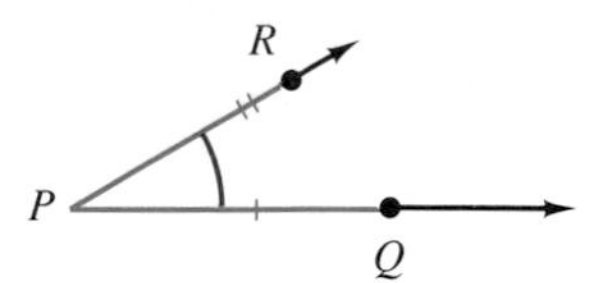

Step 3 Draw segment $\overline{QR}$; then $\triangle PQR \cong \triangle ABC$.

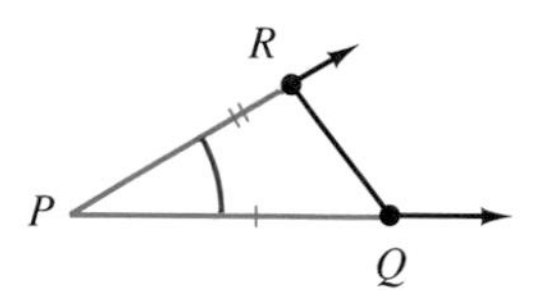

■

The procedure described in Example 11.5 uniquely determines the size and shape of $\triangle PQR$ when we are given the three parts, side-angle-side, of $\triangle ABC$. The angle had to be the **included angle**, the one containing the given sides. This property is often abbreviated as **SAS (side-angle-side).**

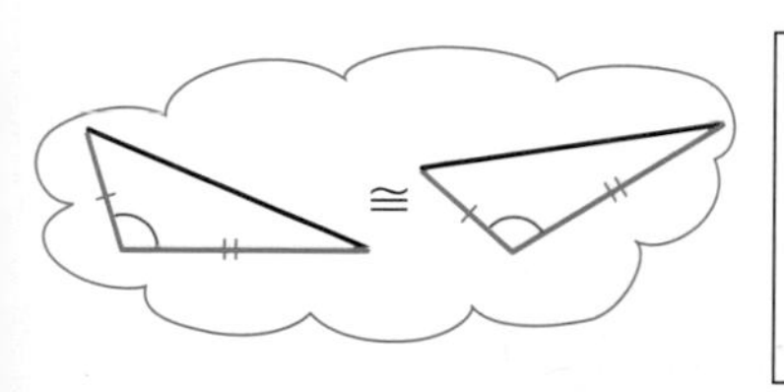

PROPERTY Side-Angle-Side (SAS)

If two sides and the included angle of one triangle are congruent to two sides and the included angle of another triangle, then the two triangles are congruent.

EXAMPLE 11.6

Using the SAS Property

Two line segments, $\overline{AB}$ and $\overline{CD}$, intersect at their common midpoint M. Show that $\overline{AD}$ and $\overline{BC}$ are parallel and have the same length.

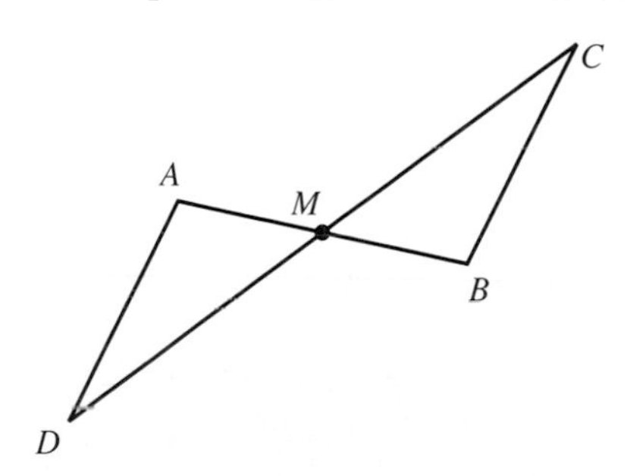

SOLUTION

It is useful to add tick marks and arcs to your drawing to summarize the given information. In this problem, M is the midpoint of $\overline{AB}$, so $AM = MB$. We indicate this on the drawing by putting a single tick mark on each of the segments $\overline{AM}$ and $\overline{MB}$. Similarly $CM = MD$ and we put double tick marks on each of the segments $\overline{CM}$ and $\overline{MD}$. We also use single arcs to indicate the congruence of the vertical angles at M formed by the intersecting segments.

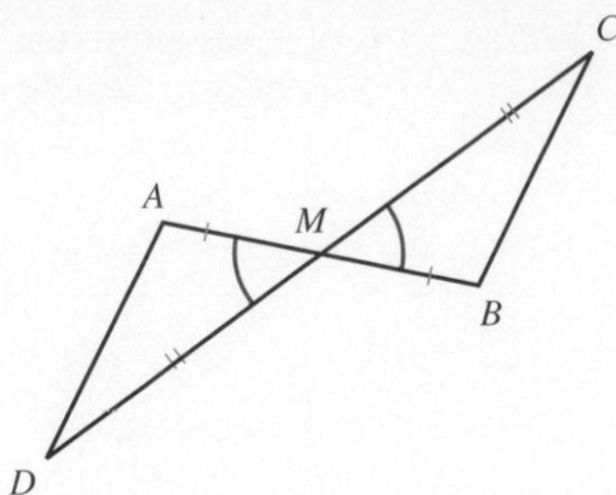

It is now apparent that the SAS property gives us the congruence $\triangle AMD \cong \triangle BMC$. It follows that the corresponding sides $\overline{AD}$ and $\overline{BC}$ are congruent, so that $AD = BC$. We also have that $\angle A \cong \angle B$, so the alternate interior angles theorem of Chapter 10 guarantees that $\overline{AD}$ and $\overline{CB}$ are parallel. ■

The following theorem about isosceles triangles is an important consequence of the SAS property.

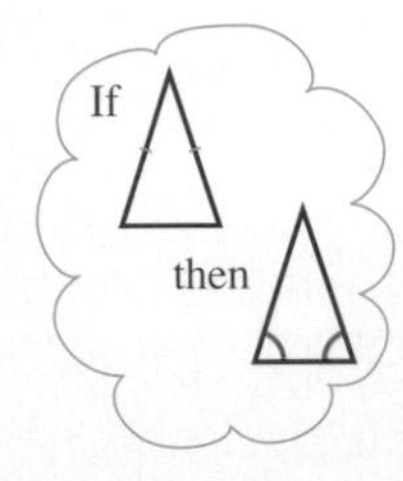

THEOREM Isosceles Triangle Theorem

The angles opposite the congruent sides of an isosceles triangle are congruent.

PROOF Let $\triangle ABC$ be isosceles, with $\overline{AB}$ and $\overline{AC}$ congruent. Consider the vertex correspondence $A \leftrightarrow A$, $B \leftrightarrow C$, $C \leftrightarrow B$, (which amounts to looking at the same triangle from the back!).

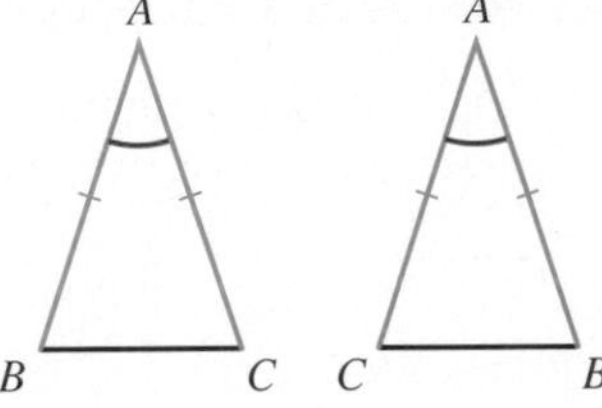

Since $\overline{AB} \cong \overline{AC}$ and $\angle A \cong \angle A$ it follows from the SAS property that $\triangle ABC \cong \triangle ACB$. But then all six corresponding parts of $\triangle ABC$ and $\triangle ACB$ are congruent, including $\angle B \cong \angle C$. ■

The isosceles triangle theorem has many uses. In particular it gives a simple way to prove **Thales' theorem.**

THEOREM Thales' Theorem

Any triangle ABC inscribed in a semi-circle with diameter $\overline{AB}$ has a right angle at point C.

PROOF Draw the radius $\overline{OC}$. This divides $\triangle ABC$ into two isosceles triangles, $\triangle AOC$ and $\triangle COB$. The isosceles triangle theorem tells us that the measures of the base angles of $\triangle AOC$ are equal, say x. Likewise, the base angles of $\triangle COB$ are equal, say y. Since the sum of the measures of the interior angles of $\triangle ABC$ is 180°, we have that $x + y + (x + y) = 180°$. But this equation tells us that $m(\angle C) = x + y = 90°$.

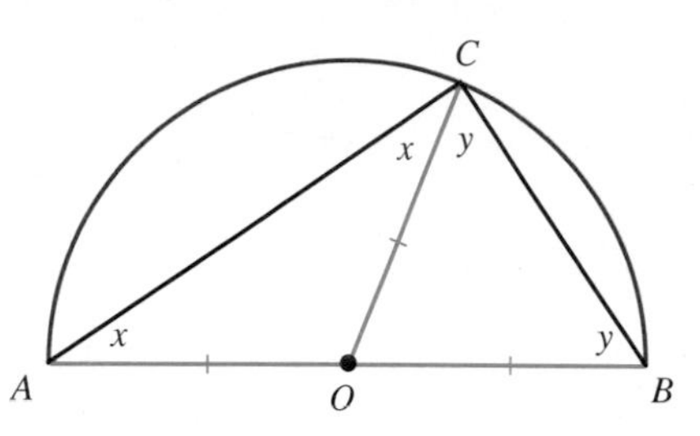

■

The Angle-Side-Angle (ASA) Property

In the next example, we suppose that two angles of a triangle, and the side included between these angles, are given. Is this information sufficient to be able to construct a congruent triangle?

EXAMPLE 11.7

Exploring the Angle-Side-Angle Property

Two angles and their **included side** are given for $\triangle ABC$, as shown. Construct a triangle PQR that is congruent to triangle ABC.

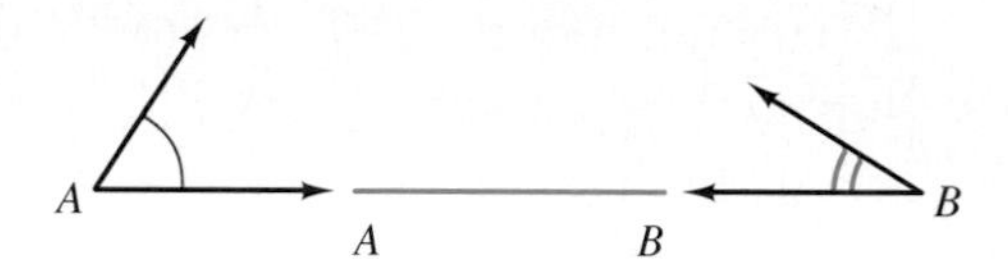

SOLUTION

Step 1 Construct $\angle P \cong \angle A$.

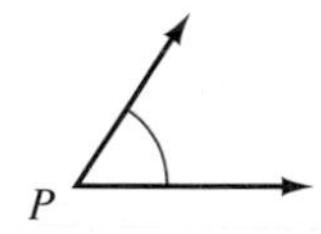

Step 2 Construct segment $\overline{PQ}$ on a side of $\angle P$ so that $PQ = AB$.

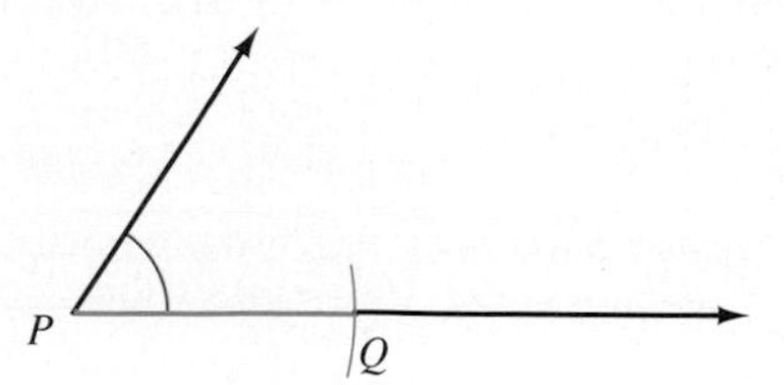

Step 3 Construct an angle congruent to $\angle B$ at vertex Q, with one side containing P and the other side intersecting $\angle P$ to determine point R. Then $\triangle PQR \cong \triangle ABC$.

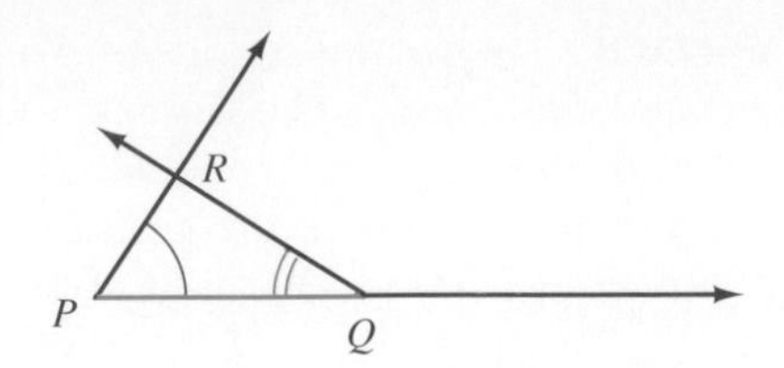

The construction just shown illustrates the **angle-side-angle property,** abbreviated as **ASA.**

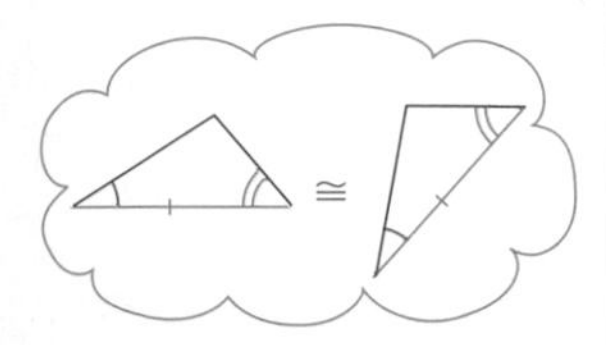

PROPERTY Angle-Side-Angle (ASA)

If two angles and the included side of one triangle are congruent to the two angles and the included side of another triangle, then the two triangles are congruent.

The ASA property allows us to prove that any triangle with two angles of the same measure is isosceles.

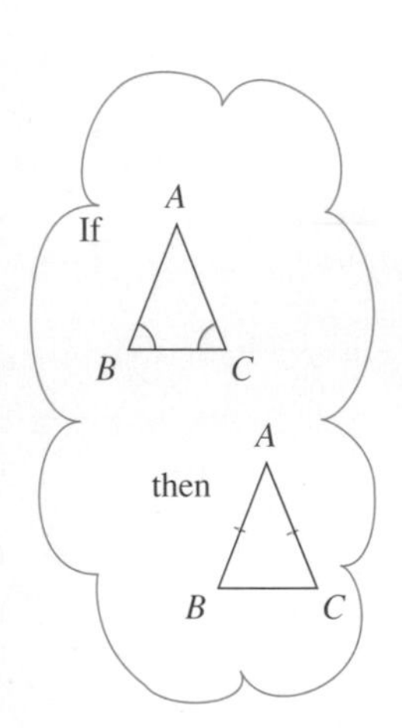

THEOREM Converse of the Isosceles Triangle Theorem

If two angles of a triangle are congruent, then the sides opposite them are congruent; that is, the triangle is isosceles.

PROOF Let $\triangle ABC$ have two congruent angles, say $\angle B \cong \angle C$. We know that $\overline{BC} \cong \overline{CB}$, since a line segment is congruent to itself. By the ASA property it follows that $\triangle ABC \cong \triangle ACB$. This means that the corresponding sides of $\triangle ABC$ and $\triangle ACB$ are congruent, so $\overline{AB} \cong \overline{AC}$. ■

The Angle-Angle-Side (AAS) Property

The side in the ASA theorem is the one included by the two angles. However if *any* two angles of one triangle are congruent to two angles of a second triangle, then all three pairs of corresponding angles are congruent. This follows from the fact that the measures of the three angles of a triangle add up to 180°, so the third angle is uniquely determined by the other two angles. This gives us the **angle-angle-side property,** abbreviated as **AAS.**

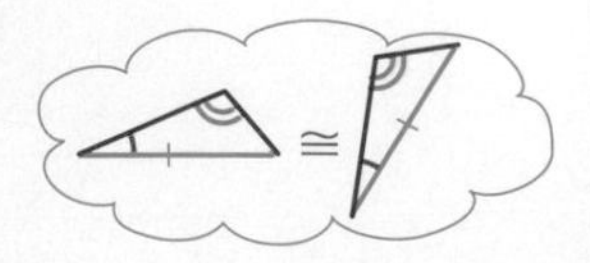

PROPERTY Angle-Angle-Side (AAS)

If two angles and a nonincluded side of one triangle are congruent respectively to two angles and the corresponding nonincluded side of a second triangle, then the two triangles are congruent.

JUST FOR FUN

Twice Around a Triangle

Draw any triangle ABC and let P be any point on side $\overline{AB}$. Draw an arc centered at B to determine the point Q on side $\overline{BC}$ for which $BP = BQ$. In the same way, draw an arc at C to locate point R on $\overline{CA}$, and then draw an arc centered at A to determine point P' on $\overline{AB}$. In three more steps, go around the triangle a second time, constructing Q', R', and finally P'' on $\overline{AB}$. What do you find interesting about P''?

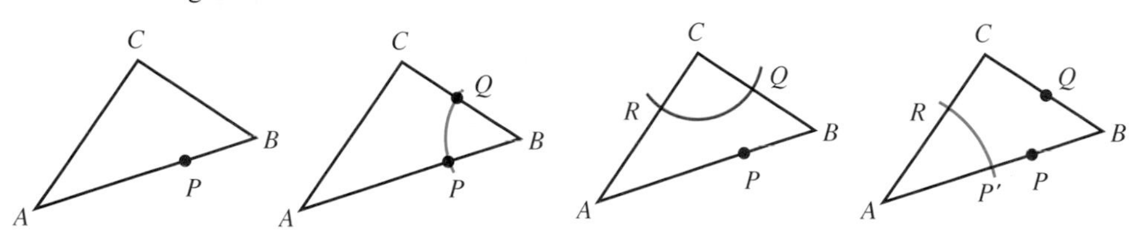

There is no "SSA" congruence property since it is possible for two noncongruent triangles to have two pairs of congruent sides and a congruent nonincluded angle. An example is shown in Figure 11.5.

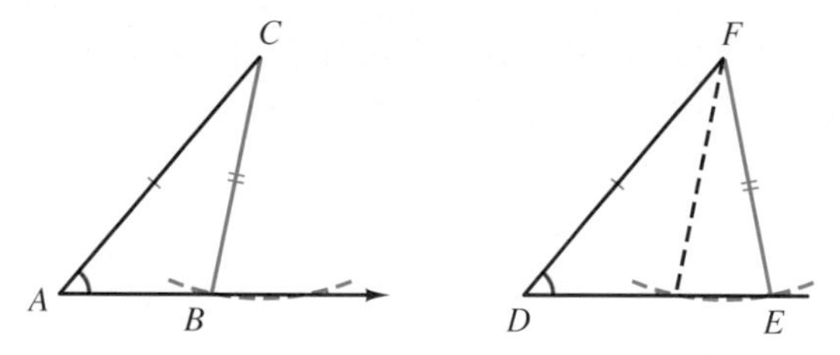

Figure 11.5
Triangles ABC and DEF are not congruent even though $AC = DF$, $BC = EF$, and $\angle A \cong \angle D$.

Congruent Right Triangles

The SSS, SAS, ASA, and AAS properties apply to all triangles, with no restriction placed on their type. The following theorem applies only to right triangles. It tells us that the shape and size of a right triangle is completely determined by the lengths of the hypotenuse and one of the legs.

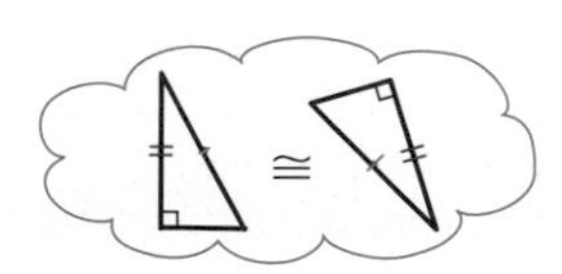

THEOREM The Hypotenuse-Leg (HL) Theorem

If the hypotenuse and a leg of one right triangle are congruent to the hypotenuse and leg of a second right triangle, then the two right triangles are congruent.

PROOF Let $\triangle ABC$ and $\triangle DEF$ be right triangles with right angles at C and F, and suppose $AB = DE$ and $AC = DF$. We are to prove that $\triangle ABC \cong \triangle DEF$.

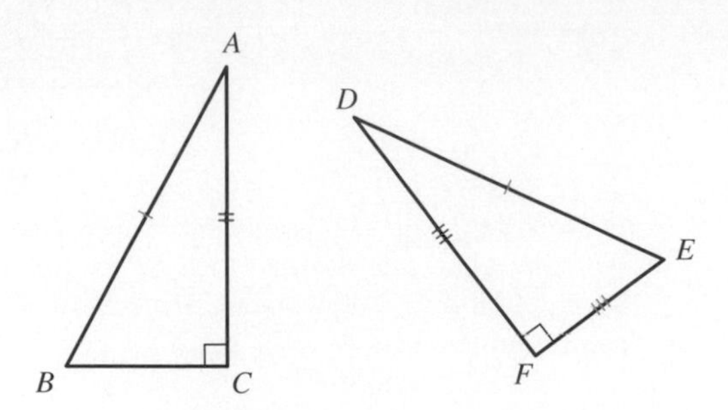

Consider the point G on line $\overleftrightarrow{BC}$ which satisfies $CG = EF$, as shown here.

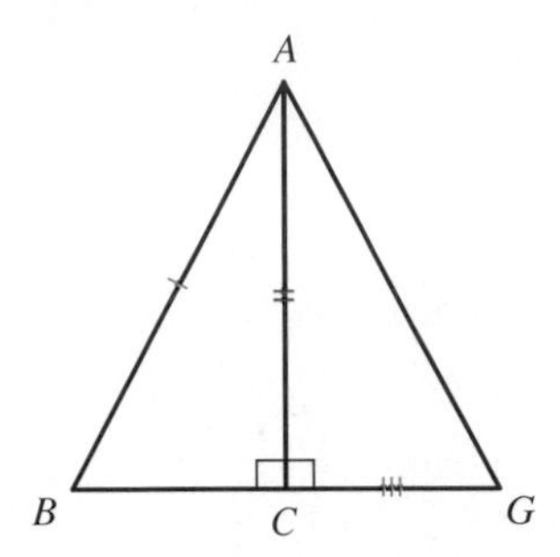

Since $\angle ACG$ and $\angle DFE$ are both right angles, and $\overline{AC} \cong \overline{DF}$, it follows that $\triangle ACG \cong \triangle DFE$ by the SAS property. Thus $AG = DE$. Since we are given that $DE = AB$, we also have $AG = AB$. This means that $\triangle ABG$ is isosceles, and so by the isosceles triangle theorem, we know that $\angle B \cong \angle G$. By the AAS property this means $\triangle ABC \cong \triangle AGC$. But $\triangle AGC \cong \triangle DEF$, and therefore $\triangle ABG \cong \triangle DEF$. ■

EXAMPLE 11.8 **Applying the HL Theorem**

The diagonals of a convex quadrilateral have the same length and one side is included between two right angles. Show that the quadrilateral is a rectangle.

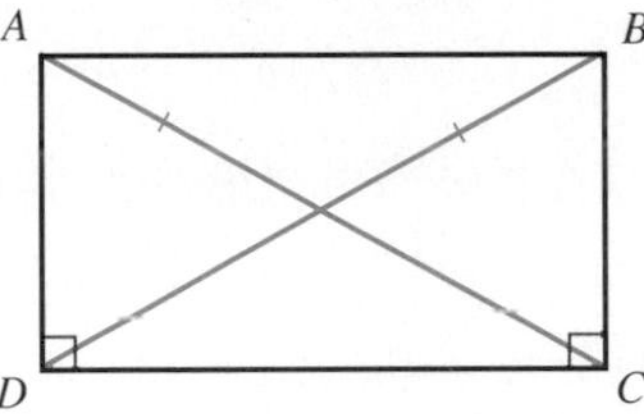

SOLUTION

The quadrilateral $ABCD$ has the properties shown in the diagram above. The right angles are at vertices C and D and $AC = BD$. It follows that $\triangle ADC$ is a right triangle with hypotenuse $\overline{AC}$ and leg $\overline{DC}$. Likewise $\triangle BCD$ is a right triangle with hypotenuse $\overline{BD}$ and leg $\overline{CD}$. Since $AC = BC$ and $DC = CD$, the HL theorem gives $\triangle ADC \cong \triangle BCD$. Therefore, $\overline{AD} \cong \overline{BC}$. Moreover $\overline{AD} \parallel \overline{BC}$ since both segments are perpendicular to $\overline{DC}$. By the alternate interior angles theorem (Chapter 10), $\angle DAC \cong \angle BCA$. Therefore, $\triangle DAC \cong \triangle BCA$ by the SAS property. It follows that $\angle CBA \cong \angle ADC$, so the angle at B in quadrilateral $ABCD$ is also a right angle. Since the four interior angles of $ABCD$ have a total measure of 360°, the angle at A is also a right angle. Thus $ABCD$ is a rectangle. ■

PROBLEM SET 11.1

Understanding Concepts

1. The two triangles shown below are congruent.

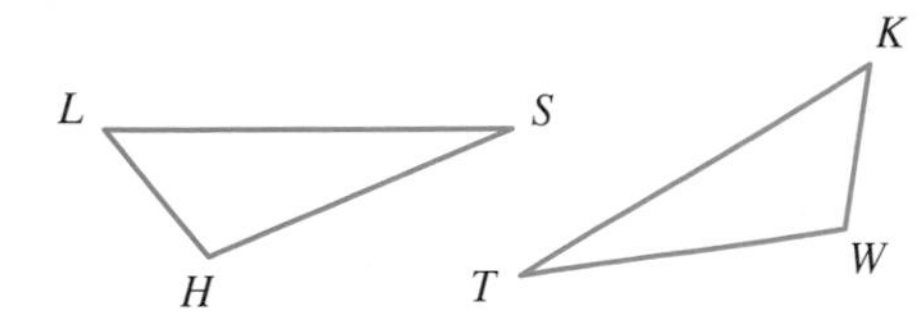

Determine the following:

(a) Corresponding vertices
$L \leftrightarrow$ _____, $H \leftrightarrow$ _____, $S \leftrightarrow$ _____

(b) Corresponding sides
$\overline{LH} \leftrightarrow$ _____, $\overline{HS} \leftrightarrow$ _____, $\overline{SL} \leftrightarrow$ _____

(c) Corresponding angles
$\angle L \leftrightarrow$ _____, $\angle H \leftrightarrow$ _____, $\angle S \leftrightarrow$ _____

(d) $\triangle LHS \cong \triangle$ _____.

2. Suppose $\triangle JKL \cong \triangle ABC$, where $\triangle ABC$ is shown below.

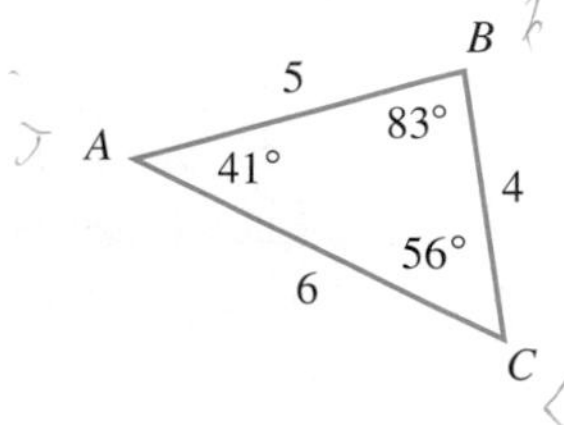

Find the following:
(a) KL **(b)** LJ **(c)** $m(\angle L)$ **(d)** $m(\angle J)$

3. Segments of length x and y are shown below.

A ——— x ——— B C ——— y ——— D

Describe procedures, using only a straightedge and a compass, to construct:

(a) a line segment $\overline{EF}$ of length $x + y$;

(b) a line segment of length $x - y$.

4. Two angles, $\angle A$ and $\angle B$ are shown below.

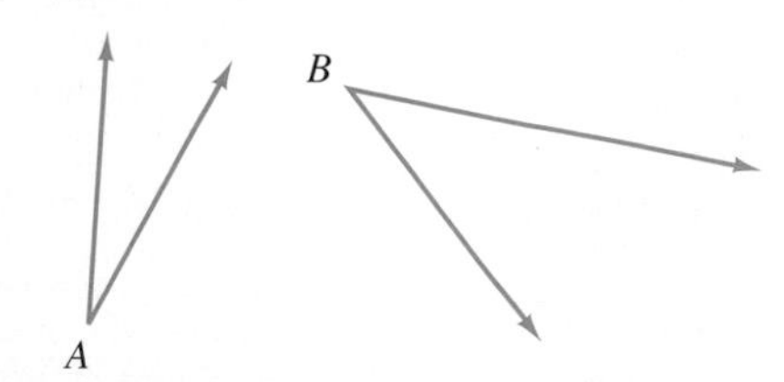

Describe procedures, using only a straightedge and a compass, to construct

(a) $\angle C$, so that $m(\angle C) = m(\angle A) + m(\angle B)$.

(b) $\angle D$, so that $m(\angle D) = m(\angle B) - m(\angle A)$.

(c) $\angle E$, so that $m(\angle E) = m(\angle A) + m(\angle B) = 180°$.

5. Use a ruler, protractor, and compass to construct, when possible, a triangle with the stated properties. If such a triangle cannot be drawn, explain why. Decide if there can be two or more noncongruent triangles with the stated properties.

(a) an isosceles triangle with two sides of length 5 cm and an apex angle of measure 28°

(b) an equilateral triangle with sides of length 6 cm

(c) a triangle with sides of length 8 cm, 2 cm, and 5 cm

(d) a triangle with angles measuring 30° and 110° and a nonincluded side of length 5 cm

(e) a right triangle with legs (the sides including the right angle) of length 6 cm and 4 cm

(f) a triangle with sides of length 10 cm and 6 cm and a nonincluded angle of 45°

(g) a triangle with sides of length 5 cm and 3 cm and an angle of 20°

6. Suppose your compass can open to a maximum radius of AB. Describe how you can construct a line segment congruent to $\overline{CD}$ if the length of CD is greater than AB.

7. Each part below shows two triangles, with arcs and tick marks identifying congruent parts. If it is possible to conclude that the triangles are congruent, describe what property or theorem you use. If you cannot be sure the triangles are congruent, state "no conclusion possible." The first one is done for you.

(a)

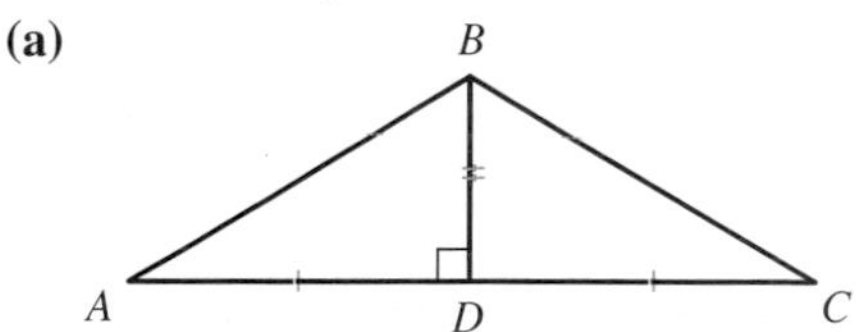

Answer $\triangle ABD \cong \triangle CBD$ by SAS.

(b)

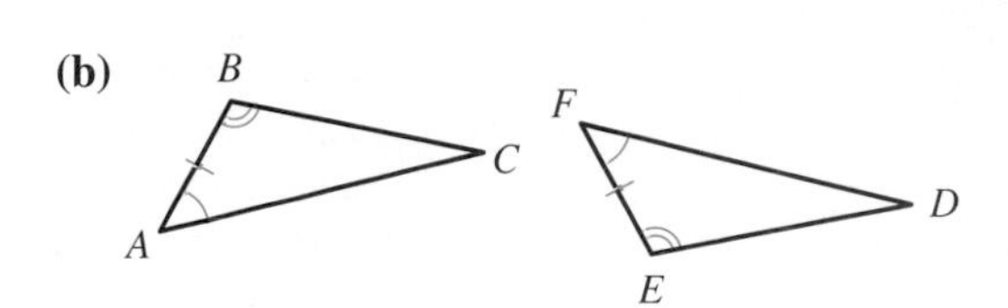

(c)

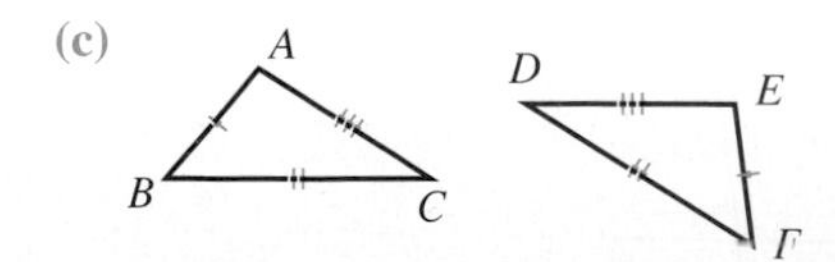

(d)

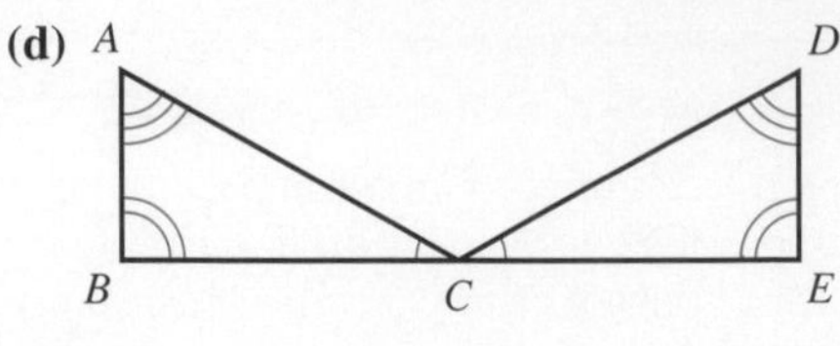

(e)

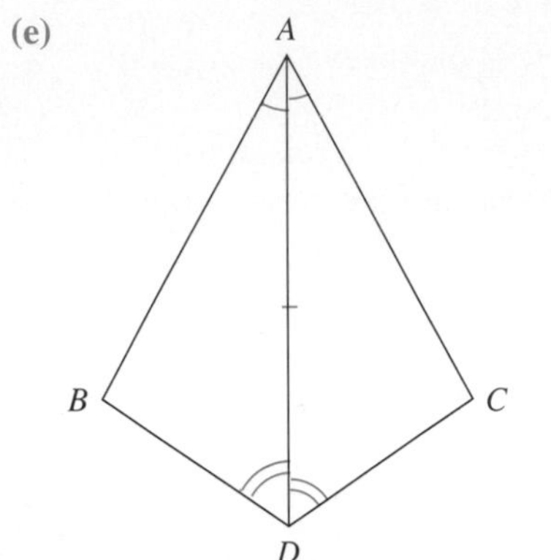

(f)

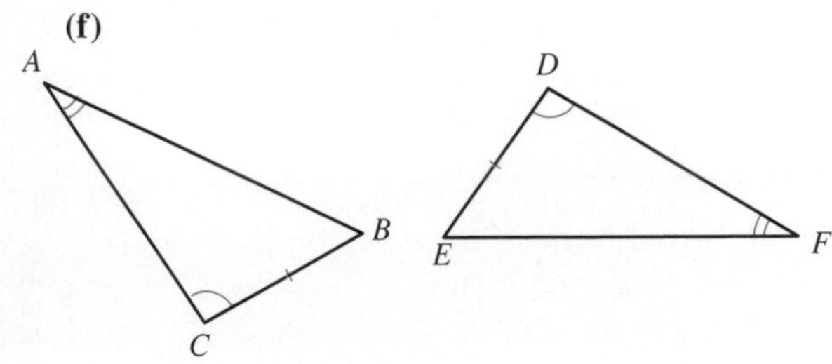

(g)

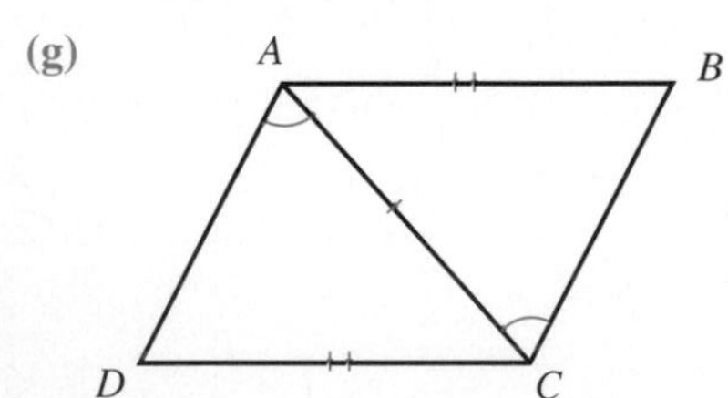

(h)

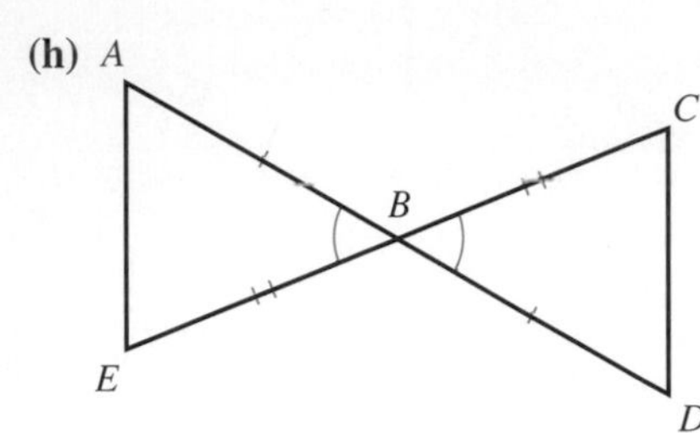

(i)

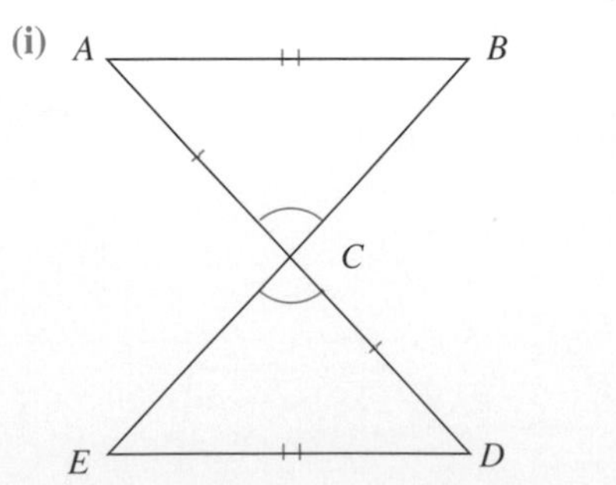

(j)

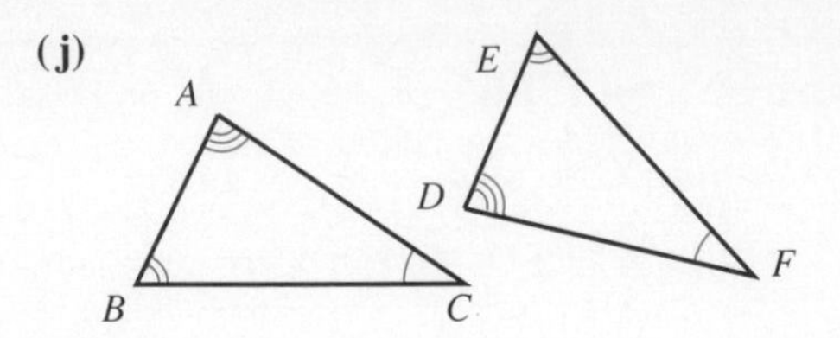

(k)

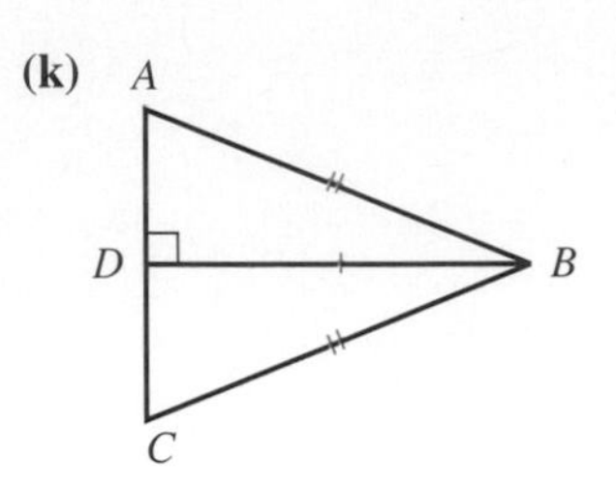

8. Prove that an equilateral triangle is equiangular.
9. Prove that an equiangular triangle is equilateral.
10. If a right triangle has an angle measuring 30° show that the side opposite this angle is half the length of the hypotenuse. (*Hint:* Construct a related equilateral triangle.)
11. Let $ABCD$ be a parallelogram.

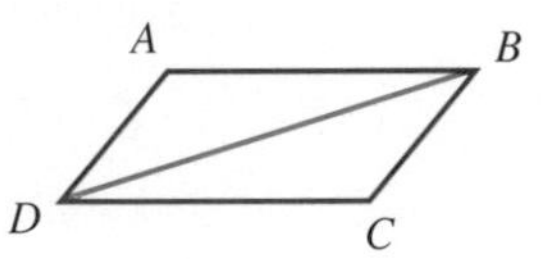

   (a) Prove that $\triangle ABD \cong \triangle CDB$. (*Hint:* Use the ASA property.)
   (b) Prove that opposite sides of a parallelogram have the same length.
   (c) Prove that opposite angles of a parallelogram have the same measure.

12. Suppose you have shown that opposite sides of a parallelogram are congruent (see part (b) of problem 11). Let M be the point of intersection of the diagonals of the parallelogram. Prove that M is the midpoint of each diagonal. (*Hint:* First prove that $\triangle ABM \cong \triangle CDM$.)

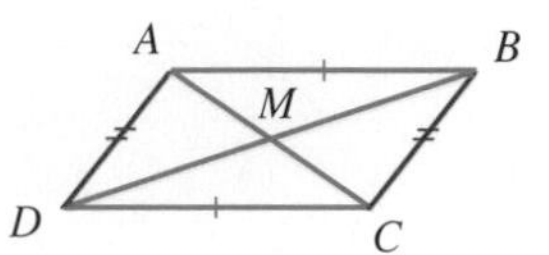

13. Suppose you know that the diagonals of any parallelogram intersect at the common midpoint of each diagonal (see problem 12).
   (a) If the diagonals of a parallelogram are perpendicular, prove that the parallelogram is a rhombus.
   (b) Prove the converse of part (a); that is, that the diagonals of a rhombus are perpendicular.

14. A triangle has sides of length 4 cm and 9 cm. What can you say about the length of the third side?

15. (a) A quadrilateral has sides of length 2 cm, 7 cm, and 5 cm. What inequality does the length of the fourth side satisfy?

(b) Let A, B, C, and D be any four points in the plane. Explain why $AD \leq AB + BC + CD$.

16. Let $\angle K$ and $\angle T$ be right angles, and suppose $EK = ET$. Prove that $KITE$ is a kite; that is, show that $KI = TI$.

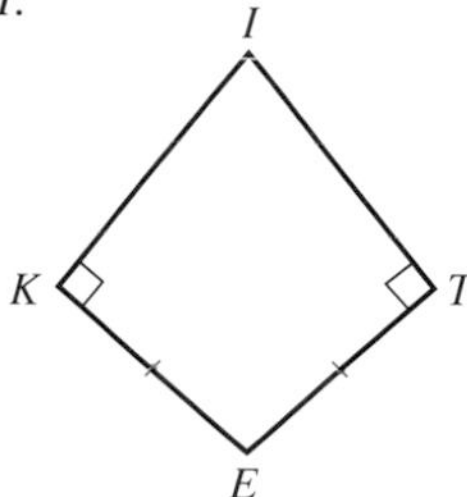

17. Suppose you drew a circle by tracing around a bowl. Explain how you can locate the center of the circle with a piece of notebook paper. (*Hint:* Use the easily proved converse to Thales' theorem: the hypotenuse of a right triangle inscribed in a circle is a diameter of the circle.)

Thinking Critically

18. Two angles and a *nonincluded* side of $\triangle ABC$ are drawn below.

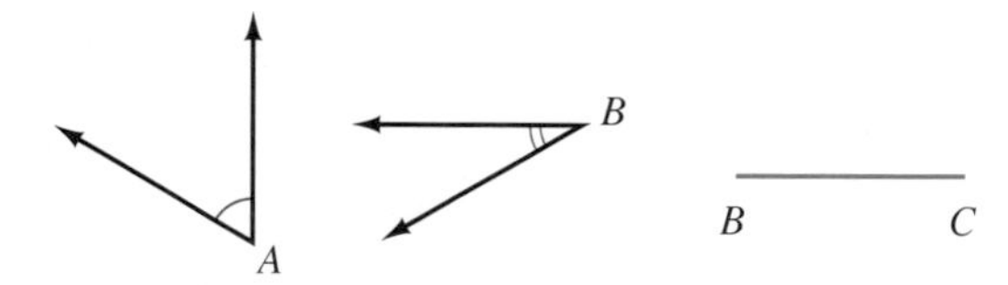

Describe and show the steps of a straightedge and compass construction of a triangle PQR that is congruent to $\triangle ABC$.

19. In the figure below, $AB = AE$ and $AC = AD$.

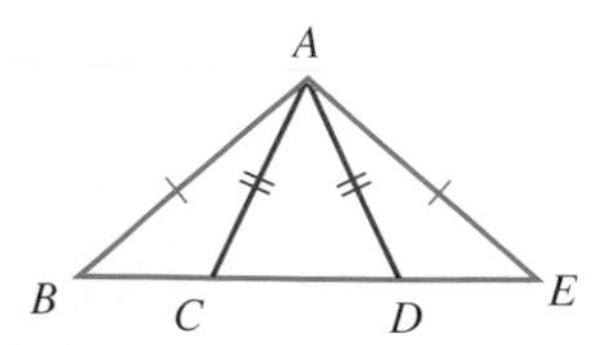

(a) Why is $\angle B \cong \angle E$?

(b) Why is $\angle ACD \cong \angle ADC$?

(c) Use the AAS property to prove that $\triangle ABC \cong \triangle AED$.

(d) Prove that $BC = DE$.

20. Recall that a trapezoid with a pair of congruent angles adjacent to one of its bases is called an isosceles trapezoid.

(a) Prove that the sides joining the bases of an isosceles trapezoid are congruent. The following figure may be helpful.

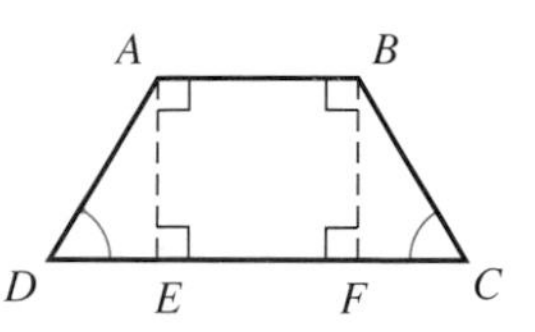

(b) Prove that the diagonals of an isosceles trapezoid are congruent.

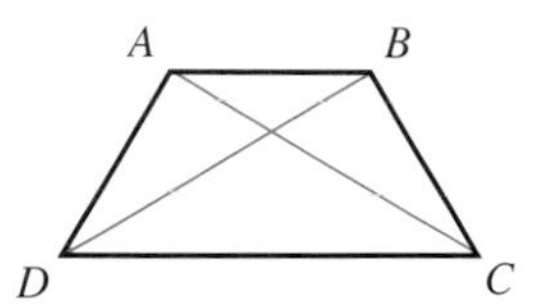

21. For each part, decide if the given conditions are sufficient to conclude that $\triangle ABC \cong \triangle ADE$. If so, give a proof; if not, draw a figure which satisfies the information but shows that $\triangle ABC$ is not congruent to $\triangle ADE$.

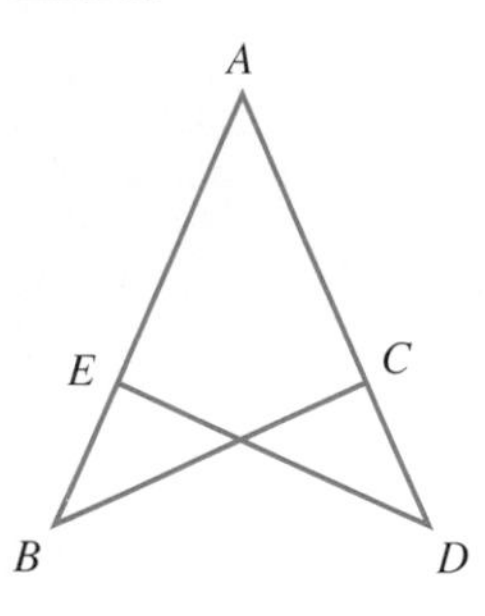

(a) $AB = AD$ and $m(\angle B) = m(\angle D)$

(b) $AB = AD$ and $BC = DE$

(c) $AB = AD$ and $AE = AC$

(d) $EB = CD$ and $BC = DE$

22. Points D, E, and F are located on the sides of equilateral triangles ABC so that $AD = BE = CF$. Prove that $\triangle DEF$ is equilateral.

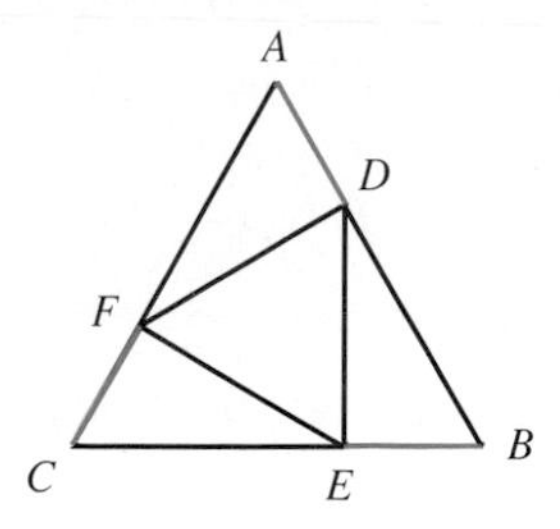

23. If E, F, G, and H are points on the sides, $\overline{AB}$, $\overline{BC}$, $\overline{CD}$, and $\overline{DA}$ of square $ABCD$ such that $AE = BF = CG = DH$, prove that $EFGH$ is a square.

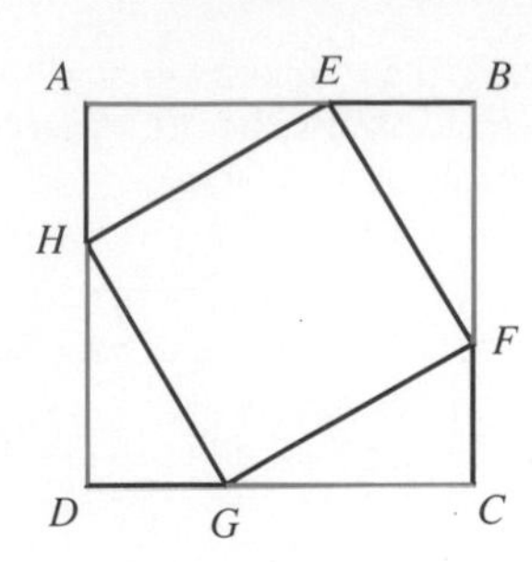

24. Each edge of a tetrahedron is congruent to its opposite edge. For example, $\overline{AB} \cong \overline{CD}$. Prove that the faces of the tetrahedron are congruent to one another.

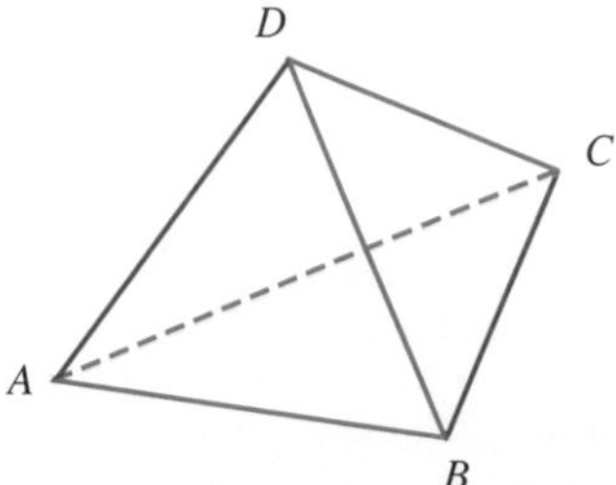

25. All six faces of a hexahedron (that is, a polyhedron with six faces) are congruent parallelograms. Prove that the faces must be congruent rhombuses.

Thinking Cooperatively

26. Let S be a point in the interior of triangle PQR. Prove that $SQ + SR < PQ + PR$. (*Hint:* Let T be the point on side $\overline{PR}$ which is intersected by the extension of $\overline{QS}$. Now use the triangle inequality twice.)

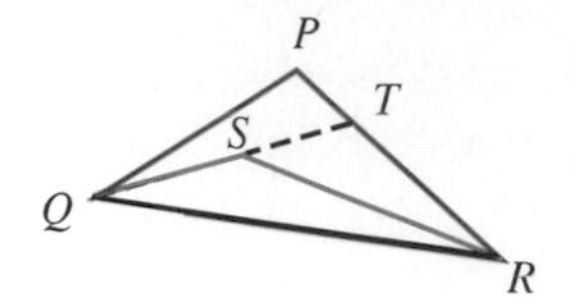

27. In Example 11.4 the four towns involved formed a convex quadrilateral. Suppose instead that Davis is interior to the triangle formed by the other three towns.

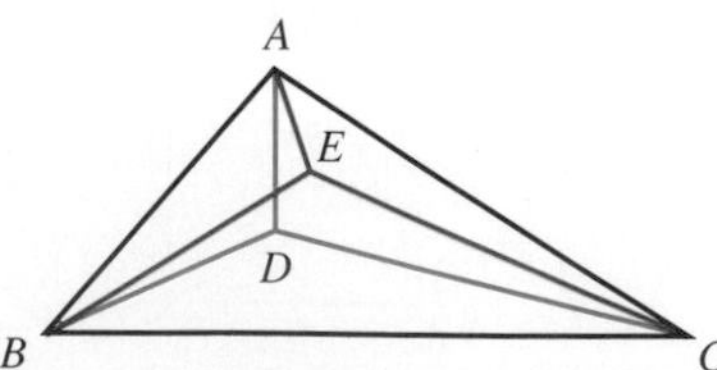

Prove that the power station should be located at Davis, since its sum of distances to the other towns is as small as possible. (*Hint:* Use the result of problem 26 to prove that $DA + DB + DC < EA + EB + EC + ED$, where E is any point other than D.)

28. Six towns are located at the vertices A, B, C, D, E, F of a regular hexagon. A power station located at Q would require $QA + QB + QC + QD + QE + QF$ miles of transmission line to serve the six communities. Describe a better location P, for the station, and prove it has the least possible sum of distances to A, B, C, D, E, and F.

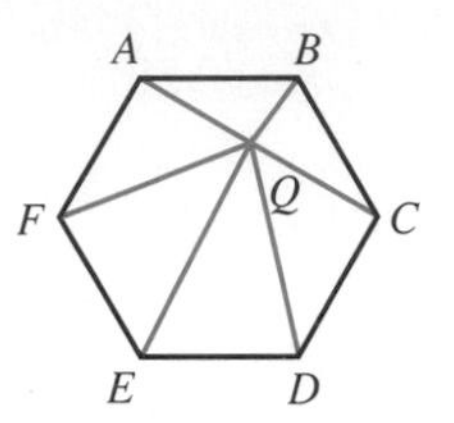

Making Connections

29. A bicycle rack is made from three pieces of metal box tubing. There are bolts at A, B, C, and D that join the tubing and attach the rack to the bumper of a car.
    (a) Why is the top of the rack likely to shift sideways?
    (b) If a fourth piece of tubing is available, where can it be attached to provide reinforcement? Explain why this works.
    (c) Would a rope from A to C reinforce the frame? How about two pieces of rope, from A to C and from B to D?

30. Carpenters construct a wall by nailing studs to a top and bottom plate, as shown below.

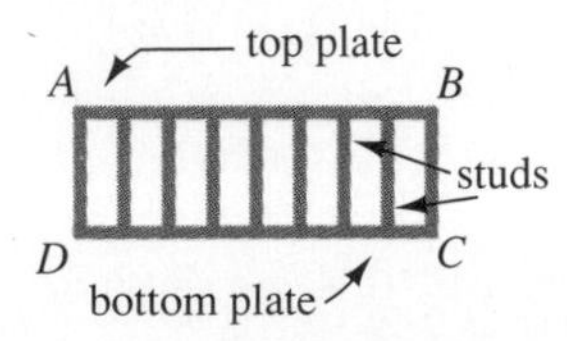

The studs are cut to the same length, and the top and bottom plates are the same length.
    (a) If the pieces are properly cut and nailed, is $ABCD$ necessarily rectangular, or are there other shapes the framework can take?
    (b) Carpenters frequently "square up" a stud wall by adjusting it so that it has diagonals of equal

length. Prove that a parallelogram with congruent diagonals is a rectangle.

(c) Once the wall is "squared up" a diagonal brace is nailed across the frame. Why is this? What geometric principle is involved?

31. A TV antenna is being erected on top of a flat roof. Three guy wires of equal length are attached to a point A near the top of the mast. If the mast is to be vertical, what is important about the location of the points C, D, and E to which the guy wires are anchored to the roof?

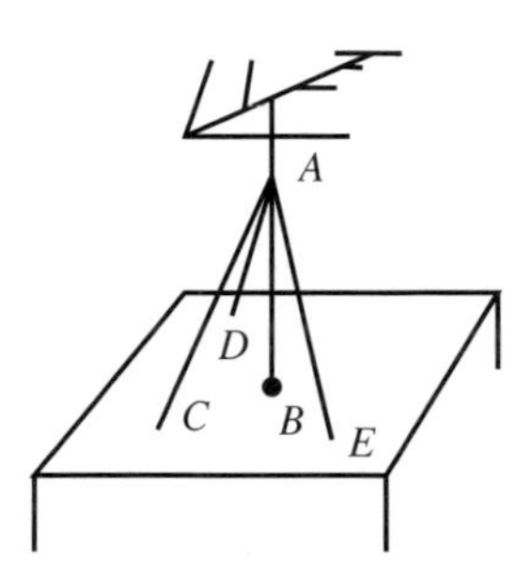

Using a Computer
Using an Automatic Drawer

32. Draw a circle, and label its center O. Next draw three points A, B, and P on the circle and draw two angles, $\angle AOB$ and $\angle APB$, that both intercept the arc of the circle between A and B.

(a) Measure $\angle AOB$ and $\angle APB$. What relationship do you observe? If your software permits, move P around the circle and investigate what happens to the measure of $\angle APB$.

(b) Make a conjecture that relates the measures of $\angle AOB$ and $\angle APB$.

(c) Justify your conjecture. (*Hint:* Draw the diameter through P and then mimic the proof of Thales' theorem.)

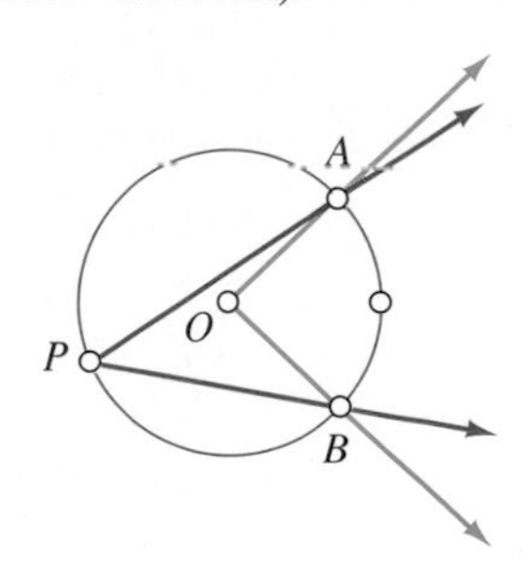

33. Draw a circle. Next draw four points A, B, C, and D on the circle, which are joined by line segments to form the inscribed quadrilateral $ABCD$.

(a) Measure $\angle A$ and $\angle C$. What relationship do you observe? If your software permits, move some of the points of your quadrilateral and see if the relationship between $\angle A$ and $\angle C$ is preserved.

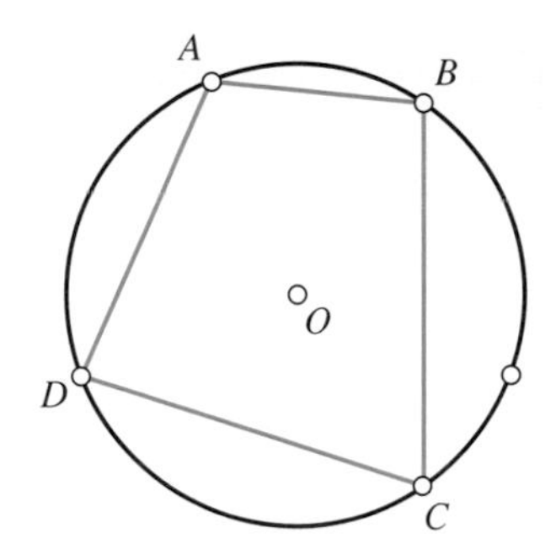

(b) Make a conjecture concerning the relationship of opposite angles of an inscribed quadrilateral.

(c) Justify your conjecture. (*Hint:* Draw the radii $\overline{OA}$, $\overline{OB}$, $\overline{OC}$, and $\overline{OD}$. This creates four isosceles triangles.)

For Review

34. Prove that the measure of an exterior angle of a triangle is greater than the measure of either of the opposite interior angles: $m(\angle 3) > m(\angle 1)$ and $m(\angle 3) > m(\angle 2)$.

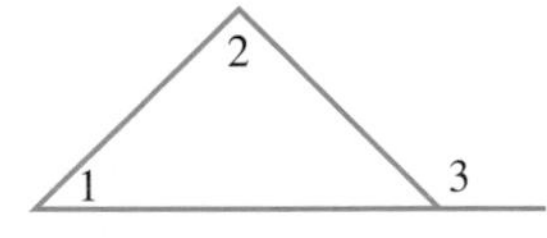

35. Use a protractor and a ruler to construct a regular heptagon. Label the measures of an interior, an exterior, and a central angle.

36. **(a)** Make a freehand (pencil only—no ruler or other drawing tools) drawing of a regular octahedron.

(b) Beside your drawing of the octahedron draw a truncated octahedron. Do this by removing the corners of the octahedron to form squares, which reduce the triangular faces of the octahedron to regular hexagonal faces.

11.2 Constructing Geometric Figures

In the last section a basic construction was described:

- Construction 1 Construct a congruent angle

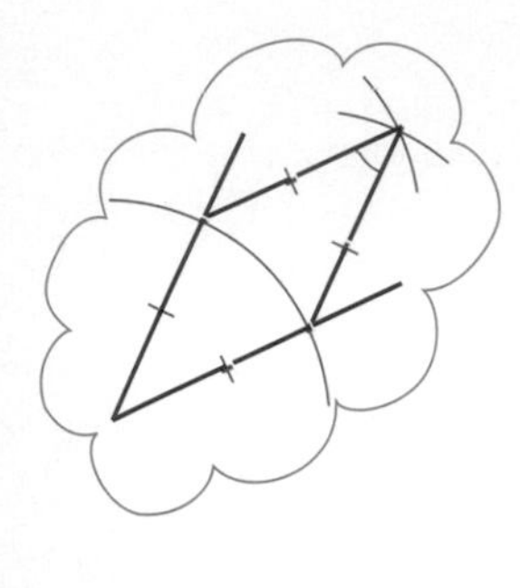

Since only the straightedge and compass were used, this is an example of an Euclidean construction. In this section we describe a number of other Euclidean constructions and explore some related applications and theorems. We also investigate constructions with the Mira and by paper folding. The constructions shown in the examples and called for in the problems can be done with an automatic drawer if available.

To be certain that a construction results in a figure that has a desired property, a proof of the validity of the construction must be given. Construction 1 of a congruent angle is a consequence of the SSS property. Many constructions can be verified by appealing to the properties of a rhombus listed in Figure 11.6. A rhombus is easily constructed by drawing intersecting arcs of circles of the same radius.

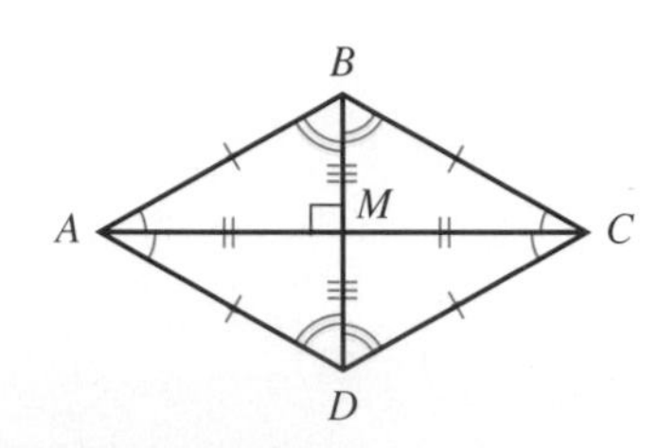

Figure 11.6
The rhombus *ABCD* has many useful properties:
- **The diagonals are angle bisectors.**
- **The diagonals are perpendicular.**
- **The diagonals intersect at their common midpoint *M*.**
- **The sides are all congruent to one another.**
- **The opposite sides are parallel.**

INTO THE CLASSROOM
Constructions in Space

Constructions using a compass and a straightedge lead to pictorial figures, confined to a sheet of paper. Interest and excitement can also be generated by constructing figures in space, using sticks, brass fasteners, paper clips, string, cut paper, multilink cubes, or indeed whatever is available. Such figures can be held and felt, literally giving students a feel for shape. In some cases the shapes can be bent or flexed to give a dynamic liveliness to figures which would remain of lesser interest when only drawn on a sheet of paper.

Books, pamphlets, and journals published by the National Council of Teachers of Mathematics and other publishing companies provide the teacher with a wide variety of ideas and resources for three-dimensional constructions and related activities. Every teacher will want to gather a personal collection of favorite hands-on constructions suitable for his or her own classroom. Here are three suggestions for constructions to do in the classroom:

- *Hinged polygons.* Strips of cardstock can be joined with brass fasteners through holes at the ends of each strip. Any triangle is rigid, demonstrating the SSS congruence property. Any quadrilateral, however, is flexible. As the quadrilateral flexes the sum of the angle measures remains at 360°, as can be checked with a protractor.

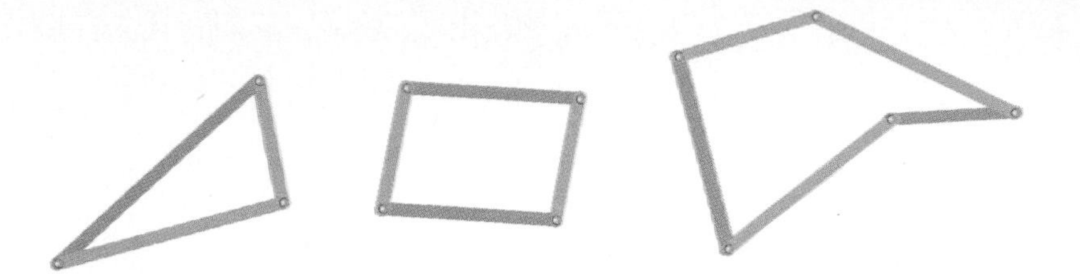

- *Space polygons and polyhedra.* Thin wooden sticks, say from a "Pick-Up-Sticks" game, can be cut to differing lengths and joined by short pieces of rubber tubing at their endpoints. More than two sticks can meet at a single vertex by inserting a length (or several lengths) of tubing through a hole punched sideways through another section of tubing. It is easy to form quadrilaterals that flex in space, and joining the midpoints of the four sides by elastic bands shows that a parallelogram is always formed. Properties of cubes, tetrahedra, and other polyhedra can also be explored with easily constructed skeletal models.

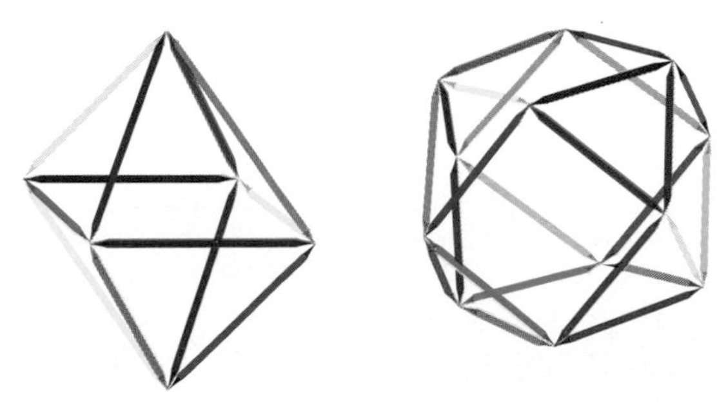

- *Geodesic domes.* Even primary grade children will enjoy constructing simple geodesic dome models with toothpicks and mini-marshmallows. The pattern for a small model is shown in its flattened position. The framework is then positioned vertically and wrapped around to insert toothpicks a and b into marshmallow A, toothpick C into marshmallow B, and the four upward pointing toothpicks into marshmallow C. The photograph shows first-graders with the domes they've made. A full discussion is found in the article "Marshmallows, Toothpicks, and Geodesic Domes" by Stacy Wahl (in *Geometry for Grades K–6: Readings from the Arithmetic Teacher,* National Council of Teachers of Mathematics, 1987).

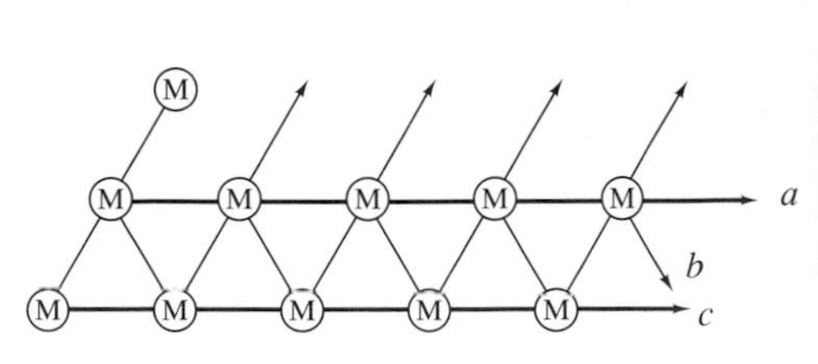

Basic Pattern, 1 *v*. 5/8 geodesic dome.

Constructing Parallel and Perpendicular Lines

If l is a given line and P is a given point not on l, it is useful to know how to construct the line through P that is either parallel or perpendicular to l. There are several procedures that can be devised and it is interesting to invent some of your own. The following construction takes advantage of the properties of a rhombus.

2 CONSTRUCTION Construct a Parallel Line

Given P and l as shown, construct a line through P that is parallel to l.

Procedure

Step 1
Draw a line through P that intersects l at a point labeled A.

Step 2
Draw an arc centered at A through P, and let B denote its intersection with l.

Step 3
With the same radius AB draw arcs with centers at P and B; let C be the intersection of the two arcs.

Step 4
Draw the line $k = \overleftrightarrow{PC}$; since $ABCP$ is a rhombus its opposite sides are parallel, and so $k \parallel l$.

A "corresponding angles" construction of a parallel line is outlined in problem 1 as an alternative to the construction just shown.

3 CONSTRUCTION Construct a Perpendicular Line Through a Point Not on the Given Line

Given line l and point P not on l as shown, construct a line through P that is perpendicular to l.

Procedure

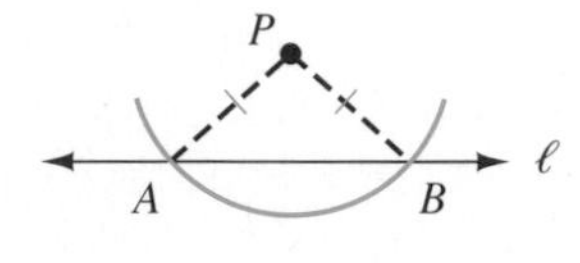

Step 1
Draw an arc at P that intersects l at two points, A and B.

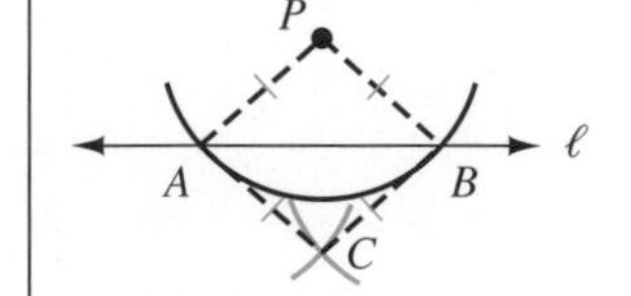

Step 2
With the compass still at radius AP, draw arcs at A and B, and let C be their point of intersection.

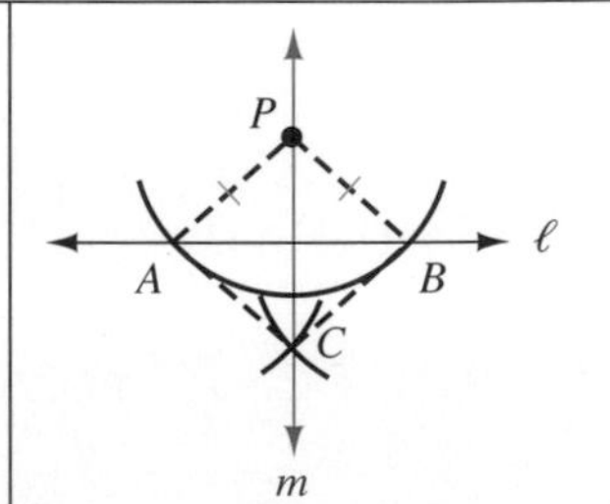

Step 3
Draw the line $\overleftrightarrow{PC}$; since $\overleftrightarrow{PC}$ is a diagonal of rhombus $APBC$ it is perpendicular to $\overline{AB}$.

The construction of perpendicular lines has several applications, as shown in Figure 11.7. The point F at which the perpendicular intersects the line is called the **foot of the perpendicular through *P*.** The length of $\overline{PF}$ is the **distance from the point *P* to the line *l*.** The point P' on the opposite side of l from P for which $PF = P'F$ is called the **point of reflection** of P across l. In a triangle, the perpendicular line segment from a vertex to the line containing the opposite side is an **altitude of the triangle.**

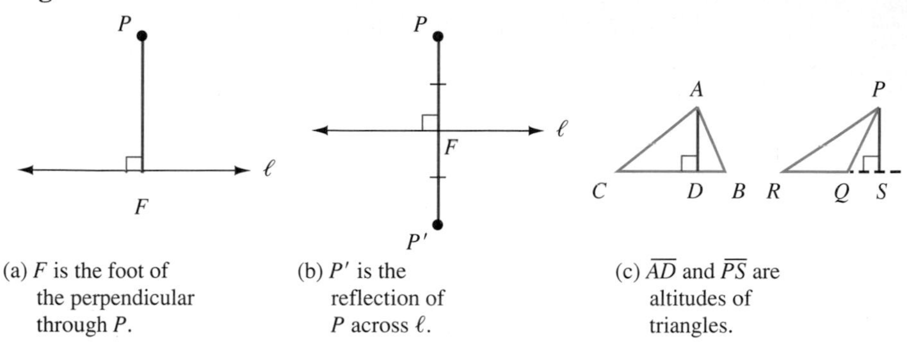

(a) F is the foot of the perpendicular through P.

(b) P' is the reflection of P across ℓ.

(c) $\overline{AD}$ and $\overline{PS}$ are altitudes of triangles.

Figure 11.7
Applications of perpendicular lines

If point P lies on line l a small modification in the second step of the procedure above is required to construct the line perpendicular to l at P.

4

CONSTRUCTION Construct the Perpendicular Line Through a Point on a Given Line

Given line l and point P on l as shown, construct the line through P perpendicular to l.

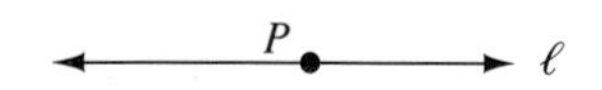

Procedure

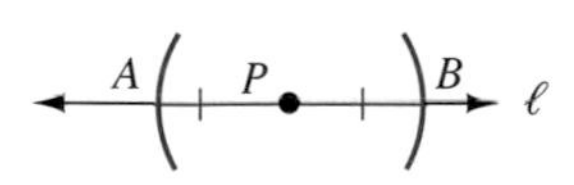

Step 1
Draw two arcs of equal radius centered at P; let A and B be their points of intersection with l.

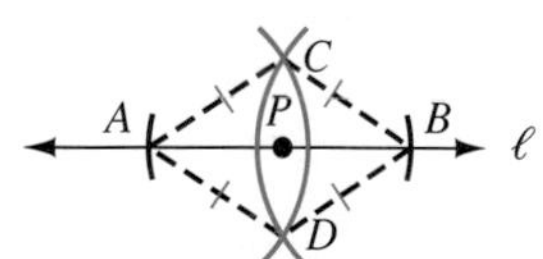

Step 2
Draw arcs centered at A and B with a radius *greater* than AP; let C and D be their points of intersection.

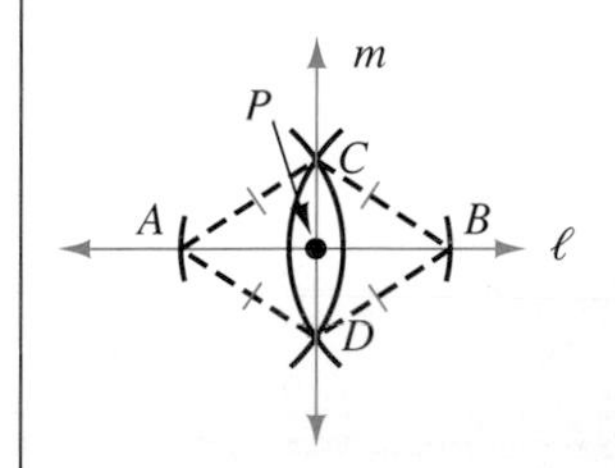

Step 3
Draw line $m = \overleftrightarrow{CD}$; since $\overline{CD}$ is a diagonal of the rhombus $ABCD$ and P is the midpoint of diagonal $\overline{AB}$, m passes through P and is perpendicular to $l = \overleftrightarrow{AB}$; that is, m is perpendicular to l.

Constructing the Midpoint and Perpendicular Bisector of a Line Segment

The line perpendicular to a segment at its midpoint is called the **perpendicular bisector** of the segment. The following construction is also justified by properties of the rhombus.

5 **CONSTRUCTION Construct the Perpendicular Bisector of a Line Segment**

Construct the perpendicular bisector of the segment $\overline{AB}$ shown.
(Note that the midpoint of the segment is not given, but will be constructed.)

A —————————— B

Procedure

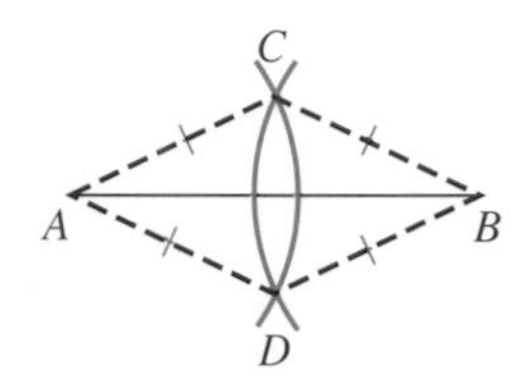 **Step 1** Draw arcs of the same radius centered at A and B and intersecting in points C and D.	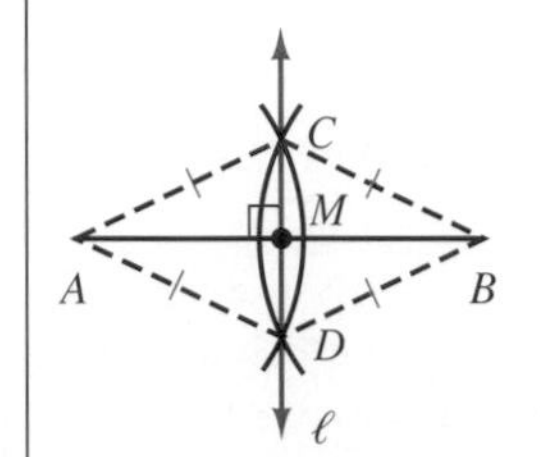 **Step 2** Draw the line $\overleftrightarrow{CD}$. Since $ADBC$ is a rhombus, $\overline{CD}$ intersects $\overline{AB}$ at a right angle at the midpoint M of $\overline{AB}$.

The midpoint of a line segment M is the same distance from A and B. Indeed, *all* points of the perpendicular bisector of a segment are equidistant from the end points of the segment.

THEOREM Equidistance Property of the Perpendicular Bisector

A point lies on the perpendicular bisector of a line segment if, and only if, the point is equidistant from the endpoints of the segment.

PROOF Let l be the perpendicular bisector of segment $\overline{AB}$. Thus, l intersects $\overline{AB}$ at right angles at the midpoint M, as shown in the left side of the diagram on the next page. Let P be any point of l. By the SAS property, $\triangle PMA \cong \triangle PMB$. This means that the corresponding sides $\overline{PA}$ and $\overline{PB}$ are congruent. Hence, $PA = PB$ as claimed.

The proof that equidistant points from A and B lie on the perpendicular bisector is similar. (See problem 24.)

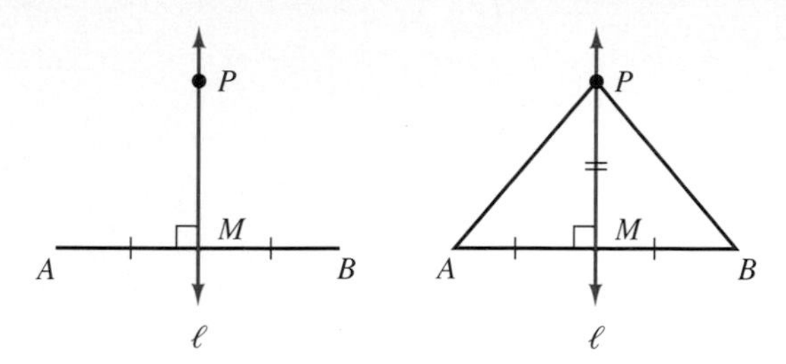

■

EXAMPLE 11.9 **Locating an Airport**

The Tri-Cities Airport Authority wishes to locate a new airport to serve their three cities, located at A, B, and C as shown. If possible, they would like a site, P, that is the same distance from A, B, and C. How can P be located?

A •

• C

B •

SOLUTION

To be equidistant from A and B, P must be on the perpendicular bisector of the segment $\overline{AB}$. Similarly, P must be on the perpendicular bisector of $\overline{BC}$. Since A, B, and C are not collinear, the perpendicular bisectors to $\overline{AB}$ and $\overline{BC}$ are not parallel and we can choose P as their point of intersection. Since $PA = PB$ and $PB = PC$ we have that $PA = PC$. Therefore, P is also equidistant from A and C, so P is also on the perpendicular bisector of $\overline{AC}$. Therefore P is equidistant from A, B, and C, as desired. ■

Point P is the center of a unique circle containing A, B, and C as shown in Figure 11.8. The circle is called the **circumscribing circle** of $\triangle ABC$, and P is called the **circumcenter** of $\triangle ABC$. Frequently the circumscribing circle is called the **circumcircle** of the triangle.

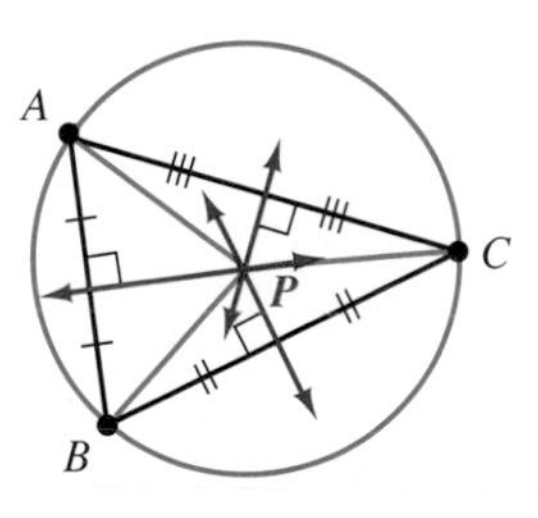

Figure 11.8
The perpendicular bisectors of the sides of $\triangle ABC$ are concurrent at a point P equidistant from A, B, and C. Point P is the center of the circumscribing circle of $\triangle ABC$.

Constructing the Angle Bisector

Given $\angle ABC$ (see Figure 11.9) we wish to construct the ray $\overrightarrow{BD}$ that forms congruent angles with the sides $\overrightarrow{BA}$ and $\overrightarrow{BC}$. If $\angle ABD \cong \angle CBD$ then $\overrightarrow{BD}$ is the **angle bisector** of $\angle ABC$.

Once again the properties of a rhombus justify the following construction.

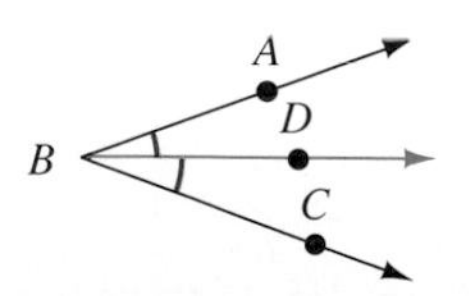

Figure 11.9
$\overrightarrow{BD}$ is the angle bisector of $\angle ABC$ if $\angle ABD \cong \angle DBC$.

JUST FOR FUN

Circle Amazement

Use a jar lid (or other handy circular object) to draw any three congruent circles through a common point P. Let A, B, and C denote the three points of intersection of pairs of your circles other than P. Now use the jar lid to discover something amazing about the circumscribing circle of $\triangle ABC$.

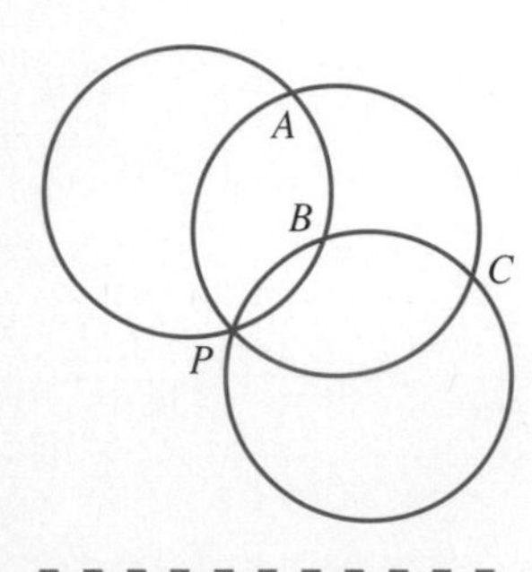

"THERE'S A GUY I KNOW NAMED NOBEL WHO WANTS TO GIVE YOU A PRIZE FOR THIS THING."

6 **CONSTRUCTION** **Construct the Angle Bisector**

Construct the angle bisector of $\angle E$ shown.

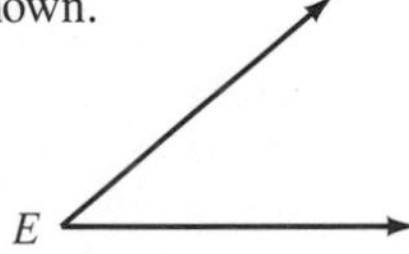

Procedure

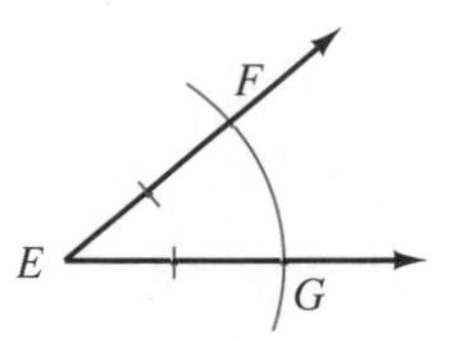 **Step 1** Draw an arc centered at E; let F and G denote the points at which the arc intersects the sides of $\angle E$.	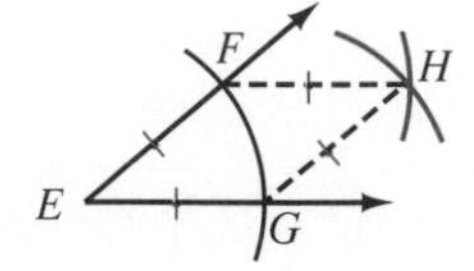 **Step 2** Draw arcs, centered at F and G, of radius EF. Let H be the point of intersection of the two arcs.	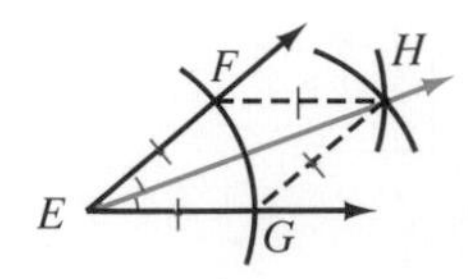 **Step 3** Draw the ray $\overrightarrow{EH}$. Since the diagonal $\overline{EH}$ forms congruent angles to the sides $\overline{EG}$ and $\overline{EF}$ of the rhombus $EGHF$, $\overrightarrow{EH}$ is the angle bisector of $\angle E$.

The following theorem tells how far the points on the angle bisector are from the sides of an angle.

THEOREM Equidistance Property of the Angle Bisector

A point lies on the bisector of an angle if, and only if, the point is equidistant from the sides of the angle.

PROOF Let P be any point on the bisector of $\angle A$ as shown. Let F and F' be the feet of the perpendiculars from P to the sides of $\angle A$.

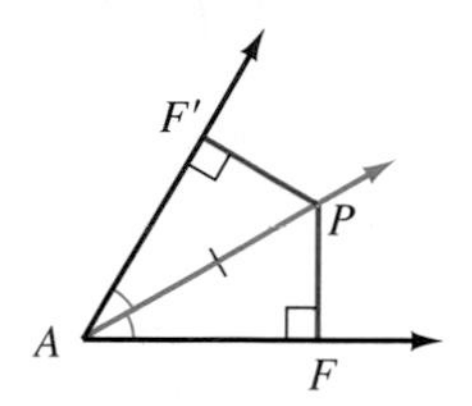

By the AAS congruence property, $\triangle PAF \cong \triangle PAF'$. Therefore, $PF = PF'$. The proof of the converse is similar and left to the reader. ■

By mimicking the solution of the airport problem of Example 11.9 it can be shown that the bisectors of the interior angles of a triangle are concurrent at the point I which is equidistant to three sides of the triangle (see Figure 11.10). Point I, called the **incenter** of $\triangle ABC$, is the center of the **inscribed circle,** or **incircle,** of $\triangle ABC$.

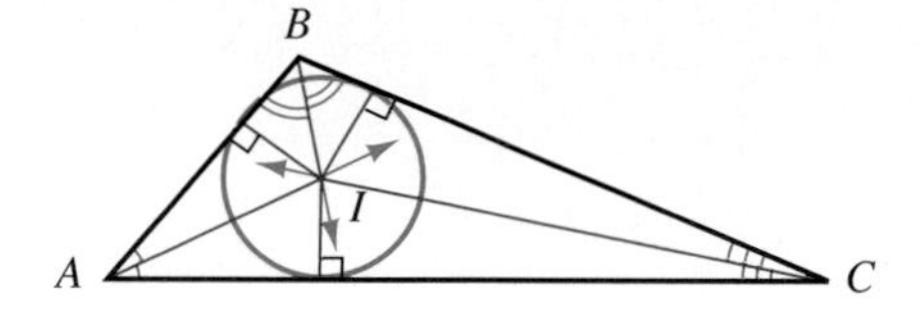

Figure 11.10
The bisectors of the interior angles of a triangle are concurrent at a point I equidistant from the sides of the triangle. I is the center of the inscribed circle of the triangle.

Constructing Regular Polygons

A square is easily constructed with compass and straightedge. For example, draw a circle at the intersection of two perpendicular lines, as shown in Figure 11.11(a). By constructing angle bisectors, the eight vertices of a regular octagon are

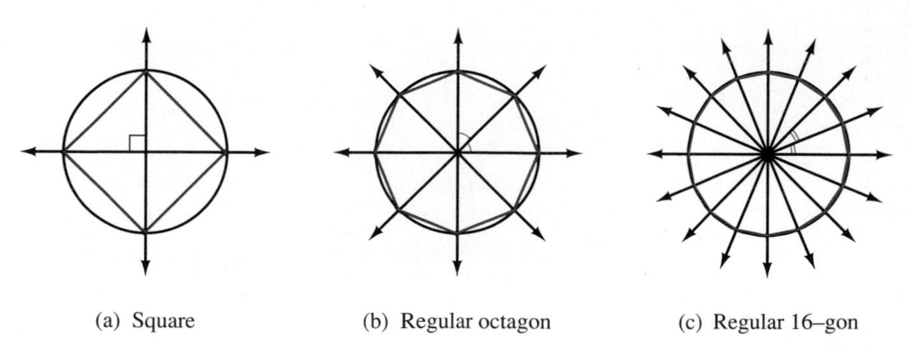

(a) Square (b) Regular octagon (c) Regular 16–gon

Figure 11.11
Constructing angle bisectors doubles the number of sides of a regular polygon inscribed in a circle.

HIGHLIGHT FROM HISTORY
Compass or Straightedge Alone Constructions

In 1797 the Italian geometer and poet Lorenzo Mascheroni (1750–1800) published *Geometria del Compasso* ("Geometry with Compasses"). The book contains the remarkable discovery that if the given and required elements of a construction are all points, and not lines, then the straightedge is unnecessary in any construction that could otherwise be done with both compass and straightedge. Compass alone constructions are frequently called Masheroni constructions, although it is now known that Masheroni's discovery was actually made much earlier by an obscure writer named Georg Mohr. Mohr included the result, and its proof, in a book appearing in 1672.

Masheroni's discovery inspired the French mathematician Jean Victor Poncelet (1788–1867) and later the Swiss-German geometer Jacob Steiner (1796–1867) to consider constructions with straightedge alone. It is not quite true that all Euclidean constructions can be done without a compass, but amazingly, if even one circle and its center are drawn in the plane of the construction, then the compass can be discarded and the straightedge alone is sufficient for any Euclidean construction.

The figure below shows a compass alone construction of the vertices of a square *AFDG* inscribed in a circle centered at *O*. The construction is attributed to Napoleon.

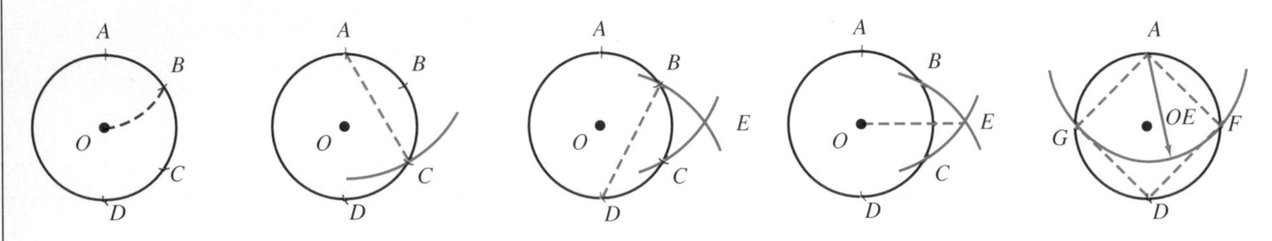

constructed. Angle bisectors can then be constructed to the octagon of Figure 11.11(b) to yield the regular 16-gon of part (c) of the figure.

When the vertices of a polygon are all points of a given circle, the polygon is called an **inscribed polygon.** A regular hexagon is particularly easy to inscribe in a given circle with a compass and straightedge: pick any point *A* on the circumference of the circle centered at *O* and then successively strike arcs of radius *OA* around the circle to locate *B*, *C*, *D*, *E*, and *F*. The hexagon *ABCDEF* is regular since joining the sides to the center *O* forms six congruent equilateral triangles *OAB*, *OBC*, . . . , *OFA*. As shown in Figure 11.12, connecting every other vertex gives a construction of the inscribed equilateral triangle *ACE*. On the other hand, constructing angle bisectors of the central angles of the hexagon produces an inscribed regular dodecagon.

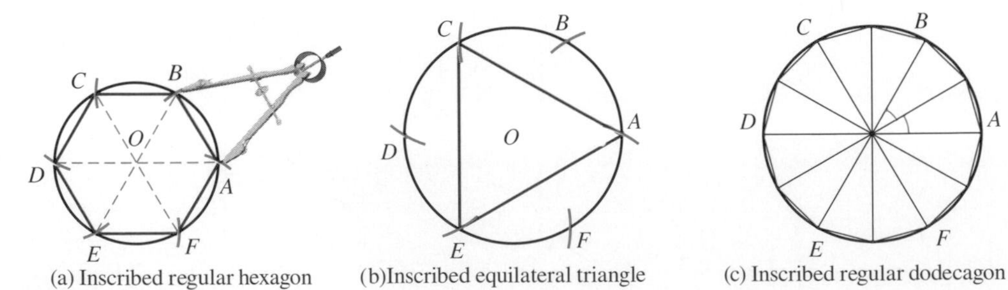

(a) Inscribed regular hexagon (b)Inscribed equilateral triangle (c) Inscribed regular dodecagon

Figure 11.12
The regular 3-, 6-, and 12-gons can be inscribed with compass and straightedge in a given circle.

A compass and straightedge construction of the regular pentagon requires more ingenuity. (One method is outlined in problem 19 at the end of this section.)

The ancient Greek geometers knew how to construct the regular polygons shown so far. They also knew that the regular 15-gon can be constructed with compass and straightedge. (The construction is outlined in problem 26 at the end of this section.) For over 2000 years, the only regular polygons known to be constructible with compass and straightedge were the ones contained in Book IV of Euclid's *Elements:* the regular 3-, 4-, 5- and 15-gons and, by angle bisection, their successive "doubles," $2^r \cdot 3$-, $2^r \cdot 4$-, $2^r \cdot 5$-, and $2^r \cdot 15$-gons, where r is any whole number.

On March 30, 1796, one month before his nineteenth birthday, Carl Friedrich Gauss (1777–1855) entered into a notebook his discovery that a number of other regular polygons were constructible, including those of 17, 257, and 65,537 sides. The numbers 3, 5, 17, 257, and 65,537 are prime numbers of the special form $F_k = 2^{2^k} + 1$, where k is a nonnegative integer. For example, $F_0 = 2^{2^0} + 1 = 2^1 + 1 = 3$, $F_1 = 2^{2^1} + 1 = 2^2 + 1 = 4 + 1 = 5$, and so on. Numbers of this form had been studied earlier by Pierre de Fermat (1601–1665), and prime numbers of the form $2^{2^k} + 1$ are known as **Fermat primes.**

Here is the remarkable theorem of Gauss. The proof of the "only if" part of the theorem is due to Pierre Wantzel (1814–1848).

THEOREM The Gauss-Wantzel Constructibility Theorem

A regular polygon of n sides is constructible with compass and straightedge if, and only if, n has one of the following forms:

1. $n = 2^r \cdot 4$, where $r = 0, 1, 2, \ldots$;
2. $n = 2^r \cdot p$, where p is Fermat prime and $r = 0, 1, \ldots$;
3. $n = 2^r \cdot p_1 \cdots p_m$, where $p_1 \cdots, p_m$ are distinct Fermat primes, and $r = 0, 1, \ldots$.

For example, the regular heptagon is not constructible since 7 is not a prime number of the form $2^{2^k} + 1$. A regular nonagon (9-gon) is also not constructible since 9 has two factors of 3. On the other hand, a regular polygon of $1020 = 2^2 \cdot 3 \cdot 5 \cdot 17$ sides is constructible since its odd prime factors are distinct Fermat primes.

Fermat believed that all numbers of the form $F_k = 2^{2^k} + 1$ were prime, but Euler proved this assertion to be false by showing that $F_5 = 2^{2^5} + 1$ is composite; in fact, $F_5 = 4{,}294{,}967{,}297 = (641)(6{,}700{,}417)$. Likewise, $F_6, F_7, \ldots, F_{21}$ are now known to be composite. It is generally believed that there are only five Fermat primes, namely 3, 5, 17, 257, and 65,537.

Mira and Paper Folding Constructions

Figure 11.13 shows a Mira, a simple yet very effective device for constructing plane geometric figures. Activities with a Mira can introduce even young students to congruence, parallelism, perpendicularity, symmetry and other basic concepts of geometry. The plastic surface of a Mira reflects an object in front and still allows an object in back to be seen through the surface. By superimposing the direct and reflected images, the line traced along the drawing edge of the Mira constructs a line of symmetry, the perpendicular bisector, the angle bisector, and so on. Constructions with a Mira are quick yet quite accurate with a little practice.

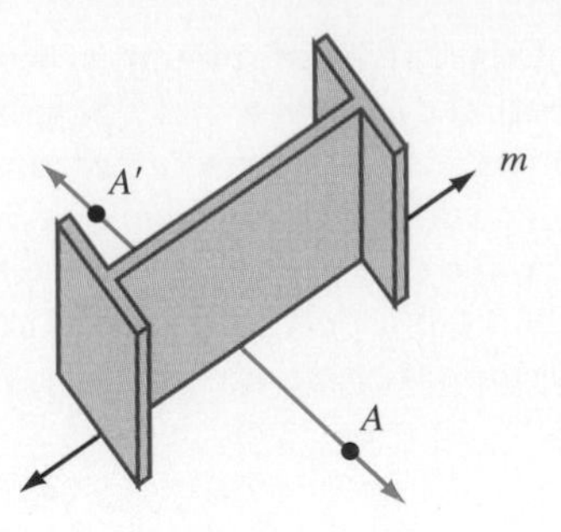

(a) The Mira

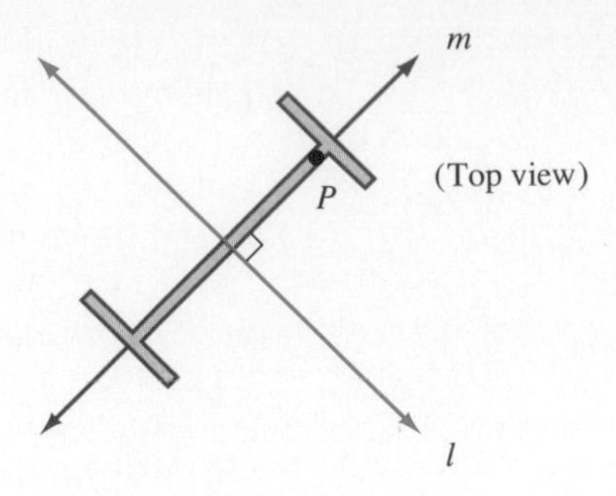

(b) Pivot the Mira about P until line l coincides with its reflection to construct the line m through P that is perpendicular to l.

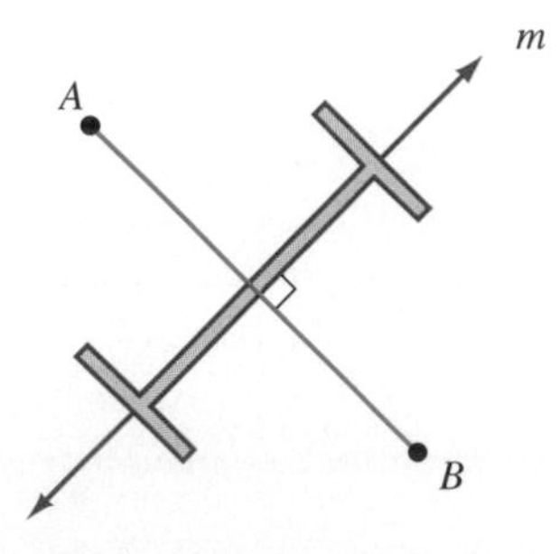

(c) When the reflection of B coincides with A, the drawing edge of the Mira determines the perpendicular bisector of $\overline{AB}$.

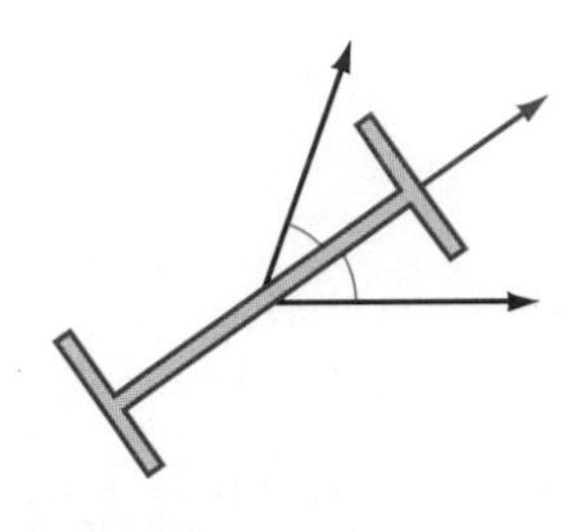

(d) Pivot the Mira about the vertex until the reflection of the near side coincides with the far side to construct the angle bisector.

Figure 11.13
The Mira and its use in three basic constructions

Mira constructions can usually be converted to equivalent paper folding procedures. The drawing line of the Mira is replaced by the crease line of a fold. It is helpful to draw lines and points very dark, so they can be seen from the reverse side of the paper. A folding construction of the perpendicular bisector is shown in Figure 11.14.

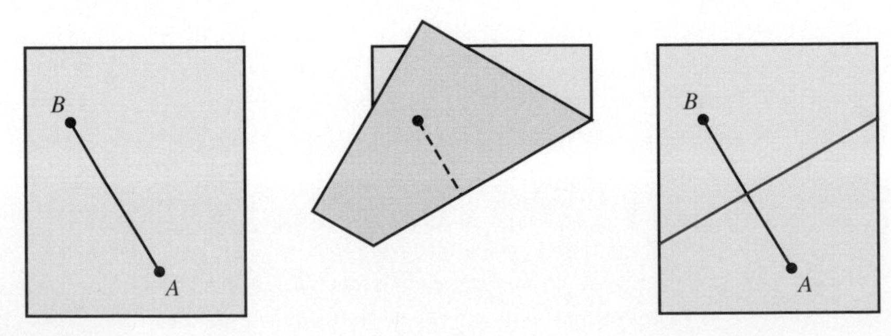

Figure 11.14
Folding point A onto point B forms a crease on the perpendicular bisector of $\overline{AB}$.

HIGHLIGHT FROM HISTORY
Three Impossible Construction Problems

The straightedge allows us to draw a line of indefinite length through any two given points, and the compass* allows us to draw a circle with a given point as its center and passing through any given distinct second point. It then becomes a challenge to find procedures to construct a figure using only these simple tools. Many important contributions to geometry were inspired by attempts to solve the following famous problems, each of which arose in antiquity.

1. *The trisection of an angle:* Divide an arbitrary given angle into three congruent angles.

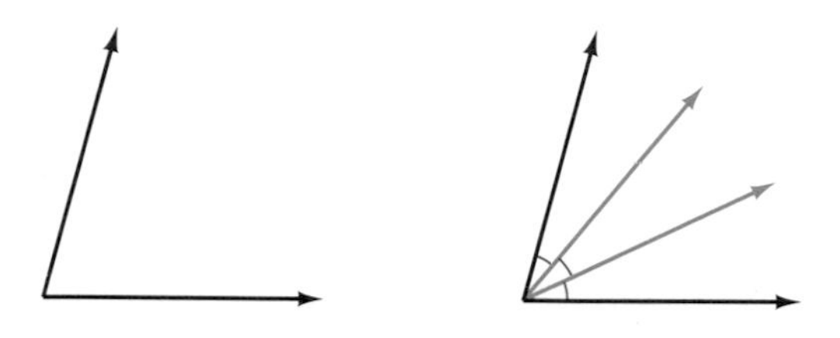

2. *The duplication of the cube:* Given a cube, construct a cube with twice the volume.

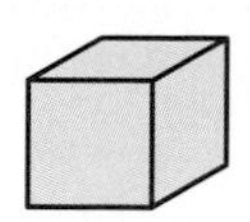
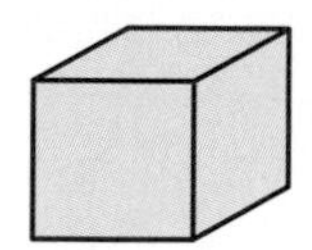

3. *The squaring of the circle:* Given a circle, construct a square of the same area as the circle.

Extensive efforts for over 2000 years failed to solve any of these problems. It was not until the early 1800s that it was shown that these problems were impossible to solve when only the straightedge and compass were allowed. It is interesting to note that methods of algebra were used to prove the impossibility of these geometric construction problems.

* The usage "the compass" is common, but some texts and authors still prefer "compasses" or even "a pair of compasses."

FROM THE NCTM STANDARDS **Constructing and Using Geometric Models**

Geometry gives children a different view of mathematics. As they explore patterns and relationships with models, blocks, geoboards, and graph paper, they learn about the properties of shapes and sharpen their intuitions and awareness of spatial concepts. Children's geometric ideas can be developed by having them sort and classify models of plane and solid figures, construct models from straws, make drawings, and create and manipulate shapes on a computer screen. Folding paper cutouts or using mirrors to investigate lines of symmetry are other ways for children to observe figures in a variety of positions, become aware of their important properties, and compare and contrast them. Related experiences help children avoid simplistic and misleading ideas about shapes, such as that implied by one child's observation, "This is an upside-down triangle."

SOURCE: From *Curriculum and Evaluation Standards for School Mathematics Grades K–4*, pp. 48–49.

PROBLEM SET 11.2

Understanding Concepts

1. The sequence of steps shown below outline the "corresponding angles" construction to draw a line k that is parallel to a line l and passes through the point P not on l.

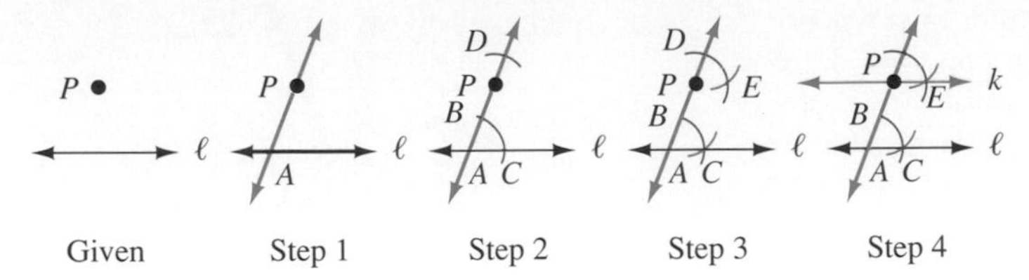

(a) Give a written description of each of the four steps.
(b) Explain why the construction gives the desired line k.

2. **(a)** Describe, in words and drawings, a Mira construction which gives the line k parallel to a line l and through a point P not on l.
(b) Answer (a), but use paper folding instead of the Mira.

3. The drafting triangle and straightedge can be used to construct parallel lines, as shown below.

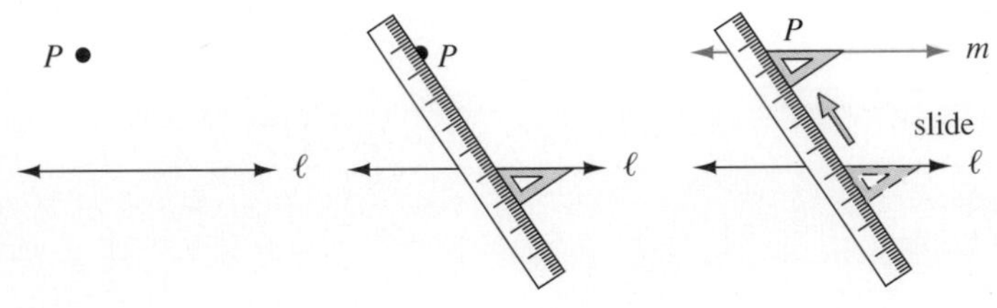

(a) What geometric principle justifies this construction?
(b) Describe, in words and drawings, a procedure using a straightedge and drafting triangle to construct the line m perpendicular to a given line l and passing through a given point P.

4. Use compass and straightedge to construct the following:
(a) Line perpendicular to l through P

(b) Line perpendicular to l through Q

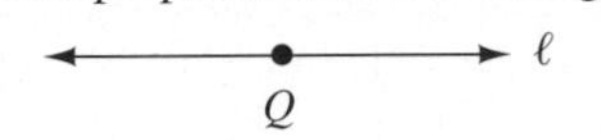

(c) Perpendicular bisector to $\overline{ST}$

S ——————— T

(d) Bisector of $\angle A$

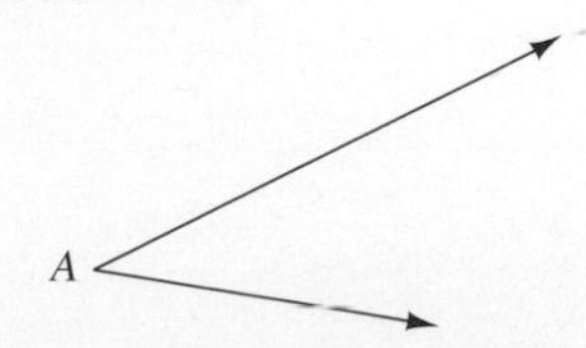

5. Repeat the constructions in problem 4 with a Mira (if available).

6. Repeat the constructions in problem 4 with paper folding.

7. Construct the circumcenters and circumscribing circles of triangles of the three types shown. Use any tools you wish to draw the perpendicular bisectors of the sides of the triangles.
(a) Acute triangle
(b) Right triangle
(c) Obtuse triangle

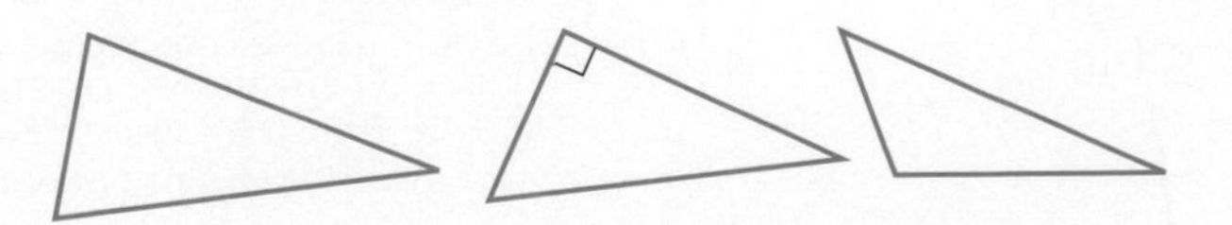

(d) Make a conjecture about where—inside, on, or outside the triangle—the circumcenter of a triangle will be located.

8. Construct the incenters and inscribed circles of the three types of triangles shown in problem 7, using

(a) a compass and straightedge.

(b) a Mira and compass (if available).

(c) paper folding (copy and cut the triangles from paper with scissors, then fold) and compass.

9. A point P is exterior to a circle centered at Q. Use Thales' theorem to justify why drawing the circle with diameter $\overline{PQ}$ gives a construction of the two lines $\overleftrightarrow{PS}$ and $\overleftrightarrow{PT}$ that are tangent to the circle at Q.

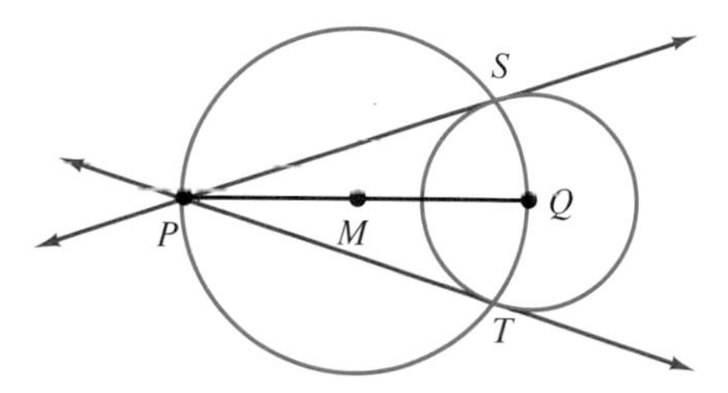

10. Draw any triangle ABC. Using any tools you wish, draw both the inscribed and the circumscribed circles of your triangle. Now choose an arbitrary point X on the larger circle and draw the chords $\overline{XY}$ and $\overline{XZ}$ that are tangent to the smaller circle. Finally, draw chord $\overline{YZ}$ of the larger circle. Start with a new point X' and again draw a triangle $X'Y'Z'$ (colored pencils are helpful). Make a conjecture about all such triangles.

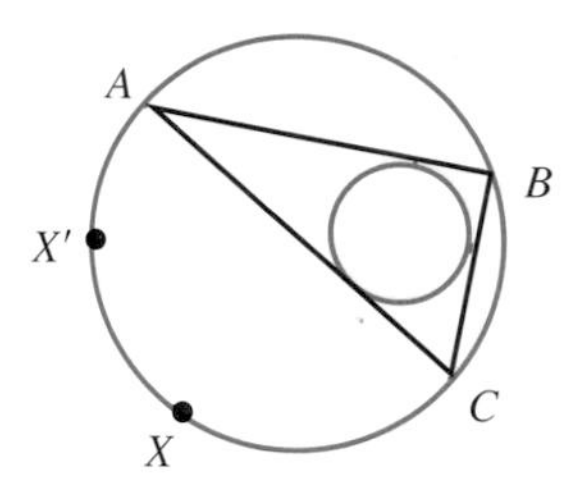

11. **(a)** Use any drawing tools you wish to construct the altitudes of the three types of triangles in problem 7.

(b) Make a conjecture about the three altitudes of an acute triangle.

(c) Make a conjecture about the three altitudes of a right triangle.

(d) Make a conjecture about the three lines containing the altitudes of an obtuse triangle.

12. Justify the following construction of an equilateral triangle inscribed in a given circle.

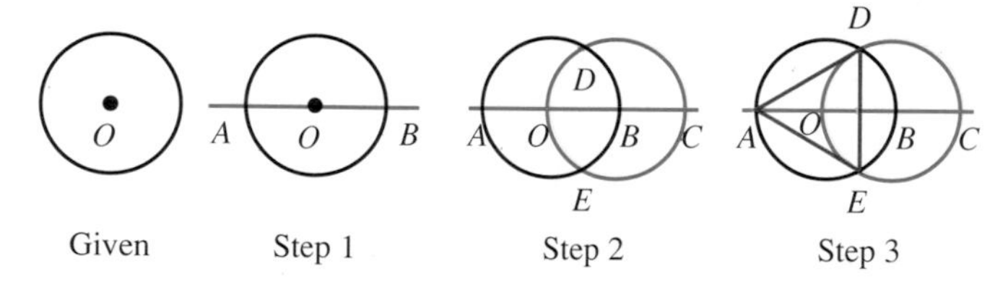

Given Step 1 Step 2 Step 3

13. A **median** of a triangle is a line segment from a vertex to the midpoint of the opposite side. Using any drawing tools you wish, draw the three medians in each of several triangles. Give a statement that describes what you observe.

14. Let the lines k and l intersect to form two pairs of vertical angles. Prove that the lines m and n that bisect the vertical angles are perpendicular.

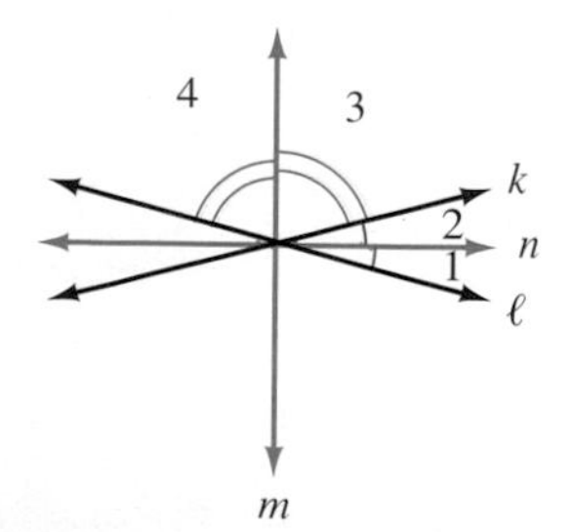

15. (a) Prove that the perpendicular bisector of any chord of a circle contains the center of the circle.

(b) Explain how to construct the center of the circle whose arc is shown below. (*Hint:* Use part (a), twice!)

(c) The three congruent circles below are centered at points A, B, and C on the large circular arc, and the circles at centers A and C contain point B. Where do the dashed lines intersect? Explain why.

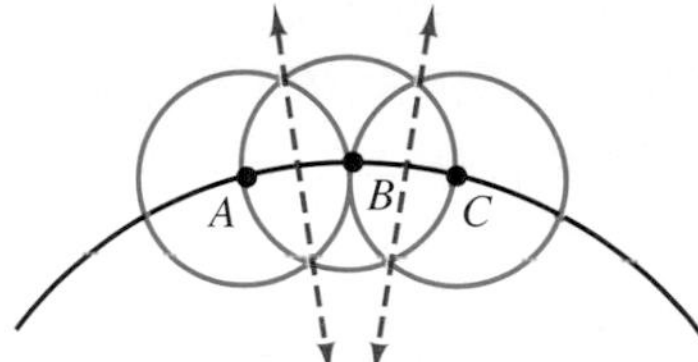

16. Prove that the angle bisectors of a triangle are concurrent, using the equidistance property of the angle bisector. The discussion in Example 11.9 can be used as a model for your proof.

17. The diagram shows how to erect an equilateral triangle ABC on a given line segment $\overline{AB}$ using a compass and straightedge.

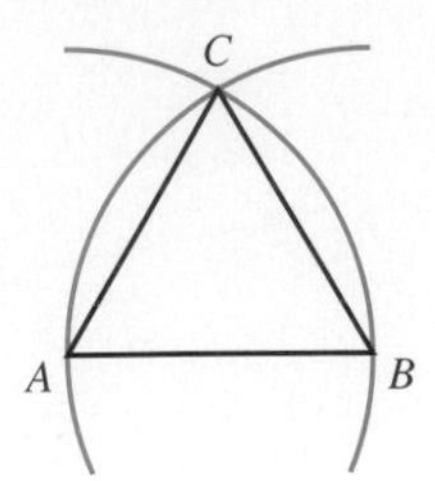

Give careful step-by-step instructions to construct these polygons erected on a given side $\overline{AB}$.
(a) a square **(b)** a regular hexagon
(c) a regular octagon

18. Reflecting point B to fall on the perpendicular bisector of $\overline{AB}$ shows how a Mira can be used to draw an equilateral triangle ABC on a given side $\overline{AB}$.

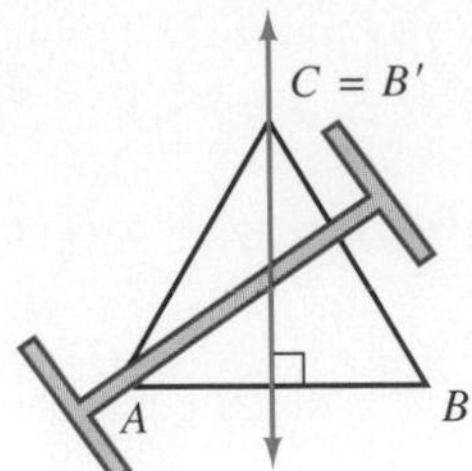

Give careful step-by-step instructions to construct these polygons on a given side with a Mira.
(a) a square **(b)** a regular hexagon
(c) a regular octagon

19. **(a)** Construct a regular pentagon inscribed in a circle by following the steps outlined.
(b) Use a ruler and protractor to check that $PENTA$ is regular.

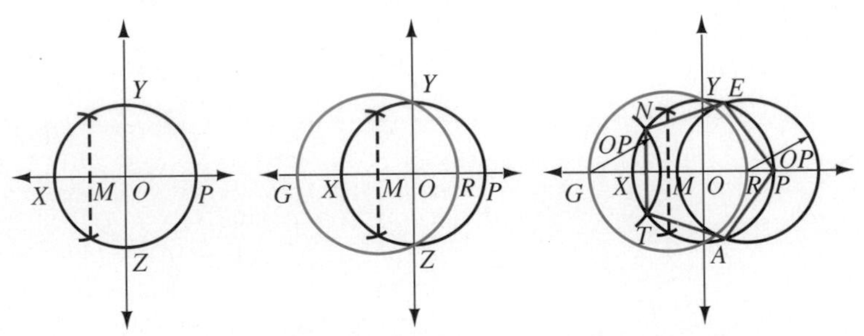

Draw perpendicular diameters, and construct midpoint M of radius $\overline{OX}$.

Draw circle at M through Y to locate G and R.

Draw circles at G and R of radius OP to locate the vertices of the regular pentagon $PENTA$.

20. **(a)** Construct a heptagon inscribed in a given circle by following the steps outlined below.

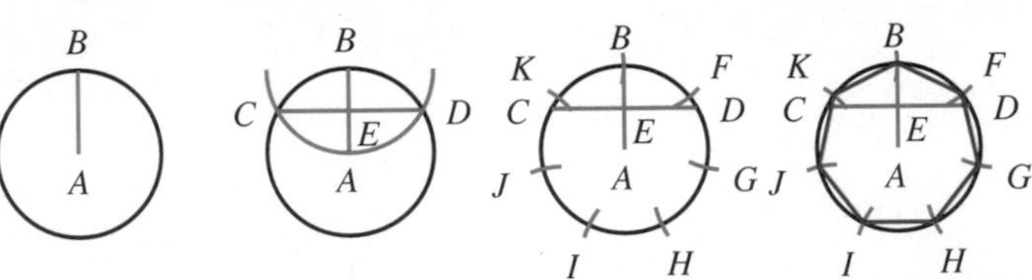

Draw a radius $\overline{AB}$.

Draw a circle at B through A and then draw chord $\overline{CD}$.

Lay off arcs of radius DE from B.

Draw the heptagon $BFGHIJK$.

(b) Is it possible for $BFGHIJK$ to be a regular heptagon, or is it just a close approximation?

21. Continue the list of constructible versus nonconstructible regular n-gons up to $n = 100$, using the Gauss-Wantzel theorem.

Constructible	3	4	5	6		8		10		12
Nonconstructible					7		9		11	

22. Someone claims that trisecting the chord $\overline{BC}$ of an arc centered at A gives a compass and straightedge trisection of $\angle A$.
How would you respond to this assertion? How well does the method appear to work on an angle of measure 150°? Make a drawing and use a protractor to measure the angles.

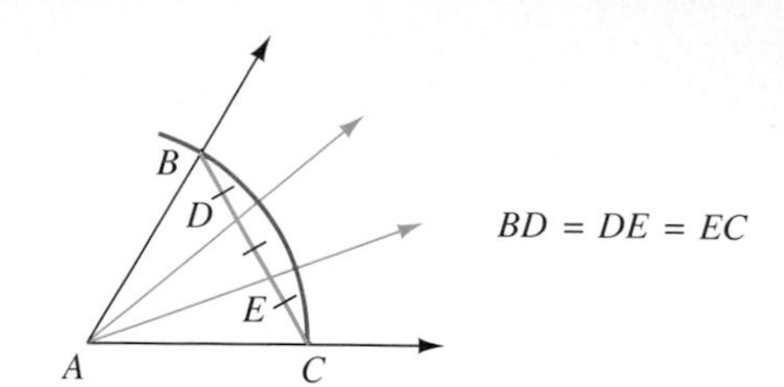

Thinking Critically

23. Justify the "kite" construction of an angle bisector shown below.

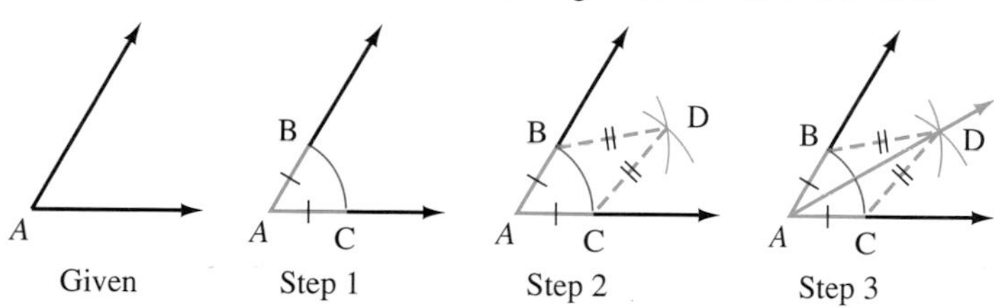

24. Complete the proof of the equidistance property of the perpendicular bisector. Do so by showing that if P is equidistant from A and B, then the line $\overleftrightarrow{PM}$ containing P and the midpoint M of $\overline{AB}$ is the perpendicular bisector of $\overline{AB}$.

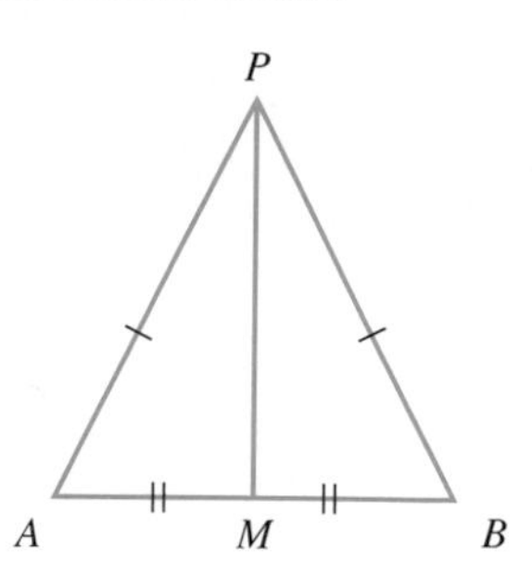

25. Draw any triangle ABC. Using any tools for construction you wish, draw the line through vertex A that is parallel to the opposite side $\overline{BC}$. Similarly, draw the lines through B and C that are parallel to their respective opposite sides. Let D, E, and F be the points of intersection of the three lines you have constructed.

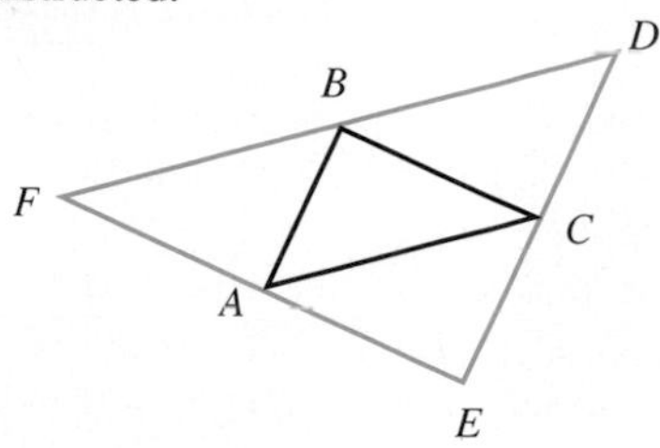

(a) Prove that your construction gives three new triangles, each congruent to $\triangle ABC$. (*Hint: ABDC* is a parallelogram.)

(b) Prove that A, B, and C are the midpoints of the sides of $\triangle DEF$.

(c) Construct the altitudes of $\triangle ABC$. Explain why they are also the perpendicular bisectors of $\triangle DEF$.

(d) Prove that the lines containing the altitudes of $\triangle ABC$ are concurrent. (This point is called the **orthocenter** of $\triangle ABC$.)

26. Draw any triangle ABC. Using any tools you wish, draw its orthocenter (that is, the point of concurrence of the three altitudes of $\triangle ABC$; see problem 25). Label the orthocenter D. Now trace the four points A, B, C, and D on a clean sheet of paper, and draw the orthocenter of $\triangle BCD$. Are you surprised? Guess, and then check what the orthocenters of $\triangle ACD$ and $\triangle ABD$ are.

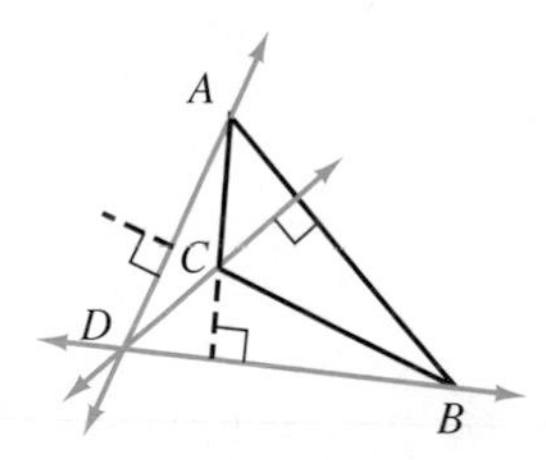

27. A regular pentagon $PENTA$ and an equilateral triangle PQR are both inscribed in the same circle centered at O, with P a common vertex. Calculate $m(\angle NOQ)$ and explain why laying off segments of length QN about the circle constructs a regular inscribed 15-gon.

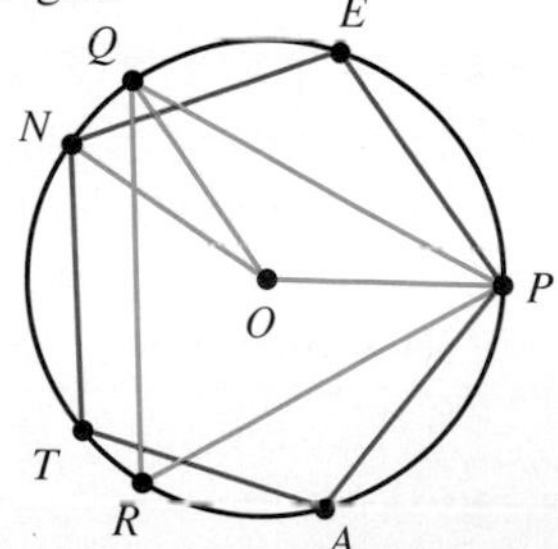

28. Archimedes proposed the following method to trisect an angle.

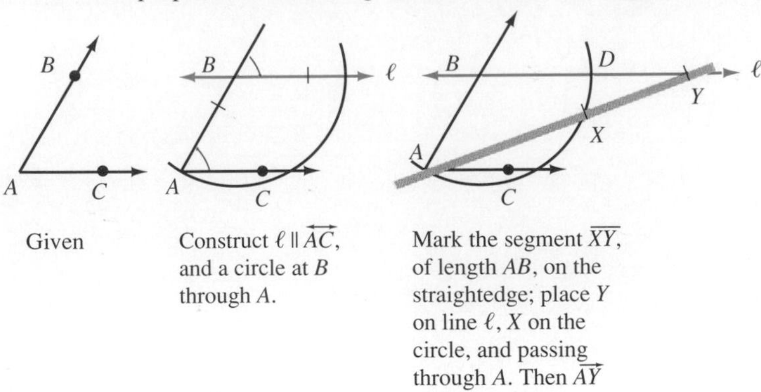

Given

Construct $\ell \parallel \overrightarrow{AC}$, and a circle at B through A.

Mark the segment $\overline{XY}$, of length AB, on the straightedge; place Y on line ℓ, X on the circle, and passing through A. Then $\overrightarrow{AY}$ trisects $\angle BAC$.

(a) Prove that Archimedes' construction is exact. The following figure should help you show that $m(\angle BAC) = 3 \cdot m(\angle YAC)$. (*Hint:* The measure of an exterior angle of a triangle is the sum of the measures of the opposite interior angles.)

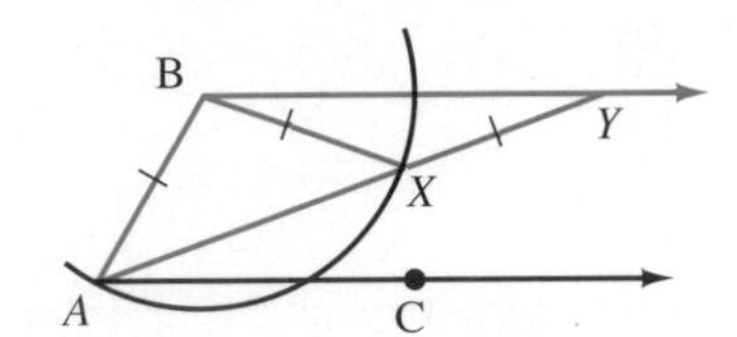

(b) Why is Archimedes' method not a compass and straightedge construction in the sense of Euclid?

Using a Calculator

29. Verify that $F_3 = 2^{2^3} + 1$ and $F_4 = 2^{2^4} + 1$ are given decimally by 257 and 65,537.

30. Verify that the Fermat number $F_5 = 2^{2^5} + 1$ is 4,294,967,297. On a calculator with an 8-digit display it may help to write

$$2^{32} = 2^{12} \cdot 2^{20} = 4096 \cdot 1048576$$
$$= 4 \cdot 1048576 \cdot 10^3 + 96 \cdot 10^6 + 96 \cdot 48576.$$

Justify each step in the expansion just shown and carefully explain how you have used both the calculator and paper and pencil addition to obtain your final answer.

31. Verify that the Fermat number $F_5 = 4{,}294{,}967{,}297$ is composite by computing (641) (6700417). You may find it useful to write 6,700,417 in the form $67 \times 10^5 + 417$.

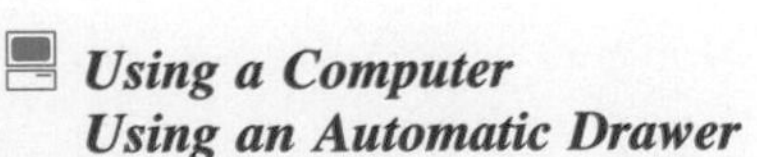

Using a Computer
Using an Automatic Drawer

32. Draw any triangle ABC. Construct the following three points:
G, the **centroid** (intersection of the medians, see problem 13);
H, the orthocenter (intersection of the altitudes, see problem 25); and P, the circumcenter (intersection of the perpendicular bisectors of the sides).

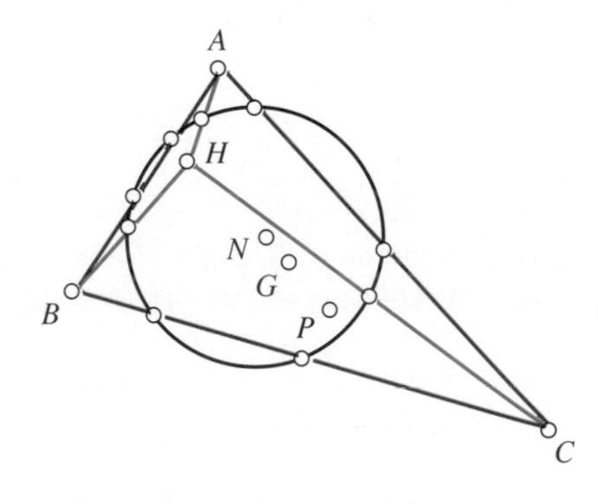

(a) Euler discovered an interesting fact about G, H, and P. What do you suppose the "Euler line" might be?

(b) Measure the distances PG and GH, and then make a conjecture concerning the ratio of these distances.

(c) Find the midpoint N of the segment $\overline{PH}$, and draw the circle centered at N which passes through the midpoint of a side of your triangle. Where does it intersect the other sides of the triangle?

(d) Describe how the circle at N intersects the segments $\overline{HA}$, $\overline{HB}$, and $\overline{HC}$.

33. Draw any triangle ABC. On each side construct outward pointing equilateral triangles BCR, CAS, and ABT. Also construct the incenters X, Y, and Z of the equilateral triangles.

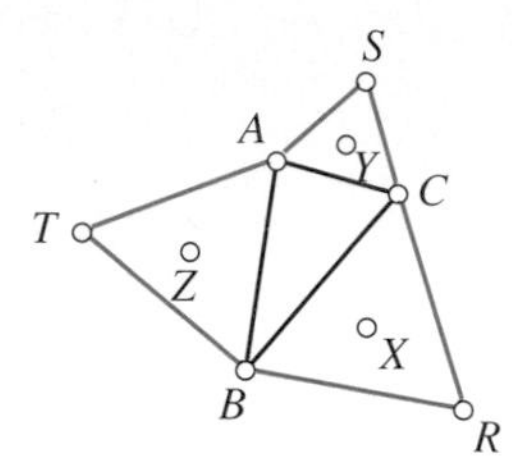

(a) What kind of triangle is XYZ? Measure lengths and angles to support your guess.

(b) Construct the segments $\overline{AR}$, $\overline{BS}$, and $\overline{CT}$, showing they are concurrent at a point F. (F is called the **Fermat point,** and in an acute triangle ABC it is the point with the smallest sum $FA + FB + FC$ of distances to the vertices of $\triangle ABC$).

(c) At what angles do the lines drawn in part (b) intersect at F? Measure to verify your guess.

(d) Draw the segments $\overline{AX}$, $\overline{BY}$, $\overline{CZ}$, showing they are concurrent at a point N. (N is called the **Napoleon point;** supposedly it was Napoleon who discovered the theorem that you likely discovered in answering part (a).)

(e) Construct the circumcenter P of $\triangle ABC$. What can you conjecture about the three points F, N, and P?

34. Draw three equilateral triangles that share a common vertex A. Let the triangles, labeled counterclockwise around their respective interiors, be $\triangle ABC$, $\triangle AB'C'$, and $\triangle AB''C''$. Draw the midpoint T of $\overline{BC''}$, as shown below. Similarly, draw the midpoint R of $\overline{B'C}$ and the midpoint I of $\overline{C'B''}$. Finally, draw the triangle TRI. What kind of a triangle does TRI seem to be? Measure TRI to check your conjecture.

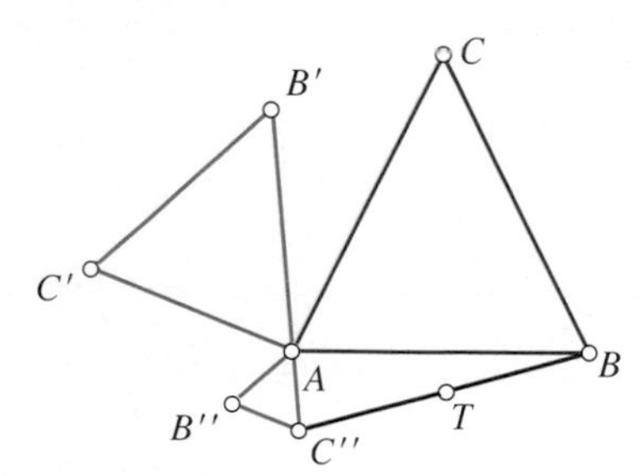

35. Draw a circle. At four points A, B, C, and D on the circle, draw tangent lines to the circle. Let Q, R, S, and T denote the points at which successive pairs of tangent lines intersect, to give a quadrilateral $QRST$ that is circumscribed about the circle.

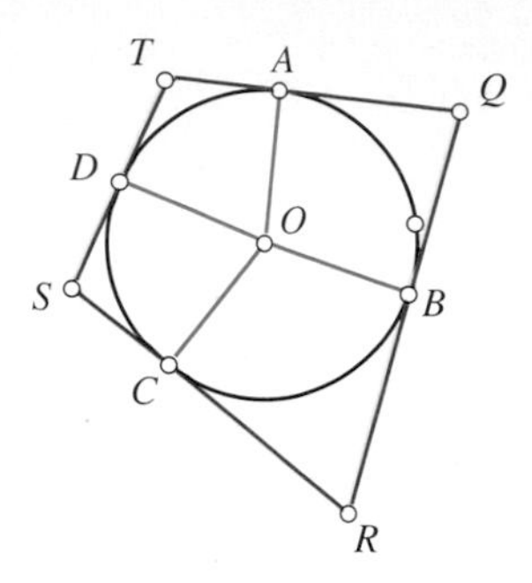

(a) Investigate how the sums of opposite lengths of the sides of $QRST$ compare. Make a conjecture about $QR + ST$ and $RS + QT$.

(b) Prove your conjecture. (*Hint:* Notice that $QA = QB$, $RB = RC$, $SC = SD$, and $TD = TA$.)

For Review

36. Consider the two triangles shown below.

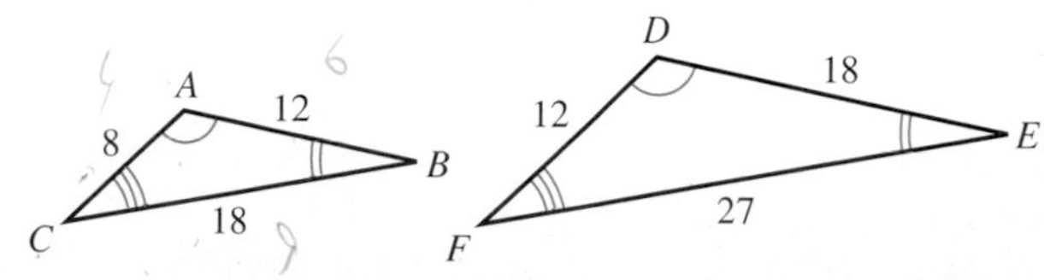

Are the following assertions *true* or *false*? Explain your answer.

(a) Five parts of $\triangle ABC$ are congruent to five parts of $\triangle DEF$.

(b) $\triangle ABC$ is congruent to $\triangle DEF$.

37. The Girl Scout troop needed to determine the width of the river. Alicia followed this plan. First she paced off equally spaced markers at A, B, and C, where A is across the river from a tree at point T. Then Alicia walked 126 feet directly away from the river until she reached point D, at which the tree at T was in line with the stake at B. What is the width of the river, and what geometric property is Alicia relying on?

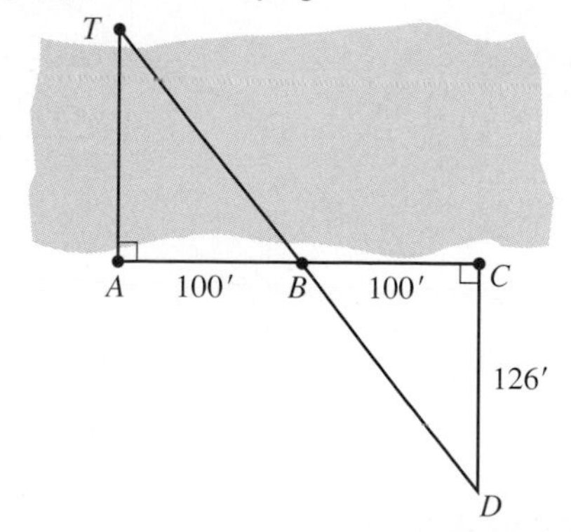

38. Two segments, $\overline{AB}$ and $\overline{CD}$, are congruent, parallel and not on the same line. Prove that the endpoints of the segments are the vertices of a parallelogram.

COOPERATIVE *Investigation* — *Exploring Quadrilaterals and Circles*

Every triangle has both an inscribed circle and a circumscribed circle, but most quadrilaterals have neither an inscribed nor circumscribed circle. Let's investigate how to construct some quadrilaterals that have at least one of these special circles. Work in pairs to discuss and compare your discoveries.

Materials Needed

Spaghetti noodles (uncooked), 3″ × 5″ cards, scissors, drawing and measuring tools (ruler, compass, Mira, and so on)

Exploring Spaghetti Quadrilaterals

Break each of two equally long spaghetti noodles into two pieces. Form a quadrilateral with the four lengths of spaghetti, with opposite sides coming from the same noodle. Put points at A, B, C, and D, and then remove the noodles and draw the quadrilateral $ABCD$.

1. Assuming that the quadrilateral $ABCD$ has an inscribed circle, discuss how you can construct its center and radius. Carry out your procedure: does it appear that $ABCD$ has an inscribed circle?
2. Make another quadrilateral of a different shape but using the same pieces of spaghetti. Does it have an inscribed circle?

Exploring Notecard Quadrilaterals

Cut a 3″ × 5″ note card diagonally into two pieces. Arrange the pieces as shown and use your ruler to extend the sides to form a quadrilateral $APBQ$.

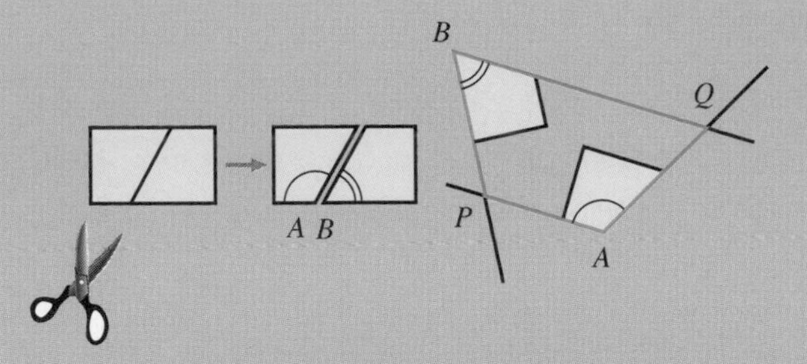

3. Assuming that the quadrilateral $APBQ$ has a circumscribed circle, discuss how you can construct its center. Carry out your procedure: does it appear that there is a circle through all the vertices of $APBQ$?
4. Rearrange the two pieces of notecard to form another quadrilateral of a different shape. Does it have a circumscribed circle?

A Further Investigation

5. Draw a quadrilateral that has both an inscribed and a circumscribed circle. Describe your method, and draw the two circles.

11.3 Similar Triangles

Two figures are **similar** if they have the same shape but not necessarily the same size. For example, an overhead projector forms an image on the screen that is similar to the figure on the transparency. The same is true of a photocopy made on a machine at the enlarge or reduce setting. Figures in space can also be similar to one another. Design engineers often build small scale models of buildings, airplanes, or ships and then make tests and measurements on the model to predict whether or not the design objectives will be met in the full scale structure.

A map or model will indicate how its size compares to actual size by giving a **scale factor.** For example, a ship model may be scaled at $1:100$, meaning that two points at a distance x on the model correspond to points on the real ship at a distance $100x$. Conversely, any length on the actual ship divided by 100 gives the corresponding length for the model. The scale factor is also called the **ratio of similitude.**

The following definition gives an exact description of similarity for triangles.

DEFINITION Similar Triangles and the Scale Factor

Triangle ABC is **similar** to triangle DEF, written $\triangle ABC \sim \triangle DEF$, if, and only if, corresponding angles are congruent and the lengths of corresponding sides have the same ratio. That is, $\triangle ABC \sim \triangle DEF$ if, and only if, $\angle A \cong \angle D$, $\angle B \cong \angle E$, $\angle C \cong \angle F$ and

$$\frac{DE}{AB} = \frac{EF}{BC} = \frac{DF}{AC}.$$

The common ratio of lengths of corresponding sides is called the **scale factor** from $\triangle ABC$ to $\triangle DEF$.

The scale factor from $\triangle DEF$ to $\triangle ABC$ is the reciprocal of the scale factor from $\triangle ABC$ to $\triangle DEF$.

Two examples of similar triangles and their scale factors are shown in Figure 11.15.

It is possible to conclude that two triangles are similar even when we have incomplete information about the sides and angles of the triangles. The most commonly used criteria for similarity are the angle-angle (AA), side-side-side (SSS), and side-angle-side (SAS) properties. *

The Angle-Angle (AA) Similarity Property

In Figure 11.16, $\triangle ABC$ and $\triangle DEF$ have corresponding angles measuring 110°, 40°, and 30°. We see that $\triangle ABC \sim \triangle DEF$. The scale factor can be determined by measuring the lengths of two corresponding sides and forming their ratio. For example, the scale factor is DE/AB.

In general, two triangles with three pairs of congruent angles are similar, an observation known as the **angle-angle-angle (AAA) property** of similar triangles.

* In some books, some or even all of these properties are proved on the basis of other assumptions, making the properties theorems. In other texts, the properties are adopted as postulates. In this text, in keeping with an informal treatment of Euclidean geometry, we will refer to the AA, SSS, and SAS criteria for similarity as *properties*.

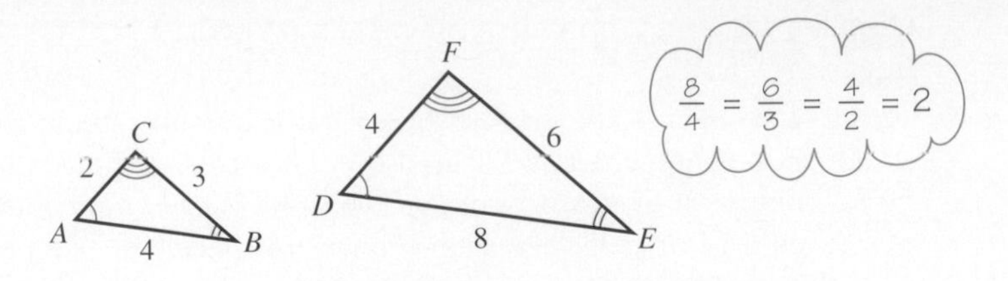

$\triangle ABC \sim \triangle DEF$
Scale factor from $\triangle ABC$ to $\triangle DEF = 2$

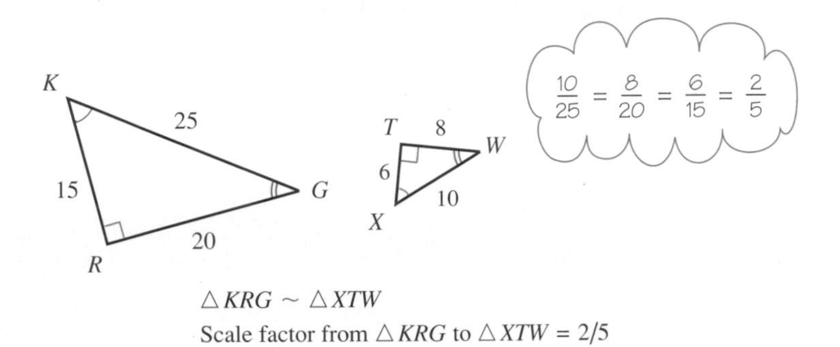

$\triangle KRG \sim \triangle XTW$
Scale factor from $\triangle KRG$ to $\triangle XTW = 2/5$

Figure 11.15
Two pairs of similar triangles

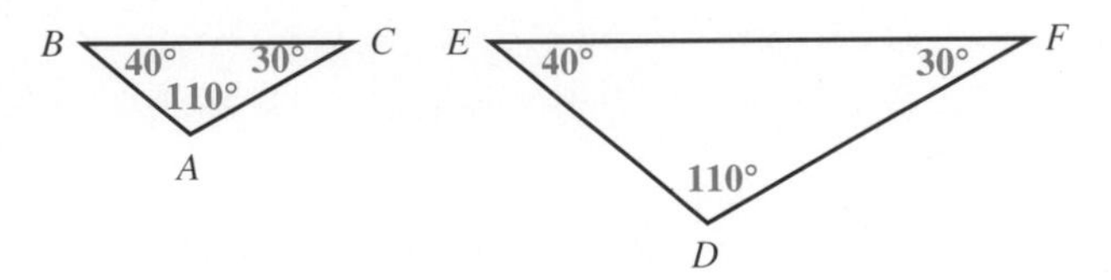

Figure 11.16
Two triangles with three congruent angles are similar by the AAA Similarity Property.

However all three angles of a triangle are determined once we know two of its angles, since the three measures of the angles add up to 180°. Therefore the property is called the **angle-angle (AA) property of similarity.**

PROPERTY The AA Similarity Property

If two angles of one triangle are congruent respectively to two angles of a second triangle then the triangles are similar.

EXAMPLE 11.10 **Making an Indirect Measurement with Similarity**

A tree at point T is in line with a stake at point L when viewed from point N. Use the information in the diagram to measure the distance across the river.

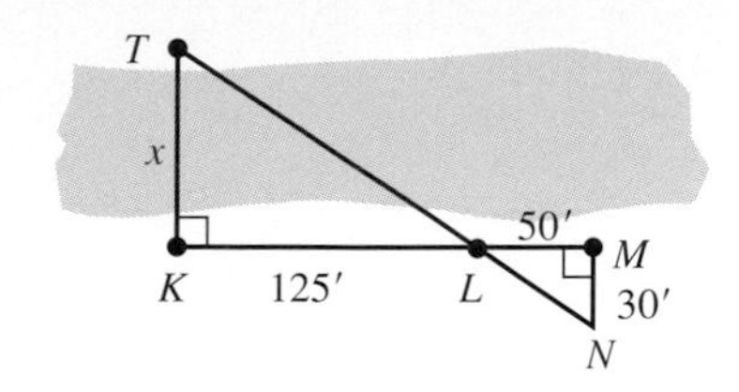

SOLUTION

By the vertical angles theorem, $\angle TLK \cong \angle NLM$. Also $\angle K \cong \angle M$ since both are right angles. By the AA similarity property, $\triangle TLK \sim \triangle NLM$. Thus $\frac{TK}{NM} = \frac{KL}{ML}$, since the lengths of corresponding sides have the same ratio. Since $TK = x$, $NM = 30'$, $KL = 125'$, and $ML = 50'$ this gives the proportion $x/30' = 125'/50'$. Therefore, $x = (30') \cdot (125')/50' = 75'$. We have found the distance across the river by an indirect measurement using similar triangles. ■

The Side-Side-Side (SSS) Similarity Property

PROPERTY The SSS Similarity Property

If the three sides of one triangle are proportional to the three sides of a second triangle, then the triangles are similar. That is, if $\frac{DE}{AB} = \frac{EF}{BC} = \frac{DF}{AC}$ then $\triangle ABC \sim \triangle DEF$.

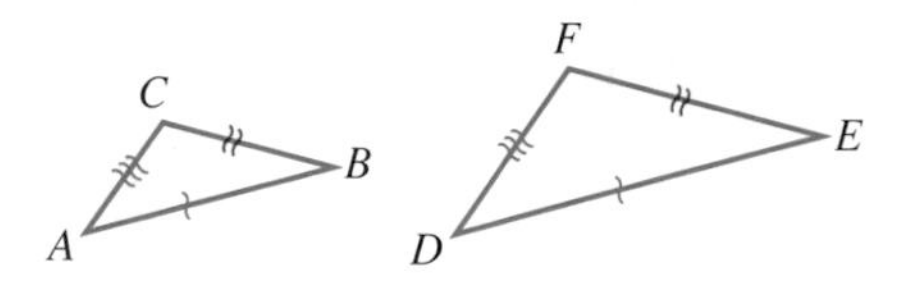

EXAMPLE 11.11

Applying the SSS Similarity Property

A contractor wishes to build an L-shaped concrete footing for a brick wall, with a 12 foot leg of the wall meeting a 10 foot leg of the wall at a right angle. The contractor knows that a 3 by 4 by 5 foot triangle has a right angle opposite the 5 foot side. How can the contractor place stakes at points X, Y, and Z to form a right angle at point Y? The wall will be built along string lines stretched from X to Y and from Y to Z.

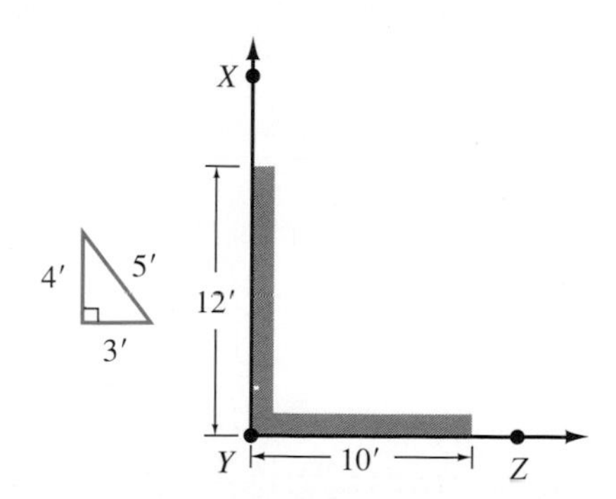

SOLUTION

By the SSS similarity property, the 3–4–5 foot right triangle can be magnified by a convenient scale factor to give an accurate right triangle. A scale factor of 4 gives a 12–16–20 foot right triangle. The contractor can place a stake at X that is 16 feet from the corner point Y. Using two measuring tapes, a stake is placed at the point Z where the 20 foot mark on the tape from X crosses the 12 foot mark on the tape from Y.

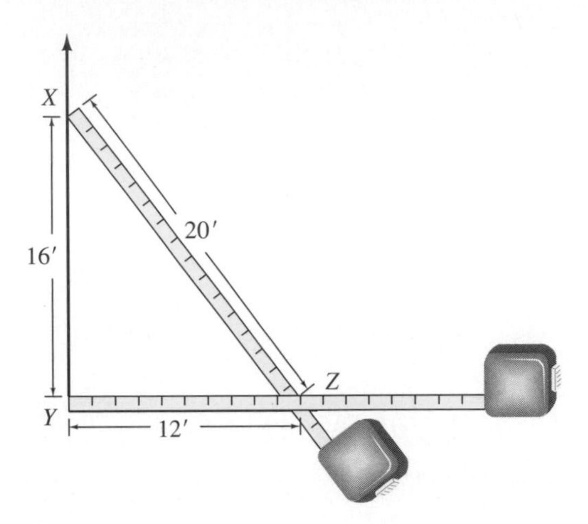

■

WORTH READING

The Navaho's Organization of Space and Time

The Navajo believe in a dynamic universe. Rather than consisting of objects and situations, the universe is made up of processes. Central to our Western mode of thought is the idea that things are separable entities that can be subdivided into smaller discrete units. For us, things that change through time do so by going from one specific state to another specific state. While we believe time to be continuous, we often even break it into discrete units or freeze it and talk about an instant or point in time. Or, we often just ignore time. For example, when we speak of a boundary line dividing a surface into two parts or a line being divided by a point, we are describing a static situation, that is, one in which time plays no role whatever. Among the Navajo, where the focus is on process, change is everpresent; interrelationship and motion are of primary significance. These incorporate and subsume space and time. . . . To us, the significant aspect of a boundary is that it is a spatial divider; to the Navajo, the significance is the processes of which the boundary is a part and how it affects and is being affected by those processes. The Navajo react quite negatively when fences are placed upon their reservation land. One major reason is their belief that space should *not* be segmented in an arbitrary and static way.

These excerpts, contrasting Western and Navajo organization of space and time, are from Marcia Ascher's book *Ethnomathematics, A Multicultural View of Mathematical Ideas.* The book gives a fascinating introduction to multicultural mathematical ideas from peoples such as the Inuit, Navajo, and Iroquois of North America; the Incas of South America; the Malekula, Warlpiri, Maori, and Caroline Islanders from Oceania; and the Tshokwe, Bushoong, and Kpelle of Africa.

SOURCE: Marcia Ascher, *Ethnomathematics: A Multicultural View of Mathematical Ideas.* Pacific Grove, CA: Brooks/Cole Publishing Company, 1991, pp. 128–29.

JUST FOR FUN

Thales' Puzzle

Thales is reputed to have calculated the height of the pyramids in Egypt by comparing the shadow cast by the pyramid to the shadow cast by a vertical stick. By similar right triangles $H/h = y/x$, where H is the height of the pyramid, h is the height of the stick, x is the length of the shadow of the stick, and $y = PB$ is the length of the shadow of the pyramid from a point P on the ground to the point B directly beneath apex A. Thus, the height of the pyramid is given by the formula $H = yh/x$. Unfortunately, Thales still has a difficulty: he can easily measure h and x, but y is not simple to measure since point B is somewhere *inside* the pyramid!

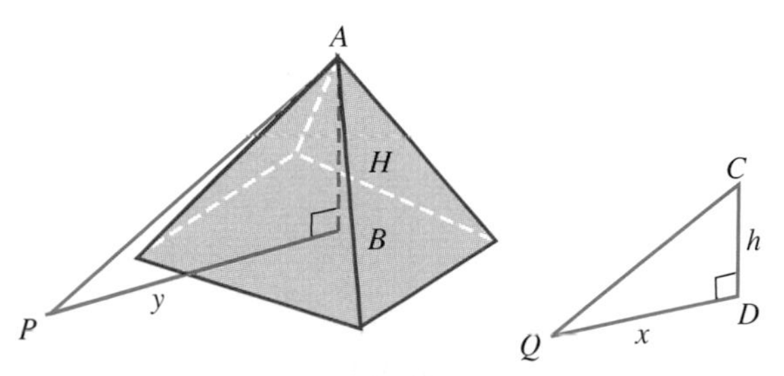

After marking points P and Q one morning, Thales returned a few hours later and realized he could now calculate the height of the pyramid. How did Thales solve his problem?

The Side-Angle-Side (SAS) Property

PROPERTY The SAS Similarity Property

If, in two triangles, the ratios of any two pairs of corresponding sides are equal and the included angles are congruent, then the two triangles are similar. That is, if $\frac{DE}{AB} = \frac{DF}{AC}$ and $\angle A \cong \angle D$ then $\triangle ABC \sim \triangle DEF$.

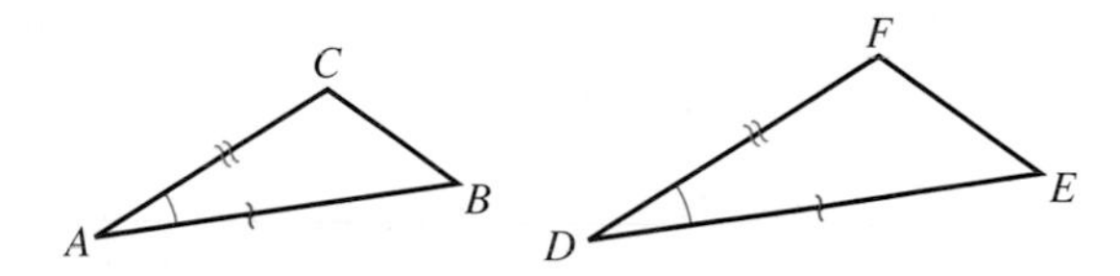

EXAMPLE 11.12

Applying the SAS Similarity Property

The sun is about 93 million miles from Earth and the moon is about 240,000 miles distant. If the diameter of the moon is 2200 miles, what is the approximate diameter of the sun? (*Hint:* From the earth, the sun and moon appear to have the same diameter.)

SOLUTION

Because the moon (M) appears to have the same diameter as the sun (S) during an eclipse, they form nearly congruent angles when viewed from

Earth (E). The above illustration is far from a true scale drawing, but it does show that $\triangle EM_1M_2 \sim \triangle ES_1S_2$ by the SAS similarity property. Thus,

$\dfrac{S_1S_2}{M_1M_2} = \dfrac{ES_1}{EM_1}$, so $S_1S_2 = \dfrac{ES_1}{EM_1} \cdot M_1M_2 = \dfrac{93{,}000{,}000}{240{,}000} \cdot 2200$ miles $= 852{,}500$ miles. This estimate compares well with the sun's actual diameter of 864,000 miles given by more accurate methods. ■

Geometric Problem Solving Using Similar Triangles

The following sequence of examples illustrates how similar triangles can be used to explore the properties of geometric figures. These examples are unified by exploring figures constructed by joining the midpoints of sides of triangles or quadrilaterals.

EXAMPLE 11.13 Exploring the Medial Triangle

If X, Y, and Z are the midpoints of the sides of $\triangle ABC$, then $\triangle XYZ$ is called the **medial triangle** of $\triangle ABC$.

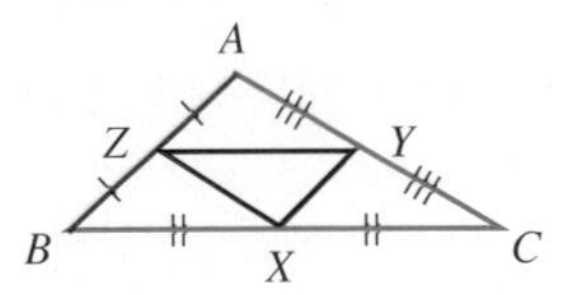

Show that:

(a) Each side of $\triangle XYZ$ is parallel to and half the length of the side of $\triangle ABC$ through the opposite vertex. For example, $\overline{XY} \parallel \overline{AB}$ and $XY = AB/2$.

(b) Show that the four triangles $\triangle XYZ$, $\triangle AZY$, $\triangle BZX$, and $\triangle YXC$ are congruent.

SOLUTION

(a) The sides $\overline{AC}$ and $\overline{BC}$ of $\triangle ABC$ are twice the length of the respective sides $\overline{YC}$ and $\overline{XC}$ of $\triangle YCX$. Since these triangles have the same included angle, $\angle C$, we conclude that $\triangle ABC \sim \triangle YXC$ by the SAS similarity property. The scale factor from $\triangle ABC$ to $\triangle YXC$ is $1/2$, so $XY = AB/2$. Also $\angle BAC \cong \angle XYC$ and so $\overline{BA} \parallel \overline{XY}$ by the corresponding angles theorem for parallel lines. The analysis just completed for side $\overline{XY}$ can be repeated for the other two sides of $\triangle XYZ$.

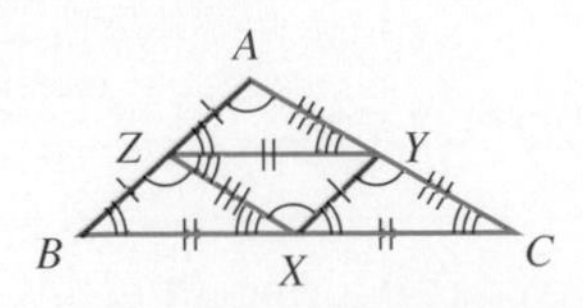

(b) The four triangles are congruent by the SSS congruence theorem. ■

EXAMPLE 11.14

Classifying the Midpoint Figure of a Quadrilateral

Two quadrilaterals are shown. It appears that joining successive midpoints of the sides of these quadrilaterals forms a parallelogram. Prove that this is indeed the case.

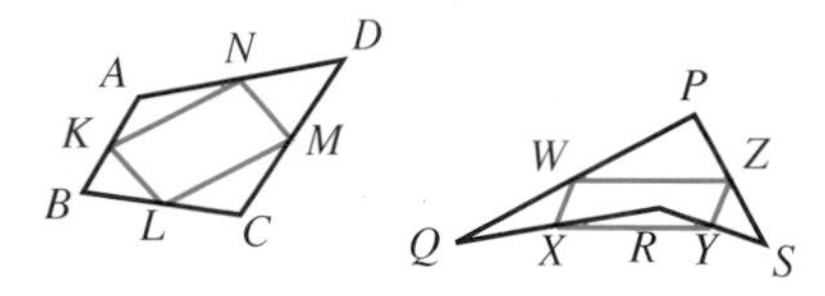

SOLUTION

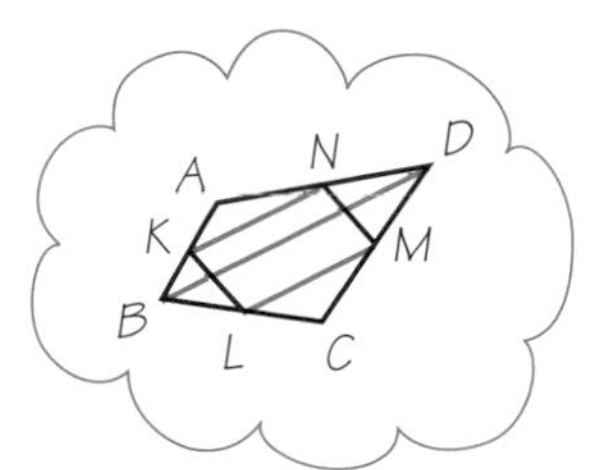

Let $KLMN$ be the midpoint figure of the quadrilateral $ABCD$. Draw the diagonal $\overline{BD}$ and consider $\triangle ABD$ and $\triangle CBD$. By Example 11.13 we know $\overline{KN}$ and $\overline{LM}$ are both parallel to $\overline{BD}$ and have length $BD/2$. But then $\overline{KN}$ and $\overline{LM}$ are congruent and parallel segments. The same reasoning shows $\overline{KL}$ and $\overline{NM}$ are congruent and parallel, since both segments are parallel to $\overline{AC}$ and have length $AC/2$. By definition, $KLMN$ is a parallelogram. ■

The proof just given remains valid for *space* quadrilaterals, for which the four vertices are not necessarily in the same plane. For example, the quadrilateral $PQRS$ shown above may be easily visualized as a nonplanar quadrilateral. However, the midpoint quadrilateral $WXYZ$ is a parallelogram, so it lies in a plane. It's interesting to confirm this result with a quadrilateral made of sticks whose midpoints are joined by elastic bands to form the midpoint parallelogram.

EXAMPLE 11.15

Discovering the Centroid of a Triangle

A **median** of a triangle is a line segment joining a vertex to the midpoint of the opposite side as shown. Prove that the three medians of a triangle are concurrent at a point G which is 2/3 of the distance along each median from the vertex toward the midpoint. G is the **centroid** or **center of gravity** of the triangle.

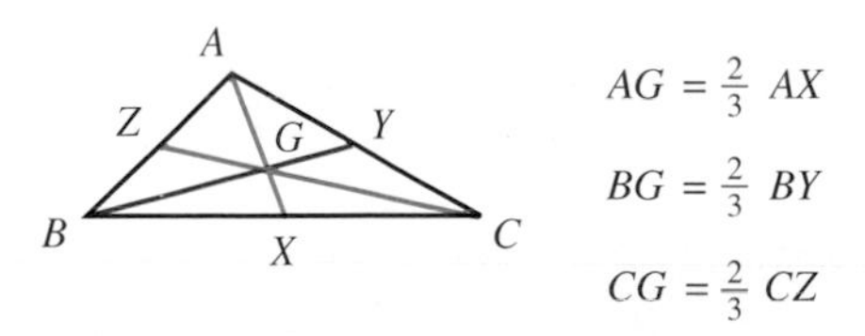

SOLUTION

Understand the problem • • • •

There are really *two* problems to be solved. First, there is a *distance* problem: if G is the point where the two medians $\overline{BY}$ and $\overline{CZ}$ intersect, then we must show that $BG = 2/3BY$, or equivalently $BG = 2GY$. Second, there is a *concurrence* problem: if G is the point of intersection of $\overline{BY}$ and $\overline{CZ}$, then we must show that the third median $\overline{AX}$ also passes through G.

Devise a plan • • • •

Since we hope to show $BG = 2GY$ and $CG = 2GZ$, it may be useful to consider the midpoints M of $\overline{BG}$ and N of $\overline{CG}$. The distance problem for medians $\overline{BY}$ and $\overline{CZ}$ will be solved if it can be shown that M and G trisect $\overline{BY}$, and N and G trisect $\overline{CZ}$. Since Z, M, N, and Y are the successive midpoints of the quadrilateral $ABGC$, we should gain important information by constructing the midpoint figure $ZMNY$, which we know is a parallelogram by Example 11.14.

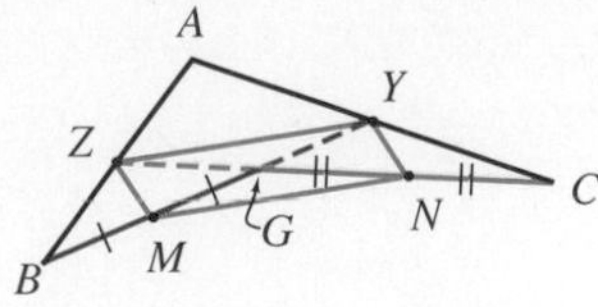

Carry out the plan • • • •

Because $ZMNY$ is a parallelogram, the point G at which the diagonals intersect is the midpoint of the diagonal $\overline{MY}$ of the parallelogram. Thus $MG = GY$. But M is the midpoint of $\overline{BG}$, so $BM = MG$. This shows that G is $2/3$ of the distance from B to Y along the median $\overline{BY}$. The same reasoning shows that G and N trisect the median $\overline{CZ}$. The concurrence problem is now solved by symmetry: if G' is the point of intersection of the medians $\overline{BY}$ and $\overline{AX}$, then G' is $2/3$ of the distance from a vertex along either median; therefore, $G = G'$.

Look back • • • •

The problem-solving strategies used in this example are likely to be helpful for other problems:

- *Divide the problem into simpler parts:* we solved a distance problem and a concurrence problem.
- *Consider a simpler problem first:* G was defined as the intersection of *two* medians, and it was to be shown G was $2/3$ of the distances along the two medians from the vertices.
- *Use a related result:* the previous example, showing that the midpoints of sides of any quadrilateral formed a parallelogram, was a key idea in the solution. ■

PROBLEM SET 11.3

Understanding Concepts

1. Which of the following pairs of triangles are similar? If they are similar, explain why, express the similarity with the ~ notation, and give the scale factor.

(a)

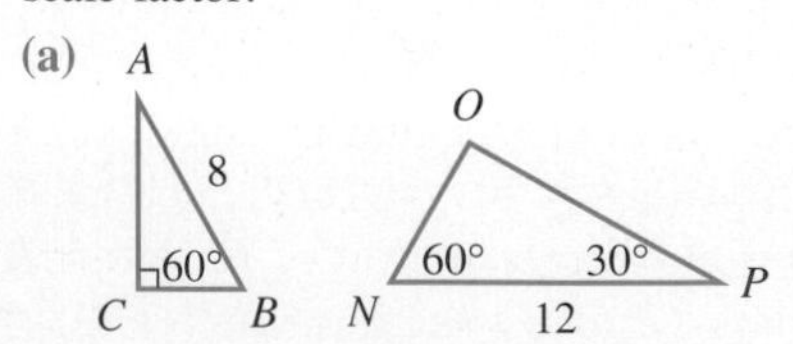

(b)

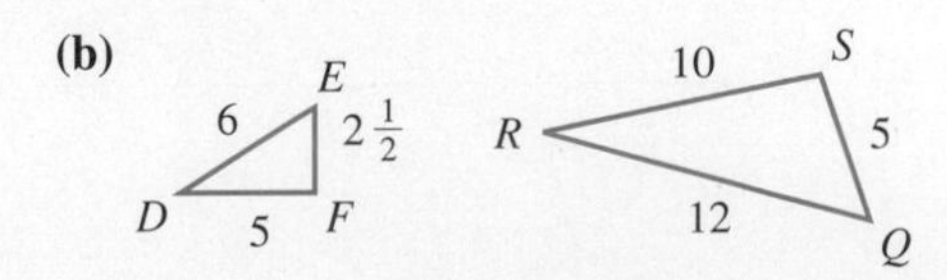

(c)

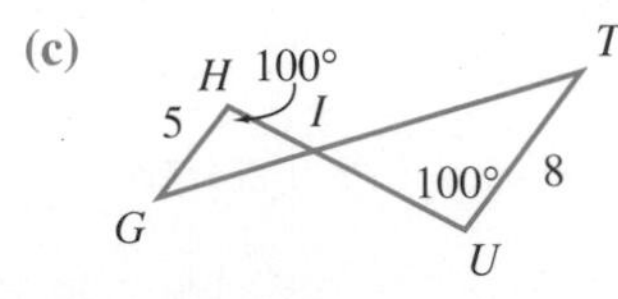

(d)

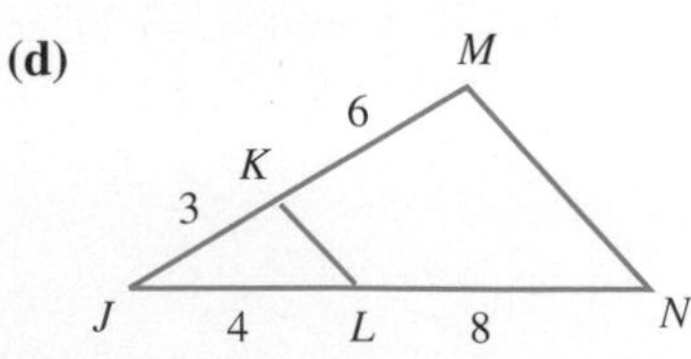

(e)

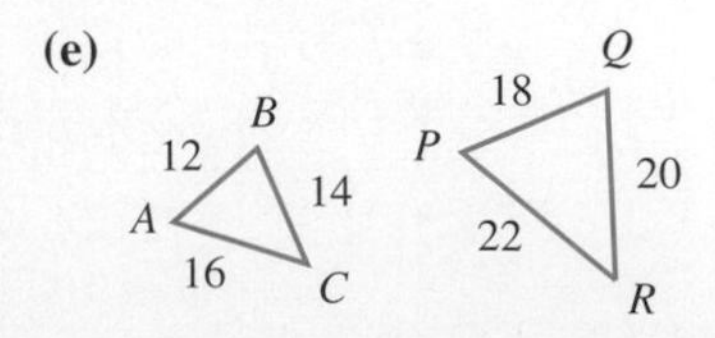

2. Are the following figures necessarily similar? If so, explain why. If not, draw an example to show why not.
   (a) Any two equilateral triangles.
   (b) Any two isosceles triangles.
   (c) Any two right triangles having an acute angle of measure 36°.
   (d) Any two isosceles right triangles.
   (e) Any two congruent triangles.
   (f) A triangle with sides of length 3 and 4 and an angle of 30° and a triangle with sides of length 6 and 8 and an angle of 30°.
3. A pair of similar triangles is shown in each part. Find the measures of the segments marked with a letter a, b, c, or d.
   (a)

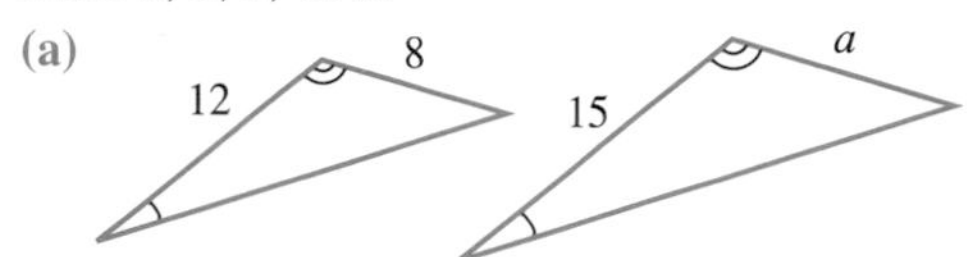

   (b)

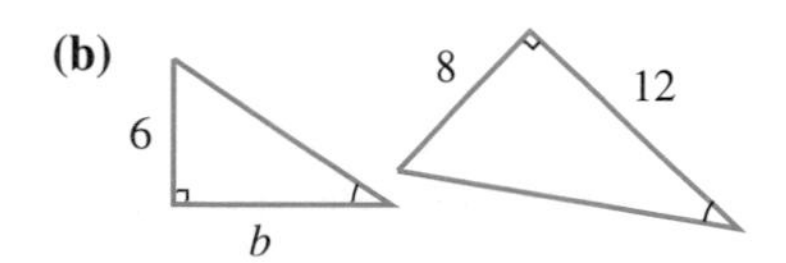

   (c) **(d)**

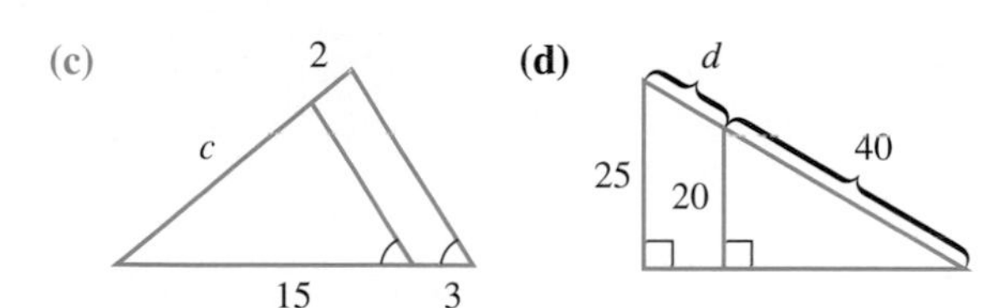

4. (a) Two convex quadrilaterals, $ABCD$ and $EFGH$, have congruent angles at their corresponding vertices: $\angle A \cong \angle E$, $\angle B \cong \angle F$, $\angle C \cong \angle G$, $\angle D \cong \angle H$. Can you conclude that the two convex quadrilaterals are similar? Explain.
   (b) Two convex quadrilaterals have corresponding sides in the same ratio. Are the quadrilaterals necessarily similar? Explain.
5. Suppose $\triangle ABC \sim \triangle DEF$, where only points D and E are shown. Find all possible locations for F, and draw the corresponding triangles DEF.

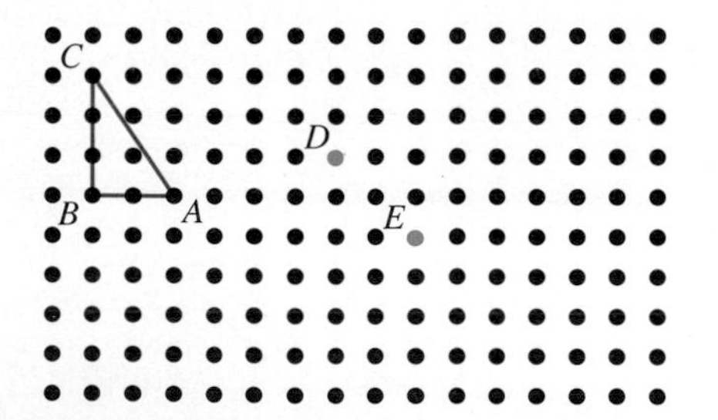

6. The diagonals of the trapezoid $ABCD$ shown below intersect at E, where $\overline{AB} \parallel \overline{CD}$.
   (a) Explain why $\triangle ABE \sim \triangle CDE$.
   (b) Determine x and y.

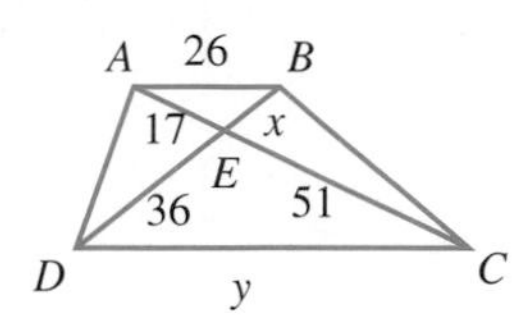

7. Let $\overline{AB}$ and $\overline{CD}$ be parallel, and let $\overline{AD}$ and $\overline{BC}$ intersect at E. Prove that $a \cdot y = x \cdot b$.

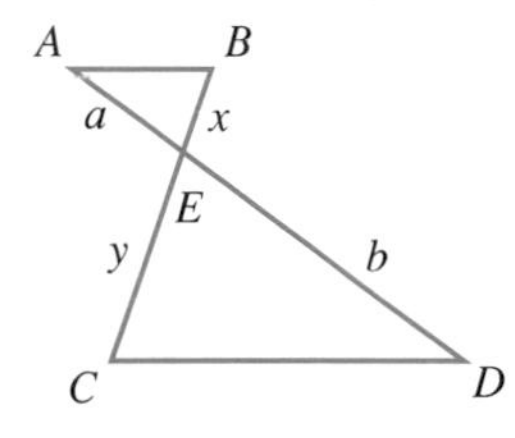

8. $\triangle ABC$ is a right triangle with altitude $\overline{CD}$.
   (a) Explain why $\triangle ADC \sim \triangle CDB$.
   (b) Find an equation for h and solve it to show that $h = \sqrt{9 \cdot 25} = 3 \cdot 5 = 15$.

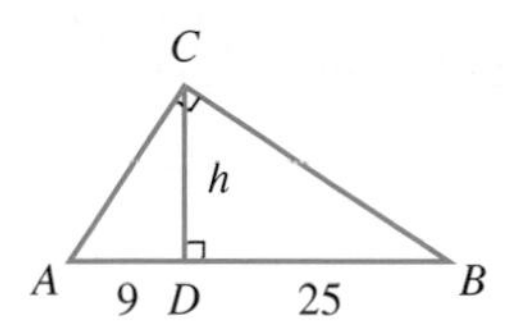

9. An isosceles triangle whose apex angle measures 36° is sometimes called a **golden triangle.**

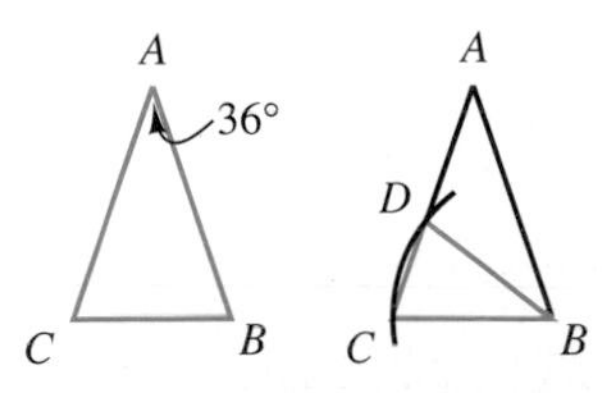

   (a) Draw an arc centered at B and passing through C. If D is the point at which the arc intersects $\overline{AC}$, prove that $\triangle BCD$ is also a golden triangle. (*Hint:* What is $m(\angle C)$?)
   (b) Construct three more golden triangles CDE, DEF, and EFG, where each contains the next.
10. Lined notebook paper provides a convenient way to subdivide a line segment into a specified number of congruent subsegments. The diagram shows how swinging an arc of radius AB subdivides the segment into five congruent parts: $\overline{AX} \cong \overline{XY} \cong \overline{YZ} \cong \overline{ZW} \cong \overline{WB}$.

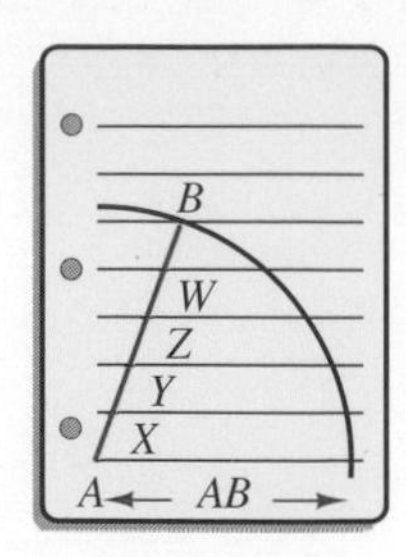

(a) Show how to subdivide a segment $\overline{AB}$ into seven congruent segments.

(b) Carefully explain why the procedure is valid.

11. (a) If $\overline{AB}$ and $\overline{XU}$ in the diagram shown are parallel, find $x = OX$ in terms of a and b.

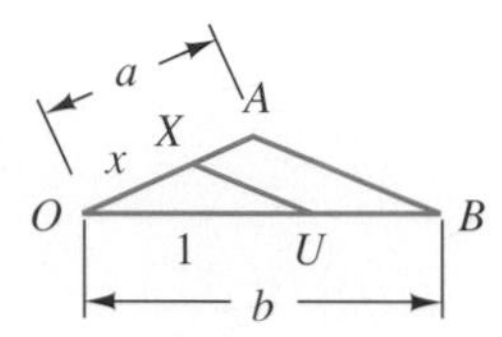

(b) Given segments of length a and b and a unit segment of length 1, describe a procedure to construct a segment of length a/b with compass and straightedge.

12. (a) Find y, in terms of a and b, where $\overline{UA}$ and $\overline{YB}$ in the diagram shown are parallel.

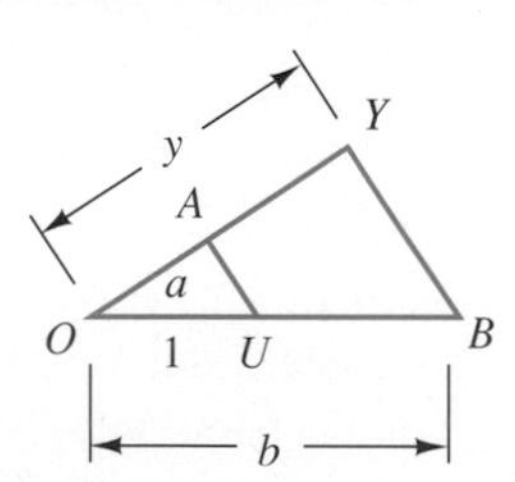

(b) Given segments of length a and b, and a unit segment of length 1, describe a procedure to construct a segment of length $a \cdot b$ with compass and straightedge.

Thinking Critically

13. Let $\overline{CD}$ be the altitude drawn to the hypotenuse of the right triangle ABC shown.

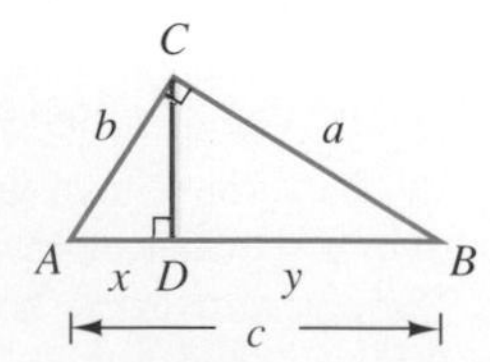

$a = BC,\quad b = AC,\quad c = AB$
$x = AD,\quad y = DB$

(a) Explain why $\triangle ABC \sim \triangle ACD$ and $\triangle ABC \sim \triangle CBD$.

(b) Explain why $\dfrac{x}{b} = \dfrac{b}{c}$ and $\dfrac{y}{a} = \dfrac{a}{c}$.

(c) Use part (b) to show that $c^2 = a^2 + b^2$. (This gives a proof of the Pythagorean theorem.)

14. Prove that the square shown inscribed in a right triangle has sides of length $x = \dfrac{ab}{a + b}$, where a and b are the lengths of the legs of the triangle.

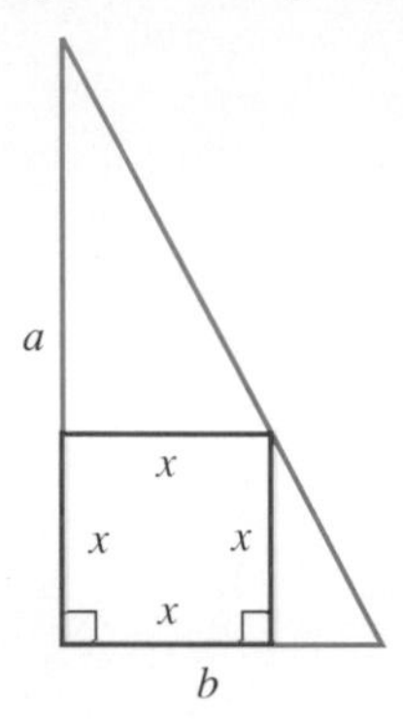

15. Let $\triangle ABC$, as illustrated in the diagram, have an angle bisector at A which intersects side $\overline{BC}$ at D. The diagram suggests that D divides the side $\overline{BC}$ opposite the angle in the same ratio as the lengths of the respective sides adjacent to the angle. That is, the angle bisector gives the proportion $DB/DC = AB/AC$. Show this is true by justifying the four statements below.

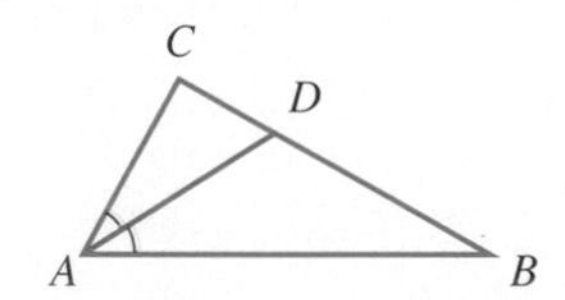

(a) Let l be the line parallel to $\overline{AD}$ passing through C, and let P be its intersection with the extension of side $\overline{AB}$. Then $\angle APC \cong \angle ACP$. (Why?)

(b) $PA = AC$ (Why?)

(c) $\dfrac{PA}{AB} = \dfrac{CD}{DB}$ (Why?)

(d) $\dfrac{DB}{DC} = \dfrac{AB}{AC}$ (Why?)

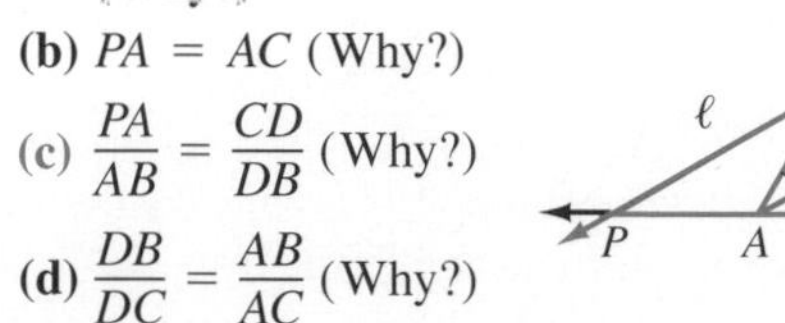

16. Let three arbitrary perpendiculars to the sides of $\triangle ABC$ of the diagram be drawn, meeting in pairs at the points P, Q, and R. Prove that $\triangle PQR \sim \triangle ABC$.

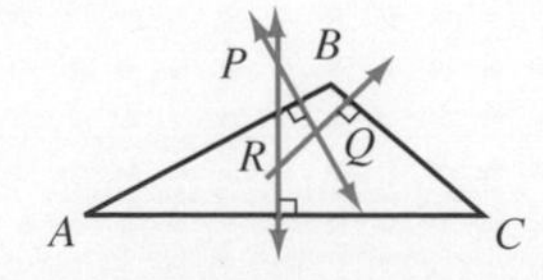

17. (a) Let $PQRS$ be a space quadrilateral, and let W, X, Y, Z be the midpoints of successive sides. Explain why $\overline{WY}$ and $\overline{XZ}$ intersect at their common midpoint M.

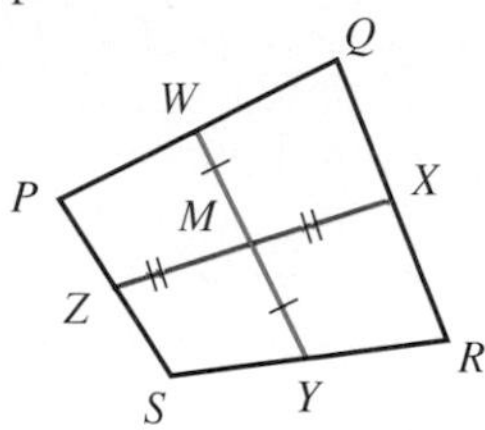

(b) A **bimedian** of a tetrahedron is segment joining the midpoint of one edge to the midpoint of the opposite (nonadjacent) edge of the tetrahedron. Explain why the three bimedians of a tetrahedron are concurrent at their midpoints.

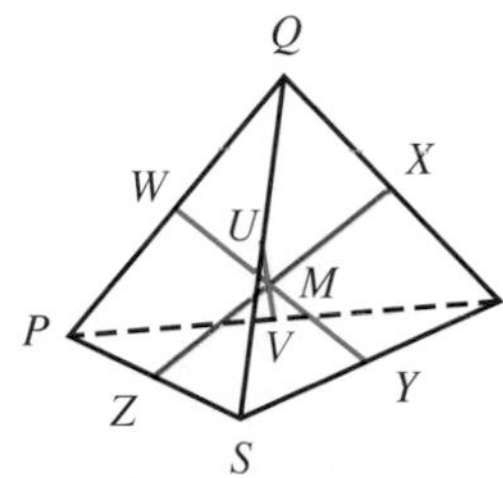

18. Let $ABCD$ be a trapezoid as illustrated, with bases of length $a = AB$ and $b = CD$. Let the diagonals $\overline{AC}$ and $\overline{BD}$ intersect at P, and suppose $\overline{EF}$ is the segment parallel to the bases that passes through P. Show that $EP = FP = \dfrac{ab}{a + b}$. (Thus $EF = \dfrac{2ab}{a + b}$, which is called the **harmonic mean** of a and b.)

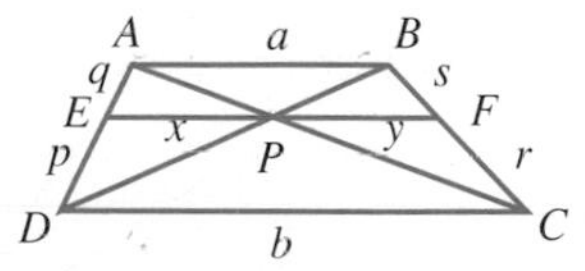

(*Hint:* Explain why $x/a = p/(p + q)$ and $x/b = q/(q + p)$. What happens when these equations are added?)

19. Let M and N be the midpoints of the sides of parallelogram $ABCD$ opposite A as shown. Show that $\overline{AM}$ and $\overline{AN}$ divide the diagonal $\overline{BD}$ into 3 congruent segments: $BP = PQ = QD$. (*Hint:* Construct $\overline{AC}$, and see Example 11.15.)

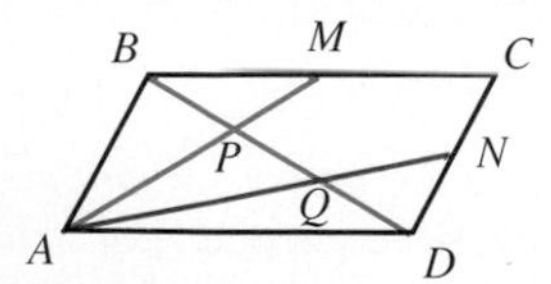

20. Let M be the midpoint of the median AD of $\triangle ABC$ as illustrated, and let the ray $\overrightarrow{CM}$ intersect $\overline{AB}$ at T. Show that T trisects $\overline{AB}$; that is, $AT = (1/3)AB$.

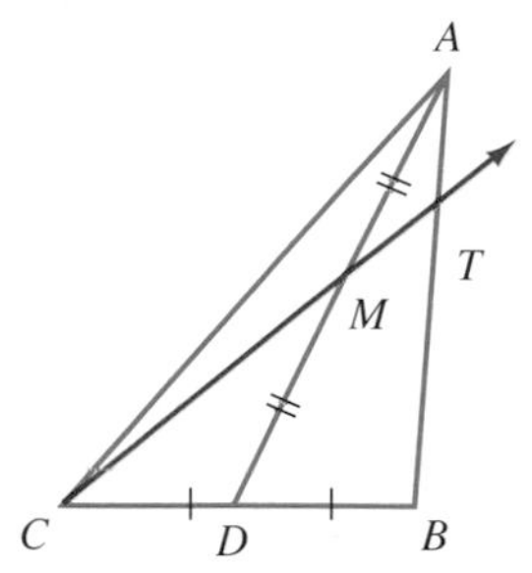

(*Hint:* Construct a line through B parallel to $\overline{AD}$, and let it intersect the extension of $\overline{CA}$ at a point labeled P. What is special about T relative to $\triangle PBC$?)

21. Use similarity to find the distances AP, BP, CP, and DP in the figure shown. The squares on the lattice have sides of unit length.

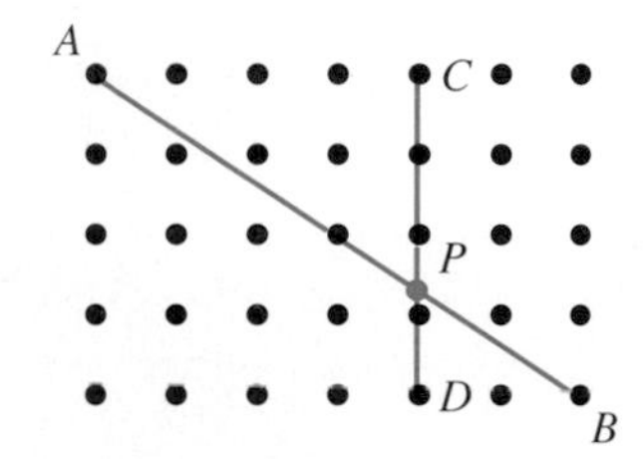

Making Connections

22. Mingxi is standing 75 feet from the base of tree. The shadow from the top of Mingxi's head coincides with the shadow from the top of the tree. If Mingxi is 5′9″ tall and his shadow is 7′ long, how tall is the tree?

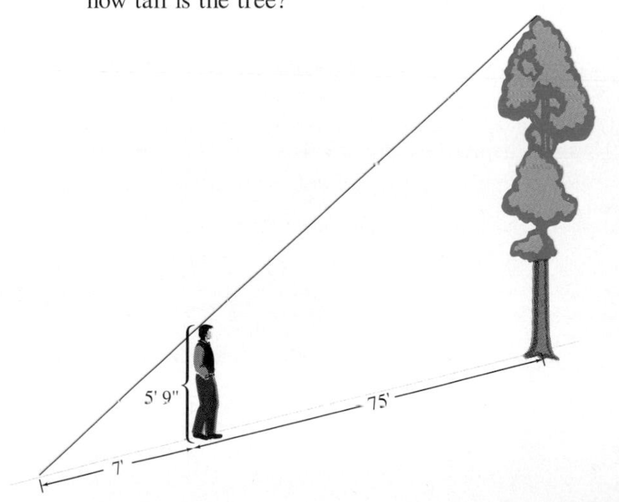

23. Mohini laid a mirror on the level ground 15 feet from the base of the pole as shown. Standing 4 feet from the mirror, she can see the top of the pole reflected in the mirror. If Mohini is 5′5″ tall, how can she estimate the height of the pole? What must she allow for in her calculation?

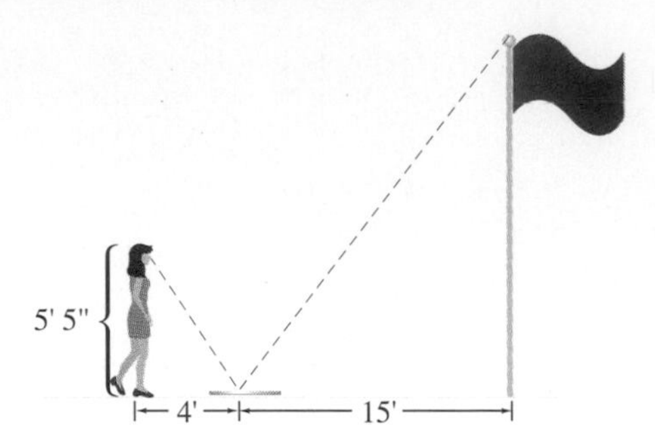

24. By holding a ruler 2 feet in front of her eyes as illustrated in the diagram, Ginny sees that the top and bottom points of a vertical cliff face line up with marks separated by 3.5″ on the ruler. According to the map, Ginny is about a half mile from the cliff. What is the approximate height of the cliff? Recall that a mile is 5280 feet.

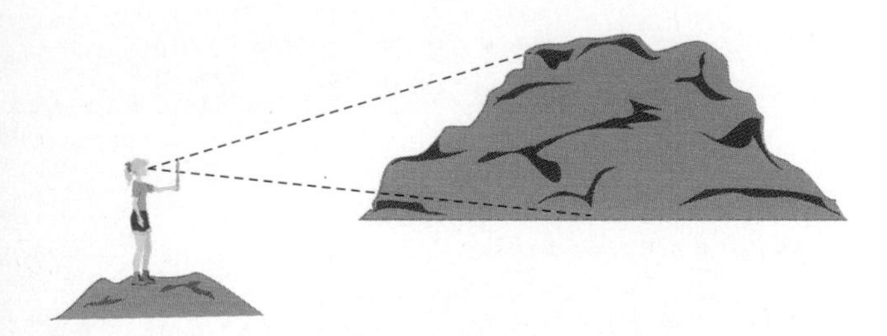

25. A vertical wall 18 feet high casts a shadow 6 feet wide on level ground. If Lisa is 5′3″ tall, how far away from the wall can she stand and still be entirely in the shade?

26. A drive from Prineville to Queenstown currently requires passing through Renton, due to a steep intervening hill. This is shown on the rough map drawn below. The two roads are level and straight, and meet at 55° at Renton.

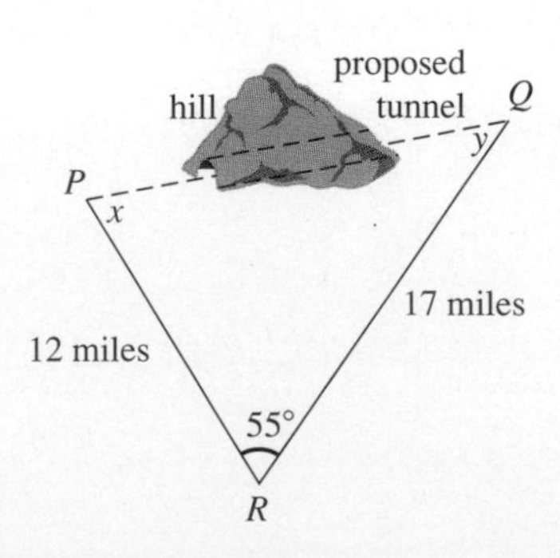

(a) How many miles would be saved by boring a tunnel through the hill? Make an accurate scale drawing, and take measurements off of your drawing.

(b) Two construction crews will dig the tunnel from opposite sides of the hill. What angle measures, x and y, will ensure that the two crews meet properly at the center of the hill? Take measurements from your scale drawing to give accurate estimates.

Using a Computer
Using an Automatic Drawer

27. Draw a circle and two chords, $\overline{AB}$ and $\overline{CD}$, that intersect at a point P. Measure the angles in the triangles BCP and DAP.

(a) What conclusion about $\triangle BCP$ and $\triangle DAP$ is suggested by your measurements?

(b) Measure the lengths of $\overline{PA}$, $\overline{PB}$, $\overline{PC}$, and $\overline{PD}$, and then compare $PA \cdot PB$ with $PC \cdot PD$. Does your answer to part (a) justify your observation about the two products?

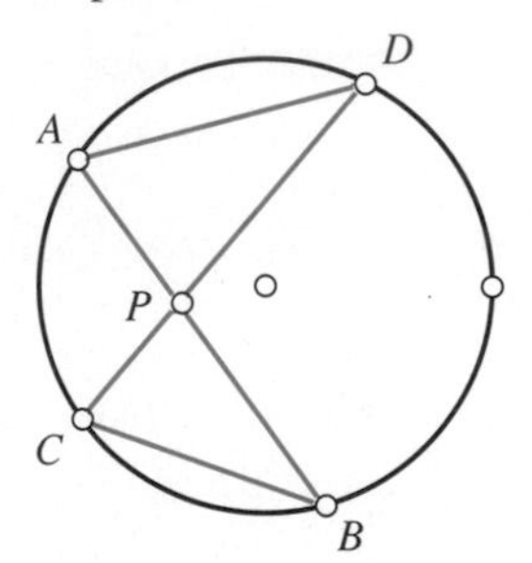

28. Draw a circle and a point P that is outside of the circle as shown. Next, draw two rays from P that intersect the circle at the points A, B, C, and D. Measure the angles in $\triangle BCP$ and $\triangle DAP$.

(a) What conclusion about $\triangle BCP$ and $\triangle DAP$ is suggested by your measurements?

(b) Measure the lengths of $\overline{PA}$, $\overline{PB}$, $\overline{PC}$, and $\overline{PD}$, and then compare $PA \cdot PB$ with $PC \cdot PD$. Does your answer to part (a) justify your observation about the two products?

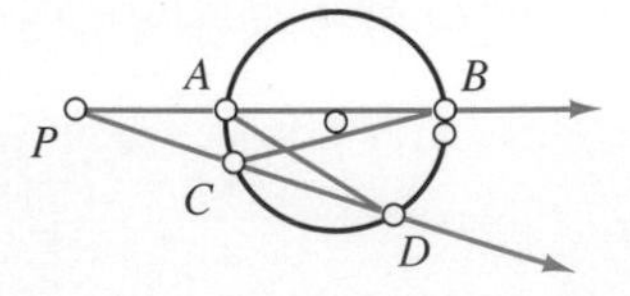

Communicating

29. (a) In Example 11.14 it was shown that joining the successive midpoints of a quadrilateral formed a parallelogram. Explore what you find by *reversing* the construction. That is, given a

parallelogram $KLMN$ can you construct a quadrilateral $ABCD$ for which $KLMN$ is the midpoint figure? Is $ABCD$ unique?

(b) Write a report discussing your results of part (a).

For Review

30. Find three pairs of congruent triangles in this figure.

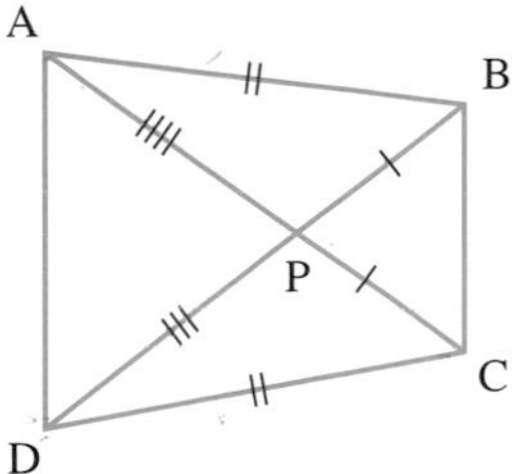

31. Find the measures of the interior angles of $\triangle RST$.

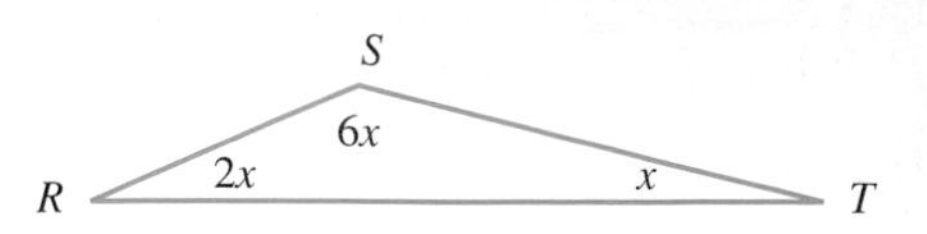

32. Let A and B be the points of tangency from P to a circle centered at C as shown. Prove that $PA = PB$. (*Hint:* Draw the segment $\overline{PC}$.)

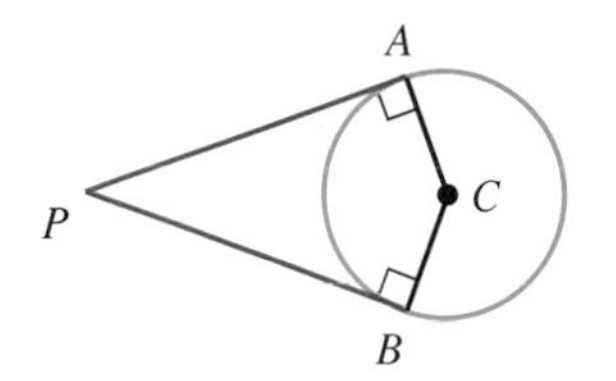

COOPERATIVE Investigation *Explorations with Centers of Gravity*

The **center of gravity,** or **centroid,** of a figure is its point of balance. A plane figure, cut from cardboard, will hang horizontally when it is suspended by a string pinned at the centroid. If pushed slowly over the edge of a table, the figure will begin to teeter just as the centroid reaches the table's edge.

Materials Needed

Heavy cardboard, scissors, ruler, pins, and strings

Directions

1. Cut a large (sides ranging from 6″ to 12″) triangle from cardboard. Hang the triangle from string pinned at some point of the triangle (alternatively, balance the figure on the upright point of a pin). Move the pin to new points to determine the centroid experimentally. Balance the triangle on the edge of a table, and see if the point you found experimentally lies directly over the table's edge.

2. Draw the medians on the triangle, and see if they intersect at the point found experimentally.
3. Cut a large convex quadrilateral from cardboard, and find its centroid experimentally with both the string and table edge methods.
4. Devise a method to *draw* the centroid of a convex quadrilateral. (*Hint:* Each diagonal of the quadrilateral forms two triangles.)
5. Use the method you've devised to draw the centroids of the cardboard quadrilaterals. Do your points coincide with the experimentally found centroids?

Archimedes, by the way, made many discoveries about the center of gravity for both plane and solid figures. He was well aware that the medians of a triangle intersected at the centroid.

EPILOGUE From Compasses to Computers

In the tradition of Euclid only two instruments were allowed when constructing a geometric figure—the straightedge and the compass. Within the strict limitations of the Euclidean rules of construction, it is still possible to draw a large number of useful figures, such as angle bisectors, parallel and perpendicular lines, midpoints, and perpendicular bisectors. In this chapter we have used compass and straightedge constructions to motivate the properties of congruent and similar triangles. We have also seen practical applications of congruence and similarity, such as to the indirect measurement of angles and distances.

In the early nineteenth century the limitations of the compass and straightedge were made clear with the results of Gauss, Wantzel, and others. It is easy to trisect any line segment, but not all angles can be trisected by Euclidean methods. Any procedure which claims otherwise is flawed: either the construction is not Euclidean, or else it is an approximation.

Non-Euclidean drawing tools, such as rulers and protractors, widen the scope of figures which can be constructed and investigated. For example, the angle trisectors of any triangle meet in pairs to form the most special triangle of all, the equilateral triangle. Since Euclid was not able to construct angle trisectors with compass and straightedge, it is not too surprising that he missed this remarkable theorem discovered in 1911 by Frank Morley.

At the dawn of the twenty-first century, traditional drawing instruments such as the ruler and protractor play an increasingly minor role in the construction of figures and images. Industries now turn to methods of computer-aided design (CAD). Astronomers and astrophysicists depend on radio telescopes and sophisticated optical telescopes to study the large scale structure of the universe. In the opposite direction, the force tunnelling microscope forms images in which single atoms are discernible. Ultrasound and CAT scan images have become invaluable in diagnostic procedures. The construction and study of shape is increasingly dependent on computers and other modern technology. Nevertheless the geometric principles involved are not entirely outside the K–12 curriculum. Indeed, it has become clear that the construction and study of geometric shape should be an integral part of the mathematics classroom beginning with the earliest grades.

CLASSIC CONUNDRUM Diagonal Diagnosis

A rectangle is inscribed in a quarter circle as shown. Given the two distances indicated, can you determine the length of the diagonal $\overline{AC}$?

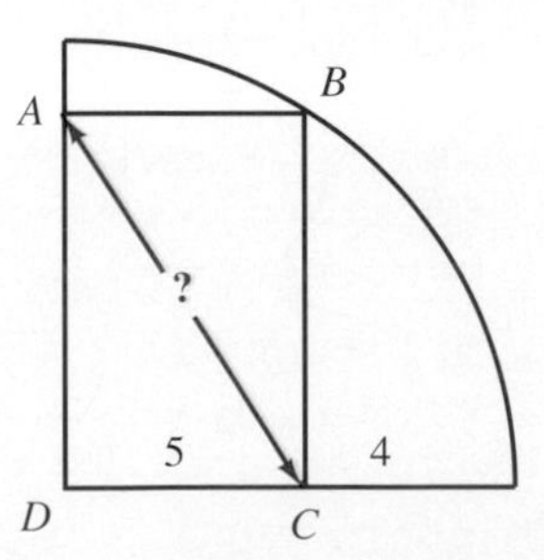

CHAPTER 11 SUMMARY

Key Concepts

1. Congruent Triangles

(a) Two figures are **congruent** if, and only if, they have the same size and shape. In particular, two triangles are congruent if, and only if, the angles and sides of one triangle are congruent to the corresponding angles and sides of the second triangle.

(b) Two triangles are congruent if they satisfy one of the following properties:

1. SSS (Side-Side-Side)
2. SAS (Side-Angle-Side)
3. ASA (Angle-Side-Angle)
4. AAS (Angle-Angle-Side)

(c) The **triangle inequality:** The length of the side of a triangle is less than the sum of the lengths of the other two sides.

(d) The **isosceles triangle theorem and its converse:** Two angles of a triangle are congruent if, and only if, their opposite sides are congruent.

(e) **Thales' theorem:** A triangle inscribed in a semicircle is a right triangle with the diameter the hypotenuse.

(f) The **HL theorem:** Two right triangles are congruent if the hypotenuse and leg of one triangle are congruent to the hypotenuse and leg of the second triangle.

2. Constructing Geometric Figures

(a) With compass and straightedge it is possible to copy a line segment; copy an angle; construct a parallel to a line through a point not on the line; construct a perpendicular from a point to a line; construct a perpendicular to a line through a point on the line; construct a perpendicular bisector; and construct an angle bisector.

(b) **Equidistance property of the perpendicular bisector:** A point is on the perpendicular bisector of a segment if, and only if, the point is equidistant from the endpoints of the segment.

(c) **Equidistance property of the angle bisector:** A point in the interior of an angle is on the angle bisector if, and only if, it is equidistant from the sides of the angle.

(d) The **Gauss-Wantzel theorem** describes which regular n-gons have compass and straightedge constructions.

3. Similar Triangles

(a) Two triangles are **similar** if corresponding angles are congruent and corresponding sides have the same ratio of lengths of sides. The common ratio of lengths of corresponding sides is the **scale factor.**

(b) Two triangles are similar if they satisfy one of the following properties:

1. AA (Angle-Angle) similarity (two pairs of congruent angles)
2. SSS (Side-Side-Side) similarity (three proportional sides)
3. SAS (Side-Angle-Side) similarity (two proportional pairs of sides include congruent angles)

Vocabulary and Notation

Section 11.1

Congruent figures
Parts of a triangle
Congruent triangles, $\triangle ABC \cong \triangle DEF$
Congruence properties of triangles: SSS, SAS, ASA, AAS

Section 11.2

Constructions: compass and straightedge, Mira, paper folding, automatic drawer
Altitude of a triangle
Perpendicular bisector
Angle bisector

Section 11.3

Similar figures
Similar triangles, $\triangle ABC \sim \triangle DEF$
Scale factor (ratio of similitude)
Similarity properties of triangles: AA, SSS, SAS

CHAPTER REVIEW EXERCISES

Section 11.1

1. In each figure, find at least one pair of congruent triangles. Express the congruence using the $\cong$ symbol, and justify why the triangles are congruent.

(a)

(b)

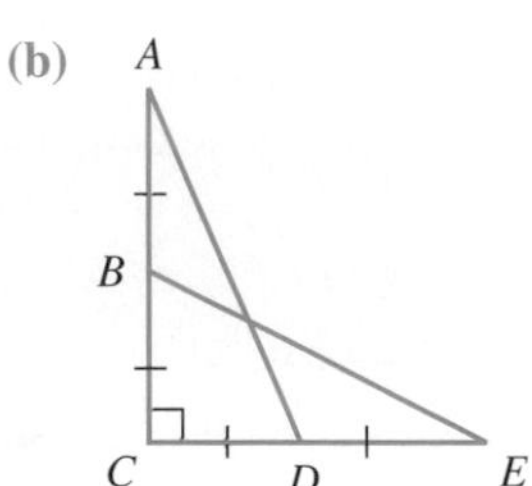

(c)

(d)

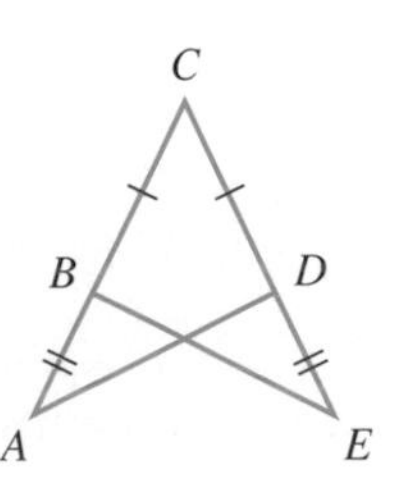

(e)

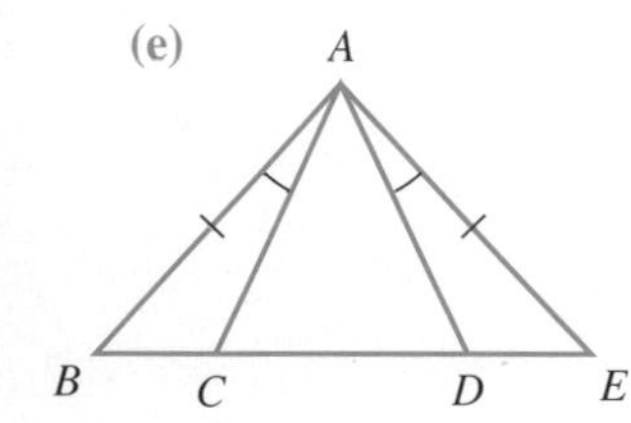

(f)

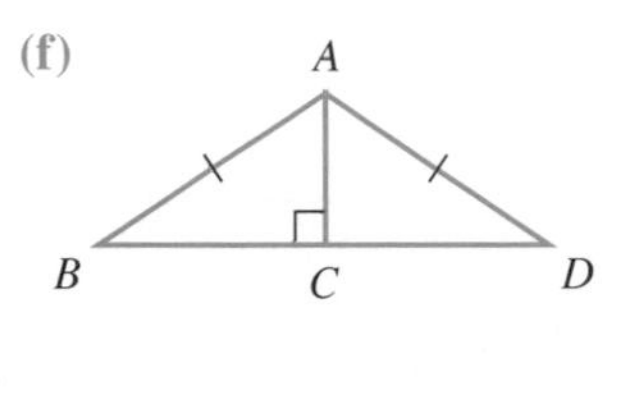

2. Without measuring, fill in the blanks below these figures.

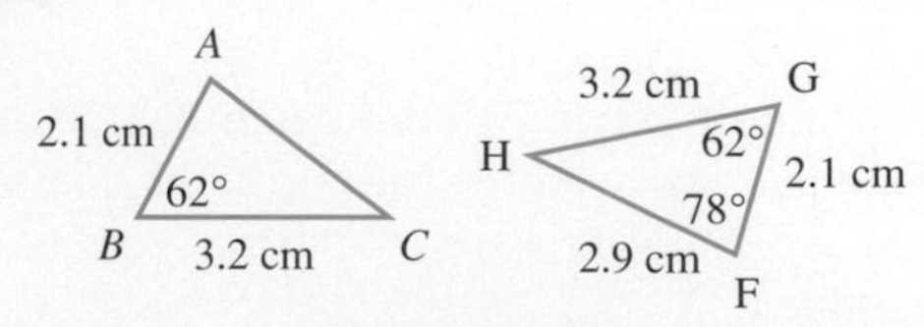

(a) $AC =$ ____ (b) $m(\angle H) =$ ____
(c) $m(\angle A) =$ ____ (d) $m(\angle C) =$ ____

3. Let D be any point of the base $\overline{BC}$ of an isosceles triangle ABC. Locate E on $\overline{AC}$ and F on $\overline{AB}$ so that $EC = BD$ and $BF = DC$. Draw the figure as it is described and then prove that $DE = DF$.

Section 11.2

4. Construct the following with a compass and a straightedge. Show and describe all of your steps.

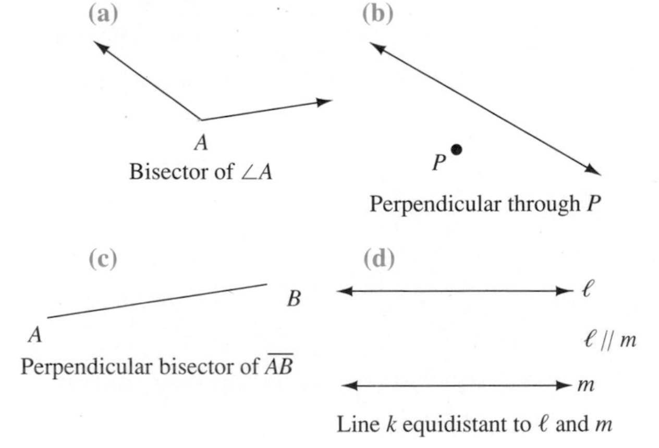

5. Show and describe the position of a Mira which performs construction of problem 4 in one step.

6. (a) Construct $\triangle ABC$, where $\angle A$, $\overline{AB}$, and $\overline{BC}$ are the parts shown. Is the shape of $\triangle ABC$ uniquely determined?
   (b) Is the shape of $\triangle ABC$ uniquely determined if $\angle C$ is obtuse?

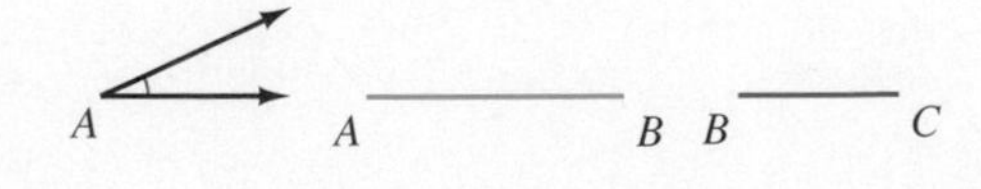

7. Construct a regular hexagon *ABCDEF*, where diagonal *AD* is given below.

Section 11.3

8. (a) If only a ruler is available, is it possible to determine if two triangles are similar?
   (b) If only a protractor is available, is it possible to determine if two triangles are similar?

9. Triangle *ABC* is given below. Construct a triangle *DEF* for which $\triangle ABC \sim \triangle DEF$ and $DE = (3/2)AB$, using a compass and a straightedge.

10. Explain why each pair of triangles is similar. State the similarity using the $\sim$ symbol and give the scale factor.

(a)

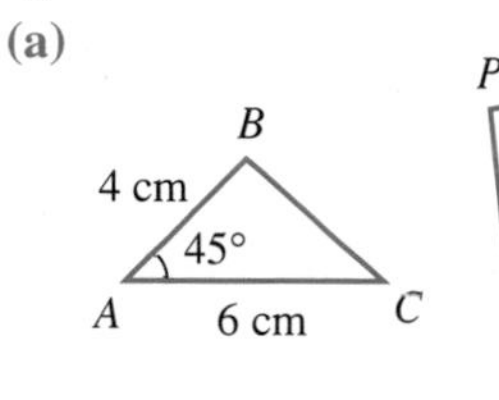

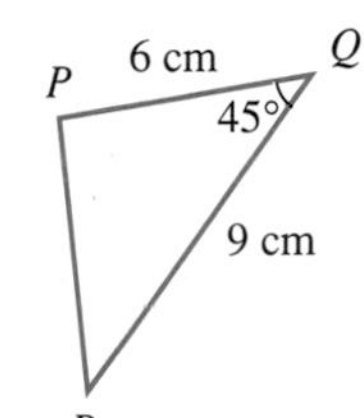

(b)

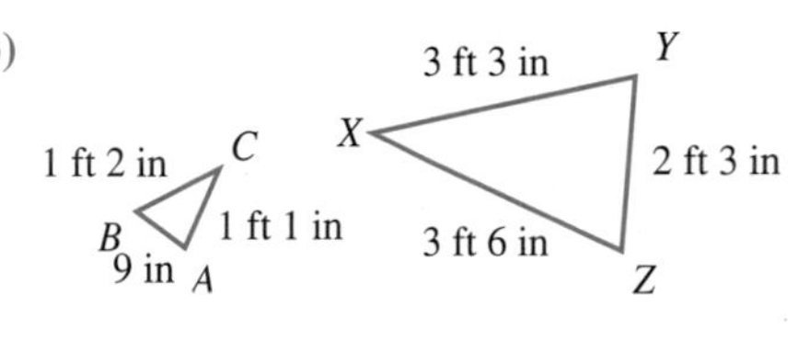

(c)

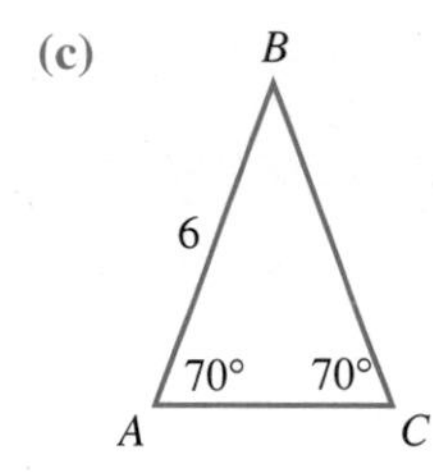

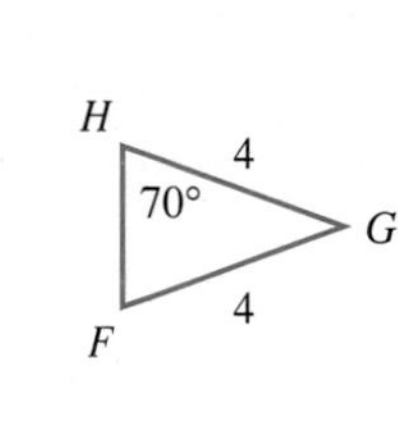

(d)

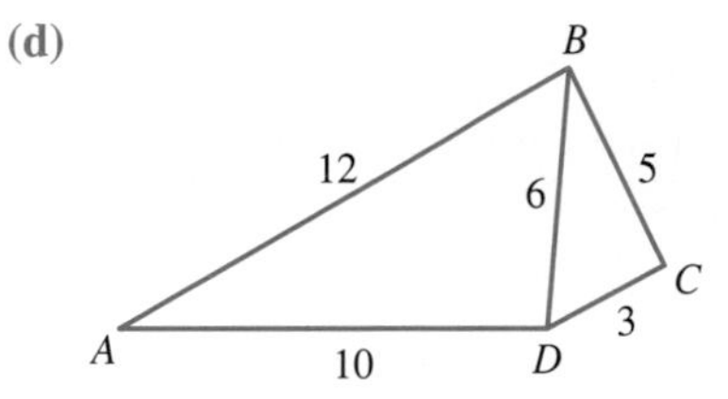

11. Lines *k*, *l*, and *m* are parallel. Find the distances *x* and *y* using similar triangles.

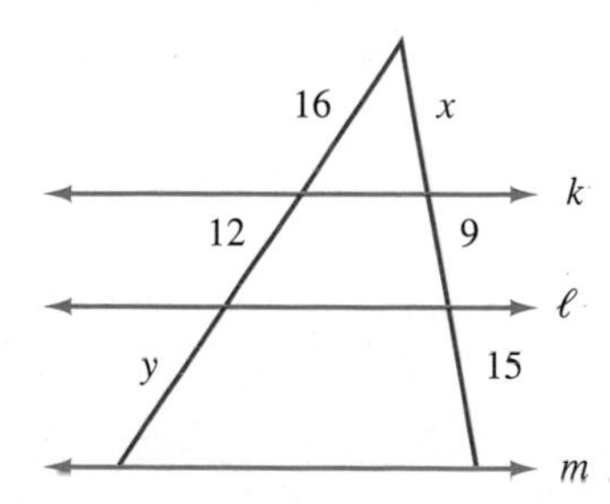

CHAPTER TEST

1. In each figure find a pair of congruent triangles. State what congruence property justifies your conclusion, and write the congruence using the congruence symbol $\cong$.

(a)

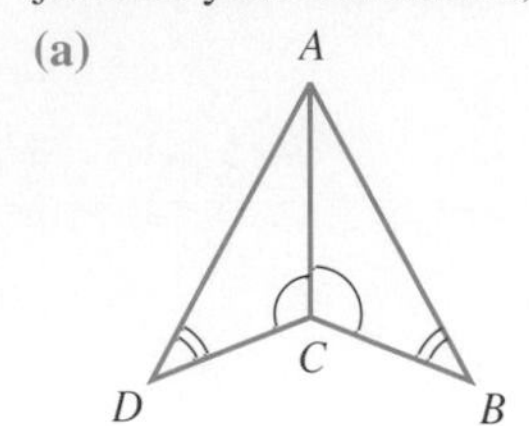

(b)

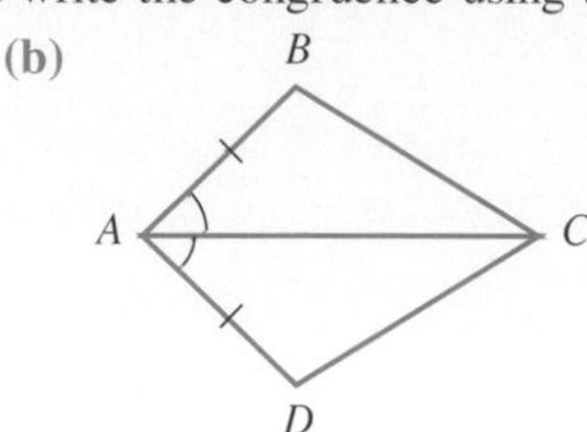

(c)

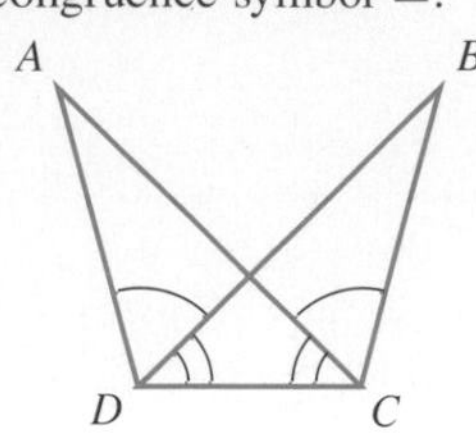

(d)

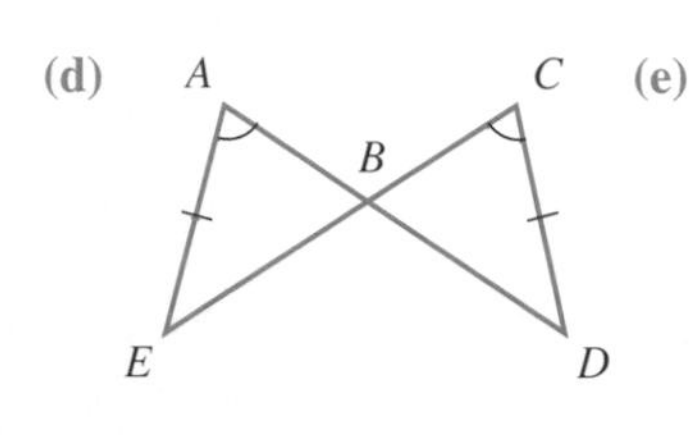

(e)

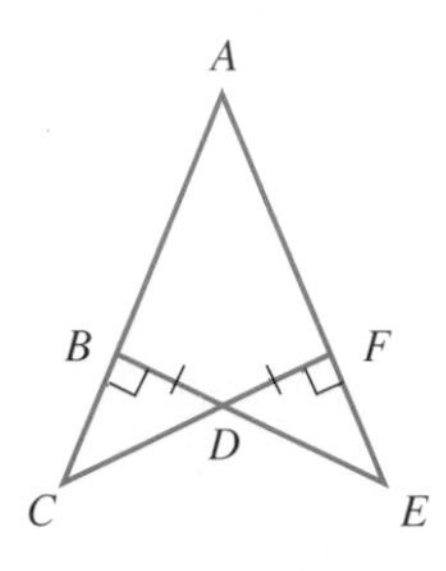

(f)

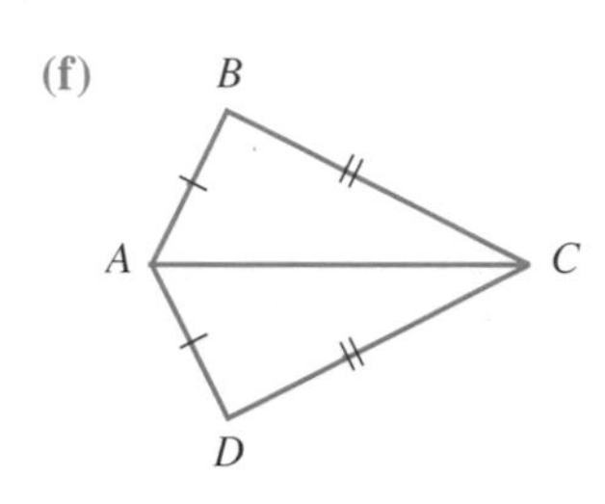

2. A triangle has sides of length 10 feet and 16 feet. What is the range of lengths of the third side?

3. Construct a triangle DEF so that $\triangle DEF \cong \triangle ABC$ and $\angle A$, $\overline{AB}$, and $\angle B$ are given as shown below. List the steps you follow.

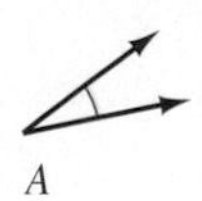

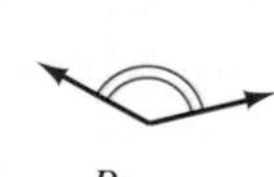

4. Decide if the following pairs of triangles are necessarily congruent. If so state why, and give the vertex correspondence. If not give a counterexample.

(a)

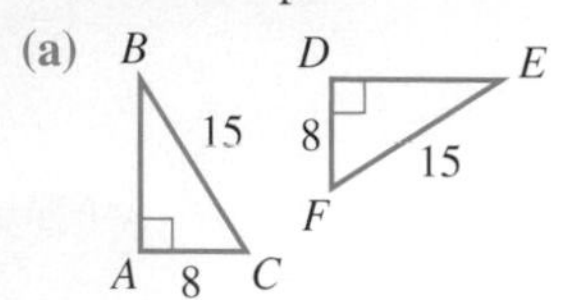

(b)

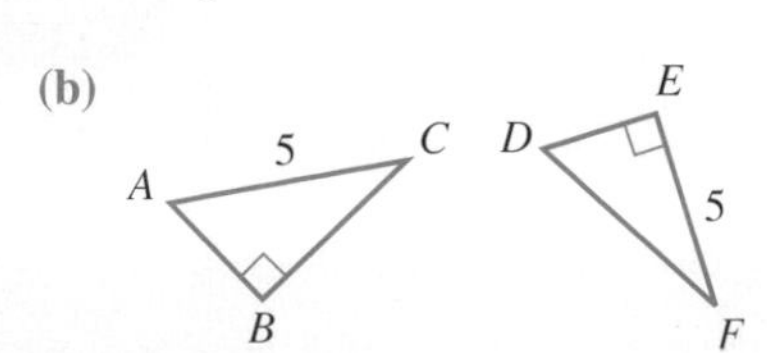

(c)

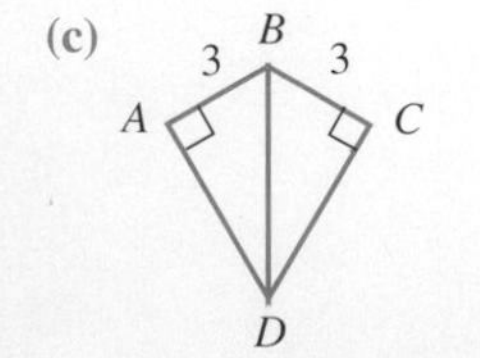

(d)

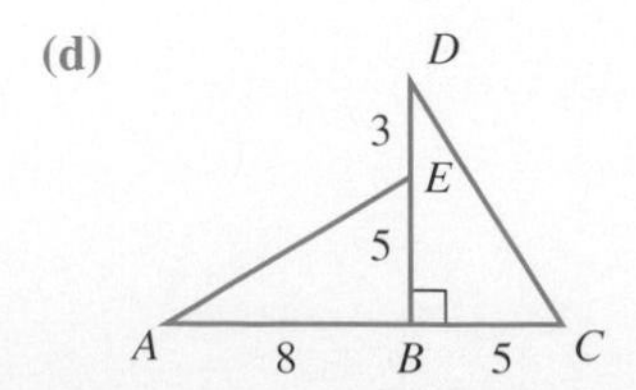

5. Let $ABCDE$ be a regular pentagon. Let $PQRST$ be inscribed so that $AP = BQ = CR = DS = ET$. Prove that $PQRST$ is a regular pentagon.

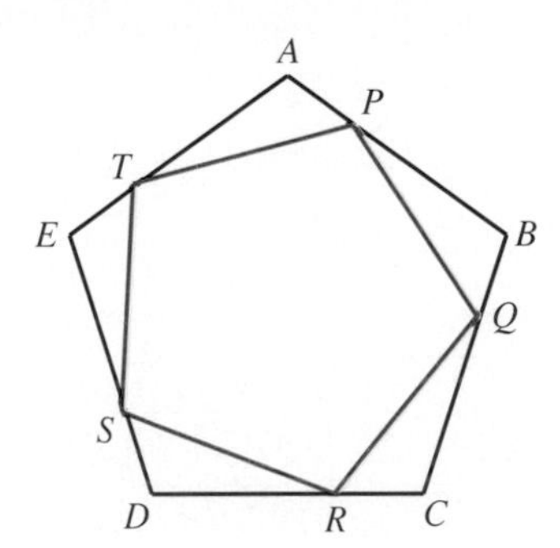

6. In $\triangle FGH$, $\angle F \cong \angle G$ and $FG = FH$. What can you conclude about the triangle?

7. Let $\overline{PQ}$ be perpendicular to line l.

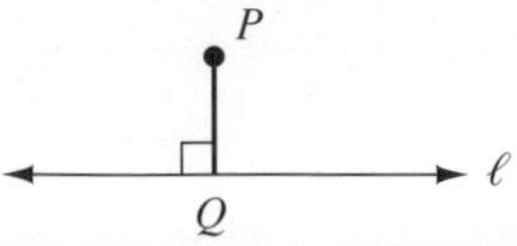

(a) Use a compass and a straightedge to construct two equilateral triangles, $\triangle PQR$ and $\triangle PQS$, each with $\overline{PQ}$ as a side.

(b) Use a compass and a straightedge to construct an equilateral triangle PTU for which $\overline{PQ}$ is an altitude. Describe your procedure.

8. Explain why each pair of triangles is similar. Express the similarity with the $\sim$ notation.

(a)

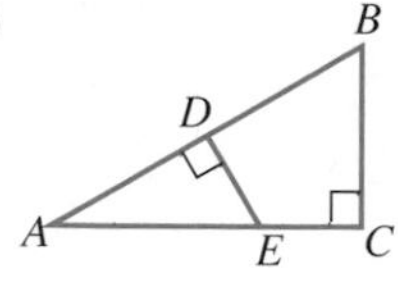

(b)

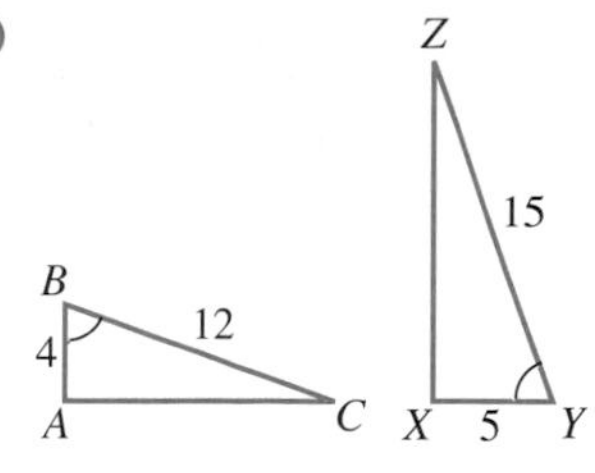

(c)

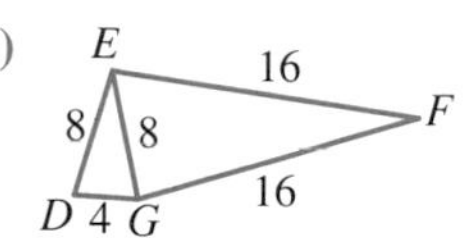

(d)

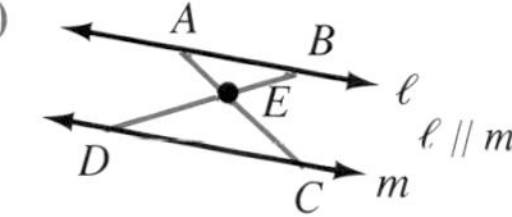

9. Fill in the blanks below, where $\triangle KLM \sim \triangle UVW$.

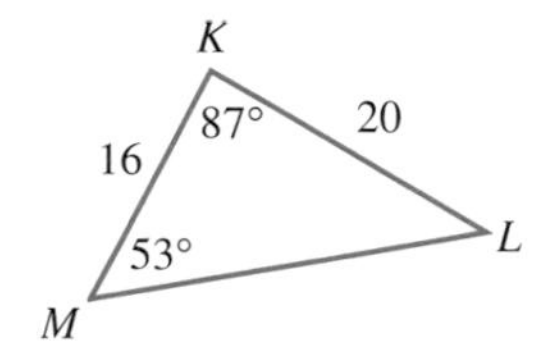

(a) $m(\angle W)$ = ________

(b) Scale factor = ________

(c) UV = ________

10. A person 6 feet tall casts a shadow 7 feet long, and a tree casts a shadow 56 feet long.

(a) What assumption can you make about the sun's rays?

(b) What other assumption can you make to conclude that $\triangle ABC \sim \triangle DEF$?

(c) How tall is the tree?

11. If $AB = 12$ and $AD = 2DC$ find AE and EB.

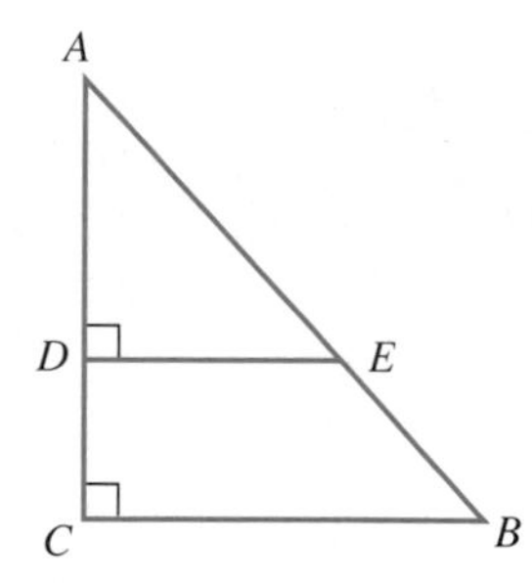

Explain how you found your answer.

12. In $\triangle ABC$ the midpoint M of side $\overline{AB}$ satisfies $MC = MA$. Prove that $\triangle ABC$ is a right triangle.

12 Measurement

HANDS ON

Measurements in Beanland

Materials Needed

Dry beans (small red kidney or white navy beans); enlarged copies of the figures shown in the two activities below

Directions

In Beanland, the lengths of curves are measured in *beanlengths*, abbreviated bl. Similarly, the areas of regions are measured in *beanareas*, abbreviated ba. The diagram below shows that the length of the curve is about 15 bl.

The next diagram shows a region bounded by a simple closed curve. By counting the beans, we see that the region has an area of about 55 ba.

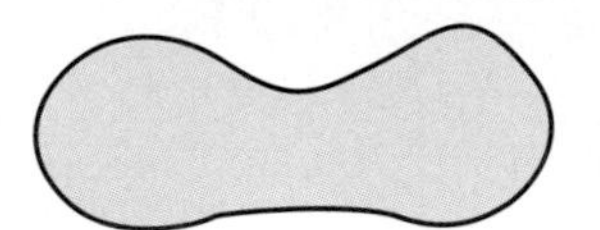

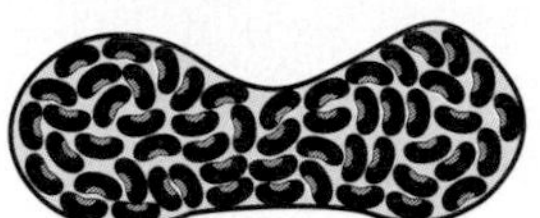

Activities

1. Work with an enlargement of the park shown below.

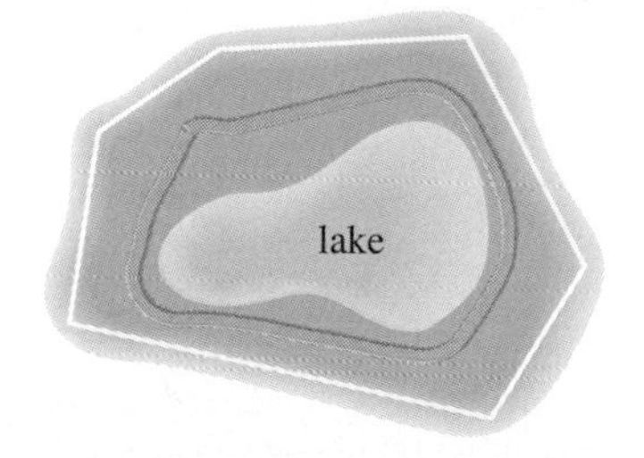

Use beanlengths and beanareas to estimate:

(a) the length of a fence needed to enclose the park
(b) the length of the footpath
(c) the area of the park
(d) the area of the lake

2. Consider this system of squares and circles. (An enlarged copy of the figure would be best to use.)

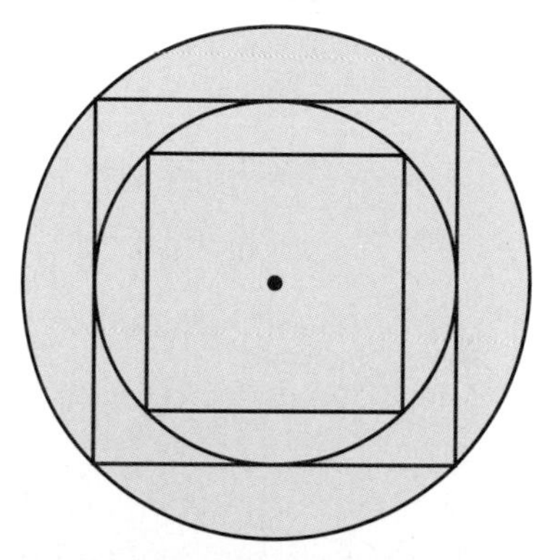

(a) Measure the ring-shaped region between the two circles, using your beans.
(b) Measure the area of the smaller circle, and then compare this area to that of the ring.
(c) Measure the area of the small square and then the area of the region between the two squares. How do these areas compare?
(d) Measure the perimeter (the distance around) of the small square in beanlengths. Next measure the length of the diagonal of the large square. How do these lengths compare?

CONNECTIONS

The Principles and Processes of Measurement

Measurement played a limited but important role in Chapters 10 and 11. Only two figures were measured: line segments, measured by the distance between their endpoints; and angles, measured by the degrees of rotation needed to turn one side to the other. In this chapter, we introduce more general notions of measurement of geometric figures. All curves, not just segments, will be given a length. Plane regions will be measured by area and perimeter. Finally, space figures will be measured by surface area and volume.

We begin by discussing the general process of measurement and the concept of a unit of measurement. The two principal systems of measurement are then described:

The U.S. Customary (English) System, used in the United States but almost no other country; and the International System (metric), used by all countries worldwide including the United States.

12.1 The Measurement Process

The geometry of the Babylonians and ancient Egyptians always had a practical purpose, and often this purpose was dependent on having knowledge of size and capacity. It was important to know the areas of fields, the volumes of granaries, and so on. Many engineering projects gave rise to geometric problems concerned with magnitudes. To be specific, suppose a canal has a given trapezoidal cross section and a known length. Knowing how much volume of earth one worker can dig in one day, how many workers are needed to excavate the canal in a given period of time?

Determining size requires that a comparison be made to a **unit.** For example, the volume of a canal could be expressed in "worker-days," where a worker-day is the volume one person can excavate in one day's labor. The worker-day is thus a unit of volume. It is analogous to the original definition of acre, which was the area of land that could be plowed in one day with one team of oxen.

In early times, units of measurement were defined more for convenience than accuracy. For example, many units of length correspond to parts of the human body, some of which are shown in Figure 12.1. The hand, span, foot, and cubit all appear in early records of Babylonia and Egypt.

Many of these units later became standardized and persist today. For example, horses are still measured in hands, where a hand is now 4 inches. Originally an inch was the length of 3 barleycorns placed end-to-end.

FROM THE NCTM STANDARDS **Measurement in K–4**

Measurement is of central importance to the curriculum because of its power to help children see that mathematics is useful in everyday life and to help them develop many mathematical concepts and skills. Measuring is a natural context in which to introduce the need for learning about fractions and decimals, and it encourages children to be actively involved in solving and discussing problems.

Instruction at the K–4 level emphasizes the importance of establishing a firm foundation in the basic underlying concepts and skills of measurement. Children need to understand the attribute to be measured as well as what it means to measure. Before they are capable of such understanding, they must first experience a variety of activities that focus on comparing objects directly, covering them with various units, and counting the units. Premature use of instruments or formulas leaves children without the understanding necessary for solving measurement problems.

Estimation should be emphasized because it helps children understand the attributes and the process of measuring as well as gain an awareness of the sizes of units. Everyday situations in which only an estimate is required should be included. Since measurements are not exact, children should realize that it is often appropriate, for example, to report a measurement as between eight and nine centimeters or about three hours.

SOURCE: From *Curriculum and Evaluation Standards for School Mathematics*, Grades K–4, p. 51,

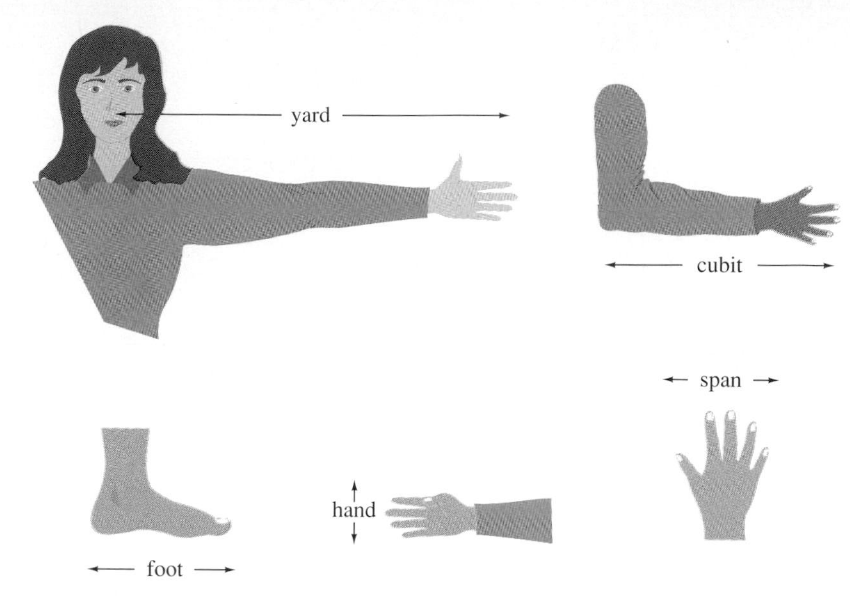

Figure 12.1
Examples of traditional units of length based on the human body

The **measurement process** can be viewed as a sequence of steps.

The Measurement Process

1. Choose the property (such as length, area, volume, capacity, temperature, time, weight), of an object or event which is to be measured.
2. Select an appropriate unit of measurement.
3. Use a measurement device to "cover," "fill," "time," or otherwise provide a comparison of the object to the unit.
4. Express the measurement as the number of units used.

EXAMPLE 12.1

Investigating Tangram Measurements

Tangrams are the seven shapes I, II, . . . , VII that form a square, as shown below. Use shape I, the small isosceles right triangle, as the unit of "one tangram area" (abbreviated 1 tga) to measure:

(a) the area of each of the tangram pieces,
(b) the area of the "fish,"
(c) the area of the circle that circumscribes the square. Are your measurements exact, or only approximate?

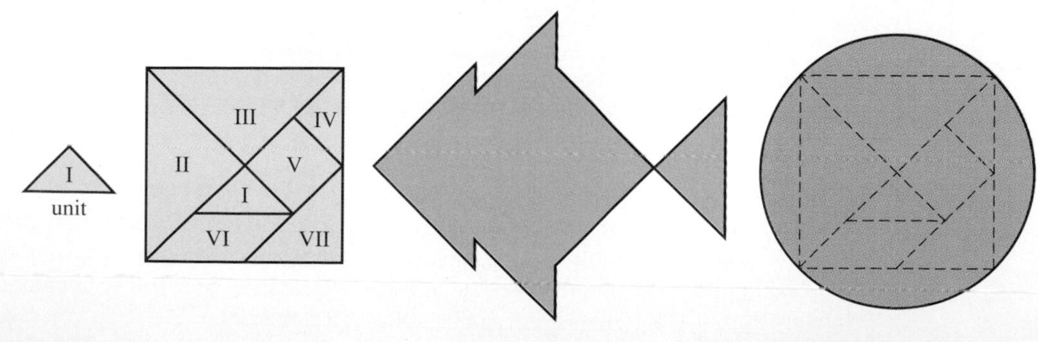

SOLUTION

(a) Each tangram piece can be covered by copies of the unit shape I, giving the exact measurements in the table.

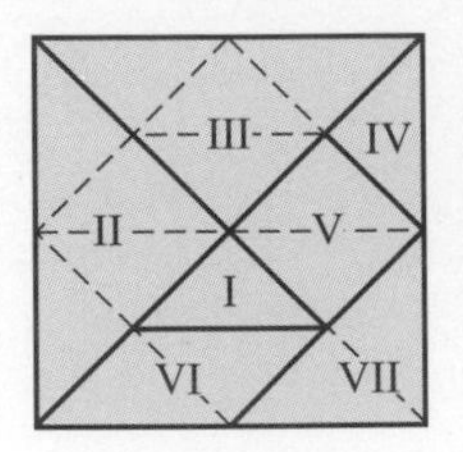

Piece	Area
I, IV	1 tga
II, III	4 tga
V, VI, VII	2 tga

(b) The fish is covered by the seven tangram pieces, so its area is exactly $(4 + 2 + 1 + 2 + 1 + 4 + 2)$ tga $=$ 16 tga.

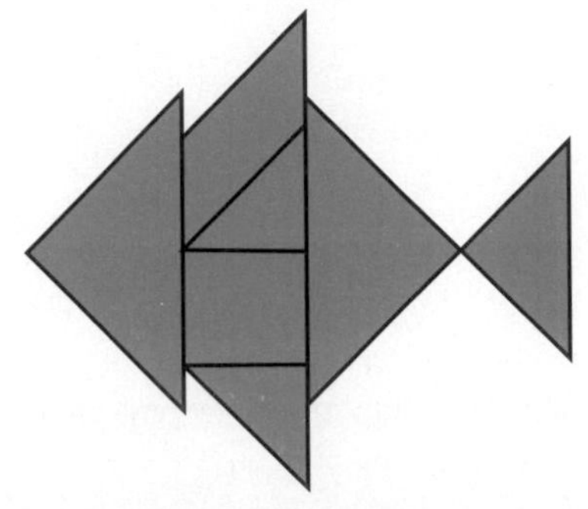

(c) The circle can be covered by the seven tangram pieces, together with 12 additional copies of the unit shape. This shows that the circle's area is between 16 tga and 28 tga. Thus we might estimate the area at about 25 tga.

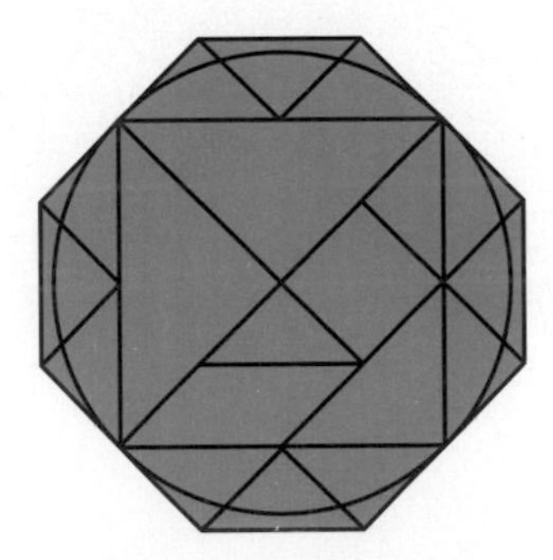

■

An important practical purpose of measurement is communication. By agreeing on common units of measurement, people are able to express and interpret information about size, quantity, capacity, and so on. Historically, as commerce developed and goods were traded over increasingly large distances, the need for a standard system of units became more and more apparent. In the seventeenth and eighteenth centuries, the rise of science and the beginnings of the industrial revolution gave further impetus to the development of universal systems of measurement.

The U.S. Customary, or "English," System of Measures

The English system arose from a hodgepodge of traditional informal units of measurement. Table 12.1 lists some of the units of **length** in this system. The ratios comparing one unit of length to another are clearly the result of accident, not planning.

TABLE 12.1 Units of Length in the Customary System

Unit	Abbreviation	Equivalent Measurement in Feet
Inch	in	$\frac{1}{12}$ ft
Foot	ft	1 ft
Yard	yd	3 ft
Rod	rd	$16\frac{1}{2}$ ft
Furlong*	fur	660 ft
Mile	mi	5280 ft

* The *furlong* is a shortening of "furrow long," revealing its origin in agriculture.

Learning the customary system requires extensive memorization, and using the system involves computations with cumbersome numerical factors.

Area is a measure of the region bounded by a plane curve. Any shape that tiles the plane could be chosen as a unit, but the square is the most common shape. The size of the square is arbitrary, but it is natural to choose the length of a side to correspond to a unit measure of length. Areas are therefore usually measured in square inches, square feet, and so on. A moderate sized house may have 1800 square feet of floor space, a living room carpet may cover 38 square yards, and a national forest may cover 642 square miles. An exception to this pattern is the acre; 640 acres have a total area of one square mile. Some common units of area are listed in Table 12.2. The notation ft^2 indicates square feet.

TABLE 12.2 Units of Area in the Customary System

Unit	Abbreviation	Equivalent Measure in Other Units
Square inch	in^2	$\frac{1}{144}$ ft^2
Square foot	ft^2	144 in^2, or $\frac{1}{9}$ yd^2
Square yard	yd^2	9 ft^2
Acre	acre	$\frac{1}{640}$ mi^2, or 43,560 ft^2
Square mile	mi^2	640 acres, or 27,878,400 ft^2

The ratios comparing one unit of area to another can be visualized, as shown in Figure 12.2 on page 766. We see that the area of a 3 ft by 3 ft square is obtained by the multiplication 3 ft $\times$ 3 ft $=$ 3 $\times$ 3 $\times$ ft $\times$ ft $=$ 9 ft^2. *In computing with dimensioned quantities it is essential to retain the units in all equations and expressions.* For example, it is correct to write 12 in $=$ 1 ft; without the dimensions this equation would be incorrect since $12 \neq 1$. Omitting the units in expressions is a common source of errors.

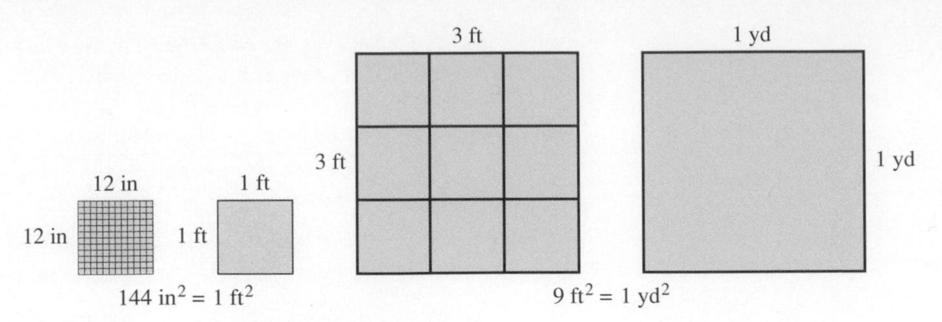

Figure 12.2
Comparing units of area measure

Volume is the measure of space taken up by a solid in three-dimensional space. The unit, as shown in Table 12.3, is the volume of a cube whose side length is one of the standard units of length. For example, the displacement of the pistons in a car engine may be 327 cubic inches, which is abbreviated 327 in^3.

TABLE 12.3 Units of Volume in the Customary System

Unit	Abbreviation	Equivalent Measure in Other Units
Cubic inch	in^3	$\frac{1}{1728}$ ft^3
Cubic foot	ft^3	1728 in^3, or $\frac{1}{27}$ yd^3
Cubic yard	yd^3	27 ft^3

The ratios comparing units of volume are illustrated in Figure 12.3.

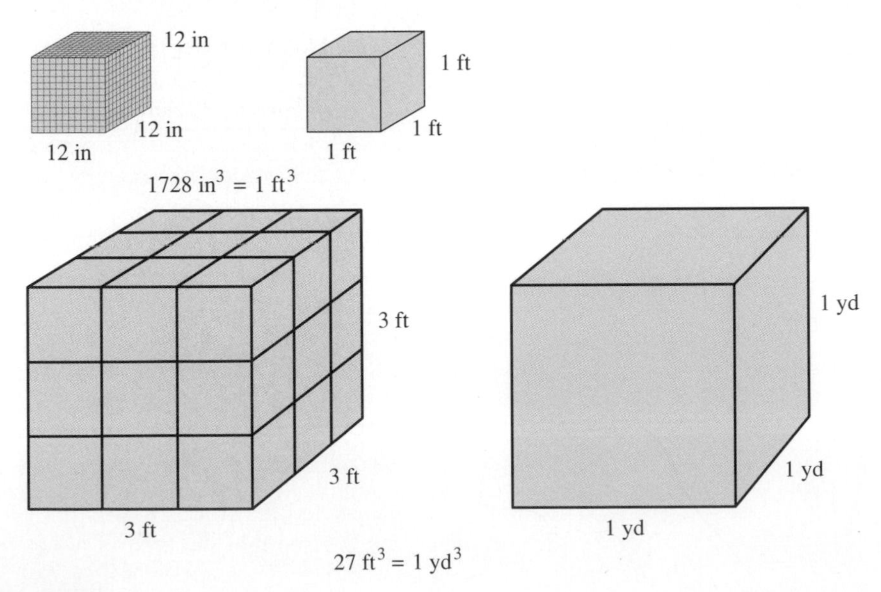

Figure 12.3
Comparing units of volume measure

Capacity is the volume which can be held in a container such as a bottle, pan, basket, tank, and so on. Capacity is often expressed in in^3, ft^3, or yd^3, but other units are also in common use, particularly for liquid measures. Examples of liquid measures include the fluid ounce, cup (8 fluid ounces), pint (2 cups), quart (2 pints), and the gallon (4 quarts). A gallon is equivalent to 231 in^3. Dry measures of capacity include the bushel (2150.42 in^3) and the peck (1/4 of a bushel).

Metric Units: The International System

The metric system of measurement originated in France shortly after the Revolution of 1789. The definitions of the units have been modified over succeeding years, taking advantage of scientific and technological advances. The system was codified in 1981 by the International Standardization Organization. The International System of Units, also called the **SI System** after its French name *Systeme International*, has now achieved worldwide acceptance. The metric system has been a legal standard since 1866 in the United States; indeed the Customary units were *defined* in terms of metric units in 1893. In the 1970s a movement to replace customary units with metric units was unsuccessful. About the same time most other English speaking countries, including Great Britain, Canada, Australia, and New Zealand, did change to metric units. Even day-to-day measurements in those countries—speed limits, distances between cities, and amounts in recipes—were replaced with metric units.

The principal advantage of the metric system—other than its universality—is the ease of comparison of units. The ratio of one unit to another is always a power of 10, which ties the metric system conveniently to the base ten numeration system. This makes it quite simple to convert a measurement in one metric unit to the equivalent measurement in another metric unit.

Each power of 10 is given a prefix which modifies the fundamental unit. For example, the factor 1000 (that is, 10^3) is expressed by the prefix *kilo*. Thus a kilometer is 1000 meters. Similarly, the factor $\frac{1}{100}$ (that is, 10^{-2}) is expressed by the prefix *centi*.

Therefore, a centimeter is $\frac{1}{100}$ of a meter. The more commonly used prefixes and their symbols are listed in Table 12.4.

TABLE 12.4 The SI Decimal Prefixes

Prefix	Factor	Symbol
kilo	$1000 = 10^3$	k
hecto	$100 = 10^2$	h
deka	$10 = 10^1$	da
(none for basic unit)	$1 = 10^0$	(none)
deci	$0.1 = 10^{-1}$	d
centi	$0.01 = 10^{-2}$	c
milli	$0.001 = 10^{-3}$	m
micro	$0.000001 = 10^{-6}$	μ (Greek *mu*)

Length

The fundamental unit of length in the SI system is the **meter,** abbreviated by the symbol m. The unit symbol is always written last in SI, so there can be no confusion with the prefix milli, which is also given the symbol m. For example, one thousandth of a meter is a millimeter, written as 1 mm. There is no space between the first and second m, and there are no periods between or after the symbols.

The most commonly used metric units of length are listed in Table 12.5.

TABLE 12.5 Metric Units of Length

Unit	Abbreviation	Multiple or Fraction of 1 Meter
1 kilometer	1 km	1000 m
1 hectometer	1 hm	100 m
1 dekameter	1 dam	10 m
1 meter	1 m	1 m
1 decimeter	1 dm	0.1 m
1 centimeter	1 cm	0.01 m
1 millimeter	1 mm	0.001 m
1 micron (micrometer)	1 μm	0.000001 m

The prefix in a metric measurement can be replaced with its corresponding numerical factor. For example,

$$251 \text{ cm} = 251 \times 10^{-2} \text{ m} = 2.51 \text{ m}.$$

Similarly, a power of 10 can be replaced with the corresponding prefix, as in

$$0.179 \text{ m} = 179 \times 10^{-3} \text{ m} = 179 \text{ mm}.$$

Some metric measurements are shown in Figure 12.4.

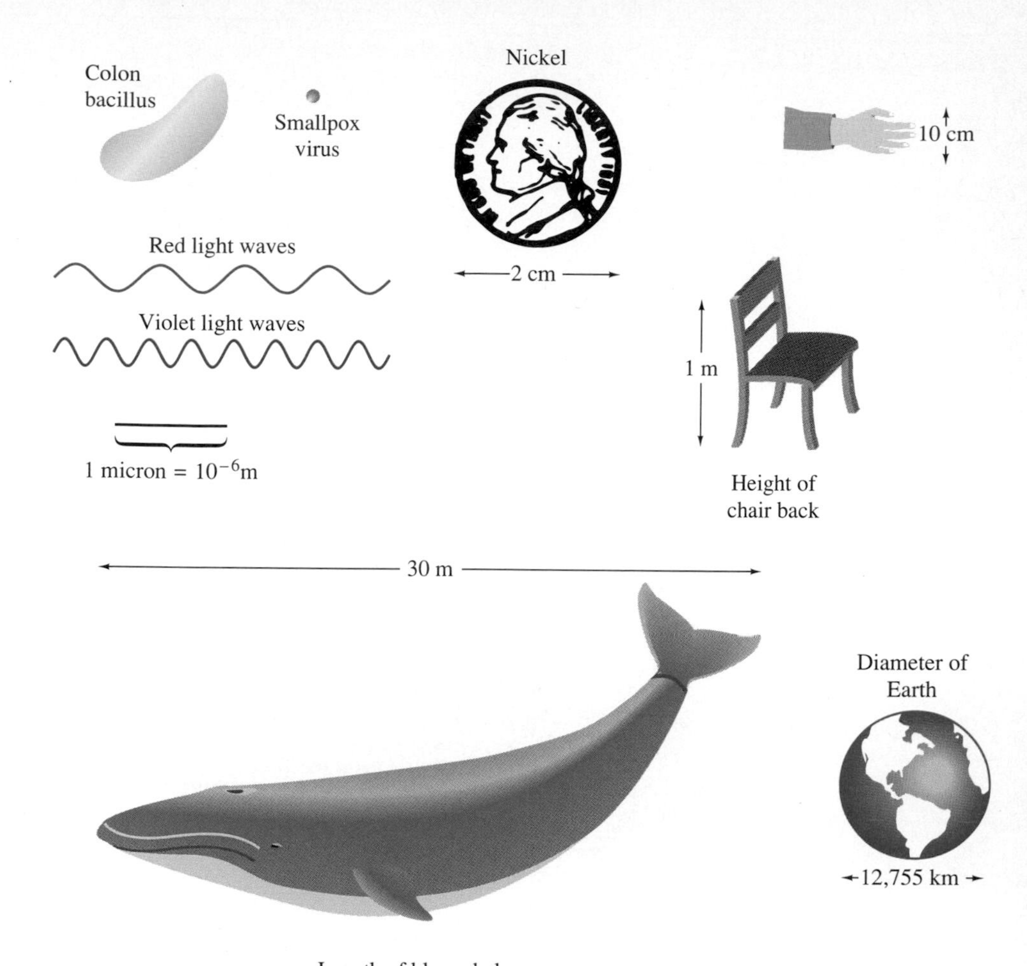

Figure 12.4
Examples of metric measurements of length

HIGHLIGHT FROM HISTORY
Redefining the Meter

The meter was originally defined in the 1790s as one ten-millionth of the distance from the North Pole to the equator. Making the measurement, however, presented impossible difficulties. Not only was the earth quite unlike a perfect sphere, but political turmoil in revolutionary France led to the arrest of government surveyors as royalist spies, who narrowly escaped the guillotine. In 1889, the meter was fixed as the distance between two marks on a platinum-iridium bar, but scientists were unsatisfied since measurements could be no more accurate than one part in a million. In 1960 the meter was again redefined, becoming 1,650,763.73 wave lengths of the reddish-orange light emitted by krypton 86, a rare atmospheric gas. The new meter was accurate to 4 parts per billion, but still was an irritant to scientists measuring continental drift and the distance to the moon. In 1983, the meter was redefined yet again, in a way that connects length to time. The new definition specifies the meter as the distance traveled by light in space in 1/299,792,458 of a second. One second of time, which can be precisely measured with atomic clocks, is defined as the duration of 9,193,631,770 vibration cycles of the Cesium 133 atom. The definition invokes a sacred tenet of physics, that the speed of light in space is a universal constant, namely 299,792,458 meters per second.

EXAMPLE 12.2 **Changing Units in the Metric System**

Convert these measurements to the unit shown.

(a) 1495 mm = ____ m
(b) 29.4 cm = ____ mm
(c) 38741 m = ____ km

SOLUTION

(a) 1495 mm $= 1495 \times 10^{-3}$ m = 1.495 m
(b) 29.4 cm $= (294 \times 10^{-1}) \times 10^{-2}$ m $= 294 \times 10^{-3}$ m = 294 mm
(c) 38,741 m $= 38.741 \times 10^{3}$ m = 38.741 km ■

Area

Area is usually expressed in square meters (m^2) or square kilometers (km^2). Another common unit is the hectare. A **hectare** (ha) is the area of a 100 m square; that is, 1 ha = 10,000 m^2. See Table 12.6.

The floorspace of a classroom might typically be about one **are** (pronounced "air"). A hectare is about 2.5 acres, so the area of farm land is measured in hectares in metric countries.

TABLE 12.6 Metric Units of Area

Unit	Abbreviation	Multiple or Fraction of 1 Square Meter
1 square centimeter	1 cm^2	0.0001 m^2
1 square meter	1 m^2	1 m^2
1 are (1 square dekameter)	1 a	100 m^2
1 hectare (1 square hectometer)	1 ha	10000 m^2
1 square kilometer	1 km^2	1,000,000 m^2

Volume and Capacity

Small volumes are measured typically in cubic centimeters (abbreviated cm^3). Large volumes are often measured in cubic meters (m^3). A convenient unit of capacity is the **liter,** defined as a cubic decimeter. Thus, the liter (abbreviated either as ℓ or L) is the volume of a cube whose sides each measure 1 dm = 10 cm; see Figure 12.5. Since 10 cm $\times$ 10 cm $\times$ 10 cm = 1000 cm^3, a liter is also 1000 cubic centimeters. Recalling that *milli* is the prefix for $\frac{1}{1000}$, a milliliter (ml) is the same as one cubic centimeter.

$$1 \text{ L} = 1 \text{ liter} = 1000 \text{ cm}^3$$
$$1 \text{ mL} = 1 \text{ milliliter} = 1 \text{ cm}^3$$

Large plastic bottles of soda usually contain 2 liters, while a typical soft drink can contain about 354 milliliters. A child's dose of cough medicine may be 3 mL. A recipe may call for 0.5 liters of water. In metric countries, gasoline is priced by the liter, and to fill a car's gas tank takes about 40 to 60 liters.

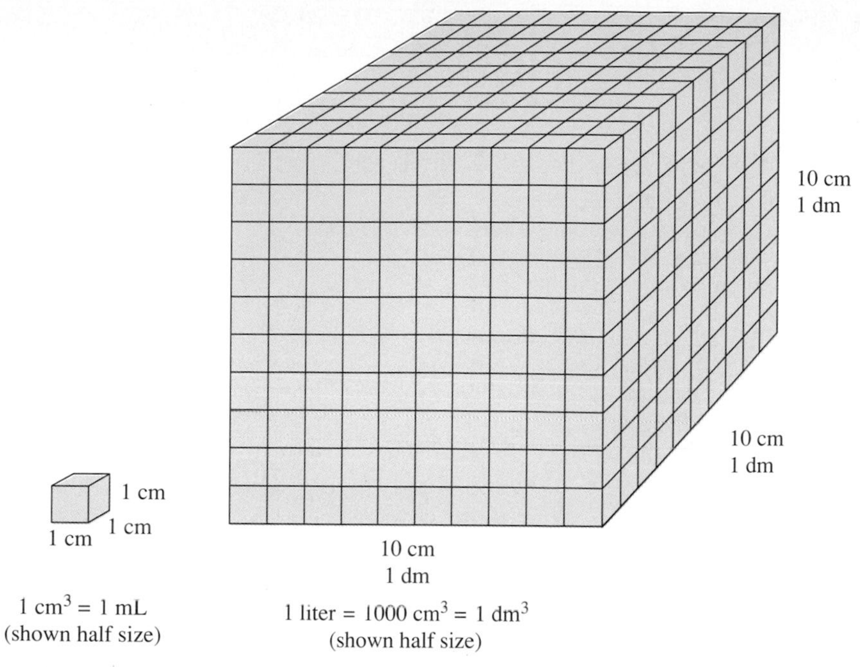

Figure 12.5
A liter is a cubic decimeter or, equivalently, 1000 cubic centimeters.

Weight and Mass

The **weight** of an object is the force exerted on the object by gravity. For example, a brick on the surface of the earth may weigh 6 pounds, but on the surface of the moon would only weigh about 1 pound. During the journey from earth to moon, the brick would weigh nearly nothing at all. Nevertheless, an astronaut would not care to be hit even by a weightless brick, since the brick never loses its **mass.**

In science, the distinction between mass and weight is very important: mass is the amount of matter of an object, and weight is the force of gravity on the object. But on the surface of the earth and in everyday life situations, the weight of an object is proportional to its mass. That is, the mass of an object is accurately estimated by weighing it.

The U.S. customary unit of weight is the familiar pound. Lighter weights are often given in ounces, and very heavy weights are given in tons:

$$16 \text{ ounces (oz)} = 1 \text{ pound (lb)}$$
$$2000 \text{ pounds} = 1 \text{ ton}$$

Table 12.7 lists some metric units of weight. The base unit of weight in the metric system is the **kilogram,** which is the weight of one liter of water.

TABLE 12.7 Metric Units of Weight

Unit	Abbreviation	Multiples of Other Metric Units
1 milligram	1 mg	0.001 g
1 gram	1 g	0.001 kg
1 kilogram	1 kg	1000 g
1 metric ton	1 t	1000 kg

SCHOOL BOOK PAGE *Measuring Temperature*

Temperature: Celsius and Fahrenheit

Build Understanding

Do you know the temperature right now? A thermometer gives the temperature. Temperature is measured in units called ***degrees***.

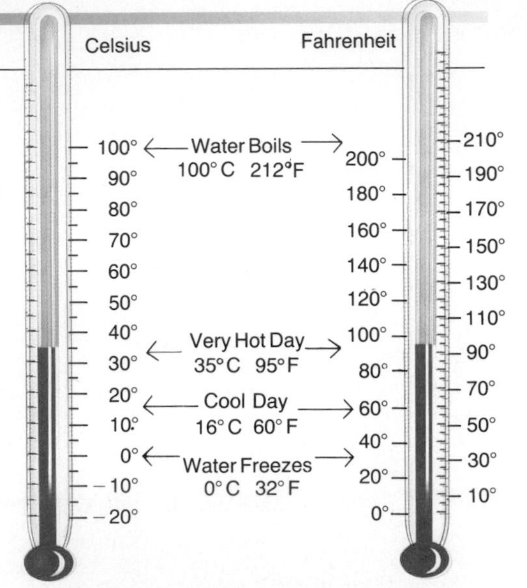

A. Look at the thermometers. The number next to the top of the liquid is the temperature. On the Celsius thermometer, it is 35 degrees (35°C).

When it is 35°C, what is the Fahrenheit temperature? Look at that thermometer. The liquid is halfway between 90 and 100. It is 95 degrees Fahrenheit (95°F).

B.

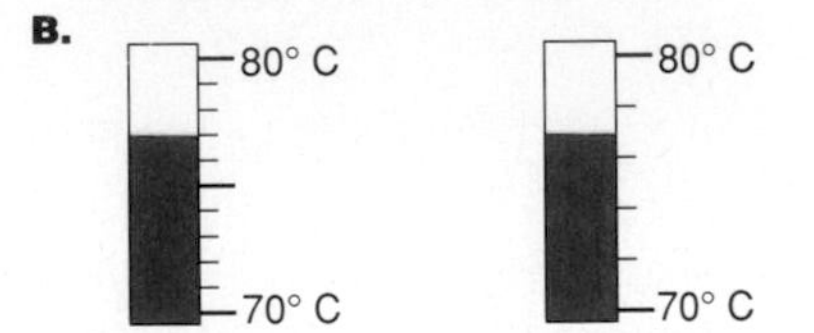

On some thermometers, every degree is marked. On others, every two degrees are marked. Both of the thermometers at the left read 77°C.

Talk About Math How are the Celsius and Fahrenheit thermometers different?

Check Understanding

For another example, see Set E, pages 134–135.

1. **Estimation** Snow is falling outside. What might the Fahrenheit temperature read?

2. A thermometer reads about 20°. Do you need to wear a heavy coat outdoors?

3. **Mental Math** If the temperature rose 4 degrees, what would the thermometer in Exercise 2 read?

4. What is the temperature in degrees Celsius?

20° C
10° C

1. Estimate the temperature, measured both in degrees Celsius and Fahrenheit, of these items:
   (a) an oven ready to bake a cake
   (b) noontime temperature on a hot summer day in Death Valley
   (c) hot tea **(d)** ice cream in a freezer

One milligram is approximately the weight of a grain of salt. It is a common measure of vitamins and medicines. A gram is approximately the weight of half a cube of sugar. Canned goods and dry packaged items at the grocery store will usually be weighed in grams. In metric countries, larger food items, such as meats, fruits, and vegetables, are priced by the kilogram. A kilogram is about 2.2 pounds.

EXAMPLE 12.3

Estimating Weights in the Metric System

Match the item in the column to the approximate weight of the item taken from the list that follows.

(a) Nickel
(b) Compact automobile
(c) Two liter bottle of soda
(d) Recommended daily allowance of vitamin B-6
(e) Size D battery
(f) Large watermelon

List of weights: 2 mg, 2 kg, 100 g, 1200 kg, 9 kg, 5 g

SOLUTION

(a) 5 g **(b)** 1200 kg **(c)** 2 kg **(d)** 2 mg **(e)** 100 g
(f) 9 kg ■

Temperature

There are two commonly used scales used to measure temperature. According to the **Fahrenheit scale,** 32°F represents the freezing point of water and 212° the boiling point of water. Thus, the Fahrenheit scale introduces 180 degrees of division between the freezing and boiling temperatures. The **Celsius scale** divides this temperature range into 100 degrees: the freezing point is 0° Celsius and the boiling point is 100° Celsius.

PROBLEM SET 12.1

Understanding Concepts

1. For each object listed below make a list of measurable properties.
   (a) A bulletin board **(b)** An extension cord
   (c) A file box **(d)** A table
2. Suppose you are designing a house. Give examples of measurements you believe are important to consider. For example, the height of the house may be needed to satisfy a zoning regulation. Discuss examples of measurements of (a) length, (b) area, and (c) volume and capacity. What units are appropriate?
3. Let a *pen* be the area of a penny.
   (a) Estimate the area of a 4″ × 6″ card in pens.
   (b) Discuss why pens are a difficult unit of area to use.
4. Arrange these solids in a list according to volume, from smallest to largest. Are there any ties?

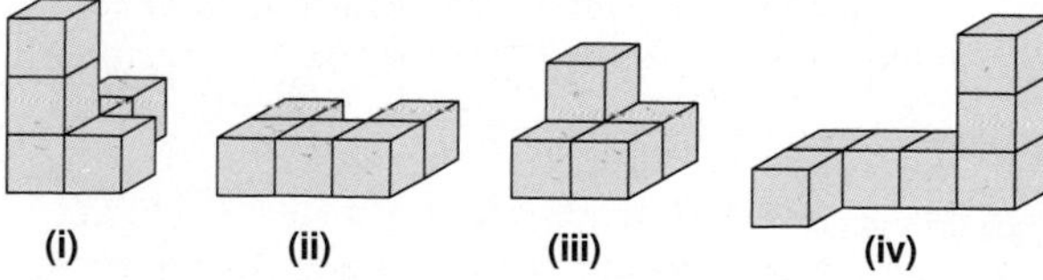

5. (a) Verify that an acre contains 43,560 square feet. Show your computation.
   (b) A square lot contains 1 acre. What is the length of each side to the nearest foot?

6. The precision of a measurement depends on the smallest subdivision of the measuring device. The measurement is given to the nearest marked subdivision. Measure the width of a one dollar bill, using the following rulers.

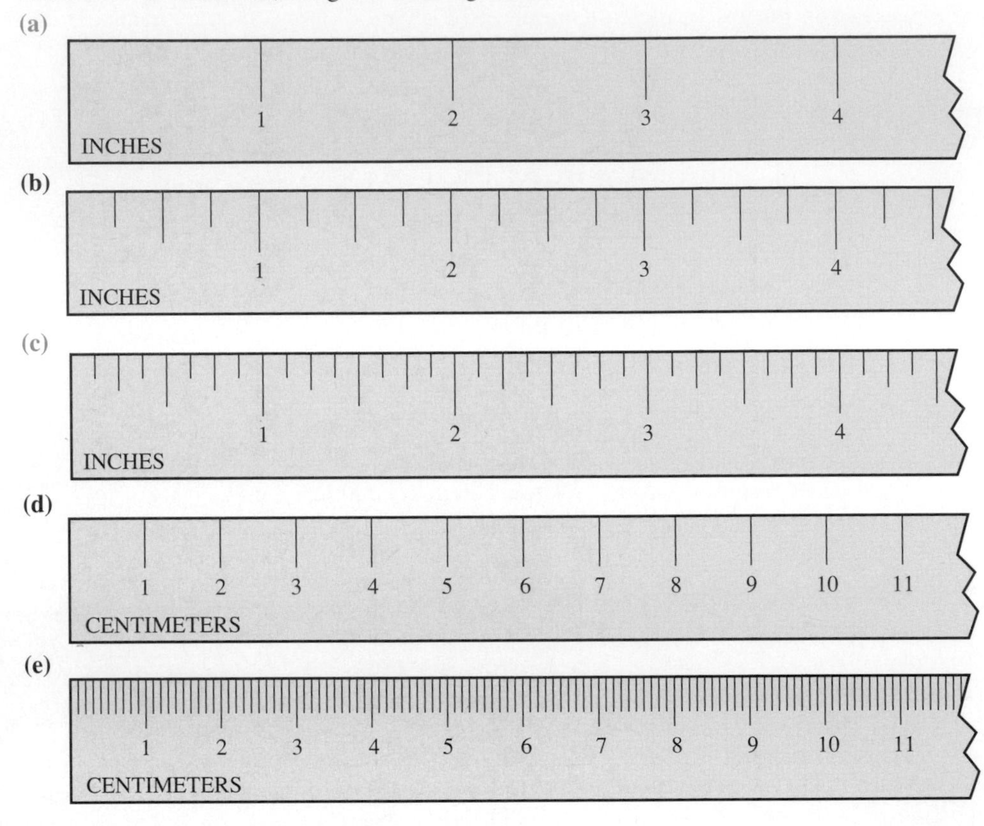

7. A measurement of 12.6 cm is assumed to be precise to the nearest last digit. That is, the true length is between 12.55 cm and 12.65 cm. Find the minimum and maximum true values for the following measurements.
   (a) A distance of 166 kilometers from Portland to Eugene
   (b) A piece of notebook paper 27.9 centimeters long
   (c) A pencil lead 0.50 millimeters in diameter
   (d) A diving board 3.5 meters over the water

8. The first inch of a ruler is often subdivided more finely than the remaining inches. Explain, in a few sentences and a sketch, how the width of a dollar bill can be measured to the nearest sixteenth of an inch.

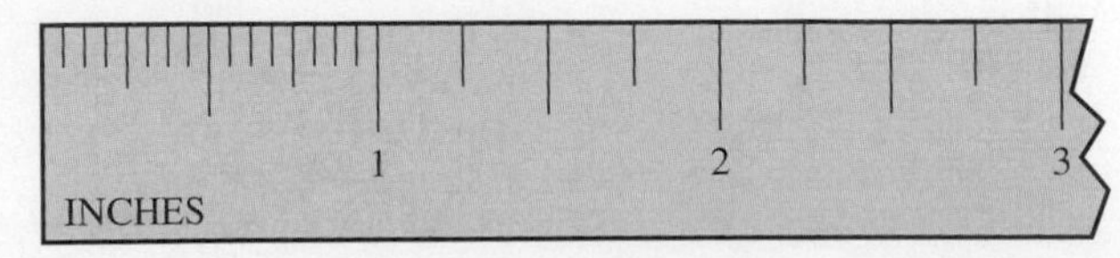

9. A small bottle of Perrier sparkling mineral water contains 33 cL.
   (a) What is the volume in milliliters?
   (b) Will three small bottles fill a 1 liter bottle?

10. Fill in the blanks.
   (a) 58,728 g = _____ kg
   (b) 632 mg = _____ g
   (c) 0.23 kg = _____ g
   (d) A cubic meter of water weighs _____ kg

11. Give the most reasonable answer listed in each part.
   (a) A newborn baby weighs about: 8.3 kg, 3.5 kg, 750 g, 1625 mg
   (b) A compact car weighs about: 5000 kg, 2000 g, 1200 kg, 50 kg.
   (c) The recommended daily allowance of vitamin C is: 250 g, 60 mg, 0.3 kg, 0.002 mg.

12. In each of the following, select the most reasonable metric measurement.

(a) The height of the typical center in the National Basketball Association is: 6.11 m, 3 m, 95 cm, 212 cm

(b) The diameter of a coffee cup is about: 50 m, 50 mm, 500 mm, 5 km

(c) A coffee cup has a capacity of about: 8 L, 8 mL, 240 mL, 500 mL

13. Use a metric ruler to measure these items.
(a) The size of a sheet of standard notebook paper
(b) The length and width of the cover of this textbook
(c) The diameter of a nickel
(d) The perimeter of (distance around) your wrist

14. The dimensions of Noah's Ark are given in the Bible as: 300 cubits long, 50 cubits wide, and 30 cubits high. Give the dimension in (a) meters and (b) feet. Use a meter stick and ruler to measure your own cubit, as shown in Figure 12.1.

Thinking Critically

15. Pints, quarts, and gallons are part of a larger "doubling" system of capacity measure.

1 jigger = 2 mouthfuls	1 pint = 2 cups
1 jack = 2 jiggers	1 quart = 2 pints
1 jill = 2 jacks	1 bottle = 2 quarts
1 cup = 2 jills	1 gallon = 2 bottles
	1 pail = 2 gallons

(a) How many mouthfuls are in a jill? A cup? A pint?

(b) Suppose one mouthful, one jigger, one jack, . . . , and one gallon are poured into an empty pail. Does the pail overflow, is it exactly filled, or is there room for more? (*Hint:* Draw an empty pail. Put one gallon in, then one bottle, and so on.)

16. In 1991 physicists at IBM used a scanning-tunneling microscope to write their corporate logo in individual xenon atoms, making it the first structure ever built one atom at a time. The image below is magnified by about five million times.

(a) Use a metric ruler to estimate the overall length of the IBM logo. Give your answer in microns, where a micron, μ, is 10^{-6} m.

(b) A typical human hair has a circumference of about 200 microns. About how many times can an "IBM" logo of the size shown be written in a circular band about one hair?

Using a Calculator

17. Verify that a hectare is about $2\frac{1}{2}$ acres. Use the approximate conversion 1.6 km = 1 mile, and show all of your steps.

18. (a) Use the conversion 1 in = 2.54 cm to calculate the number of cubic inches in a liter.
(b) Which volume is larger, 6.2 liters or 327 cubic inches?

19. A fortnight is 2 weeks. Convert a speed of 25 inches per minute to its equivalent in furlongs per fortnight.

20. Show there are about 30 million seconds in a year.

Making Connections

21. Here is a metric recipe for spaghetti sauce, except that the prefixes (if any) of some of the measures have been obliterated by some previous spills. Fill in the correct prefix, or indicate no prefix needed, for these ingredients in the recipe.

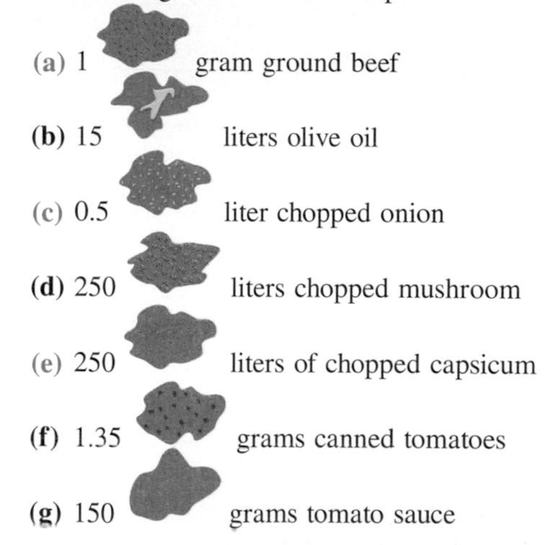

(a) 1 ___ gram ground beef
(b) 15 ___ liters olive oil
(c) 0.5 ___ liter chopped onion
(d) 250 ___ liters chopped mushroom
(e) 250 ___ liters of chopped capsicum
(f) 1.35 ___ grams canned tomatoes
(g) 150 ___ grams tomato sauce

22. In Australia, gasoline is 72.9 A¢ per liter, where A¢ is Australian cents. If the exchange rate is 1 A¢ = 0.79 U.S.¢, what is the equivalent cost expressed in United States dollars per gallon? A U.S. gallon is approximately 3.7854 liters.

23. Metric countries rate the fuel efficiency of a car by the number of liters of gasoline required to drive 100 kilometers. If a car takes 9 liters per 100 kilometers, what is its efficiency in miles per gallon? Use the conversions 1 gal $\doteq$ 3.7854 L and 1 mile $\doteq$ 1.6 km.

24. A light-year is the distance light travels in empty space in one year.

(a) Light travels at a speed of 186,000 miles per second. The star nearest the sun is Proxima Centauri, in the constellation Centaurus, whose distance is 4 light years. What is the distance to Proxima Centauri in miles?

(b) In metric measurements the speed of light is 3.00×10^8 meters per second. Verify that a light year is about 10^{16} meters.

25. An herbicide is bottled in concentrated form. A working solution is mixed by adding 1 part concentrate to 80 parts water.

(a) How many liquid ounces of concentrate should be added to 5 gallons of water?

(b) How many liters of water should be added to 65 milliliters of concentrate?

26. Lumber is measured in board-feet, where a board-foot is the volume of a piece of lumber one foot square and one inch thick.

(a) How many board feet are in a two-by-four (2″ by 4″) that is 10 feet long? The volume of a rectangular solid is length times width times height.

(b) Lumber is priced in dollars per thousand board feet. Suppose two-by-fours ten feet long are \$690 per thousand board feet. What is the cost of 144 two-by-fours, each 10 feet long?

27. How far did Captain Nemo's submarine *Nautilus* travel under the sea according to the title of Jules Verne's well-known novel? (*Hint*: Find the definition of "league.")

Communicating

28. Write a one paragraph report that explains why the words *ounce* and *inch* are related.

29. Write a one paragraph report that explains why the abbreviation for pound is *lb*.

30. Write a report in two or three pages that introduces and compares the Fahrenheit and Celsius scales of temperature measurement. Include sketches, examples, and a few problems.

For Review

31. What are the measures of the interior angles of this triangle?

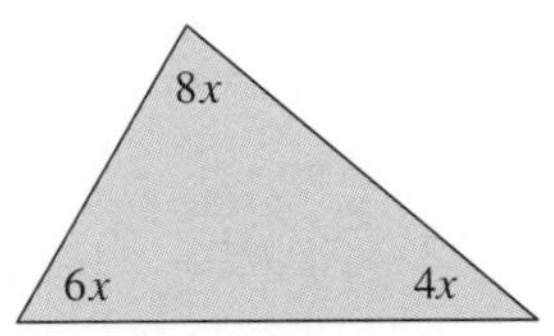

32. Explain how to construct, with a compass and straightedge, a square $ABCD$ where the diagonal $\overline{AC}$ is given.

A ———————— C

33. In each part below, information is given about ΔABC and ΔDEF. Determine if the conditions are sufficient to conclude that the triangles are congruent or similar, or if no such conclusion can be drawn.

(a) $AB = EF, \quad BC = DF, \quad \angle B \cong \angle F$

(b) $AC = EF, \quad \angle B \cong \angle D, \quad BC = DF$

(c) $AB = 2, \quad BC = 3, \quad m(\angle B) = 35°, EF = 6, ED = 9, \quad m(\angle D) = 100°, \quad m(\angle F) = 45°$

(d) $m(\angle C) = 90°, \quad AC = 5, \quad m(\angle D) = 90°, EF = 5$

12.2 Area and Perimeter

The number of units required to cover a region in the plane is the **area** of the region. Usually squares are chosen to define a **unit of area,** but any shape that tiles the plane (that is, covers the plane without gaps or overlaps) can serve equally well. Working with a nonstandard unit allows students to discover important general principles of the measurement process.

EXAMPLE 12.4 **Making Measurements in Nonstandard Units**

Find the area of each figure in terms of the unit of area shown at the right.

(a)

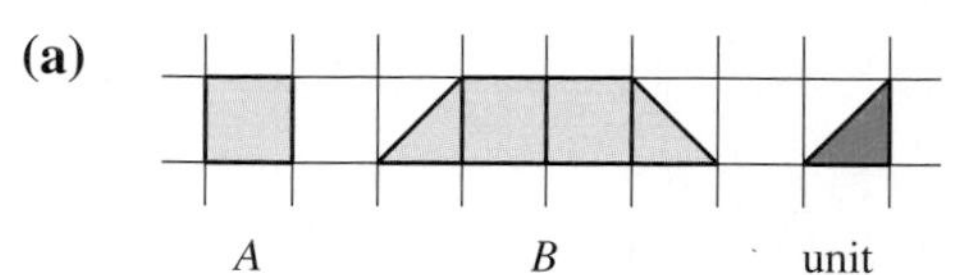

(b)

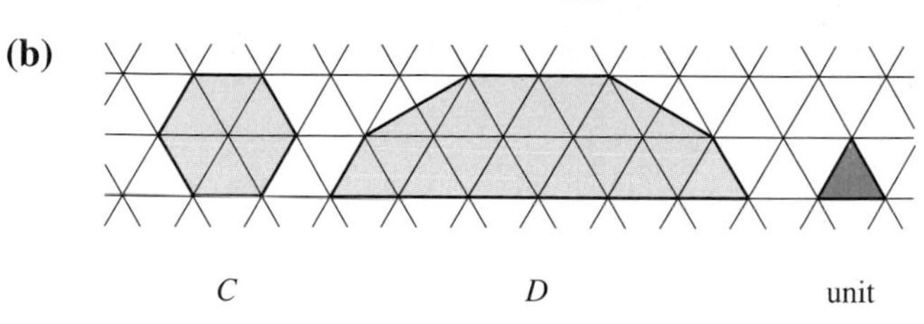

SOLUTION

(a) The full square A can be covered by 2 of the unit shapes, so area $(A) = 2$ units. Region B is covered by 6 units, so area $(B) = 6$ units.

(b) The hexagon C is covered by 6 of the triangular units, so area $(C) = 6$ units. Region D cannot be covered directly by the triangular units, although it is evident that the area of D is between 16 and 20 units. To find the exact area, remove and then rejoin a triangular piece as shown below to form a new shape D' of the same area as D. That is, area $(D') =$ area (D). Since D' can be covered by 18 triangular units it follows that area $(D) = 18$ units.

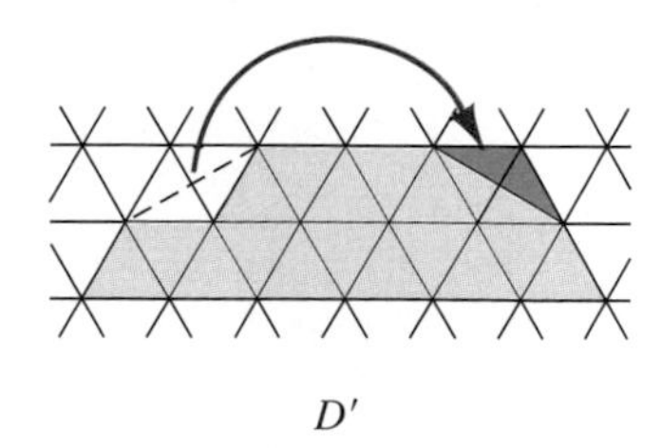

■

The solution just given uses two useful properties of area. The following **congruence and dissection properties** of area will be used repeatedly.

PROPERTY The Congruence and Dissection Properties of Area

Congruence property

If region R is congruent to region S then the two regions have the same area:

$$\text{area}(R) = \text{area}(S)$$

Dissection property

If a region R is dissected into nonoverlapping subregions $A, B, \ldots, F$, then the area of R is the sum of the areas of the subregions:

$$\text{area}(R) = \text{area}(A) + \text{area}(B) + \cdots + \text{area}(F).$$

JUST FOR FUN

How to Cover a Long Hole with a Short Board

Can you cut the board shown into just two pieces to exactly cover the 60 cm by 12 cm hole?

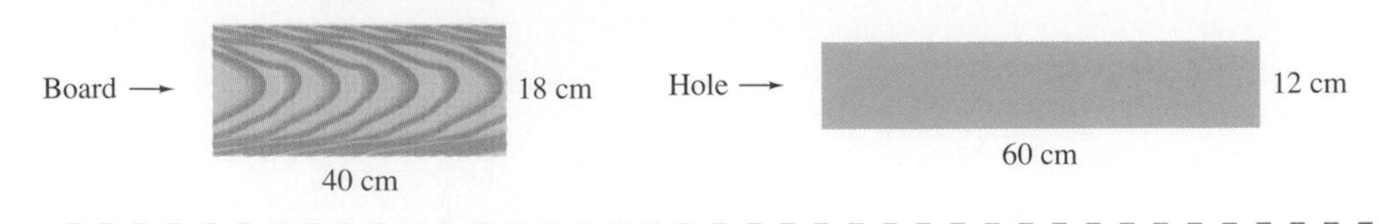

These properties are illustrated in Figure 12.6.

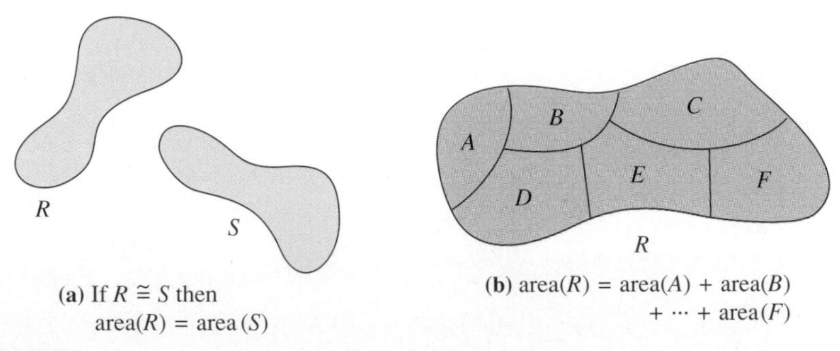

(a) If $R \cong S$ then area(R) = area(S)

(b) area(R) = area(A) + area(B) + ⋯ + area(F)

Figure 12.6
The congruence property and the dissection property of area

The congruence and dissection properties show that rearranging the pieces of a figure forms a new figure with the same area as the original figure.

EXAMPLE 12.5

Solving Leonardo's Problems

Leonardo da Vinci (1452–1519) once became absorbed in showing how the areas of certain curvilinear (curved sided) regions could be determined and compared among themselves and to rectangular regions. The pendulum and the ax are two of the examples he included in notes for his book *De Ludo Geometrico* (roughly meaning "Fun with Geometry"), which he never completed. The dots show the centers of the circular arcs which form the boundary of the region.

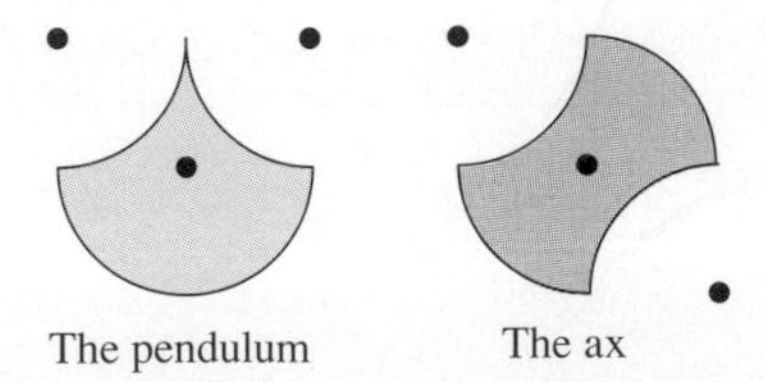

If the arcs forming the pendulum and the ax have radius 1, show that the areas of both figures are equal to that of a 1 by 2 rectangle.

SOLUTION

After inscribing the figures in a square, we then use the congruence-dissection property of area to rearrange the subregions to form a 1 by 2 rectangle.

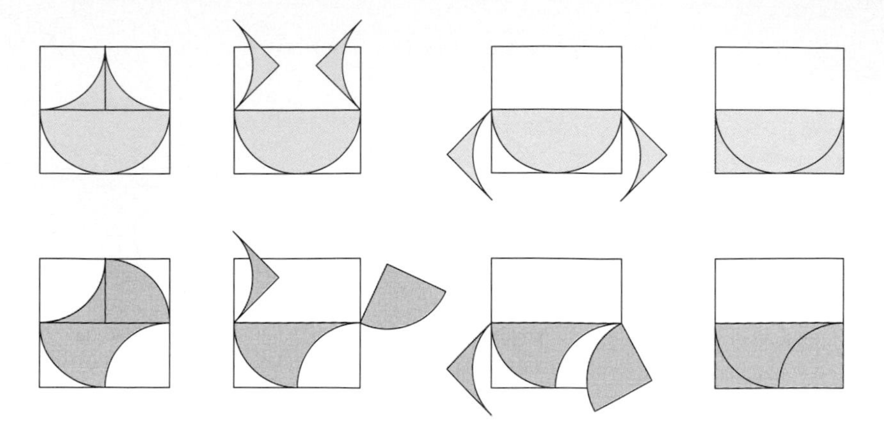

It is surprising to learn that the pendulum and ax both have an area of 2 square units. (These examples and others are described in a booklet by Herbert Wills III, *Leonardo's Dessert*, National Council of Teachers of Mathematics, Reston, Virginia, 1985.) ■

Unlike Examples 12.4 and 12.5, most area measurement problems are answered by giving a reasonable *estimate* of the area. Units of square shape are easy to dissect into smaller squares to give a more precise estimation.

EXAMPLE 12.6

Investigating the Area of a Cycloidal Arch

Imagine rolling a wheel along a straight line, with a reflector at point P on the rim. Point P traces an arch-shaped curve. In the early seventeenth century Galileo investigated this curve and named it the **cycloid**. Discover for yourself the conjecture Galileo made about how the area of the cycloidal arch compares to the area of the circle used to generate the arch.

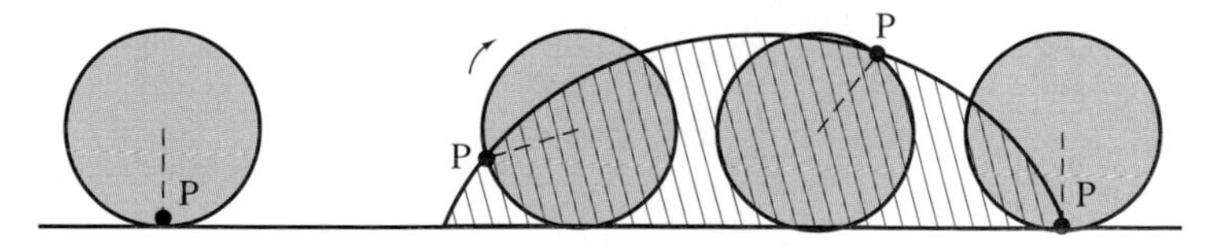

SOLUTION

Understand the problem • • • •

It is visually clear that the cycloidal arch has an area that is much larger than that of the circle. Our goal is to guess the ratio area (cycloid)/area (circle) that gives the comparison between the areas.

Devise a plan • • • •

The areas of the arch and circle must both be measured in some unit of area. For example, we can use squares of size U, where the diameter of the circle is equal to the sum of four side lengths of U. For better accuracy, we can also use small square units of size u, where eight side lengths of u equal a diameter of the circle.

Carry out the plan • • • •

The circle and arch are overlaid by a square grid, with squares of unit area U.

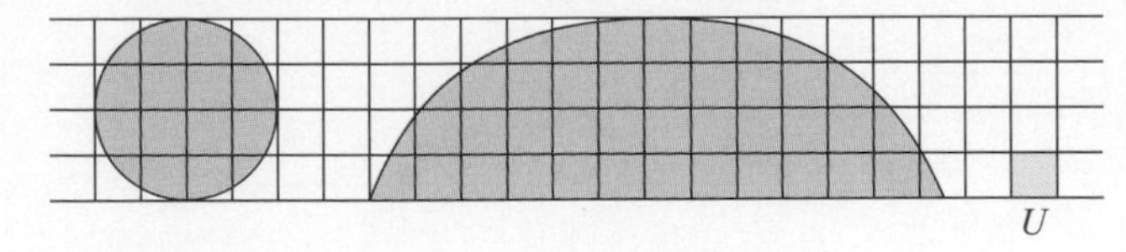

The circle is entirely within 16 squares of size U but does not entirely cover about one unit of area in each of the four corners. Thus we estimate that area (circle) $\doteq$ 12 U. Similarly, we see that area (arch) $\doteq$ 37 U is a reasonable estimate.

Better accuracy is given by the grid of squares of unit area u. The diagram below leads us to the estimates area (circle) $\doteq$ 50 u and area (arch) $\doteq$ 149 u.

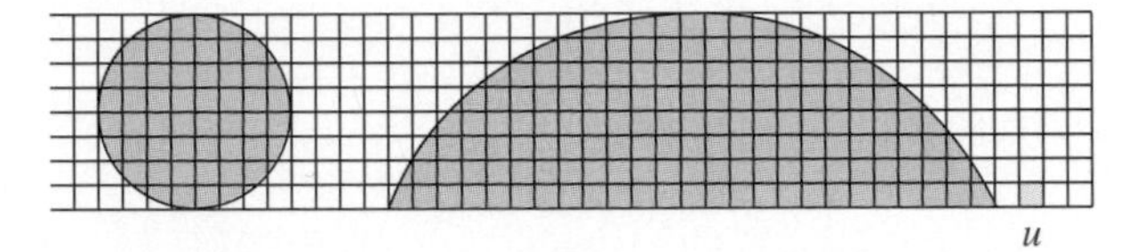

Both $\frac{37\ u}{12\ u}$ and $\frac{149\ U}{50\ U}$ are nearly 3, which in fact was Galileo's conjecture. The correctness of Galileo's conjecture was proved in 1634 by Gilles Persone de Roberval.

Look back • • • •

The finer grid of squares gave us additional precision in our measurements, but this required considerably more time and effort to obtain. The measurement process nearly always requires us to make a judgment about how to balance the conflicting needs of precision versus cost. ■

Areas of Polygons
Rectangles

A 3 cm by 5 cm rectangle can be covered by 15 unit squares when the unit square is 1 cm^2 as shown in Figure 12.7. Similarly a 2.5 cm by 3.5 cm rectangle can be covered by 6 whole units, 5 half unit squares and one quarter unit square, giving a total area of 8.75 square centimeters. This is also the product of the width and the length since 2.5 cm × 3.5 cm = 8.75 cm^2.

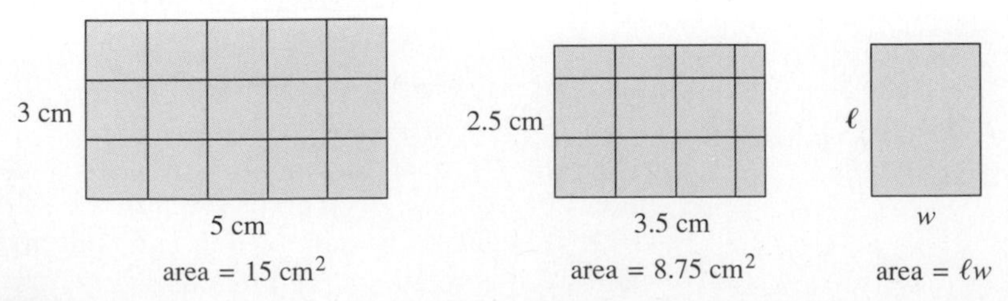

Figure 12.7
The area of a rectangle is the product of its length and width.

For any rectangle the formula for the area A is as follows.

FORMULA Area of a Rectangle

A rectangle of length l and width w has area A given by the formula $A = lw$.

Parallelograms

Suppose a parallelogram has a pair of opposite sides b units long, and these sides are h units apart; an example is shown in Figure 12.8. We say that b is the **base** of the parallelogram and h is the **altitude, or height.** (Unless the parallelogram is a rectangle, the altitude is *not* the same as the length of the other two sides of the parallelogram.) Removing and replacing a triangle T forms a rectangle of the same area as the parallelogram. The rectangle has length b and width h, so its area is bh. Therefore, the area of the parallelogram in Figure 12.8 is also bh.

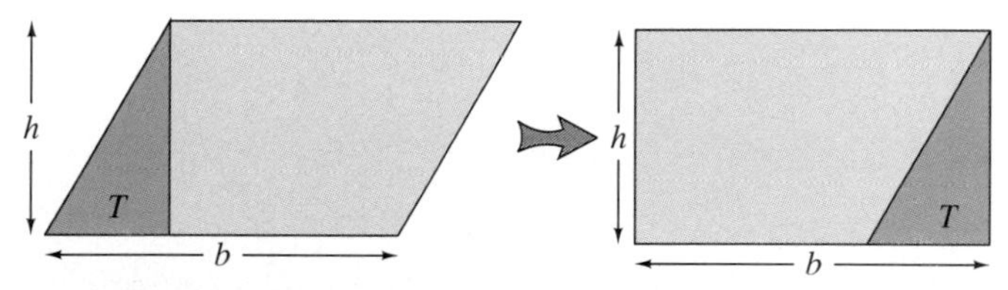

Figure 12.8
A parallelogram of base b and altitude h has the same area as a b by h rectangle. Therefore, the area of the parallelogram is bh.

Any parallelogram with base b and altitude h can be dissected and rearranged to form a rectangle of length b and width h in a similar way to that shown in Figure 12.8. (See problem 8 in the section exercises for a more general case.) This yields the following formula.

JUST FOR FUN

Tile and Smile

Cut a convex quadrilateral from card stock, locate the midpoints of its sides, and then cut along the segments joining successive midpoints to give four triangles T_1, T_2, T_3, T_4, and a parallelogram P.

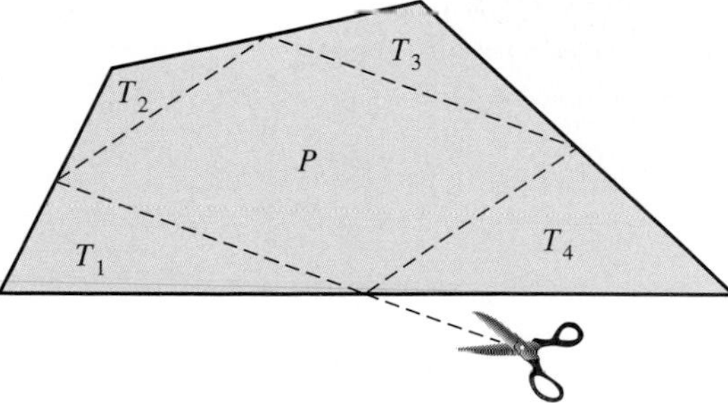

- Show that the four triangles tile the parallelogram.
- How does the area of the parallelogram compare to the area of the original quadrilateral?

FORMULA Area of a Parallelogram

A parallelogram of base b and altitude h has area A given by $A = bh$.

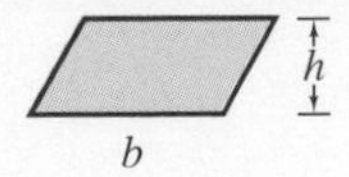

EXAMPLE 12.7

Using the Parallelogram Area Formula

Find the area of each parallelogram, and then compute the lengths x and y.

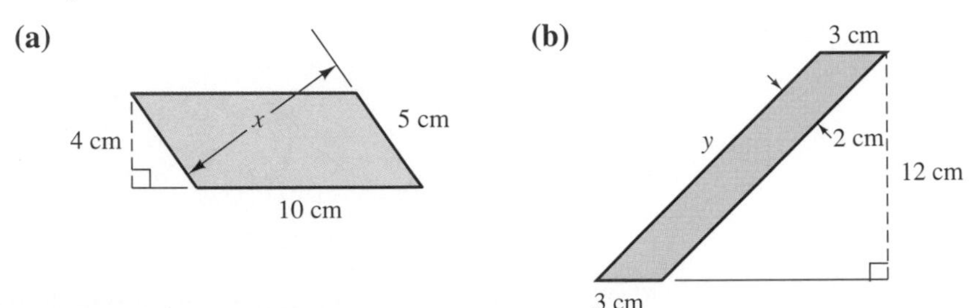

SOLUTION

(a) The parallelogram has base 10 cm and height 4 cm, so its area is $A = (10 \text{ cm})(4 \text{ cm}) = 40 \text{ cm}^2$. If the side of length 5 cm is considered the base, then x is the corresponding height and $A = (5 \text{ cm})x$. Since $A = 40 \text{ cm}^2$, we find $x = 40 \text{ cm}^2/5 \text{ cm} = 8 \text{ cm}$.

(b) The procedure for (a) is followed. The area is $A = (3 \text{ cm})(12 \text{ cm}) = 36 \text{ cm}^2$. Viewing the side of length y as a base with corresponding altitude 2 cm, we have $36 \text{ cm}^2 = y(2 \text{ cm})$. Therefore $y = 36 \text{ cm}^2/2 \text{ cm} = 18 \text{ cm}$. ■

Triangles

Figure 12.9 shows that a triangle of base b and altitude h can be dissected and rearranged to form a parallelogram of base b and altitude $\frac{1}{2}h$. The formula $\frac{1}{2}bh$ for the area of the triangle then follows from the area formula already derived for the parallelogram.

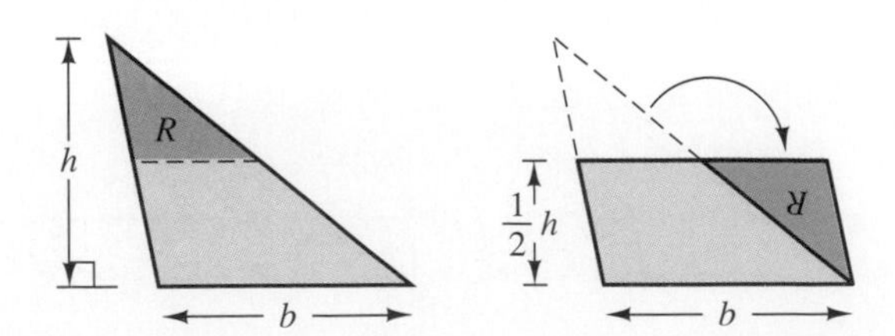

Figure 12.9
A triangle of base b and altitude h can be dissected and rearranged to form a parallelogram of base b and altitude $\frac{1}{2}h$.

FORMULA Area of a Triangle

A triangle of base b and altitude h has area $A = \frac{1}{2}bh$.

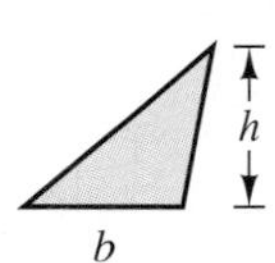

Any side of a triangle can be considered as the base so there are three pairs of bases and altitudes.

EXAMPLE 12.8

Using the Triangle Area Formula

Find the area of each triangle and the distances v and w.

(a)

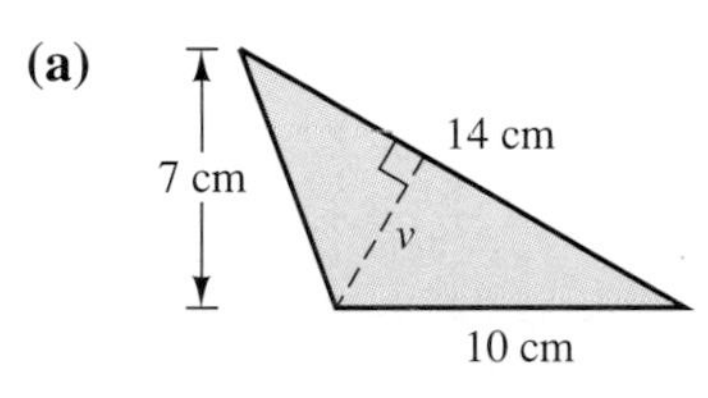

(b)

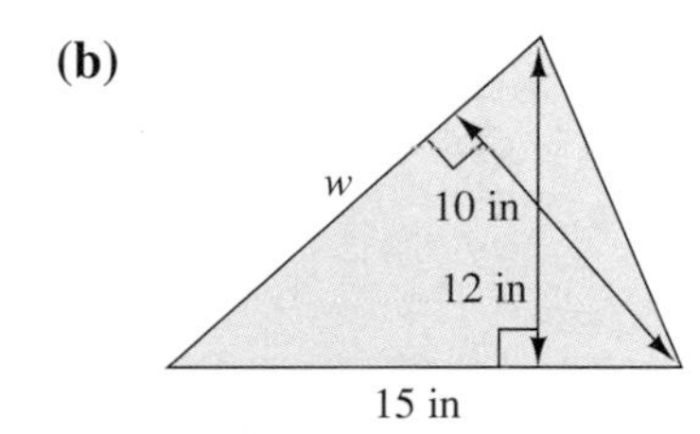

SOLUTION

(a) The formula $A = \frac{1}{2}bh$ shows the area of the triangle is $A = \frac{1}{2}(10 \text{ cm})(7 \text{ cm}) = 35 \text{ cm}^2$. If the side of length 14 cm is considered the base, then the corresponding altitude is v. Since $A = \frac{1}{2}(14 \text{ cm})(v)$, $v = A/(7 \text{ cm}) = (35 \text{ cm}^2)/(7 \text{ cm}) = 5 \text{ cm}$.

(b) $A = \frac{1}{2}(15 \text{ in})(12 \text{ in}) = 90 \text{ in}^2$ is the area of the triangle. Considering the side of length w as the base, the corresponding altitude is 10 in and $A = \frac{1}{2}w(10 \text{ in})$. Therefore, $w = (90 \text{ in}^2)/(5 \text{ in}) = 18 \text{ in}$. ■

Trapezoids

There are several ways to derive the formula for the area of a trapezoid; some suggestions are given in problem 11 at the end of the section, but you may enjoy searching for a method of your own.

FORMULA Area of a Trapezoid

A trapezoid with bases of length a and b and altitude h has area $A = \frac{1}{2}(a + b)h$.

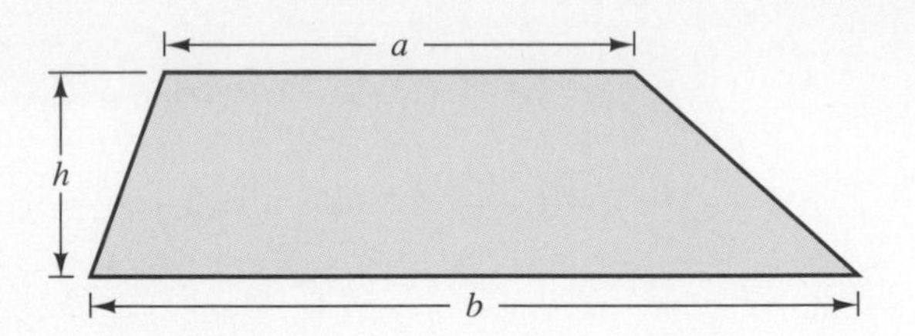

Figure 12.10
The area of a trapezoid is $\frac{1}{2}(a + b)h$.

EXAMPLE 12.9

Finding the Areas of Lattice Polygons

A polygon formed by joining the points of a square array is called a **lattice polygon.** Lattice polygons are easy to draw on dot paper, or they can be formed with rubber bands on a geoboard. Find the area of the lattice polygons shown, where the unit of area is the area of a small square of the array.

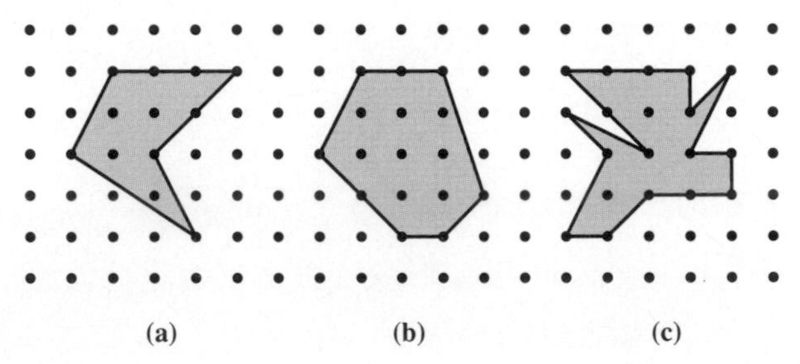

SOLUTION

(a) A horizontal line dissects the polygon into a trapezoid A of area $\frac{1}{2}(3 + 2) \cdot (2) = 5$ and a triangle B of area $\frac{1}{2}(2)(2) = 2$. The area of the polygon is therefore 7.

(b) The lattice hexagon can be dissected into trapezoids C and D and triangle E. The total area of the hexagon is therefore

$$\frac{1}{2}(2 + 3)(2) + \frac{1}{2}(3 + 1)(2) + \frac{1}{2}(4)(1) = 11.$$

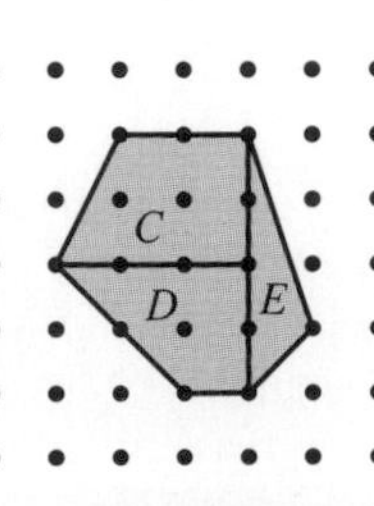

Other dissections of the hexagon can be used, but the total area will always be the same.

(c) We could solve the problem in the same way as parts (a) and (b), but there is another useful technique: construct a square about the polygon, and then subtract the areas of the regions F, G, H, I, and J. Therefore the area of the polygon is

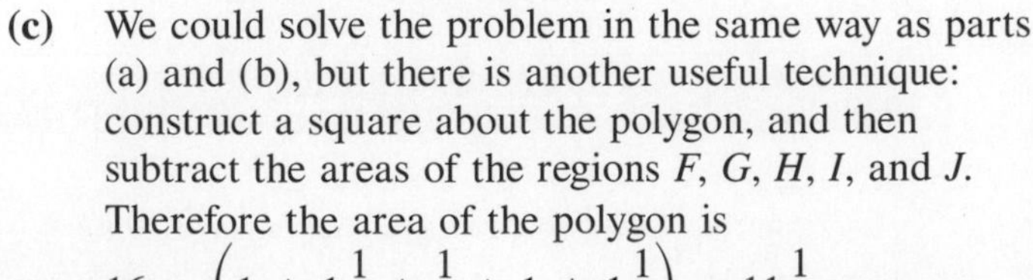

$$16 - \left(1 + 1\frac{1}{2} + \frac{1}{2} + 1 + 1\frac{1}{2}\right) = 11\frac{1}{2}.$$

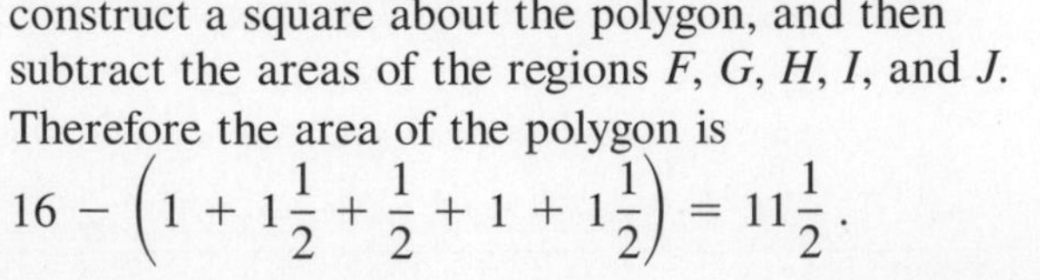

■

Length of a Curve

The length of a polygonal curve is obtained by summing the lengths of its sides. The length of a nonpolygonal curve is measured, or at least estimated, by calculating the length of an approximating polygonal curve with vertices on the given curve. The accuracy of the estimation is improved by using an approximating polygonal curve with more vertices, as shown in Figure 12.11.

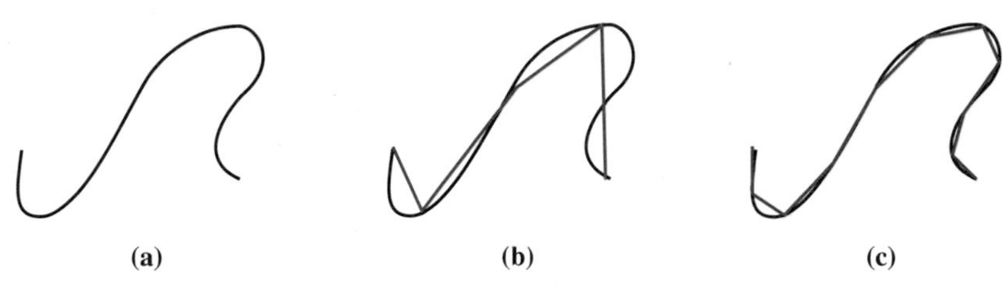

Figure 12.11
The length of a curve (a) is estimated by measuring the length of a polygonal approximation (b). Increasing the number of vertices gives an improved estimate, as in (c).

The length of a curve can also be measured by first laying a string along the curve and then straightening the string along a ruler. This is the principle that makes the flexible tape measure used for sewing so useful.

EXAMPLE 12.10

Determining the Length of a Cycloid

A circle and the cycloid it generates (see Example 12.7) are shown below. Use a marker pen and a piece of string (or thin strip of paper) to make a tape measure, where the unit of length is the diameter d of the circle.

(a) According to your tape measure, what is the length from point A to point B along the cycloid?

(b) What is the approximate length of the line segment $\overline{AB}$?

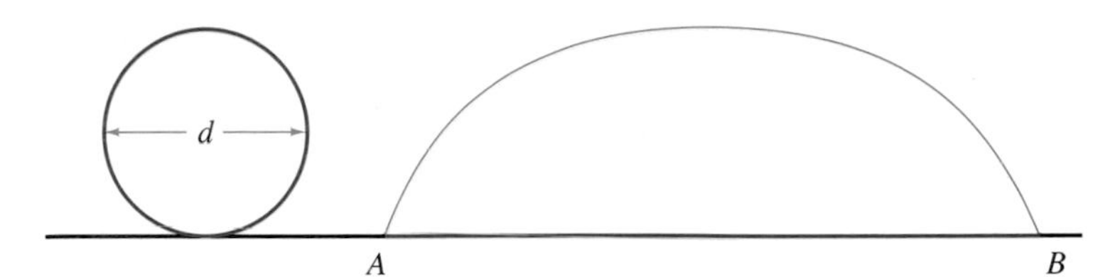

SOLUTION

(a) The tape measure shows that the length of the cycloid is very near four diameters of the circle. In 1658, Christopher Wren (1632–1723) proved that the length of a cycloid is *exactly* four diameters. Wren is perhaps best known as the architect of St. Paul's Cathedral in London.

(b) The segment $\overline{AB}$ is a bit over 3 diameters. Because $\overline{\text{AB}}$ is covered by rolling the circle once around, AB is the length around the circle; that is, AB is the circumference. ■

Perimeter

The length of a simple closed plane curve is called its **perimeter.** Therefore, perimeter is a *length* measurement and is given in centimeters, inches, feet, meters, and so on. It is important that the *area* of the region enclosed by a simple closed curve not be confused with the perimeter of the figure. Area is given in cm^2, in^2, ft^2, m^2, and so on. In summary: perimeter is the measure of the distance around a region, and area is the measure of the size of the region within a boundary.

EXAMPLE 12.11

Finding Perimeters

The following figures have been drawn on a square grid, where each square is 1 cm on a side. Give the perimeter and area of each figure.

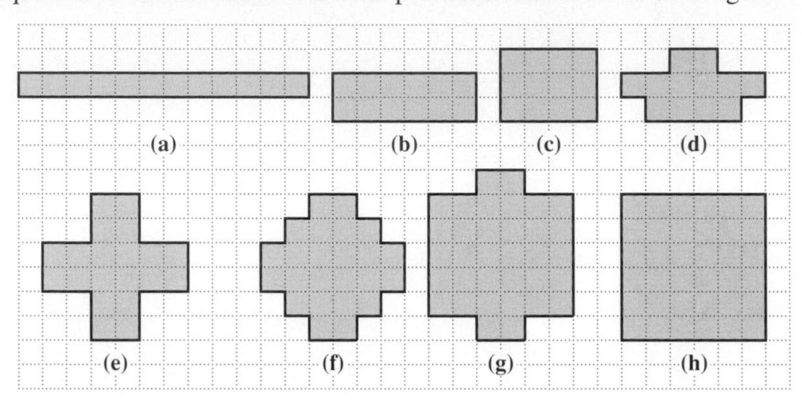

SOLUTION

Figure	(a)	(b)	(c)	(d)	(e)	(f)	(g)	(h)	
Perimeter	26	16	14	18	24	24	24	24	centimeters
Area	12	12	12	12	20	24	28	36	square centimeters

Figures (a), (b), (c), and (d) have the same area but different perimeters. Figures (e), (f), (g), and (h) have the same perimeter but different areas. ■

The Circumference of a Circle

The perimeter of a circle is called its **circumference.** Using a piece of string or a tape measure, or by rolling a disc along a line (as in Example 12.10 (b)), it is easy to rediscover a fact known even in ancient times: The ratio of the circumference of a circle to its diameter is the same for all circles. Two examples are shown in Figure 12.12. This ratio, which is somewhat larger than 3, is given by the symbol π, the lower case Greek letter *pi*.

DEFINITION π

The ratio of the circumference C to the diameter d of a circle is π. Therefore,

$$\frac{C}{d} = \pi \quad \text{and} \quad C = \pi d.$$

Since the diameter d is twice the radius r of the circle, we also have the formula $C = 2\pi r$.

In 1761 John Lambert proved that π is an irrational number, so it is impossible to express π exactly by a fraction or as a terminating or repeating decimal. The values

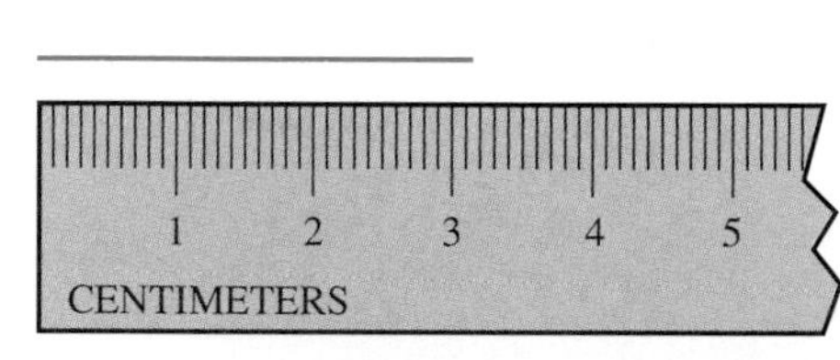

Figure 12.12
The ratio of the circumference C to the diameter d is the same for all circles: $c/d = \pi$, or $C = \pi d$.

$3\frac{1}{7}$ and 3.14 are useful approximate values, but precision measurements require using more decimal places in the unending decimal expansion $\pi = 3.1415927 \cdots$. A circle 100 feet in diameter has an *approximate* circumference of 314 feet, but the exact circumference is 100 π feet. It is acceptable to use the symbol π to express results, since this gives exact values. When an approximate numerical value is needed, an appropriate estimate of π such as 3.1416 can be used in the calculations.

EXAMPLE 12.12

Calculating the Equatorial Circumference of the Earth

The equatorial diameter of the earth is 7926 miles. Calculate the distance around the earth at the equator using the following approximations for π: (a) 3.14 (b) 3.1416.

SOLUTION

(a) (3.14) (7926 miles) $\doteq$ 24,887.64 miles

(b) (3.1416) (7926 miles) $\doteq$ 24,900.322 miles

The two different approximations of π account for the difference of about 12.7 miles in the answers. ■

The Area of a Circle

The area of a circle of radius r is given by the formula πr^2, first proved rigorously by Archimedes.

FORMULA Area of a Circle

The area A enclosed by a circle of radius r is $A = \pi r^2$.

Since π is defined as a ratio of lengths, it seems surprising to find that π also occurs in the formula for the area of a circle. A convincing, but informal, derivation of the formula, $A = \pi r^2$, is shown in Figure 12.13. The circle of radius r and circumference $C = 2\pi r$ is dissected into congruent sectors which are rearranged to form a "parallelogram" of base $\frac{1}{2}C = \pi r$ and altitude r. By the formula for the area of a parallelogram, the wavy-based "parallelogram" has area $\pi r \times r = \pi r^2$. If the number of sections is made larger and larger, the thin sectors form an increasingly exact approximation to a true parallelogram of area πr^2.

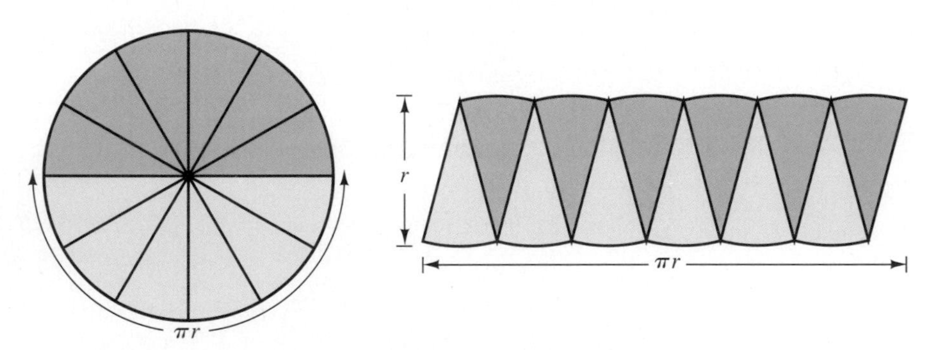

Figure 12.13
The sectors of a circle can be rearranged to approximate a parallelogram of area πr^2.

HIGHLIGHT FROM HISTORY
A Brief History of π

In the third century B.C., Archimedes showed that π is approximately $3\frac{1}{7}$. To estimate π, Archimedes inscribed a regular polygon in a circle and then calculated the ratio of the polygon's perimeter to the diameter of the circle. An inscribed hexagon shows π is about 3, but by using a 96-gon Archimedes proved that $3\frac{10}{71} < \pi < 3\frac{10}{70}$. The same idea was used by the Dutch mathematician Van Ceulen (d. 1610) who used a 32,212,254,720-gon to calculate π to 20 decimals. A century later the English mathematician John Machin took advantage of the invention of calculus to calculate π to 100 decimal places. Machin's method, with some minor variations, was used well into the twentieth century. When implemented on the ENIAC in 1949, the first electronic computer spent 70 hours to calculate π to 2037 decimal places.

In 1991 π was calculated to 2,260,321,336 decimal places by David and Gregory Chudnovsky. The two brothers, immigrants from the former Soviet Union, achieved this record feat by programming their homemade supercomputer with formulas derived from the work done in 1914 by the self-taught Indian mathematician Srinivasa Ramanujan. Gregory, who has coped with the muscular disorder myasthenia gravis since age 12, nevertheless has convenient access to the computer since it is kept in his Manhattan apartment.

EXAMPLE 12.13 **Determining the Size of a Pizza π**

A 14″ pizza has the same thickness as a 10″ pizza. How many times more ingredients are there on the larger pizza?

SOLUTION

Pizzas are measured by their diameters, so the radii of the two pizzas are 7″ and 5″. Since the thicknesses are the same, the amount of ingredients used is proportional to the areas of the pizzas. The larger pizza has area $\pi(7 \text{ in})^2 = 49\pi \text{ in}^2$, and the smaller pizza has area $\pi(5 \text{ in})^2 = 25\pi \text{ in}^2$. The ratio of areas is $49\pi \text{ in}^2/25\pi \text{ in}^2 = 1.96$, showing that the 14″ pizza has about twice the ingredients of the 10″ one. ■

PROBLEM SET 12.2

Understanding Concepts

1. A square with sides 1 meter long contains 1 square meter of area. That is, a 1 meter square has 1 m^2 of area. Now compare these areas, and decide which is larger:
   (a) a 2 meter square (that is, a square with sides 2 m long), or a square of area 2 m^2.
   (b) a $\frac{1}{2}$ meter square, or a square of area $\frac{1}{2}$ m^2.

2. Botanists often need to measure the rate at which water is lost by transpiration through the leaves of a plant. It is necessary to know the leaf area of the plant. Estimate the area of the leaf shown. It has been overlaid with a grid of squares 1 cm on a side, shown at half scale.

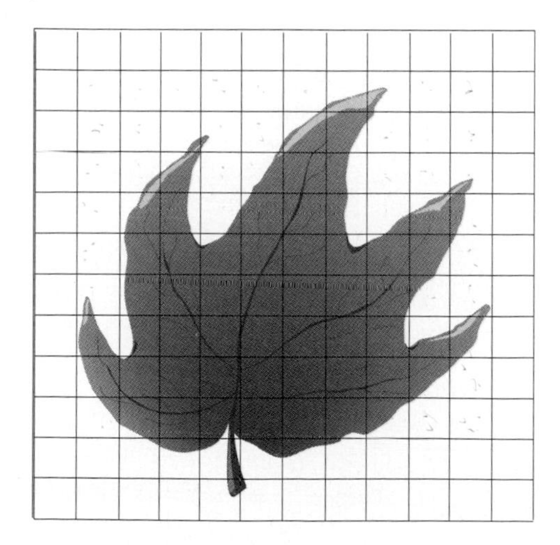

3. Measure the figure F shown below in each of the three nonstandard units of area (a), (b), and (c) shown to the right. Do so by tracing F and then tiling the region with the unit area shape.

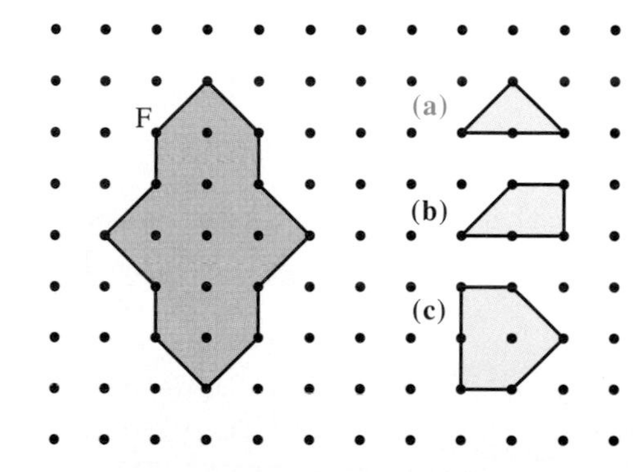

4. The Hungarian mathematician J. Kurschak (1864–1933) dissected a regular 12-gon into triangles of two shapes, as shown on the left below. If the 12-gon has a diameter of 2 as shown, what is its area? Explain your reasoning, using the right hand figure.

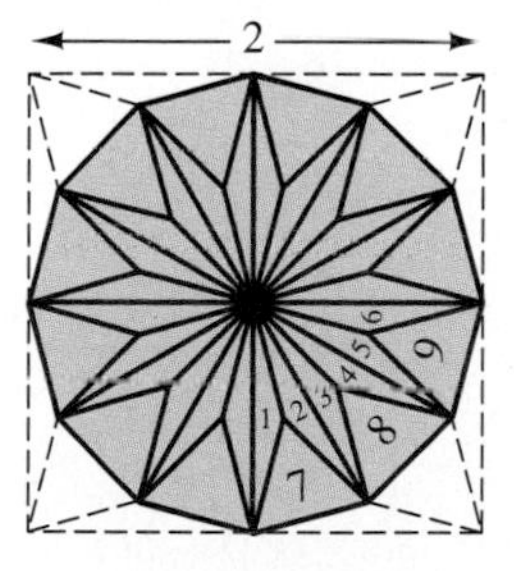

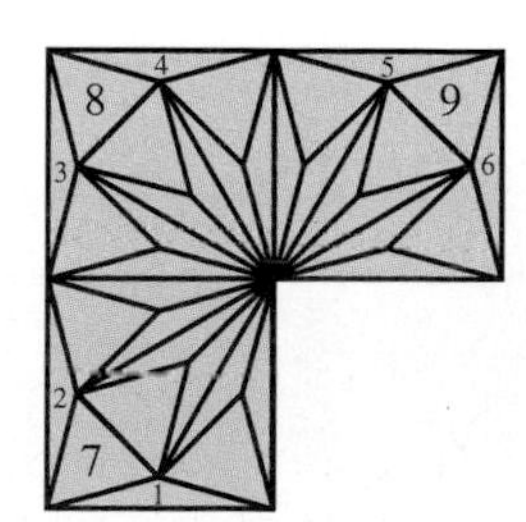

5. Four mutually tangent circles of diameter 10 cm with their centers at the vertices of a square are used to draw a vase.
   (a) Use dissection and rearrangement to form a square of the same area as the vase. (Show a sequence of steps similar to the solutions in Example 12.5.)
   (b) Show the vase has area 100 cm^2.

(c) As an extra challenge, see if you can do part **(a)** by cutting the vase into just 3 pieces.

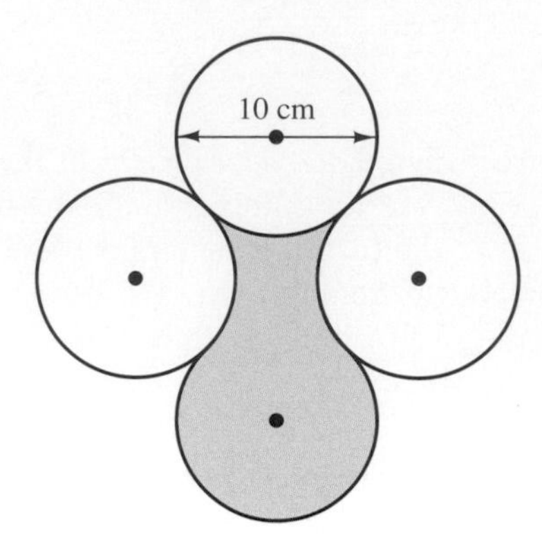

6. **(a)** A rectangle has area 36 cm^2 and width 3 cm. What is the length of the rectangle?
   (b) A rectangle has area 60 cm^2 and perimeter 38 cm. Use "guess and check" to find the length and width of the rectangle.
7. Twenty-four 1 cm by 1 cm squares are used to tile a rectangle.
   (a) Find the dimensions of all possible rectangles.
   (b) Which rectangle has the smallest perimeter?
   (c) Which rectangle has the largest perimeter?
8. Draw the parallelogram shown below on graph paper and cut it out with scissors. Then show how to dissect the parallelogram into subregions that can be rearranged into a 2 cm by 7 cm rectangle. (*Hint:* Use several vertical lines, 2 cm apart.)

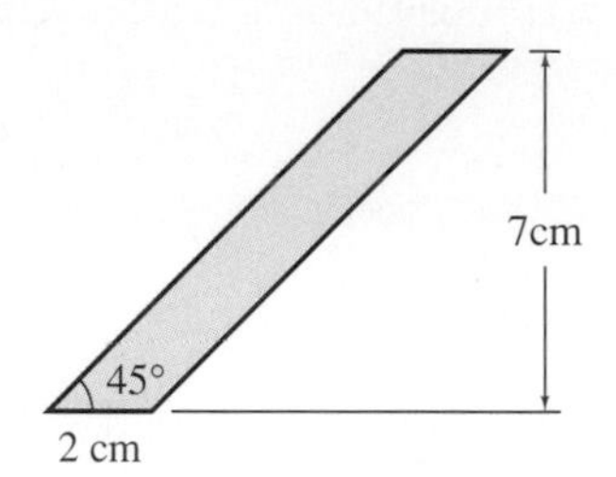

9. Cut several congruent pairs of triangles of assorted shapes from paper. By first folding a sheet of paper in half, the two triangles in each pair can be cut simultaneously.
   (a) Show that the two triangles in each pair can be arranged to form a parallelogram.
   (b) Use part (a) to explain how the area formula for a triangle follows from the area formula for a parallelogram.
10. Cut several triangles as illustrated from paper.

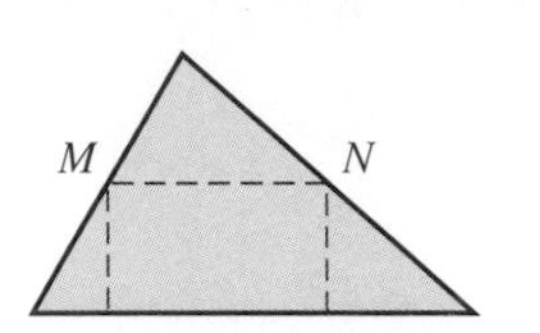

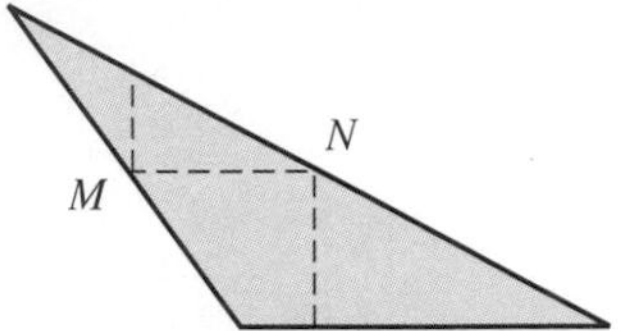

Find the midpoints, M and N, of the slanted sides (that is, the sides not the base) by folding.
   (a) Fold on the horizontal and vertical lines (shown dashed) to form a doubly covered (two layers of paper) rectangle.
   (b) If the triangle has base b and altitude h, what are lengths of the sides of the rectangle you formed by folding?
11. There are many ways to derive the formula $\frac{1}{2}(a + b)h$ for the area of a trapezoid with bases a and b and altitude h. Use words and sketches to explain how to derive the formula using each of these different methods.
   (a) Form a parallelogram with two congruent copies of the trapezoid.
   (b) Dissect the trapezoid by a line through he midpoints of the sides joining the bases and rearrange the two subregions into a parallelogram.
   (c) Dissect the trapezoid into two triangles by a diagonal.

---

12. Find the areas and perimeters of the following parallelograms. Be sure to express your answer in the appropriate units of measurement.

(a)

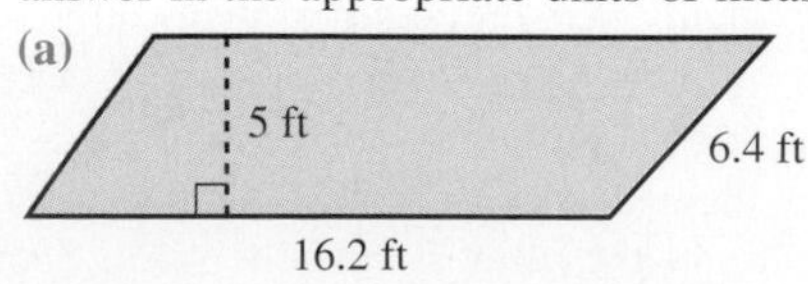

(b)

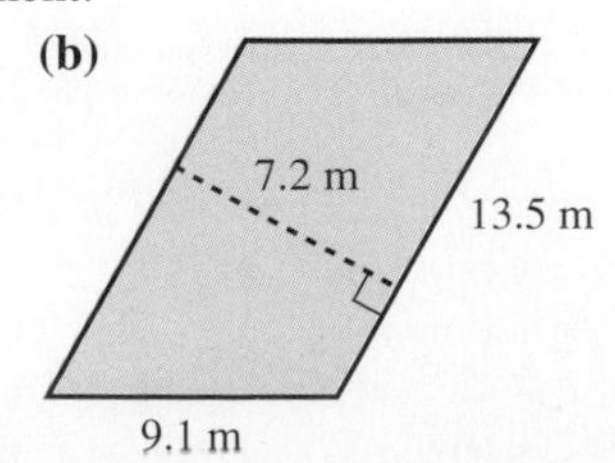

(c)

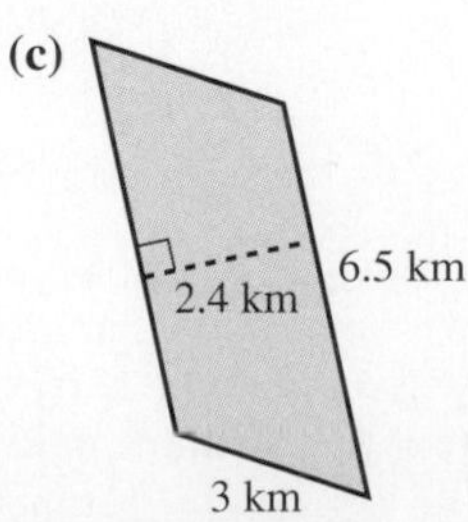

13. Find the areas and perimeters of these triangles.

(a)

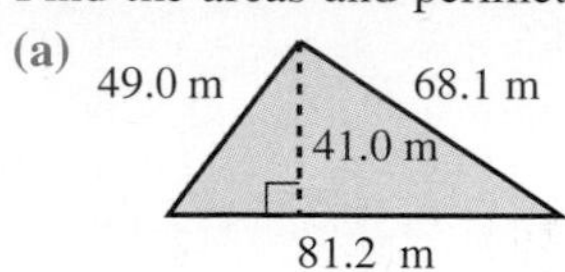

(b)

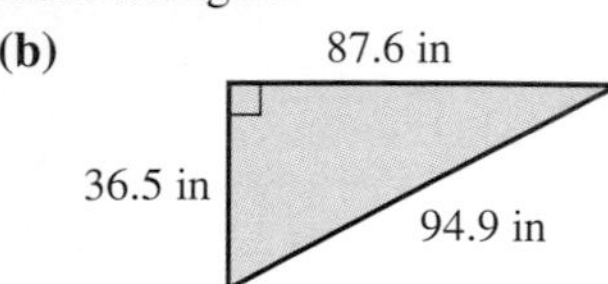

(c)

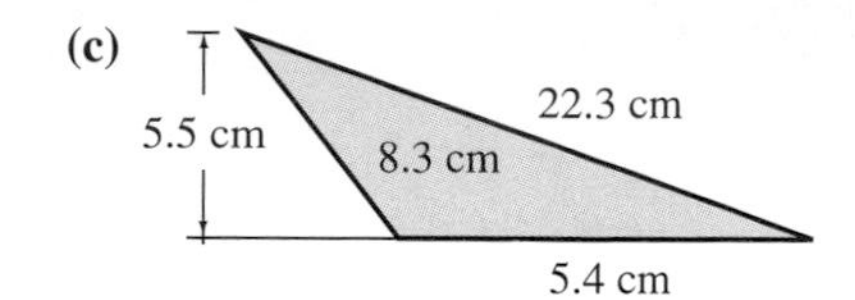

14. Find the areas of these figures. Express the area in square units.

(a)

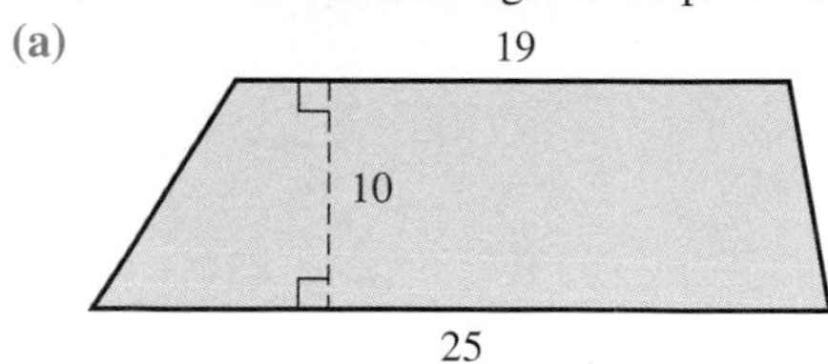

(b)

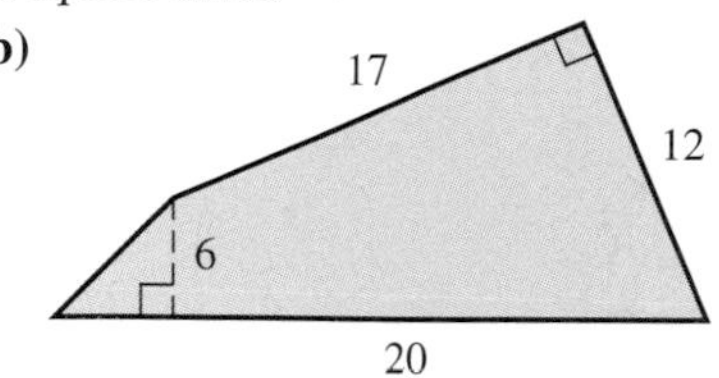

(c)

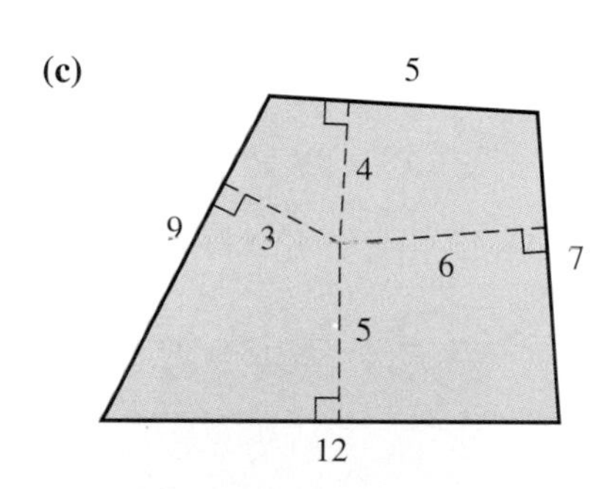

(d)

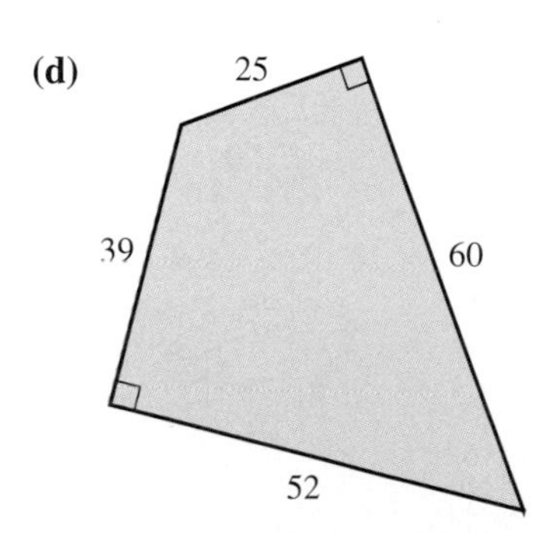

15. Lines k, l, and m are parallel to the line containing the side $\overline{AB}$ of the triangles shown here.

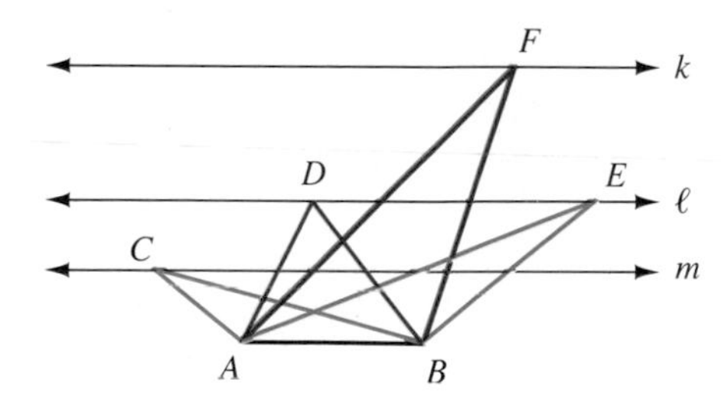

(a) What triangle has the smallest area? Why?

(b) What triangle has the largest area? Why?

(c) Which two triangles have the same area? Why?

16. Lines k, l, and m are equally spaced parallel lines. Let $ABCD$ be a parallelogram of area 12 square units.

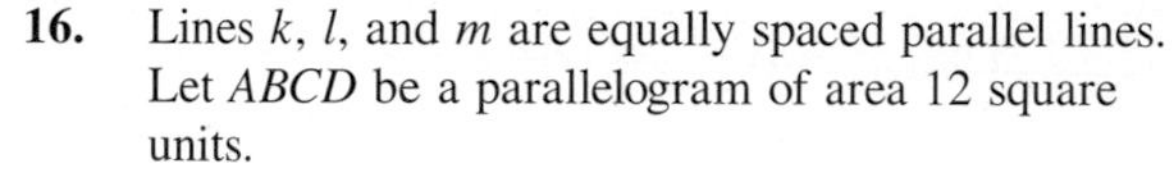

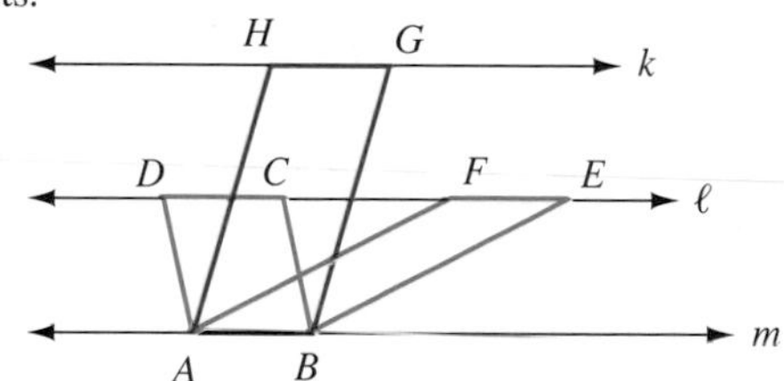

(a) What is the area of the parallelogram $ABEF$?

(b) What is the area of the parallelogram $ABGH$?

(c) If $AB = 4$ units of length, what is the distance between the parallel lines?

17. Find the area of each lattice polygon shown below.

(a)

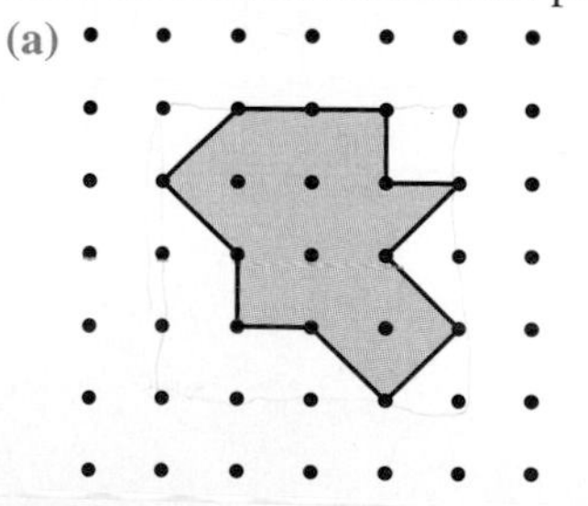

(b)

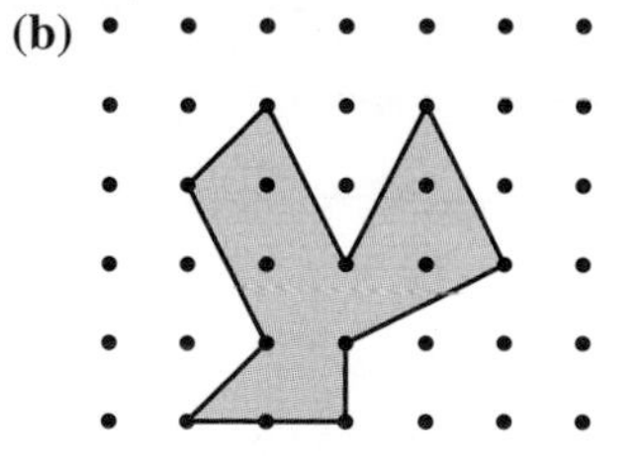

(c)

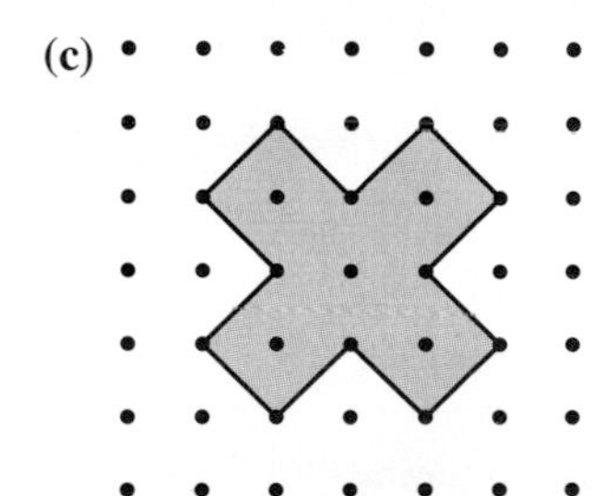

18. **(a)** Find the areas of the three similar triangles A, B, and C.

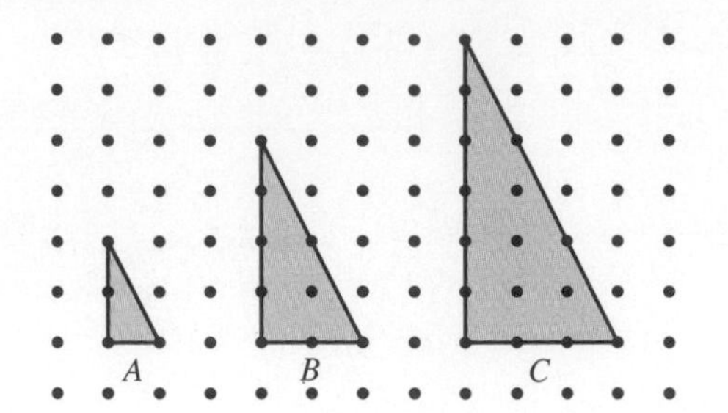

(b) The sides of triangle B are twice the length of the corresponding sides of triangle A. What is the ratio, area (B)/area (A), of areas of the triangles?

(c) The scale factor from triangle A to triangle C is 3. What is ratio of areas, area (C)/area (A)?

19. An oval track is made by erecting semicircles on each end of a 50 m by 100 m rectangle.

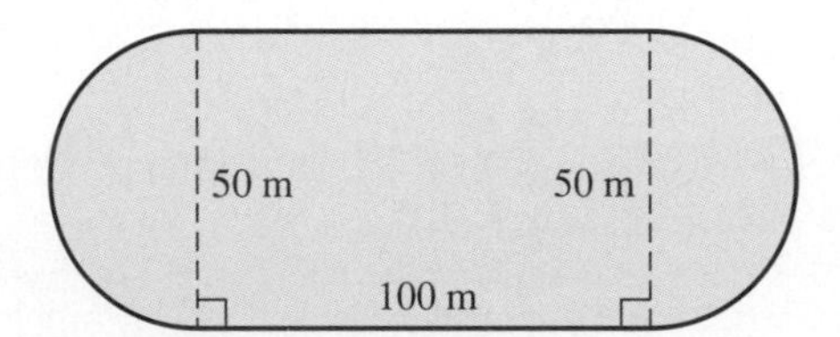

(a) What is the length of the track?

(b) What is the area of the region enclosed by the track?

20. An **annulus** is the region bounded by two concentric circles.

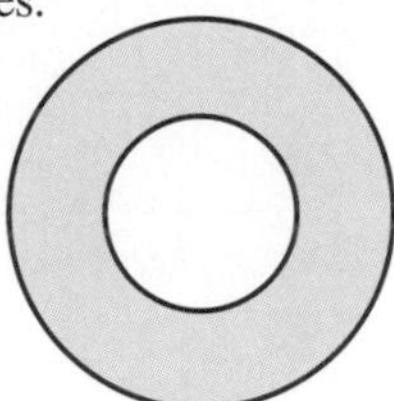

(a) If the radius of the small circle is 1 and the radius of the larger circle is 2, what is the area of the annulus?

(b) A dart board has four annular rings surrounding a bull's-eye.

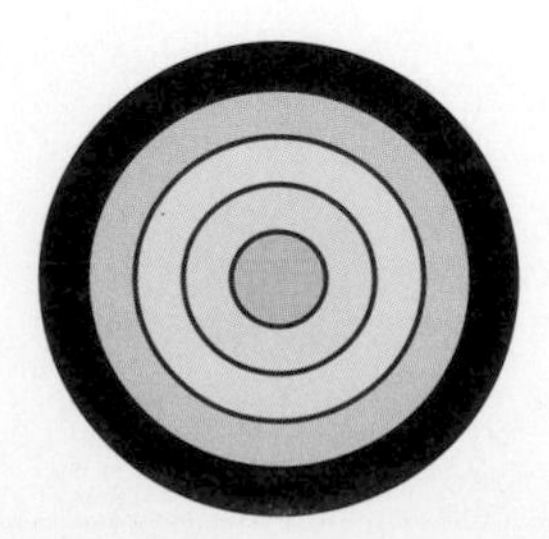

The circles have radii 1, 2, 3, 4, and 5. Suppose a dart is equally likely to hit any point of the board. Is the dart more likely to hit in the outermost ring (shown black) or inside the region consisting of the bull's-eye and the two innermost rings?

21. **(a)** A circle is inscribed in a square. What percentage of the area of the square is inside the circle?

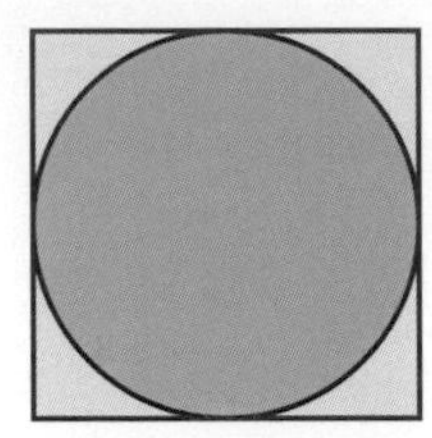

(b) A wall is covered with a pattern of 15 discs as shown. If a dart is thrown at the wall, what is the probability that the dart sticks at a point in one of the spaces between the discs?

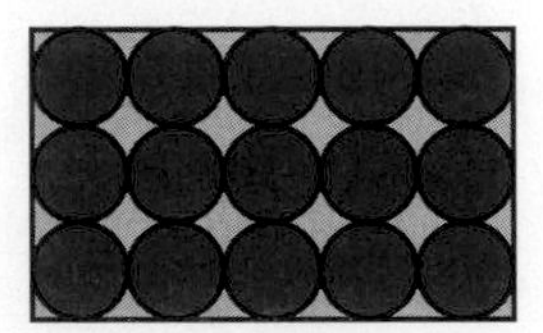

22. The meter was originally defined as one ten-millionth of the distance from the North Pole to the equator.

(a) Assuming the earth is a perfect sphere what would be the circumference of a great circle on the earth that passes through the North and South poles?

(b) The diameter of the equatorial circle of the earth is 12,755 kilometers. What is the circumference of the equator?

(c) Which is longer, the polar circle or the equator? Can you account for the difference?

Thinking Critically

23. Two regions, A and B, are cut from paper. Suppose the area of region A is 20 cm^2 larger than that of region B. If the regions are overlapped, by how much does the area of the nonoverlapped part of region A exceed the nonoverlapped part of region B? Explain carefully.

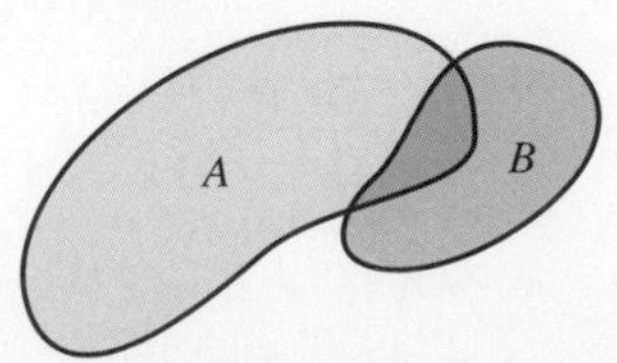

24. The four boards shown are each 1 foot wide and 12 feet long, and overlap at right angles. What is the area of the ground shaded by the boards?

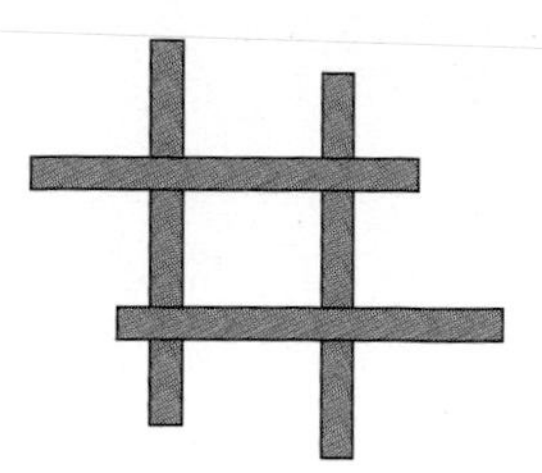

25. (a) The colored region shown below is formed by circular arcs drawn from two opposite corners of a 1 by 1 square. What is the area of the region?

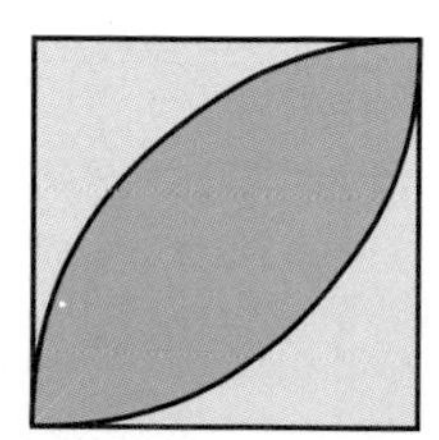

(b) Four semicircles are drawn at the midpoints of the sides of a 1 by 1 square. What is the area of the shaded region?

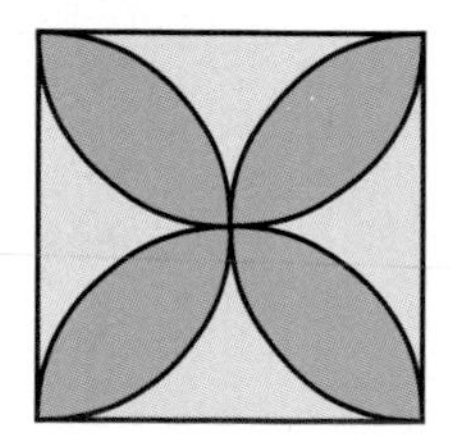

26. A square cake measures 8″ by 8″. A wedge-shaped piece is cut by two slices meeting at 90° at the cake's center. What is the area of the top of the piece? Explain your reasoning carefully.

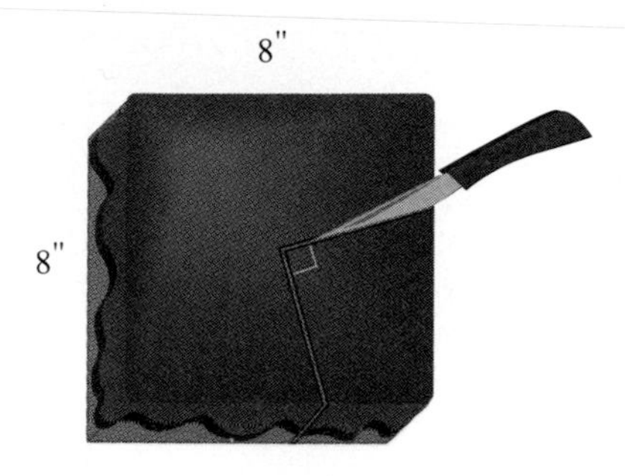

27. A sidewalk 8 feet wide surrounds the polygonally shaped garden of perimeter 300 feet as shown. The sidewalk makes circular arcs of 8 foot radius at the vertices of the polygon. Explain why the area of the area covered by the walk is $2400 + 64\pi$ square feet.

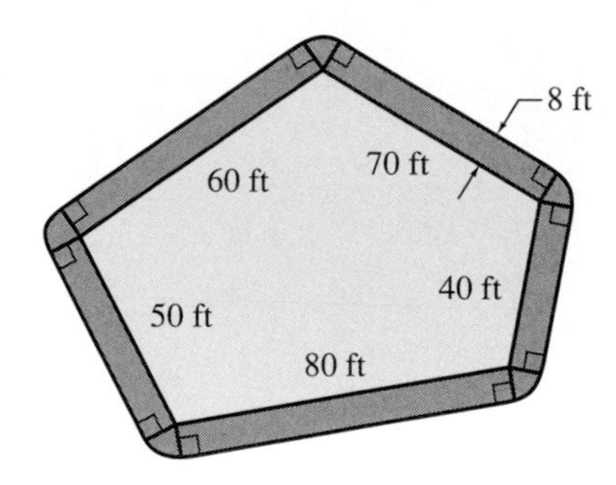

28. The commentaries of the Talmud (Tosfos Pesachim 109a, Tosfos Succah 8a, Marsha Babba Bathra 27a) present a nice approach to the formula, $A = \pi r^2$, for the area of a circle. Imagine that the interior of a circle is covered by concentric circles of yarn. Clipping the yarn circles along a vertical radius, each strand is straightened to cover an isosceles triangle. Find the area of the triangle and then explain how the area formula for a circle follows.

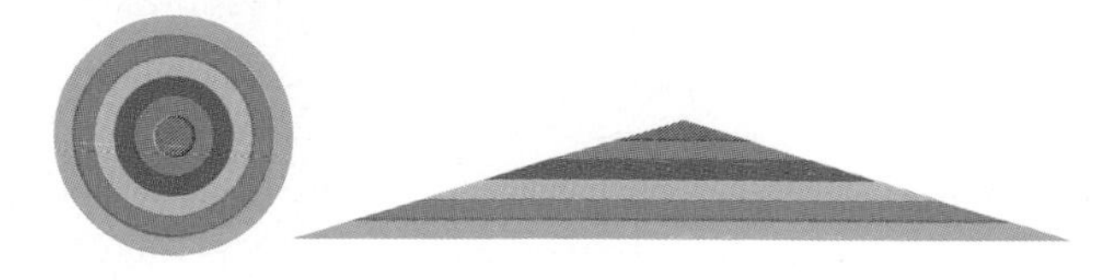

29. A board lays across tin cans of 10″ circumference as illustrated to the right. Suppose the cans all roll to the right without slipping.

(a) When the cans have made one complete revolution how far has the board moved?

(b) If the cans move to the right at 10 miles per hour, what is the speed of the board?

Thinking Cooperatively

30. (a) Find the areas of the circular sectors.

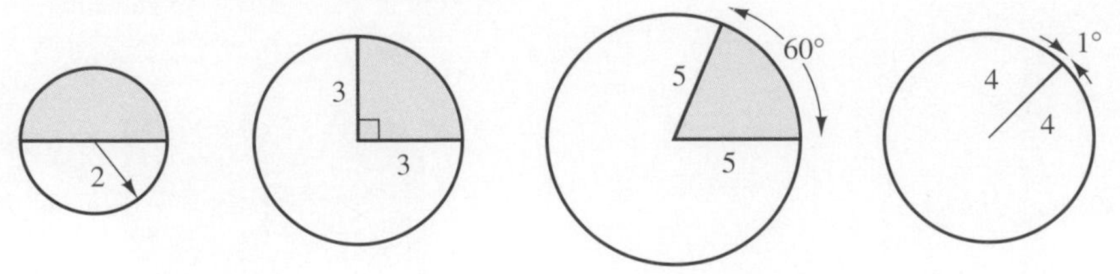

(b) Propose a formula for the area of a sector of angle $x°$ in a circle of radius r.

31. A student found the formula $x\pi r/180$ for the length of the arc intercepted by a central angle measuring $x°$ in a circle of radius r.

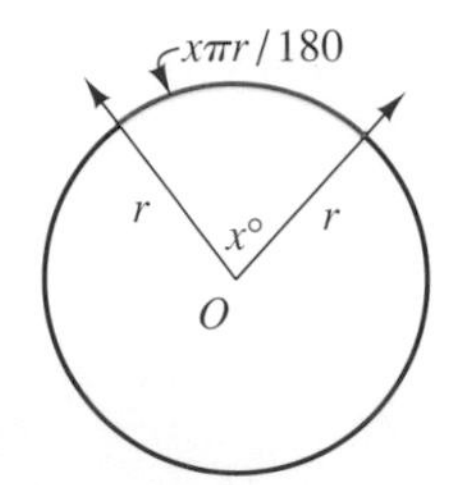

(a) Does this formula give correct results for the arcs in the four circles shown in problem 30(a)?

(b) Explain, as you would to a student, why the formula is reasonable.

32. Let P be an arbitrary point in an equilateral triangle ABC of altitude h and side s as shown. What is the sum $x + y + z$ of the distances to the sides of the triangle? (*Hint:* The areas of $\triangle ABP$, $\triangle BCP$, and $\triangle ACP$ add up to area ($\triangle ABC$).)

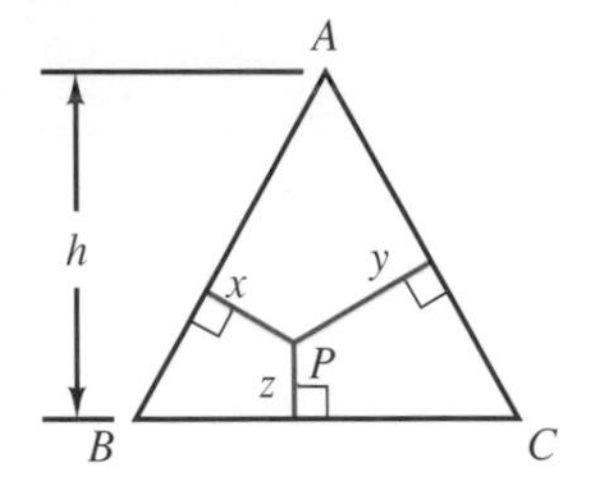

33. Joining each vertex of a triangle shown here to the midpoint of the opposite side divides the triangle into six small triangles.

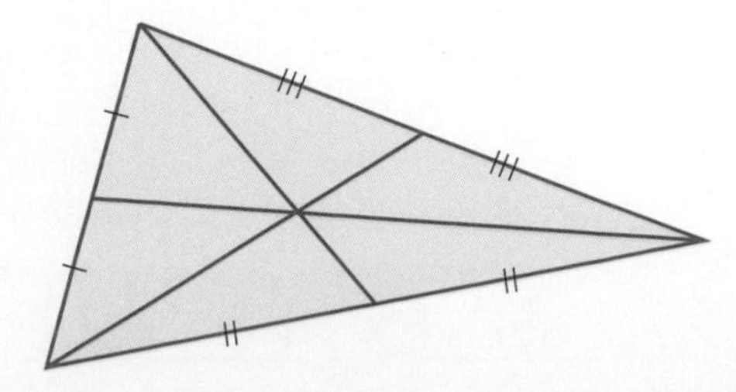

Show that all six triangles have the same area. (*Hint:* Look for pairs of triangles with the same base and height.)

Making Connections

34. Roll-ends of carpet are on sale for six dollars per square yard. To finish the rough-cut edges, edging material costing ten cents per foot is glued in place. Compute the total cost of a roll-end measuring 8 feet by 10 feet.

35. A carpet is made by sewing a one inch wide braid around and around until the final shape is an oval with semicircular ends as shown. Estimate the length of braid required. (*Hint:* Estimate the area of the carpet.)

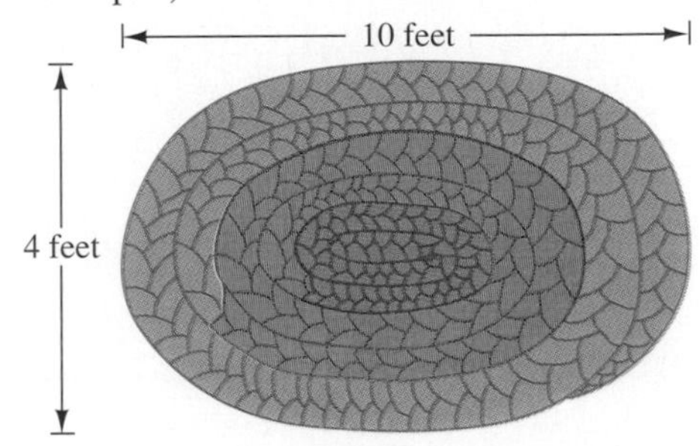

36. An L-shaped house, walkway, garage, and driveway are shown situated on a 70′ by 120′ lot. How many bags of fertilizer are needed for the lawn? Assume the bags are each 20 pounds, and one pound of fertilizer will treat 200 square feet of lawn.

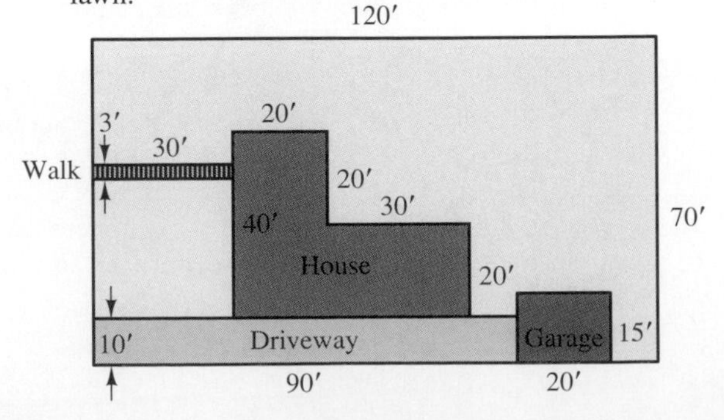

37. A 10′ by 12′ kitchen is to be tiled with 8″ square tiles. Estimate the number of tiles this will require.

38. **(a)** The normal sized tires on a truck have a 14 inch radius. If oversized tires of 15 inch radius are used how much farther does the truck travel per revolution of the wheel?

(b) If the speedometer indicates the truck is traveling at 56 miles per hour, what is the true speed when the truck is running on the oversized tires?

39. Sunaina has 600 feet of fencing. She wishes to build a corral along an existing high straight wall. She has already decided to make the corral in the shape of an isosceles triangle, with two sides each 300 feet long. What is the measure of $\angle A$ which will give Sunaina the corral of most area? (*Hint:* Consider one of the sides of length 300 feet as a base of the triangle.)

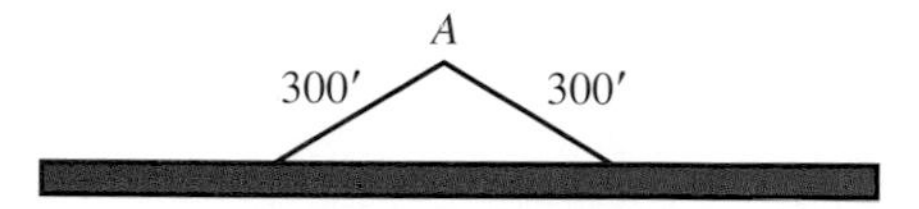

Using a Calculator

40. By the SSS congruence property the size and shape of a triangle are uniquely determined by the lengths a, b, and c of the triangle's sides. This also means that the area A of a triangle is determined from a, b, and c. Letting s denote half the perimeter of the triangle, so $s = \frac{1}{2}(a + b + c)$, the area of the triangle is given by **Hero's** (or **Heron's**) **formula:**

$$A = \sqrt{s(s - a)(s - b)(s - c)}.$$

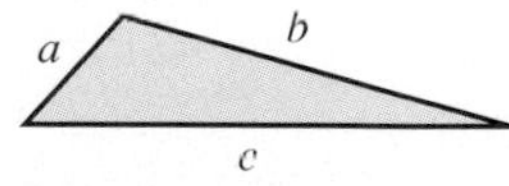

The formula is named for the Greek mathematician Hero of Alexandria, who lived about A.D. 50, but the formula was discovered much earlier by Archimedes. Use Hero's formula to compute the area of these triangles. It should be easy to check your calculation for the first two triangles.

(a)

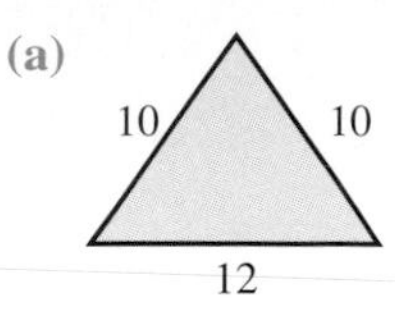

(b)

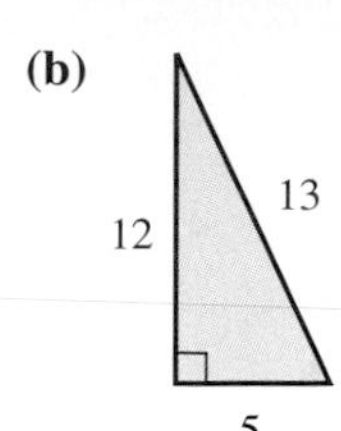

(c)

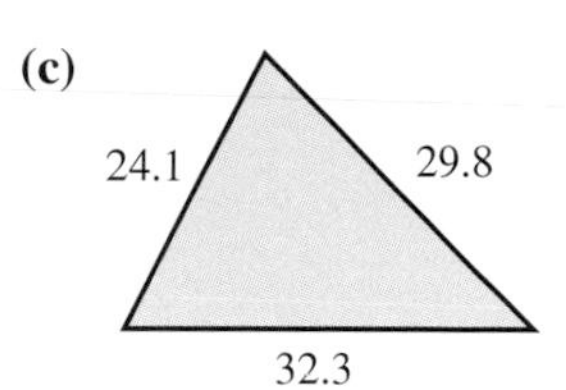

Using a Computer
Using an Automatic Drawer

41. Draw any convex quadrilateral. Next join the successive midpoints of the sides of the sides of the quadrilateral to form a parallelogram. Let the computer compute both the area of the quadrilateral and the area of the parallelogram. What relationship seems to exist between these areas?

42. Draw an equilateral triangle and its inscribed and circumscribed circles. How do the areas of the two circles compare? Investigate with your software.

43. Draw a regular hexagon and its inscribed and circumscribed circles. Use your software to compare the areas of the two circles.

For Review

44. Convert these measurements to the unit shown.

(a) 1 m = _____ cm **(b)** 352 mm = _____ cm

(c) 1 m^2 = _____ cm^2 **(d)** 1 m^3 = _____ liters

45. ΔABC and ΔXYZ have a congruent side, $\overline{AB} \cong \overline{XY}$, and a congruent angle, $\angle A \cong \angle Y$. List three additional pieces of information, each of which is sufficient to guarantee that the two triangles are congruent. Explain what congruence theorem is used in each case.

46. Prove that the diagonals of a rectangle are congruent.

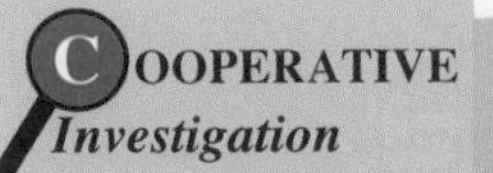

Discovering Pick's Formula

In 1899 the German mathematician Georg Pick discovered a remarkable formula for the area of a polygon drawn on square dot paper. Polygons of this special type are also known as lattice polygons. Stretching a rubber band onto a geoboard is another easy way to form lattice polygons.

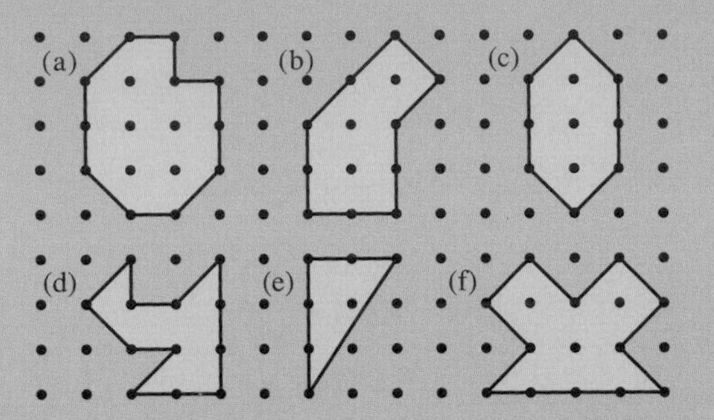

1. Complete the table of values for each polygon, where

b = number of dots on the boundary of the polygon,

i = number of dots in the interior of the polygon, and

A = area of the polygon.

The values of b, i, and A for polygon **(a)** are given as an example.

Polygon	b	i	A
(a)	11	5	$9\frac{1}{2}$
(b)			
(c)			
(d)			
(e)			
(f)			

2. Try to guess a formula for A in terms of b and i. (If you have trouble add a column of the values of b/2. You may also want to obtain more data by drawing other lattice polygons.)
3. Search for an extension of Pick's formula for regions with a hole; that is, a region inside one lattice polygon but outside a second lattice polygon.

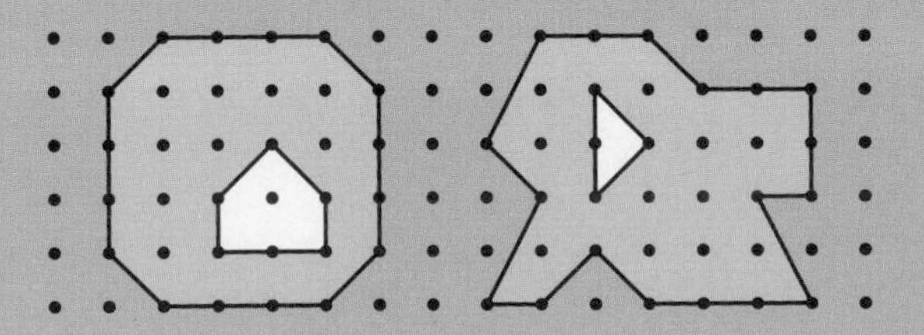

12.3 The Pythagorean Theorem

The Pythagorean theorem is the single most remarkable result in geometry. The theorem is aesthetically pleasing and also very useful in solving practical problems.

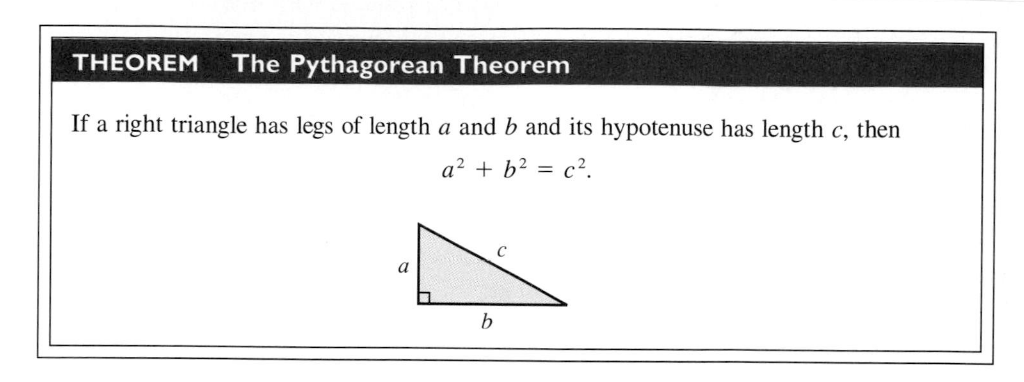

THEOREM The Pythagorean Theorem

If a right triangle has legs of length a and b and its hypotenuse has length c, then

$$a^2 + b^2 = c^2.$$

By erecting squares on the sides of a right triangle, the Pythagorean relation $a^2 + b^2 = c^2$ can be interpreted as a result about areas: *The sum of the areas of the squares on the legs of a right triangle is equal to the area of the square on the hypotenuse.* The area interpretation of the Pythagorean theorem is shown in Figure 12.14.

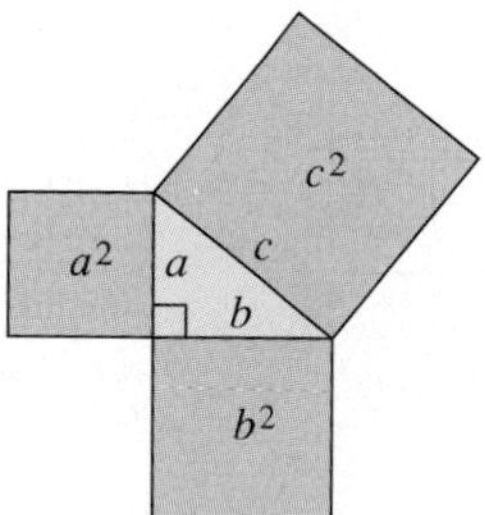

Figure 12.14
The sum of the areas of the squares on the legs of a right triangle equals the area of the square on the hypotenuse.

The area interpretation probably led to the discovery of the theorem, at least in special cases. For example, the Babylonian clay tablet shown in Figure 12.15 shows a large square subdivided by congruent isosceles right triangles. It is apparent that the area of the square on the hypotenuse equals the area in the two squares on the legs.

Figure 12.15
A Babylonian tablet suggesting a special case of the Pythagorean theorem

The demonstration depicted in Figure 12.15 is not a general proof of the Pythagorean theorem, because the right triangle is isosceles. However, a similar idea can be followed for arbitrary right triangles. In Figure 12.16 (a) we begin with any right triangle, letting a and b denote the lengths of the legs and c the length of the hypotenuse. Next consider a square with sides of length $a + b$. Four congruent copies of the right triangle are placed inside the squares in two different ways. In Figure 12.16 (b) the four triangles leave two squares uncovered, with respective areas a^2 and b^2. In Figure 12.16 (c) the four triangles leave one square of area c^2 uncovered. Since the four triangles leave the same area uncovered in both arrangements, we conclude $a^2 + b^2 = c^2$.

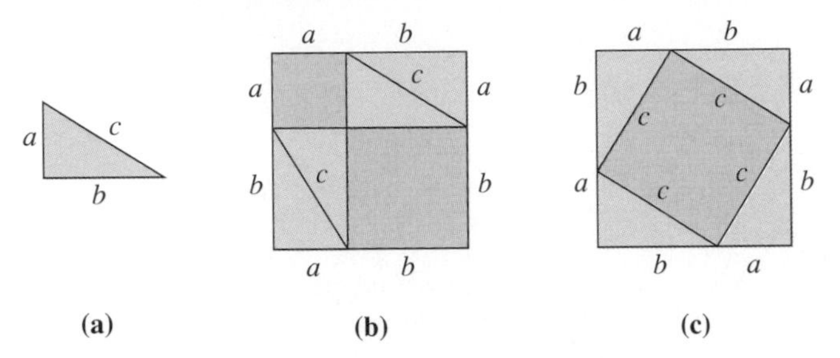

Figure 12.16
A dissection proof of the Pythagorean theorem

No records have survived to indicate what proof, if any, Pythagoras (c. 572–501 B.C.) may have offered. The dissection proof requires showing that the inner quadrilateral of Figure 12.16 (c) is actually a square (why is it?), but the Pythagorean's knowledge of angles in a right triangle was sufficient to do this. Since the time of Pythagoras a tremendous number of proofs have been devised. In the second edition of *The Pythagorean Proposition,* E. S. Loomis catalogs 370 different proofs.

EXAMPLE 12.14

Using the Pythagorean Theorem

Find the lengths x, y, and z in the following figures.

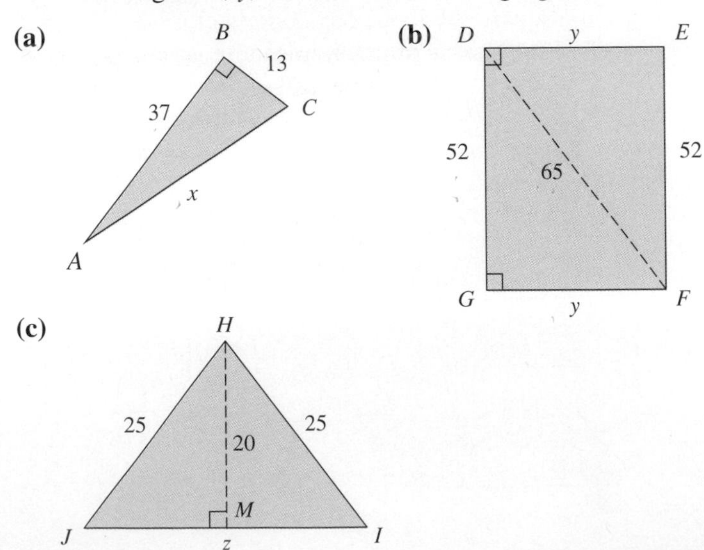

SOLUTION

(a) By the Pythagorean theorem, $x^2 = 13^2 + 37^2 = 169 + 1369 = 1538$. Therefore $x = \sqrt{1538} \doteq 39.2$.

(b) The diagonal $\overline{DF}$ of rectangle $DEFG$ is the hypotenuse of the right triangle DEF. The Pythagorean theorem, applied to $\triangle DEF$, gives $y^2 + 52^2 = 65^2$. Therefore $y^2 = 65^2 - 52^2 = 4225 - 2704 = 1521$, and $y = \sqrt{1521} = 39$.

(c) By the HL Theorem (Chapter 11, Section 1), $\triangle HIM$ and $\triangle HJM$ are congruent right triangles. Therefore $MI = MJ$, so that M is the midpoint of $\overline{IJ}$. Applying the Pythagorean theorem to $\triangle HIM$ gives $25^2 = 20^2 + MI^2$, or $MI^2 = 25^2 - 20^2 = 625 - 400 = 225$. Thus $MI = \sqrt{225} = 15$ and $z = JI = 2MI = 2(15) = 30$. ■

For many applications it is necessary to write and solve equations based on the Pythagorean relation. Here is an example.

EXAMPLE 12.15 Determining How Far You Can See

Imagine yourself on top of a mountain, or perhaps in an airplane, at a known altitude given in feet. Approximately how far away, in miles, is the horizon?

SOLUTION

Understand the problem • • • •

Altitude is a measure of the distance above the surface of the earth. The horizon is the circle of points where our line of sight is tangent to the sphere of the earth's surface. The problem is to derive a formula that expresses, or at least approximates, the distance to the horizon as it depends on the altitude of the observer. Since the altitude is given in miles, special care must be taken to handle the units of measure properly.

Devise a plan • • • •

The earth is very nearly a sphere. A line of sight to the horizon forms a leg of a right triangle, as the diagram shows. Since the radius, r, of the earth is about 4000 miles, and the altitude, s, is known, the Pythagorean theorem can be used to solve for the distance to the horizon. In the diagram all distances, including s, are expressed in miles; if h is the altitude in feet we can use the conversion formula $h = 5280\ s$.

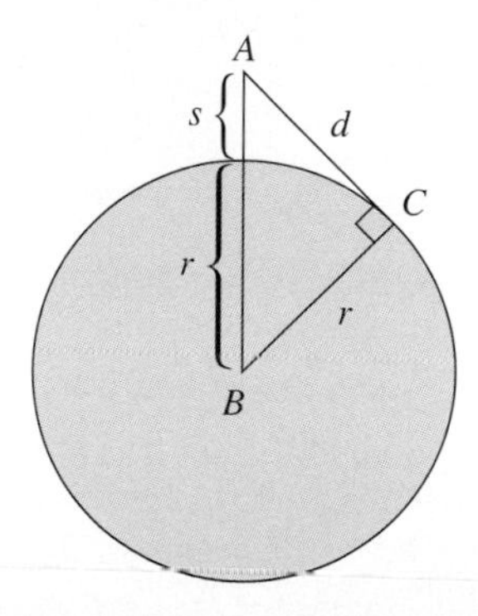

Carry out the plan • • • •

Applying the Pythagorean theorem to the right triangle ABC gives $d^2 + r^2 = (s + r)^2$. The squared term on the right can be written $s^2 + 2sr + r^2$, so $d^2 + r^2 = s^2 + 2sr + r^2$. Subtracting r^2 from both sides shows that $d^2 = s^2 + 2sr$. Therefore the exact distance d, in miles, is given by the formula

$$d = \sqrt{s^2 + 2rs},$$

where $r = 4000$ miles and s denotes the miles of altitude.

To see how d is calculated when the altitude h is given in feet we use the equation $h = 5280\ s$. Thus d, still in miles, is given by

$$d = \sqrt{\left(\frac{h}{5280}\right)^2 + \frac{2 \cdot 4000 \cdot h}{5280}}.$$

At an altitude h of 5280 feet, $d = \sqrt{1+8000}$ miles. At an altitude of 52,800 feet, $d = \sqrt{100 + 80{,}000}$. For altitudes even as high as an orbiting space shuttle the first term under the square root in our formula is much smaller than the second term. Therefore the first term can be omitted with little loss of accuracy. This gives the approximate formula

$$d \doteq \sqrt{\frac{2 \cdot 4000}{5280}\,h}.$$

Since $\sqrt{\frac{8000}{5280}} \doteq 1.2$, we obtain a simple formula for the miles to the horizon seen from an altitude of h feet:

$$d \doteq 1.2\sqrt{h}.$$

For example, the distance to the horizon as seen from an airplane flying at 40,000 feet is about $1.2\sqrt{40{,}000} = (1.2)(200) = 240$ miles.

Look back • • • •

This problem involved several steps which are typical of the way the Pythagorean theorem is used:

- draw a figure and label all the distances,
- identify all the right triangles in the drawing,
- write the Pythagorean relationships for all of the right triangles, and
- solve the Pythagorean formulas to determine unknown values needed for the solution of the problem. ■

The Converse of the Pythagorean Theorem

The numbers 5, 12, and 13 satisfy $5^2 + 12^2 = 13^2$. Is the triangle with sides of length 5, 12, and 13 a right triangle? The answer is yes, since the Pythagorean relation $a^2 + b^2 = c^2$ holds if, *and only if,* a, b, and c are the side lengths of a right triangle.

> **THEOREM Converse of the Pythagorean Theorem**
>
> Let a triangle have sides of length a, b, and c. If $a^2 + b^2 = c^2$, then the triangle is a right triangle and the angle opposite the side of length c is its right angle.

The key idea is to use the Pythagorean theorem itself to prove its own converse.

EXAMPLE 12.16 **Checking for Right Triangles**

Determine if the three lengths given can be the lengths of the sides of a right triangle.

(a) 15; 17; 8 **(b)** 10; 5; $5\sqrt{3}$ **(c)** 231; 520; 568

SOLUTION

(a) $8^2 + 15^2 = 64 + 225 = 289 = 17^2$, so 8, 15, and 17 are the lengths of the sides of a right triangle.

(b) $5^2 + (5\sqrt{3})^2 = 25 + 25 \cdot 3 = 25 + 75 = 100 = 10^2$, so 5, $5\sqrt{3}$, and 10 are the lengths of the sides of a right triangle.

(c) $231^2 + 520^2 = 53{,}361 + 270{,}400 = 323{,}761 \neq 322{,}624 = 568^2$, so 231, 520, and 568 are not the lengths of sides of a right triangle. This would be difficult to see by measuring angles with a protractor, since this triangle closely resembles the right triangle with sides of length 231, 520, and 569. ■

PROBLEM SET 12.3

Understanding Concepts

1. Find the distance x in each figure.

(a)

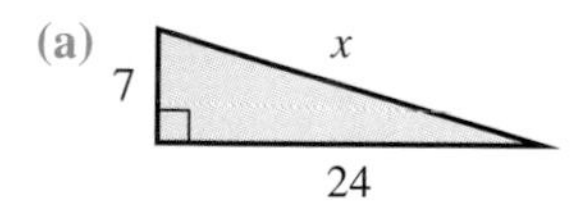

(b)

x
1
x

(c)

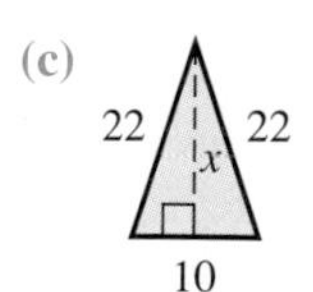

(d)

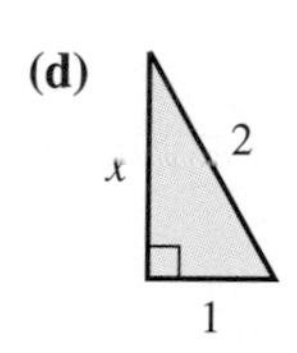

(e)

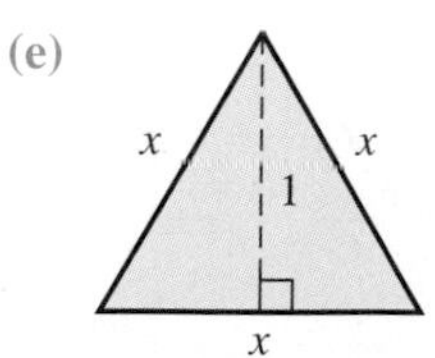

(f)

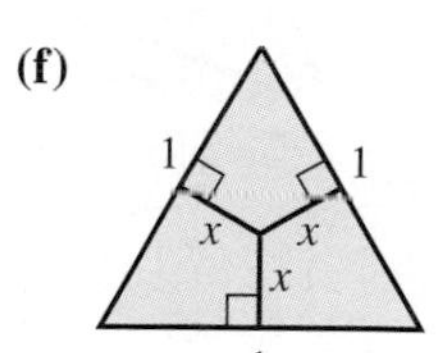

2. Find the distance x in each figure.

(a)

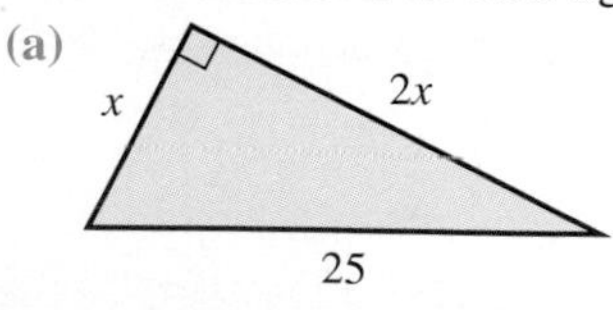

(b)

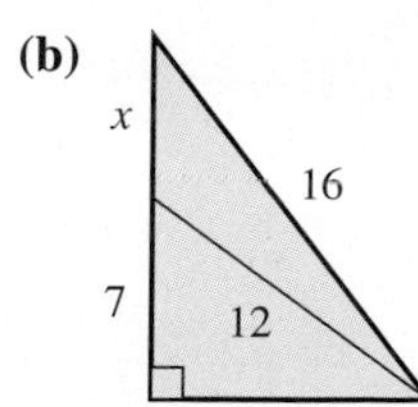

(c)

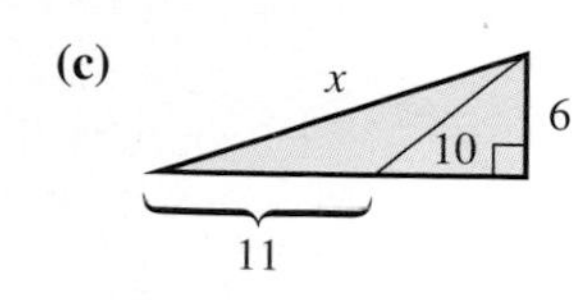

3. Find the distances x and y in the rectangular prism and the cube.

(a)

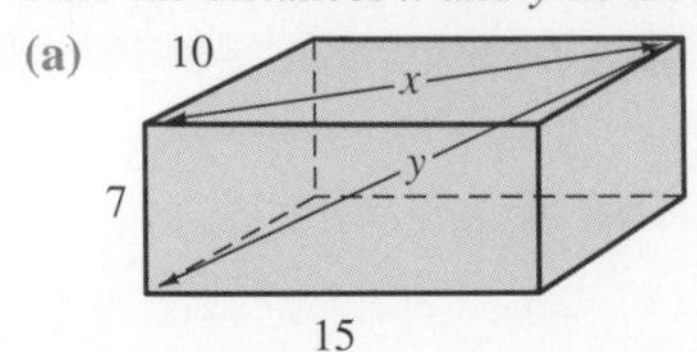

(b)

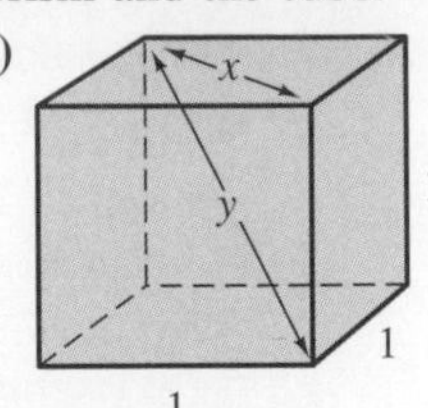

4. Find the distance x in these space figures.

(a)

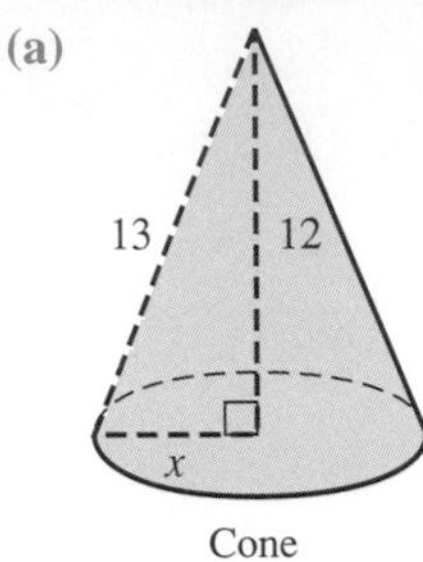

Cone

(b)

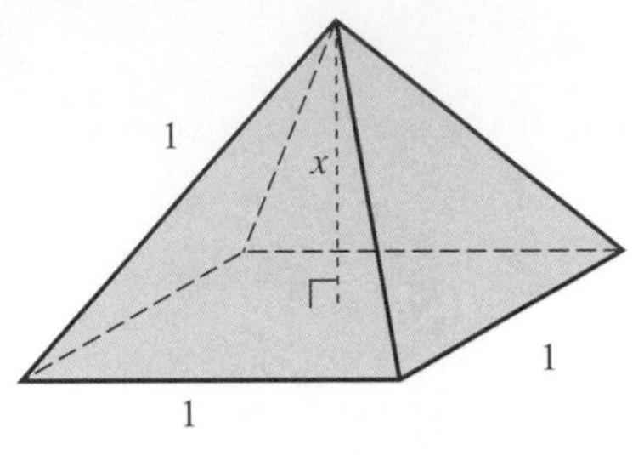

Square based pyramid, with equilateral triangle faces

(c)

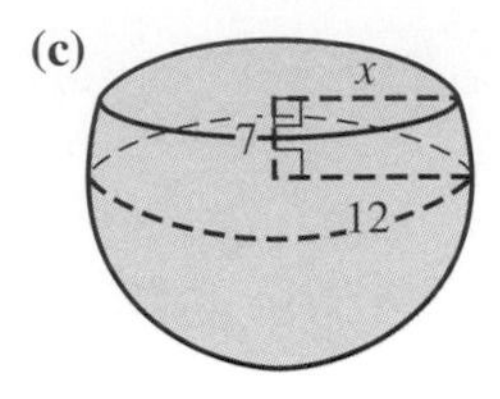

Sphere cut by plane

5. Find the areas of these figures.

(a)

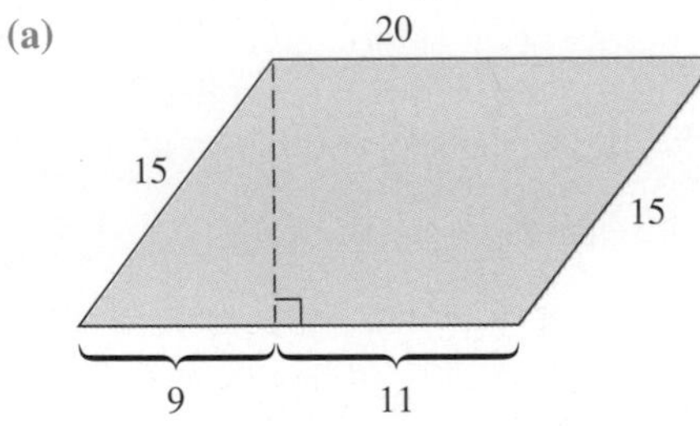

(b)

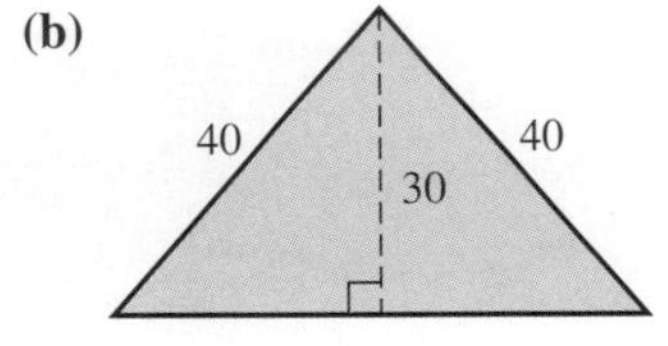

(c)

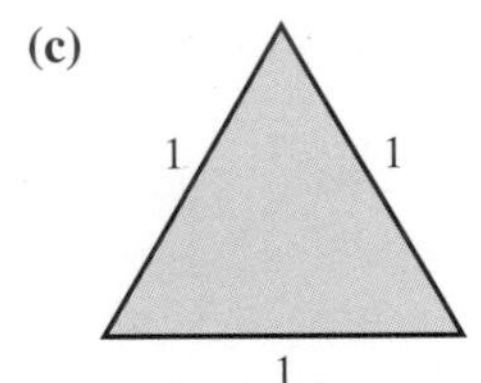

---

6. Francoise and Maurice cut across a 50 foot by 100 foot vacant lot on their way to school. How much distance do they save by not staying on the sidewalk?

7. A square with sides of length 2 is inscribed in a circle and circumscribed around another circle. Which is larger, the area of the region between the circles or the area inside the smaller circle?

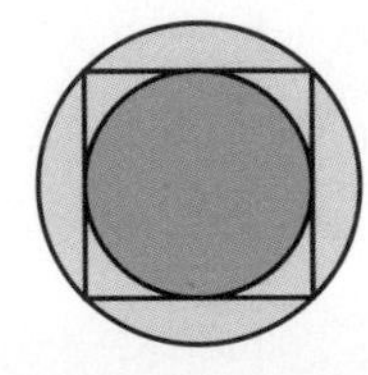

8. At noon car A left town heading due east at 50 miles per hour. At 1 P.M. car B left the same town heading due north at 40 miles per hour. How far apart were the two cars at

(a) 2 P.M. **(b)** 3:30 P.M.?

9. Find AG in this spiral of right triangles.

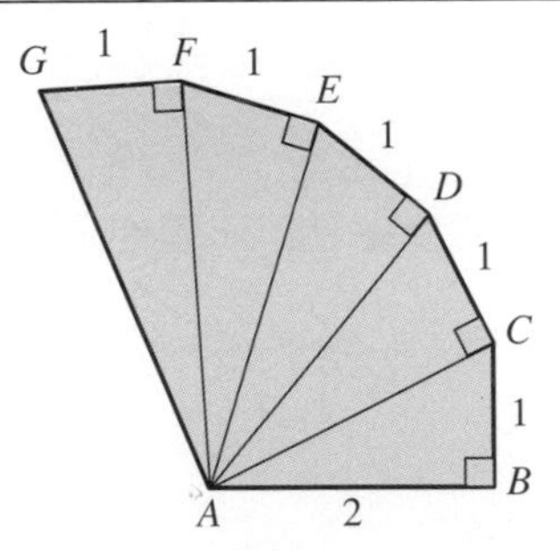

10. **(a)** What is the length of the side of a square inscribed in a circle of radius 1?

(b) What is the length of the side of a cube inscribed in a sphere of radius 1?

11. What is the distance between the centers of these circles?

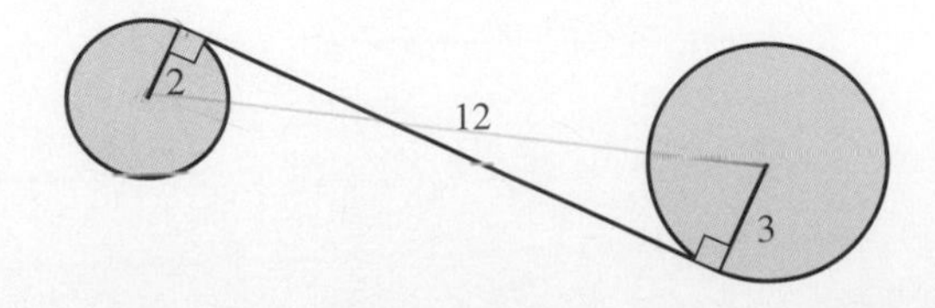

12. What is the radius of the circle shown below?

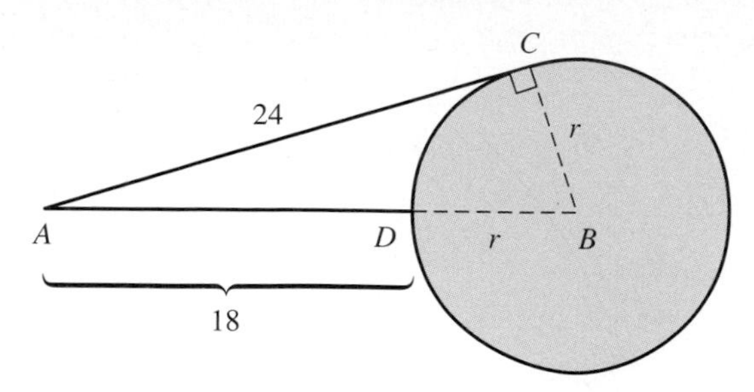

13. Which of the following can be the lengths of the sides of a right triangle?

(a) 21, 28, 35 **(b)** 9, 40, 41
(c) 12, 35, 37 **(d)** 14, 27, $\sqrt{533}$
(e) $7\sqrt{2}$, $4\sqrt{7}$, $2\sqrt{77}$ **(f)** 9.5, 16.8, 19.3

14. If a, b, and c are the lengths of the sides of a right triangle, explain why $10a$, $10b$, and $10c$ are also the lengths of the sides of a right triangle.

Thinking Critically

15. The twelfth century Hindu mathematician Bhaskara arranged four copies of a right triangle of side lengths a, b, c into a c by c square, filling in the remaining region with a small square which is shown as black in the diagram below.

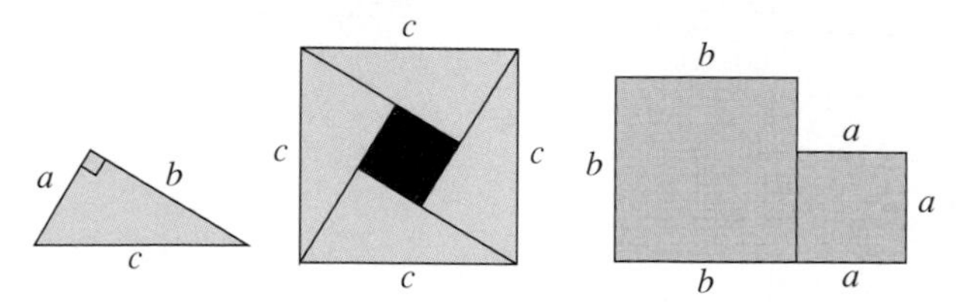

(a) Show how the five pieces in the c by c square can be arranged to fill the "double" square region at the right.

(b) Explain how the Pythagorean theorem follows from part (a).

16. Trace and cut the following five shapes from paper.

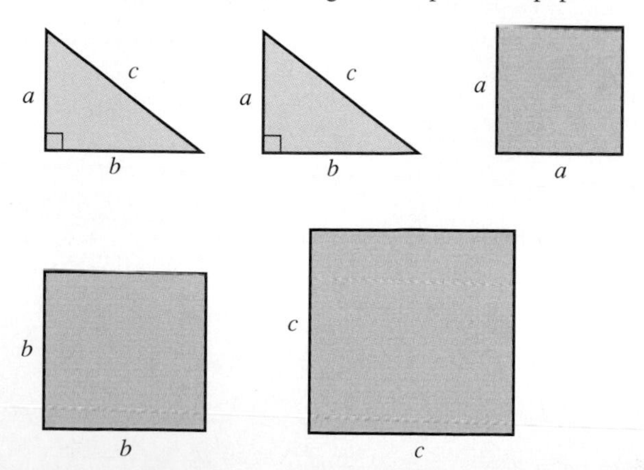

Now tile the pentagon shown below in two ways:

(a) with the two triangles and the two smaller squares;

(b) with the two triangles and the largest square.

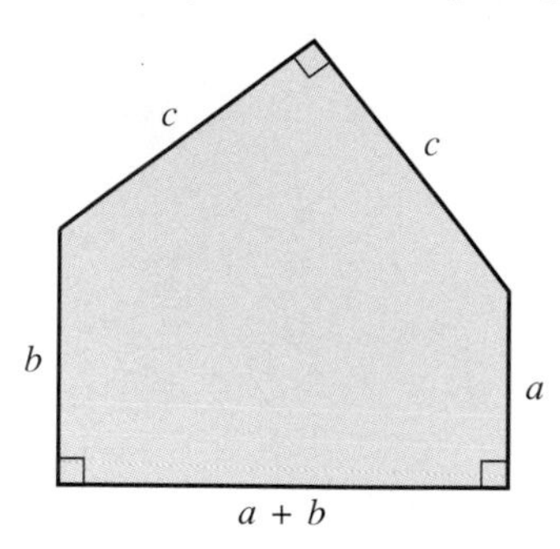

(c) Explain why the two tilings in parts (a) and (b) prove the Pythagorean theorem.

17. In the nineteenth century, Henry Perigal, a London stockbroker and amateur astronomer, discovered a beautiful scissors-and-paper demonstration of the Pythagorean theorem. Follow these steps to complete your own demonstration. Through the center of the larger square on the leg of the right triangle, draw one line perpendicular to the hypotenuse and a second line parallel to the hypotenuse. Cut along these two lines to divide the square into four congruent pieces, and then show how to arrange these four pieces, together with the square on the shorter leg, to form a square on the hypotenuse of the right triangle.

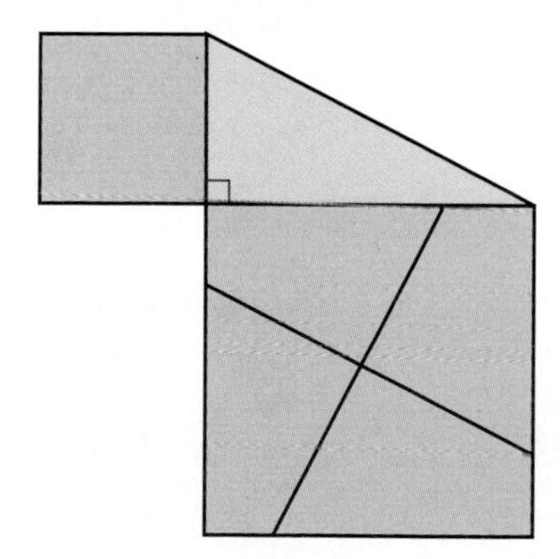

18. Justify why the shaded regions (which are all parallelograms) in the following sequence of diagrams have the same area. This provides a striking dynamical proof of the Pythagorean theorem.

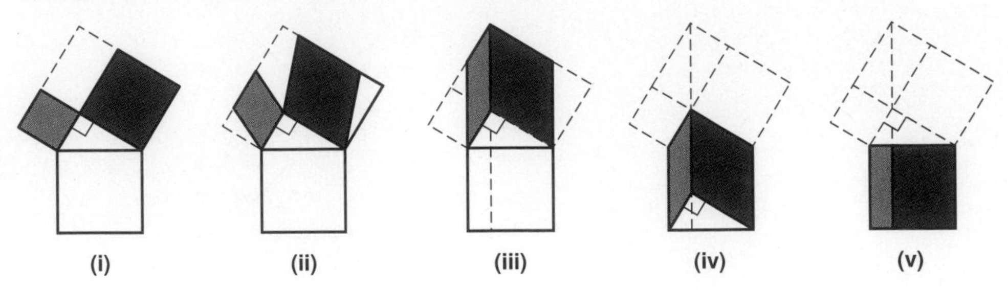

19. Prove the converse of the Pythagorean theorem by justifying the steps below. Recall that a, b, and c are the lengths of $\triangle ABC$. If $a^2 + b^2 = c^2$, you are to show $\angle C$ is a right angle.

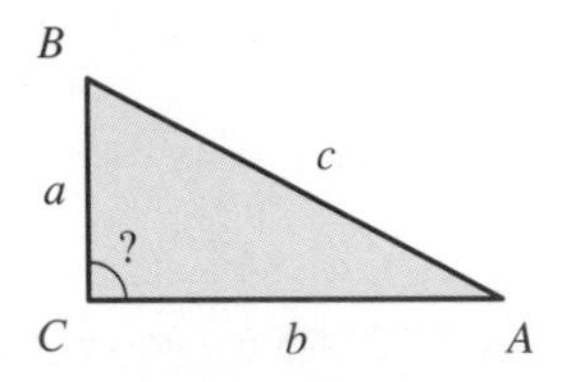

(a) Construct a right triangle DEF with legs of length a and b. If z is the length of the hypotenuse of $\triangle DEF$, why is $z^2 = a^2 + b^2$?

(b) Why is $c = z$?

(c) What congruence property proves that $\triangle DEF$ is congruent to $\triangle ABC$?

(d) Why is $\angle C$ a right angle?

20. (a) Show that the altitude h of an equilateral triangle with sides of length s is given by $h = \frac{\sqrt{3}}{2}s$.

(b) Find a formula for the area of an equilateral triangle of side length s.

(c) Find a formula for the area of a regular hexagon of side length s.

(d) Show that the area of the inscribed circle of a regular hexagon is $\frac{3}{4}$ the area of the circumscribed circle.

21. There are five different lengths between pairs of nails on a 3 nail by 3 nail geoboard:

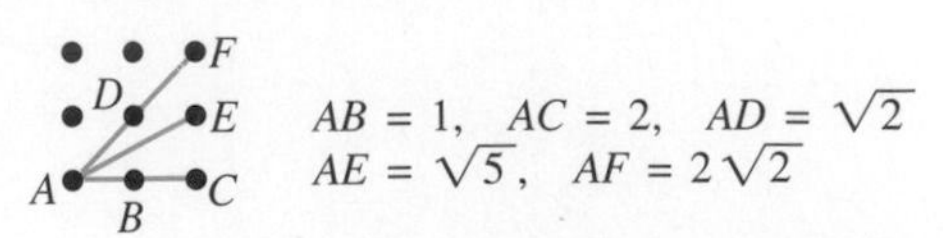

How many different lengths can you find on a 5 nail by 5 nail geoboard?

22. In Frank Baum's *The Wonderful Wizard of Oz,* Scarecrow shows off his newly acquired brain by giving his rendition of the Pythagorean theorem:

> *"The sum of the square roots of any two sides of an isosceles triangle equals the square root of the remaining side."*

(a) What errors does Scarecrow make?

(b) Is there an isosceles triangle with this property?

23. Find the length of the diagonals of this isosceles trapezoid.

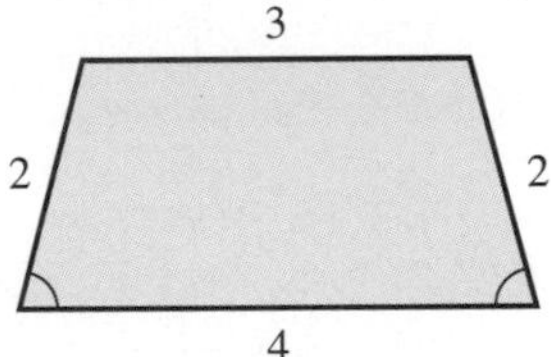

Making Connections

24. A baseball diamond is actually a square 90 feet on a side. What distance must a catcher throw the ball to pick off a runner attempting to steal second base?

25. Approximately what height can be reached from a 24 foot ladder? What assumptions have you made to arrive at your answer?

26. The ancient Egyptians squared off fields with a rope 12 units long, with knots tied to indicate each unit. Explain how such a rope could be used to form a right angle. What theorem justifies their procedure?

27. A water lily is floating in a murky pond rooted on the bottom of the pond by a stem of unknown length.

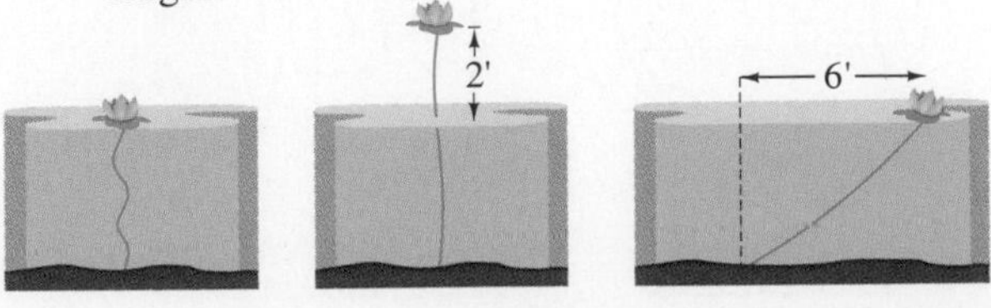

The lily can be lifted 2 feet over the water and moved 6 feet to the side. What is the depth of the pond?

28. A stop sign is to be made by cutting off triangles from the corners of a square sheet of metal 32 inches on a side. What length x will leave a regular octagon? Give your answer to the nearest eighth inch.

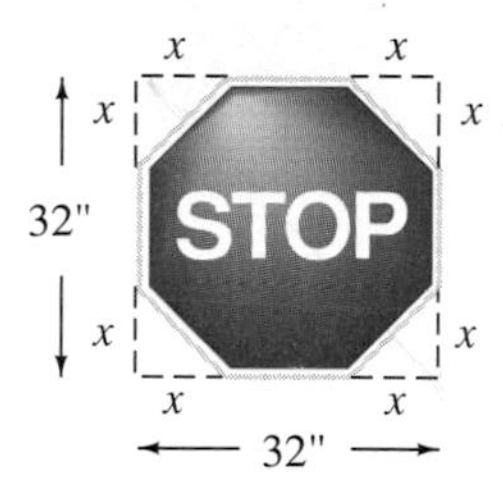

Using a Calculator

29. Example 12.15 showed that the distance (in miles) to the horizon from an altitude of s miles is exactly

$$d = \sqrt{s^2 + 8000s}$$

on a perfectly spherical earth of radius 4000 miles. Compare this exact formula to the approximate formula $d = \sqrt{8000s}$ when the altitude s is

(a) 200 miles **(b)** 400 miles **(c)** 1000 miles.

30. Use the approximate formula $d \doteq 1.2\sqrt{h}$ of Example 12.15 to answer these questions.

(a) On a cliff top 100 feet over the ocean, what is the distance to the horizon?

(b) The observation deck of the Sears Tower in Chicago is 1353 feet above ground level. How far can you see across Lake Michigan?

(c) In the Dr. Seuss book *Yurtle the Turtle,* Yurtle stands on the backs of other turtles and can see 40 miles. How high is Yurtle?

Using a Computer
Using an Automatic Drawer

31. Draw any right triangle. On each side, draw outward pointing equilateral triangles. Use your software to calculate the areas of the triangles.

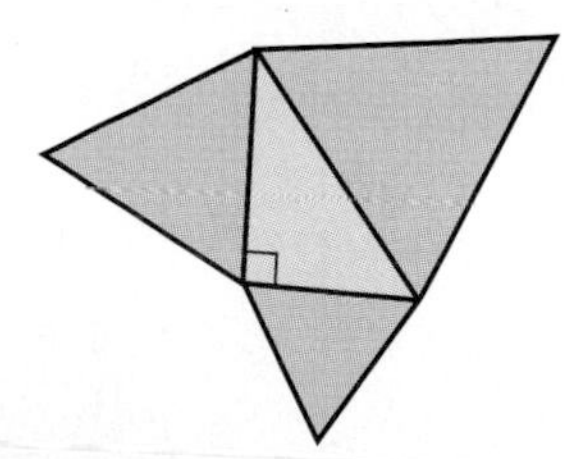

How does the sum of the areas of the triangles on the two legs compare to the area of the equilateral triangle on the hypotenuse?

32. Draw any parallelogram and its two diagonals. Next draw squares erected on each segment of your drawing, and find the areas of all six squares.

(a) Verify that the sum of the areas of the squares on the four sides of the parallelogram equals the sum of the areas of the two squares on the diagonals of the parallelogram.

(b) Explain how the result in part (a) is equivalent to the Pythagorean theorem when the parallelogram is a rectangle.

For Review

33. The length of a rectangle is increased by $33\frac{1}{3}$ percent and the width is decreased by 25 percent. By what percent does the area change?

34. Find the areas of these figures.

(a)

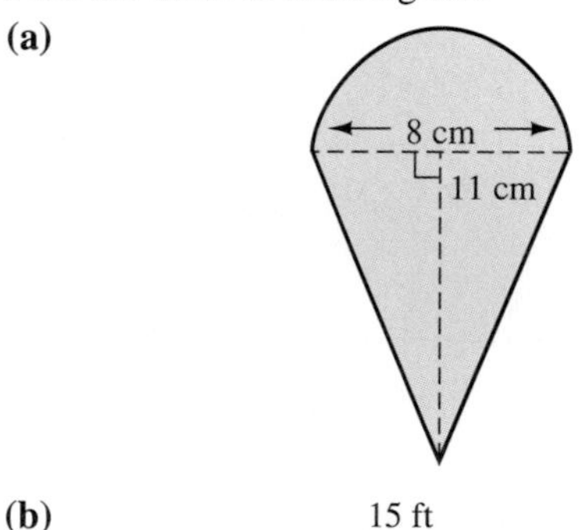

(b)

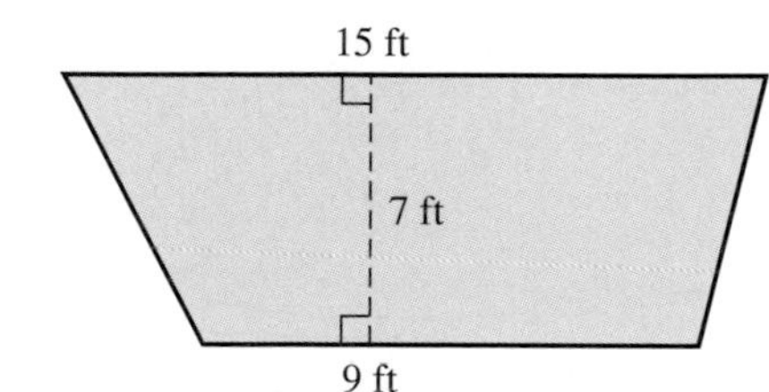

35. Two semicircular arcs, of radius 3 m and 5 m, are centered on the diameter $\overline{AB}$ of a large semicircle as shown. Which route from A to B is shorter: along the large semicircle, or along the two smaller semicircles which touch tangentially at C?

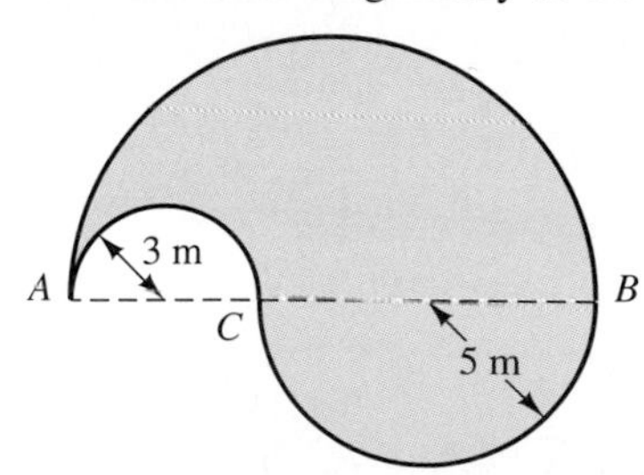

12.4 Surface Area and Volume

The **surface area** of any polyhedron is the sum of the areas of its faces. It's often useful to cut the faces apart and rearrange them in a single plane. In many cases the figure that is formed has an easily calculated area. The procedure also suggests ideas that are useful for space figures other than polyhedra.

Surface Area of Right Prisms

Figure 12.17 shows how the surface of a right prism is cut into two congruent bases, with the lateral surfaces of the prism unfolded to form a rectangle. If the right prism has height h and the perimeter of the base is p, then the rectangle has area hp.

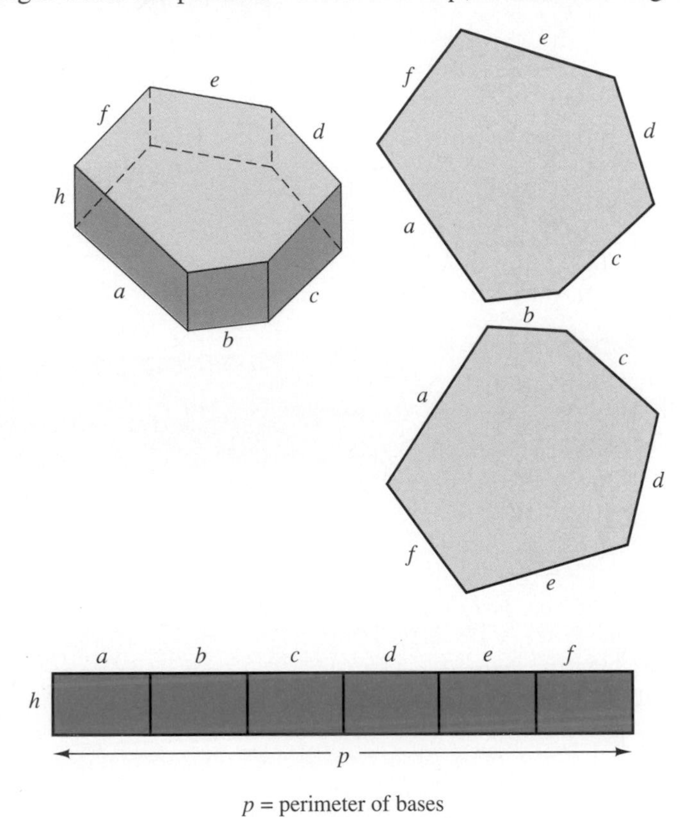

Figure 12.17
The surface of a right prism can be cut and unfolded to form two congruent bases and a rectangle.

EXAMPLE 12.17 **Finding Surface Area of a Prism-Shaped Box**

A fancy gift box is shaped as a regular six-pointed Star of David. The box is 10 cm deep and the side of each point is 8 cm long. What is the surface area of the box?

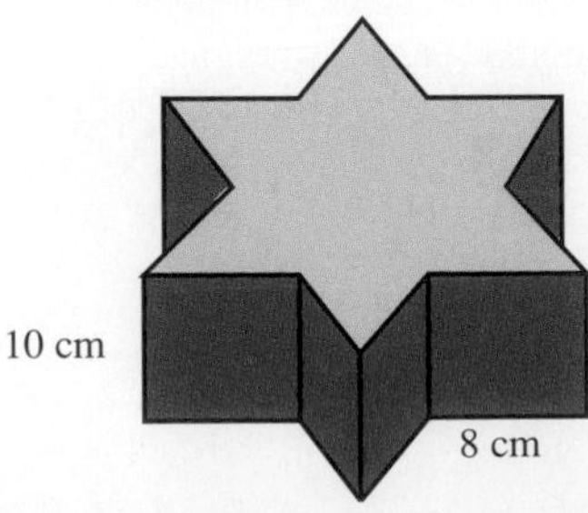

SOLUTION

The box consists of two six-pointed stars, and a lateral surface that unfolds to become a rectangle of length $12 \cdot 8$ cm = 96 cm and width 10 cm. Therefore the lateral surface area of the box is 10 cm · 96 cm = 960 cm^2.

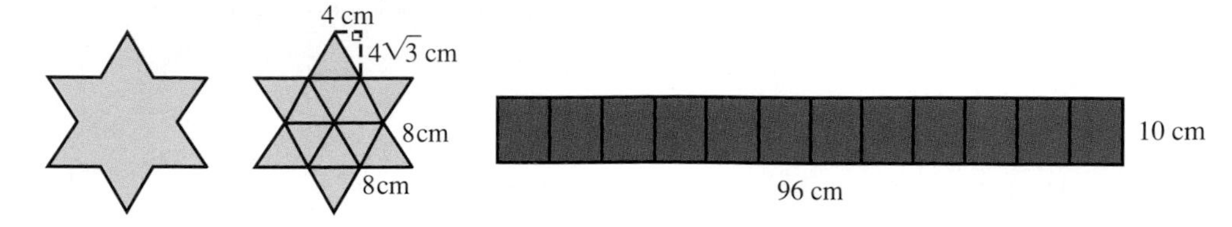

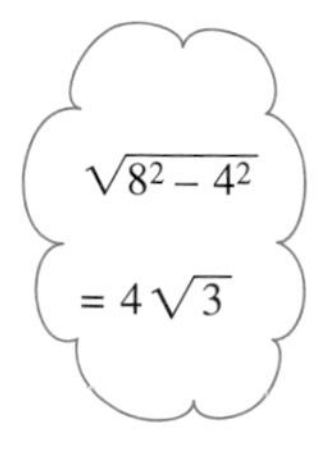

The area of each star-shaped base can be found by dissecting the star into 12 equilateral triangles, each 8 cm on a side. The altitude of the equilateral triangle is $4\sqrt{3}$ cm (use the Pythagorean theorem to see why), so the area of each triangle is $\frac{1}{2} \cdot 8 \cdot 4\sqrt{3}$ cm^2, or $16\sqrt{3}$ cm^2. The area of one base of the box is then $12 \cdot 16\sqrt{3}$ cm^2 = $192\sqrt{3}$ cm^2. The total area of the box is therefore $384\sqrt{3} + 960$ cm^2, or about 1625 cm^2. ■

Surface Area of Pyramids

The surface area of a pyramid is computed by adding the area of the base to the sum of the areas of the triangles forming the lateral surface of the pyramid. Of special importance is the **right regular pyramid,** for which the base is a regular polygon and the lateral surface is formed by congruent isosceles triangles. The altitude of the triangles is called the **slant height** of the pyramid.

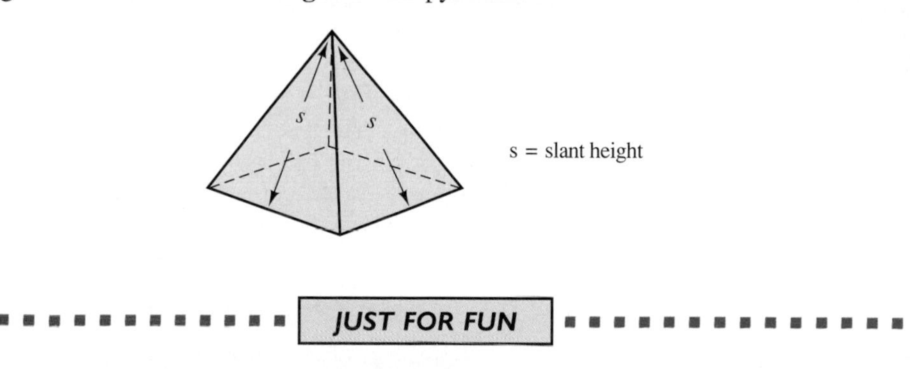

JUST FOR FUN

The Woven Cube

Cut three rectangular strips of paper, each strip a different color. Fold each along the dotted lines shown. Now weave the three strips into a rigid cube. The opposite faces should be the same color, and the end squares of the strips should be neatly tucked inside.

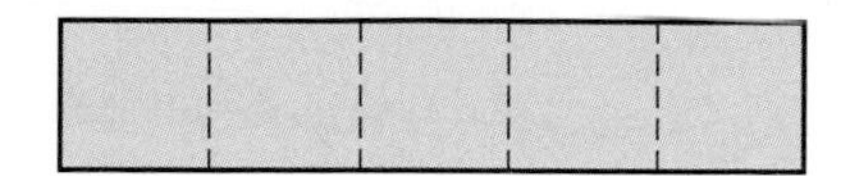

(The regular polyhedra can all be woven with suitably cut and folded paper strips. Patterns and directions may be found in *Build Your Own Polyhedra,* by Peter Hilton and Jean Pedersen, Addison-Wesley Publishing Company, 1988.)

EXAMPLE 12.18 **Finding the Surface Area of a Right Regular Pyramid**

A pyramid has a square base 10 cm on a side. The edges meeting at the apex of the pyramid are each 12 cm long. Find the slant height of the pyramid and then calculate the total surface area (including the base) of the pyramid.

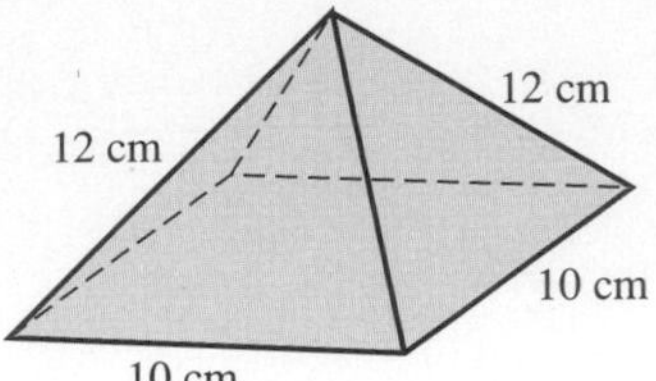

SOLUTION

The surface of the pyramid consists of a square and four isosceles triangles. The Pythagorean theorem shows that the altitude of each triangle, and therefore the slant height of the pyramid, is $\sqrt{12^2 - 5^2}$ cm $= \sqrt{119}$ cm. Thus each triangle has area $\frac{1}{2} \cdot 10 \text{ cm} \cdot \sqrt{119} \text{ cm} = 5\sqrt{119} \text{ cm}^2$. Adding the areas of the four triangles to the area of the base, we get the total area of $20\sqrt{119} + 100 \text{ cm}^2$.

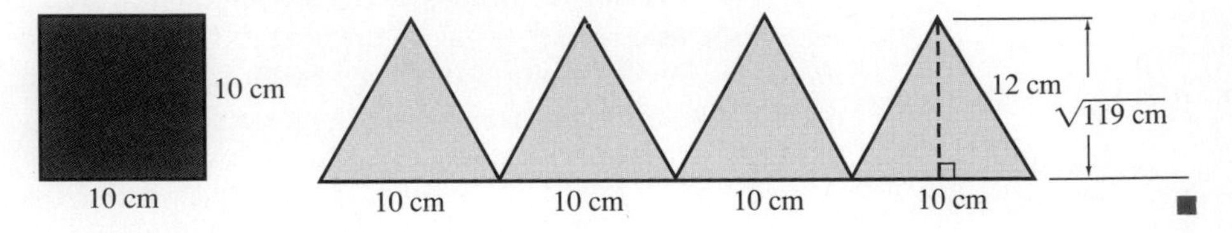

■

Surface Area of a Right Cylinder

The surface area of a right cylinder can be found by nearly the same method used earlier for the right prism. The only difference is that the lateral surface is unrolled, rather than unfolded, to form a rectangle. The width of the rectangle is the height of the cylinder and the length of the rectangle is the perimeter of the base of the cylinder.

EXAMPLE 12.19 **Finding the Surface Area of a Right Cylinder**

A small can of frozen orange juice is about 9.5 cm tall and has a diameter of about 5.5 cm. The circular ends are metal and the rest of the can is cardboard. How much metal and how much cardboard are needed to make a juice can?

SOLUTION

The rectangle below is 9.5 cm wide and 5.5π cm long, so the area of cardboard is $(9.5)\ (5.5)\pi\ \text{cm}^2 = 52.25\pi\ \text{cm}^2$, or about 164 cm^2. The circles have a radius of 2.75 cm, so each circle has area $\pi\ (2.75\ \text{cm})^2$. Twice this is $15.125\pi\ \text{cm}^2$, so the area of the two metal ends is about 47.5 cm^2.

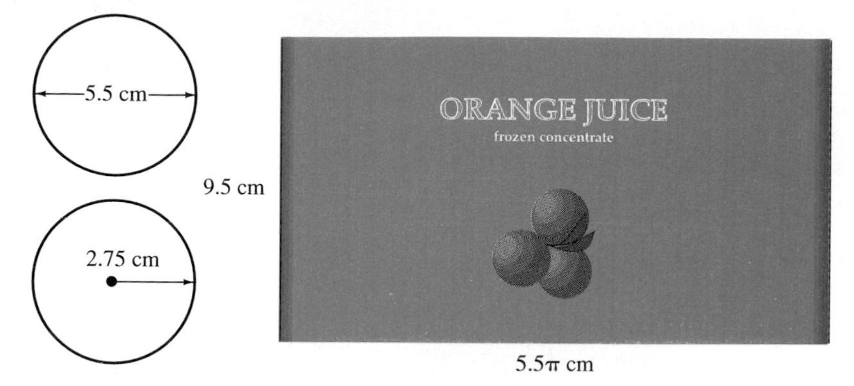

■

Lateral Surface Area of a Right Circular Cone

The lateral surface of a right circular cone can be cut and unrolled to form a sector of a circle. The area of the sector is determined by calculating the fraction of the corresponding full circle that is covered by the sector. Here is an example that illustrates the procedure.

EXAMPLE 12.20 **Finding the Surface Area of an Ice Cream Cone**

An ice cream cone has a diameter of 2.5 inches and slant height of 6 inches. What is the area of the cone?

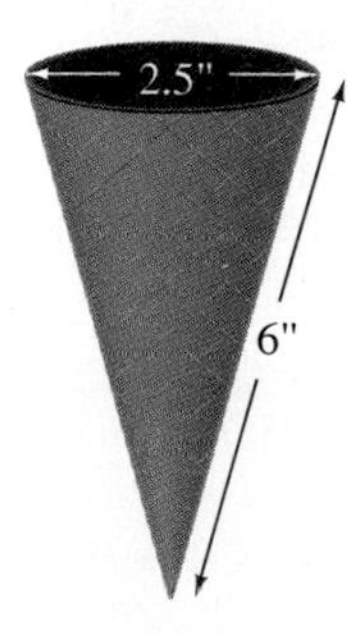

SOLUTION

Imagine making a cut from the apex of the cone to the circular rim of the cone. The cone can, in the mind's eye, be unrolled to a circular sector of a circle of radius 6 inches as shown on page 810. The rim of the cone is the circumference of a circle of diameter 2.5 inches, so the length of the arc of the sector is 2.5π inches. Since the circle of radius 6 inches has a circumference of 12π inches, we see that the sector fills in the fraction $2.5\pi/12\pi$ of the circle. The area of the circle is $\pi(6\ \text{in})^2 = 36\pi\ \text{in}^2$, so the sector has area $\frac{2.5\pi}{12\pi} \times 36\pi\ \text{in}^2 = 2.5 \times 3\pi\ \text{in}^2 \doteq 23.6\ \text{in}^2$. This is also the area of the cone.

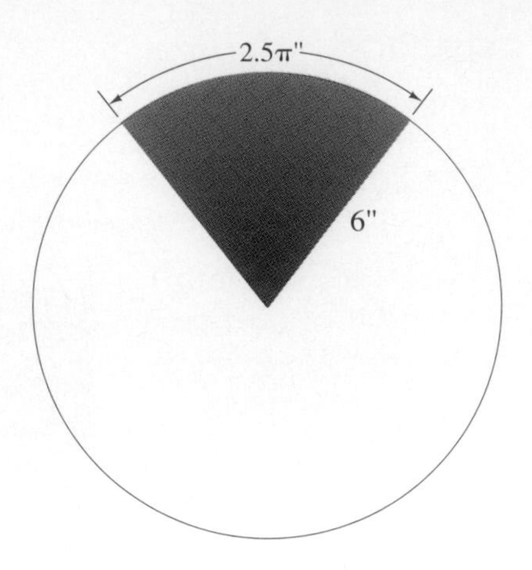

■

Volumes of Right Prisms and Right Cylinders

The volume of the rectangular box shown in Figure 12.18 is given by ℓwh, where ℓ, w, and h are, respectively, the length, width, and height of the box. Since ℓw gives the area B of the base of the box, the volume can also be written in the form $V = Bh$, where B is the area of the base and h is the height.

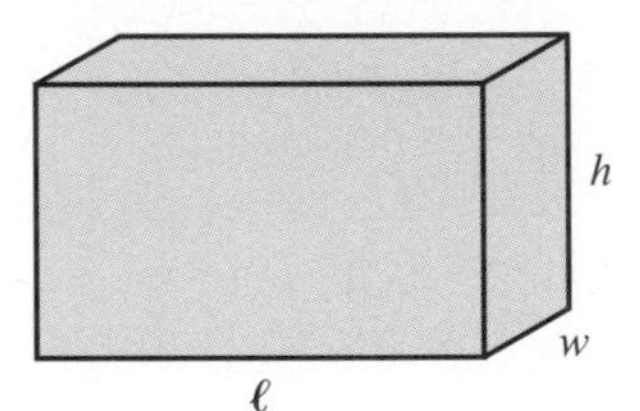

Figure 12.18
A rectangular box has volume $V = lwh$. Equivalently, $V = Bh$, where B is the area of the base and h is the height.

Figure 12.19(a) shows a solid composed of many (say n) small right rectangular prisms, all of the height h. If $B_1, B_2, \cdots, B_n$ are the areas of the bases, the total volume, v, of the prism is $B_1h + B_2h + \cdots + B_nh = (B_1 + B_2 + \cdots + B_n)$. That is, the volume V is given by $V = Bh$, where $B = B_1 + B_2 + \cdots + B_h$ is the total area of the base. The right cylinder depicted in Figure 12.19 (b) can be approximated to arbitrary accuracy by prisms of height h as shown at the left.

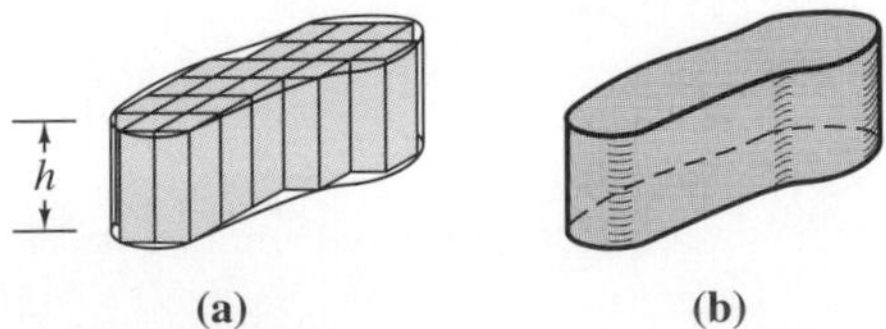

Figure 12.19
Right rectangular prisms can approximate a right prism or a right cylinder.

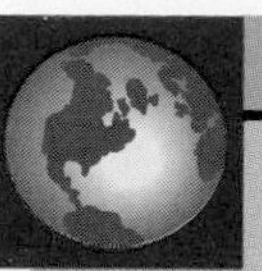

CURRENT AFFAIRS

Finding Answers in Higher-dimensional Space

Why would mathematicians want to leave the comfort of our familiar three-dimensional world? Because, curiously, by poking their heads up into higher dimensions, they can get a clearer view of complex problems—they can see relationships that look hopelessly tangled in the squashed and compacted universe of lower dimensions. Similarly, astrophysicists enter higher dimensions to see patterns in star clusters; particle physicists to look for unified theories; engineers to analyze mechanical linkages; and communications specialists to find ways to pack information into tight spaces.

There's nothing like hopping into a higher dimension to make a complex problem easier. If that sounds counterintuitive, just think about what going to a higher dimension really means. Say you're living on a one-dimensional line. You can move forward or backward, like a train on its track. But you can't move sideways. It's not only out of bounds, it's out of your universe. Now imagine that your universe suddenly spreads out into two dimensions. You can roam freely over the entire surface: east, west, north, south, or any direction in between. Or, better yet, imagine that you're a movie character, living your life on a two-dimensional screen. Add a third dimension and suddenly you can step off into the audience. You can simply walk away from that gunman about to shoot you. Thanks to that extra dimension, you have new freedom to move about.

SOURCE: From "Escape from 3-D" by K. C. Cole from *Discover*, July 1993, Vol. 14, no. 7, p. 54.

FORMULA Volume of a Right Prism or a Right Cylinder

Let a right prism or right cylinder have height h and a base of area B. Then its volume, V, is given by

$$V = Bh.$$

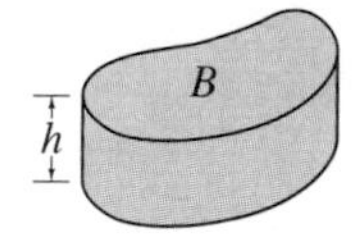

B = area of base

EXAMPLE 12.21

Computing the Volume of a Right Prism and a Right Cylinder

Find the volume of the gift box and the juice can.

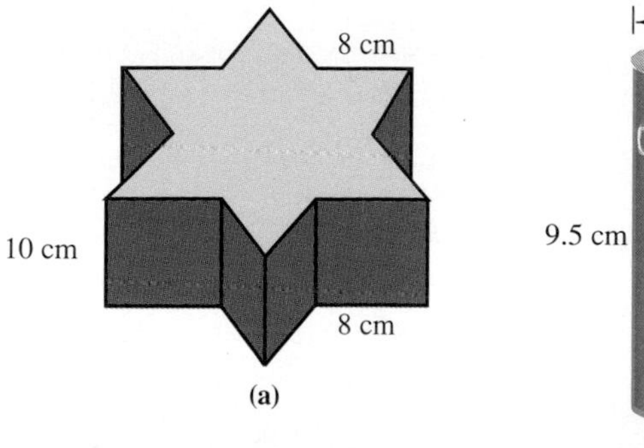

(a)

(b)

SOLUTION

(a) The base area, $B = 192\sqrt{3}\text{ cm}^2$, of the gift box was calculated in Example 12.17. The height is $h = 10$ cm, so the volume is $V = Bh = 1920\sqrt{3}\text{ cm}^3$, or about 3326 cm^3. This can also be expressed as 3.326 liters.

(b) The area of the circular base of the juice can is $\pi\,(2.75\text{ cm})^2 = 7.5625\pi\text{ cm}^2$. The height is $h = 9.5$ cm, so the volume, $V = Bh$, is $71.84375\pi\text{ cm}^3$, or about 226 cm^3. ■

Volume of Oblique Prisms and Cylinders

A deck of neatly stacked playing cards forms a right rectangular prism as shown in Figure 12.20. The total volume of the deck is the sum of the volumes of each card. If the cards slide easily on one another, it is easy to tilt the deck to form an oblique prism of the same base area B and the same height h. The solid is still made up of the same cards, so the oblique prism still has volume Bh.

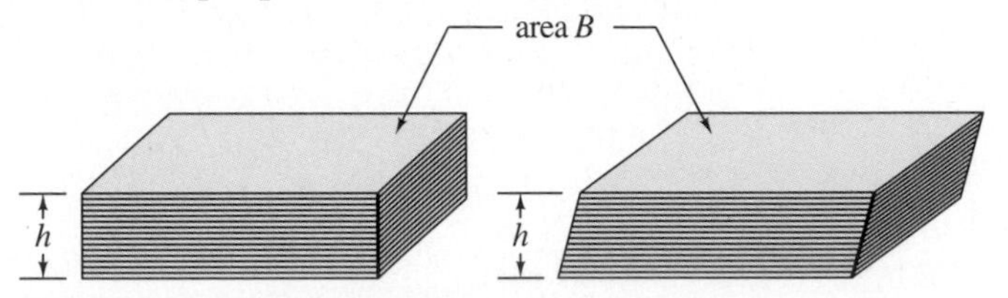

Figure 12.20
An oblique prism of base area B and height h has the same volume $V = Bh$ as the corresponding right prism.

Any oblique prism or cylinder can be imagined as a stack of very thin cards, all shaped like the base of the solid. With no change of volume, the oblique stack can be straightened to form a right prism or right cylinder of the same height h and base area B. Both the right and oblique shapes therefore have the same volume, namely $V = Bh$. This is illustrated in Figure 12.21.

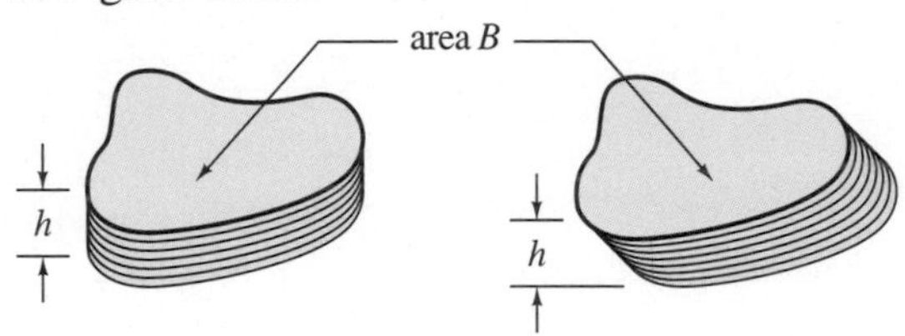

Figure 12.21
Any prism or cylinder, either right or oblique has volume $V = Bh$, where B is the base area and h is the height.

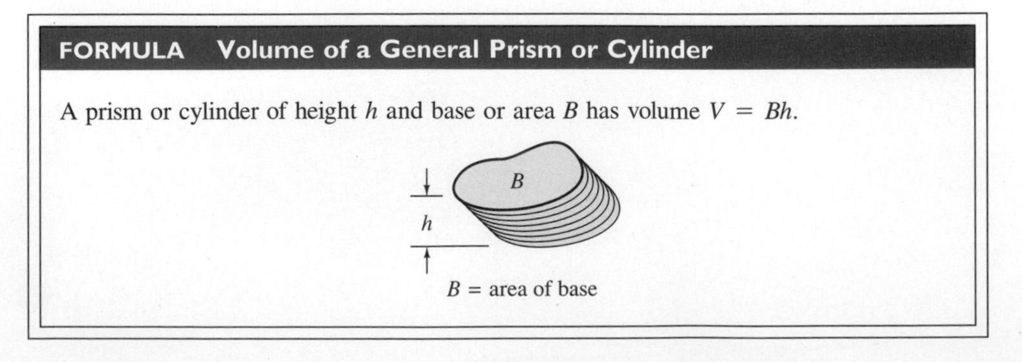
FORMULA Volume of a General Prism or Cylinder

A prism or cylinder of height h and base or area B has volume $V = Bh$.

INTO THE CLASSROOM

Problem Solving with Measurement

Each pair of students has an orange and a sheet of centimeter squared graph paper. The challenge is then given. *Find the area of the peel of the orange.* Students well-versed in the basic principles of measurement may solve the problem in a direct yet appropriate way: the orange is peeled and then the peeling is cut or torn into small pieces to tile a region of the graph paper; the region's boundary is traced and then its area, which equals that of the orange peel, is estimated by counting the number of square centimeters covered.

Problem solving with measurement reinforces both the principles and processes of measurement. On the other hand, overemphasis on exercises which require only a routine application of a formula reduces measurement to a mechanistic level. Here are two more examples illustrating the difference between a routine exercise and a problem.

Exercise: A right triangle has legs of length 6″ and 10.″ What is the area of the triangle?

Problem: Two straws, of lengths 6″ and 10,″ are joined with paperclips to form a flexible hinge. At what angle should the straws meet to form the sides of a triangle of largest possible area?

Exercise: A rectangular solid has length 6 cm, width 2 cm, and height 2 cm. What is the surface area of the solid?

Problem: The Math Manipulative Supply House sells wooden centimeter cubes in sets of 24 cubes each. What is the best shape of a box that will hold one set of cubes?

Volumes of Pyramids and Cones

In two-dimensional space (that is, in the plane), a diagonal dissects a square into two congruent right triangles. Thus the area of each triangle is one-half of that of the corresponding square.

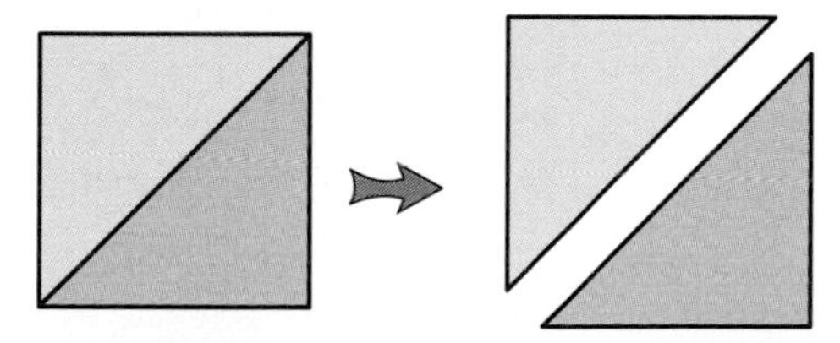

In three-dimensional space, the diagonals from one corner of a cube form the edges of three congruent pyramids that fill the cube. Therefore each pyramid has one-third the volume of the corrresponding cube as shown in Figure 12.22.

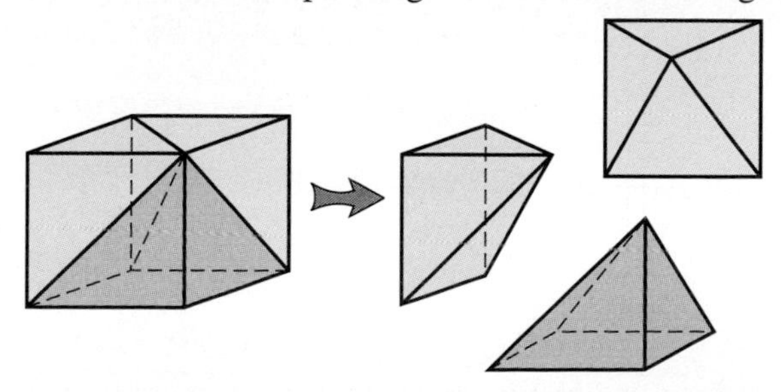

Figure 12.22
A cube can be dissected into three congruent pyramids.

Suppose next that, instead of a cube, we begin with a rectangular solid and use the diagonals from one corner to decompose the solid into three pyramids. An example is shown in Figure 12.23. In general, the three pyramids are not congruent to one another. However, it can be shown that the volumes of the three pyramids are equal. Therefore, if the prism has base area B and height h, we conclude that each pyramid has volume $\frac{1}{3}Bh$.

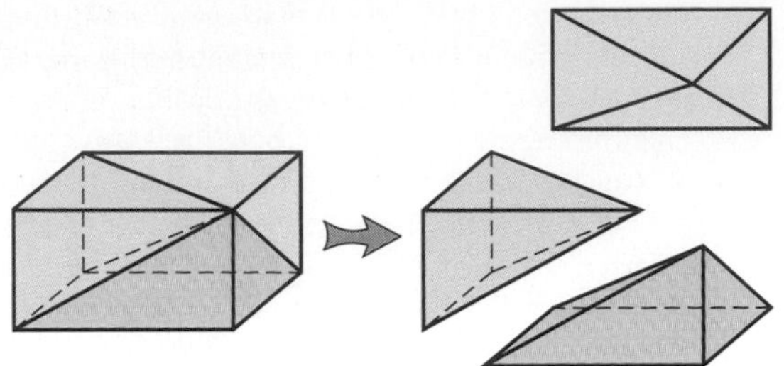

Figure 12.23
A rectangular prism can be dissected into three pyramids of equal volume.

Similar reasoning shows that *all* pyramids of base B and height h have volume $\frac{1}{3}Bh$. The base can be any polygon and the apex can be any point at distance h to the plane of the base as shown in Figure 12.24.

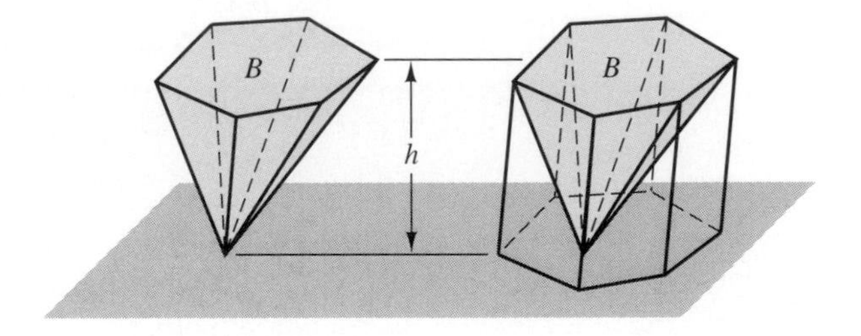

Figure 12.24
A pyramid of height h and base of area B has one-third the volume of a corresponding prism of base area B and height h. Therefore the pyramid has volume $\frac{1}{3}Bh$.

The base of a cone can be approximated to arbitrary accuracy by a polygon with sufficiently many sides, so the volume of a cone of base area B and height h is also given by the formula $\frac{1}{3}Bh$.

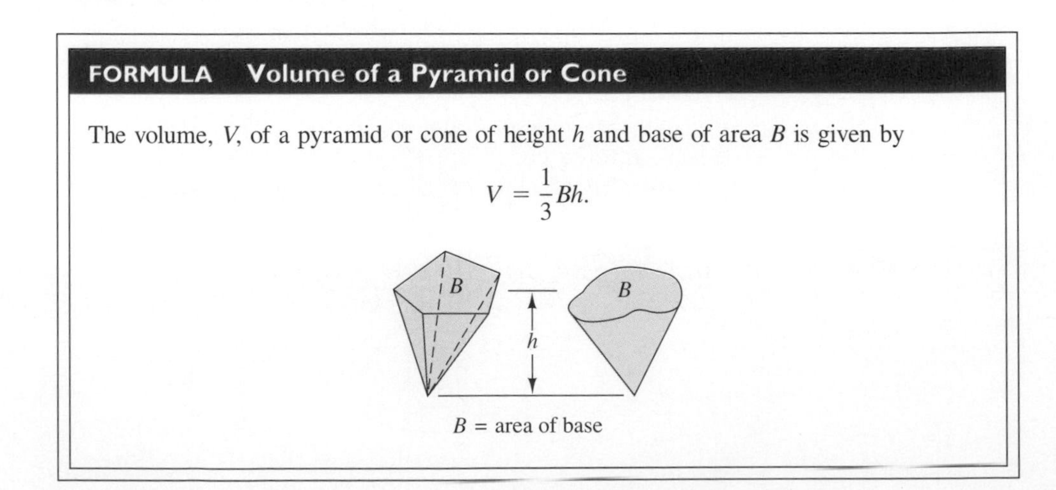

FORMULA Volume of a Pyramid or Cone

The volume, V, of a pyramid or cone of height h and base of area B is given by

$$V = \frac{1}{3}Bh.$$

EXAMPLE 12.22 **Determining the Volume of an Egyptian Pyramid**

The pyramid of Khufu is 147 m high and its square base is 231 m on each side. What is the volume of the pyramid?

SOLUTION

The area of the base is $(231 \text{ m})^2 = 53{,}361 \text{ m}^2$. Therefore, the volume is $\frac{1}{3}(53{,}361 \text{ m}^2)(147 \text{ m}) = 2{,}610{,}000 \text{ m}^3$. If the stones were stacked on a football field, a rectangular prism nearly 2000 feet high would result. For comparison, the 110 story Sears Tower in Chicago, the world's tallest building, reaches 1454 feet. ■

The Volume of a Sphere

Suppose that a solid sphere of radius r is placed in the right circular cylinder of height $2r$ that just contains it. Filling the remaining space in the cylinder with water, it is found that removing the sphere will leave the cylinder one-third full as illustrated in Figure 12.25. This means that the sphere takes up two-thirds of the volume of the cylinder. Since the volume of the cylinder is $Bh = (\pi r^2)(2r) = 2\pi r^3$, the experiment suggests that the volume of a sphere of radius r is given by $\frac{2}{3}(2\pi r^3) = \frac{4}{3}\pi r^3$. The first rigorous proof of this remarkable formula was given by Archimedes.

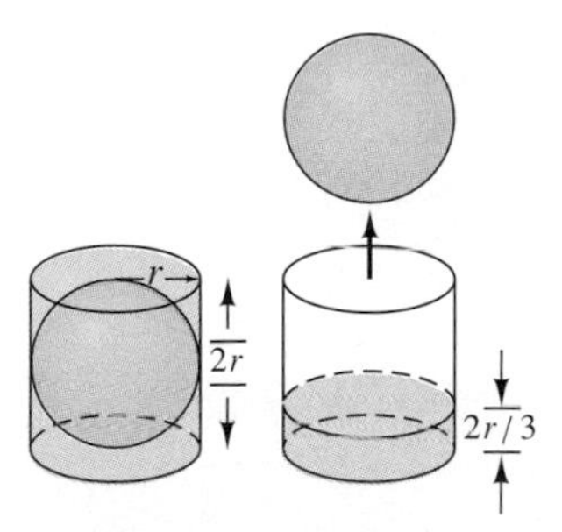

Figure 12.25
A sphere fills two-thirds of the circular cylinder containing the sphere.

FORMULA Volume of a Sphere

The volume, V, of a sphere of radius r is given by the formula

$$V = \frac{4}{3}\pi r^3.$$

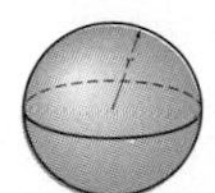

HIGHLIGHT FROM HISTORY
Sophie Germain (1776–1831)

Sophie Germain grew up in a time of social, political, and economic upheaval in France. To shield Sophie from the violence in the streets of Paris during the time of the fall of the Bastille, her wealthy parents confined their 13-year-old daughter to the family's library. Here she chanced upon J. E. Montucla's *History of Mathematics,* which recounts the legend of Archimedes' death. The story tells how a Carthaginian soldier, heedless of orders to spare the renowned mathematician, killed the unsuspecting Archimedes who remained absorbed in a geometry problem. Sophie wished to explore for herself a subject of such compelling interest.

Sophie's family initially resisted her determination to study mathematics, but eventually they gave her the freedom to follow her intellectual instincts. Since women were not permitted to enroll in the Ecole Polytechnique, which opened in Paris in 1794, Sophie resorted to collecting lecture notes from various professors at the university. The absence of a formal mathematical education was compensated by her courage to overcome strenuous challenges.

Sophie's early research was in number theory. She corresponded regularly with the great Carl Friedrich Gauss, who gave her work high praise. At the turn of the century, Sophie turned her attention increasingly to the mathematical theory of vibrating elastic surfaces. Her prize winning paper on vibrating elastic plates in 1816 placed her in the ranks of the most celebrated mathematicians of the time. Gauss recommended that she be awarded an honorary doctorate from the University of Göttingen, but unfortunately Sophie Germain's death came too soon for the awarding of the degree.

EXAMPLE 12.23 **Using the Sphere Volume Formula**

An ice cream cone is 5 inches high and has an opening 3 inches in diameter. If filled with ice cream and given a hemispherical top, how much ice cream is there?

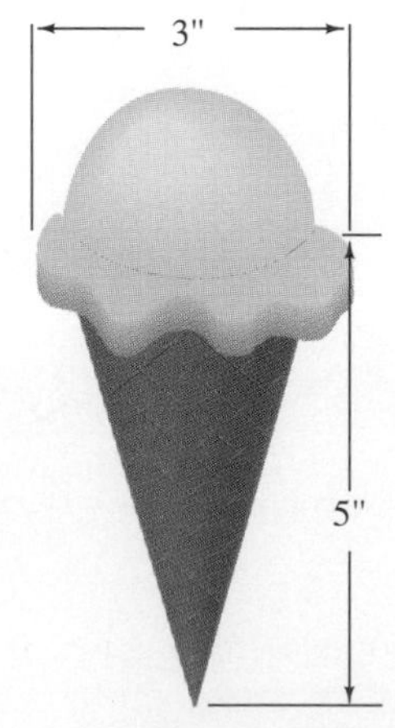

SOLUTION

The hemisphere has radius 1.5 inches, so its volume is $\frac{2}{3}\pi(1.5 \text{ in})^3 = 2.25\ \pi \text{ in}^3$. The cone has volume $\frac{1}{3}Bh = \frac{1}{3}\pi(1.5 \text{ in})^2 (5 \text{ in}) = 3.75\ \pi \text{ in}^3$. Thus the total volume is $2.25\pi \text{ in}^3 + 3.75\pi \text{ in}^3 = 6\pi \text{ in}^3$, or about 19 in^3. Since a gallon is 231 in^3, we see that the cone holds very close to a third of a quart of ice cream. ■

The Surface Area of a Sphere

A formula for the surface area of a sphere of radius r can be discovered by dividing the sphere's surface into many (say n) tiny regions of area $B_1, B_2, B_3, \cdots, B_n$. The sum $B_1 + B_2 + B_3 + \cdots + B_n$ is the surface area, S, of the sphere. Each region can also be viewed as the "base" of a pyramid-like solid whose apex is the center of the sphere. Each of the "pyramids" has height r, so the pyramids have volumes $\frac{1}{3}B_1 r, \frac{1}{3}B_2 r, \frac{1}{3}B_3 r$, and so on. An example is shown in Figure 12.26.

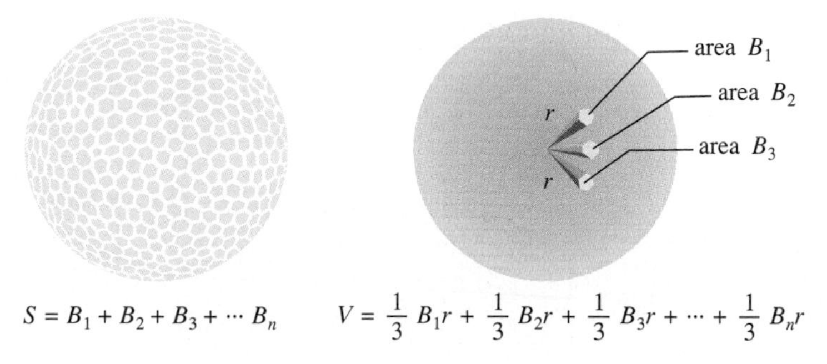

Figure 12.26
A solid sphere can be viewed as made up of pyramid-like pieces.

We have the following relationships:

$$S = B_1 + B_2 + B_3 + \cdots + B_n \qquad \text{(surface area of the sphere)}$$

and

$$V = \frac{1}{3}B_1 r + \frac{1}{3}B_2 r + \frac{1}{3}B_3 r + \cdots + \frac{1}{3}B_n r. \qquad \text{(volume of the sphere)}$$

Therefore,

$$V = \frac{r}{3}(B_1 + B_2 + B_3 + \cdots + B_n) = \frac{r}{3}S.$$

Since $V = \frac{4}{3}\pi r^3$, this gives us the equation

$$\frac{4}{3}\pi r^3 = \frac{r}{3}S.$$

Multiplying both sides by 3, and dividing both sides by r, we can solve for the surface area S.

FORMULA Surface Area of a Sphere

The surface area S of a sphere of radius r is given by the formula

$$S = 4\pi r^2.$$

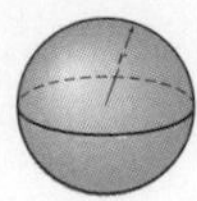

EXAMPLE 12.24

Comparing Earth to Jupiter

The diameter of Jupiter is about 11 times larger than the diameter of our planet Earth. How many times greater is (a) the surface area of Jupiter? (b) the volume of Jupiter?

SOLUTION

(a) Let r denote the radius of Earth and R the radius of Jupiter. Therefore $R = 11r$. Using the formula for the surface area of a sphere, the ratio of the surface area of Jupiter to that of Earth is

$$\frac{4\pi R^2}{4\pi r^2} = \frac{R^2}{r^2} = \left(\frac{R}{r}\right)^2 = (11)^2.$$

That is, the surface area of Jupiter is 11^2, or 121 times the surface area of Earth.

(b) Using the sphere volume formula, the ratio of volumes is

$$\frac{\frac{4}{3}\pi R^3}{\frac{4}{3}\pi r^3} = \frac{R^3}{r^3} = \left(\frac{R}{r}\right)^3 = (11)^3.$$

Therefore, the volume of Jupiter is about 11^3, or 1331 times the volume of Earth. Precise measurements of the two not quite spherical planets show that the volume ratio is 1323.3, which agrees closely with 1331. ■

Comparing Measurements of Similar Figures

Two figures are similar if they have the same shape but possibly different size. The ratio of all pairs of corresponding lengths in the two figures is a constant value called the scale factor, which we will denote by the letter k. The ratio of any *linear measurement* of the two figures—perimeter, height, diameter, slant height, and so on—is also that of the scale factor k. For example, since the diameter of Jupiter is 11 times the diameter of Earth then the scale factor is $k = 11$. We then also know that the equator of Jupiter is 11 times as long as Earth's equator.

The ratio of *areas* of similar figures is given by the *square,* k^2, of the scale factor k. The ratio of volumes is given by the *cube,* k^3, of the scale factor. This basic fact is evident for the cubes shown in Figure 12.27.

The same comparison of areas and volumes holds for any pair of similar figures, not just the cube shown in Figure 12.27. The following result is an important principle for the comparison of the measurements of similar figures.

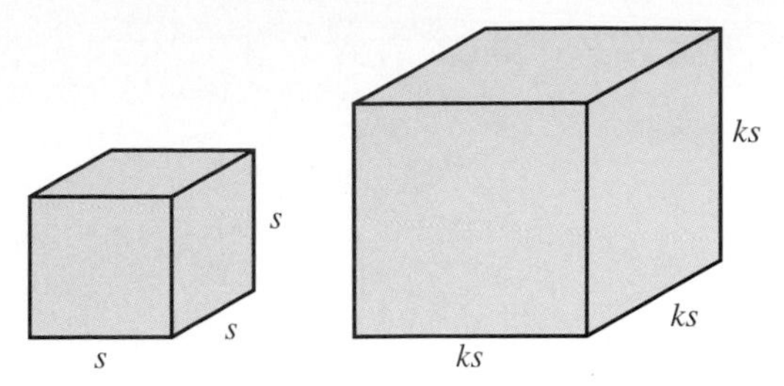

Scale factor = k

	Cube I	Cube II
Length of edge	s	ks
Area of each face	s^2	$k^2 s^2$
Volume	s^3	$k^3 s^3$

Figure 12.27
Area varies by the square, k^2, of the scale factor k, and volume varies by the cube, k^3, of the scale factor.

THEOREM The Similarity Principle of Measurement

Let Figures I and II be similar. Suppose some length dimension of Figure II is k times as the corresponding dimension of Figure I; that is, k is the scale factor. Then:

1. *any* length measurement—perimeter, diameter, height, slant height, and so on—of Figure II is k times that of the corresponding length measurement Figure I;
2. *any* area measurement—surface area, area of a base, lateral surface area, and so on—of Figure II is k^2 times that of the corresponding area measurement Figure I;
3. *any* volume measurement—total volume, capacity, half-full, and so on—of Figure II is k^3 times the corresponding volume measurement of Figure I.

HIGHLIGHT FROM HISTORY
Jonathan Swift (1667–1745)

Jonathan Swift published his best known book, *Gulliver's Travels,* in England in 1726. This book has been popular for two-and-a-half centuries because it is both a delightful story about pygmies and giants and a clever satire on many eighteenth-century institutions and scholars. *Gulliver's Travels* stresses the relative contrasts of the pygmies of Lilliput and the giants of Brobdingnag. Swift points out that "nothing is great or little otherwise than by comparison." Comparisons can be fun: if a Lilliputian is 6 inches tall, about how long would his or her shoes be? What is the approximate weight of a Lilliputian?

WORTH READING

Sizing Up the Universe

We all, children and grownups alike, are inclined to live in our own little world, in our immediate surroundings, or at any rate with our attention concentrated on those things with which we are directly in touch. We tend to forget how vast are the ranges of existing reality which our eyes cannot directly see, and our attitudes may become narrow and provincial. We need to develop a wider outlook, to see ourselves in our relative position in the great and mysterious universe in which we have been born and live.

At school we are introduced to many different spheres of existence, but they are often not connected with each other, so that we are in danger of collecting a large number of images without realizing that they all join together in one great whole. It is therefore important in our education to find the means of developing a wider and more connected view of our world and a truly cosmic view of the universe and our place in it.

This book presents a series of forty pictures composed so that they may help to develop this wider view. They really give a series of views as seen during an imaginary and fantastic journey through space—a journey in one direction, straight upward from the place where it begins. Although these views are as true to reality as they can be made with our present knowledge, they portray a wonderland as full of marvels as that which Alice saw in her dreams.

Kees Boeke, a sixth grade teacher in Holland, worked with the children in his class to make a picture book that takes the reader on an imaginary journey through the universe. Going from the picture on one page to that of the next page changes the scale of view by a factor of ten, so in 40 jumps the journey moves from the gamma ray to dots representing clumps of distant galaxies. A similar book, which more fully develops Boeke's idea, is *Power of Ten,* by Philip and Phyllis Morrison and The Office of Charles and Ray Eames (Scientific American Books, 1982).

SOURCE: From *Cosmic View: The Universe in 40 Jumps* by Kees Boeke, with an introduction by Arthur H. Compton. Copyright ©1957 by Kees Boeke. Reprinted by permission of Harper Collins Publishers, Inc.

EXAMPLE 12.25 **Using the Similarity Principle**

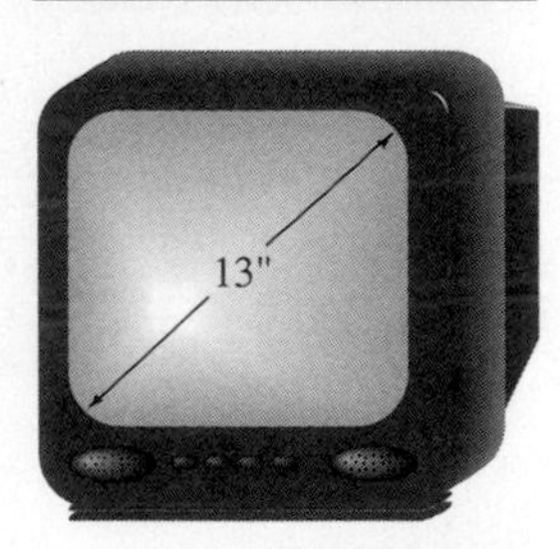

(a) Television sets are measured by the length of the diagonal of the rectangular screen. How many times larger is the screen area of a 40 inch model than a 13 inch table model?

(b) A 2″ by 4″ by 8″ rectangular brick of gold weighs about 44 pounds. What are the dimensions of a similarly shaped brick that weighs 10 pounds?

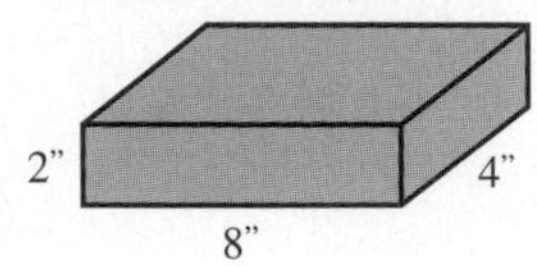

SOLUTION

(a) The scale factor, k, is $40/13 \doteq 3.08$. Since area varies by $k^2 \doteq (3.08)^2 = 9.5$, the large screen has about 9.5 times the area of the similarly shaped small screen.

(b) The weight of a gold brick is proportional to its volume, and the volume of similarly shaped bricks varies by the factor k^3, the cube of the scale factor k. Therefore $10 = k^3 44$, so $k^3 = 10/44$ and $k \doteq (10/44)^{1/3} \doteq 0.6$. Multiplying the length, width, and length of the 44 pound brick by 0.6 gives the approximate dimensions of a similarly shaped 10 pound brick of gold, namely 1.2″ by 2.4″ by 4.8″. ■

PROBLEM SET 12.4

Understanding Concepts

1. Find the surface area and volume of each figure.

(a)

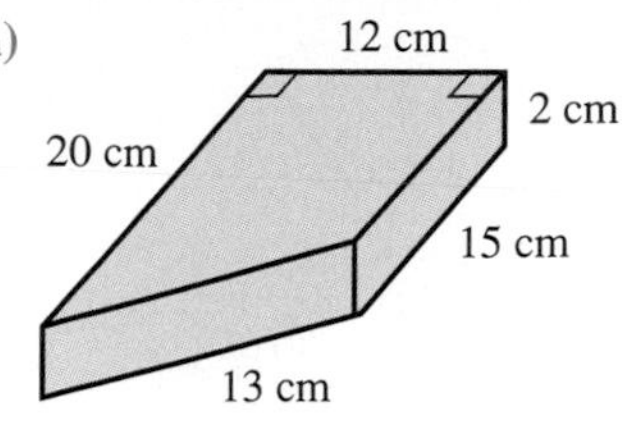

Right trapezoidal prism

(b)

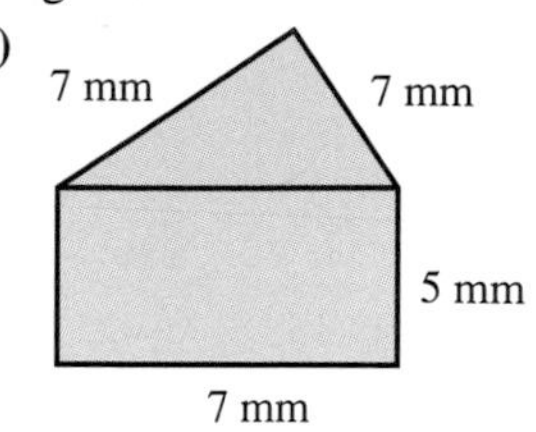

Regular right triangular prism

(c)

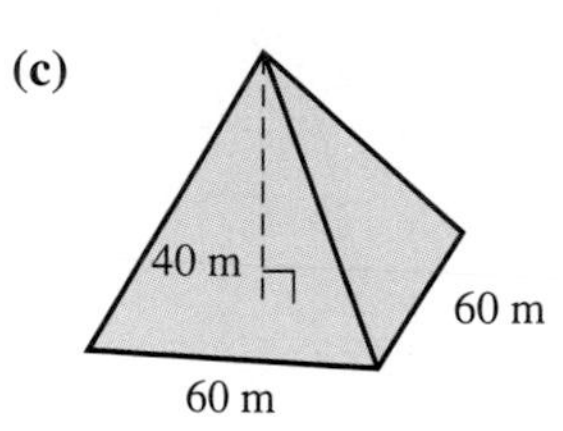

Right square pyramid

2. Find the surface area and volume of these figures.

(a)

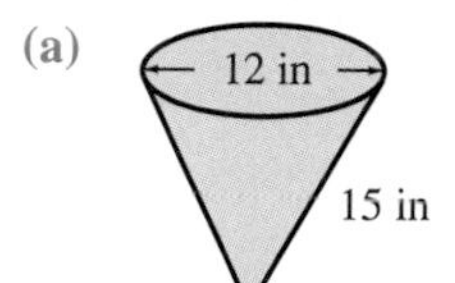

Right circular cone

(b)

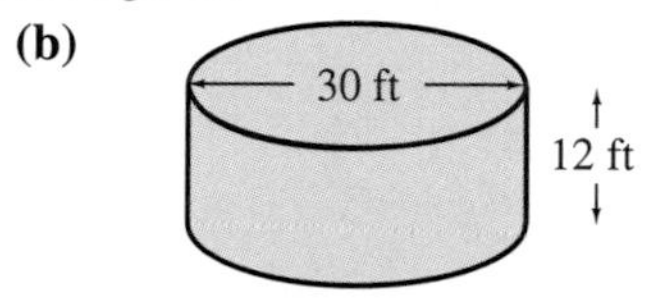

Right circular cylinder

(c)

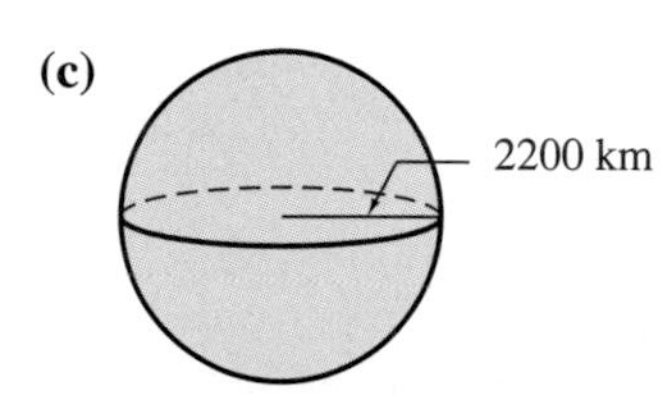

Sphere

3. Find the surface area and volume of each figure.

(a)

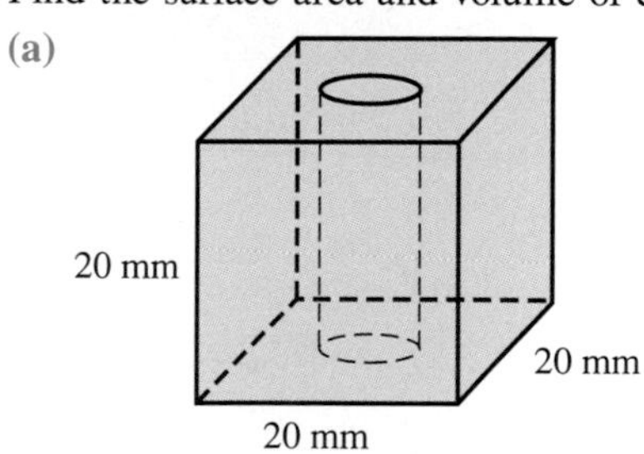

Cube with cylindrical hole of diameter 4mm

(b)

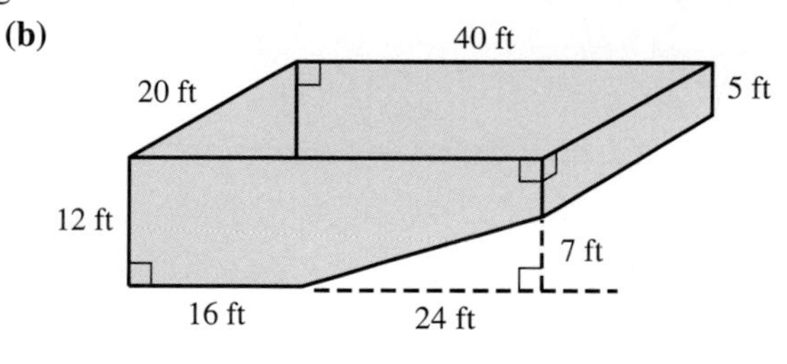

Swimming pool

4. Find the surface area and volume of these figures.

(a)

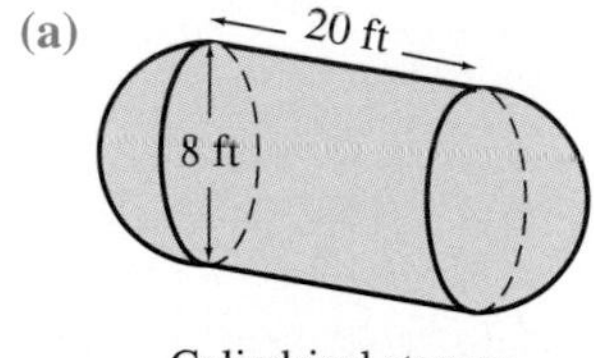

Cylindrical storage tank with hemispherical ends

(b)

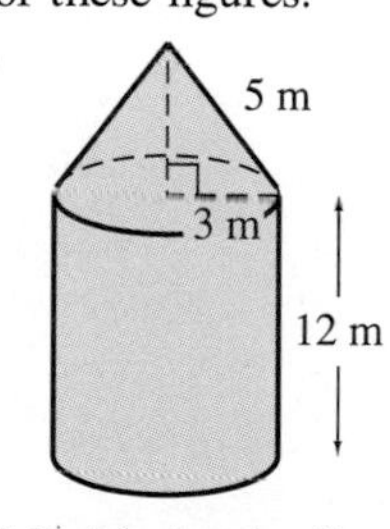

Cylindrical grain silo with conical top

5. A square right regular pyramid is formed by cutting, folding, and gluing the following pattern.

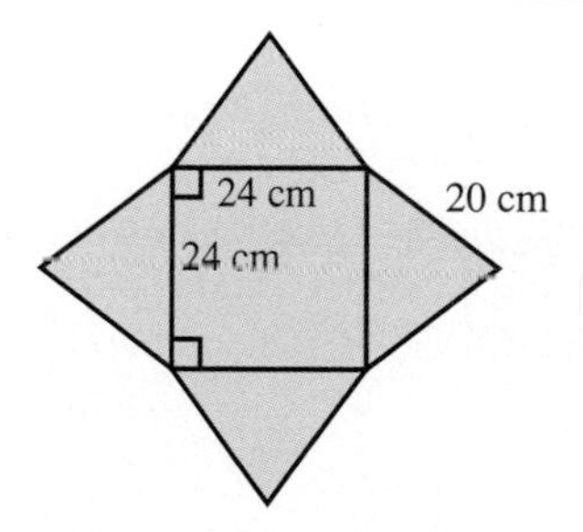

(a) What is the slant height of the pyramid?
(b) What is the lateral surface area of the pyramid?
(c) Use the Pythagorean theorem to find the height of the pyramid.
(d) What is the volume of the pyramid?

6. A pyramid is formed by joining a vertex of a unit cube to the four vertices of an opposite face.

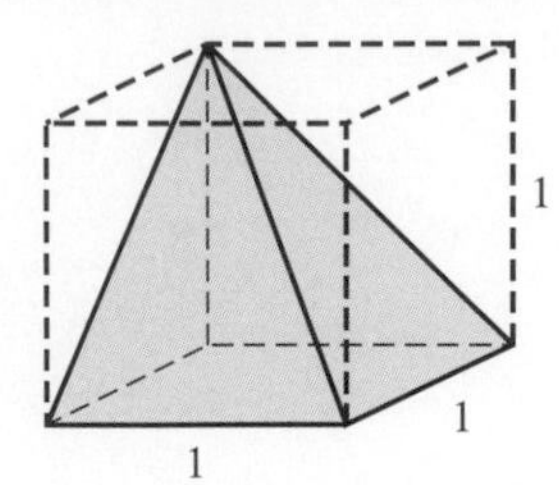

(a) Use the "cut-and-flatten" idea to view the five faces of the pyramid spread out on a plane. Label the dimensions of all of the sides of the faces.

(b) Calculate the area of each face of the pyramid, and give the total surface area of the pyramid.

7. (a) Make a pattern for the pyramid shown in problem 6.

(b) Copy your pattern onto card stock and make three congruent paper pyramids.

(c) Show that a cube can be formed with the three pyramids. Make a drawing to show this.

(d) Why does it follow from part (c) that each pyramid has volume $\frac{1}{3}$ cubic units? Explain.

8. A sheet of $8\frac{1}{2}''$ by $11''$ notebook paper can be rolled into a cylinder in either of two ways.

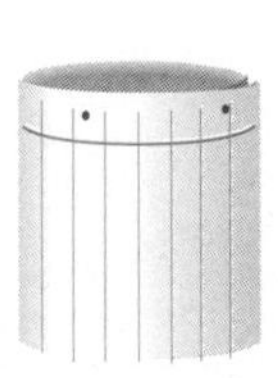

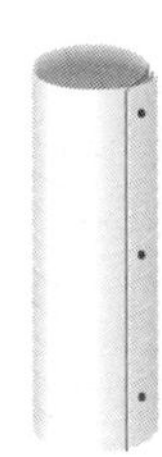

Which way encloses the largest volume? Make a prediction, then check it.

9. An aluminum soda pop can has a diameter of 6.5 cm and a height of about 11 cm. If there are 30 milliliters in a fluid ounce, verify that the capacity of the can is 12 fluid ounces as printed on the can's label.

10. The circular sector shown can be rolled into a right circular cone in which the two 4-inch radial segments are joined.
Find the following measurements.

(a) The radius of the base of the cone.

(b) The height of the cone.

(c) The volume of the cone.

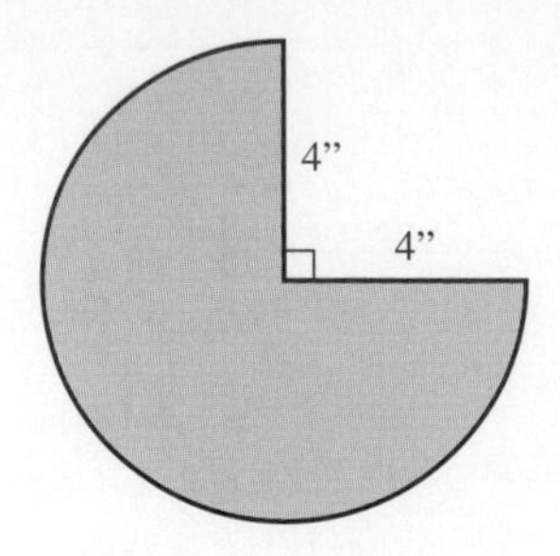

11. A semicircular sector is rolled up and joined along the two radial segments to form a cone. Show that the diameter of the cone is equal to the slant height of the cone.

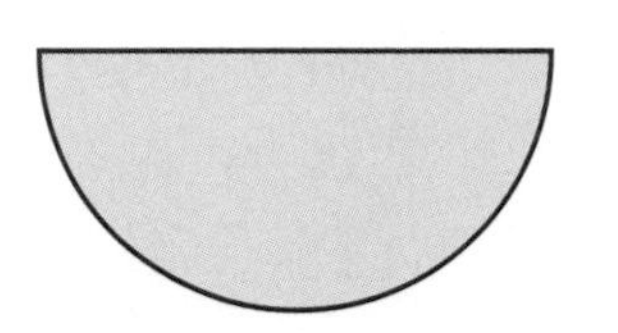

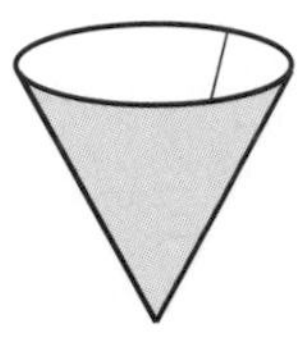

12. If it takes a quart of paint to cover the base of a hemisphere, how many quarts does it take to paint the spherical part of the same hemisphere?

13. Archimedes showed that the volume of a sphere is two-thirds the volume of the right circular cylinder just containing the sphere. Show that the area of the sphere is also two-thirds the surface area of the cylinder. (Archimedes was so pleased with these discoveries that he requested that the figure shown be placed on his tombstone.)

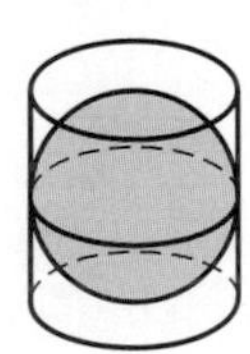

14. Find the volumes of the following solids.

(a)

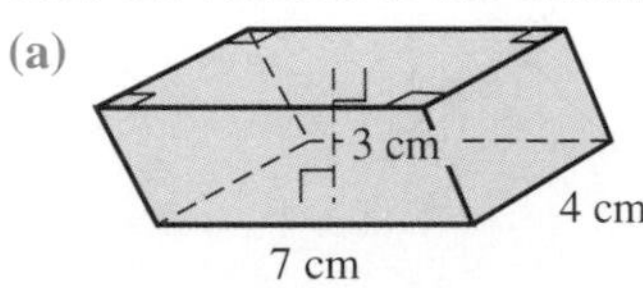

Oblique prism

(b)

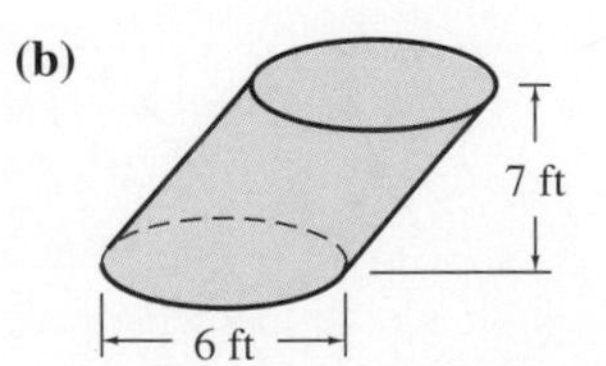

Oblique circular cylinder

(c)

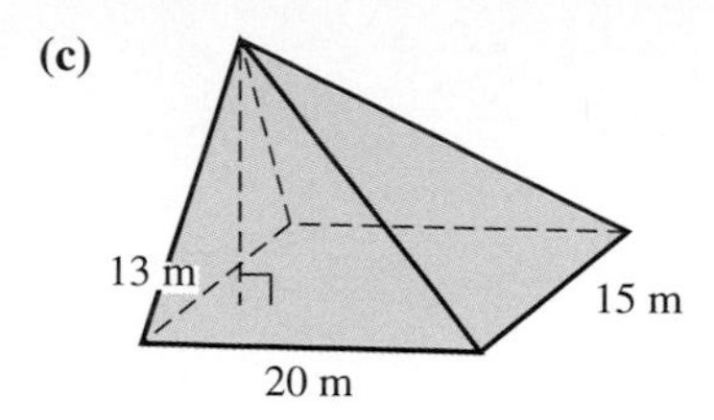

Rectangular pyramid

(d)

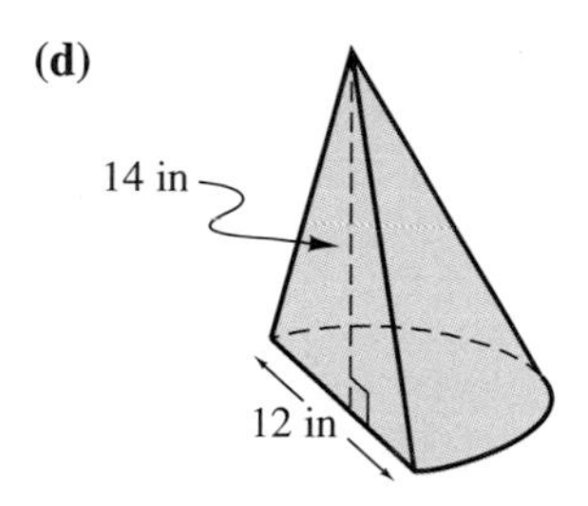

Semicircular cone

Use the similarity principle to answer problems 15, 16, 17, 18, 19, and 20. Explain carefully how the principle is used.

15. (a) An 8″ (diameter) pizza will feed one person. How many people will a 16″ pizza feed?

(b) Is it better to buy one 14″ pizza at $10 or two 10″ pizzas at $6 each? (*Hint:* 1.4^2 is about 2.0.)

16. (a) Eight spherical lead fishing sinkers are melted to form a single spherical sinker. If the small sinkers each have diameter 1/4 inch, what is the diameter of the new large one?

(b) How many small sinkers would it take to make a sinker 1 inch in diameter?

17. What fraction of the area of the large circle is shaded in each figure?

(a)

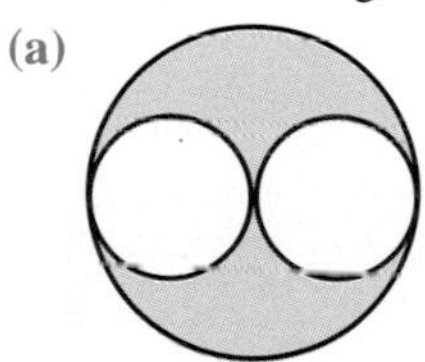

(b)

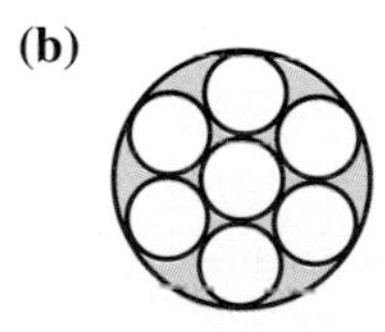

(*Hint:* First compare each unshaded circle to the large circle.)

18. Cones I, II, and III are similar to one another. Fill in the measurements left blank in the following table.

	I	II	III
Height	6	18	cm
Perimeter of base		30	15 cm
Lateral surface area	40		cm^2
Volume			10 cm^3

19. A cylindrical can holds 100 milliliters.

(a) If the radius of the base is doubled and the height halved, what is the new volume of the can?

(b) If the radius of the base is halved and height is doubled, what is the new volume of the can?

20. A cube 10 cm on a side holds 1 liter.

(a) How many liters does a cube 20 cm on a side hold?

(b) What is the length of each side of a cube that holds 2 liters?

Thinking Critically

21. A right circular cone has height r and a circular base of radius $2r$. Compare the volume of the cone to that of a sphere of radius r. Sketch both solids, using the same scale.

22. Similar figures are erected on the three sides of a right triangle. What formula relates the areas of the three figures? Explain your reasoning carefully.

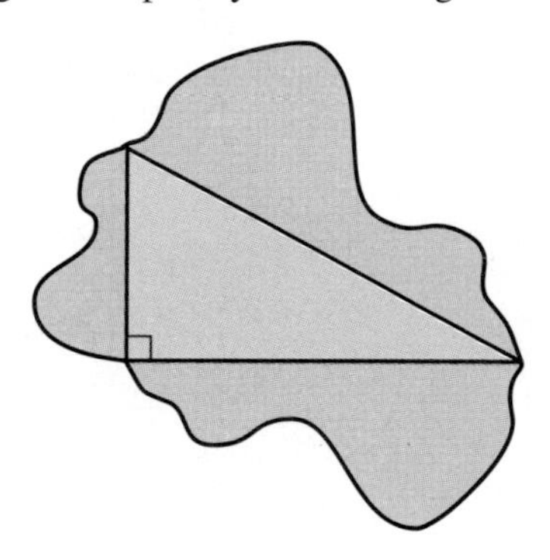

23. Find the area of each region R_1, R_2, R_3, and R_4 of the geoboard figure shown.
(*Hint:* Why are R_1 and R_3 similar? What is the scale factor?)

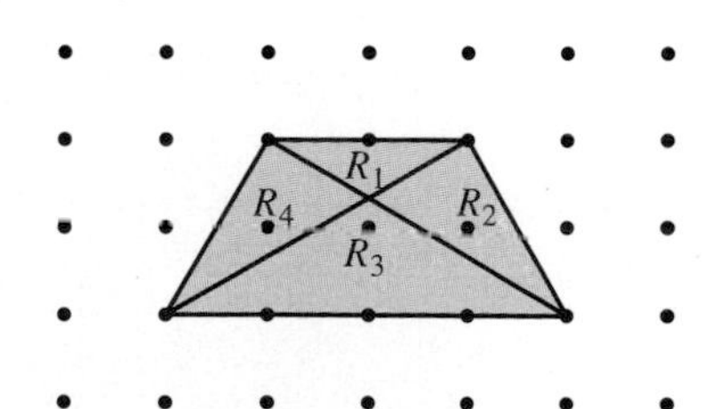

24. An ice cream soda glass is shaped like a cone of height 6 inches, and has a capacity of 16 fluid ounces when filled to the rim. Use the similarity principle to answer the following questions, using the fact that the cone of liquid is similar to the cone of the entire region inside the glass.

(a) How high is the soda in the glass when it contains 2 fluid ounces?

(b) How much soda is in the glass when it is filled to a level 1 inch below the rim?

Thinking Cooperatively

25. A birthday cake has been baked in a 7″ by 7″ by 2″ pan. Frosting covers the top and sides of the cake.

(a) What is the volume of the cake?

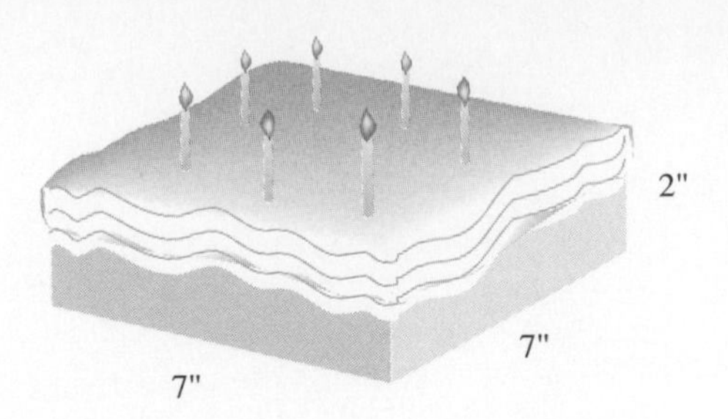

(b) What is the area covered by the frosting?

(c) Describe how to cut the cake into eight pieces, so that each piece has the same size (measured by volume) *and* the same amount of frosting (measured by area covered with frosting).

(d) Describe how to cut the cake in seven pieces, each with the same size and amount of frosting. (*Hint:* Consider a slice made by two vertical cuts from the center that intercepts 4 inches of the perimeter.)

26. A **frustum** of a pyramid or cone is obtained by slicing the top off the pyramid or cone by a plane parallel to the base. Find the volumes of these two frustums.

(a)

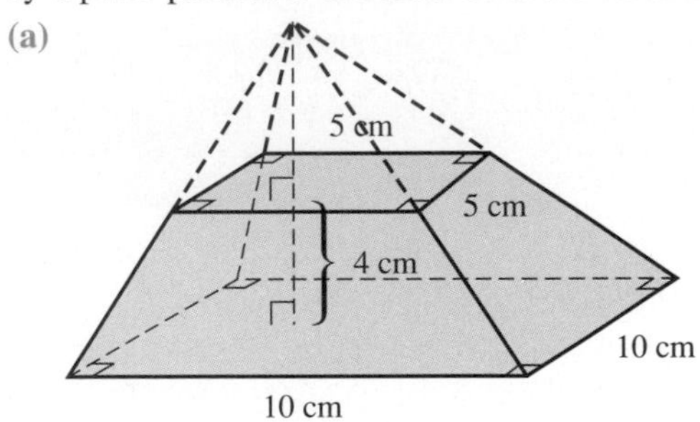

(b)

27. The cube *ABCDEFGH* with edges of length ℓ, contains the tetrahedron *ACEG*.

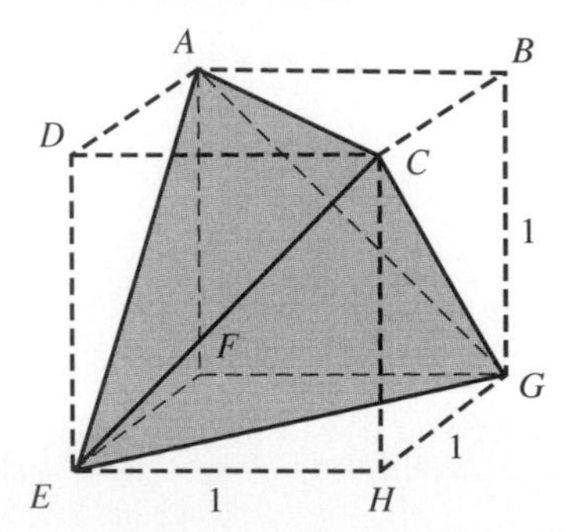

(a) Explain why *ACEG* is a regular tetrahedron with edges of length $\sqrt{2}$.

(b) Show that the volume of *ACEG* is $1/3$. (*Hint:* What is the volume of the tetrahedron *ABCG?*)

(c) Use the similarity principle to explain why a regular tetrahedron with edges of length b has volume $\sqrt{2}b^3/12$.

Making Connections

28. A napkin ring is being made of cast silver. It has the shape of a cylinder 1.25 inches high, with a cylindrical hole 1 inch in diameter and a thickness of $1/16$ inch. How many ounces of silver are required? It will help to know that silver weighs about 6 ounces per cubic inch.

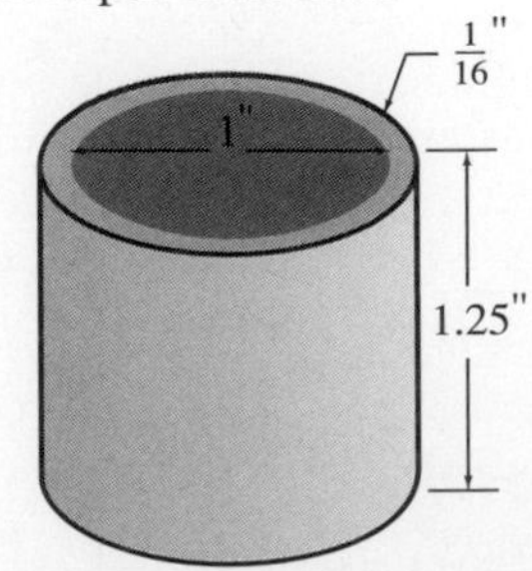

29. Give the dimensions of a rectangular aquarium 40 cm high that holds 48 liters of water.

30. A theater sells 4″ by 5″ by 8″ boxes of popcorn for \$1.75. It also sells cylindrical "tubs" of popcorn for \$3.50 where the tub is 10″ high and has a diameter of 6″. Is it better to buy one tub or two boxes of popcorn?

31. Small grapefruits of diameter 3 in are on sale at five for a dollar. The large 4 in diameter grapefruit are three for a dollar. If you are buying \$5 worth of grapefruit, should you choose small ones or large ones?

32. **World Records.** *The Guiness Book of World Records,* published annually by Facts on File, New York, contains a fascinating collection of measurements.

(a) The world's largest flawless crystal ball weighs 106.75 pounds and is 13 inches in diameter. What is the weight of a crystal ball 5 inches in diameter?

(b) The largest pyramid is the Quetzacoatl, 63 miles southeast of Mexico City. It is 177 feet tall and covers an area of 45 acres. Estimate the volume of the pyramid. By comparison, the largest Egyptian pyramid of Khufu (called Cheops by the Greeks) has a volume of 88.2 million ft^3. Recall that an acre is 43,560 ft^2.

(c) The building with the largest volume in the world is the Boeing Company's main assembly plant in Everett, Washington. The building encloses 472 million ft^3 and covers 98.3 acres. What is the size of a cube of equal volume?

33. A water pipe with an inside diameter of 3/4 inches is 50 feet long in its run from the hot water tank to the faucet. How much hot water is wasted when the water inside the pipe cools down? Give your answer in gallons, where 1 gallon = 231 in^3.

34. In Jonathan Swift's satirical novel *Gulliver's Travels,* Dr. Lemuel Gulliver encounters the tiny Lilliputians. A Lilliputian is similar to Gulliver but is 6 inches tall compared to Gulliver's 6 feet.

(a) Explain why the Lilliputians ordered 1728 rations for Gulliver's dinner.

(b) If the material from Gulliver's shirt was cut up to make shirts for the Lilliputians, how many shirts could be made?

35. In many countries the size and shape of sheets of paper is based on the metric system. An A0 sheet is a rectangle of area 1 m^2. When cut in half across its width, it forms two A1 sheets, each of which is similar to the A0 sheet. Cutting an A1 sheet forms two A2 sheets, and so on.

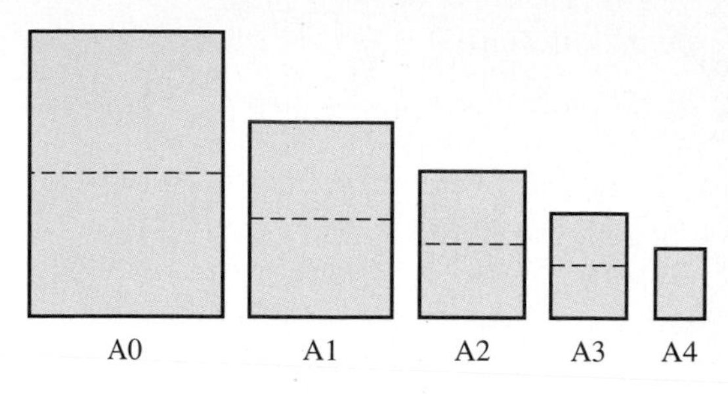

(a) What is the scale factor by which the linear dimensions of an A0 sheet are multiplied to give the corresponding dimensions of an A1 sheet?

(b) Find the width and length, in centimeters, of an A0 sheet.

(c) Find the width and length of an A4 sheet.

(d) What metric sized paper do you think is used in place of the $8\frac{1}{2}$″ by 11″ sheets in common use in the U.S.?

36. A geologist wishes to determine the density of a rock specimen: density is the weight of the rock divided by its volume. The weight is 7.34 kg, easily measured on a scale. To find the volume, the specimen is submerged in a cylindrical water tank 16 cm in diameter, causing the level of the water to rise 5.7 cm.

(a) What is the volume of the specimen?

(b) What is the density of the specimen? Give your answer in grams per cubic centimeter.

37. A rain gauge has a funnel 6 inches in diameter at the top, tapering into a plexiglass collection cylinder whose inside diameter is 2 inches. How far apart should marks be placed on the cylinder to indicate the number of inches of rainfall? (*Hint:* Use the similarity principle.)

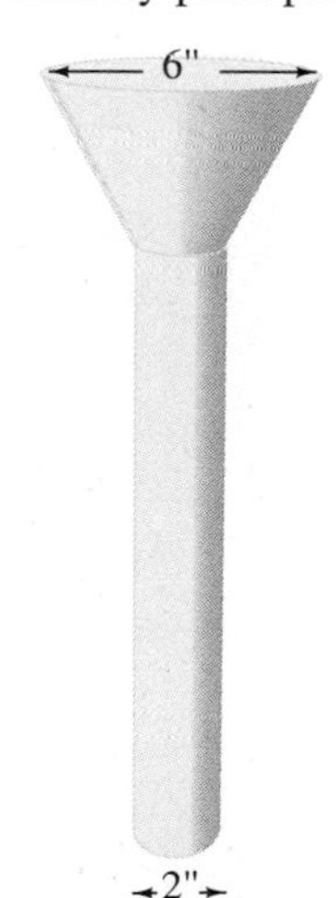

Communicating

38. Consider the following problem: *Which is a better buy, a 9″ pizza for \$5 or a 12″ pizza for \$8?*
 (a) Write your own solution to the problem, and for each of the following answers, write an explanation why the reasoning is faulty.
 (b) "They are equal. There are three more inches in the 12″ pizza but the price also increases by \$3."
 (c) "If there are eight slices per pizza then the 9″ pizza costs 62¢ per slice. The 12″ pizza costs \$1 per slice. Therefore the 9″ pizza would be the better buy."
 (d) "Since $9/5 = 1.8$ inches per dollar and $12/8 = 1.5$ inches per dollar, the 9″ pizza is the better buy."

For Review

39. Which of the following triples could be the lengths of sides of a right triangle?
 (a) 30, 72, 78 **(b)** 12, 35, 37
 (c) 2.0, 2.1, 2.9
40. Find formulas for the perimeter and area of a regular hexagon with sides of length b.
41. Find the perimeter and area of this lattice polygon.

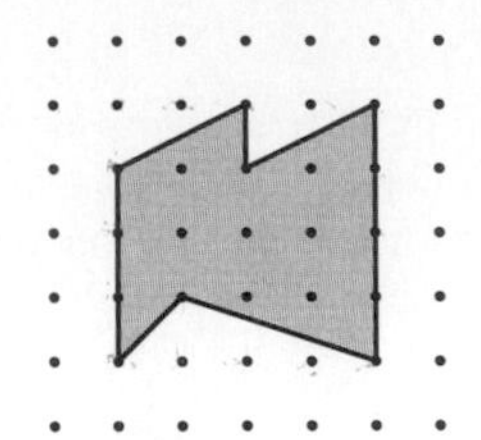

42. What is the speed equivalent to 50 inches per second when expressed in miles per hour?
43. Convert these measurements to the unit shown.
 (a) 0.278 m = ________ cm
 (b) 2.3 km^2 = ________ m^2
 (c) 68,532 cm^3 = ________ L.

EPILOGUE That's about the size of it!

The size of our everyday world spans about six orders of magnitude; that is, the longest distances of interest are about 10^6, or a million times larger than the smallest sized items we deal with. At arm's length—that is, on a scale of about a meter—we find most of the objects (chairs, beachballs, and so on) and life forms (dogs, cats, horses, and so on) familiar in daily life. A thousandfold increase—to the scale of kilometers—encompasses the distances of ordinary travel, whether across town or cross country. A thousandfold decrease—to the scale of millimeters—encompasses all that can be seen easily with the unaided eye. The dot at the end of this sentence is several tenths of a millimeter across.

To a great extent, the story of science and technology is told by the increasing number of orders of magnitude required to encompass newly discovered objects and phenomena. The first major step to the measurement of the large was Eratosthenes measurement (c. 240 B.C.) of the Earth's circumference. At noon during the summer solstice, a vertical rod at Syene (now Aswan) cast no shadow, whereas 5000 stadia to the north in Alexandria a vertical rod made an angle of "1/50 of four right angles" (that is, 7°12′). Thus the earth's circumference is 50×5000, or 250,000 stadia, a value accurate to about 6 percent. The measurement of the solar system—the sun and its planets—was much more difficult: The Copernican model was only 1/7 of the true size, and the first accurate distances to the planets (to within 10 percent) were made in 1672 by the newly created French Academy of Science. The first accurate distance to a star was given in 1837 by Friedrich Wilhelm Bessel, who found that 61 Cygni was 619,000 times as far from earth as the sun. In 1924 Edwin Hubble proved that William Herschel's "island universes" were separate galaxies far from our own Milky Way galaxy. To measure the universe, where distant quasars are 10 billion light years away, we must measure on the astonishingly large scale of about 10^{25} meters.

Recent breakthroughs in microscopy reveal new images of the very small. At 10^{-6} m we find bacteria, at 10^{-9} m a single sodium atom, at 10^{-12} m the nucleus of the sodium atom. Current thought suggests that 10^{-16} m—the scale of quarks—may present all there is to see, at least until we reach 10^{-31} m.

This chapter has introduced the basic notions of measurement. Measurement is concerned with how size is determined and communicated. Generally measurement is an approximation, calling for appropriate judgments to be made about the selection of measurement tools and the level of precision required. For some ideal shapes—triangles, prisms, pyramids, circles, and spheres, to name a few—the measurement process is supplemented by the use of formulas. Measurement, however, is not simply a collection of formulas; often we must return to the basic principles of the measurement process.

CLASSIC CONUNDRUM Strings and Balls

Imagine tying a loop of string tightly about a basketball. By cutting the string and inserting an extra piece 6 feet long, the lengthened loop forms a circle almost 1 foot above the surface of the basketball.

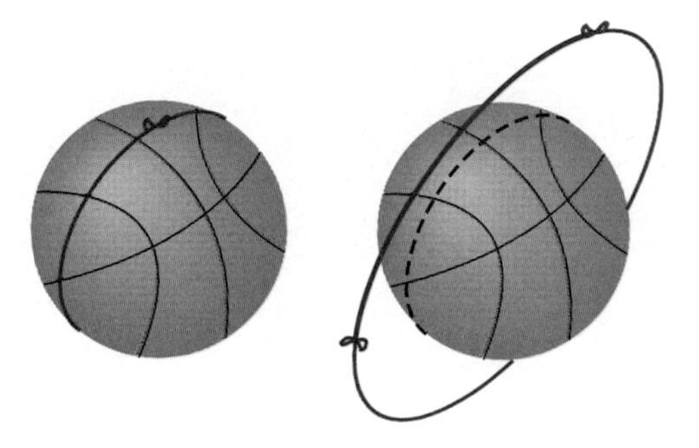

Now imagine a *very* long loop of string wrapped snuggly around the earth's equator. Once more, cut the string, insert an additional 6 feet of string, and hold the string equally distant above the equator. Could an ant crawl under the string? Could a rabbit run under the string?

CHAPTER 12 SUMMARY

Key Concepts

1. The Measurement Process

The **measurement process** can be divided into four steps: (1) choose the property (length, area, volume, and so on) to be measured; (2) select a unit of measurement; (3) compare the object to the unit by covering, filling, and so on; (4) express the measurement as the number of units used. **Measurements are not exact,** and decisions must be made when selecting measurement tools to give appropriate accuracy and precision. A standardized system of units provides a way to communicate size and magnitude. The **metric (SI) system** is used around the world. Units are related by powers of 10 described by a **prefix system.** The most commonly used prefixes are *milli* = 1/1000, *centi* = 1/100, and *kilo* = 1000. The basic unit of length is the **meter,** so that areas are measured in square meters, m^2, and volumes in cubic meters, m^3. Areas are also measured in **hectares** (1 hectare = 10,000 m^2) and capacities are measured in **liters** (1 L = 1000 cm^3 = 1/1000 m^3). In the United States the customary system is unofficial but still in use.

2. Area and Perimeter

Area is the amount of the plane covered by a plane region. The **unit of area** is arbitrary, but usually a square one unit of length on a side is chosen. To compare the area of one region to another, the **congruence** and **dissection** properties are often useful. Some polygons have areas given by the following formulas:

Rectangle of width w and length l: $A = wl$
Parallelogram of base b and height h: $A = bh$
Triangle of base b and altitude h: $A = \frac{1}{2}bh$
Trapezoid of bases a and b and altitude h: $A = \frac{1}{2}(a + b)h$

Length is the distance along a curve, and **perimeter** is the length of a simple closed curve. In particular the perimeter of a circle is called the **circumference** of the circle. The number π(**pi**), about 3.1416, is defined to be the ratio of the circumference to the diameter of a circle. Therefore a circle of radius r has circumference $2\pi r$. The **area of the circle** is given as πr^2.

3. The Pythagorean Theorem

Three numbers a, b, and c satisfy the **Pythagorean relationship** if $a^2 + b^2 = c^2$. The **Pythagorean theorem** states that the lengths of the sides of a right triangle satisfy the Pythagorean relationship. The theorem can also be viewed as a result about areas: The sum of the areas of squares on the legs equals the area of the square on the hypotenuse. The converse of the Pythagorean theorem also holds; if $a^2 + b^2 = c^2$, then a triangle with sides of length a, b, and c is a right triangle.

4. Surface Area and Volume

The surface area of a polyhedron is the sum of the areas of the plane faces. For some polyhedra, such as right prisms and right regular pyramids, it is useful to imagine that the surface is cut and unfolded onto the plane. Similarly, the surface areas of right cylinders and right circular cones can be determined by a cutting and unrolling transformation.

The volume of a prism or cylinder of base area B and height h is Bh. A pyramid or cone of base area B and height h is $\frac{1}{3}Bh$.

A sphere of radius r has surface area $4\pi r^2$ and volume $\frac{4}{3}\pi r^3$.

The **similarity principle** describes the relationship between measurements of similar figures as it depends on the scale factor k. All linear measurements are multiplied by k, all area measurements are multiplied by k^2, and all volume measurements are multiplied by k^3.

Vocabulary and Notation

Section 12.1

Measurement process

Unit of measure

U.S. Customary (English) System
Units of length (inch, foot, yard, mile), area (in^2, ft^2, yd^2, acre, mi^2), volume and capacity (in^3, ft^3, yd^3, quart, gallon)

Metric (SI) System
Prefixes: m = *milli* (1/1000), c = *centi* (1/100), k = *kilo* (1000)
Units of length (meter, centimeter, kilometer), area (m^2, cm^2, km^2, hectare), volume and capacity (m^3, cm^3, km^3, liter)

Section 12.2

Area
Unit of Area
Congruence property of area
Dissection property of area
Altitude and base of a parallelogram or triangle
Length of curve
Perimeter
Circumference of circle; π (pi)

Section 12.3

Pythagorean theorem
Converse of the Pythagorean theorem

Section 12.4

Surface area of a surface in space
Area of base, lateral surface area
Right regular pyramid
Slant height
Similarity principle of measurement

CHAPTER REVIEW EXERCISES

Section 12.1

1. Select an appropriate metric unit of measurement for each of the following.
   (a) The length of a sheet of a notebook paper.
   (b) The diameter of a camera lens.
   (c) The distance from Los Angeles to Mexico City.
   (d) The height of the Washington monument.
   (e) The area of Central Park.
   (f) The area of the state of Kentucky.
   (g) The volume of a raindrop.
   (h) The capacity of a punchbowl.
2. Give the most likely answer:
   (a) A bottle of cider contains: 30 mL, 4L, 15L
   (b) Cross-country skis have length: 190 cm, 190 km, 190 m
   (c) The living area of a house is: 2000 cm^2, 1.2 ha, 200 m^2
3. An aquarium is a rectangular prism 60 cm long, 40 cm wide, and 35 cm deep. What is the capacity of the aquarium in liters?
4. A sailfish off the coast of Florida took out 300 feet of line in 3 seconds. Estimate the speed of the fish in miles per hour.

Section 12.2

5. Let M be the midpoint of side $\overline{AD}$ of the trapezoid $ABCD$. What is the ratio of the area of triangle MBC to the area of the trapezoid? (*Hint:* Dissect the parallelogram by a horizontal line through M, and rearrange the two pieces to form a parallelogram.)

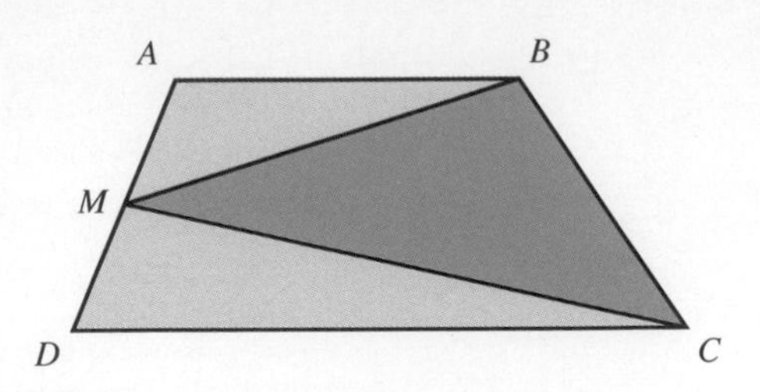

6. Find the areas of these figures.

(a)

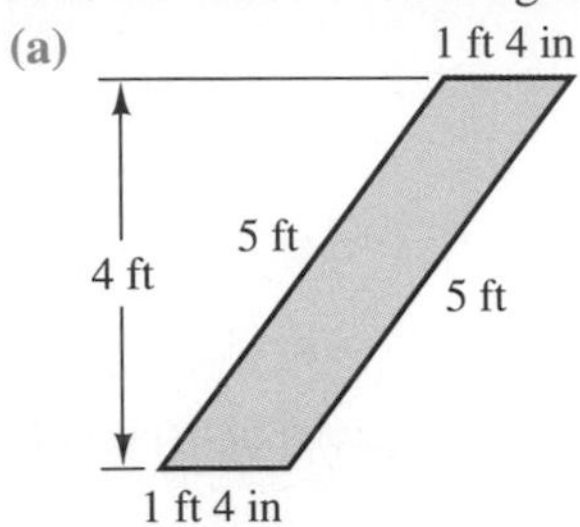

(b)

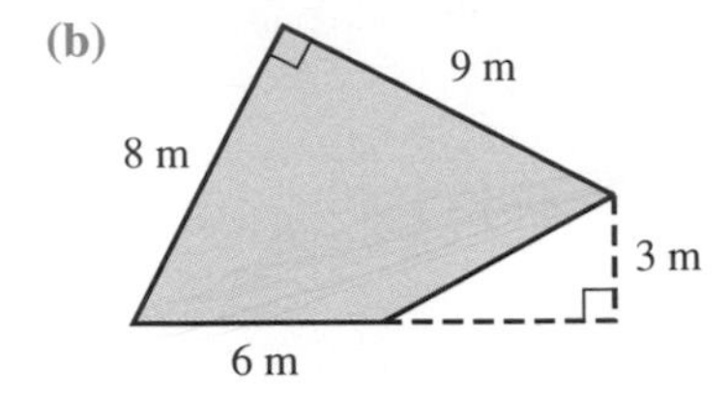

(c)

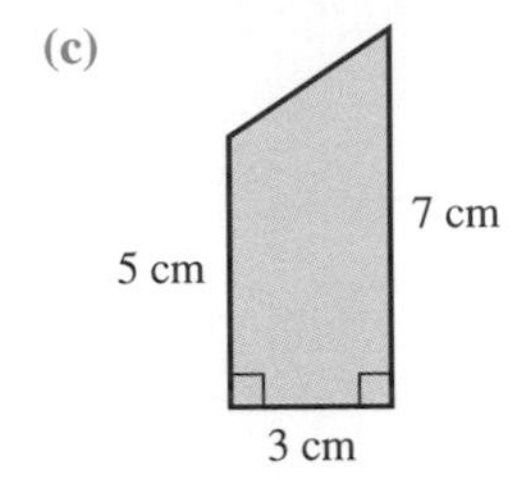

7. Find the area of each lattice polygon.

(a)

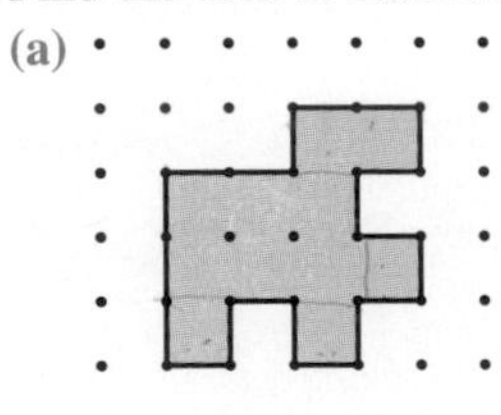

(b)

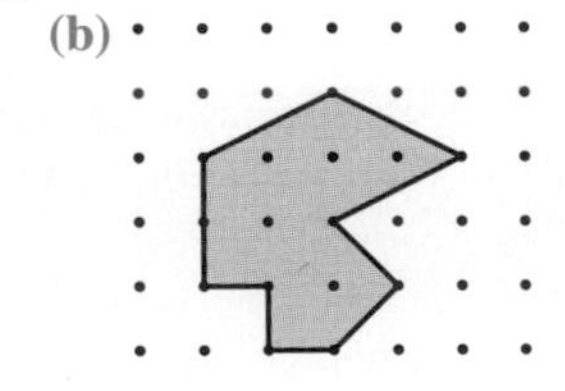

(c)

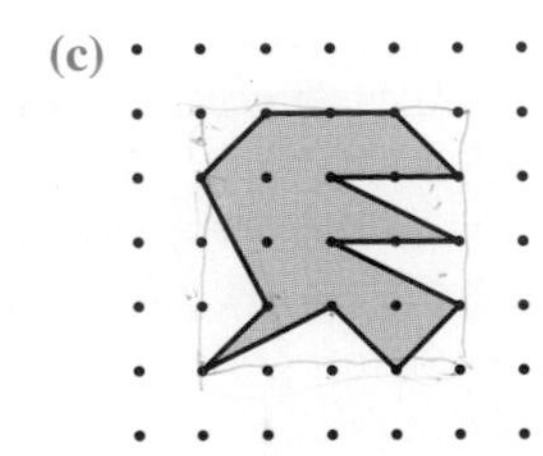

8. Find the areas and perimeters of these figures.

(a)

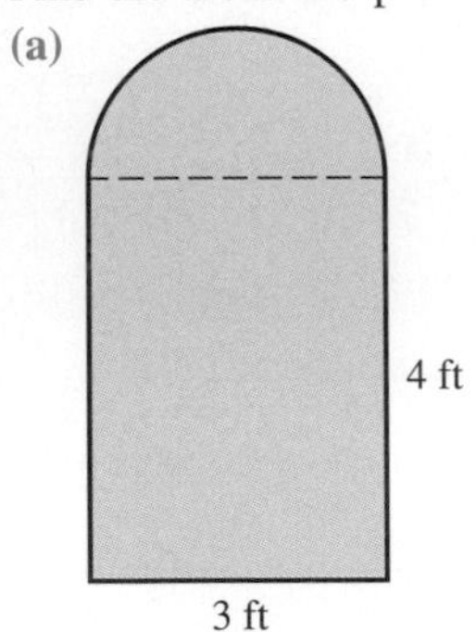

(b)

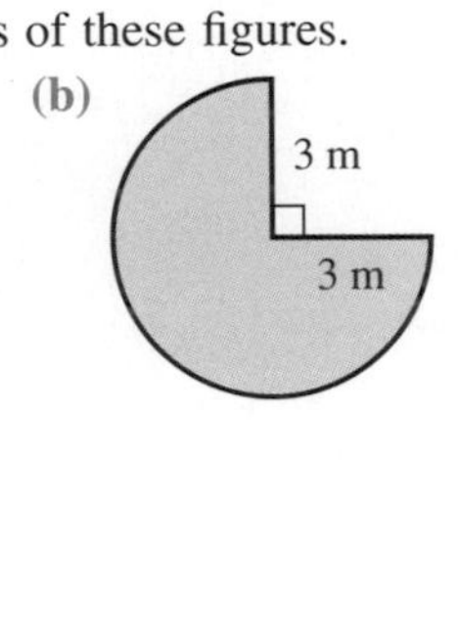

Section 12.3

9. Solve for x and y in the figure.

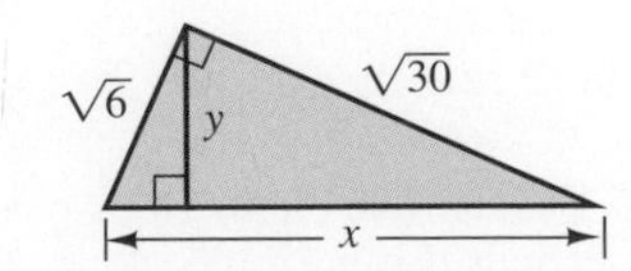

10. A right circular cone has slant height 35 cm and a base of diameter 20 cm. What is the height of the cone?

11. A rectangular box has sides of length 4 inches, 10 inches, and 12 inches. What are the lengths of each of the four diagonals of the box?

12. Find the perimeter of the following lattice polygon.

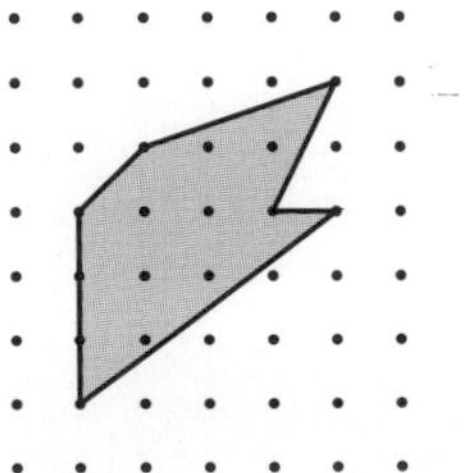

Section 12.4

13. Find the volume and surface area of these figures.

(a)

(b)

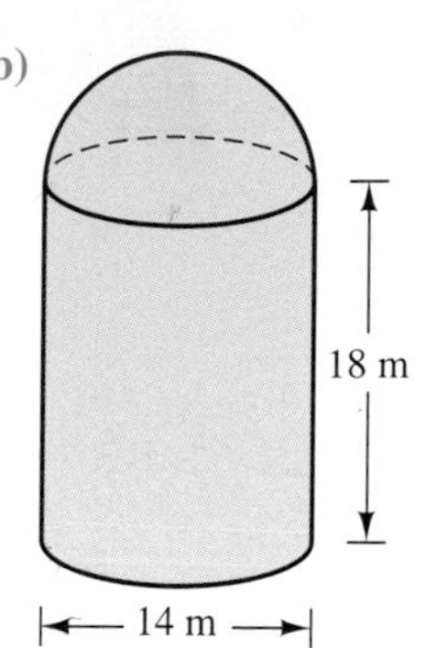

(c)

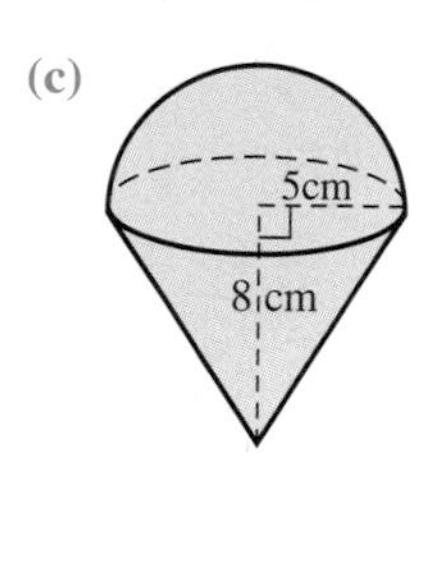

14. Which has the largest volume, a sphere of radius 10 meters or four cubes with sides of length 10 meters?

15. Heather's garden has a similar shape as Johan's, but is 75 feet long whereas Johan's is 50 feet long.

(a) Johan needed 180 feet of fencing to enclose his garden. How much fencing does Heather need?

(b) Heather used 45 pounds of fertilizer. How much will Johan use, assuming it is applied at the same number of pounds per square foot?

CHAPTER TEST

1. Fill in the blank with the metric unit of measurement that makes the statement reasonable.
   (a) The haze filter on Donna's camera has a diameter of 52 ________.
   (b) A gray whale has length 18 ________.
   (c) In the 1968 Olympics Bob Beaman had a long jump of 8.90 ________.
   (d) The Mississippi River has a length of 1450 ________.
   (e) A cup of coffee contains about 250 ________.
   (f) A fill-up at the gas station took 46 ________.

2. Fill in the blanks:
   (a) 2161 mm = ________ cm
   (b) 1.682 km = ________ cm
   (c) 0.5 m^2 = ________ cm^2
   (d) 1 ha = ________ m^2
   (e) 4719 mL = ________ L
   (f) 3.2 L = ________ cm^3

3. Complete the conversions of the measurements in the U.S. Customary system, using your calculator when convenient.
   (a) 1147 in = ______ yd
   (b) 7942 ft = ______ mi
   (c) 32.4 yd^2 = ______ ft^2
   (d) 9402 acres = ______ mi^2
   (e) 7.6 yd^3 = ______ ft^3
   (f) 5961 in^3 = ______ ft^3

4. The world's fastest growing plant grew 12 feet in 14 days. Calculate how many minutes it takes this plant to grow one inch at the same rate of growth.

5. Explain how the formula for the area of a triangle can be derived from that of a parallelogram by cutting the triangle from the midpoint of the base to the midpoint of one of the sides.

6. Find the area of each geoboard polygon. The nails are 1 cm apart.

(a)

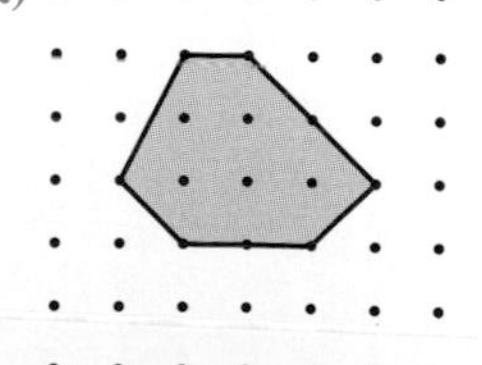

(b)

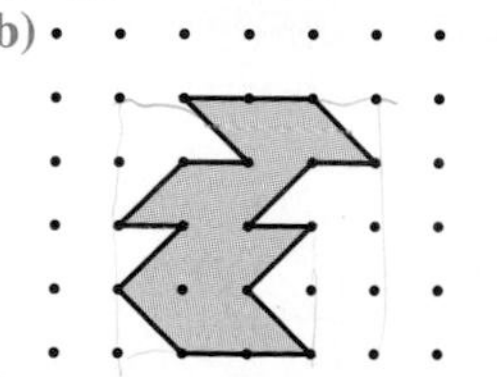

(c)

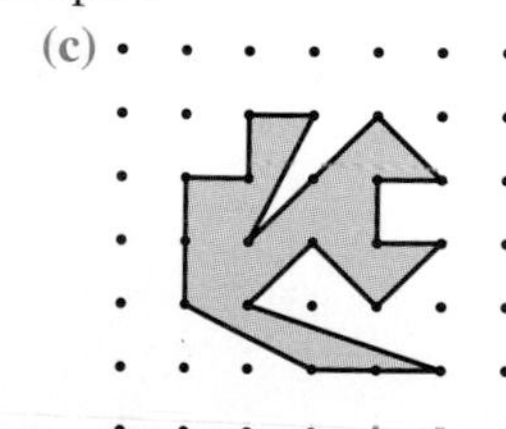

7. Find the area and perimeter of the following kite.

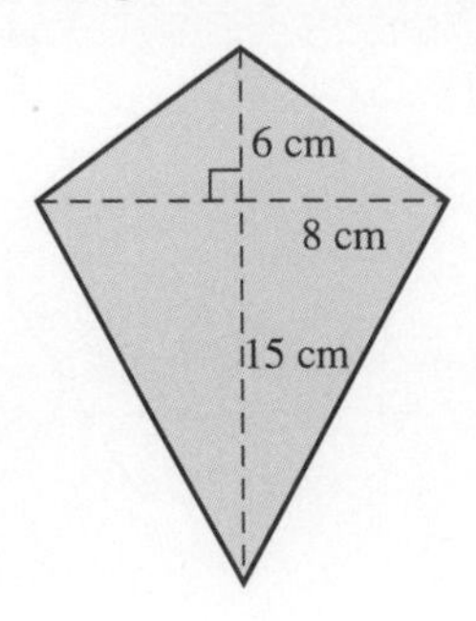

8. Find the area and perimeter of the following figure.

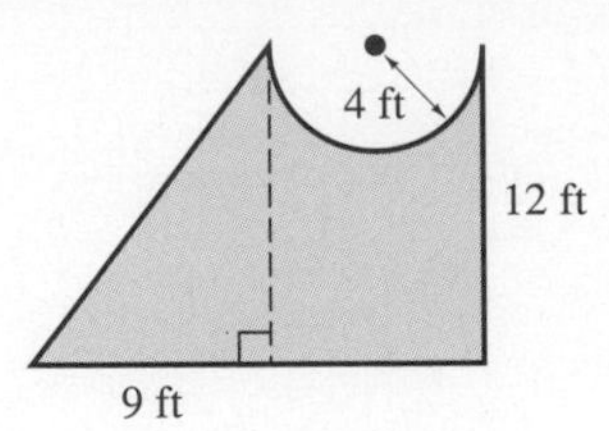

9. Find the area of each figure.

(a)

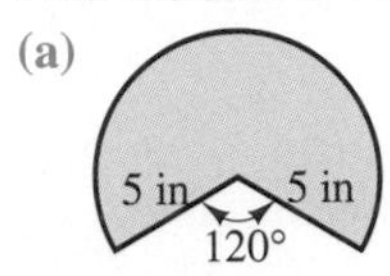

(b)

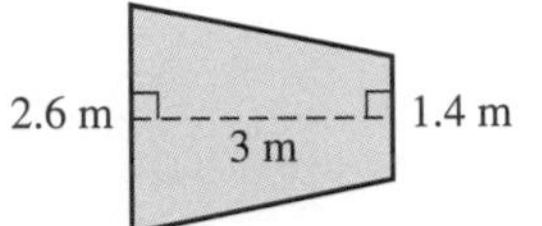

(c)

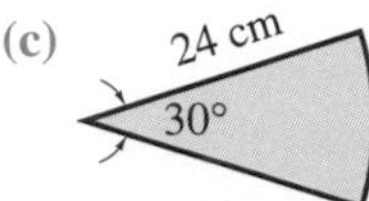

10. What is the ratio of the area of the inscribed square to the area of the square circumscribed about the same circle?

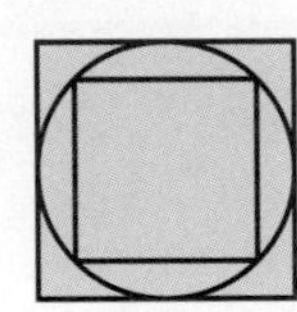

11. A lens is made by cutting a section from a sphere. If the lens has diameter 10 mm and height 4 mm, what is the radius of the sphere from which it was cut?

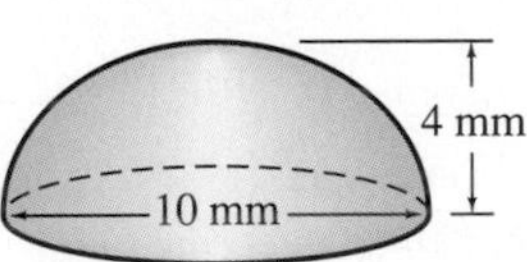

12. A ladder 15 feet long rests against a vertical wall. If the bottom of the ladder is 6 feet from the base of the wall, how high does the ladder reach?

13. (a) Find the perimeter of the geoboard triangle PQR.

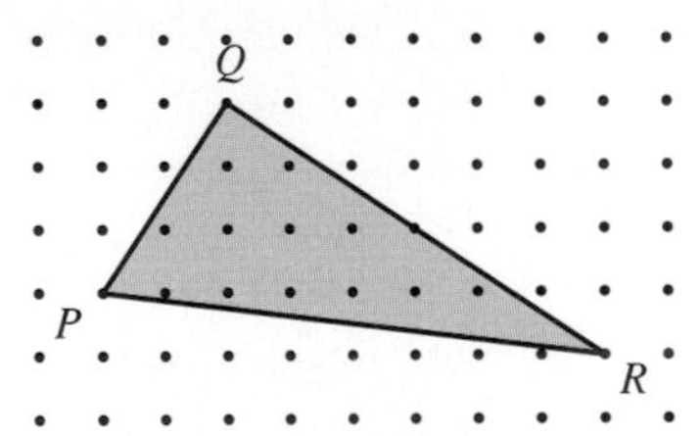

(b) Is ΔPQR a right triangle?

14. Find the surface area and volume of these figures.

(a)

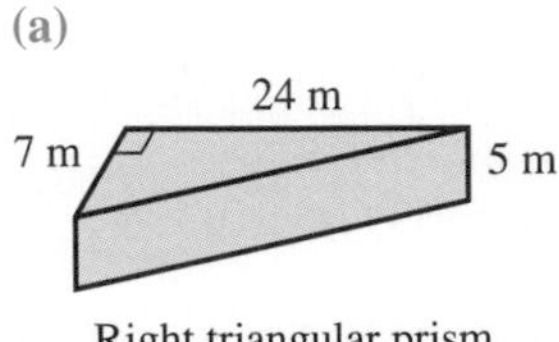

Right triangular prism

(b)

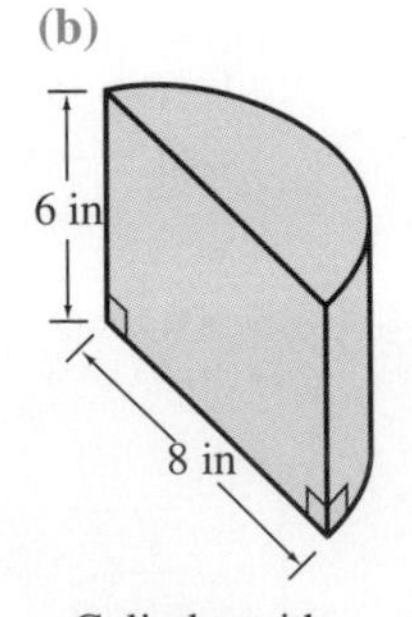

Cylinder with semicircular base

(c)

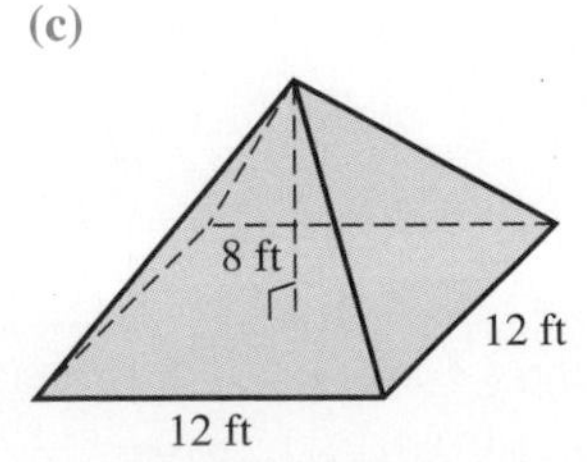

Right square pyramid

(d)

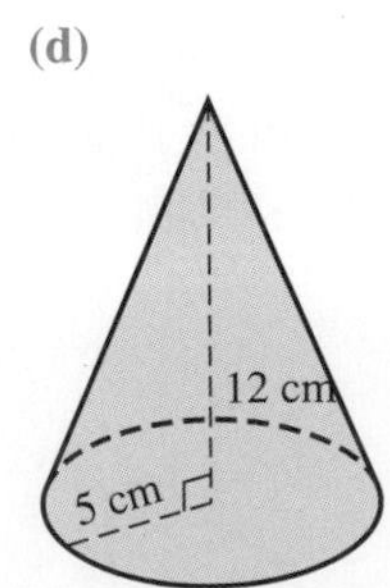

Right circular cone

15. Find the surface area and volume of each figure.

(a)

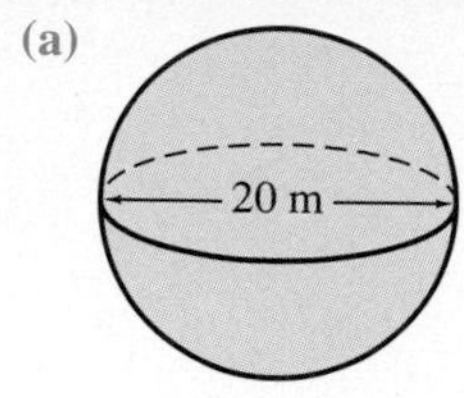

Sphere

(b)

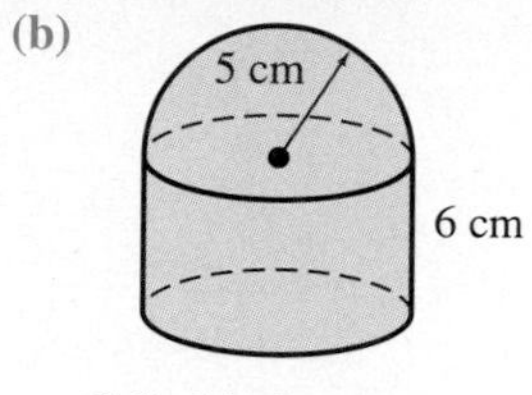

Cylindrical tank with hemispherical top

16. Papa Bear, Mama Bear, and Baby Bear have similar shapes, except Papa Bear is 5 ft tall, Mama Bear is 4 ft tall and Baby Bear is 2 ft tall. Fill in the values left blank in the following chart.

	Papa Bear	Mama Bear	Baby Bear
Length of suspenders		40 in	
Weight			30 lb
Number of fleas	6000		

17. A grapefruit has an outside diameter of 5 inches. When cut open it is discovered that the peel is 3/4 inches thick. What percentage of the grapefruit's volume is peel?

13 Geometric Transformations

HANDS ON — Exploring Reflection and Rotation Symmetry

Materials Needed

Clear acetate sheets, overhead transparency pens, tissues to clean acetate sheets for reuse, Mira (if available)

How to Check for Reflection and Rotation Symmetry

A figure has **reflection** (or **line**) **symmetry** if there is a mirror line which reflects the figure onto itself. For example the parafoil kite below has a vertical line of symmetry. The wheel cover at the right does not have reflection symmetry but it does have **rotation symmetry** since the figure turns onto itself when rotated through 72° about the center point.

Reflection symmetry can be verified by the "trace-and-flip" test. The figure and its line of symmetry is traced on an acetate sheet. The acetate is then flipped across the proposed symmetry line, turning the sheet upside down, to check that the points of the traced figure coincide with the original figure. Alternatively, reflection symmetry can be investigated by placing the drawing line of a Mira over a proposed line of symmetry of a figure.

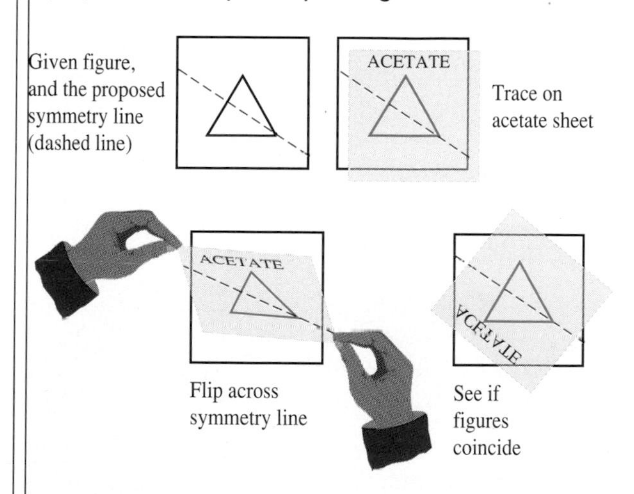

Rotation symmetry can be investigated by the "trace-and-turn" test, illustrated below. The point held fixed is the **center of rotation.** Since the tracing coincides with the original figure after a 120° turn, the "trace-and-turn" test shows that an equilateral triangle has 120° rotation symmetry.

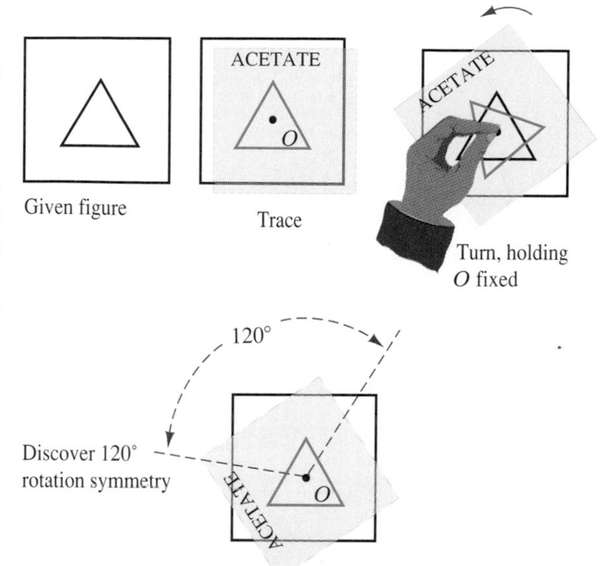

Activities

1. Use either a Mira or the "trace-and-flip" test to find all lines of symmetry of the following figures. Use dashed lines to draw the symmetry lines.

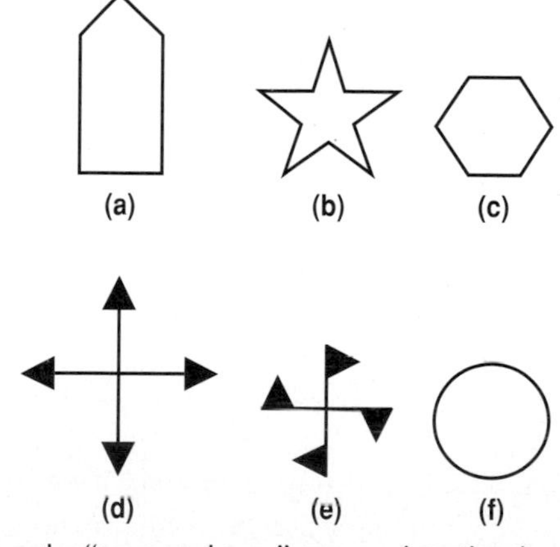

2. Use the "trace-and-turn" test to describe the rotation symmetries of the figures in Activity 1. Indicate the center of rotation and the angle measure of the rotation.
3. Sketch all lines of symmetry and describe all rotation symmetries for the following polygons.
   (a) Triangles: equilateral, isosceles, scalene
   (b) Quadrilaterals: square, rhombus, rectangle, parallelogram, trapezoid, isosceles trapezoid, kite

CONNECTIONS

Transformational Geometry, the Mathematics of Patterns and Symmetries

The Northwest Coast Indians occupy a narrow band of land along the western coasts of Washington, Canada, and southeast Alaska. The tribal groups of this region are renowned for their striking graphic art depicting whales, seals, eagles, bears, and other animal forms. Often the design is highly symmetric, such as the example of Haida art shown in Figure 13.1.

Figure 13.1
Sea lions on a Haida dance tunic

A trip to your local museum will show that symmetry is a common element found in the decorations and artwork created by all the world's cultures. Moreover, while a design from China is easily distinguished from a design from Central America, it is often evident that the two designs are based on the same underlying pattern of symmetry.

Most people, including school children, have an intuitive sense of symmetry, but usually the sense is too vague for precise classification and understanding. In this chapter we will see that the concept of a geometric transformation makes it possible to give a precise meaning to symmetry.

Section 1 of this chapter defines and investigates the rigid motions. It is discovered that any rigid motion of the plane is one of four basic types—slides, turns, flips, and glide reflections. Slides, turns, and flips are also called translations, rotations, and reflections respectively.

In Section 2, rigid motions are used as a tool to explore symmetry. *How much and what kind* of symmetry a figure possesses is discussed. Of special interest are the patterns in the plane, created by endlessly repeating a single motif.

In Section 3 another type of geometric transformation is investigated—the similarity transformation.

13.1 Rigid Motions

Imagine that each point P of the plane is "moved" to a new position P' in the same plane. Call P' the **image** of P, and call P the **preimage** of P'. If distinct points P and Q have distinct images P' and Q', and every point has a unique preimage point, then the association $P \leftrightarrow P'$ defines a one-to-one correspondence of the plane onto itself. Such a correspondence is called a **transformation of the plane.**

DEFINITION Transformation of the Plane

A one-to-one correspondence of the set of points in the plane to itself is a **transformation of the plane.** If point P corresponds to point P', then P' is called the **image** of P under the transformation. Point P is called the **preimage** of P'.

In this section and the next we will investigate a special type of transformation called a **rigid motion.** As the name implies, a rigid motion does not allow stretching or shrinking.

DEFINITION Rigid Motion of the Plane

A transformation of the plane is a **rigid motion** if, and only if, the distance between any two points P and Q equals the distance between their image points P' and Q'. That is, $PQ = P'Q'$ for all points P and Q.

A rigid motion is also called an **isometry,** meaning "same measure" (*iso* = same, *metry* = measure).

A useful physical model of a rigid motion of the plane can be realized with a sheet of clear acetate and a sheet of paper on which figures with points labeled A, B, C, . . . are drawn. The figures are traced onto the transparency and a rigid motion is modeled by moving the transparency to a new position in the plane of the paper. An example is shown in Figure 13.2, where primed letters A', B', C', . . . indicate the points in the image figure which correspond to the respective points A, B, C, . . . in the original figure. A rigid motion actually maps *all* of the points of the plane, but usually it is enough to show how a simple figure such as a triangle is moved to describe the motion. It is allowable to turn the transparency upside down before it is returned to the plane of the paper, since the definition of rigid motion is still satisfied.

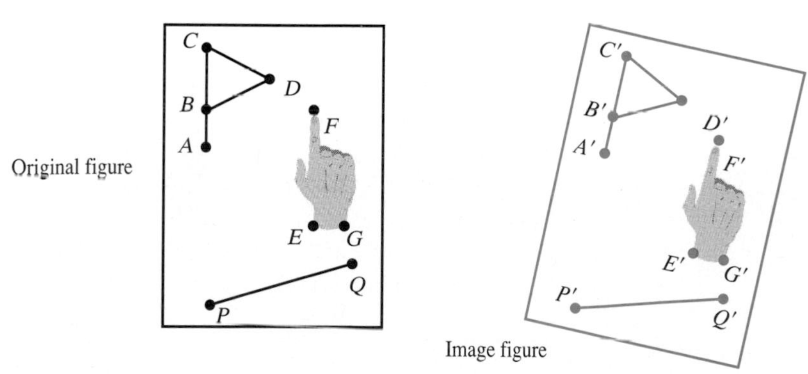

Figure 13.2
Illustrating a rigid motion of the plane

It needs to be emphasized that only the initial and final positions of a transformation are of interest. The transparency could have been taken for a roller coaster ride

JUST FOR FUN

Quadrilateral + Quadrilateral = Parallelogram?

Fold a sheet of paper in half, and then use scissors to cut a pair of congruent convex quadrilaterals. Cut one of the quadrilaterals along one of the diagonals, and cut the second quadrilateral

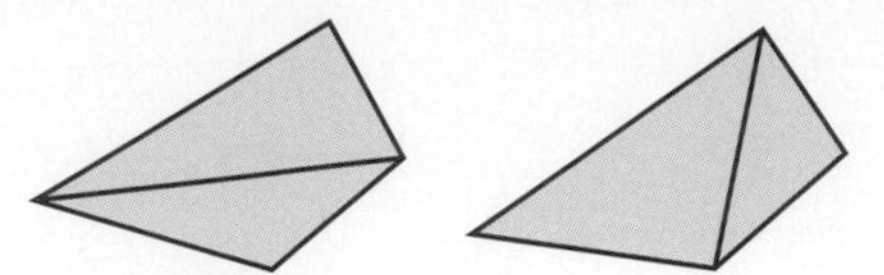

along the other diagonal. Show that the four triangles can be arranged to form a parallelogram. (This brainteaser of V. Proizvolov appeared in *Quantum, The Student Magazine of Math and Science,* September/October 1992, p. 31. See problem 28 in the section exercises for a useful idea for a solution using a translation.)

before reaching its final position. When the net outcomes of two motions are the same, the transformations are said to be **equivalent.**

Four transformations of the plane have special importance. They are the four **basic rigid motions of the plane:** translations, rotations, reflections, and glide reflections.

Translations

A **translation,** also known as a **slide,** is the rigid motion in which all points of the plane are moved in the same direction and by the same distance. A translation is illustrated by the "trace-and-slide" model in Figure 13.3. An arrow drawn from a point P to its image point P' completely specifies the two pieces of information required to define a translation: the direction of the slide is the direction of the arrow and the distance moved is the length of the arrow. The arrow is called the **slide arrow** of the translation. Another name for the slide arrow is the **translation vector.**

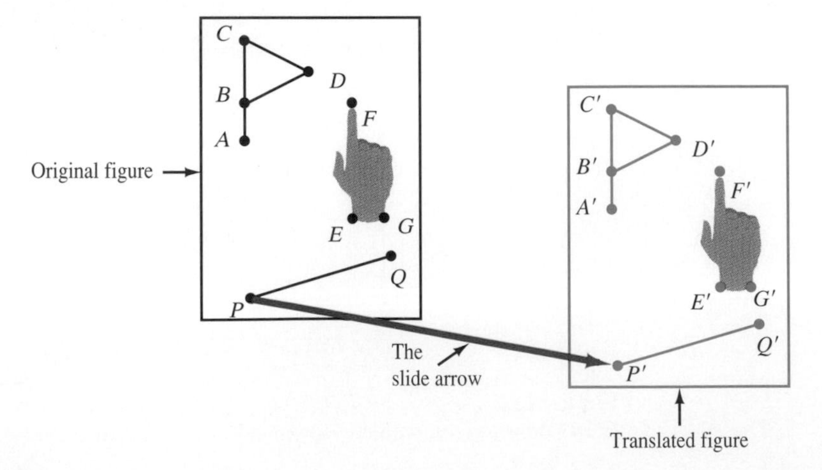

Figure 13.3
A slide, or translation, moves each point of the plane in the same direction and through the same distance.

EXAMPLE 13.1

Finding the Image Under a Translation

Find the image of the pentagon $ABCDE$ under the slide that takes the point C to C'.

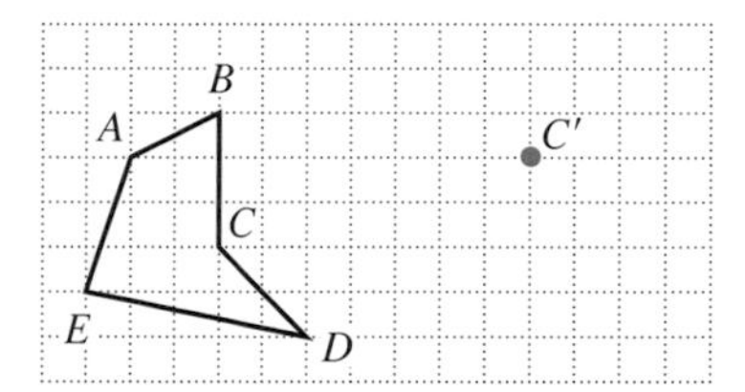

SOLUTION

The slide arrow from C to C' is 7 units to the right and 2 units up. Therefore A' is found by moving 7 units to the right of A and then 2 units up. The remaining points are found in the same way.

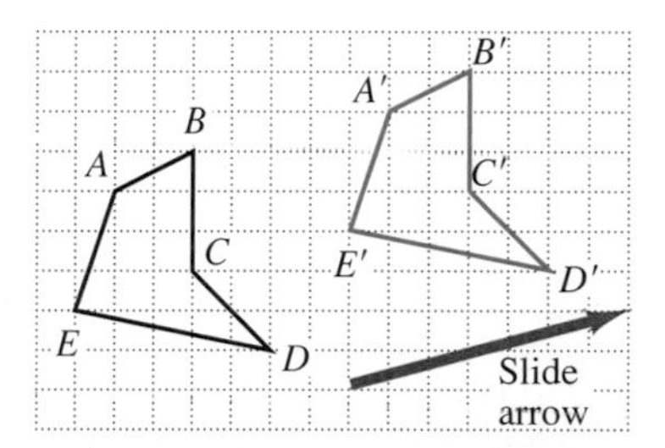

■

Rotations

The **rotation,** also called a **turn,** is another basic rigid motion. One point of the plane—called the **turn center** or the **center of rotation**—is held fixed and the remaining points are turned about the center of rotation through the same number of degrees—the **turn angle.** A counterclockwise turn about point O through 120° is shown in Figure 13.4.

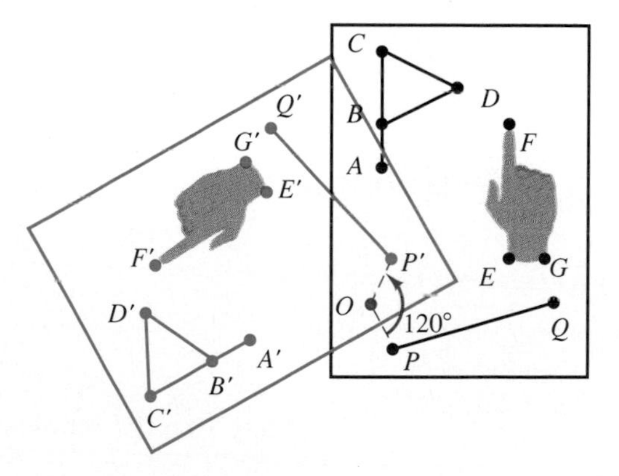

Figure 13.4
A turn, or rotation, rotates each point of the plane about a fixed point O—the turn center—through the same number of degrees and in the same direction of rotation.

A rotation is determined by giving the turn center and the directed angle corresponding to the turn angle. This information can be pictured by a **turn arrow,** as shown in Figure 13.5.

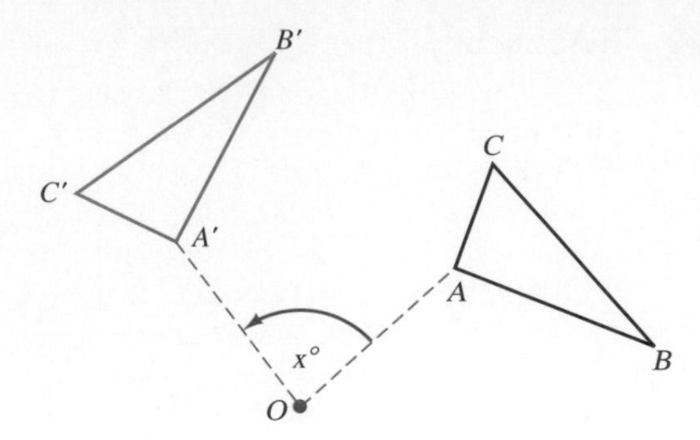

Figure 13.5
The rotation about center O by $x°$ can be indicated by a turn arrow.

Usually counterclockwise turn angles are assigned positive degree measures, whereas negative measures indicate that the rotation is clockwise. In this way a $-120°$ turn is equivalent to a $240°$ turn about the same center. Remember that only the initial and final positions are considered, not the actual physical motion.

EXAMPLE 13.2

Finding Images Under Rotations

Find the image of each figure under the indicated turn.

(a) $90°$ rotation about P **(b)** $180°$ rotation about Q **(c)** $-90°$ rotation about R

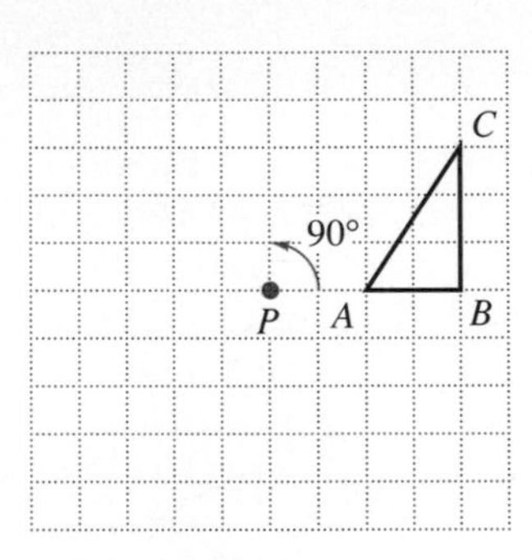

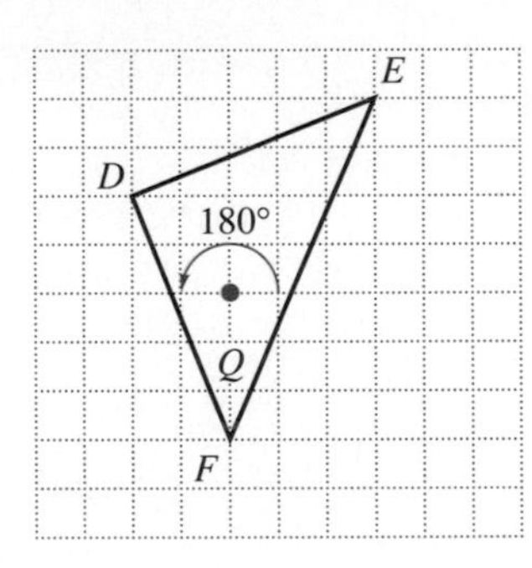

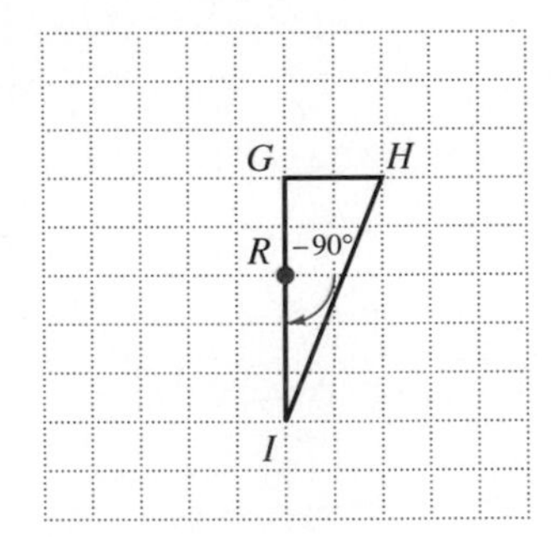

SOLUTION

(a)

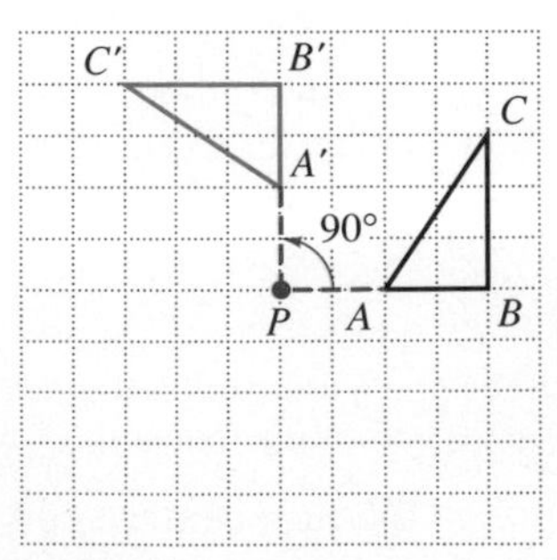

(b)

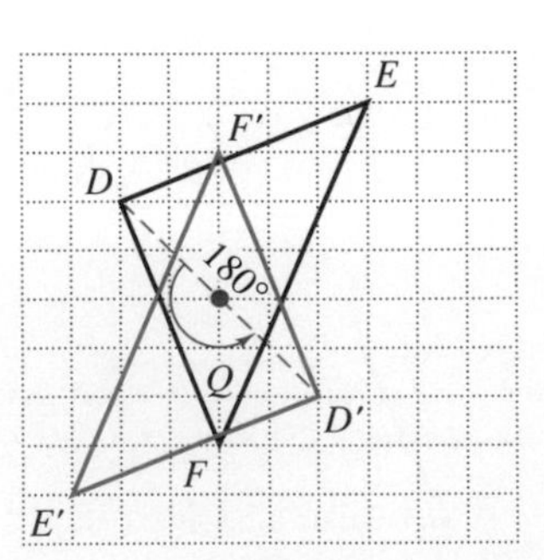

(c)

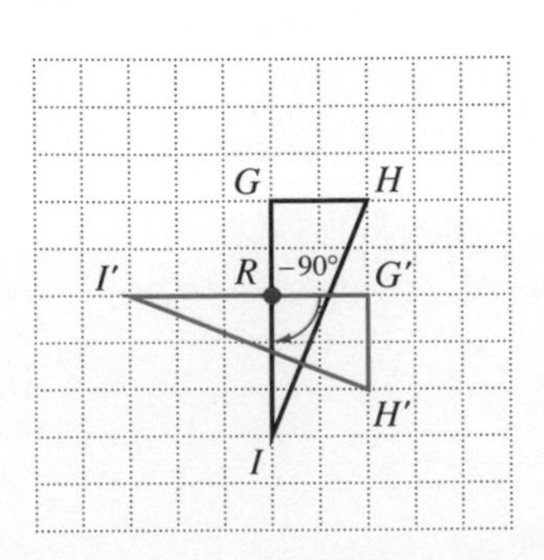

■

Reflections

The third basic rigid motion is a **reflection,** which is also called a **flip** or **mirror reflection.** A reflection is determined by a line in the plane called the **line of reflection** or the **mirror line.** Each point P of the plane is transformed to the point P' on the opposite side of the mirror line m and at the same distance from m, as shown in Figure 13.6. Note that P' is located so that m is the perpendicular bisector of $\overline{PP'}$. Every point Q on m is transformed to itself; that is, $Q' = Q$ if Q is any point on m.

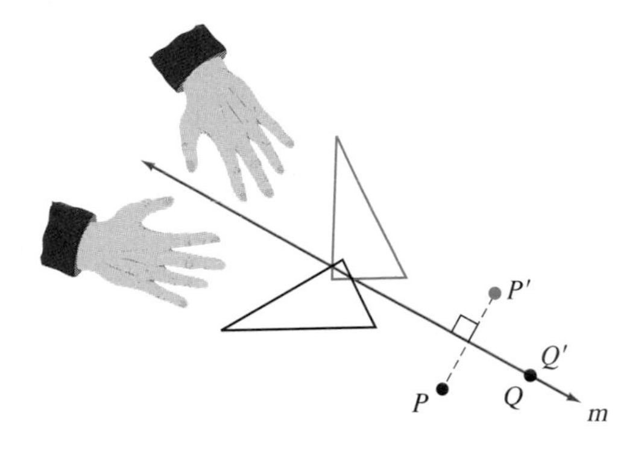

Figure 13.6
A flip, or reflection, about a line *m* transforms each point of the plane to its mirror image on the opposite side of *m*.

Reflections can be performed with a "trace-and-flip" procedure using an acetate transparency. First, the original figure is traced, including the line of reflection and a reference point (such as Q in Figure 13.6). The transparency is then turned over to perform the flip, and the points along the line of reflection are placed over their original position. The alignment of the reference point ensures that no sliding along the reflection line occurred.

The Mira is also ideally suited to perform reflections. The drawing edge of the Mira is placed along the line of reflection and the reflected image is then traced with a pencil.

EXAMPLE 13.3

Finding Images Under Reflections

Sketch the image of the "flag" under a flip across line m.

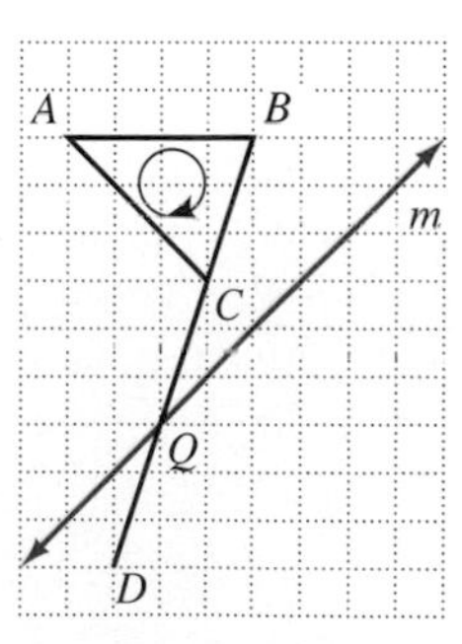

SOLUTION

We can follow the "trace-and-flip" method if an acetate sheet or tracing paper is available, or we can use a Mira. Alternatively, the point A' which is the mirror point of A across line m can be plotted. Similarly B', C', and so on, can be plotted, until the entire image can be sketched accurately.

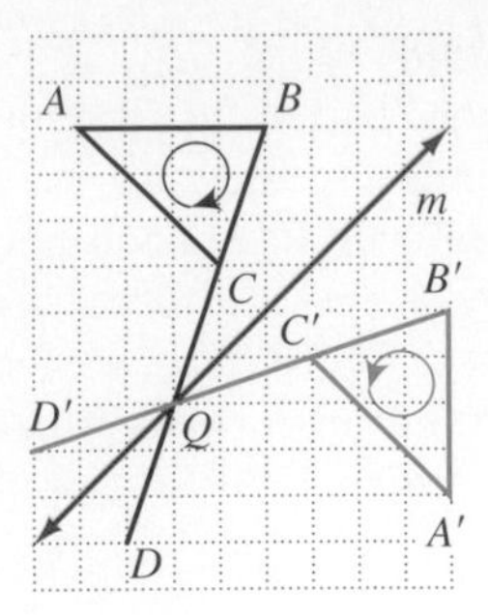

■

It's important to notice that a reflection reverses left-handed and right-handed orientations. For example, the left-pointing flag in Example 13.3 is transformed to a right-pointing flag, and the clockwise-pointing arrow on the circle becomes a counterclockwise-pointing arrow in the image. A rigid motion which interchanges "handedness" is called **orientation reversing.** Thus, a reflection is orientation reversing. Translations and rotations, since they do not reverse handedness, are examples of **orientation preserving** transformations.

Glide Reflections

The fourth and last basic rigid motion is the **glide reflection.** As the name suggests, a glide reflection combines both a slide and a reflection. The example most easily recalled is the motion which transforms a left footprint to a right footprint, as depicted in Figure 13.7. It is required that the line of reflection be parallel to the direction of the slide. In Figure 13.7 the slide came before the reflection, but if the reflection had preceded the slide the net outcome would have been the same.

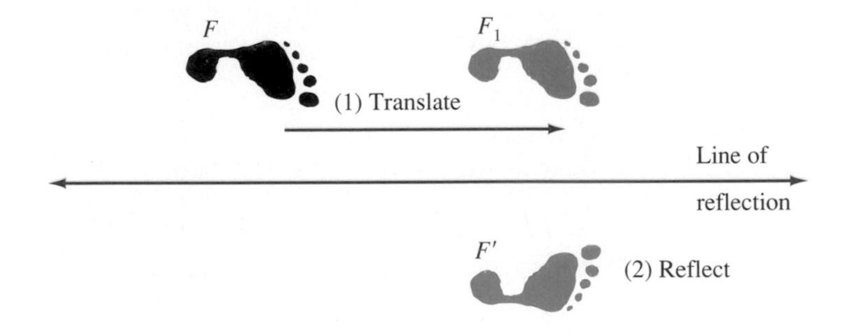

Figure 13.7
A glide reflection combines (1) a slide and (2) a reflection, where the line of reflection is parallel to the direction of the slide.

A glide reflection changes handedness, so it is an orientation reversing rigid motion. This is due to the reflection part of the motion.

To determine a glide reflection, it is useful to observe from Figure 13.8 that the midpoint M of the segment $\overline{PP'}$ lies on the mirror line of the reflection. This information is the key to solving the problem in the next example.

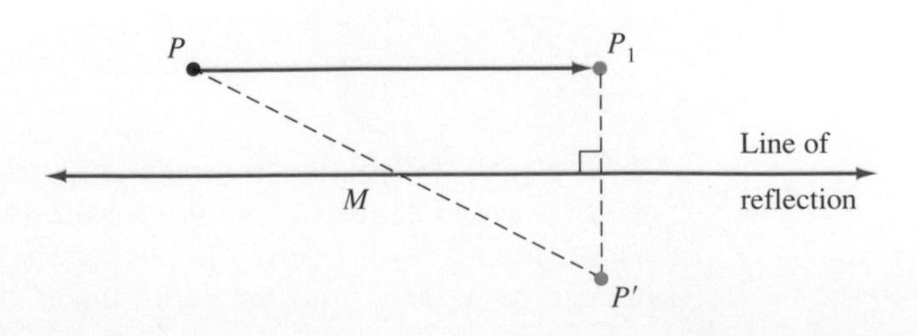

Figure 13.8
If points P and P' correspond under a glide reflection, then the midpoint M of PP' lies on the line of reflection.

SCHOOL BOOK PAGE *The Rigid Motions in Grade 6*

Transformations

Build Understanding

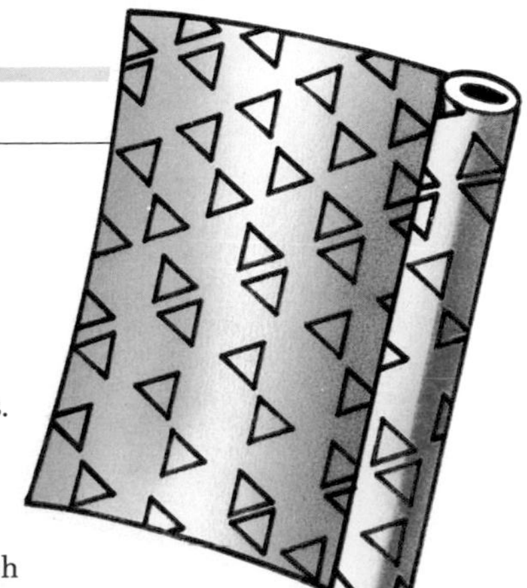

Jane Dunagan is buying wallpaper. She notices that many wallpaper patterns are made from geometric figures.

Geometric figures can be moved on a plane without changing their size or shape. Notice how the geometric figures in the wallpaper pattern all have the same size and shape, but are in various positions.

A ***slide***, ***turn***, or ***flip*** of a geometric figure is called a ***transformation***.

A. A slide moves a figure right, left, up, or down. In each diagram, figure II is a result of sliding figure I.

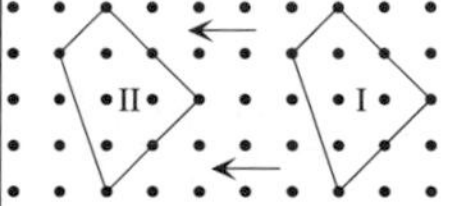

A slide 5 units left

A slide 2 units down

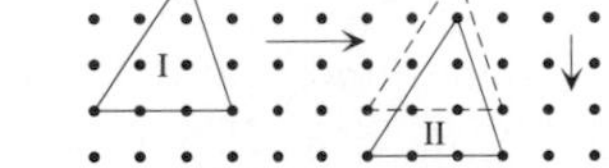

A slide 6 units right and 1 unit down

B. A flip is a mirror image of a figure. When triangle I is flipped over $\overleftrightarrow{AB}$, the result is triangle II. If you folded the dot paper at $\overleftrightarrow{AB}$, the parts of the triangles would fit exactly. Triangles I and II are ***symmetric***.

C. Figures can be turned about a given point. A $\frac{1}{4}$ clockwise turn of figure I about point P results in figure II. A $\frac{1}{4}$ turn is a 90° turn.

A 90° clockwise turn of figure II about point P results in figure III. A $\frac{1}{2}$, or 180° clockwise turn of figure I also results in figure III.

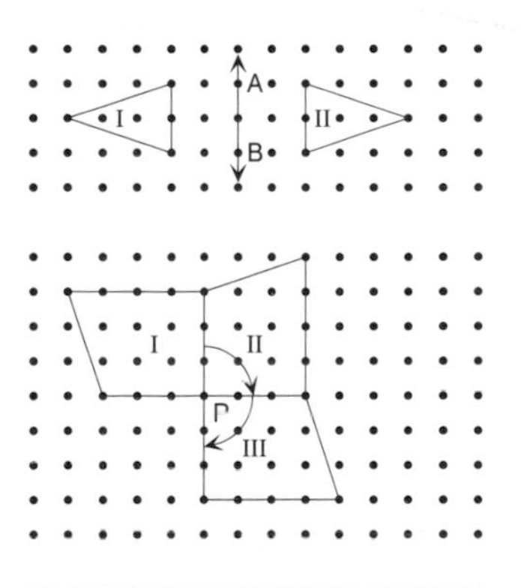

A 180° clockwise turn of a figure is not the same as a mirror image.

Talk About Math What happens if you turn a figure 180° and then turn it 180° again about the same point?

1. In part A, are the arrows shown slide arrows? Explain why not.
2. What other line, besides $\overleftrightarrow{AB}$, is a line of symmetry of both triangles I and II shown in part B?
3. What are some of the advantages and disadvantages of illustrating rigid motions on dot paper or the geoboard? Make a list, and briefly discuss each observation on your list.

EXAMPLE 13.4 Determining a Glide Reflection

A glide reflection has taken points A and B of triangle ABC to the points A' and B', as shown below. Find the line of reflection and the slide arrow of the glide reflection, and then sketch the image triangle $A'B'C'$ under the glide.

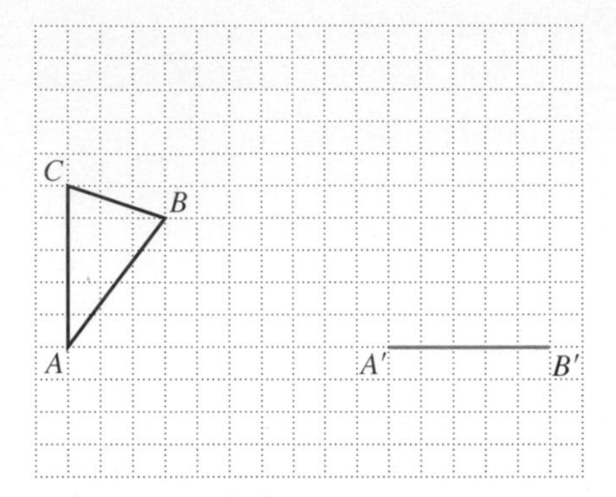

SOLUTION

The square grid makes it easy to draw the respective midpoints M and N of the line segments $\overline{AA'}$ and $\overline{BB'}$. Since both M and N lie on the mirror line, $m = \overleftrightarrow{MN}$ is the line of reflection of the glide. Reflecting A' across m determines the point A_1, and the slide arrow of the glide is drawn by connecting A to A_1. The slide arrow is 8 units to the right and 4 units up, which allows us to find C_1. Reflecting C_1 across the line of reflection locates C', and therefore $\triangle A'B'C'$ can be completed.

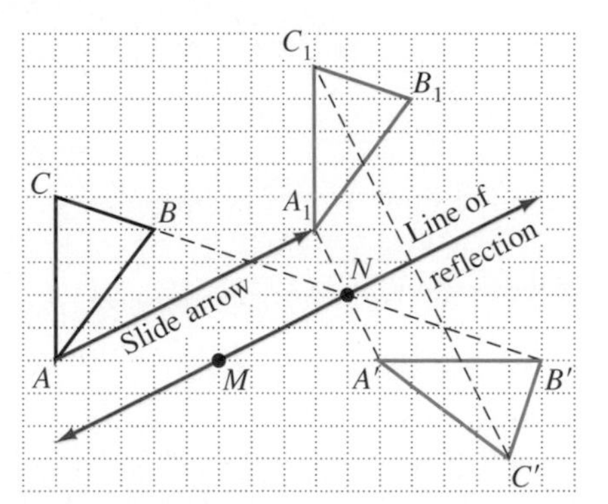

■

The Net Outcome of Two Successive Reflections

Recall that any two rigid motions which have the same net outcome are called equivalent. For example, rotations of $-120°$ and $+240°$ about the same center O are equivalent. Similarly the motion consisting of two consecutive 180° rotations about a point O is equivalent to the **identity transformation,** which is the rigid "motion" that leaves all points of the plane fixed.

Two consecutive reflections across the same line of reflection also bring each point back to its original position, so any "double flip" is also equivalent to the identity transformation. Suppose, however, that two flips are taken in succession across two *different* lines of reflection, say first over m_1 and next over m_2. There are two cases to consider, where m_1 and m_2 are parallel and where m_1 and m_2 intersect.

Table 13.1 summarizes useful information about the four basic rigid motions.

TABLE 13.1 The Four Basic Rigid Motions

Name (alternate name) and Sketch	Information Needed	Description	Orientation Property
Translation (slide)	Slide arrow, indicating distance and direction	Every point of the plane is moved the same distance in the same direction.	Orientation is preserved.
Rotation (turn)	Turn arrow, indicating turn center and turn angle.	Every point of the plane is rotated through the same directed angle about the turn center.	Orientation is preserved.
Reflection (flip)	Line of reflection.	Every point of the plane is moved to its mirror image on the opposite side of the line of reflection.	Orientation is reversed.
Glide reflection (glide)	Slide arrow and a line of reflection parallel to the slide direction.	Every point of the plane is moved by the same translation and reflected across the same line parallel to the slide direction.	Orientation is reversed.

EXAMPLE 13.5 Exploring Consecutive Reflections Across Parallel Lines

Let m_1 and m_2 be parallel lines of reflection.

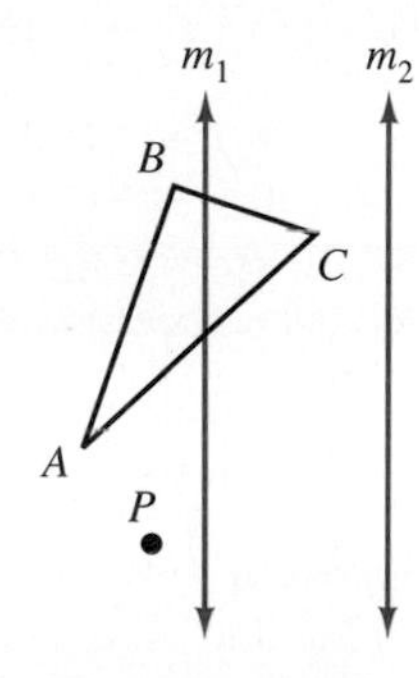

(a) Sketch the image of $\triangle ABC$ and point P under the reflection across m_1; let the image be labeled as $\triangle A_1B_1C_1$ and P_1.

(b) Sketch the image of $\triangle A_1B_1C_1$ and P_1 under reflection across line m_2; let the image be labeled $\triangle A'B'C'$ and P'.

(c) Describe the net outcome of the rigid motion consisting of the two successive reflections, first across m_1 and next across m_2.

SOLUTION

(a) and **(b)** Each reflection can be drawn with a Mira, or by the "trace-and-flip" method with an acetate sheet. Whatever method is used will result in the images shown below.

(c) If d is the directed distance from line m_1 to line m_2, we see that point P is moved a distance $2d$ in the direction perpendicular to m_1 and m_2 and pointing from m_1 toward m_2. In fact *all* points of the plane are moved in this direction through the same distance $2d$ and so the net outcome of two successive reflections across a pair of parallel lines is equivalent to the translation shown at the right below.

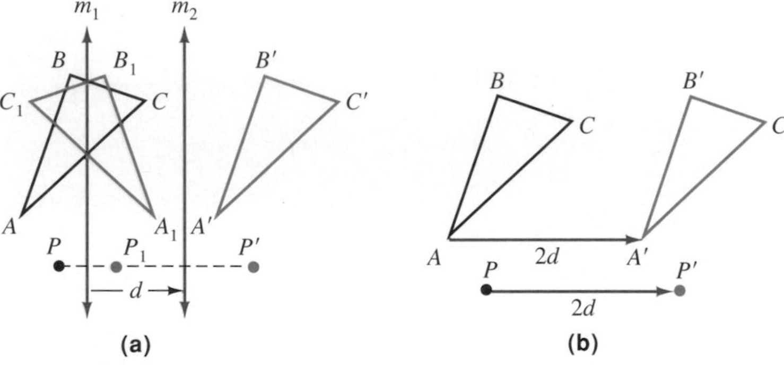

(a) Two reflections across parallel lines

(b) The equivalent translation

■

A similar investigation can be carried out for the motion consisting of two successive reflections across lines m_1 and m_2 which intersect at a point O. Most people find the result very surprising: *the net outcome of the two reflections is equivalent to a rotation about the point O of intersection of m_1 and m_2. The angle of rotation is twice the measure of the directed angle that turns line m_1 onto line m_2.* This is illustrated in Figure 13.9.

The following theorem summarizes the two possible net outcomes of a pair of successive reflections.

> **THEOREM The Net Outcome of Two Reflections**
>
> The net outcome of two successive reflections is either a translation, if the lines of reflection are parallel, or a rotation, if the lines of reflection intersect.

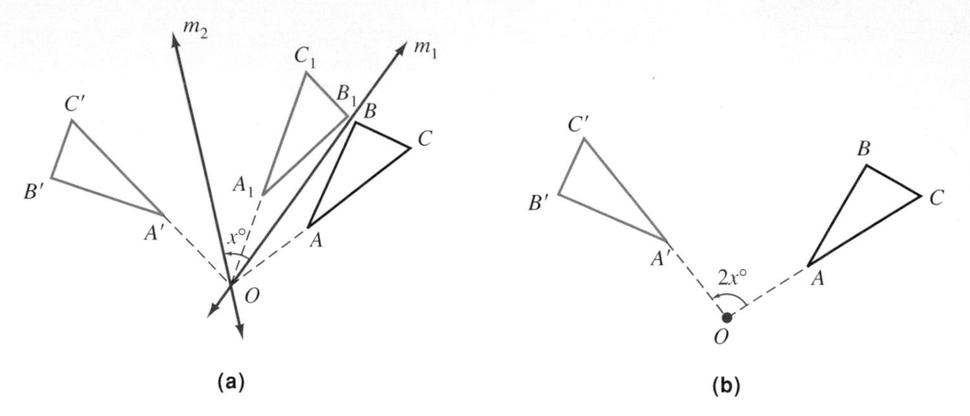

Figure 13.9
Two reflections across intersecting lines are equivalent to a rotation about the point of intersection of the two lines.

The Net Outcome of Three Successive Reflections

Three lines can be arranged in several ways in the plane. For example, the lines m_1, m_2, and m_3 in Figure 13.10(a) are parallel to one another. The successive image of $\triangle ABC$ across the lines are shown, with $\triangle A'B'C'$ the image at the completion of all three reflections. In Figure 13.10(b), we see that $\triangle ABC$ can be taken to $\triangle A'B'C'$ by a *single* reflection over the line l. Line l is the image of line m_1 under the translation that takes m_2 to m_3. Thus successive reflections across three parallel lines is equivalent to a single reflection.

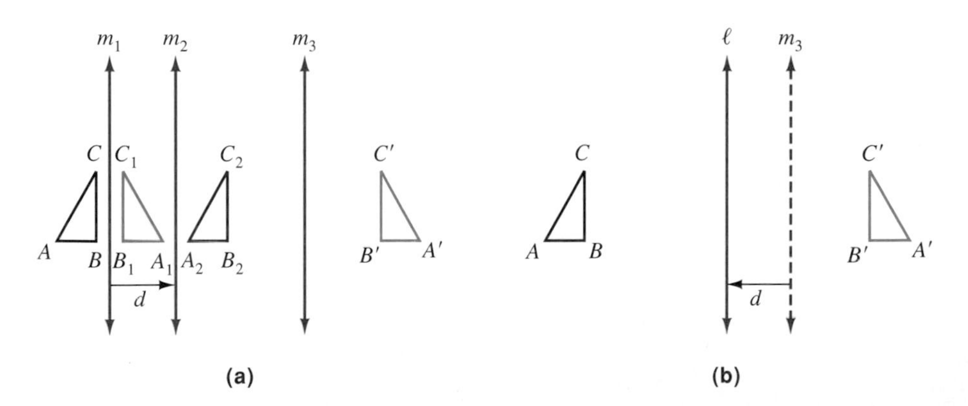

Figure 13.10
Three reflections across parallel lines m_1, m_2, m_3, is equivalent to a reflection across one line, l.

Three reflections across concurrent lines m_1, m_2, and m_3 can also be discovered to be equivalent to a single reflection across a certain line l which passes through the point O of concurrence. (See problem 23 in the section exercises.) In all other cases, where the three lines are neither parallel nor concurrent, it can be shown that the net outcome of three successive reflections is equivalent to a certain glide reflection. (See problem 24 in the section exercises.)

In summary, we have the following result.

THEOREM The Net Outcomes of Three Reflections

The net outcome of three successive reflections across lines m_1, m_2, and m_3 is equivalent to either:

- a reflection, if m_1, m_2, and m_3 are parallel or concurrent;

or

- a glide reflection, if m_1, m_2, and m_3 are neither parallel nor concurrent.

Classification of General Rigid Motions

Any rigid motion can be modeled by moving an acetate sheet to a new position. If the motion takes $\triangle ABC$ to $\triangle A'B'C'$, then the final position of the sheet is uniquely determined by aligning a copy of $\triangle ABC$, drawn on the acetate, with $\triangle A'B'C'$. Let's now see how to move the copy of $\triangle ABC$ onto $\triangle A'B'C'$ by the sequence of at most three reflections.

First, a reflection across the perpendicular bisector of $\overline{AA'}$ takes A to A'. Let B_1 and C_1 denote the images of B and C under this reflection. If $B \neq B'$ a second reflection, this time across the perpendicular bisector of $\overline{B_1B'}$, takes B_1 onto B'. Point A' remains fixed during the second reflection since A' is a point on the line of reflection. Let C_2 denote the image of C_1 under the second reflection. If $B_1 = B'$, the second reflection just described can be omitted, and we just let $C_2 = C_1$. If $C_2 = C'$, we are done: the acetate copy of $\triangle ABC$ has been moved to $\triangle A'B'C'$ in at most two reflections. If $C_2 \neq C'$ then another reflection is taken across line $\overline{A'B'}$, moving C_2 to C' and leaving A' and B' fixed. In this case, the motion is completed by a sequence of at most three reflections.

This informal reasoning can be made rigorous, proving that a general rigid motion is equivalent to either two or three successive reflections. But we already know that two reflections are equivalent to either a translation or rotation, and three reflections are equivalent to either a single reflection or a glide reflection. Thus we have a remarkable theorem.

THEOREM Classification of General Rigid Motions

Any rigid motion of the plane is equivalent to one of the four basic rigid motions: a translation, a rotation, a reflection, or a glide reflection.

Rigid motions have many application in geometry. For example, the informal definition of congruence to mean "same size and shape" can now be made precise.

DEFINITION Congruent Figures

Two figures are **congruent** if, and only if, one figure is the image of the other under a rigid motion.

The periodic drawings and block prints of M. C. Escher show how the plane can be tiled by congruent figures. Figure 13.11 illustrates a two motif pattern of fish of two sizes.

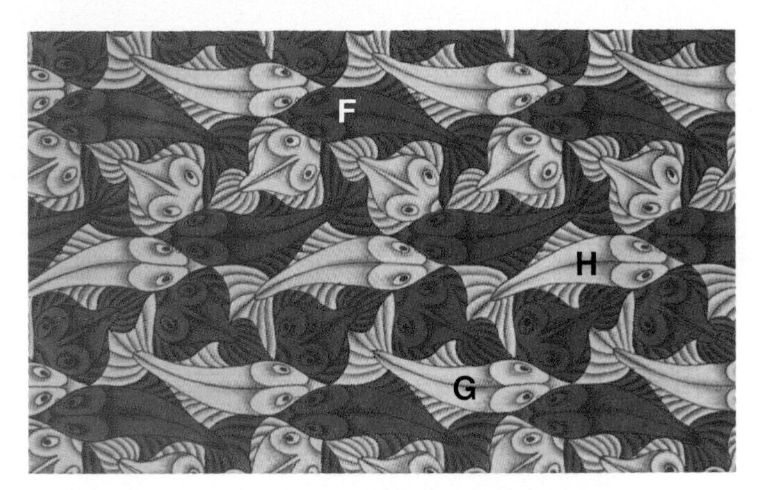

Figure 13.11
A two-motif tiling of the plane by M. C. Escher

EXAMPLE 13.6

Classifying Rigid Motions

Examine the tiling of M. C. Escher shown in Figure 13.11. Three of the large fish are labeled F, G, and H.

(a) What type of rigid motion takes F onto G ?
(b) What type of rigid motion takes F onto H ?

SOLUTION

(a) Since F and G have the same orientation (both bend the tail to the left), they are related by an orientation preserving transformation. Since the fish face in opposite directions, the motion is not a translation and so it must be a rotation. (Can you identify the turn center and the size of the angle of rotation?)
(b) F and H have opposite orientation, so either a reflection or glide reflection takes F onto H. H is not a reflection of F, so it must be a glide reflection. (Can you determine the line of reflection of the glide?) ■

In the next example, we see how rigid motions provide a problem-solving tool.

EXAMPLE 13.7

Proving Napoleon's Theorem with Rigid Motions

The following result is known as *Napoleon's Theorem: Let equilateral triangles be erected outward on the sides of an arbitrary triangle* $T = \triangle ABC$. *Then the centers* X, Y, *and* Z *of the equilateral triangles are the vertices of an equilateral triangle.*
It is doubtful Napoleon proved the theorem, but if he had known about rigid motions he could have given a proof. Show how this can be done.

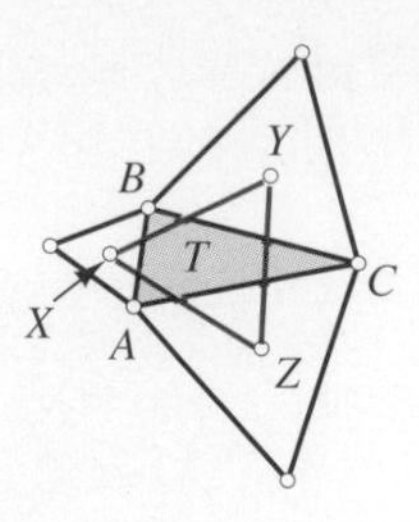

SOLUTION

Understand the problem

Nothing special is assumed about triangle T, but we must show that $\triangle XYZ$ is equilateral—the most special triangle of all.

Devise a plan

An equilateral triangle has a property no other triangle has: a rotation of $+120^\circ$ or -120° about the center of the triangle moves each vertex of the triangle to a different vertex of the same triangle. Let's investigate the effect of such rotations taken about the centers X, Y, and Z of the equilateral triangles we have erected on the sides of $\triangle ABC$.

Carry out the plan

Consider two successive 120° counterclockwise rotations, first about X and next about Y. By viewing X and Y as two adjacent points in an equilateral triangular lattice, it is readily seen (check it!) that the net outcome of the $+120^\circ$ rotations about X and Y is equivalent to a single -120° (clockwise) rotation about a point Z' for which $\triangle XYZ'$ is equilateral.

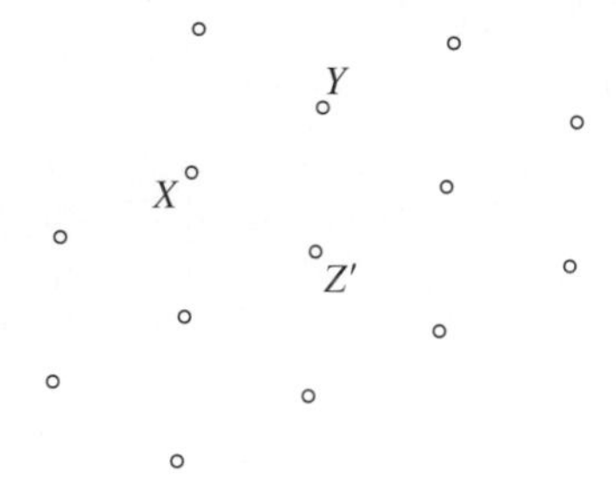

In the diagram below we see that the first 120° rotation about X takes triangle T to T_1, and then the 120° rotation about Y takes T_1 to T'. On the other hand a -120° (clockwise) rotation about Z also takes T to T'. Thus we see that $Z = Z'$, and therefore $\triangle XYZ$ is equilateral.

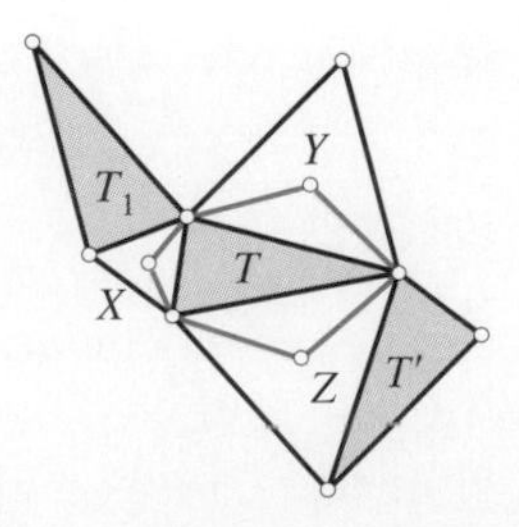

Look back

Our approach suggests that we continue applying additional 120° rotations, thereby extending the pattern of congruent copies of triangle T, each of which is surrounded by equilateral triangles. The following tiling pattern shows that Napoleon's theorem is just one aspect of an underlying tiling of the plane with equilateral triangles.

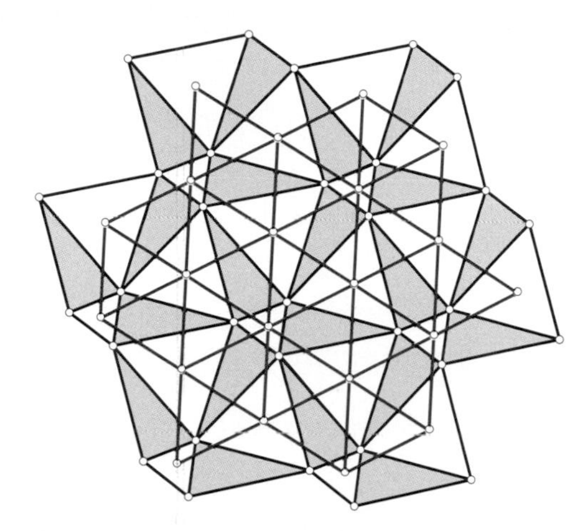

■

Rigid Motions in Logo

Logo Translations

To perform a translation in Logo, you need only lift the turtle, move it to a new starting position, turn it back to its original heading, and put the pen back down before using the original procedure. This will produce a translation of the first drawing, with sides of the new figure parallel to those of the original.

LOGO EXAMPLE 13.A

Translating a Polygon

This procedure will draw a parallelogram.

```
to   parallelogram
     repeat 2 [ fd 40 rt 100 fd 30 rt 80 ]
end
```

Write a procedure to translate the parallelogram by 80 turtle steps in the direction 60 degrees east of north.

SOLUTION

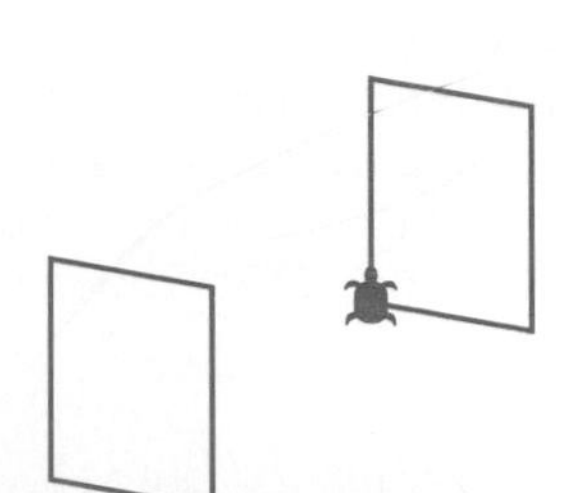

```
to   trans.par
       parallelogram
       pu rt 60 fd 80 lt 60 pd
       parallelogram
end
```

■

In the **trans.par** procedure, the parallelogram procedure can be replaced with any procedure that draws a geometric figure. The 60s may be replaced by any other number to select the angle of the translation vector, and the 80 may be replaced by any other number indicating the length of the translation vector.

Logo Rotations

A rotation in Logo is performed by turning the turtle about the center of rotation and then redrawing the figure with the new starting heading.

LOGO EXAMPLE 13.B

Rotating a Polygon

The following procedure draws a regular polygon with sides of length **:side** and with **:nsides** giving the number of sides.

```
to    polygon :side :nsides
         repeat :nsides [ fd :side rt 360 / :nsides ]
end
```

Write a procedure to rotate a regular polygon through an angle given by the variable **:angle.**

SOLUTION

```
to    spin.poly :side :nsides :angle
         polygon :side :nsides
         rt :angle
         polygon :side :nsides
end
```

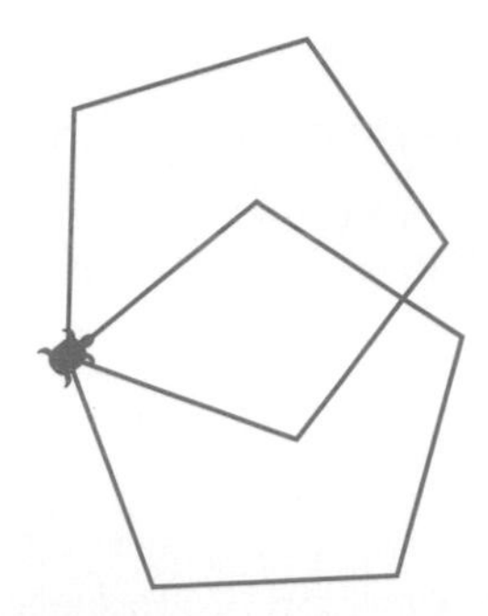

■

Interesting pictures can be drawn by repeating these rotations through a full circle. The angle of rotation is found by dividing 360 degrees by the number of times the figure is repeated. The number of repetitions in the following program is specified by the variable **:reps.**

```
to  spin.poly2 :side :nsides :reps
        repeat :reps [ polygon :side :nsides rt 360 / :reps]
end
```

The figures below were drawn using **spin.poly2.**

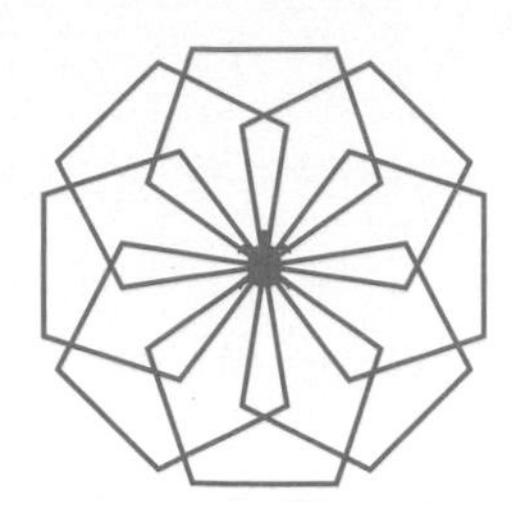

spin.poly2 40 5 8

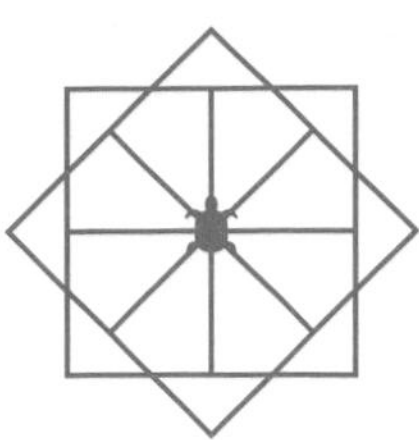

spin.poly2 40 4 8

spin.poly2 40 7 18

Logo Reflections

Reversing orientation is the key characteristic used to reflect figures in Logo. We can draw a reflection of any figure by replacing left and right by right and left, respectively, each time they occur in a procedure. The new procedure will then draw the reflection of the figure drawn by the original procedure.

LOGO EXAMPLE 13.C

Reflecting a Polygon

The following procedure draws the hexagon shown here. The turtle indicates the start and end position of the turtle in this procedure.

```
to    hexa
        lt 45 fd 10 * sqrt 2
        rt 90 fd 30 * sqrt 2
        rt 45 fd 30
        rt 135 fd 20 * sqrt 2
        lt 90 fd 20 * sqrt 2
        rt 135 fd 50 rt 90
end
```

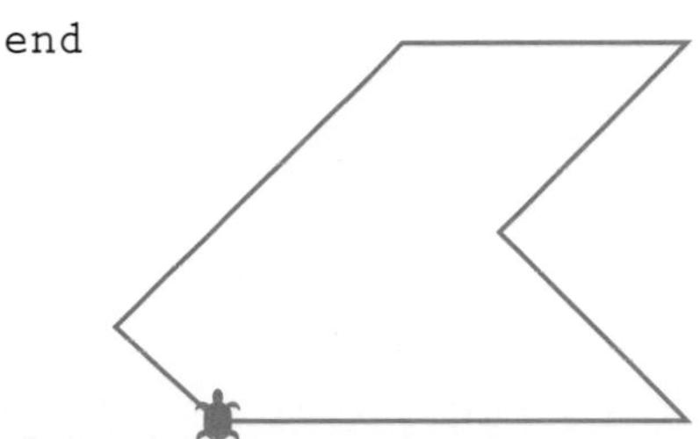

Write a procedure that draws the hexagon reflected across a vertical line drawn through the starting position.

SOLUTION

Interchanging left turns with right turns gives the procedure below. Both the original hexagon and its reflection are shown on the same screen.

```
to    refl.hexa
        rt 45 fd 10 * sqrt 2
        lt 90 fd 30 * sqrt 2
```

```
        lt 45 fd 30
        lt 135 fd 20 * sqrt 2
        rt 90 fd 20 * sqrt 2
        lt 135 fd 50 lt 90
end
```

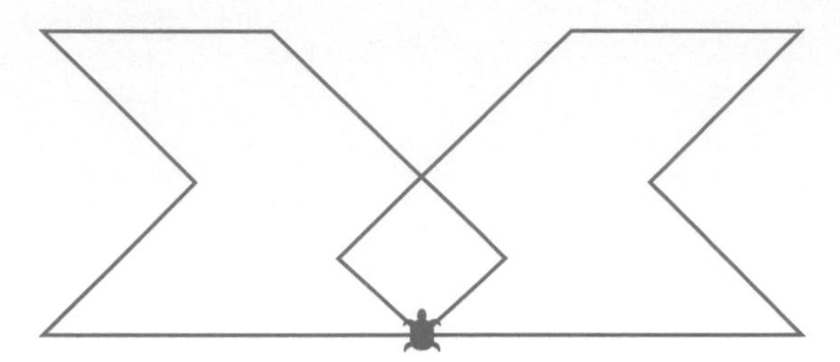

■

Logo Glide Reflections

A glide reflection is a composition of a reflection followed by a slide in the direction parallel to the line of reflection. The example below uses Logo to illustrate a glide reflection.

LOGO EXAMPLE 13.D

Glide Reflecting a Polygon

Use paper and pencil sketches to predict the figures drawn by these procedures. Enter and run the procedures on a computer to see if your predictions are accurate.

```
to  trap
      fd 70 lt 120
      fd 30 lt 60
      fd 40 lt 60
      fd 30 lt 120
end
```

```
to  refl.trap
      fd 70 rt 120
      fd 30 rt 60
      fd 40 rt 60
      fd 30 rt 120
end
```

```
to  glide.refl
      rt 90
      trap pu
      fd 100 pd
      refl.trap
      lt 90
end
```

SOLUTION

The procedure **trap** draws an isosceles trapezoid with a vertically oriented base. The procedure **refl.trap** draws a reflection of the trapezoid across the vertical line through the base drawn first. The procedure **glide.refl** draws the trapezoid with its base horizontal, and then draws the glide reflection of the trapezoid 100 turtle steps to the right.

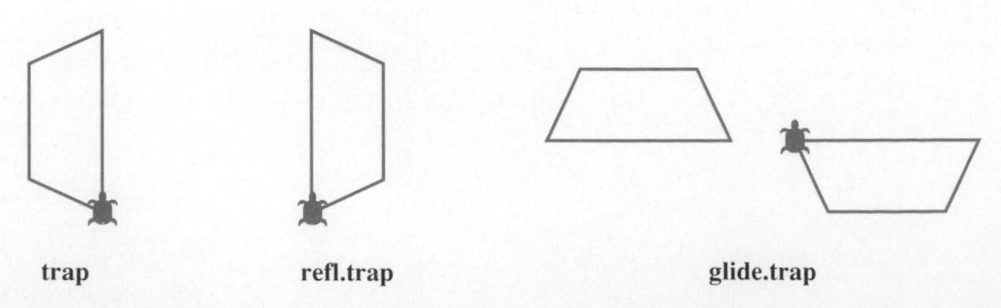

trap **refl.trap** **glide.trap**

■

PROBLEM SET 13.1

Understanding Concepts

1. Which of the following "transformations" correspond to a rigid motion? Explain the reasoning you have used to give your answer.
   (a) A deck of cards is shuffled.
   (b) A completed jigsaw puzzle is taken apart and then put back together.
   (c) A jigsaw puzzle is taken from the box, assembled, and then replaced in its box.
   (d) A painting is moved to a new position on the same wall.
   (e) Bread dough is allowed to "rise."
2. For each figure shown, find its image under the translation which takes P to P'.
   (a) (b)

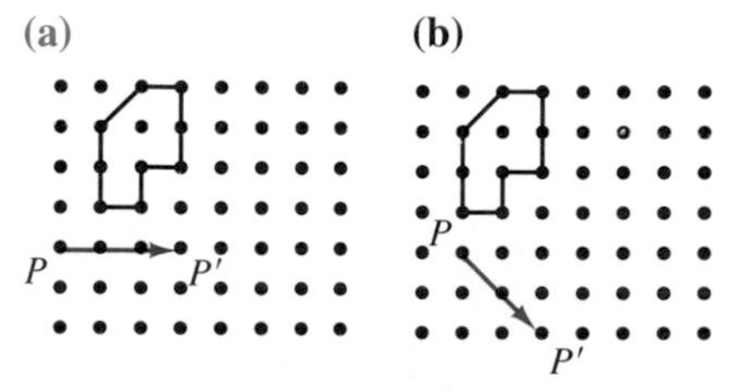

3. The translation that takes point P to P' has transformed triangle ABC (not shown) to its image $A'B'C'$ (shown below).
   (a) Draw triangle ABC.
   (b) Describe the rigid motion that transforms $\triangle A'B'C'$ to $\triangle ABC$, and compare it to the translation that takes $\triangle ABC$ to $\triangle A'B'C'$.

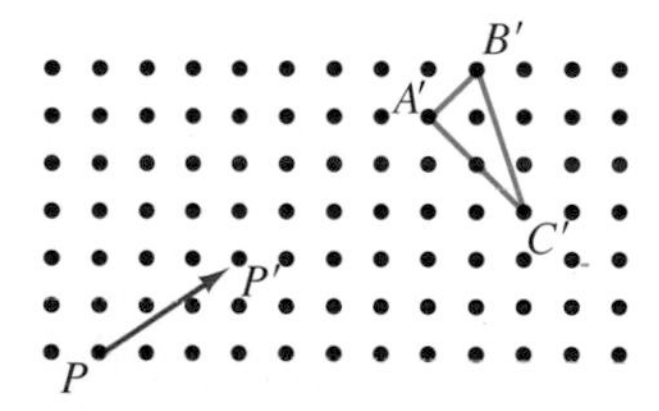

4. In each of the following give an equivalent answer between 0° and 360°.
   (a) A clockwise rotation of 60° is equivalent to a counterclockwise rotation of ________.
   (b) A clockwise rotation of 433° is equivalent to a clockwise rotation of ________.
   (c) A clockwise rotation of 3643° is equivalent to a clockwise rotation of ________.
   (d) A sequence of two consecutive clockwise rotations, first of 280° and next of 120°, is equivalent to a single clockwise rotation of ________.
   (e) A rotation of −260° is equivalent to a rotation of ________.
5. Sketch the image of $\triangle ABC$ under the given rotations.

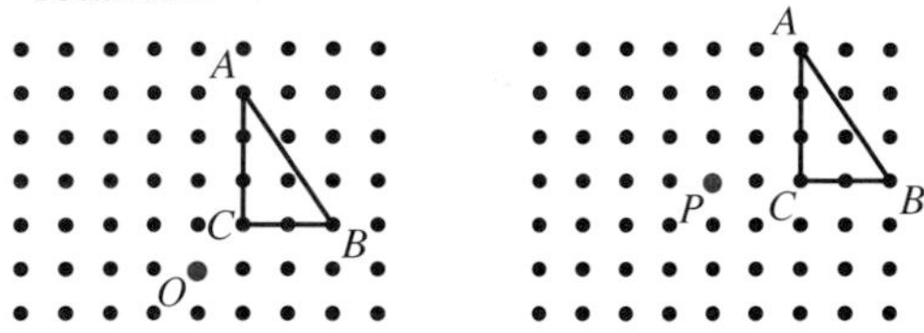

   (a) 90° counterclockwise about O
   (b) 180° about P
6. A rotation has sent A to A' and B to B', as shown below.

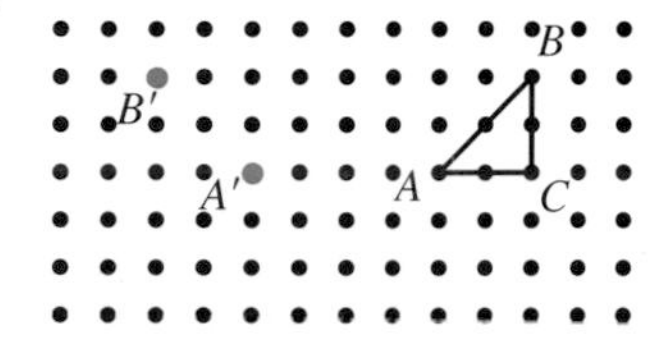

   (a) Find the center of rotation.
   (b) Find the turn angle.
   (c) Sketch the image triangle $A'B'C'$.
7. Trace the following figure, which shows $\triangle ABC$ and its image under a rotation. Use any drawing tools you wish (Mira, compass, and straightedge) to construct the center of rotation. (*Hint:* Why is the center of rotation on the perpendicular bisector of $\overline{AA'}$?)

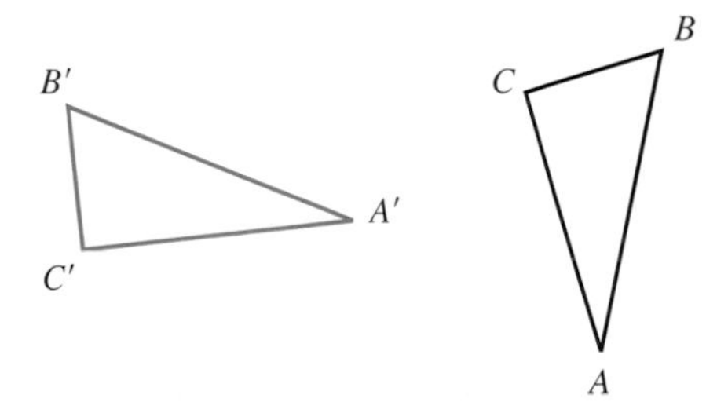

8. A **half-turn** is a rotation by a straight (180°) angle about a center point O.
   (a) Suppose P' is the image of point P under a half-turn. Describe how to find the turn center O.
   (b) Describe a simple procedure to draw the image of points A, B, and C under a half-turn centered at O, using a straightedge and compass.

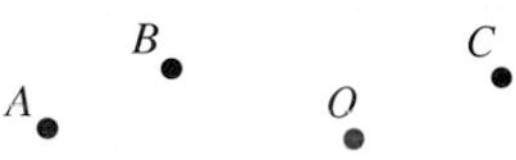

   (c) Sketch the image of the entire lower case letter P under a half-turn about point O. (See figure on next page.) What letter does the image resemble?

(d) Sketch the image of the stick-figure under a half-turn about point O.

9. What basic rigid motion is equivalent to two half-turns (that is, 180° rotations) performed in succession about two points O_1 and O_2? Explain carefully, using sketches to illustrate your discussion. It may be helpful to let x denote the distance from O_1 to O_2.

10. Redraw the figure below on squared graph paper. Then sketch the reflection of $\triangle ABC$ across the mirror line m.

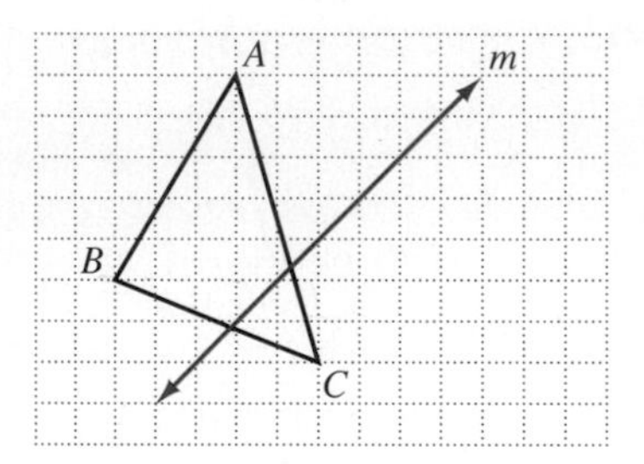

11. A reflection has sent P to P'.
   (a) Find the line of reflection.
   (b) Find the image of the polygon $PQRST$ under the reflection that takes P to P'.

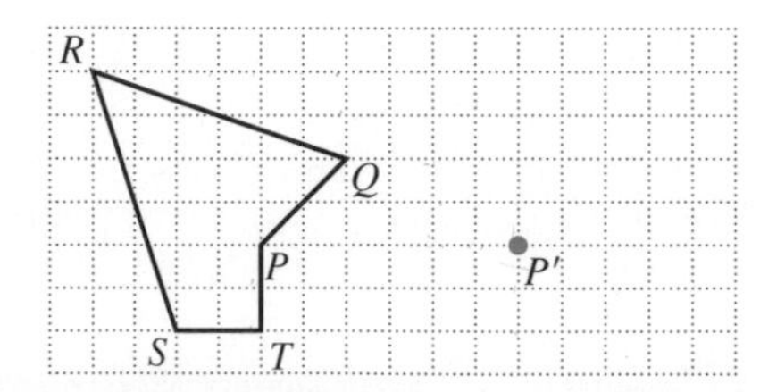

12. **(a)** A reflection across line m leaves point A fixed, so that $A' = A$. What can be said about A and m?
   (b) A reflection across line m leaves two points A and B fixed, so that $A' = A$ and $B' = B$. What can you say about A, B, and m?
   (c) A reflection takes point C to point D. Where does the reflection take point D?

13. A glide reflection is defined by the slide arrow and line of reflection m shown. Draw the following images of the polygon $ABCDE$.
   (a) The image $A_1B_1C_1D_1E_1$ under the slide.
   (b) The image $A'B'C'D'E'$ under the glide reflection.

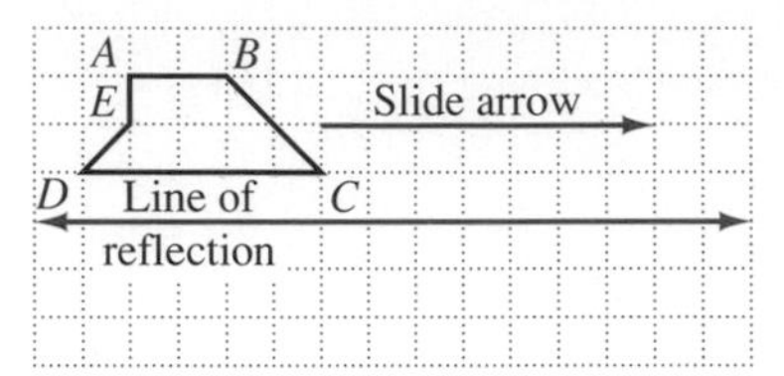

14. A glide reflection has taken B to B' and E to E'. Find
   (a) the line of reflection of the glide.
   (b) the slide arrow.
   (c) the image $A'B'C'D'E'$ of the polygon $ABCDE$.

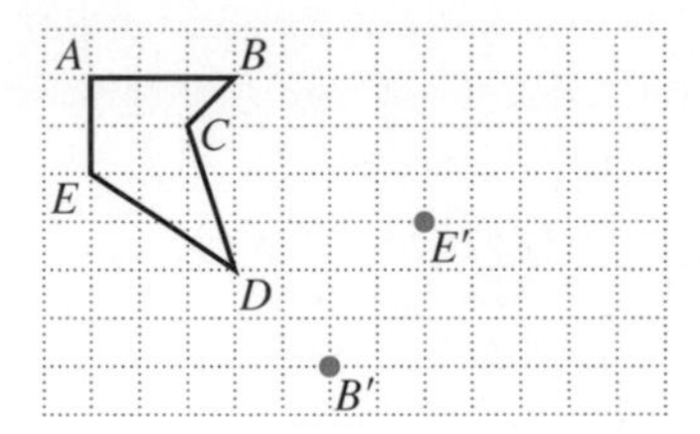

15. The slide arrow and line of reflection of a glide reflection are shown together with a point P and its image P' under the glide reflection.

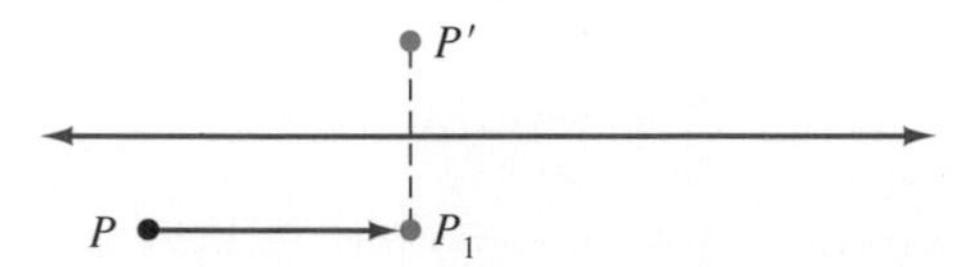

   (a) Sketch the image P'' of P which results when this glide reflection is performed twice in succession.
   (b) Describe the basic rigid motion that is equivalent to performing the same glide reflection twice in succession.
   (c) Describe the basic rigid motion that is equivalent to performing the same glide reflection three successive times.

16. In each part below draw a line m_2 so that the net outcome of successive reflections about m_1 and then m_2 is equivalent to the translation specified by the slide arrow shown.

(a)

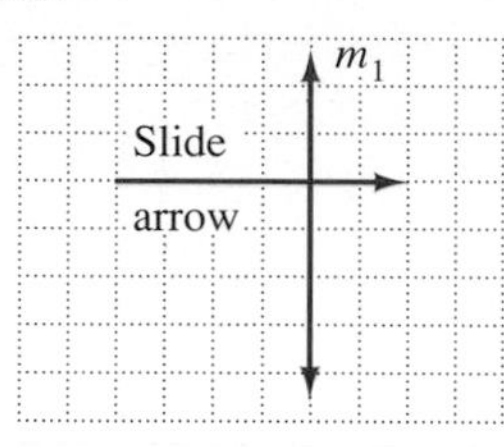

(b)

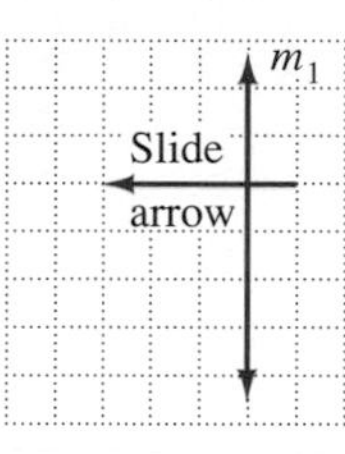

(c)

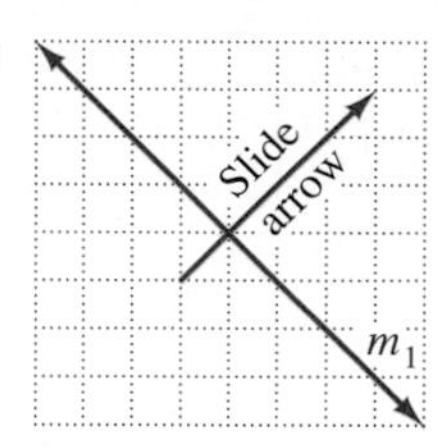

17. Copy the four vertical lines, l_1, l_2, m_1, m_2, and $\triangle ABC$ on to squared paper.

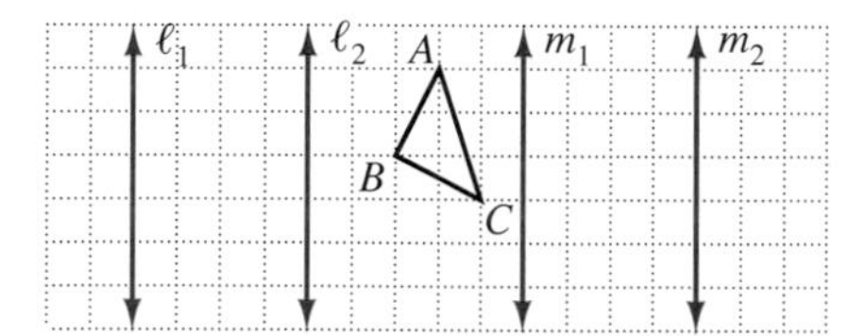

(a) Draw the image of $\triangle ABC$ obtained by reflection across line m_1. Label it $\triangle A_1B_1C_1$.

(b) Draw the image of $\triangle A_1B_1C_1$ obtained by reflection across line m_2. Label it $\triangle A'B'C'$.

(c) Draw the image of $\triangle ABC$ obtained by reflecting $\triangle ABC$ across line l_1. Label it $A_1^* B_1^* C_1^*$.

(d) Draw the image of $\triangle A_1^* B_1^* C_1^*$ under reflection across l_2.

(e) What basic rigid motion is equivalent to the net outcome of successive reflections across m_1 and m_2?

18. In each part below draw line m_2 so that the net outcome of successive reflections about m_1 and then m_2 is equivalent to the given rotation about point O. The first one is done for you.

Given Answer

(a) 90° counterclockwise rotation

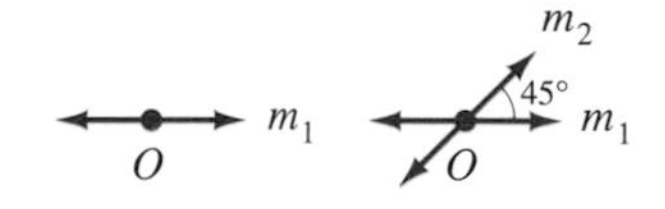

(b) 120° counterclockwise rotation

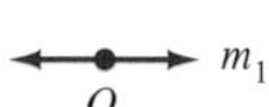

(c) Half-turn (180° rotation)

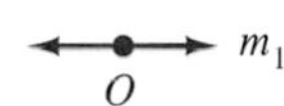

(d) Half-turn

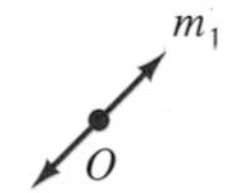

(e) 90° clockwise rotation

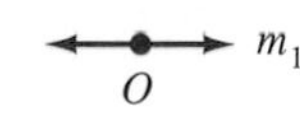

(f) 90° clockwise rotation

Thinking Critically

19. A rigid motion takes points A, B, C, and P to the respective image points A', B', C' and P', where $AB = 3$ cm, $AC = 4$ cm, $AP = 2$ cm, $BC = 2$ cm, $BP = 4$ cm, and $CP = 4$ cm. In each part use a compass and ruler to draw the smallest set of points which you know must contain the point P' when you are given:

(a) only point A'. (*Hint:* The answer is a circle.)

(b) points A' and B'.

(c) points A', B', and C'.

20. Suppose the perpendicular bisectors are drawn for all pairs of distinct points P and P' that correspond under some rigid motion other than the identity transformation. What is the basic motion if:

(a) the lines are parallel?

(b) the lines are concurrent?

(c) there is just one line?

Explain how you arrived at your answer to each part.

21. Two successive 90° rotations are taken, first about center O_1 and then about center O_2, where $O_1O_2 = 2$ cm. Describe the basic rigid motion that is equivalent to the successive rotations. Explain carefully, using words and sketches.

22. Suppose that the 90° rotations in problem 21 are each replaced with a 120° rotation. What basic rigid motion is equivalent to the net outcome of the two rotations? Explain with words and sketches.

23. (a) Find the image $\triangle A'B'C'$ of $\triangle ABC$ under the rigid motion consisting of three consecutive reflections across the concurrent lines m_1, m_2, and m_3.

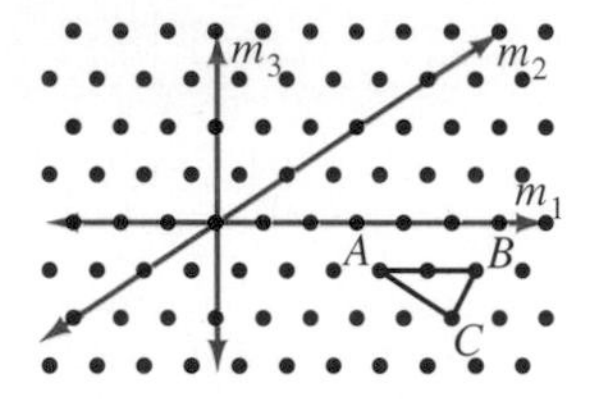

(b) Find a line l so that $\triangle ABC$ is taken onto $\triangle A'B'C'$ by one reflection across l.

24. (a) Find the image $\triangle A'B'C'$ of $\triangle ABC$ under the rigid motion consisting of three consecutive reflections across lines m_1, m_2, and m_3.

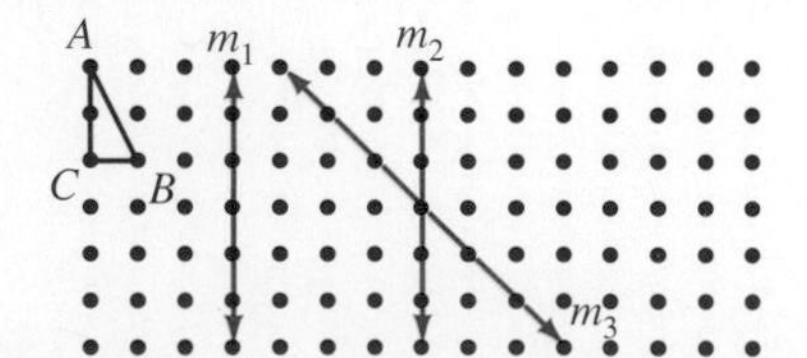

(b) Find the image of $\triangle ABC$ under the glide reflection whose slide arrow extends from P to P' and whose line of reflection is line l.

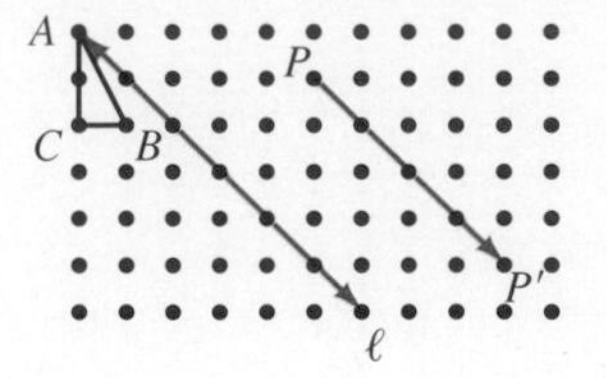

(c) What conclusion can you make about the two rigid motions described in (a) and (b)?

25. Trace the figure below, where $\triangle ABC$ is congruent to $\triangle A'B'C'$.

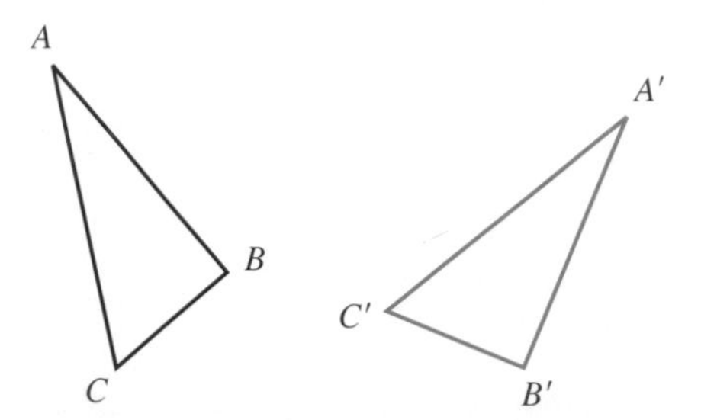

(a) Use a Mira (or other drawing tools) to draw the line m_1 across which A is reflected to A'. Also, draw the images of B and C and label them B_1 and C_1.

(b) Draw the line m_2 across which B_1 reflects to B'. What is the image of C_1 across m_2?

(c) Use the lines m_1 and m_2 to describe the basic rigid motion that takes $\triangle ABC$ to $\triangle A'B'C'$.

26. Trace the figure below, where $\triangle ABC$ is congruent to $\triangle A'B'C'$.

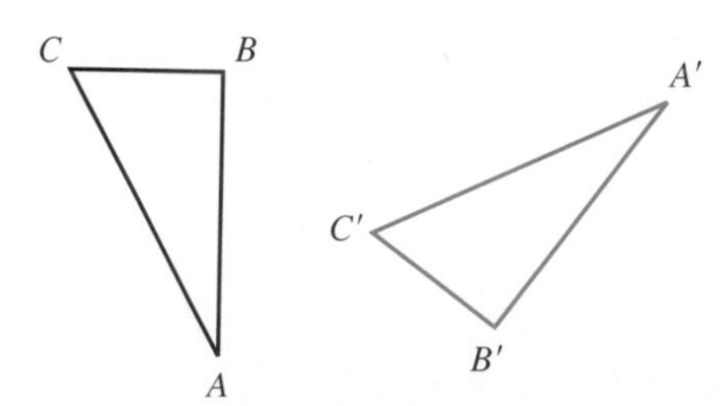

(a) Use a Mira (or other drawing tools) to draw three lines of reflection—m_1, m_2, and m_3—so that:

(i) reflection across m_1 takes A to A' (and B, C are taken to B_1, C_1);

(ii) reflection across m_2 takes B_1 to B' (and C_1 is taken to C_2);

(iii) reflection across m_3 takes C_2 to C'.

(b) Describe the type of basic rigid motion that takes $\triangle ABC$ to $\triangle A'B'C'$.

27. In each of the following parts, a complicated sequence of rigid motions is described. Explain how you know what type of basic rigid motion is equivalent to the net outcome of the motion described.

    (a) Reflections are taken across six lines and no point is taken back to its original position.

    (b) Reflections are taken across eleven lines and there are points which are taken back to their original positions.

    (c) Two different glide reflections are taken in succession, with the net outcome taking some point back to its original location.

28. A convex quadrilateral is cut along a diagonal, and a congruent quadrilateral is cut along the other diagonal. (See the Just for Fun, Quadrilateral + Quadrilateral = Parallelogram? on page 838.) Show that the four triangular regions can be arranged to form a parallelogram. (*Hint:* Translate quadrilateral $ABCD$ to $A'B'C'D'$, where $A' = C$. What do you notice about the polygon $BB'D'D$?)

29. Let C and D be the two circles on the same side of mirror m. Construct a tangent line to circle C which, after reflection, is also tangent to circle D. How many solutions can you find?

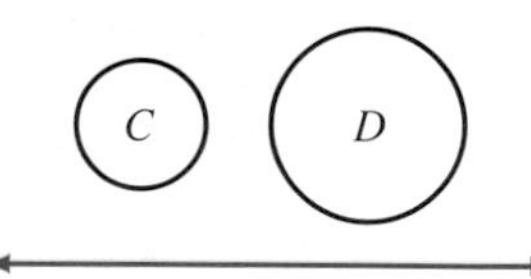

30. A tiling by M. C. Escher is shown below.

    (a) What rigid motion takes figure A onto figure B?

    (b) What rigid motion takes figure A onto figure C?

    (c) What rigid motion takes figure C onto figure D?

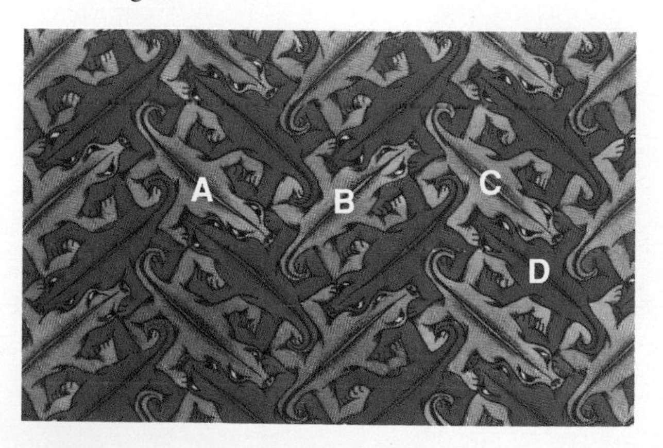

Connections

31. Estelle wants to install a "full-length" mirror on her closet door. She is 5′2″ tall, and her eyes are 6″ below the top of her head. How tall must the mirror be for Estelle to see her entire reflection? How high should the top of the mirror be off the floor? (*Hint:* Draw a picture, showing Estelle and her mirror image behind the plane of the mirror.)

32. **Fermat's Principle.** Suppose that a ray of light emanating from point P is reflected from a mirror at point R toward point S. It was known even in ancient times that the incident and reflected rays of light make congruent angles to the line m of the mirror. Pierre de Fermat (1601–1665) proposed an important principle to explain why this is so: *light follows the path of shortest distance.* According to Fermat's principle, if Q is some point on the mirror other than R, then the distance $PQ + QS$ must exceed the distance $PR + RS$ traveled by the light.

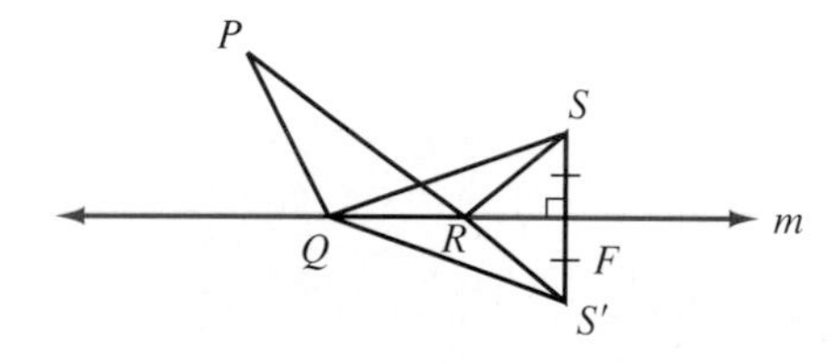

Answer the following questions to verify that $PQ + QS > PR + RS$. Let S' be the image of S under reflection across line m.

    (a) Why is $RS' = RS$ and $QS' = QS$? (*Hint:* The reflection across m is a *rigid* motion which preserves distances.)

    (b) Why is $PQ + QS' > PS'$?

    (c) How does the inequality of part (b) give the desired result that $PQ + QS > PR + RS$?

33. Let P and S be two points on the same side of mirror m. If S' is the point of reflection of point S across m, then the line drawn from P to S' intersects m at the point R of reflection (see the figure in problem 32). Suppose P and Q are between *two* mirrors, m and l, as shown on the next page.

    (a) Construct a doubly reflected light path $PQRS$ which is reflected off mirror m at Q and then off mirror l at R.

(b) Construct another doubly reflected path $PABS$ which first reflects off l at A and then off mirror m at B.

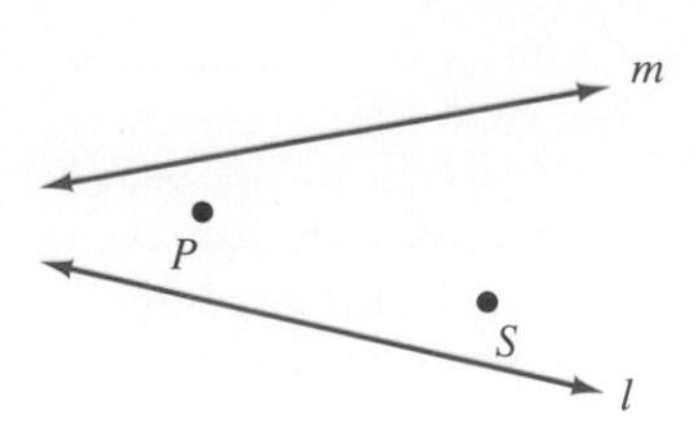

34. A **corner mirror** is formed by placing two mirrors together at a right angle. Explain why looking at yourself in a corner mirror is quite different than seeing yourself in an ordinary mirror. Use sketches to make your ideas clear.

35. A billiard ball is located at a point P along an edge of a rectangular billiard table. Show that there is a billiard shot which strikes the cushions on the other three rails (sides) of the table and then returns to bounce at P.

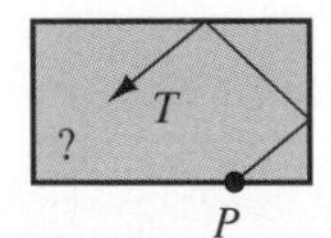

(*Hint:* Suppose the table T is reflected across its sides successively, forming the images T', T'', and T'''.) Explain how a billiard path $PQRSP$ can be found by drawing the line segment $\overline{PP'''}$. Sketch the path $PQRSP$ on the original table T.

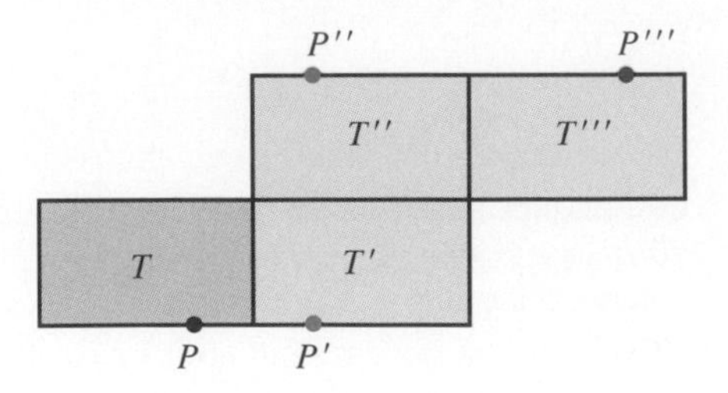

Using a Computer
Using an Automatic Drawer

36. Draw any quadrilateral $ABCD$ and locate the midpoints. M and N, of sides $\overline{CD}$ and $\overline{AD}$. Now use your software to perform the following transformations of the quadrilateral: (i) a 180° rotation about M, (ii) a 180° rotation about N, (iii) a translation with the slide arrow (vector) that takes B to D. Show that translations of the octagon that result from (i), (ii), and (iii) will tile the plane.

37. Draw any triangle. Use the transformation menu of your drawing software to create the hexagon shown below, consisting of three congruent copies of the triangle interspersed with three equilateral triangles. Now translate the hexagon to show a tiling of the plane similar to the one shown at the end of Example 13.7.

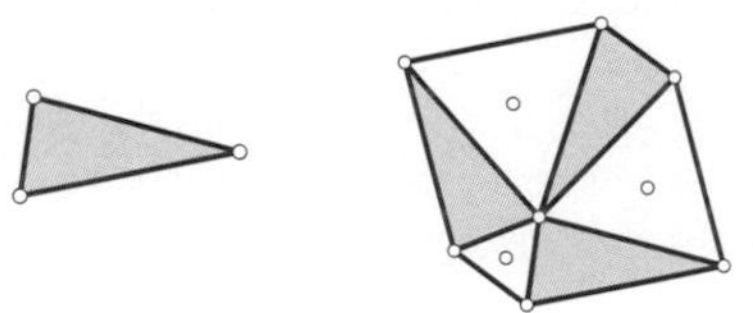

Using Logo

38. The triangle shown is drawn with this procedure.

```
to    scalene
          fd 40 lt ( 180 - 82.82)
          fd 50 lt ( 180 - 41.41)
          fd 60 lt ( 180 - 55.77)
end
```

Write a procedure to draw the figure below, using **scalene** as a subprocedure.

Note that the start position is marked by the vertical turtle. The end position is marked by the turtle facing to the right at an angle.

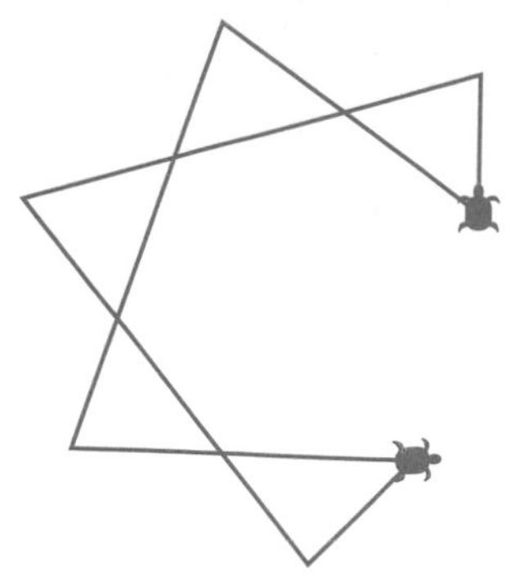

39. Use the **spin.poly2** procedure from Logo Example 13.B to create four different spin.polygons.

40. Determine the values for the variables **:nsides** and **:reps** to create the figures below using the **spin.poly2** procedure of Logo Example 13.B.

(a)

(b)

(c)

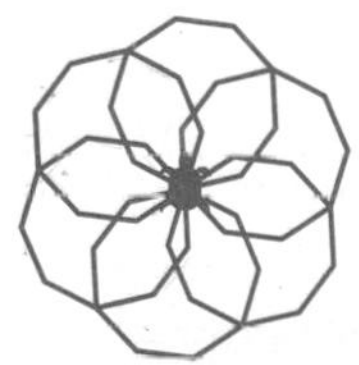

41. The following procedure draws the parallelogram shown.

```
to    parallelogram.2
        lt 48
        repeat 2 [ lt 53 fd 60 lt
                            127 fd 110 ]
        rt 48
end
```

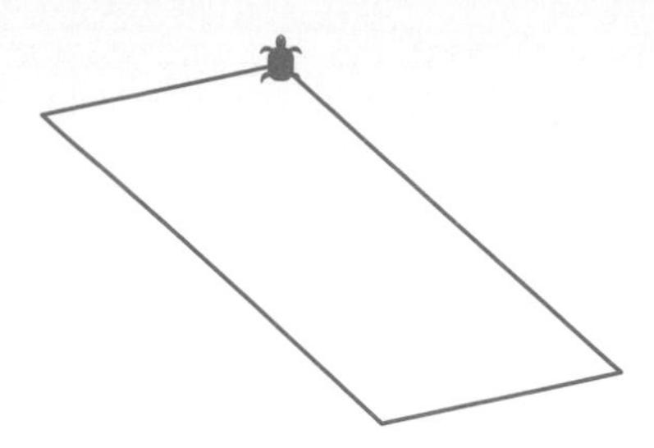

Write procedures to perform each of the following transformations:

(a) a translation of 50 turtle steps straight up;
(b) rotation through an angle of 42° to the right;
(c) reflection across a vertical line;
(d) glide reflection combining parts (a) and (c.)

Communicating

42. Explore what happens when translations are taken in succession, writing a report on your conclusions. Give both specific examples and state any general principles you find. In particular, include answers to the following questions:
   (a) Why is the net outcome of two successive translations equivalent to another translation?
   (b) If the first translation takes A to A_1 and the second translation takes A_1 to A', why is the arrow from A to A' the slide arrow of the combined motion of the two translations?
   (c) What happens to the net outcome if the order in which the two translations are taken is reversed?

For Review

43. The measures of the three angles in a triangle are $2x°$, $3x°$, and $4x°$. What are the three angle measures?

44. A triangle RST has area 87 cm^2 and $RS = 3$ cm. What is the distance from T to the line containing RS?

45. Give a counterexample that shows that the following conjecture is false: *the midpoints of the sides of a rhombus are the vertices of a square.*

46. If the word "square" in the conjecture stated in problem 45 is replaced with "rectangle" do you think the conjecture is true? Give a proof or provide a counterexample.

13.2 Patterns and Symmetries

Symmetry is a universal principle of organization and form. The circular arc of a rainbow and the hexagonal symmetry of an ice crystal are visible expressions of the symmetry of many, indeed most, of the physical processes of the universe. A sea shell and the fanned tail of the peacock are spectacular examples of biological symmetry, and indeed most life forms are highly symmetric. Symmetry is the norm of nature and natural law, not the exception.

In the human domain, all cultures of the world, even those in prehistoric times, developed a useful intuitive understanding of the basic concepts of symmetry. Decorations on pottery, walls, tools, weapons, musical instruments, and clothing are most often highly symmetric. Buildings, temples, tombs, and other structures are usually designed with an eye to symmetry and balance. Music, poetry, and dance frequently incorporate symmetry in their underlying structure.

While people have long had an informal understanding of symmetry, it is only more recently that mathematics has provided a deeper understanding of what symmetry means and how different kinds of symmetry can be described and classified. In the classroom, youngsters are drawn to this artistic and aesthetic aspect of mathematics.

What is Symmetry?

The concept of a rigid motion, which we defined and explored in the last section, makes it possible to give a precise definition of a symmetry of a geometric figure in the plane.

DEFINITION A Symmetry of a Plane Figure

A **symmetry** of a plane figure is any rigid motion of the plane that moves all the points of the figure back to points of the figure.

Thus, all points P of the figure are taken by the symmetry motion to points P' that also are points of the figure. The identity motion is a symmetry of any figure, but of more interest are figures that have symmetries other than the identity. Under a nonidentity symmetry some points in the figure move to new positions in the figure, even though the figure as a whole appears unchanged by the motion.

The classification theorem of the preceding section tells us that there are just four basic rigid motions. Therefore, any symmetry of a figure is one of these four basic types, and the symmetry properties of a figure can be fully described by listing all of the symmetries of each type.

FROM THE NCTM STANDARDS **Symmetry**

Symmetry in two and three dimensions provides rich opportunities for students to see geometry in the world of art, nature, construction, and so on. Butterflies, faces, flowers, arrangements of windows, reflections in water, and some pottery designs involve symmetry. Turning symmetry is illustrated by bicycle gears. Pattern symmetry can be observed in the multiplication table, in numbers arrayed in charts, and in Pascal's triangle.

SOURCE: From *Curriculum and Evaluation Standards for School Mathematics,* Grades K–4, p. 115.

INTO THE CLASSROOM
Mathematics in Motion

The words "transformation" and "isometry" often suggest advanced topics best left for gifted middle school students or postponed to the high school curriculum. Quite the opposite is true, however, since there are activities, games, artistic constructions, and problems, all exploring "motion geometry"—that are suitable for students at every grade level. If available, primary children can work with Miras, pattern blocks, and geoboards to create and investigate symmetric patterns. For example, each student can create a "half" figure with rubber bands on the upper half of a geoboard. Boards are then exchanged, and the student is challenged to complete a mirror image figure in the lower half of the geoboard. Older children could replace the geoboard with squared paper or dot paper, and investigate rotations and point symmetry as well as reflections and line symmetry. Logo and automatic drawing software also offer exciting possibilities for investigations in transformation geometry.

Here are three more ideas, suggesting how patterns and motions can be approached in the classroom.

- *Follow the leader.* Draw a line with a ruler down a blank sheet of paper. In pairs of students, the "leader" slowly draws a curve and simultaneously the "follower" draws the reflected curve across the line of symmetry. The students can interchange roles of leader and follower. To explore point symmetry, a prominent dot can be drawn at the center of the sheet. Some students, with a pencil in each hand, might like to attempt a solitaire game.

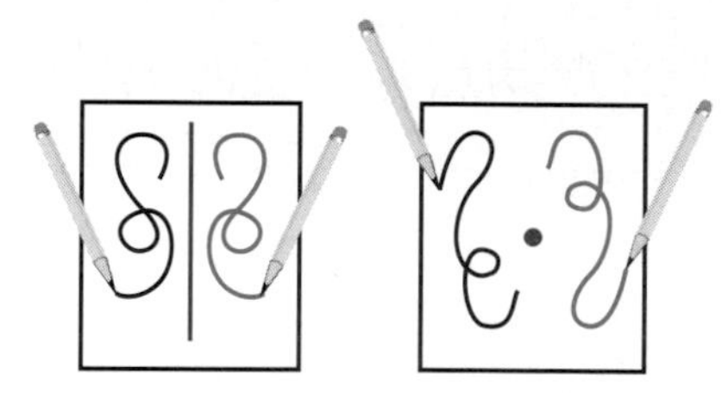

- *Punchy puzzles.* If the square shown at the far left is folded along the dashed lines, then the pattern of holes can be seen to be created with just one punch. How can a square of paper be folded and punched one time only to create the other hole patterns shown?

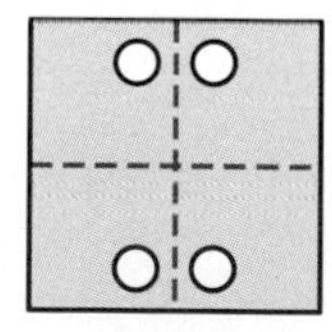
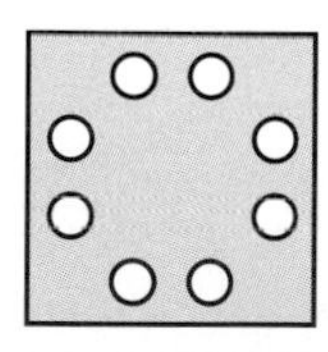
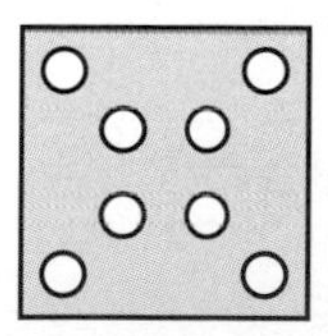
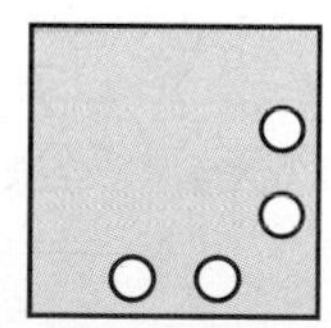
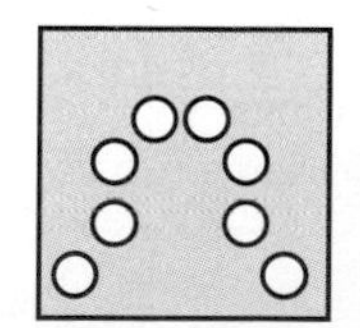

- *Stained glass window search.* Eight congruent isosceles right triangles, with four of each color, will form a square window. Three windows are shown here, but a reflection and rotation show that the first two windows are really the same. How many different window patterns are there, each with four panes of each of two colors? (You should be able to find 13 distinct patterns with no two patterns the same under either a rotation or a reflection.)

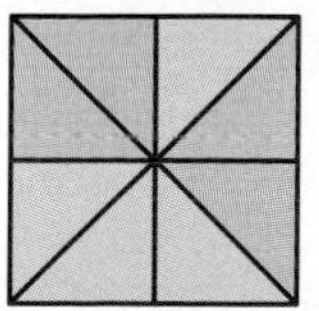
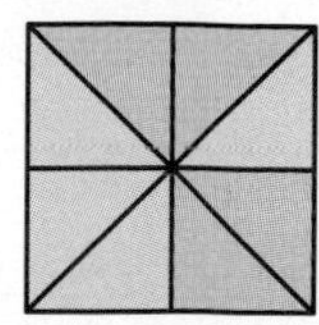
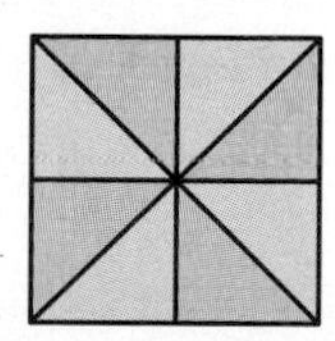

Reflection Symmetry

A figure has **reflection symmetry** if a reflection across some line is a symmetry of the figure. The line of reflection is called a **line of symmetry** of the figure. Each point P of the figure on one side of the line of symmetry is matched to a point P' of the figure on the opposite side of the line of symmetry. Some figures and their lines of symmetry (shown dashed) are shown in Figure 13.12.

Figure 13.12
Plane figures and their lines of symmetry

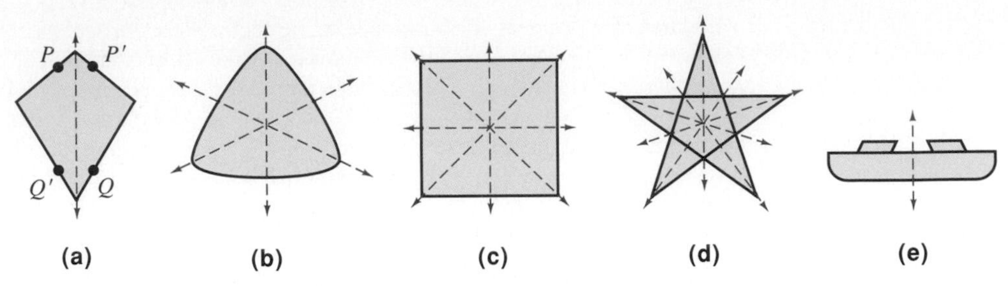

Reflection symmetry is also called **line symmetry** or **bilateral symmetry.** Bilateral symmetry is also used to describe figures in space that have a plane of symmetry. For example, ferries (as suggested in Figure 13.12(e)) are often bilaterally symmetric across midships to simplify loading and unloading their cargo of cars and trucks. Infrequent passengers on such ferries can find it very disorienting when the bow and stern are indistinguishable.

EXAMPLE 13.8

Identifying Lines of Symmetry

How many lines of symmetry does each letter shown below have?

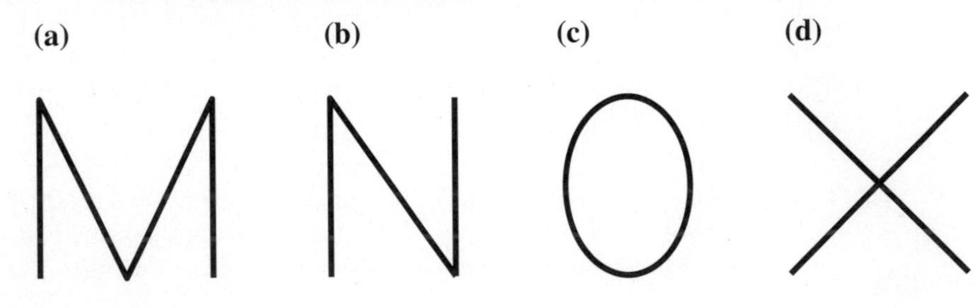

SOLUTION

(a) 1 **(b)** 0 **(c)** 2 **(d)** 4, since the segments are congruent and intersect at right angles. ■

Rotation Symmetry

A figure has **rotation symmetry**, or **turn symmetry**, if the figure comes back to itself when it is rotated through a certain angle between 0° and 360°. The center of the turn is called the **center of rotation.** Some examples of figures with rotation symmetry are shown in Figure 13.13.

Figure 13.13
Figures with rotation symmetry

(a) 90° symmetry

(b) 72° symmetry

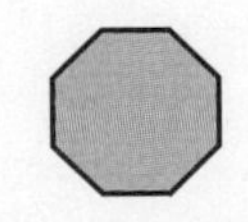

(c) 45° symmetry

(d) 180° symmetry

HIGHLIGHT FROM HISTORY

George W. Brainerd and Anna O. Shepherd, Pioneers in Mathematical Anthropology

George W. Brainerd, a North American archeologist, was the first person to use the principles of symmetry as a tool for anthropological study. His analysis of the designs on pottery of the prehistoric Anasazi of Monument Valley, Arizona, and of the Maya of the Yucatan Peninsula, led him to formulate a number of principles for the use of symmetry classifications in pattern analysis. Brainerd published his ideas in the article "Symmetry in Primitive Conventional Design," which appeared in 1942 in *American Antiquity,* the leading journal of North American anthropology. Unfortunately, Brainerd's work was almost completely neglected, although it did attract the attention of Anna O. Shepherd, a geologist at the Carnegie Institution in Washington, D.C. In her monograph *The Symmetry of Abstract Design with Special Reference to Ceramic Decoration,* published in 1948, Shepherd dicusses how certain symmetries predominate within a specific culture, and how changes within a culture can be identified by symmetry. Shepherd's work, like that of Brainerd, had to wait until the mid-1970s to be fully appreciated. Today, the anthropological significance of symmetry analysis is well established.

A figure with 90° rotation symmetry automatically has 180° and 270° rotation symmetry. For this reason it is customary to give just the *smallest* positive angle measure which turns the figure back to itself. The only exception is for figures composed of concentric circles, which turn back to themselves after *any* turn about their center. Such figures have **circular symmetry.**

EXAMPLE 13.9

Finding Angles of Rotation Symmetry

Determine the angles of measure of rotation symmetry of these figures.

(a) **(b)** **(c)** **(d)**

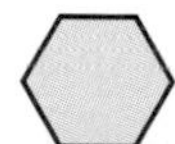

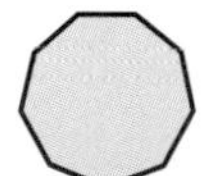

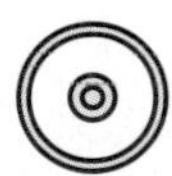

SOLUTION

(a) The smallest positive angle of rotation of a regular hexagon measures 60°, so the turn angles of all of the rotation symmetries are 60°, 120°, 180°, 240°, and 300°.

(b) The only rotation angle measures 180°.

(c) The smallest amount of turn of a regular 9-gon is 360°/9 = 40°, so the turn angles are 40°, 80°, 120°, 160°, 200°, 240°, 280°, and 320°.

(d) This figure has circular symmetry. ■

Point Symmetry

A figure has **point symmetry** if it has 180° rotation symmetry about some point O as shown in Figure 13.14. This means that a half-turn takes the figure back to itself, and every point P of the figure has a corresponding point P' of the figure that is directly opposite the turn center O.

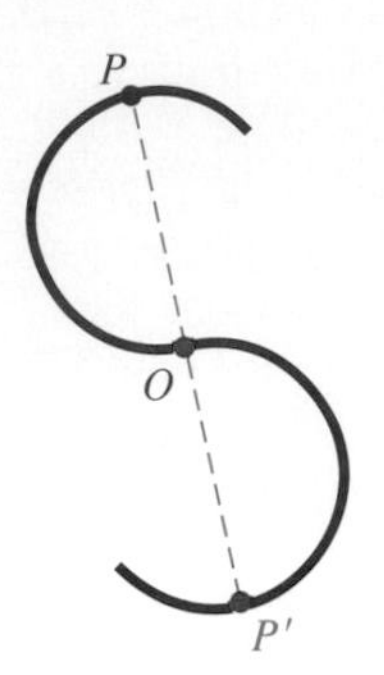

Figure 13.14
The letter *S* has point symmetry.

EXAMPLE 13.10 **Identifying Point Symmetry**

What letters, in uppercase block form, can be drawn to have point symmetry?

SOLUTION

H, I, N, O, S, X, Z. ■

Patterns: Figures with Translation Symmetries

A **pattern** is a figure with a translation symmetry. To avoid considering the whole plane as a pattern, or even just some set of horizontal lines, it is assumed that there is some minimum positive distance required to translate a pattern back onto itself. Thus a pattern must be an infinite figure (why?) with motifs repeated infinitely often. Patterns are common on wallpaper, decorative brick walls, printed and woven fabrics, ribbons, and friezes (ceiling line decorations in older buildings). Enough of the pattern must be shown to make it clear how to extend the pattern indefinitely.

There are two classes of patterns in the plane. The first are the **border patterns,** in which the motif is repeated in just *one* direction. That is, the translations taking the pattern back onto itself are all in the same direction. Some examples of border patterns are shown in Figure 13.15, where the direction of translation symmetry is horizontal for each pattern. Occasionally, a border pattern may appear with a vertical or even oblique orientation, but it is convenient to assume the pattern is viewed in a horizontal position.

The second class of plane patterns has translations in more than one direction. The pattern therefore covers the entire plane. The most common examples are found on wallpaper designs, which account for the name **wallpaper pattern.** Two wallpaper patterns, one Arabian and one Egyptian, are shown in Figure 13.16.

A pattern may have other symmetries in addition to its translation symmetries. However, the possibilities are limited. For example, the only possible rotation symmetry of a border pattern is a half-turn (why?). The fact that only certain symmetries can

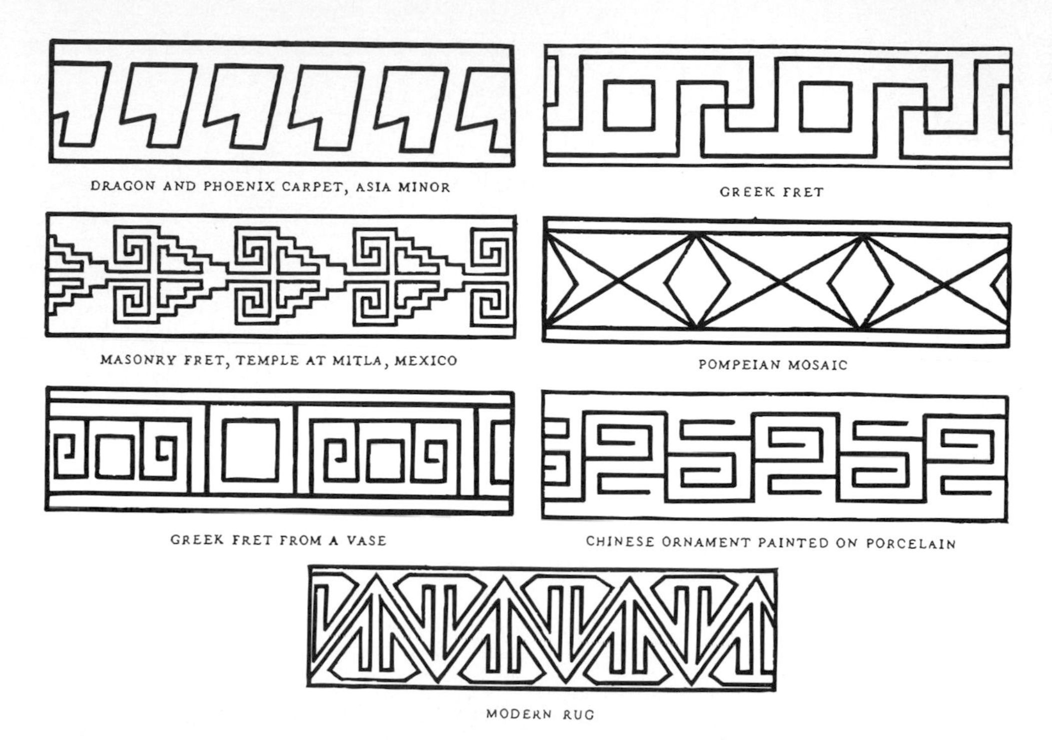

Figure 13.15
Border patterns from around the world

Figure 13.16
Examples of wallpaper patterns

coexist in a pattern makes it possible to catalog the symmetry types of patterns. In particular, it has been shown that there are just seven types of border designs, and seventeen types of wallpaper designs. A proof that no other types of patterns exist is quite difficult, but it is fairly easy to describe all of the different types of border and wallpaper designs. We begin with the classification of border patterns.

Classifying Border Patterns

We have already observed that the only possible rotation symmetry of a border pattern is a 180° rotation; that is, a half-turn. Similarly, reflection symmetry can only occur across two kinds of lines, namely a horizontal line (in the direction of the border) or possibly across certain vertical lines (perpendicular to the direction of the

WORTH READING

Symmetries of Culture

In this book we demonstrate how to use the geometric principles of crystallography to develop a descriptive classification of patterned design. Just as specific chemical assays permit objective analysis and comparison of objects, so too the description of designs by their geometric symmetries makes possible systematic study of their function and meaning within cultural contexts.

This particular type of analysis classifies the underlying structure of decorated forms; that is, the way the parts (elements, motifs, design units) are arranged in the whole design by the geometrical symmetries which repeat them. The classification emphasizes the way the design elements are repeated, not the nature of the elements themselves. The symmetry classes which this method yields, also called motion classes, can be used to describe any design whose parts are repeated in a regular fashion. On most decorated forms such repeated design, properly called pattern, is either planar or can be flattened (e.g., unrolled), so that these repeated designs can be described either as bands or strips (one-dimensional infinite) or as overall patterns (two-dimensional infinite) in a plane.

These excerpts are from the introduction to *Symmetries of Culture: Theory and Practice of Plane Pattern Analysis,* by Dorothy K. Washburn and Donald W. Crow. Nearly every page of *Symmetries of Culture* is graced by beautiful photographs and drawings that illustrate the principles of symmetry discovered and utilized by contemporary and historic cultures from around the world.

SOURCE: From *Symmetries of Culture: Theory and Practice of Plane Pattern Analysis* by Dorothy K. Washburn and Donald W. Crow, Page ix.

border). Any glide reflection necessarily has a horizontal slide arrow and a horizontal line of reflection. However, not all combinations of these symmetries can exist simultaneously. For example, if a border pattern has both a horizontal and a vertical line of symmetry, then it necessarily has a 180° rotation symmetry (why?).

It has been shown that border patterns have just seven types of symmetry. The International Crystallographic Union designates each symmetry type with a two-symbol notation, as shown in Figure 13.17. The symbols can be assigned by the following procedure, where we view the border pattern in a horizontal position.

First symbol: m, if there is a vertical line of symmetry; 1, otherwise.
Second symbol: m, if there is a horizontal line of symmetry; g, if there is a glide reflection (but no horizontal symmetry line); 2, if there is a half-turn symmetry (but no horizontal reflection or glide reflection); 1, otherwise.

The seven symmetry types are mm, mg, $m1$, $1m$, $1g$, 12, and 11. The m refers to *mirror*, g to *glide*, and 2 to a *half-turn*.

The seven types of border patterns, and their corresponding Crystallographic Union symbol, are shown in Figure 13.17.

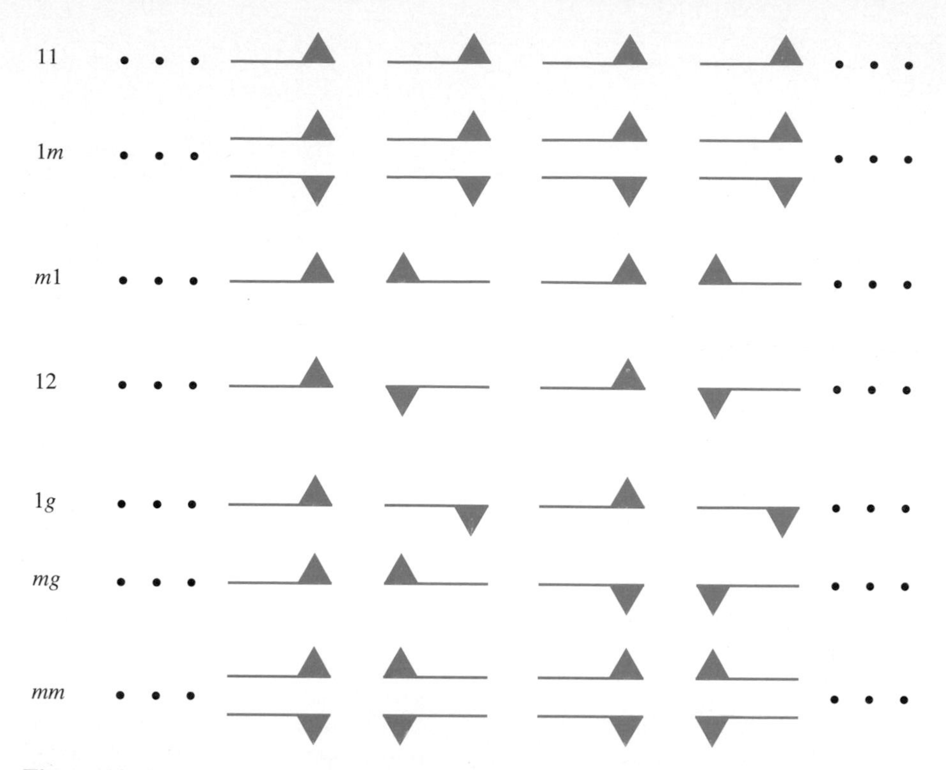

Figure 13.17
The seven symmetry types of border patterns

EXAMPLE 13.11

Classifying Border Patterns

Classify the symmetry type of the following border patterns by assigning the appropriate two-symbol notation.

(a) **(b)** **(c)**

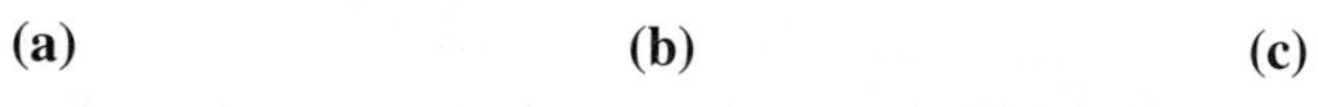

SOLUTION

It is helpful to view the patterns upside down, in a mirror, with a Mira, and so forth, since this will help you discover and verify what symmetry motions are present. A transparency copy of the pattern, if available, is an almost ideal tool to explore and classify patterns of symmetry.

(a) This border has both a vertical and horizontal line of reflection, so the symmery is of type *mm*.

(b) There is no vertical symmetry line, so the first symbol is 1. There is no horizontal symmetry line but there is a glide symmetry, so the second symbol is *g*. Altogether, the symmetry is type 1*g*.

(c) There are no lines of reflection, nor is there a glide reflection symmetry. However, there is half-turn symmetry so the symbol is 12. ■

HIGHLIGHT FROM HISTORY
Classifying Symmetry

In the years 1230 to 1354, the Moors constructed the magnificent Alhambra, a group of palatial buildings in the hills overlooking Grenada. The walls and ceilings are decorated with striking patterns formed by regularly repeated motifs. Thirteen of the 17 types of plane symmetry can be found. It is thought that the Islamic ban on human and animal motifs gave rise to the creation of such intricate abstract geometric decoration.

Similarly, artisans from other world cultures have discovered and used repeated motif designs. Indeed numerous examples representing all 17 symmetry types have been identified. It took mathematical methods, however, to *prove* that no more than 17 patterns of plane symmetry exist.

The first step toward classification was made by the Russian crystallographer Evgraf Federov in 1891. His work was made more widely available in 1924 through the work of P. Niggli and George Pólya. Fedorov, as a crystallographer, was also interested in patterns of symmetry of space figures. He was able to show that 230 patterns of symmetry exist in three-dimensional space.

Other generalizations have been investigated. For example colors may be used in a systematic way, such as the red-and-black coloration of a checkerboard. The two-colored patterns were classified by a mathematically knowledgeable textile worker in the 1930s. He discovered that there are 17 two-colored border patterns and 46 two-colored wallpaper patterns.

Classifying Wallpaper Patterns

It is not surprising that wallpaper patterns permit a larger number of symmetry combinations than are found in border patterns. For example, a checkerboard pattern of congruent squares admits quarter-turns (90° rotations). Similarly, a honeycomb pattern of regular hexagons admits 60° rotation symmetry about the center of any hexagon and 120° rotation symmetry about any vertex of a hexagon. Both the checkerboard square grid and the honeycomb hexagonal grid also admit half-turn rotations about the midpoint of any side of the tiling polygon.

These examples show that rotation symmetries of half-turns (180°), third-turns (120°), quarter-turns (90°) and one sixth-turns (60°) can exist in certain wallpaper patterns. Notice that we skipped over one fifth-turns (72°), and we have not given a pattern with a rotation symmetry of turn size $360°/n$, $n > 6$. The **crystallographic restriction** theorem tells us why: There are no wallpaper patterns which have a rotation symmetry of turn size $360°/n$ for $n = 5$ or $n > 6$. The theorem's name makes reference to the interest crystallographers have in repeated patterns.

THEOREM The Crystallographic Restriction

A rotation symmetry of a wallpaper pattern can have size 60°, 90°, 120°, or 180°, but no other size is possible.

PROOF Suppose, to the contrary, that a certain wallpaper pattern has 72° rotation symmetry about some point. Since there are translations that move the pattern back onto itself, these translations also move any turn center to new turn centers and therefore there are infinitely many turn centers. Among these turn centers, choose two—call them O_1 and O_2—that are closest together. As shown below, a clockwise 72° rotation about O_2 takes O_1 to a new turn center O'. Similarly, a counterclockwise rotation about O_1 takes O_2 to a new turn center O_2. But O_1' and O_2' are closer together than O_1 and O_2, contradicting the assumption we made earlier that O_1 and O_2 were as near to one another as turn centers can be. Thus, the assumption that a wallpaper pattern has 72° rotation symmetry leads us to a contradiction. Therefore, no wallpaper pattern can have 72° rotation symmetry.

A similar but even simpler argument shows that no wallpaper pattern can have a rotation symmetry of size smaller than 60°. (See problem 22 in the exercise section.)

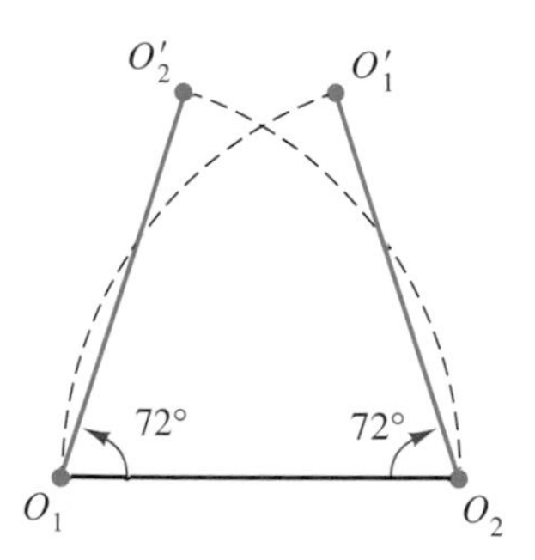

■

The crystallographic restriction provides a useful first step in the classification of wallpaper patterns. A more detailed analysis has shown that there are four types without rotation symmetry, five types with a smallest rotation symmetry of 180°, three types with a smallest rotation symmetry of 120°, three types with a smallest rotation symmetry of 90°, and two types with 60° rotation symmetry. Altogether this gives just 17 symmetry types of wallpaper patterns. Figure 13.18 shows examples of each type.

JUST FOR FUN

Mirror Magic

Amy's shop project had been coming along nicely. She had glued in all of the wood shapes in the frame and it only remained to glue in the triangular mirror. Unfortunately, she discovered that the mirror faced *inward* when fit into the triangular space left in the frame. "Oh, oh—I wasn't thinking when I traced my pattern on the backside of the mirror," said Amy. "Oh well, with two straight cuts across the mirror I can fill in the space in the frame with three outward facing pieces of mirrors. No problem!" What was Amy's solution to her difficulty? You may find it useful to reflect the triangle across the line containing the midpoints of the two shorter sides.

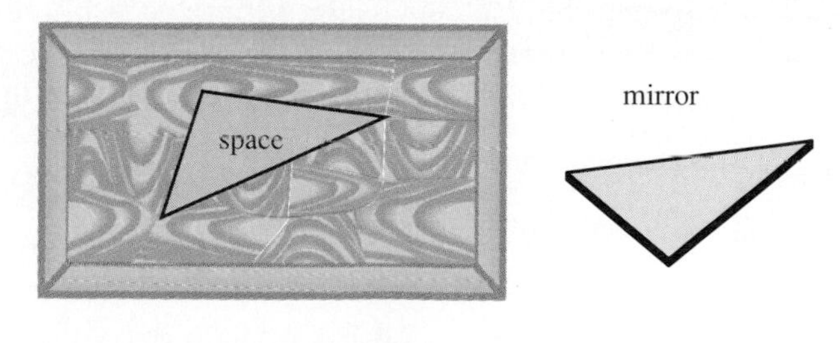

Figure 13.18
The 17 symmetry types of wallpaper patterns

Logo Symmetry

Symmetric figures can be drawn by carefully writing a procedure for the repeated motif of the figure, lifting the pen, returning to a new starting position and heading that correspond to the motion of the symmetry type, putting the pen back down, and then drawing the next part of the figure.

LOGO EXAMPLE 13.E

Creating a Symmetric Figure

The **arc.right** and **arc.left** procedures below draw arcs of a given radius and angle. (See Appendix C.)

```
to arc.right :radius :angle
   repeat :angle [ fd :radius * 2 * 3.14 / 360 rt 1 ]
end
```

```
to  arc.left :radius :angle
    repeat :angle [ fd :radius * 2 * 3.14 / 360 lt 1 ]
end
```

Use the arc procedures to draw a symmetric Valentine heart.

SOLUTION

First, we draw the right half of the heart.

```
to    half.heart.right
        pu  fd 40 pd
        rt 30 arc.right 50 180
        rt 15 fd 123
end
```

A reflection draws the symetric left half of the heart.

```
to    half.heart.left
        pu fd 40 pd
        lt 30 arc.left 50 180
        lt 15 fd 123
end
```

The entire heart is drawn by combining the two halves.

```
to    heart
        pu home pd
        half.heart.right
```

JUST FOR FUN

The Penny Game

Lynn and Kelly are playing a game with just a few simple rules. Each player takes turns placing a penny on a rectangular table top. Each new penny must be flat on the table, and cannot touch any of the pennies already on the table. The first player unable to put another penny on the table according to these rules is the loser. Lynn, who gets to make the first move, puts the first penny at the center of the table and is confident of a win. What is Lynn's strategy?

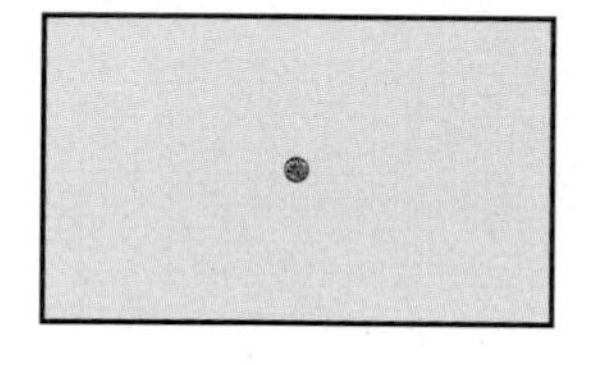

```
        pu home pd
        half.heart.left
        pu home pd
end
```

■

Rotation symmetry has already been used to write procedures to draw regular polygons and stars. Other figures with rotation symmetry are just as easy to draw.

LOGO EXAMPLE 13.F

Drawing a Figure with Rotation Symmetry

Create a figure with 40° rotation symmetry that incorporates the flag motif drawn by the following procedure.

```
to   flag
        fd 25
        repeat 2 [ fd 10 rt 90 fd 20 rt 90 ]
        bk 25
end
```

SOLUTION

The figure below is drawn with the **spin.flag 9** procedure. Nine was used as the number of repetitions because this gives a rotation of 360°/9 = 40°.

```
to  spin.flag :reps
        repeat :reps [ pu fd 20 pd flag pu bk 20 pd
        rt 360 / :reps]
end
```

■

PROBLEM SET 13.2

Understanding Concepts

1. Find all lines of symmetry of these figures.

   (a) (b) (c) (d) (e)

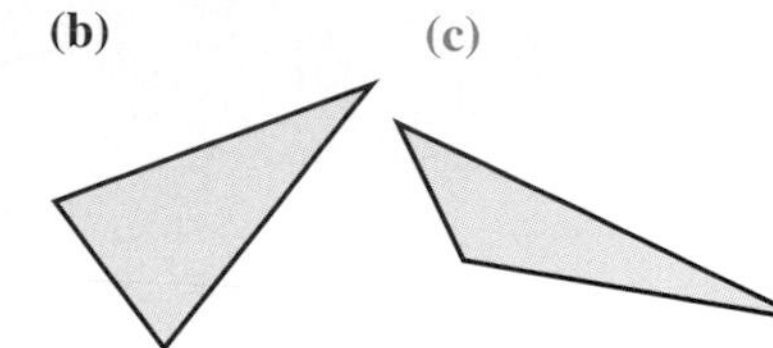

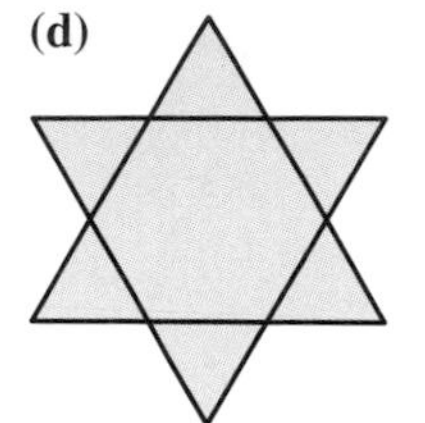

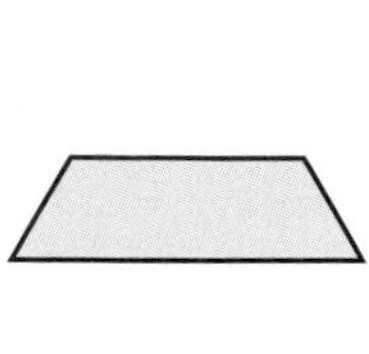

2. Find all lines of symmetry of these figures.

   (a) (b) (c) (d) (e)

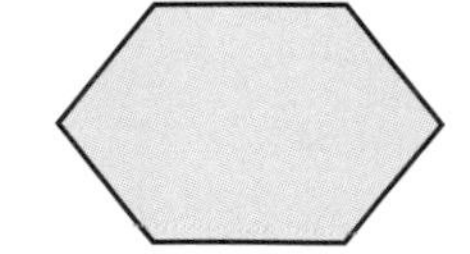

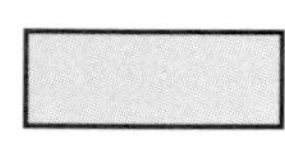

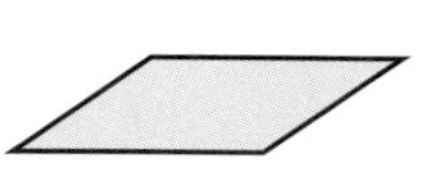

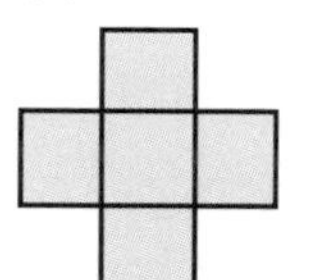

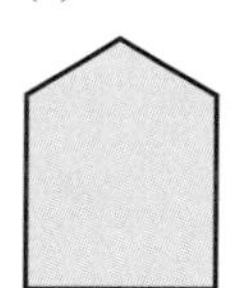

3. Draw polygons with the following symmetries, if possible.
   (a) One line of symmetry but no rotation symmetry.
   (b) Rotation symmetry but no reflection symmetry.
   (c) One line of symmetry and rotation symmetry.

4. Complete each figure to give it reflection symmetry about line m.

   (a) (b)

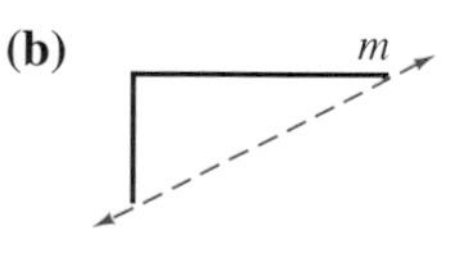

   (c) (d)

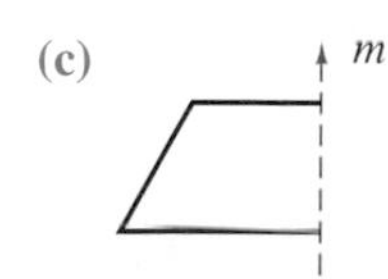

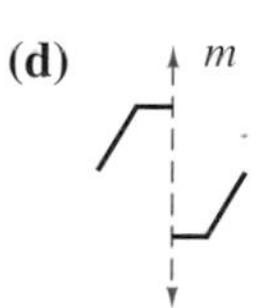

5. Complete each of these figures to give it point symmetry about point O.

   (a) (b)

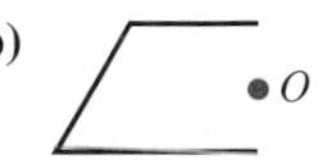

   (c) (d)

6. Copy the figures shown to the right onto graph paper. Then complete each figure to give it reflection symmetry across the dashed line.

(a)

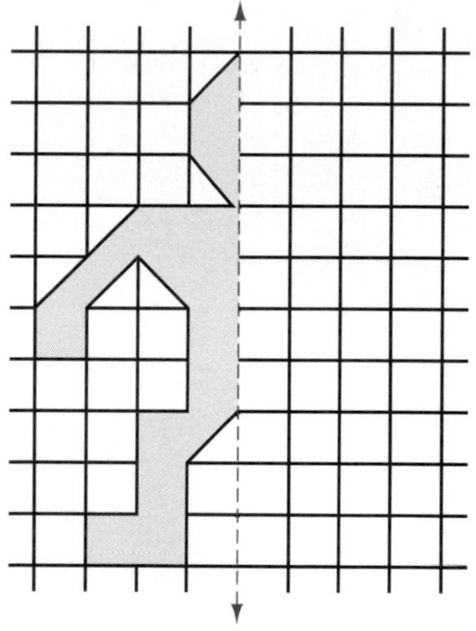

(b)

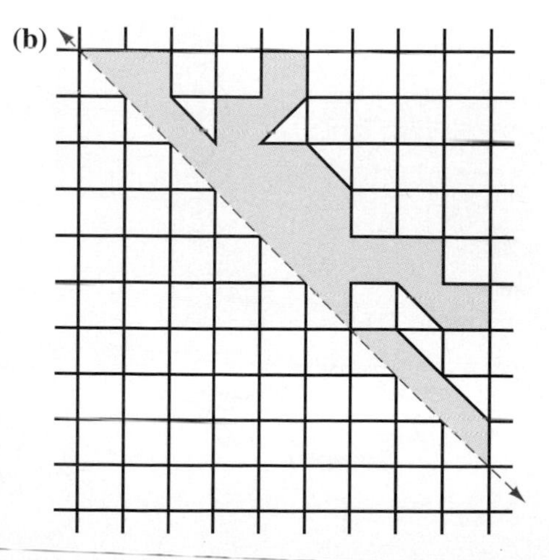

7. A valentine heart is easy to make symmetric: Cut it from a piece of construction paper folded once in half.

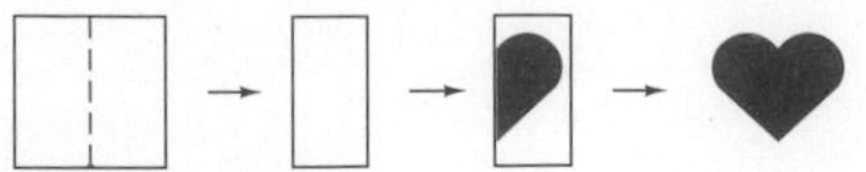

(a) Suppose the paper is folded in half twice. Sketch the shape you obtain when you unfold the cut pattern.

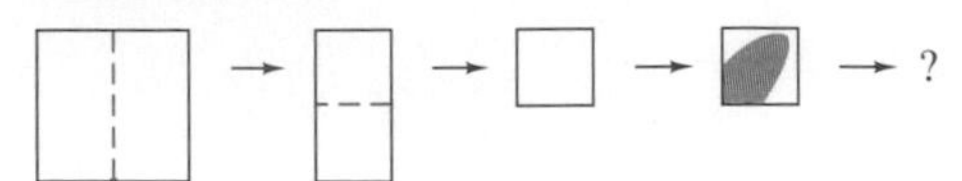

(b) Describe how to make a 6-fold symmetric snowflake by folding and cutting a sheet of paper.

8. Describe all symmetries of each of the following company logos.

(a)

(b)

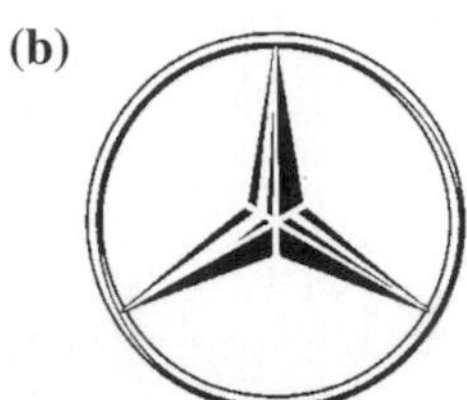

(c)

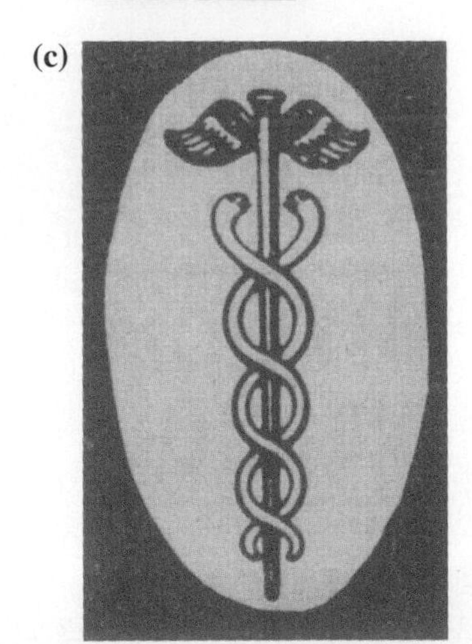

(d)

(e)

9. Describe the symmetries of the wheel covers shown.

(a) (b)

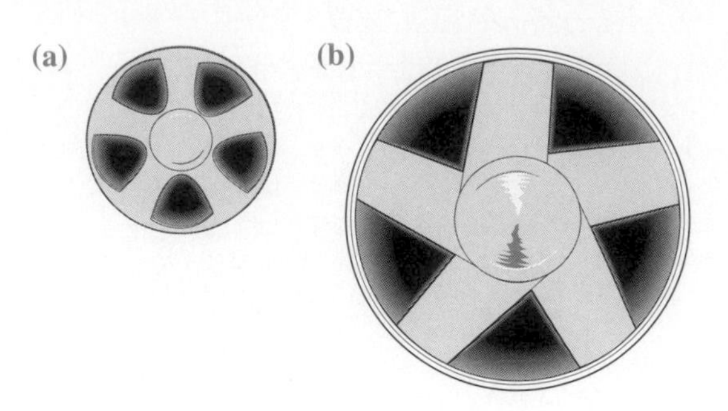

(c) **(d)**

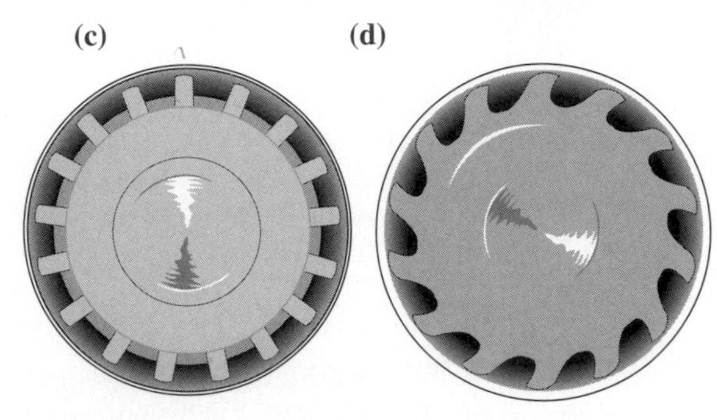

10. Draw figures, if possible, with the following symmetries.

(a) Point symmetry, but no reflection symmetry.

(b) Two or more lines of symmetry but no rotation symmetry. (*Hint:* Why must the figure be of infinite extent?)

11. (a) Complete the figure shown to give it 90° rotation symmetry about point O.

(b) Repeat part (a), but giving the resulting figure 60° symmetry.

12. Identify the regular n-gons in each part that have the given symmetries.

(a) There are exactly three lines of symmetry.

(b) There are exactly four lines of symmetry.

(c) There are exactly 19 lines of symmetry.

(d) The polygon has 10° rotation symmetry.

(e) The polygon has both 6° and 15° rotation symmetry.

13. List all the digits 0, 1, 2, 3, 4, 5, 6, 7, 8, 9 that have:
   (a) vertical reflection symmetry.
   (b) horizontal reflection symmetry.
   (c) vertical and horizontal reflection symmetry.
   (d) point symmetry.
   Write the digits in the most symmetric way you can.
14. Repeat problem 13, but for uppercase capital letters A, B, . . . , Z, written as symmetrically as possible.
15. Repeat problem 13 for the lowercase letters a, b, . . . , z, written as symmetrically as possible.
16. Describe all the symmetries of each border pattern, and classify it by the two-symbol notation used in crystallography.
   (a) . . . A A A A A A . . .
   (b) . . . B B B B B B . . .
   (c) . . . N N N N N N . . .
17. Describe all the symmetries of each border pattern, and give its two-symbol classification used by crystallographers.
   (a) . . . H O H O H O . . .
   (b) . . . M W M W M W . . .
   (c) . . . 9 6 9 6 9 6 . . .

---

18. Describe the symmetries in the following wallpaper patterns. For any rotation symmetry, give the turn center and the size of the turn angle. Give the directions of each reflection and glide symmetry in each pattern.

(a)
A A A A
A A A A
A A A A
A A A A

(b)
E E E E
E E E E
E E E E
E E E E

(c)
N N N N
N N N N
N N N N
N N N N

(d)
Z N Z N
N Z N Z
Z N Z N
N Z N Z

(e)
p q p q
d b d b
p q p q
d b d b

(f)
E E E E
  E E E E
E E E E
  E E E E

Thinking Critically

19. A *palindrome* is a word, phrase, sentence, or numeral that is the same read either forward or backward. Examples are WOW, NOON, and TOOT.
   (a) If a word written in all capital letters has a vertical line of symmetry, why must it be a palindrome?
   (b) NOON does not have a line of symmetry. Find another palindrome with no line of symmetry.
   (c) What symmetry do you see in pod?
20. Describe what symmetries you see in these statements.
   (a) "Sums are not set as a test on Erasmus"
   (b) "Is it odd how asymmetrical
   is 'symmetry'?
   'Symmetry' is asymmetrical.
   How odd it is."
   (c) "Able was I ere I saw Elba." (attributed to Napoleon)
21. Carefully explain why no border pattern has the symbol "$m2$." (*Hint:* If a border pattern has a vertical line of symmetry and 180° rotation symmetry, what other symmetry must it also have?)
22. Prove that a wallpaper pattern cannot have a rotation symmetry of size $360°/n$ for any $n > 6$. (*Hint:* Follow the same ideas used to prove no pattern has turn centers of size $360°/5$, but this time compare how close O'_2 is to O_2. The following figure should be helpful, where O_1 and O_2 are two $360°/n$ turn centers which are at the smallest distance that occurs between any two turn centers.)

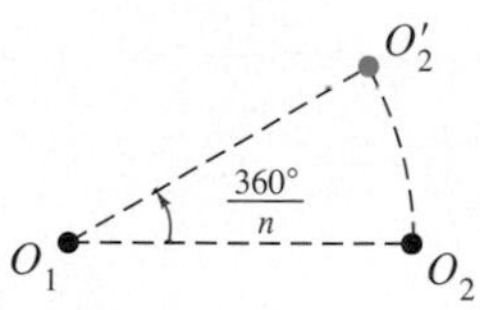

23. Why is it impossible for a wallpaper pattern to have both 90° rotation symmetry about some center O_1 and 60° rotation about some center O_2? (*Hint:* What symmetry would be created by combining a 90° clockwise rotation followed by a 60° counterclockwise rotation?)

24. Classify the following Maori rafter patterns. The Maori, the indigenous people of New Zealand, used principles of symmetry to express their belief system. Disregard the color scheme when you classify the symmetry type of the pattern.

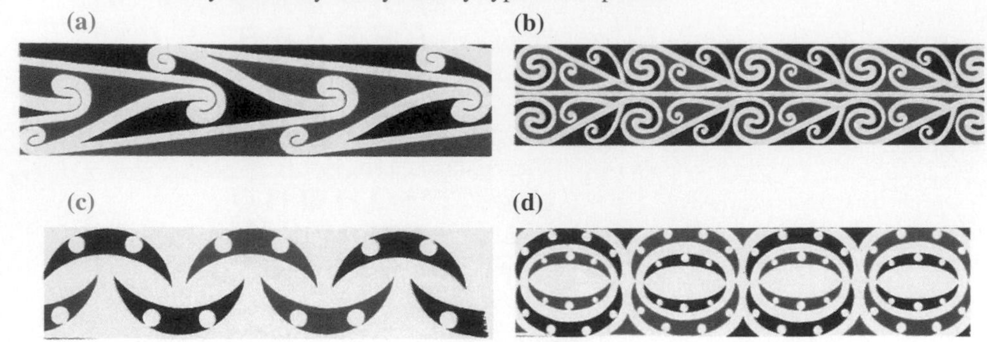

25. Classify the following Inca border patterns.

Thinking Cooperatively

26. In each strip of rectangles shown below, a certain rigid motion applied to the left most rectangle takes the figure to the next rectangle. Apply the same motion, but to the second rectangle, to draw the image of the second rectangle in the third rectangle. Continue to use the same motion to fill in the successive rectangles, and then classify the border pattern which is produced.

(a)

p	p	p			

(b)

p	q				

(c)

p	d				

(d)

p	b				

27. The following border is of type 11. For each of the other six types of border symmetry, complete the pattern to give it the corresponding symmetry type by adding as little as possible.

		p				p				p			

28. These drawings were made by George Pólya for his 1924 paper classifying the 17 pattern types.

(a)

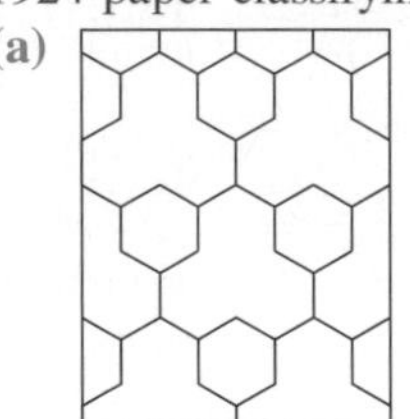

(b)

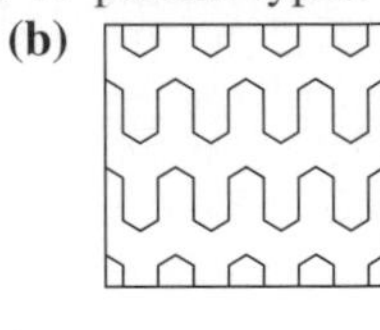

(c)

For each pattern, give
 (i) the number of directions of reflection symmetry.
 (ii) the number of directions of glide symmetry.
 (iii) the sizes of angles of rotation symmetry.

29. A three-dimensional figure has **bilateral symmetry** if there is a plane for which every point P of the figure has a mirror image P' of the figure on the opposite side of the plane. For example, a right prism with an equilateral triangular base has four planes of symmetry. Find the number of planes of symmetry of these space figures:
 (a) a 1 by 2 by 3 rectangular prism.
 (b) a $1 \times 2 \times 2$ rectangular prism.
 (c) a cube.
 (d) a square-based right regular pyramid (the apex is equidistant from the vertices of the base).

Making Connections

30. Describe what symmetries, or lack of symmetry, you see in the following forms and objects.
(a) a pair of scissors
(b) a tee shirt
(c) a dress shirt
(d) a golf club
(e) a tennis racket
(f) a crossword puzzle

31. Describe the symmetry you find in
(a) an addition table.
(b) a multiplication table.
(c) Pascal's triangle.

32. The figures below result from a famous experiment in physics, the Chladni plate. A square metal plate is supported horizontally at its center, sprinkled with fine dry sand, and then vibrated at different frequencies. The sand migrates to the *nodal lines,* where there is no movement of the plate. In the dark regions between the nodal lines, the plate is in vertical vibrational motion.

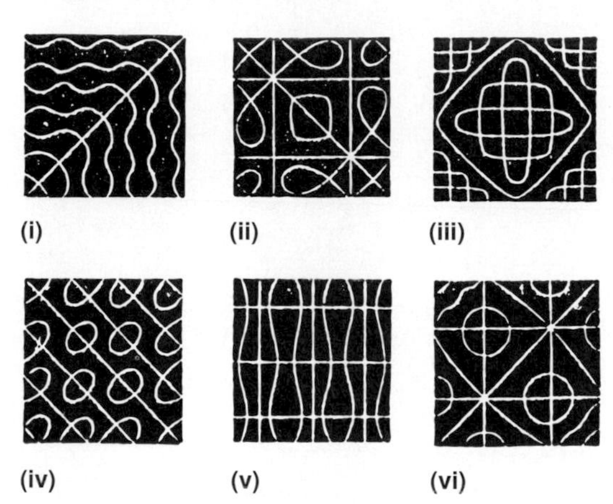

(i) (ii) (iii)

(iv) (v) (vi)

(a) Describe the symmetries of each of the six Chladni plates shown.

(b) The apparatus of the experiment has the symmetries of a square. Do the nodal lines always have the symmetries of a square, or can the vibrational pattern "break" square symmetry? (The Chladni plates shown were published in 1834 in *Of the Connexion of the Physical Science* by Mary Somerville, one of the great women mathematicians of the nineteenth century.)

Communicating

33. Go on a symmetry hunt across campus, looking for striking examples of symmetry in buildings, decorative brickwork, sculptures, gardens, or wherever you may find it. Provide photos or drawings of three or four examples that you find especially interesting. Describe and classify the types of symmetry found in your examples. Include a border pattern and a wallpaper pattern.

Using a Computer
Using Logo

34. Using Logo Example 13.E for the Valentine heart as a guide, write a procedure to draw a figure resembling the one below. Use subprocedures for the symmetric halves of the figure and a superprocedure to call upon them and move the turtle between halves.

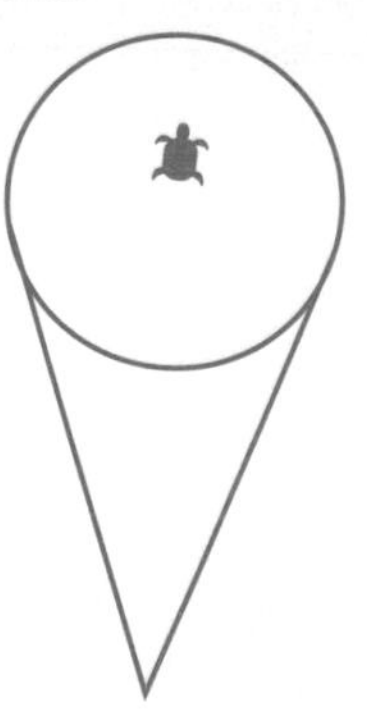

35. Write a procedure using the **side** procedure to create the figures below.

```
to side
    fd 20 rt 60 fd 30 rt 60 fd 20
end
```

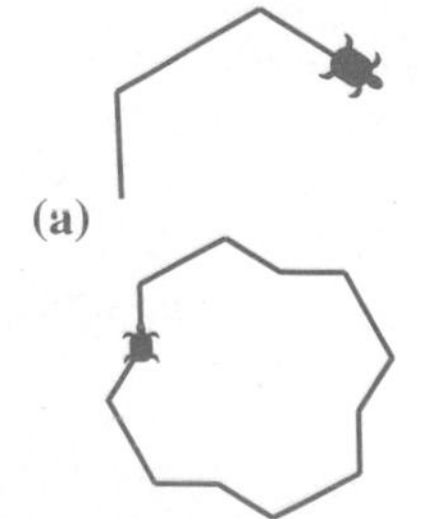

(a)

(b)

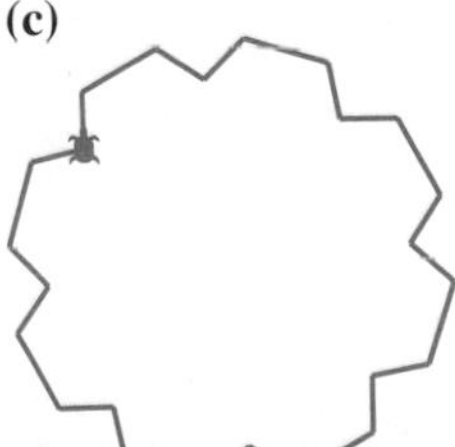

(c)

For Review

36. Some images of a rectangle are shown below. Which images could be the result of a rigid motion? If the transformation is not possibly a rigid motion, give a reason why this is so.

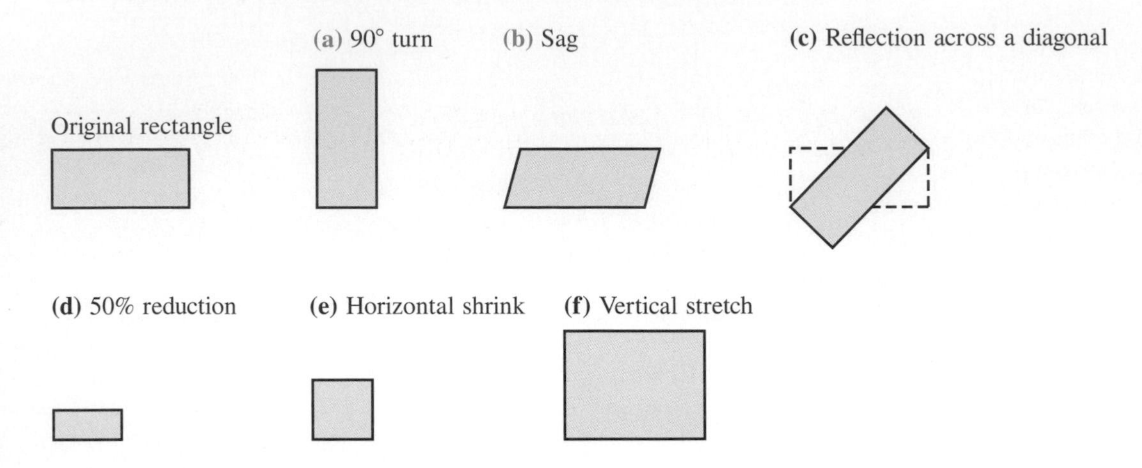

37. Fill in the blanks in the two statements below.

(a) A counterclockwise rotation of size 130° about point O followed by a counterclockwise rotation of 220° also about O is equivalent to a single counterclockwise rotation of _____° about O.

(b) The net outcome of the two rotations described in (a) is equivalent to a clockwise rotation of _____° about O.

38. (a) Triangle ABC is equilateral with sides of length 2 units. Find the images of points A, B, and C under reflection across the three successive lines m_1, m_2, and m_3 shown. Label the respective image points A', B', and C'.

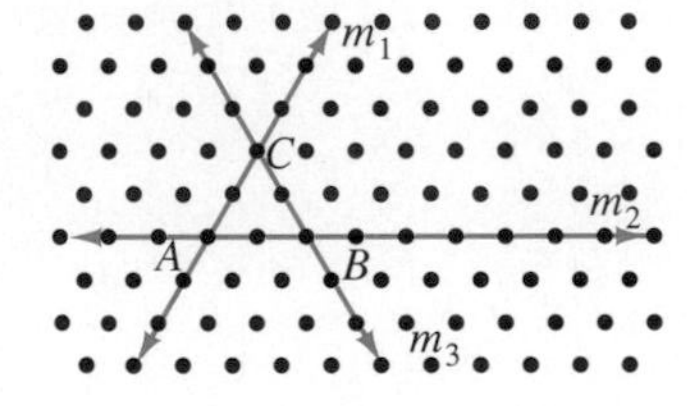

(b) Describe the basic rigid motion that takes $\triangle ABC$ onto $\triangle A'B'C'$.

13.3 Similarity Transformations

In the first section of this chapter the concept of a rigid motion was used to define the congruence of two figures. This gives precision to the informal description of congruence as "the same size and the same shape," and allows us to consider congruence for figures other than just triangles.

In this section the informal description of similarity as "figures of the same shape" is made precise in an analogous way. The key new concept is the **similarity transformation.** Our first step is to explore a special type of similarity transformation of the plane called a **size transformation.**

Size Transformations

Children often like to play shadow games, using their hands and fingers to create shapes that cast funny or scary shadows on a screen. When a child's hands are close to the light source—a light bulb or a slide projector, say—the shadow image can be made quite large. That is, there has been a size change.

A size transformation for us will be much like the shadow game although it will take place in the plane. Our definition requires that we know two things: the **center** of the size transformation and the **scale factor.**

DEFINITION Size Transformation

Let O be a point in the plane and k a positive real number. A **size transformation** with **center** O and **scale factor** k is the geometric transformation that takes each point P, $P \neq O$, of the plane to the point P' on the ray $\overrightarrow{OP}$ for which $OP' = k \cdot OP$, and takes the point O to itself.

Two examples of size transformations are shown in Figure 13.19.

Figure 13.19
Two size transformations

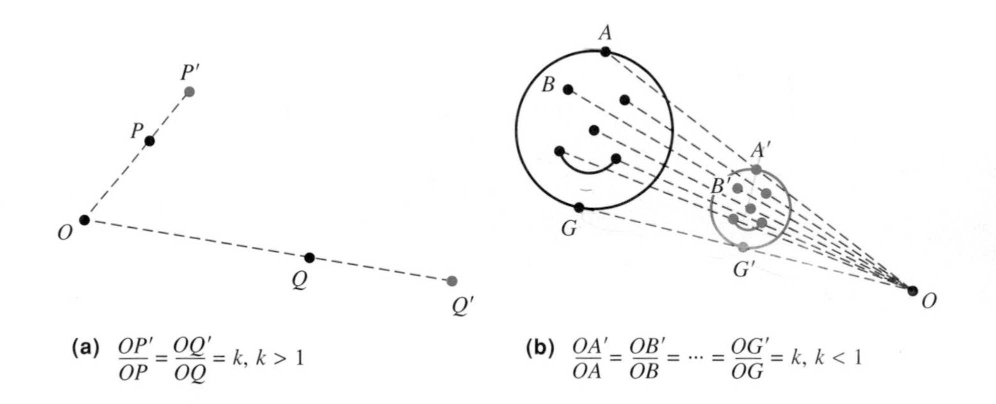

When the scale factor k is larger than 1, the image of a figure is larger than the original and the size transformation is an **expansion.** If $k < 1$ the size transformation is a **contraction.** If $k = 1$ then all points are left unmoved—that is, $P = P'$ for all P—and the size transformation is the identity transformation.

The most important fact about size transformations is contained in the following theorem. Recall that if a point P is taken to the image point P', we say that P is the preimage of P'.

THEOREM Distance Change Under a Size Transformation

Under a size transformation with scale factor k, the distance between any two image points is k times the distance between their preimages. That is, for all points P and Q, $P'Q' = k \cdot PQ$.

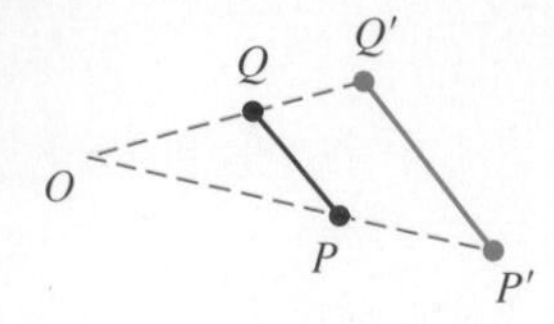

PROOF By the definition of a size transformation, it follows from the SAS similarity property that $\triangle POQ$ is similar to $\triangle P'OQ'$ and the scale factor is k. Thus, the ratio of the length of side $P'Q'$ to the corresponding side PQ is also k. That is, $P'Q'/PQ = k$, and this is equivalent to the desired result. ■

More Properties of Size Transformations

The theorem just proved shows that a size transformation changes *all* distances between pairs of points by the same factor k. On the other hand, many geometric properties of figures other than distances are left unchanged under size transformations. Such properties are said to be **invariant**. Some examples of invariant properties are illustrated in Figure 13.20. We see that the images of a line segment, line, ray, and angle are, respectively, a line segment, line, ray, and angle. Moreover, the line segments, lines, and rays have a parallel image, and the image of an angle is a congruent angle.

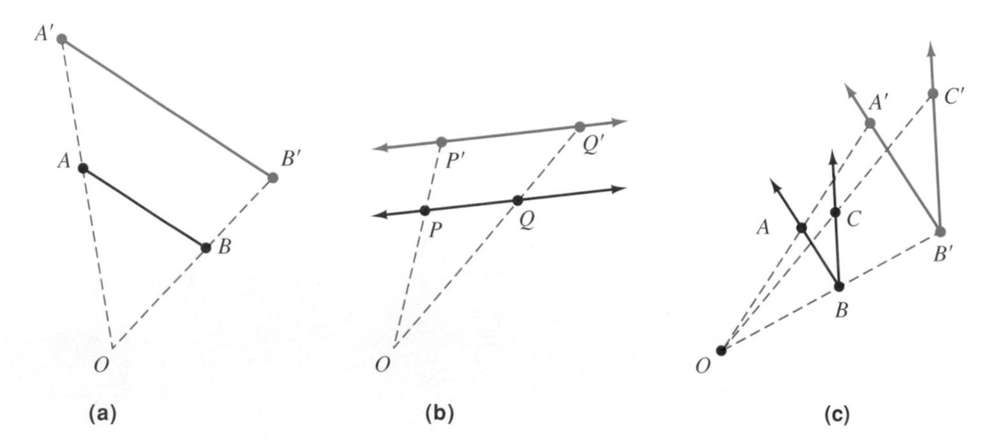

Figure 13.20
Examples of size transformations

We can summarize some of the important geometric properties that are preserved under size transformations.

THEOREM Invariance Properties of Size Transformations

1. *Line segments are preserved.* The image of a line segment $\overline{AB}$ under a size transformation is the line segment $\overline{A'B'}$. Moreover $\overline{A'B'}$ is parallel to $\overline{AB}$.
2. *Lines and rays are preserved.* If $l = \overleftrightarrow{PQ}$ is a line, then its image l' is also a line. In fact, l' is the line $\overleftrightarrow{P'Q'}$ and l' is parallel to l. Similarly, the image of ray $\overrightarrow{PQ}$ is the parallel ray $\overrightarrow{P'Q'}$
3. *Angles and their measure are preserved.* The image of an angle is also an angle. Moreover, the angle and its image are congruent.
4. *Ratios of distances are preserved.* If points A, B, C, and D are taken to A', B', C', and D' by a size transformation, then $\dfrac{A'B'}{C'D'} = \dfrac{AB}{CD}$.

Since a size transformation preserves segments and angles, it also preserves polygons. That is, the image of a polygon will also be a polygon.

CURRENT AFFAIRS

Self-similarity and Fractals

A line is one-dimensional, and a plane is two-dimensional. But what if someone asks for a geometric figure whose dimension is *between* one and two? The question may seem unanswerable. However, beginning in the late nineteenth century, mathematicians have created numerous objects of uncertain dimension. The Koch curve is an example, named after the Swedish mathematician, Helge von Koch, who first described it in 1904. To construct the Koch curve, begin with a line segment. Next replace the middle third by an equilateral triangular bump, resulting in a 4-segment polygonal curve. In the third step, a triangular bump is added to each side of the curve, yielding a curve with 16 sides.

(a) (b) (c)

Repeating the process infinitely often gives the Koch curve.

The Koch curve has many interesting properties, but of special interest is its self-similarity: if fragments of the curve are viewed with microscopes of 100 power, or 1000 power, or any power whatever, the enlargements all appear identical. This unlimited "roughness" of the Koch curve suggests that its dimension is larger than one. In 1975, Benoit B. Mandelbrot, a Fellow at IBM's Thomas J. Watson Research Center, published the first comprehensive study of the geometry of self-similar shapes such as the Koch curve, the Sierpinski carpet, and the Menger sponge. According to Mandelbrot's definition, the Koch curve has dimension D given by the equation $4 = 3^D$. That is, $D = 1.2619\ldots$. Since self-similar objects may have nonintegral dimension, Mandelbrot called such objects fractals, from the same verb *frangere* (to break) that is the source of our word fraction.

Fractals are not simply abstract creations of mathematicians. Indeed, intense ongoing scientific research suggests that many, and possibly most, of the physical and life processes are what can be described as chaotic, and the geometric shapes of chaos are fractals. (For a nontechnical introduction to self-similarity, fractals, and chaos consult James Gleick's *Chaos: Making a New Science*, Penguin Books, New York, 1987.)

EXAMPLE 13.12 **Size Transformations of a Triangle**

Find the image $\triangle ABC$ under each of the size transformations centered at O_1 and O_2, each with a scale factor of $k = 2$. How do the two images compare?

SOLUTION

The images of $\triangle ABC$ under the two size transformations are shown below.

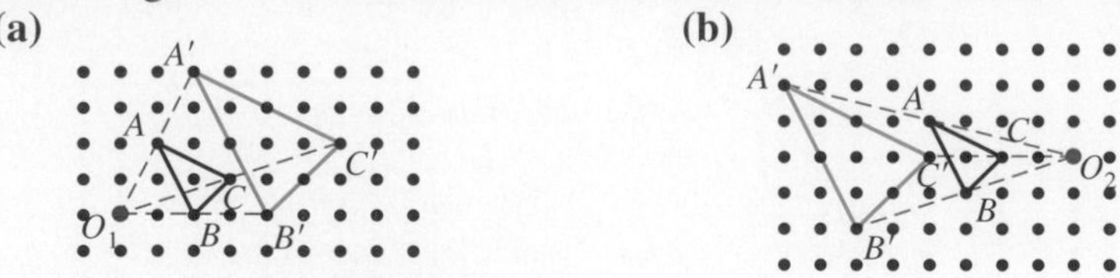

The image triangles in (a) and (b) are congruent. Indeed, one can be moved to the other by a translation. ■

Similarity Transformations

A sequence of size tranformations and rigid motions, performed in succession, will be called a **similarity transformation.**

DEFINITION Similarity Transformation

A transformation is a **similarity transformation** if, and only if, it is a sequence of size changes and rigid motions.

"WE DID THE WHOLE ROOM OVER IN FRACTALS."

An example of a similarity transformation is shown in Figure 13.21. Here a size transformation centered at O takes $\triangle ABC$ to $\triangle A_1B_1C_1$. This is followed by a reflection across line l, taking $\triangle A_1B_1C_1$ to $\triangle A_2B_2C_2$. Finally, a contractive size transformation centered at P takes $\triangle A_2B_2C_2$ to $\triangle A'B'C'$. The combination of the three transformations, taking $\triangle ABC$ to $\triangle A'B'C'$, is the similarity transformation.

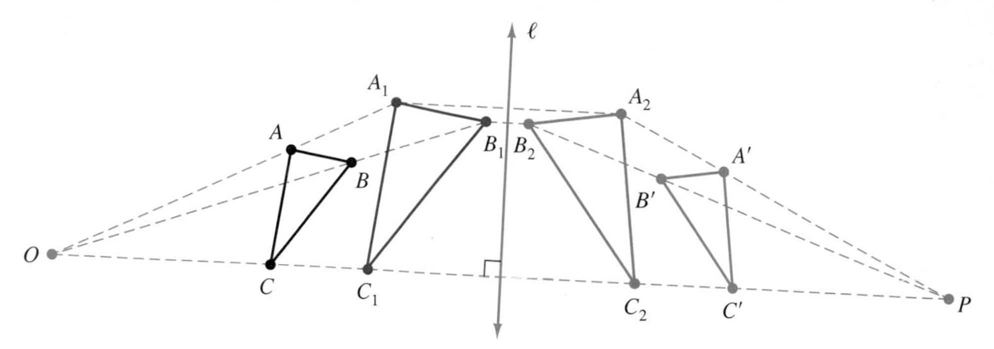

Figure 13.21
A similarity transformation—composed of a size transformation, a reflection, and a size transformation—takes $\triangle ABC$ to $\triangle A'B'C'$.

We can now give a precise definition of similarity of figures.

DEFINITION Similar Figures

Two figures F and G are **similar,** written $F \sim G$, if, and only if, there is similarity transformation that takes one figure onto the other figure.

It can be checked that two triangles F and G are similar under this definition precisely when they are also similar under our earlier definition in Chapter 11 for similar triangles.

EXAMPLE 13.13

Verifying Similarity

Show that the small letter F and the larger letter F are similar by describing a similarity transformation that takes the smaller figure onto the larger one.

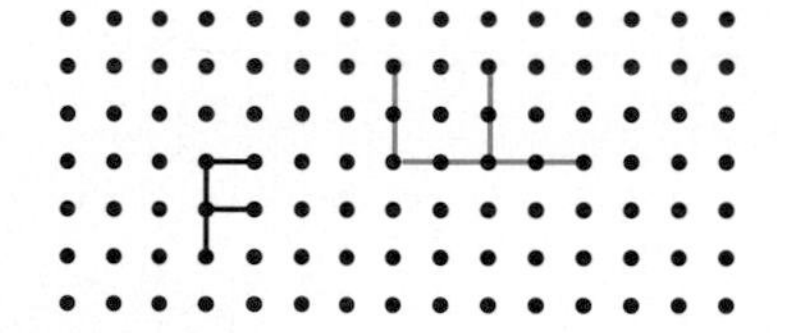

SOLUTION

We need a sequence of size transformations and rigid motions that rotate, stretch, and position the smaller letter onto the larger. This can be done in three steps: (a) make a 90° rotation, (b) do a size transformation with scale factor $k = 2$, and (c) perform a translation. The steps are shown on the next page.

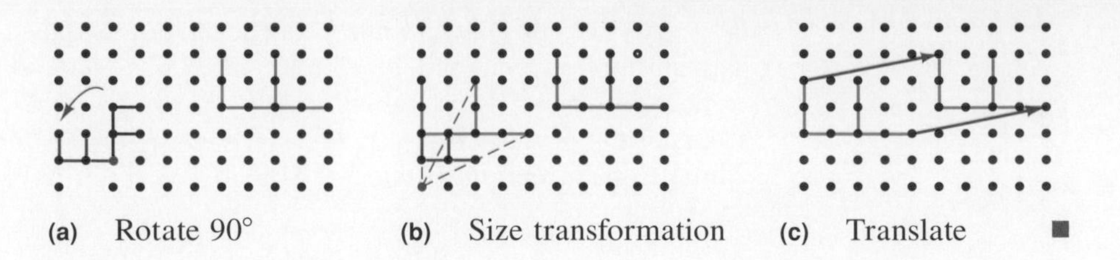

(a) Rotate 90° (b) Size transformation (c) Translate ■

These three steps are not unique. For example, a translation could have been taken first, then a 90° rotation, and finally a size transformation with a properly chosen center. Showing that *some* similarity transformation takes one figure to the other is all that is required.

Similar Figures with Logo

Similar polygons have corresponding angles congruent and corresponding sides proportional. Thus, it is easy to modify a procedure to draw a similar figure: keep all turn angles the same, and multiply all turtle step distances by the same scale factor.

LOGO EXAMPLE 13.G

Drawing Similar Polygons

Create a procedure to draw a sequence of rectangles that are each similar to the rectangle drawn by this procedure.

```
to    rectangle :length :width
          repeat 2 [ fd :length rt 90 fd :width rt 90 ]
end
```

SOLUTION

Four rectangles are drawn by the procedure below, successively enlarged by the scale factor 1.2. The **make** command is described in Appendix C.

```
to    similar.rectangle :length :width
          rt 90
          repeat 4 [ rectangle :length :width
              make "length :length * 1.2
              make "width :width * 1.2
              pu fd :length pd ]
end
```

■

PROBLEM SET 13.3

Understanding Concepts

1. Use dot or graph paper to copy $\triangle ABC$ and points O and P. Then draw the image of $\triangle ABC$ for:
   (a) the size transformation with center O and scale factor 2;
   (b) the size transformation with center P and scale factor $1/2$.

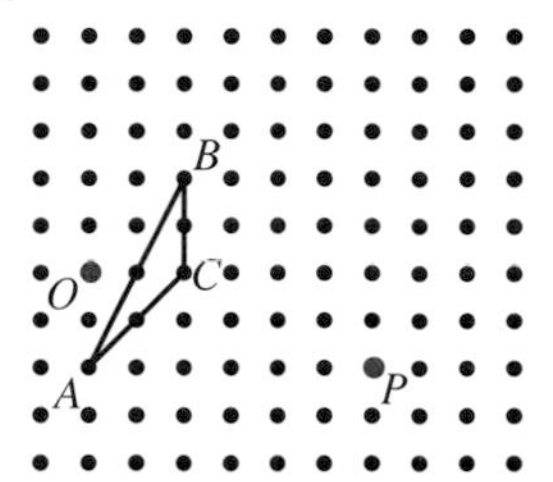

2. Copy $\triangle ABC$ and point O. Use a ruler to draw:
   (a) the figure resulting from the size transformation with center O and scale factor $3/2$;
   (b) the figure resulting from the size transformation with center B and scale factor $3/2$.

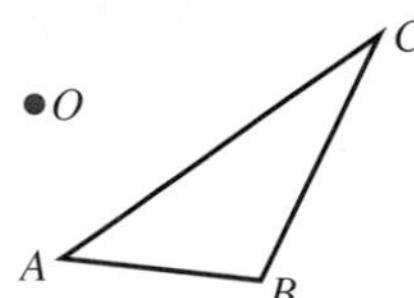

3. Use dot or graph paper to copy square $ABCD$ below. Draw the images of the square for the following transformations. Make a separate drawing for each of (a), (b), and (c).
   (a) The size transformations with center O and each of the scale factors $1/3$, $2/3$, $4/3$;
   (b) the size transformations with center A and each of the scale factors $1/3$, $2/3$, $4/3$;
   (c) the size transformations with center P and each of the scale factors $1/3$, $2/3$, $4/3$.

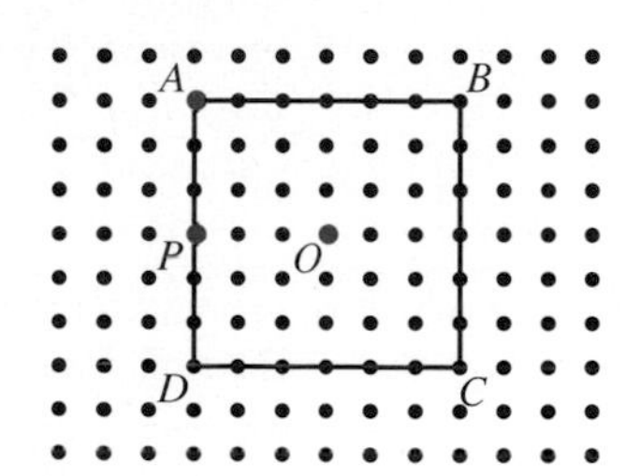

4. Quadrilateral $A'B'C'D'$ is the image of $ABCD$ under a size transformation.

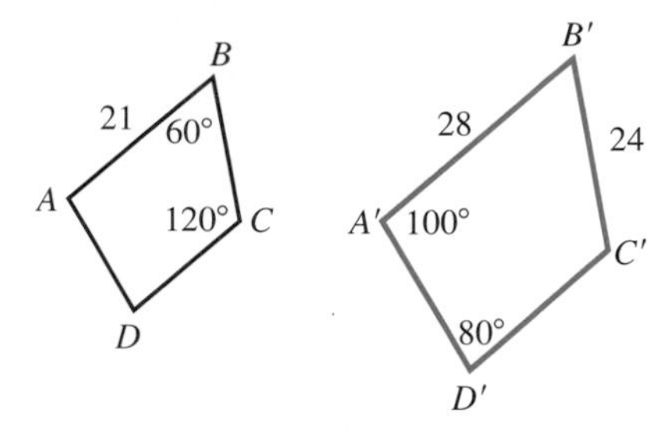

   (a) What is the scale factor of the size transformation?
   (b) Explain how the center of the size transformation can be located.
   (c) Give the measurements:
   $BC =$ ______, $m(\angle D) =$ ______.

5. Find the center and the scale factor of the size transformation that takes the line segment $\overline{PQ}$ onto the line segment $\overline{ST}$.

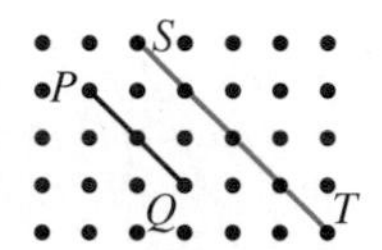

6. In the figure shown below, $\triangle DEF$ is the image of $\triangle ABC$ under the size transformation with center O and scale factor 2.
   (a) Describe the size transformation that takes $\triangle DEF$ onto $\triangle GHI$. Show the center of the transformation on a sketch and give the scale factor.
   (b) Describe the size transformation that takes $\triangle ABC$ onto $\triangle GHI$ by locating the center and giving the scale factor.
   (c) The Pythagorean theorem shows that $AB = \sqrt{5}$, so the perimeter of $\triangle ABC$ is $3 + \sqrt{5}$. Explain how to use the scale factors determined in parts (a) and (b) to obtain the perimeters of $\triangle DEF$ and $\triangle GHI$.
   (d) A size transformation with scale factor 4 takes $\triangle ABC$ onto $\triangle JKL$. What is the perimeter of $\triangle JKL$?
   (e) A size transformation with scale factor $\sqrt{5}$ takes $\triangle DEF$ onto $\triangle XYZ$. What is the perimeter of $\triangle XYZ$?

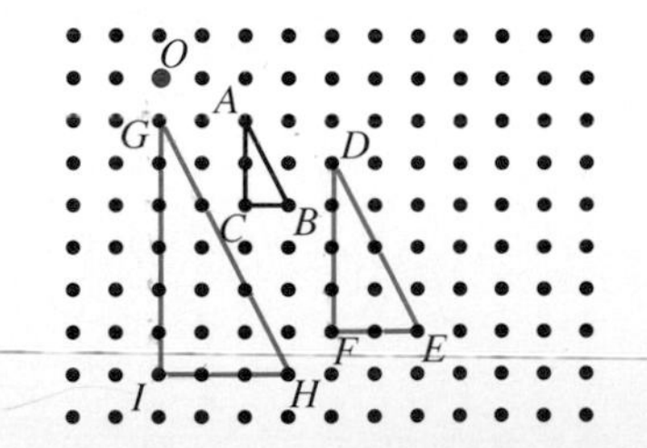

7. A similarity transformation is composed of a half-turn about point H followed by a size transformation centered at point O with scale factor 2. Draw the image of $\triangle ABC$ under the similarity transformation just described. Be sure to show the image, $\triangle A_1 B_1 C_1$, of $\triangle ABC$ under just the half-turn.

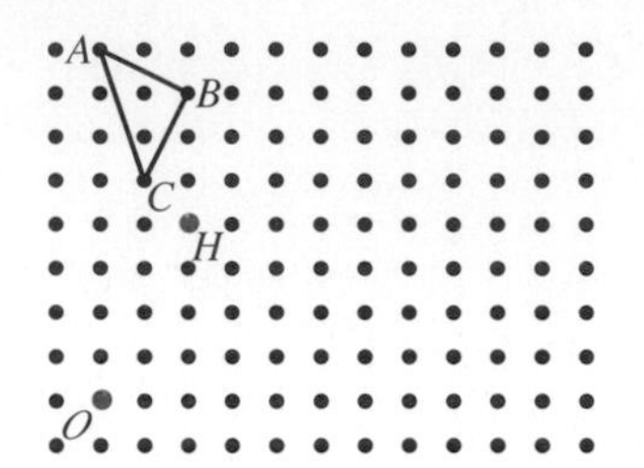

8. Sketch the image of $\triangle JKL$ under the similarity transformation composed of a size transformation centered at point P with scale factor $2/3$ followed by a reflection across line m.

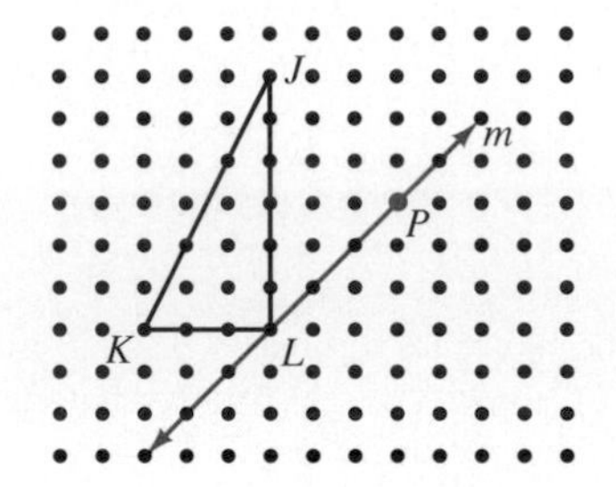

9. Describe a similarity transformation that takes quadrilateral $ABCD$ onto quadrilateral $A'B'C'D'$ as shown. Sketch the intermediate images of the size transformations and rigid motions that compose the similarity transformation.

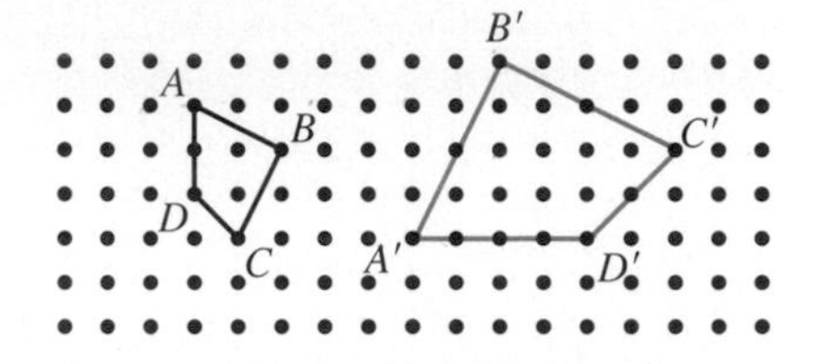

10. Copy $\triangle ABC$ and $\triangle XYZ$ onto triangular dot paper. Describe a similarity motion that takes the smaller triangle onto the larger one.

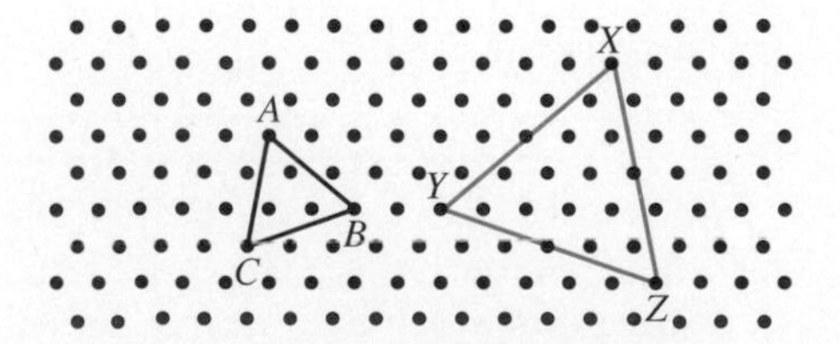

11. Quadrilateral $ABCD$ has consecutive sides of length 36, 18, 24, and 60, and quadrilateral $WXYZ$ has consecutive sides of length 12, 6, 8, and 20. Are the two quadrilaterals necessarily similar? Explain how you arrived at your answer.

12. Show that a circle centered at P of radius r is similar to a circle centered at Q of radius R.

13. Use a ruler and protractor to draw a polygon $A'B'C'D'E'F'$ that is similar to the given hexagon $ABCDEF$, where $A'B'$ is given as shown. Explain what properties of similar figures you rely on to make your drawing.

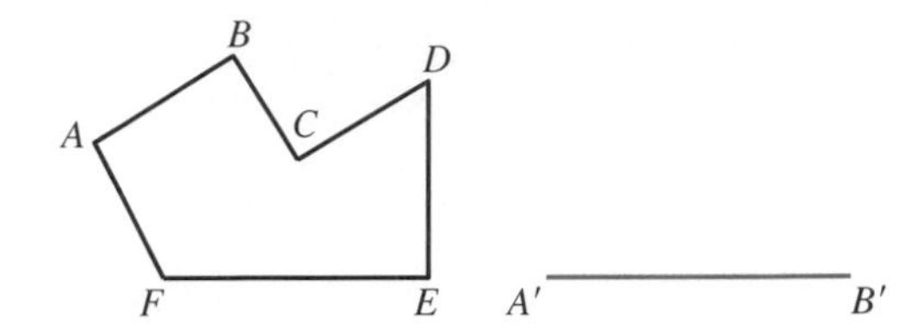

Thinking Critically

14. Let $\triangle ABC$ be a given acute triangle. Construct the square that has two vertices on side $\overline{AB}$ and one vertex on each of the other two sides of the triangle. (*Hint*: Construct several squares, such as $SQRE$, that touches all but side $\overline{BC}$, and then consider a size transformation centered at point A.)

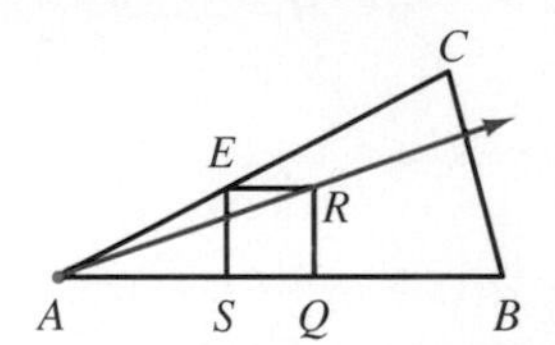

15. Construct a regular hexagon $ABCDEF$. Then give instructions how to construct a square $WXYZ$ inscribed in the hexagon, where the side $\overline{WX}$ of the square is parallel to side $\overline{BC}$ of the hexagon, as shown.

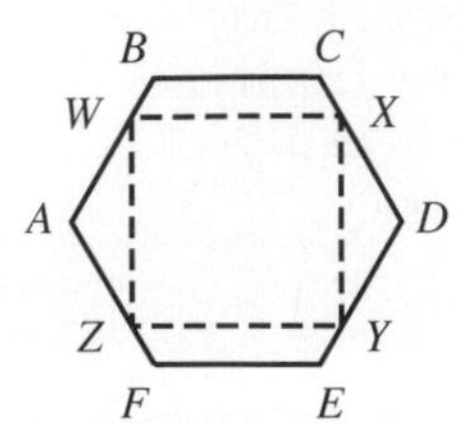

(*Hint*: Draw a certain smaller square first, and then expand by a size transformation.)

16. A **cycloid** is the curve traced by a point of the rim of a circle that rolls on a line l. For example, the circle of diameter 1 has traced the cycloid which passes through A, B, and C (but misses point P).

What is the diameter of the circle that generates a cycloid containing the points A and P? (*Hint:* Draw the ray $\overrightarrow{AP}$ and stretch the cycloid by a size transformation centered at A. How can the scale factor be expressed in terms of the distances AP and AQ?)

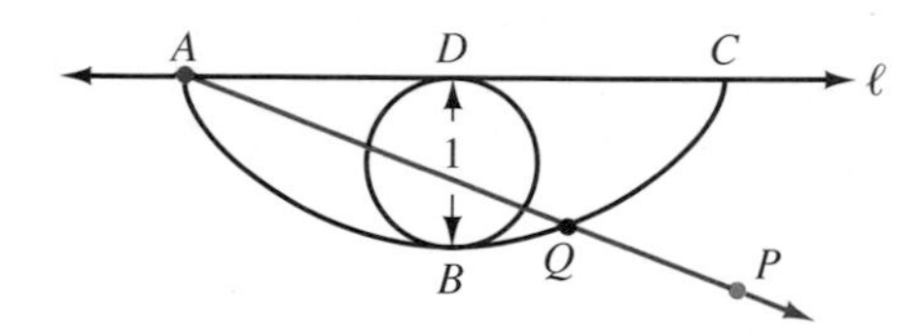

(This problem is of historic interest, since Johann Bernoulli (1667–1748) and others in 1697 showed that a bead falling along a frictionless cycloidal wire from A to P will require less time than if the bead fell along *any* other curve that joins A to P. In particular, if two beads are released simultaneously from A, with one falling along the cycloid and the other falling along the straight line segment $\overline{AP}$, then the bead following the cycloidal path arrives at P ahead of the bead following the straight path between A and P.)

17. A size transformation centered at some point O of line l has taken point P to P'. Explain how to draw (a) the center O of the transformation, and (b) the image Q' of the point Q, where Q lies on l.

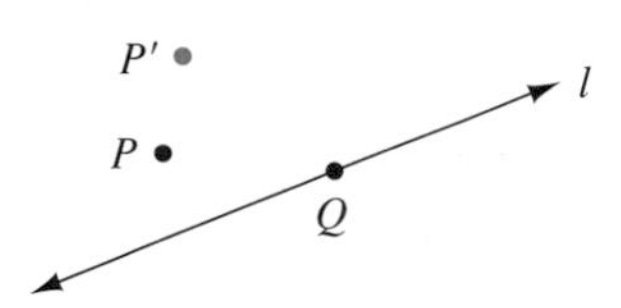

What properties of size transformations are used to verify your construction methods?

18. A size transformation takes P to P' and Q to Q', where the four points P, P', Q, Q', are collinear as shown here. Explain (a) how to draw the image R' of point R, and (b) how to locate the center O of the size transformation. (*Hint:* What line through P' must contain R'?)

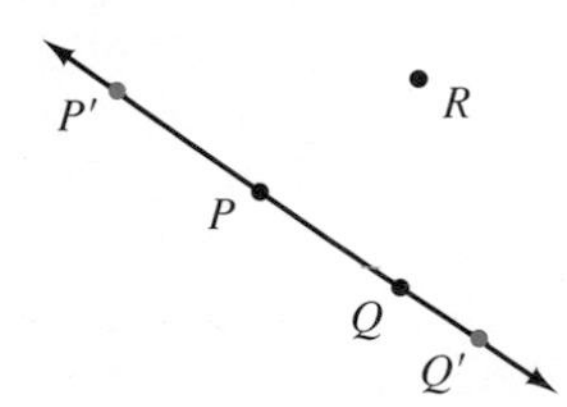

19. Consider the collinear equally spaced points A, B, C, D, E, F, G. Find the center and scale factor of:
(a) the size transformation taking $\overline{AB}$ to $\overline{AF}$.
(b) the size transformation taking $\overline{BC}$ to $\overline{CE}$.
(c) the size transformation taking $\overline{AG}$ to $\overline{BE}$.
(d) the size transformation taking $\overline{AD}$ to $\overline{EF}$.

A B C D E F G

20. Is it always possible to decompose any similarity transformation into a size transformation followed by a rigid motion? Explain how you get your answer.

Making Connections

21. The small figure of a fish has been overlaid by a grid of small squares. An enlargement can be made by drawing the figure on paper overlaid by a grid of large squares. What properties of similarity transformations show why this procedure is effective?

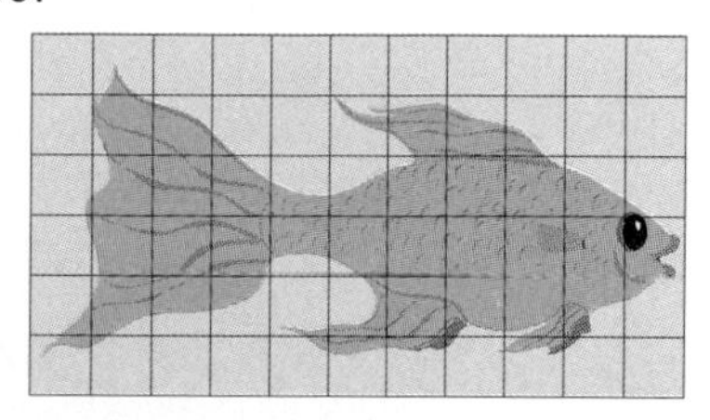

22. A **pantograph** is a mechanical device used to draw enlargements. A simple version can be constructed from cardboard strips that are hinged with brass fasteners.

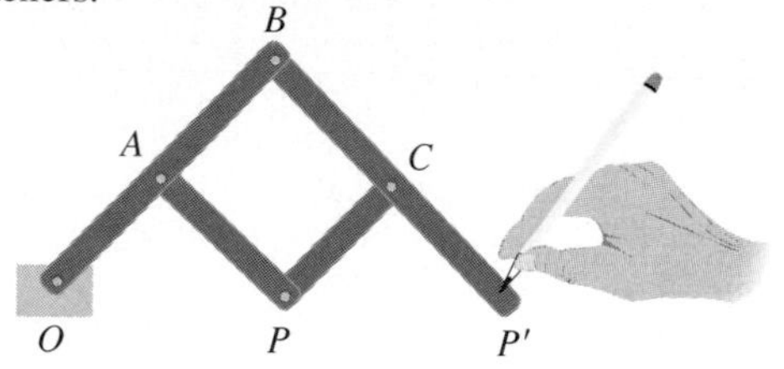

The pivot point at O is held fixed as P is moved over the figure. The pencil point at point P' traces out an enlargement.
(a) Explain why the pantograph mechanically gives a size transformation.
(b) What is the scale factor of the size transformation? Assume that all of the adjacent points along a strip are the same distance apart.

23. Suppose a knot is placed at a point along a rubber band. One end of the band is held by a pin at point O, and a pencil is placed in the loop of the band opposite O. An enlarged figure is drawn by moving the pencil so as to keep the knot directly on top of points P in the given figure. The pencil draws points P' of a new figure.
(a) What is important about the way a rubber band stretches that guarantees that the procedure gives a size transformation?

(b) If the knot is at the midpoint of the band, what is the scale factor?

(c) Where along the rubber band should a knot be tied to produce an enlargement that is three times the size of the original figure?

24. Make a list of devices in common use to produce similarity transformations. A photocopying machine is an example since many allow you to make copies that vary in size from, say, 64 percent to 142 percent of the original. That is, the scale factors vary between 0.64 and 1.42. For each example you give, try to give an estimate of the scale factors.

Using a Computer
Using an Automatic Drawer

25. **The Golden Rectangle.** Use your drawer to carry out these steps:

Draw two points, A and B.
Construct a square $ABCD$ erected on side $\overline{AB}$.
Draw the rays $\overrightarrow{BC}$ and $\overrightarrow{AD}$.
Construct the midpoint M of $\overline{BC}$.
Construct the circle at M through D.
Construct G, the intersection of the circle with ray $\overrightarrow{BC}$.
Construct the line perpendicular to $\overrightarrow{BC}$ at G.
Construct H, the intersection of the line with ray $\overrightarrow{AD}$.
Construct the rectangle $ABGH$.
Hide the rays, circle, midpoint, and vertical line, leaving just the rectangle and square.

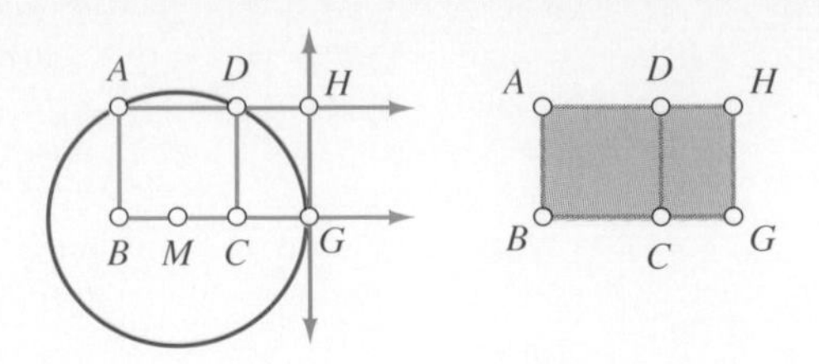

(a) Measure AB, BG, CG, and GH, and show that $AB/CG = BG/GH$.

(b) Explain why rectangles $ABGH$ and $CGHD$ are similar.

Using Logo

26. Write a Logo procedure to draw an irregular pentagon. Next write a Logo procedure to draw a line of four pentagons of varying size that are similar to the original.

For Review

27. For each whole number $n = 0, 1, 2, \ldots$, draw (if possible) a hexagon with exactly n lines of symmetry.

28. Draw a border pattern with the following symmetry properties.

(a) A vertical and a horizontal line of symmetry.

(b) No lines of symmetry, but centers of point (half-turn) symmetry.

29. A rigid motion is composed of a half-turn about point P, a reflection across line l, and a translation to the right by 6 feet. If this rigid motion takes point A to point B, describe the inverse rigid motion that returns point B to point A.

EPILOGUE The Dynamical View of Geometry

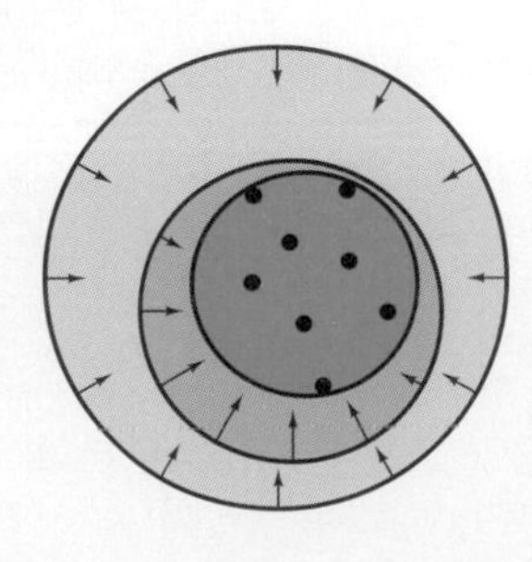

Geometry in Euclid's time presented a static view of shape, as if only still pictures of figures were to be seen in the mind's eye. The concept of a geometric transformation, introduced in the latter part of the nineteenth century, has provided a dynamic view of geometry. Figures are allowed, and even invited, to move and perhaps even change size. The mind's eye sees an animated world of shapes in action.

The dynamic viewpoint is a useful tool for problem solving and discovery. Here's an example: *Show that any finite set of points in the plane is contained inside (or on) some circle of smallest radius.* To see why, first imagine surrounding the points with a very large circle. Now let the circle shrink around the set of points as tightly as possible, allowing the centers of the circles to move as well.

As a second example, imagine that a small circle within a triangle is allowed to grow until it just touches all three sides of the triangle. This makes it clear that every triangle has an inscribed circle.

Recent developments in computer graphics, both hardware and software, provide new opportunities to explore how the properties of a figure change as the figure is altered. For example, "draw" any triangle and its three medians with automatic drawing software. Next "drag" (that is, relocate) a vertex to a new position, leaving the other vertices stationary. The medians of the triangle undergo corresponding movement, but they will be seen to remain concurrent at the new centroid.

CLASSIC CONUNDRUM Inverting the Tetractys

Ten pennies can be tightly packed to form an equilateral triangle. This pattern was called the "tetractys" by the ancient Pythagoreans, and is still familiar today in ten pin bowling alleys. The problem is to turn the tetractys upside down by sliding one penny at a time to a new position in which it touches two other pennies. It's easy to do moving four pennies, but you should be able to find a three move solution.

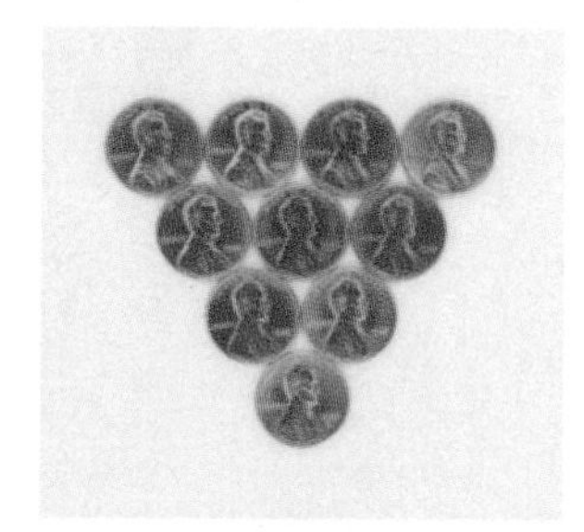

A three coin triangle can be inverted in one move, a six coin triangle in two moves, and, as noted above, a ten coin triangle takes three moves. It would be reasonable to guess that a 15 coin triangle can be inverted in four moves, but it actually takes five moves. Try it!

CHAPTER 13 SUMMARY

Key Concepts

1. **Rigid Motions**
   (a) A **transformation of the plane** is a one-to-one correspondence of the points of the plane. If point P corresponds to point P', P' is the **image** of P and P is the **preimage** of P'.
   (b) A **rigid motion,** also called an **isometry,** is a transformation of the plane that preserves distance: $PQ = P'Q'$ for all points P, Q and corresponding image points P', Q'.
   (c) There are four **basic rigid motions: translations, rotations, reflections,** and **glide reflections.** These are also called **slides, turns, flips,** and **glides.**
   (d) The **identity transformation** is the rigid motion for which each point corresponds to itself: $P = P'$ for all points P.
   (e) Two transformations are **equivalent** if both transformations take each point P to the same image point P'.
   (f) A sequence of two reflections across parallel lines is equivalent to a translation. A sequence of two reflections across intersecting lines is equivalent to a rotation.

(g) A sequence of three reflections across parallel or concurrent lines is equivalent to a reflection. Otherwise a sequence of three reflections is equivalent to a glide reflection.
(h) Every rigid motion is equivalent to one of the basic motions: a translation, a rotation, a reflection, or a glide reflection.
(i) Two figures are **congruent** if, and only if, one figure is the image of the other under a rigid motion.

2. Patterns and Symmetries

(a) A **symmetry** of a figure is a rigid motion which takes every point of the figure to an image point that is also a point of the figure.
(b) A figure has **reflection symmetry** if reflection across some line—called a **line of symmetry**—takes a figure onto itself.
(c) A figure has **rotation symmetry** if rotation about some center O takes the figure to itself.
(d) A figure has **point symmetry** about point O if it has half-turn (180°) symmetry about O.
(e) A **pattern** is a plane figure with translation symmetries, including some translation of smallest positive distance.
(f) A pattern with translations in one direction only is a **border pattern.** There are seven symmetry types of border patterns.
(g) A pattern with translations in more than one direction is a **wallpaper pattern.** There are 17 symmetry types of wallpaper patterns.
(h) The **crystallographic restriction:** Any rotation symmetry of a wallpaper has size 60°, 90°, 120°, or 180°.

3. Similarity Transformations

(a) A **size transformation** with **center** O **scale factor** k takes point P, $P \neq O$ to the point P' on ray $\overrightarrow{OP}$ for which $OP' = k \cdot OP$, and leaves O fixed.
(b) A size transformation multiplies all distances by the scale factor k, so $P'Q' = k \cdot PQ$ for all P, Q.
(c) A size transformation takes parallel lines to parallel lines, angles to congruent angles, and line segments to parallel line segments. Ratios of distances are invariant under size transformations: $AB/CD = A'B'/C'D'$ for all points A, B, C, D.
(d) A **similarity transformation** is a sequence of size transformations and rigid motions.
(e) Two figures are **similar** if, and only if, there is a similarity transformation that takes one figure onto the other.

Vocabulary and Notation

Section 13.1

Transformation of the plane
Preimage, image
Rigid motion, or isometry
Basic rigid motion: translation (slide), rotation (turn), reflection (flip), glide reflection (glide)
Slide arrow (or translation vector)
Center of rotation
Turn angle, turn arrow
Line of reflection
Orientation reversing/preserving transformation
Equivalent transformations
Identity transformation
Congruent figures, F ≅ G

Section 13.2

Symmetry of a figure
Reflection symmetry (line symmetry, bilateral symmetry)
Line of symmetry
Rotation symmetry (turn symmetry)
Center of rotation
Circular symmetry
Point symmetry
Pattern
Border pattern
Wallpaper pattern
Crystallographic restriction

Section 13.3

Size transformation
Center, O
Scale factor k ($0 < k < 1$: contraction, $k > 1$: expansion)
Invariant property
Similarity transformation
Similar figures

CHAPTER REVIEW EXERCISES

Section 13.1

1. Draw the image of $ABCDE$ under the translation that takes A to A'.

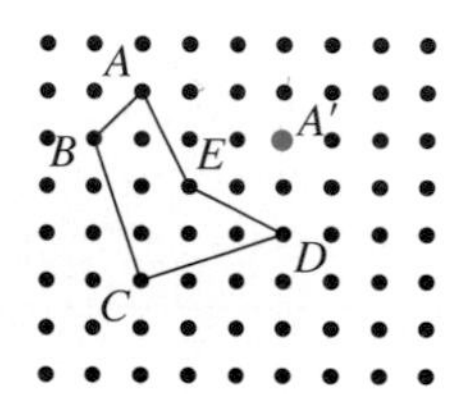

2. Determine the center and turn angle of the rotation that takes A to A' and B to B'. Use a protractor, Mira, ruler, or whatever drawing tools you wish.

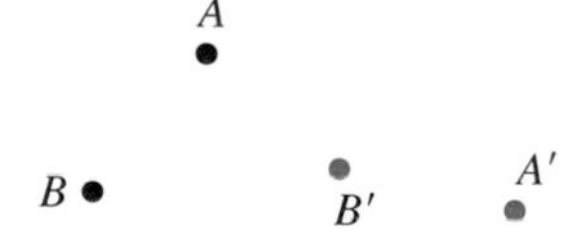

3. Describe the basic rigid motion that takes A, B, C to A', B', C'. Use any drawing tools you wish.

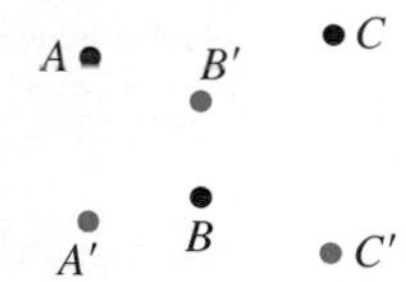

4. A glide reflection has horizontal line l as its line of reflection and translates 4 inches to the right. Draw three lines of reflection—m_1, m_2, and m_3—so that successive reflections across m_1, m_2, and m_3 result in a motion equivalent to the glide reflection.

Section 13.2

5. The geometric forms shown are from African art. How many lines of symmetry does each figure have?

(a) (b)

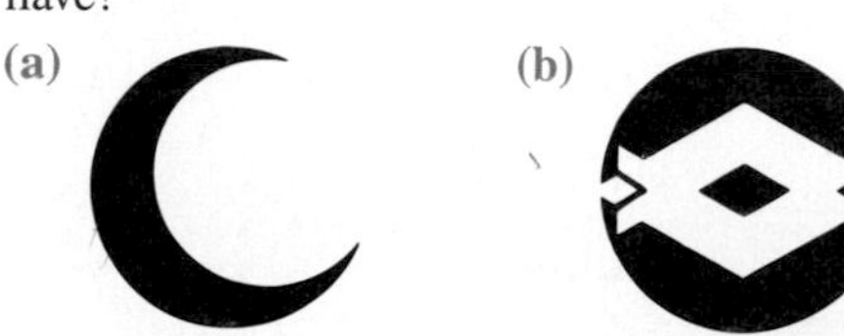

(c) (d)

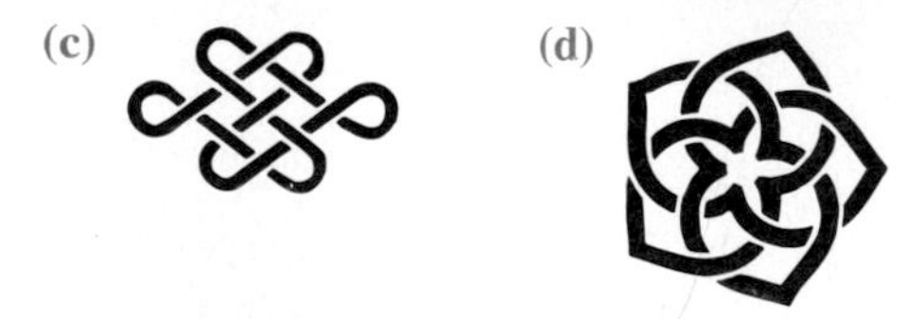

(e) (f)

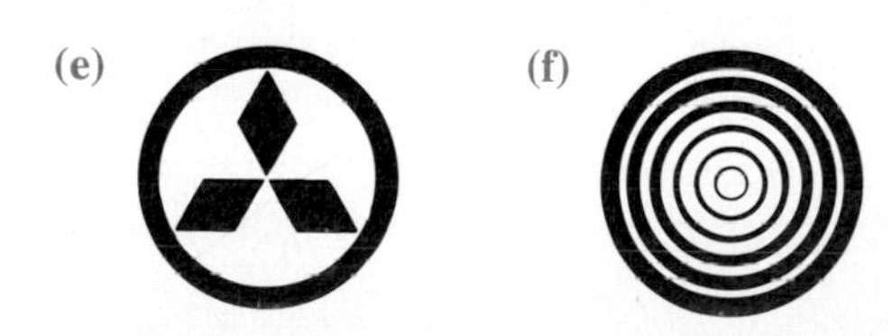

6. For each of the figures shown in problem 5, give all of the angles of rotation symmetry.

7. Describe the symmetries of each of these border patterns.

(a)

FRENCH RENAISSANCE ORNAMENT FROM CASKET

(b)

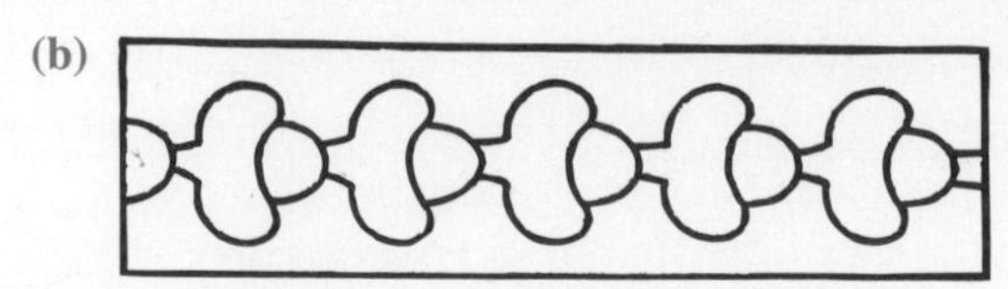

STAINED GLASS, CATHEDRAL OF BOURGES

Section 13.3

8. Find the center of a size transformation of scale factor $3/2$ that takes A to A', and then draw the image $A'B'C'D'$ of the square $ABCD$. Use a ruler, compass, or other drawing tools.

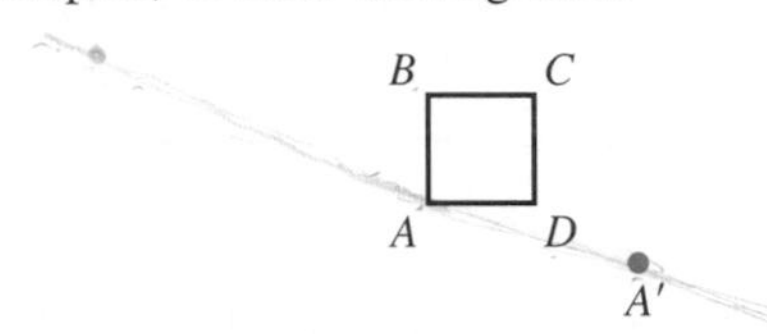

9. A size transformation with scale factor k takes figure F onto figure F_1. A second size transformation, this one with scale factor $1/k$, takes F_1 onto F'. What kind of a transformation will take F onto F'? Explain why.

10. Describe a similarity transformation that takes square $ABCD$ onto the square $JKLM$, where J, K, L, and M are the respective midpoints of the sides $\overline{AB}$, $\overline{BC}$, $\overline{CD}$, and $\overline{AD}$ of the square $ABCD$.

CHAPTER TEST

1. A translation takes points A, B, C to A', B', C'. Show the location of C' and B.

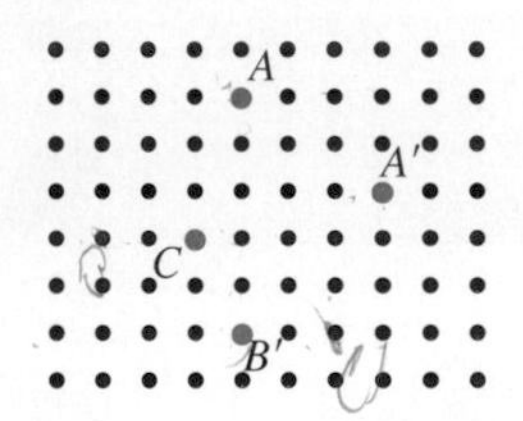

2. Copy the rectangle $ABCD$ and points A' and B' onto squared paper.
   (a) Show that A' and B' are the image of A and B under a rotation. Give the center of rotation and the size of the rotation angle.
   (b) Draw the image rectangle $A'B'C'D'$.

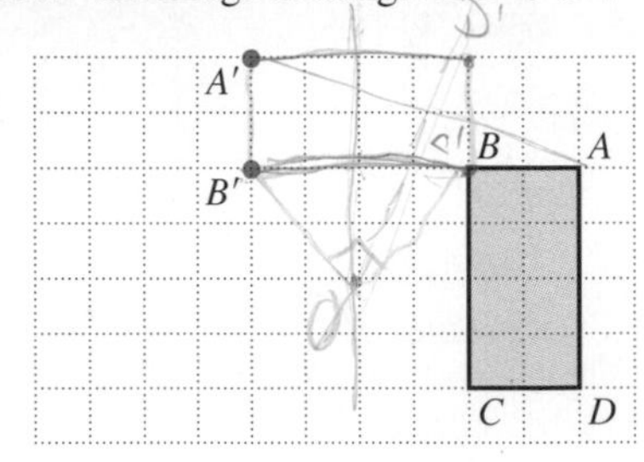

3. Trace the drawing of $\angle A$. Suppose A' is the image of A under a reflection. Explain how to draw the image of $\angle A$ under the reflection.

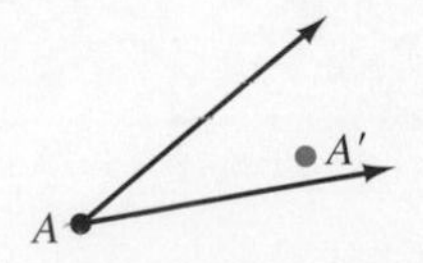

4. Find the line of reflection and slide arrow of the glide reflection that takes rectangle $ABCD$ onto $A'B'C'D'$.

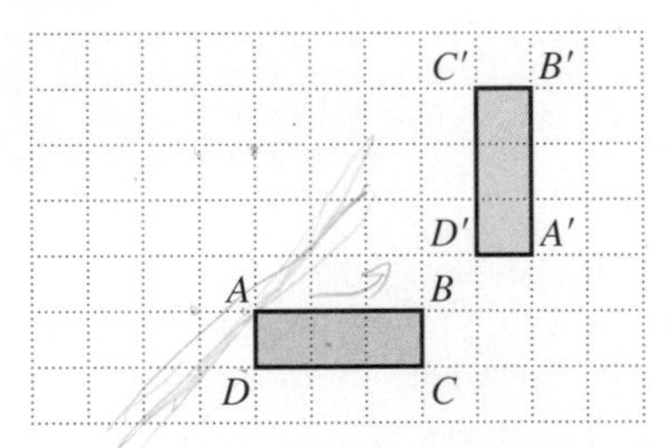

5. Draw two parallel lines, m_1 and m_2, so that the sequence of reflections across m_1 and m_2 will map point P to point P'.

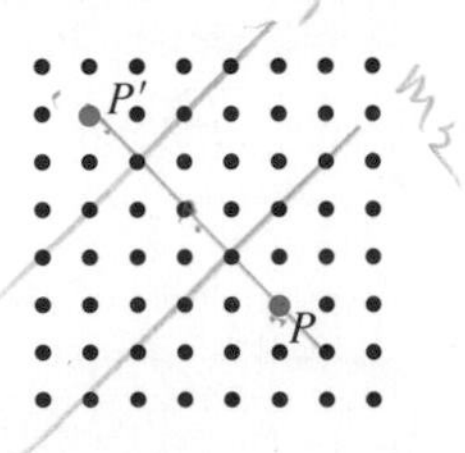

6. Draw two lines, l_1 and l_2, so that a sequence of reflections across l_1 and l_2 will rotate point Q to point Q' about the turn center O.

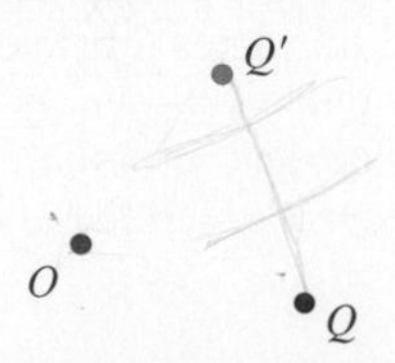

7. Three lines—m_1, m_2, and m_3 are shown in each part. What type of rigid motion is equivalent to a sequence of reflections across m_1, m_2, and m_3?

(a)

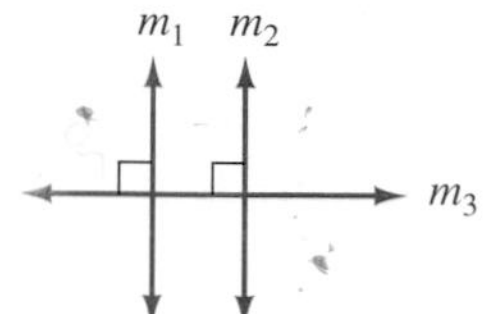

(b) Parallel lines

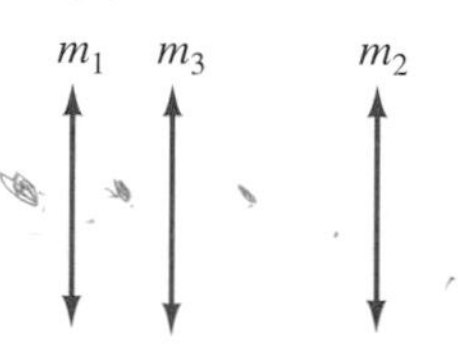

(c)

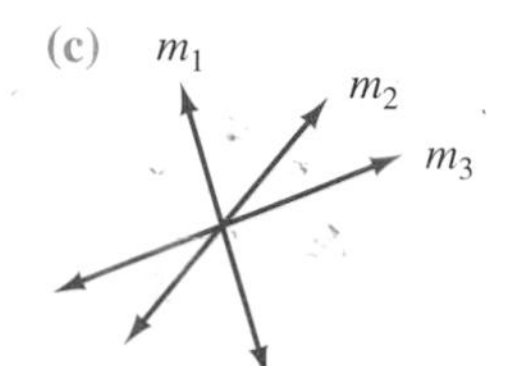

(d)

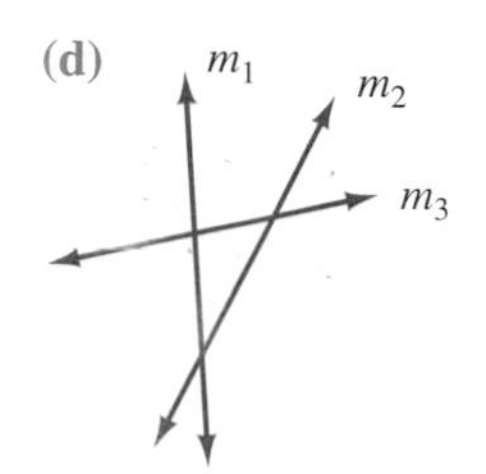

8. In the Escher tiling shown, what type of rigid motion

(a) takes figure A onto figure B?

(b) takes figure B onto figure C?

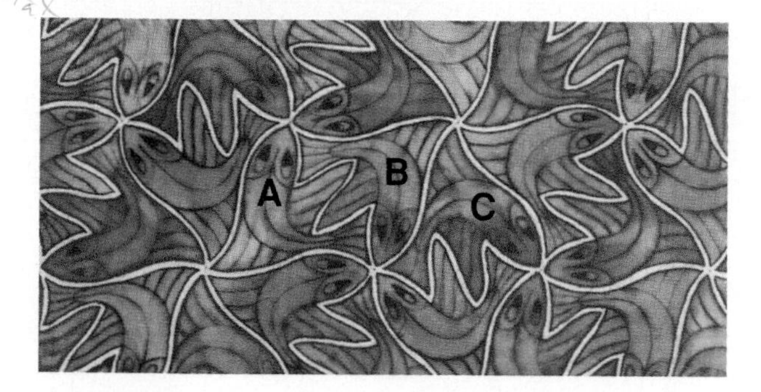

9. Draw all of the lines of symmetry for these figures.

(a)

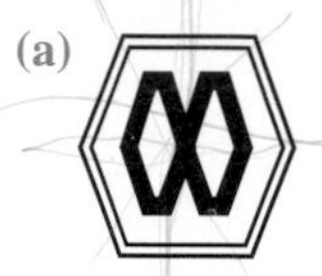

(b)

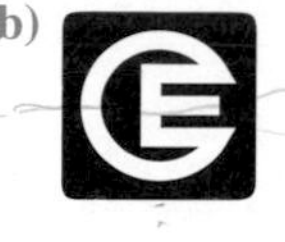

(c)

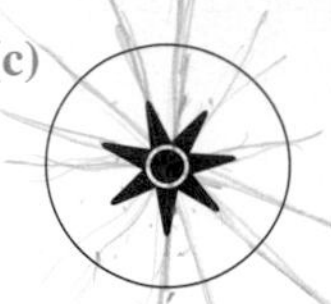

10. What is the size of the smallest positive angle of rotation symmetry in each figure shown in problem 9?

11. Two blank crossword puzzles are shown. What symmetries are found in the grid of black and white squares used by the puzzle maker?

(a)

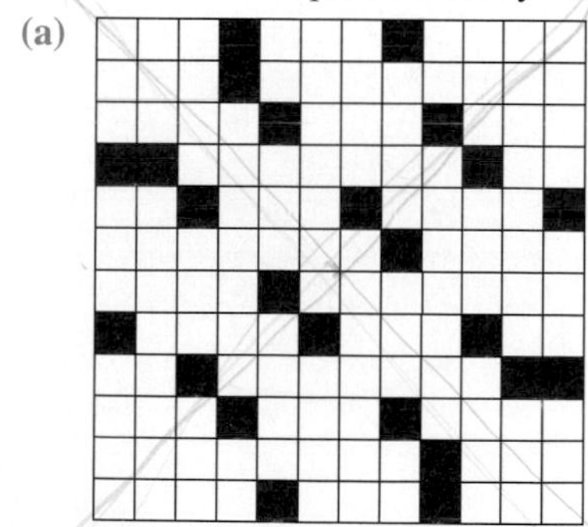

(b)

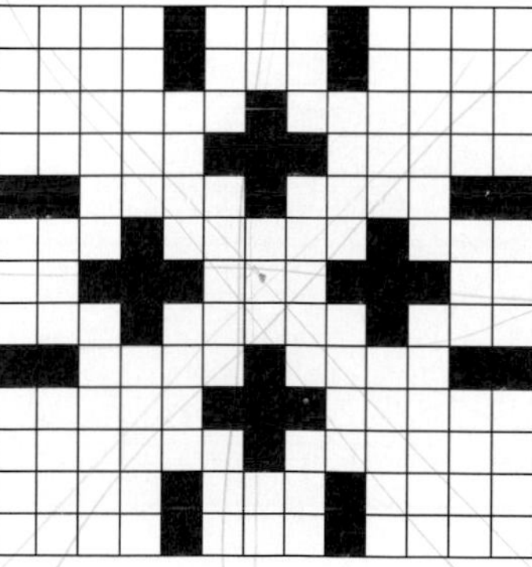

12. List the symmetries found in each of the border patterns shown.

(a)

INDIAN PAINTED LACQUER WORK

(b)

MALTESE LACE

(c)

ANCIENT GREEK SCROLL BORDER

(d)

ITALIAN DAMASK OF THE RENAISSANCE

13. What is the crystallographic restriction?

14. Describe the rotation symmetries in each of these wallpaper patterns.

(a)

CEILING DECORATION FROM THEBES

(b)

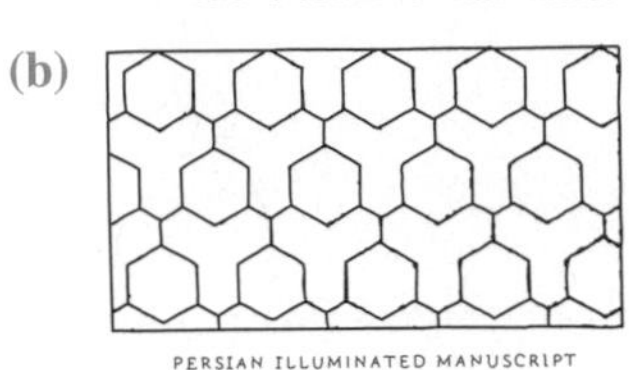

PERSIAN ILLUMINATED MANUSCRIPT

---

15. $\triangle DEF$ is a size transformation image of $\triangle ABC$ with center O, $AB = 8$, $BC = 12$, $DF = 36$, and $DE = 16$.
Find the following values.

(a) k, the scale factor of the size transformation.

(b) EF

(c) AC

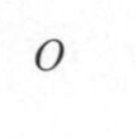

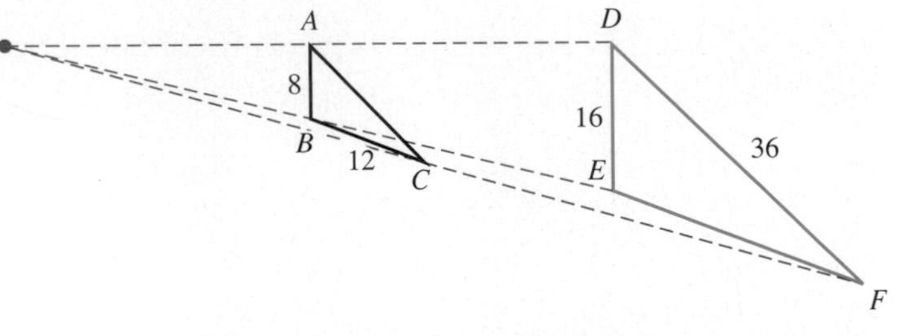

---

16. A size transformation has taken $ABCDEF$ onto $AVWXYZ$ as shown, where $AV = 8$, $VB = 4$, $BC = 20$, and $AZ = 18$.

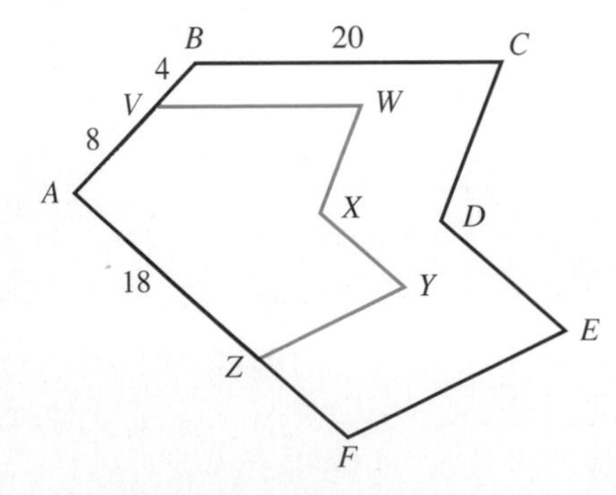

(a) What is the center of the size transformation?

(b) What is the scale factor?

(c) What is the distance VW?

(d) What is the distance ZF?

17. Describe a similarity transformation that takes the square $ABCD$ onto the square $A'B'C'D'$.

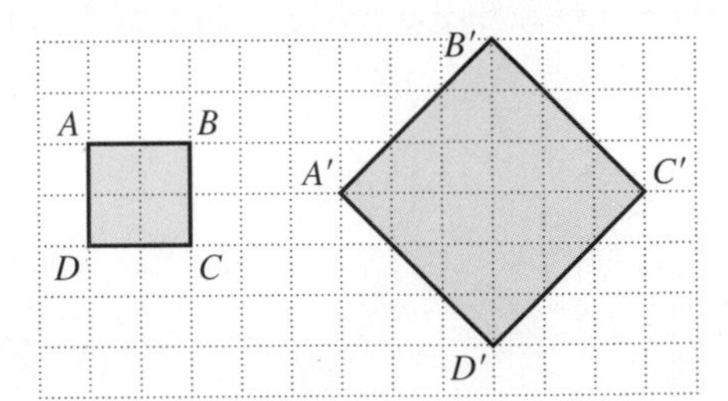

14 Coordinate Geometry

HANDS ON

Joining Lattice Points

Materials Needed

Two sheets of graph paper (preferably quarter-inch)

Directions

Step 1. Draw two intersecting number lines—one vertical and one horizontal as shown. Number each number line starting with 0 at the point of intersection. Any point where the grid lines intersect is called a **lattice point** and can be identified uniquely by an ordered pair of integers (a, b). For example, the lattice point 3 units to the right and 4 units up from the point of intersection of the number lines is identified by the pair (3, 4) and conversely. Marking a point like (3, 4) on the grid is called **plotting the point.**

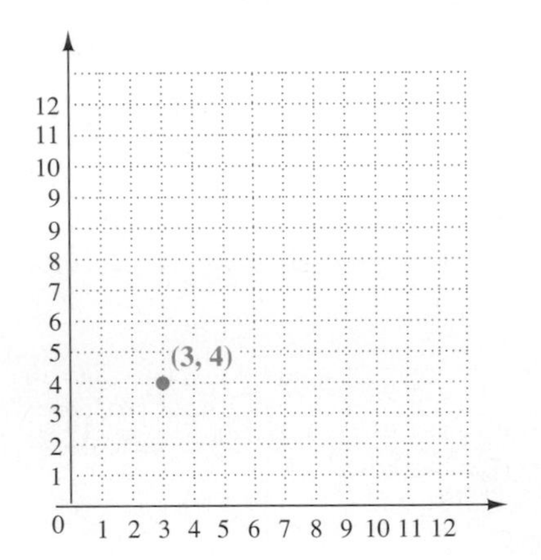

Step 2. Plot four lattice points on a grid so that none of the six line segments determined by the four points contains a lattice point other than its end points. For example, the line segment joining (3, 4) and (6, 5) is such a segment. However, the segment joining (3, 4) and (7, 2) also contains the lattice point (5, 3). Repeat this process of plotting four such sets of points several times and on a different grid each time (about four such grids can be drawn conveniently on a single sheet of quarter-inch graph paper).

Step 3. Return to each of the examples produced in Step 2. See if you can add a fifth lattice point to any of these configurations without creating a situation where at least one of the 10 line segments determined by the five lattice points contains a lattice point other than its end points. If a line segment is created that contains a lattice point other than its end points, carefully identify and describe the lattice point. Make a conjecture based on your investigations in this part of the activity.

CONNECTIONS

Synthetic and Analytic Geometry

In Chapters 10, 11, 12, and 13 we considered various aspects of geometry involving points, lines, planes, angles, lengths, and areas. We discovered many interesting and surprising geometric properties and these were "proved" or justified most frequently using notions like similarity, congruence, and measurement. Such arguments are usually called **synthetic** because they depend on the synthesis of ideas and results obtained earlier. In this chapter we consider geometry from a different point of view using numbers and equations to establish geometric properties. Geometry from this perspective is called **coordinate geometry** and is developed by a powerful blending of synthetic and algebraic methods. This approach makes it possible to discover and verify geometric properties algebraically. Conversely, we can use geometric insight to understand and visualize algebraic phenomena. Historically, this blending of algebra and geometry breathed new life into both subjects, and is carried on in such subjects as analytic geometry, linear algebra, calculus, differential equations and much of the rest of mathematics. Coordinate geometry provides an important method to solve geometric problems.

14.1 The Cartesian Coordinate System

Consider the two perpendicular number lines illustrated in Figure 14.1. Any point in the plane can be uniquely located by giving its distance to the right or left of the vertical number line and its distance above or below the horizontal number line. In Figure 14.1, the point P is 5 units to the right of the vertical number line and 3 units above the horizontal number line, and there is only one such point. Thus, P is identified by the ordered number pair (5, 3) and we sometimes write $P(5, 3)$ as shown. Other times we will just write (5, 3).

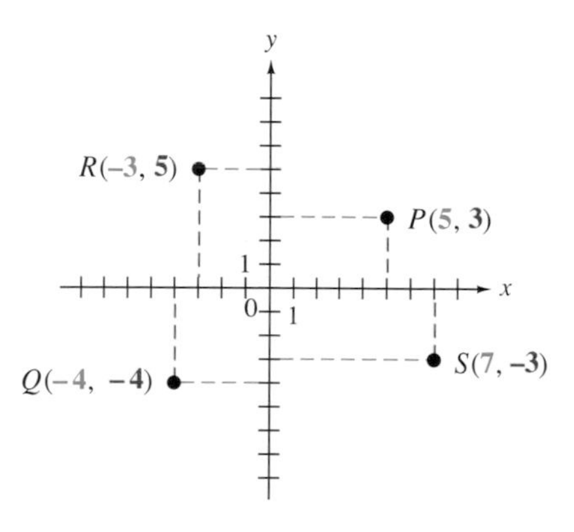

Figure 14.1
The Cartesian coordinate system

The two number lines are called **coordinate axes** or just **axes.** Typically the horizontal number line is called the ***x*-axis** and the vertical number line is called the ***y*-axis.** The two numbers in the number pair (a, b) that locate a point are called the

HIGHLIGHT FROM HISTORY
René Descartes and Coordinate Geometry

It is often difficult to ascribe the development of any particular body of mathematics to its originator. The fact is that mathematics is the cumulative result of the efforts of numerous individuals working over hundreds or even thousands of years. The invention of coordinate geometry is customarily ascribed to René Descartes (1596–1650), one of the leading seventeenth century mathematicians, who did indeed make great strides in commingling the ideas of geometry and algebra as explained in a book titled *La géométrie.* However, Descartes never thought of an ordered pair (a, b) as the coordinates of a point in the plane. Thus, the terminology, "Cartesian product" and "Cartesian coordinate system," ascribing these ideas to Descartes, is largely misplaced. The idea of coordinates goes back at least as far as the Greek, Apollonius of Perga, in the third century B.C., and was utilized by Nicole Oresme (1323?–1382), the French Bishop of Lisieux, and by

others. The idea was also known to the amateur but inspired Pierre de Fermat, a contemporary of Descartes, and was certainly popularized by the Dutch mathematician, Frans van Schooten (1615–1700) in his *Geometria a Renato Des Cartes (Geometry by René Descartes)* in 1649. It is probably not unreasonable to say that our modern ideas of coordinate geometry were inspired by Descartes but organized and popularized by Schooten.

coordinates of the point. The first number in the pair is called the ***x*-coordinate** and gives the distance of the point to the right or left of the vertical axis; that is, in the direction of the x-axis. The second number in the ordered pair is called the ***y*-coordinate** and gives the distance of the point above or below the horizontal axis; that is, in the direction of the y-axis. The axes divide the plane into four regions or **quadrants** numbered I, II, III, and IV numbering counterclockwise from the upper right-hand quadrant. A point:

- lies in quadrant I if both its coordinates are positive,
- lies in quadrant II if the first coordinate is negative and the second coordinate is positive,
- lies in quadrant III if both coordinates are negative, and
- lies in quadrant IV if the first coordinate is positive and the second is negative.

The point (0, 0) where the axes intersect is called the **origin** of the coordinate system. These notions are summarized in Figure 14.2.

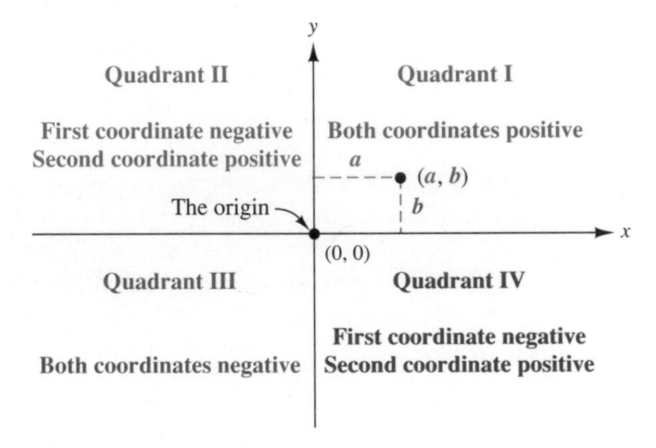

Figure 14.2
Salient features of the Cartesian coordinate system

EXAMPLE 14.1 Plotting Points

The diagram below shows the partial outline of a house.

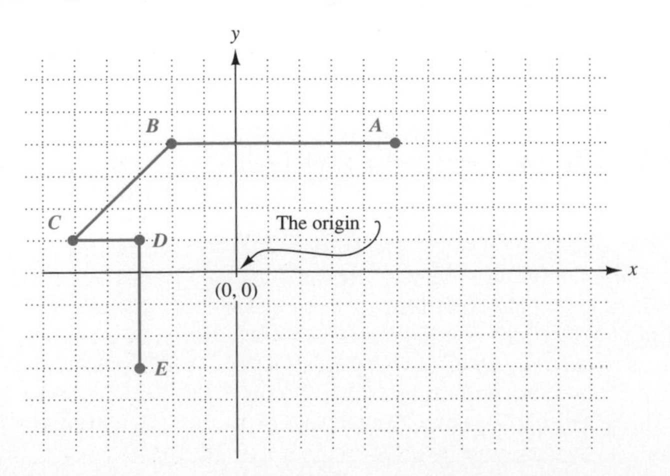

(a) Give the coordinates; that is, the ordered pair naming each of A, B, C, D, and E.

(b) Plot the points $F(8, -3)$, $G(8, 1)$, $H(10, 1)$, $I(7, 4)$, $J(6, 4)$, $K(6, 5)$, and $L(5, 5)$.

(c) Draw the segments $\overline{EF}$, $\overline{FG}$, $\overline{GH}$, $\overline{HI}$, $\overline{IJ}$, $\overline{JK}$, $\overline{KL}$, and $\overline{LA}$.

SOLUTION

(a) A is 5 units to the right of $(0, 0)$ (that is, in the x-direction from the origin) and 4 units above $(0, 0)$ (that is, in the y-direction from the origin). Thus, A is the point $(5, 4)$. The x-coordinate of A is 5 and the y-coordinate of A is 4. Similarly, we determine that the coordinates of the other points are $B(-2, 4)$, $C(-5, 1)$, $D(-3, 1)$, and $E(-3, -3)$.

(b) The point $F(8, -3)$ is 8 units to the right of $(0, 0)$ and 3 units *below* $(0, 0)$. Similarly, the other points are located as shown.

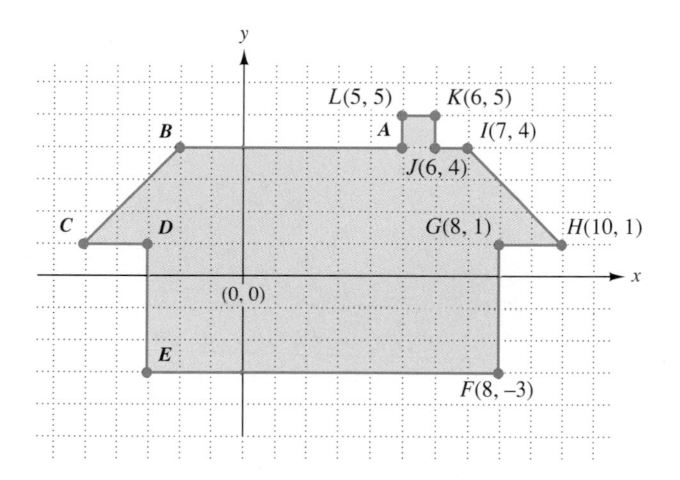

(c) Drawing the designated line segments, the completed figure forms the outline of a house. ■

The Midpoint Formula

Suppose M is the midpoint of the line segment $\overline{PQ}$. If the coordinates of P and Q are known, can we calculate the coordinates of M? Let's begin with a particular example.

EXAMPLE 14.2 **Determining the Midpoint of a Line Segment**

Compute the coordinates of the midpoint, M, of the line segment joining $P(3, 6)$ and $Q(7, 11)$.

SOLUTION

Understand the problem • • • •

We are given the coordinates of the endpoints P and Q. We must find the coordinates of the midpoint, M.

Devise a plan • • • •

A diagram will allow us to see how P, Q, and M are related on a coordinate plane. If horizontal and vertical distances can be identified, we should be able to determine the coordinates of M.

Carry out the plan • • • •

If we draw a coordinate system and plot the points P and Q as shown here, we certainly get a better feeling for the problem.

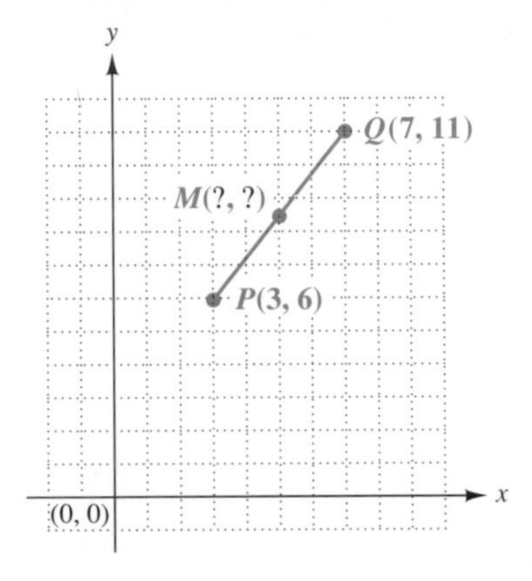

Perhaps we can add to the picture and use some of the geometric facts from Chapter 11. Draw the horizontal segments $\overline{PS}$ and $\overline{MT}$ and the vertical segments $\overline{MR}$ and $\overline{QS}$ as shown. Then, since $\overline{MT}$ and $\overline{PS}$ are parallel as are $\overline{RM}$ and $\overline{SQ}$, and $\overline{PM} \cong \overline{MQ}$, it follows that $\triangle PMR \cong \triangle MQT$. But then R is the midpoint of $\overline{PS}$ and must have coordinates $(r, 6)$ where

$$\begin{aligned} r &= 3 + \frac{1}{2}PS \\ &= 3 + \frac{1}{2}(7 - 3) \\ &= 5. \end{aligned}$$

Also, T is the midpoint of $\overline{SQ}$ and must have coordinates $(7, s)$ where

$$\begin{aligned} s &= 6 + \frac{1}{2}SQ \\ &= 6 + \frac{1}{2}(11 - 6) \\ &= 8.5. \end{aligned}$$

Finally, M has coordinates $(r, s) = (5, 8.5)$ as was to be determined.

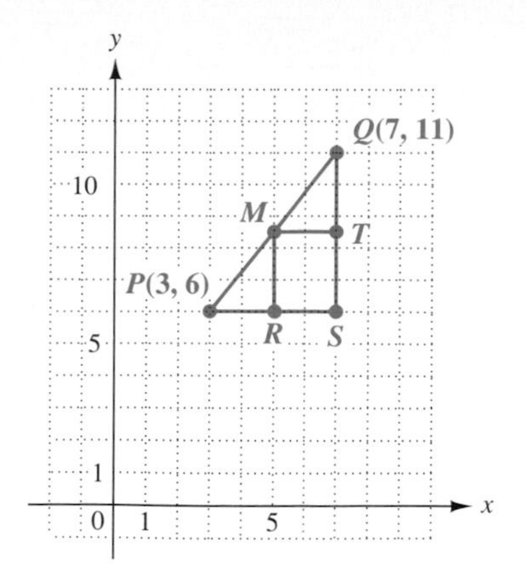

Look back • • • •

The key to this solution was drawing a picture on a coordinate plane. This made it possible to determine the coordinates of key points and, finally, the desired coordinates of the midpoint, M.

Also, observe that

$$\begin{aligned} 5 &= 3 + \frac{1}{2}(7 - 3) \\ &= 3 + \frac{1}{2} \cdot 7 - \frac{1}{2} \cdot 3 \\ &= \frac{1}{2} \cdot 3 + \frac{1}{2} \cdot 7 \\ &= \frac{3 + 7}{2} \end{aligned}$$

The average of 3 and 7 is 5.

and, similarly,

$$8.5 = \frac{6 + 11}{2}.$$

The average of 6 and 11 is 8.5.

Thus, the coordinates of the midpoint of $\overline{PQ}$ are the *average* of the x-coordinates of P and Q and the average of the y-coordinates of P and Q. ■

The preceding example is typical of the general case which is stated here as a theorem.

THEOREM The Midpoint Formula

Consider points $P(a, b)$ and $Q(c, d)$. The midpoint of the segment $\overline{PQ}$ is the point

$$M\left(\frac{a + c}{2}, \frac{b + d}{2}\right).$$

RENÉ DESCARTES EXPLAINS THE COORDINATE SYSTEM WHICH TIES TOGETHER ALGEBRA AND GEOMETRY
46°
5.1 sec.
370 FT
90 FT
S. harris

A useful notational device when working with midpoints is to write a number as twice or even four times another number. For example, we can write 3 as $2 \cdot (1.5)$, -4.5 as $2 \cdot (-2.25)$, 7 as $4 \cdot (1.75)$, and -3 as $4 \cdot (-0.75)$. In general, we can write any number in the form $2a$ or $4b$ or $3c$, and so on. This device helps avoid fractions. For example, the midpoint of the segment joining $P(2r, 2s)$ and $Q(2u, 2v)$ is

$$M\left(\frac{2r + 2u}{2}, \frac{2s + 2v}{2}\right) = M(r + u, s + v).$$

SCHOOL BOOK PAGE *Coordinate Geometry in Third Grade*

SOCIAL STUDIES CONNECTION

Locating Points on a Grid

Build Understanding

Maps and graphs are very much alike. They are both drawn on a grid.

A. Start at the town square. Walk 3 blocks east on Main Street. Where are you? Are you at the corner of Main Street and Park Avenue?

Now walk 3 blocks north on Park Avenue. Where are you? What is at the corner of Park Avenue and Sunset Road?

B. You can use ***number pairs*** to find places on a map. Use the numbers at the side and bottom. What will you find at (5, 2) on the map?

Begin at the star.

(5, 2)

Move 5 units to the right. Move 2 units up.

The store is at (5, 2) on the map.

C. What number pair tells where the theater is on the map?

Start at the theater. Follow the street down to the number on the bottom. The theater is 4 units to the right of the star.

Now start at the theater and follow the street to the number on the side. The theater is 5 units up from the star.

The theater is at (4, 5) on the map.

■ **Write About Math** Some people use a memory trick to find places on a map. They say, "Step right up." How can this sentence help you locate points on a grid?

SOURCE: From *Scott Foresman Exploring Mathematics* Grades 1–7 by L. Carey Bolster et al. Copyright © 1994 Scott, Foresman and Company. Reprinted by permission of Scott, Foresman and Company.

1. What number pair identifies the corner of Poe Street and Stadium Place?
2. List the number pairs that identify points along Market Street. What do you conclude?
3. List the number pairs that identify points along Elm Street. What do you conclude?
4. What would the coordinates of the intersection nine blocks west of the theater be?

EXAMPLE 14.3 **Proving that the Diagonals of a Rectangle Bisect Each Other**

Show that the diagonals of a rectangle bisect each other.

SOLUTION

Any rectangle $ABCD$ can be placed on a coordinate system with $\overline{AB}$ along the positive y-axis and $\overline{AD}$ along the positive x-axis as shown. Then A is the point (0, 0), B is (0, $2b$), and D is ($2d$, 0) for some positive real numbers b and d. Also C is the point ($2d$, $2b$). By the preceding theorem the midpoint of $\overline{AC}$ is

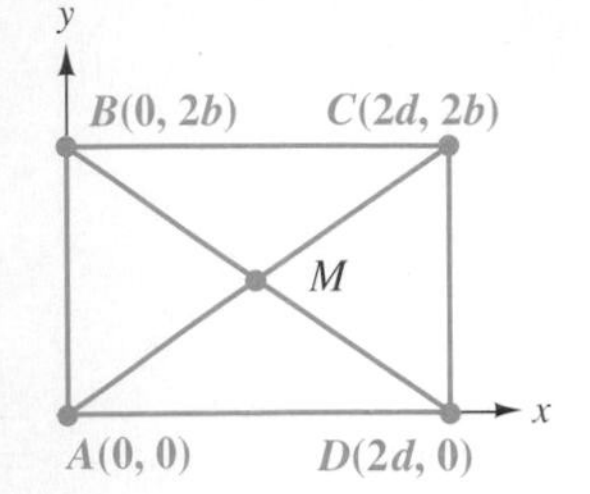

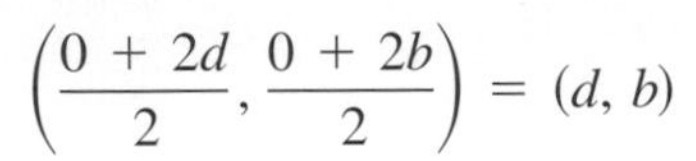

$$\left(\frac{0 + 2d}{2}, \frac{0 + 2b}{2}\right) = (d, b)$$

and the midpoint of $\overline{BD}$ is

$$\left(\frac{0 + 2d}{2}, \frac{2b + 0}{2}\right) = (d, b)$$

But this implies that the two diagonals meet at the point M which is the midpoint of each diagonal. ■

Dividing a Line Segment

We just determined the coordinates of the midpoint of a line segment. Suppose we wanted to determine M so that M is 1/3 of the way from A to B; that is, so that

$$\frac{AM}{AB} = \frac{1}{3}.$$

Consider Figure 14.3 with B above and to the right of A. Draw horizontal segments $\overline{ME}$ and $\overline{AC}$ and vertical segments $\overline{MD}$ and $\overline{BC}$. It follows that $\triangle AMD$, and $\triangle ABC$ are similar so that corresponding sides are of proportional length. Since

$$AM = \frac{1}{3}AB,$$

it follows that

$$AD = \frac{1}{3}AC \qquad \text{and} \qquad CE = \frac{1}{3}CB.$$

Thus,

$$\begin{aligned} r &= a + \frac{1}{3}(c - a) \\ &= a + \frac{1}{3}c - \frac{1}{3}a \\ &= \frac{2}{3}a + \frac{1}{3}c \end{aligned}$$

and similarly

$$s = \frac{2}{3}b + \frac{1}{3}d.$$

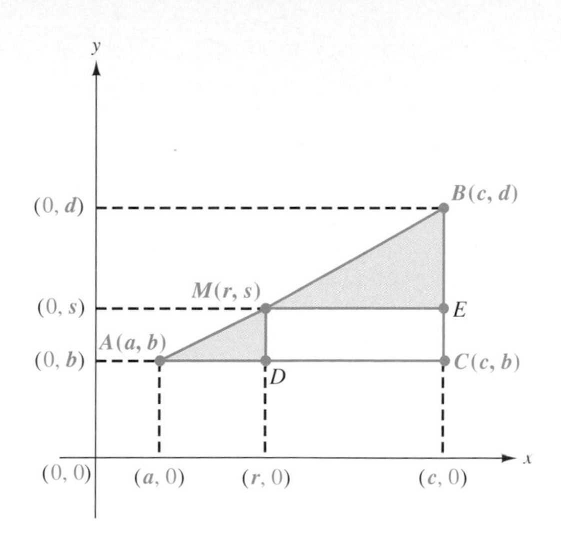

Figure 14.3
The point 1/3 of the way from A to B

The argument when B is to the right and below A is similar and the final result is the same.

EXAMPLE 14.4

Determining the Point 1/3 of the Way from (3, 6) to (12, −6)

Determine the coordinates of the point 1/3 of the way from $A(3, 6)$ to $B(12, -6)$.

SOLUTION

Let $M(r, s)$ be the desired point. Then,

$$r = \frac{2}{3}(3) + \frac{1}{3}(12) = 6$$

and

$$s = \frac{2}{3}(6) + \frac{1}{3}(-6) = 2.$$

The point M is $(6, 2)$. As a check we compute the distances AM and AB using the Pythagorean theorem from Chapter 12.

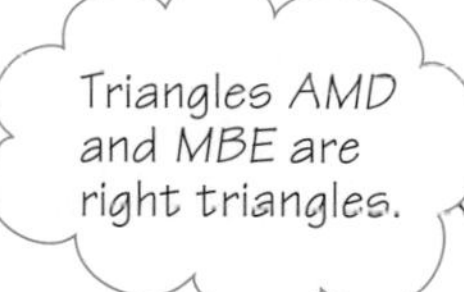

$$AM = \sqrt{(3-6)^2 + (6-2)^2} = \sqrt{9+16} = 5$$
$$AB = \sqrt{(3-12)^2 + (6-(-6))^2} = \sqrt{81+144} = 15$$

Thus, $AM/AB = 1/3$ as desired. ■

THEOREM Division of a Line Segment

The point the fraction t of the distance from $A(a, b)$ to $B(c, d)$ is

$$M((1-t)a + tc, \quad (1-t)b + td).$$

The Distance Formula

We have just discovered how to divide a line segment in a given ratio. We can also determine its length. Consider this example.

EXAMPLE 14.5

Determining the Length of a Line Segment

Compute the length of the line segment $\overline{PQ}$ where P is the point (2, 5) and Q is (7, 8).

SOLUTION

Drawing the line segment $\overline{PQ}$, we are again led to construct the right triangle PQR where $\overline{PR}$ and $\overline{QR}$ are parallel to the x- and y-axes respectively. Since $\triangle PQR$ is a right triangle, the Pythagorean theorem applies. Thus,

$$PQ^2 = PR^2 + RQ^2.$$

Hence,

$$\begin{aligned} PQ &= \sqrt{PR^2 + RQ^2} \\ &= \sqrt{(7-2)^2 + (8-5)^2} \\ &= \sqrt{25+9} = \sqrt{34} \doteq 5.8 \end{aligned}$$

as required.

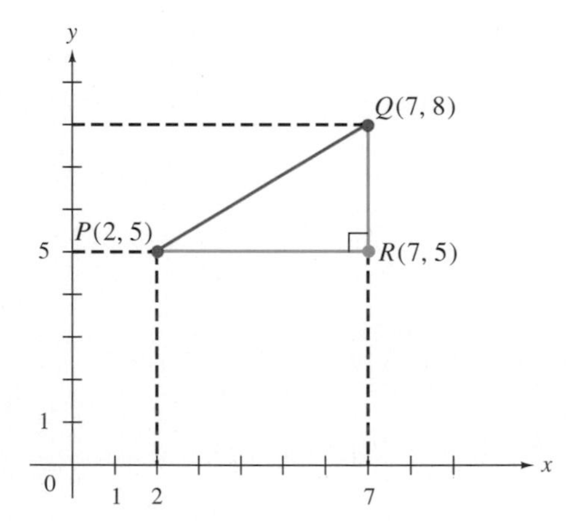

■

The preceding example suggests a formula for finding the distance between two points whose coordinates are known. As before, the proof is an exact translation of the preceding solution where we replace $P(2, 5)$ and $Q(7, 8)$ by $P(a, b)$ and $Q(c, d)$ respectively. Since the translation is exact, the proof is omitted.

THEOREM The Distance Formula

Let P and Q be the points (a, b) and (c, d). Then the distance between P and Q is

$$PQ = \sqrt{(c-a)^2 + (d-b)^2}.$$

Note that

$$(c - a)^2 = (a - c)^2 \quad \text{and} \quad (d - b)^2 = (b - d)^2$$

since the square of the negative of a number is the same as the square of the number. Thus, in the distance formula, it does not make any difference which point is chosen for P and which is chosen for Q. For example, the distance between (4, 7) and (1, 3) is given by both

$$\begin{aligned}\sqrt{(4 - 1)^2 + (7 - 3)^2} &= \sqrt{3^2 + 4^2} \\ &= \sqrt{9 + 16} = \sqrt{25} = 5\end{aligned}$$

and

$$\begin{aligned}\sqrt{(1 - 4)^2 + (3 - 7)^2} &= \sqrt{(-3)^2 + (-4)^2} \\ &= \sqrt{9 + 16} = \sqrt{25} = 5.\end{aligned}$$

EXAMPLE 14.6

Proving that a Triangle Is Isosceles

Prove that the triangle with vertices $R(1, 4)$, $S(5, 0)$, and $T(7, 6)$ is isosceles.

SOLUTION

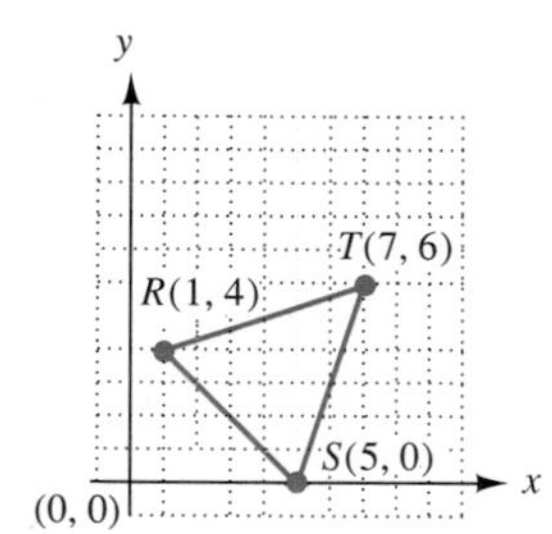

We compute the length of the three sides.

$$\begin{aligned}RS &= \sqrt{(1 - 5)^2 + (4 - 0)^2} = \sqrt{16 + 16} = \sqrt{32} \\ RT &= \sqrt{(1 - 7)^2 + (4 - 6)^2} = \sqrt{36 + 4} = \sqrt{40} \\ ST &= \sqrt{(5 - 7)^2 + (0 - 6)^2} = \sqrt{4 + 36} = \sqrt{40}\end{aligned}$$

Since $RT = ST$, it follows that $\triangle RST$ is isosceles. ■

Logo Coordinate Geometry

Although no axes appear, Logo treats the screen as a Cartesian coordinate system. The home position at the center of the screen has coordinates (0, 0). The command **home** sends the turtle to (0, 0) from its current position, drawing a straight line segment if the pen is down. More generally, the turtle can be sent along a line segment to any new position on the screen by using the **setx, sety,** and **setpos** commands. For example, to move the turtle to the position (20, −50) type the command

setpos [**20 −50**] (or, in Terrapin Logo, **setxy 20 (−50)**; parentheses are required about the negative y-coordinate to avoid confusion with $20 - 50 = -30$.)

Different versions of Logo give different coordinate sizes of the screen, which can be determined with a little experimentation. For example, observing the turtle when **setx 80**, **setx 90**, **setx 100**, and so on are executed will allow you to discover the x-coordinate of the right edge of the screen.

Any figure drawn with line segments on squared graph paper can also be drawn with a corresponding Logo procedure.

LOGO EXAMPLE 14.A

Using Logo Coordinates

Write a procedure to draw the house shown on graph paper.

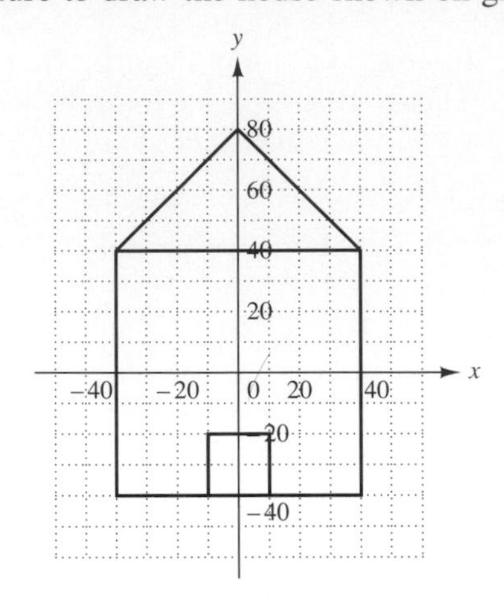

SOLUTION

The procedure below draws the house on the screen. (For Terrapin Logo, replace **setpos [ ]** with **setxy,** enclosing negative coordinate values in parentheses.)

```
to  house
        pu setpos [−40  −40] pd
        setpos [−40  −40]
        setpos [40  40]
        setpos [40  −40]
        setpos [−40  −40]
        pu setpos [−40  40]
        pd setpos [0  80]
        setpos [40  40]
        pu setpos [−10  −40]
        pd setpos [−10  −20]
        setpos [10  −20]
        setpos [10  −40]
        pu home pd
end
```

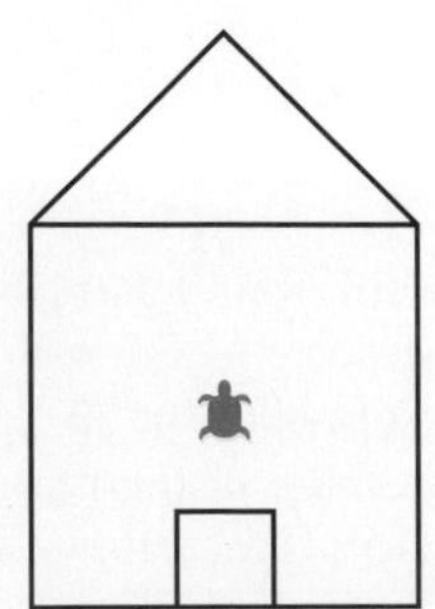

■

PROBLEM SET 14.1

Understanding Concepts

1. What can be said about the coordinates of a point P if
   (a) P lies on the x-axis?
   (b) P lies in the second quadrant?
   (c) P lies in the fourth quadrant?
   (d) P lies on the y-axis?
   (e) P is at the origin?
2. Plot and label the following points on a Cartesian coordinate system drawn on a sheet of graph paper.
   (a) $P(5, 7)$ (b) $Q(5, -7)$ (c) $R(-5, 7)$
   (d) $S(-5, -7)$ (e) $T(0, 5)$ (f) $U(7.5, 0)$
   (g) $V(0, -5.2)$ (h) $W(-7, 0)$ (i) $X(0, 0)$
3. Plot these points and connect them in order with line segments: (1, 1), (1, 11), (4, 13), (5, 15), (6, 13), (7, 12), (10, 11), (9, 10), (6, 9), (4, 7), (1, 1).
4. (a) Plot the points (5, 0), (4, 3), (3, 4), (0, 5), (−3, 4), (−4, 3), (−5, 0), (−4, −3), (−3, −4), (0, −5), (3, −4), and (4, −3).
   (b) What do you observe about the points in part (a)?
5. Let $A(0, 0)$, $B(2, 3)$, $C(r, s)$, and $D(5, 0)$ be the vertices of parallelogram $ABCD$ reading clockwise around from A. Determine r and s. (*Hint:* Draw a diagram.)
6. Let $M(r, s)$ be the midpoint of $\overline{PQ}$. Determine $M(r, s)$ for these choices of P and Q. Also, plot P, Q, and M on graph paper in each case.
   (a) $P(2, 7)$, $Q(6, 1)$ (b) $P(-1, 1)$, $Q(3, 5)$
   (c) $P(0, 7)$, $Q(-4, 1)$ (d) $P(-2, -5)$, $Q(4, 1)$
   (e) $P(2, 3)$, $Q(7, -3)$ (f) $P(3, 4)$, $Q(-1, -5)$
7. Consider the points $A(1, -7)$ and $B(-7, 9)$. Determine the coordinates of the point P
   (a) one-fourth of the way from A to B.
   (b) three fourths of the way from A to B.
   (c) five-fourths of the way from A to B.
   (d) Plot the points in parts (a), (b), and (c) on a Cartesian coordinate system.
8. Consider the convex quadrilateral $A(-2, -1)$, $B(4, 3)$, $C(6, 1)$, and $D(0, -7)$.
   (a) Determine the midpoints P, Q, R, and S of $\overline{AB}$, $\overline{BC}$, $\overline{CD}$, and $\overline{DA}$ respectively.
   (b) Show that $\overline{PR}$ and $\overline{QS}$ from part (a) bisect each other.
   (c) On graph paper, draw the quadrilaterals $ABCD$ and $PQRS$.
9. Compute the distance between these pairs of points.
   (a) (−2, 5) and (4, 13) (b) (3, −4) and (8, 8)
   (c) (0, 7) and (8, −8) (d) (3, 5) and (2, −4.3)
10. (a) Prove that $R(1, 2)$, $S(7, 10)$, and $T(5, -1)$ are the vertices of a right triangle.
    (b) Draw the triangle RST of part (a) on graph paper.
11. Consider the quadrilateral with vertices $A(2, 4)$, $B(12, 0)$, $C(10, -5)$, and $D(0, -1)$.
    (a) Show that $ABCD$ is a rectangle.
    (b) Show that $\overline{AC}$ and $\overline{BD}$ bisect one another.
    (c) Plot $ABCD$ and draw the line segments $\overline{AC}$ and $\overline{BD}$ on a coordinate system.
12. Classify the triangles with these vertices as one or more of right, acute, obtuse, equilateral, isosceles, or scalene.
    (a) (1, 1), (1, −4), (6, −4)
    (b) (−5, −4), (−1, −1), (3, −4)
    (c) (−1, 3), (2, 5), (7, −3)
    (d) (0, 2), (3, 7), (8, 2)

Thinking Critically

13. (a) Plot the set of points $P(a, b)$ with a and b integers, $a + b = 6$, and $-5 \le a \le 5$.
    (b) What do you observe about the points in part (a)?
14. (a) Plot the set of points $Q(r, s)$ with r and s being real numbers, $r - s = 6$ and $-5 \le r \le 5$.
    (b) What do you observe about the points in part (a)?
15. (a) Plot the set of points $(a, 5)$ with a being an integer and $-6 \le a \le 6$.
    (b) What can you say about the points in part (a)?
16. (a) Plot the set of points $(-2, b)$ with b being a real number and $-4 \le b \le 4$.
    (b) What seems to be true about the points in part (a)?
17. (a) Plot the set of points (x, y) with x and y being integers, $2x + 3y = 6$, and $-12 \le x \le 12$. (*Hint:* Give x integer values in the prescribed range, compute the corresponding values of y, and plot the resulting point (x, y) if y is an integer.)
    (b) Carefully describe the set of points in part (a).
    (c) The set of plotted points in part (a) is called the **graph** of the equation $2x + 3y = 6$ subject to

the given conditions. Conjecture what the graph would look like if the conditions that x and y be integers and $-12 \le x \le 12$ were removed.

18. Conjecture what the graph of the set of points (x, y) with $2x + 3y \le 6$ and x and y as integers would look like. (*Hint:* Carefully consider problem 16.) Explain with a sketch.

19. Conjecture what the graph of $2x + 3y \le 6$ with x and y as real numbers would look like. Explain with a sketch.

20. **(a)** Plot and label the points $A(3, 7)$ and $B(5, -4)$ on a Cartesian coordinate system.

(b) Plot the point $C(3(1 - t) + 5t, 7(1 - t) - 4t)$ for $t = -2, -1, 0, 1/2, 1, 2$, and 3 on the Cartesian coordinate system of part (a).

(c) What do you observe about the points of part (b)?

(d) Conjecture what the set of plotted points $(3(1 - t) + 5t, 7(1 - t) - 4t)$ would look like if t were allowed to assume all real values without restriction. Explain briefly.

21. Determine the point Q such that $\overline{PQ}$ is symmetric about the y-axis for each of the following choices for P. Also, plot P and Q and draw $\overline{PQ}$ in each case but the last.

(a) $P(3, 5)$ **(b)** $P(-2, 4)$ (c) $P(-3, -5)$
(d) $P(0, 5)$ **(e)** $P(5, 0)$ **(f)** $P(a, b)$

22. Determine the point T such that the segment $\overline{ST}$ is symmetric about the x-axis for all of these choices for S. Also, plot S and T and draw $\overline{ST}$ in each case but the last.

(a) $S(2, 7)$ **(b)** $S(-2, -4)$ (c) $S(3, 0)$
(d) $S(0, 4)$ (e) $S(-3, 5)$ **(f)** $S(u, v)$

23. Determine the point D that is the image of C under the indicated rotation about the origin in each case. Plot C and D in each case.

(a) $C(2, 4)$, 90° counterclockwise rotation.
(b) $C(3, -5)$, 90° clockwise rotation.
(c) $C(3, -5)$, 270° counterclockwise rotation.
(d) $C(2, 0)$, 60° counterclockwise rotation.
(e) $C(0, -2)$, 45° clockwise rotation.

24. Determine the point F that is the image of E under the indicated slides. Plot E and F in each case except the last.

(a) $E(3, 4)$, slide 2 to the right and 3 up.
(b) $E(-1, -3)$, slide 3 to the left and 4 up.
(c) $E(a, b)$, slide c to the right and d down, $c > 0$ and $d > 0$.

25. Determine the point H such that the segment $\overline{GH}$ is symmetric about the origin. Plot G, H, and draw $\overline{GH}$ in each case but the last.

(a) $G(3, -5)$ **(b)** $G(-2, -4)$ (c) $G(0, 3)$
(d) $G(-4, 0)$ (e) $G(-2, 7)$ **(f)** $G(a, b)$

26. (a) If $A(0, 0)$, $B(3, 5)$, $C(r, s)$, and $D(7, 0)$ are the vertices of a parallelogram determine r and s.

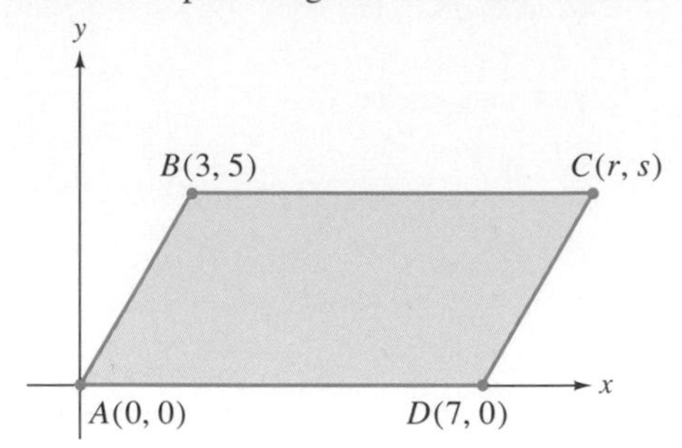

(b) Using the result of part (a), show that the diagonals of the parallelogram bisect each other.

27. **(a)** Consider the quadrilateral $A(0, 0)$, $B(2, 5)$, $C(6, 9)$, and $D(8, 0)$. Let P, Q, R, and S be the midpoints of $\overline{AB}$, $\overline{BC}$, $\overline{CD}$, and $\overline{DA}$ respectively. Show that the segments $\overline{PR}$ and $\overline{QS}$ bisect each other.

(b) Draw the quadrilateral $ABCD$ and the segments $\overline{PR}$ and $\overline{QS}$ on a coordinate system.

28. **(a)** Consider the diagram shown with P above and to the left of Q. Determine an expression for the area of the triangle in terms a, b, c, and d. (*Hint:* Draw vertical lines from P and Q to the x-axis.)

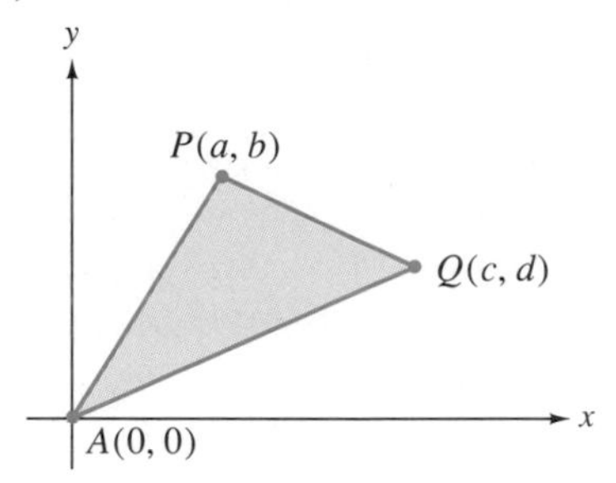

(b) Repeat part (a) with Q above and to the left of P.

(c) Express the results of (a) and (b) in a single formula using absolute value.

(d) Use your answer to (c) to compute the area of the triangle with vertices $A(0, 0)$, $B(0, b)$, and $C(c, 0)$ with $b > 0$ and $c > 0$.

Using a Calculator

29. **(a)** Reading in order and clockwise around a polygon, its vertices are $A(3, 2)$, $B(1, 6)$, $C(4, 9)$, $D(4, 7)$, $E(7, 7)$, $F(11, 4)$, and $A(3, 2)$. Determine the perimeter of this polygon to the nearest tenth.

(b) Using graph paper, carefully draw the polygon of part (a) and determine its area using Pick's formula $A = \frac{b}{2} + i - 1$ (see page 796).

(c) The formula for the area of a polygon with vertices $(x_1, y_1), (x_2, y_2), (x_3, y_3), \cdots, (x_n, y_n)$ is

$$A = \frac{1}{2}\left| x_1y_2 + x_2y_3 + x_3y_4 + \cdots + x_{n-1}y_n + x_ny_1 - x_2y_1 - x_3y_2 - x_4y_3 - \cdots - x_ny_{n-1} - x_1y_n \right|.$$

Use this formula to check the result of part (b). (*Note*: Here $n = 6$ and $(x_1, y_1) = (3, 2)$, $(x_2, y_2) = (1, 6)$, $(x_3, y_3) = (4, 9)$, $(x_4, y_4) = (4, 7)$, $(x_5, y_5) = (7, 7)$, and $(x_6, y_6) = (11, 4)$.)

Using a Computer
Using Logo

30. Determine the size of the screen in turtle steps for your version of Logo and your computer. Move the turtle around to find out the Cartesian coordinates of the four corners of the screen.

31. Use graph paper to draw a simple picture with at least 15 line segments in it. Center your picture about the origin. Write a Logo procedure using set position (or setxy) to draw the figure.

Making Connections

32. (a) Write an equation using x and y that states that the point $P(x, y)$ is 3 units from the origin.

(b) Carefully describe what the graph of the set of points in part (a) looks like.

(c) Compute the perimeter of the figure of part (a) correct to the nearest tenth.

(d) Compute the area of the figure in part (a) correct to the nearest tenth.

33. A cross-country course is laid out as shown. Determine the length of the course correct to the nearest meter.

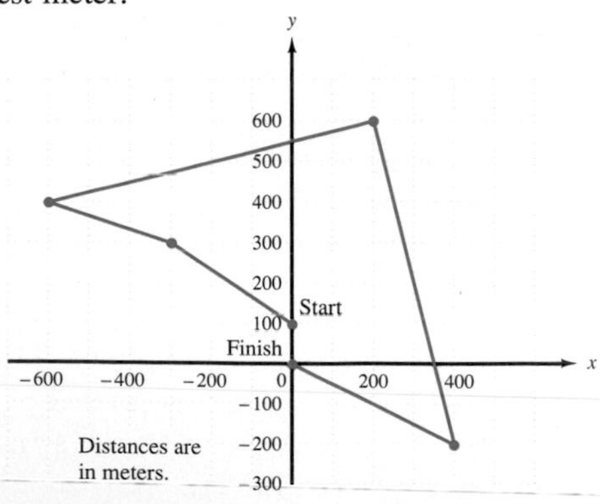

For Review

34. In the figure shown, $\overline{AB}$ and $\overline{DE}$ are parallel. Prove that $\triangle ABC \sim \triangle EDC$.

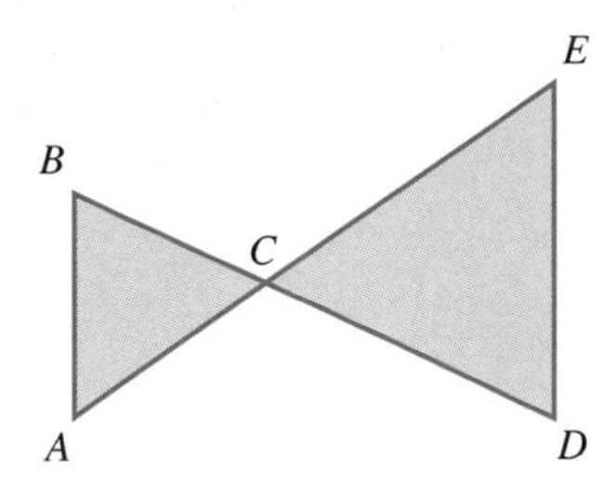

35. Determine x in the diagram shown given that $\overline{AB}$ is parallel to $\overline{CD}$.

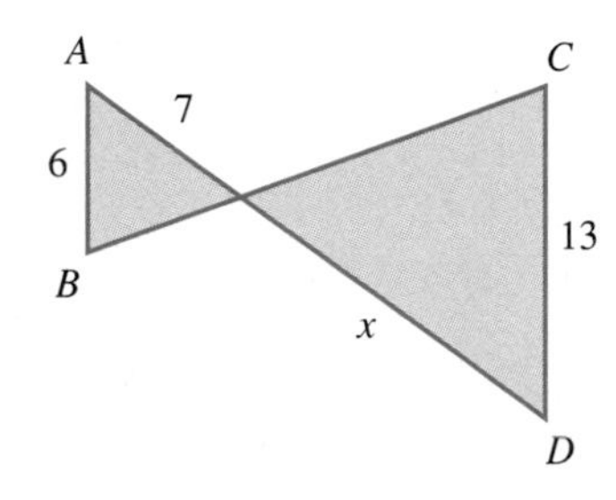

36. Determine y in the diagram shown given that $\overline{CD}$ is parallel to $\overline{AB}$.

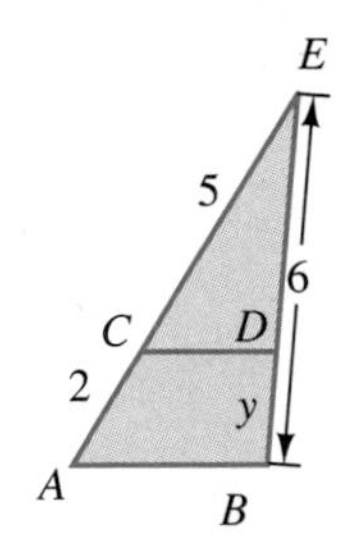

37. Determine z in the diagram shown.

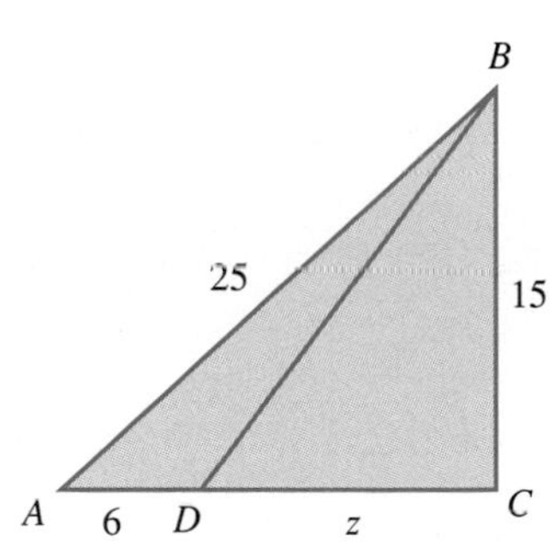

COOPERATIVE *Investigation 1*

Locating Centers of Gravity

Materials Needed

1. Tagboard or cardboard and a sheet of graph paper
2. Scissors

Directions

Step 1. Paste a sheet of graph paper onto a piece of tagboard.
Step 2. Cut out a tagboard triangle and note the coordinates of its vertices.
Step 3. Find the center of gravity of the triangle by balancing it on the edge of a table. Carefully draw a line on the triangle by holding it tight and by using the table edge as a straight edge. Rotate the triangle and repeating the balancing procedure, draw a second line. Where the two lines intersect is the *center of gravity*, that is, the triangle would hang horizontally if hung by a thread from this point.
Step 4. Determine the coordinates of the center of gravity.
Step 5. Using the data from Steps 3 and 4, conjecture a formula for the coordinates of the center of gravity in terms of the coordinates of the vertices of any triangle.

14.2 Lines and Their Graphs

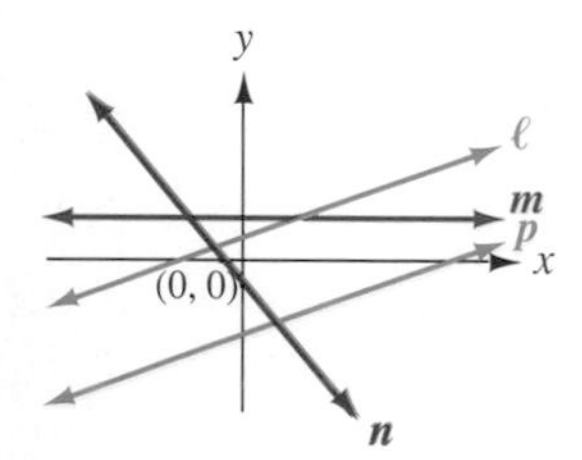

Figure 14.4
Lines in the plane

Slope

Consider the lines l, m, n, and p in Figure 14.4. The properties that distinguish between two lines are their location on the coordinate system and their direction or steepness. The direction or steepness of a line is the same as that of any segment of the line and this leads to the notion of **slope.**

Carpenters use the ratio

$$s = \frac{h}{r},$$

where h is the distance a roof rises and r is the distance over which the rise takes place, to compute the slope of a roof as shown in Figure 14.5. This is sometimes expressed by saying that the slope of a roof is "the rise over the run." Surveyors use the same idea when calculating the slope of a road. If a road rises five feet while moving forward horizontally 100 feet, the road has a slope of 0.05. In surveying, slopes are usually expressed as percents. Thus, a grade with a slope of 0.05 is said to be a 5% grade.

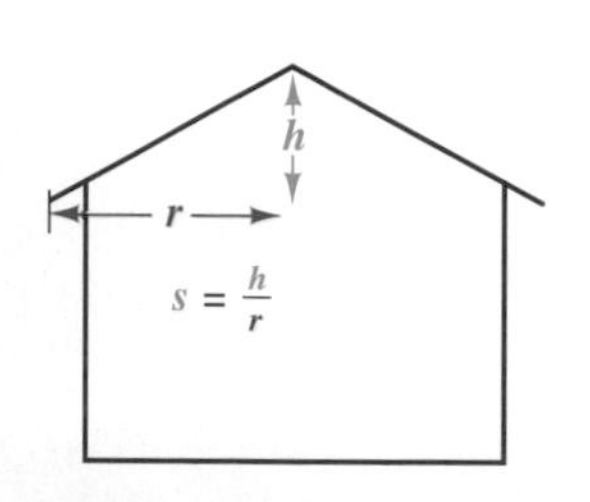

Figure 14.5
Slope on a roof

We also use the idea *rise over run* to determine the slope of a line segment. Consider the points $P(3, 5)$ and $Q(9, 7)$ shown in Figure 14.6. In moving from P to Q one moves up 2 units while moving to the right 6 units. The rise over the run gives a slope of $1/3$ indicated by the letter m. In this case,

$$m = \frac{7 - 5}{9 - 3} = \frac{2}{6} = \frac{1}{3}.$$

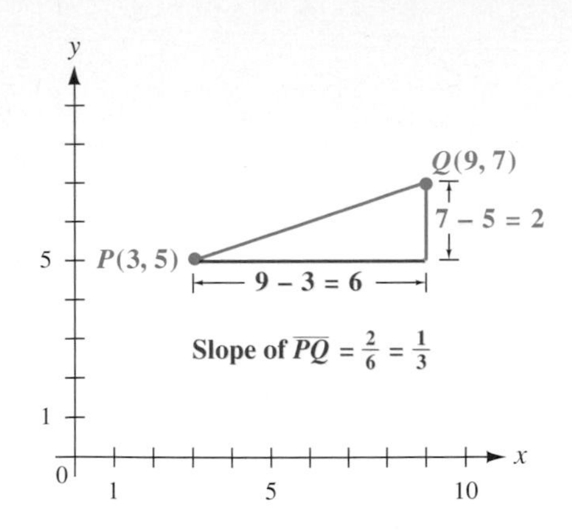

Figure 14.6
Slope of a segment

In general, the reasoning is similar and leads to this definition of slope.

DEFINITION Slope of a Line Segment

Let $P(x_1, y_1)$ and $Q(x_2, y_2)$ with $x_1 \neq x_2$ be two points. Then the **slope of the line segment** $\overline{PQ}$ is given by

$$m = \frac{y_2 - y_1}{x_2 - x_1}.$$

If $x_1 = x_2$ in the preceding definition, then $x_2 - x_1 = 0$ and $\overline{PQ}$ is vertical. Since division by zero is undefined, we must say that *a vertical segment has no slope* or, equivalently, that *the slope of a vertical segment is undefined.*

In computing the slope of a segment, it makes no difference which point is chosen as P and which is chosen as Q. However, once the choice is made, one must stick with it and always subtract *in the same direction* in both numerator and denominator. For example, in computing the slope of the segment in Figure 14.6, we identified P and Q as (3, 5) and (9, 7) respectively. But this could have been reversed to obtain

$$\frac{5-7}{3-9} = \frac{7-5}{9-3}$$

$$m = \frac{5-7}{3-9} = \frac{-2}{-6} = \frac{1}{3}$$

as before.

Finally, in Figure 14.6 the slope was positive and the segment sloped upward to the right. This is always the case for segments with positive slopes. (Why?) If $y_1 = y_2$ in the definition of slope then the slope is 0 and the segment $\overline{PQ}$ is necessarily horizontal. If the slope of a segment is negative, the segment slopes *downward to the right* as illustrated in the next example.

EXAMPLE 14.7

Determining a Negative Slope

Compute the slope of the line segment determined by $R(-3, 7)$ and $S(5, -2)$.

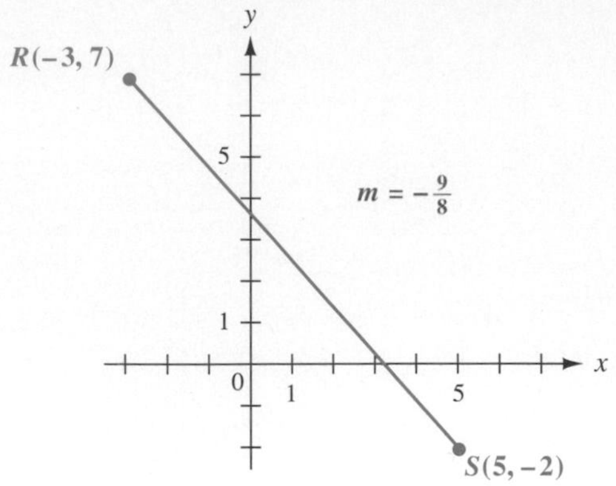

SOLUTION

As seen in the diagram, the segment slopes downward to the right. Also, from the definition,

$$m = \frac{7 - (-2)}{-3 - 5}$$
$$= \frac{9}{-8} = -\frac{9}{8}.$$

Subtracting in the other direction, we obtain

$$m = \frac{-2 - 7}{5 - (-3)}$$
$$= \frac{-9}{8} = -\frac{9}{8}$$

as well. ■

The **slope of a line** is the slope of any segment on the line as in the following definition.

DEFINITION Slope of a Line

The **slope of a line** is the slope of any segment on the line.

It makes no difference which segment (that is, which two points on the line) are used to determine the slope. Indeed, this follows immediately from what we already know about similar triangles.

Consider the segments $\overline{AB}$ and $\overline{CD}$ on the line of Figure 14.7. If we choose P and Q so that $\overline{AP}$ and $\overline{CQ}$ are parallel to the x-axis and $\overline{BP}$ and $\overline{DQ}$ are parallel to the y-axis, then $\triangle ABP \sim \triangle CDQ$. (Why?) But corresponding sides of similar triangles are of proportional lengths. Thus,

$$\frac{CQ}{AP} = \frac{DQ}{BP} = k$$

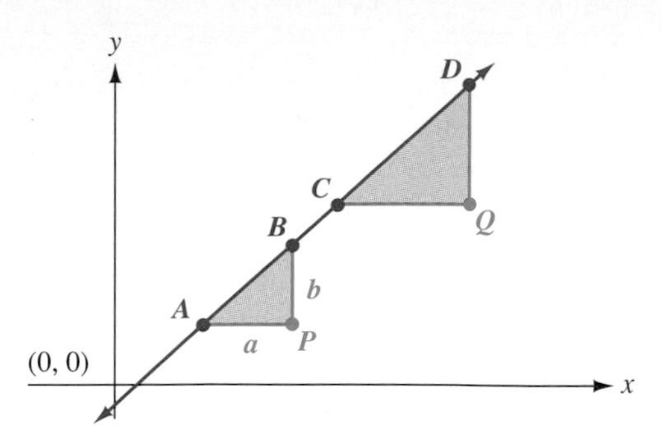

Figure 14.7
Slope of a line from different segments

for some scale factor k. It follows that if $AP = a$ and $BP = b$, then

$$CQ = k \cdot AP = ka \qquad \text{and} \qquad DQ = k \cdot BP = kb.$$

But then

$$\text{slope } \overline{AB} = \frac{BP}{AP} = \frac{b}{a} = \frac{kb}{ka} = \frac{DQ}{CQ} = \text{slope } \overline{CD}.$$

Hence, the slope of the line can be computed using either segment.

Similar considerations are used to prove the following important theorem.

> **THEOREM Condition for Parallelism**
>
> Two segments (lines) are parallel if, and only if, they have same slope.

JUST FOR FUN

The Greek Cross—I

A Greek cross is formed by adjoining five squares as shown here. Cut out a Greek cross and then cut it into four pieces along the dotted lines determined by the midpoints P, Q, R, and S of their respective segments as illustrated. Show how to reassemble the pieces into a single square.

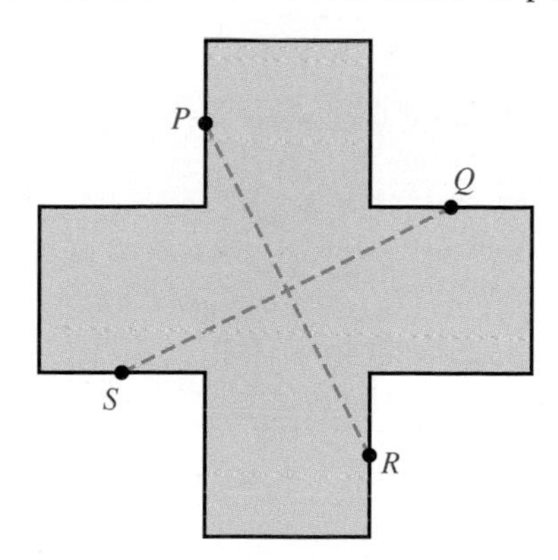

PROOF Consider parallel lines l and m as shown. Draw a horizontal line intersecting l and m at P and Q. Draw vertical lines that intersect l and m at points S and V and intersect the horizontal line at R and T as shown.

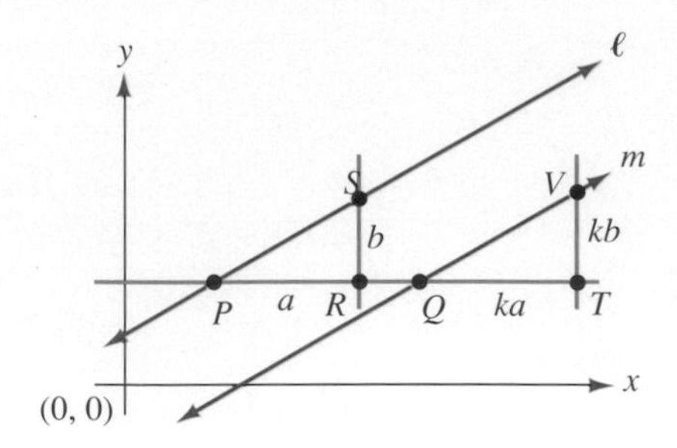

Since the corresponding sides of $\triangle PSR$ and $\triangle QVT$ are parallel, it follows that the triangles are similar. (Why?) Hence, as above

$$\frac{SR}{VT} = \frac{PR}{QT}$$

and

$$\text{slope } \overline{PS} = \frac{SR}{PR} = \frac{VT}{QT} = \text{slope } \overline{QV}.$$

Thus, parallel lines have equal slopes. The reverse argument is only slightly more difficult but we omit it here. ■

EXAMPLE 14.8

Showing that Line Segments Are Parallel

Let P, Q, R, and S be the points $(-7, -1)$, $(-1, 3)$, $(-1, -3)$, and $(8, 3)$ respectively. Show that $\overline{PQ}$ is parallel to $\overline{RS}$.

SOLUTION

By the preceding theorem, line segments are parallel if, and only if, they have the same slope. Since

$$\text{slope } \overline{PQ} = \frac{3 - (-1)}{-1 - (-7)} = \frac{4}{6} = \frac{2}{3}$$

and

$$\text{slope } \overline{RS} = \frac{3 - (-3)}{8 - (-1)} = \frac{6}{9} = \frac{2}{3},$$

it follows that the two segments are parallel. ■

EXAMPLE 14.9

Determining a Segment with a Given Slope

Let P be the point $(3, -4)$. Determine a point Q so that $\overline{PQ}$ has slope $-2/3$.

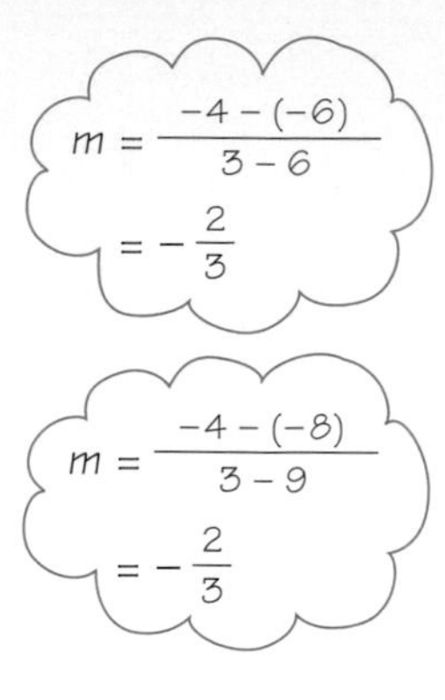

SOLUTION

Locate the point $(3, -4)$ on a coordinate system as shown. Since $\overline{PQ}$ must have slope $-2/3$, the segment slopes downward to the right. Also, the "rise" is -2 and the run is 3. Thus, we can find Q by counting down 2 units from $(3, -4)$ and to the right 3 units. Thus, Q is the point $(6, -6)$. Moreover, since the slope is a ratio, we could count down 4 units and to the right 6 units to determine $R(9, -8)$, such that $\overline{PR}$ also has slope $-2/3$. Similarly, counting backwards, we determine points, S, T, and V such that $\overline{PS}$, $\overline{PT}$, and $\overline{PV}$ all have slope $-2/3$. (Check these.) Apparently there are infinitely many points that satisfy the condition of the problem.

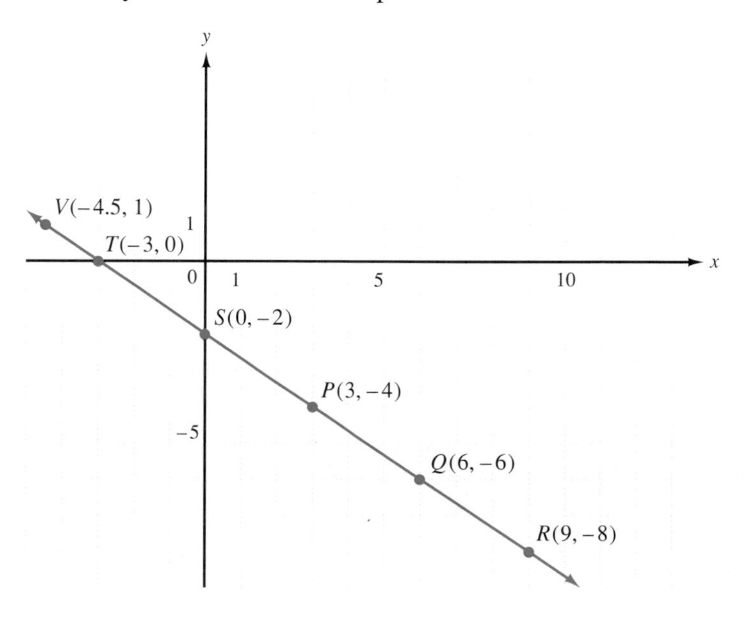

■

Two lines are parallel if, and only if, they have the same slope. What is the case when lines are perpendicular? Since parallel lines have the same slope, we can restrict our attention to perpendicular lines that meet at the origin.

Consider the diagram of Figure 14.8. Lines r and s are perpendicular and meet at the origin. Let $P(a, b)$ be a point in the first quadrant and on r, and let $\overline{PQ}$ be parallel

Figure 14.8
Slopes of perpendicular line segments

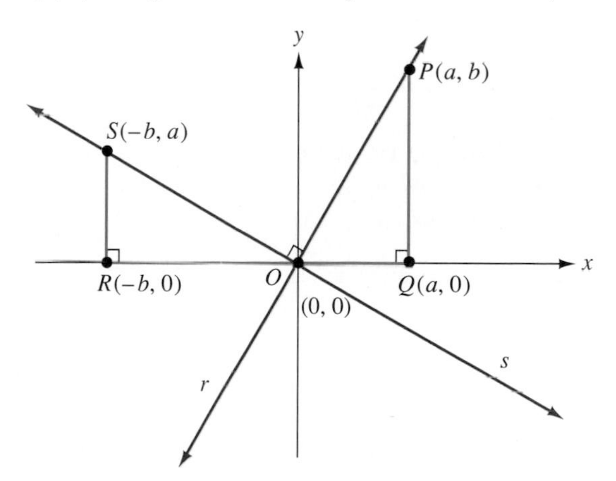

HIGHLIGHT FROM HISTORY
Maria Agnesi (1718–1799)

Maria Gaetana Agnesi was a child prodigy. Her father Pietro Gaetana, who was a professor of mathematics at the University of Bologna in Italy, recognized his daughter's talent and encouraged her studies. Before she was 13, Agnesi spoke many languages including Greek, Hebrew, and Latin. As a young woman, she gave talks on mathematics and philosophy to adult friends of her parents during parties at their home.

Agnesi's famous two-volume book, *Analytical Institutions,* took her ten years to write. It includes discussions of algebra, geometry, and calculus. In it, she describes the *Witch of Agnesi,* a curve first proposed by the French mathematician Fermat and studied extensively by Agnesi herself. Because *Analytical Institutions* was so clearly written, it was translated into French and English and used as a textbook. It is the first work of such stature that has survived that was written by a woman.

According to some accounts, Pope Benedict XIV appointed Agnesi to teach at the University of Bologna around 1750; however, it is possible that she never actually taught there. It is certain that Agnesi retired from mathematics just as her intellectual powers were at their peak. When in her forties, she decided to devote the remaining years of her life to helping the sick and poor. Maria Gaetana Agnesi died at the age of 81, leaving the world a rich scholastic and humanitarian legacy.

SOURCE: From *Portraits for Classroom Bulletin Boards: Women Mathematicians,* text by Virginia Slachman.

to the y-axis. Place R on the x-axis to the left of 0 so that $\overline{RO} \cong \overline{QP}$. Thus, R has coordinates $(-b, 0)$. Let $\overline{RS}$ be vertical and intersect s at S. It follows that $\triangle OPQ \cong \triangle OSR$. Hence, S has coordinates $(-b, a)$ as shown. But then

$$\text{slope } \overline{OP} = \frac{b-0}{a-0} = \frac{b}{a}$$

and

$$\text{slope } \overline{OS} = \frac{a-0}{-b-0} = -\frac{a}{b}.$$

This shows that the slopes of perpendicular lines have the property that one is the negative of the reciprocal of the other. Equivalently, the product of the slopes of perpendicular lines is -1 since

$$\frac{b}{a} \cdot \left(-\frac{a}{b}\right) = -1.$$

The converse of these last assertions is also true but we omit the proof. In any case, the condition for perpendicularity can be formalized as a theorem.

THEOREM Condition for Perpendicularity

Let two line segments (or lines) have slopes m_1 and m_2. Then the segments (lines) are perpendicular if, and only if, $m_1 m_2 = -1$ or, equivalently,

$$m_2 = -\frac{1}{m_1}.$$

EXAMPLE 14.10

Showing that the Diagonals of a Rhombus Are Perpendicular

Consider the quadrilateral $A(0, 0)$, $B(1, \sqrt{3})$, $C(3, \sqrt{3})$, and $D(2, 0)$.

(a) Show that the figure $ABCD$ is a rhombus.

(b) Show that the diagonals $\overline{AC}$ and $\overline{BD}$ are perpendicular.

SOLUTION

It is helpful to make a sketch to assist in the analysis of the problem.

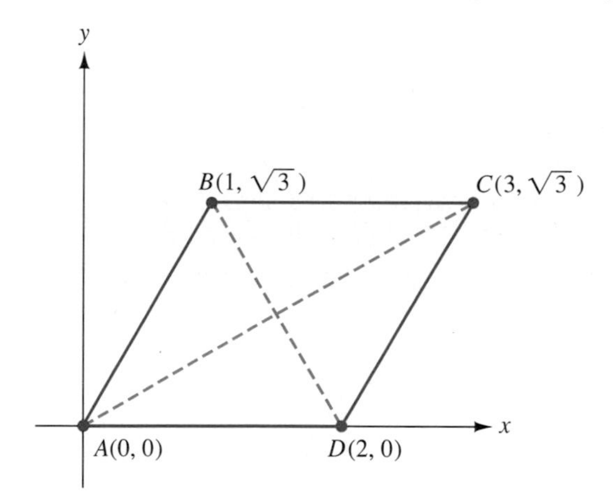

(a) To show that the figure is a rhombus, we must show that the sides have equal lengths. Clearly, $AD = BC = 2$. Moreover, by the distance formula,

$$AB = \sqrt{(1-0)^2 + (\sqrt{3}-0)^2} = \sqrt{1+3} = 2$$

and

$$DC = \sqrt{(3-2)^2 + (\sqrt{3}-0)^2} = \sqrt{1+3} = 2.$$

(b) Using the formula that defines the slope of a line segment

$$\text{slope } \overline{AC} = \frac{\sqrt{3}-0}{3-0} = \frac{\sqrt{3}}{3}$$

and

$$\text{slope } \overline{BD} = \frac{\sqrt{3}-0}{1-2} = \frac{\sqrt{3}}{-1} = -\sqrt{3}.$$

Since

$$\frac{\sqrt{3}}{3} \cdot (-\sqrt{3}) = \frac{-3}{3} = -1,$$

it follows from the preceding theorem that $\overline{AC}$ and $\overline{BD}$ are perpendicular. ■

Equations of Lines

Using the tools of coordinate geometry it is now possible to write equations whose graphs are lines. Then, just as we could use coordinates alone to prove geometric results as above, we can use algebra and the equations of lines to obtain other geometric results. Conversely, we can use geometric ideas to clarify and/or demonstrate results in algebra.

We begin by considering a particular example.

EXAMPLE 14.11 **Determining the Equation of a Line Through (2, 3) with Slope 4/3**

Derive an equation of the line through point $P(2, 3)$ and having slope $4/3$.

SOLUTION

Understand the problem • • • •

There is one, and only one, line through the point (2, 3) and having slope 4/3. One can draw the line by plotting the point $P(2, 3)$ and then plotting the point $Q(5, 7)$ 3 units to the right and 4 units above P. The line segment through these points must have slope 4/3, so the line through these points must be the desired line. Next suppose $R(x, y)$ is *any* point on the line. We must find an equation involving x and y that is satisfied by those, and only those, points that lie on the line.

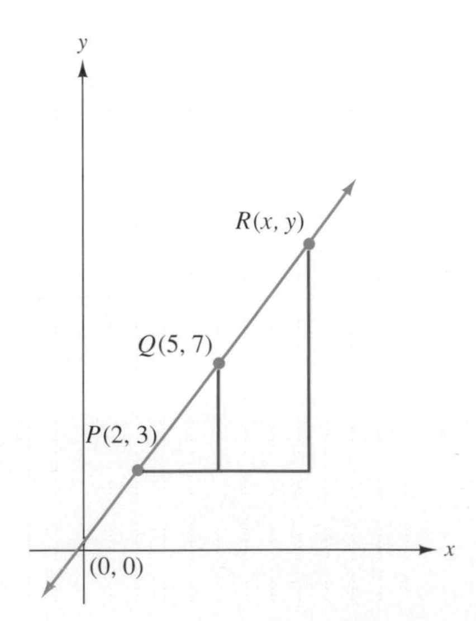

Devise a plan • • • •

Since the slope of a line can be determined by *any* two points on the line, $R(x, y)$ is on the line if, and only if,

$$\text{slope } \overline{PR} = \frac{4}{3}.$$

Perhaps we can use this fact to derive the desired equation.

Carry out the plan • • • •

Since

$$\text{slope } \overline{PR} = \frac{y-3}{x-2},$$

it follows that R is on the line in question if, and only if,

$$\frac{y-3}{x-2} = \frac{4}{3}$$

or, alternatively,

$$y - 3 = \frac{4}{3}(x - 2).$$

Hence, this must be the desired equation.

Look back • • • •

Since there is one, and only one, line through a given point and having a given slope, $R(x, y)$ is on the desired line if, and only if, slope $\overline{PR} = 4/3$. By expressing slope $\overline{PR}$ in terms of x and y, we obtained the desired equation. ■

The preceding example is typical and the result can be stated as a theorem which would be proved by exactly the same argument as that of Example 14.11.

THEOREM Point-Slope Form of the Equation of a Line

The equation of the line through $P(a, b)$ and having slope m is

$$y - b = m(x - a).$$

This is called the **point-slope form** of the equation of a line.

As this theorem suggests, there are several forms of the equation of a line. Another particularly useful form is stated in the next theorem. First we note that if a line crosses the y-axis at the point $(0, b)$, b is called the **y-intercept** of the line.

THEOREM Slope-Intercept Form of the Equation of a Line

The **slope-intercept form of the equation of a line** is

$$y = mx + b$$

where m is the slope and b is the y-intercept.

PROOF Since b is the y-intercept, the line passes through the point $(0, b)$. Also, it has slope m. Therefore, by the point-slope form of the equation of a line, the desired equation is

$$y - b = m(x - 0)$$

or, equivalently,

$$y = mx + b$$

as claimed. ■

EXAMPLE 14.12 Using the Slope-Intercept Form of the Equation of a Line

Write the equations of the following lines with the slope and y-intercept as indicated. Also, draw each line on a coordinate system.

(a) $m = -3, \quad b = 5$ **(b)** $m = 0, \quad b = -4$

SOLUTION

(a) Using the above theorem, we obtain the equation $y = -3x + 5$. To draw the line we plot the point $(0, 5)$ and the point $(1, 2)$ which is 3 units *below* and 1 unit to the right of $(0, 5)$. Then draw the line through these 2 points.

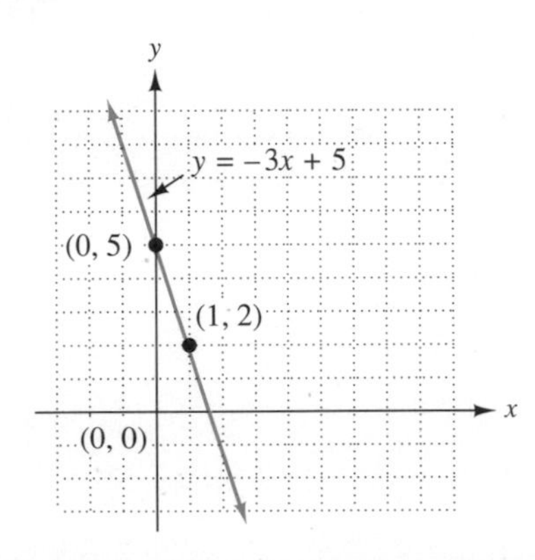

(b) This line goes through the point $(0, -4)$ and has slope 0. Therefore, using the slope-intercept form, we obtain the equation

$$y = 0x + (-4)$$

or just

$$y = -4.$$

This says that the line is horizontal and that a point is on this line if, and only if, its y-coordinate is -4. The x-coordinates of these points are unrestricted. The line is as shown.

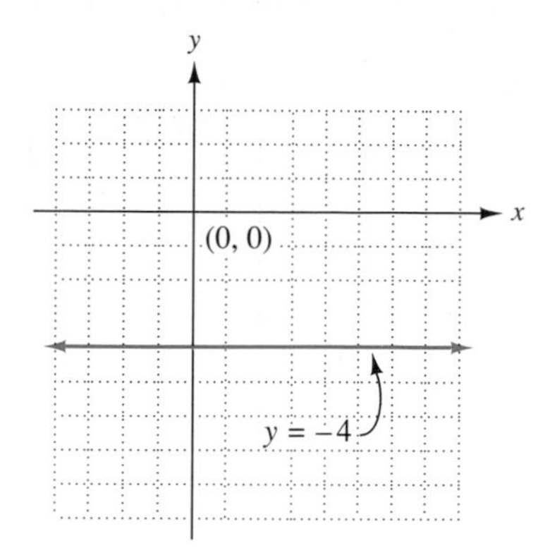

■

The preceding example suggests that *every* horizontal line has an equation of the form $y = b$; the slope is 0 and the x values of points on the lines are left unrestricted. By analogy, *every* vertical line has an equation of the form $x = a$; there is *no* slope and the y values of points on the lines are left unrestricted.

EXAMPLE 14.13

Determining the Equation of a Line Through Two Points

(a) Determine the equation of the line through $P(3, 5)$ and $Q(-2, 1)$.

(b) Determine the equation of the line through $P(3, 5)$ and perpendicular to the line in part (a).

SOLUTION

(a) The slope of the desired line is the slope of the segment $\overline{PQ}$;

$$\text{slope } \overline{PQ} = \frac{5 - 1}{3 - (-2)} = \frac{4}{5}.$$

We now use the point-slope form of the equation of the line using either P or Q. Using P we have

Multiply both sides by 5 and use the distributive property.

$$y - 5 = \frac{4}{5}(x - 3),$$
$$5y - 25 = 4x - 12,$$
$$4x - 5y + 13 = 0.$$

Subract $5y - 25$ from both sides.

Or, using Q as the point, we have

$$y - 1 = \frac{4}{5}(x - (-2))$$

which can be simplified to

$$4x - 5y + 13 = 0$$

as before. It makes no difference which of the two points is used.

(b) Since the slopes of perpendicular lines are negative reciprocals of one another, the slope of the line through P and perpendicular to the line of part (a) is $-5/4$. Hence, the desired equation is

$$y - 5 = -\frac{5}{4}(x - 3)$$

which can be simplified to

$$5x + 4y - 35 = 0.$$

■

Lastly, we observe that all equations of lines considered so far can be written in the form

$$Ax + By + C = 0$$

where A, B, and C are real numbers and not both A and B are zero. Such an equation is called a **linear equation** and every such equation is the equation of a line.

- If $A = 0$ and $B \neq 0$, the line is horizontal and crosses the y-axis at the point $(0, -C/B)$.
- If $A \neq 0$ and $B = 0$, the line is vertical and crosses the x-axis at $(-C/A, 0)$.
- If $A \neq 0$ and $B \neq 0$, the line has slope $-A/B$, which is positive if A and B have opposite signs and negative if A and B have the same sign.

Intersections of Lines

By definition and as illustrated in Figure 14.9, two distinct lines must either be parallel or they must intersect in a single point. Since the graph of a line consists of all points whose coordinates satisfy the equation of the line, if two lines intersect at a common point (a, b) that point is said to be a **simultaneous solution of the two linear equations** corresponding to the two lines. Thus, from the geometry:

- two linear equations have precisely one simultaneous solution, or
- two linear equations have no simultaneous solution, or
- two linear equations actually represent the same line and so have infinitely many simultaneous solutions.

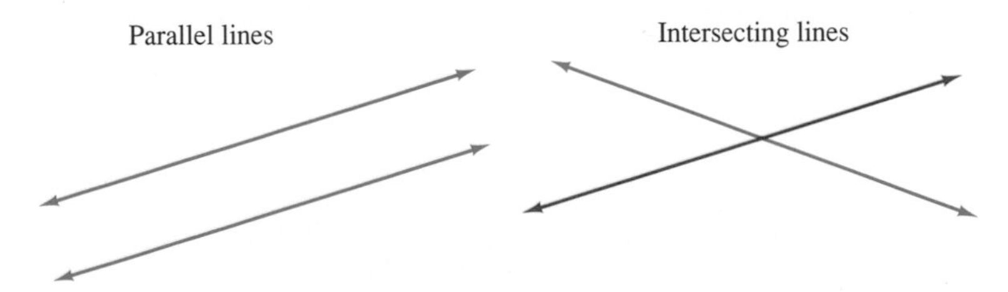

Figure 14.9
Two distinct lines must be parallel or they must intersect in a single point.

EXAMPLE 14.14 **Determining Simultaneous Solutions of Linear Equations**

(a) Show that the lines $2x + 5y = -11$ and $3x - 2y = 12$ are not parallel.

(b) Find the simultaneous solution to the equations in part (a).

SCHOOL BOOK PAGE *Graphing in Grade Five*

SCIENCE CONNECTION

Number Relationships and Ordered Pairs

Build Understanding

A. The distance in feet between you and a lightning strike can be computed if you count the seconds between the flash and the thunder. Then multiply the number of seconds by 1,100.

How far away was a lightning strike if the time between the flash and thunder was 4 seconds?

The relationship between the time and distance can be shown in a table and as ordered pairs plotted on a graph.

Time (seconds)	Distance (feet)	Ordered Pair
1	1,100	(1, 1,100)
2	2,200	(2, 2,200)
3	3,300	(3, 3,300)
4	4,400	(4, 4,400)

The lightning strike was 4,400 feet away.

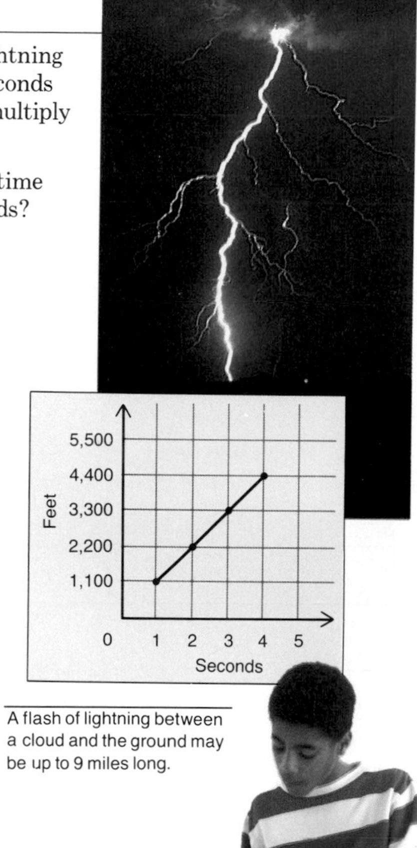

A flash of lightning between a cloud and the ground may be up to 9 miles long.

B. The relationship between numbers can be stated using a rule. In the table below, the relationship between A and B is given by the rule A + 5. The rule can be used to find more numbers or missing numbers.

Rule: A + 5

A	B	A + 5
7	12	
18	23	
26	▒	26 + 5
▒	41	n + 5

Talk About Math Can the graph in Example A be extended to show longer amounts of time? Why or why not?

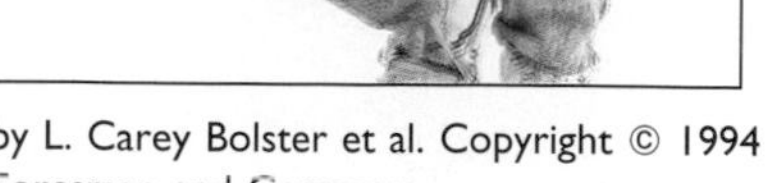

1. Discuss how you could use this page to begin to introduce students to the notion of the equation of a line.
2. What equation relates A and B in part B above?
3. Consider part A above. How far would you be from a lightning strike if it is 10 seconds between seeing the strike and hearing the thunder?

SOLUTION

(a) Solving for y in terms of x to obtain the slope-intercept forms of the equations of the lines in question, we obtain

$$y = -\frac{2}{5}x - \frac{11}{5} \quad \text{and} \quad y = \frac{3}{2}x - 6.$$

Thus, the slopes of the lines are $-2/5$ and $3/2$ and the lines are not parallel.

(b) Since the x and y values of the point of intersection of the lines must be the same, it follows from part (a) that, at this point,

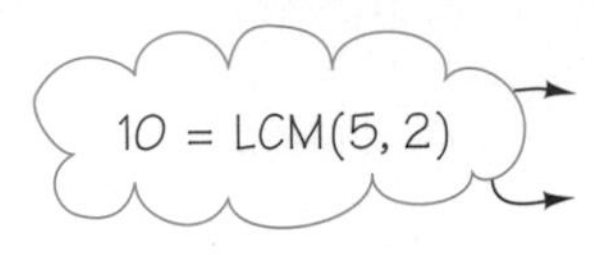

$$-\frac{2}{5}x - \frac{11}{5} = \frac{3}{2}x - 6.$$

Multiplying through by 10 to eliminate fractions we have

$$\begin{aligned} -4x - 22 &= 15x - 60, \\ -19x - 22 &= -60, \\ -19x &= -38, \end{aligned}$$

and

$$x = 2.$$

Now, substituting x for 2 in either equation in part (a), we obtain

$$y = -\frac{2}{5} \cdot 2 - \frac{11}{5} = -\frac{4}{5} - \frac{11}{5} = \frac{-15}{5} = -3.$$

Even more easily, from the other equation,

$$y = \frac{3}{2} \cdot 2 - 6 = 3 - 6 = -3.$$

Thus, the simultaneous solution is the point $(2, -3)$.

The solution can be checked by drawing the lines and reading off the coordinates of the point of intersection as shown here.

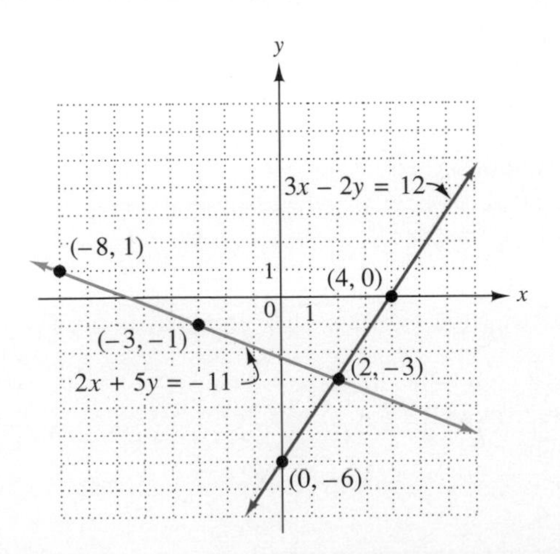

WORTH READING

In Mathematical Circles

The ingenious honey bees. Man has long shown interest in the seeming geometrical sagacity of the honey bees. The first known man to report on this mathematical acumen was the eminent Greek geometer Pappus, who flourished in Alexandria some sixteen hundred years ago. In Book V of his famous *Mathematical Collection* we find the following passage concerning the extremum properties of the cells of the bees' honeycombs:

> Presumably because they know themselves to be entrusted with the task of bringing from the gods to the accomplished portion of mankind a share of ambrosia in this form, they do not think it proper to pour it carelessly on ground or wood, or any other ugly or irregular material; but first collecting the sweets of the most beautiful flowers which grow upon the earth, they make from them for the reception of the honey, the vessels which we call honeycombs (with cells) all equal and similar, and contiguous to one another, and hexagonal in form. And that they have contrived this by virtue of a certain geometrical forethought, we may infer in this way. They would necessarily think that the figures must be such as to be contiguous to one another, that is to say, to have their sides in common in order that no foreign matter could enter into the interstices between them and so defile the purity of the produce. Now there are three rectilinear figures which are capable of fulfilling this condition, I mean regular figures which are equilateral and equiangular, for bees would have none of figures which are not uniform. . . . There being then three figures capable by themselves of exactly filling up the space about the same point, the bees by reason of their instinctive wisdom chose for their construction the figure which has most angles, because they conceived it would hold more honey than either of the other two.

This excerpt is item 4 in a collection of 360 interesting and entertaining mathematical vignettes on mathematics and mathematicians by Howard Eves. The two volume collection is titled *In Mathematical Circles* (Prindle, Weber and Schmidt, Inc., Boston, 1969) and provides a wonderful supply of tidbits that can be used to enliven any mathematics class.

Since two points determine a line, we plot two points for each equation. For the first equation, if $y = 1$, then $x = -8$, and if $y = -1$, then $x = -3$. Therefore, $(-8, 1)$ and $(-3, -1)$ determine the first line. Similarly $(4, 0)$ and $(0, -6)$ determine the second line. We read off $(2, -3)$ as the simultaneous solution as shown. ■

PROBLEM SET 14.2

Understanding Concepts

1. Compute the slopes of the line segments determined by these pairs of points. In each case tell whether the segment is vertical, horizontal, slopes upward to the right, or slopes downward to the right.
   (a) $P(1, 4)$, $Q(3, 8)$ **(b)** $R(-2, 5)$, $S(-2, -6)$
   (c) $U(-2, -3)$, $V(-4, -7)$
   (d) $C(3, 5)$, $D(-3, 5)$
   (e) $E(1, -2)$, $F(-2, -5)$
   (f) $G(-2, -2)$, $H(4, -5)$
2. Determine b so that the slope of $\overline{PQ}$ is 2 where P and Q are the points $(b, 3)$ and $(4, 7)$ respectively.
3. Determine d so that the slope of $\overline{CD}$ is undefined if C and D are the points $(d, 3)$ and $(-5, 5)$ respectively.
4. Determine e so that $\overline{EF}$ is horizontal if E and F are the points $(-3, -5)$ and $(2, e)$ respectively.
5. Determine a if the point $(a, 3)$ is on the line $2x + 3y = 18$.
6. **(a)** Determine two different points on the line $3x + 5y + 15 = 0$.
   (b) Draw the graph of the line in part (a) on a coordinate system.
7. Graph each of these lines on a single coordinate system and label each line.
   (a) $3x + 5y = 12$ **(b)** $6x = -10y + 12$
   (c) $5y - 3x = 15$ **(d)** $6x + 10y = 24$
   (e) What do you conclude about the lines of parts (a) and (d)?

8. Graph each of these lines on a coordinate system.
(a) $2y - 16 = 0$ (b) $x = -7$ (c) $x + y = 2$
(d) $x - y = 4$ (e) $y = 3x + 4$
(f) $y = -2x + 6$

9. (a) Determine two points on the line $4x + 2y = 6$.
(b) Use the points determined in part (a) to compute the slope of the line.
(c) Solve the equation in part (a) for y in terms of x and so again determine the slope of the line and also the y-intercept. (*Hint:* Solving for y in terms of x gives the slope-intercept form of the equation of a line.)

10. In each case determine k so that the line is parallel to the line $3x - 5y + 45 = 0$.
(a) $7x + ky = 21$ (b) $kx - 8y - 24 = 0$
(c) $y = kx + 5$ (d) $x = ky + 5$

11. Determine the slope and y-intercept of each of these lines.
(a) $3x - 7y + 21 = 0$ (b) $2x + 5y = 20$
(c) $y = 6$ (d) $y = 0.3x + 15$
(e) $x = -5$ (f) $2x = 3y - 18$

12. Determine whether these pairs of lines are parallel, perpendicular, or neither parallel nor perpendicular.
(a) $3x + 7y + 15 = 0$, $6x + 15y + 31 = 0$
(b) $2x - 5y = 7$, $4x = 10y + 20$
(c) $x + 4y + 6 = 0$, $8x = 2y + 13$
(d) $7x + 3y + 21 = 0$, $3x - 7y + 21 = 0$
(e) $y = 3x + 15$, $6x - 2y + 12 = 0$

13. Determine k so that $2x + ky + 6 = 0$ is
(a) parallel to $3x - 5y = 15$.
(b) perpendicular to $3x - 5y = 15$.

14. Determine if these are equations of lines that intersect. If the lines intersect, determine the coordinates of the point of intersection; that is, determine the simultaneous solution to the pair of equations.
(a) $2x - 3y = 9$, $4x - 4y = 16$
(b) $3x + 5y = 15$, $6x + 10y = 30$
(c) $4x - 3y = 12$, $8x - 6y = 0$

15. (a) Write the equation of the line perpendicular to the line $5x - 2y = 10$ and which passes through the point $(-1, 7)$.
(b) Determine the point where the two lines of part (a) intersect.

Thinking Critically

16. (a) Compute the midpoint of the segment $\overline{PQ}$ where P and Q are $(3, -5)$ and $(5, 9)$ respectively.
(b) Compute the slope of $\overline{PQ}$ from part (a).
(c) Use the results of (a) and (b) and determine the equation of the perpendicular bisector of $\overline{PQ}$.

17. A point is on the perpendicular bisector of a segment if, and only if, it is equidistant from the end points of the segment. Use this fact and the distance formula to determine the perpendicular bisector of the segment $\overline{PQ}$ in problem 16. (*Hint:* $(r - s)^2 = r^2 - 2rs + s^2$.)

18. Determine the equation of the perpendicular bisectors of the segments determined by these pairs of points.
(a) $P(3, 5)$, $Q(-1, 7)$ (b) $R(4, -6)$, $S(8, -4)$
(c) $C(3, -2)$, $D(4, 6)$ (d) $E(3, 7)$, $F(3, -4)$
(e) $G(-1, 5)$, $H(5, 5)$ (f) $I(2, 2)$, $J(-3, 5)$

19. (a) Determine the equations of the three altitudes of the triangle with vertices $A(0, 0)$, $B(3, 6)$, and $C(9, 0)$.
(b) Show that the three altitudes of part (a) meet at a common point. (*Hint:* Determine the point of intersection of each pair of equations in part (a).)
(c) Draw the triangle and graph the equations of part (a) on a coordinate system.

20. Show that the three perpendicular bisectors of the sides of the triangle in problem 19 meet at a common point.

21. Show that the medians of the triangle in problem 19 meet in a common point.

22. Show that the point where the three medians in problem 21 meet is two-thirds of the way from each vertex to the midpoint of the opposite side.

23. Show that the three points determined in problems 19, 20, and 21 are collinear. The line on which they lie is called the **Euler line.** (*Hint:* If P, Q, and R are collinear, what must be true about slope $\overline{PQ}$ and slope $\overline{PR}$?)

24. Determine the shortest distance from the point $(1, 5)$ to the line $3x - 2y = 6$.

Communicating

25. (a) On four separate coordinate systems sketch these lines.
(i) $\frac{x}{2} + \frac{y}{5} = 1$ (ii) $\frac{x}{4} + \frac{y}{-3} = 1$
(iii) $\frac{x}{-4} + \frac{y}{-3} = 1$ (iv) $\frac{x}{-5} + \frac{y}{2} = 1$
(b) On the basis of your graphs in part (a), carefully discuss the significance of a and b for the graph of $\frac{x}{a} + \frac{y}{b} = 1$.

26. If $a \neq 0$ and $b \neq 0$, discuss what can be said about the lines

$$ax + by = c \quad \text{and} \quad bx - ay = d.$$

(*Hint:* Compute the slope in each case.)

Making Connections

27. **(a)** The maximum allowable slope for a ramp for disabled persons using a wheelchair is $1/20$. If the sill of the door to a building is 3.5 feet above the level of the sidewalk, at least how far from the building must a ramp for wheelchair users begin?

3.5′

(b) How long is the shortest allowable ramp in part (a)?

28. It is customary to design highways so that the grade (slope) never exceeds 6%. If it is necessary to exceed this limit, truckers are always warned by a sign indicating how long the stretch of highway is with the grade exceeding 6%.

(a) If it is 9 miles from Lenore to Pierce and the elevation of Pierce is 3000 feet greater than that of Lenore, what is the slope of the straight line connecting the two towns? There are 5280 feet in a mile.

(b) Is it possible to build a highway from Lenore to Pierce without exceeding the 6% restriction on the grade? Explain.

---

29. Scientists often need to determine the relationship between two quantities, say x and y. This is often done by determining corresponding values of the quantities, plotting the points determined by these number pairs on a coordinate system, then drawing a line that seems to fit the plotted points most closely, and finally determining the equation of this line. The line described is called the **line of best fit** and there is a technical method for determining it. Here we do it "by eye." For example, the data points and the "line of best fit" might appear as shown.

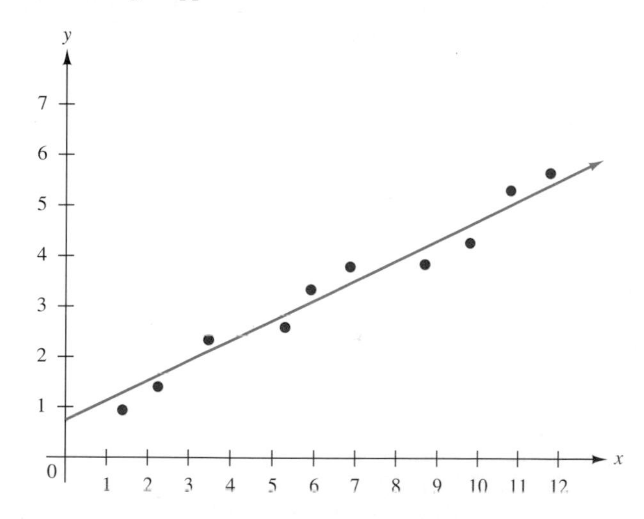

Since the points $(1, 1.1)$ and $(10, 5)$ appear to lie on the line, the equation of the line of best fit is approximately

$$y - 5 \doteq \frac{5 - 1.1}{10 - 1}(x - 10)$$

which simplifies to

$$y \doteq 0.43x + 0.67.$$

Using this equation, we could predict that, when $x = 20$, $y \doteq 9.27$. By eye, draw the line of best fit for each of the data sets illustrated, determine the equation of the line, and predict the value of y when $x = 15$.

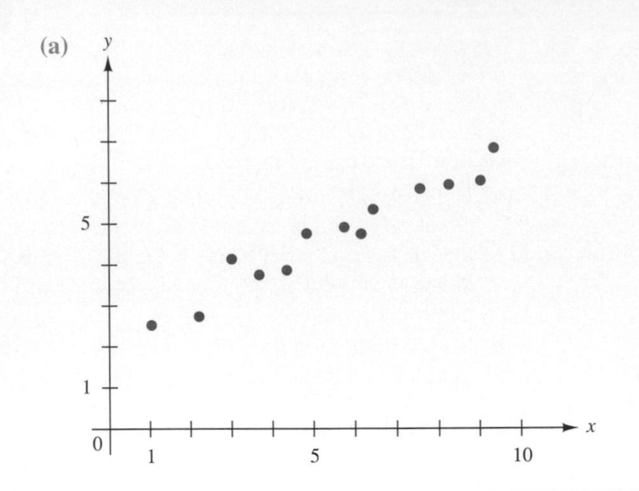

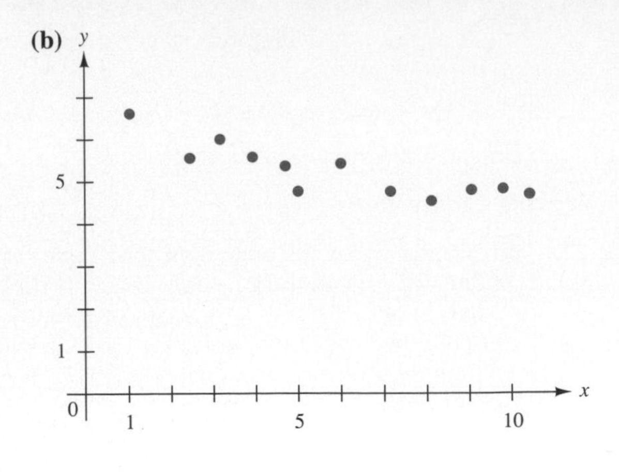

For Review

30. Consider the figures shown here.

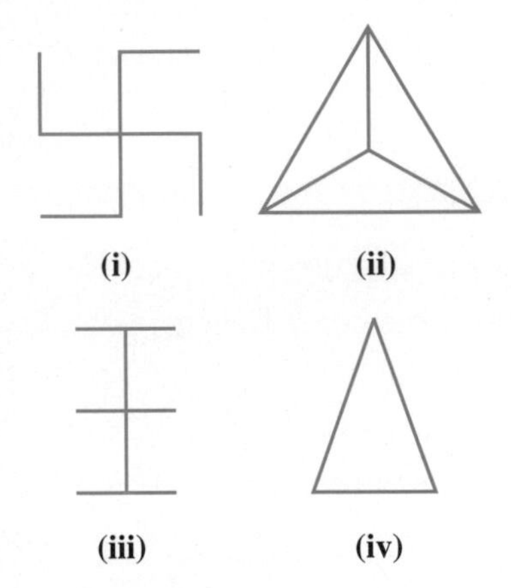

(a) Which of the figures has precisely one line of symmetry?
(b) Which of the figures has precisely two lines of symmetry?
(c) Do any of the figures have more than two lines of symmetry?
(d) Which of the figures is symmetric about a point?
(e) List those figures that have rotational symmetry and indicate the angle or angles of rotation in each case.

31. If a geometrical figure is symmetric about a point, must it have rotational symmetry? Explain briefly.

32. If a plane geometric figure has one or more rotational symmetries, must it also have symmetry about a point? Explain briefly.

33. Using graph paper, copy the figure shown and draw its reflection in the given line.

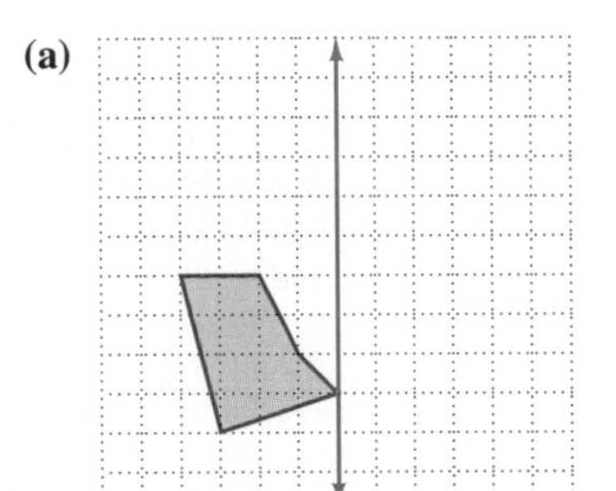

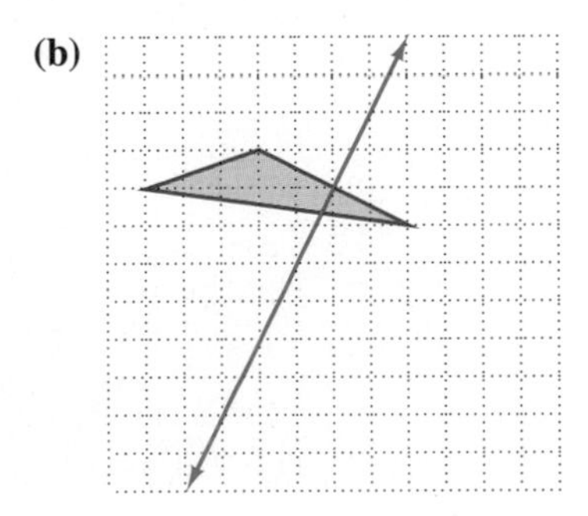

34. Using graph paper, copy the figure shown and draw the figure that results from rotating the given figure through an angle of 90° counterclockwise about the point P.

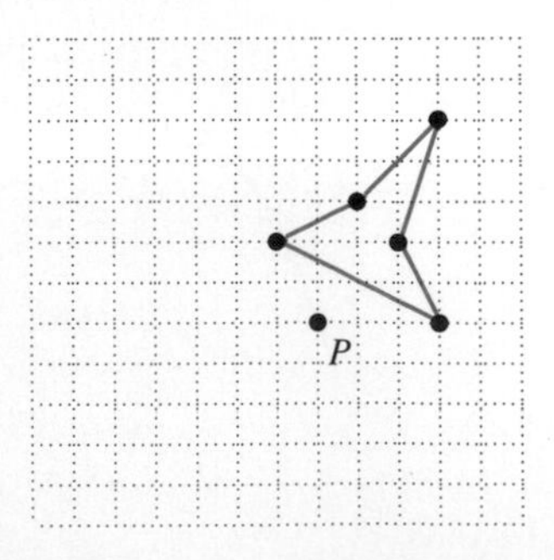

COOPERATIVE Investigation 2

Stretching a Spring

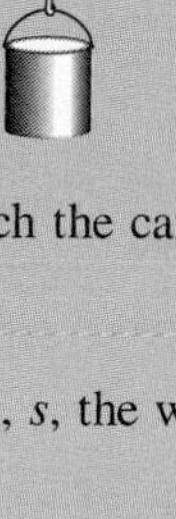

Materials Needed

1. A screen door spring about 20″ long.
2. Several objects weighing between 0.5 and 2 pounds.
3. A tin can that can be suspended from the spring.
4. A scale.
5. A meter stick.

Directions

1. Attach the spring to a support as indicated in the diagram and attach the can to the spring.
2. Weigh each object and record its weight, w.
3. Place each object in the can on the spring and measure the amount, s, the weight *stretches* the spring.
4. Plot the points (w, s) on an appropriate coordinate system.
5. Draw the line of best fit and determine its equation in terms of w and s.
6. Use the equation of the line of best fit to predict s when $w = 3$ lbs.
7. Actually, measure s when $w = 3$ lbs. and see how close your approximation is.

COOPERATIVE Investigation 3

Swinging a Pendulum

Materials Needed

1. A pendulum made by tying three or four heavy washers together on the end of a string about 150 centimeters long.
2. A stopwatch.
3. Graph paper for each student.
4. A calculator for each student or pair of students.

The Problem

The problem is to determine the relationship between the length, l, of the pendulum and the time, t, required for the pendulum to complete 10 complete swings back and forth.

Directions

Step 1. Mark the string at 10 centimeter intervals starting where it ties onto the washers.

Step 2. Hold the pendulum at lengh, $l = 10$, centimeters on a solid support (say a coat rack) and carefully note the time in seconds for 10 complete swings.

Step 3. Repeat Step 2 for $l = 20, 30, 40, 50,$ and 60 centimeters.

Step 4. Let $x = \sqrt{l}$ and plot the points (t, x) for the measurements in Steps 2 and 3.

Step 5. Draw a line of best fit for the points in Step 4 and determine an equation relating t and x and hence t and $\sqrt{l}$.

Step 6. Use the formula in Step 5 to predict how long it will take for 10 swings of the pendulum if $l = 100$ centimeters.

Step 7. Actually time 10 swings of the pendulum when $l = 100$ to check the accuracy of your prediction in Step 6.

14.3 Solving Geometric Problems Using Coordinates

Proofs Using Coordinates

The problems and examples in sections 14.1 and 14.2 suggest that many of the general theorems of geometry can be obtained using the techniques of coordinate geometry and methods of algebra. Suppose, for example, that you are asked to show that the line segment joining the midpoints of two sides of a triangle is parallel to the third side. This does not mean we are to show that the property holds for a particular triangle, or even for a certain type of triangle—equilateral, isosceles, scalene, right, or acute. We are asked to show that this is a general result, true for all triangles. This means that the argument we use must be completely general and must not depend on

INTO THE CLASSROOM

Henry Richard on Using Coordinate Geometry in His Mathematics Classes

When I teach my classes about using ordered pairs to locate points, I like to use visual aids and concrete examples. I create a bulletin board/learning center that makes a cross-curricular connection with social studies. I use an old wall-sized map of the United States and make a grid over the map with heavy black knitting yarn. I staple not only the ends but also each piece of yarn where it intersects with the other. I label the vertical pieces with consecutive letters and the horizontal pieces with consecutive numbers.

I prepare 3-in.-by-5-in. cards, each with one pair of ordered numbers. The location of each ordered pair should pinpoint a fairly large city on the map. For example, F-5 might indicate Chicago, H-6 Pittsburgh, and A-2 San Francisco, and so forth. Students match each pair of ordered pairs with a major city. After completing all the cards, students put the cities in any logical order from east to west that could be used for a trip across the United States.

Not only does the activity provide for reinforcement of the ordered-pair concept, it also affords students an opportunity to develop a more acute awareness of United States geography.

Another practical application for ordered pairs, again using a cross-curricular connection with social studies, is to demonstrate latitude and longitude as ordered pairs on a large world map. To begin the activity, I again prepare 3-in. by-5-in. cards, each listing a latitude and longitude. I divide the class into two teams. One member from Team A selects a card from the stack and finds the location of the ordered pair on the world map. If the student cannot find the location correctly, the card is placed back in the stack and Team B has a turn. The game continues until all the cards have been used, and one team is declared the winner.

To provide additional student involvement, I have each student make an ordered-pair card listing a latitude and longitude. The cards are placed in a stack, shuffled, and chosen randomly. As a variation, I sometimes have students take turns orally giving an ordered pair to members of the opposite team.

 Henry Richard teaches at Governor Wolfe School in Bethlehem, Pennsylvania.

Remember: any real number can be written as twice another real number.

special properties of some particular triangle. Consider any triangle—say the one shown in Figure 14.10. This is not any particular triangle, just one that we drew to assist our thought processes. To continue our analysis using coordinate geometry, we add axes to the drawing as in Figure 14.11 with A at the origin, C on the positive x-axis, and B above the x-axis as shown.

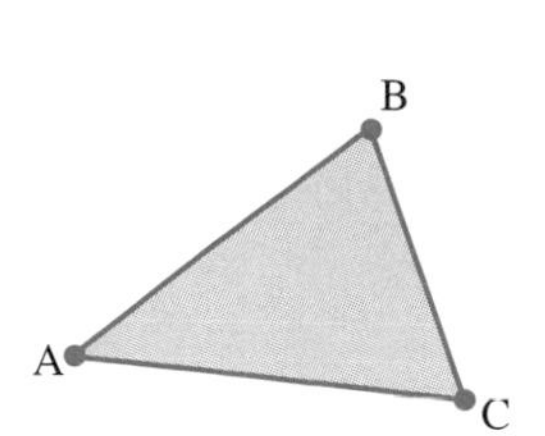

Figure 14.10
Triangle *ABC*

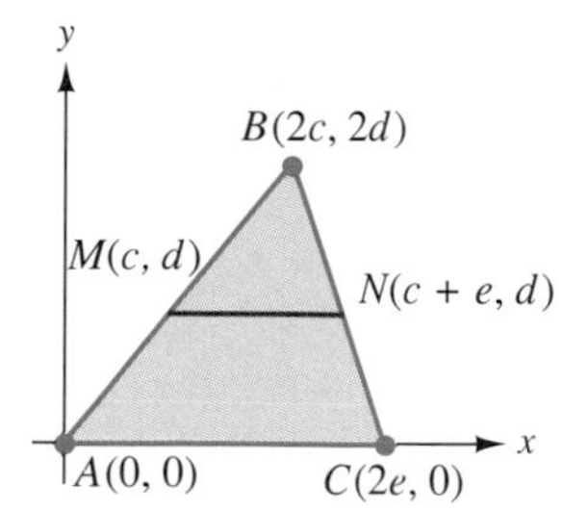

Figure 14.11
Triangle *ABC* on a coordinate system

Then A has coordinates $(0, 0)$, B has coordinates $(2c, 2d)$ and C has coordinates $(2e, 0)$ for suitable real numbers c, d, and e. Since this could be done for *any* triangle and c, d, and e are not further specified, the argument will apply to *any* triangle; it is completely general. Let M and N be the midpoints of $\overline{AB}$ and $\overline{BC}$ respectively. Then, M and N have coordinates (c, d) and $(c + e, d)$ respectively and

$$\begin{aligned} \text{slope } \overline{MN} &= \frac{d - d}{c - (c + e)} \\ &= 0. \end{aligned}$$

Since

$$\begin{aligned} \text{slope } \overline{AC} &= \frac{0 - 0}{2e - 0} \\ &= 0, \end{aligned}$$

it follows that $\overline{MN}$ is parallel to $\overline{AC}$ as we were to show.

To illustrate the method further, we consider two more examples.

EXAMPLE 14.15

Proving a Result via Coordinates

Show that the line segments joining the midpoints of the sides of a quadrilateral form a parallelogram.

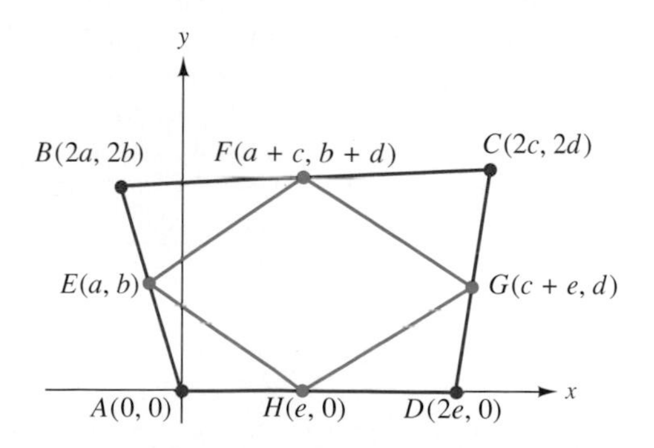

SOLUTION

Any quadrilateral can be placed on a coordinate system with one vertex at the origin, one vertex on the positive x-axis, and the other two vertices above the x-axis. Let quadrilateral $ABCD$ be so placed as shown. Then, for suitable numbers a, b, c, d, and e, with $e \neq a$, the coordinates of the vertices are $A(0, 0)$, $B(2a, 2b)$, $C(2c, 2d)$, and $D(2e, 0)$. By the midpoint formula, the midpoints of $\overline{AB}$, $\overline{BC}$, $\overline{CD}$, and $\overline{DA}$ are respectively $E(a, b)$, $F(a + c, b + d)$, $G(c + e, d)$ and $H(e, 0)$. Therefore,

$$\text{slope } \overline{EF} = \frac{(b + d) - b}{(a + c) - a} = \frac{d}{c},$$

$$\text{slope } \overline{FG} = \frac{(b + d) - d}{(a + c) - (c + e)} = \frac{b}{a - e},$$

$$\text{slope } \overline{GH} = \frac{d - 0}{(c + e) - e} = \frac{d}{c},$$

and

$$\text{slope } \overline{HE} = \frac{b - 0}{a - e} = \frac{b}{a - e}.$$

This shows that $\overline{EF}$ is parallel to $\overline{GH}$ and $\overline{FG}$ is parallel to $\overline{HE}$. Thus, $EFGH$ is a parallelogram as was to be shown. ■

EXAMPLE 14.16 Proving that the Diagonals of a Rhombus Are Perpendicular

Show that the diagonals of a rhombus are perpendicular.

SOLUTION

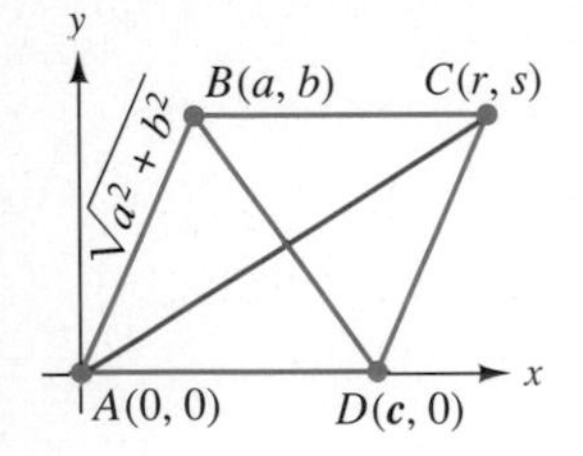

Any rhombus can be placed on a coordinate system so that one vertex is at the origin, one vertex is on the positive x-axis, and the other two vertices are in quadrant I as shown. Let $A(0, 0)$, $B(a, b)$, $C(r, s)$, and $D(c, 0)$ be the vertices of the rhombus as shown. Since a rhombus is a parallelogram, $\overline{BC}$ is parallel to $\overline{AD}$ and so is horizontal. Thus, $s = b$. Also, the sides of a rhombus are the same length so

$$r = a + c \qquad \text{and} \qquad c = \sqrt{a^2 + b^2}.$$

This implies that

$$c^2 = a^2 + b^2$$

or, equivalently, that

$$-b^2 = a^2 - c^2.$$

To show that $\overline{AC}$ and $\overline{BD}$ are perpendicular, we consider their slopes.

$$\text{slope } \overline{AC} = \frac{s - 0}{r - 0}$$

$$= \frac{b}{a + c}$$

since $s = b$ and $r = a + c$ from above

$$\text{slope } \overline{BD} = \frac{b}{a - c}$$

But then

$$
\begin{aligned}
\text{slope } \overline{AC} \cdot \text{slope } \overline{BD} &= \frac{b}{a+c} \cdot \frac{b}{a-c} \\
&= \frac{b^2}{a^2 - c^2} \\
&= \frac{b^2}{-b^2} \\
&= -1.
\end{aligned}
$$

Remember, $(a + c)(a - c) = a^2 - c^2$.

since $-b^2 = a^2 - c^2$

This shows that the diagonals of the rhombus are perpendicular as claimed. ■

The idea in each of the preceding examples was that any plane geometrical figure can be placed on a coordinate system. Moreover, this was done in such a way that the coordinates of certain key points were kept general and yet as simple as possible to facilitate later computation. Thus, the proofs were completely general and showed that the results claimed held for *all* such figures.

Proofs Using Equations and Coordinates

The preceding proofs only depended on the use of coordinates, the midpoint formula, and the notion of slope. Algebraic methods, however, make it possible to use equations of lines as well.

EXAMPLE 14.17

Showing that the Altitudes of a Triangle Are Concurrent

Show that the altitudes of a triangle are concurrent.

SOLUTION

We orient the triangle on a coordinate system with one vertex at the origin, one vertex on the positive x-axis, and one vertex in the upper half plane. Let $A(0, 0)$, $B(c, d)$, and $C(e, 0)$, with $c \neq e$, be these vertices.

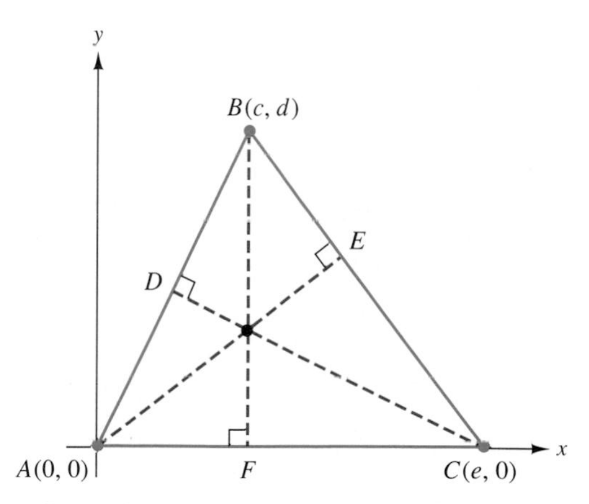

We now use the notion of slope, the fact that the slopes of perpendicular lines are negative reciprocals of one another, and the point-slope form of the equation of a line. Let $\overline{AE}$, $\overline{BF}$, and $\overline{CD}$ be the altitudes of the triangle.

Then

$$\text{slope } \overline{BC} = \frac{d}{c - e} \quad \text{so slope } \overline{AE} = \frac{e - c}{d}.$$

Therefore, the equation of $\overleftrightarrow{AE}$ is

$$y - 0 = \frac{e - c}{d}(x - 0)$$

or simply

$$dy = ex - cx.$$

Similarly,

$$\text{slope } \overline{AB} = \frac{d}{c} \text{ so slope } \overline{DC} = -\frac{c}{d}.$$

Therefore, the equation of $\overleftrightarrow{DC}$ is

$$y - 0 = -\frac{c}{d}(x - e)$$

or simply

$$dy = -cx + ce.$$

To find where $\overleftrightarrow{AE}$ and $\overleftrightarrow{DC}$ intersect, we determine the simultaneous solution of the equations of these two lines. Subtracting the equation for $\overleftrightarrow{DC}$ from the equation for $\overleftrightarrow{AE}$, we obtain

$$0 = ex - ec.$$

Therefore, the x-coordinate of the point of intersection is

$$x = \frac{ec}{e} = c.$$

Hence, without even determining the y-coordinate of the point of intersection of $\overleftrightarrow{AE}$ and $\overleftrightarrow{DC}$, it follows that the point of intersection lies on $\overleftrightarrow{BF}$ since $\overleftrightarrow{BF}$ is vertical, passes through the point $B(c, d)$, and so has equation $x = c$. Thus, the three altitudes are concurrent as was to be shown. ■

Equations of Circles

Since circles possess many interesting properties that can be treated by methods of coordinate geometry, we now consider equations of circles.

EXAMPLE 14.18 Determining Points Equidistant from a Fixed Point

(a) In words, describe the set of all points in a coordinate plane that are at distance 2 from the point (2, 3).

(b) Derive a formula that expresses the condition that $P(x, y)$ be one of the points in part (a).

SOLUTION

(a) The set of all points at distance 2 from the point (2, 3) is a circle of radius 2 with (2, 3) as its center as shown.

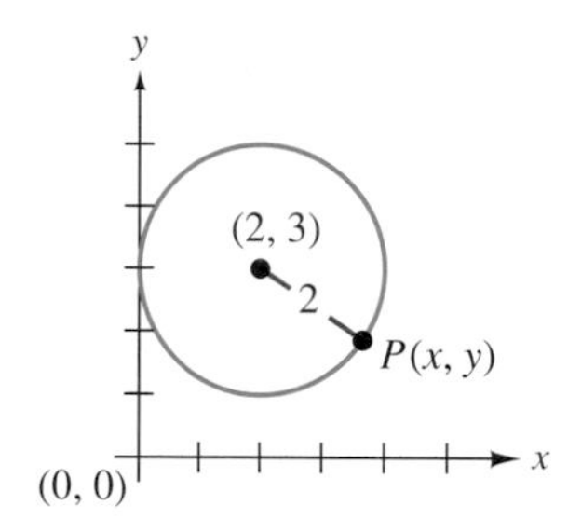

(b) The point P is on the circle if, and only if, its distance from (2, 3) is 2. Thus, using the distance formula,

$$\sqrt{(x - 2)^2 + (y - 3)^2} = 2$$

or, equivalently,

$$(x - 2)^2 + (y - 3)^2 = 4.$$

Since $P(x, y)$ lies on this curve if, and only if, its coordinates satisfy the equation, this equation is the equation of the circle described. ■

The preceding example is typical of the general case stated here as a theorem.

> **THEOREM Equation of a Circle**
>
> The equation of a circle of radius r with center $C(a, b)$ is
>
> $$(x - a)^2 + (y - b)^2 = r^2.$$
>
> In the special case where the center is the origin, the equation is $x^2 + y^2 = r^2$.

EXAMPLE 14.19

Finding the Circle with a Given Center and Passing Through a Given Point

Find the equation of the circle with center (5, 12) and which passes through the origin.

SOLUTION

If the center of the circle is (5, 12) and the circle passes through the origin, then the radius of the circle is the distance from (0, 0) to (5, 12). Thus,

$$r = \sqrt{(5 - 0)^2 + (12 - 0)^2} - \sqrt{169} = 13.$$

By the theorem, the desired equation is,

$$(x - 5)^2 + (y - 12)^2 = 169.$$

■

A **chord** of a circle is the line segment joining any two points on the circle.

JUST FOR FUN

The Greek Cross—II

In the earlier Greek cross puzzle you were asked to cut the cross into four pieces that could be reassembled into a single square. Show how to cut the cross into four pieces, different from the first time, that can still be reassembled into a single square. (*Hint:* How long must the side of the square be?)

EXAMPLE 14.20

Finding the Center of a Circle

(a) Write the equation of the circle of radius 5 and center $C(1, 2)$.

(b) Show that the points $A(-3, -1)$ and $B(4, 6)$ both lie on the circle.

(c) Determine the equation of the perpendicular bisector of chord $\overline{AB}$.

(d) Show that the center of the circle lies on the perpendicular bisector of $\overline{AB}$.

SOLUTION

(a) By the theorem, the desired equation is

$$(x - 1)^2 + (y - 2)^2 = 25.$$

(b) Since

$$(-3 - 1)^2 + (-1 - 2)^2 = (-4)^2 + (-3)^2 = 25$$

and

$$(4 - 1)^2 + (6 - 2)^2 = 3^2 + 4^2 = 25,$$

$(-3, -1)$ and $(4, 6)$ both satisfy the equation of part (a) and so lie on the circle.

(c) Let $M(c, d)$ be the midpoint of $\overline{AB}$. By the midpoint formula,

$$c = \frac{-3 + 4}{2} = \frac{1}{2} \quad \text{and} \quad d = \frac{-1 + 6}{2} = \frac{5}{2}.$$

Also, the slope of $\overline{AB}$ is

$$m = \frac{6 - (-1)}{4 - (-3)} = \frac{7}{7} = 1.$$

It follows that the slope of the perpendicular bisector is -1 and its equation is

$$y - \frac{5}{2} = -1\left(x - \frac{1}{2}\right)$$

which simplifies to

$$x + y = 3.$$

(d) Since $1 + 2 = 3$, it follows that the center, $C(1, 2)$, of the circle lies on the perpendicular bisector of the chord. ■

PROBLEM SET 14.3

Understanding Concepts

1. (a) Let $A(0, 0)$, $B(a, 0)$, $C(r, s)$, and $D(0, c)$ with $a > 0$ and $c > 0$ as the vertices of a square. Determine r, s, and c in terms of a.
   (b) Draw the figure of part (a) on a Cartesian coordinate system.
2. **(a)** Let $A(0, 0)$, $B(a, 0)$, $C(r, s)$, and $D(0, c)$ with $a > 0$ and $c > 0$ be the vertices of a rectangle that is not a square. Determine r and s in terms of a and c.
   (b) Draw the figure of part (a) on a Cartesian coordinate system.
3. (a) Let $A(0, 0)$, $B(a, b)$, $C(r, s)$, and $D(c, 0)$ with $a > 0$ and $c > 0$ be the vertices of a parallelogram that is not a rectangle or rhombus. Determine r and s in terms of a, b, and c.
   (b) Draw the figure of part (a) on a Cartesian coordinate system.
4. **(a)** Let $A(0, 0)$, $B(a, b)$, and $C(2c, 0)$ with a, b, and c positive be the vertices of an equilateral triangle. Determine a and b in terms of c.
   (b) Sketch the figure of part (a) on a Cartesian coordinate system.
5. In words, describe the set of points satisfying each of these equations.
   (a) $x^2 + y^2 = 81$ **(b)** $(x - 2)^2 + y^2 = 4$
   (c) $x^2 + (y + 5)^2 = 0$
   (d) $(x - 3)^2 + (y - 5)^2 = 36$
   (e) $(x + 2)^2 + (y + 3)^2 = 25$
   (f) $(x - 1)^2 + (y - 2)^2 = -4$
6. Plot the set of all points satisfying each of these equations.
   (a) $(x - 2)^2 + (y + 3)^2 = 49$
   (b) $(x - 3)^2 + (y - 5)^2 = -9$
7. Write the equations of the circles satisfying these conditions.
   (a) Center at $(2, 5)$, radius 3
   (b) Center at $(3, -4)$, radius 1
   (c) Center at $(-1, 2)$ and just touching the x-axis
   (d) Center at $(2, b)$ and just touching the y-axis
8. (a) Show that the point $(1, 2)$ lies on the circle $(x + 2)^2 + (y - 3)^2 = 10$.
   (b) Write the equation of the line through the point $(1, 2)$ and the center of the circle of part (a).
   (c) Write the equation of the line through the point $(1, 2)$ and perpendicular to the line in part (b).
   (d) Carefully draw the circle of part (a) and the lines in parts (b) and (c).
   (e) How many points does the line of part (c) appear to have in common with the circle of part (a)? This line is said to be **tangent** to the circle.
9. **(a)** Show that the point $(-1, 6)$ lies on the circle of problem 8.
   (b) The points $(1, 2)$ and $(-1, 6)$ both lie on the circle of problem 8. Determine the equation of the perpendicular bisector of the chord joining $(1, 2)$ and $(-1, 6)$.
   (c) Show that the center of the circle lies on the perpendicular bisector of the chord determined in part (b).

10. One, and only one, circle can be drawn through any three noncollinear points.
 (a) Determine the center of the circle through the points (7, 7), (−1, 1), and (6, 0). (*Hint:* Use the idea of problem 9, part (b).)
 (b) Determine the radius of the circle in part (a).
 (c) Determine the equation of the circle in part (a).
 (d) Plot the points of part (a) and the circle of part (b) on a Cartesian coordinate system.

Thinking Critically

11. Use coordinate geometry to show that the diagonals of a parallelogram bisect each other.
12. (a) Let $A(0, 0)$, $B(4a, 4b)$, $C(4c, 4d)$, and $D(4e, 0)$ with $b > 0$, $d > 0$, and $e > 0$ as the vertices of a quadrilateral. Let P, Q, R, and S be the midpoints of $\overline{AB}$, $\overline{BC}$, $\overline{CD}$, and $\overline{DA}$ respectively. Show that $\overline{PR}$ and $\overline{QS}$ bisect each other.
 (b) Draw a sketch to illustrate part (a).
13. Using the results of problem 4, show that the medians of an equilateral triangle are also the altitudes of the triangle.
14. Let $A(a, b)$, $B(c, d)$, and $C(e, f)$ be the vertices of any triangle.
 (a) Show that the three medians of the triangle meet at the point

$$G\left(\frac{a + c + e}{3}, \frac{b + d + f}{3}\right).$$

 The point G is called the **centroid** of the triangle. If the triangle were cut out of heavy cardboard, it would balance on a pinpoint placed under the centroid.
 (b) Show that the centroid of a triangle is two-thirds of the way from each vertex to the midpoint of the opposite side.
15. Show that the perpendicular bisectors of the sides of a triangle are concurrent using methods of coordinate geometry.
16. Use coordinate methods to show that the sum of the squares of the lengths of the diagonals equals the sum of the squares of the lengths of the sides of a parallelogram. (*Hint:* Any parallelogram can be placed on a coordinate system as shown. Let P be the point (a, b) and R the point $(c, 0)$.)

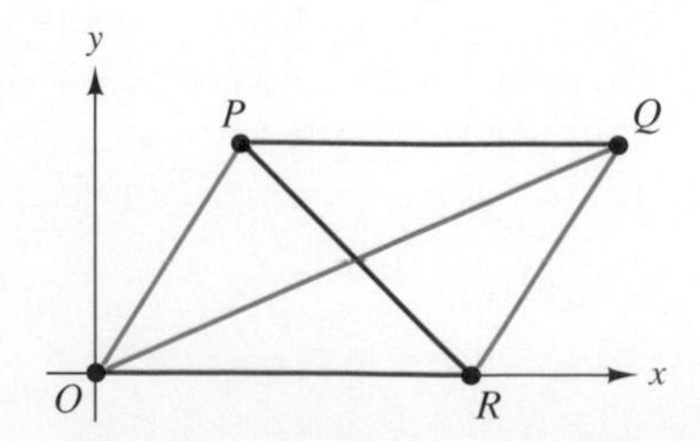

17. Consider the squares shown in this diagram. Let P, Q, R, and S be the centers of the squares as shown.
 (a) Using coordinate geometry, show that $\overline{PR} \cong \overline{QS}$.
 (b) Show that $\overline{PR}$ is perpendicular to $\overline{QS}$.

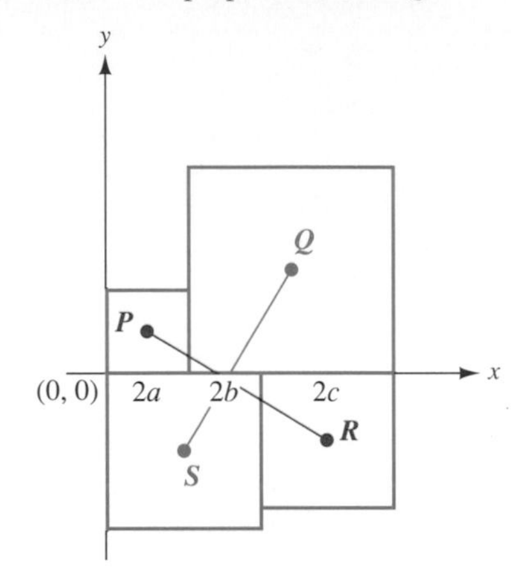

18. Consider the triangle SPQ inscribed in a semicircle as shown. Use coordinate methods to prove Thales' theorem; that is, show that $\overline{PQ}$ is perpendicular to $\overline{PS}$. (*Hint:* Recall that $x^2 + y^2 = r^2$ and $(x + r)(x - r) = x^2 - r^2$.)

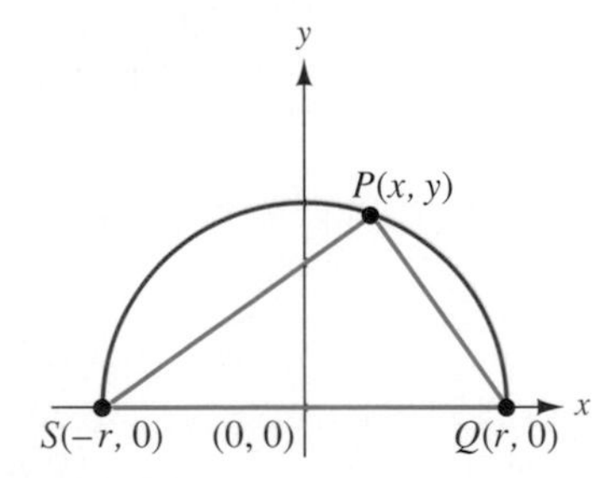

Making Connections

19. (a) Consider a ladder 12 feet long standing vertically against a wall. Let $M(x, y)$ be the midpoint of the ladder.

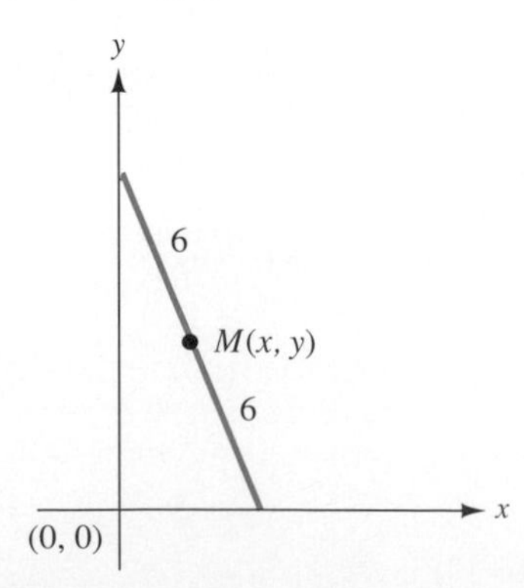

Determine the equation of the path described by M as the ladder is pulled away from the wall until it is lying flat on the ground but with the top and bottom of the ladder touching the wall and the ground respectively at all times.

(b) Copy the sketch in part (a) and draw the path described by M.

For Review

20. Determine the sum of the interior angles of a convex decagon.
21. Determine the sum of the exterior angles of a convex decagon.
22. Determine the measure of each interior angle of a regular decagon.
23. Determine the measure of each exterior angle of a regular decagon.

---

24. Determine if the point P is in the interior or the exterior of the region formed by the closed curve shown.

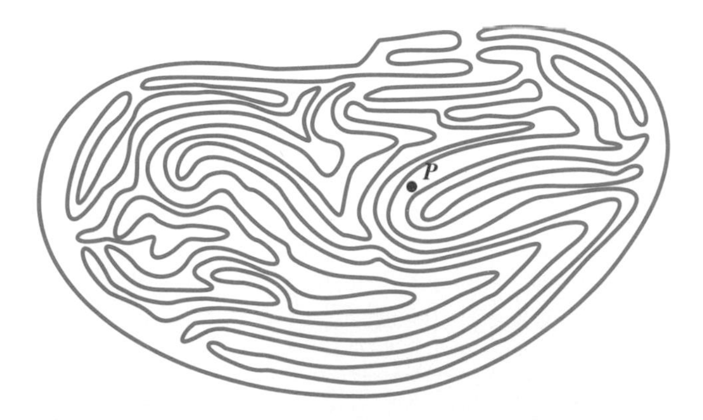

25. If possible draw an example of a nonconvex simple (a) triangle, (b) quadrilateral, (c) pentagon, (d) hexagon.
26. Determine the area of the figure shown if the area of the smallest possible square with lattice points as corners is 1.

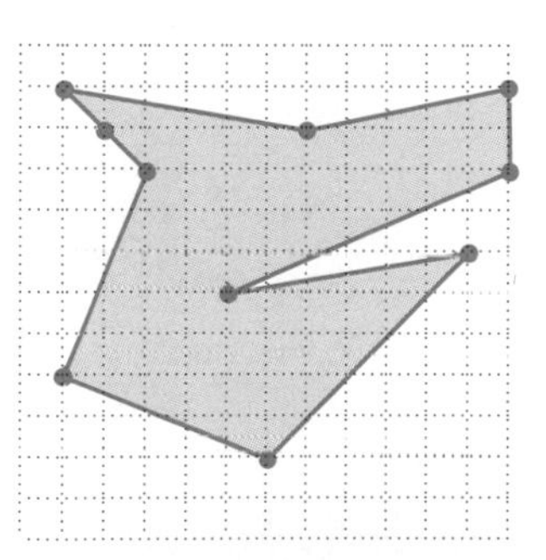

14.4 Graphing Functions and Relations

Graphing Functions

Suppose you invest \$1000 at 7% interest compounded annually. How much will your investment be worth in one year? In two years? In twenty years? In n years? The answer to these questions can be found by using the compound interest formula

$$A = 1000(1.07)^n$$

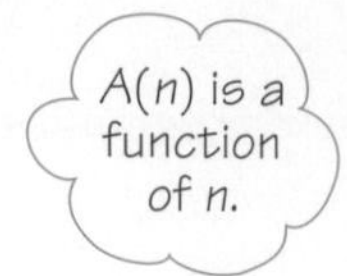

as was developed in Section 7.4. This formula gives the value of the investment at the end of n years; that is, A is a *function* of n and, to emphasize this fact, we sometimes write

$$A(n) = 1000(1.07)^n.$$

$A(1)$ is the value of the investment at the end of the first year, $A(1.5)$ is the value of the investment at the end of a year and a half, and so on.

The function $A(n)$ certainly enables one to answer the questions at the first of the preceding paragraph.* However, just looking at the function does not give most people good understanding of the way an investment grows at compound interest. A table like Table 14.1 certainly helps, but even the table does not suggest how rapidly the investment grows in later years.

TABLE 14.1 The Value at the End of Each Year of $1000 Invested at 7% Compounded Annually

n	$A(n)$	n	$A(n)$	n	$A(n)$	n	$A(n)$
1	1070.00	6	1500.73	11	2104.85	16	2952.16
2	1144.90	7	1605.78	12	2252.19	17	3158.82
3	1225.04	8	1718.19	13	2409.85	18	3379.93
4	1310.80	9	1838.46	14	2578.53	19	3616.53
5	1402.55	10	1967.15	15	2759.03	20	3869.68

This can perhaps best be conveyed by the visual impact of a graph as in Figure 14.12. The points on the graph are those with coordinates $(n, A(n))$ determined by the formula. Actually, we plotted only the points from the data in Table 14.1, but if these points are joined by a smooth curve, we can read from the graph the values of $A(n)$ for other values of n. Thus, from the graph, $A(10.5) \doteq 2030$ and, by direct computation, $A(10.5) \doteq 2034.84$. Note, in particular, that the graph shows that the investment grows steadily but modestly at first (the first half dozen points lie almost on a straight line), but that the growth is much more dramatic in later years. In fact, at the end of 40 years the investment is worth an astonishing $14,974.46! (Show this by using the $\boxed{y^x}$ key on a calculator.)

The following definition makes the notion of the graph of a function more precise.

DEFINITION The Graph of a Function

The **graph of the function $f(x)$** is the set of points (x, y) whose coordinates satisfy the equation $y = f(x)$.

---

* To compute $1000(1.07)^2$ enter this string into a calculator: $\boxed{\text{ON/AC}}$ 1.07 $\boxed{y^x}$ 2 $\boxed{\times}$ 1000 $\boxed{=}$

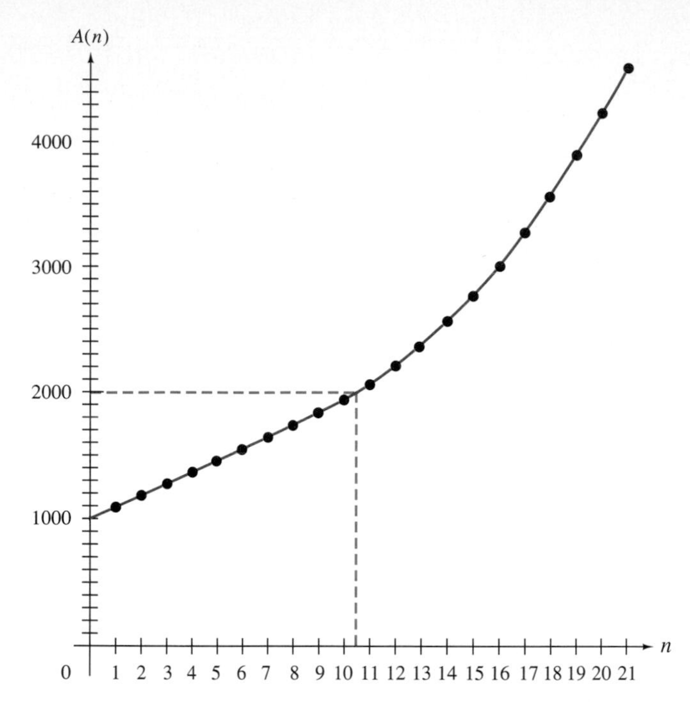

Figure 14.12
Graph of $A(n) = 1000(1.07)^n$

EXAMPLE 14.21

Drawing the Graph of a Function

Graph the function $y = \sqrt{16 - x^2}$.

SOLUTION

Since the graph is the set of all points (x, y) that satisfy the equation, we use a calculator to prepare a table of values of x and corresponding values of y as shown. However, since we can do this for only a finite set of points, we do it only for a representative set of values of x, plot the points, and then connect them with a smooth curve. Note, by the way, that the domain of this function is $\{x: -4 \leq x \leq 4\}$. For values of x outside this range $16 - x^2$ is negative and so has no real number as its square root.* The graph of the function is the semicircle of radius 4 as shown.

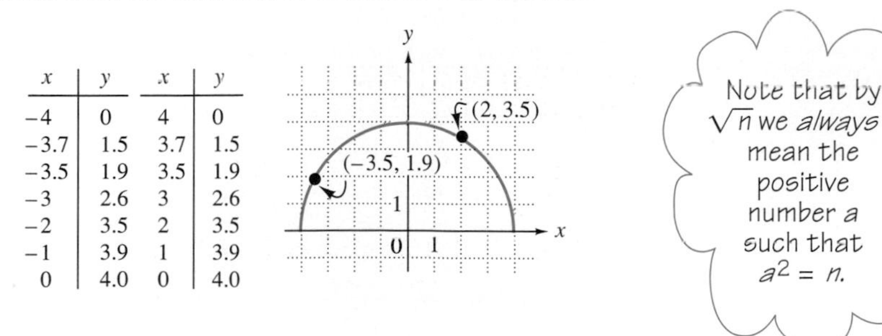

x	y	x	y
−4	0	4	0
−3.7	1.5	3.7	1.5
−3.5	1.9	3.5	1.9
−3	2.6	3	2.6
−2	3.5	2	3.5
−1	3.9	1	3.9
0	4.0	0	4.0

■

* Negative numbers do have square roots but they are complex numbers, inappropriate for this course and also for the elementary school curriculum.

HIGHLIGHT FROM HISTORY
George Dantzig and Linear Programming

The Second World War (1939–1945), presented the allies with huge logistical problems associated with moving and supplying enormous armies of millions of men both in Europe and the Far East. Necessity is the mother of invention, however, and the mathematical theory of linear programming was invented to solve the problem of moving and utilizing men and materiel most efficiently.

In the beginning, the calculations involved were of limited extent and tedious to perform. All this was changed, however, by a technique invented by George B. Dantzig, a professor at Stanford University. The technique, called the simplex method, greatly simplified the calculations and did much to enhance the allied war effort. With the advent of the computer, the process has been enormously speeded up, and it is now possible to deal with problems involving literally thousands of variables. The significance of these methods to business and industry worldwide is incalculable.

It is important to note that any vertical line cuts the graph in Example 14.21 in at most one point. This is the visual counterpart of the requirement in the definition of a function that to each x in the domain of the function there corresponds precisely one y in the range. Thus, one can tell at a glance if a graph is or is not the graph of a function.

THEOREM The Vertical Line Test

A graph is the graph of a function if, and only if, every vertical line cuts the graph in at most one point.

EXAMPLE 14.22 **Using the Vertical Line Test**

In each of the following indicate if the graph is the graph of a function. If it is not the graph of a function, tell why.

(a)

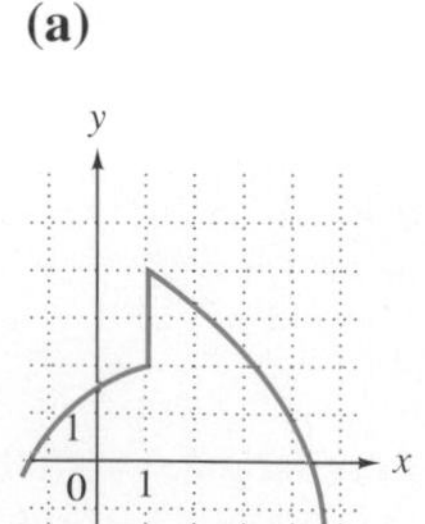

(b)

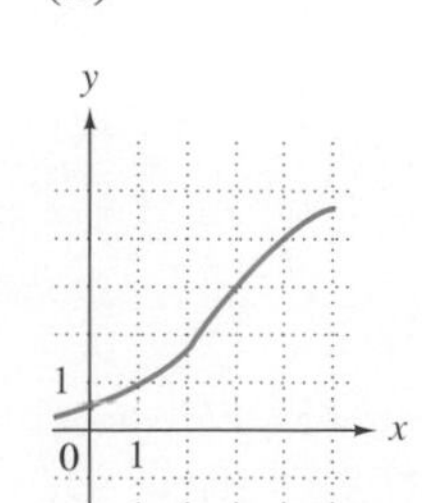

(c)

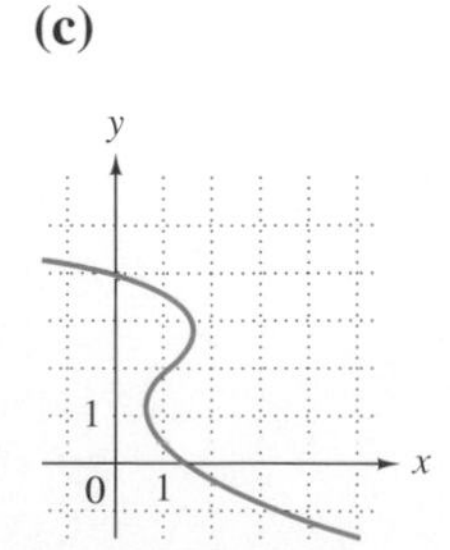

SOLUTION

(a) This graph is not the graph of a function since the line $x = 1$ cuts the graph in a whole interval of values of y.

(b) This is the graph of a function since each vertical line cuts the graph just once.

(c) This is not the graph of a function since some vertical lines cut the graph more than once. In particular, this is true of the line $x = 1$. ■

Graphs of functions can be symmetric with respect to the y-axis and also with respect to the origin. Can the graph of a function be symmetric with respect to the x-axis? Unless $f(x) = 0$ for all x, the answer is *no* since, as is apparent from Figure 14.13 such a graph necessarily fails the vertical line test.

CURRENT AFFAIRS

Narendra Karmarkar and Linear Programming

In 1984, the mathematical community was excited by the announcement that Narendra Karmarkar, at AT&T Bell Laboratories, had invented a powerful new method for solving linear programming problems that was significantly faster than the classic simplex method of George Dantzig. The importance of Karmarkar's innovation is that it may make it possible to solve significant problems in "real time." An example of the need for a really fast algorithm is the situation that arises when there is a serious storm at a major airline hub like Chicago. Many flights get delayed or canceled so that numerous crews and airplanes are in the wrong places. A real-time algorithm would make it possible to key the data into a computer and come up with an on-the-spot solution that would allow the disruption to be straightened out more quickly with a minimum of cost and passenger inconvenience.

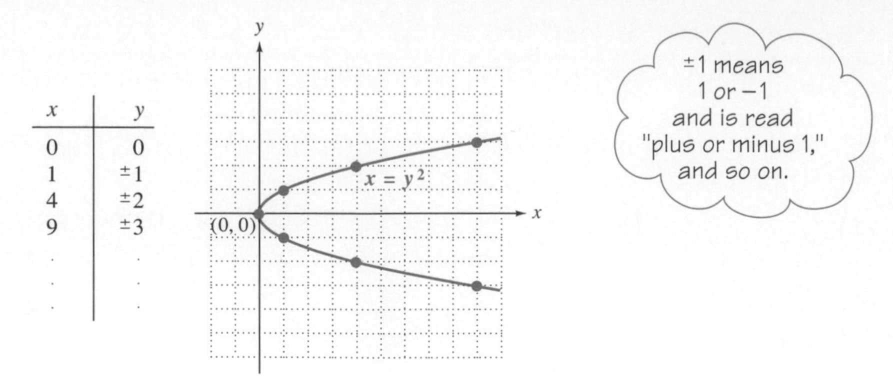

x	y
0	0
1	± 1
4	± 2
9	± 3
.	.
.	.
.	.

Figure 14.13
The graph of $y^2 = x$. Here y is not a function of x.

Maximum and Minimum Values of Functions

Companys desire to maximize profit and to minimize cost. Traffic engineers want to design systems that maximize traffic flow and minimize travel time. Telephone companies want to design networks that maximize the number of calls that can be handled. Such desires may require finding the maximum or minimum value of a function. A number of methods can be used to determine maxima and minima but, for elementary school students, one accessible way is to use a graph.

EXAMPLE 14.23

Determining the Minimum Value of $f(x) = x^2 - 4x$

Draw the graph of $f(x) = x^2 - 4x$ and determine the minimum value of the function.

SOLUTION

The desired graph is the set of points satisfying the equation

$$y = x^2 - 4x.$$

Make a table of corresponding values (x, y), plot the points, and join these with a smooth curve. The resulting graph is as shown. The minimum value of the function is the least of the y-coordinates of the points on the graph. Since the graph is symmetric about the line $x = -2$, it is apparent that the minimum value is -4 and that it occurs when $x = -2$.

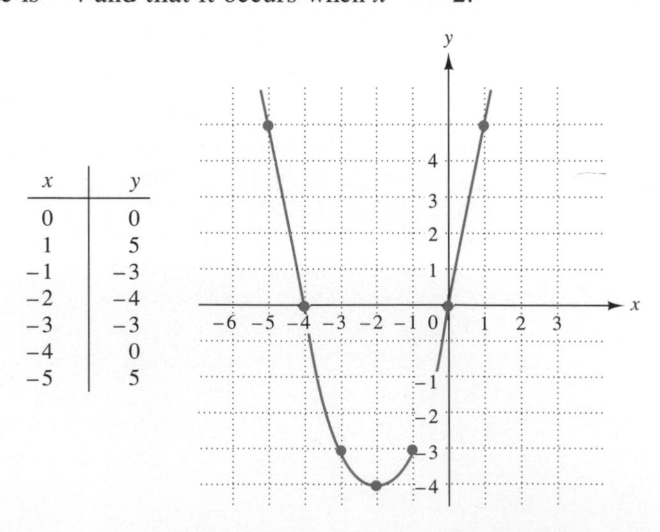

x	y
0	0
1	5
−1	−3
−2	−4
−3	−3
−4	0
−5	5

■

EXAMPLE 14.24 **Solving a Practical Problem**

A farmer has 100 yards of fencing with which to enclose a rectangular garden to be located along an existing straight fence as shown. What should the dimensions of the new garden be if it is to have the largest possible area?

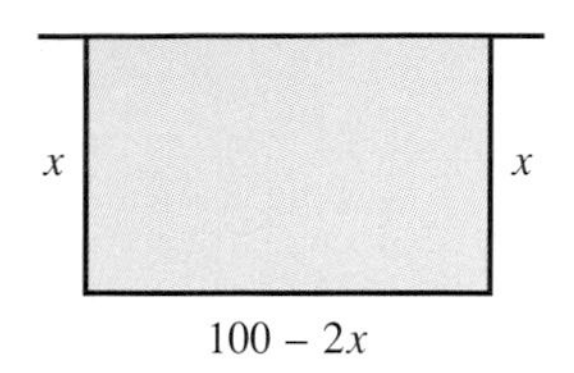

SOLUTION

Understand the problem • • • •

We are given the length of fence available to enclose the new garden. Some of the fence must be used to reach straight out from the existing fence and the remainder will run parallel to the existing fence. If the garden were to extend 50 yards out from the existing fence, it would be zero yards wide and would enclose no area. Similarly, if it were to extend 100 yards along the existing fence no area would be enclosed. For gardens between these two extremes, area *is* enclosed, and we must determine the dimensions so that this area is as large as possible.

Devise a plan • • • •

Since the desired dimensions are unknown, perhaps part of our strategy should be to represent the dimensions using a variable. Since the area of a rectangle is length times width, we can then represent the area of the new garden as a function of the variable. Since we desire the maximum area, we may then be able to proceed along the lines of the preceding example.

Carry out the plan • • • •

Let x represent the length of the two portions of fence extending out at right angles to the existing fence. Then, since the farmer has only 100 yards of fencing, the length of the fence parallel to the existing fence must be $100 - 2x$. Then the area is given by the function $A(x) = x(100 - 2x) = 100x - 2x^2$.

In the last example, we discovered the minimum of a function by graphing; perhaps the same idea will work here. Consider the table of corresponding values of x and y and the graph obtained by plotting the points (x, y) as shown on page 950. Since this graph is symmetric about the line $x = 25$, it is apparent that the maximum y value, and hence the maximum value of the function occurs when $x = 25$. Thus, the maximum area is 1250 yd^2 and it occurs when the field measures 25 yards by 50 yards with the long side of the garden along the existing fence.

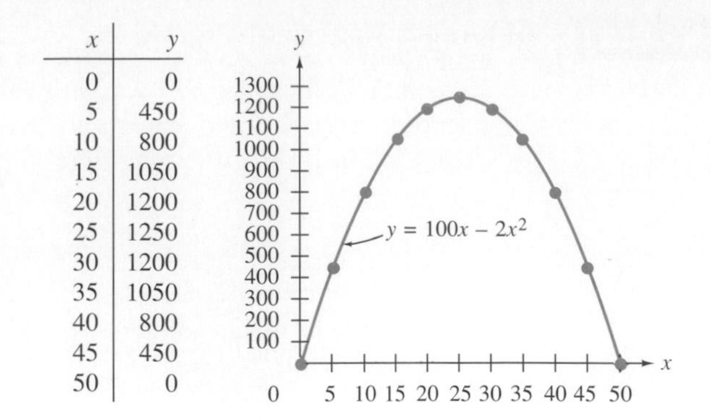

x	y
0	0
5	450
10	800
15	1050
20	1200
25	1250
30	1200
35	1050
40	800
45	450
50	0

Look back • • • •

The key to solving this problem was to use a variable to express the area of the garden as a function of x. Then, drawing a graph, we were able to determine the dimensions of the field that yielded maximum area. ■

Graphs Involving the Relations $<, >, \leq, \geq$

We have seen that the graph of every equation of the form $ax + by = c$ is a straight line. How would the graph of an inequality like $ax + by \leq c$ appear? We consider some examples.

JUST FOR FUN

Watering a Playfield

The principal of Elmwood Elementary school hired a plumber to install underground sprinklers to water the school's playfield. The plumber decided to install one large sprinkler at the middle of each side of the field—adjusting the sprinkler so that it will just reach the two adjacent corners. He shows the principal his sketch.

The principal is not convinced that the plumber's sketch is right and draws his own sketch that indicates that the middle of the field will not be watered.

Who is right—the plumber or the principal?

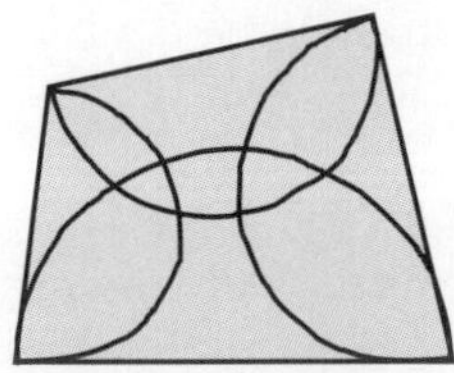

Plumber's sketch

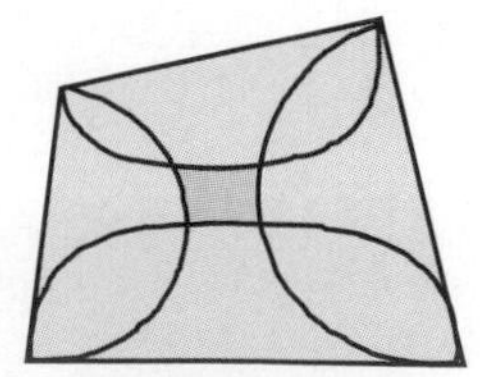

Principal's sketch

EXAMPLE 14.25 Graphing an Inequality

Draw the graph of the inequality $2x - 3y \leq 6$.

SOLUTION

The graph of the equality $2x - 3y = 6$ is the line shown in the diagram on the left. If $2a - 3b = 6$, the point $P(a, b)$ is on the line. However, if b is replaced by c for any $c > b$, then $2a - 3c < 6$ because we are subtracting more from $2a$. Also, $Q(a, c)$ is *above* $P(a, b)$ as indicated. Since this is true for any point on the line, those points (x, y) for which $2x - 3y < 6$ must lie *above* the line. Thus, the graph of $2x - 3y \leq 6$ is that portion of the plane above or on the line as indicated by the shading of the diagram on the right.

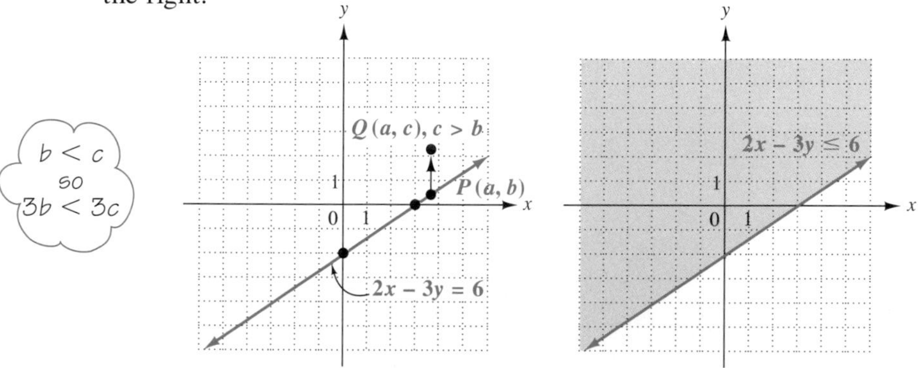

■

Note that the line itself is part of the graph in the preceding example. If we had been graphing $2x - 3y < 6$ instead of $2x - 3y \leq 6$, the graph would have been the same except we would have dashed the line to indicate that it was *not* part of the graph of the inequality.

The graph of the linear inequality in the last problem is called a **half-plane,** and this would be so even if it did not include the line itself. With this understanding, the preceding example is typical of the general case as stated in this theorem.

> **THEOREM The Graph of a Linear Inequality**
>
> The **graph of a linear inequality** is the set of all points on one side or the other of the line associated with the inequality. The line may or may not be included in the graph depending on whether equality is or is not allowed in the inequality. In either case the graph is called a **half-plane.**

EXAMPLE 14.26 Using the Idea of a Half-Plane

Graph the inequality $3x + 5y < 15$.

SOLUTION

As before, we begin by graphing the line $3x + 5y = 15$. However, this time we draw a dashed line since equality is not allowed; the inequality is strict. Since we already know that the solution is a half-plane, it only remains to determine which of the two half-planes determined by the line is the desired graph. To do this most easily, we check a single point on one side of the line, making our choice so that the arithmetic is as easy as possible. If the point checked satisfies the inequality, the half-plane containing the point is the desired graph. Otherwise, the half-plane not containing the point is the graph. Here we check the point (0, 0). Since $3 \cdot 0 + 5 \cdot 0 = 0 < 15$, (0, 0) satisfies the inequality and the graph is as shown.

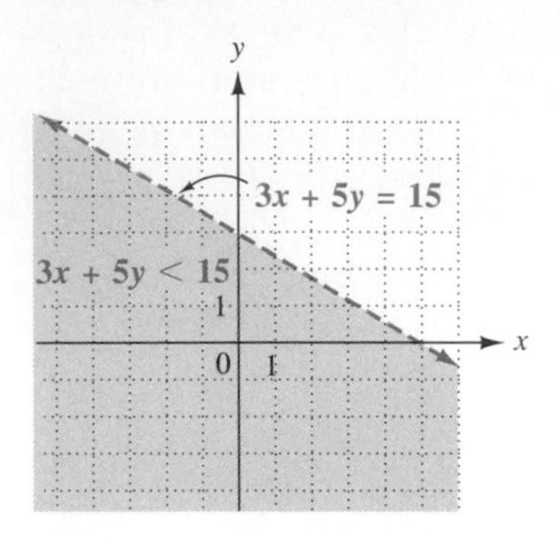

■

EXAMPLE 14.27 Graphing Simultaneous Inequalities

Graph the set of all points simultaneously satisfying $3x + 4y \geq 12$ and $-2x + 3y \leq 6$.

SOLUTION

Since equality is allowed in each of these inequalities, we begin by drawing the *solid* lines $3x + 4y = 12$, and $-2x + 3y = 6$, as shown. Each of these lines divides the plane into two half-planes, and we check the point (0, 0) to see which half-plane is the graph in each case. Since $3 \cdot 0 + 4 \cdot 0 = 0 \not\geq 12$, the graph of $3x + 4y \geq 12$, is the half-plane above and on the line $3x + 4y = 12$ as shown by the shading on the diagram. Similarly, the graph of $-2x + 3y \leq 6$ is the shaded half-plane below and on the line $-2x + 3y = 6$. It follows that the graph of the set of points satisfying both inequalities is the intersection of the two half-planes; that is, the doubly shaded region in the diagram.

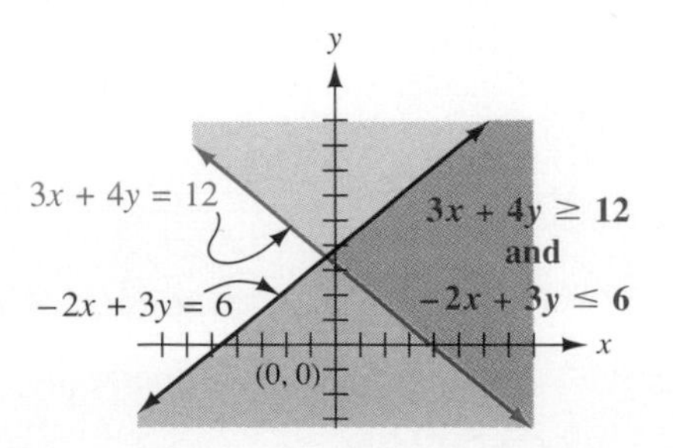

■

It is worth noting that regions defined by inequalities as in Example 14.27 play a central role in linear programming—a powerful optimization method of considerable importance in business and industry.

PROBLEM SET 14.4

Understanding Concepts

1. Draw the graphs of each of these linear functions.
   (a) $y = 2x - 3$ **(b)** $y = 0.5x + 2$
   (c) $y = -3x$
2. **(a)** On the same coordinate system draw the graphs of these three linear functions: $y = 4x$, $y = 4x + 5$, and $y = 4x - 3$.
   (b) Briefly discuss the graphs in part (a).
3. Use the vertical line test to decide which of these are graphs of functions.

(a)

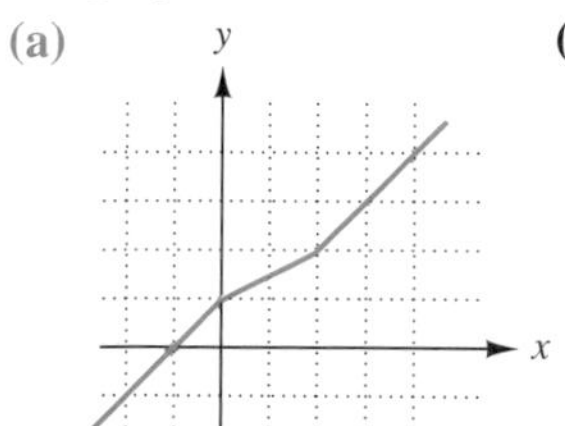

(b)

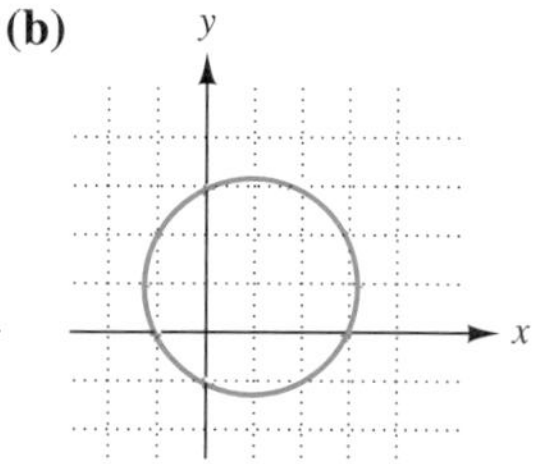

(c)

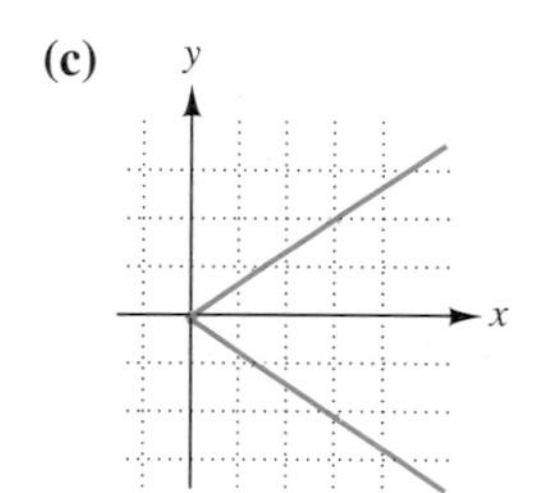

(d)

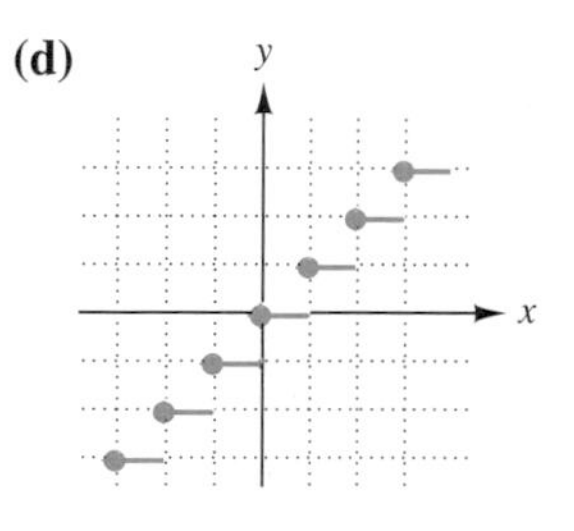

4. Could a circle be the graph of a function? Explain.
5. Draw the graph of each of these linear inequalities.
   (a) $4x - y < 8$ **(b)** $-2x + 5y \geq 10$
   (c) $x + y \leq 4$ **(d)** $3y < 2x + 6$
   (e) $3x + 5y + 15 \geq 0$ **(f)** $y \leq 7$
6. Graph the set of points that satisfy these two inequalities simultaneously:
   $3x - 4y \leq 12, \quad 2x + 5y \leq 10.$
7. Graph each of these functions.
   (a) $y = x^2$ **(b)** $y = (x - 2)^2$ **(c)** $y = (x + 3)^2$
   (d) Discuss the relationship between the graphs of parts (a), (b), and (c).
8. Graph each of these functions.
   (a) $y = x^2 - 4x + 4$
   (b) $y = x^2 + 6x + 9$ **(c)** $y = x^2 + 4x + 4$
   (d) Could you write each of these functions in a different, more concise form? (*Hint*: Consider problem 7.)
9. Use graphing techniques to find
   (a) the minimum value of the quantity $x^2 + 10x$ and the value of x at which it occurs.
   (b) the maximum value of $8x - 2x^2$ and the value of x at which it occurs.
10. Determine the minimum value of $g(x) = x^4 - 8x^2$ and for what value of x the minimum occurs.
11. Graph each of these functions.
    (a) $y = x^3$
    (b) $y = x^3 - 9x$
    (*Hint:* Be sure to consider $x = 1/2$ and $x = -1/2$.)
    (c) If the functions in parts (a) and (b) have a point of symmetry, identify it in each case.

Thinking Critically

12. (a) Graph $y = x^2 + 2x$.
    (b) If the graph of part (a) has a line of symmetry, identify it.
13. (a) Graph the set of points that satisfy $y^2 = x + 4$.
    (b) If the graph of part (a) has a line of symmetry, identify it.
    (c) Does the equation of part (a) define y as a function of x? Explain.
14. **(a)** Graph $y = \sqrt{x + 4}$. Recall that $\sqrt{n}$ is the *positive* number a such that $a^2 = n$.
    (b) Compare the graph of part (a) to that of problem 13, part (a). Explain.
    (c) Does the equation of part (a) define y as a function of x?
    (d) Discuss the graph of $y = -\sqrt{x + 4}$.
15. **(a)** Graph the function $f(x) = 1/x$. (*Hint:* Along with other values, be sure to plot points for positive and negative x values near 0.)
    (b) What is $f(0)$ in part (a)?
16. (a) Graph the function $y = |x - 1|$. Recall that $|a|$ means the absolute value of a.
    (b) If the function in part (a) has a line of symmetry, identify it.
17. The notation $\lfloor x \rfloor$ indicates the largest integer less than or equal to x. Thus, $\lfloor 3 \rfloor = 3$, $\lfloor \pi \rfloor = 3$, $\lfloor -\pi \rfloor = -4$, and so on. Graph the function $y = \lfloor x \rfloor$.
18. Graph the equation $|y| + |x| = 1$ for $-1 \leq x \leq 1$ and $-1 \leq y \leq 1$. (*Hint:* Recall that $|a| = a$ if $a \geq 0$ and $|a| = -a$ if $a < 0$ and consider the four cases **(i)** $x \geq 0, \quad y \geq 0$, **(ii)** $x \geq 0, \quad y < 0$, **(iii)** $x < 0, \quad y \geq 0$, and **(iv)** $x < 0, \quad y < 0$.)

19. The graph in problem 18 is not the graph of a function. Describe how to break it up into two functions and draw their graphs. (*Hint:* Consider problem 14 again.)

Making Connections

20. Acme Athletic Manufacturing Company has discovered that the number of running shoes they can sell per week is given by the function $s(x) = 400 + 80x - x^2$ where x is the price in dollars of a pair of shoes.

(a) What is the maximum number of shoes Acme can sell per week?

(b) What price should Acme charge in order to achieve maximum sales each week?

21. The height in feet of a ball shot upward from the ground with a velocity of 640 feet per second is given by the function $h(t)$ where $h(t) = -16t^2 + 640t$ and t is the time in seconds after the ball is shot upward.

(a) What is the height of the ball when $t = 0$?

(b) What is the height of the ball when $t = 40$?

(c) What is the maximum height reached by the ball?

(d) At what time t does the ball reach its maximum height?

22. Populations of bacteria grow because, after a suitable interval, each cell divides into two identical cells. A certain type of cell divides once every minute.

(a) If one such cell is placed in an agar dish at time $t = 0$, what function, $p(t)$, gives the population size at the end of t minutes?

(b) Graph the function $p(t)$ of part (a) using suitable scales on the vertical and horizontal axes.

Communicating

23. One of the cells described in problem 22 is placed in a bottle at 11:00 A.M. and the bottle is exactly full at noon.

(a) When is the bottle half-full?

(b) When is the bottle 1/4 full?

(c) When is the bottle 1/16 full?

(d) If you were one of the bacterium in the bottle when it was 1/16 full, do you think you would worry about having space to live? Should you worry? Explain briefly.

24. Populations grow in the same way that investments accrue at compound interest. Thus, if P is the present population of Seattle and it is growing at the rate of 5% per year, the future population is given by the formula $F(t) = P(1.05)^t$ where time t is in years.

(a) Show by direct computation using the y^x key on a calculator, that Seattle's population will essentially double in the next 14 years if the present growth rate continues.

(b) If P is the population of Seattle now, what will it be in 28 years if the present growth rate continues?

(c) Discuss in some detail what problems the growth in parts (a) and (b) pose for Seattlites in the next 40 years. For example, it is worth noting that Seattle suffered water shortages in several recent years.

For Review

25. Use the methods of coordinate geometry to show that the diagonals of a rhombus

(a) bisect each other. **(b)** are perpendicular.

26. Determine the slopes of these lines.

(a) $y = -3x + 17$ **(b)** $x = 5y + 2$

(c) $3x + 5y = 15$

27. Write the equation of the line with slope 4 that passes through the point $(3, -2)$.

28. Write the equation of the line parallel to the line $2x - 7y = 14$ and which passes through the point $(-2, -3)$.

29. Write the equation of the line perpendicular to the line $4x - 5y = 14$ and which passes through the point $(4, 2)$.

EPILOGUE The Themes of Geometry

In this chapter we have considered geometry from an algebraic point of view. That is, we have considered coordinate or analytic geometry.

In brief summary, we have seen:

- how to represent points by ordered pairs of numbers using a Cartesian coordinate system;
- how to determine the distance between two points;
- how to determine the midpoint of a line and, more generally, how to determine the coordinates of a point M on segment $\overline{PQ}$ such that $PM/PQ = t$;
- how to determine the slope of a line;
- how to write the equation of a line in various forms;
- how to find the simultaneous solution to two linear equations;
- how to write equations of circles; and
- how to use these tools to prove general results about geometric figures.

The methods of coordinate geometry are pervasive in most of mathematics that students will study after elementary school.

This concludes our study of informal geometry. Several themes have cut across these geometry chapters. These themes, which could properly form the basis for the geometry portion of the elementary school curriculum, are as follows:

- *Invariance*

  Invariance involves the idea that, while many properties of geometric figures change from figure to figure, some, often surprisingly, do not. Indeed, much of the interest in, and utility of, geometry derives from this fact. For example, the ratio of the circumference of a circle to the diameter is always $\pi = 3.1415 \cdots$; the volume of any cone is always one-third the volume of the corresponding cylinder; the sum of the exterior angles of any convex polygon is always 360°; and so on. Geometry is replete with remarkable and useful invariances.

- *Symmetry*

  There are many kinds of symmetry in geometry. There is the repetitive symmetry that manifests itself in tilings and tessellations; the symmetry of an object as if it were reflected in a mirror (that is, symmetry with respect to a line); symmetry of an object through a point; and rotational symmetry. There is also a sort of symmetry in many of the formulas of geometry. For example, the Pythagorean expression $c^2 = a^2 + b^2$ is unchanged if a and b are interchanged and similar symmetries are exhibited by the distance and slope formulas in coordinate geometry.

- *Congruence*

  The notion of congruence (that is, that one figure is exactly the same size and shape as another) is of great importance.

- *Similarity*

  This embodies the idea that two figures are the same *shape* but of different size, as if one figure is simply a magnification of the other. Perhaps the most important consequence of similarity is that corresponding lengths in similar figures are proportional.

- *Locuses*

  Though we have not used the word locus, the idea is to consider sets of points that satisfy certain conditions. For example, a circle is the set of all points in a plane that are equidistant from a fixed point C. Similarly, the set of points $P(x, y)$ that satisfy an equation like $3x - 4y = 17$ is a straight line.

- *Maxima and minima*

  Of all triangles of fixed perimeter, the equilateral triangle has maximum area. Of all rectangles of fixed area, the square has least perimeter. Such maximum and minimum questions often arise in informal geometry.

- *Homeomorphism*

  We have not previously used the word, but homeomorphism embodies the idea that objects, configurations, and so on, may look quite different but are nevertheless essentially the same. For example, the networks shown here appear quite different. However, they are actually the same in the sense that vertices of the networks can be identified so that the same vertices are connected by edges in each case.

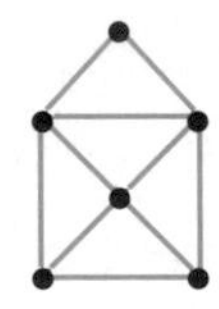
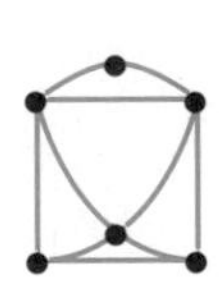
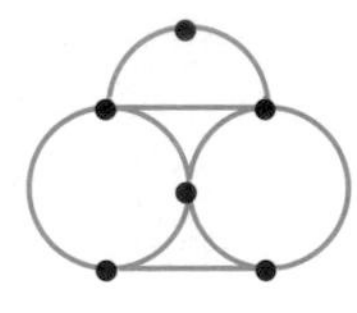

- *Limit*

  As an example of the notion of limit, as n gets larger and larger a regular n-gon more and more closely approximates a circle. Indeed, we would say that the limit of a regular n-gon as n tends to infinity *is* a circle. This notion was used in developing the formula for the area of a circle.

- *Measurement*

  This notion needs no comment. The ability to measure and communicate information concerning size and amount is basic to geometric thinking and applications of geometry in the real world.

- *Coordinates*

  This theme stresses the idea that geometric objects can be viewed as sets of points determined by ordered pairs of numbers that satisfy certain conditions. This powerful notion makes it possible to use methods of arithmetic and algebra to obtain geometric results.

- *Logical structure*

  As in the rest of mathematics, geometric ideas do not stand alone. Even in informal geometry it is important for students to see how some results follow from others and to realize that guessing or conjecturing alone is not enough.

CLASSIC CONUNDRUM Square Inch Mysteries

From a sheet of graph paper cut an 8 by 8 square into the four pieces *A*, *B*, *C*, and *D* as shown.

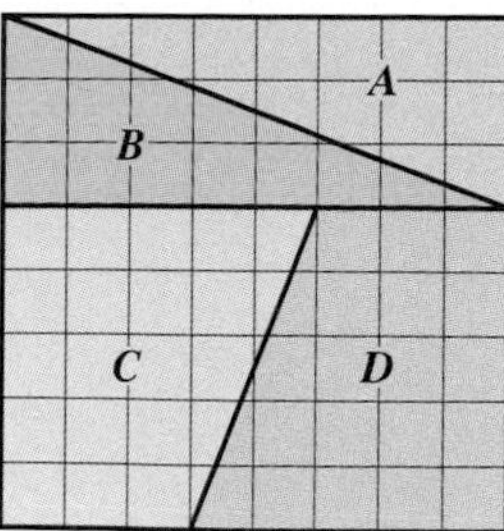

Now rearrange the four pieces into a rectangle as shown here. Does the rectangle have the same area as the square?

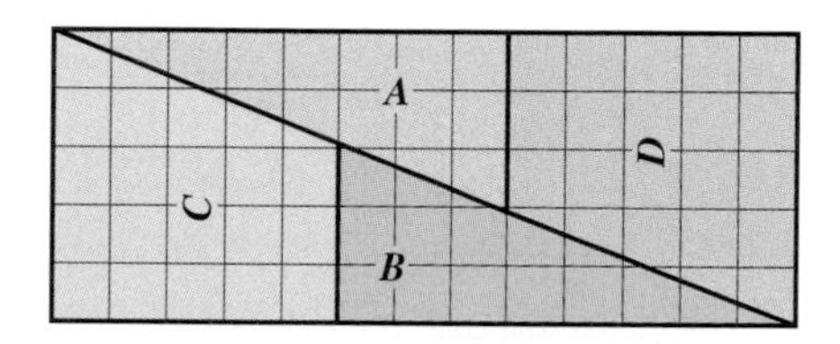

Before you run out to buy 8 by 8 bricks of gold, rearrange the four pieces into a "propeller." Can you account for the changes in area?

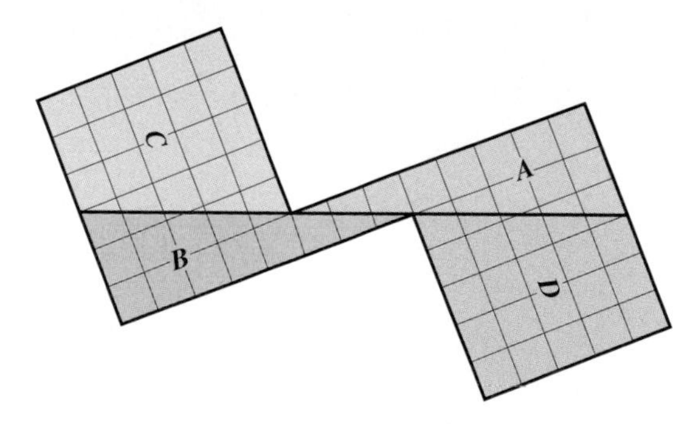

CHAPTER 14 SUMMARY

Key Concepts

The main idea of this chapter has been to introduce the idea of the Cartesian coordinate system and to show how points in the plane can be represented by ordered pairs of real numbers. In particular:

- $P(x_1, y_1)$ is the point x_1 units to the right or left of the y-axis and y_1 units above or below the x-axis.
- The midpoint of the segment $\overline{PQ}$, where P and Q are respectively the points (a, b) and (c, d), is

$$M\left(\frac{a + c}{2}, \frac{b + d}{2}\right).$$

- Let $P(a, b)$ and $Q(c, d)$ be any two points and $0 < t < 1$. The point
$$M((1 - t)a + tc, (1 - t)b + td)$$
is such that M is on PQ and
$$\frac{PM}{PQ} = t.$$
- The distance between $P(x_1, y_1)$ and $Q(x_2, y_2)$ is given by
$$PQ = \sqrt{(x_1 - x_2)^2 + (y_1 - y_2)^2}.$$
- The slope of $\overline{PQ}$ with P and Q as above and $x_1 \neq x_2$ is m, where
$$m = \frac{y_2 - y_1}{x_2 - x_1} = \frac{y_1 - y_2}{x_1 - x_2}.$$
- A segment with positive slope slopes upward to the right and one with negative slope slopes downward to the right. A segment with zero slope is horizontal and a vertical segment has no slope (it is undefined).
- The slope of a line is the slope of any segment on the line.
- Two lines are parallel if, and only if, they have the same slope.
- Two lines are perpendicular if, and only if, the product of their slopes is -1.
- The equation of the line with slope m and through the point (a, b) is
$$y - b = m(x - a).$$
This is the point-slope form of the equation of a line.
- The equation of a line with slope m through the point $(0, b)$ is
$$y = mx + b.$$
This is the slope-intercept form of the equation of a line. The number b is called the y-intercept of the line.
- A line is horizontal if, and only if, it has an equation of the form $y = b$. A line is vertical if, and only if, it has an equation of the form $x = a$.
- If two distinct lines are not parallel they must intersect and the coordinates (r, s) of their point of intersection satisfy the equation of each line and are called the simultaneous solution of the two equations.
- Every circle has an equation of the form
$$(x - a)^2 + (y - b)^2 = r^2$$
where (a, b) is the center of the circle and r is its radius.
- The preceding facts can be used to prove results about geometric figures.

Vocabulary and Notation

Section 14.1

The Cartesian coordinate system
Coordinate axes
  x-axis
  y-axis

Coordinates
  x-coordinate
  y-coordinate
  Quadrant
Origin, (0, 0)
The midpoint formula
Dividing a line segment
The distance formula

Section 14.2

Slope
  Of a line segment
  Of a line
Condition for parallelism
Condition for perpendicularity
Equations of lines
  Point-slope form
  Slope-intercept form
Linear equation
Simultaneous solution of two linear equations

Section 14.3

Proofs using coordinates
Proofs using equations and coordinates
Equation of a circle

Section 14.4

The graph of a function
The vertical line test
Maximum and minimum values of functions
Graphs involving relations
Graph of a linear inequality

CHAPTER REVIEW EXERCISES

Section 14.1

1. Plot each of these points on a Cartesian coordinate system.
   (a) $(5, -2)$ (b) $(1.4, 1.7)$
   (c) $(-3, -4.5)$ (d) $(-3, 1.5)$
2. In what quadrant does $P(a, b)$ lie
   (a) if $a > 0$ and $b < 0$?
   (b) if $a < 0$ and $b > 0$?
   (c) if $ab > 0$?
   (d) if $ab < 0$?
3. Determine a and b if $M(a, b)$ is the midpoint of the segment $\overline{PQ}$ where P and Q are the points given.
   (a) $(1, 5)$, $(3, -1)$ (b) $(3.2, 1.7)$, $(1.4, -1.5)$
   (c) $(0, 3)$, $(-5, -4)$ (d) $(0, 0)$, $(-2, -9)$
4. Determine PQ for each of the pairs of points in problem 3.
5. Show that the triangle with vertices $A(-3, 1)$, $B(0, 5)$, and $C(1, -2)$ is a right triangle.
6. Show that the triangle with vertices $R\left(-7, \frac{9}{2}\right)$, $S(3, 0)$, and $T(1, -3)$ is isosceles.

Section 14.2

7. If possible determine the slope of each of the segments with the indicated endpoints.
   (a) $M(3, -2)$, $N(2, -3)$
   (b) $P(3, -7)$, $Q(3, 5)$
   (c) $R(-2, -5)$, $S(1, -7)$
   (d) $A(2, 5)$, $B(2, -3)$
8. Determine c so that $\overline{PQ}$ has slope 5 where P and Q are respectively $(c, -2)$ and $(-2, 7)$.
9. Consider the points $C(3, 5)$, $D(2, b)$, $E(-1, 2)$, and $F(2, -7)$.
   (a) Determine b so that $\overleftrightarrow{CD}$ is parallel to $\overleftrightarrow{EF}$.
   (b) Determine b so that $\overleftrightarrow{CD}$ is perpendicular to $\overleftrightarrow{EF}$.
10. Determine if $(-4, 5)$ is on the perpendicular bisector of $\overline{PQ}$ where P and Q are respectively $(0, 5)$ and $(-3, 1)$.
11. Determine b so that the point $(3, b)$ satisfies the equation $3x - 5y = 17$.
12. Write the equation of the line through $P(3, -2)$ and $Q(-4, 7)$.
13. Write the equation of the line with slope $3/2$ and y-intercept 10.
14. Determine the slope of the line $3x - 4y = 15$.
15. Determine the equation of the line through the point $(5, 1.5)$ and perpendicular to the line of problem 14.
16. Find the simultaneous solution to the equations $3x - 5y = 19$ and $2x + 3y = 0$.

Section 14.3

17. Consider the right triangle shown, where M is the midpoint of the hypotenuse. Let D be the midpoint of the square in the second quadrant with one side $\overline{AB}$ and let E be the midpoint of the square in the fourth quadrant with one side $\overline{AC}$.
   (a) Show that $\triangle EMD$ is a right triangle.
   (b) Show that E, A, and D are collinear. (*Hint:* What must be true about the slopes of $\overline{EA}$ and $\overline{ED}$ if, and only if, E, A, and D are collinear?)

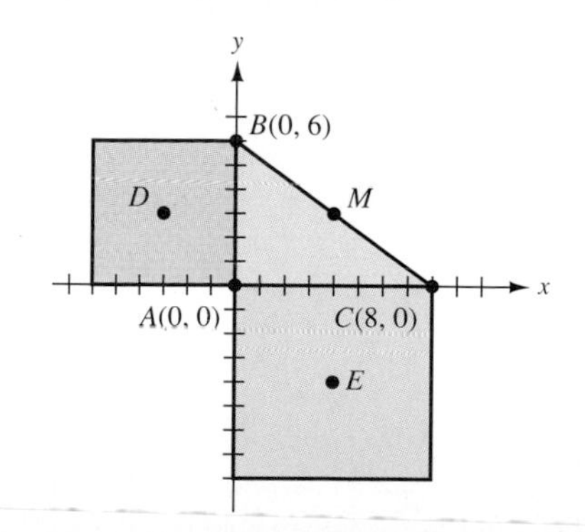

18. Consider the three squares drawn on the coordinate system as shown. Let E, G, and H be the centers of the three squares and let F be the point (6, 0).

(a) Show that $\overline{EG} \cong \overline{FH}$.

(b) Show that the line $\overleftrightarrow{EG}$ is perpendicular to the line $\overleftrightarrow{FH}$.

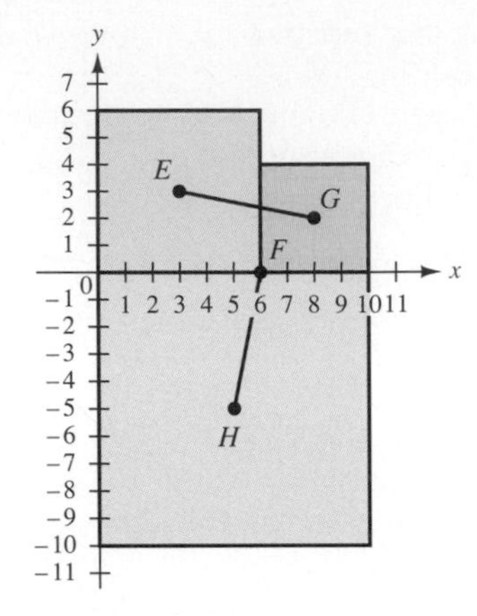

19. Consider the four squares drawn on the coordinate system as shown. Let E, F, G, and H be the midpoints of the squares.

(a) Show that $\overline{EG} \cong \overline{FH}$.

(b) Show that the line $\overleftrightarrow{EG}$ is perpendicular to the line $\overleftrightarrow{FH}$.

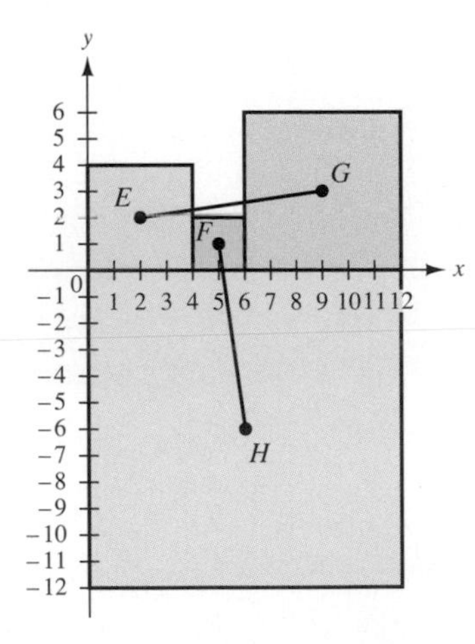

20. Consider the right triangle shown where M is the midpoint of the hypotenuse. Let D be the midpoint of the square in the second quadrant with one side $\overline{AB}$ and let E be the midpoint of the square in the fourth quadrant with one side $\overline{AC}$.

(a) Show that $\triangle EMD$ is a right triangle.

(b) Show that E, A, and D are collinear.

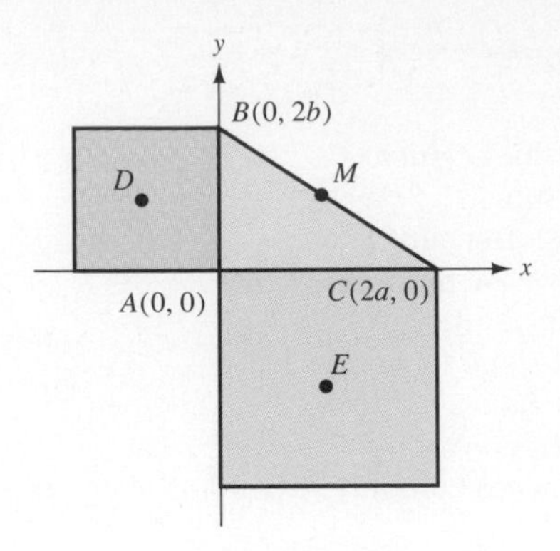

21. Consider the three squares drawn on the coordinate system. Let E, G, and H be the centers of the three squares and let F be the point $(2a, 0)$.

(a) Prove that $\overline{EG} \cong \overline{FH}$.

(b) Prove that the line $\overleftrightarrow{EG}$ is perpendicular to the line $\overleftrightarrow{FH}$.

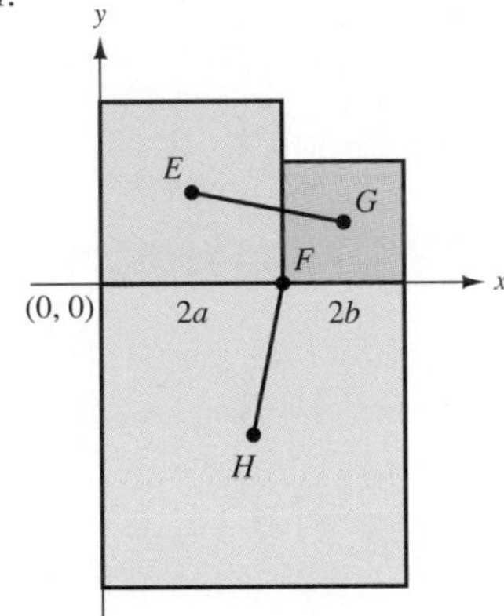

22. Consider the four squares drawn on the coordinate system and let E, F, G, and H be the midpoints of the squares as shown.

(a) Prove that $\overline{EG} \cong \overline{FH}$.

(b) Prove that line $\overleftrightarrow{EG}$ is perpendicular to line $\overleftrightarrow{FH}$.

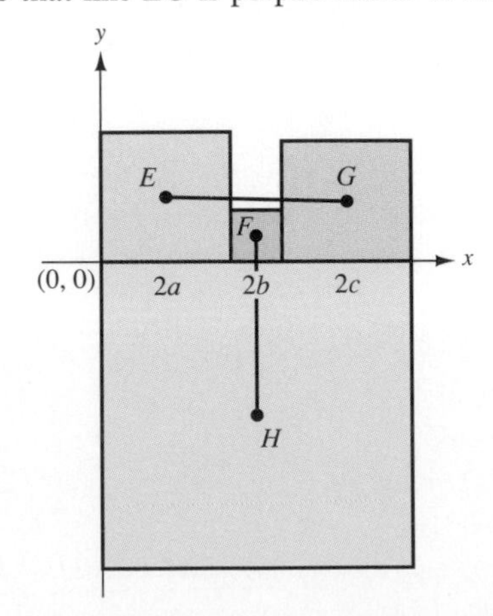

CHAPTER TEST

1. If $A(0, 0)$, $B(3, 5)$, $C(r, s)$, and $D(7, 0)$ are the vertices of a parallelogram, determine r and s.
2. Determine the point 1/4 of the way from $A(-8, 0)$ to $B(4, 12)$.
3. Determine the equation of the perpendicular bisector of $\overline{AB}$ where A and B are the points (2, 5) and (7, 4).
4. Show that the triangle with vertices $R(-3, 2)$, $S(7, 10)$, and $T(1, -3)$ is a right triangle.
5. Determine r so that the segments $\overline{PQ}$ and $\overline{RS}$ are parallel where P, Q, R, and S are the points $(2, -5)$, $(-3, 4)$, $(2, r)$, and $(4, 1)$ respectively.
6. Determine the equation of the line through $P(4, 2)$ and $Q(-1, 5)$.
7. Determine the slope of the line whose equation is $5x + 2y = 19$.
8. Determine the equation of the circle with center $(-2, 5)$ and radius 4.
9. Determine the simultaneous solution to the equations $2x - 3y = 9$ and $3x - 5y = 14$.
10. Give a general proof that the diagonals of a parallelogram bisect each other.
11. Compute the value of an investment of \$5000 at the end of 5 years if interest is computed at the rate of 7% compounded annually.
12. Consider the function $y = 2x^2 + 4x$.
    (a) Graph the function.
    (b) Determine the minimum value of the function and the value of x for which it occurs.
13. Graph the linear inequality $2x - 6y \leq 12$.

APPENDIX A
Manipulatives in the Mathematics Classroom*

What are Manipulatives?

More and more we are hearing calls for greater use of manipulatives in elementary mathematics teaching. Manipulatives are concrete materials that are used for "modeling" or representing mathematical operations or concepts. In much the same way that children make models of airplanes or clipper ships so that they can study and learn about them, your students can make and learn from models of two-digit numbers or division. The difference between the two situations is that while the airplane and boat models are smaller versions of actual concrete things, the number models or division models are concrete models of abstract concepts.

When children use bundles of Popsicle™ sticks and single sticks to represent tens and ones, or stretch rubber bands around nails on a geoboard to show squares and triangles, they are using manipulatives to "model" mathematical ideas. Technically, even when young children count on their fingers, they are concretely modeling numbers!

It is important to note that some frequently used "manipulatives" materials fail to fit this definition of manipulatives. Flashcards, for instance, even though they are certainly manipulated, are never used to model mathematical ideas. Other objects like checkers in a checker game also are manipulated but are not used to directly represent mathematical ideas.

---

*By Warren Crown, associate professor of mathematics education, Rutgers, The State University of New Jersey. Used by permission of ScottForesman Marketing Department.

What Kinds of Things Can Be Used as Manipulatives?

Good teachers use a great variety of things as manipulatives. Popsicle™ sticks, dried beans, smooth stones, egg cartons, and poker chips are all inexpensive and effective materials. Some teachers even like to use the students in their classes themselves as models for sets and numbers.

There are also, however, commercial materials that serve more specific modeling purposes. Some of these are base-ten blocks, geoboards, fraction strips or pieces, and algebra tiles. What's important about the materials you select is not their cost, but that they accurately represent the concept or operation that your students are ready to learn about.

What Does Research Say About Using Manipulatives?

For the last twenty years, research support for using manipulatives in elementary math teaching has been growing. In 1977, Suydam and Higgins were able to identify and review 23 research studies that addressed the question of the effectiveness of manipulative materials for instruction. They concluded that instruction using manipulative materials had a higher probability of producing greater mathematics achievement than nonmanipulative instruction (Suydam & Higgins, 1977).

Twelve years later, Sowell (1989) had 60 studies to work with in a similar review. These studies were conducted at all educational levels, from kindergarten to college, and focused on many different mathematical topics. After using a sophisticated statistical procedure called meta-analysis, Sowell was able to offer a much stronger conclusion than Suydam and Higgins: that manipulative materials do, indeed, have a significant positive effect on achievement, especially when they are used over a long period of time (Sowell, 1989). In addition, Sowell found that the use of concrete materials for instruction was effective in improving students' attitudes toward mathematics.

What Do the Professional Organizations Say?

Because of the growing research support, professional organizations and leading educators have been urging increased use of concrete materials in teaching. Manipulatives play a very prominent role in the National Council of Teachers of Mathematics *Curriculum and Evaluation Standards for School Mathematics*. Even though there is no single standard which says that manipulatives should be used more frequently, their use is supported for many different topics at all levels, from kindergarten through twelfth grade. In the area of Number Sense and Numeration, for example, the Standards suggest that:

> Students should be able to construct number meaning through real-world experiences and the use of physical materials. (Standards, p. 38)

In Fractions and Decimals,

> Students should be able to use models to relate fractions to decimals and to find equivalent fractions, and to explore operations on fractions and decimals. (Standards, p. 57)

Although manipulatives have traditionally been used more in the lower elementary grades than the upper grades, the Standards also make a clear statement about their applicability and desirability in grades 5–8:

> Implementation of the 5–8 standards should consider the unique characteristics of middle school students. As vast changes occur in their intellectual, psychological, social, and physical development, students in grades 5–8 begin to develop their abilities to think and reason more abstractly. Throughout this period, however, concrete experiences should continue to provide the means by which they construct knowledge. From these experiences, they abstract more complex meanings and ideas. (Standards, p. 68)

How Does Concrete Modeling Help Children Learn Mathematics?

The mechanism by which concrete modeling promotes the learning process is still somewhat of a mystery to us, but the fact that it does is becoming more certain all the time. Most research shows that the best instructional sequence to follow for the presentation of elementary mathematical material is concrete-pictorial-symbolic. Activities with concrete materials should precede those which show pictured relationships and those should, in turn, precede formal operations with symbols. Ultimately, students need to reach that final level of symbolic proficiency with many of the mathematical skills that they master, but the meanings of those symbols and abstract operations must be firmly rooted in experiences with real objects. Otherwise, their performance of the symbolic operations will simply be rote repetitions of meaningless, memorized procedures.

Concrete and pictorial models are, at best, imperfect representations of abstract mathematical ideas and concepts. Not every characteristic of the concrete model is important mathematically. For example, many teachers use a yellow, wooden rod to represent the number "5," but "yellowness" certainly has nothing to do with the "fiveness." To prevent the child from abstracting inappropriate characteristics of the model as characteristics of the mathematical idea, multiple models should be used for important concepts. Multiple models of the same idea which are perceptually very different from each other direct the student to abstract from them only what they have in common—the mathematical concept. This abstraction is what leads to meaningful mathematics learning.

When Should Manipulatives be Used?

Manipulatives and concrete models can be used almost anywhere in the elementary mathematics curriculum and should be used with all students. The most frequent occurrence of modeling in the curriculum will probably be at the point of introduction of new topics. The use of concrete materials while introducing a new topic allows students to gain that necessary foundational experience before trying to demonstrate their understandings symbolically. But intelligent use of manipulatives can also provide the most effective form of remediation. Engaging students in concrete activities related to the mathematics that they are struggling with will frequently help them to identify exactly which part of the process is causing the confusion, and then to work through it.

Manipulative use is also not just important for younger children. As stated in the NCTM Standards, older children can benefit just as much from appropriate concrete activity. The skills and concepts that they are asked to deal with are increasingly complex, and concrete introductions can frequently pave the way toward true understanding. Good concrete and pictorial models are available to help students deal with percent, ratio, geometric formulas, integers, and even solving algebraic equations.

Manipulative *Do's* and *Don'ts*

Elementary teachers who have been using manipulatives for many years have learned some rules which make their use in classrooms more effective. Don't use the materials exclusively for demonstrations and teacher explanations. The most effective way to use manipulatives is to have the children work directly with the materials. It is through touching and moving that the learning takes place. Watching the teacher do the manipulation is much less successful.

Discuss appropriate behavior before distributing the materials and allow plenty of free exploration time after. Children need to have time to do what they want with the materials before they will do what you want with them. This exploration time is also learning time. The more familiar they are with the materials, the more effective will be their later use of them.

Have children work with the materials in small groups and explain their thinking to each other as they do so. Asking children to verbalize their thoughts for others gives them the opportunity to clarify their own thinking. By listening to the discussions, you can make judgments about how well they understand the concepts.

How Can I Get More Information About Using Manipulatives?

This list of references is a good place to start. The first two citations will be helpful for the practitioner, while the second two are the research reviews mentioned earlier.

National Council of Teachers of Mathematics. *Curriculum and Evaluation Standards for School Mathematics.* Reston, Virginia: NCTM, 1989.

Reys, Robert E.; Suydam, Marilyn N.; and Lindquist, Mary Montgomery. *Helping Children Learn Mathematics,* Second Edition. Englewood Cliffs, New Jersey: Prentice Hall, Inc., 1989.

Sowell, Evelyn, J. "Effects of Manipulative Materials in Mathematics Instruction." *Journal for Research in Mathematics Education,* vol 20, no. 5, (November, 1989) pp. 498–505.

Suydam, Marilyn N., and Higgins, Jon L. *Activity-Based Learning in Elementary School Mathematics: Recommendations from Research.* Columbus, Ohio: ERIC Center for Science, Mathematics, and Environmental Education, 1977.

APPENDIX B
Logic and Mathematical Reasoning

"For a *complete* logical argument," Arthur began with admirable solemnity, "we need two prim Misses—"

"Of course!" she interrupted. "I remember that word now. And they produce_____?"

"A Delusion," said Arthur.

"Ye-es?" she said dubiously. "I don't seem to remember that so well. But what is the *whole* argument called?"

"A sillygism."

—From Lewis Carroll's *Sylvie and Bruno*

Lewis Carroll, in the quotation above, is making light of the **syllogism,** a central notion of logic in which a conclusion is drawn from the acceptance of two **premises.** Formal logic originated with the ancient Greek philosopher Aristotle, who believed the operations of valid reasoning were summed up in his list of 14 syllogisms, the most famous being "All men are mortal; all heroes are men; therefore all heroes are mortal." Medieval theologians extended Aristotle's list to 19, which constituted the verbal foundation of logic until the middle of the nineteenth century. George Boole, in his booklet *The Mathematical Analysis of Logic—Being an Essay Towards a Calculus of Deductive Reasoning* of 1847, showed that logical statements could be represented in algebraic terms, and deductions could be obtained by algebraic operations. Thus Boole originated **symbolic logic.**

In this appendix we take a middle ground, using only enough symbols to clarify the main principles of logical reasoning and operations. The main application is toward logical reasoning and problem solving in mathematics, but logic has a place in every domain of human thought. In his "Introduction to Learners" for *Symbolic Logic,*

Lewis Carroll claims that logic

> "... will be of real *use* to you in *any* subject you may take up. It will give you clearness of thought—the ability to *see your way* through a puzzle—the habit of arranging your ideas in an orderly and get-at-able form—and, more valuable than all, the power to detect *fallacies,* and to tear to pieces the flimsy illogical arguments, which you will so continually encounter in books, in newspapers, in speeches, and even in sermons, and which so easily delude those who have never taken the trouble to master this fascinating art. *Try it.* That is all I ask of you!"

Statements

Logical reasoning deals with **statements,** given by declarative sentences that are either true or false.

DEFINITION Statement

A **statement** is a declarative sentence that is either true or false, but not both.

Often a statement is symbolized by a single letter, say p or q. The **truth value** of a statement is either true, T, or false, F.

EXAMPLE B.1

Identifying Statements and Nonstatements

Classify each of these as a statement or nonstatement.

(a) $2 + 3 = 5$ **(b)** $4 + 2 = 13$ **(c)** $2x + 3 > 7$
(d) This sentence is false. **(e)** Shut the window. **(f)** Pigs can fly.

SOLUTION

(a), (b), and (f) are statements, since they are either true or false. Since x was not specified, (c) is not a statement. If we specified x however, it would become a statement. Since (e) is not declarative, it is not a statement. The self-referential statement (d) is neither true nor false and so it is not a statement. ■

There are many useful ways to combine and relate statements, including **negation, and, or, if . . . then,** and **if, and only if.**

Negation

The statements "It rained today" and "It did not rain today" are negations of one another. The statement formed from the given statement p by prefixing "It is false that," or simply "not," is called the **negation** of p, and is symbolized by $\sim p$.

DEFINITION Negation: $\sim p$

If p is a statement, the **negation** of p, written $\sim p$, is the statement "it is false that p".

Forming a statement $\sim p$ which is the negation of a given statement p is often straightforward, but special care must be taken for complex statements. This is particularly so if the statement involves such words as *all, every, some, there exists,* or *no*. These words signify a number, or quantity, of elements in some set that satisfy some condition or have some property. These words are thus known as **quantifiers** in a statement.

EXAMPLE B.2 **Forming Negations of Statements**

Write the negation of each statement below.

(a) $3 + 4 \neq 7$
(b) Every natural number is one larger than some other natural number.
(c) Some students have a calculator.
(d) No student mastered logic.
(e) All whole numbers satisfy $x(x + 1) = x^2 + x$.
(f) Some numbers in the sequence 31, 331, 3331, 33331, . . . are not prime numbers.
(g) Some numbers in the sequence 11, 111, 1111, 11111, 111111, . . . are perfect squares.

SOLUTION

(a) $3 + 4 = 7$
(b) There is some natural number that is not one larger than some other natural number.
(c) No student has a calculator.
(d) At least one student mastered logic. (Alternatively: Some students mastered logic.)
(e) There exists a whole number x for which $x(x + 1) \neq x^2 + x$.
(f) All numbers in the sequence 31, 331, 3331, 33331, . . . are prime numbers.
(g) No numbers in the sequence 11, 111, 1111, 11111, 111111, . . . are perfect squares.

■

Compound Statements

Two simple statements, p and q, can be combined by **and,** symbolized by $\wedge$, to form the compound statement $p \wedge q$ defined as follows.

DEFINITION The *and* Operation, $p \wedge q$

Let p and q be statements. Then p **and** q, written $q \wedge q$, is the statement which is true when both p and q are true and is false when one, or both, of p and q is false.

The inclusive **or** operation, symbolized with $\vee$, is defined similarly.

> **DEFINITION The *or* Operation, $p \vee q$**
>
> Let p and q be statements. Then p **or** q, written $p \vee q$, is the statement which is true if at least one of the statements of p and q is true, and is false if both p and q are false.

EXAMPLE B.3

Determining the Truth Value of Compound Statements

Let p: 7 is an even natural number
q: 4 is a square number
r: 6 is an even number
s: $10 = 2 \cdot 3$

Classify these following compound statements as *true* or *false*.

(a) $p \wedge q$ **(b)** $q \vee r$ **(c)** $p \vee s$
(d) $q \wedge r$ **(e)** $\sim q \wedge s$ **(f)** $\sim p \wedge \sim q$
(g) $\sim (p \vee q)$ **(h)** $\sim q \vee \sim s$ **(i)** $\sim(q \wedge s)$
(j) $\sim s \wedge \sim p$ **(k)** $\sim(s \wedge p)$ **(l)** $\sim p \wedge (q \vee s)$

SOLUTION

(a) F **(b)** T **(c)** F **(d)** T **(e)** F **(f)** F **(g)** F
(h) T **(i)** T **(j)** T **(k)** T **(l)** T. ■

Two statements with the same truth value are **logically equivalent.** A simple example is the logical equivalence of p and $\sim\sim p$, called the **law of double negation.** Similarly, $p \wedge q$ is logically equivalent to $q \wedge p$, and $p \vee q$ is logically equivalent to $q \vee p$.

Augustus DeMorgan (1806–1871), Professor of Mathematics at University College, London, discovered two especially useful pairs of logically equivalent statements. These were published in 1858, and have since become widely known as **DeMorgan's Laws.**

EXAMPLE B.4

Discovering DeMorgan's Laws of Logic

Explain why $\sim(p \wedge q)$ is logically equivalent to $\sim p \vee \sim q$ and why $\sim(p \vee q)$ is logically equivalent to $\sim p \wedge \sim q$.

SOLUTION

If $\sim(p \wedge q)$ is false then $p \wedge q$ is true, meaning p is true and q is true. Then $\sim p$ and $\sim q$ are both false, so $\sim p \vee \sim q$ is false. If $\sim(p \wedge q)$ is true, then $p \wedge q$ is false, meaning that one (or both) of p and q is (are) false. But then one (or both) of $\sim p$ and $\sim q$ is (are) true, so $\sim p \vee \sim q$ is also true. This shows that $\sim(p \wedge q)$ and $\sim p \vee \sim q$ are simultaneously true or false, and are therefore logically equivalent. The second DeMorgan law can be explained in the same manner. (See problems 24 and 28 in the exercises.) ■

CURRENT AFFAIRS

Fuzzy Logic May Smooth the Way in the Future

"Fuzzy" subways, air conditioners, and automatic transmissions operate more efficiently and smoothly than their conventional counterparts. Why? Fuzzy logic enables computers to simulate the ambiguities encountered in real-life situations.

A fuzzy system that controls acceleration and braking in Japan's Sendai subway provides a smoother ride than would human drivers.

This division of mathematics, called fuzzy logic, is now applied to helping computers simulate the vagueness and uncertainty inherent in our thought processes and language. Computers normally solve problems by breaking them down into a series of yes-or-no decisions, represented by ones and zeros. But fuzzy logic lets computers assign numerical values that fall somewhere between zero and one. Instead of statements or conditions being only true or false, no maybe's or sort of's, fuzzy theory sets up conditions such as slow, medium, and fast. And because it doesn't require all-or-nothing choices, and represents real-world events as continuous phenomena, fuzzy logic seems to make a variety of computer-controlled machines run more smoothly and efficiently. The Sendai subway system in Japan, for example, accelerates and brakes more evenly than a human driver (see graph). Passengers don't even bother hanging onto straps.

An American, Professor Lotfi A. Zadeh of the University of California at Berkeley, developed fuzzy-logic theory 25 years ago. "I gave quite a bit of thought to what to call it," said Zadeh. "I realized that fuzzy didn't sound good, but at the same time, I could not find a scientific term that described it as accurately as fuzzy did."

SOURCE: From "Fuzzy Logic" by J. T. Johnson from *Popular Science,* July 1990, pp. 87–88.

If-then Statements

Another way to combine two statements p and q is to form the **conditional statement** "if p, then q," symbolized by $p \to q$. The statement is true if q *must* be true when p is true. Statement p is called the **hypothesis** of the conditional statement $p \to q$, and q is called the **conclusion.**

DEFINITION The Conditional Statement "if p, then q": $p \to q$.

Let p and q be statements. Then the statement **if p, then q,** written $p \to q$, is true if q is necessarily true when p is true. The statement $p \to q$ is false only in the case that p is true and q is false.

EXAMPLE B.5

Determining When "if . . . then" Is True or False

Decide if each conditional statement is *true* or *false.*

(a) If $1 = 2$, then $2 = 3$.
(b) If $2 = -2$, then $4 = 4$.
(c) If a and b are whole numbers and $a < b$, then $a^2 < b^2$.
(d) If a and b are integers and $a < b$, then $a^2 < b^2$.

SOLUTION

(a) T
(b) T
(c) T
(d) F

In both (a) and (b) the hypothesis p is false, so the conditional statement $p \to q$ is automatically true: it does not matter if q is false, as in (a), or true, as in (b). In (c), the properties of the arithmetic operations show that if the hypothesis is true then so is the conclusion, and thus the conditional statement is true. In (d) the hypothesis can be true and the conclusion is false: take $a = -3$ and $b = 2$ for example, and notice that $-3 < 2$ but $(-3)^2 \not< 2^2$.

The truth of the implications (a) and (b) may seem very strange, and even senseless, but there is an important point being made: if you mistakenly assume something is true when it is not, then anything—true or false—can be "deduced." ■

The implication in (d) of the example above was shown to be false by exhibiting what is known as a **counterexample.** A counterexample to a conjectured implication $p \to q$ is an example for which p is true and yet q is false. Perhaps the most famous counterexample in history is the plucked chicken, proposed by Diogenes to refute Plato's definition of a human being as a featherless biped.

Many statements, although not originally in "if p, then q" form, can nevertheless be rewritten in this form. For example, "all men are mortals" is equivalent to the conditional statement "if x is a man, then x is a mortal." Other variations occur by replacing "if . . . then" by expressions which use such words as "implies," "only if," "necessary," "sufficient," and so on. Here are alternative ways to state the same conditional statement:

1. If p, then q.
2. q if p.
3. p only if q.
4. p is a sufficient condition for q.
5. q is a necessary condition for p.
6. p implies q.

"YOU WANT PROOF? I'LL GIVE YOU PROOF!"

The Contrapositive

Some other variations of the conditional statement $p \rightarrow q$ are obtained by interchanging p and q, or replacing p and q by their negations, or doing both.

> **DEFINITION The Contrapositive $\sim q \rightarrow \sim p$**
>
> Given the conditional statement "if p, then q," then "if not q, then not p" is its **contrapositive.** In symbols,
>
> conditional: $p \rightarrow q$
> contrapositive: $\sim q \rightarrow \sim p$.

EXAMPLE B.6 **Writing the Contrapositive**

Write the contrapositive for each of these conditional statements, where m denotes any whole number.

(a) If m is divisible by 4, then m is divisible by 2.
(b) If m^2 is divisible by 3, then m is divisible by 3.

SOLUTION

(a) If m is not divisible by 2, then m is not divisible by 4.
(b) If m is not divisible by 3, then m^2 is not divisible by 3. ■

The statement $p \rightarrow q$ means that q is necessarily true if p is true. Therefore, if q is false, then p must also be false. That is, if $\sim q$ is true then $\sim p$ must be true, which in symbols is written $\sim q \rightarrow \sim p$. The same reasoning shows that $\sim q \rightarrow \sim p$ is true precisely when $p \rightarrow q$ is true. We have the following result.

THEOREM Logical Equivalence of the Contrapositive

The statements "if p, then q" and its contrapositive "if not q, then not p" are logically equivalent. Symbolically,

$$p \rightarrow q \text{ is equivalent to } \sim q \rightarrow \sim p.$$

For many problems, showing that $\sim q \rightarrow \sim p$ is easier than showing $p \rightarrow q$, even though the two statements are logically equivalent.

EXAMPLE B.7

Using the Contrapositive

Let m and n be whole numbers. If mn is even show that m or n is even.

SOLUTION

Let p be the statement "mn is even" and q be the statement "m or n is even." Rather than show that $p \rightarrow q$, let's show the contrapositive $\sim q \rightarrow \sim p$. That is, let's show that if m and n are odd, then mn is odd. Since m and n are odd, we know that $m = 2k + 1$ and $n = 2h + 1$ for some whole numbers k and h. Then $mn = (2k + 1)(2h + 1) = 4kh + 2k + 2h + 1 = 2(2kh + k + h) + 1$ so mn is also an odd number. Since this shows $\sim q \rightarrow \sim p$, we also know that $p \rightarrow q$. ■

The Converse

DEFINITION The Converse Statement

Given the conditional statement "if p, then q," its **converse** is the statement "if q, then p." In symbols,

conditional: $p \rightarrow q$
converse: $q \rightarrow p$.

The converse $q \rightarrow p$ is *not* logically equivalent to $p \rightarrow q$. For example, "if m and n are even, then $m + n$ is even" is true, but its converse "if $m + n$ is even, then m and n are even" is false. Take $m = n = 1$ to see why the converse is false.

"If, and Only if" Statements

Sometimes *both* a conditional statement "if p, then q" and its converse "if q, then p" hold, in which case we say "p if, and only if, q" is true.

DEFINITION The "if, and only if" Statement $p \leftrightarrow q$

Let p and q be statements. If p is necessarily true when q is true, and q is necessarily true when p is true, then the statement "p if, and only if, q" is true. In symbols "p if, and only if, q" is written as $p \leftrightarrow q$.

Showing p "if, and only if, q" requires that *two* directions of implication be demonstrated: $p \rightarrow q$ and $q \rightarrow p$. Either of these implications may be replaced by their logically equivalent contrapositive statements. For instance, if $p \rightarrow q$ and $\sim p \rightarrow \sim q$, then $p \leftrightarrow q$.

EXAMPLE B.8 Showing an "If, and Only If"

Let m be any whole number. Show that m^2 is even if, and only if, m is even.

SOLUTION

Let p be the statement "m^2 is even" and q be the statement "m is even." The special case $m = n$ in Example B.7 shows that if m^2 is even then m is even, so we have $p \rightarrow q$. The converse $q \rightarrow p$ can be proved directly. Let m be even. Then $m = 2k$ for some whole number k and therefore $m^2 = 4k^2 = 2 \cdot (2k^2)$. This shows that if m is even then m^2 is even, so that $q \rightarrow p$. Since we've shown that $p \rightarrow q$ *and* $q \rightarrow p$, we have shown $p \leftrightarrow q$. ■

Deductive Reasoning

Once we have accepted certain statements as true, the truth of certain other statements may follow by logical reasoning. For an example, consider these statements.

1. If you wish to be a good elementary school teacher, then you need to learn some mathematics.
2. You wish to be a good elementary school teacher.
3. You need to learn some mathematics.

If statements 1 and 2 are accepted as true, then they are called the **hypotheses,** or **premises,** of the argument. **Deductive reasoning** allows us to claim that statement 3, called the **conclusion,** is also true. Correctly deducing if the conclusion follows from the hypotheses is called **valid reasoning.** If the conclusion does not follow from the premises, the reasoning is **invalid.** Invalid reasoning doesn't mean the conclusion is false; it just means that we do not know if the conclusion is true or false on the basis of the assumptions given in the hypotheses.

There are three commonly used rules which provide a valid argument for deriving a conclusion from the hypotheses. These are the **law of detachment,** the **law of contraposition,** and the **law of syllogism.**

The Law of Detachment and Direct Reasoning

This is the most common mode of deductive reasoning. We reason that if $p \to q$ is true and p is true, then q is necessarily true. This is often written symbolically this way.

Hypotheses:	$p \to q$
	p
Conclusion:	q

For example:

Hypotheses:	If the base ten representation of a whole number ends in 5, then the number is divisible by 5.
	7635 is a numeral which ends in 5.
Conclusion:	7635 is divisible by 5

Showing that a conclusion holds by the law of detachment is an example of direct reasoning in problem solving.

PROBLEM-SOLVING STRATEGY 15 • Use Direct Reasoning

To show that a statement q is true, look for a statement p such that $p \to q$ is true. The problem is solved once p is shown to be true.

EXAMPLE B.9

Using Direct Reasoning

If you have a table saw, a wooden cube 3 units on a side can be cut into 27 unit cubes by making six cuts. Similarly, a wooden cube 4 units on a side can be cut into 64 unit cubes by making nine cuts.

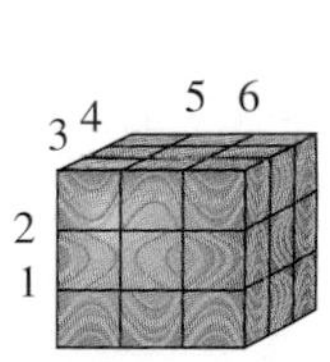

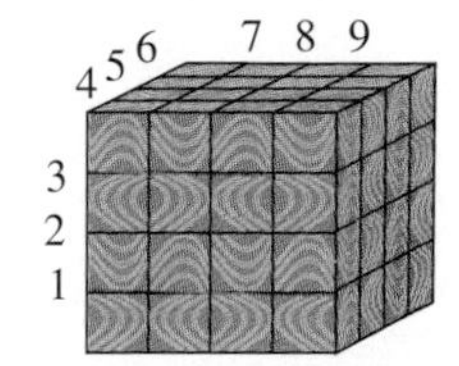

If the pieces are rearranged between cuts, is it possible to reduce the number of cuts needed to form the unit cubes?

(a) Explain why no fewer than six cuts are required to make 27 unit cubes from the $3 \times 3 \times 3$ cube.

(b) Explain how the $4 \times 4 \times 4$ cube can also be cut into unit cubes with six cuts, but no fewer.

SOLUTION

(a) The unit cube in the interior of the $3 \times 3 \times 3$ cube has six faces, and each face requires a separate cut. Thus rearranging pieces between cuts cannot reduce the number of required cuts from six.

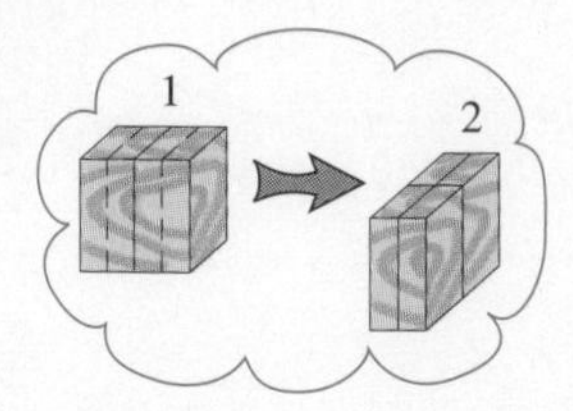

(b) Reasong as in part (a), any one of the eight interior unit cubes of the $4 \times 4 \times 4$ cube requires six cuts, so the 64 unit cubes cannot be formed with fewer than six cuts. By first cutting the cube in half and then aligning the two halves, a second cut will yield four slabs. Making the same rearrangement in the other two directions shows that six cuts in all suffice to make 64 unit cubes. ■

The Law of Contraposition and Indirect Reasoning

Since the contrapositive $\sim q \rightarrow \sim p$ is logically equivalent to $p \rightarrow q$, assuming $p \rightarrow q$ and $\sim q$ allows us to deduce $\sim p$.

$$\begin{array}{ll} \text{Hypotheses:} & p \rightarrow q \\ & \sim q \\ \hline \text{Conclusion:} & \sim p \end{array}$$

EXAMPLE B.10

Applying the Law of Contraposition

Show that 135,798,648 is *not* a square number.

SOLUTION

Since $0^2 = 0$, $1^2 = 1$, $2^2 = 4$, $3^2 = 9$, $4^2 = 16$, $5^2 = 25$, $6^2 = 36$, $7^2 = 49$, $8^2 = 64$, and $9^2 = 81$, it follows that the last digit of any square must be one of 0, 1, 4, 5, 6, or 9. Therefore, since 135,798,648 ends in 8, it is not a square. ■

A variation of the law of contraposition, called the **method of contradiction,** is often useful in proving results and solving problems. Suppose we are to prove that a statement q is true. Assume, in search of a contradiction, that q is false; or, what is the same thing, assume that $\sim q$ is true. Now suppose that $\sim q$ true implies that some statement p is false, but we already know that p is truc. Our assumption that q is false must be incorrect, and this makes q true.

In symbols, the argument looks like this.

$$\begin{array}{ll} \text{Hypotheses:} & \sim q \rightarrow \sim p \\ & p \\ \hline \text{Conclusion:} & q \end{array}$$

When applied to problem solving, the process is called **indirect reasoning.**

PROBLEM-SOLVING STRATEGY 16 • Use Indirect Reasoning

To show that a statement q is true, see if assuming that q is false leads to the contradiction of a known truth p.

EXAMPLE B.11 Using Indirect Reasoning to Solve a Coloring Problem

Suppose that we are to paint each point of the plane with one of n colors that are numbered 1, 2, . . . , n. No two points that are one inch apart can be painted with the same color. Show that this will require at least four colors. That is, show that $n \geq 4$.

SOLUTION

Understand the problem • • • •

If a point P of the plane gets, say, color 1, and a point Q is one inch from P, then color 1 cannot be assigned to Q. Thus Q must be assigned a second color, say color 2. Every point must be "painted" a color, and no two points of the same color can be one inch apart. We are to show that a coloring of all of the points of the plane which meets this condition requires at least four or more different colors.

Devise a plan • • • •

Using the strategy of indirect reasoning, let's assume that we can get by with just three colors: 1, 2, and 3. We then start painting points, and see if the condition must be violated. We need to consider new points that are one inch from already painted points, and see if we run out of colors.

Carry out the plan • • • •

Suppose the point P in the figure below is painted color 1. No point on the circle centered at P of radius 1 inch can be painted color 1, so the points Q and R one inch apart on this circle must be painted color 2 or 3. Indeed, since Q and R are one inch apart, one point (say Q) is painted color 2 and the other point is color 3. Point S, which is one inch from both Q and R, can't be painted color 2 or 3, so it must be painted color 1. The same reasoning applied to the points P, Q', R', S' shows that Q' and R' require colors 2 and 3 between them, so S' must necessarily be painted color 1. But we can also assume that S and S' are one inch apart, and since both were painted color 1, we have contradicted the condition for a valid coloring.

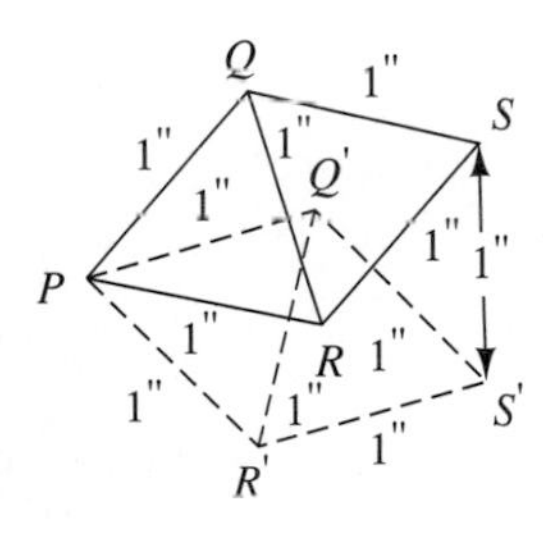

Look back • • • •

The problem asked that we show that at least four colors are required. A related question is to determine how many colors, at most, are enough to give a valid coloring. Here is a way to see that seven colors suffice. First, tile the plane with regular hexagons which are 2/5 of an inch on each side. Color a hexagon with color 1, and then color the six hexagons it touches with colors 2, 3, 4, 5, 6, and 7. The 7-hexagon coloring scheme can now be repeated over and over, as shown in this figure. Since the opposite corners on a hexagon are 4/5 inch apart, and two hexagons of the same color are more than one inch apart, we see that the conditions of the coloring are satisfied. If c is the *smallest* number of colors which suffice, we now know that $4 \leq c \leq 7$. The following drawing is at a reduced scale.

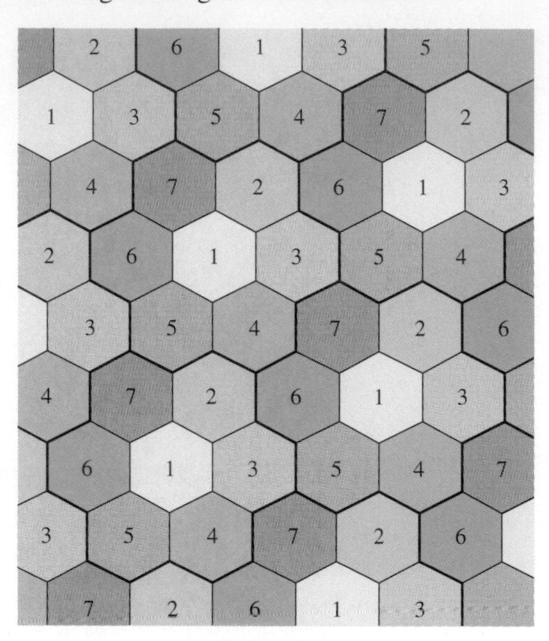

■

The Law of Syllogism

The **law of syllogism** has this form in symbols.

$$\begin{array}{ll} \text{Hypotheses:} & p \rightarrow q \\ & q \rightarrow r \\ \hline \text{Conclusion:} & p \rightarrow r \end{array}$$

The law of syllogism is also called **transitive reasoning.** Examples of the law of syllogism occur repeatedly in mathematics. We also use this method of deduction in everyday life. For example:

Hypotheses: If I eat too many cookies, then I gain weight.
If I gain weight, then my blood pressure increases.

Conclusion: If I eat too many cookies, then my blood pressure increases.

EXAMPLE B.12 **Using the Law of Syllogism**

Give a valid conclusion that follows from each set of statements.

(a) If Manuel studies hard, then he will receive high grades. Any person with high grades will be gainfully employed.

(b) If a whole number is divisible by 54, then it is divisible by 27. If a whole number is divisible by 27, then it is divisible by 9.

SOLUTION

(a) If Manuel studies hard, then he will be gainfully employed.
(b) If a whole number is divisible by 54, then it is divisible by 9. ■

PROBLEM SET B

Understanding Concepts

1. Which of these are statements?
   (a) $3 + 5 = 6$
   (b) $2 < 7$
   (c) $x = 5$
   (d) Open your book.
   (e) $2 \div 0$ is not defined.
2. The **open sentence** "$x^2 + 6 = 5x$" can be made either *true* or *false* by using different quantifiers:

   "For some whole number x, $x^2 + 6 = 5x$"

   is true, since $x = 2$ or $x = 3$ makes the equation true.

   "For all whole numbers x, $x^2 + 6 = 5x$"

   is false, since the equation is false for the whole number $x = 0$, among others. Use a quantifier—"For all . . . ", "For some . . . ", "For no . . . "—to make each of these open sentences *true,* where x is a whole number.
   (a) $x + 5 = 9$ **(b)** $x + x^2 = x(1 + x)$
   (c) $x \cdot 1 = x \cdot 2$ **(d)** $x^2 + 1 = 0$
3. Use quantifiers to make each statement in problem 2 *false.*
4. Write the negation of these statements.
   (a) $3 \cdot 1 = 3$
   (b) $4 + 6 \neq 10$
   (c) $2 < 1$
   (d) All triangles have two congruent sides.
   (e) Some rectangles are not squares.
   (f) No pig has a curly tail.
   (g) All squares have a right angle.
   (h) Some natural number is not the sum of one, two, or three triangular numbers.
   (i) No whole number whose base ten representation ends with the digit 7 is a square number.
5. Let p represent the statement "I understand the problem" and let q represent the statement "I can solve the problem." Write each of these statements in symbolic form.
   (a) I don't understand the problem.
   (b) I understand the problem and I can solve the problem.
   (c) If I don't understand the problem, then I can't solve the problem.
   (d) If I can solve the problem, then I understand the problem.
6. Let p, q, and $q \rightarrow r$ be *true* statements, and let s be a *false* statement. Determine which of these statements is *true.*
   (a) $p \vee s$ **(b)** $s \rightarrow r$
   (c) $r \wedge s$ **(d)** $\sim s \vee \sim p$
7. Decide which of the following compound statements are *true* and which are *false.*
   (a) $2 + 2 = 4$ or $3 = 5 + 1$.
   (b) 9 is even and is a square number.
   (c) 18 is divisible by 6 and 14 is divisible by 2.
8. Use DeMorgan's Laws (Example B.4) to express the negations of the following compound statements:
   (a) I own a calculator and I drive a car.
   (b) x is not a square number or x is even.
   (c) It is not a vacation and it is not expensive.
9. Decide if the following conditional statements are *true* or *false:*
   (a) If n is a natural number, then the last digit of n^4 is 0, 1, 5, or 6.
   (b) If the last digit of a natural number is 0, 1, 5, or 6, then it is a fourth power of some natural number.
   (c) n is a natural number only if $n + 1$ is a whole number.
   (d) n is a natural number if $n + 1$ is a whole number.
10. Write the converse and contrapositive of the following conditional statements, where r, s, t, and u denote whole numbers.

(a) If r is a dweeble, then r is a moosh.
(b) If $s \cdot t \neq s \cdot u$, then $t \neq u$.

11. The **inverse** of the conditional statement "if p, then q" is "if not p, then not q"; that is, $\sim p \rightarrow \sim q$ is the inverse of $p \rightarrow q$.
(a) Why is the inverse logically equivalent to the converse of $p \rightarrow q$?
(b) Write the inverse of the two statements in problem 10.

12. Let m and n be whole numbers, and consider the statement $p \rightarrow q$ given by "if $m + n$ is even, then m and n are even."
(a) Express the contrapositive, converse, and inverse of the given conditional.
(b) For the statements which are *true,* give a proof.
(c) For the statements which are *false,* give a counterexample.

13. Let n be any whole number.
(a) Show, by following the method shown in Example B.8, that n^2 is odd if, and only if, n is odd.
(b) Assuming the result of Example B.8, namely that m^2 is even if, and only if, m is even, explain why it follows that n^2 is odd if, and only if, n is odd.

14. Decide if the following arguments are valid or invalid. If valid, explain which law—detachment, contraposition, or syllogism—is used. Assume that x, y, and z are whole numbers.

(a) Hypotheses: If x, y, and z are all odd, then $x + y + z$ is odd.
$x + y + z$ is even.
Conclusion: x, y, and z are not all odd numbers.

(b) Hypotheses: If $x \neq y$, then $2xy < x^2 + y^2$.
$3 \neq 5$.
Conclusion: $2 \cdot 3 \cdot 5 < 3^2 + 5^2$.

(c) Hypotheses: If x and y are even, then $x + y$ is even.
$3 + 7$ is even.
Conclusion: 3 and 7 are even.

(d) Hypotheses: If I were rich, I would own a Rolex.
I am not rich.
Conclusion: I do not own a Rolex.

(e) Hypotheses: If x is divisible by 64, then it is divisible by 8.
If x is divisible by 32, then it is divisible by 16.
Conclusion: If x is divisible by 8, then it is divisible by 16.

(f) Hypotheses: If x and y are odd, then $x + y$ is even.
$3 + 7$ is even.
Conclusion: 3 and 7 are odd.

(g) Hypotheses: If I study, I will get a good grade.
If I get a good grade, I will get a scholarship.
Conclusion: If I study, I will get a scholarship.

15. There are three forms of invalid reasoning which occur commonly.

Fallacy of the converse
If p, then q.
q
p (invalid)

Fallacy of the inverse
If p, then q.
$\sim p$
$\sim q$ (invalid)

False transitivity
If p, then q.
If p, then r.
If q, then r. (invalid)

Which fallacies occur in these arguments?
(a) If I am a good person, nothing bad will happen to me. Nothing bad happened to me, so I am a good person.
(b) If you've got the time, I have the beer. You don't have any time, so I have no beer.
(c) If you work too hard, you'll be wealthy and die young. Therefore, if you're wealthy you'll die young.

16. What conclusion can be deduced from these sets of hypotheses? Let f stand for a statement which is *false.*
(a) Hypotheses: p or q
$\sim p$
Conclusion: ?
(b) Hypothesis: $\sim p \rightarrow f$
Conclusion: ?
(c) Hypotheses: $(p \wedge q) \rightarrow r$
p
Conclusion: ? (give a conditional statement)

17. Prove, or give a counterexample, to these statements:
(a) For all whole numbers n, $n^2 + n + 2$ is even.
(b) For all whole numbers n, $n^2 + 2n + 3$ is even.

Thinking Critically

18. Let $p \rightarrow q$, $q \rightarrow r$, $r \rightarrow s$, and r be *true* statements, and let f be a *false* statement. Classify the following statements as *true, false,* or *cannot be decided.* Explain your reasoning.
   (a) s (b) q (c) $p \rightarrow r$ (d) $r \vee t$
   (e) $q \wedge s$ (f) $\sim r \rightarrow \sim p$ (g) $p \wedge f$ (h) $r \rightarrow q$

19. Are the following arguments valid? Explain why or why not. You may wish to rephrase statements in logically equivalent ways, use symbols, and show how the laws of valid reasoning are used.
   (a) Hypotheses: If it rains, then it won't snow.
   If it is below freezing, it will snow.
   Conclusion: If it rains, then it is above freezing.
   (b) Hypotheses: It is Wednesday.
   It is not tea time.
   If it is Wednesday and cold out, then it is tea time.
   Conclusion: It is not cold out.

20. Explain why these arguments are valid.
   (a) Hypotheses: $p \rightarrow q$
   $r \rightarrow s$
   $p \vee r$
   Conclusion: $q \vee s$
   (b) Hypotheses: If p, then r.
   If q, then r.
   Conclusion: If p or q, then r.

21. Every natural number n can be written in the form $n = 2^d \cdot k$, where k is an odd number and the exponent d is some whole number. For example, $40 = 2^3 \cdot 5$, $480 = 2^5 \cdot 15$, and $27 = 2^0 \cdot 27$. Call the exponent d the *degree of evenness.* Thus, 40 has degree of evenness 3, 480 has degree of evenness 5, and 27 has degree of evenness 0.
   (a) Write the whole numbers between 1 and 24 in the form $2^d \cdot k$.
   (b) The two numbers $8 = 2^3 \cdot 1$ and $24 = 2^3 \cdot 3$ have the same degree of evenness, namely 3. Is 3 the largest degree in your list from 1 to 24?
   (c) Suppose $m = 2^d \cdot h$ and $n = 2^d \cdot k$, where $m < n$, have the same degree of evenness, d. Why do you know there is some natural number x between m and n that has a *larger* degree of evenness than d? (*Hint:* Consider an *even* number, say $2u$, which is between the odd numbers h and k, and use *direct* reasoning.)
   (d) Show that any finite list of consecutive natural numbers has just one number of the largest degree of evenness. (*Hint:* Reason *indirectly* by assuming that two numbers have the largest degree of evenness. Can you find a contradiction to this assumption?)

22. It is easy to cover the 64 squares of an 8 by 8 array with 32 domino-shaped double squares, as shown at the left. Suppose, however, that two opposite corner squares are removed, as shown at the right. Can the 62 squares remaining be covered with 31 double square pieces? (*Hint:* If the squares in the array are colored alternately red and black as on a chess board, each domino covers both a red and a black square. Now reason indirectly: If a covering by 31 dominos is possible, how many red squares are covered? How many black?) Explain carefully.

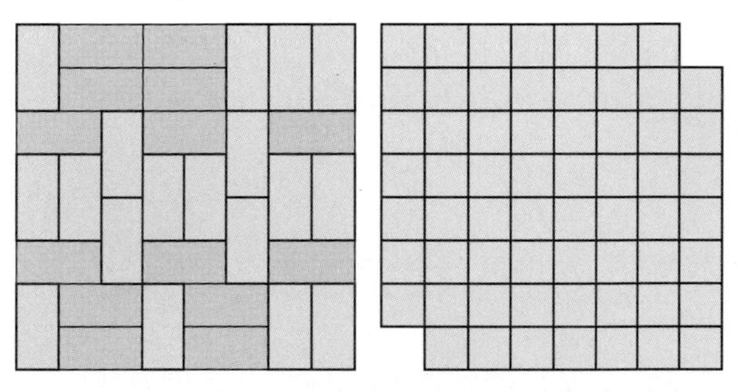

23. Solomon Golomb, a mathematician and electrical engineer, posed a tiling problem for the triangular floor shown.* There are 11 regular hexagons on each side, and 66 hexagons in all.

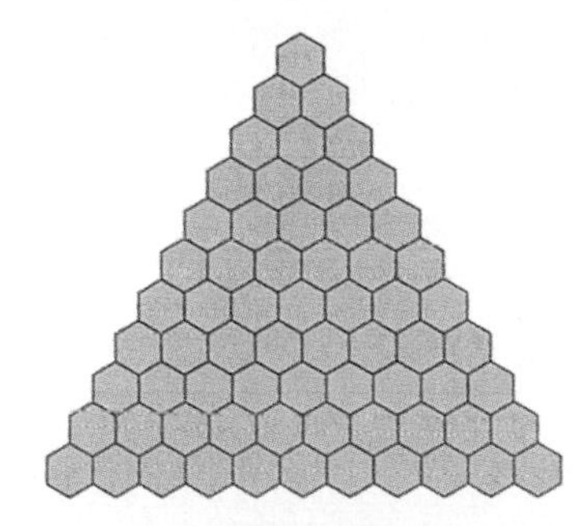

   (a) Can the floor be covered by 22 tiles of the following shape? Cutting tiles is not allowed.

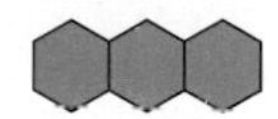

   (*Hint:* Number the tiles with 0s, 1s, 2s as shown here. The sum of the numbers covered

* Three figures of hexagonal tiles from "Brain Bogglers" by Solomon W. Golomb from *Discover Magazine,* March 1991, p. 88.

by a tile in any position is a multiple of 3. Now reason indirectly.)

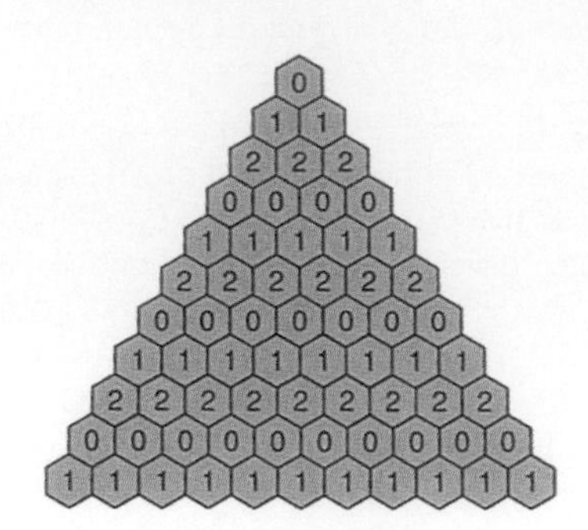

(b) Suppose that 22 tiles are available with a triangular shape as shown. Can the floor be covered? (*Hint:* Try direct reasoning.)

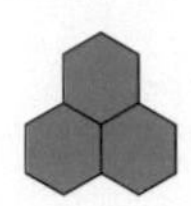

24. In Example B.4 it was claimed that DeMorgan's second law of logic could be proved in the same manner in which the first law was obtained. Show this is so by explaining why $\sim(p \vee q)$ is logically equivalent to $\sim p \wedge \sim q$.

Thinking Cooperatively

25. (a) Play the following two-person game, pitting player E ("equilateral") versus player S ("scalene"). Each player, in turn, places either a red or a blue marker at one of the eight dots in the configuration shown below. Player E scores a point for each equilateral triangle formed whose vertices are covered by markers of the same color, all red or all blue. Player S scores a point for each scalene triangle of sides a, b, and c formed which has its three vertices covered by markers of the same color. The player with the most points after all eight vertices are colored is the winner.

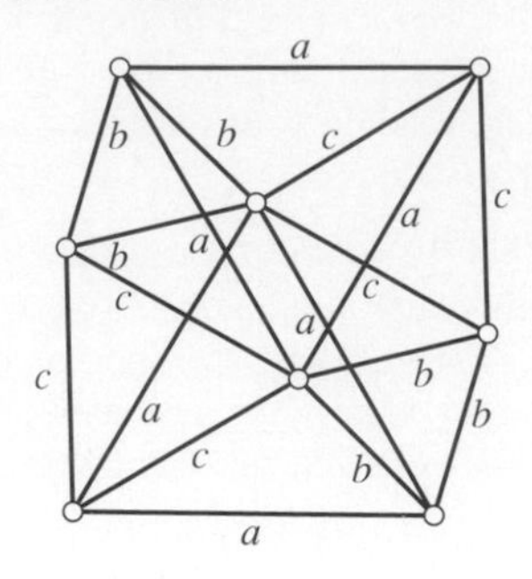

(b) Cover each of the eight points with either a red or a blue marker. Report on what your explorations reveal about these questions:

If a one-colored scalene triangle is formed, is there also a one-colored equilateral triangle? If a one-colored equilateral triangle is formed, is there also a one-colored scalene triangle?

Using a Computer

26. Example B.2 (f) considered the statement "Some numbers in the sequence 31, 331, 3331, 33331, . . . are not prime numbers." Use FACTORINTEGER to explore whether the statement is true or not. Recall that a prime number has no divisors other than itself and 1.

27. Example B.2 (g) considered the statement "some numbers in the sequence 11, 111, 1111, 11111, 111111, . . . are perfect squares." Use a computer (or calculator) to investigate the truth of the statement. If the computer search fails to prove the statement is true, look for a proof of the negation of the statement.

Making Connections

28. DeMorgan's Laws of Logic were derived in Example B.4. If we use the symbol $\equiv$ to denote the logical equivalence between statements, these laws can be written

$$\sim(p \wedge q) \equiv \sim p \vee \sim q$$

and

$$\sim(p \vee q) \equiv \sim p \wedge \sim q.$$

In problem 14 of Section 2.1 in Chapter 2, you were asked to derive DeMorgan's Laws for Sets, namely

$$\overline{A \cap B} = \overline{A} \cup \overline{B} \quad \text{and} \quad \overline{A \cup B} = \overline{A} \cap \overline{B}.$$

This suggests that concepts and operations with sets have a correspondence to logical concepts and operations.

Set Theory		Logic	
Sets	A, B	Statements	p, q
Union	$A \cup B$	Or	$p \vee q$
Intersection	$A \cap B$	And	$p \wedge q$
Complement	$\overline{A}$	Negation	$\sim p$
Equality	$A = B$	Logical equivalence	$p \equiv q$
Subset	$A \subseteq B$	Implies	$p \rightarrow q$
Universal set	U	True	T
Empty set	$\emptyset$	False	F

Translate each set identity below into a corresponding logical expression.

(a) $\overline{\overline{A}} = A$

(b) $A \cap (B \cup C) = (A \cap B) \cup (A \cap C)$

(c) $A \cap \overline{A} = \emptyset$

(d) $B \cup \overline{B} = U$

(e) If $A \subseteq B$, then $A \cap \overline{B} = \emptyset$

(f) If $A \subseteq B$, then $\overline{B} \subseteq \overline{A}$.

29. A particularly important application of logic originated with the Master's thesis of Claude Shannon in 1937. Shannon showed the connection between logical relationships and electrical circuits, which is the central idea for the design of modern digital computers. The basic logical element of a statement can be viewed as a switch in a simple circuit. If the switch p is closed (p is true), then the circuit is complete; but if the switch p is open (p is false), then the circuit is open.

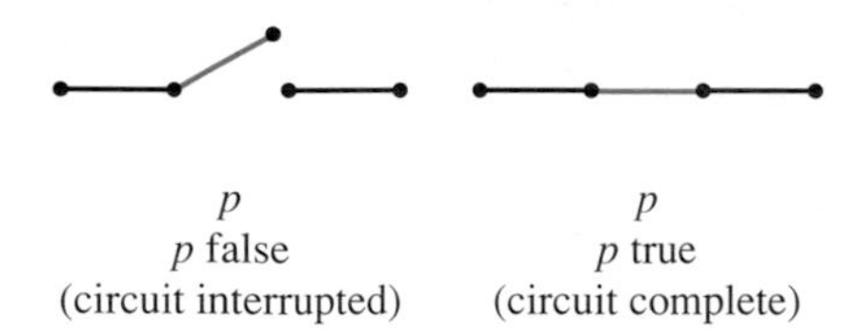

Circuits for "p and q" and "p or q" are built by combining switches in series or parallel:

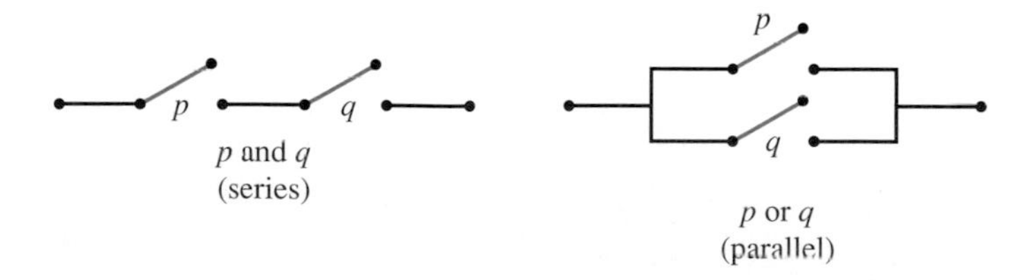

For current to flow in an "and" circuit, both switches must be closed, but current flows in an "or" circuit if either or both switches are closed.

(a) Explain why $p \rightarrow q$ is given by this circuit.

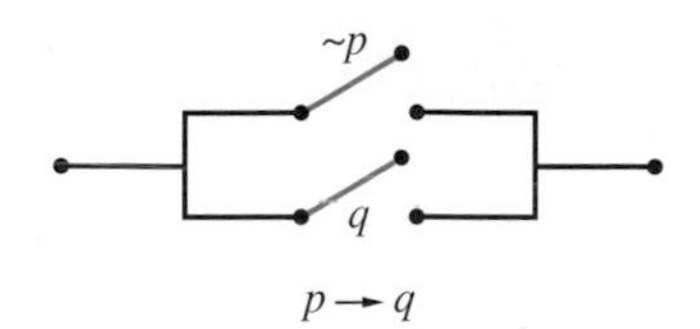

(b) Explain why the **exclusive** or (p or q but not both p and q) is given by this circuit.

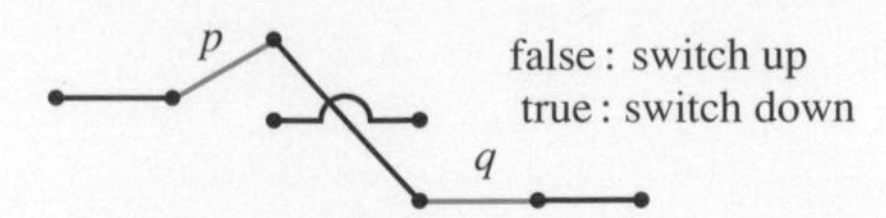

30. Here is one way to design a circuit for $(p \wedge q) \vee (p \wedge r)$.

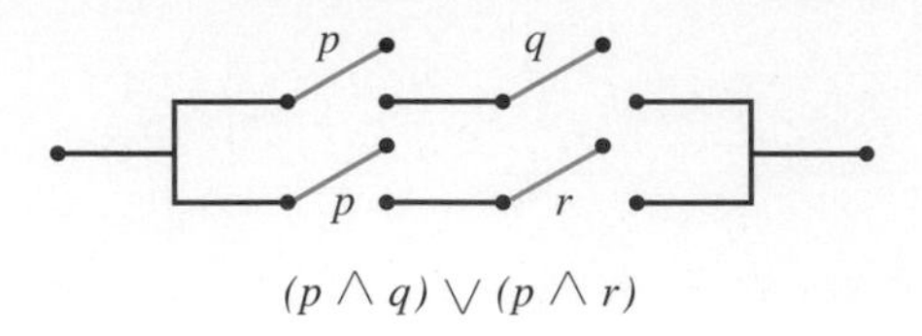

$(p \wedge q) \vee (p \wedge r)$

(a) Design a circuit which uses just one p switch instead of two, and which would be therefore cheaper to build.

(b) Design a switch for $(p \vee q) \wedge (p \vee r)$.

31. Compare these two circuits.

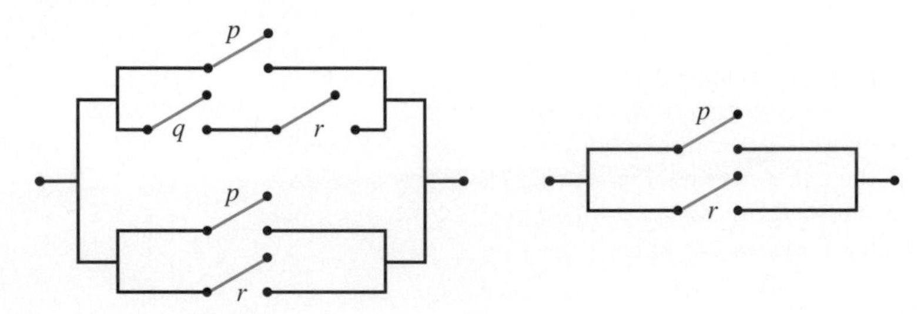

Explain why the left circuit is "redundant." Why is it important for a computer engineer to avoid redundant circuits?

Communicating

32. The word *premise* sounds much like *promise.* In what sense is a premise a promise? Consult a dictionary to see if these two words have a common etymology.

33. George Pólya, in his book *Induction and Analogy in Mathematics* (Princeton University Press, 1954), cites the dictum "When you have satisfied yourself that the theorem is true, you start proving it." Write a paragraph which elaborates this point. Why do you suppose Pólya titled a film he made "Let Us Teach Guessing"? What role do you believe guessing should have in your classroom? How would you go about implementing your response to the preceding questions in your classroom?

34. Discuss these statements, each made by a prominent mathematician.

(a) "Logic merely sanctions the conquests of the imagination." (Jacques Hadamard, 1865–1962)

(b) "Do not put the logical cart before the heuristic horse." (Max Schiffer)

APPENDIX C
Logo: A Programming Language for Learning Mathematics

Logo Philosophy

Logo is a computer language that was designed to be consistent with a constructivist philosophy of learning. Created at MIT by Seymour Papert and his team, Logo was designed to help children learn mathematics through interaction with a computer environment. This language can empower children to investigate mathematical topics and to control their learning. Logo encourages natural spontaneous learning by students through interaction with their environment, as is consistent with Piagetian theory. A key to learning in a Logo environment is the emphasis on student directed learning. Students are encouraged to investigate a problem, form a plan to resolve the problem, use logic to create programs, try them out, look at results and debug, and then repeat the process. Communication is encouraged at every step of the process. Students try what they think will work, check if it does what they expect, repair their procedures using logic, and then check again. This is a great strength of an approach encouraging students to learn mathematics by creating Logo procedures. So often, when solving textbook problems, students do the work, check their answer, then quit. However, students using Logo tend to continue to debug their programs until their procedures work. Right or wrong is no longer an answer in the book, the student can make a judgment based upon results: the procedure works or needs repair. Note that wrong is not an option in this situation. Furthermore, students can use the procedures they have created as building blocks for new, more complicated procedures. Logo provides the opportunity for students to expand their learning of mathematical concepts.

Basics in LogoWriter, Terrapin, and Apple

Turtle graphics are a key feature of the current versions of Logo. This appendix uses LogoWriter by LCSI, with instructions for Terrapin Logo and Apple Logo in parentheses ( ) where they differ. In LogoWriter, the turtle heading and position are represented on the screen by a realistic turtle shape; in Apple and Terrapin, a triangle is used. LogoWriter and Apple Logo accept either capital or lower case letters; Terrapin will accept only capital letters.

Basic commands that are intrinsic to Logo are referred to as **primitives.** This simply means that these are programs or procedures which Logo already understands. You don't have to provide instructions telling the computer how to do these. Logo primitives and their abbreviations will be listed in bold face; the primitives and their abbreviations may be used interchangeably. Other versions of Logo may vary in use of parentheses. Syntax (the programming counterpart to grammar) can be checked in the documentation that comes with the software; minor changes may be needed in the procedures used in this section. While procedures in this book are written in lower case letters, persons using older versions of Terrapin Logo should press the **Caps Lock** key before using Logo.

Screens and Modes

There are three main sections of the screen in Logo, although not all of them may be visible at the same time within a single screen view. In LogoWriter, when you have entered a new page in Logo, you will see a screen divided by a horizontal bar near the lower edge. The lower portion is the **immediate mode** section of the screen. Commands typed there are performed immediately after the **Return** or **Enter** key is pressed. The run, whether it involves graphics or printing of numbers or words, appears in the upper portion of this screen. Using a flip command, you can view the edit portion of the screen, located in the upper portion. The lower portion remains unchanged, showing any commands you may have typed in the immediate mode. Programs, called **procedures,** are typed in the edit portion of the screen. If you have a Macintosh, you can use the mouse with icon selection to move from one screen to the other. If you use an IBM or compatible, you will have to use the key combinations explained in the documentation and also found later in this appendix. While there are three main **screen sections: immediate; deferred** (or **edit,** where procedures are written); and **graphics** (or **display**), there are two **modes** in which you can use Logo. In **immediate mode,** Logo performs commands immediately after they are entered. In **deferred mode,** Logo saves the commands of a procedure as a sequence of instructions on how to do something and then actually follows those instructions when the procedure is called upon by name in the immediate mode.

For Terrapin Logo Users Only

When you enter Terrapin Logo, be sure to press the Caps Lock key. Typing **DRAW** will call up the graphics/run screen, with immediate mode in the lower portion. You should see a question mark (?) followed by a blinking square cursor. Pressing the **Control** and **T** keys simultaneously converts the screen to a text screen only, with all graphics hidden from view. **Control S** shows the split screen again. **Control F** shows the full screen as graphics output. To enter the edit mode (where procedures are written) type **EDIT** or **TO** and a procedure name.

For Apple Logo Users Only

To use Apple Logo, for immediate mode you need merely type in a command and you will be provided with a graphic screen where the command has been performed. Thus, you enter in immediate mode. The lower portion of the screen is where immediate mode commands are then typed. **Control S** (or type **SS**) shows the split screen, and **Control T** (or type **TS**) shows the text screen. To show the full graphics screen, type **FS.** The edit screen is entered by typing **EDIT** or **EDIT** followed by the name of the procedure immediately preceded by open quote (“).

Turtle Motions

Although Logo is a programming language and can be used to write procedures to do numerical computation, its most popular feature is the turtle graphics mode. Logo commands are available to move a turtle about the screen, drawing pen lines along the turtle's path.

Let's begin with the basic turtle moves: **forward, back, right,** and **left.** These are abbreviated by **fd, bk, rt, lt,** respectively. Each of these commands is followed by a space and then a number. With **forward** and **back,** the number represents the number of turtle steps the turtle moves in the direction it is currently heading. For the **right** and **left** primitives, the number represents the degree measure the turtle turns; positive measure indicates a turn in the direction stated from which the turtle is facing, and negative measure indicates a turn in the opposite direction. This means that **right -30** is the same as **left 30.** Similarly, **fd -50** and **bk 50** both move the turtle backward by 50 turtle steps.

EXAMPLE C.1

Moving the Turtle and Drawing Lines

Sketch what the computer will draw when each of the following commands or lists of commands is typed.

(a) `fd 50` **(b)** `rt 60` **(c)** `fd 40` `rt 60` `fd 30` **(d)** `bk 40 rt 90` `bk 20 lt 90` `bk 30`

SOLUTION

Each of the lists of commands below is performed separately. The **cleargraphics** command, abbreviated **cg,** returns the turtle to its home position and clears the screen before executing a new list of commands. (In Apple, use the **clearscreen** command, abbreviated **cs;** in Terrapin, use **DRAW.**) Note that in Logo it is acceptable to place two (or more) instructions on the same line.

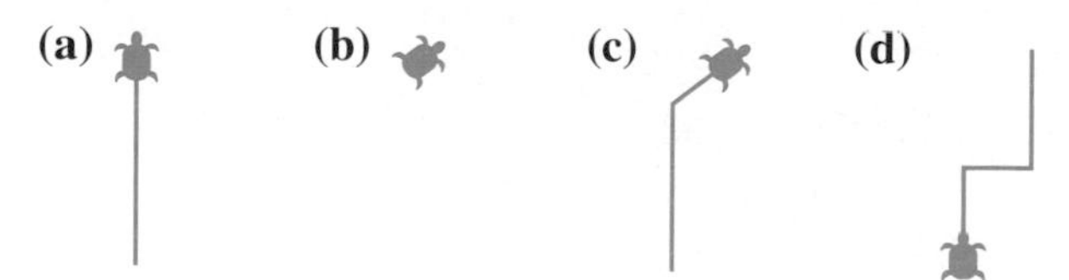

■

If you tell the turtle to move forward more than the available room on the screen, the turtle continues its path on the opposite edge of the screen, just as in many computer games; **fd 500** will move the turtle more than once through the screen in a forward direction (the number of turtle steps to make up the dimensions of the screen varies among versions of Logo and computers.)

Now let's combine the four commands **forward, back, right,** and **left** to draw a simple picture.

EXAMPLE C.2

Drawing with the fd, bk, rt, and lt Primitives

Write a list of commands that would draw the figure below. Lengths of segments are given in turtle steps and measures of angles are given in degrees. At the end of the drawing, the angle between the turtle and the line segment is 45°.

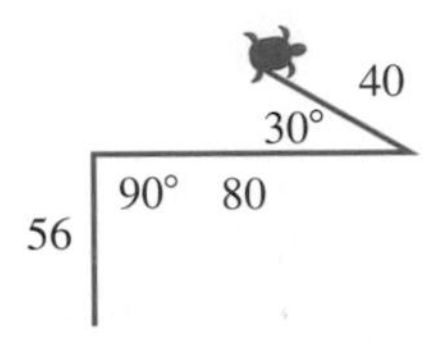

SOLUTION

There is more than one way to solve this problem. Provided below is one set of commands that will draw the figure provided.

```
fd 56
rt 90
fd 80
rt 30
bk 40
lt 45
```

■

Sometimes, when drawing a figure, you may wish the turtle to be visible; at other times you may wish it to be hidden. When you first enter Logo, the turtle is visible. Any condition that occurs when the software is first booted up is referred to as a default state. The command **hideturtle** abbreviated (**ht**) tells the turtle or triangle to hide itself; the command **showturtle** abbreviated (**st**) tells the turtle to show itself.

If you wish to clear the screen and begin again, **cg** for clear graphics (**CS** for **clearscreen**) clears the screen of all graphics that had been drawn and places the turtle in the center, facing the top of the screen. In most versions of Logo, **home** moves the turtle to the center of the screen in vertical position, but does not lift the pen. Thus, there will be a pen mark connecting the previous turtle position to the center unless the pen is up.

Some basic pen commands are:

penup (**pu**) which tells the turtle to lift its pen, thus not leaving a mark until the pen is put down again;
pendown (**pd**) which tells the turtle to put its pen down, leaving a mark wherever it travels;

penerase (**pe**) which tells the turtle to turn the pen upside down and use the eraser, removing the marks wherever the turtle travels until another command changes the pen position.*

Writing Procedures

In Logo, programs are referred to as **procedures** and are generally relatively short in length. All procedures begin with **to** and end with **end.** To write procedures in LogoWriter, you "flip" to the other side of the "page" by clicking on the F on the upper right corner of the screen on a Macintosh or by holding down the CTRL key and typing the letter F on an IBM or compatible. To return to the drawing side, repeat the same steps.

To write procedures in Apple Logo, type **to** followed by the name of the procedure and then type the instructions on the next lines, completing the procedure with the word **end** and the procedure will be defined automatically. If you wish to edit an existing procedure, type **edit** followed by the name of the procedure immediately preceded by **"**, for example **edit "square.** (In Terrapin Logo, type **EDIT** or **TO;** to return to draw mode, follow the instructions at the bottom of the edit screen.)

Once in the edit mode, you can teach the turtle. This is the process of writing procedures or programs. In Logo, you begin a procedure with the word **to** and the name of the procedure on the first line. No lines of instruction belong on this first line. The body of the instructions come next. Then, on the last line you type the word **end.** Nothing else can be included on this last line.

EXAMPLE C.3

Drawing a Square

Write a procedure to draw a square 50 turtle steps long on each side. The turtle begins in the home position or center of the graphics screen.

SOLUTION

Let's tell the turtle to walk forward 50 turtle steps and then turn right. What angle should the turtle turn? After drawing a square with paper and pencil or walking off the square like the turtle, you soon discover that the turn angle is by 90°. This is repeated for each of the four sides. The procedure is then:

```
to square
    fd 50 rt 90
    fd 50 rt 90
    fd 50 rt 90
    fd 50 rt 90
end
```

Notice that the line **fd 50 rt 90** is repeated four times within this procedure. ■

* The last primitive listed above does not exist in older versions of Terrapin Logo.

To save this procedure, we select **savepage** under **edit** (or follow the instructions for saving in the Apple or Terrapin manual). To run the **square** procedure, we type **square** and then press **return** or **enter.** Check to see that it draws a square with its lower left corner at the starting position of the turtle.

Procedure **square** works well, but the instructions seem repetitive. Procedures with repetitive instructions can be shortened by using the **repeat** command. This primitive is followed by a number, indicating the number of times a set of commands is to be repeated and immediately behind that follows a set of square brackets enclosing the instructions to be repeated.

EXAMPLE C.4 Using the Repeat Command

Use the **repeat** command to shorten the **square** procedure.

SOLUTION

If we use a slightly different name, we can keep both procedures.

```
to square2
   repeat 4 [ fd 50 rt 90 ]
end
```

To run this procedure, type **square2.** This will reproduce the same square shown in Example C.3. ■

Variables

The procedure **square2** will only draw a square with side 50 turtle steps. We could rewrite this procedure to draw a square with a different side length, but we would have to write a separate procedure for each length. To write more generally applicable programs, we introduce the use of variables. A **variable** in Logo represents a value which may be selected or computed. In Logo, a variable is always immediately preceded by a punctuation mark. If you are referring to the variable address or location in the computer's memory, use a quotation mark ("); for example, **"length.** If you are referring to the value currently stored in the variable location by the computer, use a colon (:), for example **:length** might refer to the number 50 which was the most recent value of the variable. We use a parameter, a special kind of variable, in this section. For more on variables, look at your Logo documentation or a Logo text. The parameter is introduced by a colon (:) and is first listed on the title line of a procedure.

EXAMPLE C.5 Using a Variable Input

Write a procedure called **square3** to draw a square with variable length side. Use the parameter **:side** to represent the side length.

SOLUTION

The procedure uses **:side** each time that value is referred to. Note that we use a new name again, to be able to keep the old procedures. The difference between **square3** and **square2** is the replacement of **50** with **:side.**

```
to square3 :side
   repeat 4 [ fd :side rt 90 ]
end
```

■

To draw a square with side 50 type **square3 50.** To draw a square with side length 20 type **square3 20.** Any size square may be drawn within the constraints of the screen.

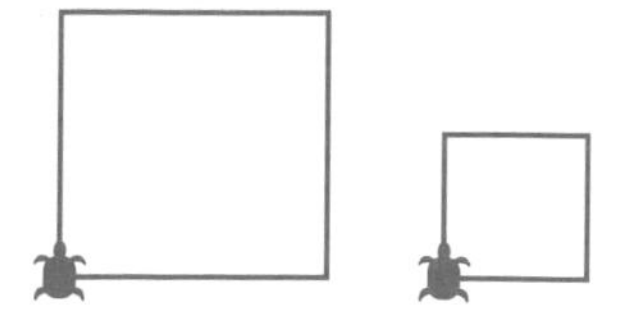

EXAMPLE C.6

Using Procedures to Draw

Write a procedure that calls upon **square3 :side** to create a picture similar to the one below. The drawing uses several pen colors, which can make the figures seen on a color monitor especially attractive.

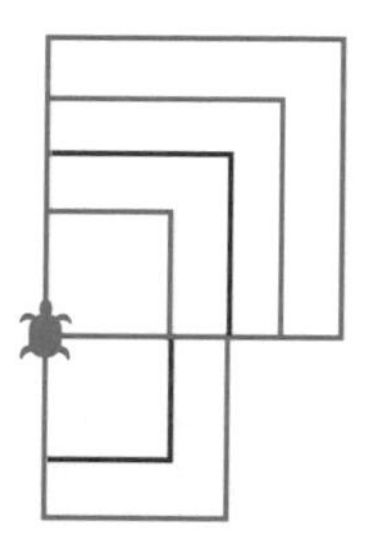

SOLUTION

In LogoWriter the pen color is selected with the **setc** command followed by the number of the color. In Apple Logo **setc** is replaced by **setpc** and by **PC** in Terrapin. You will need to change the appropriate command lines to run the program if you have a different version than LogoWriter. Color numbers vary among versions of Logo so your picture may not have the same colors as the figure shown.

```
to picture
    setc 3 square3 20
    setc 5 square3 30
    setc 4 square3 40
    setc 3 square3 50
    rt 90
    setc 5 square3 20
    setc 4 square3 30
    lt 90
end
```

■

Drawing Polygons

Now let's write procedures for an equilateral triangle, a regular pentagon, and other regular polygons. Again we will use the parameter **:side** to represent the length of a side.

EXAMPLE C.7 Drawing an Equilateral Triangle

Write a procedure to draw an equilateral triangle with variable length side.

SOLUTION

The drawing below illustrates that the turtle turns through the exterior angle at each vertex of the triangle. Note that, to draw any polygon, the turtle must turn one full revolution of 360°. This is commonly known as the Total Turtle Trip Theorem. For an equilateral triangle, each exterior angle is 360°/3, or 120°. Thus an equilateral triangle is drawn by the following procedure.

```
to triangle :side
   repeat 3 [ fd :side rt 120 ]
end
```

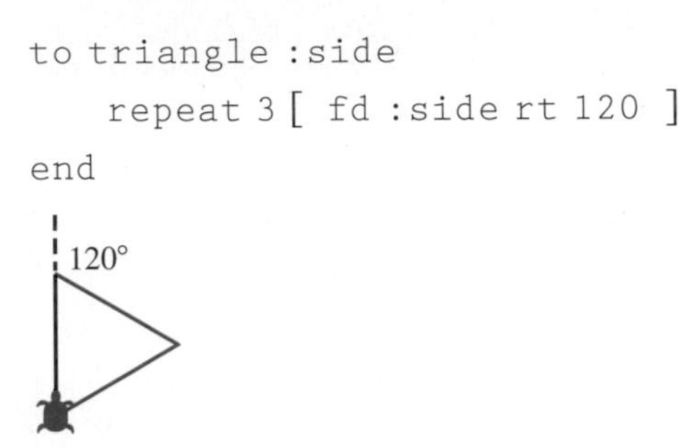

■

EXAMPLE C.8 Drawing a Regular Pentagon

Write a procedure to draw a regular pentagon with variable length side.

SOLUTION

We can use what we learned in drawing the equilateral triangle to compute the turning angle for the turtle. By the Total Turtle Trip Theorem, 360°/5 = 72°. Thus a regular pentagon is drawn by the procedure below.

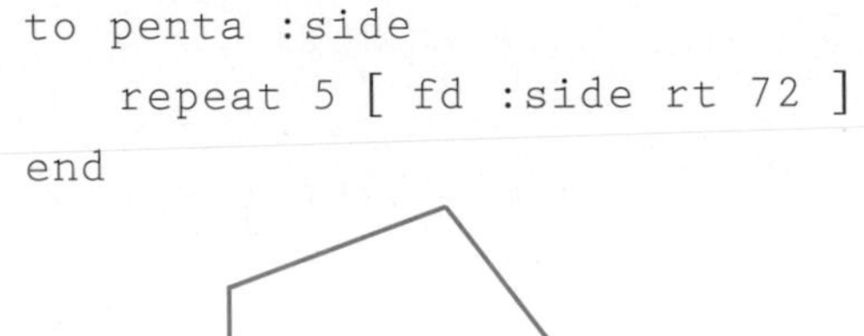

```
to penta :side
   repeat 5 [ fd :side rt 72 ]
end
```

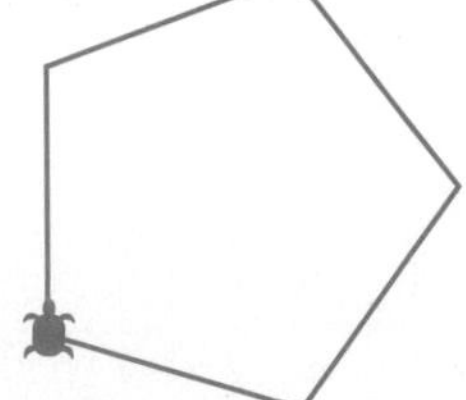

■

EXAMPLE C.9 Drawing a Regular Polygon

Write a general procedure to draw a regular polygon using **:side** to represent the length of a side and **:nsides** to represent the number of sides.

SOLUTION

We can generalize our pentagon procedure by making use of an additional variable to represent the number of sides of the regular polygon. Noting from our previous procedures that the measure of the exterior angle is found by dividing 360° by the number of sides, we write the following procedure, letting the computer do the arithmetic.

```
to polygon :nsides :side
   repeat :nsides [ fd :side rt 360 / :nsides ]
end
```

■

Take a moment now to draw regular polygons of varying number of sides and length of side. Play with it and become comfortable with the procedure, gaining understanding of each parameter and its use. Be careful that you first type in the name of the procedure, then the number of sides, and finally the length of a side. This is the order in which the computer is expecting the information. If you were to type in **polygon 30 5,** the turtle would draw a polygon with 30 sides and five units long on a side, not a pentagon with side length 30.

Examples include the figures below.

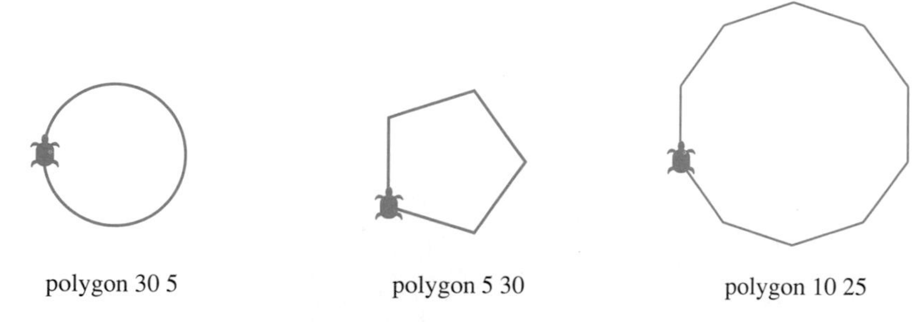

polygon 30 5 polygon 5 30 polygon 10 25

Changing the Value of a Variable

The **make** command can be used to change the value of a variable within a procedure. In the following example, we can change the square procedure into a small spiral by increasing the length of the side after each turn.

EXAMPLE C.10 **Drawing Spirals**

Write a procedure to draw the figure below. Note that the figure contains 15 sides; the length of each successive side is 10 turtle steps longer than the previous.

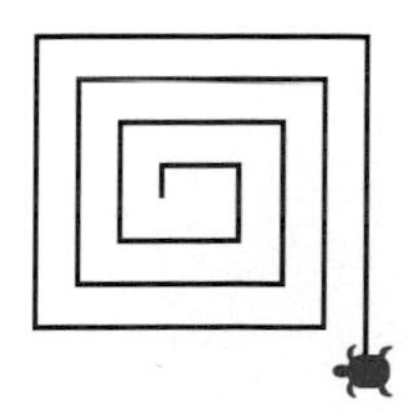

SOLUTION

```
to square.spiral :side
   repeat 15 [ fd :side rt 90 make "side :side + 10 ]
end
```

Note that the **make** command is immediately followed by the variable location or address where you want the value to be placed, then by the new value or formula for computing the new value. Look at the **square3** procedure and note the similarities. What changes were made to the **square3** procedure to create **square.spiral**? ■

EXAMPLE C.11

Using the MAKE Command to Draw Nested Polygons

Write a procedure using the **make** command to draw a triangle, square, pentagon, and hexagon where the polygons are regular and share a common side of length 30.

SOLUTION

```
to multi.poly :nsides
   lt 90
   repeat 4 [ polygon :nsides 30 make "nsides :nsides + 1]
   rt 90
end
```

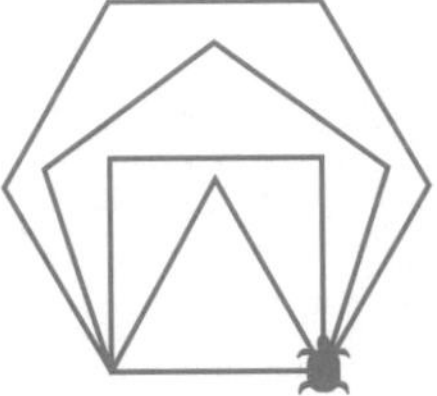

Type **multi.poly 3.** The turtle is turned to the left. The first **polygon 3 30** draws a triangle with side length 30. Then the number of sides is increased to four and the square is then drawn by **polygon 4 30.** The repeat instructions continue this process two more steps, drawing the pentagon and hexagon. The turtle turns back to its original heading and the procedure ends. ■

Drawing Circles and Arcs

After you have played with the polygon procedure and are comfortable with it, try drawing polygons with increasing number of sides. Do you notice that the sides and angles seem to almost blend together? It's difficult to identify the vertices. The polygons become more and more like circles. In the discrete situation of a computer screen, there is no discernable difference between a circle and a regular polygon with a large number of sides.

We can write a circle procedure as a 360 sided regular polygon.

```
to circle
   repeat 360 [ fd 1 rt 1 ]
end
```

The procedure above draws a circle with circumference 360 turtle steps. This demonstrates that, to Logo, a circle and a 360-gon are the same. The more sophisticated procedure below will draw a circle turning to the right with radius supplied by the user and circumference 2π times the radius. The value of π is approximated by 3.14, and the $*$ symbol indicates multiplication.

```
to circle.right :radius
   repeat 360 [ fd :radius * 2 * 3.14/360 rt 1 ]
end
```

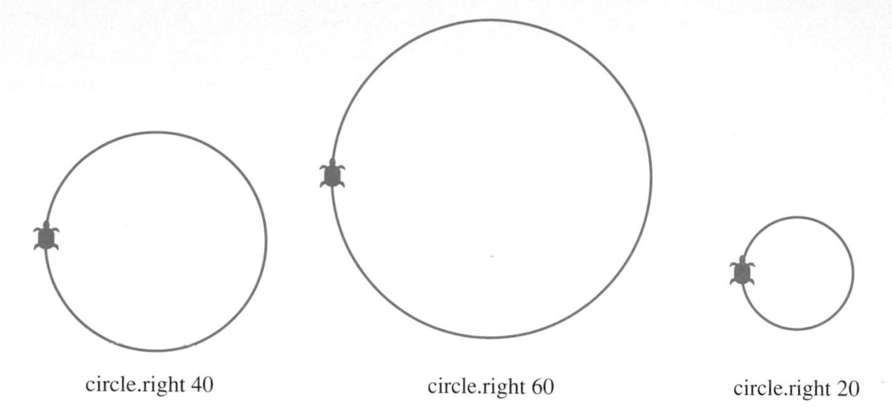

EXAMPLE C.12

Drawing Arcs

Write a procedure to draw an arc that curves to the right, using the procedure **circle.right** as a guide. Incorporate variables that allow you to select the radius and angle measure of the arc. Then write a procedure in which the turtle turns to the left.

SOLUTION

To draw an arc of degree measure **:angle** turning to the right, we adapt our circle procedure.

```
to arc.right :radius :angle
   repeat :angle [ fd :radius * 2 * 3.14/360 rt 1 ]
end
```

We need merely change right to left to draw an arc curving to the left.

```
to arc.left :radius :angle
   repeat :angle [ fd :radius * 2 * 3.14/360 lt 1 ]
end
```

Some figures using these procedures are shown below.

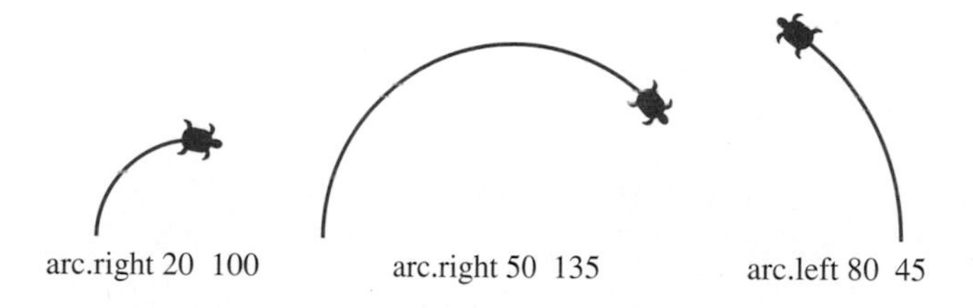

The **arc.right** and **arc.left** procedures are a useful way to incorporate curves in drawing. For example, the letters of the alphabet could be drawn using turtle lines and arc procedures. For other examples, see Chapter 13. ■

Problem Solving Using Logo: Superprocedures

Procedures may be long or short, but it is advisable to use Pólya's problem-solving strategies to write programs for complicated problems. In particular, it is often useful to break a big problem into several smaller parts and solve each part with an individual procedure. After checking that each subprocedure works as expected, a main superprocedure can be written that combines these pieces to solve the larger problem. An example follows.

EXAMPLE C.13 **Drawing a Stick Person**

Write a program to draw a stick person resembling the one shown here. Use one superprocedure to call upon a collection of subprocedures for the figure.

SOLUTION

You may use a different number and combination of procedures from the solution shown below. This set of procedures divides the figure by the part of the body and contains a separate procedure for the head, each arm, and each leg. The superprocedure combines these and hides the turtle for the finished picture. The body and turtle moves between procedures are within the superprocedure **stick.**

```
to stick
   fd 20
   head
   bk 20
   rt.arm
   lt.arm
   bk 40
   lt.leg
   rt.leg
   ht
end
```

```
to head
 rt 90
 repeat 20[ fd 4 lt 18 ]
 lt 90
end
```

```
to rt.arm
 rt 110 fd 40 bk 40 lt 110
end
to lt.arm
 lt 110 fd 40 bk 40 rt 110
end
to rt.leg
 lt 30 bk 50 fd 50 rt 30
end
to lt.leg
 rt 30 bk 50 fd 50 lt 30
end
```

■

TABLE C.1 Common Primitives and their Abbreviations

Commands	Logo Writer	Apple	Terrapin	Example
forward	FD	FD	FD	FD 50
backward	BK	BK	BK	BK 30
right	RT	RT	RT	RT 90
left	LT	LT	LT	LT 30
show turtle	ST	ST	ST	ST
hide turtle	HT	HT	HT	HT
clear screen	CG	CS	DRAW	CG
home	HOME	HOME	HOME	HOME
pen up	PU	PU	PU	PU
pen down	PD	PD	PD	PD
pen erase	PE	PE	—	PE
make	MAKE	MAKE	MAKE	MAKE "num :num + 1
flip page	FLIP	EDIT/DRAW	EDIT/DRAW	FLIP
end	END	END	END	END
repeat	REPEAT	REPEAT	REPEAT	REPEAT 3[FD 20 RT 120]
slow turtle	SLOWTURTLE	—	—	slowturtle
fast turtle	FASTTURTLE	—	—	fastturtle
set pencolor	SETPC	SETPC	PC	SETPC 3 (PC 3)
set background	SETBG	SETBG	BG	SETBG 2 (BG 2)
set position	SETPOS	SETPOS	SETXY	SETPOS[100−100] SET XY 100(−100)
set heading	SETH	SETHEADING	SETH	SETH 45
screen size	$x(-248, 248)$ $y(-110, 110)$	$x(-139, 139)$ $y(-119, 119)$	$x(-139, 139)$ $y(-120, 120)$	

—Feature does not exist in this version of Logo.

Recursion Using Logo

A powerful programming tool in Logo is **recursion,** which allows a procedure to call upon a copy of itself. Recursion is not a feature in many computer languages, including Basic. Recursion that does not create an endless loop has two key features: (1) the procedure calls upon a copy of itself as a subprocedure; and (2) there is a test or check within the procedure that stops it when a given set of conditions are met. The primItive **if** is often used in the test line.

EXAMPLE C.14 Analyzing a Recursive Procedure

Draw the picture that the recursive procedure below would create.

```
to fig
   fd 40 rt 72                (Terrapin only)
   if heading = 0 [stop]      (IF HEADING = 0 THEN STOP)
   fig
end
```

SOLUTION

This type of recursion is called tail-end recursion because the original procedure is called upon at the end of the procedure. It acts much like a repeat. The procedure **fig** moves the turtle forward 40 steps, turns it 72° to the right, and then checks the turtle's heading. The command **if** is followed by a **predicate,** which is a statement that the computer can test to determine whether it is true or false. If the predicate is true, the computer follows the instructions in brackets; if it is false, the computer skips to the next line of instructions. Thus, if the turtle is heading directly to the top of the screen, the command **stop** ends the procedure in process. If the heading is not 0, then the **stop** command is skipped and the procedure is called upon again to move the turtle and turn it and the heading check is repeated. After drawing five sides and turning the turtle 72° five times, the turtle is facing the top and the procedure will stop. Thus, the **fig** recursive procedure draws a regular pentagon and then ends.

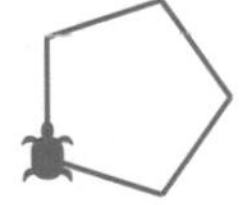

■

EXAMPLE C.15 Drawing Using Recursion

Draw the figure created by the recursive procedure below when **polygons 6** is typed.

```
to polygons :nsides                (Terrapin only)
   if :nsides < 3 [ stop ]         (IF :NSIDES < 3 THEN STOP)
   lt 90
   polygon :nsides 30
   rt 90
   make "nsides :nsides - 1
   polygons :nsides
end
```

SOLUTION

Notice that this uses **polygon :nsides :side** as a subprocedure. This procedure is tail-end recursive because the procedure calls upon a copy of itself just before the end of the procedure. The check for continuing the procedure occurs in the first line. This meets the two requirements of a recursive procedure. The drawing is shown below.

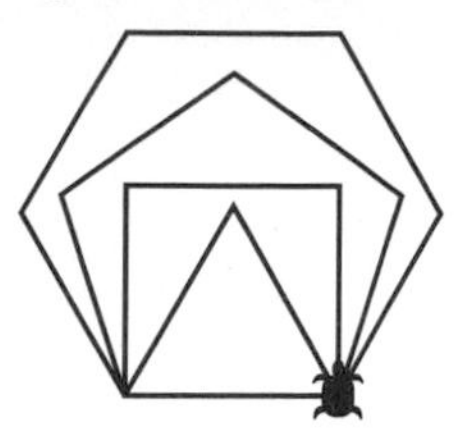

■

EXAMPLE C.16 **More Drawing with Recursion**

Draw by hand the picture created by the procedure listed below in the same order as the computer would draw it. Check by typing it into the computer.

Type **dub 1.5** and watch what the computer draws.

```
to dub :size
   repeat 18 [fd :size rt 20]         (Terrapin only)
   if :size < 20 [dub :size + 2]      (IF :SIZE < 20 THEN DUB :SIZE + 2)
   repeat 18 [fd :size lt 20]
end
```

SOLUTION

The figure is shown below. Note as you watch this drawn on a computer screen that the circles (actually, regular 18-gons) are drawn from smallest to largest on the right because the first repeat (that draws right circles) occurs before the test and the same procedure is called upon again before the second repeat is reached. This continues, until **:size** is no longer less than 20. Then, the computer completes each procedure, from the last begun until the first, thus drawing the circles from largest to smallest on the left. The order in which procedures are completed in imbedded recursion can be compared to the Russian Matrioshka dolls, the children's toy barrels, or the inventory style of last in, first out. Procedures are completed from the most nested in order to the outermost procedure.

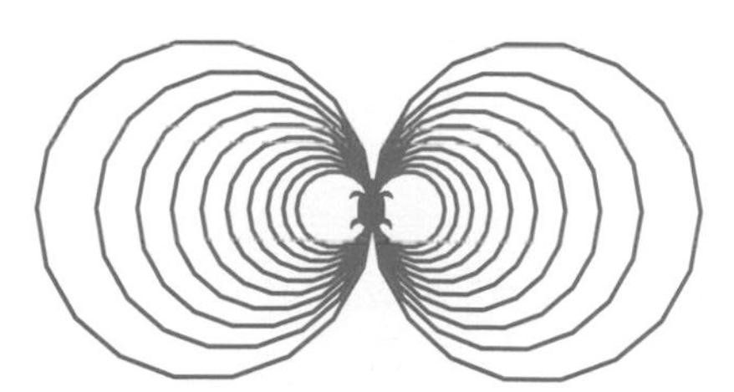

■

PROBLEM SET C

1. Write your initials on the screen. Write a procedure using those commands so that you can print your initials on the screen whenever you wish.
2. Write a set of instructions in human steps (using **fd, bk, lt, rt**) to get from your classroom to the front door of the building. Use **fd** if the person is to step forward, even if the step is up or down on a flight of stairs.
3. Use the square and triangle procedures and simple turtle lines and arcs to draw a scene from your campus.

4. Draw a maze on a transparency, tape it up on your computer screen, and try to run the turtle through the maze without touching any lines. Trade with a classmate and repeat.

5. Draw a simple map using turtle graphics. Then write a story for the turtle to stop at various locations on the map (post office, candy store, home, school, . . .). Now have a child try to move the turtle according to the trip described in the story.

6. Write a procedure to draw a right triangle. (*Hint:* You may want to draw two of the three sides and then use the **home** command.)

7. (a) Write a procedure to draw a rectangle of dimensions 50 by 80 turtle steps.

   (b) Write a procedure to draw a general rectangle. Use the line below for the title line.

   `to rectangle :length :width`

8. (a) Write a procedure to draw a parallelogram with side lengths 40 and 70 and one interior angle 38°.

   (b) Write a procedure to draw a general parallelogram using the line below to begin your procedure.

   `to parallelogram :side1 :side2 :angle`

9. (a) Write a procedure to draw a rhombus with side length 50 and one interior angle 52°.

   (b) Write a procedure to draw a general rhombus using the line below to begin your procedure.

   `to rhombus :side :angle`

10. Write a procedure to draw a simple picture using only rectangles, regular polygons, and circles.

11. Draw a house on paper using regular polygons. Write a superprocedure that uses the polygon procedure and turtle moves to draw this house on the computer screen.

---

12. The procedure below uses imbedded recursion to draw a finite representation of a fractal. The figure is shown for levels one and two.

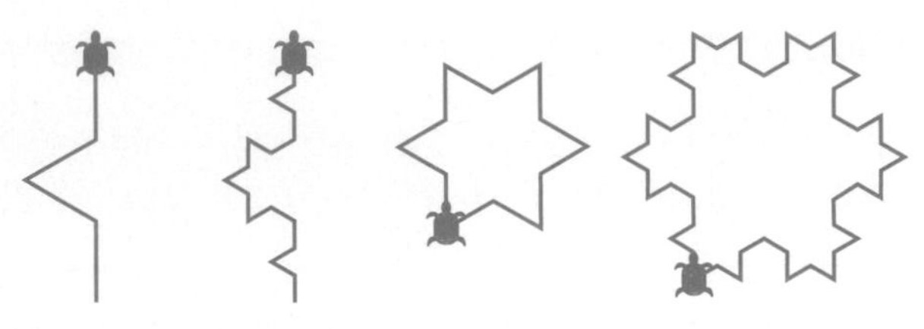

fractal 1 100   fractal 2 100   frac 1 100   frac 2 100

(a) Draw by hand the figure that will be drawn for level 3 for both **fractal** and **frac.**

```
to fractal :level :side                          (in Terrapin)
   if :level < 1 [ fd :side stop ]               (IF :LEVEL<1 THEN FD :SIDE STOP)
   fractal :level − 1 :side / 3
   lt 60
   fractal :level  −  1 :side / 3
   rt 120
   fractal :level  −  1 :side / 3
   lt 60
   fractal :level  −  1 :side / 3
end
to frac :level :side
   repeat 3 [fractal :level :side rt 120]
end
```

(b) What would change if the procedure below replaced **frac :level :side?** Draw levels 1, 2, and 3.

```
to frac2 :level :side
   repeat 5 [fractal :level :side rt 72]
end
```

(c) If you wish to view the levels on top of one another, use a procedure based upon the flakes procedure listed below.

```
to flakes :first.level :last.level
   pu setpos [−50 −50] pd
   repeat :last.level  −  :first.level + 1
   [ frac :first.level 100
   make "first.level :first.level + 1 ]
end
```

Sample runs of this program are shown below. Note that the procedure **flakes 0 3** draws levels 0 through 3 of the Koch curve drawn by the **frac** procedure. See Section 13.3 for further discussion of Koch curves.

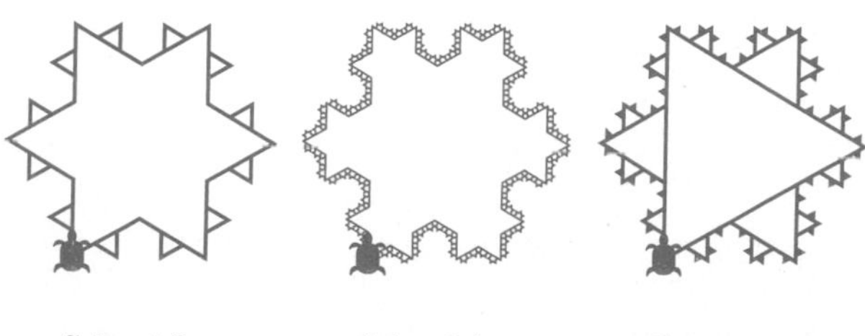

flakes 1 2 flakes 2 4 flakes 0 3

APPENDIX D
A Brief Guide to *The Geometer's Sketchpad*

Exciting new possibilities for exploring geometric concepts on the computer are available with a type of software called an **automatic drawer.** This software allows the user to construct geometric figures with both speed and precision. The software automatically gives the measures of angles, lengths of segments, and areas of regions. Once a figure is constructed, it can be manipulated to a continuum of new shapes which preserve the geometric relationships used to construct the original figure. In this way the geometric properties of the configuration can be explored in a dynamic environment not possible with traditional paper and pencil sketches.

Many automatic drawers are currently available, including the following:

Cabri (Brooks/Cole Publishing Co.)
Euclid's Toolbox (Addison-Wesley Publishing Co.)
The Geometry Inventor (Sunburst Communications, Inc.)
The Geometer's Sketchpad (Key Curriculum Press, Inc.)

This appendix provides a short introduction to *The Geometer's Sketchpad Version 2.1.* More advanced features of *Sketchpad* are described in the User Guide and Reference Manual which accompanies the software. Users of other drawing packages will need to refer to the manual pertinent to their own software. Even so, there is much common ground, and it should not be difficult to modify the procedures in the examples below to accommodate the software being used.

Using the Mouse

The mouse is used in four ways.

- **Point** Move the mouse until the tip of the arrow is over the object you want to select.

- **Click** Point, then press and quickly release the mouse button.
- **Double-click** Point, then click the mouse button twice in quick succession.
- **Drag** Point, then press and hold the mouse button. Move the pointer to a new location and then release the mouse button.

The Sketch Window

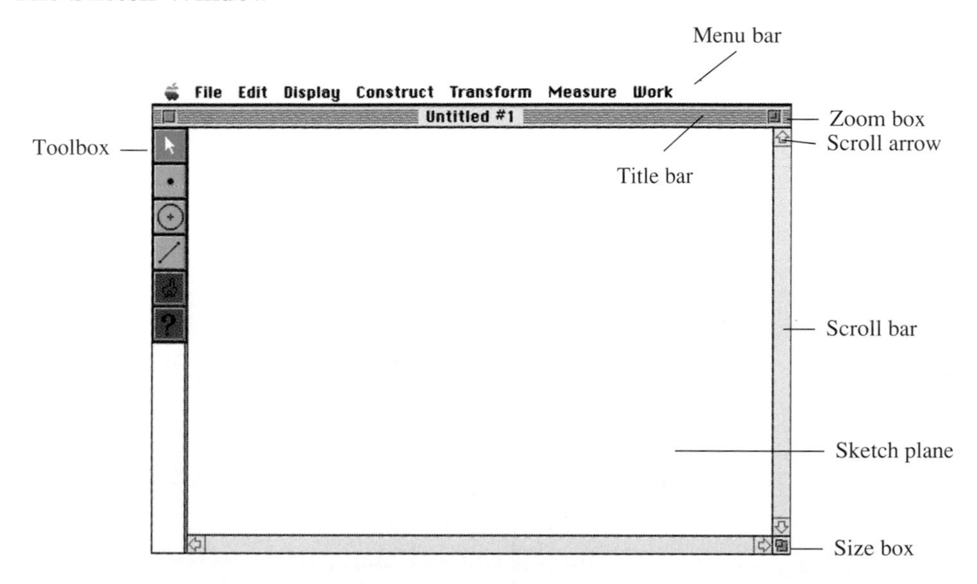

- **Menu bar** Dragging downward accesses a list of commands available to create and investigate figures.
- **Title bar** Clicking the close bar will close the window; clicking the zoom box toggles the size of the window between full size and current size; dragging on the title bar moves the window around the screen.
- **Scroll bars** Dragging the box scrolls the contents of the window vertically or horizontally.
- **Size box** Dragging this box changes the size of the window.
- **Sketch plane** The area where drawings appear.
- **Toolbox** Clicking the icon activates the corresponding tool.

Using the Toolbox

The names of the tools and their specific functions are discussed below. To activate a tool, point and click on the tool icon. You will notice that the icon becomes highlighted. The rotation and dilation tools are hidden beneath the selection tool, and similarly the ray and line tools are beneath the segment tool. To choose a hidden tool, point at the tool they are concealed under and then press and hold down the mouse button. The other tools are revealed. While holding down the mouse button, drag your pointer to the tool you wish to use and release the mouse button. Your tool can now be used. You will notice that the icon of the new tool now appears in the toolbox.

Selection Arrow Tool, Translate Tool, Rotate Tool, and Dilate Tool

There are actually three different tools represented here : the translation, rotation, and dilation tools. All of them can be used to select objects in the sketch window. The **Translate tool** is used to move an object from one position in the sketch window to another position. The **Rotate tool** is used to rotate the position of an object, and the **Dilate tool** is used to enlarge or shrink an object's size. The center of the dilation or rotation is made by selecting the point and using the **Mark Center** command in the Transform Menu.

Point Tool

This tool creates points at the cursor position in the sketch plane; each click creates a new point.

Compass Tool

This tool constructs a circle by dragging the cursor from the center to a point on the circle.

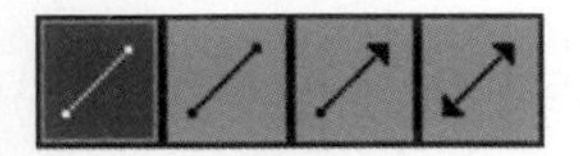

Straightedge Tools

There are three different tools concealed here: the segment, ray, and line tools. They respectively create a segment, ray, or line by clicking on one point and dragging to a second point on the object.

Text Tool

This tool enables you to label and create captions for your objects and sketches.

Object Information Tool

This tool will give you information about objects in your sketch.

EXAMPLE D.1

Constructing a Triangle with the Toolbox

Construct a triangle and investigate how it can be labeled and manipulated.

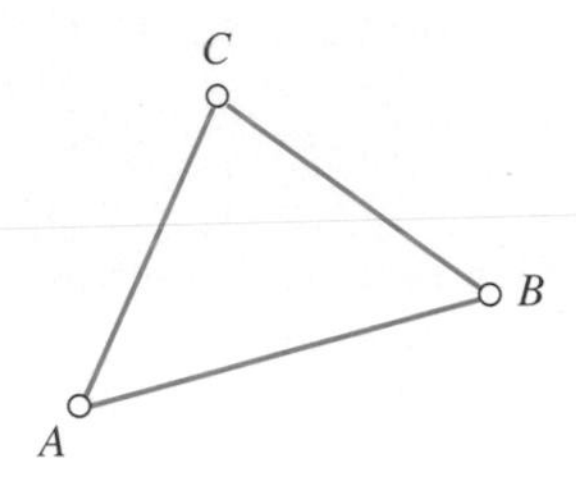

SOLUTION

Begin with a clear sketch plane. (*Note:* Dragging downward on the File menu and releasing the mouse button on the New Sketch command will create a new sketch window with a blank sketch plane.) Select the Segment tool, and move the cursor back to the sketch plane. Now click and drag from A to B, click, then drag from B to C, click, and finally drag from C back to A. This will construct triangle ABC. Next return to the toolbox, and click on the

Selection tool. Click on points and segments of your triangle to see how selected objects become highlighted. Notice that a click on another object will deselect the first object. By holding down the Shift key, notice that a sequence of objects can be selected or deselected. Also see that dragging creates a box in which all objects are selected, and clicking at a point outside your figure deselects all objects. Finally, return to the toolbox and highlight the Label tool icon. Clicking the pointer hand, when it turns black, provides a label of the vertex or side of your triangle. A second click will hide the label. Double-clicking a label allows you to edit the label. With the Label tool still activated, drag to create a box in which text can be typed. ■

The use of the toolbox can be explored in "free play" experimentation. You will quickly discover how to construct circles, rays, line segments, and polygons. However, for more sophisticated constructions you will want to use commands from the Construct and Transform Menus.

Construct

Point On Object
Point At Intersection ⌘I
Point At Midpoint ⌘M

Segment ⌘L
Perpendicular Line
Parallel Line
Angle Bisector

Circle By Center+Point
Circle By Center+Radius

Polygon Interior ⌘P
Circle Interior

Construction Help...

The Construct Menu

An alternative method to construct a line segment is to first select the desired endpoints in the sketch plane and then drag ("Pull down") the Construct Menu. Releasing the mouse button when the **Segment** command is highlighted will draw the segment between the selected points. A line is drawn similarly, but you must first highlight the line tool in the toolbox before going to the Construct Menu. Each command in the Construct Menu requires certain objects in your sketch to be selected. If the command appears a light gray color in the menu, it tells you that you do not have the proper objects selected to execute that command. Clicking on the **Construction Help . . .** command at the bottom of the menu gives a list of the required selections for each Construct Menu command.

EXAMPLE D.2

Constructing an Equilateral Triangle, the Medians, and the Centroid with the Construct Menu

Construct any two points A and B. Then draw an equilateral triangle ABC, and construct the midpoints of the sides. Draw each segment from the vertex of the triangle to the midpoint of the opposite side. These three segments, the medians, are concurrent at the centroid, G, of the triangle.

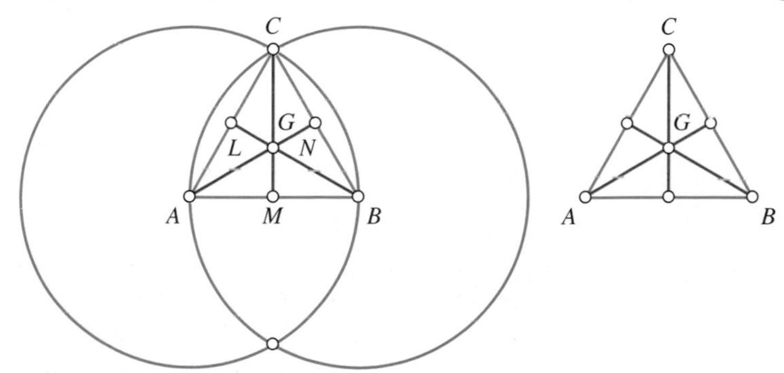

SOLUTION

Select the Point tool from the toolbox and draw two points, A and B, in the sketch plane. Choose the Select tool from the toolbox and, holding down the Shift key, select A and then B. Pull down the Construct Menu, releasing the mouse button to execute the **Circle By Center+Point** command. This will draw the circle centered at *A* that passes through *B*. In the same way, select point *B* and then point *A* to draw the circle centered at *B* that passes through point *A*. The circle just drawn will be highlighted, and the first circle drawn can also be selected by holding down the Shift key and clicking with the Select arrow. Now pull down the Construct menu and release the mouse button when the **Point At Intersection** command is highlighted. The two points at which the circles intersect will be drawn and are automatically selected. Select one of these to be point *C*. While holding down the Shift key, also select *A* and *B*. Executing the **Segment** command from the Construct Menu with three selected points will draw the sides of the desired equilateral triangle *ABC*. The three sides just drawn are automatically selected, so executing the **Point At Midpoint** command will construct the midpoints *L*, *M*, and *N* of the sides of the triangle. The centroid *G* of the triangle is now easy to construct: draw $\overline{AN}$ and $\overline{BL}$ and then construct the point, *G*, at which these two segments intersect. Selecting the circles and executing the **Hide** command from the Display Menu will eliminate the circles from view. ■

The Construct Menu permits you to construct new objects that have a specified geometric relationship to previously drawn objects in the sketch. For example, it is simple to draw parallel lines, perpendicular lines, and angle bisectors. Another way to create objects in a drawing is to turn to the Transform Menu.

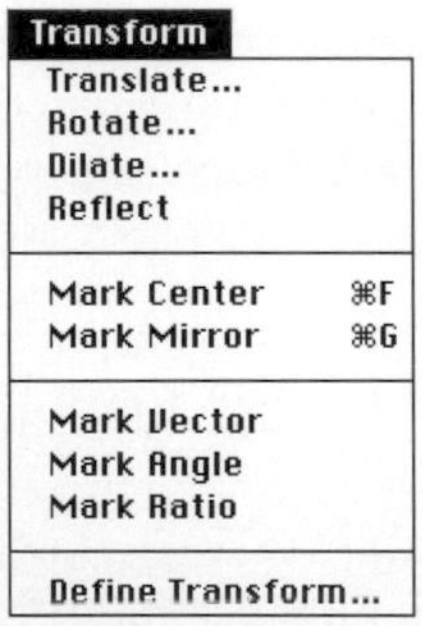

Transform
Translate...
Rotate...
Dilate...
Reflect
Mark Center ⌘F
Mark Mirror ⌘G
Mark Vector
Mark Angle
Mark Ratio
Define Transform...

The Transform Menu

The Transform Menu allows objects to be translated, rotated, dilated, and reflected. For example, to perform a reflection you would first select the desired line of reflection by clicking the Select arrow on the reflection line in the sketch plane. With this line (or ray or segment) selected, the Transform Menu is pulled down to execute the **Mark Mirror** command. Next, back in the sketch plane, select all of the objects that you wish to reflect. Now pull down the Transform Menu and execute the **Reflect** command. This will draw the reflection of the selected objects across the mirror line. Translations, rotations, and dilations are performed in a similar way, and again some free play experimentation will quickly make it clear how the commands in the Transform Menu are used in constructions and investigations.

EXAMPLE D.3 **Creating a Hexagonal Tiling with the Transform Menu**

Draw a hexagon *ABCDEF* with a pair, $\overline{AB}$ and $\overline{DE}$, of opposite sides that are parallel and congruent. Construct the midpoint, *M*, of $\overline{BC}$ and rotate the hexagon by 180° about the midpoint to create a ten-sided polygonal tile. Then show that translations of the decagon will tile the plane.

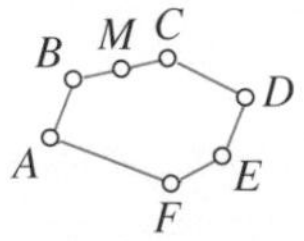

SOLUTION

Use the Segment tool to construct three sides, $\overline{AB}$, $\overline{BC}$, and $\overline{CD}$. With the Shift key held down, select points *B* and *D*, in that order, and execute the

Mark Vector command in the Transform Menu. Next, select point A and segment $\overline{AB}$ and then execute the **Translate . . . By Marked Vector** command in the Transform Menu. This will extend $ABCD$ to become $ABCDE$, where $\overline{DE}$ is parallel and the same length as $\overline{AB}$. Now complete the hexagon by constructing point F and segments $\overline{EF}$ and $\overline{FA}$. Construct the midpoint M by selecting the segment $\overline{BC}$ and executing the **Point At Midpoint** command from the Construct Menu. Now select M and execute the **Mark Center** command to choose M as the center of rotation. Next, select the hexagon (all six sides and all six vertices) and execute the **Rotate . . . By Fixed Angle** (namely by 180°) command. This creates the desired ten-sided tile. Selecting the tile and repeatedly executing the **Translate . . . By Marked Vector** command will quickly produce a partial tiling of the plane.

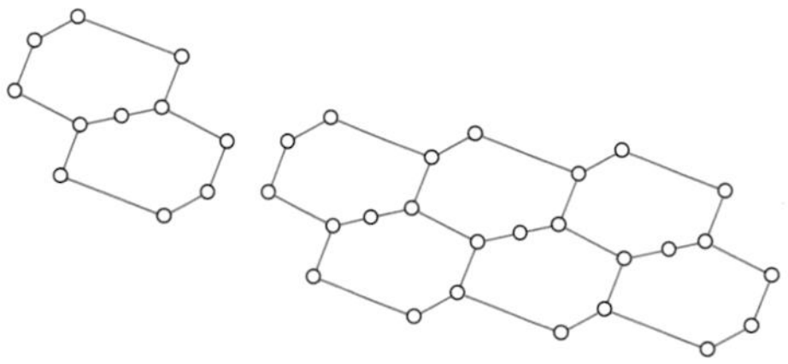

■

Measure
Distance
Length
Slope
Radius
Circumference
Area
Perimeter
Angle
Arc Angle
Arc Length
Ratio
Calculate...
Tabulate
Add Entry ⌘E

The Measure Menu

Distances, lengths, areas, perimeters, angles, and so forth, can be measured by executing commands in the Measure Menu. For example, selecting three points A, B, and C (in that order) and pulling down the Measure Menu until the **Angle** command is highlighted will display the measure of $\angle ABC$ in the sketch plane. The measurement caption can be dragged to any convenient position in the sketch plane.

Once you have displayed measurements in the sketch plane, you can select a set of measurements that you wish to include in a table. To create a table, simply pull down the Measure Menu until you have highlighted the **Tabulate** command. The current measurements of the selected objects are now displayed in the table. These entries are permanent, and will not change if the figure under investigation is manipulated. The new measurements of a manipulated figure can be added to the table by executing the **Add Entry** command in the Measure Menu. (Alternatively, double-click on the table to add a new set of measurements to the table).

Algebraic and trigonometric expressions in terms of measurements of the figure under investigation can be formed by selecting the relevant measures and then invoking the **Calculate . . .** command of the Measure Menu. The expression and its value are then displayed in the sketch plane.

EXAMPLE D.4

Exploring the Converse of the Pythagorean Theorem with the Measure Menu

Draw any triangle ABC. Measure the lengths AC, BC, and AB of the three sides and measure $\angle ACB$. Display the values of AC^2 + BC^2 − AB^2 and the measure of $\angle ACB$ in a table. What kind of a triangle is ABC when AC^2 + BC^2 − AB^2 is zero? positive? negative?

Length (Segment a) = 5.72 cm Length (Segment b) = 2.43 cm Length (Segment c) = 6.22 cm
Length (Segment a)^2 + Length (Segment b)^2 – Length (Segment c)^2 = –0.02 square cm
Angle (ACB) = 90.05°.

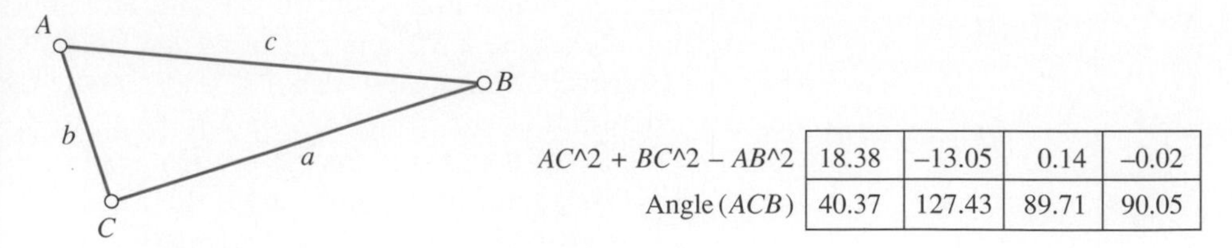

AC^2 + BC^2 – AB^2	18.38	–13.05	0.14	–0.02
Angle (ACB)	40.37	127.43	89.71	90.05

SOLUTION

We have already described how a triangle can be drawn. The length and angle measurements are displayed in a table by following the procedures just above which described the use of the Measurement Menu. Manipulating the triangle, and adding additional entries to the table should reveal that $\angle ACB$ is a right angle precisely when the expression AC^2 + BC^2 − AB^2 is zero. If AC^2 + BC^2 − AB^2 $>$ 0 the angle measure is less than 90° and triangle ABC is an acute triangle. On the other hand, if AC^2 + BC^2 − AB^2 $<$ 0 then the angle measure is greater than 90° and ABC is an obtuse triangle. ■

The File, Edit, Display, and Work Menus

The File Menu allows you to create new sketch and script windows and will open a saved sketch or script. Newly created sketches or scripts, or those that have been edited, can be saved for future use. Completed sketches can be printed.

The Edit Menu allows you to undo and redo steps in your constructions, allows you to cut, copy, and paste selected objects within or between sketch windows, and has a variety of commands that are helpful to selecting objects in the sketch window.

The Display Menu allows you to control the thickness and color of lines, the shading density and color of regions, and the size and style of the type font.

The Work Menu is a convenient way to activate an open sketch or script. Choosing a script with the Option Key held down will play the script, using the selected objects in the sketch plane as the Givens.

The Script Window

The Script Window is opened by executing the **New Script** command in the File Menu. The Script Window has buttons which mimic those of a tape recorder. Clicking on the **REC** button will record the instructions given subsequently in the Sketch Window. When the sketch is completed, the Script Window is reopened and the **STOP** button is clicked. The Script Window below the control buttons is divided into two parts. In the upper colored part is a list of the objects which need to be selected for the script to be run. The construction steps are listed in the lower white part of the window. Playing a script with the selected objects listed under the Givens will recreate the figure drawn when the script was being recorded. Scripts are especially useful for performing repeated constructions in a complex drawing.

File
New Sketch ⌘N
New Script
Open... ⌘O
Save ⌘S
Save As...
Close ⌘W
Page Setup...
Print Preview...
Print...
Share Documents...
Quit ⌘Q

Edit
Undo ⌘Z
Redo ⌘R
Cut ⌘X
Copy ⌘C
Paste ⌘V
Clear
Action Button ▶
Select All ⌘A
Select Parents ⌘U
Select Children ⌘D
Hide Toolbox
Show Clipboard
Create Publisher...

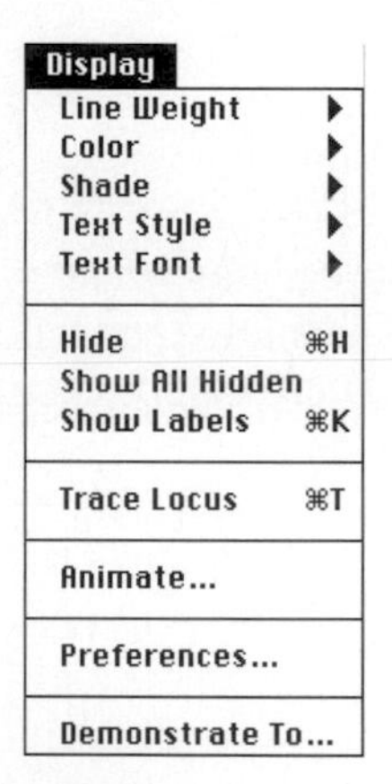

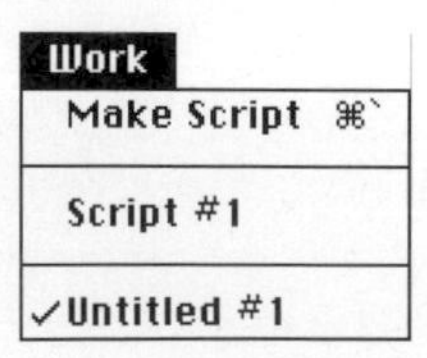

EXAMPLE D.5 Creating a Script to Construct a Square on a Given Side

Create a script that will draw a square *ABCD* and shade the enclosed region, where *A* and *B* are given points.

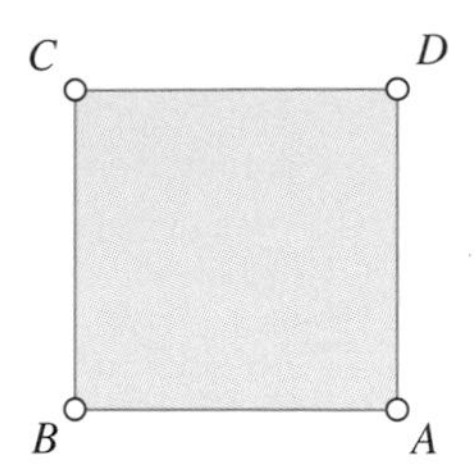

SOLUTION

The script shown below, titled "Square on a side," will construct the desired square. The command **Polygon Interior** in the Construct Menu will shade the region within the square.

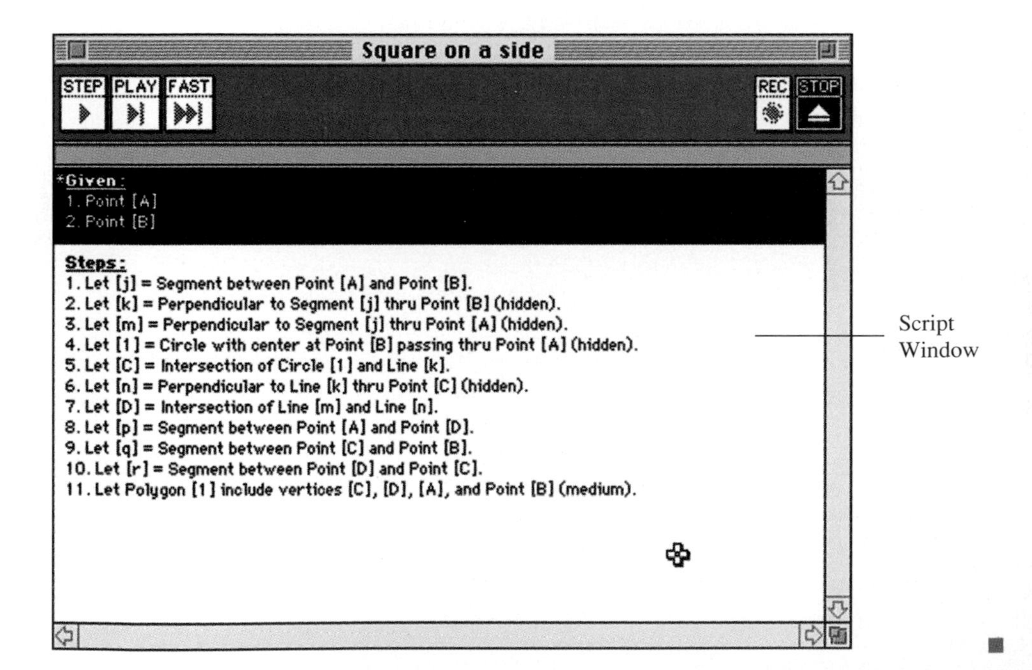

■

Advanced Features

The construction methods described and illustrated in the examples above will enable the user of *The Geometer's Sketchpad* to create a wide variety of interesting figures to investigate. However, *Sketchpad* and other automatic drawing software have other features that more experienced users will find valuable. For example, *Sketchpad* can add an exciting dynamic dimension to a sketch by incorporating animation, movement, hide/show buttons, and so on. The user's manuals for the software describes these features and how they can be used.

ANSWERS TO SELECTED PROBLEMS

Chapter 1

Problem Set 1.1 (page 7)

1. (a) $9 \times 9 = 81$
$79 \times 9 = 711$
$679 \times 9 = 6111$
$6579 \times 9 = 51{,}111$
$45{,}679 \times 9 = 411{,}111$
$345{,}679 \times 9 = 3{,}111{,}111$
$2{,}345{,}679 \times 9 = 21{,}111{,}111$
$12{,}345{,}679 \times 9 = 111{,}111{,}111$

3. (a) $1 \times 8 + 1 = 9$
$12 \times 8 + 2 = 98$
$123 \times 8 + 3 = 987$

5. (a) $67 \times 67 = 4489$
$667 \times 667 = 444{,}889$
$6667 \times 6667 = 44{,}448{,}889$

6. (b)

			⑤
6	9	8	⑤
3	5	7	⑤
2	1	4	⑤
⑤	⑤	⑤	⑤

9. (a) $1 \times 142{,}857 = 142{,}857$
$2 \times 142{,}857 = 285{,}714$
$3 \times 142{,}857 = 428{,}571$
$4 \times 142{,}857 = 571{,}428$
$5 \times 142{,}857 = 714{,}285$

11. (a) P, Q, and R appear to be collinear.
13. (a) $\overline{AD}$, $\overline{BE}$, and $\overline{CF}$ appear to be concurrent.
18. (a) Results will vary. It very likely always stops.

JUST FOR FUN For Careful Readers (page 13)

1. The second engineer was the first engineer's daughter. **2.** There was no smoke since the train was electric. **3.** The 12th rung down since the ladder rises with the ship.

Problem Set 1.2 (page 15)

1. (a) 21 bikes, 6 trikes **4.** 12 **7. (a)**

⑦
⑥ ④ ⑨
⑧

9. (a)

⑥
① ②
⑤ ③ ④

or

①
⑥ ⑤
② ④ ③

11. (d) There are no solutions. **12. (a)** 1, 2, 3, 5, 8, 13, 21 **(c)** 3, 5, 8, 13, 21, 34, 55 **(e)** 2, 1, 3, 4, 7, 11

JUST FOR FUN How Many Heaps? (page 24)

The least number each could have received is 14.

Problem Set 1.3 (page 25)

1. No. When 10 is multiplied by 5 and 13 is added, the result is 63, not 48. **3.** 3 **5. (a)** Yes. Yes. The rules are really the same. **7. (a)** **(c)** **(e)** **9.** 49

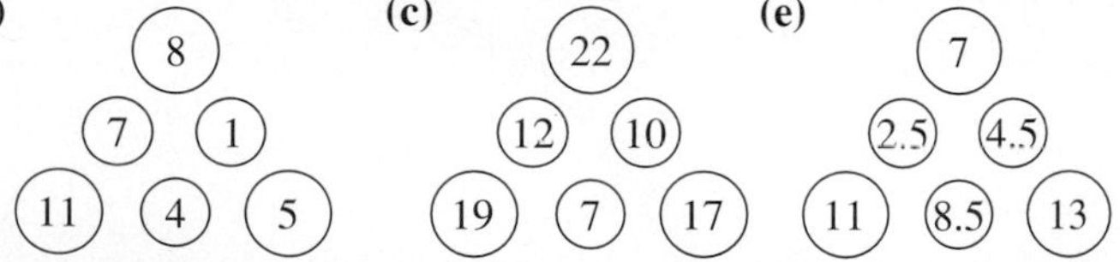

11. 1357, 1375, 1537, 1573, 1735, 1753, 3157, 3175, 3517, 3571, 3715, 3751, 5137, 5173, 5317, 5371, 5713, 5731, 7135, 7153, 7315, 7351, 7513, 7531 **13.** Assuming that the bags are identical, the possibilities are as shown.

Bag 1	Bag 2	Bag 3
23	1	1
21	3	1
19	3	3
19	5	1
17	3	5
17	7	1
15	5	5
15	3	7
15	1	9
13	9	3
13	7	5
13	11	1
11	7	7
11	9	5
11	11	3
9	9	7

15. 4 and 24, 6 and 16, 8 and 12 **17.** 9 minutes **20.** Dawkins, Chalmers, Ertl, Albright, Badgett **22.** 46 square meters.

Problem Set 1.4 (page 41)

1. (a) 14, 17, 20 **(c)** 10, 10, 15 **(e)** 162, 486, 1458 **2. (a)**

(d) The nth even number **(f)** 1,443,603 **4. (a)** 16 **5. (a)** 320 **(c)** 595

7. (a) $1 + 2 + 3 + 4 + 5 + 4 + 3 + 2 + 1 = 25$
$1 + 2 + 3 + 4 + 5 + 6 + 5 + 4 + 3 + 2 + 1 = 36$ **9. (a)** 86 **(d)** 42

11. (a) $1 - 4 + 9 - 16 + 25 = 15$
$1 - 4 + 9 - 16 + 25 - 36 = -21$
(b) $1 - 4 + 9 - 16 + 25 - 36 + 49 = 28$
$1 - 4 + 9 - 16 + 25 - 36 + 49 - 64 = -36$

(c) For even n, $1 - 4 + 9 - \cdots - n^2 = -\dfrac{n(n+1)}{2}$. For odd n, $1 - 4 + 9 - \cdots - n^2 = \dfrac{n(n+1)}{2}$.

14. (a) 6 **(c)** 4950 **17. (a)**

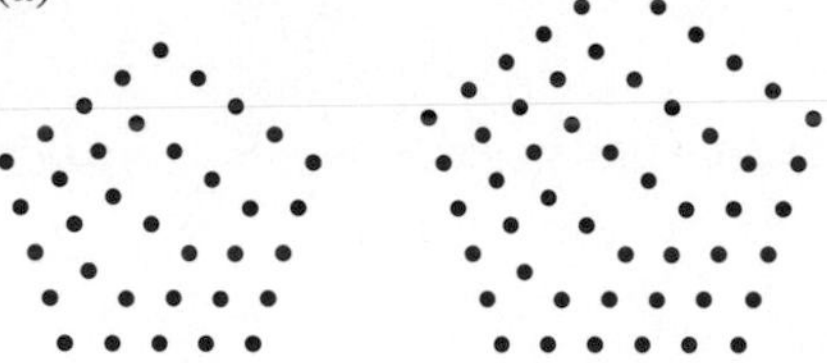

(b) 1, 5, 12, 22, 35, 51 **(c)** $1 + 4 + 7 + 10 + 13 = 35$, $1 + 4 + 7 + 10 + 13 + 16 = 51$ **(d)** 28 **(e)** 145 **(f)** $3n - 2$ **(g)** $p_n = \dfrac{n(3n-1)}{2}$ **21. (a)** 1, 2, 4, 8, 16, 32, 64, 128 **22. (a)** 1, 2, 6, 20, 70 **(b)** The sums are the numbers right down the center of Pascal's triangle. **23. (a)** 30, 2100, 29400

JUST FOR FUN How Many Pages in the Book? (page 49)

To number pages 1 through 9 takes $9 \cdot 1 = 9$ digits. To number pages 10 through 99 takes $90 \cdot 2 = 180$ digits. This leaves $867 - 180 - 9 = 678$ digits to number 3-digit pages. Thus, there are $678 \div 3 = 226$ 3-digit pages and $226 + 90 + 9 = 325$ pages in the book.

Problem Set 1.5 (page 54)

1. Yes. Answers will vary. The second player can add enough tallies to make a multiple of five at each step, forcing the first player to be the one to exceed 30. **3. (a)** \$28 **6.** Moe was wearing Hiram's coat and Joe's hat; Hiram was wearing Joe's coat and Moe's hat. **7. (a)** 25. Not all of the information was needed. **8.** Beth is the center; Jane is the guard; Mitzi is the forward. **11.** Yes. Either 5 or 7 may be placed in the center and the remaining digits paired so that the sums of the pairs are the same. The paired digits are placed opposite each other in the diagram. **12. (a)** 3 **14.** Since there are only 10 digits (0, 1, 2, . . . , 9), in any collection of 11 natural numbers there must be two which have the same units digit. The difference of these two numbers must have a zero as its units digit and is thus divisible by 10. **16.** If five points are chosen in a unit square then by the Pigeonhole Principle at least two of the points must be in or on the boundary of one of the four smaller squares shown. The farthest these two points can be from each other is $\sqrt{2}/2$, units if they are on opposite corners of the small square.

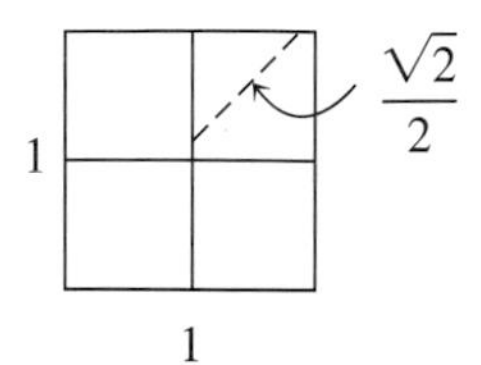

18. If the cups of marbles are arranged as described, each cup will be part of three different groups of three adjacent cups. The sum of all marbles in all groups of three adjacent cups is $3 \cdot (10 \cdot 11/2) = 165$ since each cup of marbles is counted three times. With the marble count of 165 and ten possible groups of three adjacent cups, by the Pigeonhole Principle at least one group of three adjacent cups must have 17 or more marbles since $165 \div 10 = 16.5 > 16$.
20. Case One: At least two people have no friends. Then these two people have the same number of friends.

Problem Set 1.6 (page 65)

1. 17 min, NENNEE or NENENE or NENEEN **5.** Sara should get the racket, and pay Ken \$11.25.
9. (a)

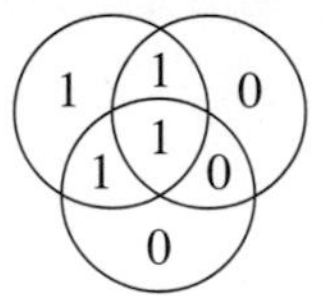

(c) Correct

10. (a)

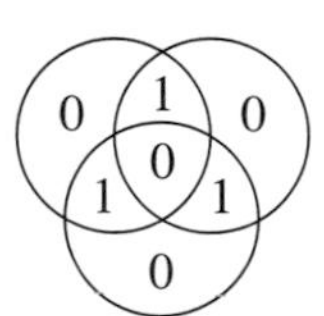

(b) Since all three circles are bad, the error must be in the intersection of all three circles.

13. (a) The warehouse should be built at store *B*.

CLASSIC CONUNDRUM 10 = 9? (page 69)

No, no. The two men in the first room were number 1 and number 2. The proprietor's scheme actually ignores the tenth man entirely!

Chapter 1 Review Exercises (page 72)

1. (a) 8; 168; 2568; 34,568; 434,568; 5,234,568; 61,234,568; 701,234,568; 7,901,234,568 **(b)** No, there is a recognizable pattern. **(c)** Yes. **(d)** No. It suggests that examples only *hint at* a pattern or result. A proof is really needed to be sure.
2. (a) The result is a 7-digit number with all digits your favorite. **(b)** 1,111,111
3. (a)

67	1	43
13	37	61
31	73	7

(b)

7	1	31
13	37	61
43	73	67

4. *P*, *Q*, and *R* appear to be collinear. **5.** 4884 **6.** 28

7. (a) Answers will vary. One solution is as follows:

$$\begin{array}{r} 179 \\ 368 \\ +\ 452 \\ \hline 999 \end{array}$$

(b) Yes. The digits in any column can be arranged in any order. **(c)** No. The hundreds column digits must add up to 8 to allow for a carry from the tens column. If the digit 1 is not in the hundreds column, the smallest this sum can be is 2 + 3 + 4 = 9. Thus, the digit 1 must be in the hundreds column. **8.** 9 **9.** 60 **10.** 88 square feet **11.** 9 **12. (a)** Multiply by 5, then subtract 2. **(b)** Answers will vary. The best strategy is to give Chanty consecutive integers starting with 0. **13. (a)** 3, 6, 12, 24, 48, 96 **(b)** 4, 8, 16, 32, 64, 128 **(c)** 1, 6, 36, 216, 1296, 7776 **(d)** 2, 10, 50, 250, 1250, 6250 **(e)** 7, 7, 7, 7, 7, 7 **14.** Kimberly is the lawyer and the painter; Terry is the engineer and the doctor; Otis is the teacher and the writer.
15. (a) $21 + 23 + 25 + 27 + 29 = 5^3$
$31 + 33 + 35 + 37 + 39 + 41 = 6^3$
$43 + 45 + 47 + 49 + 51 + 53 + 55 = 7^3$
(b) 1, 3, 7, 13, 21, 31, 43, 57, 73, 91 **(c)** $91 + 93 + 95 + 97 + 99 + 101 + 103 + 105 + 107 + 109 = 10^3$
16. (a) $14 + 16 + 18 + 20 = 4^3 + 4$
$22 + 24 + 26 + 28 + 30 = 5^3 + 5$
$32 + 34 + 36 + 38 + 40 + 42 = 6^3 + 6$
(b) $92 + 94 + 96 + 98 + 100 + 102 + 104 + 106 + 108 + 110 = 10^3 + 10$
17. (a) 25 **(b)** 1075 **18. (a)** 11th **(b)** 6141 **(c)** Duly observed.
(d) $6141 = (6 + 12 + \cdots + 3072 + 6144) - (3 + 6 + \cdots + 3072) = 6144 - 3 = 6141$. **19.** 442,865
20. (a) $\dfrac{n(n+1)}{2} + 1$ **(b)** $\dfrac{n(n-1)}{2}$ **(c)** n^2 **21.** The product of the squared entries appears to be equal to the product of the circled entries. **22. (a)** 3 **(b)** 9 **(c)** 27 **(d)** $P_0 + P_1 \cdot 2^1 + P_2 \cdot 2^2 + \cdots + P_n \cdot 2^n = 3^n$, where P_k is the kth element of the nth row of Pascal's triangle. **(e)** 4, 16, 64 **(f)** $P_0 + P_1 \cdot r^1 + P_2 \cdot r^2 + \cdots + P_n r^n = (r + 1)^n$
23. (a) 5 **(b)** 9 **(c)** 50 **24.** 17 **25. (a)** 2500 dollars **(b)** A B G I M or A B C D F J M
26. Judy gets the house; Joshua gets the motor home and the painting; JoAnn gets the automobile. Joshua should give $94,500 to Judy, $308,000 to John, and $267,500 to JoAnn. **27. (a)** 1101000 **(b)** 1000101 **(c)** 1010010 **28. (a)** 0110001 **(b)** correct **(c)** 1010010

Chapter 1 Test (page 75)

1. 21,111,111; 12,111,111; 11,211,111; 11,121,111; 11,112,111; 11,111,211; 11,111,121; 11,111,112
2. (a) $17 + 18 + 19 + 20 + 21 + 22 + 23 + 24 + 25 = 64 + 125 = 225 - 36$
$26 + 27 + 28 + 29 + 30 + 31 + 32 + 33 + 34 + 35 + 36 = 125 + 216 = 441 - 100$
(b) $82 + 83 + 84 + \cdots + 98 + 99 + 100 = 729 + 1000 = 3025 - 1296$
(c) $[(n-1)^2 + 1] + [(n-1)^2 + 2] + \cdots + n^2 = (n-1)^3 + n^3 = \left(\dfrac{n(n+1)}{2}\right)^2 - \left(\dfrac{(n-1)(n-2)}{2}\right)^2$ **3.** 36 **4.** 3
5. (a)

57	2	37
12	32	52
27	62	7

(b)

7	2	27
12	32	52
37	62	57

6. 10 days

7. (a) $2 + 5 + 8 + 11 + 8 + 5 + 2 = 41 = 3^2 + 2 \cdot 4^2$
$2 + 5 + 8 + 11 + 14 + 11 + 8 + 5 + 2 = 66 = 4^2 + 2 \cdot 5^2$
(b) $2 + 5 + 8 + \cdots + 26 + 29 + 26 + \cdots + 8 + 5 + 2 = 281 = 9^2 + 2 \cdot 10^2$
8. (a) $S_{20} = -110$, $S_{21} = 121$ **(b)** For n odd, $S_n = \left(\dfrac{n+1}{2}\right)^2$. For n even, $S_n = -\left(\dfrac{n}{2}\right)\left(\dfrac{n}{2} + 1\right)$.
9. (a) 1101000 **(b)** 0101101 **10.** 20, EENNEEN

Chapter 2

Problem Set 2.1 (page 91)

1. (a) {Arizona, California, Idaho, Oregon, Utah} **2. (a)** {1, i, s, t, h, e, m, n, a, o, y, c} **3. (a)** {8, 9, 10, 11, 12} **(c)** {3, 6, 9, 12, 15, 18} **4.** Answers will vary. **(a)** $\{x \in U \mid 11 \le x \le 14\}$ or $\{x \in U \mid 10 < x < 15\}$

(c) $\{x \in U \mid x = 4n \text{ and } n \in N\}$ **5.** Answers may vary. **(a)** $\{x \in N \mid x \text{ is even and } x > 12\}$ or $\{x \in N \mid x = 2n$ for some $n \in N$ and $n > 6\}$ **6. (a)** No, it is not clear which cities with these names are meant. **(b)** Yes, it is clear which four cities are included in the set. **7. (a)** True **(c)** True **(e)** True

8.

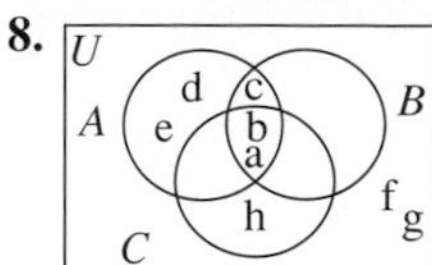

(a) $B \cup C = \{a, b, c, h\}$ **(c)** $A - B = \{d, e\}$
(e) $\overline{A} = \{f, g, h\}$ **9. (a)** $M = \{45, 90, 135, 180, 225, 270, \ldots\}$ **(b)** $L \cap M = \{90, 180, 270, \ldots\}$ = the set of simultaneous multiples of 6 and 45 = the set of multiples of 90. **(c)** 90

11. (a)

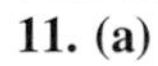

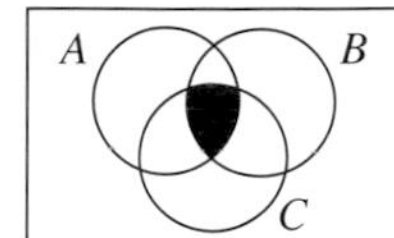

(c)

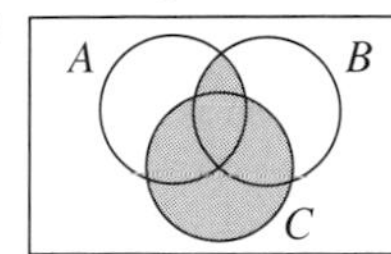

(e)

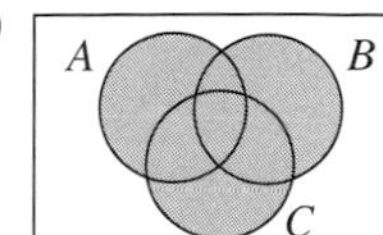

12. (a)

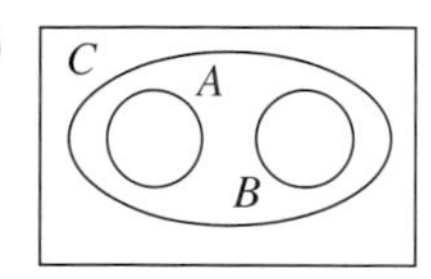

13. No, it is possible that there are elements of A that are also elements of B but not C, or C but not B. For example, Let $A = \{1, 2\}$, $B = \{2, 3\}$, $C = \{3\}$.

15. (a) $\overline{A \cap B} = \{1, 2, 3, 4, 5, 7, 8, 9, 10, 11, 13, 14, 15, 16, 17, 19, 20\}$
$\overline{A} \cup \overline{B} = \{1, 2, 3, 4, 5, 7, 8, 9, 10, 11, 13, 14, 15, 16, 17, 19, 20\}$
$\overline{A \cup B} = \{1, 5, 7, 11, 13, 17, 19\}$
$\overline{A} \cap \overline{B} = \{1, 5, 7, 11, 13, 17, 19\}$

(b) Since $\overline{A \cap B} = \overline{A} \cup \overline{B}$ and $\overline{A \cup B} = \overline{A} \cap \overline{B}$ from part (a), DeMorgan's Laws hold for these sets.

16. (a) $\{(1, a), (1, b), (2, a), (2, b)\}$ **(c)** $\{(6, t)\}$ **17. (a)** $\{d, e\} \times \{4, 6\}$ **(c)** $N \times N$ **(e)** $\varnothing \times \varnothing$, $\{2, 9\} \times \varnothing$, $\varnothing \times \{a, b, c\}$

19. (a) The first component in an ordered pair is an element of the first set and the second component is an element of the second set. For example, if $A = \{1, 2\}$ and $B = \{a, b\}$, then $A \times B = \{(1, a), (1, b), (2, a), (2, b)\}$ but $B \times A = \{(a, 1), (a, 2), (b, 1), (b, 2)\}$ **(b)** $C \times D = D \times C$ if, and only if, $C = \varnothing$ or $D = \varnothing$ or $C = D$.

20. (a)

(c)

(e)

21. (a) $L \cap T$ **(c)** $S \cup T$ **22. (a)** 8; $\varnothing$, $\{a\}$, $\{b\}$, $\{c\}$, $\{a, b\}$, $\{a, c\}$, $\{b, c\}$, $\{a, b, c\}$ **(c)** 8 **(e)** $2^{26} = 67{,}108{,}864$

23. (a)

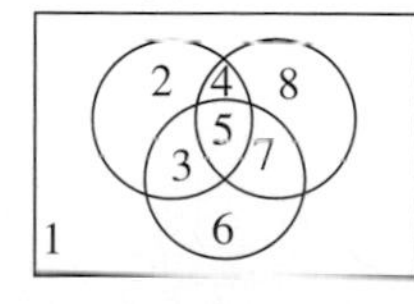

8 regions

(c) $\overline{A} \cap B \cap C \cap \overline{D}$

24. Symbolically, statements (1), (2), and (4) give $S \subseteq N$, $A \subseteq M$, $M \subseteq N$. Therefore, $A \subseteq M \subseteq N \subseteq S$. By (3), $S \cap L = \varnothing$. But $A \subseteq S$, so $A \cap S = \varnothing$. Thus, birds in this aviary live on mince pies.

26. (a) 15 **27. (a)** Only 1 way

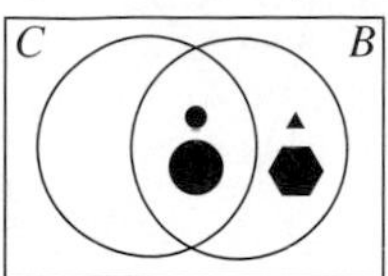

28. (a) $6 \cdot 2 \cdot 3 = 36$ **(b)** $6 \cdot 2 \cdot 3 \cdot 2 = 72$

30.

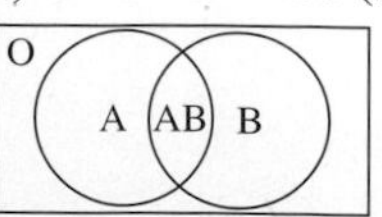

32. Answers will vary. **34.** *Understand the problem* The squares are $1^2 = 1$, $2^2 = 4$, $3^2 = 9$, and so on. The problem is to arrange the numbers 1, 2, . . . , 15 in a list so that pairs of adjacent numbers have sums that are squares. *Devise a plan* 1 can be added to 15 or 8 to give a square; 2 can be added to 14 or 7, 3 can be added to 13 or 6; and so on. However, 8 can only be added to 1 and 9 can only be added to 7. Thus, there are two possible sums for each number except for 8 and 9. Using this information, we should be able to order the list as required. *Carry out the plan* Since only one sum can contain 8 and similarly for 9, these numbers must be on the ends of our list. We start with 8 and then 1. Then the only possibility for the successor to 1 is 15, and so on. This yields the following list: 8, 1, 15, 10, 6, 3, 13, 12, 4, 5, 11, 14, 2, 7, 9 *Look back* The successful plan arose from looking at numbers in the list whose sums were squares. For all but two numbers, there were two numbers that could be added to the number being considered that would result in a square. However, there was only one possibility for 8 (and also 9) so 8 had to go on one end of the list and this determined all the remaining choices.

JUST FOR FUN Disc Discoveries (page 99)

The large blue disc is most like the other two, since it matches the small disc in color and the large pink disc in size. If the three discs shown were part of a larger four disc set, the fourth disc was likely a pink disc the size of the small blue disc.

Problem Set 2.2 (page 103)

1. (a) 13: ordinal
first: ordinal
3. Apparently, the cardinality of the set of seats in the auditorium is less than the cardinality of the set of potential audience members.
4. Answers will vary. One possibility is shown. **5. (a)** 15 **(c)** 2 **7.** Yes. $3 \leftrightarrow 5$ **8. (a)** Finite

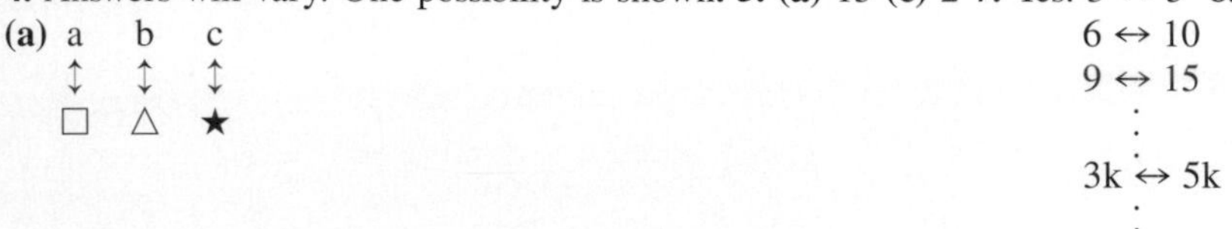

9. (a)
$$\begin{array}{ccccccc} 1 & 8 & 27 & 64 & \cdots & n^3 & \cdots \\ \updownarrow & \updownarrow & \updownarrow & \updownarrow & & \updownarrow & \\ 1^3 & 8^3 & 27^3 & 64^3 & \cdots & n^9 & \cdots \end{array}$$

Since there is a one-to-one correspondence between the set and one of its proper subsets, the set is infinite.
10. (a) For example, $Q_1 \leftrightarrow Q_2$, $Q_3 \leftrightarrow Q_4$, and so on.

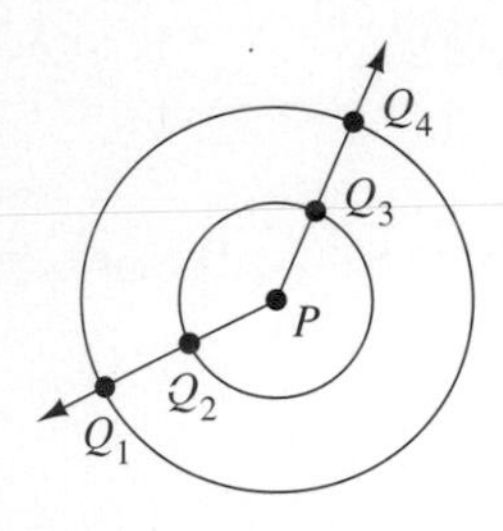

(c) For example $Q_1 \leftrightarrow Q_2$, and so on.

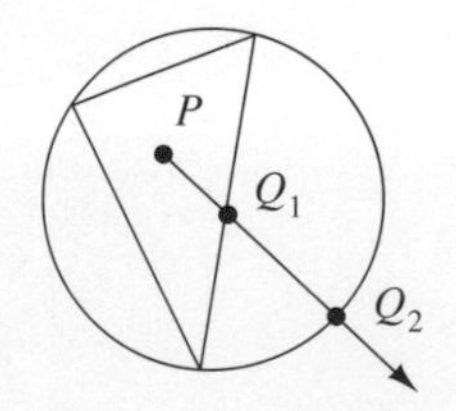

11. (a) True **(c)** True

13. Possibilities

$n(A)$	$n(B)$
1	12
2	6
3	4
4	3
6	2
12	1

14. (a) $n(A \cap B) \leq n(A)$. The set $A \cap B$ contains only the elements of A that are also elements of B. Thus, $A \cap B$ cannot have more elements than A. **15. (a)** One of four possibilities may be chosen to fill the first spot in an ordered triple. For each possibility, one of two elements may be chosen for the second position, making eight possibilities for filling the first two spots. Then one of five elements may be chosen to fill the last position, making a total of $8 \times 5 = 40$ possibilities.

16.

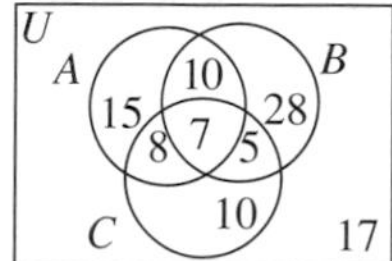

18. (a)

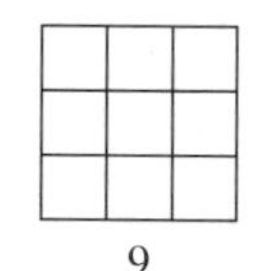

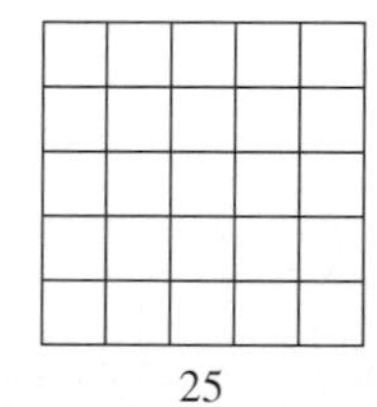

19. (a) 37, 61 **20. (a)** No. Since he lives forever, Joe will eventually earn enough to pay all his bills through any given week though he falls further and further behind. **(b)** After $(102 \cdot 50)/15 = 340$ weeks
22. (a) Let each element $a \in A$ be matched to itself: $a \leftrightarrow b$. This gives a one-to-one correspondence from A to itself, so $A \sim A$. **(b)** If $A \sim B$, there is a one-to-one correspondence that matches each element $a \in A$ to an element $b \in B$: $a \leftrightarrow b$. Since each element of B is matched exactly once, the correspondence can be reversed: $b \leftrightarrow a$. This shows that $B \sim A$, and therefore set equivalence is symmetric. **(c)** If $A \sim B$ and $B \sim C$, then there are two one-to-one correspondences: $a \leftrightarrow b$ and $b \leftrightarrow c$. Combining these correspondences gives us a one-to-one correspondence $a \leftrightarrow c$ from A to C. Thus $A \sim C$, and so set equivalence is transitive. **24.** Using a Venn diagram and guess and check, we discover that **(a)** four students have visited all three countries. **(b)** 14 students have been only to Canada.
26. (a) Making an orderly list, we obtain the following: **(b)** 24

$$a \leftrightarrow 1, \quad b \leftrightarrow 2, \quad c \leftrightarrow 3$$
$$a \leftrightarrow 1, \quad b \leftrightarrow 3, \quad c \leftrightarrow 2$$
$$a \leftrightarrow 2, \quad b \leftrightarrow 1, \quad c \leftrightarrow 3$$
$$a \leftrightarrow 2, \quad b \leftrightarrow 3, \quad c \leftrightarrow 1$$
$$a \leftrightarrow 3, \quad b \leftrightarrow 1, \quad c \leftrightarrow 2$$
$$a \leftrightarrow 3, \quad b \leftrightarrow 2, \quad c \leftrightarrow 1$$

27. (a)

r	**0**	**1**	**2**	**3**	**4**	**5**
Subsets with *r* Elements	∅	{a}, {b}, {c}, {d}	{a, b}, {a, c}, {a, d}, {b, c}, {b, d}, {c, d}	{a, b, c}, {a, b, d}, {a, c, d}, {b, c, d}	{a, b, c, d}	none
Number of Subsets with *r* Elements	1	4	6	4	1	0

29. As the Venn diagram shows, four people have type AB blood.

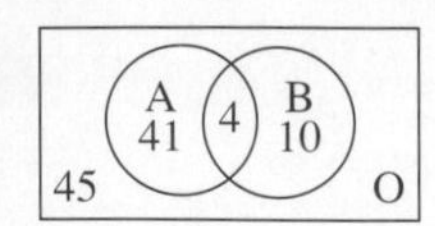

33.

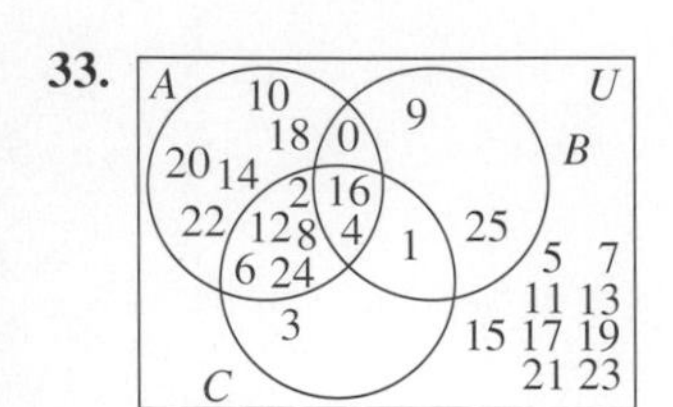

34. **(a)** $\overline{A} = \{b, c, f, g\}$ **(c)** $A \cup \overline{B} = \{a, b, d, e\}$ **(e)** $\overline{A} \cap \overline{B} = \{b\}$ **(g)** $A \times B = \{(a, c), (a, e), (a, f), (a, g), (d, c), (d, e), (d, f), (d, g), (e, c), (e, e), (e, f), (e, g)\}$

35. (a)

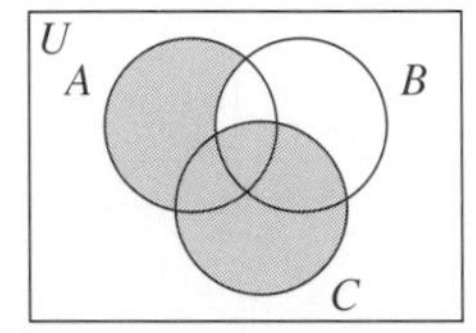

JUST FOR FUN Paper Clip Comparison (page 113)

Many people will say just one more and will be surprised when shown that the answer is two more.

Problem Set 2.3 (page 116)

1. (a) (i) 5 (ii) 5 (iii) 4 **(b)** (ii) and (iii) **3. (a)** 4, 5, 6, 7, or 8 **(b)** 4 **4. (a)**

7. (a) Closed **(c)** Closed **(e)** Closed **9.** The parentheses may be overlooked and the addends rearranged because the associative and commutative properties of addition allow the addends to be added in any order.

11. (a) $5 + 7 = 12$ $12 - 7 = 5$ **(b)** $4 + 8 = 12$ $12 - 8 = 4$
$7 + 5 = 12$ $12 - 5 = 7$ $8 + 4 = 12$ $12 - 4 = 8$

12. (a) $419 + x = 627$ **13. (a)** 9 **(c)** 0 **(e)** 0, 1, 2, 3, 4, 5 **(g)** 8, 9, 10, 11, . . . **15. (a)** Comparison **(b)** Missing addend **(c)** Take-away **(d)** Comparison **17.** $257 - 240 = 17$ pages **18. (a)** $(8 - 5) - (2 - 1) = 2$ **(c)** $((8 - 5) - 2) - 1 = 0$

19. (a)

5	2	⑦
1	2	③
⑥	④	⑦

20. (a) 1, 6, 4, ⑩, 3, 2, 5

22. $A \cup B$, A, B, $A \cap B$

$n(A \cup B) = n(A) + n(B) - n(A \cap B)$

24. (a) $1 + 2 + 3 + 4 + 5 + 6 + 7 = 28$ **25. (a)**

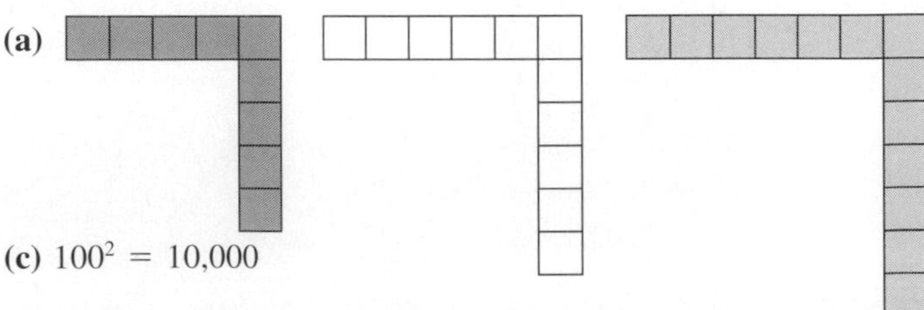

(c) $100^2 = 10{,}000$

26. (a) $d_4 = 24$, $d_5 = 32$ **27. (a)** 98 **28.** $\{0\}$

30. (a)

n	1	2	3	4	5	6	7	8	9	10	11	12	13	14	15
t_n	1	3	6	10	15	21	28	36	45	55	66	78	91	105	120

(b) $11 = 10 + 1$, $12 = 6 + 6$, $13 = 10 + 3$, $14 = 10 + 3 + 1$, $15 = 15$, $16 = 15 + 1$, $17 = 15 + 1 + 1$, $18 = 15 + 3$, $19 = 10 + 6 + 3$, $20 = 10 + 10$, $21 = 21$, $22 = 21 + 1$, $23 = 10 + 10 + 3$, $24 = 21 + 3$, $25 = 15 + 10$

31.

+	5	4	1	6	9	2	0	8	7	3
3	8	7	4	9	12	5	3	11	10	6
9	14	13	10	15	18	11	9	17	16	12
6	11	10	7	12	15	8	6	14	13	9
4	9	8	5	10	13	6	4	12	11	7
0	5	4	1	6	9	2	0	8	7	3
7	12	11	8	13	16	9	7	15	14	10
5	10	9	6	11	14	7	5	13	12	8
2	7	6	3	8	11	4	2	10	9	5
1	6	5	2	7	10	3	1	9	8	4
8	13	12	9	14	17	10	8	16	15	11

34. **(a)** 2 **(c)** 10 **(e)** 5 **36.** 5, since $3 \times 5 = 15$.

Problem Set 2.4 (page 133)

1. **(a)** $3 \times 5 = 15$ **(c)** $4 \times 10 = 40$ **(e)** $3 \times 4 = 12$ **2.** **(a)** Array model **(c)** Measurement model **4.** **(a)** Each of the a lines coming from set A intersects each of the b lines coming from set B. **(b)** $\{\square, \triangle\} \times \{♥, ♦, ♣, ♠\}$

5. **(a)** Not closed. $2 \times 2 = 4$ and 4 is not in the set. **(c)** Not closed. $2 \times 4 = 8$ and 8 is not in the set. **(e)** Closed. The product of any two odd whole numbers is always another odd whole number. **(g)** Closed. $2^m \times 2^n = 2^{m+n}$ for any whole numbers m and n. **7.** **(a)** Commutative property of multiplication **(c)** Multiplication by zero property **(e)** Associative property of multiplication **8.** **(a)** Commutative property of multiplication: $5 \times 3 = 3 \times 5$

9. **(a)**

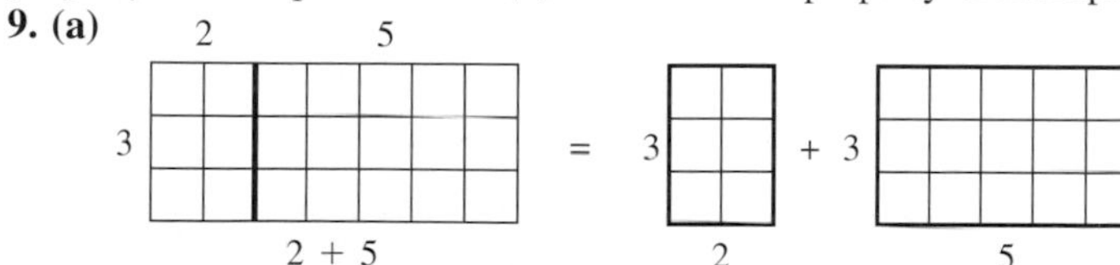

10.

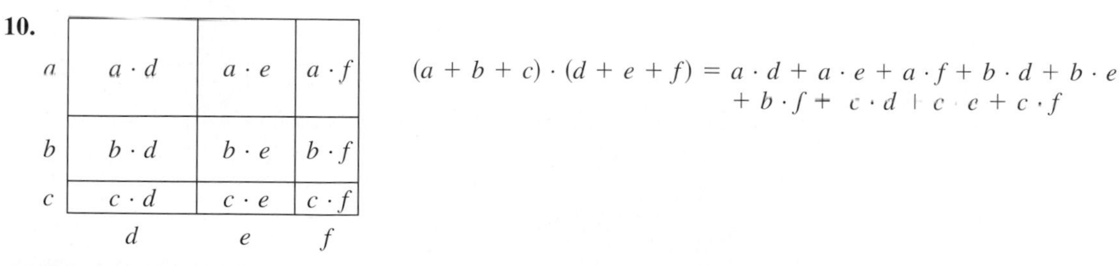

12. $18 \times 0.86 + 18 \times 0.14 = 18 \times (0.86 + 0.14) = 18 \times 1 = 18.00$ **13.** **(a)** Distributive property **14.** **(a)** $18 \div 6 = 3$ **15.** **(a)** $4 \times 8 = 32$, $8 \times 4 = 32$, $32 \div 4 = 8$, $32 \div 8 = 4$ **16.** **(a)** Repeated subtraction **17.** **(a)** 6 **(c)** 6 R 12 **18.** Answers will vary. For example, **(a)** $3 \div 2$ is not a whole number.

19. **(a)** If a/b and d/b are defined, then there exist whole numbers r and s such that $a/b = r$ and $d/b = s$. But then $a = br$ and $d = bs$ so that $a + d = br + bs = b(r + s)$. This implies that $(a + d)/b$ is defined, and also that

$$\frac{a + d}{b} = r + s = \frac{a}{b} + \frac{d}{b}.$$

20. (a) 29 **21. (a)** 3^{35} **(c)** 3^{10} **(e)** $(yz)^3$ **22. (a)** 2^3 **(c)** 2^{10} **23. (a)** $4^2 = (2^2)^2 = 2^{(2 \cdot 2)} = 2^4$ The power operation is not commutative, since $2^3 = 8$ and $3^2 = 9$ so $2^3 \neq 3^2$. **24. (a)** 4 **(c)** 10 **25.** The large square has side length $a + b$ so its area is $(a + b)^2$. But the square is divided into four regions with areas a^2, ab, ab, and b^2, showing that the total area can also be expressed by $a^2 + 2ab + b^2$.

27. The large square has area $(a + b)^2$ and the small square has area $(a - b)^2$. Thus $(a + b)^2 + (a - b)^2$ gives twice the area of the small square plus the area of the region within the large square that is outside the small square. The same area is given by adding the areas of the two overlapped squares of area a^2 and the areas of the two small squares of area b^2, giving $2a^2 + 2b^2$. Therefore $(a + b)^2 + (a - b)^2 = 2a^2 + 2b^2$.

29. (a) $4 \cdot (5 - 2) = 4 \cdot 3 = 12$, $4 \cdot 5 - 4 \cdot 2 = 20 - 8 = 12$. Therefore, $4 \cdot (5 - 2) = 4 \cdot 5 - 4 \cdot 2$. **(b)** The diagram below makes it clear that

$$a(b - c) = ab - ac$$

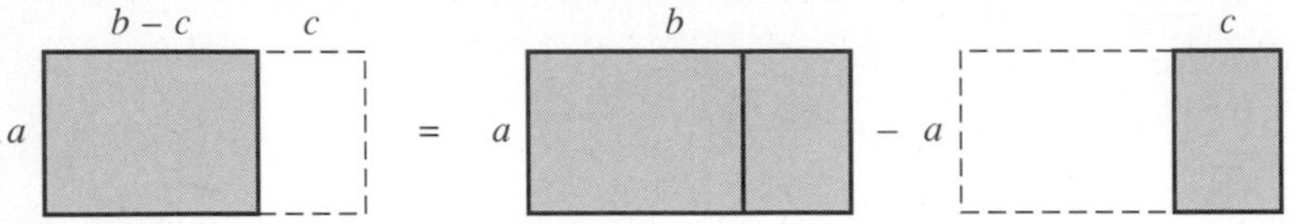

31. $2(1 + 2 + 3 + 4 + 5 + 6 + 7 + 8 + 9 + 10 + 11 + 12) = 2 \cdot \frac{12 \cdot 13}{2} = 156$ times

33.

	n	1	2	3	4	5	6	7	8	9	10	11	12	13	14	15
(a)	nth Oblong Number	2	6	12	20	30	42	56	72	90	110	132	156	182	210	240
(b)	nth Triangular Number	1	3	6	10	15	21	28	36	45	55	66	78	91	105	120

34. (a)

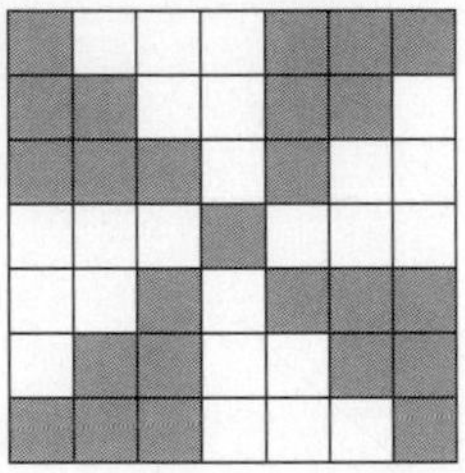

36. (b)
$$\begin{aligned}(10a + 5)^2 &= (10a + 5)(10a + 5)\\ &= (10a + 5) \cdot 10a + (10a + 5) \cdot 5\\ &= 100a^2 + 50a + 50a + 25\\ &= 100a^2 + 100a + 25\\ &= (a^2 + a)100 + 25\\ &= a \cdot (a + 1)100 + 25\end{aligned}$$

For any whole number a, $10a + 5$ is a whole number which ends in the digit 5 and conversely.

38. $2 + 7 = 9$, $7 + 2 = 9$, $9 - 7 = 2$, $9 - 2 = 7$ **40.** $2 \times 3 = 6$, $3 \times 2 = 6$

JUST FOR FUN Red and Green Jelly Beans (page 141)

They are the same. After the first move the jar labeled G contains 20 red jelly beans. Let r red and g green, with $r + g = 20$, be the number of red and green jelly beans moved back from jar G to jar R. Then jar R contains g green jelly beans and jar G contains $20 - r$ red jelly beans, but $g = 20 - r$ since $r + g = 20$.

Problem Set 2.5 (page 146)

1. (a) It depends on your point of view regarding the question, "Is a person his or her own friend?" If the answer is yes, the relation is reflexive. If the answer is no, the relation is not reflexive. Most dictionaries suggest that the answer should be no. **(c)** No, a person cannot be his own father. **(e)** No. Any person living in Washington D.C. is not related to him or herself, since the person does not live in a state. **3. (a)** Not transitive. A may be B's friend and B may be C's friend, but A and C need not be friends. Indeed they may not even know one another. **(b)** Not symmetric. Sue may be the sister of John, but John is not the sister of Sue. **(c)** Not symmetric. My dad is the father of me, but I am not the father of my dad. **(d)** Symmetric **(e)** Symmetric **4. (a)** Reflexive **(c)** Not reflexive; 0 is not related to 0 since $0 \cdot 0 \not> 0$. **5. (a)** Symmetric **(b)** Not symmetric; $2 \mid 6$ but $6 \nmid 2$. **(c)** Symmetric **7. (a)** Not symmetric, reflexive, nor transitive **8. (a)** Reflexive only

9. (a)

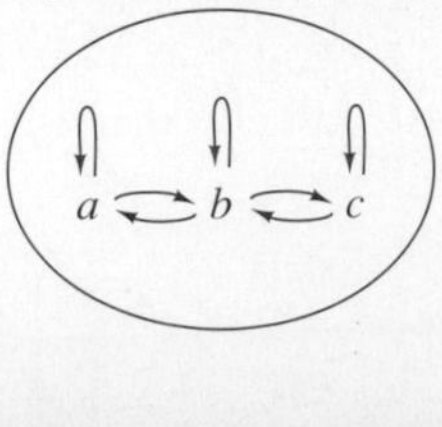

10. (a) The relation is not symmetric, so it is not an equivalence relation. **(c)** The relation is not symmetric, so it is not an equivalence relation. **11. (a)** $g(0) = 5$, $g(1) = 4$, $g(2) = 5$, $g(3) = 8$, $g(4) = 13$ **(b)** $\{4, 5, 8, 13\}$
13. (a) $A = 2x^2$ **(b)** $P = 6x$ **(c)** $D = \sqrt{5} \cdot x$ **15. (a)**

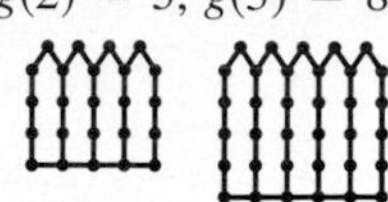

(b) 48, 63 **(c)** $d = f(n) = (n - 1)(n + 1) = n^2 - 1$ for all cases but $n = 1$.
17. (a) $19 \div 3 = 6 \text{ R } 1$ and $31 \div 3 = 10 \text{ R } 1$, so $19 \equiv 31$.
$28 \div 3 = 9 \text{ R } 1$ but $15 \div 3 = 5 \text{ R } 0$, so $28 \not\equiv 15$.
(b) Every element of E_0 is a multiple of 3 and so has a remainder of 0 when divided by 3. Every whole number that has a remainder of 0 when divided by 3 is in E_0, so E_0 is an equivalence class of the relation $\equiv$. **(c)** Every element of E_1 has a remainder of 1 when divided by 3, and every whole number that has a remainder of 1 when divided by 3 is in E_1, so E_1 is an equivalence class of the relation $\equiv$. **(d)** $E_2 = \{n \mid n = 3k + 2 \text{ and } k \in W\} = \{2, 5, 8, 11, \ldots\}$ **18. (b)** Suppose $a, b, c, d, e, f, \in W$. Then $(a, b) \square (a, b)$ since $a + b = b + a$, so the relation is reflexive. If $(a, b) \square (c, d)$, then $a + d = b + c$. But then $(c, d) = (a, b)$ and the relation is symmetric. If $(a, b) \square (c, d)$ and $(c, d) \square (e, f)$, then $a + b = b + c$ and $c + f = d + e$. Adding these equations shows that

$$a + d + c + f = b + c + d + e.$$

Subtracting d and c from both sides gives

$$a + f = b + e,$$

so $(a, b) \square (e, f)$. Thus, the relation is also transitive. Hence, the relation is an equivalence relation on $W \times W$.
(d) $E_1 = \{(1, 0), (2, 1), (3, 2), \ldots, (m, m - 1), \ldots\}$

20. (a)

n	1	2	3	4	5	6	7	8	9	10
T_n	1	2	4	7	13	24	44	81	149	274

21. (a)

Length of train, n	1	2	3	4	5
Number of trains, T_n	1	2	4	7	13

(c) $T_6 = 24$, $T_7 = 44$, $T_8 = 81$

22. (a) $S_5 = 1^2 + 2^2 + 3^2 + 4^2 + 5^2 = 55 = 10 \cdot 11 \cdot 12/24$
$S_6 = 1^2 + 2^2 + 3^2 + 4^2 + 5^2 + 6^2 = 91 = 12 \cdot 13 \cdot 14/24$
(b) $S_n = 2n(2n + 1)(2n + 2)/24$

23. (a)

Size of the Array	Number of Squares of Size 1 × 1	2 × 2	3 × 3	4 × 4	5 × 5	Total Number of Squares
1 × 1	1					1
2 × 2	4	1				5
3 × 3	9	4	1			14
4 × 4	16	9	4	1		30
5 × 5	25	16	9	4	1	55

24. (c) $(p + 1)(q + 1) + (p + q + 1)r = pq + pr + qr + p + q + r + 1$ **25.** $m = 2$, $b = 10$ **27.** $4.58\overline{3}$ sec.
29. $d = \dfrac{760t}{60 \cdot 60} \doteq \dfrac{t}{5}$ **30. (a)** \$7.15 **31. (a)** $g = 2w + 2h$ **32. (a)** 55 cents **33. (a)** \$7075 **(b)** \$28
34. (a) \$7078.98 **36.** 7501 boxes **38. (a)** 6 **(b)** 2 **(c)** 9 **(d)** 3

CLASSIC CONUNDRUM A Problem from the Dark Ages (page 153)

Give each of two of the sons five full flasks and five empty flasks and the other son ten half empty flasks. This gives each son ten half-flasks of contents and ten flasks.

Chapter 2 Review Exercises (page 155)

1. (a) $S = \{4, 9, 16, 25\}$
$P = \{2, 3, 5, 7, 11, 13, 17, 19, 23\}$
$T = \{2, 4, 8, 16\}$

(b) $\overline{P} = \{4, 6, 8, 9, 10, 12, 14, 15, 16, 18, 20, 21, 22, 24, 25\}$
$S \cap T = \{4, 16\}$
$S \cup T = \{2, 4, 8, 9, 16, 25\}$
$S - T = \{9, 25\}$

2.

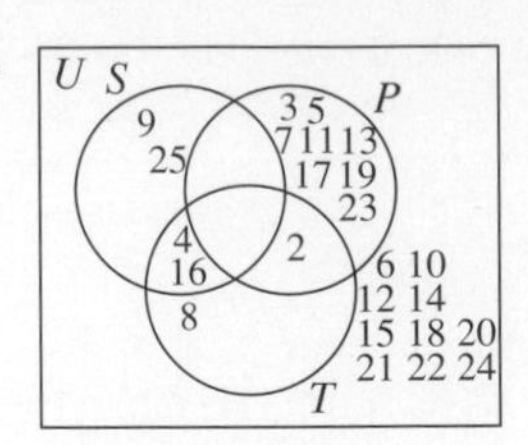

3. $S = \{$standard, legal$\}$, $C = \{$grey, tan, blue, yellow$\}$
$S \times C = \{$(standard, grey), (standard, tan), (standard, blue); (standard, yellow), (legal, grey), (legal, tan), (legal, blue), (legal, yellow)$\}$

4. $n(S) = 3$, $n(T) = 6$, $n(S \cup T) = 7$, $n(S \cap T) = 2$, $n(S - T) = 1$, $n(T - S) = 4$

5.

1	4	9	16	25	36	49	64	81	100
↕	↕	↕	↕	↕	↕	↕	↕	↕	↕
a	b	c	d	e	f	g	h	i	j

6. There is a one-to-one correspondence between the set of cubes and a proper subset. For example,

1	8	27	64	125	⋯	k^3	⋯
↕	↕	↕	↕	↕		↕	
1	64	729	4096	15,625	⋯	$(k^2)^3$	⋯

7. (a) Suppose $A = \{$a, b, c, d, e$\}$ and $B = \{$■, ★$\}$
Then $n(A) = 5$, $n(B) = 2$, $A \cap B = \varnothing$ and $n(A \cup B) = 5 + 2 = 7$.

(b)

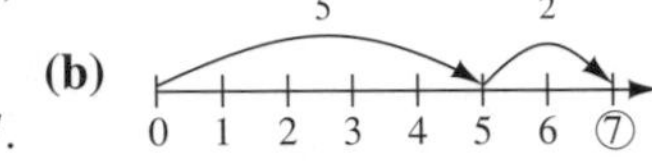

8. (a) Commutative property for addition: $7 + 3 = 3 + 7$ **(b)** Associative property for addition: $3 + (2 + 6) = (3 + 2) + 6$ **(c)** Additive identity property of zero: $7 + 0 = 7$

9. (a)

(b)

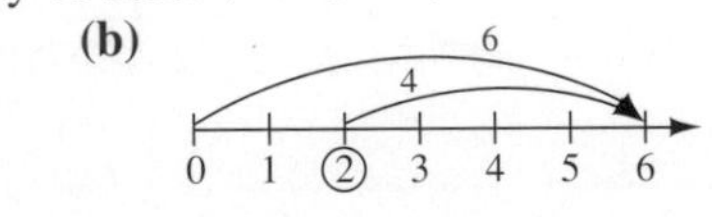

10. $6'' \times 6'' \times 8''$ **11.** eight rows, with seven full rows and eight soldiers in the back row

12. (a)

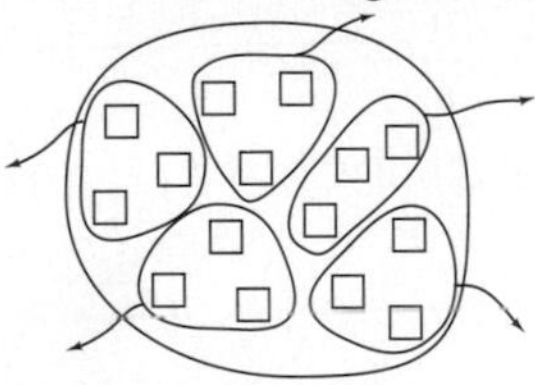

(b)

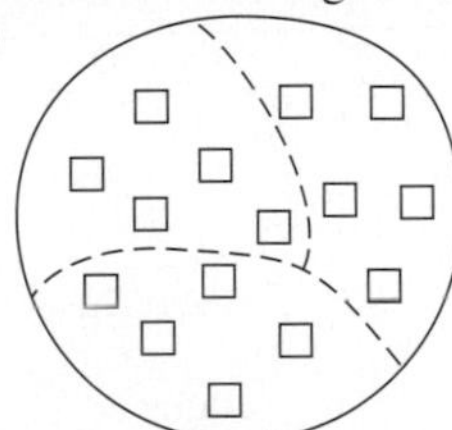

(c)

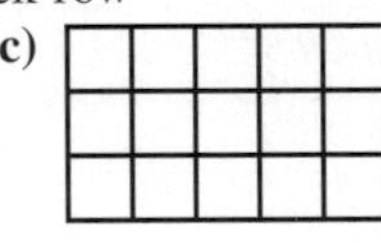

13. (a) Transitive **(b)** Reflexive, symmetric, transitive **(c)** Symmetric **14.** The relation is reflexive since any element belongs to the same subset as itself. The relation is symmetric since, if x is in the same subset as y, then y is in the same subset as x. The relation is transitive since, if x is in the same subset as y and y is in the same subset as z (and the subsets are mutually exclusive), then x must be in the same subset as z. Thus, □ is an equivalence relation on S with equivalence classes A, B, and C.

15. (a) $f(3) = 0$, $f(0.5) = -2.5$, $f(-2) = 20$ **(b)** $x = 0$ or $x = 3$ **16. (a)**

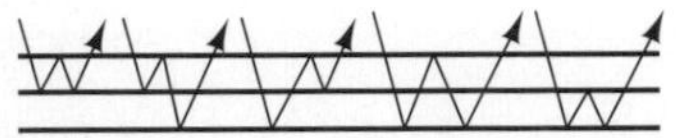

(b) Number of paths with n reflections = number of paths with $n - 1$ reflections + number of paths with $n - 2$ reflections. That is, the number of paths with n reflections is the Fibonacci number F_{n+2}.

Chapter 2 Test (page 156)

1. 15th—ordinal; 1040—nominal; \$253—cardinal

2. (a)

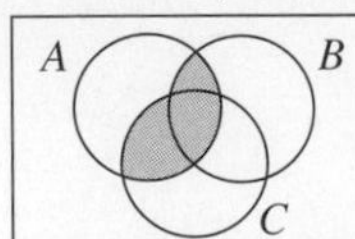

(b)

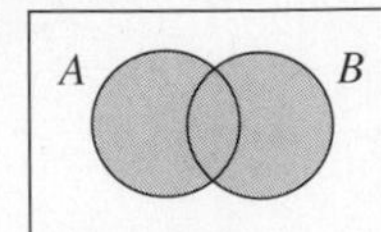

(c)

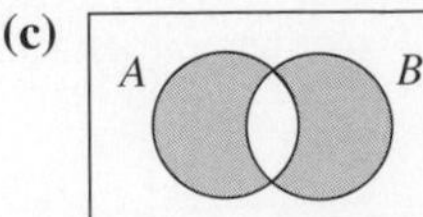

3. $A - B = \varnothing$ **4. (a)** 10 **(b)** 4 **(c)** 2 **(d)** 20
5. (a) Associative property of addition **(b)** Distributive property of multiplication over addition **(c)** Additive identity property of zero **(d)** Associative property of multiplication **6. (a)** 64 bottles **(b)** Grouping **7.** $n(A \cap B) = 5$ $n(\overline{A \cap B}) = 21$ **8.** 1, 3, 7, or 21 **9. (a)** $4 \times 2 = 8$ **(b)** $12 \div 3 = 4$ **(c)** $5 \cdot (9 + 2) = 5 \cdot 9 + 5 \cdot 2$ **(d)** $10 - 4 = 6$ **10. (a)** Yes **(b)** No

11. (a) 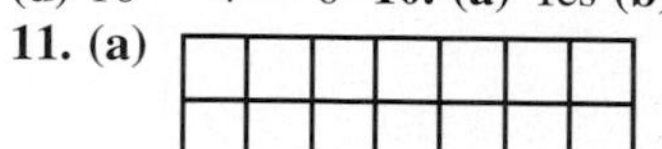 = + **(b)** 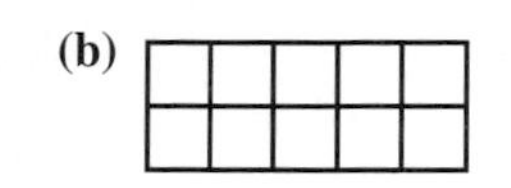 = 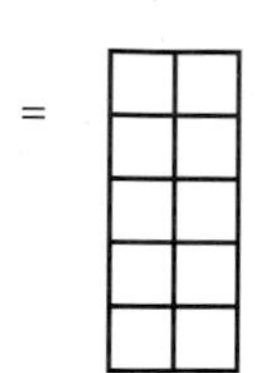

12. (a) Number line **(b)** Comparison **(c)** Missing addend **13. (a)** Let P, Q, and R be points in the plane. The relation is reflexive since $OP = OP$. The relation is symmetric since, if $OP = OQ$, then $OQ = OP$. Finally, the relation is transitive since, if $OP = OQ$ and $OQ = OR$, then $OP = OR$. Thus, □ is an equivalence relation of S. **(b)** Each equivalence class is the set of points in a circle with center at the origin O.
14. The relation is reflexive and transitive.

15. (a) 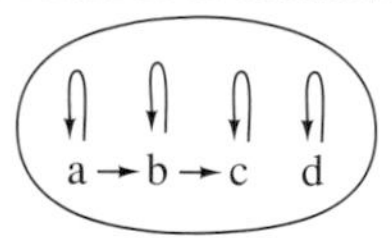

(b) **(c)** 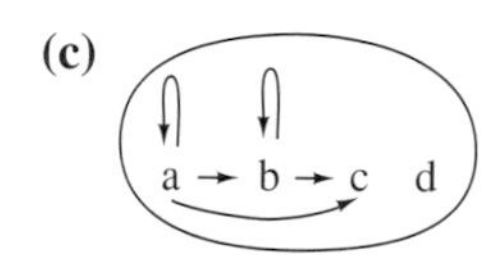

(d) 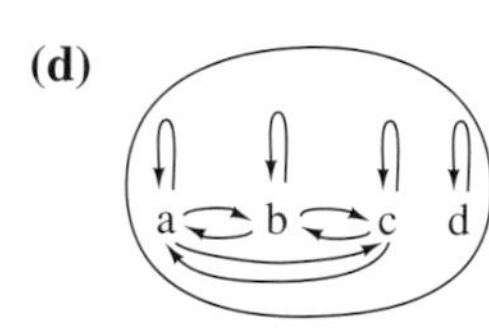

16. {1, 2, 5, 10} **17. (a)** 5 **(b)** The bee can enter cell $n + 2$ from either cell $n + 1$ or cell n. If there are $F(n + 1)$ ways to get to cell $n + 1$ and $F(n)$ ways to get to cell n, then the number of ways to get to cell $n + 2$ is $F(n + 1) + F(n)$. **(c)** 144

Chapter 3

JUST FOR FUN For Careful Readers (page 167)

1. None; it's a hole! **2.** Three inches, the difference between the radii of the two circles. **3.** A quarter and a penny. One was not a quarter but the other was.

Problem Set 3.1 (page 173)

1. (a) 2137 **(c)** 120,310 **(e)** 697 **(g)** 16.920 **2. (a)** ∩ |
3. (a) IX **4. (a)**
11.

15. 10 units will be traded for 1 strip, and 20 strips will be traded for 2 mats giving 5 mats, 5 strips, and 3 units.
18. (a) 14 **(c)** 13 **(e)** 217 **20. (a)** 413

22. (a) 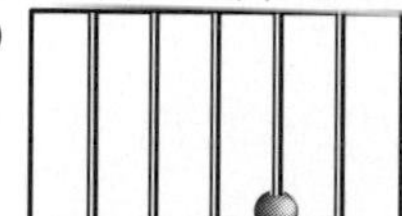 **(c)** **(e)** 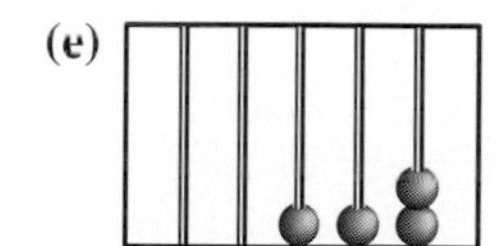 **(g)** 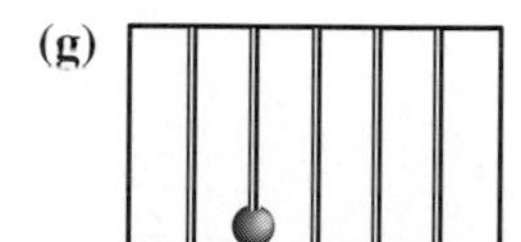

23. (a) The "fives" wire **25. (a)** Yes. No matter what number is represented, one can always be added by bringing over one more bead on the right-most wire and following the rule: If there are 5 beads showing on any wire, then they shall be moved to the back and one bead brought forward on the wire immediately to the left **27. (a)** {1, 2, 3, 4, 5, 7, 9} **(c)** {2, 4, 6, 8} **(e)** {1, 3} **(g)** $n(A) = 5$, $n(C) = 4$, $n(A \cup C) = 9$ **(i)** Since A and C are disjoint ($A \cap C = \varnothing$), $n(A \cup C) = n(A) + n(C)$. However, A and B have two elements in common ($n(A \cup B) = 2$) so the sum $n(A) + n(B)$ counts those elements twice; by subtracting $n(A \cap B)$ we get $n(A \cup B)$. **28. (a)** Distributive property of multiplication over addition **(c)** Commutative property of addition **29. (a)** $10 - 3 = 7$, $10 - 7 = 3$

Problem Set 3.2 (page 179)

1. 0 1 2 3 4 10 11 12 13 14 20 21 22 23 24 30 31 32 33 34 40 41 42 43 44 100

5. (a) 108 **(c)** 5 **(e)** 125 **6. (a)** 153 **(c)** 6 **(e)** 216 **7. (a)** 591 **(c)** 12 **(e)** 1728 **8. (a)** 2422_{five} **(c)** 10_{five} **9. (a)** 1330_{six} **(c)** 10_{six} **10. (a)** 1014_{five} **(c)** 23_{five} **11. (a)** 1707_{twelve} **(c)** 100_{twelve}
12. (a)

One Thousand Twenty-Fours	Five Hundred Twelves	Two Hundred Fifty-Sixes	One Hundred Twenty-Eights	Sixty-fours	Thirty-twos	Sixteens	Eights	Fours	Twos	Units
1024	512	256	128	64	32	16	8	4	2	1
2^{10}	2^9	2^8	2^7	2^6	2^5	2^4	2^3	2^2	2^1	2^0

(c) (i) 11000_{two} (ii) 10010_{two} (iii) 10_{two} (iv) 1000_{two} **14. (a)** $100{,}000{,}000_{\text{two}}$ **16. (a)** 000, 100, 010, 110, 001, 101, 011, 111 Append a 0 onto the end of the four 2-digit sequences, then append a 1 onto the end of the four 2-digit sequences. **(c)** 32 **17. (a)** 0, 1, 2, 3, 4, 5, 6, 7 **(c)** The whole numbers from 0 to $2^n - 1$. There are 2^n of these whole numbers, each with a different n-digit base two representation corresponding to one of the n-digit sequences of 0s and 1s.
20. (a) $51 \div 3 = 17$, $51 \div 17 = 3$ **21. (a)** $11 \times 31 = 341$, $341 \div 31 = 11$ **23. (a)** $n + 6$

JUST FOR FUN What's the difference? (page 192)

$$\begin{array}{r}91\\-\ 19\\\hline 72\end{array}\quad\begin{array}{r}95\\-\ 59\\\hline 36\end{array}\quad\begin{array}{r}42\\-\ 24\\\hline 18\end{array}\quad\begin{array}{r}62\\-\ 26\\\hline 36\end{array}\quad\begin{array}{r}74\\-\ 47\\\hline 27\end{array}$$

$$\begin{array}{r}61\\-\ 16\\\hline 45\end{array}\quad\begin{array}{r}82\\-\ 28\\\hline 54\end{array}\quad\begin{array}{r}81\\-\ 18\\\hline 63\end{array}\quad\begin{array}{r}32\\-\ 23\\\hline 9\end{array}\quad\begin{array}{r}54\\-\ 45\\\hline 9\end{array}$$

(a) All answers are evenly divisible by 9. **(b)** Yes. The first digit in the answer is 1 less than the difference between the digits in the subtrahend (the top number) in each subtraction problem and the second digit is 9 minus the first digit.
(c) In each case the sum of the digits is 9.

Problem Set 3.3 (page 195)

1. (a)

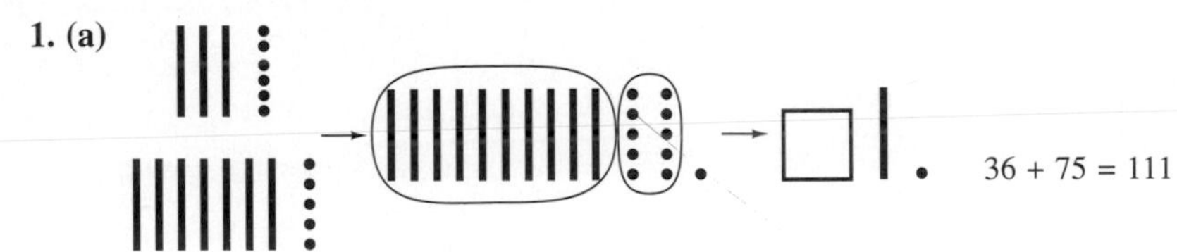

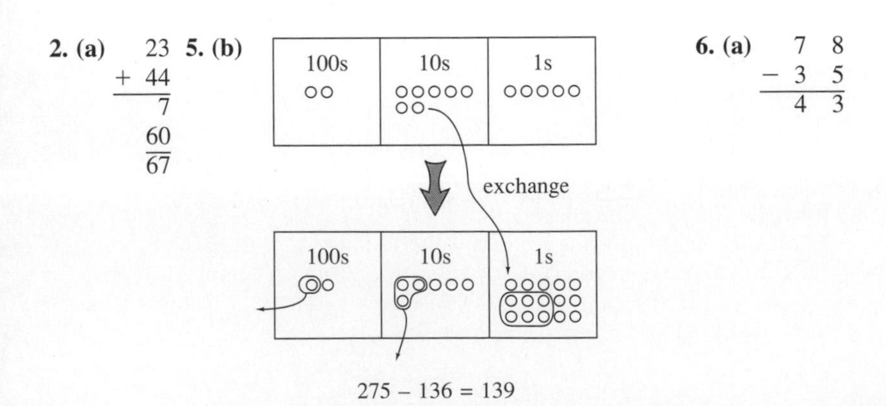

10. (a) 0, 1, 2, 3, 10, 11, 12, 13, 20, 21, 22, 23, 30, 31, 32, 33

12. (a) 1012 **(c)** 2120 **(e)** 111 **(g)** 113

13. (a)
$$\begin{array}{r} 6763 \\ +\ 5519 \\ \hline 12{,}282 \end{array}$$
(c)
$$\begin{array}{r} 881 \\ +\ 362 \\ \hline 1243 \end{array}$$
(e)
$$\begin{array}{r} 4002 \\ -\ 1843 \\ \hline 2159 \end{array}$$

14. (a)
$$\begin{array}{r} 2437 \\ 281 \\ +\ 3476 \\ \hline 6194 \end{array}$$
(c)
$$\begin{array}{r} 3891 \\ 2493 \\ +\ 5125 \\ \hline 11{,}509 \end{array}$$

15. (a)
$$\begin{array}{r} 835 \\ -\ 241 \\ \hline 594 \end{array}$$
(c)
$$\begin{array}{r} 7342 \\ -\ 6534 \\ \hline 808 \end{array}$$

16. (a) five **(c)** seven or greater **(e)** seven **(g)** twelve

19. (a)
$1 \cdot s = 19{,}998$ $6 \cdot s = 119{,}988$
$2 \cdot s = 39{,}996$ $7 \cdot s = 139{,}986$
$3 \cdot s = 59{,}994$ $8 \cdot s = 159{,}984$
$4 \cdot s = 79{,}992$ $9 \cdot s = 179{,}982$
$5 \cdot s = 99{,}990$ $10 \cdot s = 199{,}980$

(d) $14 \cdot s = 279{,}972$
$23 \cdot s = 459{,}954$
$32 \cdot s = 639{,}936$

(f) Arithmetic

21. (a) 59,341,586,218,343 **22. (a)** 41,290,291,107

23. (a) $1001 \div 91 = 11$, $1001 \div 11 = 91$, $91 \times 11 = 1001$

24. (a) $143 \times 7 = 1001$, $1001 \div 143 = 7$

26. (a) $A \cup B \cup C = \{1, 2, 3, 4, 5, 6, 7, 8\}$
$A \cap B = \{2, 4\}$
$A \cap C = \{3, 4, 5\}$
$B \cap C = \{4, 6\}$
$A \cap B \cap C = \{4\}$

JUST FOR FUN What's the Sum? (page 203)

$$\begin{array}{r} 91 \\ 19 \\ \hline 110 \end{array} \quad \begin{array}{r} 95 \\ 59 \\ \hline 154 \end{array} \quad \begin{array}{r} 42 \\ 24 \\ \hline 66 \end{array} \quad \begin{array}{r} 62 \\ 26 \\ \hline 88 \end{array} \quad \begin{array}{r} 74 \\ 47 \\ \hline 121 \end{array}$$

$$\begin{array}{r} 61 \\ 16 \\ \hline 77 \end{array} \quad \begin{array}{r} 82 \\ 28 \\ \hline 110 \end{array} \quad \begin{array}{r} 81 \\ 18 \\ \hline 99 \end{array} \quad \begin{array}{r} 32 \\ 23 \\ \hline 55 \end{array} \quad \begin{array}{r} 54 \\ 45 \\ \hline 99 \end{array}$$

(a) All the divisions come out even. **(b)** Yes. If the sum of the digits of the top number is d and $d < 10$, then the decimal representation of the sum in question is dd; that is 66, or 55, or whatever. If the sum of the two digits of the top number is the 2-digit number st, then the sum in question is $s(s + t)t$; that is $1(1 + 4)4 = 154$, $1(1 + 1)1 = 121$, and so on.

Problem Set 3.4 (page 206)

1. (a)

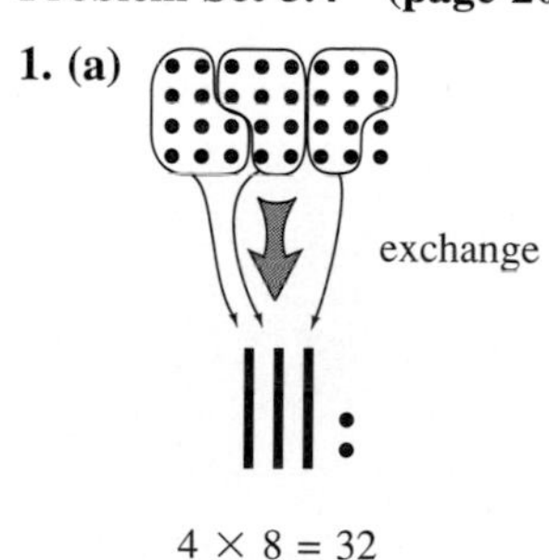

$4 \times 8 = 32$

3. (a) The number of hundreds in $30 \times 70 + 100$. **5. (a)** Distributive property of multiplication over addition **(c)** Associative property of addition

6.

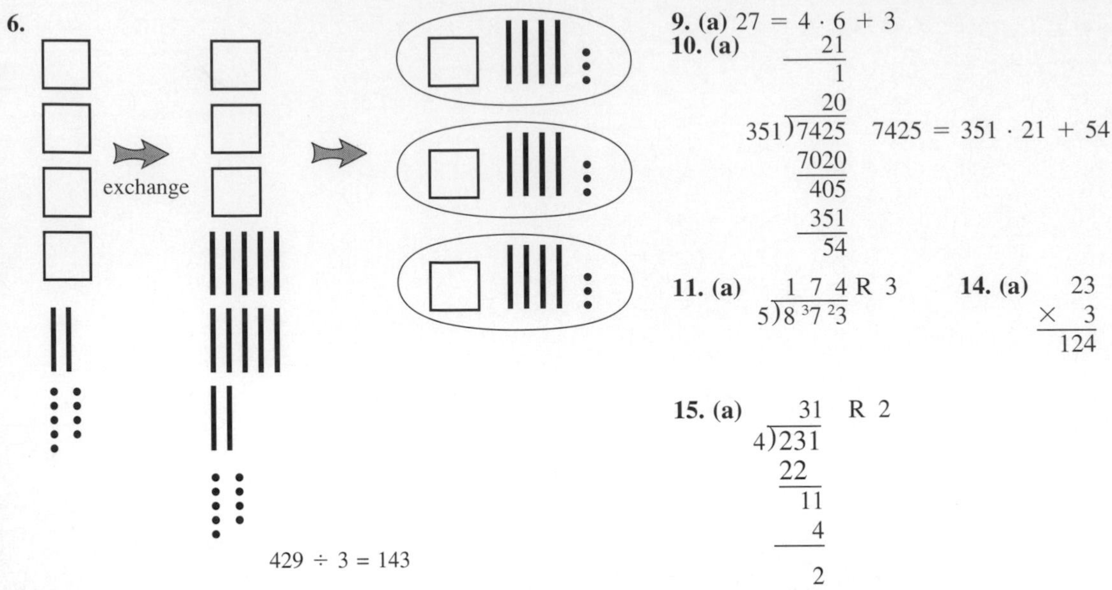

$429 \div 3 = 143$

9. (a) $27 = 4 \cdot 6 + 3$

10. (a)

$$\begin{array}{r} 21 \\ \hline 1 \\ 20 \\ 351\overline{)7425} \\ \underline{7020} \\ 405 \\ \underline{351} \\ 54 \end{array} \qquad 7425 = 351 \cdot 21 + 54$$

11. (a) $\begin{array}{r} 1\;7\;4 \text{ R } 3 \\ 5\overline{)8\;{}^{3}7\;{}^{2}3} \end{array}$

14. (a) $\begin{array}{r} 23 \\ \underline{\times\ \ 3} \\ 124 \end{array}$

15. (a)

$$\begin{array}{r} 31 \quad \text{R } 2 \\ 4\overline{)231} \\ \underline{22} \\ 11 \\ \underline{4} \\ 2 \end{array}$$

17. (a) $34 \cdot 54 = (17 \cdot 2) \cdot 54 = 17 \cdot (2 \cdot 54) = 17 \cdot 108$ since 2 evenly divides 34.

19. (a) Yes. This is simply a rearrangement of the rows in the usual algorithm. Thus, we usually write

$$\begin{array}{r} 374 \\ \underline{23} \\ 1122 \\ \underline{748\ } \\ 8602 \end{array}$$

Here we multiply by 20 first and then 3 to obtain

$$\begin{array}{r} 374 \\ \underline{\times\ \ 23} \\ 748\ \\ \underline{1122} \\ 8602 \end{array}$$

(b)

$$\begin{array}{r} 285 \\ \underline{\times\ 362} \\ 855 \\ 1710\ \\ \underline{570\ \ } \\ 103{,}170 \end{array}$$

20. (a) 1256_{seven}, $1 \cdot 7^3 + 2 \cdot 7^2 + 5 \cdot 7 + 6 = 482_{\text{ten}}$ **(c)** 111100010_{two}, $1 \cdot 2^8 + 1 \cdot 2^7 + 1 \cdot 2^6 + 1 \cdot 2^5 + 1 \cdot 2 = 482_{\text{ten}}$ **21. (a)** $\begin{array}{r} 7531 \\ \underline{\times\ \ \ \ 9} \\ 67{,}779 \end{array}$ **24. (a)** 912,000

25. (a) quotient is 541, remainder is 72

26. (a)

$$\begin{array}{rl} 634 = & 6 \text{ hundreds} + 3 \text{ tens} + 4 \text{ ones} \\ \underline{+\ 163 =} & \underline{1 \text{ hundred} + 6 \text{ tens} + 3 \text{ ones}} \\ = & 7 \text{ hundreds} + 9 \text{ tens} + 7 \text{ ones} \\ = & 797 \end{array}$$

(c)

$$\begin{array}{rl} 363 = & 3 \text{ hundreds} + 6 \text{ tens} + 3 \text{ ones} \\ \underline{+\ 532 =} & \underline{5 \text{ hundreds} + 3 \text{ tens} + 2 \text{ ones}} \\ = & 8 \text{ hundreds} + 9 \text{ tens} + 5 \text{ ones} \\ = & 895 \end{array}$$

(e)

$$\begin{array}{rl} 725 = & 7 \text{ hundreds} + 2 \text{ tens} + 5 \text{ ones} \\ \underline{-\ 413 =} & \underline{-(4 \text{ hundreds} + 1 \text{ ten} + 3 \text{ ones})} \\ = & 3 \text{ hundreds} + 1 \text{ ten} + 2 \text{ ones} \\ = & 312 \end{array}$$

27. (a)
$$\begin{array}{rl} 374 & = 3 \text{ hundreds} + 7 \text{ tens} + 4 \text{ ones} \\ +\ 483 & = 4 \text{ hundreds} + 8 \text{ tens} + 3 \text{ ones} \\ \hline & = 7 \text{ hundreds} + 15 \text{ tens} + 7 \text{ ones} \\ & = 8 \text{ hundreds} + 5 \text{ tens} + 7 \text{ ones} \\ & = 857 \end{array}$$

(c)
$$\begin{array}{rl} 724 & = 7 \text{ hundreds} + 2 \text{ tens} + 4 \text{ ones} \\ +\ 532 & = 5 \text{ hundreds} + 3 \text{ tens} + 2 \text{ ones} \\ \hline & = 12 \text{ hundreds} + 5 \text{ tens} + 6 \text{ ones} \\ & = 1 \text{ thousand} + 2 \text{ hundreds} + 5 \text{ tens} + 6 \text{ ones} \\ & = 1256 \end{array}$$

(e)
$$\begin{array}{rl} 367 & = 3 \text{ hundreds} + 6 \text{ tens} + 7 \text{ ones} \\ -\ 249 & = -(2 \text{ hundreds} + 4 \text{ tens} + 9 \text{ ones}) \\ \hline 367 & = 3 \text{ hundreds} + 5 \text{ tens} + 17 \text{ ones} \\ -\ 249 & = -(2 \text{ hundreds} + 4 \text{ tens} + 9 \text{ ones}) \\ \hline & = 1 \text{ hundred} + 1 \text{ ten} + 8 \text{ ones} \\ & = 118 \end{array}$$

28. (a)
$$\begin{array}{rl} 213 & = 2 \text{ twenty-fives} + 1 \text{ five} + 3 \text{ ones} \\ +\ 131 & = 1 \text{ twenty-five} + 3 \text{ fives} + 1 \text{ one} \\ \hline & = 3 \text{ twenty-fives} + 4 \text{ fives} + 4 \text{ ones} \\ & = 344_{\text{five}} \end{array}$$

(c)
$$\begin{array}{rl} 142 & = 1 \text{ twenty-five} + 4 \text{ fives} + 2 \text{ ones} \\ +\ 123 & = 1 \text{ twenty-five} + 2 \text{ fives} + 3 \text{ ones} \\ \hline & = 2 \text{ twenty-fives} + 6 \text{ fives} + 5 \text{ ones} \\ & = 2 \text{ twenty-fives} + 1 \text{ twenty-five} + 1 \text{ five} + \\ & \quad 1 \text{ five} + 0 \text{ ones} \\ & = 3 \text{ twenty-fives} + 2 \text{ fives} + 0 \text{ ones} \\ & = 320_{\text{five}} \end{array}$$

(e)
$$\begin{array}{rl} 344 & = 3 \text{ twenty-fives} + 4 \text{ fives} + 4 \text{ ones} \\ -\ 232 & = -(2 \text{ twenty-fives} + 3 \text{ fives} + 2 \text{ ones}) \\ \hline & = 1 \text{ twenty-five} + 1 \text{ five} + 2 \text{ ones} \\ & = 112_{\text{five}} \end{array}$$

30. (a) 112 **(c)** 241 **(e)** 233 **(g)** 12 R 13

Problem Set 3.5 (page 224)

1. (c) 92 **(e)** 240 **2. (c)** 138 **(e)** 576 **3. (a)** 787 **(e)** 1026 **4. (a)** 240,000 **5. (a)** 900 **(c)** 27,000,000 **6. (c)** $100 < 678 - 431 < 300$ **(e)** $140{,}000 < 7403 \cdot 28 < 240{,}000$ **8. (a)** 3330 **9. (a)** 52,000 **(c)** 48,000 **(e)** 13,000 **10. (a)** 90,000 **11. (a)** 750 **13. (a)** 3 **14. (a)** 27,451 **16. (a)** $(24) \cdot (678) = 16{,}272$ **(c)** $(2467) \cdot (8) = 19{,}736$ **17. (a)** $(88) + 8 + 8 + 8 + 8 + 8 + 8 = 136$ **(c)** $(888) + 8 + 8 + 8 + 8 + 8 = 928$ **18. (a)** $(844{,}422) \div 1 = 844{,}422$ **(c)** $(84) \div (44 \div (22 \div 1)) = 42$

21. (a)
$$\begin{array}{r} 2742 \\ 415 \\ 6943 \\ +\ 2718 \\ \hline 12{,}818 \end{array}$$
22. (a)
$$\begin{array}{r} 2734 \\ -\ 2643 \\ \hline 91 \end{array}$$
23. (a)
$$\begin{array}{r} 347 \\ 42 \\ \hline 694 \\ 1388 \\ \hline 14{,}574 \end{array}$$

Problem Set 3.6 (page 238)

1. (a) 641 **(c)** 101,388 **(e)** 770 **(g)** 770 **(i)** 237 **2. (a)** 98,915 **(c)** 32 **3. (a)** 49 **(c)** 13 **4. (a)** 841 **(c)** 33

5. (a) $\boxed{\text{ON/AC}}\ \boxed{(}\ 784\ \boxed{\sqrt{}}\ \boxed{-}\ 91\ \boxed{\div}\ 13\ \boxed{)}\ \boxed{\div}\ \boxed{(}\ 8\ \boxed{\times}\ 49\ \boxed{-}\ 11\ \boxed{\times}\ 35\ \boxed{)}\ \boxed{=}$

7. (a) $1831 - (17 \times 28) + 34$

13. 390 **16. (a)** $\dfrac{\left(\dfrac{\sqrt{5}+1}{2}\right)^3}{\sqrt{5}}$ **17. (a)** 1 **(c)** 2 **(e)** 5 **18. (a)** 1 **(c)** 4 **(e)** 11 **21. (b)** 141,422,324; some calculators may run out of display space. In this case, L_{37} and L_{38} may be added by hand to obtain L_{39}.

23. $17 + 18 = 35$; $35 - 17 = 18$; $35 - 18 = 17$ **25.** $11 \cdot 27 = 297$; $297 \div 27 = 11$; $297 \div 11 = 27$

27.

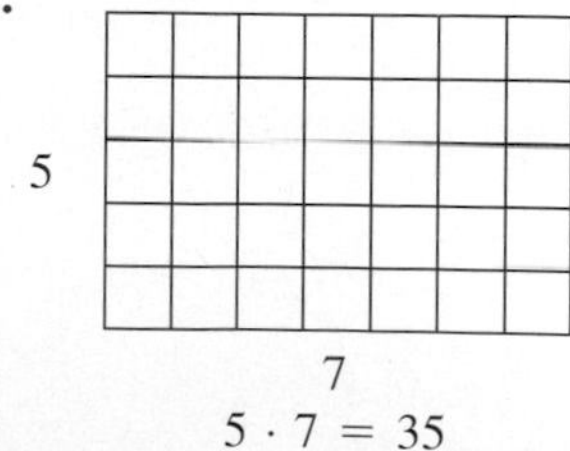

$5 \cdot 7 = 35$

CLASSIC CONUNDRUM A Cryptarithm (page 241)

$$\begin{array}{r} \text{FOOD} \\ +\ \text{FAD} \\ \hline \text{DIETS} \end{array}$$

Note that O is "oh" and not necessarily zero. We will use Ø to represent zero.
Since the sum of two different digits is at most 17 and D is a carry, it follows that D = 1 and S = 2. Also, since the carry to the last column comes from the fourth column, it must be the case that F = 9 and I = Ø. This gives

$$\begin{array}{r} 9\text{OO}1 \\ +\ 9\text{A}1 \\ \hline 1\text{ØET}2 \end{array}$$

If there is a carry from the second to the third column, then O = E and this is a contradiction. Therefore, A + O must be at most 9. Also, the only remaining digits from which to choose the remaining letter values are 3, 4, 5, 6, 7, and 8. Thus, if O = 3, A must be less than 7. If O = 4, A must be less than 6. If O = 5, A must be less than 5. If O = 6, A must be 3; and if O > 6, there is no possibility for A. We consider the cases O = 6, 5, 4, and 3 separately.

Case 1: O = 6, A = 3, $\begin{array}{r} 9661 \\ +\ 931 \\ \hline 1\text{ØE}92 \end{array}$

This gives T = 9 = F and so is impossible.

Case 2: O = 5, A = 3 or 4, $\begin{array}{r} 9551 \\ +\ 931 \\ \hline 1Ø482, \end{array}$ $\begin{array}{r} 9551 \\ +\ 941 \\ \hline 1Ø492 \end{array}$

If A = 4, then T = 9 = D, an impossibility. If A = 3, we obtain a solution.

Case 3: O = 4, A = 3 or 5, $\begin{array}{r} 9441 \\ +\ 931 \\ \hline 1Ø372, \end{array}$ $\begin{array}{r} 9441 \\ +\ 951 \\ \hline 1Ø392 \end{array}$

In the first case A = 3 = E and in the second case T = 9 = F. Since both are impossible, this case yields no solution.

Case 4: O = 3, A = 4, 5, or 6, $\begin{array}{r} 9331 \\ +\ 941 \\ \hline 1Ø272, \end{array}$ $\begin{array}{r} 9331 \\ +\ 951 \\ \hline 1Ø282, \end{array}$ $\begin{array}{r} 9331 \\ +\ 961 \\ \hline 1Ø292 \end{array}$

All possibilities here result in E = 2 = S. Since this is impossible, Case 4 yields no solution. Thus, the only solution is F = 9, O = 5, D = 1, A = 3, I = Ø, E = 4, T = 8, and S = 2 as noted in Case 2.

Chapter 3 Review Exercises (page 243)

1. **(a)** 2353 **(b)** 58,331 **(c)** 1998
2.

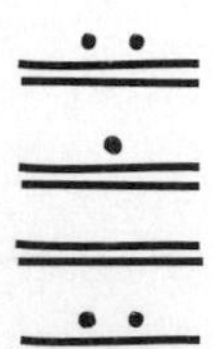

3. Exchange 30 units for 3 strips, then exchange all 30 strips for 3 mats. The result is 8 mats, 0 strips, and 2 units.
4. **(a)** 45_{ten} **(b)** 181_{ten} **(c)** 417_{ten} **5.** **(a)** 2122_{five} **(b)** 100011111_{two} **(c)** 560_{seven}

6.

$$\begin{array}{r} 47 \\ +25 \\ \hline 72 \end{array}$$

7. (a) $\begin{array}{r} 42 \\ +\ 54 \\ \hline 6 \\ 90 \\ \hline 96 \end{array}$ **(b)** $\begin{array}{r} 47 \\ +\ 35 \\ \hline 12 \\ 70 \\ \hline 82 \end{array}$ **(c)** $\begin{array}{r} 59 \\ +\ 63 \\ \hline 12 \\ 110 \\ \hline 122 \end{array}$

8. (a)

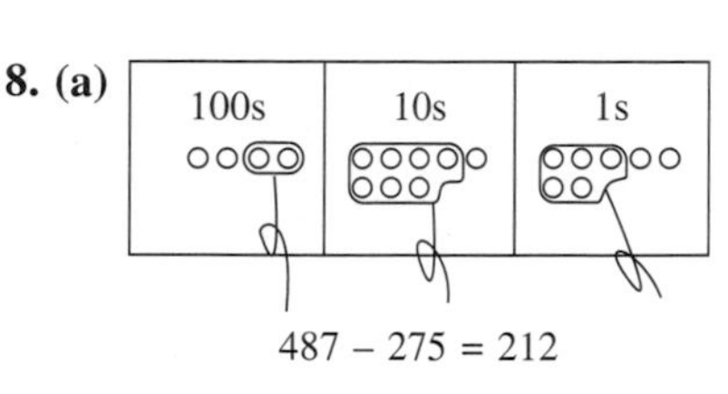

487 – 275 = 212

(b)

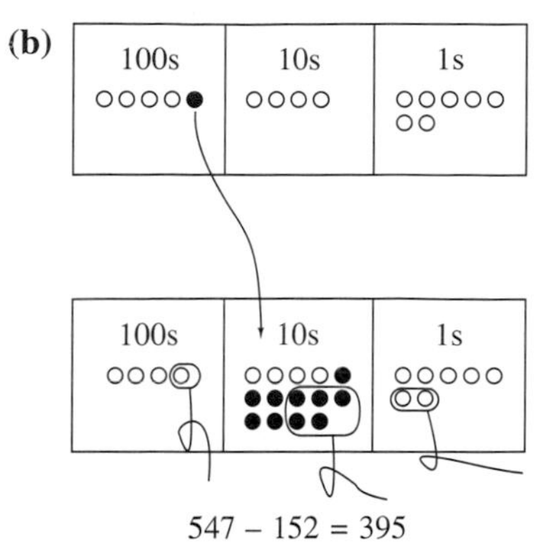

547 – 152 = 395

9. (a) $\begin{array}{r} 2433 \\ +\ 141 \\ \hline 3124 \end{array}$ **(b)** $\begin{array}{r} 2433 \\ -\ 141 \\ \hline 2242 \end{array}$ **(c)** $\begin{array}{r} 243 \\ \times\ 42 \\ \hline 1041 \\ 21320 \\ \hline 22{,}411 \end{array}$

10. (a) $\begin{array}{r} 357 \\ \times\ 4 \\ \hline 28 \\ 200 \\ 1200 \\ \hline 1428 \end{array}$ **(b)** $\begin{array}{r} 642 \\ \times\ 27 \\ \hline 14 \\ 280 \\ 4200 \\ 40 \\ 800 \\ 12000 \\ \hline 17{,}334 \end{array}$

11. (a) $\begin{array}{r} 127 \\ \hline 7 \\ 20 \\ 100 \\ 7\overline{)895} \\ 700 \\ \hline 195 \\ 140 \\ \hline 55 \\ 49 \\ \hline 6 \end{array}$ **(b)** $\begin{array}{r} 79 \\ \hline 9 \\ 70 \\ 347\overline{)27{,}483} \\ 24{,}290 \\ \hline 3193 \\ 3123 \\ \hline 70 \end{array}$

12. (a) $5\overline{)27{,}436}$ = 5487 R 1 **(b)** $8\overline{)39{,}584}$ = 4948 R 0

13. (a) 2121_{five} **(b)** 2023221_{five} **14.**

~~42~~	~~35~~
21	70
~~10~~	~~140~~
5	280
~~2~~	~~560~~
1	1120
	1470

15. (a) 300,000 **(b)** 270,000 **(c)** 275,000 **16. (a)** $1000 < 657 + 439 < 1200$ **(b)** $100 < 657 - 439 < 300$ **(c)** $240{,}000 < 657 \cdot 439 < 350{,}000$ **(d)** $50 < 1657 \div 23 < 90$ **17. (a)** 1100 **(b)** 300 **(c)** 280,000 **(d)** 100 **18.** 2 **19. (a)** 8088 **(b)** 49,149 **20. (a)** 11, 18, 29, 47 **(b)** The integer nearest $[(1 + \sqrt{5})/2] \cdot L_n = L_{n+1}$ for all integers n. **(c)** No. It does not hold for L_1, L_2, and L_3. **(d)** The nearest integer to $[(1 + \sqrt{5})/2] \cdot L_n$ is L_{n+1} for $n \geq 4$.

Chapter 3 Test (page 244)

1. (a) 197_{ten} **(b)** 207_{ten} **(c)** 558_{ten}
2. (a) 2111_{five} **(b)** 100011001_{two} **(c)** $1E5_{\text{twelve}}$ **3. (a)** 340 **(b)** 144 **(c)** 23111

4.

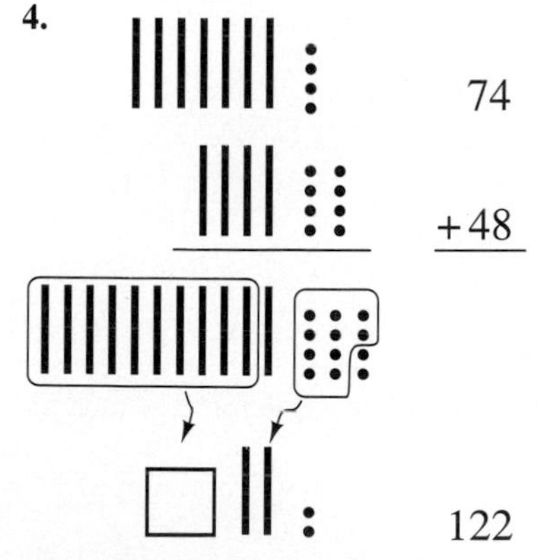

5.

6. $\begin{array}{r} 2837 \\ +\ 7224 \\ \hline 10{,}061 \end{array}$ **7.** $\begin{array}{r} 8236 \\ -\ 3542 \\ \hline 4694 \end{array}$

8. (a) 3,000,000 **(b)** 3,400,000 **(c)** 3,380,000 **(d)** 3,377,000

9. 4800 **10.** $\begin{array}{r} 751 \\ \times\ 93 \\ \hline 69{,}843 \end{array}$ **11.** $\begin{array}{r} 468 \\ \times\ 20 \\ \hline 9360 \end{array}$

12. (a) 1, 2, 5, 12 **(b)** 29, 70 **(c)** $f_n = 2f_{n-1} + f_{n-2}$ for $n \geq 3$.

13. (a) 1 and 1
9 and 9
36 and 36
100 and 100

(b) 1, 3, 6, 10 **(c)** $\left(\frac{n(n+1)}{2}\right)^2$ **14.** 1575 **15.** 1,464,843

16. Answers may vary. One possibility is as follows:

[ON/AC] 2 [M+] 5 [M+] [x⇄M] [M+] [x⇄M] [M+] ...

Chapter 4

JUST FOR FUN A Problem of Punctuation (page 260)

That that is, is; that that is not, is not.

Problem Set 4.1 (page 260)

1. (a) 4 × 9 rectangle

$36 = 4 \cdot 9$
4 divides 36

3. (a) 0, 8, 16, 24, 32, 40, 48, 56, 64, 72

5. (a)

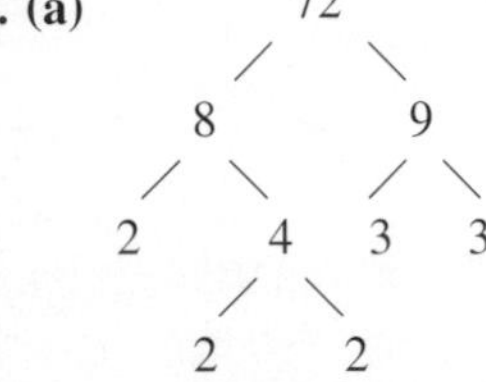

(c) 264 → 8, 33; 8 → 2, 4; 4 → 2, 2; 33 → 3, 11

6. (a)
$5\overline{)25}$ = 5
$2\overline{)50}$
$2\overline{)100}$
$7\overline{)700}$
$700 = 7 \cdot 2 \cdot 2 \cdot 5 \cdot 5$

(c)
$3\overline{)6}$ = 2
$3\overline{)18}$
$5\overline{)90}$
$5\overline{)450}$
$450 = 5 \cdot 5 \cdot 3 \cdot 3 \cdot 2$

7. (a) 1, 2, 3, 4, 6, 8, 12, 16, 24, 48 **8. (a)** $48 = 2^4 \cdot 3^1$ **(c)** $2250 = 2^1 \cdot 3^2 \cdot 5^3$ **9. (a)** Yes, because $28 = 2^2 \cdot 7^1$ so all the prime factors of 28 appear in a and to at least as high a power. **(c)** $2^1 \cdot 7^2$

12. No. For example, $10 = 2 \cdot 5$ with 2 and 5 both primes. Yet $5 > 3.162 \cdots = \sqrt{10}$.

13. (a) True. $n \cdot 0 = 0$ for every natural number n. **(c)** True. $1 \cdot n = n$ for every natural number n. **(e)** False. $0 \div 0 = q$ if, and only if, $0 \cdot q = 0$ for a unique integer q. However, this is true for *every* integer q.

15. Yes. If $p \mid bc$ then p must appear in the prime factorization of the product bc and hence in the prime factorization of b or c. But then $p \mid b$ or $p \mid c$ as claimed. **18.** If $m \mid n$, then $n = mk$ for some natural number k. Also, if $n \mid m$ then $m = nq$ for some natural number q. Hence, $n = mk = nqk$ so $qk = 1$. Since q and k are both natural numbers, $q = 1$ and $k = 1$. Thus, $m = n$. **21.** No. For example, $2 \mid (3 + 5)$ but $2 \nmid 3$ and $2 \nmid 5$. **23. (a)** $N_4 = N_2 \cdot 101$, $N_6 = N_2 \cdot 10101$, $N_8 = N_2 \cdot 1010101$ **25. (a)** No. **(c)** Yes. **26. (a)** 89, 144, 233, 377, 610, 987, 1597, 2584, 4181, 6765

(b) F_3, F_6, F_9, F_{12}, F_{15}, and F_{18}. It appears that $F_3 \mid F_n$ if, and only if, $3 \mid n$. Indeed, since $F_3 = 2$ and the sum of two odd numbers is even and the sum of an even and an odd number is odd, it follows that the Fibonacci sequence follows the pattern odd, odd, even, odd, odd, even, and so on without end. Thus, every third number in the sequence is divisible by 2 and conversely. **29. (a)** $2^2 \cdot 137^1$ **(c)** $2^1 \cdot 137^1$ **(e)** $(2^1 \cdot 137^1) \mid (2^2 \cdot 137^1)$ and $(2^3 \cdot 3^2 \cdot 13^1) \mid (2^3 \cdot 3^2 \cdot 7^2 \cdot 13^1)$
30. (a) $2^2 \cdot 3^2 \cdot 7^2 \cdot 13^2$ **(c)** $3^4 \cdot 5^6$ **(e)** The exponents in the prime power representation are even.
31. (a) False. The sum of two odd natural numbers is an even natural number. **(c)** True. $a + b = b + a$ if a and b are any natural numbers. **(e)** True. $a + (b + c) = (a + b) + c$ if a, b, and c are any natural numbers. **(g)** True. $a(b + c) = ab + ac$ if a, b, and c are any natural numbers. **(i)** True. 1 is an element of S.
32. (a) 299 **33. (a)** 19, 23, 27 **(c)** 15, 21, 28 **(e)** 48, 96, 192 **34. (a)** 1378

Problem Set 4.2 (page 269)

1. (a) Divisible by 2 and 3 **(c)** Divisible by 5 **2. (a)** 1554 **(c)** None **3. (a)** Divisible by 7 and 11 **(c)** Divisible by 7 and 13 **4. (a)** 539 **(c)** None **6. (a)** True **(c)** True **(e)** False **8. (a)** For any palindromc with an even number of digits, the digits in the odd positions are the same as the digits in the even positions, but with the order reversed. Thus, the difference of the sums of the digits in the even and odd position is zero which is divisible by 11.
9. When testing for divisibility by 7, 11, and 13 the digits of the number are broken up into 3-digit groups. If a number has form *abc,abc*, then the difference in the sums of the 3-digit numbers in odd positions and even positions will be zero, which is divisible by each of 7, 11, and 13. **12.** No. He or she could have made other kinds of errors that by chance resulted in the record being out of balance by an amount that is a multiple of 9. **13. (e)** No.
14. (a) $126 \div 9 = 14$
$10{,}206 \div 9 = 1134$
$1{,}002{,}006 \div 9 = 111{,}334$
$100{,}020{,}006 \div 9 = 11{,}113{,}334$
$10{,}000{,}200{,}006 \div 9 = 1{,}111{,}133{,}334$
$1{,}000{,}002{,}000{,}006 \div 9 = 111{,}111{,}333{,}334$
16. Suppose d divides b and c then $b = dm$ and $c = dn$ for some natural numbers m and n. But $a = b + c = dm + dn = d(m + n)$. So d also divides a. Now suppose d divides a and b. (This is the same as if d divided a and c.) Then $a = dq$ and $b = dr$ for some natural numbers q and r. If $a = b + c$, then $c = a - b = dq - dr = d(q - r)$. So d also divides c. **18. (a)** $2^7 \cdot 3^2 \cdot 7^1$

19. (a)

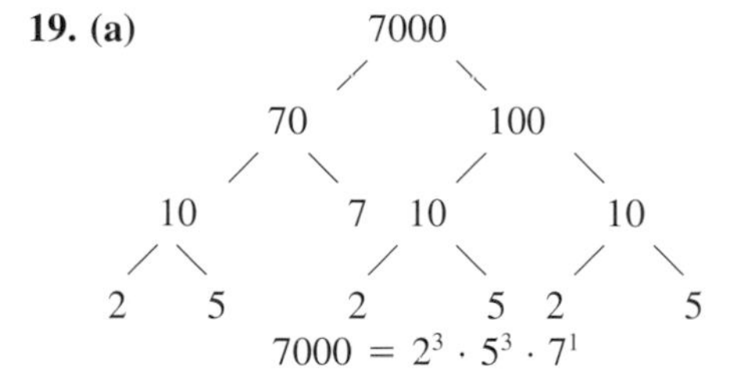

$7000 = 2^3 \cdot 5^3 \cdot 7^1$

21. $2^2 \cdot 3^2 \cdot 5^1 \cdot 7^1$

JUST FOR FUN Making a Chain (page 271)

Four. Open all four of the links in one section of chain. These can then be used to connect the remaining five sections of chain into a single chain of 24 links.

JUST FOR FUN A Weighty Matter (page 274)

If the basketball weighs 21 ounces plus half its own weight, then 21 ounces must be half the weight of the ball. Therefore, the basketball weighs 42 ounces.

Problem Set 4.3 (page 284)

1. (a)

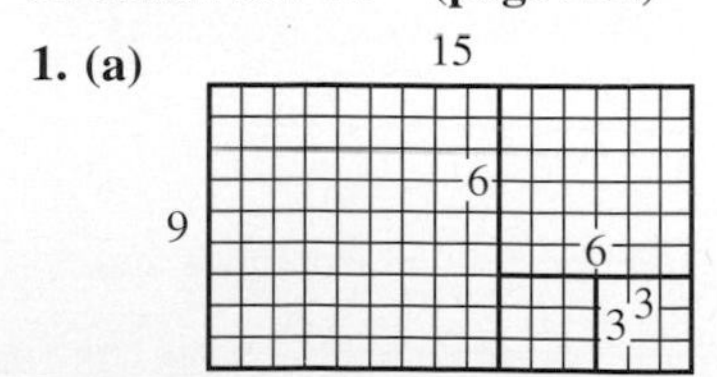

The 3 by 3 square is the largest square that will tile a 9 by 15 rectangle.

(b)

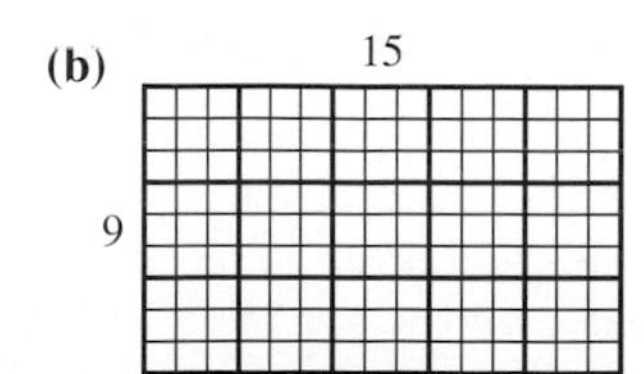

3.

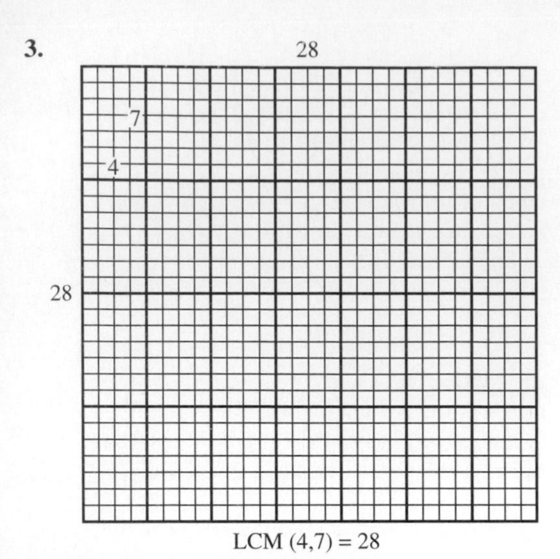

5. **(a)** 3 **6.** **(a)** 216 **7.** **(a)** GCD(24, 27) · LCM(24, 27) = 3 · 216 = 648 = 24 · 27
8. **(a)** GCD(r, s) = $2^1 \cdot 3^1 \cdot 5^2$ = 150, LCM(r, s) = $2^2 \cdot 3^3 \cdot 5^3$ = 13,500
10. **(a)** GCD(a, b, c) = $2^0 \cdot 3^1 \cdot 5^1 \cdot 7^0$ = 15, LCM(a, b, c) = $2^2 \cdot 3^3 \cdot 5^3 \cdot 7^1$ = 94,500
11. **(a)** D_{18} = {1, 2, 3, 6, 18}
D_{24} = {1, 2, 3, 4, 6, 8, 12, 24}
D_{12} = {1, 2, 3, 4, 6, 12}
GCD (18, 24, 12) = 6
M_{18} = {0, 18, 36, 54, 72, 90, . . .}
M_{24} = {0, 24, 48, 72, 96, . . .}
M_{12} = {0, 12, 24, 36, 48, 60, 72, 84, . . .}
LCM (18, 24, 12) = 72
13. **(a)** GCD(24, 18) = 6, GCD(GCD(24, 18), 12) = GCD(6, 12) = 6
LCM(24, 18) = 72, LCM(LCM(24, 18), 12) = LCM(72, 12) = 72
The final results are the same.
14. **(a)** The 1, 3, and 9 rods **(d)** The 1, 2, 3, 6, 9, and 18 trains. **15.** **(a)** 12 **(b)** 12
19. **(a)** 18; 76; 1364 **(c)** No. F_{19} = 4181 = 37 · 113 **(h)** If GCD(F_{16}, F_{20}) = 4, then the conjecture must be false.
21. **(a)** 1224 sec, 1224 = LCM(72, 68) **23.** **(a)** LCM(220, 264) = 1320; LCM(220, 275) = 1100; LCM(264, 275) = 6600 **(c)** GCD(220, 264) = 44; GCD(220, 275) = 55; GCD(264, 275) = 11
24. **(a)** 21 **25.** **(a)** 2646 **26.** **(a)** 3 **27.** **(a)** 720 **28.** **(a)** 205,800 = $2^3 \cdot 3^1 \cdot 5^2 \cdot 7^3$, 31,460 = $2^2 \cdot 5^1 \cdot 11^2 \cdot 13^1$, 25,840 = $2^4 \cdot 5^1 \cdot 17^1 \cdot 19^1$
30. **(a)** $A \cap (B \cup C)$ = {a, b, c, d, e, f, g} ∩ {a, b, c, d, e, g}
= {a, b, c, d, e, g}
$(A \cap B) \cup (A \cap C)$ = {a, c, d, e, g} ∪ {a, b, c, d}
= {a, b, c, d, e, g}
(c) $\overline{A \cup B}$ = $\overline{\{a, b, c, d, e, f, g\}}$ = {h, i, j, k, l, m, n, o, p, q, r, s, t, u, v, w, x, y, z}
$\overline{A} \cap \overline{B}$ = {h, i, j, k, l, m, n, o, p, q, r, s, t, u, v, w, x, y, z} ∩ {b, f, h, i, j, k, l, m, n, o, p, q, r, s, t, u, v, w, x, y, z}
= {h, i, j, k, l, m, n, o, p, q, r, s, t, u, v, w, x, y, z}

31. **(a)**

n	$n^2 - 81n + 1681$	Prime?
1	1601	yes
2	1523	yes
3	1447	yes
4	1373	yes
5	1301	yes

(c) No. If a conjecture is true for several cases, it does not mean it is true for *every* case. **(e)** The conjecture may seem more probable, but we still cannot say if it is *always* true.
32. **(a)** No. If $p \mid a$ and $p \nmid b$ then $p \nmid (a + b)$. **33.** **(a)** No. If $p \mid a$ and $p \nmid b$ then $p \nmid (a + b)$.

Problem Set 4.4 (page 304)

1. (a) 2 **(c)** 12 **(e)** 8 **2. (a)** 2 **(c)** 9 **(e)** 6 **3. (a)** 2 **(c)** 8 **(e)** 3 **4. (a)** 5 **(c)** 3
5. (a) $9 -_{12} 7 = 9 +_{12} 5 = 2$ **(c)** $5 -_{12} 9 = 5 +_{12} 3 = 8$ **(e)** $2 -_{12} 11 = 2 +_{12} 1 = 3$
6. (a) 11 **(c)** 12 **(e)** 12 **7. (a)** 11 **(c)** undefined **(e)** 9 **8. (a)** 1, 5, 7, 11 **10. (a)** 2 **(c)** 3 **(e)** 5 **(g)** 4 **(i)** 5 **(k)** undefined
11. (a) 5; $n +_5 5 = n$ for $n = 1, 2, 3, 4, 5$. **12. (a)** 4, 3, 2, 1, 5 **13. (a)** 1, 2, 3, 4 **14. (a)** $4 \div_{12} 7 = 4$ **(c)** $3 \div_5 2 = 4$ **(e)** $2 \div_5 4 = 3$ **(g)** $4 \times_{12} 7^{-1} = 4 \times_{12} 7 = 4$ **(i)** $3 \times_5 2^{-1} = 3 \times_5 3 = 4$ **(k)** $2 \times_5 4^{-1} = 2 \times_5 4 = 3$
15. (c)

$$\begin{aligned} y +_{12} 2 &= 3 \times_{12} 11 \\ y +_{12} 2 &= 9 \\ y &= 9 +_{12} 10 \\ y &= 7 \end{aligned}$$

16. (a)

17. (a) The bar code gives the digits: 1057486537. This code is incorrect because the sum of the digits is 46—not a multiple of 10. **20. (a)** 3 **(c)** 3 **(e)** 4 **(g)** 4 **21. (a)** 5 **25. (a)** 2, 3, 4, 6, 8, 9, 10 **26.** $y = 2, 3, 6,$ or 11
27. (a) (i) is incorrect. (ii) is correct. **28. (a)** 1 **33. (a)** $D_{60} = \{1, 2, 3, 4, 5, 6, 10, 12, 15, 20, 30, 60\}$
$D_{150} = \{1, 2, 3, 5, 6, 10, 15, 25, 30, 50, 75, 150\}$
So GCD(60, 150) = 30.

34. (a)

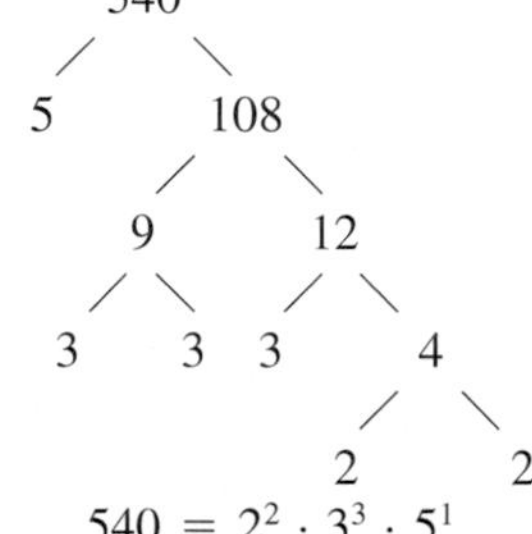

$540 = 2^2 \cdot 3^3 \cdot 5^1$

35. (a) $540 = 2^2 \cdot 3^3 \cdot 5^1$, $600 = 2^3 \cdot 3^1 \cdot 5^2$

CLASSIC CONUNDRUM The Chinese Remainder Problem (page 308)

We seek the first common term in the arithmetic progressions

$$2, 7, 12, 17, \cdots, \quad 3, 10, 17, 24, \cdots, \quad \text{and} \quad 4, 13, 22, 31, \cdots.$$

By repeated addition using the constant function of a calculator, we find that 157 is the desired number.

Chapter 4 Review Exercises (page 310)

1.

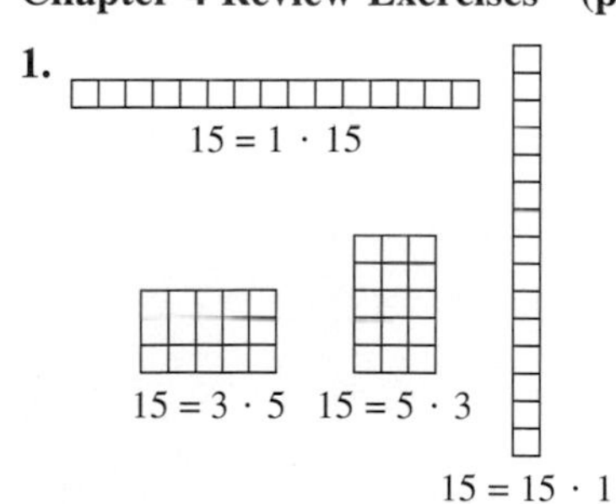

2.

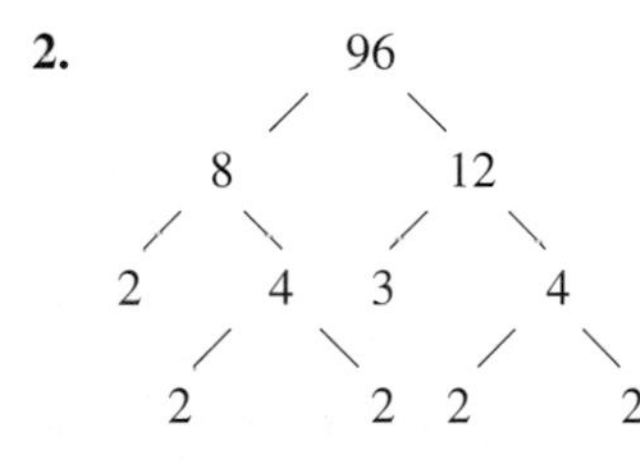

3. (a) $D_{60} = \{1, 2, 3, 4, 5, 6, 10, 12, 15, 20, 30, 60\}$
(b) $D_{72} = \{1, 2, 3, 4, 6, 8, 9, 12, 18, 24, 36, 72\}$
(c) $D_{60} \cap D_{72} = \{1, 2, 3, 4, 6, 12\}$, so GCD(60, 72) = 12.

4. (a) $1200 = 2^4 \cdot 3^1 \cdot 5^2$
(b) $2940 = 2^2 \cdot 3^1 \cdot 5^1 \cdot 7^2$
(c) $\text{GCD}(1200, 2940) = 2^2 \cdot 3^1 \cdot 5^1 \cdot 7^0 = 60$
$\text{LCM}(1200, 2940) = 2^4 \cdot 3^1 \cdot 5^2 \cdot 7^2 = 58{,}800$

5. Composite; $847 = 11 \times 77$. **6. (a)** Answers will vary. For example, $15 = 3 \cdot 5$; $5 > \sqrt{15}$. **(b)** Yes. $3 \leq \sqrt{15}$.
7. Answers will vary. For example, $8 \mid 16$ and $4 \mid 16$ but $32 \nmid 16$. **8.** Since $n = 2536$, the prime, 2, divides $3 \cdot 5 \cdot 7 + 11 \cdot 13 \cdot 17$.

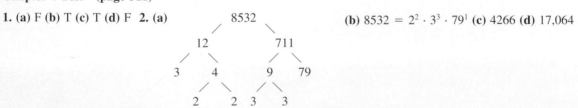

9. **(a)** Divisible by 2, 5 **(b)** Divisible by 3, 11 **(c)** Divisible by 5 **(d)** Divisible by 5, 11
10. **(a)** Divisible by 11 **(b)** Divisible by 7, 13 **(c)** Divisible by 11, 13 **11.** **(a)** False **(b)** True **(c)** False **(d)** True
12. **(a)** $(1 + 4)(1 + 2) = 15$ **(b)** 1, 3, 7, 9, 21, 27, 49, 63, 81, 147, 189, 441, 567, 1323, 3969
13. $d = 5$ **14.** **(a)** $2^3 \cdot 3^5 \cdot 7^3 \cdot 11^3 \cdot 13^1 = 11{,}537{,}501{,}976$ **(b)** $2^2 \cdot 3^5 \cdot 7^2 \cdot 11^1 = 523{,}908$ **15.** **(a)** 28 by 28 **(b)** 28
16. **(a)** 168 by 168 **(b)** 168
17. **(a)** $D_{63} = \{1, 3, 7, 9, 21, 63\}$
$D_{91} = \{1, 7, 13, 91\}$
$D_{63} \cap D_{91} = \{1, 7\}$, so GCD(91, 63) = 7.
(b) $M_{63} = \{0, 63, 126, 189, 252, 315, 378, 441, 504, 567, 630, 693, 756, 819, 882, 945, 1008, \ldots\}$
$M_{91} = \{0, 91, 182, 273, 364, 455, 546, 637, 728, 819, 910, \ldots\}$
$M_{63} \cap M_{91} = \{0, 819, 1638, \ldots\}$, so LCM(63, 91) = 819.
(c) $7 \cdot 819 = 5733 = 63 \cdot 91$
18. **(a)** $2 \cdot 11^2 = 242$ **(b)** $2^3 \cdot 3^2 \cdot 5^2 \cdot 7^1 \cdot 11^3 = 16{,}770{,}600$
19. **(a)** $\begin{array}{r} 9 \text{ R } 10{,}890 \\ 12{,}100\overline{)119{,}790} \end{array}$ $\qquad \begin{array}{r} 1 \text{ R } 1210 \\ 10{,}890\overline{)12{,}100} \end{array}$

$\begin{array}{r} 9 \text{ R } 0 \\ 1210\overline{)10{,}890} \end{array}$

Thus, GCD(119,790, 12,100) = 1210. **(b)** LCM(119,790, 12,100) = $119{,}790 \cdot 12{,}100/1210 = 1{,}197{,}900$. **20.** 2192
21. **(a)** 1 **(b)** 5 **(c)** 12 **(d)** Undefined **(e)** 5 **(f)** 12 **(g)** Undefined **(h)** 3 **(i)** 3
22. **(a)** 4 **(b)** 1 **(c)** 2 **(d)** 4 **23.** 2, 4, 5, 6, 8, 10 **24.** **(a)** 9 **(b)** 3 **25.** **(a)** 80321 − 1589 **(b)** 60648 − 9960
26. **(a)**

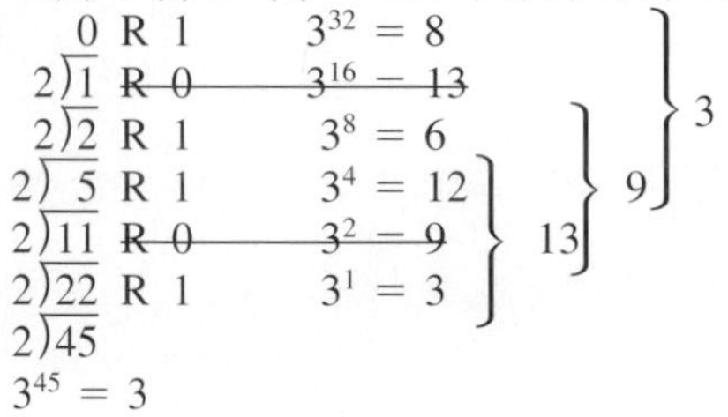
0 R 1 $3^{32} = 8$
$2\overline{)1}$ R 0 $3^{16} = 13$
$2\overline{)2}$ R 1 $3^8 = 6$
$2\overline{)5}$ R 1 $3^4 = 12$
$2\overline{)11}$ R 0 $3^2 = 9$
$2\overline{)22}$ R 1 $3^1 = 3$
$2\overline{)45}$
$3^{45} = 3$
13
9
3

(b)

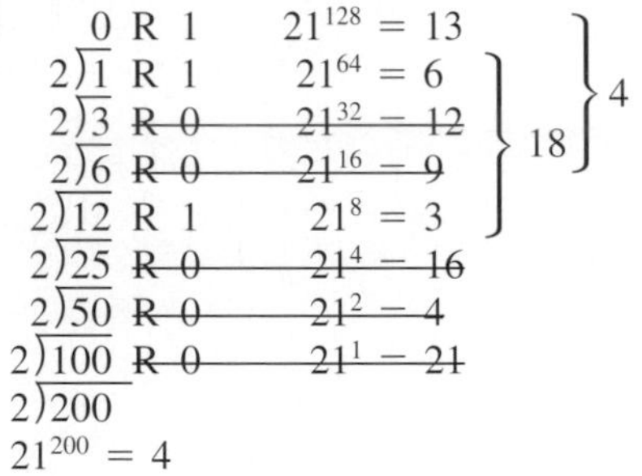
0 R 1 $21^{128} = 13$
$2\overline{)1}$ R 1 $21^{64} = 6$
$2\overline{)3}$ R 0 $21^{32} = 12$
$2\overline{)6}$ R 0 $21^{16} = 9$
$2\overline{)12}$ R 1 $21^8 = 3$
$2\overline{)25}$ R 0 $21^4 = 16$
$2\overline{)50}$ R 0 $21^2 = 4$
$2\overline{)100}$ R 0 $21^1 = 21$
$2\overline{)200}$
$21^{200} = 4$
18
4

(c)

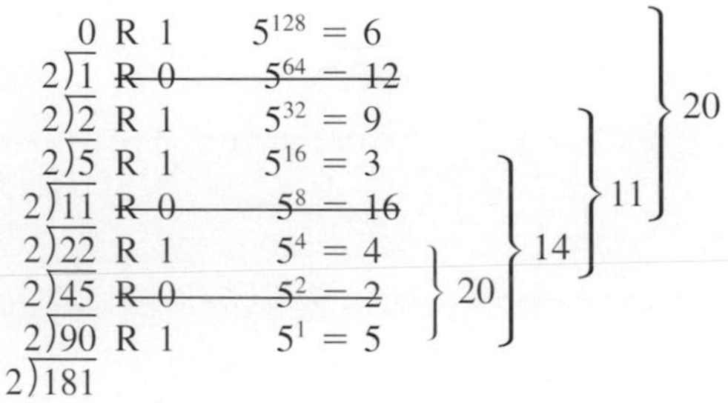
0 R 1 $5^{128} = 6$
$2\overline{)1}$ R 0 $5^{64} = 12$
$2\overline{)2}$ R 1 $5^{32} = 9$
$2\overline{)5}$ R 1 $5^{16} = 3$
$2\overline{)11}$ R 0 $5^8 = 16$
$2\overline{)22}$ R 1 $5^4 = 4$
$2\overline{)45}$ R 0 $5^2 = 2$
$2\overline{)90}$ R 1 $5^1 = 5$
$2\overline{)181}$
20
14
11
20

$5^{181} = 20$ **27.** PZBYY LUKLY **28.** $k = 21$, I WANT TO BE FREE XY

Chapter 4 Test (page 311)

1. **(a)** F **(b)** T **(c)** T **(d)** F **2.** **(a)**

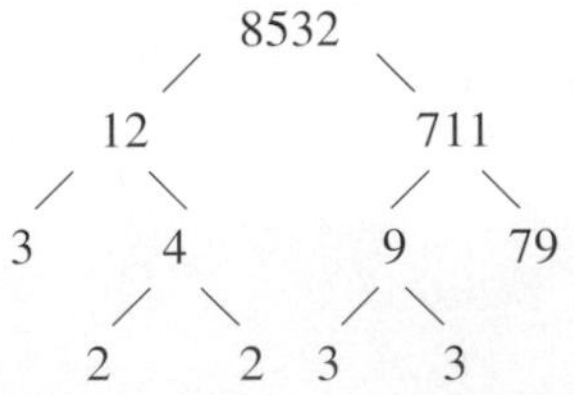
8532
12
711
3
4
9
79
2
2
3
3

(b) $8532 = 2^2 \cdot 3^3 \cdot 79^1$ **(c)** 4266 **(d)** 17,064

3. **(a)** Divisible by 2, 3 **(b)** Divisible by 3

4. (a)

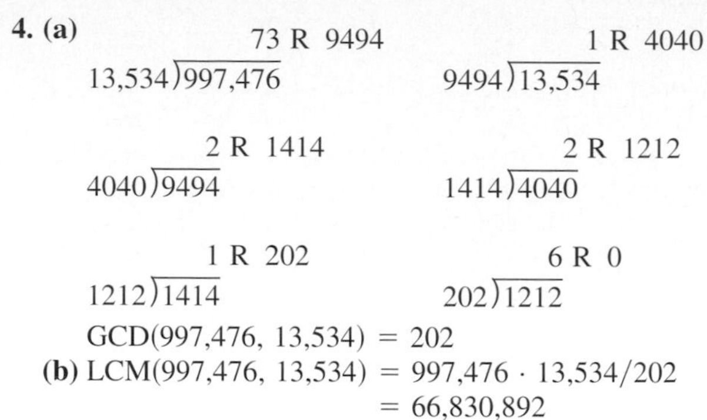

$13{,}534\overline{)997{,}476}$ = 73 R 9494 $9494\overline{)13{,}534}$ = 1 R 4040

$4040\overline{)9494}$ = 2 R 1414 $1414\overline{)4040}$ = 2 R 1212

$1212\overline{)1414}$ = 1 R 202 $202\overline{)1212}$ = 6 R 0

GCD(997,476, 13,534) = 202

(b) LCM(997,476, 13,534) $= 997{,}476 \cdot 13{,}534/202$
$= 66{,}830{,}892$

5. (a) No. The prime power representation of r contains two 7s, but the prime power representation of m contains only one 7, so $r \nmid m$. **(b)** $(3 + 1)(2 + 1)(1 + 1)(4 + 1) = 120$ **(c)** $2^2 \cdot 5^0 \cdot 7^1 \cdot 11^3 = 37{,}268$ **(d)** $2^3 \cdot 5^2 \cdot 7^2 \cdot 11^4 = 143{,}481{,}800$ **6.**

7. (a) 4 **(b)** 12 **(c)** 5 **(d)** 3 **(e)** 3 **(f)** 7 **8.** ANHYT WD **9.** $k = 21$; I FEEL FINE X

Chapter 5

Problem Set 5.1 (page 322)

1. Answers will vary. Two possibilities are shown in each case. **(a)**

(c)

2. (a) **(c)** (no counters) **3. (a)** **(c)** opp 17

4. (a) At mailtime, you are delivered a check for \$14. **6. (a)** -15 **7. (a–e)**

8. (a) 4 **(c)** 6 **9. (a)**

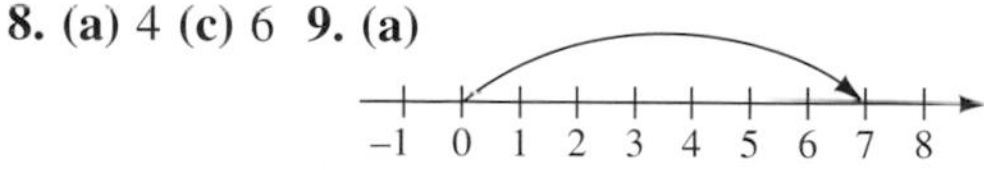

(c)

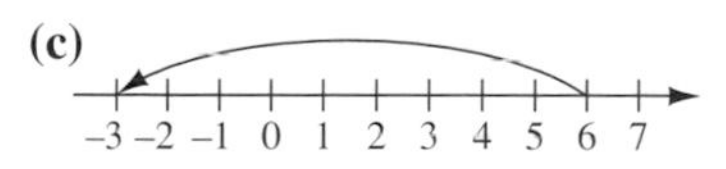

12. (a) 34 **(c)** 76 **13. (a)** 13, -13 **14. (a)** 4 red counters **15. (a)** $-12, -10, -8, -6, -4, -2, 0, 2, 4, 6, 8, 10, 12$ **17. (a)** $\{n \mid n \text{ is an integer and } -12 \leq n \leq 12\}$ **18. (a)** 2^{20} **19. (a)** 100 black, 110 red **20. (a)** $2^5 = 32$ **21. (a)** 90 **(c)** 3,402,000 **22. (a)** No. The prime power representation of c contains more 3s than the prime power representation of a. **23. (a)** $1400 = 2^3 \cdot 5^2 \cdot 7^1$

24. (a)

$4554\overline{)5445}$ = 1 R 891 $891\overline{)4554}$ = 5 R 99

$99\overline{)891}$ = 9 R 0

GCD(4554, 5445) = 99

JUST FOR FUN Choosing the Right Box (page 332)

Note that there is a symmetry in the problem between apples and oranges. Thus, it probably does not make sense to select the piece of fruit from the box labeled "apples" or the box labeled "oranges." We select from the box labeled "apples and oranges." Since the box is mislabeled, if we obtain an apple it contains only apples and should be so labeled. The remaining boxes contain oranges only and a mixture of apples and oranges. But each box is mislabeled. Thus, the box labeled "oranges" should be relabeled "apples and oranges" and the box labeled "apples and oranges" should be relabeled "oranges." A similar analysis holds if, on our first selection we obtain an orange. Then the boxes should be relabeled as shown.

apples and oranges → oranges
apples → apples and oranges
apples and oranges → apples

JUST FOR FUN Fun with a Flow Chart (page 338)

We start with 27 and x simultaneously and see what results are obtained.

27	x
54	$2x$
58	$2x + 4$
290	$10x + 20$
302	$10x + 32$
3020	$100x + 320$
2700	$100x$
27	x

We obtain the original number whether we start with a 2-digit number or not.

Problem Set 5.2 (page 341)

1. (a)

$8 + (-3) = 5$

(c)

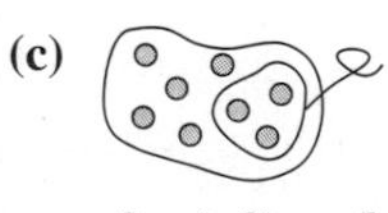

$-8 - (-3) = -5$

(e)

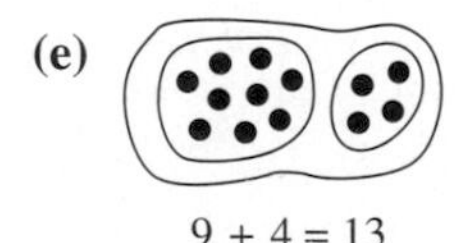

$9 + 4 = 13$

(g)

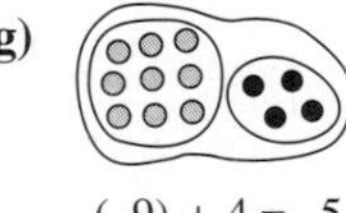

$(-9) + 4 = -5$

2. (a) At mailtime you receive a bill for \$27 and a bill for \$13. $(-27) + (-13) = -40$ **(c)** The mail carrier brings you a check for \$27 and a check for \$13. $27 + 13 = 40$. **(e)** At mailtime you receive a bill for \$41 and a check for \$13. $(-41) + 13 = -28$ **(g)** At mailtime you receive a bill for \$13 and a check for \$41. $(-13) + 41 = 28$

3. (a)

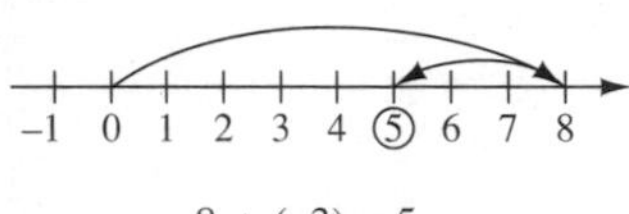

$8 + (-3) = 5$

(c)

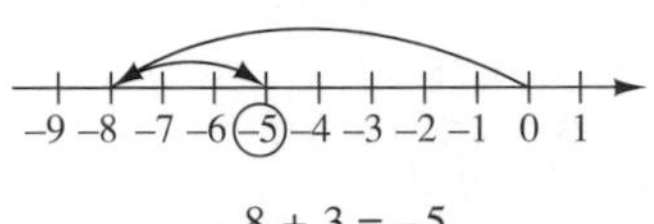

$-8 + 3 = -5$

(e)

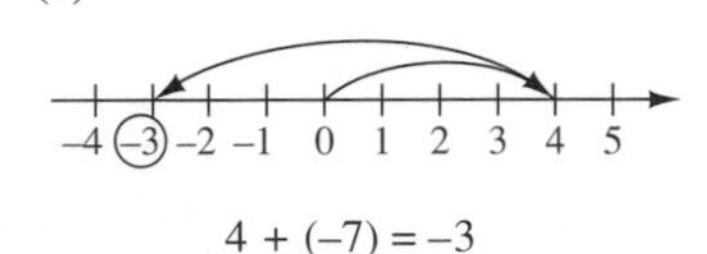

$4 + (-7) = -3$

(g)

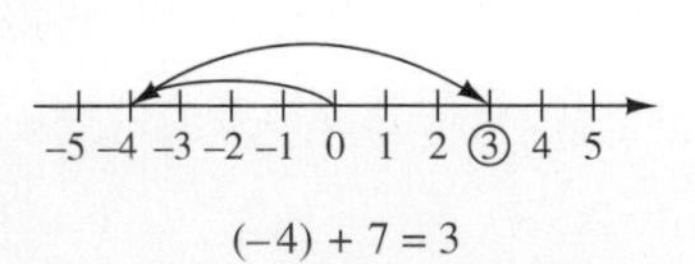

$(-4) + 7 = 3$

4. (a) $13 + (-7)$ **(c)** $(-13) + (-7)$ **(e)** $3 + (-8)$ **(g)** $(-8) + (-13)$ **5. (a)** 40 **(c)** -27 **(e)** -135 **(g)** -135 **9. (a)** More; by \$106. **11. (a)** $-117 < -24$ **(c)** $18 > 12$ **(e)** $-5 < 1$ **13. (a)** True **(c)** True **15.** No. If $a \geq b$, then it is possible that $a = b$ and so $a > b$ is false. **17.** $-6, -5, -4, -3, -2, -1, 0, 1, 2, 3, 4, 5, 6$

19. **(a)** **(i)** 6 **(iii)** 15 **(b)** **(i)**

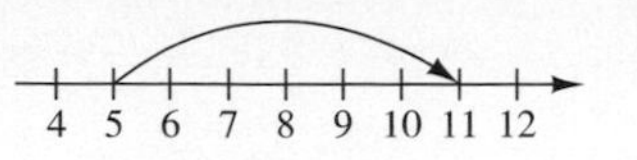

distance is 6

(iii)

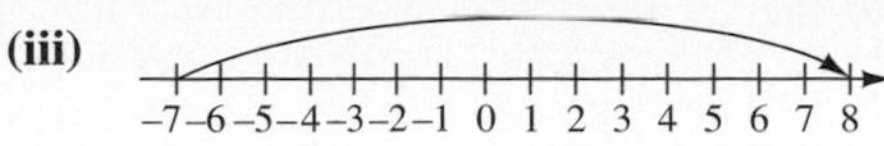

distance is 15

20. **(a)** **(i)** $|7 + 2| = 9, |7| + |2| = 9$ **(iii)** $|7 + (-6)| = 1, |7| + |-6| = 13$ **(v)** $|6 + 0| = 6, |6| + |0| = 6$

22. **(a)**

3	−4	1
−2	0	2
−1	4	−3

23. **(a)**

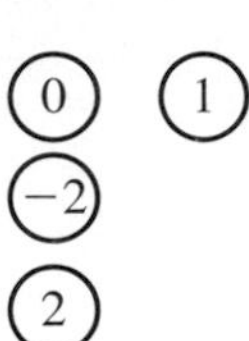

24. **(a)** $7 - (-3) = 10, (-3) - 7 = 10$ **27.** **(a)** 50 **28.** **(a)** 1575 **(c)** −5909 **(e)** 2053
30. **(a)** $F_1 + F_3 + F_5 + F_7 = 1 + 2 + 5 + 13 = 21 = F_8$
$F_1 + F_3 + F_5 + F_7 + F_9 = 1 + 2 + 5 + 13 + 34 = 55 = F_{10}$
33. **(a)** Yes. $F_0 + F_1 = 0 + 1 = 1 = F_2$ **34.** **(a)** $101 + 3 = 104$

37. **(a)**

t	h
0	0
1	80
2	128
3	144
4	128
5	80
6	0
7	−112

41. **(a)** Divisible by 3 **42.** **(a)** Divisible by 7, 13

JUST FOR FUN Three on a Bike (page 351)

The problem is that 27 + 2 should *not* add to 30. Of the \$30, \$25 is in the till, \$2 is in the helper's pocket, and \$3 has been returned to the bikers. Alternatively, of the \$27, \$25 is in the till and \$2 is in the helper's pocket.

Problem Set 5.3 (page 356)

1. **(a)** 77 **(c)** −77 **(e)** 108 **(g)** −108 **(i)** 0 **2.** **(a)** 4 **(c)** −4 **(e)** −13 **(g)** 16 **(i)** 36
3. $(-25{,}753) \cdot (-11) = 283{,}283$
$283{,}283 \div (-11) = -25{,}753$
$283{,}283 \div (-25{,}753) = -11$

5. **(a)**

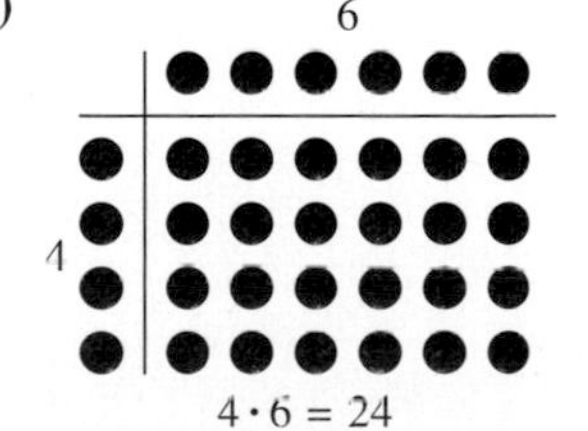

7.

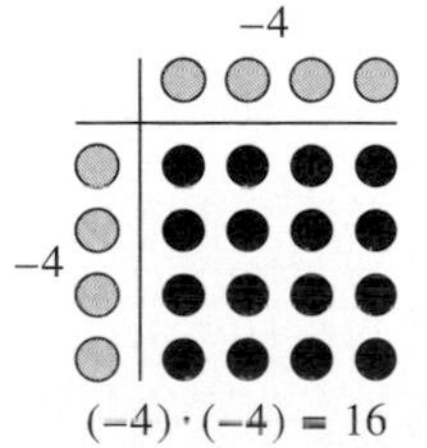

9. **(a)** Richer by \$78; $6 \cdot 13 = 78$. **10.** **(a)** $6 \cdot 3 = 18$ **12.** **(a)** $3 \cdot (-1) = -3, 3(-2) = -6, 3(-3) = -9$
14. **(a)** Multiplicative property of zero; Distributive property of multiplication over addition; Definition of the additive inverse.

15. **(a)**

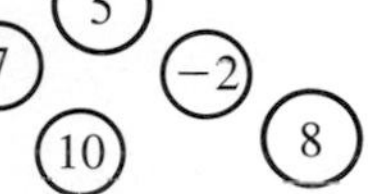

(c)

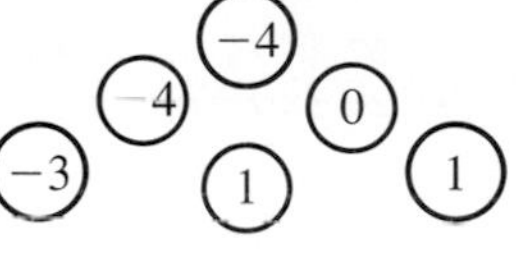

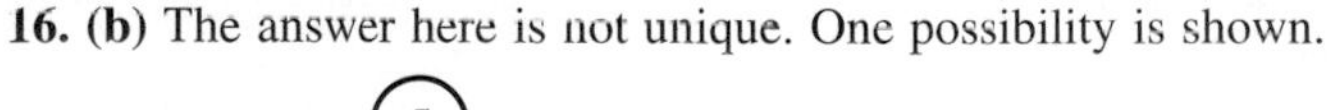

16. **(b)** The answer here is not unique. One possibility is shown.

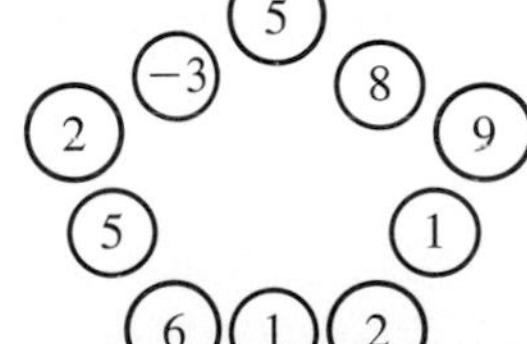

21. **(a)** \$7 **(b)** $(-105) \div 15 = -7$. A loss of \$105, shared among 15 people results in each person losing \$7.
24. **(b)** 59 **25.** **(a), (b)** The sequence of entries

ON/AC 1 x^2 M+ = − 2 x^2 M+ = + 3 x^2 M+ = . . .

successively computes $s_1, s_2, s_3, \ldots$ and $a_1, a_2, a_3, \ldots$. After each = entry, the calculator will show the appropriate entry in the a-sequence and will have the appropriate entry in the s-sequence in memory. The see the s entry press x⇄M right after = . Then, to get on with the calculation, press x⇄M again and continue as before. We thus compute the a and s entries in the following table. The $3s \div a$ entries are computed separately afterward.

n	a_n	s_n	$3s_n \div a_n$
1	1	1	3
2	−3	5	−5
3	6	14	7
4	−10	30	−9
5	15	55	11
6	−21	91	−13

27. **(b)** 2,621,440 **28.** **(a)** 5, 12, 29, 70

CLASSIC CONUNDRUM On Stealing Apples (page 364)

Working backwards, after passing the various watchmen the thief had 1, 2(1 + 2) = 6, 2(6 + 2) = 16, 2(16 + 2) = 36 apples. He originally stole 16 apples.

Chapter 5 Review Exercises (page 366)

1. **(a)** −1 **(b)** 5 **(c)** −15, −13, −11, . . . , 11, 13, 15 **2.** **(a)** Richer by \$12; 12 **(b)** Poorer by \$37; −37 **3.** **(a)** 12 **(b)** −24 **4.** **(a)** Answers will vary. Any "drop" that shows 5 more red counters than black counters represents the integer −5. **(b)** Any "drop" that shows 6 more black counters than red counters represents the integer 6.
5. Answers will vary. **(a)** At mailtime you receive a bill for \$114 and a check for \$29. **(b)** The mail carrier brings you a bill for \$19 and a check for \$66. **6.** **(a)** −44 **(b)** 61 **7.** $2 + (-4) = -2$ **8.** $(-1) - (-3) = 2$ **9.** **(a)** $45 + (-68) = -23$ **(b)** $45 - (-68) = 113$ **10.** **(a)** −2 **(b)** −22 **(c)** −32 **(d)** 12 **(e)** 20 **(f)** −4 **11.** **(a)** 27° below zero **(b)** $(-15) - 12 = -27$ **12.** **(a)** \$25 **(b)** $(-12) + 37 = 25$ **13.** **(a)** $2 \cdot 2 = 4$ **(b)** $(-3) \cdot (-5) = 15$
14. **(a)** The diagram must have 7 rows: **(b)**

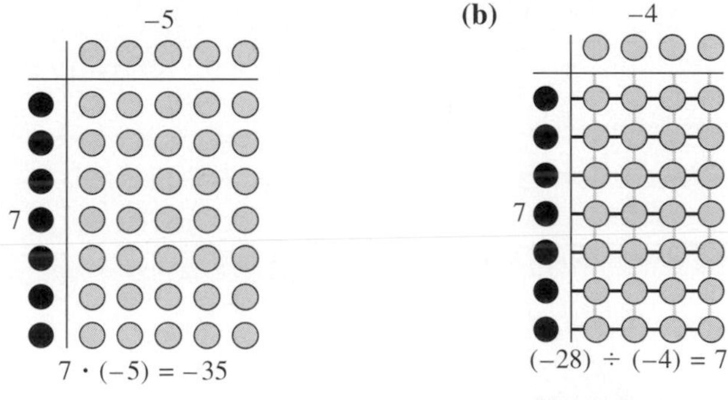

$7 \cdot (-5) = -35$

$(-28) \div (-4) = 7$

28 red counters

15. **(a)**

−4
−8

$(-8) \cdot (-4) = 32$

(b)

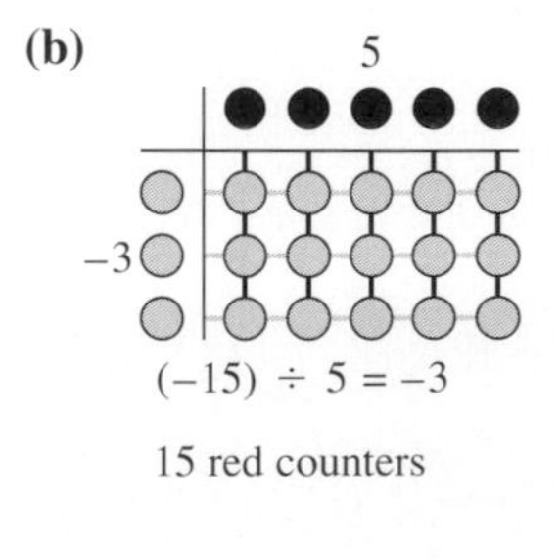

$(-15) \div 5 = -3$

15 red counters

16. **(a)** **(b)**

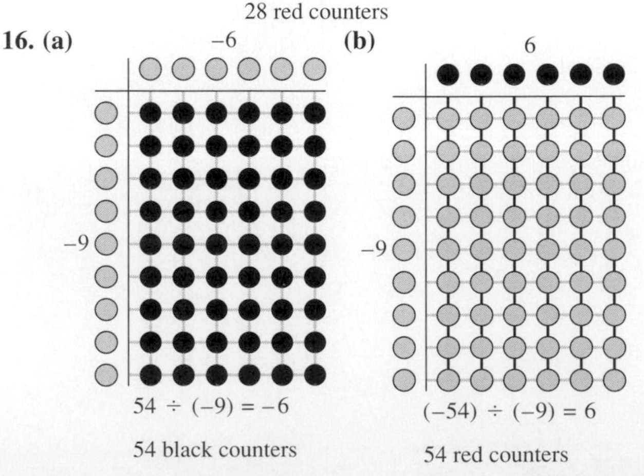

$54 \div (-9) = -6$

54 black counters

$(-54) \div (-9) = 6$

54 red counters

17. (a) 56 **(b)** -56 **(c)** -56 **(d)** -7 **(e)** -12 **(f)** 12 **18. (a)** At mailtime you receive 7 checks, each for \$12. **(b)** The mail carrier takes away 7 checks, each for \$13. **(c)** The mail carrier takes away 7 bills, each for \$13. **19. (a)** 3 **(b)** 11
20. By the division algorithm, n must be of one of these forms: $6q$, $6q + 1$, $6q + 2$, $6q + 3$, $6q + 4$, or $6q + 5$. If n is not divisible by 2, however, then n cannot be of any of the forms $6q$, $6q + 2$ or $6q + 4$. Likewise, if n is not divisible by 3, n cannot be of the form $6q + 3$ either. Thus, there must be an integer q such that either $n = 6q + 1$ or $n = 6q + 5$.

Case 1

$$\begin{aligned} n &= 6q + 1 \\ n^2 - 1 &= (6q + 1)^2 - 1 \\ &= (36q^2 + 12q + 1) - 1 \\ &= 36q^2 + 12q \\ &= 12q(3q + 1) \end{aligned}$$

If q is even, then $24 \mid 12q$ and $n^2 - 1$ is divisible by 24. If q is odd, then $3q + 1$ is even, so $24 \mid 12q(3q + 1)$ and hence $n^2 - 1$ is divisible by 24.

Case 2

$$\begin{aligned} n &= 6q + 5 \\ n^2 - 1 &= (6q + 5)^2 - 1 \\ &= (36q^2 + 60q + 25) - 1 \\ &= 36q^2 + 60q + 24 \\ &= 12(3q^2 + 5q + 2) \\ &= 12(3q + 2)(q + 1) \end{aligned}$$

If q is even, $3q + 2$ is even. If q is odd, $q + 1$ is even. In either case, it follows that $n^2 - 1$ is divisible by 24.

Chapter 5 Test (page 367)

1.

5

−5

$(-25) \div -5 = 5$

2. (a) -26 **(b)** 12 **(c)** 26 **(d)** -12 **(e)** -361 **(f)** -408 **(g)** -864 **(h)** 105 **(i)** 0

3. 2160, the same as LCM(240,54) **4.** Richer by \$135; $(-5) \cdot (-27) = 135$

5. (a)

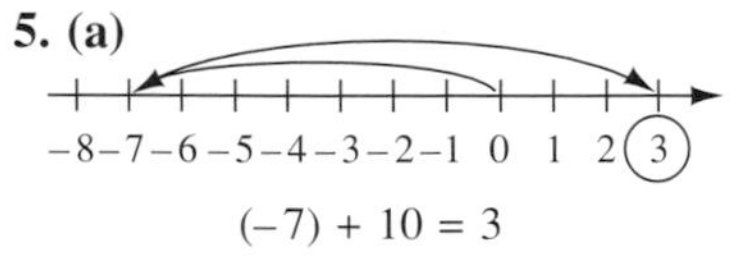

$(-7) + 10 = 3$

(b)

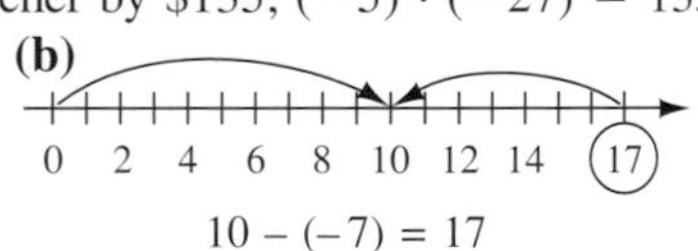

$10 - (-7) = 17$

(c)

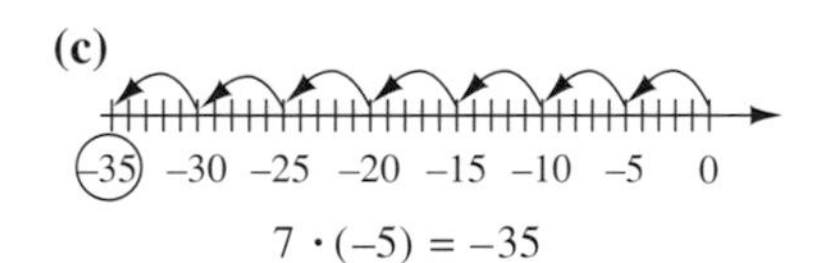

$7 \cdot (-5) = -35$

6. Answers may vary.

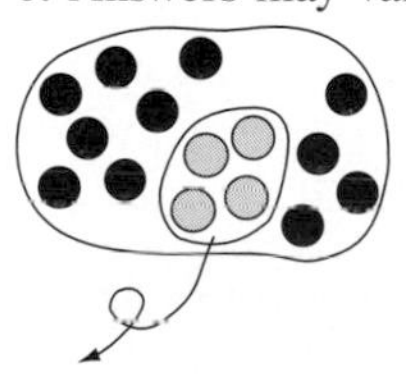

7. \$381; $129 + 341 - 13 - 47 - 29 = 381$ **8.** Poorer by \$9; $(-27) \div 3 = -9$
9. (a) $-5, -3, -8, -11, -19, -30$ **(b)** $7, -5, 2, -3, -1, -4$ **(c)** $6, -8, -2, -10, -12, -22$
10. (a) $-18, -17, -16, \ldots, -3, -2, -1, 0, 1, 2, 3, \ldots, 16, 17, 18.$
(b) Yes.

Chapter 6

JUST FOR FUN Suspicious Simplifications (page 378)

It works, but it's just lucky. For example,

$$\frac{18}{84} \neq \frac{1}{4}, \frac{27}{76} \neq \frac{2}{6},$$

and so on. A few examples that work out only suggest a property or result. Just one counterexample, like either of the two above, definitely shows that the property does not always hold. In fact, except for numbers like

$$\frac{11}{11}, \frac{22}{22}, \frac{33}{33}, \frac{44}{44}, \ldots$$

the only other fraction with two digit numerals with this property is 26/65.

Problem Set 6.1 (page 382)

1. **(a)** $\frac{1}{6}$ **(c)** $\frac{0}{1}$ **(e)** $\frac{2}{6}$ **2.** **(a)** 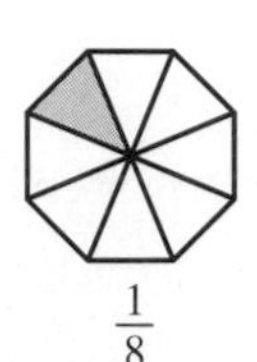 $\frac{1}{8}$ **(c)** 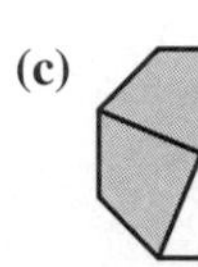 $\frac{3}{4}$

3. **(a)** A: $\frac{1}{4}$, B: $\frac{3}{4}$, C: $\frac{3}{2}$ or $\frac{6}{4}$ **(c)** G: $\frac{0}{5}$, H: $\frac{2}{5}$, I: $\frac{8}{5}$

4. **(a)** 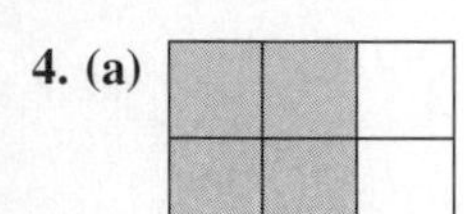 **5.** **(a)** $\frac{1}{3}$ **(c)** $\frac{5}{7}$ **(e)** $\frac{1}{4}$ **(g)** $\frac{2}{3}$ **7.** **(a)** $\frac{3}{6} = \frac{1}{2}$ **9.** **(a)** 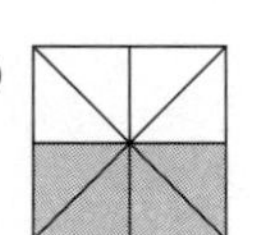

10. **(a)** 24 **(c)** -140 **11.** **(a)** Equivalent **(c)** Not Equivalent **12.** **(a)** Equivalent **13.** **(a)** Yes **(c)** Yes **14.** **(a)** $\frac{7}{12}$ **(c)** $\frac{-31}{43}$ **15.** **(a)** $\frac{96}{288} = \frac{2^5 \cdot 3^1}{2^5 \cdot 3^2} = \frac{1}{3}$ **16.** Answers will vary. Possible answers are shown. **(a)** $\frac{15}{55}$ and $\frac{22}{55}$ **(c)** $\frac{32}{24}, \frac{15}{24}$, and $\frac{4}{24}$ **17.** **(a)** $\frac{9}{24}$ and $\frac{20}{24}$ **(c)** $\frac{136}{96}$ and $\frac{21}{96}$ **18.** **(a)** $\frac{7}{12}, \frac{2}{3}$ **(c)** $\frac{29}{36}, \frac{5}{6}$ **19.** **(a)** True. Given two fractions, two equivalent fractions with a common denominator may be found by finding a common multiple of the two original denominators. Once these fractions, say $\frac{a}{c}$ and $\frac{b}{c}$ are found, infinitely many more pairs of equivalent fractions can be found; namely, $\frac{a \cdot n}{c \cdot n}$ and $\frac{b \cdot n}{c \cdot n}$ for any integer n other than 0. **(c)** False. Given any positive fraction, a smaller positive fraction may be found by multiplying the denominator by 2. So there cannot be a least positive fraction. **20.** **(a)** $\left\{\frac{a}{b} \mid a = 3n, b = 5n \text{ for any integer } n \neq 0\right\}$ **(c)** $\left\{\frac{0}{n} \mid n \text{ is any integer}, n \neq 0\right\}$

22. **(a)** $\frac{4}{8}$ **(d)** 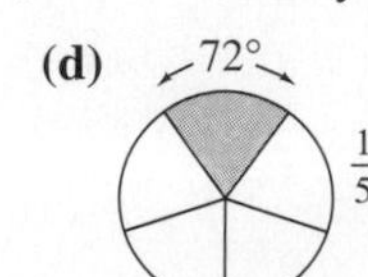

$\frac{1}{5}$ **(f)** 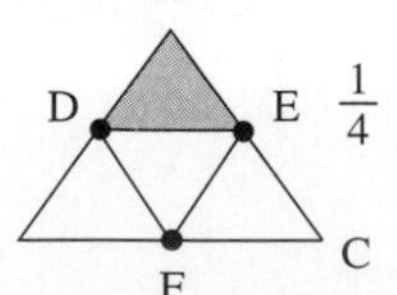

$\frac{1}{4}$

23. **(a)** $\frac{1}{1}, \frac{3}{2}, \frac{8}{5}, \frac{5}{3}, \frac{2}{1}$ **(b)** Between $\frac{8}{5}$ and $\frac{5}{3}$ **24.** **(a)** $\frac{3+4}{4+5} = \frac{7}{9}$; $\frac{3}{4} < \frac{7}{9} < \frac{4}{5}$ since $3 \cdot 9 < 4 \cdot 7$ and $7 \cdot 5 < 9 \cdot 4$.

25. **(a)** Let $\frac{a}{b} < \frac{c}{d}$ with b and d positive so that $ad < bc$. But then $ad + mcd < bc + mcd$ where m is any natural number. Therefore, $(a + mc)d < (b + md)c$ and this implies that

$$\frac{a + mc}{b + md} < \frac{c}{d}.$$

Also, if $ad < bc$, then $mad < mbc$. But then $ab + mad < ab + mbc$ and $a(b + md) < b(a + mc)$. This last statement implies that

$$\frac{a}{b} < \frac{a + mc}{b + md}.$$

So

$$\frac{a}{b} < \frac{a + mc}{b + md} < \frac{c}{d}.$$

(b) Given any two distinct rational numbers, $\frac{a}{b}$ and $\frac{c}{d}$, $\frac{a + mc}{b + md}$ is a rational number between $\frac{a}{b}$ and $\frac{c}{d}$ for any natural number m. Since there are infinitely many choices for m, and each choice gives a distinct rational number there are infinitely many rational numbers between $\frac{a}{b}$ and $\frac{c}{d}$. (Indeed, $\frac{a}{b} < \frac{a + c}{b + d} < \frac{a + 2c}{b + 2d} < \cdots < \frac{a + mc}{b + md} < \cdots < \frac{c}{d}$.

26. **(a)** Let a, b, and c be positive integers, and suppose $a < c$. Then $ab < bc$ and so $\frac{a}{b} < \frac{c}{b}$.

27. Answers to each of (a) and (b) will vary but in each case one-third of the rhombuses will have each of the three possible orientations.

28. **(a)** $\mathcal{F}_5$: $\frac{0}{1}, \frac{1}{5}, \frac{1}{4}, \frac{1}{3}, \frac{2}{5}, \frac{1}{2}, \frac{3}{5}, \frac{2}{3}, \frac{3}{4}, \frac{4}{5}, \frac{1}{1}$

$\mathcal{F}_6$: $\frac{0}{1}, \frac{1}{6}, \frac{1}{5}, \frac{1}{4}, \frac{1}{3}, \frac{2}{5}, \frac{1}{2}, \frac{3}{5}, \frac{2}{3}, \frac{3}{4}, \frac{4}{5}, \frac{5}{6}, \frac{1}{1}$

29. **(a)** 18 gal **30.** **(a)** 3/5 **(c)** 1/4 **31.** Yes. Lakeside won $\frac{19}{25}$ of their games while Shorecrest won $\frac{16}{21}$ of their games, and $\frac{16}{21} > \frac{19}{25}$. **32.** **(a)** Carol; Carol **33.** **(a)** $\frac{1}{2}$ **(c)** $\frac{3}{6}$ **35.** **(a)** 5, −8, 13, −21, 34, −55, 89, −144, 233, −377 **36.** **(a)** $m = 7, n = 5$ **37.** $D = 5$

JUST FOR FUN The Sultan's Estate (page 389)

Since $(1/2) + (1/3) + (1/9) = 17/18$, the canny old Sultan did not leave all his horses to his sons. The uncle, realizing this, saw that if he included his horse before the division, the sons could each receive a whole number of horses and the uncle could then take back his own horse.

JUST FOR FUN Bookworm Math (page 392)

The bookworm only needs to eat through the front cover of Volume 1 and the back cover of Volume 2 for a total of $1/8 + 1/8 = 1/4$ inch.

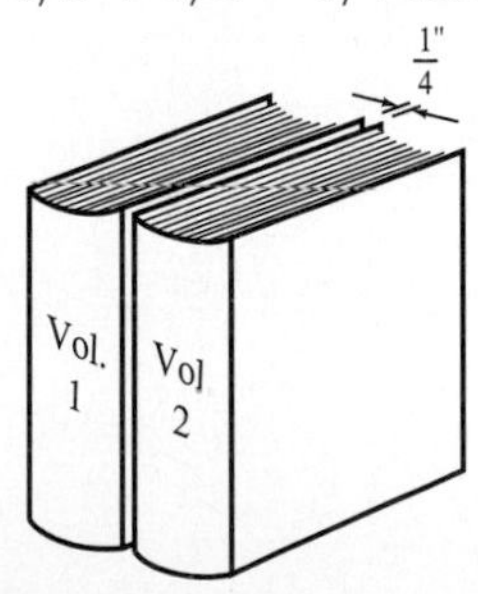

Problem Set 6.2 (page 402)

1. **(a)** $\frac{1}{3} + \frac{1}{2} = \frac{5}{6}$ **2.** **(a)** $\frac{2}{5} + \frac{6}{5} = \frac{8}{5}$ **3.** **(a)** $\frac{3}{4} + \frac{-2}{4} = \frac{1}{4}$

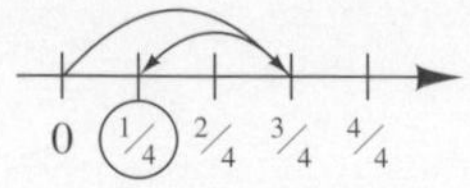

4. **(a)** $\frac{5}{7}$ **(c)** $\frac{5}{6}$ **(e)** $\frac{19}{15}$ **(g)** $\frac{73}{100}$ **5.** **(a)** $2\frac{1}{4}$ **(c)** $4\frac{19}{23}$ **6.** **(a)** $\frac{19}{8}$ **(c)** $\frac{557}{5}$ **7.** **(a)** $\frac{5}{6} - \frac{1}{4} = \frac{7}{12}$ **8.** **(a)** $\frac{3}{8}$ **(c)** $\frac{4}{3}$ **(e)** $\frac{1}{3}$ **(g)** $\frac{625}{642}$

9. **(a)** $3 \times \frac{5}{2} = \frac{15}{2}$ **10.** **(a)** $2 \times \frac{3}{5} = \frac{6}{5}$ **12.** **(a)** $\frac{8}{3}$ **(c)** $\frac{4}{9}$ **(e)** $\frac{1}{5}$ **13.** $\frac{b}{a} > 1$. $\frac{d}{c}$ is also close to 1.

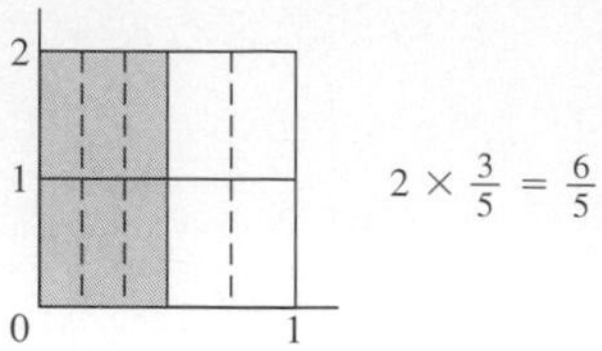

15. **(a)** $\frac{8}{15}$ **(c)** $\frac{10}{11}$ **(e)** $\frac{4}{7}$ **17.** **(a)** $\frac{7}{6}$ **18.** **(a)** $\frac{2}{3} + \frac{1}{12} = \frac{3}{4}$

19. $\frac{17}{18}$, since 17 of the total of 18 horses are given to the sons.

20. **(a)**

$\frac{1}{2}$	$\frac{1}{12}$	$\frac{5}{12}$
$\frac{1}{4}$	$\frac{1}{3}$	$\frac{5}{12}$
$\frac{1}{4}$	$\frac{7}{12}$	$\frac{1}{6}$

21. **(a)** $S_5 = \frac{137}{60}$, $S_6 = \frac{49}{20}$, $S_7 = \frac{363}{140}$, $S_8 = \frac{761}{280}$

22. The mixed number $A\frac{b}{c}$ is equal to $A + \frac{b}{c}$. But

$$\begin{aligned} A + \frac{b}{c} &= A \times \frac{c}{c} + \frac{b}{c} \\ &= \frac{Ac}{c} + \frac{b}{c} \\ &= \frac{Ac + b}{c}. \end{aligned}$$

24. **(a)** Simply add. **25.** **(a)** $\frac{2}{9} = \frac{1}{5} + \frac{1}{45}$ $\quad \frac{2}{11} = \frac{1}{6} + \frac{1}{66}$

26. **(a)** $a_1 = \frac{1}{2} = 1 - \frac{1}{2}$, $a_2 = \frac{1}{2} + \frac{1}{4} = \frac{3}{4} = 1 - \frac{1}{4}$, $a_3 = \frac{1}{2} + \frac{1}{4} + \frac{1}{8} = \frac{7}{8} = 1 - \frac{1}{8}$

27. **(a)** It terminates.

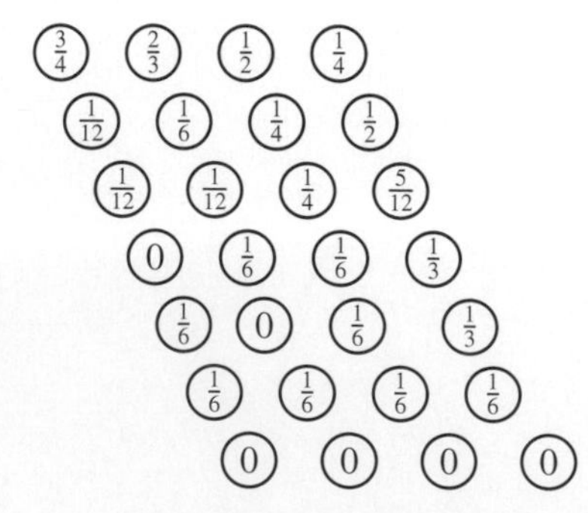

28. **(a)** The process terminates when one row consists of all ones.

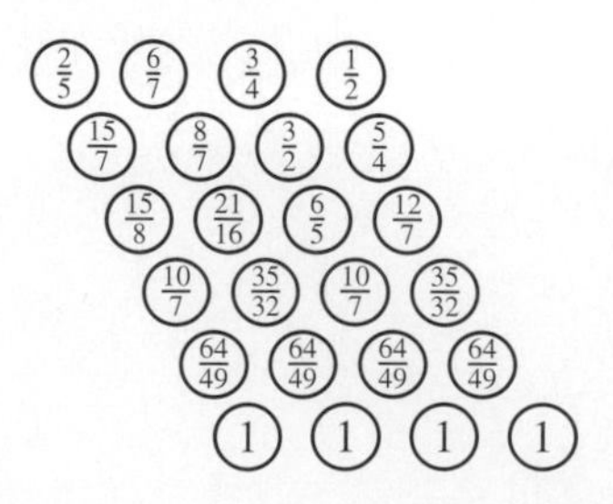

29. 3, 3, 4, 5, 6, 6 **31.** 23 bows **33.** \$42.00 **37.** **(a)** $\frac{7}{2}$ **39.** 180, $\frac{203}{180}$

Problem Set 6.3 (page 418)

1. Commutative and associative properties for addition: $(3 + 2 + 8) + \left(\frac{1}{5} + \frac{2}{5} + \frac{1}{5}\right)$

3. (a) $\frac{-4}{5}$

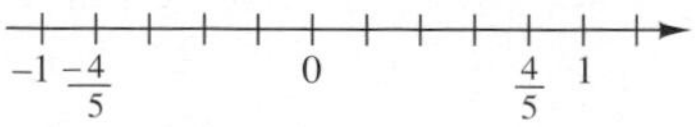

(c) $\frac{8}{3}$

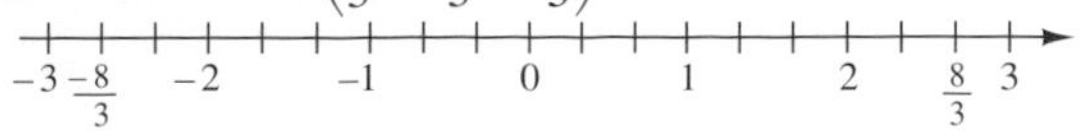

4. (a) $\frac{-1}{2}$ **(c)** $\frac{11}{8}$ The most useful property here is equivalence of fractions.

5. (a) $\frac{-7}{20}$ **(c)** $\frac{7}{24}$ **(e)** $\frac{-41}{12}$ The most useful property here is equivalence of fractions.

6. (a) $\frac{7}{8}$ **(c)** $\frac{1}{2}$ **(e)** 8 The most useful property here is equivalence of fractions.

7. (a) $\frac{2}{3}$

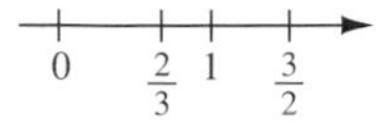

(c) $\frac{11}{4}$

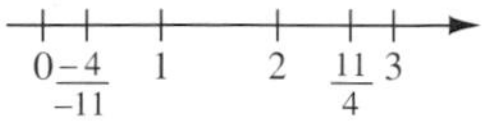

(e) $\frac{-1}{2}$

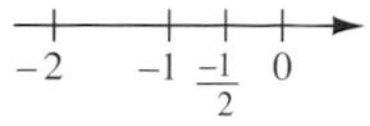

8. (a) $\frac{2}{3}$ **(c)** 0 **9. (a)** Addition of rational numbers—definition,

10. $\frac{7}{6}$, since $\frac{a}{b} = \frac{2}{3} \div \frac{4}{7} = \frac{2}{3} \cdot \frac{7}{4} = \frac{14}{12} = \frac{7}{6}$. **11. (a)** $x = \frac{-3}{4}$ **(c)** $x = \frac{-6}{5}$

12. (a) Closure property for subtraction and the existence of a multiplicative inverse.

13. (a) $\frac{-1}{5}, \frac{2}{5}, \frac{4}{5}$

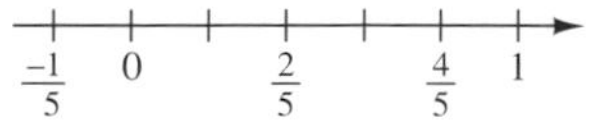

(c) $\frac{3}{8}, \frac{1}{2}, \frac{3}{4}$

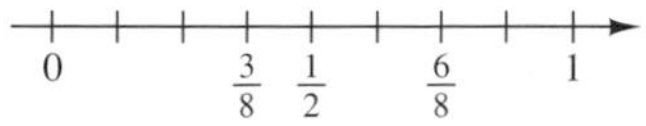

14. (a) $-4 \cdot 4 = -16 < -15 = 5(-3)$

15. (a)
$$x + \frac{2}{3} > -\frac{1}{3}$$
$$x > -\frac{1}{3} - \frac{2}{3} = -1$$

(c)
$$\frac{3}{4}x < -\frac{1}{2}$$
$$x < -\frac{1}{2} \div \frac{3}{4}$$
$$x < -\frac{1}{2} \cdot \frac{4}{3}$$
$$x < -\frac{2}{3}$$

16. (a) Answers can vary. One answer is $\frac{1}{2}$, since $\frac{4}{9} < \frac{1}{2} < \frac{6}{11}$. **(c)** Answers can vary. Since $\frac{7}{12} = \frac{14}{24} = \frac{28}{48}$ and $\frac{14}{23} = \frac{28}{46}$, one answer is $\frac{28}{47}$. **18. (a)** $\frac{1}{4}$ **(c)** -1 **19. (a)** 9 **20. (a)** $\frac{3}{2}$ **(c)** $\frac{3}{5}$ **(e)** 40 **(g)** $-2\frac{1}{8}$ **21. (a)** 1 $\left(\text{about } \frac{1}{3} + \frac{2}{3}\right)$; $\frac{44{,}155}{41{,}866} = 1\frac{2289}{41{,}866}$ **(c)** $-1\left(\text{about } \frac{5}{8} \cdot \frac{-8}{5}\right)$; $\frac{-169}{156} = \frac{-13}{12}$ **22. (a)** $-\frac{3}{125}$ **24.** $\frac{1260}{250} = 5\frac{1}{25}$ **25. (a)** 8 **(c)** $\frac{3}{5}$ **(e)** $\frac{56}{9}$

26. (a)
$$s_1 = \frac{1}{1 \cdot 2} = \frac{1}{2} = 1 - \frac{1}{2}$$
$$s_2 = \frac{1}{1 \cdot 2} + \frac{1}{2 \cdot 3}$$
$$= \frac{1}{2} + \frac{1}{6}$$
$$= \left(1 - \frac{1}{2}\right) + \left(\frac{1}{2} - \frac{1}{3}\right)$$
$$s_3 = \frac{1}{1 \cdot 2} + \frac{1}{2 \cdot 3} + \frac{1}{3 \cdot 4}$$
$$= \frac{1}{2} + \frac{1}{6} + \frac{1}{12}$$
$$= \left(1 - \frac{1}{2}\right) + \left(\frac{1}{2} - \frac{1}{3}\right) + \left(\frac{1}{3} - \frac{1}{4}\right)$$

27. If $\frac{a}{b} < \frac{c}{d}$, then $\frac{a}{b} + \frac{m}{n} = \frac{c}{d}$ for some rational number $\frac{m}{n} > 0$. If $\frac{c}{d} < \frac{e}{f}$, then $\frac{c}{d} + \frac{r}{s} = \frac{e}{f}$ for some rational number $\frac{r}{s} > 0$. But then,

$$\begin{aligned}\frac{e}{f} &= \frac{c}{d} + \frac{r}{s}\\ &= \left(\frac{a}{b} + \frac{m}{n}\right) + \frac{r}{s}\\ &= \frac{a}{b} + \left(\frac{m}{n} + \frac{r}{s}\right).\end{aligned}$$

Hence $\frac{a}{b} < \frac{e}{f}$ since $\frac{m}{n} + \frac{r}{s}$ is a positive rational number. **29. (a)** The mediant becomes $\frac{4}{7}$, which is not equal to $\frac{3}{5}$.
30. (a) $z = 1, y = \frac{4}{5}, x = z - y = \frac{1}{5}$ $L = x + \frac{1}{2} = \frac{1}{5} + \frac{1}{2} = \frac{7}{10}$
31. (a) Change $= \frac{-3}{4}$ **(c)** Wednesday close $= 47$ **33.** Cut the 8 inch side in half and the 10 inch side in thirds to get six $3\frac{1}{3}$ by 4 inch rectangles. **35. (a)** 2 yd^2 **36.** \$230 to \$345

CLASSIC CONUNDRUM The Slice Is Right? (page 425)

Let the 100 loaves be given in the shares s, $s + d$, $s + 2d$, $s + 3d$, and $s + 4d$ to the five men. Then $5s + 10d = 100$, or $s + 2d = 20$. Therefore the shares, again in increasing order, are $20 - 2d$, $20 - d$, 20, $20 + d$, and $20 + 2d$. The other condition of the problem tells us that $40 - 3d = \frac{1}{7}(60 + 3d)$, which can be solved to show that $d = \frac{55}{6} = 9\frac{1}{6}$.

Thus the shares are $1\frac{2}{3}$, $10\frac{5}{6}$, 20, $29\frac{1}{6}$, and $38\frac{1}{3}$ loaves.

Chapter 6 Review Exercises (page 427)

1. (a) $\frac{2}{4}$ **(b)** $\frac{6}{6}$ **(c)** $\frac{0}{4}$ **(d)** $\frac{5}{3}$ **2.**

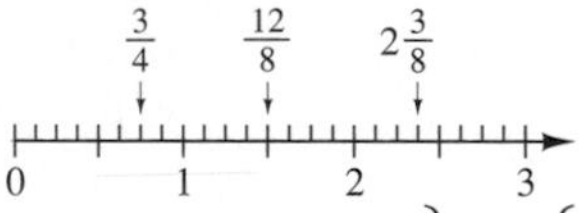

3. (a) $\frac{1}{3}$ **(b)** $\frac{4}{33}$ **(c)** $\frac{21}{4}$ **(d)** $\frac{297}{7}$ **4.** $\frac{13}{30}, \frac{13}{27}, \frac{1}{2}, \frac{25}{49}, \frac{26}{49}$

5. (a) $\left\{\frac{a}{b} \,\middle|\, a = 5n, b = 8n \text{ for any integer } n \neq 0\right\}$ **(b)** $\left\{\frac{c}{d} \,\middle|\, c = -4d, \text{ where } c \text{ and } d \text{ are integers, } d \neq 0\right\}$

(c) $\left\{\frac{e}{f} \,\middle|\, e = 3k, f = 4k \text{ for any integer } k \neq 0\right\}$

6.

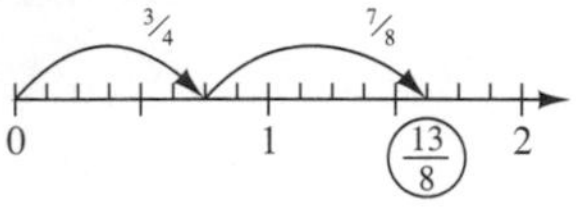

7.

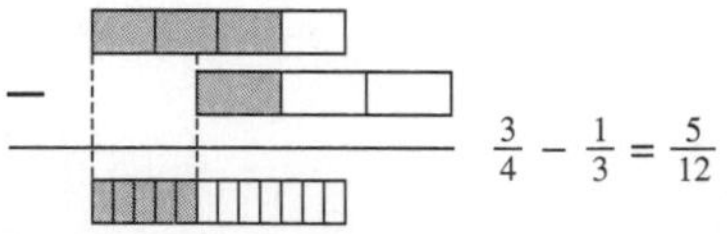

8. (a) $\frac{5}{8}$ **(b)** $\frac{-7}{36}$
(c) $\frac{2}{15}$ **(d)** $\frac{41}{12}$ or $3\frac{5}{12}$

9. (a)

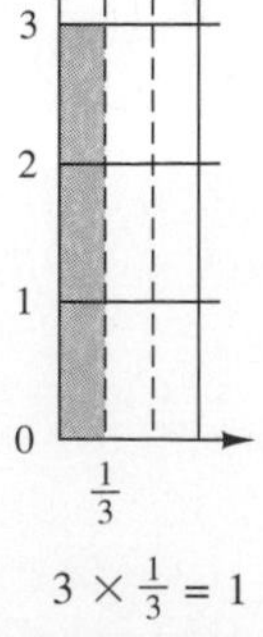

(b)

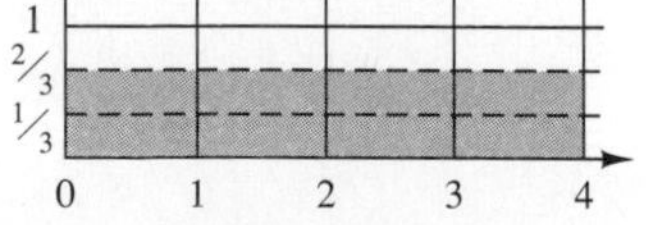

(c)

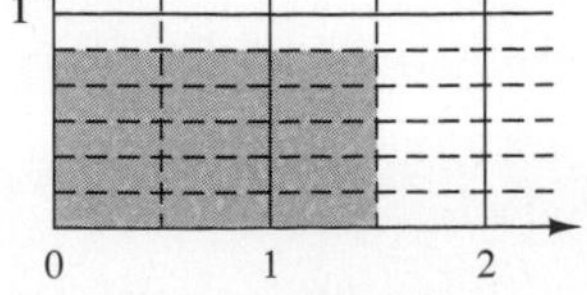

10. $\frac{57}{10}$ miles $= 5\frac{7}{10}$ miles **11.** **(a)** $\frac{-1}{8}$ **(b)** $\frac{3}{2}$ **(c)** $\frac{-1}{2}$ **(d)** $\frac{-7}{10}$ **12.** **(a)** $x = 2$ **(b)** $x = \frac{-1}{6}$ **(c)** $x = \frac{5}{18}$ **(d)** $x = \frac{9}{16}$

13. Write $\frac{5}{6} = \frac{55}{66}$ and $\frac{10}{11} = \frac{60}{66}$. Then $\frac{56}{66}$ and $\frac{57}{66}$ are between $\frac{5}{6}$ and $\frac{10}{11}$. Other answers may be given.

14. **(a)** 4 **(b)** 5 **(c)** 20

Chapter 6 Test (page 428)

1. **(a)**

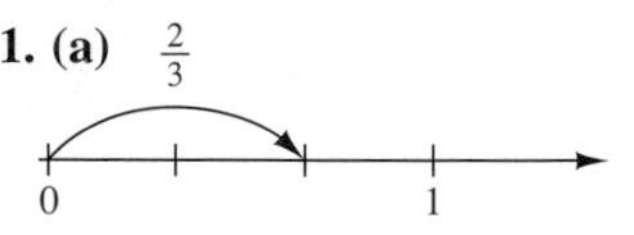

(b)

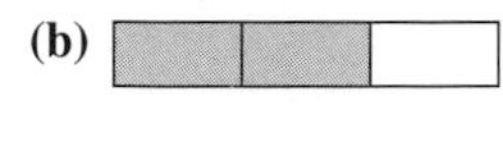

(c)

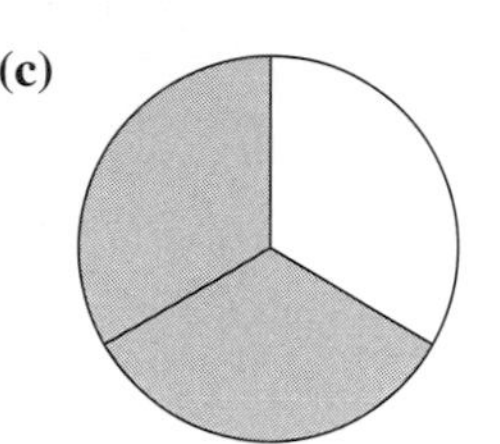

(d)

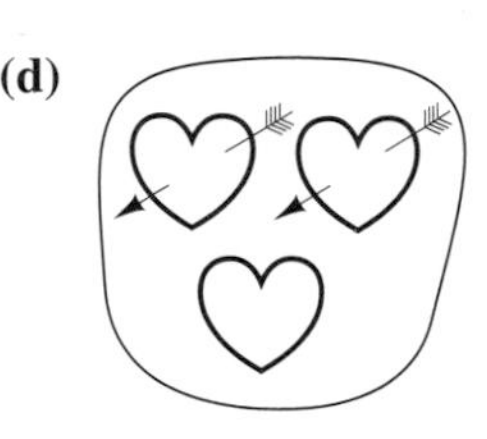

2. Answers will vary. Possibilities include $\frac{-6}{8}, \frac{-9}{12}, \frac{-12}{16}$. **3.** $-3, -1\frac{1}{2}, 0, \frac{5}{8}, \frac{2}{3}, 3, \frac{16}{5}$ **4.** **(a)** $\frac{1}{8}$ **(b)** $\frac{-7}{9}$ **(c)** 3 **(d)** 1

5. **(a)** If $\frac{a}{b}$ and $\frac{c}{d}$ are rational numbers with $\frac{c}{d} \neq 0$, then $\frac{a}{b} \div \frac{c}{d} = \frac{e}{f}$ if, and only if, $\frac{a}{b} = \frac{c}{d} \cdot \frac{e}{f}$. **(b)** If $\frac{a}{b} \div \frac{c}{d} = \frac{e}{f}$, then $\frac{a}{b} = \frac{c}{d} \cdot \frac{e}{f} = \frac{e}{f} \cdot \frac{c}{d}$. So

$$\frac{a}{b} \cdot \frac{d}{c} = \frac{e}{f} \cdot \frac{c}{d} \cdot \frac{d}{c} = \frac{e}{f}.$$

Thus, $\frac{a}{b} \div \frac{c}{d} = \frac{a}{b} \cdot \frac{d}{c}$.

6. **(a)** Answers will vary. **(b)** Answers will vary. **7.** **(a)** $x > \frac{-3}{2}$ **(b)** $x = \frac{-2}{9}$ **(c)** $x > \frac{-4}{15}$ **(d)** $x > \frac{-1}{12}$ **8.** $\frac{a+b}{b} = \frac{c+d}{d}$ if, and only if, $(a+b)d = b(c+d)$; that is, if, and only if, $ad + bd = bc + bd$. But this is so if, and only if, $ad = bc$. And this is so if, and only if, $\frac{a}{b} = \frac{c}{d}$.

9. **(a)** 40 acres **(b)** $\frac{1}{4}$ mile **10.** **(a)** If $\frac{a}{b}$ and $\frac{c}{d}$ are two rational numbers with $\frac{a}{b} < \frac{c}{d}$, then there is a rational number $\frac{e}{f}$ such that $\frac{a}{b} < \frac{e}{f} < \frac{c}{d}$. **(b)** Write $\frac{3}{5} = \frac{18}{30}$ and $\frac{2}{3} = \frac{20}{30}$ to see that $\frac{19}{30}$ is between $\frac{3}{5}$ and $\frac{2}{3}$. **11.** **(a)** $11\frac{1}{2}$ **(b)** 48 **(c)** 23

12. **(a)** If $\frac{a}{b}$ is a rational number, then its additive inverse is the rational number $\frac{-a}{b}$. **(b)** $\frac{-3}{4}, \frac{7}{4}, \frac{8}{2}$

13. **(a)** If $\frac{a}{b}$ is a rational number with $a \neq 0$, then its multiplicative inverse is the rational number $\frac{b}{a}$. **(b)** $\frac{2}{3}, -\frac{5}{4}, \frac{-1}{5}$

14. **(a)** 0, since $\frac{2}{3} + \frac{-4}{6} = 0$. **(b)** 2, since $\frac{5}{6} \cdot \frac{36}{15} = \frac{5}{15} \cdot \frac{36}{6} = \frac{1}{3} \cdot 6 = 2$. **(c)** 1, since $\frac{9}{5} - \frac{1}{5} = \frac{8}{5}$ is the reciprocal of $\frac{5}{8}$. **(d)** $\frac{1}{3}$, since $\frac{2}{\cancel{3}} \cdot \frac{\cancel{3}}{\cancel{4}} \cdot \frac{\cancel{4}}{\cancel{5}} \cdot \frac{\cancel{5}}{6} = \frac{2}{6} = \frac{1}{3}$.

Chapter 7

Problem Set 7.1 (page 449)

1. **(a)** $273.412 = 200 + 70 + 3 + \frac{4}{10} + \frac{1}{100} + \frac{2}{1000}$; $273.412 = 2 \cdot 10^2 + 7 \cdot 10^1 + 3 \cdot 10^0 + 4 \cdot 10^{-1} + 1 \cdot 10^{-2} + 2 \cdot 10^{-3}$ **2.** **(a)** $\frac{81}{250}$; $250 = 2^1 \cdot 5^3$ **3.** **(a)** 0.35 **(c)** 0.04 **4.** **(a)** $\frac{107}{333}$ **(d)** 1 **(e)** $\frac{1}{4}$ **(g)** $\frac{1}{7}$

5. **(a)** Answers will vary. For example, $2/3 = 0.666\ldots$. **6.** **(a)** $0.007, 0.017, 0.01\overline{7}, 0.027$ **(c)** $0.35, 0.\overline{35}, 0.36 = \frac{9}{25}, \frac{10}{25}$

8. Assume $3 - \sqrt{2}$ is rational, then $3 - \sqrt{2} = q$ where q is rational. This implies that $\sqrt{2} = 3 - q$. But $3 - q$ is rational since the rational numbers are closed under subtraction. This can't be true since we know $\sqrt{2}$ is irrational. Thus, the assumption that $3 - \sqrt{2}$ is rational must be false. So $3 - \sqrt{2}$ is irrational.

12. (a) $0.1 = 1/10$ **13 (a)** $0.\overline{09}$ **(c)** $0.\overline{0009}$ **14. (a)** $\frac{74}{99}$ **15. (a)** $0.\overline{5}$ **(d)** $0.\overline{51}$ **16.** Answers will vary. **(a)** One example is $\sqrt{2} + (3 - \sqrt{2}) = 3$ since $\sqrt{2}$ and $(3 - \sqrt{2})$ are both irrational. **(b)** One example is $\sqrt{3} + \sqrt{3} = 2\sqrt{3}$ since $\sqrt{3}$ and $2\sqrt{3}$ are both irrational.

19. Answers will vary. For example, $\frac{\sqrt{2}}{2\sqrt{2}} = \frac{1}{2}$ and $\frac{1}{2}$ is rational. **25. (a)** Answers will vary. For example, if $a = 1$ and $b = 4$, the sequence is 1, 4, 5, 9, 14, 23, 37, 60, 97, 157, 254, 411, 665, 1076, 1741, 2817, 4558, 7375, 11933, 19308.

27. (a) The mediant of $\frac{3}{8}$ and $\frac{17}{24}$ is $\frac{3 + 17}{8 + 24} = \frac{20}{32}$.

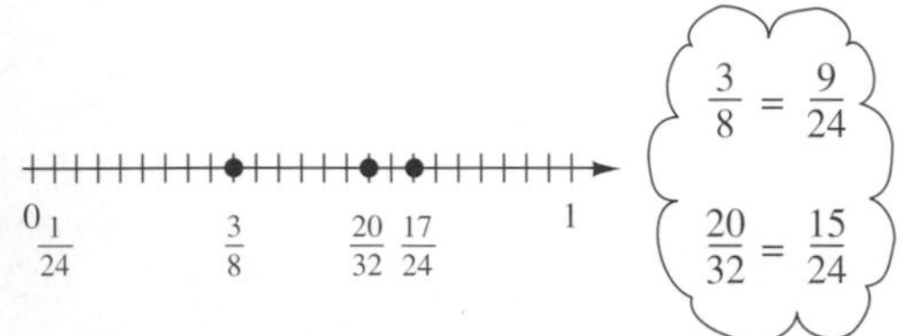

28. (a) $\frac{7}{6}$ **(c)** $\frac{1}{3}$ **(e)** $\frac{6}{5}$

Problem Set 7.2 (page 461)

1. (a) 403.674 **(c)** 1.137 **2. (a)** 174.37 **(c)** 3.12 **3. (a)** 35.412 **(c)** 128.7056 **4. (a)** $0.8\overline{3}$ **5. (a)** 2.77×10^{8} **6. (a)** 1.05×10^{-10} **(c)** 1.29×10^{-13} **7. (a)** 1.53×10^{10} **8.**

0.984	0.123	0.738
0.369	0.615	0.861
0.492	1.107	0.246

11. (c)

2.374 0.041 5.267
2.415 5.308
7.723

12. (a) 3.4, 4.3, 5.2, 6.1, 7.0, 7.9 **(c)** 0.0114, 0.1144, 0.2174, 0.3204, 0.4234, 0.5264
13. (a) 2.11, 2.327, 2.5663, 2.8302, 3.1213
15. (a) **(c)** **16. (a)**

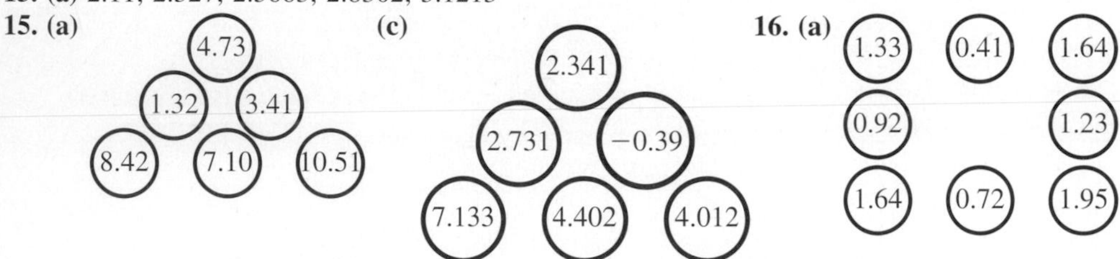

(c) Answers vary. A solution must have 5.01 in the lower right corner. **17.** \$29.16 **20. (a)** 210.375 in^2
24. (a)

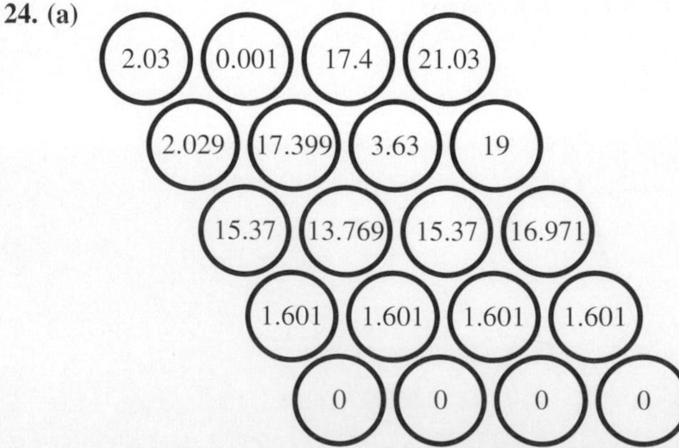

(d) 5, 9, 17, 31

25. (a) Working to nine decimal places, this yields 1, 1, 1, 1 after five steps. **26. (a)** XXIV **27. (a)** 29
28. (a) **29. (a)** 56,525 **30. (a)**

31. (a) 2303

JUST FOR FUN Who Shaves Francisco? (page 469)

This is a logical paradox. If Francisco doesn't shave himself, then he must shave himself. However, if he shaves himself then he does not shave himself. This type of self-reference often leads to a paradox and has had to be ruled out in logical discourse.

Problem Set 7.3 (page 472)

1. (a) $\frac{7}{5}$ **(c)** $\frac{7}{12}$ **(e)** $\frac{12}{5}$ **2. (a)** Yes **(c)** No **(e)** No **3. (a)** $r = 9$ **(c)** $t = 57$ **4. (a)** 3 to 2 **(c)** 2 to 3
5. (a) \$19.25 **(c)** $\frac{3.5}{19.25} = 0.\overline{18}, \frac{5}{27.5} = 0.\overline{18}$ **8.** 19 ft **10. (a)** $\begin{aligned} \frac{a}{b} &= \frac{c}{d} \\ ad &= bc \\ da &= cb \\ \frac{d}{c} &= \frac{b}{a} \end{aligned}$ **(c)** $\begin{aligned} \frac{a}{b} &= \frac{c}{d} \\ ad &= bc \\ ac + ad &= ac + bc \\ a(c + d) &= (a + b)c \\ \frac{a}{a + b} &= \frac{c}{c + d} \end{aligned}$

11. (a) $y = 108$ **12. (a)** $y = 4$ **15.** 59 to 58 **17. (a)** 32 ounces of cheese for 90¢ **21. (a)** Yes **23.** \$13.01 **26.** \$139.93
29. (a) -3 **(c)** -12 **(e)** 12 **(g)** -4 **(i)** 7 **(k)** -3 **31.** 72 **33.** $2b = 2^1 \cdot 3^2 \cdot 5^1 \cdot 11^3$

Problem Set 7.4 (page 483)

1. (a) 18.75% **(c)** 92.5% **(e)** $36.\overline{36}$% **(g)** 23.81% **2. (a)** 19% **(c)** 215% **3. (a)** $\frac{1}{10}$ **(c)** $\frac{5}{8}$ **4. (a)** 196 **(c)** 285.38 **(e)** 5.4962
5. (a) 420 **(c)** 6 **(e)** 112 **6. (a)** 25% **(c)** 50% **8.** 34.29% **11.** 42% **13. (a)** 60%
15. \$17,380 **17. (a)** \$62.40 **19. (a)** \$3576.80 **22.** \$11,956.13 **24. (a)** 15 **(c)** 6 **25. (a)** 6,400,000 **28. (a)** $\frac{17}{23}$ **29. (a)** $\frac{1}{3}$
(c) $\frac{7}{45}$ **(e)** $\frac{41}{224}$ **30. (a)** $\frac{1}{6}$ **(c)** $\frac{27}{10}$ **(e)** $\frac{17}{48}$ **31. (a)** $r = 10$

CLASSIC CONUNDRUM Some Curious Fractions (Page 487)

$$1, 2, \frac{3}{2}, \frac{5}{3}, \frac{8}{5}, \frac{13}{8}$$

The numbers appear to be quotients of consecutive Fibonacci numbers. Indeed, $1 = F_2/F_1$, $2 = F_3/F_2$, $3/2 = F_4/F_3$, $5/3 = F_5/F_4$, $8/5 = F_6/F_5$, and $13/8 = F_7/F_6$. It appears that the nth term in the sequence is F_{n+1}/F_n.

Chapter 7 Review Exercises (page 489)

1. (a) $2 \cdot 10^2 + 7 \cdot 10^1 + 3 \cdot 10^0 + 4 \cdot 10^{-1} + 2 \cdot 10^{-2} + 5 \cdot 10^{-3}$ **(b)** $3 \cdot 10^{-4} + 5 \cdot 10^{-5} + 4 \cdot 10^{-6}$
2. (a) 0.056 **(b)** 0.08 **(c)** 0.1375 **3. (a)** $\frac{63}{200}$ **(b)** $\frac{603}{500}$ **(c)** $\frac{2001}{10{,}000}$ **4.** $\frac{2}{66}$, 0.33, $\frac{4}{12}$, 0.3334, $\frac{5}{13}$ **5. (a)** $\frac{3451}{333}$ **(b)** $\frac{707}{330}$
6. Irrational. The decimal expansion of a does not have a repeating sequence of digits and it does not terminate.
7. (a) $\frac{2}{9}$ **(b)** $\frac{36}{99}$ **8. (a)** 96.1885 **(b)** 20.581 **(c)** 83.898 **(d)** 7.0 **9. (a)** 34.9437 **(b)** 27.999 **(c)** 109.23237 **(d)** 6.0
10. (a) About 60, 60.384 **(b)** About 40, 39.813 **(c)** About 500, 620.5815 **(d)** About 3, 3.5975
11. (a) 2.473×10^7 **(b)** 1.247×10^{-7} **12. (a)** 8.52×10^9 **(b)** 8.81×10^8 **13.** Suppose $3 - \sqrt{2} = r$ where r is rational. Then $3 - r = \sqrt{2}$. But this implies that $\sqrt{2}$ is rational since the rationals are closed under subtraction. This is a contradiction since $\sqrt{2}$ is irrational. Therefore, by contradiction, $3 - \sqrt{2}$ is irrational. **14.** $(3 - \sqrt{2}) + \sqrt{2} = 3$

15. The decimal expansion of an irrational number has no repeating sequence of digits and is nonterminating.
16. (a) 928.125 ft^2 **(b)** 8.4375 qts **17.** Answers will vary. For example, 4123/9997 is such a fraction.
18. (a) $5/18 = 0.2\overline{7}$; the period starts in the second decimal place. $41/333 = 1.\overline{123}$; the period starts right after the decimal point. $11/36 = 0.30\overline{5}$; the period starts in the third decimal place. $7/45 = 0.1\overline{5}$; the period starts in the second decimal place. $13/80 = 0.1625$; this decimal is terminating. **(b)** Consider the prime factor representation of the denominator in each of the above. If the highest power of 2 and/or 5 appearing in this prime factor representation is r, the period begins in the $r + 1$ decimal place. **19.** 11 to 9 **20. (a)** Yes **(b)** No **(c)** Yes **21.** $7.88 **22.** 13 gal **23.** $y = \frac{35}{3}$
24. 43.2 feet **25. (a)** 62.5% **(b)** 211.5% **(c)** 1.5% **26. (a)** 0.28 **(b)** 0.0105 **(c)** $0.\overline{3}$ **27.** $3.53 **28.** 8% **29.** 55%
30. $3514.98

Chapter 7 Test (page 490)

1. (a) 0.48 **(b)** $0.\overline{24}$ **(c)** $0.\overline{63}$ **2. (a)** $\frac{5}{11}$ **(b)** $\frac{284}{9}$ **(c)** $\frac{7}{20}$ **3.** 5 **4.** 7.30×10^{-13} **5.** Answers will vary. A suitable choice is 125/999. **6. (a)** 17 to 15 **(b)** 53.125% **7.** 15% **8.** 17 yrs **9.** 5 years ago **10.** $1196.41

Chapter 8

JUST FOR FUN A Matter of Speed (page 496)

To drive 300 miles at an average speed of 60 miles per hour requires $300 \div 60 = 5$ hours. Since it takes 5 hours to drive the first 250 miles at 50 miles per hour, it is impossible to average 60 miles per hour for the entire trip.

Problem Set 8.1 (page 505)

3.

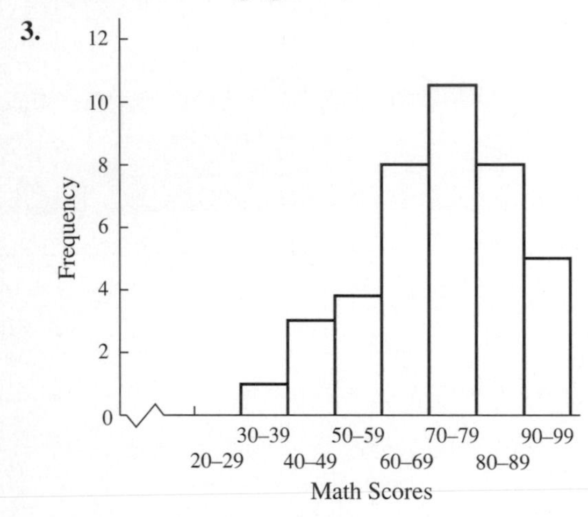

7. (a)

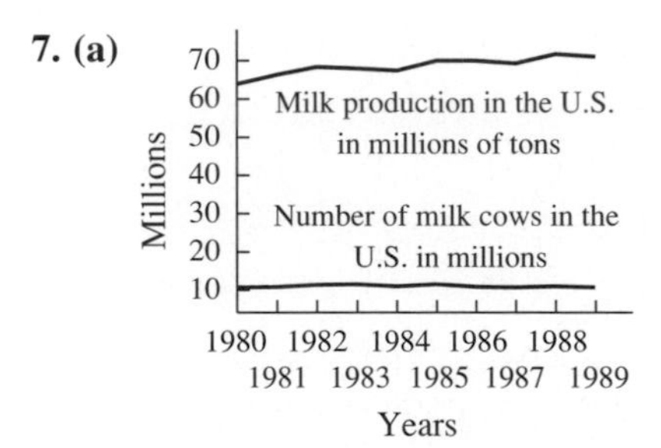

(b) 5.9 tons per cow in 1980 7.2 tons per cow in 1989
(c) It is no doubt due to better production practices on the part of dairy farmers—primarily better feeding practices, better health care for cows, and perhaps to the use of hormones.

9.

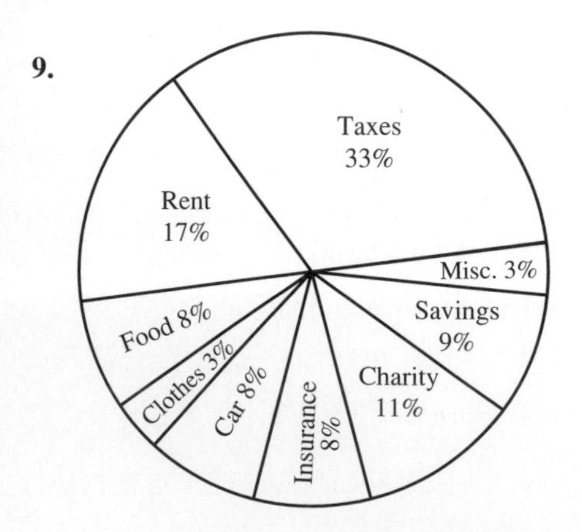

11. (a) 12 **12. (a)** $\left(\frac{5}{7}\right) \cdot (100\%) \doteq 71\%$ **16. (a)** If the curve continues to bend upward as shown, the number aged 65 and older should be about 139 million. **17. (a)** About 310 arrests per 100,000 juveniles. **21. (a)** Histogram (A) emphasizes the changes in the daily Dow Jones average by its choice of vertical scale. The changes appear to be large in histogram (A) thus exaggerating the report of stock activity on the evening news. Histogram (B) makes it clear that the changes are minimal. **(b)** $\left(\frac{36}{3086}\right) \cdot (100\%) \doteq 1.2\%$ which an investor probably would not worry about.

24. (b) First can: $V = \pi(4 \text{ cm})^2 (8 \text{ cm}) \doteq 402 \text{ cm}^3$
Second can: $V = \pi(3 \text{ cm})^2 (14 \text{ cm}) \doteq 396 \text{ cm}^3$

25. (b) Jan.–Mar., about 3% per month
April–June, about 3.5% per month
July–Sept., about 6% per month
Oct.–Dec., about 7% and 10% per month

26. (a) Expenditures per elementary or secondary school student

1960: approximately $\dfrac{\$80{,}000{,}000{,}000}{40{,}000{,}000} \doteq \$2000$

1990: approximately $\dfrac{\$230{,}000{,}000{,}000}{46{,}000{,}000} \doteq \$5000$

(d) The graph of the data in current dollars exaggerates the increase in spending since it also shows growth due to inflation. That is, if a dollar is worth less in 1990 than in 1960, you have to spend more in 1990 just to stay at the same relative level of spending. The graph of the data in constant 1990 dollars is adjusted for inflation and so gives a more realistic picture of the increase in spending. **27. (a)** This includes the categories "some college but no degree," "vocational-technical certificate," and "associate," so about 65%. **28. (a)** From the "total" column, students drinking alcohol. **36. (a)** 0.125 **(c)** 0.068 **37. (a)** $0.\overline{1}$ **(c)** $0.\overline{2142857}$ **38. (a)** $\frac{3}{8}$ **(c)** $\frac{111}{250}$ **39. (a)** $\frac{337}{90}$ **(c)** $\frac{1}{45}$ **(e)** $\frac{47{,}267}{9990}$

JUST FOR FUN How Many Are Single? (page 516)

17/41 of the population are single.

Problem Set 8.2 (page 527)

1. $\bar{x} = 20.6$, $\hat{x} = 19$, mode $= 18$ **3. (a)** $Q_L = 70$, $Q_U = 80$ **(b)** 64 – 70 – 77 – 80 – 86
5. (a) 20.56 **(b)** 4.44 **(c)** 69% **(d)** 94% **(e)** 100% **7. (a)** $3^2 + 4^2 + 5^2 + 6^2 + 7^2 + 8^2$
8. (a) $\frac{1}{1(1+1)} = \frac{1}{2}$ **(b)** $\frac{1}{1(1+1)} + \frac{1}{2(2+1)} = \frac{1}{2} + \frac{1}{6} = \frac{2}{3}$ **9. (a)** $\left(\sum_{i=1}^{5} i\right)^2 = 225$
10. (a) $\sum_{i=1}^{5} (2i + 1)$ or $\sum_{j=0}^{4} (2j + 3)$ Other answers are also possible. **(c)** $\sum_{i=0}^{6} 2 \cdot 3^i$
12. (a) $\bar{x} = 27$, $s \doteq 8.4$ **(b)** $\bar{x} = 32$, $s \doteq 8.4$ **15. (a)** $\bar{x} \doteq 36{,}900$, $\hat{x} = 30{,}000$, mode $= 22{,}000$
16. Many examples satisfy these conditions. **(a)** 5, 7, 10, 14, 14; $\bar{x} = 10$, $\hat{x} = 10$, mode $= 14$
17. (a) If $s = 0$, then all the data values are equal. **19. (a)** All 3s **20. (a)** 32 **(c)** 32 **23.** The total of data values in A is $30 \cdot 45 = 1350$. The total of data values in B is $40 \cdot 65 = 2600$. For the combined data, $\bar{x} = \frac{1350 + 2600}{30 + 40} \doteq 56.4$.
27. (a) Estimate values from the bar graph.

Month	0	N	D	J	F	M	A	M	J	J	A	S
Deficit (in billions)	7.2	7.8	7.0	7.7	7.9	10.4	10.0	8.4	12.1	10.4	10.0	10.9

$\bar{x} = 9.15$ billion **29. (a)** $\bar{x} = 69.43$ **(b)** $\hat{x} = 69$ **(c)** mode $= 69$ **(d)** $s \doteq 1.9$ **(e)** $Q_L = 68$ **(f)** $Q_U = 70.5$
31. There are many possible answers. One pair would be $\frac{11}{16}$ and $\frac{21}{32}$. **33.** Assume that $\sqrt{3} + r = s$ where s is rational. Then $\sqrt{3} = s - r$, and $s - r$ is rational since the rational numbers are closed under subtraction. But this says $\sqrt{3}$ is rational and is a contradiction. Therefore, $\sqrt{3} + r$ is irrational.

JUST FOR FUN A Matter of Sums (page 534)

We guess and check.

$$1 + 23 + 4 + 5 + 67 + 8 + 9 = 117$$
$$12 + 3 + 4 + 5 + 67 + 8 + 9 = 108$$

This reduces the sum by 9. If we could reduce it by 9 more, we would be done. Perhaps we can use the same idea; that is, replace 5 + 67 by 56 + 7. This gives

$$12 + 3 + 4 + 56 + 7 + 8 + 9 = 99$$

as desired.

Notice that any number we obtain in this way must differ from 117 by a multiple of 9. An interesting study would be to determine *all* numbers that can be obtained in the way described.

Problem Set 8.3 (page 540)

1. (a) All freshmen in U.S. colleges and universities in 1995 **(c)** All people in the U.S. **2.** Yes. Many poorer people cannot afford telephones, many people are irritated by telephone surveys and sales pitches, and so on. These factors could certainly bias a sample. **5. (a)** This is surely a poor sampling procedure. The sample is clearly not random. The selection of the colleges or universities could easily reflect biases of the investigators. The choices of the faculty to be included in the study almost surely also reflects the bias of the administrators of the chosen schools. **(b)** Presumably the population is all college and university faculty. But the opinions of faculty at large research universities are surely vastly different from those of their colleagues at small liberal arts colleges. Indeed, there are almost surely four distinct populations here.
9. (a) Suppose the container contains n beans. Then the number of marked beans is $25/n$. This fraction would be approximated by the fraction a/b of marked beans in the handful. Thus, approximately,

$$\frac{25}{n} = \frac{a}{b} \quad \text{and} \quad n = \frac{25b}{a}$$

is approximately the number of beans in the container. **10. (a)** Between 21.8 and 27.2 **11.** Yes, since all sides of the die are equally likely to come up, all sequences of 0s and 1s are equally likely to appear. **15. (b)** 4 **16. (a)** Her time going upstream is $4 \div 2 = 2$ and her time going downstream is $4 \div 4 = 1$. Thus, the average speed is $8 \div 3 = 2.\overline{6}$ miles per hour. **(b)** $\dfrac{2}{\frac{1}{4} + \frac{1}{2}} = \dfrac{8}{3}$ **23. (a)** $\bar{x} = 4.43$ **(b)** $s \doteq 0.68$ **27. (a)** 23 **(b)** 23

CLASSIC CONUNDRUM A Magic Magic Magic Square (page 544)

There are many patterns that all add to 34. Some not already shown include the following.

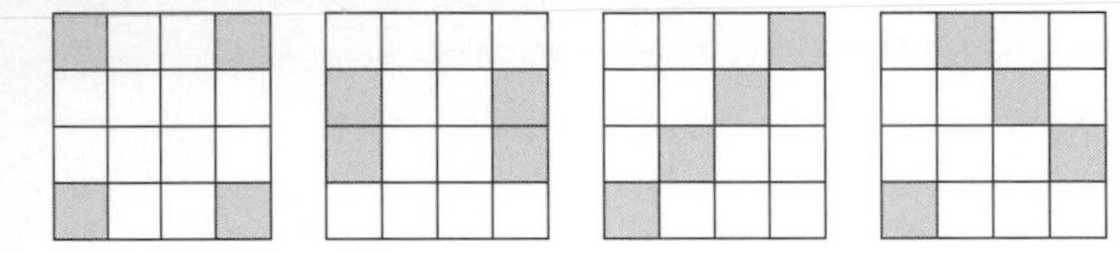

Chapter 8 Review Exercises (page 545)

1. (a)

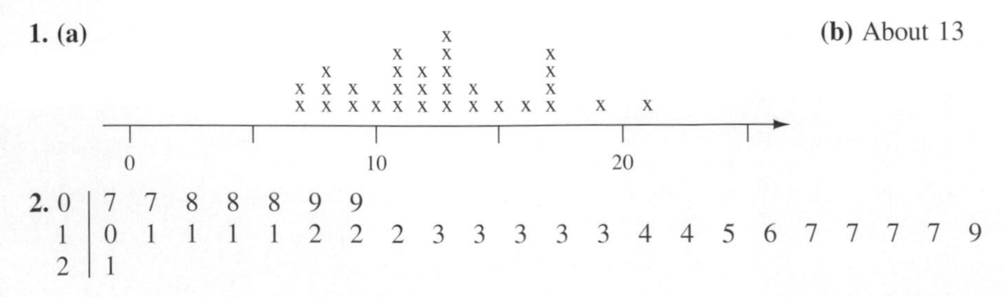

(b) About 13

2.

0	7 7 8 8 8 9 9
1	0 1 1 1 1 2 2 2 3 3 3 3 3 4 4 5 6 7 7 7 7 9
2	1

3.

4.

Mrs. Karnes		Ms. Stevens
9 9 8 8 8 7 7	0	6 6 6 8 8 8 8 8 9 9 9 9 9
9 7 7 7 7 6 5 4 4 3 3 3 3 3 2 2 2 1 1 1 1 0	1	1 1 1 1 1 1 1 1 1 2 3
1	2	

5. (a)

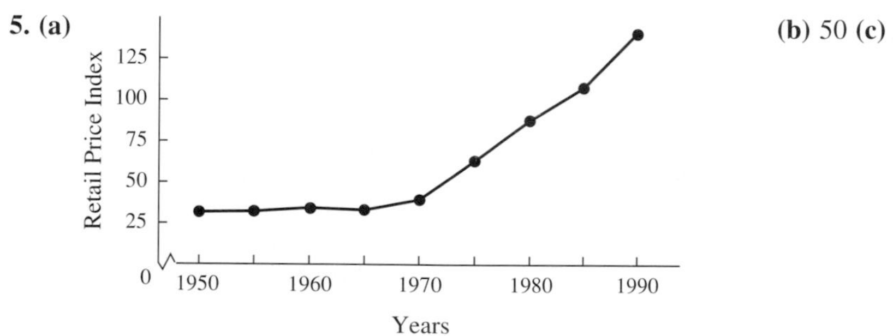

(b) 50 **(c)** About 190

6. Many examples exist. One example is 1, 2, 3, 4, 90 with a mean of 20. **7.**

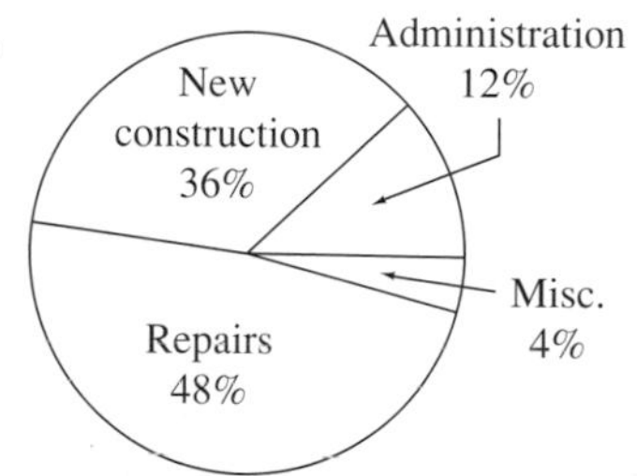

8. (a) The volume of the larger box is about double that of the smaller box. But the length of a side of the larger box is less than double the length of a side of the smaller box, suggesting that the change was less than doubling. **(b)** The volume of the larger is about twice that of the smaller, which defends the pictograph. **9.** How is medical doctor defined? Does this include all specialists? osteopaths? naturopaths? chiropractors? acupuncturists? How was the sampling done to determine the stated average?

10. $\bar{x} \doteq 12.6$, $\hat{x} = 12.5$, mode $= 13$, $s \doteq 3.5$ **11. (a)** $Q_L = 10$, $Q_U = 15$ **(b)** 7 – 10 – 12.5 – 15 – 21 **(c)** $\hat{x} = 9$, $Q_L = 8$, $Q_U = 11$ **(d)** 6 – 8 – 9 – 11 – 13 **(e)**

12. (a) $\bar{x} \doteq 27.8$, $s \doteq 2.0$ **(b)** $\bar{x} = 27.3$, $s \doteq 1.6$ **(c)** The means are about the same for the two sets of data, but the standard deviation is smaller for the second histogram since the data is less spread out from the mean. **13.** There are $21 \cdot 77 = 1617$ points for the 21 students. So there are 1839 points for all 24 students. Thus, the average is $1839 \div 24 \doteq 76.6$. **14.** There are $27 \cdot 75 = 2025$ points for the second period students and $30 \cdot 78 = 2340$ points for the fourth period students. Thus, the average for all students is $(2340 + 2025)/57 \doteq 76.6$. **15.** No. The sample only represents the population of students at State University, not university students nationwide. **16.** You might want to limit your sample to persons 20 years old and older who want to work. Alternatively, you might want to define several populations and determine figures for each: persons 20 years old and older who want to work, teenagers who want full-time employment, teenagers who want part-time employment, adults 20 years old and older who want part-time employment, and so on. **17.** Telephone polls sample only persons sufficiently affluent to own a telephone. They are also biased by the fact that many people do not like telephone polls or telephone commercial solicitations and so refuse to respond or respond inaccurately because of anger. **18.** Voluntary responses to mailed questionnaires tend to come primarily from those who feel strongly (either positively or negatively) about an issue or who represent narrow special interest groups. They are rarely representative of the population as a whole. **19.** One way is to number the students in alphabetical order and select the sample using a spinner or a random number generator on a computer. Alternatively, one might print the names of all students on slips of paper, place them in a container, mix them well, and have someone close their eyes and select from the container the names of those to be in the sample. **20.** Yes. Just continue taking samples until one finally shows up with eight out of the ten in the sample preferring WHITO toothpaste.

Chapter 8 Test (page 547)

1. (a)

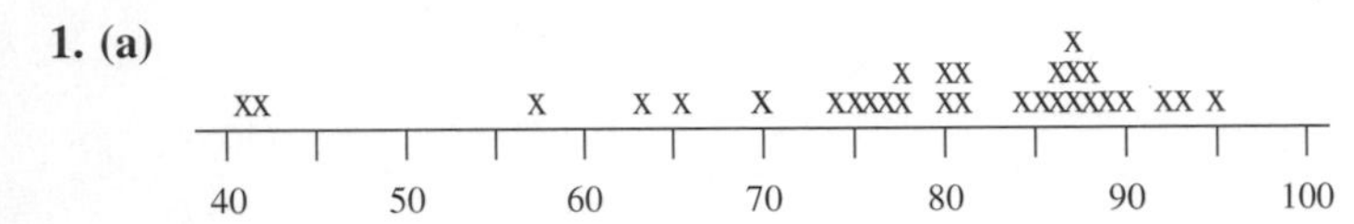

(b) 41 and 42 are certainly outliers, since they are so far removed from the major collection of data.
2. (a) $\bar{x} \doteq 78.5$ **(b)** $\hat{x} = 81$ **(c)** mode $= 87$ **(d)** $s \doteq 13.2$

3.

Stem	Leaves
4	1 2
5	7
6	3 6
7	0 4 5 6 7 8 8
8	0 0 1 1 4 5 6 6 7 7 7 8 8 9
9	0 2 3 5

4. 41 – 75 – 81 – 87 – 95

5. $(0.40) \cdot (360°) = 144°$ **6.** A random sample is one chosen in such a way that every subset of the population has an equal chance of being included. **7.** To average 80% on all five tests, she must score at least 400. Thus, the total score for her last two tests should be $400 - (77 + 79 + 72) = 172$. **8.** With a normal distribution, 68% of the data will be within one standard deviation of the mean, 95% of the data will be within two standard deviations of the mean, and 99.7% of the data will be within three standard deviations of the mean. **9.** One way would be to number the students and select 200 using random numbers from a random number generator to determine which students are included in the sample. **10.** If the sample is not chosen at random it is quite likely to reflect bias—bias of the sampler, bias reflecting the group from which the sample was actually chosen (views of teamsters, or AARP members), and so on.

Chapter 9

JUST FOR FUN Three-Card Monte (page 558)

(a) There are three possibilities for the three cards: BB, BW, and WW. If a card is dealt at random and a black side comes up, then the card is either the BB card or the BW card. Thus, it might appear that the probability that the other side is black is 1/2. However, the BB card can come up either way and still show B. Thus, in two out of three cases, a card showing B is B on the other side as well. Thus, the probability in question is 2/3. **(b)** We did 15 trials, obtaining 11 black on the second side for a probability of $P_e = 0.73$. This tends to confirm the above guess.

Problem Set 9.1 (page 561)

1. $P_e = 4/35 \doteq 0.11$ **4. (b)** 1 **7. (a)** iv **(c)** ii **(e)** ii **9. (b)** 2 + 6, 3 + 5, 4 + 4, 5 + 3, 6 + 2. **11. (a)** Answers will vary. We got $P_e(A) = 12/20 = 0.6$, $P_e(B) = 6/20 = 0.3$, and $P_e(C) = 2/20 = 0.1$. **(b)** About right since there are 360° in a complete revolution and $240/360 = 0.6\overline{7}$,

$90/360 = 0.25$, and $30/360 = .08\overline{3}$. **13.** Answers will vary. We obtained P_e (first 6 on fourth roll) $\doteq 0.2$. **16.** Answers will vary. When we did the experiment, we obtained four hearts including one face card, seven diamonds including three face cards, five spades including two face cards, and four clubs with no face cards. This yielded the following results. **(a)** $P_e(\text{R}) = 11/20 = 0.55$ **(b)** $P_e(F) = 6/20 = 0.3$ **(c)** $P_e(R \text{ or } F) = 13/20 = 0.65$ **(d)** $P_e(R \text{ and } F) = 4/20 = 0.2$ **(e)** $P_e(R) + P_e(F) - P_e(R \text{ and } F) = 0.55 + 0.3 - 0.2 = 0.65$ **(f)** This suggests that $P_e(R \text{ or } F) = P_e(R) + P_e(F) - P_e(R \text{ and } F)$ unlike the result suggested by problem 15. **(g)** The only apparent difference is that 8 and 7 in problem 15 cannot occur simultaneously while the events R and F in this problem can occur simultaneously. This means that counting the number of red cards and counting the number of face cards counts red face cards twice. To adjust the count properly we must subtract the count of those cards that are red and are face cards. **18.** Answers will vary. When we did the experiment, H occurred all told 12 times, 5 occurred all told three times, and 5 and H occurred together two times. This yielded the following results. **(a)** $P_e(\text{H}) = 12/20 = 0.6$ **(b)** $P_e(5) = 3/20 = 0.15$ **(c)** P_e (H and 5) $= 2/20 = 0.1$ **(d)** $P_e(H) \cdot P_e(5) = (0.6)(0.15) = 0.09$ **(e)** Yes. Since the events H and 5 are independent, the number of simultaneous occurrences of H and 5 should be about $P_e(\text{H}) \cdot$ (the number of occurrences of 5). But then

$$P_e(\text{H and } 5) \doteq \frac{P_e(\text{H}) \cdot (\text{the number of occurrences of } 5)}{20}$$
$$= P_e(\text{H}) \cdot P_e(5).$$

24. (e) Area of region in (a) is 0.1
Area of region in (b) is 0.12
Area of region in (c) is 0.50
The results compare very favorably for region (c) and less so for (a) and (b). With more points the comparison would be much closer. **25. (a)** P_e (a person 20 years old lives to be at least 80) $= 2{,}626{,}372/9{,}664{,}994 \doteq 0.27$.
30. $s \doteq 2.3$ **32.** Since this range extends two standard deviations on either side of the mean,

$$P(\text{an individual falls in the range}) = 0.95.$$

JUST FOR FUN Social Security Numbers (page 566)

Over 90 percent of the numbers will have an even digit in this position.

Problem Set 9.2 (page 581)

1. $4 = 1 + 3 = 2 + 2 = 3 + 1$
$6 = 1 + 5 = 2 + 4 = 3 + 3 = 4 + 2 = 5 + 1$ So there are 10 ways to score 4 or 6.
3. (a) 3 ways; H and 2, H and 4, H and 6 **5.** The Venn diagram shows that four students speak both English and Japanese

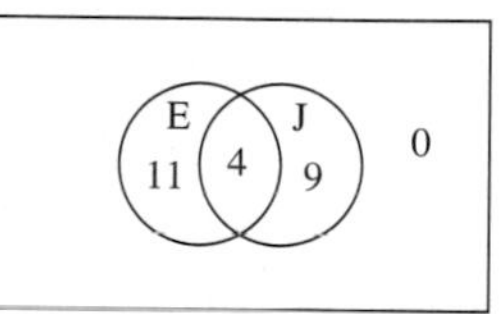

7. Let R be the set of red face cards and let A be the set of black aces. Since $R \cap A = \varnothing$, the second principle of counting applies and $n(R \text{ or } A) = n(R) + n(A) = 6 + 2 = 8$. **8. (a)** $6 \cdot 5 \cdot 4 \cdot 3 = 360$ **(b)** $3 \cdot 5 \cdot 4 \cdot 3 = 180$ **(c)** $3 \cdot 2 \cdot 4 \cdot 3 = 72$
10.

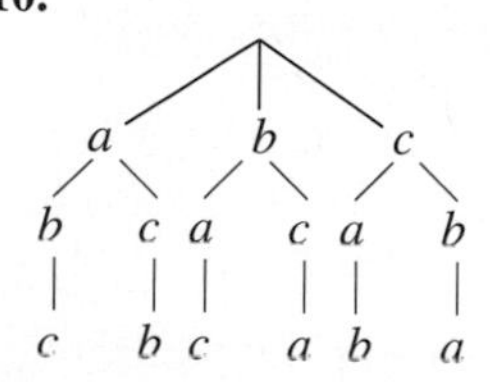

The code words are *abc*, *acb*, *bac*, *bca*, *cab*, and *cba*.

12. (a) 5040 **(c)** 72 **(e)** 35,280 **13. (a)** $13 \cdot 12 \cdot 11 \cdot 10 \cdot 9 \cdot 8 \cdot 7 \cdot 6 = 51{,}891{,}840$ **(c)** $15 \cdot 14 = 210$ **(e)** $\frac{15!}{15!} = 1$ **15. (a)** $5 \cdot 5 \cdot 5 \cdot 5 \cdot 5 = 3125$ **17. (a)** $\frac{4!}{2!\,2!} = 6$ **18. (a)** To obtain an even sum, two odd faces or two even faces must come up. This can happen in $3 \cdot 3 + 3 \cdot 3 = 18$ ways. **21. (a)** If we think of *ab* as a single entity, then there are four things to put in order. This can be done in $4! = 24$ ways. **(b)** There are 24 with *b* immediately following *a* and, by the same reasoning as in part (a), 24 with *a* immediately following *b*. Therefore, there are 48 with *a* and *b* adjacent. **(c)** By symmetry, in half the possible arrangements *a* would precede *e* and in half *e* would precede *a*. Therefore, *a* precedes *e* in $5!/2 = 60$ arrangements. Alternatively, if *a* is first, *a* precedes *e* in $4! = 24$ arrangements. If *a* is second, *a* precedes *e* in $3 \cdot 3! = 18$ arrangements. If *a* is third, *a* precedes *e* in $2 \cdot 3! = 12$ arrangements and if *a* is fourth, it precedes *e* in $3! = 6$ arrangements. Adding, we again obtain 60 as before. **26. (a)** $C(13, 3) \cdot C(11, 3) = \frac{13 \cdot 12 \cdot 11}{1 \cdot 2 \cdot 3} \cdot \frac{11 \cdot 10 \cdot 9}{1 \cdot 2 \cdot 3} =$ 47,190 ways **32.** It is incorrect since the trials are completely independent. What happens on one trial has no affect on any other trial. She may continue to lose all evening! **34. (a)**

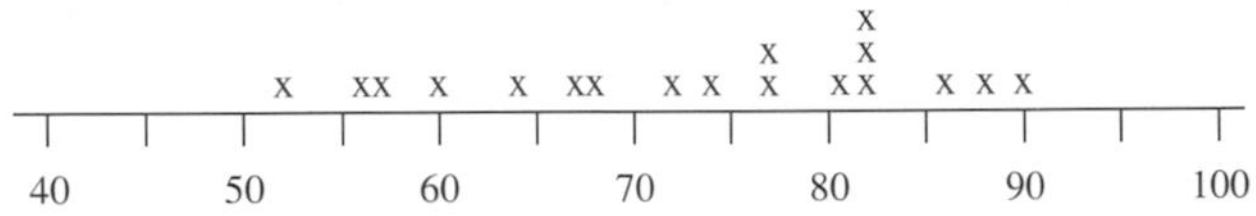

(c) 53 − 64 − 75.5 − 82 − 90

JUST FOR FUN It's a Girl! (page 585)

The possibilities in order of age are boy, boy; boy, girl; girl, boy; girl, girl. In three cases there is at least one boy and in two of these the other child is a girl. Thus, the desired probability is 2/3.

JUST FOR FUN A Probability Paradox (page 586)

For the three tables the probabilities are as follows:

$$\text{Table } A\text{: } P(R \mid G) = 5/11 \doteq 0.45,\ P(R \mid B) = 3/7 \doteq 0.43$$
$$\text{Table } B\text{: } P(R \mid G) = 6/9 \doteq 0.67,\ P(R \mid B) = 9/14 \doteq 0.64$$
$$\text{Table } C\text{: } P(R \mid G) = 11/20 = 0.55,\ P(R \mid B) = 12/21 \doteq 0.57$$

This is quite surprising since $P(R \mid G) > P(R \mid B)$ on each of tables *A* and *B*, but $P(R \mid G) < P(R \mid B)$ on table *C* where the balls in the hats were obtained by combining the balls in the hats on tables *A* and *B*. This is another example of Simpson's paradox.

Problem Set 9.3 (page 597)

1. (a) Listed in the order penny, nickel, dime, quarter, the possibilities are:

HHHH	HTHH	THHH	TTHH
HHHT	HTHT	THHT	TTHT
HHTH	HTTH	THTH	TTTH
HHTT	HTTT	THTT	TTTT

(c) $P(\text{2 heads and 2 tails}) = 6/16 = 0.375$ **3. (a)** $4/22 \doteq 0.18$ **(c)** $6/22 \doteq 0.27$ **4. (a)** $4/12 \doteq 0.33$ **5. (a)** $7/20 = 0.35$ **6. (a)** 0.19 **(c)** 0.33 **(e)** 0.25 **7. (a)** $4/50 = 0.08$ **(c)** $0/50 = 0$ **9.** $P(3\text{H}) = C(13, 3)/C(52, 3) = \frac{13 \cdot 12 \cdot 11}{1 \cdot 2 \cdot 3} \div \frac{52 \cdot 51 \cdot 50}{1 \cdot 2 \cdot 3} \doteq 0.013$ **11.** These answers are determined by ratios of angular measures of appropriate regions. **(a)** $P(\text{shaded area}) = 3/10 = 0.30$ **(c)** $P(10 \text{ or } 6) = 2/10 = 0.20$ **(e)** $P(8 \mid \text{shaded area}) = 1/3$ **(g)** $P(\text{vowel or an odd numbered region}) = 7 \cdot 7/10 = 0.70$ **12.** Here the probabilities are ratios of areas. **(a)** $P(1) = 16/25 = 0.64$ **(c)** $P(5) = 1/25 = 0.04$ **13. (a)** $P(b) = 1/20$ **(b)** $P(a \text{ or } b \text{ or } c \text{ or } d \mid a \text{ or } b \text{ or } c \text{ or } d \text{ or } e) = 4/5$ **14. (a)** 5 : 31 or, equivalently, 5/31 **16.** $P(A \text{ or } C) = P(A) + P(C) = \frac{1}{2} + \frac{1}{6} = \frac{4}{6} = \frac{2}{3}$. Thus, the odds in favor of *A* or *C* are $\frac{\frac{2}{3}}{1 - \frac{2}{3}} = \frac{2}{1}$ or 2 : 1.

18. $E = \$4 \cdot \frac{1}{36} + \$6 \cdot \frac{2}{36} + \$8 \cdot \frac{3}{36} + \$10 \cdot \frac{4}{36} + \$20 \cdot \frac{5}{36} + \$40 \cdot \frac{6}{36} + \$20 \cdot \frac{5}{36} + \$10 \cdot \frac{4}{36} + \$8 \cdot \frac{3}{36}$ $+ \$6 \cdot \frac{2}{36} + \$4 \cdot \frac{1}{36} = \frac{\$600}{36} = \$16.67$ to the nearest penny.

19. (a) $P(\text{both white}) = \frac{C(6, 2)}{C(14, 2)} = \frac{\frac{6 \cdot 5}{1 \cdot 2}}{\frac{14 \cdot 13}{1 \cdot 2}} \doteq 0.16$ **20. (a)** $P(\text{all four red}) = \frac{C(8, 4)}{C(19, 4)} = \frac{\frac{8 \cdot 7 \cdot 6 \cdot 5}{1 \cdot 2 \cdot 3 \cdot 4}}{\frac{19 \cdot 18 \cdot 17 \cdot 16}{1 \cdot 2 \cdot 3 \cdot 4}} \doteq 0.02$

21. (a) $P(\text{a code word begins with a}) = \frac{25 \cdot 24 \cdot 23 \cdot 22}{26 \cdot 25 \cdot 24 \cdot 23 \cdot 22} \doteq 0.04$

24. (a) P(a five card hand contains exactly two aces)

$$= \frac{C(4, 2) \cdot C(48, 3)}{C(52, 5)}$$

$$= \frac{\frac{4 \cdot 3}{1 \cdot 2} \cdot \frac{48 \cdot 47 \cdot 46}{1 \cdot 2 \cdot 3}}{\frac{52 \cdot 51 \cdot 50 \cdot 49 \cdot 48}{1 \cdot 2 \cdot 3 \cdot 4 \cdot 5}}$$

$$\doteq 0.04$$

25. The seven numbers with two alike in a fixed order appear with probability $(1/6)^7$. To compute the number of ways seven such numbers can appear note that we can choose the number to appear twice in six ways *and* each of the other numbers in only one way *and* we can then order these numbers in $7!/2!$ ways. Thus, the desired probability is

$$6 \cdot \frac{7!}{2!} \cdot \left(\frac{1}{6}\right)^7 \doteq 0.05.$$

26. (a) $P(A \text{ and } B) = P(A)P(B \mid A) = P(B)P(A \mid B)$ Therefore, $P(B \mid A) = P(B)P(A \mid B)/P(A)$.
(c) Replacing $P(A)$ in part (a) by its equal from part (b), we obtain

$$P(B \mid A) = \frac{P(B)P(A \mid B)}{P(B)P(A \mid B) + P(\overline{B})P(A \mid \overline{B})}.$$

29. $C(7, 4) + C(8, 4) = \frac{7 \cdot 6 \cdot 5 \cdot 4}{1 \cdot 2 \cdot 3 \cdot 4} + \frac{8 \cdot 7 \cdot 6 \cdot 5}{1 \cdot 2 \cdot 3 \cdot 4} = 105$ **31.** $5 \cdot 21 \cdot 4 \cdot 20 \cdot 3 + 21 \cdot 5 \cdot 20 \cdot 4 \cdot 19 = 184{,}800$

CLASSIC CONUNDRUM Should the Contestant Switch? (page 601)

If the contestant picks the "rich" door the host reveals the "lesser" or "joke" door. If the contestant picks the "lesser" door the host reveals the "joke" door leaving the "rich" door closed. If the contestant picks the "joke" door, the host reveals the "lesser" door leaving the "rich" door. Thus, in two of the three cases the door selected by the contestant is not the "rich" door. The contestant should switch.

Chapter 9 Review Exercises (page 603)

1. Answers will vary. When we did the experiment, three heads and a tail occurred five times so that $P_e = 5/20 = 0.25$.
2. Answers will vary. When we did the experiment we obtained the following:

2	3	4	5	6	7	8	9	10	11	12
I	I	II	II	III	III	IIII	I	II	I	

Using this data **(a)** $P_e(3 \text{ or } 4) = 3/20 = 0.15$ **(b)** $P_e(\text{score at least } 5) = 16/20 = 0.80$ **3.** Answers will vary. Using the above data, we obtain $P_e(5 \text{ or } 6 \text{ or } 7 \mid 5 \text{ or } 6 \text{ or } 7 \text{ or } 8 \text{ or } 9) = 8/13 \doteq 0.62$ **4.** Answers will vary. In our study, seven out of 20 chose chocolate so that $P_e = 7/20 = 0.35$. **5.** Answers will vary. When we did the experiment we obtained the data shown on page A-46.

Point up	5	4	3	5	2	4	3	2	2	3	2	3	3	1	2	0	4	4	3	2
Head up	0	1	2	0	3	1	2	3	3	2	3	2	2	4	3	5	1	1	2	3

Using this data **(a)** P_e(3 tacks land point up) $= 6/20 = 0.30$ **(b)** P_e(2 or 3 tacks land point up) $= 12/20 = 0.60$
6. Answers will vary. Doing the experiment, we obtain the following data with the results indicated in (a) and (b).

Number of trials to get a 5 or 6	1	2	3	4	5	6	7
			||	||||	|||		|

(a) Average number of rolls required is

$$\frac{3 + 3 + 4 + 4 + 4 + 4 + 5 + 5 + 5 + 7}{10} = 4.4$$

We guess that it should take 4 or 5 rolls to get a 5 or 6.

(b) P_e(it takes precisely five rolls to obtain 5 or 6) $= 3/10 = 0.30$ **7.** Answers will vary. Doing the experiment, we obtained the following data yielding the results in (a), (b), and (c).

	Heart	**Nonheart**
Ace	|	||
Nonace	|||| |||	|||| ||||

(a) P_e (ace or heart) $= 11/20 = 0.55$ **(b)** P_e (ace and heart) $= 1/20 = 0.05$ **(c)** P_e (ace | heart) $= 1/9 \doteq 0.11$
8. P_e (person aged 20 will live to age 90) $= 468{,}174/9{,}664{,}994 \doteq 0.05$
9. (a) HHH HTT **(b)** 3
HHT THT
HTH TTH
THH TTT

10. (a) $C(15, 9) = \dfrac{15 \cdot 14 \cdot 13 \cdot 12 \cdot 11 \cdot 10 \cdot 9 \cdot 8 \cdot 7}{1 \cdot 2 \cdot 3 \cdot 4 \cdot 5 \cdot 6 \cdot 7 \cdot 8 \cdot 9} = 5005$

(b) $C(2, 1) \cdot C(3, 1) \cdot C(13 \cdot 7) = \dfrac{2}{1} \cdot \dfrac{3}{1} \cdot \dfrac{13 \cdot 12 \cdot 11 \cdot 10 \cdot 9 \cdot 8 \cdot 7}{1 \cdot 2 \cdot 3 \cdot 4 \cdot 5 \cdot 6 \cdot 7} = 10{,}296$

11. Fill in the Venn diagram from the inside out.

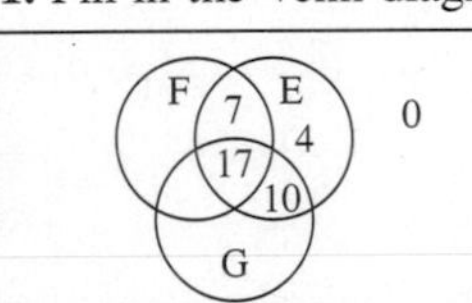

(a) $90 - 4 = 86$ **(b)** 7 **12. (a)** $C(13, 2) = \dfrac{13 \cdot 12}{1 \cdot 2} = 78$ **(b)** $C(12, 2) = \dfrac{12 \cdot 11}{1 \cdot 2} = 66$
(c) $C(13, 2) + C(12, 2) - C(3, 2) = 78 + 66 - 3 = 141$
13. (a) $26 \cdot 25 \cdot 24 \cdot 23 \cdot 22 = 7{,}893{,}600$ **(b)** $5 \cdot 4 \cdot 24 \cdot 23 \cdot 22 = 242{,}880$ **(c)** $24 \cdot 23 \cdot 22 = 12{,}144$
14. (a) $\dfrac{7!}{2!\,2!\,2!\,1!} = 630$ **(b)** $\dfrac{6!}{2!\,2!\,1!\,1!} = 180$ **(c)** $\dfrac{6!}{2!\,2!\,1!\,1!} = 180$ **15. (a)** HH1, HH2, HH3, HH4, HH5, HH6
HT1, HT2, HT3, HT4, HT5, HT6
TH1, TH2, TH3, TH4, TH5, TH6
TT1, TT2, TT3, TT4, TT5, TT6

(b) $P(\text{TT5}) = 1/24 \doteq 0.04$ **16.** $P(5 \mid \text{TT}) = 1/6 \doteq 0.17$
17. $P(\text{sum at most } 11) = 1 - P(\text{sum is } 12) = 1 - \dfrac{1}{36} \doteq 0.97$

18. (a) $[C(5, 2) + C(6, 2) + C(4, 2)]/C(15, 2) = \left(\dfrac{5 \cdot 4}{1 \cdot 2} + \dfrac{6 \cdot 5}{1 \cdot 2} + \dfrac{4 \cdot 3}{1 \cdot 2}\right) \Big/ \dfrac{15 \cdot 14}{1 \cdot 2} \doteq 0.30$

(b) $P(\text{both white}) = \dfrac{C(5, 2)}{C(15, 2)} = \dfrac{\frac{5 \cdot 4}{1 \cdot 2}}{\frac{15 \cdot 14}{1 \cdot 2}} \doteq 0.10$

(c) $P(\text{both white} \mid \text{both the same}) = \dfrac{C(5, 2)}{C(5, 2) + C(6, 2) + C(4, 2)}$

$$= \frac{10}{10 + 15 + 6}$$

$$\doteq 0.32$$

19. (a) $P(\text{b and } 8) = 0$ **(b)** $P(\text{b or } 8) = P(\text{b}) + P(8) = \dfrac{90}{360} + \dfrac{30}{360} = \dfrac{1}{3} \doteq 0.33$

(c) $P(\text{b} \mid 8) = 0$ **(d)** $P(\text{b and } 2) = \dfrac{30}{360} \doteq 0.08$

(e) $P(\text{b or } 2) = P(\text{b}) + P(2) - P(\text{b and } 2)$

$$= \frac{90}{360} + \frac{30}{360} - \frac{30}{360}$$

$$= 0.25$$

(f) $P(2 \mid \text{b}) = 30/90 \doteq 0.33$ **20. (a)** 3 : 5 or 3/5 **(b)** 1 : 7 or 1/7 **21. (a)** $P(A)/[1 - P(A)]$; that is,

$$\frac{0.85}{1 - 0.85} = \frac{0.85}{0.15} = \frac{17}{3} \quad \text{or} \quad 17 : 3$$

(b) $P(A) = 17/25 = 0.68$ **22. (a)** E = \$5 · (.50) + \$10(.25) + \$20 · (.10) = \$7 **(b)** No. On average you expect to lose \$3 per game. **23.** Since the sexes of children are independent, $P(\text{other two children are boys}) = \dfrac{1}{2} \cdot \dfrac{1}{2} = 0.25$.

Chapter 9 Test (page 604)

1. Prepare a card as shown and ask a number of people to choose a number. Calculate the empirical probability of choosing 3 as the number of times 3 is chosen divided by the number of people questioned. **2.** This would be an empirical probability obtained by keeping records for a large number of trials of treating strep throat with penicillin. **3. (a)** 5040 **(b)** 504 **(c)** 40,320 **(d)** 5040 **(e)** 6720 **(f)** 40,320 **(g)** 84 **(h)** 1 **4. (a)** Fill in the regions in the Venn diagram starting with the innermost region.

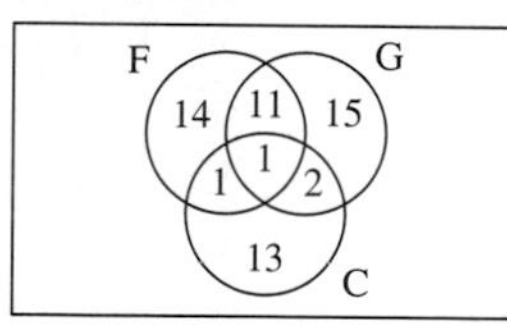

Adding all the counts, there are 57 students in all. **(b)** 13 students **(c)** 11 students

5. (a) $C(8, 5) = \dfrac{8 \cdot 7 \cdot 6 \cdot 5 \cdot 4}{1 \cdot 2 \cdot 3 \cdot 4 \cdot 5} = 56$ **(b)** 0 ways. Only 5 balls are being selected.

(c) $C(5, 5) + C(8, 5) = \dfrac{5 \cdot 4 \cdot 3 \cdot 2 \cdot 1}{1 \cdot 2 \cdot 3 \cdot 4 \cdot 5} + \dfrac{8 \cdot 7 \cdot 6 \cdot 5 \cdot 4}{1 \cdot 2 \cdot 3 \cdot 4 \cdot 5} = 57$

6. $P(\text{2 yellow balls} \mid \text{3 green balls}) = C(5, 2)/C(14, 2) \doteq 0.11$
7. 5 to 12 **8.** 7 to 13 **9. (a)** $7^4 = 2401$ **(b)** $7 \cdot 6 \cdot 5 \cdot 4 = 840$ **10. (a)** $2 \cdot 6 \cdot 5 \cdot 4 = 240$ **(b)** Choose cd as a single unit in 1 way, and choose two more letters in $C(5, 2) = 10$ ways, and arrange these three items in order in $3! = 6$ ways. Similarly for dc. Therefore, the desired number is $2 \cdot 1 \cdot 10 \cdot 6 = 120$.

Chapter 10

JUST FOR FUN Arranging Points at Integer Distances (page 613)

The isosceles trapezoid with bases 3″ and 4″ has 2″ sides and 4″ diagonals.

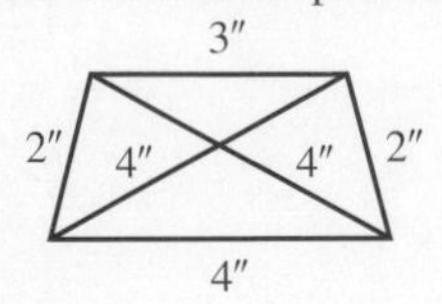

Problem Set 10.1 (page 624)

1. (a) Answers will vary; the p should extend into the lowest row and leave the upper 2 or 3 rows blank.
2. (a) E U **3. (a)** A B

4. (a) 10 angles. $\angle APB$, $\angle APC$, $\angle APD$, $\angle APE$, $\angle BPC$, $\angle BPD$, $\angle BPE$, $\angle CPD$, $\angle CPE$, $\angle DPE$.
5. $m(\angle AXB) = m(\angle AXE) - m(\angle BXE) = 180° - 132° = 48°$
$m(\angle CXD) = m(\angle BXD) - m(\angle BXC) = 90° - 35° = 55°$
$m(\angle DXE) = m(\angle BXE) - m(\angle BXD) = 132° - 90° = 42°$
7. The three lines are concurrent. **9. (a)** Opposite angles are supplementary: $m(\angle A) + m(\angle C) = 180°$, $m(\angle B) + m(\angle D) = 180°$. **10. (a)** Angle measurements will be different for different drawings. You should find $m(\angle APB) = m(\angle AQB) = m(\angle ARB)$ and this measure is half the measure of $\angle AOB$.
11. (a) $58°\ 36'\ 45'' = 58° + \left(\frac{36}{60}\right)^{\circ} + \left(\frac{45}{3600}\right)^{\circ} = 58.6125°$ **(c)** $71.32° = 71° + (0.32)(60') = 71° + 19.2' = 71° + 19' + (0.2)(60'') = 71°\ 19'\ 12''$
12. Ten times; once between 1 and 2, once between 2 and 3, . . . , and once between 10 and 11.
13. (a) 360° **(d)** 30° **14. (a)** 90° **(c)** The minute hand is on the 6 and the hour hand is halfway between the 4 and the 5. The angle between two consecutive numbers is $\frac{1}{12}$ of a revolution, or 30°. So the angle is $(1.5)(30°) = 45°$.
16. Draw a horizontal ray $\overrightarrow{PQ}$ at P, in the opposite direction of $\overrightarrow{AB}$ and $\overrightarrow{CD}$. The opposite interior angles theorem then gives $m(\angle APQ) = 130°$, $m(\angle CPQ) = 140°$. Thus $m(\angle P) = 360° - 130° - 140° = 90°$.
17. (a) 40°, since the interior angles of a triangle add up to 180° **(c)** 49°, since the interior angles of a triangle add up to 180° and a right angle has measure 90° **18. (a)** $x + x + 30° = 180°$ so $x = 75°$. The interior angles measure 75°, 75°, and 30°. **19. (a)** No, because an obtuse angle has measure greater than 90° and adding two such measures would exceed 180° which is the sum of all three interior angle measures for any triangle.
20. (a) Zero intersection points if the five lines are parallel to each other. **21. (a)** 6 lines
22. Notice that adjacent angles are not all composed from adjacent rays. But a pair of adjacent angles is determined by three rays and each collection of three rays determines a different pair of adjacent angles. So an equivalent question is: How many ways can a subset of three rays be chosen from ten rays? $C(10, 3) = 120$ ways.
24. The pencil turns through each interior angle of the triangle. Since the pencil faces the opposite direction when it returns to the starting side, it has turned a total of 180°. This demonstrates that the sum of measures of the interior angles of a triangle is 180°. **25. (a)** By trial and error, ten taxi segments. Alternatively, this is number of sequences of 3 E(east) and 2 N(north) segments, and there are ten ways to form such sequences: NNEEE, NENEE, . . . , EEENN. **26.** The four searchlights, placed anywhere in the plane, can always be turned to illuminate the whole plane. Here's one way to do this. Consider two lights that are farthest south. Point the eastern most one of these to the northwest, and the western most to the northeast. Similarly, point the remaining two lights southeast and southwest.

28. (a) 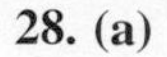

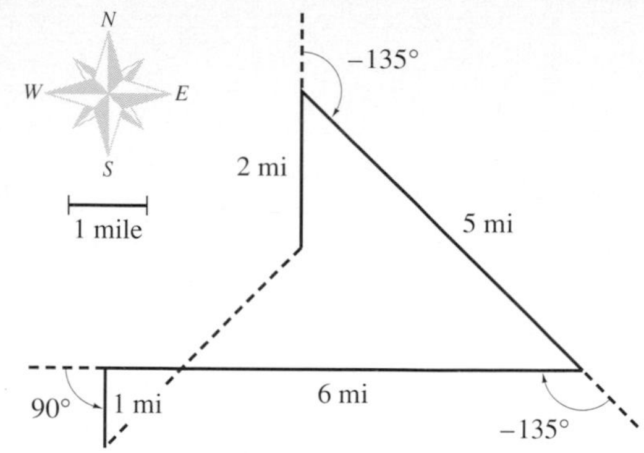

30. (a) A full revolution takes 24 hours, so the earth turns $\frac{1}{24}$ of a revolution, or 15°, in one hour.
31. The angle of latitude is equal to the angle of elevation to Polaris.
33. (a) Q, R, and S are collinear **34. (a)** The lines $\overleftrightarrow{BB'}$ and $\overleftrightarrow{DD'}$ intersect at a right angle at P.

JUST FOR FUN Triangle Pick-Up-Sticks (page 633)

It is not possible to remove nine matchsticks and leave two triangles, but the other cases are possible.

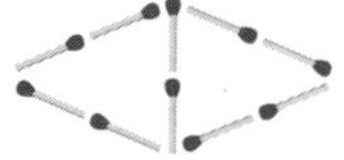
Remove 6

Remove 7

Remove 8

Remove 10

Remove 11

Problem Set 10.2 (page 644)

1.

	(a)	(b)	(c)	(d)
Simple curve		√		
Closed curve		√	√	√
Polygonal curve	√	√		
Polygon		√		

2. (a) An example is 

(c) An example is 

4. (a) Convex **(b)** Nonconvex **5. (a)** 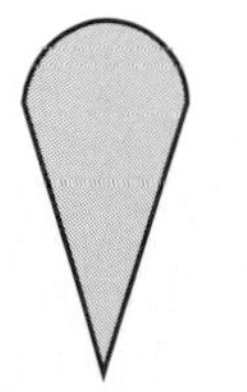 **(c)** 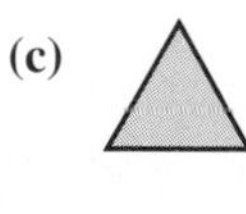

6. (a) 6 **(c)** 8 **7.** $2x + 5x + 5x + 5x + 5x + 2x = (6 - 2)(180°)$ or $24x = 720°$, so $x = 30°$. The angles measure 60°, 150°, 150°, 150°, 150°, and 60°. **9. (a)** $(5 - 2)(180°) = 540°$ **(c)** $(6 - 2)(180°) = 720°$
10. Pictures will vary. **(a)** Triangle **(c)** Decagon **12. (a)** 360° **(c)** 0°
14.

	(a)	(c)
Interior angle	$128\frac{4}{7}°$	108°
Exterior angle	$51\frac{3}{7}°$	72°
Central angle	$51\frac{3}{7}°$	72°

16. (a) $\frac{360°}{n} = 15°$ so $n = 24$. **18. (a)** 8 regions

19. (a)

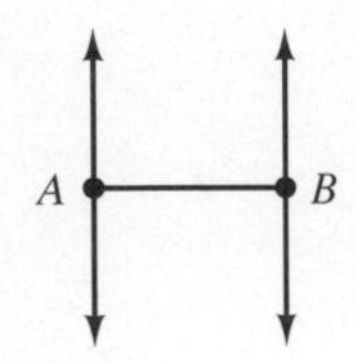

C could be any point (other than A or B) on either of the two lines drawn through A and through B which are perpendicular to $\overline{AB}$.

(b)

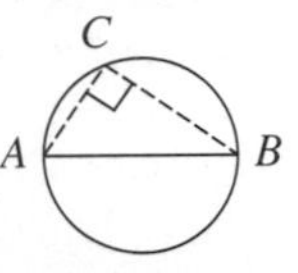

C could be any point (other than A or B) on the circle with $\overline{AB}$ as diameter.

20. At a vertex, the interior angle and the conjugate angle add up to 360°. For an n-gon, the sum of all interior and all conjugate angles is $n \cdot 360°$. All the interior angles add up to $(n - 2) \cdot 180°$, so all the conjugate angles add up to $360°n - (n - 2)\ 180° = 360°n - 180°n + 360° = 180°n + 360° = (n + 2) \cdot 180°$.

22. Such a point S allows for n triangles to be formed, all with the vertex S. The sum of all the interior angles of these n triangles is $n \cdot 180°$ which is equal to the sum of all the interior angles of the n-gon plus 360° for the angles that surround the point S. Thus the sum of the interior angles of the n-gon is $n \cdot 180° - 360° = (n - 2) \cdot 180°$.

24. (a), (b)

	Number of Points or Lines Removed							
	0	**1**	**2**	**3**	**4**	**5**	**6**	**7**
Number of Pieces of Line	1	2	3	4	5	6	7	8
Number of Pieces of Plane	1	2	4	7	11	16		

(c) The pattern is to add the number of pieces of the line and the number of pieces of the plane for a value n to get the number of pieces of the plane with $n + 1$ lines. For example, $5 + 11$ gives 16. Then $6 + 16$ gives 22, and so on. Continuing in this fashion, 10 lines determine 56 pieces of the plane. **(d)** $1 + n\ (n + 1)/2$

25. Each new circle creates a new region each time it intersects a previously drawn circle. Since the new circle intersects each of the old circles in two points, this creates the following pattern:

Number of Circles	Number of Regions
1	2
2	$2 + 2 \cdot 1 = 4$
3	$4 + 2 \cdot 2 = 8$
4	$8 + 2 \cdot 3 = 14$
5	$14 + 2 \cdot 4 = 22$

Number of Circles	Number of Regions
6	$22 + 2 \cdot 5 = 32$
7	$32 + 2 \cdot 6 = 44$
8	$44 + 2 \cdot 7 = 58$
9	$58 + 2 \cdot 8 = 74$
10	$74 + 2 \cdot 9 = 92$

27. $n = 4$ The vertices of a rhombus. Alternatively, the vertices of an equilateral triangle plus the center point. or

$n = 5$ The vertices of a regular pentagon.

$n = 6$ The vertices of a regular pentagon plus the center point.

29. **(a)** The sum of the interior angles is $360° = m(\angle P) + m(\angle Q) + m(\angle R) + m(\angle S)$. We also know $m(\angle P) = m(\angle R)$ and $m(\angle Q) = m(\angle S)$, so $360° = m(\angle P) + m(\angle Q) + m(\angle P) + m(\angle Q)$, giving us $180° = m(\angle P) + m(\angle Q)$. **(b)** $m(\angle Q) + m(\angle q) = 180°$ and $m(\angle P) + m(\angle Q) = 180°$ so $m(\angle q) = m(\angle P)$. $\angle q$ and $\angle P$ are corresponding angles, so segments $\overline{PS}$ and $\overline{QR}$ are parallel. $m(\angle q) = m(\angle P)$ and $m(\angle P) = m(\angle R)$ so $m(\angle q) = m(\angle R)$. $\angle q$ and $\angle R$ are alternate interior angles, so their congruence gives $\overline{PQ}$ parallel to $\overline{SR}$. Hence the figure is a parallelogram.

P Q
$\angle P$ $\angle Q$ $\angle q$
$\angle S$ $\angle R$
S R

31. **(a)** Drawings will vary. **(b)** After drawing the black curve, various regions are determined. The colored curve will begin in some region and cut through the black curve into another region and then into another region and so on, eventually returning to the starting place to close off the curve. Each crossing switches the colored curve from the interior to the exterior region formed by the black curve, or vice versa. So the crossings must be even in number since the colored curve gets back to where it started. **33.** The square cover could fall through the hole if it were on edge and slightly rotated from its position when in place. **35.** **(a)** *SQRE* is a square **(b)** No **37.** **(a), (c), (e), (g)** Use the procedure **polygon :nsides :side** with appropriate choices of the values of the two variables. **(i)** A regular 360-gon will draw a good approximation of a circle.

38. **(a)**
```
to openstar.5
     repeat 5 [fd 30 lt 36 fd 30 rt 108]
end
```

40.

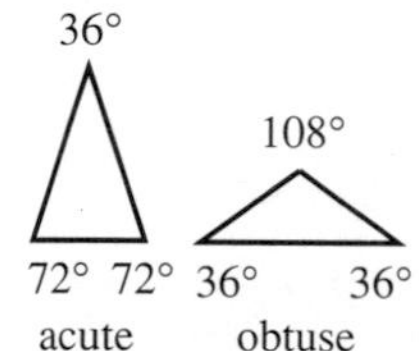

42. One angle is 90°; call the other two angles *A* and *B*. The interior angles of a triangle add up to 180°, so $90° + m(\angle A) + m(\angle B) = 180°$, or $m(\angle A) + m(\angle B) = 90°$.

JUST FOR FUN Dissection Delights (page 657)

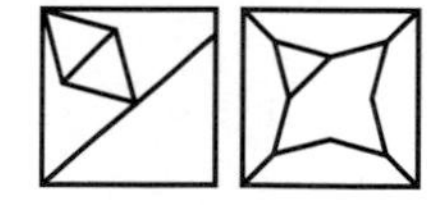

Problem Set 10.3 (page 663)

1. Many different tilings can be formed. For example:
(a)

2. Yes, tilings are possible. For example:
(a)

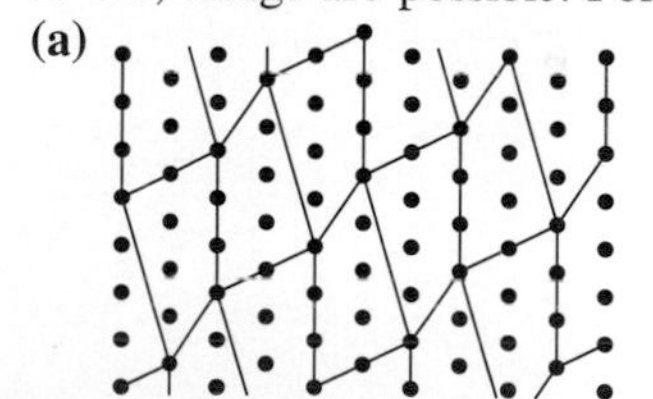

3. **(a)** Squares, pentagons, hexagons, heptagons, octagons
(b) Many vertex figures that appear in the tiling cannot correspond to regular polygons. For example, regular 5-, 6-, and 8-gons have interior angles of measure 108°, 120°, and 135°. These add up to $108° + 120° + 135° = 363° \neq 360°$.

5. The interior angle of a square is 90° and for a pentagon is $\dfrac{(5-2)(180°)}{5} = 108°$ and for a 20-gon is $\dfrac{(20-2)(180°)}{20} = 162°$. Then, $90° + 108° + 162° = 360°$. **7.** Various tilings can be formed.

9. **(a)**

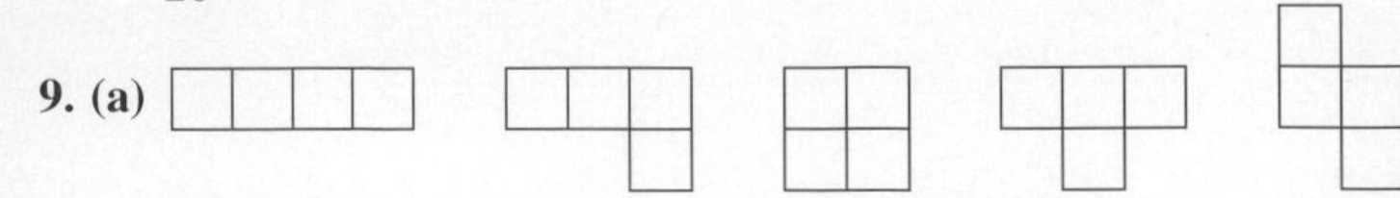

(b) The five tetrominoes cannot tile a 4 by 5 rectangle. To see why, color the rectangle in a checkerboard pattern of ten red and ten black squares. All but the T-shaped tetromino cover two red and two black squares. The T-tetromino covers three of one color and one of the other color.
11. The hexagon will tile using translations only, with no rotation necessary.
13. Only the semiregular tiling in which each vertex is surrounded by a hexagon and four equilateral triangles will appear to be different. To see why, shade the six large equilateral triangles, each made up of four small equilateral triangles about any hexagon. The "saw blade" formed has an orientation that is clearly reversed in a mirror image.

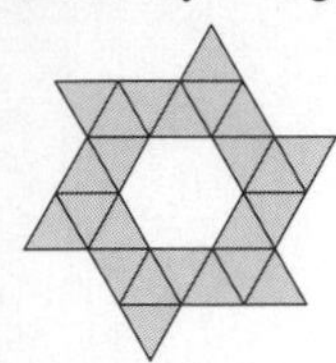

17. Here's one way to make a tiling 12-gon, by modifying a square tile.

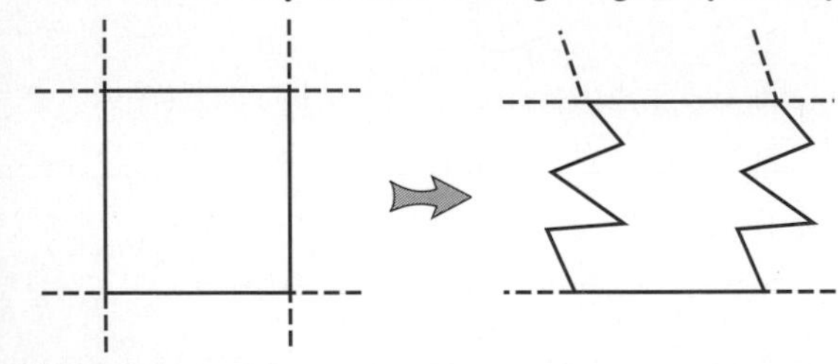

Similarly, modifying a pentagon with opposite parallel congruent sides will form a 13-gon that tiles as shown.

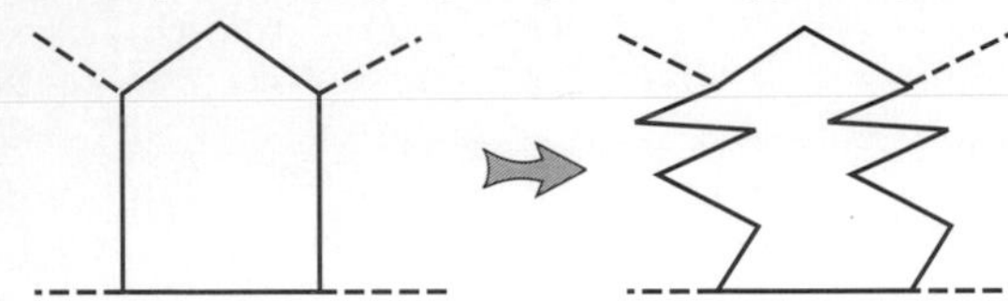

The same idea can be used for any $n \geq 6$, using a modified square for even n, and a modified pentagon for odd n.

18. **(a)**

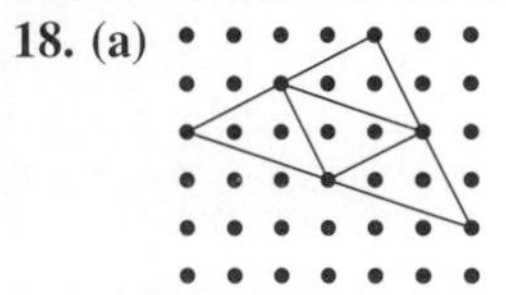

(c)

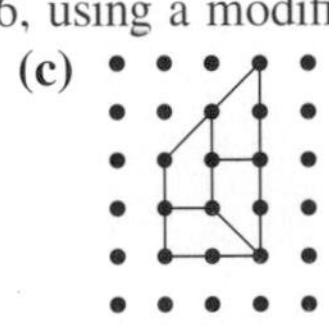

20. **(a)**

21. **(a)** A second generation Chevron is shown that uses 25 original Chevrons. It is even easier to construct a second generation that uses 36 of the original Chevrons.

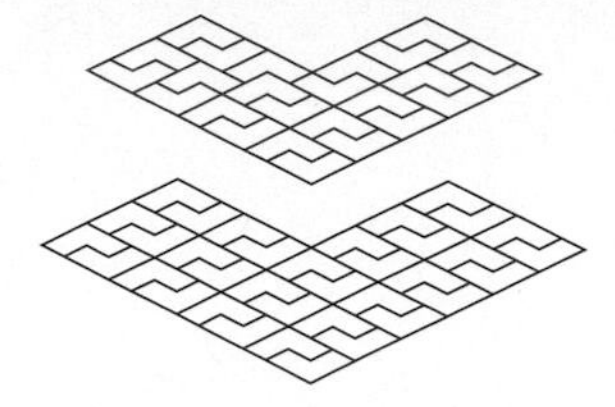

23. **(a)** The translates of V form a square grid. **(b)** These points form a square grid.

24. **(a)**
```
to square :side
    repeat 4 [fd :side rt 90]
end

to sq.tile :side
    pu bk 100 lt 90 bk 100 pd
    repeat 3 [ repeat 10 [square :side fd :side]
        rt 90 fd 2* :side rt 90
        repeat 10 [square :side fd :side]
        lt 180]
end
```

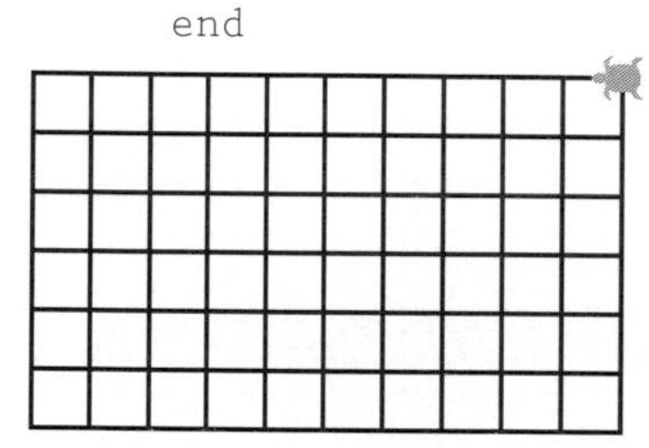

25. **(a)**
```
to tile.par
    pu fd 50 lt 90 fd 100 rt 180 pd
    repeat 2 [ repeat 12 [parallelogram fd 30]
        rt 110 fd 80 rt 70
            repeat 12 [parallelogram fd 30] rt 180]
end
```

28. Letters above the line are formed with straight pen strokes while letters below the line involved curved pen strokes.

29. Let A, B, and C be the vertices of an equilateral triangle, and D the center (centroid) of the triangle.

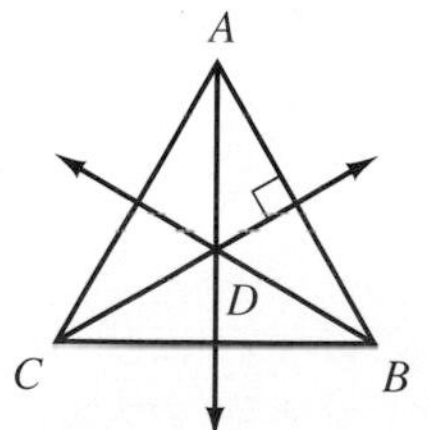

Problem Set 10.4 (page 674)

1. **(a)** Yes. *AHGFEDCBHFDBA* is one Euler path. **(c)** No **(e)** Yes. *CEAEFBFDBACD*

2. **(a)** $D = 22$, $E = 11$ for (b); $D = 24$, $E = 12$ for (c); $D = 30$, $E = 15$ for (d); $D = 22$, $E = 11$ for (e); $D = 30$, $E = 15$ for (f)

3. **(a)** Yes, since exactly two vertices are odd. **(b)** $D = 2 + 2 + 4 + 8 + 3 + 6 + 6 + 1 = 32$, so $E = 16$.

5. (a)

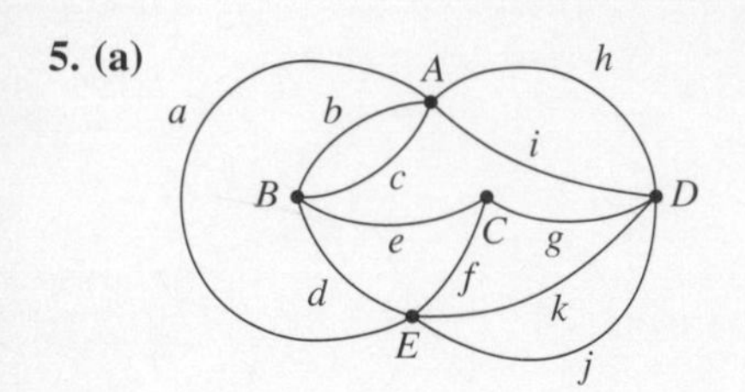

6. (a)

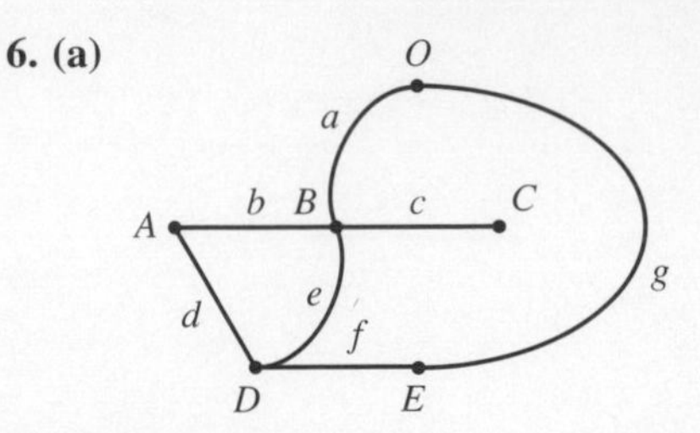

7. (a) $V = 6, R = 7, E = 11$ **9. (a)** No. Vertex 1 is of degree 5 but no vertex in the other network has degree 5. **(c)** Yes. Associate $A \leftrightarrow 3$, $B \leftrightarrow 4$, $C \leftrightarrow 5$, $D \leftrightarrow 1$, $E \leftrightarrow 2$, or $A \leftrightarrow 2$, $B \leftrightarrow 4$, $C \leftrightarrow 1$, $D \leftrightarrow 5$, $E \leftrightarrow 3$.
10. (a) Combining these two paths gives a closed path with distinct edges. **11.** The trees with no branches. **13.** Consider a network with a vertex for each person and an edge between two vertices for each time the corresponding persons have shaken hands. By the result stated in problem 12, there are an even number of people with odd numbers of handshakes. **15.** Requiring that each edge is traced exactly twice is equivalent to duplicating each edge in the network and asking if the new "doubled edge" network is traceable. Since doubling the edges at each vertex always creates an even vertex, the "doubled edge" network is always traceable. **17. (a)** 14, as can be counted from a diagram.
18. (a) **19.** One collection of edges is *HA, AB, BG, GF, FD, DC, DE.*

21. A square. **23.** $x = 30°$, $y = 45°$, $z = 135°$

JUST FOR FUN Space Out for Success! (page 682)

Form a tetrahedron with the six matches.

Problem Set 10.5 (page 689)

1. (a) Polyhedron **(c)** Polyhedron **(e)** Not a polyhedron
2. (a) Pentagonal prism **(c)** Oblique circular cone **(e)** Right rectangular prism **3. (a)** 4 **(c)** *A, B, C, D*
5. (a)

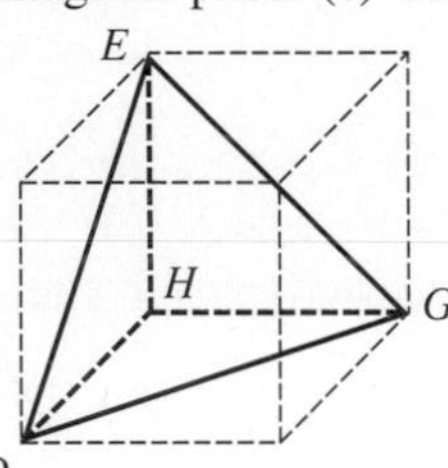

6. (a)

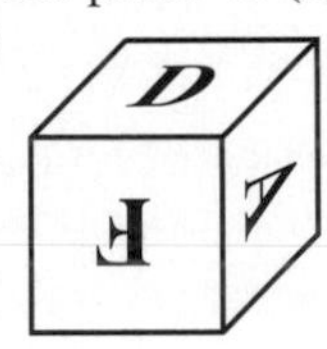

7. (a) 45° since the dihedral angle between the adjacent sides of the cube is 90° and another pyramid would fit into the gap between the original pyramid and a vertical side of the cube. **(b)** Filling the cube with six such pyramids, one sees that three pyramids sit around an edge formed by the lateral sides of a pyramid. Thus three copies of the dihedral angle give a full revolution of 360° around this edge, so the dihedral angle measures 120°. **9. (a)** Triangle **10. (a)** Possible, by having the plane cut off congruent isosceles right triangles from the three faces at a chosen corner of the cube.
(c) Possible, by cutting off along one edge a right triangular prism shape. **(e)** Possible; for example, the plane through the center of the cube that is perpendicular to a diagonal of the cube will form a regular hexagon. **11. (a)** A cube is a prism.
(b) A tetrahedron is a pyramid. **13. (a)** Pentagonal double pyramid: $V = 7$, $F = 10$, $E = 15$ $7 + 10 = 15 + 2$
Hexagonal antiprism: $V = 12$, $F = 14$, $E = 24$ $12 + 14 = 24 + 2$

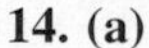

14. (a)

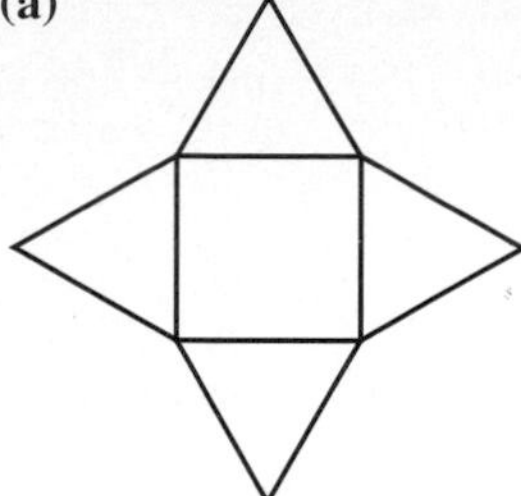

15. (a) $V = 10$, $E = 20$, $F = 12$. $V + F = 22$ and $E + 2 = 22$. Euler's formula holds.
16. Consider the plane P through the points A, B, and C. Since $\overline{CD}$ is below $\overline{AB}$, this means D is below plane P. The segment $\overline{DE}$, starting at D, is seen to first go above P as it crosses $\overline{AB}$, and then return below P as it crosses beneath $\overline{BC}$. Since no line can cross back and forth through a plane, the space "pentagon" shown is an impossible figure.
17. (a) (other examples are possible)

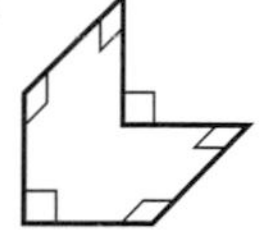

19. (a) Suppose the faces of a polyhedron consist of a p-gon, a q-gon, an r-gon, and so on. Since each edge of the polyhedron borders two faces, the sum $p + q + r + \cdots$ is twice the number of edges. That is, $p + q + r + \cdots = 2E$. Since there are F faces and $p, q, r, \ldots$ are all three or greater, we get $2E \geq 3 + 3 + \cdots = 3F$.
(b) Each of the V vertices of a polyhedron is the endpoint of three or more edges that meet at the vertex. Thus, $3V$ is less than or equal to the total number of ends of the edges. But each of the E edges has two ends, so there are $2E$ ends of edges. We see that $3V \leq 2E$. **(c)** Adding $3V \leq 2E$ and $3F \leq 2E$ shows that $3V + 3F \leq 4E$. But $V + F = E + 2$ (Euler's formula), so $3V + 3F = 3E + 6$. Comparing this to the inequality, we see that $3E + 6 \leq 4E$. Subtracting $3E$ from both sides shows that $6 \leq E$. **(d)** Suppose $E = 7$. Since $3F \leq 2E = 14$, we see that F is no larger than 4 ($F \geq 5$ would give $3F \geq 15$). Similarly, $3V \leq 2E = 14$ means that $V \leq 4$. Since both $V \leq 4$ and $F \leq 4$, then $V + F \leq 8$. But $V + F = E + 2$ (Euler's formula), and $E = 7$, so $V + F = 9$. This contradicts $V + F \leq 8$, so our assumption $E = 7$ is not possible. **(e)** A pyramid with a base of $3, 4, 5, \ldots, n, \ldots$ sides has $6, 8, 10 \ldots, 2n, \ldots$ edges, respectively. Slicing off a tiny corner at one vertex somewhere on the base of the pyramid adds three new edges, giving us polyhedra with $9, 11, 13, \ldots, 2n + 3, \ldots$ edges. Altogether, the pyramids and pyramids with a truncated base corner give us polyhedra with 6, 8, 9, 10, 11, . . . edges.
21. (a)

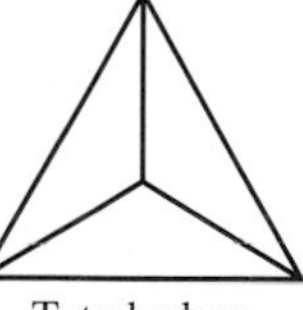

Tetrahedron

(b)

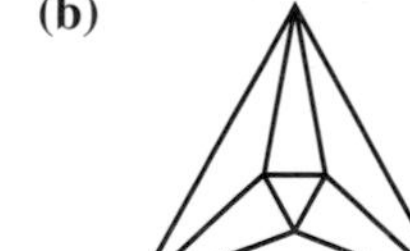

Octahedron

(c)

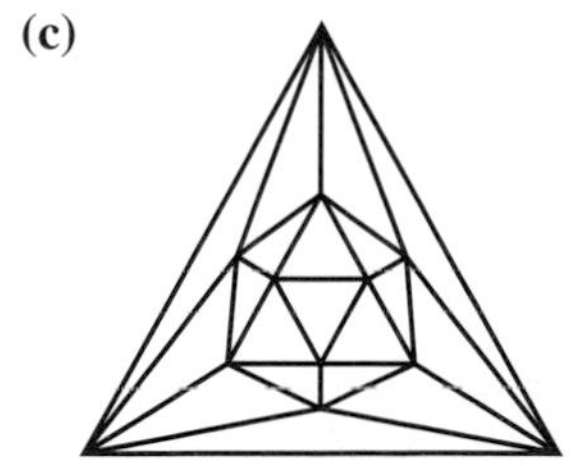

Icosahedron

23. (a) The base is a pentagon.

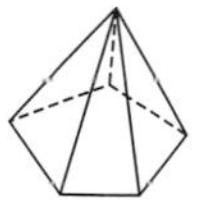

(b) The center polygon is a triangle.

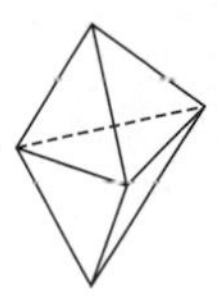

(c) The pentagonal based pyramid, viewed from above, would look like the left hand figure below. The "cube" type, with all quadrilateral faces, is the next example shown. It should be possible to give similar visualizations for the remaining hexahedra.

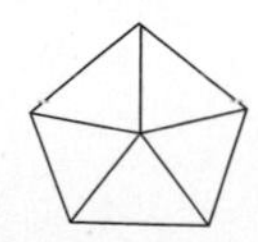
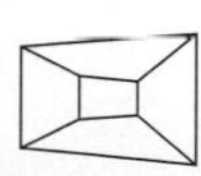
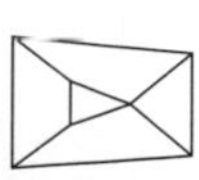
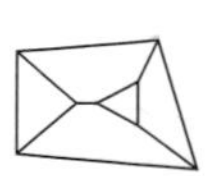
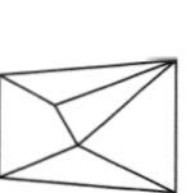
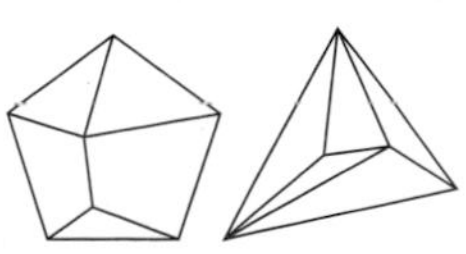

25. (a) When folded, these edges coincide and so must be the same length in the net. The dashed segment $\overline{AP}$ is perpendicular to edge $\overline{BH}$ because folding along $\overline{BH}$ moves point A along a circle that is in a plane perpendicular to the axis $\overline{BH}$ of the fold. **26.** The axis of all the hinges on a door must be along the intersection of the planes of the wall and the plane of the opened door. Since planes meet in a line, the axes of the hinges must be along a single line.
29. $m(\angle x) = 50°$, $m(\angle z) = 130°$, $m(\angle y) = 50°$
30. (a)

n	3	4	5	6	7	8
Diagonals	0	2	5	9	14	20

(c) 54 **31. (a)** Many polygons are possible.

CLASSIC CONUNDRUM An Unexpected Bisector (page 694)

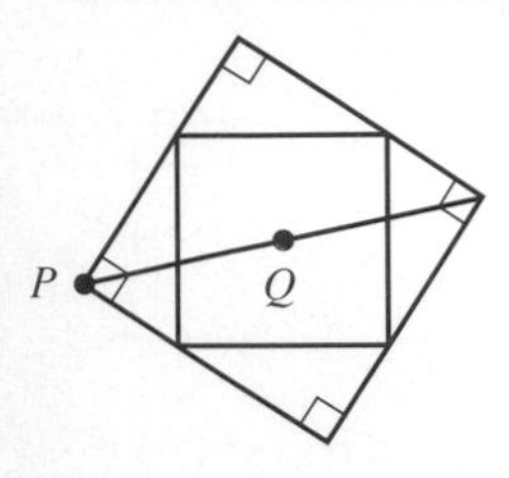

Chapter 10 Review Exercises (page 696)

1. (a) $\overleftrightarrow{AC}$ **(b)** $\overline{BD}$ **(c)** AD **(d)** $\angle ABC$ or $\angle CBA$ **(e)** $m(\angle BCD)$ **(f)** $\overrightarrow{DC}$
2. (a) $\angle BAD$ **(b)** $\angle BCD$ **(c)** $\angle ABC$, $\angle ADC$ **3. (a)** 143° **(b)** 53°
4. $p = 55°$, $r = 55°$, $s = 125°$, $q = 35°$ **5.** $x = 45°$, $y = 33°$, $z = 147°$
6. (a) iv **(b)** i **(c)** vi **(d)** v **(e)** ii **(f)** iii **7. (a)** No, because obtuse angles have measure greater than 90° and the sum of the three interior angles of a triangle is 180°. **(b)** Yes, try angles of 100°, 100°, 100°, and 60°. **(c)** No, because acute angles have measure less than 90° and the sum of the interior angles must be 360°. **8.** The interior angles add up to $(6 - 2)(180)° = 720°$. So $16x = 720°$ and $x = 45°$. The angles are 135°, 135°, 135°, 45°, 225°, 45°. **9.** 360°
10. The angles are 60° for the triangle, 90° for the square, and 120° for the hexagon. The angles of the four polygons must add up to 360°, so the fourth angle is 90°, and therefore the fourth polygon is a square.
11.

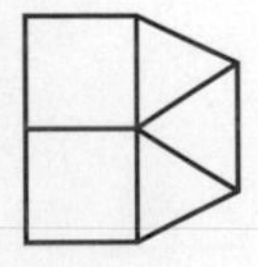

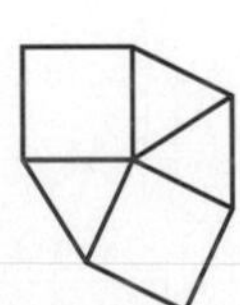

12. Use 180° rotations of the tile about the midpoints of the sides. **13. (a)** It has four odd vertices.
(b) An edge between any two of vertices A, B, C, and E. Many Euler paths are possible.
14. Construct a network with vertices A, B, C, D, and E and edges corresponding to bridges. Since just two vertices, A and D, have odd degree, there is an Euler path. The Euler path corresponds to a walking path which crosses each bridge exactly once. **15.** $V = 11$, $R = 7$, $E = 16$. $V + R = 18$ and $E + 2 = 18$, so $V + R = E + 2$ holds.
16. (a) 6 **(b)** $\overline{CD}$, $\overline{EF}$, $\overline{GH}$ **(c)** $\overline{DH}$, $\overline{GC}$, $\overline{EH}$, $\overline{FG}$ **(d)** 45° **17.** Square right prism; triangular pyramid or tetrahedron; oblique circular cylinder; sphere; hexagonal right prism.
18. (a)

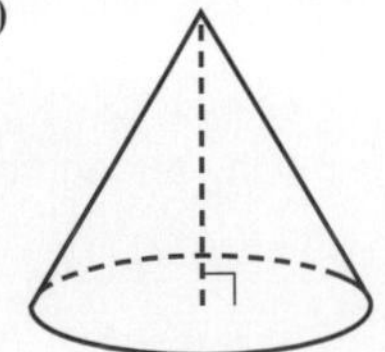

(b)

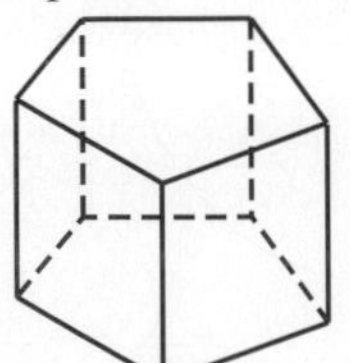

(c)

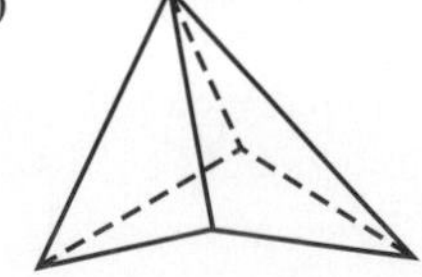

19 (a) See Figure 10.44. **(b)** $V = 6$, $F = 8$, $E = 12$. Thus $V + F = 14$ and $E + 2 = 14$, so Euler's formula holds.
20. By Euler's formula, $V + 14 = 24 + 2$, so $V = 12$.

Chapter 10 Test (page 699)

1.

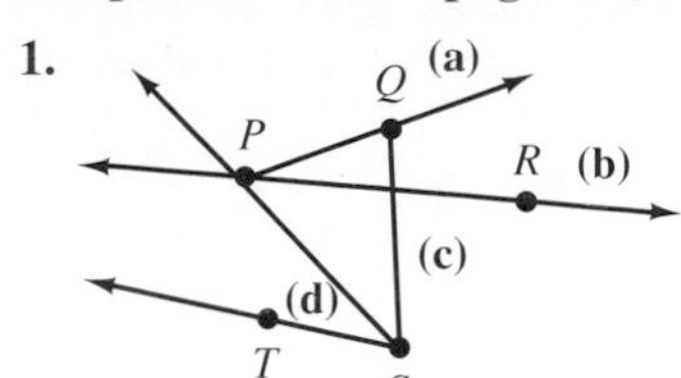

2. (a) C, A **(b)** F **(c)** D **(d)** E **(e)** B

3. Many examples of each are possible.
(a) 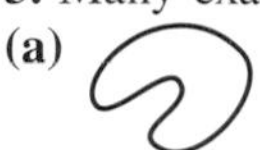 **(b)** 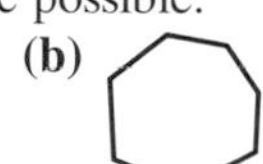 **(c)**  **(d)** 

4. (a) True **(b)** False **(c)** True **(d)** True **5. (a)** Decagon **(b)** Octagon **(c)** Nonagon
6. $t = 40°$, $r = 80°$, $s = 100°$ **7.** a, b, c, d, e **8.** A vertex figure uses two octagons and one square.
9. The total turn angle is 1080°, so the turn at each of the seven vertices is $1080°/8 = 135°$. This means that the interior angle at each point is $180° - 135° = 45°$.
10. The average interior angle measure for an n-gon is $\dfrac{(n - 2)(180°)}{n}$. We want this value to be 174°, so $\dfrac{180(n - 2)}{n} = 174$. This gives $n = 60$. **11. (a)** (iii) **(b)** (iv) **(c)** (i), (ii), (iv) **(d)** (i) add an edge between vertices of degree 3 and 7; (ii) add an edge between the two vertices of degree 3; (iii) add an edge between any two of the four odd vertices.
12. (a) By Euler's formula, $V + 7 = 11 + 2$, so $V = 6$. **(b)** Many such networks can be drawn.
13. The same as an interior angle of a regular pentagon, which is $\dfrac{3 \cdot 180°}{5} = 108°$.
14. (a) 24. Each edge borders a square face and a triangular face. So just count the edges of all the square faces (or all the edges of the triangular faces). **(b)** 12. Either count in the diagram or use Euler's formula.
15. (a) Diagrams will vary. **(b)** $V = 16$, $F = 21$, $E = 35$ **(c)** $V + F = 37$ and $E + 2 = 37$, so Euler's formula holds.

Chapter 11

JUST FOR FUN Twice Around a Triangle (page 715)

P″ coincides with the original point *P*

Problem Set 11.1 (page 717)

1. (a) $L \leftrightarrow K$, $H \leftrightarrow W$, $S \leftrightarrow T$ **(b)** $\overline{KW}$, $\overline{WT}$, $\overline{TK}$ **(c)** $\angle K$, $\angle W$, $\angle T$ **(d)** $\triangle KWT$
3. (a) Draw a line segment and mark point E. Set your compass to AB and determine a point G on your line segment with an arc centered at E. Set your compass to CD and draw an arc centered at G to determine F, away from E.
(b) Begin as in (a). Set compass to CD and draw an arc centered at G to determine F, back toward E.
5. (a) One such triangle. **(c)** Impossible by the triangle inequality, since $2 + 5 < 8$. **(e)** One such triangle.
7. (c) $\triangle ABC \cong \triangle EFD$ by SSS **(e)** $\triangle ABD \cong \triangle ACD$ by ASA **(g)** No conclusion possible. **(i)** No conclusion possible.
8. Let $\triangle ABC$ be equilateral. Since $AB = BC$, it follows from the isosceles triangle theorem that $\angle A \cong \angle C$. In the same way, since $BC = CA$ it follows that $\angle B \cong \angle A$. Thus all three angles are congruent.
10. Construct the equilateral triangle shown. Since $\triangle ABM$ is congruent to $\triangle CBM$ by the HL theorem, we see M is the midpoint of AC. Since $\triangle ABC$ is equilateral we know that $AB = AC$. Thus, $AM = \frac{1}{2}AC = \frac{1}{2}AB$.

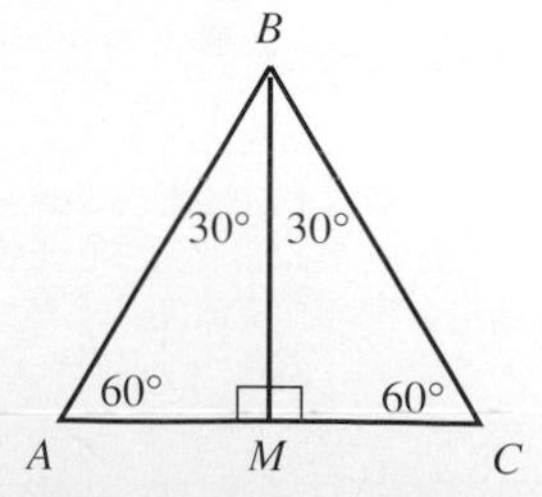

11. (a) $\angle ABD \cong \angle CDB$, as alternate interior angles between parallel lines. Likewise, $\angle ADB \cong \angle CBD$, and $DB = BD$. By ASA, $\triangle ABD \cong \triangle CDB$. **13. (a)** Refer to the diagram in problem 12. $AM = MC$ and $BM = MD$. All four angles at M are right angles, so by SAS, $\triangle AMB \cong \triangle CMB \cong \triangle CMD \cong \triangle AMD$. Thus $\overline{AB} \cong \overline{CB} \cong \overline{CD} \cong \overline{AD}$.
14. It is longer than 5 cm and shorter than 13 cm (remember the triangle inequality). **15. (a)** $0 < s < 14$ cm, where s is the length of the fourth side. **17.** Place a corner of the rectangular sheet of paper at a point C on the circle and mark the points A and B where the edges of the paper cross the circle. By the converse of Thales' theorem, $\overline{AB}$ is a diameter of the circle. Repeating the procedure at a second point C' will allow you to construct a second diameter $\overline{A'B'}$. The center of the circle is where the two diameters intersect.

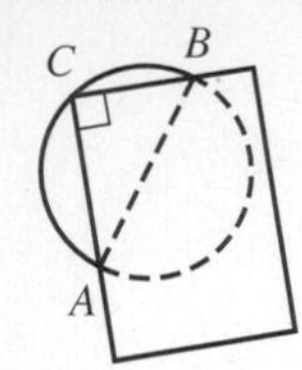

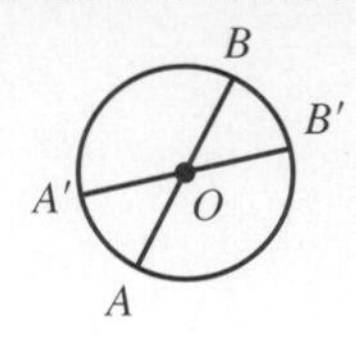

19. (a) $\triangle ABE$ is isosceles, with $AB = BE$, so the base angles are congruent by the isosceles triangle theorem. **(b)** $\triangle ACD$ is isosceles, so its base angles are congruent. **(c)** $\angle ACB$ and $\angle ACD$ are supplementary, as are $\angle ADC$ and $\angle ADE$. Since $\angle ACD \cong \angle ADC$, then we get $\angle ACB \cong \angle ADE$. Using $\angle B \cong \angle E$ and $AC = AD$, the AAS property gives $\triangle ABC \cong \triangle AED$. **(d)** From part (c), $\overline{BC} \cong \overline{ED}$. **21. (a)** Yes, using ASA. **(c)** Yes, using SAS. **22.** Since $\triangle ABC$ is equilateral, $\angle A$, $\angle B$, and $\angle C$ are congruent 60° angles, and $AB = BC = CA$. Since we are given $AD = BE = CF$, we also see that $AF = BD = CE$. By the SAS property, $\triangle ADF \cong \triangle BED \cong \triangle CFE$. Thus $FD = DE = EF$, so $\triangle DEF$ is equilateral. **24.** If $AB = CD = a$, $BC = AD = b$, and $AC = BD = c$, then each face of the tetrahedron is a triangle with sides of length a, b, and c. By the SSS property, the triangles are congruent to one another. **26.** By the triangle inequality, $PQ + PT > QT$ and $ST + TR > SR$. Note that $QT = QS + ST$. Combining inequalities, $PQ + PT + ST + TR > QT + SR$, or $PQ + PR + ST > QS + ST + SR$, and so $PQ + PR > QS + SR$.
29. (a) The angles at the vertices of a quadrilateral can change even though the lengths of the sides are fixed. (There is no "SSSS congruence property" for a quadrilateral.) **30. (a)** The framework forms a parallelogram, but not necessarily a rectangle. **32. (a), (b)** $m(\angle AOB) = 2\,m(\angle APB)$ **(c)** Draw the diameter $\overline{PQ}$. There are two cases to consider.

Case 1: *A and B lie on opposite sides of* $\overline{PQ}$. We see that $\triangle POA$ is isosceles, so the base angles are congruent. That is, $m(\angle APO) = m\,(\angle OAP) = x$. Therefore, $m(\angle AOQ) = 2x = 2m(\angle APO)$. By the same reasoning, $m(\angle BOQ) = 2y = 2m(\angle BPO)$. Thus, $m(\angle AOB) = 2x + 2y = 2m(\angle APB)$.

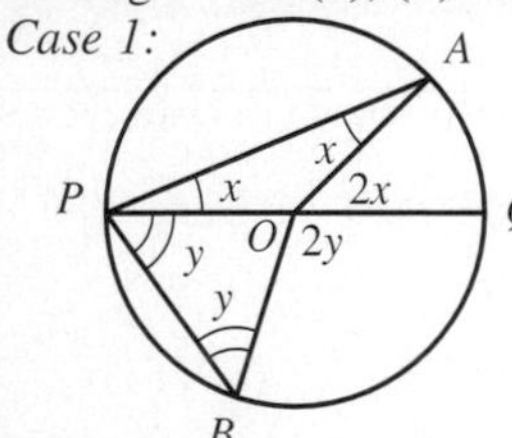

Case 2: *A and B are on the same side of* $\overline{PQ}$. Nearly the same analysis holds, but now $m(\angle AOB) = 2x - 2y = 2m(\angle APB)$.

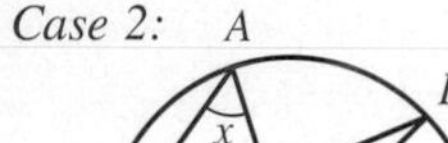
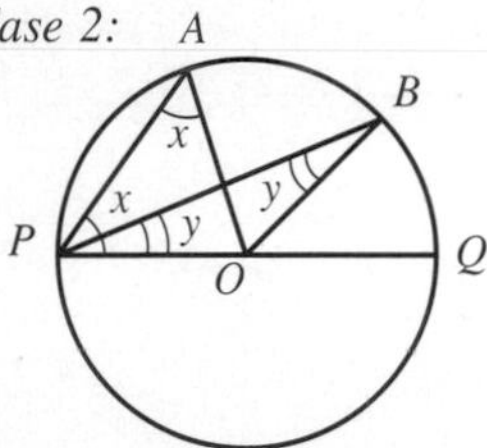

34. Since $m(\angle 3) = m(\angle 1) + m(\angle 2)$, we see that $m(\angle 3) > m(\angle 1)$ and $m(\angle 3) > m(\angle 2)$.

JUST FOR FUN Circle Amazement (page 728)

The jar lid will also give the circle through A, B, and C. When it is drawn, each of the four points A, B, C, and P is the intersection of three of the circles.

Problem Set 11.2 (page 734)

1. (a) Step 1: Draw a line through P that intersects l. Label the intersection point A. Step 2: Draw arcs of equal radius at A and P. Label, with B and C, the intersection points of the arc at A, and label as D the intersection of the arc at P with $\overleftrightarrow{AP}$. Step 3: Set the compass to radius BC, and draw an arc at D. Let E denote the intersection with the arc drawn at P.

Step 4: Construct the line k through P and E. **(b)** The construction gives the congruence of the corresponding angles, $\angle PAC \cong \angle DPE$. Therefore $k \parallel l$ by the corresponding angles property. **3. (a)** The corresponding angles property guarantees that k is parallel to l. **(b)** Align the ruler with the line; slide the drafting triangle, with one leg of the right triangle on the ruler, until the second leg of the triangle meets point P. **7. (a)** The circumcenter will be inside the acute triangle. **(b)** The circumcenter will be at the midpoint of the hypotenuse of a right triangle. **(c)** The circumcenter will be outside an obtuse triangle. **(d)** The circumcenter is inside, on, or outside a triangle if, and only if, the triangle is acute, right, or obtuse, respectively. **9.** Since $\triangle PQS$ is inscribed in the circle with diameter $\overline{PQ}$, it has a right angle at S by Thales' theorem. Thus $\overline{PS} \perp \overline{SQ}$. Similarly, $\overline{PT} \perp \overline{TQ}$. **12.** By Thales' theorem, $\angle ADB$ is a right angle. We also see that $\triangle ODB$ is an equilateral triangle, since all sides have the length of the radius. Moreover, $ODBE$ is a rhombus, so the side $\overline{DE}$ is a bisector of the 60° angle $\angle ODB$. Thus $m(\angle ADE) = m(\angle ADB) - m(\angle EDB) = 90° - 30° = 60°$. Similarly, $m(\angle AED) = 60°$. Therefore all angles of $\triangle ADE$ have measure 60°, so $\triangle ADE$ is equilateral. **14.** Since $m(\angle 1) + m(\angle 2) + m(\angle 3) + m(\angle 4) = 180°$, $m(\angle 1) = m(\angle 2)$, and $m(\angle 3) = m(\angle 4)$, it follows that $m(\angle 2) + m(\angle 3) = 180°/2 = 90°$.
15. (a) Suppose the perpendicular bisector of chord $\overline{AB}$ intersects the circle at a point C. Then the circle is the circumscribing circle of $\triangle ABC$. The center of the circumscribing circle is the point of concurrence of the perpendicular bisectors of all three sides of $\triangle ABC$. In particular, the perpendicular bisector of side $\overline{AB}$ contains the center of the circle.
17. (a) Extend $\overline{AB}$, and construct the line at A that is perpendicular to $\overline{AB}$. Set the compass to radius AB and mark off this distance on the perpendicular line to determine a point C for which $AC = AB$. Similarly, construct a perpendicular line at B to $\overline{AB}$, and determine a point D (on the same side of $\overline{AB}$ as C) on this perpendicular so $BD = AB$. $ABDC$ is a square with given side $\overline{AB}$. (Other constructions also work.) **18. (a)** Extend $\overline{AB}$ to a longer segment. Erect perpendicular rays to $\overline{AB}$ at both A and B, to the same side of $\overline{AB}$. Bisect the right angle at A, and let its intersection with the ray at B determine point C. Erect the perpendicular at C to $\overline{BC}$, and let D be the intersection with the ray constructed at A. Then $ABCD$ is a square erected on the given side $\overline{AB}$.
21. Constructible: 3, 4, 5, 6, 8, 10, 12, 15, 16, 17, 20, 24, 30, 32, 34, 40, 48, 51, 60, 64, 68, 80, 85, 96.
23. By SSS, $\triangle ABD \cong \triangle ACD$. Therefore, $\angle BAD \cong \angle CAD$. **25. (a)** $ABDC$ is a parallelogram. So (as has been shown in previous exercises) $AB = CD$ and $BD = AC$. By SSS, $\triangle ABC \cong \triangle DCB$. Similarly, $\triangle ABC \cong \triangle BAF$ and $\triangle ABC \cong \triangle CEA$. **(c)** The altitude of $\triangle ABC$ at A is perpendicular to $\overline{BC}$. But $\overline{FE}$, by construction, is parallel to $\overline{BC}$. Thus, the altitude at A is also perpendicular to $\overline{FE}$. Since A is the midpoint of $\overline{FE}$, this means the altitude of $\triangle ABC$ at A is also the perpendicular bisector of $\overline{FE}$. The same reasoning applies to the altitudes at B and C. **27.** $m(\angle NOQ) = m(\angle PON) - m(\angle POQ) = 144° - 120° = 24°$, since a central angle for a regular pentagon is 72° and is 120° for an equilateral triangle. A regular 15-gon has central angle $\frac{360°}{15} = 24°$, so laying off segments of length QN would give 15 equally spaced points around the circle. **30.** Use the laws of exponents and the distributive property of multiplication over addition to show that $F_5 = 2^{2^5} + 1 = 2^{32} + 1 = 2^{12+20} + 1 = 2^{12} \cdot 2^{20} + 1 = (4096) \cdot (1{,}048{,}576) + 1 = 4 \cdot (1{,}048{,}576) \cdot 10^3 + 96 \cdot (10^6 + 48{,}576) + 1 = 4{,}194{,}304{,}000 + 96{,}000{,}000 + 4{,}663{,}296 + 1 = 4{,}294{,}967{,}297$.
32. (a) G, H, and P are collinear. The Euler line passes through G, H, and P. **(b)** $GH/GP = 2$. Thus G is one-third of the distance from P to H along the Euler line. **(c)** The circle intersects all sides of $\triangle ABC$ at their midpoints. **(d)** The circle bisects each of the segments $\overline{AH}$, $\overline{BH}$, and $\overline{CH}$. **34.** $\triangle TRI$ is equilateral. **36. (a)** True. Three pairs of angles and two pairs of sides are congruent. **(b)** False. After pairing up congruent angles, the sides with equal lengths are not corresponding sides in the triangles, so the two triangles are not congruent.

JUST FOR FUN Thales' Puzzle (page 745)

Later in the day, the points A and C will cast new shadows, say at P' and Q'. Since both P and P' are away from the pyramid, Thales can easily measure PP' and QQ', and calculate the scale factor $PP'/QQ' = s$. The height of the pyramid is therefore sh, where h is the height of the vertical stick.

Problem Set 11.3 (page 748)

1. (a) First notice that $m(\angle O) = 180° - 60° - 30° = 90°$. Therefore, by the AA similarity property, $\triangle ABC \sim \triangle PNO$. The scale factor from $\triangle ABC$ to $\triangle PNO$ is $\frac{12}{8} = \frac{3}{2}$. **(c)** By the AA similarity property, $\triangle GHI \sim \triangle TUI$, with scale factor 8/5. **2. (a)** Yes, by AA: all angles are the same, 60°. **(c)** Yes, by the AA similarity property. **(e)** Yes, by the SSS similarity property, or AA, or SAS. **3. (a)** $\frac{12}{15} = \frac{8}{a}$, so $a = 10$. **(c)** $\frac{c}{15} = \frac{c+2}{18}$; $18c = 15c + 30$; $c = 10$.
4. (a) No. A square and a nonsquare rectangle are convex quadrilaterals with congruent angles, yet are not similar quadrilaterals. **6. (a)** $\overline{AB}$ is parallel to $\overline{CD}$ so, by alternate interior angles, $\angle BAE \cong \angle DCE$. Also, $\angle AEB \cong \angle CED$, being vertical angles, so the AA similarity property gives $\triangle ABE \sim \triangle CDE$. **(b)** By similarity, $\frac{x}{36} = \frac{17}{51}$, so $x = 12$.

Also, $\frac{26}{y} = \frac{17}{51}$, so $y = 78$. **8. (a)** $\angle CAD \cong \angle BAC$, since they are the same angle, and $m(\angle ADC) = m(\angle ACB) = 90°$. By the AA similarity property, $\triangle ADC \sim \triangle ACB$. Likewise, $\triangle CDB \sim \triangle ACB$. Thus, $\triangle ADC \sim \triangle CDB$. **10. (a)** Draw an arc of large enough radius so that point B is on the seventh line above the line with point A. **11. (a)** By corresponding angles for parallel lines, $\angle OXU \cong \angle XAB$ and $\angle OUX \cong \angle UBA$. So by the AA similarity property, $\triangle OXU \sim \triangle OAB$. Thus, $\frac{x}{a} = \frac{1}{b}$, and $x = \frac{a}{b}$. **(b)** Construct a triangle $\triangle OAB$ with $OA = a$ and $OB = b$. Mark the point U at length 1 from point O on segment $\overline{OB}$ or its extension. Construct a line through U which is parallel to $\overline{AB}$, calling the point where it crosses $\overline{OA}$ the point X. Then $OX = \frac{a}{b}$. **13. (a)** Use the AA similarity property. **(b)** Using $\triangle ACD \sim \triangle ABC$, then $\frac{AD}{AC} = \frac{AC}{AB}$ or $\frac{x}{b} = \frac{b}{c}$. Similarly, $\triangle CBD \sim \triangle ABC$ gives $\frac{BD}{BC} = \frac{CB}{AB}$ or $\frac{y}{a} = \frac{a}{c}$. **(c)** $x = \frac{b^2}{c}$ and $y = \frac{a^2}{c}$. Also, $x + y = c$, so $c = \frac{b^2}{c} + \frac{a^2}{c}$, or $c^2 = a^2 + b^2$. **15. (a)** $\angle APC$ and $\angle BAD$ are corresponding angles to the parallel segments $\overline{PC}$ and $\overline{AD}$, so $\angle APC \cong \angle BAD$. But $\angle BAD \cong \angle DAC$ since $\overline{AD}$ bisects $\angle BAC$, and $\angle DAC \cong \angle ACP$ since they are alternate interior angles of $\overline{PC}$ and $\overline{AD}$. Thus, $\angle APC \cong \angle BAD \cong \angle DAC \cong \angle ACP$. **(c)** Since $\triangle ABD \sim \triangle PBC$ by the AA similarity property, it follows that $\frac{PB}{AB} = \frac{CB}{DB}$. But $PB = PA + AB$ and $CB = CD + DB$, so $\frac{PA + AB}{AB} = \frac{CD + DB}{DB}$. Thus, $\frac{PA}{AB} + \frac{AB}{AB} = \frac{CD}{DB} + \frac{DB}{DB}$, or $\frac{PA}{AB} = \frac{CD}{DB}$, which is equivalent to the stated proportion.
17. (a) The midpoints W, X, Y, and Z are the vertices of a parallelogram, and we know that the diagonals of any parallelogram intersect at their common midpoints. **(b)** By part (a), we conclude that $\overline{WY}$ and $\overline{XZ}$ intersect at their common midpoint M. Now consider the space quadrilateral $PSQR$, whose midpoints form the parallelogram $ZUXV$. By the reasoning of part (a), we know that the diagonals $\overline{XZ}$ and $\overline{UV}$ intersect at their common midpoint. But the midpoint of $\overline{XZ}$ is M, so all three bimedians intersect at their common midpoints. **19.** Let L denote the midpoint of $\overline{AC}$, which is also the midpoint of $\overline{BD}$. By Example 11.15, P is the centroid of $\triangle ABC$ and $BP = \frac{2}{3}BL$. Since $BL = \frac{1}{2}BD$, this shows $BP = \frac{2}{3} \cdot \frac{1}{2}BD = \frac{1}{3}BD$. By the same reasoning, Q is the centroid of $\triangle ADC$ and $QD = \frac{1}{3}BD$. Finally, $PQ = BD - BP - QD = \left(1 - \frac{1}{3} - \frac{1}{3}\right)BD = \frac{1}{3}BD$.
21. The right triangles, $\triangle ACP$ and $\triangle BDP$, have congruent vertical angles at P. Thus, $\triangle ACP \sim \triangle BDP$ by AA similarity. Since $AC/BD = 4/2$, the scale factor is 2. Therefore $CP = 2DP$. Since $CD = 4$ and $CD = CP + DP$, we see that $CP = \frac{2}{3}(4) = 8/3$ and $DP = \frac{1}{3}(4) = 4/3$. **23.** The right triangles also have a congruent angle at the vertex at the mirror, so the triangles are similar by the AA property. Assuming Mohini's eyes are 5″ beneath the top of her head, this gives the proportion $h/5' = 15'/4'$, making the pole $h = (5')(15'/4') = 18'9''$ high. **25.** By similar triangles, $(6 - x)/6 = 5.25'/18'$. Therefore, $x = 6'(1 - 5.25'/18') = 4.25' = 4'3''$.

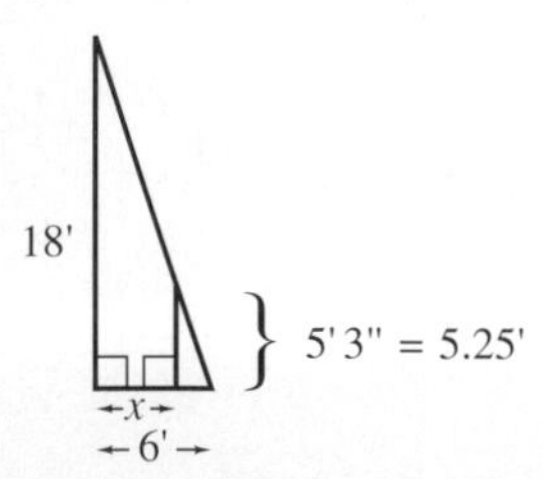

27. (a) $\triangle BCP \sim \triangle DAP$ **(b)** $PA/PC = PD/PB$, so $PA \cdot PB = PC \cdot PD$
30. $\triangle APB \cong \triangle DPC$; $\triangle ABC \cong \triangle DCB$; $\triangle ABD \cong \triangle DCA$ **32.** By the hypotenuse-leg theorem, $\triangle ACP \cong \triangle BCP$, so $PA = PB$.

CLASSIC CONUNDRUM Diagonal Diagnosis (page 754)

$AC = DB$ = radius of the circle = 5 + 4 =9.

Chapter 11 Review Exercises (page 756)

1. (a) $\triangle ACD \cong \triangle ACB$ by SSS. **(b)** $\triangle ACD \cong \triangle ECB$ by SAS. **(c)** $\triangle ADF \cong \triangle BEC$ by AAS. **(d)** $\triangle ACD \cong \triangle ECB$ by SAS. **(e)** $\angle B \cong \angle E$ since $\triangle ABE$ is isosceles. Therefore, $\triangle ABC \cong \triangle AED$ by ASA and $\triangle ABD \cong \triangle AEC$ by ASA. **(f)** $\triangle ABC \cong \triangle ADC$ by the HL (hypotenuse-leg) theorem. **2. (a)** 2.9 cm **(b)** 40° **(c)** 78° **(d)** 40° **3.** $\angle B \cong \angle C$ since $\triangle ABC$ is isosceles. By construction, $BF = DC$ and $BD = EC$. Therefore, $\triangle BDF \cong \triangle CED$ by SAS, so $DE = DF$. **4. (a), (b), (c)** Standard constructions as in Section 11.2. **(d)** Draw a circle at any point A on line m, and let it intersect line l at B and C. Construct $\overline{AB}$ and $\overline{AC}$. Draw circles of the same radius at B and C to determine the respective midpoints M and N of $\overline{AB}$ and $\overline{AC}$. Then $k = \overleftrightarrow{MN}$ is the desired line. Alternatively, construct a perpendicular line to l at a point on l. This determines a perpendicular segment between l and m. The perpendicular bisector of the segment is the desired line k.

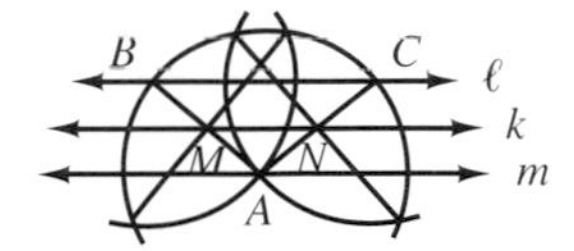

5. (a) Reflect one side of $\angle A$ to the other.

(b) Pivot Mira about P until the line reflects to itself.

(c) Reflect A onto B.

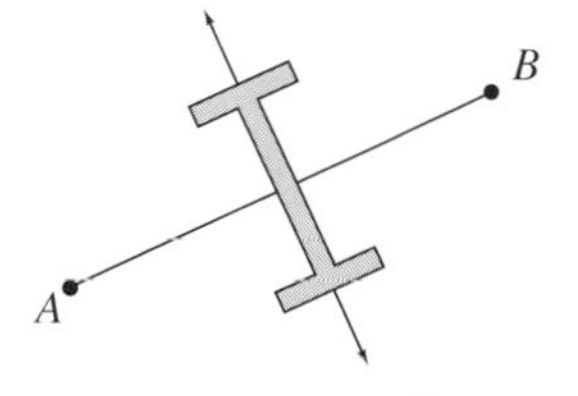

(d) Reflect line m onto line l.

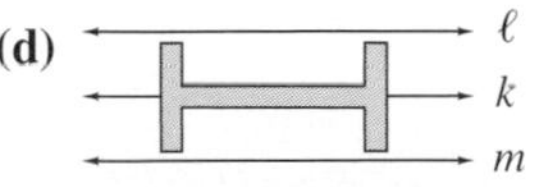

6. (a) Construct $\angle A$, lay off length AB, and draw a circle at B of radius BC. The circle intersects the other ray from A at two points, C_1 and C_2, giving two triangles $\triangle ABC_1$ and $\triangle ABC_2$.

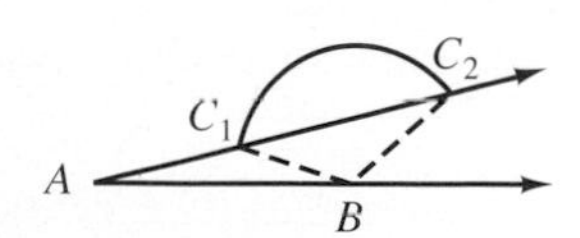

(b) Only $\triangle ABC_1$ has $\angle C = \angle C_1$ obtuse.

7. Find the midpoint M of $\overline{AD}$. Then draw circles of radius AM centered at A, D, and M.

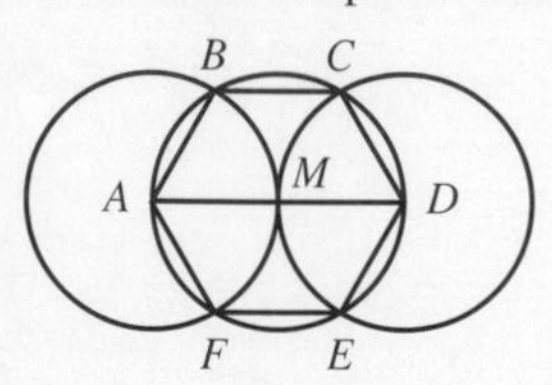

8. (a) Yes, using the SSS similarity property. **(b)** Yes, using the AA similarity property. **9.** Bisect sides $\overline{AB}$ and $\overline{AC}$ to determine their midpoints, M and N. Extend $\overline{AB}$ beyond B and draw the circle at B through M. Let E be the intersection with the extension. Similarly, extend $\overline{AC}$ beyond C, draw the circle at C through N, and let F be the intersection of this circle with the extension. Choosing $D = A$, the SAS similarity property guarantees that $\triangle ABC \sim \triangle DEF$.
10. (a) $\triangle BAC \sim \triangle PQR$ by SAS similarity. The scale factor is 6 cm/4 cm = 3/2.
(b) $\triangle ABC \sim \triangle YZX$ by SSS similarity. The scale factor is $42''/14'' = 3$. **(c)** $\triangle ABC \sim \triangle HGF$ by AA similarity. The scale factor is $4/6 = 2/3$. **(d)** $\triangle ADB \sim \triangle BCD$ by SSS similarity. The scale factor is $5/10 = 1/2$.
11. Draw additional line segments parallel to the given transversals. This creates similar triangles from which it follows that $x/9 = 16/12$, so $x = 9(16/12) = 12$ and $y/12 = 15/9$, so $y = 12 \cdot (15/9) = 20$.

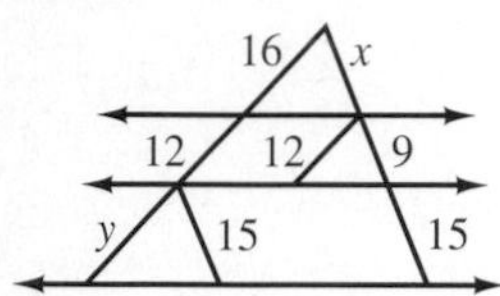

Chapter 11 Test (page 758)

1. (a) $\triangle ADC \cong \triangle ABC$ by AAS. **(b)** $\triangle ABC \cong \triangle ADC$ by SAS. **(c)** $\triangle ADC \cong \triangle BCD$ by ASA. **(d)** $\triangle ABE \cong \triangle CBD$ by AAS. **(e)** $\triangle BDC \cong \triangle FDE$ by ASA and $\triangle ABE \cong \triangle AFC$ by ASA. **(f)** $\triangle ABC \cong \triangle ADC$ by SSS. **2.** The third side is greater than 6 feet and less than 26 feet by the triangle inequality. **3.** Construct a segment $\overline{DE}$ that is congruent to $\overline{AB}$. Construct rays at D and E to the same side of $\overline{DE}$ that are respectively congruent to $\angle A$ and $\angle B$. Let F be a point of intersection of the rays. Then $\triangle DEF \cong \triangle ABC$. **4. (a)** $\triangle ABC \cong \triangle DEF$ by the hypotenuse-leg theorem. **(b)** Not congruent.

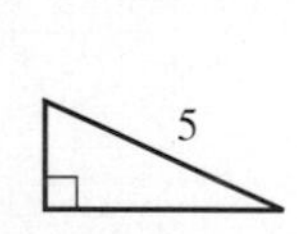

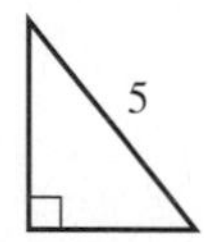

(c) $\triangle ABD \cong \triangle CBD$ by the hypotenuse-leg theorem. **(d)** $\triangle ABE \cong \triangle DBC$ by SAS. **5.** The small triangles are congruent to one another by the SAS congruence property, so $PQRST$ is equilateral. Let x and y be the measures of the acute angles in $\triangle APT$. Then each interior angle of $PQRST$ has measure $180° - x - y$, so $PQRST$ is equiangular. Altogether, $PQRST$ is regular. **6.** $\triangle FGH$ is equilateral.

7. (a) Draw circles of radius PQ, one centered at P and one centered at Q. The circles intersect at points R and S for which $\triangle PQR$ and $\triangle PQS$ are equilateral. **(b)** Construct the angle bisectors of $\angle QPR$ and $\angle QPS$, and denote their intersection with line l as T and U. $\triangle PTU$ is the desired equilateral triangle. **8. (a)** $\triangle ADE \sim \triangle ACB$ by the AA similarity property, since both triangles contain $\angle A$ and a right angle. **(b)** $\triangle ABC \sim \triangle XYZ$ by the SAS similarity property since $\frac{5}{4} = \frac{15}{12}$. **(c)** $\triangle DEG \sim \triangle EFG$ by the SSS similarity property, since $\frac{16}{8} = \frac{16}{8} = \frac{8}{4}$. **(d)** $\triangle AEB \sim \triangle CED$ by the AA similarity property since $\angle AEB \cong \angle CED$ being vertical angles and $\angle EBA \cong \angle EDC$ being alternate interior angles between parallel lines.

9. (a) $m(\angle W) = 180° - 87° - 40° = 53°$. **(b)** $\frac{20}{16} = \frac{5}{4}$ **(c)** $UV = \left(\frac{5}{4}\right)KL = \left(\frac{5}{4}\right)20 = 25$.
10. (a) That they form the same angle relative to the ground, namely the angle of elevation, since the distant sun casts shadow rays that are essentially parallel. **(b)** That the person and the tree stand at the same angle with the ground; for example, both vertical. Then $\triangle ABC \sim \triangle DEF$ by the AA similarity property. **(c)** $\frac{DE}{6'} = \frac{56'}{7'}$, so $DE = 48'$.

11. $\triangle ADE \sim \triangle ACB$ by the AA similarity property, since both triangles contain $\angle A$ and a right angle. Thus $AE/AB = AD/AC = AD/(AD + DC) = 2DC/(2DC + DC) = 2/3$. Then $AE = \frac{2}{3}(AB) = \frac{2}{3}(12) = 8$ and $EB = AB - AE = 12 - 8 = 4$. **12.** Draw the circle with radius MC centered at M. The circle passes through A, B, and C. Thus, by Thales' theorem, $\triangle ABC$ is a right triangle because it is inscribed in a semicircle of diameter $\overline{AB}$.

Chapter 12

Problem Set 12.1 (page 773)

1. (a) Height, length, thickness, area, diagonal, weight **(c)** Height, length, depth **3. (a)** Answers will vary, anywhere from 40 to 54. **(b)** The circular portions of area don't fit well together, but leave gaps between them.
5. (a) $1 \text{ acre} = \frac{1}{640} \text{ mi}^2$, so $1 \text{ acre} = \left(\frac{1}{640} \text{ mi}^2\right)\left(\frac{5280 \text{ ft}}{1 \text{ mi}}\right)^2 = 43{,}560 \text{ ft}^2$. **6. (a)** 3 in **(c)** $2\frac{5}{8}$ in
7. (a) Between 165.5 km and 166.5 km **(c)** Between 0.495 mm and 0.505 mm **9. (a)** $33 \text{ cL} = 33 \cdot 10^{-2} \text{ L} = 0.33 \text{ L} = 330 \cdot 10^{-3} \text{ L} = 330 \text{ mL}$. **(b)** Not quite, since 1 liter = 1000 mL and $3 \cdot 33 \text{ cL} = 990 \text{ mL}$.
10. (a) 58.728 kg **(c)** 230 g **11. (a)** 3.5 kg **13. (a)** About 28 cm by 22 cm **(c)** About 2 cm **15. (a)** 8, 16, 32
17. $1 \text{ ha} = (10{,}000 \text{ m}^2)\left(\frac{1 \text{ km}}{1000 \text{ m}}\right)^2\left(\frac{1 \text{ mi}}{1.6 \text{ km}}\right)^2\left(\frac{640 \text{ acre}}{1 \text{ mi}^2}\right) \doteq 2.5$ acres **19.** $\frac{25 \text{ in}}{1 \text{ min}} \cdot \frac{60 \text{ min}}{1 \text{ hr}} \cdot \frac{24 \text{ hr}}{1 \text{ day}} \cdot \frac{14 \text{ day}}{1 \text{ fortnight}} \cdot \frac{1 \text{ ft}}{12 \text{ in}} \cdot \frac{1 \text{ furlong}}{660 \text{ ft}} = 63.6$ furlong/fortnight **21. (a)** kilo **(c)** no prefix **(e)** milli **23.** $\frac{100 \text{ km}}{9 \text{ L}} \cdot \frac{3.7854 \text{ L}}{1 \text{ gal}} \cdot \frac{1 \text{ mi}}{1.6 \text{ km}} \doteq 26.3 \frac{\text{mi}}{\text{gal}}$
25. (a) $5 \text{ gal} \cdot \frac{4 \text{ qt}}{1 \text{ gal}} \cdot \frac{32 \text{ ounces}}{1 \text{ qt}} = 640$ ounces. Since $640 \div 80 = 8$, add 8 liquid ounces of concentrate. **(b)** Add $80 \times 65 \text{ mL} = 5200 \text{ mL} = 5.2 \text{ L}$ of water. **27.** A league varied from time to time in history, but was usually close to 3 miles. Thus the Nautilus traveled about 60,000 miles. **31.** Interior angle measures of a triangle add up to 180°, so $8x + 6x + 4x = 180°$, and $x = 10°$. Angles are of measure 80°, 60°, and 40°. **33. (a)** $\triangle ABC \cong \triangle EFD$ by the SAS congruence property. **(c)** $\triangle ABC \sim \triangle FED$ by the SAS similarity property.

JUST FOR FUN How to Cover a Long Hole with a Short Board (page 778)

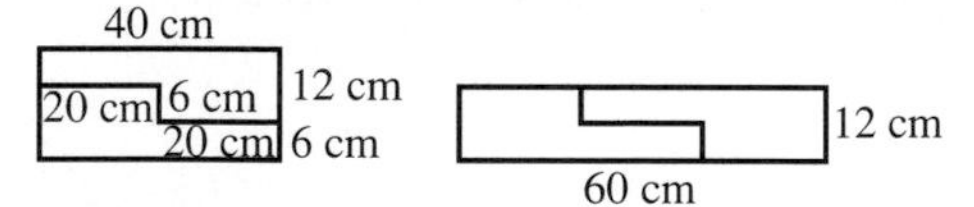

JUST FOR FUN Tile and Smile (page 781)

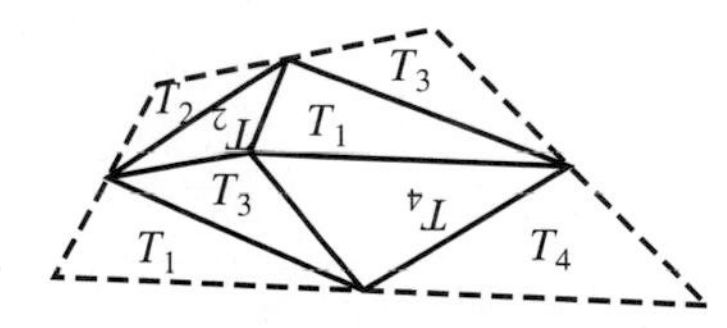

The parallelogram has one-half of the area of the quadrilateral.

Problem Set 12.2 (page 789)

1. (a) A 2 meter square has area 4 m², which is larger than 2 m². **(b)** A square of area $\frac{1}{2}$ m² is larger than the area, $\frac{1}{4}$ m², of a square with sides $\frac{1}{2}$ m. **3. (a)** 12 units

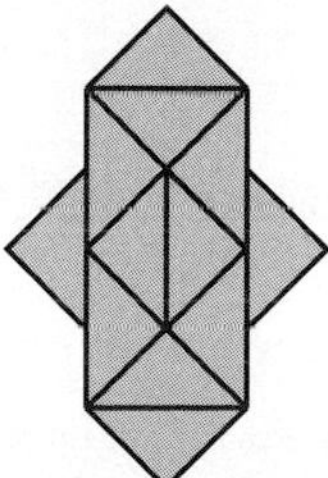

4. Rearrange the triangles numbered 1–9 in the gap between the 12-gon and the square and 3/4 of the square is covered. Hence the area of the 12-gon is 3/4 (4 units) = 3 units of area.
6. (a) 12 cm **7. (a)** 1 cm by 24 cm; 2 cm by 12 cm; 3 cm by 8 cm; 4 cm by 6 cm. The dimensions can also be given in the opposite order. **9. (a)** Starting with the two triangles on top of each other, rotate the top one 180° about the midpoint of any side to form a parallelogram. **(b)** Two triangles combine to form a parallelogram with base length the base of the triangle and height the height of the triangle. Therefore 2 · area (triangle) = area (parallelogram) = bh, so area (triangle) $= \frac{1}{2}bh$. **11. (a)**

2 · area (trapezoid) = area (parallelogram) $= h(a + b)$, so area (trapezoid) $= \frac{1}{2}(a + b)h$

12. (a) 81 ft^2, 45.2 ft **13. (a)** 1664.6 m^2, 198.3 m **14. (a)** 220 square units **15. (a)** $\triangle ABC$. All triangles have the same base and $\triangle ABC$ has the smallest height. **(b)** $\triangle ABF$. It has the largest height. **(c)** $\triangle ABD$ and $\triangle ABE$. They have equal heights and the same base. **17. (a)** 9 square units **19. (a)** 100 m + 100 m + $2\pi \cdot 50$ m = $(200 + 100\pi)$m $\doteq$ 514 m **(b)** (50 m)(100 m) + $\pi(25 \text{ m})^2 = (5000 + 625\pi)$ m$^2 \doteq 6963$ m^2 **20. (a)** $\pi(2)^2 - \pi(1)^2 = 3\pi$ square units $\doteq$ 9.4 square units. **22. (a)** 40,000,000 m **(b)** $12{,}755\pi$ km $\doteq$ 40,071,000 m **(c)** Equator is longer, since the earth bulges slightly at the equator and is slightly flattened at the poles. **23.** 20 cm^2. The common overlap reduces the area of both regions by the same amount, so the difference in area is unchanged.

25. (a) The unshaded region at the left has area $1^2 - \frac{1}{4}\pi(1^2) = 1 - \pi/4$.

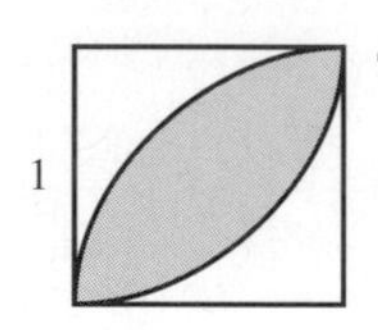

The shaded region at the left has area $1 - (1 - \pi/4) - (1 - \pi/4) = (\pi/2) - 1$.

27. The areas of the rectangular portions of sidewalk total 2400 ft^2. The pieces formed with circular areas have total turning of 360°, so when placed together form a circle with radius 8 ft of area of $\pi(8 \text{ ft})^2 = 64\pi$ ft^2. Total area is $(2400 + 64\pi)$ ft^2. **29. (a)** The length of twice the circumference, or 20″.

30. (a) $\frac{1}{2}\pi(2^2) = 2\pi$; $\frac{1}{4}\pi(3^2) = 9\pi/4$; $\frac{1}{6}\pi(5^2) = 25\pi/6$; $\frac{1}{360}\pi(4^2) = 16\pi/360$ **(b)** $\left(\frac{x}{360}\right)\pi(r^2)$

32. Draw $\overline{AP}$, $\overline{BP}$, $\overline{CP}$. Then area ($\triangle ABC$) $= \frac{1}{2}sh$ = area ($\triangle ABP$) + area ($\triangle BPC$) + area ($\triangle CPA$) =

$\frac{1}{2}sx + \frac{1}{2}sz + \frac{1}{2}sy = \frac{1}{2}s(x + y + z)$. Therefore, $\frac{1}{2}sh = \frac{1}{2}s(x + y + z)$, and $h = x + y + z$.
Alternate visual proof:

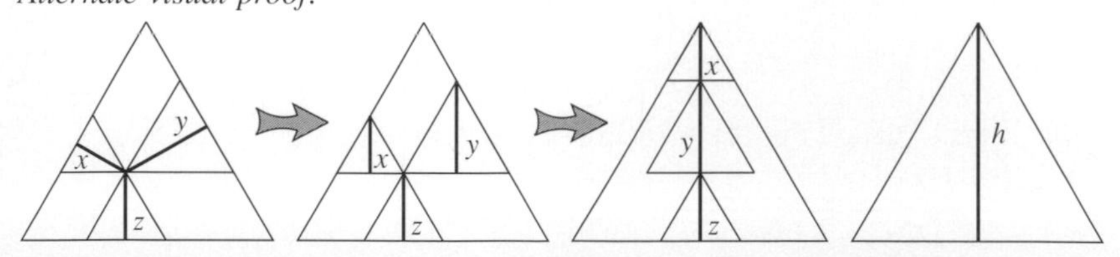

35. Consider the carpet as a 6 ft by 4 ft rectangle with semicircular ends of radius 2′. Then the carpet's area is (6 ft)(4 ft) + $\pi(2 \text{ ft})^2 \doteq 36.57$ ft$^2 \doteq 5266$ in^2. The carpet contains about 5266 inches of braid, or about 439 ft.

37. Area = 120 ft^2 = 17,280 in^2, so $\dfrac{17{,}280 \text{ in}^2}{64 \text{ in}^2} = 270$ tiles are needed. Some extra tiles should also be ordered to account for mistakes, wastage, and so on. **39.** 90°, since viewing one side of length 300′, the altitude of the triangle is greatest if the angle is 90°. **40. (a)** 48 square units. **42.** The circumscribed circle has four times the area of the inscribed circle. **44. (a)** 100 cm **(c)** 10,000 cm^2 **46.** For rectangle $ABCD$, $\triangle ABC \cong \triangle DCB$ by SAS. Therefore $\overline{AC} \cong \overline{BD}$.

Problem Set 12.3 (page 801)

1. (a) $x^2 = 7^2 + 24^2 = 625$, so $x = 25$. **(c)** $x^2 + 5^2 = 22^2$, so $x = \sqrt{459}$.
(e) $x^2 = 1^2 + (x/2)^2$, so $\left(\frac{3}{4}\right)x^2 = 1$, $x^2 = 4/3$, $x = 2/\sqrt{3}$ (or $2\sqrt{3}/3$).
2. (a) $x^2 + (2x)^2 = 25^2$, $5x^2 = 625$, $x = \sqrt{125} = 5\sqrt{5}$. **3. (a)** $x^2 = 10^2 + 15^2 = 325$, so $x = \sqrt{325}$; $y^2 = x^2 + 7^2 = 325 + 49 = 374$, so $y = \sqrt{374}$. **4. (a)** $x = \sqrt{13^2 - 12^2} = 5$. **5. (a)** Height $= \sqrt{15^2 - 9^2} = 12$, so area $= (20)(12) = 240$ square units. **7.** The areas are equal. The small circle of radius 1 has area π, and the large circle of radius $\sqrt{2}$ has area $\pi(\sqrt{2})^2 = 2\pi$, so the area between the circles is $2\pi - \pi = \pi$. **9.** $AG = 3$, since $AC = \sqrt{5}$, $AD = \sqrt{6}$, $AE = \sqrt{7}$, $AF = \sqrt{8}$, and $AG = \sqrt{9}$.
11.

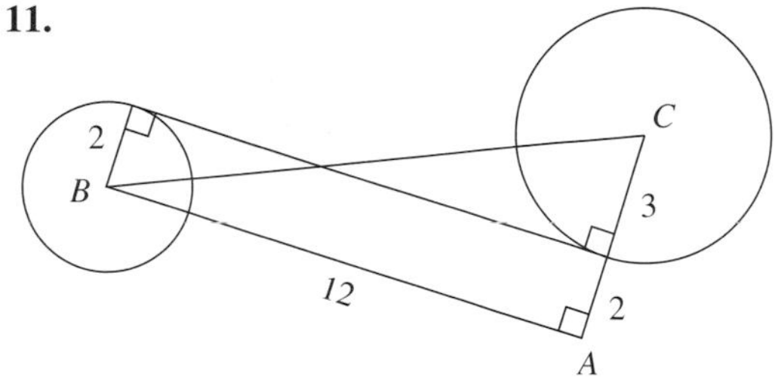

The distance between centers, B and C, is $\sqrt{12^2 + 5^2} = 13$.
13. (a) $(21)^2 + (28)^2 = 1225 = (35)^2$, yes **(c)** $(12)^2 + (35)^2 = 1369 = (37)^2$, yes
(e) $(7\sqrt{2})^2 + (4\sqrt{7})^2 = 210 \neq 308 = (2\sqrt{77})^2$, no
15. (a)

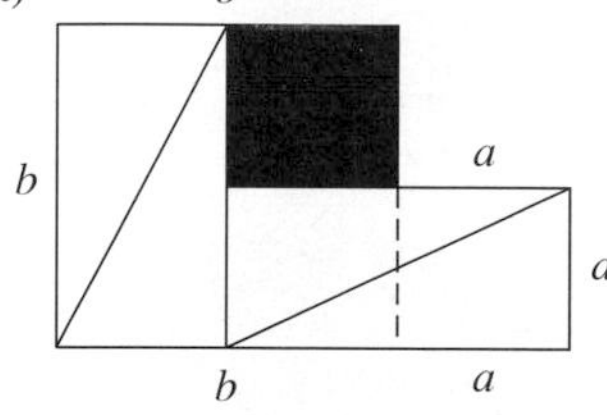

(b) Since the square and "double square" are tiled by the same five shapes, their areas are equal. The respective areas are c^2 and $a^2 + b^2$, so $c^2 = a^2 + b^2$. **17.**

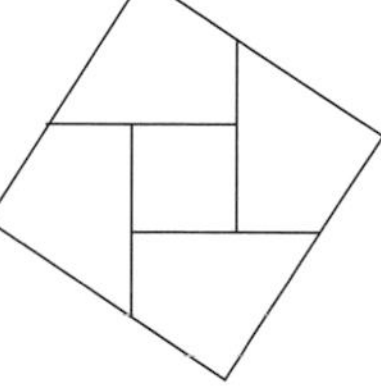

19. (a) By the Pythagorean theorem **(b)** $a^2 + b^2 = c^2$ and $a^2 + b^2 = z^2$, so $z^2 = c^2$, and therefore $z = c$.
(c) SSS property **(d)** $\angle C$ is the angle corresponding to the right angle in the congruent triangle $\triangle DEF$.
20. (a)

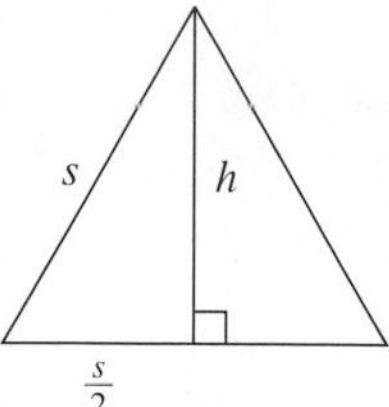

$\left(\frac{s}{2}\right)^2 + (h)^2 = s^2$, by the Pythagorean theorem. Therefore $h = \frac{\sqrt{3}}{2}s$.
22. (a) Squares (not square roots), right (not isosceles) triangle, and only the squares on the legs are added (not any of the sides). **(b)** No. Suppose an isosceles triangle has sides of length x, x, and y. By the triangle inequality, $x + x > y$. Therefore $2x > y$ and $\sqrt{2}\sqrt{x} > \sqrt{y}$, which means it is impossible for $\sqrt{x} + \sqrt{x} = \sqrt{y}$.

25. Answers will vary, but your ladder cannot be vertical, so the height is less than 24 feet. If the base is 7 feet from the wall the top of the ladder is still nearly 23 feet off of the ground. **27.** Let d be the depth of the pond. The stem length is $d + 2$ (in feet). Held to the side, a right triangle is formed with legs of length 6 and d and hypotenuse of length $(d + 2)$. Then $6^2 + d^2 = (d + 2)^2$, so $d = 8$ feet. **29. (a)** Exact $\doteq$ 1281 miles, approx $\doteq$ 1265 miles **30. (a)** $d \doteq 1.\ 2\sqrt{100} = 12$ miles **31.** The sum of the areas of the equilateral triangles on the legs equals the area of the equilateral triangle on the hypotenuse. **33.** Let the original dimensions be l and w. The changed dimensions are $\frac{4}{3}l$ and $\frac{3}{4}w$. Therefore, area $= \left(\frac{4}{3}l\right)\left(\frac{3}{4}w\right) = lw$, the same as the original area.

35. Along large semicircle: $\frac{1}{2}(2 \cdot \pi \cdot 8 \text{ m}) = 8\pi$ m. Along the two smaller semicircles: $\frac{1}{2}(2 \cdot \pi \cdot 3\text{m}) + \frac{1}{2}(2 \cdot \pi \cdot 5\ m) = 8\pi$ m. The distances are the same.

Problem Set 12.4 (page 821)

1. (a) $S = 2 \cdot \frac{1}{2}(20 \text{ cm} + 15 \text{ cm})(12 \text{ cm}) + (2 \text{ cm})(60 \text{ cm}) = 540 \text{ cm}^2$

$V = \frac{1}{2}(20 \text{ cm} + 15 \text{ cm})(12 \text{ cm})(2 \text{ cm}) = 420 \text{ cm}^3$

2. (a) The lateral surface area of the cone is $\frac{2\pi(6 \text{ in})}{2\pi(15 \text{ in})} = \frac{2}{5}$ of a circle of radius 15 in, so $S = \frac{2}{5}(\pi)(15 \text{ in})^2 + \pi(6 \text{ in})^2 = 126\pi \text{ in}^2 \doteq 396 \text{ in}^2$. Height of cone is $\sqrt{189}$ in, so $V = \frac{1}{3}\pi(6 \text{ in})^2(\sqrt{189} \text{ in}) \doteq 518 \text{ in}^3$.

3. (a) $S = 6 \cdot (20 \text{ mm})^2 - 2 \cdot \pi(2 \text{ mm})^2 + (20 \text{ mm})\, 2\pi(2 \text{ mm}) \doteq 2626 \text{ mm}^2$
$V = (20 \text{ mm})^3 - \pi(2 \text{ mm})^2(20 \text{ mm}) \doteq 7749 \text{ mm}^3$

4. (a) $S = 4\pi(4 \text{ ft})^2 + (20 \text{ ft})(2\pi)(4 \text{ ft}) = 224\pi \text{ ft}^2 \doteq 704 \text{ ft}^2$
$V = \frac{4}{3}\pi(4 \text{ ft})^3 + \pi(4 \text{ ft})^2(20 \text{ ft}) \doteq 1273 \text{ ft}^3$

5. (a) 16 cm **6. (a)**

8. $V_1 = \pi(8.\ 5''/2\pi)^2(11'') = 63.24$ cubic inches. $V_2 = \pi(11''/2\pi)^2(8.\ 5'') = 81.85$ cubic inches. The shorter cylinder has the greater volume. (Note that $V_1/V_2 = 8.5/11$.)

10. (a) Circumference of the cone is $\frac{3}{4} \cdot 2 \cdot \pi(4 \text{ in}) = 6\pi$ in, so the radius is 3 in. **11.** Let s be the radius of the semicircle. Then the slant height of the cone is s. Let d be the diameter of the cone. Then $\pi d = \frac{1}{2}(2\pi)(s)$, so $d = s$. **13.** Area (sphere) $= 4\pi^2$ square units. Area (cylinder) $= 2 \cdot \pi \cdot r^2 + 2\pi r \cdot (2r) = 6\pi r^2$ square units. Thus, $4\pi r^2/6\pi r^2 = 2/3$. **14. (a)** 84 cm³ **15. (a)** 4, considering area **(b)** One 14″ pizza is nearly the same amount of pizza, but will save \$2. **17. (a)** $1 - \frac{1}{4} - \frac{1}{4} = \frac{1}{2}$ (the small circles have one-half the diameter, so 1/4 the area) **19. (a)** 200 ml (doubling the radius increases volume by factor of 4; halving height halves the volume) **22.** Suppose the area of the similar figure with straight side of length 1 is A. By the similarity principle, the areas of the figures erected on the sides of the triangle are then a^2A, b^2A, and c^2A, since a, b, and c are the respective scale factors. Since $a^2 + b^2 = c^2$ (Pythagorean theorem), we get $a^2A + b^2A = c^2A$, showing that the sum of the areas of the figures erected on the legs is equal to the area of the figure erected on the hypotenuse. **24. (a)** 3 in, since $2/16 = 1/8 = (1/2)^3$ and $\frac{1}{2} \cdot 6'' = 3''$.

26. (a) By similarity, the height of the large complete pyramid is 8 cm. The frustum's volume is found by subtracting the volume of the removed pyramid from the volume of the complete pyramid: $V = \frac{1}{3}(10 \text{ cm})^2(8 \text{ cm}) - \frac{1}{3}(5 \text{ cm})^2(4 \text{ cm}) = \frac{700}{3} \text{ cm}^3 \doteq 233 \text{ cm}^3$.

28. $V(\text{ring}) = \pi\left(\frac{9}{16}\text{ in}\right)^2\left(\frac{5}{4}\text{ in}\right) - \pi\left(\frac{1}{2}\text{ in}\right)^2\left(\frac{5}{4}\text{ in}\right) \doteq 0.26\text{ in}^3$, so about 1.56 ounces.
30. $V(\text{box}) = 160\text{ in}^3$ and $V(\text{tub}) = \pi(3\text{ in})^2(10\text{ in}) \doteq 283\text{ in}^3$. Two boxes is a better buy.
32. (a) $\left(\frac{5\text{ in}}{13\text{ in}}\right)^3(106.75\text{ pounds}) \doteq 6.07$ pounds
34. (a) The scale factor is 12 and volume would be increased by the factor $12^3 = 1728$.
35. (a) $s = 1/\sqrt{2}$, since then $s^2 = 1/2$ showing that the area of an A1 sheet is half that of an A0 sheet.
(b) 

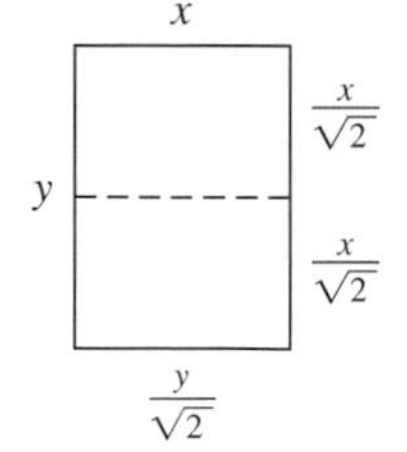

Using part (a), we see that the width x of an A0 sheet is $y/\sqrt{2}$, where y is the length. Then $xy = y^2/\sqrt{2} = 1\text{ m}^2$, so $y = 2^{1/4}\text{ m} \doteq 1.2\text{ m}$ and $x = 1/2^{1/4}\text{ m} \doteq 0.84\text{ m}$. **37.** The circular cross sections are related by a scale factor of 3, so 9 inches of water in the cylinder corresponds to 1 inch of rainfall. **38. (b)** The 3″ and \$3 are not scale factors. The reasoning seems to suggest that a 6″ pizza should cost \$2, and a 3″ pizza is given away together with \$1. **39. (a)** Yes, since $30^2 + 72^2 = 6084 = 78^2$ **41.** $P \doteq 8 + 2\sqrt{5} + \sqrt{2} + \sqrt{10} \doteq 17$. $A = 12$ square units. **42.** $\frac{50\text{ in}}{\text{sec}} \cdot \frac{1\text{ ft}}{12\text{ in}} \cdot \frac{1\text{ mile}}{5280\text{ ft}} \cdot \frac{60\text{ sec}}{1\text{ min}} \cdot \frac{60\text{ min}}{1\text{ hr}} \doteq 2.84$ miles per hour. **43. (a)** 27. 8 cm

CLASSIC CONUNDRUM Strings and Balls (page 827)

A string of length s will tightly encircle a sphere of radius r given by $r = s/2\pi$, since $s = 2\pi r$ is the circumference. A longer string of length S will make a circle of radius $R = S/2\pi$. Subtracting we see that $R - r = (S - s)/2\pi$. If S is 6 feet larger than s, then $S - s = 6$ feet and $R - r = 6'/2\pi \doteq 1'$. That is, an extra 6 feet of string will increase the radius of *any* circle by about a foot. The rabbit can easily run under the 6 foot longer string that is everywhere 1 foot above the equator.

Chapter 12 Review Exercises (page 829)

1. (a) Centimeters **(b)** Millimeters **(c)** Kilometers **(d)** Meters **(e)** Hectares **(f)** Square kilometers **(g)** Milliliters **(h)** Liters
2. (a) 4 L **(b)** 190 cm **(c)** 200 m² **3.** 84 L **4.** $\frac{300\text{ ft}}{3\text{ sec}} \cdot \frac{1\text{ mile}}{5280\text{ ft}} \cdot \frac{60\text{ sec}}{1\text{ min}} \cdot \frac{60\text{ min}}{1\text{ hr}} \doteq 68$ miles per hour
5.

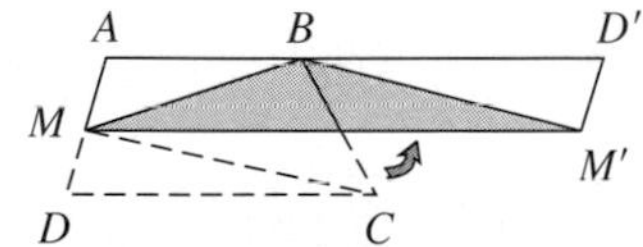

The triangle has half the area of the parallelogram $AD'M'M$, so it also has half the area of the trapezoid $ABCD$.

6. (a) $768\text{ in}^2 = 5\frac{1}{3}\text{ ft}^2$ **(b)** $\frac{1}{2}(8\text{ m})(9\text{ m}) + \frac{1}{2}(3\text{ m})(6\text{ m}) = 45\text{ m}^2$ **(c)** $\frac{1}{2}(5\text{ cm} + 7\text{ cm})(3\text{ cm}) = 18\text{ cm}^2$

7. (a) 11 square units **(b)** 9 square units **(c)** $9\frac{1}{2}$ square units **8. (a)** $A = (3\text{ ft})(4\text{ ft}) + \frac{1}{2}\pi(1.5\text{ ft})^2 \doteq 15.5\text{ ft}^2$, $P = 11\text{ ft} + \pi(1.5\text{ ft}) \doteq 15.7\text{ ft}$ **(b)** $A = \frac{3}{4}\pi(3\text{ m})^2 \doteq 21.2\text{ m}^2$, $P = \frac{3}{4}\cdot 2\pi(3\text{ m}) + 6\text{ m} \doteq 20.1\text{ m}$
9. $x = 6$, $y = \sqrt{5}$ (*Methods to solve:* Find the area two ways, or use similar triangles.) **10.** $\sqrt{1125}\text{ cm} \doteq 33.5\text{ cm}$
11. $\sqrt{116}$ in, $\sqrt{160}$ in, $\sqrt{244}$ in, $\sqrt{260}$ in **12.** $9 + \sqrt{2} + \sqrt{10} + \sqrt{5} \doteq 15.8$ units

13. **(a)** $V = [(10\text{ ft})(20\text{ ft}) + \frac{1}{2}(8\text{ ft} + 20\text{ ft})(8\text{ ft})](30\text{ ft}) = 9360\text{ ft}^3$

$S = 2 \cdot \frac{1}{2}(8\text{ ft} + 20\text{ ft})(8\text{ ft}) + 2 \cdot (10\text{ ft})(20\text{ ft}) + 2 \cdot (10\text{ ft})(30\text{ ft}) + 2 \cdot (10\text{ ft})(30\text{ ft}) +$
$(8\text{ ft})(30\text{ ft}) + (20\text{ ft})(30\text{ ft}) = 2664\text{ ft}^2$

(b) $V = \pi(7\text{m})^2(18\text{ m}) + \frac{1}{2} \cdot \frac{4}{3}\pi(7\text{ m})^3 \doteq 3489\text{ m}^3$

$S = \frac{1}{2} \cdot 4\pi(7\text{ m})^2 + 2\pi(7\text{ m})(18\text{ m}) + \pi(7\text{ m})^2 \doteq 1253\text{ m}^2$

(c) $V = \frac{1}{3}\pi(5\text{ cm})^2(8\text{ cm}) + \frac{1}{2} \cdot \frac{4}{3}\pi(5\text{ cm})^3 \doteq 471\text{ cm}^3$

$S = \frac{1}{2} \cdot 4\pi(5\text{ cm})^2 + \frac{2\pi(5\text{ cm})}{2\pi(8\text{ cm})} \cdot \pi(8\text{ cm})^2 = 90\,\pi\text{ cm}^2 \doteq 283\text{ cm}^2$

14. $V(\text{sphere}) = \frac{4}{3}\pi(10\text{ m})^3$ and V (four cubes) $= 4(10\text{ m})^3$. Since $\pi > 3$, then $\frac{4}{3}\pi > 4$, showing that the sphere has larger volume. **15.** **(a)** $(180\text{ ft})(1.5) = 270$ ft, since $k = 1.5$ is the scale factor. **(b)** $\frac{45\text{ pounds}}{(1.5)^2} = 20$ pounds, since Johan's garden area is $(1/1.5)^2$ times that of Heather's.

Chapter 12 Test (page 831)

1. **(a)** mm **(b)** m **(c)** m **(d)** km **(e)** mL **(f)** liters **2.** **(a)** 216.1 cm **(b)** 168,200 cm **(c)** 5000 cm^2 **(d)** 10,000 m^2 **(e)** 4.719 l **(f)** 3200 cm^3 **3.** **(a)** Approximately 31.86 yd **(b)** Approximately 1.5 mi **(c)** 291.6 ft^2 **(d)** Approximately 14.69 mi^2 **(e)** 205.2 ft^3 **(f)** Approximately 3.45 ft^3

4. $\frac{14\text{ day}}{12\text{ ft}} \cdot \frac{24\text{ hours}}{1\text{ day}} \cdot \frac{60\text{ min}}{\text{hour}} \cdot \frac{1\text{ ft}}{12\text{ in}} = 140\,\frac{\text{min}}{\text{in}}$, so 140 minutes to grow one inch.

5.

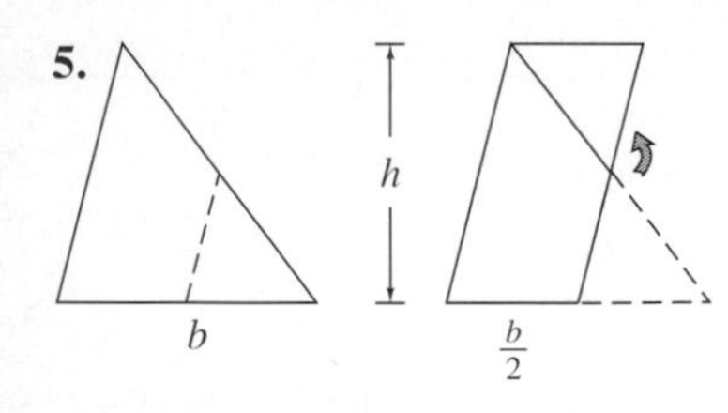

area (triangle) = area (parallelogram) = $(b/2)(h) = \frac{1}{2}bh$.

6. **(a)** 8 cm^2 **(b)** 8 cm^2 **(c)** 8.5 cm^2

7. $A = \frac{1}{2} \cdot (6\text{ cm})(16\text{ cm}) + \frac{1}{2}(15\text{ cm})(16\text{ cm}) = 168\text{ cm}^2$. Sides are 10 cm and 17 cm by the Pythagorean theorem, so $P = 54$ cm.

8. $A = \frac{1}{2}(9\text{ ft})(12\text{ ft}) + (8\text{ ft})(12\text{ ft}) - \frac{1}{2}\pi(4\text{ ft})^2 \doteq 125\text{ ft}^2$

$P = 15\text{ ft} + 9\text{ ft} + 8\text{ ft} + 12\text{ft} + \frac{1}{2} \cdot 2\pi(4\text{ ft}) \doteq 56.6\text{ ft}$

9. **(a)** $\frac{2}{3}\pi(5\text{ m})^2 \doteq 52.4\text{ in}^2$ **(b)** $\frac{1}{2}(2.6\text{ m} + 1.4\text{ m})(3\text{ m}) = 6.0\text{ m}^2$ **(c)** $\frac{1}{12} \cdot 3(24\text{ cm})^2 \doteq 151\text{ cm}^2$

10. Let the radius of the circle be r. Then the circumscribed square has sides of length $2r$ and the inscribed square has sides of length $\sqrt{2}\,r$.

$$\frac{\text{area (inscribed)}}{\text{area (circumscribed)}} = \frac{(\sqrt{2}r)^2}{(2r)^2} = \frac{2r^2}{4r^2} = \frac{1}{2}.$$

Alternate solution: The scale factor of the large to the small square is $1/\sqrt{2}$, so the small square has $(1/\sqrt{2})^2 = 1/2$ the area of the large square.

11. Cross-sectional view:

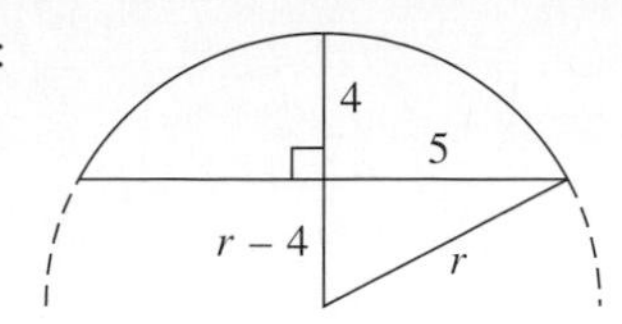

$(r-4)^2 + (5)^2 = 4^2$, or $r^2 - 8r + 16 + 25 = r^2$. Then $r = (16+25)/8$ mm $=$ 5.125 mm.

12. $\sqrt{189}$ ft $\doteq$ 13.7 ft, by the Pythagorean theorem.

13. **(a)** $P = \sqrt{52} + \sqrt{13} + \sqrt{65} \doteq 18.9$ units **(b)** Yes, $(\sqrt{52})^2 + (\sqrt{13})^2 = 65 = (\sqrt{65})^2$.

14. **(a)** $S = 2 \cdot \frac{1}{2}(7\text{ m})(24\text{ m}) + (7\text{ m} + 24\text{ m} + 25\text{ m})(5\text{ m}) = 448\text{ m}^2$, since the diagonal is 25 m.

$$V = \frac{1}{2}(7\text{ m})(24\text{ m})(5\text{ m}) = 420\text{ m}^3$$

(b) $S = 2 \cdot \frac{1}{2}\pi(4'')^2 + \frac{1}{2} \cdot 2\pi(4'')(6'') + (6'')(8'') \doteq 173.7\text{ in}^2$

$$V = \frac{1}{2}\pi(4'')^2(6'') \doteq 150.8\text{ in}^3$$

(c) Slant height is 10 ft. $S = 4 \cdot \frac{1}{2}(12\text{ ft})(10\text{ ft}) + (123\text{ ft})^2 = 384\text{ ft}^2$

$$V = \frac{1}{3}(12\text{ ft})^2(8\text{ ft}) = 384\text{ ft}^3$$

(d) Slant height is 13 cm. $S = \pi(5\text{ cm})^2 + \dfrac{2\pi(5\text{ cm})}{2\pi(13\text{ cm})} \cdot \pi(13\text{ cm})^2 = 90\pi\text{ cm}^2 \doteq 283\text{ cm}^2$

$$V = \frac{1}{3}\pi(5\text{ cm})^2(12\text{ cm}) = 100\pi\text{ cm}^3 \doteq 314\text{ cm}^3$$

15. **(a)** $S = 4\pi(10\text{ m})^2 = 400\pi\text{ m}^2 \doteq 1257\text{ m}^2$

$$V = \frac{4}{3}\pi(10\text{ m})^3 \doteq 4189\text{ m}^3$$

(b) $S = \pi(5\text{ cm})^2 + 2\pi(5\text{ cm})(6\text{ cm}) + \frac{1}{2} \cdot 4\pi(5\text{ cm})^2 = 135\pi\text{ cm}^2 \doteq 424\text{ cm}^2$

$$V = \pi(5\text{ cm})^2(6\text{ m}) + \frac{1}{2} \cdot \frac{4}{3}\pi(5\text{ cm})^3 \doteq 733\text{ cm}^3$$

16.

	Papa	**Mama**	**Baby**
Length of suspenders	50	40	20
Weight	468.75	240	30
Number of fleas	6000	3840	960

Scale Factors	
PB to MB	4/5
MB to BB	1/2

17. $V(\text{peel}) = \frac{4}{3}\pi(2.5\text{ m})^3 - \frac{4}{3}\pi(1.75\text{ in})^3 \doteq 43.0\text{ in}^3$, and $V(\text{grapefruit}) = \frac{4}{3}\pi(2.5\text{ in})^3 \doteq 65.4\text{ in}^3$. About 66 percent is peel. *Alternate solution:* The scale factor is $\left(2\frac{1}{2} - 3/4\right)\Big/2\frac{1}{2} = 0.7$. Since $(0.7)^{3'} = 0.343$, it follows that 34.3 percent of the grapefruit is not peel, and 65.7 percent is peel.

Chapter 13

JUST FOR FUN Quadrilateral + Quadrilateral = Parallelogram? (page 838)

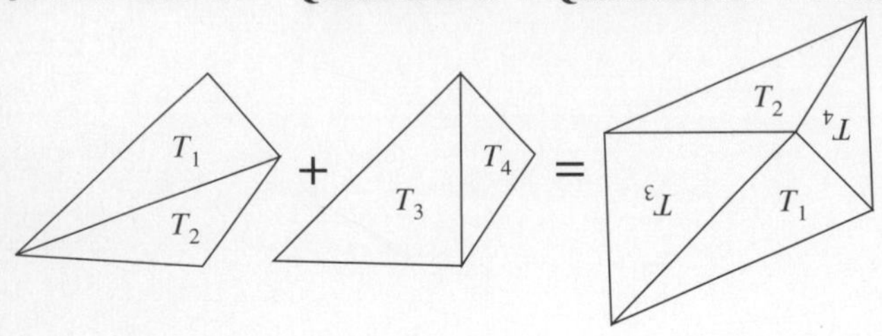

Problem Set 13.1 (page 855)

1. (a) Not a rigid motion. Distances between particular cards will change. **(c)** No, distances between particular pieces almost certainly will have changed. **2. (a)**

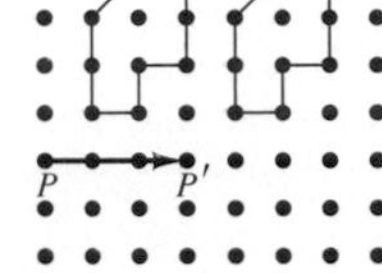

4. (a) 300° **(c)** 43° **5. (a)**

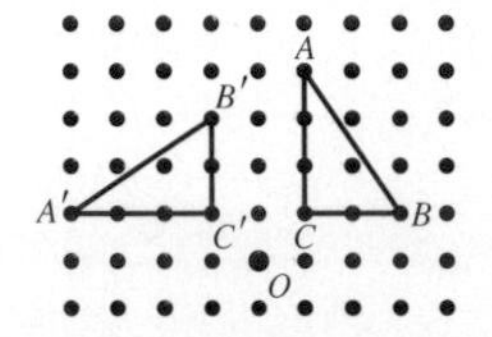

6. (a) Center O is the intersection of $\overleftrightarrow{AB}$ and $\overleftrightarrow{A'B'}$. **8. (a)** O is the midpoint of segment $\overline{PP'}$. **(b)** Draw the line $\overleftrightarrow{AO}$. Set the compass to distance AO and mark off this distance from O on $\overleftrightarrow{AO}$ away from A. This is A'. Likewise for points B and C. **9.** Let $x = O_1O_2$. A translation of distance $2x$ in the direction of the ray $\overrightarrow{O_1O_2}$. **11. (a)** Draw the vertical line through the midpoint of PP'.

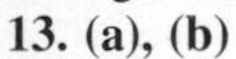

13. (a), (b)

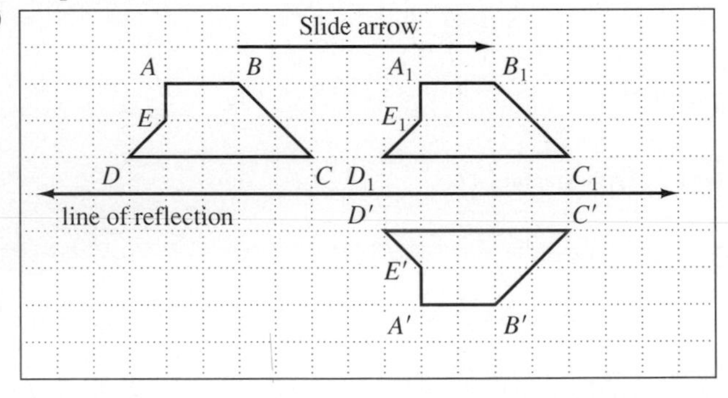

15. (a)

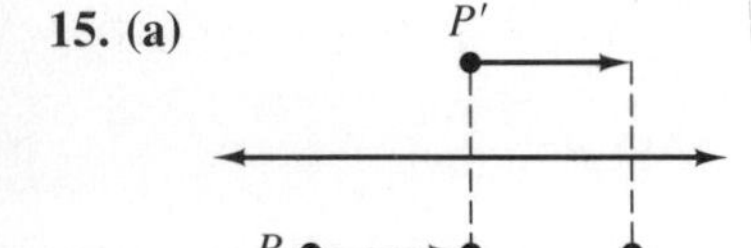

16. (a)

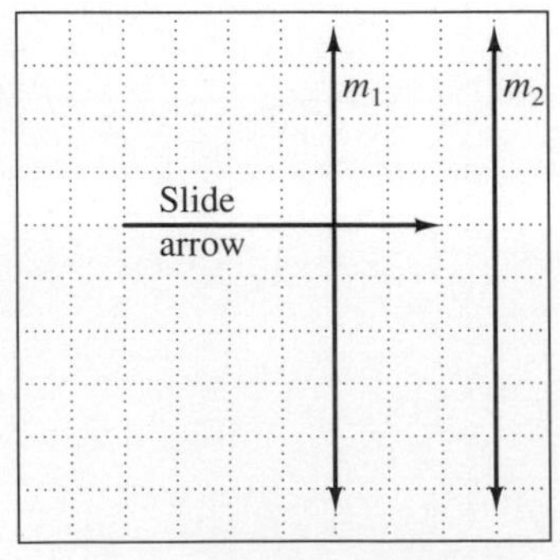

18. (c)

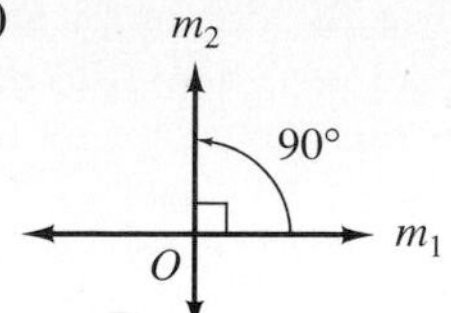

(e)

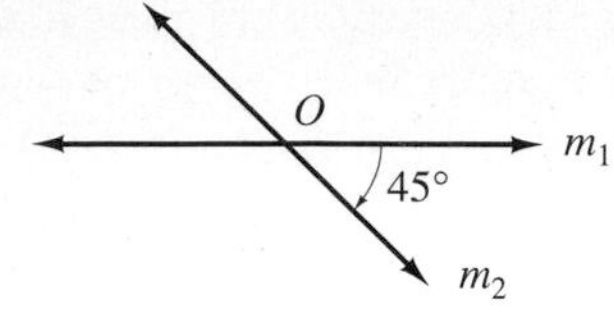

19. (a)

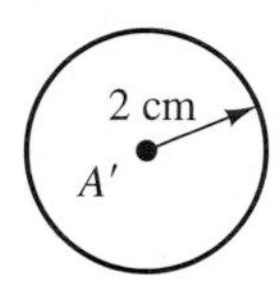

P' is some point of the circle at A' of radius 2 cm.

20. (a) Translation

21. On a 1 cm square grid, the two 90° rotations take O_1 to O_1' and O_2 to O_2'. as shown. This motion is equivalent to the 180° rotation about the point O.

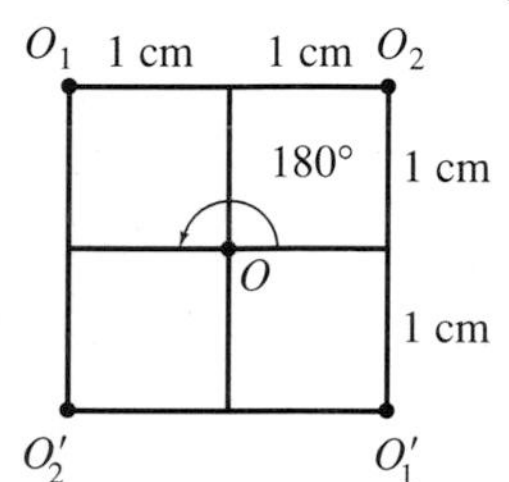

23. (a), (b)

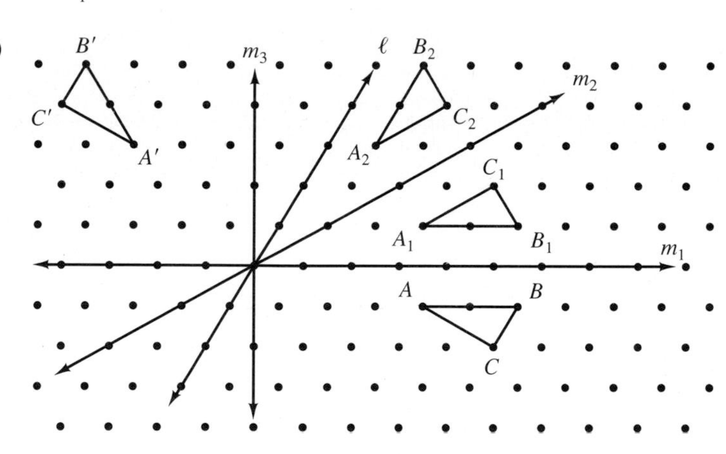

25. (a)

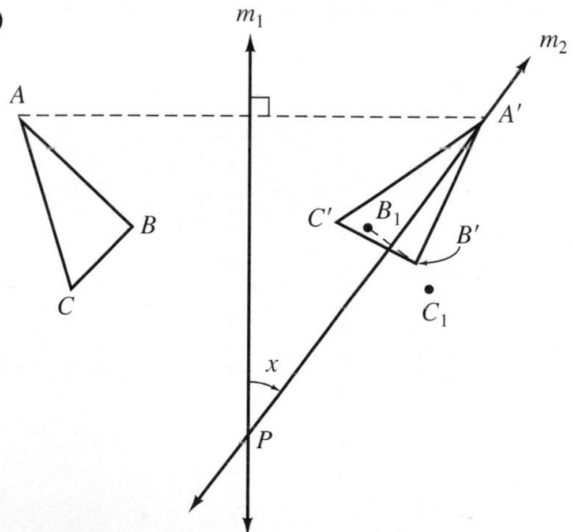

(b) C'

(c) Reflection first across line m_1 and then across line m_2 is equivalent to a rotation about the point P of intersection of m_1 and m_2, through an angle twice the measure x of the directed angle from line m_1 toward line m_2.

27. (a) A translation. Six reflections give an orientation preserving rigid motion, so it is either a rotation or a translation. Since a rotation has a fixed point (namely the rotation center), the motion is a translation. **28.** Translating $ABCD$ with the slide arrow A to C gives a parallelogram $BB'D'D$ which is clearly subdivided into triangles that are congruent to $\triangle ABC$, $\triangle BCD$, $\triangle CDA$, and $\triangle DAB$.

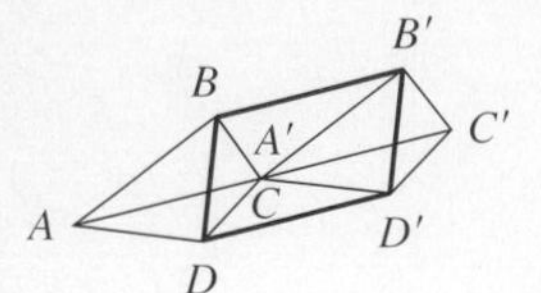

30. (a) A glide reflection

31. The mirror only needs to be 31″ tall, half of Estelle's height of 62″. The top of the mirror must be 3″ lower than the top of Estelle's head; that is, the top is 59″ off of the floor. It does not matter how far away Estelle stands; her reflection is always twice the distance she stands from the mirror.

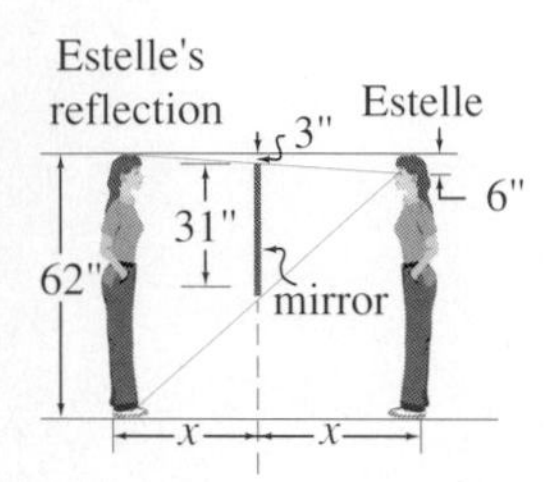

32. (a) $\overline{RS'}$ is the reflection of $\overline{RS}$ across m, so $RS' = RS$ since a reflection preserves all distances. Similarly, $QS' = QS$. **34.** A single mirror reverses orientation, so the double reflection seen in a corner mirror preserves orientation. The corner mirror reflection of your right hand will appear as a right hand.

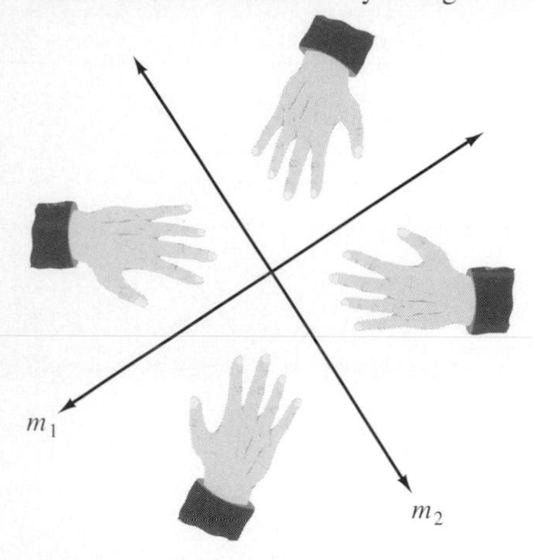

36. See the quadrilateral tilings illustrated in Example 10.14.
38.
```
to spin.scalene
    repeat 5 [scalene lt 55.77 fd 60]
end
```
40. (a) 6 6

41. (a)
```
to translate
     parallelogram.2
     pu fd 50 pd
     parallelogram.2
end
```

43. $2x + 3x + 4x = 180$, so $x = 20$, giving angles of measure 40°, 60°, and 80°.

45.

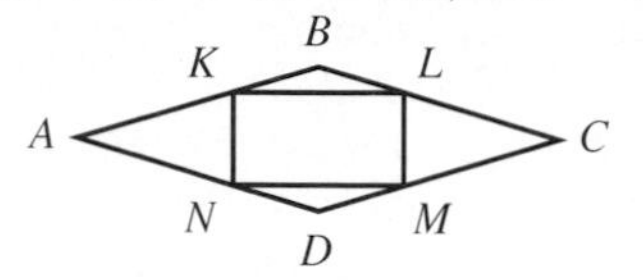

Midpoints form a nonsquare rectangle in this example.

JUST FOR FUN Mirror Magic (page 871)

Reflecting the mirror across the line through the midpoints of the shorter sides determine where Amy can make two straight cuts. Each of the three resulting pieces of mirror has a line of symmetry, and so the pieces can be turned over and glued into place.

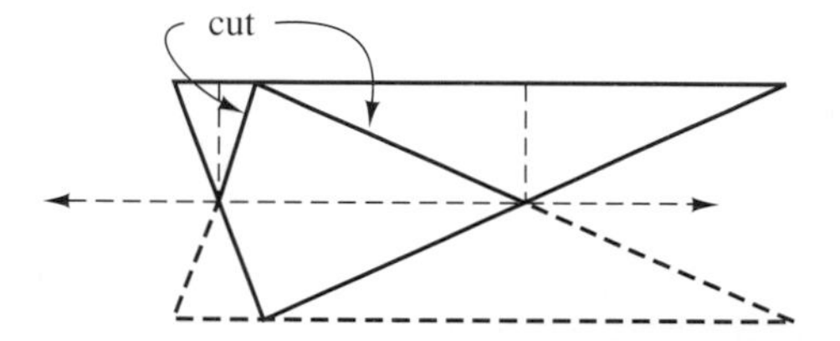

JUST FOR FUN The Penny Game (page 873)

Lynn can always place a penny at the position that is point symmetric to Kelly's last move. Since Lynn can always make a move, Kelly will be the first player unable to find space for an additional penny on the table.

Problem Set 13.2 (page 875)

1. (a)

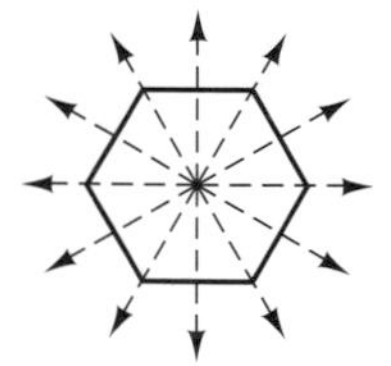

(c)

none

3. (a) Many figures are possible.

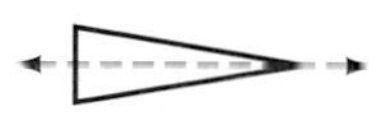

4. (a)

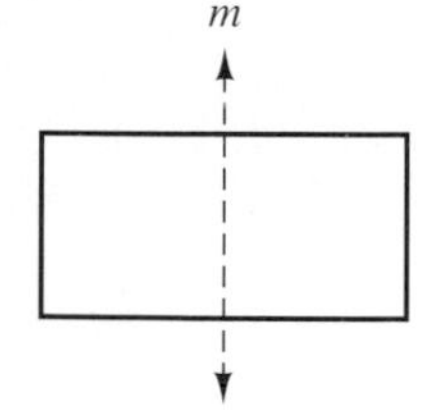

(c)

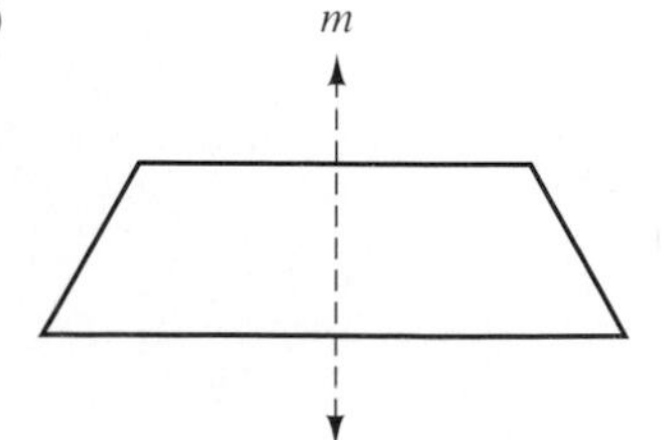

7. (a)

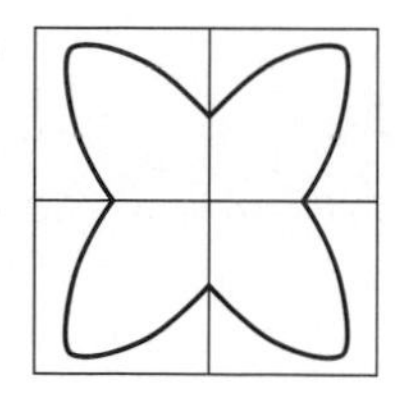

8. (a) One line of symmetry **9. (a)** Five lines of symmetry and 72° rotation symmetry **(b)** 72° rotation symmetry

11. (a)

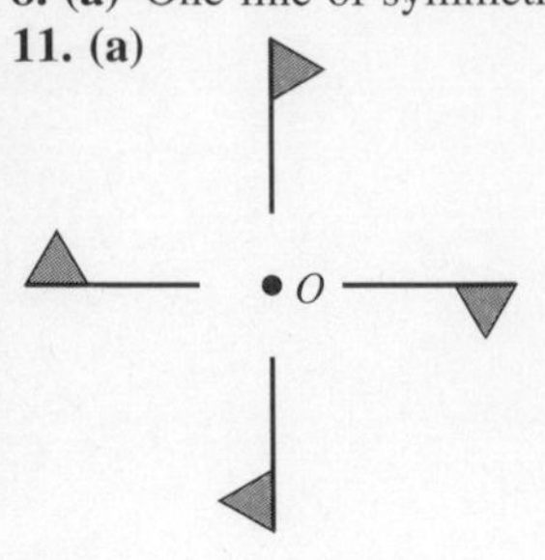

12. (a) Equilateral triangle **(b)** Square **13. (a)** 0, 1, 8 **(b)** 0, 1, 3, 8 **(c)** 0, 1, 8 **(d)** 0, 1, 8 **16. (a)** mL **17. (a)** There are vertical lines of symmetry through the centers of each letter, and there is a horizontal line of symmetry. The symbol type is *mm*. **18. (a)** Vertical line and glide symmetry across lines through a column of As or midway between columns of As. **(c)** 180° rotation symmetry about a point which is the center of an N, or midway between the centers of any two Ns. **(e)** 180° rotation centers at midpoints of vertically adjacent letters, reflection symmetry through vertical lines midway to adjacent columns, glide symmetry through horizontal lines midway to adjacent rows. **19. (a)** No letter or digit reflects vertically into a different letter or digit, so it must reflect into itself. **21.** The pattern must also have a horizontal line of symmetry, so it would be an mm pattern. **24. (a)** 1g **(c)** mg

26. (a) 11

p	p	p	p	p	p

(c) 12

p	d	p	d	p	d

28. (a) Three directions of reflection symmetry; three directions of glide symmetry; 120° rotation symmetry. **29. (a)** 3 **30. (a)** As left-handed people know well, scissors are not symmetric. **(c)** A man's dress shirt is not quite symmetric, since it buttons right handed. **(e)** Tennis rackets have two planes of bilateral symmetry. **31. (a)** Across the line of diagonal entries

35. (a)
```
to fig4
    repeat 4 [side lt 30]
end
```
36. (a) Yes **(b)** No, the distance between opposite corners has changed. **37. (a)** 350°

38. (a)

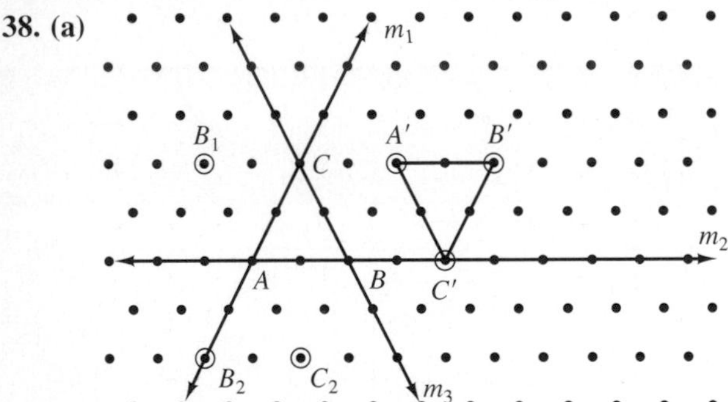

(b) Glide reflection, 3 units right and reflect across the line l parallel to $\overline{AB}$ and midway between C and $\overline{AB}$.

Problem Set 13.3 (page 887)

1. (a)

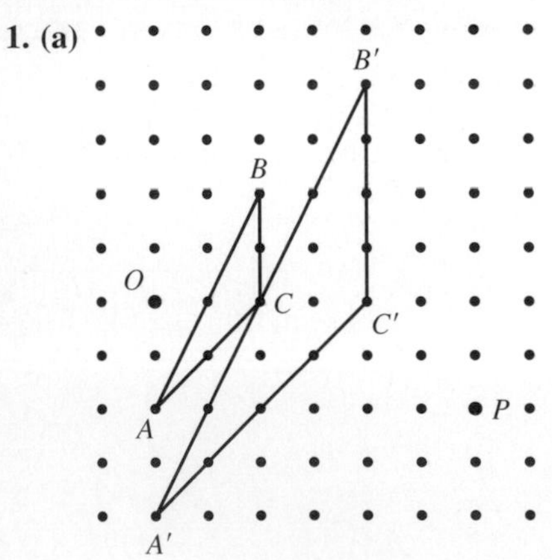

(b)

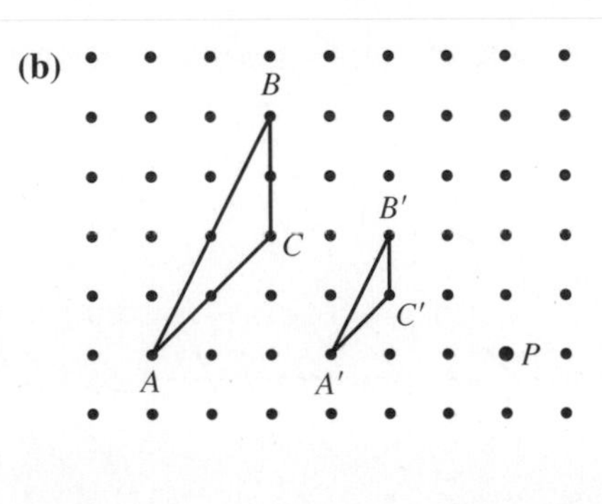

3. (a)

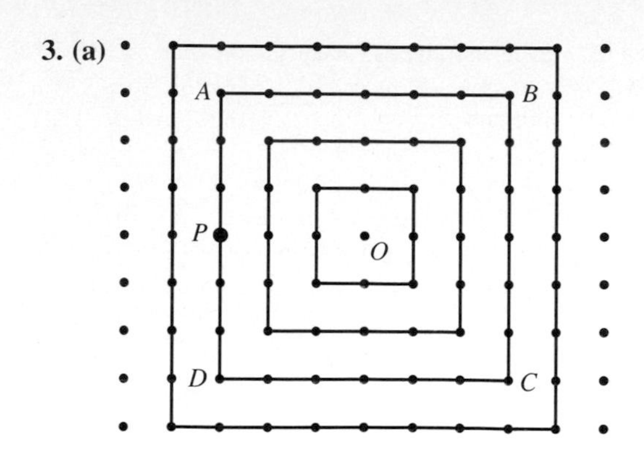

4. (a) $\frac{4}{3}$ **(b)** Find the intersection of two lines such as $\overleftrightarrow{AA'}$ and $\overleftrightarrow{BB'}$. **(c)** 18, 80°

6. (a) Center *P*, scale factor 3/2

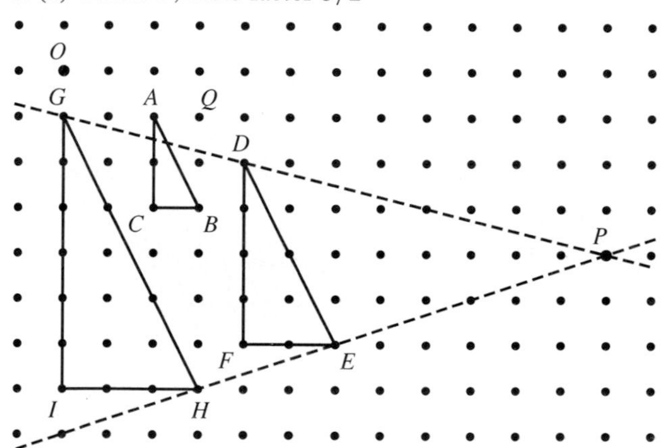

7.

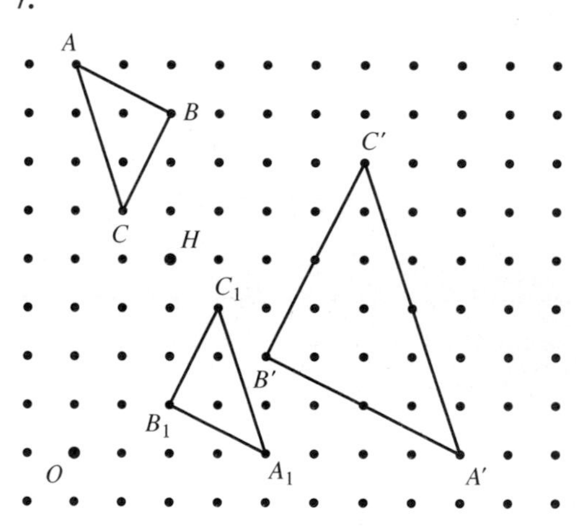

9.

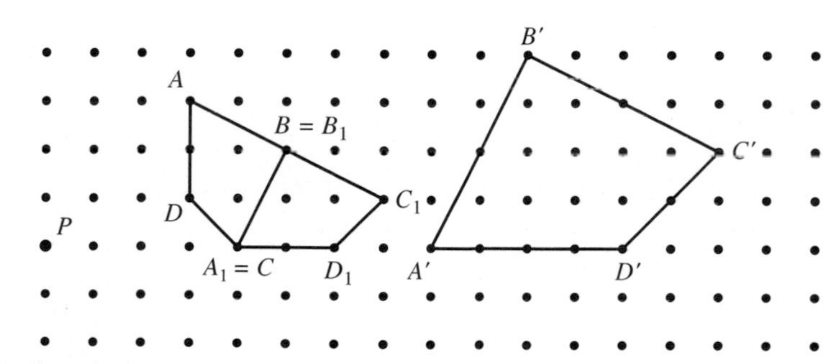

Rotate 90° counterclockwise about *B*. Then perform a size transformation centered at *P* with scale factor 2. (Other sequences will also work.)

11. No. A quadrilateral with sides of those lengths can be in many shapes with different angles. **14.** The vertices corresponding to R all lie on the ray $\overrightarrow{AR}$, so the intersection of this ray with side $\overline{BC}$ gives the desired square.

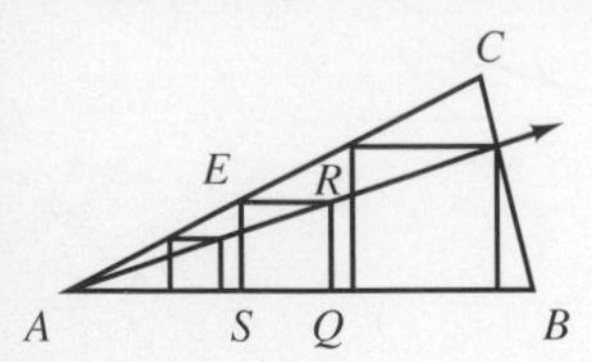

16. A size transformation about A with scale factor AP/AQ will yield a cycloid passing through P. So the diameter of the circle would be $\frac{AP}{AQ} \cdot 1 = \frac{AP}{AQ}$.
17. The line $\overleftrightarrow{PP'}$ passes through O, so constructing this line determines point O. Since $\overline{PQ}$ and $\overline{P'Q'}$ are parallel, constructing the line through P' that is parallel to $\overline{PQ}$ will determine Q'.

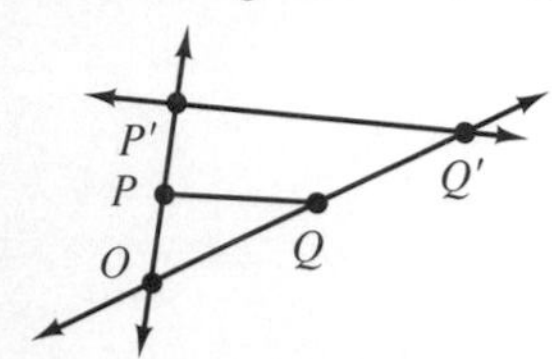

19. (a) Center A, scale factor $AF/AB = 5/1 = 5$. **21.** The larger grid is a size transformation that can easily be drawn. The short curves within each square can be drawn with good accuracy. **23. (a)** All parts of the rubber band stretch by the same factor. **(b)** 2 **(c)** $\frac{1}{3}$ of the way from the pinned point to the other end of the loop.

27.

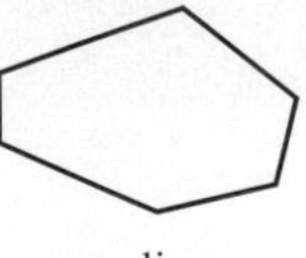

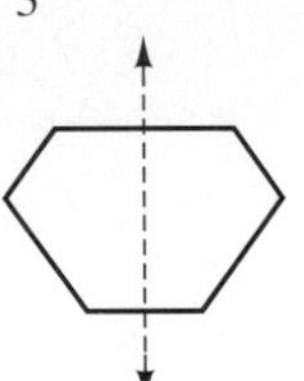

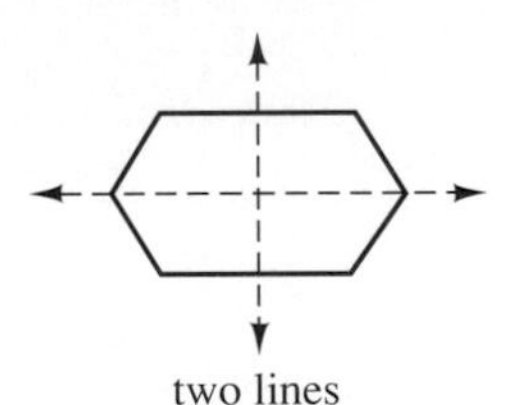

three lines

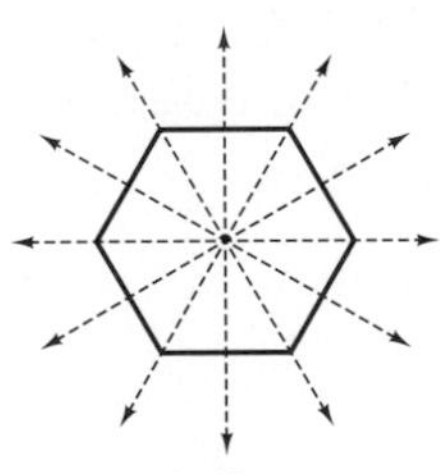

Hexagons with four, five, and seven or more lines of symmetry are not possible.
29. Translate left by 6 feet, reflect across l, then half-turn about P.

CLASSIC CONUNDRUM Inverting the Tetractys (page 891)

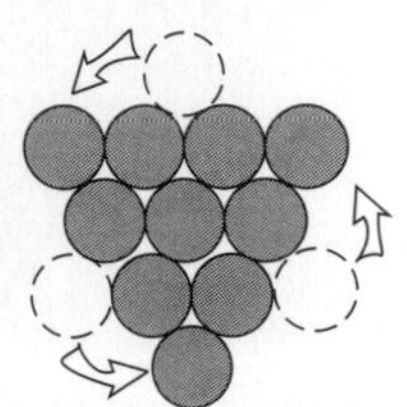

Chapter 13 Review (page 893)

1.

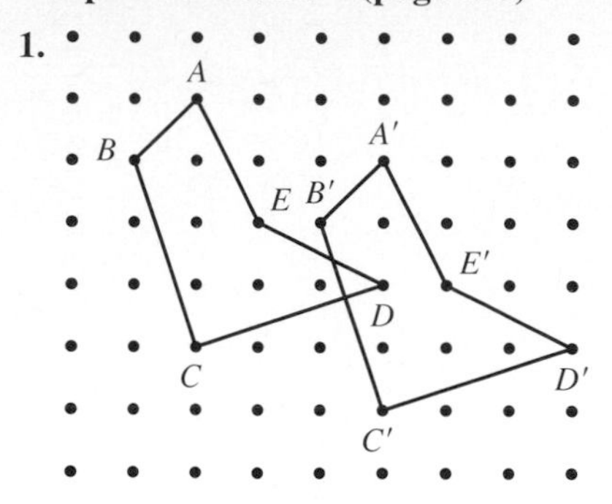

2. Find the perpendicular bisectors of segments $\overline{AA'}$ and $\overline{BB'}$. Their intersection is the turn center O, and the measure of $\angle AOA'$ is the turn angle.

3. Reflection across line l.

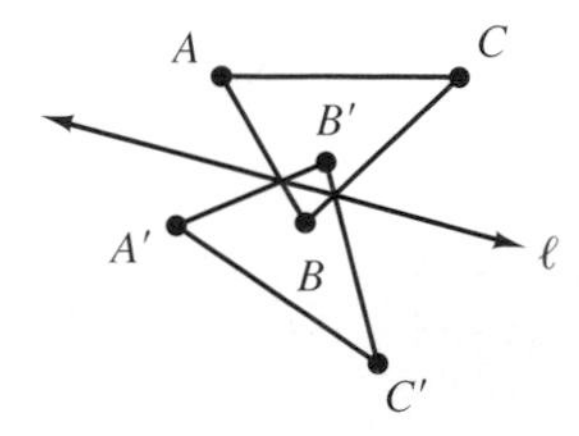

4. Draw any two vertical lines two inches apart. There are then three ways to choose the successive lines of reflection.

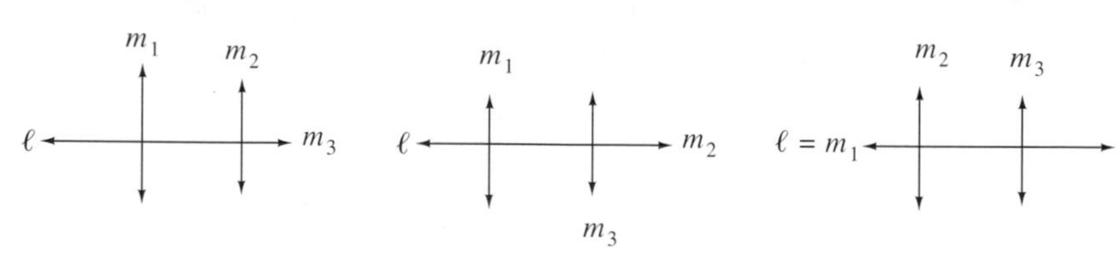

5. **(a)** 1 **(b)** 2 **(c)** 0 **(d)** 0 **(e)** 3 **(f)** All lines through the center point, since the figure has circular symmetry. **6.** **(a)** 0° **(b)** 0°, 180° **(c)** 0°, 180° **(d)** 0°, 72°, 144°, 216°, 288° **(e)** 0°, 120°, 240° **(f)** Any angle **7.** **(a)** Vertical line of symmetry **(b)** Horizontal line of symmetry **8.** O is the point on the line $\overleftrightarrow{AA'}$ for which $OA = 2AA'$ as shown.

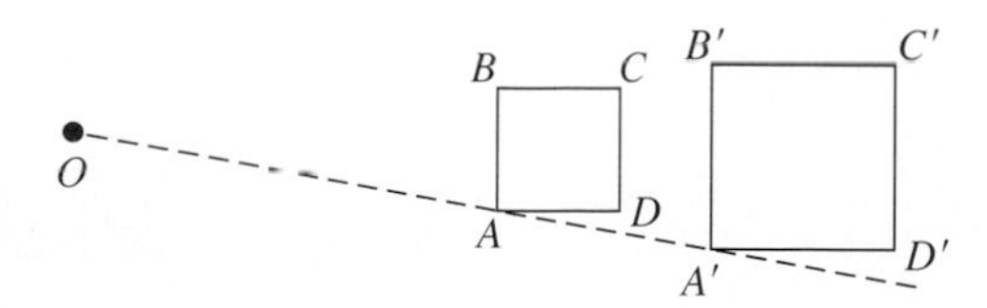

9. Since F and F' are similar figures with scale factor $k \cdot (1/k) = 1$, F and F' are congruent. Thus, some rigid motion will transform F to F'. Since no rotation or reflection takes place, this rigid motion must be a translation. **10.** Rotate $ABCD$ 45° about the center point P of the square. Then do a size transformation about P with scale factor $\frac{\sqrt{2}}{2}$.

Chapter 13 Test (page 894)

1. C' is 2 units right of B'; B is 2 units left of C.

2. **(a)** Center P and rotation angle 90° **(b)**

3. Construct the perpendicular bisector of $\overline{AA'}$, and suppose it intersects the sides of $\angle A$ at points labeled B and C. Then $\angle BA'C$ is the desired angle.

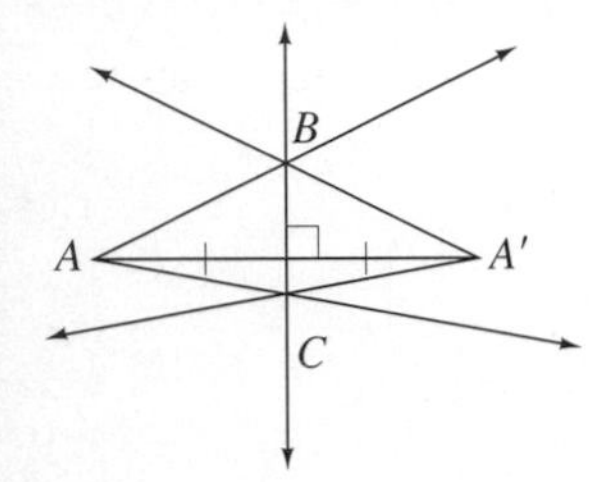

4.

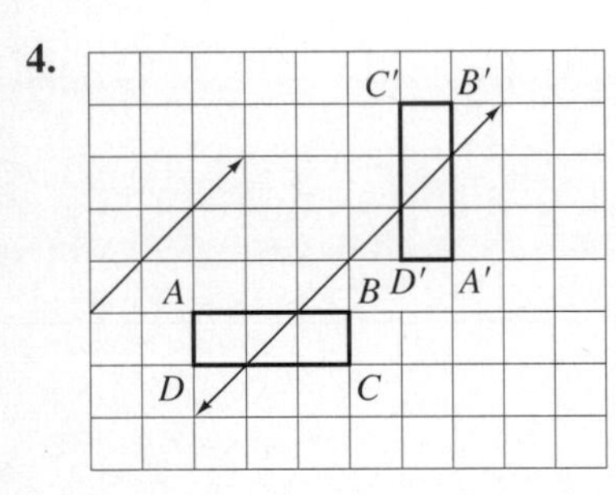

5. Various pairs of lines are possible. The distance between the lines must be half the distance between P and P'.

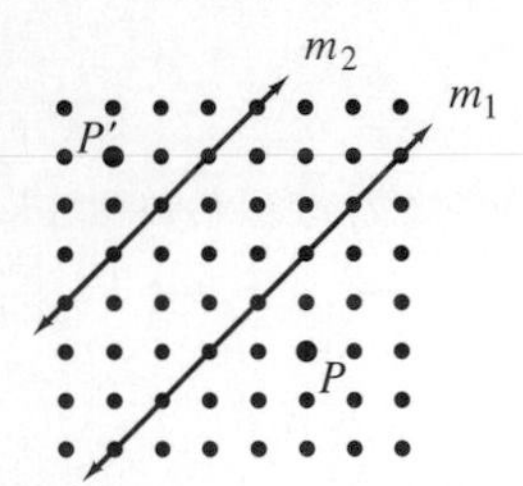

6. Many pairs of lines are possible. The two lines need to intersect at point O and the directed angle between the lines should be half the measure of $\angle QOQ'$.

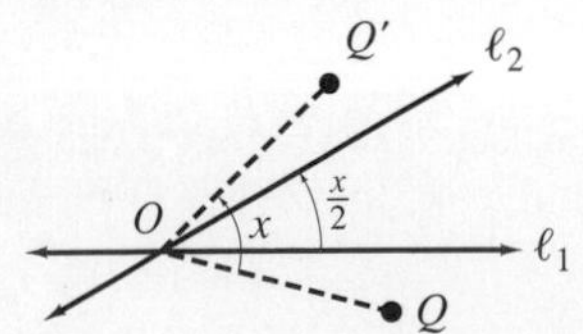

7. **(a)** Glide reflection **(b)** Reflection **(c)** Reflection **(d)** Glide reflection

8. **(a)** Rotation **(b)** Rotation

9. (a)

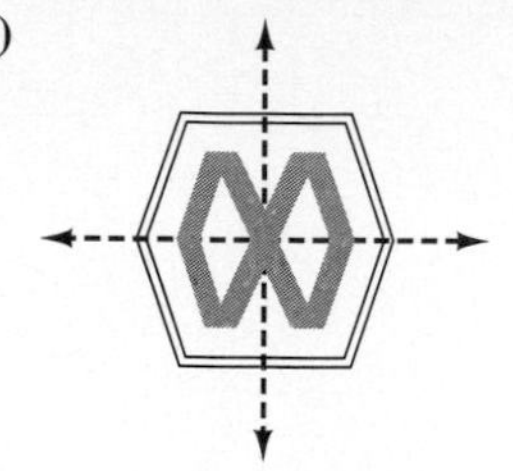

(b)

(c)

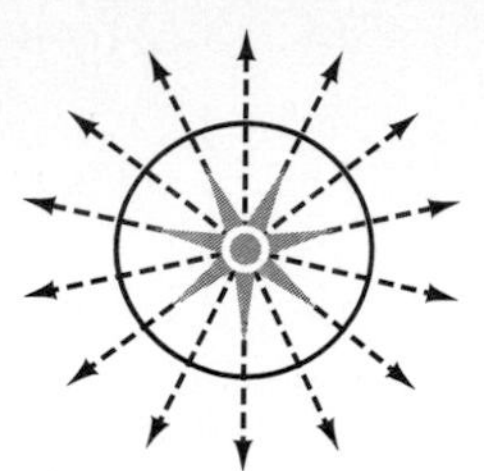

10. (a) 180° **(b)** 360° (only the identity rotation) **(c)** 360°/7 **11. (a)** Point symmetry, and two diagonal lines of symmetry **(b)** Four lines of symmetry, and 90° rotation symmetry **12. (a)** Glide reflection (1g) **(b)** Vertical and horizontal lines of symmetry (mm), and glide and half-turn symmetries **(c)** Half-turn symmetry (12) **(d)** Half-turn symmetry, vertical line of symmetry, glide reflection (mg) **13.** In wallpaper patterns the only possible rotation symmetries are 60°, 90°, 120°, and 180°. **14. (a)** 90° rotation symmetry **(b)** 120° rotation symmetry **15. (a)** 2 **(b)** 24 **(c)** 18 **16. (a)** A **(b)** $\frac{2}{3}$ **(c)** 40/3 **(d)** 9 **17.** Many transformations are possible. Here is one sequence: translate the square so A is taken to A'; rotate about A' by 45°; perform a size transformation about A' with scale factor $3\sqrt{2}/2$ (since $A'B' = 3\sqrt{2}$ and $AB = 2$).

Chapter 14

Problem Set 14.1 (page 911)

1. (a) The second coordinate is 0. **(c)** The first coordinate is positive and the second coordinate is negative.
2. (a), (c), (e), (g), (i)

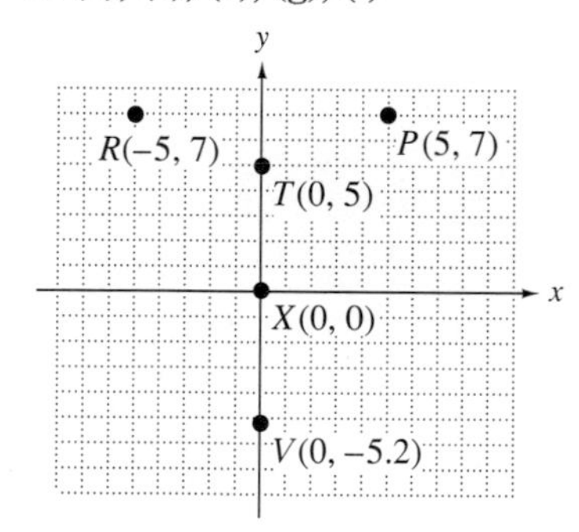

6. (a)

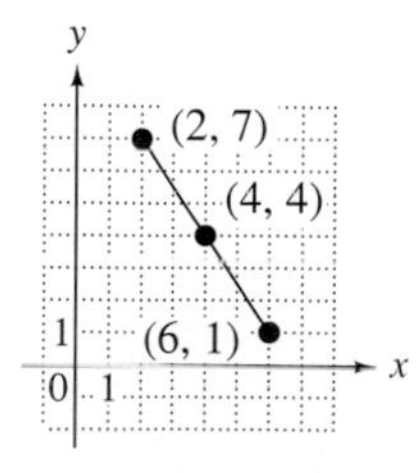

(c)

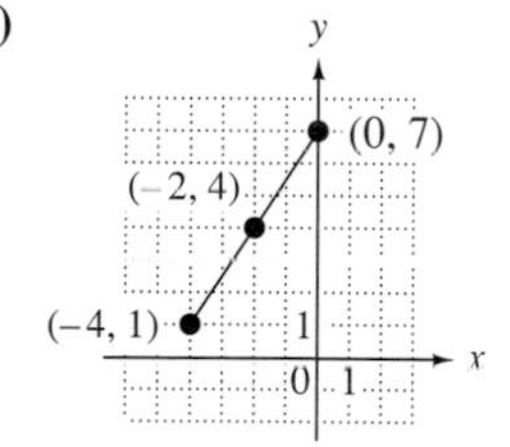

(e)

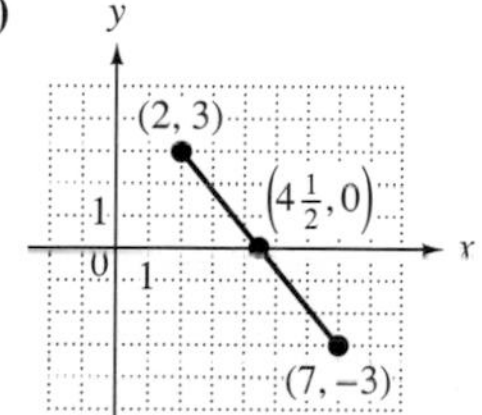

7.

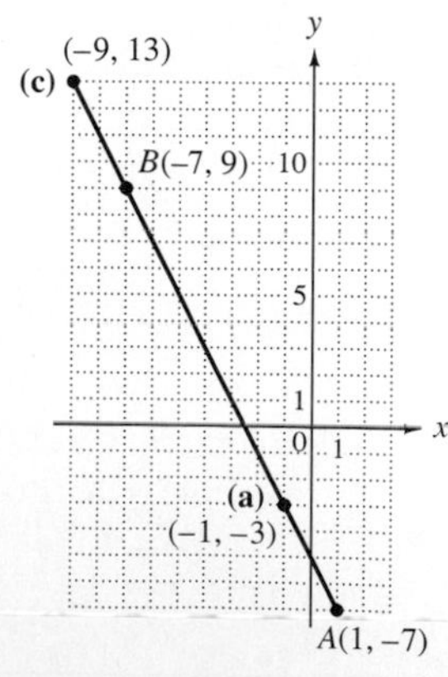

9. (a) $\sqrt{(4-(-2))^2+(13-5)^2}=\sqrt{36+64}=\sqrt{100}=10$
(c) $\sqrt{(8-0)^2+(-8-7)^2}=\sqrt{64+225}=\sqrt{289}=17$
10. (a) $(RS)^2=(\sqrt{(7-1)^2+(10-2)^2})^2=(\sqrt{36+64})^2=(\sqrt{100})^2=100$
$(RT)^2=(\sqrt{(5-1)^2+(-1-2)^2})^2=(\sqrt{16+9})^2=(\sqrt{25})^2=25$
$(ST)^2=(\sqrt{(7-5)^2+(10-(-1))^2})^2=(\sqrt{4+121})^2=(\sqrt{125})^2=125$
Since $(RS)^2+(RT)^2=(ST)^2$, by the Pythagorean theorem $\triangle RST$ is a right triangle.
12. By plotting vertices and calculating the lengths of sides we determine the following. **(a)** Right, isosceles **(c)** Acute, scalene

13. (a)

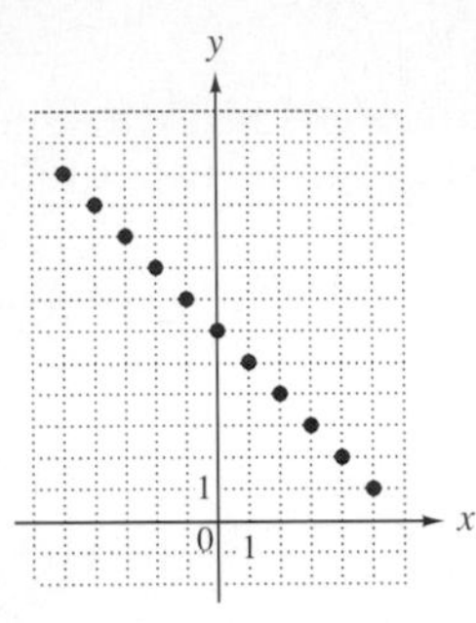

15. (a)

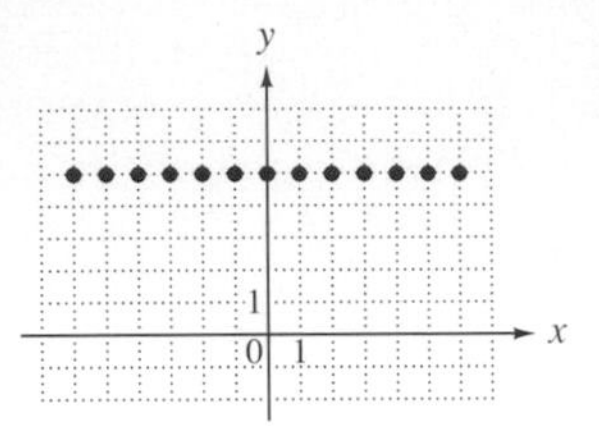

21. (a)

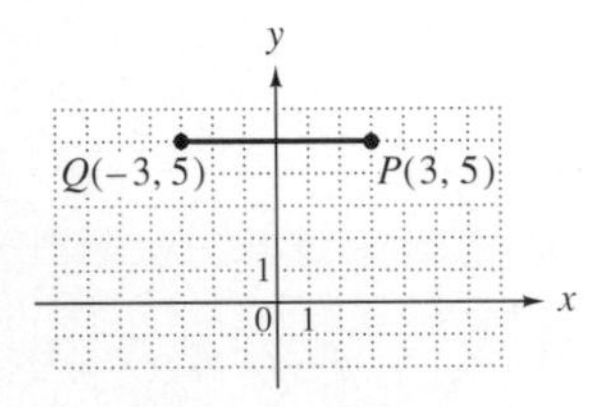

(c)

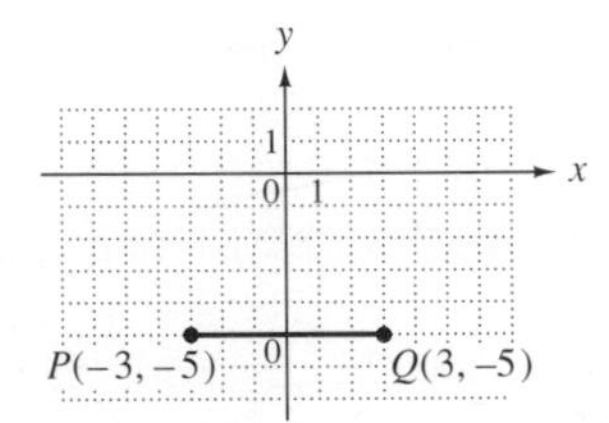

(e)

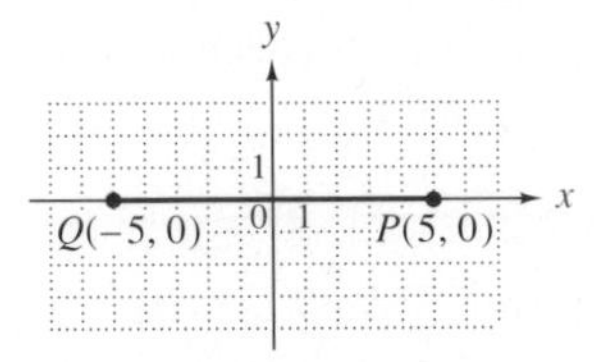

22. (a)

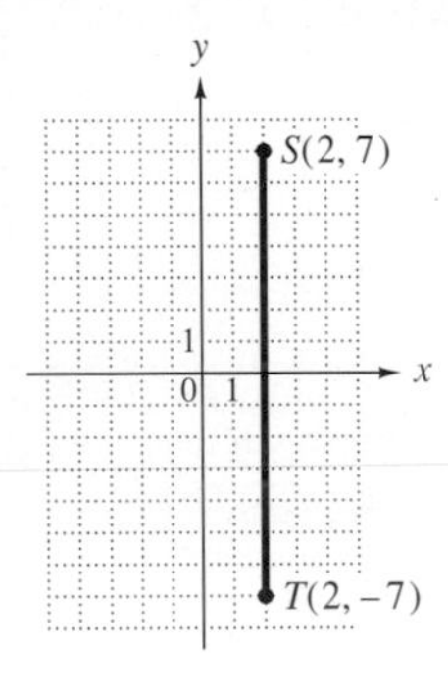

(c)

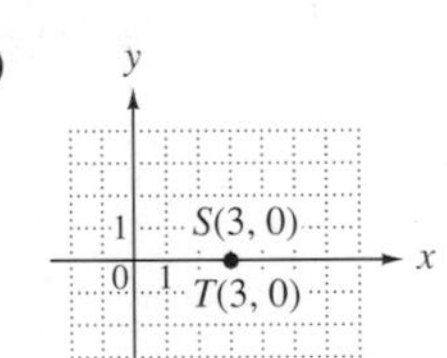

(e)

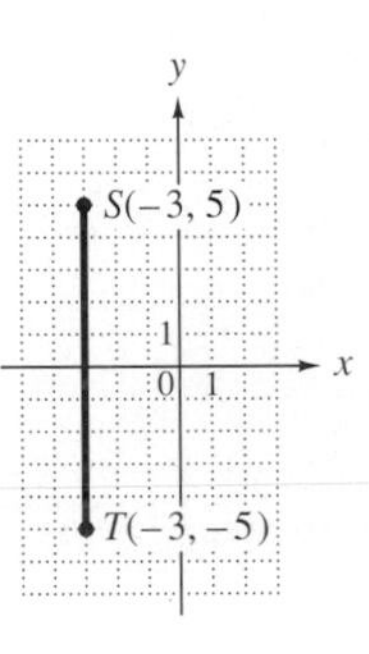

23. (a)

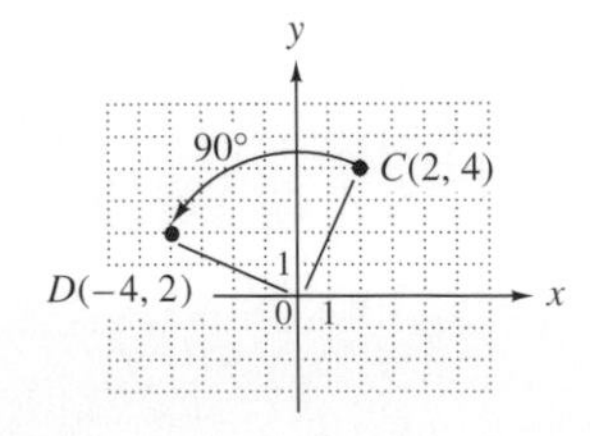

(c)

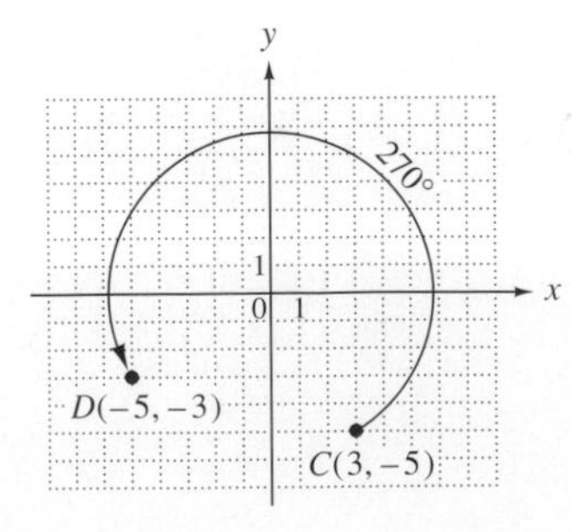

25. (a)

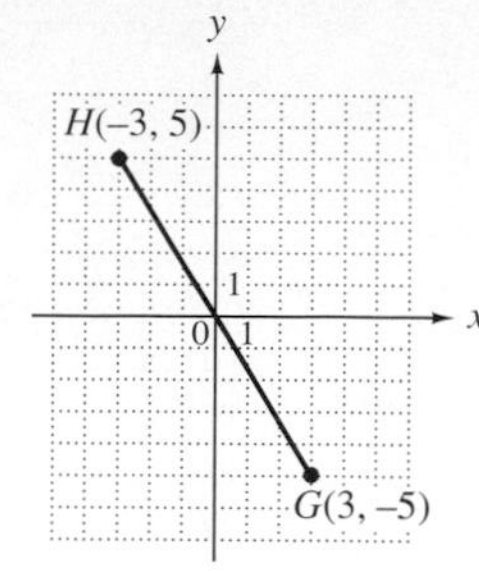

(c)

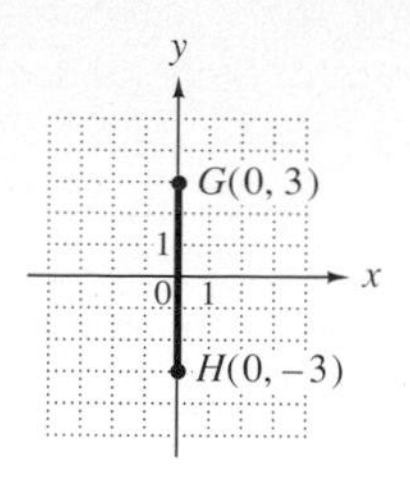

(e)

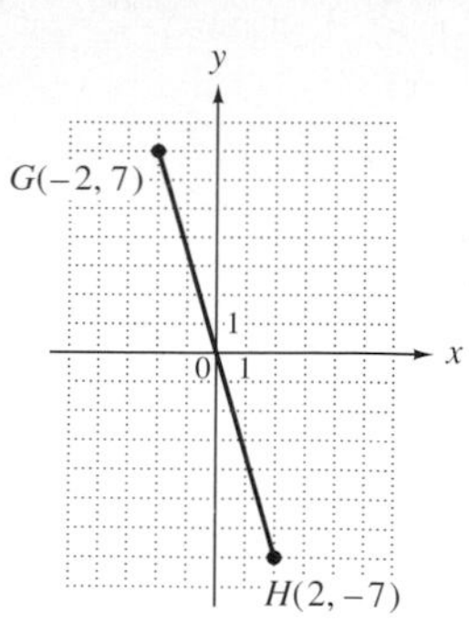

26. (a) $r = 10$, $s = 5$ **(b)** The midpoint of $\overline{AC}$ is $\left(\frac{10+0}{2}, \frac{5+0}{2}\right) = \left(5, \frac{5}{2}\right)$. The midpoint of $\overline{BD}$ is $\left(\frac{3+7}{2}, \frac{5+0}{2}\right) = \left(5, \frac{5}{2}\right)$. Therefore, the diagonals bisect each other.
32. (a) $\sqrt{(x-0)^2 + (y-0)^2} = 3$ or $x^2 + y^2 = 9$ **(b)** A circle of radius 3 and center at the origin.
34. $\angle BCA \cong \angle ECD$ as vertical angles. $\angle ABC \cong \angle CDE$ as alternate interior angles when parallel lines are cut by a third line. Similarly, $\angle BAC \cong \angle CED$. So, by the AA similarity property, $\triangle ABC \sim \triangle EDC$. **36.** $\angle CED \cong \angle AEB$. Since $\overline{CD}$ is parallel to $\overline{AB}$, the corresponding angles, $\angle CDE$ and $\angle ABE$, are congruent. Hence, by the AA similarity property, $\triangle CDE \sim \triangle ABE$. Since corresponding sides are proportional, we have:

$$\frac{5}{7} = \frac{6-y}{6}$$
$$30 = 42 - 7y$$
$$-12 = -7y$$
$$\frac{12}{7} = y$$

JUST FOR FUN The Greek Cross–I (page 917)

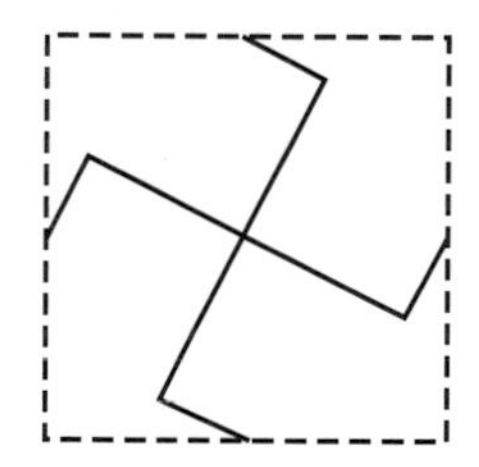

Problem Set 14.2 (page 929)

1. (a) 2, upward **(c)** 2, upward **(e)** 1, upward **2.** 2 **5.** $a = \frac{9}{2}$

7.

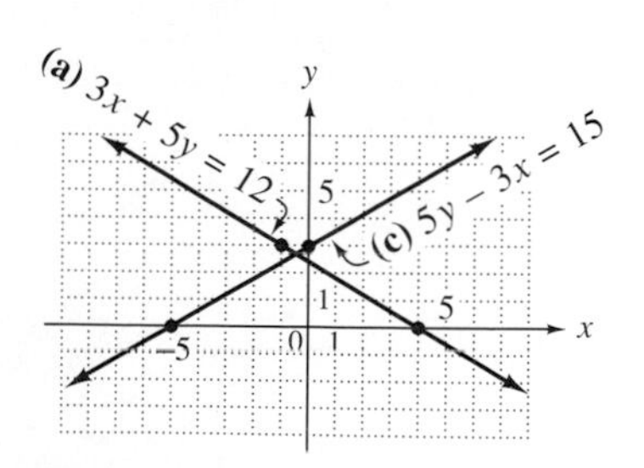

8.

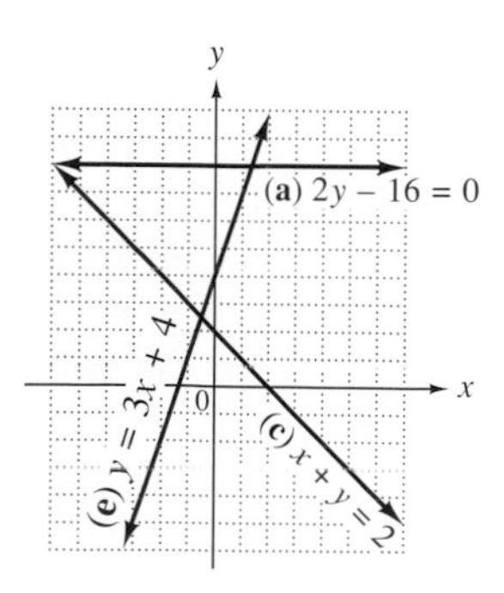

10. (a) $\frac{-35}{3}$ **(c)** $\frac{3}{5}$ **11. (a)** $m = \frac{3}{7}$, $b = 3$ **(c)** $m = 0$, $b = 6$ **(e)** Slope undefined, no y-intercept

12. (a) Neither **(c)** Perpendicular **13. (a)** $\frac{-10}{3}$ **14. (a)** $(3, -1)$ **16. (a)** $(4, 2)$ **18. (a)** $y - 6 = 2(x - 1)$

(c) $y - 2 = -\frac{1}{8}\left(x - \frac{7}{2}\right)$ **(e)** $x = 2$ **24.** The shortest distance is along the line $y - 5 = -\frac{2}{3}(x - 1)$ through $(1, 5)$ and perpendicular to $3x - 2y = 6$. These two lines meet at $(4, 3)$ so the shortest distance is $\sqrt{(1-4)^2 + (5-3)^2} = \sqrt{13}$.
29. (a) Many lines are possible. One line is $y - 6 = \frac{1}{2}(x - 8)$ or $y = \frac{1}{2}x + 2$. When $x = 15$, $y = \frac{1}{2}(15) + 2 = \frac{19}{2}$.
30. (a) iv **(c)** Yes, ii has three lines of symmetry. **32.** No. The figure shown has 120° rotation symmetry, but no symmetry about a point.

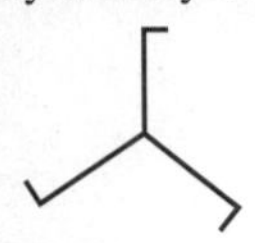

JUST FOR FUN The Greek Cross—II (page 940)

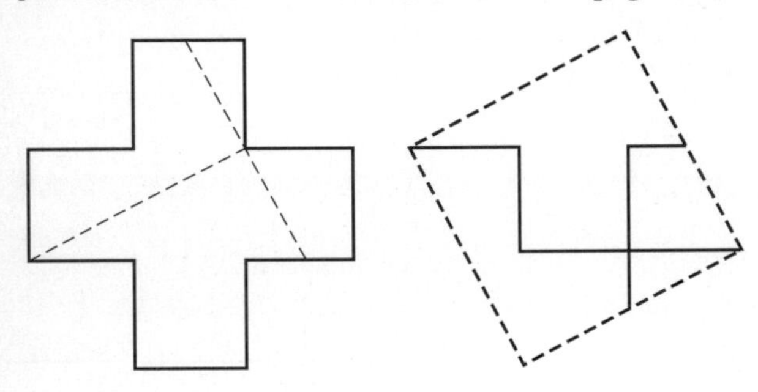

Problem Set 14.3 (page 941)

1. (a) $r = a$, $s = a$, $c = a$ **3. (a)** $r = a + c$, $s = b$ **5. (a)** Circle centered at $(0, 0)$ with radius 9
(c) The single point $(0, -5)$ **(e)** Circle centered at $(-2, -3)$ with radius 5
6. (a)

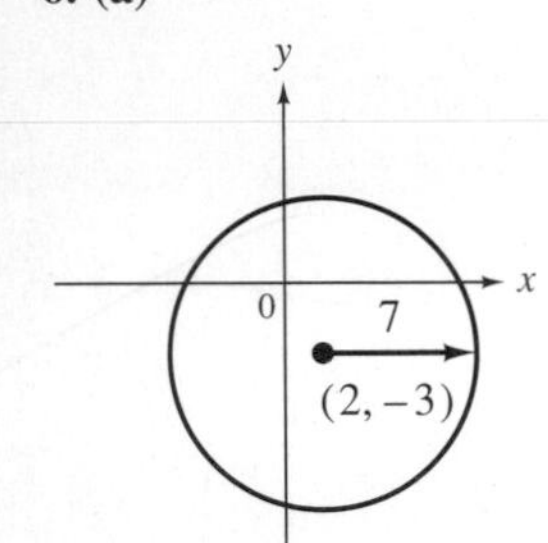

7. (a) $(x - 2)^2 + (y - 5)^2 = 9$ **(c)** $(x + 1)^2 + (y - 2)^2 = 4$ **8. (a)** $(1 + 2)^2 + (2 - 3)^2 = 3^2 + (-1)^2 = 9 + 1 = 10$ so the coordinates of the point satisfy the equation. **10. (a)** Find the equations of the perpendicular bisectors of two chords. Their point of intersection is the center of the circle. The three perpendicular bisectors are $y - 4 = \frac{-4}{3}(x - 3)$, $y - \frac{1}{2} = 7\left(x - \frac{5}{2}\right)$, and $y - \frac{7}{2} = \frac{-1}{7}\left(x - \frac{13}{2}\right)$. The center is $(3, 4)$.

11. A general parallelogram can be drawn as shown.

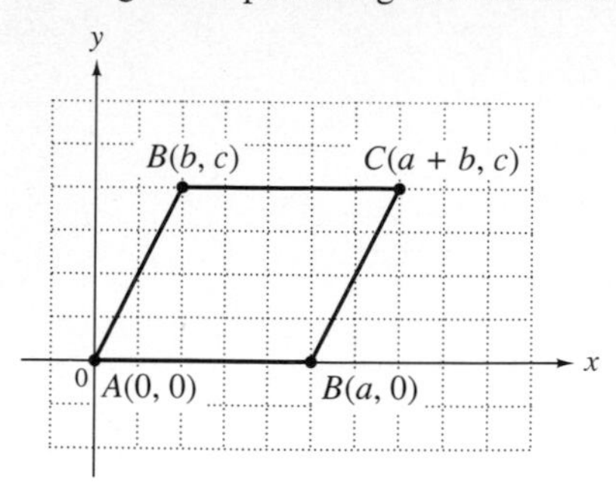

The midpoint of $\overline{AC}$ is $\left(\frac{a+b}{2}, \frac{c}{2}\right)$ and the midpoint of $\overline{BD}$ is $\left(\frac{a+b}{2}, \frac{c}{2}\right)$ so the diagonals bisect each other.

14. (a) The medians are

$$y - b = \left(\frac{d + f - 2b}{c + e - 2a}\right)(x - a),$$
$$y - d = \left(\frac{b + f - 2d}{a + e - 2c}\right)(x - c),$$

and

$$y - f = \left(\frac{b + d - 2f}{a + c - 2e}\right)(x - e).$$

Each pair of medians intersect in the point $\left(\frac{a + c + e}{3}, \frac{b + d + f}{3}\right)$.

17. (a) The points are $P(a, a)$, $Q(2a + b + c, b + c)$, $R(2a + 2b + c, -c)$ and $S(a + b, -(a + b))$. Thus,

$$PR = \sqrt{(a + 2b + c)^2 + (-c - a)^2} = \sqrt{(a + 2b + c)^2 + (c + a)^2}$$

and

$$SQ = \sqrt{((2a + b + c) - (a + b))^2 + ((b + c) - (-(a + b)))^2}$$
$$= \sqrt{(a + c)^2 + (a + 2b + c)^2}.$$

Therefore, $\overline{PR} \cong \overline{SQ}$.

(b) Slope $\overline{PR} = \dfrac{-c - a}{(2a + 2b + c) - a} = -\dfrac{a + c}{a + 2b + c}$

Slope $\overline{SQ} = \dfrac{(b + c) - (-(a + b))}{(2a + b + c) - (a + b)} = \dfrac{a + 2b + c}{a + c}$

Since the slope of $\overline{SQ}$ is the negative of the reciprocal of the slope of $\overline{PR}$, $\overline{PR} \perp \overline{SQ}$. **20.** 1440° **22.** 144°

JUST FOR FUN Watering a Playfield (page 950)

The plumber is right. Is he ever just *barely* right?

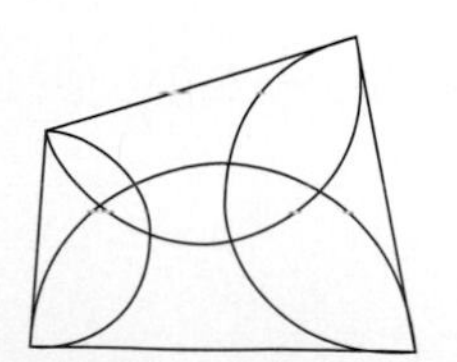

Problem Set 14.4 (page 953)

1.

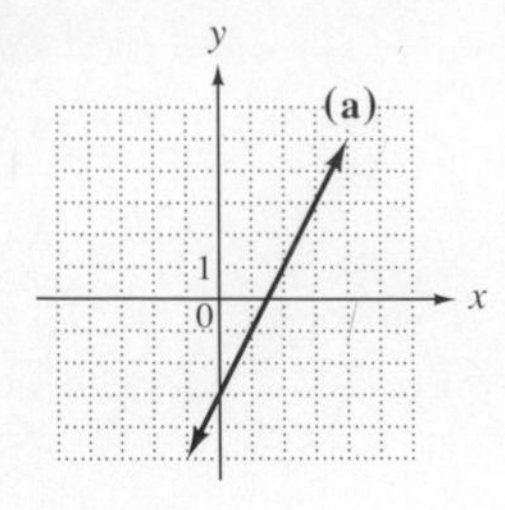

3. (a) Yes, this is a graph of a function.

5. (a)

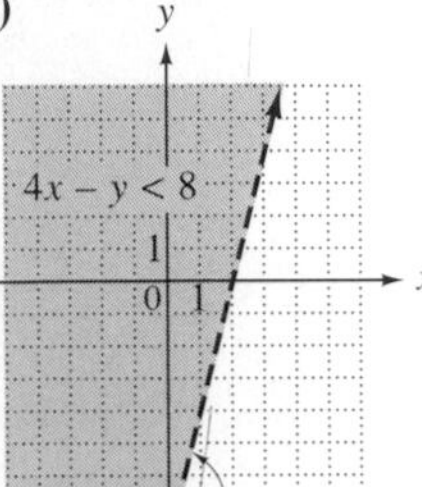

(c)

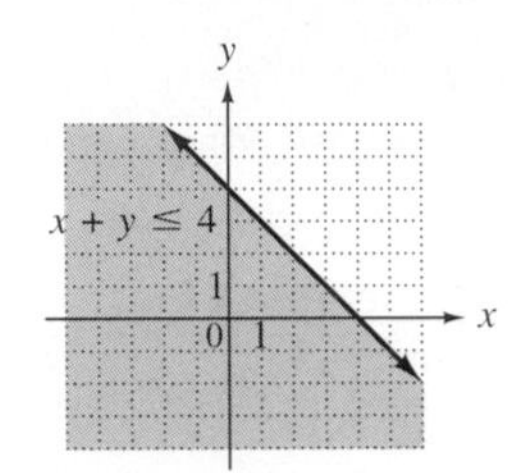

(e)

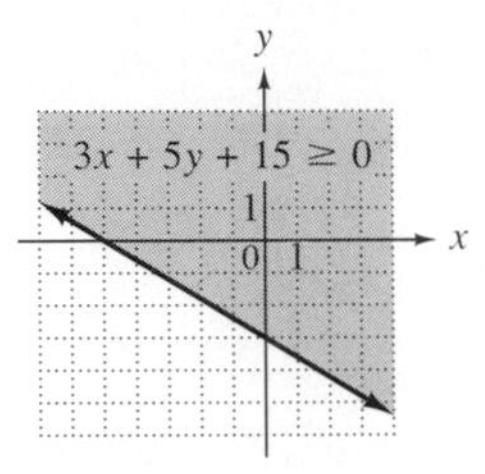

7.

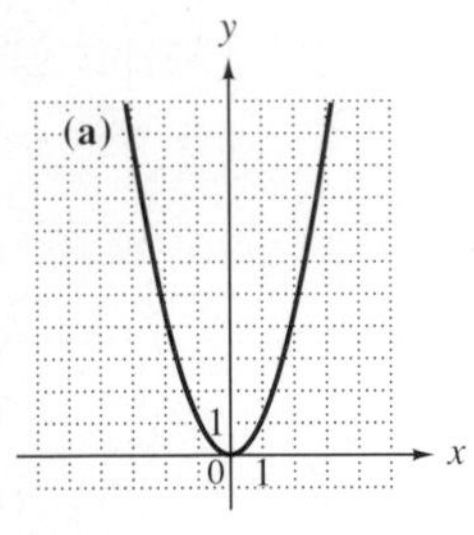

8.

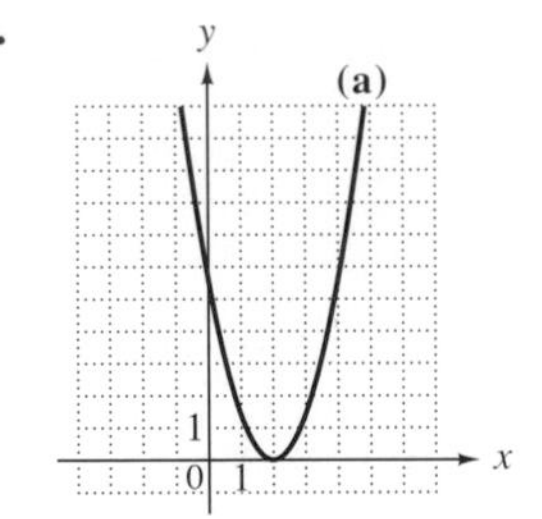

9. (a) From the graph, the minimum of $x^2 + 10x$ is -25.

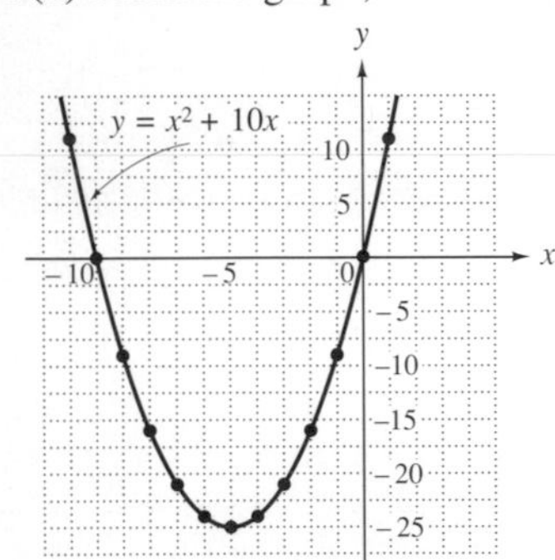

12. (a)

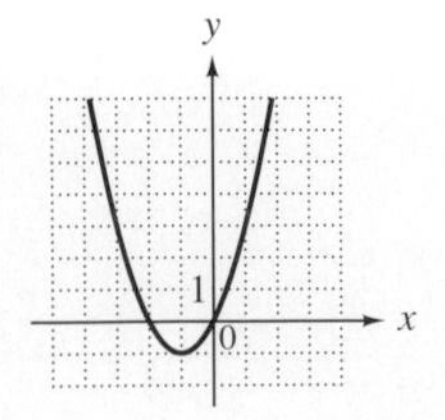

(b) $x = -1$

13. (a)

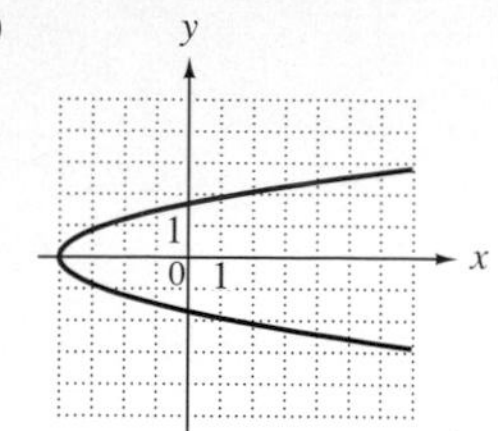

16. (a)

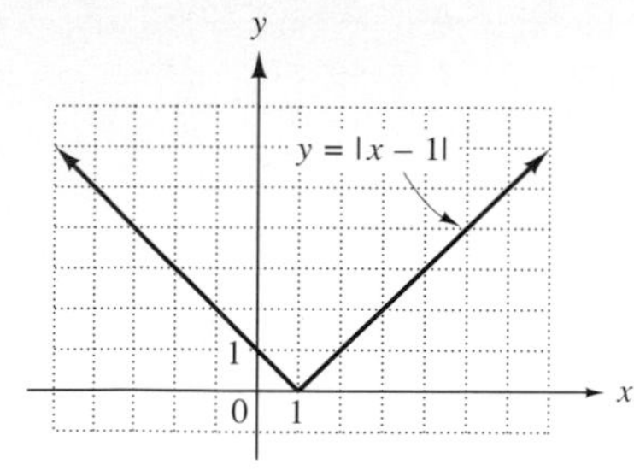

20. (a) 2000 **23. (a)** 11:59 A.M. **25. (a)** In the diagram below D is a units from A. If D has coordinates (b, c), then by the Pythagorean theorem, $b^2 + c^2 = a^2$. Hence, $c^2 = b^2 - a^2$ and $c = \sqrt{a^2 - b^2}$. Since $DC = a$, C has coordinates $(a + b, \sqrt{a^2 - b^2})$. Thus, the midpoint of AC is

$$\left(\frac{a+b}{2}, \frac{\sqrt{a^2 - b^2}}{2}\right)$$

and the midpoint of $\overline{BD}$ is also

$$\left(\frac{a+b}{2}, \frac{\sqrt{a^2 - b^2}}{2}\right).$$

Thus, the diagonals bisect each other.

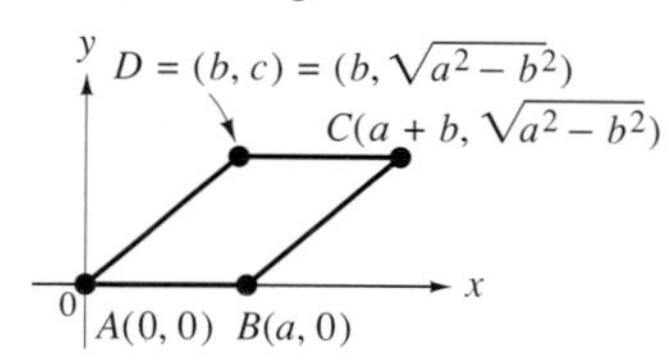

26. (a) -3 **28.** $y + 3 = \frac{2}{7}(x + 2)$

CLASSIC CONUNDRUM Square Inch Mysteries (page 957)

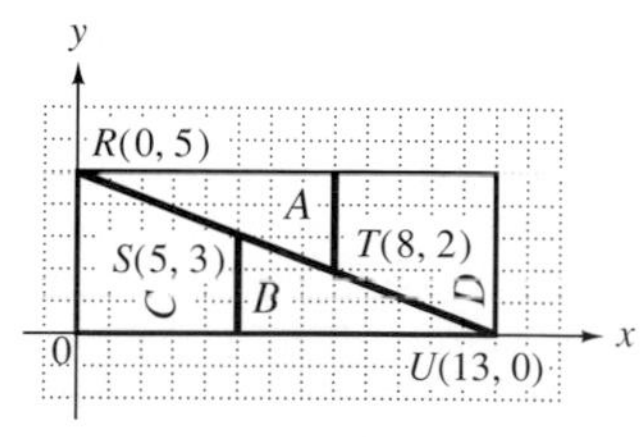

Place the rectangle on a coordinate system and label the points $R(0, 5)$, $S(5, 3)$, $T(8, 2)$, and $U(13, 0)$ as shown. Since

$$\text{slope } \overline{RS} = \frac{3 - 5}{5 - 0} = \frac{-2}{5} = -0.40$$

and

$$\text{slope } \overline{SU} = \frac{0 - 3}{13 - 5} = \frac{-3}{8} = -0.375,$$

R, S, and U do *not* lie on a straight line. Similarly,

$$\text{slope } \overline{RT} = \frac{2-5}{8-0} = -\frac{3}{8} = -0.375$$

and

$$\text{slope } \overline{TU} = \frac{0-2}{13-8} = \frac{-2}{5} = -0.40,$$

so R, T, and U do *not* lie on a straight line. Indeed, as the slopes show, $RTUS$ is a very thin parallelogram—a fact obscured by the heavy lines used in the diagram. It turns out that this parallelogram has area one and this accounts for the extra unit of area.

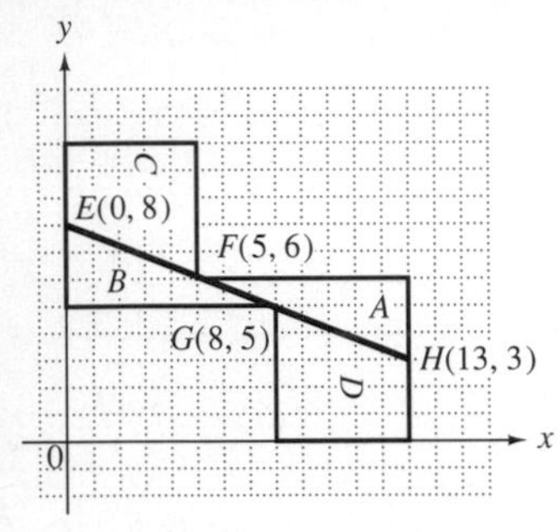

Now place the "propeller" on a coordinate system and label the points $E(0, 8)$, $F(5, 6)$, $G(8, 5)$, and $H(13, 3)$ as shown. Here

$$\text{slope } \overline{EF} = \frac{6-8}{5-0} = -\frac{2}{5} = -0.40$$

and

$$\text{slope } \overline{FH} = \frac{3-6}{13-5} = \frac{-3}{8} = -0.375,$$

so E, F, and H do not lie on a straight line. Similarly,

$$\text{slope } \overline{EG} = \frac{5-8}{8-0} = \frac{-3}{8} = -0.375$$

and

$$\text{slope } \overline{GH} = \frac{3-5}{13-8} = \frac{-2}{5} = -0.40,$$

so E, G, and H also do not lie on a straight line. Indeed, as the slopes show, $EFHG$ forms a small parallelogram where the pieces of the puzzle overlap—a fact again obscured by the heavy lines used in making the drawing. It turns out that this parallelogram has area one, and this accounts for the missing one unit of area.

Chapter 14 Review Exercises (page 959)

1.

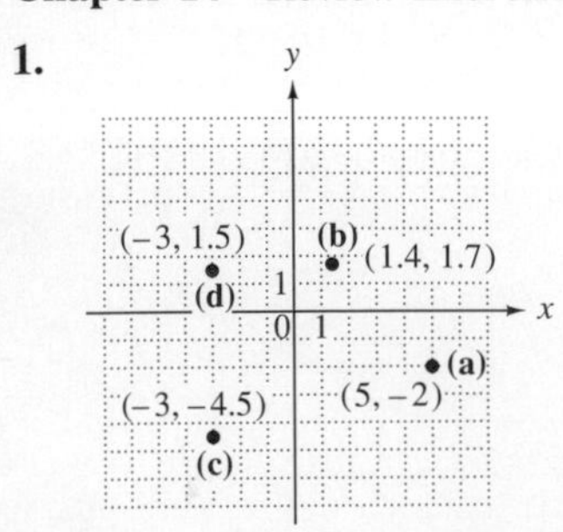

2. **(a)** IV **(b)** II **(c)** I or III **(d)** II or IV **3.** **(a)** $a = 2, b = 2$ **(b)** $a = 2.3, b = 0.1$ **(c)** $a = -2.5, b = -0.5$ **(d)** $a = -1, b = -4.5$ **4.** **(a)** $\sqrt{40} = 2\sqrt{10}$ **(b)** $\sqrt{13.48} \doteq 3.67$ **(c)** $\sqrt{74}$ **(d)** $\sqrt{85}$ **5.** $AB = 5$, $BC = \sqrt{50}$, $AC = 5$. $(AB)^2 + (AC)^2 = 5^2 + 5^2 = 50 = (\sqrt{50})^2 = BC^2$. Thus, by the Pythagorean theorem, the triangle is a right triangle.
6. $RS = \sqrt{\dfrac{481}{4}} = RT$ so the triangle is isosceles. **7.** **(a)** 1 **(b)** Undefined **(c)** $-2/3$ **(d)** Undefined **8.** $-19/5$
9. **(a)** $b = 8$ **(b)** $b = 14/3$ **10.** $R(-4, 5)$ is on the perpendicular bisector of $\overline{PQ}$ if, and only if, it is equidistant from P and Q. Since

$$RP = \sqrt{(-4 - 0)^2 + (5 - 5)^2} = 4$$

and

$$RQ = \sqrt{(-4 - (-3))^2 + (1 - 5)^2} = \sqrt{17} \doteq 4.12,$$

R is *not* on the perpendicular bisector of $\overline{PQ}$. **11.** $b = -8/5$ **12.** Since the slope $\overline{PQ} = -9/7$, the equation of the line is $y + 2 = -\frac{9}{7}(x - 3)$ or, equivalently, $y - 7 = -\frac{9}{7}(x + 4)$. **13.** $y = \frac{3}{2}x + 10$ **14.** Solving for y in terms of x, we obtain $y = (3/4)x - 15/4$. Thus, the slope is $3/4$. **15.** $y - 1.5 = (-4/3)(x - 5)$ **16.** $(3, -2)$ **17.** **(a)** The coordinates of M, D, and E are $M(4, 3)$, $D(-3, 3)$, and $E(4, -4)$. Thus, $\overline{DM}$ is horizontal, $\overline{ME}$ is vertical and $\angle DME$ is a right angle. **(b)** slope $\overline{EA} = \dfrac{0 - (-4)}{0 - 4} = -1$, slope $\overline{ED} = \dfrac{3 - (-4)}{-3 - 4} = -1$ Since the slopes are the same, E, A, and D are collinear. **18.** **(a)** Since the coordinates of E, G, and H are $E(3, 3)$, $G(8, 2)$, and $H(5, -5)$, $EG = \sqrt{26} = FH$ and $\overline{EG} \cong \overline{FH}$. **(b)** Since the slope of $\overline{EG}$ is $-1/5$ and the slope of $\overline{FH}$ is 5, it follows that $\overline{EG}$ and $\overline{FH}$ are perpendicular. **19.** **(a)** Since the coordinates of E, F, G, and H are $E(2, 2)$, $F(5, 1)$, $G(9, 3)$, and $H(6, -6)$, $EG = \sqrt{50} = FH$ and $\overline{EG} \cong \overline{FH}$. **(b)** Since the slope of $\overline{EG}$ is $1/7$ and the slope of $\overline{FH}$ is -7, it follows that $\overline{EG}$ and $\overline{FH}$ are perpendicular. **20.** **(a)** Since the coordinates of M, D, and E are $M(a, a)$, $D(-b, b)$, and $E(a, -a)$, $\overline{MD}$ is horizontal and $\overline{ME}$ is vertical. Thus, $\angle EMD$ is a right angle and $\triangle EMD$ is a right triangle. **(b)** Since slope $\overline{EA} = -1 =$ slope $\overline{ED}$, it follows that E, A, and D are collinear. **21.** **(a)** The coordinates of E, F, G, and H are $E(a, a)$, $F(2a, 0)$, $G(2a + b, b)$, and $H(a + b, -a - b)$. Therefore,

$$EG = \sqrt{(a + b)^2 + (b - a)^2} = \sqrt{2a^2 + 2b^2}$$

and

$$FH = \sqrt{(b - a)^2 + (-a - b)^2} = \sqrt{2a^2 + 2b^2},$$

so $\overline{EG} \cong \overline{FH}$. **(b)** Since

$$(\text{slope } \overline{EG}) \cdot (\text{slope } \overline{FH}) = \frac{b - a}{a + b} \cdot \frac{a + b}{a - b} = -1,$$

$\overline{EG}$ and $\overline{FH}$ are perpendicular. **22.** **(a)** The coordinates of E, F, G, and H are $E(a, a)$, $F(2a + b, b)$, $G(2a + 2b + c, c)$, and $H(a + b + c, -(a + b + c))$ Therefore,

$$EG = \sqrt{(a + 2b + c)^2 + (c - a)^2} = \sqrt{(a + 2b + c)^2 + (a - c)^2}$$

and

$$FH = \sqrt{(a - c)^2 + (a + 2b + c)^2},$$

so $\overline{EG} \cong \overline{FH}$. **(b)** Since $(\text{slope } \overline{EG}) \cdot (\text{slope } \overline{FH}) = \dfrac{c - a}{a + 2b + c} \cdot \dfrac{a + 2b + c}{a - c} = -1$, it follows that $\overline{EG}$ and $\overline{FH}$ are perpendicular.

Chapter 14 Test **(page 961)**

1. $r = 10$, $s = 5$ **2.** The desired point is

$$\left(\left(1 - \frac{1}{4}\right) \cdot (-8) + \frac{1}{4} \cdot 4, \left(1 - \frac{1}{4}\right) \cdot 0 + \frac{1}{4} \cdot 12\right) = (-5, 3).$$

3. The midpoint of $\overline{AB}$ is $(9/2, 9/2)$ and the slope is $-1/5$. Therefore, the equation of the perpendicular bisector is

$$y - \frac{9}{2} = 5\left(x - \frac{9}{2}\right).$$

4. $RS = \sqrt{164}$, $ST = \sqrt{205}$, and $RT = \sqrt{41}$. Therefore, $(RS)^2 + (RT)^2 = 164 + 41 = 205 = (ST)^2$ and so $\triangle RST$ is a right triangle. **5.** $r = 23/5$ **6.** Slope $\overline{PQ} = \dfrac{2-5}{4-(-1)} = \dfrac{-3}{5}$, equation is $y - 2 = -\dfrac{3}{5}(x-4)$ or, equivalently, $y - 5 = -\dfrac{3}{5}(x+1)$. **7.** $-5/2$ **8.** $(x+2)^2 + (y-5)^2 = 16$ **9.** $(3, -1)$

10.

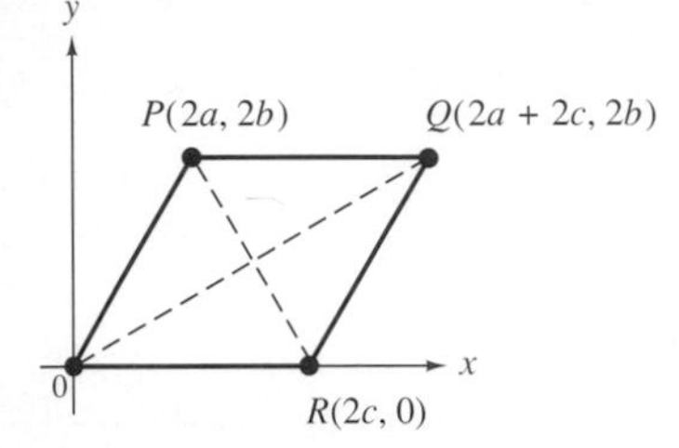

Any parallelogram can be placed on a coordinate axes as shown and with vertices with coordinates as indicated with a, b, and c all positive. The midpoint of $\overline{PR}$ is $((2a+2c)/2, (2b+0)/2) = (a+c, b)$ and the midpoint of $\overline{OQ}$ is $((2a+2c)/2, (2b+0)/2) = (a+c, b)$ also. Thus, the diagonals bisect one another. **11.** $\$5000 \cdot (1.07)^5 \doteq \$7012.76$

12. (a)

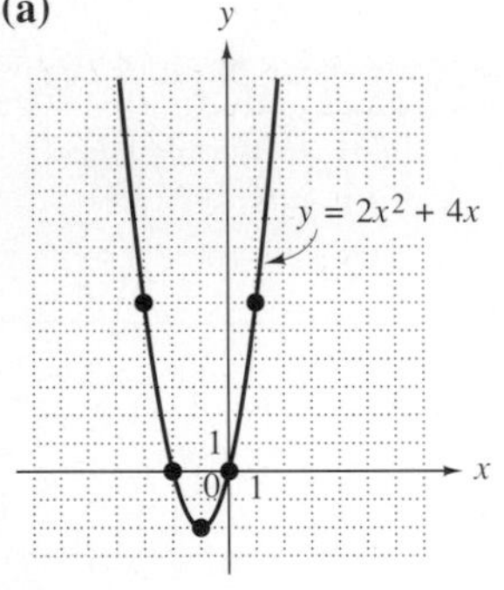

(b) Minimum value is -2 when $x = -1$.

13.

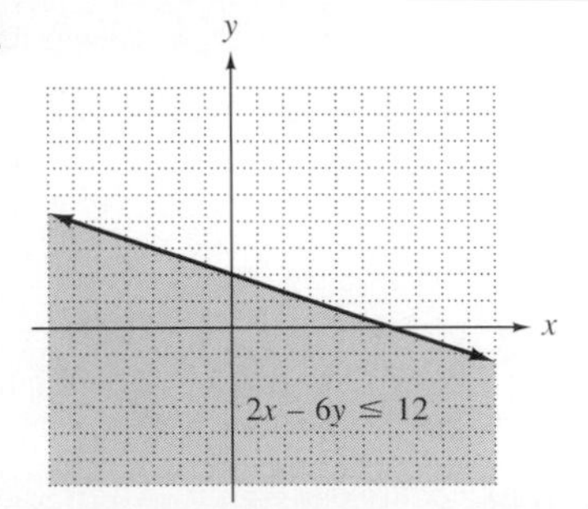

Appendix B (page 979)

1. (a) Statement **(c)** Not a statement **2. (a)** For some **(c)** For some **4. (a)** $3 \cdot 1 \neq 3$ **(c)** $2 \geq 1$ **(e)** All rectangles are squares. **(g)** Some squares do not have a right angle. **5. (a)** $\sim p$ **(c)** $\sim p \rightarrow \sim q$ **6. (a)** T **(c)** F **7. (a)** T **8. (a)** I do not own a calculator or I do not drive a car. **10. (a)** Converse: If r is a moosh, then r is a dweeble. Contrapositive: If r is not a moosh, then r is not a dweeble. **11. (a)** The contrapositive of the inverse is logically equivalent to the inverse. Therefore $\sim p \rightarrow \sim q$ is equivalent to $\sim(\sim q) \rightarrow \sim(\sim p)$, so $\sim p \rightarrow \sim q$ is equivalent to $q \rightarrow p$ by the law of double negation. **(b)** If r is not a dweeble, then r is not a moosh. If $s \cdot t = s \cdot u$, then $t = u$. **13. (a)** Let p be the statement "n^2 is odd" and q be "n is odd." First prove $q \rightarrow p$. Let n be any odd whole number, so $n = 2i + 1$ for some whole number

i. Then $n^2 = (2i + 1)^2 = 4i^2 + 4i + 1 = 2(2i^2 + 2i) + 1$ so n^2 is odd. Prove $p \to q$ by showing the contrapositive $\sim q \to \sim p$ is true. Let n be any whole number which is not odd. Thus n must be even, and $n = 2j$ for some whole number j. Then $n^2 = (2j)^2 = 4j^2 = 2(2j^2)$ so n^2 is even, and n^2 is not odd. This proves $\sim q \to \sim p$, so $p \to q$ is true. Together, we have $p \leftrightarrow q$. **(b)** Take the contrapositive of the implications.
14. (a) Valid, contraposition **(c)** Not valid **(e)** Not valid **15. (a)** Fallacy of the converse **16. (a)** q **18. (a)** T, law of detachment **(c)** T, law of syllogism **(e)** Cannot be decided **(g)** F
19. (a) Valid by law of syllogism since

$$\begin{array}{l} p \to q \\ r \to \sim q \\ \hline p \to \sim r \end{array}$$

and $r \to \sim q$ is logically equivalent to $q \to \sim r$
21. (a)

$1 = 2^0 \cdot 1$	$13 = 2^0 \cdot 13$
$2 = 2^1 \cdot 1$	$14 = 2^1 \cdot 7$
$3 = 2^0 \cdot 3$	$15 = 2^0 \cdot 15$
$4 = 2^2 \cdot 1$	$16 = 2^4 \cdot 1$
$5 = 2^0 \cdot 5$	$17 = 2^0 \cdot 17$
$6 = 2^1 \cdot 3$	$18 = 2^1 \cdot 9$
$7 = 2^0 \cdot 7$	$19 = 2^0 \cdot 19$
$8 = 2^3 \cdot 1$	$20 = 2^2 \cdot 5$
$9 = 2^0 \cdot 9$	$21 = 2^0 \cdot 21$
$10 = 2^1 \cdot 5$	$22 = 2^1 \cdot 11$
$11 = 2^0 \cdot 11$	$23 = 2^0 \cdot 23$
$12 = 2^2 \cdot 3$	$24 = 2^3 \cdot 3$

(b) 4 is the largest degree of evenness in the list, occurring for $16 = 2^4$. **(c)** h and k are odd numbers, and $h < k$. Let $2u$ be an even number between h and k. Then $m < 2^d \cdot 2u < n$, and $2^d \cdot 2u = 2^{d+1}u$, so the degree of $2^d \cdot 2u$ is at least $d + 1$. **(d)** Assume that the list has two numbers with the largest degree of evenness for the list. Then using the method in part (c) we find a number between these two with a larger degree of evenness. This contradicts the assumption that our original two numbers have the largest degree of evenness in the list, so we cannot have two numbers with the largest degree of evenness. **23. (a)** Assume you can do the tiling. Each tile covers numbers which add to a multiple of 3, so the sum of all numbers covered by the tiling is a multiple of 3. Adding the numbers in the triangle gives 62, which is not a multiple of 3. Since we have a contradiction, the tiling is not possible. **25. (a)** The game always ends in a tie. **(b)** The number of scalene and equilateral triangles is the same. By placing the eight-vertex configuration at an arbitrary position in the plane, this proves an interesting theorem: *Let each point of the plane be colored red or blue, and let T be any scalene triangle of sides a, b, and c. Then there is a triangle in the plane congruent to T with vertices of one color if, and only if, there is an equilateral triangle of side a, b, or c with vertices of one color.* **27.** The statement is false. If the last digit of n^2 is 1, then the last digit of n must be either 1 or 9. If d is the tens digit of n, squaring $10d + 1$ and $10d + 9$ shows that the tens digit of n^2 is always an even digit. In particular, it is not 1, so no perfect square can end in 11 as the final two digits. **28. (a)** $\sim(\sim p) \equiv p$, law of double negation **(c)** $p \wedge (\sim p) \equiv F$ **(e)** $(p \to q) \to (p \vee \overline{q}) \equiv F$)
29. (a) Current flows when p is false and q is either true or false and when p is true and q is true, which are precisely the conditions for $p \to q$ to be true.

Appendix C (page 999)

1. See sample solution that draws LOGO.

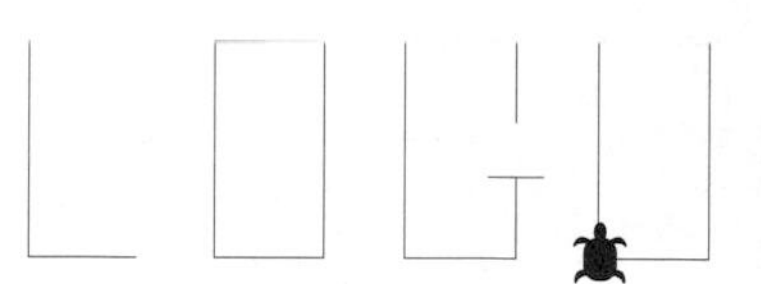

```
to  LOGO
    pu lt 90 fd 150 rt 90 pd
    fd 80 bk 80 rt 90 fd 40
    pu fd 30 lt 90 pd
    repeat 2 [fd 80 rt 90 fd 40 rt 90]
    pu rt 90 fd 70 lt 90 pd
    fd 80 rt 90 fd 40 rt 90 fd 30
    bk 30 lt 90 bk 40 lt 90 bk 80
    rt 90 fd 40 lt 90 fd 30 lt 90 fd 10 bk 20
    fd 10 rt 90 bk 30
    pu rt 90 fd 30 lt 90 pd
    repeat 2 [fd 80 rt 90 fd 40 rt 90]
    end
```

6.
```
to  right.triangle  :leg1 :leg2
  fd  :leg1  rt  90
  fd  :leg2
  home
end
```

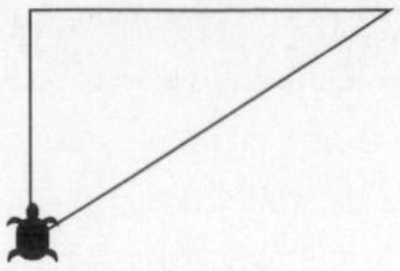

right. triangle 60 90

7. (a)
```
to  rectangle1
   repeat 2 [ fd 50 rt 90 fd 80 rt 90 ]
end
```

(b)
```
to  rectangle  :length  :width
   repeat 2 [ fd :length rt 90 fd :width rt 90 ]
end
```

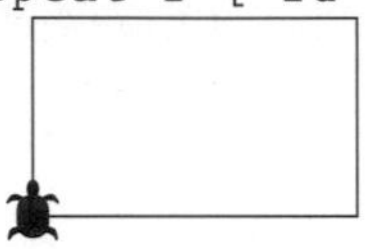

rectangle 50 80

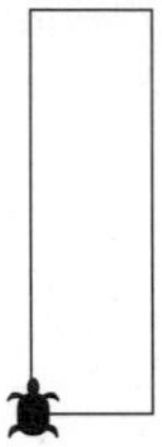

rectangle 100 30

rectangle 60 80

8. (a)
```
to  parallelogram1
   repeat 2 [ fd 40 rt 142 fd 70 rt 38 ]
end
```

(b)
```
to  parallelogram  :side1  :side2  :angle
   repeat 2 [ fd :side1 rt 180 - :angle fd :side2 rt :angle ]
```

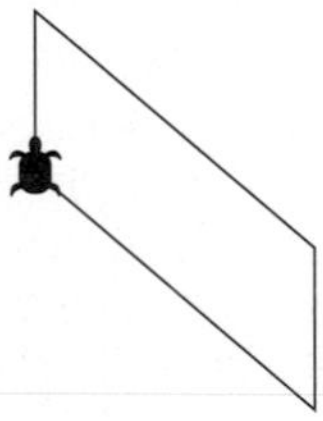

parallelogram 30 70 48

parallelogram 50 30 79

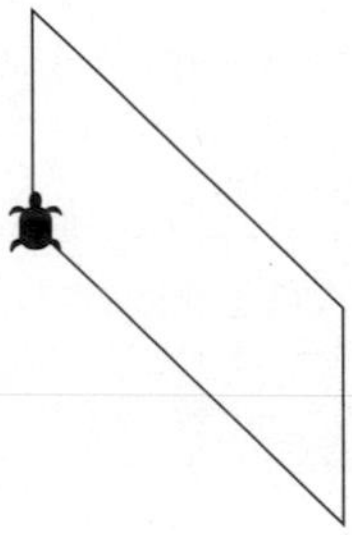

parallelogram 40 80 45

9. (a)
```
to  rhombus1
   repeat 2 [ fd 50 rt 128 fd 50 rt 52 ]
end
```

(b)
```
to  rhombus  :side  :angle
   repeat 2 [ fd :side rt 180 - :angle fd :side rt :angle ]
end
```

rhombus 50 45

rhombus 50 75

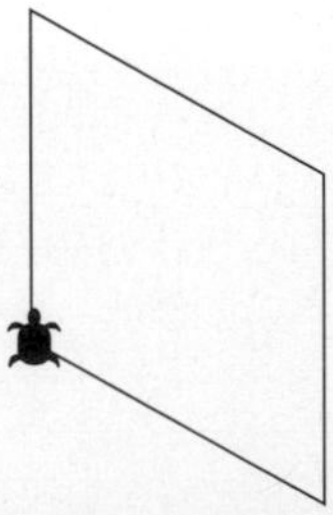

rhombus 90 60

ACKNOWLEDGMENTS

Unless otherwise acknowledged, all photographs are the property of Scott, Foresman and Company. Page abbreviations are as follows: (T)top, (C)center, (B)bottom, (L)left, (R)right.

xxx, CALVIN AND HOBBES 1990 copyright Watterson. Dist. by UNIVERSAL PRESS SYNDICATE. Reprinted with permission. All Rights Reserved. **9,** Jim Davis. Copyright 1989/Reprinted by permission of United Feature Syndicate, Inc. **17,** Stanford University News Service **24,** Focus On Sports **33,** Culver Pictures **52,** G.I.Bernard/Oxford Scientific Films/ANIMALS ANIMALS **57,** Jeff Guerrant Photography **59,** Library of Congress **79L,** Bridgeman/Art Resource, New York **79R,** Museo Nacional de Antropologia y Arqueologia in Lima, Peru **81,** Baveria-Verlag **89R,** Bettmann Archive **150,** Comstock Inc. **212,** CALVIN AND HOBBES copyright Watterson. Dist. by UNIVERSAL PRESS SYNDICATE. Reprinted with permission. All Rights Reserved. **219,** Prof.Gottried E. Noether **252,** Sidney Harris **276,** American Mathematical Society **281,** Denise Applewhite/Sygma **319,** Yale University Library, New Haven, CT **327,** Library of Congress **348,** Bryn Mawr College Archives **371,** Sidney Harris **443,** Sidney Harris **446,** Bettmann Archive **532,** UPI/Bettmann **533,** Sidney Harris **536,** From R. A. Fisher, "THE LIFE OF A SCIENTIST" **539,** Bettmann Archive **552,** Museum fur Volkskunde von Schweizerishches **594,** Arthur Tilley/Tony Stone Images **595,** By permission of Johnny Hart and Creators Syndicate **610TR,** John Bova/Photo Researchers **610CR,** SS/Photo Researchers **610BR,** Ray E. Ellis/Photo Researchers **610TL,** Manfred P. Kage **610CL,** ©1995 M.C.Escher/Cordon Art—Baarn—Holland. Haags-Gementemuseum **610BL,** "Tail of the Seahorse" Mandelbrat set from H. O. PEITGEN and P.H.RICHTER, "The Beauty of Fractals," Heidelberg, Springer-Verlag, 1986 **612,** By permission of Johnny Hart and Creators Syndicate **641T,** Courtesy National Baha'i Headquarters **651TL,** ISUMI **651CL,** Israel National Museum, Jerusalem **651CR,** Courtesy St.Marks Cathedral, Venice **651BL,** SIMOES **651BR,** Dye, Daniel Sheets, A GRAMMAR OF CHINESE LATTICE, Figs. C9b,S12a. Harvard-Yenching Institute Monograph V. Cambridge, 1937 **658T,** from Martin Gardner's PENROSE TILES TO TRAPDOOR CIPHERS, W. H. Freeman 1989 **658B,** ©1995, M.C.Escher/Cordon Art—Baarn—Holland. Private Collection **663L,** from ALTAIR DESIGN coloring book E.Holiday, (London: Pantheon, 1970). **667 and 674ALL,** From MATHEMATICS: AN INTRODUCTION TO ITS SPIRIT AND USE, W. H. Freeman and Company **678TL,** S.Camazine/Photo Researchers **678BL,** Thompson, D'Arcy W., ON GROWTH AND FORM, New edition, Cambridge University Press, Cambridge and New York, 1948. **678BR,** from ERNST HAECKEL, CHALLENGER, Monograph, 1887 **678CL,** Dr. Giuseppe Mazza **678CR,** Rare Book Room/New York Public Library, Astor, Lenox and Tilden Foundations **687L,** Murray & Assoc./Tony Stone Images **687R,** From Thompson D'Arcy, ON GROWTH AND FORM **693,** H.Chaumeton/Photo Researchers **708,** Washington University **710,** From NATIONAL CYCLOPEDIA OF AMERICAN BIOGRAPHY **728,** Sidney Harris **767,** Sidney Harris **775,** Courtesy IBM Research **787,** NASA **797,** Copyright the British Museum **815,** Catherine Koehler **816,** Ecole Sophie Germain, Paris **819,** National Portrait Gallery, London **836,** Thomas Burke Memorial/Washington State Museum **849,** ©1995 M. C.Escher/Cordon Art—Baarn—Holland. **859,** © 1995 M.C.Escher/Cordon Art—Baarn—Holland. **865L,** Courtesy Southwest Museum, Los Angeles **865R,** Carnegie Institute, Washington, DC **867T,** From AESTHETIC MEASURE by George D. Birkhoff, ©1933/Harvard University Press, Cambridge, MA **868T,** Musee Royal de l'Afrique Centrale, Tervuren, Belgium **872ALL,** From AESTHETIC MEASURE by George Birkhoff, ©1933/Harvard University Press, Cambridge, MA **876,** Mercedes-Benz emblem, Courtesy Mercedes-Benz of North America, Inc. **876,** "The Chevrolet Bow Tie emblem and the Oldsmobile Rocket emblem are trademarks of the Chevrolet Motor Division and Oldsmobile Motor Division, respectively, General Motors Corporation, used with permission. General Motors Corporation, however, does not endorse or assume any responsibility for the text, errors, or omissions of Harper Collins Publishers, its officers, agents, employees, or other representatives." **876,** Sterling Savings Assn. emblem, Sterling Savings Assn., Colfax, Washington **879,** From FEARFUL SYMMETRY, copyright © 1992 Ian Stewart and Martin Golubitsky, Printed by Blackwell Publishers **884,** Sidney Harris **893,** From AESTHETIC MEASURE by George D. Birkhoff, ©1933/Harvard University Press, Cambridge, MA **895,** © 1995, M.C.Escher /Cordon Art—Baarn—Holland. Private Collection **896ALL,** From AESTHETIC MEASURE by George D. Birkhoff ©1933/Harvard University Press, Cambridge, MA **904,** Sidney Harris **920,** Bettmann Archive **947,** Courtesy A. T. & T., Bell Laboratories **972,** Sidney Harris

MATHEMATICAL LEXICON

Many of the words, prefixes, and suffixes forming the vocabulary of mathematics are derived from words and word roots from Latin, Greek, and other languages. Some of the most common terms are listed below, to serve as an aid to learning and understanding the terminology of mathematics.

acute from Latin *acus* "needle" by way of *acutus* "pointed, sharp"

algorithm distortion of Arabic name *al-Khowarazmi* "the man from Khwarazm," whose book on the use of Hindu-Arabic numeration was translated into Latin as *Liber Algorismi* meaning "Book of al-Khowarazmi"

angle from Latin *angulus*, "corner, angle"

apex from Latin word meaning "tip, peak"

area Latin *area* "vacant piece of ground, plot of ground, open court"

associative from Latin *ad* "to" and *socius* "partner, companion"

axis from Latin word meaning "axle, pivot"

bi- from Latin prefix derived from *dui-* "two"; *bi*nary, *bi*nomial, *bi*sect

calculate from Latin *calc* "chalk, limestone" and diminutive suffix *-ulus* [a *calculus* was a small pebble; *calculare* meant "to use pebbles" = to do arithmetic]

cent- from Latin *centum* "hundred"; *cent*imeter, per*cent*

circum- from Latin *circum* "around"; *circum*ference, *circum*scribe

co-, col-, from Old Latin *com* "together with, beside, near"; *com*mutative,
com-, con- *col*linear, *com*plement, *con*gruent

commutative from Latin *co-* "together with" and *mutare* "to move"

concurrent from Latin *co-*, "together with" and *currere* "to run"

conjecture from Latin *co-* "together with" and *iactus* "to throw" [conjecture = throw (ideas) together]

cylinder from Greek *kulindros* "a roller"

de- Latin preposition *de* "from, down from, away from, out of "; *de*nominator, *de*duction

deca-, deka- from Greek deka- "ten"; *deca*gon, do*deca*hedron, *deka*meter

deci- from Latin *decimus* "tenth"; *deci*mal, *deci*meter

diagonal from Greek *dia-* "through, across" and *gon-* "angle"

diameter from Greek *dia-* "through, across" and *metron* "measure"

digit from Latin *digitus* "a finger"

distribute from Latin prefix *dis-* "apart, away" and Latin *tribu* "a tribe [of Romans]"

empirical from Latin *empiricus* "a physician whose art is founded solely on practice"

equal from Latin *æquus*, "even, level"

equilateral from Latin *æquus*, "even, level" and *latus* "side"

equivalent from Latin *æquus*, "even, level" and *valere* "to have value"

exponent from Latin *ex* "away" and *ponent-* = present participial stem of *ponere* "to put"

figure from Latin *figura* "shape, form, figure"

fraction from Latin *fractus*, past participle of *frangere* "to break"

geometry from Greek *geo-* "earth" and *metron* "measure"

-gon from Greek *gonia* "angle, corner"; poly*gon*, penta*gon*

-hedron from Greek *hedra* "base, seat", poly*hedron*, tetra*hedron*

hept-, sept- Greek *hept*, from prehistoric Greek *sept*, meaning "seven"; *hepta*gon

heuristic from Greek *heuriskein* "to find, discover"

hex- from Greek *hex* (prehistoric Greek *sex*) "six"; *hex*agon, *hex*omino

icosahedron from Greek *eikosi* "twenty" and *hedra* "bases, seat"

inch from Latin *uncia*, a unit of weight equal to one twelfth of the *libra*, or Roman pound

inscribe from Latin *in* "in" and *scribere* "to scratch" hence "to write"
integer from Latin *in-* "not" and Indo-European root *tag-* "to touch" [an integer is untouched, hence "whole"]
inverse from Latin *in* "in" and *versus*, past participle of *vertere* "to turn"
isosceles from Greek *isos* "equal" and *skelos* "leg"
kilo- from Greek *khiloi* "thousand"; *kilo*gram, *kilo*meter
lateral from Latin *latus* "side"
lb. abbreviation for pound, from *libra*, the Roman unit of weight
line from Latin *linum* "flax" [The Romans made *linea*, linen thread, from flax.]
median from Latin *medius* "in the middle"
meter from Greek *metron* "measure, length"
milli- from Latin *mille* "one thousand"
multiply from Latin *multi* "many" and Indo-European *pel* "to fold"
nonagon from Latin *nonus* "ninth" and Greek *gon* "angle"
number from Latin *numerus* "number"
obtuse from Latin *ob* "against, near, at" and *tusus* "to strike, to beat" [*obtusus* = beaten down to the point of being dull]
oct- from Greek *octo* "eight"; *oct*agon, *oct*ahedron
parallel from Greek *para* "alongside" and *allenon* "one another"
pent- Greek *pent* "five"; *pent*agon, *pent*agram, *pent*omino
percent from Latin *per* "for" and *centum* "hundred" [percent = for (each) hundred]
peri- from Greek *peri* "around"; *peri*meter
plane from Latin *planus "flat"*
poly- from Greek *polus* "many"; *poly*gon, *poly*hedra, *poly*omino
prism from Greek *prisma* "something that has been sawed"
quadr- Latin *quadr-* "four"; *quandr*ant, *quadr*ilateral
rectangle from Latin *rectangulus* "right-angled"
-sect from Latin *sectus*, past participle of *secui* "to cut"; bi*sect*, inter*sect*,
surface from Latin *super* "over" and *facies* "form, shape"
symmetric from Greek *sun-* "together with" and *metron* "measure"
tetra- Greek *tetra-* "four"; *tetra*hedron, *tetr*omino
trans- Latin *trans* "across"; *trans*itive, *trans*lation, *trans*versal
tri- from Latin *tri* "three"; *tri*angle, *tri*sect
vertex from Latin verb *vertere* "to turn"
zero from Arabic *çifr* "empty"

INDEX AND PRONUNCIATION GUIDE

a	**a**ct, b**a**t	j	**j**ust, fu**dg**e	œ	as in German sch**ö**n or in French f**eu**
ā	c**a**pe, w**ay**	k	**k**eep, to**k**en	R	rolled r as in French **r**ouge or in German **r**ot
â	d**a**re, M**a**ry	KH	as in Scottish lo**ch** on in German i**ch**	sh	**sh**oe, fi**sh**
ä	**a**lms, c**a**lm	N	as in French bo**n** or u**n**	th	**th**in, pa**th**
ch	**ch**ild, bea**ch**	o	**o**x, w**a**sp	u	**u**p, l**o**ve
e	s**e**t, m**e**rry	ō	**o**ver, n**o**	û	**u**rge, b**u**rn
ē	**e**qual, b**ee**	o͝o	b**oo**k, p**oo**r	y	**y**es, oni**o**n
e	like a in **a**lone or e in syst**e**m	*oo*	**oo**ze, f**oo**l	z	**z**eal, la**z**y
g	**g**ive, be**g**	ô	**ou**ght, r**aw**	zh	trea**s**ure, mira**g**e
i	**i**f, b**i**g	oi	**oi**l, j**oy**		
ī	**i**ce, b**i**te	ou	**ou**t, c**ow**		

A

D

E

F

G

H

I

N

Q

R

S

T

U

V

W

Z

CURRICULUM STANDARDS FOR GRADES 5-8

STANDARD 1: Mathematics as Problem Solving

In grades 5-8, the mathematics curriculum should include numerous and varied experiences with problem solving as a method of inquiry and application so that students can—
- use problem-solving approaches to investigate and understand mathematical content;
- formulate problems from situations within and outside mathematics;
- develop and apply a variety of strategies to solve problems, with emphasis on multistep and nonroutine problems;
- verify and interpret results with respect to the original problem situation;
- generalize solutions and strategies to new problem situations;
- acquire confidence in using mathematics meaningfully.

STANDARD 2: Mathematics as Communication

In grades 5-8, the study of mathematics should include opportunities to communicate so that students can—
- model situations using oral, written, concrete, pictorial, graphical, and algebraic methods;
- reflect on and clarify their own thinking about mathematical ideas and situations;
- develop common understandings of mathematical ideas, including the role of definitions;
- use the skills of reading, listening, and viewing to interpret and evaluate mathematical ideas;
- discuss mathematical ideas and make conjectures and convincing arguments;
- appreciate the value of mathematical notation and its role in the development of mathematical ideas.

STANDARD 3: Mathematics as Reasoning

In grades 5-8, reasoning shall permeate the mathematics curriculum so that students can—
- recognize and apply deductive and inductive reasoning;
- understand and apply reasoning processes, with special attention to spatial reasoning and reasoning with proportions and graphs;
- make and evaluate mathematical conjectures and arguments;
- validate their own thinking;
- appreciate the pervasive use and power of reasoning as a part of mathematics.

STANDARD 4: Mathematical Connections

In grades 5-8, the mathematics curriculum should include the investigation of mathematical connections so that students can—
- see mathematics as an integrated whole;
- explore problems and describe results using graphical, numerical, physical, algebraic, and verbal mathematical models or representations;
- use a mathematical idea to further their understanding of other mathematical ideas;
- apply mathematical thinking and modeling to solve problems that arise in other disciplines, such as art, music, psychology, science, and business;
- value the role of mathematics in our culture and society.

STANDARD 5: Number and Number Relationships

In grades 5-8, the mathematics curriculum should include the continued development of number and number relationships so that students can—
- understand, represent, and use numbers in a variety of equivalent forms (integer, fraction, decimal, percent, exponential, and scientific notation) in real-world and mathematical problem situations;
- develop number sense for whole numbers, fractions, decimals, integers, and rational numbers;
- understand and apply ratios, proportions, and percents in a wide variety of situations;
- investigate relationships among fractions, decimals, and percents;
- represent numerical relationships in one- and two-dimensional graphs.

STANDARD 6: Number Systems and Number Theory

In grades 5-8, the mathematics curriculum should include the study of number systems and number theory so that students can—
- understand and appreciate the need for numbers beyond the whole numbers;
- develop and use order relations for whole numbers, fractions, decimals, integers, and rational numbers;
- extend their understanding of whole number operations to fractions, decimals, integers, and rational numbers;
- understand how the basic arithmetic operations are related to one another;
- develop and apply number theory concepts (e.g., primes, factors, and multiples) in real-world and mathematical problem situations.

STANDARD 7: Computation and Estimation

In grades 5-8, the mathematics curriculum should develop the concepts underlying computation and estimation in various contexts so that students can—
- compute with whole numbers, fractions, decimals, integers and rational numbers;
- develop, analyze, and explain procedures for computation and techniques for estimation;
- develop, analyze, and explain methods for solving proportions;